MOORE

Clinically Oriented ANATOMY

Seventh Edition

Keith L. Moore, M.Sc., Ph.D., D.Sc. (Hon), F.I.A.C., F.R.S.M., F.A.A.A.
Professor Emeritus in Division of Anatomy, Department of Surgery
Former Chair of Anatomy and Associate
Dean for Basic Medical Sciences
Faculty of Medicine, University of Toronto
Toronto, Ontario, Canada

Arthur F. Dalley II, Ph.D.
Professor, Department of Cell and Developmental Biology
Adjunct Professor, Department of Orthopaedics and Rehabilitation
Director, Programs in Medical Gross Anatomy and Anatomical Donations Program
Vanderbilt University School of Medicine
Adjunct Professor for Anatomy
Belmont University School of Physical Therapy
Nashville, Tennessee, U.S.A.

Anne M. R. Agur, B.Sc. (OT), M.Sc., Ph.D.
Professor, Division of Anatomy, Department of Surgery, Faculty of Medicine
Department of Physical Therapy
Department of Occupational Science & Occupational Therapy
Division of Physiatry, Department of Medicine
Division of Biomedical Communications
Institute of Medical Science
Graduate Department of Rehabilitation Science
Graduate Department of Dentistry
University of Toronto
Toronto, Ontario, Canada

Wolters Kluwer | Lippincott Williams & Wilkins
Health
Philadelphia · Baltimore · New York · London
Buenos Aires · Hong Kong · Sydney · Tokyo

Acquisitions Editor: Crystal Taylor
Product Manager: Julie Montalbano
Marketing Manager: Joy Fisher Williams
Art Director, Digital Content: Jennifer Clements
Artists: Imagineeringart.com, lead artist Natalie Intven, MSc, BMC
Designer: Terry Mallon
Compositor: SPi Global

7th Edition

DISCLAIMER
Care has been taken to confirm the accuracy of the information presented and to describe generally accepted practices. However, the authors, editors, and publisher are not responsible for errors or omissions or for any consequences from application of the information in this book and make no warranty, expressed or implied, with respect to the currency, completeness, or accuracy of the contents of the publication. Application of this information in a particular situation remains the professional responsibility of the practitioner; the clinical treatments described and recommended may not be considered absolute and universal recommendations.

The authors, editors, and publisher have exerted every effort to ensure that drug selection and dosage set forth in this text are in accordance with the current recommendations and practice at the time of publication. However, in view of ongoing research, changes in government regulations, and the constant flow of information relating to drug therapy and drug reactions, the reader is urged to check the package insert for each drug for any change in indications and dosage and for added warnings and precautions. This is particularly important when the recommended agent is a new or infrequently employed drug.

Some drugs and medical devices presented in this publication have Food and Drug Administration (FDA) clearance for limited use in restricted research settings. It is the responsibility of the healthcare provider to ascertain the FDA status of each drug or device planned for use in their clinical practice.

To purchase additional copies of this book, call our customer service department at **(800) 638-3030** or fax orders to **(301) 223-2320**. International customers should call **(301) 223-2300.**

Visit Lippincott Williams & Wilkins on the Internet: http://www.lww.com. Lippincott Williams & Wilkins customer service representatives are available from 8:30 am to 6:00 pm, EST.

The publishers have made every effort to trace the copyright holders for borrowed material. If they have inadvertently overlooked any, they will be pleased to make the necessary arrangements at the first opportunity.

CCS0113

Clinically Oriented ANATOMY

Seventh Edition

In Loving Memory of Marion

My best friend, wife, colleague, mother of our five children and grandmother of our nine grandchildren for her love, unconditional support, and understanding. Wonderful memories keep you in our hearts and minds. • (KLM)

To Pam and Ron

I am grateful to my eldest daughter Pam, who assumed the office duties her mother previously carried out. She is also helpful in many other ways. I am also grateful to my son-in-law Ron Crowe whose technical skills have helped me prepare the manuscript for this book. • (KLM)

To My Grandchildren

Melissa, Kristin, Alecia, Lauren, Mitchel, Caitin, Jayme, Courtney and Brooke. With best wishes for your future endeavours. Love, Grandpa • (KLM)

To Muriel

My bride, best friend, counselor, and mother of our sons; and to our family—Tristan, Lana, Elijah, Finley and Sawyer; Denver, and Skyler—with love and great appreciation for their support, understanding, good humor, and—most of all—patience. • (AFD)

To my husband, Enno, and my children, Erik and Kristina, for their support and encouragement. • (AMRA)

To Our Students

You will remember some of what you hear, much of what you read, more of what you see, and almost all of what you experience and understand fully.

To Anatomical Donors

With sincere appreciation to all those who donate their bodies for anatomical study and research, without whom anatomical textbooks and atlases, and anatomical study in general would not be possible.

**Keith L. Moore, Ph.D.,
D.Sc. (Hon)., F.I.A.C.,
F.R.S.M., F.A.A.A.**

Dr. Moore has been the recipient of many prestigious awards and recognitions. He has received the highest awards for excellence in human anatomy education at the medical, dental, graduate, and undergraduate levels—and for his remarkable record of textbook publications in clinically oriented anatomy and embryology—from both the American Association of Anatomists (AAA: Distinguished Educator Award, 2007) and the American Association of Clinical Anatomists (AACA: Honored Member Award, 1994). In 2008 Dr. Moore was inducted as a Fellow of the American Association of Anatomists. The rank of Fellow honors distinguished members who have demonstrated excellence in science and their overall contributions to the medical sciences. In 2012, Dr. Moore received an honorary Doctor of Science degree from The Ohio State University, the Queen Elizabeth II Diamond Jubilee Medal honoring significant contributions and achievements by Canadians, and the R. Benton Adkins, Jr. Distinguished Service Award for his outstanding record of service to the American Association of Clinical Anatomists.

Arthur F. Dalley II

Arthur F. Dalley II, Ph.D.

**Anne M. R. Agur, B.Sc.
(OT), M.Sc., Ph.D.**

Preface

A third of a century has passed since the first edition of *Clinically Oriented Anatomy* appeared on bookstore shelves. Although the factual basis of anatomy is remarkable among basic sciences for its longevity and consistency, this book has evolved markedly since its inception. This is a reflection of changes in the clinical application of anatomy, new imaging technologies that reveal living anatomy in new ways, and improvements in graphic and publication technology that enable superior demonstration of this information. Efforts continue to make this book even more student friendly and authoritative. The seventh edition has been thoroughly reviewed by students, anatomists, and clinicians for accuracy and relevance and revised with significant new changes and updates.

KEY FEATURES

Clinically Oriented Anatomy has been widely acclaimed for the relevance of its clinical correlations. As in previous editions, the seventh edition places clinical emphasis on anatomy that is important in physical diagnosis for primary care, interpretation of diagnostic imaging, and understanding the anatomical basis of emergency medicine and general surgery. Special attention has been directed toward assisting students in learning the anatomy they will need to know in the twenty-first century, and to this end new features have been added and existing features updated.

Extensive art program. The seventh edition is distinguished by an extensive revision of the art program. Working with a team of artists from Imagineering, every illustration has been revised, improving accuracy and consistency and giving classical art derived from *Grant's Atlas of Anatomy* a fresh, vital, new appearance. An effort has been made to ensure that all the anatomy presented and covered in the text is also illustrated. The text and illustrations have been developed to work together for optimum pedagogical effect, aiding the learning process and markedly reducing the amount of searching required to find structures. The great majority of the clinical conditions are supported by photographs and/or color illustrations; multipart illustrations often combine dissections, line art, and medical images; tables are accompanied by illustrations to aid the student's understanding of the structures described.

Clinical correlations. Popularly known as "blue boxes," the clinical information sections have grown, and many of them are supported by photographs and/or dynamic color illustrations to help with understanding the practical value of anatomy. In response to our readers' suggestions, the blue boxes have been grouped together within chapters, enabling presentation of topics with less interruption of the running text.

Bottom line summaries. Frequent "bottom line" boxes summarize the preceding information, ensuring that primary concepts do not become lost in the many details necessary for thorough understanding. These summaries provide a convenient means of ongoing review and underscore the big picture point of view.

Anatomy described in a practical, functional context. A more realistic approach to the musculoskeletal system emphasizes the action and use of muscles and muscle groups in daily activities, emphasizing gait and grip. The eccentric contraction of muscles, which accounts for much of their activity, is now discussed along with the concentric contraction that is typically the sole focus in anatomy texts. This perspective is important to most health professionals, including the growing number of physical and occupational therapy students using this book.

Surface anatomy and medical imaging. Surface anatomy and medical imaging, formerly presented separately, are now integrated into the chapter, presented at the time each region is being discussed, clearly demonstrating anatomy's relationship to physical examination and diagnosis. Both natural views of unobstructed surface anatomy and illustrations superimposing anatomical structures on surface anatomy photographs are components of each regional chapter. Medical images, focusing on normal anatomy, include plain and contrast radiographic, MRI, CT, and ultrasonography studies, often with correlative line art as well as explanatory text, to help prepare future professionals who need to be familiar with diagnostic images.

Case studies, accompanied by clinico-anatomical problems and board review-style multiple-choice questions. Interactive case studies and multiple-choice questions are available to our readers online at http://thePoint.lww.com, providing a convenient and comprehensive means of self-testing and review.

Terminology. The terminology fully adheres to *Terminologia Anatomica* (1998), approved by the International Federation of Associations of Anatomists (IFAA). Although the official English-equivalent terms are used throughout the book, when new terms are introduced, the Latin form, used in Europe, Asia, and other parts of the world, is also provided.

The roots and derivations of terms are provided to help students understand meaning and increase retention. Eponyms, although not endorsed by the IFAA, appear in parentheses in this edition—for example, sternal angle (angle of Louis)—to assist students who will hear eponymous terms during their clinical studies. The terminology is now available online at http://www.unifr.ch/ifaa.

RETAINED AND IMPROVED FEATURES

Students and faculty have told us what they want and expect from *Clinically Oriented Anatomy*, and we listened:

- A *comprehensive text* enabling students to fill in the blanks, as time allotted for lectures continues to decrease, laboratory guides become exclusively instructional, and multiauthored lecture notes develop inconsistencies in comprehension, fact, and format.
- A *resource capable of supporting areas of special interest and emphasis* within specific anatomy courses that *serves the anatomy needs of students during both the basic science and the clinical phases of their studies.*
- A *thorough Introduction* that covers important systemic information and concepts basic to the understanding of the anatomy presented in the subsequent regional chapters. Students from many countries and backgrounds have written to express their views of this book—gratifyingly, most are congratulatory. Health professional students have more diverse backgrounds and experiences than ever before. Curricular constraints often result in unjustified assumptions concerning the prerequisite information necessary for many students to understand the presented material. The Introduction includes efficient summaries of functional systemic anatomy. Students' comments specifically emphasized the need for a systemic description of the nervous system and the peripheral autonomic nervous system (ANS) in particular.
- Routine facts (such as muscle attachments, innervations, and actions) presented in *tables organized to demonstrate shared qualities and illustrated to demonstrate the provided information. Clinically Oriented Anatomy* provides more tables than any other anatomy textbook.
- *Illustrated clinical correlations* that not only describe but also *show anatomy as it is applied clinically.*
- *Illustrations that facilitate orientation.* Many orientation figures have been added, along with arrows to indicate the locations of the inset figures (areas shown in close-up views) and viewing sequences. Labels have been placed to minimize the distance between label and object, with leader lines running the most direct course possible.

- Blue boxes are classified by the following icons to indicate the type of clinical information covered:

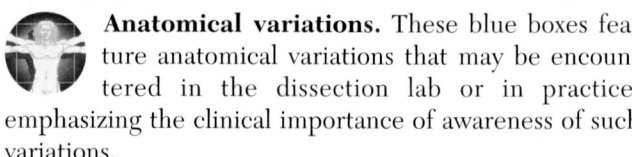 **Anatomical variations.** These blue boxes feature anatomical variations that may be encountered in the dissection lab or in practice, emphasizing the clinical importance of awareness of such variations.

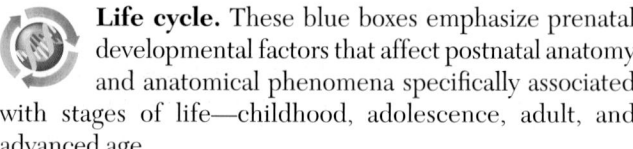 **Life cycle.** These blue boxes emphasize prenatal developmental factors that affect postnatal anatomy and anatomical phenomena specifically associated with stages of life—childhood, adolescence, adult, and advanced age.

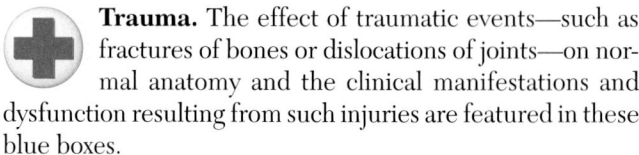 **Trauma.** The effect of traumatic events—such as fractures of bones or dislocations of joints—on normal anatomy and the clinical manifestations and dysfunction resulting from such injuries are featured in these blue boxes.

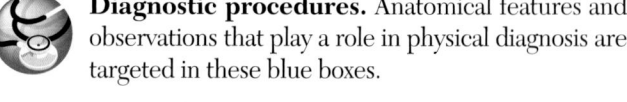 **Diagnostic procedures.** Anatomical features and observations that play a role in physical diagnosis are targeted in these blue boxes.

Surgical procedures. These blue boxes address such topics as the anatomical basis of surgical procedures, such as the planning of incisions, and the anatomical basis of regional anesthesia.

Pathology. The effect of disease on normal anatomy, such as cancer of the breast, and anatomical structures or principles involved in the confinement or dissemination of disease within the body are the types of topics covered in these blue boxes.

- **Boldface type** indicates the main entries of anatomical terms, when they are introduced and defined. In the index, the page numbers of these main entries also appear in boldface type, so that the main entries can be easily located. Boldface type is also used to introduce clinical terms in the clinical correlation (blue) boxes.
- *Italic type* indicates anatomical terms important to the topic and region of study or labeled in an illustration that is being referenced.
- Useful content outlines appear at the beginning of every chapter.
- Instructor's resources and supplemental materials are available at http://thePoint.lww.com.

Anne M. R. Agur, Ph.D., joined Keith Moore and Arthur Dalley as a co-author for the sixth edition. From the outset, *Clinically Oriented Anatomy* has utilized materials from *Grant's Atlas*, for which Anne has had responsibility since 1991. Anne made significant contributions to previous editions of *Clinically Oriented Anatomy* beyond the sharing of materials from *Grant's Atlas,* and has been involved in—and been an asset to—every stage of the development of this and the previous editions.

COMMITMENT TO EDUCATING STUDENTS

This book is written for health science students, keeping in mind those who may not have had a previous acquaintance with anatomy. We have tried to present the material in an interesting way so that it can be easily integrated with what will be taught in more detail in other disciplines such as physical diagnosis, medical rehabilitation,

and surgery. We hope this text will serve two purposes: to educate and to excite. If students develop enthusiasm for clinical anatomy, the goals of this book will have been fulfilled.

Keith L. Moore
University of Toronto
Faculty of Medicine

Arthur F. Dalley II
Vanderbilt University
School of Medicine

Anne M. R. Agur
University of Toronto
Faculty of Medicine

ABBREVIATIONS

a., aa.	artery, arteries	lev.	levator
ant.	anterior	lit.	literally
B.C.E.	before the Common (Christian) era	M	male
C	cervical	m., mm.	muscle, muscles
C.E.	Common (Christian) era	Mediev.	medieval
Co	coccygeal	Mod.	modern
dim.	diminutive	post.	posterior
e.g.	for example	S	sacral
et al.	and others	sup.	superior, superioris
F	female	supf.	superficial
Fr.	French	T	thoracic
G.	Greek	TA	*Terminologia Anatomica*
i.e.	that is	TE	Terminologia Embryologica
inf.	inferior	TH	Terminologia Histologica
L	liter, lumbar	v., vv.	vein, veins
L.	Latin	vs.	versus

Acknowledgments

We wish to thank the following colleagues who were invited by the publisher to assist with the development of the seventh edition through their critical analysis.

- Ernest Adeghate, PhD, Chair, Department of Anatomy, United Arab Emirates University
- Christopher Bise, PT, MS, DPT, OCS, Assistant Professor, University of Pittsburgh; Physical Therapist at UPMC Children's Hospital of Pittsburgh
- Christopher Briggs, PhD, Assistant Professor, Department of Anatomy and Neuroscience, University of Melbourne, Australia
- Anjanette Clifford, BS, MS, DC, Assistant Professor of Master of Science Degree in Sports Science and Rehabilitation, Logan College of Chiropractic University
- James Culberson, PhD, Professor, Department of Neurobiology and Anatomy, West Virginia University
- Terry Dean, MD, PhD, Post-Doctoral Fellow, Department of Cell and Molecular Biology, University of Pennsylvania
- Robert Frysztak, PhD, Professor, Department of Physiology, National University Health Sciences
- Tom Gillingwater, MD, Professor, Department of Biological and Clinical Lab Sciences, University of Edinburgh, United Kingdom
- Daniel Graney, PhD, Professor, Department of Anatomy and Embryology, University of Washington School of Medicine
- Robert Hage, MD, PhD, DLO, MBA, Professor, Department of Anatomy, St.George's University, West Indies
- Beth Jones, PT, DPT, MS, OCS, Assistant Professor, Physical Therapy Program, Department of Orthopaedics and Rehabilitation, School of Medicine, University of New Mexico; Assistant Professor, Department of Cell Biology and Physiology, School of Medicine, University of New Mexico
- Randy Kulesza, PhD, Associate Professor, Department of Anatomy, Assistant Dean- Post Baccalaureate and MS in Biomedical Sciences Programs, Lake Erie College of Osteopathic Medicine
- Octavian Lucaciu, MD, PhD, Associate Professor, Department of Anatomy, Canadian Memorial Chiropractic College
- Linda McLoon, PhD, Professor, Department of Ophthalmology and Visual Neuroscience, University of Minnesota
- Gary Nieder, PhD, Professor, Department of Neuroscience, Cell Biology & Physiology, Wright State University
- El Sayed Nosair, PhD, Assistant Professor, Sharjah University, United Arab Emirates
- James Walker, PhD, Associate Professor, Department of Basic Medical Sciences, Purdue University; Associate Director, Department of Human Anatomy, Indiana University School of Medicine-Lafayette

Several students were also invited by the publisher to review the textbook:

- Shloka Ananthanarayana, Mount Sinai School of Medicine
- Jennifer Gulas, Universidad Autonoma de Guadalajara, Mexico
- Paul Johnson, Michigan State University
- Liying Low, University of Glasgow, Scotland
- Kristina Medhus, University of Texas Medical School-Houston
- Vanessa Scott, Albert Einstein College of Medicine
- Sara Thorp, Ohio University College of Osteopathic Medicine
- Ryckie Wade, University of East Anglia, United Kingdom

In addition to reviewers, many people—some of them unknowingly—helped us by perusing, discussing, or contributing to parts of the manuscript and/or providing constructive criticism of the text and illustrations in this and previous editions:

- Dr. Peter Abrahams, Prof. of Clinical Anatomy, Medical Teaching Centre, Institute of Clinical Education, Warwick Medical School, University of Warwick, Coventry, UK
- Dr. Robert D. Acland, Professor of Surgery/Microsurgery, Division of Plastic and Reconstructive Surgery, University of Louisville, Louisville, Kentucky
- Dr. Edna Becker, Associate Professor of Medical Imaging, University of Toronto Faculty of Medicine, Toronto
- Dr. Donald R. Cahill, Professor of Anatomy (retired; former Chair), Mayo Medical School; former Editor-in-Chief of Clinical Anatomy, Rochester, Minnesota
- Dr. Joan Campbell, Assistant Professor of Medical Imaging, University of Toronto Faculty of Medicine, Toronto
- Dr. Stephen W. Carmichael, Professor Emeritus, Mayo Medical School, Editor-in-Chief of Clinical Anatomy, Rochester, Minnesota
- Dr. Carmine D. Clemente, Professor of Anatomy and Orthopedic Surgery, University of California, Los Angeles School of Medicine, Los Angeles
- Dr. James D. Collins, Professor of Radiological Sciences, University of California, Los Angeles School of Medicine/Center for Health Sciences, Los Angeles

- Dr. Raymond F. Gasser, Emeritus Professor of Anatomy, Louisiana State University School of Medicine, New Orleans
- Dr. Ralph Ger (Deceased), Professor of Anatomy and Structural Biology, Albert Einstein College of Medicine, Bronx, New York
- Dr. Paul Gobee, Assistant Professor, Developer Anatomical E-Learning, Department of Anatomy & Embryology, Leiden University Medical Center, Leiden, Netherlands
- Dr. Douglas J. Gould, Professor of Biomedical Sciences, Oakland University William Beaumont School of Medicine, Detroit, Michigan
- Dr. Daniel O. Graney, Professor of Biological Structure, University of Washington School of Medicine, Seattle
- Dr. David G. Greathouse, former Professor and Chair, Belmont University School of Physical Therapy, Nashville, Tennessee
- Scott L. Hagan, 4th year Medical Student, Vanderbilt University School of Medicine, Nashville, TN
- Dr. Masoom Haider, Assistant Professor of Medical Imaging, University of Toronto Faculty of Medicine, Toronto
- Dr. John S. Halle, Professor and Chair, Belmont University School of Physical Therapy, Nashville, Tennessee
- Dr. Jennifer L. Halpern, Assistant Professor, Orthopedic Surgery—Oncology, Vanderbilt University School of Medicine, Nashville, Tennessee
- Dr. Nicole Herring, Assistant Professor, Anatomical Sciences and Neurobiology, University of Louisville School of Medicine, Louisville, Kentucky
- Dr. Walter Kuchareczyk, Professor and Neuroradiologist Senior Scientist, Department of Medical Resonance Imaging, University Health Network, Toronto, Ontario, Canada.
- Dr. Nirusha Lachman, Professor of Anatomy, Mayo Medical School, Rochester, Minnesota
- Dr. H. Wayne Lambert, Professor, Neurobiology and Anatomy Department, West Virginia University School of Medicine, Morgantown, West Virginia
- Dr. Michael von Lüdinghausen, University Professor, Anatomy Institute, University of Würzburg, Würzburg, Germany (retired)
- Dr. Shirley McCarthy, Director of MRI, Department of Diagnostic Radiology, Yale University School of Medicine, New Haven, Connecticut
- Dr. Lillian Nanney, Professor of Plastic Surgery and Cell and Developmental Biology, Vanderbilt University School of Medicine, Nashville, Tennessee
- Dr. Todd R. Olson, Professor of Anatomy and Structural Biology, Albert Einstein College of Medicine, Bronx, New York
- Dr. Wojciech Pawlina, Professor and Chair of Anatomy, Mayo Medical School, Rochester, Minnesota
- Dr. T. V. N. Persaud, Professor Emeritus of Human Anatomy and Cell Science, Faculties of Medicine and Dentistry, University of Manitoba, Winnipeg, Manitoba, Canada. Professor of Anatomy and Embryology, St. George's University, Granada, West Indies
- Dr. Cathleen C. Pettepher, Professor of Cancer Biology, Vanderbilt University School of Medicine, Nashville, Tennessee
- Dr. Thomas H. Quinn, Professor of Biomedical Sciences, Creighton University School of Medicine, Omaha, Nebraska
- Dr. Christopher Ramnanan, Assistant Professor, Department of Cellular and Functional Anatomy, Division of Clinical and Functional Anatomy, University of Ottawa Faculty of Medicine, Ottawa, Ontario, Canada
- Dr. David Resuehr, Assistant Professor, Department of Cellular, Developmental and Integrative Biology, University of Alabama at Birmingham School of Medicine, Birmingham, Alabama
- Dr. George E. Salter, Professor of Anatomy (retired), Department of Cell Biology, University of Alabama, Birmingham
- Dr. Ryan Splittgerber, Assistant Professor, Department of Genetics, Cell Biology and Anatomy, University of Nebraska Medical Center, College of Medicine, Omaha, Nebraska
- Dr. Tatsuo Sato, Professor and Head (retired), Second Department of Anatomy, Tokyo Medical and Dental University Faculty of Medicine, Tokyo
- Professor Colin P. Wendell-Smith, Department of Anatomy and Physiology, University of Tasmania, Hobart, Tasmania, Australia (retired)
- Dr. Andreas H. Weiglein, Professor, Institut fur Anatomie, Medical University Graz, Graz, Austria
- Dr. David G. Whitlock, Professor of Anatomy, University of Colorado Medical School, Denver

Art plays a major role in facilitating learning. We extend our sincere gratitude and appreciation for the skills, talents, and timely work of our medical illustrators for this edition. Wynne Auyeung and Natalie Intven from Imagineering did a superb job of managing a team of talented artists to revise all of the illustrations for a more consistent, vibrant art program. Photographs taken during a major surface anatomy photography project for the fifth edition continue to be a tremendous asset. E. Anne Rayner, Senior Photographer, Vanderbilt Medical Art Group did an excellent job photographing the surface anatomy models, working in association with authors Arthur Dalley and Anne Agur. We greatly appreciate the contribution the models made to the quality of both the previous and the current edition. Although the number of illustrations from *Grant's Atlas of Anatomy* continues to be reduced and replaced by new art, we gratefully acknowledge the excellence of Professor J. C. B. Grant's dissections and the excellent art done by the following: Dorothy Foster Chubb,

Elizabeth Blackstock, Nancy Joy, Nina Kilpatrick, David Mazierski, Stephen Mader, Bart Vallecoccia, Sari O'Sullivan, Kam Yu, Caitlin Duckwall, and Valerie Oxorn.

Many thanks also to those at Lippincott Williams & Wilkins who participated in the development of this edition: Crystal Taylor, Acquisitions Editor; Jennifer Clements, Art Director; and Julie Montalbano, Product Manager. Finally, thanks are due to the sales and marketing division at LWW, which has played a key role in the continued success of this book.

Keith L. Moore
Arthur F. Dalley II
Anne M. R. Agur

Contents

List of Clinical Blue Boxes

3   Pelvis and Perineum

7   Head

8   Neck

9   Summary of Cranial Nerves

Figure Credits

INTRODUCTION

Fig. I.20 Based on Hamill JH, Knutzan K: Biochemical Basis of Human Movement. 2nd ed. Baltimore: Williams & Wilkins, 2003.

Fig. I.22C Based on Silverthorn. Human Physiology. 4th ed. Tappan, NJ: Pearson Education, 2007. P. 459.

Fig. I.50 Daffner RH: Clinical Radiology: The Essentials. 2nd ed. Baltimore: Williams & Wilkins, 1998.

Fig. I.53 Wicke L: Atlas of Radiologic Anatomy. 6th English ed. Ed and trans: Taylor AN. Baltimore: Williams & Wilkins, 1998. [Wicke L: Roentgen-Anatomie Normalbefunde. 5th ed. Munich: Urban and Schwarzenberg, 1995.]

Fig. I.54 Atlas of Radiologic Anatomy. 6th English ed.

Fig. I.55 Atlas of Radiologic Anatomy. 6th English ed.

Fig. I.56 Atlas of Radiologic Anatomy. 6th English ed.

Fig. BI.1 Reprinted with permission from van de Graaff K: Human Anatomy. 4th ed. Dubuque, IA: WC Brown, 1995.

Fig. BI.2 Rassner G: Atlas of Dermatology. 3rd ed. Trans: Burgdorf WHC. Philadelphia: Lea & Febiger, 1994 (photo); Stedman's Medical Dictionary. 27th ed. Baltimore: Lippincott Williams & Wilkins, 2000. (artist: Neil O. Hardy, Westport, CT).

Fig. BI.4 Based on Stedman's Medical Dictionary. 27th ed. (artist: Neil O. Hardy, Westport, CT).

Fig. BI.5 Based on Stedman's Medical Dictionary. 27th ed. (artist: Neil O. Hardy, Westport, CT).

Fig. BI.8 Based on Willis MC: Medical Terminology: The Language of Health Care. Baltimore: Lippincott Williams & Wilkins, 1995.

Fig. BI.9 Reprinted with permission from Roche Lexikon Medizin. 4th ed. Munich: Urban & Schwarzenberg, 1998.

CHAPTER 1

Fig. 1.20. B&C. Based on Clay JH, Pounds DM: Basic Clinical Massage Therapy: Integrating Anatomy and Treatment. 2nd ed. Baltimore: Lippincott Williams & Wilkins, 2002.

Fig. 1.24C Based on Stedman's Medical Dictionary. 27th ed. (artist: Michael Schenk, Jackson, MS).

Fig. 1.30A Dean D, Herbener TE: Cross-Sectional Anatomy. Baltimore: Lippincott Williams & Wilkins, 2000.

Fig. 1.48 Adapted with permission from Moore KL, Persaud TVN. The Developing Human: Clinically Oriented Embryology. 7th ed. Philadelphia: Saunders, 2003. Fig. 8.5A, B, & D, p. 150.

Fig. 1.50 Adapted with permission from Torrent-Guasp F, Buckberg GD, Clemente C, et al.: The structure and function of the helical heart and its buttress wrapping. The normal macroscopic structure of the heart. Semin Thoracic Cardiovasc Surg 2001;13:30.

Fig. B1.6A Based on Bickley LS, Szilagyi PG: Bates' Guide to Physical Examination. 8th ed. Baltimore: Lippincott Williams & Wilkins, 2002.

Fig. B1.6B Brant WE, Helms CA: Fundamentals of Diagnostic Radiology. 2nd ed. Baltimore: Williams & Wilkins, 1999.

Fig. B1.7 Based on Hall-Craggs ECB: Anatomy as the Basis of Clinical Medicine. 3rd ed. Baltimore: Williams & Wilkins, 1995

Fig. B1.9 Based on Stedman's Medical Dictionary. 27th ed. (artist: Neil O. Hardy, Westport, CT).

Fig. B1.11A Based on Stedman's Medical Dictionary. 27th ed. Baltimore: Lippincott Williams & Wilkins, 2000. (artist: Mikki Senkarik, San Antonio, TX).

Fig. B1.11B Olympus America, Inc., Melville, NY.

Fig. B1.12A Based on Stedman's Medical Dictionary. 27th ed. (artist: Neil O. Hardy, Westport, CT).

Fig. B1.12B Bates' Guide to Physical Examination, 10th ed., p. 300.

Fig. B1.13 Stedman's Medical Dictionary. 27th ed. (artist: Neil O. Hardy, Westport, CT); photographs of bronchus, carina, and trachea—Feinsilver SH, Fein A: Textbook of Bronchoscopy. Baltimore: Williams & Wilkins, 1995; photograph of bronchoscopy procedure—courtesy of Temple University Hospital, Philadelphia.

Fig. B1.14 Clinical Radiology: The Essentials. 2nd ed.

Fig. B1.18 Based on Stedman's Medical Dictionary. 27th ed. (artist: Neil O. Hardy, Westport, CT).

Fig. B1.19 With permission from The Developing Human: Clinically Oriented Embryology. 7th ed. Figs. 14.15 and 14.14, p. 345–346.

Fig. B1.23 Siemens Medical Solutions USA, Inc.

Fig. B1.26 Based on figure provided by the Anatomical Chart Company.

Fig. B1.28 Based on Stedman's Medical Dictionary. 27th ed. (artist: Neil O. Hardy, Westport, CT).

Fig. B1.29 Based on Stedman's Medical Dictionary. 27th ed. (artist: Neil O. Hardy, Westport, CT); photograph—courtesy of Quinton Cardiology, Inc.

Fig. B1.34 Based on Clinical Radiology: The Essentials. 2nd ed.

Fig. B1.37 Cross-Sectional Anatomy. P. 25.

Fig. B1.38B-E Madden ME. Introduction to Sectional Anatomy. Baltimore: Lippincott Williams & Wilkins, 2000.

CHAPTER 2

Fig. 2.5 Basic Clinical Massage Therapy: Integrating Anatomy and Treatment. 2nd ed.

Fig. 2.6B Based on Bates' Guide to Physical Examination. 10th ed., p 415; Fig. 2.6B(slices) Based on Grants' Atlas of Anatomy 12th ed. fig 2.6DE

Fig. 2.18 Based on Sauerland EK: Grant's Dissector. 12th ed. Baltimore: Williams & Wilkins, 1999.

Fig. 2.33B Cormack DH, Clinically Integrated Histology, Baltimore: Lippincott Williams and Wilkins, 1998. (Fig. 8.8 p. 191)

Fig. 2.36B Based on Agur AMR: Grant's Method of Anatomy. 9th ed. Baltimore: Williams & Wilkins, 1975.

Fig. 2.43C Based on Stedman's Medical Dictionary. 27th ed. (artist: Neil O. Hardy, Westport, CT).

Fig. 2.47 Based on McConnell TH, Hull K: Human Form, Human Function: Essentials of Anatomy and Physiology. 1st ed. Baltimore: Lippincott Williams & Wilkins, 2011. fig. 14.16a, p. 565.

Fig. 2.48B Based on Grant's Dissector. 12th ed.

Fig. 2.57A Based on Grant's Method of Anatomy. 9th ed.

Fig. 2.57B Based on Grant's Method of Anatomy. 9th ed.

Fig. 2.58C Based on Stedman's Medical Dictionary. 27th ed. (artist: Neil O. Hardy, Westport, CT)

Fig. 2.59D Gartner LP, Hiatt JL: Color Atlas of Histology. 3rd ed. Baltimore: Lippincott Williams & Wilkins, 2001.

Fig. 2.62 Based on Bates' Guide to Physical Examination. 8th ed.

Fig. 2.64E Based on Grant's Dissector. 12th ed.

Fig. 2.67B–E Reprinted with permission from Karaliotas C. et al: Liver and Biliary Tract Surgery: Embryological Anatomy to 3D-Imaging and Transplant Innovations. Vienna: Springer, 2007.

Fig. 2.90 Based on Rosse C, Gaddum-Rosse P: Hollinshead's Textbook of Anatomy. 5th ed. Philadelphia, Lippincott-Raven, 1997.

Fig. 2.91A Based on Basic Clinical Massage Therapy: Integrating Anatomy and Treatment. 2nd ed.

Fig. 2.102B & C Cross-Sectional Anatomy.

Fig. B2.2 Lockhart RD, Hamilton GF, Fyfe FW: Anatomy of the Human Body. Philadelphia: Lippincott, 1959

Fig. B2.3 ACD Based on Tank W, Gest TR: LWW Atlas of Anatomy. Baltimore: Lippincott Williams & Wilkins, 2008. Plate 5.11A

Fig. B2.7 Based on Stedman's Medical Dictionary. 27th ed.; photograph—courtesy of Mission Hospital, Mission Viejo, CA.

Fig. B2.8 Fundamentals of Diagnostic Radiology. 2nd ed

Fig. B2.9 Based on Stedman's Medical Dictionary. 27th ed. (artist: Neil O. Hardy, Westport, CT).

Fig. B2.10 Based on Stedman's Medical Dictionary. 28th ed. Baltimore: Lippincott Williams & Wilkins, 2006 (artist: Mikki Senkarik, San Antonio, TX).

Fig. B2.11 Based on Stedman's Medical Dictionary. 27th ed. (artist: Neil O. Hardy, Westport, CT).

Fig. B2.12 Based on Bates, 10th ed., p. 429

Fig. B2.15B Based on Stedman's Medical Dictionary. 27th ed. (artist: Neil O. Hardy)

Fig. B2.16 Based on Stedman's Medical Dictionary. 27th ed. (artist: Neil O. Hardy, Westport, CT).

Fig. B2.17 Photograph of colonoscope—Olympus America, Inc; photograph of diverticulosis—Schiller, KFR et al. Colour Atlas of Endoscopy. Chapman and Hall, London, 1986, Springer Science and Business Media; drawings—Stedman's Medical Dictionary. 27th ed. (diverticulosis—artist: Neil O. Hardy, Westport, CT; colonoscopy—artist: Mikki Senkarik, San Antonio, TX).

Fig. B2.18 Based on Cohen BS. Medical Terminology. 4th ed. Baltimore: Lippincott Williams & Wilkins, 2003. Fig. 12.8.

Fig. B2.19A Bates 10th ed., p. 444.

Fig. B2.22 Based on Bates' Guide to Physical Examination and History Taking. 8th ed. Baltimore: Lippincott Williams and Wilkins, 2003.

Fig. B2.27 Rubin et al., Rubin's Pathology: Clinicopathologic Foundations of Medicine. 4th ed. Baltimore: Lippincott Williams & Wilkins: 2004.

Fig. B2.29 Based on Stedman's Medical Dictionary. 27th ed. (artist: Neil O. Hardy, Westport, CT).

Fig. B2.30 (inset) Based on Stedman's Medical Dictionary. 28th ed.

Fig. B2.33 Stedman's Dictionary for Health Professionals and Nursing. 5th ed. Baltimore: Lippincott Williams & Wilkins, 2005. P. 987.

Fig. B2.34A Reprinted with permission from Moore KL, Persaud TVN: Before We Are Born. 7th ed., Saunders (Elsevier), Philadelphia, 2008. Fig. 9.10; courtesy of Dr. Nathan E. Wiseman, Professor of Surgery, Children's Hospital, University of Manitoba, Winnipeg, Manitoba, Canada.

Fig. B2.34B Reprinted with permission from Moore KL, Persaud TVN: The Developing Human. 8th ed., Saunders (Elsevier), Philadelphia 2008. Fig. 8.12C; courtesy of Dr. Prem S. Sahni, formerly of Department of Radiology, Children's Hospital, Winnipeg, Manitoba, Canada.

Fig. B2.35 Reprinted with permission from Medscape Gastroenterology 6 (1), 2004. http://www.medscape.com/viewarticle/474658 ©2004, Medscape.

Fig. 2.36 Based on Hardin, DMJr: Acute Appendicitis: Review and Update. American Family Physician. 60(7):2027-34 (1999) Fig. 1B©Floyd E. Hosmer

Fig. B 2.37B Based on Eckert, P et al.: Fibrinklebung, Indikation und Anwendung. München: Urban & Schwarzenberg, 1986.

Fig. TB 2.1 Based on LWW Atlas of Anatomy Plates 5.10B, 5.11B, and 5.11C.

CHAPTER 3

Fig. 3.14B Based on DeLancey JO. Structure support of the urethra as it relates to stress urinary incontinence: The hammock hypothesis. Am J Obstet Gynecol 1994;170:1713–1720

Fig. 3.38A Left—Based on Dauber W: Pocket Atlas of Human Anatomy. Rev. 5th ed. New York: Thieme: 2007. P. 195.

Fig. 3.51B Based on Clemente, CD: Anatomy: A Regional Atlas of the Human Body. 5th ed. Baltimore: Lippincott Williams & Wilkins, 2006. Fig. 272.1.

Fig. 3.61E Based on Das Lexicon der Gesundheit. Munich: Urban & Schwarzenberg Verlag, 1996 (artist: Jonathan Dimes), p. 3.

Fig. 3.72D Lee JKT, Sagel SS, Stanley RJ, et al.: Computed Body Tomography with MRI Correlation. 3rd ed. Baltimore: Lippincott Williams & Wilkins, 1998.

Fig. B3.2B Based on Anatomy as the Basis of Clinical Medicine. 3rd ed.

Fig. B3.7A & B Reprinted with permission from LearningRadiology. com.

Fig. B3.9 Based on Stedman's Medical Dictionary. 27th ed.

Fig. B3.10 Based on Hartwig W: Fundamental Anatomy. Baltimore: Lippincott Williams & Wilkins; 2007. P. 176.

Fig. B3.11 Based on Stedman's Medical Dictionary. 27th ed.

Fig. B3.14A & B Based on Beckmann CR. Obstetrics and Gynecology. 4th ed. Baltimore: Lippincott Williams and Wilkins, 2002.

Fig. B3.16 Reprinted with permission from Stuart GCE, Reid DF. Diagnostic studies. In: Copeland LJ, ed. Textbook of Gynecology. Philadelphia: Saunders, 1993.

Fig. B3.17A–D Based on Stedman's Medical Dictionary. 27th ed.

Fig. B3.18A & B Based on Fuller J, Schaller-Ayers J: A Nursing Approach. 2nd ed. Philadelphia: Lippincott, 1994. Fig. B3.11 (artist: Larry Ward, Salt Lake City, UT).

Fig. B3.20A & C–E Based on Stedman's Medical Dictionary. 27th ed.

Fig. B3.22 Based on Obstetrics and Gynecology. 4th ed.

Fig. B3.23 Based on A Nursing Approach. 2nd ed.

Fig. B3.24 Based on Stedman's Medical Dictionary. 27th ed.

Fig. B3.26A Based on Stedman's Medical Dictionary. 28th ed.

Fig. B3.26B With permission from Bristow RE, Johns Hopkins School of Medicine, Baltimore, MD.

Fig. 3.27A-C Based on LWW Atlas Plate 6.19A, p. 276

Fig. B3.28 Based on Obstetrics and Gynecology. 4th ed.

Fig. B3.29A and B Based on Stedman's Medical Dictionary. 27th ed.

Fig. B3.32 Based on Stedman's Medical Dictionary. 27th ed.

Fig. B3.33A Based on Stedman's Medical Dictionary. 27th ed. (artist: Neil O. Hardy, Westport, CT) and Clinically Oriented Anatomy, 7th ed., fig. B3.30B

Fig. B3.33B Edwards L, ed: Atlas of Genital Dermatology. Baltimore: Lippincott Williams & Wilkins, 2004.

CHAPTER 4

Fig. 4.1C Based on Olson TR: Student Atlas of Anatomy. Baltimore: Williams & Wilkins, 1996.

Fig. 4.4 Based on Pocket Atlas of Human Anatomy. Rev. 5th ed. Fig. B, p. 49.

Fig. 4.7D Becker RF, Wilson JW, Gehweiler JA: Anatomical Basic of Medical Practice. Baltimore: Williams & Wilkins, 1974.

Fig. 4.30 Based on Student Atlas of Anatomy.

Fig. 4.31 Based on Student Atlas of Anatomy.

Fig. B4.3A–E Based on Clark CR: The Cervical Spine. 3rd ed. Philadelphia: Lippincott Williams & Wilkins, 1998.

Fig. B4.3F & G Computed Body Tomography with MRI Correlation. 3rd ed.

Fig. B4.10 Van de Graff. Human Anatomy. 4th ed. Dubuque: WC Brown. P. 163.

Fig. B4.11 Median MRI ©LUHS2008. Loyola University Health System, Maywood, IL. transverse MRI—Choi S-J et al. The use of MRI to predict the clinical outcome of non-surgical treatment for lumbar I-V disc herniation. Korean J Radiol 2007;8:156–163:5a.
Fig. B4.13B GE Healthcare, www.medcyclo.com.
Fig. B4.13C Cross-sectional Human Anatomy.
Fig. B4.13D LearningRadiology.com.
Fig. B4.13E LearningRadiology.com.
Fig. B4.15C Based on The Cervical Spine. 3rd ed.
Fig. B4.16B Based on eMedicine.com, 2008/ http://www.emedicine.com/sports/TOPIC71.HTM.
Fig. B4.16C Based on Drake R et al.: Gray's Atlas of Anatomy. New York: Churchill Livingstone, 2004. P. 30.
Fig. B4.17F Science Photo Library/Custom Medical Stock Photo, Inc.
Fig. B4.17G Princess Margaret Rose Orthopaedic Hospital/Science Photo Library/Photo Researchers, Inc.; right—Anatomical Basic of Medical Practice.

CHAPTER 5

Fig. 5.5A Atlas of Radiologic Anatomy. 6th English ed.
Fig. 5.21E Based on Rose J, Gamble JG. Human Walking. 2nd ed. Baltimore: Williams & Wilkins, 1994.
Fig. 5.21G Based on Basic Clinical Massage Therapy: Integrating Anatomy and Treatment. 2nd ed.
Fig. 5.22C Based on Melloni, JL: Melloni's Illustrated Review of Human Anatomy: By Structures—Arteries, Bones, Muscles, Nerves, Veins. Lippincott Williams & Wilkins, 1988.
Fig. 5.34A & B Based on Basic Clinical Massage Therapy: Integrating Anatomy and Treatment. 2nd ed.
Fig. 5.40A Based on Basic Clinical Massage Therapy: Integrating Anatomy and Treatment. 2nd ed.
Fig. 5.40F Based on Basic Clinical Massage Therapy: Integrating Anatomy and Treatment. 2nd ed. Fig. 9.12, p. 342.
Fig. 5.40H Based on Basic Clinical Massage Therapy, 2nd. Ed. Fig. 9.14, p. 344.
Fig. 5.42(left) Based on Basic Clinical Massage Therapy: Integrating Anatomy and Treatment. 2nd ed.
Fig. 5.42(right) Based on Melloni's Illustrated Review of Human Anatomy: By Structures—Arteries, Bones, Muscles, Nerves, Veins. P. 173.
Fig. 5.55C–F Based on Basic Clinical Massage Therapy: Integrating Anatomy and Treatment. 2nd ed.
Fig. 5.60F–K Based on Basic Clinical Massage Therapy: Integrating Anatomy and Treatment. 2nd ed.
Fig. 5.68A & H Based on Basic Clinical Massage Therapy: Integrating Anatomy and Treatment. 2nd ed.
Fig. 5.68C Based on Basic Clinical Massage Therapy and Grant's Atlas of Anatomy, 13th ed., Fig. 78, p. 453
Fig. 5.69A,C Based on Grant's Atlas of Anatomy. 13th ed.
Fig. 5.69B Based on Grant's Atlas of Anatomy, 13th ed., Fig. 6.78C
Fig. 5.69D Based in part on Grant's Atlas of Anatomy, 13th ed., Fig. 5.80C, p. 455
Fig. 5.73 Based on Basmajian JV, Slonecker CE: Grant's Method of Anatomy: A Clinical Problem-Solving Approach. 11 ed. Baltimore: Williams & Wilkins, 1989.
Fig. 5.76A Based on Basic Clinical Massage Therapy: Integrating Anatomy and Treatment. 2nd ed.
Fig. 5.79A & B Based on Kapandji, IA. The Physiology of the Joints. Vol. 2: Lower Limb. 5th ed. Edinburgh, UK, Churchill Livingstone, 1987.
Fig. 5.79C Based on Basic Clinical Massage Therapy: Integrating Anatomy and Treatment. 2nd ed.
Fig. 5.80B Atlas of Radiologic Anatomy. 6th English ed.
Fig. 5.85B Atlas of Radiologic Anatomy. 6th English ed.
Fig. 5.92B Based on Student Atlas of Anatomy.
Fig. 5.95A Atlas of Radiologic Anatomy. 6th English ed.

Fig. B5.1A Yochum TR, Rowe LJ. Essentials of Skeletal Radiology, Vol. 1, 2nd ed., Baltimore: Lippincott Williams & Wilkins, 1996. Fig. 9.85, p. 707.
Fig. B5.1B Brunner, LC, Kuo TY: Hip fractures in adults. Am Fam Phys 2003;67(3):Fig. 2.
Fig. B5.1D Rossi F, Dragoni S. Acute avulsion fractures of the pelvis in adolescent competitive athletes. Skel Radiol 2001;30(3):Fig. 7.
Fig. B5.3D Yochum TR, Rowe LJ. Essentials of Skeletal Radiology, 3rd Ed. Baltimore: Lippincott Williams & Wilkins, 2005.
Fig. B5.4 Essentials of Skeletal Radiology, 3rd ed.
Fig. B5.5 ©eMedicine.com, 2008.
Fig. B5.8D Hatch RL et al.: Diagnosis and management of metatarsal fractures. Am Fam Phys 2007;76(6):217.
Fig. B5.8E Essentials of Skeletal Radiology, Vol. 1, 2nd edition, Fig. 9.104A, p. 737.
Fig. B5.9 Davies M. The os trigonum syndrome. Foot 2004;14(3):Fig. 2.
Fig. B5.10 Doda P, Peh W: Woman with possible right toe fracture. Asia Pacific J Fam Med 2006;5(3):50.
Fig. B5.11A Reprinted with permission from Roche Lexikon Medizin. 4th ed.
Fig. B5.11B–D Stedman's Medical Dictionary. 28th ed. (artist: Neil O. Hardy, Westport, CT), p. 2090.
Fig. B5.12 LearningRadiology.com.
Fig. B5.13B Kavanagh EC et al.: MRI findings in bipartite patella. Skel Radiol 2007;36(3):Fig. 1a.
Fig. B5.14 Bates 10e, p. 699 upper fig.
Fig. B5.22 Bates 10e, p. 700, upper fig.
Fig. B5.25 Bates 10e, p. 485, posterior tibial pulse
Fig. B5.26(top) www.xray200.co.uk
Fig. B5.27 Bates 10e, p. 485, dorsalis pedis pulse.
Fig. B5.28 Essentials of Skeletal Radiology. 2nd ed.
Fig. B5.30 Drawings—Willis MC: Medical Terminology: A Programmed Learning Approach to the Language of Health Care. Baltimore: Lippincott Williams & Wilkins, 2002; radiograph—Clinical Radiology—The Essentials.
Fig. B5.32A–C Modified from Palastanga NP, Field DG, Soames R: Anatomy and Human Movement. 4th ed. Oxford, UK: Butterworth-Heinemann, 2002.
Fig. B5.32D Clinical Radiology—The Essentials.
Fig. B5.34 Based on Roche Lexikon Medizin. 4th ed.
Fig. B5.35C Stedman's Medical Dictionary. 28th ed, p. 1184.

CHAPTER 6

Fig. 6.17 LWW Atlas of Anatomy. Baltimore: Pl. 2.53, p. 82.
Fig. 6.32 Basic Clinical Massage Therapy: Integrating Anatomy and Treatment, 2nd ed. Fig. 4.28, p. 147.
Fig. 6.49B–D, F, & G Basic Clinical Massage Therapy: Integrating Anatomy and Treatment. 2nd ed. Figs. 5.1, 5.12, 5.3, 5.6, and 5.10, pgs. 193, 201, 195, 197, and 199.
Fig. 6.53 Based on Hoppenfeld, S, de Boer P. Surgical Exposures in Orthopaedics, 3rd ed. Baltimore: Lippincott Williams & Wilkins, 2003. Fig. 2.27, p. 89.
Fig. 6.60B & C Basic Clinical Massage Therapy, 2nd ed. Fig. 5.5, p. 186.
Fig. 6.92 Modified from Biomechanical Basis of Human Motion. Fig. 5.8, p. 153.
Fig. 6.93 Platzer W. Color Atlas of Human Anatomy. Vol. 1: Locomotor System. 4th ed. New York: Thieme, 1992, p. 147 and 149.
Fig. 6.102 Based on LWW Atlas of Anatomy Plate 2.43
Fig. 6.103 B and C Based on from Anatomy as the Basis of Clinical Medicine. 3rd ed.
Fig. 6.109B Grant's Method of Anatomy: A Clinical Problem-Solving Approach. 11th ed.
Fig. B6.5 Rowland LP: Merritt's Textbook of Neurology. 9th ed. Baltimore: Williams & Wilkins, 1995.
Fig. B6.9 Left—Meschan I. An Atlas of Anatomy Basic to Radiology. Philadelphia: Saunders, 1975; right—Salter RB. Textbook of Disorders

and Injuries of the Musculoskeletal System. 3rd ed. Baltimore: Williams & Wilkins, 1998.

Fig. B6.13 Bates, 10th ed, p. 697 bottom left.

Fig. B6.14 Based on Anderson MK, Hall SJ, Martin M: Foundations of Athletic Training. 3rd ed. Baltimore: Lippincott Williams & Wilkins, 1995.

Fig. B6.31 www.xray200.co.uk.

Fig. B6.37A John Sleezer/MCT/Landov.

Fig. B6.37B Basic Clinical Massage Therapy: Integrating Anatomy and Treatment, second edition. Fig. 5-35, p. 223.

Fig. B6.38 Textbook of Disorders and Injuries of the Musculoskeletal System. 3rd ed.

CHAPTER 7

Fig. 7.16 LWW Atlas of Anatomy. Plate 7.29, p. 324

Fig. 7.25 Based on LWW Atlas of Anatomy Plate 7.73, p. 368

Fig. 7.26 Based on LWW Atlas of Anatomy Plate 7.74, p. 369

Fig. 7.28A Based on LWW Atlas of Anatomy Plate 7.50B, p. 345.

Fig. 7.31C Based on LWW Atlas of Anatomy Plate 760B, p. 365.

Fig. 7.44A Based on Anatomy as the Basis of Clinical Medicine. 3rd ed.

Fig. 7.46A Based on LWW Atlas of Anatomy. Plate 7.58B, p. 353.

Fig. 7.51A Based on Melloni's Illustrated Review of Human Anatomy: By Structures—Arteries, Bones, Muscles, Nerves, Veins, p. 149.

Fig. 7.51B Based on Human Anatomy. 4th ed. Fig. 15.18, p. 419.

Fig. 7.52 Welch Allyn, Inc., Skaneateles Falls, NY.

Fig. 7.53 Based on Human Anatomy. 4th ed. Fig. 15.17.

Fig. 7.54B Melloni's Illustrated Review of Human Anatomy: By Structures—Arteries, Bones, Muscles, Nerves, Veins. P. 141.

Fig. 7.54C Based on Melloni's Illustrated Review of Human Anatomy: By Structures—Arteries, Bones, Muscles, Nerves, Veins. P. 143.

Fig. 7.56A–D Based on Girard, Louis: Anatomy of the Human Eye. II. The Extra-ocular Muscles. Teaching Films, Inc. Houston, TX.

Fig. 7.57 Based on Melloni's Illustrated Review of Human Anatomy: By Structures—Arteries, Bones, Muscles, Nerves, Veins. P. 189.

Fig. 7.65 Based in part on Clemente C. Atlas of Anatomy 6th ed., Baltimore: Lippincott Williams and Wilkins, 2011. figs. 529-531.

Fig. 7.71 Based on Paff, GH: Anatomy of the Head & Neck. Philadelphia: WB Saunders Co., 1973. Fig. 122.3, p. 62-63.

Fig. 7.72 Based on Basic Clinical Massage Therapy: Integrating Anatomy and Treatment, 2nd ed. Figs. 3.15, 3.16, and 3.19, p. 82, 84, and 86.

Fig. 7.88 Based on LWW Atlas of Anatomy Plate 7.39A

Fig. 7.90A Based on LWW Atlas of Anatomy Plate 7-40A, p. 335

Fig. 7.90C Based on LWW Atlas of Anatomy Plate 7-38C., p. 333

Fig. 7.98 Based on Anatomy of the Head & Neck. Figs. 238–240, p. 142–143 and Grant's Atlas of Anatomy, 13th ed., Fig. 7.78B, p. 705.

Fig. 7.100B & C Based on Hall-Craggs ECB: Anatomy as the Basis of Clinical Medicine. 2nd ed. Baltimore: Williams & Wilkins, 1990. Fig. 9.100, p. 536.

Fig. 7.112 Based on LWW Atlas of Anatomy. Pl. 7.66B& C.

Fig. 7.120 Seeley RR, Stephens TR, and Tate P: Anatomy & Physiology. 6th ed. New York: McGraw-Hill 2003. Fig. 15.28, p. 532.

Fig. B7.12 Ger R, Abrahams P, Olson T: Essentials of Clinical Anatomy. 3rd ed. New York: Parthenon, 1996. Fig. B7.12.

Fig. B7.14 ©LUHS2008. Loyola University Health System, Maywood, IL.

Fig. B7.15 Skin Cancer Foundation.

Fig. B7.20A Visuals Unlimited.

Fig. B7.20B Courtesy of Dr. Gerald S. Smyser, Altru Health System, Grand Forks, ND.

Fig. B7.23 Stedman's Medical Dictionary. 28th ed. (artist: Neil O. Hardy, Westport, CT).

Fig. B7.24 Mann IC: The Development of the Human Eye. New York: Grune & Stratton. 1974.

Fig. B7.25 Welch Allyn, Inc., Skaneateles Falls, NY.

Fig. B7.26 Medical Terminology. 4th ed.

Fig. B7.27 Digital Reference of Ophthalmology, Edward S. Harkness Eye Institute, Department of Ophthalmology of Columbia University.

Fig. B7.28 Stedman's Medical Dictionary. 28th ed. (artist: Neil O. Hardy, Westport, CT).

Fig. B7.29 Mehrle G, Augenheikunde fur Krankenpfegeberufe S aufl. Munchen: Urban & Fischer, 1991.

Fig. B7.32 The Developing Human: Clinically Oriented Embryology. 7th ed.

Fig. B7.33A – D Stedman's Medical Dictionary, 28th ed. (artist: Neil O. Hardy, Westport, CT).

Fig. B7.39 Courtesy of Eugene Kowaluk Photography.

Fig. B7.41 Turner, JS: An overview of head and neck. In Walker HK, Hall WD, Hurst JW, eds: Clinical Methods—The History, Physical and Laboratory Examinations. Butterworths, 1990. Figs. 119.1 and 119.2.

Fig. B7.42 Anatomy as the Basis of Clinical Medicine. 3rd ed.

Fig. B7.43 Bechara Y. Ghorayeb MD, Houston, TX.

Fig. B7.44 Welch Allyn, Inc., Skaneateles Falls, NY.

Fig. B7.45 Stedman's Medical Dictionary. 28th ed. (artist: Neil O. Hardy, Westport, CT).

CHAPTER 8

Fig. 8.1 Based on LWW Atlas of Anatomy Plate 7.13, p. 308 and Tank, PW, Grant's Dissector, 15th ed., Baltimore: Lippincott Williams and Wilkins, 2012, Fig 7.6, p. 209 and 7.10, p. 214

Fig. 8.4A Based on LWW Atlas of Anatomy Plate 7.10A&B, p. 305

Fig. 8.7A Based on Basic Clinical Massage Therapy: Integrating Anatomy and Treatment. 2nd ed., Fig. 3.28

Fig. 8.7B&F Based on Basic Clinical Massage Therapy: Integrating Anatomy and Treatment. 2nd ed., Fig. 6.24

Fig. 8.7CDE Based on Basic Clinical Massage Therapy: Integrating Anatomy and Treatment. 2nd ed. Fig. 8.7 CDE

Fig. 8.12 Based on Grant's Dissector, 15e, Fig. 7-5, p. 208

Fig. 8.20 Based on LWW Atlas of Anatomy Plate 7.73, p 368.

Fig. 8.24A Based on Grant's Dissector. 15th ed., Fig. 7.11

Fig. 8.32C Based on Pocket Atlas of Human Anatomy. 5th ed. P. 169, Fig. C, p. 169.

Fig. 8.43 Based on LWW Atlas of Anatomy Plate 7.10

Fig. 8.44A Abrahams P: The Atlas of the Human Body. San Diego, CA: Thunder Bay Press, 2002. P. 66.

Fig. 8.46B Based on LWW Atlas of Anatomy Plate 7-21, p. 316

Fig. B8.1 Based on Merritt's Textbook of Neurology. 9th ed.

Fig. B8.3 Based on Siemens Medical Solutions USA, Inc.

Fig. B8.6 Based on Sadler TW. Langman's Medical Embryology. 7th ed. Baltimore: Williams & Wilkins, 1995.

Fig. B8.7 Leung AKC, Wong AI, Robson WLLM: Ectopic thyroid gland simulating a thyroglossal duct cyst. Can J Surg 1995;38:87. ©1995 Canadian Medical Association.

Fig. B8.9 Klima: Schilddrüsen-Sonographie. München: Urban & Schwarzenberg Verlag, 1989.

Fig. B8.11 Based on Rohen JW et al.: Color Atlas of Anatomy: A Photographic Study of the Human Body. 5th ed. Baltimore: Lippincott Williams & Wilkins, 2002.

Fig. B8.12 Based on Stedman's Medical Dictionary. 27th ed.

CHAPTER 9

No credits

Note: *Credits for figures based on* Grant's Atlas of Anatomy *and* Essential Clinical Anatomy *illustrations are available at* thePoint.lww.com.

Introduction to Clinically Oriented Anatomy

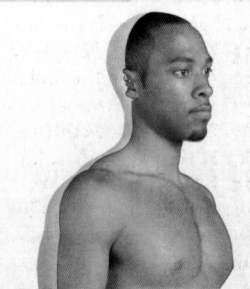

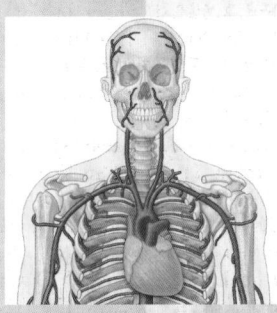

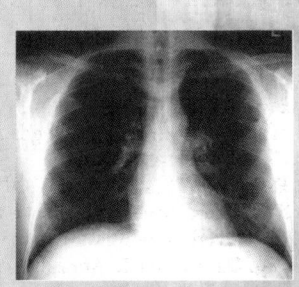

APPROACHES TO STUDYING ANATOMY

Anatomy is the setting (structure) in which the events (functions) of life occur. This book deals mainly with functional *human gross anatomy*—the examination of structures of the human that can be seen without a microscope. The three main approaches to studying anatomy are regional, systemic, and clinical (or applied), reflecting the body's organization and the priorities and purposes for studying it.

Regional Anatomy

Regional anatomy (topographical anatomy) considers the organization of the human body as major parts or segments (Fig. I.1): a main body, consisting of the head, neck, and trunk (subdivided into thorax, abdomen, back, and pelvis/perineum), and paired upper limbs and lower limbs. All the major parts may be further subdivided into areas and regions. Regional anatomy is the method of studying the body's structure by focusing attention on a specific part (e.g., the head), area (the face), or region (the orbital or eye region); examining the arrangement and relationships of the various systemic structures (muscles, nerves, arteries, etc.) within it; and then usually continuing to study adjacent regions in an ordered sequence. Outside of this Introduction, the regional approach is followed in this book, with each chapter addressing the anatomy of a major part of the body. This is the approach usually followed in anatomy courses that have a laboratory component involving dissection. When studying anatomy by this approach, it is important to routinely put the regional anatomy into the context of that of adjacent regions, parts, and of the body as a whole.

Regional anatomy also recognizes the body's organization by layers: skin, subcutaneous tissue, and deep fascia covering the deeper structures of muscles, skeleton, and cavities, which contain *viscera* (internal organs). Many of these deeper structures are partially evident beneath the body's outer covering and may be studied and examined in living individuals via surface anatomy.

Surface anatomy is an essential part of the study of regional anatomy. It is specifically addressed in this book in "surface anatomy sections" (orange background) that provide knowledge of what lies under the skin and what structures are perceptible to touch (palpable) in the living body at rest and in action. We can learn much by observing the external form and surface of the body and by observing or feeling the superficial aspects of structures beneath its surface. The aim of this method is to *visualize* (recall distinct mental images of) structures that confer contour to the surface or are palpable beneath it and, in clinical practice, to distinguish any unusual or abnormal findings. In short, surface anatomy requires a thorough understanding of the anatomy of the structures beneath the surface. In people with stab wounds, for example, a physician must be able to visualize the deep structures that may be injured. Knowledge of surface anatomy can also decrease the need to memorize facts because the body is always available to observe and palpate.

Physical examination is the clinical application of surface anatomy. **Palpation** is a clinical technique, used with **observation** and **listening** for examining the body. *Palpation of arterial pulses*, for instance, is part of a physical examination. Students of many of the health sciences will learn to use instruments to facilitate examination of the body (such as an *ophthalmoscope* for observation of features of the eyeballs) and to listen to functioning parts of the body (a *stethoscope* to auscultate the heart and lungs).

Regional study of deep structures and abnormalities in a living person is now also possible by means of radiographic and sectional imaging and endoscopy. *Radiographic and sectional imaging (radiographic anatomy)* provides useful information about normal structures in living individuals, demonstrating the effect of muscle tone, body fluids and pressures, and gravity that cadaveric study does not. *Diagnostic radiology* reveals the effects of trauma, pathology, and aging on normal structures. In this book, most radiographic and many sectional

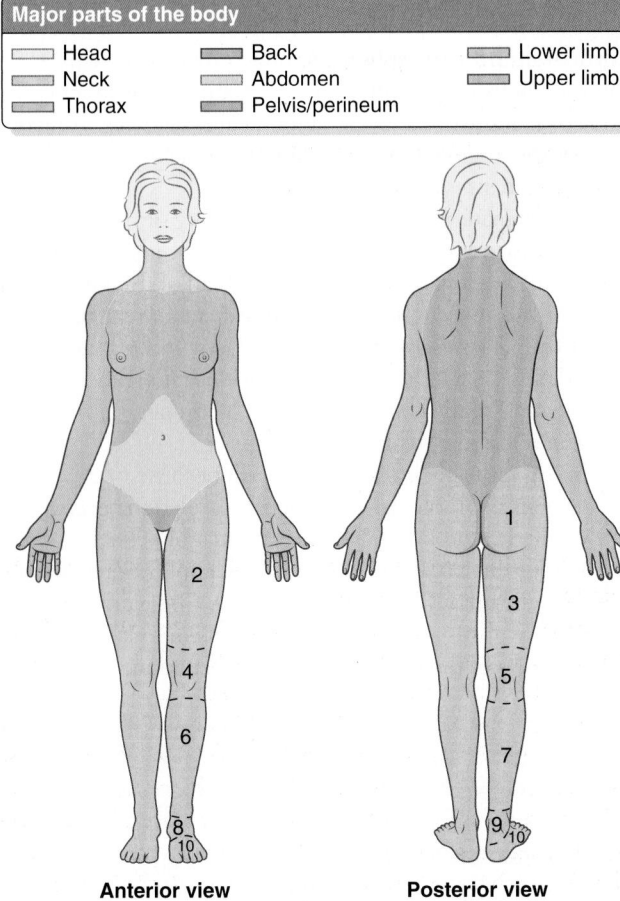

Major parts of the body

Head	Back	Lower limb
Neck	Abdomen	Upper limb
Thorax	Pelvis/perineum	

Anterior view   **Posterior view**

Regions of lower limb

1 = Gluteal region	6 = Anterior leg region
2 = Anterior thigh region	7 = Posterior leg region
3 = Posterior thigh region	8 = Anterior talocrural (ankle) region
4 = Anterior knee region	9 = Posterior talocrural region
5 = Posterior knee region	10 = Foot region

FIGURE I.1. **Major parts of the body and regions of the lower limb.** Anatomy is described relative to the anatomical position illustrated here.

images are integrated into the chapters where appropriate. The medical imaging sections at the end of each chapter provide an introduction to the techniques of radiographic and sectional imaging and include series of sectional images that apply to the chapter. *Endoscopic techniques* (using a insertable flexible fiber optic device to examine internal structures, such as the interior of the stomach) also demonstrate living anatomy. The detailed and thorough learning of the three-dimensional anatomy of deep structures and their relationships is best accomplished initially by dissection. In clinical practice, surface anatomy, radiographic and sectional images, endoscopy, and your experience from studying anatomy will combine to provide you with knowledge of your patient's anatomy.

The computer is a useful adjunct in teaching regional anatomy because it facilitates learning by allowing interactivity and manipulation of two- and three-dimensional graphic

models. **Prosections,** carefully prepared dissections for the demonstration of anatomical structures, are also useful. However, learning is most efficient and retention is highest when didactic study is combined with the experience of first-hand **dissection**—that is, learning by doing. During dissection you observe, palpate, move, and sequentially reveal parts of the body. In 1770, Dr. *William Hunter,* a distinguished Scottish anatomist and obstetrician, stated: "Dissection alone teaches us where we may cut or inspect the living body with freedom and dispatch."

Systemic Anatomy

Systemic anatomy is the study of the body's organ systems that work together to carry out complex functions. The basic systems and the field of study or treatment of each (in parentheses) are:

- The **integumentary system** (*dermatology*) consists of the skin (L. *integumentum,* a covering) and its appendages—hairs, nails, and sweat glands, for example—and the subcutaneous tissue just beneath it. The skin, an extensive sensory organ, forms the body's outer, protective covering and container.
- The **skeletal system** (*osteology*) consists of bones and cartilage; it provides our basic shape and support for the body and is what the muscular system acts on to produce movement. It also protects vital organs such as the heart, lungs, and pelvic organs.
- The **articular system** (*arthrology*) consists of joints and their associated ligaments, connecting the bony parts of the skeletal system and providing the sites at which movements occur.
- The **muscular system** (*myology*) consists of skeletal muscles that act (contract) to move or position parts of the body (e.g., the bones that articulate at joints), or smooth and cardiac muscle that propels, expels, or controls the flow of fluids and contained substance.
- The **nervous system** (*neurology*) consists of the *central nervous system* (brain and spinal cord) and the *peripheral nervous system* (nerves and ganglia, together with their motor and sensory endings). The nervous system controls and coordinates the functions of the organ systems, enabling the body's responses to and activities within its environment. The sense organs, including the olfactory organ (sense of smell), eye or visual system (*ophthalmology*), ear (sense of hearing and balance—*otology*), and gustatory organ (sense of taste), are often considered with the nervous system in systemic anatomy.
- The **circulatory system** (*angiology*) consists of the cardiovascular and lymphatic systems, which function in parallel to transport the body's fluids.
  - The **cardiovascular system** (*cardiology*) consists of the heart and blood vessels that propel and conduct blood through the body, delivering oxygen, nutrients, and hormones to cells and removing their waste products.

- The **lymphatic system** is a network of lymphatic vessels that withdraws excess tissue fluid (lymph) from the body's interstitial (intercellular) fluid compartment, filters it through lymph nodes, and returns it to the bloodstream.
- The **alimentary** or **digestive system** (*gastroenterology*) consists of the digestive tract from the mouth to the anus, with all its associated organs and glands that function in ingestion, mastication (chewing), deglutition (swallowing), digestion, and absorption of food and the elimination of the solid waste (feces) remaining after the nutrients have been absorbed.
- The **respiratory system** (*pulmonology*) consists of the air passages and lungs that supply oxygen to the blood for cellular respiration and eliminate carbon dioxide from it. The diaphragm and larynx control the flow of air through the system, which may also produce tone in the larynx that is further modified by the tongue, teeth, and lips into speech.
- The **urinary system** (*urology*) consists of the kidneys, ureters, urinary bladder, and urethra, which filter blood and subsequently produce, transport, store, and intermittently excrete urine (liquid waste).
- The **genital (reproductive) system** (*gynecology* for females; *andrology* for males) consists of the gonads (ovaries and testes) that produce oocytes (eggs) and sperms, the ducts that transport them, and the genitalia that enable their union. After conception, the female reproductive tract nourishes and delivers the fetus.
- The **endocrine system** (*endocrinology*) consists of specialized structures that secrete hormones, including discrete ductless endocrine glands (such as the thyroid gland), isolated and clustered cells of the gut and blood vessel walls, and specialized nerve endings. **Hormones** are organic molecules that are carried by the circulatory system to distant effector cells in all parts of the body. The influence of the endocrine system is thus as broadly distributed as that of the nervous system. Hormones influence metabolism and other processes, such as the menstrual cycle, pregnancy, and parturition (childbirth).

None of the systems functions in isolation. The passive skeletal and articular systems and the active muscular system collectively constitute a *supersystem,* the **locomotor system** or **apparatus** (*orthopedics*), because they must work together to produce locomotion of the body. Although the structures directly responsible for locomotion are the muscles, bones, joints, and ligaments of the limbs, other systems are indirectly involved as well. The brain and nerves of the nervous system stimulate them to act; the arteries and veins of the circulatory system supply oxygen and nutrients to and remove waste from these structures; and the sensory organs (especially vision and equilibrium) play important roles in directing their activities in a gravitational environment.

In this Introduction, an overview of several systems significant to all parts and regions of the body will be provided before Chapters 1 through 8 cover regional anatomy in detail. Chapter 9 also presents systemic anatomy in reviewing the cranial nerves.

Clinical Anatomy

Clinical anatomy (applied anatomy) emphasizes aspects of bodily structure and function important in the practice of medicine, dentistry, and the allied health sciences. It incorporates the regional and systemic approaches to studying anatomy and stresses clinical application.

Clinical anatomy often involves inverting or reversing the thought process typically followed when studying regional or systemic anatomy. For example, instead of thinking, "The action of this muscle is to . . . ," clinical anatomy asks, "How would the absence of this muscle's activity be manifest?" Instead of noting, "The . . . nerve provides innervation to this area of skin," clinical anatomy asks, "Numbness in this area indicates a lesion of which nerve?"

Clinical anatomy is exciting to learn because of its role in solving clinical problems. The clinical correlation boxes (popularly called "blue boxes," appearing on a blue background) throughout this book describe practical applications of anatomy. "Case studies," such as those on the Clinically Oriented Anatomy website (http://thePoint.lww.com/COA7e), are integral parts of the clinical approach to studying anatomy.

The Bottom Line

STUDYING ANATOMY

Anatomy is the study of the structure of the human body. ◆ Regional anatomy considers the body as organized into segments or parts. ◆ Systemic anatomy sees the body as organized into organ systems. ◆ Surface anatomy provides information about structures that may be observed or palpated beneath the skin. ◆ Radiographic, sectional, and endoscopic anatomy allows appreciation of structures in living people, as they are affected by muscle tone, body fluids and pressures, and gravity. ◆ Clinical anatomy emphasizes application of anatomical knowledge to the practice of medicine.

ANATOMICOMEDICAL TERMINOLOGY

Anatomical terminology introduces and makes up a large part of medical terminology. To be understood, you must express yourself clearly, using the proper terms in the correct way. Although you are familiar with common, colloquial terms for parts and regions of the body, you must learn the *international anatomical terminology* (e.g., axillary fossa instead of armpit and clavicle instead of collarbone) that enables precise communication among healthcare professionals and scientists worldwide. Health professionals must also know the common and colloquial terms people are likely to use when they describe their complaints. Furthermore, you must be able to use terms people will understand when explaining their medical problems to them.

The terminology in this book conforms to the new *International Anatomical Terminology. Terminologia Anatomica* (TA) and *Terminologia Embryologica (TE)* list terms both in Latin and as English equivalents (e.g., the common shoulder muscle is *musculus deltoideus* in Latin and *deltoid* in English). Most terms in this book are English equivalents. Official terms are available at www.unifr.ch/ifaa. Unfortunately, the terminology commonly used in the clinical arena may differ from the official terminology. Because this discrepancy may be a source of confusion, this text clarifies commonly confused terms by placing the unofficial designations in parentheses when the terms are first used—for example, *pharyngotympanic tube* (auditory tube, eustachian tube) and *internal thoracic artery* (internal mammary artery). *Eponyms,* terms incorporating the names of people, are not used in the new terminology because they give no clue about the type or location of the structures involved. Further, many eponyms are historically inaccurate in terms of identifying the original person to describe a structure or assign its function, and do not conform to an international standard. Notwithstanding, commonly used eponyms appear in parentheses throughout the book when these terms are first used—such as *sternal angle* (angle of Louis)—since you will surely encounter them in your clinical years. Note that eponymous terms do not help to locate the structure in the body. The Clinically Oriented Anatomy website (http://thePoint.lww.com/COA7e) provides a list of eponymous terms.

Structure of terms. Anatomy is a descriptive science and requires names for the many structures and processes of the body. Because most terms are derived from Latin and Greek, medical language may seem difficult at first; however, as you learn the origin of terms, the words make sense. For example, the term *gaster* is Latin for stomach or belly. Consequently, the esophagogastric junction is the site where the esophagus connects with the stomach, gastric acid is the digestive juice secreted by the stomach, and a digastric muscle is a muscle divided into two bellies.

Many terms provide information about a structure's shape, size, location, or function or about the resemblance of one structure to another. For example, some muscles have descriptive names to indicate their main characteristics. The *deltoid muscle,* which covers the point of the shoulder, is triangular, like the symbol for *delta,* the fourth letter of the Greek alphabet. The suffix *-oid* means "like"; therefore, *deltoid* means like delta. *Biceps* means two-headed and *triceps* means three-headed. Some muscles are named according to their shape—the *piriformis muscle,* for example, is pear shaped (L. *pirum,* pear + L. *forma,* shape or form). Other muscles are named according to their location. The *temporal muscle* is in the temporal region (temple) of the cranium (skull). In some cases, actions are used to describe muscles—for example, the *levator scapulae* elevates the scapula (L. shoulder blade). Anatomical terminology applies logical reasons for the names of muscles and other parts of the body,

and if you learn their meanings and think about them as you read and dissect, it will be easier to remember their names.

Abbreviations. Abbreviations of terms are used for brevity in medical histories and in this and other books, such as in tables of muscles, arteries, and nerves. Clinical abbreviations are used in discussions and descriptions of signs and symptoms. Learning to use these abbreviations also speeds note taking. Common anatomical and clinical abbreviations are provided in this text when the corresponding term is introduced—for example, temporomandibular joint (TMJ). The Clinically Oriented Anatomy website (http://thePoint.lww.com/COA7e) provides a list of commonly used anatomical abbreviations. More extensive lists of common medical abbreviations may be found in the appendices of comprehensive medical dictionaries (e.g., *Stedman's Medical Dictionary,* 28th ed.).

Anatomical Position

All anatomical descriptions are expressed in relation to one consistent position, ensuring that descriptions are not ambiguous (Figs. I.1 and I.2). One must visualize this position in the mind when describing patients (or cadavers), whether they are lying on their sides, supine (recumbent, lying on the back, face upward), or prone (lying on the abdomen, face downward). The **anatomical position** refers to the body position as if the person were standing upright with the:

- head, gaze (eyes), and toes directed anteriorly (forward),
- arms adjacent to the sides with the palms facing anteriorly, and
- lower limbs close together with the feet parallel.

This position is adopted globally for anatomicomedical descriptions. By using this position and appropriate terminology, you can relate any part of the body precisely to any other part. It should also be kept in mind, however, that gravity causes a downward shift of internal organs (viscera) when the upright position is assumed. Since people are typically examined in the supine position, it is often necessary to describe the position of the affected organs when supine, making specific note of this exception to the anatomical position.

Anatomical Planes

Anatomical descriptions are based on four imaginary planes (median, sagittal, frontal, and transverse) that intersect the body in the anatomical position (Fig. I.2):

- The **median plane** (median sagittal plane), the vertical plane passing longitudinally through the body, divides the body into right and left halves. The plane defines the midline of the head, neck, and trunk where it intersects the surface of the body. *Midline* is often erroneously used as a synonym for the median plane.
- **Sagittal planes** are vertical planes passing through the body *parallel to the median plane. Parasagittal* is commonly used but is unnecessary because any plane parallel to and on either side of the median plane is sagittal

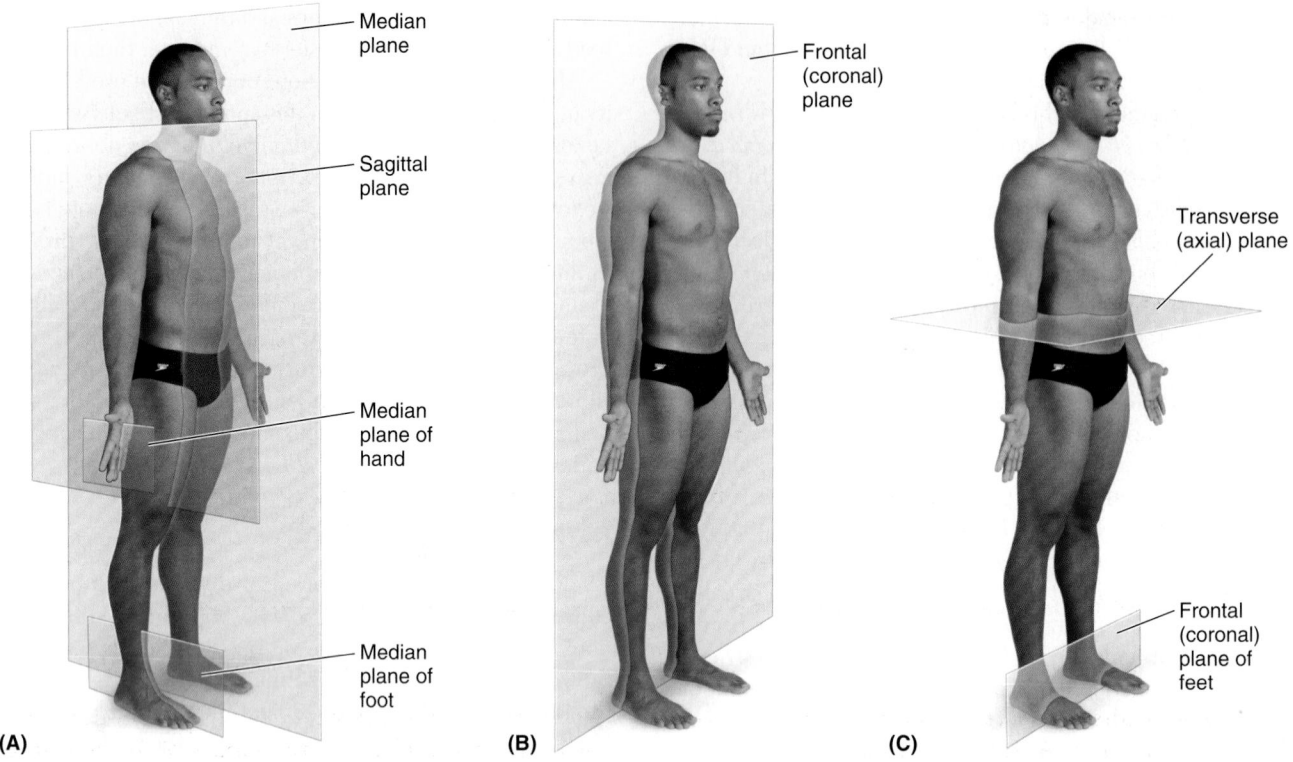

FIGURE I.2. Anatomical planes. The main planes of the body are illustrated.

by definition. However, a plane parallel and near to the median plane may be referred to as a *paramedian plane*.

- **Frontal (coronal) planes** are vertical planes passing through the body *at right angles to the median plane,* dividing the body into anterior (front) and posterior (back) parts.
- **Transverse planes** are horizontal planes passing through the body *at right angles to the median and frontal planes,* dividing the body into superior (upper) and inferior (lower) parts. Radiologists refer to transverse planes as *transaxial,* which is commonly shortened to *axial planes.*

Since the number of sagittal, frontal, and transverse planes is unlimited, a reference point (usually a visible or palpable landmark or vertebral level) is necessary to identify the location or level of the plane, such as a "transverse plane through the umbilicus" (Fig. I.2C). Sections of the head, neck, and trunk in precise frontal and transverse planes are symmetrical, passing through both the right and left members of paired structures, allowing some comparison.

The main use of anatomical planes is to describe *sections* (Fig. I.3):

- **Longitudinal sections** run lengthwise or parallel to the long axis of the body or of any of its parts, and the term applies regardless of the position of the body. Although median, sagittal, and frontal planes are the standard (most commonly used) longitudinal sections, there is a 180° range of possible longitudinal sections.
- **Transverse sections,** or cross sections, are slices of the body or its parts that are cut at right angles to the longitudinal axis of the body or of any of its parts. Because

the long axis of the foot runs horizontally, a transverse section of the foot lies in the frontal plane (Fig. I.2C).

- **Oblique sections** are slices of the body or any of its parts that are not cut along the previously listed anatomical planes. In practice, many radiographic images and anatomical sections do not lie precisely in sagittal, frontal, or transverse planes; often they are slightly oblique.

Anatomists create sections of the body and its parts anatomically, and clinicians create them by planar imaging technologies, such as computerized tomography (CT), to describe and display internal structures.

Terms of Relationship and Comparison

Various adjectives, arranged as pairs of opposites, describe the relationship of parts of the body or compare the position of two structures relative to each other (Fig. I.4). Some of these terms are specific for comparisons made in the anatomical position, or with reference to the anatomical planes:

Superior refers to a structure that is nearer the **vertex,** the topmost point of the cranium (Mediev. L., skull). **Cranial** relates to the cranium and is a useful directional term, meaning toward the head or cranium. **Inferior** refers to a structure that is situated nearer the sole of the foot. **Caudal** (L. *cauda,* tail) is a useful directional term that means toward the feet or tail region, represented in humans by the coccyx (tail bone), the small bone at the inferior (caudal) end of the vertebral column.

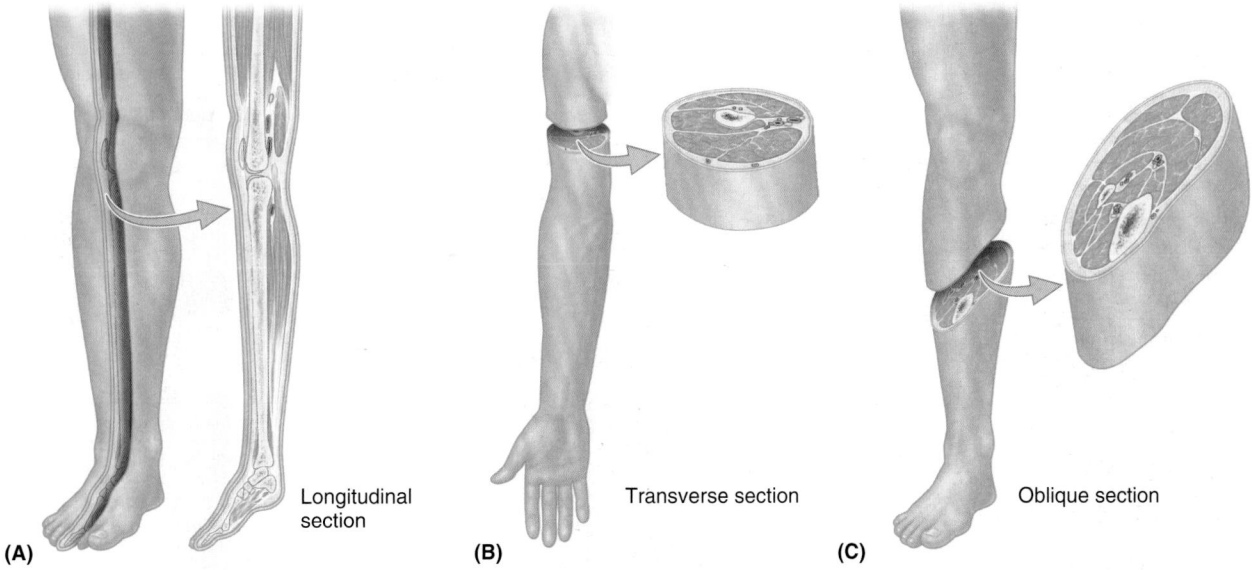

Longitudinal
section

(A)

Transverse section

(B)

Oblique section

(C)

FIGURE I.3. **Sections of the limbs.** Sections may be obtained by anatomical sectioning or medical imaging techniques.

Posterior (dorsal) denotes the back surface of the body or nearer to the back. **Anterior** (ventral) denotes the front surface of the body. **Rostral** is often used instead of anterior when describing parts of the brain; it means toward the rostrum (L. for beak); however, in humans it denotes nearer the anterior part of the head (e.g., the frontal lobe of the brain is rostral to the cerebellum).

Medial is used to indicate that a structure is nearer to the median plane of the body. For example, the 5th digit of the hand (little finger) is medial to the other digits. Conversely, **lateral** stipulates that a structure is farther away from the median plane. The 1st digit of the hand (thumb) is lateral to the other digits.

Dorsum usually refers to the superior aspect of any part that protrudes anteriorly from the body, such as the dorsum of the tongue, nose, penis, or foot. It is also used to describe the posterior surface of the hand, opposite the **palm.** Because the term dorsum may refer to both superior and posterior surfaces in humans, the term is easier to understand if one thinks of a quadripedal plantigrade animal that walks on its palms and soles, such as a bear. The **sole** is the inferior aspect or bottom of the foot, opposite the dorsum, much of which is in contact with the ground when standing barefoot. The surface of the hands, the feet, and the digits of both corresponding to the dorsum is the **dorsal surface,** the surface of the hand and fingers corresponding to the palm is the **palmar surface,** and the surface of the foot and toes corresponding to the sole is the **plantar surface.**

Combined terms describe intermediate positional arrangements: **inferomedial** means nearer to the feet and median plane—for example, the anterior parts of the ribs run inferomedially; **superolateral** means nearer to the head and farther from the median plane.

Other terms of relationship and comparisons are independent of the anatomical position or the anatomical planes, relating primarily to the body's surface or its central core:

Superficial, intermediate, and **deep** describe the position of structures relative to the surface of the body or the relationship of one structure to another underlying or overlying structure.

External means outside of or farther from the center of an organ or cavity, while **internal** means inside or closer to the center, independent of direction.

Proximal and **distal** are used when contrasting positions nearer to or farther from the attachment of a limb or the central aspect of a linear structure, respectively.

Terms of Laterality

Paired structures having right and left members (e.g., the kidneys) are **bilateral,** whereas those occurring on one side only (e.g., the spleen) are **unilateral.** Designating whether you are referring specifically to the right or left member of bilateral structures can be critical, and is a good habit to begin at the outset of one's training to become a health professional. Something occurring on the same side of the body as another structure is **ipsilateral;** the right thumb and right great (big) toe are ipsilateral, for example. **Contralateral** means occurring on the opposite side of the body relative to another structure; the right hand is contralateral to the left hand.

Terms of Movement

Various terms describe movements of the limbs and other parts of the body (Fig. I.5). Most movements are defined in relationship to the anatomical position, with movements occurring within, and around axes aligned with, specific anatomical planes. While most movements occur at joints where two or more bones or cartilages articulate with one another, several non-skeletal structures exhibit movement (e.g., tongue, lips, eyelids). Terms of movement may also be considered in pairs of opposing movements:

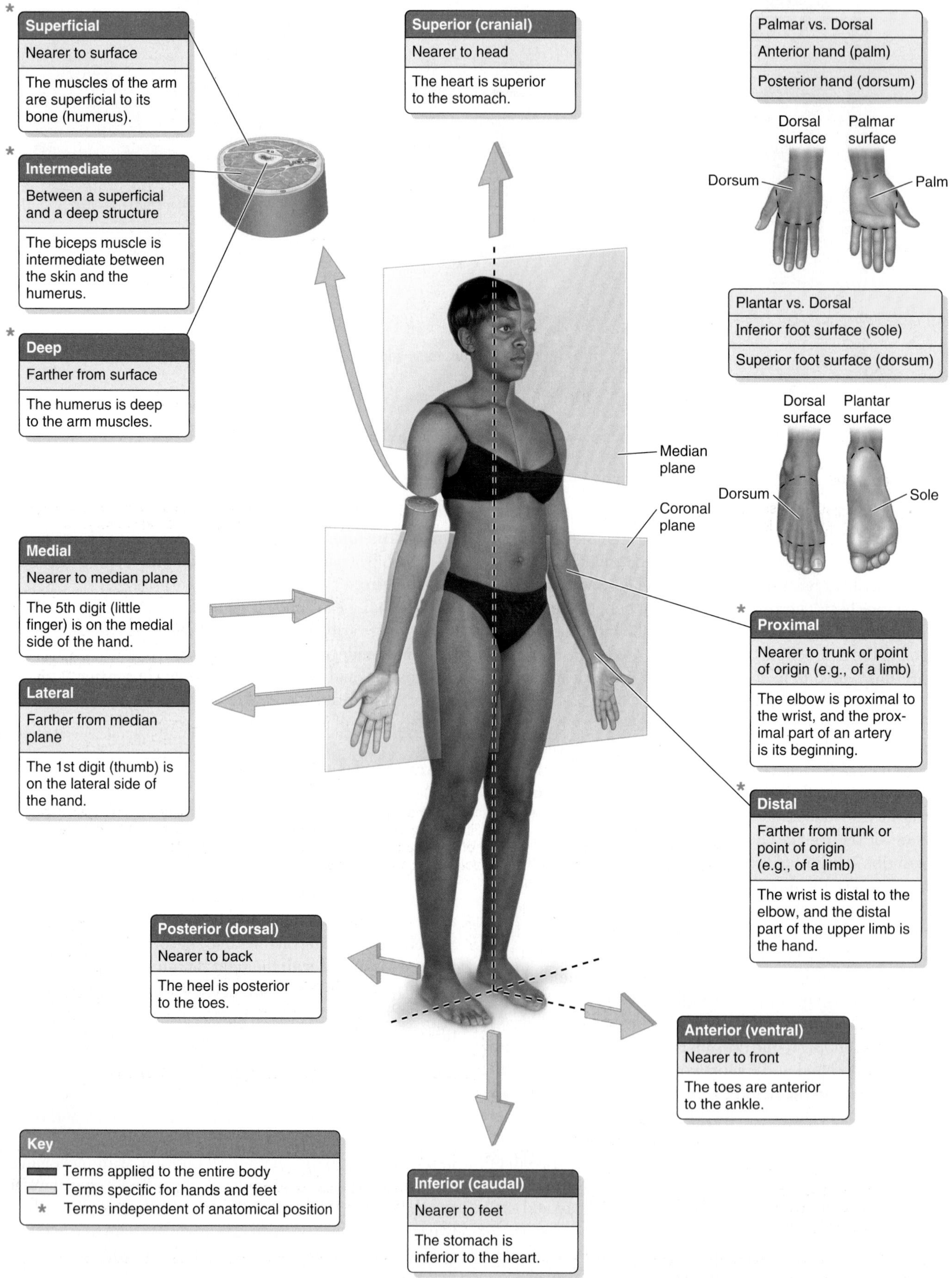

Superficial

Nearer to surface

The muscles of the arm are superficial to its bone (humerus).

Intermediate

Between a superficial and a deep structure

The biceps muscle is intermediate between the skin and the humerus.

Deep

Farther from surface

The humerus is deep to the arm muscles.

Superior (cranial)

Nearer to head

The heart is superior to the stomach.

Palmar vs. Dorsal

Anterior hand (palm)

Posterior hand (dorsum)

Dorsal surface Palmar surface

Dorsum Palm

Plantar vs. Dorsal

Inferior foot surface (sole)

Superior foot surface (dorsum)

Dorsal surface Plantar surface

Dorsum Sole

Median plane

Coronal plane

Medial

Nearer to median plane

The 5th digit (little finger) is on the medial side of the hand.

Lateral

Farther from median plane

The 1st digit (thumb) is on the lateral side of the hand.

Proximal

Nearer to trunk or point of origin (e.g., of a limb)

The elbow is proximal to the wrist, and the proximal part of an artery is its beginning.

Distal

Farther from trunk or point of origin (e.g., of a limb)

The wrist is distal to the elbow, and the distal part of the upper limb is the hand.

Posterior (dorsal)

Nearer to back

The heel is posterior to the toes.

Anterior (ventral)

Nearer to front

The toes are anterior to the ankle.

Key

Terms applied to the entire body
Terms specific for hands and feet
* Terms independent of anatomical position

Inferior (caudal)

Nearer to feet

The stomach is inferior to the heart.

FIGURE I.4. Terms of relationship and comparison. These terms describe the position of one structure relative to another.

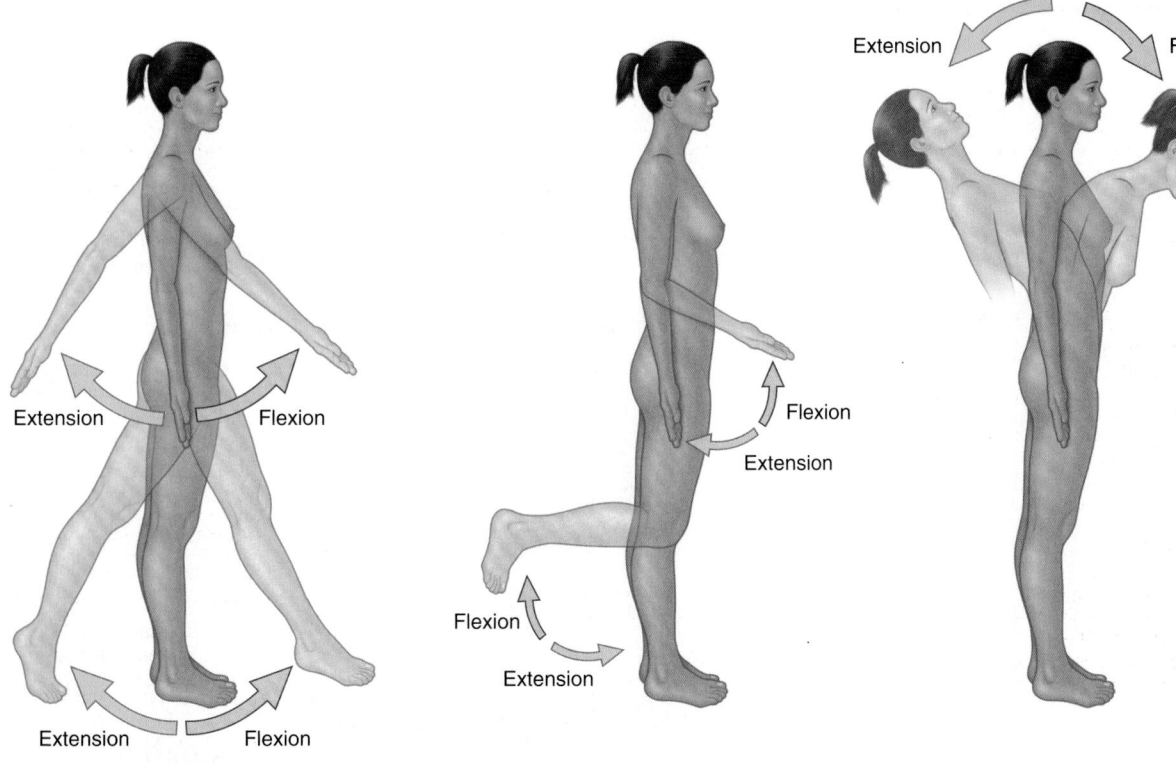

(A) Flexion and extension of upper limb at shoulder joint and lower limb at hip joint

Flexion and extension of forearm at elbow joint and of leg at knee joint

Flexion and extension of vertebral column at intervertebral joints

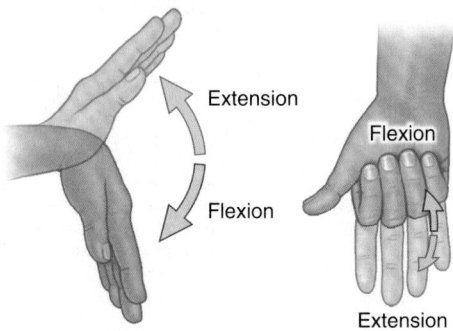

(B) Flexion and extension of hand at wrist joint

Flexion and extension of digits (fingers) at metacarpophalangeal and interphalangeal joints

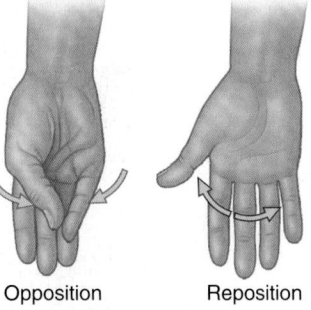

(C) Opposition and reposition of thumb and little finger at carpometacarpal joint of thumb combined with flexion at metacarpophalangeal joints

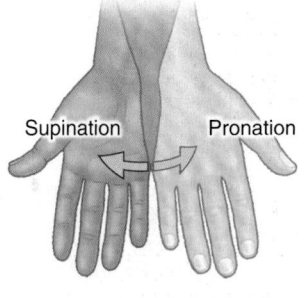

(D) Pronation and supination of forearm at radio-lnar joints

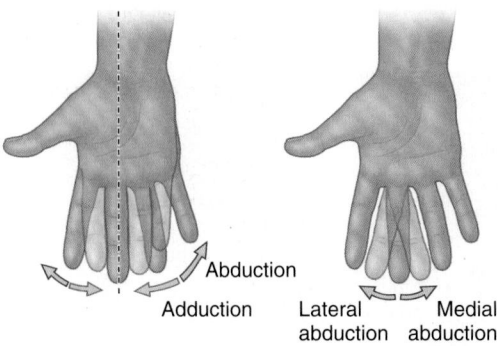

(E) Abduction and adduction of 2nd, 4th, and 5th digits at metacarpo-phalangeal joints

Abduction of 3rd digit at metacarpophalangeal joint

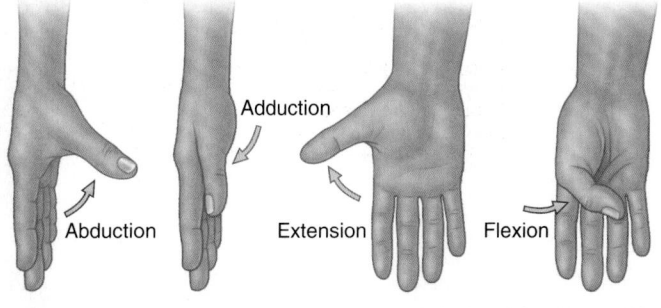

(F) The thumb is rotated 90° relative to other structures. Abduction and adduction at metacarpophalangeal joint occurs in a sagittal plane; flexion and extension at metacarpophalangeal and interphalangeal joints occurs in frontal planes, opposite to these movements at all other joints.

FIGURE I.5. Terms of movement. These terms describe movements of the limbs and other parts of the body; most movements take place at joints, where two or more bones or cartilages articulate with one another.

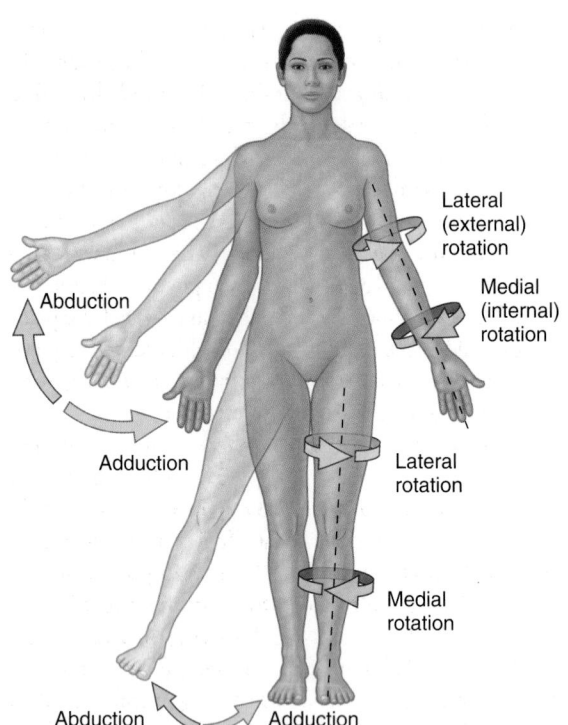

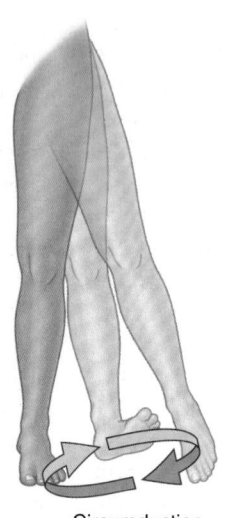

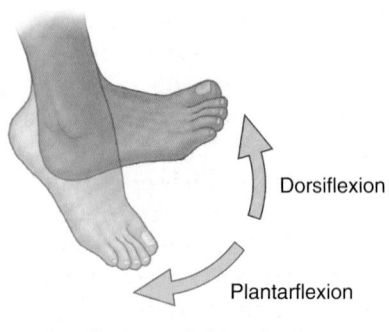

Dorsiflexion and plantarflexion of foot at ankle joint

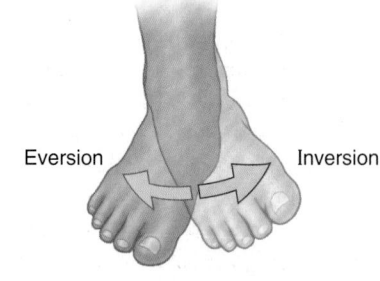

(G) Abduction and adduction of right limbs and rotation of left limbs at glenohumeral and hip joints, respectively

(H) Circumduction (circular movement) of lower limb at hip joint

(I) Inversion and eversion of foot at subtalar and transverse tarsal joints

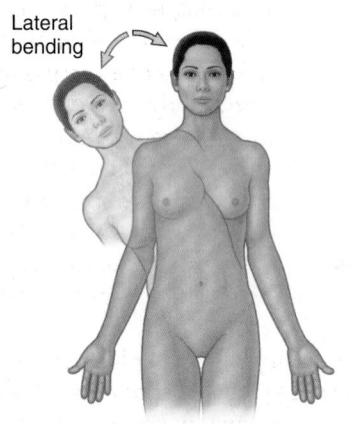

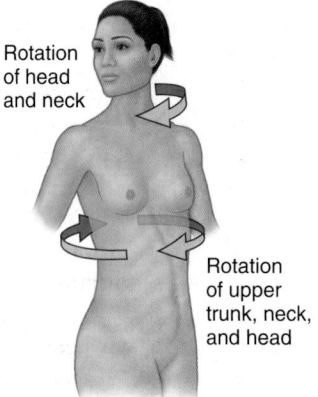

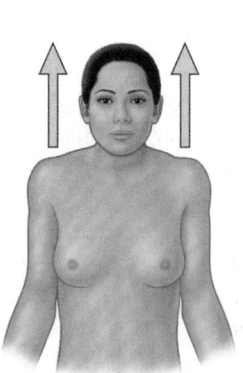

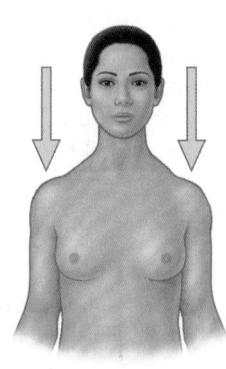

(K) Elevation and depression of shoulders

(J) Lateral bending (lateral flexion) of trunk and rotation of upper trunk, neck, and head

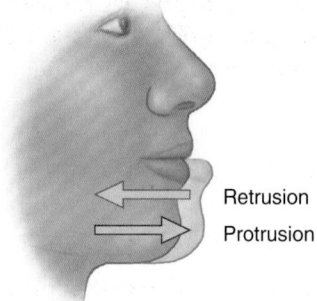

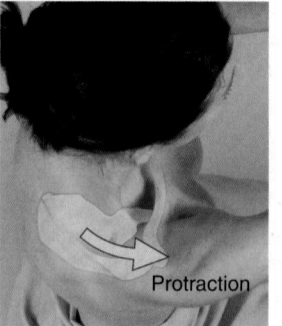

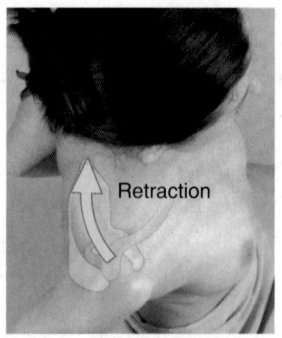

(L) Protrusion and retrusion of jaw at temporomandibular joints

(M) Protraction and retraction of scapula on thoracic wall

FIGURE I.5. *(Continued)*

Flexion and extension movements generally occur in sagittal planes around a transverse axis (Fig. I.5A & B). **Flexion** indicates bending or decreasing the angle between the bones or parts of the body. For most joints (e.g., elbow), flexion involves movement in an anterior direction. **Extension** indicates straightening or increasing the angle between the bones or parts of the body. Extension usually occurs in a posterior direction. The knee joint, rotated 180° to other joints, is exceptional in that flexion of the knee involves posterior movement and extension involves anterior movement. **Dorsiflexion** describes flexion at the ankle joint, as occurs when walking uphill or lifting the front of the foot and toes off the ground (Fig. I.5I). **Plantarflexion** bends the foot and toes toward the ground, as when standing on your toes. Extension of a limb or part beyond the normal limit—**hyperextension** (overextension)—can cause injury, such as "whiplash" (i.e., hyperextension of the neck during a rear-end automobile collision).

Abduction and adduction movements generally occur in a frontal plane around an anteroposterior axis (Fig. I.5E & G). Except for the digits, **abduction** means moving away from the median plane (e.g., when moving an upper limb laterally away from the side of the body) and **adduction** means moving toward it. In *abduction of the digits* (fingers or toes), the term means spreading them apart—moving the other fingers away from the neutrally positioned 3rd (middle) finger or moving the other toes away from the neutrally positioned 2nd toe. The 3rd finger and 2nd toe *medially* or *laterally abduct* away from the neutral position. *Adduction of the digits* is the opposite—bringing the spread fingers or toes together, toward the neutrally positioned 3rd finger or 2nd toe. Right and left lateral flexion (lateral bending) are special forms of abduction for only the neck and trunk (Fig. I.5J). The face and upper trunk are directed anteriorly as the head and/or shoulders tilt to the right or left side, causing the midline of the body itself to become bent sideways. This is a compound movement occurring between many adjacent vertebrae.

As you can see by noticing the way the thumbnail faces (laterally instead of posteriorly in the anatomical position), the thumb is rotated 90° relative to the other digits (Fig. I.5F). Therefore, the thumb flexes and extends in the frontal plane and abducts and adducts in the sagittal plane.

Circumduction is a circular movement that involves sequential flexion, abduction, extension, and adduction (or in the opposite order) in such a way that the distal end of the part moves in a circle (Fig. I.5H). Circumduction can occur at any joint at which all the above-mentioned movements are possible (e.g., the shoulder and hip joints).

Rotation involves turning or revolving a part of the body around its longitudinal axis, such as turning one's head to face sideways (Fig. I.5G). *Medial rotation* (internal rotation) brings the anterior surface of a limb closer to the median plane, whereas *lateral rotation* (external rotation) takes the anterior surface away from the median plane.

Pronation and supination are the rotational movements of the forearm and hand that swing the distal end of the radius (the lateral long bone of the forearm) medially and laterally around and across the anterior aspect of the ulna (the other long bone of the forearm) while the proximal end of the radius rotates in place (Fig. I.5D). **Pronation** rotates the radius medially so that the palm of the hand faces posteriorly and its dorsum faces anteriorly. When the elbow joint is flexed, pronation moves the hand so that the palm faces inferiorly (e.g., placing the palms flat on a table). **Supination** is the opposite rotational movement, rotating the radius laterally and uncrossing it from the ulna, returning the pronated forearm to the anatomical position. When the elbow joint is flexed, supination moves the hand so that the palm faces superiorly. (*Memory device:* You can hold *soup* in the palm of your hand when the flexed forearm is *sup*inated but are *prone* [likely] to spill it if the forearm is then *pron*ated!)

Eversion moves the sole of the foot away from the median plane, turning the sole laterally (Fig. I.5I). When the foot is fully everted it is also dorsiflexed. **Inversion** moves the sole of the foot toward the median plane (facing the sole medially). When the foot is fully inverted it is also plantarflexed. *Pronation of the foot* actually refers to a combination of eversion and abduction that results in lowering of the medial margin of the foot (the feet of an individual with flat feet are pronated), and *supination of the foot* generally implies movements resulting in raising the medial margin of the foot, a combination of inversion and adduction.

Opposition is the movement by which the pad of the 1st digit (thumb) is brought to another digit pad (Fig. I.5C). This movement is used to pinch, button a shirt, and lift a teacup by the handle. **Reposition** describes the movement of the 1st digit from the position of opposition back to its anatomical position.

Protrusion is a movement anteriorly (forward) as in protruding the mandible (chin), lips, or tongue (Fig. I.5L). **Retrusion** is a movement posteriorly (backward), as in retruding the mandible, lips, or tongue. The similar terms **protraction** and **retraction** are used most commonly for anterolateral and posteromedial movements of the scapula on the thoracic wall, causing the shoulder region to move anteriorly and posteriorly (Fig. I.5M).

Elevation raises or moves a part superiorly, as in elevating the shoulders when shrugging, the upper eyelid when opening the eye, or the tongue when pushing it up against the palate (roof of mouth) (Fig. I.5K). **Depression** lowers or moves a part inferiorly, as in depressing the shoulders when standing at ease, the upper eyelid when closing the eye, or pulling the tongue away from the palate.

The Bottom Line

ANATOMICOMEDICAL TERMINOLOGY

Anatomical terms are descriptive terms standardized in an international reference guide, *Terminologia Anatomica* (TA). These terms, in English or Latin, are used worldwide. ♦ Colloquial terminology is used by—and to communicate with—lay people. ♦ Eponyms are often used in clinical settings but are not recommended because they do not provide anatomical context and are not standardized. ♦ Anatomical directional terms are based on the body in the anatomical position. ♦ Four anatomical planes divide the body, and sections divide the planes into visually useful and descriptive parts. ♦ Other anatomical terms describe relationships of parts of the body, compare the positions of structures, and describe laterality and movement.

The Bottom Line

ANATOMICAL VARIATIONS

Anatomical variations are common and students should expect to encounter them during dissection. It is important to know how such variations may influence physical examinations, diagnosis, and treatment.

ANATOMICAL VARIATIONS

Anatomy books describe (initially, at least) the structure of the body as it is usually observed in people—that is, the most common pattern. However, occasionally a particular structure demonstrates so much variation within the normal range that the most common pattern is found less than half the time! Beginning students are frequently frustrated because the bodies they are examining or dissecting do not conform to the atlas or text they are using (Bergman et al., 1988). Often students ignore the variations or inadvertently damage them by attempting to produce conformity. Therefore, you should *expect anatomical variations* when you dissect or inspect prosected specimens.

In a random group of people, individuals differ from each other in physical appearance. The bones of the skeleton vary not only in their basic shape but also in lesser details of surface structure. A wide variation is found in the size, shape, and form of the attachments of muscles. Similarly, considerable variation exists in the patterns of branching of veins, arteries, and nerves. *Veins vary the most and nerves the least.* Individual variation must be considered in physical examination, diagnosis, and treatment.

Most descriptions in this text assume a normal range of variation. However, the frequency of variation often differs among human groups, and variations collected in one population may not apply to members of another population. Some variations, such as those occurring in the origin and course of the cystic artery to the gallbladder, are clinically important (see Chapter 2), and any surgeon operating without knowledge of them is certain to have problems. Clinically significant variations are described in clinical correlation (blue) boxes identified with an Anatomical Variation icon (at left).

Apart from racial and sexual differences, humans exhibit considerable genetic variation, such as polydactyly (extra digits). Approximately 3% of newborns show one or more significant birth defects (Moore et al, 2012). Other defects (e.g., atresia or blockage of the intestine) are not detected until symptoms occur. Discovering variations and congenital anomalies in cadavers is actually one of the many benefits of firsthand dissection, because it enables students to develop an awareness of the occurrence of variations and a sense of their frequency.

INTEGUMENTARY SYSTEM

Because the skin (L. *integumentum,* a covering) is readily accessible and is one of the best indicators of general health, careful observation of it is important in physical examinations. It is considered in the differential diagnosis of almost every disease. The skin provides:

- *Protection* of the body from environmental effects, such as abrasions, fluid loss, harmful substances, ultraviolet radiation, and invading microorganisms.
- *Containment* for the body's structures (e.g., tissues and organs) and vital substances (especially extracellular fluids), preventing dehydration, which may be severe when extensive skin injuries (e.g., burns) are experienced.
- *Heat regulation* through the evaporation of sweat and/or the dilation or constriction of superficial blood vessels.
- *Sensation* (e.g., pain) by way of superficial nerves and their sensory endings.
- *Synthesis and storage* of vitamin D.

The **skin,** the body's largest organ, consists of the epidermis, a superficial cellular layer, and the dermis, a deep connective tissue layer (Fig. I.6).

The **epidermis** is a *keratinized epithelium*—that is, it has a tough, horny *superficial layer* that provides a protective outer surface overlying its regenerative and pigmented deep or *basal layer.* The epidermis has no blood vessels or lymphatics. The *avascular epidermis* is nourished by the underlying *vascularized dermis.* The dermis is supplied by arteries that enter its deep surface to form a cutaneous plexus of anastomosing arteries. The skin is also supplied with afferent nerve endings that are sensitive to touch, irritation (pain), and temperature. Most nerve terminals are in the dermis, but a few penetrate the epidermis.

The **dermis** is a dense layer of interlacing *collagen* and *elastic fibers.* These fibers provide skin tone and account for the strength and toughness of skin. The dermis of animals is removed and tanned to produce leather. Although the bundles of collagen fibers in the dermis run in all directions to produce a tough felt-like tissue, in any specific location most fibers run in the same direction. The predominant pattern of collagen fibers determines the characteristic tension and wrinkle lines in the skin.

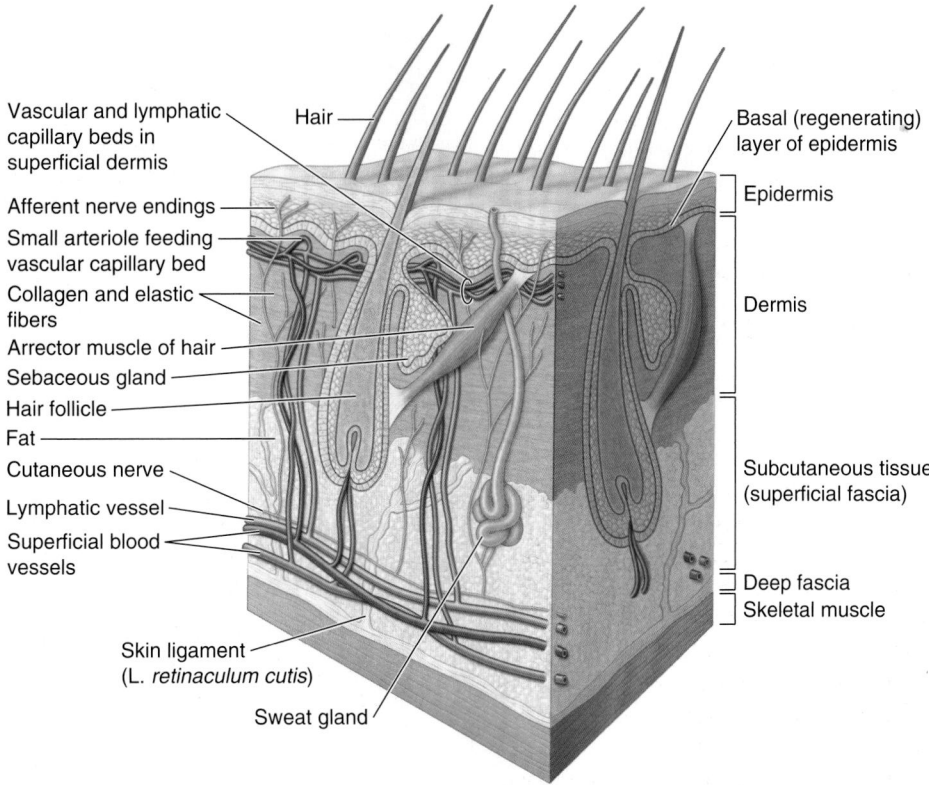

Vascular and lymphatic capillary beds in superficial dermis

Hair

Basal (regenerating) layer of epidermis

Epidermis

Afferent nerve endings

Small arteriole feeding vascular capillary bed

Collagen and elastic fibers

Dermis

Arrector muscle of hair

Sebaceous gland

Hair follicle

Fat

Cutaneous nerve

Lymphatic vessel

Superficial blood vessels

Subcutaneous tissue (superficial fascia)

Deep fascia

Skeletal muscle

Skin ligament (L. *retinaculum cutis*)

Sweat gland

FIGURE I.6. Skin and some of its specialized structures.

The **tension lines** (also called cleavage lines or Langer lines) tend to spiral longitudinally in the limbs and run transversely in the neck and trunk (Fig. I.7). Tension lines at the elbows, knees, ankles, and wrists are parallel to the transverse creases that appear when the limbs are flexed. The elastic fibers of the dermis deteriorate with age and are not replaced; consequently, in older people, the skin wrinkles and sags as it loses its elasticity.

The skin also contains many specialized structures (Fig. I.6). The deep layer of the dermis contains *hair follicles*, with associated smooth arrector muscles and sebaceous glands. Contraction of the **arrector muscles of hairs** (L. *musculi arrector pili*) erects the hairs, causing goose bumps. Hair follicles are generally slanted to one side, and several *sebaceous glands* lie on the side the hair is directed toward ("points to") as it emerges from the skin. Thus, contraction of the arrector muscles causes the hairs to stand up straighter, thereby compressing the sebaceous glands and helping them secrete their oily product onto the skin surface. The evaporation of the watery secretion (sweat) of the *sweat glands* from the skin provides a *thermoregulatory mechanism* for heat loss (cooling). Also involved in the loss or retention of body heat are the small arteries (arterioles) within the dermis. They dilate to fill *superficial capillary beds* to radiate heat (skin appears red) or constrict to minimize surface heat loss (skin, especially of the lips and fingertips, appears blue). Other skin structures or derivatives include the nails (fingernails, toenails), the mammary glands, and the enamel of teeth.

Located between the overlying skin (dermis) and underlying deep fascia, the **subcutaneous tissue** (superficial fascia) is composed mostly of *loose connective tissue* and *stored fat* and contains sweat glands, *superficial blood vessels, lymphatic vessels,* and *cutaneous nerves* (Fig. I.6). The neurovascular structures course in the subcutaneous tissue, distributing only their terminal branches to the skin.

The subcutaneous tissue provides for most of the body's fat storage, so its thickness varies greatly, depending on the person's nutritional state. In addition, the distribution of subcutaneous tissue varies considerably in different sites in the same individual. Compare, for example, the relative abundance of subcutaneous tissue evident by the thickness of the fold of skin that can be pinched at the waist or thighs with the anteromedial part of the leg (the shin, the anterior border of the tibia) or the back of the hand, the latter two being nearly devoid of subcutaneous tissue. Also consider the distribution of subcutaneous tissue and fat between the sexes: In mature females, it tends to accumulate in the breasts and thighs, whereas in males, subcutaneous fat accumulates in the lower abdominal wall.

Subcutaneous tissue participates in thermoregulation, functioning as insulation, retaining heat in the body's core. It also provides padding that protects the skin from compression by bony prominences, such as those in the buttocks.

Skin ligaments (L. *retinacula cutis*), numerous small fibrous bands, extend through the subcutaneous tissue and attach the deep surface of the dermis to the underlying

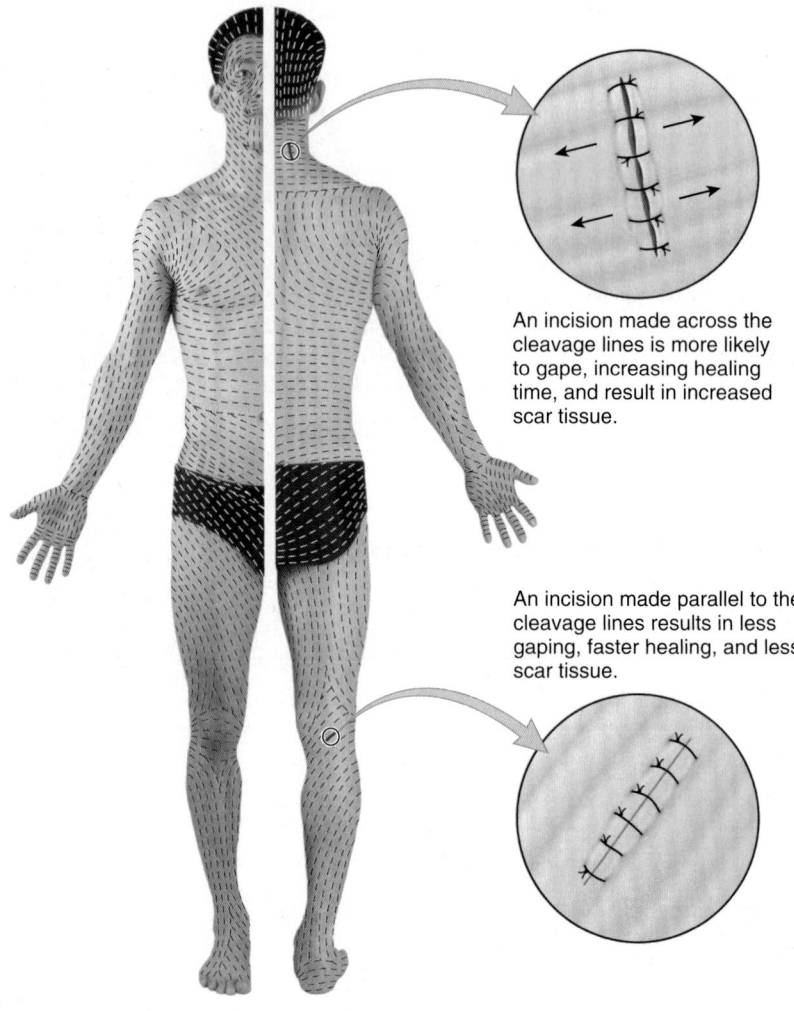

An incision made across the cleavage lines is more likely to gape, increasing healing time, and result in increased scar tissue.

An incision made parallel to the cleavage lines results in less gaping, faster healing, and less scar tissue.

FIGURE I.7. Tension lines in the skin. The *dashed lines* indicate the predominant direction of the collagen fibers in the dermis.

Anterior view   Posterior view

deep fascia (Fig. I.6). The length and density of these ligaments determines the mobility of the skin over deep structures. Where skin ligaments are longer and sparse, the skin is more mobile, such as on the back of the hand (Fig. I.8A & B). Where ligaments are short and abundant, the skin is firmly attached to the underlying deep fascia, such as in the palms and soles (Fig. I.8C). In dissection, removal of skin where the skin ligaments are short and abundant requires use of a sharp scalpel. The skin ligaments are long but particularly well developed in the breasts, where they form weight-bearing *suspensory ligaments* (see Chapter 1).

INTEGUMENTARY SYSTEM

Skin Color Signs in Physical Diagnosis

Blood flow through the superficial capillary beds of the dermis affects the color of skin and can provide important clues for diagnosing certain clinical conditions. When the blood is not carrying enough oxygen from the lungs, such as in a person who has stopped breathing or in a person having a defective circulation that is sending an inadequate amount of blood through the lungs, the skin can appear bluish (*cyanotic*). This occurs because the oxygen-carrying hemoglobin of blood is bright red when carrying oxygen (as it does in arteries and usually does in capillaries), and appears deep, purplish blue when depleted of oxygen, as it does in veins. Cyanosis is especially evident where skin is thin, such as the lips, eyelids, and deep to the transparent nails. Skin injury, exposure to excess heat, infection, inflammation, or allergic reactions may cause the superficial capillary beds to become engorged, making the skin look abnormally red, a sign called *erythema*. In certain liver disorders, a yellow pigment called bilirubin builds up in the blood, giving a yellow appearance to the whites of the eyes and skin, a condition called *jaundice*. Skin color changes are most readily observed in people with light-colored skin and may be difficult to discern in people with dark skin.

Skin Incisions and Scarring

The skin is always under tension. In general, lacerations or incisions that parallel the tension lines usually heal well with little scarring because there is minimal disruption of fibers. The uninterrupted fibers tend to retain the cut edges in place. However, a laceration or incision across the tension lines disrupts more collagen fibers. The disrupted lines of force cause the wound to gape and it may heal with excessive (keloid) scarring. When other considerations, such as adequate exposure and access or avoidance of nerves, are not of greater importance, surgeons attempting to minimize scarring for cosmetic reasons may use surgical incisions that parallel the tension lines.

Stretch Marks in Skin

The collagen and elastic fibers in the dermis form a tough, flexible meshwork of tissue. Because the skin can distend considerably, a relatively small incision can be made during surgery compared with the much larger incision required to attempt the same procedure in an embalmed cadaver, which no longer exhibits elasticity. The skin can stretch and grow to accommodate gradual increases in size. However, marked and relatively fast size increases, such as the abdominal enlargement and weight gain accompanying pregnancy, can stretch the skin too much, damaging the collagen fibers in the dermis (Fig. BI.1). Bands of thin wrinkled skin, initially red but later becoming purple, and white stretch marks (L. *striae gravidarum*) appear on the abdomen, buttocks, thighs, and breasts during pregnancy. *Stretch marks* (L. *striae cutis distensae*) also form in obese individuals and in certain diseases (e.g., hypercortisolism or Cushing syndrome); they occur along with distension and loosening of the deep fascia due to protein breakdown leading to reduced cohesion between the collagen fibers. Stretch marks generally fade after pregnancy and weight loss, but they never disappear completely.

Skin Injuries and Wounds

Lacerations. Accidental cuts and skin tears are superficial or deep. Superficial lacerations violate the epidermis and perhaps the superficial layer of the dermis; they bleed but do not interrupt the continuity of the dermis. Deep lacerations penetrate the deep layer of the dermis, extending into the subcutaneous tissue or beyond; they gape and require approximation of the cut edges of the dermis (by suturing, or stitches) to minimize scarring.

Burns are caused by thermal trauma, ultraviolet or ionizing radiation, or chemical agents. Burns are classified, in increasing order of severity, based on the depth of skin injury (Fig. BI.2):

* *1st-degree (superficial) burn* (e.g., sunburn): damage is limited to the epidermis; symptoms are *erythema* (hot red skin), pain, and *edema* (swelling); *desquamation* (peeling) of the superficial layer usually occurs several days later, but the layer is quickly replaced from the basal layer of the epidermis without significant scarring.

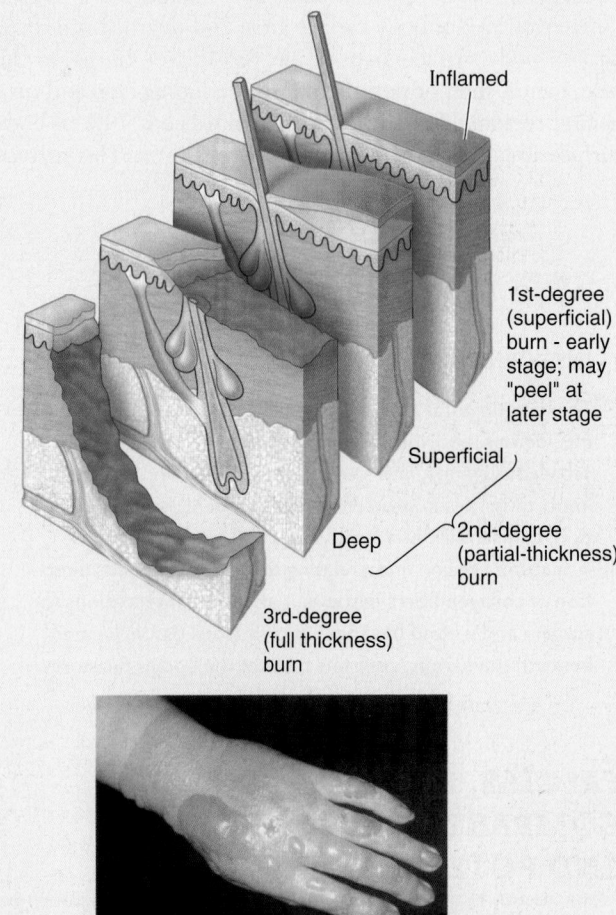

Inflamed

1st-degree (superficial) burn - early stage; may "peel" at later stage

Superficial

Deep

2nd-degree (partial-thickness) burn

3rd-degree (full thickness) burn

2nd-degree (partial-thickness) burn

FIGURE BI.2.

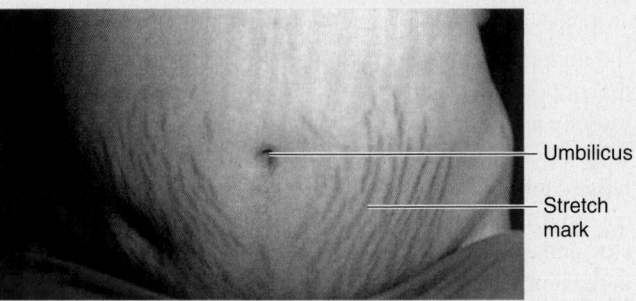

Umbilicus

Stretch mark

FIGURE BI.1.

- *2nd-degree (partial-thickness) burn:* epidermis and superficial dermis are damaged with blistering (superficial 2nd degree) or loss (deep 2nd degree); nerve endings are damaged, making this variety the most painful; except for their most superficial parts, the sweat glands and hair follicles are not damaged and can provide the source of replacement cells for the basal layer of the epidermis along with cells from the edges of the wound; healing will occur slowly (3 weeks to several months), leaving scarring and some contracture, but it is usually complete.
- *3rd-degree (full-thickness) burn:* the entire thickness of the skin is damaged and perhaps underlying muscle. There is marked edema and the burned area is numb since sensory endings are destroyed. A minor degree of healing may occur at the edges, but the open, ulcerated portions require skin grafting: dead material (*eschar*) is removed and replaced (*grafted*) over the burned area with skin *harvested* (taken) from a non-burned location (*autograft*) or using skin from human cadavers or pigs, or cultured or artificial skin.

The extent of a burn (percent of total body surface affected) is generally more significant than the degree (severity in terms of depth) in estimating the effect on the well-being of the victim. According to the American Burn Association's classification of burn injury, a major burn includes 3rd-degree burns over 10% of body surface area; 2nd-degree burns over 25% of body surface area; or any 3rd-degree burns on the face, hands, feet, or perineum (area including anal and urogenital regions). When the burn area exceeds 70% of body surface area, the mortality rate exceeds 50%. The surface area affected by a burn in an adult can be estimated by applying the "Rule of Nines" in which the body is divided into areas that are approximately 9% or multiples of 9% of the total body surface (Fig. BI.3).

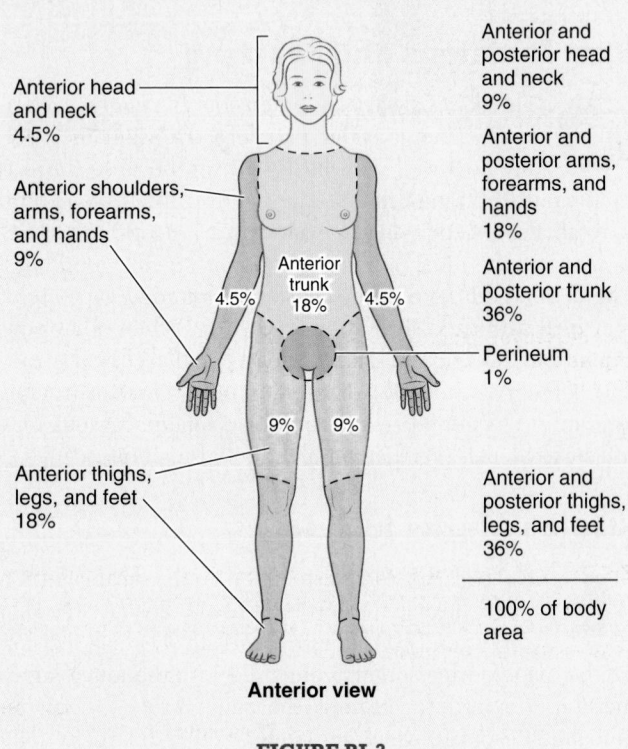

Anterior view

FIGURE BI.3.

The Bottom Line

INTEGUMENTARY SYSTEM

The integumentary system (the skin) consists of the epidermis, dermis, and specialized structures (hair follicles, sebaceous glands, and sweat glands). The skin: ♦ plays important roles in protection, containment, heat regulation, and sensation; ♦ synthesizes and stores vitamin D; ♦ features tension lines, relating to the predominant direction of collagen fibers in the skin, that have implications for surgery and wound healing. Subcutaneous tissue, located beneath the dermis, contains most of the body's fat stores.

FASCIAS, FASCIAL COMPARTMENTS, BURSAE, AND POTENTIAL SPACES

Fascias (L. *fasciae*) constitute the wrapping, packing, and insulating materials of the deep structures of the body. Underlying the *subcutaneous tissue* (superficial fascia) almost everywhere is the deep fascia (Fig. I.9). The **deep fascia** is a dense, organized connective tissue layer, devoid of fat, that covers most of the body parallel to (deep to) the skin and subcutaneous tissue. Extensions from its internal surface invest deeper structures, such as individual muscles and neurovascular bundles, as **investing fascia.** Its thickness varies widely. For example, in the face, distinct layers of deep fascia are absent.

In the limbs, groups of muscles with similar functions sharing the same nerve supply are located in **fascial compartments,** separated by thick sheets of deep fascia, called **intermuscular septa,** that extend centrally from the surrounding fascial sleeve to attach to bones. These compartments may contain or direct the spread of an infection or a tumor.

In a few places, the deep fascia gives attachment (origin) to the underlying muscles (although it is not usually included in lists or tables of origins and insertions); but in most places, the muscles are free to contract and glide deep to it. However, the deep fascia itself never passes freely over bone; where deep fascia contacts bone, it blends firmly with the *periosteum* (bone covering). The relatively unyielding deep fascia investing muscles, and especially that surrounding the fascial compartments in the limbs, limits the outward expansion of the bellies of contracting skeletal muscles. Blood is thus pushed out as the veins of the muscles and compartments are compressed. *Valves*

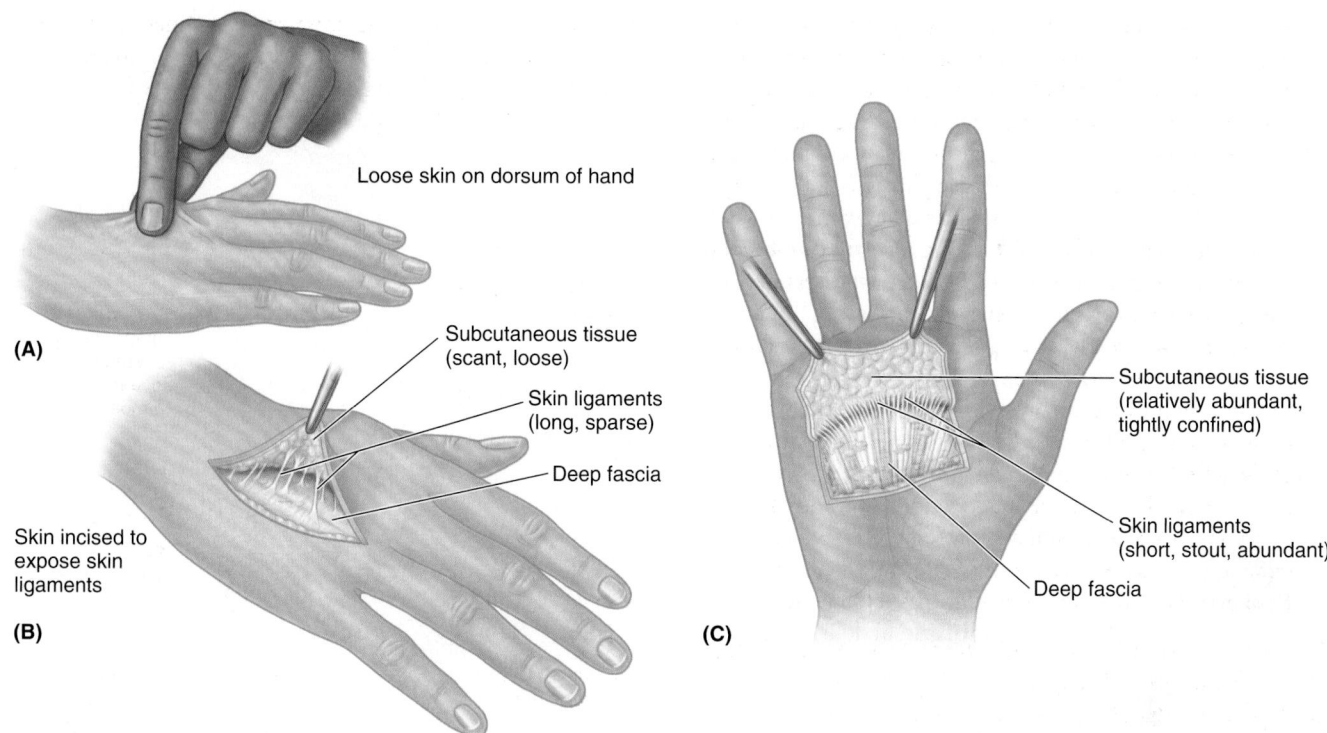

Loose skin on dorsum of hand

(A)

Subcutaneous tissue
(scant, loose)

Skin ligaments
(long, sparse)

Deep fascia

Skin incised to
expose skin
ligaments

(B)

Subcutaneous tissue
(relatively abundant,
tightly confined)

Skin ligaments
(short, stout, abundant)

Deep fascia

(C)

FIGURE I.8. Skin ligaments in subcutaneous tissue. A. The thickness of subcutaneous tissue can be estimated as being approximately half that of a pinched fold of skin (i.e., a fold of skin includes a double thickness of subcutaneous tissue). The dorsum of the hand has relatively little subcutaneous tissue. **B.** Long, relatively sparse skin ligaments allow the mobility of the skin demonstrated in part **A. C.** The skin of the palm (like that of the sole) is firmly attached to the underlying deep fascia.

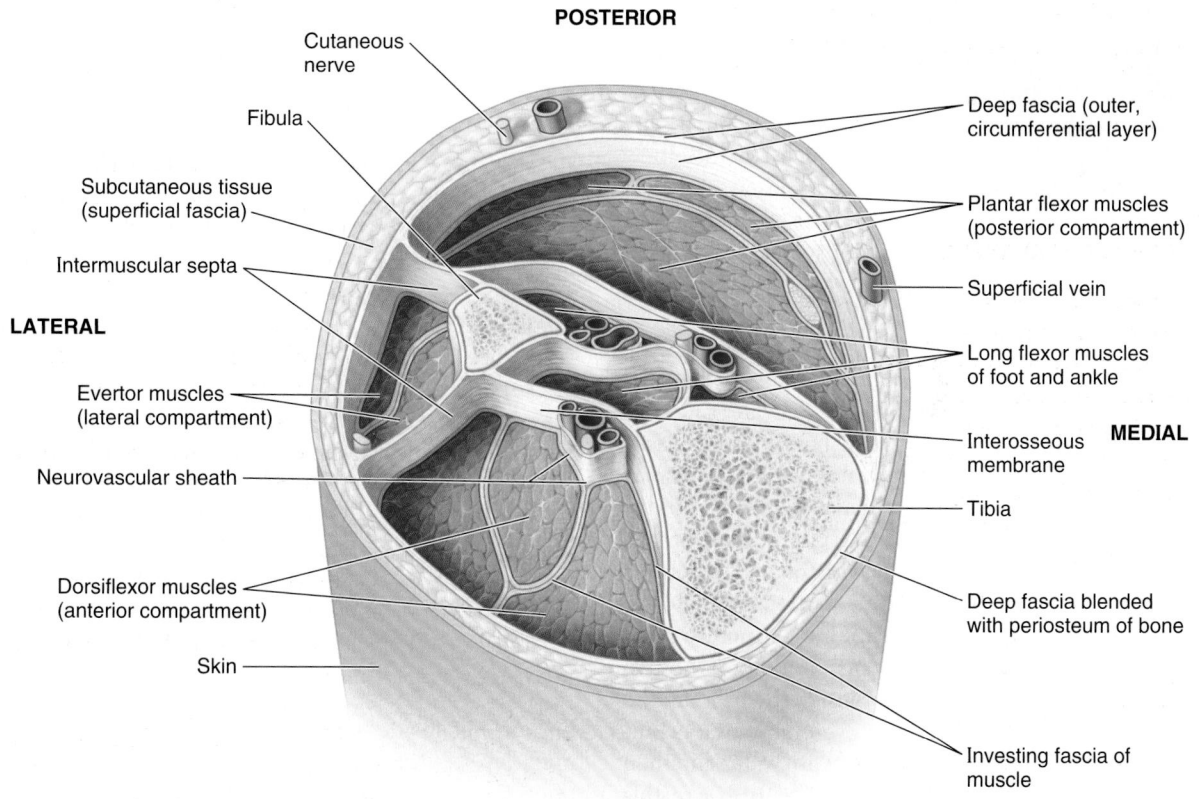

POSTERIOR

Cutaneous nerve

Fibula

Subcutaneous tissue
(superficial fascia)

Intermuscular septa

LATERAL

Evertor muscles
(lateral compartment)

Neurovascular sheath

Dorsiflexor muscles
(anterior compartment)

Skin

Deep fascia (outer,
circumferential layer)

Plantar flexor muscles
(posterior compartment)

Superficial vein

Long flexor muscles
of foot and ankle

Interosseous
membrane **MEDIAL**

Tibia

Deep fascia blended
with periosteum of bone

Investing fascia of
muscle

Anterosuperior view of right leg

FIGURE I.9. Excavated section of the leg demonstrating the deep fascia and fascial formations.

within the veins allow the blood to flow only in one direction (toward the heart), preventing the backflow that might occur as the muscles relax. Thus deep fascia, contracting muscles, and venous valves work together as a *musculovenous pump* to return blood to the heart, especially in the lower limbs where blood must move against the pull of gravity (Fig. I.26).

Near certain joints (e.g., wrist and ankle), the deep fascia becomes markedly thickened, forming a **retinaculum** (plural = retinacula) to hold tendons in place where they cross the joint during flexion and extension, preventing them from taking a shortcut, or *bow stringing,* across the angle created (Fig. I.19).

Subserous fascia, with varying amounts of fatty tissue, lies between the internal surfaces of the musculoskeletal walls and the serous membranes lining the body cavities. These are the *endothoracic, endoabdominal,* and *endopelvic fascias;* the latter two may be referred to collectively as *extraperitoneal fascia.*

Bursae (singular = bursa; Mediev. L., a purse) are closed sacs or envelopes of **serous membrane** (a delicate connective tissue membrane capable of secreting fluid to lubricate a smooth internal surface). Bursae are normally collapsed. Unlike three-dimensional realized or actual spaces, these potential spaces have no depth; their walls are apposed with only a thin film of lubricating fluid between them that is secreted by the enclosing membranes. When the wall is interrupted at any point, or when a fluid is secreted or formed within them in excess, they become realized spaces; however, this condition is abnormal or pathological.

Usually occurring in locations subject to friction, bursae enable one structure to move more freely over another. **Subcutaneous bursae** occur in the subcutaneous tissue between the skin and bony prominences, such as at the elbow or knee; **subfascial bursae** lie beneath deep fascia; and **subtendinous bursae** facilitate the movement of tendons over bone. **Synovial tendon sheaths** are a specialized type of elongated bursae that wrap around tendons, usually enclosing them as they traverse osseofibrous tunnels that anchor the tendons in place (Fig. I.10A).

Bursae occasionally communicate with the synovial cavities of joints. Because they are formed by delicate, transparent serous membranes and are collapsed, bursae are not easily noticed or dissected in the laboratory. It is possible to display bursae by injecting and distending them with colored fluid.

Collapsed bursal sacs surround many important organs (e.g., the heart, lungs, and abdominal viscera) and structures (e.g., portions of tendons). This configuration is much like wrapping a large but empty balloon around a structure, such as a fist (Fig. I.10B). The object is surrounded by the two layers of the empty balloon but is not inside the balloon; the balloon itself remains empty. For an even more exact comparison, the balloon should first be filled with water and then emptied, leaving the empty balloon wet inside. In exactly this way, the heart is surrounded by—but is not inside—the pericardial sac. Each lung is surrounded by—but is not inside—a *pleural sac,* and the abdominal viscera are surrounded by—

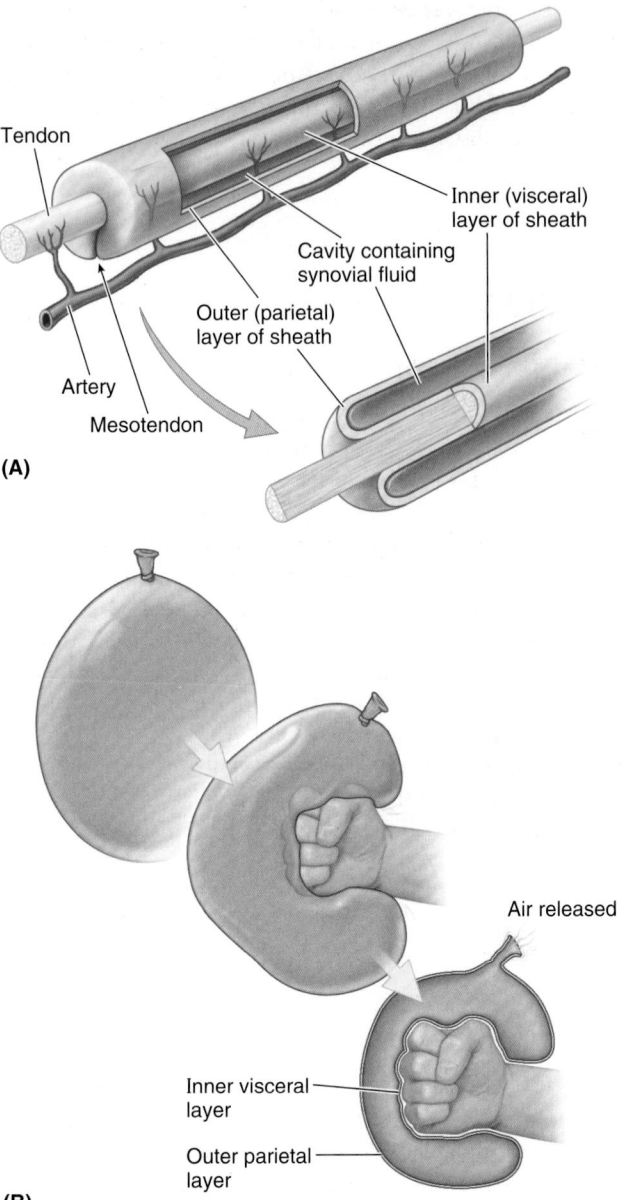

(A)

(B)

FIGURE I.10. Synovial tendon sheaths and bursal sacs. A. Synovial tendon sheaths are longitudinal bursae that surround tendons as they pass deep to retinacula or through fibrous digital sheaths. **B.** Bursal sacs enclose several structures, such as the heart, lungs, abdominal viscera, and tendons, much like this collapsed balloon encloses the fist. A thin film of lubricating fluid between the parietal and visceral layers confers mobility to the structure surrounded by the bursa within a confined compartment. The transitional folds of synovial membrane between the continuous parietal and visceral layers surrounding the connecting stalks (the wrist in this example) and/or neurovascular structures serving the surrounded mass are called mesenteries. In the case of a synovial tendon sheath, the mesentery is called a mesotendon.

but are not inside—the *peritoneal sac.* In such cases, the inner layer of the balloon or serous sac (the one adjacent to the fist, viscus, or viscera) is called the **visceral layer;** the outer layer of the balloon (or the one in contact with the body wall) is called the **parietal layer.** Such a surrounding double layer of membranes, moistened on their apposed surfaces, confers freedom of movement on the surrounded structure when it is contained within a confined space, such as the heart within its surrounding fibrous sac (*pericardium*) or flexor tendons within the fibrous tunnels that hold the tendons against the bones of the fingers.

FASCIAS

Fascial Planes and Surgery

In living people, fascial planes (*interfascial* and *intrafascial*) are potential spaces between adjacent fascias or fascia-lined structures, or within loose areolar fascias, such as the *subserous fascias*. Surgeons take advantage of these interfascial planes, separating structures to create spaces that allow movement and access to deeply placed structures. In some procedures, surgeons use extrapleural or extraperitoneal fascial planes, which allow them to operate outside the membranes lining the body cavities, minimizing the potential for contamination, the spread of infection, and consequent formation of adhesions (adherences) within the cavities. Unfortunately, these planes are often fused and difficult to establish or appreciate in embalmed cadavers.

The Bottom Line

FASCIAS AND BURSAE

Deep fascia is an organized connective tissue layer that completely envelops the body beneath the subcutaneous tissue underlying the skin. Extensions and modifications of the deep fascia: ♦ divide muscles into groups (intermuscular septa), ♦ invest individual muscles and neurovascular bundles (investing fascia), ♦ lie between musculoskeletal walls and the serous membranes lining body cavities (subserous fascia), and ♦ hold tendons in place during joint movements (retinacula). Bursae are closed sacs formed of serous membrane that occur in locations subject to friction; they enable one structure to move freely over another.

SKELETAL SYSTEM

The skeletal system may be divided into two functional parts (Fig. I.11):

- The **axial skeleton** consists of the bones of the head (*cranium* or *skull*), neck (*hyoid bone* and *cervical vertebrae*), and trunk (*ribs, sternum, vertebrae*, and *sacrum*).

- The **appendicular skeleton** consists of the *bones of the limbs,* including those forming the pectoral (shoulder) and pelvic girdles.

Cartilage and Bones

The skeleton is composed of cartilages and bones. **Cartilage** is a resilient, semirigid form of connective tissue that forms parts of the skeleton where more flexibility is required—for example, where the *costal cartilages* attach the ribs to the sternum. Also, the **articulating surfaces** (bearing surfaces) of bones participating in a synovial joint are capped with **articular cartilage** that provides smooth, low-friction, gliding surfaces for free movement (Fig. I.16A). Blood vessels do not enter cartilage (i.e., it is *avascular*); consequently, its cells obtain oxygen and nutrients by diffusion. The proportion of bone and cartilage in the skeleton changes as the body grows; the younger a person is, the more cartilage he or she has. The bones of a newborn are soft and flexible because they are mostly composed of cartilage.

Bone, a living tissue, is a highly specialized, hard form of connective tissue that makes up most of the skeleton. Bones of the adult skeleton provide:

- support for the body and its vital cavities; it is the chief supporting tissue of the body.
- protection for vital structures (e.g., the heart).
- the mechanical basis for movement (leverage).
- storage for salts (e.g., calcium).
- a continuous supply of new blood cells (produced by the marrow in the medullary cavity of many bones).

A fibrous connective tissue covering surrounds each skeletal element like a sleeve, except where articular cartilage occurs; that surrounding bones is **periosteum** (Fig. I.15), whereas that around cartilage is **perichondrium.** The periosteum and perichondrium nourish the external aspects of the skeletal tissue. They are capable of laying down more cartilage or bone (particularly during fracture healing) and provide the interface for attachment of *tendons* and *ligaments.*

The two types of bone are **compact bone** and **spongy** (trabecular) **bone.** They are distinguished by the relative amount of solid matter and by the number and size of the spaces they contain (Fig. I.12). All bones have a superficial thin layer of compact bone around a central mass of spongy bone, except where the latter is replaced by a *medullary (marrow) cavity.* Within the medullary cavity of adult bones, and between the **spicules** (trabeculae) of spongy bone, *yellow* (fatty) or *red* (blood cell and platelet forming) *bone marrow*—or a combination of both—is found.

The architecture and proportion of compact and spongy bone vary according to function. Compact bone provides strength for weight bearing. In *long bones* designed for rigidity and attachment of muscles and ligaments, the amount of compact bone is greatest near the middle of the *shaft* where the bones are liable to buckle. In addition, long bones have elevations (e.g., *ridges, crests,* and *tubercles*) that serve as *buttresses*

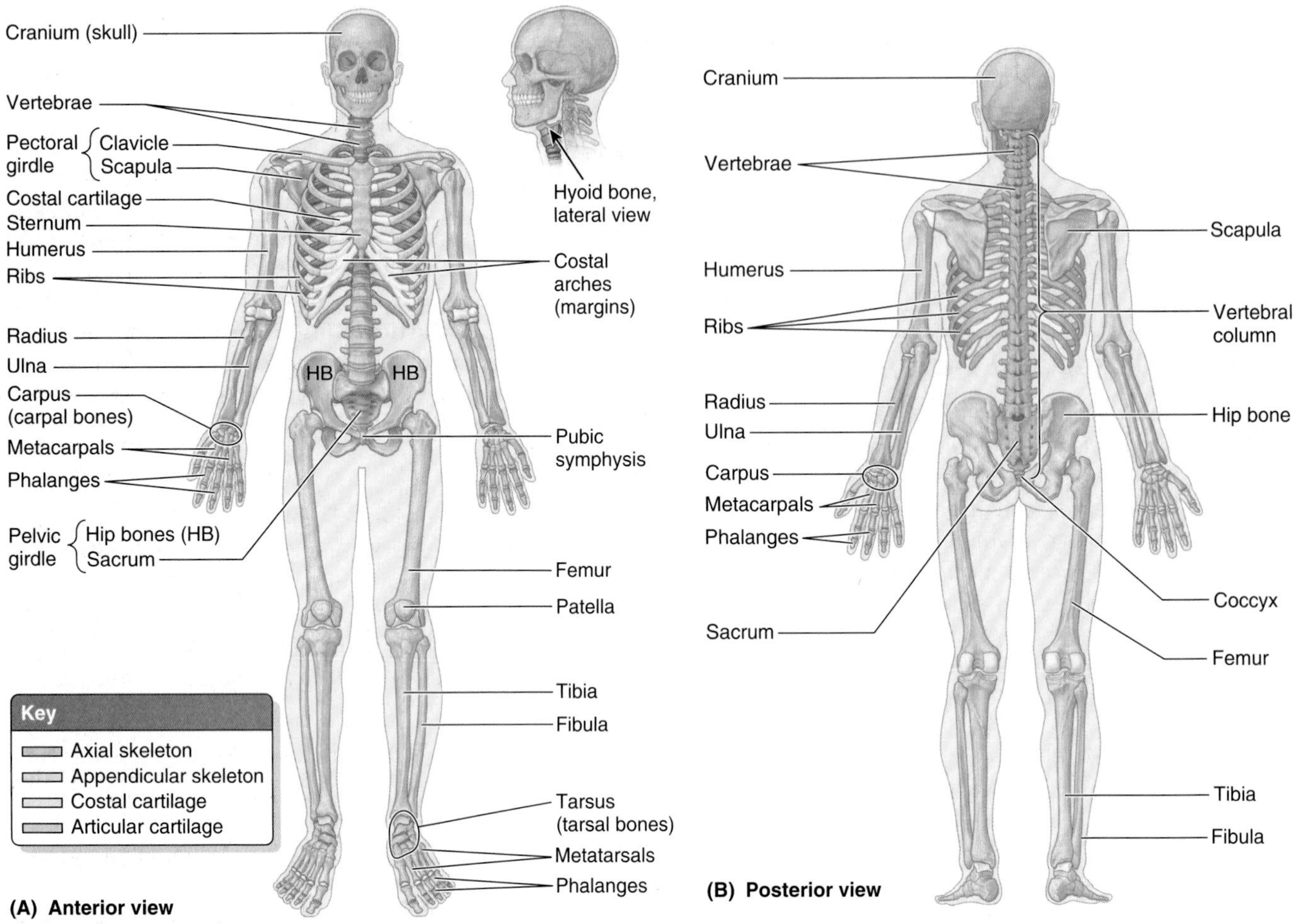

Cranium (skull)

Vertebrae

Pectoral { Clavicle
girdle   { Scapula

Costal cartilage

Sternum

Humerus

Ribs

Hyoid bone,
lateral view

Costal
arches
(margins)

Radius

Ulna

Carpus
(carpal bones)

Metacarpals

Phalanges

HB   HB

Pubic
symphysis

Pelvic { Hip bones (HB)
girdle { Sacrum

Femur

Patella

Tibia

Fibula

Key
- Axial skeleton
- Appendicular skeleton
- Costal cartilage
- Articular cartilage

Tarsus
(tarsal bones)

Metatarsals

Phalanges

(A) Anterior view

Cranium

Vertebrae

Humerus

Ribs

Scapula

Vertebral
column

Radius

Ulna

Carpus

Metacarpals

Phalanges

Hip bone

Sacrum

Coccyx

Femur

Tibia

Fibula

(B) Posterior view

FIGURE I.11. Skeletal system.

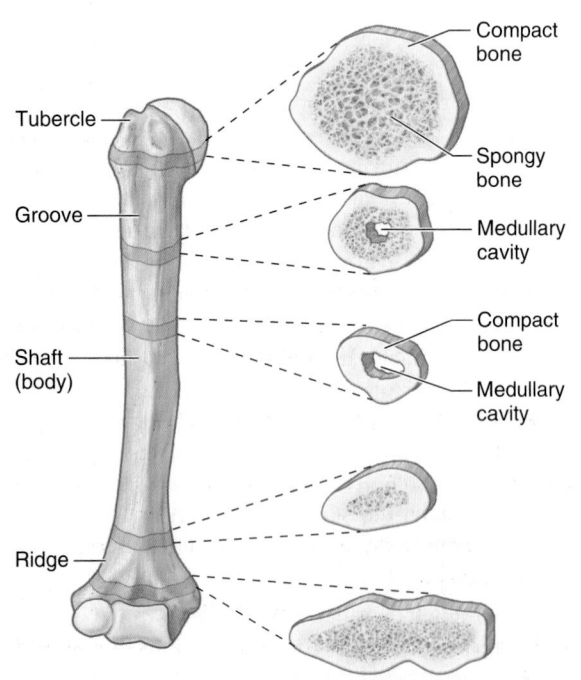

Tubercle

Groove

Shaft
(body)

Ridge

Compact
bone

Spongy
bone

Medullary
cavity

Compact
bone

Medullary
cavity

FIGURE I.12. Transverse sections of the humerus. The shaft of a living bone is a tube of compact bone that surrounds a medullary cavity.

(supports) where large muscles attach. Living bones have some *elasticity* (flexibility) and great *rigidity* (hardness).

CLASSIFICATION OF BONES

Bones are classified according to their shape.

- *Long bones* are tubular (e.g., the humerus in the arm).
- *Short bones* are cuboidal and are found only in the tarsus (ankle) and carpus (wrist).
- *Flat bones* usually serve protective functions (e.g., the flat bones of the cranium protect the brain).
- *Irregular bones* have various shapes other than long, short, or flat (e.g., bones of the face).
- *Sesamoid bones* (e.g., the patella or knee cap) develop in certain tendons and are found where tendons cross the ends of long bones in the limbs; they protect the tendons from excessive wear and often change the angle of the tendons as they pass to their attachments.

Bone Markings and Formations

Bone markings appear wherever tendons, ligaments, and fascias are attached or where arteries lie adjacent to or enter

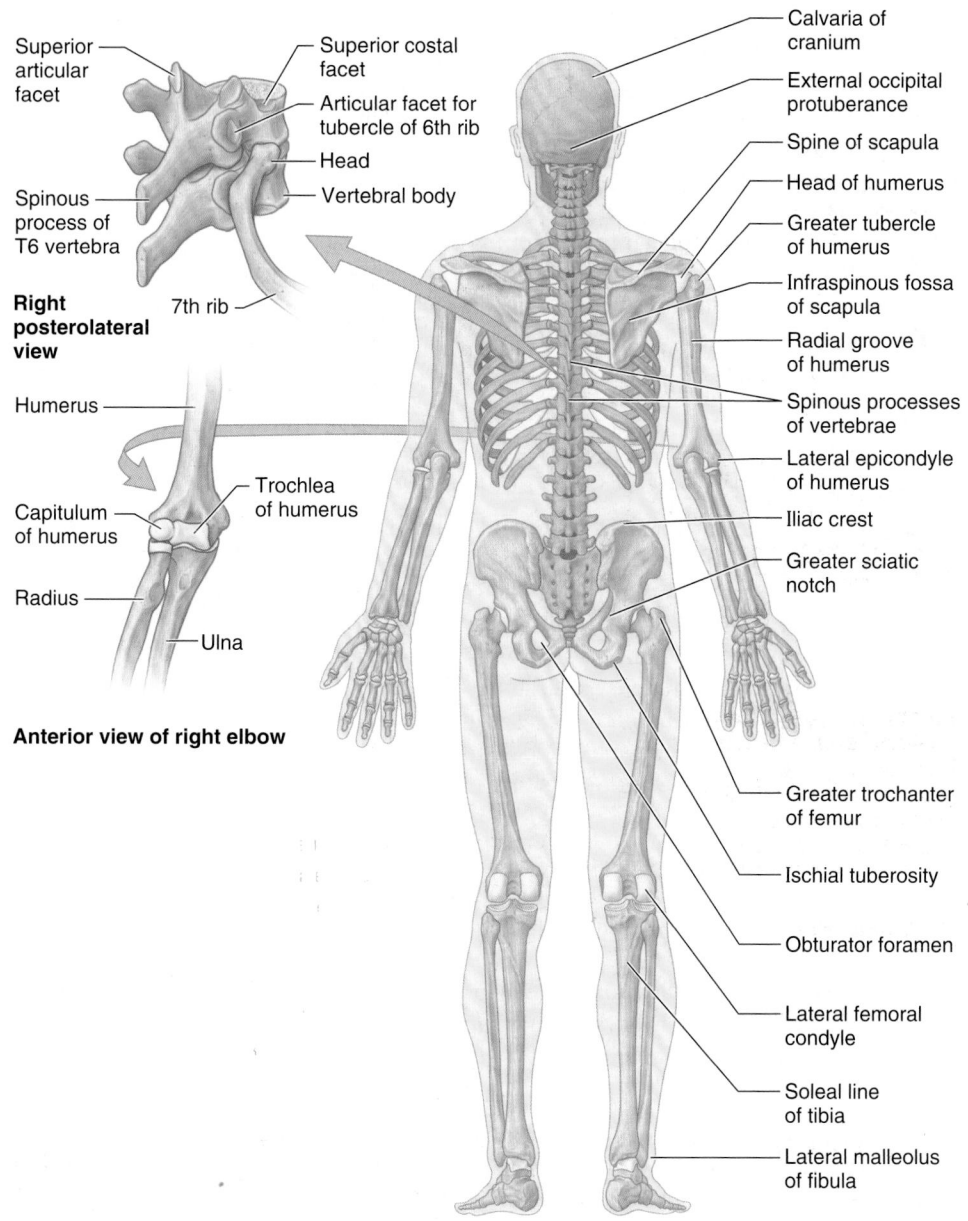

Superior articular facet

Superior costal facet

Articular facet for tubercle of 6th rib

Head

Vertebral body

Spinous process of T6 vertebra

Right posterolateral view

7th rib

Humerus

Trochlea of humerus

Capitulum of humerus

Radius

Ulna

Anterior view of right elbow

Calvaria of cranium

External occipital protuberance

Spine of scapula

Head of humerus

Greater tubercle of humerus

Infraspinous fossa of scapula

Radial groove of humerus

Spinous processes of vertebrae

Lateral epicondyle of humerus

Iliac crest

Greater sciatic notch

Greater trochanter of femur

Ischial tuberosity

Obturator foramen

Lateral femoral condyle

Soleal line of tibia

Lateral malleolus of fibula

Posterior view

FIGURE I.13. Bone markings and formations. Markings appear on bones wherever tendons, ligaments, and fascia attach. Other formations relate to joints, the passage of tendons, and the provision of increased leverage.

bones. Other formations occur in relation to the passage of a tendon (often to direct the tendon or improve its leverage) or to control the type of movement occurring at a joint. Some of the various markings and features of bones are (Fig. I.13):

- *Capitulum:* small, round, articular head (e.g., capitulum of the humerus).
- *Condyle:* rounded, knuckle-like articular area, often occurring in pairs (e.g., the lateral and medial femoral condyles).
- *Crest:* ridge of bone (e.g., the iliac crest).
- *Epicondyle:* eminence superior or adjacent to a condyle (e.g., lateral epicondyle of the humerus).
- *Facet:* smooth flat area, usually covered with cartilage, where a bone articulates with another bone (e.g., superior

costal facet on the body of a vertebra for articulation with a rib).

- *Foramen:* passage through a bone (e.g., obturator foramen).
- *Fossa:* hollow or depressed area (e.g., infraspinous fossa of the scapula).
- *Groove:* elongated depression or furrow (e.g., radial groove of the humerus).
- *Head* (L. *caput*): large, round articular end (e.g., head of the humerus).
- *Line:* linear elevation (e.g., soleal line of the tibia).
- *Malleolus:* rounded process (e.g., lateral malleolus of the fibula).
- *Notch:* indentation at the edge of a bone (e.g., greater sciatic notch).

- *Protuberance:* projection of bone (e.g., external occipital protuberance).
- *Spine:* thorn-like process (e.g., the spine of the scapula).
- *Spinous process:* projecting spine-like part (e.g., spinous process of a vertebra).
- *Trochanter:* large blunt elevation (e.g., greater trochanter of the femur).
- *Trochlea:* spool-like articular process or process that acts as a pulley (e.g., trochlea of the humerus).
- *Tubercle:* small raised eminence (e.g., greater tubercle of the humerus—Fig. I.13).
- *Tuberosity:* large rounded elevation (e.g., ischial tuberosity).

BONE DEVELOPMENT

Most bones take many years to grow and mature. The humerus (arm bone), for example, begins to ossify at the end of the embryonic period (8 weeks); however, ossification is not complete until age 20. All bones derive from *mesenchyme* (embryonic connective tissue) by two different processes: intramembranous ossification (directly from mesenchyme) and endochondral ossification (from cartilage derived from mesenchyme). The histology (microscopic structure) of a bone is the same by either process (Ross et al., 2011).

- In **intramembranous ossification** (membranous bone formation), mesenchymal models of bones form during the embryonic period, and direct ossification of the mesenchyme begins in the fetal period.
- In **endochondral ossification** (cartilaginous bone formation), cartilage models of the bones form from mesenchyme during the fetal period, and bone subsequently replaces most of the cartilage.

A brief description of endochondral ossification helps explain how long bones grow (Fig. I.14). The mesenchymal cells condense and differentiate into *chondroblasts,* dividing cells in growing cartilage tissue, thereby forming a *cartilaginous bone model.* In the midregion of the model, the cartilage *calcifies* (becomes impregnated with calcium salts), and *periosteal capillaries* (capillaries from the fibrous sheath surrounding the model) grow into the calcified cartilage of the bone model and supply its interior. These blood vessels, together with associated *osteogenic (bone-forming) cells,* form a *periosteal bud* (Fig. I.14A). The capillaries initiate the **primary ossification center,** so named because the bone tissue it forms replaces most of the cartilage in the main body of the bone model. The shaft of a bone ossified from the primary ossification center is the **diaphysis,** which grows as the bone develops.

Most **secondary ossification centers** appear in other parts of the developing bone after birth; the parts of a bone ossified from these centers are **epiphyses.** The chondrocytes in the middle of the epiphysis hypertrophy, and the *bone matrix* (extracellular substance) between them

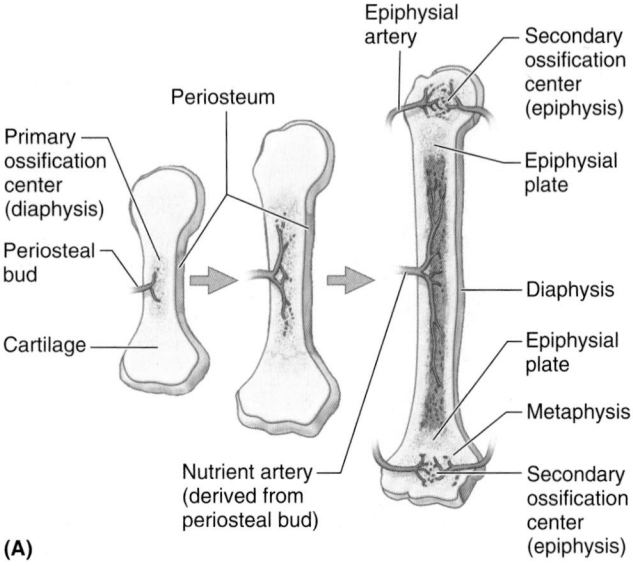

(A)

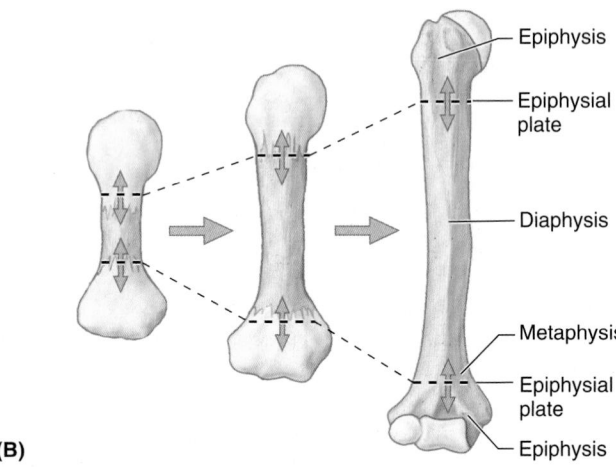

(B)

FIGURE I.14. Development and growth of a long bone. A. The formation of primary and secondary ossification centers is shown. **B.** Growth in length occurs on both sides of the cartilaginous epiphysial plates (*double-headed arrows*). The bone formed from the primary center in the diaphysis does not fuse with that formed from the secondary centers in the epiphyses until the bone reaches its adult size. When growth ceases, the depleted epiphysial plate is replaced by a synostosis (bone-to-bone fusion), observed as an epiphysial line in radiographs and sectioned bone.

calcifies. *Epiphysial arteries* grow into the developing cavities with associated osteogenic cells. The flared part of the diaphysis nearest the epiphysis is the **metaphysis.** For growth to continue, the bone formed from the primary center in the diaphysis does not fuse with that formed from the secondary centers in the epiphyses until the bone reaches its adult size. Thus, during growth of a long bone, cartilaginous **epiphysial plates** intervene between the diaphysis and epiphyses (Fig. I.14B). These growth plates are eventually replaced by bone at each of its two sides, diaphysial and epiphysial. When this occurs, bone growth ceases and the diaphysis fuses with the epiphyses.

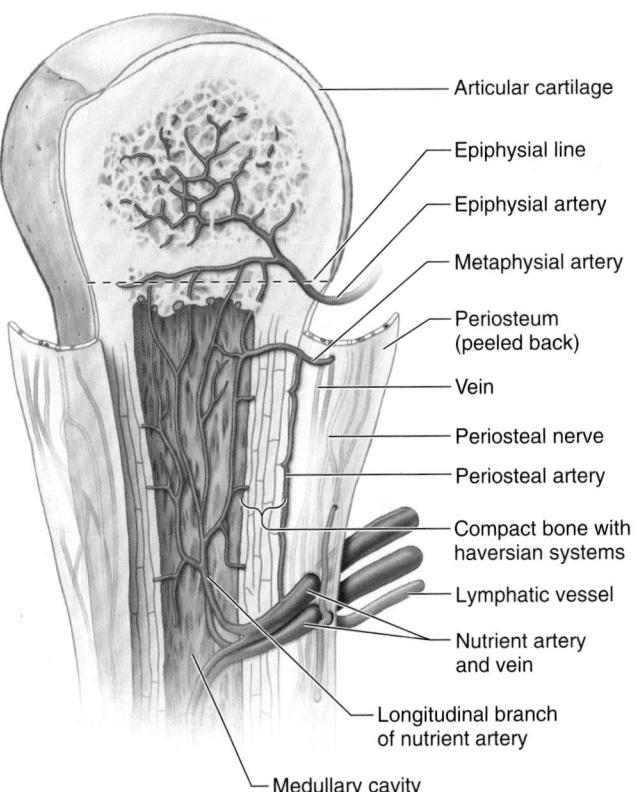

Articular cartilage

Epiphysial line

Epiphysial artery

Metaphysial artery

Periosteum (peeled back)

Vein

Periosteal nerve

Periosteal artery

Compact bone with haversian systems

Lymphatic vessel

Nutrient artery and vein

Longitudinal branch of nutrient artery

Medullary cavity

FIGURE I.15. Vasculature and innervation of a long bone.

The seam formed during this fusion process (*synostosis*) is particularly dense and is recognizable in sectioned bone or radiographs as an **epiphysial line** (Fig. I.15). The epiphysial fusion of bones occurs progressively from puberty to maturity. *Ossification of short bones* is similar to that of the primary ossification center of long bones, and only one short bone, the calcaneus (heel bone), develops a secondary ossification center.

VASCULATURE AND INNERVATION OF BONES

Bones are richly supplied with blood vessels. Most apparent are the **nutrient arteries** (one or more per bone) that arise as independent branches of adjacent arteries outside the periosteum and pass obliquely through the compact bone of the shaft of a long bone via **nutrient foramina.** The nutrient artery divides in the medullary cavity into longitudinal branches that proceed toward each end, supplying the bone marrow, spongy bone, and deeper portions of the compact bone (Fig. I.15). However, many small branches from the periosteal arteries of the periosteum are responsible for nourishment of most of the compact bone. Consequently, a bone from which the periosteum has been removed dies.

Blood reaches the *osteocytes* (bone cells) in the compact bone by means of **haversian systems** or *osteons* (microscopic canal systems) that house small blood vessels. The ends of the bones are supplied by metaphysial and epiphysial arteries that arise mainly from the arteries that supply the joints. In the limbs, these arteries are typically part of a *periarticular arterial plexus,* which surrounds the joint, ensuring blood flow distal to the joint regardless of the position assumed by the joint.

Veins accompany arteries through the nutrient foramina. Many large veins also leave through foramina near the articular ends of the bones. Bones containing red bone marrow have numerous large veins. *Lymphatic vessels* are also abundant in the periosteum.

Nerves accompany blood vessels supplying bones. The periosteum is richly supplied with sensory nerves—**periosteal nerves**—that carry pain fibers. The periosteum is especially sensitive to tearing or tension, which explains the acute pain from bone fractures. Bone itself is relatively sparsely supplied with sensory endings. Within bones, **vasomotor nerves** cause constriction or dilation of blood vessels, regulating blood flow through the bone marrow.

BONES

Accessory (Supernumerary) Bones

Accessory (supernumerary) bones develop when additional ossification centers appear and form extra bones. Many bones develop from several centers of ossification, and the separate parts normally fuse. Sometimes one of these centers fails to fuse with the main bone, giving the appearance of an extra bone. Careful study shows that the apparent extra bone is a missing part of the main bone. Circumscribed areas of bone are often seen along the sutures of the cranium where the flat bones abut, particularly those related to the parietal bone (see Chapter 7). These small, irregular, worm-like bones are **sutural bones** (wormian bones). It is important to know that accessory bones are common in the foot, to avoid mistaking them for bone fragments in radiographs and other medical images.

Heterotopic Bones

Bones sometimes form in soft tissues where they are not normally present (e.g., in scars). Horse riders often develop *heterotopic bones* in their thighs (*rider's bones*), probably because of chronic muscle strain resulting in small hemorrhagic (bloody) areas that undergo calcification and eventual ossification.

Trauma to Bone and Bone Changes

Bones are living organs that cause pain when injured, bleed when fractured, remodel in relationship to stresses placed on them, and change

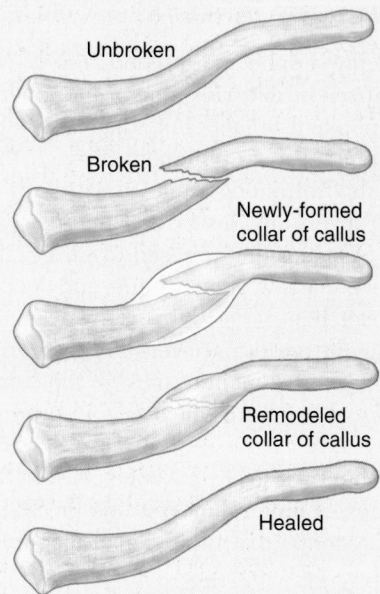

Unbroken

Broken

Newly-formed
collar of callus

Remodeled
collar of callus

Healed

FIGURE BI.4.

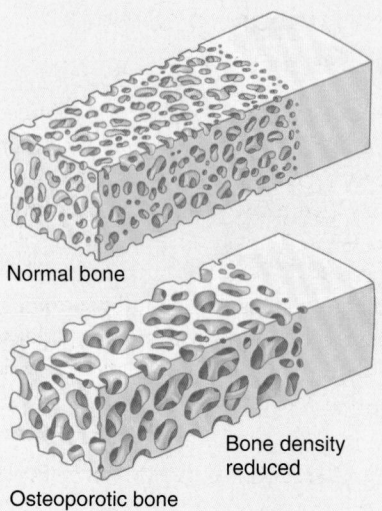

Normal bone

Bone density
reduced

Osteoporotic bone

FIGURE BI.5.

with age. Like other organs, bones have blood vessels, lymphatic vessels, and nerves, and they may become diseased. Unused bones, such as in a paralyzed limb, *atrophy* (decrease in size). Bone may be absorbed, which occurs in the mandible when teeth are extracted. Bones *hypertrophy* (enlarge) when they support increased weight for a long period.

Trauma to a bone may break it. For the fracture to heal properly, the broken ends must be brought together, approximating their normal position. This is called *reduction of a fracture*. During bone healing, the surrounding *fibroblasts* (connective tissue cells) proliferate and secrete collagen, which forms a *collar of callus* to hold the bones together (Fig. BI.4). Bone remodeling occurs in the fracture area, and the callus calcifies. Eventually, the callus is resorbed and replaced by bone. After several months, little evidence of the fracture remains, especially in young people. Fractures are more common in children than in adults because of the combination of their slender, growing bones and carefree activities. Fortunately, many of these breaks are *greenstick fractures* (incomplete breaks caused by bending of the bones). Fractures in growing bones heal faster than those in adult bones.

Osteoporosis

During the aging process, the organic and inorganic components of bone both decrease, often resulting in *osteoporosis*, a reduction in the quantity of bone, or atrophy of skeletal tissue (Fig. BI.5). Hence, the bones become brittle, lose their elasticity, and fracture easily. *Bone scanning* is an imaging method used to assess normal and diminished bone mass (see the section on Medical Imaging Techniques at the end of this Introduction).

Sternal Puncture

Examination of bone marrow provides valuable information for evaluating hematological (blood) diseases. Because it lies just beneath the skin (i.e., is subcutaneous) and is easily accessible, the sternum (breast bone) is a commonly used site for harvesting bone marrow. During a *sternal puncture,* a wide-bore (large diameter) needle is inserted through the thin cortical bone into the spongy bone. A sample of *red bone marrow* is aspirated with a syringe for laboratory examination. *Bone marrow transplantation* is sometimes performed in the treatment of leukemia.

Bone Growth and Assessment of Bone Age

Knowledge of the sites where ossification centers occur, the times of their appearance, the rates at which they grow, and the times of fusion of the sites (times when synostosis occurs) is important in clinical medicine, forensic science, and anthropology. A general index of growth during infancy, childhood, and adolescence is indicated by *bone age,* as determined from radiographs (negative images on X-ray films). The age of a young person can be determined by studying the ossification centers in bones. The main criteria are: (1) the appearance of calcified material in the diaphysis and/or epiphyses and (2) the disappearance of the radiolucent (dark) line representing the epiphysial plate (absence of this line indicates that epiphysial fusion has occurred; fusion occurs at specific times for each epiphysis). The fusion of epiphyses with the diaphysis occurs 1 to 2 years earlier in girls than in boys. Determining bone age can be helpful in predicting adult height in early- or late-maturing adolescents. Assessment of bone age also helps establish the approximate age of human skeletal remains in medicolegal cases.

Effects of Disease and Diet on Bone Growth

 Some diseases produce early epiphysial fusion (ossification time) compared with what is normal for the person's chronological age; other diseases result in delayed fusion. The growing skeleton is sensitive to relatively slight and transient illnesses and to periods of malnutrition. Proliferation of cartilage at the metaphyses slows down during starvation and illness, but the degeneration of cartilage cells in the columns continues, producing a dense line of provisional calcification. These lines later become bone with thickened trabeculae, or *lines of arrested growth*.

Displacement and Separation of Epiphyses

 Without knowledge of bone growth and the appearance of bones in radiographic and other diagnostic images at various ages, a *displaced epiphysial plate*

could be mistaken for a fracture, and *separation of an epiphysis* could be interpreted as a displaced piece of a fractured bone. Knowing the patient's age and the location of epiphyses can prevent these anatomical errors. The edges of the diaphysis and epiphysis are smoothly curved in the region of the epiphysial plate. Bone fractures always leave a sharp, often uneven edge of bone. An injury that causes a fracture in an adult usually causes the displacement of an epiphysis in a child.

Avascular Necrosis

 Loss of arterial supply to an epiphysis or other parts of a bone results in the death of bone tissue—*avascular necrosis* (G. *nekrosis*, deadness). After every fracture, small areas of adjacent bone undergo necrosis. In some fractures, avascular necrosis of a large fragment of bone may occur. A number of clinical disorders of epiphyses in children result from avascular necrosis of unknown etiology (cause). These disorders are referred to as *osteochondroses*.

The Bottom Line

CARTILAGE AND BONES

The skeletal system can be divided into the axial (bones of the head, neck, and trunk) and appendicular skeletons (bones of the limbs). The skeleton itself is composed of several types of tissue: ♦ cartilage, a semirigid connective tissue; ♦ bone, a hard form of connective tissue that provides support, protection, movement, storage (of certain electrolytes), and synthesis of blood cells; ♦ periosteum, which surrounds bones, and perichondrium, which surrounds cartilage, provide nourishment for these tissues and are the sites of new cartilage and bone formation. Two types of bone, spongy and compact, are distinguished by the amount of solid matter and the size and number of spaces they contain. Bones can be classified as long, short, flat, irregular, or sesamoid. Standard terms for specific bone markings and features are used when describing the structure of individual bones.

Most bones take many years to grow. Bones grow through the processes of: ♦ intramembraneous ossification, in which mesenchymal bone models are formed during the embryonic and prenatal periods, and ♦ endochondral ossification, in which cartilage models are formed during the fetal period, with bone subsequently replacing most of the cartilage after birth.

Joints

Joints (articulations) are unions or junctions between two or more bones or rigid parts of the skeleton. Joints exhibit a variety of forms and functions. Some joints have no movement, such as the epiphysial plates between the epiphysis and diaphysis of a growing long bone; others allow only slight movement, such as teeth within their sockets; and some are freely movable, such as the glenohumeral (shoulder) joint.

CLASSIFICATION OF JOINTS

Three classes of joints are described, based on the manner or type of material by which the articulating bones are united.

1. The articulating bones of **synovial joints** are united by a **joint** (articular) **capsule** (composed of an outer **fibrous layer** lined by a serous **synovial membrane**) spanning and enclosing an articular cavity. The **joint cavity** of a synovial joint, like the knee, is a potential space that contains a small amount of lubricating **synovial fluid,** secreted by the synovial membrane. Inside the capsule, articular cartilage covers the articulating surfaces of the bones; all other internal surfaces are covered by synovial membrane. The bones in Figure I.16A, normally closely apposed, have been pulled apart for demonstration, and the joint capsule has been inflated. Consequently the normally potential joint cavity is exaggerated. The *periosteum* investing the participating bones external to the joint blends with the fibrous layer of the joint capsule.

2. The articulating bones of **fibrous joints** are united by fibrous tissue. The amount of movement occurring at a fibrous joint depends in most cases on the length of the fibers uniting the articulating bones. The *sutures of the cranium* are examples of fibrous joints (Fig. I.16B). These bones are close together, either interlocking along a wavy line or overlapping. A **syndesmosis** type of fibrous joint unites the bones with a sheet of fibrous tissue, either a ligament or a fibrous membrane. Consequently, this type of

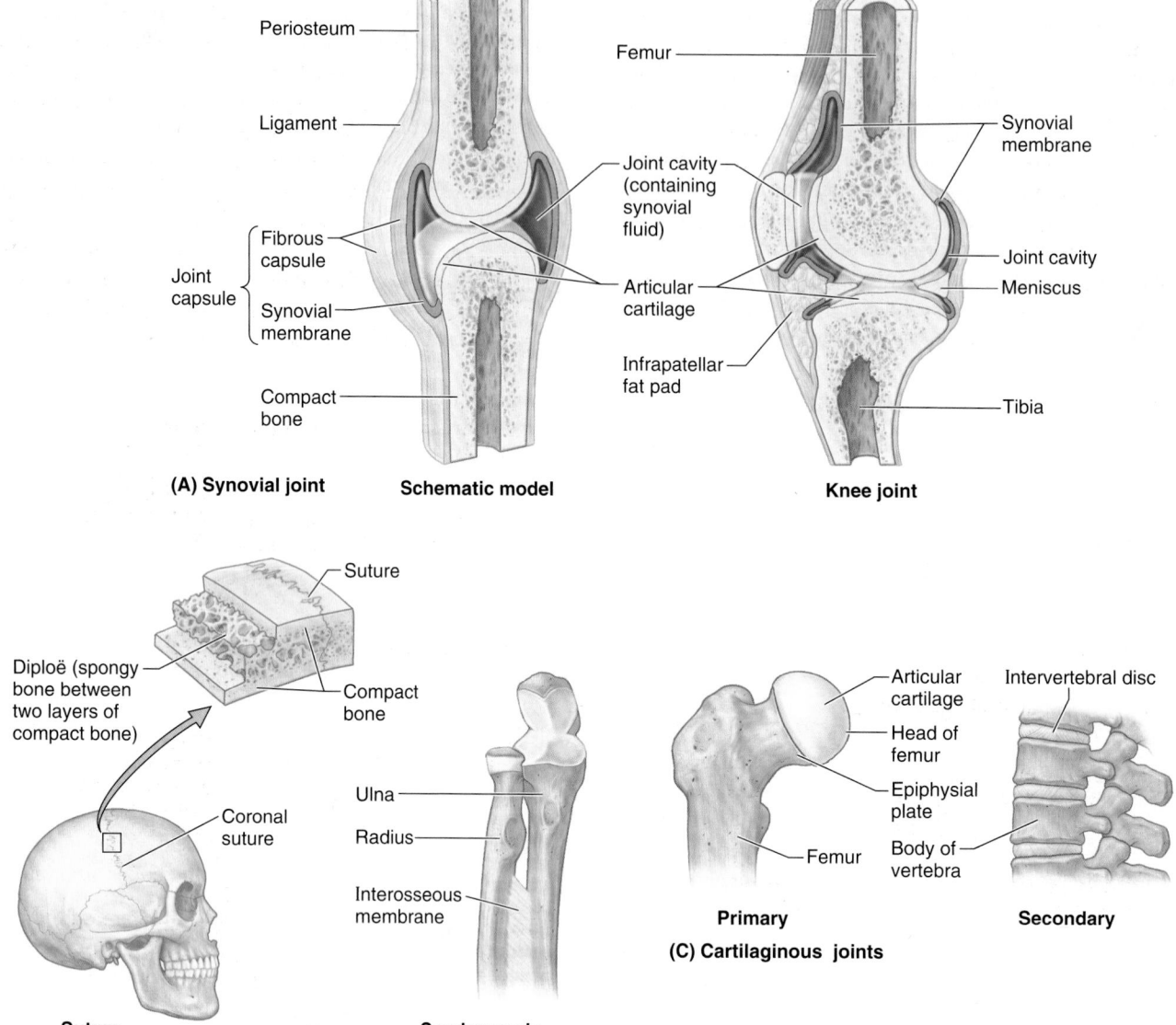

FIGURE I.16. Three classes of joint. Examples of each class are shown. **A.** Two models demonstrating basic features of a synovial joint.

joint is partially movable. The interosseous membrane in the forearm is a sheet of fibrous tissue that joins the radius and ulna in a syndesmosis. A **dento-alveolar syndesmosis** (gomphosis or socket) is a fibrous joint in which a peg-like process fits into a socket articulation between the root of the tooth and the alveolar process of the jaw. Mobility of this joint (a loose tooth) indicates a pathological state affecting the supporting tissues of the tooth. However, microscopic movements here give us information (via the sense of proprioception) about how hard we are biting or clenching our teeth and whether we have a particle stuck between our teeth.

3. The articulating structures of **cartilaginous joints** are united by hyaline cartilage or fibrocartilage. In **primary cartilaginous joints,** or synchondroses, the bones are united by hyaline cartilage, which permits slight bending during early life. Primary cartilaginous joints are usually temporary unions, such as those present during the development of a long bone (Figs. I.14 and I.16C), where the bony epiphysis and the shaft are joined by an epiphysial plate. Primary cartilaginous joints permit growth in the length of a bone. When full growth is achieved, the epiphysial plate converts to bone and the epiphyses fuse with the diaphysis. **Secondary cartilaginous joints,** or symphyses, are strong, slightly movable joints united by fibrocartilage. The fibrocartilaginous *intervertebral discs* (Fig. I.16C) between the vertebrae consist of binding connective tissue that joins the vertebrae together. Cumulatively, these joints provide strength and shock absorption as well as considerable flexibility to the vertebral column (spine).

Synovial joints, the most common type of joint, provide free movement between the bones they join; they are joints of locomotion, typical of nearly all limb joints. Synovial joints are usually reinforced by **accessory ligaments** that are either separate (*extrinsic*) or are a thickening of a portion of the joint capsule (*intrinsic*). Some synovial joints have other distinguishing features, such as a fibrocartilaginous *articular disc* or meniscus, which are present when the articulating surfaces of the bones are incongruous (Fig. I.16A).

The six major types of synovial joints are classified according to the shape of the articulating surfaces and/or the type of movement they permit (Fig. I.17):

1. **Plane joints** permit gliding or sliding movements in the plane of the articular surfaces. The opposed surfaces of the bones are flat or almost flat, with movement limited by their tight joint capsules. Plane joints are numerous and are nearly always small. An example is the *acromioclavicular joint* between the acromion of the scapula and the clavicle.

2. **Hinge joints** permit flexion and extension only, movements that occur in one plane (sagittal) around a single axis that runs transversely; thus hinge joints are *uniaxial joints.* The joint capsule of these joints is thin and lax anteriorly and posteriorly where movement occurs; however, the bones are joined by strong, laterally placed collateral ligaments. The *elbow joint* is a hinge joint.

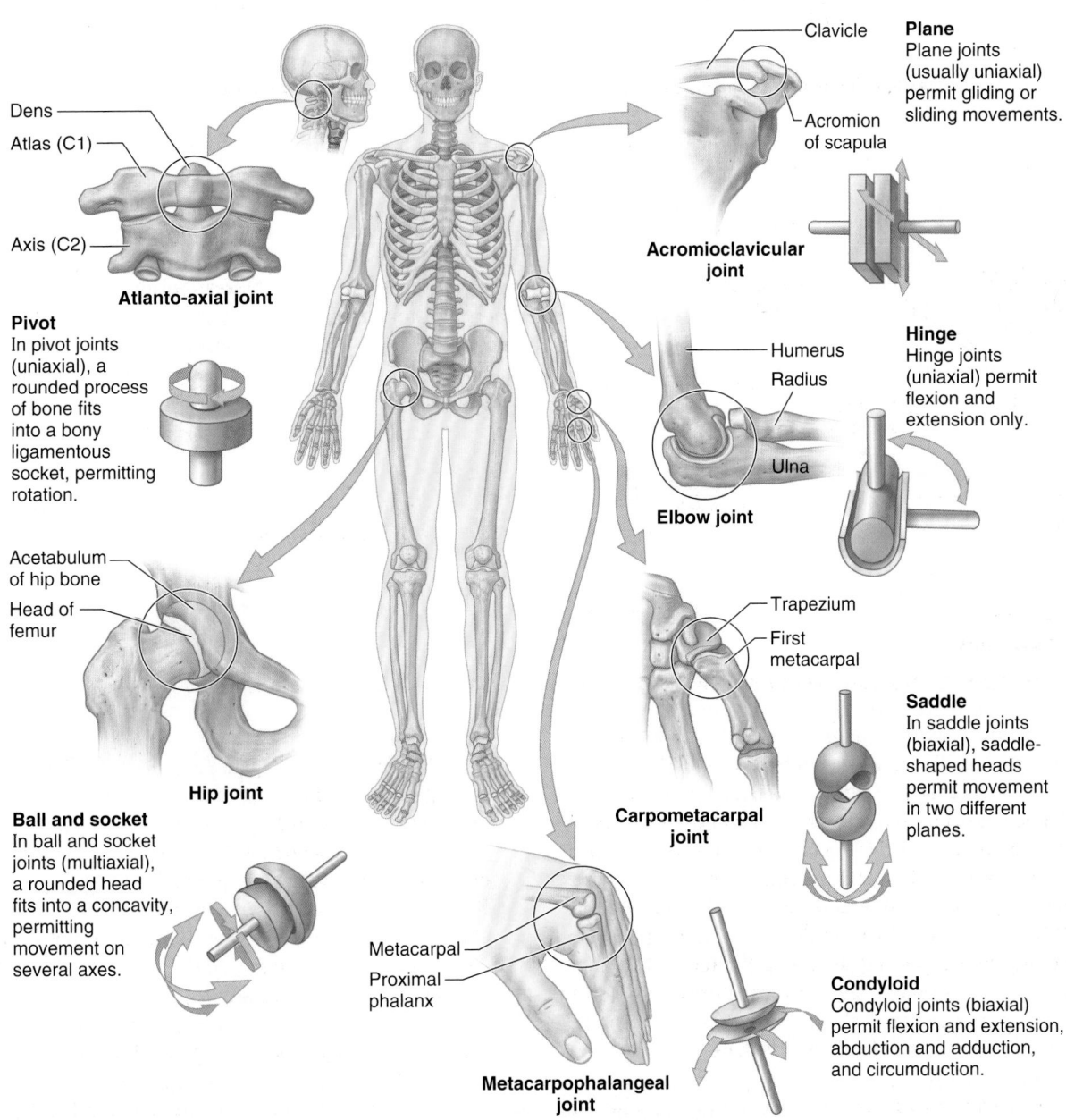

FIGURE I.17. **The six types of synovial joints.** Synovial joints are classified according to the shape of their articulating surfaces and/or the type of movement they permit.

3. **Saddle joints** permit abduction and adduction as well as flexion and extension, movements occurring around two axes at right angles to each other; thus saddle joints are *biaxial joints* that allow movement in two planes, sagittal and frontal. The performance of these movements in a circular sequence (circumduction) is also possible. The opposing articular surfaces are shaped like a saddle (i.e., they are reciprocally concave and convex). The *carpometacarpal joint* at the base of the 1st digit (thumb) is a saddle joint (Fig. I.17).

4. **Condyloid joints** permit flexion and extension as well as abduction and adduction; thus condyloid joints are also biaxial. However, movement in one plane (sagittal) is usually greater (freer) than in the other. Circumduction, more restricted than that of saddle joints, is also possible. The *metacarpophalangeal joints* (knuckle joints) are condyloid joints.

5. **Ball and socket joints** allow movement in multiple axes and planes: flexion and extension, abduction and adduction, medial and lateral rotation, and circumduction; thus ball and socket joints are *multiaxial joints.* In these highly mobile joints, the spheroidal surface of one bone moves within the socket of another. The *hip joint* is a ball and socket joint in which the spherical *head of the femur* rotates within the socket formed by the *acetabulum* of the hip bone.

6. **Pivot joints** permit rotation around a central axis; thus they are uniaxial. In these joints, a rounded process of bone rotates within a sleeve or ring. The *median atlantoaxial joint* is a pivot joint in which the atlas (C1 vertebra) rotates around a finger-like process, the dens of the axis (C2 vertebra), during rotation of the head.

JOINT VASCULATURE AND INNERVATION

Joints receive blood from **articular arteries** that arise from the vessels around the joint. The arteries often *anastomose* (communicate) to form networks (*peri-articular arterial anastomoses*) to ensure a blood supply to and across the joint in the various positions assumed by the joint. **Articular veins** are communicating veins that accompany arteries (L. *venae comitantes*) and, like the arteries, are located in the joint capsule, mostly in the synovial membrane.

Joints have a rich nerve supply provided by **articular nerves** with sensory nerve endings in the joint capsule. In the distal parts of the limbs (hands and feet), the articular nerves are branches of the cutaneous nerves supplying the overlying skin. However, most articular nerves are branches of nerves that supply the muscles that cross and therefore move the joint. The **Hilton law** states that the nerves supplying a joint also supply the muscles moving the joint and the skin covering their distal attachments.

Articular nerves transmit sensory impulses from the joint that contribute to the sense of **proprioception**, which provides an awareness of movement and position of the parts of the body. The synovial membrane is relatively insensitive. Pain fibers are numerous in the fibrous layer of the joint capsule and the accessory ligaments, causing considerable pain when the joint is injured. The sensory nerve endings respond to the twisting and stretching that occurs during sports activities.

JOINTS

Joints of Newborn Cranium

The bones of the *calvaria* (skullcap) of a newborn infant's cranium do not make full contact with each other (Fig. BI.6). At these sites, the sutures form wide areas of fibrous tissue called **fontanelles**. The *anterior fontanelle* is the most prominent; laypeople call it the "soft spot." The fontanelles in a newborn are often felt as ridges because of the overlapping of the cranial bones by molding of the calvaria as it passes through the birth canal. Normally, the anterior fontanelle is flat. A bulging fontanelle may indicate increased intracranial pressure; however, the fontanelle normally bulges during crying. Pulsations of the fontanelle reflect the pulse of cerebral arteries. A depressed fontanelle may be observed when the baby is dehydrated (Swartz, 2001).

Degenerative Joint Disease

Synovial joints are well designed to withstand wear, but heavy use over several years can cause degenerative changes. Some destruction is inevitable during such activities as jogging, which wears away the articular cartilages and sometimes erodes the underlying articulating

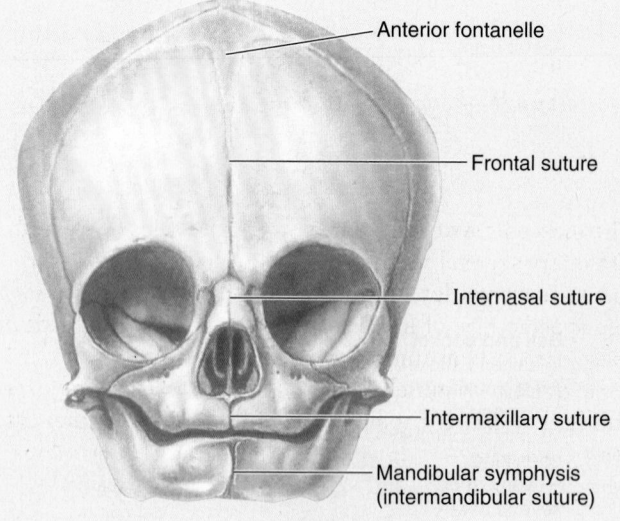

Anterior fontanelle

Frontal suture

Internasal suture

Intermaxillary suture

Mandibular symphysis (intermandibular suture)

FIGURE BI.6.

surfaces of the bones. The normal aging of articular cartilage begins early in adult life and progresses slowly thereafter, occurring on the ends of the articulating bones, particularly those of the hip, knee, vertebral column, and hands (Salter, 1998). These irreversible degenerative changes in joints result in the articular cartilage becoming a less effective shock absorber and lubricated surface. As a result, the articulation becomes increasingly vulnerable to the repeated friction that occurs during joint movements. In some people, these changes do not produce significant symptoms; in others, they cause considerable pain.

Degenerative joint disease or *osteoarthritis* is often accompanied by stiffness, discomfort, and pain. Osteoarthritis is common in older people and usually affects joints that support the weight of their bodies (e.g., the hips and knees). Most substances in the bloodstream, normal or pathological, easily enter the joint cavity. Similarly, traumatic infection of a joint may be followed by *arthritis,* inflammation of a joint, and *septicemia,* blood poisoning.

The Bottom Line

JOINTS

A joint is a union between two or more bones or rigid parts of the skeleton. Three general types of joints are recognized: fibrous, cartilaginous, and synovial. Freely moveable synovial joints: ♦ are the most common type; ♦ can be classified into plane, hinge, saddle, condyloid, ball and socket, and pivot; ♦ receive their blood supply from articular arteries that often form networks; ♦ are drained by articular veins originating in the synovial membrane; ♦ are richly innervated by articular nerves that transmit the sensation of proprioception, an awareness of movement and position of parts of the body.

MUSCLE TISSUE AND MUSCULAR SYSTEM

The muscular system consists of all the muscles of the body. Voluntary skeletal muscles constitute the great majority of the named muscles. All skeletal muscles are composed of one specific type of muscle tissue. However, other types of muscle tissue constitute a few named muscles (e.g., the ciliary and detrusor muscles, and the arrector muscles of hairs) and form important components of the organs of other systems, including the cardiovascular, alimentary, genitourinary, integumentary, and visual systems.

Types of Muscle (Muscle Tissue)

Muscle cells, often called *muscle fibers* because they are long and narrow when relaxed, are specialized contractile cells. They are organized into tissues that move body parts or temporarily alter the shape (reduce the circumference of all or part) of internal organs. Associated connective tissue conveys nerve fibers and capillaries to the muscle cells as it binds

Arthroscopy

 The cavity of a synovial joint can be examined by inserting a cannula and an arthroscope (a small telescope) into it. This surgical procedure—*arthroscopy*—enables orthopedic surgeons to examine joints for abnormalities, such as torn menisci (partial articular discs of the knee joint). Some surgical procedures can also be performed during arthroscopy (e.g., by inserting instruments through small puncture incisions). Because the opening in the joint capsule for inserting the arthroscope is small, healing is more rapid after this procedure than after traditional joint surgery.

them into bundles or fascicles. Three types of muscle are described based on distinct characteristics relating to:

- Whether it is normally willfully controlled (*voluntary* vs. *involuntary*).
- Whether it appears striped or unstriped when viewed under a microscope (*striated* vs. *smooth* or *unstriated*).
- Whether it is located in the body wall (*soma*) and limbs or makes up the hollow organs (*viscera,* e.g., the heart) of the body cavities or blood vessels (*somatic* vs. *visceral*).

There are three muscle types (Table I.1):

1. **Skeletal striated muscle** is voluntary somatic muscle that makes up the gross *skeletal muscles* that compose the muscular system, moving or stabilizing bones and other structures (e.g., the eyeballs).
2. **Cardiac striated muscle** is involuntary visceral muscle that forms most of the walls of the heart and adjacent parts of the great vessels, such as the aorta, and pumps blood.
3. **Smooth muscle** (unstriated muscle) is involuntary visceral muscle that forms part of the walls of most vessels and hollow organs (viscera), moving substances through them by coordinated sequential contractions (pulsations or peristaltic contractions).

Skeletal Muscles

FORM, FEATURES, AND NAMING OF MUSCLES

All skeletal muscles, commonly referred to simply as "muscles," have fleshy, reddish, contractile portions (one or more heads or bellies) composed of skeletal striated muscle. Some muscles are fleshy throughout, but most also have white non-contractile portions (tendons), composed mainly of organized collagen bundles, that provide a means of attachment (Fig. I.18).

When referring to the length of a muscle, both the belly and the tendons are included. In other words, a muscle's length is the distance between its attachments. Most skeletal muscles are attached directly or indirectly to bones, cartilages, ligaments, or fascias or to some combination of these structures. Some muscles are attached to organs (the eyeball, for example), skin (such as facial muscles), and mucous

TABLE I.1. TYPES OF MUSCLE (MUSCLE TISSUE)

Muscle Type	Location	Appearance of Cells	Type of Activity	Stimulation
Skeletal striated muscle Striation Muscle fiber Nucleus Satellite cell	Composes gross, named muscles (e.g., biceps of arm) attached to skeleton and fascia of limbs, body wall, and head/neck	Large, very long, unbranched, cylindrical fibers with transverse striations (stripes) arranged in parallel bundles; multiple, peripherally located nuclei	Intermittent (phasic) contraction above a baseline tonus; acts primarily to produce movement (isotonic contraction) through shortening (concentric contraction) or controlled relaxation (eccentric contraction), or to maintain position against gravity or other resisting force without movement (isometric contraction)	Voluntary (or reflexive) by somatic nervous system
Cardiac striated muscle Nucleus Intercalated disc Striation Muscle fiber	Muscle of heart (myocardium) and adjacent portions of great vessels (aorta, vena cava)	Branching and anastomosing shorter fibers with transverse striations (stripes) running parallel and connected end to end by complex junctions (intercalated discs); single, central nucleus	Strong, quick, continuous rhythmic contraction; acts to pump blood from heart	Involuntary; intrinsically (myogenically) stimulated and propagated; rate and strength of contraction modified by autonomic nervous system
Smooth (unstriated or unstriped) muscle Smooth muscle fiber Nucleus	Walls of hollow viscera and blood vessels, iris, and ciliary body of eye; attached to hair follicles of skin (arrector muscle of hair)	Single or agglomerated small, spindle-shaped fibers without striations; single central nucleus	Weak, slow, rhythmic, or sustained tonic contraction; acts mainly to propel substances (peristalsis) and to restrict flow (vasoconstriction and sphincteric activity)	Involuntary by autonomic nervous system

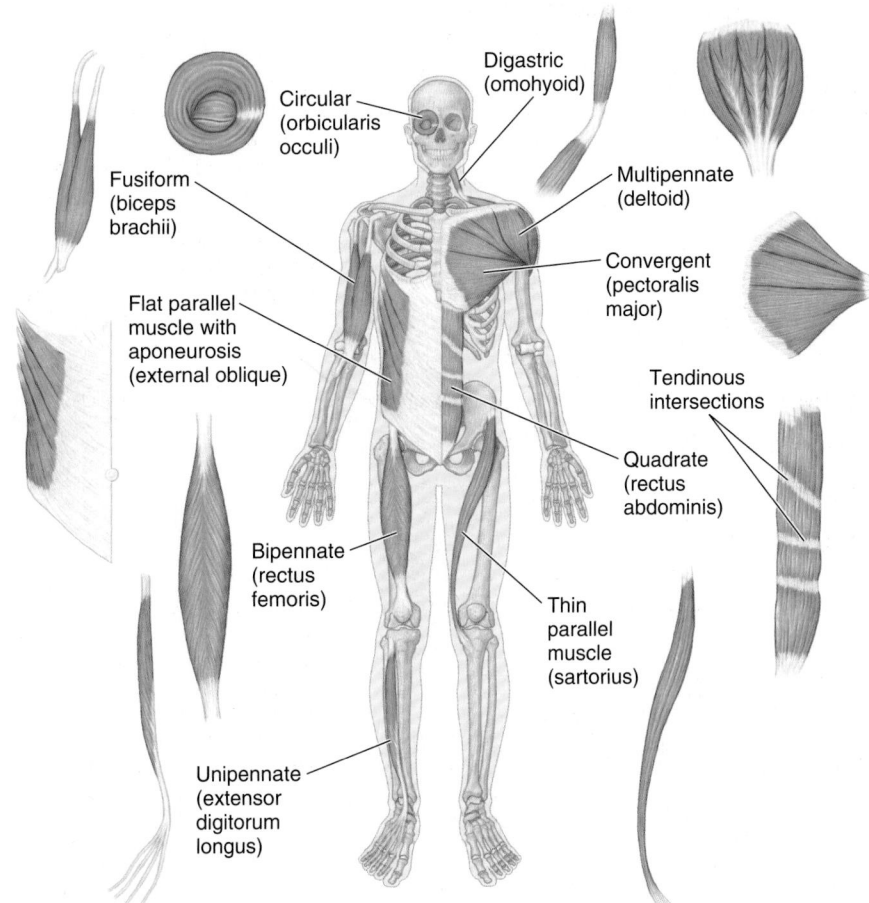

FIGURE I.18. Architecture and shape of skeletal muscles. The architecture and shape of a skeletal muscle depend on the arrangement of its fibers.

membranes (intrinsic tongue muscles). Muscles are organs of locomotion (movement), but they also provide static support, give form to the body, and provide heat. Figure I.19 identifies the skeletal muscles that lie most superficially. The deep muscles are identified when each region is studied.

The architecture and shape of muscles vary (Fig. I.18). The tendons of some muscles form flat sheets, or **aponeuroses,** that anchor the muscle to the skeleton (usually a ridge or a series of spinous processes) and/or to deep fascia (such as the latissimus dorsi muscle of the back), or to the aponeurosis of another muscle (such as the oblique muscles of the anterolateral abdominal wall). Most muscles are named on the basis of their function or the bones to which they are attached. The abductor digiti minimi muscle, for example, abducts the little finger. The sternocleidomastoid muscle (G. *kleidos*, bolt or bar, clavicle) attaches inferiorly to the sternum and clavicle and superiorly to the mastoid process of the temporal bone of the cranium. Other muscles are named on the basis of their position (medial, lateral, anterior, posterior) or length (brevis, short; longus, long). Muscles may be described or classified according to their shape, for which a muscle may also be named:

- **Flat muscles** have parallel fibers often with an aponeurosis—for example, the external oblique (broad flat muscle). The sartorius is a narrow flat muscle with parallel fibers.
- **Pennate muscles** are feather-like (L. *pennatus,* feather) in the arrangement of their fascicles, and may be *unipennate, bipennate,* or *multipennate*—for example, extensor digitorum longus (unipennate), rectus femoris (bipennate), and deltoid (multipennate).
- **Fusiform muscles** are spindle shaped with a round, thick belly (or bellies) and tapered ends—for example, biceps brachii.
- **Convergent muscles** arise from a broad area and converge to form a single tendon—for example, pectoralis major.
- **Quadrate muscles** have four equal sides (L. *quadratus,* square)—for example, the rectus abdominis, between its tendinous intersections.
- **Circular** or **sphincteral muscles** surround a body opening or orifice, constricting it when contracted—for example, orbicularis oculi (closes the eyelids).
- **Multiheaded** or **multibellied muscles** have more than one head of attachment or more than one contractile belly, respectively. *Biceps muscles* have two heads of attachment (e.g., biceps brachii), *triceps muscles* have three heads (e.g., triceps brachii), and the digastric and gastrocnemius muscles have two bellies. (Those of the former are arranged in tandem; those of the latter lie parallel.)

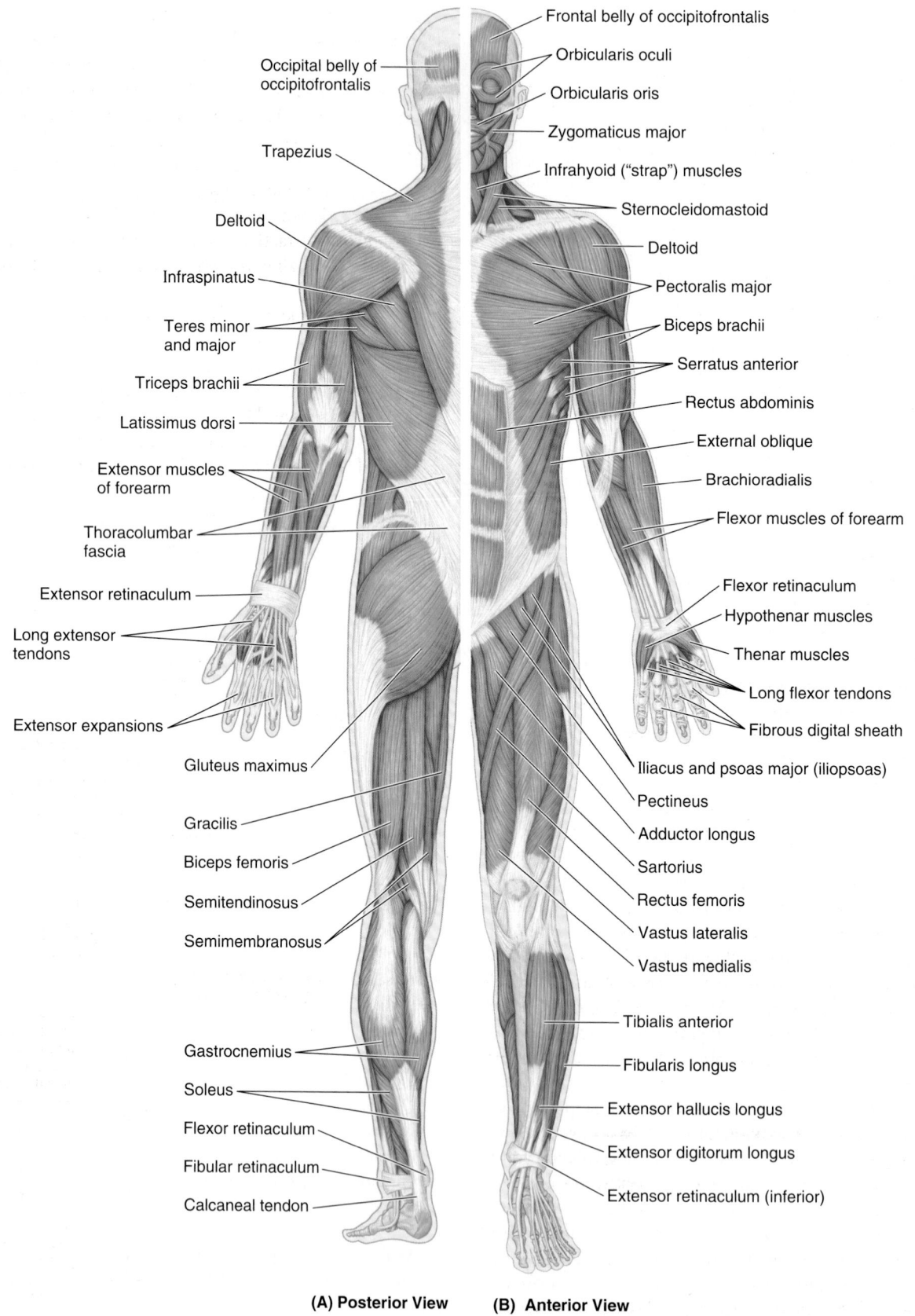

Occipital belly of occipitofrontalis

Trapezius

Deltoid

Infraspinatus

Teres minor and major

Triceps brachii

Latissimus dorsi

Extensor muscles of forearm

Thoracolumbar fascia

Extensor retinaculum

Long extensor tendons

Extensor expansions

Gluteus maximus

Gracilis

Biceps femoris

Semitendinosus

Semimembranosus

Gastrocnemius

Soleus

Flexor retinaculum

Fibular retinaculum

Calcaneal tendon

Frontal belly of occipitofrontalis

Orbicularis oculi

Orbicularis oris

Zygomaticus major

Infrahyoid ("strap") muscles

Sternocleidomastoid

Deltoid

Pectoralis major

Biceps brachii

Serratus anterior

Rectus abdominis

External oblique

Brachioradialis

Flexor muscles of forearm

Flexor retinaculum

Hypothenar muscles

Thenar muscles

Long flexor tendons

Fibrous digital sheath

Iliacus and psoas major (iliopsoas)

Pectineus

Adductor longus

Sartorius

Rectus femoris

Vastus lateralis

Vastus medialis

Tibialis anterior

Fibularis longus

Extensor hallucis longus

Extensor digitorum longus

Extensor retinaculum (inferior)

(A) Posterior View   **(B) Anterior View**

FIGURE I.19. Superficial skeletal muscles. Most of the muscles shown move the skeleton for locomotion, but some muscles—especially those of the head—move other structures (e.g., the eyeballs, scalp, eyelids, skin of face, and tongue). The sheath of the left rectus abdominis, formed by aponeuroses of the flat abdominal muscles, has been removed to reveal the muscle. Retinaculae are deep fascial thickenings that tether tendons to underlying bones as they cross joints.

CONTRACTION OF MUSCLES

Skeletal muscles function by contracting; they pull and never push. However, certain phenomena—such as "popping of the ears" to equalize air pressure and the *musculovenous pump* (pp. 41–42, Fig. I.26)—take advantage of the expansion of muscle bellies during contraction. When a muscle contracts and shortens, one of its attachments usually remains fixed while the other (more mobile) attachment is pulled toward it, often resulting in movement. Attachments of muscles are commonly described as the origin and insertion; the *origin* is usually the proximal end of the muscle, which remains fixed during muscular contraction, and the *insertion* is usually the distal end of the muscle, which is movable. However, this is not always the case. Some muscles can act in both directions under different circumstances. For example, when doing push-ups, the distal end of the upper limb (the hand) is fixed (on the floor) and the proximal end of the limb and the trunk (of the body) are being moved. Therefore, this book usually uses the terms *proximal* and *distal* or *medial* and *lateral* when describing most muscle attachments. Note that if the attachments of a muscle are known, the action of the muscle can usually be deduced (rather than memorized). When studying muscle attachments, act out the action; you are more likely to learn things you have experienced.

Reflexive Contraction. Although skeletal muscles are also referred to as voluntary muscles, certain aspects of their activity are automatic (**reflexive**) and therefore not voluntarily controlled. Examples are the respiratory movements of the diaphragm, controlled most of the time by reflexes stimulated by the levels of oxygen and carbon dioxide in the blood (although we can willfully control it within limits), and the myotatic reflex, which results in movement after a muscle stretch produced by tapping a tendon with a reflex hammer.

Tonic Contraction. Even when "relaxed," the muscles of a conscious individual are almost always slightly contracted. This slight contraction, called **muscle tone** (tonus), does not produce movement or active resistance (as phasic contraction does) but gives the muscle a certain firmness, assisting the stability of joints and the maintenance of posture, while keeping the muscle ready to respond to appropriate stimuli. Muscle tone is usually absent only when unconscious (as during deep sleep or under general anesthesia) or after a nerve lesion resulting in paralysis.

Phasic Contraction. There are two main types of **phasic** (active) **muscle contractions**: (1) **isotonic contractions,** in which the muscle changes length in relationship to the production of movement, and (2) **isometric contractions,** in which muscle length remains the same—no movement occurs, but the force (muscle tension) is increased above tonic levels to resist gravity or other antagonistic force (Fig. I.20). The latter type of contraction is important in maintaining upright posture and when muscles act as fixators or shunt muscles as described below.

There are two types of isotonic contractions. The type we most commonly think of is **concentric contraction,** in which movement occurs as a result of the muscle shortening—for example, when lifting a cup, pushing a door, or striking a blow. The ability to apply exceptional force by means of concentric contraction often is what distinguishes an athlete from an amateur. The other type of isotonic contraction is **eccentric contraction,** in which a contracting muscle lengthens—that is, it undergoes a controlled and gradual relaxation while continually exerting a (diminishing) force, like playing out a rope. Although people are generally not as aware of them, eccentric contractions are as important as concentric contractions for coordinated, functional movements such as walking, running, and setting objects (or one's self) down.

Often, when the main muscle of a particular movement (the *prime mover*) is undergoing a concentric contraction, its antagonist is undergoing a coordinated eccentric contraction. In walking, we contract concentrically to pull our center of gravity forward and then, as it passes ahead of the limb, we contract eccentrically to prevent a lurching during the transfer of weight to the other limb. Eccentric contractions

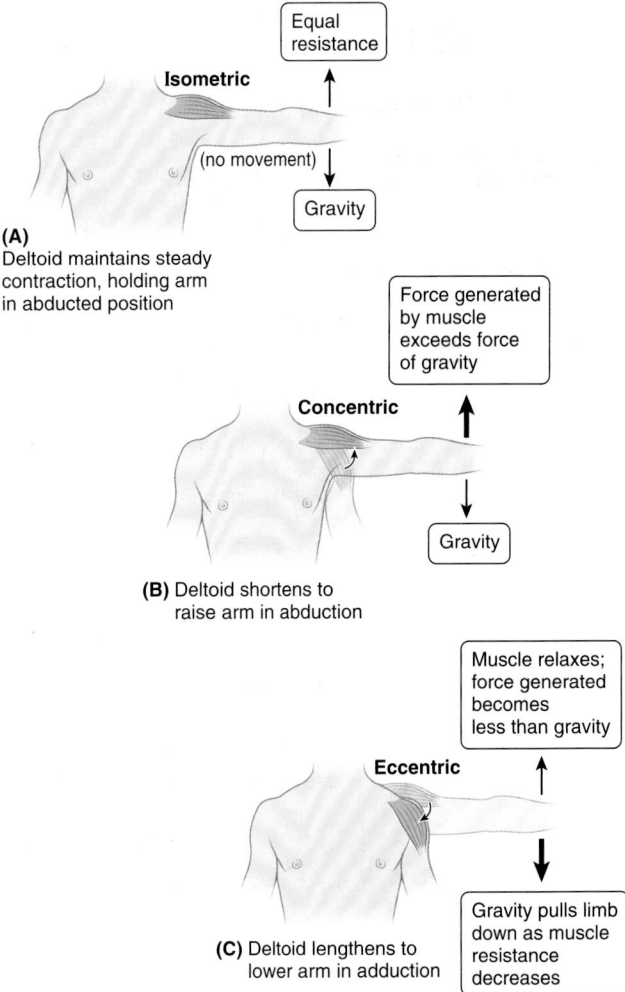

(A)
Deltoid maintains steady contraction, holding arm in abducted position

(B) Deltoid shortens to raise arm in abduction

(C) Deltoid lengthens to lower arm in adduction

FIGURE I.20. Isometric and isotonic contractions. Isometric contraction sustains the position of a joint without producing movement. Concentric and eccentric contractions are isotonic contractions in which the muscle changes length: concentric contractions by shortening and eccentric contractions by actively controlled lengthening (relaxation).

require less metabolic energy at the same load but, with a maximal contraction, are capable of generating much higher tension levels than concentric contractions—as much as 50% higher (Marieb, 2004).

Whereas the **structural unit** of a muscle is a skeletal striated muscle fiber, the **functional unit** of a muscle is a **motor unit,** consisting of a motor neuron and the muscle fibers it controls (Fig. I.21). When a *motor neuron* in the spinal cord is stimulated, it initiates an impulse that causes all the muscle fibers supplied by that motor unit to contract simultaneously. The number of muscle fibers in a motor unit varies from one to several hundred. The number of fibers varies according to the size and function of the muscle. Large motor units, in which one neuron supplies several hundred muscle fibers, are in the large trunk and thigh muscles. In smaller eye and hand muscles, where precision movements are required, the motor units include only a few muscle fibers. Movement (phasic contraction) results from the activation of an increasing number of motor units, above the level required to maintain muscle tone.

FUNCTIONS OF MUSCLES

Muscles serve specific functions in moving and positioning the body.

- A **prime mover** (agonist) is the main muscle responsible for producing a specific movement of the body. It contracts concentrically to produce the desired movement, doing most of the work (expending most of the energy) required. In most movements, there is a single prime mover, but some movements involve two prime movers working in equal measure.
- A **fixator** steadies the proximal parts of a limb through isometric contraction while movements are occurring in distal parts.

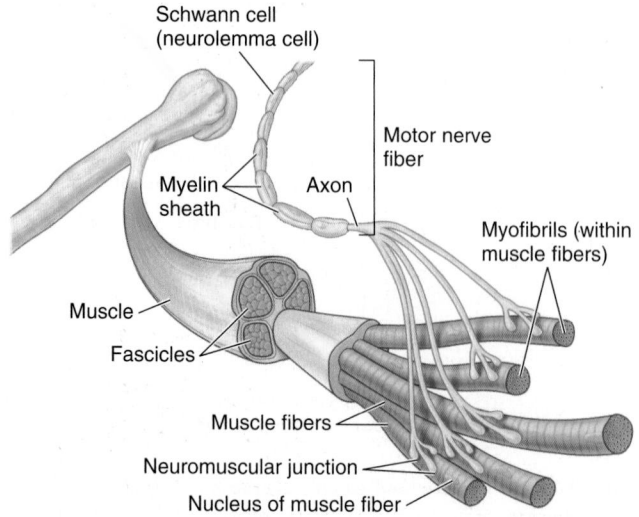

FIGURE I.21. Motor unit. A motor unit consists of a single motor neuron and the muscle fibers innervated by it.

- A **synergist** complements the action of a prime mover. It may directly assist a prime mover, providing a weaker or less mechanically advantaged component of the same movement, or it may assist indirectly, by serving as a fixator of an intervening joint when a prime mover passes over more than one joint, for example. It is not unusual to have several synergists assisting a prime mover in a particular movement.
- An **antagonist** is a muscle that opposes the action of another muscle. A primary antagonist directly opposes the prime mover, but synergists may also be opposed by secondary antagonists. As the active movers concentrically contract to produce a movement, antagonists eccentrically contract, relaxing progressively in coordination to produce a smooth movement.

The same muscle may act as a prime mover, antagonist, synergist, or fixator under different conditions. Note also that the actual prime mover in a given position may be gravity. In such cases, a paradoxical situation may exist in which the prime mover usually described as being responsible for the movement is inactive (passive), while the controlled relaxation (eccentric contraction) of the antigravity antagonist(s) is the active (energy requiring) component in the movement. An example is lowering (adducting) the upper limbs from the abducted position (stretched out laterally at 90° to the trunk) when standing erect (Fig. I.20C). The prime mover (adductor) is gravity; the muscles described as the prime movers for this movement (pectoralis major and latissimus dorsi) are inactive or passive; and the muscle being actively innervated (contracting eccentrically) is the deltoid (an abductor, typically described as the antagonist for this movement).

When a muscle's pull is exerted along a line that parallels the axis of the bones to which it is attached, it is at a disadvantage for producing movement. Instead it acts to maintain contact between the articular surfaces of the joint it crosses (i.e., it resists dislocating forces); this type of muscle is a **shunt muscle.** For example, when the arms are at one's sides, the deltoid functions as a shunt muscle. The more oblique a muscle's line of pull is oriented to the bone it moves (i.e., the *less* parallel the line of pull is to the long axis of the bone, e.g., the biceps brachii when the elbow is flexed), the more capable it is of rapid and effective movement; this type of muscle is a **spurt muscle.** The deltoid becomes increasingly effective as a spurt muscle after other muscles have initiated abduction of the arm.

NERVES AND ARTERIES TO MUSCLES

Variation in the nerve supply of muscles is rare; it is a nearly constant relationship. In the limb, muscles of similar actions are generally contained within a *common fascial compartment* and share innervation by the same nerves (Fig. I.9); therefore, you should learn the innervation of limb muscles in terms of the functional groups, making it necessary to memorize only the exceptions. Nerves supplying skeletal muscles (**motor nerves**) usually enter the fleshy portion of the muscle (vs. the tendon), almost always from the deep aspect (so the nerve is protected by the muscle it supplies). The few exceptions will be pointed out later in the text. When a nerve pierces a muscle, by passing through its fleshy portion or between its two heads of attachment, it usually supplies that muscle. Exceptions are the sensory branches that innervate the skin of the back after penetrating the superficial muscles of the back.

The blood supply of muscles is not as constant as the nerve supply and is usually multiple. Arteries generally supply the structures they contact. Thus you should learn the course of the arteries and deduce that a muscle is supplied by all the arteries in its vicinity.

SKELETAL MUSCLES

Muscle Dysfunction and Paralysis

From the clinical perspective, it is important not only to think in terms of the action normally produced by a given muscle but also to consider what loss of function would occur if the muscle failed to function (paralysis). How would the dysfunction of a given muscle or muscle group be manifest (i.e., what are the visible signs)?

Absence of Muscle Tone

Although a gentle force, muscle tone can have important effects: the tonus of muscles in the lips helps keep the teeth aligned, for instance. When this gentle but constant pressure is absent (due to paralysis or a short lip that leaves the teeth exposed), teeth migrate, becoming everted ("buck teeth").

The absence of muscle tone in an unconscious patient (e.g., under a general anesthetic) may allow joints to be dislocated as he or she is being lifted or positioned. When a muscle is denervated (loses its nerve supply), it becomes *paralyzed* (flaccid, lacking both its tonus and its ability to contract phasically on demand or reflexively). In the absence of a muscle's normal tonus, that of opposing (*antagonist*) muscle(s) may cause a limb to assume an abnormal resting position. In addition, the denervated muscle will become fibrotic and lose its elasticity, also contributing to the abnormal position at rest.

Muscle Soreness and "Pulled" Muscles

Eccentric contractions that are either excessive or associated with a novel task are often the cause of delayed-onset *muscle soreness*. Thus walking down many flights of stairs would actually result in more soreness, owing to the eccentric contractions, than walking up the same flights of stairs. The muscle stretching that occurs during the lengthening type of eccentric contraction appears to be more likely to produce microtears in the muscles and/or periosteal irritation than that associated with concentric contraction (shortening of the muscle belly).

Skeletal muscles are limited in their ability to lengthen. Usually muscles cannot elongate beyond one third of their resting length without sustaining damage. This is reflected in their attachments to the skeleton, which usually do not permit excessive lengthening. An exception is the hamstring muscles of the posterior thigh. When the knee is extended, the hamstrings typically reach their maximum length before the hip is fully flexed (i.e., flexion at the hip is limited by the hamstring's ability to elongate). Undoubtedly this, as well as forces related to their eccentric contraction, explains why hamstring muscles are "pulled" (sustain tears) more commonly than other muscles (Fig. BI.7).

Growth and Regeneration of Skeletal Muscle

Skeletal striated muscle fibers cannot divide, but they can be replaced individually by new muscle fibers derived from **satellite cells of skeletal muscle**

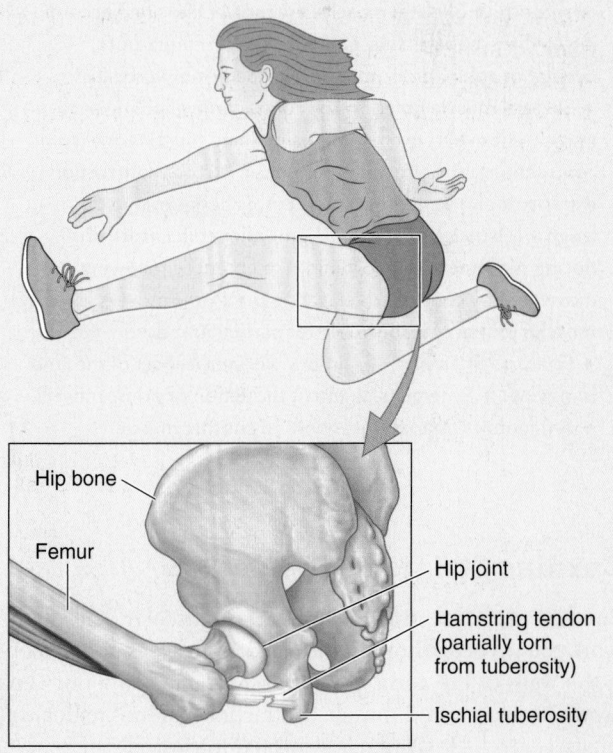

Hip bone

Femur

Hip joint

Hamstring tendon (partially torn from tuberosity)

Ischial tuberosity

FIGURE BI.7.

(see skeletal muscle figure, Table I.1). Satellite cells represent a potential source of *myoblasts,* precursors of muscle cells, which are capable of fusing with each other to form new skeletal muscle fibers if required (Ross et al., 2011). The number of new fibers that can be produced is insufficient to compensate for major muscle degeneration or trauma. Instead of becoming regenerated effectively, the new skeletal muscle is composed of a disorganized mixture of muscle fibers and fibrous scar tissue. Skeletal muscles are able to grow larger in response to frequent strenuous exercise, such as body building. This growth results from *hypertrophy of existing fibers,* not from the addition of new muscle fibers. Hypertrophy lengthens and increases the *myofibrils* within the muscle fibers (Fig. I.21), thereby increasing the amount of work the muscle can perform.

Muscle Testing

 Muscle testing helps examiners diagnose nerve injuries. There are two common testing methods:

- The person performs movements that resist those of the examiner. For example, the person keeps the forearm flexed while the examiner attempts to extend it. This technique enables the examiner to gauge the power of the person's movements.
- The examiner performs movements that resist those of the person. When testing flexion of the forearm, the examiner asks the person to flex his or her forearm while the examiner resists the efforts. Usually muscles are tested in bilateral pairs for comparison.

Electromyography (EMG), the electrical stimulation of muscles, is another method for testing muscle action. The examiner places surface electrodes over a muscle, asks the person to perform certain movements, and then amplifies and records the differences in electrical action potentials of the muscles. A normal resting muscle shows only a baseline activity (muscle tone), which disappears only during deep sleep, during paralysis, and when under anesthesia. Contracting muscles demonstrate variable peaks of phasic activity. EMG makes it possible to analyze the activity of an individual muscle during different movements. EMG may also be part of the treatment program for restoring the action of muscles.

The Bottom Line

SKELETAL MUSCLES

Muscles are categorized as skeletal striated, cardiac striated, or smooth. ♦ Skeletal muscles are further classified according to their shape as flat, pennate, fusiform, quadrate, circular or sphincteral, and multiheaded or multi-bellied. ♦ Skeletal muscle functions by contracting, enabling automatic (reflexive) movements, maintaining muscle tone (tonic contraction), and providing for phasic (active) contraction with (isotonic) or without (isometric) change in muscle length. ♦ Isotonic movements are either concentric (producing movement by shortening) or eccentric (allowing movement by controlled relaxation). ♦ Prime movers are the muscles primarily responsible for particular movements. ♦ Fixators "fix" a part of a limb while another part of the limb is moving. ♦ Synergists augment the action of prime movers. ♦ Antagonists oppose the actions of another muscle.

Cardiac Striated Muscle

Cardiac striated muscle forms the muscular wall of the heart, the **myocardium.** Some cardiac muscle is also present in the walls of the aorta, pulmonary vein, and superior vena cava. Cardiac striated muscle contractions are not under voluntary control. Heart rate is regulated intrinsically by a *pacemaker,* an impulse-conducting system composed of specialized cardiac muscle fibers; they, in turn, are influenced by the autonomic nervous system (ANS) (discussed later in this chapter).

Cardiac striated muscle has a distinctly striped appearance under microscopy (Table I.1). Both types of striated muscle—skeletal and cardiac—are further characterized by the immediacy, rapidity, and strength of their contractions. *Note:* Even though the trait applies to both skeletal and cardiac striated muscle, in common usage the terms *striated* and *striped* are used to designate voluntary skeletal striated muscle.

As demonstrated in Table I.1, cardiac striated muscle is distinct from skeletal striated muscle in its location, appearance, type of activity, and means of stimulation. To support its continuous level of high activity, the blood supply to cardiac striated muscle is twice as rich as that to skeletal striated muscle.

Smooth Muscle

Smooth muscle, named for the absence of striations in the appearance of the muscle fibers under microscopy, forms a large part of the middle coat or layer (*tunica media*) of the walls of blood vessels (above the capillary level) (Fig. I.23; Table I.1). Consequently, it occurs in all vascularized tissue. It also makes up the muscular parts of the walls of the alimentary tract and ducts. Smooth muscle is found in skin, forming the *arrector muscles of hairs* associated with hair follicles (Fig. I.6), and in the eyeball, where it controls lens thickness and pupil size.

Like cardiac striated muscle, smooth muscle is *involuntary muscle;* however, it is directly innervated by the ANS. Its contraction can also be initiated by hormonal stimulation or by local stimuli, such as stretching. Smooth muscle responds more slowly than striated muscle and with a delayed and more leisurely contraction. It can undergo partial contraction for long periods and has a much greater ability than striated muscle to elongate without suffering paralyzing injury. Both of these factors are important in regulating the size of sphincters and the caliber of the *lumina* (interior spaces) of tubular structures (e.g., blood vessels or intestines). In the walls of the alimentary tract, uterine tubes, and ureters, smooth muscle cells are responsible for peristalsis, rhythmic contractions that propel the contents along these tubular structures.

CARDIAC AND SMOOTH MUSCLE

Hypertrophy of the Myocardium and Myocardial Infarction

In *compensatory hypertrophy,* the myocardium responds to increased demands by increasing the size of its fibers. When cardiac striated muscle fibers are damaged by loss of their blood supply during a heart attack, the tissue becomes necrotic (dies) and the fibrous scar tissue that develops forms a *myocardial infarct* (MI), an area of *myocardial necrosis* (pathological death of cardiac tissue). Muscle cells that degenerate are not replaced, because cardiac muscle cells do not divide. Furthermore, there is no equivalent to the satellite cells of skeletal muscle that can produce new cardiac muscle fibers.

Hypertrophy and Hyperplasia of Smooth Muscle

Smooth muscle cells undergo *compensatory hypertrophy* in response to increased demands. Smooth muscle cells in the uterine wall during pregnancy increase not only in size but also in number (*hyperplasia*) because these cells retain the capacity for cell division. In addition, new smooth muscle cells can develop from incompletely differentiated cells (*pericytes*) that are located along small blood vessels (Ross et al., 2011).

The Bottom Line

CARDIAC AND SMOOTH MUSCLE

Cardiac muscle is a striated muscle type found in the walls of the heart, or myocardium, as well as in some major blood vessels. ♦ Contraction of cardiac muscle is not under voluntary control but is instead activated by specialized cardiac muscle fibers forming the pacemaker, the activity of which is regulated by the autonomic nervous system (ANS). ♦ Smooth muscle does not have striations. It occurs in all vascular tissues and in the walls of the alimentary tract and other organs. ♦ Smooth muscle is directly innervated by the ANS and thus is not under voluntary control.

CARDIOVASCULAR SYSTEM

The circulatory system transports fluids throughout the body; it consists of the cardiovascular and lymphatic systems. The heart and blood vessels make up the blood transportation network, *the cardiovascular system.* Through this system, the heart pumps blood through the body's vast system of blood vessels. The blood carries nutrients, oxygen, and waste products to and from the cells.

Vascular Circuits

The *heart* consists of two muscular pumps that, although adjacently located, act in series, dividing the circulation into two components: the *pulmonary* and *systemic circulations* or circuits (Fig. I.22A & B). The *right ventricle* propels low-oxygen blood returning from the systemic circulation into the lungs via the *pulmonary arteries.* Carbon dioxide is exchanged for oxygen in the capillaries of the lungs, and then the oxygen-rich blood is returned via the *pulmonary veins* to the heart's *left atrium.* This circuit, from the right ventricle through the lungs to the left atrium, is the **pulmonary circulation.** The *left ventricle* propels the oxygen-rich blood returned to the heart from the pulmonary circulation through **systemic arteries** (the *aorta* and its branches), exchanging oxygen and nutrients for carbon dioxide in the remainder of the body's capillaries. Low-oxygen blood returns to the heart's right atrium via **systemic veins** (tributaries of the *superior* and *inferior vena cavae*). This circuit, from left ventricle to right atrium, is the **systemic circulation.**

The systemic circulation actually consists of many parallel circuits serving the various regions and/or organ systems of the body (Fig. I.22C).

Blood Vessels

There are three types of blood vessels: *arteries, veins,* and *capillaries* (Fig. I.23). Blood under high pressure leaves the heart and is distributed to the body by a branching system of thick-walled arteries. The final distributing vessels, *arterioles,* deliver oxygen-rich blood to capillaries. Capillaries form a *capillary bed,* where the interchange of oxygen, nutrients, waste products, and other substances with the extracellular fluid occurs. Blood from the capillary bed passes into thin-walled *venules,* which resemble wide capillaries. Venules drain into small veins that open into larger veins. The largest veins, the superior and inferior venae cavae, return low-oxygen blood to the heart.

Most vessels of the circulatory system have three coats, or tunics:

- **Tunica intima,** an inner lining consisting of a single layer of extremely flattened epithelial cells, the **endothelium,** supported by delicate connective tissue. Capillaries consist only of this tunic, with blood capillaries also having a supporting basement membrane.

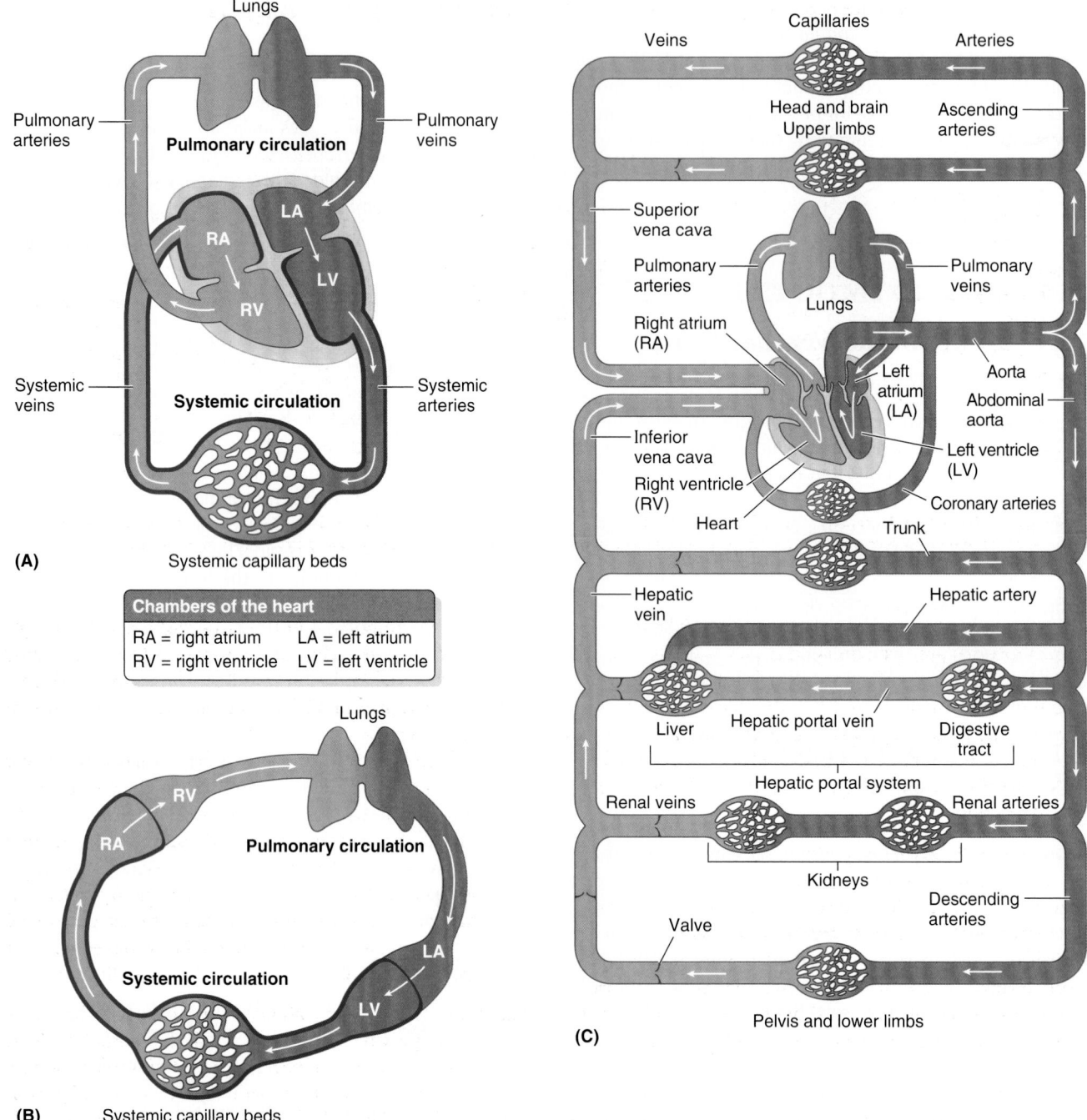

FIGURE I.22. The circulation. A. Schematic illustration of the anatomic arrangement of the two muscular pumps (right and left heart) serving the pulmonary and systemic circulations. **B.** Schematic illustration of the body's circulation, with the right and left heart depicted as two pumps in series. The pulmonary and systemic circulations are actually serial components of one continuous loop. **C.** A more detailed schematic illustration demonstrating that the systemic circulation actually consists of many parallel circuits serving the various organs and regions of the body.

- **Tunica media,** a middle layer consisting primarily of smooth muscle.
- **Tunica adventitia,** an outer connective tissue layer or sheath.

The tunica media is the most variable coat. Arteries, veins, and lymphatic ducts are distinguished by the thickness of this layer relative to the size of the lumen, its organization, and,

in the case of arteries, the presence of variable amounts of elastic fibers.

ARTERIES

Arteries are blood vessels that carry blood under relatively high pressure (compared to the corresponding veins) from the heart and distribute it to the body (Fig. I.24A). The blood

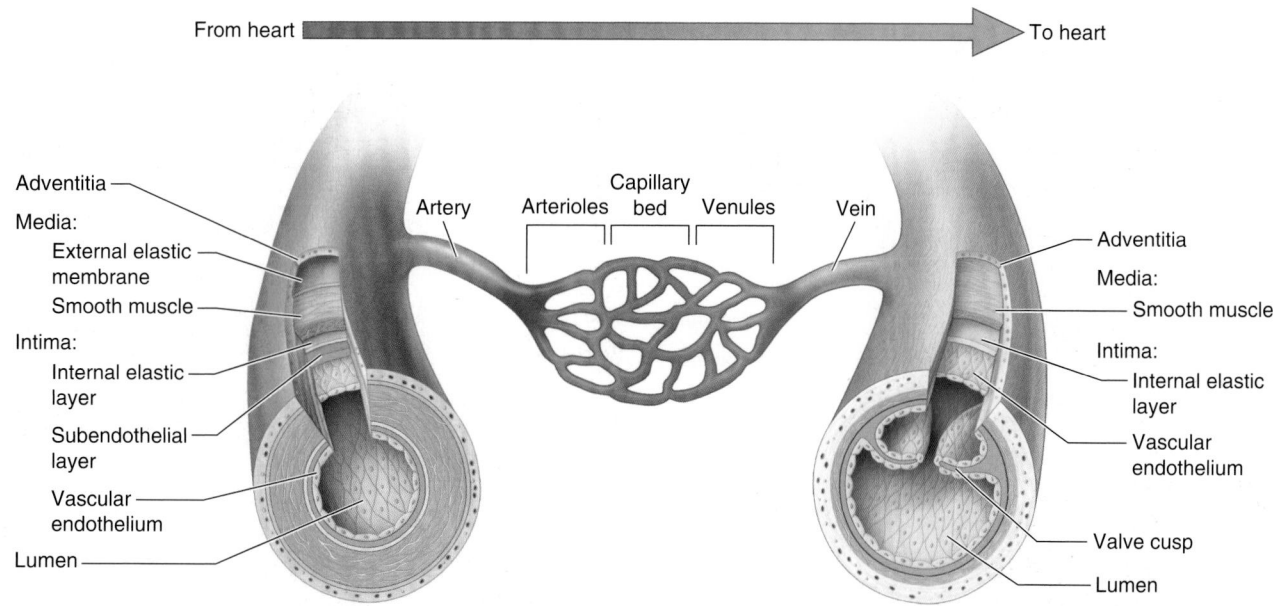

FIGURE I.23. Blood vessel structure. The walls of most blood vessels have three concentric layers of tissue, called tunics (L. *tunicae,* coats). With less muscle, veins are thinner walled than their companion arteries and have wide lumens (L. *luminae*) that usually appear flattened in tissue sections.

passes through arteries of decreasing caliber. The different types of arteries are distinguished from each other on the basis of overall size, relative amounts of elastic tissue or muscle in the tunica media (Fig. I.23), the thickness of the wall relative to the lumen, and function. Artery size and type is a continuum—that is, there is a gradual change in morphological characteristics from one type to another. There are three types of arteries:

- **Large elastic arteries** (conducting arteries) have many elastic layers (sheets of elastic fibers) in their walls. These large arteries initially receive the cardiac output. Their elasticity enables them to expand when they receive the cardiac output from the ventricles, minimizing the pressure change, and return to normal size between ventricular contractions, as they continue to push the blood into the medium arteries downstream. This maintains the blood pressure in the arterial system between cardiac contractions (at a time when ventricular pressure falls to zero). Overall, this minimizes the ebb in blood pressure as the heart contracts and relaxes. Examples of large elastic arteries are the aorta, the arteries that originate from the arch of the aorta (brachiocephalic trunk, subclavian and carotid arteries), and the pulmonary trunk and arteries (Fig. I.24A).
- **Medium muscular arteries** (distributing arteries) have walls that consist chiefly of circularly disposed smooth muscle fibers. Their ability to decrease their diameter (*vasoconstrict*) regulates the flow of blood to different parts of the body as required by circumstance (e.g., activity, thermoregulation). Pulsatile contractions of their muscular walls (regardless of lumen caliber) temporarily and rhythmically constrict their lumina in progressive

sequence, propelling and distributing blood to various parts of the body. Most of the named arteries, including those observed in the body wall and limbs during dissection such as the brachial or femoral arteries, are medium muscular arteries.

- **Small arteries** and **arterioles** have relatively narrow lumina and thick muscular walls. The degree of filling of the capillary beds and level of arterial pressure within the vascular system are regulated mainly by the degree of tonus (firmness) in the smooth muscle of the arteriolar walls. If the tonus is above normal, *hypertension* (high blood pressure) results. Small arteries are usually not named or specifically identified during dissection, and arterioles can be observed only under magnification.

Anastomoses (communications) between multiple branches of an artery provide numerous potential detours for blood flow in case the usual pathway is obstructed by compression due to the position of a joint, pathology, or surgical ligation. If a main channel is occluded, the smaller alternate channels can usually increase in size over a period of time, providing a **collateral circulation** that ensures the blood supply to structures distal to the blockage. However, collateral pathways require time to open adequately; they are usually insufficient to compensate for sudden occlusion or ligation.

There are areas, however, where collateral circulation does not exist, or is inadequate to replace the main channel. Arteries that do not anastomose with adjacent arteries are **true** (anatomic) **terminal arteries** (end arteries). Occlusion of an end artery interrupts the blood supply to the structure or segment of an organ it supplies. True terminal arteries supply the retina, for example, where occlusion will result in blindness. While not true terminal arteries, *functional*

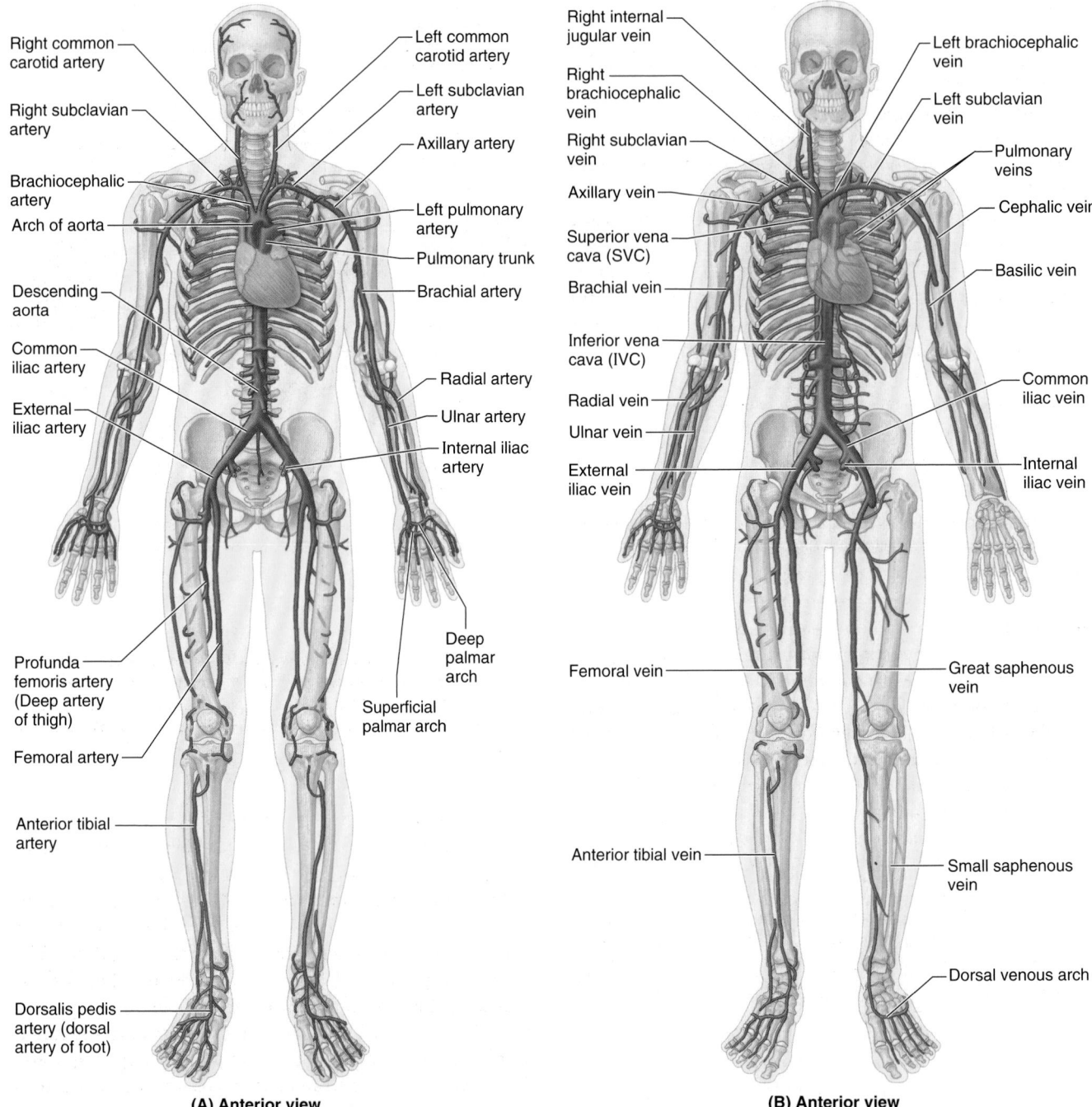

(A) Anterior view

(B) Anterior view

FIGURE I.24. Systemic portion of the cardiovascular system. The arteries and veins shown here carry oxygen-rich blood from the heart to the systemic capillary beds and return low-oxygen blood from the capillary beds to the heart, respectively, constituting the systemic circulation. Although commonly depicted and considered as single vessels, as shown here, the deep veins of the limbs usually occur as pairs of accompanying veins.

terminal arteries (arteries with ineffectual anastomoses) supply segments of the brain, liver, kidneys, spleen, and intestines; they may also exist in the heart.

VEINS

Veins generally return low-oxygen blood from the capillary beds to the heart, which gives the veins a dark blue appearance (Fig. I.24B). The large pulmonary veins are atypical in

that they carry oxygen-rich blood from the lungs to the heart. Because of the lower blood pressure in the venous system, the walls (specifically, the tunica media) of veins are thinner than those of their companion arteries (Fig. I.23). Normally, veins do not pulsate and do not squirt or spurt blood when severed. There are three sizes of veins:

- **Venules** are the smallest veins. Venules drain capillary beds and join similar vessels to form small veins.

Magnification is required to observe venules. Small veins are the tributaries of larger veins that unite to form **venous plexuses,** such as the dorsal venous arch of the foot (Fig. I.24B). Small veins are unnamed.

- **Medium veins** drain venous plexuses and accompany medium arteries. In the limbs, and in some other locations where the flow of blood is opposed by the pull of gravity, the medium veins have **venous valves,** passive flap valves that permit blood to flow toward the heart but not in the reverse direction (Fig. I-26). Examples of medium veins include the named superficial veins (cephalic and basilic veins of the upper limbs and great and small saphenous veins of the lower limbs) and the accompanying veins that are named according to the artery they accompany (Fig. I-24B).
- **Large veins** are characterized by wide bundles of longitudinal smooth muscle and a well-developed tunica adventitia. An example is the superior vena cava.

Veins are more abundant than arteries. Although their walls are thinner, their diameters are usually larger than those of the corresponding artery. The thin walls allow veins to have a large capacity for expansion, and do so when blood return to the heart is impeded by compression or internal pressures (e.g., after taking a large breath and holding it; this is the *Valsalva maneuver*).

Since the arteries and veins make up a circuit, it might be expected that half the blood volume would be in the arteries and half in the veins. Because of the veins' larger diameter and ability to expand, typically only 20% of the blood occupies arteries, whereas 80% is in the veins.

Although often depicted as single vessels in illustrations for simplicity, veins tend to be double or multiple. Those that accompany deep arteries—**accompanying veins** (L. *venae comitantes*)—surround them in an irregular branching network (Fig. I.25). This arrangement serves as a *countercurrent heat exchanger*, the warm arterial blood warming the cooler venous blood as it returns to the heart from a cold limb. The accompanying veins occupy a relatively unyielding fascial **vascular sheath** with the artery they accompany. As a

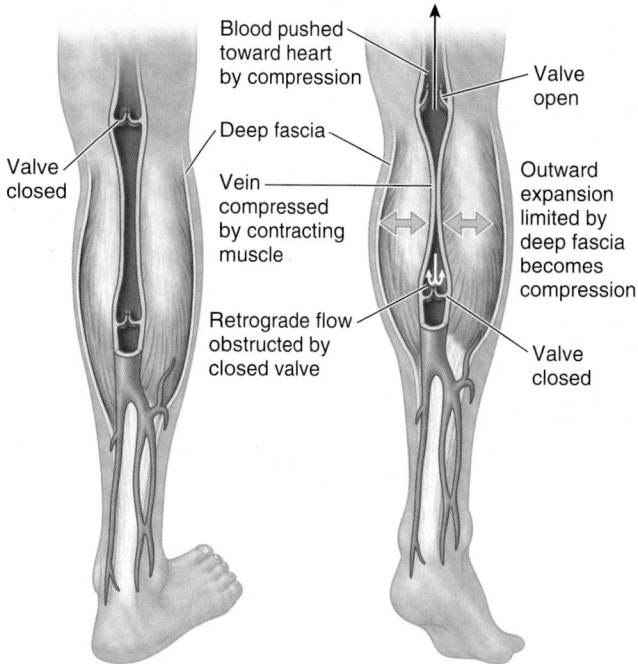

FIGURE I.26. Musculovenous pump. Muscular contractions in the limbs function with the venous valves to move blood toward the heart. The outward expansion of the bellies of contracting muscles is limited by deep fascia and becomes a compressive force, propelling the blood against gravity.

result, they are stretched and flattened as the artery expands during contraction of the heart, which aids in driving venous blood toward the heart—an *arteriovenous pump.*

Systemic veins are more variable than arteries, and *venous anastomoses*—natural communications, direct or indirect, between two veins—occur more often between them. The outward expansion of the bellies of contracting skeletal muscles in the limbs, limited by the deep fascia, compresses the veins, "milking" the blood superiorly toward the heart; another (musculovenous) type of venous pump (Fig. I.26). The valves of the veins break up the columns of blood, thus relieving the more dependent parts of excessive pressure, allowing venous blood to flow only toward the heart. The venous congestion that hot and tired feet experience at the end of a busy day is relieved by resting the feet on a footstool that is higher than the trunk (of the body). This position of the feet also helps the veins return blood to the heart.

BLOOD CAPILLARIES

For the oxygen and nutrients carried by the arteries to benefit the cells that make up the tissues of the body, they must leave the transporting vessels and enter the *extravascular space* between the cells, the extracellular (intercellular) space in which the cells live. **Capillaries** are simple

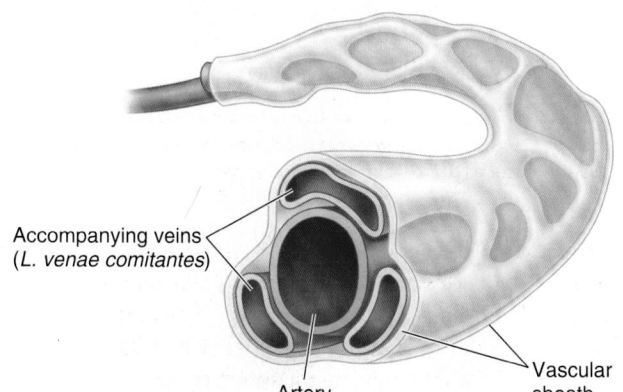

FIGURE I.25. Accompanying veins. Although most veins of the trunk occur as large single vessels, veins in the limbs occur as two or more smaller vessels that accompany an artery in a common vascular sheath.

endothelial tubes connecting the arterial and venous sides of the circulation that allow the exchange of materials with the interstitial or **extracellular fluid** (ECF). Capillaries are generally arranged in **capillary beds,** networks that connect the arterioles and venules (Fig. I.23). The blood enters the capillary beds through arterioles that control the flow and is drained from them by venules.

As the hydrostatic pressure in the arterioles forces blood into and through the capillary bed, it also forces fluid containing oxygen, nutrients, and other cellular materials out of the blood at the arterial end of the capillary bed (upstream) into the extracellular spaces, allowing exchange with cells of the surrounding tissue. Capillary walls are relatively impermeable, however, to plasma proteins. Downstream, at the venous end of the capillary bed, most of this ECF—now containing waste products and carbon dioxide—is reabsorbed into the blood as a result of the osmotic pressure from the higher concentrations of proteins within the capillary.

(Although firmly established, this principle is referred to as the *Starling hypothesis*.)

In some regions, such as in the fingers, there are direct connections between the small arterioles and venules proximal to the capillary beds they supply and drain. The sites of such communications—**arteriolovenular** (arteriovenous) **anastomoses** (AVAs)—permit blood to pass directly from the arterial to the venous side of the circulation without passing through capillaries. AV shunts are numerous in the skin, where they have an important role in conserving body heat.

In some situations, blood passes through two capillary beds before returning to the heart; a venous system linking two capillary beds constitutes a **portal venous system.** The venous system by which nutrient-rich blood passes from the capillary beds of the alimentary tract to the capillary beds or sinusoids of the liver—the *hepatic portal system*—is the major example (Fig. I.22C).

CARDIOVASCULAR SYSTEM

Arteriosclerosis: Ischemia and Infarction

The most common acquired disease of arteries—and a common finding in cadaver dissection—is *arteriosclerosis* (hardening of the arteries), a group of diseases characterized by thickening and loss of elasticity of the arterial walls. A common form, *atherosclerosis,* is associated with the buildup of fat (mainly cholesterol) in the arterial walls. A calcium deposit forms an *atheromatous plaque* (*atheroma*)—well-demarcated, hardened yellow areas or swellings on the intimal surfaces of arteries (Fig. BI.8A). The consequent arterial narrowing and surface irregularity may result in *thrombosis* (formation of a local intravascular clot, or *thrombus*), which may occlude the artery or be flushed into the bloodstream and block smaller vessels distally as an *embolus* (a plug occluding a vessel) (Fig. BI.8B). The consequences of atherosclerosis include *ischemia* (reduction of

blood supply to an organ or region) and *infarction* (local death, or necrosis, of an area of tissue or an organ resulting from reduced blood supply). These consequences are particularly significant in regard to the heart (*ischemic heart disease* and *myocardial infarction* [MI] or heart attack), brain (*stroke*), and distal parts of limbs (*gangrene*).

Varicose Veins

When the walls of veins lose their elasticity, they become weak. A weakened vein dilates under the pressure of supporting a column of blood against gravity. This results in *varicose veins*—abnormally swollen, twisted veins—most often seen in the legs (Fig. BI.9). Varicose veins have a caliber greater than normal, and their valve cusps do not meet or have been destroyed by

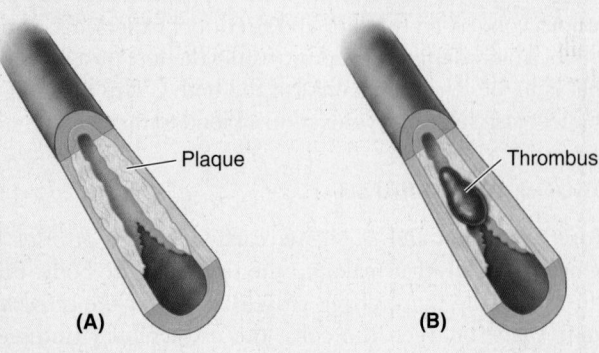

FIGURE BI.8.

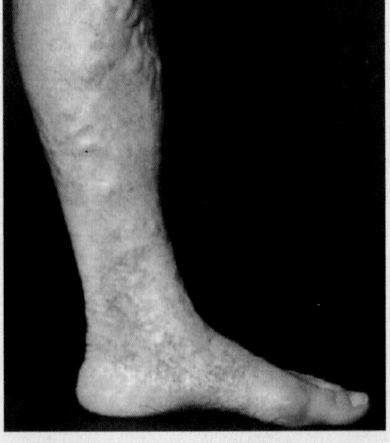

FIGURE BI.9.

inflammation. Varicose veins have *incompetent valves;* thus the column of blood ascending toward the heart is unbroken, placing increased pressure on the weakened walls, further exacerbating the varicosity problem. Varicose veins also occur in the presence of degenerated deep fascia. *Incompetent fascia* is incapable of containing the expansion of contracting muscles; thus the (musculofascial) musculovenous pump is ineffective.

The Bottom Line

CARDIOVASCULAR SYSTEM

The cardiovascular system consists of the heart and blood vessels—the arteries, veins, and capillaries. ♦ Arteries and veins (and lymphatics) have three coats or tunics—tunica intima, tunica media, and tunica adventitia. ♦ Arteries have both elastic and muscle fibers in their walls, which allow them to propel blood throughout the cardiovascular system. ♦ Veins have thinner walls than arteries and are distinguished by valves, which prevent backflow of blood. ♦ As simple endothelial tubes, capillaries are the smallest blood vessels and provide the linkage between the smallest arteries (arterioles) and veins (venules).

LYMPHOID SYSTEM

Although widely distributed throughout most of the body, most of the lymphoid (lymphatic) system is not apparent in the cadaver, yet it is essential to survival. Knowledge of the anatomy of the lymphoid system is important for clinicians. The *Starling hypothesis* (see "Blood Capillaries," p. 41) explains how most of the fluid and electrolytes entering the extracellular spaces from the blood capillaries is also reabsorbed by them. However, as much as 3 L each day fails to be reabsorbed by the blood capillaries. Furthermore, some plasma protein leaks into the extracellular spaces, and material originating from the tissue cells that cannot pass through the walls of blood capillaries, such as cytoplasm from disintegrating cells, continually enters the space in which the cells live. If this material were to accumulate in the extracellular spaces, a reverse osmosis would occur, bringing even more fluid and resulting in **edema** (an excess of interstitial fluid, manifest as swelling). However, the amount of interstitial fluid remains fairly constant under normal conditions, and proteins and cellular debris normally do not accumulate in the extracellular spaces because of the lymphoid system.

The lymphoid system thus constitutes a sort of "overflow" system that provides for the drainage of surplus tissue fluid and leaked plasma proteins to the bloodstream, as well as for the removal of debris from cellular decomposition and infection. The important components of the lymphoid system are (Fig. I.27):

- **Lymphatic plexuses,** networks of lymphatic capillaries that originate blindly in the extracellular (intercellular) spaces of most tissues. Because they are formed of a highly attenuated endothelium lacking a basement membrane, along with surplus tissue fluid, plasma proteins, bacteria, cellular debris, and even whole cells (especially lymphocytes) can readily enter them.
- **Lymphatic vessels** (lymphatics), a nearly bodywide network of thin-walled vessels that have abundant lymphatic valves. In living individuals, the vessels bulge where each of the closely spaced valves occur, giving lymphatics a beaded appearance. Lymphatic capillaries and vessels occur almost everywhere blood capillaries are found, except for example, teeth, bone, bone marrow, and the entire central nervous system. (Excess tissue fluid here drains into the cerebrospinal fluid.)
- **Lymph** (L. *lympha,* clear water), the tissue fluid that enters lymph capillaries and is conveyed by lymphatic vessels. Usually clear, watery, and slightly yellow, lymph is similar in composition to blood plasma.
- **Lymph nodes,** small masses of lymphatic tissue located along the course of lymphatic vessels through which lymph is filtered on its way to the venous system (Fig. I.27B).
- **Lymphocytes,** circulating cells of the immune system that react against foreign materials.
- **Lymphoid organs,** parts of the body that produce lymphocytes, such as the thymus, red bone marrow, spleen, tonsils, and the solitary and aggregated lymphoid nodules in the walls of the alimentary tract and appendix.

Superficial lymphatic vessels, more numerous than veins in the subcutaneous tissue and anastomosing freely, converge toward and follow the venous drainage. These vessels eventually drain into **deep lymphatic vessels** that accompany the arteries and also receive the drainage of internal organs. It is likely that the deep lymphatic vessels are also compressed by the arteries they accompany, milking the lymph along these valved vessels in the same manner described earlier for accompanying veins. Both superficial and deep lymphatic vessels traverse lymph nodes (usually several sets) as they course proximally, becoming larger as they merge with vessels draining adjacent regions. Large lymphatic vessels enter large collecting vessels, called lymphatic trunks, which unite to form either the right lymphatic duct or the thoracic duct (Fig. I.27A):

- The **right lymphatic duct** drains lymph from the body's right upper quadrant (right side of head, neck, and thorax plus the right upper limb). At the root of the neck, it enters the junction of the right internal jugular and right subclavian veins, the **right venous angle.**
- The **thoracic duct** drains lymph from the remainder of the body. The *lymphatic trunks* draining the lower half

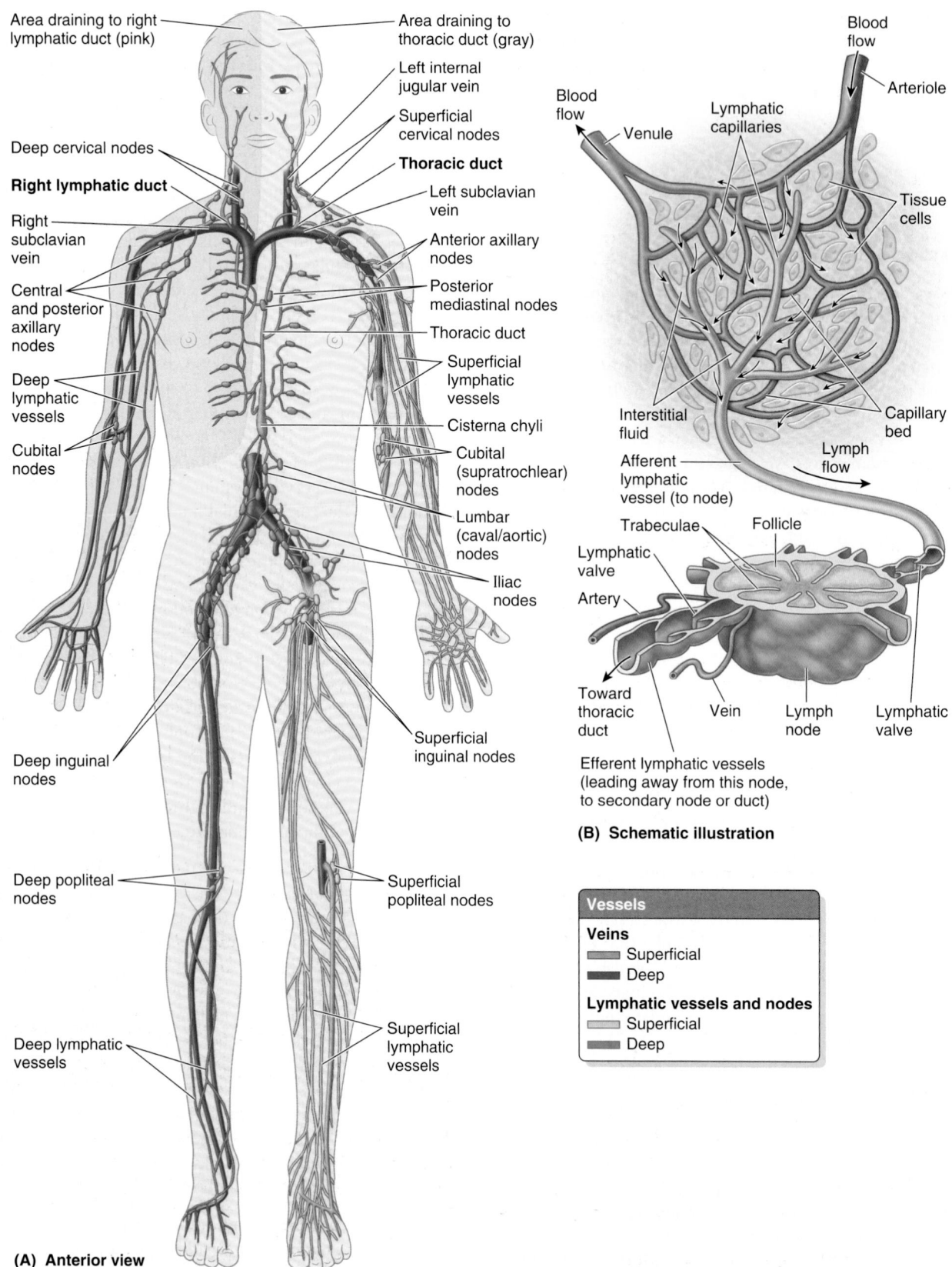

(A) Anterior view

(B) Schematic illustration

Vessels
Veins
▬ Superficial
▬ Deep
Lymphatic vessels and nodes
▬ Superficial
▬ Deep

FIGURE I.27. Lymphoid system. A. Pattern of lymphatic drainage. Except for the right superior quadrant of the body (pink), lymph ultimately drains into the left venous angle via the thoracic duct. The right superior quadrant drains to the right venous angle, usually via a right lymphatic duct. Lymph typically passes through several sets of lymph nodes, in a generally predictable order, before it enters the venous system. **B.** Schematic illustration of lymph flow from extracellular spaces through a lymph node. *Small black arrows* indicate the flow (leaking) of interstitial fluid out of blood capillaries and (absorption) into the lymphatic capillaries.

of the body merge in the abdomen, sometimes forming a dilated collecting sac, the cisterna chyli. From this sac (if present), or from the merger of the trunks, the thoracic duct ascends into and then through the thorax to enter the **left venous angle** (junction of left internal jugular and left subclavian veins).

Although this is the typical drainage pattern of most lymph, lymphatic vessels communicate with veins freely in many parts of the body. Consequently, ligation of a lymphatic trunk or even the thoracic duct itself may have only a transient effect as a new pattern of drainage is established through more peripheral lymphaticovenous—and later interlymphatic—anastomoses.

Additional functions of the lymphoid system include:

- *Absorption and transport of dietary fat.* Special lymphatic capillaries, called *lacteals* (L. *lacteus*, milk), receive all lipid and lipid-soluble vitamins absorbed by the intestine. Visceral lymphatics then convey the milky fluid, *chyle* (G. *chylos*, juice), to the thoracic duct and into the venous system.
- *Formation of a defense mechanism for the body.* When foreign protein drains from an infected area, antibodies specific to the protein are produced by immunologically competent cells and/or lymphocytes and dispatched to the infected area.

LYMPHOID SYSTEM

Spread of Cancer

Cancer invades the body by *contiguity* (growing into adjacent tissue) or by *metastasis* (the dissemination of tumor cells to sites distant from the original or primary tumor). Metastasis occurs three ways:

1. *Direct seeding* of serous membranes of body cavities.
2. *Lymphogenous spread* (via lymphatic vessels).
3. *Hematogenous spread* (via blood vessels).

It is surprising that often even a thin fascial sheet or serous membrane deflects tumor invasion. However, once a malignancy penetrates into a potential space, the direct seeding of cavities—that is, of its serous membranes—is likely.

Lymphogenous spread of cancer is the most common route for the initial dissemination of *carcinomas* (epithelial tumors), the most common type of cancer. Cells loosened from the primary cancer site enter and travel via lymphatics. The lymph-borne cells are filtered through and trapped by lymph nodes, which thus become secondary (metastatic) cancer sites.

The pattern of cancerous lymph node involvement follows the natural routes of lymph drainage. Thus when removing a potentially metastatic tumor, surgeons *stage the metastasis* (determine the degree to which cancer has spread) by removing and examining lymph nodes that receive lymph from the organ or region in the order the lymph normally passes through them. Therefore, it is important for physicians to literally know the lymphatic drainage "backward and forward"—that is, (1) to know what nodes are likely to be affected when a tumor is identified in a certain site or organ (and the order in which they receive lymph) and (2) to be able to determine likely sites of primary cancer sites (sources of metastasis) when an enlarged node is detected. *Cancerous nodes* enlarge as the tumor cells within them increase; however, unlike swollen infected nodes, cancerous nodes are not usually painful when compressed.

Hematogenous spread of cancer is the most common route for the metastasis of the less common (but more malignant) *sarcomas* (connective tissue cancers). Because veins are more abundant and have thinner walls that offer less resistance, metastasis occurs more often by venous than arterial routes. Since the blood-borne cells follow venous flow, the liver and lungs are the most common sites of secondary sarcomas. Typically, the treatment or removal of a primary tumor is not difficult, but the treatment or removal of all the affected lymph nodes or other secondary (metastatic) tumors may be impossible (Cotran et al., 1999).

Lymphangitis, Lymphadenitis, and Lymphedema

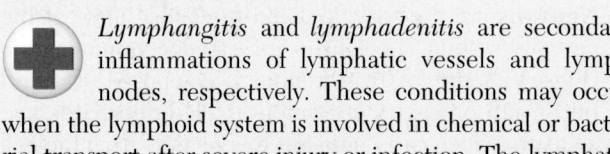

Lymphangitis and *lymphadenitis* are secondary inflammations of lymphatic vessels and lymph nodes, respectively. These conditions may occur when the lymphoid system is involved in chemical or bacterial transport after severe injury or infection. The lymphatic vessels, not normally evident, may become apparent as red streaks in the skin, and the nodes become painfully enlarged. This condition is potentially dangerous because the uncontained infection may lead to *septicemia* (blood poisoning). *Lymphedema*, a localized type of edema, occurs when lymph does not drain from an area of the body. For instance, if cancerous lymph nodes are surgically removed from the axilla (compartment superior to the armpit), lymphedema of the limb may occur. Solid cell growths may permeate lymphatic vessels and form minute *cellular emboli* (plugs), which may break free and pass to regional lymph nodes. In this way, further lymphogenous spread to other tissues and organs may occur.

The Bottom Line

NERVOUS SYSTEM

The nervous system enables the body to react to continuous changes in its internal and external environments. It also controls and integrates the various activities of the body, such as circulation and respiration. For descriptive purposes, the nervous system is divided:

- Structurally into the *central nervous system* (CNS), consisting of the brain and spinal cord, and the *peripheral nervous system* (PNS), the remainder of the nervous system outside of the CNS.
- Functionally into the *somatic nervous system* (SNS) and the *autonomic nervous system* (ANS).

Nervous tissue consists of two main cell types: neurons (*nerve cells*) and neuroglia (*glial cells*), which support the neurons.

- **Neurons** are the structural and functional units of the nervous system specialized for rapid communication (Figs. I.28 and I.29). A neuron is composed of a **cell body** with processes (extensions) called **dendrites** and an **axon,** which carry impulses to and away from the cell body, respectively. *Myelin,* layers of lipid, and protein substances form a **myelin sheath** around some axons, greatly increasing the velocity of impulse conduction. Two types of neurons constitute the majority of neurons comprising the nervous system (and the peripheral nervous system in particular) (Fig. I.28):
  1. **Multipolar motor neurons** have two or more dendrites and a single axon that may have one or more collateral branches. They are the most common type of neuron in the nervous system (CNS and PNS). All of the motor neurons that control skeletal muscle and those comprising the ANS are multipolar neurons.
  2. **Pseudounipolar sensory neurons** have a short, apparently single (but actually double) process

extending from the cell body. This common process separates into a peripheral process, conducting impulses from the receptor organ (touch, pain, or temperature sensors in the skin, for example) toward the cell body, and a central process that continues from the cell body into the CNS. The cell bodies of pseudounipolar neurons are located outside the CNS in sensory ganglia and are thus part of the PNS.

Neurons communicate with each other at **synapses,** points of contact between neurons (Fig. I.29). The communication occurs by means of *neurotransmitters,* chemical agents released or secreted by one neuron, which may excite or inhibit another neuron, continuing or terminating the relay of impulses or the response to them.

- **Neuroglia** (glial cells or glia), approximately five times as abundant as neurons, are non-neuronal, non-excitable cells that form a major component of nervous tissue, supporting, insulating, and nourishing the neurons. In the CNS, neuroglia include *oligodendroglia, astrocytes, ependymal cells,* and *microglia* (small glial cells). In the

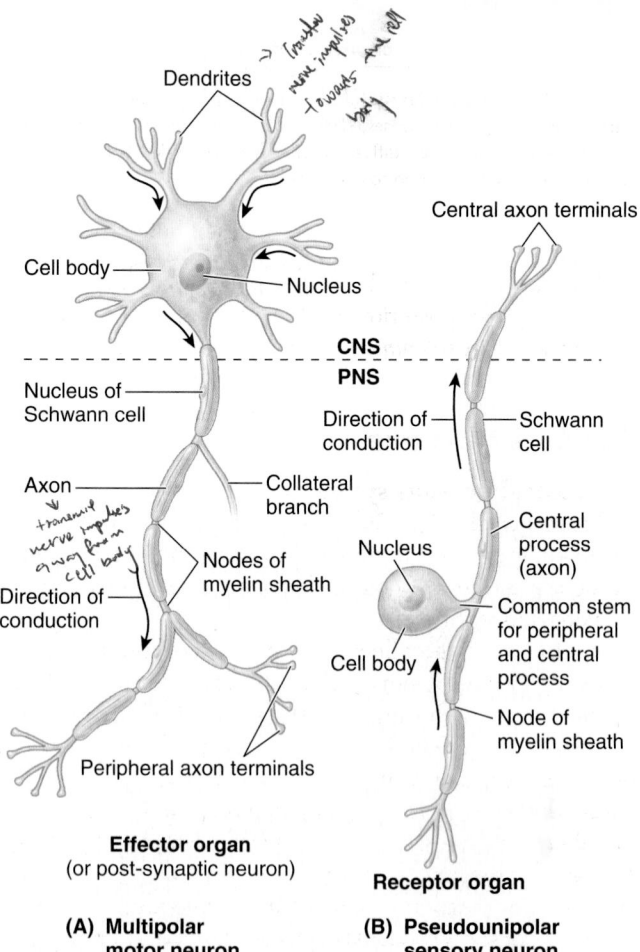

FIGURE I.28. Neurons. The most common types of neurons are shown. **A.** Multipolar motor neurons. All of the motor neurons that control skeletal muscle and those comprising the ANS are multipolar neurons. **B.** Except for some of the special senses (for example, olfaction and vision), all sensory neurons of the PNS are pseudounipolar neurons with cell bodies located in sensory ganglia.

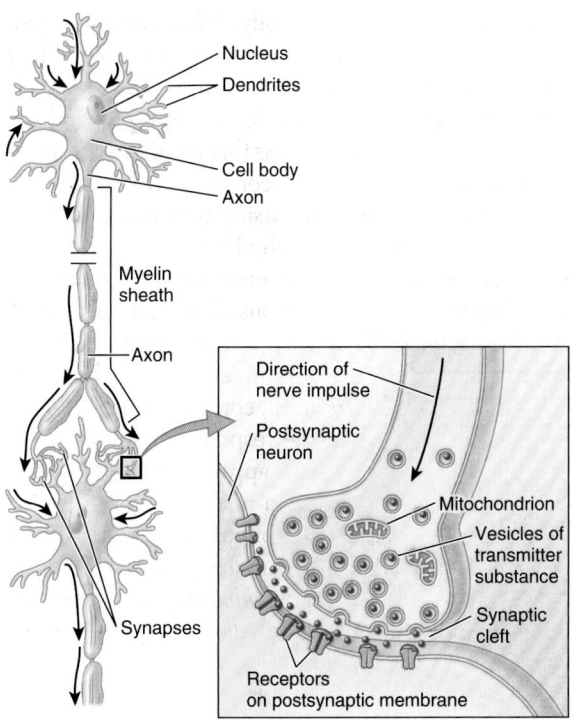

FIGURE I.29. Multipolar motor neurons synapsing. A neuron influences other neurons at synapses. *Inset:* Detailed structure of an axodendritic synapse. Neurotransmitters diffuse across the synaptic cleft between the two cells and become bound to receptors.

PNS, neuroglia include *satellite cells* around the neurons in the spinal (posterior root) and autonomic ganglia and Schwann (*neurolemma*) cells (Figs. I.28 and I.29).

Central Nervous System

The **central nervous system** (CNS) consists of the brain and spinal cord (Fig. I.30). The principal roles of the CNS are to integrate and coordinate incoming and outgoing neural signals and to carry out higher mental functions, such as thinking and learning.

A **nucleus** is a collection of nerve cell bodies in the CNS. A bundle of nerve fibers (axons) within the CNS connecting neighboring or distant nuclei of the cerebral cortex is a **tract.** The brain and spinal cord are composed of gray matter and white matter. The nerve cell bodies lie within and constitute the **gray matter;** the interconnecting fiber tract systems form the **white matter** (Fig. I.31). In transverse sections of the spinal cord, the gray matter appears roughly as an H-shaped area embedded in a matrix of white matter. The struts (supports) of the H are **horns;** hence there are right and left posterior (dorsal) and anterior (ventral) gray horns.

Three membranous layers—pia mater, arachnoid mater, and dura mater—collectively constitute the **meninges.** The meninges and the **cerebrospinal fluid** (CSF) surround and protect the CNS. The brain and spinal cord are intimately

covered on their outer surface by the innermost meningeal layer, a delicate, transparent covering, the **pia mater.** The CSF is located between the pia mater and the **arachnoid mater.** External to the pia mater and arachnoid mater is the thick, tough **dura mater**. The *dura mater of the brain is* intimately related to the internal aspect of the bone of the surrounding *neurocranium* (braincase); *the dura mater of the spinal cord* is separated from the surrounding bone of the vertebral column by a fat-filled *epidural space.*

Peripheral Nervous System

The **peripheral nervous system** (PNS) consists of nerve fibers and cell bodies outside the CNS that conduct impulses to or away from the CNS (Fig. I.30). The PNS is organized into nerves that connect the CNS with peripheral structures.

A **nerve fiber** consists of an axon, its neurolemma (G. *neuron*, nerve + G. *lemma*, husk), and surrounding endoneurial connective tissue (Figs. I.32). The **neurolemma** consists of the cell membranes of Schwann cells that immediately surround the axon, separating it from other axons. In the PNS, the neurolemma may take two forms, creating two classes of nerve fibers:

1. The neurolemma of myelinated nerve fibers consists of Schwann cells specific to an individual axon, organized into a continuous series of enwrapping cells that form myelin.
2. The neurolemma of unmyelinated nerve fibers is composed of Schwann cells that do not make up such an apparent series; multiple axons are separately embedded within the cytoplasm of each cell. These Schwann cells do not produce myelin. Most fibers in cutaneous nerves (nerves supplying sensation to the skin) are unmyelinated.

A **nerve** consists of:

- a bundle of nerve fibers outside the CNS (or a "bundle of bundled fibers," or *fascicles*, in the case of a larger nerve),
- the connective tissue coverings that surround and bind the nerve fibers and fascicles together, and
- the blood vessels (*vasa nervorum*) that nourish the nerve fibers and their coverings (Fig. I.33).

Nerves are fairly strong and resilient because the nerve fibers are supported and protected by three connective tissue coverings:

1. **Endoneurium,** delicate connective tissue immediately surrounding the neurilemma cells and axons.
2. **Perineurium,** a layer of dense connective tissue that encloses a fascicle of nerve fibers, providing an effective barrier against penetration of the nerve fibers by foreign substances.
3. **Epineurium,** a thick connective tissue sheath that surrounds and encloses a bundle of fascicles, forming the outermost covering of the nerve; it includes fatty tissue, blood vessels, and lymphatics.

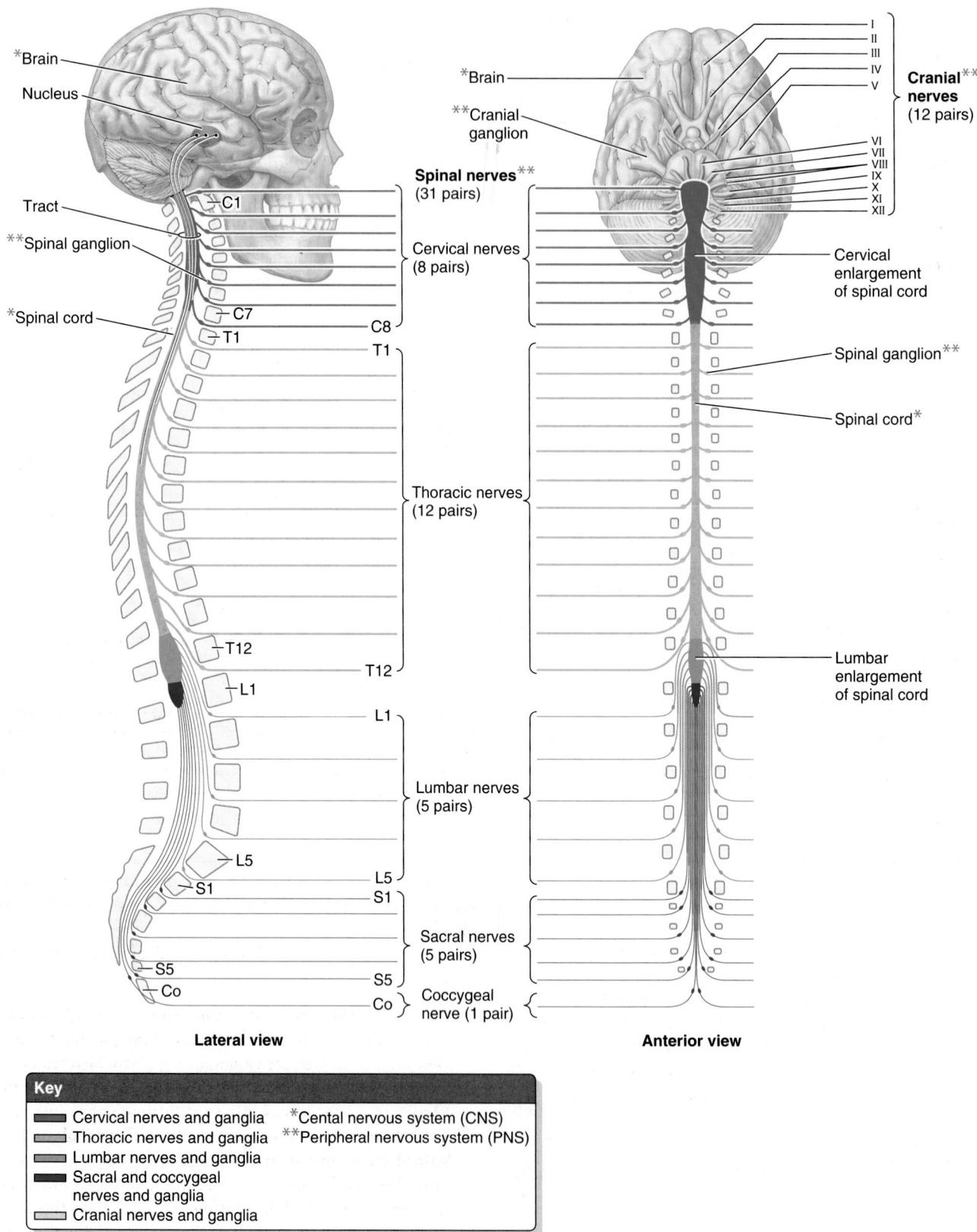

Lateral view

Anterior view

*Brain

Nucleus

Tract

**Spinal ganglion

*Spinal cord

C1

C7

T1

T12

L1

L5

S1

S5

Co

Spinal nerves
(31 pairs)

Cervical nerves
(8 pairs)

C8

T1

T12

L1

L5

S1

S5

Co

Thoracic nerves
(12 pairs)

Lumbar nerves
(5 pairs)

Sacral nerves
(5 pairs)

Coccygeal
nerve (1 pair)

*Brain

**Cranial
ganglion

I
II
III
IV
V

VI
VII
VIII
IX
X
XI
XII

Cranial
nerves
(12 pairs)

Cervical
enlargement
of spinal cord

Spinal ganglion**

Spinal cord*

Lumbar
enlargement
of spinal cord

Key

Cervical nerves and ganglia   *Cental nervous system (CNS)
Thoracic nerves and ganglia   **Peripheral nervous system (PNS)
Lumbar nerves and ganglia
Sacral and coccygeal
nerves and ganglia
Cranial nerves and ganglia

FIGURE I.30. Basic organization of the nervous system. The CNS consists of the brain and spinal cord. The PNS consists of nerves and ganglia. Nerves are either cranial nerves or spinal (segmental) nerves, or derivatives of them. Except in the cervical region, each spinal nerve bears the same letter–numeral designation as the vertebra forming the superior boundary of its exit from the vertebral column. In the cervical region, each spinal nerve bears the same letter–numeral designation as the vertebra forming its inferior boundary. Spinal nerve C8 exits between vertebrae C7 and T1. The cervical and lumbar enlargements of the spinal cord occur in relationship to the innervation of the limbs.

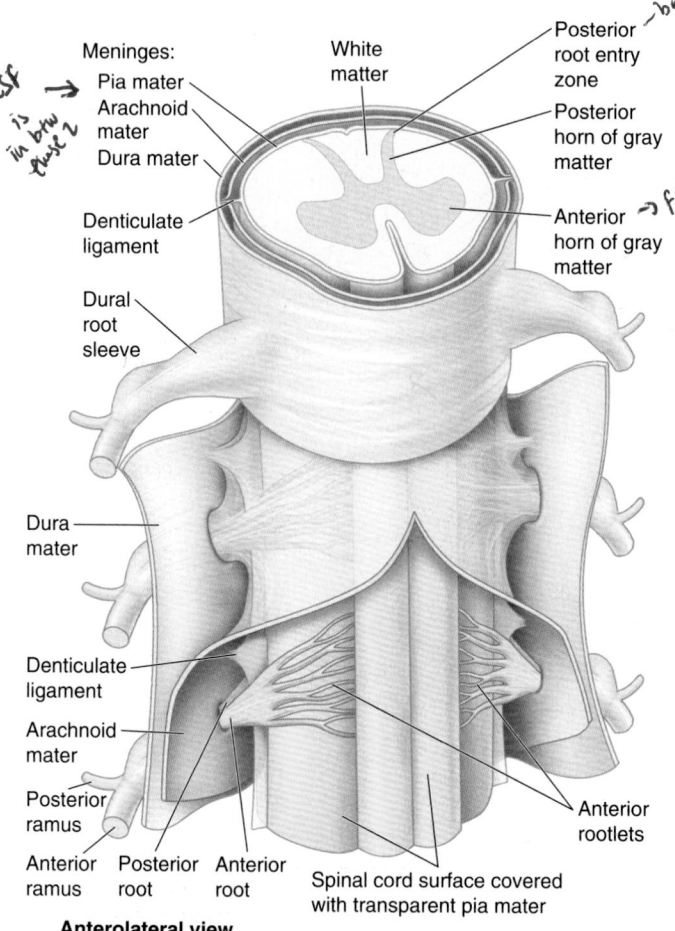

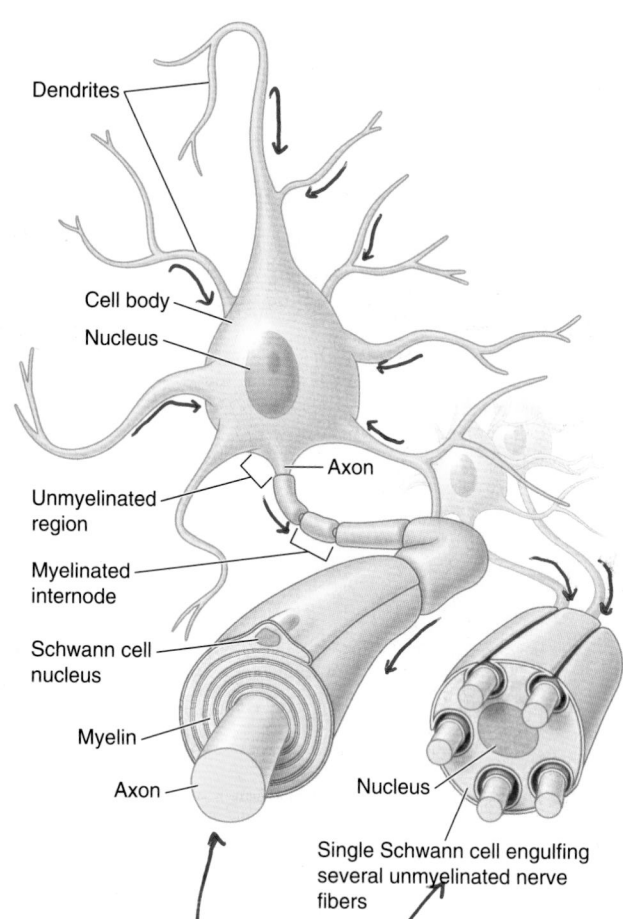

[handwritten margin notes: "CSF is in btw these 2"]

Meninges:
Pia mater
Arachnoid mater
Dura mater

White matter

Posterior root entry zone *[handwritten: → back]*

Posterior horn of gray matter

Denticulate ligament

Anterior horn of gray matter *[handwritten: → front]*

Dural root sleeve

Dura mater

Denticulate ligament

Arachnoid mater

Posterior ramus

Anterior ramus   Posterior root   Anterior root

Anterior rootlets

Spinal cord surface covered with transparent pia mater

Anterolateral view

Dendrites

Cell body

Nucleus

Axon

Unmyelinated region

Myelinated internode

Schwann cell nucleus

Myelin

Axon

Nucleus

Single Schwann cell engulfing several unmyelinated nerve fibers

FIGURE I.31. Spinal cord and spinal meninges. The dura mater and arachnoid mater are incised and reflected to show the posterior and anterior roots and the denticulate ligament (a bilateral, longitudinal, toothed thickening of the pia mater that anchors the cord in the center of the vertebral canal). The spinal cord is sectioned to show its horns of gray matter. The meninges extend along the nerve roots then blend with the epineurium at the point where the posterior and anterior roots join, forming the dural root sleeves that enclose the sensory (posterior root) ganglia.

FIGURE I.32. Myelinated and unmyelinated nerve fibers. Myelinated nerve fibers have a sheath composed of a continuous series of neurilemma (Schwann) cells that surround the axon and form a series of myelin segments. Multiple unmyelinated nerve fibers are individually engulfed within a single neurilemma cell that does not produce myelin.

[handwritten annotations: "Rem. Nerve fibers → axons along its axolemma & surrounding endoneural connective tissue."]

Nerves are organized much like a telephone cable: The axons are like individual wires insulated by the neurolemma and endoneurium; the insulated wires are bundled by the perineurium, and the bundles are surrounded by the epineurium forming the cable's outer wrapping (Fig. I.33). It is important to distinguish between *nerve fibers* and *nerves*, which are sometimes depicted diagrammatically as being one and the same.

A collection of neuron cell bodies outside the CNS is a **ganglion.** There are both motor (autonomic) and sensory ganglia. *[handwritten: → Pseudounipolar sensory neuron]*

[handwritten margin: ⊘ 10/5/14]

[handwritten: Nerve is a → Nerve fibers bundled up]

TYPES OF NERVES

The PNS is anatomically and operationally continuous with the CNS (Fig. I.30). Its **afferent (sensory) fibers** convey neural impulses to the CNS from the sense organs (e.g., the eyes) and from sensory receptors in various parts of the body (e.g., in the skin). Its **efferent (motor) fibers** convey neural impulses from the CNS to *effector organs* (muscles and glands).

Nerves are either cranial nerves or spinal nerves, or derivatives of them.

- **Cranial nerves** exit the cranial cavity through foramina (openings) in the cranium (G. *kranion*, skull) and are identified by a descriptive name (e.g., "trochlear nerve") or a Roman numeral (e.g., "CN IV"). Only 11 of the 12 pairs of cranial nerves arise from the brain; the other pair (CN XI) arises from the superior part of the spinal cord.
- **Spinal (segmental) nerves** exit the vertebral column (spine) through intervertebral foramina (Fig. I.30). Spinal nerves arise in bilateral pairs from a specific segment of the spinal cord. The 31 spinal cord segments and the 31 pairs of nerves arising from them are identified by a letter and number (e.g., "T4") designating the region of the spinal cord and their superior-to-inferior order (*C*, cervical; *T*, thoracic; *L*, lumbar; *S*, sacral; *Co*, coccygeal).

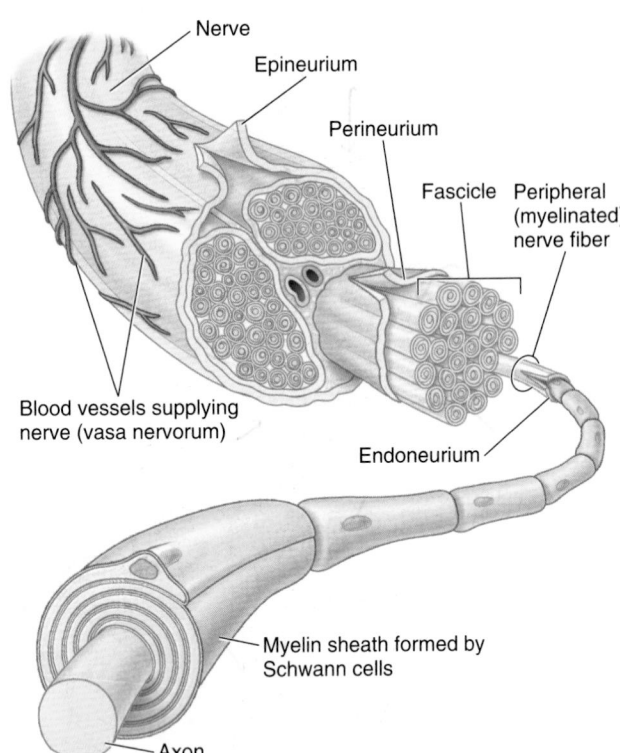

FIGURE I.33. Arrangement and ensheathment of myelinated nerve fibers. Nerves consist of the bundles of nerve fibers, the layers of connective tissue binding them together, and the blood vessels (vasa nervorum) that serve them. All but the smallest nerves are arranged in bundles called fascicles.

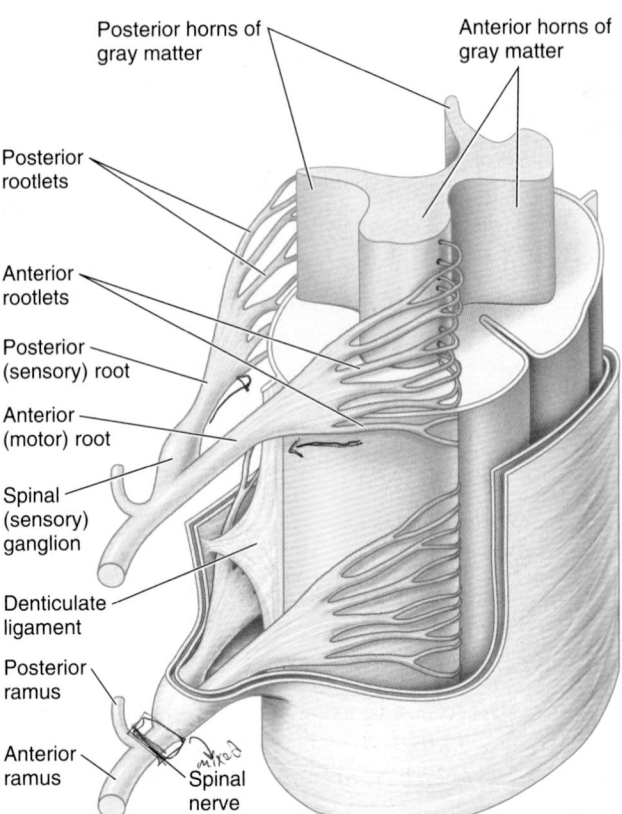

FIGURE I.34. Spinal cord gray matter, spinal roots, and spinal nerves. The meninges are incised and reflected to show the H-shaped gray matter in the spinal cord and the posterior and anterior rootlets and roots of two spinal nerves. The posterior and anterior rootlets enter and leave the posterior and anterior gray horns, respectively. The posterior and anterior nerve roots unite distal to the spinal ganglion to form a mixed spinal nerve, which immediately divides into posterior and anterior rami.

Spinal Nerves. Spinal nerves initially arise from the spinal cord as *rootlets* (a detail commonly omitted from diagrams for the sake of simplicity); the rootlets converge to form two *nerve roots* (Fig. I.34):

1. An **anterior (ventral) nerve root,** consisting of motor (efferent) fibers passing from nerve cell bodies in the anterior horn of spinal cord gray matter to effector organs located peripherally.
2. A **posterior (dorsal) nerve root,** consisting of sensory (afferent) fibers from cell bodies in the spinal (sensory) or posterior (dorsal) root ganglion (commonly abbreviated in clinical use as "DRG") that extend peripherally to sensory endings and centrally to the posterior horn of spinal cord gray matter.

The posterior and anterior nerve roots unite, within or just proximal to the intervertebral foramen, to form a *mixed* (both motor and sensory) *spinal nerve,* which immediately divides into two *rami* (L., branches): a *posterior (dorsal) ramus* and an *anterior (ventral) ramus.* As branches of the mixed spinal nerve, the posterior and anterior rami carry both motor and sensory fibers, as do all their subsequent branches. The terms *motor nerve* and *sensory nerve* are almost always relative terms, referring to the *majority* of fiber types conveyed

by that nerve. Nerves supplying muscles of the trunk or limbs (motor nerves) also contain about 40% sensory fibers, which convey pain and proprioceptive information. Conversely, cutaneous (sensory) nerves contain motor fibers, which serve sweat glands and the smooth muscle of blood vessels and hair follicles.

The unilateral area of skin innervated by the sensory fibers of a single spinal nerve is called a **dermatome;** the unilateral muscle mass receiving innervation from the fibers conveyed by a single spinal nerve is a **myotome** (Fig. I.35). From clinical studies of lesions of the posterior roots or spinal nerves, dermatome maps have been devised to indicate the typical pattern of innervation of the skin by specific spinal nerves (Fig. I.36). However, a lesion of a single posterior root or spinal nerve would rarely result in numbness over the area demarcated for that nerve in these maps because the fibers conveyed by adjacent spinal nerves overlap almost completely as they are distributed to the skin, providing a type of double coverage. The lines indicating dermatomes on dermatome maps would thus

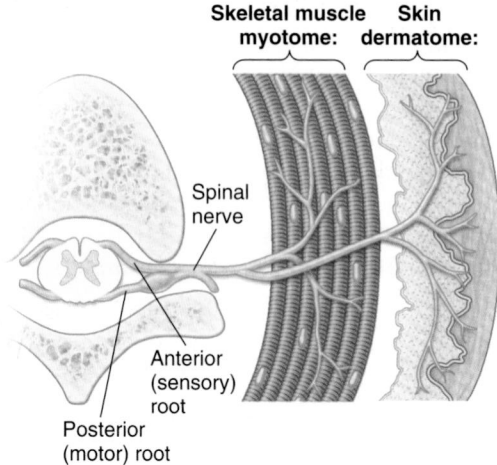

Skeletal muscle myotome: **Skin dermatome:**

Spinal nerve

Anterior (sensory) root

Posterior (motor) root

FIGURE I.35. Dermatomes and myotomes. Schematic representation of a dermatome (the unilateral area of skin) and myotome (the unilateral portion of skeletal muscle) receiving innervation from a single spinal nerve.

be better represented by smudges or gradations of color. Generally, at least two adjacent spinal nerves (or posterior roots) must be interrupted to produce a discernible area of numbness.

As they emerge from the intervertebral foramina, spinal nerves are divided into two rami (Fig. I.37):

1. **Posterior (primary) rami of spinal nerves** supply nerve fibers to the synovial joints of the vertebral column, deep muscles of the back, and the overlying skin in a segmental pattern. As a general rule, the posterior rami remain separate from each other (do not merge to form major somatic nerve plexuses). [networks?]

2. **Anterior (primary) rami of spinal nerves** supply nerve fibers to the much larger remaining area, consisting of the anterior and lateral regions of the trunk and the upper and lower limbs. The anterior rami that are distributed exclusively to the trunk generally remain separate from each other, also innervating muscles and skin in a segmental pattern (Figs. I.38 and I.39). However, primarily in relationship to the innervation of the limbs, the majority of anterior rami merge with one or more adjacent anterior rami, forming the major *somatic nerve plexuses* (networks) in which their fibers intermingle and from which a new set of *multisegmental peripheral nerves* emerges (Figs. I.39 and I.40A & B). The anterior rami of spinal nerves participating in plexus formation contribute fibers to multiple peripheral nerves arising from the plexus (Fig. I.40A); conversely, most peripheral nerves arising from the plexus contain fibers from multiple spinal nerves (Fig. I.40B).

Although the spinal nerves lose their identity as they split and merge in the plexus, the fibers arising from a specific

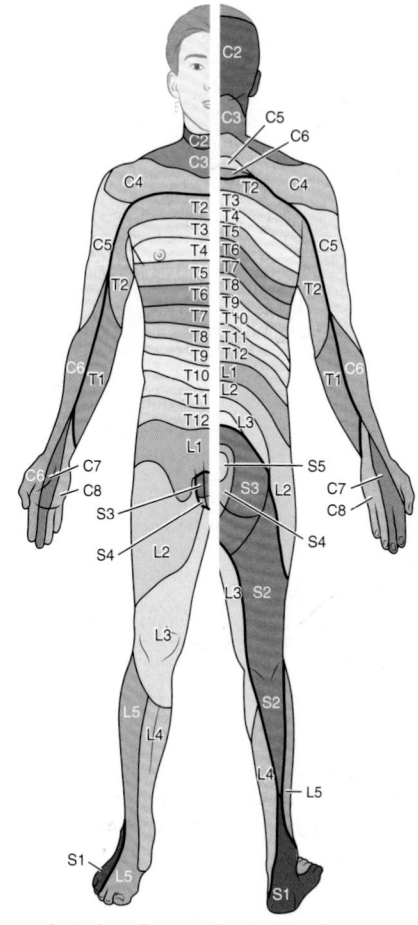

Anterior view　**Posterior view**

FIGURE I.36. Dermatomes (segmental cutaneous innervation). Dermatome maps of the body are based on an accumulation of clinical findings following spinal nerve injuries. This map is based on the studies of Foerster (1933) and reflects both anatomical (actual) distribution or segmental innervation and clinical experience. Another popular but more schematic map is that of Keegan and Garrett (1948), which is appealing for its regular, more easily extrapolated pattern. Spinal nerve C1 lacks a significant afferent component and does not supply the skin; therefore, no C1 dermatome is depicted. Note that in the Foerster map, C5–T1 and L3–S1 are distributed almost entirely in the limbs (i.e., have little or no representation on the trunk).

spinal cord segment and conveyed from it by a single spinal nerve are ultimately distributed to one segmental dermatome, although they may reach it by means of a multisegmental peripheral nerve arising from the plexus that also conveys fibers to all or parts of other adjacent dermatomes (Fig. I.40C).

It is therefore important to distinguish between the distribution of the fibers carried by spinal nerves (*segmental innervation* or *distribution*—i.e., dermatomes and myotomes labeled with a letter and a number, such as "T4") and of the fibers carried by branches of a plexus (*peripheral nerve innervation* or *distribution*, labeled with the names of peripheral nerves, such as "the median

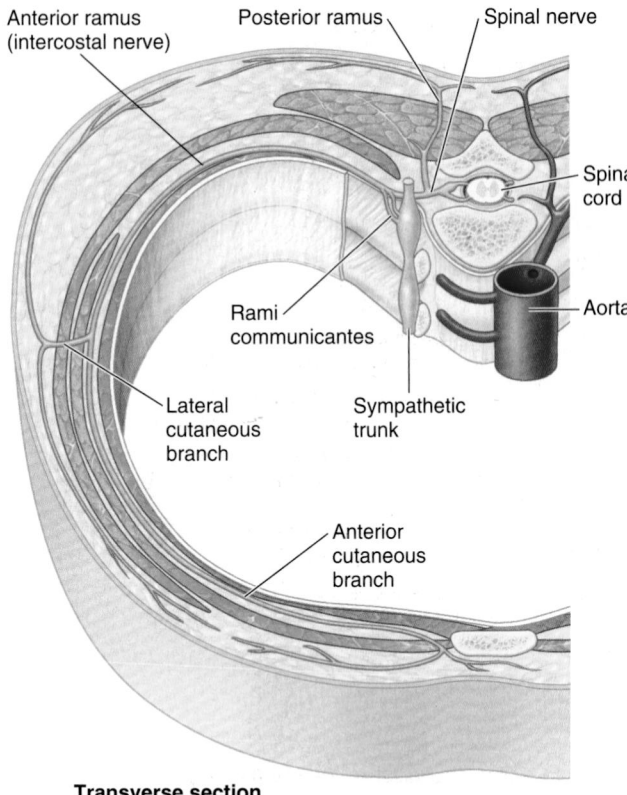

Anterior ramus
(intercostal nerve)

Posterior ramus

Spinal nerve

Spinal
cord

Aorta

Rami
communicantes

Lateral
cutaneous
branch

Sympathetic
trunk

Anterior
cutaneous
branch

Transverse section

FIGURE I.37. Distribution of spinal nerves. Almost as soon as they are formed by the merging of posterior and anterior roots, spinal nerves divide into anterior and posterior (primary) rami. Posterior rami are distributed to the synovial joints of the vertebral column, deep muscles of the back, and the overlying skin. The remaining anterolateral body wall, including the limbs, is supplied by anterior rami. Posterior rami and the anterior rami of spinal nerves T2–T12 generally do not merge with the rami of adjacent spinal nerves to form plexuses.

nerve") (Figs. I.36 and I.38). Mapping segmental innervation (dermatomes, determined by clinical experience) and mapping the distribution of peripheral nerves (determined by dissecting the branches of a named nerve distally) produce entirely different maps, except for most of the trunk where, in the absence of plexus formation, segmental and peripheral distributions are the same. The overlapping in the cutaneous distribution of nerve fibers conveyed by adjacent spinal nerves also occurs in the cutaneous distribution of nerve fibers conveyed by adjacent peripheral nerves.

Cranial Nerves. As they arise from the CNS, some cranial nerves convey only sensory fibers, some only motor fibers, and some carry a mixture of both types of fibers (Fig. I.41). Communication occurs between cranial nerves, and between cranial nerves and upper cervical (spinal) nerves; thus a nerve that initially conveys only motor fibers may receive sensory fibers distally in its course, and vice

versa. Except for the first two (those involved in the senses of smell and sight), cranial nerves that convey sensory fibers into the brain bear sensory ganglia (similar to spinal or posterior root ganglia), where the cell bodies of the pseudounipolar fibers are located. Although, by definition, the term *dermatome* applies only to spinal nerves, similar areas of skin supplied by single cranial nerves can be identified and mapped. Unlike dermatomes, however, there is little overlap in the innervation of zones of skin supplied by cranial nerves.

21/5/13

SOMATIC AND VISCERAL FIBERS

The types of fibers conveyed by cranial or spinal nerves are as follows (Fig. I.41):

- Somatic fibers
  - **General sensory fibers** (*general somatic afferent [GSA] fibers*) transmit sensations from the body to the CNS; they may be *exteroceptive sensations* from the skin (pain, temperature, touch, and pressure) or pain and *proprioceptive sensations* from muscles, tendons, and joints. Proprioceptive sensations are usually subconscious, providing information regarding joint position and the tension of tendons and muscles. This information is combined with input from the vestibular apparatus of the internal ear, resulting in awareness of the orientation of the body and limbs in space, independent of visual input.
  - **Somatic motor fibers** (*general somatic efferent [GSE] fibers*) transmit impulses to skeletal (voluntary) muscles.
- Visceral fibers (Autonomic)
  - **Visceral sensory fibers** (*general visceral afferent [GVA] fibers*) transmit pain or subconscious visceral *reflex sensations* (information concerning distension, blood gas, and blood pressure levels, for example) from hollow organs and blood vessels to the CNS.
  - **Visceral motor fibers** (*general visceral efferent [GVE] fibers*) transmit impulses to smooth (involuntary) muscle and glandular tissues. Two varieties of fibers, *presynaptic* and *postsynaptic,* work together to conduct impulses from the CNS to smooth muscle or glands.

Both types of sensory fibers—visceral sensory and general sensory—are processes of pseudounipolar neurons with cell bodies located outside of the CNS in spinal or cranial sensory ganglia (Figs. I.41 and I.42). The motor fibers of nerves are axons of multipolar neurons. The cell bodies of somatic motor and presynaptic visceral motor neurons are located in the gray matter of the spinal cord. Cell bodies of postsynaptic motor neurons are located outside the CNS in autonomic ganglia.

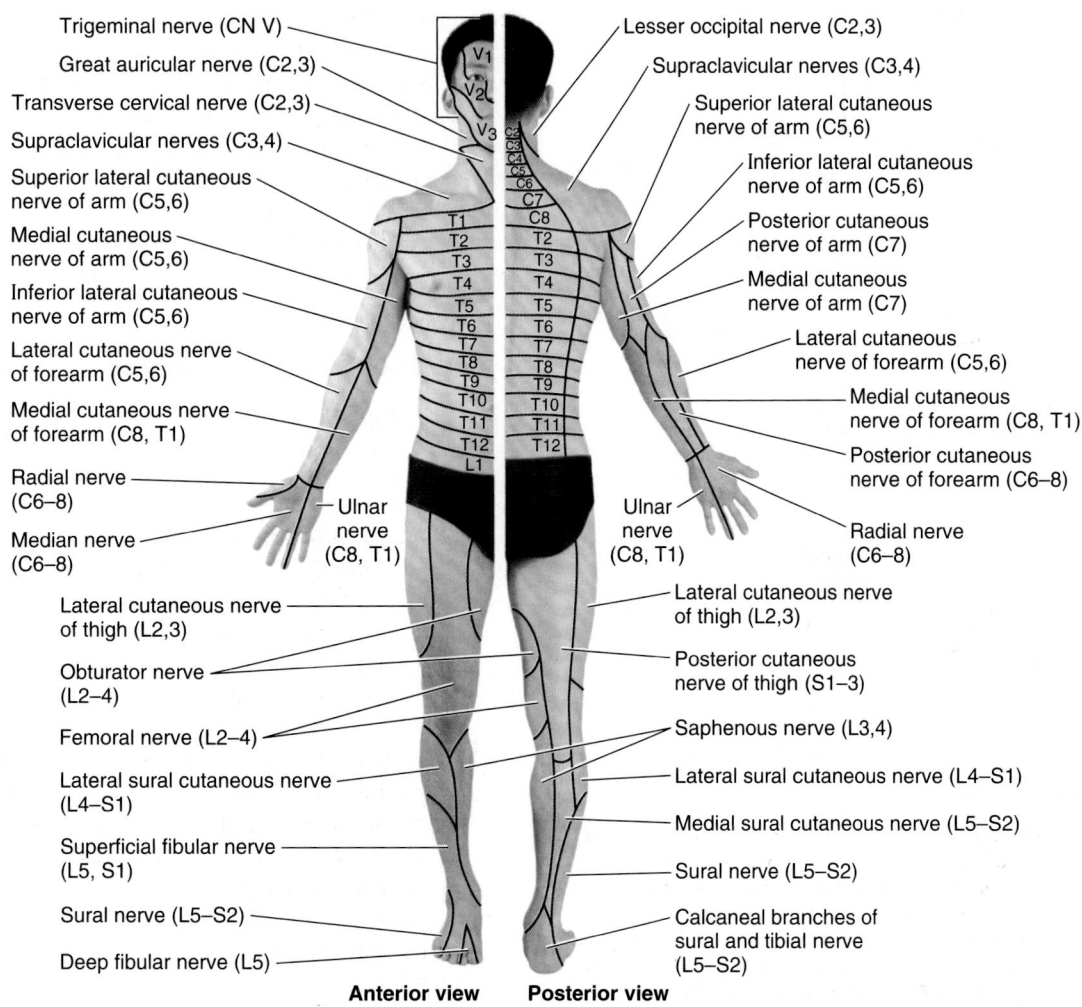

Anterior view   Posterior view

FIGURE I.38. Distribution of peripheral cutaneous nerves. Maps of the cutaneous distribution of peripheral nerves are based on dissection and supported by clinical findings.

In addition to the fiber types listed above, some cranial nerves also convey **special sensory fibers** for the special senses (smell, sight, hearing, balance, and taste). On the basis of the embryologic/phylogenetic derivation of certain muscles of the head and neck, some motor fibers conveyed by cranial nerves to striated muscle have traditionally been classified as "special visceral"; however, since the designation is confusing and not applied clinically, that term will not be used here. These fibers are occasionally designated as *branchial motor*, referring to muscle tissue derived from the pharyngeal arches in the embryo.

CENTRAL AND PERIPHERAL NERVOUS SYSTEMS

Damage to the CNS

When the *brain or spinal cord is damaged*, the injured axons do not recover in most circumstances. Their proximal stumps begin to regenerate, sending sprouts into the area of the lesion; however, this growth is blocked by *astrocyte proliferation* at the injury site, and the axonal sprouts are soon retracted. As a result, permanent disability follows destruction of a tract in the CNS.

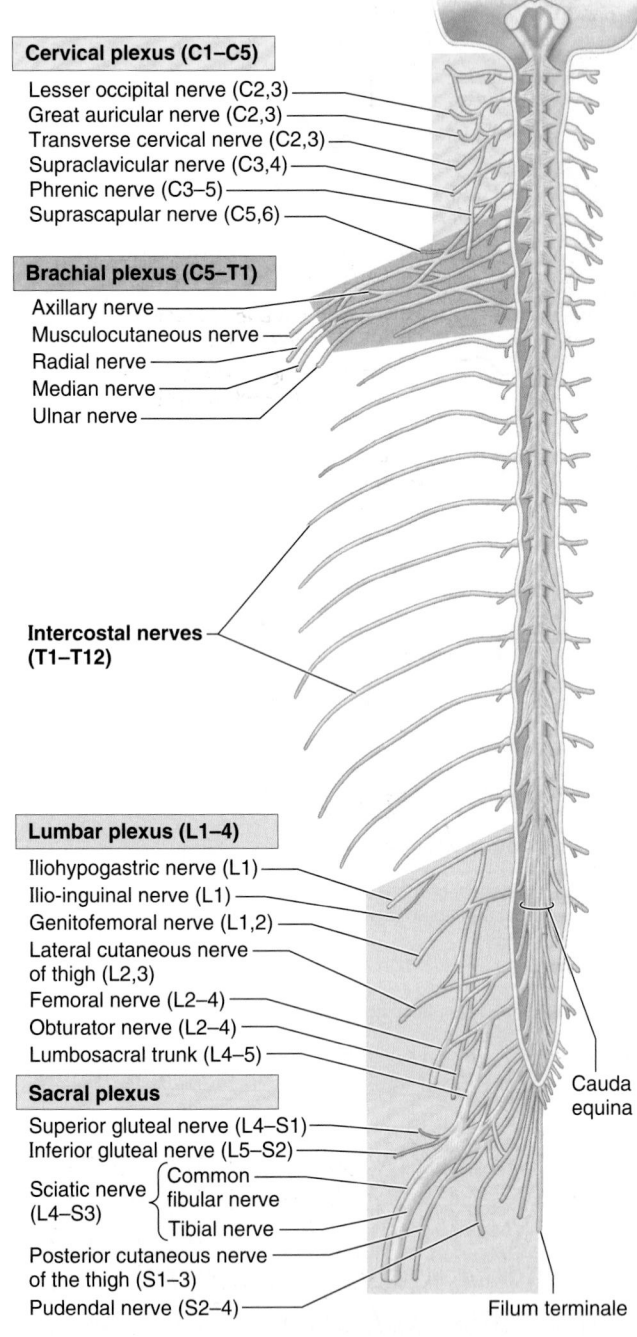

Cervical plexus (C1–C5)

Lesser occipital nerve (C2,3)
Great auricular nerve (C2,3)
Transverse cervical nerve (C2,3)
Supraclavicular nerve (C3,4)
Phrenic nerve (C3–5)
Suprascapular nerve (C5,6)

Brachial plexus (C5–T1)

Axillary nerve
Musculocutaneous nerve
Radial nerve
Median nerve
Ulnar nerve

Intercostal nerves (T1–T12)

Lumbar plexus (L1–4)

Iliohypogastric nerve (L1)
Ilio-inguinal nerve (L1)
Genitofemoral nerve (L1,2)
Lateral cutaneous nerve of thigh (L2,3)
Femoral nerve (L2–4)
Obturator nerve (L2–4)
Lumbosacral trunk (L4–5)

Sacral plexus

Superior gluteal nerve (L4–S1)
Inferior gluteal nerve (L5–S2)
Sciatic nerve (L4–S3) { Common fibular nerve / Tibial nerve }
Posterior cutaneous nerve of the thigh (S1–3)
Pudendal nerve (S2–4)

Cauda equina

Filum terminale

Posterior view

FIGURE I.39. Anterior rami of spinal nerves and their participation in plexus formation. Although the posterior rami (not shown) generally remain separate from each other and follow a distinctly segmental pattern of distribution, most anterior rami (20 of 31 pairs) participate in the formation of plexuses, which are primarily involved in the innervation of the limbs. The anterior rami distributed only to the trunk generally remain separate, however, and follow a segmental distribution similar to that of the posterior rami.

Rhizotomy

The posterior and anterior roots are the only sites where the motor and sensory fibers of a spinal nerve are segregated. Therefore, only at these locations can the surgeon selectively section either functional element for the relief of intractable pain or spastic paralysis (*rhizotomy*).

Nerve Degeneration and Ischemia of Nerves

Neurons do not proliferate in the adult nervous system, except those related to the sense of smell in the olfactory epithelium. Therefore, neurons destroyed through disease or trauma are not replaced (Hutchins et al., 2002). When nerves are stretched, crushed, or severed, their axons degenerate mainly distal to the lesion because they depend on their nerve cell bodies for survival. If the axons are damaged but the cell bodies are intact, regeneration and return of function may occur. The chance of survival is best when a nerve is compressed. Pressure on a nerve commonly causes *paresthesia*, the pins-and-needles sensation that occurs when one sits too long with the legs crossed, for example.

A *crushing nerve injury* damages or kills the axons distal to the injury site; however, the nerve cell bodies usually survive, and the nerve's connective tissue coverings remain intact. No surgical repair is needed for this type of nerve injury because the intact connective tissue coverings guide the growing axons to their destinations. Regeneration is less likely to occur in a severed nerve. Sprouting occurs at the proximal ends of the axons, but the growing axons may not reach their distal targets. A *cutting nerve injury* requires surgical intervention because regeneration of the axon requires apposition of the cut ends by sutures through the epineurium. The individual nerve bundles are realigned as accurately as possible. *Anterograde (wallerian) degeneration* is the degeneration of axons detached from their cell bodies. The degenerative process involves the axon and its myelin sheath, even though this sheath is not part of the injured neuron.

Compromising a nerve's blood supply for a long period by *compression of the vasa nervorum* (Fig. I.33) can also cause nerve degeneration. Prolonged *ischemia* (inadequate blood supply) of a nerve may result in damage no less severe than that produced by crushing or even cutting the nerve. The Saturday night syndrome, named after an intoxicated individual who "passes out" with a limb dangling across the arm of a chair or the edge of a bed, is an example of a more serious, often permanent, paresthesia. This condition can also result from the sustained use of a tourniquet during a surgical procedure. If the ischemia is not too prolonged, temporary numbness or paresthesia results. *Transient paresthesias* are familiar to anyone who has had an injection of anesthetic for dental repairs.

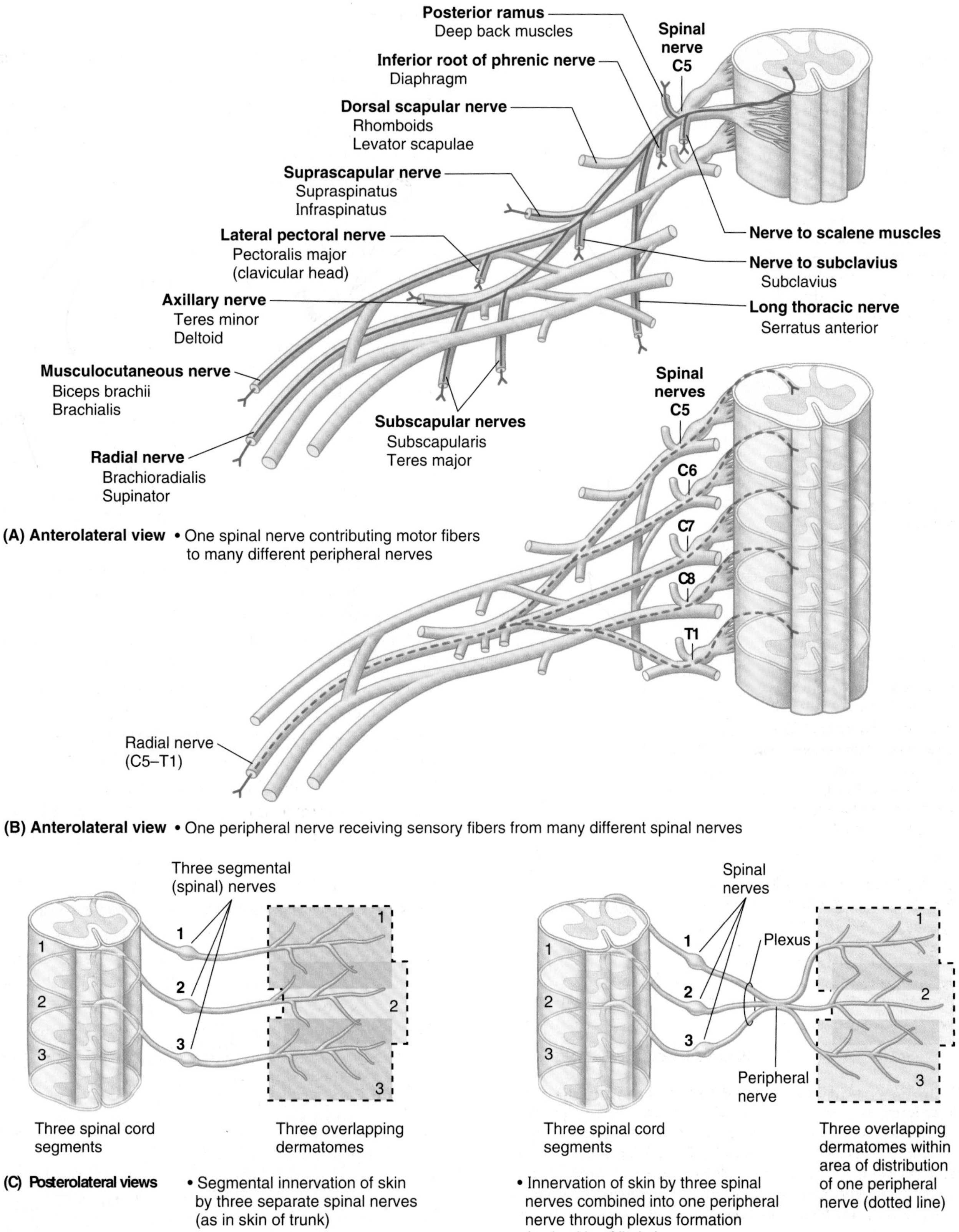

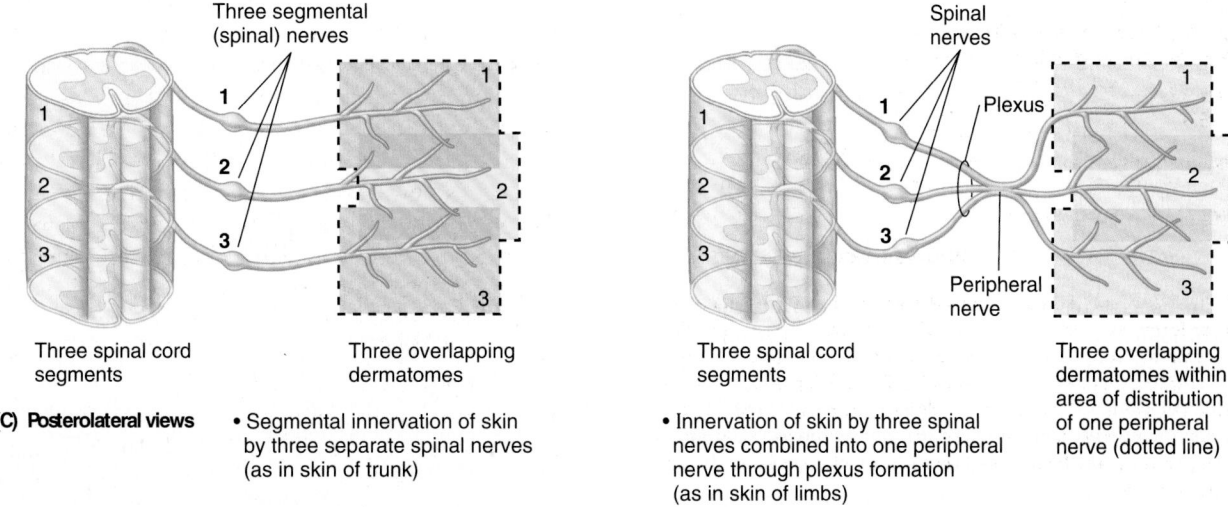

Posterior ramus — Deep back muscles

Inferior root of phrenic nerve — Diaphragm

Dorsal scapular nerve — Rhomboids / Levator scapulae

Suprascapular nerve — Supraspinatus / Infraspinatus

Lateral pectoral nerve — Pectoralis major (clavicular head)

Axillary nerve — Teres minor / Deltoid

Musculocutaneous nerve — Biceps brachii / Brachialis

Radial nerve — Brachioradialis / Supinator

Subscapular nerves — Subscapularis / Teres major

Spinal nerve C5

Nerve to scalene muscles

Nerve to subclavius — Subclavius

Long thoracic nerve — Serratus anterior

Spinal nerves C5

(A) Anterolateral view • One spinal nerve contributing motor fibers to many different peripheral nerves

Radial nerve (C5–T1)

Spinal nerves C5, C6, C7, C8, T1

(B) Anterolateral view • One peripheral nerve receiving sensory fibers from many different spinal nerves

Three segmental (spinal) nerves

1, 2, 3

Three spinal cord segments

Three overlapping dermatomes

(C) Posterolateral views • Segmental innervation of skin by three separate spinal nerves (as in skin of trunk)

Spinal nerves

Plexus

Peripheral nerve

Three spinal cord segments

Three overlapping dermatomes within area of distribution of one peripheral nerve (dotted line)

• Innervation of skin by three spinal nerves combined into one peripheral nerve through plexus formation (as in skin of limbs)

FIGURE I.40. Plexus formation. Adjacent anterior rami merge to form plexuses in which their fibers are exchanged and redistributed, forming a new set of multisegmental peripheral (named) nerves. **A.** The fibers of a single spinal nerve entering the plexus are distributed to multiple branches of the plexus. **B.** The peripheral nerves derived from the plexus contain fibers from multiple spinal nerves. **C.** Although segmental nerves merge and lose their identity when plexus formation results in multisegmental peripheral nerves, the segmental (dermatomal) pattern of nerve fiber distribution remains.

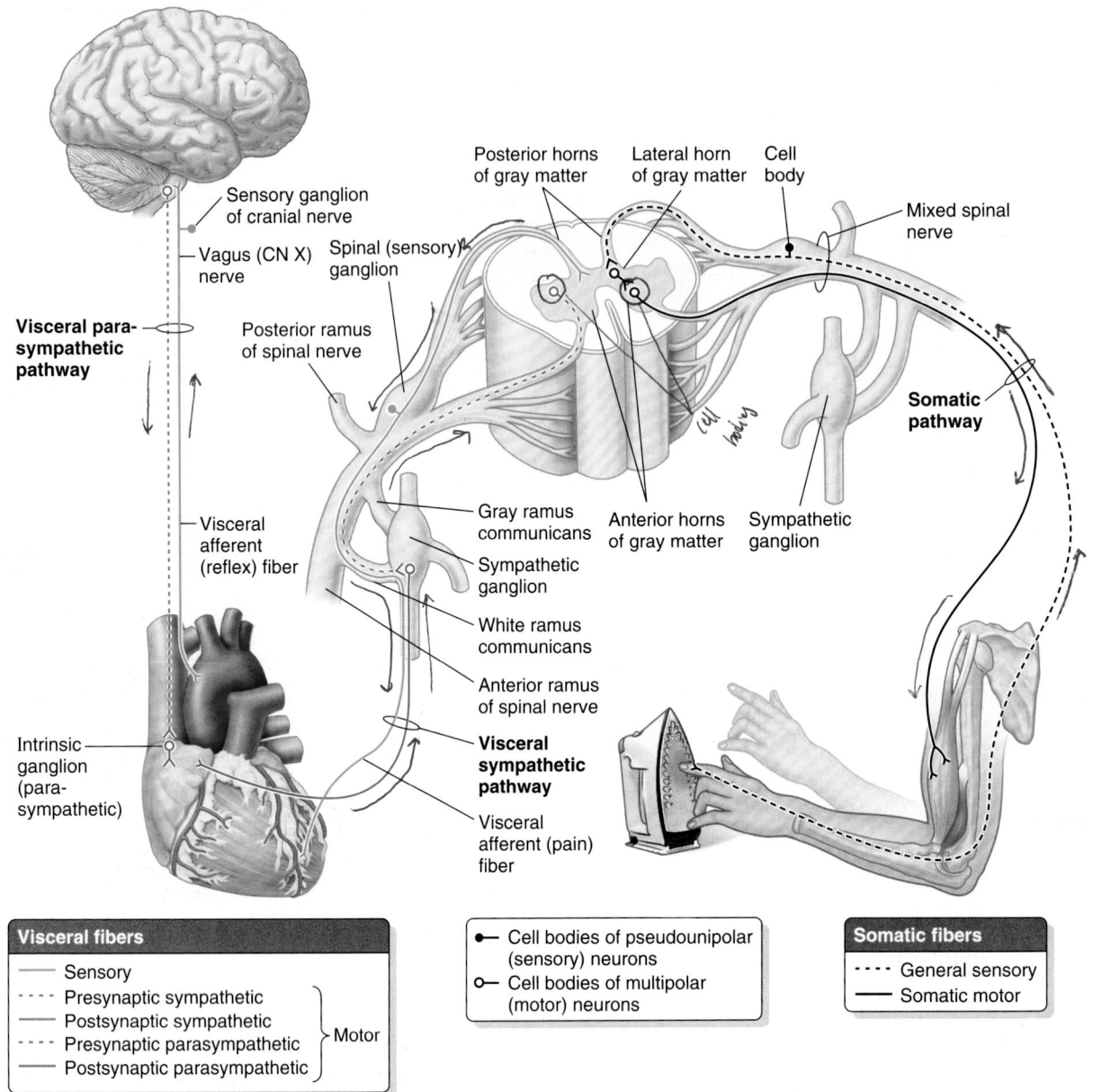

Posterior horns of gray matter

Lateral horn of gray matter

Cell body

Mixed spinal nerve

Sensory ganglion of cranial nerve

Vagus (CN X) nerve

Spinal (sensory) ganglion

Visceral para-sympathetic pathway

Posterior ramus of spinal nerve

Somatic pathway

Visceral afferent (reflex) fiber

Gray ramus communicans

Sympathetic ganglion

White ramus communicans

Anterior ramus of spinal nerve

Anterior horns of gray matter

Sympathetic ganglion

Visceral sympathetic pathway

Intrinsic ganglion (para-sympathetic)

Visceral afferent (pain) fiber

Visceral fibers	
——— Sensory	
- - - - Presynaptic sympathetic	⎫
——— Postsynaptic sympathetic	⎬ Motor
- - - - Presynaptic parasympathetic	
——— Postsynaptic parasympathetic	⎭

- ● Cell bodies of pseudounipolar (sensory) neurons
- ○ Cell bodies of multipolar (motor) neurons

Somatic fibers	
- - - - General sensory	
——— Somatic motor	

FIGURE I.41. Somatic and visceral innervation via spinal, splanchnic, and cranial nerves. The somatic motor system permits voluntary and reflexive movement caused by contraction of skeletal muscles, such as occurs when one touches a hot iron.

The Bottom Line

CENTRAL AND PERIPHERAL NERVOUS SYSTEMS

The nervous system can be functionally divided into the central nervous system (CNS), which consists of the brain and spinal cord, and the peripheral nervous system (PNS), which consists of the nerve fibers and their nerve cell bodies that reside outside the CNS. ♦ Neurons are the functional units of the nervous system. They are composed of a cell body, dendrites, and axons. ♦ The neuronal axons (nerve fibers) transmit impulses to other neurons or to a target organ or muscle, or in the case of sensory nerves, transmit impulses to the CNS from peripheral sensory organs. ♦ Neuroglia are non-neuronal, supporting cells of the nervous system.

♦ Within the CNS, a collection of nerve cell bodies is called a nucleus; in the PNS, nerve cell body aggregations (or even solitary nerve cell bodies) constitute a ganglion. ♦ In the CNS, a bundle of nerve fibers that connect the nuclei is called a tract; in the PNS, a bundle of nerve fibers, the connective tissue holding it together, and the blood vessels serving it (vasa nervorum) constitute a nerve. ♦ Nerves exiting the cranium are cranial nerves; those exiting the vertebral column (formerly, the spine), are spinal nerves. ♦ Although some cranial nerves convey a single type of fiber, most nerves convey a variety of visceral or somatic and sensory or motor fibers.

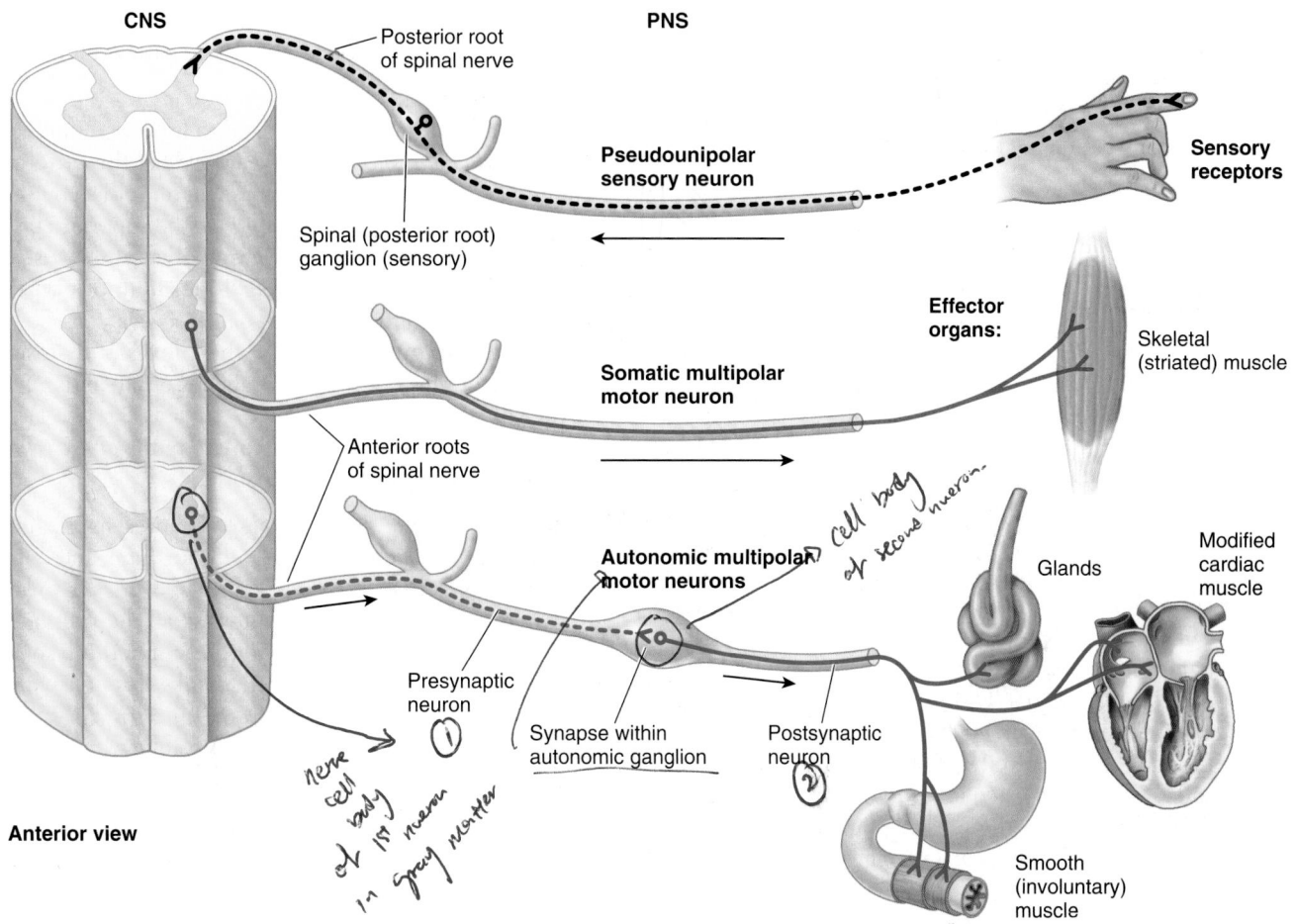

FIGURE I.42. Neurons of the PNS. Note the types of neurons involved in the somatic and visceral nervous systems, the general location of their cell bodies in relation to the CNS, and their receptors or effector organs.

Somatic Nervous System

The somatic nervous system (SNS), composed of somatic parts of the CNS and PNS, provides sensory and motor innervation to all parts of the body (G. *soma*), except the viscera in the body cavities, smooth muscle, and glands (Figs. I.41 and I.42). The **somatic sensory system** transmits sensations of touch, pain, temperature, and position from sensory receptors. Most of these sensations reach conscious levels (i.e., we are aware of them). The **somatic motor system** innervates only skeletal muscle, stimulating voluntary and reflexive movement by causing the muscle to contract, as occurs in response to touching a hot iron.

Autonomic Nervous System (ANS)

The **autonomic nervous system** (ANS), classically described as the *visceral nervous system* or *visceral motor system* (Figs. I.41 and I.42), consists of motor fibers that stimulate smooth (involuntary) muscle, modified cardiac muscle (the intrinsic stimulating and conducting tissue of the heart), and glandular (secretory) cells. However, the visceral efferent fibers of the ANS are accompanied by visceral afferent

fibers. As the afferent component of autonomic reflexes and in conducting visceral pain impulses, these visceral afferent fibers also play a role in the regulation of visceral function.

The efferent nerve fibers and ganglia of the ANS are organized into two systems or divisions: the *sympathetic (thoracolumbar) division* and the *parasympathetic (craniosacral) division*. Unlike sensory and somatic motor innervation, in which the passage of impulses between the CNS and the sensory ending or effector organ involves a single neuron, in both divisions of the ANS, conduction of impulses from the CNS to the effector organ involves a series of two *multipolar neurons* (Fig. I.42). The nerve cell body of the first **presynaptic (preganglionic) neuron** is located in the gray matter of the CNS. Its fiber (axon) synapses only on the cell body of a **postsynaptic (postganglionic) neuron,** the second neuron in the series. The cell bodies of these second neurons are located outside the CNS in autonomic ganglia, with fibers terminating on the effector organ (smooth muscle, modified cardiac muscle, or glands).

The anatomical distinction between the sympathetic and parasympathetic divisions of the ANS is based primarily on:

1. The location of the presynaptic cell bodies, and
2. Which nerves conduct the presynaptic fibers from the CNS.

A functional distinction of pharmacological importance for medical practice is that the postsynaptic neurons of the two divisions generally liberate different neurotransmitter substances: *norepinephrine* by the sympathetic division (except in the case of sweat glands) and *acetylcholine* by the parasympathetic division.

SYMPATHETIC (THORACOLUMBAR) DIVISION OF ANS

The cell bodies of the presynaptic neurons of the sympathetic division of the ANS are found in only one location: the **intermediolateral cell columns** (IMLs) or nuclei of the spinal cord (Fig. I.43). The paired (right and left) IMLs are a part of the gray matter of the thoracic (T1–12) and the upper lumbar (L1–L2 or 3) segments of the spinal cord (hence the alternate name "thoracolumbar" for the division). In transverse sections of this part of the spinal cord, the IMLs appear as small lateral horns of the H-shaped gray matter, looking somewhat like an extension of the cross-bar of the H between the posterior and the anterior horns. The IMLs are organized *somatotopically* (i.e., arranged like the body, the cell bodies involved with innervation of the head located superiorly, and those involved with innervation of the pelvic viscera and lower limbs located inferiorly). Thus it is possible to deduce the location of the presynaptic sympathetic cell bodies involved in innervation of a specific part of the body.

The cell bodies of postsynaptic neurons of the sympathetic nervous system occur in two locations, the paravertebral and prevertebral ganglia (Fig. I.44):

- **Paravertebral ganglia** are linked to form right and left *sympathetic trunks (chains)* on each side of the vertebral column and extend essentially the length of this column. The *superior paravertebral ganglion* (the superior cervical ganglion of each sympathetic trunk) lies at the base of the cranium. The *ganglion impar* forms inferiorly where the two trunks unite at the level of the coccyx.
- **Prevertebral ganglia** are in the plexuses that surround the origins of the main branches of the abdominal aorta (for which they are named), such as the two large *celiac ganglia* that surround the origin of the *celiac trunk* (a major artery arising from the aorta).

Because they are motor fibers, the axons of presynaptic neurons leave the spinal cord through anterior roots and enter the anterior rami of spinal nerves T1–L2 or 3 (Figs. I.45 and I.46). Almost immediately after entering, all the presynaptic sympathetic fibers leave the anterior rami of these spinal nerves and pass to the sympathetic trunks through **white rami communicantes** (*communicating branches*). Within the sympathetic trunks, presynaptic fibers follow one of four possible courses:

- Ascend in the sympathetic trunk to synapse with a postsynaptic neuron of a higher paravertebral ganglion.

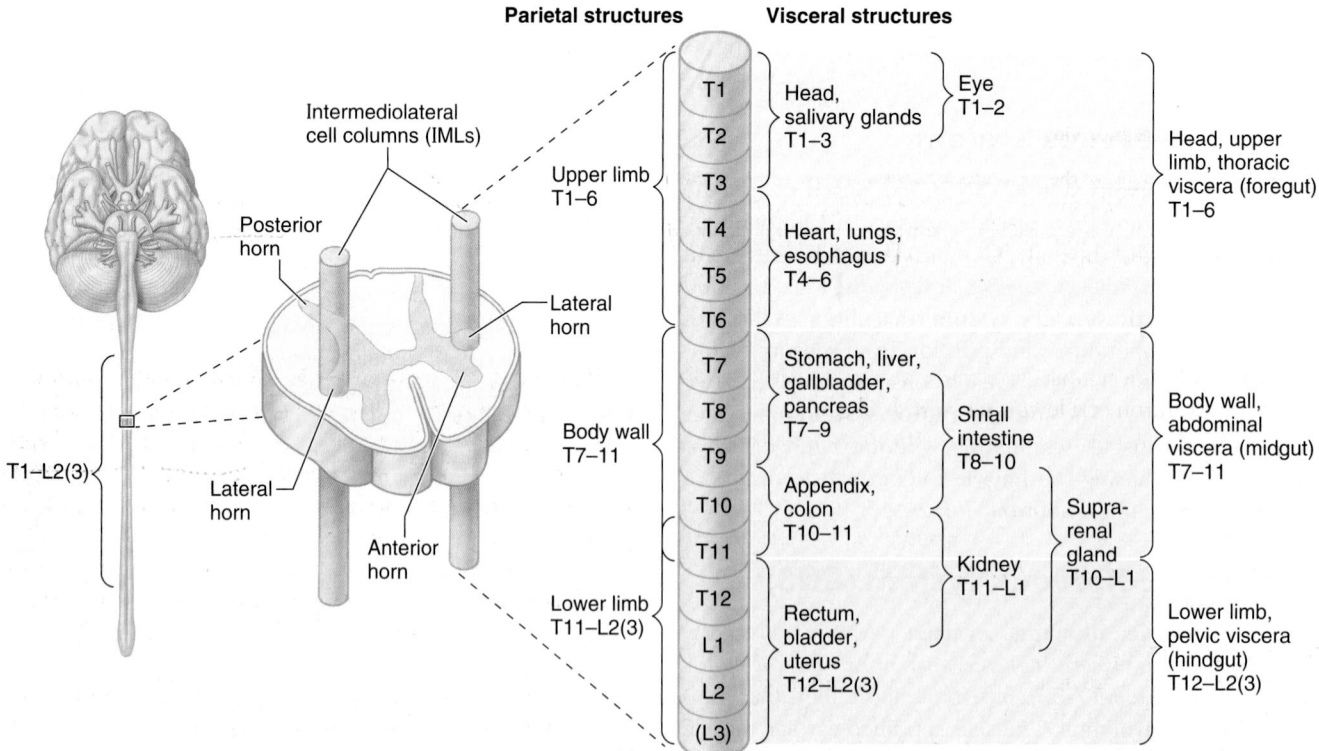

FIGURE I.43. Intermediolateral cell columns. Each IML or nucleus constitutes the lateral horn of gray matter of spinal cord segments T1–L2 or 3 and consists of the cell bodies of the presynaptic neurons of the sympathetic nervous system, which are somatotopically arranged.

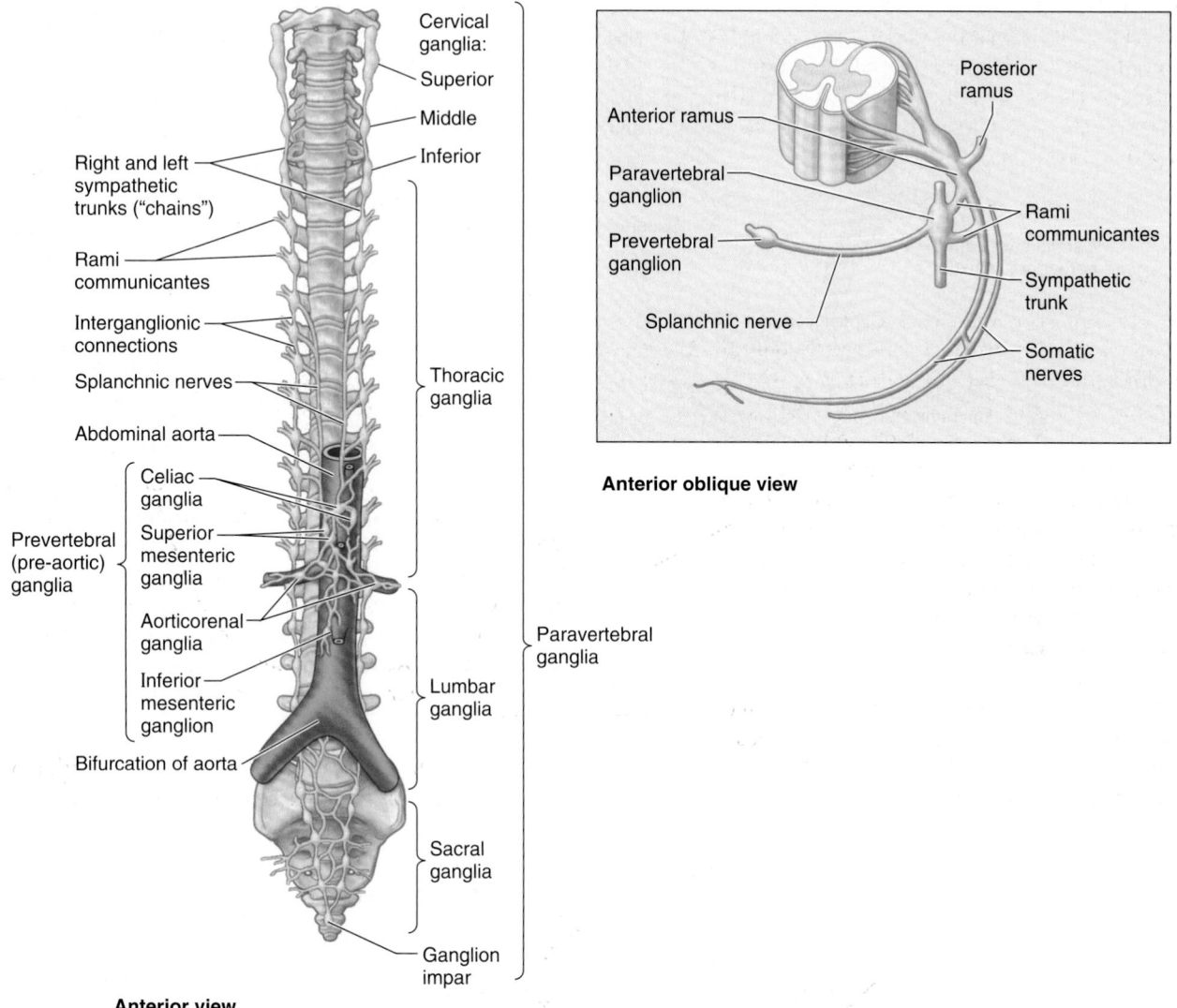

Anterior view

Anterior oblique view

FIGURE I.44. Ganglia of the sympathetic nervous system. In the sympathetic nervous system, cell bodies of postsynaptic neurons occur either in the paravertebral ganglia of the sympathetic trunks or in the prevertebral ganglia that occur mainly in relationship to the origins of the main branches of the abdominal aorta. Prevertebral ganglia are specifically involved in the innervation of abdominopelvic viscera. The cell bodies of postsynaptic neurons distributed to the remainder of the body occur in the paravertebral ganglia.

- Descend in the sympathetic trunk to synapse with a post-synaptic neuron of a lower paravertebral ganglion.
- Enter and synapse immediately with a postsynaptic neuron of the paravertebral ganglion at that level.
- Pass through the sympathetic trunk without synapsing, continuing through an abdominopelvic splanchnic nerve (a branch of the trunk involved in innervating abdominopelvic viscera) to reach the prevertebral ganglia.

Presynaptic sympathetic fibers that provide autonomic innervation within the head, neck, body wall, limbs, and thoracic cavity follow one of the first three courses, synapsing within the paravertebral ganglia. Presynaptic sympathetic fibers innervating viscera within the abdominopelvic cavity follow the fourth course.

Postsynaptic sympathetic fibers greatly outnumber the presynaptic fibers; each presynaptic sympathetic fiber synapses with 30 or more postsynaptic fibers. Those postsynaptic sympathetic fibers, destined for distribution within the neck, body wall, and limbs, pass from the paravertebral ganglia of the sympathetic trunks to adjacent anterior rami of spinal nerves through **gray rami communicantes** (Fig. I.46). By this means, they enter all branches of all 31 pairs of spinal nerves, including the posterior rami.

The postsynaptic sympathetic fibers stimulate contraction of the blood vessels (*vasomotion*) and arrector muscles associated with hairs (*pilomotion*, resulting in "goose bumps"), and to cause sweating (*sudomotion*). Postsynaptic sympathetic fibers that perform these functions in the

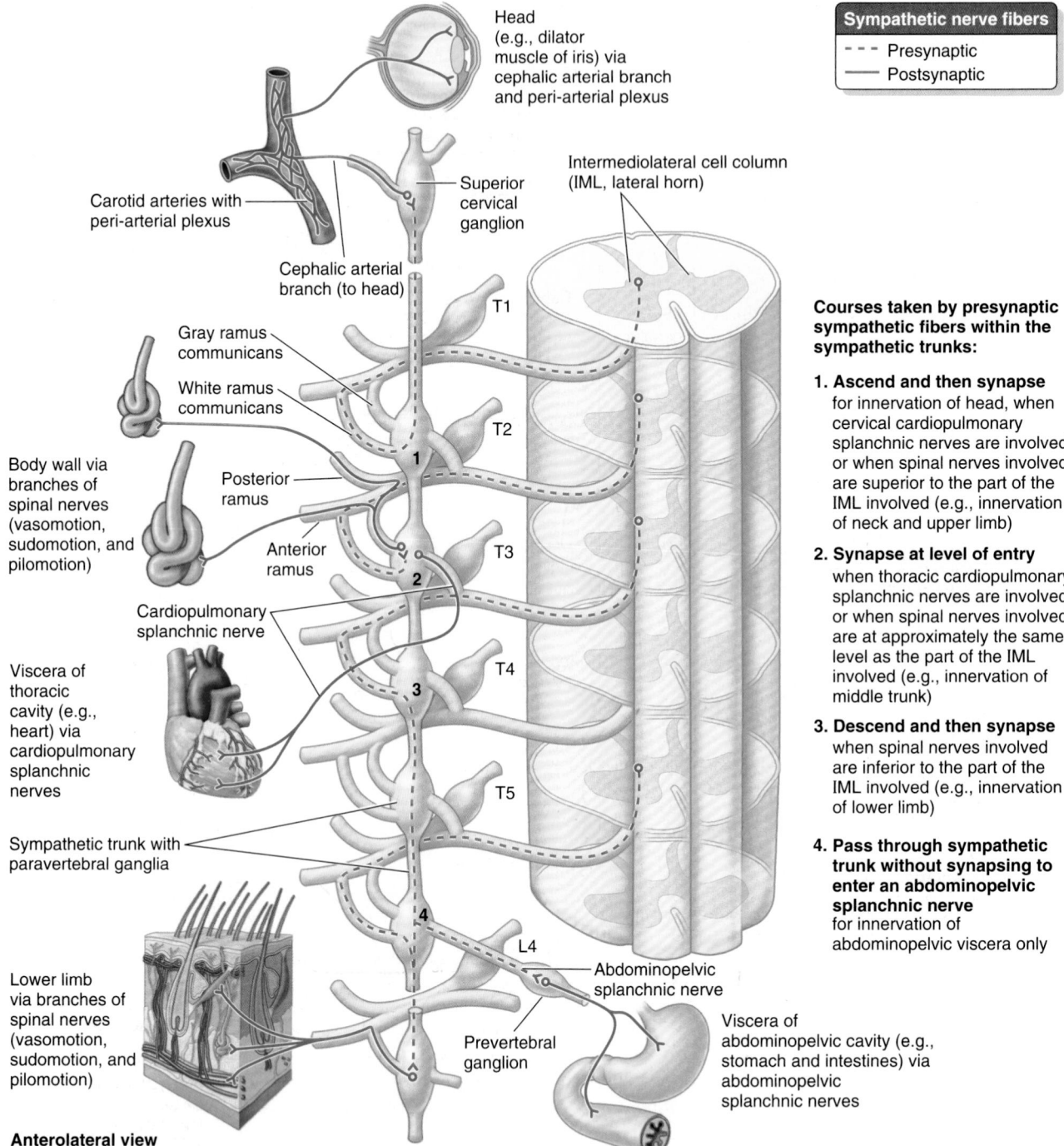

Sympathetic nerve fibers
- - - Presynaptic
——— Postsynaptic

Head
(e.g., dilator
muscle of iris) via
cephalic arterial branch
and peri-arterial plexus

Intermediolateral cell column
(IML, lateral horn)

Carotid arteries with
peri-arterial plexus

Superior
cervical
ganglion

Cephalic arterial
branch (to head)

T1

Gray ramus
communicans

White ramus
communicans

T2

Body wall via
branches of
spinal nerves
(vasomotion,
sudomotion, and
pilomotion)

Posterior
ramus

Anterior
ramus

T3

Cardiopulmonary
splanchnic nerve

T4

Viscera of
thoracic
cavity (e.g.,
heart) via
cardiopulmonary
splanchnic
nerves

T5

Sympathetic trunk with
paravertebral ganglia

Lower limb
via branches of
spinal nerves
(vasomotion,
sudomotion, and
pilomotion)

L4

Abdominopelvic
splanchnic nerve

Prevertebral
ganglion

Viscera of
abdominopelvic cavity (e.g.,
stomach and intestines) via
abdominopelvic
splanchnic nerves

Anterolateral view

Courses taken by presynaptic sympathetic fibers within the sympathetic trunks:

1. **Ascend and then synapse**
   for innervation of head, when cervical cardiopulmonary splanchnic nerves are involved, or when spinal nerves involved are superior to the part of the IML involved (e.g., innervation of neck and upper limb)

2. **Synapse at level of entry**
   when thoracic cardiopulmonary splanchnic nerves are involved, or when spinal nerves involved are at approximately the same level as the part of the IML involved (e.g., innervation of middle trunk)

3. **Descend and then synapse**
   when spinal nerves involved are inferior to the part of the IML involved (e.g., innervation of lower limb)

4. **Pass through sympathetic trunk without synapsing to enter an abdominopelvic splanchnic nerve**
   for innervation of abdominopelvic viscera only

FIGURE I.45. Courses taken by sympathetic motor fibers. All presynaptic fibers follow the same course until they reach the sympathetic trunks. In the trunks, they follow one of four possible courses. Fibers involved in providing sympathetic innervation to the body wall and limbs or viscera above the level of the diaphragm follow paths 1–3 to synapse in the paravertebral ganglia of the sympathetic trunks. Fibers involved in innervating abdominopelvic viscera follow path 4 to prevertebral ganglion via abdominopelvic splanchnic nerves.

Parietal distribution   **Visceral distribution**

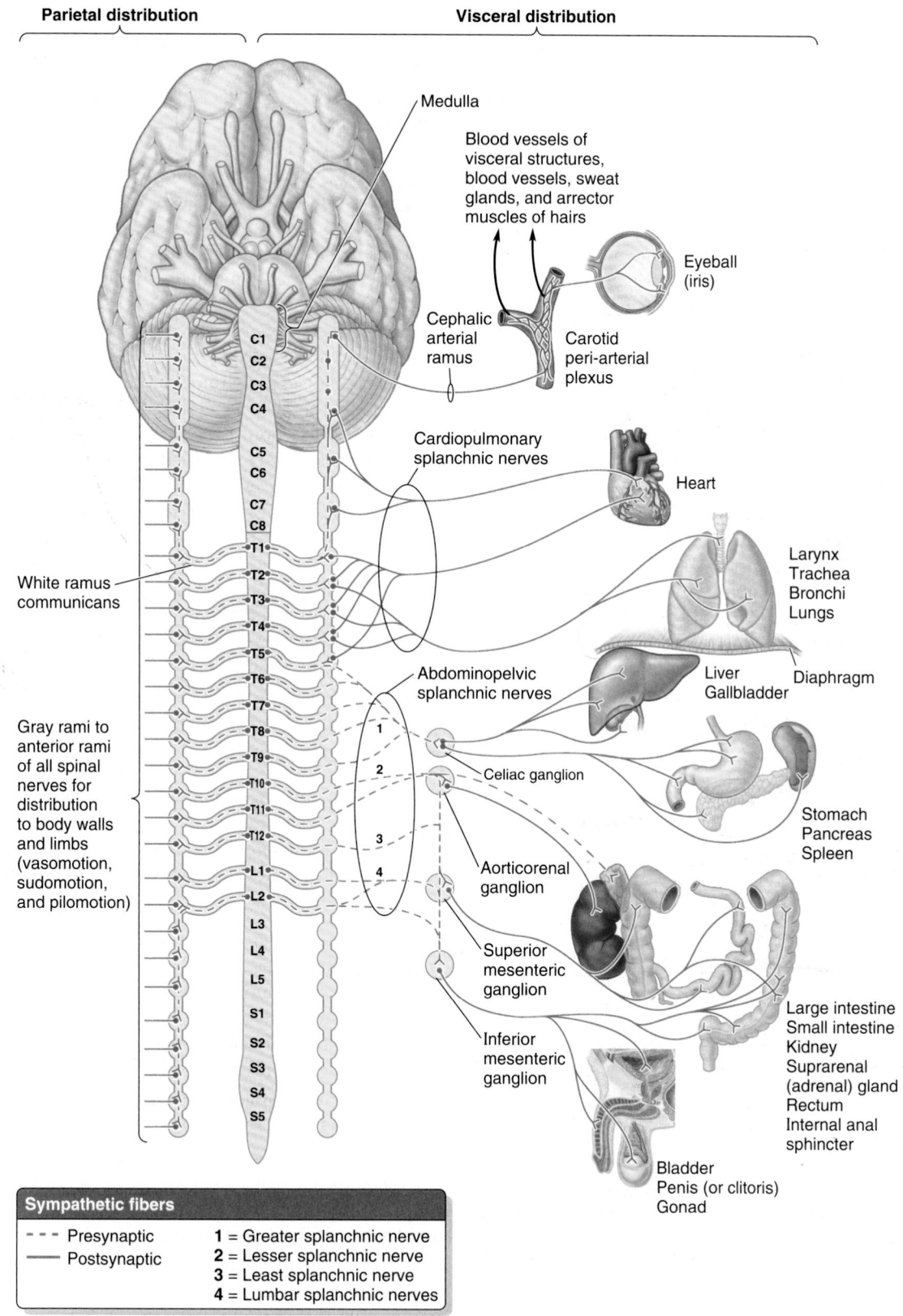

Sympathetic fibers

- - - Presynaptic   **1** = Greater splanchnic nerve
——— Postsynaptic   **2** = Lesser splanchnic nerve
                     3 = Least splanchnic nerve
                     4 = Lumbar splanchnic nerves

FIGURE I.46. The sympathetic (thoracolumbar) division of ANS. Postsynaptic sympathetic fibers exit from the sympathetic trunks by different means, depending on their destination: Those destined for parietal distribution within the neck, body wall, and limbs pass from the sympathetic trunks to adjacent anterior rami of all spinal nerves through gray communicating branches (L. rami communicantes); those destined for the head pass from cervical ganglia by means of cephalic arterial rami to form a carotid peri-arterial plexus; and those destined for viscera of the thoracic cavity (e.g., the heart) pass through cardiopulmonary splanchnic nerves. Presynaptic sympathetic fibers involved in the innervation of viscera of the abdominopelvic cavity (e.g., the stomach) pass through the sympathetic trunks to the prevertebral ganglia by means of abdominopelvic splanchnic nerves. Postsynaptic fibers from the prevertebral ganglia form peri-arterial plexuses, which follow branches of the abdominal aorta to reach their destination.

head (plus innervation of the dilator muscle of the iris—dilator pupillae) all have their cell bodies in the *superior cervical ganglion* at the superior end of the sympathetic trunk. They pass from the ganglion by means of a **cephalic arterial branch** to form *peri-arterial plexuses of nerves,* which follow the branches of the carotid arteries, or they may pass directly to nearby cranial nerves, to reach their destination in the head (Maklad et al., 2001).

Splanchnic nerves convey visceral efferent (autonomic) and afferent fibers to and from the viscera of the body cavities. Postsynaptic sympathetic fibers destined for the viscera of the thoracic cavity (e.g., the heart, lungs, and esophagus) pass through **cardiopulmonary splanchnic nerves** to enter the cardiac, pulmonary, and esophageal plexuses (Figs. I.45 and I.46). The presynaptic sympathetic fibers involved in the innervation of viscera of the abdominopelvic cavity (e.g., the stomach and intestines) pass to the prevertebral ganglia through **abdominopelvic splanchnic nerves** (including the greater, lesser, least thoracic, and lumbar splanchnic nerves) (Figs. I.45–I.47). All presynaptic sympathetic fibers of the abdominopelvic splanchnic nerves, except those involved in innervating the suprarenal (adrenal) glands, synapse in prevertebral ganglia. The postsynaptic fibers from the prevertebral ganglia form peri-arterial plexuses, which follow branches of the abdominal aorta to reach their destination.

Some presynaptic sympathetic fibers pass through the celiac prevertebral ganglia without synapsing, continuing to terminate directly on cells of the medulla of the suprarenal gland (Fig. I.47). The suprarenal medullary cells function as a special type of postsynaptic neuron that, instead of releasing their neurotransmitter substance onto the cells of a specific effector organ, release it into the bloodstream to circulate throughout the body, producing a widespread sympathetic response. Thus the sympathetic innervation of this gland is exceptional.

As described earlier, postsynaptic sympathetic fibers are components of virtually all branches of all spinal nerves. By this means and via peri-arterial plexuses, they extend to and innervate all the body's blood vessels (the sympathetic system's primary function) as well as sweat glands, arrector muscles of hairs, and visceral structures. Thus the sympathetic nervous system reaches virtually all parts of the body, with the rare exception of such avascular tissues as cartilage and nails. Because the two sets of sympathetic ganglia (para- and prevertebral) are centrally placed in the body and are close to the midline (hence relatively close to the spinal cord), in this division the presynaptic fibers are relatively short, whereas the postsynaptic fibers are relatively long, having to extend to all parts of the body.

PARASYMPATHETIC (CRANIOSACRAL) DIVISION OF ANS

Presynaptic parasympathetic nerve cell bodies are located in two sites within the CNS, and their fibers exit by two routes. This arrangement accounts for the alternate name "craniosacral" for the parasympathetic division of the ANS (Fig. I.48):

- In the gray matter of the brainstem, the fibers exit the CNS within cranial nerves III, VII, IX, and X; these fibers constitute the **cranial parasympathetic outflow**.
- In the gray matter of the sacral segments of the spinal cord (S2–4), the fibers exit the CNS through the anterior roots of sacral spinal nerves S2–4 and the pelvic splanchnic nerves that arise from their anterior rami; these fibers constitute the **sacral parasympathetic outflow.**

Not surprisingly, the cranial outflow provides parasympathetic innervation of the head, and the sacral outflow provides the parasympathetic innervation of the pelvic viscera. However, in terms of the innervation of thoracic and abdominal viscera, the cranial outflow through the vagus nerve (CN X) is dominant. It provides innervation to all thoracic viscera and most of the gastrointestinal (GI) tract from the esophagus through most of the large intestine (to its left colic flexure).

The sacral outflow to the GI tract supplies only the descending and sigmoid colon and rectum.

Regardless of the extensive influence of its cranial outflow, the parasympathetic system is much more restricted

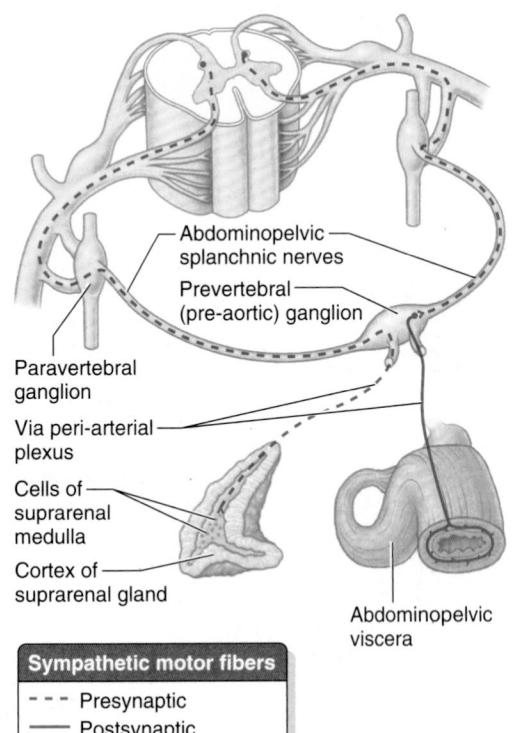

FIGURE I.47. Sympathetic supply to medulla of suprarenal (adrenal) gland. The sympathetic supply to the suprarenal gland is exceptional. The secretory cells of the medulla are postsynaptic sympathetic neurons that lack axons or dendrites. Consequently, the suprarenal medulla is supplied directly by presynaptic sympathetic neurons. The neurotransmitters produced by medullary cells are released into the bloodstream to produce a widespread sympathetic response.

Labels in figure:
- Abdominopelvic splanchnic nerves
- Prevertebral (pre-aortic) ganglion
- Paravertebral ganglion
- Via peri-arterial plexus
- Cells of suprarenal medulla
- Cortex of suprarenal gland
- Abdominopelvic viscera

Sympathetic motor fibers
- - - Presynaptic
- —— Postsynaptic

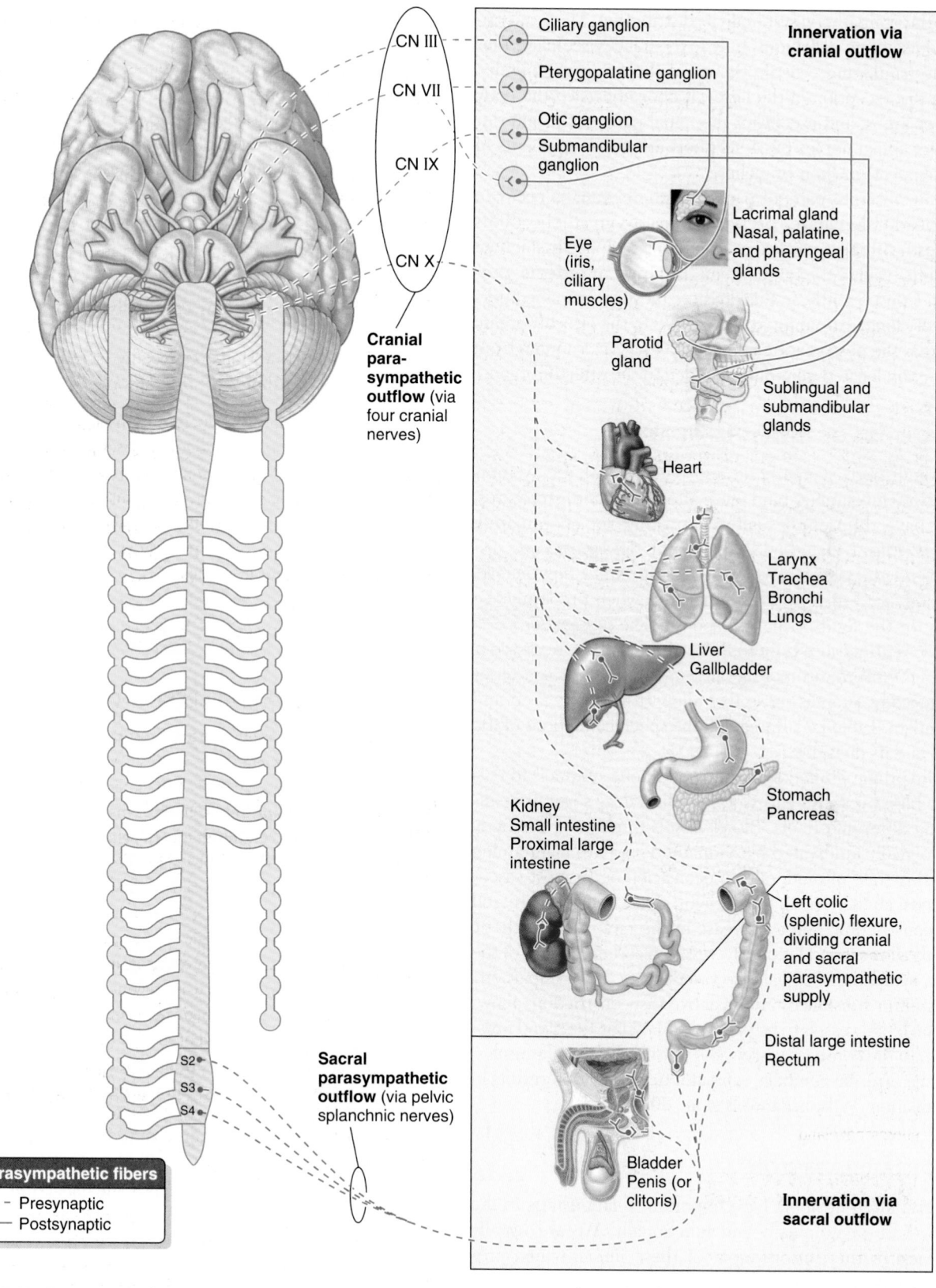

CN III
Ciliary ganglion

CN VII
Pterygopalatine ganglion

CN IX
Otic ganglion
Submandibular ganglion

CN X

Innervation via cranial outflow

Eye (iris, ciliary muscles)

Lacrimal gland
Nasal, palatine, and pharyngeal glands

Parotid gland

Sublingual and submandibular glands

Heart

Larynx
Trachea
Bronchi
Lungs

Liver
Gallbladder

Stomach
Pancreas

Kidney
Small intestine
Proximal large intestine

Left colic (splenic) flexure, dividing cranial and sacral parasympathetic supply

Distal large intestine
Rectum

Bladder
Penis (or clitoris)

Innervation via sacral outflow

Cranial parasympathetic outflow (via four cranial nerves)

S2
S3
S4

Sacral parasympathetic outflow (via pelvic splanchnic nerves)

Parasympathetic fibers
- - - Presynaptic
——— Postsynaptic

FIGURE I.48. Parasympathetic (craniosacral) division of ANS. Presynaptic parasympathetic nerve cell bodies are located in opposite ends of the CNS and their fibers exit by two different routes: (1) in the gray matter of the brainstem, with fibers exiting the CNS within cranial nerves III, VII, IX, and X (these fibers constitute the cranial parasympathetic outflow), and (2) in the gray matter of the sacral (S2–4) segments of the spinal cord, with fibers exiting the CNS via the anterior roots of spinal nerves S2–4 and the pelvic splanchnic nerves that arise from their anterior rami. (These fibers constitute the sacral parasympathetic outflow.) The cranial outflow provides parasympathetic innervation of the head, neck, and most of the trunk; the sacral outflow provides the parasympathetic innervation of the pelvic viscera.

than the sympathetic system in its distribution. The parasympathetic system distributes only to the head, visceral cavities of the trunk, and erectile tissues of the external genitalia. With the exception of the latter, it does not reach the body wall or limbs, and except for the initial parts of the anterior rami of spinal nerves S2–4, its fibers are not components of spinal nerves or their branches.

Four discrete pairs of parasympathetic ganglia occur in the head. Elsewhere, presynaptic parasympathetic fibers synapse with postsynaptic cell bodies, which occur singly in or on the wall of the target organ (**intrinsic** or **enteric ganglia**). Consequently, in this division, most presynpatic fibers are very long, extending from the CNS to the effector organ, whereas the postsynaptic fibers are very short, running from a ganglion located near or embedded in the effector organ.

FUNCTIONS OF DIVISIONS OF ANS

Although both sympathetic and parasympathetic systems innervate involuntary (and often affect the same) structures, they have different, usually contrasting yet coordinated, effects (Figs. I.46 and I.48). In general, the *sympathetic system* is a **catabolic** (energy-expending) **system** that enables the body to deal with stresses, such as when preparing the body for the fight-or-flight response. The *parasympathetic system* is primarily a **homeostatic** or **anabolic** (energy-conserving) **system,** promoting the quiet and orderly processes of the body, such as those that allow the body to feed and assimilate. Table I.2 summarizes the specific functions of the ANS and its divisions.

The primary function of the sympathetic system is to regulate blood vessels. This is accomplished by several means having different effects. Blood vessels throughout the body are tonically innervated by sympathetic nerves, maintaining a resting state of moderate vasoconstriction. In most vascular beds, an increase in sympathetic signals causes increased vasoconstriction, and a decrease in the rate of sympathetic signals allows vasodilation. However, in certain regions of the body, sympathetic signals are vasodilatory (i.e., sympathetic transmitter substances inhibit active vasoconstriction, allowing the blood vessels to be passively dilated by the blood pressure). In the coronary vessels, the vessels of skeletal muscles, and the external genitalia, sympathetic stimulation results in vasodilation (Wilson-Pauwels et al., 2010).

VISCERAL SENSATION

Visceral afferent fibers have important relationships to the ANS, both anatomically and functionally. We are usually unaware of the sensory input of these fibers, which provides information about the condition of the body's internal environment. This information is integrated in the CNS, often triggering visceral or somatic reflexes or both. Visceral reflexes regulate blood pressure and chemistry by altering such functions as heart and respiratory rates and vascular resistance.

Visceral sensation that reaches a conscious level is generally perceived as pain that is either poorly localized or felt as cramps or that may convey a feeling of hunger, fullness, or nausea. Surgeons operating on patients who are under local anesthesia may handle, cut, clamp, or even burn (cauterize) visceral organs without evoking conscious sensation. However, adequate stimulation, such as the following, may elicit visceral pain:

- Sudden distension.
- Spasms or strong contractions.
- Chemical irritants.
- Mechanical stimulation, especially when the organ is active.
- Pathological conditions (especially ischemia) that lower the normal thresholds of stimulation.

Normal activity usually produces no sensation, but it may do so when the blood supply is inadequate (ischemia). Most visceral reflex (unconscious) sensation and some pain travel in visceral afferent fibers that accompany the parasympathetic fibers retrograde (backward). Most visceral pain impulses (from the heart and most organs of the peritoneal cavity) travel centrally along visceral afferent fibers accompanying sympathetic fibers.

The Bottom Line

AUTONOMIC NERVOUS SYSTEM (ANS)

The autonomic nervous system is a subdivision of the motor nervous system that controls functions of the body not under conscious control. ♦ Two neurons, a presynaptic and a postsynaptic fiber, connect the CNS with an end organ, consisting of smooth muscle, gland, or modified cardiac muscle. ♦ Based on the location of the cell body of the presynaptic fibers, the ANS can be subdivided into two divisions: the sympathetic and parasympathetic. ♦ Presynaptic cell bodies of the sympathetic division are found only in the intermediolateral cell columns of gray matter in the thoracolumbar spinal cord, which are organized somatotopically. ♦ The presynaptic sympathetic nerve fibers terminate in sympathetic ganglia formed of the cell bodies of postsynaptic sympathetic neurons. ♦ Sympathetic ganglia are in the sympathetic trunks (paravertebral ganglia) or around the roots of the major branches of the abdominal aorta (prevertebral ganglia). ♦ Cell bodies of the presynaptic neurons of the parasympathetic division are in the gray matter of the brainstem and sacral segments of the spinal cord. ♦ Cell bodies of postsynaptic parasympathetic neurons of the trunk are located in or on the structure being innervated, whereas those in the head are organized into discrete ganglia. ♦ The sympathetic and parasympathetic divisions usually have opposite but coordinated effects.

♦ The sympathetic system primarily regulates blood vessels and facilitates emergency (flight-or-fight) responses.

always opposites

TABLE I.2. FUNCTIONS OF AUTONOMIC NERVOUS SYSTEM (ANS)

Organ, Tract, or System		Effect of Sympathetic Stimulation[a]	Effect of Parasympathetic Stimulation[b]
Eyes	Pupil	Dilates pupil (admits more light for increased acuity at a distance)	Constricts pupil (protects pupil from excessively bright light)
	Ciliary body		Contracts ciliary muscle, allowing lens to thicken for near vision (accommodation)
Skin	Arrector muscles of hair	Causes hairs to stand on end ("gooseflesh" or "goose bumps")	No effect (does not reach)[c]
	Peripheral blood vessels	Vasoconstricts (blanching of skin, lips, and turning fingertips blue)	No effect (does not reach)[c]
	Sweat glands	Promotes sweating[d]	No effect (does not reach)[c]
Other glands	Lacrimal glands	Slightly decreases secretion[e]	Promotes secretion
	Salivary glands	Secretion decreases, becomes thicker, more viscous[e]	Promotes abundant, watery secretion
Heart		Increases the rate and strength of contraction; inhibits the effect of parasympathetic system on coronary vessels, allowing them to dilate[e]	Decreases the rate and strength of contraction (conserving energy); constricts coronary vessels in relation to reduced demand
Lungs		Inhibits effect of parasympathetic system, resulting in bronchodilation and reduced secretion, allowing for maximum air exchange	Constricts bronchi (conserving energy) and promotes bronchial secretion
Digestive tract		Inhibits peristalsis, and constricts blood vessels to digestive tract so that blood is available to skeletal muscle; contracts internal anal sphincter to aid fecal continence	Stimulates peristalsis and secretion of digestive juices; Contracts rectum, inhibits internal anal sphincter to cause defecation
Liver and gallbladder		Promotes breakdown of glycogen to glucose (for increased energy)	Promotes building/conservation of glycogen; increases secretion of bile
Urinary tract		Vasoconstriction of renal vessels slows urine formation; internal sphincter of bladder contracted to maintain urinary continence	Inhibits contraction of internal sphincter of bladder, contracts detrusor muscle of the bladder wall causing urination
Genital system		Causes ejaculation and vasoconstriction resulting in remission of erection	Produces engorgement (erection) of erectile tissues of the external genitals
Suprarenal medulla		Release of adrenaline into blood	No effect (does not innervate)

[a]In general, the effects of sympathetic stimulation are catabolic, preparing body for the fight-or-flight response.

[b]In general, the effects of parasympathetic stimulation are anabolic, promoting normal function and conserving energy.

[c]The parasympathetic system is restricted in its distribution to the head, neck, and body cavities (except for erectile tissues of genitalia); otherwise, parasympathetic fibers are never found in the body wall and limbs. Sympathetic fibers, by comparison, are distributed to all vascularized portions of the body.

[d]With the exception of the sweat glands, glandular secretion is parasympathetically stimulated.

[e]With the exception of the coronary arteries, vasoconstriction is sympathetically stimulated; the effects of sympathetic stimulation on glands (other than sweat glands) are the indirect effects of vasoconstriction.

♦ The parasympathetic system—distributed only to the viscera of the head, neck, and cavities of the trunk and the erectile tissues of the genitalia—is primarily concerned with body conservation, often reversing the effects of sympathetic stimulation. ♦ Some nerves distributing autonomic nerve fibers to the body cavities also convey visceral sensory nerve fibers from the viscera that conduct impulses for pain or reflexes.

MEDICAL IMAGING TECHNIQUES

Radiologic anatomy is the study of the structure and function of the body using medical imaging techniques. It is an important part of clinical anatomy and is the anatomic basis of radiology, the branch of medical science dealing with the use of radiant energy in the diagnosis and treatment of disease. Being able to identify normal structures on radiographs (X-rays) makes it easier to recognize the changes caused by disease and injury. Familiarity with medical imaging techniques commonly used in clinical settings enables one to recognize congenital anomalies, tumors, and fractures. The most commonly used medical imaging techniques are

- Conventional radiography (X-ray images).
- Computerized tomography (CT).
- Ultrasonography (US).
- Magnetic resonance imaging (MRI).
- Nuclear medicine imaging.

Although the techniques differ, each is based on the receipt of attenuated beams of energy that have been passed through, reflected off of, or generated by the body's tissues. Medical imaging techniques permit the observation of anatomical structures in living people and the study of their movements in normal and abnormal activities (e.g., the heart and stomach).

Conventional Radiography

Conventional radiographic studies, in which special techniques such as contrast media have not been used, is referred to clinically as *plain film studies* (Fig. I.49), although today most images are produced and viewed digitally on monitors instead of film. In a radiologic examination, a highly penetrating beam of X-rays transilluminates the patient, showing tissues of differing densities of mass within the body as images of differing intensities (areas of relative light and dark) on the film or monitor (Fig. I.50). A tissue or organ that is relatively dense in mass (e.g., compact bone) absorbs or reflects more X-rays than does a less dense tissue (e.g., spongy bone). Consequently, a dense tissue or organ produces a somewhat transparent area on the X-ray film or bright area on a monitor because fewer X-rays reach the film or detector. A dense substance is *radiopaque*, whereas a substance of less density is *radiolucent*.

Many of the same principles that apply to making a shadow apply to conventional radiography. When making a shadow of your hand on a wall, the closer your hand is to the wall, the sharper the shadow produced. The farther your hand is from the wall (and therefore the closer to the light source), the more the shadow is magnified. Radiographs are made with the part of the patient's body being studied close to the X-ray film or detector to maximize the clarity of the image and minimize magnification artifacts. In basic radiologic nomenclature, *postero-anterior (PA) projection* refers to a radiograph in which the X-rays traversed the patient from posterior (P) to anterior (A); the X-ray tube was posterior to the patient and the X-ray film or detector was anterior (Fig. I.51A). A radiograph using *anteroposterior (AP) projection* radiography is the opposite. Both PA and AP projection radiographs are viewed as if you and the patient were facing each other (the patient's right side is opposite your left); this is referred to as an *anteroposterior (AP) view*. (Thus the standard chest X-ray, taken to examine the heart and lungs, is an AP view of

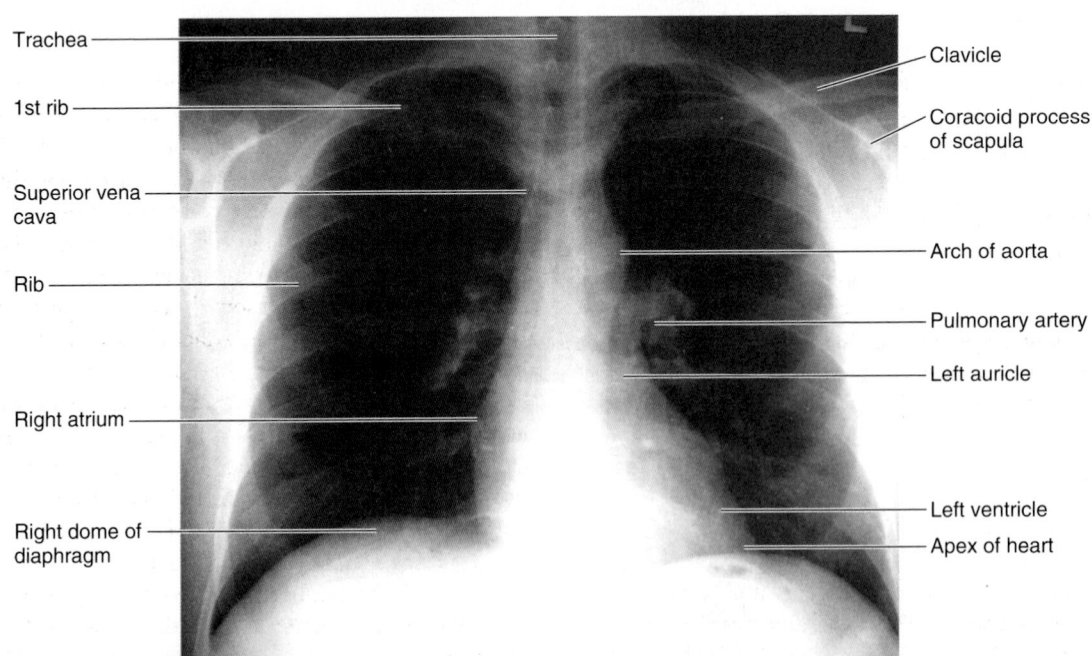

FIGURE I.49. Radiograph of thorax (chest). AP view of a PA projection radiograph demonstrates the arch of the aorta, parts of the heart, and domes of the diaphragm. Note that the dome of the diaphragm is higher on the right side. (Courtesy of Dr. E. L. Lansdown, Professor of Medical Imaging, University of Toronto, Toronto, ON, Canada.)

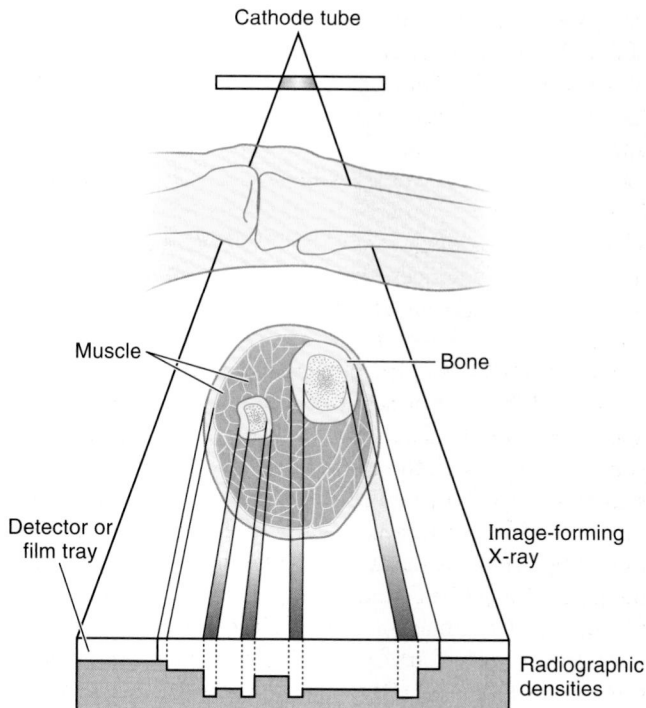

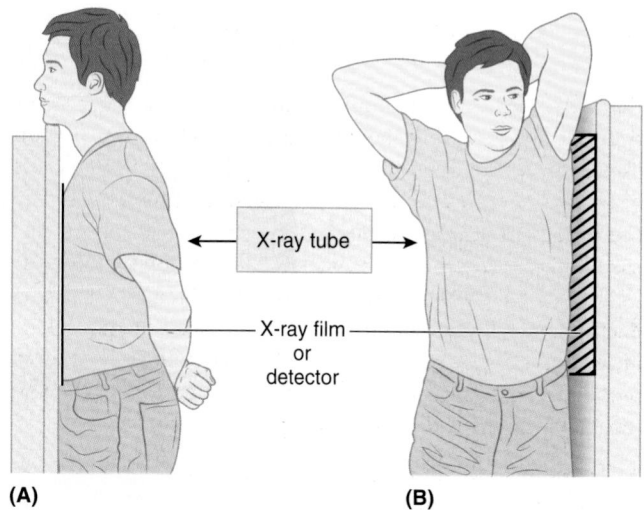

FIGURE I.51. Orientation of patient's thorax during radiography.
A. When taking a PA projection, the X-rays from the X-ray tube pass through the thorax from the back to reach the X-ray film or detector anterior to the person. **B.** When taking a lateral projection, the X-rays pass through the thorax from the side to reach the X-ray film adjacent to the person's other side.

FIGURE I.50. Principles of X-ray image formation. Portions of the beam of X-rays traversing the body become attenuated to varying degrees based on tissue thickness and density. The beam is diminished by structures that absorb or reflect it, causing less reaction on the film or by the detector compared with areas that allow the beam to pass relatively uninterrupted.

a PA projection.) For lateral radiographs, radiopaque letters (R or L) are used to indicate the side placed closest to the film or detector, and the image is viewed from the same direction that the beam was projected (Fig. I.51B).

The introduction of contrast media (radiopaque fluids such as iodine compounds or barium) allows the study of various luminal or vascular organs and potential or actual spaces—such as the digestive tract, blood vessels, kidneys, synovial cavities, and the subarachnoid space—that are not visible in plain films (Fig. I.52). Most radiologic examinations are performed in at least two projections at right angles to each other. Because each radiograph presents a two-dimensional representation of a three-dimensional structure, structures sequentially penetrated by the X-ray beam overlap each other. Thus more than one view is usually necessary to detect and localize an abnormality accurately.

Computed Tomography

In computed tomography (CT), the scans show radiographic images of the body that resemble transverse anatomical sections (Fig. I.53). In this technique, a beam of X-rays passes through the body as the X-ray tube and detector rotate around the axis of the body. Multiple overlapping radial energy absorptions are measured, recorded, and compared by a computer to determine the radiodensity of each volumetric pixel (*voxel*) of the chosen body plane. The radiodensity of

(amount of radiation absorbed by) each voxel is determined by factors that include the amount of air, water, fat, or bone in that element. The computer maps the voxels into a planar image (slice) that is displayed on a monitor or printout. CT images relate well to conventional radiographs, in that areas of great absorption (e.g., bone) are relatively transparent (white) and those with little absorption are black (Fig. I.53). CT scans are always displayed as if the viewer were standing at a supine patient's feet—i.e., from an inferior view.

Ultrasonography

Ultrasonography (US) is a technique that visualizes superficial or deep structures in the body by recording pulses of ultrasonic waves reflecting off the tissues (Fig. I.54). US has the advantage of a lower cost than CT and MRI, and the machine is portable. The technique can be performed virtually anywhere, including the clinic examination room, bedside, or on the operating table. A transducer in contact with the skin generates high-frequency soundwaves that pass through the body and reflect off tissue interfaces between tissues of differing characteristics, such as soft tissue and bone. Echoes from the body reflect into the transducer and convert to electrical energy. The electrical signals are recorded and displayed on a monitor as a cross-sectional image, which can be viewed in real time and recorded as a single image or on videotape.

A major advantage of US is its ability to produce real-time images, demonstrating motion of structures and flow within blood vessels. In *Doppler ultrasonography*, the shifts in frequency between emitted ultrasonic waves and their echoes are used to measure the velocities of moving objects. This technique is based on the principle of the *Doppler effect*. Blood flow through vessels is displayed in color, superimposed on the two-dimensional cross-sectional image.

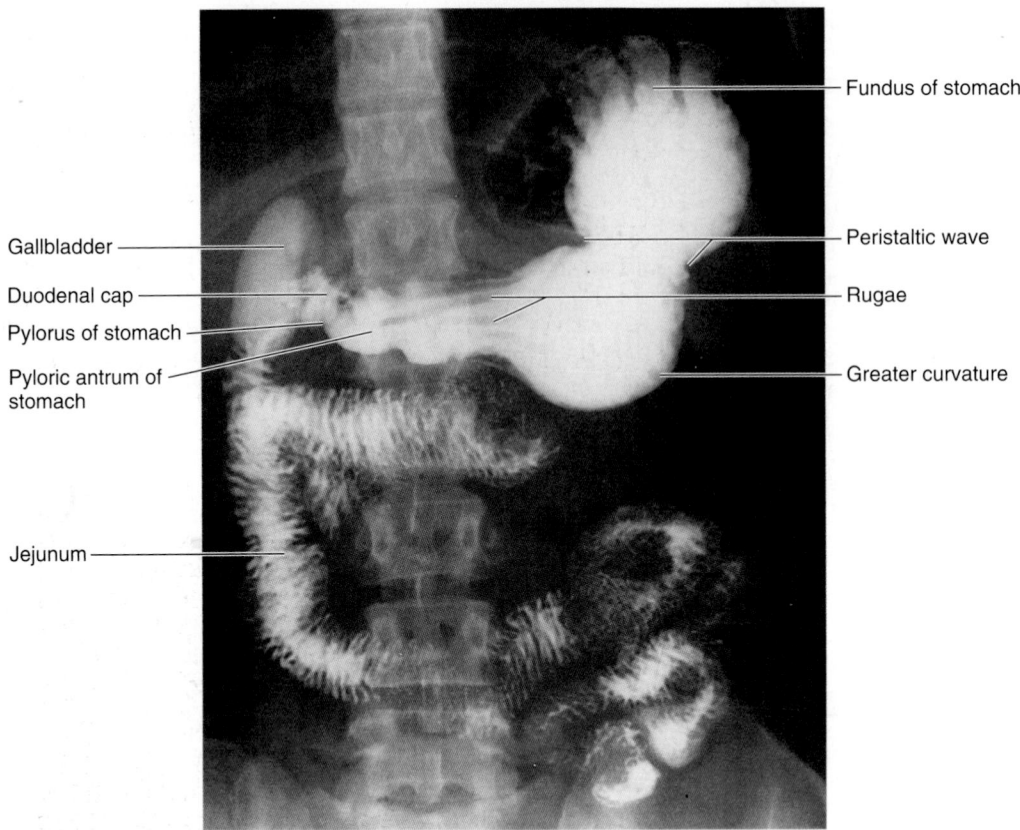

FIGURE I.52. Radiograph of stomach, small intestine, and gallbladder. Observe the gastric folds, or rugae (longitudinal folds of the mucous membrane). Also note the peristaltic wave that is moving the gastric contents toward the duodenum, which is closely related to the gallbladder. (Courtesy of Dr. J. Heslin, Toronto, ON, Canada.)

Scanning of the pelvic viscera from the surface of the abdomen requires a fully distended bladder. The urine serves as an "acoustical window," transmitting soundwaves to and from the posteriorly placed pelvic viscera with minimal attenuation. The distended bladder also displaces gas-filled intestinal loops out of the pelvis. *Transvaginal sonography* permits the positioning of the transducer closer to the organ of interest (e.g., the ovary) and avoids fat and gas, which absorb or reflect soundwaves. Bone reflects nearly all ultrasound waves, whereas air conducts them poorly. Consequently, US is not generally used for examining the CNS and aerated lungs of adults.

The appeal of ultrasonography in obstetrics is that it is a non-invasive procedure that does not use radiation; it can yield useful information about the pregnancy, such as determining whether it is intra-uterine or extra-uterine (ectopic) and whether the embryo or fetus is living. It has also become a standard method of evaluating the growth and development of the embryo and fetus.

Magnetic Resonance Imaging

Magnetic resonance imaging (MRI) provides images of the body similar to those of CT scans, but MRI is better for tissue differentiation. MRI studies closely resemble ana-tomical sections, especially of the brain (Fig. I.55). The person is placed in a scanner with a strong magnetic field, and the body is pulsed with radiowaves. Signals subsequently emitted from the patient's tissues are stored in a computer and reconstructed into various images of the body. The appearance of tissues on the generated images can be varied by controlling how radiofrequency pulses are sent and received.

Free protons in the tissues that become aligned by the surrounding magnetic field are excited (flipped) with a radiowave pulse. As the protons flip back, minute but measurable energy signals are emitted. Tissues that are high in proton density, such as fat and water, emit more signals than tissues that are low in proton density. The tissue signal is based primarily on three properties of protons in a particular region of the body. These are referred to as T1 and T2 relaxation (producing T1- and T2-weighted images), and proton density. Although liquids have a high density of free protons, the excited free protons in moving fluids such as blood tend to move out of the field before they flip and give off their signal and are replaced by unexcited protons. Consequently, moving fluids appear black in T1-weighted images.

Computers associated with MRI scanners have the capacity to reconstruct tissues in any plane from the data acquired: transverse, median, sagittal, frontal, and even arbitrary oblique planes. The data may also be used to generate

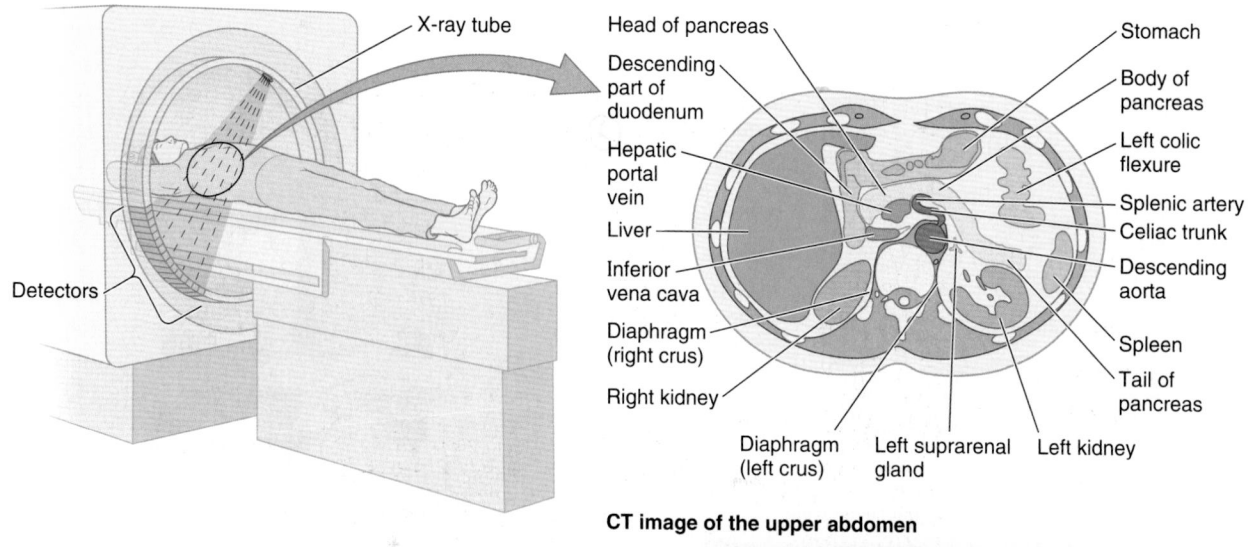

Head of pancreas
Descending part of duodenum
Hepatic portal vein
Liver
Inferior vena cava
Diaphragm (right crus)
Right kidney

Stomach
Body of pancreas
Left colic flexure
Splenic artery
Celiac trunk
Descending aorta
Spleen
Tail of pancreas

Diaphragm (left crus) Left suprarenal gland Left kidney

CT image of the upper abdomen

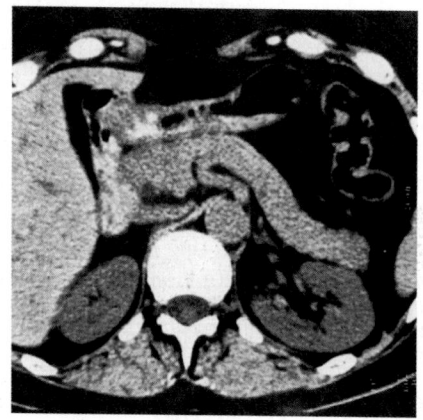

FIGURE I.53. Technique for producing a CT scan. The X-ray tube rotates around the person in the CT scanner and sends a fan-shaped beam of X-rays through the upper abdomen from a variety of angles. X-ray detectors on the opposite side of the body measure the amount of radiation that passes through a horizontal section. A computer reconstructs the images from several scans, and a CT scan is produced. The scan is oriented so it appears the way an examiner would view it when standing at the foot of the bed and looking toward a supine person's head.

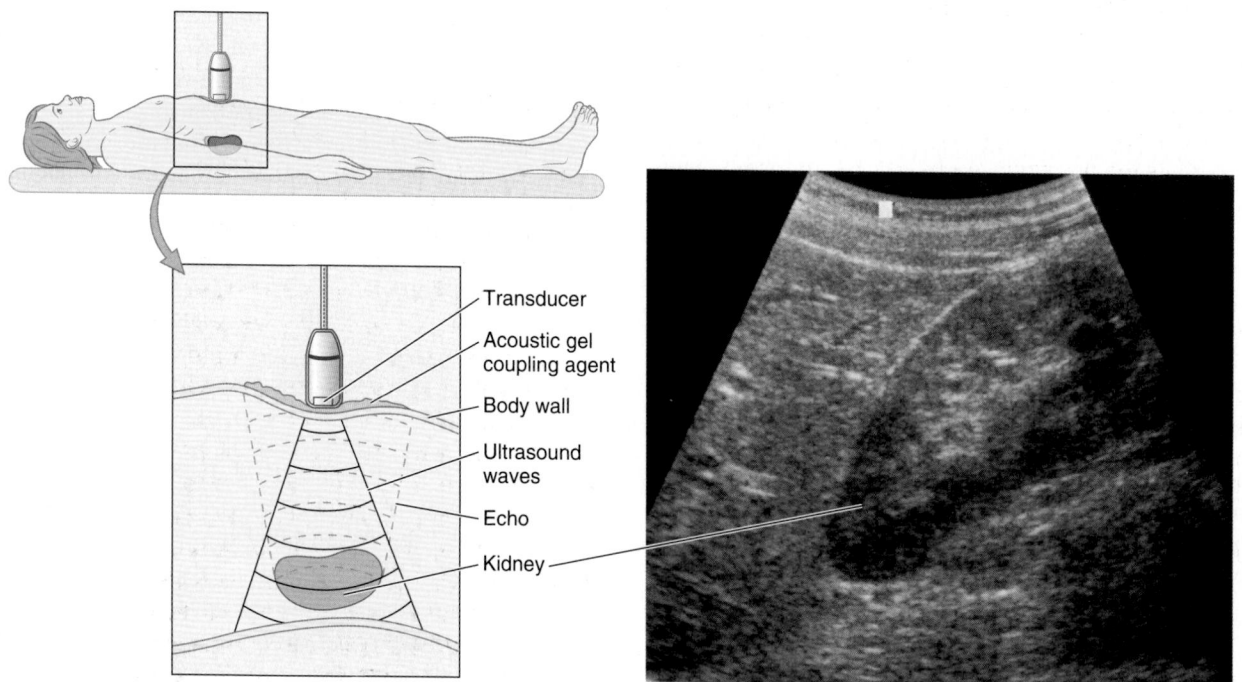

Transducer
Acoustic gel coupling agent
Body wall
Ultrasound waves
Echo
Kidney

FIGURE I.54. Technique for producing an ultrasound image of the upper abdomen. The image results from the echo of ultrasound waves from abdominal structures of different densities. The image of the right kidney is displayed on a monitor.

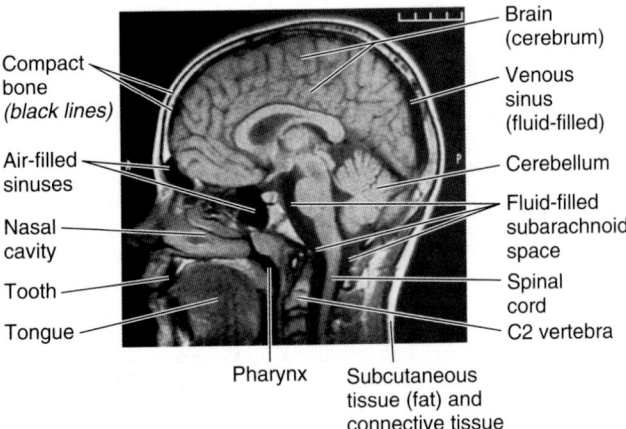

Compact bone (black lines)

Air-filled sinuses

Nasal cavity

Tooth

Tongue

Brain (cerebrum)

Venous sinus (fluid-filled)

Cerebellum

Fluid-filled subarachnoid space

Spinal cord

C2 vertebra

Pharynx

Subcutaneous tissue (fat) and connective tissue

FIGURE I.55. **Median MRI of head.** Many details of the CNS and structures in the nasal and oral cavities and upper neck are seen in this study. The black low-signal areas superior to the anterior and posterior aspects of the nasal cavity are the air-filled frontal and sphenoidal sinuses.

three-dimensional reconstructions. MRI scanners produce good images of soft tissues without the use of ionizing radiation. Motion made by the patient during long scanning sessions created problems for early-generation scanners, but fast scanners now in use can be gated or paced to visualize moving structures, such as the heart and blood flow, in real time.

Nuclear Medicine Imaging

Nuclear medicine imaging techniques provide information about the distribution or concentration of trace amounts of radioactive substances introduced into the body. Nuclear medicine scans show images of specific organs after intravenous (IV) injection of a small dose of radioactive material. The radionuclide is tagged to a compound that is selectively taken up by an organ, such as technetium-99m methylene diphosphonate (^{99m}Tc-MDP) for bone scanning (Fig. I.56).

Positron emission tomography (PET) scanning uses cyclotron-produced isotopes of extremely short half-life that emit positrons. PET scanning is used to evaluate the physiologic

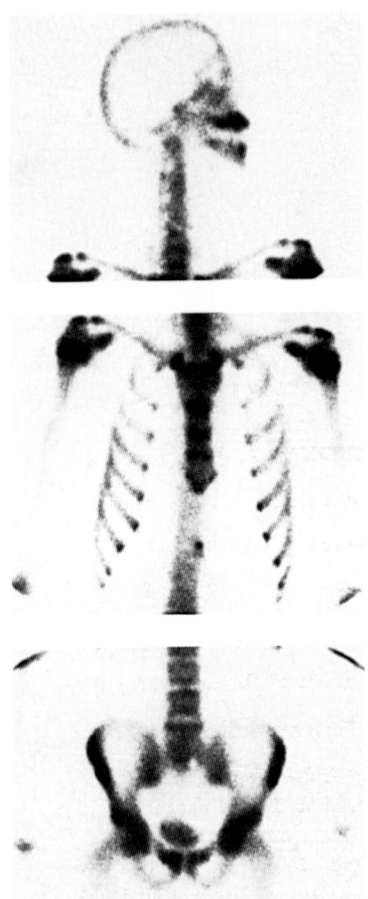

FIGURE I.56. **Bone scans of head and neck, thorax, and pelvis.** These nuclear medicine images can be viewed as a whole or in cross section.

function of organs, such as the brain, on a dynamic basis. Areas of increased brain activity will show selective uptake of the injected isotope. Images can be viewed as the whole organ or in cross sections. Single-photon emission computed tomography (SPECT) scans are similar but use longer lasting tracers. They are less costly, but require more time and have lower resolution.

The Bottom Line

MEDICAL IMAGING TECHNIQUES

Medical imaging techniques enable the visualization of anatomy in living people. These techniques enable structures to be examined with their normal tonus, fluid volumes, internal pressures, etc., which are not present in the cadaver. The primary goal of medical imaging is, of course, to detect pathology. However, a sound knowledge of radiologic anatomy is required to distinguish pathologies and abnormalities from normal anatomy.

Thorax

CHAPTER 1

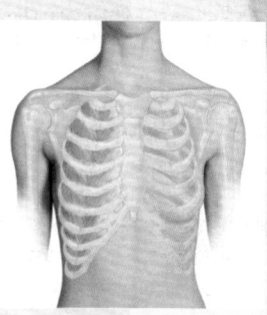
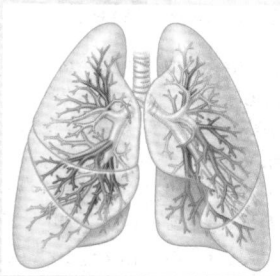
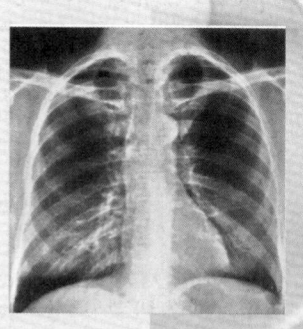

OVERVIEW OF THORAX

The **thorax** is the part of the body between the neck and abdomen. Commonly the term *chest* is used as a synonym for thorax, but the chest is much more extensive than the thoracic wall and cavity contained within it. The **chest** is generally conceived as the superior part of the trunk that is broadest superiorly owing to the presence of the *pectoral (shoulder) girdle* (clavicles and scapulae), with much of its girth accounted for by the pectoral and scapular musculature and, in adult females, the breasts.

The **thoracic cavity** and its wall have the shape of a truncated cone, being narrowest superiorly, with the circumference increasing inferiorly, and reaching its maximum size at the junction with the abdominal portion of the trunk. The *wall of the thoracic cavity* is relatively thin, essentially as thick as its skeleton. The thoracic skeleton takes the form of a domed birdcage. The **thoracic cage** (rib cage), with the horizontal bars formed by ribs and costal cartilages, is also supported by the vertical sternum (breastbone) and thoracic vertebrae (Fig. 1.1). Furthermore, the floor of the thoracic cavity (thoracic diaphragm) is deeply invaginated inferiorly (i.e., is pushed upward) by viscera of the abdominal cavity. Consequently, nearly the lower half of the thoracic wall surrounds and protects abdominal rather than thoracic viscera (e.g., liver). Thus the thorax and its cavity are much smaller than one might expect based on external appearances of the chest.

The thorax includes the primary organs of the respiratory and cardiovascular systems. The thoracic cavity is divided into three major spaces: the central compartment or *mediastinum* that houses the thoracic viscera except for the lungs and, on each side, the *right* and *left pulmonary cavities* housing the lungs.

The majority of the thoracic cavity is occupied by the lungs, which provide for the exchange of oxygen and carbon dioxide between the air and blood. Most of the remainder of the thoracic cavity is occupied by the heart and structures involved in conducting the air and blood to and from the lungs. Additionally, nutrients (food) traverse the thoracic cavity via the esophagus, passing from the site of entry in the head to the site of digestion and absorption in the abdomen.

Although in terms of function and development the mammary glands are most related to the reproductive system, the breasts are located on and are typically dissected with the thoracic wall; thus they are included in this chapter.

THORACIC WALL

The true thoracic wall includes the thoracic cage and the muscles that extend between the ribs as well as the skin, subcutaneous tissue, muscles, and fascia covering its anterolateral aspect. The same structures covering its posterior aspect are considered to belong to the back. The mammary glands of the breasts lie within the subcutaneous tissue of the thoracic wall. The anterolateral axio-appendicular muscles (see Chapter 6) that overlie the thoracic cage and form the bed of the breast are encountered in the thoracic wall and may be

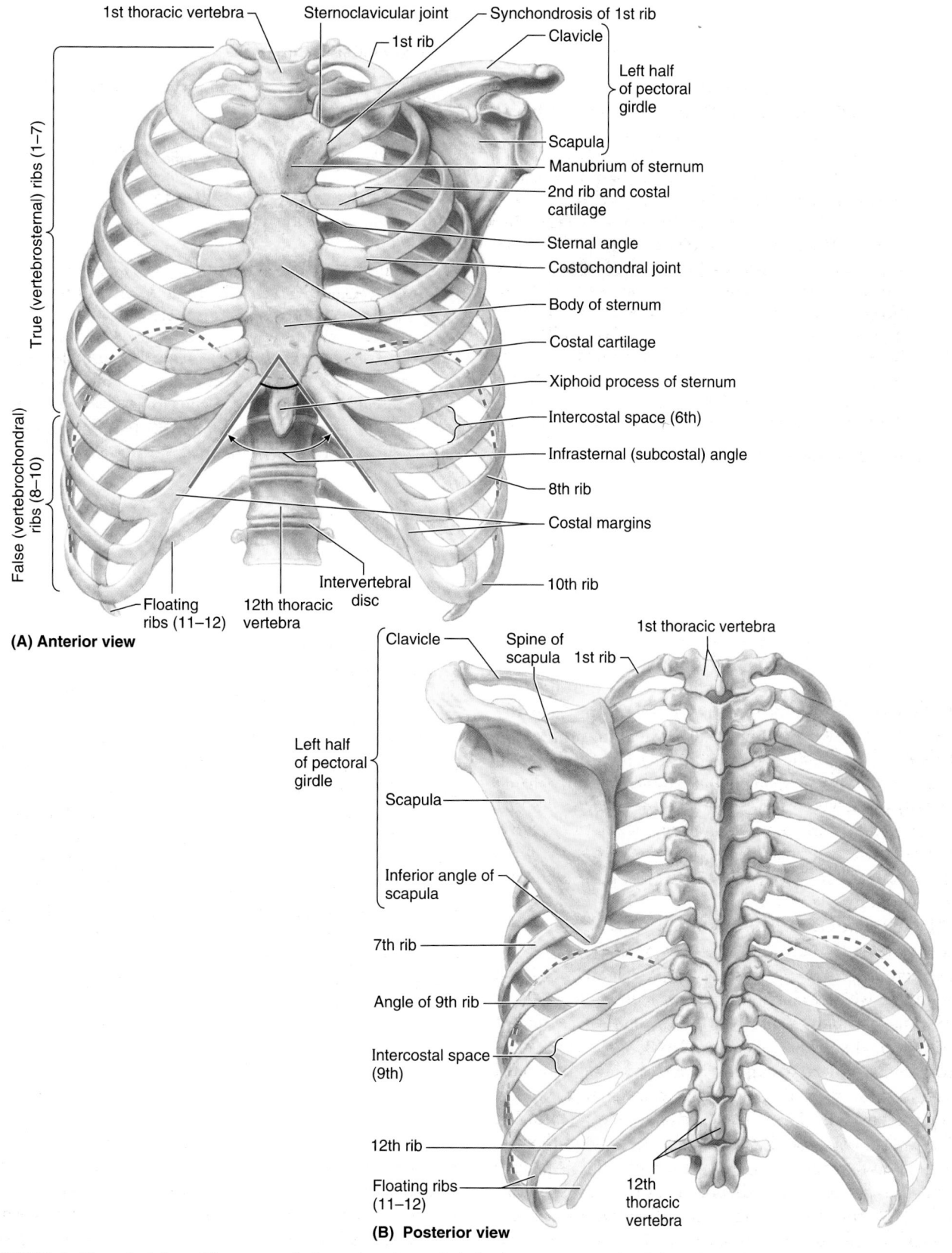

1st thoracic vertebra

Sternoclavicular joint

Synchondrosis of 1st rib

1st rib

Clavicle

Left half of pectoral girdle

Scapula

Manubrium of sternum

2nd rib and costal cartilage

Sternal angle

Costochondral joint

Body of sternum

Costal cartilage

Xiphoid process of sternum

Intercostal space (6th)

Infrasternal (subcostal) angle

8th rib

Costal margins

10th rib

True (vertebrosternal) ribs (1–7)

False (vertebrochondral) ribs (8–10)

Floating ribs (11–12)

12th thoracic vertebra

Intervertebral disc

(A) Anterior view

Clavicle

Spine of scapula

1st rib

1st thoracic vertebra

Left half of pectoral girdle

Scapula

Inferior angle of scapula

7th rib

Angle of 9th rib

Intercostal space (9th)

12th rib

Floating ribs (11–12)

12th thoracic vertebra

(B) Posterior view

FIGURE 1.1. Thoracic skeleton. The osteocartilaginous thoracic cage includes the sternum, 12 pairs of ribs and costal cartilages, and 12 thoracic vertebrae and intervertebral discs. The clavicles and scapulae form the pectoral (shoulder) girdle, one side of which is included here to demonstrate the relationship between the thoracic (axial) and upper limb (appendicular) skeletons. The red dotted line indicates the position of the diaphragm, which separates the thoracic and abdominal cavities.

considered part of it, but are distinctly upper limb muscles based on function and innervation. They will be mentioned only briefly here.

The domed shape of the thoracic cage provides remarkable rigidity, given the light weight of its components, enabling it to:

- Protect vital thoracic and abdominal organs (most air or fluid filled) from external forces.
- Resist the negative (sub-atmospheric) internal pressures generated by the elastic recoil of the lungs and inspiratory movements.
- Provide attachment for and support the weight of the upper limbs.
- Provide the anchoring attachment (origin) of many of the muscles that move and maintain the position of the upper limbs relative to the trunk, as well as provide the attachments for muscles of the abdomen, neck, back, and respiration.

Although the shape of the thoracic cage provides rigidity, its joints and the thinness and flexibility of the ribs allow it to absorb many external blows and compressions without fracture and to change its shape for respiration. Because the most important structures within the thorax (heart, great vessels, lungs, and trachea), as well as its floor and walls, are constantly in motion, the thorax is one of the most dynamic regions of the body. With each breath, the muscles of the thoracic wall—working in concert with the diaphragm and muscles of the abdominal wall—vary the volume of the thoracic cavity, first by expanding the capacity of the cavity, thereby causing the lungs to expand and draw air in and then, due to lung elasticity and muscle relaxation, decreasing the volume of the cavity and causing them to expel air.

The Bottom Line

OVERVIEW OF THORAX

The thorax, consisting of the thoracic cavity, its contents, and the wall that surrounds it, is the part of the trunk between the neck and abdomen. ♦ The shape and size of the thoracic cavity and thoracic wall are different from that of the chest (upper trunk or torso) because the latter includes some upper limb bones and muscles and, in adult females, the breasts. ♦ The thorax includes the primary organs of the respiratory and cardiovascular systems. ♦ The thoracic cavity is divided into three compartments: the central mediastinum, occupied by the heart and structures transporting air, blood, and food; and the right and left pulmonary cavities, occupied by the lungs.

Skeleton of Thoracic Wall

The **thoracic skeleton** forms the osteocartilaginous *thoracic cage* (Fig. 1.1), which protects the thoracic viscera and some abdominal organs. The thoracic skeleton includes 12 pairs of ribs and associated costal cartilages, 12 thoracic vertebrae and the intervertebral (IV) discs interposed between them, and the sternum. The ribs and costal cartilages form the largest part of the thoracic cage; both are identified numerically, from the most superior (1st rib or costal cartilage) to the most inferior (12th).

RIBS, COSTAL CARTILAGES, AND INTERCOSTAL SPACES

Ribs (L. *costae*) are curved, flat bones that form most of the thoracic cage (Figs. 1.1 and 1.2). They are remarkably light in weight yet highly resilient. Each rib has a spongy interior containing *bone marrow* (hematopoietic tissue), which forms blood cells. There are *three types of ribs* that can be classified as typical or atypical:

1. **True (vertebrosternal) ribs** (1st–7th ribs): They attach directly to the sternum through their own costal cartilages.
2. **False (vertebrochondral) ribs** (8th, 9th, and usually 10th ribs): Their cartilages are connected to the cartilage of the rib above them; thus their connection with the sternum is indirect.
3. **Floating (vertebral, free) ribs** (11th, 12th, and sometimes 10th ribs): The rudimentary cartilages of these ribs do not connect even indirectly with the sternum; instead they end in the posterior abdominal musculature.

Typical ribs (3rd–9th) have the following components:

- **Head:** wedge-shaped and has two facets, separated by the **crest of the head** (Figs. 1.2 and 1.3); one facet for articulation with the numerically corresponding vertebra and one facet for the vertebra superior to it.
- **Neck:** connects the head of the rib with the body at the level of the tubercle.
- **Tubercle:** located at the junction of the neck and body; a smooth *articular part* articulates with the corresponding transverse process of the vertebra, and a rough *nonarticular part* provides attachment for the costotransverse ligament (see Fig. 1.8B).
- **Body** (shaft): thin, flat, and curved, most markedly at the **costal angle** where the rib turns anterolaterally. The angle also demarcates the lateral limit of attachment of the deep back muscles to the ribs (see Table 4.6 in Chapter 4). The concave internal surface of the body has a **costal groove** paralleling the inferior border of the rib, which provides some protection for the intercostal nerve and vessels.

Atypical ribs (1st, 2nd, and 10th–12th) are dissimilar (Fig. 1.3):

- The **1st rib** is the broadest (i.e., its body is widest and nearly horizontal), shortest, and most sharply curved of

Neck
Head
Superior facet
Crest of head
Inferior facet
Articular facet
Nonarticular part
Tubercle
Costal angle
Body (shaft)
Costal groove
6th rib
Site of articulation with costal cartilage
8th rib
External surface
Internal surface
Compact bone
Bone marrow (hematopoietic tissue)
Costal groove
(A)
Posterior views
(B)

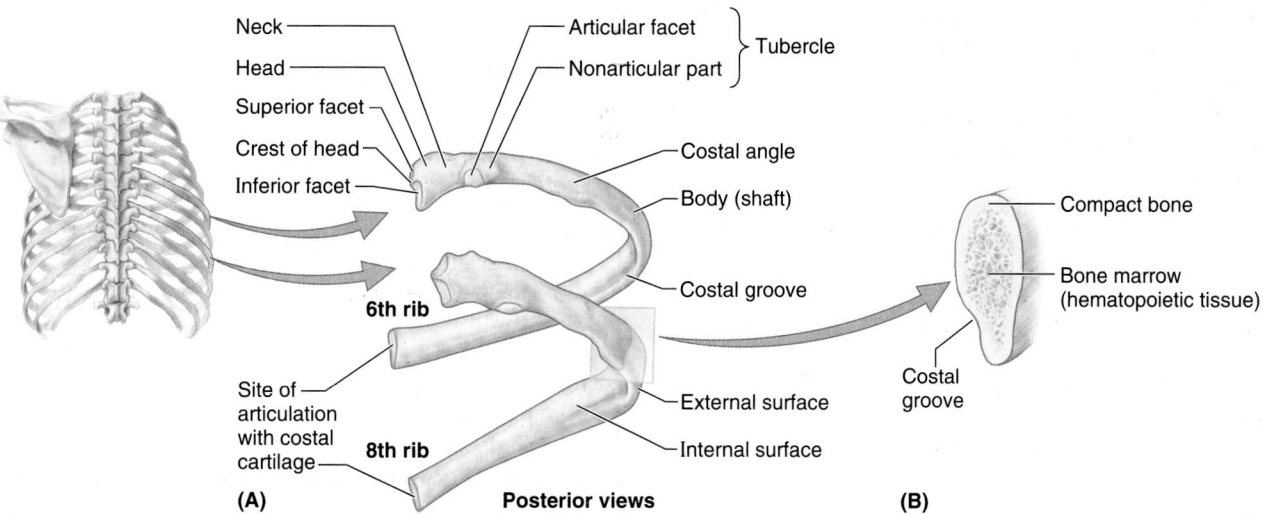

FIGURE 1.2. Typical ribs. A. The 3rd–9th ribs have common characteristics. Each rib has a head, neck, tubercle, and body (shaft). **B.** Cross section of the midbody of a rib.

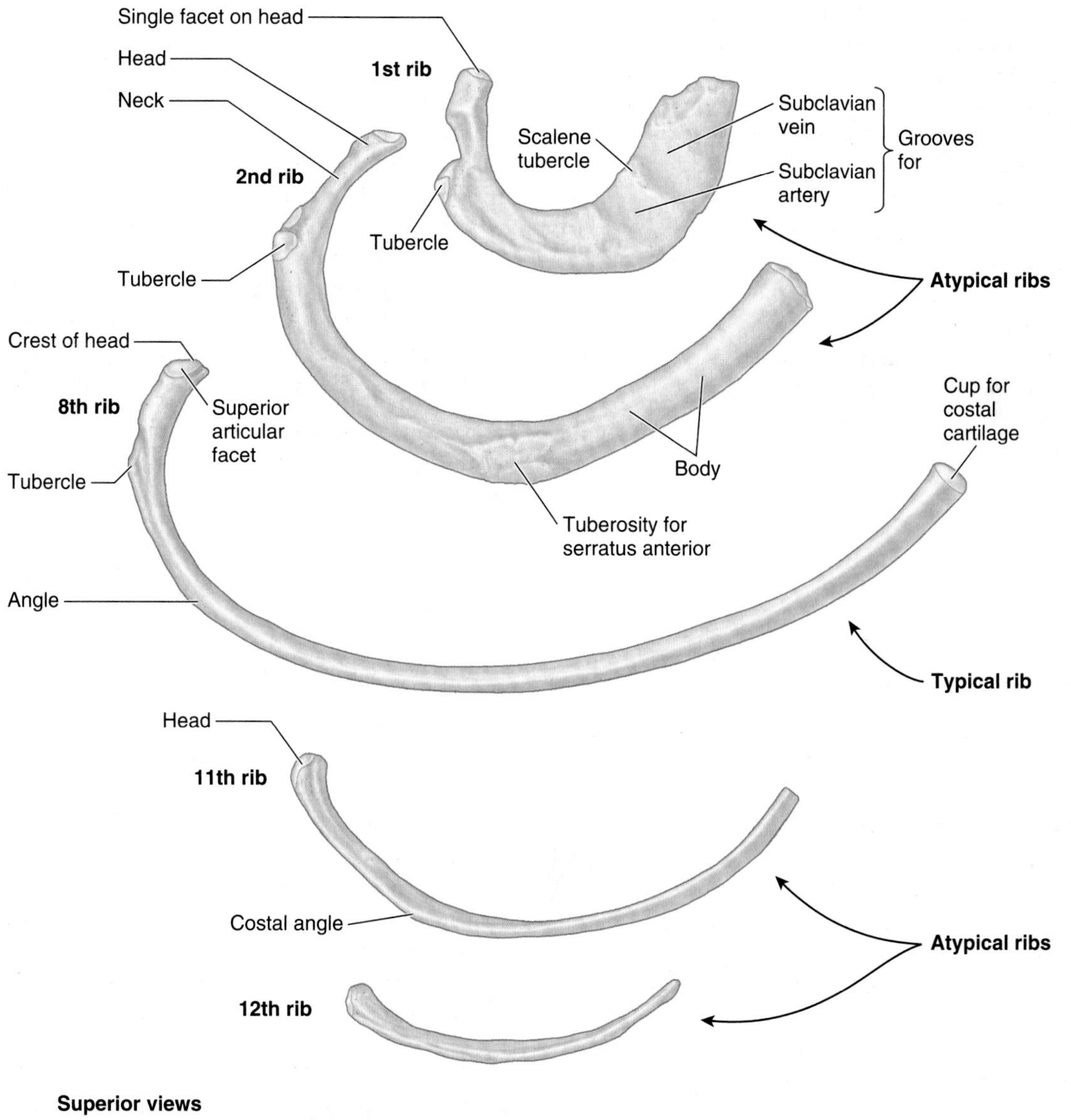

Single facet on head
Head
Neck
1st rib
2nd rib
Scalene tubercle
Subclavian vein
Subclavian artery
Grooves for
Tubercle
Tubercle
Atypical ribs
Crest of head
8th rib
Superior articular facet
Tuberosity for serratus anterior
Body
Cup for costal cartilage
Angle
Typical rib
Head
11th rib
Costal angle
Atypical ribs
12th rib
Superior views

FIGURE 1.3. Atypical ribs. The atypical 1st, 2nd, 11th, and 12th ribs differ from typical ribs (e.g., the 8th rib, shown in center).

the seven true ribs. It has a single facet on its head for articulation with the T1 vertebra only and two transversely directed grooves crossing its superior surface for the subclavian vessels; the grooves are separated by a **scalene tubercle** and ridge, to which the anterior scalene muscle is attached.

- The **2nd rib** is has a thinner, less curved body and is substantially longer than the 1st rib. Its head has two facets for articulation with the bodies of the T1 and T2 vertebrae; its main atypical feature is a rough area on its upper surface, the **tuberosity for serratus anterior,** from which part of that muscle originates.
- The *10th–12th ribs,* like the 1st rib, have only one facet on their heads and articulate with a single vertebra.
- The *11th and 12th ribs* are short and have no neck or tubercle.

Costal cartilages prolong the ribs anteriorly and contribute to the elasticity of the thoracic wall, providing a flexible attachment for their anterior ends (tips). The cartilages increase in length through the first 7 and then gradually decrease. The first 7 costal cartilages attach directly and independently to the sternum; the 8th, 9th, and 10th articulate with the costal cartilages just superior to them, forming a continuous, articulated, cartilaginous **costal margin** (Fig. 1.1A; see also Fig. 1.13). The 11th and 12th costal cartilages form caps on the anterior ends of the corresponding ribs and do not reach or attach to any other bone or cartilage. The costal cartilages of ribs 1–10 clearly anchor the anterior end of the rib to the sternum, limiting its overall movement as the posterior end rotates around the transverse axis of the rib (Fig. 1.5).

Intercostal spaces separate the ribs and their costal cartilages from one another (Fig. 1.1A). The spaces are named according to the rib forming the superior border of the space—for example, the 4th intercostal space lies between ribs 4 and 5. There are 11 intercostal spaces and 11 intercostal nerves. Intercostal spaces are occupied by intercostal muscles and membranes, and two sets (main and collateral) of intercostal blood vessels and nerves, identified by the same number assigned to the space. The space below the 12th rib does not lie between ribs and thus is referred to as the **subcostal space,** and the anterior ramus (branch) of spinal nerve T12 is the subcostal nerve. The intercostal spaces are widest anterolaterally, and they widen further with inspiration. They can also be further widened by extension and/or lateral flexion of the thoracic vertebral column to the contralateral side.

THORACIC VERTEBRAE

Most **thoracic vertebrae** are typical in that they are independent, have bodies, vertebral arches, and seven processes for muscular and articular connections (Figs. 1.4 and 1.5). Characteristic features of thoracic vertebrae include:

- Bilateral *costal facets (demifacets) on the vertebral bodies,* usually occurring in inferior and superior pairs, for articulation with the heads of ribs.

- *Costal facets on the transverse processes* for articulation with the tubercles of ribs, except for the inferior two or three thoracic vertebrae.
- Long, inferiorly slanting *spinous processes.*

Superior and **inferior costal facets,** most of which are small *demifacets,* occur as bilaterally paired, planar surfaces on the superior and inferior posterolateral margins of the bodies of typical thoracic vertebrae (T2–T9). Functionally, the facets are arranged in pairs on adjacent vertebrae, flanking an interposed IV disc, an inferior (demi)facet of the superior vertebra and a superior (demi)facet of the inferior vertebra. Typically, two demifacets paired in this manner and the posterolateral margin of the IV disc between them form a single socket to receive the head of the rib of the same identifying number as the inferior vertebra (e.g., head of rib 6 with the superior costal facet of vertebra T6). Atypical thoracic vertebrae bear whole costal facets in place of demifacets:

- The superior costal facets of vertebra T1 are not demifacets because there are no demifacets on the C7 vertebra above, and rib 1 articulates only with vertebra T1. T1 has a typical inferior costal (demi)facet.
- T10 has only one bilateral pair of (whole) costal facets, located partly on its body and partly on its pedicle.
- T11 and T12 also have only a single pair of (whole) costal facets, located on their pedicles.

The *spinous processes* projecting from the vertebral arches of typical thoracic vertebrae (e.g., vertebrae T6 or T7) are long and slope inferiorly, usually overlapping the vertebra below (Figs. 1.4D and 1.5). They cover the intervals between the *laminae* of adjacent vertebrae, thereby preventing sharp objects such as a knife from entering the *vertebral canal* and injuring the spinal cord. The convex superior articular facets of the *superior articular processes* face mainly posteriorly and slightly laterally, whereas the concave inferior articular facets of the *inferior articular processes* face mainly anteriorly and slightly medially. The bilateral joint planes between the respective articular facets of adjacent thoracic vertebrae define an arc, centering on an axis of rotation within the vertebral body (Fig. 1.4A–C). Thus small rotatory movements are permitted between adjacent vertebrae, limited by the attached rib cage.

THE STERNUM

The **sternum** (G. *sternon,* chest) is the flat, elongated bone that forms the middle of the anterior part of the thoracic cage (Fig. 1.6). It directly overlies and affords protection for mediastinal viscera in general and much of the heart in particular. The sternum consists of three parts: manubrium, body, and xiphoid process. In adolescents and young adults, the three parts are connected together by cartilaginous joints (*synchondroses*) that ossify during middle to late adulthood.

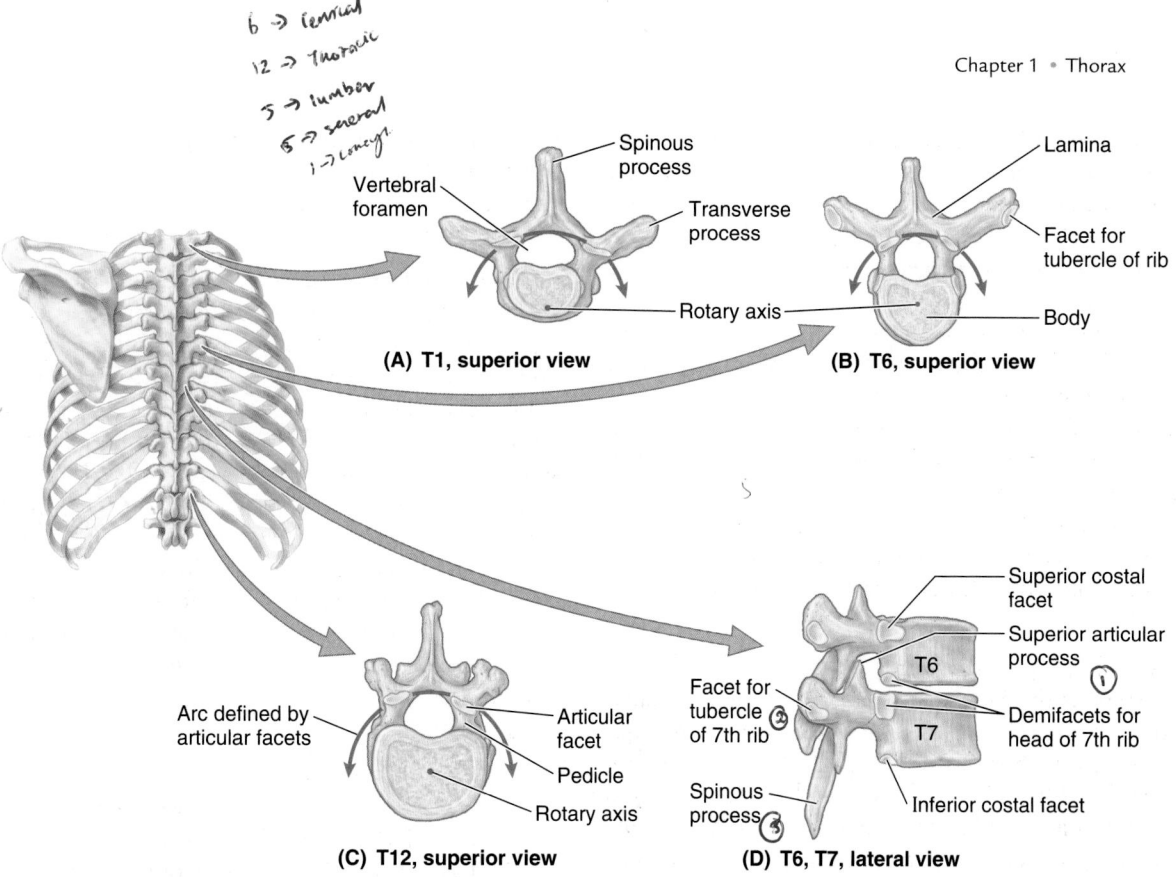

Handwritten notes (top left):
6 → cervical
12 → Thoracic
5 → lumbar
5 → sacral
1 → coccyx.

(A) T1, superior view

- Spinous process
- Vertebral foramen
- Transverse process
- Rotary axis

(B) T6, superior view

- Lamina
- Facet for tubercle of rib
- Body

(C) T12, superior view

- Arc defined by articular facets
- Articular facet
- Pedicle
- Rotary axis

(D) T6, T7, lateral view

- Superior costal facet
- Superior articular process ①
- Facet for tubercle of 7th rib ②
- Demifacets for head of 7th rib
- Spinous process ③
- Inferior costal facet
- T6
- T7

FIGURE 1.4. Thoracic vertebrae. A. T1 has a vertebral foramen and body similar in size and shape to a cervical vertebra. **B.** T5–T9 vertebrae have typical characteristics of thoracic vertebrae. **C.** T12 has bony processes and a body size similar to a lumbar vertebra. The planes of the articular facets of thoracic vertebrae define an arc (*red arrows*) that centers on an axis traversing the vertebral bodies vertically. **D.** Superior and inferior costal facets (demifacets) on the vertebral body and costal facets on the transverse processes. Long sloping spinous processes are characteristic of thoracic vertebrae. ①

② ③

IV disc

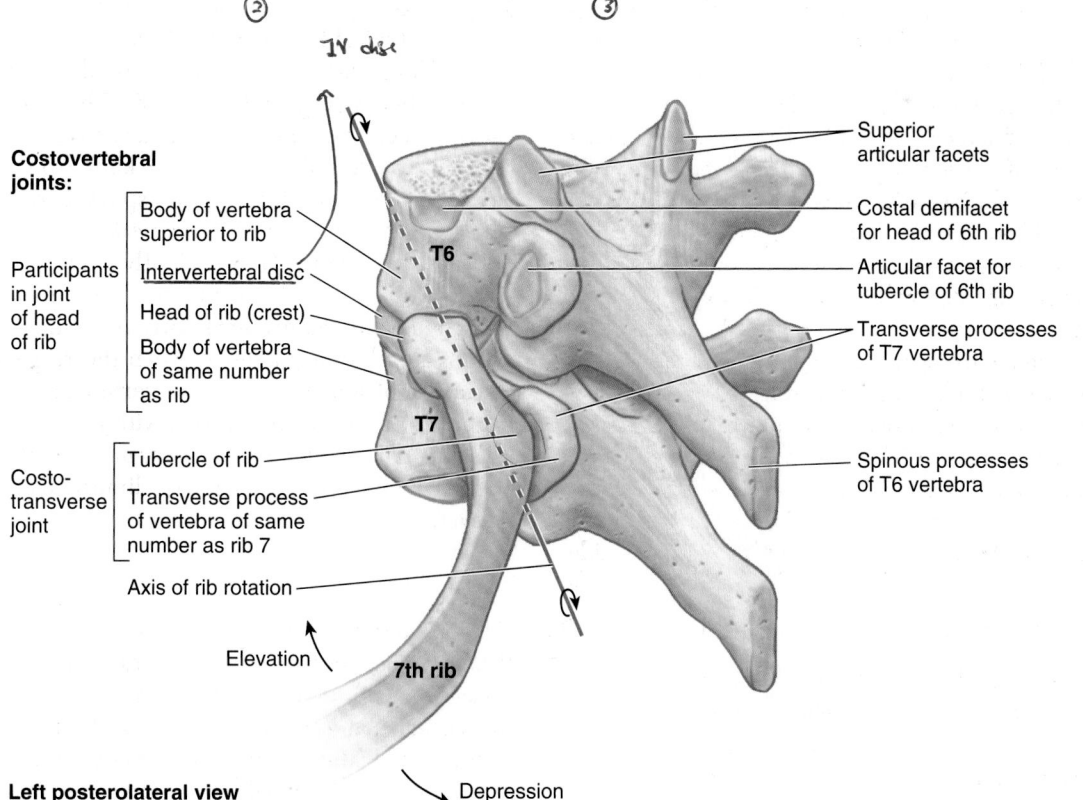

Costovertebral joints:

Participants in joint of head of rib
- Body of vertebra superior to rib
- Intervertebral disc
- Head of rib (crest)
- Body of vertebra of same number as rib

Costo-transverse joint
- Tubercle of rib
- Transverse process of vertebra of same number as rib 7

- Axis of rib rotation
- Elevation
- 7th rib
- Depression

- Superior articular facets
- Costal demifacet for head of 6th rib
- Articular facet for tubercle of 6th rib
- Transverse processes of T7 vertebra
- Spinous processes of T6 vertebra

T6

T7

Left posterolateral view

FIGURE 1.5. Costovertebral articulations of a typical rib. The costovertebral joints include the joint of head of rib, in which the head articulates with two adjacent vertebral bodies and the intervertebral disc between them, and the costotransverse joint, in which the tubercle of the rib articulates with the transverse process of a vertebra. The rib moves (elevates and depresses) around an axis that traverses the head and neck of the rib (*arrows*).

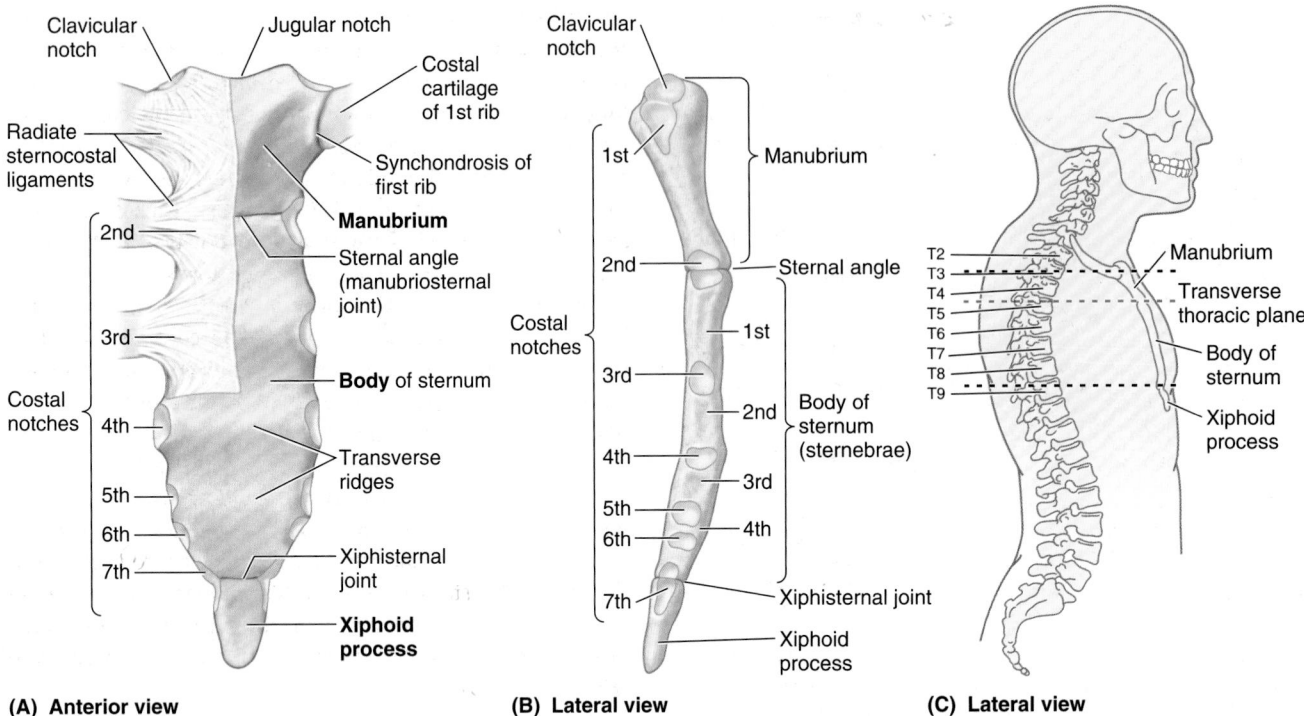

FIGURE 1.6. Sternum. A. The thin, broad membranous bands of the radiate sternocostal ligaments pass from the costal cartilages to the anterior and posterior surfaces of the sternum—is shown on the upper right side. **B.** Observe the thickness of the superior third of the manubrium between the clavicular notches. **C.** The relationship of the sternum to the vertebral column is shown.

The **manubrium** (L. handle, as in the handle of a sword, with the sternal body forming the blade) is a roughly trapezoidal bone. The manubrium is the widest and thickest of the three parts of the sternum. The easily palpated concave center of the superior border of the manubrium is the **jugular notch (suprasternal notch)**. The notch is deepened by the medial (sternal) ends of the clavicles, which are much larger than the relatively small **clavicular notches** in the manubrium that receive them, forming the *sternoclavicular (SC) joints* (Fig. 1.1A). Inferolateral to the clavicular notch, the costal cartilage of the 1st rib is tightly attached to the lateral border of the manubrium—the **synchondrosis of the first rib** (Figs. 1.1A and 1.6A). The manubrium and body of the sternum lie in slightly different planes superior and inferior to their junction, the **manubriosternal joint** (Fig. 1.6A & B); hence, their junction forms a projecting **sternal angle** (of Louis).

The **body of the sternum,** is longer, narrower, and thinner than the manubrium, and is located at the level of the T5–T9 vertebrae (Fig. 1.6A–C). Its width varies because of the scalloping of its lateral borders by the **costal notches.** In young people, four *sternebrae* (primordial segments of the sternum) are obvious. The sternebrae articulate with each other at primary cartilaginous joints (*sternal synchondroses*). These joints begin to fuse from the inferior end between puberty (sexual maturity) and age 25. The nearly flat anterior surface of the body of the sternum is marked in adults by three variable **transverse ridges** (Fig. 1.6A), which represent the lines of fusion (*synostosis*) of its four originally separate sternebrae.

The **xiphoid process,** the smallest and most variable part of the sternum, is thin and elongated. Its inferior end lies at the level of T10 vertebra. Although often pointed, the process may be blunt, bifid, curved, or deflected to one side or anteriorly. It is cartilaginous in young people but more or less ossified in adults older than age 40. In elderly people, the xiphoid process may fuse with the sternal body.

The xiphoid process is an important landmark in the median plane because

- Its junction with the sternal body at the **xiphisternal joint** indicates the inferior limit of the central part of the thoracic cavity projected onto the anterior body wall; this joint is also the site of the **infrasternal angle** (subcostal angle) formed by the right and left costal margins (Fig. 1.1A).
- It is a midline marker for the superior limit of the liver, the central tendon of the diaphragm, and the inferior border of the heart.

Thoracic Apertures

While the thoracic cage provides a complete wall peripherally, it is open superiorly and inferiorly. The much smaller superior opening (aperture) is a passageway that allows communication with the neck and upper limbs. The larger inferior opening provides the ring-like origin of the diaphragm, which completely occludes the opening. Excursions of the diaphragm primarily control the volume/internal pressure of the thoracic cavity, providing the basis for tidal respiration (air exchange).

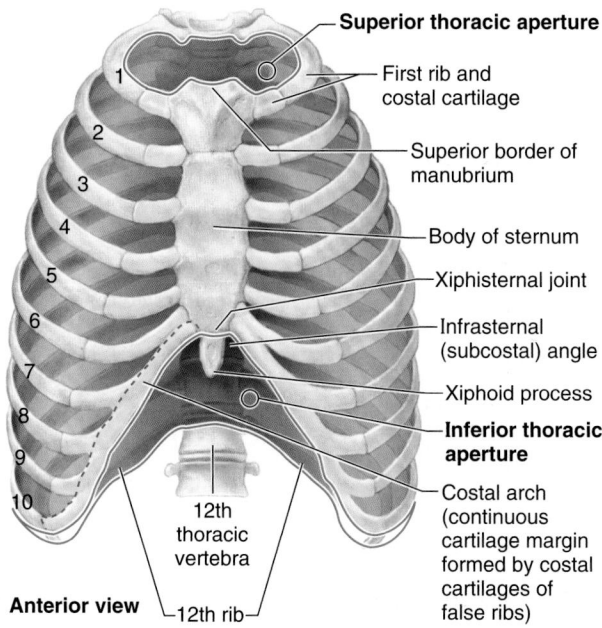

Superior thoracic aperture

First rib and costal cartilage

Superior border of manubrium

Body of sternum

Xiphisternal joint

Infrasternal (subcostal) angle

Xiphoid process

Inferior thoracic aperture

Costal arch (continuous cartilage margin formed by costal cartilages of false ribs)

1 2 3 4 5 6 7 8 9 10

12th thoracic vertebra

12th rib

Anterior view

FIGURE 1.7. Thoracic apertures. The superior thoracic aperture is the "doorway" between the thoracic cavity and the neck and upper limb. The inferior thoracic aperture provides attachment for the diaphragm, which protrudes upward so that upper abdominal viscera (e.g., liver) receive protection from the thoracic cage. The continuous cartilaginous bar formed by the articulated cartilages of the 7th–10th (false) ribs forms the costal margin.

SUPERIOR THORACIC APERTURE

The **superior thoracic aperture** is bounded (Fig. 1.7):

- Posteriorly, by vertebra T1, the body of which protrudes anteriorly into the opening.
- Laterally, by the 1st pair of ribs and their costal cartilages.
- Anteriorly, by the superior border of the manubrium.

Structures that pass between the thoracic cavity and the neck through the oblique, kidney-shaped superior thoracic aperture include the trachea, esophagus, nerves, and vessels that supply and drain the head, neck, and upper limbs.

The adult superior thoracic aperture measures approximately 6.5 cm anteroposteriorly and 11 cm transversely. To visualize the size of this opening, note that this is slightly larger than necessary to allow the passage of a 2-inch × 4-inch piece of lumber. Because of the obliquity of the 1st pair of ribs, the aperture slopes antero-inferiorly.

INFERIOR THORACIC APERTURE

The **inferior thoracic aperture,** the anatomical *thoracic outlet,* is bounded as follows:

- Posteriorly, by the 12th thoracic vertebra, the body of which protrudes anteriorly into the opening.
- Posterolaterally, by the 11th and 12th pairs of ribs.
- Anterolaterally, by the joined costal cartilages of ribs 7–10, forming the costal margins.
- Anteriorly, by the xiphisternal joint.

The inferior thoracic aperture is much more spacious than the superior thoracic aperture and is irregular in outline. It is also oblique because the posterior thoracic wall is much longer than the anterior wall. By closing the inferior thoracic aperture, the diaphragm separates the thoracic and abdominal cavities almost completely. Structures passing from or to the thorax to or from the abdomen pass through openings that traverse the diaphragm (e.g., esophagus and inferior vena cava), or pass posterior to it (e.g., aorta).

Just as the size of the thoracic cavity (or its contents) is often overestimated, its inferior extent (corresponding to the boundary between the thoracic and abdominal cavities) is often incorrectly estimated because of the discrepancy between the inferior thoracic aperture and the location of the diaphragm (floor of the thoracic cavity) in living persons. Although the diaphragm takes origin from the structures that make up the inferior thoracic aperture, the domes of the diaphragm rise to the level of the 4th intercostal space, and abdominal viscera, including the liver, spleen, and stomach, lie superior to the plane of the inferior thoracic aperture, within the thoracic wall (Fig. 1.1A & B).

30/8/14

Joints of Thoracic Wall

Although movements of the joints of the thoracic wall are frequent—for example, in association with normal respiration—the range of movement at the individual joints is relatively small. Nonetheless, any disturbance that reduces the mobility of these joints interferes with respiration. During deep breathing, the excursions of the thoracic cage (anteriorly, superiorly, or laterally) are considerable. Extending the vertebral column further increases the anteroposterior (AP) diameter of the thorax. The joints of the thoracic wall are illustrated in Figure 1.8. The type, participating articular surfaces, and ligaments of the joints of the thoracic wall are provided in Table 1.1.

The *intervertebral joints* between the bodies of adjacent vertebrae are joined by longitudinal ligaments and *intervertebral discs*. These joints are discussed with the back in Chapter 4; the sternoclavicular joints are discussed in Chapter 6.

COSTOVERTEBRAL JOINTS

A typical rib articulates posteriorly with the vertebral column at two joints, the joints of heads of ribs and costotransverse joints (Fig. 1.5).

Joints of Heads of Ribs. The *head of the rib* articulates with the *superior costal facet* of the corresponding (same-numbered) vertebra, the *inferior costal facet* of the vertebra superior to it, and the adjacent intervertebral (IV) disc uniting the two vertebrae (Figs. 1.4 and 1.8A). For example, the head of the 6th rib articulates with the superior costal facet of the body of the T6 vertebra, the inferior costal facet of T5, and the IV disc between these vertebrae.

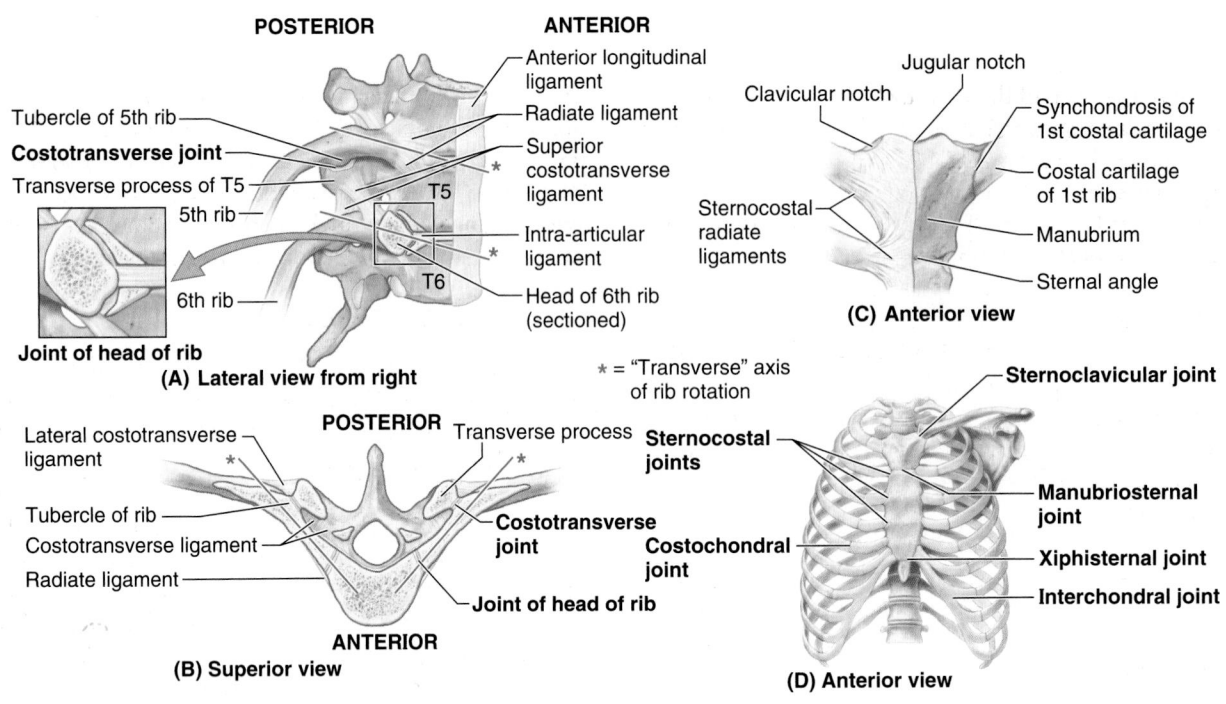

FIGURE 1.8. Joints of thoracic wall.

TABLE 1.1. JOINTS OF THORACIC WALL

Joint	Type	Articulation	Ligaments	Comments
Intervertebral (of vertebrae T1–T12)	Symphysis (secondary cartilaginous)	Adjacent vertebral bodies bound together by IV disc	Anterior and posterior longitudinal	Movement mostly limited to small degrees of rotation
Costovertebral Joints of head of rib	Synovial plane joint	Head of each rib with superior demi or costal facet of vertebral body of same number and inferior demi- or costal facet of vertebral body superior to it and IV disc between them	Radiate and intra-articular ligaments of head of rib	Heads of 1st, 11th, and 12th ribs (sometimes 10th) articulate only with vertebral body of same number
Costotransverse		Tubercle of rib with transverse process of vertebra of same number	Costotransverse; lateral and superior costotransverse	11th and 12th ribs do not articulate with transverse process of vertebrae of same number
Costochondral	Primary (hyaline) cartilaginous joint	Lateral end of costal cartilage with sternal end of rib	Cartilage and bone bound together by periosteum	No movement normally occurs at this joint; costal cartilage provides flexibility
Interchondral	Synovial plane joint	Between costal cartilages of 6th and 7th, 7th and 8th, and 8th and 9th ribs	Interchondral ligaments	Articulation between costal cartilages of 9th and 10th ribs is fibrous
Sternocostal	1st: primary cartilaginous joint (synchondrosis)	Articulation of 1st costal cartilages with manubrium of sternum		
	2nd–7th: synovial plane joint	Articulation of the 2nd–7th pairs of costal cartilages with sternum	Anterior and posterior radiate sternocostal; intra-articular	Articular cavities often absent; fibrocartilage covers articular surfaces
Sternoclavicular	Saddle type of synovial joint	Sternal end of clavicle with manubrium of sternum and 1st costal cartilage	Anterior and posterior sternoclavicular; costo-clavicular	This joint is divided into two compartments by an articular disc
Manubriosternal	Secondary cartilaginous joint (symphysis)	Articulation between manubrium and body of sternum		These joints often fuse and become synostoses in older individuals
Xiphisternal	Primary cartilaginous joint (synchondrosis)	Articulation between xiphoid process and body of sternum		

IV, intervertebral.

The crest of the head of the rib attaches to the IV disc by an **intra-articular ligament of head of rib** within the joint, dividing the enclosed space into two synovial cavities.

The *fibrous layer of the joint capsule* is strongest anteriorly, where it forms a **radiate ligament of head of rib** that fans out from the anterior margin of the head of the rib to the sides of the bodies of two vertebrae and the IV disc between them (Fig. 1.8A & B). The heads of the ribs connect so closely to the vertebral bodies that only slight gliding movements occur at the (demi)facets (pivoting around the intra-articular ligament) of the joints of the heads of ribs; however, even slight movement here may produce a relatively large excursion of the distal (sternal or anterior) end of a rib.

Costotransverse Joints. Abundant ligaments lateral to the posterior parts (vertebral arches) of the vertebrae provide strength to and limit the movements of these joints, which have only thin joint capsules. A **costotransverse ligament** passing from the neck of the rib to the transverse process and a **lateral costotransverse ligament** passing from the tubercle of the rib to the tip of the transverse process strengthen the anterior and posterior aspects of the joint, respectively. A **superior costotransverse ligament** is a broad band that joins the crest of the neck of the rib to the transverse process superior to it. The aperture between this ligament and the vertebra permits passage of the spinal nerve and the posterior branch of the intercostal artery. The superior costotransverse ligament may be divided into a strong *anterior costotransverse ligament* and a weak *posterior costotransverse ligament*.

The strong costotransverse ligaments binding these joints limit their movements to slight gliding. However, the articular surfaces on the tubercles of the superior 6 ribs are convex and fit into concavities on the transverse processes (Fig. 1.9). As a result, rotation occurs around a mostly transverse axis that traverses the intra-articular ligament and the head and neck of the rib (Fig. 1.8A & B). This results in elevation and depression movements of the sternal ends of the

ribs and the sternum in the sagittal plane (*pump-handle movement*) (Fig. 1.10A & C). Flat articular surfaces of tubercles and transverse processes of the 7th–10th ribs allow gliding (Fig. 1.9), resulting in elevation and depression of the lateral-most portions of these ribs in the transverse plane (*bucket-handle movement*) (Fig. 1.10B & C).

STERNOCOSTAL JOINTS

The 1st pair of costal cartilages articulates with the manubrium by means of a thin dense layer of tightly adherent fibrocartilage interposed between the cartilage and manubrium, the **synchondrosis of the 1st rib.** The 2nd–7th pairs of costal cartilages articulate with the sternum at synovial joints with fibrocartilaginous articular surfaces on both the chondral and sternal aspects, allowing movement during respiration. The weak joint capsules of these joints are thickened anteriorly and posteriorly to form **radiate sternocostal ligaments.** These continue as thin, broad membranous bands passing from the costal cartilages to the anterior and posterior surfaces of the sternum, forming a felt-like covering for this bone.

Movements of Thoracic Wall

Movements of the thoracic wall and the diaphragm during inspiration produce increases in the intrathoracic volume and diameters of the thorax (Fig. 1.10D & F). Consequent pressure changes result in air being alternately drawn into the lungs (*inspiration*) through the nose, mouth, larynx, and trachea and expelled from the lungs (*expiration*) through the same passages. During passive expiration, the diaphragm, intercostal muscles, and other muscles relax, decreasing intrathoracic volume and increasing the *intrathoracic pressure* (Fig. 1.10E & C). Concurrently, *intra-abdominal pressure* decreases and abdominal viscera are decompressed. This allows the stretched elastic tissue of the lungs to recoil, expelling most of the air.

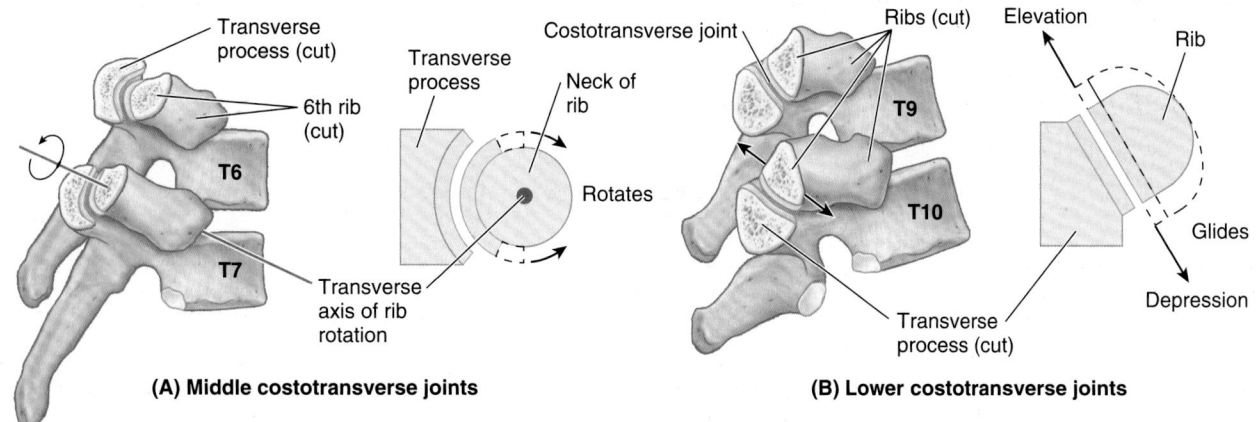

(A) Middle costotransverse joints

(B) Lower costotransverse joints

Right lateral views

FIGURE 1.9. Costotransverse joints. Conformation of articular surfaces, revealed in sagittal sections of the costotransverse joints, demonstrates how the 1st–7th ribs rotate about an axis that runs longitudinally through the neck of the rib (**A**), whereas the 8th–10th ribs glide (**B**).

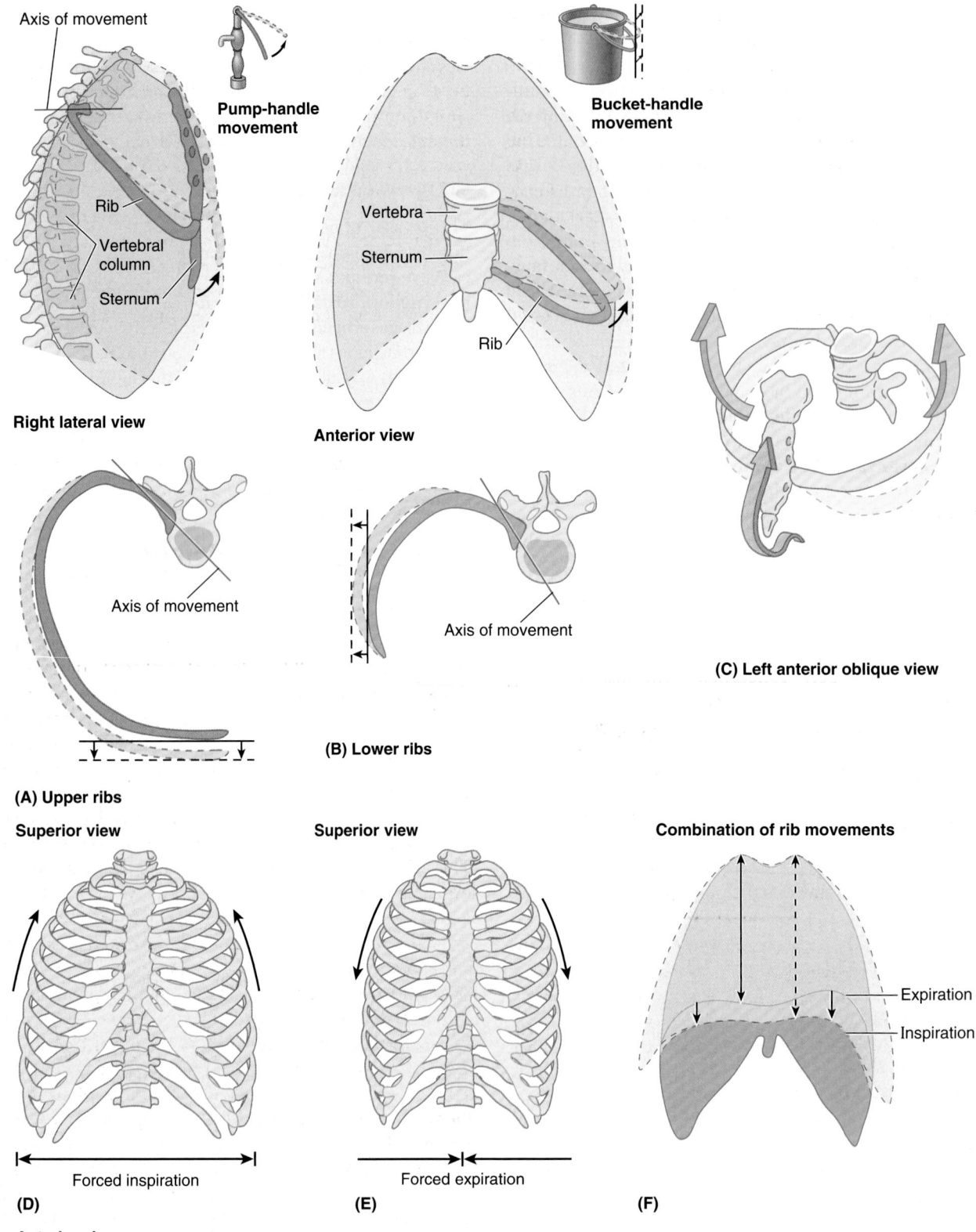

FIGURE 1.10. Movements of thoracic wall. A. When the upper ribs are elevated, the AP dimension of the thorax is increased (pump-handle movement), with a greater excursion (increase) occurring inferiorly, at the end of the pump handle. **B.** The middle parts of the lower ribs move laterally when they are elevated, increasing the transverse dimension (bucket-handle movement). **C.** The combination of rib movements (*arrows*) that occur during forced inspiration increase the AP and transverse dimensions of the thoracic cage. **D.** The thorax widens during forced inspiration as the ribs are elevated (*arrows*). **E.** The thorax narrows during expiration as the ribs are depressed (*arrows*). **F.** The primary movement of inspiration (resting or forced) is contraction of the diaphragm, which increases the vertical dimension of the thoracic cavity (*arrows*). When the diaphragm relaxes, decompression of the abdominal viscera pushes the diaphragm upward, reducing the vertical dimension for expiration.

The *vertical dimension* (height) of the central part of the thoracic cavity increases during inspiration as contraction of the diaphragm causes it to descend, compressing the abdominal viscera (Fig. 1.10F). During expiration, the vertical dimension returns to the neutral position as the elastic recoil of the lungs produces sub-atmospheric pressure in the pleural cavities, between the lungs and the thoracic wall. As a result of this and the absence of resistance to the previously compressed viscera, the domes of the diaphragm ascend, diminishing the vertical dimension.

The *AP dimension* of the thorax increases considerably when the intercostal muscles contract. Movement of the ribs (primarily 2nd–6th) at the costovertebral joints around an axis passing through the necks of the ribs causes the anterior ends of the ribs to rise—*the pump-handle movement* (Fig. 1.10A & C). Because the ribs slope inferiorly, their elevation also results in anterior–posterior movement of the sternum, especially its inferior end, with slight movement occurring at the manubriosternal joint in young people, in whom this joint has not yet synostosed (united).

The *transverse dimension* of the thorax also increases slightly when the intercostal muscles contract, raising the middle (lateral-most parts) of the ribs (especially the lower ones)—*the bucket-handle movement* (Fig. 1.10B & C). The combination of all these movements moves the thoracic cage anteriorly, superiorly, and laterally (Fig. 1.10C & F).

THORACIC WALL

Chest Pain

Although *chest pain* can result from pulmonary disease, it is probably the most important symptom of cardiac disease (Swartz, 2009). However, chest pain may also occur in intestinal, gallbladder, and musculoskeletal disorders. When evaluating a patient with chest pain, the examination is largely concerned with discriminating between serious conditions and the many minor causes of pain. People who have had a *heart attack* usually describe the associated pain as a "crushing" sub-sternal pain (deep to the sternum) that does not disappear with rest.

Rib Fractures

The short, broad 1st rib, posteroinferior to the clavicle, is rarely fractured because of its protected position (it cannot be palpated). When it is broken, however, structures crossing its superior aspect may be injured, including the brachial plexus of nerves and subclavian vessels that serve the upper limb. The middle ribs are most commonly fractured. *Rib fractures* usually result from blows or crushing injuries. The weakest part of a rib is just anterior to its angle; however, direct violence may fracture a rib anywhere, and its broken end may injure internal organs such as a lung and/or the spleen. Fractures of the lower ribs may tear the diaphragm and result in a *diaphragmatic hernia* (see Chapter 2). Rib fractures are painful because the broken parts move during respiration, coughing, laughing, and sneezing.

Flail Chest

Multiple rib fractures may allow a sizable segment of the anterior and/or lateral thoracic wall to move freely. The loose segment of the wall moves paradoxically (inward on inspiration and outward on expiration). *Flail chest* is an extremely painful injury and impairs ventilation, thereby affecting oxygenation of the blood. During treatment, the loose segment may be fixed by hooks and/or wires so that it cannot move.

Thoracotomy, Intercostal Space Incisions, and Rib Excision

The surgical creation of an opening through the thoracic wall to enter a pleural cavity is a *thoracotomy* (Fig. B1.1). An *anterior thoracotomy* may involve making H-shaped cuts through the perichondrium of one or more costal cartilages and then shelling out segments of costal cartilage to gain entrance to the thoracic cavity (see Fig 1.13, right side).

The posterolateral aspects of the 5th–7th intercostal spaces are important sites for *posterior thoracotomy incisions*. In general, a lateral approach is most satisfactory for entry through the thoracic cage (Fig. B1.1). With the patient lying on the contralateral side, the upper limb is fully abducted, placing the forearm beside the patient's head. This elevates and laterally rotates the inferior angle of scapula, allowing access as high as the 4th intercostal space.

Surgeons use an H-shaped incision to incise the superficial aspect of the periosteum that ensheaths the rib, strip the periosteum from the rib, and then *excise a wide segment of the rib* to gain better access, as might be required to enter the thoracic cavity and remove a lung (*pneumonectomy*), for example. In the rib's absence, entry into the thoracic cavity can be made through the deep aspect of the periosteal sheath, sparing the adjacent intercostal muscles. After the operation, the missing pieces of ribs regenerate from the intact periosteum, although imperfectly.

Supernumerary Ribs

People usually have 12 ribs on each side, but the number is increased by the presence of *cervical* and/or *lumbar ribs*, or decreased by failure of

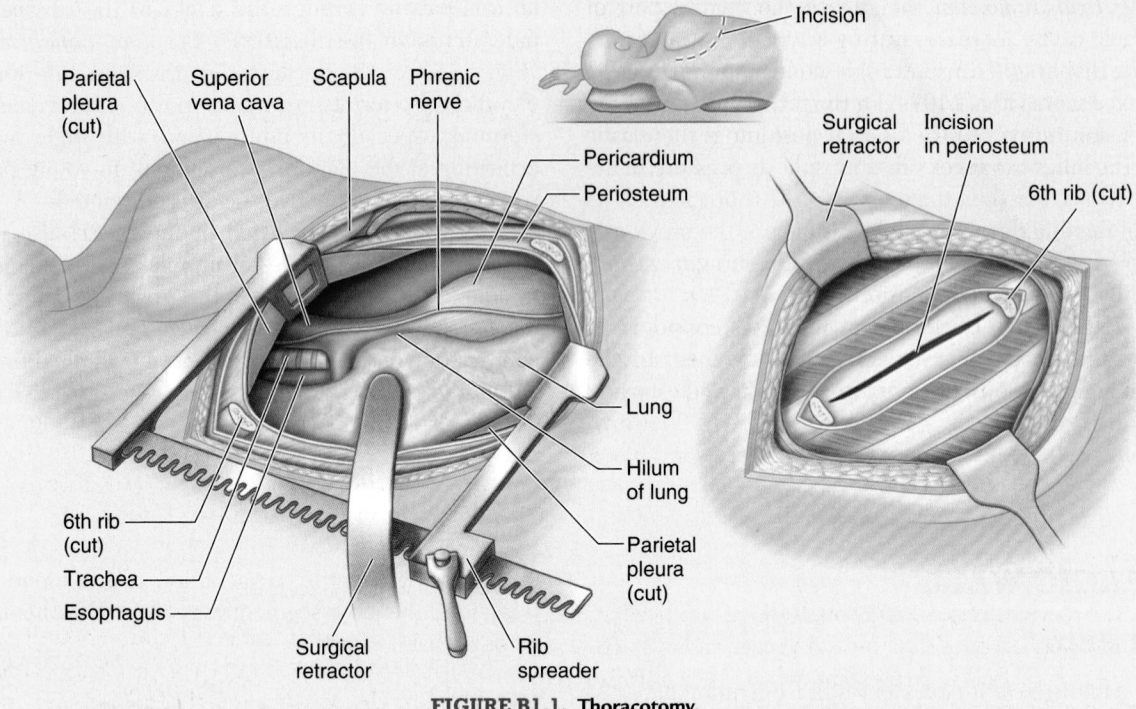

FIGURE B1.1. Thoracotomy.

the 12th pair to form. Cervical ribs are relatively common (0.5–2%) and may interfere with neurovascular structures exiting the superior thoracic aperture. Lumbar ribs are less common. *Supernumerary* (extra) *ribs* also have clinical significance in that they may confuse the identification of vertebral levels in radiographs and other diagnostic images.

Protective Function and Aging of Costal Cartilages

Costal cartilages provide resilience to the thoracic cage, preventing many blows from fracturing the sternum and/or ribs. Because of the remarkable elasticity of the ribs and costal cartilages in children, chest compression may produce injury within the thorax even in the absence of a rib fracture. In elderly people, the costal cartilages lose some of their elasticity and become brittle; they may undergo calcification, making them radiopaque (i.e., appear white in radiographs).

Ossified Xiphoid Process

Many people in their early 40s suddenly become aware of their partly *ossified xiphoid process* and consult their physician about the hard lump in the "pit of their stomach" (*epigastric fossa*). Never having been aware of their xiphoid process before, they fear they have developed a tumor.

Sternal Fractures

Despite the subcutaneous location of the sternum, *sternal fractures* are not common. Crush injuries can occur after traumatic compression of the thoracic wall in automobile accidents when the driver's chest is forced into the steering column, for example. The installation and use of air bags in vehicles has reduced the number of sternal fractures. A fracture of the sternal body is usually a *comminuted fracture* (a break resulting in several pieces). Displacement of the bone fragments is uncommon because the sternum is invested by *deep fascia* (fibrous continuities of radiate sternocostal ligaments; Fig. 1.6A) and the sternal attachments of the pectoralis major muscles. The most common site of sternal fracture in elderly people is at the sternal angle, where the manubriosternal joint has fused. The fracture results in *dislocation of the manubriosternal joint*.

The concern in sternal injuries is not primarily for the fracture itself, but for the likelihood of heart injury (myocardial contusion, cardiac rupture, tamponade) or lung injury. The mortality (death rate) associated with sternal fractures is 25–45%, largely owing to these underlying injuries. Patients with *sternal contusion* should be evaluated for underlying visceral injury (Marx et al., 2009).

Median Sternotomy

To gain access to the thoracic cavity for surgical operations in the mediastinum—such as *coronary artery bypass grafting*, for example—the sternum is divided (split) in the median plane and retracted. The flexibility of ribs and costal cartilages enables spreading of the halves of the sternum during procedures requiring *median sternotomy*. Such "sternal splitting" also gives good exposure for removal of tumors in the superior lobes of the lungs. After surgery, the halves of the sternum are usually joined using wire sutures or clips.

Sternal Biopsy

The sternal body is often used for *bone marrow needle biopsy* because of its breadth and subcutaneous position. The needle pierces the thin cortical bone and enters the vascular spongy bone. Sternal biopsy is commonly used to obtain specimens of marrow for transplantation and for detection of metastatic cancer and *blood dyscrasias* (abnormalities).

Sternal Anomalies

The sternum develops through the fusion of bilateral, vertical condensations of precartilaginous tissue, *sternal bands* or *bars*. The halves of the sternum of the fetus may not fuse. *Complete sternal cleft* is an uncommon anomaly through which the heart may protrude (*ectopia cordis*). Partial clefts involving the manubrium and superior half of the body are V- or U-shaped and can be repaired during infancy by direct apposition and fixation of the sternal halves. Sometimes a perforation (*sternal foramen*) remains in the sternal body because of incomplete fusion. It is not clinically significant; however, one should be aware of its possible presence so that it will not be misinterpreted in chest X-ray, as a being an unhealed bullet wound for example. A receding (*pectus excavatum,* or funnel chest) or projecting (*pectus cavinatum,* or pigeon breast) sternum are anomalous variations that may become evident or more pronounced during childhood.

The xiphoid process is commonly perforated in elderly persons because of age-related changes; this perforation is also not clinically significant. Similarly, an anteriorly-protruding xiphoid process in neonates is not unusual; when it occurs, it does not usually require correction.

Thoracic Outlet Syndrome

Anatomists refer to the superior thoracic aperture as the *thoracic inlet* because non-circulating substances (air and food) may enter the thorax only through this aperture. When clinicians refer to the superior thoracic aperture as the *thoracic outlet,* they are emphasizing the arteries and T1 spinal nerves that emerge from the thorax through this aperture to enter the lower neck and upper limbs. Hence, various types of *thoracic outlet syndrome* (TOS) exist in which emerging structures are affected by obstructions of the superior thoracic aperture (Rowland and Pedley, 2010). Although TOS implies a thoracic location, the obstruction actually occurs outside the aperture in the root of the neck, and the manifestations of the syndromes involve the upper limb (see Chapters 6 and 8).

Dislocation of Ribs

Rib dislocation ("slipping rib" syndrome) is the displacement of a costal cartilage from the sternum—*dislocation of a sternocostal joint* or the displacement of the interchondral joints. Rib dislocations are common in body-contact sports; complications may result from pressure on or damage to nearby nerves, vessels, and muscles. *Displacement of interchondral joints* usually occurs unilaterally and involves ribs 8, 9, and 10. Trauma sufficient to displace these joints often injures underlying structures, such as the diaphragm and/or liver, causing severe pain, particularly during deep inspiratory movements. The injury produces a lump-like deformity at the displacement site.

Separation of Ribs

"Rib separation" refers to *dislocation of the costochondral junction* between the rib and its costal cartilage. In separations of the 3rd–10th ribs, tearing of the perichondrium and periosteum usually occurs. As a result, the rib may move superiorly, overriding the rib above and causing pain.

Paralysis of Diaphragm

Paralysis of half of the diaphragm (one dome or hemidiaphragm) because of injury to its motor supply from the *phrenic nerve* does not affect the other half because each dome has a separate nerve supply. One can detect *paralysis of the diaphragm* radiographically by noting its paradoxical movement. Instead of descending as it normally does during inspiration owing to diaphragmatic contraction (Fig. B1.2A), the paralyzed dome ascends as it is pushed superiorly by the abdominal viscera that are being compressed by the active contralateral dome (Fig. B1.2B). Instead of ascending during expiration, the paralyzed dome descends in response to the positive pressure in the lungs.

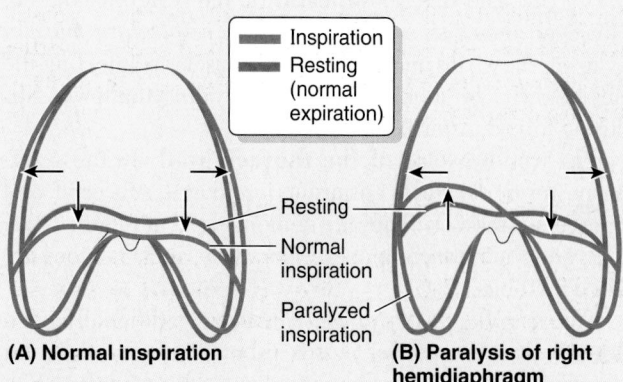

FIGURE B1.2. Normal and paradoxical movements of diaphragm.

The Bottom Line

SKELETON, APERTURES, JOINTS, AND MOVEMENTS OF THORACIC WALL

Skeleton of thoracic wall: The thoracic wall (1) protects the contents of the thoracic cavity; (2) provides the mechanics for breathing; and (3) provides for attachment of neck, back, upper limb, and abdominal musculature. ♦ The domed shape of the thoracic cage gives it strength, and its osteocartilaginous elements and joints give it flexibility. ♦ Posteriorly the thoracic cage consists of a column of 12 thoracic vertebrae and interposed IV discs. ♦ Laterally and anteriorly the cage consists of 12 ribs that are continued anteriorly by costal cartilages. Anteriorly, the 3-part sternum protects the central thoracic viscera.

Apertures of thoracic wall: Although the thoracic cage is complete peripherally, it is open superiorly and inferiorly. ♦ The superior thoracic aperture is a small passageway for the transmittal of structures to and from the neck and upper limbs. ♦ The large inferior thoracic aperture provides a rim to which the diaphragm is attached. Structures passing between the thorax and abdomen traverse openings in the diaphragm (e.g., esophagus) or pass posterior to it (e.g., aorta).

Joints of thoracic wall: The joints enable and determine movements of the thoracic wall. ♦ Posteriorly, ribs articulate with the semiflexible thoracic vertebral column via costovertebral joints. These include joints of heads of ribs and costotransverse joints, both strongly supported by multiple ligaments. ♦ Anteriorly, ribs articulate with costal cartilages via costochondral joints. ♦ Costal cartilages 1–7 articulate directly and costal cartilages 8–10 articulate indirectly with the sternum via the synchondrosis of the 1st rib, synovial sternocostal joints, and interchondral joints.

Movements of thoracic wall: The movements of most ribs occur around a generally transverse axis that passes through the head, neck, and tubercle of the rib. ♦ This axis, plus the slope and curvature of the ribs, results in pump-handle-type movements of the upper ribs that alter the AP diameter of the thorax and bucket-handle-type movements of lower ribs that alter its transverse diameter. ♦ Contraction and relaxation of the superiorly convex diaphragm alters its vertical dimensions. ♦ Increasing dimensions produce inhalation, and decreasing dimensions produce exhalation.

Muscles of Thoracic Wall

Some muscles attached to and/or covering the thoracic cage are primarily involved in serving other regions. **Axio-appendicular muscles** extend from the thoracic cage (axial skeleton) to bones of the upper limb (appendicular skeleton). Similarly, some muscles of the anterolateral abdominal wall, back, and neck muscles have attachments to the thoracic cage (Fig. 1.11). The axio-appendicular muscles act primarily on the upper limbs (see Chapter 6). But several of them, including the *pectoralis major* and *pectoralis minor* and the inferior part of the *serratus anterior,* may also function as accessory muscles of respiration, helping elevate the ribs to expand the thoracic cavity when inspiration is deep and forceful (e.g., after a 100-m dash). The *scalene muscles* of the neck, which descend from vertebrae of the neck to the 1st and 2nd ribs, act primarily on the vertebral column. However, they also serve as *accessory respiratory muscles* by fixing these ribs and enabling the muscles connecting the ribs below to be more effective in elevating the lower ribs during forced inspiration.

The true **muscles of the thoracic wall** are the serratus posterior, levatores costarum, intercostal, subcostal, and transversus thoracis. They are demonstrated in Figure 1.12A & B, and their attachments, innervations, and functions are listed in Table 1.2.

The *serratus posterior muscles* have traditionally been described as inspiratory muscles, but this function is not supported by electromyography or other evidence. On the basis of its attachments and disposition, the **serratus posterior superior** was said to elevate the superior four ribs, thus increasing the AP diameter of the thorax and raising the sternum. On the basis of its attachments and disposition, the **serratus posterior inferior** was said to depress the inferior ribs, preventing them from being pulled superiorly by the diaphragm. However, recent studies (Vilensky et al., 2001) suggest that these muscles, which span the superior and inferior thoracic apertures as well as the transitions from the relatively inflexible thoracic vertebral column to the much more flexible cervical and lumbar segments of the column, may not be primarily motor in function. Rather, they may have a proprioceptive function. These muscles, particularly the serratus posterior superior, have been implicated as a source of chronic pain in myofascial pain syndromes.

The **levatores costarum muscles** (L. *levator,* a lifter) are 12 fan-shaped muscles that elevate the ribs (Fig. 1.17), but their role, if any, in normal inspiration is uncertain. They may play a role in vertebral movement and/or proprioception.

The *intercostal muscles* occupy the intercostal spaces (Figs. 1.11–1.14; Table 1.2). The superficial layer is formed by the external intercostals, the inner layer by the internal intercostals. The deepest fibers of the latter, lying internal to the intercostal vessels, are somewhat artificially designated as a separate muscle, the *innermost intercostals.*

- The **external intercostal muscles** (11 pairs) occupy the intercostal spaces from the tubercles of the ribs posteriorly to the costochondral junctions anteriorly (Figs. 1.11–1.13, and 1.15). Anteriorly, the muscle fibers are replaced

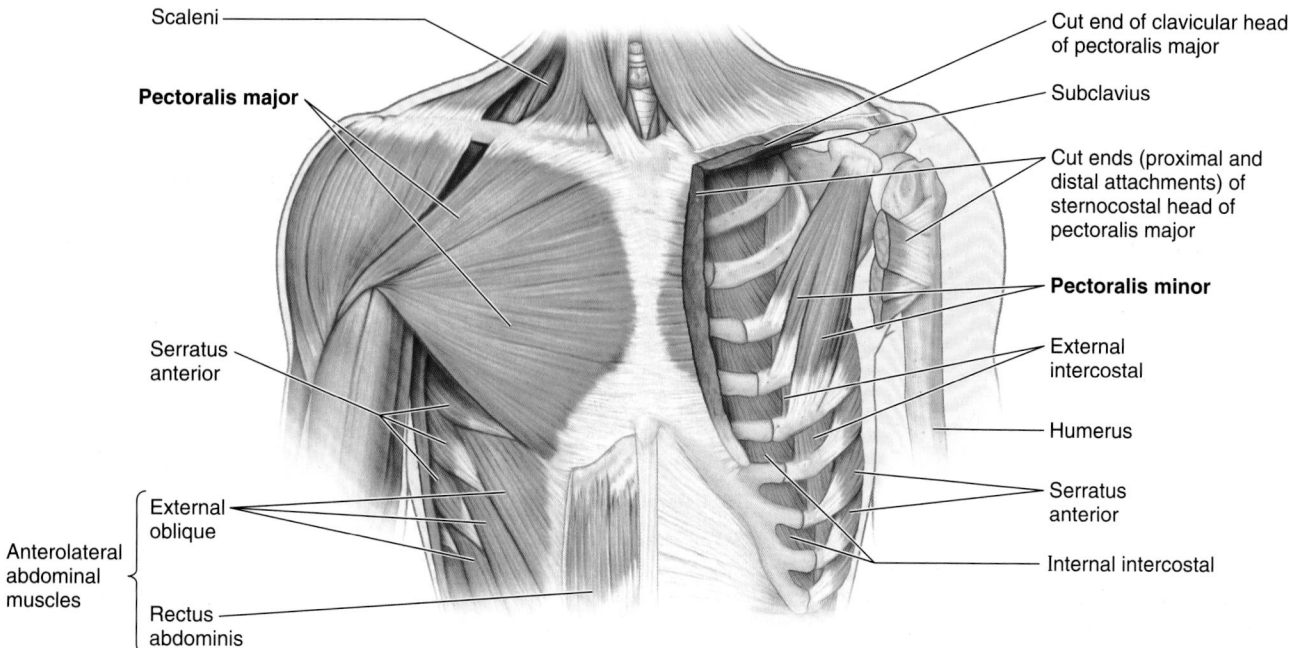

Scaleni

Pectoralis major

Serratus
anterior

Anterolateral
abdominal
muscles

External
oblique

Rectus
abdominis

Cut end of clavicular head
of pectoralis major

Subclavius

Cut ends (proximal and
distal attachments) of
sternocostal head of
pectoralis major

Pectoralis minor

External
intercostal

Humerus

Serratus
anterior

Internal intercostal

Anterior view

FIGURE 1.11.   Axio-appendicular, neck, and anterolateral abdominal muscles overlying thoracic wall. The pectoralis major has been removed on the left side to expose the pectoralis minor, subclavius, and external intercostal muscles. When the upper limb muscles are removed, the superiorly tapering domed shape of the thoracic cage is revealed.

by the **external intercostal membranes** (Fig. 1.15A). These muscles run infero-anteriorly from the rib above to the rib below. Each muscle attaches superiorly to the inferior border of the rib above and inferiorly to the superior border of the rib below (Fig. 1.15C). These muscles are continuous inferiorly with the *external oblique muscles* in the anterolateral abdominal wall. The external intercostals are most active during inspiration.

- The **internal intercostal muscles** (11 pairs) run deep to and at right angles to the external intercostals (Figs. 1.12B, 1.14, and 1.15C). Their fibers run inferoposteriorly from the floors of the costal grooves to the superior borders of the ribs inferior to them. The internal intercostals attach to the bodies of the ribs and their costal cartilages as far anteriorly as the sternum and as far posteriorly as the angles of the ribs (Fig. 1.16). Between the ribs posteriorly, medial to the angles, the internal intercostals are replaced by the **internal intercostal membranes** (Fig. 1.15A). The inferior internal intercostal muscles are continuous with the *internal oblique muscles* in the anterolateral abdominal wall. The internal intercostals—weaker than the external intercostal muscles—are most active during expiration—especially their interosseous (vs. interchondral) portions.

- The **innermost intercostal muscles** are similar to the internal intercostals and are essentially their deeper parts. The innermost intercostals are separated from the internal intercostals by intercostal nerves and vessels (Figs. 1.15A & B and 1.16). These muscles pass between the internal

surfaces of adjacent ribs and occupy the lateral-most parts of the intercostal spaces. It is likely (but undetermined) that their actions are the same as those of the internal intercostal muscles.

The **subcostal muscles** are variable in size and shape, usually being well developed only in the lower thoracic wall. These thin muscular slips extend from the internal surface of the angle of one rib to the internal surface of the second or third rib inferior to it. Crossing one or two intercostal spaces, the subcostals run in the same direction as the internal intercostals and blend with them (Fig. 1.15B).

The **transversus thoracis muscles** consist of four or five slips that radiate superolaterally from the posterior aspect of the inferior sternum (Figs. 1.13–1.15A). The transversus thoracis muscles are continuous inferiorly with the *transversus abdominis muscles* in the anterolateral body wall. These muscles appear to have a weak expiratory function and may also provide proprioceptive information.

Although the external and internal intercostals are active during inspiration and expiration, respectively, most activity is isometric (increases tonus without producing movement); the role of these muscles in producing movement of the ribs appears to be related mainly to *forced respiration*. The diaphragm is the primary muscle of inspiration. Expiration is passive unless one is exhaling against resistance (e.g., inflating a balloon) or trying to expel air more rapidly than usual (e.g., coughing, sneezing, blowing one's nose, or shouting). The elastic recoil of the lungs and decompression of

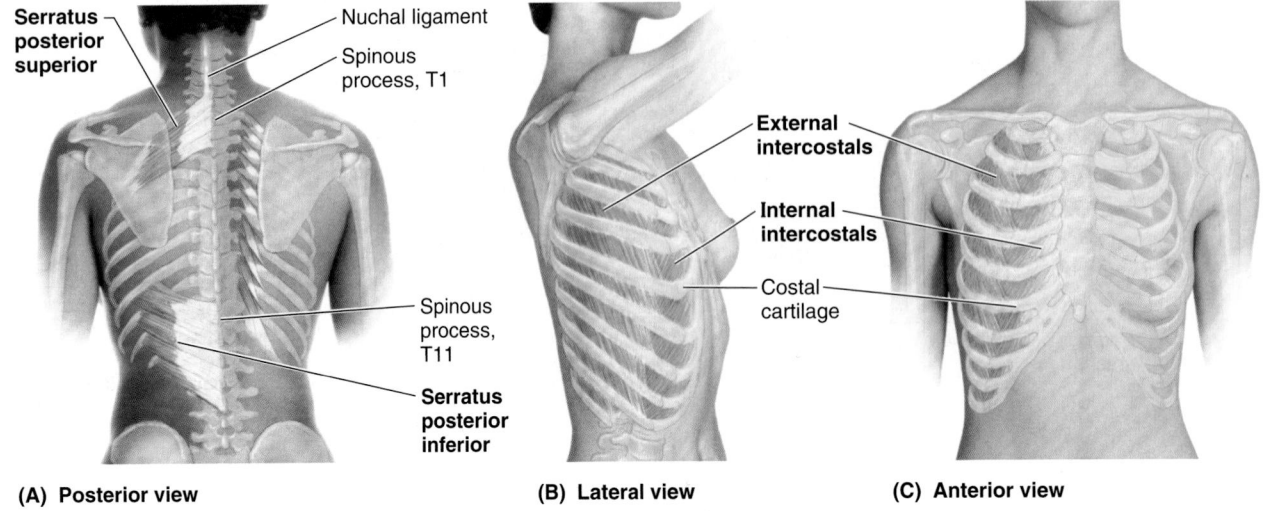

FIGURE 1.12. Muscles of thoracic wall.

TABLE 1.2. MUSCLES OF THORACIC WALL

Muscle	Superior Attachment	Interior Attachment	Innervation	Main Action	
Serratus posterior superior	Nuchal ligament, spinous processes of C7 to T3 vertebrae	Superior borders of 2nd to 4th ribs	2nd to 5th intercostal nerves	Proprioception (elevate ribs)[b]	
Serratus posterior inferior	Spinous processes of T11 to L2 vertebrae	Inferior borders of 8th to 12th ribs near their angles	Anterior rami to T9 to T12 thoracic spinal nerves	Proprioception (depress ribs)[b]	
Levator costarum	Transverse processes of T7–11	Subjacent ribs between tubercle and angle	Posterior primary rami of C8–T11 nerves	Elevate ribs	
External intercostal	Inferior border of ribs	Superior border of ribs below	Intercostal nerve	Elevate ribs during forced inspiration[a]	
Internal intercostal				Interosseous part: depresses ribs	During active (forced) respiration[a]
Innermost intercostal				Interchondral part: elevates ribs	
Subcostal	Internal surface of lower ribs near their angles	Superior borders of 2nd or 3rd ribs below		Probably act in same manner as internal intercostal muscles	
Transversus thoracis	Posterior surface of lower sternum	Internal surface of costal cartilages 2–6		Weakly depress ribs[b] Proprioception?	

[a]All intercostal muscles keep intercostal spaces rigid, thereby preventing them from bulging out during expiration and from being drawn in during inspiration. The role of individual intercostal muscles and accessory muscles of respiration in moving the ribs is difficult to interpret despite many electromyographic studies.

[b]Action traditionally assigned based on attachments; appear to be largely proprioceptive in function.

abdominal viscera expel previously inhaled air. The primary role of the intercostal muscles in respiration is to support (increase the tonus or rigidity of) the intercostal space, resisting paradoxical movement especially during inspiration when internal thoracic pressures are lowest (most negative). This is most apparent following a high spinal cord injury, when there is an initial flaccid paralysis of the entire trunk but the diaphragm remains active. In these circumstances, the vital capacity is markedly compromised by the paradoxical incursion of the thoracic wall during inspiration. Several weeks later, the paralysis becomes spastic; the thoracic wall stiffens and vital capacity rises (Standring, 2008).

The mechanical action of the intercostal muscles in rib movement, especially during forced respiration, can be

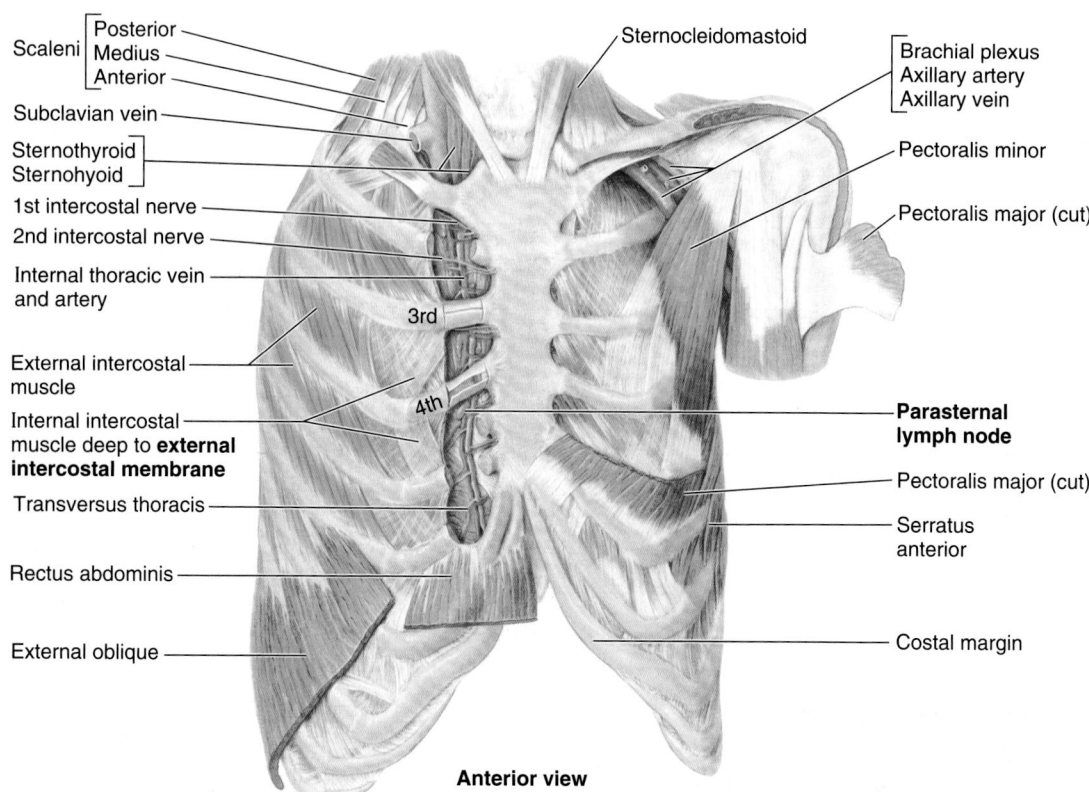

Scaleni { Posterior — Medius — Anterior — }

Subclavian vein —

Sternothyroid } Sternohyoid }

1st intercostal nerve —

2nd intercostal nerve —

Internal thoracic vein and artery —

External intercostal muscle —

Internal intercostal muscle deep to **external intercostal membrane** —

Transversus thoracis —

Rectus abdominis —

External oblique —

Sternocleidomastoid

Brachial plexus Axillary artery Axillary vein

Pectoralis minor

Pectoralis major (cut)

3rd

4th

Parasternal lymph node

Pectoralis major (cut)

Serratus anterior

Costal margin

Anterior view

FIGURE 1.13. Dissection of anterior aspect of anterior thoracic wall. The external intercostal muscles are replaced by membranes between costal cartilages. The H-shaped cuts through the perichondrium of the 3rd and 4th costal cartilages are used to shell out pieces of cartilage, as was done with the 4th costal cartilage. It is not uncommon for the 8th rib to attach to the sternum, as in this specimen. The internal thoracic vessels and parasternal lymph nodes (*green*) lie inside the thoracic cage lateral to the sternum.

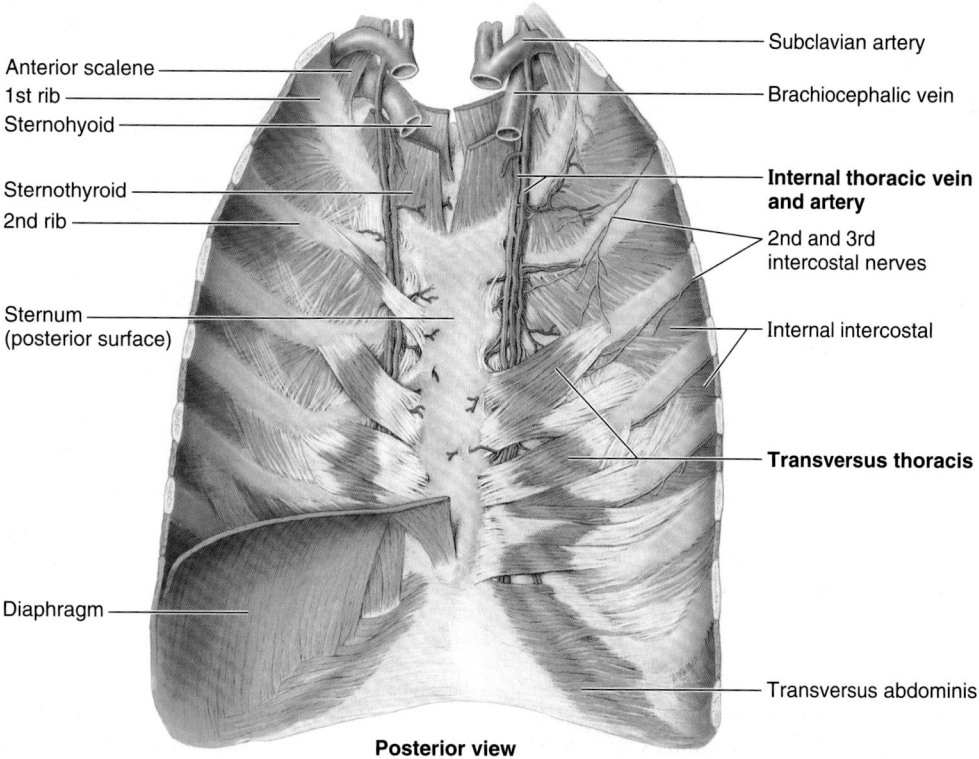

Anterior scalene —

1st rib —

Sternohyoid —

Sternothyroid —

2nd rib —

Sternum (posterior surface) —

Diaphragm —

Subclavian artery

Brachiocephalic vein

Internal thoracic vein and artery

2nd and 3rd intercostal nerves

Internal intercostal

Transversus thoracis

Transversus abdominis

Posterior view

FIGURE 1.14. Posterior aspect of anterior thoracic wall. The internal thoracic arteries arise from the subclavian arteries and have paired accompanying veins (L. *venae comitantes*) inferiorly. Superior to the 2nd costal cartilage, there is only a single internal thoracic vein on each side, which drains into the brachiocephalic vein. The continuity of the transversus thoracis muscle with the transversus abdominis muscle becomes apparent when the diaphragm is removed, as has been done here on the right side.

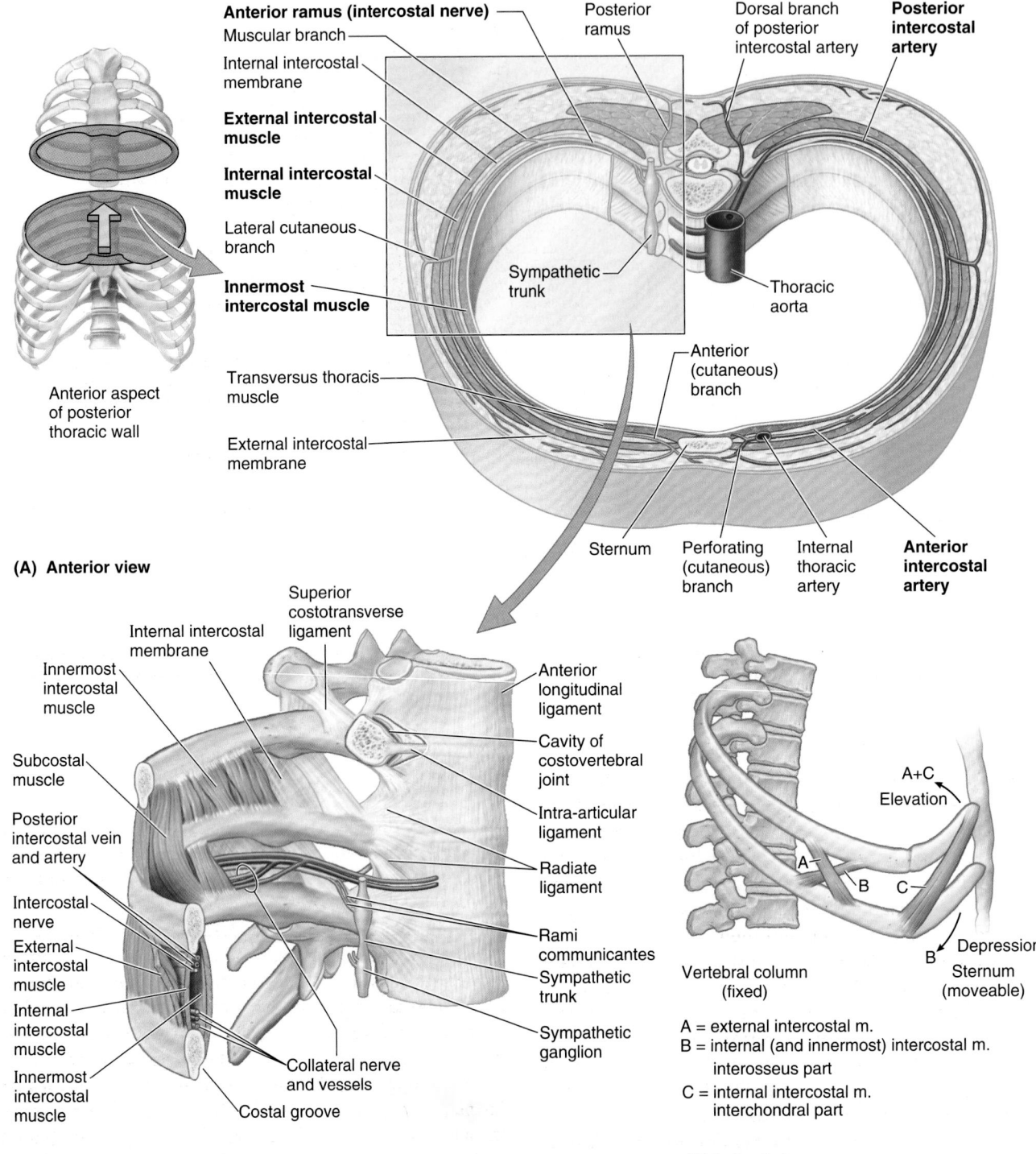

Anterior ramus (intercostal nerve)
Muscular branch
Internal intercostal membrane
External intercostal muscle
Internal intercostal muscle
Lateral cutaneous branch
Innermost intercostal muscle

Posterior ramus
Dorsal branch of posterior intercostal artery
Posterior intercostal artery
Sympathetic trunk
Thoracic aorta

Anterior aspect of posterior thoracic wall

Transversus thoracis muscle
External intercostal membrane

(A) Anterior view

Anterior (cutaneous) branch
Sternum
Perforating (cutaneous) branch
Internal thoracic artery
Anterior intercostal artery

Superior costotransverse ligament
Internal intercostal membrane
Innermost intercostal muscle
Subcostal muscle
Posterior intercostal vein and artery
Intercostal nerve
External intercostal muscle
Internal intercostal muscle
Innermost intercostal muscle
Collateral nerve and vessels
Costal groove

Anterior longitudinal ligament
Cavity of costovertebral joint
Intra-articular ligament
Radiate ligament
Rami communicantes
Sympathetic trunk
Sympathetic ganglion

(B) Anterolateral view

A+C Elevation
A
B   C
B   Depression
Sternum (moveable)

Vertebral column (fixed)

A = external intercostal m.
B = internal (and innermost) intercostal m.
    interosseus part
C = internal intercostal m.
    interchondral part

(C) Lateral view

FIGURE 1.15. Contents of an intercostal space. A. This transverse section shows nerves (*right side*) and arteries (*left side*) in relation to the intercostal muscles. **B.** The posterior part of an intercostal space is shown. The joint capsule (radiate ligament) of one costovertebral joint has been removed. Innermost intercostal muscles bridge one intercostal space; subcostal muscles bridge two. The mnemonic for the order of the neurovascular structures in the intercostal space from superior to inferior is VAN—vein, artery, and nerve. Communicating branches (L. *rami communicantes*) extend between the intercostal nerves and the sympathetic trunk. **C.** A simple model of the action of the intercostal muscles is shown. Contraction of the muscle fibers that most closely parallel the slope of the ribs at a given point (fibers *A* and *C*) will elevate the ribs and sternum; contraction of muscle fibers that are approximately perpendicular to the slope of the ribs (fibers *B*) will depress the ribs.

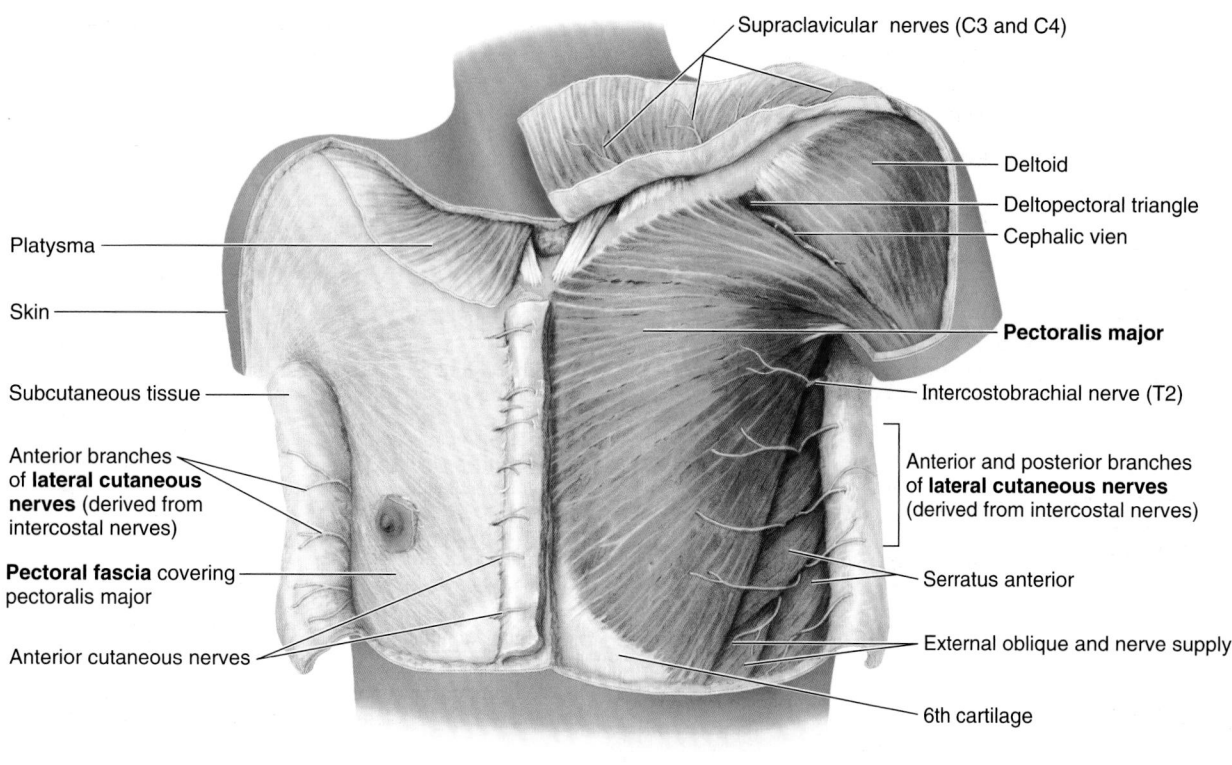

Supraclavicular nerves (C3 and C4)

Deltoid

Deltopectoral triangle

Cephalic vien

Platysma

Skin

Pectoralis major

Subcutaneous tissue

Intercostobrachial nerve (T2)

Anterior branches of **lateral cutaneous nerves** (derived from intercostal nerves)

Anterior and posterior branches of **lateral cutaneous nerves** (derived from intercostal nerves)

Pectoral fascia covering pectoralis major

Serratus anterior

Anterior cutaneous nerves

External oblique and nerve supply

6th cartilage

Anterior view (slightly oblique)

FIGURE 1.16. Superficial dissection of male pectoral region. The platysma is cut short on the right side and is reflected on the left side, together with the underlying supraclavicular nerves. Filmy pectoral fascia covers the right pectoralis major. The fascia has been removed on the left side. The cutaneous branches of the intercostal nerves that supply the breast are shown.

appreciated by means of a simple model (Fig. 1.15C). A pair of curved levers, representing the ribs bordering an intercostal space, are hinged posteriorly to a fixed vertebral column and anteriorly to a moveable sternum. The ribs (and intervening intercostal space) descend as they run anteriorly, reaching their low point approximately at the costochondral junction, and then ascend to the sternum. Muscles with fibers that most closely approximate the slope of the ribs at their attachments (external intercostal and interchondral portion of the internal intercostal muscles) rotate the ribs superiorly at their posterior axes, elevating the ribs and sternum. Muscle with fibers that are approximately perpendicular to the slope of the ribs at their attachment (interosseous part of internal intercostal muscles) rotate the ribs inferiorly at their posterior axes, depressing the ribs and sternum (Slaby et al., 1994).

The (thoracic) *diaphragm* is a shared wall (actually floor/ceiling) separating the thorax and abdomen. Although it has functions related to both compartments of the trunk, its most important (vital) function is serving as the *primary muscle of inspiration*. The detailed description of the diaphragm appears in Chapter 2 because the attachments of its crura occur at abdominal levels (i.e., to lumbar vertebrae) and all of its attachments are best observed from its inferior (abdominal) aspect.

Fascia of Thoracic Wall

Each part of the deep fascia is named for the muscle it invests or for the structure(s) to which it is attached. Consequently, a large portion of the deep fascia overlying the anterior thoracic wall is called **pectoral fascia** for its association with the pectoralis major muscles (Fig. 1.16). In turn, much of the pectoral fascia forms a major part of the *bed of the breast* (structures against which the posterior surface of the breast lies). Deep to the pectoralis major and its fascia is another layer of deep fascia suspended from the clavicle and investing the pectoralis minor muscle, the *clavipectoral fascia*.

The thoracic cage is lined internally with **endothoracic fascia** (see Fig. 1.30C). This thin fibro-areolar layer attaches the adjacent portion of the lining of the lung cavities (costal parietal pleura) to the thoracic wall. It becomes more fibrous over the apices of the lungs (*suprapleural membrane*).

Nerves of Thoracic Wall

The 12 pairs of *thoracic spinal nerves* supply the thoracic wall. As soon as they leave the IV foramina in which they are formed, the mixed thoracic spinal nerves divide into anterior and posterior (primary) rami or branches (Fig. 1.15A and 1.17). The *anterior rami of nerves T1–T11* form the **intercostal nerves** that run along the extent of the intercostal

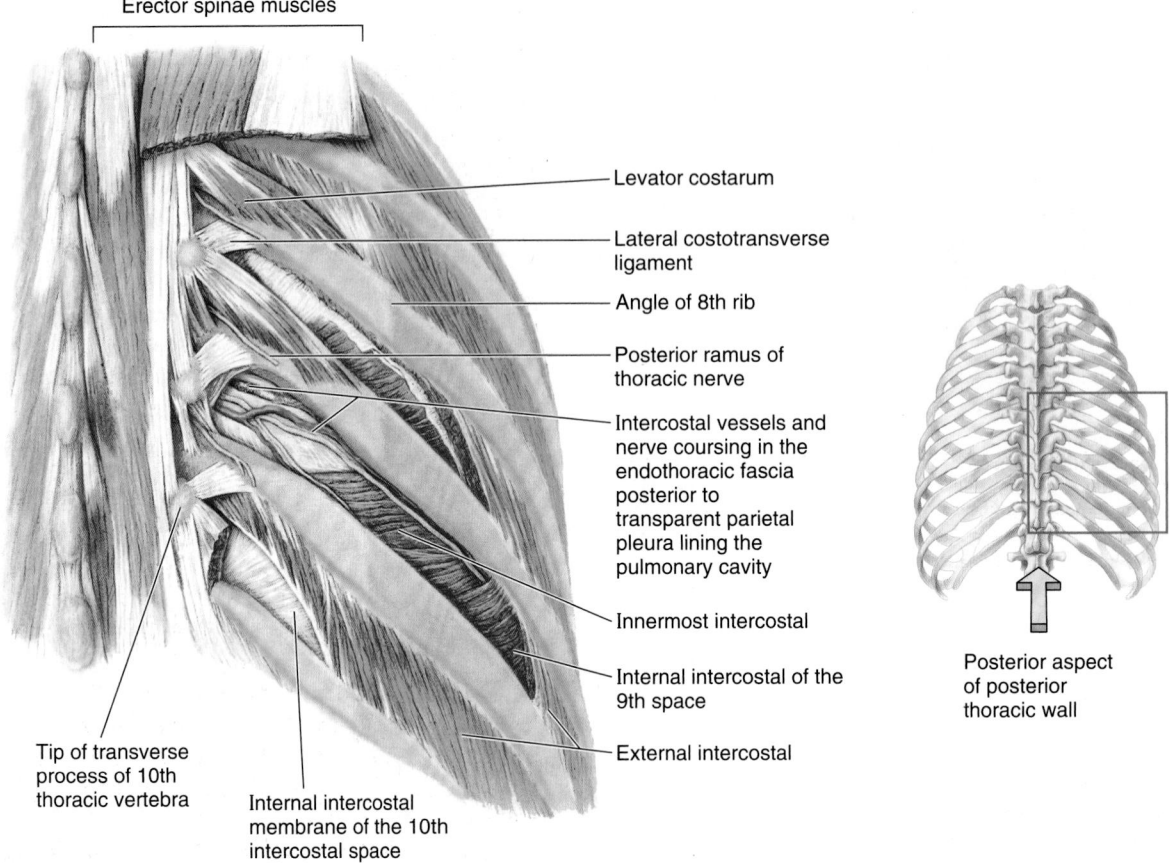

Erector spinae muscles

Levator costarum

Lateral costotransverse ligament

Angle of 8th rib

Posterior ramus of thoracic nerve

Intercostal vessels and nerve coursing in the endothoracic fascia posterior to transparent parietal pleura lining the pulmonary cavity

Innermost intercostal

Internal intercostal of the 9th space

External intercostal

Posterior aspect of posterior thoracic wall

Tip of transverse process of 10th thoracic vertebra

Internal intercostal membrane of the 10th intercostal space

Posterior view

FIGURE 1.17. Dissection of posterior aspect of thoracic wall. Most of the deep muscles of the back have been removed to expose the levatores costarum muscles. In the 8th and 10th intercostal spaces, varying parts of the external intercostal muscle have been removed to expose the underlying internal intercostal membrane, which is continuous with the internal intercostal muscle. In the 9th intercostal space, the levator costorum has been removed to expose the intercostal vessels and nerve.

spaces. The *anterior ramus of nerve T12,* coursing inferior to the 12th rib, is the **subcostal nerve.** The *posterior rami of thoracic spinal nerves* pass posteriorly, immediately lateral to the articular processes of the vertebrae, to supply the joints, deep back muscles, and skin of the back in the thoracic region.

TYPICAL INTERCOSTAL NERVES

The 3rd–6th intercostal nerves enter the medial-most parts of the posterior intercostal spaces, running initially within the endothoracic fascia between the parietal pleura (serous lining of pulmonary cavity) and the internal intercostal membrane nearly in the middle of the intercostal spaces (Figs. 1.15A & B and 1.17). Near the angles of the ribs, the nerves pass between the internal intercostal and the innermost intercostal muscles. At this point, the intercostal nerves pass to and then continue to course in or just inferior to the *costal grooves,* running inferior to the intercostal arteries (which, in turn, run inferior to the intercostal veins). The neurovascular bundles (especially the vessels) are thus sheltered by the inferior margins of the overlying ribs. Collateral branches of

these nerves arise near the angles of the ribs and run along the superior border of the rib below. The nerves continue anteriorly between the internal and innermost intercostal muscles, supplying these and other muscles and giving rise to lateral cutaneous branches in approximately the midaxillary line (MAL). Anteriorly, the nerves appear on the internal surface of the internal intercostal muscle. Near the sternum, the nerves turn anteriorly, passing between the costal cartilages to become anterior cutaneous branches.

Through its posterior ramus and the lateral and anterior cutaneous branches of its anterior ramus, most thoracic spinal nerves (T2–T12) supply a strip-like *dermatome* of the trunk extending from the posterior median line to the anterior median line (Fig. 1.18). The group of muscles supplied by the posterior ramus and anterior ramus (intercostal nerve) of each pair of thoracic spinal nerves constitutes a *myotome.* The myotomes of most thoracic spinal nerves (T2–T11) include the intercostal, subcostal, transversus thoracis, levatores costarum, and serratus posterior muscles associated with the intercostal space that includes the anterior ramus (intercostal nerve) of the specific spinal nerve, plus the overlying portion of the deep muscles of the back.

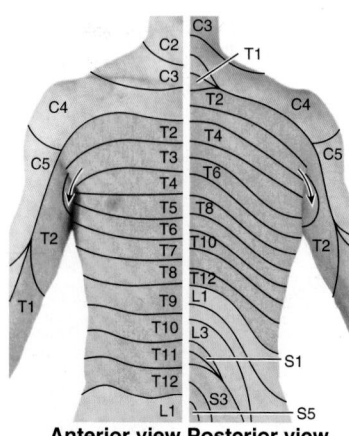

FIGURE 1.18. Segmental innervation (dermatomes) of thoracic wall (after Foerster). Dermatomes C5–T1 are located mostly in the upper limbs and are not represented significantly on the body wall. Since the anterior rami of spinal nerves T2–T12 are not involved in plexus formation, there is no difference between the dermatomes and the zones of peripheral nerve distribution here. Dermatome T4 includes the nipple; dermatome T10 includes the umbilicus.

The *branches of a typical intercostal nerve* are (Fig. 1.15A & B):

- **Rami communicantes,** or communicating branches, that connect each intercostal nerve to the ipsilateral sympathetic trunk. Presynaptic fibers leave the initial portions of the anterior ramus of each thoracic (and upper lumbar) spinal nerve by means of a white communicating ramus and pass to the *sympathetic trunk.* Postsynaptic fibers distributed to the body wall and limbs pass from the ganglia of the sympathetic trunk via gray rami to join the anterior ramus of the nearest spinal nerve, including all intercostal nerves. Sympathetic nerve fibers are distributed through all branches of all spinal nerves (anterior and posterior rami) to reach the blood vessels, sweat glands, and smooth muscle of the body wall and limbs.
- **Collateral branches** that arise near the angles of the ribs and descend to course along the superior margin of the lower rib, helping supply intercostal muscles and parietal pleura.
- **Lateral cutaneous branches** that arise near the MAL, pierce the internal and external intercostal muscles, and divide in turn into *anterior and posterior branches.* These terminal branches supply the skin of the lateral thoracic and abdominal walls.
- **Anterior cutaneous branches** pierce the muscles and membranes of the intercostal space in the parasternal line and divide into *medial and lateral branches.* These terminal branches supply the skin on the anterior aspect of the thorax and abdomen.
- **Muscular branches** that supply the intercostal, subcostal, transversus thoracis, levatores costarum, and serratus posterior muscles.

ATYPICAL INTERCOSTAL NERVES

Although the anterior ramus of most thoracic spinal nerves is simply the intercostal nerve for that level, the *anterior ramus*

of the 1st thoracic (T1) spinal nerve first divides into a large superior and a small inferior part. The superior part joins the *brachial plexus,* the nerve plexus supplying the upper limb, and the inferior part becomes the 1st intercostal nerve. Other atypical features of specific intercostal nerves include the following:

- The *1st and 2nd intercostal nerves* course on the internal surface of the 1st and 2nd ribs, instead of along the inferior margin in costal grooves (Fig. 1.14).
- The *1st intercostal nerve* has no anterior cutaneous branch and often no lateral cutaneous branch. When there is a lateral cutaneous branch, it supplies the skin of the axilla and may communicate with either the intercostobrachial nerve or the medial cutaneous nerve of the arm.
- The *2nd* (and sometimes 3rd) *intercostal nerve* gives rise to a large lateral cutaneous branch, the **intercostobrachial nerve;** it emerges from the 2nd intercostal space at the MAL, penetrates the serratus anterior, and enters the axilla and arm. The intercostobrachial nerve usually supplies the floor—skin and subcutaneous tissue—of the axilla and then communicates with the *medial cutaneous nerve of the arm* to supply the medial and posterior surfaces of the arm. The lateral cutaneous branch of the 3rd intercostal nerve frequently gives rise to a second intercostobrachial nerve.
- The *7th–11th intercostal nerves,* after giving rise to lateral cutaneous branches, cross the costal margin posteriorly and continue on to supply abdominal skin and muscles. No longer being between ribs (intercostal), they now become *thoraco-abdominal nerves* of the anterior abdominal wall (see Chapter 2). Their anterior cutaneous branches pierce the rectus sheath, becoming cutaneous close to the median plane.

Vasculature of Thoracic Wall

In general, the pattern of vascular distribution in the thoracic wall reflects the structure of the thoracic cage—that is, it runs in the intercostal spaces, parallel to the ribs.

ARTERIES OF THORACIC WALL

The arterial supply to the thoracic wall (Fig. 1.19; Table 1.3) derives from the:

- *Thoracic aorta,* through the posterior intercostal and subcostal arteries.
- *Subclavian artery,* through the internal thoracic and supreme intercostal arteries.
- *Axillary artery,* through the superior and lateral thoracic arteries.

The **intercostal arteries** course through the thoracic wall between the ribs. With the exception of the 10th and 11th intercostal spaces, each intercostal space is supplied by three arteries: a large posterior intercostal artery (and its collateral branch) and a small pair of anterior intercostal arteries.

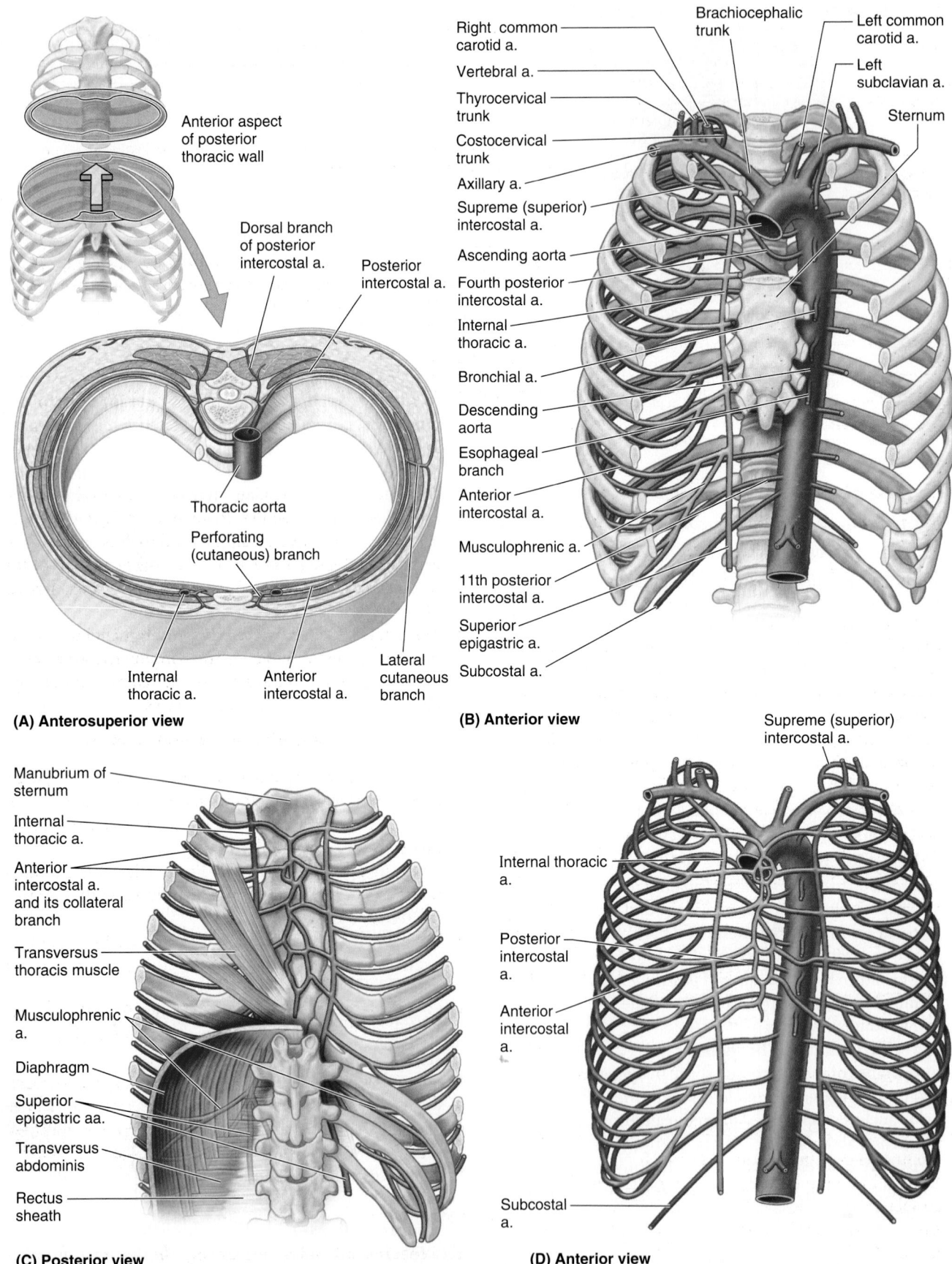

Anterior aspect of posterior thoracic wall

Dorsal branch of posterior intercostal a.

Posterior intercostal a.

Thoracic aorta

Perforating (cutaneous) branch

Internal thoracic a.

Anterior intercostal a.

Lateral cutaneous branch

(A) Anterosuperior view

Right common carotid a.

Vertebral a.

Thyrocervical trunk

Costocervical trunk

Axillary a.

Supreme (superior) intercostal a.

Ascending aorta

Fourth posterior intercostal a.

Internal thoracic a.

Bronchial a.

Descending aorta

Esophageal branch

Anterior intercostal a.

Musculophrenic a.

11th posterior intercostal a.

Superior epigastric a.

Subcostal a.

Brachiocephalic trunk

Left common carotid a.

Left subclavian a.

Sternum

(B) Anterior view

Manubrium of sternum

Internal thoracic a.

Anterior intercostal a. and its collateral branch

Transversus thoracis muscle

Musculophrenic a.

Diaphragm

Superior epigastric aa.

Transversus abdominis

Rectus sheath

(C) Posterior view

Supreme (superior) intercostal a.

Internal thoracic a.

Posterior intercostal a.

Anterior intercostal a.

Subcostal a.

(D) Anterior view

FIGURE 1.19. Arteries of thoracic wall. The arterial supply to the thoracic wall derives from the thoracic aorta through the posterior intercostal and subcostal arteries (**A, B,** and **D**), from the axillary artery (**B**), and from the subclavian artery through the internal thoracic (**C**) and supreme intercostal arteries (**B**). Connections (anastomoses) between the arteries permit collateral circulation pathways to develop (**D**).

TABLE 1.3. ARTERIAL SUPPLY OF THORACIC WALL

Artery	Origin	Course	Distribution
Posterior intercostals	Superior intercostal artery (intercostal spaces 1 and 2) and thoracic aorta (remaining intercostal spaces)	Pass between internal and innermost intercostal muscles	Intercostal muscles, overlying skin, and parietal pleura
Anterior intercostals	Internal thoracic (intercostal spaces 1–6) and musculophrenic arteries (intercostal spaces 7–9)		
Internal thoracic	Subclavian artery	Passes inferiorly and lateral to sternum between costal cartilages and transversus thoracic muscle to divide into superior epigastric and musculophrenic arteries	By way of anterior intercostal arteries to intercostal spaces 1–6 and musculophrenic artery (lateral terminal branch)
Subcostal	Thoracic aorta	Courses along inferior border of 12th rib	Muscles of anterolateral abdominal wall

The **posterior intercostal arteries:**

- Of the 1st and 2nd intercostal spaces arise from the **supreme (superior) intercostal artery,** a branch of the costocervical trunk of the subclavian artery.
- Of the 3rd–11th intercostal spaces (and the subcostal arteries of the subcostal space) arise posteriorly from the thoracic aorta (Fig. 1.19). Because the aorta is slightly to the left of the vertebral column, the right 3rd–11th intercostal arteries cross the vertebral bodies, running a longer course than those on the left side (Fig. 1.19B).
- All give off a posterior branch that accompanies the posterior ramus of the spinal nerve to supply the spinal cord, vertebral column, back muscles, and skin.
- Give rise to a small collateral branch that crosses the intercostal space and runs along the superior border of the rib.
- Accompany the intercostal nerves through the intercostal spaces. Close to the angle of the rib, the arteries enter the costal grooves, where they lie between the intercostal vein and nerve. At first the arteries run in the endothoracic fascia between the parietal pleura and the internal intercostal membrane (Fig. 1.17); then they run between the innermost intercostal and internal intercostal muscles.
- Have terminal and collateral branches that anastomose anteriorly with anterior intercostal arteries (Fig. 1.19A).

The **internal thoracic arteries** (historically, the internal mammary arteries):

- Arise in the root of the neck from the inferior surfaces of the first parts of the *subclavian arteries.*
- Descend into the thorax posterior to the clavicle and 1st costal cartilage (Figs. 1.13, 1.14, and 1.19).
- Are crossed near their origins by the ipsilateral phrenic nerve. (crosses underneath the artery)
- Descend on the internal surface of the thorax slightly lateral to the sternum and posterior to the upper six costal

cartilages and intervening internal intercostal muscles. After descending past the 2nd costal cartilage, the internal thoracic arteries run anterior to the transversus thoracis muscle (Figs. 1.15A and 1.19C). Between slips of the transversus thoracis muscle, the arteries contact parietal pleura posteriorly.
- Terminate in the 6th intercostal space by dividing into the *superior epigastric* and the **musculophrenic arteries.**
- Directly give rise to the anterior intercostal arteries supplying the superior 6 intercostal spaces.

Ipsilateral pairs of **anterior intercostal arteries:**

- Supply the anterior parts of the upper 9 intercostal spaces.
- Pass laterally in the intercostal space, one near the inferior margin of the superior rib and the other near the superior margin of the inferior rib.
- Of the first 2 intercostal spaces lie initially in the endothoracic fascia between the parietal pleura and the internal intercostal muscles.
- Supplying the 3rd–6th intercostal spaces are separated from the pleura by slips of the transversus thoracis muscle.
- Of the 7th–9th intercostal spaces derive from the *musculophrenic arteries,* also branches of the internal thoracic arteries.
- Supply the intercostal muscles and send branches through them to supply the pectoral muscles, breasts, and skin.
- Are absent from the inferior two intercostal spaces; these spaces are supplied only by the posterior intercostal arteries and their collateral branches.

VEINS OF THORACIC WALL

The **intercostal veins** accompany the intercostal arteries and nerves and lie most superior in the costal grooves (Figs. 1.15B and 1.20). There are 11 **posterior intercostal**

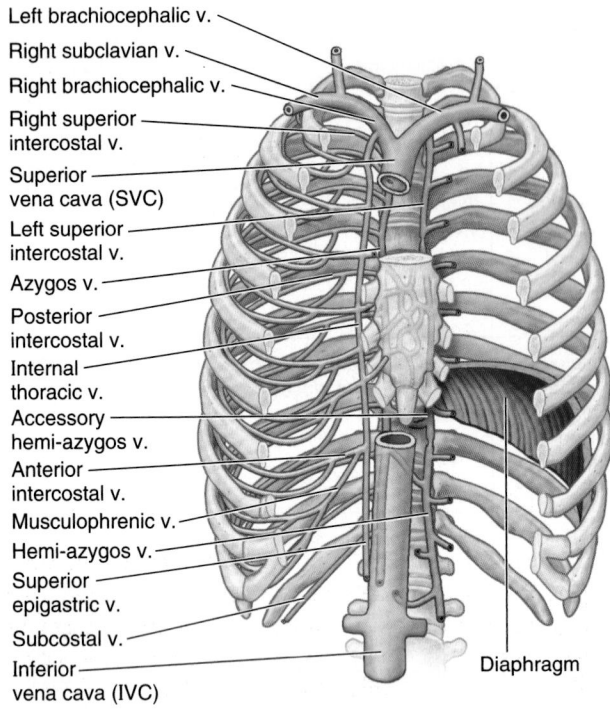

Left brachiocephalic v.
Right subclavian v.
Right brachiocephalic v.
Right superior intercostal v.
Superior vena cava (SVC)
Left superior intercostal v.
Azygos v.
Posterior intercostal v.
Internal thoracic v.
Accessory hemi-azygos v.
Anterior intercostal v.
Musculophrenic v.
Hemi-azygos v.
Superior epigastric v.
Subcostal v.
Inferior vena cava (IVC)
Diaphragm

Anterior view

FIGURE 1.20. Veins of thoracic wall. Although depicted here as continuous channels, the anterior and posterior intercostal veins are separate vessels, normally draining in opposite directions, the tributaries of which communicate (anastomose) in approximately the anterior axillary line. Because these veins lack valves, however, flow can be reversed.

veins and one **subcostal vein** on each side. The posterior intercostal veins anastomose with the **anterior intercostal veins** (tributaries of internal thoracic veins). As they approach the vertebral column, the posterior intercostal veins receive a *posterior branch*, which accompanies the posterior ramus of the spinal nerve of that level, and an *intervertebral vein* draining the vertebral venous plexuses associated with the vertebral column. Most posterior intercostal veins (4–11) end in the *azygos/hemi-azygos venous system*, which conveys venous blood to the superior vena cava (SVC). The posterior intercostal veins of the 1st intercostal space usually enter directly into the right and left brachiocephalic veins. The posterior intercostal veins of the 2nd and 3rd (and occasionally 4th) intercostal spaces unite to form a trunk, the *superior intercostal vein* (Fig. 1.20).

The **right superior intercostal vein** is typically the final tributary of the *azygos vein*, before it enters the SVC. The **left superior intercostal vein,** however, usually empties into the *left brachiocephalic vein*. This requires the vein to pass anteriorly along the left side of the superior mediastinum, specifically across the arch of the aorta or the root of the great vessels arising from it, and between the vagus and phrenic nerves (see Fig. 1.70B). It usually receives the left bronchial veins and may receive the left pericardiacophrenic vein as well. Typically, it communicates inferiorly with the *accessory hemi-azygos vein*. The **internal thoracic veins** are the companion veins (L. *venae comitantes*) of the internal thoracic arteries.

MUSCLES AND NEUROVASCULATURE OF THORACIC WALL

Dyspnea: Difficult Breathing

When people with respiratory problems (e.g., *asthma*) or with *heart failure* have difficulty breathing (*dypsnia*), they use their accessory respiratory muscles to assist the expansion of their thoracic cavity. They lean on their knees or on the arms of a chair to fix their pectoral girdle, so these muscles are able to act on their rib attachments and expand the thorax.

Extrapleural Intrathoracic Surgical Access

Fixation makes it difficult to appreciate in the embalmed cadaver, but in surgery the relatively loose nature of the thin endothoracic fascia provides a natural cleavage plane, allowing the surgeon to separate the costal parietal pleura lining the lung cavity from the thoracic wall. This allows *intrathoracic access to extrapleural structures* (e.g., lymph nodes) and instrument placement without opening and perhaps contaminating the potential space (pleural cavity) that surrounds the lungs.

Herpes Zoster Infection of Spinal Ganglia

A *herpes zoster infection* causes a classic, dermatomally distributed skin lesion—*shingles*—an agonizingly painful condition (Fig. B1.3). Herpes zoster is primarily a viral disease of spinal ganglia, usually a reactivation of the varicella-zoster virus (VZV), or

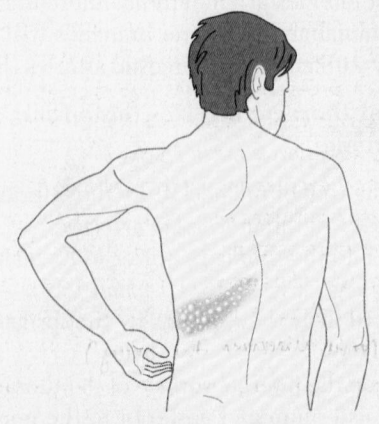

FIGURE B1.3. Herpes zoster.

chickenpox virus. After invading a ganglion, the virus produces a sharp burning pain in the dermatome supplied by the involved nerve (Fig. 1.18). The affected skin area becomes red, and vesicular eruptions appear. The pain may precede or follow the skin eruptions. Although primarily a *sensory neuropathy* (pathological change in the nerve), weakness from motor involvement occurs in 0.5–5.0% of people, commonly in elderly cancer patients (Rowland, 2010). Muscular weakness usually occurs in the same myotomal distribution, as do the dermatomal pain and vesicular eruptions.

Intercostal Nerve Block

 Local anesthesia of an intercostal space is produced by injecting an anesthetic agent around the intercostal nerves between the paravertebral line and the area of required anesthesia. This procedure, an *intercostal nerve block*, involves infiltration of the anesthetic around the intercostal nerve trunk and its collateral branches (Fig. B1.4). The term *block* indicates that the nerve endings in the skin and transmission of impulses through the sensory nerves carrying information about pain are interrupted (blocked)

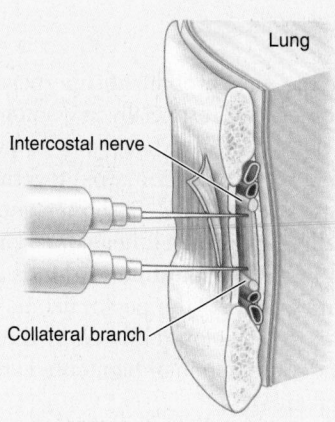

FIGURE B1.4. Intercostal nerve block.

before the impulses reach the spinal cord and brain. Because any particular area of skin usually receives innervation from two adjacent nerves, considerable overlapping of contiguous dermatomes occurs. Therefore, complete loss of sensation usually does not occur unless two or more intercostal nerves are anesthetized.

The Bottom Line

MUSCLES AND NEUROVASCULATURE OF THORACIC WALL

Muscles of thoracic wall: The thorax is overlapped by the axio-appendicular muscles of the upper limb as well as some neck, back, and abdominal muscles. ♦ Most of these muscles can affect deep respiration when the pectoral girdle is fixed and account for many of the surface features of the thoracic region. The muscles that are truly thoracic, however, provide few if any surface features. ♦ The serratus posterior muscles are thin with small bellies that may be proprioceptive organs. ♦ The costal muscles can move the ribs during forced respiration. The costal muscles function primarily to support (provide tonus for) the intercostal spaces, resisting negative and positive intrathoracic pressures. ♦ The diaphragm is the primary muscle of respiration, responsible for most of inspiration (normally expiration is mostly passive). ♦ Deep fascia overlies and invests the muscles of the thoracic wall, as it does elsewhere. ♦ Where the fleshy portions of the intercostal muscles are absent, their fascia is continued as intercostal membranes so that the wall is complete. ♦ The endothoracic fascia is a thin, fibro-areolar layer between the internal aspect of the thoracic cage and the lining of the pulmonary cavities, which can be opened surgically to gain access to intrathoracic structures.

Neurovasculature of thoracic wall: The pattern of distribution of neurovascular structures to the thoracic wall reflects the construction of the thoracic cage. ♦ These neurovascular structures course in the intercostal spaces, parallel to the ribs, and serve the intercostal muscles as well as the integument and parietal pleura on their superficial and deep aspects. ♦ Because plexus formation does not occur in relationship to the thoracic wall, the pattern of peripheral and segmental (dermatomal) innervation is identical in this region. ♦ The intercostal nerves run a posterior to anterior course along the length of each intercostal space, and the anterior and posterior intercostal arteries and veins converge toward and anastomose in approximately the anterior axillary line. ♦ The posterior vessels arise from the thoracic aorta and drain to the azygos venous system. ♦ The anterior vessels arise from the internal thoracic artery, branches, and tributaries and drain to the internal thoracic vein, branches, and tributaries.

Breasts

The breasts are the most prominent superficial structures in the anterior thoracic wall, especially in women. The **breasts** (L. *mammae*) consist of glandular and supporting fibrous tissue embedded within a fatty matrix, together with blood vessels, lymphatics, and nerves. Both men and women have breasts; normally they are well developed only in women (Figs. 1.21 and 1.22). The **mammary glands** are in the subcutaneous tissue overlying the pectoralis major and minor muscles. At the greatest prominence of the breast is the *nipple,* surrounded by a circular pigmented area of skin, the **areola** (L. small area).

The mammary glands within the breasts are accessory to reproduction in women. They are rudimentary and functionless in men, consisting of only a few small ducts or epithelial cords. Usually, the fat present in male breasts is not different from that of subcutaneous tissue elsewhere, and the glandular system does not normally develop.

FEMALE BREASTS

The amount of fat surrounding the glandular tissue determines the size of non-lactating breasts. The roughly circular body of the female breast rests on a **bed of the breast** that extends transversely from the lateral border of the sternum to the midaxillary line and vertically from the 2nd through 6th ribs. Two thirds of the bed are formed by the *pectoral fascia* overlying the pectoralis major; the other third, by the fascia covering the serratus anterior. Between the breast and the pectoral fascia is a loose subcutaneous tissue plane or

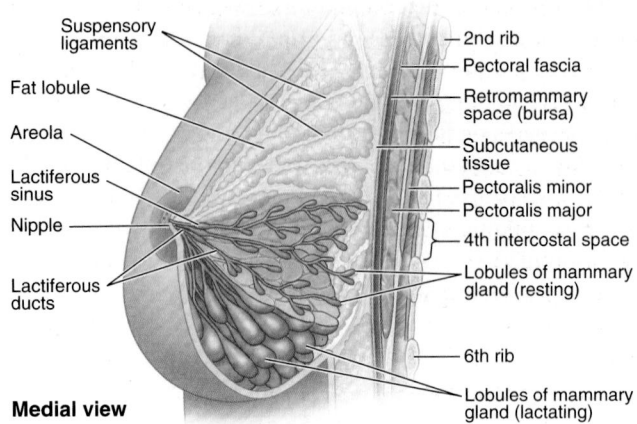

Medial view

FIGURE 1.22. Sagittal section of female breast and anterior thoracic wall. The superior two thirds of the figure demonstrates the suspensory ligaments and alveoli of the breast with resting lobules of mammary gland; the inferior part shows lactating lobules of mammary gland.

potential space—the **retromammary space** (bursa). This plane, containing a small amount of fat, allows the breast some degree of movement on the pectoral fascia. A small part of the mammary gland may extend along the inferolateral edge of the pectoralis major toward the axillary fossa (armpit), forming an **axillary process** or tail (of Spence). Some women discover this (especially when it may enlarge during a menstrual cycle) and become concerned that it may be a lump (tumor) or enlarged lymph nodes.

The mammary glands are firmly attached to the dermis of the overlying skin, especially by substantial skin ligaments (L. *retinacula cutis*), the **suspensory ligaments** (of Cooper). These condensations of fibrous connective tissue, particularly well developed in the superior part of the gland, help support the *lobes* and *lobules of the mammary gland.*

During puberty (ages 8–15 years), the breasts normally enlarge, owing in part to glandular development but primarily from increased fat deposition. The areolae and nipples also enlarge. Breast size and shape are determined in part by genetic, ethnic, and dietary factors. The **lactiferous ducts** give rise to buds that develop into 15–20 **lobules of the mammary gland,** which constitute the **parenchyma** (functional substance) of the mammary gland. Thus each lobule is drained by a lactiferous duct, all of which converge to open independently. Each duct has a dilated portion deep to the areola, the **lactiferous sinus,** in which a small droplet of milk accumulates or remains in the nursing mother. As the neonate begins to suckle, compression of the areola (and lactiferous sinus beneath it) expresses the accumulated droplets and encourages the neonate to continue nursing as the hormonally mediated *let-down reflex* ensues. The mother's milk is secreted into—not sucked from the gland by—the baby's mouth.

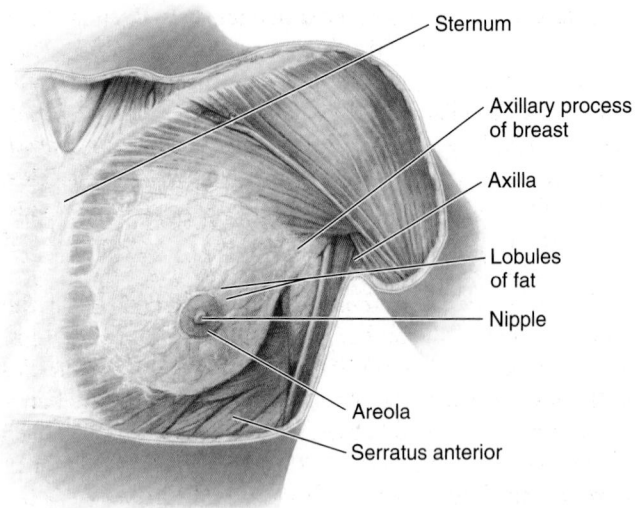

Anterior view

FIGURE 1.21. Superficial dissection of female pectoral region. The pectoral fascia has been removed, except where it lies deep to the breast. The bed of the breast extends from the 2nd through the 6th ribs. The axillary process of the breast extends toward or into the axillary fossa.

The areolae contain numerous **sebaceous glands,** which enlarge during pregnancy and secrete an oily substance that provides a protective lubricant for the areola and nipple. The areola and nipple are particularly subject to chaffing and irritation as mother and baby begin the nursing experience. The **nipples** are conical or cylindrical prominences in the centers of the areolae. The nipples have no fat, hair, or sweat glands. The tips of the nipples are fissured with the lactiferous ducts opening into them. The nipples are composed mostly of circularly arranged smooth muscle fibers that compress the lactiferous ducts during lactation and erect the nipples in response to stimulation, as when a baby begins to suckle.

The mammary glands are modified sweat glands; therefore, they have no capsule or sheath. The rounded contour and most of the volume of the breasts are produced by subcutaneous fat, except during pregnancy when the mammary glands enlarge and new glandular tissue forms. The milk-secreting **alveoli** (L. small hollow spaces) are arranged in grape-like clusters. In most women, the breasts enlarge slightly during the menstrual period from increased release of gonadotropic hormones—follicle-stimulating hormone (FSH) and luteinizing hormone (LH)—on the glandular tissue.

VASCULATURE OF BREAST

The *arterial supply of the breast* (Fig. 1.23A & B) derives from the:

- **Medial mammary branches of perforating branches** and *anterior intercostal branches of the internal thoracic artery,* originating from the subclavian artery.
- *Lateral thoracic* and *thoraco-acromial arteries,* branches of the axillary artery.
- *Posterior intercostal arteries,* branches of the thoracic aorta in the 2nd, 3rd, and 4th intercostal spaces.

The *venous drainage of the breast* is mainly to the *axillary vein,* but there is some drainage to the internal thoracic vein (Fig. 1.23C).

The *lymphatic drainage of the breast* is important because of its role in the metastasis of cancer cells. Lymph passes from the nipple, areola, and lobules of the gland to the **subareolar lymphatic plexus** (Fig. 1.24A & B). From this plexus:

- Most lymph (>75%), especially from the lateral breast quadrants, drains to the *axillary lymph nodes,* initially to the *anterior* or *pectoral nodes* for the most part. However, some lymph may drain directly to other axillary nodes or even to interpectoral, deltopectoral, supraclavicular, or inferior deep cervical nodes. (The axillary lymph nodes are covered in detail in Chapter 6.)
- Most of the remaining lymph, particularly from the medial breast quadrants, drains to the *parasternal lymph nodes* or to the opposite breast, whereas lymph from the inferior quadrants may pass deeply to abdominal lymph nodes (subdiaphragmatic *inferior phrenic lymph nodes*).

Lymph from the skin of the breast, except the nipple and areola, drains into the ipsilateral axillary, inferior deep cervical, and infraclavicular lymph nodes and into the parasternal lymph nodes of both sides.

Lymph from the axillary nodes drains into *clavicular* (*infraclavicular* and *supraclavicular*) *lymph nodes* and from them into the *subclavian lymphatic trunk,* which also drains lymph from the upper limb. Lymph from the parasternal nodes enters the *bronchomediastinal lymphatic trunks,* which also drain lymph from the thoracic viscera. The termination of the lymphatic trunks varies; traditionally, these trunks are described as merging with each other and with the *jugular lymphatic trunk,* draining the head and neck to form a short *right lymphatic duct* on the right side or entering the termination at the *thoracic duct* on the left side. However, in many (perhaps most) cases, the trunks open independently into the junction of the internal jugular and subclavian veins, the *right or left venous angles,* that form the *right* and *left brachiocephalic veins* (Fig. 1.24C). In some cases, they open into both contributing veins immediately prior to the angle.

NERVES OF BREAST

The *nerves of the breast* derive from *anterior and lateral cutaneous branches of the 4th–6th intercostal nerves* (Fig. 1.15). The branches of the intercostal nerves pass through the pectoral fascia covering the pectoralis major to reach overlying subcutaneous tissue and skin of the breast. The branches of the intercostal nerves convey sensory fibers from the skin of the breast and sympathetic fibers to the blood vessels in the breasts and smooth muscle in the overlying skin and nipple.

Surface Anatomy of Thoracic Wall

The *clavicles* (collar bones) lie subcutaneously, forming bony ridges at the junction of the thorax and neck (Fig. 1.25). They can be palpated easily throughout their length, especially where their medial ends articulate with the manubrium of the sternum. The clavicles demarcate the superior division between zones of lymphatic drainage: above the clavicles, lymph flows ultimately to inferior jugular lymph nodes; below them, parietal lymph (that from the body wall and upper limbs) flows to the axillary lymph nodes.

The *sternum* (breastbone) lies subcutaneously in the anterior median line and is palpable throughout its length. Between the prominences of the medial ends of the clavicles at the sternoclavicular joints, the *jugular notch* in the manubrium can be palpated between the prominent medial ends of the clavicles. The notch lies at the level of the inferior border of the body of T2 vertebra and the space between the 1st and 2nd thoracic spinous processes.

The *manubrium,* approximately 4 cm long, lies at the level of the bodies of T3 and T4 vertebrae (Fig. 1.26). The *sternal angle* is palpable and often visible in young people because of the slight movement that occurs at the manubriosternal joint during forced respiration. The sternal angle lies at the level

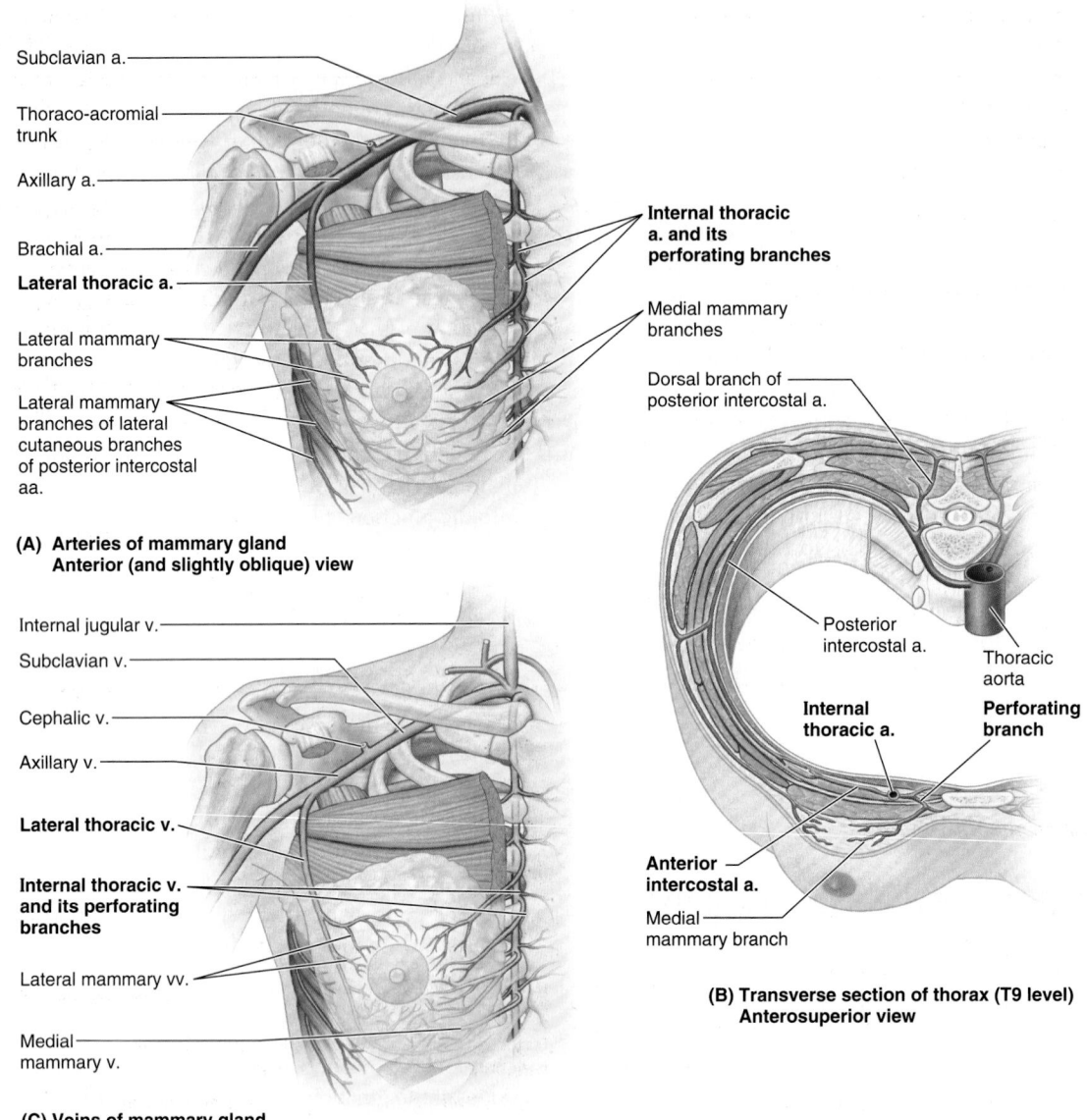

Subclavian a.

Thoraco-acromial trunk

Axillary a.

Brachial a.

Lateral thoracic a.

Lateral mammary branches

Lateral mammary branches of lateral cutaneous branches of posterior intercostal aa.

Internal thoracic a. and its perforating branches

Medial mammary branches

Dorsal branch of posterior intercostal a.

**(A) Arteries of mammary gland
Anterior (and slightly oblique) view**

Posterior intercostal a.

Thoracic aorta

Internal thoracic a.

Perforating branch

Internal jugular v.

Subclavian v.

Cephalic v.

Axillary v.

Lateral thoracic v.

Internal thoracic v. and its perforating branches

Lateral mammary vv.

Medial mammary v.

Anterior intercostal a.

Medial mammary branch

**(B) Transverse section of thorax (T9 level)
Anterosuperior view**

**(C) Veins of mammary gland
Anterior (and slightly oblique) view**

FIGURE 1.23. Vasculature of breast. A. The mammary gland is supplied from its medial aspect mainly by perforating branches of the internal thoracic artery and by several branches of the axillary artery (principally the lateral thoracic artery) superiorly and laterally. **B.** The breast is supplied deeply by branches arising from the intercostal arteries. **C.** Venous drainage is to the axillary vein (mainly) and the internal thoracic veins. [a., artery; v., vein; vv. veins]

of the T4–T5 IV disc and the space between the 3rd and 4th thoracic spinous processes. The sternal angle marks the level of the 2nd pair of costal cartilages. The left side of the manubrium is anterior to the arch of the aorta, and its right side directly overlies the merging of the brachiocephalic veins to form the *superior vena cava* (SVC) (Fig. 1.24C). Because it is common clinical practice to insert catheters into the SVC for intravenous feeding of extremely ill patients and for other purposes, it is essential to know the surface anatomy of this large vein. The SVC passes inferiorly deep to the manubrium and manubriosternal junction but projects as much as a fingerbreadth to the right of the margin of the manubrium. The SVC enters the right atrium of the heart opposite the right 3rd costal cartilage.

The *body of the sternum*, approximately 10 cm long, lies anterior to the right border of the heart and vertebrae T5–T9 (Fig. 1.26). The **intermammary cleft** (midline depression or cleavage between the mature female breasts) overlies the sternal body (Figs. 1.25 & 1.29). The *xiphoid process* lies in a slight depression, the **epigastric fossa.** This fossa is used as a guide in cardiopulmonary resuscitation (CPR) to properly position the hand on the inferior part of the sternum. The *xiphisternal joint* is palpable and is often seen as a ridge, at the level of the inferior border of T9 vertebra.

The *costal margins*, formed by the joined costal cartilages of the 7th–10th ribs, are easily palpable because they extend inferolaterally from the xiphisternal joint. The converging right and left costal margins form the *infrasternal angle*.

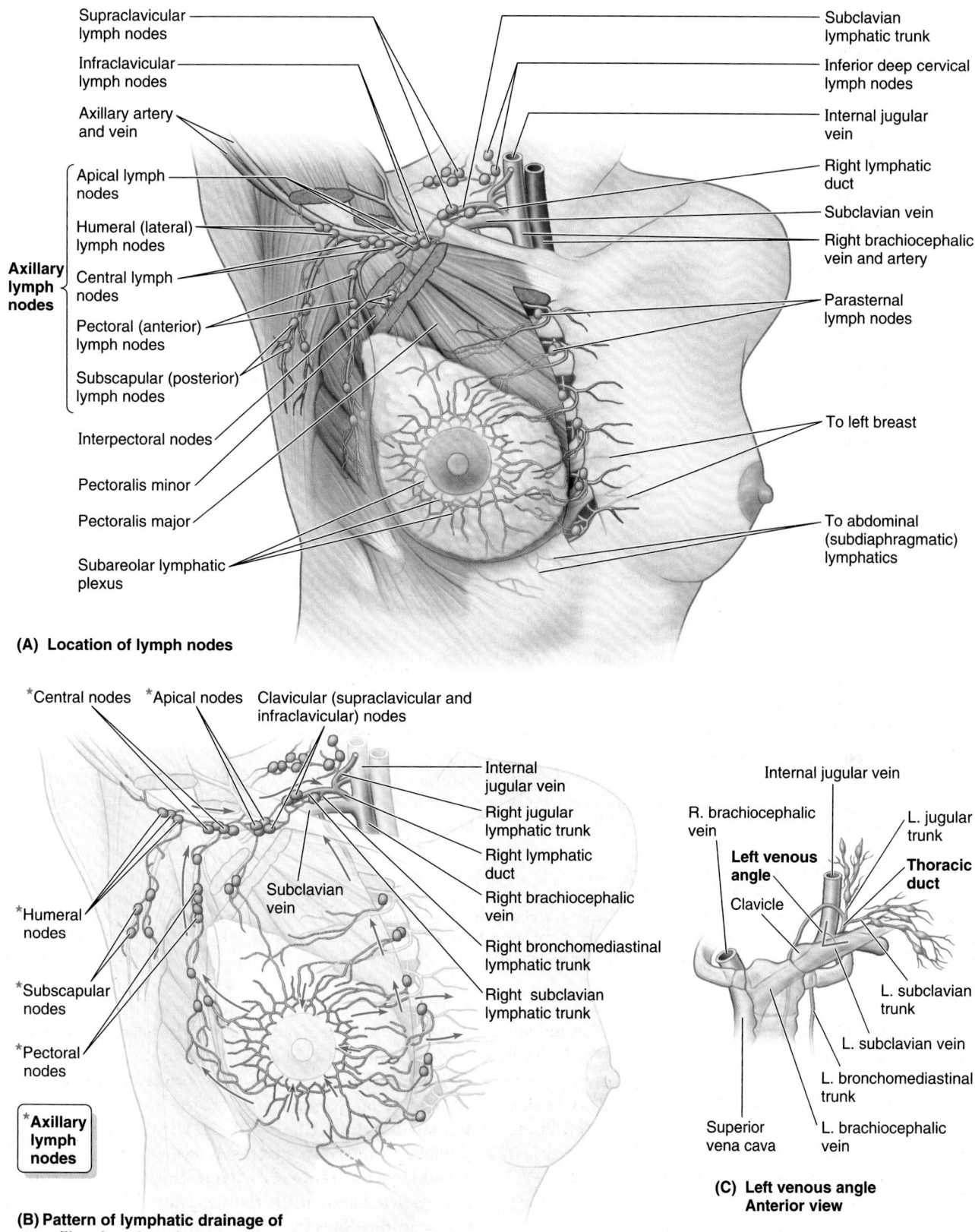

(A) **Location of lymph nodes**

Supraclavicular lymph nodes

Infraclavicular lymph nodes

Axillary artery and vein

Axillary lymph nodes
- Apical lymph nodes
- Humeral (lateral) lymph nodes
- Central lymph nodes
- Pectoral (anterior) lymph nodes
- Subscapular (posterior) lymph nodes

Interpectoral nodes

Pectoralis minor

Pectoralis major

Subareolar lymphatic plexus

Subclavian lymphatic trunk

Inferior deep cervical lymph nodes

Internal jugular vein

Right lymphatic duct

Subclavian vein

Right brachiocephalic vein and artery

Parasternal lymph nodes

To left breast

To abdominal (subdiaphragmatic) lymphatics

(B) **Pattern of lymphatic drainage of axillary lymph nodes**

*Central nodes   *Apical nodes   Clavicular (supraclavicular and infraclavicular) nodes

*Humeral nodes

*Subscapular nodes

*Pectoral nodes

*Axillary lymph nodes

Internal jugular vein

Right jugular lymphatic trunk

Right lymphatic duct

Right brachiocephalic vein

Right bronchomediastinal lymphatic trunk

Right subclavian lymphatic trunk

Subclavian vein

Anterior (and slightly oblique) view

(C) **Left venous angle Anterior view**

Internal jugular vein

R. brachiocephalic vein

Left venous angle

Clavicle

L. jugular trunk

Thoracic duct

L. subclavian trunk

L. subclavian vein

L. bronchomediastinal trunk

L. brachiocephalic vein

Superior vena cava

FIGURE 1.24. Lymphatic drainage of breast. A. The lymph nodes receiving drainage from the breast. **B.** The *red arrows* indicate lymph flow from the right breast. Most lymph, especially that from the superior lateral quadrant and center of the breast, drains to the axillary lymph nodes, which, in turn, are drained by the subclavian lymphatic trunk. On the right side, it enters the venous system via the right lymphatic duct. **C.** Most lymph from the left breast returns to the venous system via the thoracic duct.

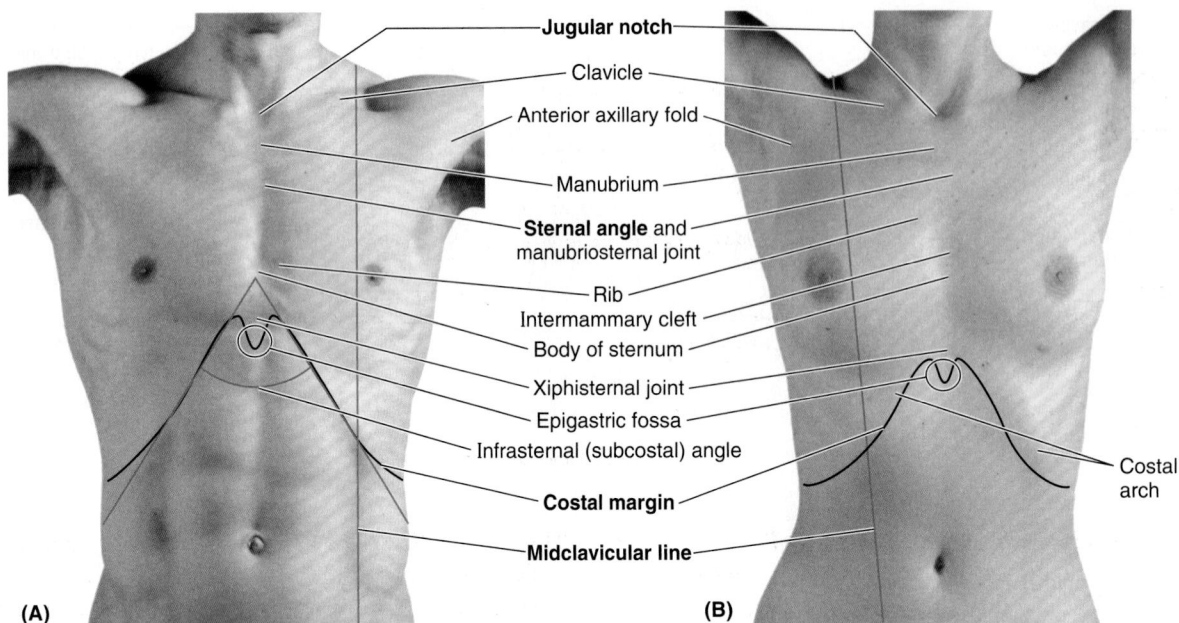

FIGURE 1.25. Surface features of anterior thoracic wall.

The *ribs* and *intercostal spaces* provide a basis for locating or describing the position of structures or sites of trauma or pathology on or deep to the thoracic wall. Because the 1st rib is not palpable, rib counting in physical examinations starts with the 2nd rib adjacent to the subcutaneous and easily palpated sternal angle. To count the ribs and intercostal spaces anteriorly, slide the fingers (digits) laterally from the sternal angle onto the 2nd costal cartilage and begin counting the ribs and spaces by moving the fingers from here. The 1st intercostal space is that superior to the 2nd costal cartilage—that is, intercostal spaces are numbered according to the rib forming their superior boundary. Generally, it is more reliable to count intercostal spaces, since the fingertip tends to rest in (slip into) the gaps between the ribs. One finger should remain in place while another is used to locate the next space. Using all the fingers, it is possible to locate four spaces at a time. The spaces are widest anterolaterally (approximately in the midclavicular line). If the fingers are removed from the thoracic wall while counting spaces, the finger may easily be returned to the same space, mistaking it for the one below. Posteriorly, the medial end of the spine of the scapula overlies the 4th rib.

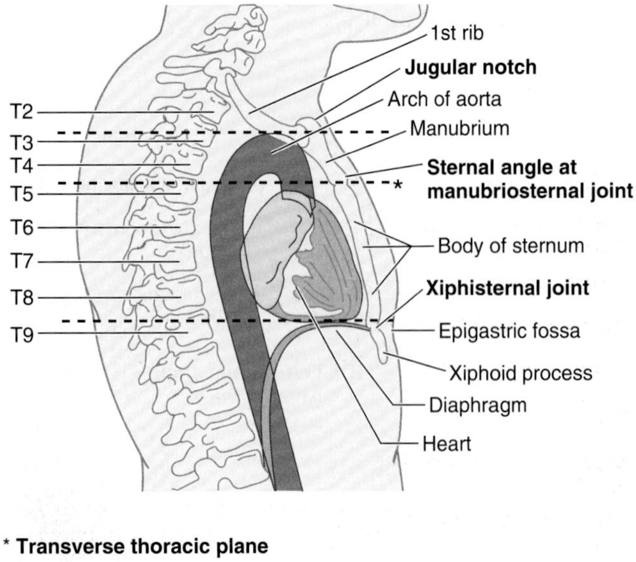

While the ribs and/or intercostal spaces provide the "latitude" for navigation and localization on the thoracic wall, several imaginary lines facilitate anatomical and clinical descriptions by providing "longitude." The following lines are extrapolated over the thoracic wall based on visible or palpable superficial features:

- The **anterior median (midsternal) line** (AML) indicates the intersection of the median plane with the anterior thoracic wall (Fig. 1.27A).
- The **midclavicular line** (MCL) passes through the midpoint of the clavicle, parallel to the AML.
- The **anterior axillary line** (AAL) runs vertically along the anterior axillary fold that is formed by the inferolateral border of the pectoralis major as it spans from the thoracic cage to the humerus in the arm (Figs. 1.27B).
- The **midaxillary line** (MAL) runs from the apex (deepest part) of the axillary fossa (armpit), parallel to the AAL.
- The **posterior axillary line** (PAL), also parallel to the AAL, is drawn vertically along the posterior axillary fold formed by the latissimus dorsi and teres major muscles as they span from the back to the humerus.

* Transverse thoracic plane

FIGURE 1.26. Vertebral levels of sternum and transverse thoracic plane.

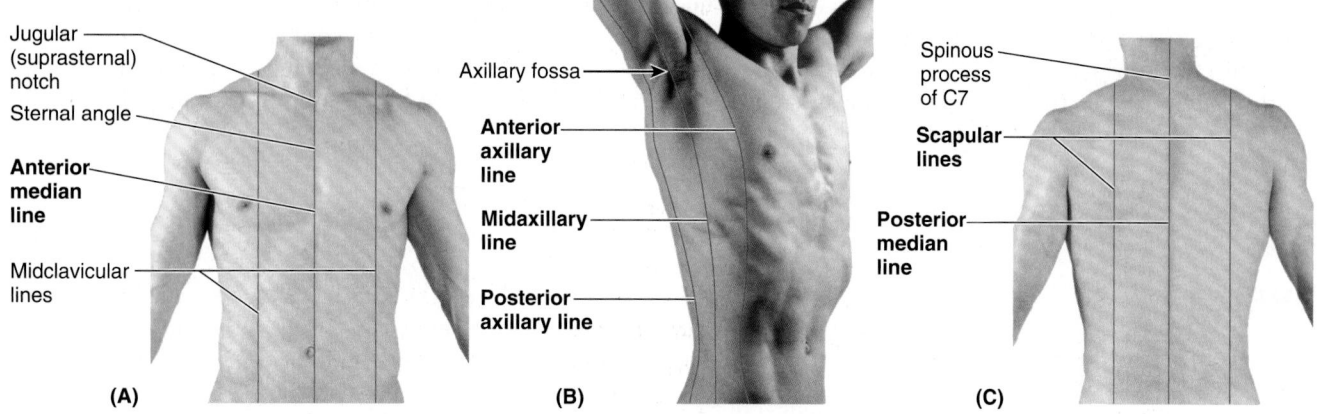

Jugular (suprasternal) notch
Sternal angle
Anterior median line
Midclavicular lines

Axillary fossa
Anterior axillary line
Midaxillary line
Posterior axillary line

Spinous process of C7
Scapular lines
Posterior median line

(A) **(B)** **(C)**

FIGURE 1.27. Vertical lines of thoracic wall.

- The **posterior median (midvertebral) line** (PML) is a vertical line along the tips of the spinous processes of the vertebrae (Fig. 1.27C).
- The **scapular lines** (SLs) are parallel to the posterior median line and intersect the inferior angles of the scapula.

Additional lines (not illustrated) are extrapolated along the borders of palpable bony formations such as the sternum and vertebral column, such as the parasternal and paravertebral lines (G. *para,* alongside of, adjacent to).

Breasts are the most prominent surface features of the anterior thoracic wall, especially in women. Except when there is an overabundance of subcutaneous tissue, the breasts in men are mostly an accentuation of the contour of the *pectoralis major muscles,* highlighted by the presence of the nipple in the 4th intercostal space, lateral to the MCL (Fig. 1.28). In moderately athletic individuals, the contour of the pectoralis major muscles is apparent, separated in the midline by the intermammary cleft overlying the sternum, with the lateral border forming the *anterior axillary fold* (Fig. 1.25). Inferolaterally, finger-like slips, or **digitations of the serratus anterior,** have a serrated (sawtooth) appearance as they attach to the ribs and interdigitate with the *external oblique* (Fig. 1.28).

The *inferior ribs* and *costal margins* are often apparent, especially when the abdominal muscles are contracted to "pull the belly in." The intercostal musculature is not normally evident; however, in (rare) cases in which there is an absence or atrophy of the intercostal musculature, the intercostal spaces become apparent with respiration: during inspiration, they are concave; during expiration, they protrude.

The female breasts vary in the size, shape, and symmetry—even in the same person. Their flattened superior surfaces show no sharp demarcation from the anterior surface of the thoracic wall, but laterally and inferiorly, their borders are well defined (Fig. 1.29). A venous pattern over the breasts is often visible, especially during pregnancy.

The *nipple* is surrounded by the slightly raised and circular pigmented *areola,* the color of which depends on the woman's complexion. The areola usually darkens during pregnancy and retains the darkened pigmentation thereafter. The areola is normally dotted with the papular (small elevated) openings of the *areolar glands* (sebaceous glands in the skin of the areola). On occasion, one or both nipples are inverted (retracted); this minor congenital anomaly may make breast feeding difficult.

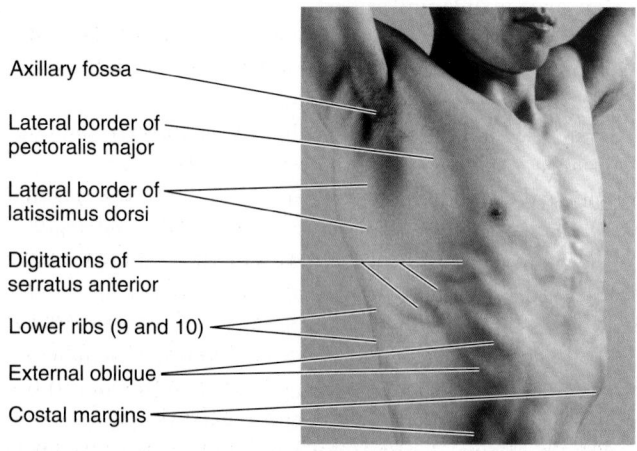

Axillary fossa
Lateral border of pectoralis major
Lateral border of latissimus dorsi
Digitations of serratus anterior
Lower ribs (9 and 10)
External oblique
Costal margins

FIGURE 1.28. Surface anatomy of thoracic wall musculature.

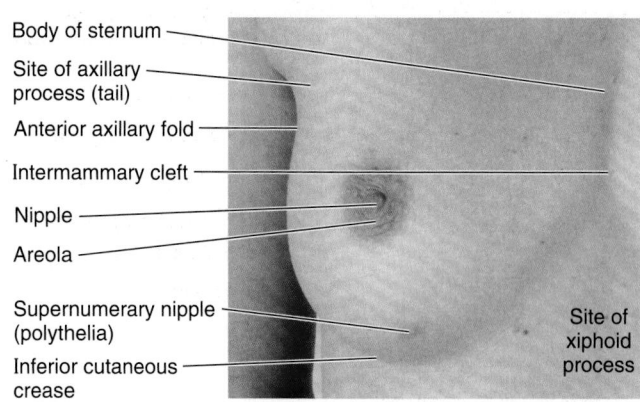

Body of sternum
Site of axillary process (tail)
Anterior axillary fold
Intermammary cleft
Nipple
Areola
Supernumerary nipple (polythelia)
Inferior cutaneous crease
Site of xiphoid process

FIGURE 1.29. Surface anatomy of female breast.

In men and young *nulliparous women*—those who have never borne a viable child—with moderate breast size, the nipple lies anterior to the 4th intercostal space, approximately 10 cm from the AML. Usually, however, the position of nipples varies considerably with breast size, especially in *multiparous women*—those who have given birth to two or more children. Consequently, because of variations in size and shape, the nipples are not a reliable guide to the 4th intercostal spaces in adult females.

BREASTS

Changes in Breasts

Changes in breast tissue, such as branching of the lactiferous ducts, occur during menstrual periods and pregnancy. Although mammary glands are prepared for secretion by midpregnancy, they do not produce milk until shortly after the baby is born. *Colostrum*, a creamy white to yellowish premilk fluid, may secrete from the nipples during the last trimester of pregnancy and during initial episodes of nursing. Colostrum is believed to be especially rich in protein, immune agents, and a growth factor affecting the infant's intestines. In multiparous women (those who have given birth two or more times), the breasts often become large and pendulous. The breasts in elderly women are usually small because of the decrease in fat and the atrophy of glandular tissue.

Breast Quadrants

For the anatomical location and description of tumors and cysts, the surface of the breast is divided into *four quadrants* (Fig. B1.5). For example, a physician's record might state: "A hard irregular mass was felt in the superior medial quadrant of the breast at the 2 o'clock position, approximately 2.5 cm from the margin of the areola."

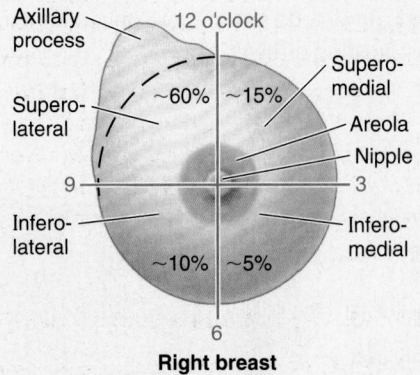

FIGURE B1.5. Breast quadrants.

Carcinoma of the Breast

Understanding the lymphatic drainage of the breasts is of practical importance in predicting the metastases (dispersal) of cancer cells from a *carci-noma of the breast* (breast cancer). Carcinomas of the breast are malignant tumors, usually *adenocarcinomas* (glandular cancer) arising from the epithelial cells of the lactiferous ducts in the mammary gland lobules (Fig. B1.6A). Metastatic cancer cells that enter a lymphatic vessel usually pass through two or three groups of lymph nodes before entering the venous system.

Interference with the lymphatic drainage by cancer may cause *lymphedema* (edema, excess fluid in the subcutaneous tissue), which in turn may result in deviation of the nipple and a thickened, leather-like appearance of the skin. Prominent "puffy" skin between dimpled pores give it an orange-peel appearance (*peau d'orange sign*). Larger dimples (fingertip size or bigger) result from cancerous invasion of the glandular tissue and fibrosis (fibrous degeneration), which causes shortening or places traction on the suspensory ligaments. *Subareolar breast cancer* may cause retraction of the nipple by a similar mechanism involving the lactiferous ducts.

Breast cancer typically spreads by means of lymphatic vessels (*lymphogenic metastasis*), which carry cancer cells from the breast to the lymph nodes, chiefly those in the axilla. The cells lodge in the nodes, producing nests of tumor cells (*metastases*). Abundant communications among lymphatic pathways and among axillary, cervical, and parasternal nodes may also cause metastases from the breast to develop in the supraclavicular lymph nodes, the opposite breast, or the abdomen (Fig. 1.24A & B). Because most of lymphatic drainage of the breast is to the *axillary lymph nodes*, they are the most common site of metastasis from a breast cancer. Enlargement of these palpable nodes suggests the possibility of breast cancer and may be key to early detection. However, the absence of enlarged axillary lymph nodes is no guarantee that metastasis from a breast cancer has not occurred; the malignant cells may have passed to other nodes, such as the infraclavicular and supraclavicular lymph nodes.

The posterior intercostal veins drain into the *azygos/hemi-azygos system of veins* alongside the bodies of the vertebrae (see Fig. 1.38B) and communicate with the internal vertebral venous plexus surrounding the spinal cord. Cancer cells can also spread from the breast by these venous routes to the vertebrae, and from there to the cranium and brain. Cancer also spreads by contiguity (invasion of adjacent tissue). When breast cancer cells invade the *retromammary space* (Fig. 1.22), attach to or invade the pectoral fascia overlying the pectoralis major, or metastasize to the interpectoral nodes, the breast elevates when the muscle contracts. This movement is a clinical sign of advanced cancer of the breast.

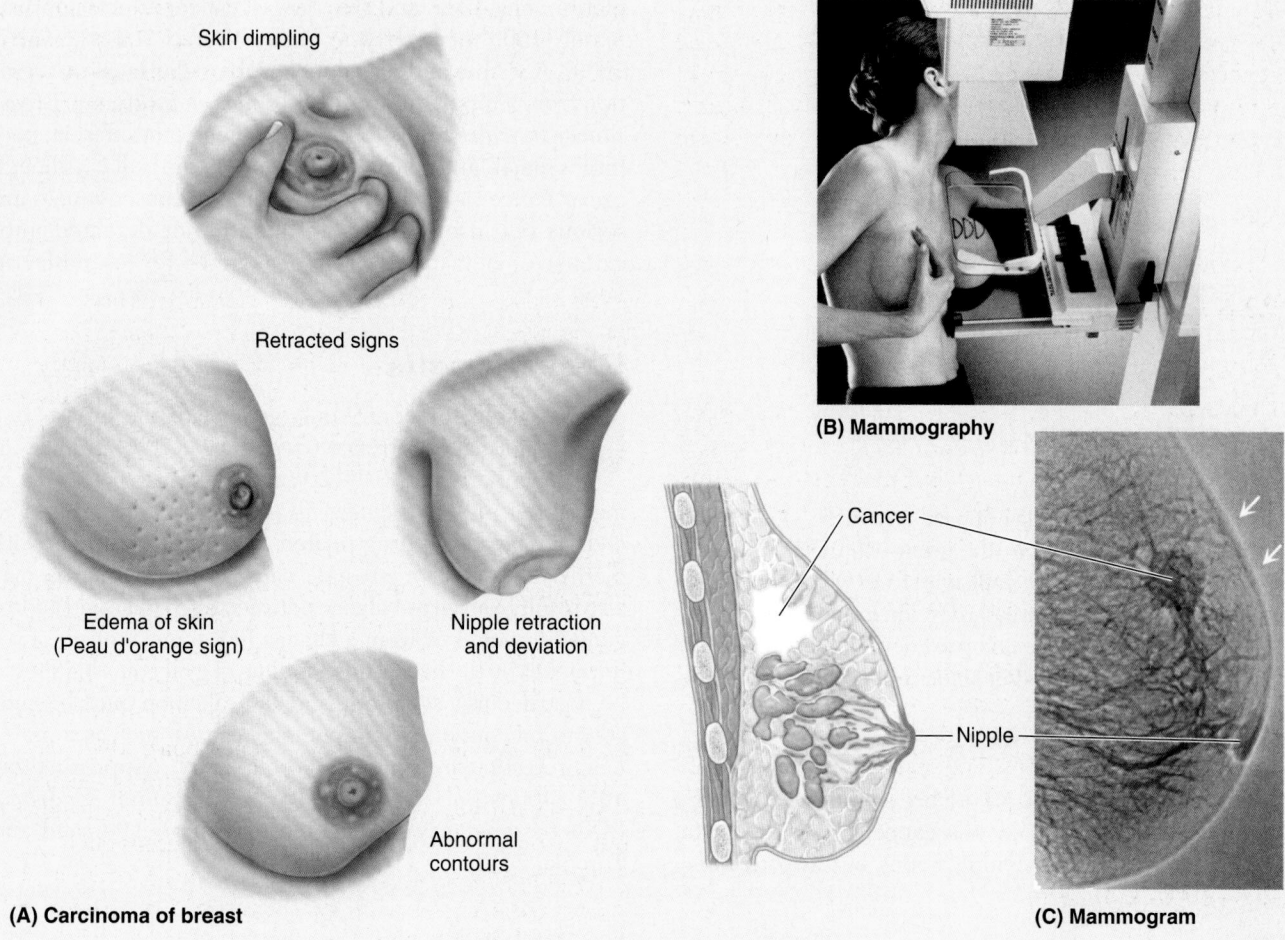

Skin dimpling

Retracted signs

Edema of skin
(Peau d'orange sign)

Nipple retraction
and deviation

Abnormal
contours

(A) Carcinoma of breast

(B) Mammography

Cancer

Nipple

(C) Mammogram

FIGURE B1.6. **Detection of breast cancer.**

To observe this upward movement, the physician has the patient place her hands on her hips and press while pulling her elbows forward to tense her pectoral muscles.

Mammography

Radiographic examination of the breasts, *mammography,* is one of the techniques used to detect breast masses (Fig. B1.6B). A carcinoma appears as a large, jagged density in the *mammogram.* The skin is thickened over the tumor (*upper two arrows* in Fig. B1.6C) and the nipple is depressed. Surgeons use mammography as a guide when removing breast tumors, cysts, and abscesses.

Surgical Incisions of Breast

Surgical incisions are made in the inferior breast quadrants when possible because these quadrants are less vascular than the superior ones. The transition between the thoracic wall and breast is most abrupt inferiorly, producing a line, crease, or deep skin fold—the *inferior cutaneous crease* (Fig. 1.29). Incisions made along this crease will be least evident and may be hidden by overlap of the breast. Incisions that must be made near the areola, or on the breast itself, are usually directed radially to either side of the nipple (Langer tension lines run transversely here) or circumferentially (Fig. I.7 in Introduction).

Mastectomy (breast excision) is not as common as it once was as a treatment for breast cancer. In *simple mastectomy,* the breast is removed down to the retromammary space. *Radical mastectomy,* a more extensive surgical procedure, involves removal of the breast, pectoral muscles, fat, fascia, and as many lymph nodes as possible in the axilla and pectoral region. In current practice, often only the tumor and surrounding tissues are removed—a *lumpectomy* or *quadrantectomy* (known as *breast-conserving surgery,* a wide local excision)—followed by radiation therapy (Goroll, 2009).

Polymastia, Polythelia, and Amastia

Polymastia (supernumerary breasts) or *polythelia* (accessory nipples) may occur superior or inferior to the normal pair, occasionally developing in the axillary fossa or anterior abdominal wall (Figs. 1.29 and B1.7). Supernumerary breasts usually consist of only a rudimentary nipple and areola, which may be mistaken for a mole (nevus) until they change pigmentation with the normal nipples during pregnancy. However, glandular tissue may also be present and further develop with lactation. These small

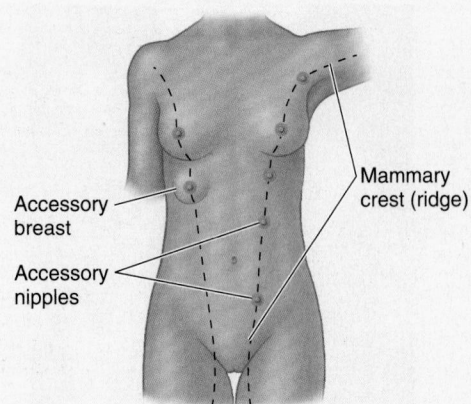

Accessory breast

Accessory nipples

Mammary crest (ridge)

FIGURE B1.7. Polymastia and polythelia.

supernumerary breasts may appear anywhere along a line extending from the axilla to the groin—the location of the *embryonic mammary crest* (milk line) from which the breasts develop, and along which breasts develop in animals with multiple breasts. There may be no breast development (*amastia*), or there may be a nipple and/or areola, but no glandular tissue.

Breast Cancer in Men

 Approximately 1.5% of breast cancers occur in men. As in women, the cancer usually metastasizes to axillary lymph nodes but also to bone, pleura, lung, liver, and skin. Breast cancer affects approximately 1000 men per year in the United States (Swartz, 2009). A visible and/or palpable subareolar mass or secretion from a nipple may indicate a *malignant tumor. Breast cancer in males* tends to infiltrate the pectoral fascia, pectoralis major, and apical lymph nodes in the axilla. Although breast cancer is uncommon in men, the consequences are serious because they are frequently not detected until extensive metastases have occurred—for example, in bone.

Gynecomastia

 Slight temporary enlargement of the breasts is a normal occurrence (frequency = 70%) in males at puberty (age 10–12 years). Breast hypertrophy in males after puberty (*gynecomastia*) is relatively rare (<1%) and may be age or drug related (e.g., after treatment with diethylstilbestrol for prostate cancer). Gynecomastia may also result from an imbalance between estrogenic and androgenic hormones or from a change in the metabolism of sex hormones by the liver. Thus a finding of gynecomastia should be regarded as a symptom, and an evaluation must be initiated to rule out important potential causes, such as suprarenal or testicular cancers (Goroll, 2009). Approximately 40% of postpubertal males with *Klinefelter syndrome* (XXY trisomy) have gynecomastia (Moore, Persaud and Torchia, 2012).

The Bottom Line

BREASTS AND SURFACE ANATOMY OF THORACIC WALL

Breasts: The mammary glands are in the subcutaneous tissue of the breast, overlying the pectoralis major and serratus anterior muscles and associated deep fascia (bed of the breast). ♦ Lobules of glandular tissue converge toward the nipple, each having its own lactiferous duct, which opens there. ♦ The superior lateral quadrant of the breast has the most glandular tissue, largely owing to an extension toward or into the axilla (axillary process) and, therefore, is the site of most tumors. ♦ The breast is served by the internal thoracic and lateral thoracic vessels and the 2nd–6th intercostal vessels and nerves. Most lymph from the breast drains to the axillary lymph nodes; this is significant when treating breast cancer.

♦ Because the mammary glands and axillary lymph nodes are superficial, the ability to palpate primary and metastatic tumors during routine breast examination enables early detection and treatment.

Surface anatomy of thoracic wall: The thoracic wall is especially well provided with visible and/or palpable features useful in examining the wall and underlying visceral features. ♦ Ribs and intercostal spaces, counted from the 2nd rib at the level of the sternal angle, provide latitude. ♦ Clavicles, nipples, axillary folds, scapulae, and the vertebral column provide longitude. ♦ The breasts are prominent features and, in males, the nipples demarcate the 4th intercostal space.

VISCERA OF THORACIC CAVITY

When sectioned transversely, it is apparent that the *thoracic cavity* is kidney shaped: a transversely ovoid space deeply indented posteriorly by the thoracic vertebral column and the heads and necks of the ribs that articulate with it (Fig. 1.30A). The thoracic cavity is divided into three compartments (Figs. 1.30A & C):

• Right and left **pulmonary cavities,** bilateral compartments that contain the lungs and *pleurae* (lining membranes) and occupy the majority of the thoracic cavity.

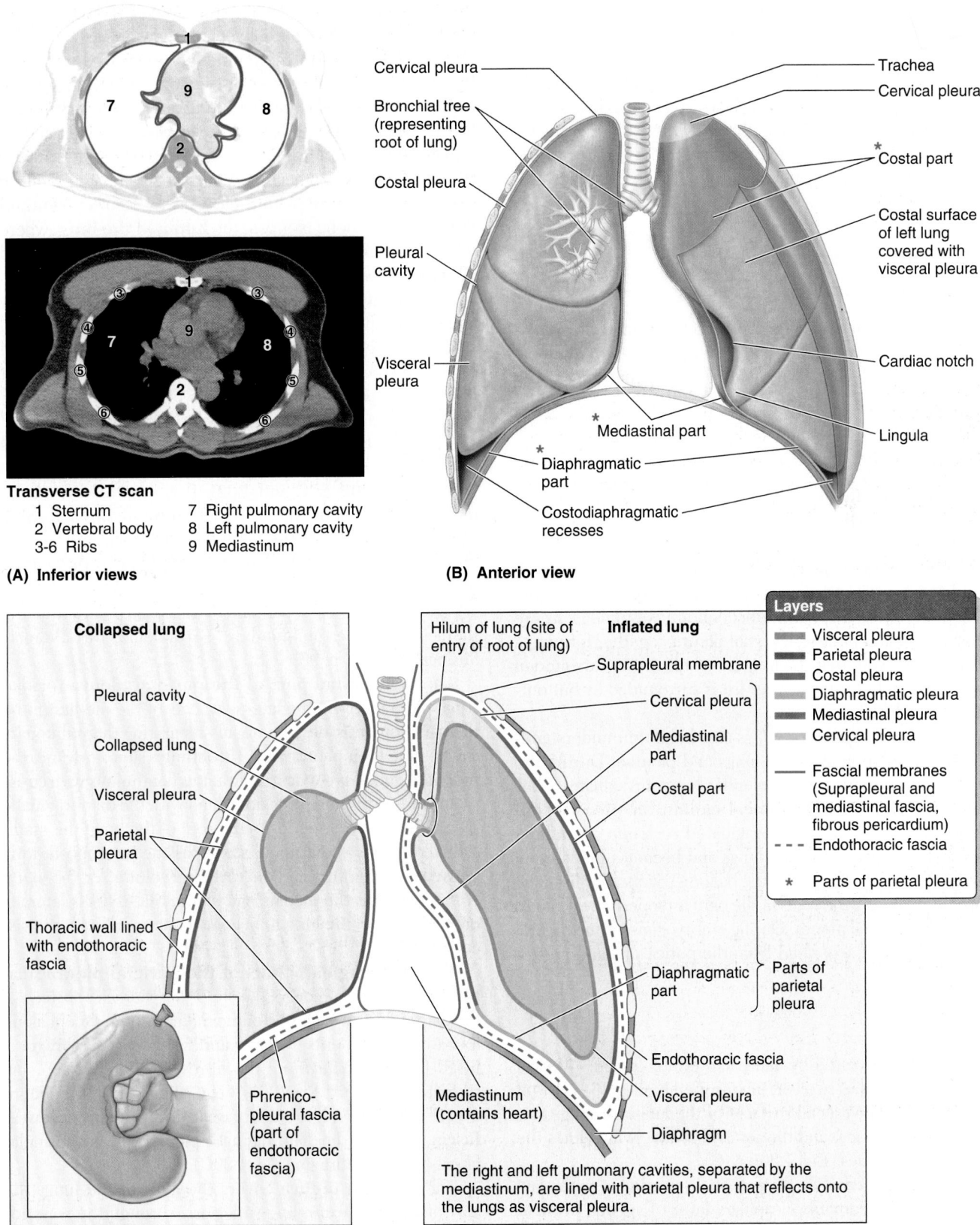

Transverse CT scan

1 Sternum	7 Right pulmonary cavity
2 Vertebral body	8 Left pulmonary cavity
3-6 Ribs	9 Mediastinum

(A) Inferior views

(B) Anterior view

(C) Anterior view

FIGURE 1.30. Divisions of thoracic cavity and lining of pulmonary cavities. A. The CT scan and interpretive diagram above it are transverse cross-sectional views of the thoracic cavity demonstrating its kidney-like shape, resulting from the protruding vertebral bodies, and division into three compartments. The dimensional (**B**) and coronal cross-sectional (**C**) diagrams demonstrate the linings of the pleural cavities and lungs (pleurae). Each lung is invested by the inner layer of a closed sac that has been invaginated by the lung. **Inset:** A fist invaginating an underinflated balloon demonstrates the relationship of the lung (represented by the fist) to walls of the pleural sac (parietal and visceral layers of pleura).

- A central **mediastinum,** a compartment intervening between and completely separating the two pulmonary cavities, which contains essentially all other thoracic structures—the heart, thoracic parts of the great vessels, thoracic part of the trachea, esophagus, thymus, and other structures (e.g., lymph nodes). It extends vertically from the superior thoracic aperture to the diaphragm and anteroposteriorly from the sternum to the thoracic vertebral bodies.

Pleurae, Lungs, and Tracheobronchial Tree

Each *pulmonary cavity* (right and left) is lined by a *pleural membrane* (**pleura**) that also reflects onto and covers the external surface of the lungs occupying the cavities (Fig. 1.30B & C). To visualize the relationship of the pleurae and lungs, push your fist into an underinflated balloon (Fig. 1.30C). The inner part of the balloon wall (adjacent to your fist, which represents the lung), is comparable to the *visceral pleura;* the remaining outer wall of the balloon represents the *parietal pleura.* The cavity between the layers of the balloon, here filled with air, is analogous to the pleural cavity, although the pleural cavity contains only a thin film of fluid. At your wrist (representing the root of the lung), the inner and outer walls of the balloon are continuous, as are the visceral and parietal layers of pleura, together forming a *pleural sac.* Note that the lung is outside of but surrounded by the pleural sac, just as your fist is surrounded by but outside of the balloon.

The inset in Figure 1.30C is also helpful in understanding the development of the lungs and pleurae. During the embryonic period, the developing lungs invaginate (grow into) the **pericardioperitoneal canals,** the precursors of the pleural cavities. The invaginated *coelomic epithelium* covers the primordia of the lungs and becomes the visceral pleura in the same way that the balloon covers your fist. The epithelium lining the walls of the pericardioperitoneal canals forms the parietal pleura. During embryogenesis, the pleural cavities become separated from the pericardial and peritoneal cavities.

PLEURAE

Each lung is invested by and enclosed in a serous **pleural sac** that consists of two continuous membranes: the *visceral pleura,* which invests all surfaces of the lungs forming their shiny outer surface, and the *parietal pleura,* which lines the pulmonary cavities (Fig. 1.30B & C).

The **pleural cavity**—the potential space between the layers of pleura—contains a capillary layer of **serous pleural fluid,** which lubricates the pleural surfaces and allows the layers of pleura to slide smoothly over each other during respiration. The surface tension of the pleural fluid provides the cohesion that keeps the lung surface in contact with the thoracic wall; consequently, the lung expands and fills with air

when the thorax expands while still allowing sliding to occur, much like a film of water between two glass plates.

The **visceral pleura** (pulmonary pleura) closely covers the lung and adheres to all its surfaces, including those within the horizontal and oblique fissures (Figs. 1.30B & C and 1.31A). In cadaver dissection, the visceral pleura cannot usually be dissected from the surface of the lung. It provides the lung with a smooth slippery surface, enabling it to move freely on the parietal pleura. The visceral pleura is continuous with the parietal pleura at the **hilum of the lung,** where structures making up the *root of the lung* (e.g., bronchus and pulmonary vessels) enter and leave the lung (Fig. 1.30C).

The **parietal pleura** lines the pulmonary cavities, thereby adhering to the thoracic wall, mediastinum, and diaphragm. It is thicker than the visceral pleura, and during surgery and cadaver dissections, it may be separated from the surfaces it covers. The parietal pleura consists of three *parts—costal, mediastinal,* and *diaphragmatic*—and the *cervical pleura.*

The **costal part of the parietal pleura** (costovertebral or costal pleura) covers the internal surfaces of the thoracic wall (Figs. 1.30B & C and 1.32). It is separated from the internal surface of the thoracic wall (sternum, ribs and costal cartilages, intercostal muscles and membranes, and sides of thoracic vertebrae) by *endothoracic fascia.* This thin, extrapleural layer of loose connective tissue forms a natural cleavage plane for surgical separation of the costal pleura from the thoracic wall (see the blue box "Extrapleural Intrathoracic Surgical Access," p. 96).

The **mediastinal part of the parietal pleura** (mediastinal pleura) covers the lateral aspects of the mediastinum, the partition of tissues and organs separating the pulmonary cavities and their pleural sacs. It continues superiorly into the root of the neck as cervical pleura. It is continuous with costal pleura anteriorly and posteriorly, and with the diaphragmatic pleura inferiorly. Superior to the root of the lung, the mediastinal pleura is a continuous sheet passing anteroposteriorly between the sternum and the vertebral column. At the *hilum of the lung,* it is the mediastinal pleura that reflects laterally onto the root of the lung to become continuous with the visceral pleura.

The **diaphragmatic part of the parietal pleura** (diaphragmatic pleura) covers the superior (thoracic) surface of the diaphragm on each side of the mediastinum, except along its costal attachments (origins) and where the diaphragm is fused to the *pericardium;* the fibroserous membrane surrounding the heart (Figs. 1.30B & C and 1.32). A thin, more elastic layer of endothoracic fascia, the **phrenicopleural fascia,** connects the diaphragmatic pleura with the muscular fibers of the diaphragm (Fig. 1.30C).

The **cervical pleura** covers the apex of the lung (the part of the lung extending superiorly through the superior thoracic aperture into the root of the neck—Figs. 1.30B & C and 1.31A). It is a superior continuation of the costal and mediastinal parts of the parietal pleura. The cervical pleura forms a cup-like dome (**pleural cupula**) over the apex that reaches its summit 2–3 cm superior to the level of the

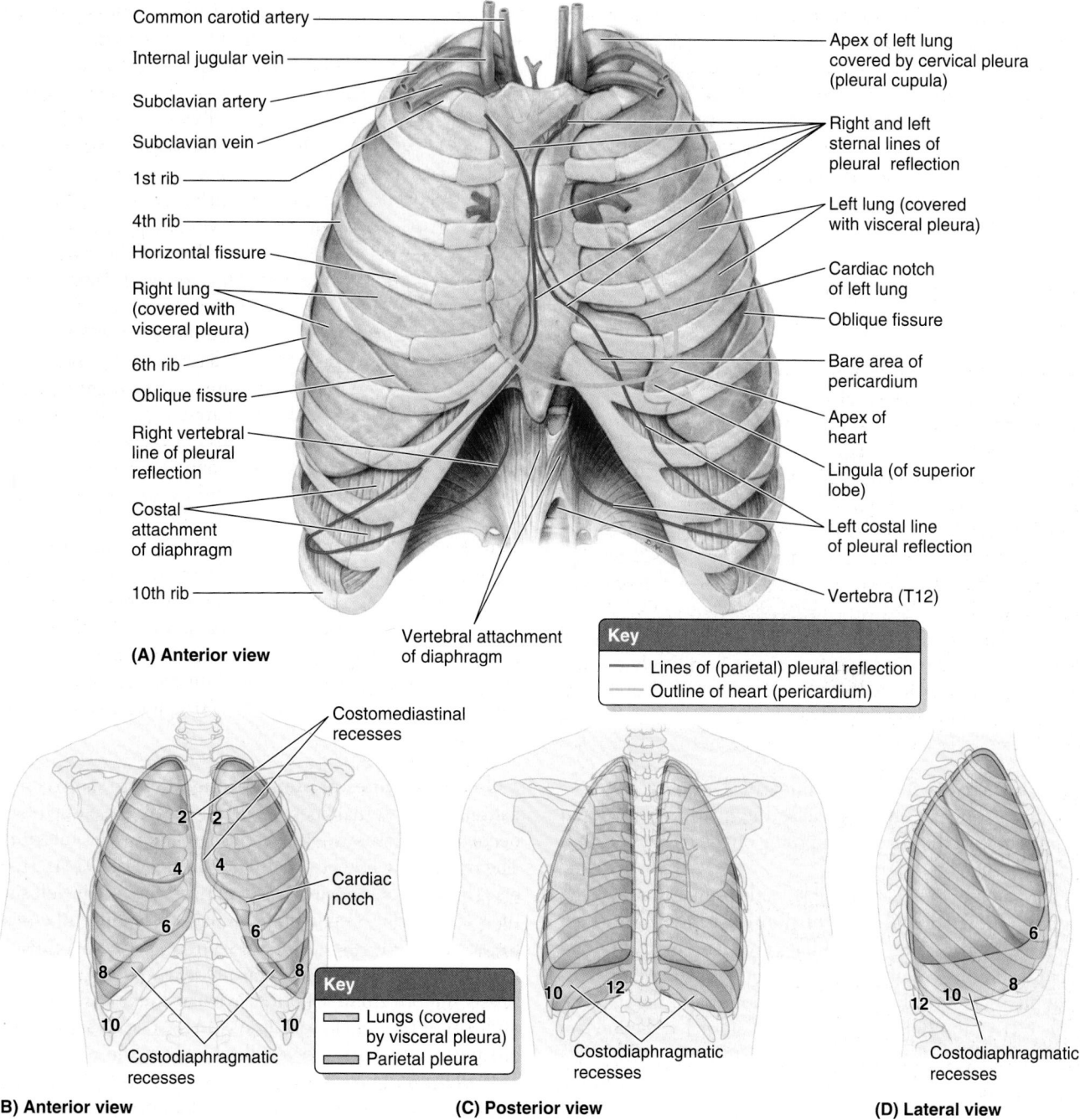

Common carotid artery

Internal jugular vein

Subclavian artery

Subclavian vein

1st rib

4th rib

Horizontal fissure

Right lung
(covered with
visceral pleura)

6th rib

Oblique fissure

Right vertebral
line of pleural
reflection

Costal
attachment
of diaphragm

10th rib

Apex of left lung
covered by cervical pleura
(pleural cupula)

Right and left
sternal lines of
pleural reflection

Left lung (covered
with visceral pleura)

Cardiac notch
of left lung

Oblique fissure

Bare area of
pericardium

Apex of
heart

Lingula (of superior
lobe)

Left costal line
of pleural reflection

Vertebra (T12)

Vertebral attachment
of diaphragm

(A) Anterior view

Key
— Lines of (parietal) pleural reflection
— Outline of heart (pericardium)

Costomediastinal
recesses

2 2

4 4

6 6

8 8

10 10

Cardiac
notch

Key
☐ Lungs (covered
by visceral pleura)
☐ Parietal pleura

Costodiaphragmatic
recesses

(B) Anterior view

10 12

Costodiaphragmatic
recesses

(C) Posterior view

6

10 8

12

Costodiaphragmatic
recesses

(D) Lateral view

FIGURE 1.31.   Relationship of thoracic contents and linings of thoracic cage. A. The apices of the lungs and cervical pleura extend into the neck. The left sternal reflection of parietal pleura and anterior border of the left lung deviate from the median plane, circumventing the area where the heart is, lies adjacent to the anterior thoracic wall. In this "bare area" the pericardial sac is accessible for needle puncture with less risk of puncturing the pleural cavity or lung. **B–D.** The shapes of the lungs and the larger pleural sacs that surround them during quiet respiration are demonstrated. The costodiaphragmatic recesses, not occupied by lung, are where pleural exudate accumulates when the body is erect. The outline of the horizontal fissure of the right lung clearly parallels the 4th rib. The ribs are identified by number.

medial third of the clavicle at the level of the neck of the 1st rib. The cervical pleura is reinforced by a fibrous extension of the endothoracic fascia, the **suprapleural membrane** (Sibson fascia). The membrane attaches to the internal border of the 1st rib and the transverse process of C7 vertebra (Fig. 1.30C).

The relatively abrupt lines along which the parietal pleura changes direction as it passes (reflects) from one wall of the pleural cavity to another are the *lines of pleural reflection* (Figs. 1.31 and 1.32). Three lines of pleural reflection outline the extent of the pulmonary cavities on each side: *sternal, costal,* and *diaphragmatic.* The outlines of the right and

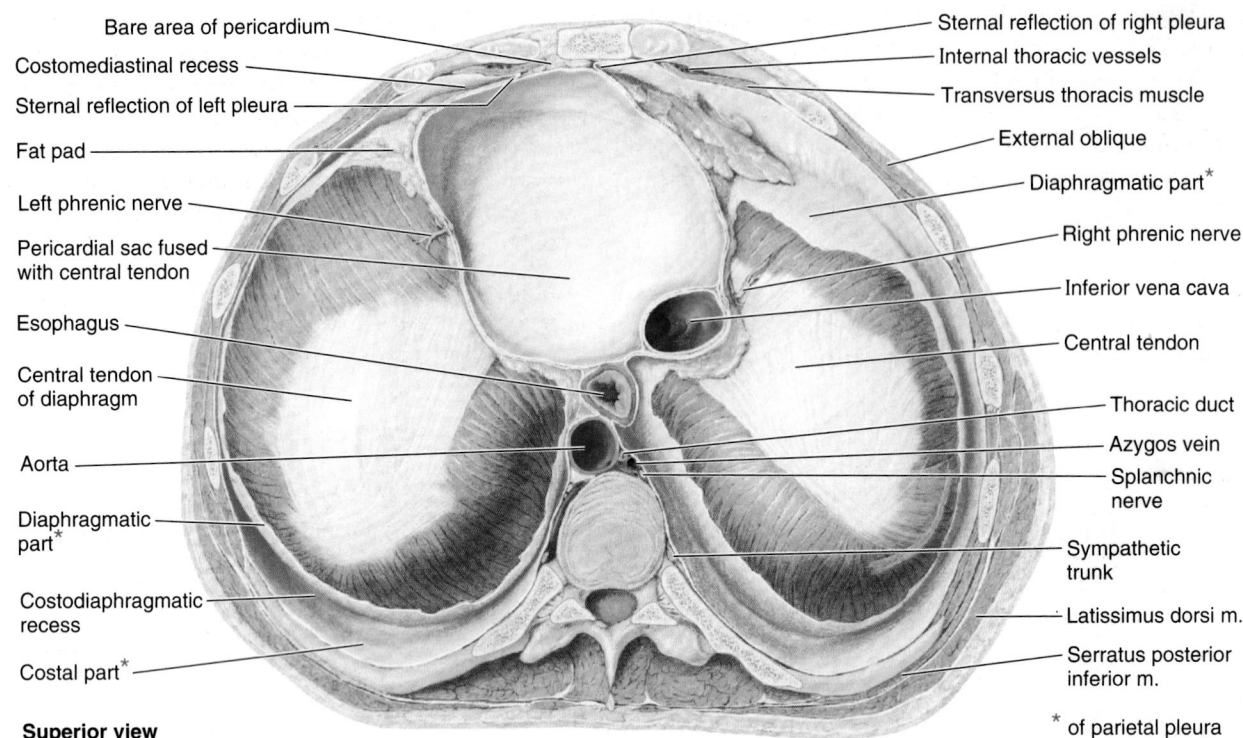

Bare area of pericardium
Costomediastinal recess
Sternal reflection of left pleura
Fat pad
Left phrenic nerve
Pericardial sac fused with central tendon
Esophagus
Central tendon of diaphragm
Aorta
Diaphragmatic part*
Costodiaphragmatic recess
Costal part*

Superior view

Sternal reflection of right pleura
Internal thoracic vessels
Transversus thoracis muscle
External oblique
Diaphragmatic part*
Right phrenic nerve
Inferior vena cava
Central tendon
Thoracic duct
Azygos vein
Splanchnic nerve
Sympathetic trunk
Latissimus dorsi m.
Serratus posterior inferior m.

* of parietal pleura

FIGURE 1.32. Diaphragm, base of pulmonary cavities and mediastinum, and costodiaphragmatic recesses. Most of the diaphragmatic pleura has been removed. At this level, the mediastinum consists of the pericardial sac (middle mediastinum) and the posterior mediastinum, mainly containing the esophagus and aorta. The deep groove surrounding the convexity of the diaphragm is the costodiaphragmatic recess, lined with parietal pleura. Anteriorly at this level, the pericardium and costomediastinal recesses and, between the sternal reflections of pleura, an area of pericardium only (the bare area) lie between the heart and the thoracic wall.

left pulmonary cavities are asymmetrical (i.e., are not mirror images of each other) because the heart is turned and extends toward the left side, imposing on the left cavity more markedly than on the right.

Deviation of the heart to the left side primarily affects the **right** and **left sternal lines of pleural reflection,** which are asymmetrical. The sternal lines are sharp or abrupt and occur where the costal pleura is continuous with the mediastinal pleura anteriorly. Starting superiorly from the cupulae (Fig. 1.31A), the right and left lines of sternal reflection run inferomedially, passing posterior to the sternoclavicular joints to meet at the anterior median line (AML), posterior to the sternum at the level of its sternal angle. Between the levels of costal cartilages 2–4, the right and left lines descend in contact. The pleural sacs may even slightly overlap each other.

The *sternal line of pleural reflection on the right side* continues to pass inferiorly in the AML to the posterior aspect of the xiphoid process (level of the 6th costal cartilage), where it turns laterally (Fig. 1.31). The *sternal line of reflection on the left side,* however, descends in the AML only to the level of the 4th costal cartilage. Here it passes to the left margin of the sternum and continues inferiorly to the 6th costal cartilage, creating a shallow notch as it runs lateral to an area of direct contact between the pericardium (heart sac) and the anterior thoracic wall. This shallow notch in the pleural sac, and the "bare area" of pericardial contact with the anterior wall, are important for *pericardiocentesis* (see blue box, "Pericardiocentesis," later in this chapter).

The **costal lines of pleural reflection** are sharp continuations of the sternal lines, occurring where the costal pleura becomes continuous with diaphragmatic pleura inferiorly. The right costal line proceeds laterally from the AML. However, because of the bare area of pericardium on the left side, the left costal line begins at the midclavicular line; otherwise, the right and left costal lines are symmetrical as they proceed laterally, posteriorly, and then medially, passing obliquely across the 8th rib in the midclavicular line (MCL) and the 10th rib in the midaxillary line (MAL), becoming continuous posteriorly with the vertebral lines at the necks of the 12th ribs inferior to them.

The **vertebral lines of pleural reflection** are much rounder, gradual reflections and occur where the costal pleura becomes continuous with the mediastinal pleura posteriorly. The vertebral lines of pleural reflection parallel the vertebral column, running in the paravertebral planes from vertebral level T1 through T12, where they become continuous with the costal lines.

The lungs do not fully occupy the pulmonary cavities during expiration; thus the peripheral diaphragmatic pleura is in contact with the lowermost parts of the costal pleura. The potential pleural spaces here are the **costodiaphragmatic recesses,** pleura-lined "gutters," which surround the upward convexity of the diaphragm inside the thoracic wall (Figs. 1.30B, 1.32, and 1.33C). Similar but smaller pleural recesses are located posterior to the sternum where the costal pleura is in contact with the mediastinal pleura. The potential pleural spaces here are the **costomediastinal recesses.** The left recess is larger (less occupied) because the *cardiac notch* in the left lung is more pronounced than the corresponding notch in the pleural sac. The *inferior borders of the lungs* move farther into the pleural recesses during deep inspiration and retreat from them during expiration (Fig. 1.33B & C).

LUNGS

The **lungs** are the vital organs of respiration. Their main function is to oxygenate the blood by bringing inspired air

into close relation with the venous blood in capillaries. Although cadaveric lungs may be or hard, and discolored, healthy lungs in livin normally light, soft, and spongy, and fully occu e pulmonary cavities. They are also elastic and recoil to approximately one third their size when the thoracic cavity is opened (Fig. 1.30C). The lungs are separated from each other by the mediastinum. Each lung has (Figs. 1.33 and 1.34):

- An **apex,** the blunt superior end of the lung ascending above the level of the 1st rib into the root of the neck; the apex is covered by cervical pleura.

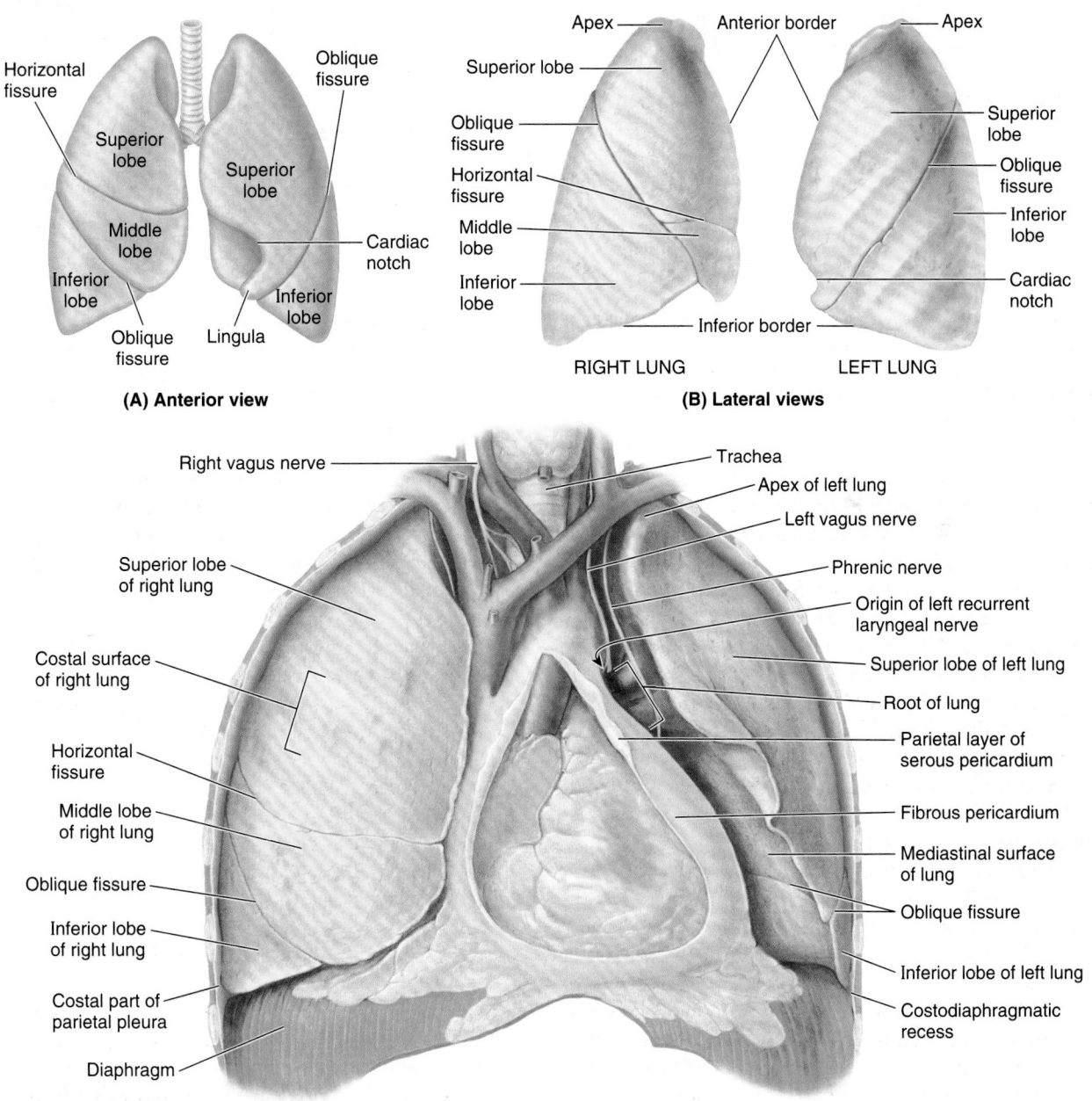

FIGURE 1.33. Costal surfaces of lungs. The lungs are shown in isolation in anterior (**A**) and lateral views (**B**), demonstrating lobes and fissures. **C.** The heart and lungs are shown in situ. The left lung is retracted from the heart (covered by fibrous pericardium) revealing the phrenic nerve as it passes anterior to the root of the lung, while the vagus nerve (CN X) passes posterior to the root. The superior lobe of the left lung in **C** is a variation that has neither a marked cardiac notch nor a lingula.

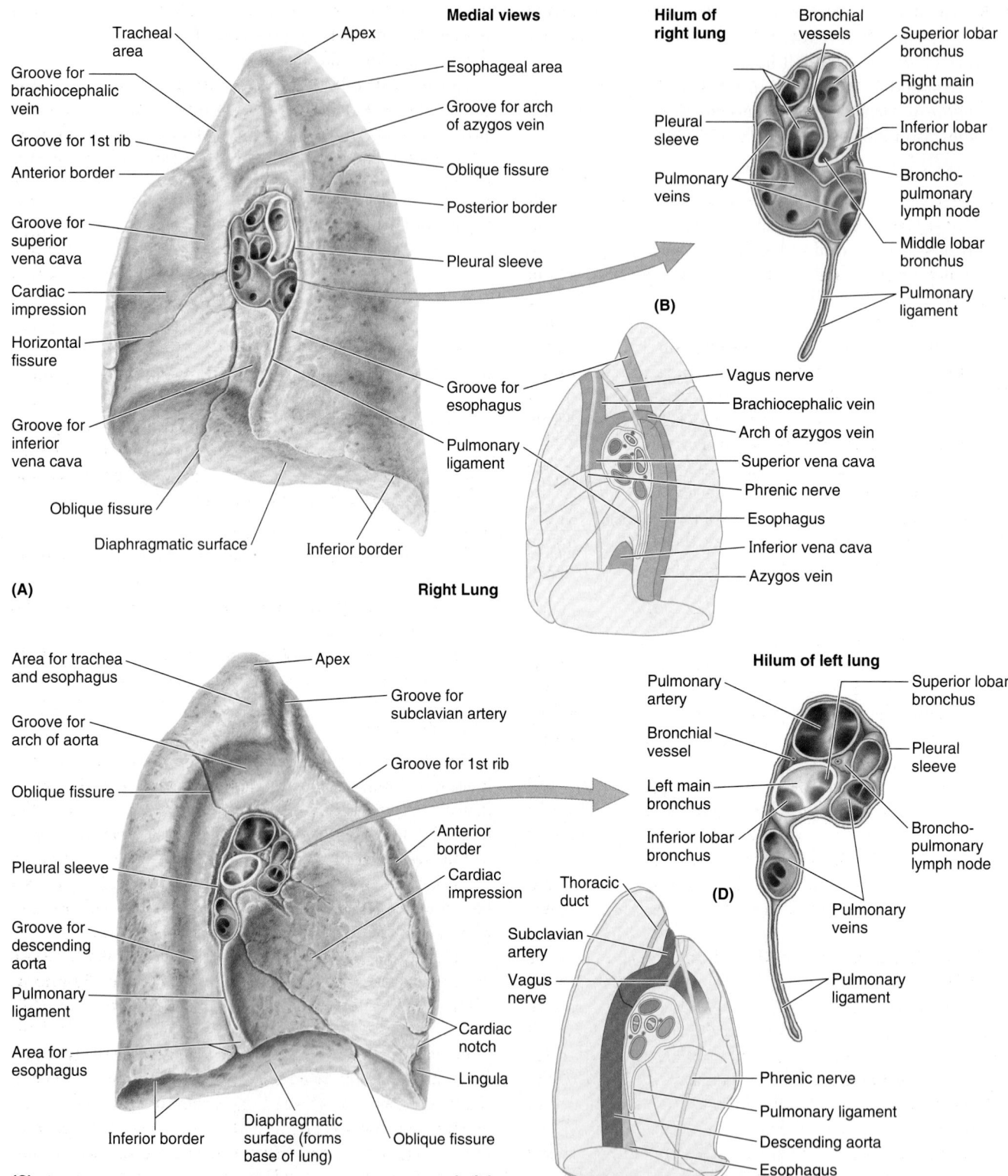

Medial views

(A) Right Lung

Tracheal area
Groove for brachiocephalic vein
Groove for 1st rib
Anterior border
Groove for superior vena cava
Cardiac impression
Horizontal fissure
Groove for inferior vena cava
Oblique fissure
Diaphragmatic surface
Inferior border
Apex
Esophageal area
Groove for arch of azygos vein
Oblique fissure
Posterior border
Pleural sleeve
Groove for esophagus
Pulmonary ligament

Hilum of right lung

(B)

Bronchial vessels
Superior lobar bronchus
Right main bronchus
Inferior lobar bronchus
Broncho-pulmonary lymph node
Middle lobar bronchus
Pulmonary ligament
Pleural sleeve
Pulmonary veins

Vagus nerve
Brachiocephalic vein
Arch of azygos vein
Superior vena cava
Phrenic nerve
Esophagus
Inferior vena cava
Azygos vein

(C) Left Lung

Area for trachea and esophagus
Groove for arch of aorta
Oblique fissure
Pleural sleeve
Groove for descending aorta
Pulmonary ligament
Area for esophagus
Inferior border
Diaphragmatic surface (forms base of lung)
Oblique fissure
Apex
Groove for subclavian artery
Groove for 1st rib
Anterior border
Cardiac impression
Cardiac notch
Lingula

Hilum of left lung

(D)

Pulmonary artery
Superior lobar bronchus
Bronchial vessel
Pleural sleeve
Left main bronchus
Inferior lobar bronchus
Broncho-pulmonary lymph node
Pulmonary veins
Pulmonary ligament

Thoracic duct
Subclavian artery
Vagus nerve
Phrenic nerve
Pulmonary ligament
Descending aorta
Esophagus

FIGURE 1.34. Mediastinal surfaces and hila of lungs. Impressions are formed in embalmed lungs by contact with adjacent structures. Superior to the root of the right lung (**A**), a groove is formed as the arch of the azygos vein courses anteriorly to enter the superior vena cava (SVC); whereas in the left lung (**C**), a similar but larger groove is formed superior to the root as the aorta arches posteriorly and descends as the thoracic aorta. The hilum of each lung is centered in the mediastinal surface. At the hilum (**B, D**), the root of each lung is surrounded by a pleural sleeve that descends inferior to the root as the pulmonary ligament. The pulmonary veins are the most anterior and inferior in the root, with the bronchi central and posteriorly placed.

- A **base,** the concave inferior surface of the lung, opposite the apex, resting on and accommodating the ipsilateral dome of the diaphragm.
- Two or three *lobes,* created by one or two *fissures.*
- Three *surfaces* (costal, mediastinal, and diaphragmatic).
- Three *borders* (anterior, inferior, and posterior).

The **right lung** features **right oblique** and **horizontal fissures** that divide it into three **right lobes: superior, middle,** and **inferior.** The right lung is larger and heavier than the left, but it is shorter and wider because the right dome of the diaphragm is higher and the heart and pericardium bulge more to the left. The *anterior border of the right lung* is relatively straight. The **left lung** has a single **left oblique fissure** dividing it into two **left lobes, superior** and **inferior.** The *anterior border of the left lung* has a deep **cardiac notch,** an indentation consequent to the deviation of the apex of the heart to the left side. This notch primarily indents the antero-inferior aspect of the superior lobe. This indentation often shapes the most inferior and anterior part of the superior lobe into a thin, tongue-like process, the **lingula** (L. dim. of *lingua,* tongue), which extends below the cardiac notch and slides in and out of the costomediastinal recess during inspiration and expiration (Figs. 1.30B, 1.31A, and 1.34C).

The lungs of an embalmed cadaver, usually firm to the touch, demonstrate impressions formed by structures adjacent to them, such as the ribs, heart, and great vessels (Figs. 1.33A and 1.34A & C). These markings provide clues to the relationships of the lungs; however, only the cardiac impressions are evident during surgery or in fresh cadaveric or postmortem specimens.

The **costal surface of the lung** is large, smooth, and convex. It is related to the costal pleura, which separates it from the ribs, costal cartilages, and innermost intercostal muscles (Fig. 1.33C). The posterior part of the costal surface is related to the bodies of the thoracic vertebrae and is sometimes referred to as the *vertebral part of the costal surface.*

The **mediastinal surface of the lung** is concave because it is related to the middle mediastinum, which contains the pericardium and heart (Fig. 1.34). The mediastinal surface includes the *hilum,* which receives the root of the lung. If embalmed, there is a *groove for the esophagus* and a *cardiac impression for the heart* on the mediastinal surface of the right lung. Because two thirds of the heart lies to the left of the midline, the cardiac impression on the mediastinal surface of the left lung is much larger. This surface of the left lung also features a prominent, continuous groove for the *arch of the aorta* and the *descending aorta* as well as a smaller *area for the esophagus* (Fig. 1.34C).

The **diaphragmatic surface of the lung,** which is also concave, forms the **base of the lung,** which rests on the dome of the diaphragm. The concavity is deeper in the right lung because of the higher position of the right dome, which overlies the liver. Laterally and posteriorly, the diaphragmatic surface is bounded by a thin, sharp margin (inferior border) that projects into the *costodiaphragmatic recess* of the pleura (Figs. 1.33C and 1.34).

The **anterior border of the lung** is where the costal and mediastinal surfaces meet anteriorly and overlap the heart. The cardiac notch indents this border of the left lung. The **inferior border of the lung** circumscribes the diaphragmatic surface of the lung and separates this surface from the costal and mediastinal surfaces. The **posterior border of the lung** is where the costal and mediastinal surfaces meet posteriorly; it is broad and rounded and lies in the cavity at the side of the thoracic region of the vertebral column.

The lungs are attached to the mediastinum by the **roots of the lungs**—that is, the bronchi (and associated bronchial vessels), pulmonary arteries, superior and inferior pulmonary veins, the pulmonary plexuses of nerves (sympathetic, parasympathetic, and visceral afferent fibers), and lymphatic vessels (Fig. 1.34). If the lung root is sectioned before the (medial to) branching of the main (primary) bronchus and pulmonary artery, its general arrangement is

- *Pulmonary artery:* superiormost on left (the superior lobar or "eparterial" bronchus may be superiormost on the right).
- *Superior and inferior pulmonary veins:* anteriormost and inferiormost, respectively.
- *Main bronchus:* against and approximately in the middle of the posterior boundary, with the bronchial vessels coursing on its outer surface (usually on posterior aspect at this point).

The **hilum of the lung** is a wedge-shaped area on the mediastinal surface of each lung through which the structures forming the root of the lung pass to enter or exit the lung (Fig. 1.34B & D). The hilum ("doorway") can be likened to the area of earth where a plant's roots enter the ground. Medial to the hilum, the lung root is enclosed within the area of continuity between the parietal and the visceral layers of pleura—the **pleural sleeve** (mesopneumonium).

Inferior to the root of the lung, this continuity between parietal and visceral pleura forms the **pulmonary ligament,** extending between the lung and the mediastinum, immediately anterior to the esophagus (Fig. 1.34A-D). The pulmonary ligament consists of a double layer of pleura separated by a small amount of connective tissue. When the root of the lung is severed and the lung is removed, the pulmonary ligament appears to hang from the root. To visualize the root of the lung, the pleural sleeve surrounding it, and the pulmonary ligament hanging from it, put on an extra-large lab coat and abduct your upper limb. Your forearm is comparable to the root of the lung, and the coat sleeve represents the pleural sleeve surrounding it. The pulmonary ligament is comparable to the slack of the sleeve as it hangs from your forearm; and your wrist, hand, and abducted fingers represent the branching structures of the root—the bronchi and pulmonary vessels.

TRACHEOBRONCHIAL TREE

Beginning at the *larynx*, the walls of the airway are supported by horseshoe- or C-shaped rings of hyaline cartilage. The sub-laryngeal airway constitutes the **tracheobronchial tree.** The *trachea* (described with the *superior mediastinum,* later in this chapter), located within the superior mediastinum, constitutes the trunk of the tree. It bifurcates at the level of the transverse thoracic plane (or sternal angle) into **main bronchi,** one to each lung, passing inferolaterally to enter the lungs at the hila (singular = hilum) (Fig. 1.35E).

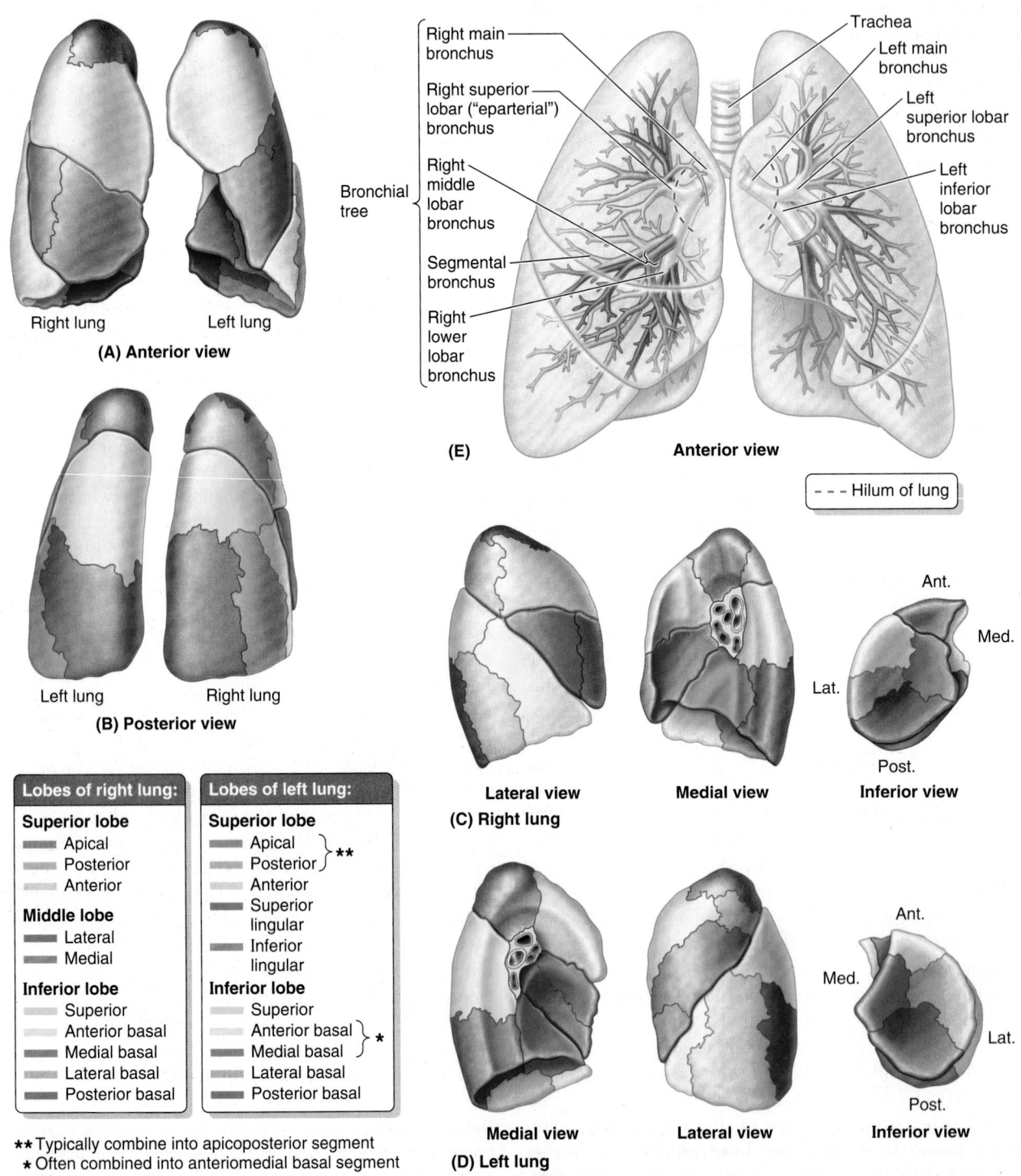

Lobes of right lung:

Superior lobe
- Apical
- Posterior
- Anterior

Middle lobe
- Lateral
- Medial

Inferior lobe
- Superior
- Anterior basal
- Medial basal
- Lateral basal
- Posterior basal

Lobes of left lung:

Superior lobe
- Apical
- Posterior }**
- Anterior
- Superior lingular
- Inferior lingular

Inferior lobe
- Superior
- Anterior basal }
- Medial basal }*
- Lateral basal
- Posterior basal

** Typically combine into apicoposterior segment
* Often combined into anteriomedial basal segment

FIGURE 1.35. **Tracheobronchial tree and bronchopulmonary segments. A–D.** The bronchopulmonary segments are demonstrated after injection of a different color latex into each tertiary segmental bronchus, as shown in **E.** The right main bronchus gives off the right superior lobar (lobe) bronchus before entering the hilum of the lung.

- The **right main bronchus** is wider, shorter, and runs more vertically than the left main bronchus as it passes directly to the hilum of the lung.
- The **left main bronchus** passes inferolaterally, inferior to the arch of the aorta and anterior to the esophagus and thoracic aorta, to reach the hilum of the lung.

Within the lungs, the bronchi branch in a constant fashion to form the branches of the tracheobronchial tree. Note that the *branches* of the tracheobronchial tree are components of the *root* of each lung (consisting of branches of the pulmonary artery and veins as well as the bronchi).

Each main (primary) bronchus divides into secondary **lobar bronchi,** two on the left and three on the right, each of which supplies a lobe of the lung. Each lobar bronchus divides into several tertiary **segmental bronchi** that supply the bronchopulmonary segments (Figs. 1.35 and 1.36).

The **bronchopulmonary segments** are:

- The largest subdivisions of a lobe.
- Pyramidal-shaped segments of the lung, with their apices facing the lung root and their bases at the pleural surface.
- Separated from adjacent segments by connective tissue septa.
- Supplied independently by a segmental bronchus and a tertiary branch of the pulmonary artery.
- Named according to the segmental bronchi supplying them.
- Drained by intersegmental parts of the pulmonary veins that lie in the connective tissue between and drain adjacent segments.

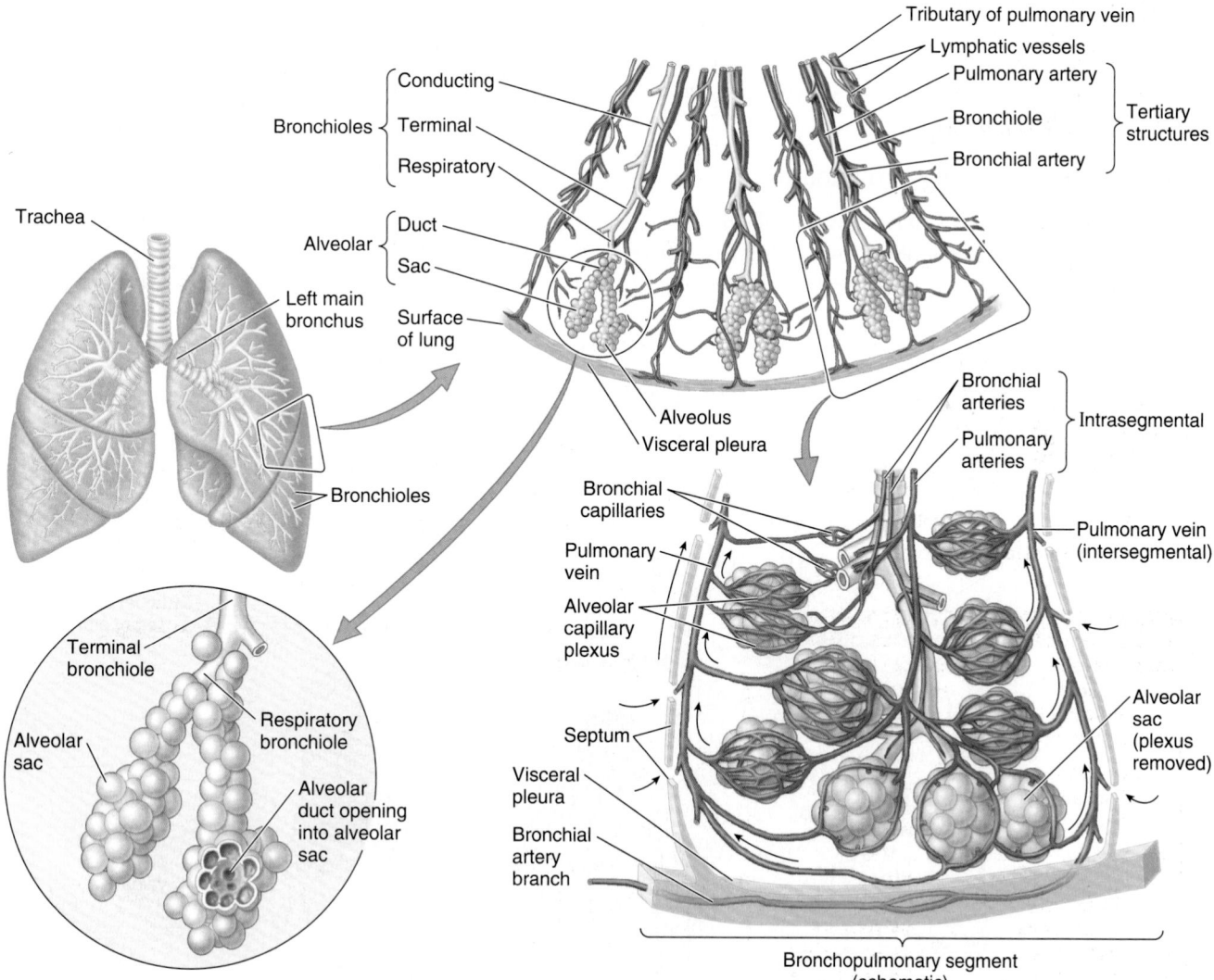

FIGURE 1.36. Internal structure and organization of lungs. Within the lungs, the bronchi and pulmonary arteries are paired and branch in unison. Tertiary segmental (tertiary) branches supply the bronchopulmonary segments. Each intrasegmental pulmonary artery, carrying poorly oxygenated blood, ends in a capillary plexus in the walls of the alveolar sacs and alveoli, where oxygen and carbon dioxide are exchanged. The intersegmental pulmonary veins arise from the pulmonary capillaries, carrying well-oxygenated blood to the heart. Bronchial arteries are distributed along and supply the bronchial tree. Their distalmost branches supply capillary beds drained by the pulmonary veins, such as those of the visceral pleura. A very small amount of low-oxygen blood thus drains into the otherwise oxygen-rich blood conveyed by the pulmonary veins.

- Usually 18–20 in number (10 in the right lung; 8–10 in the left lung, depending on the combining of segments).
- Surgically resectable.

Beyond the tertiary segmental bronchi (Fig. 1.35B), there are 20 to 25 generations of branching *conducting bronchioles* that eventually end as **terminal bronchioles,** the smallest conducting bronchioles (Fig. 1.36). **Bronchioles** lack cartilage in their walls. **Conducting bronchioles** transport air but lack glands or alveoli. Each terminal bronchiole gives rise to several generations of **respiratory bronchioles,** characterized by scattered, thin-walled outpocketings (alveoli) that extend from their lumens. The **pulmonary alveolus** is the basic structural unit of gas exchange in the lung. Due to the presence of the alveoli, the respiratory bronchioles are involved both in air transportation and gas exchange. Each respiratory bronchiole gives rise to 2–11 *alveolar ducts,* each of which gives rise to 5–6 *alveolar sacs.* **Alveolar ducts** are elongated airways densely lined with alveoli, leading to common spaces, the **alveolar sacs,** into which clusters of alveoli open. New alveoli continue to develop until about age 8 years, by which time there are approximately 300 million alveoli.

VASCULATURE OF LUNGS AND PLEURAE

Each lung has a pulmonary artery supplying blood to it and two pulmonary veins draining blood from it (Fig. 1.37). The **right** and **left pulmonary arteries** arise from the *pulmonary trunk* at the level of the *sternal angle;* they carry low-oxygen ("venous") blood to the lungs for oxygenation. (They are usually colored blue, like veins, or bluish-purple in anatomical illustrations.) Each pulmonary artery becomes part of the root of the corresponding lung and divides secondary **lobar arteries.** The *right* and *left superior lobar arteries* to the superior lobes arise first, before entering the hilum. Continuing into the lung, the artery descends posterolateral to the main bronchus as the *inferior lobar artery of the left lung* and as an intermediate artery that will divide into *middle* and *inferior lobar arteries of the right lung.* Lobar arteries divide into tertiary **segmental arteries.** The arteries and bronchi are paired in the lung, branching simultaneously and running parallel courses. Consequently, a paired secondary lobar artery and bronchus serves each lobe. Likewise, a paired tertiary segmental artery and bronchus supply each bronchopulmonary segment of the lung. Usually, the artery is located on the anterior aspect of the corresponding bronchus.

Two *pulmonary veins,* a **superior** and an **inferior pulmonary vein** on each side, carry oxygen-rich ("arterial") blood from corresponding lobes of each lung to the left atrium of the heart. The **middle lobe vein** is a tributary of the right superior pulmonary vein. (Pulmonary veins are commonly colored red, like arteries, or reddish-purple in anatomical illustrations.) The pulmonary veins run independently of the arteries and bronchi in the lung, coursing between and receiving blood from adjacent bronchopulmonary segments

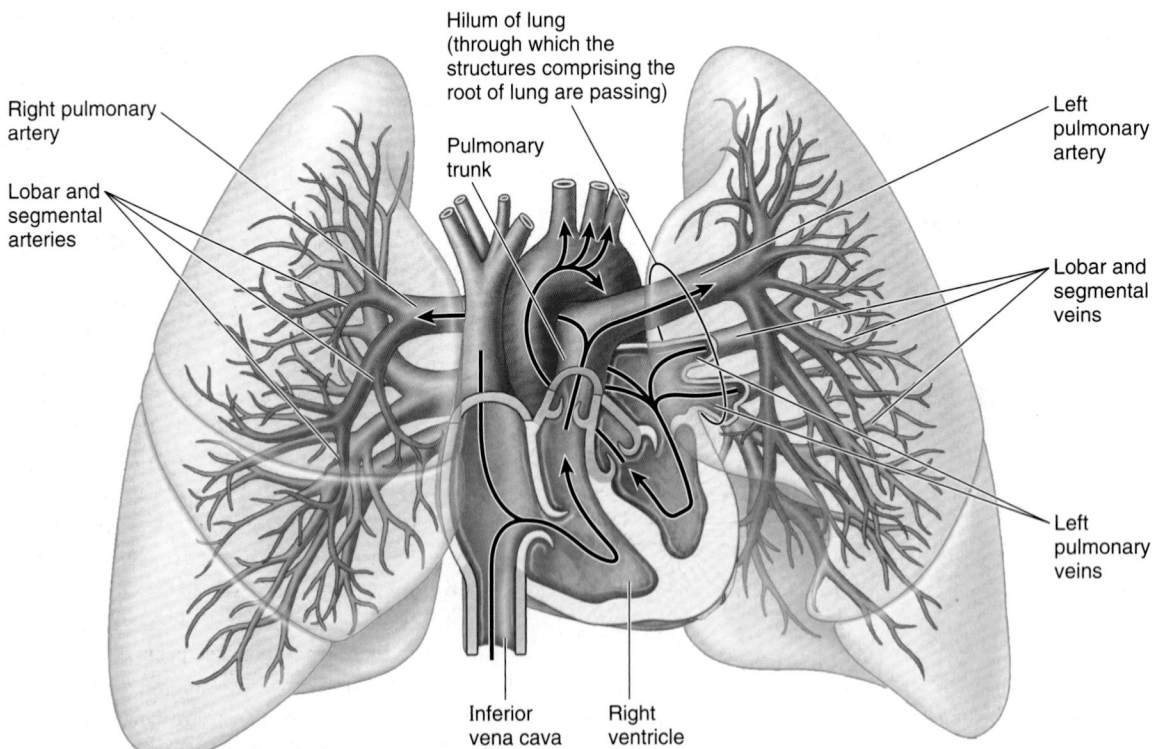

FIGURE 1.37. Pulmonary circulation. Although the intrapulmonary relationships are accurately demonstrated, the separation of the vessels of the root of the lung has been exaggerated in the hilar region to show them as they enter and leave the lung. Note that the right pulmonary artery passes under the arch of the aorta to reach the right lung and that the left pulmonary artery lies completely to the left of the arch.

as they run toward the hilum. Except in the central, perihilar region of the lung, the veins from the visceral pleura and the bronchial venous circulation drain into the pulmonary veins, the relatively small volume of low-oxygen blood entering the large volume of oxygen-rich blood returning to the heart. Veins from the parietal pleura join systemic veins in adjacent parts of the thoracic wall.

Bronchial arteries supply blood for nutrition of the structures making up the root of the lungs, the supporting

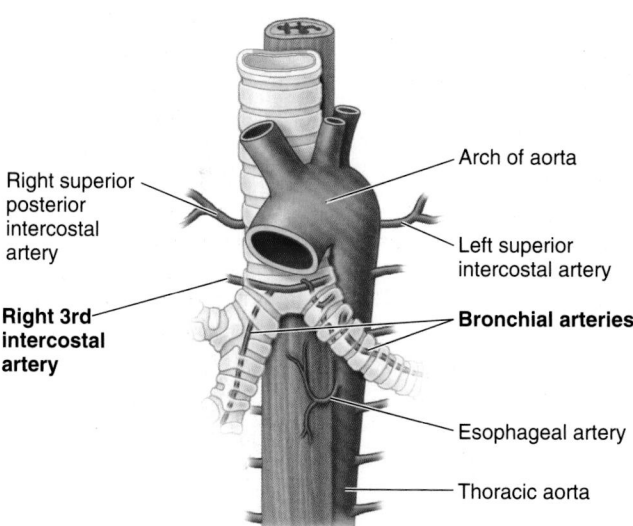

(A) Bronchial arteries

Right superior posterior intercostal artery

Right 3rd intercostal artery

Arch of aorta

Left superior intercostal artery

Bronchial arteries

Esophageal artery

Thoracic aorta

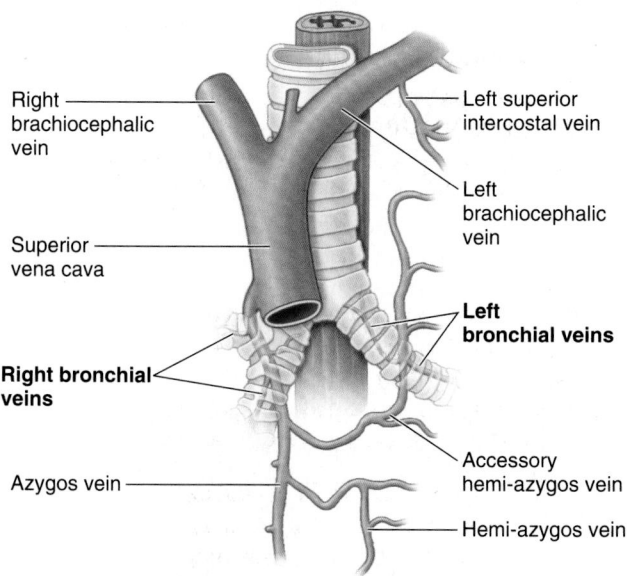

(B) Bronchial veins

Right brachiocephalic vein

Superior vena cava

Right bronchial veins

Azygos vein

Left superior intercostal vein

Left brachiocephalic vein

Left bronchial veins

Accessory hemi-azygos vein

Hemi-azygos vein

FIGURE 1.38. Bronchial arteries and veins. A. The bronchial arteries supply the supporting tissues of the lungs and visceral pleura. **B.** The bronchial veins drain the more proximal capillary beds supplied by the bronchial arteries; the rest is drained by the pulmonary veins.

tissues of the lungs, and the visceral pleura (Fig. 1.38A). The two **left bronchial arteries** usually arise directly from the thoracic aorta. The single **right bronchial artery** may also arise directly from the aorta; however it commonly arises indirectly, either by way of the proximal part of one of the upper posterior intercostal arteries (usually the right 3rd posterior intercostal artery), or from a common trunk with the left superior bronchial artery.

The small bronchial arteries provide branches to the upper esophagus. Then they typically pass along the posterior aspects of the main bronchi, supplying them and their branches as far distally as the respiratory bronchioles. (However, see the variation shown in Figs. 1.63 and 1.69, drawn from a cadaver dissection.) The most distal branches of the bronchial arteries anastomose with branches of the pulmonary arteries in the walls of the bronchioles and in the visceral pleura. The parietal pleura is supplied by the arteries that supply the thoracic wall.

The **bronchial veins** (Fig. 1.38B) drain only part of the blood supplied to the lungs by the bronchial arteries, primarily that distributed to or near the more proximal part of the roots of the lungs. The remainder of the blood is drained by the pulmonary veins, specifically the blood returning from the visceral pleura, the more peripheral regions of the lung, and the distal components of the root of the lung. The *right bronchial vein* drains into the *azygos vein*, and the *left bronchial vein* drains into the *accessory hemi-azygos vein* or the *left superior intercostal vein*. Bronchial veins also receive some blood from esophageal veins.

The **pulmonary lymphatic plexuses** communicate freely (Fig. 1.39). The superficial **subpleural\lymphatic plexus** lies deep to the visceral pleura and drains the lung parenchyma (tissue) and visceral pleura. Lymphatic vessels from this superficial plexus drain into the **bronchopulmonary lymph nodes** (hilar lymph nodes) in the region of the lung hilum.

The deep **bronchopulmonary lymphatic plexus** is located in the submucosa of the bronchi and in the peribronchial connective tissue. It is largely concerned with draining the structures that form the root of the lung. Lymphatic vessels from this deep plexus drain initially into the intrinsic **pulmonary lymph nodes,** located along the lobar bronchi. Lymphatic vessels from these nodes continue to follow the bronchi and pulmonary vessels to the hilum of the lung, where they also drain into the bronchopulmonary lymph nodes. From them, lymph from both the superficial and deep lymphatic plexuses drains to the **superior** and **inferior tracheobronchial lymph nodes,** superior and inferior to the bifurcation of the trachea and main bronchi, respectively. The right lung drains primarily through the consecutive sets of nodes on the right side, and the superior lobe of the left lung drains primarily through corresponding nodes of the left side. Many, but not all, of the lymphatics from the lower lobe of the left lung, however, drain to the right superior tracheobronchial nodes; the lymph then continues to follow the right-side pathway.

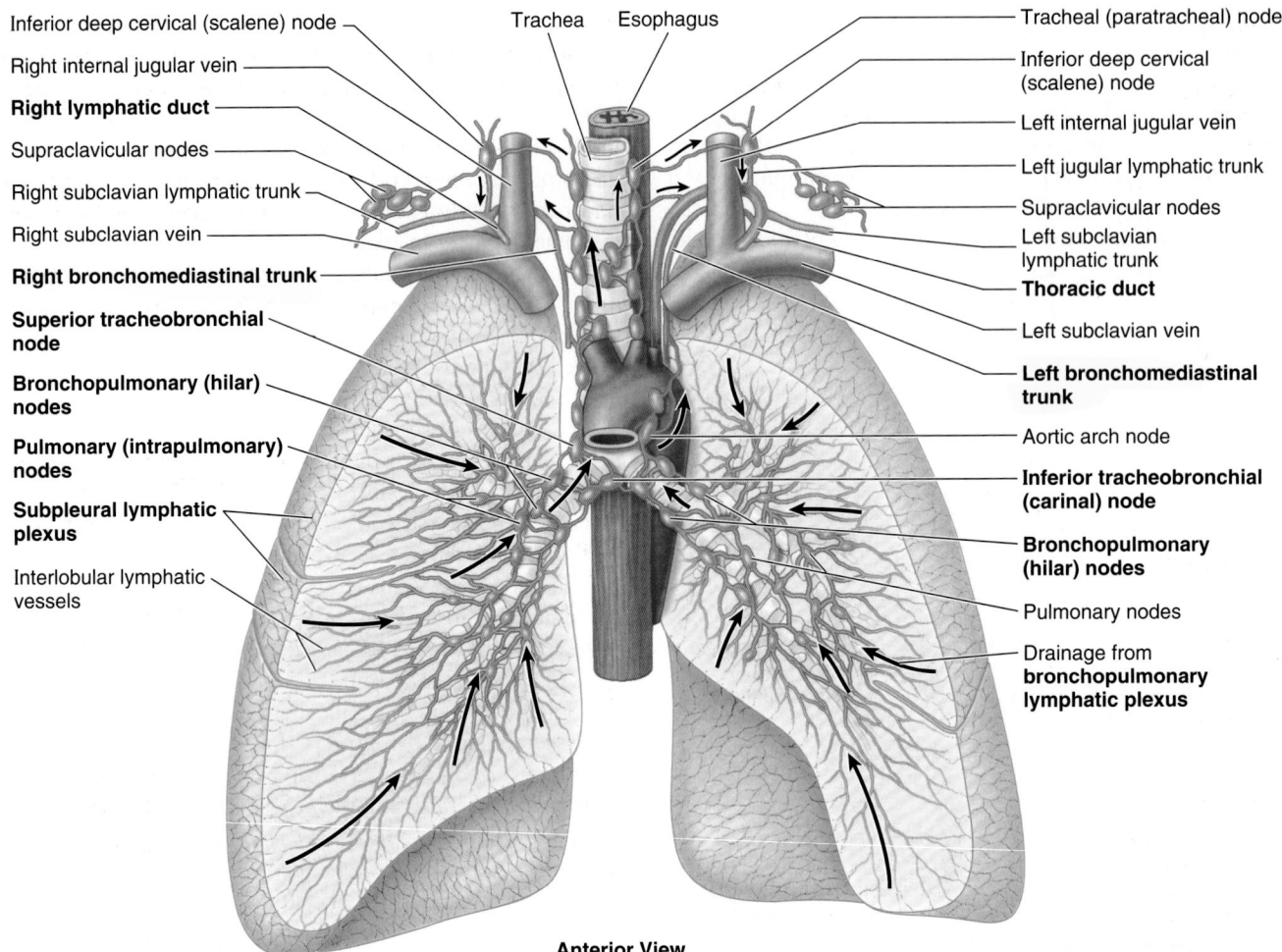

FIGURE 1.39. Lymphatic drainage of lungs. The lymphatic vessels originate from superficial subpleural and deep lymphatic plexuses. All lymph from the lung leaves along the root of the lung and drains to the inferior or superior tracheobronchial lymph nodes. The inferior lobe of *both* lungs drains to the centrally placed inferior tracheobronchial (carinal) nodes, which primarily drain to the right side. The other lobes of each lung drain primarily to the ipsilateral superior tracheobronchial lymph nodes. From here the lymph traverses a variable number of paratracheal nodes and enters the bronchomediastinal trunks.

Lymph from the tracheobronchial lymph nodes passes to the **right** and **left bronchomediastinal lymph trunks,** the major lymph conduits draining the thoracic viscera. These trunks usually terminate on each side at the *venous angles* (junctions of the subclavian and internal jugular veins); however, the right bronchomediastinal trunk may first merge with other lymphatic trunks, converging here to form the short *right lymphatic duct.* The left bronchomediastinal trunk may terminate in the *thoracic duct.* Lymph from the parietal pleura drains into the lymph nodes of the thoracic wall (intercostal, parasternal, mediastinal, and phrenic). A few lymphatic vessels from the cervical parietal pleura drain into the axillary lymph nodes.

NERVES OF LUNGS AND PLEURAE

The nerves of the lungs and visceral pleura are derived from the **pulmonary plexuses** anterior and (mainly) posterior to the roots of the lungs (Fig. 1.40). These nerve networks contain parasympathetic, sympathetic, and visceral afferent fibers.

The *parasympathetic fibers* conveyed to the pulmonary plexus are presynaptic fibers from the vagus nerve (CN X). They synapse with *parasympathetic ganglion cells* (cell bodies of postsynaptic neurons) in the pulmonary plexuses and along the branches of the bronchial tree. The parasympathetic fibers are motor to the smooth muscle of the bronchial tree (*bronchoconstrictor*), inhibitory to the pulmonary vessels (*vasodilator*), and secretory to the glands of the bronchial tree (*secretomotor*).

The *sympathetic fibers* of the pulmonary plexuses are postsynaptic fibers. Their cell bodies (*sympathetic ganglion cells*) are in the *paravertebral sympathetic ganglia of the sympathetic trunks.* The sympathetic fibers are inhibitory to the bronchial muscle (*bronchodilator*), motor to the pulmonary vessels (*vasoconstrictor*), and inhibitory to the alveolar glands of the bronchial tree—type II secretory epithelial cells of the alveoli (Fig. 1.36).

The *visceral afferent fibers* of the pulmonary plexuses are either reflexive (conducting subconscious sensations associated with reflexes that control function) or nociceptive

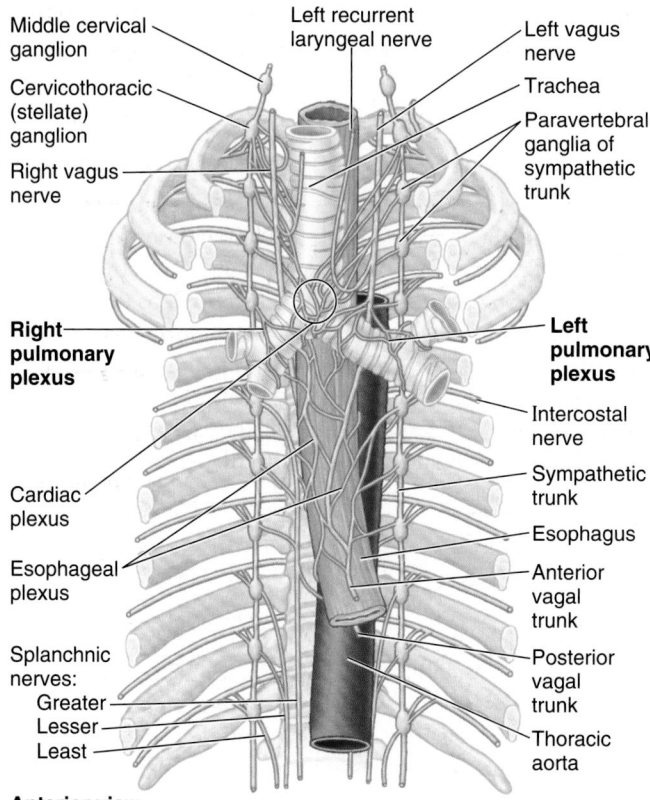

FIGURE 1.40. Nerves of lungs and visceral pleura. The right and left pulmonary plexuses, anterior and posterior to the roots of the lungs, receive sympathetic contributions from the right and left sympathetic trunks and parasympathetic contributions from the right and left vagus nerves (CN X). After contributing to the posterior pulmonary plexus, the vagus nerves continue inferiorly and become part of the esophageal plexus, often losing their identity and then reforming as anterior and posterior vagal trunks. Branches of the pulmonary plexuses accompany pulmonary arteries and especially bronchi to and within the lungs.

(conducting pain impulses generated in response to painful or injurious stimuli, such as chemical irritants, ischemia, or excessive stretch). *Reflexive visceral afferent fibers* with cell bodies in the sensory ganglion of the vagus nerve (CN X) accompany the parasympathetic fibers, conveying impulses centrally from nerve endings associated with the:

- Bronchial mucosa, probably in association with tactile sensation for cough reflexes.
- Bronchial muscles, possibly involved in stretch reception.
- Interalveolar connective tissue, in association with *Hering-Breuer reflexes* (a mechanism that tends to limit respiratory excursions).
- Pulmonary arteries, serving pressor receptors (receptors sensitive to blood pressure).
- Pulmonary veins, serving chemoreceptors (receptors sensitive to blood gas levels).

Nociceptive afferent fibers from the visceral pleura and bronchi accompany the sympathetic fibers through the

sympathetic trunk to the sensory ganglia of upper thoracic spinal nerves, whereas those from the trachea accompany the parasympathetic fibers to the sensory ganglion of the vagus nerve (CN X).

The *nerves of the parietal pleura* derive from the intercostal and phrenic nerves. The costal pleura and the peripheral part of the diaphragmatic pleura are supplied by the *intercostal nerves.* They mediate the sensations of touch and pain. The central part of the diaphragmatic pleura and the mediastinal pleura are supplied by the *phrenic nerves* (Figs. 1.32 and 1.34B & D).

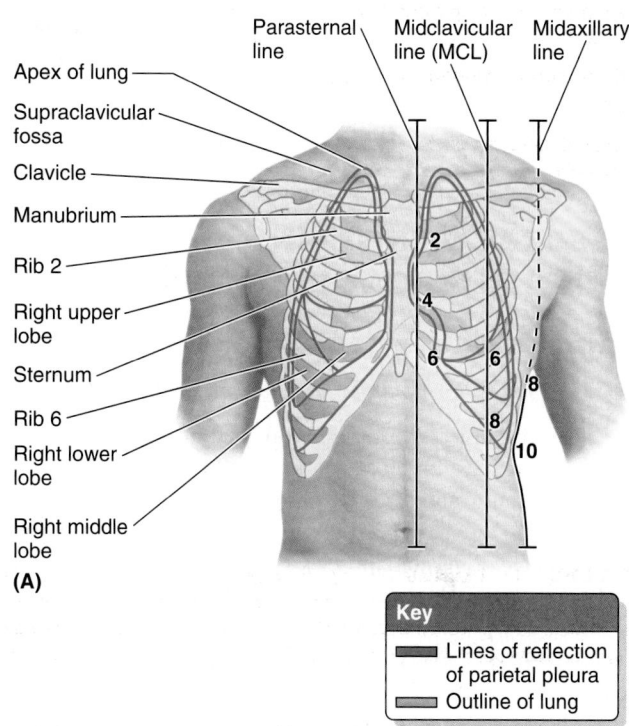

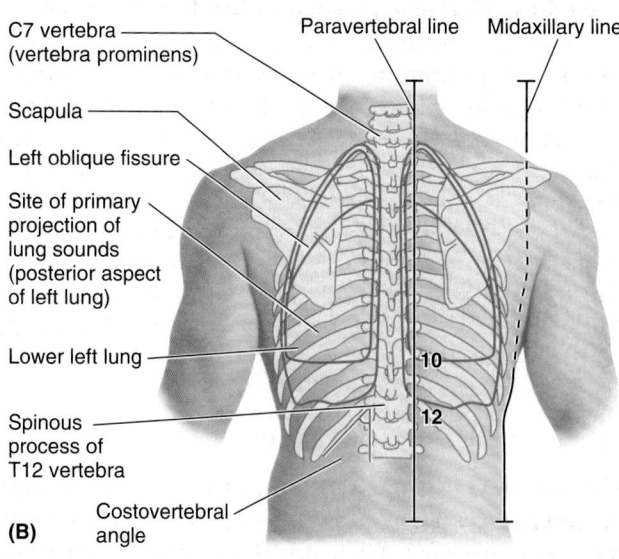

FIGURE 1.41. Surface anatomy of pleurae and lungs.

SURFACE ANATOMY OF PLEURAE AND LUNGS

The cervical pleurae and apices of the lungs pass through the superior thoracic aperture deep to the **supraclavicular fossae,** depressions located posterior and superior to the clavicles and lateral to the tendons of the sternocleidomastoid muscles (Fig. 1.41). The anterior borders of the lungs lie adjacent to the anterior line of reflection of the parietal pleura between the 2nd and 4th costal cartilages. Here, the margin of the left pleural reflection moves laterally and then inferiorly at the cardiac notch to reach the 6th costal cartilage. The anterior border of the left lung is more deeply indented by its cardiac notch. On the right side, the pleural reflection continues inferiorly from the 4th to the 6th costal cartilage, paralleled closely by the anterior border of the right lung. Both pleural reflections and anterior lung borders pass laterally at the 6th costal cartilages. The pleural reflections reach the mid-clavicular line (MCL) at the level of the 8th costal cartilage, the 10th rib at the midaxillary line (MAL), and the 12th rib at the scapular line (SL); however, the inferior margins of the lungs reach the MCL at the level of the 6th rib, the MAL at the 8th rib, and the SL at the 10th rib, proceeding toward the spinous process of T10 vertebra. They then proceed toward the spinous process of T12 vertebra. Thus the parietal pleura generally extends approximately two ribs inferior to the lung.

The *oblique fissure* of the lungs extends from the level of the spinous process of T2 vertebra posteriorly to the 6th costal cartilage anteriorly, which coincides approximately with the medial border of the scapula when the upper limb is elevated above the head (causing the inferior angle to rotate laterally). The *horizontal fissure* of the right lung extends from the oblique fissure along the 4th rib and costal cartilage anteriorly.

PLEURAE, LUNGS, AND TRACHEOBRONCHIAL TREE

Injuries of Cervical Pleura and Apex of Lung

Because of the inferior slope of the 1st pair of ribs and the superior thoracic aperture they form, the cervical pleura and apex of the lung project through this opening into the neck, posterior to the inferior attachments of the sternocleidomastoid muscles. Consequently, the lungs and pleural sacs may be injured in *wounds to the base of the neck* resulting in a *pneumothorax,* the presence of air (G. *pneuma*) in the pleural cavity. The cervical pleura reaches a relatively higher level in infants and young children because of the shortness of their necks. Consequently, the cervical pleura is especially vulnerable to injury during infancy and early childhood.

Injury to Other Parts of Pleurae

The pleurae descend inferior to the costal margin in three regions, where an abdominal incision might inadvertently enter a pleural sac: the right part of the infrasternal angle (Fig. 1.25) and right and left costovertebral angles (Fig. 1.41B). The small areas of pleura exposed in the costovertebral angles inferomedial to the 12th ribs are posterior to the superior poles of the kidneys. The pleura is in danger here (i.e., a pneumothorax may occur) from an incision in the posterior abdominal wall when surgical procedures expose a kidney, for example.

Pulmonary Collapse

The lungs (more specifically, the air sacs that collectively make up the lung) are comparable to an inflated balloon when they are distended. If the distension is not maintained, their inherent elasticity will cause them to collapse (*atelectasis: secondary atelectasis* is the collapse of a previously inflated lung; *primary atelectasis* refers to the failure of a lung to inflate at birth). An inflated balloon remains distended only as long as its outlet is closed because its walls are free to fully contract. Normal lungs in situ remain distended even when the airway passages are open because the outer surfaces of the lungs (visceral pleura) adhere to the inner surface of the thoracic walls (parietal pleura) as a result of the surface tension provided by the pleural fluid. The elastic recoil

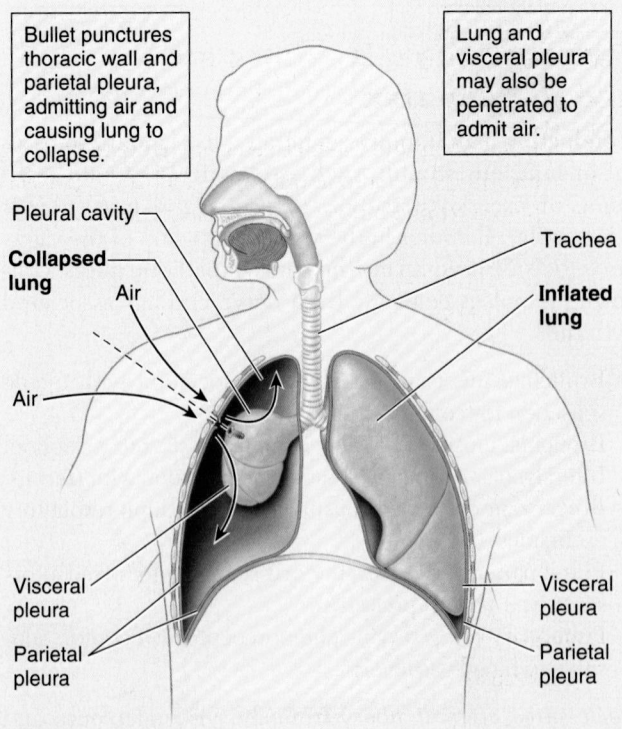

Bullet punctures thoracic wall and parietal pleura, admitting air and causing lung to collapse.

Lung and visceral pleura may also be penetrated to admit air.

Pleural cavity

Collapsed lung

Air

Air

Trachea

Inflated lung

Visceral pleura

Visceral pleura

Parietal pleura

Parietal pleura

FIGURE B1.8. Pulmonary collapse.

of the lungs causes the pressure in the pleural cavities to be sub-atmospheric. The pressure is usually about −2 mm Hg; during inspiration, it drops to about −8 mm Hg.

If a penetrating wound opens through the thoracic wall or the surface of the lungs, air will be sucked into the pleural cavity because of the negative pressure (Fig. B1.8). The surface tension adhering visceral to parietal pleura (lung to thoracic wall) will be broken, and the lung will collapse, expelling most of its air because of its inherent elasticity (elastic recoil). When a lung collapses, the pleural cavity (normally a potential space) becomes a real space.

The pleural sacs do not normally communicate; thus one lung may be collapsed after surgery, for example, without the other lung collapsing. Laceration or rupture of the surface of a lung (and its visceral pleura) or penetration of the thoracic wall (and its parietal pleura) results in hemorrhage and the entrance of air into the pleural cavity. The amount of blood and air that accumulates determines the extent of pulmonary collapse.

When a lung collapses, it occupies less volume within the pulmonary cavity and the pulmonary cavity does not increase in size (in fact, it may decrease in size) during inspiration. This reduction in size will be evident radiographically on the affected side by elevation of the diaphragm above its usual levels, intercostal space narrowing (ribs closer together), and displacement of the mediastinum (*mediastinal shift;* most evident via the air-filled trachea within it) toward the affected side. In addition, the collapsed lung will usually appear more dense (whiter) surrounded by more radiolucent (blacker) air.

In *open-chest surgery,* respiration and lung inflation must be maintained by intubating the trachea with a cuffed tube and using a positive-pressure pump, varying the pressure to alternately inflate and deflate the lungs.

Pneumothorax, Hydrothorax, and Hemothorax

Entry of air into the pleural cavity (*pneumothorax*), resulting from a penetrating wound of the parietal pleura from a bullet, for example, or from rupture

of a pulmonary lesion into the pleural cavity (*bronchopulmonary fistula*), results in collapse of the lung (Fig. B1.8). Fractured ribs may also tear the visceral pleura and lung, thus producing pneumothorax. The accumulation of a significant amount of fluid in the pleural cavity (*hydrothorax*) may result from *pleural effusion* (escape of fluid into the pleural cavity). With a chest wound, blood may also enter the pleural cavity (*hemothorax*) (Fig. B1.9). Hemothorax results more commonly from injury to a major intercostal or internal thoracic vessel than from laceration of a lung. If both air and fluid (*hemopneumothorax*, if the fluid is blood) accumulate in the pleural cavity, an *air–fluid level* or interface (sharp line, horizontal regardless of the patient's position, indicating the upper surface of the fluid) will be seen on a radiograph.

Thoracentesis

Sometimes it is necessary to insert a hypodermic needle through an intercostal space into the pleural cavity (*thoracentesis*) to obtain a sample of fluid or to remove blood or pus (Fig. B1.10). To avoid damage to the intercostal nerve and vessels, the needle is inserted superior to the rib, high enough to avoid the collateral branches. The needle passes through the intercostal muscles and costal parietal pleura into the pleural cavity. When the patient is in the upright position, intrapleural fluid accumulates in the costodiaphragmatic recess. Inserting the needle into the 9th intercostal space in the midaxillary line during expiration will avoid the inferior border of the lung. The needle should be angled upward, to avoid penetrating the deep side of the recess (a thin layer of diaphragmatic parietal pleura and diaphragm overlying the liver).

Insertion of a Chest Tube

Major amounts of air, blood, serous fluid, pus, or any combination of these substances in the pleural cavity are typically removed by *insertion of a chest*

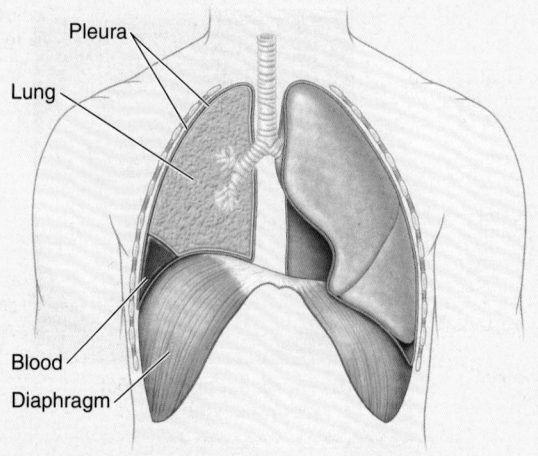

FIGURE B1.9. Hemothorax in right pleural cavity.

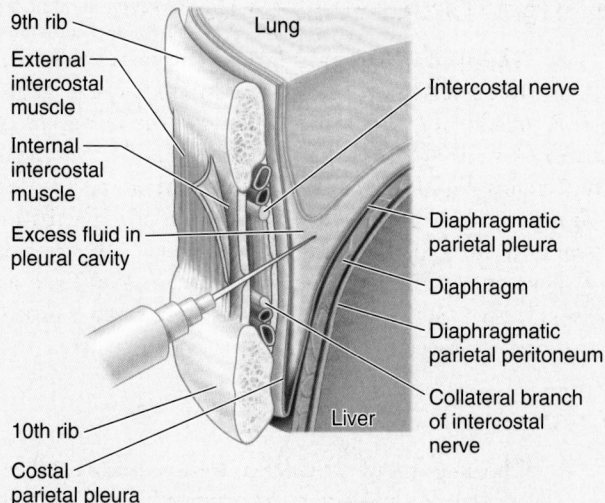

FIGURE B1.10. Technique for midaxillary thoracentesis.

tube. A short incision is made in the 5th or 6th intercostal space in the midaxillary line (which is approximately at nipple level). The tube may be directed superiorly (toward the cervical pleura [Fig. 1.31A]) for air removal, or inferiorly (toward the costodiaphragmatic recess) for fluid drainage. The extracorporeal end of the tube (i.e., the end that is outside of the body) is connected to an underwater drainage system, often with controlled suction, to prevent air from being sucked back into the pleural cavity. Removal of air allows reinflation of a collapsed lung. Failure to remove fluid may cause the lung to develop a resistant fibrous covering that inhibits expansion unless it is peeled off (*lung decortication*).

Pleurectomy and Pleurodesis

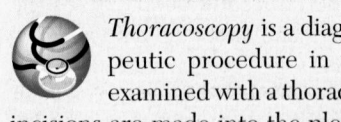

Obliteration of a pleural cavity by disease, such as *pleuritis* (inflammation of pleura), or during surgery (*pleurectomy*, or excision of a part of the pleura, for example) (Fig. B1.11A) does not cause appreciable functional consequences; however, it may produce pain during exertion. In other procedures, adherence of the parietal and visceral layers of pleura is induced by covering the apposing layers of pleura with an irritating powder or sclerosing agent (*pleurodesis*). Pleurectomy and pleurodesis are performed to prevent recurring *spontaneous secondary atelectasis* (spontaneous lung collapse) caused by chronic pneumothorax or malignant effusion resulting from lung disease (Shields et al., 2009).

Thoracoscopy

Thoracoscopy is a diagnostic and sometimes therapeutic procedure in which the pleural cavity is examined with a thoracoscope (Fig. B1.11B). Small incisions are made into the pleural cavity via an intercostal space. In addition to observation, biopsies can be taken and some thoracic conditions can be treated (e.g., disrupting adhesions or removing plaques).

Pleuritis (Pleurisy)

During inspiration and expiration, the sliding of normally smooth, moist pleurae makes no detectable sound during *auscultation of the lungs* (listening to breath sounds); however, inflammation of the pleura, *pleuritis (pleurisy)*, makes the lung surfaces rough. The resulting friction (*pleural rub*) is detectable with a stethoscope. It sounds like a clump of hairs being rolled between the fingers. The inflamed surfaces of pleura may also cause the parietal and visceral layers of pleura to adhere (*pleural adhesion*). Acute pleuritis is marked by sharp, stabbing pain, especially on exertion, such as climbing stairs, when the rate and depth of respiration may be increased even slightly.

Variations in Lobes of Lung

Variations should be anticipated in the form of the lungs. The oblique and horizontal fissures may be incomplete or absent in some specimens, with consequent reductions in the number or distinctiveness of lobes. Occasionally an extra fissure divides a lung. Consequently, the left lung sometimes has three lobes and the right lung only two. The superior left lobe may not feature a lingula (Fig. 1.33A & B). The most common "accessory" lobe is the *azygos lobe*, which appears in the right lung in approximately 1% of people. The usually small accessory lobe appears superior to the hilum of the right lung, separated from the rest of the lung by a deep groove lodging the arch of the azygos vein. A less common large azygos lobe may appear as a bifurcated apex.

Appearance of Lungs and Inhalation of Carbon Particles and Irritants

The lungs are light pink in healthy children and people who are non-smokers and live in a clean environment. The lungs are commonly dark and mottled in

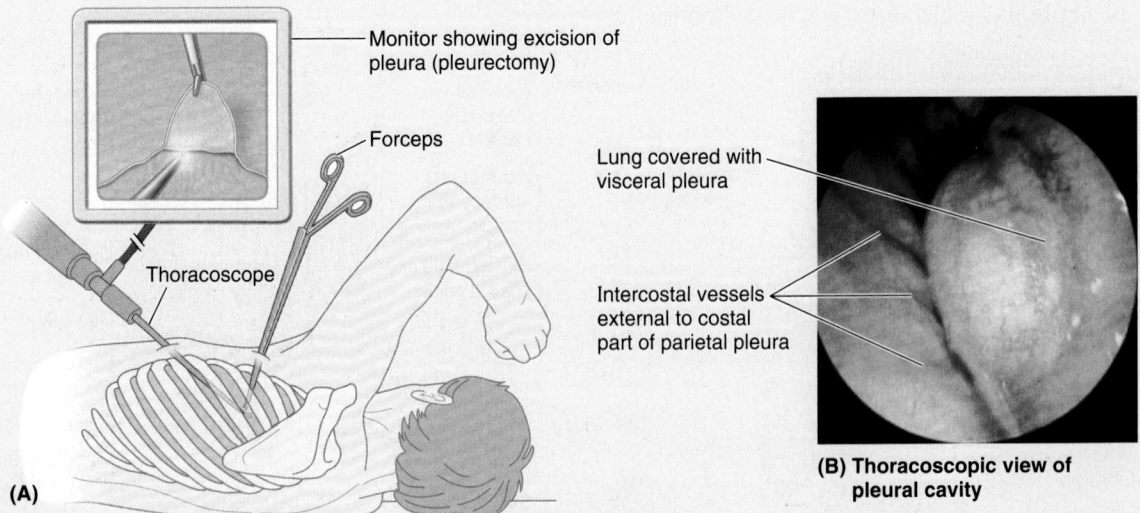

Monitor showing excision of pleura (pleurectomy)

Forceps

Thoracoscope

(A)

Lung covered with visceral pleura

Intercostal vessels external to costal part of parietal pleura

(B) Thoracoscopic view of pleural cavity

FIGURE B1.11. Pleurectomy.

most adults who live in either urban or agricultural areas, especially those who smoke, because of the accumulation of carbon and dust particles in the air and irritants in tobacco that are inhaled. *Smoker's cough* results from the inhalation of these irritants. However the lungs are capable of accumulating a considerable amount of carbon without being adversely affected. Lymph from the lungs carries special cells (*phagocytes*) that remove carbon from the gas-exchanging surfaces and deposit it in the "inactive" connective tissue, which supports the lung, or in lymph nodes receiving lymph from the lungs.

Auscultation of Lungs and Percussion of Thorax

Auscultation of the lungs (listening to their sounds with a stethoscope) and *percussion of the thorax* (tapping on fingers pressed firmly on the thoracic wall over the lungs to detect sounds in the lungs) (Fig. B1.12A) are important techniques used during physical examinations. Auscultation assesses airflow through the tracheobronchial tree into the lobes of the lung. Percussion helps establish whether the underlying tissues are air filled (resonant sound), fluid filled (dull sound), or solid (flat sound). An awareness of normal anatomy, particularly the projection of the lungs and the portions that are overlapped by bone (e.g., the scapula) with associated muscles, enable the examiner to know where flat and resonant sounds should be expected (Fig. B1.12B). Auscultation of the lungs and percussion of the thorax should always include the root of the neck where the apices of the lungs are located (Fig. 1.41A). When clinicians refer to "auscultating the base of the lung," they are not usually referring to its diaphragmatic surface or anatomical base. They are usually referring to the inferoposterior part of the inferior lobe. To auscultate this area, the clinician applies a stethoscope to the posterior thoracic wall at the level of the 10th thoracic vertebra.

Aspiration of Foreign Bodies

Because the right main bronchus is wider and shorter and runs more vertically than the left main bronchus, *aspirated foreign bodies* or food is more likely to enter and lodge in it or one of its branches. A potential hazard encountered by dentists is an aspirated foreign body, such as a piece of tooth or filling material, that is likely to enter the right main bronchus. To maintain a more sterile environment and avoid aspiration of foreign objects, some dentists insert a thin rubber dam into the oral cavity before performing certain procedures.

Bronchoscopy

As a *bronchoscope* proceeds down the trachea to enter a main bronchus, a keel-like ridge, the **carina** (L. keel of a boat), is observed between the orifices of the right and left main bronchi (Fig. B1.13). A cartilaginous projection of the last tracheal ring, the carina normally lies in a sagittal plane and has a fairly definite edge. If the tracheobronchial lymph nodes in the angle between the main bronchi are enlarged because cancer cells have metastasized from a *bronchogenic carcinoma,* for example, the carina is distorted, widened posteriorly, and immobile. Hence, morphological changes in the carina are important diagnostic signs to bronchoscopists in assisting with the differential diagnosis of respiratory disease.

The mucous membrane covering the carina is one of the most sensitive areas of the tracheobronchial tree and is associated with the *cough reflex.* For example, when someone

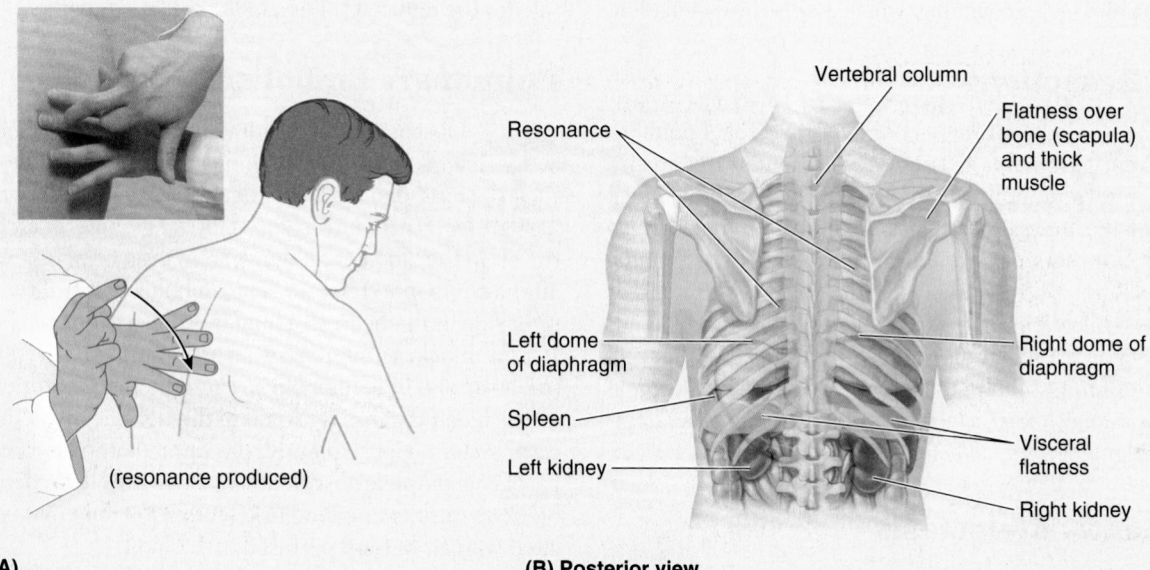

(A)   **(B) Posterior view**

FIGURE B1.12.   Percussion of the lungs. A. Bimanual percussion. **B.** Areas of flatness (yellow shading) and resonance (unshaded) are shown.

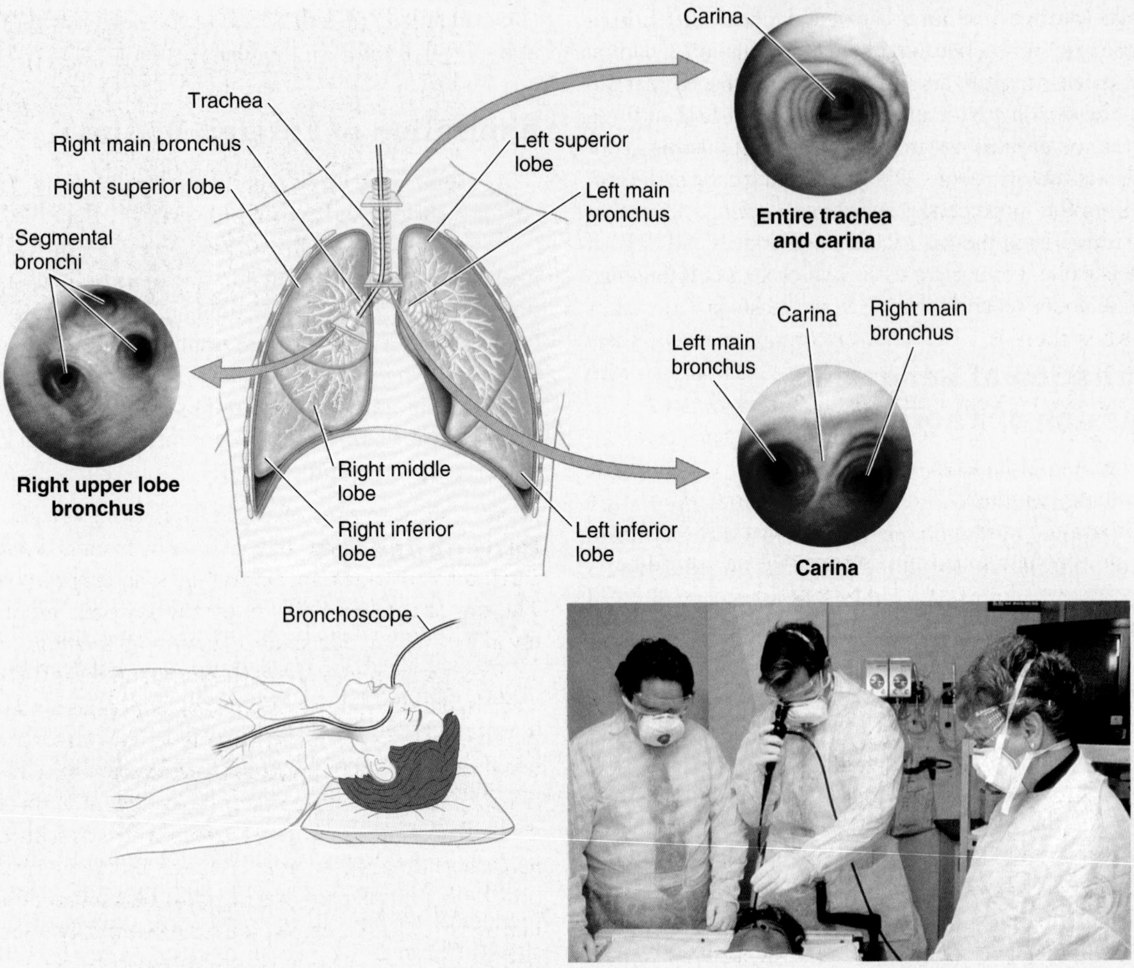

Carina

Trachea

Right main bronchus

Left superior
lobe

Right superior lobe

Left main
bronchus

**Entire trachea
and carina**

Segmental
bronchi

Carina   Right main
bronchus

Left main
bronchus

**Right upper lobe
bronchus**

Right middle
lobe

Right inferior
lobe

Left inferior
lobe

Carina

Bronchoscope

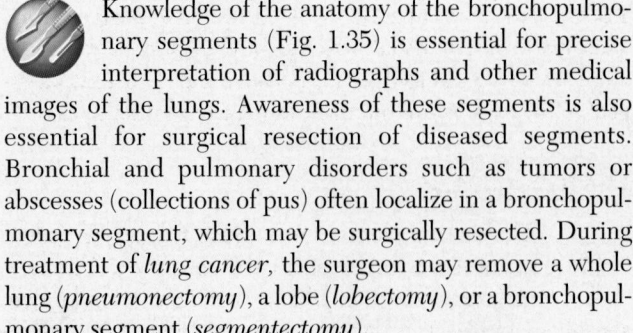

FIGURE B1.13. Bronchoscopy.

aspirates a peanut, they choke and cough. Once the peanut passes the carina, coughing usually stops. If the choking victim is inverted to make use of gravity to expel the foreign body (postural drainage of the lungs), lung secretions passing the carina also cause coughing, which assists the expellation.

Lung Resections

Knowledge of the anatomy of the bronchopulmonary segments (Fig. 1.35) is essential for precise interpretation of radiographs and other medical images of the lungs. Awareness of these segments is also essential for surgical resection of diseased segments. Bronchial and pulmonary disorders such as tumors or abscesses (collections of pus) often localize in a bronchopulmonary segment, which may be surgically resected. During treatment of *lung cancer*, the surgeon may remove a whole lung (*pneumonectomy*), a lobe (*lobectomy*), or a bronchopulmonary segment (*segmentectomy*).

Segmental Atelectasis

Blockage of a segmental bronchus (Fig. 1.35E, e.g., by an aspirated object) will prevent air from reaching the bronchopulmonary segment it supplies. The air

in the blocked segment will gradually be absorbed into the blood, and the segment will collapse. A segmental collapse does not require compensating space in the pleural cavity to be realized because adjacent segments will expand to compensate for the reduced volume of the collapsed segment.

Pulmonary Embolism

Obstruction of a pulmonary artery by a blood clot (*embolus*) is a common cause of morbidity (sickness) and mortality (death). An embolus in a pulmonary artery forms when a blood clot, fat globule, or air bubble travels in the blood to the lungs from a leg vein, for example, after a compound fracture. The embolus passes through the right side of the heart to a lung through a pulmonary artery. It may block a pulmonary artery—*pulmonary embolism (PE)*—or one of its branches. The pulmonary arteries carry all the blood that has returned to the right heart via the vena caval system. Consequently, the immediate result of PE is partial or complete obstruction of blood flow to the lung. The blockage results in a lung or a sector of a lung that is ventilated with air but not perfused with blood.

When a large embolus occludes a pulmonary artery, the patient suffers *acute respiratory distress* because of a major decrease in the oxygenation of blood, owing to blockage of blood flow through the lung. Conversely, the right side of the heart may become acutely dilated because the volume of blood arriving from the systemic circuit cannot be pushed through the pulmonary circuit (*acute cor pulmonale*). In either case, death may occur in a few minutes. A medium-size embolus may block an artery supplying a bronchopulmonary segment, producing a *pulmonary infarct*, an area of necrotic (dead) lung tissue.

In physically active people, a collateral circulation—an indirect, accessory blood supply—often exists and develops further when there is a PE so that infarction is not as likely to occur, or at least not be as devastating. Anastomoses with branches of the bronchial arteries abound in the region of the terminal bronchioles. In ill people with impaired circulation in the lung, such as *chronic congestion*, PE commonly results in lung infarction. When an area of visceral pleura is also deprived of blood, it becomes inflamed (*pleuritis*) and irritates or becomes fused to the sensitive parietal pleura, resulting in pain. Pain from the parietal pleura is referred to the cutaneous distribution of the intercostal nerves to the thoracic wall, or, in the case of inferior nerves, to the anterior abdominal wall.

Lymphatic Drainage and Pleural Adhesion

If the parietal and visceral layers of pleura adhere (*pleural adhesion*), the lymphatic vessels in the lung and visceral pleura may *anastomose* (join) with parietal lymphatic vessels that drain into the axillary lymph nodes. The presence of carbon particles in these nodes is presumptive evidence of pleural adhesion.

Hemoptysis

Spitting of blood or blood-stained sputum derived from the lungs and/or bronchial tree is due to bronchial or pulmonary hemorrhage. In about 95% of cases, the bleeding is from branches of the bronchial arteries. The most common causes include *bronchitis* (inflammation of the bronchi), *lung cancer, pneumonia, bronchiectasis, pulmonary embolism,* and *tuberculosis.*

Bronchogenic Carcinoma

The term *bronchogenic carcinoma* was once a specific designation for cancer arising in a bronchus—usually squamous- (oat) or small-cell carcinoma (cancer)—but now the term refers to any lung cancer. *Lung cancer* (carcinoma—CA) is mainly caused by cigarette smoking; most cancers arise in the mucosa of the large bronchi and produce a persistent, productive cough or *hemoptysis* (spitting of blood). Malignant (cancer) cells can be detected in the sputum (saliva-borne matter). The primary tumor, observed radiologically as an enlarging lung mass (Fig. B1.14), metastasizes early to the bronchopulmonary lymph nodes and subsequently to other thoracic lymph

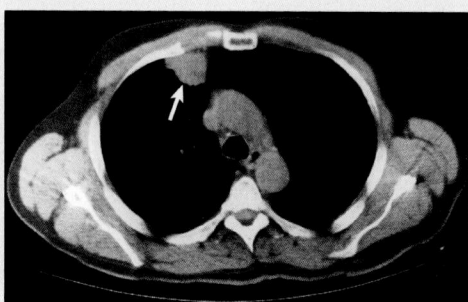
Lung carcinoma: CT scan shows a mass in anterior segment of right upper lobe (arrow) adjacent to pleura.
FIGURE B1.14. Lung cancer.

nodes. Common sites of *hematogenous metastases* (spreading through the blood) of cancer cells from a bronchogenic carcinoma are the brain, bones, lungs, and suprarenal glands. The tumor cells probably enter the systemic circulation by invading the wall of a sinusoid or venule in a lung. It is then transported through the pulmonary veins, left heart, and aorta to these structures. Often the lymph nodes superior to the clavicle—the *supraclavicular lymph nodes*—are enlarged when bronchogenic carcinoma develops owing to metastases of cancer cells from the tumor. Consequently, the supraclavicular lymph nodes were once referred to as *sentinel lymph nodes* because their enlargement alerted the physician to the possibility of malignant disease in the thoracic and/or abdominal organs. More recently, the term sentinel lymph node has been applied to a node or nodes that first receive lymph draining from a cancer-containing area, regardless of location, following injection of blue dye containing radioactive tracer (technetium-99).

Lung Cancer and Mediastinal Nerves

Lung cancer involving a phrenic nerve may result in paralysis of one half of the diaphragm (hemidiaphragm). Because of the intimate relationship of the *recurrent laryngeal nerve* to the apex of the lung (Fig. 1.33C), this nerve may be involved in *apical lung cancers.* This involvement usually results in hoarseness owing to paralysis of a vocal fold (cord) because the recurrent laryngeal nerve supplies all but one of the laryngeal muscles.

Pleural Pain

The visceral pleura is insensitive to pain because it receives no nerves of general sensation. The parietal pleura (particularly the costal part) is extremely sensitive to pain. The parietal pleura is richly supplied by branches of the intercostal and phrenic nerves. Irritation of the parietal pleura may produce local pain or referred pain projected to dermatomes supplied by the same spinal (posterior root) ganglia and segments of the spinal cord. Irritation of the costal and peripheral parts of the diaphragmatic pleura results in local pain and referred pain to the dermatomes of

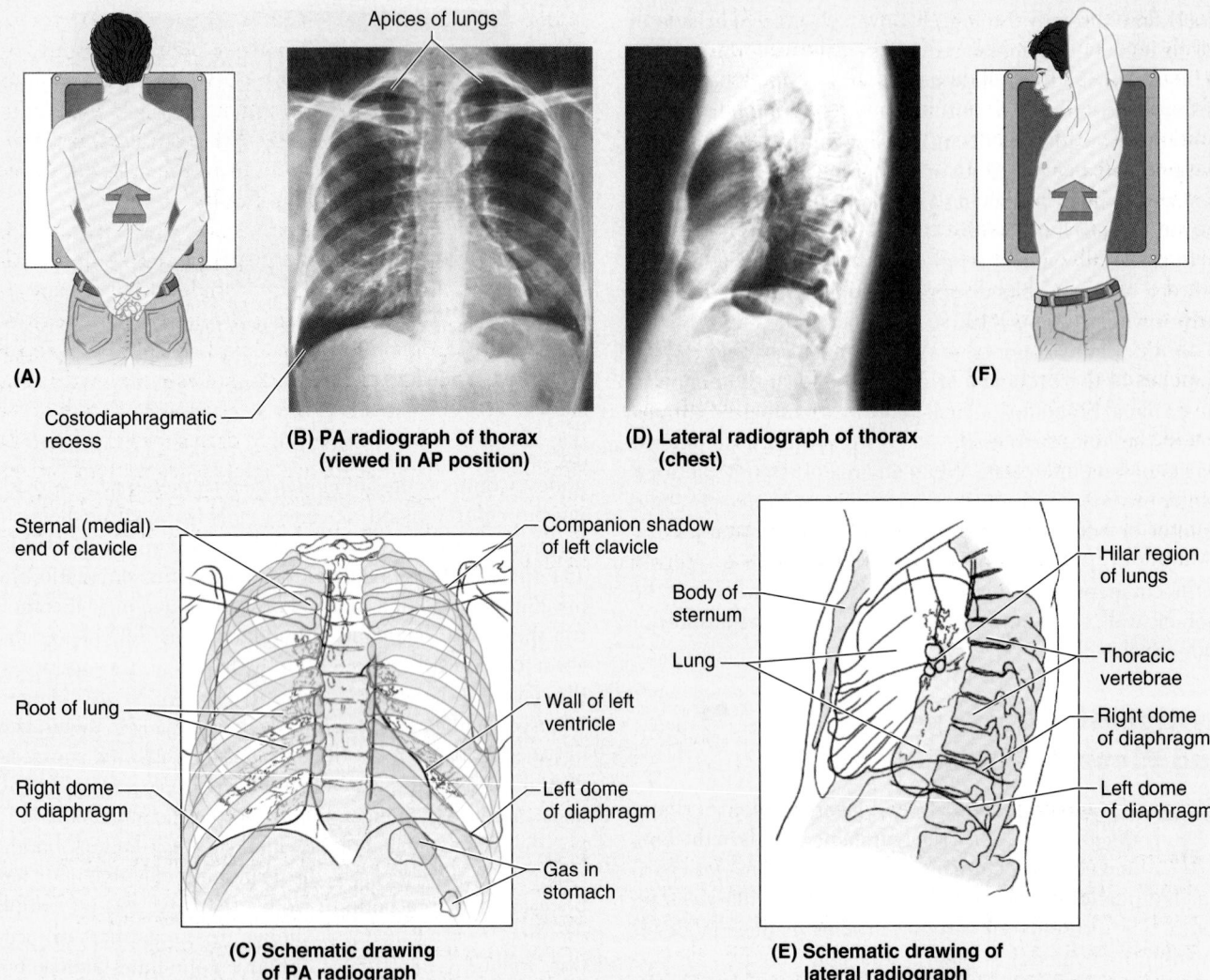

Apices of lungs

(A)

Costodiaphragmatic recess

(B) PA radiograph of thorax (viewed in AP position)

(D) Lateral radiograph of thorax (chest)

(F)

Sternal (medial) end of clavicle

Companion shadow of left clavicle

Root of lung

Wall of left ventricle

Right dome of diaphragm

Left dome of diaphragm

Gas in stomach

(C) Schematic drawing of PA radiograph

Body of sternum

Lung

Hilar region of lungs

Thoracic vertebrae

Right dome of diaphragm

Left dome of diaphragm

(E) Schematic drawing of lateral radiograph

FIGURE B1.15. Thoracic radiography. A. Orientation for PA projection (*arrow* = X-ray beam) **B.** PA radiograph of thorax (viewed in AP position). **C.** Schematic drawing of PA radiograph. **D.** Lateral radiograph of the thorax (chest). **E.** Schematic drawing of lateral radiograph. **F.** Orientation for a lateral radiograph (*arrow* = X-ray beam).

the thoracic and abdominal walls. Irritation of the mediastinal and central diaphragmatic areas of parietal pleura results in referred pain to the root of the neck and over the shoulder (C3–C5 dermatomes).

Chest X-ray

The most common radiographic study of the thorax is the *posteroanterior (PA) projection* (Fig. B1.15A), producing a **PA radiograph** (Fig. B1.15B) used primarily to examine the thoracic respiratory and cardiovascular structures, as well as the thoracic wall. The radiologist or technician places the anterior aspect of the patient's thorax against the X-ray detector or film cassette and rotates the shoulders anteriorly to move the scapulae away from the superior parts of the lungs (Fig. B1.15A). The person takes a deep breath and holds it. The deep inspiration causes the diaphragmatic domes to descend, filling the lungs with air

(increasing their radiolucency), and moving the inferior margins of the lungs into the costodiaphragmatic recesses. The inferior margins should appear as sharp, acute angles. Pleural effusions accumulating here do not allow the inferior margin to descend into the recess and the usual radiolucent air density here is replaced with a hazy radiopacity. Lobar disease, such as pneumonia, appears as localized, relatively radiodense areas that contrast with the radiolucency of the remainder of the lung.

A PA radiograph, which is viewed as if you were facing the patient (an *anteroposterior [AP] view*), is a composite of the images cast by the soft tissues and bones of the thoracic wall. Soft tissues, including those of the breasts, cast shadows of varying density, depending on their composition and thickness. Paralleling the superior margins of the clavicles are shadows cast by the skin and subcutaneous tissues covering these bones. The clavicles, ribs, and inferior cervical and superior thoracic vertebrae are visible. In PA radiographs, most ribs stand out clearly against the background of the relatively lucent lungs (Fig. B1.15B & C). The inferior ribs tend to be obscured by the diaphragm and the superior contents of the abdomen (e.g., liver), depending on the phase of respiration when the radiograph was taken. Usually only lateral margins of the manubrium are visible in these projections. The lower thoracic vertebrae are more or less obscured by the sternum and mediastinum. Uncommonly, cervical ribs, missing ribs, forked ribs, and fused ribs are visible. Occasionally, the costal cartilages are calcified in older people (especially the inferior cartilages).

In PA projections, the right and left domes of the diaphragm are separated by the central tendon, which is obscured by the heart. The right dome of the diaphragm, formed by the underlying liver, is usually approximately half an intercostal space higher than the left dome. The lungs, because of their low density, are relatively lucent compared with surrounding structures. The lungs exhibit a radiodensity similar to that of air and, therefore, produce paired radiolucent areas. In PA projections, the lungs are obscured inferior to the domes of the diaphragm and anterior and posterior to the mediastinum. The pulmonary arteries are visible in the hilar region of each lung. Intrapulmonary vessels are slightly larger in caliber in the inferior lobes. Transverse sections of the air-filled bronchi have lucent centers and thin walls.

The areas obscured in PA projections are usually visible in lateral radiographs. In **lateral projections,** the middle and inferior thoracic vertebrae are visible, although they are partially obscured by the ribs (Fig. B1.15D & E). The three parts of the sternum are also visible. **Lateral radiographs** allow better viewing of a lesion or anomaly confined to one side of the thorax. In a lateral projection, both domes of the diaphragm are often visible as they arch superiorly from the sternum. A lateral radiograph is made using a lateral projection, with the side of the thorax against the film cassette or X-ray detector and the upper limbs elevated over the head (Fig. B1.15F).

The Bottom Line

PLEURAE, LUNGS, AND TRACHEOBRONCHIAL TREE

Pleurae: The thoracic cavity is divided into three compartments: two bilateral pulmonary cavities that are entirely separated by the central mediastinum. ♦ The pulmonary cavities are completely lined by membranous parietal pleura that reflects onto the lungs at their roots, becoming the visceral pleura that intimately invests the lungs' outer surface. ♦ The pleural cavity between the two layers of the pleural sac is empty, except for a lubricating film of pleural fluid. The pleural fluid prevents the lungs from collapse and causes the lungs to expand when the thorax expands for inhalation. ♦ Most of the parietal pleura is named for the structures it covers: costal, mediastinal, and diaphragmatic parts. ♦ The cervical pleura extends into the root of the neck forming a dome above the anterior aspect of the 1st rib and clavicle. ♦ Parietal pleura is sensitive, being innervated by the phrenic and intercostal nerves. ♦ Because the lungs do not completely fill the pulmonary cavities, and because of the protrusion of the diaphragm and underlying abdominal viscera into the inferior thoracic aperture, a peripheral gutter—the costodiaphragmatic recess—is formed. Extrapulmonary fluids (exudates) accumulate in this space when the trunk is erect.

Lungs: The lungs are the vital organs of respiration in which venous blood exchanges oxygen and carbon dioxide with a tidal airflow. ♦ Air and blood are delivered to each lung via its root, consisting of a pulmonary artery and vein and a main bronchus and their branches/tributaries that enter the lung at its hilum. ♦ Both lungs are pyramidal, having an apex, a base, three surfaces, and three borders. ♦ The right lung has three lobes that are separated by horizontal and oblique fissures. ♦ The left lung has two lobes, separated by an oblique fissure, and features a marked cardiac notch in its anterior border owing to the asymmetrical placement of the heart.

Tracheobronchial tree: The tracheobronchial tree is distinguished grossly by cartilage in its walls. ♦ The bifurcation of the trachea (at the level of the sternal angle) is asymmetrical: the right main bronchus is more vertical and of greater caliber than the left. ♦ The bronchi and pulmonary arteries course and branch together: the main bronchi/arteries each serve a lung, second-order lobar branches supply two left and three right lobes, and third-order segmental branches supply the 8–10 bronchopulmonary segments of each lung. ♦ The bronchopulmonary segment is the smallest resectable division of the lung. ♦ The pulmonary veins run independent intersegmental courses, draining adjacent bronchopulmonary segments. ♦ The structures of the root of the lung and supporting tissues (and part of the esophagus) are supplied by bronchial arteries. ♦ The lymphatic drainage of the lungs follows a mostly predictable course, with most of the right lung and the superior lobe of the left lung following ipsilateral pathways to the right lymphatic trunk and thoracic duct. However, most of the drainage from the left inferior lobe passes to the right pathway. Nerve fibers of the pulmonary plexuses are autonomic (bronchioconstrictive and secretomotor vagal parasympathetic fibers; inhibitory and vasoconstrictive sympathetic fibers) and visceral afferent (reflex and pain).

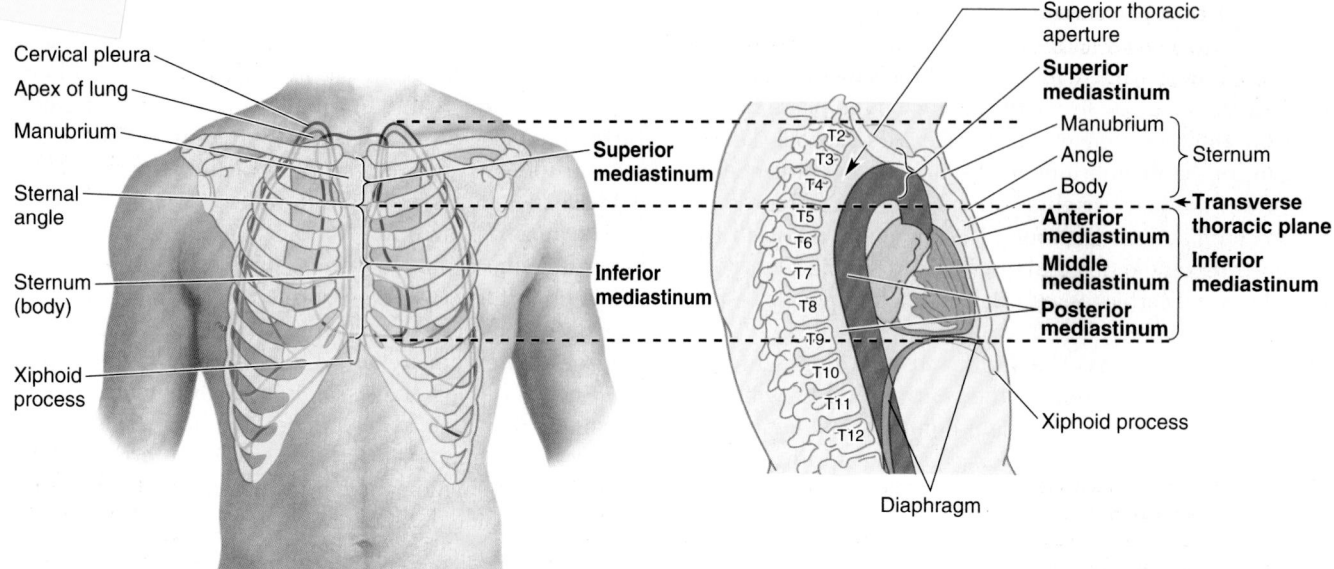

Cervical pleura
Apex of lung
Manubrium
Sternal angle
Sternum (body)
Xiphoid process

Superior mediastinum
Inferior mediastinum

Superior thoracic aperture
Superior mediastinum
Manubrium
Angle — Sternum
Body
Anterior mediastinum
Middle mediastinum
Posterior mediastinum
← Transverse thoracic plane
Inferior mediastinum

Xiphoid process

Diaphragm

FIGURE 1.42. Subdivisions and levels of mediastinum. The subdivisions of the mediastinum are demonstrated as if the person were in the supine position. The level of the viscera relative to the subdivisions as defined by thoracic cage landmarks depends on the individual's position because the soft tissue of the mediastinum sags with the pull of gravity.

Overview of Mediastinum

The **mediastinum** (Mod. L. middle septum), occupied by the mass of tissue between the two pulmonary cavities, is the central compartment of the thoracic cavity (Fig. 1.42). It is covered on each side by *mediastinal pleura* and contains all the thoracic viscera and structures except the lungs. The mediastinum extends from the superior thoracic aperture to the diaphragm inferiorly, and from the sternum and costal cartilages anteriorly to the bodies of the thoracic vertebrae posteriorly. Unlike the rigid structure observed in the embalmed cadaver, the mediastinum in living people is a highly mobile region because it consists primarily of hollow (liquid- or air-filled) visceral structures united only by loose connective tissue, often infiltrated with fat. The major structures in the mediastinum are also surrounded by blood and lymphatic vessels, lymph nodes, nerves, and fat.

The looseness of the connective tissue and the elasticity of the lungs and parietal pleura on each side of the mediastinum enable it to accommodate movement as well as volume and pressure changes in the thoracic cavity; for example, those resulting from movements of the diaphragm, thoracic wall, and tracheobronchial tree during respiration, contraction (beating) of the heart and pulsations of the great arteries, and passage of ingested substances through the esophagus. The connective tissue becomes more fibrous and rigid with age; hence the mediastinal structures become less mobile. The mediastinum is divided into superior and inferior parts for purposes of description (Fig. 1.42).

The **superior mediastinum** extends inferiorly from the superior thoracic aperture to the horizontal plane that includes the *sternal angle* anteriorly and passes approximately through the *junction (IV disc) of T4 and T5 vertebrae* posteriorly, often referred to as the **transverse thoracic plane.** The **inferior mediastinum**—between the transverse thoracic plane and the diaphragm—is further subdivided by the pericardium into anterior, middle, and posterior parts. The pericardium and its contents (heart and roots of its great vessels) constitute the **middle mediastinum.** Some structures, such as the esophagus, pass vertically through the mediastinum and therefore lie in more than one mediastinal compartment.

Pericardium

The middle mediastinum includes the pericardium, heart, and roots of its great vessels (Fig. 1.34)—ascending aorta, pulmonary trunk, and SVC—passing to and from the heart.

The **pericardium** is a fibroserous membrane that covers the heart and the beginning of its great vessels (Figs. 1.33B and 1.43). The pericardium is a closed sac composed of two layers. The tough external layer, the **fibrous pericardium,** is continuous with the central tendon of the diaphragm (Fig. 1.32). The internal surface of the fibrous pericardium is lined with a glistening serous membrane, the **parietal layer of serous pericardium.** This layer is reflected onto the heart at the great vessels (aorta, pulmonary trunk and veins, and superior and inferior venae cavae) as the **visceral layer of serous pericardium.** The **serous pericardium** is composed mainly of mesothelium, a single layer of flattened cells forming an epithelium that lines both the internal surface of the fibrous pericardium and the external surface of the heart. The fibrous pericardium is:

- Continuous superiorly with the *tunica adventitia* (perivascular connective tissue) of the great vessels entering and leaving the heart and with the pretracheal layer of deep cervical fascia.
- Attached anteriorly to the posterior surface of the sternum by the **sternopericardial ligaments,** which are highly variable in their development.

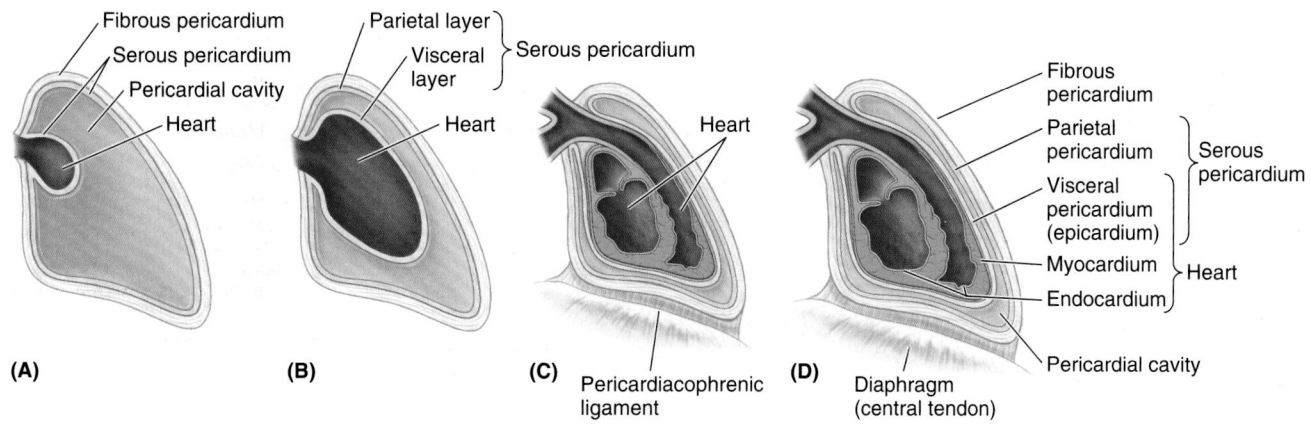

FIGURE 1.43. Pericardium and heart. A. The heart occupies the middle mediastinum and is enclosed by pericardium, composed of two parts. The tough, outer fibrous pericardium stabilizes the heart and helps prevent it from overdilating. Between the fibrous pericardium and the heart is a "collapsed" sac, the serous pericardium. The embryonic heart invaginates the wall of the serous sac (**B**) and soon practically obliterates the pericardial cavity (**C**), leaving only a potential space between the layers of serous pericardium. **C.** and **D.** The pericardiacophrenic ligament is the continuity of the fibrous pericardium with the central tendon of the diaphragm.

- Bound posteriorly by loose connective tissue to structures in the posterior mediastinum.
- Continuous inferiorly with the central tendon of the diaphragm (Fig. 1.43C & D).

The inferior wall (floor) of the fibrous pericardial sac is firmly attached and confluent (partially blended) centrally with the central tendon of the diaphragm. The site of continuity has been referred to as the **pericardiacophrenic ligament;** however, the fibrous pericardium and central tendon are not separate structures that fused together secondarily, nor are they separable by dissection. As a result of the attachments just described, the heart is relatively well tethered in place inside this fibrous sac. The pericardium is influenced by movements of the heart and great vessels, the sternum, and the diaphragm.

The heart and roots of the great vessels within the pericardial sac lie posterior to the sternum, costal cartilages, and anterior ends of the 3rd–5th ribs on the left side (Fig. 1.44). The heart and pericardial sac are situated obliquely, approximately two thirds to the left and one third to the right of the median plane. If you turn your face to the left about 45° without rotating your shoulders, the rotation of your head approximates that of the heart relative to the trunk.

The fibrous pericardium protects the heart against sudden overfilling because it is so unyielding and closely related to the great vessels that pierce it superiorly. The ascending aorta carries the pericardium superiorly beyond the heart to the level of the sternal angle.

The **pericardial cavity** is the potential space between opposing layers of the parietal and visceral layers of serous pericardium. It normally contains a thin film of fluid that enables the heart to move and beat in a frictionless environment.

The *visceral layer of serous pericardium* forms the *epicardium*, the outermost of three layers of the heart wall. It extends onto the beginning of the great vessels, becoming continuous with the parietal layer of serous pericardium (1) where the aorta and pulmonary trunk leave the heart and (2) where the SVC,

inferior vena cava (IVC), and pulmonary veins enter the heart. The **transverse pericardial sinus** is a transversely running passage within the pericardial cavity between these two groups of vessels and the reflections of serous pericardium around them. The reflection of the serous pericardium around the second group of vessels defines the *oblique pericardial sinus*. The pericardial sinuses form during development of the heart as a consequence of the folding of the primordial heart tube. As the heart tube folds, its venous end moves posterosuperiorly (Fig. 1.45) so that the venous end of the tube lies adjacent to the arterial end, separated only by the transverse pericardial sinus (Fig. 1.46). Thus the transverse sinus is posterior to the intrapericardial parts of the pulmonary trunk and ascending aorta, anterior to the SVC, and superior to the atria of the heart.

As the veins of the heart develop and expand, a pericardial reflection surrounding them forms the **oblique pericardial sinus,** a wide pocket-like recess in the pericardial cavity posterior to the base (posterior aspect) of the heart, formed by the left atrium (Figs. 1.45 and 1.46). The oblique sinus is bounded laterally by the pericardial reflections surrounding the pulmonary veins and IVC and posteriorly by the pericardium overlying the anterior aspect of the esophagus. The oblique sinus can be entered inferiorly and will admit several fingers; however, they cannot pass around any of these structures because the sinus is a blind sac (cul-de-sac).

The **arterial supply of the pericardium** (Fig. 1.47) is mainly from a slender branch of the internal thoracic artery, the **pericardiacophrenic artery,** that often accompanies or at least parallels the phrenic nerve to the diaphragm. Smaller contributions of blood come from the:

- *Musculophrenic artery,* a terminal branch of the internal thoracic artery.
- *Bronchial, esophageal,* and *superior phrenic arteries,* branches of the thoracic aorta.
- *Coronary arteries* (visceral layer of serous pericardium only), the first branches of the aorta.

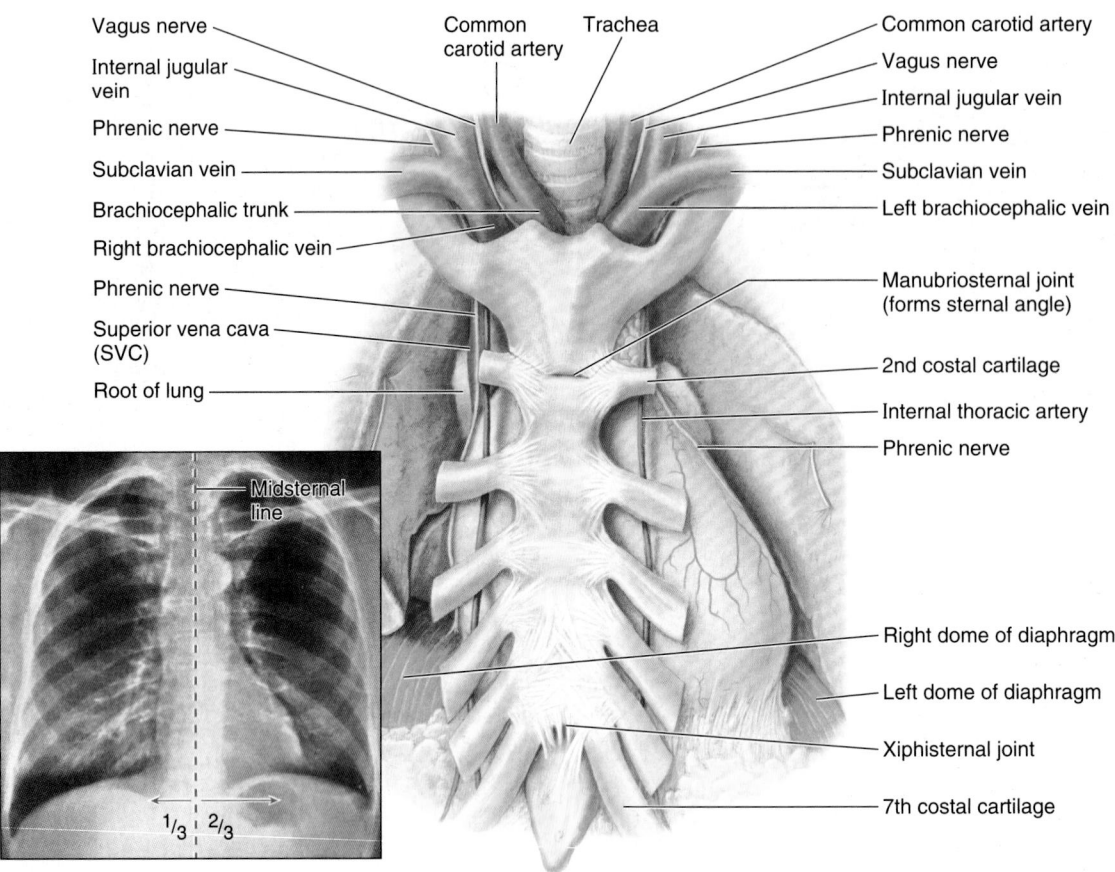

Anterior views

FIGURE 1.44. Pericardial sac in relation to sternum and phrenic nerves. This dissection exposes the pericardial sac posterior to the body of the sternum from just superior to the sternal angle to the level of the xiphisternal joint. The pericardial sac (and therefore the heart) lies approximately one third to the right of the midsternal line and two thirds to the left (inset).

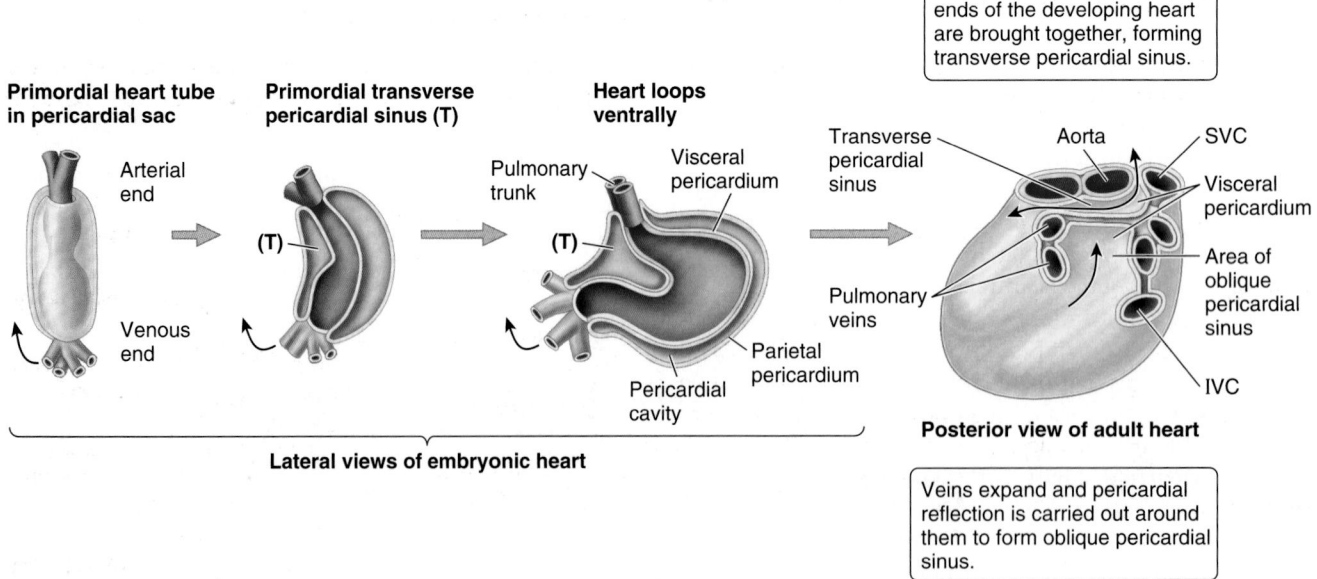

FIGURE 1.45. Development of heart and pericardium. The longitudinal embryonic heart tube invaginates the double-layered pericardial sac (somewhat like placing a wiener in a hot dog bun). The primordial heart tube then "loops" ventrally, bringing the primordial arterial and venous ends of the heart together and creating the primordial transverse pericardial sinus (*T*) between them. With growth of the embryo, the veins expand and spread apart, inferiorly and laterally. The pericardium reflected around them forms the boundaries of the oblique pericardial sinus. *IVC*, inferior vena cava; *SVC*, superior vena cava.

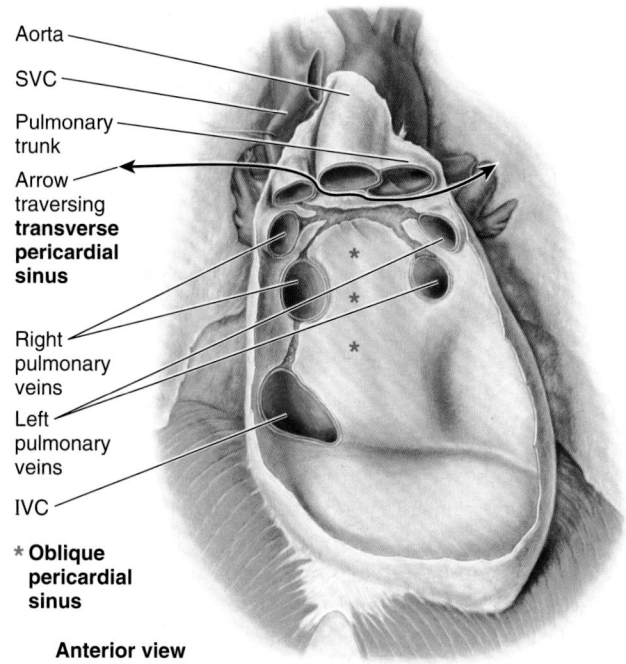

Anterior view

FIGURE 1.46. **Interior of pericardial sac.** To remove the heart from the sac, the eight vessels piercing the sac were severed. The oblique pericardial sinus is circumscribed by five veins. The superior vena cava (*SVC*), pulmonary trunk, and especially the aorta have intrapericardial portions. The peak of the pericardial sac occurs at the junction of the ascending aorta and the arch of the aorta. The transverse pericardial sinus is bounded anteriorly by the serous pericardium covering the posterior aspect of the pulmonary trunk and ascending aorta, posteriorly by that covering the SVC, and inferiorly by the visceral pericardium covering the atria of the heart. *IVC,* inferior vena cava.

The **venous drainage of the pericardium** is from the:

• *Pericardiacophrenic veins,* tributaries of the brachioce-phalic (or internal thoracic) veins.
• Variable tributaries of the *azygos venous system* (discussed later in this chapter).

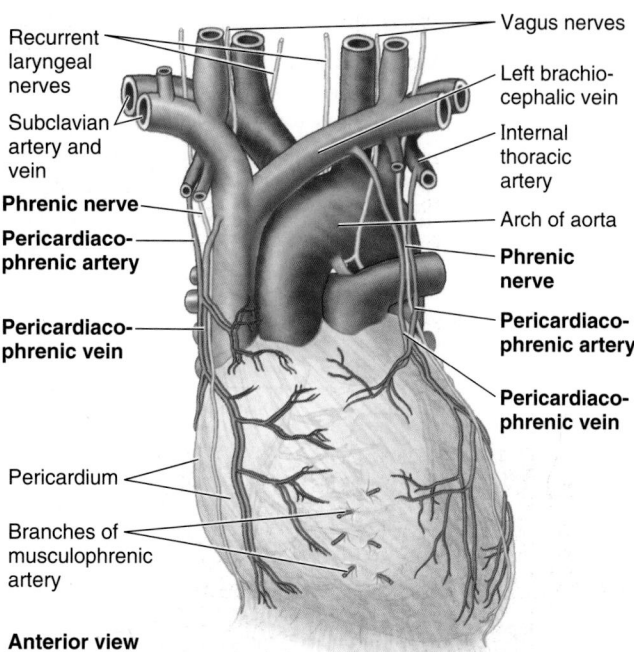

Anterior view

FIGURE 1.47. **Arterial supply and venous drainage of pericardium.** The arteries of the pericardium derive primarily from the internal thoracic arteries with minor contributions from their musculophrenic branches and the thoracic aorta. The veins are tributaries of the brachiocephalic veins.

The **nerve supply of the pericardium** is from the:

• *Phrenic nerves* (C3–C5), primary source of sensory fibers; pain sensations conveyed by these nerves are commonly referred to the skin (C3–C5 dermatomes) of the ipsilateral supraclavicular region (top of the shoulder of the same side).
• *Vagus nerves,* function uncertain.
• *Sympathetic trunks,* vasomotor.

The innervation of the pericardium by the phrenic nerves, and the course of these *somatic nerves* between the heart and the

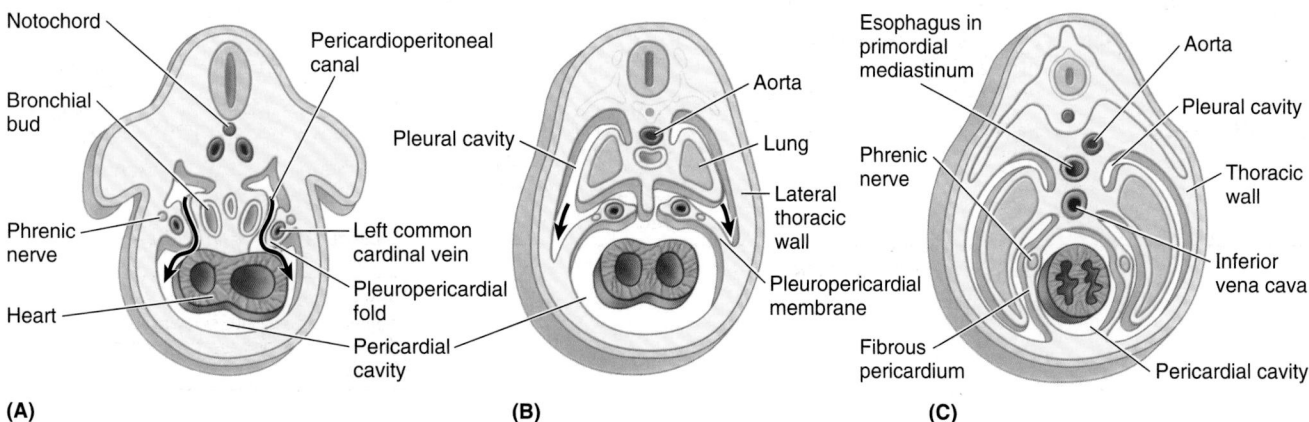

FIGURE 1.48. **Development of fibrous pericardium and relocation of phrenic nerve.** Exuberant growth of the lungs into the primordial pleura cavities (pleuroperitoneal canals) cleaves the pleuropericardial folds from the body wall, creating the pleuropericardial membranes. The membranes include the phrenic nerve and become the fibrous pericardium that encloses the heart and separates the pleural and pericardial cavities.

lungs, make little sense unless the development of the fibrous pericardium is considered. A membrane (*pleuropericardial membrane*) that includes the phrenic nerve is split or separated from the developing body wall by the developing pleural cavities, which extend to accommodate the rapidly growing lungs (Fig. 1.48). The lungs develop within the pericardioperitoneal canals that run on both sides of the foregut, connecting the thoracic and abdominal cavities on each side of the septum transversum. The canals (primordial pleural cavities) are too small to accommodate the rapid growth of the lungs, and they begin to invade the mesenchyme of the body wall posteriorly, laterally, and anteriorly, splitting it into two layers: an outer layer that becomes the definitive thoracic wall (ribs and intercostal muscles) and an inner or deep layer (the pleuropericardial membranes) that contains the phrenic nerves and forms the fibrous pericardium (Moore, Persaud, and Torchia, 2012). Thus the pericardial sac can be a source of pain just as the rib cage or parietal pleura can be, although the pain tends to be referred to dermatomes of the body wall—areas from which we more commonly receive sensation.

MEDIASTINUM OVERVIEW AND PERICARDIUM

Levels of Viscera Relative to Mediastinal Divisions

The division between the superior and inferior mediastinum (*transverse thoracic plane*) is defined in terms of bony body wall structures and is mostly independent of gravitational effects. The level of the viscera relative to the subdivisions of the mediastinum depends on the position of the person (i.e., gravity). When a person is supine or when a cadaver is being dissected, the viscera are positioned higher (more superior) relative to the subdivisions of the mediastinum than when the person is upright (Figs. 1.42 and B1.16A). In other words, gravity pulls the viscera downward when we are vertical.

Anatomical descriptions traditionally describe the level of the viscera as if the person were in the *supine position*—that is, lying face upward in bed or on the operating or dissection table. In this position, the abdominal viscera spread horizontally, pushing the mediastinal structures superiorly. However, when the individual is standing or sitting erect, the levels of the viscera are as shown in Figure B1.16B. This occurs because the soft structures in the mediastinum, especially the pericardium and its contents, the heart and great vessels, and the abdominal viscera supporting them, sag inferiorly under the influence of gravity.

In the supine position (Fig. B1.16A), the:

- Arch of the aorta lies superior to the transverse thoracic plane.
- Bifurcation of the trachea is transected by the transverse thoracic plane.
- Central tendon of the diaphragm (or the diaphragmatic surface or inferior extent of the heart) lies at the level of the xiphisternal junction and vertebra T9.

When *standing or sitting upright* (Fig. B1.16B), the:

- Arch of the aorta is transected by the transverse thoracic plane.
- Tracheal bifurcation lies inferior to the transverse thoracic plane.
- Central tendon of the diaphragm may fall to the level of the middle of the xiphoid process and T9–T10 IV discs.

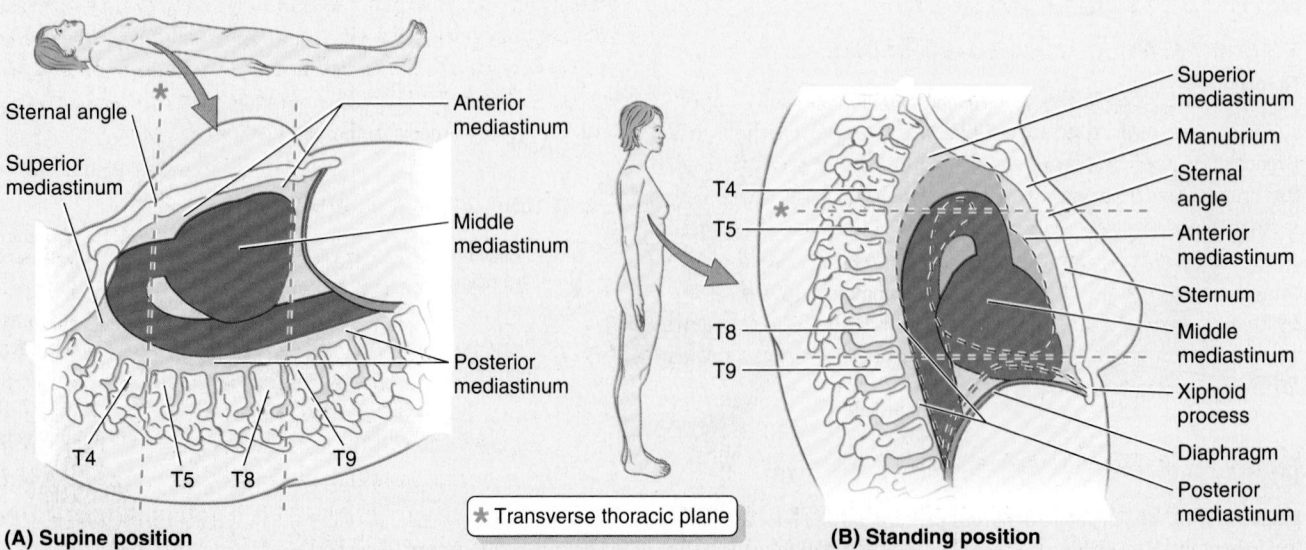

(A) Supine position   ★ Transverse thoracic plane   **(B) Standing position**

FIGURE B1.16. Position of thoracic viscera in supine and standing positions.

This vertical movement of mediastinal structures must be considered during physical and radiological examinations in the erect and supine positions. In addition, when lying on one's side, the mediastinum sags toward the lower side under the pull of gravity.

Mediastinoscopy and Mediastinal Biopsies

Using an endoscope (*mediastinoscope*), surgeons can see much of the mediastinum and conduct minor surgical procedures. They insert the endoscope through a small incision at the root of the neck, just superior to the jugular notch of the manubrium, into the potential space anterior to the trachea. During *mediastinoscopy,* surgeons can view or biopsy mediastinal lymph nodes to determine if cancer cells have metastasized to them from a bronchogenic carcinoma, for example. The mediastinum can also be explored and biopsies taken through an *anterior thoracotomy* (removing part of a costal cartilage; see the blue box "Thoracotomy, Intercostal Space Incisions, and Rib Excision," p. 83).

Widening of Mediastinum

Radiologists and emergency physicians sometimes observe widening of the mediastinum when viewing chest radiographs. Any structure in the mediastinum may contribute to pathological widening. It is often observed after trauma resulting from a head-on collision, for example, which produces hemorrhage into the mediastinum from lacerated great vessels, such as the aorta or SVC. Frequently, *malignant lymphoma* (cancer of lymphatic tissue) produces massive enlargement of mediastinal lymph nodes and widening of the mediastinum. *Hypertrophy* (enlargement) *of the heart* (often occurring due to *congestive heart failure,* in which venous blood returns to the heart at a rate that exceeds cardiac output) is a common cause of widening of the inferior mediastinum.

Surgical Significance of Transverse Pericardial Sinus

The *transverse pericardial sinus* is especially important to cardiac surgeons. After the pericardial sac is opened anteriorly, a finger can be passed through the transverse pericardial sinus posterior to the ascending aorta and pulmonary trunk (Fig. B1.17). By passing a surgical clamp or a ligature around these large vessels, inserting the tubes of a coronary bypass machine, and then tightening the ligature, surgeons can stop or divert the circulation of blood in these arteries while performing cardiac surgery, such as *coronary artery bypass grafting.*

Exposure of Venae Cavae

After ascending through the diaphragm, the entire thoracic part of the IVC (approximately 2 cm) is enclosed by the pericardium. Consequently, the pericardial sac must be opened to expose this terminal part

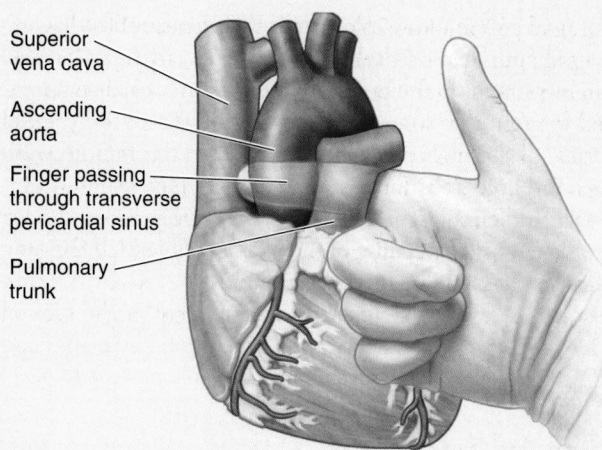

FIGURE B1.17. **Transverse pericardial sinus.**

of the IVC. The same is true for the terminal part of the SVC, which is partly inside and partly outside the pericardial sac.

Pericarditis, Pericardial Rub, and Pericardial Effusion

The pericardium may be involved in several disease processes. Inflammation of the pericardium (*pericarditis*) usually causes chest pain. It may also make the serous pericardium rough. Usually the smooth opposing layers of serous pericardium make no detectable sound during auscultation. If there is pericarditis, friction of the roughened surfaces may sound like the rustle of silk when listening with a stethoscope over the left sternal border and upper ribs (*pericardial friction rub*). A chronically inflamed and thickened pericardium may calcify, seriously hampering cardiac efficiency. Some inflammatory diseases produce *pericardial effusion* (passage of fluid from pericardial capillaries into the pericardial cavity, or an accumulation of pus). As a result, the heart becomes compressed (unable to expand and fill fully) and ineffective. Noninflammatory pericardial effusions often occur with *congestive heart failure*, in which venous blood returns to the heart at a rate that exceeds cardiac output, producing right cardiac hypertension (elevated pressure in the right side of the heart).

Cardiac Tamponade

The fibrous pericardium is a tough, inelastic, closed sac that contains the heart, normally the only occupant other than a thin lubricating layer of pericardial fluid. If extensive pericardial effusion exists, the compromised volume of the sac does not allow full expansion of the heart, limiting the amount of blood the heart can receive, which in turn reduces cardiac output. *Cardiac tamponade* (heart compression), is a potentially lethal condition because heart volume is increasingly compromised by the fluid outside the heart but inside the pericardial cavity.

Blood in the pericardial cavity, *hemopericardium*, likewise produces cardiac tamponade. Hemopericardium may result from perforation of a weakened area of heart muscle owing to a previous *myocardial infarction (MI)* or heart attack, from bleeding into the pericardial cavity after cardiac operations, or from stab wounds. This situation is especially lethal because of the high pressure involved and the rapidity with which the fluid accumulates. The heart is increasingly compressed and circulation fails. The veins of the face and neck become engorged because of the backup of blood, beginning where the SVC enters the pericardium.

In patients with *pneumothorax*—air or gas in the pleural cavity—the air may dissect along connective tissue planes and enter the pericardial sac, producing a *pneumopericardium.*

Pericardiocentesis

Drainage of fluid from the pericardial cavity, *pericardiocentesis,* is usually necessary to relieve cardiac tamponade. To remove the excess fluid, a wide-bore needle may be inserted through the left 5th or 6th intercostal space near the sternum. This approach to the pericardial sac is possible because the cardiac notch in the left lung and the shallower notch in the left pleural sac leaves part of the pericardial sac exposed—the bare area of the pericardium (Figs. 1.31A and 1.32). The pericardial sac may also be reached via the infrasternal angle by passing the needle superoposteriorly (Fig. B1.18). At this site, the needle avoids the lung and pleurae and enters the pericardial cavity; however, care must be taken not to puncture the internal thoracic artery or its terminal branches. In acute cardiac tamponade from hemopericardium, an emergency thoracotomy may be performed (the thorax is rapidly opened) so that the pericardial sac may be incised to immediately relieve the tamponade and establish *stasis of the hemorrhage* (stop the escape of blood) from the heart (see blue box "Thoracotomy, Intercostal Space Incisions, and Rib Excision," earlier in this chapter).

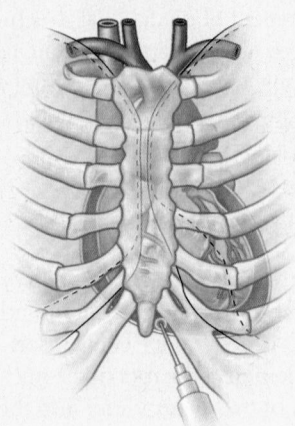

FIGURE B1.18. Pericardiocentesis.

Positional Abnormalities of the Heart

Abnormal folding of the embryonic heart may cause the position of the heart to be completely reversed so that the apex is misplaced to the right instead of the left—*dextrocardia.* This congenital anomaly is the most common positional abnormality of the heart, but it is still relatively uncommon. Dextrocardia is associated with mirror image positioning of the great vessels and arch of the aorta. This anomaly may be part of a general transposition of the thoracic and abdominal viscera (*situs inversus*), or the transposition may affect only the heart (*isolated dextrocardia*). In dextrocardia with situs inversus, the incidence of accompanying cardiac defects is low, and the heart usually functions normally. In isolated dextrocardia, however, the congenital anomaly is complicated by severe cardiac anomalies, such as *transposition of the great arteries.*

The Bottom Line

MEDIASTINUM OVERVIEW AND PERICARDIUM

Mediastinum overview: The mediastinum is the central compartment of the thoracic cavity and contains all thoracic viscera except the lungs. ♦ Occupying structures are hollow (fluid or air filled), and although contained between bony formations anteriorly and posteriorly, lie between "pneumatic packing," inflated to constantly changing volumes on each side. ♦ The mediastinum is a pliable, dynamic structure that is moved by structures contained within it (e.g., heart) and surrounding it (the diaphragm and other movements of respiration), as well as by the effect of gravity and body position. ♦ The superior mediastinum (above the transverse thoracic plane) is occupied by the trachea and upper parts of the great vessels. ♦ The middle part (most) of the inferior mediastinum is occupied by the heart. ♦ Most of the posterior mediastinum is occupied by structures vertically traversing all or much of the thorax.

Pericardium: The pericardium is a fibroserous sac, invaginated by the heart and roots of the great vessels, that encloses the serous cavity surrounding the heart. ♦ The fibrous pericardium is inelastic, attached anteriorly and inferiorly to the sternum and diaphragm, and blends with the adventitia of the great vessels as they enter or leave the sac. Thus it holds the heart in its middle mediastinal position and limits expansion (filling) of the heart. ♦ If fluid or a tumor occupies the pericardial space, the capacity of the heart is compromised. ♦ The serous pericardium lines the fibrous pericardium and the exterior of the heart. This glistening lubricated surface allows the heart (attached only by its afferent and efferent vessels and related reflections of serous membrane) the free movement required for its "wringing-out" motions during contraction. ♦ The parietal layer of the serous pericardium is sensitive. Pain impulses conducted from it by the somatic phrenic nerves result in referred pain sensations.

Heart

The **heart,** slightly larger than one's loosely clenched fist, is a double, self-adjusting suction and pressure pump, the parts of which work in unison to propel blood to all parts of the body. The right side of the heart (*right heart*) receives poorly oxygenated (venous) blood from the body through the SVC and IVC and pumps it through the pulmonary trunk and arteries to the lungs for oxygenation (Fig. 1.49A). The left side of the heart (*left heart*) receives well-oxygenated (arterial) blood from the lungs through the pulmonary veins and pumps it into the aorta for distribution to the body.

The heart has four chambers: **right** and **left atria** and **right** and **left ventricles.** The atria are receiving chambers that pump blood into the ventricles (the discharging chambers). The synchronous pumping actions of the heart's

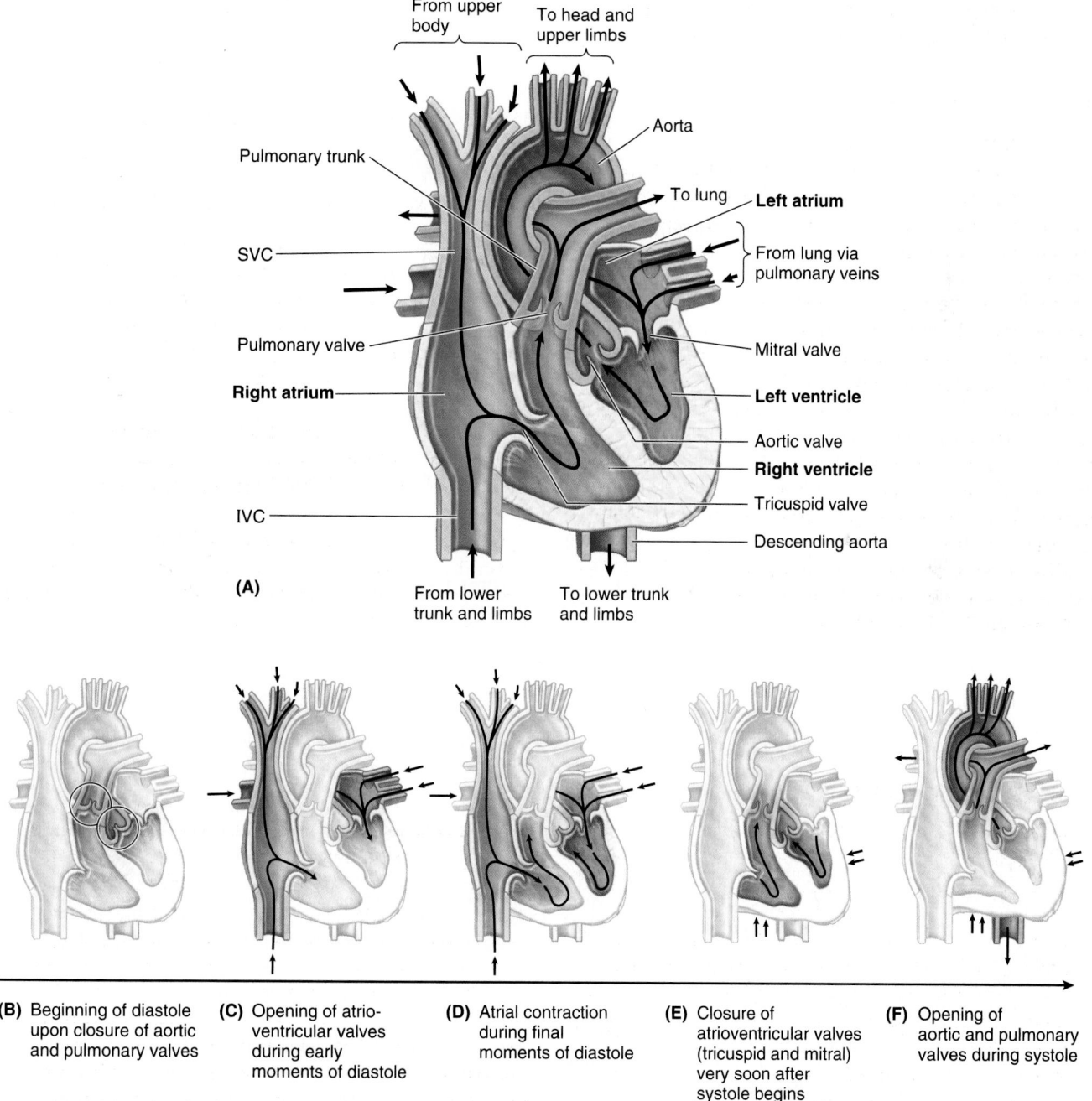

(B) Beginning of diastole upon closure of aortic and pulmonary valves

(C) Opening of atrioventricular valves during early moments of diastole

(D) Atrial contraction during final moments of diastole

(E) Closure of atrioventricular valves (tricuspid and mitral) very soon after systole begins

(F) Opening of aortic and pulmonary valves during systole

Anterior views

FIGURE 1.49. Cardiac cycle. The cardiac cycle describes the complete movement of the heart or heartbeat and includes the period from the beginning of one heartbeat to the beginning of the next one. The cycle consists of diastole (ventricular relaxation and filling) and systole (ventricular contraction and emptying). The right heart (blue side) is the pump for the pulmonary circuit; the left heart (red side) is the pump for the systemic circuit.

two atrioventricular (AV) pumps (right and left chambers) constitute the **cardiac cycle** (Fig. 1.49B–F). The cycle begins with a period of ventricular elongation and filling (**diastole**) and ends with a period of ventricular shortening and emptying (**systole**).

Two **heart sounds** are heard with a stethoscope: a *lub* (1st) sound as the blood is transferred from the atria into the ventricles, and a *dub* (2nd) sound as the ventricles expel blood from the heart. The heart sounds are produced by the snapping shut of the oneway valves that normally keep blood from flowing backward during contractions of the heart.

The wall of each heart chamber consists of three layers, from superficial to deep (Fig. 1.43):

1. **Endocardium,** a thin internal layer (endothelium and subendothelial connective tissue) or lining membrane of the heart that also covers its valves.
2. **Myocardium,** a thick, helical middle layer composed of cardiac muscle.
3. **Epicardium,** a thin external layer (mesothelium) formed by the visceral layer of serous pericardium. *⟶ part of the serous membrane of the heart (visceral layer)*

The walls of the heart consist mostly of myocardium, especially in the ventricles. When the ventricles contract, they produce a wringing motion because of the double helical orientation of the cardiac muscle fibers (Torrent-Guasp et al., 2001) (Fig. 1.50). This motion initially ejects the blood from the ventricles as the outer (basal) spiral contracts, first narrowing and then shortening the heart, reducing the volume of the ventricular chambers. Continued sequential contraction of the inner (apical) spiral elongates the heart, followed by widening as the myocardium briefly relaxes, increasing the volume of the chambers to draw blood from the atria.

The cardiac muscle fibers are anchored to the **fibrous skeleton of the heart** (Fig. 1.51). This is a complex framework of dense collagen forming four **fibrous rings** (L. *anuli fibrosi*) that surround the orifices of the valves, a right and left **fibrous trigone** (formed by connections between rings), and the membranous parts of the interatrial and interventricular septa. The fibrous skeleton of the heart:

- Keeps the orifices of the AV and semilunar valves patent and prevents them from being overly distended by an increased volume of blood pumping through them.
- Provides attachments for the leaflets and cusps of the valves.
- Provides attachment for the myocardium, which, when uncoiled, forms a continuous **ventricular myocardial band** that originates primarily from the fibrous ring of the pulmonary valve and inserts primarily into the fibrous ring of the aortic valve (Fig. 1.50).
- Forms an electrical "insulator," by separating the myenterically conducted impulses of the atria and ventricles so that they contract independently and by surrounding and providing passage for the initial part of the AV bundle of the conducting system of the heart (discussed later in this chapter).

Externally, the atria are demarcated from the ventricles by the **coronary sulcus** (**atrioventricular groove**), and

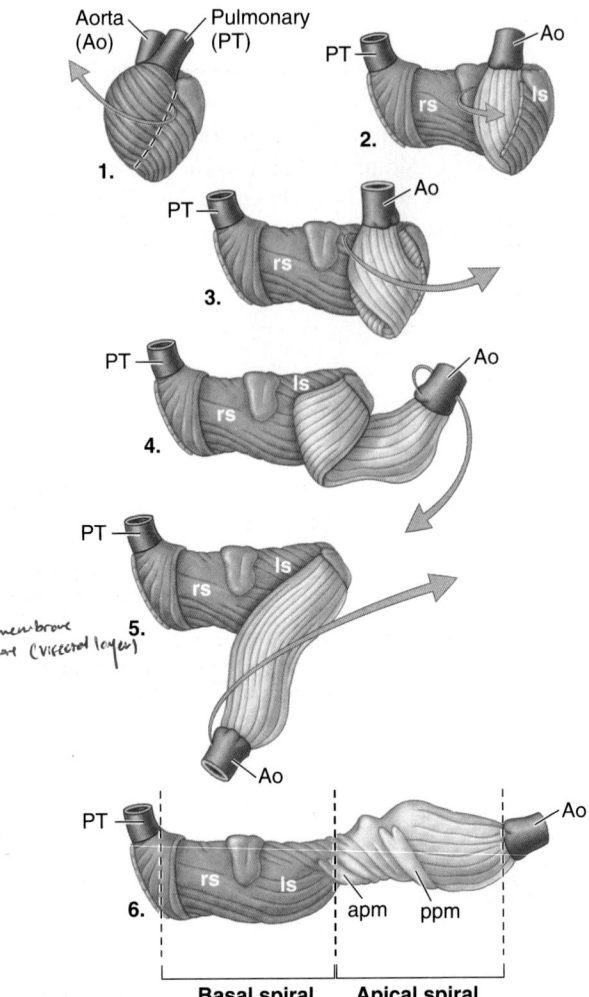

FIGURE 1.50. Arrangement of the myocardium and the fibrous skeleton of heart. The helical (double spiral) arrangement of the myocardium (modified from Torrent-Guasp et al., 2001). When the superficial myocardium is incised along the anterior interventricular groove (*dashed red line*) and peeled back starting at its origin from the fibrous ring of the pulmonary trunk (*PT*), the thick double spirals of the ventricular myocardial band are revealed. The ventricular myocardial band is progressively unwrapped. A band of nearly horizontal fibers forms an outer basal spiral (*dark brown*) that comprises the outer wall of the right ventricle (right segment; *rs*) and external layer of the outer wall of the left ventricle (left segment; *ls*). The deeper apical spiral (*light brown*), comprises the internal layer of the outer wall of the left ventricle. Its criss crossing fibers make up the interventricular septum. Thus the septum, like the outer wall of the left ventricle, is also double layered. The sequential contraction of the myocardial band enables the ventricles to function as parallel, sucking and propelling pumps; on contraction, the ventricles do not merely collapse inward but rather wring themselves out. *apm,* anterior papillary muscles; *ppm,* posterior papillary muscles.

the right and left ventricles are demarcated from each other by **anterior** and **posterior interventricular (IV) sulci** (grooves) (Fig. 1.52B & D). The heart appears trapezoidal from an anterior or posterior view (Fig. 1.52A), but in three dimensions it is shaped like a tipped-over pyramid with its apex (directed anteriorly and to the left), a base (opposite the apex, facing mostly posteriorly), and four sides.

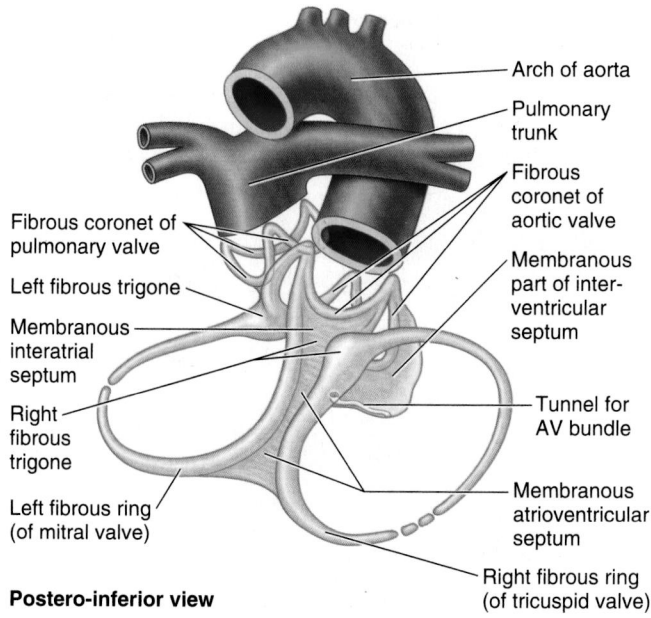

Arch of aorta

Pulmonary trunk

Fibrous coronet of aortic valve

Membranous part of inter-ventricular septum

Tunnel for AV bundle

Membranous atrioventricular septum

Right fibrous ring (of tricuspid valve)

Fibrous coronet of pulmonary valve

Left fibrous trigone

Membranous interatrial septum

Right fibrous trigone

Left fibrous ring (of mitral valve)

Postero-inferior view

FIGURE 1.51. Fibrous skeleton of heart. The isolated fibrous skeleton is composed of four fibrous rings (or two rings and two "coronets"), each encircling a valve; two trigones; and the membranous portions of the interatrial, interventricular, and atrioventricular septa.

The **apex of the heart** (Fig. 1.52B):

- Is formed by the inferolateral part of the left ventricle.
- Lies posterior to the left 5th intercostal space in adults, usually approximately 9 cm (a hand's breadth) from the median plane.
- Remains motionless throughout the cardiac cycle.
- Is where the sounds of mitral valve closure are maximal (**apex beat**); the apex underlies the site where the heart-beat may be auscultated on the thoracic wall.

The **base of the heart** (Fig. 1.52C & D):

- Is the heart's posterior aspect (opposite the apex).
- Is formed mainly by the left atrium, with a lesser contribution by the right atrium.
- Faces posteriorly toward the bodies of vertebrae T6–T9 and is separated from them by the pericardium, oblique pericardial sinus, esophagus, and aorta.
- Extends superiorly to the bifurcation of the pulmonary trunk and inferiorly to the coronary sulcus.
- Receives the pulmonary veins on the right and left sides of its left atrial portion, and the superior and inferior venae cavae at the superior and inferior ends of its right atrial portion.

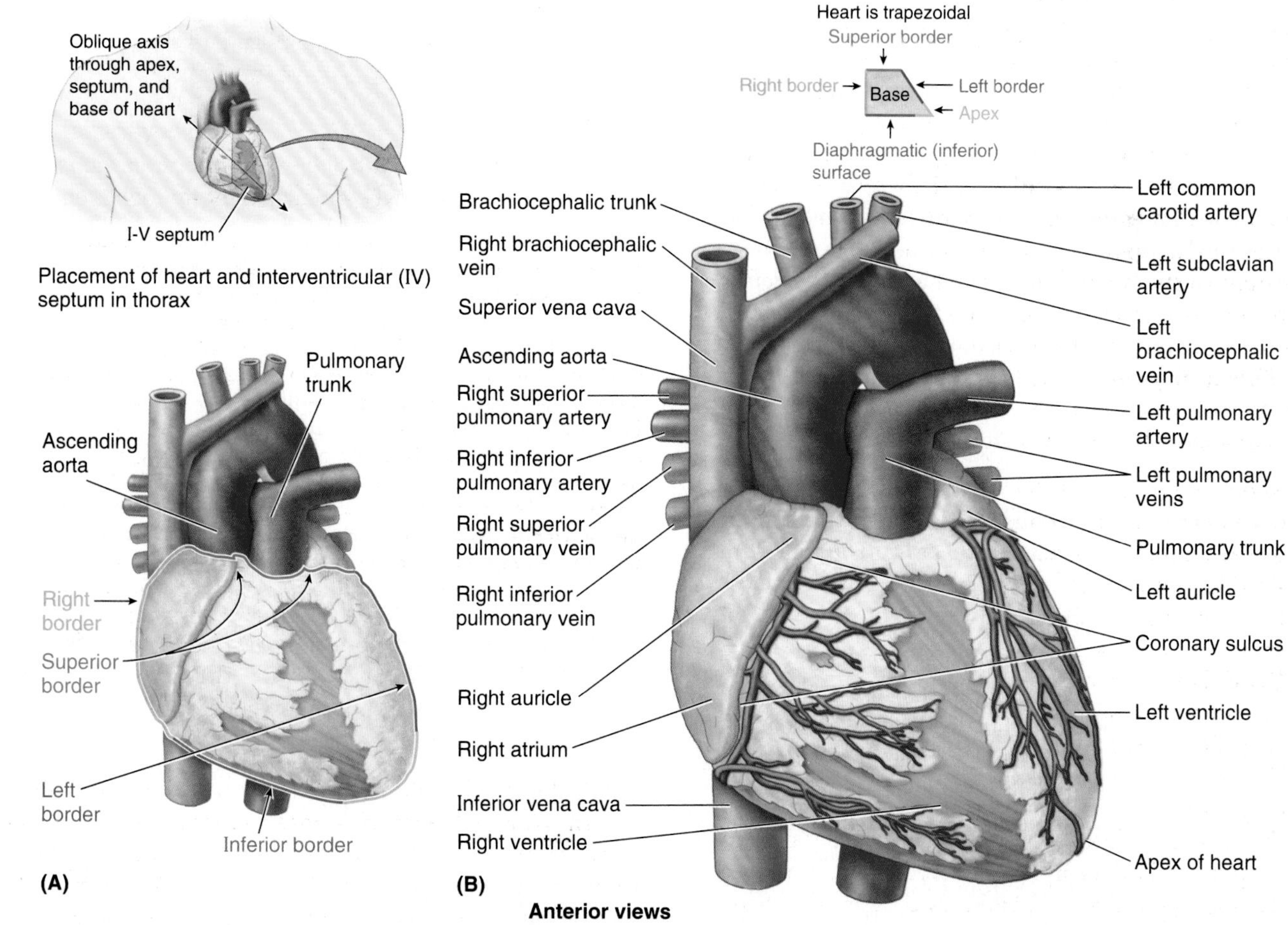

Oblique axis through apex, septum, and base of heart

I-V septum

Placement of heart and interventricular (IV) septum in thorax

Heart is trapezoidal

Superior border

Right border → **Base** ← **Left border**

← **Apex**

Diaphragmatic (inferior) surface

(A)

Pulmonary trunk

Ascending aorta

Right border

Superior border

Left border

Inferior border

(B)

Brachiocephalic trunk

Right brachiocephalic vein

Superior vena cava

Ascending aorta

Right superior pulmonary artery

Right inferior pulmonary artery

Right superior pulmonary vein

Right inferior pulmonary vein

Right auricle

Right atrium

Inferior vena cava

Right ventricle

Left common carotid artery

Left subclavian artery

Left brachiocephalic vein

Left pulmonary artery

Left pulmonary veins

Pulmonary trunk

Left auricle

Coronary sulcus

Left ventricle

Apex of heart

Anterior views

FIGURE 1.52. Shape, orientation, surfaces, and borders of heart. A, B. The sternocostal surface of the heart and the relationship of the great vessels are shown. The ventricles dominate this surface (two thirds right ventricle, one third left ventricle).

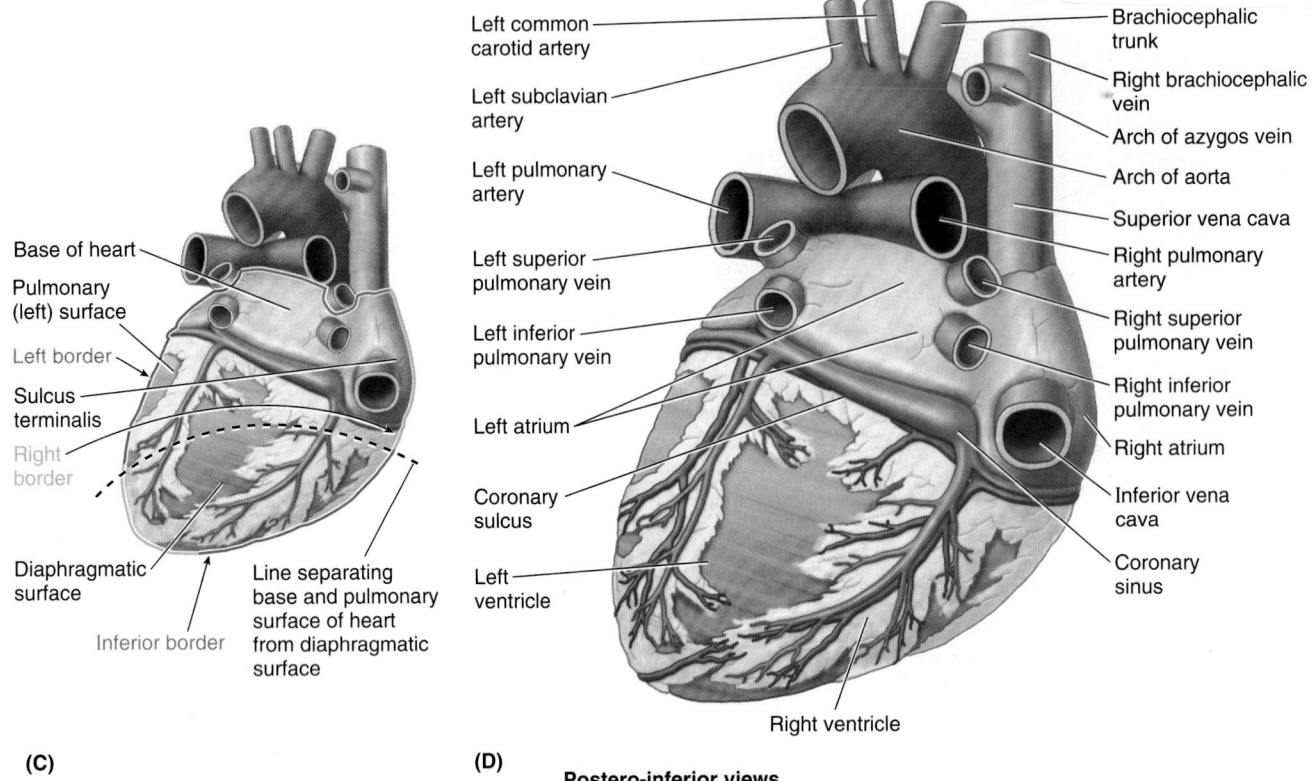

FIGURE 1.52. *(Continued)* **C, D.** The pulmonary (left) and diaphragmatic (inferior) surfaces and the base of the heart are shown as well as the relationship of the great vessels.

The four *surfaces of the heart* (Fig. 1.52A–D) are the:

1. **Anterior (sternocostal) surface,** formed mainly by the right ventricle.
2. **Diaphragmatic (inferior) surface,** formed mainly by the left ventricle and partly by the right ventricle; it is related mainly to the central tendon of the diaphragm.
3. **Right pulmonary surface,** formed mainly by the right atrium.
4. **Left pulmonary surface,** formed mainly by the left ventricle; it forms the cardiac impression in the left lung.

The heart appears trapezoidal in both anterior (Fig. 1.52A & B) and posterior views (Fig. 1.52C & D). The four *borders of the heart* are the:

1. **Right border** (slightly convex), formed by the right atrium and extending between the SVC and the IVC.
2. **Inferior border** (nearly horizontal), formed mainly by the right ventricle and slightly by the left ventricle.
3. **Left border** (oblique, nearly vertical), formed mainly by the left ventricle and slightly by the left auricle.
4. **Superior border,** formed by the right and left atria and auricles in an anterior view; the ascending aorta and pulmonary trunk emerge from this border and the SVC enters its right side. Posterior to the aorta and pulmonary trunk and anterior to the SVC, this border forms the inferior boundary of the transverse pericardial sinus.

The **pulmonary trunk,** approximately 5 cm long and 3 cm wide, is the arterial continuation of the right ventricle and divides into *right* and *left pulmonary arteries*. The pulmonary trunk and arteries conduct low-oxygen blood to the lungs for oxygenation (Figs. 1.49A and 1.52B).

RIGHT ATRIUM

The right atrium forms the right border of the heart and receives venous blood from the SVC, IVC, and coronary sinus (Fig. 1.52B & D). The ear-like **right auricle** is a conical muscular pouch that projects from this chamber like an add-on room, increasing the capacity of the atrium as it overlaps the ascending aorta.

The *interior of the right atrium* (Fig. 1.53A & B) has a:

- Smooth, thin-walled, posterior part (the **sinus venarum**) on which the venae cavae (SVC and IVC) and coronary sinus open, bringing poorly oxygenated blood into the heart.
- Rough, muscular anterior wall composed of pectinate muscles (L. *musculi pectinati*).
- Right AV orifice through which the right atrium discharges the poorly oxygenated blood it has received into the right ventricle.

The smooth and rough parts of the atrial wall are separated externally by a shallow vertical groove, the **sulcus terminalis**

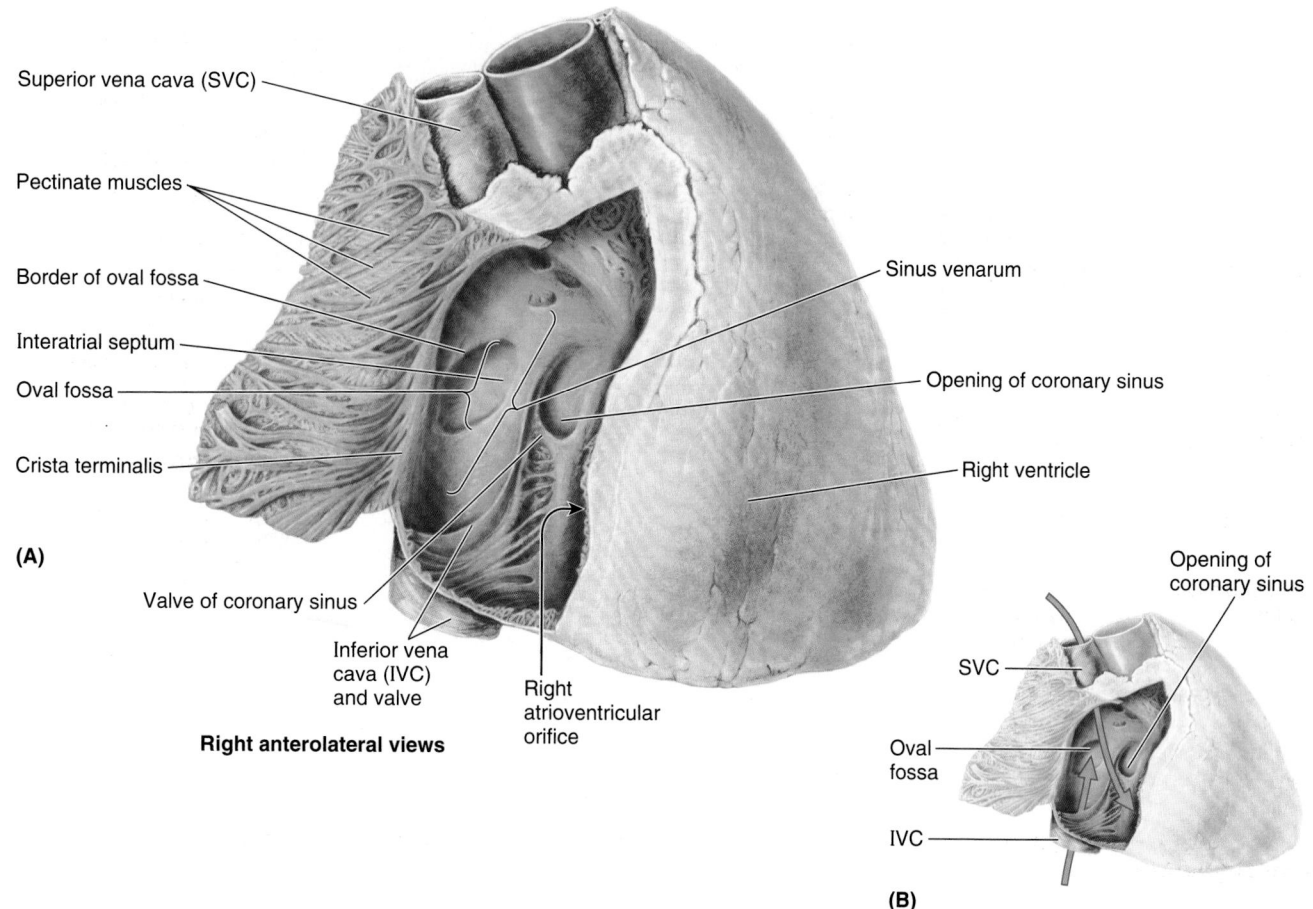

FIGURE 1.53. Right atrium of heart. A. The outer wall of the right atrium has been incised from the right auricle to the diaphragmatic surface. The wall has been retracted to reveal the smooth-walled part of the atrium, the sinus venarum, derived from absorption of the venous sinus of the embryonic heart. All of the venous structures entering the right atrium (superior and inferior vena cavae and coronary sinus) open into the sinus venarum. The shallow oval fossa is the site of fusion of the embryonic valve of the oval foramen with the interatrial septum. **B.** The inflow from the superior vena cava (*SVC*) is directed toward the right atrioventricular orifice, whereas blood from the inferior vena cava (*IVC*) is directed toward the oval fossa, as it was before birth.

or terminal groove (Fig. 1.52C), and internally by a vertical ridge, the **crista terminalis** or terminal crest (Fig. 1.53A). The SVC opens into the superior part of the right atrium at the level of the right 3rd costal cartilage. The IVC opens into the inferior part of the right atrium almost in line with the SVC at approximately the level of the 5th costal cartilage.

The **opening of the coronary sinus,** a short venous trunk receiving most of the cardiac veins, is between the right AV orifice and the IVC orifice. The **interatrial septum** separating the atria has an oval, thumbprint-size depression, the **oval fossa** (L. *fossa ovalis*), which is a remnant of the **oval foramen** (L. *foramen ovale*) and its valve in the fetus. Full understanding of the features of the right atrium requires an awareness of the development of the heart; see the blue box "Embryology of the Right Atrium," p. 151.

RIGHT VENTRICLE

The right ventricle forms the largest part of the anterior surface of the heart, a small part of the diaphragmatic surface, and almost the entire inferior border of the heart (Fig. 1.52B).

Superiorly it tapers into an arterial cone, the **conus arteriosus** (infundibulum), which leads into the pulmonary trunk (Fig. 1.54). The interior of the right ventricle has irregular muscular elevations (**trabeculae carneae**). A thick muscular ridge, the **supraventricular crest,** separates the ridged muscular wall of the inflow part of the chamber from the smooth wall of the conus arteriosus, or outflow part. The inflow part of the ventricle receives blood from the right atrium through the **right AV (tricuspid) orifice** (Fig. 1.55A), located posterior to the body of the sternum at the level of the 4th and 5th intercostal spaces. The right AV orifice is surrounded by one of the fibrous rings of the *fibrous skeleton of the heart* (Fig. 1.51). The fibrous ring keeps the caliber of the orifice constant (large enough to admit the tips of three fingers), resisting the dilation that might otherwise result from blood being forced through it at varying pressures.

The **tricuspid valve** (Figs. 1.54 and 1.55) guards the right AV orifice. The bases of the valve cusps are attached to the fibrous ring around the orifice. Because the fibrous ring maintains the caliber of the orifice, the attached valve cusps contact each other in the same way with each heartbeat. **Tendinous**

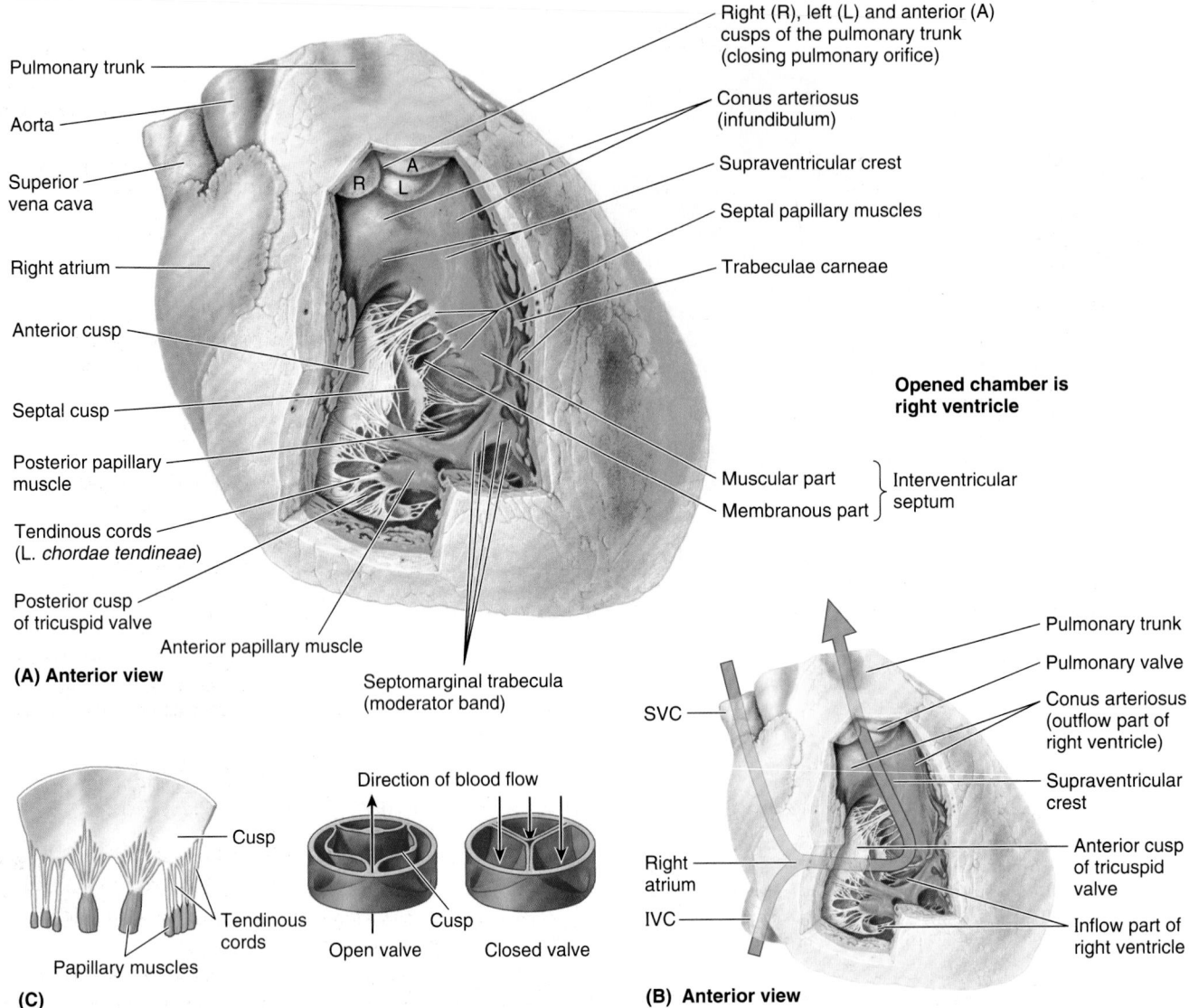

Pulmonary trunk

Aorta

Superior
vena cava

Right atrium

Anterior cusp

Septal cusp

Posterior papillary
muscle

Tendinous cords
(L. *chordae tendineae*)

Posterior cusp
of tricuspid valve

Anterior papillary muscle

(A) Anterior view

Septomarginal trabecula
(moderator band)

Right (R), left (L) and anterior (A)
cusps of the pulmonary trunk
(closing pulmonary orifice)

Conus arteriosus
(infundibulum)

Supraventricular crest

Septal papillary muscles

Trabeculae carneae

**Opened chamber is
right ventricle**

Muscular part ⎫ Interventricular
Membranous part ⎬ septum

Cusp

Tendinous
cords

Papillary muscles

(C)

Direction of blood flow

Cusp

Open valve   Closed valve

SVC

Right
atrium

IVC

Pulmonary trunk

Pulmonary valve

Conus arteriosus
(outflow part of
right ventricle)

Supraventricular
crest

Anterior cusp
of tricuspid
valve

Inflow part of
right ventricle

(B) Anterior view

FIGURE 1.54. Interior of right ventricle of heart. The sternocostal wall of the right ventricle has been excised. **A.** The tricuspid valve at the entrance to the ventricle (right atrioventricular [AV] orifice) is open and the pulmonic valve at the exit to the pulmonary trunk is closed, as they would be during ventricular filling (diastole). The smooth funnel-shaped conus arteriosus is the outflow tract of the chamber. **B.** The inflow of blood enters the chamber from its posterior and inferior aspect, flowing anteriorly and to the left (toward the apex); the outflow of blood to the pulmonary trunk leaves superiorly and posteriorly. *IVC*, inferior vena cava; *SVC*, superior vena cava.

cords (L. *chordae tendineae*) attach to the free edges and ventricular surfaces of the anterior, posterior, and septal cusps, much like the cords attaching to a parachute (Fig. 1.54A). The tendinous cords arise from the apices of **papillary muscles,** which are conical muscular projections with bases attached to the ventricular wall. The papillary muscles begin to contract before contraction of the right ventricle, tightening the tendinous cords and drawing the cusps together. Because the cords are attached to adjacent sides of two cusps, they prevent separation of the cusps and their inversion when tension is applied to the tendinous cords and maintained throughout ventricular contraction (*systole*)—that is, the cusps of the tricuspid valve are prevented from prolapsing (being driven into the right atrium) as ventricular pressure rises. Thus, regurgitation of

blood (backward flow of blood) from the right ventricle back into the right atrium is blocked during ventricular systole by the valve cusps (Fig. 1.55C).

Three papillary muscles in the right ventricle correspond to the cusps of the tricuspid valve (Fig. 1.54A):

1. The **anterior papillary muscle,** the largest and most prominent of the three, arises from the anterior wall of the right ventricle; its tendinous cords attach to the anterior and posterior cusps of the tricuspid valve.
2. The **posterior papillary muscle,** smaller than the anterior muscle, may consist of several parts; it arises from the inferior wall of the right ventricle, and its tendinous cords attach to the posterior and septal cusps of the tricuspid valve.

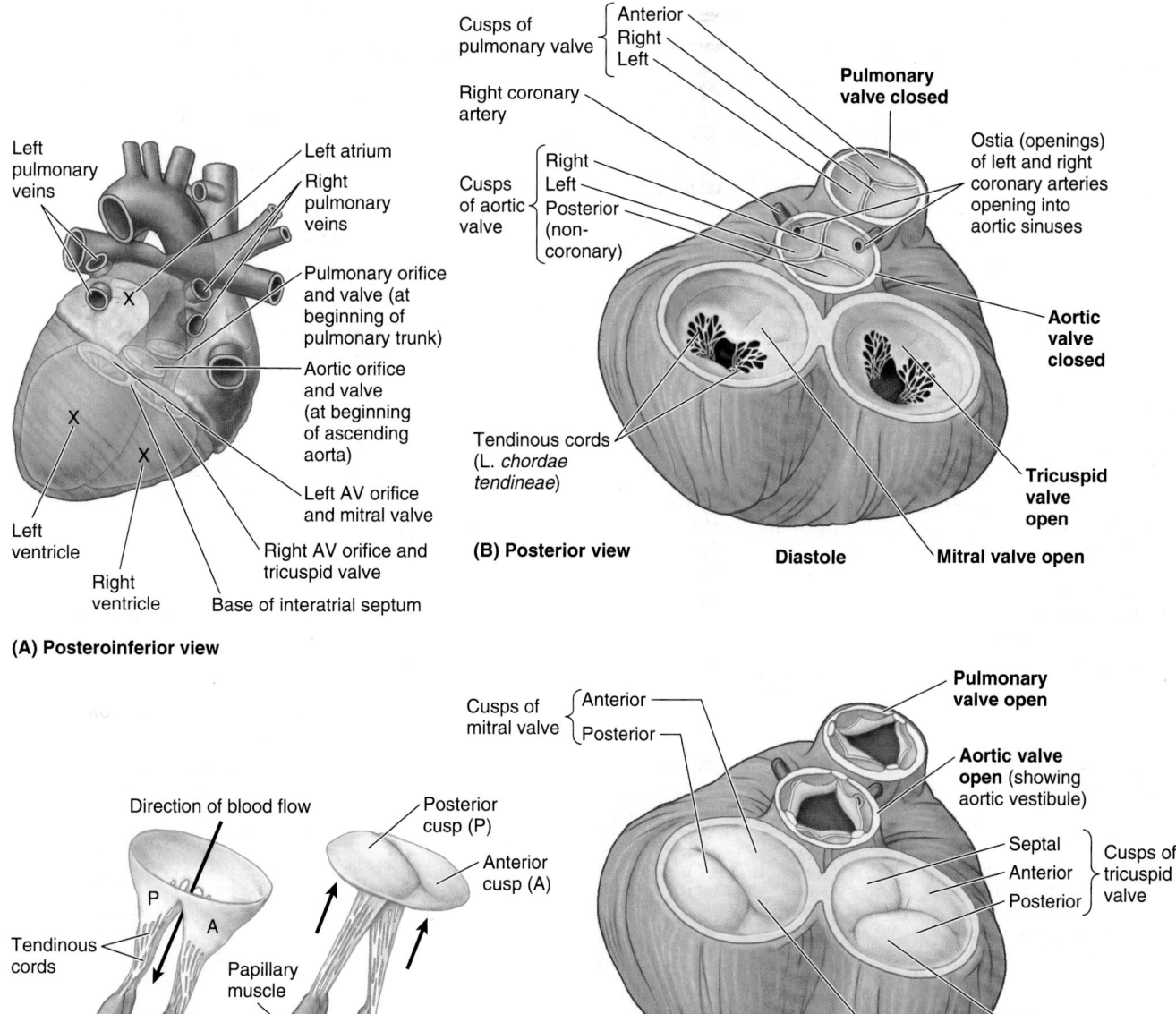

FIGURE 1.55. Valves of heart and great vessels. A. The coronary valves are shown in situ. *AV,* atrioventricular. **B.** At the beginning of diastole (ventricular relaxation and filling), the aortic and pulmonary valves are closed; shortly thereafter, the tricuspid and mitral valves open (also shown in Fig. 1.49). **C.** Shortly after systole (ventricular contraction and emptying) begins, the tricuspid and mitral valves close and the aortic and pulmonary valves open.

3. The **septal papillary muscle** arises from the interventricular septum, and its tendinous cords attach to the anterior and septal cusps of the tricuspid valve.

The **interventricular septum (IVS),** composed of muscular and membranous parts, is a strong, obliquely placed partition between the right and left ventricles (Fig. 1.54A and 1.57), forming part of the walls of each. Because of the much higher blood pressure in the left ventricle, the **muscular part of the IVS,** which forms the majority of the septum, has the thickness of the remainder of the wall of the left ventricle (two to three times as thick as the wall of the right ventricle)

and bulges into the cavity of the right ventricle. Superiorly and posteriorly, a thin membrane, part of the fibrous skeleton of the heart (Fig. 1.51), forms the much smaller **membranous part of the IVS.** On the right side, the septal cusp of the tricuspid valve (Fig. 1.54) is attached to the middle of this membranous part of the fibrous skeleton. This means that inferior to the cusp, the membrane is an interventricular septum, but superior to the cusp it is an *atrioventricular septum,* separating the right atrium from the left ventricle.

The **septomarginal trabecula (moderator band)** is a curved muscular bundle that traverses the right ventricular chamber from the inferior part of the IVS to the base of the anterior papillary muscle. This trabecula is important because it carries part of the **right branch of the AV bundle,** a part of the conducting system of the heart to the anterior papillary muscle (see "Stimulating and Conducting System of the Heart," pp. 148–149). This "shortcut" across the chamber seems to facilitate conduction time, allowing coordinated contraction of the anterior papillary muscle.

The right atrium contracts when the right ventricle is empty and relaxed; thus blood is forced through this orifice into the right ventricle, pushing the cusps of the tricuspid valve aside like curtains. The inflow of blood into the right ventricle (*inflow tract*) enters posteriorly; and when the ventricle contracts, the outflow of blood into the pulmonary trunk (*outflow tract*) leaves superiorly and to the left (Fig. 1.54B). Consequently, the blood takes a U-shaped path through the right ventricle, changing direction about 140°. This change in direction is accommodated by the **supraventricular crest,** which deflects the incoming flow into the main cavity of the ventricle, and the outgoing flow into the conus arteriosus toward the pulmonary orifice. The inflow (AV) orifice and outflow (pulmonary) orifice are approximately 2 cm apart. The **pulmonary valve** (Figs. 1.54B and 1.55) at the apex of the conus arteriosus is at the level of the left 3rd costal cartilage.

LEFT ATRIUM

The left atrium forms most of the base of the heart (Fig. 1.52C & D). The valveless pairs of right and left pulmonary veins enter the smooth-walled atrium (Fig. 1.56). In the embryo, there is only one common pulmonary vein, just as there is a single pulmonary trunk. The wall of this vein and four of its tributaries were incorporated into the wall of the left atrium, in the same way that the sinus venosus was incorporated into the right atrium. The part of the wall derived from the embryonic pulmonary vein is smooth walled. The tubular, muscular **left auricle,** its wall trabeculated with *pectinate muscles,* forms the superior part of the left border of the heart and overlaps the root of the *pulmonary trunk* (Fig. 1.52A & B). It represents the remains of the left part of the primordial atrium. A *semilunar depression* in the interatrial septum indicates the floor of the oval fossa (Fig. 1.56A); the surrounding ridge is the valve of the oval fossa (L. *valvulae foramen ovale*).

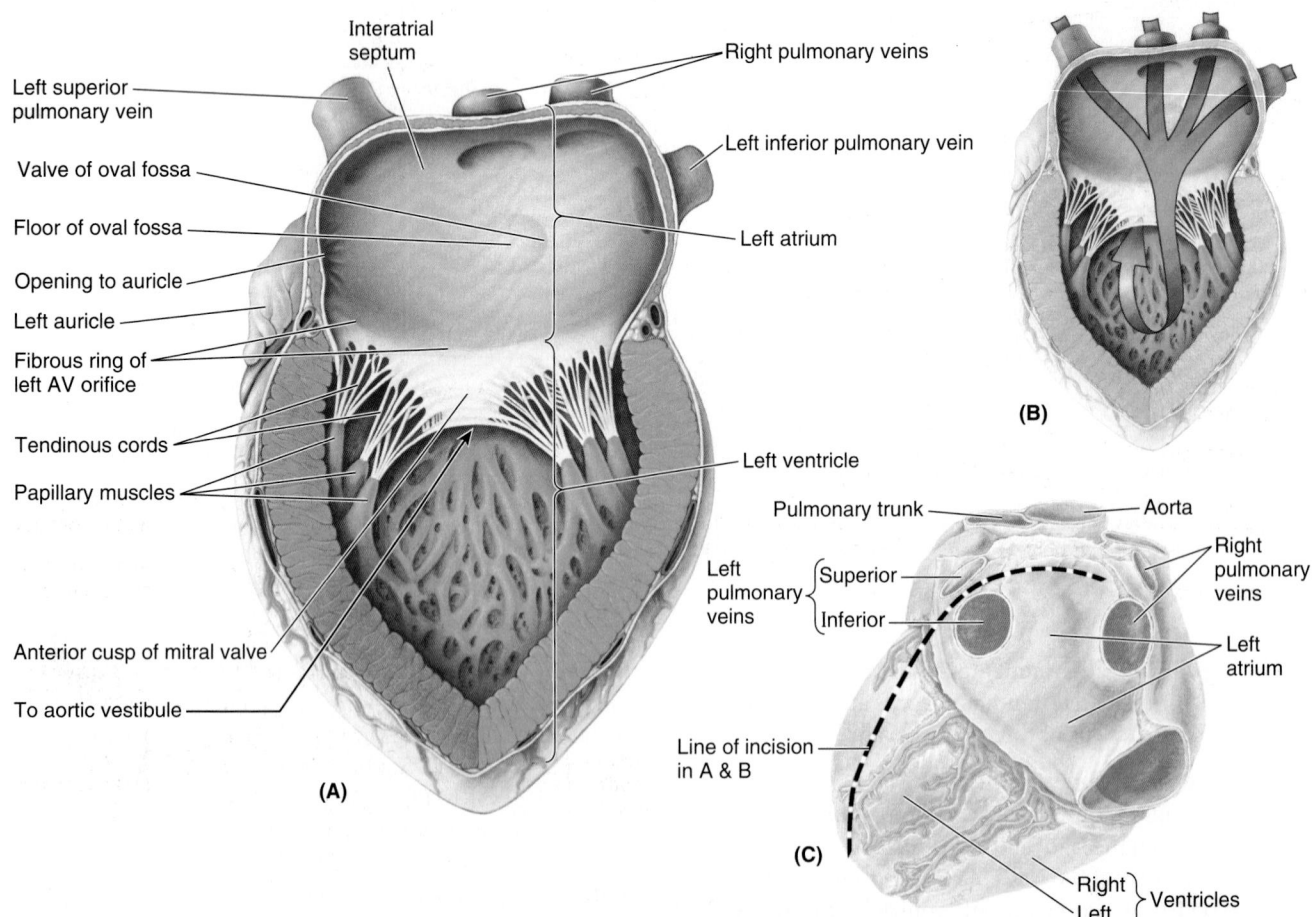

FIGURE 1.56. Interior of left atrium and ventricle of heart. A. The features of the internal aspects of the left atrium and the inflow tract of the left ventricle are shown. *AV,* atrioventricular. **B.** The pattern of blood flow through the left side of the heart. **C.** For parts **A** and **B,** the heart was incised vertically along its left border and then transversely across the superior part of its base, passing between the superior and the inferior left pulmonary veins.

The *interior of the left atrium* has

- A larger smooth-walled part and a smaller muscular auricle containing pectinate muscles.
- Four pulmonary veins (two superior and two inferior) entering its smooth posterior wall (Fig. 1.56A–C).
- A slightly thicker wall than that of the right atrium.
- An interatrial septum that slopes posteriorly and to the right.
- A left AV orifice through which the left atrium discharges the oxygenated blood it receives from the pulmonary veins into the left ventricle (Fig. 1.56B).

LEFT VENTRICLE

The left ventricle forms the apex of the heart, nearly all its left (pulmonary) surface and border, and most of the diaphragmatic surface (Figs. 1.52 and 1.57). Because arterial pressure is much higher in the systemic than in the pulmonary circulation, the left ventricle performs more work than the right ventricle.

The *interior of the left ventricle* has (Fig. 1.57)

- Walls that are two to three times as thick as those of the right ventricle.
- Walls that are mostly covered with a mesh of *trabeculae carneae* that are finer and more numerous than those of the right ventricle.
- A conical cavity that is longer than that of the right ventricle.
- Anterior and posterior *papillary muscles* that are larger than those in the right ventricle.
- A smooth-walled, non-muscular, supero-anterior outflow part, the **aortic vestibule,** leading to the aortic orifice and *aortic valve.*
- A double-leaflet *mitral valve* that guards the left AV orifice (Figs. 1.55 and 1.57A).
- An **aortic orifice** that lies in its right posterosuperior part and is surrounded by a fibrous ring to which the right posterior, and left cusps of the aortic valve are attached; the ascending aorta begins at the aortic orifice.

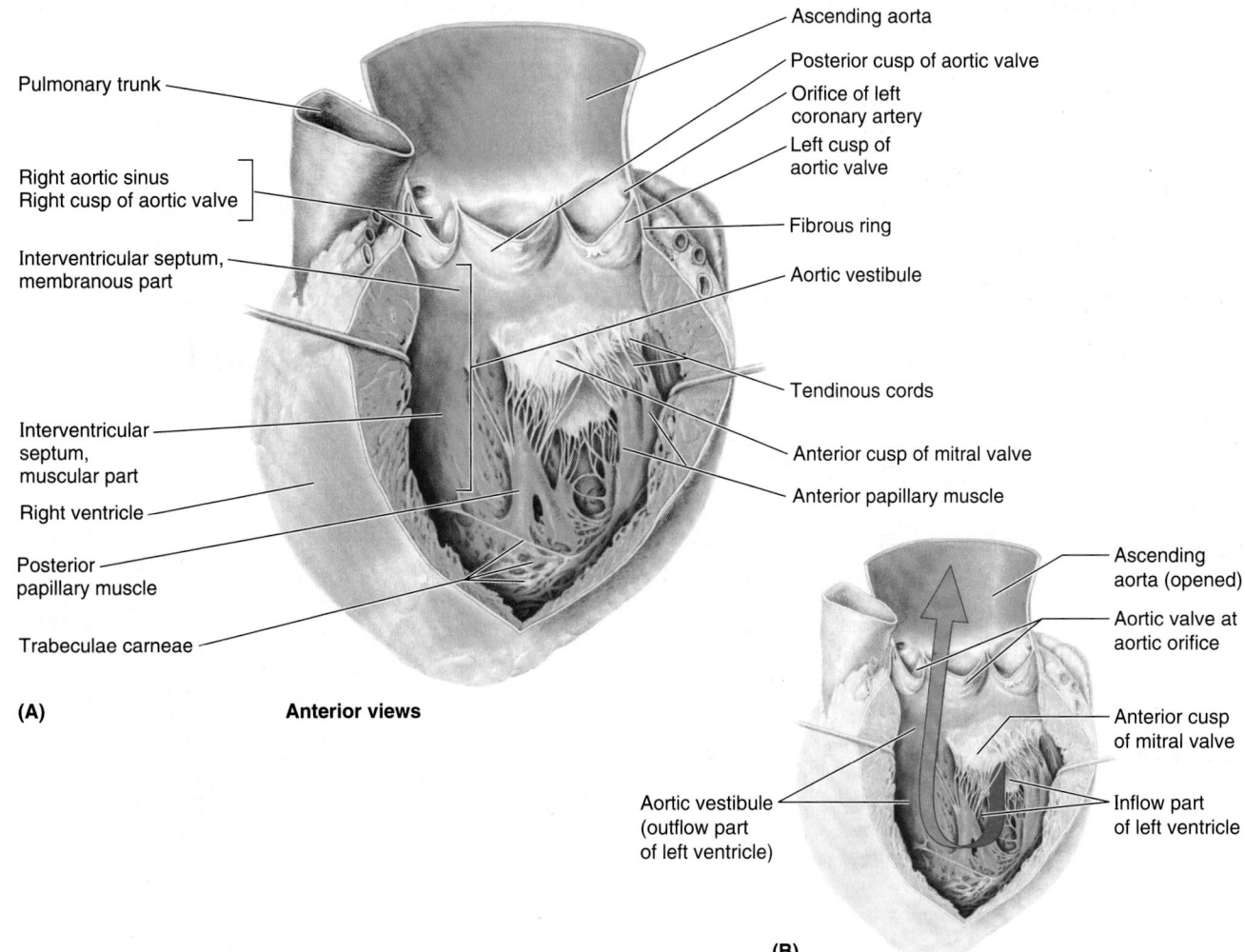

FIGURE 1.57. **Interior and outflow tract of left ventricle of heart. A, B.** The anterior surface of the left ventricle has been incised parallel to the interventricular groove, with the right margin of the incision retracted to the right, revealing an anterior view of the chamber. **B.** The left atrioventricular orifice and mitral valve are located posteriorly, and the aortic vestibule leads superiorly and to the right to the aortic valve.

The **mitral valve** has two cusps, anterior and posterior. The adjective *mitral* derives from the valve's resemblance to a bishop's miter (headdress). The mitral valve is located posterior to the sternum at the level of the 4th costal cartilage. Each of its cusps receives tendinous cords from more than one papillary muscle. These muscles and their cords support the mitral valve, allowing the cusps to resist the pressure developed during contractions (pumping) of the left ventricle (Fig. 1.57A). The cords become taut just before and during systole, preventing the cusps from being forced into the left atrium. As it traverses the left ventricle, the bloodstream undergoes two right angle turns, which together result in a 180° change in direction. This reversal of flow takes place around the anterior cusp of the mitral valve (Fig. 1.57B).

The semilunar **aortic valve,** between the left ventricle and the ascending aorta, is obliquely placed (Fig. 1.55). It is located posterior to the left side of the sternum at the level of the 3rd intercostal space.

SEMILUNAR VALVES

Each of three **semilunar cusps of the pulmonary valve** (*anterior, right,* and *left*), like the **semilunar cusps of the aortic valve** (*posterior, right,* and *left*), is concave when viewed superiorly (Figs. 1.55B and 1.57A). (See the blue box "Basis for Naming Cusps of the Aortic and Pulmonary Valves," p. 153.) Semilunar cusps do not have tendinous cords to support them. They are smaller in area than the cusps of the AV valves, and the force exerted on them is less than half that exerted on the cusps of the tricuspid and mitral valves. The cusps project into the artery but are pressed toward (and not against) its walls as blood leaves the ventricle (Figs. 1.55C and 1.58B). After relaxation of the ventricle (*diastole*), the elastic recoil of the wall of the pulmonary trunk or aorta

forces the blood back toward the heart. However, the cusps snap closed like an umbrella caught in the wind as they catch the reversed blood flow (Figs. 1.55B and 1.58C). They come together to completely close the orifice, supporting each other as their edges abut (meet), and preventing any significant amount of blood from returning to the ventricle.

The edge of each cusp is thickened in the region of contact, forming the **lunule;** the apex of the angulated free edge is thickened further as the **nodule** (Fig. 1.58A). Immediately superior to each semilunar cusp, the walls of the origins of the pulmonary trunk and aorta are slightly dilated, forming a sinus. The **aortic sinuses** and **sinuses of the pulmonary trunk** (pulmonary sinuses) are the spaces at the origin of the pulmonary trunk and ascending aorta between the dilated wall of the vessel and each cusp of the semilunar valves (Figs. 1.55B and 1.57A). The blood in the sinuses and the dilation of the wall prevent the cusps from sticking to the wall of the vessel, which might prevent closure.

The mouth of the right coronary artery is in the **right aortic sinus,** the mouth of the left coronary artery is in the **left aortic sinus,** and no artery arises from the **posterior aortic (non-coronary) sinus** (Figs. 1.57A and 1.58).

VASCULATURE OF HEART

The blood vessels of the heart comprise the coronary arteries and cardiac veins, which carry blood to and from most of the myocardium (Figs. 1.59 and 1.61). The endocardium and some subendocardial tissue located immediately external to the endocardium receive oxygen and nutrients by diffusion or microvasculature directly from the chambers of the heart. The blood vessels of the heart, normally embedded in fat, course across the surface of the heart just deep to the epicardium. Occasionally, parts of the vessels become embedded within

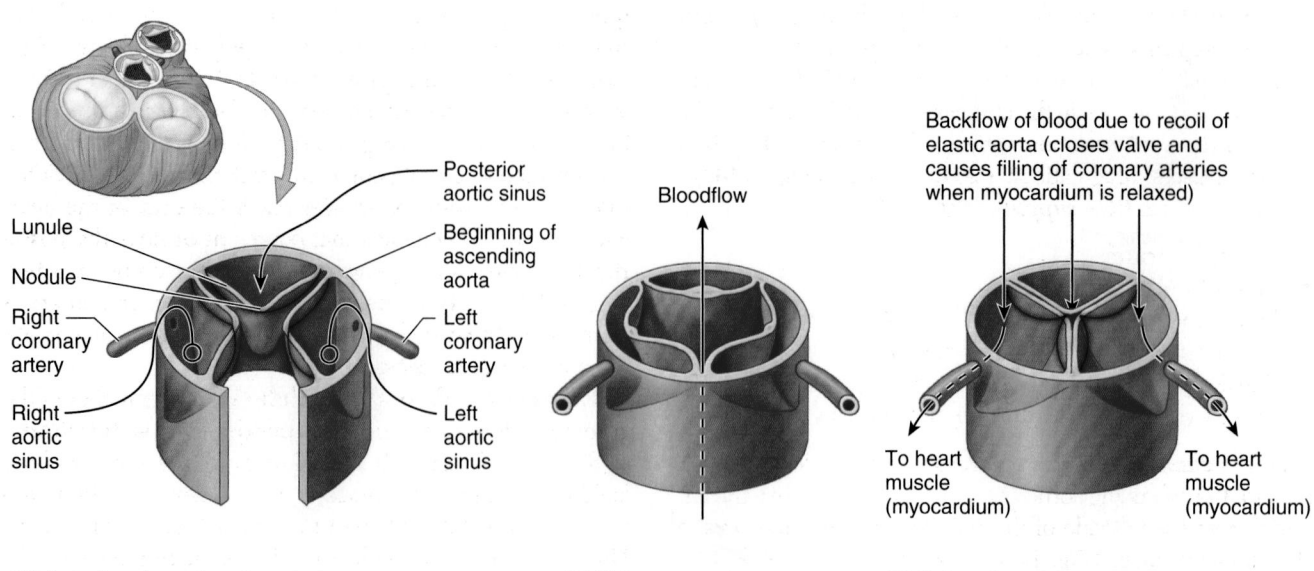

(A) Anterior view of aortic valve   **(B) Valve open**   **(C) Valve closed**

FIGURE 1.58. Aortic valve, aortic sinuses, and coronary arteries. A. Like the pulmonary valve, the aortic valve has three semilunar cusps: right, posterior, and left. **B.** Blood ejected from the left ventricle forces the cusps apart. **C.** When the valve closes, the nodules and lunules meet in the center.

the myocardium. The blood vessels of the heart are affected by both sympathetic and parasympathetic innervation.

Arterial Supply of Heart. The *coronary arteries,* the first branches of the aorta, supply the myocardium and epicardium. The right and left coronary arteries arise from the corresponding aortic sinuses at the proximal part of the ascending aorta, just superior to the aortic valve, and pass around opposite sides of the pulmonary trunk (Figs. 1.58 and 1.59; Table 1.4). The coronary arteries supply both the atria and the ventricles; however, the atrial branches are usually small and not readily apparent in the cadaveric heart. The ventricular distribution of each coronary artery is not sharply demarcated.

The **right coronary artery** (RCA) arises from the right aortic sinus of the ascending aorta and passes to the right side of the pulmonary trunk, running in the coronary sulcus (Figs. 1.58 and 1.59A). Near its origin, the RCA usually gives off an ascending **sinu-atrial nodal branch,** which supplies the SA node. The RCA then descends in the coronary sulcus and gives off the **right marginal branch,** which supplies the right border of the heart as it runs toward (but does not reach) the apex of the heart. After giving off this branch, the RCA turns to the left and continues in the coronary sulcus to the posterior aspect of the heart. At the posterior aspect of the **crux** (L. cross) **of the heart**—the junction of the interatrial and interventricular (IV) septa between the four heart chambers—the RCA gives rise to the **atrioventricular nodal branch,** which supplies the AV node (Fig. 1.59A–C). The SA and AV nodes are part of the *conducting system of the heart* (p. 148).

Dominance of the coronary arterial system is defined by which artery gives rise to the posterior interventricular (IV) branch (posterior descending artery). Dominance of the right coronary artery is typical (approximately 67%) (Fig. 1.59A); the right coronary artery gives rise to the large **posterior interventricular branch,** which descends in the posterior IV groove toward the apex of the heart. This branch supplies adjacent areas of both ventricles and sends perforating **interventricular septal branches** into the IV septum (Fig. 1.59C). The terminal (left ventricular) branch of the RCA then continues for a short distance in the coronary sulcus (Fig. 1.59A & B). Thus, in the most common pattern of distribution, the RCA supplies the diaphragmatic surface of the heart (Fig. 1.59D).

Typically, *the RCA supplies* (Figs. 1.59):

- The right atrium.
- Most of right ventricle.
- Part of the left ventricle (the diaphragmatic surface).
- Part of the IV septum, usually the posterior third.
- The SA node (in approximately 60% of people).
- The AV node (in approximately 80% of people).

The **left coronary artery** (LCA) arises from the *left aortic sinus* of the ascending aorta (Fig. 1.58), passes between the left auricle and the left side of the pulmonary trunk, and runs in the coronary sulcus (Fig. 1.59A & B). In approximately 40% of people, the **SA nodal branch** arises from the circumflex branch of the LCA and ascends on the posterior surface of the left atrium to the SA node. As it enters the coronary sulcus, at the superior end of the anterior IV groove, the LCA divides into two branches, the *anterior IV branch* (clinicians continue to use *LAD*, the abbreviation for the former term "left anterior descending" artery) and the *circumflex branch* (Fig. 1.59A & C).

The **anterior IV branch** passes along the IV groove to the apex of the heart. Here it turns around the inferior border of the heart and commonly anastomoses with the posterior IV branch of the right coronary artery (Fig. 1.59B). The anterior IV branch supplies adjacent parts of both ventricles and, via IV septal branches, the anterior two thirds of the IVS (Fig. 1.59C). In many people, the anterior IV branch gives rise to a **lateral branch** (diagonal artery), which descends on the anterior surface of the heart (Fig. 1.59A).

The smaller **circumflex branch of the LCA** follows the coronary sulcus around the left border of the heart to the posterior surface of the heart. The **left marginal** branch of the circumflex branch follows the left margin of the heart and supplies the left ventricle. Most commonly, the circumflex branch of the LCA terminates in the coronary sulcus on the posterior aspect of the heart before reaching the crux of the heart (Fig. 1.59B), but in approximately one third of hearts it continues to supply a branch that runs in or adjacent to the posterior IV groove (Fig. 1.60B).

Typically, *the LCA supplies* (Fig. 1.59):

- The left atrium.
- Most of the left ventricle.
- Part of the right ventricle.
- Most of the IVS (usually its anterior two thirds), including the AV bundle of the conducting system of the heart, through its perforating IV septal branches.
- The SA node (in approximately 40% of people).

Variations of the Coronary Arteries. Variations in the branching patterns and distribution of the coronary arteries are common. In the most common *right dominant pattern,* present in approximately 67% of people, the RCA and LCA share about equally in the blood supply of the heart (Figs. 1.59 and 1.60A). In approximately 15% of hearts, the LCA is dominant in that the posterior IV branch is a branch of the circumflex artery (Fig. 1.60B). There is codominance in approximately 18% of people, in which branches of both the right and left coronary arteries reach the crux of the heart and give rise to branches that course in or near the posterior IV groove. A few people have only one coronary artery (Fig. 1.60C). In other people, the circumflex branch arises from the right aortic sinus (Fig. 1.60D). Approximately 4% of people have an accessory coronary artery.

Coronary Collateral Circulation. The branches of the coronary arteries are generally considered to be **functional end arteries** (arteries that supply regions of the myocardium lacking sufficient anastomoses from other large branches to maintain viability of the tissue should occlusion occur). However, anastomoses do exist between branches of the coronary arteries, subepicardial or myocardial, and between these arteries and extracardiac vessels such as thoracic vessels (Standring, 2008). Anastomoses exist between the

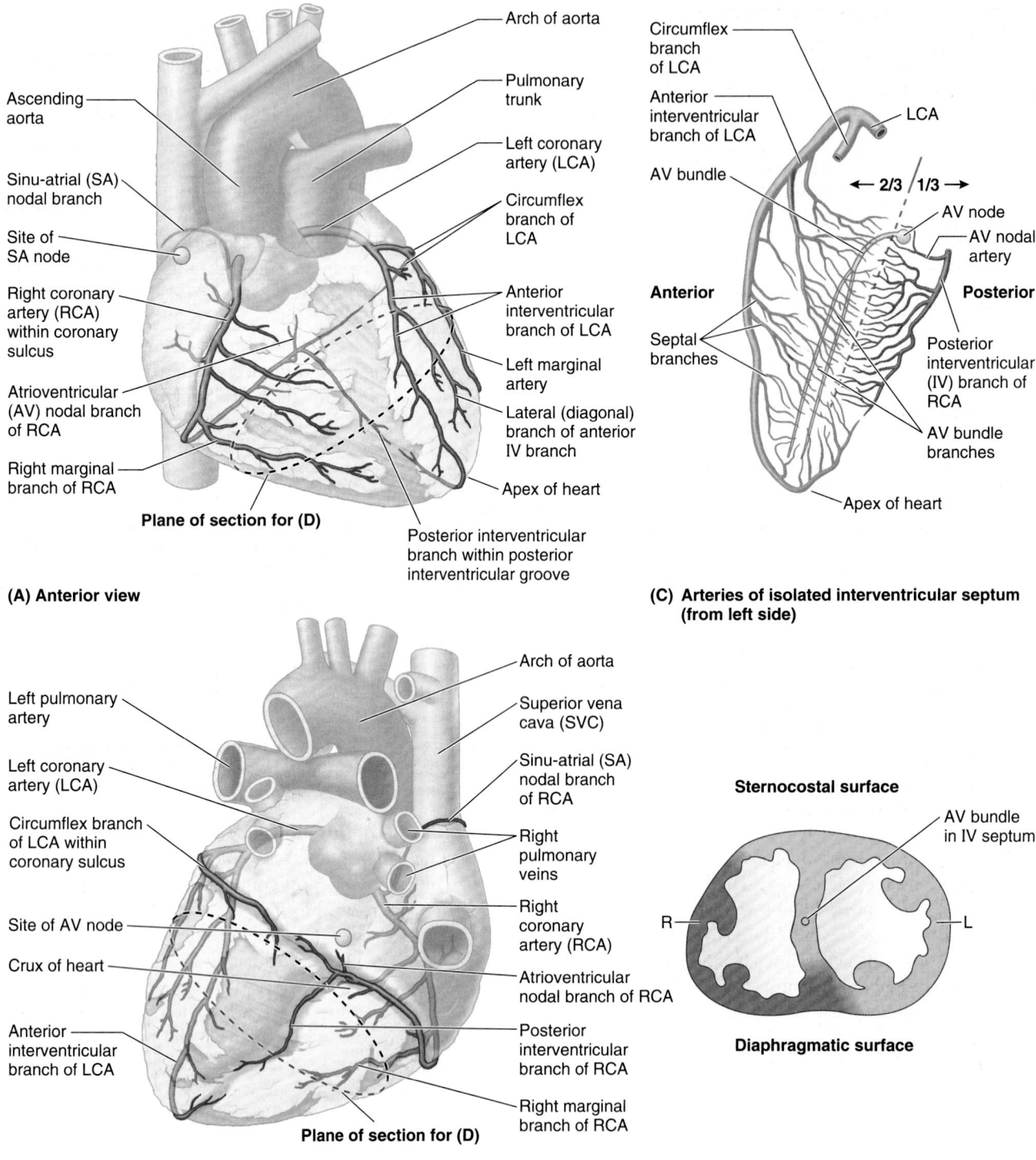

(A) Anterior view

Arch of aorta

Ascending aorta

Sinu-atrial (SA) nodal branch

Site of SA node

Right coronary artery (RCA) within coronary sulcus

Atrioventricular (AV) nodal branch of RCA

Right marginal branch of RCA

Plane of section for (D)

Pulmonary trunk

Left coronary artery (LCA)

Circumflex branch of LCA

Anterior interventricular branch of LCA

Left marginal artery

Lateral (diagonal) branch of anterior IV branch

Apex of heart

Posterior interventricular branch within posterior interventricular groove

(C) Arteries of isolated interventricular septum (from left side)

Circumflex branch of LCA

Anterior interventricular branch of LCA

LCA

AV bundle

← 2/3 / 1/3 →

AV node

AV nodal artery

Anterior

Posterior

Septal branches

Posterior interventricular (IV) branch of RCA

AV bundle branches

Apex of heart

(B) Postero-inferior view

Left pulmonary artery

Left coronary artery (LCA)

Circumflex branch of LCA within coronary sulcus

Site of AV node

Crux of heart

Anterior interventricular branch of LCA

Plane of section for (D)

Arch of aorta

Superior vena cava (SVC)

Sinu-atrial (SA) nodal branch of RCA

Right pulmonary veins

Right coronary artery (RCA)

Atrioventricular nodal branch of RCA

Posterior interventricular branch of RCA

Right marginal branch of RCA

(D) Inferior view of cross-section through ventricles at plane indicated in (A) and (B)

Sternocostal surface

AV bundle in IV septum

R

L

Diaphragmatic surface

FIGURE 1.59. Coronary arteries. A, B. In the most common pattern of distribution, the RCA anastomoses with the circumflex branch of the LCA (anastomoses are not shown) after the RCA has given rise to the posterior interventricular (IV) artery. **A–C.** The anterior IV artery (also called the left anterior descending branch) hooks around the apex of the heart to anastomose with the posterior IV artery. **C.** Arteries of the interventricular septum (*IVS*) are shown. The RCA branch to the AV node is the first of many septal branches of the posterior IV artery. The septal branches of the anterior interventricular branch of the LCA supply the anterior two thirds of the IVS. Because the AV bundle and bundle branches are centrally placed in and on the IVS, the LCA typically provides most blood to this conducting tissue. **D.** A cross-section of the right and left ventricles demonstrates the most common pattern of distribution of blood from the RCA (*red*) and LCA (*orange*) to the ventricular walls and IVS.

TABLE 1.4. ARTERIAL SUPPLY TO HEART

Artery/Branch	Origin	Course	Distribution	Anastomoses
Right coronary (RCA)	Right aortic sinus	Follows coronary (AV) sulcus between atria and ventricles	Right atrium, SA and AV nodes, and posterior part of IVS	Circumflex and anterior IV branches of LCA
SA nodal	RCA near its origin (in 60%)	Ascends to SA node	Pulmonary trunk and SA node	
Right marginal	RCA	Passes to inferior margin of heart and apex	Right ventricle and apex of heart	IV branches
Posterior interventricular	RCA (in 67%)	Runs in posterior IV groove to apex of heart	Right and left ventricles and posterior third of IVS	Anterior IV branch of LCA (at apex)
AV nodal	RCA near origin of posterior IV artery	Passes to AV node	AV node	
Left coronary (LCA)	Left aortic sinus	Runs in AV groove and gives off anterior IV and circumflex branches	Most of left atrium and ventricle, IVS, and AV bundles; may supply AV node	RCA
SA nodal	Circumflex branch of LCA (in 40%)	Ascends on posterior surface of left atrium to SA node	Left atrium and SA node	
Anterior interventricular	LCA	Passes along anterior IV groove to apex of heart	Right and left ventricles and anterior two thirds of IVS	Posterior IV branch of RCA (at apex)
Circumflex	LCA	Passes to left in AV sulcus and runs to posterior surface of heart	Left atrium and left ventricle	RCA
Left marginal	Circumflex branch of LCA	Follows left border of heart	Left ventricle	IV branches
Posterior interventricular	LCA (in 33%)	Runs in posterior IV groove to apex of heart	Right and left ventricles and posterior third of IVS	Anterior IV branch of L CA (at apex)

AV, atrioventricular; *IV*, interventricular; *IVS*, interventricular septum; *SA*, sinu-atrial.

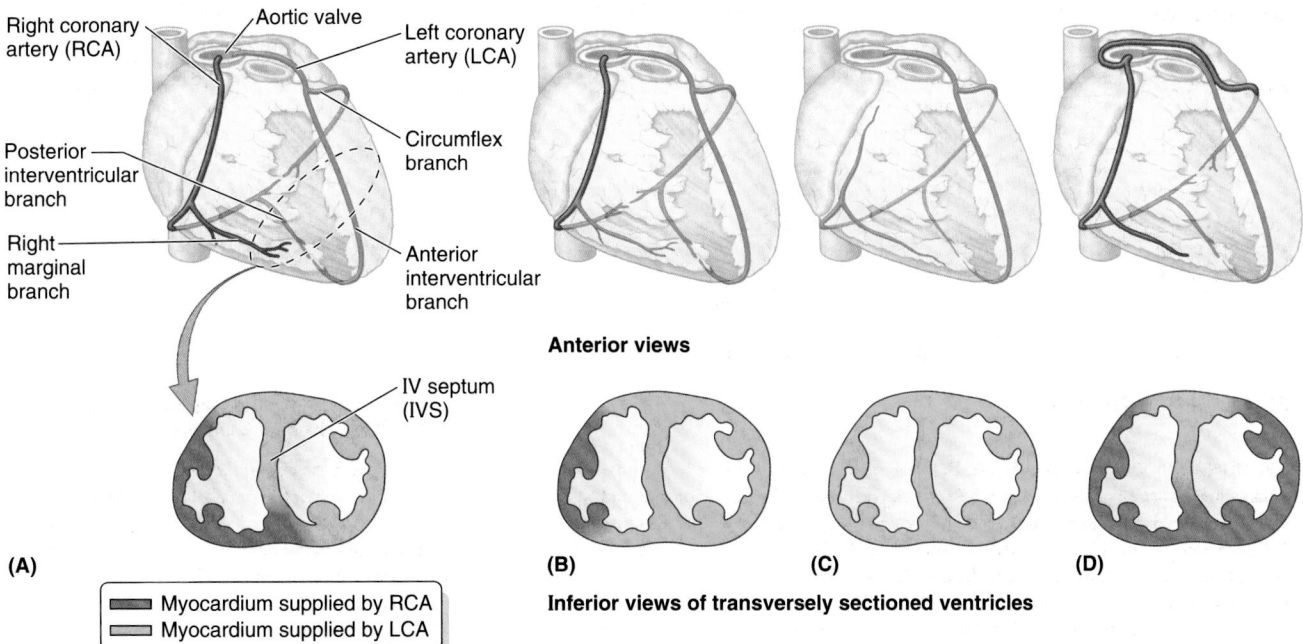

FIGURE 1.60. Variations in distribution of coronary arteries. A. In the most common pattern (67%), the RCA is dominant, giving rise to the posterior interventricular branch. **B, C.** The LCA gives rise to the posterior interventricular branch in approximately 15% of individuals. **D.** Many other variations occur.

terminations of the right and the left coronary arteries in the coronary sulcus and between the IV branches around the apex in approximately 10% of apparently normal hearts. The potential for development of collateral circulation probably exists in most if not all hearts.

Venous Drainage of the Heart. The heart is drained mainly by veins that empty into the coronary sinus and partly by small veins that empty into the right atrium (Fig. 1.61). The **coronary sinus,** the main vein of the heart, is a wide venous channel that runs from left to right in the posterior part of the coronary sulcus. The coronary sinus receives the great cardiac vein at its left end and the middle cardiac vein and small cardiac veins at its right end. The left posterior ventricular vein and left marginal vein also open into the coronary sinus.

The **great cardiac vein** is the main tributary of the coronary sinus. Its first part, the **anterior interventricular vein,** begins near the apex of the heart and ascends with the *anterior IV branch of the LCA.* At the coronary sulcus it turns left, and its second part runs around the left side of the heart with the circumflex branch of the LCA to reach the coronary sinus. (An unusual situation is occurring here: Blood is flowing in the same direction within a paired artery and vein!) The great cardiac vein drains the areas of the heart supplied by the LCA.

The **middle cardiac vein (posterior IV vein)** accompanies the *posterior interventricular branch* (usually arising from the RCA). A **small cardiac vein** accompanies *the right marginal branch of the RCA.* Thus these two veins drain most of the areas commonly supplied by the RCA. The **oblique vein of the left atrium** (of Marshall) is a small vessel, relatively unimportant postnatally, that descends over the posterior wall of the left atrium and merges with the great cardiac vein to form the *coronary sinus* (defining the beginning of the sinus). The oblique vein is the remnant of the embryonic left SVC, which usually atrophies during the fetal period, but occasionally persists in adults, replacing or augmenting the right SVC.

Some cardiac veins do not drain via the coronary sinus. Several small **anterior cardiac veins** begin over the anterior surface of the right ventricle, cross over the coronary sulcus, and usually end directly in the right atrium; sometimes they enter the small cardiac vein. The **smallest cardiac veins** (L. *venae cordis minimae*) are minute vessels that begin in the capillary beds of the myocardium and open directly into the chambers of the heart, chiefly the atria. Although called veins, they are valveless communications with the capillary beds of the myocardium and may carry blood from the heart chambers to the myocardium.

Lymphatic Drainage of the Heart. Lymphatic vessels in the myocardium and subendocardial connective tissue pass to the **subepicardial lymphatic plexus.** Vessels from this plexus pass to the coronary sulcus and follow the coronary arteries. A single lymphatic vessel, formed by the union of various lymphatic vessels from the heart, ascends between the pulmonary trunk and left atrium and ends in the **inferior tracheobronchial lymph nodes,** usually on the right side.

STIMULATING, CONDUCTING, AND REGULATING SYSTEMS OF HEART

Stimulating and Conducting System of the Heart. In the ordinary sequence of events in the cardiac cycle, the atrium and ventricle work together as one pump. The **conducting system of the heart** (Fig. 1.62) generates and transmits the impulses that produce the coordinated contractions of the *cardiac cycle* (discussed earlier in this chapter). The conducting system consists of *nodal tissue* that initiates the heartbeat and coordinates contractions of the four heart chambers, and highly specialized *conducting fibers* for conducting them rapidly to the different areas of the heart. The impulses are then propagated by the cardiac striated muscle cells so that the chamber walls contract simultaneously.

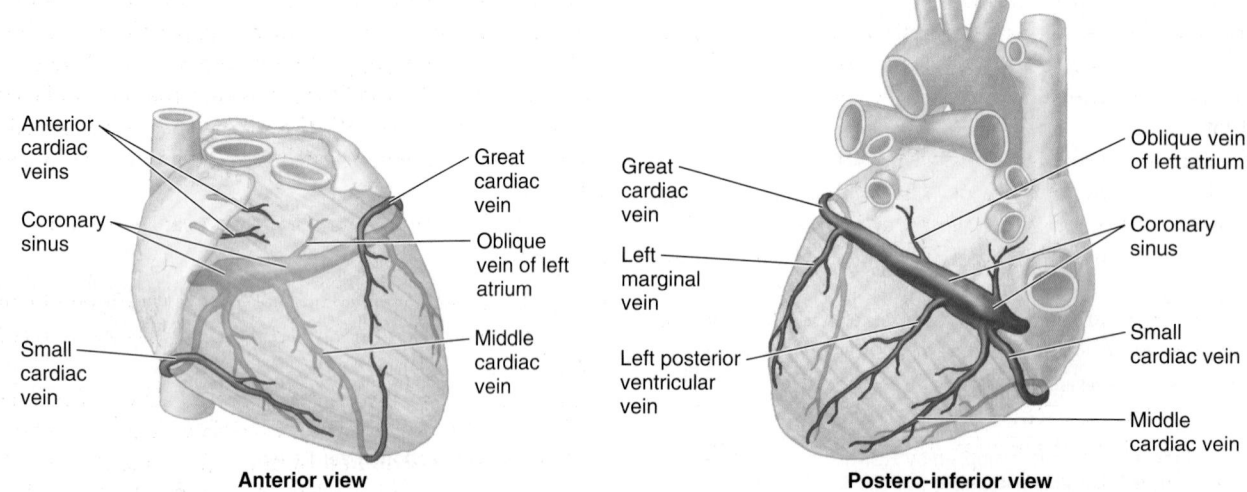

Anterior view

Postero-inferior view

FIGURE 1.61. Cardiac veins. The great, middle, and small cardiac veins; the oblique vein of the left atrium; and the left posterior ventricular vein are the main vessels draining into the coronary sinus. The coronary sinus, in turn, empties into the right atrium. The anterior cardiac veins drain directly into the auricle of the right atrium.

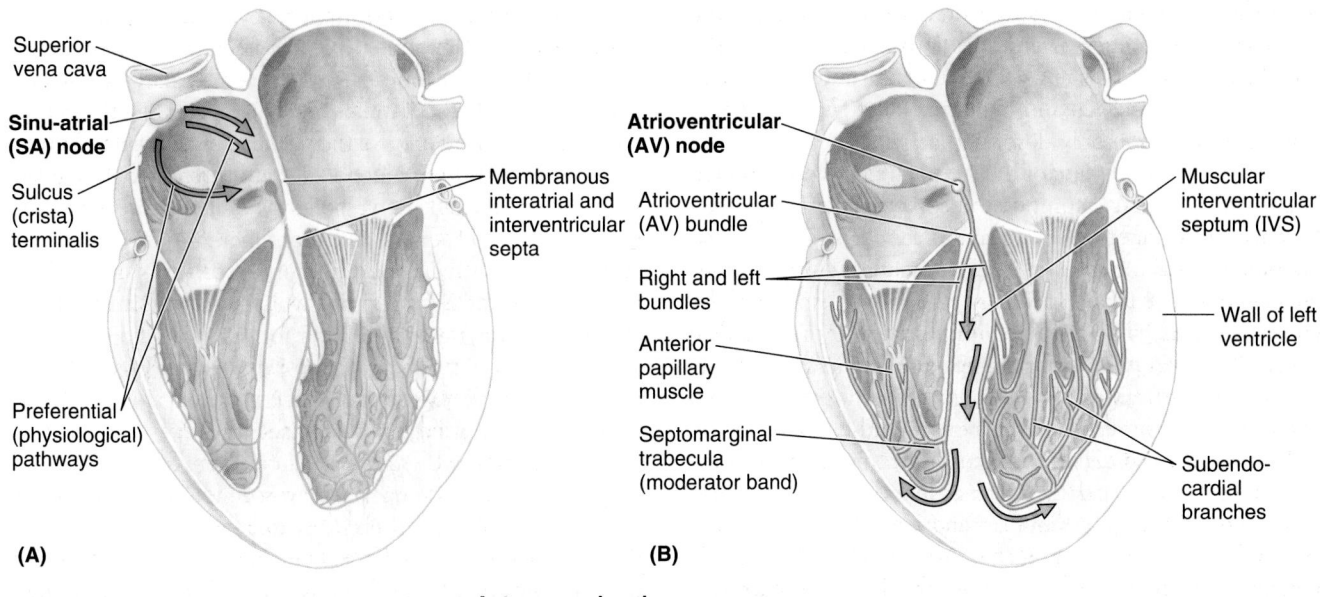

FIGURE 1.62. Conducting system of heart. A. Impulses (*arrows*) initiated at the SA node, located at the superior end of the sulcus (internally, crista) terminalis, are propagated through the atrial musculature to the AV node. **B.** Impulses (*arrows*) received by the AV node, in the inferior part of the interatrial septum, are conducted through the AV bundle and its branches to the myocardium. The AV bundle begins at the AV node and divides into right and left bundles at the junction of the membranous and muscular parts of the IVS.

The **sinu-atrial (SA) node** is located anterolaterally just deep to the epicardium at the junction of the SVC and right atrium, near the superior end of the *sulcus terminalis* (Figs. 1.59A and 1.62A). The SA node—a small collection of nodal tissue, specialized cardiac muscle fibers, and associated fibroelastic connective tissue—is the *pacemaker of the heart. The SA node initiates and regulates the impulses for the contractions of the heart*, giving off an impulse approximately 70 times per minute in most people most of the time. The contraction signal from the SA node spreads myogenically (through the musculature) of both atria. The SA node is supplied by the **sinu-atrial nodal artery,** which usually arises as an atrial branch of the RCA (in 60% of people), but it often arises from the LCA (in 40%). The SA node is stimulated by the sympathetic division of the autonomic nervous system to accelerate the heart rate and is inhibited by the parasympathetic division to return to or approach its basal rate.

The **atrioventricular (AV) node** is a smaller collection of nodal tissue than the SA node. The AV node is located in the posteroinferior region of the interatrial septum near the opening of the coronary sinus (Figs. 1.59A–C and 1.62B). The signal generated by the SA node passes through the walls of the right atrium, propagated by the cardiac muscle (**myogenic conduction**), which transmits the signal rapidly from the SA node to the AV node. The AV node then distributes the signal to the ventricles through the **AV bundle** (Fig. 1.62B). Sympathetic stimulation speeds up conduction, and parasympathetic stimulation slows it down. The AV bundle, the only bridge between the atrial and ventricular myocardium, passes from the AV node through the *fibrous skeleton of the heart* (see Fig. 1.51) and along the membranous part of the IVS.

At the junction of the membranous and muscular parts of the IVS, the AV bundle divides into **right** and **left bundles** (Fig. 1.62B). These branches proceed on each side of the muscular IVS deep to the endocardium and then ramify into **subendocardial branches** (Purkinje fibers), which extend into the walls of the respective ventricles. The subendocardial branches of the *right bundle* stimulate the muscle of the IVS, the anterior papillary muscle through the septomarginal trabecula (moderator band), and the wall of the right ventricle. The *left bundle* divides near its origin into approximately six smaller tracts, which give rise to subendocardial branches that stimulate the IVS, the anterior and posterior papillary muscles, and the wall of the left ventricle.

The AV node is supplied by the **AV nodal artery,** the largest and usually the first IV septal branch of the posterior IV artery, a branch of the RCA in 80% of people (Fig. 1.59A–C). Thus the arterial supply to both the SA and AV nodes is usually derived from the RCA. However, the AV bundle traverses the center of the IVS, the anterior two thirds of which is supplied by the septal branches of the anterior IV branch of the LCA (Fig. 1.59C & D).

Impulse generation and conduction can be summarized as follows:

- The SA node initiates an impulse that is rapidly conducted to cardiac muscle fibers in the atria, causing them to contract (Fig. 1.62A).
- The impulse spreads by myogenic conduction, which rapidly transmits the impulse from the SA node to the AV node.
- The signal is distributed from the AV node through the AV bundle and its branches (the right and left bundles), which pass on each side of the IVS to supply subendocardial branches to the papillary muscles and the walls of the ventricles (Fig. 1.62B).

Innervation of the Heart. The heart is supplied by autonomic nerve fibers from the **cardiac plexus** (Fig. 1.63; see also Fig. 1.68B & C), which is often quite artificially divided into superficial and deep portions. This nerve network is most commonly described as lying on the anterior surface of the bifurcation of the *trachea* (a respiratory structure), since it is most commonly observed in dissection after removal of the ascending aorta and the bifurcation of the pulmonary trunk. However, its primary relationship is to the posterior aspect of the latter two structures, especially the ascending aorta. The cardiac plexus is formed of both sympathetic and parasympathetic fibers en route to the heart, as well as visceral afferent fibers conveying reflexive and nociceptive fibers from the heart. Fibers extend from the plexus along and to the coronary vessels and to components of the conducting system, particularly the SA node.

The **sympathetic supply** is from presynaptic fibers, with cell bodies in the intermediolateral cell columns (IMLs) of the superior five or six thoracic segments of the spinal cord, and postsynaptic sympathetic fibers, with cell bodies in the cervical and superior thoracic paravertebral ganglia of the sympathetic trunks. The postsynaptic fibers traverse *cardiopulmonary splanchnic nerves* and the cardiac plexus to end in the SA and AV nodes and in relation to the terminations of parasympathetic fibers on the coronary arteries. *Sympathetic stimulation*

causes increased heart rate, impulse conduction, force of contraction, and, at the same time, increased blood flow through the coronary vessels to support the increased activity. Adrenergic stimulation of the SA node and conducting tissue increases the rate of depolarization of the pacemaker cells while increasing atrioventricular conduction. Direct *adrenergic stimulation* from the sympathetic nerve fibers, as well as indirect suprarenal (adrenal) hormone stimulation, increases atrial and ventricular contractility. Most adrenergic receptors on coronary blood vessels are b2-receptors, which, when activated, cause relaxation (or perhaps inhibition) of vascular smooth muscle and, therefore, dilation of the arteries (Wilson-Pauwels et al., 1997). This supplies more oxygen and nutrients to the myocardium during periods of increased activity.

The *parasympathetic supply* is from presynaptic fibers of the vagus nerves. Postsynaptic parasympathetic cell bodies (intrinsic ganglia) are located in the atrial wall and interatrial septum near the SA and AV nodes and along the coronary arteries. *Parasympathetic stimulation* slows the heart rate, reduces the force of the contraction, and constricts the coronary arteries, saving energy between periods of increased demand. Postsynaptic parasympathetic fibers release acetylcholine, which binds with *muscarinic receptors* to slow the rates of depolarization of the pacemaker cells and atrioventricular conduction and decrease atrial contractility.

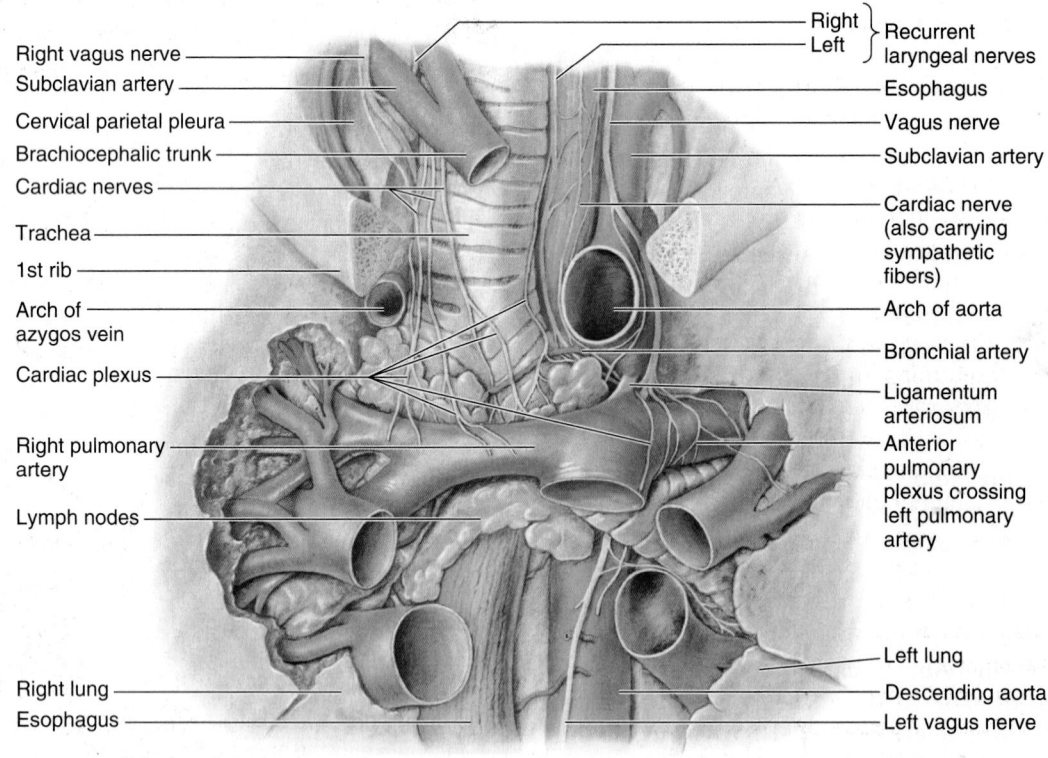

Anterior view

FIGURE 1.63. Cardiac nerves and plexus. This dissection of the superior and posterior mediastina demonstrates cardiac branches of the vagus nerve (CN X) and sympathetic trunks running down the sides of the trachea to form the cardiac plexus. Although shown lying anterior to the tracheal bifurcation here, the primary relationship of the cardiac plexus is to the ascending aorta and pulmonary trunk, the former having been removed to expose the plexus.

HEART

Cardiac Catheterization

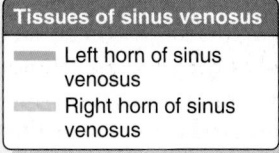

In *cardiac catheterization*, a radiopaque catheter is inserted into a peripheral vein (e.g., the femoral vein) and passed under fluoroscopic control into the right atrium, right ventricle, pulmonary trunk, and pulmonary arteries, respectively. Using this technique, intracardiac pressures can be recorded and blood samples may be removed. If a radiopaque contrast medium is injected, it can be followed through the heart and great vessels using serially exposed X-ray films. Alternatively, *cineradiography* or **cardiac ultrasonography** can be performed to observe the flow of dye in real time. Both techniques permit study of the circulation through the functioning heart and are helpful in the study of congenital cardiac defects.

Embryology of the Right Atrium

The primordial atrium is represented in the adult by the right auricle. The definitive atrium is enlarged by incorporation of most of the embryonic *sinus venosus* (Fig. B1.19A–C). The coronary sinus is also a derivative of this venous sinus. The part of the venous sinus incorporated into the primordial atrium becomes the smooth-walled sinus venarum of the adult right atrium (see Fig. 1.53A) into which all the veins drain, including the coronary sinus. The line of fusion of the primordial atrium (the adult auricle) and the sinus venarum (the derivative of the venous sinus) is indicated internally by the *crista terminalis* and externally by the *sulcus terminalis*. The sinu-atrial (SA) node (discussed earlier in this chapter) is located just in front of the opening of the SVC at the superior end of the crista terminalis—that is, in the border between the primordial atrium and the sinus venosus; hence its name.

Before birth, the valve of the IVC directs most of the oxygenated blood returning from the placenta in the umbilical vein and IVC toward the oval foramen in the interatrial septum, through which it passes into the left atrium (Fig. B1.19D). The oval foramen has a flap-like valve that permits a right to left shunt of blood but prevents a left to right shunt. At birth, when the baby takes its first breath, the lungs expand with air and pressure in the right atrium falls below that in the left atrium (Fig. B1.19E). Consequently, the oval foramen closes for its first and last time, and its valve usually fuses with the interatrial septum. The closed oval foramen is represented in the postnatal interatrial septum by the depressed oval fossa. The **border of the oval fossa** (L. *limbus fossae ovalis*) surrounds the fossa. The floor of the fossa is formed by the valve of the oval foramen. The rudimentary **IVC valve,** a semilunar crescent of tissue, has no function after birth; it varies considerably in size and is occasionally absent.

Tissues of sinus venosus
- Left horn of sinus venosus
- Right horn of sinus venosus

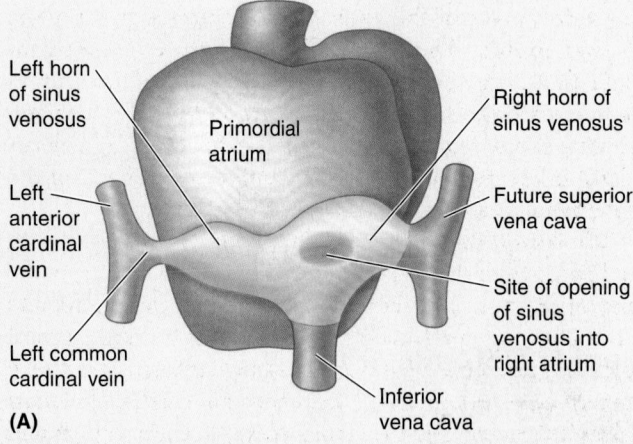

(A)

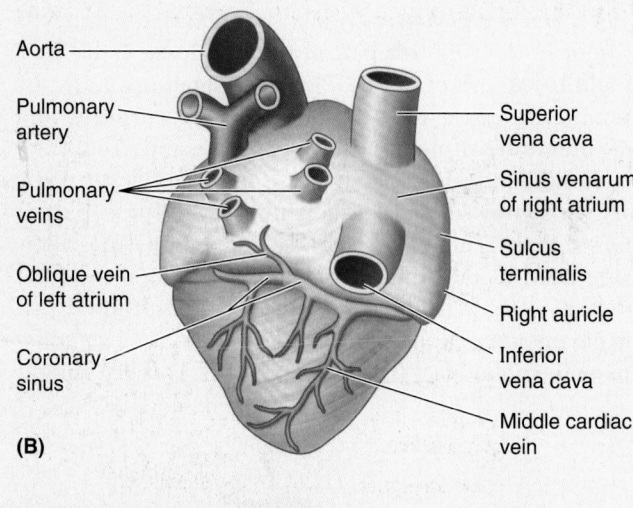

(B)

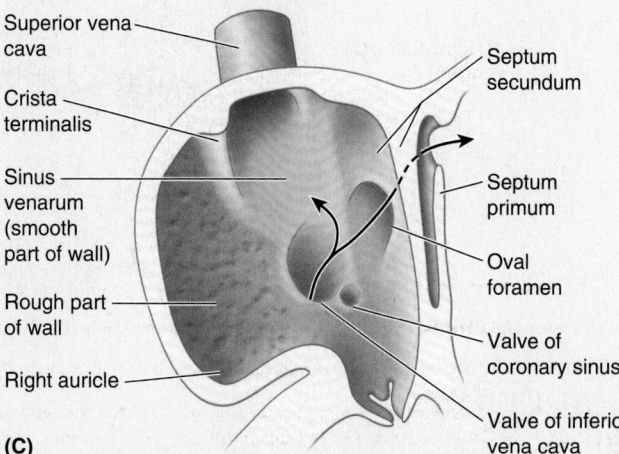

(C)

FIGURE B1.19. Development of features of right atrium.

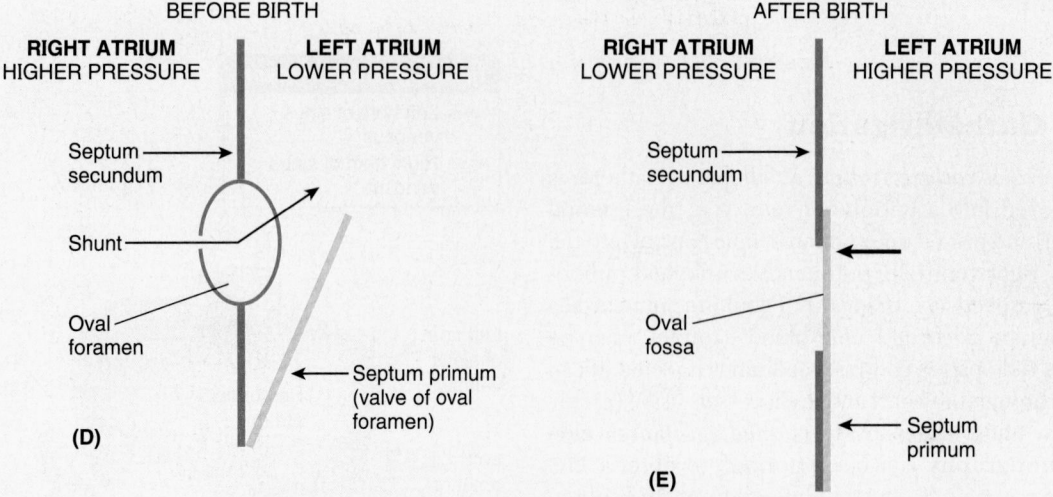

BEFORE BIRTH

RIGHT ATRIUM
HIGHER PRESSURE

LEFT ATRIUM
LOWER PRESSURE

Septum secundum

Shunt

Oval foramen

Septum primum (valve of oval foramen)

(D)

AFTER BIRTH

RIGHT ATRIUM
LOWER PRESSURE

LEFT ATRIUM
HIGHER PRESSURE

Septum secundum

Oval fossa

Septum primum

(E)

FIGURE B1.19. *(Continued)* **Development of features of right atrium.**

Septal Defects

ATRIAL SEPTAL DEFECTS

A congenital anomaly of the interatrial septum, usually incomplete closure of the oval foramen, is an *atrial septal defect (ASD).* A probe-size patency is present in the superior part of the oval fossa in 15–25% of adults (Moore et al., 2012). These small openings, by themselves, cause no hemodynamic abnormalities and are, therefore, of no clinical significance and should not be considered forms of ASDs. *Clinically significant ASDs* vary widely in size and location and may occur as part of more complex congenital heart disease. Large ASDs allow oxygenated blood from the lungs to be shunted from the left atrium through the ASD into the right atrium, causing enlargement of the right atrium and ventricle and dilation of the pulmonary trunk (Fig B1.20A). This left to right shunt of blood overloads the pulmonary vascular system, resulting in *hypertrophy of the right atrium and ventricle* and pulmonary arteries.

VENTRICULAR SEPTAL DEFECTS

The membranous part of the IVS develops separately from the muscular part and has a complex embryological origin. Consequently, this part is the common site of *ventricular septal defects (VSDs),* although defects also occur in the muscular part (Fig. B1.20B). *VSDs rank first on all lists of cardiac defects.* Isolated VSDs account for approximately 25% of all forms of congenital heart disease. The size of the defect varies from 1 to 25 mm. A VSD causes a left to right shunt of blood through the defect. A large shunt increases pulmonary blood flow, which causes severe pulmonary disease (*hypertension,* or increased blood pressure) and may cause *cardiac failure.* The much less common VSD in the muscular part of the septum frequently closes spontaneously during childhood (Green et al., 2009).

Percussion of Heart

Percussion defines the density and size of the heart. The classical percussion technique is to create vibration by tapping the chest with a finger while listening and feeling for differences in soundwave conduction. *Cardiac percussion* is performed at the 3rd, 4th, and 5th intercostal spaces from the left anterior axillary line to the right anterior axillary line (Fig. B1.21) Normally, the percussion note changes from resonance to dullness (because of the presence of the heart) approximately 6 cm lateral to the left border of the sternum.

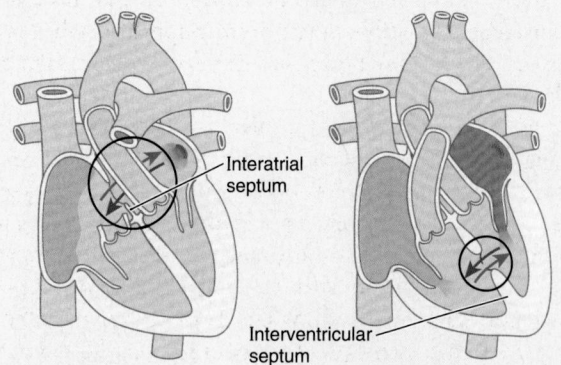

Interatrial septum

Interventricular septum

(A) Atrial septal defect (ASD)   **(B) Ventricular septal defect (VSD)**

FIGURE B1.20. Septal defects. A. Atrial septal defect (ASD). **B.** Ventricular septal defect (VSD).

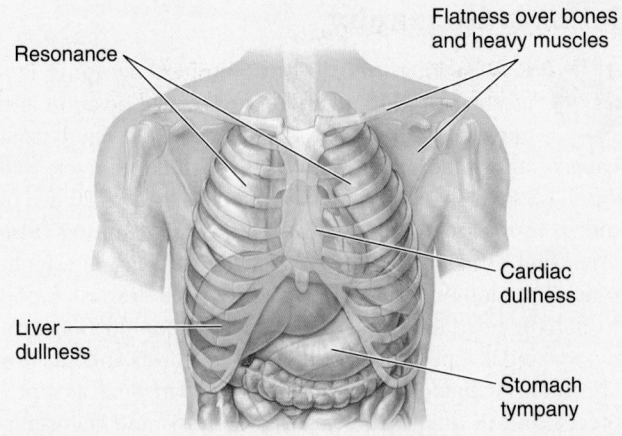

FIGURE B1.21. Areas of flatness (yellow) and resonance (unshaded) of thorax.

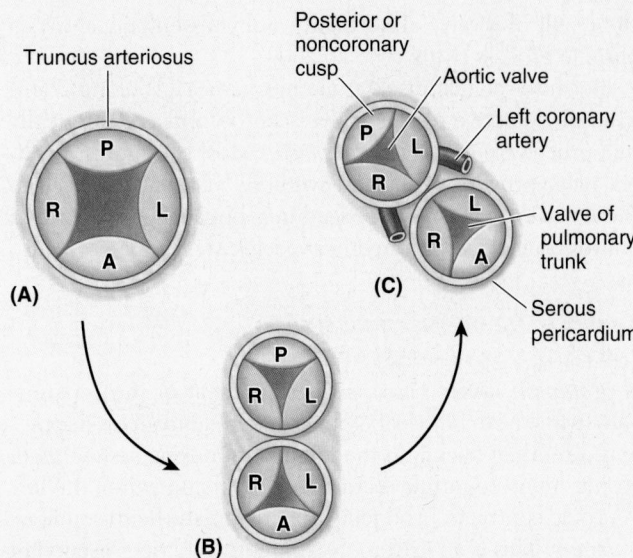

FIGURE B1.22. Developmental basis for names of valve cusps.

Stroke or Cerebrovascular Accident

Thrombi (clots) form on the walls of the left atrium in certain types of heart disease. If these thrombi detach, or pieces break off from them, they pass into the systemic circulation and occlude peripheral arteries. Occlusion of an artery supplying the brain results in a *stroke* or *cerebrovascular accident (CVA),* which may affect vision, cognition, or the motor function of parts of the body previously controlled by the now-damaged (ischemic) area of the brain.

Basis for Naming Cusps of the Aortic and Pulmonary Valves

The following account explains the *embryological basis for naming the pulmonary and aortic valves.*

The **truncus arteriosus,** the common arterial trunk from both ventricles of the embryonic heart, has four cusps (Fig. B1.22A). The truncus arteriosus divides into two vessels, each with its own three-cusp valve (pulmonary and aortic) (Fig. B1.22B). The heart undergoes partial rotation so that its apex becomes directed to the left, resulting in the arrangement of cusps as shown in Figure B1.22C. Consequently, the cusps are named according to their embryological origin, not their postnatal anatomical position. Thus the pulmonary valve has right, left, and anterior cusps, and the aortic valve has right, left, and posterior cusps. Similarly, the aortic sinuses are named right, left, and posterior.

This terminology also agrees with the coronary arteries. Note that the right coronary artery arises from the right aortic sinus, superior to the right cusp of the aortic valve, and that the left coronary has a similar relation to the left cusp and sinus. The posterior cusp and sinus do not give rise to a coronary artery; thus they are also referred to as a "noncoronary" cusp and sinus.

Valvular Heart Disease

Disorders involving the valves of the heart disturb the pumping efficiency of the heart. *Valvular heart disease* produces either stenosis (narrowing) or insufficiency. *Stenosis* is the failure of a valve to open fully, slowing blood flow from a chamber. *Insufficiency* or *regurgitation,* on the other hand, is failure of the valve to close completely, usually owing to nodule formation on (or scarring and contraction of) the cusps so that the edges do not meet or align. This allows a variable amount of blood (depending on the severity) to flow back into the chamber it was just ejected from. Both stenosis and insufficiency result in an increased workload for the heart.

Restriction of high-pressure blood flow (stenosis) or passage of blood through a narrow opening into a larger vessel or chamber (stenosis and regurgitation) produces turbulence. Turbulence sets up *eddies* (small whirlpools) that produce vibrations that are audible as *murmurs.* Superficial vibratory sensations (*thrills*) may be felt on the skin over an area of turbulence.

The clinical significance of a valvular dysfunction ranges from slight and physiologically insignificant to severe and rapidly fatal. Factors such as degree, duration, and etiology (cause) affect secondary changes in the heart, blood vessels, and other organs, both proximal and distal to the valve lesion. Valvular disorders may be congenital or acquired. Insufficiency may result from pathology of the valve itself or its supporting structures (anulus, tendinous cords, dilation of chamber wall, etc.). It may occur *acutely* (suddenly—for example, from a rupture of the cords) or *chronically* (over a relatively long time—for example, scarring and retraction). Valvular stenosis, on the other hand, is almost always

[handwritten margin notes:]
- stenosis usually presents with hypertrophy.
- insufficiency or incompetence presents as murmurs.

the result of a valve abnormality and is essentially always a chronic process (Kumar et al., 2009).

Because valvular diseases are mechanical problems, damaged or defective cardiac valves can be replaced surgically in a procedure called *valvuloplasty*. Most commonly, artificial *valve prostheses* made of synthetic materials are used in these valve-replacement procedures, but xenografted valves (valves transplanted from other species, such as pigs) are also used.

MITRAL VALVE INSUFFICIENCY (MITRAL VALVE PROLAPSE)

A *prolapsed mitral valve* is an insufficient or incompetent valve with one or both leaflets enlarged, redundant or "floppy," and extending back into the left atrium during systole. As a result, blood regurgitates into the left atrium when the left ventricle contracts, producing a characteristic heart sound or *murmur*. This is an extremely common condition, occurring in up to 1 in every 20 people, most often in young females. Usually, it is an incidental finding on physical examination; but it is of clinical importance in a small fraction of those affected, with the patient suffering chest pain and fatigue.

PULMONARY VALVE STENOSIS

In *pulmonary valve stenosis*, the valve cusps are fused, forming a dome with a narrow central opening. In *infundibular pulmonary stenosis*, the conus arteriosus is underdeveloped. Both types of pulmonary stenoses produce a restriction of right ventricular outflow and may occur together. The degree of hypertrophy of the right ventricle is variable.

PULMONARY VALVE INCOMPETENCE

If the free margins (*lunules*) of the cusps of a semilunar valve thicken and become inflexible or are damaged by disease, the valve will not close completely. An *incompetent pulmonary valve* results in a backrush of blood under high pressure into the right ventricle during diastole. Pulmonic regurgitation may be heard through a stethoscope as a *heart murmur,* an abnormal sound from the heart, produced in this case by damage to the cusps of the pulmonary valve.

AORTIC VALVE STENOSIS

Aortic valve stenosis is the most frequent valve abnormality. For those born in the early and mid-20th century, rheumatic fever was a common cause but now accounts for <10% of cases of aortic stenosis. The great majority of aortic stenoses is a result of degenerative calcification and comes to clinical attention in the 6th decade of life or later. Aortic stenosis causes extra work for the heart, resulting in *left ventricular hypertrophy*.

AORTIC VALVE INSUFFICIENCY

Insufficiency of the aortic valve results in aortic regurgitation (backrush of blood into the left ventricle), producing a heart murmur and a *collapsing pulse* (forcible impulse that rapidly diminishes).

Echocardiography

Echocardiography (ultrasonic cardiography) is a method of graphically recording the position and motion of the heart by the echo obtained from beams of ultrasonic waves directed through the thoracic wall (Fig. B1.23). This technique may detect as little as 20 mL of fluid in the pericardial cavity, such as that resulting from pericardial effusion. *Doppler echocardiography* is a technique that demonstrates and records the flow of blood through the heart and great vessels by Doppler ultrasonography, making it especially useful in the diagnosis and analysis of problems with blood flow through the heart, such as septal defects, and in delineating valvular stenosis and regurgitation, especially on the left side of the heart.

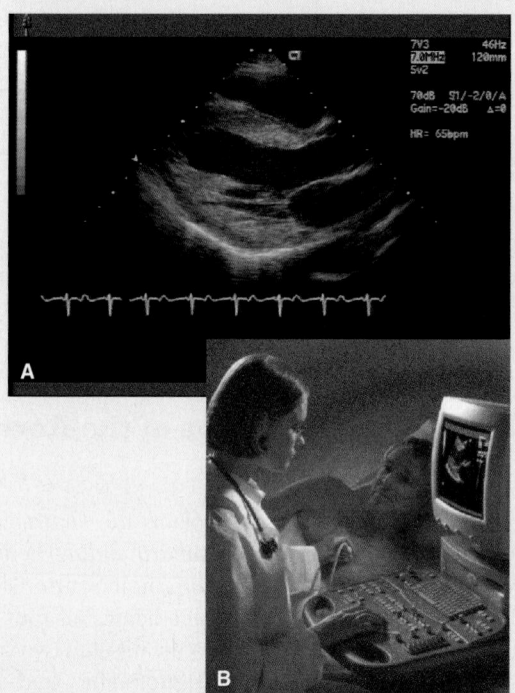

FIGURE B1.23. Echocardiography. A. Normal echocardiogram. **B.** Sonographer placing transducer in a left intercostal space in the parasternal line, overlying the heart.

Coronary Angiography

Using *coronary angiography*, the coronary arteries can be visualized with *coronary arteriograms* (Fig. B1.24). A long, narrow catheter is passed into the ascending aorta via the femoral artery in the inguinal region. Under fluoroscopic control, the tip of the catheter is placed just inside the opening of a coronary artery. A small injection of radiopaque contrast material is made, and cineradiographs are taken to show the lumen of the artery and its branches, as well as any stenotic areas that may be present.

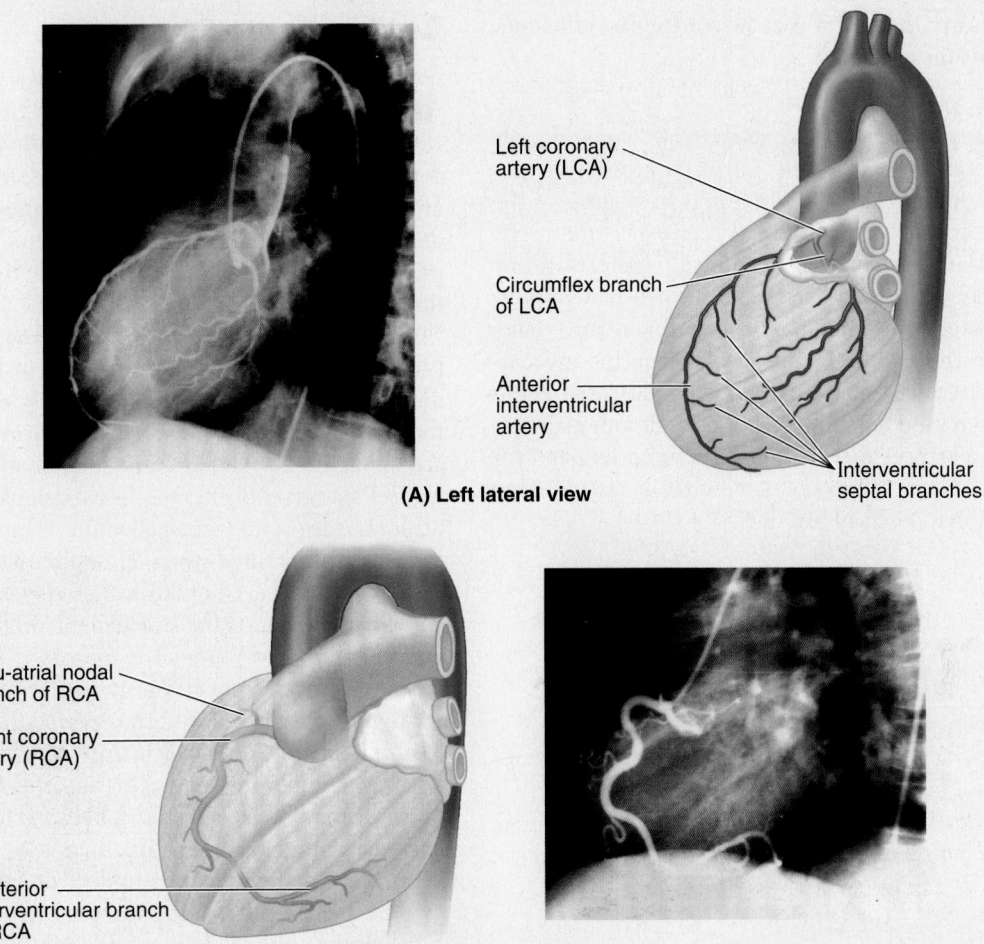

(A) Left lateral view

Left coronary
artery (LCA)

Circumflex branch
of LCA

Anterior
interventricular
artery

Interventricular
septal branches

Sinu-atrial nodal
branch of RCA

Right coronary
artery (RCA)

Posterior
interventricular branch
of RCA

(B) Left anterior oblique view

FIGURE B1.24. Coronary arteriograms.

Coronary Artery Disease or Coronary Heart Disease

Coronary artery disease (CAD) is one of the leading causes of death. It has many causes, all of which result in a reduced blood supply to the vital myocardial tissue.

MYOCARDIAL INFARCTION

With sudden occlusion of a major artery by an embolus (G. *embolos,* plug), the region of myocardium supplied by the occluded vessel becomes *infarcted* (rendered virtually bloodless) and undergoes *necrosis* (pathological tissue death). The three most common sites of coronary artery occlusion and the percentage of occlusions involving each artery are (Fig. B1.25A & B) the:

1. Anterior IV (LAD) branch of the LCA (40–50%).
2. RCA (30–40%).
3. Circumflex branch of the LCA (15–20%).

An area of myocardium that has undergone necrosis constitutes a *myocardial infarction* (MI). The most common cause

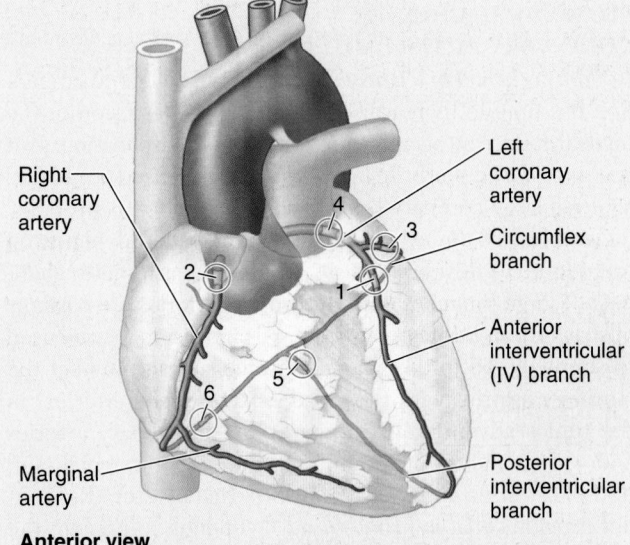

Right
coronary
artery

Left
coronary
artery

Circumflex
branch

Anterior
interventricular
(IV) branch

Posterior
interventricular
branch

Marginal
artery

Anterior view

Sites of coronary artery occlusion, in order of frequency (1–6)

FIGURE B1.25. Sites 1–3 account for at least 85% of all occlusions.

[handwritten margin notes: "① Reduced blood flow ② Ischemia"]

of *ischemic heart disease* is coronary artery insufficiency resulting from atherosclerosis.

CORONARY ATHEROSCLEROSIS

The *atherosclerotic process*, characterized by lipid deposits in the intima (lining layer) of the coronary arteries, begins during early adulthood and slowly results in stenosis of the lumina of the arteries (Fig. B1.26). As *coronary atherosclerosis* progresses, the collateral channels connecting one coronary artery with the other expand, which may initially permit adequate perfusion of the heart during relative inactivity. Despite this compensatory mechanism, the myocardium may not receive enough oxygen when the heart needs to perform increased amounts of work. Strenuous exercise, for example, increases the heart's activity and its need for oxygen. Insufficiency of blood supply to the heart (*myocardial ischemia*) may result in MI.

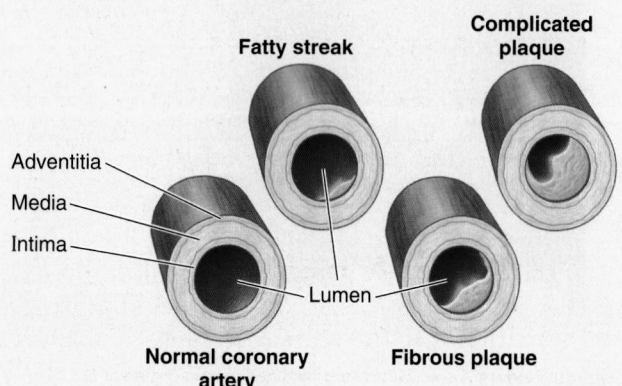

FIGURE B1.26. **Atherosclerosis:** stages of development in a coronary artery.

SLOWLY PROGRESSIVE CORONARY ARTERY DISEASE

In *slow occlusion of a coronary artery*, the collateral circulation has time to increase so that adequate perfusion of the myocardium can occur when a potentially ischemic event occurs. Consequently, MI may not result. On sudden blockage of a large coronary branch, some infarction is probably inevitable, but the extent of the area damaged depends on the degree of development of collateral anastomotic channels. If large branches of both coronary arteries are partially obstructed, an extracardiac collateral circulation may be used to supply blood to the heart. These collaterals connect the coronary arteries with the vasa vasorum (small arteries) in the tunica adventitia of the aorta and pulmonary arteries and with branches of the internal thoracic, bronchial, and phrenic arteries. Clinical studies show that anastomoses cannot provide collateral routes quickly enough to prevent the effects of *sudden coronary artery occlusion*. The functional value of these anastomoses thus appears to be more effective in slowly progressive CAD in individuals that are physically active.

Angina Pectoris

Pain that originates in the heart is called *angina* or *angina pectoris* (L. *angina,* strangling pain + L. *pectoris,* of the chest). Individuals with angina commonly describe the transient (15 sec to 15 min) but moderately severe constricting pain as tightness in the thorax, deep to the sternum. The pain is the result of ischemia of the myocardium that falls short of inducing the cellular necrosis that defines infarction.

Most often, angina results from narrowed coronary arteries. The reduced blood flow results in less oxygen being delivered to the cardiac striated muscle cells. As a result of the limited anaerobic metabolism of the myocytes, lactic acid accumulates and the pH is reduced in affected areas of the heart. Pain receptors in muscle are stimulated by lactic acid. Strenuous exercise (especially after a heavy meal), sudden exposure to cold, and stress all require increased activity on the part of the heart, but the occluded vessels cannot provide it. When food enters the stomach, blood flow to it and other parts of the digestive tract is increased. As a result, some blood is diverted from other organs, including the heart.

Anginal pain is relieved by a period of rest (1–2 min are often adequate). *Sublingual nitroglycerin* (medication placed or sprayed under the tongue for absorption through the oral mucosa) may be administered because it dilates the coronary (and other) arteries. This increases blood flow to the heart, while decreasing the workload and the heart's need for oxygen because the heart is pumping against less resistance. Furthermore, the dilated vessels accommodate more of the blood volume, so less blood arrives in the heart, relieving heart congestion. Thus the angina is usually relieved. Such angina provides a warning that the coronary arteries are compromised and that there is a need for a change of lifestyle, a healthcare intervention, or both.

The pain resulting from MI is usually more severe than with angina pectoris, and the pain resulting from the infarction does not disappear after 1–2 min of rest.

Coronary Bypass Graft

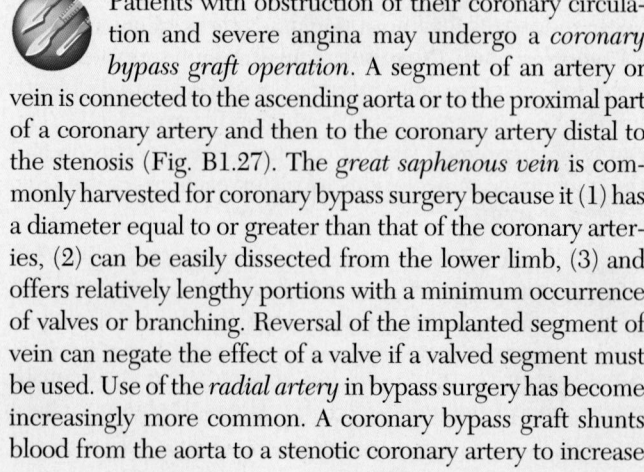

Patients with obstruction of their coronary circulation and severe angina may undergo a *coronary bypass graft operation*. A segment of an artery or vein is connected to the ascending aorta or to the proximal part of a coronary artery and then to the coronary artery distal to the stenosis (Fig. B1.27). The *great saphenous vein* is commonly harvested for coronary bypass surgery because it (1) has a diameter equal to or greater than that of the coronary arteries, (2) can be easily dissected from the lower limb, (3) and offers relatively lengthy portions with a minimum occurrence of valves or branching. Reversal of the implanted segment of vein can negate the effect of a valve if a valved segment must be used. Use of the *radial artery* in bypass surgery has become increasingly more common. A coronary bypass graft shunts blood from the aorta to a stenotic coronary artery to increase

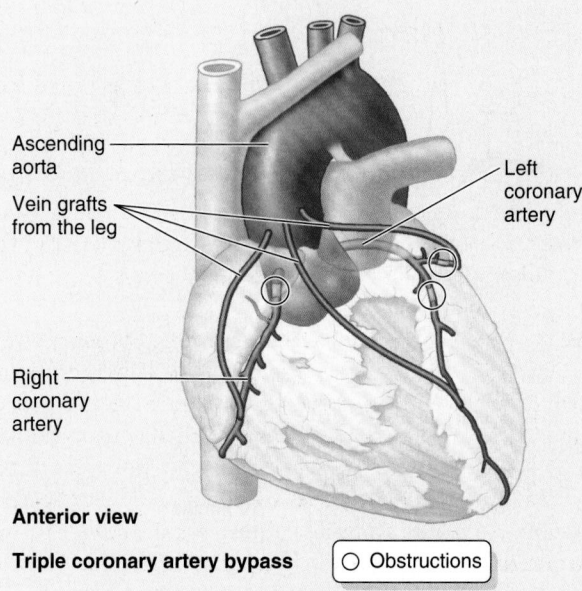

Ascending aorta

Vein grafts from the leg

Left coronary artery

Right coronary artery

Anterior view

Triple coronary artery bypass   ○ Obstructions

FIGURE B1.27. Triple coronary artery bypass.

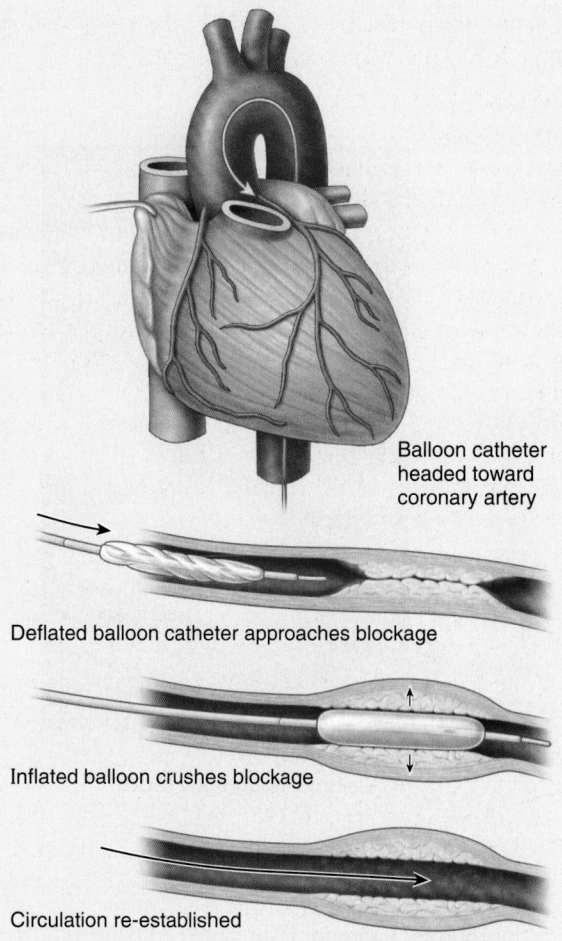

Balloon catheter headed toward coronary artery

Deflated balloon catheter approaches blockage

Inflated balloon crushes blockage

Circulation re-established

FIGURE B1.28. Percutaneous transluminal angioplasty.

the flow distal to the obstruction. Simply stated, it provides a detour around the stenotic area (arterial stenosis) or blockage (arterial atresia). Revascularization of the myocardium may also be achieved by surgically anastomosing an internal thoracic artery with a coronary artery.

Coronary Angioplasty

 In selected patients, surgeons use *percutaneous transluminal coronary angioplasty* in which they pass a catheter with a small inflatable balloon attached to its tip into the obstructed coronary artery (Fig. B1.28). When the catheter reaches the obstruction, the balloon is inflated, flattening the atherosclerotic plaque against the vessel's wall. The vessel is stretched to increase the size of the lumen, thus improving blood flow. In other cases, *thrombokinase* is injected through the catheter; this enzyme dissolves the blood clot. Intraluminal instruments with rotating blades and lasers have also been employed. After dilation of the vessel, an *intravascular stent* may be introduced to maintain the dilation. Intravascular stents are composed of rigid or semirigid tubular meshes, collapsed during introduction. Once in place, they expand or are expanded with a balloon catheter, to maintain luminal patency.

Collateral Circulation via the Smallest Cardiac Veins

 Reversal of flow in the *anterior* and *smallest cardiac veins* may bring **luminal blood** (blood from the heart chambers) to the capillary beds of the myocardium in some regions, providing some additional collateral circulation. However, unless these collaterals have dilated in response to pre-existing ischemic heart disease, especially in conjunction with physical conditioning, they are

unlikely to be able to supply sufficient blood to the heart during an acute event and thus prevent MI.

Electrocardiography

 The passage of impulses over the heart from the SA node can be amplified and recorded as an *electrocardiogram (ECG or EKG)* (Fig. B1.29). Functional testing of the heart includes *exercise tolerance tests* (treadmill stress tests), primarily to check the consequences of possible coronary artery disease. Exercise tolerance tests are of considerable importance in detecting the cause of heartbeat irregularities. Heart rate, ECG, and blood pressure readings are monitored as the patient does increasingly demanding exercise on a treadmill. The results show the maximum effort a patient's heart can safely tolerate.

Coronary Occlusion and Conducting System of Heart

Damage to the conducting system of the heart, often resulting from ischemia caused by coronary artery disease, produces disturbances of cardiac muscle

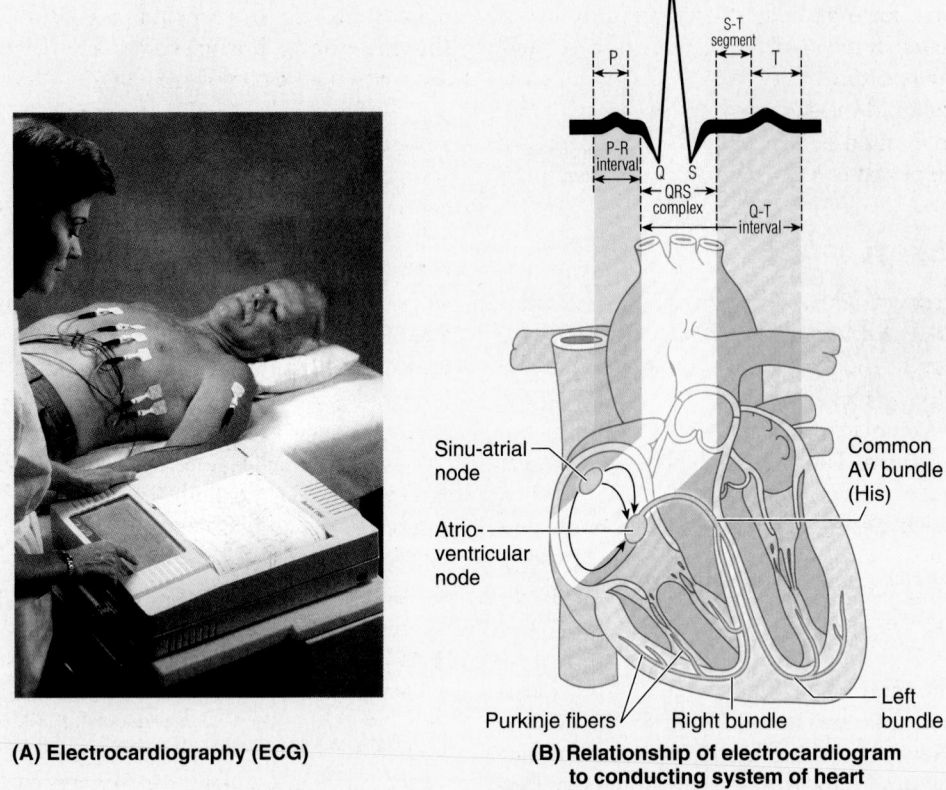

(A) Electrocardiography (ECG)

(B) Relationship of electrocardiogram to conducting system of heart

FIGURE B1.29. A. Electrocardiography (ECG). B. Relationship of electrocardiogram to conducting system of heart.

contraction. Since the anterior IV branch (LAD) gives rise to the septal branches supplying the AV bundle in most people, and branches of the RCA supply both the SA and AV nodes (Figs. B1.30 and 1.59C), parts of the conducting system of the heart are likely to be affected by their occlusion, and a *heart block* may occur. In this case (if the patient survives the initial stages),

Anterior view

FIGURE B1.30. **Blood supply of conducting system of heart.** *AV,* atrioventricular; *SA,* sinu-atrial.

the ventricles will begin to contract independently at their own rate: 25–30 times per minute (much slower than the slowest normal rate (40–45 times per minute). The atria continue to contract at the normal rate if the SA node has been spared, but the impulse generated by the SA node no longer reaches the ventricles.

Damage to one of the bundle branches results in a *bundle branch block,* in which excitation passes along the unaffected branch and causes a normally timed systole of that ventricle only. The impulse then spreads to the other ventricle via *myogenic* (muscle propagated) conduction, producing a late asynchronous contraction. In these cases, a *cardiac pacemaker* (artificial heart regulator) may be implanted to increase the ventricular rate of contraction to 70–80 per minute.

With a VSD, the AV bundle usually lies in the margin of the VSD. Obviously, this vital part of the conducting system must be preserved during surgical repair of the defect. Destruction of the AV bundle would cut the only physiological link between the atrial and ventricular musculature, also producing a heart block as described above.

Artificial Cardiac Pacemaker

In some people with a heart block, an *artificial cardiac pacemaker* (approximately the size of a pocket watch) is inserted subcutaneously. The pacemaker consists of a pulse generator or battery pack, a wire (lead), and an electrode. Pacemakers produce electrical impulses that initiate ventricular contractions at a predetermined rate. An electrode with a catheter connected to it is inserted into a vein and its progression through the venous pathway is followed with a *fluoroscope,* a device for examining deep structures in real time (as motion occurs) by means of radiographs. The terminal of the electrode is passed through the SVC to the right atrium and through the tricuspid valve into the right ventricle. Here the electrode is firmly fixed to the trabeculae carneae in the ventricular wall and placed in contact with the endocardium.

Restarting Heart

 In most cases of cardiac arrest, first-aid workers perform *cardiopulmonary resuscitation* (CPR) to restore cardiac output and pulmonary ventilation. By applying firm pressure to the thorax over the inferior part of the sternal body (external or closed chest massage), the sternum moves posteriorly 4–5 cm. The increased intrathoracic pressure forces blood out of the heart into the great arteries. When the external pressure is released and the intrathoracic pressure falls, the heart again fills with blood. If the heart stops beating (*cardiac arrest*) during heart surgery, the surgeon attempts to restart it using internal or open chest heart massage.

Fibrillation of Heart

 Fibrillation is multiple, rapid, circuitous contractions or twitchings of muscular fibers, including cardiac muscle. In *atrial fibrillation,* the normal regular rhythmical contractions of the atria are replaced by rapid irregular and uncoordinated twitchings of different parts of the atrial walls. The ventricles respond at irregular intervals to the dysrhythmic impulses received from the atria, but usually circulation remains satisfactory. In *ventricular fibrillation,* the normal ventricular contractions are replaced by rapid, irregular twitching movements that do not pump (i.e., they do not maintain the systemic circulation, including the coronary circulation). The damaged conducting system of the heart does not function normally. As a result, an irregular pattern of uncoordinated contractions occurs in the ventricles, except in those areas that are infarcted. Ventricular fibrillation is the most disorganized of all *dysrhythmias,* and in its presence no effective cardiac output occurs. The condition is fatal if allowed to persist.

Defibrillation of Heart

 A *defibrillating electric shock* may be given to the heart through the thoracic wall via large electrodes (paddles). This shock causes cessation of all cardiac movements and a few seconds later the heart may begin to beat more normally. As coordinated contractions and hence pumping of the heart is re-established, some degree of systemic (including coronary) circulation results.

Cardiac Referred Pain

 The heart is insensitive to touch, cutting, cold, and heat; however, ischemia and the accumulation of metabolic products stimulate pain endings in the myocardium. The afferent pain fibers run centrally in the middle and inferior cervical branches and especially in the thoracic cardiac branches of the sympathetic trunk. The axons of these primary sensory neurons enter spinal cord segments T1 through T4 or T5, especially on the left side.

Cardiac referred pain is a phenomenon whereby noxious stimuli originating in the heart are perceived by a person as pain arising from a superficial part of the body—the skin on the left upper limb, for example. *Visceral referred pain* is transmitted by visceral afferent fibers accompanying sympathetic fibers and is typically referred to somatic structures or areas such as a limb having afferent fibers with cell bodies in the same spinal ganglion, and central processes that enter the spinal cord through the same posterior roots (Hardy and Naftel, 2005).

Anginal pain is commonly felt as radiating from the substernal and left pectoral regions to the left shoulder and the medial aspect of the left upper limb (Fig. B1.31A). This part of the limb is supplied by the medial cutaneous nerve of the arm. Often the lateral cutaneous branches of the 2nd and 3rd intercostal nerves (the intercostobrachial nerves) join or overlap in their distribution with the medial cutaneous nerve of the arm. Consequently, cardiac pain is referred to the upper limb because the spinal cord segments of these cutaneous nerves (T1–T3) are also common to the visceral afferent terminations for the coronary arteries. Synaptic contacts may also be made with commissural (connector) neurons, which conduct impulses to neurons on the right side of comparable areas of the spinal cord. This occurrence explains why pain of cardiac origin, although usually referred to the left side, may be referred to the right side, both sides, or the back (Fig. B1.31B & C).

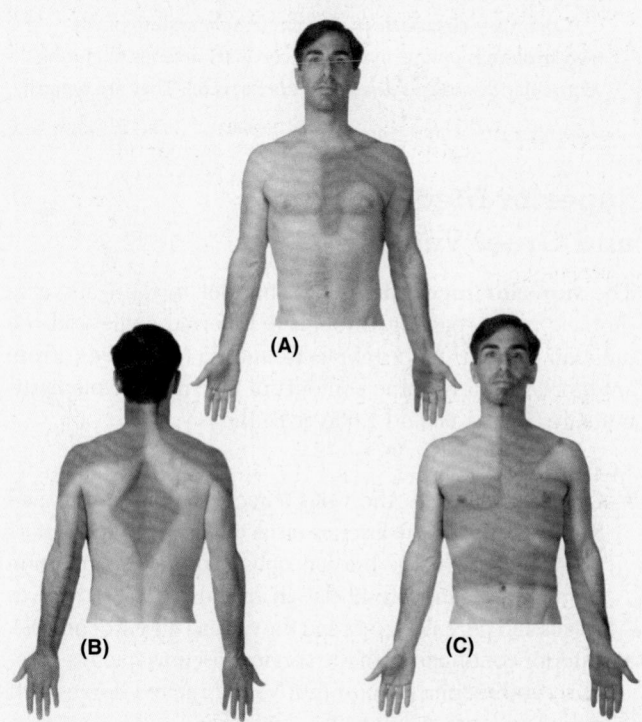

FIGURE B1.31. Areas of cardiac referred pain (*red*).

The Bottom Line

HEART

Heart: The heart is a dual suction and pressure pump that propels blood through the infinite double loop formed by the pulmonary and systemic circuits. ◆ The right heart serves the former and the left heart the latter. ◆ The heart is shaped like a tipped-over pyramid, with the apex directed anteroinferiorly and to the left and the base opposite the apex (posterior). ◆ Each side of the heart includes a receiving chamber (atrium) and a suction-compression-expulsion chamber (ventricle). ◆ The bilateral chambers (and thus the high-pressure systemic and lower-pressure pulmonary circuits) are separated by a cardiac septum that is largely muscular but partly membranous. ◆ AV valves are placed between unilateral chambers to facilitate two-stage (accumulate and then eject) pumping. ◆ Oneway semilunar valves (pulmonic and aortic) placed at the exit on each side prevent backflow (except that which fills the coronary arteries) and maintains the diastolic pressure of the arteries. ◆ The chambers have a glistening endothelial lining, the endocardium; a muscular wall or myocardium, the thickness of which is proportional to the internal pressures occurring within the specific chamber; and a glistening outer covering (the visceral layer of serous pericardium, or epicardium). ◆ The myocardium of the atria and ventricles (and the myogenic propagation of contracting stimuli through it) is attached to and separated by the connective tissue of the fibrous skeleton of the heart. ◆ The fibrous skeleton consists of four fibrous rings, two trigones, and the membranous parts of the cardiac septa. ◆ Only specialized muscle conducting contractile impulses from the atria to the ventricles penetrates the fibrous skeleton at defined sites. ◆ The fibrous skeleton provides attachment for the myocardium and cusps of valves and maintains the integrity of the orifices.

Coronary circulation: The circulatory system of the myocardium is unique in that the coronary arteries fill during ventricular diastole as a result of aortic recoil. They are typically (but not necessarily) functional end arteries. ◆ The right coronary artery (RCA) and circumflex branch of the left coronary artery (LCA) supply the walls of the atria via small branches. ◆ The RCA typically supplies the SA and AV nodes, the myocardium of the external wall of the right ventricle (except its anterior surface), the diaphragmatic surface of the left ventricle, and the posterior third of the IVS. ◆ The LCA typically supplies the anterior two thirds of the IVS (including the AV bundle of conductive tissue), the anterior wall of the right ventricle, and the external wall of the left ventricle (except the diaphragmatic surface). ◆ The capillary beds of the myocardium drain primarily into the right atrium via veins emptying into the coronary sinus. However, the vein also may enter directly into the chambers via the smallest cardiac veins. Both pathways lack valves.

Conducting, stimulating, and regulating system of heart: The conducting system of the heart consists of specialized intrinsic nodes that rhythmically generate stimuli and bundles of modified cardiac muscle that conduct the impulses. The result is the coordinated contraction of the atria and ventricles. ◆ The rate of generation and speed of conductivity are increased by the sympathetic division and inhibited by the parasympathetic division of the ANS to meet demands or conserve energy. ◆ The impulse-generating sinu-atrial (SA) node and the relaying atrioventricular (AV) node are typically supplied by nodal branches of the RCA. The atrioventricular bundle and its branches are primarily supplied by septal branches of the LCA. ◆ Occlusion of either coronary artery with subsequent infarction of nodal or conductive tissue may require placement of an artificial cardiac pacemaker. ◆ The effect of the ANS on the coronary arteries is paradoxical. Sympathetic stimulation produces vasodilation and parasympathetic stimulation produces vasoconstriction.

Superior Mediastinum and Great Vessels

The **superior mediastinum** is superior to the transverse thoracic plane, passing through the sternal angle and the junction (IV disc) of vertebrae T4 and T5 (Fig. 1.64). From anterior to posterior, the contents of the superior mediastinum are (Figs. 1.65 and 1.66A & B) the:

- Thymus.
- Great vessels, with the veins (brachiocephalic veins and SVC) anterior to the arteries (arch of aorta and roots of its major branches—the brachiocephalic trunk, left common carotid artery, and left subclavian artery) and related nerves (vagus and phrenic nerves and the cardiac plexus of nerves).
- Inferior continuation of the cervical viscera (trachea anteriorly and esophagus posteriorly) and related nerves (left recurrent laryngeal nerve).
- Thoracic duct and lymphatic trunks.

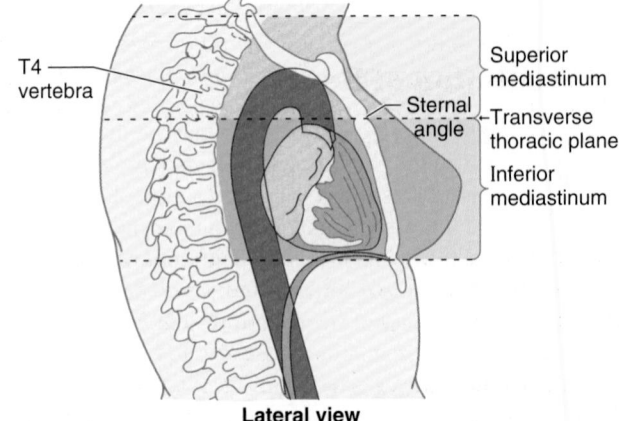

Lateral view

FIGURE 1.64. Boundaries of superior mediastinum. The superior mediastinum extends inferiorly from the superior thoracic aperture to the transverse thoracic plane.

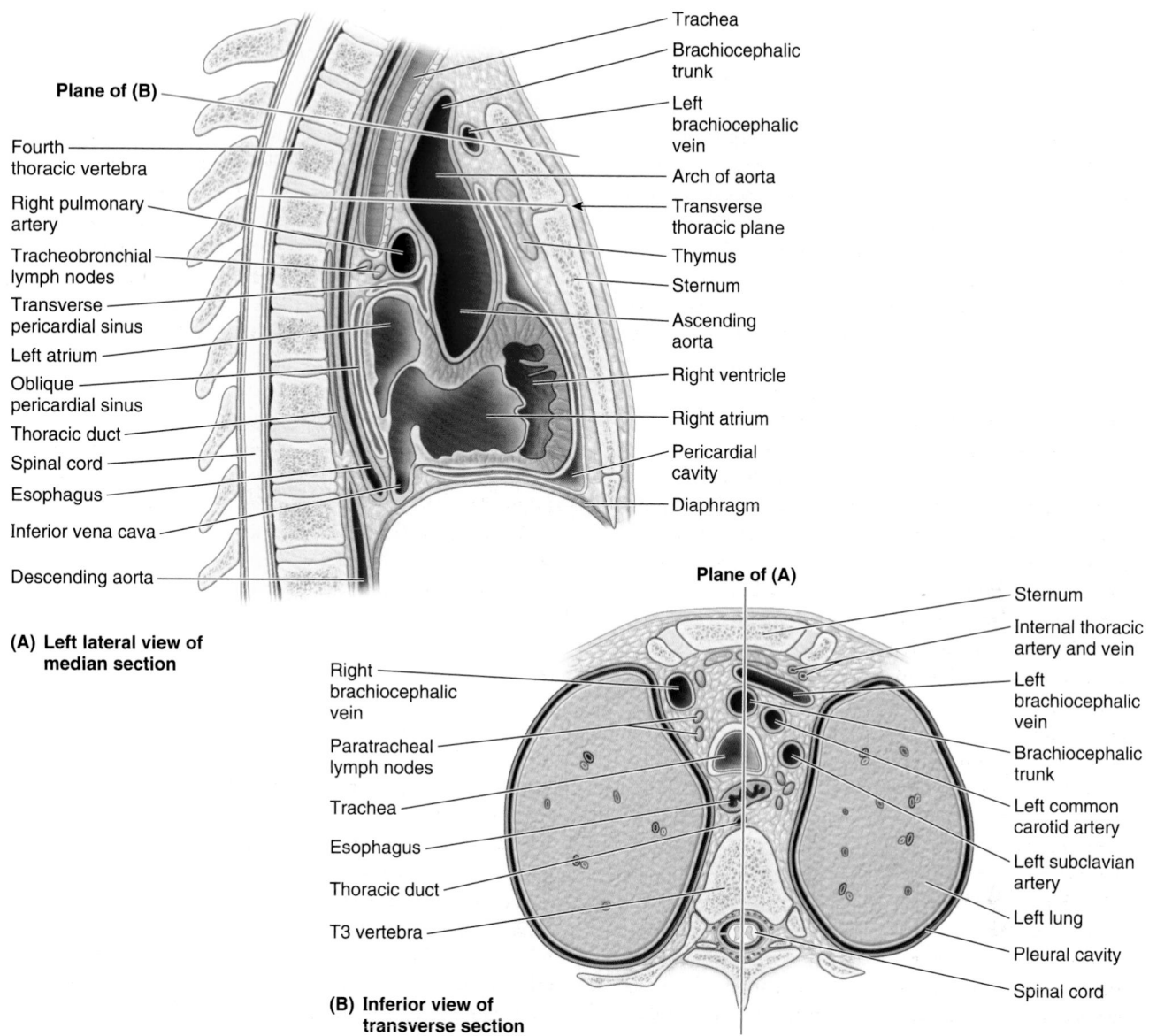

Plane of (B)

Fourth thoracic vertebra

Right pulmonary artery

Tracheobronchial lymph nodes

Transverse pericardial sinus

Left atrium

Oblique pericardial sinus

Thoracic duct

Spinal cord

Esophagus

Inferior vena cava

Descending aorta

(A) Left lateral view of median section

Trachea

Brachiocephalic trunk

Left brachiocephalic vein

Arch of aorta

Transverse thoracic plane

Thymus

Sternum

Ascending aorta

Right ventricle

Right atrium

Pericardial cavity

Diaphragm

Plane of (A)

Right brachiocephalic vein

Paratracheal lymph nodes

Trachea

Esophagus

Thoracic duct

T3 vertebra

(B) Inferior view of transverse section

Sternum

Internal thoracic artery and vein

Left brachiocephalic vein

Brachiocephalic trunk

Left common carotid artery

Left subclavian artery

Left lung

Pleural cavity

Spinal cord

FIGURE 1.65. Relationships of structures in superior mediastinum. The order of systemic structures in the superior mediastinum, from anterior to posterior, is demonstrated in both views: thymus, veins, arteries, airway (trachea), alimentary tract (esophagus), lymphatic ducts, vertebral bodies/intervertebral discs, and spinal cord.

To summarize systemically, *the order of the major structures in the superior mediastinum, from anterior to posterior, is: (1) thymus, (2) veins, (3) arteries, (4) airway, (5) alimentary tract, and (6) lymphatic trunks.*

THYMUS

The **thymus,** a primary lymphoid organ, is located in the inferior part of the neck and the anterior part of the superior mediastinum (Figs. 1.65 and 1.66A). It is a flat gland with flask-shaped lobes that lies posterior to the manubrium and extends into the anterior mediastinum, anterior to the fibrous pericardium. After puberty, the thymus undergoes gradual involution and is largely replaced by fat. The rich *arterial supply of the thymus* is derived mainly from the ante-

rior intercostal and **anterior mediastinal branches of the internal thoracic arteries.** The *veins of the thymus* end in the left brachiocephalic, internal thoracic, and inferior thyroid veins. The *lymphatic vessels of the thymus* end in the parasternal, brachiocephalic, and tracheobronchial lymph nodes.

GREAT VESSELS

The **right** and **left brachiocephalic veins** are formed posterior to the sternoclavicular (SC) joints by the union of the internal jugular and subclavian veins. At the level of the inferior border of the 1st right costal cartilage, the brachiocephalic veins unite to form the SVC (Figs. 1.65B and 1.66B). The *left brachiocephalic vein* is more than twice as long as

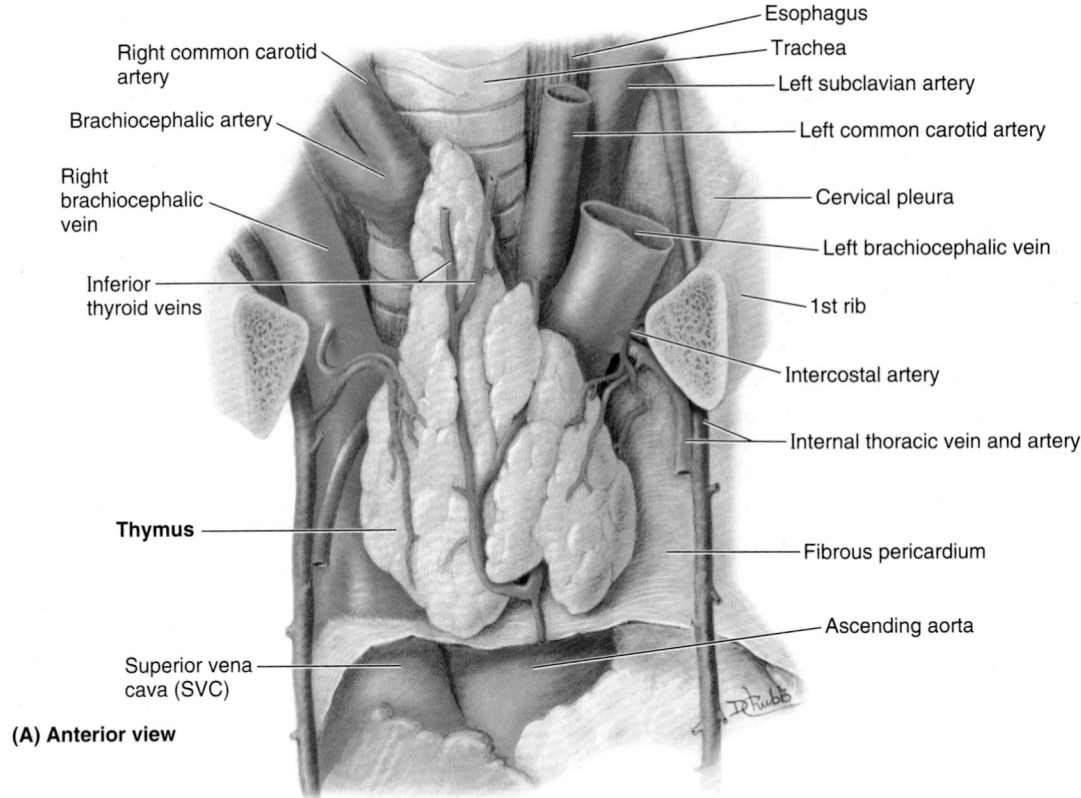

Right common carotid artery

Brachiocephalic artery

Right brachiocephalic vein

Inferior thyroid veins

Thymus

Superior vena cava (SVC)

Esophagus

Trachea

Left subclavian artery

Left common carotid artery

Cervical pleura

Left brachiocephalic vein

1st rib

Intercostal artery

Internal thoracic vein and artery

Fibrous pericardium

Ascending aorta

(A) Anterior view

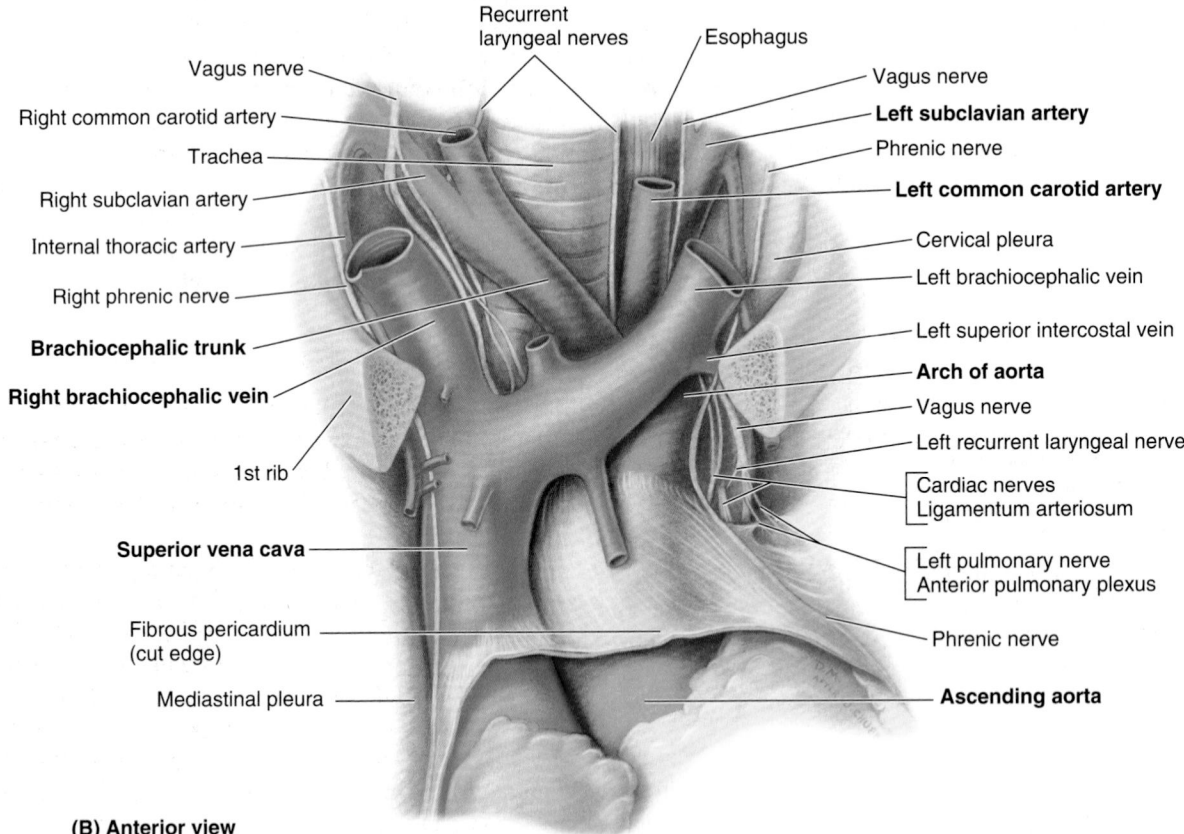

Recurrent laryngeal nerves

Esophagus

Vagus nerve

Right common carotid artery

Trachea

Right subclavian artery

Internal thoracic artery

Right phrenic nerve

Brachiocephalic trunk

Right brachiocephalic vein

1st rib

Superior vena cava

Fibrous pericardium (cut edge)

Mediastinal pleura

Vagus nerve

Left subclavian artery

Phrenic nerve

Left common carotid artery

Cervical pleura

Left brachiocephalic vein

Left superior intercostal vein

Arch of aorta

Vagus nerve

Left recurrent laryngeal nerve

Cardiac nerves
Ligamentum arteriosum

Left pulmonary nerve
Anterior pulmonary plexus

Phrenic nerve

Ascending aorta

(B) Anterior view

FIGURE 1.66. Dissections of superior mediastinum. A. In this superficial dissection of the mediastinum, the sternum and ribs have been excised and the overlapping parietal pleurae removed. It is unusual to see such a distinct thymus in an adult; usually it is impressive during puberty but subsequently regresses and becomes largely replaced by fat and fibrous tissue. **B.** In this deep dissection of the root of the neck and superior mediastinum, the thymus has been removed. The right vagus nerve (CN X) crosses anterior to the right subclavian artery and gives off the right recurrent laryngeal nerve, which passes medially to reach the trachea and esophagus. The left recurrent laryngeal nerve passes inferior and then posterior to the arch of the aorta and ascends between the trachea and esophagus to the larynx.

the right brachiocephalic vein because it passes from the left to the right side, anterior to the roots of the three major branches of the arch of the aorta (Fig. 1.66B). The brachiocephalic veins shunt blood from the head, neck, and upper limbs to the right atrium.

The **superior vena cava** (**SVC**) returns blood from all structures superior to the diaphragm, except the lungs and heart. It passes inferiorly and ends at the level of the 3rd costal cartilage, where it enters the right atrium of the heart. The SVC lies in the right side of the superior mediastinum, anterolateral to the trachea and posterolateral to the ascending aorta. The right phrenic nerve lies between the SVC and the mediastinal pleura. The terminal half of the SVC is in the middle mediastinum, where it lies beside the ascending aorta and forms the posterior boundary of the transverse pericardial sinus (Fig. 1.46).

The **ascending aorta,** approximately 2.5 cm in diameter, begins at the aortic orifice. Its only branches are the coronary arteries, arising from the aortic sinuses (Fig. 1.55B). The ascending aorta is intrapericardial (Figs. 1.66A & B); for this reason, and because it lies inferior to the transverse thoracic plane, it is considered a content of the middle mediastinum (part of inferior mediastinum).

The **arch of the aorta** (aortic arch), the curved continuation of the ascending aorta (Figs. 1.65A and 1.67; see Table 1.5), begins posterior to the 2nd right sternocostal (SC) joint at the level of the sternal angle. It arches superiorly, posteriorly and to the left, and then inferiorly. The arch ascends anterior to the right pulmonary artery and the bifurcation of the trachea, reaching its apex at the left side of the trachea and esophagus as it passes over the root of the left lung. The arch descends posterior to the left root of the lung beside the T4 vertebra. The arch ends by becoming the **thoracic (descending) aorta** posterior to the 2nd left sternocostal joint.

The **arch of the azygos vein** occupies a position corresponding to the aorta on the right side of the trachea over the root of the right lung, although the blood is flowing in the opposite direction (Fig. 1.63). The **ligamentum arteriosum,** the remnant of the fetal ductus arteriosus, passes from the root of the left pulmonary artery to the inferior surface of the arch of the aorta. The usual branches of the arch are the *brachiocephalic trunk, left common carotid artery,* and *left subclavian artery* (Figs. 1.67 and 1.68A).

The **brachiocephalic trunk,** the first and largest branch of the arch of the aorta, arises posterior to the manubrium, where it is anterior to the trachea and posterior to the left brachiocephalic vein (Figs. 1.65A & B, 1.66B, and 1.68A). The trunk ascends superolaterally to reach the right side of the trachea and the right SC joint, where it divides into the right common carotid and right subclavian arteries.

The **left common carotid artery,** the second branch of the arch of the aorta, arises posterior to the manubrium, slightly posterior and to the left of the brachiocephalic trunk. It ascends anterior to the left subclavian artery and is at first anterior to the trachea and then to its left. It enters the neck by passing posterior to the left SC joint.

The **left subclavian artery,** the third branch of the arch of the aorta, arises from the posterior part of the arch, just posterior to the left common carotid artery. It ascends lateral to the trachea and left common carotid artery through the superior mediastinum; it has no branches in the mediastinum. As it leaves the thorax and enters the root of the neck, it passes posterior to the left SC joint.

NERVES IN THE SUPERIOR MEDIASTINUM

The *vagus nerves* exit the cranium and descend through the neck posterolateral to the common carotid arteries (Fig. 1.68A; see Table 1.6). Each vagus nerve enters the superior mediastinum posterior to the respective SC joint and brachiocephalic vein.

The **right vagus nerve** (RVN) enters the thorax anterior to the right subclavian artery, where it gives rise to the **right recurrent laryngeal nerve** (Fig. 1.68A–C). The right recurrent laryngeal nerve hooks around the right subclavian artery and ascends between the trachea and esophagus to supply the larynx. The RVN runs posteroinferiorly through the superior mediastinum on the right side of the trachea. The RVN then passes posterior to the right brachiocephalic vein, SVC, and root of the right lung. Here it divides into many branches, which contribute to the **right pulmonary plexus** (Fig. 1.68C). Usually, the RVN leaves this plexus as a single nerve and passes to the esophagus, where it again breaks up and contributes fibers to the **esophageal (nerve) plexus.** The RVN also gives rise to nerves that contribute to the *cardiac plexus.*

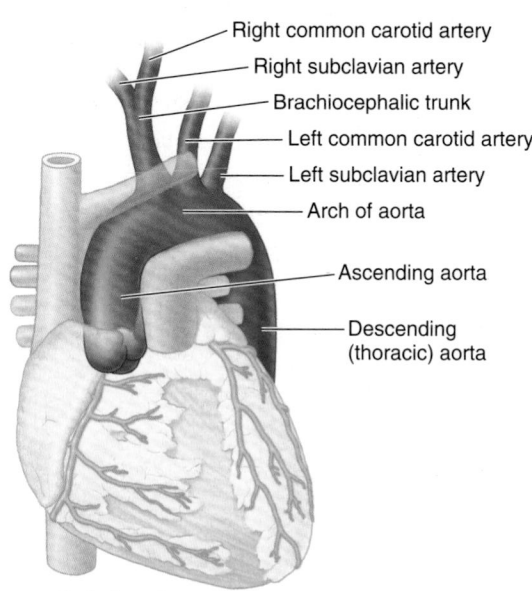

- Right common carotid artery
- Right subclavian artery
- Brachiocephalic trunk
- Left common carotid artery
- Left subclavian artery
- Arch of aorta
- Ascending aorta
- Descending (thoracic) aorta

Anterior view

FIGURE 1.67. Common pattern of branches of arch of aorta. The pattern shown is present in approximately 65% of people. The largest branch (brachiocephalic trunk) arises from the beginning of the arch, the next artery (left common carotid artery) arises from the superior part of the arch, and the third branch (left subclavian artery) arises from the arch approximately 1 cm distal to the left common carotid.

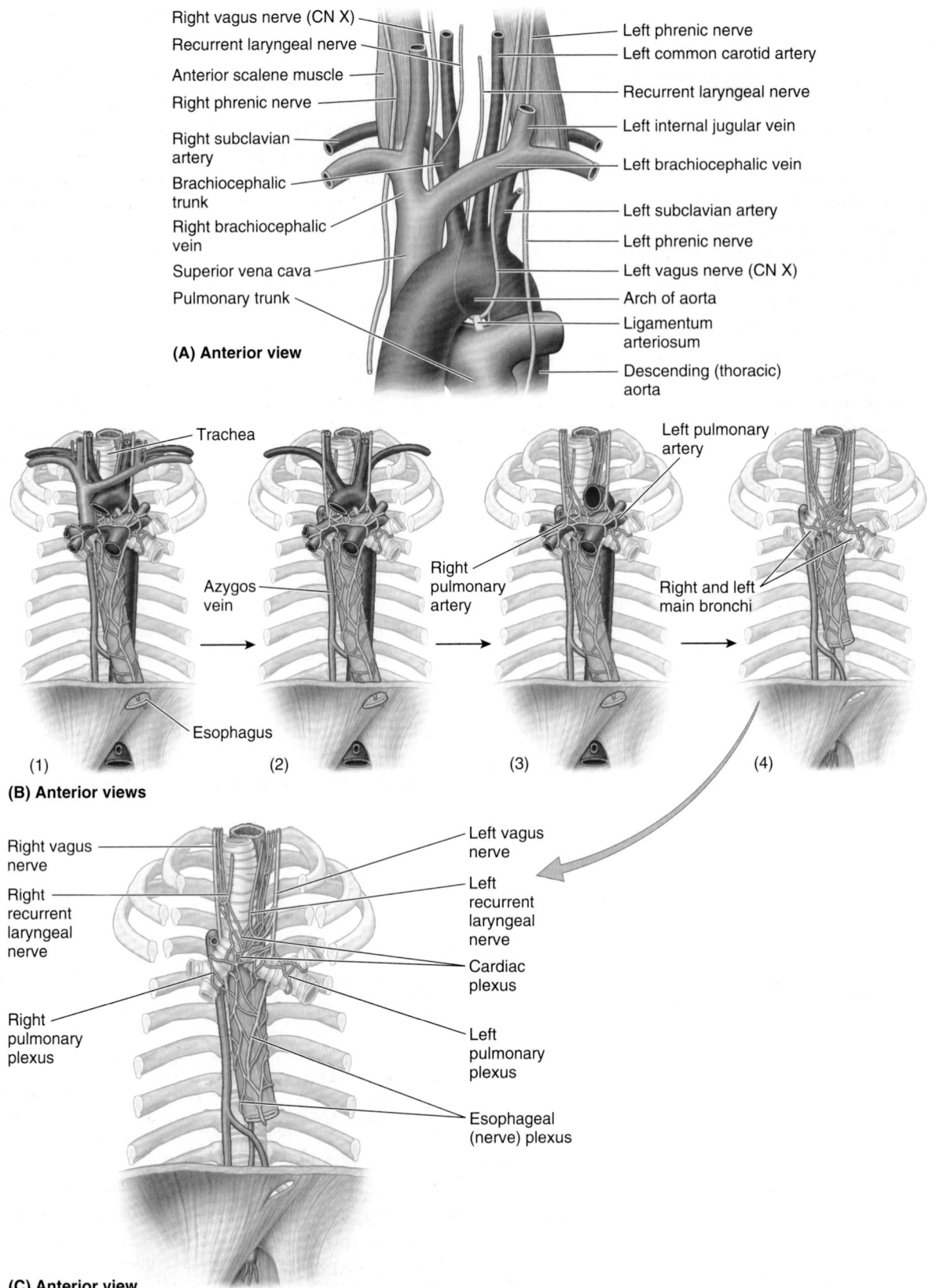

Right vagus nerve (CN X)
Recurrent laryngeal nerve
Anterior scalene muscle
Right phrenic nerve

Right subclavian artery
Brachiocephalic trunk
Right brachiocephalic vein
Superior vena cava
Pulmonary trunk

(A) Anterior view

Left phrenic nerve
Left common carotid artery
Recurrent laryngeal nerve
Left internal jugular vein
Left brachiocephalic vein
Left subclavian artery
Left phrenic nerve
Left vagus nerve (CN X)
Arch of aorta
Ligamentum arteriosum
Descending (thoracic) aorta

Trachea

Azygos vein

Esophagus

(1)   (2)   (3)   (4)

(B) Anterior views

Left pulmonary artery
Right pulmonary artery
Right and left main bronchi

Right vagus nerve
Right recurrent laryngeal nerve
Right pulmonary plexus

Left vagus nerve
Left recurrent laryngeal nerve
Cardiac plexus
Left pulmonary plexus
Esophageal (nerve) plexus

(C) Anterior view

FIGURE 1.68. Great vessels and nerves. A. The relationships of the vessels and nerves in the superior mediastinum are shown. The ligamentum arteriosum is the remnant of the fetal shunt (ductus arteriosus) that bypasses the prefunctional lungs. **B.** The relationships at the bifurcation of the trachea from superficial to deep are shown. (1) Most anterior. The left brachiocephalic vein passes across the roots of the three major branches of the arch of the aorta. (2) The ascending aorta and arch pass anterior and superior, respectively, to the right pulmonary artery. (3) The bifurcation of the pulmonary trunk and right pulmonary artery lie directly anterior to the bifurcation of the trachea. (4) The cardiac plexus remains on the anterior aspect of the tracheal bifurcation after removal of the pulmonary trunk and arteries, the ascending aorta, and the arch of the aorta, to which the plexus is primarily related. **C.** The nerves in the superior and posterior mediastina are seen after the viscera that lie anterior to the trachea and esophagus have been removed.

The **left vagus nerve** (LVN) descends in the neck posterior to the left common carotid artery (Fig. 1.68A). It enters the mediastinum between the left common carotid artery and left subclavian artery. When the LVN reaches the left side of the arch of the aorta, it diverges posteriorly from the left phrenic nerve. The LVN is separated laterally from the phrenic nerve by the left superior intercostal vein. As the LVN curves medially at the inferior border of the arch of the aorta, it gives off the **left recurrent laryngeal nerve.** The left recurrent laryngeal nerve passes inferior to the arch of the aorta, immediately lateral to the ligamentum arteriosum, and ascends to the larynx in the groove between the trachea and the esophagus (Figs. 1.63, 1.66B, 1.68A–C, and 1.69). The LVN passes posterior to the root of the left lung, where it breaks up into many branches that contribute to the **left pulmonary plexus.** The LVN leaves this plexus as a single trunk and passes to the esophagus, where it joins fibers from the right vagus in the *esophageal (nerve) plexus* (Fig. 1.68B & C).

The *phrenic nerves* (Fig. 1.68A) supply the diaphragm with motor and sensory fibers, the latter accounting for approximately one third of the nerve's fibers. The phrenic nerves also supply sensory fibers to the pericardium and mediastinal pleura. Each phrenic nerve enters the superior mediastinum between the subclavian artery and the origin of the brachiocephalic vein (see Table 1.6). The fact that the phrenic nerves pass anterior to the roots of the lungs provides an important means of distinguishing them from the vagus nerves, which pass posterior to the roots.

The **right phrenic nerve** passes along the right side of the right brachiocephalic vein, SVC, and the pericardium over the right atrium. It also passes anterior to the root of the right lung and descends on the right side of the IVC to the diaphragm, which it pierces near the caval opening (Fig. 1.70A).

The **left phrenic nerve** descends between the left subclavian and left common carotid arteries. It crosses the left surface of the arch of the aorta anterior to the left vagus nerve and passes over the left superior intercostal vein. The left phrenic nerve then descends anterior to the root of the left lung and runs along the fibrous pericardium, superficial to the left atrium and ventricle of the heart, where it pierces the diaphragm to the left of the pericardium (Fig. 1.70B). Most branching of the phrenic nerves for distribution to the diaphragm occurs on the diaphragm's inferior (abdominal) surface.

TRACHEA

The **trachea** descends anterior to the esophagus and enters the superior mediastinum, inclining a little to the right of

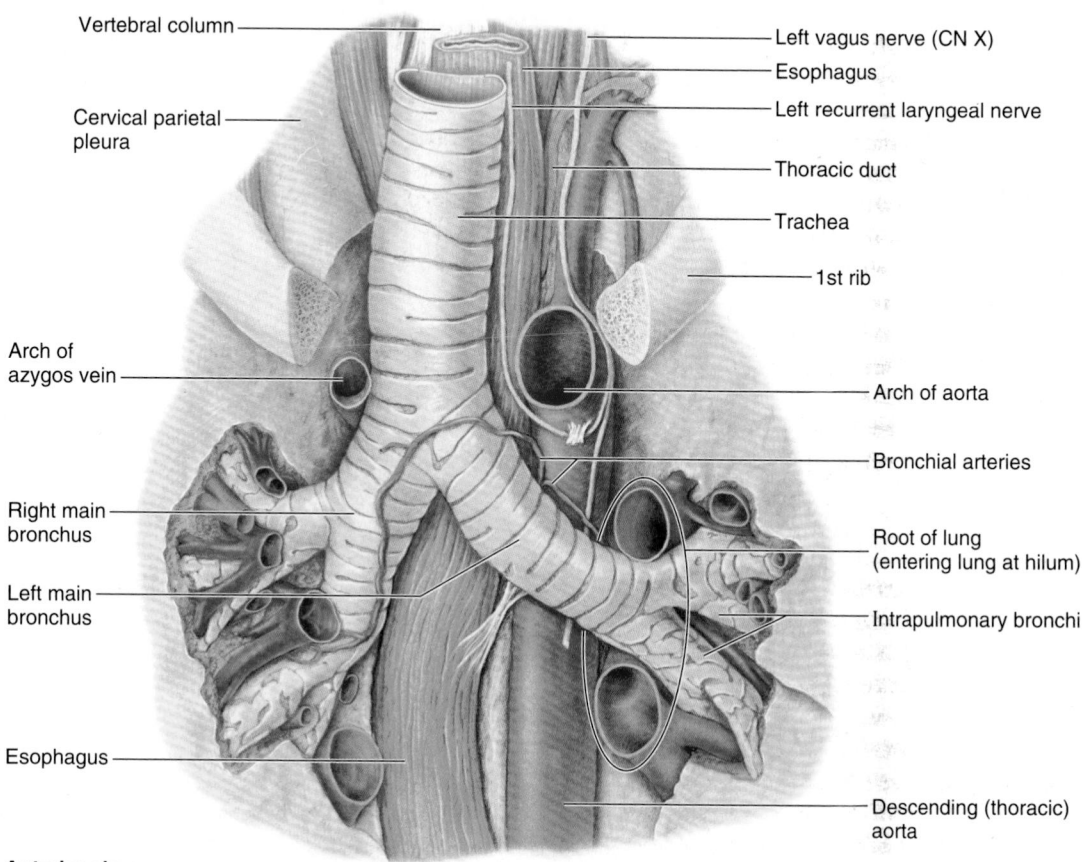

Vertebral column

Cervical parietal pleura

Arch of azygos vein

Right main bronchus

Left main bronchus

Esophagus

Left vagus nerve (CN X)

Esophagus

Left recurrent laryngeal nerve

Thoracic duct

Trachea

1st rib

Arch of aorta

Bronchial arteries

Root of lung (entering lung at hilum)

Intrapulmonary bronchi

Descending (thoracic) aorta

Anterior view

FIGURE 1.69. Deep dissection of superior mediastinum. Four structures run parallel as they traverse the superior thoracic aperture: trachea, esophagus, left recurrent laryngeal nerve, and thoracic duct. The right main bronchus is more vertical, shorter, and wider than the left main bronchus. The course of the right bronchial artery shown here is aberrant; usually it passes posterior to the bronchus.

the median plane (Figs. 1.68B & C and 1.69). The posterior surface of the trachea is flat where it is applied to the esophagus (Fig. 1.65B). The trachea ends at the level of the sternal angle by dividing into the right and left main bronchi (Figs. 1.65A and 1.69). The trachea terminates superior to the level of the heart and is not a component of the posterior mediastinum.

ESOPHAGUS

The **esophagus** is a fibromuscular tube that extends from the pharynx to the stomach (Figs. 1.65A & B, 1.68B & C, 1.69, 1.70A, and 1.71). The esophagus enters the superior mediastinum between the trachea and vertebral column, where it lies anterior to the bodies of the T1–T4 vertebrae. The esophagus is usually flattened anteroposteriorly. Initially, it inclines to the left but is pushed back to the median plane by the arch of the aorta. It is then compressed anteriorly by the root of the left lung. In the superior mediastinum, the *thoracic duct* usually lies on the left side of the esophagus,

deep (medial) to the arch of the aorta (Figs. 1.69 and 1.70B). Inferior to the arch, the esophagus again inclines to the left as it approaches and passes through the **esophageal hiatus** in the diaphragm (Fig. 1.71).

Posterior Mediastinum

The **posterior mediastinum** (the posterior part of the inferior mediastinum) is located inferior to the transverse thoracic plane, anterior to the T5–T12 vertebrae, posterior to the pericardium and diaphragm, and between the parietal pleura of the two lungs (Figs. 1.65A and 1.68C). The posterior mediastinum contains the thoracic aorta, thoracic duct and lymphatic trunks, posterior mediastinal lymph nodes, azygos and hemi-azygos veins, and esophagus and esophageal nerve plexus. Some authors also include the thoracic sympathetic trunks and thoracic splanchnic nerves; however, these structures lie lateral to the vertebral bodies and are not within the posterior mediastinal compartment or space per se.

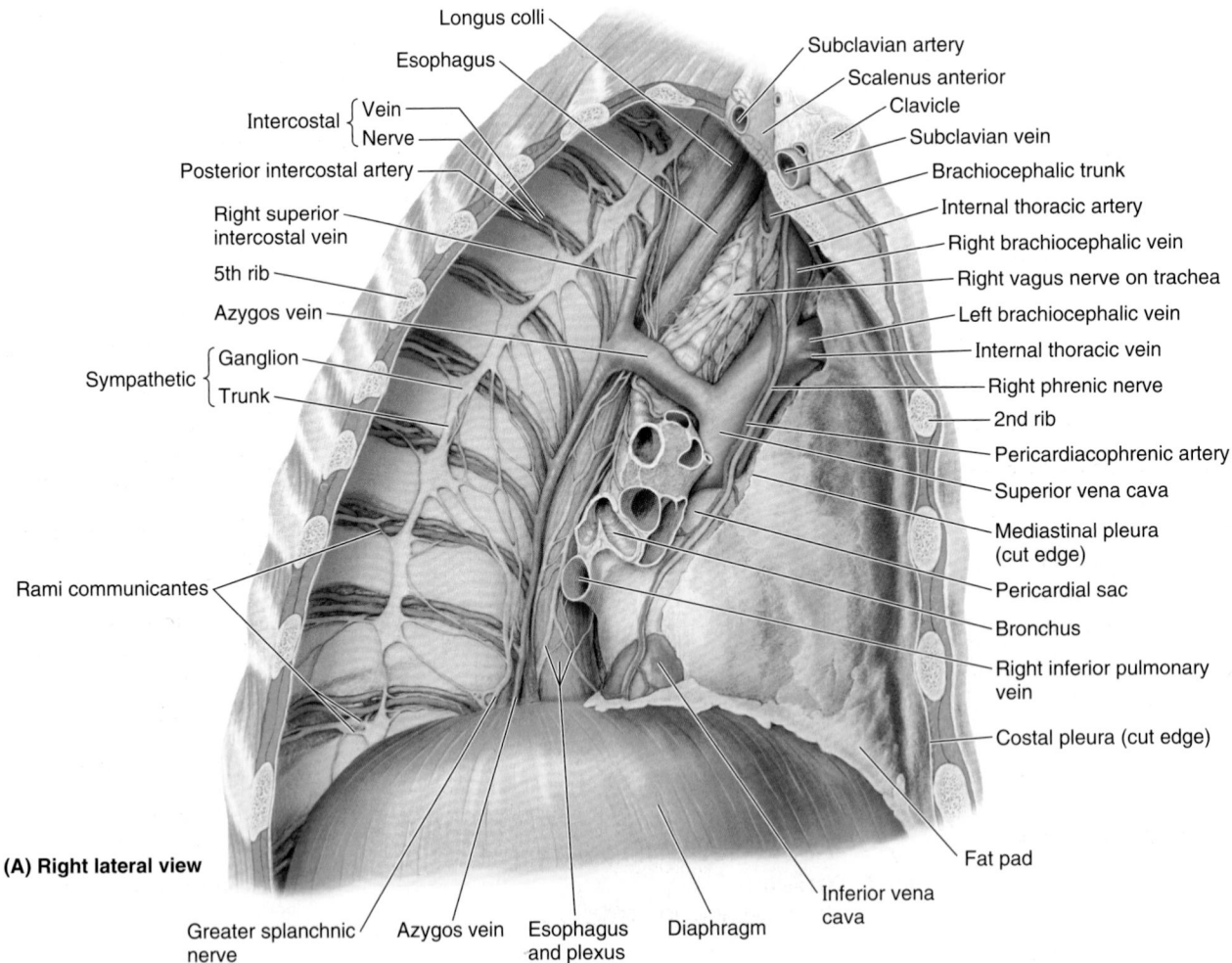

FIGURE 1.70. Lateral exposures of mediastinum. A. In this right side view, most of the costal and mediastinal pleura has been removed to expose the underlying structures. This side of the mediastinum, the "blue side," is dominated by venous structures: the azygos vein and arch, superior vena cava, right atrium, and inferior vena cava.

THORACIC AORTA

The **thoracic aorta** is the continuation of the arch of the aorta). (Figs. 1.69, 1.71, and 1.72; Table 1.5). It begins on the left side of the inferior border of the body of the T4 vertebra and descends in the posterior mediastinum on the left sides of the T5–T12 vertebrae. As it descends, the thoracic aorta approaches the median plane and displaces the esophagus to the right. The **thoracic aortic plexus** (Fig. 1.70B), an autonomic nerve network, surrounds it. The thoracic aorta lies posterior to the root of the left lung (Figs. 1.69 and 1.70B), pericardium, and esophagus. It terminates (with a name change to *abdominal aorta*) anterior to the inferior border of the T12 vertebra and enters the abdomen through the **aortic hiatus** in the diaphragm (Fig. 1.71). The thoracic duct and azygos vein ascend on its right side and accompany it through this hiatus.

In a pattern that will be more evident in the abdomen, the *branches of the descending aorta* arise and course within three "vascular planes" (Fig. 1.72):

- An *anterior, midline plane* of *unpaired visceral branches* to the gut (embryonic digestive tube) and its derivatives (A in Fig. 1.72 inset).
- *Lateral planes* of *paired visceral branches* serving viscera other than the gut and its derivatives (B).
- *Posterolateral planes* of *paired (segmental) parietal branches* to the body wall (C).

In the thorax, the *unpaired visceral branches* of the anterior vascular plane are the **esophageal arteries**—usually two, but there may be as many as five (Fig. 1.72; Table 1.5). The *paired visceral branches* of the lateral plane are represented in the thorax by the *bronchial arteries* (Fig. 1.69). Although the right and left bronchial arteries may arise directly from the aorta, most commonly only the paired left bronchial arteries do so; the right bronchial arteries arise indirectly as branches of a right posterior intercostal artery (usually the 3rd). The *paired parietal branches* of the thoracic aorta that arise posterolaterally are the nine *posterior intercostal arteries* that supply all but the upper two intercostal spaces and

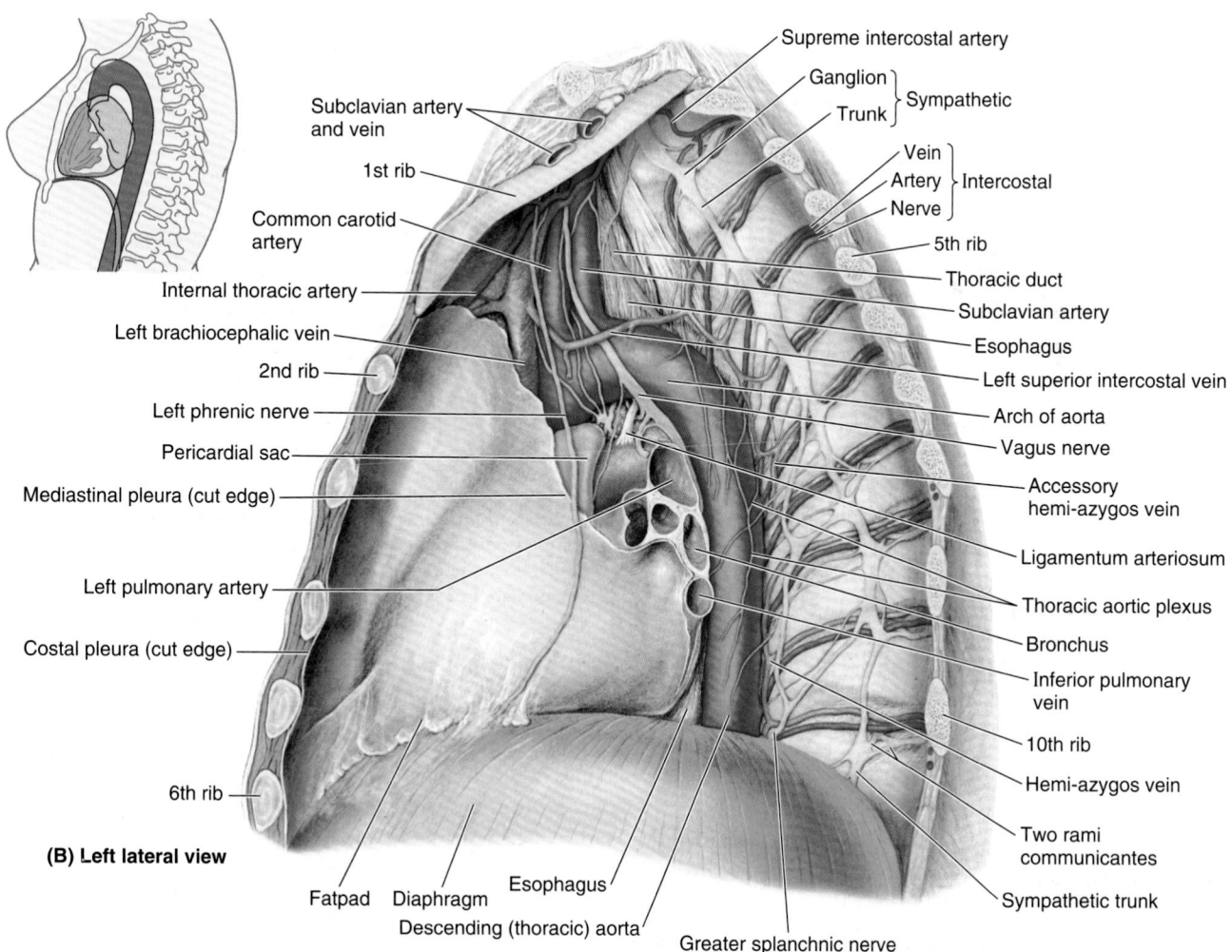

FIGURE 1.70. *(Continued)* **B.** The left side of the mediastinum, the "red side," is dominated by arterial structures: the arch of aorta and thoracic aorta, left common carotid and subclavian arteries, and left ventricle (plus the pulmonary trunk and left pulmonary artery). At the thoracic and superior lumbar levels, the sympathetic trunk is attached to intercostal nerves by paired (white and gray) rami communicantes. The left superior intercostal vein, draining the upper two to three intercostal spaces, passes anteriorly to enter the left brachiocephalic vein.

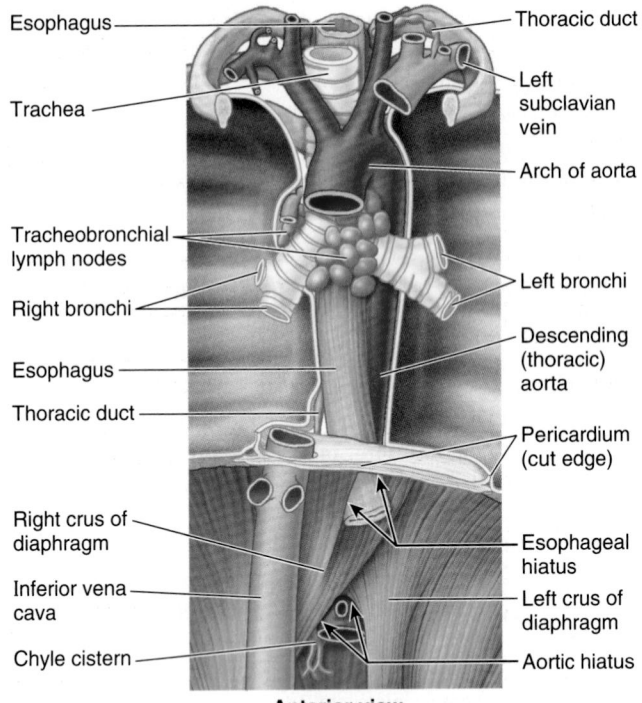

Anterior view

FIGURE 1.71. Anterior view of esophagus, trachea, bronchi, and aorta. The arch of aorta curves posteriorly on the left side of the trachea and esophagus. Enlargement of the inferior tracheobronchial (carinal) nodes may widen the angle between the main bronchi. In this specimen, the thoracic duct enters the left subclavian vein.

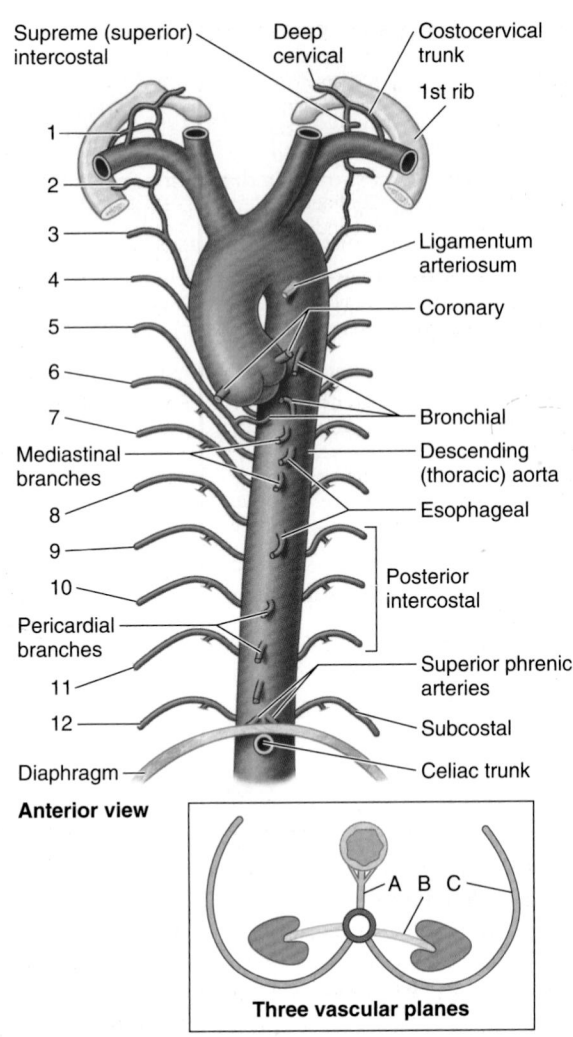

Anterior view

Three vascular planes

Superior view

FIGURE 1.72. Branches of thoracic aorta. Branches of the thoracic aorta tend to arise within three vascular planes (*inset*). Esophageal and pericardial branches represent unpaired visceral branches (*A* in inset) arising anteriorly; the bronchial arteries represent paired lateral visceral branches (*B* in inset); posterior intercostal and subcostal arteries (*1–12*) represent paired, segmental parietal branches that mostly arise posterolaterally (*C* in inset). The paired superior phrenic arteries arising from the inferior part of the thoracic aorta that supply the diaphragm are exceptions to the pattern; they are paired parietal branches that have migrated anteriorly.

the *subcostal arteries* (Fig. 1.72). The latter vessels arise from the thoracic aorta but course below the diaphragm. They are in series with the posterior intercostal arteries.

Exceptions to this pattern include the:

- **Superior phrenic arteries,** paired parietal branches that pass anterolaterally to the superior surface of the diaphragm (which is actually facing posteriorly at this level owing to the convexity of the diaphragm), where they anastomose with the musculophrenic and pericardiacophrenic branches of the internal thoracic artery.
- **Pericardial branches,** unpaired branches that arise anteriorly but, instead of passing to the gut, send twigs to the pericardium. The same is true for the small **mediastinal arteries** that supply the lymph nodes and other tissues of the posterior mediastinum.

ESOPHAGUS

The **esophagus** descends into the posterior mediastinum from the superior mediastinum, passing posterior to and to the right of the arch of the aorta (Figs. 1.68C, 1.69, and 1.71) and posterior to the pericardium and left atrium. The esophagus constitutes the primary posterior relationship of the base of the heart. It then deviates to the left and passes through the *esophageal hiatus* in the diaphragm at the level of the T10 vertebra, anterior to the aorta.

The esophagus may have three impressions, or "constrictions," in its thoracic part. These may be observed as narrowings of the lumen in oblique chest radiographs that are taken as barium is swallowed. The esophagus is compressed by three structures: (1) the arch of the aorta, (2) the left main bronchus, and (3) the diaphragm. The first two impressions occur in close proximity. The aortic arch compression is most evident in a postero-anterior (PA) radiograph after a barium swallow, and the bronchial impression is more evident in lateral views. No constrictions are visible in the empty esophagus; however, as it expands during filling, the structures noted above compress its walls.

TABLE 1.5. AORTA AND ITS BRANCHES IN THORAX

Artery	Origin	Course	Branches
Ascending aorta	Aortic orifice of left ventricle	Ascends approximately 5 cm to sternal angle where it becomes arch of aorta	Right and left coronary arteries
Arch of aorta	Continuation of ascending aorta	Arches posteriorly on left side of trachea and esophagus and superior to left main bronchus	Brachiocephalic, left common carotid, left subclavian
Thoracic (descending) aorta	Continuation of arch of aorta	Descends in posterior mediastinum to left of vertebral column; gradually shifts to right to lie in median plane at aortic hiatus	Posterior intercostal arteries, subcostal, some phrenic arteries and visceral branches (e.g., esophageal)
Posterior intercostal	Posterior aspect of thoracic aorta	Pass laterally and then anteriorly parallel to ribs	Lateral and anterior cutaneous branches
Bronchial (1–2 branches)	Anterior aspect of aorta or posterior intercostal artery	Run with the tracheobronchial tree	Bronchial and peribronchial tissue, visceral pleura *left usually has 2 arteries & right has 1*
Esophageal (4–5 branches)	Anterior aspect of thoracic aorta	Run anteriorly to esophagus *T12*	To esophagus
Superior phrenic (vary in number)	Anterior aspects of thoracic aorta	Arise at aortic hiatus and pass to superior aspect of diaphragm	To diaphragm

THORACIC DUCT AND LYMPHATIC TRUNKS

The **thoracic duct** is the largest lymphatic channel in the body. In the posterior mediastinum, it lies on the anterior aspect of the bodies of the inferior 7 thoracic vertebrae (Fig. 1.73). The thoracic duct conveys most lymph of the body to the venous system: that from the lower limbs; pelvic cavity; abdominal cavity; left upper limb; and left side of the thorax, head, and neck—that is, all lymph except that from the right superior quadrant (see the overview of the lymphatic system in the Introduction chapter).

The thoracic duct originates from the **cisterna chyli** (chyle cistern) in the abdomen and ascends through the aortic hiatus in the diaphragm (Fig. 1.71). The duct is usually thin walled and dull white. Often it is beaded because of its numerous valves. It ascends in the posterior mediastinum among the thoracic aorta on its left, the azygos vein on its right, the esophagus anteriorly, and the vertebral bodies posteriorly. At the level of the T4, T5, or T6 vertebra, the thoracic duct crosses to the left, posterior to the esophagus, and ascends into the superior mediastinum.

The thoracic duct receives branches from the middle and superior intercostal spaces of both sides through several collecting trunks. It also receives branches from posterior mediastinal structures. Near its termination, the thoracic duct often receives the *jugular, subclavian,* and *bronchomediastinal lymphatic trunks* (although any or all these vessels may terminate independently). The thoracic duct usually empties into the venous system near the union of the left internal jugular and subclavian veins—the left *venous angle* or origin of the left brachiocephalic vein (Figs. 1.72 and 1.73A)—but it may open into the left subclavian vein as shown in Figure 1.71.

VESSELS AND LYMPH NODES
OF POSTERIOR MEDIASTINUM

The thoracic aorta and its branches have been discussed earlier. **Posterior mediastinal lymph nodes** (Fig. 1.73A & B) lie posterior to the pericardium, where they are related to the esophagus and thoracic aorta. There are several nodes posterior to the inferior part of the esophagus and more (up to eight) anterior and lateral to it. The posterior mediastinal lymph nodes receive lymph from the esophagus, the posterior aspect of the pericardium and diaphragm, and the middle posterior intercostal spaces. Lymph from the nodes drains to the right or left venous angles via the right lymphatic duct or the thoracic duct.

The **azygos system of veins,** on each side of the vertebral column, drains the back and thoracoabdominal walls (Figs. 1.73A and 1.74A & B) and mediastinal viscera. The azygos system exhibits much variation in its origin, course, tributaries, and anastomoses. The *azygos vein* (G., *azygos,* unpaired) and its main tributary, the *hemi-azygos vein,* usually arise from "roots" arising from the posterior aspect of the IVC and/or renal vein, respectively, which merge with the ascending *lumbar veins.*

The **azygos vein** forms a collateral pathway between the SVC and IVC and drains blood from the posterior walls of the thorax and abdomen. It ascends in the posterior mediastinum, passing close to the right sides of the bodies of the inferior 8 thoracic vertebrae. It arches over the superior aspect of the root of the right lung to join the SVC, similar to the way the arch of the aorta passes over the root of the left lung. In addition to the *posterior intercostal veins,* the azygos vein communicates with the vertebral venous plexuses that drain the back, vertebrae, and structures in the vertebral canal.

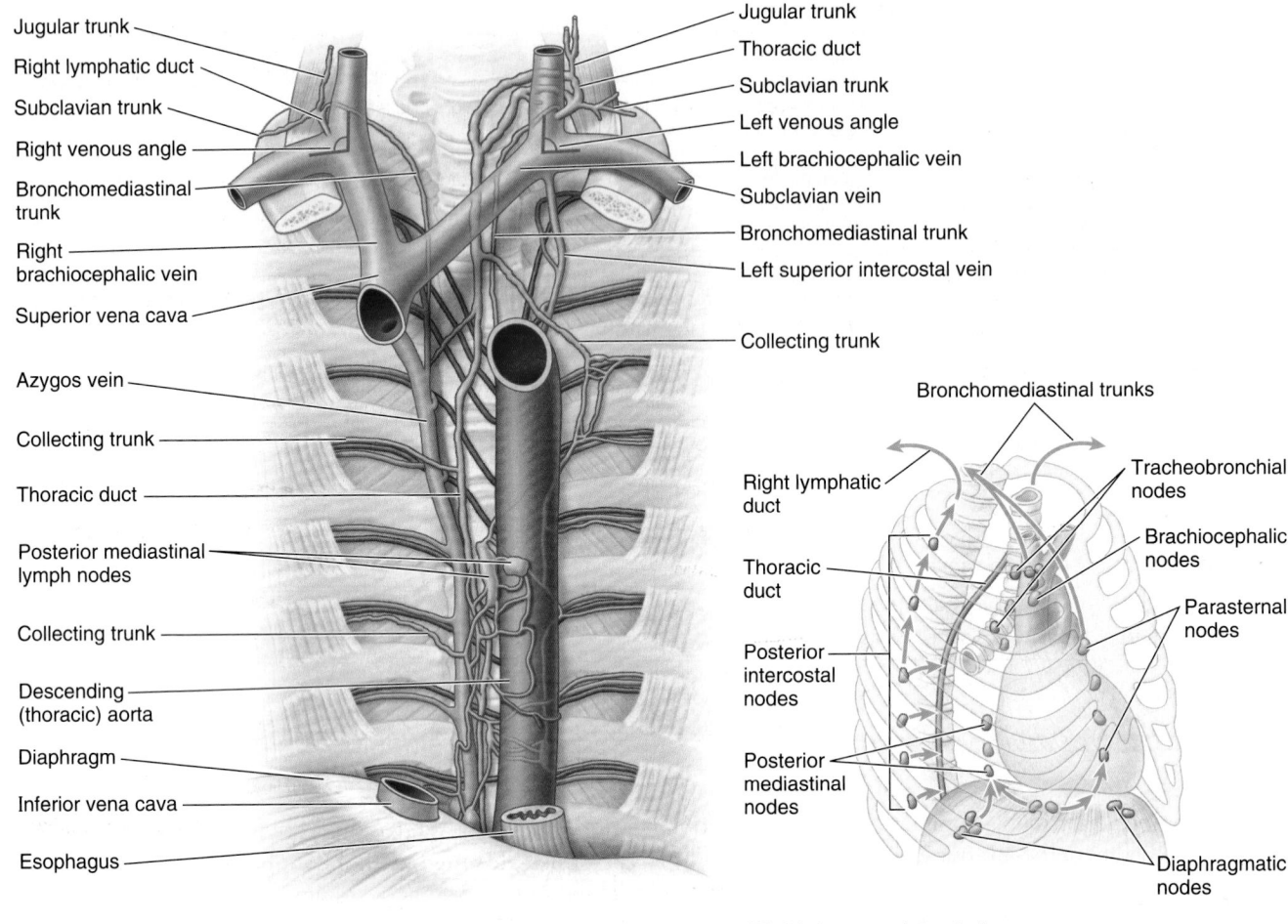

Jugular trunk
Right lymphatic duct
Subclavian trunk
Right venous angle
Bronchomediastinal trunk
Right brachiocephalic vein
Superior vena cava
Azygos vein
Collecting trunk
Thoracic duct
Posterior mediastinal lymph nodes
Collecting trunk
Descending (thoracic) aorta
Diaphragm
Inferior vena cava
Esophagus

Jugular trunk
Thoracic duct
Subclavian trunk
Left venous angle
Left brachiocephalic vein
Subclavian vein
Bronchomediastinal trunk
Left superior intercostal vein
Collecting trunk

Bronchomediastinal trunks
Tracheobronchial nodes
Brachiocephalic nodes
Parasternal nodes
Right lymphatic duct
Thoracic duct
Posterior intercostal nodes
Posterior mediastinal nodes
Diaphragmatic nodes

(A) Anterior view

(B) Right anterolateral view

FIGURE 1.73. Thoracic duct and bronchomediastinal trunks. A. The thoracic aorta has been pulled slightly to the left and the azygos vein slightly to the right to expose the thoracic duct. At approximately the transverse thoracic plane (sternal angle, T4–T5 intervertebral disc level), the thoracic duct passes to the left and continues its ascent to the neck where it arches laterally to enter the left venous angle. The right lymphatic duct is formed by the union of the contralateral partners of the ducts that join the termination of the thoracic duct. **B.** Lymph nodes and pathways that provide lymphatic drainage of the thoracic cavity.

The azygos vein also receives the mediastinal, esophageal, and bronchial veins (Fig. 1.74).

The **hemi-azygos vein** arises on the left side by the junction of the left subcostal and ascending lumbar veins. It ascends on the left side of the vertebral column, posterior to the thoracic aorta as far as the T9 vertebra. Here it crosses to the right, posterior to the aorta, thoracic duct, and esophagus, and joins the azygos vein. The hemi-azygos vein receives the inferior three posterior intercostal veins, the inferior esophageal veins, and several small mediastinal veins. The **accessory hemi-azygos vein** begins at the medial end of the 4th or 5th intercostal space and descends on the left side of the vertebral column from T5 through T8. It receives tributaries from veins in the 4th–8th intercostal spaces and sometimes from the left bronchial veins. It crosses over the T7 or T8 vertebra, posterior to the thoracic aorta and thoracic duct, where it joins the azygos vein. Sometimes the accessory hemi-azygos vein joins the hemi-azygos vein and opens with it into the azygos vein. The accessory hemi-azygos

is frequently connected to the left superior intercostal vein, as shown in Figure 1.74. The left superior intercostal vein, which drains the 1st–3rd intercostal spaces, may communicate with the accessory hemi-azygos vein; however, it drains primarily into the left brachiocephalic vein.

NERVES OF POSTERIOR MEDIASTINUM

The sympathetic trunks and their associated ganglia form a major portion of the autonomic nervous system (Fig. 1.75; Table 1.6). The **thoracic sympathetic trunks** are in continuity with the cervical and lumbar sympathetic trunks. The thoracic trunks lie against the heads of the ribs in the superior part of the thorax, the costovertebral joints in the midthoracic level, and the sides of the vertebral bodies in the inferior part of the thorax. The **lower thoracic splanchnic nerves**—also known as greater, lesser, and least splanchnic nerves—are part of the *abdominopelvic splanchnic nerves* because they supply viscera inferior to the diaphragm. They consist of presynaptic

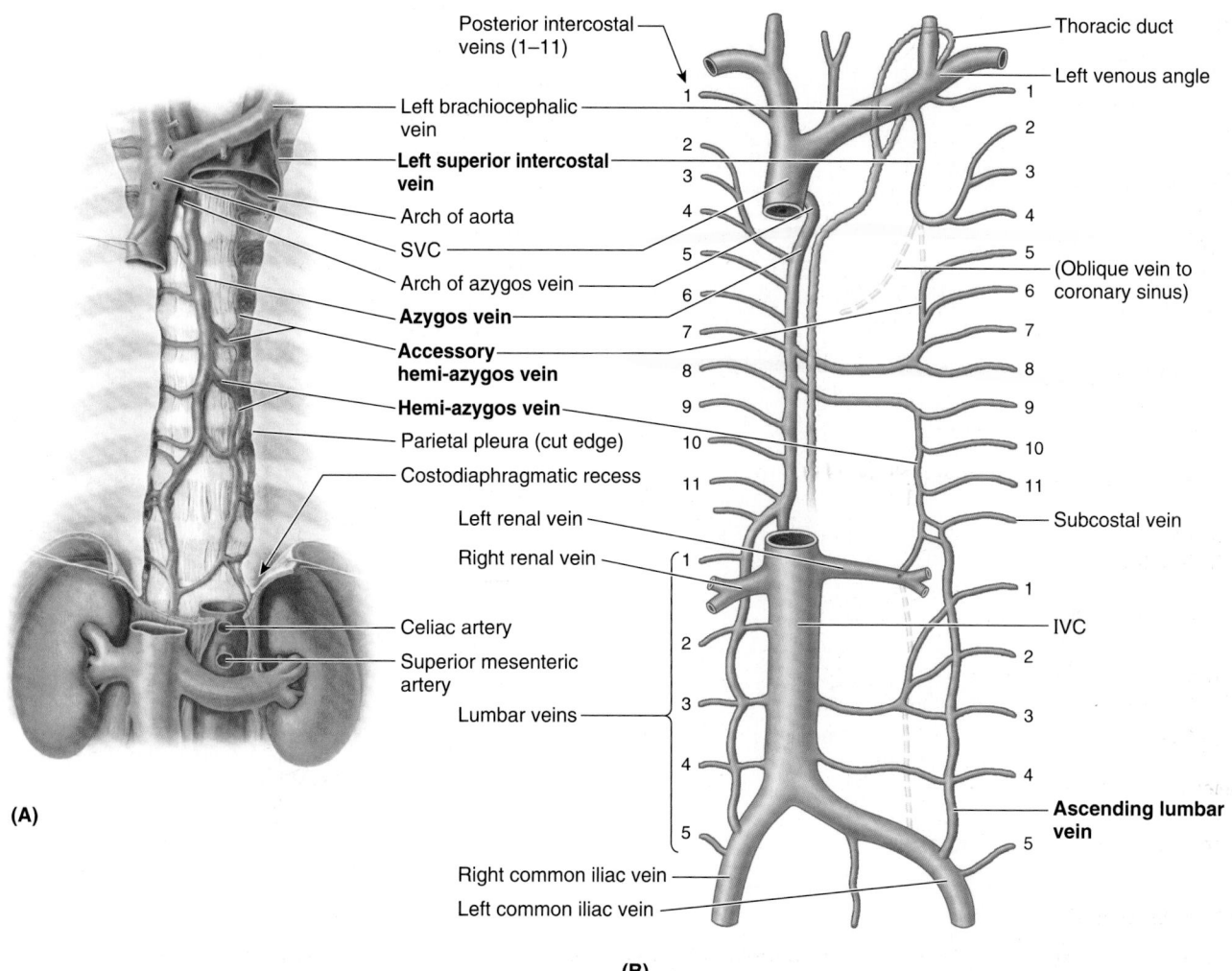

FIGURE 1.74. Azygos system of veins. The azygos vein forms a direct connection between the inferior vena cava (*IVC*) and the superior vena cava (*SVC*). The azygos and hemi-azygos veins are also continuous inferiorly (below diaphragm) with the ascending lumbar veins.

fibers from the 5th through the 12th sympathetic ganglia, which pass through the diaphragm and synapse in prevertebral ganglia in the abdomen. They supply sympathetic innervation for most of the abdominal viscera. These splanchnic nerves are discussed further in Chapter 2.

Anterior Mediastinum

The **anterior mediastinum,** the smallest subdivision of the mediastinum (Fig. 1.42), lies between the body of the sternum and the transversus thoracis muscles anteriorly and the pericardium posteriorly. It is continuous with the superior mediastinum at the sternal angle and is limited inferiorly by the diaphragm. The anterior mediastinum consists of loose connective tissue (**sternopericardial ligaments**), fat, lymphatic vessels, a few lymph nodes, and branches of the internal thoracic vessels. In infants and children, the anterior mediastinum contains the inferior part of the thymus. In unusual cases, this lymphoid organ may extend to the level of the 4th costal cartilages.

Surface Anatomy of Heart and Mediastinal Viscera

The heart and great vessels are approximately in the middle of the thorax, surrounded laterally and posteriorly by the lungs and anteriorly by the sternum and the central part of the thoracic cage (Fig. 1.76). The borders of the heart are variable and depend on the position of the diaphragm and the build and physical condition of the person. The outline of the heart can be traced on the anterior surface of the thorax by using the following guidelines (Fig. 1.76C):

- The *superior border* corresponds to a line connecting the inferior border of the 2nd left costal cartilage to the superior border of the 3rd right costal cartilage.
- The *right border* corresponds to a line drawn from the 3rd right costal cartilage to the 6th right costal cartilage; this border is slightly convex to the right.
- The *inferior border* corresponds to a line drawn from the inferior end of the right border to a point in the 5th

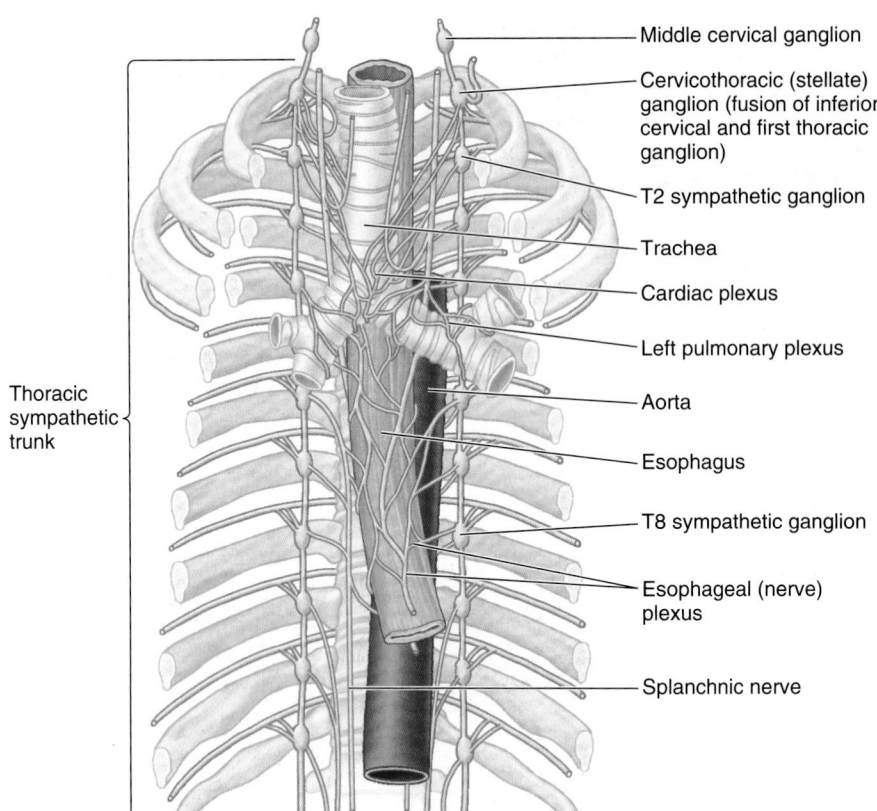

Middle cervical ganglion

Cervicothoracic (stellate) ganglion (fusion of inferior cervical and first thoracic ganglion)

T2 sympathetic ganglion

Trachea

Cardiac plexus

Left pulmonary plexus

Aorta

Esophagus

T8 sympathetic ganglion

Esophageal (nerve) plexus

Splanchnic nerve

Thoracic sympathetic trunk

FIGURE 1.75. Autonomic nerves of superior and posterior mediastina.

TABLE 1.6. NERVES OF THORAX

Nerve	Origin	Course	Distribution
Vagus (CN X)	8–10 rootlets from medulla of brainstem	Enters superior mediastinum posterior to sternoclavicular joint and brachiocephalic vein; gives rise to recurrent laryngeal nerve; continues into abdomen	Pulmonary plexus, esophageal plexus, and cardiac plexus
Phrenic	Anterior rami of C3–C5 nerves	Passes through superior thoracic aperture and runs between mediastinal pleura and pericardium	Central portion of diaphragm
Intercostals (1–11)	Anterior rami of T1–T11 nerves	Run in intercostal spaces between internal and innermost layers of intercostal muscles	Muscles in and skin over intercostal space; lower nerves supply muscles and skin of anterolateral abdominal wall
Subcostal	Anterior ramus of T12 nerve	Follows inferior border of 12th rib and passes into abdominal wall	Abdominal wall and skin of gluteal region
Recurrent laryngeal	Vagus nerve	Loops around subclavian artery on right; on left runs around arch of aorta and ascends in tracheo-esophageal groove	Intrinsic muscles of larynx (except cricothyroid); sensory inferior to level of vocal folds
Cardiac plexus	Cervical and cardiac branches of vagus nerve and sympathetic trunk	From arch of aorta and posterior surface of heart, fibers extend along coronary arteries and to sinu-atrial node	Impulses pass to sinu-atrial node; parasympathetic fibers slow rate, reduce force of heartbeat, and constrict coronary arteries; sympathetic fibers have the opposite effect
Pulmonary plexus	Vagus nerve and sympathetic trunk	Forms on root of lung and extends along bronchial subdivisions	Parasympathetic fibers constrict bronchioles; sympathetic fibers dilate them; afferents convey reflexes
Esophageal plexus	Vagus nerve, sympathetic ganglia, greater splanchnic nerve	Distal to tracheal bifurcation; vagus and sympathetic nerves form a plexus around esophagus	Vagal and sympathetic fibers to smooth muscle and glands of inferior two thirds of esophagus

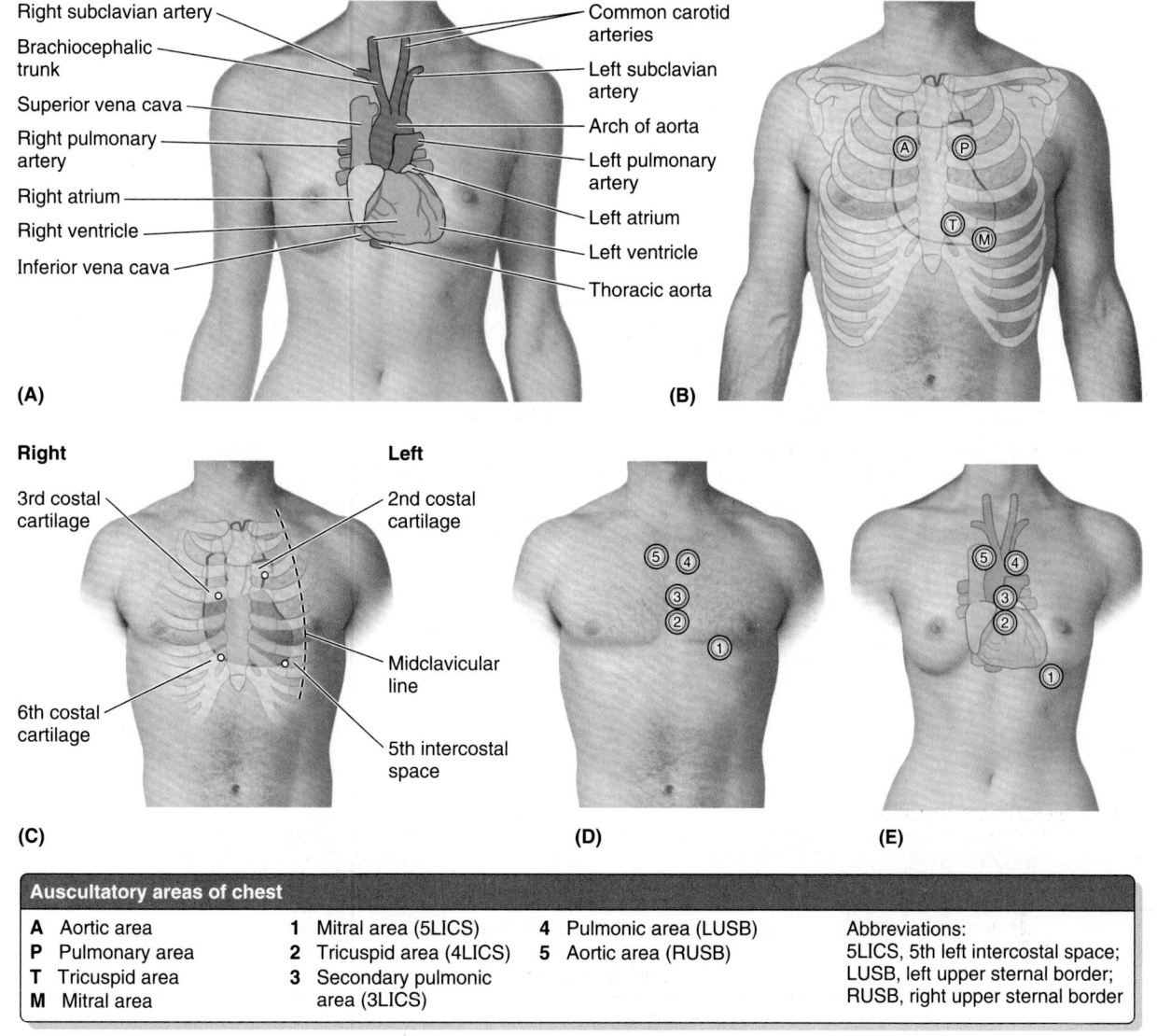

FIGURE 1.76. Surface anatomy of mediastinal viscera.

intercostal space close to the left MCL; the left end of this line corresponds to the location of the apex of the heart and the apex beat.

- The *left border* corresponds to a line connecting the left ends of the lines representing the superior and inferior borders.
- The valves are located posterior to the sternum; however, the sounds produced by them are projected to the areas shown in Figure 1.76—pulmonary (*P*), aortic (*A*), mitral (*M*), and tricuspid (*T*)—where the stethoscope may be placed to avoid intervening bone.

The **apex beat** is the impulse that results from the apex of the heart being forced against the anterior thoracic wall when the left ventricle contracts. The location of the apex beat (mitral area, M) varies in position and may be located

in the 4th or 5th intercostal spaces, 6–10 cm from the AML (anterior median line).

Auscultatory Areas

Clinicians' interest in the surface anatomy of the heart and cardiac valves results from their need to listen to valve sounds. The five areas (two areas are for the pulmonary valve) are as wide apart as possible so that the sounds produced at any given valve may be clearly distinguished from those produced at other valves (Fig. 1.76D & E). Blood tends to carry the sound in the direction of its flow; consequently, each area is situated superficial to the chamber or vessel into which the blood has passed and in a direct line with the valve orifice.

SUPERIOR, POSTERIOR, AND ANTERIOR MEDIASTINUM

Variations of Great Arteries

BRANCHES OF ARCH OF AORTA

The usual pattern of branches of the arch of the aorta is present in approximately 65% of people (Fig. 1.64). *Variations in the origin of the branches of the arch* are fairly common (Fig. B1.32A). In approximately 27% of people, the left common carotid artery originates from the brachiocephalic trunk. A brachiocephalic trunk fails to form in approximately 2.5% of people; in these cases each of the four arteries (right and left common carotid and subclavian arteries) originate independently from the arch of the aorta. The left vertebral artery originates from the arch of the aorta in approximately 5% of people. Both right and left brachiocephalic trunks originate from the arch in approximately 1.2% of people (Bergman et al., 1988).

A *retro-esophageal right subclavian artery* sometimes arises as the last (most-left-sided) branch of the arch of the aorta (Fig. B1.32B). The artery crosses posterior to the esophagus to reach the right upper limb and may compress the esophagus, causing difficulty in swallowing (*dysphagia*). An accessory artery to the thyroid gland, the **thyroid ima artery** (L. *arteria thyroidea ima*), may arise from the arch of the aorta or the brachiocephalic artery.

ANOMALIES OF ARCH OF AORTA

The most superior part of the arch of the aorta is usually approximately 2.5 cm inferior to the superior border of the manubrium, but it may be more superior or inferior. Sometimes the arch curves over the root of the right lung and passes inferiorly on the right side, forming a **right arch of the aorta.** In some cases, the abnormal arch, after passing over the root of the right lung, passes posterior to the esophagus to reach its usual position on the left side. Less frequently, a **double arch of the aorta** forms a vascular ring around the esophagus and trachea (Fig. B1.32C). A trachea that is compressed enough to affect breathing may require surgical division of the vascular ring.

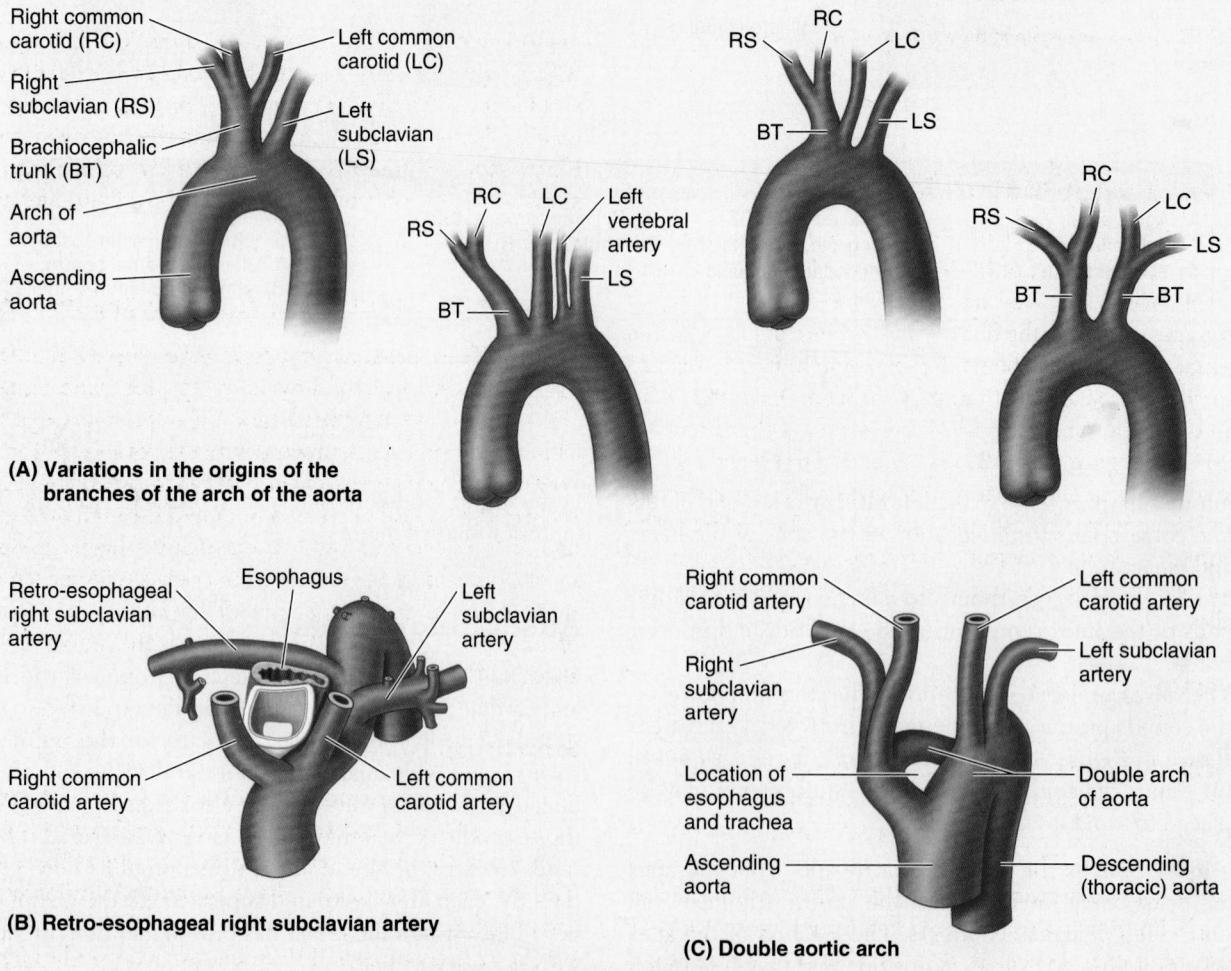

(A) Variations in the origins of the branches of the arch of the aorta

Right common carotid (RC)
Right subclavian (RS)
Brachiocephalic trunk (BT)
Arch of aorta
Ascending aorta
Left common carotid (LC)
Left subclavian (LS)

RC — RS — LC — Left vertebral artery — RS — LS — BT

RS — RC — LC — LS — BT

RS — RC — LC — LS — BT — BT

(B) Retro-esophageal right subclavian artery

Retro-esophageal right subclavian artery
Esophagus
Left subclavian artery
Right common carotid artery
Left common carotid artery

(C) Double aortic arch

Right common carotid artery
Left common carotid artery
Right subclavian artery
Left subclavian artery
Location of esophagus and trachea
Double arch of aorta
Ascending aorta
Descending (thoracic) aorta

FIGURE B1.32. Variations and anomalies of branches of aortic arch.

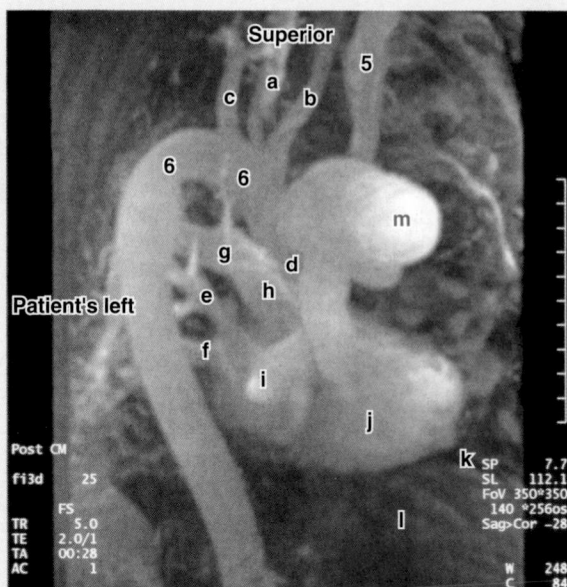

Aneurysm of the aortic arch: MR angiogram in a sagittal oblique plane; injected contrast medium shows vascular structures as bright; (5) superior vena cava, (6) aortic arch, (a) left common carotid artery, (b) brachiocephalic artery, (c) left subclavian artery, (d) ascending aorta, (e) right pulmonary vein, (f) left pulmonary vein, (g) left pulmonary artery (h) pulmonary trunk, (i) left atrium, (j) left ventricle, (k) diaphragm, (l) liver, (m) large saccular aneurysm arising from the ascending aorta.

FIGURE B1.33. **Aneurysm of the aortic arch** (magnetic resonance [*MR*] angiogram).

Aneurysm of Ascending Aorta

The distal part of the ascending aorta receives a strong thrust of blood when the left ventricle contracts. Because its wall is not yet reinforced by fibrous pericardium (the fibrous pericardium blends with the aortic adventitia at the beginning of the arch; Fig. 1.66B), an *aneurysm* (localized dilation) may develop. An aortic aneurysm is evident on a chest film (radiograph of the thorax) or an MR angiogram (Fig. B1.33) as an enlarged area of the ascending aorta silhouette. Individuals with an aneurysm usually complain of chest pain that radiates to the back. The aneurysm may exert pressure on the trachea, esophagus, and recurrent laryngeal nerve, causing difficulty in breathing and swallowing.

Coarctation of Aorta

In *coarctation of the aorta*, the arch of the aorta or thoracic aorta has an abnormal narrowing (stenosis) that diminishes the caliber of the aortic lumen, producing an obstruction to blood flow to the inferior part of the body (Fig. B1.34). The most common site for a coarctation is near the ligamentum arteriosum (Fig. 1.63). When the coarctation is inferior to this site (*postductal coarctation*), a good collateral circulation usually develops between the proximal and distal parts of the aorta through the intercostal and inter-

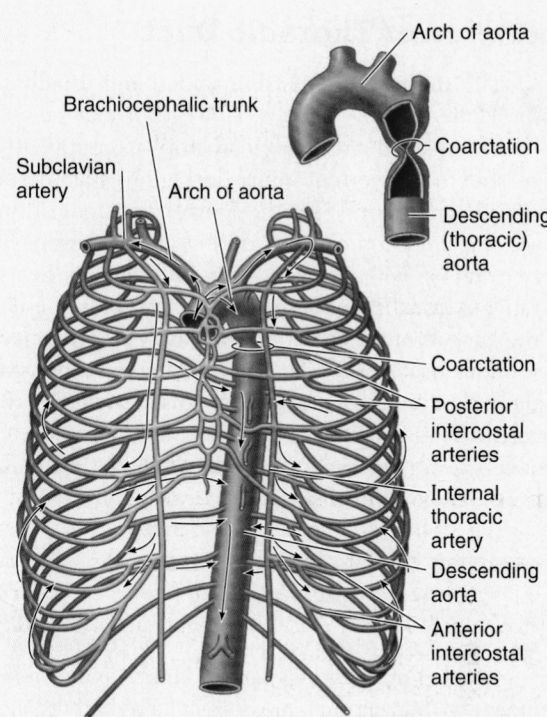

FIGURE B1.34. **Coarctation of aorta.**

nal thoracic arteries. This type of coarctation is compatible with many years of life because the collateral circulation carries blood to the thoracic aorta inferior to the stenosis. The collateral vessels may become so large that they cause notable pulsation in the intercostal spaces and erode the adjacent surfaces of the ribs, which is visible in radiographs of the thorax.

Injury to Recurrent Laryngeal Nerves

The recurrent laryngeal nerves supply all intrinsic muscles of the larynx, except one. Consequently, any investigative (diagnostic) procedure (e.g., *mediastinotomy*) or disease process in the superior mediastinum may injure these nerves and affect the voice. Because the left recurrent laryngeal nerve winds around the arch of the aorta and ascends between the trachea and esophagus, it may be involved in a bronchogenic or esophageal carcinoma, enlargement of mediastinal lymph nodes, or an *aneurysm of the arch of the aorta*. In the latter condition, the nerve may be stretched by the dilated arch.

Blockage of Esophagus

The impressions produced in the esophagus by adjacent structures are of clinical interest because of the slower passage of substances at these sites. The impressions indicate where swallowed foreign objects are most likely to lodge and where a stricture may develop, e.g., after the accidental drinking of a caustic liquid such as lye.

Laceration of Thoracic Duct

 The thoracic duct is thin walled and usually dull white in living persons. However it may be colorless, making it difficult to identify. Consequently, it is vulnerable to inadvertent injury during investigative and/or surgical procedures in the posterior mediastinum. *Laceration of the thoracic duct* during an accident or lung surgery results in lymph escaping into the thoracic cavity at rates ranging from 75 to 200 mL per hour. Lymph or chyle from the lacteals of the intestine may also enter the pleural cavity, producing *chylothorax*. This fluid may be removed by a needle tap or by thoracentesis; in some cases it may be necessary to ligate (tie off) the thoracic duct. The lymph then returns to the venous system by other lymphatic channels that join the thoracic duct superior to the ligature.

Variations of Thoracic Duct

 Variations of the thoracic duct are common because the superior part of the duct represents the original left member of a pair of lymphatic vessels in the embryo. Sometimes two thoracic ducts are present for a short distance.

Alternate Venous Routes to Heart

 The azygos, hemi-azygos, and accessory hemi-azygos veins offer alternate means of venous drainage from the thoracic, abdominal, and back regions when *obstruction of the IVC* occurs. In some people, an accessory azygos vein parallels the azygos vein on the right side. Other people have no hemi-azygos system of veins. A clinically important variation, although uncommon, is when the azygos system receives all the blood from the IVC except that from the liver. In these people, the azygos system drains nearly all the blood inferior to the diaphragm, except from the digestive tract. If *obstruction of the SVC* occurs superior to the entrance of the azygos vein, blood can drain inferiorly into the veins of the abdominal wall and return to the right atrium through the azygos venous system and the IVC.

Age Changes in Thymus

The thymus is a prominent feature of the superior mediastinum during infancy and childhood. In some infants, the thymus may compress the trachea. The thymus plays an important role in the development and maintenance of the immune system. As puberty is reached, the thymus begins to diminish in relative size. By adulthood, it is usually replaced by adipose tissue and is often scarcely recognizable; however, it continues to produce T-lymphocytes.

Aortic Angiography

To radiographically visualize the arch of the aorta and the branches arising from it, a long, narrow catheter is passed into the ascending

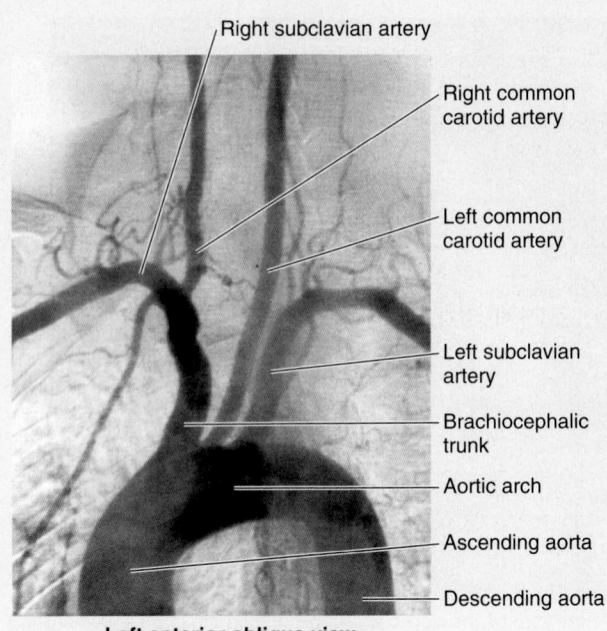

Right subclavian artery
Right common carotid artery
Left common carotid artery
Left subclavian artery
Brachiocephalic trunk
Aortic arch
Ascending aorta
Descending aorta

Left anterior oblique view

FIGURE B1.35. Aortic angiogram (aortogram).

aorta via the femoral or brachial artery in the inguinal or elbow region, respectively. Under fluoroscopic control, the tip of the catheter is placed just inside the opening of a coronary artery; an **aortic angiogram** can be made by injecting radiopaque contrast material into the aorta and into openings of the arteries arising from the arch of the aorta (Fig. B1.35).

Radiography of Mediastinum

 The heart casts most of the central radiopaque shadow in PA projections (Fig. B1.36), but the separate chambers of the heart are not distinguishable. Knowledge of the structures forming the **cardiovascular shadow** or silhouette is important because changes in the shadow may indicate anomalies or functional disease (Fig. B1.36A). In PA radiographs (AP views), the borders of the cardiovascular shadow are as follows:

- *Right border,* right brachiocephalic vein, SVC, right atrium, and IVC.
- *Left border,* terminal part of the arch of aorta, pulmonary trunk, left auricle, and left ventricle.

The left inferior part of the cardiovascular shadow presents the region of the apex. The typical anatomical apex, if present, is often inferior to the shadow of the diaphragm. Three main types of cardiovascular shadows occur, depending primarily on body type or habitus (Fig. B1.36B):

- *Transverse type,* observed in obese persons, pregnant women, and infants.
- *Oblique type,* characteristic of most people.
- *Vertical type,* present in people with narrow chests.

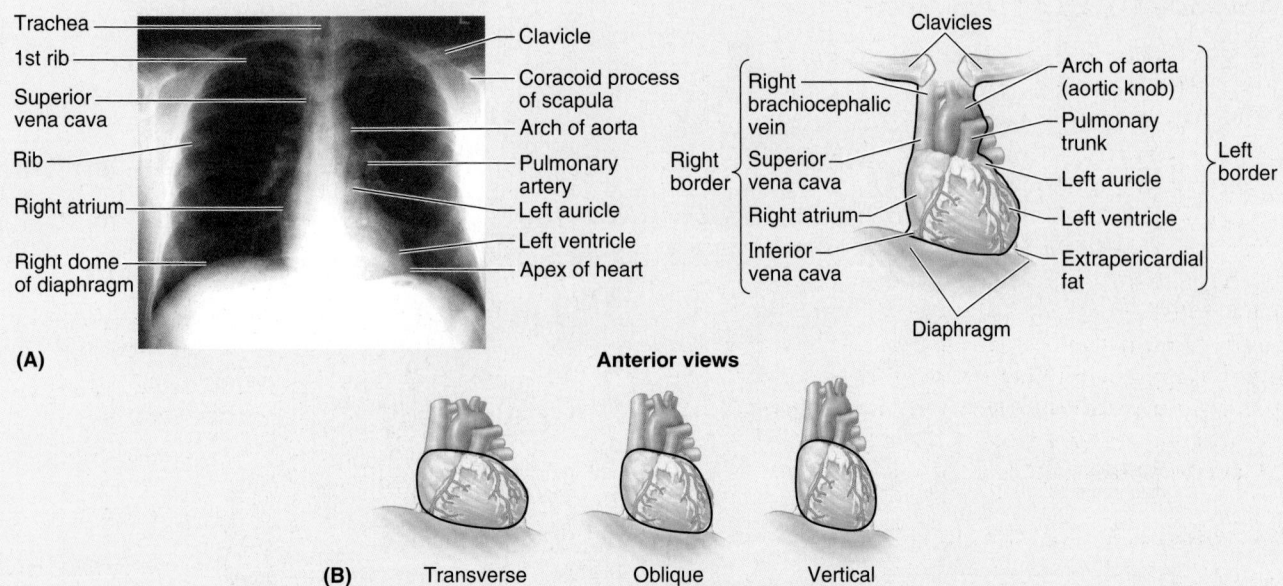

Trachea
1st rib
Superior vena cava
Rib
Right atrium
Right dome of diaphragm

Clavicle
Coracoid process of scapula
Arch of aorta
Pulmonary artery
Left auricle
Left ventricle
Apex of heart

(A)

Clavicles

Right brachiocephalic vein
Right border
Superior vena cava
Right atrium
Inferior vena cava

Arch of aorta (aortic knob)
Pulmonary trunk
Left auricle
Left ventricle
Extrapericardial fat
Left border

Diaphragm

Anterior views

(B)   Transverse   Oblique   Vertical

FIGURE B1.36. Cardiovascular shadows (mediastinal silhouettes). A. Composition of the margins of the cardiovascular shadow. **B.** Common types of cardiovascular shadow.

CT and MRI of Mediastinum

CT and MRI are commonly used to examine the thorax. CT is sometimes combined with mammography to examine the breasts (Fig. B1.37). Before CT scans are taken, an iodide contrast material is given intravenously. Because breast cancer cells have an unusual affinity for iodide, they become recognizable. MRI is usually better for detecting and delineating soft tissue lesions. It is especially useful for examining the viscera and lymph nodes of the mediastinum and roots of the lungs, by means of both planar (Fig. B1.38) and reconstructed (Fig. B1.39) images. Transverse (axial) CT and MR scans are always oriented to show how a horizontal section of a patient's body lying on an examination table would appear to the physician who is at the patient's feet. Therefore, the top of the image is anterior, and the left lateral edge of the image represents the right lateral surface of the patient's body. Data from CT and MR scans can be graphically reconstructed by the computer as transverse, sagittal, oblique, or coronal sections of the body.

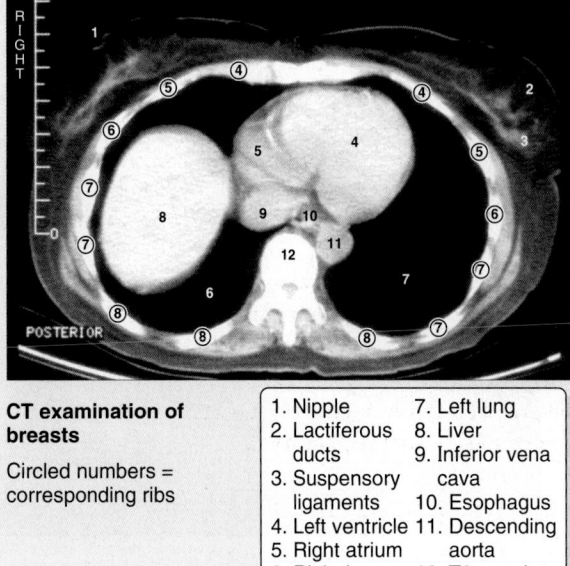

CT examination of breasts

Circled numbers = corresponding ribs

1. Nipple
2. Lactiferous ducts
3. Suspensory ligaments
4. Left ventricle
5. Right atrium
6. Right lung
7. Left lung
8. Liver
9. Inferior vena cava
10. Esophagus
11. Descending aorta
12. T9 vertebra

FIGURE B1.37. CT examination of breasts.

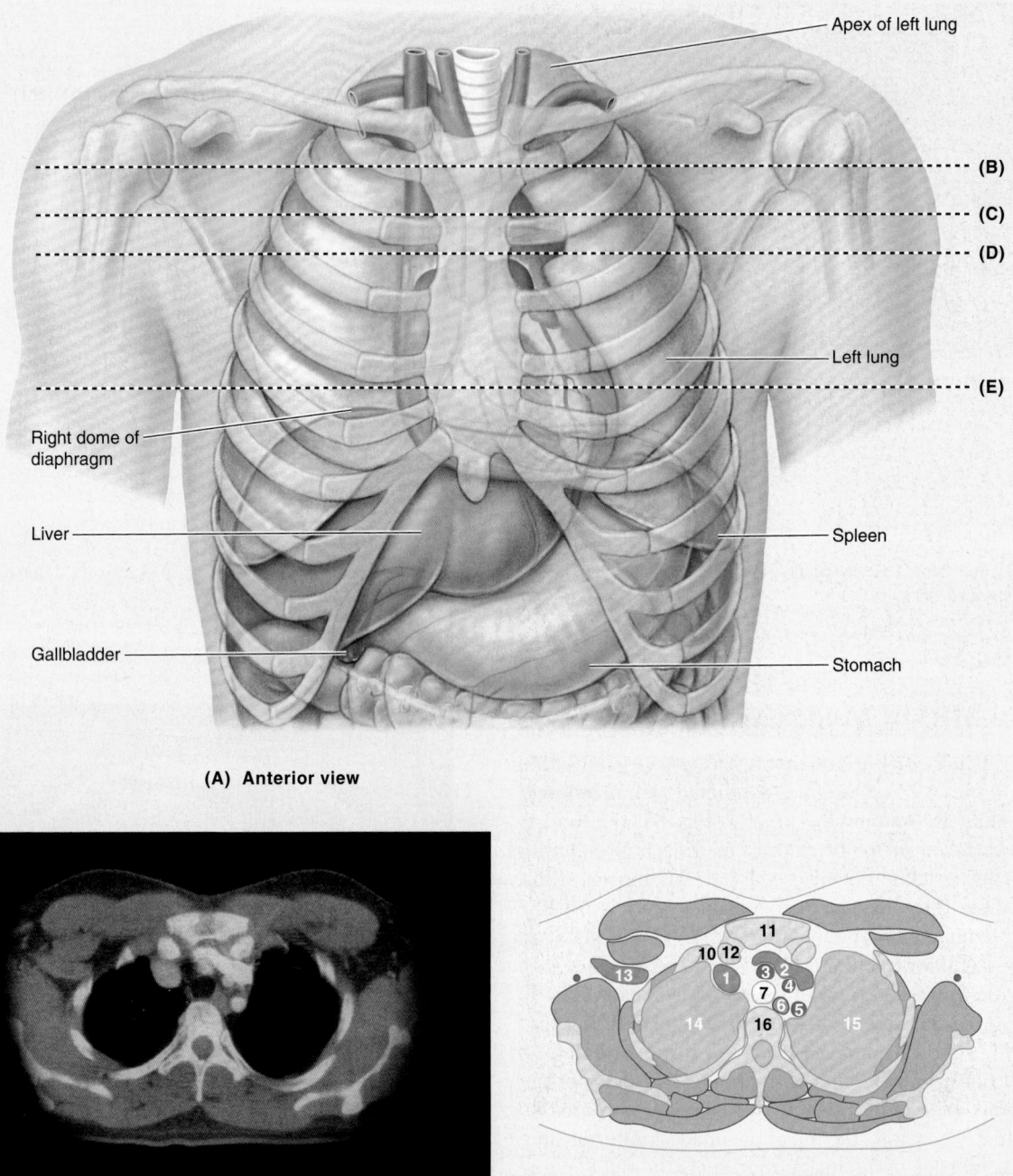

(A) Anterior view

FIGURE B1.38. Serial transverse CT scans of thorax. A. The level of each scan is indicated (broken lines). **B.** At the level of the sternoclavicular joints, the left brachiocephalic vein (*2*) *crosses the midline anterior to the three branches of the arch of the aorta* (*3, 4,* and *5*) to join the right brachiocephalic vein (*1*), forming the superior vena cava [SVC] (*22*) at a more inferior level.

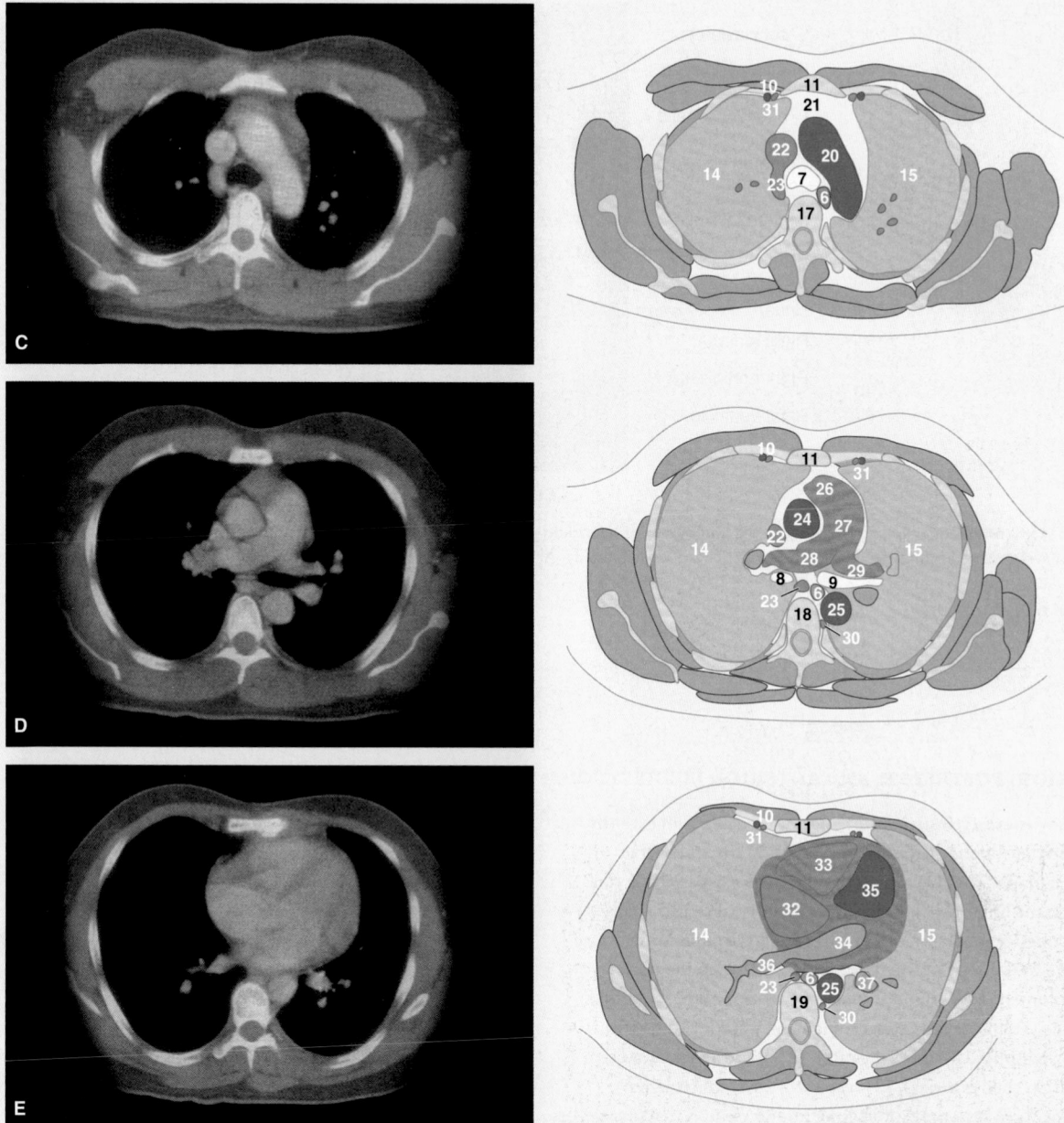

FIGURE B1.38. *(Continued)* **C.** The arch of the aorta (*20*) is obliquely placed (more sagittal than transverse), with the ascending end anteriorly in the midline, and the descending end posteriorly and to the left of the vertebral bodies (*17*). The SVC (*22*) on the right side receives the arch of the azygos vein (*23*) from its posterior aspect. **D.** The pulmonary trunk (*27*) forms the stem of an inverted Y, with the arms formed by the right (*28*) and left (*29*) pulmonary arteries. The right pulmonary artery (*28*) passes beneath the arch of the aorta [between ascending (*24*) and descending (*25*) aortae]. **E.** A scan at the level of the maximum diameter of the heart demonstrates all four chambers (*32–35*) and the diagonal slant of the interventricular septum (between *33* and *35*).

Key to Structures in Transverse CT Scans of Thorax

1 Right brachiocephalic vein	**14** Right lung	**26** Conus arteriosus
2 Left brachiocephalic vein	**15** Left lung	**27** Pulmonary trunk
3 Brachiocephalic artery	**16** T4 vertebral body	**28** Right pulmonary artery
4 Left common carotid artery	**17** T5 vertebral body	**29** Left pulmonary artery
5 Left subclavian artery	**18** T6 vertebral body	**30** Hemi-azygos vein
6 Esophagus	**19** T8 vertebral body	**31** Internal thoracic vessels
7 Trachea	**20** Arch of aorta	**32** Right atrium
8 Right main bronchus	**21** Anterior mediastinum	**33** Right ventricle
9 Left main bronchus	(region of thymic remnant)	**34** Left atrium
10 Costal cartilage	**22** Superior vena cava	**35** Left ventricle
11 Sternum	**23** Arch of azygos vein	**36** Inferior right pulmonary vein
12 Clavicle	**24** Ascending aorta	**37** Inferior left pulmonary vein
13 Axillary vein	**25** Descending aorta	

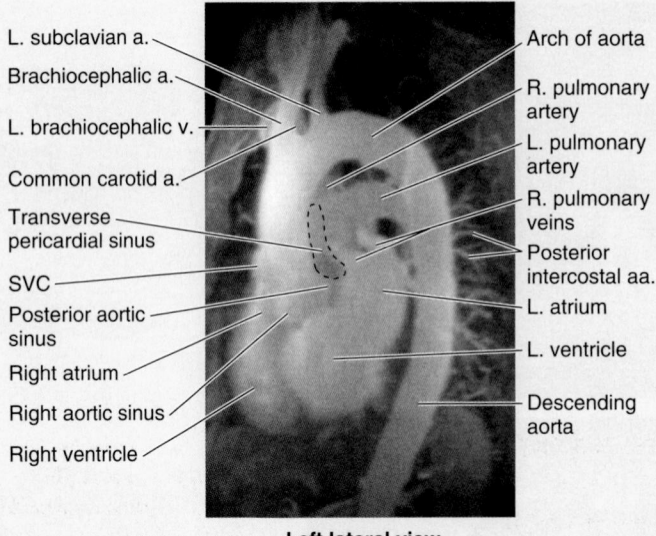

L. subclavian a.
Brachiocephalic a.
L. brachiocephalic v.
Common carotid a.
Transverse pericardial sinus
SVC
Posterior aortic sinus
Right atrium
Right aortic sinus
Right ventricle

Arch of aorta
R. pulmonary artery
L. pulmonary artery
R. pulmonary veins
Posterior intercostal aa.
L. atrium
L. ventricle
Descending aorta

Left lateral view

FIGURE B1.39. Magnetic resonance reconstructed angiogram of heart and great vessels. Lateral view (from left and slightly anterior). Reconstructed from data generated and accumulated by spiral magnetic resonance imaging. All chambers of the heart and great vessels are clearly visible. *SVC*, superior vena cava.

The Bottom Line

SUPERIOR, POSTERIOR, AND ANTERIOR MEDIASTINUM; SURFACE ANATOMY OF THORACIC VISCERA

Superior mediastinum: The superior mediastinum extends between the superior thoracic aperture and the transverse thoracic plane. The only organ that belongs exclusively to this region is the adult thymus. ♦ The remainder of the structures in the superior mediastinum pass through the superior thoracic aperture to the root of the neck, or pass between the neck and abdomen. ♦ Within the superior mediastinum, structures occur in systemic layers, proceeding from anterior to posterior: (1) lymphoid system (thymus), (2) blood vascular system (veins first, then arteries), (3) respiratory system (trachea), (4) alimentary system (esophagus), and (5) lymph vascular system. ♦ The nervous system does not have its own layer in the superior mediastinum, but it is integrated with layer 2 (phrenic and vagus nerves) and lies between layers 3 and 4 (recurrent laryngeal nerves). ♦ The pattern of the branches of the arch of the aorta is atypical in approximately 35% of people.

Posterior mediastinum: The posterior mediastinum is the narrow passageway that lies posterior to the heart and diaphragm and between the lungs. It contains structures passing from thorax to abdomen or vice versa. ♦ Contents include the esophagus and esophageal nerve plexus, thoracic aorta, thoracic duct and lymphatic trunks, posterior mediastinal lymph nodes, and azygos and hemi-azygos veins. ♦ Branches of the thoracic aorta occur primarily within three vascular planes. ♦ The azygos/hemi-azygos venous system constitutes the venous counterpart to the thoracic aorta and its posterior mediastinal branches. ♦ The thoracic portion of the sympathetic trunks and thoracic splanchnic nerves may or may not be considered components of the posterior mediastinum.

Anterior mediastinum: The smallest subdivision of the mediastinum, between sternum and transversus thoracis muscles, significant primarily as a surgical plane, contains primarily loose connective tissue and, in infants and children, the inferior extend of the thymus.

Surface anatomy of thoracic viscera: The heart and great vessels are in the central thorax, surrounded laterally and posteriorly by the lungs, and are overlapped anteriorly by the lines of pleural reflection and anterior borders of the lungs, sternum and the central part of the thoracic cage. ♦ The position of the mediastinal viscera depends on position relative to gravity, phase of respiration, and the build and physical condition of the person. ♦ Apical portions of pleurae and lungs lie posterior to the supraclavicular fossa. ♦ The transverse thoracic plane intersects the sternal angle and demarcates the great vessels superiorly from the pericardium/heart. ♦ The xiphisternal junction provides an indication of the central tendon of the diaphragm.

Abdomen

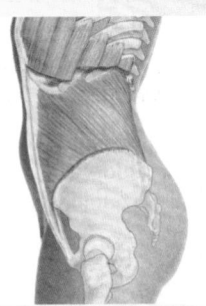

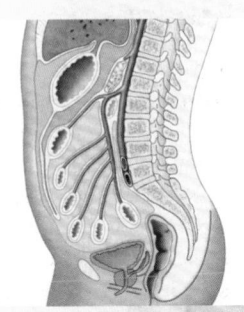

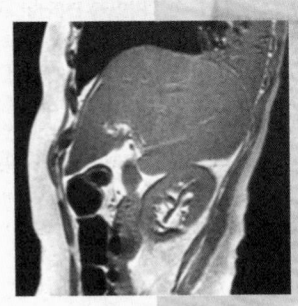

The **abdomen** is the part of the trunk between the thorax and the pelvis (Fig. 2.1). It is a flexible, dynamic container, housing most of the organs of the alimentary system and part of the urogenital system. Containment of the abdominal organs and their contents is provided by *musculo-aponeurotic walls* anterolaterally, the diaphragm superiorly, and the muscles of the pelvis inferiorly. The anterolateral musculo-aponeurotic walls are suspended between and supported by two bony rings (the inferior margin of the thoracic skeleton superiorly and the pelvic girdle inferiorly) linked by a semirigid lumbar vertebral column in the posterior abdominal wall. Interposed between the more rigid thorax and pelvis, this arrangement enables the abdomen to enclose and protect its contents while providing the flexibility required by respiration, posture, and locomotion.

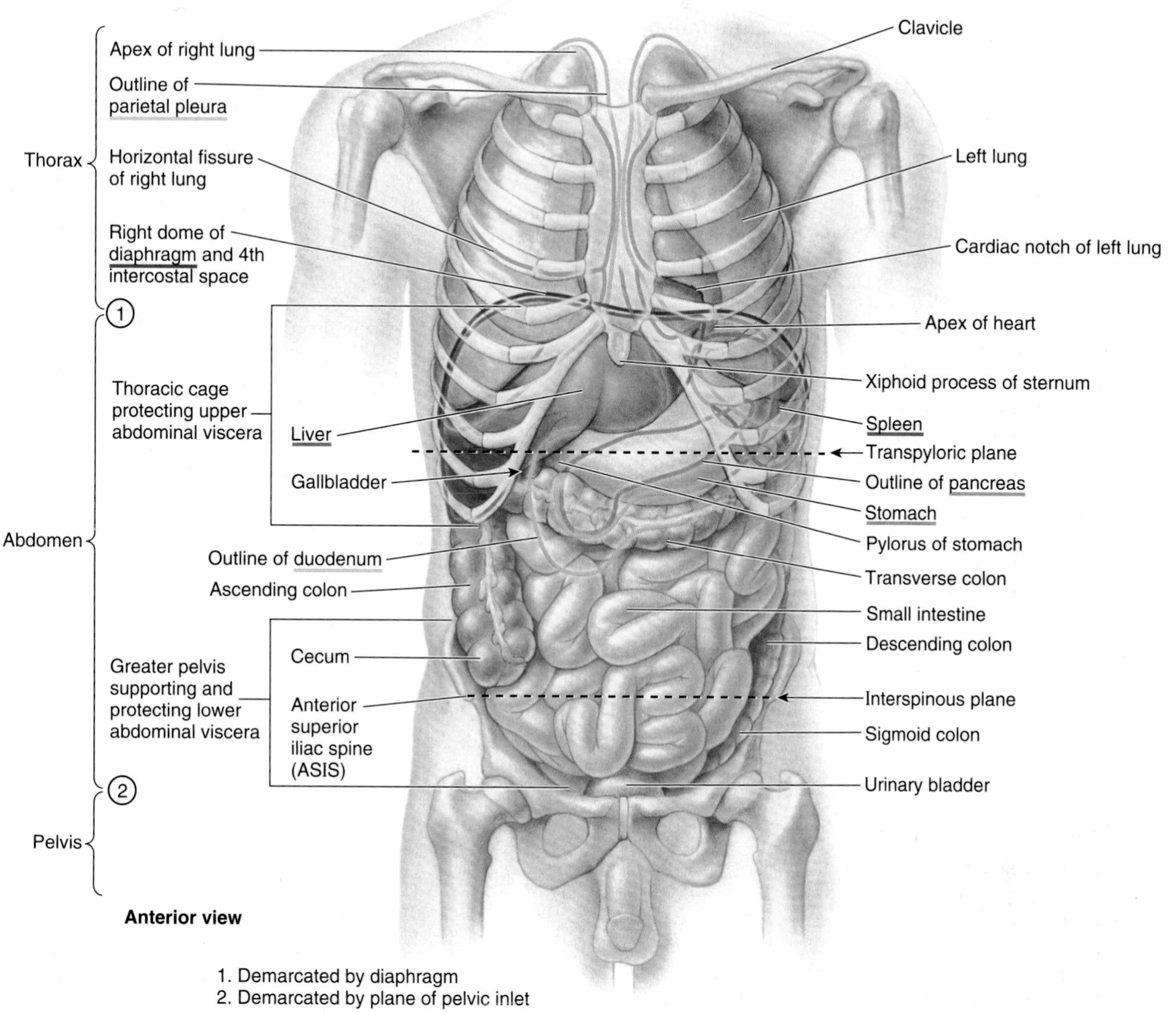
Apex of right lung
Outline of parietal pleura
Thorax
Horizontal fissure of right lung
Right dome of diaphragm and 4th intercostal space
1
Thoracic cage protecting upper abdominal viscera
Liver
Gallbladder
Abdomen
Outline of duodenum
Ascending colon
Greater pelvis supporting and protecting lower abdominal viscera
Cecum
Anterior superior iliac spine (ASIS)
2
Pelvis
Anterior view
Clavicle
Left lung
Cardiac notch of left lung
Apex of heart
Xiphoid process of sternum
Spleen
Transpyloric plane
Outline of pancreas
Stomach
Pylorus of stomach
Transverse colon
Small intestine
Descending colon
Interspinous plane
Sigmoid colon
Urinary bladder
1. Demarcated by diaphragm
2. Demarcated by plane of pelvic inlet

FIGURE 2.1. Overview of viscera of thorax and abdomen *in situ*.

Through voluntary or reflexive contraction, its muscular roof, anterolateral walls, and floor can raise internal (intra-abdominal) pressure to aid expulsion of air from the thoracic cavity (lungs and bronchi) or of fluid (e.g., urine or vomitus), flatus, feces, or fetuses from the abdominopelvic cavity.

OVERVIEW: WALLS, CAVITIES, REGIONS, AND PLANES

The dynamic, multi-layered, musculo-aponeurotic **abdominal walls** not only contract to increase intra-abdominal pressure, but also distend considerably, accommodating expansions caused by ingestion, pregnancy, fat deposition, or pathology.

The *anterolateral abdominal wall* and several organs lying against the *posterior wall* are covered on their internal aspects with a serous membrane or *peritoneum* (serosa) that *reflects* (turns sharply and continues) onto the **abdominal viscera** (L., soft parts, internal organs), such as the stomach, intestine, liver, and spleen. Thus, a bursal sac or lined potential space (*peritoneal cavity*) is formed between the walls and the viscera that normally contains only enough extracellular (parietal) fluid to lubricate the membrane covering most of the surfaces of the structures forming or occupying the abdominal cavity. Visceral movement associated with digestion occurs freely, and the double-layered reflections of peritoneum passing between the walls and the viscera provide passage for the blood vessels, lymphatics, and nerves. Variable amounts of fat may also occur between the walls and viscera and the peritoneum lining them.

The **abdominal cavity:**

- forms the superior and major part of the **abdominopelvic cavity** (Fig. 2.2), the continuous cavity that extends between the *thoracic diaphragm* and *pelvic diaphragm.*
- has no floor of its own because it is continuous with the *pelvic cavity.* The plane of the *pelvic inlet* (superior pelvic aperture) arbitrarily, but not physically, separates the abdominal and the pelvic cavities.
- extends superiorly into the osseocartilaginous *thoracic cage* to the 4th intercostal space (Fig. 2.1). Consequently, the more superiorly placed abdominal organs (spleen, liver, part of the kidneys, and stomach) are protected by the thoracic cage. The *greater pelvis* (expanded part of the pelvis superior to the pelvic inlet) supports and partly protects the lower abdominal viscera (part of the ileum, cecum, appendix, and sigmoid colon).
- is the location of most digestive organs, parts of the urogenital system (kidneys and most of the ureters), and the spleen.

Nine regions of the abdominal cavity are used to describe the location of abdominal organs, pains, or pathologies (Table 2.1A & B). The regions are delineated by four planes: two sagittal (vertical) and two transverse (horizontal) planes. The two sagittal planes are usually the *midclavicular planes* that pass from the midpoint of the clavicles (approximately 9 cm

from the midline) to the **midinguinal points,** midpoints of the lines joining the *anterior superior iliac spine* (ASIS) and the *pubic tubercles* on each side.

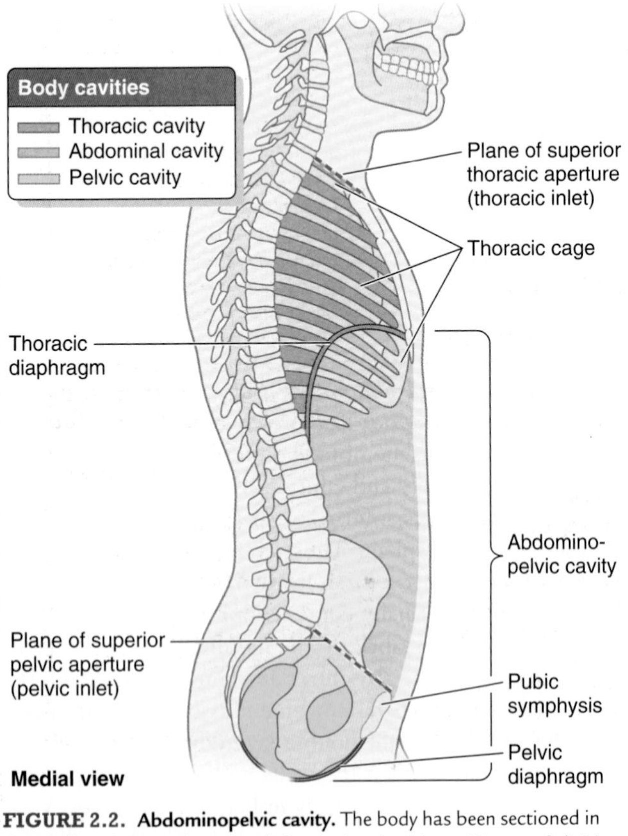

Most commonly, the transverse planes are the **subcostal plane,** passing through the inferior border of the 10th costal cartilage on each side, and the **transtubercular plane,** passing through the iliac tubercles (approximately 5 cm posterior to the ASIS on each side) and the body of the L5 vertebra. Both of these planes have the advantage of intersecting palpable structures.

Some clinicians use the transpyloric and interspinous planes to establish the nine regions. The **transpyloric plane,** extrapolated midway between the superior borders of the manubrium of the sternum and the pubic symphysis (typically the L1 vertebral level), commonly transects the *pylorus* (the distal, more tubular part of the stomach) when the patient is recumbent (supine or prone) (Fig. 2.1). Because the viscera sag with the pull of gravity, the pylorus usually lies at a lower level when the individual is standing erect. The transpyloric plane is a useful landmark because it also transects many other important structures: the fundus of the gallbladder, neck of the pancreas, origins of the superior mesenteric artery (SMA) and hepatic portal vein, root of the transverse mesocolon, duodenojejunal junction, and hila of the kidneys. The **interspinous plane** passes through the easily palpated ASIS on each side (Table 2.1B).

For more general clinical descriptions, four quadrants of the abdominal cavity (right and left upper and lower quadrants) are defined by two readily defined planes: (1) the transverse **transumbilical plane,** passing through the umbilicus (and the intervertebral [IV] disc between the L3 and L4 vertebrae), dividing it into upper and lower halves, and (2) the vertical *median plane,* passing longitudinally through the body, dividing it into right and left halves (Table 2.1C).

It is important to know what organs are located in each abdominal region or quadrant so that one knows where to auscultate, percuss, and palpate them (Table 2.1), and to record the locations of findings during a physical examination.

ANTEROLATERAL ABDOMINAL WALL

Although the abdominal wall is continuous, it is subdivided into the *anterior wall, right and left lateral walls,* and *posterior wall* for descriptive purposes (Fig. 2.3). The wall is musculo-aponeurotic, except for the posterior wall, which includes the lumbar region of the vertebral column. The boundary between the anterior and lateral walls is indefinite, therefore the term **anterolateral abdominal wall** is often used. Some structures, such as muscles and cutaneous nerves, are in both the anterior and lateral walls. The anterolateral abdominal wall extends from the thoracic cage to the pelvis.

The anterolateral abdominal wall is bounded superiorly by the cartilages of the 7th–10th ribs and the xiphoid process of the sternum, and inferiorly by the inguinal ligament and the

Body cavities
- Thoracic cavity
- Abdominal cavity
- Pelvic cavity

Plane of superior thoracic aperture (thoracic inlet)

Thoracic cage

Thoracic diaphragm

Abdomino-pelvic cavity

Plane of superior pelvic aperture (pelvic inlet)

Pubic symphysis

Pelvic diaphragm

Medial view

FIGURE 2.2. Abdominopelvic cavity. The body has been sectioned in the median plane to show the abdominal and pelvic cavities as subdivisions of the continuous abdominopelvic cavity.

TABLE 2.1. ABDOMINAL REGIONS (A), REFERENCE PLANES (B), AND QUADRANTS (C)

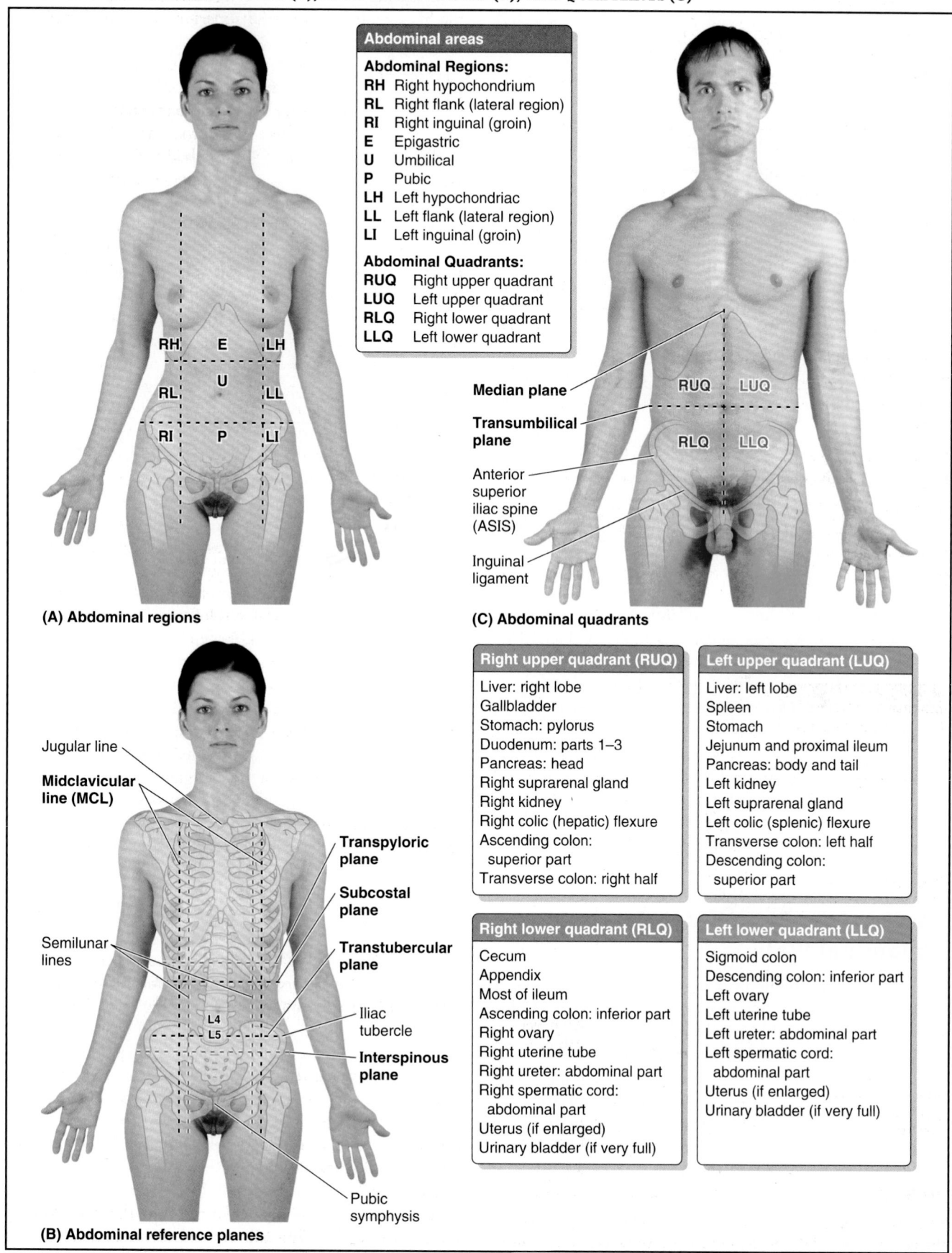

Abdominal areas

Abdominal Regions:
- **RH** Right hypochondrium
- **RL** Right flank (lateral region)
- **RI** Right inguinal (groin)
- **E** Epigastric
- **U** Umbilical
- **P** Pubic
- **LH** Left hypochondriac
- **LL** Left flank (lateral region)
- **LI** Left inguinal (groin)

Abdominal Quadrants:
- **RUQ** Right upper quadrant
- **LUQ** Left upper quadrant
- **RLQ** Right lower quadrant
- **LLQ** Left lower quadrant

(A) Abdominal regions

(C) Abdominal quadrants

Median plane
Transumbilical plane
Anterior superior iliac spine (ASIS)
Inguinal ligament

Jugular line
Midclavicular line (MCL)
Semilunar lines
Transpyloric plane
Subcostal plane
Transtubercular plane
Iliac tubercle
Interspinous plane
L4
L5
Pubic symphysis

(B) Abdominal reference planes

Right upper quadrant (RUQ)	Left upper quadrant (LUQ)
Liver: right lobe	Liver: left lobe
Gallbladder	Spleen
Stomach: pylorus	Stomach
Duodenum: parts 1–3	Jejunum and proximal ileum
Pancreas: head	Pancreas: body and tail
Right suprarenal gland	Left kidney
Right kidney	Left suprarenal gland
Right colic (hepatic) flexure	Left colic (splenic) flexure
Ascending colon: superior part	Transverse colon: left half
Transverse colon: right half	Descending colon: superior part

Right lower quadrant (RLQ)	Left lower quadrant (LLQ)
Cecum	Sigmoid colon
Appendix	Descending colon: inferior part
Most of ileum	Left ovary
Ascending colon: inferior part	Left uterine tube
Right ovary	Left ureter: abdominal part
Right uterine tube	Left spermatic cord: abdominal part
Right ureter: abdominal part	Uterus (if enlarged)
Right spermatic cord: abdominal part	Urinary bladder (if very full)
Uterus (if enlarged)	
Urinary bladder (if very full)	

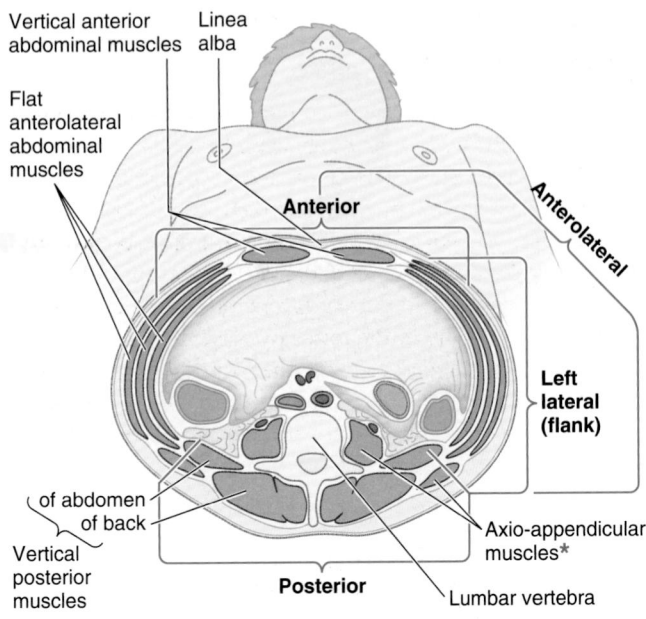

FIGURE 2.3. Subdivisions of abdominal wall. A transverse section of the abdomen demonstrates various aspects of the wall and its components. [*The relatively superficial latissimus dorsi and deeper psoas major muscles are axioappendicular muscles that attach distally in the upper and lower limbs, respectively.]

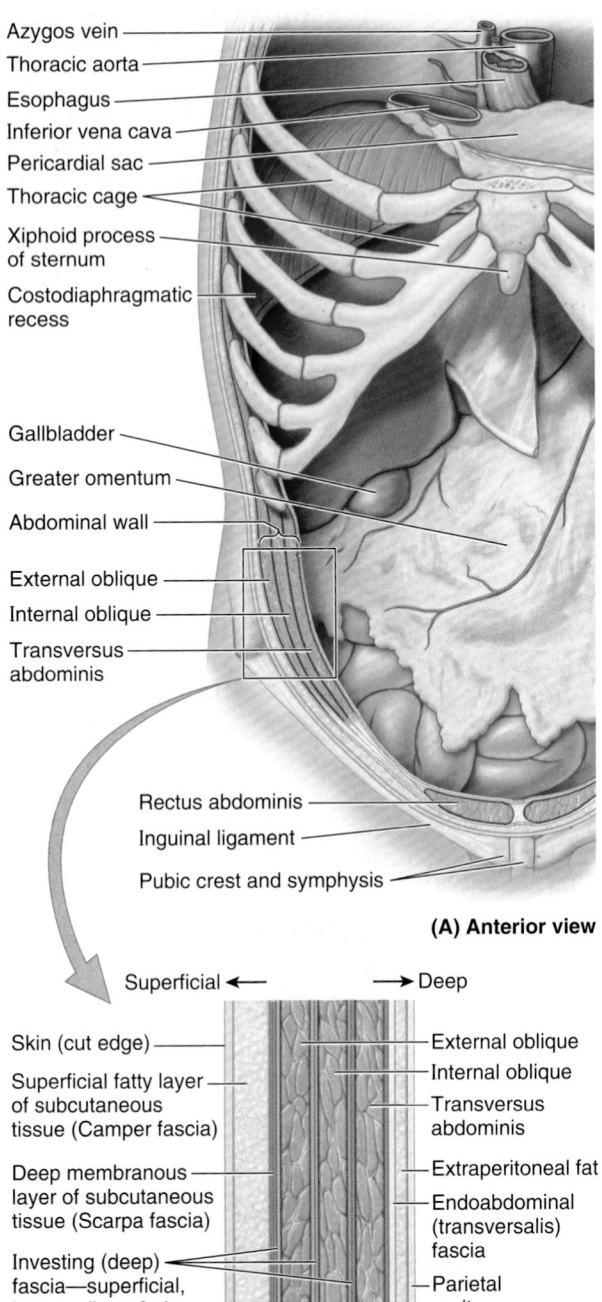

(A) Anterior view

(B) Longitudinal section

FIGURE 2.4. Abdominal contents, undisturbed, and layers of antero-lateral abdominal wall. A. The anterior abdominal wall and soft tissues of the anterior thoracic wall have been removed. Most of the intestine is covered by the apron-like greater omentum, a peritoneal fold hanging from the stomach. B. Layers of the anterolateral abdominal wall, including the trilaminar flat muscles, are shown.

superior margins of the anterolateral aspects of the pelvic girdle (iliac crests, pubic crests, and pubic symphysis) (Fig. 2.4A).

The anterolateral abdominal wall consists of skin and subcutaneous tissue (superficial fascia) composed mainly of fat, muscles and their aponeuroses and deep fascia, extraperitoneal fat, and parietal peritoneum (Fig. 2.4B). The skin attaches loosely to the subcutaneous tissue, except at the umbilicus, where it adheres firmly. Most of the anterolateral wall includes three musculotendinous layers; the fiber bundles of each layer run in different directions. This three-ply structure is similar to that of the intercostal spaces in the thorax.

Fascia of the Anterolateral Abdominal Wall

The subcutaneous tissue over most of the wall includes a variable amount of fat. It is a major site of fat storage. Males are especially susceptible to subcutaneous accumulation of fat in the lower anterior abdominal wall. In *morbid obesity,* the fat is many inches thick, often forming one or more sagging folds (L. *panniculi;* singular = *panniculus,* apron).

Superior to the umbilicus, the subcutaneous tissue is consistent with that found in most regions. Inferior to the umbilicus, the deepest part of the subcutaneous tissue is reinforced by many elastic and collagen fibers, so it has two layers: the **superficial fatty layer** (Camper fascia) and the **deep membranous layer** (Scarpa fascia) **of subcutaneous tissue.** The membranous layer continues inferiorly into

the perineal region as the superficial perineal fascia (Colles fascia), but not into the thighs.

Superficial, intermediate, and deep layers of investing fascia cover the external aspects of the three muscle layers of the anterolateral abdominal wall and their *aponeuroses* (flat expanded tendons) and cannot be easily separated from them.

The investing fascias here are extremely thin, being represented mostly by the *epimysium* (outer fibrous connective tissue layer surrounding all muscles—see Introduction) superficial to or between muscles. The internal aspect of the abdominal wall is lined with membranous and areolar sheets of varying thickness constituting **endoabdominal fascia.** Although continuous, different parts of this fascia are named according to the muscle or aponeurosis it is lining. The portion lining the deep surface of the transversus abdominis muscle and its aponeurosis is the **transversalis fascia.** The glistening lining of the abdominal cavity, the *parietal peritoneum,* is formed by a single layer of epithelial cells and supporting connective tissue. The parietal peritoneum is internal to the transversalis fascia and is separated from it by a variable amount of **extraperitoneal fat.**

Muscles of Anterolateral Abdominal Wall

There are five (bilaterally paired) muscles in the anterolateral abdominal wall (Fig. 2.3): three flat muscles and two vertical muscles. Their attachments are demonstrated in Figure 2.5 and listed, along with their nerve supply and main actions, in Table 2.2.

The *three flat muscles* are the *external oblique, internal oblique,* and *transversus abdominis.* The muscle fibers of these three concentric muscle layers have varying orientations, with the fibers of the outer two layers running diagonally and perpendicular to each other for the main part, and the fibers of the deep layer running transversely. All three flat muscles are continued anteriorly and medially as strong,

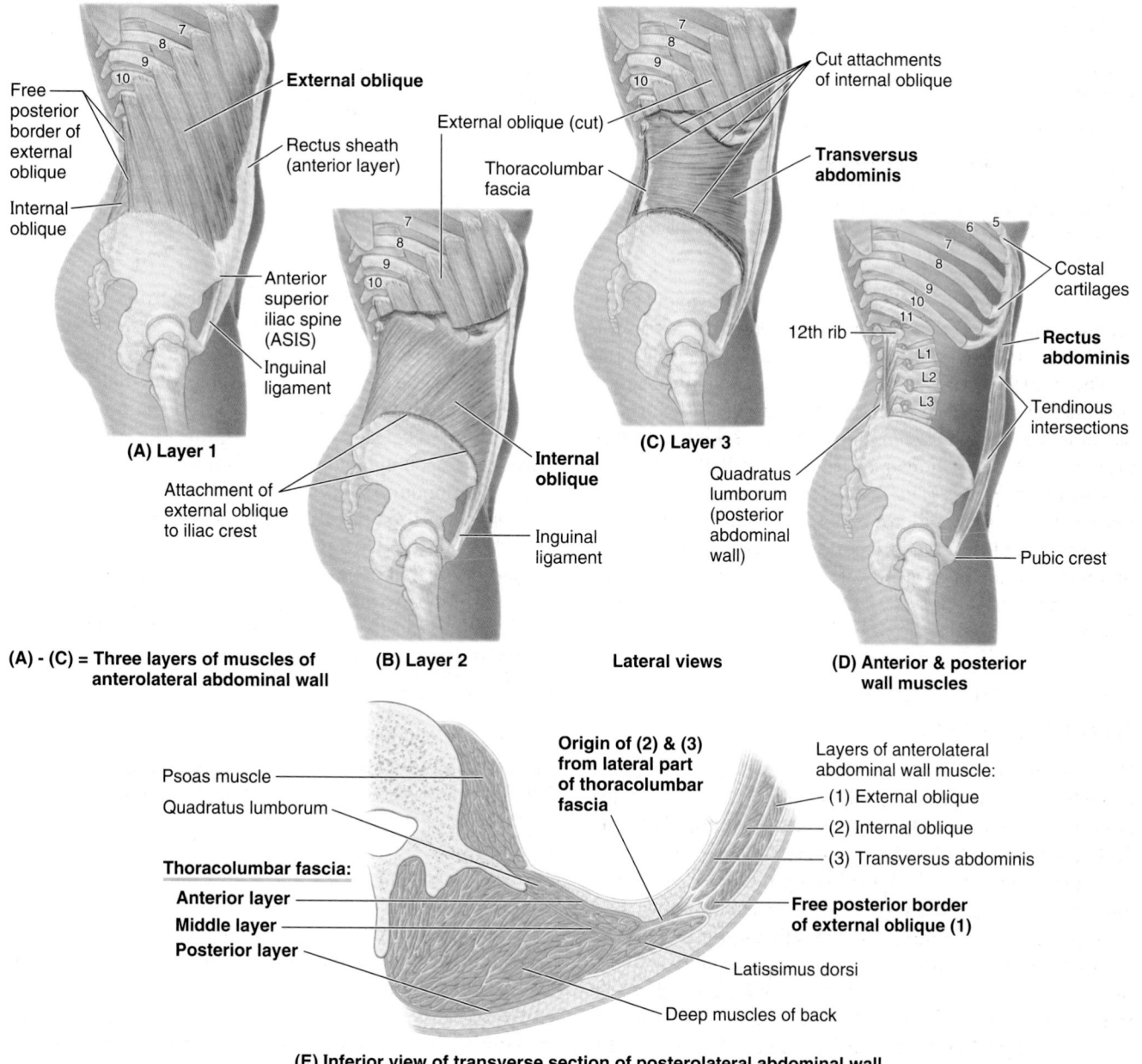

FIGURE 2.5. Muscles of anterolateral abdominal wall.

TABLE 2.2. MUSCLES OF ANTEROLATERAL ABDOMINAL WALL

Muscle	Origin	Insertion	Innervation	Main Action[a]
External oblique (A)	External surfaces of 5th–12th ribs	Linea alba, pubic tubercle, and anterior half of iliac crest	Thoraco-abdominal nerves (T7–T11 spinal nerves) and subcostal nerve	Compresses and supports abdominal viscera,[b] flexes and rotates trunk
Internal oblique (B)	Thoracolumbar fascia, anterior two thirds of iliac crest, and connective tissue deep to lateral third of inguinal ligament	Inferior borders of 10th–12th ribs, linea alba, and pecten pubis via conjoint tendon	Thoraco-abdominal nerves (anterior rami of T6–T12 spinal nerves) and first lumbar nerves	
Transversus abdominis (C)	Internal surfaces of 7th–12th costal cartilages, thoracolumbar fascia, iliac crest, and connective tissue deep to lateral third of inguinal ligament	Linea alba with aponeurosis of internal oblique, pubic crest, and pecten pubis via conjoint tendon		Compresses and supports abdominal viscera[b]
Rectus abdominis (D)	Pubic symphysis and pubic crest	Xiphoid process and 5th–7th costal cartilages	Thoraco-abdominal nerves (anterior rami of T6–T12 spinal nerves)	Flexes trunk (lumbar vertebrae) and compresses abdominal viscera;[b] stabilizes and controls tilt of pelvis (antilordosis)

[a]Approximately 80% of people have an insignificant muscle, the *pyramidalis,* which is located in the rectus sheath anterior to the most inferior part of the rectus abdominis. It extends from the pubic crest of the hip bone to the linea alba. This small muscle draws down on the linea alba.

[b]In so doing, these muscles act as antagonists of the diaphragm to produce expiration.

sheet-like aponeuroses (Fig. 2.6A). Between the midclavicular line (MCL) and the midline, the aponeuroses form the tough, aponeurotic, tendinous *rectus sheath* enclosing the rectus abdominis muscle (Fig. 2.6B). The aponeuroses then interweave with their fellows of the opposite side, forming a midline raphe (G. *rhaphe,* suture, seam), the **linea alba** (L. white line), which extends from the xiphoid process to the pubic symphysis. The decussation and interweaving of the aponeurotic fibers here is not only between right and left sides but also between superficial and intermediate and intermediate and deep layers.

The *two vertical muscles* of the anterolateral abdominal wall, contained within the rectus sheath, are the large *rectus abdominis* and the small *pyramidalis.*

EXTERNAL OBLIQUE MUSCLE

The **external oblique muscle** is the largest and most superficial of the three flat anterolateral abdominal muscles (Fig. 2.7). The attachments of the external oblique are demonstrated in Figure 2.5A, and listed, along with the nerve supply, and main actions, in Table 2.2. In contrast to the two deeper layers, the external oblique does not originate posteriorly from the thoracolumbar fascia; its posteriormost fibers (the thickest part of the muscle) have a free edge where they span between its costal origin and the iliac crest (Fig 2.5D & E). The fleshy part of the muscle contributes primarily to the lateral part of the abdominal wall. Its aponeurosis contributes to the anterior part of the wall.

Although the posteriormost fibers from rib 12 are nearly vertical as they run to the iliac crest, more anterior fibers fan out, taking an increasingly medial direction so that most of the fleshy fibers run inferomedially—in the same direction as the fingers do when the hands are in one's side pockets—with the most anterior and superior fibers approaching a horizontal course. The muscle fibers become aponeurotic approximately at the MCL medially and at the **spino-umbilical line** (line running from the umbilicus to the ASIS) inferiorly, forming a sheet of tendinous fibers that decussate at the linea alba, most becoming continuous with tendinous fibers of the contralateral internal oblique (see Fig. 2.6A). Thus, the contralateral external and internal oblique muscles together form a "digastric muscle," a two-bellied muscle sharing a common central tendon that works as a unit (see Introduction chapter). For example, the right external oblique and left internal oblique work together when flexing and rotating to bring the right shoulder toward the left hip (torsional movement of trunk).

Inferiorly, the external oblique aponeurosis attaches to the *pubic crest* medial to the *pubic tubercle.* The inferior margin of the external oblique aponeurosis is thickened as an undercurving fibrous band with a free posterior edge that spans between the ASIS and the pubic tubercle as the *inguinal ligament* (Poupart ligament) (Figs. 2.7B and 2.8).

Palpate your inguinal ligament by pressing deeply into the center of the crease between the thigh and trunk and moving the fingertips up and down. Inferiorly the

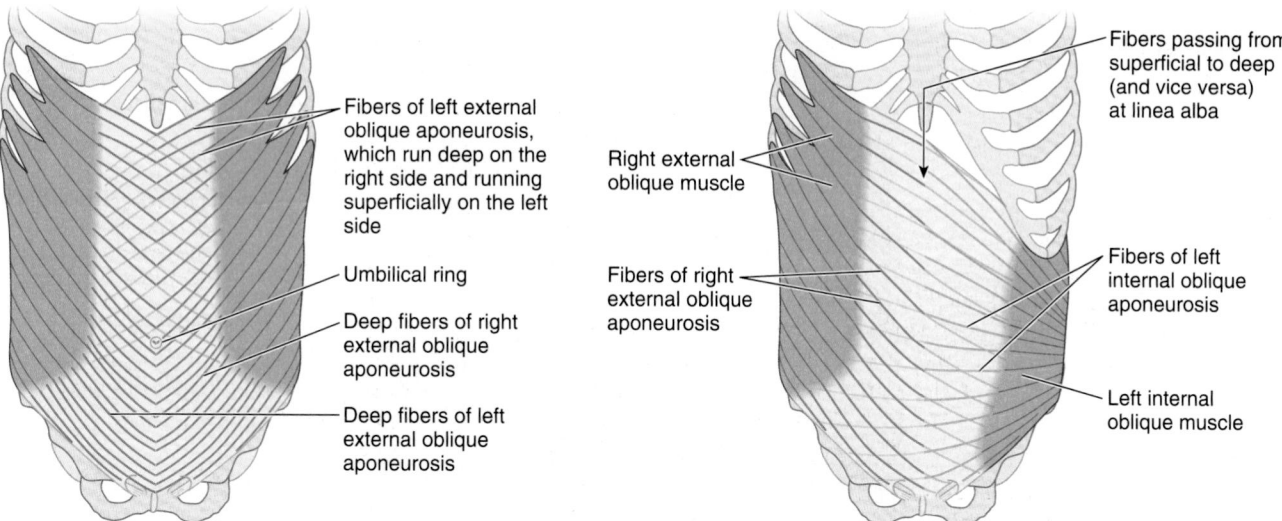

Intramuscular exchange of superficial and deep fibers within aponeuroses of contralateral external oblique muscles.

Intermuscular exchange of fibers between aponeuroses of contralateral external and internal oblique muscles.

(A) Anterior views

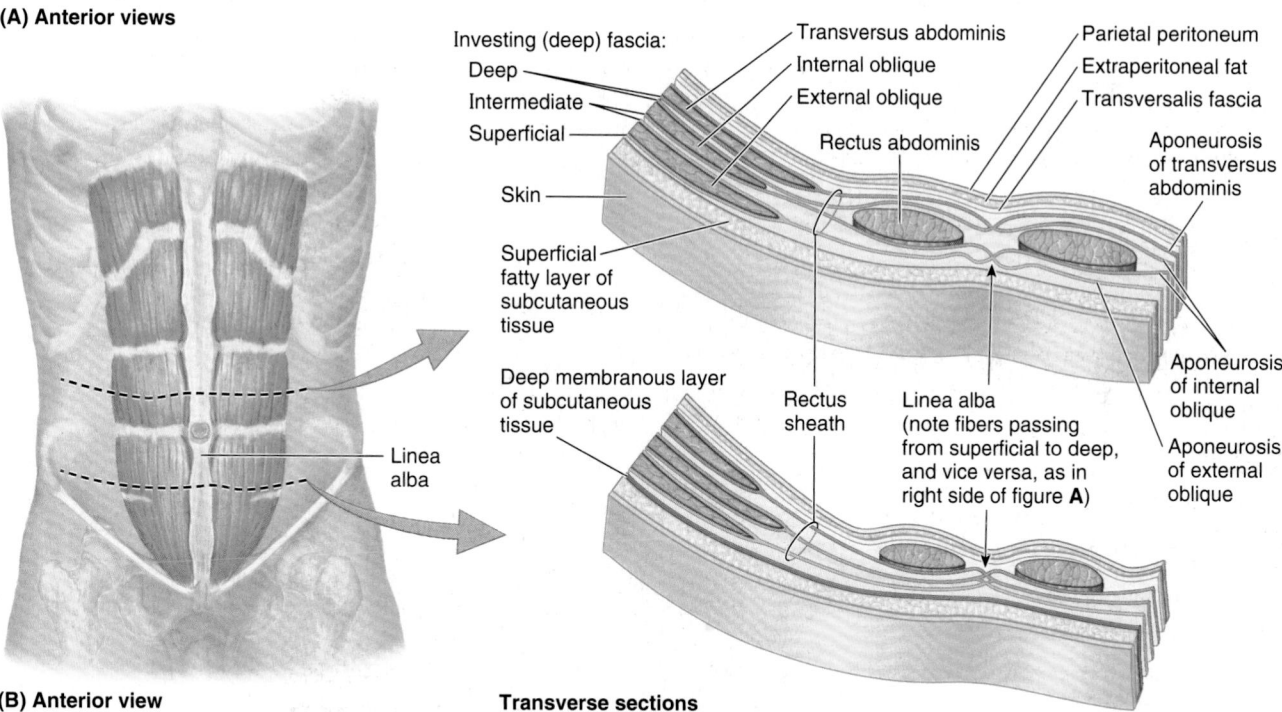

(B) Anterior view

Transverse sections

FIGURE 2.6. Structure of anterolateral abdominal wall. A. Intramuscular and intermuscular fiber exchanges within the bilaminar aponeuroses of the external and internal oblique muscles are shown. **B.** Transverse sections of the wall superior and inferior to the umbilicus show the makeup of the rectus sheath.

inguinal ligament is continuous with the deep fascia of the thigh. The inguinal ligament is therefore not a free-standing structure, although—as a useful landmark—it is frequently depicted as such. It serves as a *retinaculum* (retaining band) for the muscular and neurovascular structures passing deep to it to enter the thigh. The inferior parts of the two deeper anterolateral abdominal muscles arise in relationship to the lateral portion of the inguinal ligament. The complex modifications and attachments of the inguinal ligament, and of the inferomedial portions

of the aponeuroses of the anterolateral abdominal wall muscles, are discussed in detail with the inguinal region (later in this chapter).

INTERNAL OBLIQUE MUSCLE

The intermediate of the three flat abdominal muscles, the **internal oblique** is a thin muscular sheet that fans out anteromedially (Figs. 2.5B, 2.8, and 2.9A). Except for its lowermost fibers, which arise from the lateral half of the

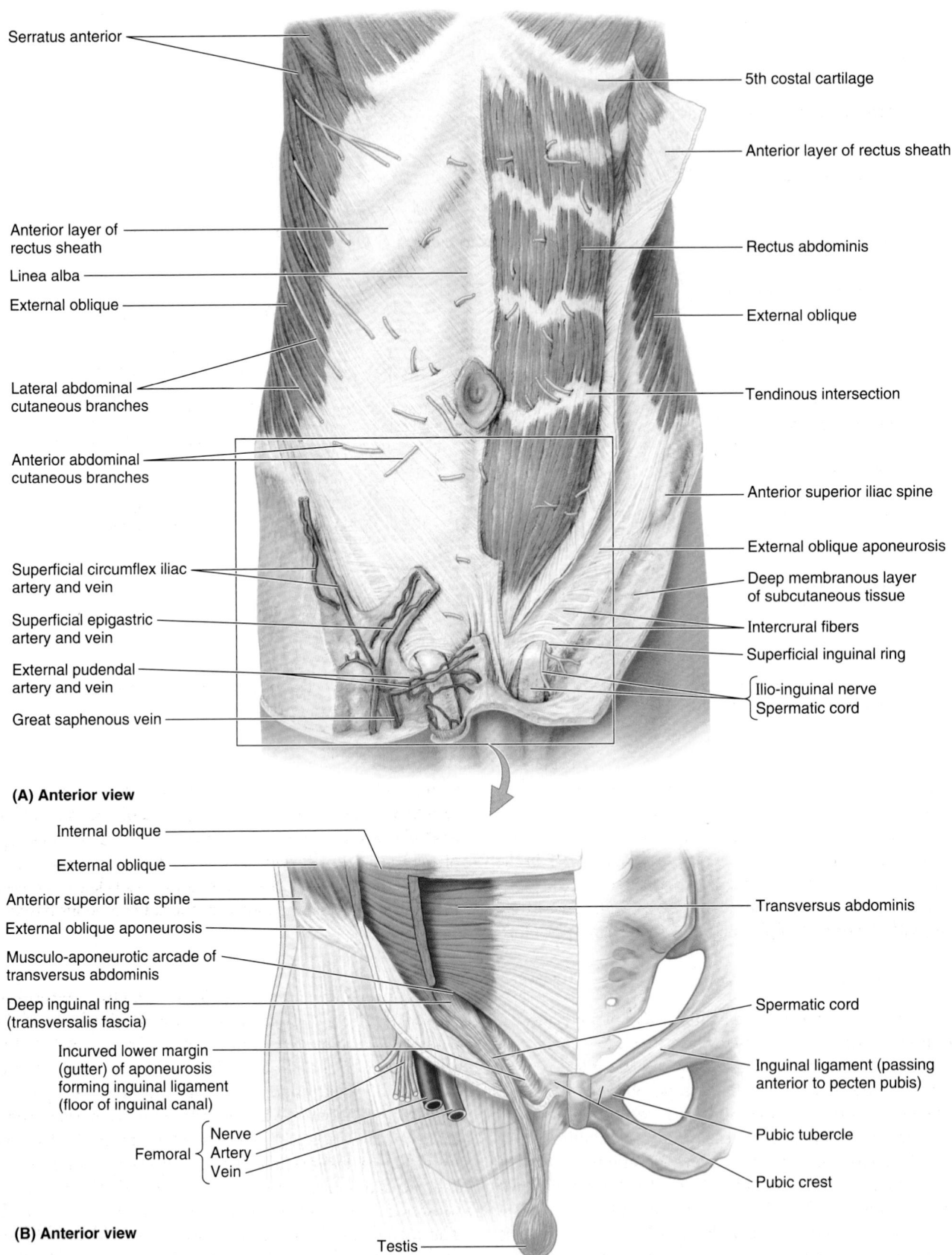

Serratus anterior

5th costal cartilage

Anterior layer of rectus sheath

Anterior layer of
rectus sheath

Rectus abdominis

Linea alba

External oblique

External oblique

Tendinous intersection

Lateral abdominal
cutaneous branches

Anterior abdominal
cutaneous branches

Anterior superior iliac spine

External oblique aponeurosis

Superficial circumflex iliac
artery and vein

Deep membranous layer
of subcutaneous tissue

Superficial epigastric
artery and vein

Intercrural fibers

Superficial inguinal ring

External pudendal
artery and vein

Ilio-inguinal nerve
Spermatic cord

Great saphenous vein

(A) Anterior view

Internal oblique

External oblique

Transversus abdominis

Anterior superior iliac spine

External oblique aponeurosis

Musculo-aponeurotic arcade of
transversus abdominis

Deep inguinal ring
(transversalis fascia)

Spermatic cord

Incurved lower margin
(gutter) of aponeurosis
forming inguinal ligament
(floor of inguinal canal)

Inguinal ligament (passing
anterior to pecten pubis)

Nerve
Femoral ⟨ Artery
Vein

Pubic tubercle

Pubic crest

(B) Anterior view

Testis

FIGURE 2.7. Anterolateral abdominal wall. A. In this superficial dissection, the anterior layer of the rectus sheath is reflected on the left side. Observe the anterior cutaneous nerves (T7–T12) piercing the rectus abdominis and the anterior layer of the rectus sheath. **B.** The three flat abdominal muscles and the formation of the inguinal ligament are demonstrated.

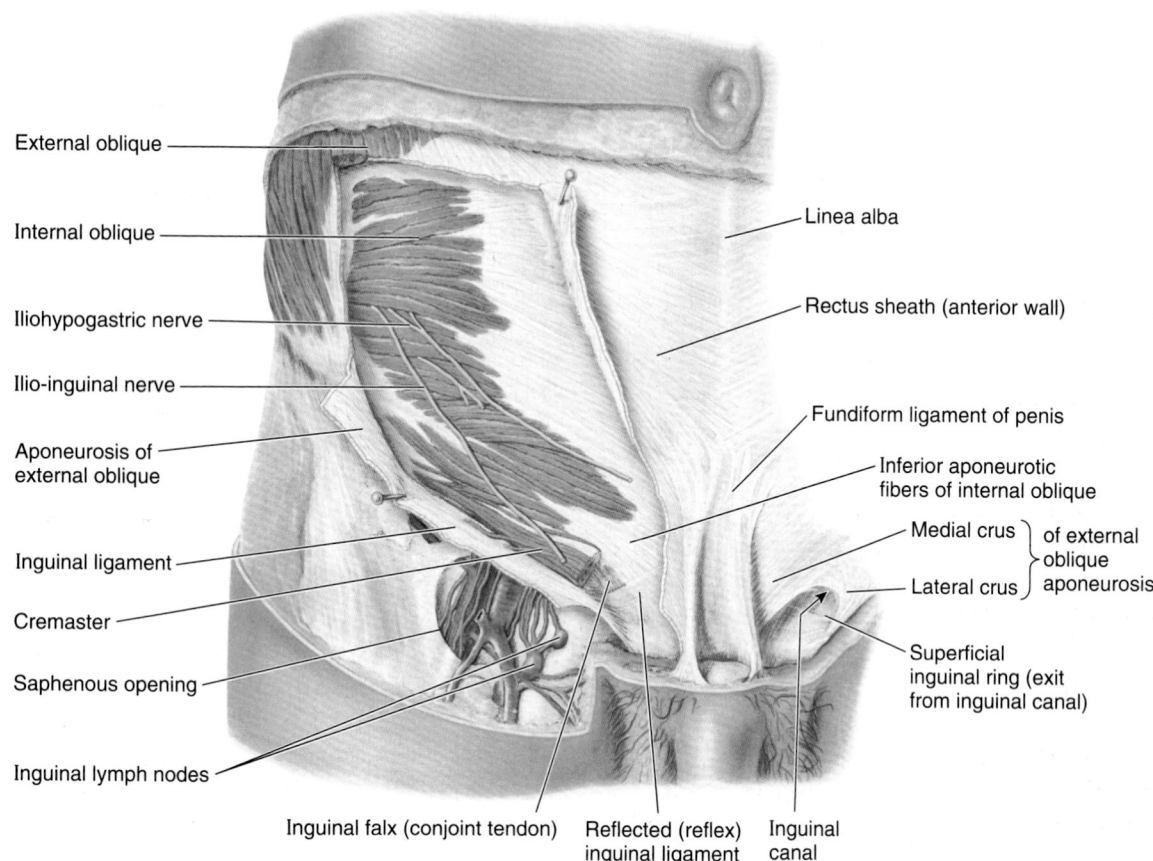

External oblique

Internal oblique

Iliohypogastric nerve

Ilio-inguinal nerve

Aponeurosis of
external oblique

Inguinal ligament

Cremaster

Saphenous opening

Inguinal lymph nodes

Linea alba

Rectus sheath (anterior wall)

Fundiform ligament of penis

Inferior aponeurotic
fibers of internal oblique

Medial crus ⎫ of external
⎬ oblique
Lateral crus ⎭ aponeurosis

Superficial
inguinal ring (exit
from inguinal canal)

Inguinal falx (conjoint tendon)   Reflected (reflex)   Inguinal
inguinal ligament   canal

FIGURE 2.8.   Inferior abdominal wall and inguinal region of a male. The aponeurosis of the external oblique is partly cut away, and the spermatic cord has been cut and removed from the inguinal canal.

inguinal ligament, its fleshy fibers run perpendicular to those of the external oblique, running superomedially (like your fingers when the hand is placed over your chest). Its fibers also become aponeurotic at the MCL and participate in the formation of the rectus sheath. The attachments of the internal oblique are demonstrated in Figure 2.5B, and listed, along with the nerve supply and main actions, in Table 2.2.

TRANSVERSUS ABDOMINIS MUSCLE

The fibers of the **transversus abdominis,** the innermost of the three flat abdominal muscles (see Figs. 2.5C and 2.7B), run more or less transversally, except for the inferior ones, which run parallel to those of the internal oblique. This transverse, circumferential orientation is ideal for compressing the abdominal contents, increasing intra-abdominal pressure. The fibers of the transversus abdominis muscle also end in an aponeurosis, which contributes to the formation of the rectus sheath (Fig. 2.9). The attachments of the transversus abdominis are demonstrated in Figure 2.5C, and listed, along with the nerve supply and main actions, in Table 2.2.

Between the internal oblique and the transversus abdominis muscles is a *neurovascular plane,* which corresponds with a similar plane in the intercostal spaces. In both regions, the

plane lies between the middle and deepest layers of muscle (Fig. 2.9A). The **neurovascular plane of the anterolateral abdominal wall** contains the nerves and arteries supplying the anterolateral abdominal wall. In the anterior part of the abdominal wall, the nerves and vessels leave the neurovascular plane and lie mostly in the subcutaneous tissue.

RECTUS ABDOMINIS MUSCLE

A long, broad, strap-like muscle, the **rectus abdominis** (L. *rectus,* straight) is the principal vertical muscle of the anterior abdominal wall (Figs. 2.5D, 2.6A, and 2.8B). The attachments of the rectus abdominis are demonstrated in Figure 2.5D, and listed, along with the nerve supply and main actions, in Table 2.2. The paired rectus muscles, separated by the linea alba, lie close together inferiorly. The rectus abdominis is three times as wide superiorly as inferiorly; it is broad and thin superiorly and narrow and thick inferiorly. Most of the rectus abdominis is enclosed in the rectus sheath. The rectus muscle is anchored transversely by attachment to the anterior layer of the rectus sheath at three or more **tendinous intersections** (see Figs. 2.5D and 2.7A). When tensed in muscular people, the areas of muscle between the tendinous intersections bulge outward. The intersections,

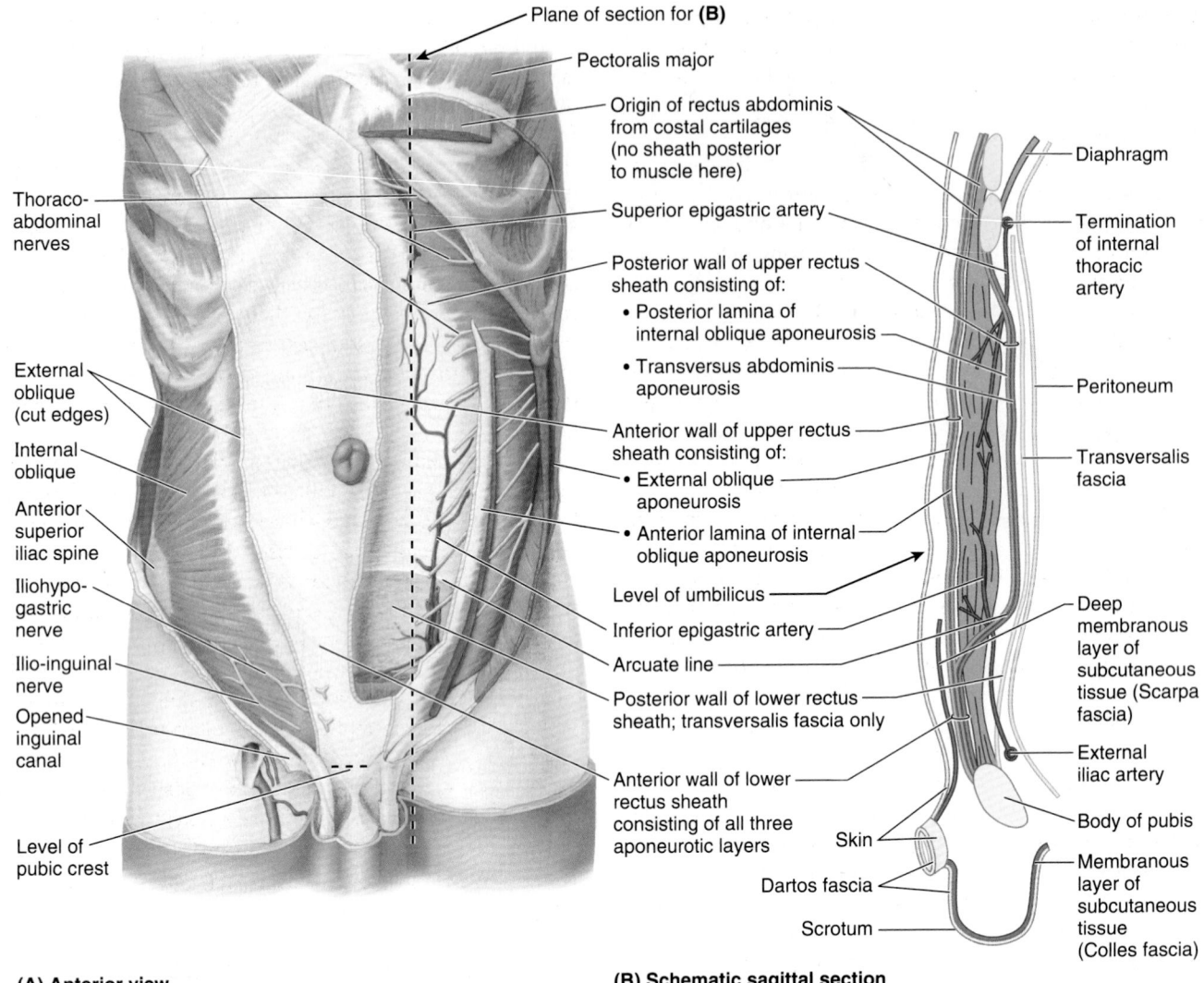

Plane of section for **(B)**

Pectoralis major

Origin of rectus abdominis from costal cartilages (no sheath posterior to muscle here)

Superior epigastric artery

Posterior wall of upper rectus sheath consisting of:
- Posterior lamina of internal oblique aponeurosis
- Transversus abdominis aponeurosis

Anterior wall of upper rectus sheath consisting of:
- External oblique aponeurosis
- Anterior lamina of internal oblique aponeurosis

Level of umbilicus

Inferior epigastric artery

Arcuate line

Posterior wall of lower rectus sheath; transversalis fascia only

Anterior wall of lower rectus sheath consisting of all three aponeurotic layers

Thoraco-abdominal nerves

External oblique (cut edges)

Internal oblique

Anterior superior iliac spine

Iliohypo-gastric nerve

Ilio-inguinal nerve

Opened inguinal canal

Level of pubic crest

Diaphragm

Termination of internal thoracic artery

Peritoneum

Transversalis fascia

Deep membranous layer of subcutaneous tissue (Scarpa fascia)

External iliac artery

Body of pubis

Skin

Dartos fascia

Scrotum

Membranous layer of subcutaneous tissue (Colles fascia)

(A) Anterior view

(B) Schematic sagittal section

FIGURE 2.9. Formation of rectus sheath and neurovascular structures of anterolateral abdominal wall. A. In this deep dissection, the fleshy portion of the external oblique is excised on the right side, but its aponeurosis and the anterior wall of the rectus sheath are intact. The anterior wall of the sheath and the rectus abdominis are removed on the left side so that the posterior wall of the sheath may be seen. Lateral to the left rectus sheath, the fleshy part of the internal oblique has been cut longitudinally; the edges of the cut are retracted to reveal the thoraco-abdominal nerves coursing in the neurovascular plane between the internal oblique and the transversus abdominis. **B.** Sagittal section through the rectus sheath of the anterior abdominal wall.

indicated by grooves in the skin between the muscular bulges, usually occur at the level of the xiphoid process, umbilicus, and halfway between these structures.

PYRAMIDALIS

The **pyramidalis** is a small, insignificant triangular muscle that is absent in approximately 20% of people. It lies anterior to the inferior part of the rectus abdominis and attaches to the anterior surface of the pubis and the anterior pubic ligament. It ends in the linea alba, which is especially thickened for a variable distance superior to the pubic symphysis. The pyramidalis tenses the linea alba. When present, surgeons use the attachment of the pyramidalis to the linea alba as a landmark for median abdominal incision (Skandalakis et al., 2009).

RECTUS SHEATH, LINEA ALBA, AND UMBILICAL RING

The **rectus sheath** is the strong, incomplete fibrous compartment of the rectus abdominis and pyramidalis muscles (Figs. 2.7–2.9). Also found in the rectus sheath are the superior and inferior epigastric arteries and veins, lymphatic vessels, and distal portions of the thoraco-abdominal nerves (abdominal portions of the anterior rami of spinal nerves T7–T12).

The rectus sheath is formed by the decussation and interweaving of the aponeuroses of the flat abdominal muscles (Fig. 2.6B). The external oblique aponeurosis contributes to the anterior wall of the sheath throughout its length. The superior two thirds of the internal oblique aponeurosis splits into two layers (laminae) at the lateral border of the rectus abdominis; one lamina passing anterior to the muscle and the other passing posterior to it. The anterior lamina joins the aponeurosis of the external oblique to form the anterior layer of the rectus sheath. The posterior lamina joins the aponeurosis of the transversus abdominis to form the posterior layer of the rectus sheath.

Beginning approximately one third of the distance from the umbilicus to the pubic crest, the aponeuroses of the three flat muscles pass anterior to the rectus abdominis to form the anterior layer of the rectus sheath, leaving only the relatively thin transversalis fascia to cover the rectus abdominis posteriorly. A crescentic **arcuate line** (Fig. 2.9) demarcates the transition between the aponeurotic posterior wall of the sheath covering the superior three quarters of the rectus and the transversalis fascia covering the inferior quarter. Throughout the length of the sheath, the fibers of the anterior and posterior layers of the sheath interlace in the anterior median line to form the complex *linea alba*.

The posterior layer of the rectus sheath is also deficient superior to the costal margin because the transversus abdominis is continued superiorly as the transversus thoracis, which lies internal to the costal cartilages (see Fig. 1.14, p. 89), and the internal oblique attaches to the costal margin. Hence, superior to the costal margin, the rectus abdominis lies directly on the thoracic wall (Fig. 2.9B).

The linea alba, running vertically the length of the anterior abdominal wall and separating the bilateral rectus sheaths (Fig. 2.7A), narrows inferior to the umbilicus to the width of the pubic symphysis and widens superiorly to the width of the xiphoid process. The linea alba transmits small vessels and nerves to the skin. In thin muscular people, a groove is visible in the skin overlying the linea alba. At its middle, underlying the umbilicus, the linea alba contains the **umbilical ring,** a defect in the linea alba through which the fetal umbilical vessels passed to and from the umbilical cord and placenta. All layers of the anterolateral abdominal wall fuse at the umbilicus. As fat accumulates in the subcutaneous tissue postnatally, the skin becomes raised around the umbilical ring and the umbilicus becomes depressed. This occurs 7–14 days after birth, when the atrophic umbilical cord "falls off."

FUNCTIONS AND ACTIONS OF ANTEROLATERAL ABDOMINAL MUSCLES

The muscles of the anterolateral abdominal wall:

- Form a strong expandable support for the anterolateral abdominal wall.
- Support the abdominal viscera and protect them from most injuries.
- Compress the abdominal contents to maintain or increase the intra-abdominal pressure and, in so doing, oppose the diaphragm (increased intra-abdominal pressure facilitates expulsion).
- Move the trunk and help to maintain posture.

The oblique and transverse muscles, acting together bilaterally, form a muscular girdle that exerts firm pressure on the abdominal viscera. The rectus abdominis participates little, if at all, in this action. Compressing the abdominal viscera and increasing intra-abdominal pressure elevates the relaxed diaphragm to expel air during respiration, and more forcibly for coughing, sneezing, nose blowing, voluntary eructation (burping), and yelling or screaming. When the diaphragm contracts during inspiration, the anterolateral abdominal wall expands as its muscles relax to make room for the organs, such as the liver, that are pushed inferiorly. The combined actions of the anterolateral muscles also produce the force required for defecation (discharge of feces), micturition (urination), vomiting, and parturition (childbirth). Increased intra-abdominal (and intrathoracic) pressure is also involved in heavy lifting, the resulting force sometimes producing a hernia.

The anterolateral abdominal muscles are also involved in movements of the trunk at the lumbar vertebrae and in controlling the tilt of the pelvis when standing for maintenance of posture (resisting lumbar lordosis). Consequently, strengthening the anterolateral abdominal wall musculature improves standing and sitting posture. The rectus abdominis is a powerful flexor of the thoracic and especially lumbar regions of the vertebral column, pulling the anterior costal margin and pubic crest toward each other. The oblique abdominal muscles also assist in movements of the trunk, especially lateral flexion and rotation of the lumbar and lower thoracic vertebral column. The transversus abdominis probably has no appreciable effect on the vertebral column (Standring, 2008).

Neurovasculature of Anterolateral Abdominal Wall

DERMATOMES OF ANTEROLATERAL ABDOMINAL WALL

The map of dermatomes of the anterolateral abdominal wall is almost identical to the map of peripheral nerve distribution (Fig. 2.10). This is because the anterior rami of spinal nerves T7–T12, which supply most of the abdominal wall, do not participate in plexus formation. The exception occurs at the L1 level, where the L1 anterior ramus bifurcates into two named peripheral nerves. Each dermatome begins posteriorly overlying the intervertebral foramen by which the spinal nerve exits the vertebral column and follows the slope of the ribs around the trunk. Dermatome T10 includes the umbilicus, whereas dermatome L1 includes the inguinal fold.

NERVES OF ANTEROLATERAL ABDOMINAL WALL

The skin and muscles of the anterolateral abdominal wall are supplied mainly by the following nerves (Fig. 2.9A and 2.10; Table 2.3):

- **Thoraco-abdominal nerves:** these are the distal, abdominal parts of the anterior rami of the inferior six thoracic spinal nerves (T7–T11); they are the former inferior intercostal nerves distal to the costal margin.

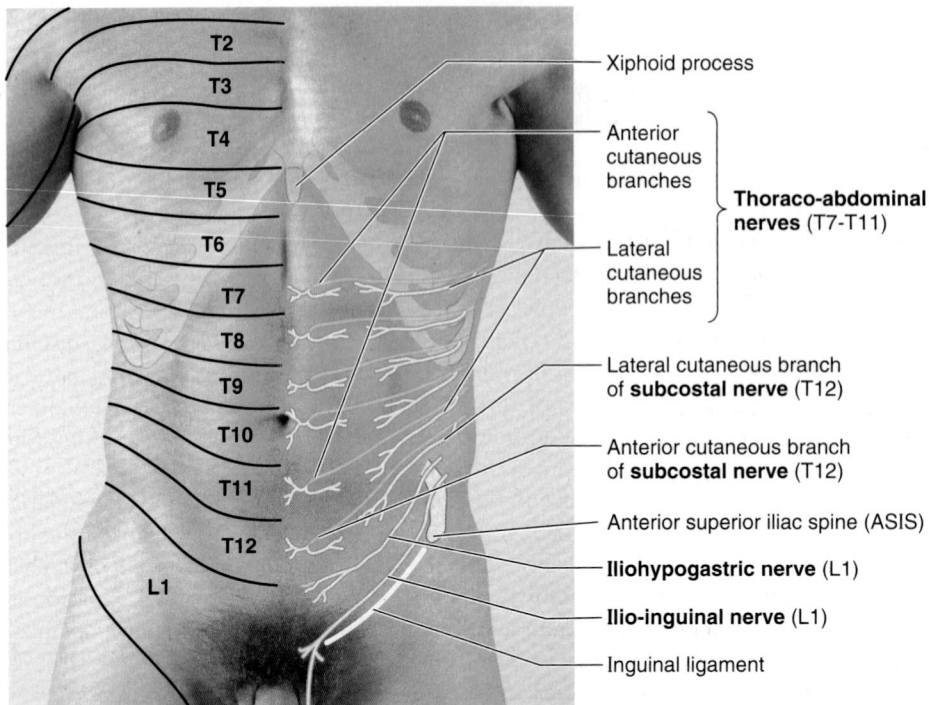

FIGURE 2.10. Dermatomes and nerves of anterolateral abdominal wall.

TABLE 2.3. NERVES OF ANTEROLATERAL ABDOMINAL WALL

Nerves	Origin	Course	Distribution
Thoraco-abdominal (T7–T11)	Continuation of lower (7th–11th) intercostal nerves distal to costal margin	Run between second and third layers of abdominal muscles; branches enter subcutaneous tissue as lateral cutaneous branches of T10–T11 (in anterior axillary line) and anterior cutaneous branches of T7–T11 (parasternal line)	Muscles of anterolateral abdominal wall and overlying skin
7th–9th lateral cutaneous branches	7th–9th intercostal nerves (anterior rami of spinal nerves T7–T9)	Anterior divisions continue across costal margin in subcutaneous tissue	Skin of right and left hypochondriac regions
Subcostal (anterior ramus of T12)	Spinal nerve T12	Runs along inferior border of 12th rib; then passes onto subumbilical abdominal wall between second and third layers of abdominal muscles	Muscles of anterolateral abdominal wall (including most inferior slip of external oblique) and overlying skin, superior to iliac crest and inferior to umbilicus
Iliohypogastric (L1)	As superior terminal branch of anterior ramus of spinal nerve L1	Pierces transversus abdominis muscle to course between second and third layers of abdominal muscles; branches pierce external oblique aponeuroses of most inferior abdominal wall	Skin overlying iliac crest, upper inguinal, and hypogastric regions; internal oblique and transversus abdominis muscles
Ilio-inguinal (L1)	As inferior terminal branch of anterior ramus of spinal nerve L1	Passes between second and third layers of abdominal muscles; then traverses inguinal canal	Skin of lower inguinal region, mons pubis, anterior scrotum or labium majus, and adjacent medial thigh; inferiormost internal oblique and transversus abdominis

- **Lateral (thoracic) cutaneous branches:** of the thoracic spinal nerves T7–T9 or T10.
- **Subcostal nerve:** the large anterior ramus of spinal nerve T12.
- **Iliohypogastric** and **ilio-inguinal nerves:** terminal branches of the anterior ramus of spinal nerve L1.

The *thoraco-abdominal nerves* pass inferoanteriorly from the intercostal spaces and run in the neurovascular plane between the internal oblique and the transversus abdominis muscles to supply the abdominal skin and muscles. The *lateral cutaneous branches* emerge from the musculature of the anterolateral wall to enter the subcutaneous tissue along the anterior axillary line (as anterior and posterior divisions), whereas the anterior abdominal cutaneous branches pierce the rectus sheath to enter the subcutaneous tissue a short distance from the median plane. *Anterior abdominal cutaneous branches of thoraco-abdominal nerves* (Fig. 2.10; Table 2.3):

- *T7–T9* supply the skin superior to the umbilicus.
- *T10* supplies the skin around the umbilicus.
- *T11,* plus the cutaneous branches of the subcostal (T12), iliohypogastric, and ilio-inguinal (L1), supply the skin inferior to the umbilicus.

During their course through the anterolateral abdominal wall, the thoraco-abdominal, subcostal, and iliohypogastric nerves communicate with each other.

VESSELS OF ANTEROLATERAL ABDOMINAL WALL

The skin and subcutaneous tissue of the abdominal wall are served by an intricate subcutaneous venous plexus, draining superiorly to the internal thoracic vein medially and the lateral thoracic vein laterally and inferiorly to the superficial and inferior epigastric veins, tributaries of the femoral and external iliac veins, respectively (Fig. 2.11). Cutaneous veins surrounding the umbilicus anastomose with *para-umbilical veins,* small tributaries of the *hepatic portal vein* that parallel the *obliterated umbilical vein* (round ligament of the liver). A relatively direct lateral superficial anastomotic channel, the **thoraco-epigastric vein,** may exist or develop (as a result of

altered venous flow) between the *superficial epigastric vein* (a femoral vein tributary) and the *lateral thoracic vein* (an axillary vein tributary). The deeper veins of the anterolateral abdominal wall accompany the arteries, bearing the same name. A deeper, medial venous anastomosis may exist or develop between the *inferior epigastric vein* (an external iliac vein tributary) and the *superior epigastric/internal thoracic veins* (subclavian vein tributaries). The superficial and deep anastomoses may afford collateral circulation during blockage of either vena cava.

The primary blood vessels (arteries and veins) of the anterolateral abdominal wall are the

- *Superior epigastric vessels* and branches of the *musculophrenic vessels* from the internal thoracic vessels.
- *Inferior epigastric* and *deep circumflex iliac vessels* from the external iliac vessels.
- *Superficial circumflex iliac* and *superficial epigastric vessels* from the femoral artery and greater saphenous vein, respectively.
- *Posterior intercostal vessels* of the 11th intercostal space and the anterior branches of *subcostal vessels.*

The *arterial supply to the anterolateral abdominal wall* is demonstrated in Figure 2.12 and summarized in Table 2.4. The distribution of the deep abdominal blood vessels reflects the arrangement of the muscles: The vessels of the anterolateral abdominal wall have an oblique, circumferential pattern (similar to the intercostal vessels; Fig. 2.11), whereas the vessels of the central anterior abdominal wall are oriented more vertically.

The **superior epigastric artery** is the direct continuation of the *internal thoracic artery*. It enters the rectus sheath superiorly through its posterior layer and supplies the superior part of the rectus abdominis and anastomoses with the inferior epigastric artery approximately in the umbilical region (see Fig. 2.9, Table 2.4)

The **inferior epigastric artery** arises from the *external iliac artery* just superior to the inguinal ligament. It runs superiorly in the transversalis fascia to enter the rectus sheath below the arcuate line. It enters the lower rectus abdominis and anastomoses with the superior epigastric artery (Fig. 2.9).

Lymphatic drainage of the anterolateral abdominal wall follows the following patterns (Fig. 2.11):

- *Superficial lymphatic vessels* accompany the subcutaneous veins; those superior to the transumbilical plane drain mainly to the **axillary lymph nodes;** however, a few drain to the **parasternal lymph nodes.** Superficial lymphatic vessels inferior to the transumbilical plane drain to the **superficial inguinal lymph nodes.**
- *Deep lymphatic vessels* accompany the deep veins of the abdominal wall and drain to the *external iliac, common iliac,* and *right* and *left lumbar (caval* and *aortic) lymph nodes.*

For an overview of superficial and deep lymphatic drainage, see the Introduction to Clinically Oriented Anatomy.

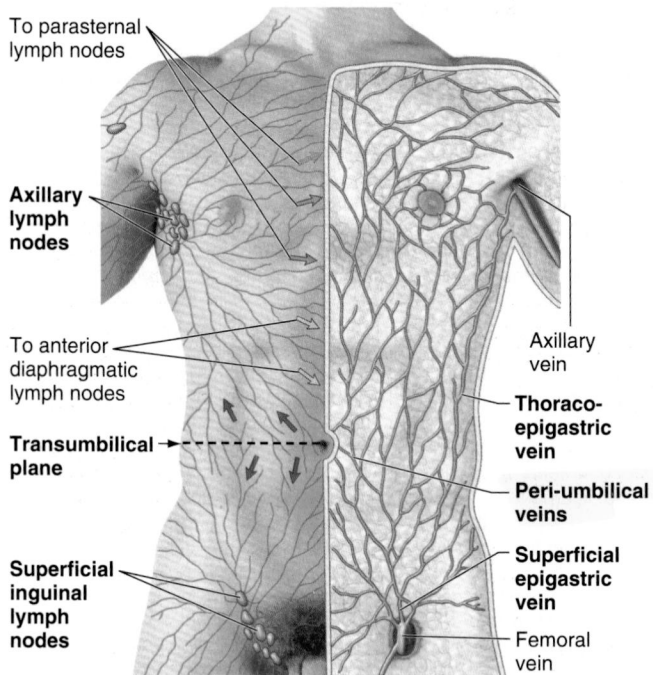

Anterior view

FIGURE 2.11. Lymphatics and superficial veins of anterolateral abdominal wall.

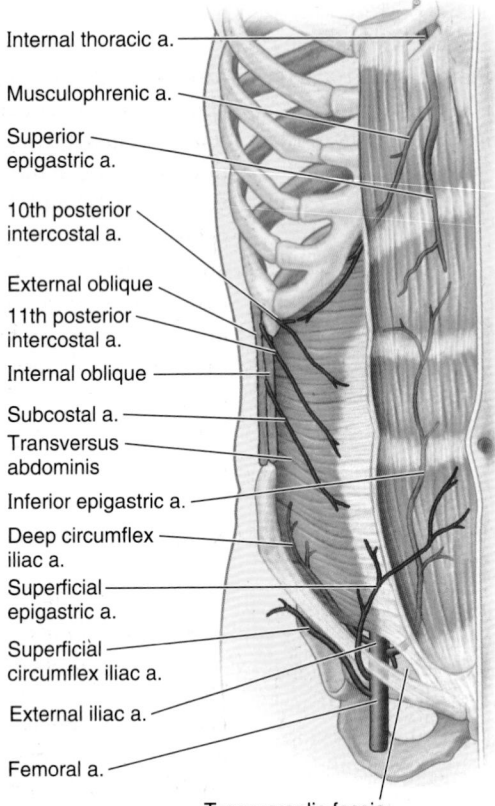

Internal thoracic a.

Musculophrenic a.

Superior
epigastric a.

10th posterior
intercostal a.

External oblique

11th posterior
intercostal a.

Internal oblique

Subcostal a.

Transversus
abdominis

Inferior epigastric a.

Deep circumflex
iliac a.

Superficial
epigastric a.

Superficial
circumflex iliac a.

External iliac a.

Femoral a.

Transversalis fascia

FIGURE 2.12. Arteries of anterolateral abdominal wall.

TABLE 2.4. ARTERIES OF ANTEROLATERAL ABDOMINAL WALL

Artery	Origin	Course	Distribution
Musculophrenic	Internal thoracic artery	Descends along costal margin	Superficial and deep abdominal wall of hypochondriac region; anterolateral diaphragm
Superior epigastric		Descends in rectus sheath deep to rectus abdominis	Rectus abdominis muscle; superficial and deep abdominal wall of epigastric and upper umbilical regions
10th and 11th posterior intercostal arteries	Aorta	Arteries continue beyond ribs to descend in abdominal wall between internal oblique and transversus abdominis muscles	Superficial and deep abdominal wall of lateral (lumbar or flank) region
Subcostal artery			
Inferior epigastric	External iliac artery	Runs superiorly and enters rectus sheath; runs deep to rectus abdominis	Rectus abdominis muscle; deep abdominal wall of pubic and inferior umbilical regions
Deep circumflex iliac		Runs on deep aspect of anterior abdominal wall, parallel to inguinal ligament	Iliacus muscle and deep abdominal wall of inguinal region; iliac fossa
Superficial circumflex iliac	Femoral artery	Runs in subcutaneous tissue along inguinal ligament	Superficial abdominal wall of inguinal region and adjacent anterior thigh
Superficial epigastric		Runs in subcutaneous tissue toward umbilicus	Superficial abdominal wall of pubic and inferior umbilic regions

FASCIA AND MUSCLES OF ANTEROLATERAL ABDOMINAL WALL

Clinical Significance of Fascia and Fascial Spaces of Abdominal Wall

 Liposuction is a surgical method for removing unwanted subcutaneous fat using a percutaneously placed suction tube and high vacuum pressure. The tubes are inserted subdermally through small skin incisions.

When closing abdominal skin incisions inferior to the umbilicus, surgeons include the membranous layer of subcutaneous tissue when suturing because of its strength. Between this layer and the deep fascia covering the rectus abdominis and external oblique muscles is a potential space where fluid may accumulate (e.g., urine from a ruptured urethra). Although there are no barriers (other than gravity) to prevent fluid from spreading superiorly from this space, it cannot spread inferiorly into the thigh because the deep membranous layer of subcutaneous tissue fuses with the deep fascia of the thigh (*fascia lata*) along a line approximately 2.5 cm inferior and parallel to the inguinal ligament.

The endoabdominal fascia is of special importance in surgery. It provides a plane that can be opened, enabling the surgeon to approach structures on or in the anterior aspect of the posterior abdominal wall, such as the kidneys or bodies of lumbar vertebrae, without entering the membranous peritoneal sac containing the abdominal viscera. Thus, the risk of contamination is minimized. An anterolateral part of this potential space between the transversalis fascia and the parietal peritoneum (*space of Bogros*) is used for placing prostheses (Gore-Tex mesh, for example) when repairing inguinal hernias (Skandalakis et al., 1996) (see Fig. 2.15A & B).

Protuberance of Abdomen

 A prominent abdomen is normal in infants and young children because their gastrointestinal tracts contain considerable amounts of air. In addition, their anterolateral abdominal cavities are enlarging and their abdominal muscles are gaining strength. An infant's and young child's relatively large liver also accounts for some bulging.

Abdominal muscles protect and support the viscera most effectively when they are well toned; thus the well-conditioned adult of normal weight has a flat or scaphoid (lit. boat shaped; i.e., hollowed or concave) abdomen when in the supine position.

The six common causes of *abdominal protrusion* begin with the letter *F:* food, fluid, fat, feces, flatus, and fetus. Eversion of the umbilicus may be a sign of increased intra-abdominal pressure, usually resulting from ascites (abnormal accumulation of serous fluid in the peritoneal cavity), or a large mass (e.g., a tumor, a fetus, or an enlarged organ such as the liver).

Excess fat accumulation owing to overnourishment most commonly involves the subcutaneous fatty layer; however, there may also be excessive depositions of extraperitoneal fat in some types of obesity.

Tumors and organomegaly (organ enlargement such as splenomegaly—enlargement of the spleen) also produce abdominal enlargement. When the anterior abdominal muscles are underdeveloped or become atrophic, as a result of old age or insufficient exercise, they provide insufficient tonus to resist the increased weight of a protuberant abdomen on the anterior pelvis. The pelvis tilts anteriorly at the hip joints when standing (the pubis descends and the sacrum ascends) producing excessive *lordosis* (sway back) of the lumbar region.

Abdominal Hernias

 The anterolateral abdominal wall may be the site of *abdominal hernias*. Most hernias occur in the inguinal, umbilical, and epigastric regions (see the blue box "Inguinal Hernias," p. 212). *Umbilical hernias* are common in neonates because the anterior abdominal wall is relatively weak in the umbilical ring, especially in low-birth-weight infants. Umbilical hernias are usually small and result from increased intra-abdominal pressure in the presence of weakness and incomplete closure of the anterior abdominal wall after ligation of the umbilical cord at birth. Herniation occurs through the umbilical ring. *Acquired umbilical hernias* occur most commonly in women and obese people. Extraperitoneal fat and/or peritoneum protrude into the hernial sac. The lines along which the fibers of the abdominal aponeuroses interlace are also potential sites of herniation (see Fig. 2.6B). Occasionally, gaps exist where these fiber exchanges occur—for example, in the midline or in the transition from aponeurosis to rectus sheath. These gaps may be congenital, the result of the stresses of obesity and aging, or the consequence of surgical or traumatic wounds.

An *epigastric hernia*, a hernia in the epigastric region through the linea alba, occurs in the midline between the xiphoid process and the umbilicus. *Spigelian hernias* are those occurring along the semilunar lines (see Table 2.1B). These types of hernia tend to occur in people older than 40 years and are usually associated with obesity. The hernial sac, composed of peritoneum, is covered with only skin and fatty subcutaneous tissue.

NEUROVASCULATURE OF ANTEROLATERAL ABDOMINAL WALL

Palpation of Anterolateral Abdominal Wall

 Warm hands are important when palpating the abdominal wall because cold hands make the anterolateral abdominal muscles tense, producing involuntary spasms of the muscles, known as *guarding*. Intense guarding, board-like reflexive muscular rigidity that cannot be willfully suppressed, occurs during palpation when an organ (such as the appendix) is inflamed and in itself constitutes a clinically significant sign of *acute abdomen*. The involuntary muscular spasms attempt to protect the viscera from pressure, which is painful when an abdominal infection is present. The common nerve supply of the skin and muscles of the wall explains why these spasms occur.

Palpation of abdominal viscera is performed with the patient in the supine position with thighs and knees semi-flexed to enable adequate relaxation of the anterolateral abdominal wall. Otherwise, the deep fascia of the thighs pulls on the membranous layer of abdominal subcutaneous tissue, tensing the abdominal wall. Some people tend to place their hands behind their heads when lying supine, which also tightens the muscles and makes the examination difficult. Placing the upper limbs at the sides and putting a pillow under the person's knees tends to relax the anterolateral abdominal muscles.

Superficial Abdominal Reflexes

 The abdominal wall is the only protection most of the abdominal organs have. Consequently, the wall will react if an organ is diseased or injured. With the person supine and the muscles relaxed, the *superficial abdominal reflex* is elicited by quickly stroking horizontally, lateral to medial, toward the umbilicus. Usually, contraction of the abdominal muscles is felt; this reflex may not be observed in obese people. Similarly, any injury to the abdominal skin results in a rapid reflex contraction of the abdominal muscles.

Injury to Nerves of Anterolateral Abdominal Wall

The inferior thoracic spinal nerves (T7–T12) and the iliohypogastric and ilio-inguinal nerves (L1) approach the abdominal musculature separately to provide the multi-segmental innervation of the abdominal muscles. Thus, they are distributed across the anterolateral abdominal wall, where they run oblique but mostly horizontal courses. They are susceptible to injury in surgical incisions or from trauma at any level of the abdominal wall. *Injury to nerves of the anterolateral abdominal wall* may result in weakening of the muscles. In the inguinal region, such a weakness may predispose an individual to development of an inguinal hernia (see the blue box "Inguinal Hernias" on p. 212).

Abdominal Surgical Incisions

 Surgeons use various *abdominal surgical incisions* to gain access to the abdominal cavity. When possible, the incisions follow the *cleavage lines* (Langer lines)

in the skin (see Introduction for discussion of these lines). The incision that allows adequate exposure, and secondarily, the best possible cosmetic effect, is chosen. The location of the incision also depends on the type of operation, the location of the organ(s) the surgeon wants to reach, bony or cartilaginous boundaries, avoidance of (especially motor) nerves, maintenance of blood supply, and minimizing injury to muscles and fascia of the abdominal wall while aiming for favorable healing. Thus, before making an incision, the surgeon considers the direction of the muscle fibers and the location of the aponeuroses and nerves. Consequently, a variety of incisions are routinely used, each having specific advantages and limitations.

Instead of transecting muscles, causing irreversible necrosis (death) of muscle fibers, the surgeon splits them in the direction of (and between) their fibers. The rectus abdominis is an exception; it can be transected because its muscle fibers run short distances between tendinous intersections, and the segmental nerves supplying it enter the lateral part of the rectus sheath where they can be located and preserved. Generally, incisions are made in the part of the anterolateral abdominal wall that gives the freest access to the targeted organ with the least disturbance to the nerve supply to the muscles. Muscles and viscera are retracted toward, not away from, their neurovascular supply.

Cutting a motor nerve paralyzes the muscle fibers supplied by it, thereby weakening the anterolateral abdominal wall. However, because of overlapping areas of innervation between nerves, one or two small branches of nerves may usually be cut without a noticeable loss of motor supply to the muscles or loss of sensation to the skin.

LONGITUDINAL INCISIONS
Longitudinal incisions, such as median and paramedian incisions (Fig. B2.1), are preferred for exploratory operations because they offer good exposure of and access to the viscera, and can be extended as necessary with minimal complication.

Median or *midline incisions* can be made rapidly without cutting muscle, major blood vessels, or nerves. Median incisions can be made along any part or the length of the linea alba from the xiphoid process to pubic symphysis. Because the linea alba transmits only small vessels and nerves to the skin, a midline incision is relatively bloodless, and avoids major nerves; however, incisions in some people may reveal abundant and well-vascularized fat. Conversely, because of its relatively poor blood supply, the linea alba may undergo necrosis and subsequent degeneration after incision if its edges are not aligned properly during closure.

Paramedian incisions (lateral to the median plane) are made in a sagittal plane and may extend from the costal margin to the pubic hairline. After the incision passes through the anterior layer of the rectus sheath, the muscle is freed and retracted laterally to prevent tension and injury to the vessels and nerves. The posterior layer of the rectus sheath and the peritoneum are then incised to enter the peritoneal cavity.

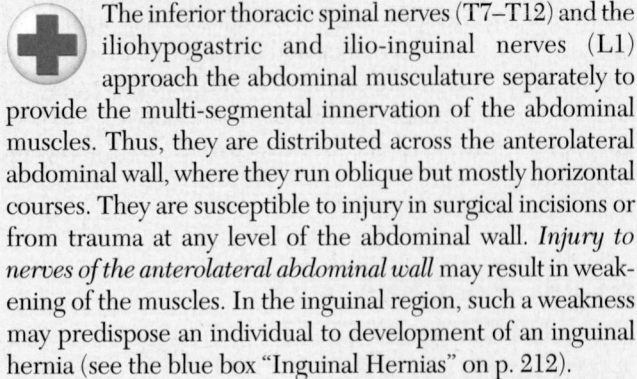

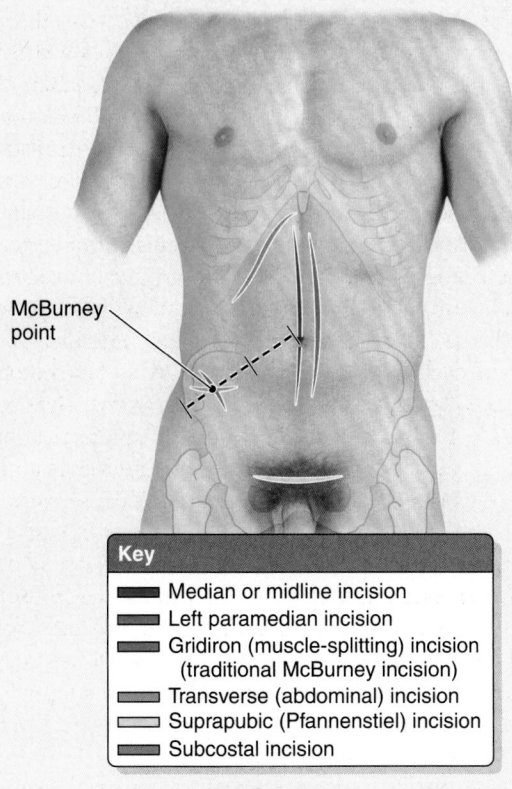

McBurney
point

Key
- ■ Median or midline incision
- ■ Left paramedian incision
- ■ Gridiron (muscle-splitting) incision
  (traditional McBurney incision)
- ■ Transverse (abdominal) incision
- ▢ Suprapubic (Pfannenstiel) incision
- ■ Subcostal incision

FIGURE B2.1.

OBLIQUE AND TRANSVERSE INCISIONS

The direction of *oblique* and *transverse incisions* is related to muscle fiber orientation, nearby hard tissue (costal margin or iliac or pubic crest), or minimizing potential nerve damage. *Gridiron (muscle-splitting) incisions* are often used for an appendectomy. The oblique *McBurney incision* is made at the *McBurney point*, approximately 2.5 cm superomedial to the ASIS on the *spino-umbilical line*. The external oblique aponeurosis is incised inferomedially in the direction of its fibers and retracted. The musculoaponeurotic fibers of the internal oblique and transversus abdominis are then split in the line of their fibers and retracted. The iliohypogastric nerve, running deep to the internal oblique, is identified and preserved. Carefully made, the entire exposure cuts no musculo-aponeurotic fibers; therefore, when the incision is closed, the muscle fibers move together and the abdominal wall is as strong after the operation as it was before.

Suprapubic (Pfannenstiel) incisions ("bikini" incisions) are made at the pubic hairline. These incisions—horizontal with a slight convexity—are used for most gynecological and obstetrical operations (e.g., for cesarean section). The linea alba and anterior layers of the rectus sheaths are transected and resected superiorly, and the rectus muscles are retracted laterally or divided through their tendinous parts allowing reattachment without muscle fiber injury.

The iliohypogastric and ilio-inguinal nerves are identified and preserved.

Transverse incisions through the anterior layer of the rectus sheath and rectus abdominis provide good access and cause the least possible damage to the nerve supply of the rectus abdominis. This muscle may be divided transversely without serious damage because a new transverse band forms when the muscle segments are rejoined. Transverse incisions are not made through the tendinous intersections because cutaneous nerves and branches of the superior epigastric vessels pierce these fibrous regions of the muscle. Transverse incisions can be increased laterally as needed to increase exposure but are not utilized for exploratory procedures because superior and inferior extension is difficult.

Subcostal incisions provide access to the gallbladder and biliary ducts on the right side and the spleen on the left. The incision is made parallel but at least 2.5 cm inferior to the costal margin to avoid the 7th and 8th thoracic spinal nerves (Table 2.3).

HIGH-RISK INCISIONS

High-risk incisions include pararectus and inguinal incisions. *Pararectus incisions* along the lateral border of the rectus sheath are undesirable because they may cut the nerve supply to the rectus abdominis. *Inguinal incisions* for repairing hernias may injure the ilio-inguinal nerve.

INCISIONAL HERNIA

An *incisional hernia* is a protrusion of *omentum* (a fold of peritoneum) or an organ through a surgical incision. If the muscular and aponeurotic layers of the abdomen do not heal properly, an incisional hernia can result.

MINIMALLY INVASIVE (ENDOSCOPIC) SURGERY

Many abdominopelvic surgical procedures (e.g., removal of the gallbladder) are now performed using an *endoscope*, in which tiny perforations of the abdominal wall allow the entry of remotely operated instruments, replacing the larger conventional incisions. Thus, the potential for nerve injury, incisional hernia, or contamination through the open wound and the time required for healing are minimized.

Reversal of Venous Flow and Collateral Pathways of Superficial Abdominal Veins

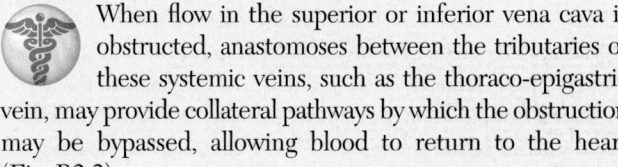 When flow in the superior or inferior vena cava is obstructed, anastomoses between the tributaries of these systemic veins, such as the thoraco-epigastric vein, may provide collateral pathways by which the obstruction may be bypassed, allowing blood to return to the heart (Fig. B2.2).

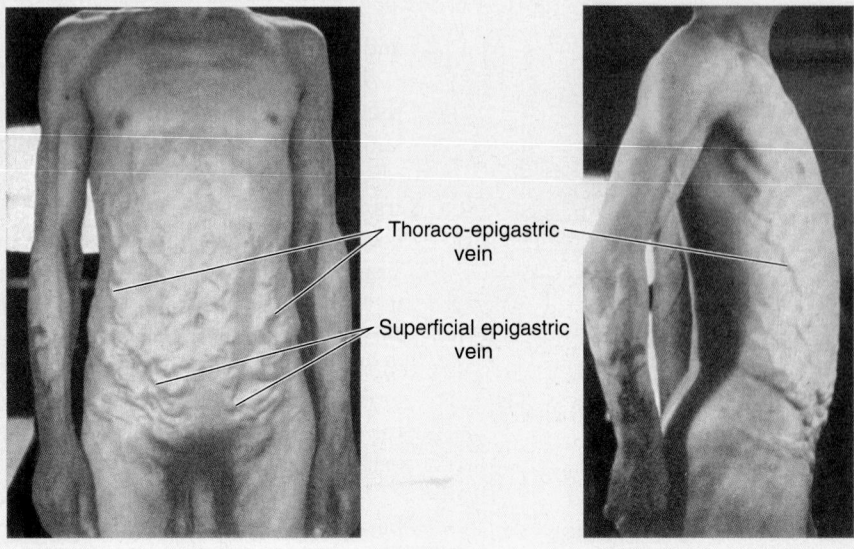

Thoraco-epigastric
vein

Superficial epigastric
vein

FIGURE B2.2.

The Bottom Line

FASCIA, MUSCLES, AND NEUROVASCULATURE OF ANTEROLATERAL ABDOMINAL WALL

Fascia: The fascia of the anterolateral abdominal wall consists of subcutaneous (superficial), investing (deep), and endoabdominal portions. ♦ The subcutaneous layer is modified inferior to the umbilicus to include a superficial fatty layer and a deep membranous layer. ♦ The superficial fatty layer is specialized here, particularly in males, for lipid storage, whereas the deep membranous layer is sufficiently complete to compartmentalize extravasated fluids (blood or urine) and allow placement of sutures during surgery. ♦ The investing layer is typical of deep fascias ensheathing voluntary muscles, and here reflects the trilaminar arrangement of the flat abdominal muscles and their aponeuroses. ♦ The endoabdominal fascia is of particular importance in surgery, enabling the establishment of an extraperitoneal space that allows anterior access to retroperitoneal structures (e.g., kidneys, ureters, and bodies of lumbar vertebra) without entering the peritoneal cavity.

Muscles: The anterolateral abdominal muscles consist of concentric, flat muscles located anterolaterally and vertical muscles placed anteriorly adjacent to the midline. ♦ A trilaminar arrangement of the flat muscles, like that in the thorax, also occurs here; however, other than their innervation by multiple but separate segmental nerves, the metamerism (segmentation) characteristic of the thoracic intercostal musculature is not apparent in the abdomen. ♦ The fleshy portions of the flat muscles become aponeurotic anteriorly. The fibers of the aponeuroses interlace in the midline, forming the linea alba, and continue into the aponeuroses of the contralateral muscles. ♦ The aponeurotic fibers of the external obliques are also continuous across the midline with those of the contralateral

internal oblique muscles. ♦ Three layers of flat, bilateral digastric muscles encircle the trunk, forming oblique and transverse girdles that enclose the abdominal cavity. ♦ In the upper two thirds of the abdominal wall, the aponeurotic layers separate on each side of the linea alba to form longitudinal sheaths that contain the rectus muscles. This brings them into a functional relationship with the flat muscles in which the vertical muscles brace the girdles anteriorly. ♦ In the lower third of the anterior abdominal wall, the aponeuroses of all three layers of flat muscles pass anterior to the rectus muscles (recti). ♦ As flexors of the trunk, the recti are the antagonistic partners of the deep (extensor) muscles of the back. Balance in the development and tonus of these partners affects posture (and thus weakness of the abdominal muscles may result in excessive lumbar lordosis—an abnormally convex curvature of the lower vertebral column). ♦ The special arrangements of the anterolateral abdominal muscles enable them to provide flexible containing walls for the abdominal contents, to increase intra-abdominal pressure or decrease abdominal volume for expulsion, and to produce anterior and lateral flexion and torsional (rotatory) movements of the trunk.

Nerves: The anterolateral abdominal muscles receive multi-segmental innervation via the anterior rami of lower thoracic (T7–T12) and the L1 spinal nerves. ♦ The rami pass separately to the muscles as five thoraco-abdominal nerves (T7–T11), a subcostal nerve (T12), and iliohypogastric and ilio-inguinal nerves (L1) that course in a plane between the second and third layers. ♦ Lateral cutaneous branches supply the overlying abdominal skin lateral to the MCL. ♦ Anterior cutaneous branches supply skin medial to the MCL. ◆ Except for L1, the maps of the abdominal dermatomes and of the peripheral nerves are thus identical. ◆ Landmark dermatomes are dermatome T10, which includes the umbilicus, and dermatome L1, which includes the inguinal fold.

Vessels: ◆ The skin and subcutaneous tissue of the abdominal wall drain superiorly (ultimately to the superior vena caval system) via the internal thoracic vein medially and the lateral thoracic vein laterally, and inferiorly (ultimately to the inferior vena caval system) via the superficial and inferior epigastric veins. ◆ Cutaneous veins surrounding the umbilicus anastomose with small tributaries of the hepatic portal vein.
◆ The distribution of the deeper abdominal blood vessels reflects the arrangement of the muscles: an oblique, circumferential pattern (similar to the intercostal vessels above)

over the anterolateral abdominal wall and a vertical pattern anteriorly. ◆ The circumferential vessels of the anterolateral wall are continuations of the 7th–11th posterior intercostal vessels, subcostal vessels, and deep circumflex iliac vessels.
◆ Vertical vessels include an anastomosis between the superior and the inferior epigastric vessels within the rectus sheath.
◆ A superficial anastomotic channel, the thoraco-epigastric vein, and the deeper medial pathway between the inferior and the superior epigastric veins afford collateral circulation during blockage of superior or inferior vena cava. ◆ Superficial abdominal lymphatic vessels superior to the transumbilical plane drain primarily to the axillary lymph nodes; those inferior to the plane drain to the superficial inguinal lymph nodes. ◆ Deep lymphatic vessels accompany deep veins of the abdominal wall to the iliac and right and left lumbar (caval and aortic) lymph nodes.

Internal Surface of Anterolateral Abdominal Wall

The internal (posterior) surface of the anterolateral abdominal wall is covered with transversalis fascia, a variable amount of extraperitoneal fat, and parietal peritoneum (Fig. 2.13).

The infraumbilical part of this surface exhibits five *umbilical peritoneal folds* passing toward the umbilicus, one in the median plane and two on each side:

- The **median umbilical fold** extends from the apex of the urinary bladder to the umbilicus and covers the **median**

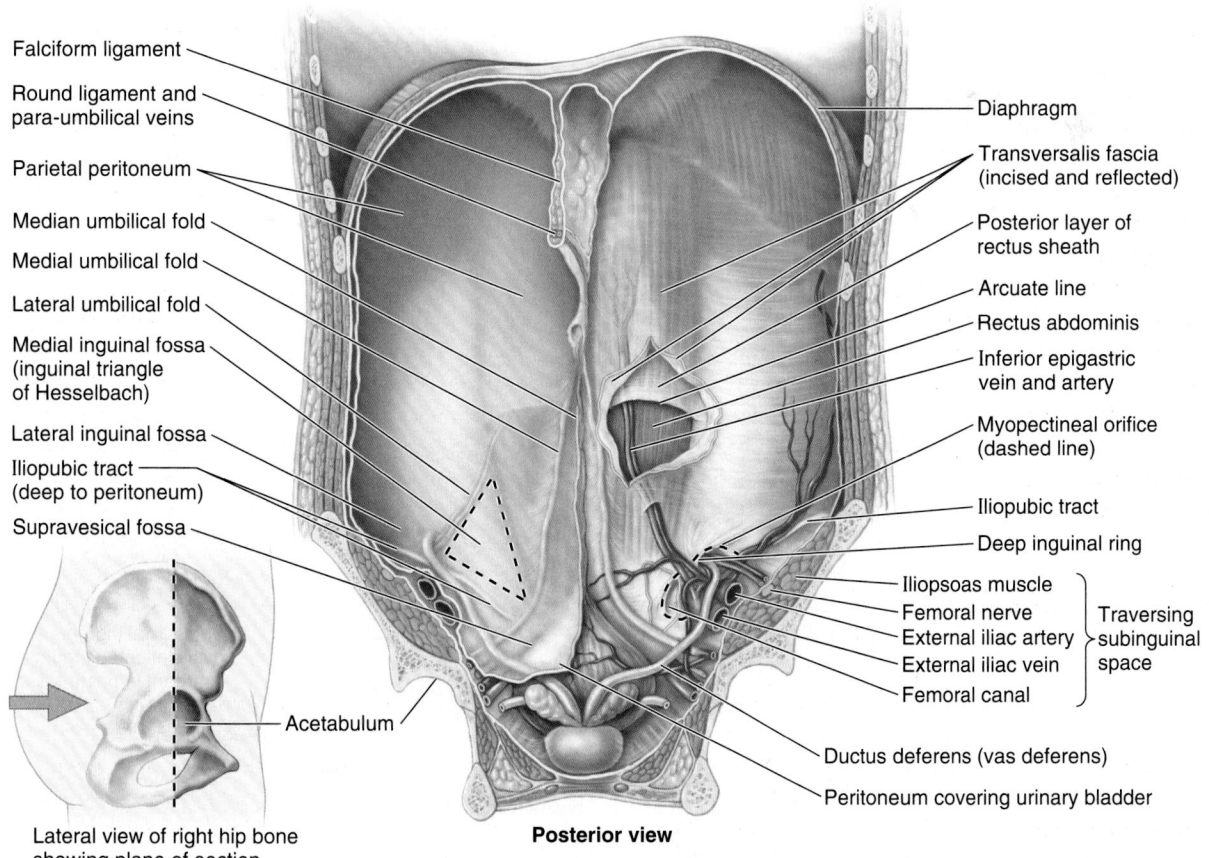

Falciform ligament
Round ligament and para-umbilical veins
Parietal peritoneum
Median umbilical fold
Medial umbilical fold
Lateral umbilical fold
Medial inguinal fossa (inguinal triangle of Hesselbach)
Lateral inguinal fossa
Iliopubic tract (deep to peritoneum)
Supravesical fossa

Acetabulum

Diaphragm
Transversalis fascia (incised and reflected)
Posterior layer of rectus sheath
Arcuate line
Rectus abdominis
Inferior epigastric vein and artery
Myopectineal orifice (dashed line)
Iliopubic tract
Deep inguinal ring
Iliopsoas muscle
Femoral nerve
External iliac artery
External iliac vein
Femoral canal
} Traversing subinguinal space
Ductus deferens (vas deferens)
Peritoneum covering urinary bladder

Lateral view of right hip bone showing plane of section in figure at right

Posterior view

FIGURE 2.13. Posterior aspect of anterolateral abdominal wall of a male. The peritoneal ligaments, folds, and fossae are the main features of this aspect.

umbilical ligament, a fibrous remnant of the *urachus* that joined the apex of the fetal bladder to the umbilicus.

- Two **medial umbilical folds,** lateral to the median umbilical fold, cover the **medial umbilical ligaments,** formed by *occluded parts of the umbilical arteries.*
- Two **lateral umbilical folds,** lateral to the medial umbilical folds, cover the *inferior epigastric vessels* and therefore bleed if cut.

The depressions lateral to the umbilical folds are the *peritoneal fossae,* each of which is a potential site for a hernia. The location of a hernia in one of these fossae determines how the hernia is classified. The shallow fossae between the umbilical folds are the:

- **Supravesical fossae** between the median and the medial umbilical folds, formed as the peritoneum reflects from the anterior abdominal wall onto the bladder. The level of the supravesical fossae rises and falls with filling and emptying of the bladder.
- **Medial inguinal fossae** between the medial and the lateral umbilical folds, areas also commonly called **inguinal triangles** (Hesselbach triangles), which are potential sites for the less common direct inguinal hernias.
- **Lateral inguinal fossae,** lateral to the lateral umbilical folds, include the *deep inguinal rings* and are potential sites for the most common type of hernia in the lower abdominal wall, the *indirect inguinal hernia* (see clinical correlation [blue] box "Inguinal Hernias," on p. 212).

The supra-umbilical part of the internal surface of the anterior abdominal wall has a sagittally oriented peritoneal reflection, the **falciform ligament,** that extends between the superior anterior abdominal wall and the liver. It encloses the *round ligament of the liver* (L. *ligamentum teres hepatis*) and *para-umbilical veins* in its inferior free edge. The round ligament is a fibrous remnant of the *umbilical vein,* which passed from the umbilicus to the liver prenatally (Fig. 2.13).

Inguinal Region

The **inguinal region** (groin) extends between the ASIS and pubic tubercle. It is an important area anatomically and clinically: anatomically because it is a region where structures exit and enter the abdominal cavity, and clinically because the pathways of exit and entrance are potential sites of herniation.

Although the testis is located in the perineum postnatally, the male gonad originally forms in the abdomen. Its relocation out of the abdomen into the perineum through the inguinal canal accounts for many of the structural features of the region. Traditionally, the testis and scrotum are dissected and studied in relation to the anterior abdominal wall and the inguinal region. For this reason, male anatomy receives greater emphasis in this section.

INGUINAL LIGAMENT AND ILIOPUBIC TRACT

Thickened fibrous bands, or *retinacula,* occur in relationship to many joints that have a wide range of movement to retain structures against the skeleton during the various positions of the joint (see Introduction). The *inguinal ligament* and *iliopubic tract,* extending from the ASIS to the *pubic tubercle,* constitute a bilaminar anterior (flexor) retinaculum of the hip joint (Figs. 2.13 and 2.14). The retinaculum spans the **subinguinal**

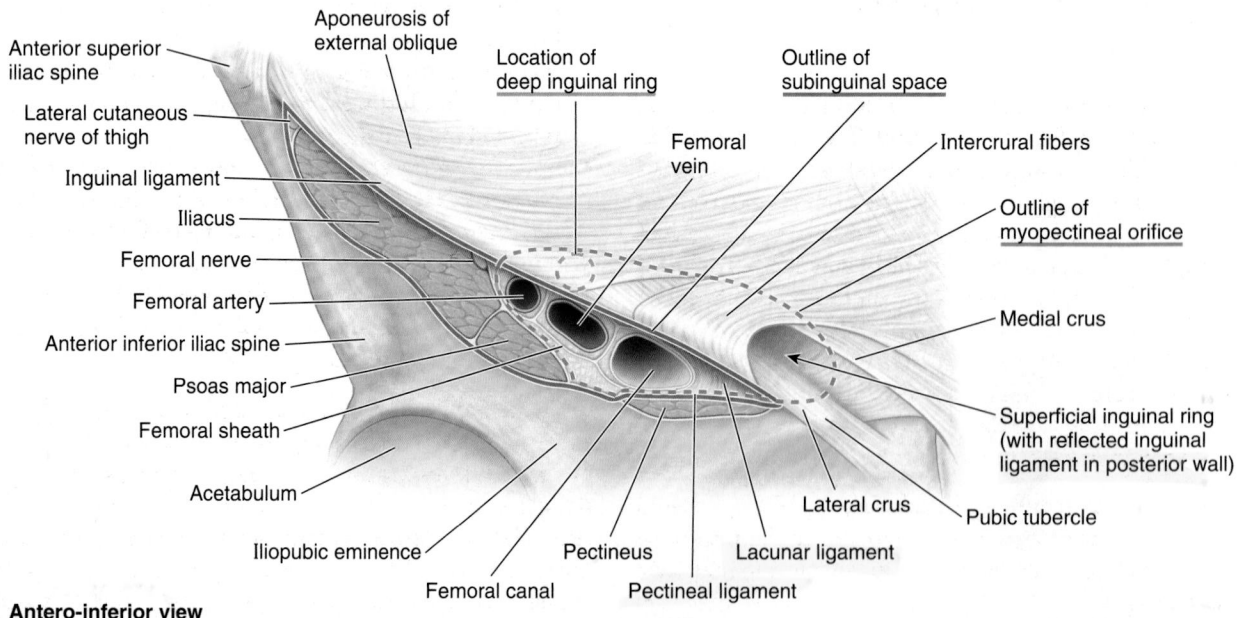

Anterior superior iliac spine
Lateral cutaneous nerve of thigh
Inguinal ligament
Iliacus
Femoral nerve
Femoral artery
Anterior inferior iliac spine
Psoas major
Femoral sheath
Acetabulum
Iliopubic eminence
Femoral canal
Pectineus
Pectineal ligament
Lacunar ligament
Lateral crus
Pubic tubercle
Superficial inguinal ring (with reflected inguinal ligament in posterior wall)
Medial crus
Outline of myopectineal orifice
Intercrural fibers
Outline of subinguinal space
Femoral vein
Location of deep inguinal ring
Aponeurosis of external oblique

Antero-inferior view

FIGURE 2.14. Formations of inguinal region. The inguinal ligament is the thickened, underturned, inferior margin of the aponeurosis of the external oblique, forming a retinaculum that bridges the subinguinal space. A slit-like gap between the medial and the lateral crura of the external oblique aponeurosis, bridged by intercrural fibers, forms the superficial inguinal ring.

space, through which pass the flexors of the hip and neurovascular structures serving much of the lower limb. These fibrous bands are the thickened inferolateral-most portions of the external oblique and aponeurosis and the inferior margin of the transversalis fascia. They are major landmarks of the region.

The **inguinal ligament** is a dense band constituting the inferiormost part of the external oblique aponeurosis. Although most fibers of the ligament's medial end insert into the pubic tubercle, some follow other courses (Fig. 2.14):

- Some of the deeper fibers pass posteriorly to attach to the *superior pubic ramus* lateral to the tubercle, forming the arching **lacunar ligament** (of Gimbernat), which forms the medial boundary of the subinguinal space. The most lateral of these fibers continue to run along the *pecten pubis* as the **pectineal ligament** (of Cooper).
- Some of the more superior fibers fan upward, bypassing the *pubic tubercle* and crossing the linea alba to blend with the lower fibers of the contralateral external oblique aponeurosis. These fibers form the **reflected inguinal ligament** (Figs. 2.8, 2.14, and 2.15A).

The **iliopubic tract** is the thickened inferior margin of the transversalis fascia, which appears as a fibrous band running parallel and posterior (deep) to the inguinal ligament (Figs. 2.13 and 2.15B). The iliopubic tract, seen in the place of the inguinal ligament when the inguinal region is viewed from its internal (posterior) aspect (e.g., during laparoscopy), reinforces the posterior wall and floor of the inguinal canal as it bridges the structures traversing the subinguinal space.

The inguinal ligament and iliopubic tract span an area of innate weakness in the body wall in the inguinal region called the **myopectineal orifice** (Fruchaud, 1956). This weak area, occurring in relation to structures traversing the body wall, is the site of direct and indirect inguinal and femoral hernias.

INGUINAL CANAL

The **inguinal canal** is formed in relation to the relocation of the testis during fetal development. The inguinal canal in adults is an oblique passage, approximately 4 cm long, directed inferomedially through the inferior part of the anterolateral abdominal wall. It lies parallel and superior to the medial half of the inguinal ligament (see Figs. 2.14 and 2.15). The main occupant of the inguinal canal is the spermatic cord in males and the round ligament of the uterus in females. These are functionally and developmentally distinct structures that occur in the same location. The inguinal canal also contains blood and lymphatic vessels and the ilio-inguinal nerve in both sexes. The inguinal canal has an opening at each end:

- The **deep (internal) inguinal ring** is the entrance to the inguinal canal. It is located superior to the middle of the inguinal ligament and lateral to the inferior epigastric artery (Fig. 2.14). It is the beginning of an evagination in the transversalis fascia that forms an opening like the entrance to a cave (Figs. 2.7B, 2.13, and 2.15). Through this opening, the extraperitoneal ductus deferens (vas deferens) and testicular vessels in males (or round ligament

of the uterus in females) pass to enter the inguinal canal. The transversalis fascia itself continues into the canal, forming the innermost covering (internal fascia) of the structures traversing the canal.

- The **superficial (external) inguinal ring** is the exit by which the spermatic cord in males, or the round ligament in females, emerges from the inguinal canal (Figs. 2.7A, 2.14, and 2.15). The superficial ring is a split that occurs in the diagonal, otherwise parallel fibers of the external oblique aponeurosis just superolateral to the pubic tubercle. The parts of the aponeurosis that lie lateral and medial to, and form the margins of, the superficial ring are *crura* (L. leg-like parts). The **lateral crus** attaches to the pubic tubercle, and the **medial crus** attaches to the pubic crest. Fibers of the superficial layer of investing (deep) fascia overlying the external oblique muscle and aponeurosis, running perpendicular to the fibers of the aponeurosis, pass from one crus to the other across the superolateral part of the ring. These **intercrural fibers** help prevent the crura from spreading apart (i.e., they keep the "split" in the aponeurosis from expanding).

The *inguinal canal* is normally collapsed anteroposteriorly against the structures it conveys. Between its two openings (rings), the inguinal canal has two walls (anterior and posterior), as well as a roof and floor (Figs. 2.14 and 2.15A & B). The structures forming these boundaries are listed in Table 2.5.

The inguinal canal has two walls (anterior and posterior), a roof, and a floor (Figs. 2.8 and 2.15A & B):

- *Anterior wall:* formed by the external oblique aponeurosis throughout the length of the canal; its lateral part is reinforced by muscle fibers of the internal oblique.
- *Posterior wall:* formed by the transversalis fascia; its medial part is reinforced by pubic attachments of the internal oblique and transversus abdominis aponeuroses that frequently merge to variable extents into a common tendon—the **inguinal falx** (conjoint tendon)—and the reflected inguinal ligament.
- *Roof:* formed laterally by the transversalis fascia, centrally by musculo-aponeurotic arches of the internal oblique and transversus abdominis, and medially by the medial crus of the external oblique aponeurosis.
- *Floor:* formed laterally by the iliopubic tract, centrally by gutter formed by the infolded inguinal ligament, and medially by the lacunar ligament.

As the inguinal ligament and iliopubic tract span the *myopectineal orifice* (Fig. 2.13), they demarcate the inferior boundaries of the inguinal canal and its openings. The inguinal triangle separates these formations from the structures of the femoral sheath (femoral vessels and femoral canal) that traverse the medial part of the subinguinal space. Most groin hernias in males pass superior to the iliopubic tract (inguinal hernias), whereas most pass inferior to it in females (femoral hernias). Because of its relative weakness, the myopectineal orifice is overlaid with prosthetic mesh placed in the extraperitoneal retro-inguinal space ("space of Bogros") in many hernia repairs.

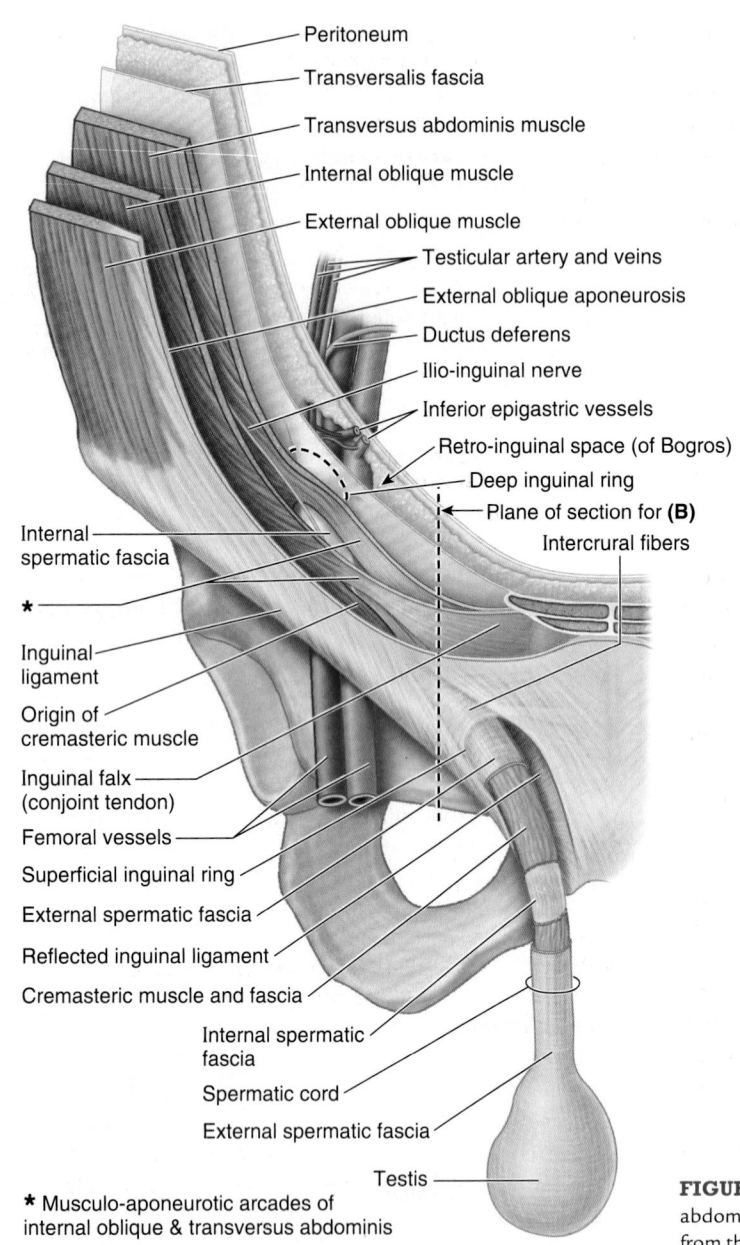

Peritoneum
Transversalis fascia
Transversus abdominis muscle
Internal oblique muscle
External oblique muscle
Testicular artery and veins
External oblique aponeurosis
Ductus deferens
Ilio-inguinal nerve
Inferior epigastric vessels
Retro-inguinal space (of Bogros)
Deep inguinal ring
Plane of section for **(B)**
Intercrural fibers

Internal spermatic fascia
*
Inguinal ligament
Origin of cremasteric muscle
Inguinal falx (conjoint tendon)
Femoral vessels
Superficial inguinal ring
External spermatic fascia
Reflected inguinal ligament
Cremasteric muscle and fascia
Internal spermatic fascia
Spermatic cord
External spermatic fascia
Testis

* Musculo-aponeurotic arcades of internal oblique & transversus abdominis

(A) Anterior view

Aponeuroses of:
external ⎫ oblique
internal ⎭
Skin
Fatty layer
Inguinal canal
Spermatic cord
Membranous layer of sub-cutaneous tissue
Anterior wall of inguinal canal (intercrural fibers)
Ductus deferens
Transversus abdominis
Transversalis fascia
Retro-inguinal space (of Bogros)
Peritoneum
Inguinal falx (conjoint tendon) forming posterior wall of canal
Iliopubic tract
Inguinal ligament forming "gutter" (floor of inguinal canal)
Superior ramus of pubis
Fascia lata of thigh

(B) Schematic sagittal section of inguinal canal

FIGURE 2.15. Inguinal canal and spermatic cord. A. The layers of the abdominal wall and the coverings of the spermatic cord and testis derived from them are shown. **B.** Sagittal section of the anterior abdominal wall and inguinal canal at the plane shown in part **A.**

TABLE 2.5. BOUNDARIES OF INGUINAL CANAL

Boundary	Lateral Third/Deep Ring	Middle Third	Medial Third/Superficial Ring
Posterior wall	Transversalis fascia	Transversalis fascia	Inguinal falx (conjoint tendon) plus reflected inguinal ligament
Anterior wall	Internal oblique plus lateral crus of aponeurosis of external oblique	Aponeurosis of external oblique (lateral crus and intercrural fibers)	Aponeurosis of external oblique (intercrural fibers), with fascia of external oblique continuing onto cord as external spermatic fascia
Roof	Transversalis fascia	Musculo-aponeurotic arches of internal oblique and transverse abdominal	Medial crus of aponeurosis of external oblique
Floor	Iliopubic tract	Inguinal ligament	Lacunar ligament

Development of Inguinal Canal. The testes develop in the extraperitoneal connective tissue in the superior lumbar region of the posterior abdominal wall (Fig. 2.16A). The **male gubernaculum** is a fibrous tract connecting the **primordial testis** to the anterolateral abdominal wall at the site of the future deep ring of the inguinal canal. A peritoneal diverticulum, the **processus vaginalis,** traverses the developing inguinal canal, carrying muscular and fascial layers of the anterolateral abdominal wall before it as it enters the **primordial scrotum.** By the 12th week, the testis is in the pelvis, and by 28 weeks (7th month), it lies close to the developing deep inguinal ring (Fig. 2.16B). The testis begins to pass through the inguinal canal during the 28th week and takes approximately 3 days to traverse it. Approximately 4 weeks later, the testis enters the scrotum (Fig. 2.16C). As the testis, its duct (*ductus deferens*), and its vessels and nerves relocate, they are ensheathed by musculofascial extensions of the anterolateral abdominal wall, which account for the presence of their derivatives in the adult scrotum: the internal and external *spermatic fasciae* and cremaster muscle (Fig. 2.15). The stalk of the processus vaginalis normally degenerates; however, its distal saccular part forms the *tunica vaginalis,* the serous sheath of the testis and epididymis (Moore et al., 2012).

The *ovaries* also develop in the superior lumbar region of the posterior abdominal wall and relocate to the lateral wall of the pelvis (Fig. 2.17). The processus vaginalis of the peritoneum traverses the transversalis fascia at the site of the deep inguinal ring, forming the inguinal canal as in the male, and protrudes into the developing labium majus.

The **female gubernaculum,** a fibrous cord connecting the ovary and primordial uterus to the developing labium majus, is represented postnatally by the **ovarian ligament,** between the ovary and uterus, and the **round ligament of the uterus** (L. *ligamentum teres uteri*), between the uterus and labium majus. Because of the attachment of the ovarian ligaments to the uterus, the ovaries do not relocate to the inguinal region; however, the round ligament passes through the inguinal canal and attaches to the subcutaneous tissue of the labium majus (Fig. 2.17B & C).

Except for its most inferior part, which becomes a serous sac engulfing the testis, the *tunica vaginalis,* the processus vaginalis obliterates by the 6th month of fetal development. The inguinal canals in females are narrower than those in males, and the canals in infants of both sexes are shorter and much less oblique than in adults. The superficial inguinal rings in infants lie almost directly anterior to the deep inguinal rings.

Inguinal Canal and Increased Intra-Abdominal Pressure. The deep and superficial inguinal rings in the adult do not overlap because of the oblique path of the inguinal canal. Consequently, increases in intra-abdominal pressure act on the inguinal canal, forcing the posterior wall of the canal against the anterior wall and strengthening this wall, thereby decreasing the likelihood of herniation until the pressures overcome the resistant effect of this mechanism.

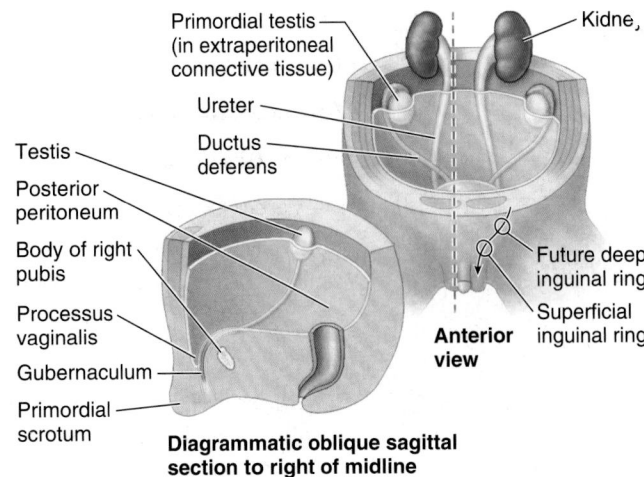

(A) Seventh week

Primordial testis (in extraperitoneal connective tissue)
Kidney
Ureter
Testis
Ductus deferens
Posterior peritoneum
Body of right pubis
Processus vaginalis
Gubernaculum
Primordial scrotum
Future deep inguinal ring
Superficial inguinal ring
Anterior view

Diagrammatic oblique sagittal section to right of midline

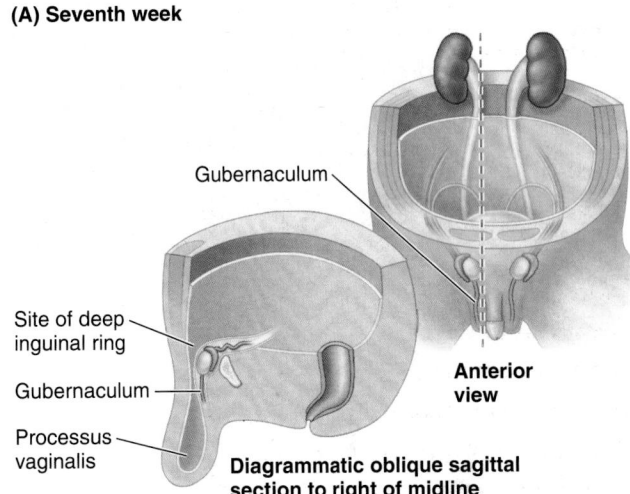

(B) Seventh month

Gubernaculum
Site of deep inguinal ring
Gubernaculum
Processus vaginalis
Anterior view

Diagrammatic oblique sagittal section to right of midline

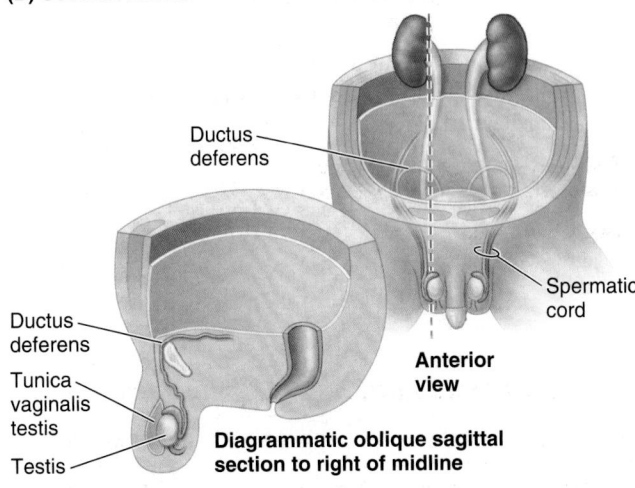

(C) Ninth month

Ductus deferens
Ductus deferens
Tunica vaginalis testis
Testis
Spermatic cord
Anterior view

Diagrammatic oblique sagittal section to right of midline

FIGURE 2.16. Formation of inguinal canals and relocation of testes. A. In a 7-week embryo, the testis is attached to the posterior abdominal wall. **B.** A fetus at 28 weeks (seventh month) shows the processus vaginalis and testis passing through the inguinal canal. The testis passes posterior to the processus vaginalis, not through it. **C.** In a newborn infant, obliteration of the stalk of the processus vaginalis has occurred. The remains of the processus vaginalis have formed the tunica vaginalis of the testis. The remnant of the gubernaculum has disappeared.

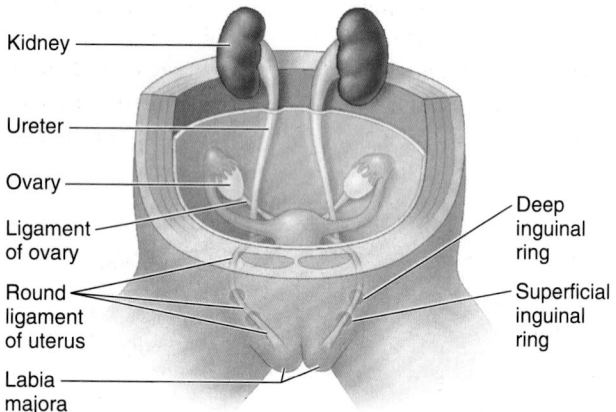

(A) 2 months

mesonephric duct

Developing kidney

Mesonephric duct

Upper gubernaculum (inguinal fold— becomes ligament of ovary)

Lower gubernaculum (becomes round ligament of uterus)

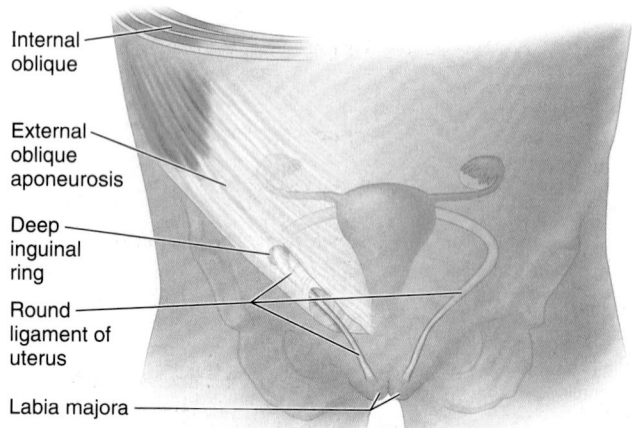

(B) 15 weeks

Kidney

Ureter

Ovary

Ligament of ovary

Round ligament of uterus

Labia majora

Deep inguinal ring

Superficial inguinal ring

Internal oblique

External oblique aponeurosis

Deep inguinal ring

Round ligament of uterus

Labia majora

(C) Mature

Anterior view

FIGURE 2.17. Formation of inguinal canals in females. A. At 2 months the undifferentiated gonads (primordial ovaries) are located on the dorsal abdominal wall. **B.** At 15 weeks the ovaries have descended into the greater pelvis. The processus vaginalis (not shown) passes through the abdominal wall, forming the inguinal canal on each side as in the male fetus. The round ligament passes through the canal and attaches to the subcutaneous tissue of the labium majus. **C.** In the mature female, the processus vaginalis has degenerated but the round ligament persists and passes through the inguinal canal.

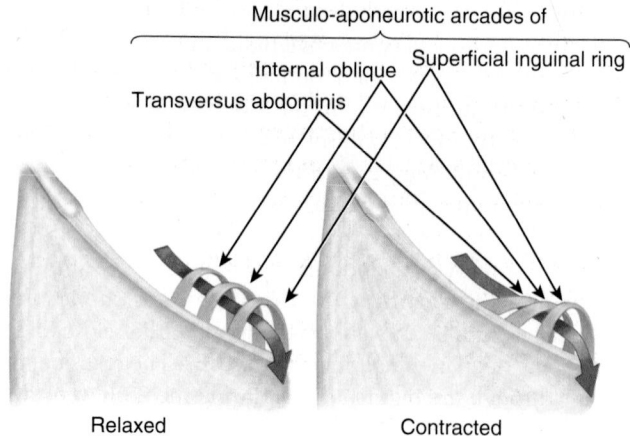

Musculo-aponeurotic arcades of

Transversus abdominis

Internal oblique

Superficial inguinal ring

Relaxed

Contracted

Anterior view

FIGURE 2.18. Arcades of inguinal canal. The inguinal canal consists of a series of three musculo-aponeurotic arcades traversed by the spermatic cord or round ligament of the uterus (*arrow*). The muscular contraction that increases intra-abdominal pressure also causes the roof of the canal to descend, narrowing the canal as it is simultaneously collapsed anteroposteriorly by the increased internal pressure.

Simultaneously, contraction of the external oblique approximates the anterior wall of the canal to the posterior wall. It also increases tension on the medial and lateral crura, resisting enlargement (dilation) of the superficial inguinal ring. Contraction of the musculature that forms the lateral part of the arcades of the internal oblique and transversus abdominis muscles makes the roof of the canal descend, constricting the canal (Fig. 2.18).

Spermatic Cord, Scrotum, and Testes

SPERMATIC CORD

The **spermatic cord** contains structures running to and from the testis and suspends the testis in the scrotum (see Fig. 2.19; Table 2.6). The spermatic cord begins at the deep inguinal ring lateral to the inferior epigastric vessels, passes through the inguinal canal, exits at the superficial inguinal ring, and ends in the scrotum at the posterior border of the testis. Fascial coverings derived from the anterolateral abdominal wall during prenatal development surround the spermatic cord. The *coverings of the spermatic cord* include the following:

- **Internal spermatic fascia:** derived from the transversalis fascia.
- **Cremasteric fascia:** derived from the investing fascia of both the superficial and deep surfaces of the internal oblique muscle.
- **External spermatic fascia:** derived from the external oblique aponeurosis and its investing fascia.

The cremasteric fascia contains loops of **cremaster muscle,** which is formed by the lowermost fascicles of the internal oblique muscle arising from the inguinal ligament (Figs. 2.8 and 2.15A). The cremaster muscle reflexively draws the testis superiorly in the scrotum, particularly in response to cold. In a warm environment, such as a hot bath, the cremaster relaxes and the testis descends deeply in the scrotum. Both responses occur in an attempt to regulate the temperature of the testis for **spermatogenesis** (formation of sperms), which requires a constant temperature approximately one degree cooler than core temperature, or during sexual activity as a protective response. The cremaster typically acts coincidentally with the **dartos muscle,** smooth muscle of the fat-free subcutaneous tissue of the scrotum (*dartos fascia*), which inserts into the skin, assisting testicular elevation as it produces contraction of the skin of the scrotum in response to the same stimuli. The cremaster muscle is innervated by the *genital branch of the genitofemoral nerve* (L1, L2), a derivative of the *lumbar plexus* (Fig. 2.19). The cremaster is striated muscle receiving somatic innervation, whereas the dartos is smooth muscle receiving autonomic innervation. Although less well developed and usually indistinct, the round ligament of the female receives similar contributions from the layers of the abdominal wall as it traverses the inguinal canal. The constituents of the spermatic cord are the following (Figs. 2.19 and 2.21; Table 2.6):

- *Ductus deferens* (vas deferens): a muscular tube approximately 45 cm long that conveys sperms from the epididymis to the ejaculatory duct.
- *Testicular artery*: arising from the aorta and supplying the testis and epididymis.
- *Artery of ductus deferens*: arising from the inferior vesical artery.
- *Cremasteric artery*: arising from the inferior epigastric artery.
- *Pampiniform venous plexus*: a network formed by up to 12 veins that converge superiorly as right or left testicular veins.
- Sympathetic *nerve fibers* on arteries and sympathetic and parasympathetic nerve fibers on the ductus deferens.
- *Genital branch of the genitofemoral nerve*: supplying the cremaster muscle.
- *Lymphatic vessels*: draining the testis and closely associated structures and passing to the lumbar lymph nodes.
- **Vestige of processus vaginalis:** may be seen as a fibrous thread in the anterior part of the spermatic cord extending between the abdominal peritoneum and the tunica vaginalis; it may not be detectable.

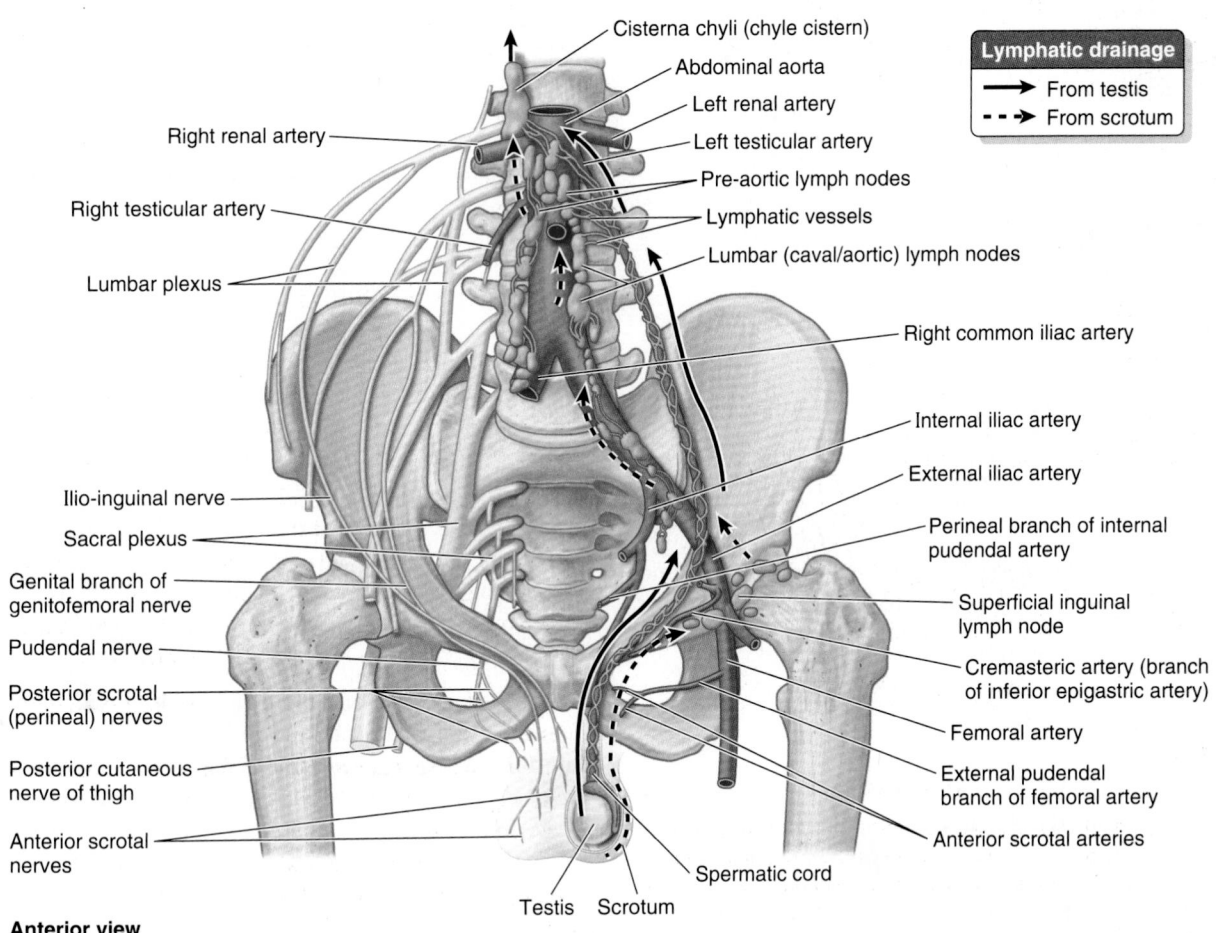

Anterior view

FIGURE 2.19. Arterial supply and lymphatic drainage of testis and scrotum; innervation of scrotum. Lymph draining from the testis and scrotum follows different courses. The lumbar plexus provides innervation to the anterolateral aspect of the scrotum; the sacral plexus provides innervation to the postero-inferior aspect.

TABLE 2.6. CORRESPONDING LAYERS OF ANTERIOR ABDOMINAL WALL, SCROTUM, AND SPERMATIC CORD

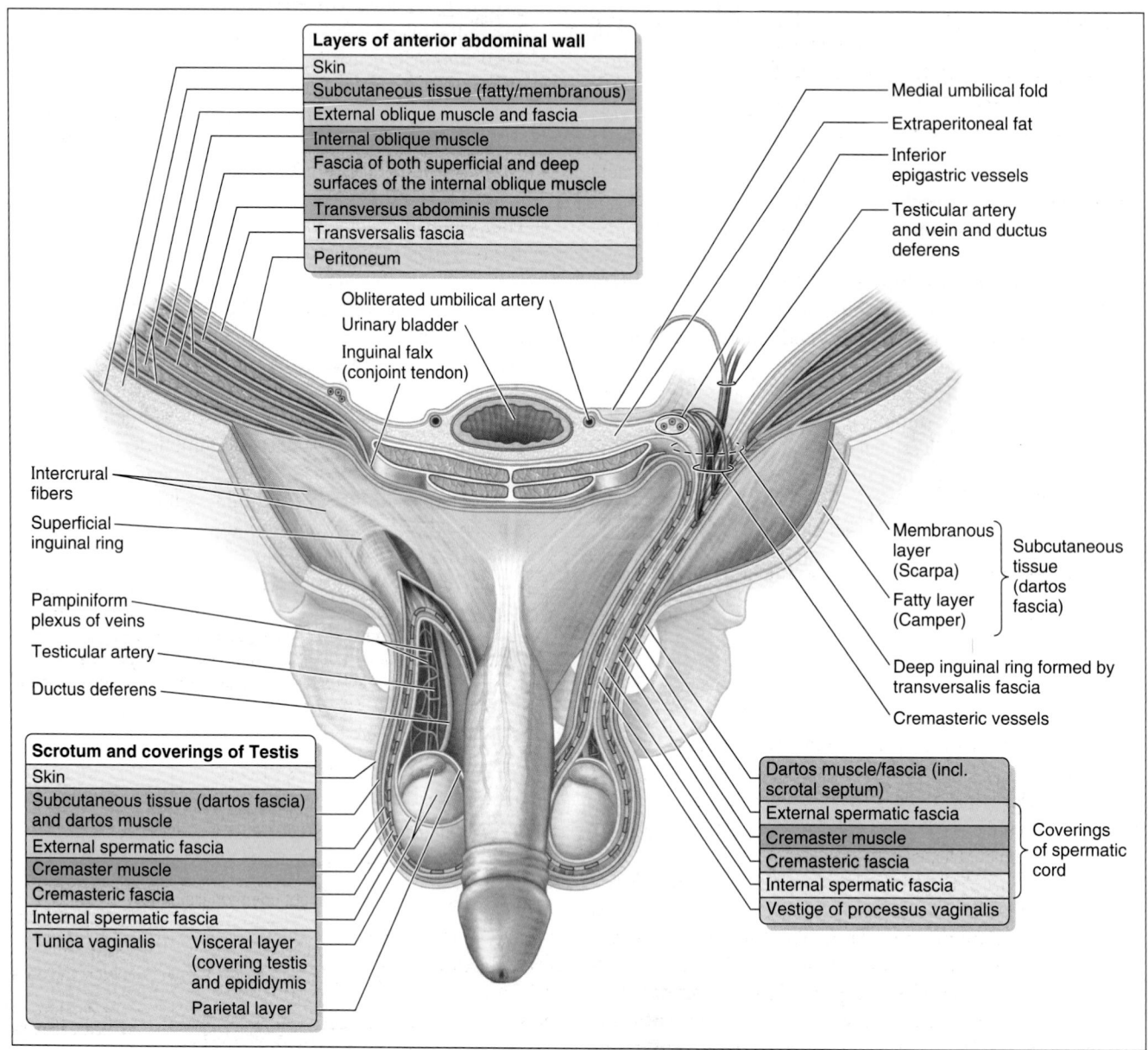

Because it is not a homolog of the spermatic cord, the round ligament does not contain comparable structures. It includes only vestiges of the lower part of the ovarian gubernaculum and the processus vaginalis.

SCROTUM

The **scrotum** is a cutaneous sac consisting of two layers: heavily *pigmented skin* and the closely related **dartos fascia,** a fat-free fascial layer including smooth muscle fibers (*dartos muscle*) responsible for the rugose (wrinkled) appearance of the scrotum (see Fig. 2.9B; Table 2.6). Because the dartos muscle attaches to the skin, its contraction causes the scrotum to wrinkle when cold, thickening the integumentary layer while reducing scrotal surface area and assisting the cremaster muscles in holding the testes closer to the body, all of which reduces heat loss.

The scrotum is divided internally by a continuation of the dartos fascia, the *septum of the scrotum,* into right and left compartments. The septum is demarcated externally by the *scrotal raphe* (see Chapter 3), a cutaneous ridge marking the line of fusion of the embryonic *labioscrotal swellings.* The superficial dartos fascia is devoid of fat and is continuous anteriorly with the *membranous layer of subcutaneous tissue of the abdomen* (Scarpa fascia) and posteriorly with the *membranous layer of subcutaneous tissue of the perineum* (Colles fascia) (see Fig. 2.9B).

The development of the scrotum is closely related to the formation of the inguinal canals. The scrotum develops from the *labioscrotal swellings,* two cutaneous outpouchings of the anterior abdominal wall that fuse to form a pendulous cutaneous pouch. Late in the fetal period, the testes and spermatic cords enter the scrotum.

The arterial supply of the scrotum (Fig. 2.19) is from the:

- **Posterior scrotal branches of the perineal artery:** a branch of the internal pudendal artery.
- **Anterior scrotal branches of the deep external pudendal artery:** a branch of the femoral artery.
- **Cremasteric artery:** a branch of the inferior epigastric artery.

Scrotal veins accompany the arteries. The lymphatic vessels of the scrotum drain into the superficial inguinal lymph nodes.

The nerves of the scrotum (Fig. 2.19) include branches of the lumbar plexus to the anterolateral surface, and branches of the sacral plexus to the posterior and inferior surfaces:

- *Genital branch of the genitofemoral nerve* (L1, L2): supplying the anterolateral surface.
- **Anterior scrotal nerves:** branches of the ilio-inguinal nerve (L1) supplying the anterior surface.
- **Posterior scrotal nerves:** branches of the perineal branch of the *pudendal nerve* (S2–S4) supplying the posterior surface.
- **Perineal branches of the posterior cutaneous nerve of thigh** (S2, S3): supplying the postero-inferior surface.

TESTES

The **testes** (testicles) are the male gonads—paired ovoid reproductive glands that produce **sperms (spermatozoa)** and male hormones, primarily testosterone (Fig. 2.20). The testes are suspended in the scrotum by the spermatic cords, with the left testis usually suspended (hanging) more inferiorly than the right testis.

The surface of each testis is covered by the **visceral layer of the tunica vaginalis,** except where the testis attaches to the epididymis and spermatic cord. The **tunica vaginalis** is a closed peritoneal sac partially surrounding the testis, which represents the closed-off distal part of the embryonic processus vaginalis. The visceral layer of the tunica vaginalis is closely applied to the testis, epididymis, and inferior part of the ductus deferens. The slit-like recess of the tunica vaginalis, the **sinus of the epididymis,** is between the body of the epididymis and the posterolateral surface of the testis.

The **parietal layer of the tunica vaginalis,** adjacent to the internal spermatic fascia, is more extensive than the visceral layer and extends superiorly for a short distance onto the distal part of the spermatic cord. The small amount of fluid in the cavity of the tunica vaginalis separates the visceral and parietal layers, allowing the testis to move freely in the scrotum.

The testes have a tough fibrous outer surface, the **tunica albuginea,** that thickens into a ridge on its internal, posterior aspect as the **mediastinum of the testis** (Fig. 2.21). From this internal ridge, fibrous septa extend inward between lobules of minute but long and highly coiled **seminiferous tubules** in which the sperms are produced. The seminiferous tubules are joined by **straight tubules** to the **rete testis** (L. *rete*, a net), a network of canals in the mediastinum of the testis.

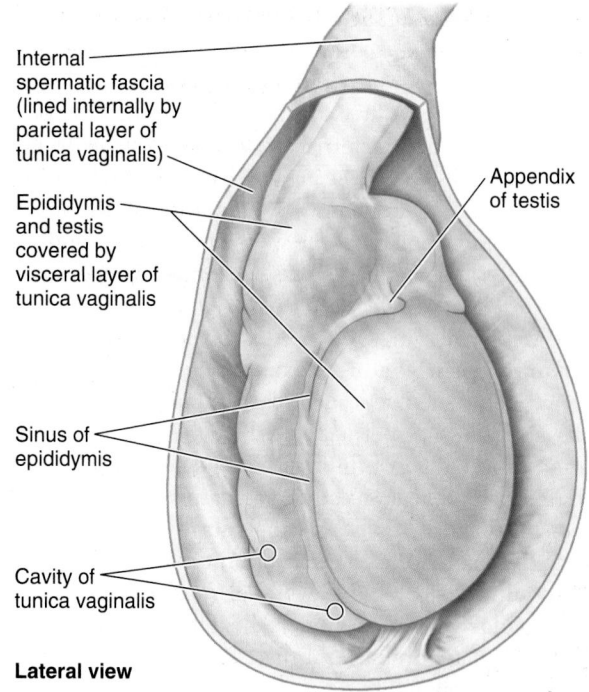

FIGURE 2.20. Tunica vaginalis (opened). The distal part of the contents of the spermatic cord, the epididymis, and most of the testis are surrounded by a collapsed sac, the tunica vaginalis. Consequently, the testis and epididymis—directly covered by the tunica's visceral layer—are mobile within the scrotum. The outer parietal layer lines the peritesticular continuation of the internal spermatic fascia.

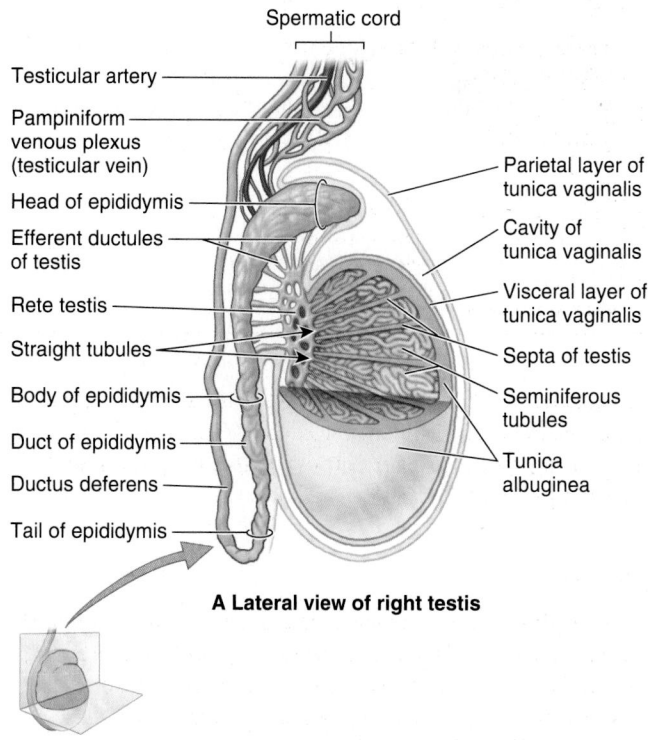

FIGURE 2.21. Structures of testis and epididymis. The coverings and a quarter-section of the testis has been removed to demonstrate the contents of the distal spermatic cord, features of the epididymis, and internal structural details of the testis. The cavity of the tunica vaginalis—actually a potential space—is highly exaggerated.

The long **testicular arteries** arise from the anterolateral aspect of the *abdominal aorta* just inferior to the *renal arteries* (Fig. 2.19). They pass retroperitoneally (posterior to the peritoneum) in an oblique direction, crossing over the ureters and the inferior parts of the external iliac arteries to reach the deep inguinal rings. They enter the inguinal canals through the deep rings, pass through the canals, exit them through the superficial inguinal rings, and enter the spermatic cords to supply the testes. The testicular artery or one of its branches anastomoses with the *artery of the ductus deferens*.

The veins emerging from the testis and epididymis form the **pampiniform venous plexus,** a network of 8–12 veins lying anterior to the ductus deferens and surrounding the testicular artery in the spermatic cord (Fig. 2.21). The pampiniform plexus is part of the thermoregulatory system of the testis (along with the cremasteric and dartos muscles) helping to keep this gland at a constant temperature. The veins of each pampiniform plexus converge superiorly, forming a **right testicular vein,** which enters the inferior vena cava (IVC), and a **left testicular vein,** which enters the left renal vein.

The lymphatic drainage of the testis follows the testicular artery and vein to the *right* and *left lumbar (caval/aortic)* and *pre-aortic lymph nodes* (see Fig. 2.19). The autonomic nerves of the testis arise as the **testicular plexus of nerves** on the testicular artery, which contains vagal parasympathetic and visceral afferent fibers and sympathetic fibers from the T10(–T11) segment of the spinal cord.

EPIDIDYMIS

The **epididymis** is an elongated structure on the posterior surface of the testis (Fig. 2.20). **Efferent ductules** of the testis transport newly developed sperms to the epididymis from the rete testis. The epididymis is formed by minute convolutions of the **duct of the epididymis,** so tightly compacted that they appear solid (Fig. 2.21). The duct becomes progressively smaller as it passes from the head of the epididymis on the superior part of the testis to its tail. At the **tail of the epididymis,** the ductus deferens begins as the continuation of the epididymal duct. In the lengthy course of this duct, the sperms are stored and continue to mature. The epididymis consists of the:

- **Head of the epididymis:** the superior expanded part that is composed of lobules formed by the coiled ends of 12–14 *efferent ductules.*
- **Body of the epididymis:** major part consisting of the tightly convoluted duct of the epididymis.
- **Tail of the epididymis:** tapering continuation with the ductus deferens, the duct that transports the sperms from the epididymis to the ejaculatory duct for expulsion via the urethra during ejaculation (see Chapter 3).

Surface Anatomy of Anterolateral Abdominal Wall

The **umbilicus** is an obvious feature of the anterolateral abdominal wall and is the reference point for the *transumbilical plane* (Fig. 2.22). This puckered indentation of skin in the center of the anterior abdominal wall is typically at the level of the IV disc between the L3 and L4 vertebrae. However, its position varies with the amount of subcutaneous fat present. The umbilicus indicates the level of the T10 dermatome. The **epigastric fossa** (pit of the stomach) is a slight depression in the epigastric region, just inferior to the *xiphoid process.* This fossa is particularly noticeable when a person is in the supine position because the abdominal organs spread out, drawing the anterolateral abdominal wall posteriorly in this region. The pain caused by pyrosis ("heartburn," result-

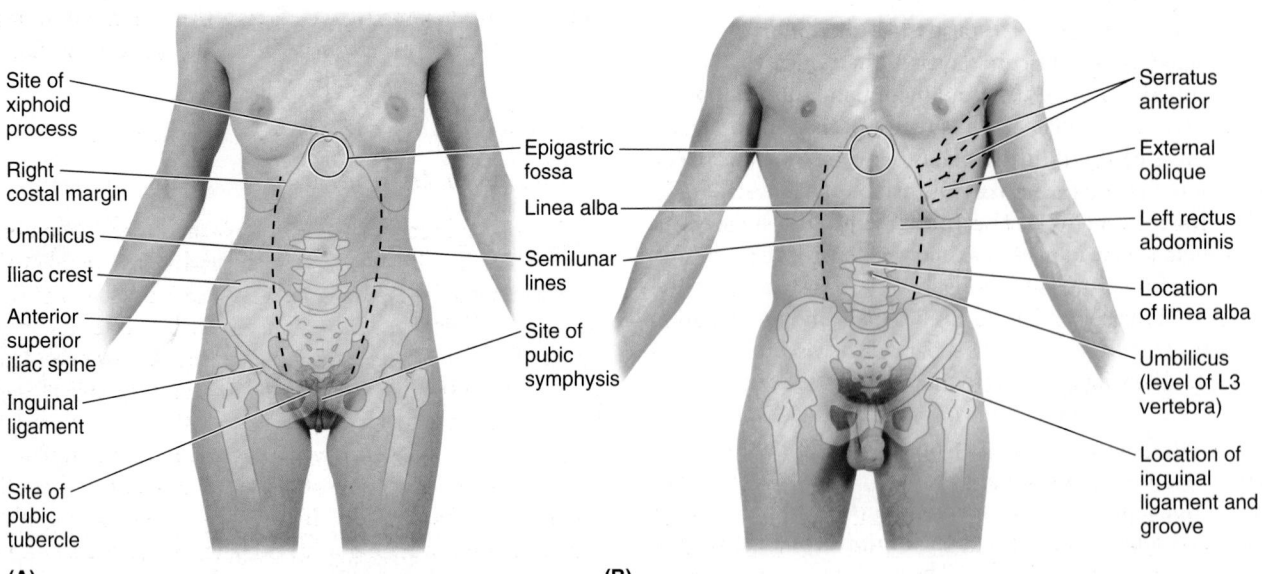

FIGURE 2.22. Surface anatomy of the anterolateral abdominal wall.

ing from reflux of gastric acid into the esophagus) is often felt at this site. The 7th–10th costal cartilages unite on each side of the epigastric fossa, their medial borders forming the *costal margin.* Although the abdominal cavity extends higher, the costal margin is the demarcation between the thoracic and abdominal portions of the body wall. When a person is in the supine position, observe the rise and fall of the abdominal wall with respiration: superiorly with inspiration and inferiorly with expiration. The rectus abdominis muscles can be palpated and observed when a supine person is asked to raise their head and shoulders against resistance.

The location of the *linea alba* is visible in lean individuals because of the vertical skin groove superficial to this raphe. The groove is usually obvious because the linea alba is approximately 1 cm wide between the two parts of the rectus abdominis superior to the umbilicus. Inferior to the umbilicus, the linea alba is not indicated by a groove. Some pregnant women, especially those with dark hair and a dark complexion, have a heavily pigmented line, the **linea nigra,** in the midline skin external to the linea alba. After pregnancy, the color of this line fades.

The upper margins of the pubic bones (*pubic crest*) and the cartilaginous joint that unite them (*pubic symphysis*) can be felt at the inferior end of the linea alba. The inguinal fold is a shallow oblique groove overlying the inguinal ligament as it extends between the *anterior superior iliac spine* (ASIS) and the *pubic tubercle.* The bony *iliac crest* at the level of L4 vertebra can be easily palpated as it extends posteriorly from the ASIS. The pubic crest, inguinal folds, and iliac crests demarcate the inferior limit of the anterior abdominal wall, distinguishing it from the perineum centrally and the lower limbs (thighs) laterally.

The **semilunar lines** (L. *lineae semilunares*) are slightly curved, linear impressions in the skin that extend from the inferior costal margin near the 9th costal cartilages to the pubic tubercles. These semilunar skin grooves (5–8 cm from the midline) are clinically important because they are parallel with the lateral edges of the rectus sheath.

Skin grooves also overlie the *tendinous intersections* of the rectus abdominis, which are clearly visible in persons with well-developed rectus muscles. The interdigitating bellies of the *serratus anterior* and *external oblique* muscles are also visible.

The site of the inguinal ligament is indicated by the **inguinal groove,** a skin crease that is parallel and just inferior to the inguinal ligament. This groove is readily visualized by having the person drop one leg to the floor while lying supine on an examining table. The inguinal groove marks the division between the anterolateral abdominal wall and the thigh.

INTERNAL SURFACE OF ANTEROLATERAL ABDOMINAL WALL AND INGUINAL REGION

Undescended (Cryptorchid) Testis

The testes are undescended in approximately 3% of full-term and 30% of premature infants (Moore, Persaud, and Torchia, 2012). About 95% of undescended testes occur unilaterally. If a testis has not descended or is not retractable (capable of being drawn down), the condition is *cryptorchidism* (G. *orchis,* testis + L. from G. *kryptos,* hidden). The *undescended testis* usually lies somewhere along the normal path of its prenatal descent, commonly in the inguinal canal. The importance of cryptorchidism is a greatly increased risk for developing malignancy in the undescended testis, particularly problematic because it is not palpable and is not usually detected until cancer has progressed.

External Supravesical Hernia

An *external supravesical hernia* leaves the peritoneal cavity through the supravesical fossa (Fig. 2.13). The site of the hernia is medial to that of a direct inguinal hernia (see the blue box "Inguinal Hernias," on p. 212). The iliohypogastric nerve is in danger of injury during the repair of this rare type of hernia.

Postnatal Patency of Umbilical Vein

Before the birth of a fetus, the umbilical vein carries well-oxygenated, nutrient-rich blood from the placenta to the fetus. Although reference is often made to the "occluded" umbilical vein forming the *round ligament of the liver,* this vein is patent for some time after birth, and is used for *umbilical vein catheterization* for exchange transfusion during early infancy—for example, in infants with *erythroblastosis fetalis* or *hemolytic disease* of the newborn (Kliegman et al., 2011).

Metastasis of Uterine Cancer to Labium Majus

Lymphogenous metastasis of cancer most commonly occurs along lymphatic pathways that parallel the venous drainage of the organ that is the site of the primary tumor. This is also true of the uterus, the veins and lymph vessels of which mostly drain via deep routes. However, some lymphatic vessels follow the course of the round ligament through the inguinal canal. Thus, while occurring less often, metastatic uterine cancer cells (especially from tumors adjacent to the proximal attachment of the round ligament) can spread from the uterus to the labium majus (the developmental homolog of the scrotum and site of distal attachment of the round ligament), and from there to the superficial inguinal nodes, which receive lymph from the skin of the perineum (including the labia).

SPERMATIC CORD, SCROTUM, AND TESTES

Inguinal Hernias

The majority of abdominal hernias occur in the inguinal region. Inguinal hernias account for 75% of abdominal hernias. These herniations occur in both sexes, but most inguinal hernias (approximately 86%) occur in males because of the passage of the spermatic cord through the *inguinal canal.*

An *inguinal hernia* is a protrusion of parietal peritoneum and viscera, such as the small intestine, through a normal or abnormal opening from the cavity in which they belong. Most hernias are reducible, meaning they can be returned to their normal place in the peritoneal cavity by appropriate manipulation. The two types of inguinal hernia are *direct* and *indirect inguinal hernias.* More than two thirds are indirect hernias. Characteristics of direct and indirect inguinal hernias are listed and illustrated in Table B2.1, with related anatomy illustrated in Figure B2.3A–C.

Normally, most of the processus vaginalis obliterates before birth, except for the distal part that forms the tunica vaginalis of the testis (Table 2.6). The peritoneal part of the hernial sac of an indirect inguinal hernia is formed by the persisting processus vaginalis. If the entire stalk of the processus vaginalis persists, the hernia extends into the scrotum superior to the testis, forming a complete indirect inguinal hernia (Table B2.1).

The *superficial inguinal ring* is palpable superolateral to the pubic tubercle by invaginating the skin of the upper scrotum with the index finger (Fig. B2.3D). The examiner's finger follows the spermatic cord superolaterally to the superficial inguinal ring. If the ring is dilated, it may admit the finger without causing pain. Should a hernia be present, a sudden impulse is felt against either the tip or pad of the examining finger when the patient is asked to cough (Swartz, 2009). However, because both inguinal hernia types exit the superficial ring, palpation of an impulse at this site does not discriminate type.

With the palmar surface of the finger against the anterior abdominal wall, the *deep inguinal ring* may be felt as a skin depression superior to the inguinal ligament, 2–4 cm superolateral to the pubic tubercle. Detection of an impulse at the superficial ring and a mass at the site of the deep ring suggests an indirect hernia.

Palpation of a direct inguinal hernia is performed by placing the palmar surface of the index and/or middle finger over the inguinal triangle and asking the person to cough or bear down (strain). If a hernia is present, a forceful impulse is felt against the pad of the finger. The finger can also be placed in the superficial inguinal ring; if a direct hernia is present, a sudden impulse is felt medial to the finger when the person coughs or bears down.

Cremasteric Reflex

Contraction of the cremaster muscle is elicited by lightly stroking the skin on the medial aspect of the superior part of the thigh with an applicator stick or tongue depressor. The *ilio-inguinal nerve* supplies this area of skin. The rapid elevation of the testis on the same side is the *cremasteric reflex.* This reflex is extremely active in children; consequently, hyperactive cremasteric reflexes may simulate undescended testes. A hyperactive reflex can be abolished by having the child sit in a cross-legged, squatting position; if the testes are descended, they can then be palpated in the scrotum.

Cysts and Hernias of Canal of Nuck

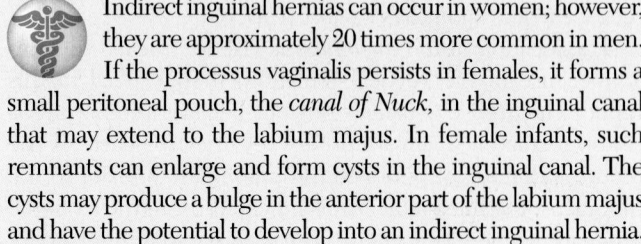

Indirect inguinal hernias can occur in women; however, they are approximately 20 times more common in men.
If the processus vaginalis persists in females, it forms a small peritoneal pouch, the *canal of Nuck,* in the inguinal canal that may extend to the labium majus. In female infants, such remnants can enlarge and form cysts in the inguinal canal. The cysts may produce a bulge in the anterior part of the labium majus and have the potential to develop into an indirect inguinal hernia.

Hydrocele of Spermatic Cord and/or Testis

A *hydrocele* is the presence of excess fluid in a *persistent processus vaginalis.* This congenital anomaly may be associated with an indirect inguinal hernia. The fluid accumulation results from secretion of an abnormal amount of serous fluid from the visceral layer of the tunica vaginalis. The size of the hydrocele depends on how much of the processus vaginalis persists.

A *hydrocele of the testis* is confined to the scrotum and distends the tunica vaginalis (Fig. B2.4A). A *hydrocele of the spermatic cord* is confined to the spermatic cord and distends the persistent part of the stalk of the processus vaginalis (Fig. B2.4B). A congenital hydrocele of the cord and testis may communicate with the peritoneal cavity.

Detection of a hydrocele requires *transillumination,* a procedure during which a bright light is applied to the side of the scrotal enlargement in a darkened room. The transmission of light as a red glow indicates excess serous fluid in the scrotum. Newborn male infants often have residual peritoneal fluid in their tunica vaginalis; however, this fluid is usually absorbed during the 1st year of life. Certain pathological conditions, such as injury and/or inflammation of the epididymis, may also produce a hydrocele in adults.

Hematocele of Testis

A *hematocele of the testis* is a collection of blood in the tunica vaginalis that results, for example, from rupture of branches of the testicular artery by trauma to the testis (Fig. B2.4C). Trauma may produce a scrotal and/or testicular *hematoma* (accumulation of blood, usually clotted, in any extravascular location). Blood does not transilluminate; therefore, transillumination can differentiate a hematocele or hematoma from a hydrocele. A hematocele of the testis may be associated with a *scrotal hematocele,* resulting from effusion of blood into the scrotal tissues.

TABLE B2.1. CHARACTERISTICS OF INGUINAL HERNIAS

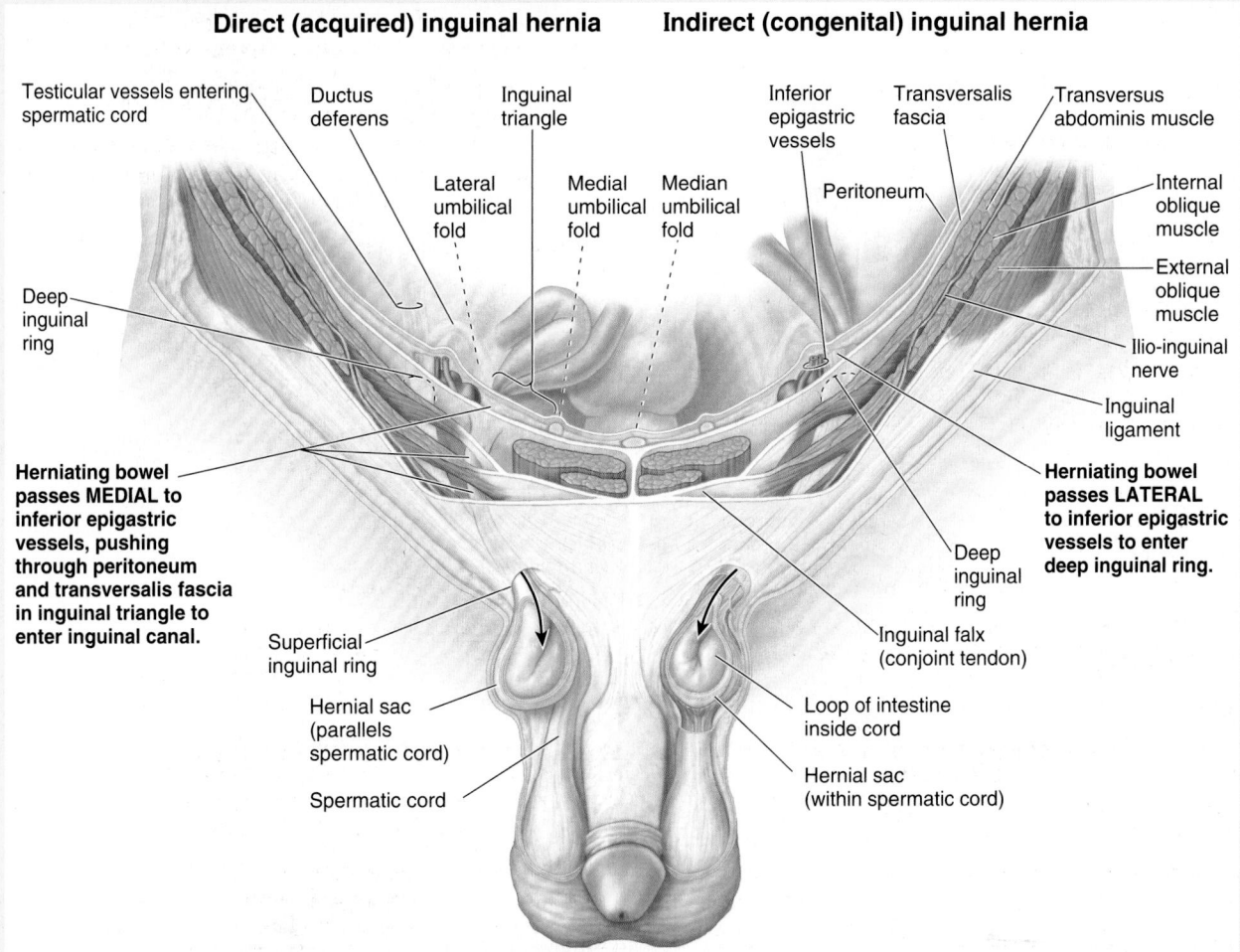

Direct (acquired) inguinal hernia   Indirect (congenital) inguinal hernia

Characteristic	Direct (Acquired)	Indirect (Congenital)
Predisposing factors	Weakness of anterior abdominal wall in inguinal triangle (e.g., owing to distended superficial ring, narrow inguinal falx, or attenuation of aponeurosis in males >40 years of age)	Patency of processus vaginalis (complete or at least superior part) in younger persons, the great majority of which are males
Frequency	Less common (one third to one quarter of inguinal hernias)	More common (two thirds to three quarters) of inguinal hernias
Exit from abdominal cavity (Fig. B2.3A & B)	Peritoneum plus transversalis fascia (lies outside inner one or two fascial coverings of cord)	Peritoneum of persistent processus vaginalis plus all three fascial coverings of cord/round ligament
Course (Fig. B2.3C)	Passes through or around inguinal canal, usually traversing only medial third of canal, external and parallel to vestige of processus vaginalis	Traverses inguinal canal (entire canal if it is of sufficient size) within processus vaginalis
Exit from anterior abdominal wall	Via superficial ring, lateral to cord; rarely enters scrotum	Via superficial ring inside cord, commonly passing into scrotum/labium majus

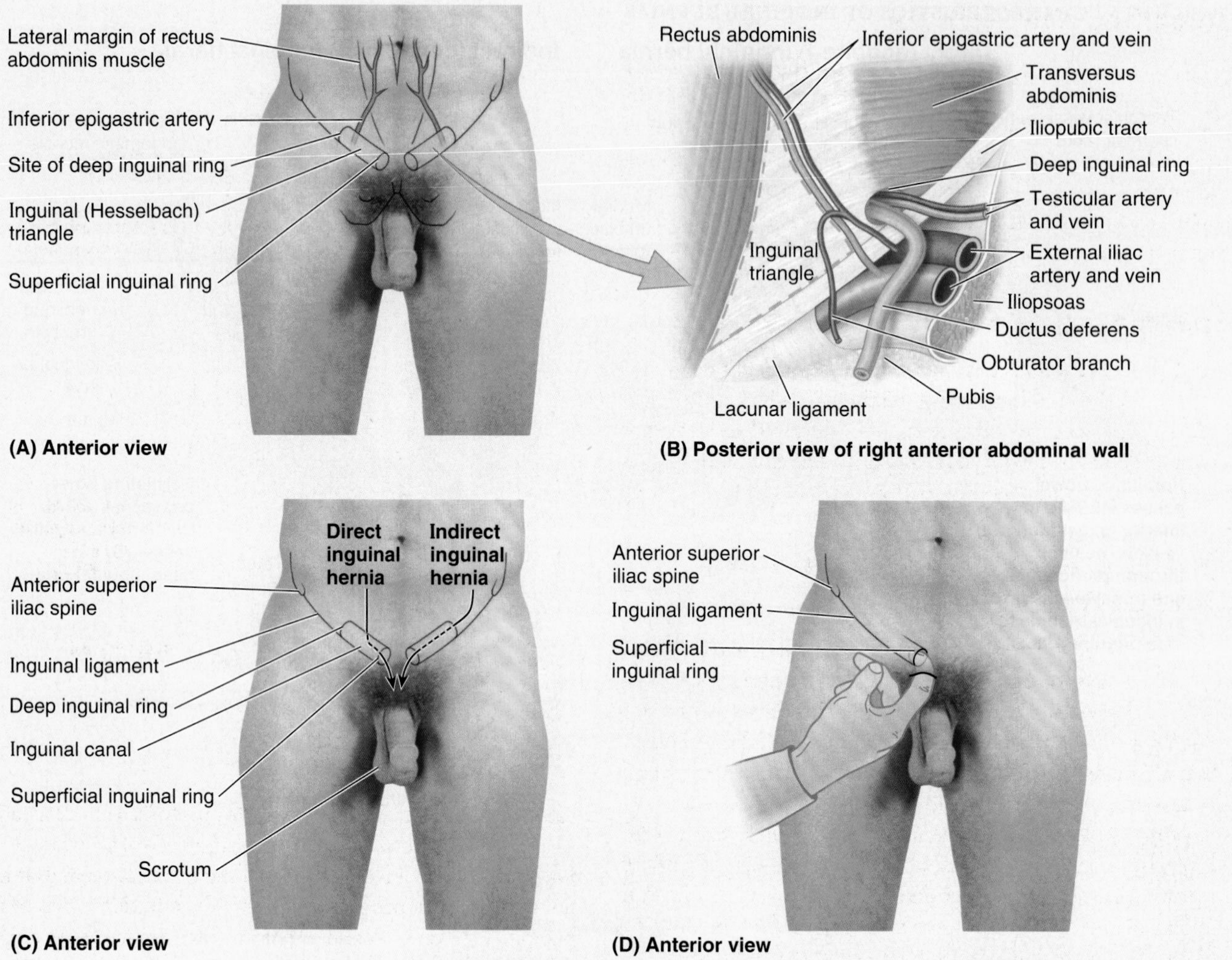

Lateral margin of rectus abdominis muscle

Inferior epigastric artery

Site of deep inguinal ring

Inguinal (Hesselbach) triangle

Superficial inguinal ring

(A) Anterior view

Rectus abdominis

Inferior epigastric artery and vein

Transversus abdominis

Iliopubic tract

Deep inguinal ring

Testicular artery and vein

External iliac artery and vein

Iliopsoas

Ductus deferens

Obturator branch

Pubis

Inguinal triangle

Lacunar ligament

(B) Posterior view of right anterior abdominal wall

Anterior superior iliac spine

Inguinal ligament

Deep inguinal ring

Inguinal canal

Superficial inguinal ring

Scrotum

Direct inguinal hernia   **Indirect inguinal hernia**

(C) Anterior view

Anterior superior iliac spine

Inguinal ligament

Superficial inguinal ring

(D) Anterior view

FIGURE B2.3.

Torsion of Spermatic Cord

Torsion of the spermatic cord is a surgical emergency because necrosis (pathologic death) of the testis may occur. The torsion (twisting) obstructs the venous drainage, with resultant edema and hemorrhage, and subsequent arterial obstruction. The twisting usually occurs just above the upper pole of the testis (Fig. B2.4D). To prevent recurrence or occurrence on the contralateral side, which is likely, both testes are surgically fixed to the scrotal septum.

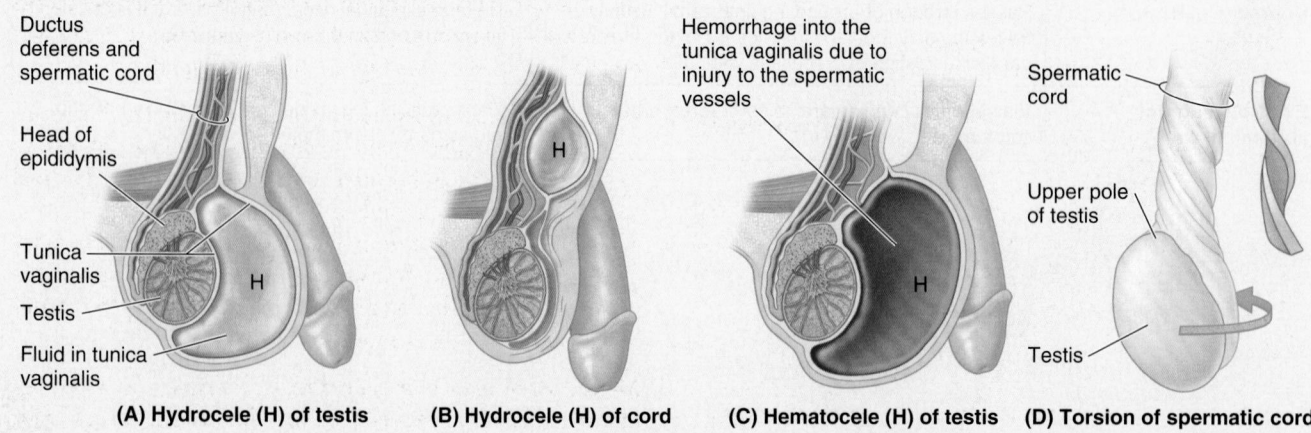

Ductus deferens and spermatic cord

Head of epididymis

Tunica vaginalis

Testis

Fluid in tunica vaginalis

(A) Hydrocele (H) of testis

(B) Hydrocele (H) of cord

Hemorrhage into the tunica vaginalis due to injury to the spermatic vessels

(C) Hematocele (H) of testis

Spermatic cord

Upper pole of testis

Testis

(D) Torsion of spermatic cord

FIGURE B2.4.

Anesthetizing Scrotum

Since the anterolateral surface of the scrotum is supplied by the lumbar plexus (primarily L1 fibers via the ilio-inguinal nerve) and the postero-inferior aspect is supplied by the sacral plexus (primarily S3 fibers via the pudendal nerve), a spinal anesthetic agent must be injected more superiorly to anesthetize the anterolateral surface of the scrotum than is necessary to anesthetize its postero-inferior surface.

Spermatocele and Epididymal Cyst

A *spermatocele* is a retention cyst (collection of fluid) in the epididymis (Fig. B2.5A), usually near its head. Spermatoceles contain a milky fluid and are generally asymptomatic. An *epididymal cyst* is a collection of fluid anywhere in the epididymis (Fig. B2.5B).

Vestigial Remnants of Embryonic Genital Ducts

When the tunica vaginalis is opened, rudimentary structures may be observed at the superior aspects of the testes and epididymis (Fig. B2.6). These structures are small remnants of genital ducts in the embryo. They are rarely observed unless pathological changes occur. The **appendix of the testis** is a vesicular remnant of the cranial end of the *paramesonephric (müllerian) duct,* the embryonic genital duct that in the female forms half of the uterus. It is attached to the upper pole of the testis. The **appendices of the epididymis** are remnants of the cranial end of the *mesonephric (wolffian) duct,* the embryonic genital duct that in the male forms part of the ductus deferens. The appendices are attached to the head of the epididymis.

Varicocele

The vine-like *pampiniform plexus of veins* may become dilated (varicose) and tortuous, producing a *varicocele,* which is usually visible only when the man is standing or straining. The enlargement usually disappears when the person lies down, particularly if the scrotum is ele-

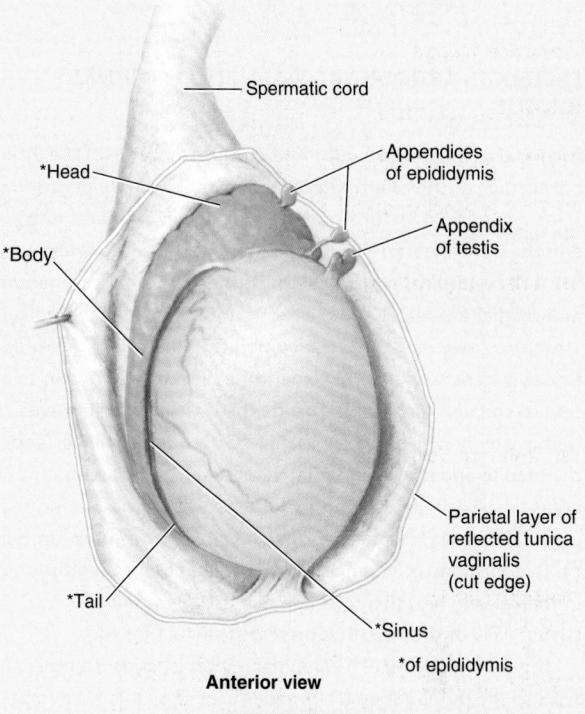

Anterior view

FIGURE B2.6.

vated while supine, allowing gravity to empty the veins. Palpating a varicocele can be likened to feeling a bag of worms. Varicoceles may result from defective valves in the testicular vein, but kidney or renal vein problems can also result in distension of the pampiniform veins. Varicocele occurs predominantly on the left side, probably because the acute angle at which the right vein enters the IVC is more favorable to flow than the nearly 90° angle at which the left testicular vein enters the left renal vein, making it more susceptible to obstruction or reversal of flow.

Cancer of Testis and Scrotum

Lymphogenous metastasis is common to all testicular tumors, so a knowledge of lymphatic drainage is helpful in treatment (Kumar et al., 2009). Because the testes relocate from the posterior abdominal wall to the scrotum during fetal development, their lymphatic drainage differs from that of the scrotum, which is an outpouching of anterolateral abdominal skin (see Fig. 2.15). Consequently:

- *Cancer of the testis:* metastasizes initially to the retroperitoneal *lumbar lymph nodes,* which lie just inferior to the renal veins. Subsequent spread may be to mediastinal and supraclavicular nodes.
- *Cancer of the scrotum:* metastasizes to the *superficial inguinal lymph nodes,* which lie in the subcutaneous tissue inferior to the inguinal ligament and along the terminal part of the great saphenous vein.

Metastasis of testicular cancer may also occur by hematogenous spread of cancer cells (via the blood) to the lungs, liver, brain, and bone.

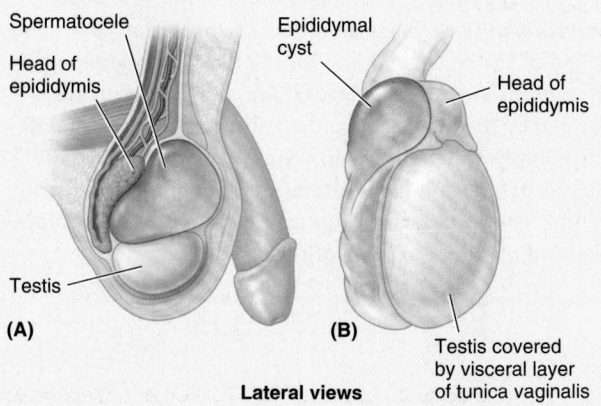

FIGURE B2.5. **A.** Spermatocele. **B.** Epididymal cyst.

The Bottom Line

INTERNAL ABDOMINAL WALL AND INGUINAL REGION

Internal abdominal wall: The primary features of the internal surface of the anterolateral abdominal wall are peritoneal folds overlying structures radiating from the umbilical ring and the peritoneal fossae formed in relation to the folds. ♦ Of the umbilical peritoneal folds, the central three (median and medial umbilical folds) cover remnants of embryological structures, whereas the lateral umbilical folds cover the inferior epigastric vessels. ♦ The peritoneal fossae formed in relation to the umbilical folds include the transitional supravesical fossae, the height of which changes with bladder filling, and the medial and lateral inguinal fossa, overlying potential weak areas in the anterior abdominal wall where direct and indirect inguinal hernias, respectively, may occur. ♦ The supra-umbilical falciform ligament encloses the remnant of the embryonic umbilical vein and the accompanying para-umbilical veins (tributaries of the hepatic portal vein) in its free edge.

Inguinal region: The inguinal region extends from the ASIS to the pubic tubercle; its superficial inguinal fold demarcates the abdomen from the lower limb. It lies within the L1 dermatome. ♦ Most structures and formations in the inguinal region relate to a double (bilaminar) retinaculum formed by the inguinal ligament and iliopubic tract as they extend between the two bony points. These two bands are thickenings of the inferior margins of the external oblique aponeurosis and transversalis fascia of the abdominal wall, respectively.

To allow the testis to descend prenatally to a subcutaneous location that will be cooler postnatally (a requirement for the development of sperms), the inguinal canal traverses the abdominal wall superior and parallel to the medial half of the inguinal ligament. ♦ In females, only the inferior portion of the gubernaculum traverses the canal, becoming the round ligament of the uterus. ♦ The inguinal canal itself consists of a deep ring internally, a superficial ring externally, and two musculo-aponeurotic arches in between. ♦ The oblique passageway through the offset rings and arches collapses when intra-abdominal pressure increases. ♦ Collapse of the canal, combined with the prenatal occlusion of the peritoneal evagination (processus vaginalis) and the contraction of the arches, normally resists the tendency for abdominal contents to herniate (protrude through) the canal. ♦ Failure of the processus vaginalis to occlude, or defective anatomy, or degeneration of tissues, may result in the development of inguinal hernias.

SPERMATIC CORD, SCROTUM, AND TESTES

Spermatic cord: In their passage through the inguinal canal, the processus vaginalis, testis, ductus deferens, and neurovascular structures of the testis (or processus vaginalis and lower ovarian gubernaculum of the female) become engulfed by fascial extensions derived from most (three of four) of the layers traversed. This results in a trilaminar covering. ♦ The transversalis fascia, internal oblique, and external oblique layers contribute the internal spermatic fascia, cremasteric muscle and fascia, and external spermatic fascia, respectively, to the spermatic cord. ♦ Although the portion of the processus vaginalis within the spermatic cord occludes, that adjacent to the testis remains patent as the tunica vaginalis testis. ♦ The contents of the spermatic cord are the ductus deferens and neurovascular structures, which trailed the testis as it relocated from the posterior abdominal wall during development.

Scrotum: The scrotum is the integumentary sac formed from the labioscrotal swellings of the male to house the testes after their relocation. The fatty layer of subcutaneous tissue of the abdominal wall is replaced in the scrotum by the smooth dartos muscle, whereas the membranous layer is continued as the dartos fascia and scrotal septum. ♦ The scrotum receives anterior scrotal arteries from the thigh (via the external pudendal artery), posterior scrotal arteries from the perineum (internal pudendal artery), and internally cremasteric arteries from the abdomen (inferior epigastric artery). ♦ Anterior scrotal nerves are derived from the lumbar plexus (via the genitofemoral and ilio-inguinal nerves), and posterior scrotal nerves from the sacral plexus (via the pudendal nerve).

Testes: The testes are the male gonads, shaped and sized like large olives or small plums, that produce sperms and male hormones. ♦ Each testis is engulfed, except posteriorly and superiorly, by a double-layered serous sac, the tunica vaginalis, derived from the peritoneum. ♦ The outer surface of the testis is covered with the fibrous tunica albuginea, which is thickened internally and posteriorly as the mediastinum of the testis from which septa radiate. ♦ Between the septa are loops of fine seminiferous tubules in which the sperms develop. The tubules converge and empty into the rete testis in the mediastinum, which is connected in turn to the epididymis by the efferent ductules. ♦ The innervation, blood vasculature, and lymphatic drainage all reflect the posterior abdominal origin of the testes and are, for the main part, independent of the surrounding scrotal sac. ♦ The epididymis is formed by the highly convoluted and compacted duct of the epididymis leading from the efferent ductules to the ductus deferens. It is the site of sperm storage and maturation. The epididymis clings to the more protected superior and posterior aspects of the testis.

PERITONEUM AND PERITONEAL CAVITY

The **peritoneum** is a continuous, glistening, and slippery transparent serous membrane. It lines the abdominopelvic cavity and invests the viscera (Fig. 2.23). The peritoneum consists of two continuous layers: the *parietal peritoneum,* which lines the internal surface of the abdominopelvic wall, and the *visceral peritoneum,* which invests viscera such as the stomach and intestines. Both layers of peritoneum consist of *mesothelium,* a layer of simple squamous epithelial cells.

The **parietal peritoneum** is served by the same blood and lymphatic vasculature and the same somatic nerve supply, as is the region of the wall it lines. Like the overlying skin, the peritoneum lining the interior of the body wall is sensitive to pressure, pain, heat and cold, and laceration. Pain from the parietal peritoneum is generally well localized, except for that on the inferior surface of the central part of the diaphragm, where innervation is provided by the phrenic nerves (discussed later in this chapter); irritation here is often referred to the C3–C5 dermatomes over the shoulder.

The **visceral peritoneum** and the organs it covers are served by the same blood and lymphatic vasculature and visceral nerve supply. The visceral peritoneum is insensitive to touch, heat and cold, and laceration; it is stimulated primarily by stretching and chemical irritation. The pain produced is poorly localized, being referred to the dermatomes of the spinal ganglia providing the sensory fibers, particularly to midline portions of these dermatomes. Consequently, pain

from foregut derivatives is usually experienced in the epigastric region, that from midgut derivatives in the umbilical region, and that from hindgut derivatives in the pubic region.

The peritoneum and viscera are in the abdominopelvic cavity. The relationship of the viscera to the peritoneum is as follows:

- *Intraperitoneal organs* are almost completely covered with visceral peritoneum (e.g., the stomach and spleen). *Intraperitoneal* in this case does *not* mean inside the peritoneal cavity (although the term is used clinically for substances injected into this cavity). Intraperitoneal organs have conceptually, if not literally, invaginated into the closed sac, like pressing your fist into an inflated balloon (see the discussion of potential spaces in the Introduction).
- *Extraperitoneal, retroperitoneal,* and *subperitoneal organs* are also outside the peritoneal cavity—external to the parietal peritoneum—and are only partially covered with peritoneum (usually on just one surface). Retroperitoneal organs such as the kidneys are between the parietal peritoneum and the posterior abdominal wall and have parietal peritoneum only on their anterior surfaces (often with a variable amount of intervening fat). Similarly, the subperitoneal urinary bladder has parietal peritoneum only on its superior surface.

The **peritoneal cavity** is within the abdominal cavity and continues inferiorly into the pelvic cavity. The peritoneal cavity is a potential space of capillary thinness between the parietal and visceral layers of peritoneum. It contains no organs but contains a thin film of **peritoneal fluid,** which is composed of

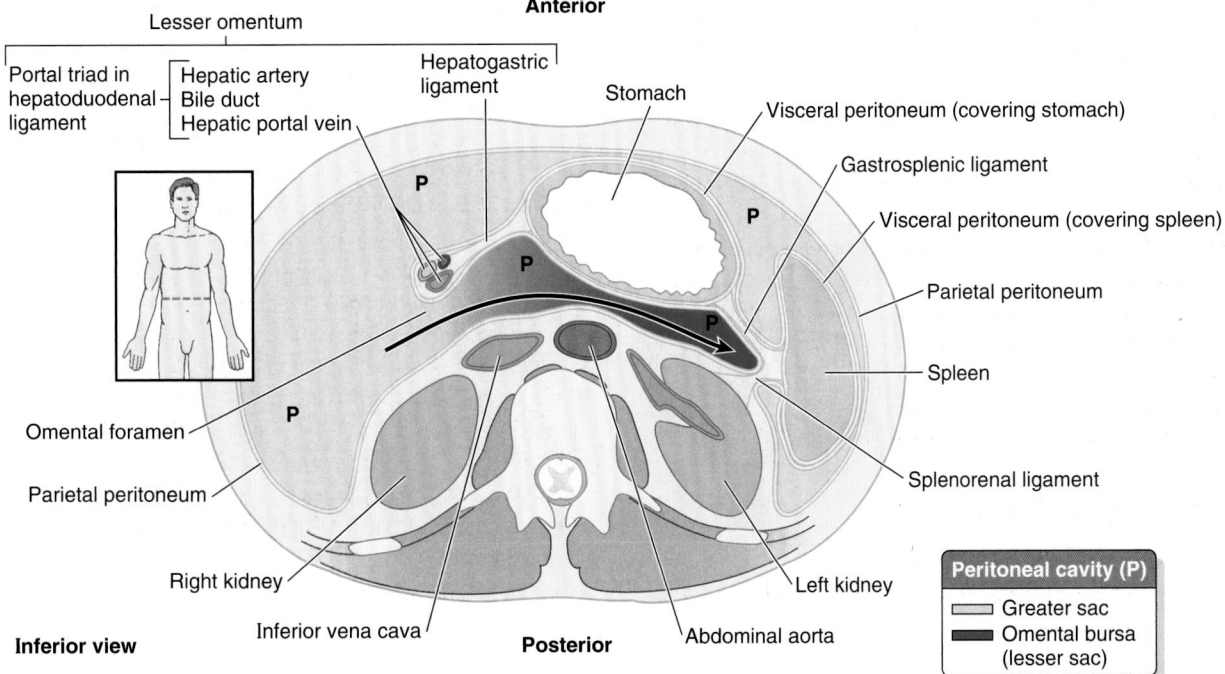

FIGURE 2.23. Transverse section of abdomen at level of omental bursa. The orientation figure (inset) indicates the level of the section superficially. The *dark arrow* passes from the greater sac of the peritoneal cavity (*P*) through the omental (epiploic) foramen and across the full extent of the omental bursa (lesser sac).

water, electrolytes, and other substances derived from interstitial fluid in adjacent tissues. Peritoneal fluid lubricates the peritoneal surfaces, enabling the viscera to move over each other without friction, and allowing the movements of digestion. In addition to lubricating the surfaces of the viscera, the peritoneal fluid contains leukocytes and antibodies that resist infection. Lymphatic vessels, particularly on the inferior surface of the constantly active diaphragm, absorb the peritoneal fluid. The peritoneal cavity is completely closed in males. However, there is a communication pathway in females to the exterior of the body through the uterine tubes, uterine cavity, and vagina. This communication constitutes a potential pathway of infection from the exterior.

Embryology of Peritoneal Cavity

When it is initially formed, the gut is the same length as the developing body. It undergoes exuberant growth, however, to provide the large absorptive surface required by nutrition. By the end of the 10th week of development, the gut is much longer than the body that contains it. For this increase in length to occur, the gut must gain freedom of movement relative to the body wall at an early stage, while still maintaining the connection with it necessary for innervation and blood supply. This growth (and later, the activity of the gut) is accommodated by the development of a serous cavity within the trunk that houses the increasingly lengthy and convoluted gut in a relatively compact space. The rate of growth of the gut initially outpaces the development of adequate space within the trunk (body), and for a time the rapidly lengthening gut extends outside the developing anterior body wall (see "Brief Review of the Embryological Rotation of the Midgut," on p. 258).

Early in its development, the embryonic body cavity (*intraembryonic coelom*) is lined with *mesoderm*, the primordium of the peritoneum. At a slightly later stage, the primordial abdominal cavity is lined with *parietal peritoneum* derived from

mesoderm, which forms a closed sac. The lumen of the peritoneal sac is the *peritoneal cavity*. As the organs develop, they invaginate (protrude) to varying degrees into the peritoneal sac, acquiring a peritoneal covering, the *visceral peritoneum*. A viscus (organ) such as the kidney protrudes only partially into the peritoneal cavity; hence, it is primarily retroperitoneal, always remaining external to the peritoneal cavity and posterior to the peritoneum lining the abdominal cavity. Other viscera, such as the stomach and spleen, protrude completely into the peritoneal sac and are almost completely invested by visceral peritoneum—that is, they are *intraperitoneal*.

These viscera are connected to the abdominal wall by a *mesentery* of variable length, which is composed of two layers of peritoneum with a thin layer of loose connective tissue between them. Generally, viscera that vary relatively little in size and shape, such as the kidneys, are retroperitoneal, whereas viscera that undergo marked changes in shape owing to filling, emptying, and peristalsis, such as the stomach, are invested with visceral peritoneum. Intraperitoneal viscera with a mesentery, such as most of the small intestine, are mobile, the degree of which varies with the length of the mesentery. Although the liver and spleen do not change shape as a result of intrinsic activity (although they may slowly change in size when engorged with blood), their need for a covering of visceral peritoneum is dictated by the need to accommodate passive changes in position imposed by the adjacent, highly active diaphragm.

As organs protrude into the peritoneal sac, their vessels, nerves, and lymphatics remain connected to their extraperitoneal (usually retroperitoneal) sources or destinations so that these connecting structures lie between the layers of the peritoneum forming their mesenteries. Initially, the entire primordial gut is suspended in the center of the peritoneal cavity by a posterior mesentery attached to the midline of the posterior body wall. As the organs grow, they gradually reduce the size of the peritoneal cavity until it is only a potential space

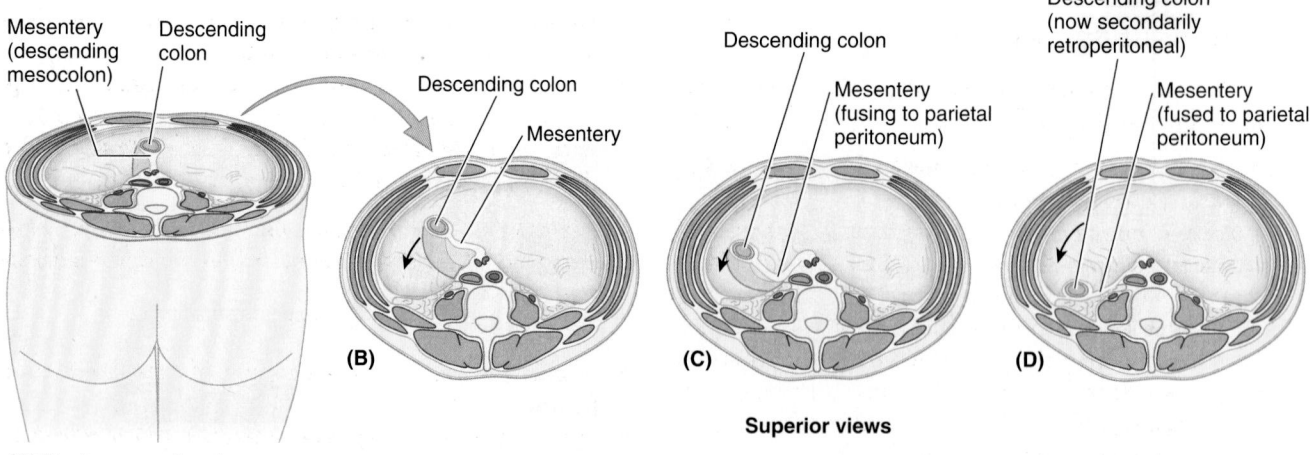

Superior views

(A) Posterosuperior view

FIGURE 2.24. Migration and fusion of descending mesocolon. Starting from the primordial position, suspended from the midline of the posterior abdominal wall (**A**), the mesocolon shifts to the left (**B**) and gradually fuses with the left posterior parietal peritoneum (**C**). **D.** The descending colon has become secondarily retroperitoneal. The *arrow* indicates the left paracolic gutter, the site where an incision is made during mobilization of the colon during surgery. Sometimes the descending colon retains a short mesentery, similar to the stage shown in **C**, especially where the colon is in the iliac fossa.

between the parietal and visceral layers of peritoneum. As a consequence, several parts of the gut come to lie against the posterior abdominal wall, and their posterior mesenteries become gradually reduced because of pressure from overlying organs (Fig. 2.24). For example, during development, the growing coiled mass of small intestine pushes the part of the gut that will become the descending colon to the left side and presses its mesentery against the posterior abdominal wall. The mesentery is held there until the layer of peritoneum that formed the left side of the mesentery and the part of the visceral peritoneum of the colon lying against the body wall fuse with the parietal peritoneum of the body wall. As a result, the colon becomes fixed to the posterior abdominal wall on the left side with peritoneum covering only its anterior aspect. The descending colon (as well as the ascending colon on the right side) has thus become *secondarily retroperitoneal,* having once been intraperitoneal (Moore et al., 2012).

The layers of peritoneum that fused now form a **fusion fascia,** a connective tissue plane in which the nerves and vessels of the descending colon continue to lie. Thus, the descending colon of the adult can be freed from the posterior body wall (surgically mobilized) by incising the peritoneum along the lateral border of the descending colon and then bluntly dissecting along the plane of the fusion fascia, elevating the neurovascular structures from the posterior body wall until the midline is reached. The ascending colon can be similarly mobilized on the right side.

Several parts of the gastrointestinal tract and associated organs become secondarily retroperitoneal (e.g., most of the duodenum and pancreas as well as the ascending and descending parts of the colon). They are covered with peritoneum only on their anterior surface. Other parts of the viscera (e.g., the sigmoid colon and spleen) retain a relatively short mesentery. However, the roots of the short mesenteries do not remain attached to the midline but shift to the left or right by a fusion process like that described for the descending colon.

Peritoneal Formations

The *peritoneal cavity* has a complex shape. Some of the facts relating to this include the following:

- The peritoneal cavity houses a great length of gut, most of which is covered with peritoneum.
- Extensive continuities are required between the parietal and visceral peritoneum to convey the necessary neurovascular structures from the body wall to the viscera.
- Although the volume of the abdominal cavity is a fraction of the body's volume, the parietal and visceral peritoneum lining the peritoneal cavity within it have a much greater surface area than the body's outer surface (skin); therefore, the peritoneum is highly convoluted.

Various terms are used to describe the parts of the peritoneum that connect organs with other organs, or to the abdominal wall, and the compartments and recesses that are formed as a consequence.

A **mesentery** is a double layer of peritoneum that occurs as a result of the invagination of the peritoneum by an organ and constitutes a continuity of the visceral and parietal peritoneum. It provides a means for neurovascular communications between the organ and the body wall (Fig. 2.25A & E). A mesentery connects an intraperitoneal organ to the body wall—usually the posterior abdominal wall (e.g., mesentery of the small intestine).

The **small intestine mesentery** is usually referred to simply as "the mesentery"; however, mesenteries related to other specific parts of the alimentary tract are named accordingly—for example, the *transverse* and *sigmoid mesocolons* (Fig. 2.25B), *mesoesophagus*, *mesogastrium*, and *mesoappendix*. Mesenteries have a core of connective tissue containing blood and lymphatic vessels, nerves, lymph nodes, and fat (see Fig. 2.48A).

An **omentum** is a double-layered extension or fold of peritoneum that passes from the stomach and proximal part of the duodenum to adjacent organs in the abdominal cavity (Fig. 2.25).

- The **greater omentum** is a prominent, four-layered peritoneal fold that hangs down like an apron from the greater curvature of the stomach and the proximal part of the duodenum (Fig. 2.25A, C, & E). After descending, it folds back and attaches to the anterior surface of the transverse colon and its mesentery.
- The **lesser omentum** is a much smaller, double-layered peritoneal fold that connects the lesser curvature of the stomach and the proximal part of the duodenum to the liver (Fig. 2.25B & D). It also connects the stomach to a triad of structures that run between the duodenum and liver in the free edge of the lesser omentum (Fig. 2.23).

A **peritoneal ligament** consists of a double layer of peritoneum that connects an organ with another organ or to the abdominal wall.

The liver is connected to the:

- Anterior abdominal wall by the *falciform ligament* (Fig. 2.26).
- Stomach by the **hepatogastric ligament,** the membranous portion of the lesser omentum.
- Duodenum by the **hepatoduodenal ligament,** the thickened free edge of the lesser omentum, which conducts the *portal triad:* portal vein, hepatic artery, and bile duct (Figs. 2.23 and 2.26).

The hepatogastric and hepatoduodenal ligaments are continuous parts of the lesser omentum and are separated only for descriptive convenience.

The stomach is connected to the:

- Inferior surface of the diaphragm by the **gastrophrenic ligament.**
- Spleen by the **gastrosplenic ligament,** which reflects to the hilum of the spleen.
- Transverse colon by the **gastrocolic ligament,** the apron-like part of the greater omentum, which descends from the greater curvature, turns under, and then ascends to the transverse colon.

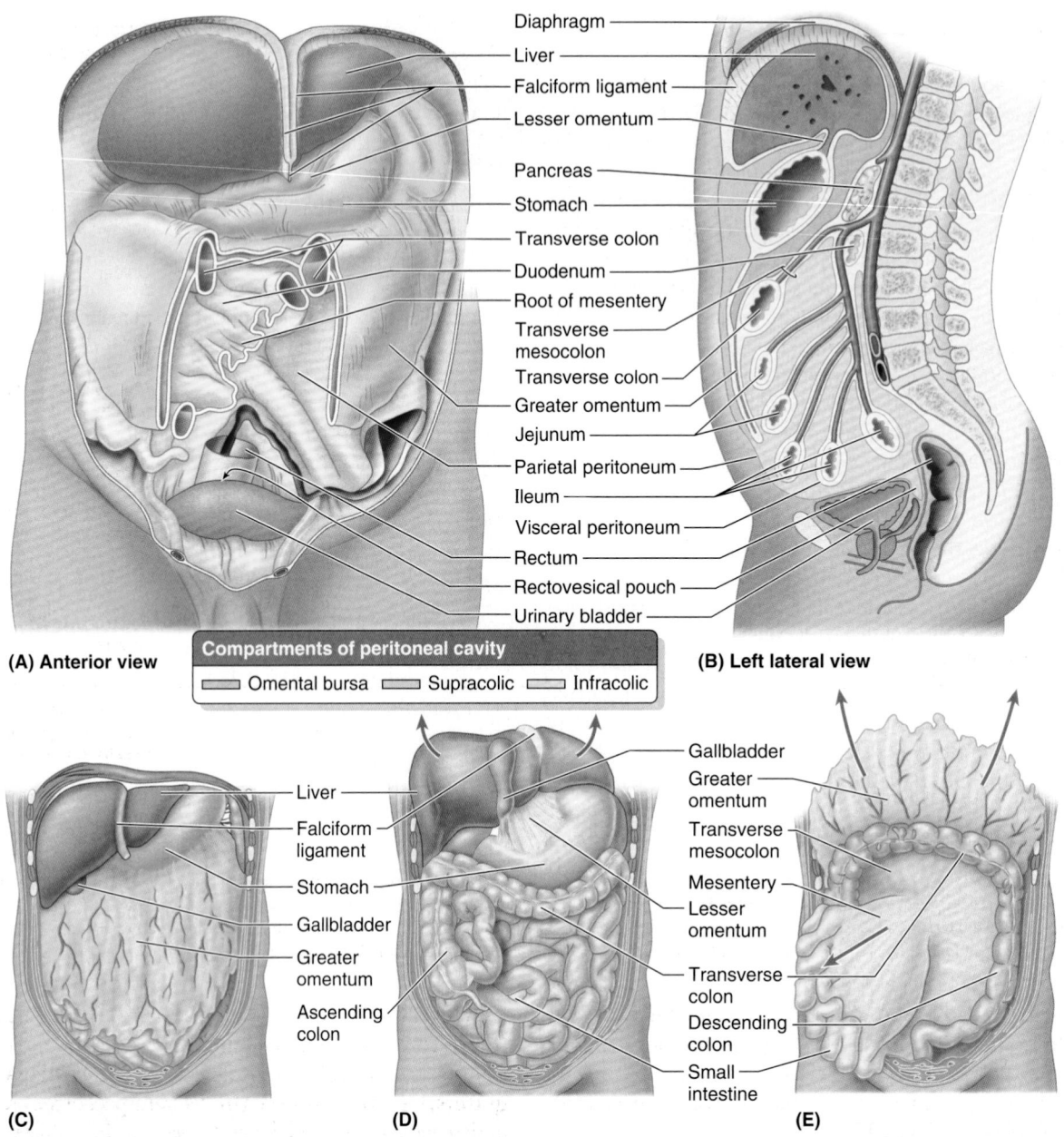

FIGURE 2.25. Principal formations of peritoneum. A. In this opened peritoneal cavity, parts of the greater omentum, transverse colon, and the small intestine and its mesentery have been cut away to reveal deep structures and the layers of the mesenteric structures. The mesentery of the jejunum and ileum (small intestine) and sigmoid mesocolon have been cut close to their parietal attachments. **B.** This median section of the abdominopelvic cavity of a male shows the relationships of the peritoneal attachments. **C.** The greater omentum is shown in its "normal" position, covering most of the abdominal viscera. **D.** The lesser omentum, attaching the liver to the lesser curvature of the stomach, is shown by reflecting the liver and gallbladder superiorly. The greater omentum has been removed from the greater curvature of the stomach and transverse colon to reveal the intestines. **E.** The greater omentum has been reflected superiorly, and the small intestine has been retracted to the right side to reveal the mesentery of the small intestine and the transverse mesocolon.

All these structures have a continuous attachment along the greater curvature of the stomach, and are all part of the greater omentum, separated only for descriptive purposes.

Although intraperitoneal organs may be almost entirely covered with visceral peritoneum, every organ must have an area that is not covered to allow the entrance or exit of neurovascular structures. Such areas are called **bare areas,** formed in relation to the attachments of the peritoneal formations to the organs, including mesenteries,

omenta, and ligaments that convey the neurovascular structures.

A **peritoneal fold** is a reflection of peritoneum that is raised from the body wall by underlying blood vessels, ducts, and ligaments formed by obliterated fetal vessels (e.g., the *umbilical folds* on the internal surface of the anterolateral abdominal wall, Fig. 2.13). Some peritoneal folds contain blood vessels and bleed if cut, such as the lateral umbilical folds, which contain the inferior epigastric arteries.

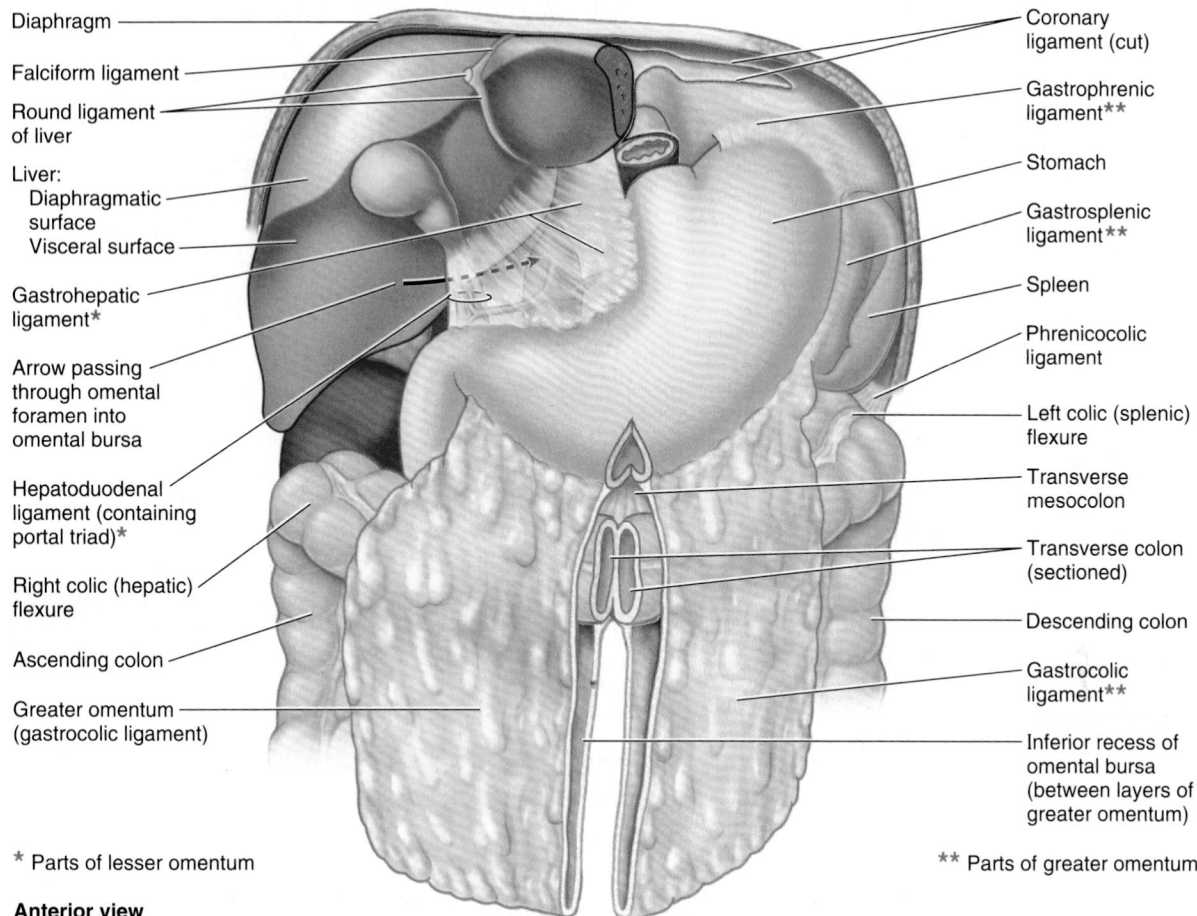

Diaphragm

Falciform ligament

Round ligament
of liver

Liver:
  Diaphragmatic
  surface
  Visceral surface

Gastrohepatic
ligament*

Arrow passing
through omental
foramen into
omental bursa

Hepatoduodenal
ligament (containing
portal triad)*

Right colic (hepatic)
flexure

Ascending colon

Greater omentum
(gastrocolic ligament)

Coronary
ligament (cut)

Gastrophrenic
ligament**

Stomach

Gastrosplenic
ligament**

Spleen

Phrenicocolic
ligament

Left colic (splenic)
flexure

Transverse
mesocolon

Transverse colon
(sectioned)

Descending colon

Gastrocolic
ligament**

Inferior recess of
omental bursa
(between layers of
greater omentum)

* Parts of lesser omentum   ** Parts of greater omentum

Anterior view

FIGURE 2.26.   Parts of greater and lesser omenta. The liver and gallbladder have been reflected superiorly. The central part of the greater omentum has been cut out to show its relation to the transverse colon and mesocolon. The term *greater omentum* is often used as a synonym for the gastrocolic ligament, but it actually also includes the gastrosplenic and gastrophrenic ligaments, all of which have a continuous attachment to the greater curvature of the stomach. The hepatoduodenal ligament (free edge of lesser omentum) conveys the *portal triad*: hepatic artery, bile duct, and portal vein.

A **peritoneal recess,** or **fossa,** is a pouch of peritoneum that is formed by a peritoneal fold (e.g., the inferior recess of the omental bursa between the layers of the greater omentum, and the supravesical and umbilical fossae between the umbilical folds; see Fig. 2.13).

Subdivisions of Peritoneal Cavity

After the rotation and development of the greater curvature of the stomach during development (see the blue box "Brief Review of the Embryological Rotation of Midgut," p. 258), the peritoneal cavity is divided into the greater and lesser peritoneal sacs (Fig. 2.27A). The *greater sac* is the main and larger part of the peritoneal cavity. A surgical incision through the anterolateral abdominal wall enters the greater sac. The *omental bursa* (*lesser sac*) lies posterior to the stomach and lesser omentum.

The **transverse mesocolon** (mesentery of the transverse colon) divides the abdominal cavity into a **supracolic compartment,** containing the stomach, liver, and spleen, and an **infracolic compartment,** containing the small intestine and ascending and descending colon. The infracolic compartment lies posterior to the greater omentum and is divided into **right**

and **left infracolic spaces** by the *mesentery of the small intestine* (Fig. 2.27B). Free communication occurs between the supracolic and the infracolic compartments through the **paracolic gutters,** the grooves between the lateral aspect of the ascending or descending colon and the posterolateral abdominal wall.

The **omental bursa** is an extensive sac-like cavity that lies posterior to the stomach, lesser omentum, and adjacent structures (Figs. 2.23, 2.27A, and 2.28). The omental bursa has a *superior recess,* limited superiorly by the diaphragm and the posterior layers of the coronary ligament of the liver, and an *inferior recess* between the superior parts of the layers of the greater omentum (Figs. 2.26 and 2.28A).

The omental bursa permits free movement of the stomach on the structures posterior and inferior to it because the anterior and posterior walls of the omental bursa slide smoothly over each other. Most of the inferior recess of the bursa becomes sealed off from the main part posterior to the stomach after adhesion of the anterior and posterior layers of the greater omentum (Fig. 2.28B).

The omental bursa communicates with the greater sac through the **omental foramen** (epiploic foramen), an opening situated posterior to the free edge of the lesser omentum

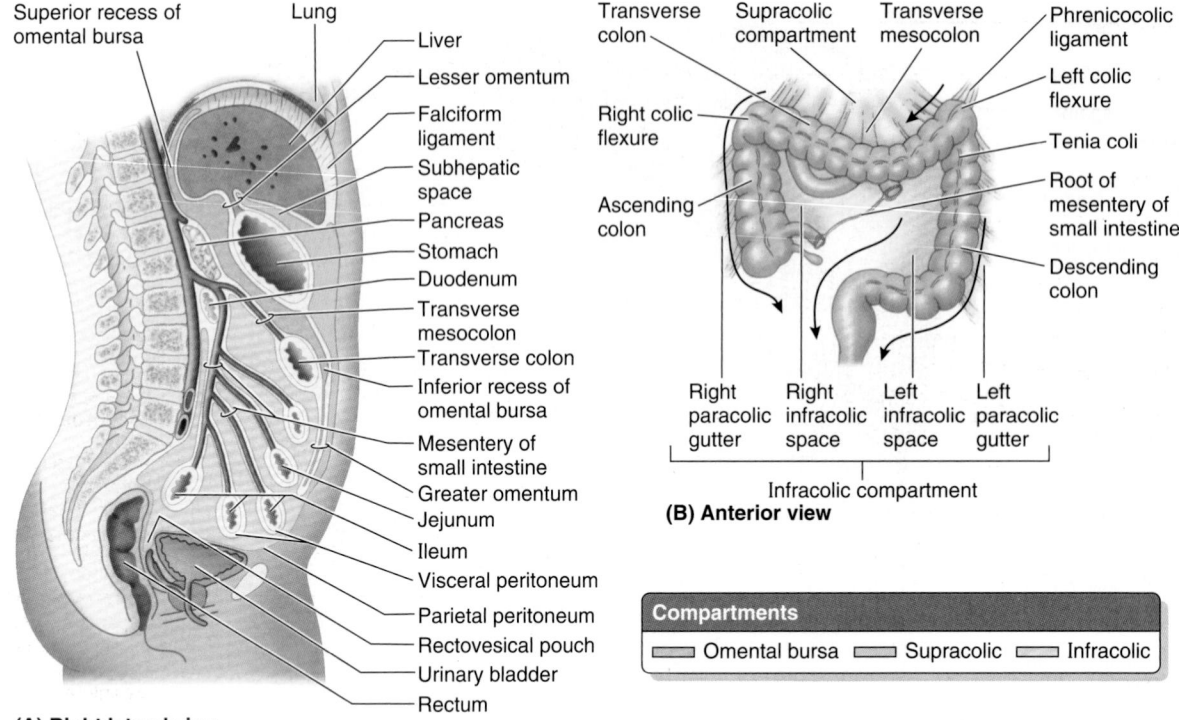

(A) Right lateral view

(B) Anterior view

Compartments

⬜ Omental bursa   ⬜ Supracolic   ⬜ Infracolic

FIGURE 2.27. **Subdivisions of peritoneal cavity. A.** This median section of the abdominopelvic cavity shows the subdivisions of the peritoneal cavity. **B.** The supracolic and infracolic compartments of the greater sac are shown after removal of the greater omentum. The infracolic spaces and paracolic gutters determine the flow of ascitic fluid (*arrows*) when inclined or upright.

(hepatoduodenal ligament). The omental foramen can be located by running a finger along the gallbladder to the free edge of the lesser omentum (Fig. 2.29). The omental foramen usually admits two fingers. The *boundaries of the omental foramen* are

- *Anteriorly:* the hepatoduodenal ligament (free edge of lesser omentum), containing the hepatic portal vein, hepatic artery, and bile duct (Figs. 2.23 and 2.26).

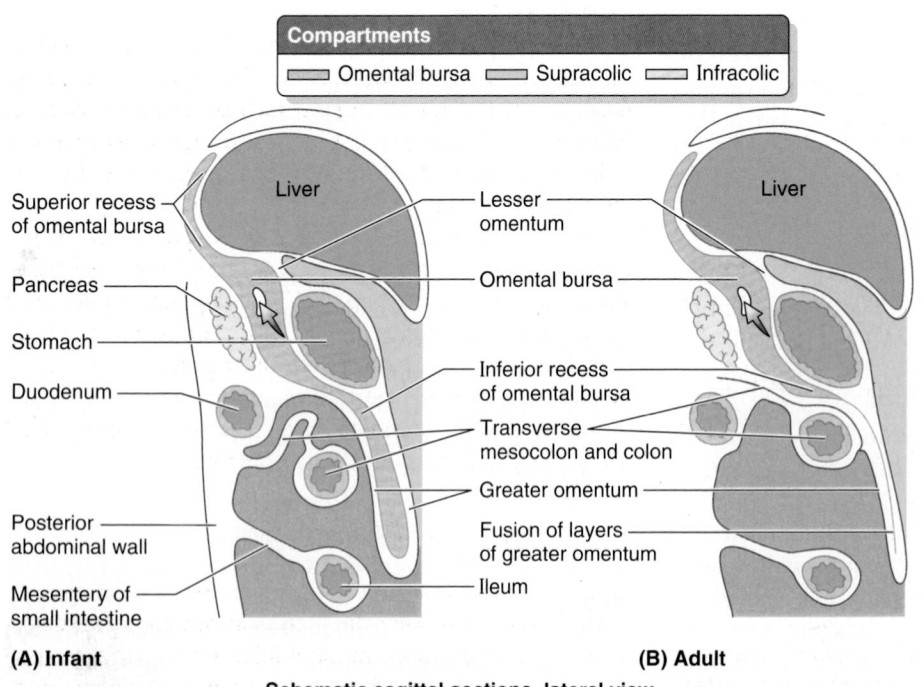

Compartments

⬜ Omental bursa   ⬜ Supracolic   ⬜ Infracolic

(A) Infant   **(B) Adult**

Schematic sagittal sections, lateral view

FIGURE 2.28. **Walls and recesses of omental bursa. A.** This section shows that the omental bursa is an isolated part of the peritoneal cavity, lying dorsal to the stomach and extending superiorly to the liver and diaphragm (superior recess) and inferiorly between the layers of the greater omentum (inferior recess). **B.** This section shows the abdomen after fusion of the layers of the greater omentum. The inferior recess now extends inferiorly only as far as the transverse colon. The red arrows pass from the greater sac through the omental foramen into the omental bursa.

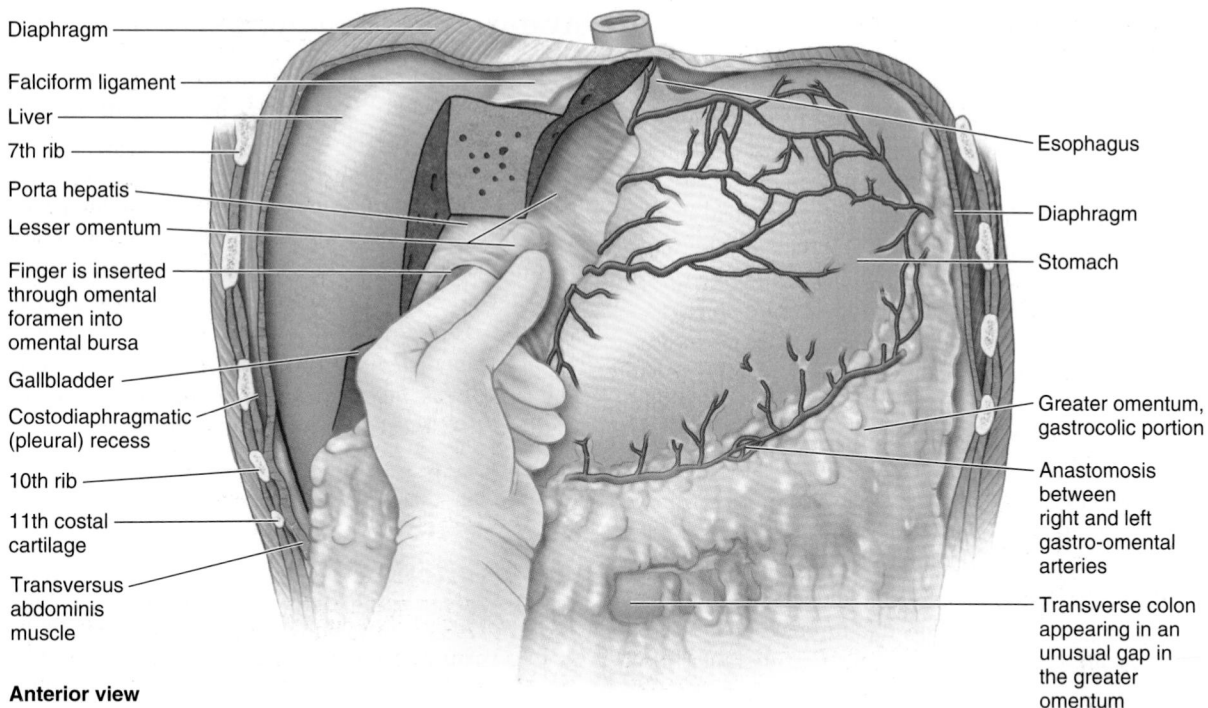

Diaphragm
Falciform ligament
Liver
7th rib
Porta hepatis
Lesser omentum
Finger is inserted through omental foramen into omental bursa
Gallbladder
Costodiaphragmatic (pleural) recess
10th rib
11th costal cartilage
Transversus abdominis muscle

Esophagus
Diaphragm
Stomach
Greater omentum, gastrocolic portion
Anastomosis between right and left gastro-omental arteries
Transverse colon appearing in an unusual gap in the greater omentum

Anterior view

FIGURE 2.29. Omental (epiploic) foramen and omental bursa. The index finger is passing from the greater sac through the omental foramen into the omental bursa (lesser sac). The hepatoduodenal ligament is being pinched between thumb and index finger, which would compress the structures of the portal triad (portal vein, hepatic artery, and bile duct).

- *Posteriorly:* the IVC and a muscular band, the right crus of the diaphragm, covered anteriorly with parietal peritoneum. (They are retroperitoneal.)

- *Superiorly:* the liver, covered with visceral peritoneum (Figs. 2.28 and 2.29).
- *Inferiorly:* the superior or first part of the duodenum.

PERITONEUM AND PERITONEAL CAVITY

Patency and Blockage of Uterine Tubes

While theoretically it is possible for organisms to enter the female peritoneal cavity directly via the uterine tubes, such primary peritonitis is rare, bearing testimony to the effectiveness of the protective mechanisms of the female reproductive tract. A primary mechanism in preventing such infection is a *mucous plug* that effectively blocks the external os (opening) of the uterus to most pathogens, but not to sperms. The *patency of the uterine tubes* can be tested clinically by means of a technique in which air or radiopaque dye is injected into the uterine cavity, from which it normally flows through the uterine tubes into the peritoneal cavity (*hysterosalpingography;* see Chapter 3 for more details).

The Peritoneum and Surgical Procedures

Because the peritoneum is well innervated, patients undergoing abdominal surgery experience more pain with large, invasive, open incisions of the peritoneum

(*laparotomy*) than they do with small laparoscopic incisions or vaginal operations.

It is the covering of peritoneum (often referred to clinically as the *serosa*) that makes watertight end-to-end anastomoses of intraperitoneal organs, such as the small intestine, relatively easy to achieve. It is more difficult to achieve watertight anastomoses of extraperitoneal structures that have an outer adventitial layer, such as the thoracic esophagus.

Because of the high incidence of complications such as peritonitis and adhesions (see the blue box "Peritoneal Adhesions and Adhesiotomy," p. 225) after operations in which the peritoneal cavity is opened, efforts are made to remain outside the peritoneal cavity whenever possible (e.g., translumbar or extraperitoneal anterior approach to the kidneys). When opening the peritoneal cavity is necessary, great effort is made to avoid contamination of the cavity.

Peritonitis and Ascites

When bacterial contamination occurs during laparotomy, or when the gut is traumatically penetrated or ruptured as the result of infection and inflammation (e.g., appendicitis), allowing gas, fecal matter, and bacteria to enter the peritoneal cavity, the result is infection and inflammation of the peritoneum—*peritonitis*. Exudation of serum, fibrin, cells, and pus into the peritoneal cavity occurs, accompanied by pain in the overlying skin and an increase in the tone of the anterolateral abdominal muscles. Given the extent of the peritoneal surfaces and the rapid absorption of material, including bacterial toxins, from the peritoneal cavity, when a peritonitis becomes *generalized* (widespread in the peritoneal cavity), the condition is dangerous and perhaps lethal. In addition to the severe abdominal pain, tenderness, nausea and/or vomiting, fever, and constipation are present.

General peritonitis also occurs when an ulcer perforates the wall of the stomach or duodenum, spilling acid content into the peritoneal cavity. Excess fluid in the peritoneal cavity is called *ascitic fluid*. The clinical condition in which one has ascitic fluid is referred to as *ascites*. Ascites may also occur as a result of mechanical injury (which may also produce internal bleeding) or other pathological conditions, such as portal hypertension (venous congestion), widespread metastasis of cancer cells to the abdominal viscera, and starvation (when plasma proteins fail to be produced, altering concentration gradients and producing a paradoxically protuberant abdomen). In all these cases, the peritoneal cavity may be distended with several liters of abnormal fluid, interfering with movements of the viscera.

Rhythmic movements of the anterolateral abdominal wall normally accompany respirations. If the abdomen is drawn in as the chest expands (*paradoxical abdominothoracic rhythm*) and muscle rigidity is present, either peritonitis or pneumonitis (inflammation of the lungs) may be present. Because the intense pain worsens with movement, people with peritonitis commonly lie with their knees flexed to relax their anterolateral abdominal muscles. They also breath shallowly (and hence more rapidly), reducing the intra-abdominal pressure and pain.

Peritoneal Adhesions and Adhesiotomy

If the peritoneum is damaged, by a stab wound for example, or infected, the peritoneal surfaces become inflamed, making them sticky with *fibrin*. As healing occurs, the fibrin may be replaced with fibrous tissue, forming abnormal attachments between the visceral peritoneum of adjacent viscera, or between the visceral peritoneum of an organ and the parietal peritoneum of the adjacent abdominal wall. *Adhesions* (scar tissue) may also form after an abdominal operation (e.g., owing to a ruptured appendix) and limit the normal movements of the viscera. This tethering may cause chronic pain or emergency complications such as intestinal obstruction when the intestine becomes twisted around an adhesion (*volvulus*).

Adhesiotomy refers to the surgical separation of adhesions. Adhesions are often found during dissection of cadavers (see the adhesion binding the spleen to the diaphragm in Fig. 2.39B, for example).

Abdominal Paracentesis

Treatment of generalized peritonitis includes removal of the ascitic fluid and, in the presence of infection, administration of large doses of antibiotics. Occasionally, more localized accumulations of fluid may have to be removed for analysis. Surgical puncture of the peritoneal cavity for the aspiration or drainage of fluid is called *paracentesis*. After injection of a local anesthetic agent, a needle or trocar and a cannula are inserted through the anterolateral abdominal wall into the peritoneal cavity through the linea alba, for example. The needle is inserted superior to the empty urinary bladder, in a location that avoids the inferior epigastric artery.

Intraperitoneal Injection and Peritoneal Dialysis

The peritoneum is a semipermeable membrane with an extensive surface area, much of which (subdiaphragmatic portions in particular) overlies blood and lymphatic capillary beds. Therefore, fluid injected into the peritoneal cavity is absorbed rapidly. For this reason, anesthetic agents, such as solutions of barbiturate compounds, may be injected into the peritoneal cavity by *intraperitoneal (I.P.) injection*.

In *renal failure,* waste products such as urea accumulate in the blood and tissues and ultimately reach fatal levels. *Peritoneal dialysis* may be performed in which soluble substances and excess water are removed from the system by transfer across the peritoneum, using a dilute sterile solution that is introduced into the peritoneal cavity on one side and then drained from the other side. Diffusible solutes and water are transferred between the blood and the peritoneal cavity as a result of concentration gradients between the two fluid compartments. Peritoneal dialysis is usually employed only temporarily, however. For the long term, it is preferable to use direct blood flow through a renal dialysis machine.

Functions of Greater Omentum

The greater omentum, large and fat laden, prevents the visceral peritoneum from adhering to the parietal peritoneum. It has considerable mobility and moves around the peritoneal cavity with peristaltic movements of the viscera. It often forms adhesions adjacent to an inflamed organ, such as the appendix, sometimes walling it off and thereby protecting other viscera from it. Thus, it is common when entering the abdominal cavity, in either dissection or surgery, to find the omentum markedly displaced from the "normal" position in which it is almost always depicted in anatomical illustrations. The greater omentum also cushions the abdominal organs against injury and forms insulation against loss of body heat.

Abscess Formation

 Perforation of a duodenal ulcer, rupture of the gallbladder, or perforation of the appendix may lead to the formation of a an *abscess* (circumscribed collection of purulent exudate, i.e., pus) in the subphrenic recess. The abscess may be walled inferiorly by adhesions (see the blue box "Subphrenic Abscesses," p. 283).

Spread of Pathological Fluids

 Peritoneal recesses are of clinical importance in connection with the spread of pathological fluids such as pus, a product of inflammation. The recesses determine the extent and direction of the spread of fluids that may enter the peritoneal cavity when an organ is diseased or injured.

Flow of Ascitic Fluid and Pus

 The *paracolic gutters* are of clinical importance because they provide pathways for the flow of ascitic fluid and the spread of intraperitoneal infections (Fig. 2.27B). Purulent material (consisting of or containing pus) in the abdomen can be transported along the paracolic gutters into the pelvis, especially when the person is upright. Thus, to facilitate the flow of exudate into the pelvic cavity where absorption of toxins is slow, patients with peritonitis are often placed in the sitting position (at least a 45° angle). Conversely, infections in the pelvis may extend superiorly to a subphrenic recess situated under the diaphragm (see the blue box "Subphrenic Abscesses," p. 283), especially when the person is supine. Similarly, the paracolic gutters provide pathways for the spread of cancer cells that have sloughed from the ulcerated surface of a tumor and entered the peritoneal cavity.

Fluid in Omental Bursa

 Perforation of the posterior wall of the stomach results in the passage of its fluid contents into the omental bursa. An inflamed or injured pancreas can also result in the passage of pancreatic fluid into the bursa, forming a *pancreatic pseudo-cyst*.

Intestine in Omental Bursa

 Although uncommon, a loop of small intestine may pass through the omental foramen into the omental bursa and be strangulated by the edges of the foramen. As none of the boundaries of the foramen can be incised because each contains blood vessels, the swollen intestine must be decompressed using a needle so it can be returned to the greater sac of the peritoneal cavity through the omental foramen.

Severance of Cystic Artery

 The cystic artery must be ligated or clamped and then severed during *cholecystectomy,* removal of the gallbladder. Sometimes, however, the artery is accidentally severed before it has been adequately ligated. The surgeon can control the hemorrhage by compressing the hepatic artery as it traverses the hepatoduodenal ligament. The index finger is placed in the omental foramen and the thumb on its anterior wall (Fig. 2.29). Alternate compression and release of pressure on the hepatic artery allows the surgeon to identify the bleeding artery and clamp it.

The Bottom Line

PERITONEUM, PERITONEAL CAVITY, AND PERITONEAL FORMATIONS

Peritoneum and peritoneal cavity: The peritoneum is a continuous, serous membrane that lines the abdominopelvic cavity (the parietal peritoneum) and the contained viscera (the visceral peritoneum). ♦ The collapsed peritoneal cavity between the parietal and visceral peritoneum normally contains only enough peritoneal fluid (about 50 mL) to lubricate the inner surface of the peritoneum. This arrangement allows the gut the freedom of movement required for alimentation (digestion). ♦ Adhesions formed as a result of infection or injury interfere with these movements. ♦ The parietal peritoneum is a sensitive, semipermeable membrane, with blood and lymphatic capillary beds especially abundant deep to its subdiaphragmatic surface.

Peritoneal formations and subdivisions of peritoneal cavity: Continuities and connections between the visceral and parietal peritoneum occur where the gut enters and exits the abdominopelvic cavity. ♦ Parts of the peritoneum also occur as double folds (mesenteries and omenta, and subdivisions called ligaments) that convey neurovascular structures and the ducts of accessory organs to and from the viscera. ♦ Peritoneal ligaments are named for the particular structures connected by them. ♦ As a result of the rotation and exuberant growth of the intestine during development, the disposition of the peritoneal cavity becomes complex. The main part of the peritoneal cavity (greater sac) is divided by the transverse mesocolon into supracolic and infracolic compartments. ♦ A smaller part of the peritoneal cavity, the omental bursa (lesser sac) lies posterior to the stomach, separating it from retroperitoneal viscera on the posterior wall. It communicates with the greater sac via the omental foramen. ♦ The complex disposition of the peritoneal cavity determines the flow and pooling of excess (ascitic) fluid occupying the peritoneal cavity during pathological conditions.

ABDOMINAL VISCERA

Overview of Abdominal Viscera and Digestive Tract

The viscera of the abdomen comprise the majority of the alimentary system: are the terminal part of the esophagus and the stomach, intestines, spleen, pancreas, liver, gallbladder, kidneys, and suprarenal glands (Figs. 2.30 and 2.31). When the abdominal cavity is opened to study these organs, it becomes evident that the liver, stomach, and spleen almost fill the domes of the diaphragm. Because they indent the thoracic cavity, they receive protection from the lower thoracic cage. It is also evident that the *falciform ligament* normally attaches along a continuous line to the anterior abdominal wall as far inferiorly as the umbilicus. It divides the liver superficially into right and left lobes. The fat-laden *greater omentum,* when in its typical

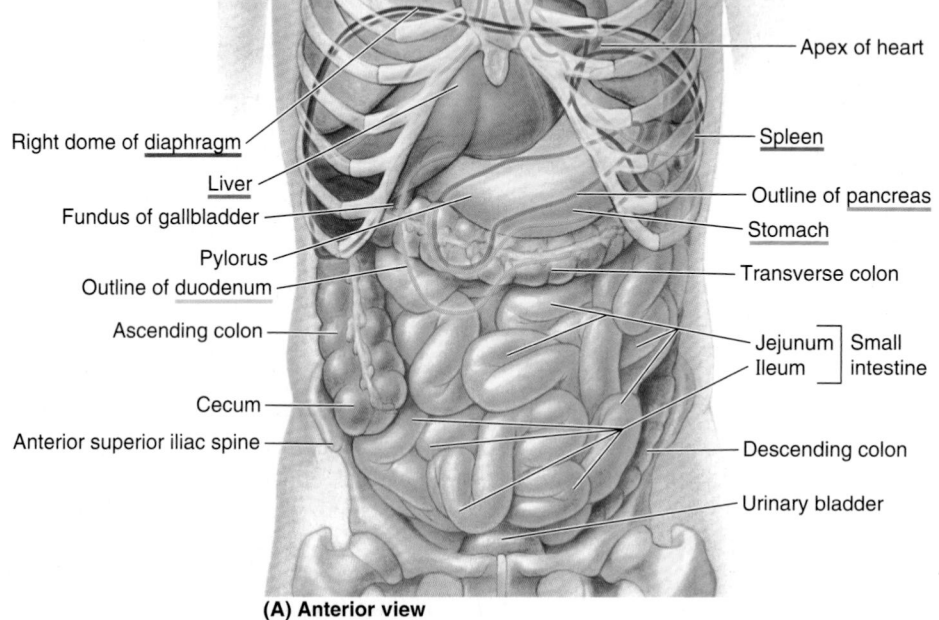

Apex of heart
Right dome of diaphragm
Liver
Fundus of gallbladder
Pylorus
Outline of duodenum
Ascending colon
Cecum
Anterior superior iliac spine
Spleen
Outline of pancreas
Stomach
Transverse colon
Jejunum ⎤ Small
Ileum ⎦ intestine
Descending colon
Urinary bladder

(A) Anterior view

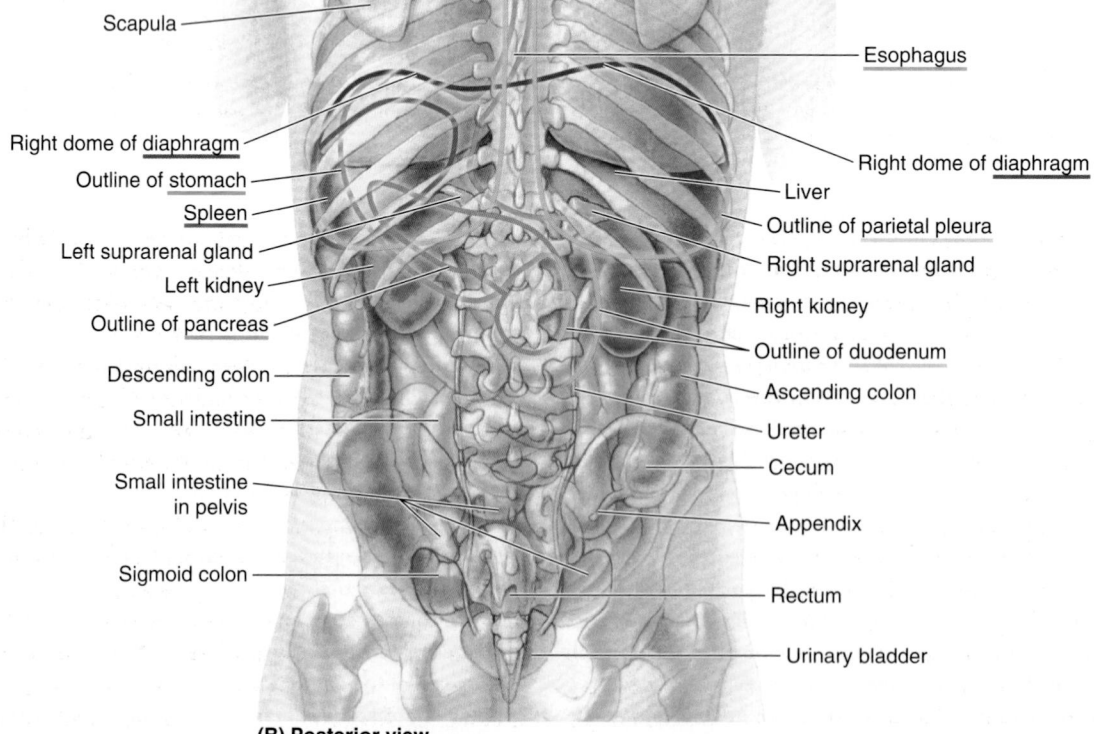

Scapula
Right dome of diaphragm
Outline of stomach
Spleen
Left suprarenal gland
Left kidney
Outline of pancreas
Descending colon
Small intestine
Small intestine in pelvis
Sigmoid colon
Esophagus
Right dome of diaphragm
Liver
Outline of parietal pleura
Right suprarenal gland
Right kidney
Outline of duodenum
Ascending colon
Ureter
Cecum
Appendix
Rectum
Urinary bladder

(B) Posterior view

FIGURE 2.30. Overview of thoracic and abdominal viscera. A and B. Some abdominal organs extend superiorly into the thoracic cage and are protected by it. Partially protected by the lowest ribs, the right kidney is lower than the left kidney, owing to the mass of the liver on the right side. A large part of the small intestine is in the pelvis.

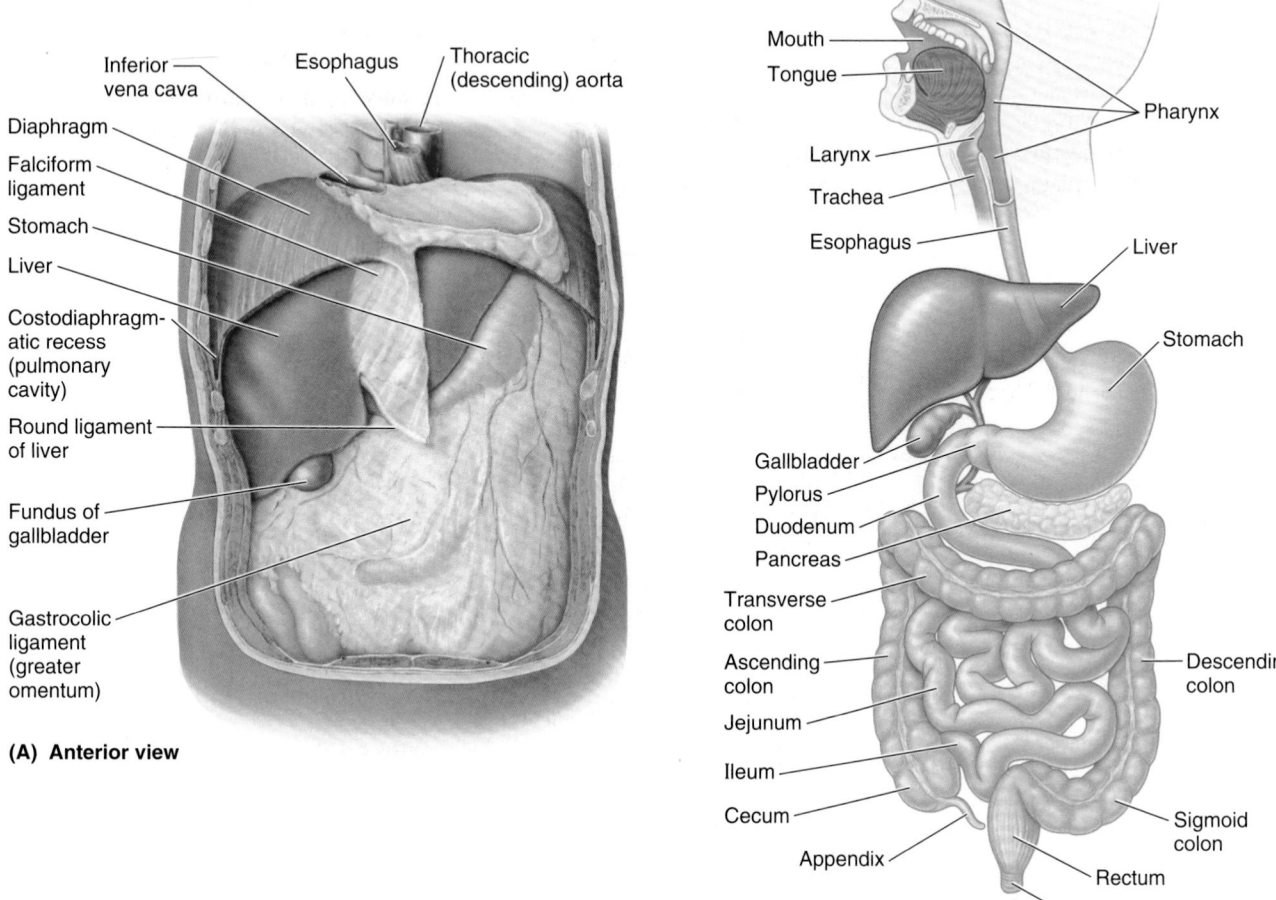

(A) Anterior view

Inferior vena cava
Esophagus
Thoracic (descending) aorta
Diaphragm
Falciform ligament
Stomach
Liver
Costodiaphragmatic recess (pulmonary cavity)
Round ligament of liver
Fundus of gallbladder
Gastrocolic ligament (greater omentum)

Mouth
Tongue
Pharynx
Larynx
Trachea
Esophagus
Liver
Stomach
Gallbladder
Pylorus
Duodenum
Pancreas
Transverse colon
Ascending colon
Descending colon
Jejunum
Ileum
Cecum
Sigmoid colon
Appendix
Rectum
Anal canal

(B) Diagrammatic anterior view; medial view of bisected head

FIGURE 2.31. **Abdominal contents *in situ* and in relation to alimentary system. A.** The undisturbed abdominal contents are shown. The anterior abdominal and thoracic walls are cut away. The falciform ligament is severed at its attachment to the anterior abdominal wall. **B.** Overview of **alimentary system,** consisting of the digestive tract from the mouth to the anus, with all of its accessory glands and organs.

position, conceals almost all of the intestine. The gallbladder projects inferior to the sharp border of the liver (Fig. 2.31A).

Food passes from the mouth and pharynx through the *esophagus* to the *stomach,* where it mixes with gastric secretions (Fig. 2.31B). Digestion mostly occurs in the stomach and duodenum. **Peristalsis,** a series of ring-like contraction waves, begins around the middle of the stomach and moves slowly toward the *pylorus.* It is responsible for mixing the masticated (chewed) food mass with gastric juices and for emptying the contents of the stomach into the duodenum.

Absorption of chemical compounds occurs principally in the *small intestine,* a coiled 5- to 6-m-long tube (shorter in life, when tonus is present, than in the cadaver) consisting of the *duodenum, jejunum,* and *ileum.* Peristalsis also occurs in the jejunum and ileum; however, it is not forceful unless an obstruction is present. The stomach is continuous with the duodenum, which receives the openings of the ducts from the *pancreas* and *liver,* the major glands of the alimentary system.

The *large intestine* consists of the *cecum* (which receives the terminal part of the ileum), *appendix, colon* (ascending, transverse, descending, and sigmoid), *rectum,* and *anal canal.* Most reabsorption of water occurs in the ascending colon. Feces form in the descending and sigmoid colon and accumulate in the rectum before defecation. The esophagus, stomach, and small and large intestines constitute the **gastrointestinal tract** and are derived from the *primordial foregut, midgut,* and *hindgut.*

The arterial supply to the abdominal part of the alimentary system is from the *abdominal aorta.* The three major branches of the aorta supplying it are the *celiac trunk* and the *superior* and *inferior mesenteric arteries* (Fig. 2.32A).

The *hepatic portal vein* is formed by the union of the *superior mesenteric* and *splenic veins* (Fig. 2.32B). It is the main channel of the *portal venous system,* which collects blood from the abdominal part of the alimentary tract, pancreas, spleen, and most of the gallbladder, and carries it to the liver.

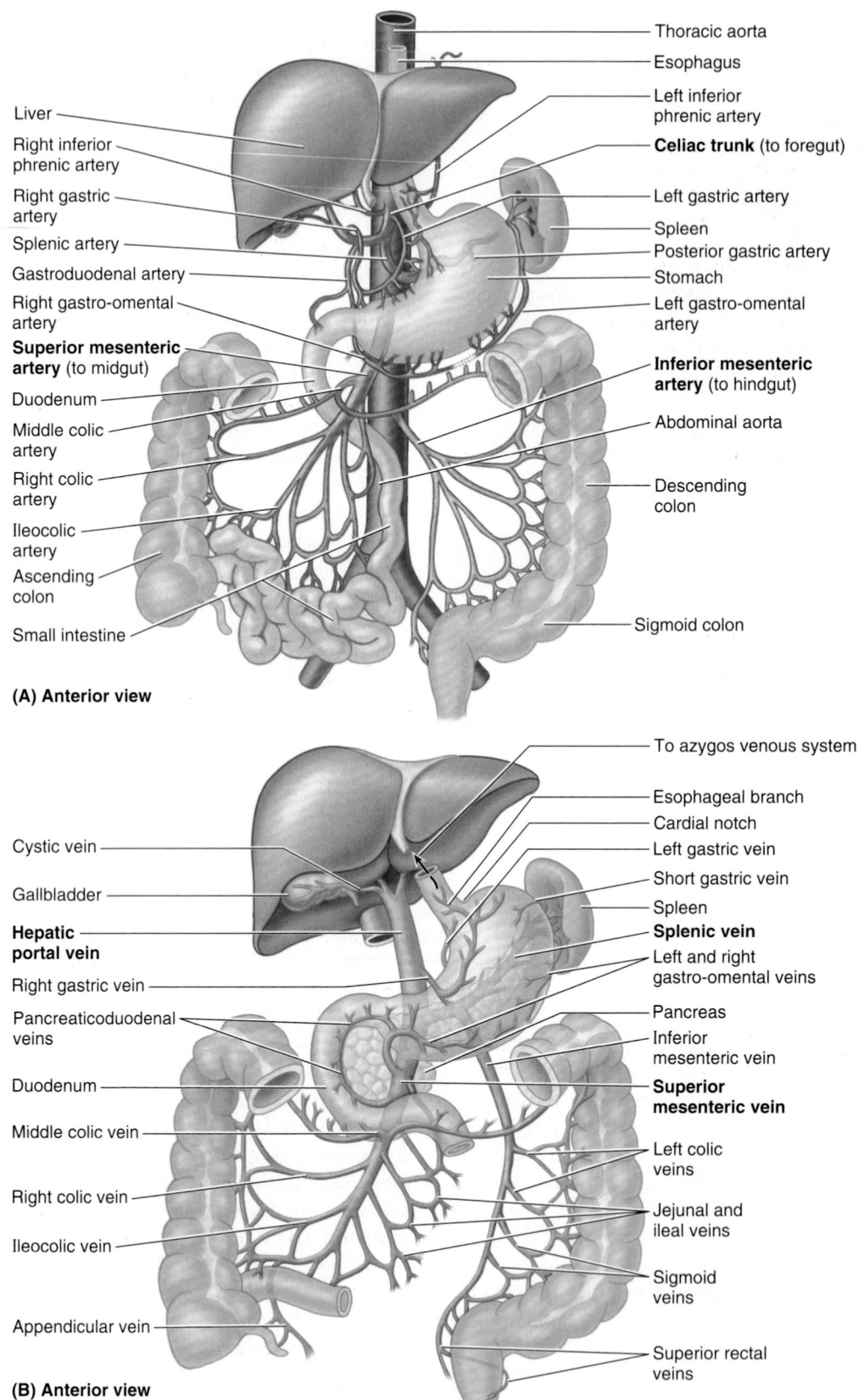

(A) Anterior view

(B) Anterior view

FIGURE 2.32. Arterial supply and venous drainage of abdominal parts of alimentary system. A. The arterial supply is demonstrated. The three unpaired branches of the abdominal aorta supply, in succession, the derivatives of the foregut, midgut, and hindgut. **B.** The venous drainage is shown. The nutrient-rich blood from the gastrointestinal tract and that from the spleen, pancreas, and gallbladder all drain to the liver via the portal vein. The *black arrow* indicates the communication of the esophageal vein with the azygos (systemic) venous system.

Esophagus

The **esophagus** is a muscular tube (approximately 25 cm [10 in] long) with an average diameter of 2 cm that conveys food from the pharynx to the stomach (Fig. 2.33A). As seen during *fluoroscopy* (x-ray, using a fluoroscope) after a barium swallow (Fig. 2.34), the esophagus normally has three constrictions where adjacent structures produce impressions:

- **Cervical constriction (upper esophageal sphincter):** at its beginning at the **pharyngoesophageal junction,** approximately 15 cm from the incisor teeth; caused by the *cricopharyngeus muscle* (see Chapter 8).
- **Thoracic (broncho-aortic) constriction:** a compound constriction where it is first crossed by the arch of the aorta, 22.5 cm from the incisor teeth, and then where it is crossed by the left main bronchus, 27.5 cm from the

incisor teeth; the former is seen in anteroposterior views, the latter in lateral views.

- **Diaphragmatic constriction:** where it passes through the *esophageal hiatus* of the diaphragm, approximately 40 cm from the incisor teeth (Fig. 2.33A).

Awareness of these constrictions is important when passing instruments through the esophagus into the stomach, and when viewing radiographs of patients who are experiencing *dysphagia* (difficulty in swallowing).

The esophagus:

- Follows the curve of the vertebral column as it descends through the neck and mediastinum—the median partition of the thoracic cavity (Fig. 2.33A).
- Has internal *circular* and external *longitudinal layers of muscle* (Fig. 2.33B). In its superior third, the external layer consists of voluntary striated muscle; the inferior

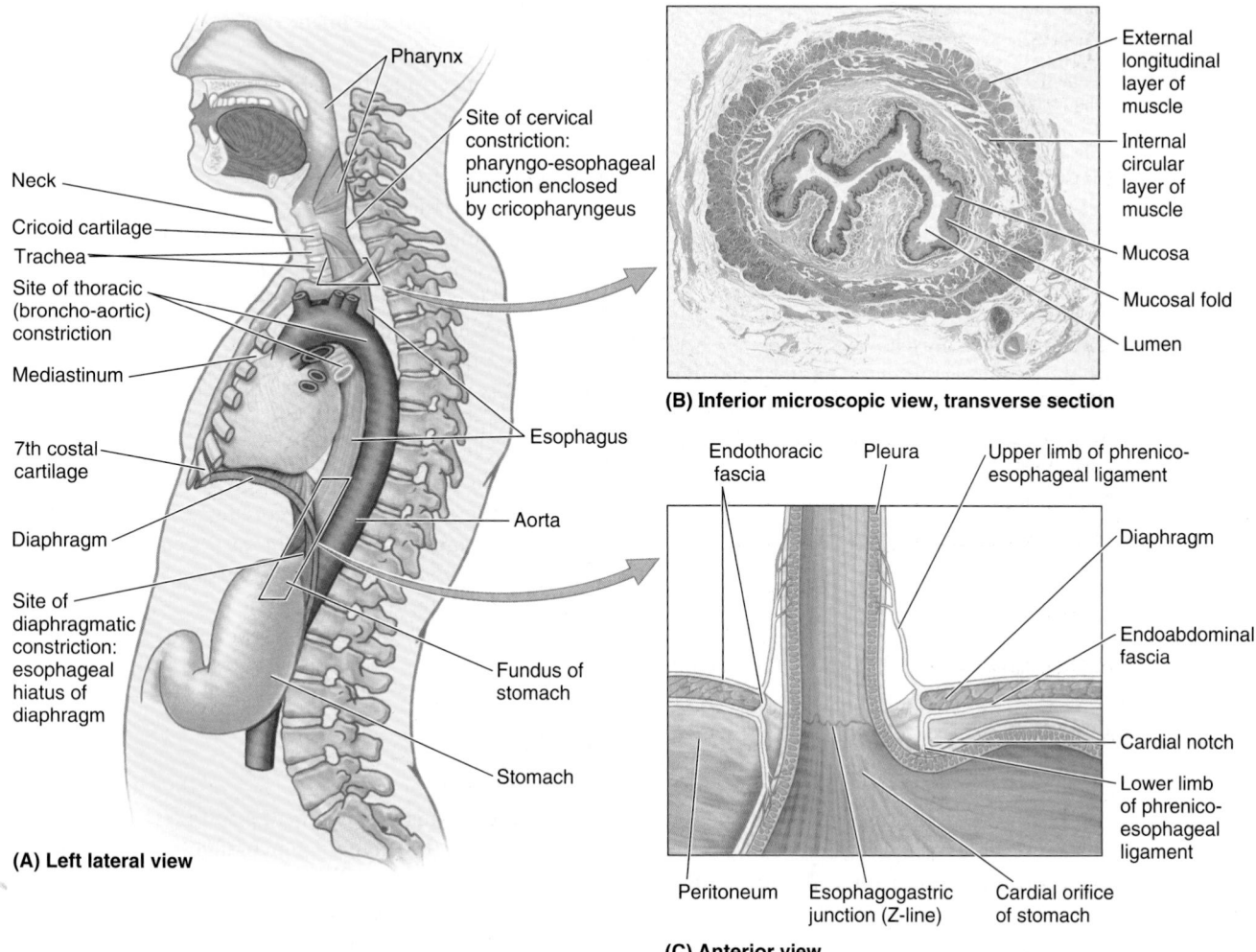

(A) Left lateral view

(B) Inferior microscopic view, transverse section

(C) Anterior view

FIGURE 2.33. The esophagus and its relationships. A. This view shows the full length of the esophagus and structures related to it. The esophagus begins at the level of the cricoid cartilage and descends posterior to the trachea. It leaves the thorax through the esophageal hiatus of the diaphragm. **B.** The transverse section of the esophagus shows the double muscular and plicated mucosal layers of its wall. **C.** A coronal section of the inferior esophagus, diaphragm, and superior stomach is shown. The phrenicoesophageal ligament connects the esophagus flexibly to the diaphragm; it limits upward movement of the esophagus while permitting some movement during swallowing and respiration.

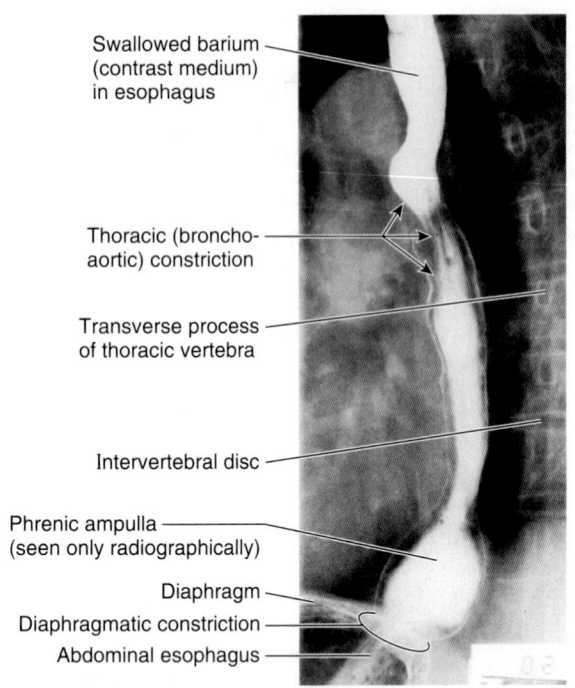

Swallowed barium
(contrast medium)
in esophagus

Thoracic (broncho-
aortic) constriction

Transverse process
of thoracic vertebra

Intervertebral disc

Phrenic ampulla
(seen only radiographically)

Diaphragm
Diaphragmatic constriction
Abdominal esophagus

FIGURE 2.34. Radiograph of esophagus after swallowing barium meal. This left posterior oblique (LPO) view demonstrates two of the three normal "constrictions" (impressions) caused by the arch of the aorta and left main bronchus. The phrenic ampulla, which is seen only radiographically, is the distensible part of the esophagus superior to the diaphragm. (Courtesy of Dr. E. L. Lansdown, Professor of Medical Imaging, University of Toronto, Toronto, ON, Canada.)

third is composed of smooth muscle, and the middle third is made up of both types of muscle.

- Passes through the elliptical *esophageal hiatus* in the muscular *right crus of the diaphragm*, just to the left of the median plane at the level of the T10 vertebra.
- Terminates by entering the stomach at the *cardial orifice of the stomach* (Fig. 2.33C) to the left of the midline at the level of the 7th left costal cartilage and T11 vertebra.
- Is encircled by the esophageal (nerve) plexus distally (Fig. 2.35).

Food passes through the esophagus rapidly because of the peristaltic action of its musculature, aided by but not dependent on gravity (one can still swallow if inverted). The esophagus is attached to the margins of the esophageal hiatus in the diaphragm by the **phrenico-esophageal ligament** (Fig. 2.33C), an extension of inferior diaphragmatic fascia. This ligament permits independent movement of the diaphragm and esophagus during respiration and swallowing.

The trumpet-shaped **abdominal part of the esophagus,** only 1.25 cm long, passes from the esophageal hiatus in the right crus of the diaphragm to the *cardial orifice of the stomach*, widening as it approaches, passing anteriorly and to the left as it descends inferiorly. Its anterior surface is covered with peritoneum of the greater sac, continuous with that covering the anterior surface of the stomach. It fits into a groove on the posterior (visceral) surface of the liver.

The posterior surface of the abdominal part of the esophagus is covered with peritoneum of the omental bursa, continuous with that covering the posterior surface of the stomach. The right border of the abdominal esophagus is continuous with the lesser curvature of the stomach; however, its left border is separated from the fundus of the stomach by the *cardial notch* between the esophagus and fundus (Fig. 2.37A).

The **esophagogastric junction** lies to the left of the T11 vertebra on the horizontal plane that passes through the tip of the xiphoid process. Surgeons and endoscopists designate the **Z-line** (Fig. 2.33C), a jagged line where the mucosa abruptly changes from esophageal to gastric mucosa, as the junction. Immediately superior to this junction, the diaphragmatic musculature forming the esophageal hiatus functions as a physiological *inferior esophageal sphincter* that contracts and relaxes. Radiologic studies show that food stops here momentarily and that the sphincter mechanism is normally efficient in preventing reflux of gastric contents into the esophagus. When one is not eating, the lumen of the esophagus is normally collapsed superior to this level to prevent food or stomach juices from regurgitating into the esophagus.

Details concerning the neurovasculature of the cervical and thoracic portions of the esophagus are provided in Chapters 1 and 8. The arterial supply of the abdominal part of the esophagus is from the *left gastric artery*, a branch of the celiac trunk, and the *left inferior phrenic artery* (Fig. 2.32A). The venous drainage from the *submucosal veins* of this part of the esophagus is both to the *portal venous system* through the *left gastric vein* (Fig. 2.32B) and into the *systemic venous system* through **esophageal veins** entering the *azygos vein*.

The lymphatic drainage of the abdominal part of the esophagus is into the *left gastric lymph nodes* (Fig. 2.35); efferent lymphatic vessels from these nodes drain mainly to *celiac lymph nodes*.

The esophagus is innervated by the **esophageal plexus,** formed by the *vagal trunks* (becoming anterior and posterior gastric branches), and the *thoracic sympathetic trunks* via the *greater (abdominopelvic) splanchnic nerves* and *periarterial plexuses* around the left gastric and inferior phrenic arteries. (See also "Summary of Innervation of Abdominal Viscera," p. 300.)

Stomach

The **stomach** is the expanded part of the digestive tract between the esophagus and small intestine (Fig. 2.31B). It is specialized for the accumulation of ingested food, which it chemically and mechanically prepares for digestion and passage into the duodenum. The stomach acts as a food blender and reservoir; its chief function is enzymatic digestion. The *gastric juice* gradually converts a mass of food into a semiliquid mixture, *chyme* (G. juice), which passes fairly quickly into the duodenum. An empty stomach is only of slightly larger caliber than the large intestine; however, it is capable of considerable expansion and can hold 2–3 L of food.

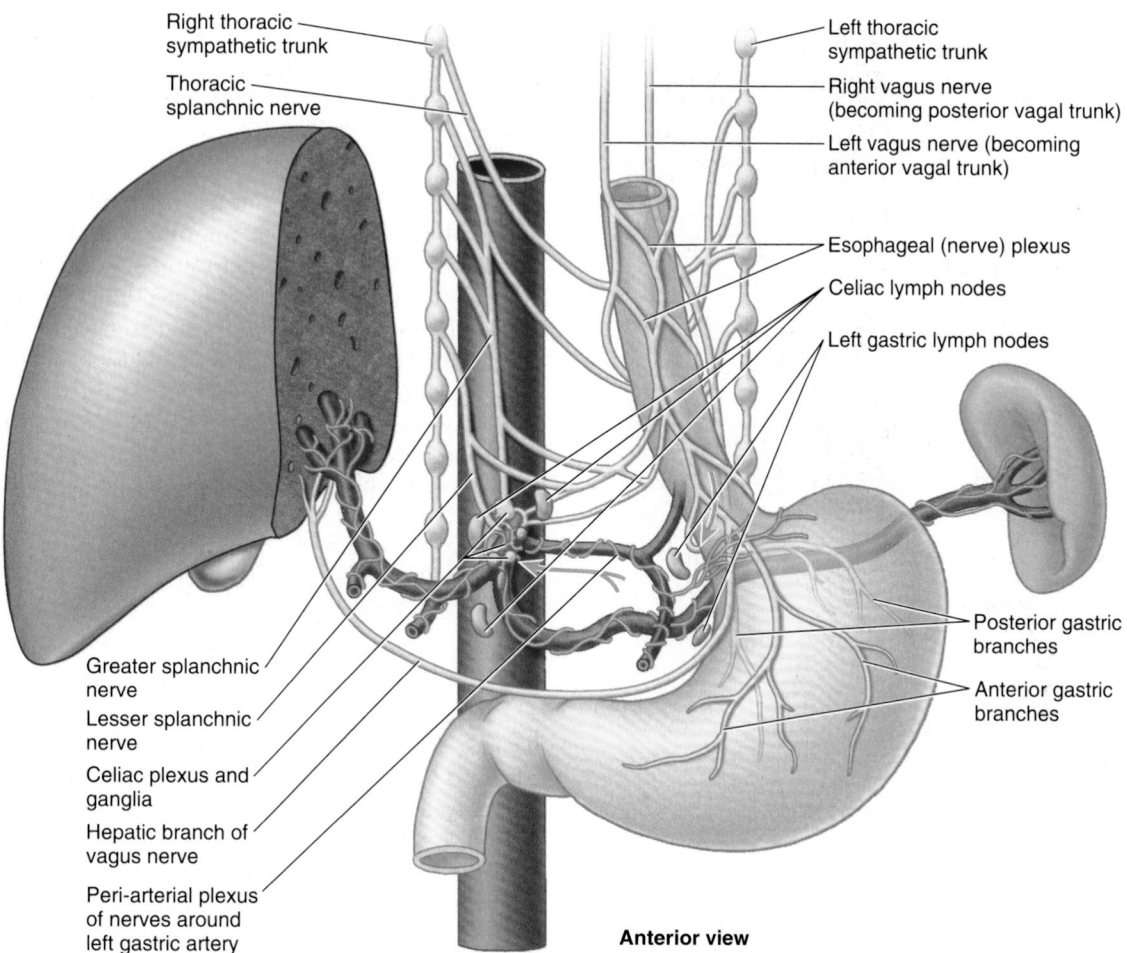

Right thoracic sympathetic trunk

Thoracic splanchnic nerve

Left thoracic sympathetic trunk

Right vagus nerve (becoming posterior vagal trunk)

Left vagus nerve (becoming anterior vagal trunk)

Esophageal (nerve) plexus

Celiac lymph nodes

Left gastric lymph nodes

Posterior gastric branches

Anterior gastric branches

Greater splanchnic nerve

Lesser splanchnic nerve

Celiac plexus and ganglia

Hepatic branch of vagus nerve

Peri-arterial plexus of nerves around left gastric artery

Anterior view

FIGURE 2.35. Nerves and lymphatics of abdominal esophagus and stomach. The vagus nerves (CN X) divide into branches that form the esophageal (nerve) plexus around the inferior esophagus. Anterior and posterior gastric branches of the plexus accompany the esophagus through the esophageal hiatus for distribution to the anterior and posterior aspects of the stomach. The anterior branches also extend to the pylorus and liver. Postsynaptic sympathetic nerve fibers from the celiac plexus are distributed to these organs through peri-arterial plexuses. The lymphatic vessels of the stomach follow a pattern similar to that of the arteries, although the flow is in the opposite direction. Thus, lymph from the stomach and abdominal part of the esophagus drains to the gastric and then celiac lymph nodes.

POSITION, PARTS, AND SURFACE ANATOMY OF STOMACH

The size, shape, and position of the stomach can vary markedly in persons of different body types (bodily habitus) and may change even in the same individual as a result of diaphragmatic movements during respiration, the stomach's contents (empty vs. after a heavy meal), and the position of the person. In the supine position, the stomach commonly lies in the right and left upper quadrants, or epigastric, umbilical, and left hypochondrium and flank regions (Fig. 2.36A). In the erect position, the stomach moves inferiorly. In asthenic (thin, weak) individuals, the body of stomach may extend into the pelvis (Fig. 2.36B).

The stomach has four parts (Figs. 2.36A and 2.37A–C):

- **Cardia:** the part surrounding the **cardial orifice** (opening), the superior opening or inlet of the stomach. In the supine position, the cardial orifice usually lies posterior to

the 6th left costal cartilage, 2–4 cm from the median plane at the level of the T11 vertebra.
- **Fundus:** the dilated superior part that is related to the left dome of the diaphragm, limited inferiorly by the horizontal plane of the cardial orifice. The **cardial notch** is between the esophagus and the fundus. The fundus may be dilated by gas, fluid, food, or any combination of these. In the supine position, the fundus usually lies posterior to the left 6th rib in the plane of the MCL (Fig. 2.36A).
- **Body:** the major part of the stomach between the fundus and pyloric antrum.
- **Pyloric part:** the funnel-shaped outflow region of the stomach; its wider part, the **pyloric antrum,** leads into the **pyloric canal,** its narrower part (Fig. 2.37A–E). The **pylorus** (G., gatekeeper) is the distal, sphincteric region of the pyloric part. It is a marked thickening of the circular layer of smooth muscle that controls discharge of the stomach contents through the **pyloric orifice** (infe-

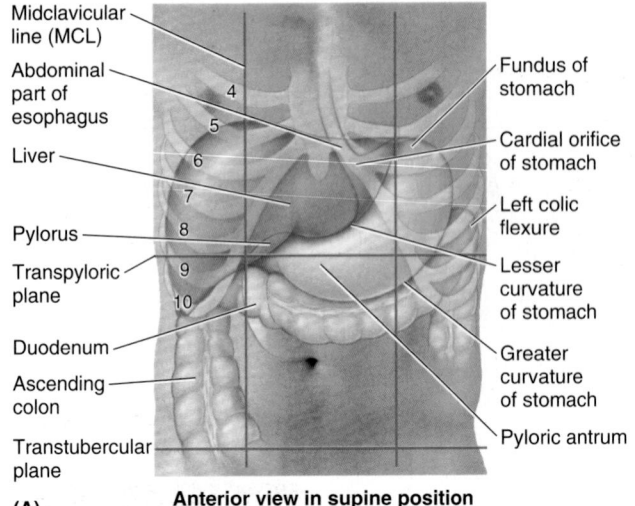

Midclavicular line (MCL)
Abdominal part of esophagus
Liver
Pylorus
Transpyloric plane
Duodenum
Ascending colon
Transtubercular plane

4
5
6
7
8
9
10

Fundus of stomach
Cardial orifice of stomach
Left colic flexure
Lesser curvature of stomach
Greater curvature of stomach
Pyloric antrum

(A)   **Anterior view in supine position**

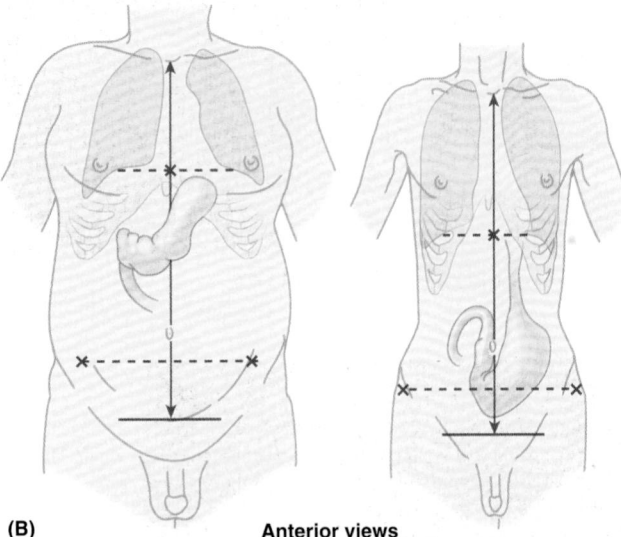

(B)   **Anterior views**

FIGURE 2.36. Surface anatomy and effect of body type (bodily habitus) on disposition and shape of stomach. A. The most common position of the stomach in a person of medium build in the supine or prone position. **B.** A heavily built hyperasthenic individual with a short thorax and long abdomen is likely to have a stomach that is placed high and more transversely disposed. In people with a slender asthenic physique, the stomach is likely to be low and vertical.

rior opening or outlet of the stomach) into the duodenum (Fig. 2.37D). Intermittent emptying of the stomach occurs when intragastric pressure overcomes the resistance of the pylorus. The pylorus is normally tonically contracted so that the pyloric orifice is reduced, except when emitting *chyme* (semifluid mass). At irregular intervals, *gastric peristalsis* pushes the chyme through the pyloric canal and orifice into the small intestine for further mixing, digestion, and absorption.

In the supine position, the pyloric part of the stomach lies at the level of the **transpyloric plane,** midway between the jugular notch superiorly and the pubic crest inferiorly

(Fig. 2.36A). The plane transects the 8th costal cartilages and the L1 vertebra. When erect its location varies from the L2 through L4 vertebra. The pyloric orifice is approximately 1.25 cm right of the midline.

The stomach also features two curvatures (Fig. 2.37A–C):

- **Lesser curvature:** forms the shorter concave right border of the stomach. The **angular incisure** (notch), the most inferior part of the curvature, indicates the junction of the body and pyloric part of the stomach (Fig. 2.37A & B). The angular incisure lies just to the left of the midline.
- **Greater curvature:** forms the longer convex left border of the stomach. It passes inferiorly to the left from the junction of the 5th intercostal space and MCL, then curves to the right, passing deep to the 9th or 10th left cartilage as it continues medially to reach the pyloric antrum.

Because of the unequal lengths of the lesser curvature on the right and the greater curvature on the left, in most people the shape of the stomach resembles the letter J.

INTERIOR OF STOMACH

The smooth surface of the gastric mucosa is reddish brown during life, except in the pyloric part, where it is pink. In life, it is covered by a continuous mucous layer that protects its surface from the gastric acid the stomach's glands secrete. When contracted, the gastric mucosa is thrown into longitudinal ridges or wrinkles called **gastric folds** (gastric rugae) (Fig. 2.38A & B); they are most marked toward the pyloric part and along the greater curvature. During swallowing, a temporary groove or furrow-like **gastric canal** forms between the longitudinal gastric folds along the lesser curvature. It can be observed radiographically and endoscopically. The gastric canal forms because of the firm attachment of the gastric mucosa to the muscular layer, which does not have an oblique layer at this site. Saliva and small quantities of masticated food and other fluids drain along the gastric canal to the pyloric canal when the stomach is mostly empty. The gastric folds diminish and obliterate as the stomach is distended (fills).

RELATIONS OF STOMACH

The stomach is covered by visceral peritoneum, except where blood vessels run along its curvatures and in a small area posterior to the cardial orifice (Fig. 2.36A). The two layers of the lesser omentum extend around the stomach and leave its greater curvature as the greater omentum (Figs. 2.28, 2.31, and 2.37A). Anteriorly, the stomach is related to the diaphragm, left lobe of liver, and anterior abdominal wall. Posteriorly, the stomach is related to the omental bursa and pancreas; the posterior surface of the stomach forms most of the anterior wall of the omental bursa (Fig. 2.39A). The transverse colon is related inferiorly and laterally to the stomach as it courses along the greater curvature of the stomach to the left colic flexure.

The **bed of the stomach,** on which the stomach rests in the supine position, is formed by the structures forming the posterior wall of the omental bursa. From superior to inferior,

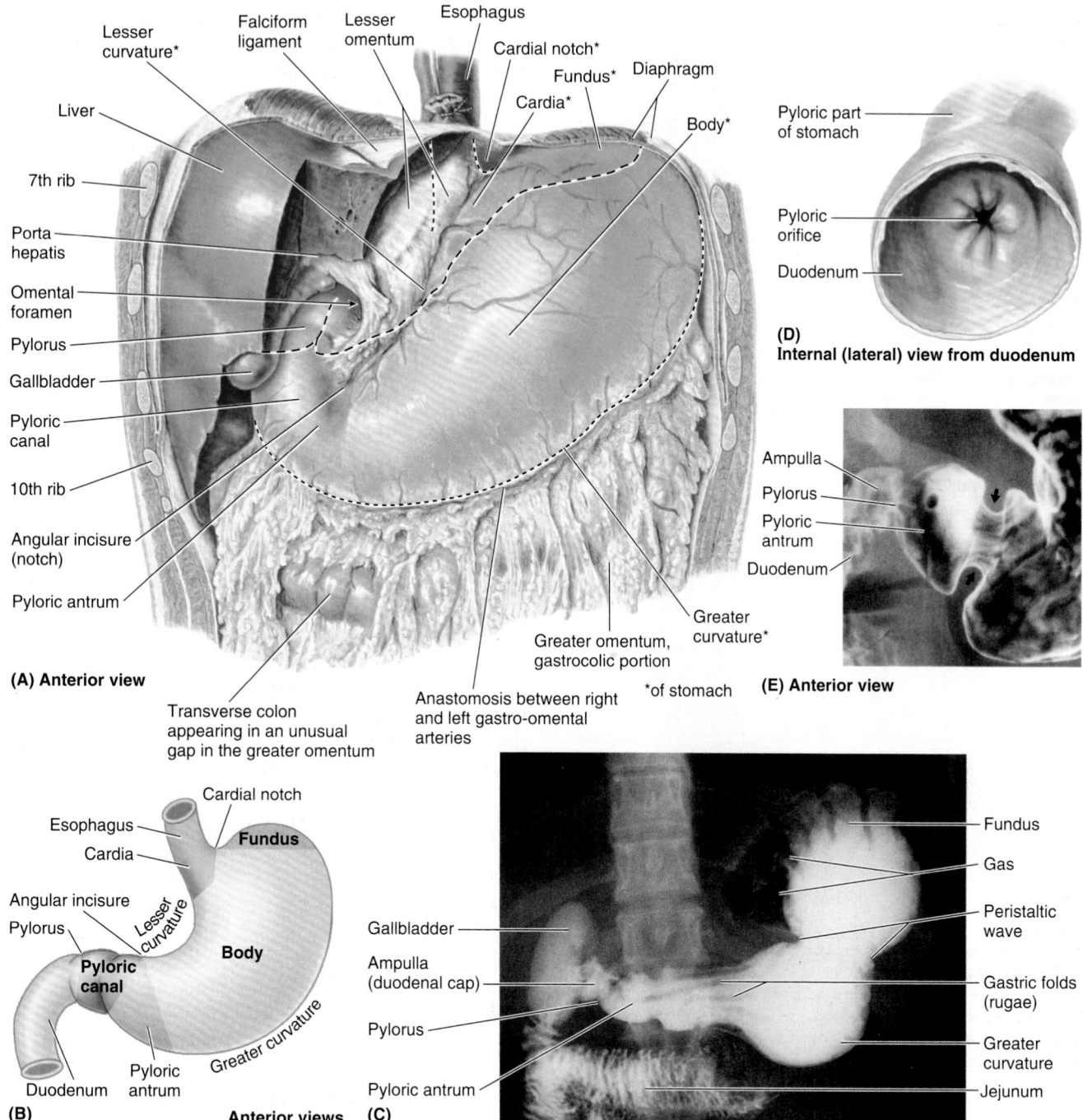

(A) Anterior view

Lesser curvature*
Liver
7th rib
Porta hepatis
Omental foramen
Pylorus
Gallbladder
Pyloric canal
10th rib
Angular incisure (notch)
Pyloric antrum

Falciform ligament
Lesser omentum
Esophagus
Cardial notch*
Fundus*
Cardia*
Diaphragm
Body*

Transverse colon appearing in an unusual gap in the greater omentum
Anastomosis between right and left gastro-omental arteries
Greater omentum, gastrocolic portion
Greater curvature*
*of stomach

Pyloric part of stomach
Pyloric orifice
Duodenum

(D) Internal (lateral) view from duodenum

Ampulla
Pylorus
Pyloric antrum
Duodenum

(E) Anterior view

(B)
Esophagus
Cardia
Angular incisure
Pylorus
Pyloric canal
Duodenum
Pyloric antrum
Cardial notch
Fundus
Lesser curvature
Body
Greater curvature
Anterior views

(C)
Gallbladder
Ampulla (duodenal cap)
Pylorus
Pyloric antrum
Fundus
Gas
Peristaltic wave
Gastric folds (rugae)
Greater curvature
Jejunum

FIGURE 2.37. Abdominal part of esophagus and stomach. A. The stomach is inflated with air. The left part of the liver is cut away so that the lesser omentum and omental foramen can be seen. The extent of the intact liver is indicated by the longer *dotted lines*. **B.** Parts of the stomach. **C.** Radiograph of the stomach after a barium meal. Circular peristaltic waves begin in the body of the stomach and sweep toward the pyloric canal, as shown in **E** *(arrowheads)*, where they stop. Gas can be seen in the cardia and fundus of this supine patient. **D.** The pylorus is the significantly constricted terminal part of the stomach. The pyloric orifice is the opening of the pyloric canal into the duodenum. **E.** Radiograph demonstrating the pyloric region of the stomach and the superior part of the duodenum. (**C** and **E** courtesy of Dr. E. L. Lansdown, Professor of Medical Imaging, University of Toronto, Toronto, ON, Canada.)

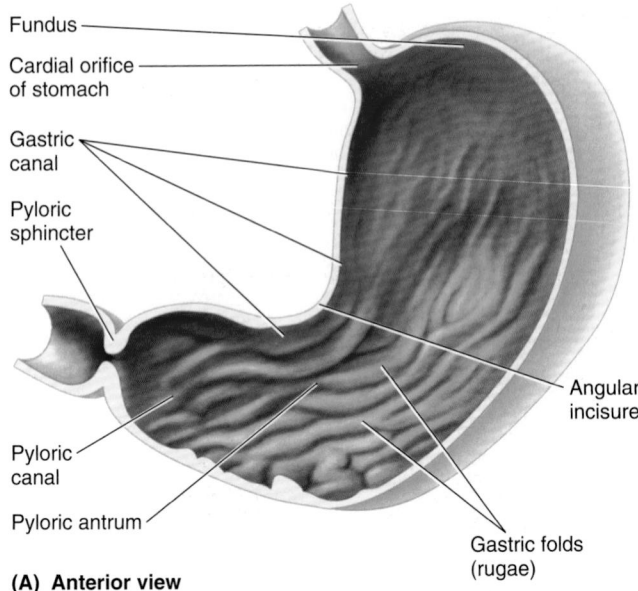

Fundus

Cardial orifice
of stomach

Gastric
canal

Pyloric
sphincter

Pyloric
canal

Pyloric antrum

Angular
incisure

Gastric folds
(rugae)

(A) Anterior view

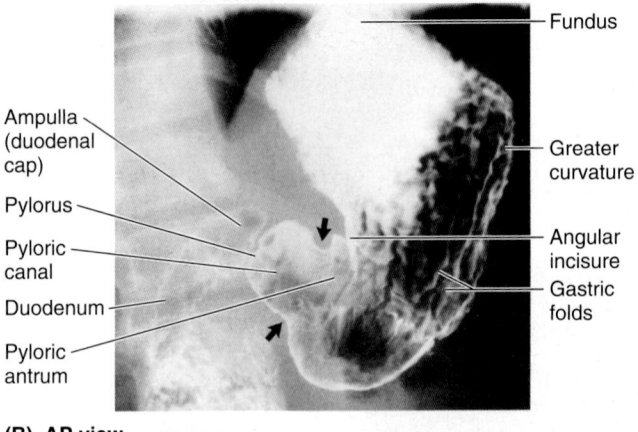

Ampulla
(duodenal
cap)

Pylorus

Pyloric
canal

Duodenum

Pyloric
antrum

Fundus

Greater
curvature

Angular
incisure

Gastric
folds

(B) AP view

FIGURE 2.38. The internal surface of the stomach. A. The anterior wall of the stomach has been removed to demonstrate its interior. The longitudinal gastric folds disappear on distension. Along the lesser curvature, several longitudinal mucosal folds extend from the esophagus to the pylorus, making up the gastric canal along which ingested liquids pass. **B.** Radiograph of stomach after a barium meal. Observe the peristaltic wave in the stomach and the longitudinal gastric folds (rugae) of mucous membrane. (**A** courtesy of Dr. J. Helsin, Toronto, ON, Canada.)

the bed of the stomach is formed by the left dome of the diaphragm, spleen, left kidney and suprarenal gland, splenic artery, pancreas, and transverse mesocolon (Fig. 2.39B).

VESSELS AND NERVES OF STOMACH

The rich *arterial supply of the stomach* arises from the celiac trunk and its branches (Fig. 2.40; Table 2.7). Most blood is supplied by anastomoses formed along the lesser curvature by the **right** and **left gastric arteries,** and along the greater curvature by the **right** and **left gastro-omental (gastro-epiploic) arteries.** The fundus and upper body receive blood from the **short** and **posterior gastric arteries.**

The *veins of the stomach* parallel the arteries in position and course (Fig. 2.41). The **right** and **left gastric veins** drain into the *hepatic portal vein;* the **short gastric veins** and **left gastro-omental veins** drain into the *splenic vein,* which joins the superior mesenteric vein (SMV) to form the hepatic portal vein. The **right gastro-omental vein** empties in the SMV. A **prepyloric vein** ascends over the pylorus to the right gastric vein. Because this vein is obvious in living persons, surgeons use it for identifying the pylorus.

The **gastric lymphatic vessels** (Fig. 2.42A) accompany the arteries along the greater and lesser curvatures of the stomach. They drain lymph from its anterior and posterior surfaces toward its curvatures, where the **gastric** and **gastro-omental lymph nodes** are located. The efferent vessels from these nodes accompany the large arteries to the *celiac lymph nodes.* The following is a summary of the *lymphatic drainage of the stomach:*

- Lymph from the superior two thirds of the stomach drains along the right and left gastric vessels to the **gastric lymph nodes;** lymph from the fundus and superior part of the body of the stomach also drains along the short gastric arteries and left gastro-omental vessels to the *pancreaticosplenic lymph nodes.*
- Lymph from the right two thirds of the inferior third of the stomach drains along the right gastro-omental vessels to the **pyloric lymph nodes.**
- Lymph from the left one third of the greater curvature drains to the **pancreaticoduodenal lymph nodes,** which are located along the short gastric and splenic vessels.

The *parasympathetic nerve supply of the stomach* (Fig. 2.42B) is from the anterior and posterior vagal trunks and their branches, which enter the abdomen through the esophageal hiatus.

The *anterior vagal trunk,* derived mainly from the left vagus nerve (CN X), usually enters the abdomen as a single branch that lies on the anterior surface of the esophagus. It runs toward the lesser curvature of the stomach, where it gives off hepatic and duodenal branches, which leave the stomach in the hepatoduodenal ligament. The rest of the anterior vagal trunk continues along the lesser curvature, giving rise to anterior gastric branches.

The larger *posterior vagal trunk,* derived mainly from the right vagus nerve, enters the abdomen on the posterior surface of the esophagus and passes toward the lesser curvature of the stomach. The posterior vagal trunk supplies branches to the anterior and posterior surfaces of the stomach. It gives off a celiac branch, which passes to the *celiac plexus,* and then continues along the lesser curvature, giving rise to posterior gastric branches.

The *sympathetic nerve supply of the stomach,* from the T6 through T9 segments of the spinal cord, passes to the celiac plexus through the *greater splanchnic nerve* and is distributed through the plexuses around the gastric and gastroomental arteries. (See also "Summary of Innervation of Abdominal Viscera," p. 301).

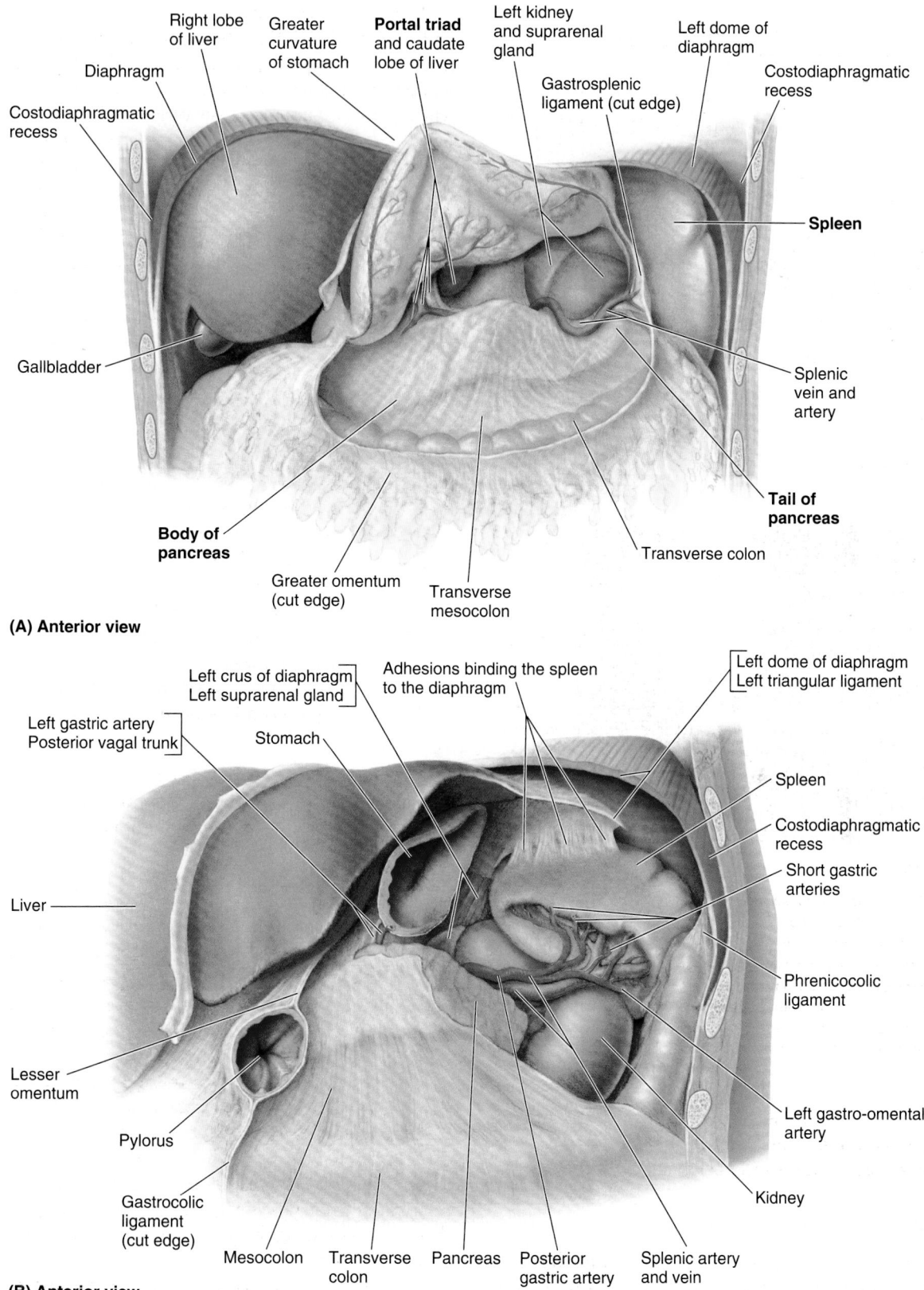

(A) Anterior view

(B) Anterior view

FIGURE 2.39. Omental bursa and stomach bed. A. The omental bursa, the greater omentum, and gastrosplenic ligament have been cut along the greater curvature of the stomach, and the stomach has been reflected superiorly to open the bursa anteriorly. At the right end of the bursa, two of the boundaries of the omental foramen can be seen: the inferior root of the hepatoduodenal ligament (containing the portal triad) and caudate lobe of the liver. **B.** The stomach and most of the lesser omentum have been excised, and the peritoneum of the posterior wall of the omental bursa covering the stomach bed is largely removed to reveal the organs in the bed. Although adhesions, such as those binding the spleen to the diaphragm here, are common postmortem findings, they are not normal anatomy.

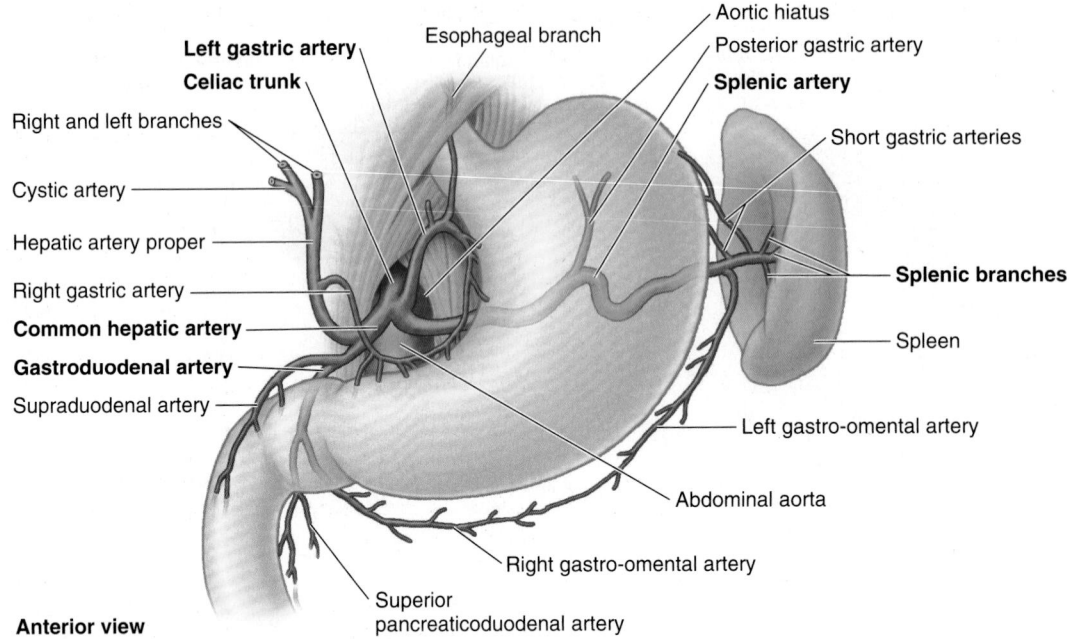

FIGURE 2.40. Arteries of stomach, duodenum, and spleen. The arterial supply of the abdominal part of the esophagus, stomach, upper (superior and upper descending parts) duodenum, and spleen is from the celiac artery. The direct branches of the celiac trunk appear in boldface.

TABLE 2.7. ARTERIAL SUPPLY TO ABDOMINAL FOREGUT DERIVATIVES: ESOPHAGUS, STOMACH, LIVER, GALLBLADDER, PANCREAS, AND SPLEEN

Artery	Origin	Course	Distribution
Celiac trunk	Abdominal aorta (at level of aortic hiatus)	After short antero-inferior course, bifurcates into splenic and common hepatic arteries	Esophagus, stomach, proximal duodenum, liver/biliary apparatus, pancreas
Left gastric	Celiac trunk	Ascends retroperitoneally to esophageal hiatus, giving rise to an esophageal branch; then descending along lesser curvature to anastomose with right gastric artery	Distal (mostly abdominal) part of esophagus and lesser curvature of stomach
Splenic		Runs retroperitoneally along superior border of pancreas; traverses splenorenal ligament to hilum of spleen	Body of pancreas, spleen, and greater curvature and posterior stomach body
Posterior gastric	Splenic artery posterior to stomach	Ascends retroperitoneally along posterior wall of lesser omental bursa to enter gastrophrenic ligament	Posterior wall and fundus of stomach
Left gastro-omental (left gastroepiploic)	Splenic artery in hilum of spleen	Passes between layers of gastrosplenic ligament to stomach, then along greater curvature in greater omentum to anastomose with right gastro-omental artery	Left portion of greater curvature of stomach
Short gastric ($n = 4$–5)		Passes between layers of gastrosplenic ligament to fundus of stomach	Fundus of stomach
Hepatic[a]	Celiac trunk	Passes retroperitoneally to reach hepatoduodenal ligament; passing between layers to porta hepatis; bifurcates into right and left hepatic arteries	Liver, gallbladder and biliary ducts, stomach, duodenum, pancreas, and respective lobes of liver

TABLE 2.7. ARTERIAL SUPPLY TO ABDOMINAL FOREGUT DERIVATIVES: ESOPHAGUS, STOMACH, LIVER, GALLBLADDER, PANCREAS, AND SPLEEN (Continued)

Artery	Origin	Course	Distribution
Cystic	Right hepatic artery	Arises within hepatoduodenal ligament (in cysto-hepatic triangle of Calot)	Gallbladder and cystic duct
Right gastric	Hepatic artery	Runs along lesser curvature of stomach to anastomose with left gastric artery	Right portion of lesser curvature of stomach
Gastroduodenal		Descends retroperitoneally, posterior to gastro-duodenal junction	Stomach, pancreas, first part of duodenum, and distal part of bile duct
Right gastro-omental (right gastroepiploic)	Gastroduodenal artery	Passes between layers of greater omentum along greater curvature of stomach to anastomose with left gastro-omental artery	Right portion of greater curvature of stomach
Superior pancreaticoduodenal		Divides into anterior and posterior arteries that descend on each side of pancreatic head, anastomosing with similar branches of inferior pancreaticoduodenal artery	Proximal portion of duodenum and superior part of head of pancreas
Inferior pancreaticoduodenal	Superior mesenteric artery	Divides into anterior and posterior arteries that ascend on each side of pancreatic head, anastomosing with similar branches of superior pancreaticoduodenal artery	Distal portion of duodenum and head of pancreas

[a]For descriptive purposes, the hepatic artery is often divided into the *common hepatic artery*, from its origin to the origin of the gastroduodenal artery, and *hepatic artery proper*, made up of the remainder of the vessel.

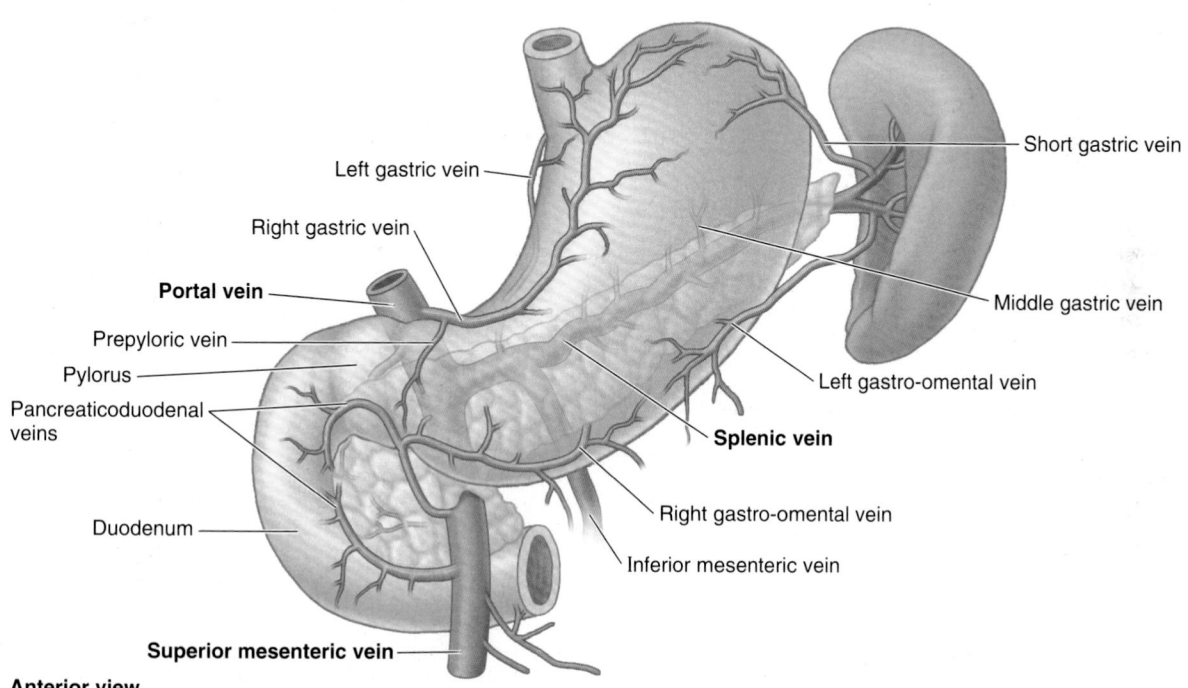

FIGURE 2.41. Veins of stomach, duodenum, and spleen. Venous drainage from the abdominal part of the esophagus, stomach, upper (superior and upper descending parts) duodenum, pancreas, and spleen is into the portal vein, either directly or indirectly via the splenic or superior mesenteric vein (SMV). The gastric veins parallel the arteries in position and course.

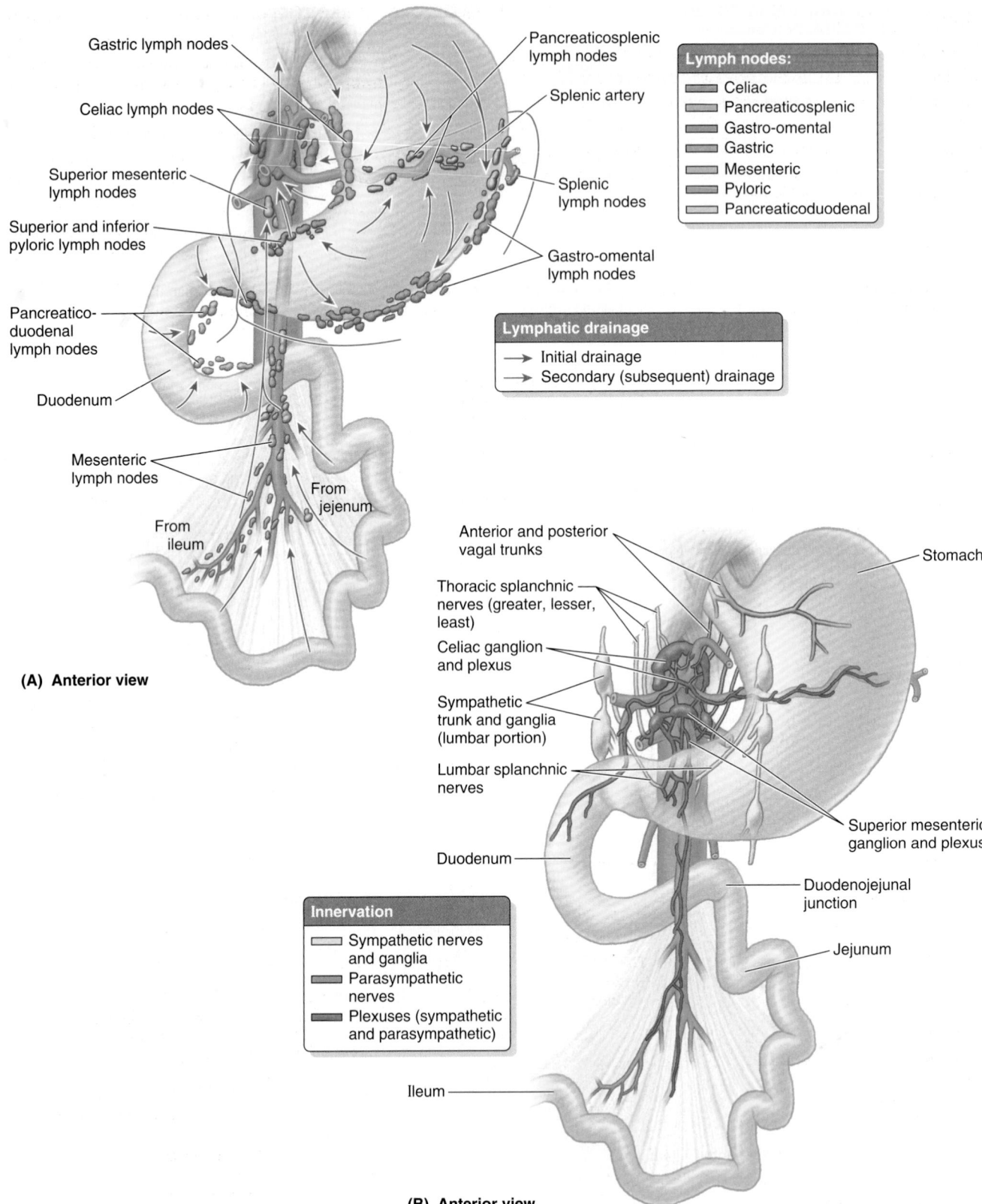

Gastric lymph nodes

Celiac lymph nodes

Superior mesenteric
lymph nodes

Superior and inferior
pyloric lymph nodes

Pancreatico-
duodenal
lymph nodes

Duodenum

Mesenteric
lymph nodes

From
ileum

From
jejenum

Pancreaticosplenic
lymph nodes

Splenic artery

Splenic
lymph nodes

Gastro-omental
lymph nodes

Lymph nodes:
- Celiac
- Pancreaticosplenic
- Gastro-omental
- Gastric
- Mesenteric
- Pyloric
- Pancreaticoduodenal

Lymphatic drainage
→ Initial drainage
→ Secondary (subsequent) drainage

(A) Anterior view

Anterior and posterior
vagal trunks

Thoracic splanchnic
nerves (greater, lesser,
least)

Celiac ganglion
and plexus

Sympathetic
trunk and ganglia
(lumbar portion)

Lumbar splanchnic
nerves

Duodenum

Stomach

Superior mesenteric
ganglion and plexus

Duodenojejunal
junction

Jejunum

Ileum

Innervation
- Sympathetic nerves
  and ganglia
- Parasympathetic
  nerves
- Plexuses (sympathetic
  and parasympathetic)

(B) Anterior view

FIGURE 2.42. Lymphatic drainage and innervation of stomach and small intestine. A. The arrows indicate the direction of lymph flow to the lymph nodes. **B.** Innervation of the stomach is both parasympathetic, from the vagus nerves (CN X) via the esophageal plexus, and sympathetic, via the greater (abdominopelvic) splanchnic, the celiac plexus, and peri-arterial plexuses.

Small Intestine

The **small intestine,** consisting of the duodenum, jejunum, and ileum (Fig. 2.43), is the primary site for absorption of nutrients from ingested materials. It extends from the pylorus to the ileocecal junction where the ileum joins the cecum (the first part of the large intestine). The pyloric part of the stomach empties into the duodenum, duodenal admission being regulated by the pylorus.

DUODENUM

The **duodenum** (L. breadth of 12 fingers), the first and shortest (25 cm) part of the small intestine, is also the widest and most fixed part. The duodenum pursues a C-shaped course around the head of the pancreas (Figs. 2.43C and 2.44A & C). It begins at the pylorus on the right side and ends at the **duodenojejunal flexure (junction)** on the left side (Fig. 2.44B & C). This junction occurs approximately at the level of the L2 vertebra, 2–3 cm to the left of the midline. The junction usually takes the form of an acute angle, the **duodenojejunal flexure.** Most of the duodenum is fixed by peritoneum to structures on the posterior abdominal wall and is considered partially retroperitoneal. The duodenum is divisible into four parts (Figs. 2.44C and 2.45; Table 2.8):

- *Superior* (first) *part:* short (approximately 5 cm) and lies anterolateral to the body of the L1 vertebra.

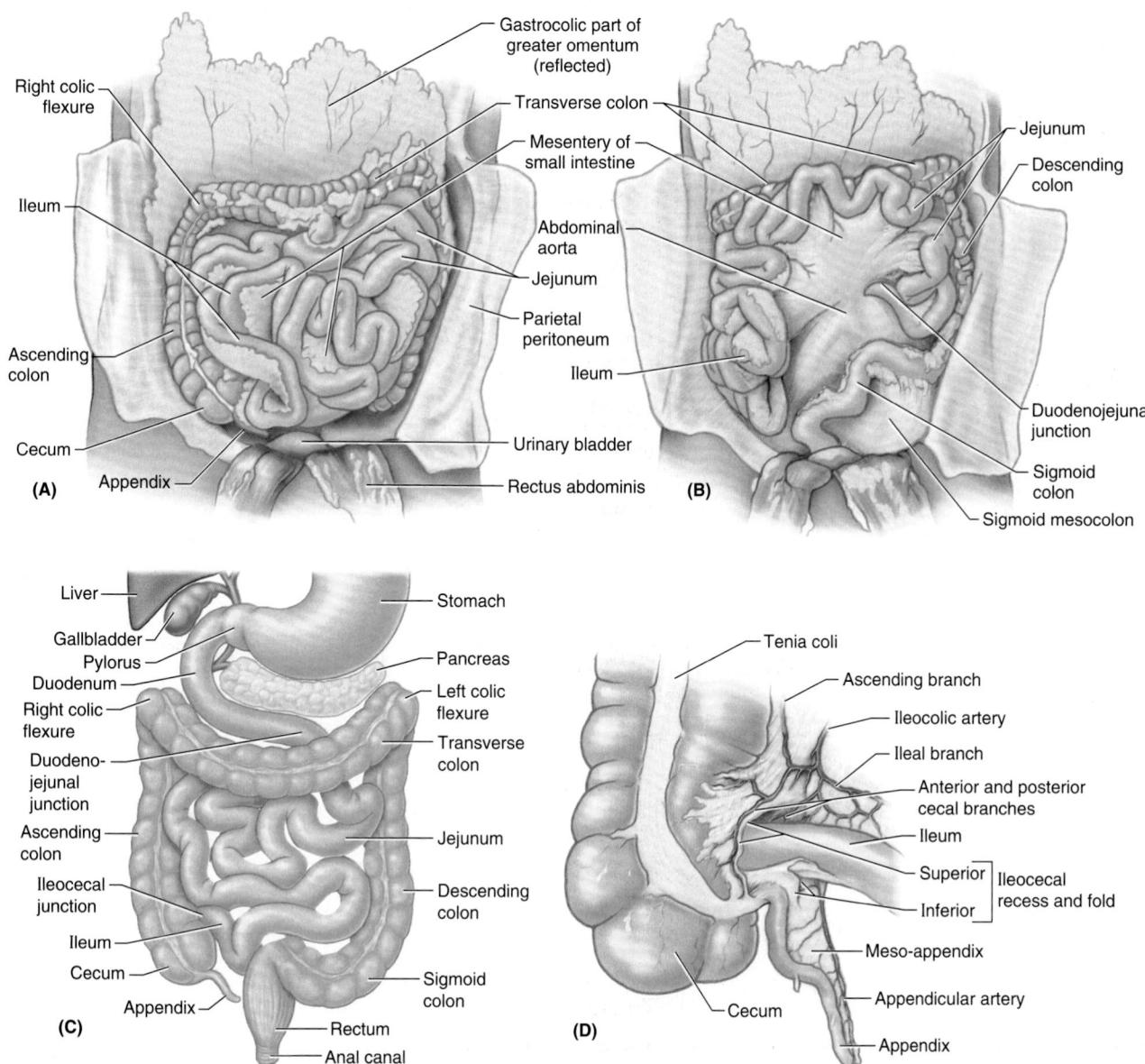

Anterior views

FIGURE 2.43. Small and large intestines. A. Note the convolutions of small intestine in situ, encircled on three sides by the large intestine and revealed by elevating the greater omentum. **B.** The convolutions of small intestine have been retracted superiorly to demonstrate the mesentery. **C.** This orientation drawing of the alimentary system indicates the general position and relationships of the intestines. **D.** The blood supply of the ileocecal region is shown.

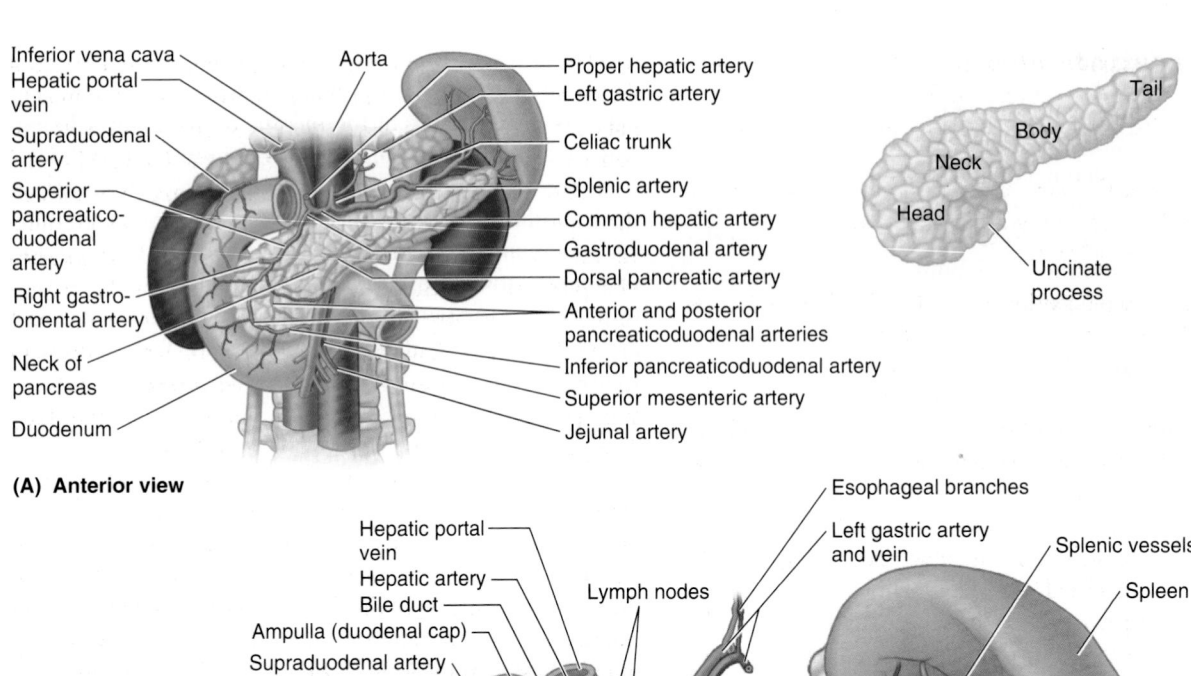

Inferior vena cava
Hepatic portal vein
Supraduodenal artery
Superior pancreatico-duodenal artery
Right gastro-omental artery
Neck of pancreas
Duodenum

Aorta
Proper hepatic artery
Left gastric artery
Celiac trunk
Splenic artery
Common hepatic artery
Gastroduodenal artery
Dorsal pancreatic artery
Anterior and posterior pancreaticoduodenal arteries
Inferior pancreaticoduodenal artery
Superior mesenteric artery
Jejunal artery

(A) Anterior view

Tail
Body
Neck
Head
Uncinate process

Esophageal branches
Left gastric artery and vein
Splenic vessels
Spleen

Hepatic portal vein
Hepatic artery
Bile duct
Ampulla (duodenal cap)
Supraduodenal artery
Lymph nodes

Pylorus of stomach
Gastroduodenal artery
Right gastro-omental artery
Anterior superior pancreaticoduodenal artery
Superior mesenteric vein and artery
Anterior inferior pancreatico-duodenal artery
Right colic artery and vein
Ileocolic vein and artery

2
4
3

Pancreas
Duodenojejunal junction (flexure)
Middle colic artery
Mesentery (cut edge)

(B) Anterior view

Left gastric artery and vein
Greater pancreatic artery
Splenic vein and artery
Union of splenic vein with SMV, posterior to neck of pancreas
Inferior mesenteric vein
Superior mesenteric artery and vein
Duodenojejunal flexure
Common stem of posterior inferior and anterior inferior pancreaticoduodenal arteries

Celiac trunk
Hepatic artery
Hepatic portal vein
Bile duct

1
2
4
3

Lymph node
Uncinate process

(C) Posterior view

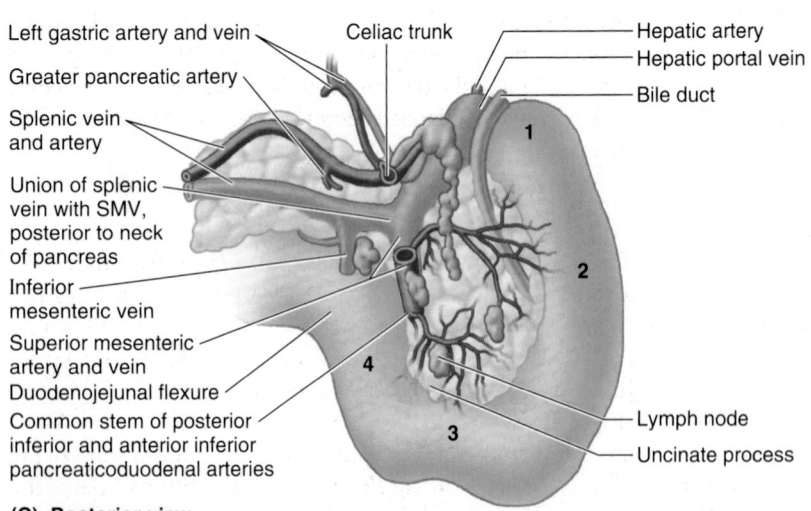

Parts of duodenum:
1 Superior
2 Descending
3 Horizontal
4 Ascending

FIGURE 2.44. Duodenum, pancreas, and spleen.
A. The duodenum, pancreas, and spleen and their blood supply are revealed by removal of the stomach, transverse colon, and peritoneum. **B.** The anterior aspect of the duodenum, pancreas, and related vasculature is shown. The duodenum is molded around the head of the pancreas. **C.** The posterior aspect of the duodenum and pancreas is shown. The abdominal aorta and inferior vena cava occupy the vertical concavity posterior to the head of the pancreas and third part of the duodenum. The uncinate process is the extension of the head of the pancreas that passes posterior to the superior mesenteric vessels. The bile duct is descending in a fissure (opened up) in the posterior part of the head of the pancreas. *SMV*, superior mesenteric vein.

- *Descending* (second) *part:* longer (7–10 cm) and descends along the right sides of the L1–L3 vertebrae.
- *Inferior* (third) *part:* 6–8 cm long and crosses the L3 vertebra.
- *Ascending* (fourth) *part:* short (5 cm) and begins at the left of the L3 vertebra and rises superiorly as far as the superior border of the L2 vertebra.

The first 2 cm of the superior part of the duodenum, immediately distal to the pylorus, has a mesentery and is mobile. This free part, called the **ampulla** (duodenal cap), has an appearance distinct from the remainder of the duodenum when observed radiographically using contrast medium (Fig. 2.37C & E). The distal 3 cm of the superior part and the other three parts of the duodenum have no mesentery and are immobile because they are retroperitoneal. The principal relationships of the duodenum are illustrated in Figures 2.44 and 2.45 and summarized in Table 2.8.

The **superior part of the duodenum** ascends from the pylorus and is overlapped by the liver and gallbladder. Peritoneum covers its anterior aspect, but it is bare of peritoneum posteriorly, except for the ampulla. The proximal part has the *hepatoduodenal ligament* (part of the lesser omentum) attached superiorly and the greater omentum attached inferiorly (see Fig. 2.26).

The **descending part of the duodenum** runs inferiorly, curving around the head of the pancreas (Figs. 2.44 and 2.45; Table 2.8). Initially, it lies to the right of and parallel to the IVC. The *bile* and *main pancreatic ducts* enter its posteromedial wall. These ducts usually unite to form the **hepatopancreatic ampulla,** which opens on an eminence, called the **major duodenal papilla,** located posteromedially in the descending duodenum. The descending part of the duodenum is entirely retroperitoneal. The anterior surface of its proximal and distal thirds is covered with peritoneum; however, the peritoneum reflects from its middle third to form the double-layered mesentery of the transverse colon, the transverse mesocolon.

The **inferior (horizontal) part of the duodenum** runs transversely to the left, passing over the IVC, aorta, and L3 vertebra. It is crossed by the superior mesenteric artery and vein and the root of the mesentery of the jejunum and ileum. Superior to it is the head of the pancreas and its uncinate process. The anterior surface of the inferior part is covered with peritoneum, except where it is crossed by the superior mesenteric vessels and the root of the mesentery. Posteriorly it is separated from the vertebral column by the right psoas major, IVC, aorta, and the right testicular or ovarian vessels.

The **ascending part of the duodenum** runs superiorly and along the left side of the aorta to reach the inferior border of the body of the pancreas. Here it curves anteriorly to join the jejunum at the duodenojejunal flexure, supported by the attachment of a **suspensory muscle of the duodenum** (ligament of Treitz). This muscle is composed of a slip of skeletal muscle from the diaphragm and a fibromuscular band of smooth muscle from the third and fourth parts of the

duodenum. Contraction of this muscle widens the angle of the duodenojejunal flexure, facilitating movement of the intestinal contents. The suspensory muscle passes posterior to the pancreas and splenic vein and anterior to the left renal vein.

The *arteries of the duodenum* arise from the celiac trunk and the superior mesenteric artery (Fig. 2.44). The celiac trunk, via the **gastroduodenal artery** and its branch, the **superior pancreaticoduodenal artery,** supplies the duodenum proximal to the entry of the bile duct into the descending part of the duodenum. The superior mesenteric artery, through its branch, the **inferior pancreaticoduodenal artery,** supplies the duodenum distal to the entry of the bile duct. The pancreaticoduodenal arteries lie in the curve between the duodenum and the head of the pancreas and supply both structures. The anastomosis of the superior and inferior pancreaticoduodenal arteries (i.e., between the celiac and superior mesenteric arteries) occurs between the entry of the bile duct and the junction of the descending and inferior parts of the duodenum. An important transition in the blood supply of the digestive tract occurs here: proximally, extending *orad* (toward the mouth) to and including the abdominal part of the esophagus, the blood is supplied to the digestive tract by the celiac trunk; distally, extending *aborad* (away from the mouth) to the left colic flexure, the blood is supplied by the SMA. The basis of this transition in blood supply is embryological; this is the junction of the foregut and midgut.

The *veins of the duodenum* follow the arteries and drain into the *hepatic portal vein,* some directly and others indirectly, through the superior mesenteric and splenic veins (Fig. 2.41).

The *lymphatic vessels of the duodenum* follow the arteries. The *anterior lymphatic vessels* drain into the *pancreaticoduodenal lymph nodes,* located along the superior and inferior pancreaticoduodenal arteries, and into the *pyloric lymph nodes,* which lie along the gastroduodenal artery (Fig. 2.46). The *posterior lymphatic vessels* pass posterior to the head of the pancreas and drain into the **superior mesenteric lymph nodes.** Efferent lymphatic vessels from the duodenal lymph nodes drain into the *celiac lymph nodes.*

The *nerves of the duodenum* derive from the *vagus* and *greater* and *lesser (abdominopelvic) splanchnic nerves* by way of the celiac and superior mesenteric plexuses. The nerves are next conveyed to the duodenum via peri-arterial plexuses extending to the pancreaticoduodenal arteries (see also "Summary of the Innervation of the Abdominal Viscera," p. 300).

JEJUNUM AND ILEUM

The second part of the small intestine, the **jejunum,** begins at the duodenojejunal flexure where the gastrointestinal tract resumes an intraperitoneal course. The third part of the small intestine, the **ileum,** ends at the **ileocecal junction,** the union of the terminal ileum and the cecum (Figs. 2.43C and 2.47). Together, the jejunum and ileum are 6–7 m long, the jejunum constituting approximately two fifths and the ileum approximately three fifths of the intraperitoneal section of the small intestine.

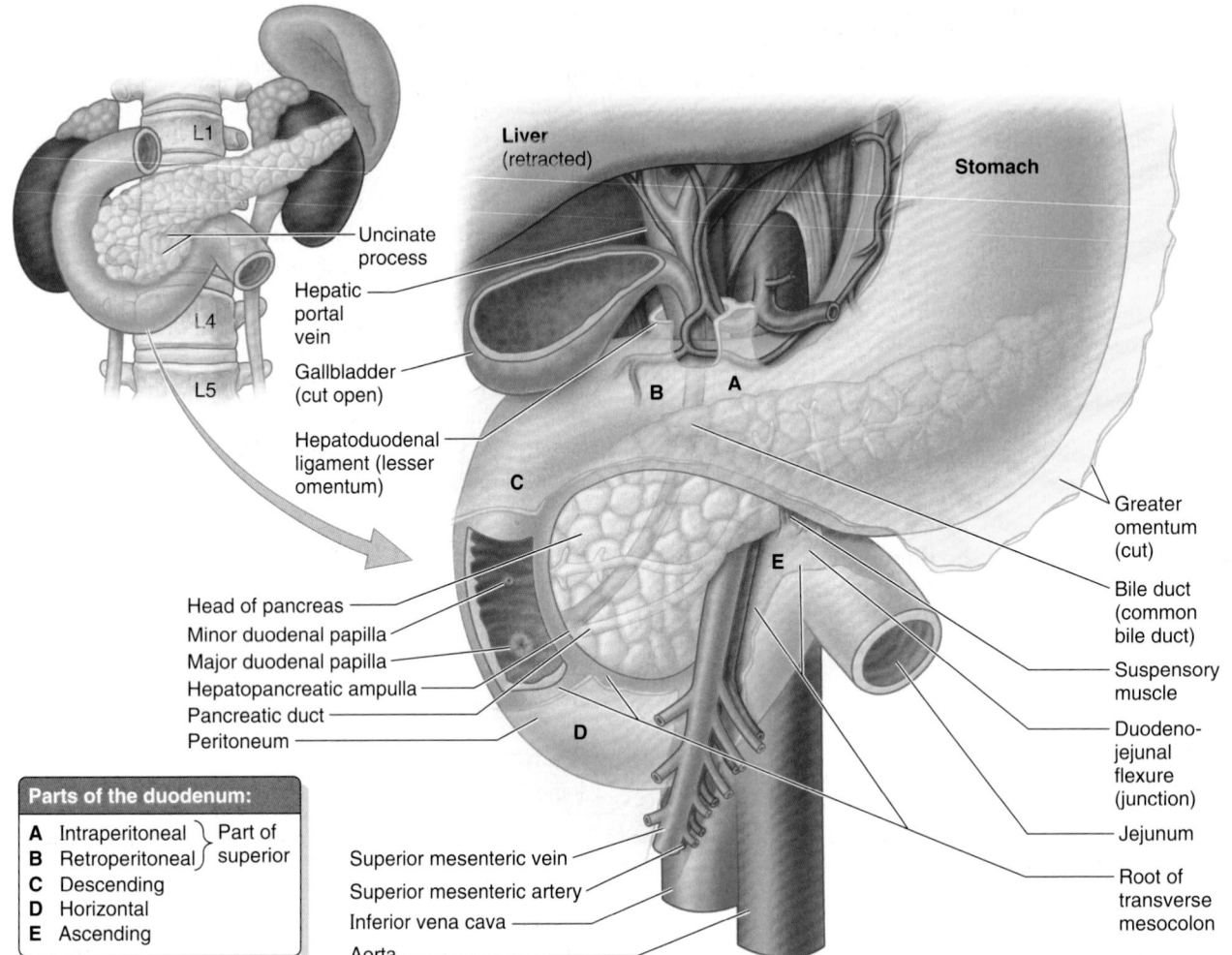

FIGURE 2.45. Relationships of duodenum. The duodenum pursues a C-shaped course around the head of the pancreas.

Parts of the duodenum:

A Intraperitoneal ⎫ Part of
B Retroperitoneal ⎬ superior
C Descending
D Horizontal
E Ascending

TABLE 2.8. RELATIONSHIPS OF DUODENUM

Part of Duodenum	Anterior	Posterior	Medial	Superior	Inferior	Vertebral Level
Superior (1st part) **(A & B)**	Peritoneum Gallbladder Quadrate lobe of liver	Bile duct Gastroduodenal artery Hepatic portal vein IVC	Pylorus	Neck of gallbladder	Neck of pancreas	Anterolateral to L1 vertebra
Descending (2nd part) **(C)**	Transverse colon Transverse mesocolon Coils of small intestine	Hilum of right kidney Renal vessels Ureter Psoas major	Head of pancreas Pancreatic duct Bile duct	Superior part of duodenum	Inferior part of duodenum	Right of L2–L3 vertebrae
Inferior (horizontal) (3rd part) **(D)**	SMA SMV Coils of small intestine	Right psoas major IVC Aorta Right ureter		Head and uncinate Process of pancreas Superior mesenteric vessels	Coils of small intestine (ilium)	Anterior to L3 vertebra
Ascending (4th part) **(E)**	Beginning of root of mesentery Coils of jejunum	Left psoas major Left margin of aorta	SMA, SMV, uncinate process of pancreas	Body of pancreas	Coils of jejunum	Left of L3 vertebra

IVC, inferior vena cava; *SMA*, superior mesenteric artery; *SMV*, superior mesenteric vein.

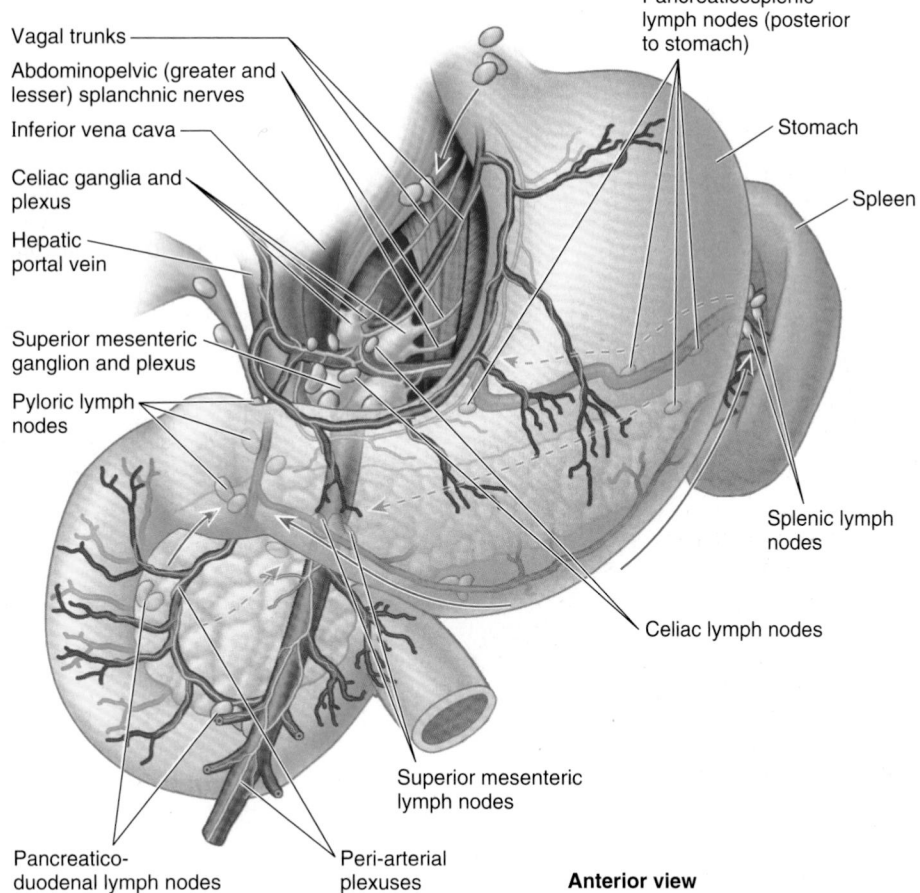

FIGURE 2.46. Lymphatic drainage and innervation of duodenum, pancreas, and spleen. The close positional relationship of these organs results in sharing of blood vessels, lymphatic vessels, and nerve pathways, in whole or in part.

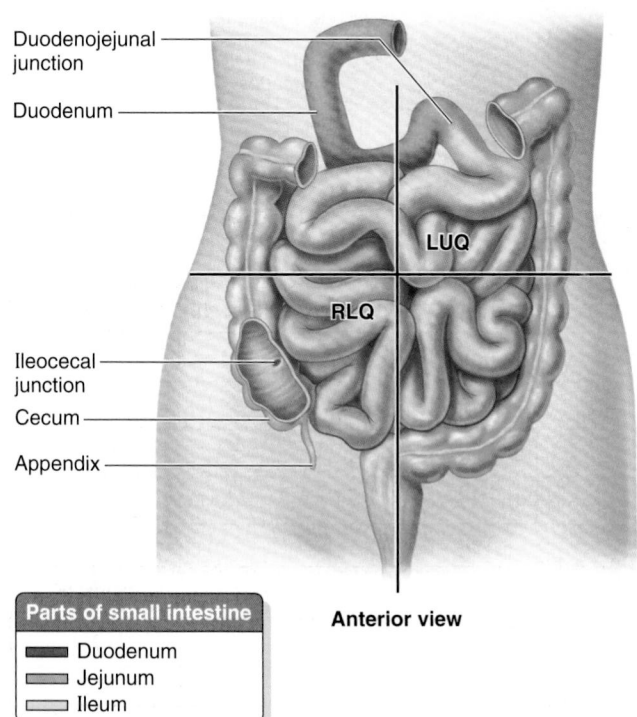

FIGURE 2.47. Jejunum and ileum. The jejunum begins at the duodeno-jejunal flexure and the ileum ends at the cecum. The combined term *jejuno-ileum* is sometimes used as an expression of the fact that there is no clear external line of demarcation between the jejunum and the ileum. *LUQ,* left upper quadrant; *RLQ,* right lower quadrant.

Most of the jejunum lies in the left upper quadrant (LUQ) of the infracolic compartment, whereas most of the ileum lies in the right lower quadrant (RLQ). The terminal ileum usually lies in the pelvis from which it ascends, ending in the medial aspect of the cecum. Although no clear line of demarcation between the jejunum and ileum exists, they have distinctive characteristics that are surgically important (Fig. 2.48B–E; Table 2.9).

The **mesentery** is a fan-shaped fold of peritoneum that attaches the jejunum and ileum to the posterior abdominal wall (Fig. 2.43B and 2.48A). The origin or **root of the mesentery** (approximately 15 cm long) is directed obliquely, inferiorly, and to the right (Fig. 2.49A). It extends from the duodenojejunal junction on the left side of vertebra L2 to the ileocolic junction and the right sacro-iliac joint. The average length of the mesentery from its root to the intestinal border is 20 cm. The root of the mesentery crosses (successively) the ascending and inferior parts of the duodenum, abdominal aorta, IVC, right ureter, right psoas major, and right testicular or ovarian vessels. Between the two layers of the mesentery are the superior mesenteric vessels, lymph nodes, a variable amount of fat, and autonomic nerves.

The **superior mesenteric artery** (SMA) supplies the jejunum and ileum via **jejunal** and **ileal arteries** (Fig. 2.49B).

The SMA usually arises from the abdominal aorta at the level of the L1 vertebra, approximately 1 cm inferior to the

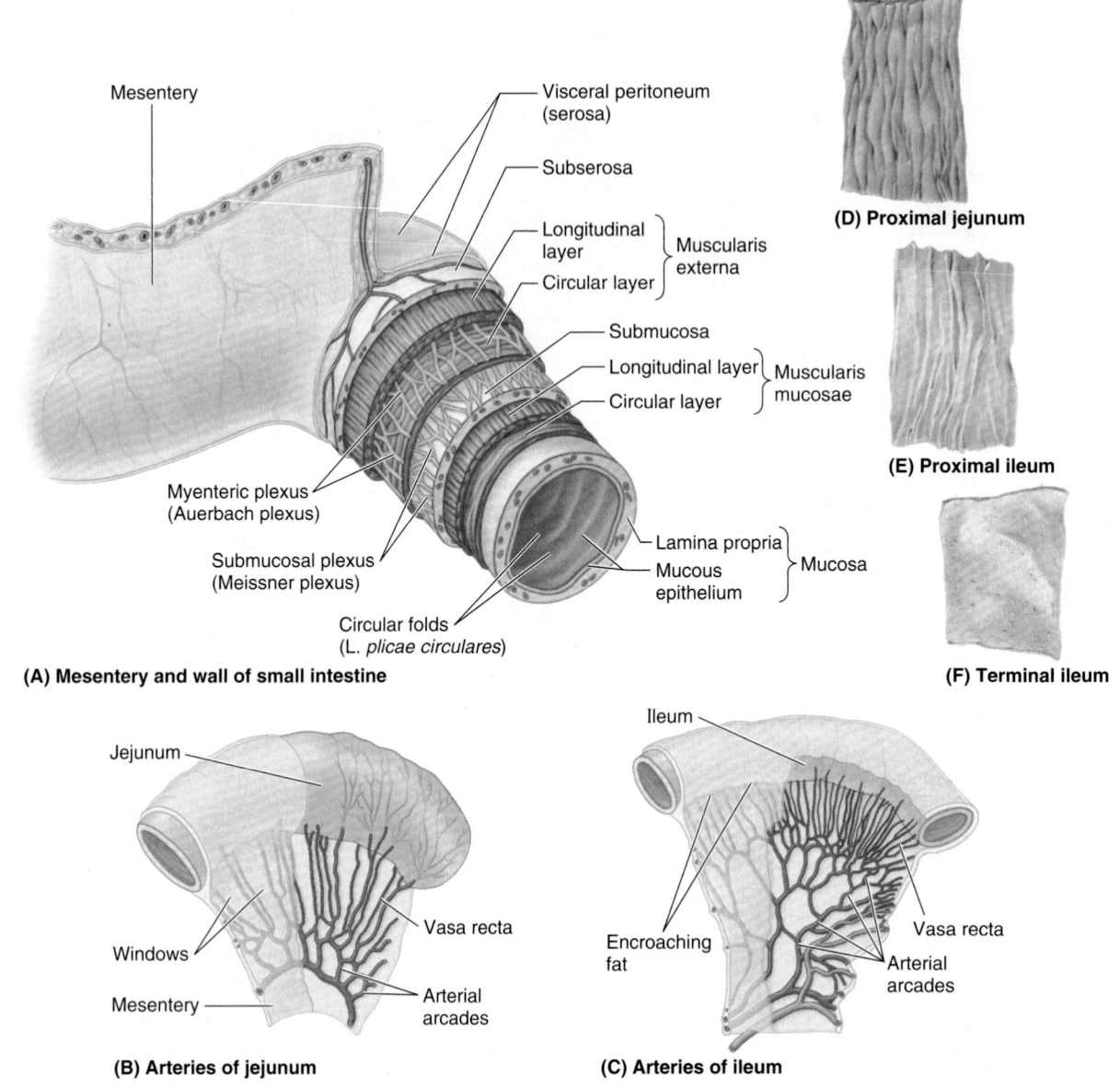

(D) Proximal jejunum

(E) Proximal ileum

(F) Terminal ileum

(A) Mesentery and wall of small intestine

Mesentery

Visceral peritoneum (serosa)

Subserosa

Longitudinal layer ⎱ Muscularis
Circular layer ⎰ externa

Submucosa

Longitudinal layer ⎱ Muscularis
Circular layer ⎰ mucosae

Myenteric plexus (Auerbach plexus)

Submucosal plexus (Meissner plexus)

Circular folds (L. *plicae circulares*)

Lamina propria
Mucous epithelium ⎱ Mucosa

(B) Arteries of jejunum

Jejunum

Windows

Mesentery

Vasa recta

Arterial arcades

(C) Arteries of ileum

Ileum

Encroaching fat

Vasa recta

Arterial arcades

FIGURE 2.48. Structure of mesentery and small intestine: distinctive features of jejunum and ileum. A. The mesentery is a double-layered fold of visceral peritoneum that suspends the gut and conducts neurovasculature from the posterior body wall. **B–E.** Distinctive features of the jejunum and ileum, outlined in Table 2.9, are illustrated.

TABLE 2.9. DISTINGUISHING CHARACTERISTICS OF JEJUNUM AND ILEUM IN LIVING BODY (FIG. 2.48)

Characteristic	Jejunum (B & D)*		Ileum (C, E & F)*	
Color	Deeper red		Paler pink	
Caliber	2–4 cm		2–3 cm	
Wall	Thick and heavy		Thin and light	
Vascularity	Greater		Less	
Vasa recta	Long	(B)	Short	(C)
Arcades	A few large loops		Many short loops	
Fat in mesentery	Less		More	
Circular folds (L. *plicae circulares*)	Large, tall, and closely packed (**D**)		Low and sparse (**E**); absent in distal part (**F**)	
Lymphoid nodules (Peyer patches)	Few		Many (**F**)	

*Letters in parentheses refer to individual figures in Fig. 2.48.

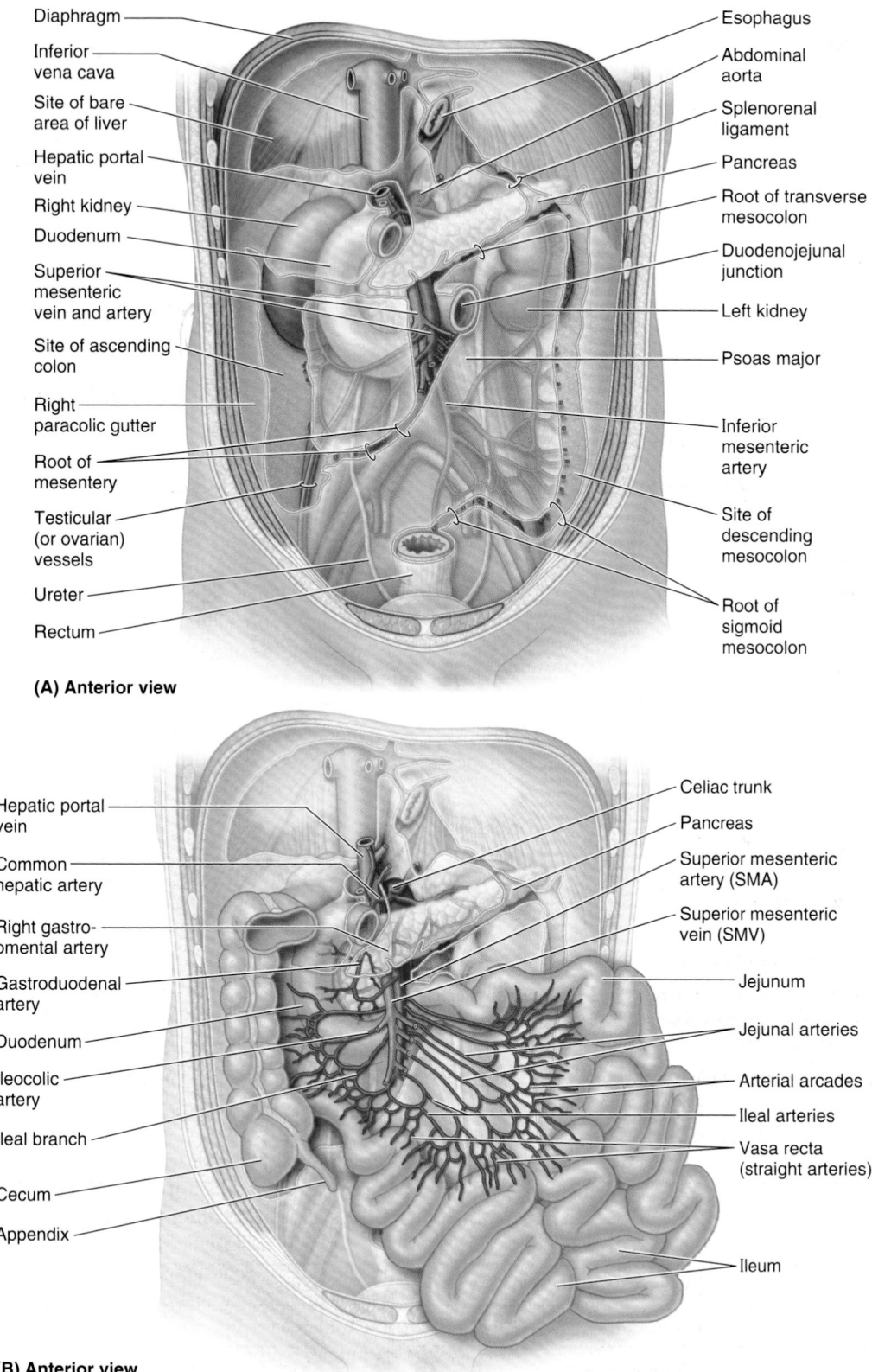

Diaphragm

Inferior vena cava

Site of bare area of liver

Hepatic portal vein

Right kidney

Duodenum

Superior mesenteric vein and artery

Site of ascending colon

Right paracolic gutter

Root of mesentery

Testicular (or ovarian) vessels

Ureter

Rectum

Esophagus

Abdominal aorta

Splenorenal ligament

Pancreas

Root of transverse mesocolon

Duodenojejunal junction

Left kidney

Psoas major

Inferior mesenteric artery

Site of descending mesocolon

Root of sigmoid mesocolon

(A) Anterior view

Hepatic portal vein

Common hepatic artery

Right gastro-omental artery

Gastroduodenal artery

Duodenum

Ileocolic artery

Ileal branch

Cecum

Appendix

Celiac trunk

Pancreas

Superior mesenteric artery (SMA)

Superior mesenteric vein (SMV)

Jejunum

Jejunal arteries

Arterial arcades

Ileal arteries

Vasa recta (straight arteries)

Ileum

(B) Anterior view

FIGURE 2.49. Arterial supply and mesenteries of intestines. A. Arterial supply of the large intestine. The transverse and sigmoid mesocolons and the mesentery of the jejunum and ileum have been cut at their roots. The ileocolic and right colic arteries on the right side and the left colic and sigmoid arteries on the left side originally coursed within mesenteries (ascending and descending mesocolons) that later fused to the posterior wall; they can be re-established surgically. **B.** Arterial supply and venous drainage of the small intestine. Except for the proximal duodenum, all of the intestine depicted in **B** is supplied by the SMA (as is most of the transverse colon, not shown). The SMV drains blood from the same portions of the intestine into the hepatic portal vein.

celiac trunk, and runs between the layers of the mesentery, sending 15–18 branches to the jejunum and ileum (see also Figs. 2.54 and 2.55). The arteries unite to form loops or arches, called **arterial arcades,** which give rise to straight arteries, called **vasa recta** (Figs. 2.48B and 2.49B).

The *superior mesenteric vein* drains the jejunum and ileum (Fig. 2.49B). It lies anterior and to the right of the SMA in the root of the mesentery (Fig. 2.49A). The SMV ends posterior to the neck of the pancreas, where it unites with the splenic vein to form the hepatic portal vein (Fig. 2.44C).

Specialized lymphatic vessels in the **intestinal villi** (tiny projections of the mucous membrane) that absorb fat are called **lacteals.** They empty their milk-like fluid into the lymphatic plexuses in the walls of the jejunum and ileum. The lacteals drain in turn into lymphatic vessels between the layers of the mesentery. Within the mesentery, the lymph passes sequentially through three groups of lymph nodes (Fig. 2.50):

- **Juxta-intestinal lymph nodes:** located close to the intestinal wall.
- **Mesenteric lymph nodes:** scattered among the arterial arcades.
- **Superior central nodes:** located along the proximal part of the SMA.

Efferent lymphatic vessels from the mesenteric lymph nodes drain to the *superior mesenteric lymph nodes.* Lymphatic vessels from the terminal ileum follow the ileal branch of the ileocolic artery to the **ileocolic lymph nodes.**

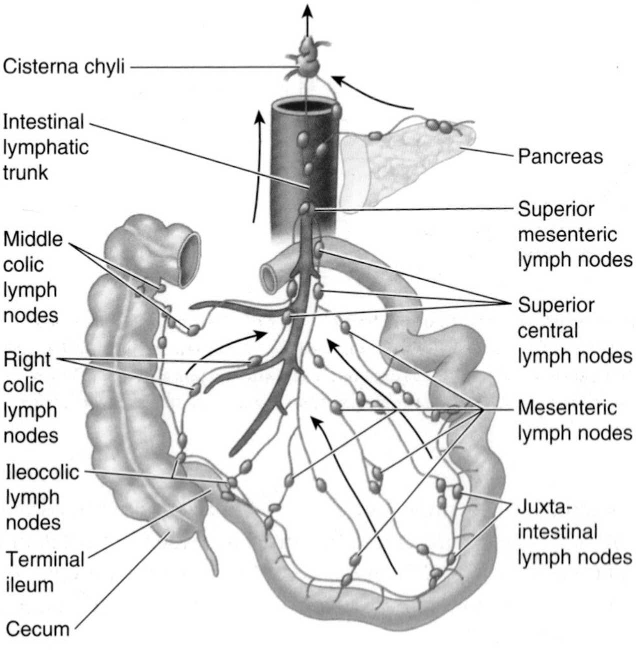

The SMA and its branches are surrounded by a *periarterial nerve plexus* through which the nerves are conducted to the parts of the intestine supplied by this artery (Fig. 2.51). The sympathetic fibers in the nerves to the jejunum and ileum originate in the T8–T10 segments of the spinal cord and reach the **superior mesenteric nerve plexus** through the *sympathetic trunks* and *thoracic abdominopelvic* (*greater, lesser,* and *least*) *splanchnic nerves.* The presynaptic sympathetic fibers synapse on cell bodies of postsynaptic sympathetic neurons in the *celiac* and *superior mesenteric* (*prevertebral*) *ganglia.* The parasympathetic fibers in the nerves to the jejunum and ileum derive from the *posterior vagal trunks.* The presynaptic parasympathetic fibers synapse with postsynaptic parasympathetic neurons in the *myenteric* and *submucosal plexuses* in the intestinal wall (see also "Summary of Innervation of Abdominal Viscera," p. 301).

Sympathetic stimulation reduces peristaltic and secretory activity of the intestine and acts as a vasoconstrictor, reducing or stopping digestion and making blood (and energy) available for "fleeing or fighting." Parasympathetic stimulation increases peristaltic and secretory activity of the intestine, restoring the digestion process following a sympathetic reaction. The small intestine also has sensory (visceral afferent) fibers. The intestine is insensitive to most pain stimuli, including cutting and burning; however, it is sensitive to distension that is perceived as *colic* (spasmodic abdominal pains or "intestinal cramps").

Large Intestine

The **large intestine** is where water is absorbed from the indigestible residues of the liquid chyme, converting it into semisolid stool or feces that is stored temporarily and allowed to accumulate until defecation occurs. The large intestine consists of the *cecum; appendix; ascending, transverse, descending,* and *sigmoid colon; rectum;* and *anal canal* (Fig. 2.52). The large intestine can be distinguished from the small intestine by:

- **Omental appendices:** small, fatty, omentum-like projections.
- **Teniae coli:** three distinct longitudinal bands: (1) **mesocolic tenia,** to which the transverse and sigmoid mesocolons attach; (2) **omental tenia,** to which the omental appendices attach; and (3) **free tenia** (L. *t. libera*), to which neither mesocolons nor omental appendices are attached.
- **Haustra:** sacculations of the wall of the colon between the teniae
- A much greater caliber (internal diameter).

The *teniae coli* (thickened bands of smooth muscle representing most of the longitudinal coat) begin at the base of the appendix as the thick longitudinal layer of the appendix separates into three bands. The teniae run the length of the large intestine, abruptly broadening and merging with each other again at the rectosigmoid junction into a continuous

FIGURE 2.50. Mesenteric lymph nodes. The superior nodes form a system in which the central nodes, at the root of the superior mesenteric artery, receive lymph from the mesenteric, ileocolic, right colic, and middle colic nodes, which in turn receive lymph from juxta-intestinal lymph nodes. The juxta-intestinal nodes adjacent to the intestines are most abundant. Fewer occur along the arteries.

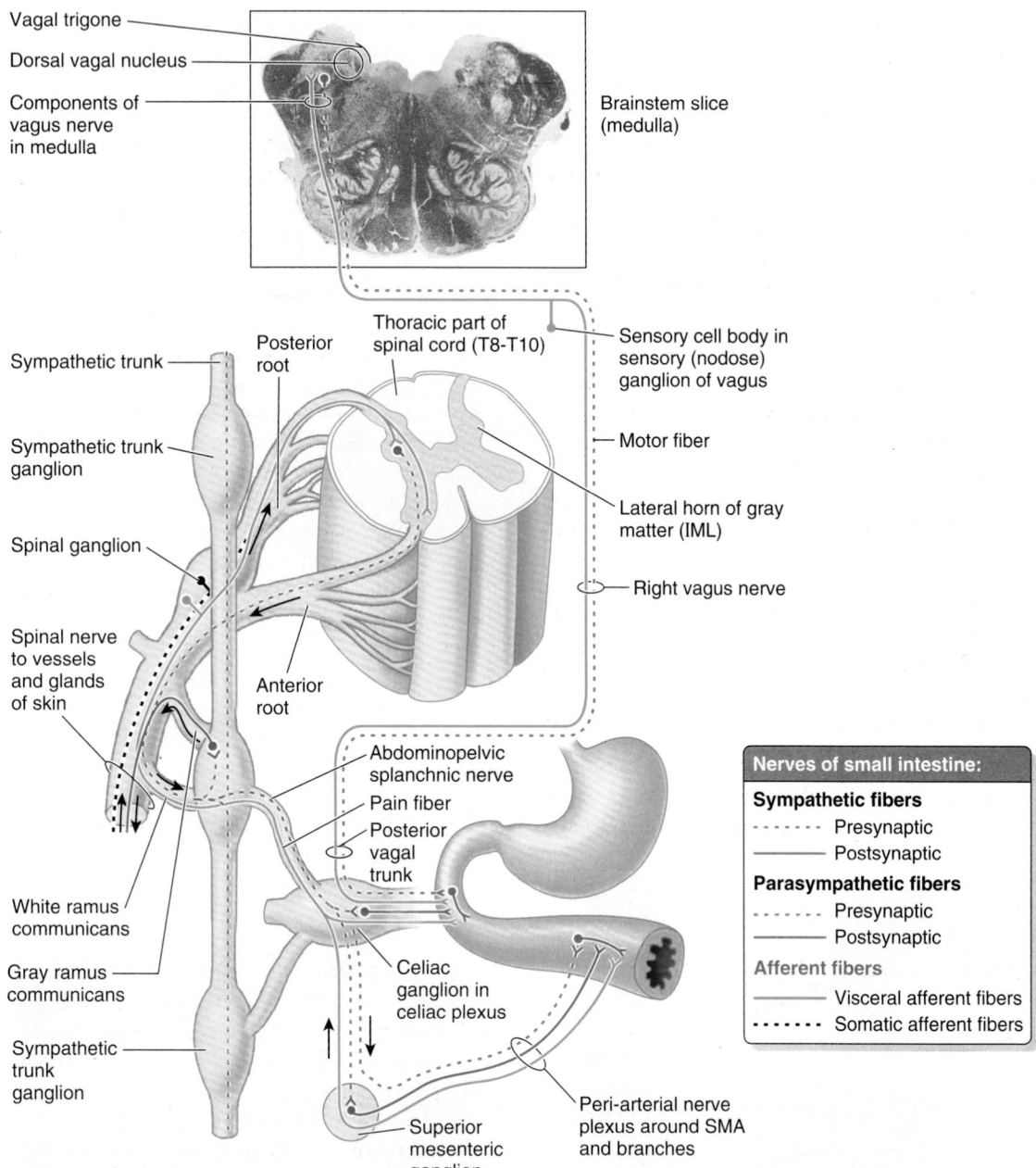

Vagal trigone
Dorsal vagal nucleus
Components of vagus nerve in medulla
Brainstem slice (medulla)

Sympathetic trunk
Posterior root
Thoracic part of spinal cord (T8-T10)
Sensory cell body in sensory (nodose) ganglion of vagus

Sympathetic trunk ganglion
Motor fiber
Lateral horn of gray matter (IML)

Spinal ganglion
Right vagus nerve

Spinal nerve to vessels and glands of skin
Anterior root

White ramus communicans
Abdominopelvic splanchnic nerve
Pain fiber
Posterior vagal trunk

Gray ramus communicans
Celiac ganglion in celiac plexus

Sympathetic trunk ganglion
Superior mesenteric ganglion
Peri-arterial nerve plexus around SMA and branches

Nerves of small intestine:

Sympathetic fibers
- - - - - Presynaptic
———— Postsynaptic

Parasympathetic fibers
- - - - - Presynaptic
———— Postsynaptic

Afferent fibers
———— Visceral afferent fibers
• • • • • Somatic afferent fibers

FIGURE 2.51. Innervation of small intestine. Presynaptic sympathetic nerve fibers originate in the T8 or T9 through T10 or T11 segments of the spinal cord and reach the celiac plexus through the sympathetic trunks and greater and lesser (abdominopelvic) splanchnic nerves. After synapsing in the celiac and superior mesenteric ganglia, postsynaptic nerve fibers accompany the arteries to the intestine. Afferent fibers are concerned with reflexes and pain. Presynaptic parasympathetic (vagus) nerves originate in the medulla (oblongata) and pass to the intestine via the posterior vagal trunk. They synapse with intrinsic postsynaptic neurons located in the intestinal wall. *SMA,* superior mesenteric artery.

longitudinal layer around the rectum. Because their tonic contraction shortens the part of the wall with which they are associated, the colon becomes sacculated or "baggy" between the teniae, forming the haustra.

CECUM AND APPENDIX

The **cecum** is the first part of the large intestine; it is continuous with the ascending colon. The cecum is a blind intestinal pouch, approximately 7.5 cm in both length and

breadth. It lies in the iliac fossa of the right lower quadrant of the abdomen, inferior to the junction of the terminal ileum and cecum (Figs. 2.52 and 2.53). If distended with feces or gas, the cecum may be palpable through the anterolateral abdominal wall.

The cecum usually lies within 2.5 cm of the inguinal ligament; it is almost entirely enveloped by peritoneum and can be lifted freely. However, the cecum has no mesentery. Because of its relative freedom, it may be displaced from the iliac fossa, but it is commonly bound to the lateral abdominal wall by one

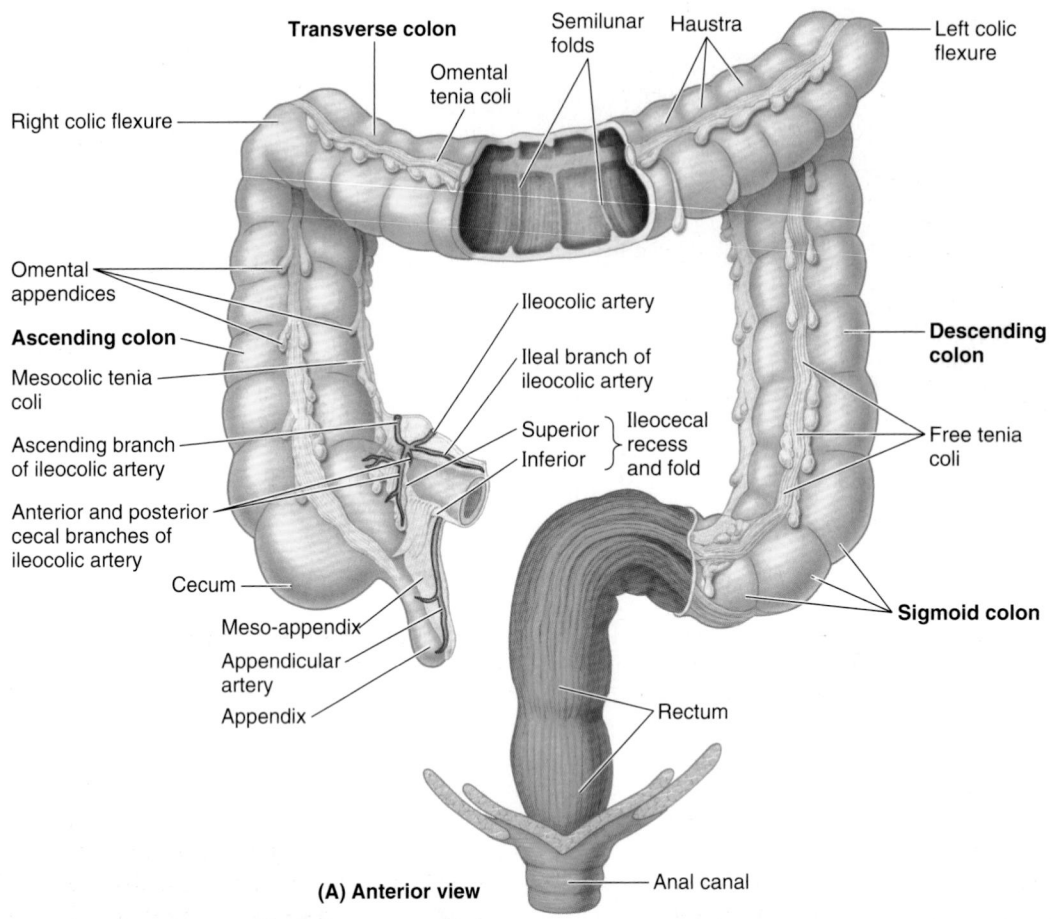

Transverse colon

Semilunar folds

Haustra

Left colic flexure

Omental tenia coli

Right colic flexure

Omental appendices

Ascending colon

Mesocolic tenia coli

Ascending branch of ileocolic artery

Anterior and posterior cecal branches of ileocolic artery

Ileocolic artery

Ileal branch of ileocolic artery

Superior ⎱ Ileocecal recess and fold
Inferior ⎰

Descending colon

Free tenia coli

Cecum

Meso-appendix

Appendicular artery

Appendix

Sigmoid colon

Rectum

Anal canal

(A) Anterior view

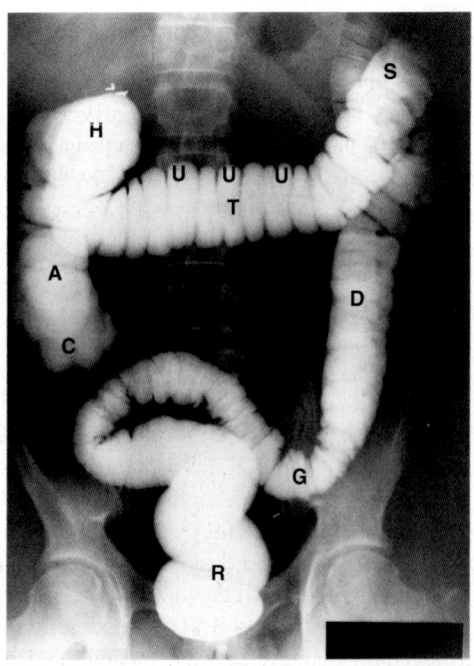

(B) Anteroposterior view (with contrast medium)

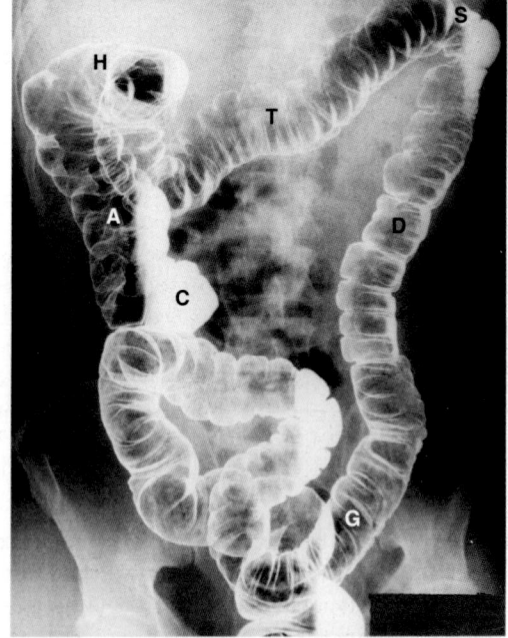

(C) Anteroposterior view (double contrast study)

FIGURE 2.52. Terminal ileum and large intestine (including appendix). A. Teniae, haustra, and fatty omental appendices (characteristic of the colon) are not associated with the rectum. **B.** To examine the colon, a barium enema has been given after the bowel is cleared of fecal material by a cleansing enema. Single contrast barium studies demonstrate the semilunar folds demarcating the haustra. **C.** Following the single contrast study, the patient has evacuated the barium and the colon was distended with air for this double-contrast study. The luminal surface remains coated with a thin layer of barium. *A,* ascending colon; *C,* cecum; *D,* descending colon; *G,* sigmoid colon; *H,* hepatic or right colic flexure; *R,* rectum; *S,* splenic or left colic flexure; *T,* transverse colon; *U,* haustra. (**B** courtesy of Dr. C. S. Ho, Professor of Medical Imaging, University of Toronto, Toronto, ON, Canada; **C** courtesy of Dr. E. L. Lansdown, Professor of Medical Imaging, University of Toronto, Toronto, ON, Canada.)

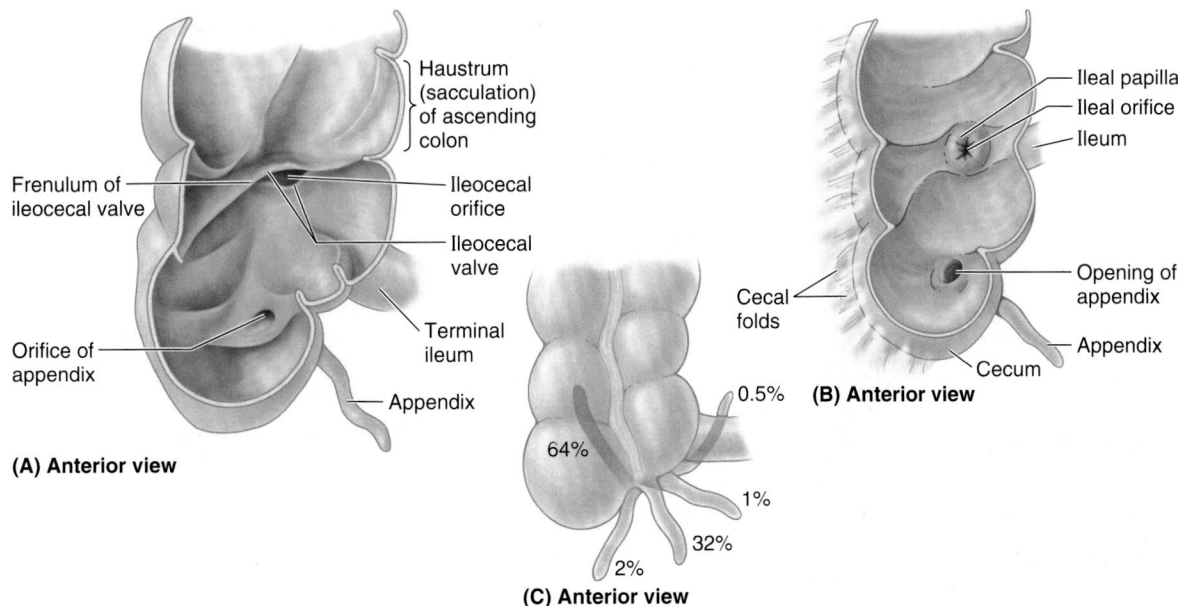

FIGURE 2.53. Terminal ileum, cecum, and appendix. A. The cecum was filled with air until dry and then opened. Observe the ileocecal valve and ileal orifice. The frenulum is a fold (more evident in cadavers) that runs from the ileocecal valve along the wall at the junction of the cecum and ascending colon. **B.** The interior of the cecum showing the endoscopic (living) appearance of the ileocecal valve. **C.** The approximate incidences of various locations of the appendix, based on an analysis of 10,000 cases, are shown.

or more **cecal folds** of peritoneum (Fig. 2.53B). The terminal ileum enters the cecum obliquely and partly invaginates into it.

In dissection, the **ileal orifice** enters the cecum between **ileocolic lips** (superior and inferior), folds that meet laterally forming ridges called the **frenula of the ileal orifice** (Fig. 2.53A). It was believed that when the cecum is distended or when it contracts, the lips and frenula actively tighten, closing the valve to prevent reflux from the cecum into the ileum. However, direct observation by endoscopy in living persons does not support this description. The circular muscle is poorly developed around the orifice; therefore, the valve is unlikely to have any sphincteric action that controls passage of the intestinal contents from the ileum into the cecum. The orifice is usually closed by tonic contraction, however, appearing as an **ileal papilla** on the cecal side (Fig. 2.53B). The papilla probably serves as a relatively passive flap valve, preventing reflux from the cecum into the ileum as contractions occur to propel contents up the ascending colon and into the transverse colon (Magee and Dalley, 1986).

The **appendix** (vermiform appendix; L. *vermis,* wormlike) is a blind intestinal diverticulum (6–10 cm in length) that contains masses of lymphoid tissue. It arises from the posteromedial aspect of the cecum inferior to the ileocecal junction. The appendix has a short triangular mesentery, the **meso-appendix,** which derives from the posterior side of the mesentery of the terminal ileum (Fig. 2.52A). The meso-appendix attaches to the cecum and the proximal part of the appendix. The position of the appendix is variable, but it is usually retrocecal (Fig. 2.53C). Clinical correlations involving the appendix are included in the blue box on pp. 259–260.

The *arterial supply of the cecum* is from the **ileocolic artery,** the terminal branch of the SMA (Figs. 2.54 and 2.55; Table 2.10). The **appendicular artery,** a branch of the ileocolic artery, supplies the appendix. *Venous drainage* from the

cecum and appendix flow through a tributary of the SMV, the **ileocolic vein** (Fig. 2.56A).

Lymphatic drainage of the cecum and appendix passes to lymph nodes in the meso-appendix and to the *ileocolic lymph nodes* that lie along the ileocolic artery (Fig. 2.56B). Efferent lymphatic vessels pass to the *superior mesenteric lymph nodes.*

The *nerve supply to the cecum and appendix* derives from the sympathetic and parasympathetic nerves from the *superior mesenteric plexus* (Fig. 2.56C). The *sympathetic nerve fibers* originate in the lower thoracic part of the spinal cord, and the *parasympathetic nerve fibers* derive from the vagus nerves. Afferent nerve fibers from the appendix accompany the sympathetic nerves to the T10 segment of the spinal cord (see also "Summary of Innervation of Abdominal Viscera," p. 300).

COLON

The **colon** has four parts—ascending, transverse, descending, and sigmoid—that succeed one another in an arch (Figs. 2.43C and 2.52). The colon encircles the small intestine, the ascending colon lying to the right of the small intestine, the transverse colon superior and/or anterior to it, the descending colon to the left if it, and the sigmoid colon inferior to it.

The **ascending colon** is the second part of the large intestine. It passes superiorly on the right side of the abdominal cavity from the cecum to the right lobe of the liver, where it turns to the left at the **right colic flexure** (*hepatic flexure*). This flexure lies deep to the 9th and 10th ribs and is overlapped by the inferior part of the liver.

The ascending colon is narrower than the cecum and is secondarily retroperitoneal along the right side of the posterior abdominal wall. The ascending colon is usually covered by peritoneum anteriorly and on its sides; however, in

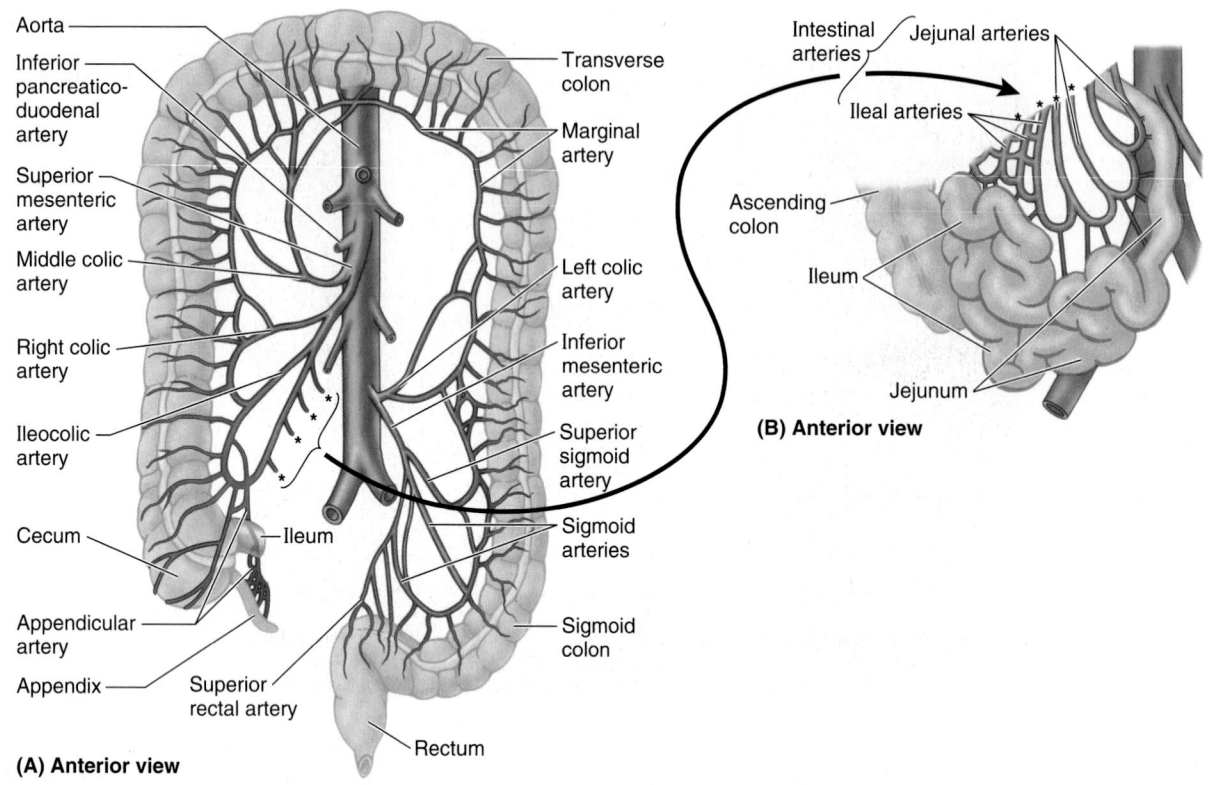

FIGURE 2.54. Arterial supply to intestines.

TABLE 2.10. ARTERIAL SUPPLY TO INTESTINES

Artery	Origin	Course	Distribution
Superior mesenteric	Abdominal aorta	Runs in root of mesentery to ileocecal junction	Part of gastrointestinal tract derived from midgut
Intestinal (jejunal and ileal) (n=15–18)	Superior mesenteric artery	Passes between two layers of mesentery	Jejunum and ileum
Middle colic		Ascends retroperitoneally and passes between layers of transverse mesocolon	Transverse colon
Right colic		Passes retroperitoneally to reach ascending colon	Ascending colon
Ileocolic	Terminal branch of superior mesenteric artery	Runs along root of mesentery and divides into ileal and colic branches	Ileum, cecum, and ascending colon
Appendicular	Ileocolic artery	Passes between layers of meso-appendix	Appendix
Inferior mesenteric	Abdominal aorta	Descends retroperitoneally to left of abdominal aorta	Supplies part of gastrointestinal tract derived from hindgut
Left colic	Inferior mesenteric artery	Passes retroperitoneally toward left to descending colon	Descending colon
Sigmoid (n = 3–4)		Passes retroperitoneally toward left to descending colon	Descending and sigmoid colon
Superior rectal	Terminal branch of inferior mesenteric artery	Descends retroperitoneally to rectum	Proximal part of rectum
Middle rectal	Internal iliac artery	Passes retroperitoneally to rectum	Midpart of rectum
Inferior rectal	Internal pudendal artery	Crosses ischioanal fossa to reach rectum	Distal part of rectum and anal canal

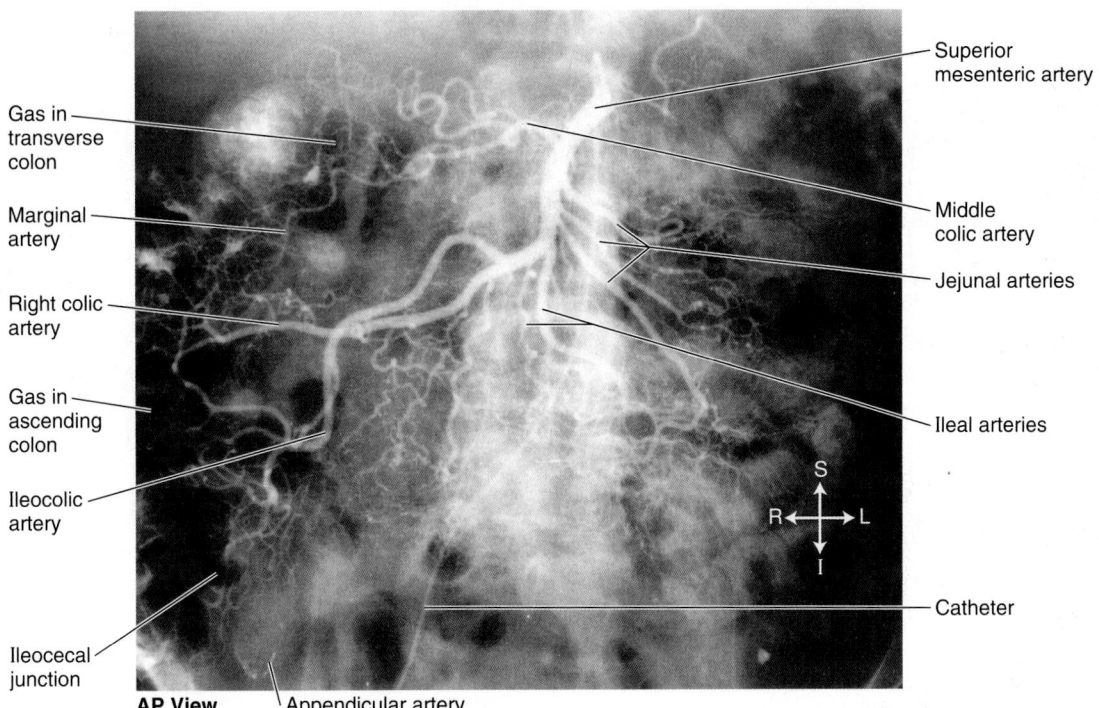

Gas in transverse colon

Marginal artery

Right colic artery

Gas in ascending colon

Ileocolic artery

Ileocecal junction

AP View

Appendicular artery

Superior mesenteric artery

Middle colic artery

Jejunal arteries

Ileal arteries

Catheter

FIGURE 2.55. Superior mesenteric arteriogram. Radiopaque dye was injected into the bloodstream by means of the catheter introduced into the femoral artery and advanced through the iliac arteries and aorta to the opening of the superior mesenteric artery. (Courtesy of Dr. E. L. Lansdown, Professor of Medical Imaging, University of Toronto, Toronto, ON, Canada.)

approximately 25% of people, it has a short mesentery. The ascending colon is separated from the anterolateral abdominal wall by the greater omentum. A deep vertical groove lined with parietal peritoneum, the *right paracolic gutter*, lies between the lateral aspect of the ascending colon and the adjacent abdominal wall (see Fig. 2.49A).

The *arterial supply to the ascending colon and right colic flexure* is from branches of the SMA, the *ileocolic* and **right colic arteries** (Figs. 2.54 and 2.55; Table 2.10). These arteries anastomose with each other and with the *right branch of the middle colic artery*, the first of a series of anastomotic arcades that is continued by the left colic and sigmoid arteries to form a continuous arterial channel, the **marginal artery** (*juxtacolic artery*). This artery parallels and extends the length of the colon close to its mesenteric border.

Venous drainage from the ascending colon flows through tributaries of the SMV, the *ileocolic* and **right colic veins** (Fig. 2.56A). The *lymphatic drainage* passes first to the **epicolic** and **paracolic lymph nodes,** next to the *ileocolic* and intermediate **right colic lymph nodes,** and from them to the *superior mesenteric lymph nodes* (Fig. 2.56B). The *nerve supply to the ascending colon* is derived from the *superior mesenteric nerve plexus* (Fig. 2.56C).

The **transverse colon** is the third, longest, and most mobile part of the large intestine (Fig. 2.52). It crosses the abdomen from the *right colic flexure* to the *left colic flexure*, where it turns inferiorly to become the descending colon. The **left colic flexure** (splenic flexure) is usually more superior, more acute, and less mobile than the right colic flexure. It lies

anterior to the inferior part of the left kidney and attaches to the diaphragm through the *phrenicocolic ligament* (see Fig. 2.26). The transverse colon and its mesentery, the *transverse mesocolon*, loops down, often inferior to the level of the iliac crests (Fig. 2.57B). The mesentery is adherent to or fused with the posterior wall of the omental bursa. The **root of the transverse mesocolon** (see Fig. 2.49A) lies along the inferior border of the pancreas and is continuous with the parietal peritoneum posteriorly. Being freely movable, the transverse colon is variable in position, usually hanging to the level of the umbilicus (L3 vertebral level) (Fig. 2.57A). However, in tall thin people, the transverse colon may extend into the pelvis (Fig. 2.57B).

The *arterial supply of the transverse colon* is mainly from the **middle colic artery** (Figs. 2.54 and 2.55; Table 2.10), a branch of the SMA. However, the transverse colon may also receive arterial blood from the *right and left colic arteries* via anastomoses, part of the series of anastomotic arcades that collectively form the marginal artery (juxtacolic artery).

Venous drainage of the transverse colon is through the SMV (Fig. 2.56A). The *lymphatic drainage* of the transverse colon is to the **middle colic lymph nodes,** which in turn drain to the *superior mesenteric lymph nodes* (Fig. 2.56B).

The *nerve supply of the transverse colon* is from the *superior mesenteric nerve plexus* via the peri-arterial plexuses of the right and middle colic arteries (Fig. 2.56C). These nerves transmit sympathetic, parasympathetic (vagal), and visceral afferent nerve fibers (see also "Summary of Innervation of Abdominal Viscera," p. 300).

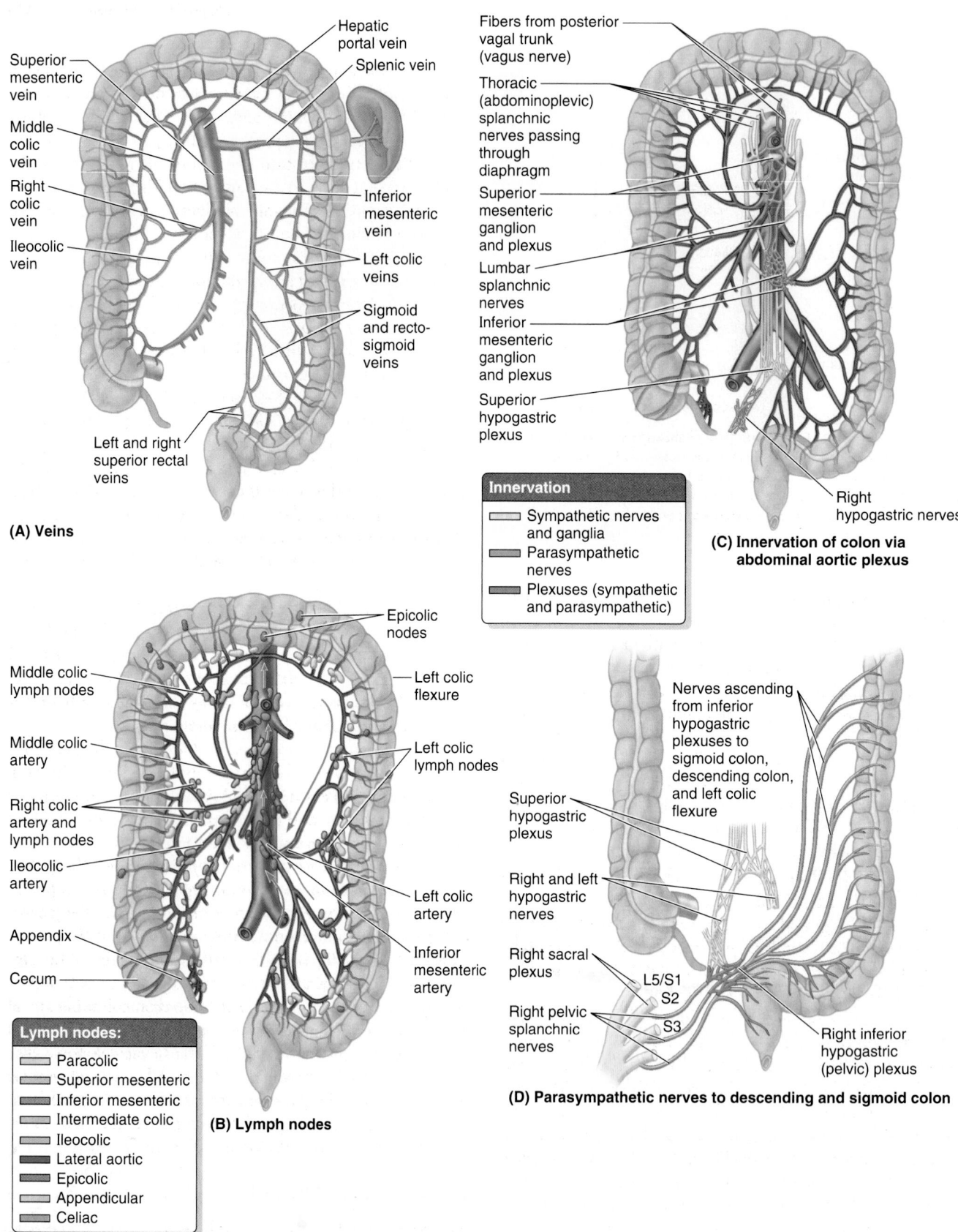

(A) Veins

Superior mesenteric vein
Middle colic vein
Right colic vein
Ileocolic vein
Hepatic portal vein
Splenic vein
Inferior mesenteric vein
Left colic veins
Sigmoid and recto-sigmoid veins
Left and right superior rectal veins

(C) Innervation of colon via abdominal aortic plexus

Fibers from posterior vagal trunk (vagus nerve)
Thoracic (abdominoplevic) splanchnic nerves passing through diaphragm
Superior mesenteric ganglion and plexus
Lumbar splanchnic nerves
Inferior mesenteric ganglion and plexus
Superior hypogastric plexus
Right hypogastric nerves

Innervation
- Sympathetic nerves and ganglia
- Parasympathetic nerves
- Plexuses (sympathetic and parasympathetic)

(B) Lymph nodes

Middle colic lymph nodes
Middle colic artery
Right colic artery and lymph nodes
Ileocolic artery
Appendix
Cecum
Epicolic nodes
Left colic flexure
Left colic lymph nodes
Left colic artery
Inferior mesenteric artery

Lymph nodes:
- Paracolic
- Superior mesenteric
- Inferior mesenteric
- Intermediate colic
- Ileocolic
- Lateral aortic
- Epicolic
- Appendicular
- Celiac

(D) Parasympathetic nerves to descending and sigmoid colon

Nerves ascending from inferior hypogastric plexuses to sigmoid colon, descending colon, and left colic flexure
Superior hypogastric plexus
Right and left hypogastric nerves
Right sacral plexus
Right pelvic splanchnic nerves
L5/S1
S2
S3
Right inferior hypogastric (pelvic) plexus

FIGURE 2.56. Veins, lymph nodes, and nerves of large intestine. A. Venous drainage by the superior mesenteric vein and inferior mesenteric vein corresponds to the pattern of the superior mesenteric artery and inferior mesenteric artery. **B.** Lymph from the large intestine flows sequentially to epicolic nodes (on the gut), paracolic nodes (along mesenteric border), intermediate colic nodes (along the colic arteries), and then to the superior or inferior mesenteric nodes and the intestinal trunks. **C.** Innervation of the colon occurs by means of mixed peri-aterial plexuses extending from the superior and inferior mesenteric ganglia along the respective arteries. **D.** Parasympathetic fibers from S2–S4 spinal cord levels ascend independently from the inferior hypogastric (pelvic) plexuses to reach the sigmoid colon, descending colon, and distal transverse colon.

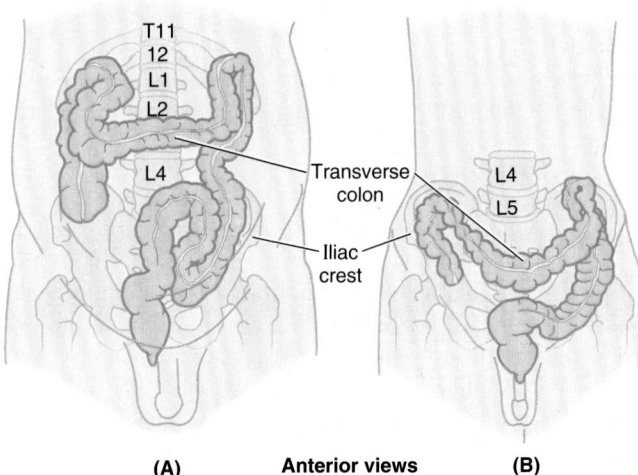

FIGURE 2.57. Effect of body type (body habitus) on disposition of the transverse colon. **A.** A heavily built hyperesthenic individual with a short thorax and a long abdomen is likely to have a transverse colon that is placed high. **B.** Individuals with a slender asthenic physique are likely to have a transverse colon that dips down toward or into the pelvis.

The **descending colon** occupies a secondarily retroperitoneal position between the left colic flexure and the left iliac fossa, where it is continuous with the sigmoid colon (Fig. 2.52). Thus, peritoneum covers the colon anteriorly and laterally and binds it to the posterior abdominal wall. Although retroperitoneal, the descending colon, especially in the iliac fossa, has a short mesentery in approximately 33% of people; however, it is usually not long enough to cause volvulus (twisting) of the colon. As it descends, the colon passes anterior to the lateral border of the left kidney. As with the ascending colon, the descending colon has a *paracolic gutter* (the left one) on its lateral aspect (see Fig. 2.49A).

The **sigmoid colon,** characterized by its S-shaped loop of variable length, links the descending colon and the rectum (Fig. 2.52). The sigmoid colon extends from the iliac fossa to the third sacral (S3) vertebra, where it joins the rectum. The termination of the teniae coli, approximately 15 cm from the anus, indicates the *rectosigmoid junction.*

The sigmoid colon usually has a long mesentery—the **sigmoid mesocolon**—and therefore has considerable freedom of movement, especially its middle part. (See the blue box "Volvulus of Sigmoid Colon," p. 261.) The **root of the sigmoid mesocolon** has an inverted V-shaped attachment, extending first medially and superiorly along the external iliac vessels and then medially and inferiorly from the bifurcation of the common iliac vessels to the anterior aspect of the sacrum. The *left ureter* and the division of the left common iliac artery lie retroperitoneally, posterior to the apex of the root of the sigmoid mesocolon. The *omental appendices of the sigmoid colon* are long (Fig. 2.52A); they disappear when the sigmoid mesentery terminates. The teniae coli also disappear as the longitudinal muscle in the wall of the colon broadens to form a complete layer in the rectum.

The *arterial supply of the descending and sigmoid colon* is from the *left colic* and **sigmoid arteries,** branches of the inferior mesenteric artery (Fig. 2.54; Table 2.10). Thus, at approximately the left colic flexure, a second transition occurs in the blood supply of the abdominal part of the alimentary canal: the SMA supplying blood to that part orad (proximal) to the flexure (derived from the embryonic midgut), and the IMA supplying blood to the part aborad (distal) to the flexure (derived from the embryonic hindgut). The sigmoid arteries descend obliquely to the left, where they divide into ascending and descending branches. The superior branch of the most superior sigmoid artery anastomoses with the descending branch of the left colic artery, thereby forming a part of the *marginal artery. Venous drainage from the descending colon and sigmoid colon* is provided by the *inferior mesenteric vein,* flowing usually into the splenic vein and then the hepatic portal vein on its way to the liver (see Figs. 2.56A and 2.75B).

Lymphatic drainage from the descending colon and sigmoid colon is conducted through vessels passing to the epicolic and paracolic nodes, and then through the **intermediate colic lymph nodes** along the left colic artery (Fig. 2.56B). Lymph from these nodes passes to the **inferior mesenteric lymph nodes** that lie around the IMA. However, lymph from the left colic flexure may also drain to the *superior mesenteric lymph nodes.*

Orad (toward the mouth, or proximal) to the left colic flexure, sympathetic and parasympathetic fibers travel together from the abdominal aortic plexus via peri-arterial plexuses to reach the abdominal part of the alimentary tract (Fig. 2.56C); however, aborad (away from the mouth, or distal) to the flexure, they follow separate routes.

The *sympathetic nerve supply of the descending and sigmoid colon* is from the lumbar part of the sympathetic trunk via lumbar (abdominopelvic) splanchnic nerves, the superior mesenteric plexus, and the peri-arterial plexuses following the inferior mesenteric artery and its branches.

The *parasympathetic nerve supply* is from the *pelvic splanchnic nerves* via the inferior hypogastric (pelvic) plexus and nerves, which ascend retroperitoneally from the plexus, independent of the arterial supply to this part of the gastrointestinal tract (Fig. 2.56D). Orad to the middle of the sigmoid colon, visceral afferents conveying pain sensation pass retrogradely with sympathetic fibers to thoracolumbar spinal sensory ganglia, whereas those carrying reflex information travel with the parasympathetic fibers to vagal sensory ganglia. Aborad to the middle of the sigmoid colon, all visceral afferents follow the parasympathetic fibers retrogradely to the sensory ganglia of spinal nerves S2–S4 (see also "Summary of Innervation of Abdominal Viscera," p. 300).

RECTUM AND ANAL CANAL

The **rectum** is the fixed (primarily retroperitoneal and subperitoneal) terminal part of the large intestine. It is continuous with the sigmoid colon at the level of S3 vertebra. The junction is at the inferior end of the mesentery of the sigmoid colon (Fig. 2.52). The rectum is continuous inferiorly with the *anal canal.* These parts of the large intestine are described with the pelvis in Chapter 3.

ESOPHAGUS AND STOMACH

Esophageal Varices

 Because the submucosal veins of the inferior esophagus drain to both the portal and systemic venous systems, they constitute a portosystemic anastomosis. In *portal hypertension* (an abnormally increased blood pressure in the portal venous system), blood is unable to pass through the liver via the hepatic portal vein, causing a reversal of flow in the esophageal tributary. The large volume of blood causes the submucosal veins to enlarge markedly, forming *esophageal varices* (Fig. B2.7). These distended collateral channels may rupture and cause severe hemorrhage that is life-threatening and difficult to control surgically. Esophageal varices commonly develop in persons who have developed *alcoholic cirrhosis* (fibrous scarring) of the liver (see the blue box "Cirrhosis of the Liver," p. 285).

Pyrosis

 Pyrosis (G., burning), or "heartburn," is the most common type of esophageal discomfort or substernal pain. This burning sensation in the abdominal part of the esophagus is usually the result of regurgitation of small amounts of food or gastric fluid into the lower esophagus (*gastroesophageal reflux disorder; GERD*). Pyrosis may also be associated with *hiatal hernia* (see "Hiatal Hernia"). As indicated by its common name, heartburn, pyrosis is commonly perceived as a "chest" (vs. abdominal) sensation.

Displacement of Stomach

 Pancreatic pseudo-cysts and abscesses in the omental bursa may push the stomach anteriorly. This displacement is usually visible in lateral radiographs of the stomach and other diagnostic images, such as computed tomography (CT). Following *pancreatitis* (inflammation of the pancreas), the posterior wall of the stomach may adhere to the part of the posterior wall of the omental bursa that covers the pancreas. This adhesion occurs because of the close relationship of the posterior wall of the stomach to the pancreas.

Hiatal Hernia

 A *hiatal (hiatus) hernia* is a protrusion of part of the stomach into the mediastinum through the esophageal hiatus of the diaphragm. The hernias occur most often in people after middle age, possibly because of weakening of the muscular part of the diaphragm and widening of the esophageal hiatus. Although clinically there are several types of hiatal hernia, the two main types are paraesophageal hiatal hernia and sliding hiatal hernia (Skandalakis et al., 1996).

In the less common *para-esophageal hiatal hernia*, the cardia remains in its normal position (Fig. B2.8A). However, a pouch of peritoneum, often containing part of the fundus of the stomach, extends through the esophageal hiatus anterior to the esophagus. In these cases, usually no regurgitation of gastric contents occurs because the cardial orifice is in its normal position.

In the common *sliding hiatal hernia*, the abdominal part of the esophagus, the cardia, and parts of the fundus of the stomach slide superiorly through the esophageal hiatus into the thorax, especially when the person lies down or bends over (Fig. B2.8B). Some regurgitation of stomach contents into the esophagus is possible because the clamping action of the right crus of the diaphragm on the inferior end of the esophagus is weak.

Pylorospasm

 Spasmodic contraction of the pylorus sometimes occurs in infants, usually between 2 and 12 weeks of age. *Pylorospasm* is characterized by failure of the smooth muscle fibers encircling the pyloric canal to relax normally. As a result, food does not pass easily from the stomach into the duodenum and the stomach becomes overly full, usually resulting in discomfort and vomiting.

Congenital Hypertrophic Pyloric Stenosis

 Congenital hypertrophic pyloric stenosis is a marked thickening of the smooth muscle (hypertrophy) in the pylorus that affects approximately 1 of every 150 male infants and 1 of every 750 female infants (Moore, Persaud, and Torchia, 2012). Normally, gastric peristalsis pushes chyme through the pyloric canal and orifice into the small intestine at irregular intervals (Fig. B2.9A). In neonates with pyloric stenosis, the elongated overgrown pylorus is hard and the pyloric canal is narrow (Fig. B2.9B), resisting gastric emptying. Proximally, the stomach may become secondarily dilated because of the pyloric stenosis (narrowing). Although the cause of congenital hypertrophic pyloric stenosis is

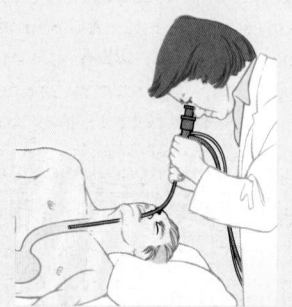
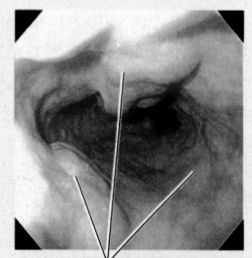

View through esophagoscope of esophageal varices

FIGURE B2.7.

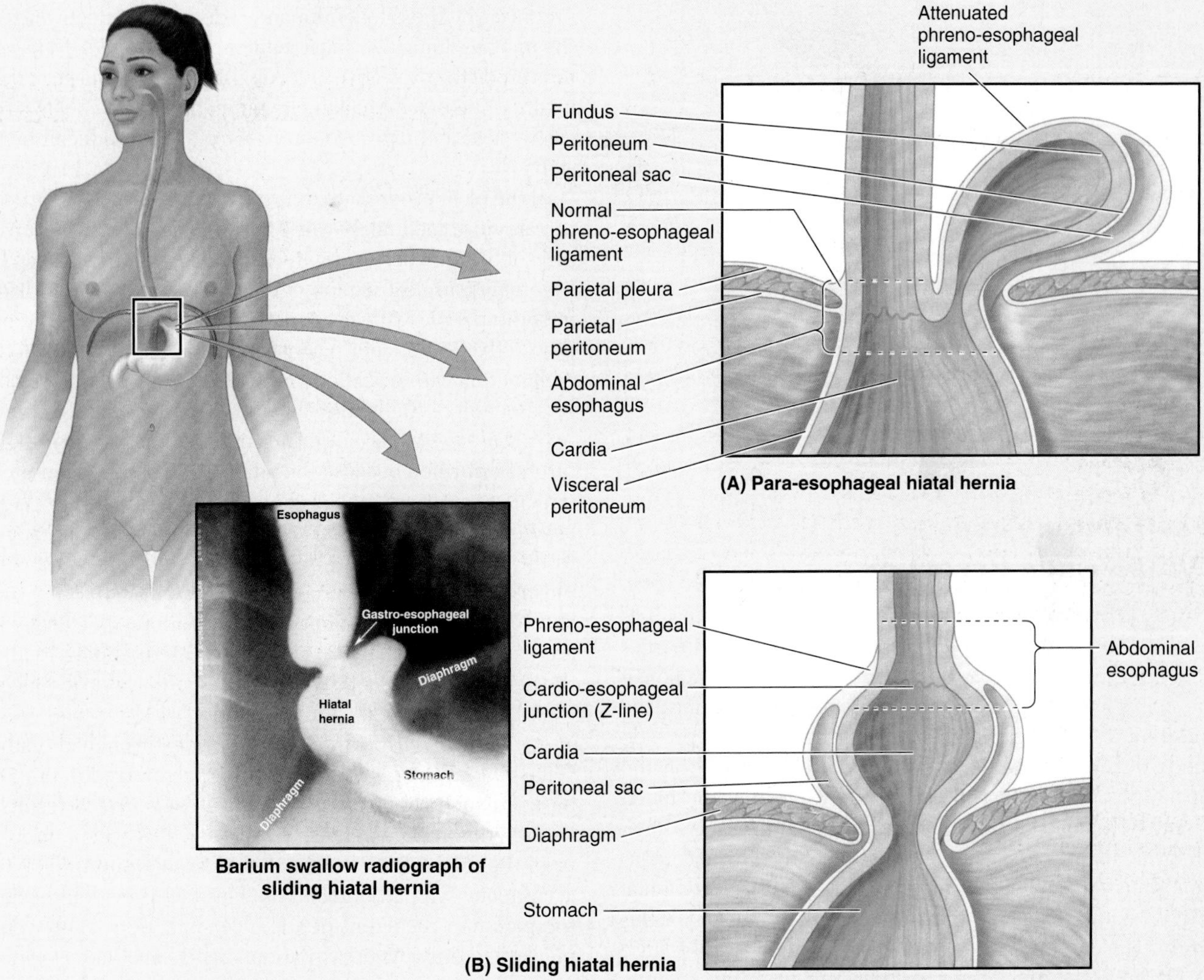

Attenuated phreno-esophageal ligament

Fundus

Peritoneum

Peritoneal sac

Normal phreno-esophageal ligament

Parietal pleura

Parietal peritoneum

Abdominal esophagus

Cardia

Visceral peritoneum

(A) Para-esophageal hiatal hernia

Esophagus

Gastro-esophageal junction

Diaphragm

Hiatal hernia

Stomach

Diaphragm

Barium swallow radiograph of sliding hiatal hernia

Phreno-esophageal ligament

Cardio-esophageal junction (Z-line)

Cardia

Peritoneal sac

Diaphragm

Stomach

Abdominal esophagus

(B) Sliding hiatal hernia

FIGURE B2.8.

unknown, genetic factors appear to be involved because of this condition's high incidence in infants of monozygotic twins.

Carcinoma of Stomach

When the body or pyloric part of the stomach contains a malignant tumor, the mass may be palpable. Using a *gastroscope*, physicians can inspect the mucosa of the air-inflated stomach, enabling them to observe gastric lesions and take biopsies (Fig. B2.10). The extensive lymphatic drainage of the stomach and the impossibility of removing all the lymph nodes create a surgical problem. The nodes along the splenic vessels can be excised by removing the spleen, gastrosplenic and splenorenal ligaments, and the body and tail of the pancreas. Involved nodes along the gastro-omental vessels can be removed by resecting the greater omentum; however, removal of the aortic and celiac nodes and those around the head of the pancreas is difficult.

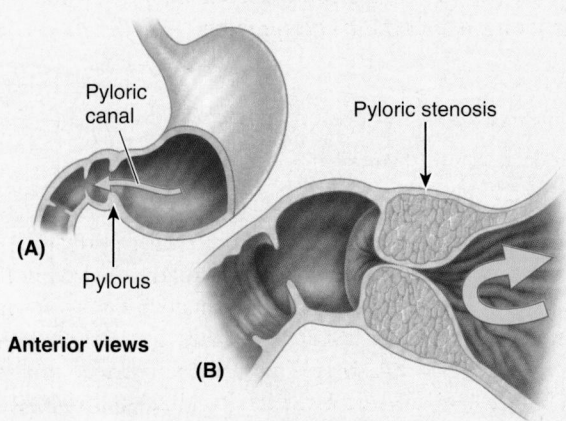

Pyloric canal

Pyloric stenosis

(A)

Pylorus

Anterior views

(B)

FIGURE B2.9. Congenital hypertrophic pyloric stenosis. A. Normal passage through the pyloric sphincter is shown. **B.** Stoppage of flow owing to stenosis is demonstrated.

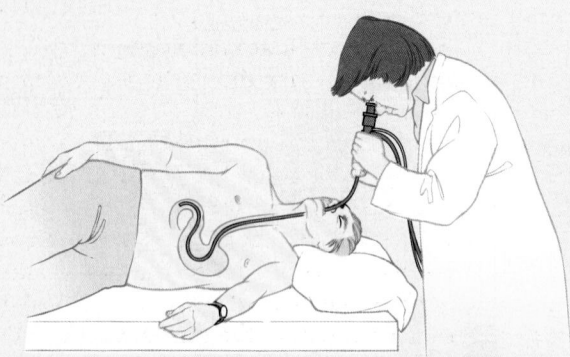

Gastroscope is introduced nasally or orally and led slowly down the esophagus and gastrointestinal tract until desired level is reached.

FIGURE B2.10.

Gastrectomy and Lymph Node Resection

Total gastrectomy (removal of the entire stomach) is uncommon. *Partial gastrectomy* (removal of part of the stomach) may be performed to remove a region of the stomach involved by a carcinoma, for example. Because the anastomoses of the arteries supplying the stomach provide good collateral circulation, one or more arteries may be ligated during this procedure without seriously affecting the blood supply to the part of the stomach remaining in place. When removing the pyloric antrum, for example, the greater omentum is incised parallel and inferior to the right gastro-omental artery, requiring ligation of all the omental branches of this artery. The omentum does not degenerate, however, because of anastomoses with other arteries, such as the omental branches of the left gastro-omental artery, which are still intact. Partial gastrectomy to remove a carcinoma usually also requires removal of all involved regional lymph nodes. Because cancer frequently occurs in the pyloric region, removal of the *pyloric lymph nodes* as well as the right *gastro-omental lymph nodes* also receiving lymph drainage from this region is especially important. As stomach cancer becomes more advanced, the lymphogenous dissemination of malignant cells involves the *celiac lymph nodes,* to which all gastric nodes drain.

Gastric Ulcers, Peptic Ulcers, *Helicobacter pylori*, and Vagotomy

Gastric ulcers are open lesions of the mucosa of the stomach, whereas *peptic ulcers* are lesions of the mucosa of the pyloric canal or, more often, the duodenum. Most ulcers of the stomach and duodenum are associated with an infection of a specific bacterium, *Helicobacter pylori* (*H. pylori*). People experiencing severe chronic anxiety are most prone to the development of peptic ulcers. They often have gastric acid secretion rates that are markedly higher than normal between meals. It is thought that the high acid in the stomach and duodenum overwhelms the bicarbonate normally

produced by the duodenum and reduces the effectiveness of the mucous lining, leaving it vulnerable to *H. pylori.* The bacteria erode the protective mucous lining of the stomach, inflaming the mucosa and making it vulnerable to the effects of the gastric acid and digestive enzymes (pepsin) produced by the stomach.

If the ulcer erodes into the gastric arteries, it can cause life-threatening bleeding. Because the secretion of acid by parietal cells of the stomach is largely controlled by the vagus nerves, *vagotomy* (surgical section of the vagus nerves) is performed in some people with chronic or recurring ulcers to reduce the production of acid. Vagotomy may also be performed in conjunction with resection of the ulcerated area (*antrectomy,* or resection of the pyloric antrum) to reduce acid secretion. A *truncal vagotomy* (surgical section of the vagal trunks) is rarely performed because the innervation of other abdominal structures is also sacrificed (Fig. B2.11A). In *selective gastric vagotomy*, the stomach is denervated but the vagal branches to the pylorus, liver and biliary ducts, intestines, and celiac plexus are preserved (Fig. B2.11B). A *selective proximal vagotomy* attempts to denervate even more specifically the area in which the parietal cells are located, hoping to affect the acid-producing cells while sparing other gastric function (motility) stimulated by the vagus nerve (Fig. B2.11C).

A *posterior gastric ulcer* may erode through the stomach wall into the pancreas, resulting in referred pain to the back. In such cases, *erosion of the splenic artery* results in severe hemorrhage into the peritoneal cavity. Pain impulses from the stomach are carried by visceral afferent fibers that accompany sympathetic nerves. This fact is evident because the pain of a recurrent peptic ulcer may persist after complete vagotomy, whereas patients who have had a bilateral

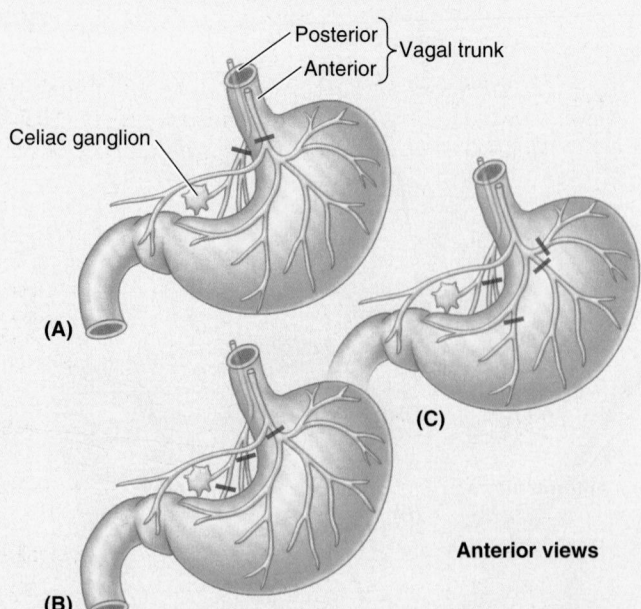

FIGURE B2.11. Vagotomy. Truncal (**A**), selective gastric (**B**), and selective proximal (**C**) vagotomy are shown. *Red bars* indicate site of surgical section of nerve.

sympathectomy may have a perforated peptic ulcer and experience no pain.

Visceral Referred Pain

 Pain is an unpleasant sensation associated with actual or potential tissue damage and mediated by specific nerve fibers to the brain, where its conscious appreciation may be modified. *Organic pain* arising from an organ such as the stomach varies from dull to severe; however, the pain is poorly localized. It radiates to the dermatome level, which receives visceral afferent fibers from the organ concerned. *Visceral referred pain* from a gastric ulcer, for example, is referred to the epigastric region because the stomach is supplied by pain afferents that reach the T7 and T8 spinal sensory ganglia and spinal cord segments through the greater splanchnic nerve (Fig. B2.12). The brain interprets the pain as though the irritation occurred in the skin of the epigastric region, which is also supplied by the same sensory ganglia and spinal cord segments.

Pain arising from the parietal peritoneum is of the somatic type and is usually severe. The site of its origin can be localized. The anatomical basis for this localization of pain is that the parietal peritoneum is supplied by somatic sensory fibers through thoracic nerves, whereas a viscus such as the appendix is supplied by visceral afferent fibers in the lesser splanchnic nerve. Inflamed parietal peritoneum is extremely sensitive to stretching. When digital pressure is applied to the anterolateral abdominal wall over the site of inflammation, the parietal peritoneum is stretched. When the fingers are suddenly removed, extreme localized pain is usually felt, known as *rebound tenderness.*

SMALL AND LARGE INTESTINE

Duodenal Ulcers

 Duodenal ulcers (peptic ulcers) are inflammatory erosions of the duodenal mucosa. Most (65%) duodenal ulcers occur in the posterior wall of the superior part of the duodenum within 3 cm of the pylorus. Occasionally, an ulcer perforates the duodenal wall, permitting the contents to enter the peritoneal cavity and causing *peritonitis.* Because the superior part of the duodenum closely relates to the liver, gallbladder, and pancreas, any of these structures may become adherent to the inflamed duodenum. They may also become ulcerated as the lesion continues to erode the tissue that surrounds it. Although bleeding from duodenal ulcers commonly occurs, *erosion of the gastroduodenal artery* (a posterior relation of the superior part of the duodenum) by a duodenal ulcer results in severe hemorrhage into the peritoneal cavity and subsequent peritonitis.

Developmental Changes in Mesoduodenum

 During the early fetal period, the entire duodenum has a mesentery; however, most of it fuses with the posterior abdominal wall because of pressure from the overlying transverse colon. Because the attachment of the mesoduodenum to the wall is secondary (occurred through formation of a *fusion fascia;* discussed under "Embryology of Peritoneal Cavity," p. 218), the duodenum and the closely associated pancreas can be separated (surgically mobilized) from the underlying retroperitoneal viscera during surgical operations involving the duodenum without endangering the blood supply to the kidney or the ureter.

Paraduodenal Hernias

There are two or three inconstant folds and fossae (recesses) around the duodenojejunal flexure (Fig. B2.13). The *paraduodenal fold and fossa* are large and lie to the left of the ascending part of the duodenum. If a loop of intestine enters this fossa, it may strangulate. During repair of a *paraduodenal hernia,* care must be taken not to injure the branches of the inferior mesenteric artery

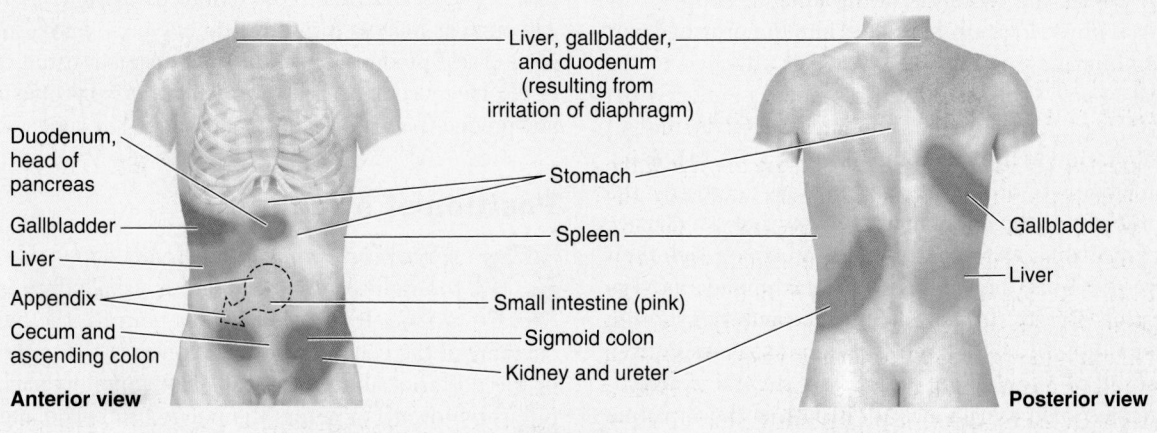

FIGURE B2.12.

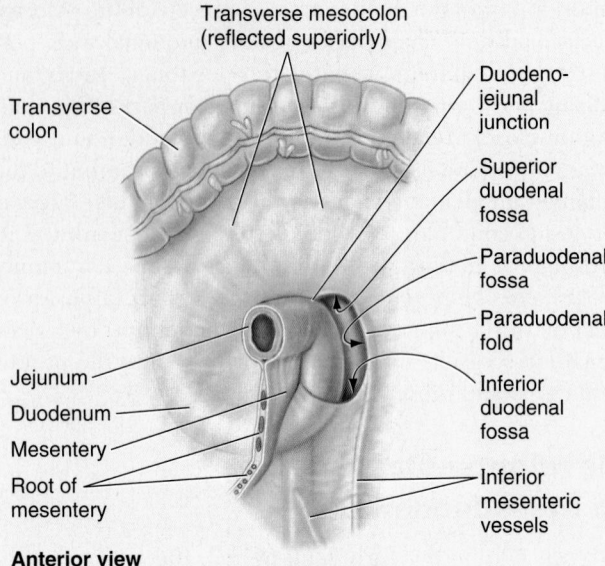

Transverse mesocolon
(reflected superiorly)

Transverse
colon

Duodeno-
jejunal
junction

Superior
duodenal
fossa

Paraduodenal
fossa

Paraduodenal
fold

Jejunum

Duodenum

Mesentery

Root of
mesentery

Inferior
duodenal
fossa

Inferior
mesenteric
vessels

Anterior view

FIGURE B2.13.

and vein or the ascending branches of the left colic artery, which are closely related to the paraduodenal fold and fossa.

Brief Review of Embryological Rotation of Midgut

 An understanding of the rotation of the midgut clarifies the adult arrangement of the intestines. The primordial gut comprises the foregut, midgut, and hindgut. Pain arising from foregut derivatives—esophagus, stomach, pancreas, duodenum, liver, and biliary ducts—localizes in the *epigastric region*. Pain arising from midgut derivatives—the small intestine distal to bile duct, cecum, appendix, ascending colon, and most of the transverse colon—localizes in the *peri-umbilical region*. Pain arising from hindgut derivatives—the distal part of the transverse colon, descending colon, sigmoid colon, and rectum—localizes in the *hypogastric region* (see Table 2.1).

For 4 weeks, the rapidly growing midgut, supplied by the SMA, is physiologically herniated into the proximal part of the umbilical cord (Fig. B2.14A). It is attached to the umbilical vesicle (yolk sac) by the omphalo-enteric duct (yolk stalk). As it returns to the abdominal cavity, the midgut rotates 270° around the axis of the SMA (Fig. B2.14B & C). As the relative size of the liver and kidneys decreases, the midgut returns to the abdominal cavity as increased space becomes available. As the parts of the intestine reach their definitive positions, their mesenteric attachments undergo modification (Fig. B2.14D & E). Some mesenteries shorten and others disappear (e.g., most of duodenal mesentery). Malrotation of the midgut results in several congenital anomalies such as volvulus (twisting) of the intestine (Moore et al., 2012).

Navigating Small Intestine

 When portions of the small intestine have been delivered through a surgical wound, the proximal (orad—toward the mouth) and distal (aborad—away from the mouth) ends of a loop of intestine are not apparent. If you try to follow the intestine in a particular direction (e.g., attempting to follow the ileum to the ileocecal junction), it is important to know which end is which. Normal peristalsis may not be present to provide a clue. Place your hands on each side of the intestine and its mesentery and then follow the mesentery with your fingers to its root (its attachment to the posterior abdominal wall), untwisting the loop of intestine as necessary. Once the mesentery and intestine are straightened to match the direction of the root, the cranial end must be the orad end, and the caudal end the aborad end.

Ischemia of Intestine

 Occlusion of the vasa recta (see Fig. 2.48B) by emboli (e.g., blood clots) results in *ischemia* of the part of the intestine concerned. If the ischemia is severe, *necrosis* (tissue death) of the involved segment results and *ileus* (obstruction of the intestine) of the paralytic type occurs. Ileus is accompanied by a severe colicky pain, along with abdominal distension, vomiting, and often fever and dehydration. If the condition is diagnosed early (e.g., using a *superior mesenteric arteriogram*), the obstructed part of the vessel may be cleared surgically.

Ileal Diverticulum

 An *ileal diverticulum* (or Meckel diverticulum) is a congenital anomaly that occurs in 1–2% of the population. A remnant of the proximal part of the embryonic omphaloenteric duct (yolk stalk), the diverticulum usually appears as a finger-like pouch (Fig. B2.15A). It is always at the site of attachment of the omphaloenteric duct on the antimesenteric border (border opposite the mesenteric attachment) of the ileum. The diverticulum is usually located 30–60 cm from the ileocecal junction in infants and 50 cm in adults. It may be free (74%) or attached to the umbilicus (26%) (Fig. B2.15B). Although its mucosa is mostly ileal in type, it may also include areas of acid-producing gastric tissue, pancreatic tissue, or jejunal or colonic mucosa. An ileal diverticulum may become inflamed and produce pain mimicking that produced by appendicitis.

Position of Appendix

 A *retrocecal appendix* extends superiorly toward the right colic flexure and is usually free (see Fig. 2.53C). It occasionally lies beneath the peritoneal covering of the cecum, where it is often fused to the cecum or the posterior abdominal wall. The appendix may project inferiorly toward or across the pelvic brim. The anatomical position of the appendix determines the symptoms and the

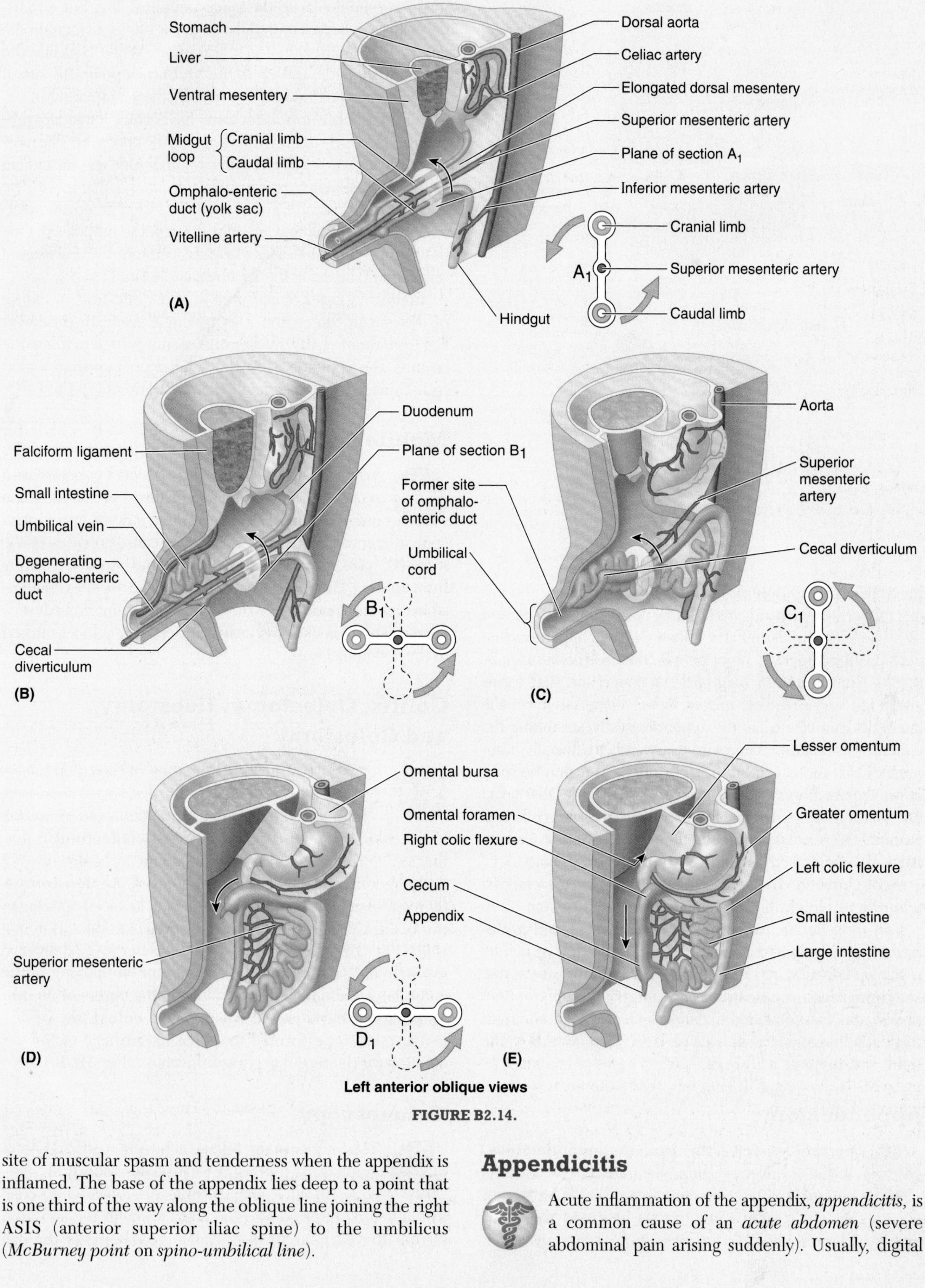

(A)

Stomach
Liver
Ventral mesentery
Midgut loop { Cranial limb, Caudal limb }
Omphalo-enteric duct (yolk sac)
Vitelline artery

Dorsal aorta
Celiac artery
Elongated dorsal mesentery
Superior mesenteric artery
Plane of section A₁
Inferior mesenteric artery
Cranial limb
Superior mesenteric artery
Caudal limb
Hindgut

A₁

(B)

Falciform ligament
Small intestine
Umbilical vein
Degenerating omphalo-enteric duct
Cecal diverticulum

Duodenum
Plane of section B₁

B₁

(C)

Former site of omphalo-enteric duct
Umbilical cord

Aorta
Superior mesenteric artery
Cecal diverticulum

C₁

(D)

Superior mesenteric artery

Omental bursa

D₁

(E)

Omental foramen
Right colic flexure
Cecum
Appendix

Lesser omentum
Greater omentum
Left colic flexure
Small intestine
Large intestine

Left anterior oblique views

FIGURE B2.14.

site of muscular spasm and tenderness when the appendix is inflamed. The base of the appendix lies deep to a point that is one third of the way along the oblique line joining the right ASIS (anterior superior iliac spine) to the umbilicus (*McBurney point* on *spino-umbilical line*).

Appendicitis

Acute inflammation of the appendix, *appendicitis*, is a common cause of an *acute abdomen* (severe abdominal pain arising suddenly). Usually, digital

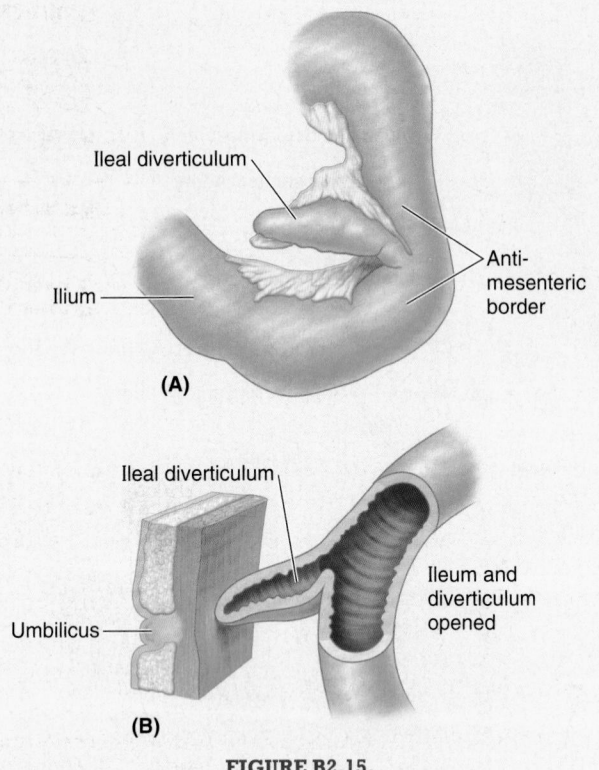

(A)

(B)

FIGURE B2.15.

pressure over the McBurney point registers maximum abdominal tenderness. Appendicitis in young people is usually caused by hyperplasia of lymphatic follicles in the appendix that occludes the lumen. In older people, the obstruction usually results from a *fecalith* (*coprolith*), a concretion that forms around a center of fecal matter. When secretions from the appendix cannot escape, the appendix swells, stretching the visceral peritoneum. The pain of appendicitis usually commences as a vague pain in the peri-umbilical region because afferent pain fibers enter the spinal cord at the T10 level. Later, severe pain in the right lower quadrant results from irritation of the parietal peritoneum lining the posterior abdominal wall. Extending the thigh at the hip joint elicits pain.

Acute infection of the appendix may result in *thrombosis* (clotting of blood) in the appendicular artery, which often results in ischemia, gangrene (death of tissue), and perforation of an inflamed appendix. *Rupture of the appendix* results in infection of the peritoneum (*peritonitis*), increased abdominal pain, nausea and/or vomiting, and *abdominal rigidity* (stiffness of abdominal muscles). Flexion of the right thigh ameliorates the pain because it causes relaxation of the right psoas muscle, a flexor of the thigh.

Appendectomy

Surgical removal of the appendix (**appendectomy**) may be performed through a transverse or gridiron (muscle-splitting) incision centered at the McBurney point in the right lower quadrant (see the blue box "Abdominal Surgical Incisions," p. 198). Traditionally, a gridiron incision is

made perpendicular to the spino-umbilical line, but a transverse incision is also commonly used. The choice of incision site and type is at the surgeon's discretion. While typically the inflamed appendix is deep to the McBurney point, the site of maximal pain and tenderness indicates the actual location.

Laparoscopic appendectomy has become a standard procedure selectively utilized for removing the appendix. The peritoneal cavity is first inflated with carbon dioxide gas, distending the abdominal wall, to provide viewing and working space. The laparoscope is passed through a small incision in the anterolateral abdominal wall (e.g., near or through the umbilicus). One or two other small incisions ("portals") are required for surgical (instrument) access to the appendix and related vessels.

In unusual cases of *malrotation of the intestine*, or failure of descent of the cecum, the appendix is not in the lower right quadrant (LRQ). When the cecum is high (*subhepatic cecum*), the appendix is in the right hypochondriac region (see Table 2.1) and the pain localizes there, not in the LRQ.

Mobile Ascending Colon

When the inferior part of the ascending colon has a mesentery, the cecum and proximal part of the colon are abnormally mobile. This condition, present in approximately 11% of individuals, may cause *volvulus of the colon* (L. *volvo*, to roll), an obstruction of the intestine resulting from twisting. *Cecopexy* (fixation) may avoid volvulus and possible obstruction of the colon. In this anchoring procedure, a tenia coli of the cecum and proximal ascending colon is sutured to the abdominal wall.

Colitis, Colectomy, Ileostomy, and Colostomy

  Chronic inflammation of the colon (*ulcerative* **colitis,** *Crohn disease*) is characterized by severe inflammation and ulceration of the colon and rectum. In some cases, a **colectomy** is performed, during which the terminal ileum and colon, as well as the rectum and anal canal, are removed. An **ileostomy** is then constructed to establish a *stoma*, an artificial opening of the ileum through the skin of the anterolateral abdominal wall (Fig. B2.16A). The terminating ileum is delivered through and sutured to the periphery of an opening in the anterolateral abdominal wall, allowing the egress of its contents. Following a partial colectomy, a **colostomy** or *sigmoidostomy* is performed to create an artificial cutaneous opening for the terminal part of the colon (Fig. B2.16B).

Colonoscopy

The interior of the colon can be observed and photographed in a procedure called **colonoscopy** or *coloscopy*, using a long, flexible fiberoptic endoscope (*colonoscope*) inserted into the colon through the anus and rectum (Fig. B2.17A). Small instruments can be passed through

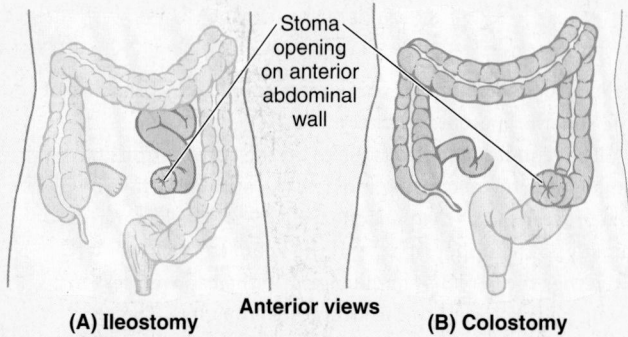

FIGURE B2.16. **A.** Ileostomy. **B.** Colostomy.

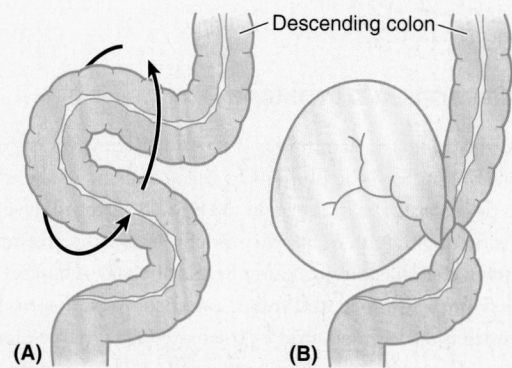

FIGURE B2.18. Volvulus of sigmoid colon.

the colonoscope and used to facilitate minor operative procedures, such as biopsies or removal of polyps. Most tumors of the large intestine occur in the sigmoid colon and rectum; they often appear near the rectosigmoid junction. The interior of the sigmoid colon is observed with a *sigmoidoscope,* a shorter endoscope, in a procedure called *sigmoidoscopy.*

Diverticulosis

Diverticulosis is a disorder in which multiple false *diverticula* (external evaginations or out-pocketings of the mucosa of the colon) develop along the intestine. It primarily affects middle-aged and elderly people. Diverticulosis is commonly found in the sigmoid colon (Fig. B2.17C & D). *Colonic diverticula* are not true diverticula because they are formed from protrusions of mucous membrane only, evaginated through weak points (separations) developed between muscle fibers rather than involving the whole wall of the colon. They occur most commonly on the

mesenteric side of the two nonmesenteric teniae coli, where nutrient arteries perforate the muscle coat to reach the submucosa. Diverticula are subject to infection and rupture, leading to diverticulitis, which can distort and erode the nutrient arteries, leading to hemorrhage. Diets high in fiber have proven beneficial in reducing the occurrence of diverticulosis.

Volvulus of Sigmoid Colon

Rotation and twisting of the mobile loop of the sigmoid colon and mesocolon—**volvulus of the sigmoid colon** (Fig. B2.18)—results in obstruction of the lumen of the descending colon and any part of the sigmoid colon proximal to the twisted segment. *Constipation* (inability of the stool to pass) and *ischemia* (absence of blood flow) of the looped part of the sigmoid colon result, which may progress to *fecal impaction* (an immovable collection of compressed or hardened feces) of the colon and possible *necrosis* (tissue death) of the involved segment if untreated.

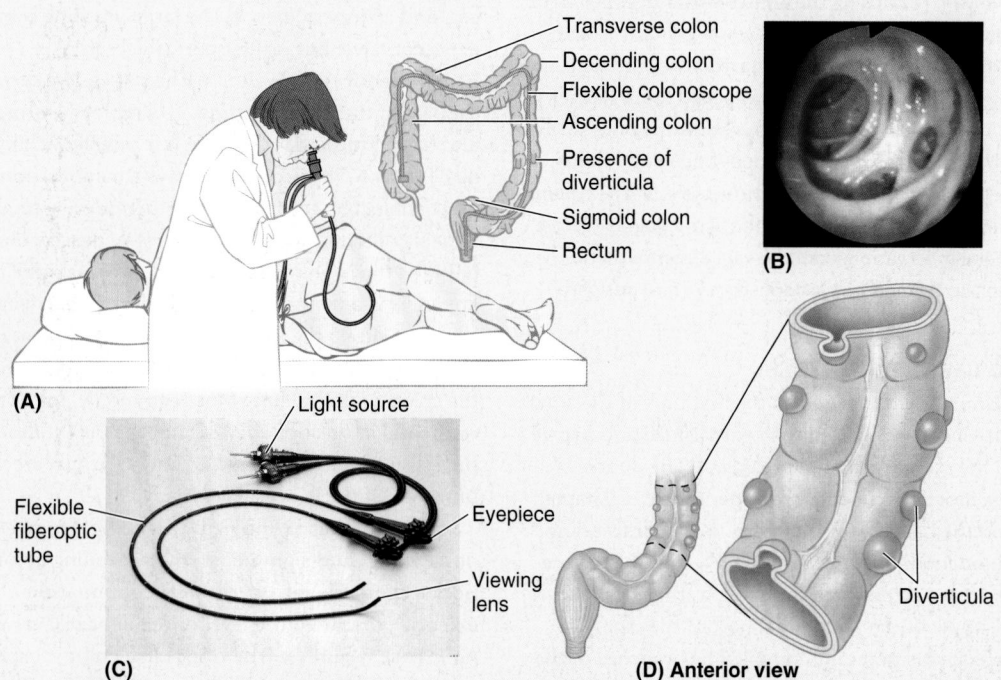

FIGURE B2.17. **Examination of large intestine. A.** The colonoscopic procedure. **B.** Diverticulosis of the colon, as photographed through a colonoscope. **C.** The parts of a colonoscope. Photographs can be taken by a camera attached to the colonoscope. **D.** Diverticula in the sigmoid colon are shown.

The Bottom Line

ESOPHAGUS AND STOMACH

Esophagus: The esophagus is a tubular conveyer of food, delivering it from the pharynx to the stomach. ♦ The esophagus penetrates the diaphragm at the T10 vertebral level, passing through its right crus, which decussates around it to form the physiological inferior esophageal sphincter. ♦ The trumpet-shaped abdominal part, composed entirely of smooth muscle innervated by the esophageal nerve plexus, enters the cardial part of the stomach. ♦ The abdominal part of the esophagus receives blood from esophageal branches of the left gastric artery (from the celiac trunk). ♦ Submucosal veins drain to both the systemic and portal venous systems and thus constitute portocaval anastomoses that may become varicose in the presence of portal hypertension. ♦ Internally, in living people, the esophagus is demarcated from the stomach by an abrupt mucosal transition, the Z-line.

Stomach: The stomach is the dilated portion of the alimentary tract between the esophagus and the duodenum, specialized to accumulate ingested food and prepare it chemically and mechanically for digestion. ♦ The stomach lies asymmetrically in the abdominal cavity, to the left of the midline and usually in the upper left quadrant. However, the position of the stomach can vary markedly in persons of different body types. ♦ The abdominal portion of the esophagus enters its cardial portion, and its pyloric part leads to the exit to the duodenum. ♦ Gastric emptying is controlled by the pylorus. ♦ In life, the internal surface of the stomach is covered with a protective layer of mucus, overlying gastric folds that disappear with distension. ♦ The stomach is intraperitoneal, with the lesser omentum (enclosing the anastomoses between right and left gastric vessels) attached to its lesser curvature, and the greater omentum (enclosing the anastomoses between right and left gastro-omental vessels) attached to its greater curvature. ♦ The vessels of its curvatures serve the body and pyloric antrum of the stomach. The upper body and fundus are served by short and posterior gastric vessels. ♦ The trilaminar smooth muscle of the stomach and gastric glands receive parasympathetic innervation from the vagus; sympathetic innervation to the stomach is vasoconstrictive and antiperistaltic.

SMALL AND LARGE INTESTINES

Small intestine: The duodenum is the first part of the small intestine, receiving chyme mixed with gastric acid and pepsin directly from the stomach via the pylorus. ♦ The duodenum follows a mostly secondarily retroperitoneal, C-shaped course around the head of the pancreas. ♦ The descending part of the duodenum receives both the bile and the pancreatic ducts. ♦ At or just distal to this level, a transition occurs in the blood supply of the abdominal part of the digestive tract. Proximal to this point it is supplied by branches of the celiac trunk; distal to this point, it is supplied by branches of the superior mesenteric artery.

The jejunum and ileum make up the convolutions of the small intestine occupying most of the infracolic division of the greater sac of the peritoneal cavity. ♦ The jejunum is mostly to the upper right, and the ileum to the lower left. Together they are 3–4 m in length (in the cadaver; less in living people owing to the structures' tonicity). The orad (proximal relative to the mouth) two fifths is jejunum and the aborad (distal) three fifths is ileum, although there is no clear line of transition. The diameter of the small intestine becomes increasingly smaller as the semifluid chyme progresses through it. ♦ Its blood vessels also become smaller, but the number of tiers of arcades increases while the length of the vasa recta decreases. ♦ The fat in which the vessels are embedded within the mesentery increases, making these features more difficult to see. ♦ The ileum is characterized by an abundance of lymphoid tissue, aggregated into lymphoid nodules (Peyer patches). ♦ The intraperitoneal portion of the small intestine (jejunum and ileum) is suspended by the mesentery, the root of which extends from the duodenojejunal junction to the left of the midline at the L2 level to the ileocecal junction in the right iliac fossa. ♦ An ileal diverticulum is a congenital anomaly present in 1–2% of the population. It is 3–6 cm in length and is typically located 50 cm from the ileocecal junction in adults.

Large intestine: The large intestine consists of the cecum; appendix; ascending, transverse, descending, and sigmoid colon; rectum; and anal canal. ♦ The large intestine is characterized by teniae coli, haustra, omental appendices, and a large caliber. ♦ The large intestine begins at the ileocecal valve; but its first part, the cecum, is a pocket that hangs inferior to the valve. ♦ The pouch-like cecum, the widest part of the large intestine, is completely intraperitoneal and has no mesentery, so that it is mobile within the right iliac fossa. ♦ The ileocecal valve is a combination valve and weak sphincter, actively opening periodically to allow entry of ileal contents and forming a largely passive one-way valve between the ileum and the cecum, preventing reflux. ♦ The appendix is an intestinal diverticulum, rich in lymphoid tissue, that enters the medial aspect of the cecum, usually deep to the junction of the lateral third and medial two thirds of the spino-umbilical line. Most commonly, the appendix is retrocecal in position, but 32% of the time it descends into the lesser pelvis. ♦ The cecum and appendix are supplied by branches of the ileocecal vessels.

The colon has four parts: ascending, transverse, descending, and sigmoid. ♦ The ascending colon is a superior, secondarily retroperitoneal continuation of the cecum, extending between the level of the ileocecal valve and the right colic flexure. ♦ The transverse colon, suspended by the transverse mesocolon between the right and left flexures, is the longest and most mobile part of the large intestine. The level to which it descends depends largely on body type (habitus). ◆ The descending colon occupies a secondarily retroperitoneal position between the left colic flexure and left iliac fossa, where it is continuous with the sigmoid colon. ◆ The S-shaped sigmoid colon, suspended by the sigmoid mesocolon, is highly variable in length and disposition, ending at the rectosigmoid junction. The teniae, haustra, and omental appendices cease at the junction, located anterior to the third sacral segment.

The part of large intestine orad (proximal) to the left colic flexure (cecum, appendix, and ascending and transverse colons) is served by branches of the superior mesenteric vessels. Aborad (distal) to the flexure, most of the remainder of the large intestine (descending and sigmoid colons and superior rectum) is served by mesenteric vessels. ◆ The left colic flexure also marks the divide between cranial (vagal) and sacral (pelvic splanchnic) parasympathetic innervation of the alimentary tract. ◆ Sympathetic fibers are conveyed to the large intestine via abdominopelvic (lesser and lumbar) splanchnic nerves via the prevertebral (superior and inferior mesenteric) ganglia and peri-arterial plexuses. ◆ The middle of the sigmoid colon marks a divide in the sensory innervation of the abdominal alimentary tract: orad, visceral afferents for pain travel retrogradely with sympathetic fibers to spinal sensory ganglia, whereas those conveying reflex information travel with parasympathetic fibers to vagal sensory ganglia; aborad, both types of visceral afferent fibers travel with parasympathetic fibers to spinal sensory ganglia.

Spleen

The **spleen** is an ovoid, usually purplish, pulpy mass about the size and shape of one's fist. It is relatively delicate and considered the most vulnerable abdominal organ. The spleen is located in the superolateral part of the left upper quadrant (LUQ), or hypochondrium of the abdomen, where it enjoys protection of the inferior thoracic cage (Fig. 2.58A & B). As the largest of the lymphatic organs, it participates in the body's defense system as a site of lymphocyte (white blood cell) proliferation and of immune surveillance and response.

Prenatally, the spleen is a hematopoietic (blood-forming) organ, but after birth it is involved primarily in identifying, removing, and destroying expended red blood cells (RBCs) and broken-down platelets, and in recycling iron and globin. The spleen serves as a blood reservoir, storing RBCs and platelets, and, to a limited degree, can provide a sort of "self-transfusion" as a response to the stress imposed by hemorrhage. In spite of its size and the many useful and important functions it provides, it is not a vital organ (not necessary to sustain life).

To accommodate these functions, the spleen is a soft, vascular (sinusoidal) mass with a relatively delicate fibroelastic capsule (Fig. 2.58E). The thin capsule is covered with a layer of visceral peritoneum that entirely surrounds the spleen except at the **splenic hilum,** where the splenic branches of the splenic artery and vein enter and leave (Fig. 2.58D). Consequently, it is capable of marked expansion and some relatively rapid contraction.

The spleen is a mobile organ although it normally does not descend inferior to the costal (rib) region; it rests on the *left colic flexure* (Fig. 2.58A & B). It is associated posteriorly with the left 9th–11th ribs (its long axis is roughly parallel to the 10th rib) and separated from them by the diaphragm and the *costodiaphragmatic recess*—the cleft-like extension of the pleural cavity between the diaphragm and the lower part of the thoracic cage. The **relations of the spleen** are:

- Anteriorly, the stomach
- Posteriorly, the left part of the diaphragm, which separates it from the pleura, lung, and ribs 9–11
- Inferiorly, the left colic flexure
- Medially, the left kidney.

The spleen varies considerably in size, weight, and shape; however, it is usually approximately 12 cm long and 7 cm wide. (A nonmetric memory device exploits odd numbers: the spleen is 1 inch thick, 3 inches wide, 5 inches long, and weighs 7 ounces.)

The **diaphragmatic surface of the spleen** is convexly curved to fit the concavity of the diaphragm and curved bodies of the adjacent ribs (Fig. 2.58A–C). The close relationship of the spleen to the ribs that normally protect it can be a detrimental one in the presence of rib fractures (see the blue box "Rupture of Spleen," p. 281). The **anterior** and **superior borders of the spleen** are sharp and often notched, whereas its **posterior (medial) end** and **inferior border** are rounded (Fig. 2.58D). Normally, the spleen does not extend inferior to the left costal margin; thus it is seldom palpable through the anterolateral abdominal wall unless it is enlarged. When it is hardened and enlarged to approximately three times its normal size, it moves inferior to the left costal margin, and its **superior (notched) border** lies inferomedially (see the blue box "Splenectomy and Splenomegaly," p. 281). The notched border is helpful when palpating an enlarged spleen because when the person takes a deep breath, the notches can often be palpated.

The spleen normally contains a large quantity of blood that is expelled periodically into the circulation by the action of the smooth muscle in its capsule and trabeculae. The large size of the splenic artery (or vein) indicates the volume of

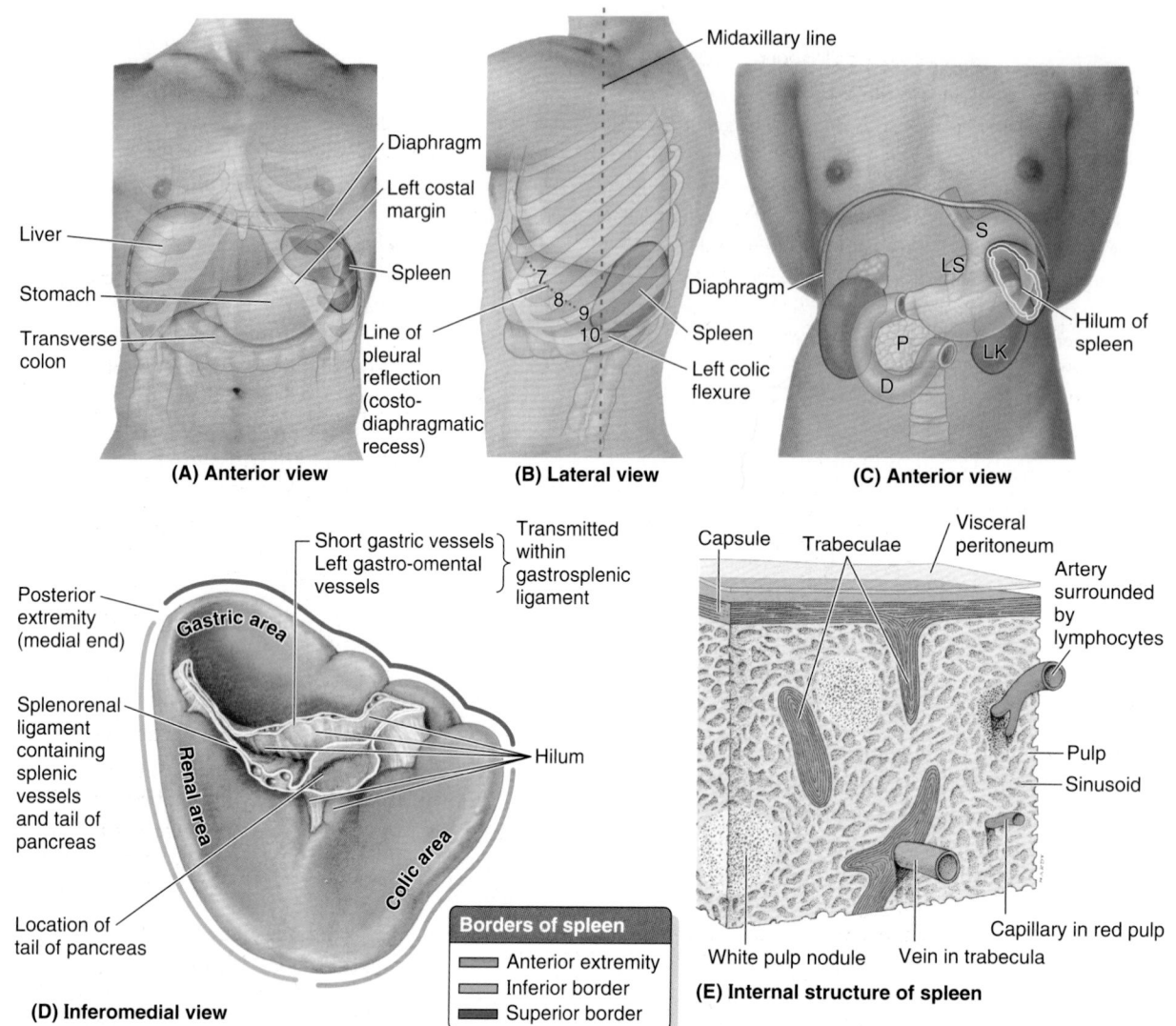

(A) Anterior view

(B) Lateral view

(C) Anterior view

(D) Inferomedial view

(E) Internal structure of spleen

Borders of spleen
- Anterior extremity
- Inferior border
- Superior border

FIGURE 2.58. **The spleen. A and B.** Surface anatomy of spleen relative to the rib cage, anterior abdominal organs, and thoracic viscera and costophrenic pleural recess. **C.** Surface anatomy of the spleen and pancreas relative to the diaphragm and posterior abdominal viscera. *D*, duodenum, *LK*, left kidney, *LS*, left suprarenal gland, *P*, pancreas, *S*, stomach. **D.** The visceral surface of the spleen. Notches are characteristic of the superior border. Concavities on the visceral surface are impressions formed by the structures in contact with the spleen. **E.** The internal structure of the spleen.

blood that passes through the spleen's capillaries and sinuses. The thin **fibrous capsule of the spleen** is composed of dense, irregular, fibroelastic connective tissue that is thickened at the splenic hilum (Fig. 2.58E). Internally the **trabeculae** (small fibrous bands), arising from the deep aspect of the capsule, carry blood vessels to and from the parenchyma or **splenic pulp,** the substance of the spleen.

The spleen contacts the posterior wall of the stomach and is connected to its greater curvature by the *gastrosplenic ligament,* and to the left kidney by the **splenorenal ligament.** These ligaments, containing splenic vessels, are attached to the hilum of the spleen on its medial aspect (Fig. 2.58D). The splenic hilum is often in contact with the tail of the pancreas and constitutes the left boundary of the omental bursa.

The *arterial supply of the spleen* is from the **splenic artery,** the largest branch of the celiac trunk (Fig. 2.59A). It follows a tortuous course posterior to the omental bursa,

anterior to the left kidney, and along the superior border of the pancreas. Between the layers of the splenorenal ligament, the splenic artery divides into five or more branches that enter the hilum. The lack of anastomosis of these arterial vessels within the spleen results in the formation of *vascular segments of the spleen:* two in 84% of spleens and three in the others, with relatively avascular planes between them, enabling subtotal splenectomy (see the blue box, "Splenectomy and Splenomegaly," p. 281).

Venous drainage from the spleen flows via the **splenic vein,** formed by several tributaries that emerge from the hilum (Figs. 2.59A and 2.60B). It is joined by the IMV and runs posterior to the body and tail of the pancreas throughout most of its course. The splenic vein unites with the SMV posterior to the neck of the pancreas to form the *hepatic portal vein.*

The *splenic lymphatic vessels* leave the lymph nodes in the splenic hilum and pass along the splenic vessels to the

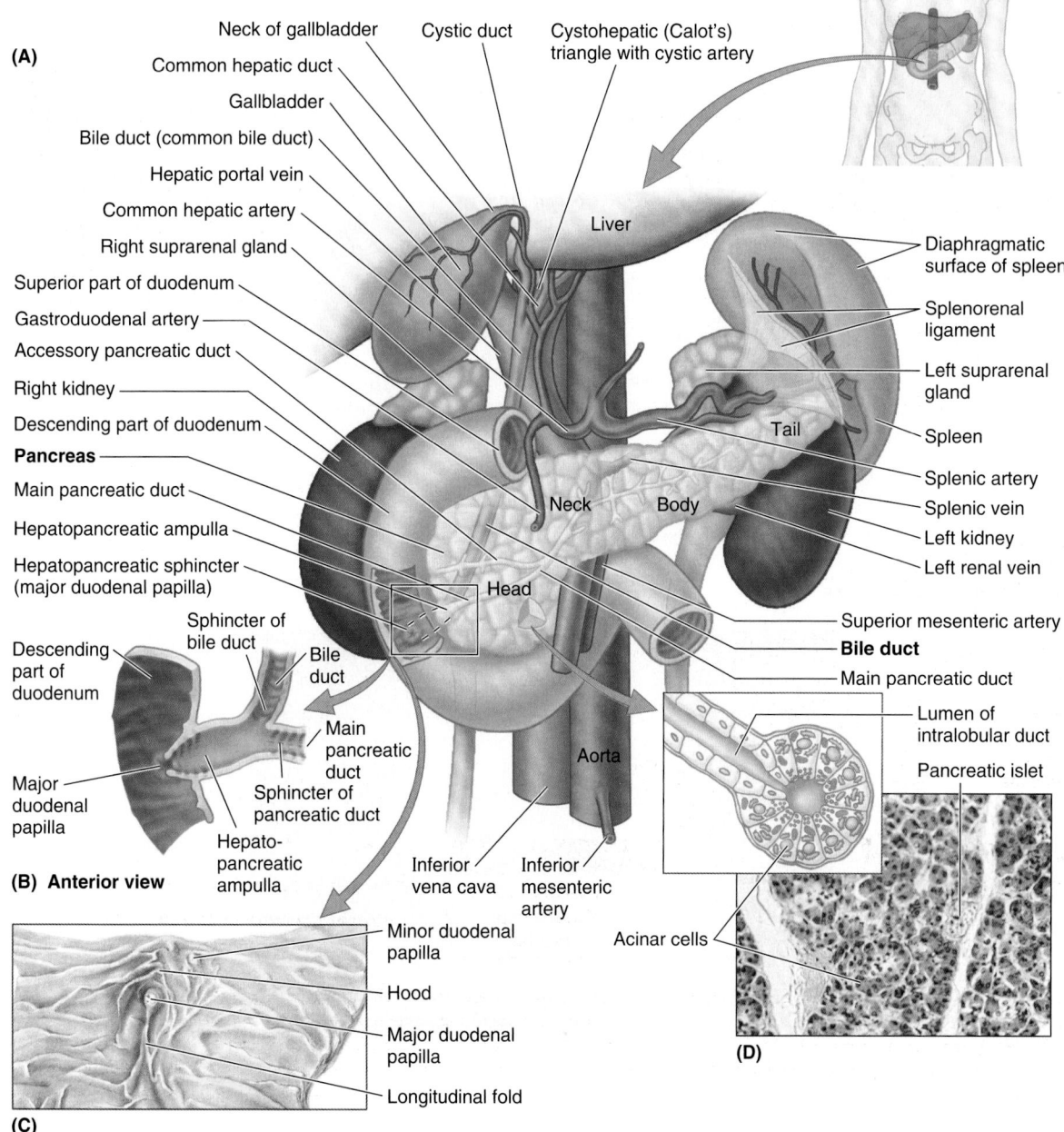

(A)

Neck of gallbladder
Common hepatic duct
Gallbladder
Bile duct (common bile duct)
Hepatic portal vein
Common hepatic artery
Right suprarenal gland
Superior part of duodenum
Gastroduodenal artery
Accessory pancreatic duct
Right kidney
Descending part of duodenum
Pancreas
Main pancreatic duct
Hepatopancreatic ampulla
Hepatopancreatic sphincter
(major duodenal papilla)

Cystic duct
Cystohepatic (Calot's)
triangle with cystic artery

Liver

Diaphragmatic
surface of spleen
Splenorenal
ligament
Left suprarenal
gland
Tail
Spleen
Neck   Body
Splenic artery
Splenic vein
Left kidney
Left renal vein
Head
Superior mesenteric artery
Bile duct
Main pancreatic duct

Sphincter of
bile duct
Descending
part of
duodenum
Bile
duct

Major
duodenal
papilla
Sphincter of
pancreatic duct
Main
pancreatic
duct
Hepato-
pancreatic
ampulla

Aorta

Lumen of
intralobular duct
Pancreatic islet

(B) Anterior view

Inferior
vena cava
Inferior
mesenteric
artery

Acinar cells

Minor duodenal
papilla
Hood
Major duodenal
papilla
Longitudinal fold

(D)

(C)

FIGURE 2.59. Spleen, pancreas, duodenum, and biliary ducts. A. Relationships of spleen, pancreas, and extrahepatic biliary ducts to other retroperitoneal viscera. **B.** The entry of the bile duct and pancreatic duct into the duodenum through the hepatopancreatic ampulla. **C.** The interior of the descending part of the duodenum reveals the major and minor duodenal papillae. **D.** The structure of the acinar (enzyme-producing) tissue is demonstrated. The photomicrograph of the pancreas displays secretory acini and a pancreatic islet.

pancreaticosplenic lymph nodes en route to the *celiac nodes* (Fig. 2.61A). The pancreaticosplenic nodes relate to the posterior surface and superior border of the pancreas.

The *nerves of the spleen*, derived from the *celiac plexus* (Fig. 2.61B), are distributed mainly along branches of the splenic artery, and are vasomotor in function.

Pancreas

The **pancreas** is an elongated, *accessory digestive gland* that lies retroperitoneally, overlying and transversely crossing the bodies of the L1 and L2 vertebra (the level of the transpy-

loric plane) on the posterior abdominal wall (Fig. 2.58C). It lies posterior to the stomach between the duodenum on the right and the spleen on the left (Fig. 2.59A). The transverse mesocolon attaches to its anterior margin (see Fig. 2.39A). The pancreas produces:

- an exocrine secretion (*pancreatic juice* from the acinar cells) that enters the duodenum through the main and accessory pancreatic ducts.
- endocrine secretions (*glucagon* and *insulin* from the *pancreatic islets* [of Langerhans]) that enter the blood (Fig. 2.59D).

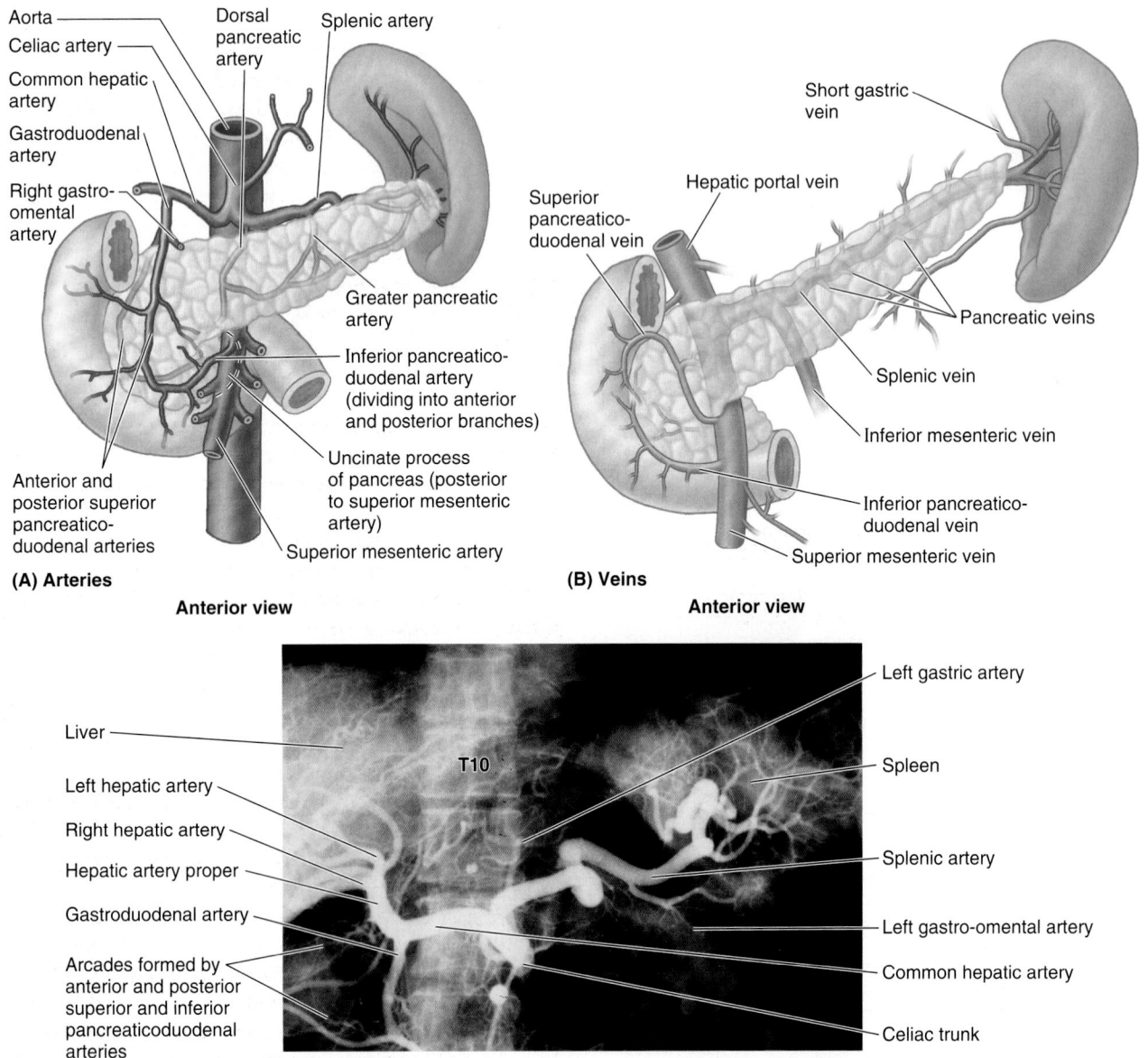

FIGURE 2.60. Arterial supply and venous drainage of pancreas. Because of the close relationship of the pancreas and duodenum, their blood vessels are the same in whole or in part. **A.** Arteries. Except for the inferior part of the pancreatic head (including uncinate process), the spleen and pancreas receive blood from the celiac artery. **B.** Venous drainage. **C.** Celiac arteriogram. Radiopaque dye was selectively injected into the lumen of the celiac artery.

For descriptive purposes, the pancreas is divided into four parts: *head, neck, body,* and *tail.*

The **head of the pancreas** is the expanded part of the gland that is embraced by the C-shaped curve of the duodenum to the right of the superior mesenteric vessels just inferior to the transpyloric plane. It firmly attaches to the medial aspect of the descending and horizontal parts of the duodenum. The **uncinate process,** a projection from the inferior part of the pancreatic head, extends medially to the left, posterior to the SMA (Fig. 2.60A). The pancreatic head rests posteriorly on the IVC, right renal artery and vein, and left renal vein. On its way to opening into the descending part of the duodenum, the *bile duct* lies in a groove on the posterosuperior surface of the head or

is embedded in its substance (Fig. 2.59A & B; see also Fig. 2.45).

The **neck of the pancreas** is short (1.5–2 cm) and overlies the superior mesenteric vessels, which form a groove in its posterior aspect (see Fig. 2.44B & C). The anterior surface of the neck, covered with peritoneum, is adjacent to the *pylorus of the stomach.* The SMV joins the splenic vein posterior to the neck to form the hepatic portal vein (Fig. 2.60).

The **body of the pancreas** continues from the neck and lies to the left of the superior mesenteric vessels, passing over the aorta and L2 vertebra, continuing just above the transpyloric plane posterior to the omental bursa. The anterior surface of the body of the pancreas is covered with peritoneum and lies

The **tail of the pancreas** lies anterior to the left kidney, where it is closely related to the splenic hilum and the left colic flexure. The tail is relatively mobile and passes between the layers of the splenorenal ligament with the splenic vessels (Fig. 2.58D).

The **main pancreatic duct** begins in the tail of the pancreas and runs through the parenchyma of the gland to the pancreatic head: here it turns inferiorly and is closely related to the bile duct (Fig. 2.59A & B). The main pancreatic duct and bile duct usually unite to form the short, dilated **hepatopancreatic ampulla** (of Vater), which opens into the descending part of the duodenum at the summit of the *major duodenal papilla* (Fig. 2.59B & C). At least 25% of the time, the ducts open into the duodenum separately.

The **sphincter of the pancreatic duct** (around the terminal part of the pancreatic duct), the **sphincter of the bile duct** (around the termination of the bile duct), and the **hepatopancreatic sphincter** (of Oddi)—around the hepatopancreatic ampulla—are smooth muscle sphincters that control the flow of bile and pancreatic juice into the ampulla and prevent reflux of duodenal content into the ampulla.

The **accessory pancreatic duct** (Fig. 2.59A) opens into the duodenum at the summit of the *minor duodenal papilla* (Fig. 2.59C). Usually, the accessory duct communicates with the main pancreatic duct. In some cases, the main pancreatic duct is smaller than the accessory pancreatic duct and the two may not be connected. In such cases, the accessory duct carries most of the pancreatic juice.

The *arterial supply of the pancreas* is derived mainly from the branches of the markedly tortuous *splenic artery*. Multiple **pancreatic arteries** form several arcades with pancreatic branches of the *gastroduodenal* and *superior mesenteric arteries* (Fig. 2.60A). As many as 10 branches may pass from the splenic artery to the body and tail of the pancreas. The *anterior* and *posterior superior pancreaticoduodenal arteries,* branches of the gastroduodenal artery, and the *anterior* and *posterior inferior pancreaticoduodenal arteries,* branches of the SMA, form anteriorly and posteriorly placed arcades that supply the head of the pancreas.

Venous drainage from the pancreas occurs via corresponding *pancreatic veins,* tributaries of the splenic and superior mesenteric parts of the hepatic portal vein; most empty into the *splenic vein* (Fig. 2.60B).

The *pancreatic lymphatic vessels* follow the blood vessels (Fig. 2.61A). Most vessels end in the *pancreaticosplenic lymph nodes,* which lie along the splenic artery. Some vessels end in the *pyloric lymph nodes.* Efferent vessels from these nodes drain to the *superior mesenteric lymph nodes* or to the *celiac lymph nodes* via the *hepatic lymph nodes.*

The *nerves of the pancreas* are derived from the *vagus* and *abdominopelvic splanchnic nerves* passing through the diaphragm (Fig. 2.61B). The parasympathetic and sympathetic fibers reach the pancreas by passing along the arteries from the *celiac plexus* and **superior mesenteric plexus** (see also "Summary of Innervation of Abdominal Viscera," p. 300). In addition to sympathetic fibers that pass to blood vessels, sympathetic and parasympathetic fibers are distributed to

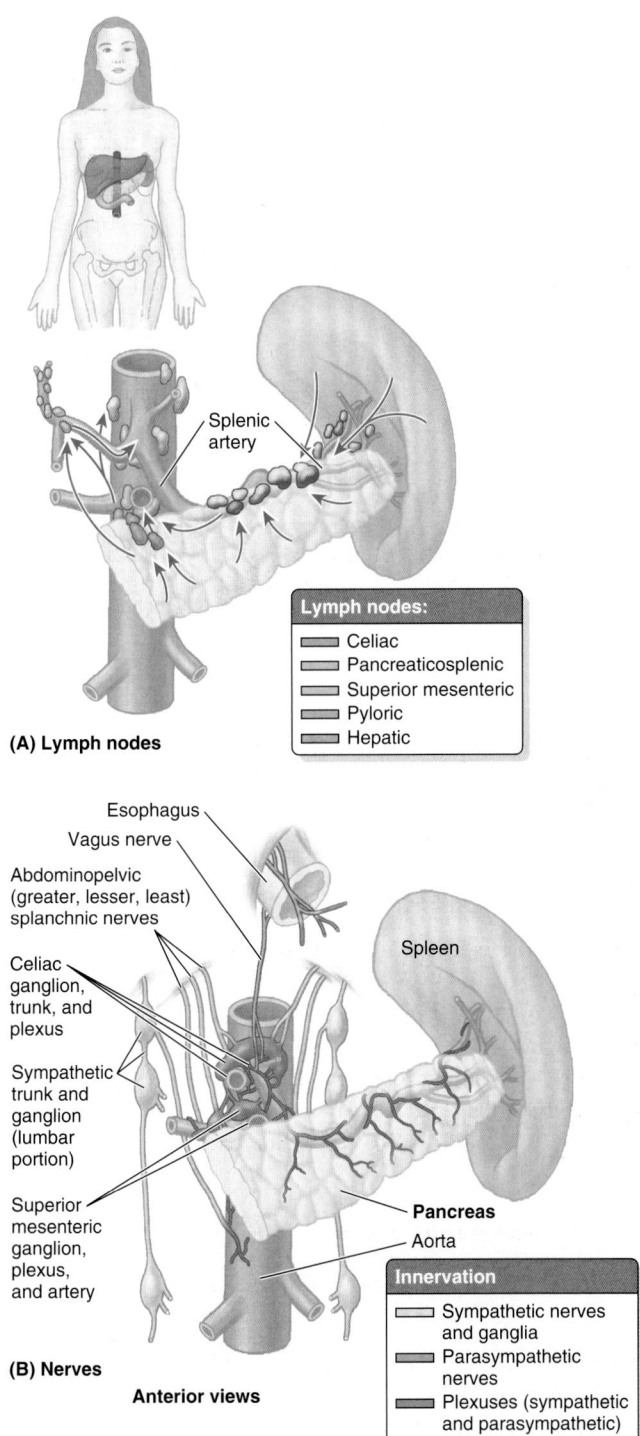

(A) Lymph nodes

Splenic artery

Lymph nodes:
- Celiac
- Pancreaticosplenic
- Superior mesenteric
- Pyloric
- Hepatic

(B) Nerves

Esophagus
Vagus nerve
Abdominopelvic (greater, lesser, least) splanchnic nerves
Celiac ganglion, trunk, and plexus
Sympathetic trunk and ganglion (lumbar portion)
Superior mesenteric ganglion, plexus, and artery
Spleen
Pancreas
Aorta

Innervation
- Sympathetic nerves and ganglia
- Parasympathetic nerves
- Plexuses (sympathetic and parasympathetic)

Anterior views

FIGURE 2.61. Lymphatic drainage and innervation of pancreas and spleen. A. The *arrows* indicate lymph flow to the lymph nodes. **B.** The nerves of the pancreas are autonomic nerves from the celiac and superior mesenteric plexuses. A dense network of nerve fibers passes from the celiac plexus along the splenic artery to the spleen. Most are postsynaptic sympathetic fibers to smooth muscle of the splenic capsule, trabeculae, and intrasplenic vessels.

in the floor of the omental bursa and forms part of the stomach bed (see Fig. 2.39A & B). The posterior surface of the body is devoid of peritoneum and is in contact with the aorta, SMA, left suprarenal gland, left kidney, and renal vessels (Fig. 2.59A).

pancreatic acinar cells and islets. The parasympathetic fibers are secretomotor, but pancreatic secretion is primarily mediated by secretin and cholecystokinin, hormones formed by the epithelial cells of the duodenum and proximal intestinal mucosa under the stimulus of acid contents from the stomach.

Liver

The **liver** is the largest gland in the body and, after the skin, the largest single organ. It weighs approximately 1500 g and accounts for approximately 2.5% of adult body weight. In a mature fetus—when it serves as a hematopoietic organ—it is proportionately twice as large (5% of body weight).

Except for fat, all nutrients absorbed from the gastrointestinal tract are initially conveyed to the liver by the portal venous system. In addition to its many metabolic activities, the liver stores glycogen and secretes **bile,** a yellow-brown or green fluid that aids in the emulsification of fat.

Bile passes from the liver via the *biliary ducts—right* and *left hepatic ducts—*that join to form the *common hepatic duct,* which unites with the *cystic duct* to form the *(common) bile duct.* The liver produces bile continuously; however, between meals it accumulates and is stored in the gallbladder, which also concentrates the bile by absorbing water and salts. When food arrives in the duodenum, the gallbladder sends concentrated bile through the biliary ducts to the duodenum.

SURFACE ANATOMY, SURFACES, PERITONEAL REFLECTIONS, AND RELATIONSHIPS OF LIVER

The *liver* lies mainly in the right upper quadrant of the abdomen, where it is protected by the *thoracic (rib) cage* and the *diaphragm* (Fig. 2.62). The normal liver lies deep to ribs 7–11 on the right side and crosses the midline toward the left nipple. The liver occupies most of the right hypochondrium and upper epigastrium and extends into the left hypochondrium. The liver moves with the

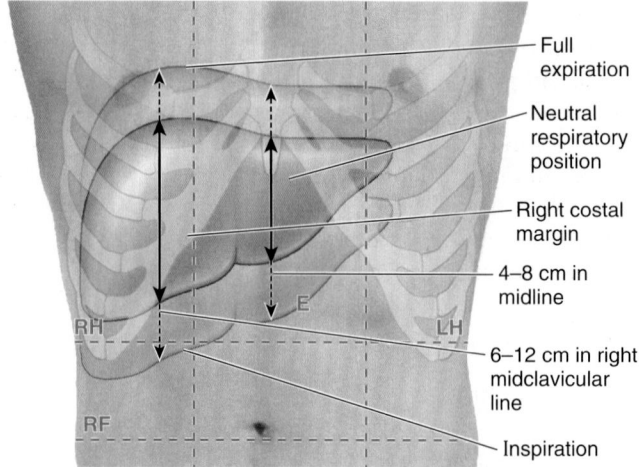

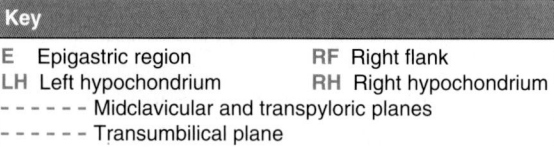

Vertical dimensions and range of movement of liver

Key		
E Epigastric region		**RF** Right flank
LH Left hypochondrium		**RH** Right hypochondrium
- - - - - - Midclavicular and transpyloric planes		
- - - - - - Transumbilical plane		

FIGURE 2.62. Surface anatomy of the liver. The liver's location, extent, relationship to the thoracic cage, and range of movements with change of position and diaphragmatic excursion are demonstrated.

excursions of the diaphragm and is located more inferiorly when one is erect because of gravity. This mobility facilitates palpation (see the blue box "Palpation of Liver" on p. 283).

The liver has a convex *diaphragmatic surface* (anterior, superior, and some posterior) and a relatively flat or even concave *visceral surface* (postero-inferior), which are separated anteriorly by its sharp *inferior border* that follows the right costal margin. inferior to the diaphragm (Fig. 2.63A).

The **diaphragmatic surface of the liver** is smooth and dome shaped, where it is related to the concavity of

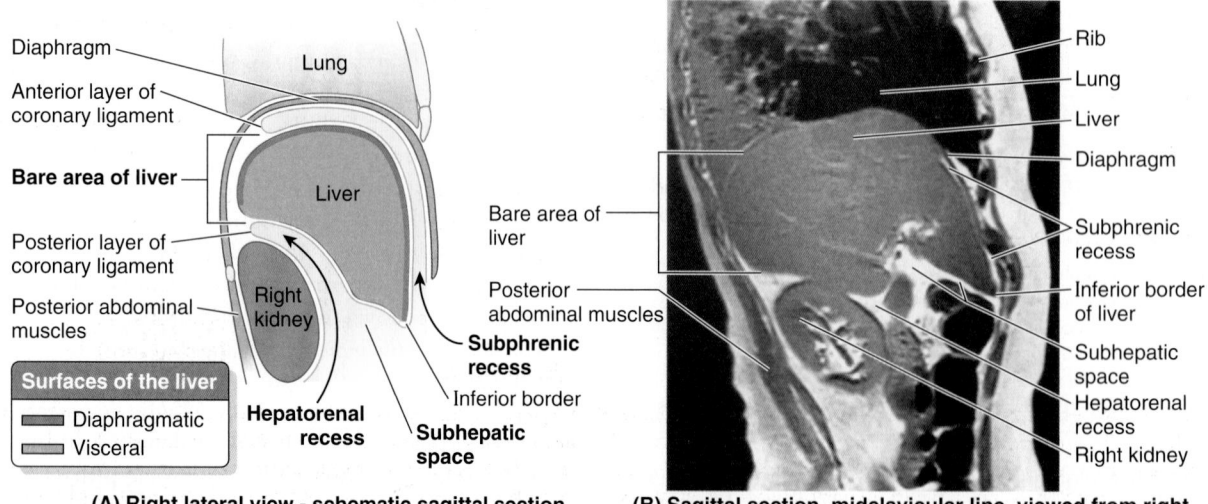

(A) Right lateral view - schematic sagittal section

(B) Sagittal section, midclavicular line, viewed from right

FIGURE 2.63. Surfaces of the liver and related potential spaces. A. This schematic sagittal section through the diaphragm, liver, and right kidney demonstrates the two surfaces of the liver and related peritoneal recesses. **B.** Sagittal magnetic resonance imaging section demonstrating the relationships featured in **A** in a living person.

the inferior surface of the diaphragm, which separates it from the pleurae, lungs, pericardium, and heart (Fig. 2.63A & B). **Subphrenic recesses**—superior extensions of the peritoneal cavity (greater sac)—exist between diaphragm and the anterior and superior aspects of the diaphragmatic surface of the liver. The subphrenic recesses are separated into right and left recesses by the *falciform ligament*, which extends between the liver and the anterior abdominal wall. The portion of the supracolic compartment of the perito-

neal cavity immediately inferior to the liver is the **subhepatic space.**

The **hepatorenal recess** (hepatorenal pouch; Morison pouch) is the posterosuperior extension of the subhepatic space, lying between the right part of the visceral surface of the liver and the right kidney and suprarenal gland. The hepatorenal recess is a gravity-dependent part of the peritoneal cavity in the supine position; fluid draining from the omental bursa flows into this recess (Fig. 2.64B & E). The hepatorenal recess communicates

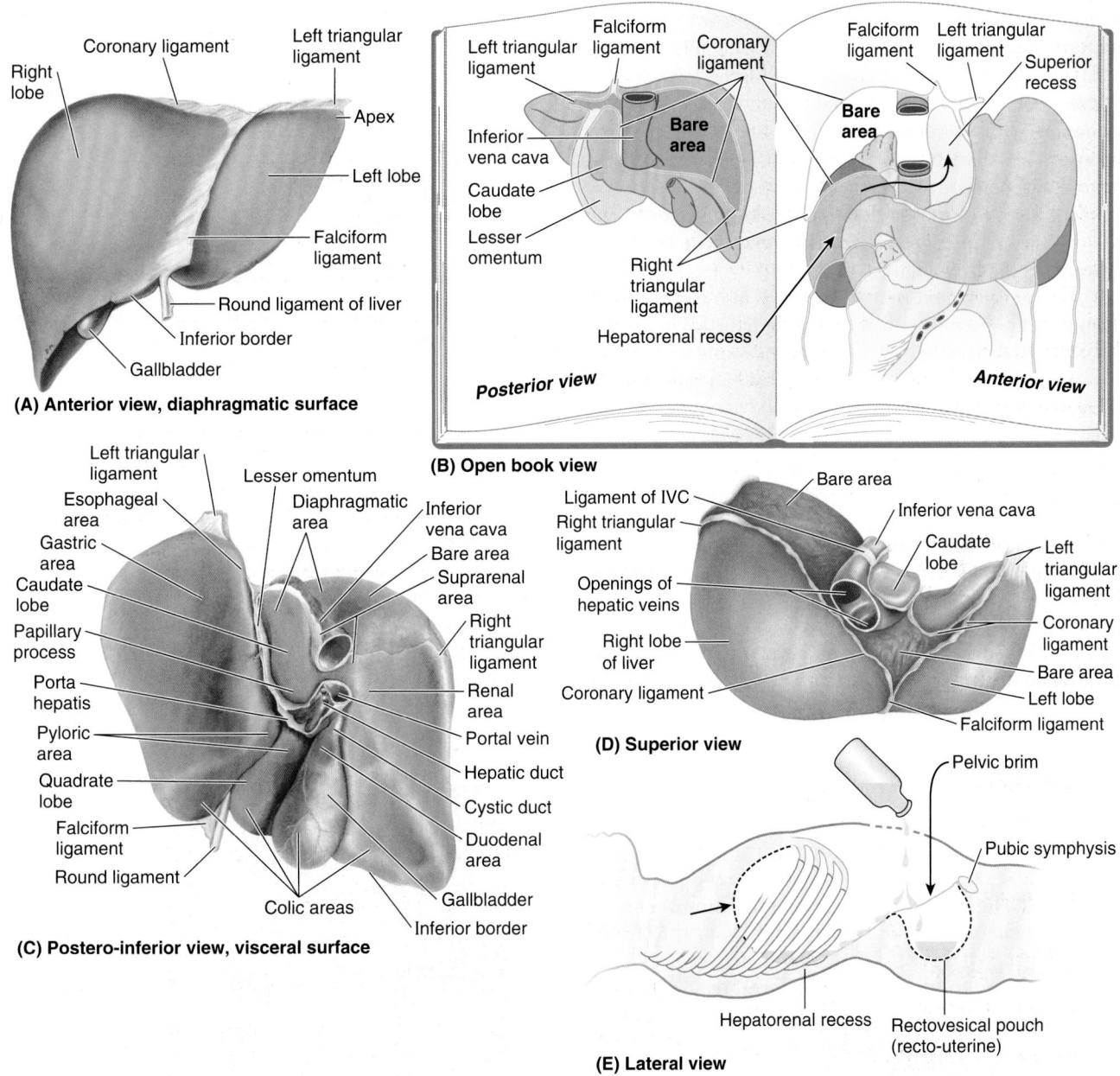

FIGURE 2.64. Peritoneal and visceral relationships of liver. A. The domed diaphragmatic surface of the liver conforms to the inferior surface of the diaphragm. This surface is divided into right and left lobes of the liver by the falciform and coronary ligaments (see also **D**). **B.** The peritoneal reflections (ligaments) and cavity related to the liver are shown diagrammatically. The attachments of the liver are cut through, and the liver is removed from its site, placed on the specimen's right, and turned posteriorly, as when turning the page of a book. **C.** In the anatomical position, the visceral surface of the liver is directed inferiorly, posteriorly, and to the left. In embalmed specimens, impressions remain where this surface is contacted by adjacent structures. **D.** The two layers of peritoneum forming the falciform ligament separate over the superior aspect of the liver to form the anterior layer of the coronary ligament, leaving the bare area of the liver without a peritoneal covering. *IVC,* inferior vena cava. **E.** Of the two gravity-dependent recesses of the abdominopelvic cavity in the supine position, the hepatorenal recess is the upper one, receiving drainage from the omental bursa and upper abdominal (supracolic) portions of the greater sac.

anteriorly with the right subphrenic recess (Fig. 2.63A & B). Recall that normally all recesses of the peritoneal cavity are potential spaces only, containing just enough peritoneal fluid to lubricate the adjacent peritoneal membranes.

The diaphragmatic surface of the liver is covered with visceral peritoneum, except posteriorly in the **bare area of the liver** (Fig. 2.64B–D), where it lies in direct contact with the diaphragm. The bare area is demarcated by the reflection of peritoneum from the diaphragm to it as the anterior (upper) and posterior (lower) layers of the **coronary ligament** (Fig. 2.63A). These layers meet on the right to form the **right triangular ligament** and diverge toward the left to enclose the triangular bare area (Fig. 2.64A–D). The anterior layer of the coronary ligament is continuous on the left with the right layer of the falciform ligament, and the posterior layer is continuous with the right layer of the lesser omentum. Near the **apex** (the left extremity) of the wedge-shaped liver, the anterior and posterior layers of the left part of the coronary ligament meet to form the **left triangular ligament.** The IVC traverses a deep **groove for the vena cava** within the bare area of the liver (Fig. 2.64B–D).

The **visceral surface of the liver** is also covered with visceral peritoneum (Fig. 2.64C), except in the **fossa for the gallbladder** (Fig. 2.65B) and the **porta hepatis**—a transverse fissure where the vessels (hepatic portal vein, hepatic artery, and lymphatic vessels), the hepatic nerve plexus, and hepatic ducts that supply and drain the liver enter and leave it. In contrast to the smooth diaphragmatic surface, the visceral surface bears multiple fissures and impressions from contact with other organs.

Two sagittally oriented fissures, linked centrally by the transverse *porta hepatis,* form the letter H on the visceral surface (Fig. 2.65A). The **right sagittal fissure** is the continuous groove formed anteriorly by the fossa for the gallbladder and posteriorly by the groove for the vena cava. The **umbilical (left sagittal) fissure** is the continuous groove formed anteriorly by the **fissure for the round ligament** and posteriorly by the **fissure for the ligamentum venosum.** The **round ligament of the liver** (L. *ligamentum teres hepatis*) is the fibrous remnant of the *umbilical vein,* which carried well-oxygenated and nutrient-rich blood from the placenta to the fetus (Fig. 2.65B). The round ligament and small *para-umbilical veins* course in the free edge of the falciform ligament. The **ligamentum venosum** is the fibrous remnant of the fetal *ductus venosus,* which shunted blood from the umbilical vein to the IVC, short-circuiting the liver.

The *lesser omentum,* enclosing the **portal triad** (bile duct, hepatic artery, and hepatic portal vein) passes from the liver to the lesser curvature of the stomach and the first 2 cm of the superior part of the duodenum (Fig. 2.66A). The thick, free edge of the lesser omentum extends between the porta hepatis and the duodenum (the *hepatoduodenal ligament*) and encloses the structures that pass through the porta hepatis. The sheet-like remainder of the lesser omentum, the *hepatogastric ligament,* extends between the groove for the ligamentum venosum and the lesser curvature of the stomach.

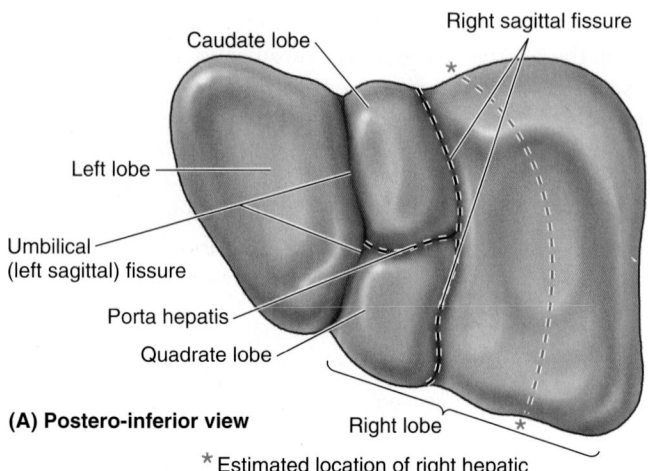

(A) Postero-inferior view

* Estimated location of right hepatic vein = right portal fissure

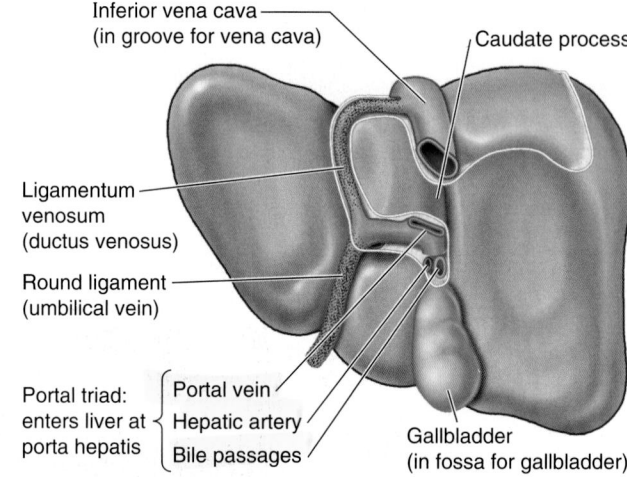

(B) Postero-inferior view

FIGURE 2.65. Visceral surface of the liver. A. The four anatomical lobes of the liver are defined by external features (peritoneal reflections and fissures). **B.** Structures forming and occupying the fissures of the visceral surface are shown.

In addition to the fissures, impressions on (areas of) the visceral surface (Fig. 2.64C) reflect the liver's relationship to the:

- Right side of the anterior aspect of the stomach (*gastric* and *pyloric areas*).
- Superior part of the duodenum (*duodenal area*).
- Lesser omentum (extends into the fissure for the ligamentum venosum).
- Gallbladder (fossa for gallbladder).
- Right colic flexure and right transverse colon (*colic area*).
- Right kidney and suprarenal gland (*renal* and *suprarenal areas*) (Fig. 2.66B).

ANATOMICAL LOBES OF LIVER

Externally, the liver is divided into two anatomical lobes and two accessory lobes by the reflections of peritoneum from its surface, the fissures formed in relation to those reflections

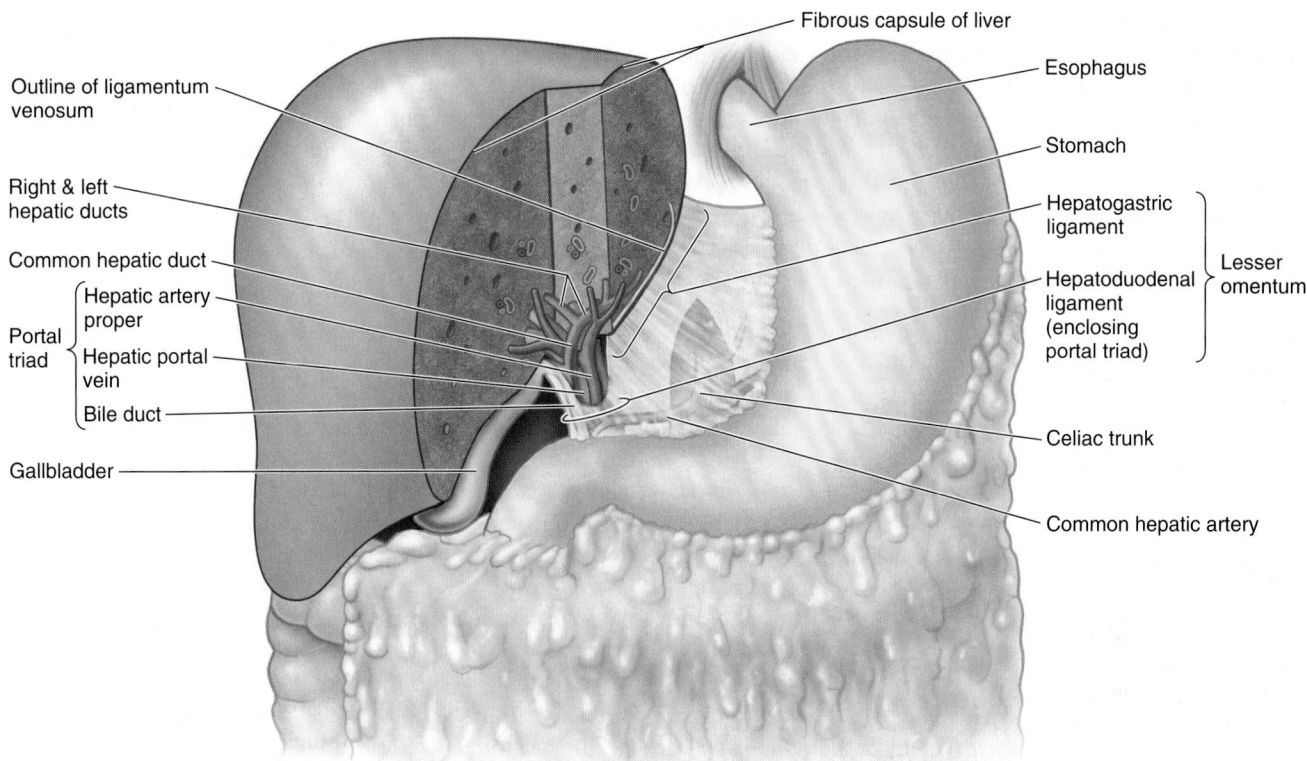

(A) Anterior view

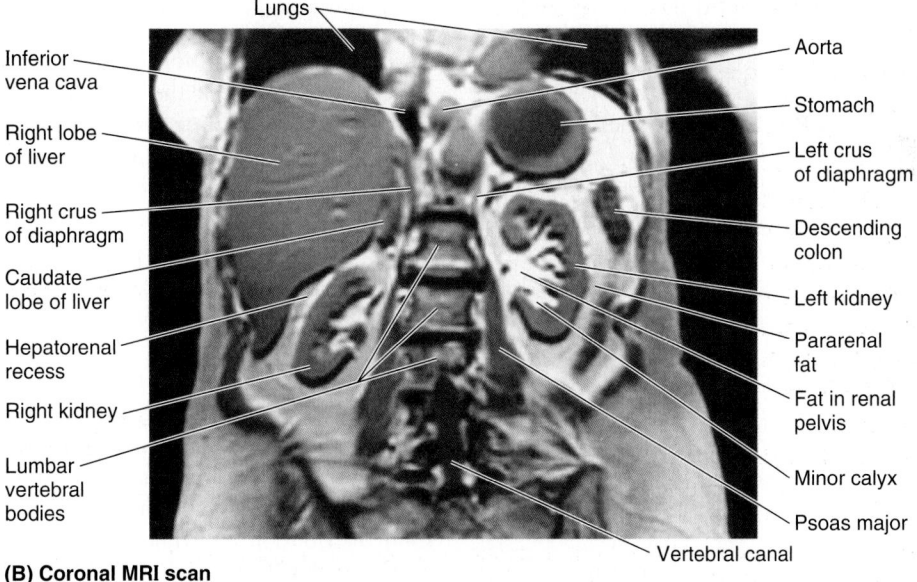

(B) Coronal MRI scan

FIGURE 2.66. **Relationships of liver to other abdominal viscera, lesser omentum, and portal triad. A.** The anterior sagittal cut is made in the plane of the fossa for the gallbladder, and the posterior sagittal cut is in the plane of the fissure for the ligamentum venosum. These cuts have been joined by a narrow coronal cut in the plane of the porta hepatis. The relationship of the liver to anterior (intraperitoneal) abdominal viscera is demonstrated. The portal triad passes between the layers of the hepatoduodenal ligament to enter the liver at the porta hepatis. The common hepatic artery passes between the layers of the hepatogastric ligament. **B.** Coronal magnetic resonance imaging (*MRI*) scan of lower thorax and abdomen demonstrating the liver's relationship to posterior (retroperitoneal) abdominal viscera (**B** courtesy of Dr. W. Kucharczyk, Professor of Medical Imaging, University of Toronto, Toronto, ON Canada.)

and the vessels serving the liver and the gallbladder. These superficial "lobes" are not true lobes as the term is generally used in relation to glands and are only secondarily related to the liver's internal architecture. The essentially midline plane defined by the attachment of the falciform ligament

and the left sagittal fissure separates a large **right lobe** from a much smaller **left lobe** (Figs. 2.64A, C, & D and 2.65). On the slanted visceral surface, the right and left sagittal fissures course on each side of—and the transverse porta hepatis separates—two accessory lobes (parts of the anatomic right lobe): the **quadrate lobe** anteriorly and inferiorly and the **caudate lobe** posteriorly and superiorly. The caudate lobe was so-named not because it is caudal in position (it is not) but because it often gives rise to a "tail" in the form of an elongated **papillary process** (Fig. 2.64C). A **caudate process** extends to the right, between the IVC and the porta hepatis, connecting the caudate and right lobes (Fig. 2.65B).

FUNCTIONAL SUBDIVISION OF LIVER

Although not distinctly demarcated internally, where the parenchyma appears continuous, the liver has functionally independent **right** and **left livers** (parts or portal lobes) that are much more equal in size than the anatomical lobes; however, the right liver is still somewhat larger (Figs. 2.67 and 2.68; Table 2.11). Each part receives its own *primary branch* of the hepatic artery and hepatic portal vein and is drained by its own hepatic duct. The *caudate lobe* may in fact be considered a third liver; its vascularization is independent of the bifurcation of the portal triad (it receives vessels from both bundles) and is drained by one or two small hepatic veins, which enter directly into the IVC distal to the main hepatic veins. The liver can be further subdivided into four *divisions* and then into eight surgically resectable *hepatic segments,* each served independently by a *secondary* or *tertiary branch* of the portal triad, respectively (Fig. 2.67).

Hepatic (Surgical) Segments of Liver. Except for the caudate lobe (*segment I*), the liver is divided into right and left livers based on the *primary (1°) division of the portal triad* into right and left branches, the plane between the right and the left livers being the **main portal fissure** in which the middle hepatic vein lies (Fig. 2.67A–C). On the visceral surface, this plane is demarcated by the *right sagittal fissure.* The plane is demarcated on the diaphragmatic surface by extrapolating an imaginary line—the *Cantlie line* (Cantlie, 1898)—from the notch for the fundus of the gallbladder to the IVC (Figs. 2.67B and 2.68A & C). The right and left livers are subdivided vertically into *medial and lateral divisions* by the *right portal and umbilical fissures,* in which the right and left hepatic veins lie (Figs. 2.67A, D, & E and 2.68). The right portal fissure has no external demarcation (Fig. 2.65A). Each of the four divisions receives a secondary (2°) branch of the portal triad (Fig. 2.67A). (Note: the medial division of the left liver—*left medial division*—is part of the right anatomical lobe; the *left lateral division* is the same as the left anatomical lobe.) A *transverse hepatic plane* at the level of the horizontal parts of the right and left branches of the portal triad subdivides three of the four divisions (all but the left medial division), creating six *hepatic segments,* each receiving tertiary branches of the triad. The left medial division is also counted as a hepatic segment, so that the main part of the liver has seven segments (**segments II–VIII,** numbered clockwise), which have also been given a descriptive name (Figs. 2.67A, D, & E and 2.68). The caudate lobe (**segment I,** bringing the total number of segments to eight) is supplied by branches of both divisions and is drained by its own minor hepatic veins.

While the pattern of segmentation described here is the most common pattern, the segments vary considerably in size and shape as a result of individual variation in the branching of the hepatic and portal vessels. The clinical significance of hepatic segments is explained in the blue box "Hepatic Lobectomies and Segmentectomy" on p. 283.

BLOOD VESSELS OF LIVER

The liver, like the lungs, has a dual blood supply (afferent vessels): a dominant venous source and a lesser arterial one (Fig. 2.67A). The *hepatic portal vein* brings 75–80% of the blood to the liver. Portal blood, containing about 40% more oxygen than blood returning to the heart from the systemic circuit, sustains the liver parenchyma (liver cells or *hepatocytes*) (Fig. 2.69). The hepatic portal vein carries virtually all of the nutrients absorbed by the alimentary tract to the sinusoids of the liver. The exception is lipids, which are absorbed into and bypass the liver via the lymphatic system. Arterial blood from the *hepatic artery,* accounting for only 20–25% of blood received by the liver, is distributed initially to nonparenchymal structures, particularly the intrahepatic bile ducts.

The *hepatic portal vein,* a short, wide vein, is formed by the superior mesenteric and splenic veins posterior to the neck of the pancreas. It ascends anterior to the IVC as part of the portal triad in the hepatoduodenal ligament (Fig. 2.66A). The **hepatic artery,** a branch of the celiac trunk, may be divided into the **common hepatic artery,** from the celiac trunk to the origin of the gastroduodenal artery, and the **hepatic artery proper,** from the origin of the gastroduodenal artery to the bifurcation of the hepatic artery (see Fig. 2.60A & C). At or close to the porta hepatis, the hepatic artery and hepatic portal vein terminate by dividing into right and left branches; these primary branches supply the right and left livers, respectively (Fig. 2.67). Within the right and left livers, the simultaneous secondary branchings of the hepatic portal vein and hepatic artery supply the medial and lateral divisions of the right and left liver, with three of the four secondary branches undergoing further (tertiary) branchings to supply independently seven of the eight hepatic segments.

Between the divisions are the **right, intermediate (middle),** and **left hepatic veins,** which are intersegmental in their distribution and function, draining parts of adjacent segments. The hepatic veins, formed by the union of *collecting veins* that in turn drain the *central veins* of the hepatic parenchyma (Fig. 2.69), open into the IVC just inferior to the diaphragm. The attachment of these veins to the IVC helps hold the liver in position.

LYMPHATIC DRAINAGE AND INNERVATION OF LIVER

The liver is a major lymph-producing organ. Between one quarter and one half of the lymph entering the thoracic duct comes from the liver.

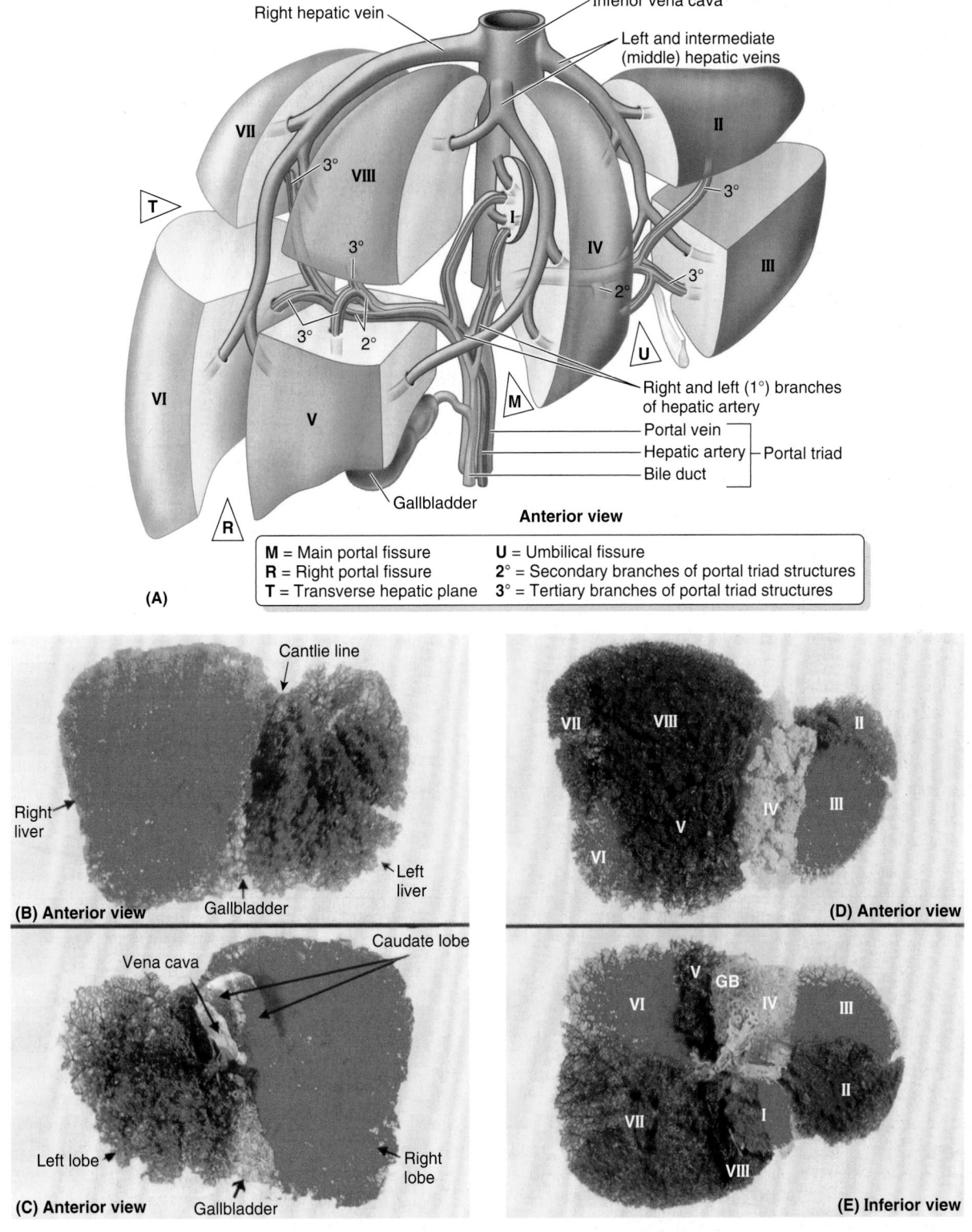

FIGURE 2.67. Hepatic segmentation. A. The right, intermediate, and left hepatic veins course within three planes or fissures [right portal (*R*), main portal (*M*), and umbilical (*U*)] that divide the liver into four vertical divisions, each served by a secondary (2°) branch of the portal triad. Three divisions are further subdivided at the transverse portal plane (*T*) into hepatic segments, each supplied by tertiary (3°) branches of the triad. The left medial division and caudate lobe are also considered hepatic segments, bringing the total to eight surgically resectable hepatic segments (segments I–VIII, each also given a name as demonstrated in Fig. 2.67 and Table 2.11). Each segment has its own intrasegmental blood supply and biliary drainage. The hepatic veins are intersegmental, draining the portions of multiple segments adjacent to them. **B and C.** Injection of latex into the right (red) and left (blue) portal veins demonstrates the right and left livers, and the Cantlie line that demarcates them on the diaphragmatic surface. **D and E.** Injection of different colors of latex into the secondary (segments IV, V, and VIII) and tertiary branches of the portal vein demonstrates the divisions of the liver and hepatic segments I–VIII.

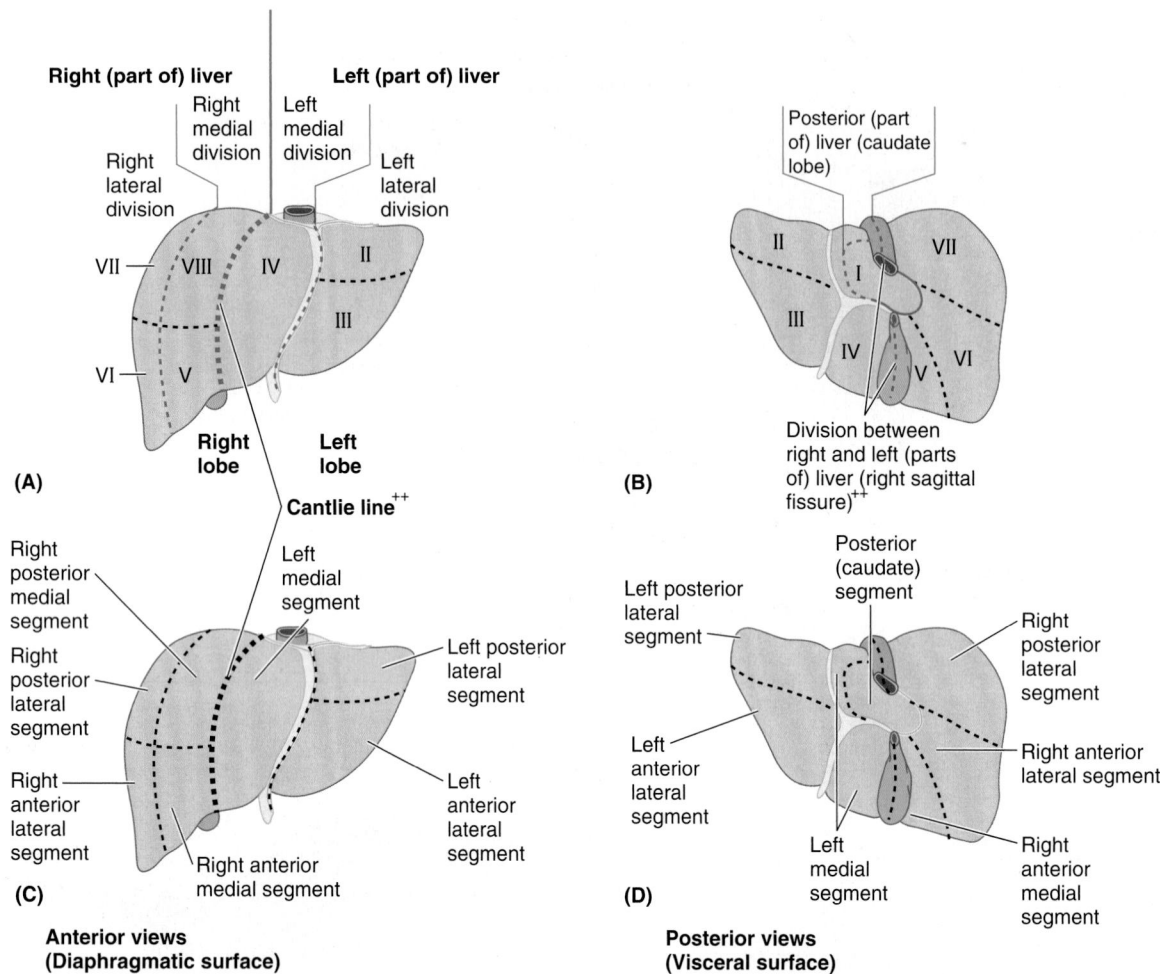

FIGURE 2.68. Parts, divisions, and segments of liver. Each part, division, and segment has an identifying name; segments are also identified by Roman numerals.

TABLE 2.11. TERMINOLOGY FOR SUBDIVISIONS OF LIVER

Anatomical Term	Right Lobe				Left Lobe		Caudate Lobe	
	Right (part of) liver [Right portal lobe*]			Left (part of) liver [Left portal lobe⁺]			Posterior (part of) liver	
	Right lateral division	Right medial division		Left medial division	Left lateral division		[Right caudate lobe*]	[Left caudate lobe⁺]
Functional/ Surgical Term**	Posterior lateral segment **Segment VII** [Posterior superior area]	Posterior lateral segment **Segment VIII** [Anterior superior area]	[Medial superior area] Left medial segment		Lateral segment **Segment II** [Lateral superior area]		Posterior segment	
	Right anterior lateral segment **Segment VI** [Posterior inferior area]	Anterior medial segment **Segment V** [Anterior inferior area]	**Segment IV** [Medial inferior area = quadrate lobe]		Left lateral anterior segment **Segment III** [Lateral inferior area]		**Segment I**	

**The labels in the table and figures above reflect the new *Terminologia Anatomica: International Anatomical Terminology (1998)*. Previous terminology is in brackets.
Under the schema of the previous terminology, the caudate lobe was divided into right and left halves, and
*The right half of the caudate lobe was considered a subdivision of the right portal lobe.
⁺The left half of the caudate lobe was considered a subdivision of the left portal lobe.
⁺⁺Cantlie line and the right sagittal fissure are surface markings defining the main portal fissure.

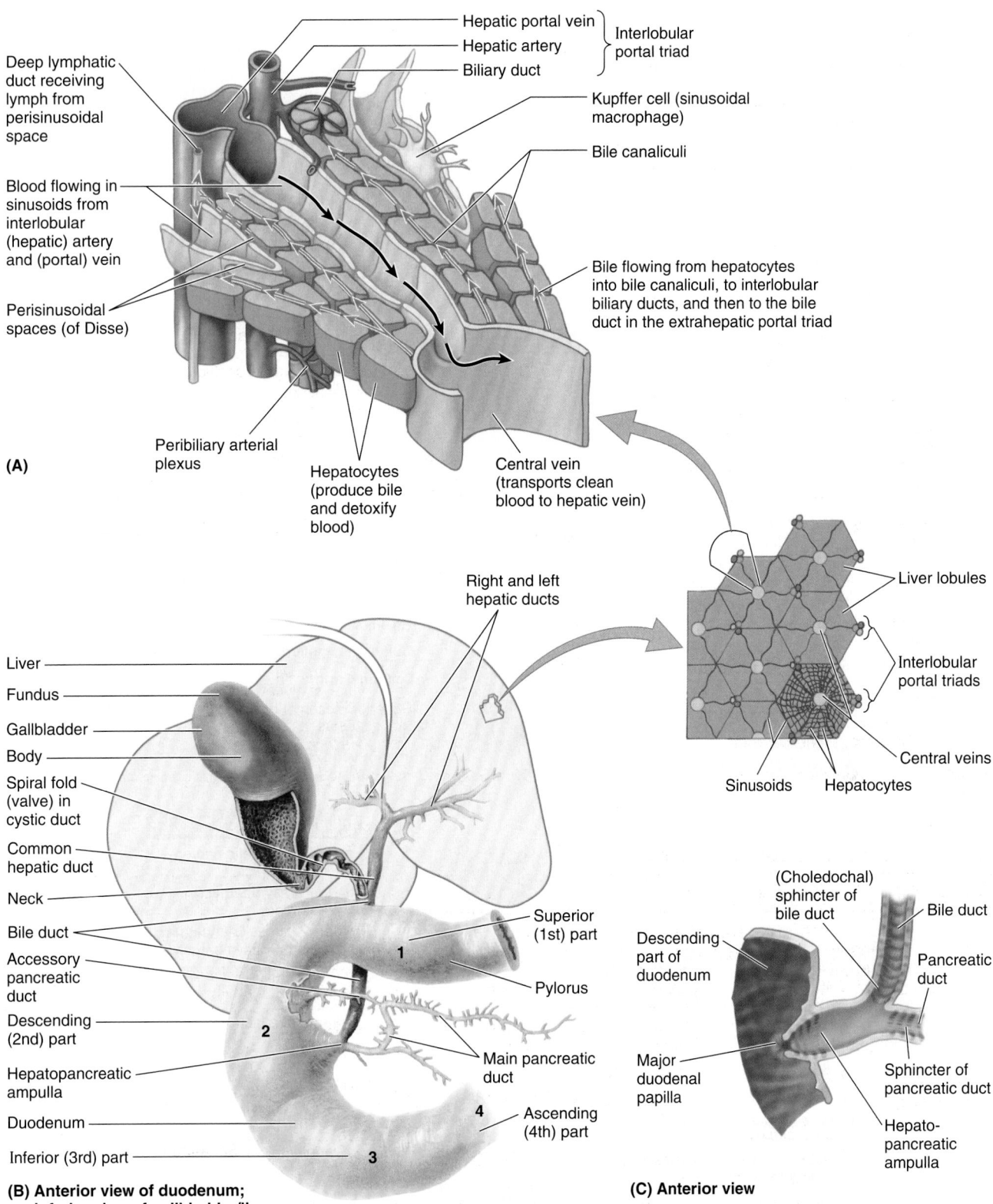

(A)

Hepatic portal vein
Hepatic artery
Biliary duct
} Interlobular portal triad

Deep lymphatic duct receiving lymph from perisinusoidal space

Kupffer cell (sinusoidal macrophage)

Bile canaliculi

Blood flowing in sinusoids from interlobular (hepatic) artery and (portal) vein

Bile flowing from hepatocytes into bile canaliculi, to interlobular biliary ducts, and then to the bile duct in the extrahepatic portal triad

Perisinusoidal spaces (of Disse)

Peribiliary arterial plexus

Hepatocytes (produce bile and detoxify blood)

Central vein (transports clean blood to hepatic vein)

Right and left hepatic ducts

Liver lobules
Interlobular portal triads
Central veins
Sinusoids   Hepatocytes

Liver
Fundus
Gallbladder
Body
Spiral fold (valve) in cystic duct
Common hepatic duct
Neck
Bile duct
Accessory pancreatic duct
Descending (2nd) part
Hepatopancreatic ampulla
Duodenum
Inferior (3rd) part

Superior (1st) part
Pylorus
Main pancreatic duct
Ascending (4th) part

(B) Anterior view of duodenum; inferior view of gallbladder/liver

(Choledochal) sphincter of bile duct
Bile duct
Descending part of duodenum
Pancreatic duct
Major duodenal papilla
Sphincter of pancreatic duct
Hepato-pancreatic ampulla

(C) Anterior view

FIGURE 2.69. Flow of blood and bile in liver. A. This view of a small part of a liver lobule illustrates the components of the interlobular portal triad and the positioning of the sinusoids and bile canaliculi. The enlarged view of the surface of a block of parenchyma removed from the liver in **B** shows the hexagonal pattern of lobes and the place of **A** within that pattern. **B.** Extrahepatic bile passages, gallbladder, and pancreatic ducts are demonstrated. **C.** The bile duct and pancreatic duct enter the hepatopancreatic ampulla, which opens into the descending part of the duodenum.

The lymphatic vessels of the liver occur as *superficial lymphatics* in the subperitoneal **fibrous capsule of the liver** (Glisson capsule), which forms its outer surface (Fig. 2.66A), and as *deep lymphatics* in the connective tissue, which accompany the ramifications of the portal triad and hepatic veins (Fig. 2.69A). Most lymph is formed in the **perisinusoidal spaces** (of Disse) and drains to the deep lymphatics in the surrounding **intralobular portal triads.**

Superficial lymphatics from the anterior aspects of the diaphragmatic and visceral surfaces of the liver, and deep lymphatic vessels accompanying the portal triads, converge toward the porta hepatis. The superficial lymphatics drain to the **hepatic lymph nodes** scattered along the hepatic vessels and ducts in the lesser omentum (Fig. 2.70A). Efferent lymphatic vessels from the hepatic nodes drain into *celiac lymph nodes*, which in turn drain into the *cisterna chyli* (chyle cistern), a dilated sac at the inferior end of the thoracic duct (see Fig. 2.100).

Superficial lymphatics from the posterior aspects of the diaphragmatic and visceral surfaces of the liver drain toward the bare area of the liver. Here they drain into **phrenic**

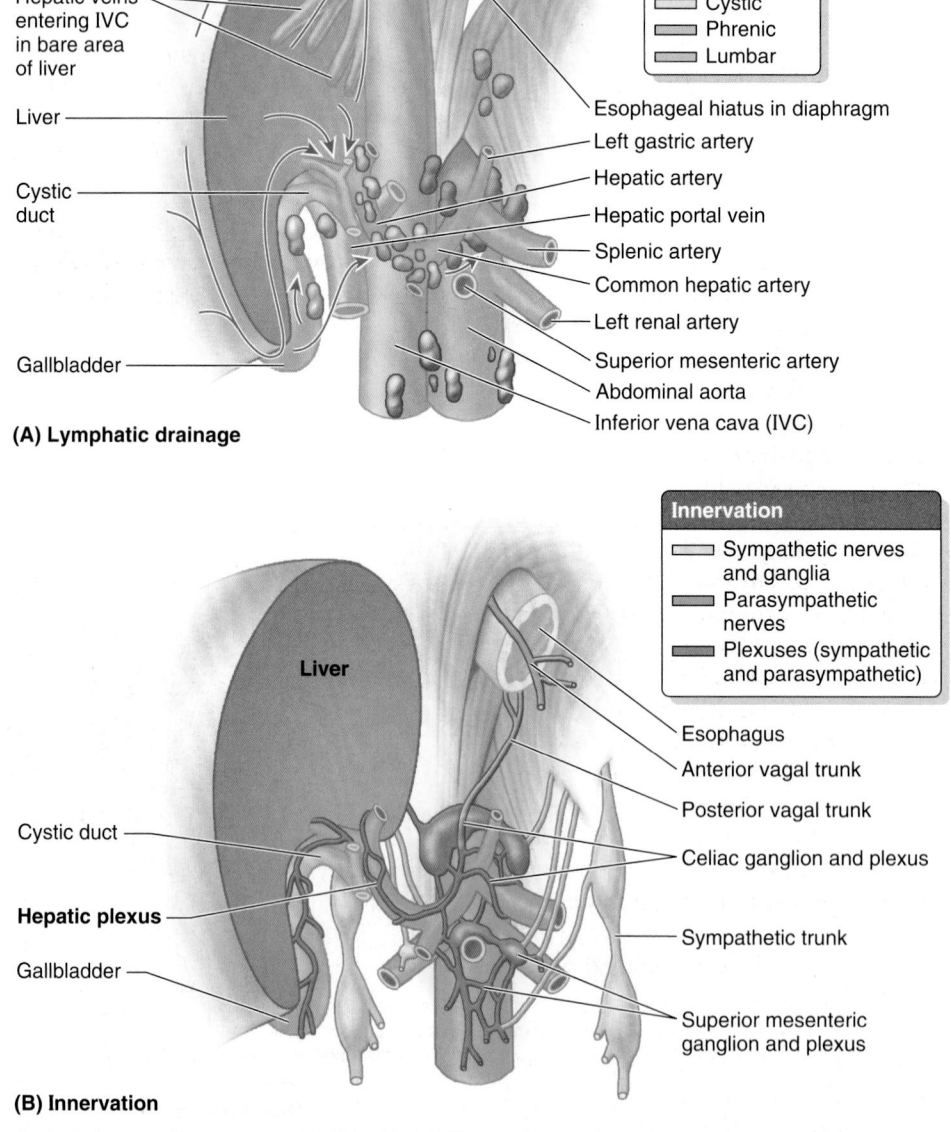

(A) Lymphatic drainage

Lymph nodes:
- Left gastric
- Mediastinal
- Celiac
- Hepatic
- Cystic
- Phrenic
- Lumbar

Caval opening in diaphragm
Hepatic veins entering IVC in bare area of liver
Liver
Cystic duct
Gallbladder

Esophageal hiatus in diaphragm
Left gastric artery
Hepatic artery
Hepatic portal vein
Splenic artery
Common hepatic artery
Left renal artery
Superior mesenteric artery
Abdominal aorta
Inferior vena cava (IVC)

(B) Innervation

Innervation
- Sympathetic nerves and ganglia
- Parasympathetic nerves
- Plexuses (sympathetic and parasympathetic)

Liver
Cystic duct
Hepatic plexus
Gallbladder

Esophagus
Anterior vagal trunk
Posterior vagal trunk
Celiac ganglion and plexus
Sympathetic trunk
Superior mesenteric ganglion and plexus

FIGURE 2.70. Lymphatic drainage and innervation of liver. A. The liver is a major lymph-producing organ. Lymph from the liver flows in two directions: that from the upper liver flows to lymph nodes located superiorly in the thorax; that from the lower liver flows to nodes located inferiorly in the abdomen. **B.** The hepatic plexus, the largest derivative of the celiac plexus, accompanies the branches of the hepatic artery to the liver conveying sympathetic and parasympathetic fibers.

lymph nodes, or join deep lymphatics that have accompanied the hepatic veins converging on the IVC, and pass with this large vein through the diaphragm to drain into the **posterior mediastinal lymph nodes.** Efferent lymphatic vessels from these nodes join the right lymphatic and thoracic ducts. A few lymphatic vessels follow different routes:

- From the posterior surface of the left lobe of the liver toward the esophageal hiatus of the diaphragm to end in the *left gastric lymph nodes.*
- From the anterior central diaphragmatic surface along the falciform ligament to the *parasternal lymph nodes.*
- Along the round ligament of the liver to the umbilicus and lymphatics of the anterior abdominal wall.

The *nerves of the liver* are derived from the **hepatic plexus** (Fig. 2.70B), the largest derivative of the celiac plexus. The hepatic plexus accompanies the branches of the hepatic artery and hepatic portal vein to the liver. This plexus consists of sympathetic fibers from the celiac plexus and parasympathetic fibers from the anterior and posterior vagal trunks. Nerve fibers accompany the vessels and biliary ducts of the portal triad. Other than vasoconstriction, their function is unclear.

Biliary Ducts and Gallbladder

The **biliary ducts** convey bile from the liver to the duodenum. Bile is produced continuously by the liver and stored and concentrated in the gallbladder, which releases it intermittently when fat enters the duodenum. Bile emulsifies the fat so that it can be absorbed in the distal intestine.

Normal hepatic tissue, when sectioned, is traditionally described as demonstrating a pattern of hexagonal-shaped *liver lobules* (Fig. 2.69A) when viewed under low magnification. Each lobule has a **central vein** running through its center from which **sinusoids** (large capillaries) and plates of **hepatocytes** (liver cells) radiate toward an imaginary perimeter extrapolated from surrounding **interlobular portal triads** (terminal branches of the hepatic portal vein and hepatic artery and initial branches of the biliary ducts). Although commonly said to be the anatomical units of the liver, hepatic "lobules" are not structural entities; instead, the lobular pattern is a physiological consequence of pressure gradients and is altered by disease. Because the bile duct is not central, the hepatic lobule does not represent a functional unit like acini of other glands. However, the hepatic lobule is a firmly established concept and is useful for descriptive purposes.

The hepatocytes secrete bile into the **bile canaliculi** formed between them. The canaliculi drain into the small *interlobular biliary ducts* and then into large *collecting bile ducts* of the intrahepatic portal triad, which merges to form the hepatic ducts (Fig. 2.69B). The **right** and **left hepatic ducts** drain the right and left (parts of the) liver, respectively. Shortly after leaving the porta hepatis, these hepatic ducts unite to form the **common hepatic duct,** which is joined on the right side by the *cystic duct* to form the *bile duct* (part of the extrahepatic portal triad of the lesser omentum), which conveys the bile to the duodenum.

BILE DUCT

The **bile duct** (formerly called the common bile duct) forms in the free edge of the lesser omentum by the union of the *cystic duct* and *common hepatic duct* (Figs. 2.65 and 2.69B). The length of the bile duct varies from 5 to 15 cm, depending on where the cystic duct joins the common hepatic duct.

The bile duct descends posterior to the superior part of the duodenum and lies in a groove on the posterior surface of the head of the pancreas. On the left side of the descending part of the duodenum, the bile duct comes into contact with the *main pancreatic duct.* These ducts run obliquely through the wall of this part of the duodenum, where they unite, forming a dilation, the *hepatopancreatic ampulla* (Fig. 2.69C). The distal end of the ampulla opens into the duodenum through the *major duodenal papilla* (see Fig. 2.45C). The circular muscle around the distal end of the bile duct is thickened to form the **sphincter of the bile duct** (L. *ductus choledochus*) (Fig. 2.69C). When this sphincter contracts, bile cannot enter the ampulla and the duodenum; hence, bile backs up and passes along the cystic duct to the gallbladder for concentration and storage.

The *arterial supply of the bile duct* is from the (Fig. 2.71):

- *Cystic artery:* supplying the proximal part of the duct.
- *Right hepatic artery:* supplying the middle part of the duct.
- *Posterior superior pancreaticoduodenal artery* and *gastroduodenal artery:* supplying the retroduodenal part of the duct.

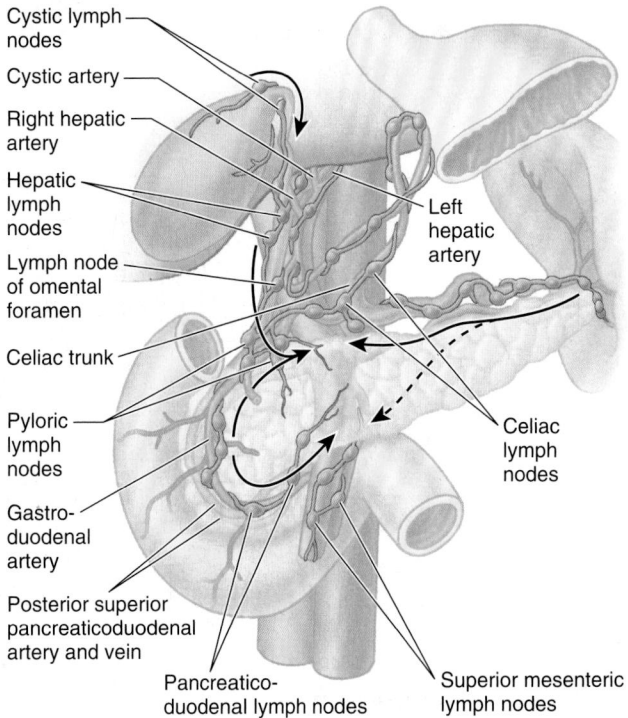

Cystic lymph nodes
Cystic artery
Right hepatic artery
Hepatic lymph nodes
Lymph node of omental foramen
Celiac trunk
Pyloric lymph nodes
Gastro-duodenal artery
Posterior superior pancreaticoduodenal artery and vein
Pancreatico-duodenal lymph nodes
Left hepatic artery
Celiac lymph nodes
Superior mesenteric lymph nodes

Anterior view

FIGURE 2.71. Arteries supplying the biliary duct and lymphatic drainage of gallbladder and bile duct. The lymphatic vessels of the gallbladder and biliary passages anastomose superiorly with those of the liver and inferiorly with those of the pancreas; most drainage flows to the celiac lymph nodes.

The *venous drainage from the proximal part of the bile duct and the hepatic ducts* usually enter the liver directly (Fig. 2.72). The *posterior superior pancreaticoduodenal vein* drains the distal part of the bile duct and empties into the hepatic portal vein or one of its tributaries.

The *lymphatic vessels from the bile duct* pass to the **cystic lymph nodes** near the neck of the gallbladder, the **lymph node of the omental foramen,** and the *hepatic lymph nodes* (Figs. 2.70 and 2.71). Efferent lymphatic vessels from the bile duct pass to the *celiac lymph nodes.*

GALLBLADDER

The **gallbladder** (7–10 cm long) lies in the *fossa for the gallbladder* on the visceral surface of the liver (Figs. 2.65B and 2.72). This shallow fossa lies at the junction of the right and left (parts of the) liver.

The relationship of the gallbladder to the duodenum is so intimate that the superior part of the duodenum is usually stained with bile in the cadaver (Fig. 2.73B). Because the liver and gallbladder must be retracted superiorly to expose the gallbladder (Fig. 2.69B) during an open anterior surgical approach (and atlases often depict it in this position), it is easy to forget that, in its natural position, the body of the gallbladder lies anterior to the superior part of the duodenum, and its neck and cystic duct are immediately superior to the duodenum (see Figs. 2.37A and 2.73B).

The pear-shaped gallbladder can hold up to 50 mL of bile. Peritoneum completely surrounds the fundus of the gallbladder and binds its body and neck to the liver. The hepatic surface of the gallbladder attaches to the liver by connective tissue of the fibrous capsule of the liver.

The gallbladder has three parts (Figs. 2.69B, 2.72, and 2.73), the:

- **Fundus:** the wide blunt end that usually projects from the inferior border of the liver at the tip of the right 9th costal cartilage in the MCL (see Figs. 2.30A and 2.31A).
- **Body:** main portion that contacts the visceral surface of the liver, transverse colon, and superior part of the duodenum.
- **Neck:** narrow, tapering end, opposite the fundus and directed toward the porta hepatis; it typically makes an S-shaped bend and joins the cystic duct (Fig. 2.72).

The **cystic duct** (3–4 cm long) connects the neck of the gallbladder to the common hepatic duct (Fig. 2.73B & C). The mucosa of the neck spirals into the **spiral fold** (spiral valve) (Fig. 2.69B). The spiral fold helps keep the cystic duct open; thus bile can easily be diverted into the gallbladder when the distal end of the bile duct is closed by the *sphincter of the bile duct* and/or hepatopancreatic sphincter, or bile can pass to the duodenum as the gallbladder contracts. The spiral fold also offers additional resistance to sudden dumping of bile when the sphincters are closed, and intra-abdominal

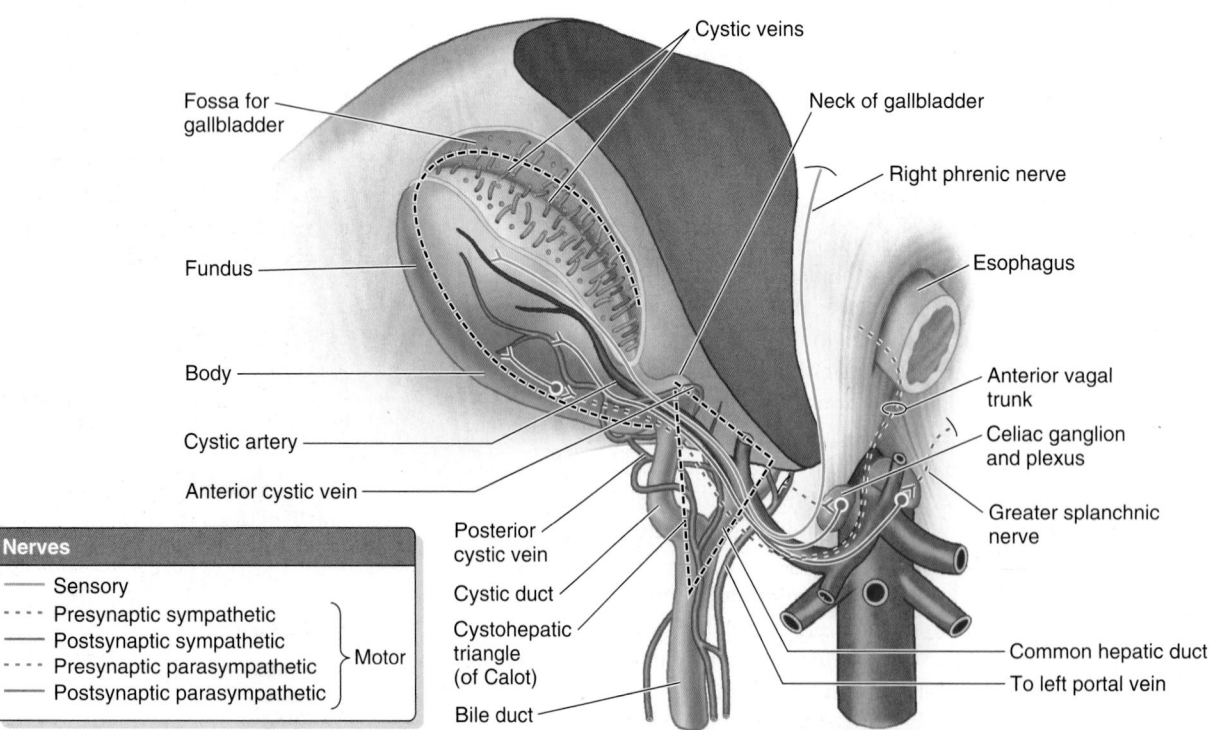

Nerves

——	Sensory
- - - -	Presynaptic sympathetic
——	Postsynaptic sympathetic
- - - -	Presynaptic parasympathetic
——	Postsynaptic parasympathetic

Motor

Anterior view, with gallbladder and liver retracted superiorly

FIGURE 2.72. Nerves and veins of liver and biliary system. Nerves are prominent along the hepatic artery and bile duct and their branches. The sympathetic nerve supply is vasomotor in the liver and biliary system. The veins of the gallbladder neck communicate with cystic veins along the cystic and biliary ducts. Small cystic veins pass from the adherent portion of the gallbladder into the sinusoids of the liver.

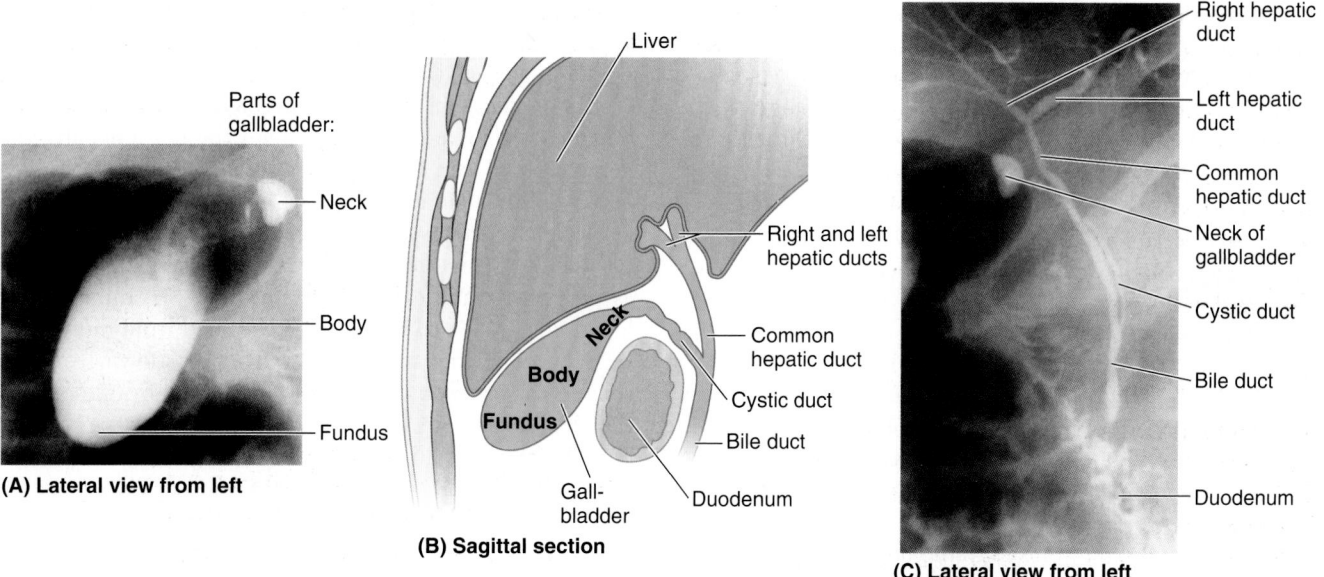

FIGURE 2.73. Normal disposition of gallbladder and extrahepatic biliary ducts. A. Gallbladder demonstrated by endoscopic retrograde cholangiography. **B.** Schematic sagittal section showing relationships to superior part of duodenum. **C.** Endoscopic retrograde cholangiograph of the bile passages. Most often the cystic duct lies anterior to the common hepatic duct. (**A** and **C** courtesy of Dr. G. B. Haber, University of Toronto, Toronto, ON, Canada.)

pressure is suddenly increased, as during a sneeze or cough. The cystic duct passes between the layers of the lesser omentum, usually parallel to the common hepatic duct, which it joins to form the bile duct.

The *arterial supply of the gallbladder and cystic duct* is from the **cystic artery** (Figs. 2.71, 2.72, and 2.74A). The cystic artery commonly arises from the *right hepatic artery* in the triangle between the common hepatic duct, cystic duct, and visceral surface of the liver, the **cystohepatic triangle** (of Calot) (Fig. 2.72). Variations occur in the origin and course of the cystic artery (Fig. 2.74B & C).

The *venous drainage from the neck of the gallbladder and cystic duct* flows via the **cystic veins.** These small and usually multiple veins enter the liver directly or drain through the hepatic portal vein to the liver, after joining the veins

draining the hepatic ducts and proximal bile duct (Fig. 2.72). The *veins from the fundus and body of the gallbladder* pass directly into the visceral surface of the liver and drain into the hepatic sinusoids. Because this is drainage from one capillary (sinusoidal) bed to another, it constitutes an additional (parallel) portal system.

The *lymphatic drainage of the gallbladder* is to the *hepatic lymph nodes* (Fig. 2.71), often through *cystic lymph nodes* located near the neck of the gallbladder. Efferent lymphatic vessels from these nodes pass to the *celiac lymph nodes.*

The *nerves to the gallbladder and cystic duct* (Fig. 2.72) pass along the cystic artery from the *celiac (nerve) plexus* (sympathetic and visceral afferent [pain] fibers), the *vagus nerve* (parasympathetic), and the *right phrenic nerve* (actually *somatic* afferent fibers). Parasympathetic stimulation

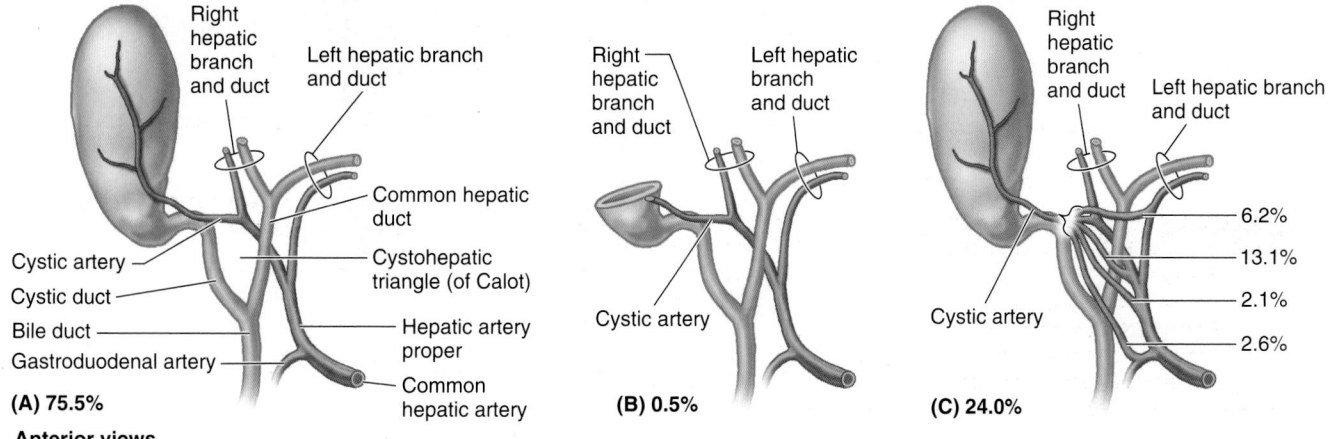

FIGURE 2.74. Variation in origin and course of cystic artery. A. The cystic artery usually arises from the right hepatic artery in the cystohepatic triangle (of Calot), bounded by the cystic duct, common hepatic duct, and visceral surface of the right liver. **B and C.** Variations in the origin and course of the cystic artery occur in 24.5% of people (Daseler et al., 1947), which is of clinical significance during cholecystectomy—surgical removal of the gallbladder.

causes contractions of the gallbladder and relaxation of the sphincters at the hepatopancreatic ampulla. However, these responses are generally stimulated by the hormone *cholecystokinin* (CCK), produced by the duodenal walls (in response to the arrival of a fatty meal), and circulated through the bloodstream.

HEPATIC PORTAL VEIN AND PORTAL–SYSTEMIC ANASTOMOSES

The **hepatic portal vein** (HPV) is the main channel of the portal venous system (Fig. 2.75A & B). It is formed anterior to the IVC and posterior to the neck of the pancreas (close to the level of the L1 vertebra and the transpyloric plane) by the union of the superior mesenteric and splenic veins. In approximately one third of individuals, the IMV joins the confluence of the superior mesenteric and splenic veins; hence, all three veins form the hepatic portal vein. In

most people, the IMV enters the splenic vein (60%—see Fig. 2.56A) or the SMV (40%).

Although the HPV is a large vessel, it runs a short course (7–8 cm), most of which is contained within the hepatoduodenal ligament. As it approaches the porta hepatis, the hepatic portal vein divides into right and left branches. The hepatic portal vein collects blood with reduced oxygenation but rich in nutrients from the abdominal part of the alimentary system, including the gallbladder and pancreas, as well as the spleen, and carries it to the liver. Streaming of the blood flow is said to occur in which blood from the splenic vein, carrying the products of RBC breakdown from the spleen, passes mostly to the left liver. Blood from the SMV, rich in absorbed nutrients from the intestines, passes mostly to the right liver. Within the liver, its branches are distributed in a segmental pattern (see "Blood Vessels of Liver," p. 272) and end in expanded capillaries, the *venous sinusoids of the liver* (see Fig. 2.69A).

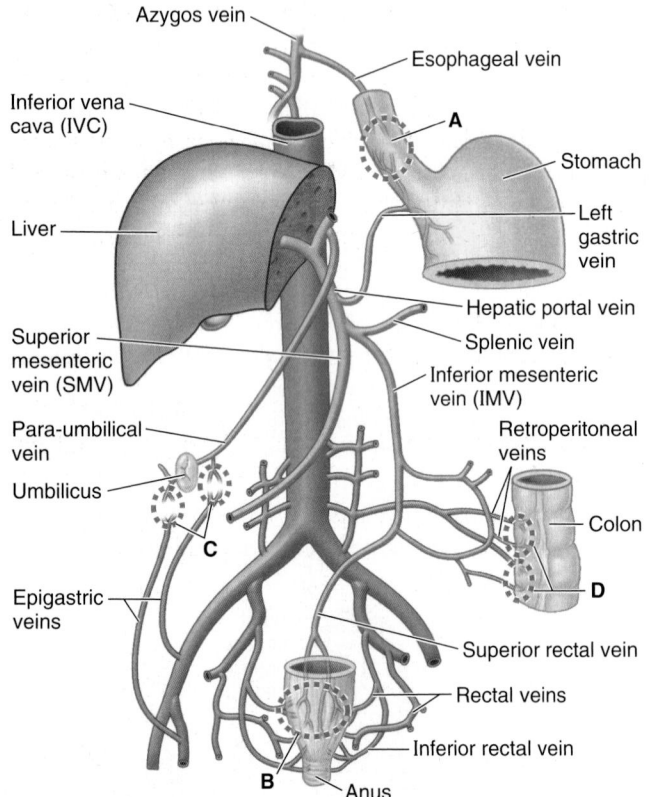

(A) Anterior view

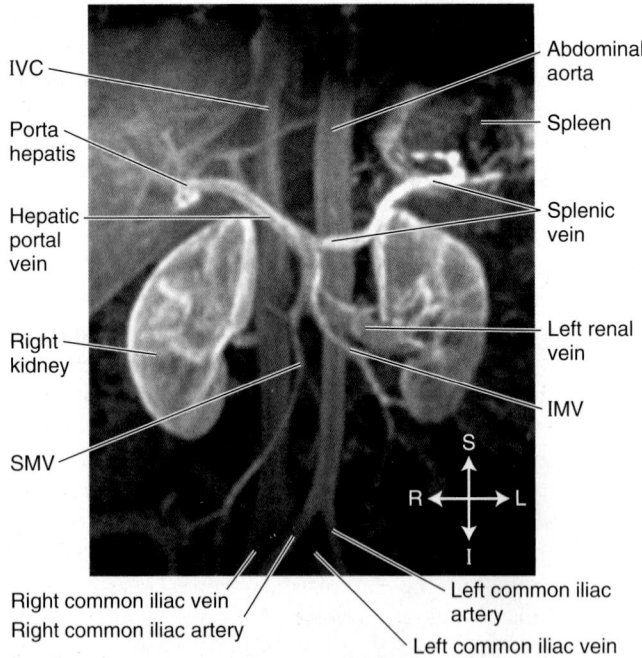

(B) Anteroposterior view, reconstructed MR angiogram

FIGURE 2.75. Tributaries of hepatic portal vein and portal–systemic anastomoses. A. Anastomoses provide a collateral circulation in cases of obstruction in the liver or portal vein. Here, the portal tributaries are darker blue and systemic tributaries are lighter blue. *A–D* indicate sites of anastomoses. *A* is between the submucosal esophageal veins draining into either the azygos vein (systemic) or the left gastric vein (portal); when dilated these are esophageal varices. *B* is between the inferior and middle rectal veins draining into the inferior vena cava (systemic) and the superior rectal vein, continuing as the inferior mesenteric vein (portal). The submucosal veins involved are normally dilated (varicose in appearance), even in newborns. When the mucosa containing them prolapses, they form *hemorrhoids*. (The varicose appearance of the veins and the occurrence of hemorrhoids are not typically related to portal hypertension, as is commonly stated.) *C* shows para-umbilical veins (portal) anastomosing with small epigastric veins of the anterior abdominal wall (systemic); this may produce the "caput medusae" (Fig. B2.24). *D* is on the posterior aspects (bare areas) of secondarily retroperitoneal viscera, or the liver, where twigs of visceral veins—for example, the colic vein, splenic veins, or the portal vein itself (portal system)—anastomose with retroperitoneal veins of the posterior abdominal wall or diaphragm (systemic system). **B.** Magnetic resonance (MR) angiogram (portal venogram) demonstrating the tributaries and formation of the portal vein in a living person.

Portal–systemic anastomoses, in which the portal venous system communicates with the systemic venous system, are formed in the submucosa of the inferior esophagus, in the submucosa of the anal canal, in the peri-umbilical region, and on the posterior aspects (bare areas) of secondarily retroperitoneal viscera, or the liver (Fig. 2.75; see legend for details). When portal circulation through the liver is diminished or obstructed because of liver disease or physical pressure from a tumor, for example, blood from the gastrointestinal tract can still reach the right side of the heart through the IVC by way of these collateral routes. These alternate routes are available because the hepatic portal vein and its tributaries have no valves; hence blood can flow in a reverse direction to the IVC. However, the volume of blood forced through the collateral routes may be excessive, resulting in potentially fatal *varices* (abnormally dilated veins) (see the blue box "Portal Hypertension" on p. 288) if the obstruction is not surgically bypassed (see the blue box "Portosystemic Shunts" on p. 288).

SPLEEN AND PANCREAS

Rupture of Spleen

Although well protected by the 9th–12th ribs (Fig. 2.30B), the spleen is the most frequently injured organ in the abdomen. The close relationship of the spleen to the ribs that normally protect it can be detrimental when there are rib fractures. Severe blows on the left side may fracture one or more of these ribs, resulting in sharp bone fragments that may lacerate the spleen. In addition, blunt trauma to other regions of the abdomen that cause a sudden, marked increase in intra-abdominal pressure (e.g., by impalement on the handlebars of a motorcycle) can cause the thin fibrous capsule and overlying peritoneum of the spleen to rupture, disrupting its soft pulp (*ruptured spleen*). If ruptured, there is profuse bleeding (*intraperitoneal hemorrhage*) and shock.

Splenectomy and Splenomegaly

Repair of a ruptured spleen is difficult; consequently, *splenectomy* (removal of the spleen) is often performed to prevent the person from bleeding to death. *Sub-total (partial) splenectomy*, when possible, is followed by rapid regeneration. Even *total splenectomy* usually does not produce serious effects, especially in adults, because most of its functions are assumed by other reticuloendothelial organs (e.g., liver and bone marrow), but there is a greater susceptibility to certain bacterial infections. When the spleen is diseased, resulting from, for example, granulocytic leukemia (high leukocyte and white blood cell count), it may enlarge to 10 or more times its normal size and weight (*splenomegaly*). Spleen engorgement sometimes accompanies hypertension (high blood pressure). The spleen is not usually palpable in the adult. Generally, if its lower edge can be detected when palpating below the left costal margin at the end of inspiration (Fig. B2.19A), it is enlarged about three times its "normal" size. Splenomegaly also occurs in some forms of hemolytic or granulocytic anemias, in which RBCs or white blood cells (WBCs), respectively, are destroyed at abnormally high rates (Fig. B2.19B). In such cases, a splenectomy may be life-saving.

Accessory Spleen(s)

One or more small **accessory spleens** may develop prenatally near the splenic hilum. They may be embedded partly or wholly in the tail of the pancreas, between the layers of the gastrosplenic ligament, in the infracolic compartment, in the mesentery, or in close proximity to an ovary or testis (Fig. B2.20). In most affected individuals, only one accessory spleen is present. Accessory spleens are relatively common, are usually small (approximately 1 cm in diameter, and range from 0.2 to 10 cm), and may resemble a lymph node. Awareness of the possible presence of an accessory spleen is important because if not removed during a splenectomy, the symptoms that indicated removal of the spleen (e.g., *splenic anemia*) may persist.

Splenic Needle Biopsy and Splenoportography

The relationship of the costodiaphragmatic recess of the pleural cavity to the spleen is clinically important (see Fig. 2.31A). This potential space descends to the

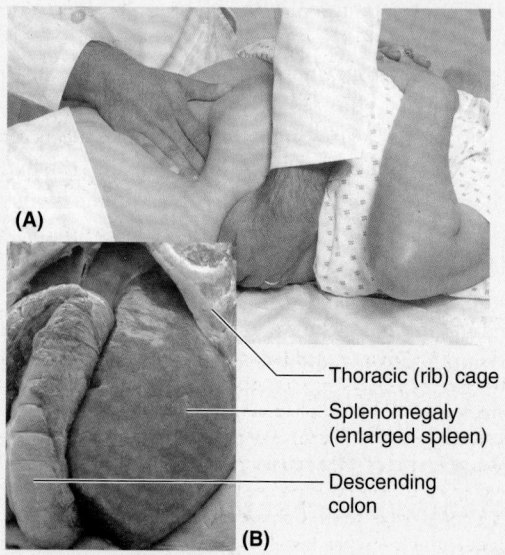

(A)

(B)

Thoracic (rib) cage

Splenomegaly (enlarged spleen)

Descending colon

FIGURE B2.19. Examination of spleen. A. Palpation of the spleen is demonstrated. **B.** This 4200-g spleen was seen at autopsy.

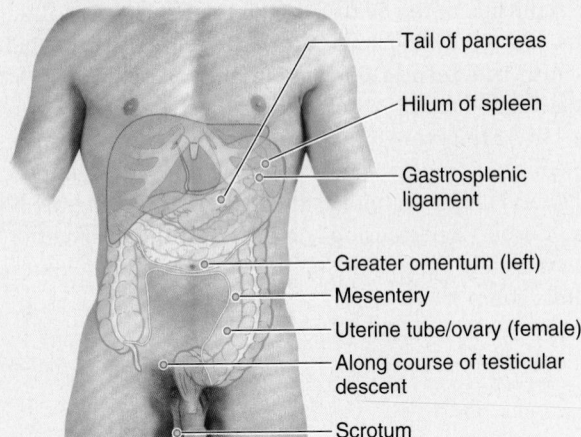

FIGURE B2.20. **Potential sites of accessory spleens.** The dots indicate where small accessory spleens may be located.

level of the 10th rib in the midaxillary line. Its existence must be kept in mind when doing a *splenic needle biopsy*, or when injecting radiopaque material into the spleen for visualization of the hepatic portal vein (*splenoportography*). If care is not exercised, this material may enter the pleural cavity, causing *pleuritis* (inflammation of the pleura).

Blockage of Hepatopancreatic Ampulla and Pancreatitis

 Because the main pancreatic duct joins the bile duct to form the hepatopancreatic ampulla and pierces the duodenal wall, a *gallstone* passing along the extrahepatic bile passages may lodge in the constricted distal end of the ampulla, where it opens at the summit of the major duodenal papilla (see Fig. 2.59A & B). In this case, both the biliary and pancreatic duct systems are blocked and neither bile nor pancreatic juice can enter the duodenum. However, bile may back up and enter the pancreatic duct, usually resulting in *pancreatitis* (inflammation of the pancreas). A similar reflux of bile sometimes results from *spasms of the hepatopancreatic sphincter*. Normally, the sphincter of the pancreatic duct prevents reflux of bile into the pancreatic duct; however, if the hepatopancreatic ampulla is obstructed, the weak pancreatic duct sphincter may be unable to withstand the excessive pressure of the bile in the hepatopancreatic ampulla. If an accessory pancreatic duct connects with the main pancreatic duct and opens into the duodenum, it may compensate for an obstructed main pancreatic duct or spasm of the hepatopancreatic sphincter.

Endoscopic Retrograde Cholangiopancreatography

 Endoscopic retrograde cholangiopancreatography (ERCP) has become a standard procedure for the diagnosis of both pancreatic and biliary disease

(Fig. B2.21). First, a fiberoptic endoscope is passed through the mouth, esophagus, and stomach. Then the duodenum is entered and a cannula is inserted into the major duodenal papilla and advanced under fluoroscopic control into the duct of choice (bile duct or pancreatic duct) for injection of radiographic contrast medium.

Accessory Pancreatic Tissue

 It is not unusual for ectopic *accessory pancreatic tissue* to develop in the stomach, duodenum, ileum, or an ileal diverticulum; however, the stomach and duodenum are the most common sites. The accessory pancreatic tissue may contain pancreatic islet cells that produce glucagon and insulin.

Pancreatectomy

For the treatment of chronic pancreatitis, most of the pancreas may be removed—a procedure called *pancreatectomy*. The anatomical relationships and blood supply of the head of the pancreas, bile duct, and duodenum make it impossible to remove the entire head of the pancreas (Skandalakis et al., 1995). Usually a rim of the pancreas is retained along the medial border of the duodenum to preserve the duodenal blood supply.

Rupture of Pancreas

The pancreas is centrally located within the body. Consequently it is not palpable and is well protected from all but the most severe penetrating trauma. The pancreas, like the liver, has a considerable functional reserve.

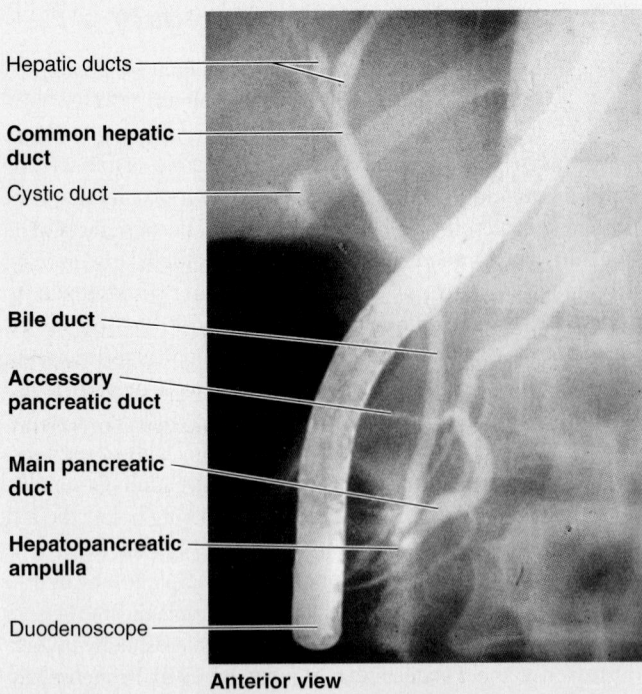

Anterior view

FIGURE B2.21. **Endoscopic retrograde cholangiopancreatogram.**

For all these reasons, the pancreas, as an exocrine organ, is not commonly a primary cause of clinical problems (discounting *diabetes,* an endocrine disorder of the islet cells). Most exocrine pancreatic problems are secondary to biliary problems. Pancreatic injury can result from sudden, severe, forceful compression of the abdomen, such as the force of impalement on a steering wheel in an automobile accident. Because the pancreas lies transversely, the vertebral column acts as an anvil, and the traumatic force may rupture the friable pancreas.

Rupture of the pancreas frequently tears its duct system, allowing pancreatic juice to enter the parenchyma of the gland and to invade adjacent tissues. Digestion of pancreatic and other tissues by pancreatic juice is very painful.

Pancreatic Cancer

Cancer involving the pancreatic head accounts for most cases of extrahepatic obstruction of the biliary ducts. Because of the posterior relationships of the pancreas, cancer of the head often compresses and obstructs the bile duct and/or the hepatopancreatic ampulla. This causes obstruction, resulting in the retention of bile pigments, enlargement of the gallbladder, and obstructive jaundice. *Jaundice* (Fr. *jaune,* yellow) is the yellow staining of most body tissues, skin, mucous membranes, and conjunctiva by circulating bile pigments.

Most people with pancreatic cancer have *ductular adenocarcinoma.* Severe pain in the back is frequently present. Cancer of the neck and body of the pancreas may cause hepatic portal or inferior vena caval obstruction because the pancreas overlies these large veins (see Fig. 2.60B). The pancreas's extensive drainage to relatively inaccessible lymph nodes, and the fact that pancreatic cancer typically metastasizes to the liver early, via the hepatic portal vein, make surgical resection of the cancerous pancreas nearly futile.

LIVER, BILIARY DUCTS, AND GALLBLADDER

Palpation of Liver

The liver may be palpated in a supine person because of the inferior movement of the diaphragm and liver that accompanies deep inspiration (see Fig. 2.62). One method of palpating the liver is to place the left hand posteriorly behind the *lower rib cage* (Fig. B2.22). Then, put the right hand on the person's right upper quadrant, lateral to the *rectus abdominis* and inferior to the *costal margin.* The person is asked to take a deep breath as the examiner presses posterosuperiorly with the right hand and pulls anteriorly with the left hand (Bickley, 2009).

Subphrenic Abscesses

Peritonitis may result in the formation of localized *abscesses* (collections of purulent exudate, or pus) in various parts of the peritoneal cavity. A common site

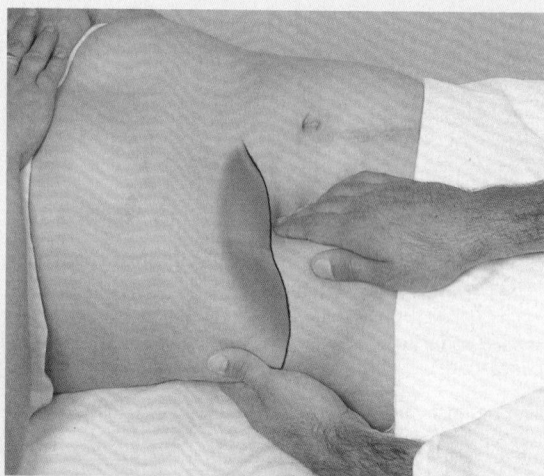

Palpation of inferior margin of liver

FIGURE B2.22. Palpation of inferior margin of liver.

for pus to collect is in the right or left subphrenic recess or space. *Subphrenic abscesses* are more common on the right side because of the frequency of ruptured appendices and perforated duodenal ulcers. Because the right and left subphrenic recesses are continuous with the hepatorenal recess (the lowest [most gravity dependent] parts of the peritoneal cavity when supine), pus from a subphrenic abscess may drain into one of the hepatorenal recesses (Fig. 2.64E), especially when patients are bedridden.

A subphrenic abscess is often drained by an incision inferior to, or through, the bed of the 12th rib (Ellis, 2010), making it unnecessary to create an opening in the pleura or peritoneum. An anterior subphrenic abscess is often drained through a subcostal incision located inferior and parallel to the right costal margin.

Hepatic Lobectomies and Segmentectomy

When it was discovered that the right and left hepatic arteries and ducts, as well as branches of the right and left hepatic portal veins, do not communicate, it became possible to perform *hepatic lobectomies,* removal of the right or left (part of the) liver, without excessive bleeding.

Most injuries to the liver involve the right liver. More recently, especially since the advent of the cauterizing scalpel and laser surgery, it has become possible to perform *hepatic segmentectomies.* This procedure makes it possible to remove (resect) only those segments that have sustained a severe injury or are affected by a tumor. The right, intermediate, and left hepatic veins serve as guides to the planes (fissures) between the hepatic divisions (Fig. B2.23); however, they also provide a major source of bleeding with which the surgeon must contend. While the pattern of branching demonstrated in Figure 2.67A is the most common

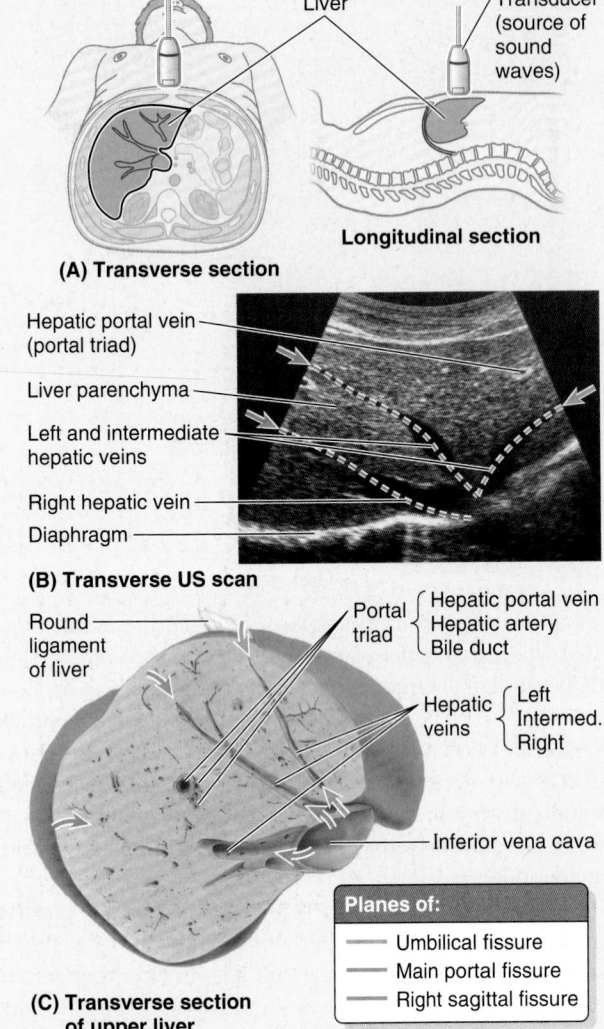

(A) Transverse section

Liver
Transducer (source of sound waves)
Longitudinal section

Hepatic portal vein (portal triad)
Liver parenchyma
Left and intermediate hepatic veins
Right hepatic vein
Diaphragm

(B) Transverse US scan

Round ligament of liver
Portal triad { Hepatic portal vein / Hepatic artery / Bile duct }
Hepatic veins { Left / Intermed. / Right }
Inferior vena cava

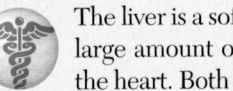

Planes of:
— Umbilical fissure
— Main portal fissure
— Right sagittal fissure

(C) Transverse section of upper liver

FIGURE B2.23. Ultrasonography (US) of hepatic veins.

pattern, the segments vary considerably in size and shape as a result of individual variation in the branching of the hepatic and portal vessels. Each hepatic resection is empirical, requiring ultrasonography, injection of dye, or balloon catheter occlusion to establish the patient's segmental pattern (Cheng et al., 1997). A more extensive injury that is likely to leave large areas of the liver devascularized may still require lobectomy.

Rupture of Liver

The liver is easily injured because it is large, fixed in position, and friable (easily crumbled). Often a fractured rib that perforates the diaphragm tears the liver. Because of the liver's great vascularity and friability, liver lacerations often cause considerable hemorrhage and right upper quadrant pain. In such cases, the surgeon must decide whether to remove foreign material and the contaminated or devitalized tissue by dissection or to perform a segmentectomy.

Aberrant Hepatic Arteries

The more common variety of right or left hepatic artery that arises as a terminal branch of the hepatic artery proper (Fig. B2.24A) may be replaced in part or entirely by an aberrant (accessory or replaced) artery arising from another source. The most common source of an *aberrant right hepatic artery* is the SMA (Fig. B2.24B). The most common source of an *aberrant left hepatic artery* is the left gastric artery (Fig. B2.24C).

Variations in Relationships of Hepatic Arteries

In most people, the right hepatic artery crosses anterior to the hepatic portal vein (Fig. B2.24D); however, in some people, the artery crosses posterior to the hepatic portal vein (Fig. B2.24E). In most people, the right hepatic artery runs posterior to the common hepatic duct (Fig. B2.24G). In some individuals, the right hepatic artery crosses anterior to the common hepatic duct (Fig. B2.24F), or the right hepatic artery arises from the SMA and so does not cross the common hepatic duct at all (Fig. B2.24H).

Hepatomegaly

The liver is a soft highly vascular organ that receives a large amount of blood immediately before it enters the heart. Both the IVC and hepatic veins lack valves. Any rise in central venous pressure is directly transmitted to the liver, which enlarges as it becomes engorged with blood. Marked temporary engorgement stretches the fibrous capsule of the liver, producing pain around the lower ribs, particularly in the right hypochondrium. This engorgement, particularly in conjunction with increased or sustained diaphragmatic activity, may be an underlying cause of "runner's stitch," perhaps explaining why it is a right-sided phenomenon.

In addition to diseases that produce hepatic engorgement such as congestive heart failure, bacterial and viral diseases such as *hepatitis* cause *hepatomegaly* (liver enlargement). When the liver is massively enlarged, its inferior edge may be readily palpated below the right costal margin and may even reach the pelvic brim in the right lower quadrant of the abdomen.

Tumors also enlarge the liver. The liver is a common site of *metastatic carcinoma* (secondary cancers spreading from organs drained by the portal system of veins, e.g., the large intestine). Cancer cells may also pass to the liver from the thorax, especially from the right breast, because of the communications between thoracic lymph nodes and the lymphatic vessels draining the bare area of the liver. Metastatic tumors form hard, rounded nodules within the hepatic parenchyma.

Cirrhosis of Liver

The liver is the primary site for detoxification of substances absorbed by the gastrointestinal tract; thus it is vulnerable to cellular damage and

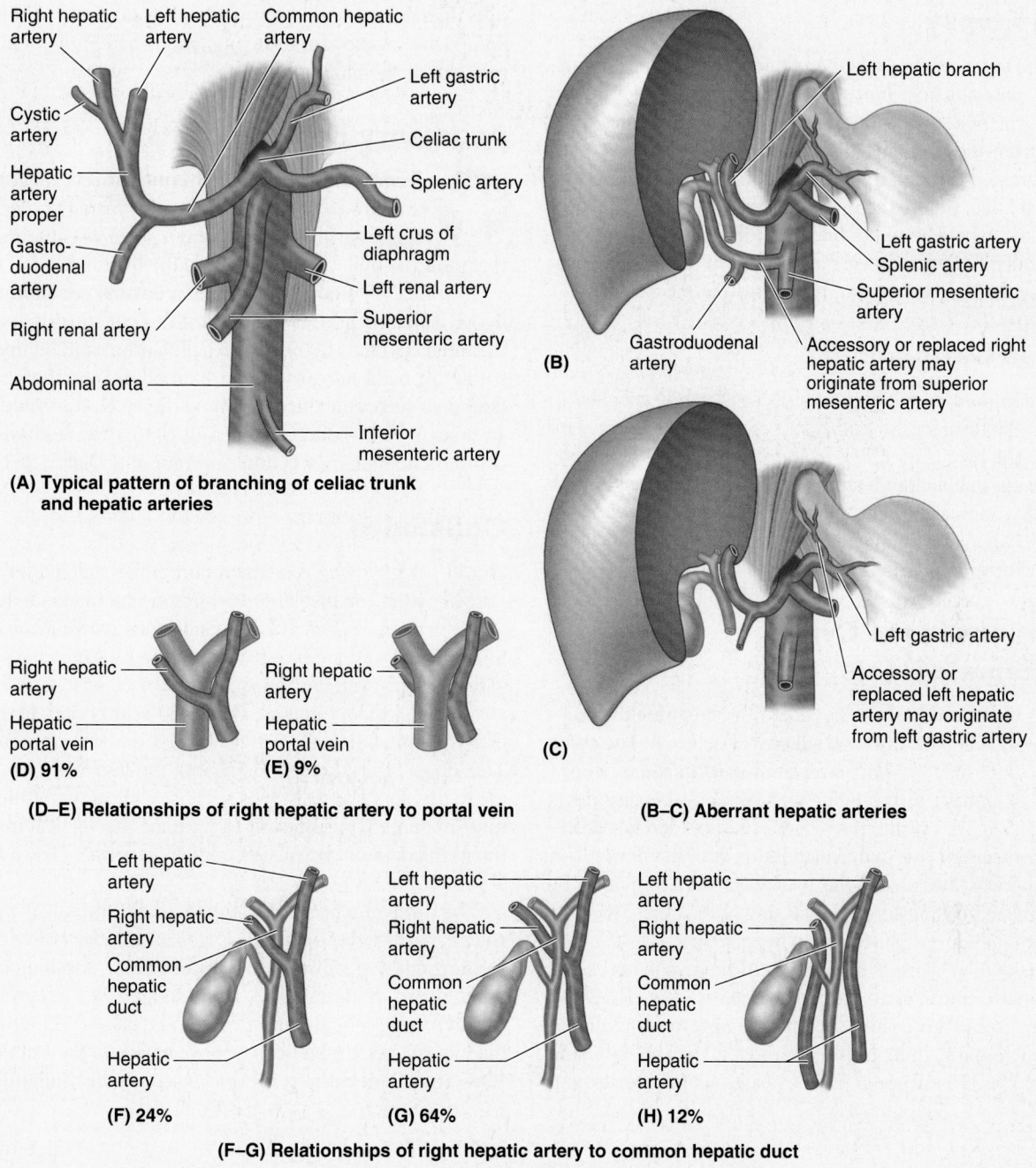

FIGURE B2.24. Variations of right and left hepatic arteries.

consequent scarring, accompanied by regenerative nodules. There is progressive destruction of hepatocytes (Fig. 2.69) in *hepatic cirrhosis* and replacement of these cells by fat and fibrous tissue. Although many industrial solvents, such as carbon tetrachloride, produce cirrhosis, the condition develops most frequently in persons suffering from chronic alcoholism.

Alcoholic cirrhosis, the most common of many causes of *portal hypertension*, is characterized by hepatomegaly and a "hobnail" appearance of the liver surface (see Fig. B2.30B

the blue box "Portosystemic Shunts" on p. 288) resulting from fatty changes and fibrosis. The liver has great functional reserve; therefore the metabolic evidence of liver failure is late to appear. Fibrous tissue surrounds the intrahepatic blood vessels and biliary ducts, making the liver firm, and impeding the circulation of blood through it (*portal hypertension*). The treatment of advanced hepatic cirrhosis may include the surgical creation of a *portosystemic* or *portocaval shunt*, anastomosing the portal and systemic venous systems (see the blue box "Portosystemic Shunts," p. 288).

Liver Biopsy

Hepatic tissue may be obtained for diagnostic purposes by *liver biopsy*. Because the liver is located in the right hypochondriac region where it receives protection from the overlying thoracic cage, the needle is commonly directed through the right 10th intercostal space in the midaxillary line. Before the physician takes the biopsy, the person is asked to hold his or her breath in full expiration to reduce the costodiaphragmatic recess and to lessen the possibility of damaging the lung and contaminating the pleural cavity.

Mobile Gallbladder

In most people the gallbladder is closely attached to the fossa for the gallbladder on the visceral surface of the liver (Fig. 2.72). In approximately 4% of people, however, the gallbladder is suspended from the liver by a short mesentery, increasing its mobility. *Mobile gallbladders* are subject to vascular torsion and infarction (sudden insufficiency of arterial or venous blood supply).

Variations in the Cystic and Hepatic Ducts

Occasionally the cystic duct runs alongside the common hepatic duct and adheres closely to it. The cystic duct may be short or even absent. In some people, there is low union of the cystic and common hepatic ducts (Fig. B2.25A). As a result, the bile duct is short and lies posterior to the superior part of the duodenum, or even inferior to it. When there is low union, the two ducts may be joined by fibrous tissue, making surgical clamping of the cystic duct difficult without injuring the common hepatic duct.

Occasionally, there is high union of the cystic and common hepatic ducts near the porta hepatis (Fig. B2.25B). In other cases, the cystic duct spirals anteriorly over the common hepatic duct before joining it on the left side (Fig. B2.25C). Awareness of the variations in arteries and bile duct formation is important for surgeons when they ligate the cystic duct during *cholecystectomy* (surgical removal of the gallbladder).

Accessory Hepatic Ducts

Accessory (aberrant) hepatic ducts are common and are in positions of danger during cholecystectomy. An accessory duct is a normal segmental duct that joins the biliary system outside the liver instead of within it (Fig. B2.26). Because it drains a normal segment of the liver, it leaks bile if inadvertently cut during surgery (Skandalakis et al., 2009). Of 95 gallbladders and biliary ducts studied, 7 had accessory ducts: 4 joined the common hepatic duct near the cystic ducts, 2 joined the cystic duct, and 1 was an anastomosing duct connecting the cystic duct with the common hepatic duct (Grant, in Agur and Dalley, 2013).

Gallstones

A *gallstone* is a concretion in the gallbladder, cystic duct, or bile duct composed chiefly of cholesterol crystals (Fig. B2.27). Gallstones (*cholelithiasis*) are much more common in females and their incidence increases with age. However, in approximately 50% of persons, gallstones are "silent" (asymptomatic). Over a 20-year period, two thirds of asymptomatic people with gallstones remain symptom free. The longer stones remain quiescent, the less likely symptoms are to develop. For gallstones to cause clinical symptoms, they must obtain a size sufficient to produce mechanical injury to the gallbladder or obstruction of the biliary tract (Townsend et al., 2012).

The distal end of the hepatopancreatic ampulla is the narrowest part of the biliary passages and is the common site for impaction of gallstones. Gallstones may also lodge in the hepatic and cystic ducts. A stone lodged in the cystic duct causes *biliary colic* (intense, spasmodic pain). When the gallbladder relaxes, the stone may move back into the gallbladder. If the stone blocks the cystic duct, *cholecystitis* (inflammation

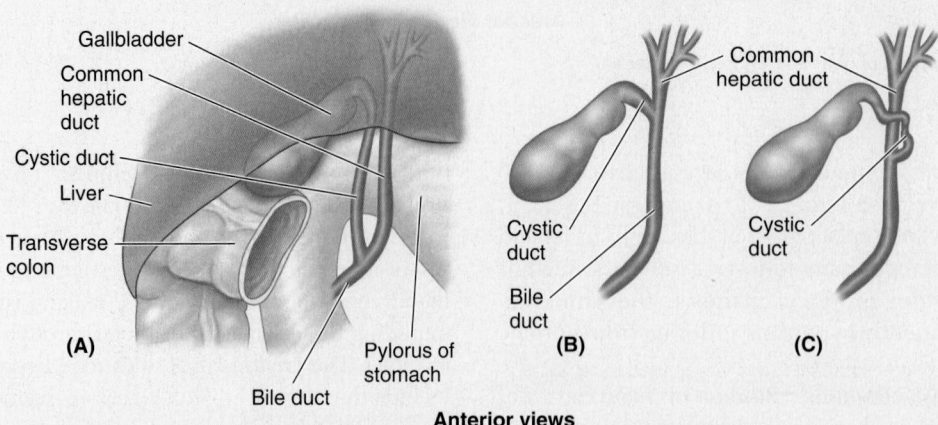

Anterior views

FIGURE B2.25. Union of cystic and common hepatic ducts. A. Low union. **B.** High union. **C.** Swerving course.

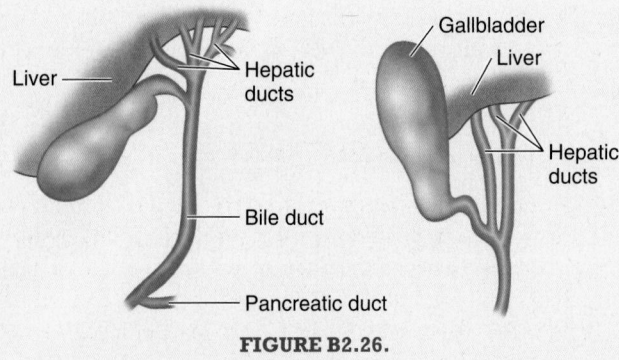

FIGURE B2.26.

of the gallbladder) occurs because of bile accumulation, causing enlargement of the gallbladder.

Another common site for impaction of gallstones is in an abnormal sacculation (Hartmann pouch), that appears in diseased states at the junction of the neck of the gallbladder and the cystic duct. When this pouch is large, the cystic duct arises from its upper left aspect, not from what appears to be the apex of the gallbladder. Gallstones commonly collect in the pouch. If a peptic duodenal ulcer ruptures, a false passage may form between the pouch and the superior part of the duodenum, allowing gallstones to enter the duodenum. (See the following blue box, "Gallstones in Duodenum.")

Pain from an *impaction of the gallbladder* develops in the epigastric region and later shifts to the right hypochondriac region at the junction of the 9th costal cartilage and the lateral border of the rectus sheath. *Inflammation of the gallbladder* may cause pain in the posterior thoracic wall, or right shoulder owing to irritation of the diaphragm. If bile cannot leave the gallbladder, it enters the blood and causes *jaundice* (see the blue box "Pancreatic Cancer," p. 283). Ultrasound and CT scans are common non-invasive techniques for locating stones.

Gallstones in Duodenum

A gallbladder that is dilated and inflamed owing to an impacted gallstone in its duct may develop adhesions with adjacent viscera. Continued inflammation may break down (ulcerate) the tissue boundaries between the gallbladder and a part of the gastrointestinal tract adherent to it, resulting in a *cholecysto-enteric fistula* (Fig. B2.28). Because of their proximity to the gallbladder, the superior part of the duodenum and the transverse colon are most likely to develop a fistula of this type. The fistula would enable a large gallstone, incapable of passing though the cystic duct, to enter the gastrointestinal tract. A large gallstone entering the small intestine in this way may become trapped at the ileocecal valve, producing a bowel obstruction (*gallstone ileus*). A cholecysto-enteric fistula also permits gas from the gastrointestinal tract to enter the gallbladder, providing a diagnostic radiographic sign.

Cholecystectomy

People with severe *biliary colic* usually have their gallbladders removed. *Laparoscopic cholecystectomy* often replaces the open-incision surgical method. The cystic artery most commonly arises from the right hepatic artery in the **cystohepatic triangle** (Calot triangle) (see Figs. 2.72 and 2.74A). In current clinical use, the cystohepatic triangle is defined inferiorly by the cystic duct,

Cholelithiasis. The gallbladder has been opened to reveal numerous yellow cholesterol gallstones.

FIGURE B2.27. Gallstones (cholelithiasis). The gallbladder has been opened to reveal numerous yellow cholesterol gallstones.

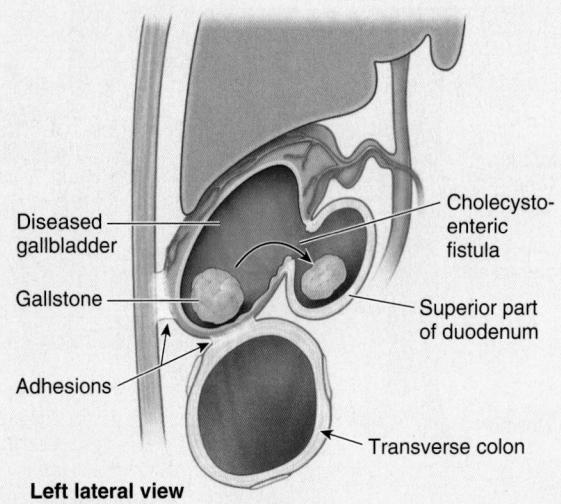

Left lateral view

FIGURE B2.28.

medially by the common hepatic duct, and superiorly by the inferior surface of the liver. Careful dissection of the cystohepatic triangle early during cholecystectomy safeguards these important structures should there be anatomical variations. Errors during gallbladder surgery commonly result from failure to appreciate the common variations in the anatomy of the biliary system, especially its blood supply. Before dividing any structure and removing the gallbladder, surgeons identify all three biliary ducts, as well as the cystic and hepatic arteries. It is usually the right hepatic artery that is in danger during surgery and must be located before ligating the cystic artery.

Portal Hypertension

When scarring and fibrosis from cirrhosis obstruct the hepatic portal vein in the liver, pressure rises in the vein and its tributaries, producing *portal hypertension.* The large volume of blood flowing from the portal system to the systemic system at the sites of portal–systemic anastomoses may produce *varicose veins,* especially in the lower esophagus. The veins may become so dilated that their walls rupture, resulting in hemorrhage (see Fig. B2.7).

Bleeding from esophageal varices (abnormally dilated veins) at the distal end of the esophagus is often severe and may be fatal. In severe cases of portal obstruction, the veins of the anterior abdominal wall (normally caval tributaries) that anastomose with the para-umbilical veins (normally portal tributaries) may become varicose and look somewhat like small snakes radiating under the skin around the umbilicus. This condition is referred to as *caput medusae* because of its resemblance to the serpents

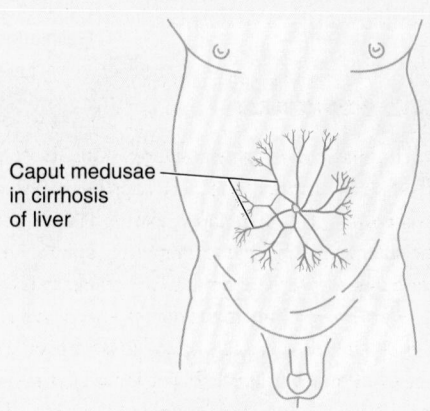

FIGURE B2.29.

on the head of Medusa, a character in Greek mythology (Fig. B2.29).

Portosystemic Shunts

A common method for reducing portal hypertension is to divert blood from the portal venous system to the systemic venous system by creating a communication between the hepatic portal vein and the IVC. This *portocaval anastomosis* or *portosystemic shunt* may be done where these vessels lie close to each other posterior to the liver (Fig. B2.30A–C). Another way of reducing portal pressure is to join the splenic vein to the left renal vein, after splenectomy (*splenorenal anastomosis* or *shunt*) (Fig. B2.30B–D) (Skandalakis et al., 2009).

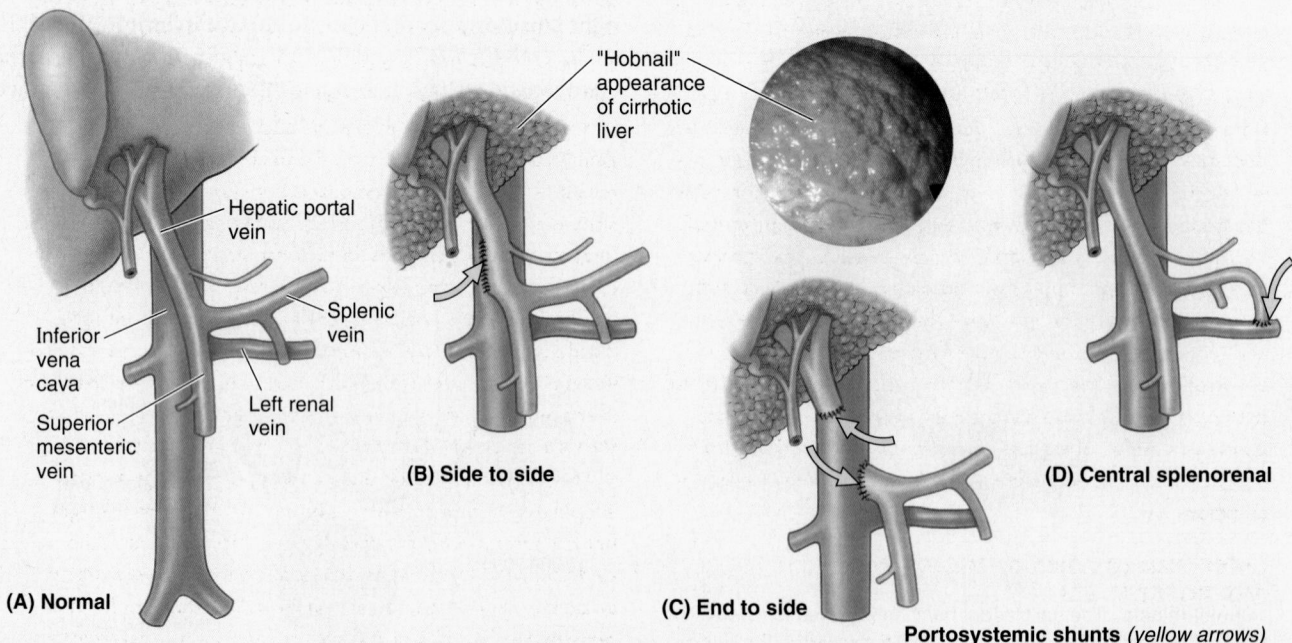

FIGURE B2.30. Portosystemic shunts (*red arrows*).

The Bottom Line

SPLEEN AND PANCREAS

Spleen: The spleen is a highly vascular (sinusoidal), pulpy mass surrounded by a delicate fibroelastic capsule. ♦ The spleen is completely covered by peritoneum, except at the splenic hilum, where the splenorenal ligament (conveying splenic vessels to the spleen) and gastrosplenic ligament (conveying the short gastric and left gastro-omental vessels to the stomach) attach. ♦ The average spleen is about the size of one's fist, within a considerable range of normal variation. ♦ The spleen is the largest of the lymphoid organs, but it is not vital. ♦ As a blood reservoir, it is normally capable of considerable temporary expansion and contraction, but it may undergo much more marked, chronic enlargement when diseased. ♦ Although it receives protection from the overlying left 9th–11th ribs, the relatively delicate spleen is the abdominal organ most vulnerable to indirect trauma. ♦ Strong blows to the abdomen can cause a sudden increase in intra-abdominal pressure that may rupture the spleen, resulting in profuse intraperitoneal hemorrhage.

Pancreas: The pancreas is both an exocrine gland, producing pancreatic juice that is secreted into the duodenum for digestion, and an endocrine gland, producing insulin and glucagon that are released as hormones into the blood. ♦ The secondarily retroperitoneal pancreas consists of a head, an uncinate process, neck, body, and tail. ♦ The head, to the right of the SMA, is embraced by the C-shaped duodenum and penetrated by the termination of the bile duct, whereas its extension, the uncinate process, passes posterior to the SMA. ♦ The neck passes anterior to the SMA and SMV, the latter merging there with the splenic vein to form the hepatic portal vein. ♦ The body lies to the left of the SMA, running transversely across the posterior wall of the omental bursa and crossing anteriorly over the L2 vertebral body and abdominal aorta. ♦ The tail enters the splenorenal ligament as it approaches the splenic hilum. ♦ The splenic vein runs parallel and posterior to the tail and body of the pancreas as it runs from the spleen to the hepatic portal vein. ♦ The main pancreatic duct runs a similar course within the pancreas, continuing transversely through the head to merge with the bile duct to form the hepatopancreatic ampulla, which enters the descending part of the duodenum. ♦ As an endocrine gland, the pancreas receives an abundant blood supply from the pancreaticoduodenal and splenic arteries. ♦ Although it receives vasomotor sympathetic and secretomotor parasympathetic nerve fibers, regulation of pancreatic secretion is primarily hormonal. ♦ The pancreas is well protected by its central location in the abdomen. The exocrine pancreas is seldom the cause of clinical problems, although diabetes, involving the endocrine pancreas, has become increasingly common.

LIVER, BILIARY DUCTS, GALLBLADDER, AND PORTAL VEIN

Liver: The liver has multiple functions. ♦ It is our major metabolic organ, initially receiving all absorbed foodstuffs, except fats. ♦ It is also our largest gland, functioning as an extrinsic intestinal gland in producing bile. ♦ The liver occupies essentially all of the right dome of the diaphragm and extends to the apex of the left dome. Consequently, it enjoys protection from the lower thoracic cage and moves with respiratory excursions. ♦ The liver is divided superficially by the falciform ligament and a groove for the ligamentum venosum into a large anatomical right lobe and a much smaller left one; formations on its visceral surface demarcate caudate and quadrate lobes. ♦ The liver is covered with peritoneum except for the bare area, demarcated by peritoneal reflections comprising the coronary ligaments. ♦ Based on interdigitated branchings of the portal triad (hepatic portal vein, hepatic artery, and intrahepatic bile ducts) and hepatic veins, the continuous parenchyma of the liver can be divided into right and left livers (plus the caudate lobe). ♦ The liver can be further subdivided into four divisions and then eight surgically resectable hepatic segments. ♦ The liver, like the lungs, receives a dual blood supply, with 75–80% of the blood arriving via the hepatic portal vein, meeting the nutritional needs of the hepatic parenchyma; 20–25% arrives via the hepatic artery, delivered primarily to the non-parenchymal elements. The hepatic portal vein and hepatic artery enter the liver via the porta hepatis, where the hepatic ducts exit. ♦ Three large hepatic veins drain directly to the IVC as it is embedded in the liver's bare area. ♦ The liver is also the body's largest lymph-producing organ. The liver's visceral aspect drains via an abdominal route, and its diaphragmatic aspect drains via a thoracic route.

Biliary ducts and gallbladder: Right and left hepatic ducts drain the bile produced by the right and left livers into the common hepatic duct, which thus carries all bile from the liver. ♦ The common hepatic duct merges with the cystic duct to form the bile duct, which conveys the bile to the descending part of the duodenum. ♦ When the sphincter of the bile duct is closed, bile backs up in the bile and cystic ducts, filling the gallbladder, where bile is stored and concentrated between meals. ♦ Although parasympathetic innervation can open the sphincter of the bile duct (and the weaker sphincter of the hepatopancreatic ampulla) and contract the gallbladder, typically these are hormonally regulated responses to fat entering the duodenum, emptying the accumulated bile into the duodenum. ♦ The pear-shaped gallbladder is attached to the visceral surface of the liver, with its fundus projecting from the liver's inferior border against the anterior abdominal wall at the intersection of the transpyloric plane and the right MCL. ♦ The gallbladder, cystic duct, and uppermost bile duct are supplied by the cystic artery, a branch arising from the right hepatic artery within the cystohepatic triangle. ♦ In addition to drainage via cystic veins that accompany the cystic artery and enter the hepatic portal vein, veins from the fundus and body make up a mini-portal system that drains directly into the hepatic sinusoids deep to the visceral surface of the liver.

c portal vein: The large but short hepatic portal vein, formed posterior to the neck of the pancreas by the merger of the SMV and splenic vein, conveys all venous blood and bloodborne nutrients from the gastrointestinal tract to the liver. ♦ The hepatic portal vein terminates at the porta hepatis, bifurcating into the right and left portal veins, distributed in a segmental pattern to the right and left livers. ♦ The hepatic portal vein traverses the hepatoduodenal ligament (free edge of lesser omentum and anterior boundary of omental foramen) as part of an extrahepatic portal triad (hepatic portal vein, hepatic artery, bile duct). ♦ Portal–systemic anastomoses provide a potential collateral pathway by which blood can return to the heart when there is an obstruction of the hepatic portal vein or disease of the liver. However, when the collateral pathways must convey large volumes, potentially lethal esophageal varices may develop.

Kidneys, Ureters, and Suprarenal Glands

The *kidneys* produce urine that is conveyed by the *ureters* to the *urinary bladder* in the pelvis. The superomedial aspect of each kidney normally contacts a *suprarenal gland*. A weak fascial septum separates the glands from the kidneys; thus they are not actually attached to each other (Fig. 2.76). The suprarenal glands function as part of the endocrine system, completely separate in function from the kidneys. The superior urinary organs (kidneys and ureters), their vessels, and the suprarenal glands are primary retroperitoneal structures

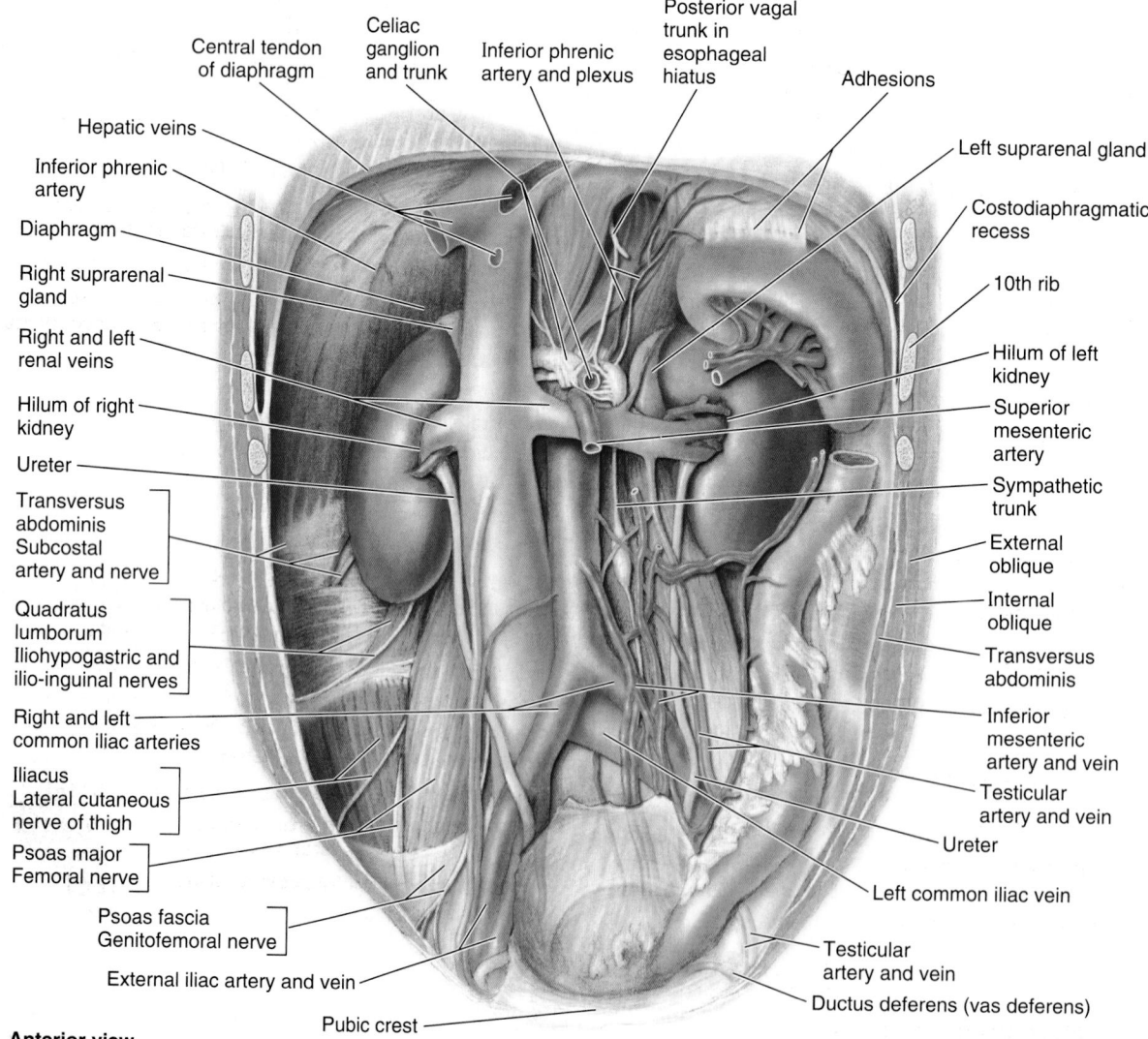

Anterior view

FIGURE 2.76. Posterior abdominal wall showing great vessels, kidneys, and suprarenal glands. Most of the fascia has been removed in this view. The ureter crosses the external iliac artery just beyond the common iliac bifurcation. The gonadal arteries (testicular arteries, as in this male, or ovarian arteries in females) cross anterior to the ureters and provide ureteric branches to them. The renal arteries are not seen here because they lie posterior to the renal veins. The superior mesenteric artery arises superior to and runs anteriorly across the left renal vein, compressing the vein against the abdominal aorta posteriorly.

on the posterior abdominal wall (Fig. 2.76)—that is, they were originally formed as and remain retroperitoneal viscera.

Perinephric fat (perirenal fat capsule) surrounds the kidneys and their vessels as it extends into their hollow centers, the **renal sinuses** (Fig. 2.77). The kidneys, suprarenal glands, and the fat surrounding them are enclosed (except inferiorly) by a condensed, membranous layer of **renal fascia,** which continues medially to ensheath the renal vessels, blending with the vascular sheaths of the latter. Inferomedially, a delicate extension of the renal fascia is prolonged along the ureter as the **peri-**

ureteric fascia. External to the renal fascia is **paranephric fat** (pararenal fat body), the extraperitoneal fat of the lumbar region that is most obvious posterior to the kidney. The renal fascia sends collagen bundles through the paranephric fat.

The collagen bundles, renal fascia, and perinephric and paranephric fat, along with the tethering provided by the renal vessels and ureter, hold the kidneys in a relatively fixed position. However, movement of the kidneys occurs during respiration and when changing from the supine to the erect position, and vice versa. Normal renal mobility is

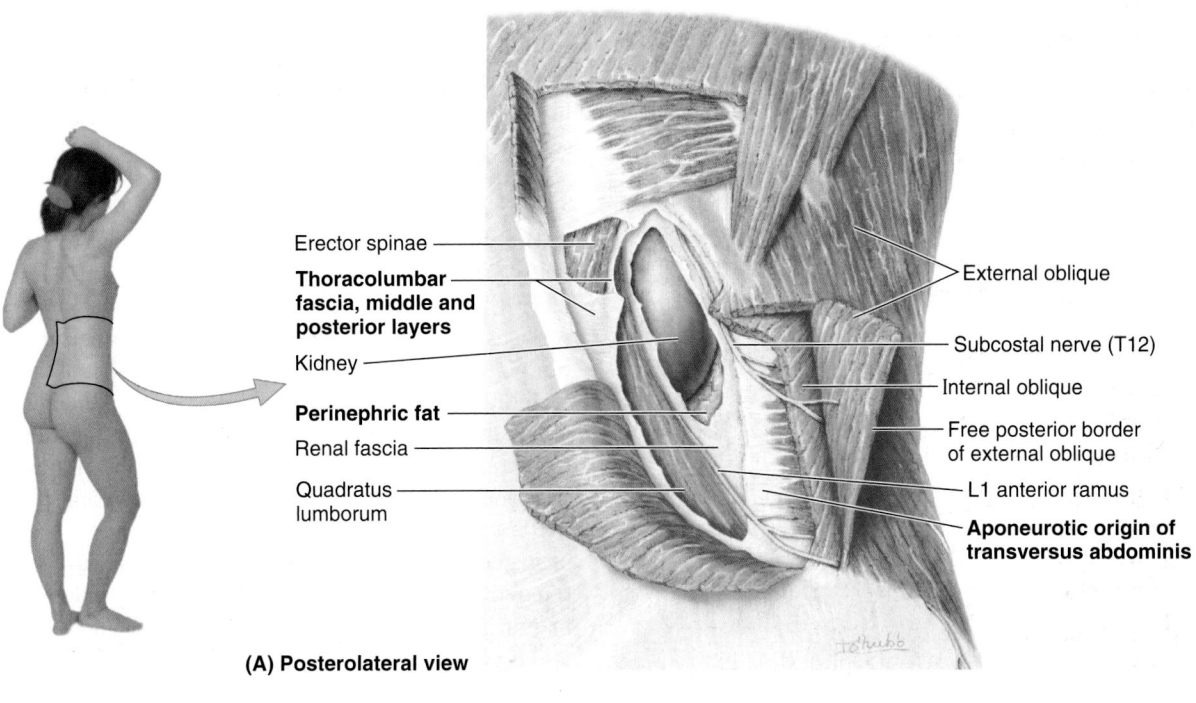

(A) Posterolateral view

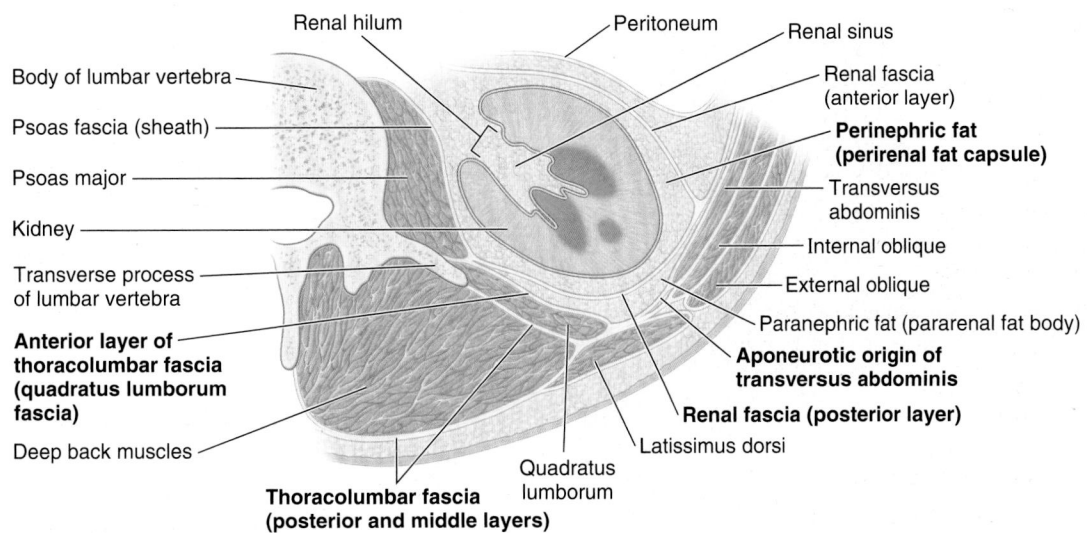

(B) Transverse section, inferior view

FIGURE 2.77. Lumbar approach to kidney and musculofascial relationships of kidney. A. The external aspect of the right posterior abdominal wall is shown. The posterolateral abdominal wall has been incised between the muscles of the anterolateral abdominal wall and the muscles of the back. The kidney and perinephric fat surrounding it inside the renal fascia are exposed. See Figure 2.95A for an earlier stage of this dissection. B. This transverse section of the kidney shows the relationships of the muscles and fascia. Because the renal fascia surrounds the kidney as a separate sheath, it must be incised in any surgical operation on the kidney, whether from an anterior or a posterior approach.

approximately 3 cm, approximately the height of one vertebral body. Superiorly, the renal fascia is continuous with the fascia on the inferior surface of the diaphragm (diaphragmatic fascia); thus the primary attachment of the suprarenal glands is to the diaphragm. Inferiorly, the anterior and posterior layers of renal fascia are only loosely united, if attached at all. (See the blue boxes "Perinephric Abscess" and "Nephroptosis" on p. 298.)

KIDNEYS

The ovoid **kidneys** remove excess water, salts, and wastes of protein metabolism from the blood while returning nutrients and chemicals to the blood. They lie retroperitoneally on the posterior abdominal wall, one on each side of the vertebral column at the level of the T12–L3 vertebrae (Fig. 2.76).

At the concave medial margin of each kidney is a vertical cleft, the **renal hilum** (Figs. 2.76 and 2.77B). The renal hilum is the entrance to a space within the kidney, the **renal sinus.** Structures that serve the kidneys (vessels, nerves, and structures that drain urine from the kidney) enter and exit the renal sinus through the renal hilum. The hilum of the left kidney lies near the transpyloric plane, approximately 5 cm from the median plane (Fig. 2.78). The transpyloric plane passes through the superior pole of the right kidney, which is approximately 2.5 cm lower than the left pole, probably due to the presence of the liver. Posteriorly, the superior parts of the kidneys lie deep to the 11th and 12th ribs. The levels of the kidneys change during respiration and with changes in posture. Each kidney moves 2–3 cm in a vertical direction during the movement of the diaphragm that occurs with deep breathing. Because the usual surgical approach to the kidneys is through the posterior abdominal

wall, it is helpful to know that the inferior pole of the right kidney is approximately a finger's breadth superior to the iliac crest.

During life, the kidneys are reddish brown and measure approximately 10 cm in length, 5 cm in width, and 2.5 cm in thickness. Superiorly, the posterior aspects of the kidneys are associated with the diaphragm, which separates them from the pleural cavities and the 12th pair of ribs (Fig. 2.76). More inferiorly, the posterior surfaces of the kidney are related to the psoas major muscles medially and the quadratus lumborum muscle (Figs. 2.76 and 2.77). See the blue box "Pain in Pararenal Region," p. 298. The subcostal nerve and vessels and the iliohypogastric and ilio-inguinal nerves descend diagonally across the posterior surfaces of the kidneys. The liver, duodenum, and ascending colon are anterior to the right kidney (Figs. 2.75B and 2.79). This kidney is separated from the liver by the *hepatorenal recess.* The left kidney is related to the stomach, spleen, pancreas, jejunum, and descending colon.

At the hilum, the *renal vein* is anterior to the *renal artery,* which is anterior to the renal pelvis (Figs. 2.76 and 2.80A). Within the kidney, the *renal sinus* is occupied by the renal pelvis, calices, vessels, and nerves, and a variable amount of fat (Fig. 2.80C & D). Each kidney has anterior and posterior surfaces, medial and lateral margins, and superior and inferior poles. However, because of the protrusion of the lumbar vertebral column into the abdominal cavity, the kidneys are obliquely placed, lying at an angle to each other (Fig. 2.77B). Consequently, the transverse diameter of the kidneys is foreshortened in anterior views (Fig. 2.76A) and anteroposterior (AP) radiographs (Fig. 2.81). The lateral margin of each kidney is convex, and the medial margin is concave where the renal sinus and renal pelvis are located. The indented medial margin gives the kidney a somewhat bean-shaped appearance.

The **renal pelvis** is the flattened, funnel-shaped expansion of the superior end of the ureter (Figs. 2.80B–D, 2.81, and 2.82). The **apex of the renal pelvis** is continuous with the ureter. The renal pelvis receives two or three **major calices** (calyces), each of which divides into two or three **minor calices.** Each minor calyx is indented by a **renal papilla,** the apex of the *renal pyramid,* from which the urine is excreted. In living persons, the renal pelvis and its calices are usually collapsed (empty). The pyramids and their associated cortex form the lobes of the kidney. The lobes are visible on the external surfaces of the kidneys in fetuses, and evidence of the lobes may persist for some time after birth.

URETERS

The **ureters** are muscular ducts (25–30 cm long) with narrow lumina that carry urine from the kidneys to the urinary bladder (Figs. 2.76 and 2.82). They run inferiorly from the apices of the renal pelves at the hila of the kidneys, passing over the pelvic brim at the bifurcation of the common iliac arteries. They then run along the lateral wall of the pelvis and enter the urinary bladder.

The abdominal parts of the ureters adhere closely to the parietal peritoneum and are retroperitoneal throughout their

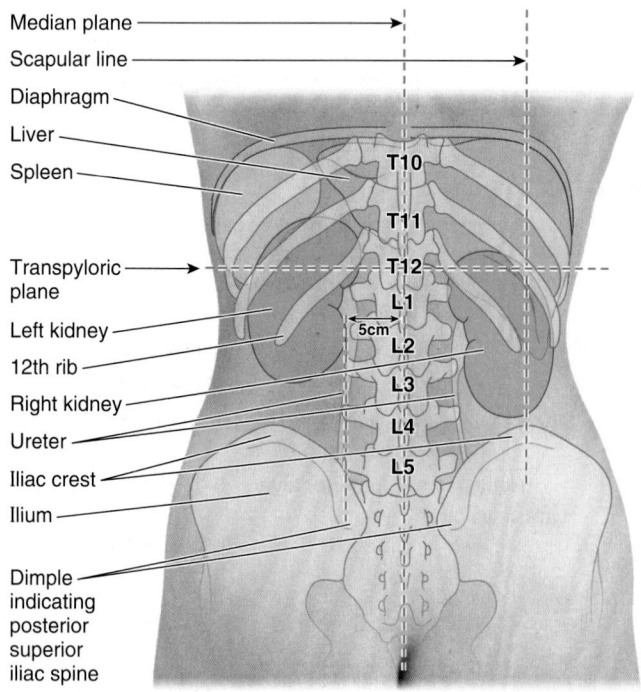

Median plane
Scapular line
Diaphragm
Liver
Spleen
Transpyloric plane
Left kidney
12th rib
Right kidney
Ureter
Iliac crest
Ilium
Dimple indicating posterior superior iliac spine

T10
T11
T12
L1
5cm
L2
L3
L4
L5

FIGURE 2.78. Surface anatomy of the kidneys and abdominal part of ureters.

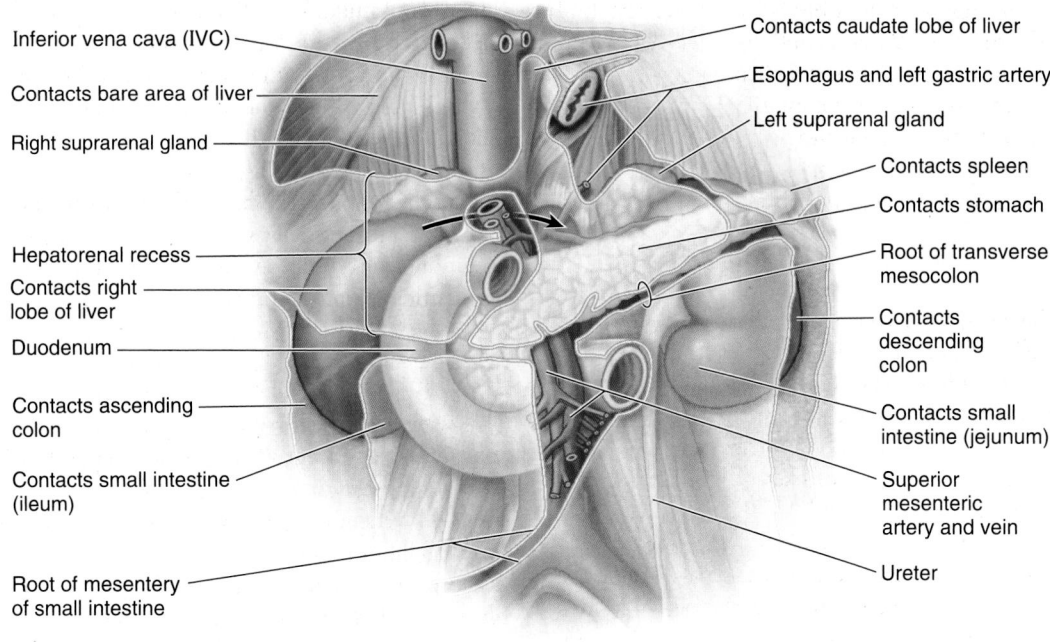

Inferior vena cava (IVC)

Contacts bare area of liver

Right suprarenal gland

Hepatorenal recess

Contacts right lobe of liver

Duodenum

Contacts ascending colon

Contacts small intestine (ileum)

Root of mesentery of small intestine

Contacts caudate lobe of liver

Esophagus and left gastric artery

Left suprarenal gland

Contacts spleen

Contacts stomach

Root of transverse mesocolon

Contacts descending colon

Contacts small intestine (jejunum)

Superior mesenteric artery and vein

Ureter

Anterior view

FIGURE 2.79. Relationships of kidneys, suprarenal glands, pancreas, and duodenum. The right suprarenal gland is at the level of the omental foramen (*black arrow*).

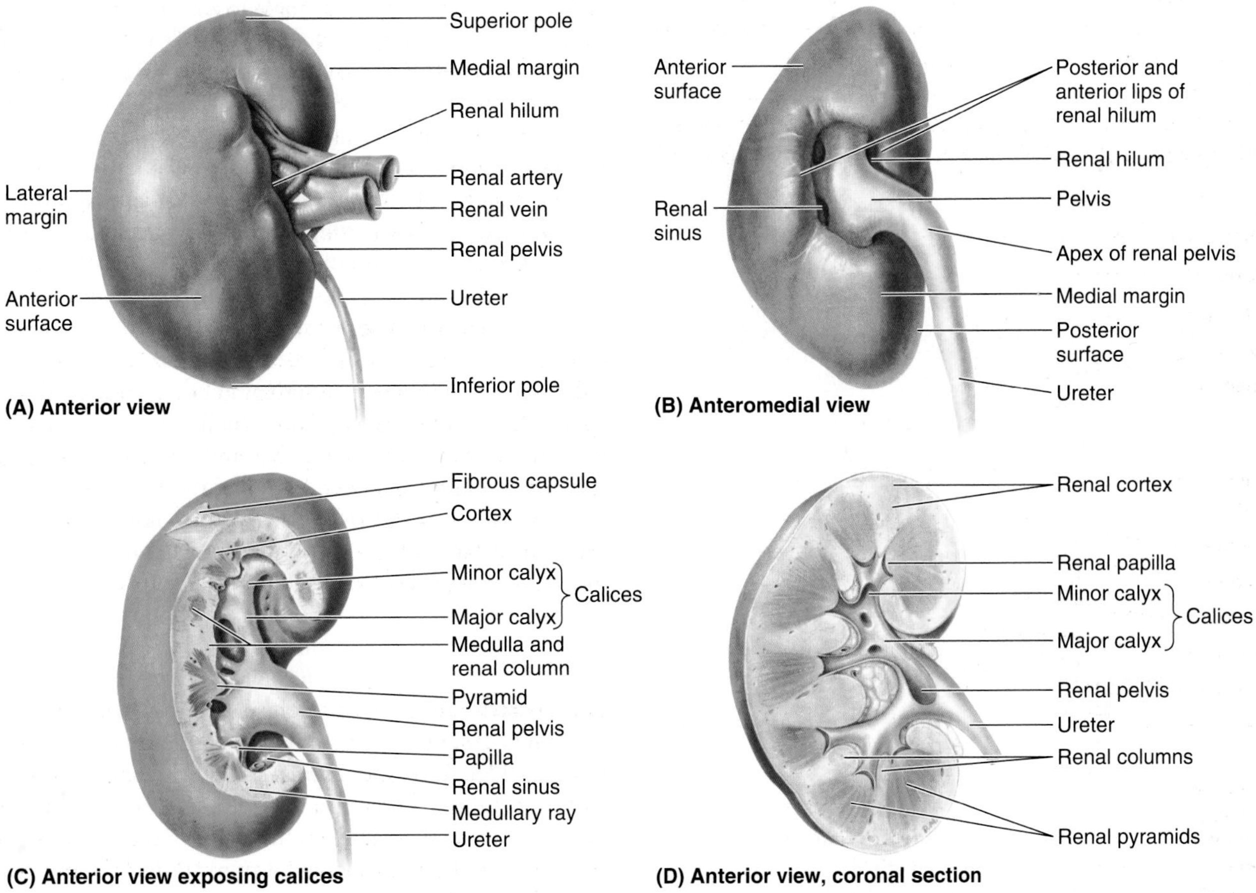

(A) Anterior view

Superior pole
Medial margin
Renal hilum
Renal artery
Renal vein
Renal pelvis
Ureter
Inferior pole
Lateral margin
Anterior surface

(B) Anteromedial view

Anterior surface
Renal sinus
Posterior and anterior lips of renal hilum
Renal hilum
Pelvis
Apex of renal pelvis
Medial margin
Posterior surface
Ureter

(C) Anterior view exposing calices

Fibrous capsule
Cortex
Minor calyx ⎱
Major calyx ⎰ Calices
Medulla and renal column
Pyramid
Renal pelvis
Papilla
Renal sinus
Medullary ray
Ureter

(D) Anterior view, coronal section

Renal cortex
Renal papilla
Minor calyx ⎱
Major calyx ⎰ Calices
Renal pelvis
Ureter
Renal columns
Renal pyramids

FIGURE 2.80. External and internal appearance of kidneys. A. The right kidney. **B.** Renal sinus, as seen through the renal hilum. **C.** The anterior lip of the renal hilum has been cut away to expose the renal pelvis and calices within the renal sinus. **D.** This coronal section of the kidney shows the organ's internal structure. The renal pyramids contain the collecting tubules and form the medulla of the kidney. The renal cortex contains the renal corpuscles.

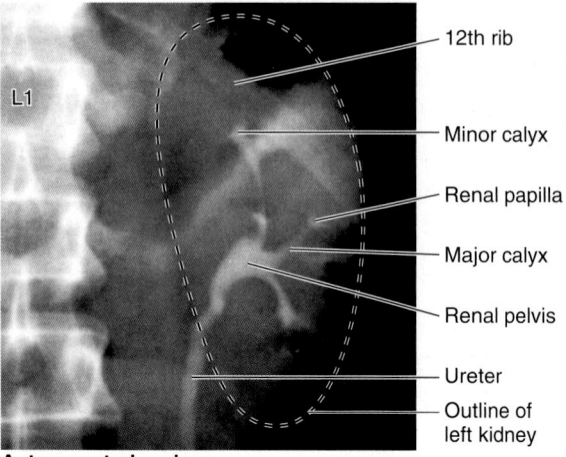

Anteroposterior view

12th rib

Minor calyx

Renal papilla

Major calyx

Renal pelvis

Ureter

Outline of left kidney

L1

FIGURE 2.81. Intravenous urogram (pyelogram). The contrast medium was injected intravenously and was concentrated and excreted by the kidneys. This AP projection shows the calices, renal pelvis, and ureter outlined by the contrast medium filling their lumina. (Courtesy of Dr. John Campbell, Department of Medical Imaging, Sunnybrook Medical Centre, University of Toronto, Toronto, ON, Canada.)

course. From the back, the surface marking of the ureter is a line joining a point 5 cm lateral to the L1 spinous process and the posterior superior iliac spine (Fig. 2.78). The ureters occupy a sagittal plane that intersects the tips of the transverse processes of the lumbar vertebrae. When examining

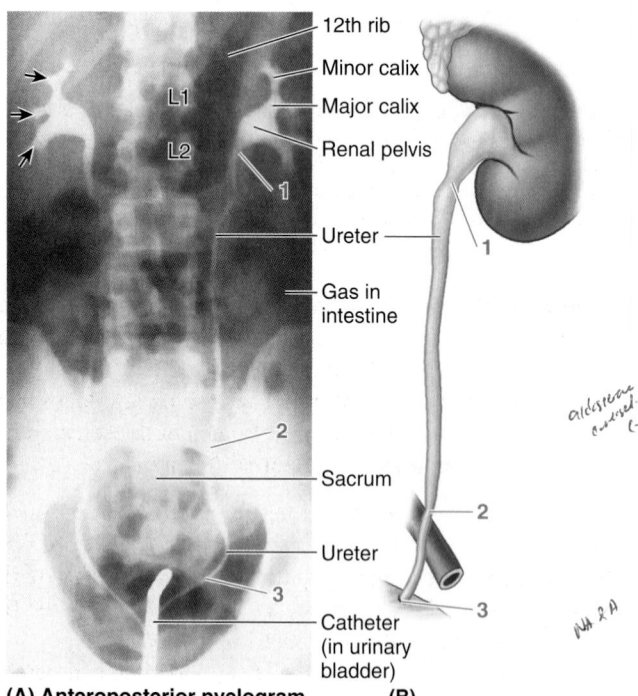

12th rib

Minor calix

Major calix

Renal pelvis

L1

L2

1

Ureter

Gas in intestine

2

Sacrum

2

Ureter

3

3

Catheter (in urinary bladder)

(A) Anteroposterior pyelogram   **(B)**

FIGURE 2.82. Normal constrictions of ureters demonstrated by retrograde pyelogram. A. Contrast medium was injected into the ureters from a flexible endoscope (urethroscope) in the bladder. The *arrows* represent papillae bulging into the minor calices. **B.** Sites at which relative constrictions in the ureters normally appear: (1) at the ureteropelvic junction, (2) crossing the external iliac artery and/or pelvic brim, and (3) as the ureter traverses the bladder wall.

the ureters radiographically using contrast medium (Figs. 2.81 and 2.82), the ureters normally demonstrate relative constrictions in three places: (1) at the junction of the ureters and renal pelves, (2) where the ureters cross the brim of the pelvic inlet, and (3) during their passage through the wall of the urinary bladder (Fig. 2.82). These constricted areas are potential sites of obstruction by ureteric stones (calculi).

Congenital anomalies of the kidneys and ureters are fairly common. (See the blue box "Congenital Anomalies of Kidneys and Ureters" on p. 299.)

SUPRARENAL GLANDS

The **suprarenal (adrenal) glands,** yellowish in living persons, are located between the superomedial aspects of the kidneys and the diaphragm (Fig. 2.83), where they are surrounded by connective tissue containing considerable perinephric fat. The suprarenal glands are enclosed by renal fascia by which they are attached to the *crura of the diaphragm.* Although the name "suprarenal" implies that the kidneys are their primary relationship, their major attachment is to the diaphragmatic crura. They are separated from the kidneys by a thin septum (part of the renal fascia—see the blue box "Renal Transplantation," p. 298).

The shape and relations of the suprarenal glands differ on the two sides. The *pyramidal right gland* is more apical (situated over the superior pole) relative to the left kidney, lies anterolateral to the right crus of the diaphragm, and makes contact with the IVC anteromedially (Fig. 2.79) and the liver anterolaterally. The *crescent-shaped left gland* is medial to the superior half of the left kidney and is related to the spleen, stomach, pancreas, and the left crus of the diaphragm.

Each gland has a *hilum,* where the veins and lymphatic vessels exit the gland, whereas, the arteries and nerves enter the glands at multiple sites. The medial borders of the suprarenal glands are 4–5 cm apart. In this area, from right to left, are the IVC, right crus of the diaphragm, celiac ganglion, celiac trunk, SMA, and the left crus of the diaphragm.

Each suprarenal gland has two parts: the *suprarenal cortex* and *suprarenal medulla* (Fig. 2.83, *inset*); these parts have different embryological origins and different functions.

The **suprarenal cortex** derives from mesoderm and secretes corticosteroids and androgens. These hormones cause the kidneys to retain sodium and water in response to stress, increasing the blood volume and blood pressure. They also affect muscles and organs such as the heart and lungs.

The **suprarenal medulla** is a mass of nervous tissue permeated with capillaries and sinusoids that derives from *neural crest cells* associated with the sympathetic nervous system (see Fig. 2.87). The *chromaffin cells* of the medulla are related to sympathetic ganglion (postsynaptic) neurons in both derivation (neural crest cells) and function. These cells secrete catecholamines (mostly epinephrine) into the bloodstream in response to signals from presynaptic neurons. Powerful medullary hormones, epinephrine (adrenaline) and norepinephrine (noradrenaline), activate the body to a flight-or-fight status in response to traumatic stress. They also increase heart

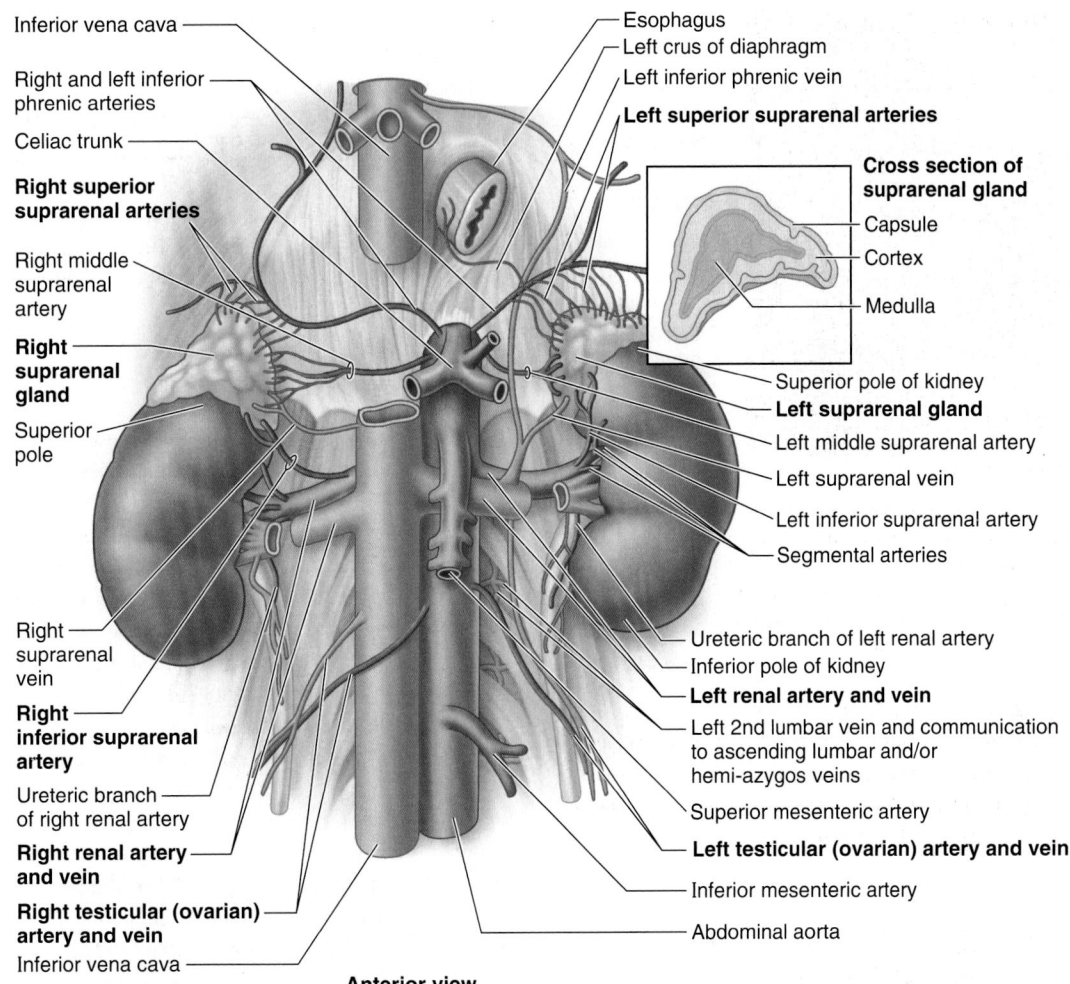

Inferior vena cava

Right and left inferior phrenic arteries

Celiac trunk

Right superior suprarenal arteries

Right middle suprarenal artery

Right suprarenal gland

Superior pole

Right suprarenal vein

Right inferior suprarenal artery

Ureteric branch of right renal artery

Right renal artery and vein

Right testicular (ovarian) artery and vein

Inferior vena cava

Esophagus

Left crus of diaphragm

Left inferior phrenic vein

Left superior suprarenal arteries

Cross section of suprarenal gland

Capsule

Cortex

Medulla

Superior pole of kidney

Left suprarenal gland

Left middle suprarenal artery

Left suprarenal vein

Left inferior suprarenal artery

Segmental arteries

Ureteric branch of left renal artery

Inferior pole of kidney

Left renal artery and vein

Left 2nd lumbar vein and communication to ascending lumbar and/or hemi-azygos veins

Superior mesenteric artery

Left testicular (ovarian) artery and vein

Inferior mesenteric artery

Abdominal aorta

Anterior view

FIGURE 2.83. Blood vessels of suprarenal glands, kidneys, and superior parts of ureters. The celiac plexus of nerves and ganglia that surrounds the celiac trunk has been removed. The inferior vena cava has been transected, and its superior part has been elevated from its normal position to reveal the arteries that pass posterior to it. The renal veins have been cut so that the kidneys could be moved laterally. For the normal relationships of the kidneys and suprarenal glands with the great vessels, see Figure 2.76. A cross section of the suprarenal gland (*inset*) shows that it is composed of two distinct parts: the cortex and medulla, which are two separate endocrine glands that became closely related during embryonic development.

rate and blood pressure, dilate the bronchioles, and change blood flow patterns, preparing for physical exertion.

VESSELS AND NERVES OF KIDNEYS, URETERS, AND SUPRARENAL GLANDS

Renal Arteries and Veins. The *renal arteries* arise at the level of the IV disc between the L1 and L2 vertebrae (Figs. 2.83 and 2.84). The longer **right renal artery** passes posterior to the IVC. Typically, each artery divides close to the hilum into five *segmental arteries* that are end arteries (i.e., they do not anastomose significantly with other segmental arteries, so that the area supplied by each segmental artery is an independent, surgically resectable unit or **renal segment**). Segmental arteries are distributed to the renal segments as follows (Fig. 2.85):

- The superior (apical) segment is supplied by the **superior (apical) segmental artery;** the anterosuperior and anteroinferior segments are supplied by the **anterosuperior segmental** and **antero-inferior segmental arter-**

ies; and the inferior segment is supplied by the **inferior segmental artery.** These arteries originate from the anterior branch of the renal artery.

- The **posterior segmental artery,** which originates from a continuation of the posterior branch of the renal artery, supplies the posterior segment of the kidney.

Multiple renal arteries are common and usually enter the hilum of the kidney (Fig. 2.84). Extrahilar renal arteries from the renal artery or aorta may enter the external surface of the kidney, commonly at their poles ("polar arteries"—see the blue box "Accessory Renal Vessels" on p. 298).

Several *renal veins* drain each kidney and unite in a variable fashion to form the **right** and **left renal veins;** these veins lie anterior to the right and left renal arteries. The longer left renal vein receives the *left suprarenal vein,* the *left gonadal (testicular or ovarian) vein,* and a communication with the *ascending lumbar vein;* it then traverses the acute angle between the SMA anteriorly and the aorta posteriorly (see the blue box "Renal Vein Entrapment Syndrome" on p. 298). Each renal vein drains into the IVC.

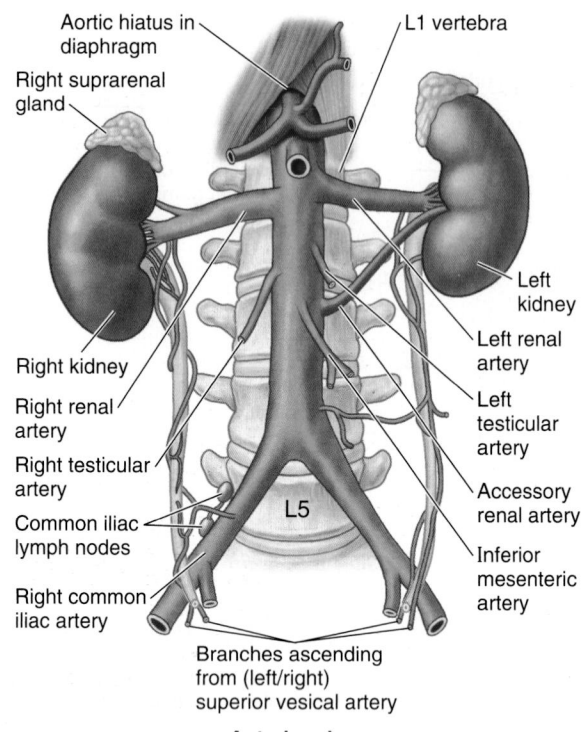

FIGURE 2.84. Arterial supply of kidneys and ureters. The abdominal aorta lies anterior to the L1–L4 vertebral bodies, usually immediately to the left of the midline. An accessory left renal artery is present.

Arterial Supply and Venous Drainage of Ureters. *Arterial branches to the abdominal portion of the ureter* arise consistently from the *renal arteries*, with less constant branches arising from the *testicular or ovarian arteries*, the *abdominal aorta,* and the *common iliac arteries* (Fig. 2.84). The branches approach the ureters medially and divide into ascending and descending branches, forming a longitudinal anastomosis on the ureteric wall. However, ureteric branches are small and relatively delicate, and disruption may lead to ischemia in spite of the continuous anastomotic channel formed. In operations in the posterior abdominal region, surgeons pay special attention to the location of ureters and are careful not to retract them laterally or unnecessarily. The arteries supplying the pelvic portion of the ureter are discussed in Chapter 3.

Veins draining the abdominal part of the ureters drain into the renal and gonadal (testicular or ovarian) veins (see Fig. 2.83).

Suprarenal Arteries and Veins. The endocrine function of the suprarenal glands makes their abundant blood supply necessary. The suprarenal arteries branch freely before entering each gland so that 50–60 branches penetrate the capsule covering the entire surface of the glands. Suprarenal arteries arise from three sources (Fig. 2.83):

- **Superior suprarenal arteries** (6–8) from the *inferior phrenic arteries.*
- **Middle suprarenal arteries** (£ 1) from the *abdominal aorta* near the level of origin of the SMA.
- **Inferior suprarenal arteries** (£ 1) from the renal arteries.

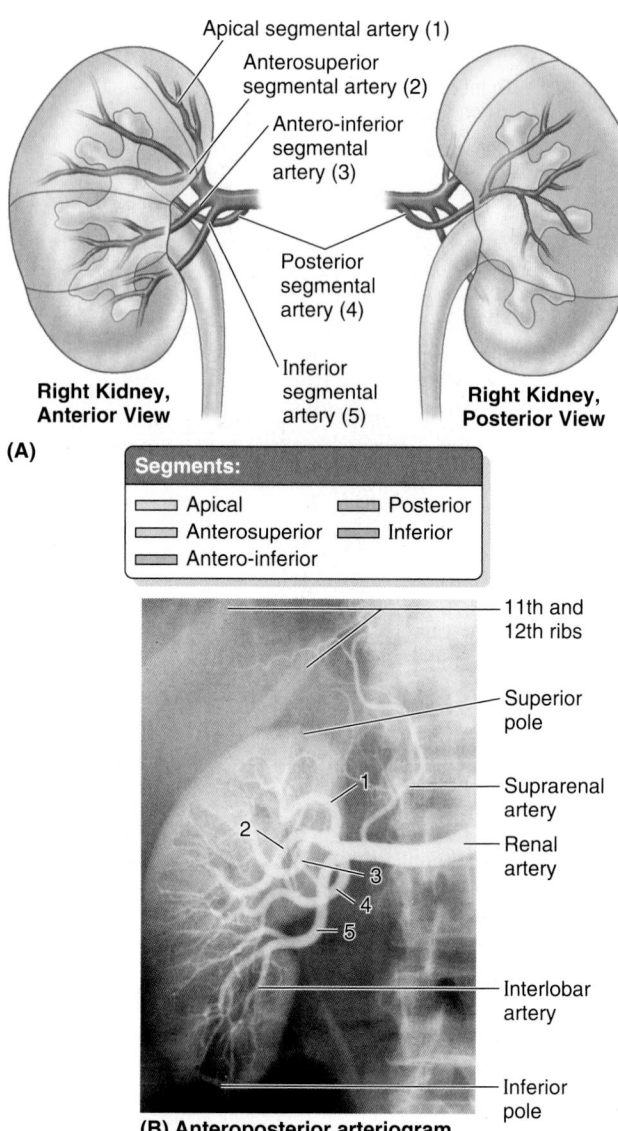

(B) Anteroposterior arteriogram

FIGURE 2.85. Renal segments and segmental arteries. A. The five renal segments and segmental renal arteries. (Numbers in parentheses identify arteries in **B**.) **B.** Renal arteriogram (*1–5*, segmental renal arteries). Although the veins of the kidney anastomose freely, segmental arteries are end arteries. (**B** courtesy of Dr. E. L. Lansdown, Professor of Medical Imaging, University of Toronto, Toronto, ON, Canada.)

The *venous drainage of the suprarenal glands* occurs via large *suprarenal veins.* The short **right suprarenal vein** drains into the IVC, whereas the longer **left suprarenal vein,** often joined by the *inferior phrenic vein,* empties into the left renal vein.

Lymphatics of Kidneys, Ureters, and Suprarenal Glands. The *renal lymphatic vessels* follow the renal veins and drain into the *right* and *left lumbar* (*caval* and *aortic*) *lymph nodes* (Fig. 2.86). Lymphatic vessels from the superior part of the ureter may join those from the kidney or pass directly to the lumbar nodes. Lymphatic vessels from the middle part of the ureter usually drain into the *common iliac lymph nodes,* whereas vessels from its inferior part drain into the *common, external,* or *internal iliac lymph nodes.*

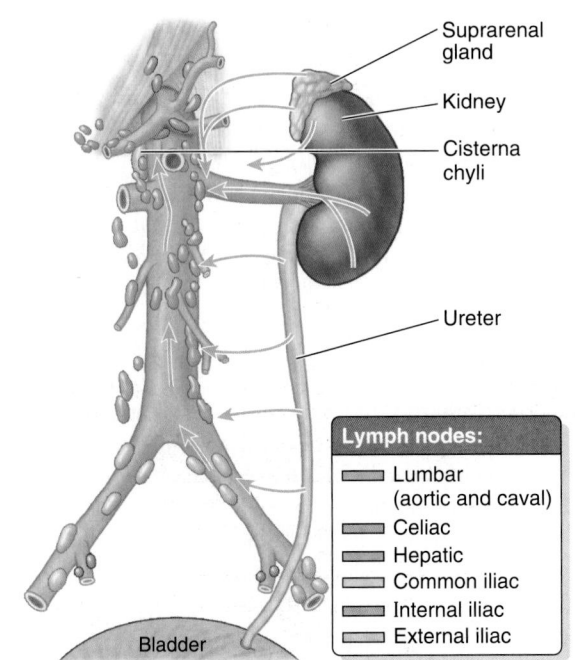

FIGURE 2.86. Lymphatics of kidneys and suprarenal glands. The lymphatic vessels of the kidneys form three plexuses: one in the substance of the kidney, one under the fibrous capsule, and one in the perirenal fat. Four or five lymphatic trunks leave the renal hilum and are joined by vessels from the capsule (*arrows*). The lymphatic vessels follow the renal vein to the lumbar (caval and aortic) lymph nodes. Lymph from the suprarenal glands also drains to the lumbar nodes. Lymphatic drainage of the ureters is also illustrated. The lumbar lymph nodes drain through the lumbar lymphatic trunks to the cisterna chyli.

The *suprarenal lymphatic vessels* arise from a plexus deep to the capsule of the gland and from one in its medulla. The lymph passes to the *lumbar lymph nodes.* Many lymphatic vessels leave the suprarenal glands.

Nerves of Kidneys, Ureters, and Suprarenal Glands. The *nerves to the kidneys* arise from the renal nerve plexus and consist of sympathetic and parasympathetic fibers (Fig. 2.87B). The **renal nerve plexus** is supplied by fibers from the abdominopelvic (especially the least) splanchnic nerves. The *nerves of the abdominal part of the ureters* derive from the renal, abdominal aortic, and superior hypogastric plexuses (Fig. 2.87A). Visceral afferent fibers conveying pain sensation (e.g., resulting from obstruction and consequent distension) follow the sympathetic fibers retrograde to spinal ganglia and cord segments T11–L2. Ureteric pain is usually referred to the ipsilateral lower quadrant of the anterior abdominal wall and especially to the groin (see the blue box "Renal and Ureteric Calculi," p. 300).

The rich *nerve supply of the suprarenal glands* is from the *celiac plexus* and abdominopelvic (greater, lesser, and least) splanchnic nerves. Myelinated presynaptic sympathetic fibers—mainly derived from the intermediolateral cell column (IML), or lateral horn, of gray matter of the spinal cord segments T10–L1—traverse both the paravertebral and the prevertebral ganglia, without synapse, to be distributed to the chromaffin cells in the suprarenal medulla (Fig. 2.87B).

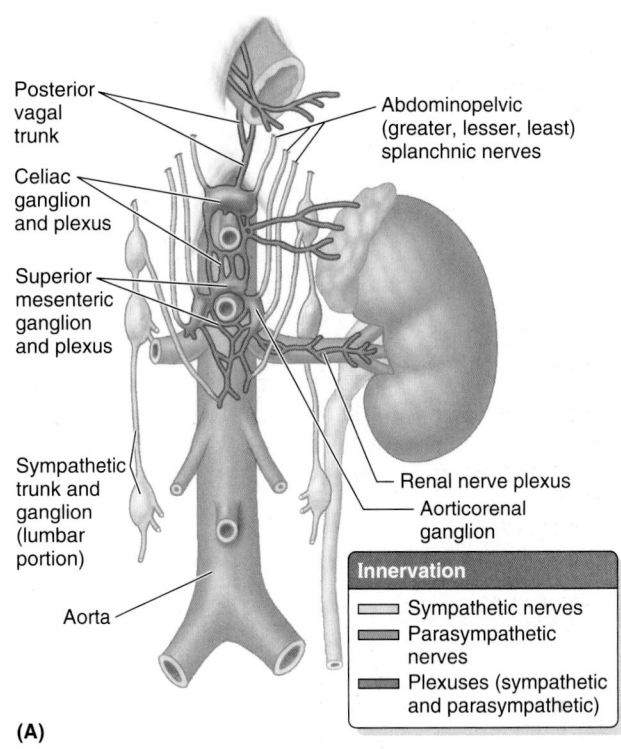

(A)

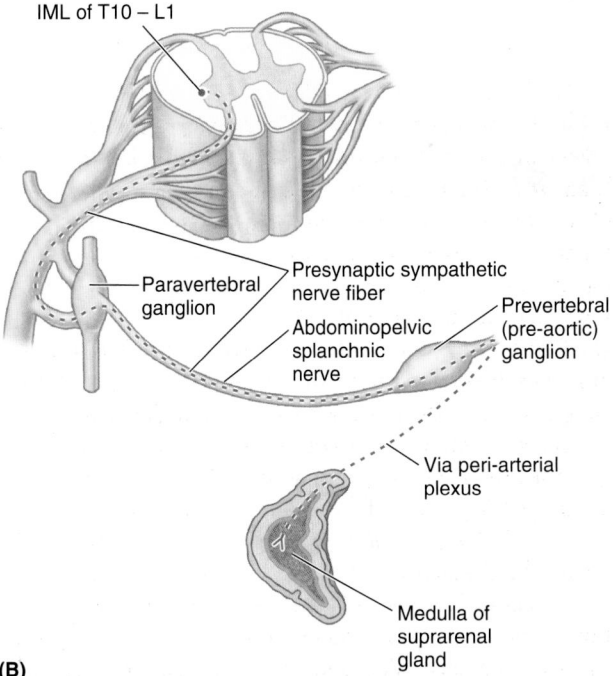

(B)

FIGURE 2.87. Nerves of the kidneys and suprarenal glands. A. The nerves of the kidneys and suprarenal glands are derived from the celiac plexus, abdominopelvic (lesser and least) splanchnic nerves, and the aorticorenal ganglion. The main efferent innervation of the kidney is vasomotor, autonomic nerves supplying the afferent and efferent arterioles. B. Exclusively in the case of the suprarenal medulla, the presynaptic sympathetic fibers pass through both the paravertebral and prevertebral ganglia without synapsing to end directly on the secretory cells of the suprarenal medulla. *IML,* intermediolateral cell column.

KIDNEYS, URETERS, AND SUPRARENAL GLANDS

Palpation of Kidneys

The kidneys are often impalpable. In lean adults, the inferior pole of the right kidney is palpable by bimanual examination as a firm, smooth, somewhat rounded mass that descends during inspiration. Palpation of the right kidney is possible because it is 1–2 cm inferior to the left one. To palpate the kidneys, press the *flank* (side of the trunk between the 11th and 12th ribs and the iliac crest) anteriorly with one hand while palpating deeply at the *costal margin* with the other. The left kidney is usually not palpable unless it is enlarged or a retroperitoneal mass has displaced it inferiorly.

Perinephric Abscess

 The attachments of the renal fascia determine the path of extension of a *perinephric abscess* (pus around the kidney). For example, fascia at the renal hilum attaches to the renal vessels and ureter, usually preventing the spread of pus to the contralateral side. However, pus from an abscess (or blood from an injured kidney) may force its way into the pelvis between the loosely attached anterior and posterior layers of the renal fascia.

Nephroptosis

 Because the layers of renal fascia do not fuse firmly inferiorly to offer resistance, abnormally mobile kidneys may descend more than the normal 3 cm when the body is erect. When kidneys descend, the suprarenal glands remain in place because they lie in a separate fascial compartment and are firmly attached to the diaphragm. *Nephroptosis* (dropped kidney) is distinguished from an *ectopic kidney* (congenital misplaced kidney) by a ureter of normal length that has loose coiling or kinks because the distance to the bladder has been reduced. The kinks do not seem to be of significance. Symptoms of intermittent pain in the renal region, relieved by lying down, appear to result from traction on the renal vessels. The lack of inferior support for the kidneys in the lumbar region is one of the reasons transplanted kidneys are placed in the iliac fossa of the greater pelvis. Other reasons for this placement are the availability of major blood vessels and convenient access to the nearby bladder.

Renal Transplantation

 Renal transplantation is the preferred treatment for selected cases of chronic renal failure. The kidney can be removed from the donor without damaging the suprarenal gland because of the weak septum of renal fascia that separates the kidney from this gland. The site for transplanting a kidney is in the iliac fossa of the greater pelvis. This site supports the transplanted kidney, so that traction is not placed on the surgically anastomosed vessels. The renal artery and vein are joined to the external iliac artery and vein, respectively, and the ureter is sutured into the urinary bladder.

Renal Cysts

 Cysts in the kidney, multiple or solitary, are common findings during ultrasound examinations and dissection of cadavers. Adult *polycystic disease of the kidneys* is an important cause of renal failure; it is inherited as an autosomal dominant trait. The kidneys are markedly enlarged and distorted by cysts as large as 5 cm.

Pain in Pararenal Region

 The close relationship of the kidneys to the psoas major muscles explains why extension of the hip joints may increase pain resulting from inflammation in the pararenal areas. These muscles flex the thighs at the hip joints.

Accessory Renal Vessels

 During their "ascent" to their final site, the embryonic kidneys receive their blood supply and venous drainage from successively more superior vessels. Usually the inferior vessels degenerate as superior ones take over. Failure of these vessels to degenerate results in *accessory renal arteries* (Fig. 2.84) and *veins*. Some accessory arteries ("polar arteries") enter/exit the poles of the kidneys. An inferior polar artery crosses the ureter and may obstruct it. Variations in the number and position of these vessels occur in approximately 30% of people.

Renal Vein Entrapment Syndrome

 In crossing the midline to reach the IVC, the longer left renal vein traverses an acute angle between the SMA anteriorly and the abdominal aorta posteriorly (Fig. B2.31). Downward traction on the SMA may compress the left renal vein (and perhaps the third part of the duodenum) resulting in a *renal vein entrapment syndrome* (mesoaortic compression of the left renal vein), also known as "nutcracker syndrome" based on the appearance of the vein in the acute arterial angle in a sagittal view. The syndrome may include hematuria or proteinuria (blood or protein in the urine), abdominal (left flank) pain, nausea and vomiting (indicating compression of the duodenum), and left testicular pain in men (related to the left testicular vein draining into the left renal vein proximal to the compression). Uncommonly, a left-sided varicocele may occur.

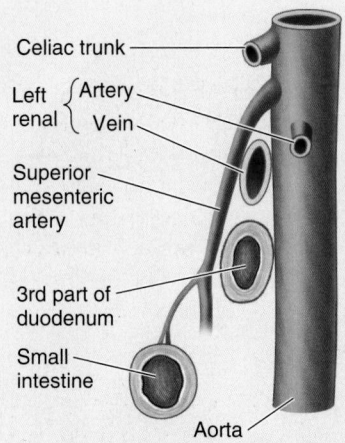

Celiac trunk

Left renal { Artery / Vein }

Superior mesenteric artery

3rd part of duodenum

Small intestine

Aorta

Lateral view (from left)

FIGURE B2.31.

Congenital Anomalies of Kidneys and Ureters

Bifid renal pelvis and ureter are fairly common (Fig. B2.32A & B). These anomalies result from division of the *ureteric bud* (metanephric diverticulum), the primordium of the renal pelvis and ureter. The extent of ureteral duplication depends on the completeness of embryonic division of the ureteric bud. The bifid renal pelvis and/or ureter may be unilateral or bilateral; however, separate openings into the bladder are uncommon. Incomplete division of the ureteric bud results in a bifid ureter; complete division results in a *supernumerary kidney* (Moore, Persaud and Torchia, 2012).

An uncommon anomaly is a *retrocaval ureter* (Fig. B2.32C), which leaves the kidney and passes posterior to the IVC.

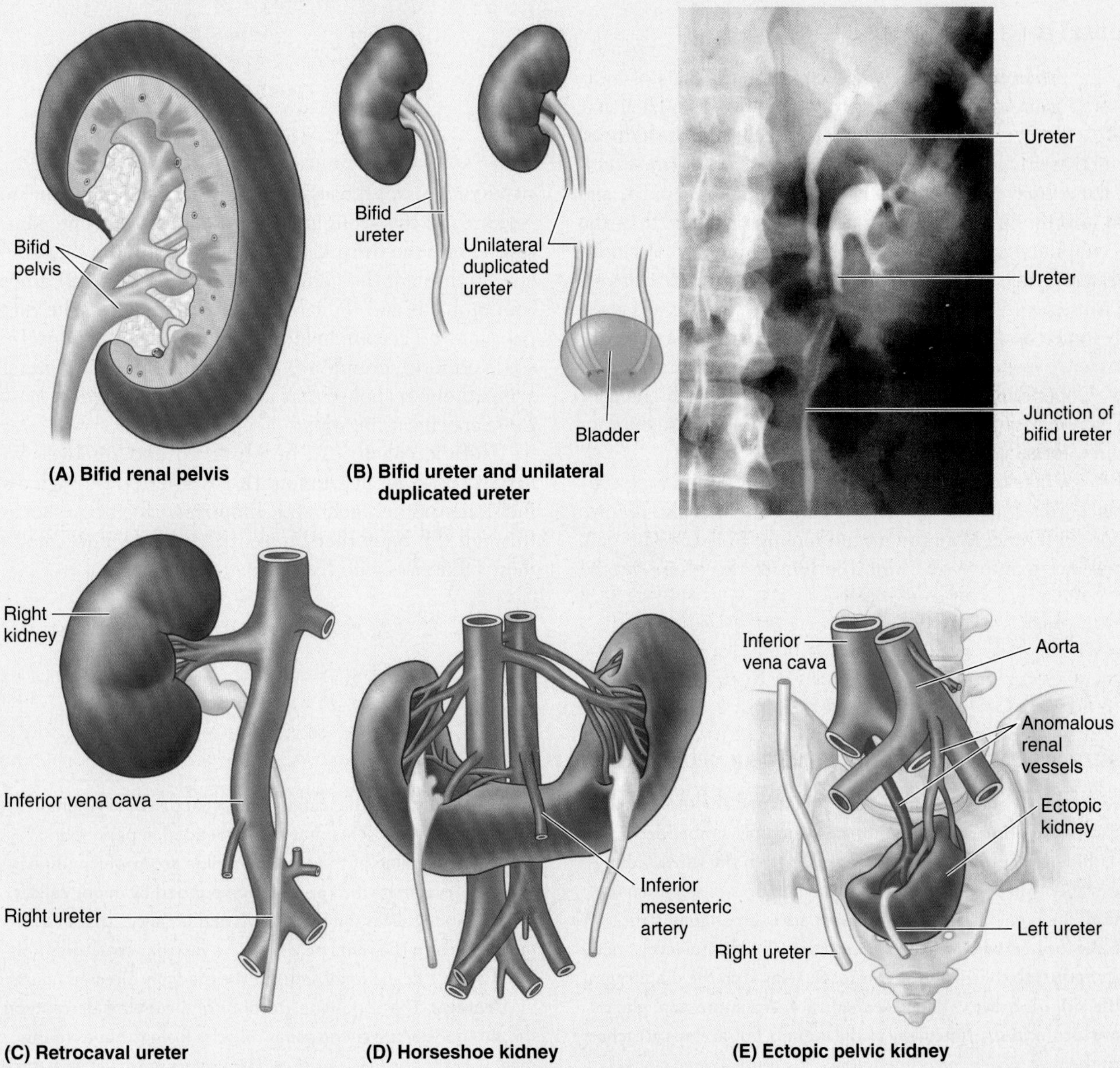

Bifid pelvis

(A) Bifid renal pelvis

Bifid ureter

Unilateral duplicated ureter

Bladder

(B) Bifid ureter and unilateral duplicated ureter

Ureter

Ureter

Junction of bifid ureter

Right kidney

Inferior vena cava

Right ureter

(C) Retrocaval ureter

Inferior mesenteric artery

Right ureter

(D) Horseshoe kidney

Inferior vena cava

Aorta

Anomalous renal vessels

Ectopic kidney

Left ureter

(E) Ectopic pelvic kidney

FIGURE B2.32.

The kidneys are close together in the embryonic pelvis. In approximately 1 in 600 fetuses, the inferior poles (rarely, the superior poles) of the kidneys fuse to form a *horseshoe kidney* (Fig. B2.32D). This U-shaped kidney usually lies at the level of L3–L5 vertebrae because the root of the *inferior mesenteric artery* prevented normal relocation of the kidneys. Horseshoe kidney usually produces no symptoms; however, associated abnormalities of the kidney and renal pelvis may be present, obstructing the ureter.

Sometimes the embryonic kidney on one or both sides fails to enter the abdomen and lies anterior to the sacrum. Although uncommon, awareness of the possibility of an *ectopic pelvic kidney* (Fig. B2.32E) should prevent it from being mistaken for a pelvic tumor and removed. A pelvic kidney in a woman also can be injured by or cause obstruction during childbirth. Pelvic kidneys usually receive their blood supply from the aortic bifurcation or a common iliac artery.

Renal and Ureteric Calculi

Calculi (L. pebbles) are composed of salts of inorganic or organic acids or of other materials. They may form and become located in the calices of the kidneys, ureters, or urinary bladder (Fig. B2.33). A *renal calculus* (kidney stone) may pass from the kidney into the renal pelvis and then into the ureter. If the stone is sharp or larger than the normal lumen of the ureter (approximately 3 mm), it causes excessive distension of this muscular tube, the *ureteric calculus* will cause severe intermittent pain (*ureteric colic*) as it is gradually forced down the ureter by waves of contraction. The calculus may cause complete or intermittent obstruction of urinary flow. Depending on the level of obstruction, which changes, the pain may be referred to the lumbar or inguinal regions, or to the external genitalia and/or testis.

The pain is referred to the cutaneous areas innervated by spinal cord segments and sensory ganglia, which also receive visceral afferents from the ureter, mainly T11–L2. The pain passes inferoanteriorly "from the loin to the groin" as the

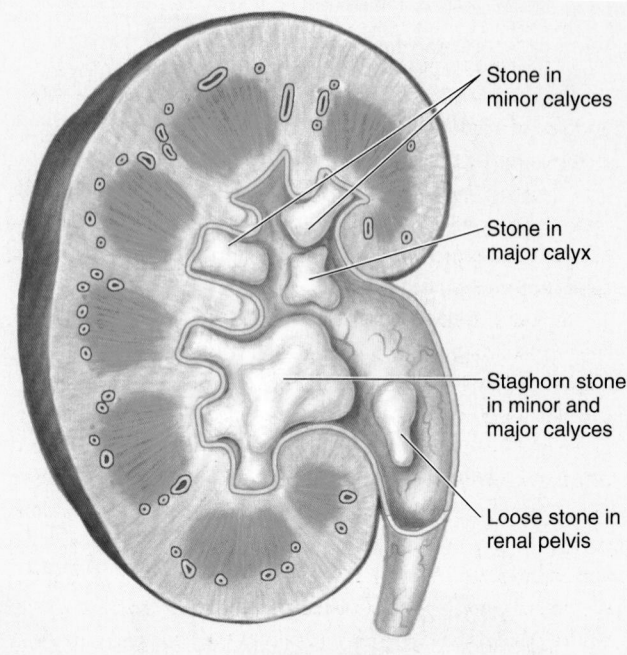

Stone in minor calyces

Stone in major calyx

Staghorn stone in minor and major calyces

Loose stone in renal pelvis

FIGURE B2.33.

stone progresses through the ureter. (The loin is the lumbar region, and the groin is the inguinal region.) The pain may extend into the proximal anterior aspect of the thigh by projection through the genitofemoral nerve (L1, L2), the scrotum in males and the labia majora in females. The extreme pain may be accompanied by marked digestive upset (nausea, vomiting, cramping, and diarrhea) and a generalized sympathetic response that may to various degrees mask the more specific symptoms.

Ureteric calculi can be observed and removed with a *nephroscope,* an instrument that is inserted through a small incision. Another technique, *lithotripsy,* focuses a shockwave through the body that breaks the calculus into small fragments that pass with the urine.

The Bottom Line

RETROPERITONEAL VISCERA AND THEIR NEUROVASCULATURE

Kidneys: The abdominal urinary organs and the suprarenal glands are primary retroperitoneal structures, embedded within perinephric fat that is separated from the surrounding extraperitoneal paranephric fat by a membranous condensation, the renal fascia. ♦ The kidneys are bean-shaped structures located between the T12 and the L3 vertebral levels, deep (anterior) to the 12th ribs. ♦ Closely related to the diaphragm, the kidneys move with its excursions. ♦ The suprarenal glands are located superomedially to the kidneys but are not attached to them. ♦ The kidneys are hollow. The central renal sinus

is occupied by the renal calices and renal pelvis, segmental arteries, and renal veins that are embedded in perinephric fat. ♦ The papillae of the renal pyramids, from which urine is excreted, evaginate into and are surrounded by minor calices. ♦ The minor calices merge to form major calices that in turn merge to form the renal pelvis. ♦ The vascular structures and renal pelvis exit the renal sinus at the medially directed hilum.

Ureters: The abdominal portions of the ureters descend on the anterior surface of the psoas muscles from the apex of the renal pelvis to the pelvic brim. ♦ The ureters normally have three sites of relative constriction, where kidney stones may lodge: the ureteropelvic junction, pelvic brim, and bladder wall. ♦ The abdominal portions of the ureters receive multiple, relatively delicate ureteric branches from the renal, testicular or ovarian, and common iliac arteries and from the abdominal aorta, which approach the ureters medially. ♦ A vertical line 5 cm lateral to the lumbar spinous processes, intersecting the posterior superior iliac spine, approximates the position of the ureter.

Suprarenal glands: The suprarenal glands are located superomedially to the kidneys but are attached primarily to the diaphragmatic crura by the surrounding renal fascia. ♦ Each suprarenal gland is actually two endocrine glands of different origin and function: suprarenal cortex and suprarenal medulla (the latter surrounded by the former). ♦ The suprarenal cortex derives from mesoderm and secretes corticosteroids and androgens; the suprarenal medulla derives from neural crest cells and secretes catecholamines (mostly epinephrine). ♦ The right suprarenal gland is more pyramidal in shape and apical in position relative to the right kidney, whereas the left gland is more crescentic and lies more medial to the superior half of the kidney.

Neurovasculature: The renal arteries arise from the abdominal aorta at the level of the L1–L2 IV disc. They lie anterior to the renal veins, with the right renal artery being longer than the left, and the left renal vein being longer than the right. ♦ Both renal veins receive renal and superior ureteric veins

and drain into the IVC, but the long left vein also receives the left suprarenal vein, the left gonadal vein, and a communication with the left ascending lumbar vein. ♦ Near the hilum, the renal arteries divide into anterior and posterior branches, the anterior branches giving rise to four segmental renal arteries. ♦ The segmental renal arteries are end arteries, each supplying a surgically resectable renal segment.

Suprarenal arteries arise from three sources: superior suprarenal arteries from the inferior phrenic arteries, middle suprarenal arteries from the abdominal aorta, and inferior suprarenal arteries from the renal arteries. ♦ The suprarenal glands drain via one large suprarenal vein, the right entering the IVC and the left entering the left renal vein.

Lymphatics from the suprarenal glands, kidneys, and upper ureters follow the venous drainage to the right or left lumbar (caval or aortic) lymph nodes.

Visceral afferent fibers (accompanying the sympathetic fibers) conduct pain sensation from the ureters to spinal cord segments T11–L2, with sensation referred to the corresponding dermatomes overlying the loin and groin. The suprarenal glands receive a rich nerve supply via presynaptic sympathetic fibers originating in the IMLs of the T10–L1 spinal cord segments. These fibers traverse both the paravertebral (sympathetic trunks) and prevertebral (celiac) ganglia without synapsing. They terminate directly on the chromaffin cells of the suprarenal medulla.

Summary of Innervation of Abdominal Viscera

For autonomic innervation of the abdominal viscera, several different splanchnic nerves and one cranial nerve (the vagus, CN X) deliver presynaptic sympathetic and parasympathetic fibers, respectively, to the abdominal aortic plexus and its associated sympathetic ganglia (Figs. 2.88 and 2.89; Table 2.12). The peri-arterial extensions of these plexuses deliver postsynaptic sympathetic fibers and the continuations of parasympathetic fibers to the abdominal viscera, where intrinsic parasympathetic ganglia occur.

SYMPATHETIC INNERVATION

The sympathetic part of the autonomic innervation of the abdominal viscera consists of the:

- Abdominopelvic splanchnic nerves from the thoracic and abdominal sympathetic trunks.
- Prevertebral sympathetic ganglia.
- Abdominal aortic plexus and its extensions, the peri-arterial plexuses.

The nerve plexuses are mixed, shared with the parasympathetic nervous system and visceral afferent fibers.

The **abdominopelvic splanchnic nerves** convey presynaptic sympathetic fibers to the abdominopelvic cavity. The fibers arise from cell bodies in the IMLs (or lateral horns) of

the gray matter of spinal cord segments T5–L2 or L3. The fibers pass successively through the anterior roots, anterior rami, and white communicating branches of thoracic and upper lumbar spinal nerves to reach the sympathetic trunks. They pass through the paravertebral ganglia of the trunks without synapsing to enter the abdominopelvic splanchnic nerves, which convey them to the prevertebral ganglia of the abdominal cavity. The abdominopelvic splanchnic nerves include the:

- *Lower thoracic splanchnic nerves* (greater, lesser, and least): from the thoracic part of the sympathetic trunks.
- *Lumbar splanchnic nerves:* from the lumbar part of the sympathetic trunks.

The *lower thoracic splanchnic nerves* are the main source of presynaptic sympathetic fibers serving abdominal viscera. The **greater splanchnic nerve** (from the sympathetic trunk at T5 through T9 or T10 vertebral levels), **lesser splanchnic nerve** (from T10 and T11 levels), and **least splanchnic nerve** (from the T12 level) are the specific abdominopelvic splanchnic nerves that arise from the thoracic part of the sympathetic trunks. They pierce the corresponding crus of the diaphragm to convey presynaptic sympathetic fibers to the celiac, superior mesenteric, and aorticorenal (prevertebral) sympathetic ganglia, respectively.

The **lumbar splanchnic nerves** arise from the abdominal part of the sympathetic trunks. Medially, the lumbar sympathetic trunks give off three to four lumbar splanchnic nerves, which pass to the *intermesenteric, inferior mesenteric,* and *superior*

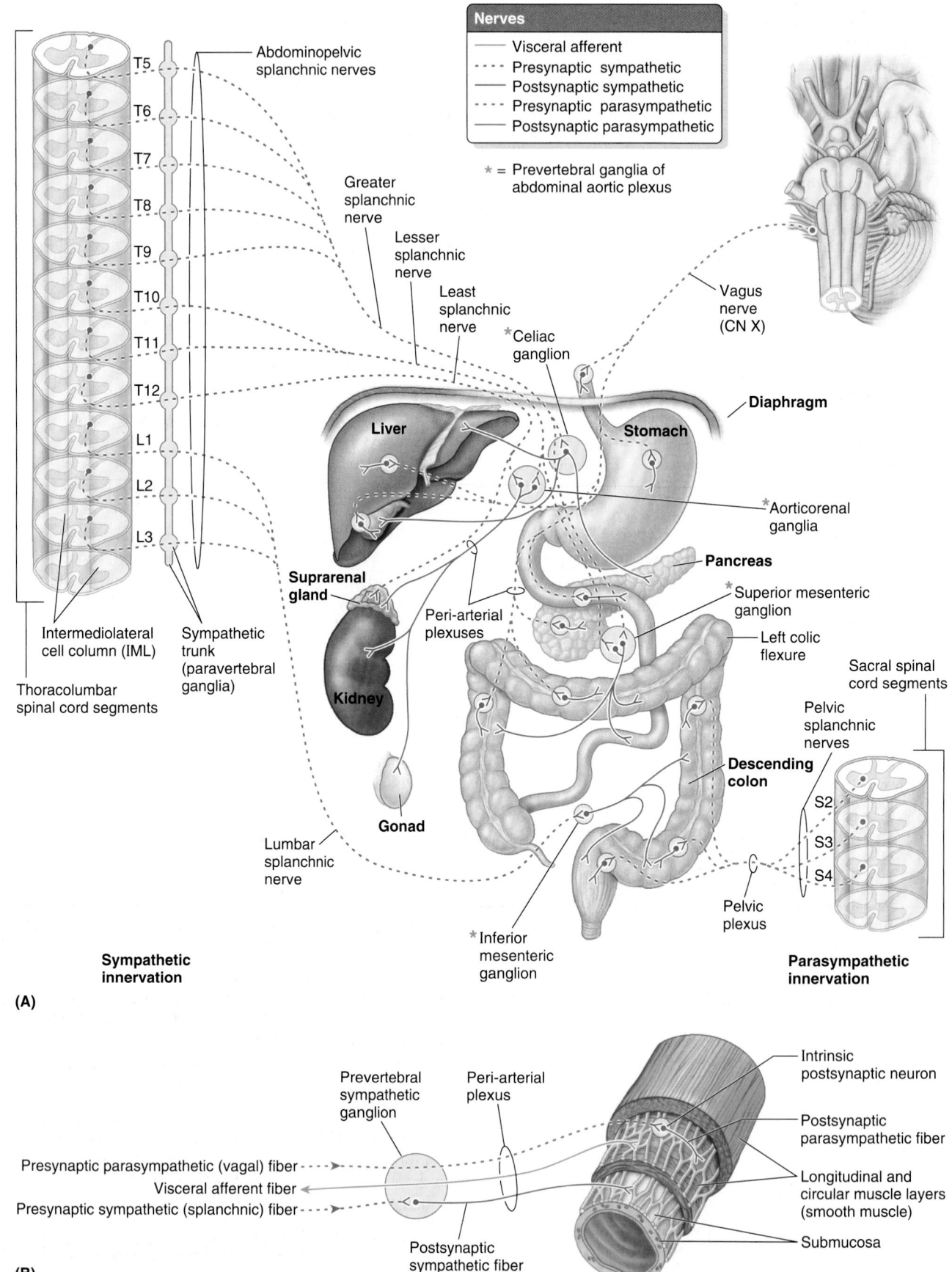

FIGURE 2.88. Autonomic nerves of posterior abdominal wall. A. Origin and distribution of presynaptic and postsynaptic sympathetic and parasympathetic fibers, and the ganglia involved in supplying abdominal viscera are shown. **B.** The fibers supplying the intrinsic plexuses of abdominal viscera are demonstrated.

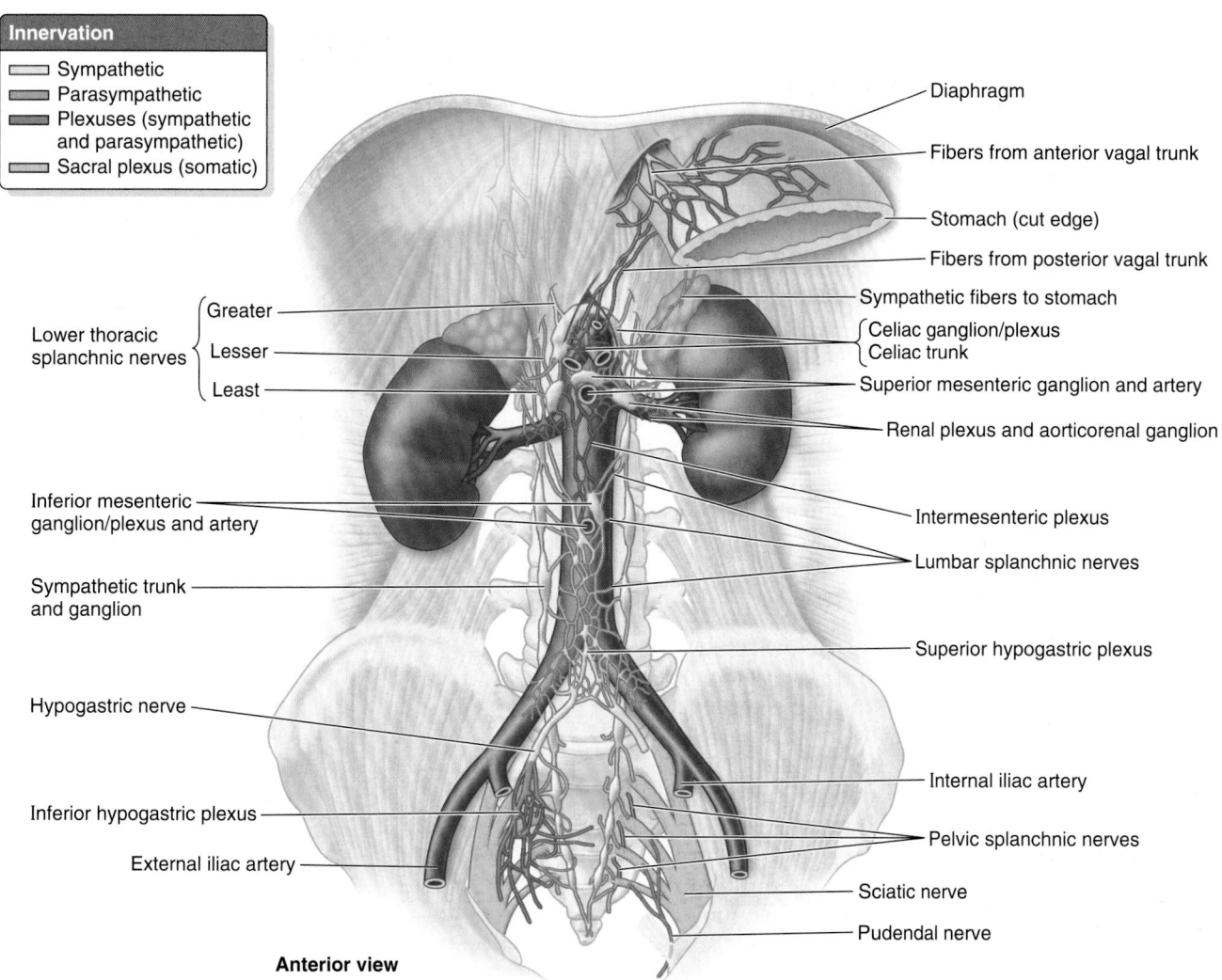

Innervation
- ☐ Sympathetic
- ☐ Parasympathetic
- ☐ Plexuses (sympathetic and parasympathetic)
- ☐ Sacral plexus (somatic)

Diaphragm

Fibers from anterior vagal trunk

Stomach (cut edge)

Fibers from posterior vagal trunk

Sympathetic fibers to stomach

Celiac ganglion/plexus
Celiac trunk

Superior mesenteric ganglion and artery

Renal plexus and aorticorenal ganglion

Intermesenteric plexus

Lumbar splanchnic nerves

Superior hypogastric plexus

Internal iliac artery

Pelvic splanchnic nerves

Sciatic nerve

Pudendal nerve

Lower thoracic splanchnic nerves { Greater / Lesser / Least }

Inferior mesenteric ganglion/plexus and artery

Sympathetic trunk and ganglion

Hypogastric nerve

Inferior hypogastric plexus

External iliac artery

Anterior view

FIGURE 2.89. Splanchnic nerves, nerve plexuses, and sympathetic ganglia in the abdomen.

TABLE 2.12.   AUTONOMIC INNERVATION OF ABDOMINAL VISCERA (SPLANCHNIC NERVES)

Splanchnic Nerves	Autonomic Fiber Type[a]	System	Origin	Destination
A. Cardiopulmonary (cervical and upper thoracic)	Postsynaptic		Cervical and upper thoracic sympathetic trunk	Thoracic cavity (viscera superior to level of diaphragm)

B. Abdominopelvic
 	
1. Lower thoracic
  a. Greater
  b. Lesser
  c. Least
2. Lumbar
3. Sacral
	Presynaptic	Sympathetic	Lower thoracic and abdomino-pelvic sympathetic trunk:
			
			1. Thoracic sympathetic trunk:
			  a. T5–T9 or T10 level
			  b. T10–T11 level
			  c. T12 level
			2. Abdominal sympathetic trunk
			3. Pelvic (sacral) sympathetic trunk
				Abdominopelvic cavity (prevertebral ganglia serving viscera and suprarenal glands inferior to level of diaphragm)
				
				1. Abdominal prevertebral ganglia:
				  a. Celiac ganglia
				  b. Aorticorenal ganglia
				  c. and 2. Other abdominal prevertebral ganglia (superior and inferior mesenteric, and of intermesenteric/hypogastric plexuses)
				3. Pelvic prevertebral ganglia

C. Pelvic	Presynaptic	Parasympathetic	Anterior rami of S2–S4 spinal nerves	Intrinsic ganglia of descending and sigmoid colon, rectum, and pelvic viscera

[a]Splanchnic nerves also convey visceral afferent fibers, which are not part of the autonomic nervous system.

hypogastric plexuses, conveying presynaptic sympathetic fibers to the associated prevertebral ganglia of those plexuses.

The cell bodies of postsynaptic sympathetic neurons constitute the major prevertebral ganglia that cluster around the roots of the major branches of the abdominal aorta: the **celiac, aorticorenal, superior mesenteric,** and **inferior mesenteric ganglia.** Minor, unnamed prevertebral ganglia occur within the intermesenteric and superior hypogastric plexuses. With the exception of the innervation of the suprarenal medulla (discussed on p. 294), the synapse between presynaptic and postsynaptic sympathetic neurons occurs in the prevertebral ganglia (Fig. 2.88B). Postsynaptic sympathetic nerve fibers pass from the prevertebral ganglia to the abdominal viscera by means of the peri-arterial plexuses associated with the branches of the abdominal aorta. Sympathetic innervation in the abdomen, as elsewhere, is primarily involved in producing vasoconstriction. With regard to the gastrointestinal tract, it acts to inhibit (slow down or stop) peristalsis.

VISCERAL SENSORY INNERVATION

Visceral afferent fibers conveying pain sensations accompany the sympathetic (visceral motor) fibers. The pain impulses pass retrogradely to those of the motor fibers along the splanchnic nerves to the sympathetic trunk, through white communicating branches to the anterior rami of the spinal nerves. Then they pass into the posterior root to the spinal sensory ganglia and spinal cord. Progressively lower spinal sensory ganglia and spinal cord segments are involved in innervating the abdominal viscera as the tract proceeds caudally. The stomach (foregut) receives innervation from the T6 to T9 levels, small intestine through transverse colon (midgut) from the T8 to 12 levels, and descending colon (hindgut) from the T12 to L2 levels (Fig. 2.90). Starting from the midpoint of the sigmoid colon, visceral pain fibers run with parasympathetic fibers, the sensory impulses being conducted to S2–S4 sensory ganglia and spinal cord levels. These are the same spinal cord segments involved in the sympathetic innervation of those portions of alimentary tract.

Visceral afferent fibers conveying reflex sensations (that generally do not reach levels of consciousness) accompany the parasympathetic (visceral motor) fibers.

PARASYMPATHETIC INNERVATION

The parasympathetic part of the autonomic innervation of the abdominal viscera (Figs. 2.88 and 2.89) consists of the following:

- *Anterior* and *posterior vagal trunks.*
- *Pelvic splanchnic nerves.*
- *Abdominal* (para-aortic) *autonomic plexuses* and their extensions, the peri-arterial plexuses.
- *Intrinsic* (enteric) *parasympathetic ganglia.*

The nerve plexuses are mixed, shared with the sympathetic nervous system and visceral afferent fibers.

The *anterior* and *posterior vagal trunks* are the continuation of the left and right vagus nerves that emerge from the

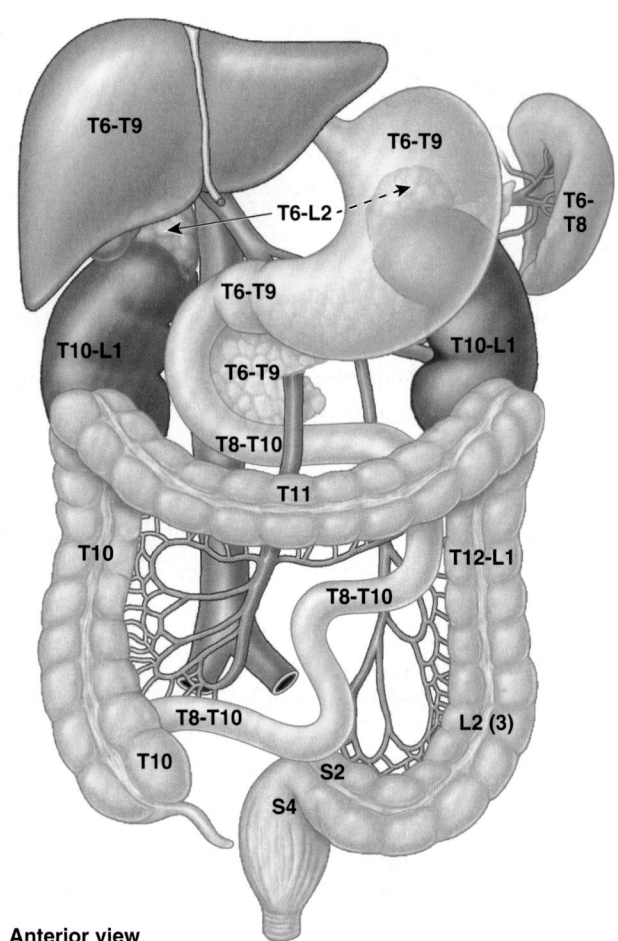

Anterior view

FIGURE 2.90. Segmental innervation of abdominal viscera. Approximate spinal cord segments and spinal sensory ganglia involved in sympathetic and visceral afferent (pain) innervation of abdominal viscera are shown.

esophageal plexus and pass through the esophageal hiatus on the anterior and posterior aspects of the esophagus and stomach (see Figs. 2.35 and 2.88A). The vagus nerves convey presynaptic parasympathetic and visceral afferent fibers (mainly for unconscious sensations associated with reflexes) to the abdominal aortic plexuses and the peri-arterial plexuses, which extend along the branches of the aorta.

The **pelvic splanchnic nerves** are distinct from other splanchnic nerves (Table 2.12) in that they:

- Have nothing to do with the sympathetic trunks.
- Derive directly from anterior rami of spinal nerves S2–S4.
- Convey presynaptic parasympathetic fibers to the inferior hypogastric (pelvic) plexus.

Presynaptic fibers terminate on the isolated and widely scattered cell bodies of postsynaptic neurons lying on or within the abdominal viscera, constituting *intrinsic (or, in the case of the GI tract, enteric) ganglia* (Fig. 2.88B).

The presynaptic parasympathetic and visceral afferent reflex fibers conveyed by the vagus nerves extend to intrinsic ganglia of the lower esophagus, stomach, small intestine, including the duodenum, ascending colon, and most of the transverse colon (see Fig. 2.88A). The fibers conveyed by the pelvic splanchnic nerves supply the descending and sigmoid parts of the colon, rectum, and pelvic organs. Thus, in terms of the gastrointestinal tract, the vagus nerves provide parasympathetic innervation of the smooth muscle and glands of the gut as far as the left colic flexure; the pelvic splanchnic nerves provide the remainder.

AUTONOMIC PLEXUSES

The **abdominal autonomic plexuses** are nerve networks consisting of both sympathetic and parasympathetic fibers, which surround the abdominal aorta and its major branches (Figs. 2.88 and 2.89). The celiac, superior mesenteric, and inferior mesenteric plexuses are interconnected. The *prevertebral sympathetic ganglia* are scattered among the celiac and mesenteric plexuses. The *intrinsic (enteric) parasympathetic ganglia,* such as the **myenteric plexus** (Auerbach plexus) in the muscular coat of the stomach and intestine, are in the walls of the viscera (see Figs. 2.48A and 2.88B).

The **celiac plexus,** surrounding the root of the celiac (arterial) trunk, contains irregular right and left **celiac ganglia** (approximately 2 cm long) that unite superior and inferior to the celiac trunk (Figs. 2.88A and 2.89). The *parasympathetic root* of the celiac plexus is a branch of the *posterior vagal trunk*, which contains fibers from the right and left vagus nerves. The *sympathetic roots* of the plexus are the greater and lesser splanchnic nerves.

The **superior mesenteric plexus** and ganglion or ganglia surround the origin of the SMA. The plexus has one median and two lateral roots. The median root is a branch of the celiac plexus, and the lateral roots arise from the lesser

and least splanchnic nerves, sometimes with a contribution from the first lumbar ganglion of the sympathetic trunk.

The **inferior mesenteric plexus** surrounds the inferior mesenteric artery and gives offshoots to its branches. It receives a medial root from the intermesenteric plexus and lateral roots from the lumbar ganglia of the sympathetic trunks. An **inferior mesenteric ganglion** may also appear just inferior to the root of the inferior mesenteric artery.

The **intermesenteric plexus** is part of the aortic plexus of nerves between the superior and the inferior mesenteric arteries. It gives rise to renal, testicular or ovarian, and ureteric plexuses.

The **superior hypogastric plexus** is continuous with the intermesenteric plexus and the inferior mesenteric plexus and lies anterior to the inferior part of the abdominal aorta at its bifurcation (Table 2.12). Right and left **hypogastric nerves** join the superior hypogastric plexus to the inferior hypogastric plexus. The superior hypogastric plexus supplies *ureteric* and *testicular plexuses* and a plexus on each common iliac artery.

The **inferior hypogastric plexus** is formed on each side by a hypogastric nerve from the superior hypogastric plexus. The right and left plexuses are situated on the sides of the rectum, cervix of the uterus, and urinary bladder. The plexuses receive small branches from the superior sacral sympathetic ganglia and the sacral parasympathetic outflow from S2 through S4 sacral spinal nerves (*pelvic [parasympathetic] splanchnic nerves*). Extensions of the inferior hypogastric plexus send autonomic fibers along the blood vessels, which form visceral plexuses on the walls of the pelvic viscera (e.g., *rectal* and *vesical plexuses*).

The Bottom Line

INNERVATION OF ABDOMINAL VISCERA

Sympathetic innervation: Presynaptic sympathetic nerve fibers involved in innervating abdominal viscera arise from cell bodies in the lower two thirds of the IMLs (T5–T6 to L2–L3 spinal cord levels) and travel via spinal nerves, anterior rami, and white communicating branches to the sympathetic trunks. ♦ The fibers traverse the paravertebral ganglia of the trunks without synapsing, continuing as components of abdominopelvic splanchnic nerves. These nerves convey them to the abdominal aortic plexus, where they are joined by presynaptic parasympathetic fibers delivered by the vagus nerve. ♦ The sympathetic fibers pass to prevertebral ganglia, most of which are clustered around the major branches of the abdominal aorta. After synapsing within the ganglia, the postsynaptic sympathetic fibers join the presynaptic parasympathetic fibers, traveling via peri-arterial plexuses around the branches of the abdominal aorta to reach the viscera. ♦ A continuation of the abdominal aortic plexus inferior to the aortic bifurcation (the superior and inferior hypogastric plexuses) convey sympathetic innervation to most of the pelvic viscera. The sympathetic fibers mainly innervate the blood vessels of abdominal viscera and are inhibitory to the parasympathetic stimulation.

♦ The parasympathetic fibers synapse on or in the walls of the viscera with intrinsic postsynaptic parasympathetic neurons, which terminate on the smooth muscle or glands of the viscera.

Parasympathetic innervation: The vagus nerves supplies parasympathetic fibers to the digestive tract from the esophagus through the transverse colon. ♦ Pelvic splanchnic nerves supply the descending and sigmoid colon and rectum. ♦ Parasympathetic stimulation promotes peristalsis and secretion (although much of the latter is usually hormonally regulated).

Sensory innervation: Visceral afferent fibers follow the autonomic fibers retrograde to sensory ganglia. ♦ Afferent fibers conveying pain sensation from abdominal viscera orad (proximal) to the middle of the sigmoid colon run with the sympathetic fibers to thoracolumbar spinal sensory ganglia; all other visceral afferent fibers run with the parasympathetic fibers. Thus, visceral afferent fibers conveying reflex information from the gut orad to the middle of the sigmoid colon pass to vagal sensory ganglia; fibers conveying both pain and reflex information from the gut aborad (distal) to the middle of the sigmoid colon pass to spinal sensory ganglia S2–S4.

DIAPHRAGM

The **diaphragm** is a double-domed, musculotendinous partition separating the thoracic and abdominal cavities. Its mainly convex superior surface faces the thoracic cavity, and its concave inferior surface faces the abdominal cavity (Fig. 2.91A & B). The diaphragm is the chief muscle of inspiration (actually, of respiration altogether, because expiration is largely passive). It descends during inspiration; however, only its central part moves because its periphery, as the fixed origin of the muscle, attaches to the inferior margin of the thoracic cage and the superior lumbar vertebrae.

The *pericardium*, containing the heart, lies on the central part of the diaphragm, depressing it slightly (Fig. 2.92A). The diaphragm curves superiorly into **right** and **left domes;** normally the right dome is higher than the left dome owing to the presence of the liver. During expiration, the right dome reaches as high as the 5th rib and the left dome ascends to the 5th intercostal space. The level of the domes of the diaphragm varies according to the:

- Phase of respiration (inspiration or expiration).
- Posture (e.g., supine or standing).
- Size and degree of distension of the abdominal viscera.

The **muscular part of the diaphragm** is situated peripherally with fibers that converge radially on the trifoliate central aponeurotic part, the **central tendon** (see Fig. 2.91). The central tendon has no bony attachments and is incompletely divided into three leaves, resembling a wide cloverleaf (Fig. 2.91B). Although it lies near the center of the diaphragm, the central tendon is closer to the anterior part of the thorax.

The *caval opening* (vena caval foramen), through which the terminal part of the IVC passes to enter the heart, perforates the central tendon. The surrounding muscular part of the diaphragm forms a continuous sheet; however, for descriptive purposes it is divided into three parts, based on the peripheral attachments:

- **Sternal part:** consisting of two muscular slips that attach to the posterior aspect of the xiphoid process; this part is not always present.
- **Costal part:** consisting of wide muscular slips that attach to the internal surfaces of the inferior six costal cartilages and their adjoining ribs on each side; the costal parts form the right and left domes.
- **Lumbar part:** arising from two aponeurotic arches, the *medial* and *lateral arcuate ligaments,* and the three superior lumbar vertebrae; the lumbar part forms right and left muscular crura that ascend to the central tendon.

The **crura of the diaphragm** are musculotendinous bands that arise from the anterior surfaces of the bodies of the superior three lumbar vertebrae, the anterior longitudinal ligament, and the IV discs. The **right crus,** larger and longer than the left crus, arises from the first three or four lumbar vertebrae. The **left crus** arises from the first two or three lumbar vertebrae. Because it lies to the left of the midline, it is surprising to find that the *esophageal hiatus* is a formation

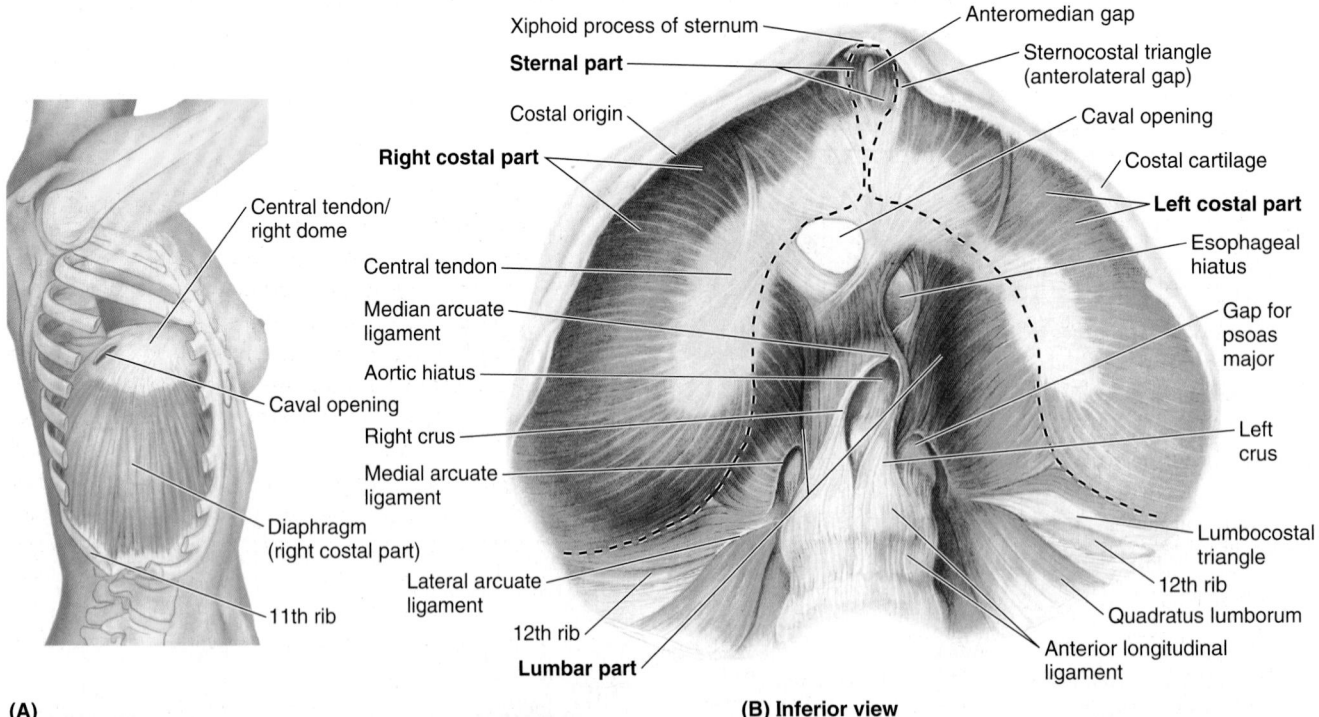

(A)

(B) Inferior view

FIGURE 2.91. Attachments, disposition, and features of the abdominal aspect of the diaphragm. A. The thoracic wall and cage have been removed to demonstrate the attachments and convexity of the right dome of the diaphragm. **B.** The fleshy sternal, costal, and lumbar parts of the diaphragm (outlined with broken lines) attach centrally to the trefoil-shaped central tendon, the aponeurotic insertion of the diaphragmatic muscle fibers.

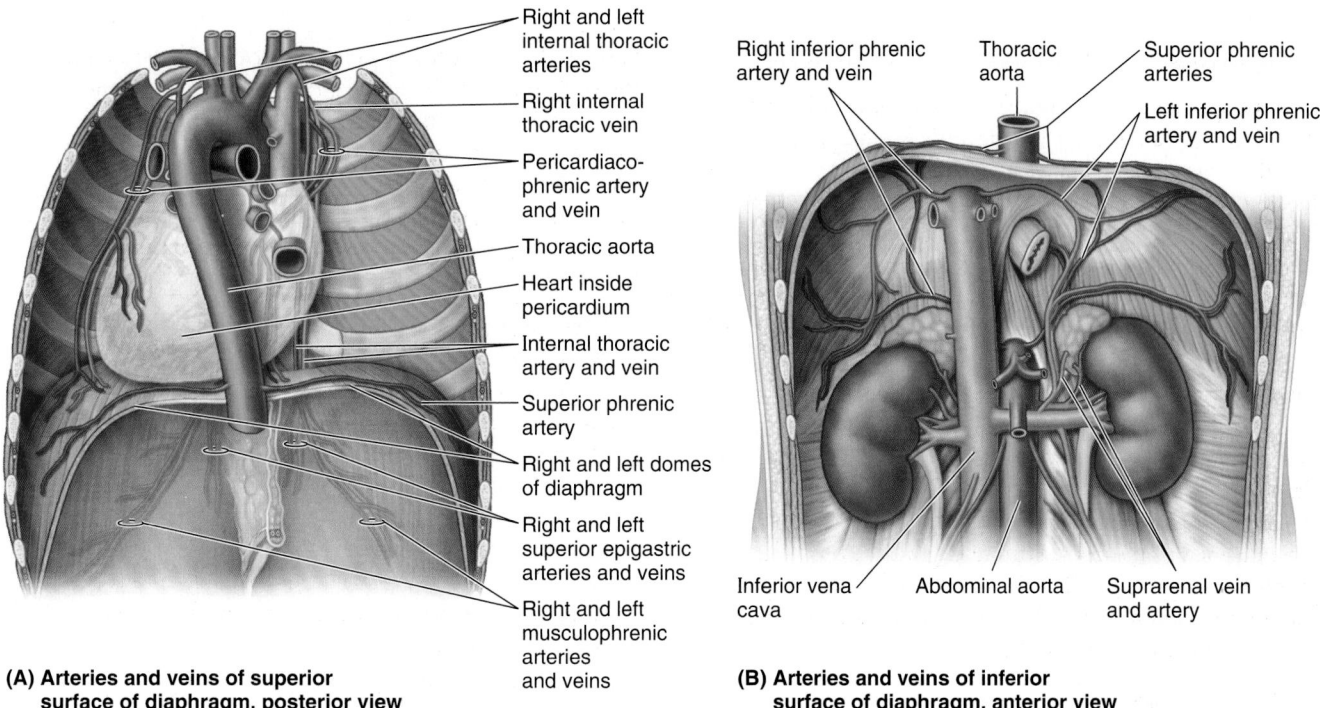

(A) Arteries and veins of superior surface of diaphragm, posterior view

Right and left internal thoracic arteries

Right internal thoracic vein

Pericardiacophrenic artery and vein

Thoracic aorta

Heart inside pericardium

Internal thoracic artery and vein

Superior phrenic artery

Right and left domes of diaphragm

Right and left superior epigastric arteries and veins

Right and left musculophrenic arteries and veins

(B) Arteries and veins of inferior surface of diaphragm, anterior view

Right inferior phrenic artery and vein

Thoracic aorta

Superior phrenic arteries

Left inferior phrenic artery and vein

Inferior vena cava

Abdominal aorta

Suprarenal vein and artery

FIGURE 2.92.   Blood vessels of diaphragm. A. The arteries and veins of the superior surface of the diaphragm. **B.** The arteries and veins of the inferior surface of the diaphragm.

in the right crus; however, if the muscular fibers bounding each side of the hiatus are traced inferiorly, it will be seen that they pass to the right of the aortic hiatus.

The right and left crura and the fibrous **median arcuate ligament,** which unites them as it arches over the anterior aspect of the aorta, form the *aortic hiatus.* The diaphragm is also attached on each side to the *medial* and *lateral arcuate ligaments.* The **medial arcuate ligament** is a thickening of the fascia covering the psoas major, spanning between the lumbar vertebral bodies and the tip of the transverse process of L1. The **lateral arcuate ligament** covers the quadratus lumborum muscles, continuing from the L12 transverse process to the tip of the 12th rib.

The superior aspect of the central tendon of the diaphragm is fused with the inferior surface of the fibrous pericardium, the strong, external part of the fibroserous *pericardial sac* that encloses the heart.

Vessels and Nerves of Diaphragm

The arteries of the diaphragm form a branch-like pattern on both its superior (thoracic) and inferior (abdominal) surfaces. The *arteries supplying the superior surface of the diaphragm* (Fig. 2.92; Table 2.13) are the *pericardiacophrenic* and *musculophrenic arteries,* branches of the internal thoracic artery, and the **superior phrenic arteries,** arising from the thoracic aorta. The *arteries supplying the inferior surface of the diaphragm* are the **inferior phrenic arteries,** which typically are the first branches of the *abdominal aorta;* however, they may arise from the celiac trunk.

The *veins draining the superior surface of the diaphragm* are the **pericardiacophrenic** and **musculophrenic veins,** which empty into the *internal thoracic veins* and, on the right side, a **superior phrenic vein,** which drains into the IVC. Some veins from the posterior curvature of the diaphragm drain into the *azygos and hemi-azygos veins* (see Chapter 1). The *veins draining the inferior surface of the diaphragm* are the *inferior phrenic veins.* The **right inferior phrenic vein** usually opens into the IVC, whereas the **left inferior phrenic vein** is usually double, with one branch passing anterior to the esophageal hiatus to end in the IVC and the other, more posterior branch usually joining the left suprarenal vein. The right and left phrenic veins may anastomose with each other.

The *lymphatic plexuses on the superior and inferior surfaces of the diaphragm* communicate freely (Fig. 2.93A). The **anterior** and **posterior diaphragmatic lymph nodes** are on the superior surface of the diaphragm. Lymph from these nodes drains into the *parasternal, posterior mediastinal,* and *phrenic lymph nodes.* Lymphatic vessels from the inferior surface of the diaphragm drain into the anterior diaphragmatic, phrenic, and superior *lumbar (caval/aortic) lymph nodes.* Lymphatic capillaries are dense on the inferior surface of the diaphragm, constituting the primary means for absorption of peritoneal fluid and substances introduced by intraperitoneal (I.P.) injection.

The entire *motor supply to the diaphragm* is from the *right* and *left phrenic nerves,* each of which arises from the anterior rami of C3–C5 segments of the spinal cord and is distributed to the ipsilateral half of the diaphragm from its

TABLE 2.13. NEUROVASCULAR STRUCTURES OF DIAPHRAGM

Vessels and Nerves	Superior Surface of Diaphragm	Inferior Surface of Diaphragm
Arterial supply	Superior phrenic arteries from thoracic aorta Musculophrenic and pericardiophrenic arteries from internal thoracic arteries	Inferior phrenic arteries from abdominal aorta
Venous drainage	Musculophrenic and pericardiacophrenic veins drain into internal thoracic veins; superior phrenic vein (right side) drains into IVC	Inferior phrenic veins; right vein drains into IVC; left vein is doubled and drains into IVC and suprarenal vein
Lymphatic drainage	Diaphragmatic lymph nodes to phrenic nodes, then to parasternal and posterior mediastinal nodes	Superior lumbar lymph nodes; lymphatic plexuses on superior and inferior surfaces communicate freely
Innervation	Motor supply: phrenic nerves (C3–C5) Sensory supply: centrally by phrenic nerves (C3–C5), peripherally by intercostal nerves (T5–T11) and subcostal nerves (T12)	

IVC, inferior vena cava.

inferior surface (Fig. 2.93B). *Sensory innervation (pain and proprioception) to the diaphragm* is also mostly from the *phrenic nerves.* Peripheral parts of the diaphragm receive their sensory nerve supply from the *intercostal nerves* (lower six or seven) and the subcostal nerves.

Diaphragmatic Apertures

The **diaphragmatic apertures** (openings, hiatus) permit structures (vessels, nerves, and lymphatics) to pass between the thorax and abdomen (Figs. 2.91, 2.92, and 2.94). There are three large apertures for the IVC, esophagus, and aorta and a number of small ones.

CAVAL OPENING

The **caval opening** is an aperture in the central tendon primarily for the IVC. Also passing through the caval opening are terminal branches of the right phrenic nerve and a few lymphatic vessels on their way from the liver to the middle phrenic and mediastinal lymph nodes. The caval opening is located to the right of the median plane at the junction of the central tendon's right and middle leaves. The most superior of the three large diaphragmatic apertures, the caval opening lies at the level of the IV disc between the T8 and T9 vertebrae. The IVC is adherent to the margin of the opening; consequently, when the diaphragm contracts during inspiration,

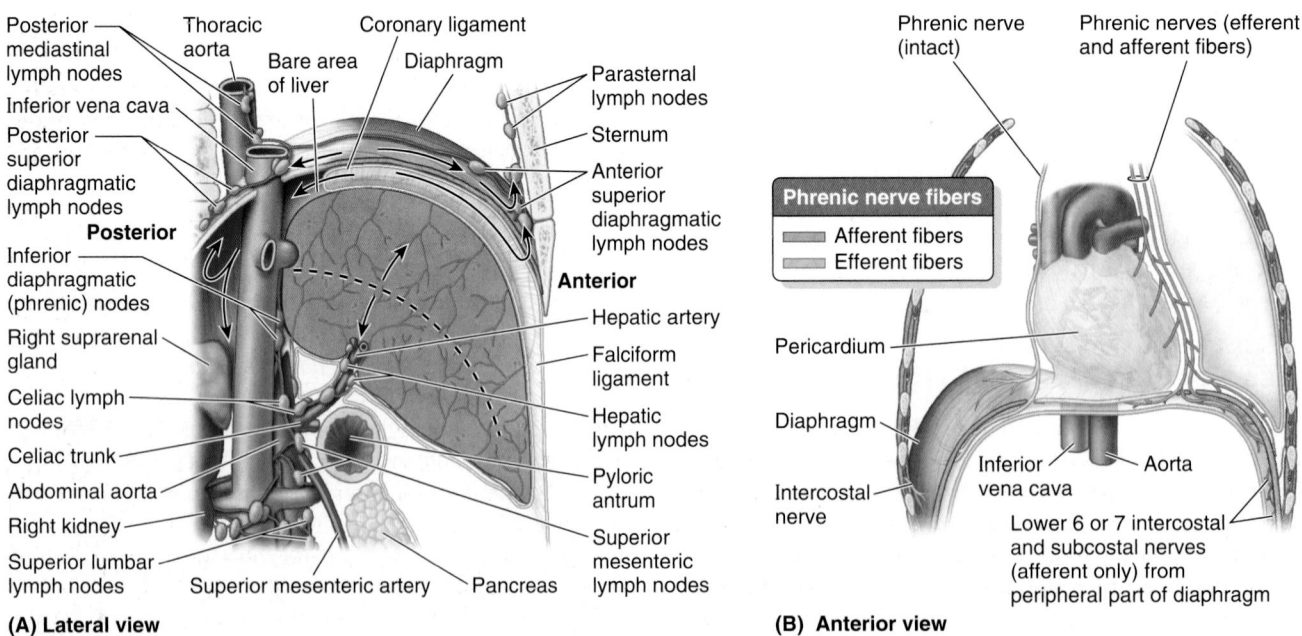

(A) Lateral view

(B) Anterior view

FIGURE 2.93. Lymphatic vessels, lymph nodes, and nerves of diaphragm. A. Lymphatic vessels are formed in two plexuses, one on the superior surface of the diaphragm and the other on its inferior surface; the plexuses communicate freely. **B.** The phrenic nerves supply all of the motor and most of the sensory innervation to the diaphragm. The lower six or seven intercostal and subcostal nerves provide sensory innervation peripherally.

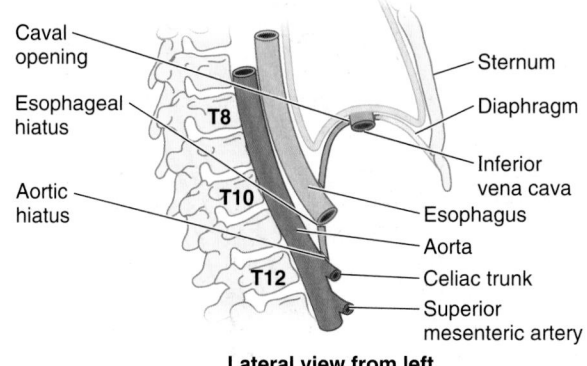

Lateral view from left

FIGURE 2.94. Apertures of diaphragm. "8-10-12" is a convenient memory device, referring to the thoracic vertebral levels at which the inferior vena cava, esophagus, and aorta penetrate the diaphragm.

it widens the opening and dilates the IVC. These changes facilitate blood flow through this large vein to the heart.

ESOPHAGEAL HIATUS

The **esophageal hiatus** is an oval opening for the esophagus in the muscle of the right crus of the diaphragm at the level of the T10 vertebra. The esophageal hiatus also transmits the anterior and posterior vagal trunks, esophageal branches of the left gastric vessels, and a few lymphatic vessels. The fibers of the right crus of the diaphragm decussate (cross one another) inferior to the hiatus, forming a muscular sphincter for the esophagus that constricts it when the diaphragm contracts. The esophageal hiatus is superior to and to the left of the aortic hiatus. In most individuals (70%), both margins of the hiatus are formed by muscular bundles of the right crus. In others (30%), a superficial muscular bundle from the left crus contributes to the formation of the right margin of the hiatus.

AORTIC HIATUS

The **aortic hiatus** is the opening posterior in the diaphragm for the descending aorta. Because the aorta does not pierce the diaphragm, movements of the diaphragm do not affect blood flow through the aorta during respiration. The aorta passes between the crura of the diaphragm posterior to the median arcuate ligament, which is at the level of the inferior border of the T12 vertebra. The aortic hiatus also transmits the thoracic duct and sometimes the azygos and hemi-azygos veins.

SMALL OPENINGS IN DIAPHRAGM

In addition to the three main apertures, there is a small opening, the **sternocostal triangle** (foramen), between the sternal and costal attachments of the diaphragm. This triangle transmits lymphatic vessels from the diaphragmatic surface of the liver and the superior epigastric vessels. The sympathetic trunks pass deep to the medial arcuate ligament, accompanied by the least splanchnic nerves. There are two small apertures in each crus of the diaphragm; one transmits the greater splanchnic nerve and the other the lesser splanchnic nerve.

Actions of Diaphragm

When the diaphragm contracts, its domes are pulled inferiorly so that the convexity of the diaphragm is somewhat flattened. Although this movement is often described as the "descent of the diaphragm," only the domes of the diaphragm descend. The diaphragm's periphery remains attached to the ribs and cartilages of the inferior six ribs. As the diaphragm descends, it pushes the abdominal viscera inferiorly. This increases the volume of the thoracic cavity and decreases the intrathoracic pressure, resulting in air being taken into the lungs. In addition, the volume of the abdominal cavity decreases slightly and intra-abdominal pressure increases somewhat.

Movements of the diaphragm are also important in circulation because the increased intra-abdominal pressure and decreased intrathoracic pressure help return venous blood to the heart. When the diaphragm contracts, compressing the abdominal viscera, blood in the IVC is forced superiorly into the heart.

The diaphragm is at its most superior level when a person is supine (with the upper body lowered, the *Trendelenburg position*). In this position, the abdominal viscera push the diaphragm superiorly in the thoracic cavity. When a person lies on one side, the hemidiaphragm rises to a more superior level because of the greater push of the viscera on that side. Conversely, the diaphragm assumes an inferior level when a person is sitting or standing. For this reason, people with dyspnea (difficult breathing) prefer to sit up, not lie down; non-tidal (reserve) lung volume is increased, and the diaphragm is working with gravity rather than opposing it.

POSTERIOR ABDOMINAL WALL

The posterior abdominal wall (Figs. 2.95–2.97) is mainly composed of the:

- Five lumbar vertebrae and associated IV discs (centrally).
- Posterior abdominal wall muscles, including the psoas, quadratus lumborum, iliacus, transversus abdominis, and oblique muscles (laterally).
- Diaphragm, which contributes to the superior part of the posterior wall.
- Fascia, including the thoracolumbar fascia.
- Lumbar plexus, composed of the anterior rami of lumbar spinal nerves.
- Fat, nerves, vessels (e.g., aorta and IVC), and lymph nodes.

If observing the anatomy of the posterior abdominal wall in only two-dimensional diagrams, such as Figure 2.97, it would be easy to suppose that it is flat. In observing a dissected cadaver or a transverse cross section such as that in Figures 2.95A & B, it is apparent that the lumbar vertebral column is a marked central prominence in the posterior wall, creating two paravertebral "gutters" on each side. The deepest (most posterior) part of these gutters is occupied by the kidneys and their surrounding fat. The abdominal aorta lies on the anterior aspect of the anteriorly protruding vertebral column. It is usually surprising

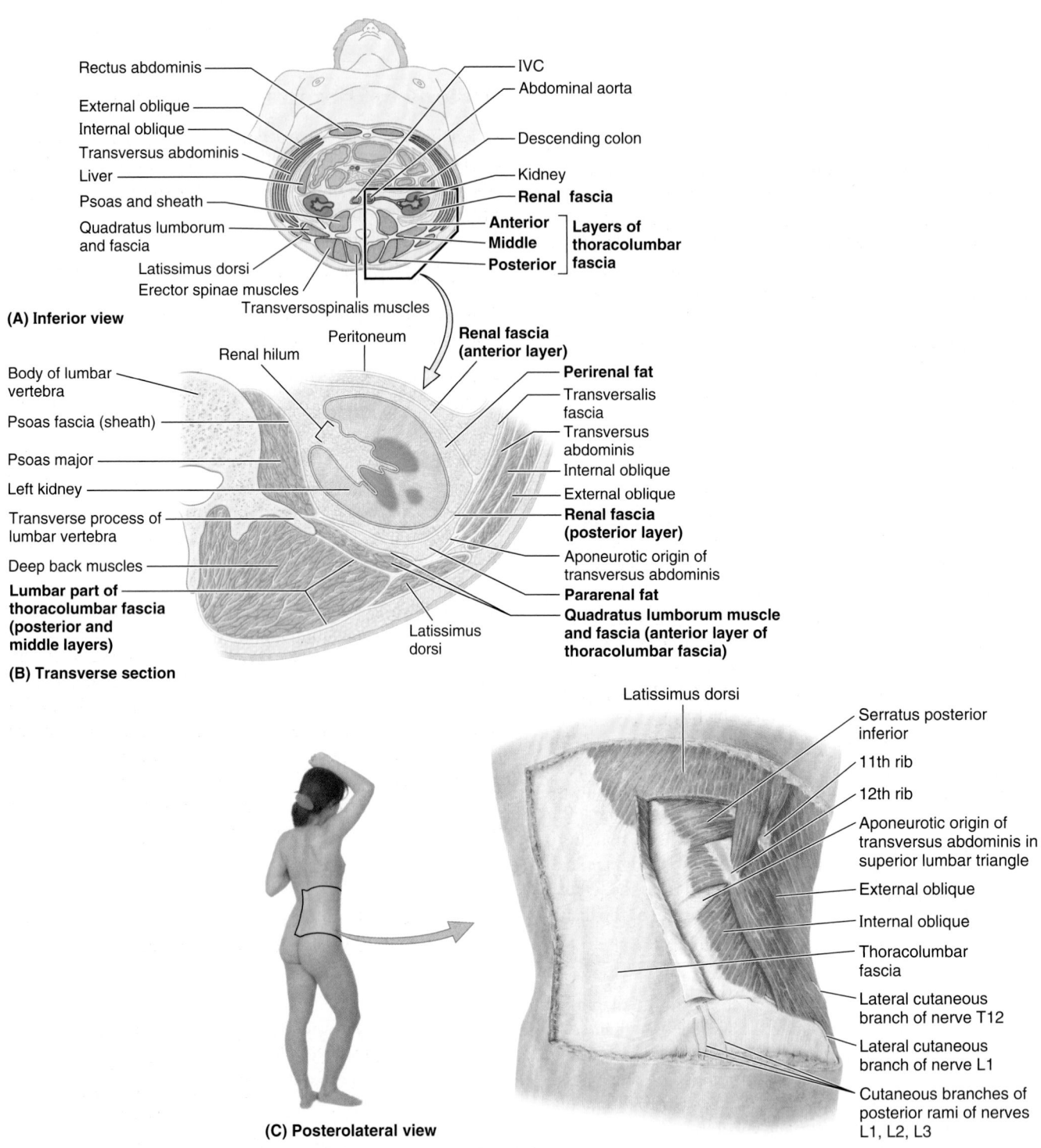

(A) Inferior view

Rectus abdominis
External oblique
Internal oblique
Transversus abdominis
Liver
Psoas and sheath
Quadratus lumborum and fascia
Latissimus dorsi
Erector spinae muscles
Transversospinalis muscles

IVC
Abdominal aorta
Descending colon
Kidney
Renal fascia
Anterior
Middle
Posterior
Layers of thoracolumbar fascia

(B) Transverse section

Body of lumbar vertebra
Psoas fascia (sheath)
Psoas major
Left kidney
Transverse process of lumbar vertebra
Deep back muscles
Lumbar part of thoracolumbar fascia (posterior and middle layers)

Peritoneum
Renal hilum
Renal fascia (anterior layer)
Perirenal fat
Transversalis fascia
Transversus abdominis
Internal oblique
External oblique
Renal fascia (posterior layer)
Aponeurotic origin of transversus abdominis
Pararenal fat
Quadratus lumborum muscle and fascia (anterior layer of thoracolumbar fascia)
Latissimus dorsi

(C) Posterolateral view

Latissimus dorsi
Serratus posterior inferior
11th rib
12th rib
Aponeurotic origin of transversus abdominis in superior lumbar triangle
External oblique
Internal oblique
Thoracolumbar fascia
Lateral cutaneous branch of nerve T12
Lateral cutaneous branch of nerve L1
Cutaneous branches of posterior rami of nerves L1, L2, L3

FIGURE 2.95. Fascia and aponeuroses of abdominal wall at level of renal hila. A. The relationships of the muscles, aponeurotic muscle sheaths, and fascia of the abdominal wall are demonstrated in transverse section. The three flat abdominal muscles forming the lateral walls span between complex anterior and posterior aponeurotic formations that ensheath vertically disposed muscles. The thin anterolateral walls (appearing disproportionately thick here) are distensible. Although flexible, the posterior abdominal wall is weight bearing and so is reinforced by the vertebral column and muscles that act on it; thus it is not distensible. *IVC,* inferior vena cava. **B.** Details of the disposition of the aponeurotic and fascial layers of the posterior abdominal wall. For details concerning those of the anterior abdominal wall, see Figure 2.5B. **C.** Dimensional view of the region demonstrated in section in **B.**

to find how close the lower abdominal aorta lies to the anterior abdominal wall in lean individuals (see Fig. B2.37C). Of course, many structures lie anterior to the aorta (SMA, parts of the duodenum, pancreas and left renal vein, etc.) and so these "posterior abdominal structures" may approach the anterior abdominal wall closer than might be expected in thin persons, especially when they are in the supine position.

Fascia of Posterior Abdominal Wall

The posterior abdominal wall is covered with a continuous layer of *endoabdominal fascia* that lies between the parietal peritoneum and the muscles (Fig. 2.95B). The fascia lining the posterior abdominal wall is continuous with the transversalis fascia that lines the transversus abdominis muscle. It is customary to name the fascia according to the structure it covers.

The *psoas fascia* covering the psoas major muscle (*psoas sheath*) is attached medially to the lumbar vertebrae and pelvic brim. The psoas fascia (sheath) is thickened superiorly to form the *medial arcuate ligament* (Fig. 2.91). The psoas fascia fuses laterally with the *quadratus lumborum* and *thoracolumbar fascias* (Fig. 2.95B). Inferior to the iliac crest, the psoas fascia is continuous with the part of the iliac fascia covering the iliacus.

The **thoracolumbar fascia** is an extensive fascial complex attached to the vertebral column medially that, in the lumbar region, has posterior, middle, and anterior layers with muscles enclosed between them (Fig. 2.95B & C). It is thin and transparent where it covers the thoracic parts of the deep muscles, but it is thick and strong in the lumbar region. The enclosure of the vertical deep back muscles (erector spinae) by the **posterior** and **middle layers of the thoracolumbar fascia** on the posterior aspect of the trunk is comparable to the enclosure of the rectus abdominis by the rectus sheath on the anterior aspect (Fig. 2.95A). This posterior sheath is even more formidable than the rectus sheath, however, because of the thickness of its posterior layer and the central attachment to the lumbar vertebrae, as opposed to the rectus sheaths, which lack bony support where they fuse to each other at the linea alba. The lumbar part of this posterior sheath, extending between the 12th rib and the iliac crest, attaches laterally to the internal oblique and transversus abdominis muscles, as does the rectus sheath. However, in contrast to the rectus sheath, the thoracolumbar fascia is not attached to the external oblique; it is attached to the latissimus dorsi (Fig. 2.95B & C).

The **anterior layer of the thoracolumbar fascia** (quadratus lumborum fascia), covering the anterior surface of the quadratus lumborum—a thinner, more transparent layer than the other two layers—attaches to the anterior surfaces

of the transverse processes of the lumbar vertebrae, the iliac crest, and the 12th rib (Figs. 2.95B and 2.97). The anterior layer is continuous laterally with the aponeurotic origin of the transversus abdominis muscle. It thickens superiorly to form the *lateral arcuate ligament* and is adherent inferiorly to the **iliolumbar ligaments** (Fig. 2.97).

Muscles of Posterior Abdominal Wall

The main paired muscles in the posterior abdominal wall (Fig. 2.96; Table 2.14) are the:

- *Psoas major:* passing inferolaterally.
- *Iliacus:* lying along the lateral sides of the inferior part of the psoas major.
- *Quadratus lumborum:* lying adjacent to the transverse processes of the lumbar vertebrae and lateral to superior parts of the psoas major.

The attachments, nerve supply, and main actions of these muscles are summarized in Table 2.14.

PSOAS MAJOR

The long, thick, fusiform **psoas major** lies lateral to the lumbar vertebrae (Figs. 2.96A and 2.97). *Psoas* is a Greek word meaning "muscle of the loin." (Butchers refer to the psoas of animals as the *tenderloin*.) The psoas major passes inferolaterally, deep to the inguinal ligament to reach the lesser trochanter of the femur. The lumbar plexus of nerves is embedded in the posterior part of the psoas major, anterior to the lumbar transverse processes.

ILIACUS

The **iliacus** is a large triangular muscle that lies along the lateral side of the inferior part of the psoas major. Most of

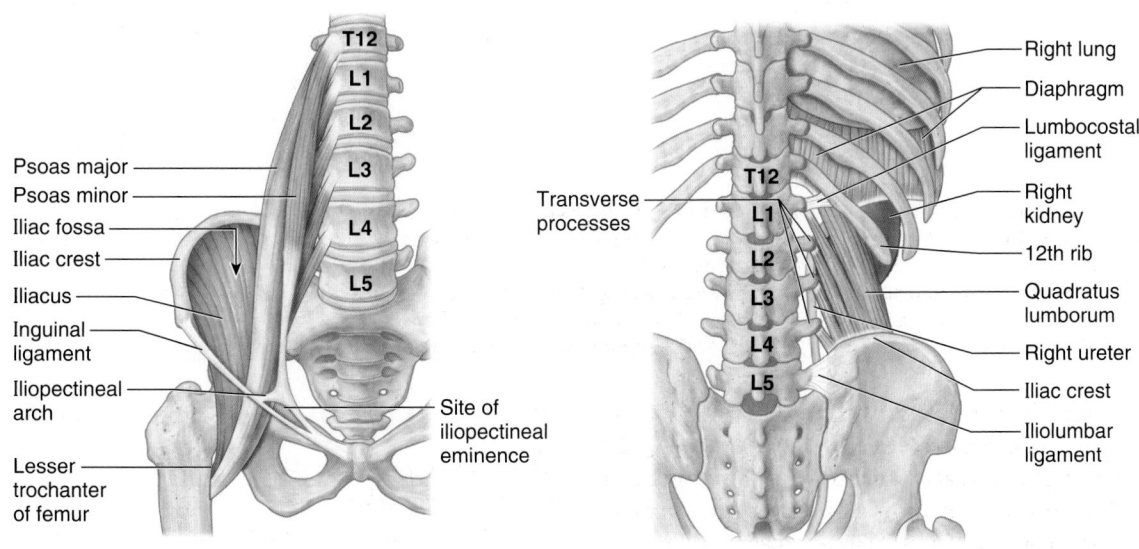

Anterior view **Posterior view**

FIGURE 2.96. Muscles of posterior abdominal wall.

TABLE 2.14. MUSCLES OF POSTERIOR ABDOMINAL WALL

Muscle	Superior Attachment	Inferior Attachment	Innervation	Main Action
Psoas major*	Transverse processes of lumbar vertebrae; sides of bodies of T12–L5 vertebrae and intervening intervertebral discs	By a strong tendon to lesser trochanter of femur	Anterior rami of lumbar nerves **L1**, **L2**, L3	Acting inferiorly with iliacus, it flexes thigh; acting superiorly it flexes vertebral column laterally; it is used to balance the trunk; when sitting it acts inferiorly with iliacus to flex trunk
Iliacus*	Superior two thirds of iliac fossa, ala of sacrum, and anterior sacro-iliac ligaments	Lesser trochanter of femur and shaft inferior to it, and to psoas major tendon	Femoral nerve (L2–L4)	Flexes thigh and stabilizes hip joint; acts with psoas major
Quadratus lumborum	Medial half of inferior border of 12th ribs and tips of lumbar transverse processes	Iliolumbar ligament and internal lip of iliac crest	Anterior branches of T12 and L1–L4 nerves	Extends and laterally flexes vertebral column; fixes 12th rib during inspiration

*Psoas minor and iliacus muscles merge inferiorly; collectively form iliopsoas muscle.

its fibers join the tendon of the psoas major. Together the psoas and iliacus form the **iliopsoas,** the chief flexor of the thigh. It is also a stabilizer of the hip joint and helps maintain the erect posture at this joint. The psoas and iliacus share in hip flexion; however, only the psoas can produce movement (flexion or lateral bending) of the lumbar vertebral column.

QUADRATUS LUMBORUM

The quadrilateral **quadratus lumborum** forms a thick muscular sheet in the posterior abdominal wall (Figs. 2.94A & B, 2.96, and 2.97). It lies adjacent to the lumbar transverse

processes and is broader inferiorly. Close to the 12th rib, the *lateral arcuate ligament* crosses the quadratus lumborum. The *subcostal nerve* passes posterior to this ligament and runs inferolaterally on the quadratus lumborum. Branches of the *lumbar plexus* run inferiorly on the anterior surface of this muscle.

Nerves of Posterior Abdominal Wall

Components of both the somatic and autonomic (visceral) nervous systems are associated with the posterior abdominal wall.

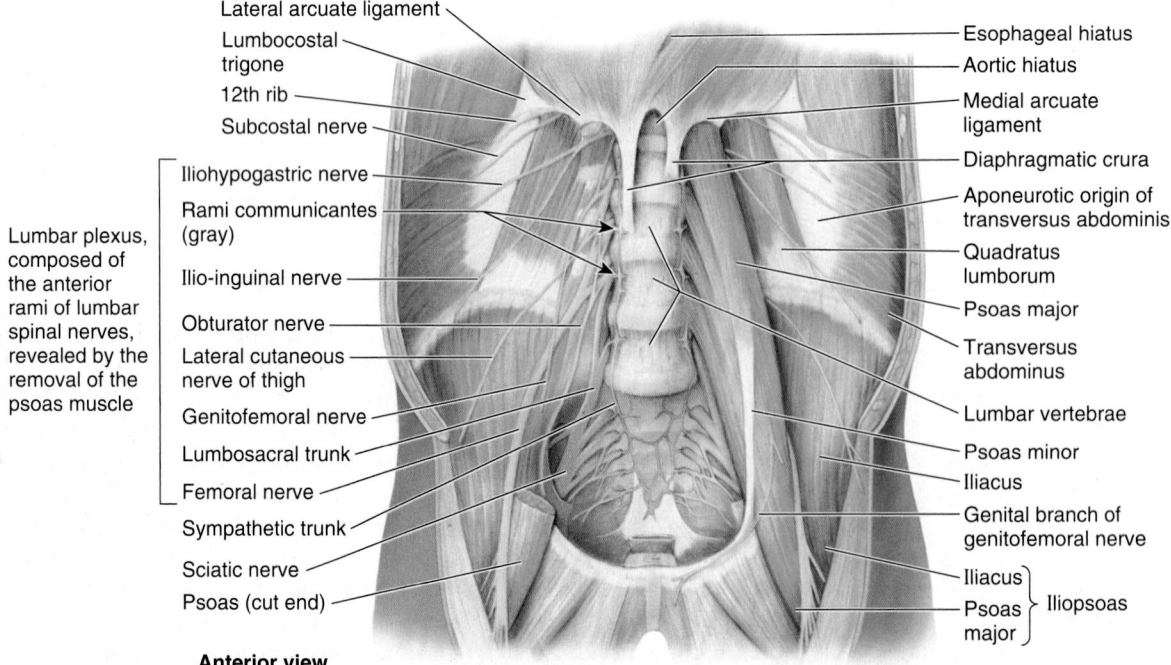

Anterior view

FIGURE 2.97. Muscles and nerves of posterior abdominal wall. Most of the right psoas major has been removed to show that the lumbar plexus of nerves is formed by the anterior rami of the first four lumbar spinal nerves and that it lies in the substance of the psoas major.

The *subcostal nerves* (anterior rami of T12) arise in the thorax, pass posterior to the lateral arcuate ligaments into the abdomen, and run inferolaterally on the anterior surface of the quadratus lumborum (Fig. 2.97). They pass through the transversus abdominis and internal oblique muscles to supply the external oblique and skin of the anterolateral abdominal wall.

The **lumbar spinal nerves** (L1–L5) pass from the spinal cord through the IV foramina inferior to the corresponding vertebrae, where they divide into posterior and anterior rami. Each ramus contains sensory and motor fibers. The posterior rami pass posteriorly to supply the muscles of the back and overlying skin, whereas the anterior rami pass laterally and inferiorly, to supply the skin and muscles of the inferiormost trunk and lower limb. The initial portions of the anterior rami of the L1, L2, and occasionally L3 spinal nerves give rise to *white communicating branches* (L. *rami communicantes*), which convey presynaptic sympathetic fibers to the lumbar sympathetic trunks.

The *abdominal part of the sympathetic trunks (lumbar sympathetic trunks)*, consisting of four lumbar *paravertebral sympathetic ganglia* and the *interganglionic branches* that connect them, are continuous with the thoracic part of the trunks deep to the medial arcuate ligaments of the diaphragm. The lumbar trunks descend on the anterolateral aspects of the bodies of the lumbar vertebrae in a groove formed by the adjacent psoas major. Inferiorly, they cross the sacral promontory and continue inferiorly into the pelvis as the sacral part of the trunks.

For the innervation of the abdominal wall and lower limbs, synapses between the presynaptic and postsynaptic fibers occur in the sympathetic trunks. Postsynaptic sympathetic fibers travel from the lateral aspect of the trunks via *gray communicating branches* to the anterior rami. They become the thoraco-abdominal and subcostal nerves, and the lumbar plexus (somatic nerves) that stimulate vasomotion, sudomotion, and pilomotion in the lowermost trunk and lower limb. *Lumbar splanchnic nerves* arising from the medial aspect of the lumbar sympathetic trunks convey presynaptic sympathetic fibers for the innervation of pelvic viscera.

The **lumbar plexus of nerves** is formed anterior to the lumbar transverse processes, within the proximal attachment of the psoas major. This nerve network is composed of the anterior rami of L1 through L4 nerves. The following nerves are branches of the lumbar plexus; the three largest are listed first:

- The *femoral nerve* (L2–L4) emerges from the lateral border of the psoas major and innervates the iliacus and passes deep to the inguinal ligament/iliopubic tract to the anterior thigh, supplying the flexors of the hip and extensors of the knee.
- The *obturator nerve* (L2–L4) emerges from the medial border of the psoas major and passes into the lesser pelvis, passing inferior to the superior pubic ramus (through the obturator foramen) to the medial thigh, supplying the adductor muscles.
- The *lumbosacral trunk* (L4, L5) passes over the ala (wing) of the sacrum and descends into the pelvis to participate

in the formation of the sacral plexus with the anterior rami of S1–S4 nerves.

- The **ilio-inguinal** and **iliohypogastric nerves** (L1) arise from the anterior ramus of L1, entering the abdomen posterior to the medial arcuate ligament and passing inferolaterally, anterior to the quadratus lumborum. They run superior and parallel to the iliac crest, piercing the transversus abdominis near the ASIS. They then pass through the internal and external obliques to supply the abdominal muscles and skin of the inguinal and pubic regions. The division of the L1 anterior ramus may occur as far distally as the ASIS, so that often only one nerve (L1) crosses the posterior abdominal wall instead of two.
- The *genitofemoral nerve* (L1, L2) pierces the psoas major and runs inferiorly on its anterior surface, deep to the psoas fascia; it divides lateral to the common and external iliac arteries into femoral and genital branches.
- The *lateral cutaneous nerve of the thigh,* or lateral femoral cutaneous nerve (L2, L3), runs inferolaterally on the iliacus and enters the thigh deep to the inguinal ligament/iliopubic tract, just medial to the ASIS; it supplies skin on the anterolateral surface of the thigh.
- An **accessory obturator nerve** (L3, L4) is present almost 10% of the time. It parallels the medial border of the psoas, anterior to the obturator nerve, crossing superior to the superior pubic ramus in close proximity to the femoral vein.

Although the larger branches (femoral, obturator, and lumbosacral trunk) are consistent in their placement, variation should be anticipated in the disposition of the smaller branches of the lumbar plexus.

Vessels of Posterior Abdominal Wall

The major neurovascular bundle of the inferior trunk, including the abdominal aorta, the inferior vena cava, and the aortic peri-arterial nerve plexus, courses in the midline of the posterior abdominal wall, anterior to the bodies of the lumbar vertebrae (see Figs. 2.70B and 2.89).

ABDOMINAL AORTA

Most arteries supplying the posterior abdominal wall arise from the abdominal aorta (Fig. 2.98A; Table 2.15). The *subcostal arteries* arise from the thoracic aorta and distribute inferior to the 12th rib. The **abdominal aorta** is approximately 13 cm in length. It begins at the *aortic hiatus* in the diaphragm at the level of the T12 vertebra and ends at the level of the L4 vertebra by dividing into the *right and left common iliac arteries*. The abdominal aorta may be represented on the anterior abdominal wall by a band (approximately 2 cm wide) extending from a median point, approximately 2.5 cm superior to the transpyloric plane to a point slightly (2–3 cm) inferior to and to the left of the umbilicus at the level of the *supracristal plane* (plane of the highest points of the iliac crests) (Fig. 2.98B). In children and lean adults, the lower

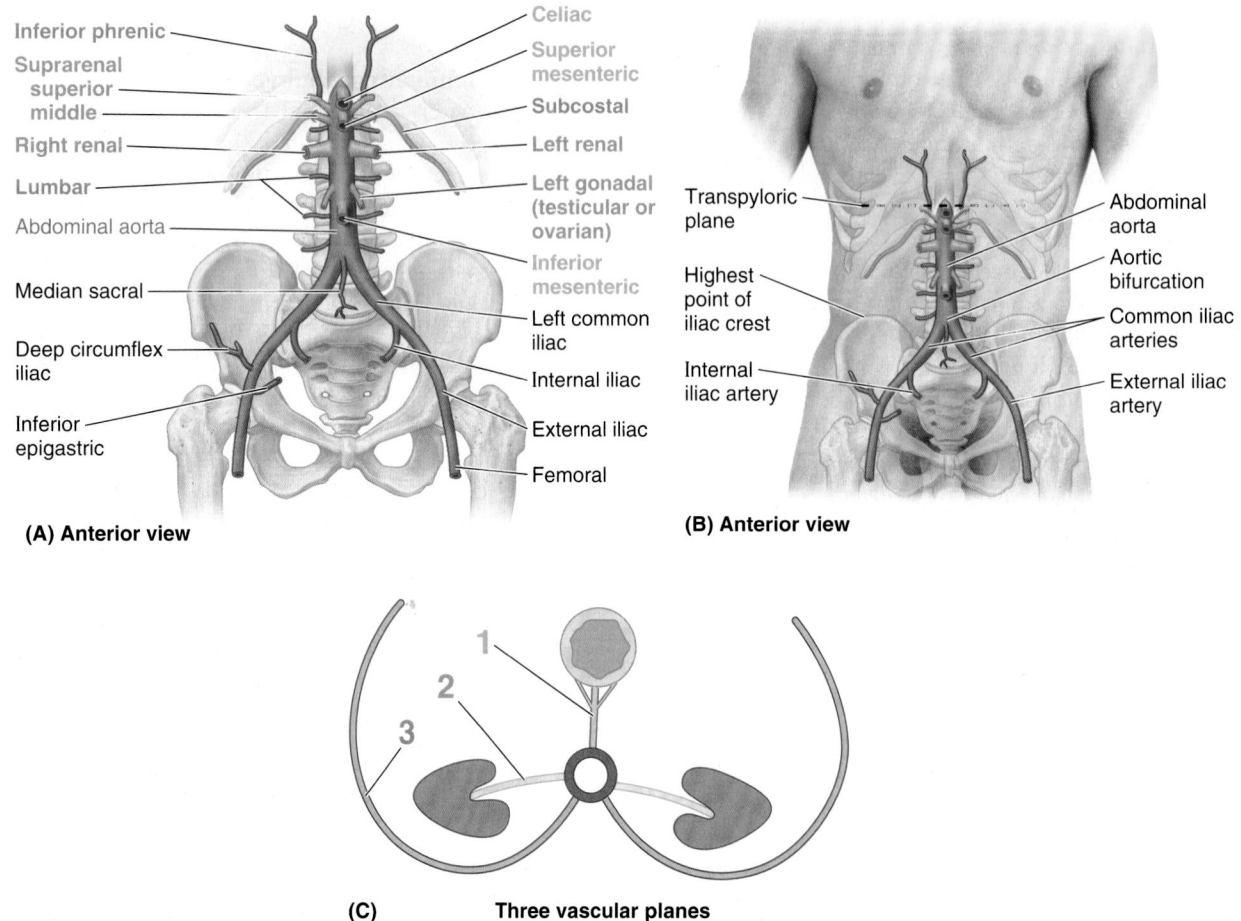

FIGURE 2.98. Arteries of posterior abdominal wall—branches of the aorta. A. Branches of abdominal aorta. **B.** Branches of upper abdominal aorta. **C.** Vascular planes in which branches of abdominal aorta are distributed.

TABLE 2.15. BRANCHES OF ABDOMINAL AORTA

Vascular Plane	Class	Distribution	Abdominal Branches (Arteries)	Vertebral Level
1. Anterior midline	Unpaired visceral	Digestive tract	Celiac	T12
			Superior mesenteric	L1
			Inferior mesenteric	L3
2. Lateral	Paired visceral	Urogenital and endocrine organs	Suprarenal	L1
			Renal	L1
			Gonadal (testicular or ovarian)	L2
3. Posterolateral	Paired parietal (segmental)	Diaphragm; body wall	Subcostal	L2
			Inferior phrenic	T12
			Lumbar	L1–L4

abdominal aorta is sufficiently close to the anterior abdominal wall that its pulsations may be detected or apparent when the wall is relaxed (see the blue box "Pulsations of Aorta and Abdominal Aortic Aneurysm" on p. 319).

The **common iliac arteries** diverge and run inferolaterally, following the medial border of the psoas muscles to the pelvic brim. Here each common iliac artery divides into the *internal* and *external iliac arteries.* The internal iliac artery enters the pelvis. (Its course and branches are described in Chapter 3.) The external iliac artery follows the iliopsoas muscle. Just before leaving the abdomen, the external iliac artery gives rise to the *inferior epigastric* and **deep circumflex iliac arteries**, which supply the anterolateral abdominal wall.

Relations of Abdominal Aorta. From superior to inferior, the important anterior relations of the abdominal aorta are the:

- *Celiac plexus* and *ganglion* (see Figs. 2.55B and 2.71).
- Body of the *pancreas* and *splenic vein* (see Fig. 2.71).
- *Left renal vein* (see Figs. 2.83 and 2.92B).
- Horizontal part of the *duodenum.*
- Coils of *small intestine.*

The abdominal aorta descends anterior to the bodies of the T12–L4 vertebrae (Fig. 2.98A). The left lumbar veins pass posterior to the aorta to reach the IVC (Fig. 2.99). On the right, the aorta is related to the azygos vein, cisterna chyi, thoracic duct, right crus of the diaphragm, and right celiac ganglion. On the left, the aorta is related to the left crus of the diaphragm and the left celiac ganglion.

Branches of the Abdominal Aorta. The branches of the descending (thoracic and abdominal) aorta may be described as arising and coursing in three "vascular planes" and can be classified as being visceral or parietal and paired or unpaired (Fig. 2.98A & C; Table 2.15). *Paired parietal branches* of the aorta serve the diaphragm and posterior abdominal wall.

The **median sacral artery,** an *unpaired parietal branch,* may be said to occupy a fourth (posterior) plane because it arises from the posterior aspect of the aorta just proximal to its bifurcation. Although markedly smaller, it could also be considered a midline "continuation" of the aorta, in which case its lateral branches, the **small lumbar arteries** and **lateral sacral branches,** would also be included as part of the paired parietal branches.

VEINS OF POSTERIOR ABDOMINAL WALL

The veins of the posterior abdominal wall are tributaries of the IVC, except for the *left testicular* or *ovarian vein,* which enters the left renal vein instead of entering the IVC (Fig. 2.99). The IVC, the largest vein in the body, has no valves except for a variable, non-functional one at its orifice in the right atrium of the heart. The IVC returns poorly oxygenated blood from the lower limbs, most of the back, the abdominal walls, and the abdominopelvic viscera. Blood from the abdominal viscera passes through the *portal venous system* and the liver before entering the IVC via the *hepatic veins.*

The **inferior vena cava** (IVC) begins anterior to the L5 vertebra by the union of the common iliac veins. The union occurs approximately 2.5 cm to the right of the median plane, inferior to the aortic bifurcation and posterior to the proximal part of the right common iliac artery (see Fig. 2.76). The IVC ascends on the right side of the bodies of the L3–L5 vertebrae and on the right psoas major to the right of the aorta.

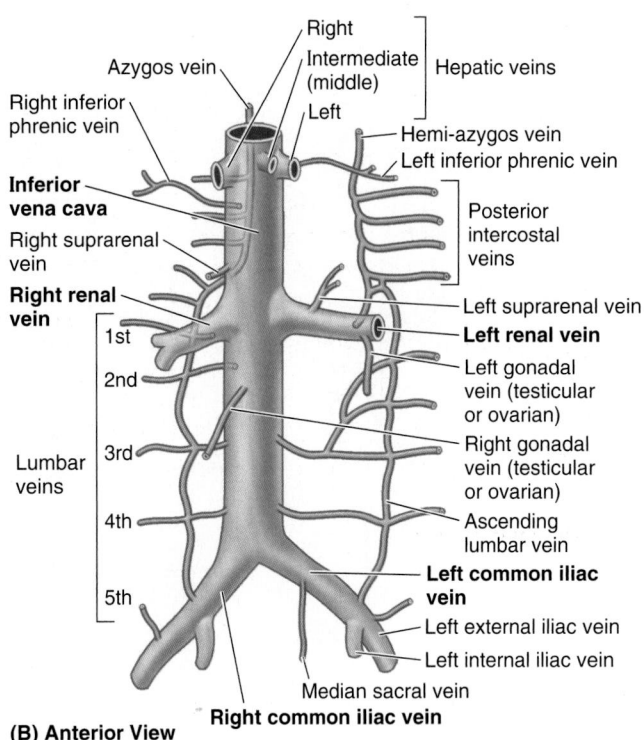

(B) Anterior View

FIGURE 2.99. Inferior vena cava and its tributaries. The asymmetry in the renal and common iliac veins reflects the placement of the IVC to the right of the midline.

The IVC leaves the abdomen by passing through the *caval opening* in the diaphragm and enters the thorax at the T8 vertebral level. Because it is formed one vertebral level inferior to the aortic bifurcation, and traverses the diaphragm four vertebral levels superior to the aortic hiatus, the overall length of the IVC is 7 cm greater than that of the abdominal aorta, although most of the additional length is intrahepatic. The IVC collects poorly oxygenated blood from the lower limbs and non-portal blood from the abdomen and pelvis. Almost all the blood from the gastrointestinal tract is collected by the hepatic portal system and passes through the hepatic veins to the IVC.

The tributaries of the IVC correspond to the paired visceral and parietal branches of the abdominal aorta. The veins that correspond to the unpaired visceral branches of the aorta are instead tributaries of the hepatic portal vein. The blood they carry does ultimately enter the IVC via the hepatic veins, after traversing the liver.

The branches corresponding to the paired visceral branches of the abdominal aorta include the right suprarenal vein, the right and left renal veins, and the right gonadal (testicular or ovarian) vein. The left suprarenal and gonadal veins drain indirectly into the IVC because they are tributaries of the left renal vein.

Paired parietal branches of the IVC include the inferior phrenic veins, the 3rd (L3) and 4th (L4) lumbar veins, and the common iliac veins. The ascending lumbar and azygos veins connect the IVC and SVC, either directly or indirectly

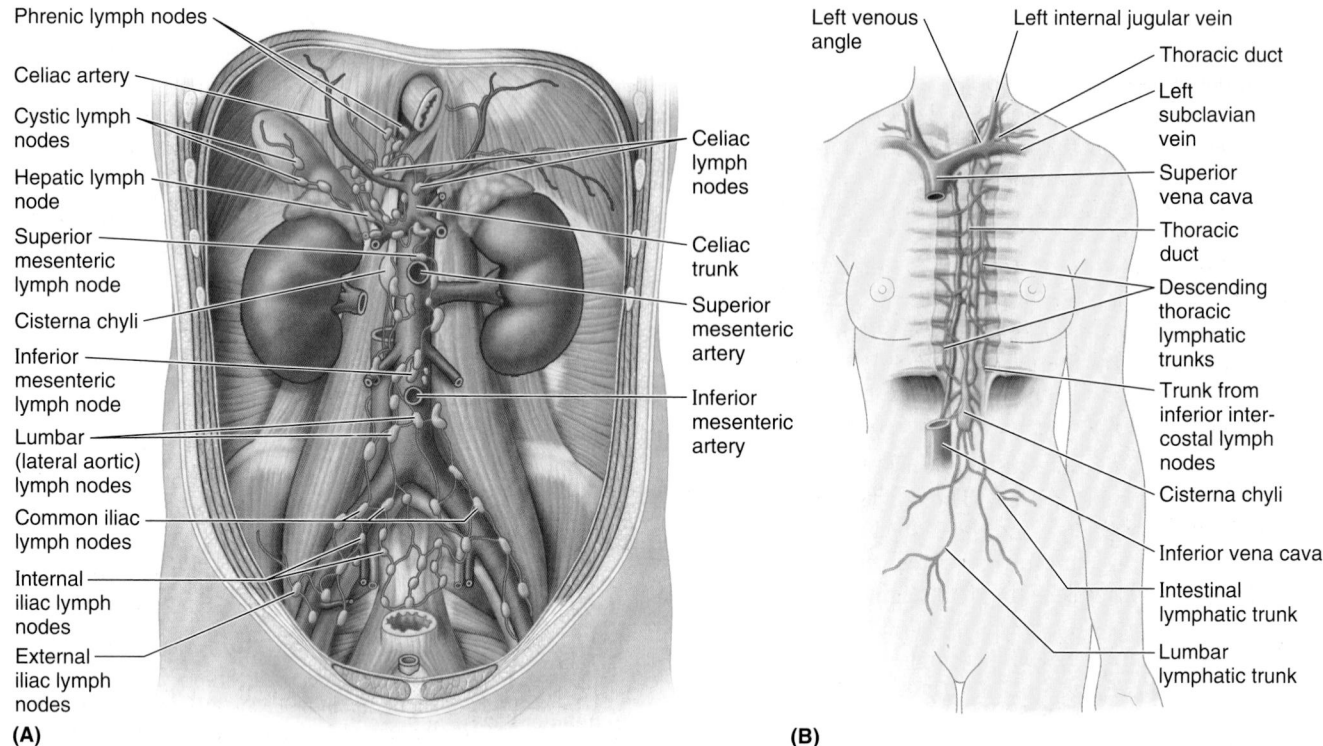

Phrenic lymph nodes
Celiac artery
Cystic lymph nodes
Hepatic lymph node
Superior mesenteric lymph node
Cisterna chyli
Inferior mesenteric lymph node
Lumbar (lateral aortic) lymph nodes
Common iliac lymph nodes
Internal iliac lymph nodes
External iliac lymph nodes

Celiac lymph nodes
Celiac trunk
Superior mesenteric artery
Inferior mesenteric artery

(A)

Left venous angle
Left internal jugular vein
Thoracic duct
Left subclavian vein
Superior vena cava
Thoracic duct
Descending thoracic lymphatic trunks
Trunk from inferior intercostal lymph nodes
Cisterna chyli
Inferior vena cava
Intestinal lymphatic trunk
Lumbar lymphatic trunk

(B)

FIGURE 2.100. Lymphatic vessels and lymph nodes of posterior abdominal wall and lymphatic trunks of abdomen. A. The parietal lymph nodes are shown. **B.** The abdominal lymphatic trunks are shown. All lymphatic drainage from the lower half of the body converges in the abdomen to enter the beginning of the thoracic duct.

providing collateral pathways (see the blue box "Collateral Routes for Abdominopelvic Venous Blood" on p. 319).

LYMPHATIC VESSELS AND LYMPH NODES OF POSTERIOR ABDOMINAL WALL

Lymphatic vessels and lymph nodes lie along the aorta, IVC, and iliac vessels (Fig. 2.100A). The *common iliac lymph nodes* receive lymph from the external and internal iliac lymph nodes. Lymph from the common iliac lymph nodes passes to the right and left *lumbar lymph nodes*. Lymph from the alimentary tract, liver, spleen, and pancreas passes along the celiac and superior and inferior mesenteric arteries to the *pre-aortic lymph nodes* (*celiac* and *superior* and *inferior mesenteric nodes*) scattered around the origins of these arteries from the aorta. Efferent vessels from these nodes form the **intestinal lymphatic trunks,** which may be single or multiple, and participate in the confluence of lymphatic trunks that gives rise to the thoracic duct (Fig. 2.100B).

The *right* and *left lumbar* (*caval* and *aortic*) *lymph nodes* lie on both sides of the IVC and aorta. These nodes receive lymph directly from the posterior abdominal wall, kidneys, ureters, testes or ovaries, uterus, and uterine tubes. They also receive lymph from the descending colon, pelvis, and lower limbs through the *inferior mesenteric* and **common iliac lymph nodes.** Efferent lymphatic vessels from the

large lumbar lymph nodes form the right and left **lumbar lymphatic trunks.**

The inferior end of the *thoracic duct* lies anterior to the bodies of the L1 and L2 vertebrae between the right crus of the diaphragm and the aorta. The thoracic duct begins with the convergence of the main lymphatic ducts of the abdomen, which in only a small proportion of individuals takes the form of the commonly depicted, thin-walled sac or dilation, the **cisterna chyli** (chyle cistern) (Fig. 2.100B). Cisterna chyli vary greatly in size and shape. More often there is merely a simple or plexiform convergence at this level of the right and left lumbar lymphatic trunks, the intestinal lymph trunk(s), and a pair of **descending thoracic lymphatic trunks,** which carry lymph from the lower six intercostal spaces on each side. Consequently, essentially all the lymphatic drainage from the lower half of the body (deep lymphatic drainage inferior to the level of the diaphragm and all superficial drainage inferior to the level of the umbilicus) converges in the abdomen to enter the beginning of the thoracic duct.

The thoracic duct ascends through the aortic hiatus in the diaphragm into the posterior mediastinum, where it collects more parietal and visceral drainage, particularly from the left upper quadrant of the body. The duct ultimately ends by entering the venous system at the junction of the *left subclavian* and *internal jugular veins* (the *left venous angle*).

DIAPHRAGM

Hiccups

Hiccups (hiccoughs) are involuntary, spasmodic contractions of the diaphragm, causing sudden inhalations that are rapidly interrupted by spasmodic closure of the glottis (aperture of the larynx) that checks the inflow of air and produces the characteristic sound. Hiccups result from irritation of afferent or efferent nerve endings, or of medullary centers in the brainstem that control the muscles of respiration, particularly the diaphragm. Hiccups have many causes, such as indigestion, diaphragm irritation, alcoholism, cerebral lesions, and thoracic and abdominal lesions, all which disturb the phrenic nerves.

Section of a Phrenic Nerve

Section of a phrenic nerve in the neck results in complete paralysis and eventual atrophy of the muscular part of the corresponding half of the diaphragm, except in persons who have an accessory phrenic nerve (see Chapter 8). *Paralysis of a hemidiaphragm* can be recognized radiographically by its permanent elevation and paradoxical movement. See the blue box "Paralysis of Diaphragm," p. 85.

Referred Pain from Diaphragm

Pain from the diaphragm radiates to two different areas because of the difference in the sensory nerve supply of the diaphragm (Table 2.12). Pain resulting from irritation of the diaphragmatic pleura or the diaphragmatic peritoneum is referred to the shoulder region, the area of skin supplied by the C3–C5 segments of the spinal cord (see the blue box "Visceral Referred Pain," p. 257). These segments also contribute anterior rami to the phrenic nerves. Irritation of peripheral regions of the diaphragm, innervated by the inferior intercostal nerves, is more localized, being referred to the skin over the costal margins of the anterolateral abdominal wall.

Rupture of Diaphragm and Herniation of Viscera

Rupture of the diaphragm and herniation of viscera can result from a sudden large increase in either the intrathoracic or intra-abdominal pressure. The common cause of this injury is severe trauma to the thorax or abdomen during a motor vehicle accident. Most diaphragmatic ruptures are on the left side (95%) because the substantial mass of the liver, intimately associated with the diaphragm on the right side, provides a physical barrier.

A non-muscular area of variable size called the *lumbocostal triangle* usually occurs between the costal and lumbar parts of the diaphragm (see Figs. 2.91 and 2.97). This part of the diaphragm is normally formed only by fusion of the superior and inferior fascias of the diaphragm. When a *traumatic diaphragmatic hernia* occurs, the stomach, small intestine and mesentery, transverse colon, and spleen may herniate through this area into the thorax.

Hiatal (hiatus) hernia, a protrusion of part of the stomach into the thorax through the esophageal hiatus, was discussed earlier in this chapter. The structures that pass through the esophageal hiatus (vagal trunks, left inferior phrenic ves-

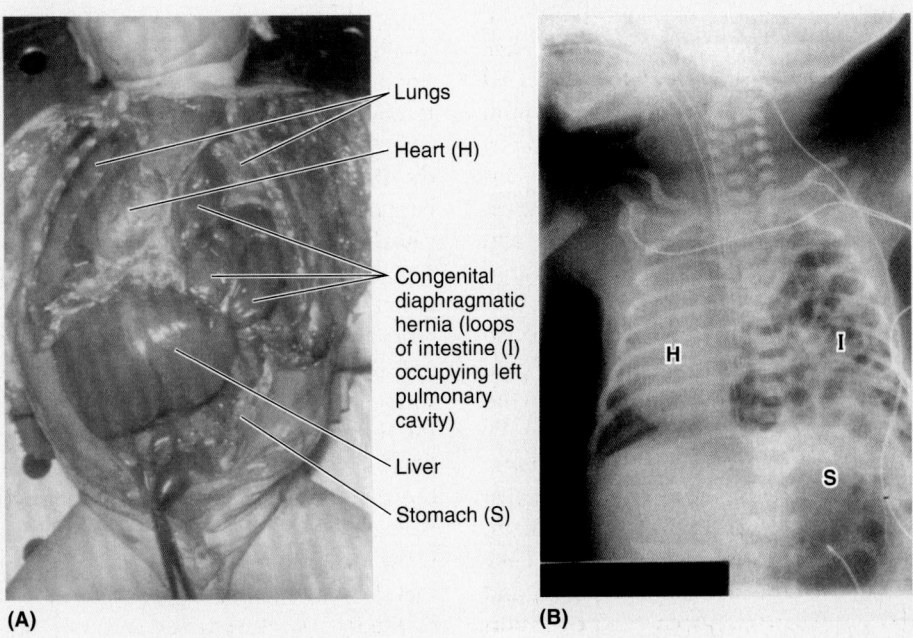

(A)   **(B)**

FIGURE B2.34. Congenital diaphragmatic hernia (CDH). A. As seen on autopsy. **B.** As seen radiographically in a newborn.

sels, esophageal branches of the left gastric vessels) may be injured in surgical procedures on the esophageal hiatus (e.g., repair of a hiatus hernia).

Congenital Diaphragmatic Hernia

In **congenital diaphragmatic hernia** (**CDH**), part of the stomach and intestine herniate through a large posterolateral defect (foramen of Bochdalek) in the region of the lumbocostal trigone of the diaphragm (Fig. B2.34). Herniation almost always occurs on the left owing to the presence of the liver on the right. This type of hernia results from the complex development of the diaphragm. *Posterolateral defect of the diaphragm* is the only relatively common congenital anomaly of the diaphragm, occurring approximately once in 2200 newborn infants (Moore, Persaud, and Torchia, 2012). With abdominal viscera in the limited space of the prenatal pulmonary cavity, one lung (usually the left lung) does not have room to develop normally or to inflate after birth. Because of the consequent *pulmonary hypoplasia*, the mortality rate in these infants is high (approximately 76%).

POSTERIOR ABDOMINAL WALL

Psoas Abscess

Although the prevalence of *tuberculosis* (TB) has been greatly reduced, there is currently a resurgence of TB, especially in Africa and Asia, sometimes in pandemic proportions, owing to AIDS and drug resistance. TB of the vertebral column is quite common. An infection may spread through the blood to the vertebrae (*hematogenous spread*), particularly during childhood. An abscess resulting from tuberculosis in the lumbar region tends to spread from the vertebrae into the psoas fascia (sheath), where it produces a *psoas abscess* (Fig. B2.35). As a consequence, the psoas fascia thickens to form a strong stocking-like tube. Pus from the psoas abscess passes inferiorly along the psoas muscle within this fascial tube over the pelvic brim and deep to the inguinal ligament. The pus usually surfaces in the superior part of the thigh.

Pus can also reach the psoas fascia by passing from the posterior mediastinum when the thoracic vertebrae are diseased.

The inferior part of the *iliac fascia* is often tense and raises a fold that passes to the internal aspect of the iliac crest. The superior part of this fascia is loose and may form a pocket, the *iliacosubfascial fossa,* posterior to the above-mentioned fold. Part of the large intestine, such as the cecum and/or appendix on the right side and the sigmoid colon on the left side, may become trapped in this fossa, causing considerable pain.

Posterior Abdominal Pain

The iliopsoas muscle has extensive, clinically important relations to the kidneys, ureters, cecum, appendix, sigmoid colon, pancreas, lumbar lymph nodes, and nerves of the posterior abdominal wall. When any of these structures is diseased, movement of the iliopsoas usually causes pain. When intra-abdominal inflammation is suspected, the *iliopsoas test* is performed. The person is asked to lie on the unaffected side and extend the thigh on the affected side against the resistance of the examiner's hand (Bickley, 2009). The elicitation of pain with this maneuver is a *positive psoas sign.* An acutely inflamed appendix, for example, will produce a positive right psoas sign (Fig. B2.36).

Because the psoas lies along the vertebral column and the iliacus crosses the sacro-iliac joint, disease of the intervertebral and sacro-iliac joints may cause *spasm of the iliopsoas,* a protective reflex. *Adenocarcinoma of the pancreas* in advanced stages invades the muscles and nerves of the posterior abdominal wall, producing excruciating pain because of the close relationship of the pancreas to the posterior abdominal wall.

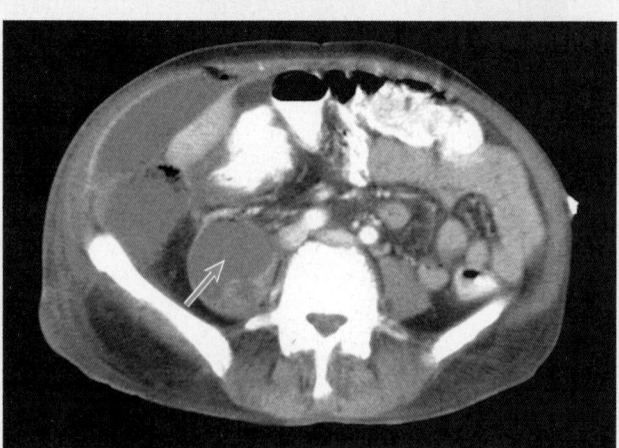

FIGURE B2.35. Psoas abscess (*arrow*).

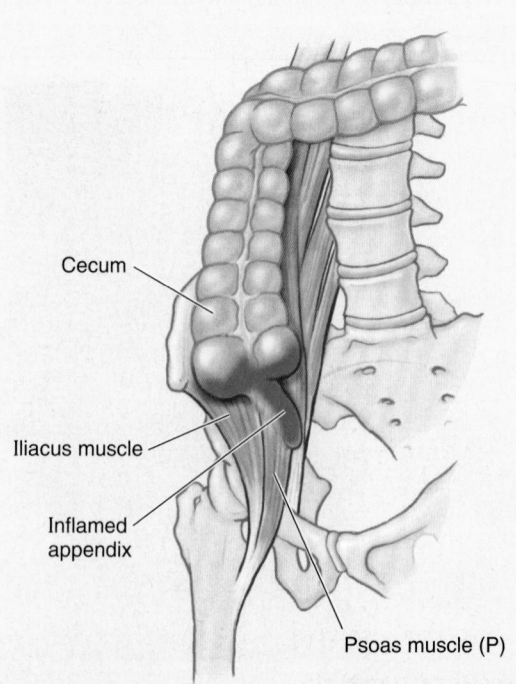

Cecum

Iliacus muscle

Inflamed appendix

Psoas muscle (P)

FIGURE B2.36. Anatomical basis of psoas sign.

Partial Lumbar Sympathectomy

The treatment of some patients with arterial disease in the lower limbs may include a *partial lumbar sympathectomy,* the surgical removal of two or more lumbar sympathetic ganglia by division of their rami communicantes. Surgical access to the sympathetic trunks is commonly through a lateral extraperitoneal approach because the sympathetic trunks lie retroperitoneally in the extraperitoneal fatty tissue (Fig. 2.97). The surgeon splits the muscles of the anterior abdominal wall and moves the peritoneum medially and anteriorly to expose the medial edge of the psoas major, along which the sympathetic trunk lies. The left trunk is often overlapped slightly by the aorta. The right sympathetic trunk is covered by the IVC. The intimate relationship of the sympathetic trunks to the aorta and IVC also makes these large vessels vulnerable to injury during lumbar sympathectomy. Consequently, the surgeon carefully retracts them to expose the sympathetic trunks that usually lie in the groove between the psoas major laterally and the lumbar vertebral bodies medially. These trunks are often obscured by fat and lymphatic tissue. Knowing that identification of the sympathetic trunks is not easy, great care is taken not to remove inadvertently part of the genitofemoral nerve, lumbar lymphatics, or ureter.

Pulsations of Aorta and Abdominal Aortic Aneurysm

Because the aorta lies posterior to the pancreas and stomach, a tumor of these organs may transmit pulsations of the aorta that could be mistaken for an *abdominal aortic aneurysm,* a localized enlargement of the aorta (Fig. B2.37A & B). Deep palpation of the midabdomen can detect an aneurysm, which usually results from a congenital or acquired weakness of the arterial wall (Fig. B2.37C & D). Pulsations of a large aneurysm can be detected to the left of the midline; the pulsatile mass can be moved easily from side to side. Medical imaging can confirm the diagnosis in doubtful cases.

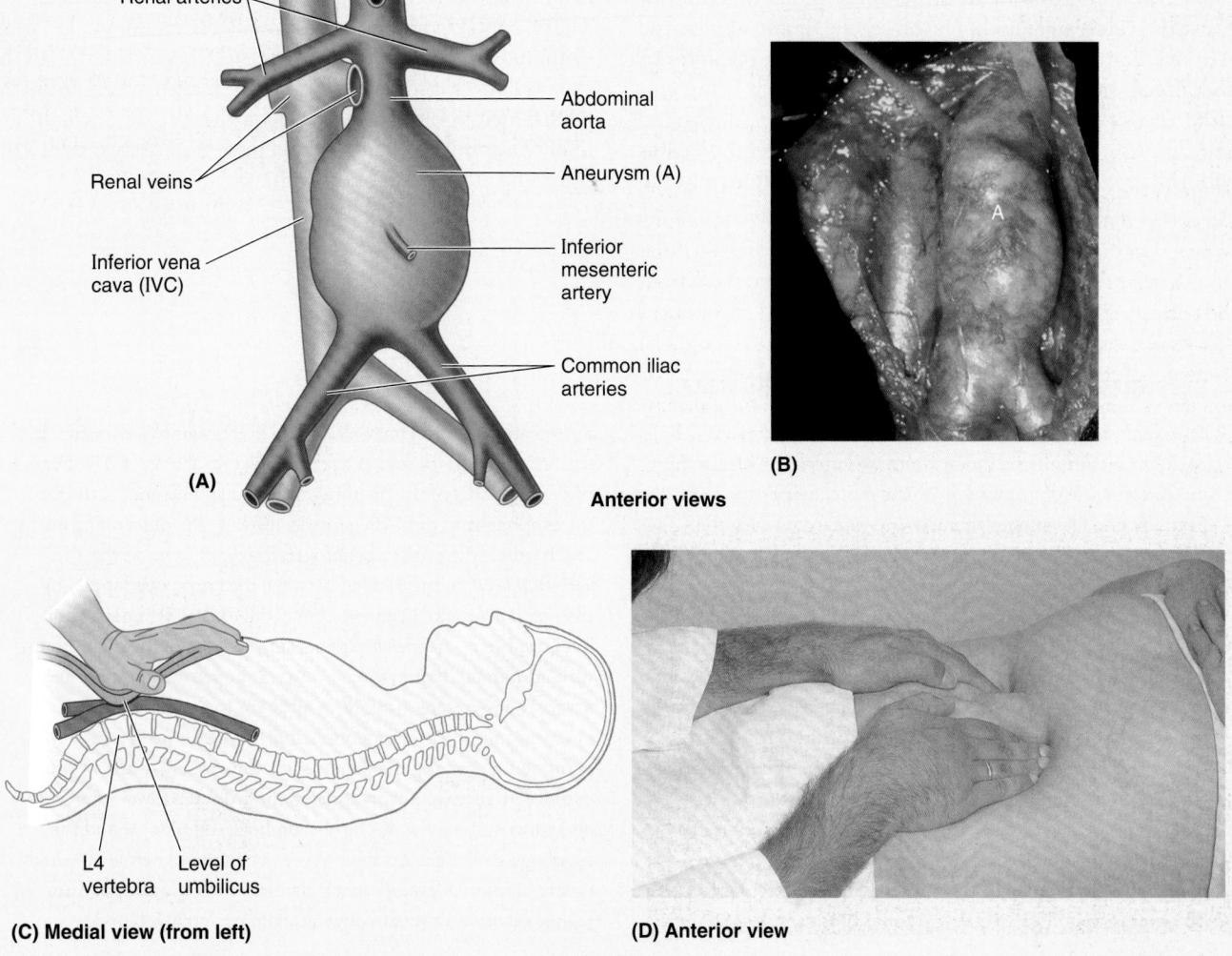

(A)

(B)

Anterior views

(C) **Medial view (from left)**

(D) **Anterior view**

FIGURE B2.37.

Acute rupture of an abdominal aortic aneurysm is associated with severe pain in the abdomen or back. If unrecognized, such an aneurysm has a mortality rate of nearly 90% because of heavy blood loss (Swartz, 2009). Surgeons can repair an aneurysm by opening it, inserting a prosthetic graft, and sewing the wall of the aneurysmal aorta over the graft to protect it. Many vascular problems formerly treated with open repair, including aneurysm repair, are now being treated by means of endovascular catheterization procedures.

When the anterior abdominal wall is relaxed, particularly in children and thin adults, the inferior part of the abdominal aorta may be compressed against the body of the L4 vertebra by firm pressure on the anterior abdominal wall, over the umbilicus (Fig. B2.37C & D). This pressure may be applied to control bleeding in the pelvis or lower limbs.

Collateral Routes for Abdominopelvic Venous Blood

Three collateral routes, formed by valveless veins of the trunk, are available for venous blood to return to the heart when the IVC is obstructed or ligated. Two of these routes (one involving the superior and inferior epigastric veins, and another involving the thoraco-epigastric vein) were discussed earlier in this chapter with the anterior abdominal wall. The third collateral route involves the *epidural venous plexus* inside the vertebral column (illustrated and discussed in Chapter 4—Back), which communicates with the *lumbar veins* of the inferior caval system, and the

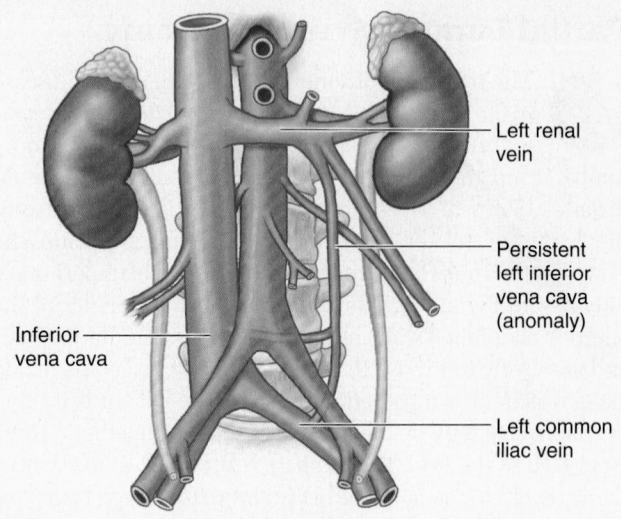

FIGURE B2.38.

Left renal vein

Persistent left inferior vena cava (anomaly)

Inferior vena cava

Left common iliac vein

tributaries of the *azygos system of veins*, which is part of the superior caval system.

The inferior part of the IVC has a complicated developmental history because it forms from parts of three sets of embryonic veins (Moore, Persaud, and Torchia, 2012). Therefore, IVC anomalies are relatively common, and most of them, such as a *persistent left IVC*, occur inferior to the renal veins (Fig. B2.38). These anomalies result from the persistence of embryonic veins on the left side, which normally disappear. If a left IVC is present, it may cross to the right side at the level of the kidneys.

The Bottom Line

DIAPHRAGM AND POSTERIOR ABDOMINAL WALL

The diaphragm is the double-domed, musculotendinous partition separating the thoracic and abdominal cavities and is the chief muscle of inspiration. ♦ The muscular portion arises from the ring-like inferior thoracic aperture from which the diaphragm rises steeply, invaginating the thoracic cage and forming a common central tendon. ♦ The right dome (higher because of the underlying liver) rises nearly to the level of the nipple, whereas the left dome is slightly lower. ♦ The central portion of the diaphragm is slightly depressed by the heart within the pericardium and is fused to the mediastinal surface of the central tendon. In the neutral respiratory position, the central tendon lies at the level of the T8–T9 IV disc and the xiphisternal joint. ♦ When stimulated by the phrenic nerves, the domes are pulled downward (descend), compressing the abdominal viscera. When stimulation ceases and the diaphragm relaxes, the diaphragm is pushed upward (ascends) by the combined decompression of the viscera and tonus of the muscles of the anterolateral abdominal wall. ♦ The diaphragm is perforated by the IVC and phrenic nerves at the T8 vertebral level. ♦ The fibers of the right crus of the diaphragm form a sphincteric hiatus for the esophagus at the T10 vertebral level. ♦ The descending aorta and thoracic duct pass posterior to the diaphragm at the T12 vertebral level, in the midline between the crura, overlapped by the median arcuate ligament connecting them. ♦ Superior and inferior phrenic arteries and veins supply most of the diaphragm, with additional drainage occurring via the musculophrenic and azygos/hemi-azygos veins. ♦ In addition to exclusive motor innervation, the phrenic nerves supply most of the pleura and peritoneum covering the diaphragm. ♦ Peripheral parts of the diaphragm receive sensory innervation from the lower intercostal and subcostal nerves. ♦ The left lumbocostal triangle and the esophageal hiatus are potential sites of acquired hernias through the diaphragm. Developmental defects in the left lumbocostal region account for most congenital diaphragmatic hernias.

Fascia and muscles: Large, complex aponeurotic formations cover the central parts of the trunk both anteriorly and posteriorly, forming dense sheaths centrally that house vertical muscles and attach laterally to the flat muscles of the anterolateral abdominal wall. ♦ The thoracolumbar fascia is the posterior aponeurotic formation. In addition to ensheathing the erector spinae between its posterior and middle layers, it encloses the quadratus lumborum between its middle and anterior layers. ♦ The anterior layer, part of the endoabdominal fascia, is continuous medially with the psoas fascia (enclosing the psoas) and laterally with the transversalis fascia (lining the transversus abdominis). ♦ The tube-like psoas fascia provides a potential pathway for the spread of infections between the vertebral column and hip joint. ♦ The endoabdominal fascia covering the anterior aspects of both the quadratus lumborum and psoas is thickened over the superiormost aspects of the muscles, forming the lateral and medial arcuate ligaments, respectively. ♦ A highly variable layer of extraperitoneal fat intervenes between the endoabdominal fascia and peritoneum. It is especially thick in the paravertebral gutters of the lumbar region, comprising the paranephric fat (pararenal fat body). ♦ The muscles of the posterior abdominal wall are the quadratus lumborum, psoas major, and iliacus.

Nerves: The lumbar sympathetic trunks deliver postsynaptic sympathetic fibers to the lumbar plexus for distribution with somatic nerves, and presynaptic parasympathetic fibers to the abdominal aortic plexus, the latter ultimately innervating pelvic viscera. ♦ With the exception of the subcostal nerve (T12) and lumbosacral trunk (L4–L5), the somatic nerves of the posterior abdominal wall are products of the lumbar plexus, formed by the anterior rami of L1–L4 deep to the psoas. ♦ Only the subcostal nerve and derivatives of the anterior ramus of L1 (iliohypogastric and ilio-inguinal nerves) have an abdominal distribution—to the muscles and skin of the inguinal and pubic regions. All other nerves pass to the muscles and skin of the lower limb.

Arteries: Except for the subcostal arteries, the arteries supplying the posterior abdominal wall arise from the abdominal aorta. ♦ The abdominal aorta descends from the aortic hiatus, coursing on the anterior aspects of the T12–L4 vertebra, immediately left of the midline, and bifurcates into the common iliac arteries at the level of the supracristal plane. ♦ Branches of the aorta arise and course in three vascular planes: anterior (unpaired visceral branches), lateral (paired visceral branches), and posterolateral (paired parietal). ♦ The median sacral artery may be considered a diminutive continuation of the aorta, which continues to give rise to paired parietal branches to the lower lumbar vertebrae and sacrum.

Veins: The veins of the posterior abdominal wall are mostly direct tributaries of the IVC, although some enter indirectly via the left renal vein. ♦ The IVC: • is the largest vein and lacks valves; • formed at the L5 vertebral level by the union of the common iliac veins; • ascends to the T8 vertebral level, passing through the caval opening of the diaphragm and entering the heart almost simultaneously; • drains poorly oxygenated blood from the body inferior to the diaphragm; and • receives the venous drainage of the abdominal viscera indirectly via the hepatic portal vein, liver, and hepatic veins. ♦ Except for the hepatic veins, the tributaries of the IVC mostly correspond to the lateral paired visceral and posterolateral paired parietal branches of the abdominal aorta. ♦ Three collateral routes (two involving the anterior abdominal wall, and one involving the vertebral canal) are available to return blood to the heart when the IVC is obstructed.

Lymph vessels and lymph nodes: Lymphatic drainage from the abdominal viscera courses retrograde along the ramifications of the three unpaired visceral branches of the abdominal aorta. ♦ Lymphatic drainage from the abdominal wall merges with that from the lower limbs, both pathways following the arterial supply retrograde from those parts. ♦ Ultimately, all lymphatic drainage from structures inferior to the diaphragm, plus that draining from the lower six intercostal spaces via the descending thoracic lymphatic trunks, enters the beginning of the thoracic duct at the T12 level, posterior to the aorta. ♦ The origin of the thoracic duct may take the form of a saccular cisterna chyli (chyle cistern).

SECTIONAL MEDICAL IMAGING OF ABDOMEN

Ultrasound, CT scans, and **MRIs** are used to examine the abdominal viscera (Figs. 2.101–2.104). Because MRIs provide better differentiation between soft tissues, its images are more revealing. An image in virtually any plane can be reconstructed after scanning is completed. Abdominal angiographic studies may also now be performed using MRA (magnetic resonance angiography) (Fig. 2.104C).

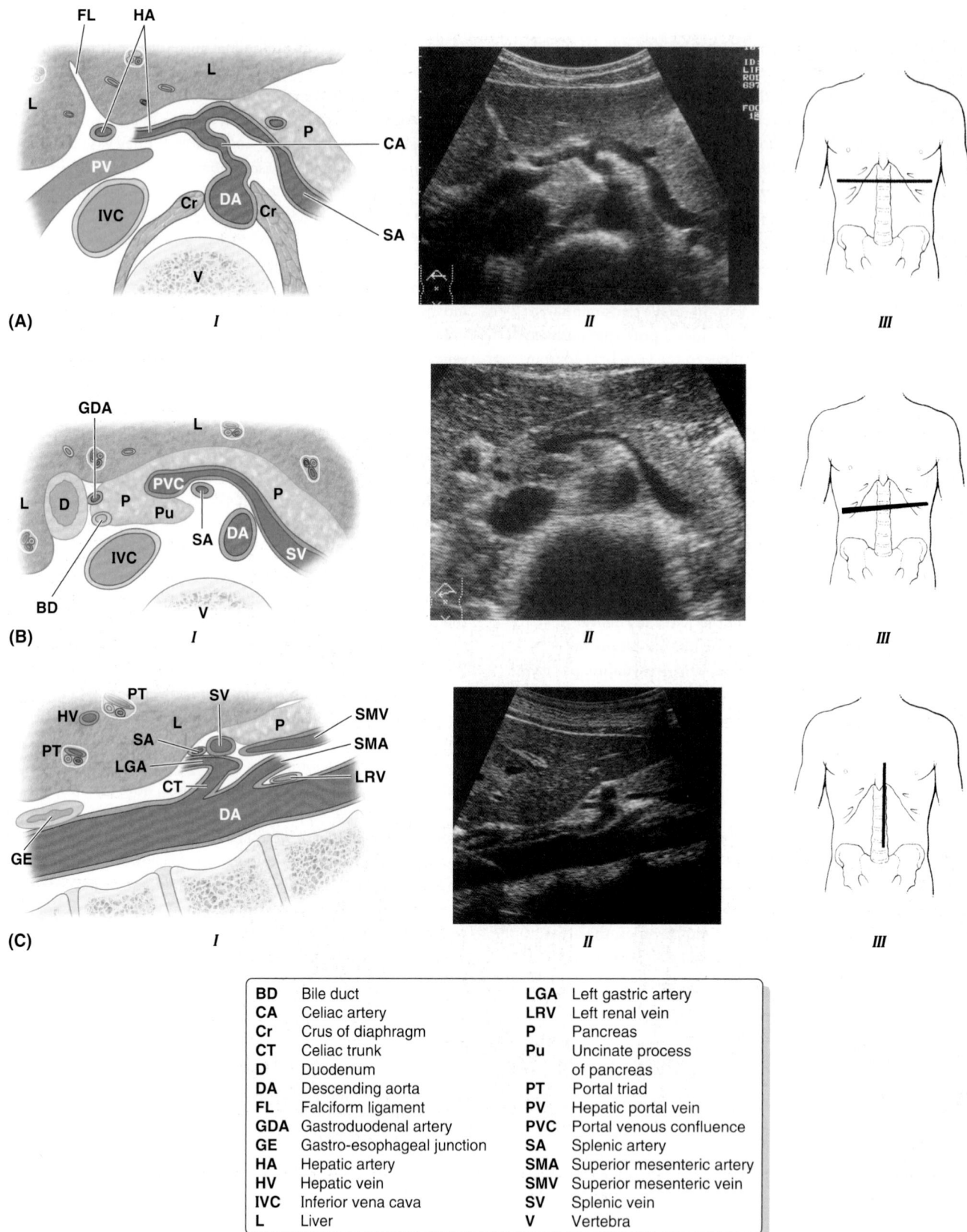

BD	Bile duct	**LGA**	Left gastric artery
CA	Celiac artery	**LRV**	Left renal vein
Cr	Crus of diaphragm	**P**	Pancreas
CT	Celiac trunk	**Pu**	Uncinate process
D	Duodenum		of pancreas
DA	Descending aorta	**PT**	Portal triad
FL	Falciform ligament	**PV**	Hepatic portal vein
GDA	Gastroduodenal artery	**PVC**	Portal venous confluence
GE	Gastro-esophageal junction	**SA**	Splenic artery
HA	Hepatic artery	**SMA**	Superior mesenteric artery
HV	Hepatic vein	**SMV**	Superior mesenteric vein
IVC	Inferior vena cava	**SV**	Splenic vein
L	Liver	**V**	Vertebra

FIGURE 2.101. Ultrasound scans of abdomen. A. A transverse scan through the celiac trunk is shown. **B.** A transverse scan through the pancreas is shown. **C.** A sagittal scan through the aorta is shown. (Courtesy of Dr. A. M. Arenson, Assistant Professor of Medical Imaging, University of Toronto, Toronto, ON, Canada.)

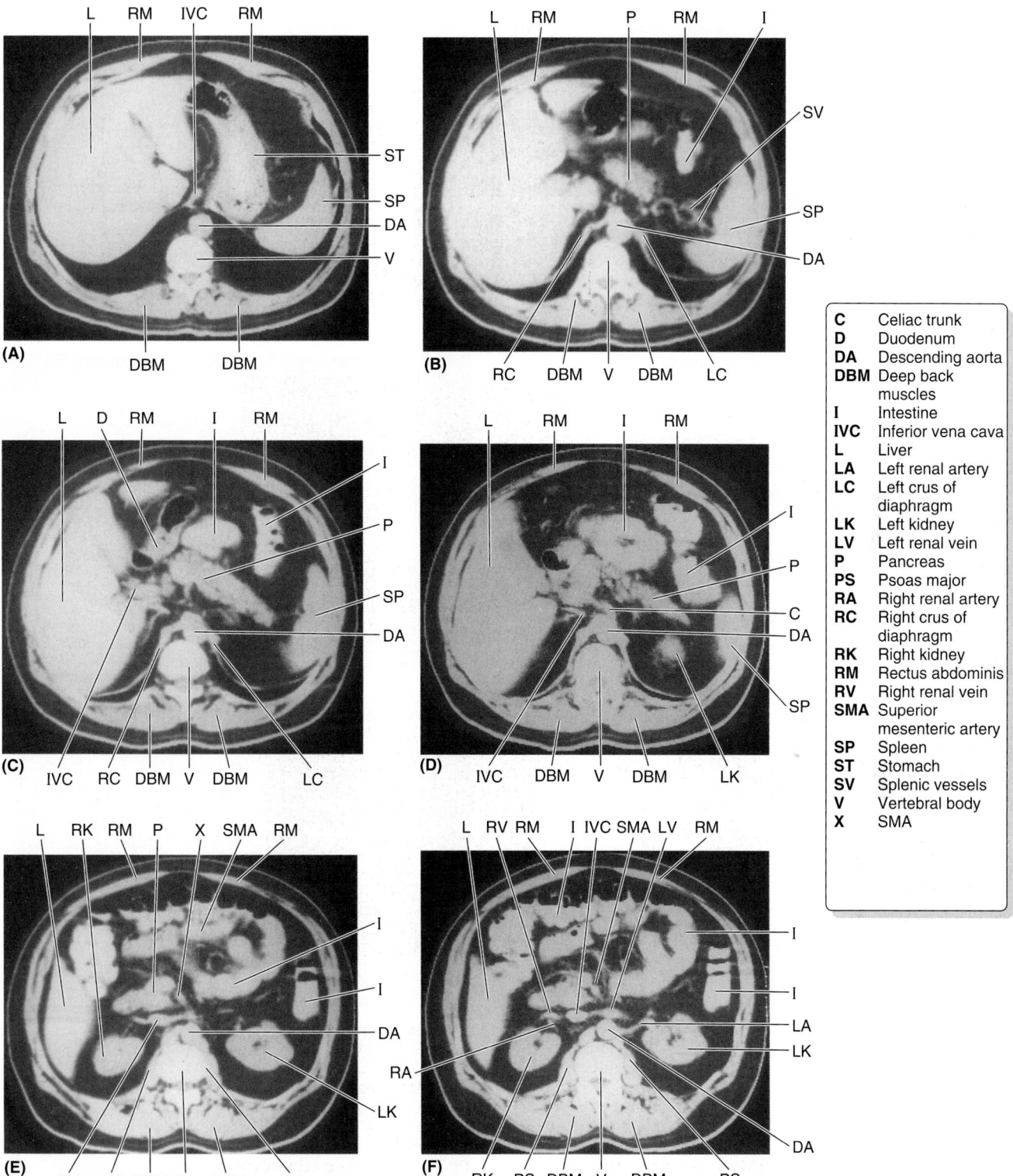

FIGURE 2.102. **Computed tomographic (CT) scans of abdomen at progressively lower levels showing viscera and blood vessels.** (Courtesy of Dr. Tom White, Department of Radiology, The Health Sciences Center, University of Tennessee, Memphis, TN.)

BAo	Bifurcation of aorta	**IM**	Iliacus muscle	**RHV**	Right hepatic vein
CC	Costal cartilage	**IVC**	Inferior vena cava	**RK**	Right kidney
CO	Cardiac orifice of stomach	**L**	Liver	**RLL**	Right lobe of liver
D	Duodenum	**LC**	Left crus	**RPC**	Right pleural cavity
DA	Descending aorta	**LHV**	Left hepatic vein		
DBM	Deep back muscles	**LK**	Left kidney	**RRV**	Right renal vein
		LPC	Left pleural cavity	**SC**	Spinal cord
F	Fat	**LRA**	Inferior vena cava	**Sp**	Spleen
FS	Fundus of stomach	**LRV**	Left renal vein	**SpV**	Spinous process of vertebra
GB	Gallbladder	**P**	Pancreas	**Sv**	Splenic vein
GM	Gluteus medius muscle	**PC**	Portal confluence	**SV**	Splenic vessels
		PF	Perirenal fat	**TC**	Transverse colon
		PS	Psoas muscle	**VB**	Vertebral body
I	Intestine	**PV**	Hepatic portal vein (triad)	**VC**	Vertebral canal
Il	Ilium	**R**	Rib	**XP**	Xiphoid process

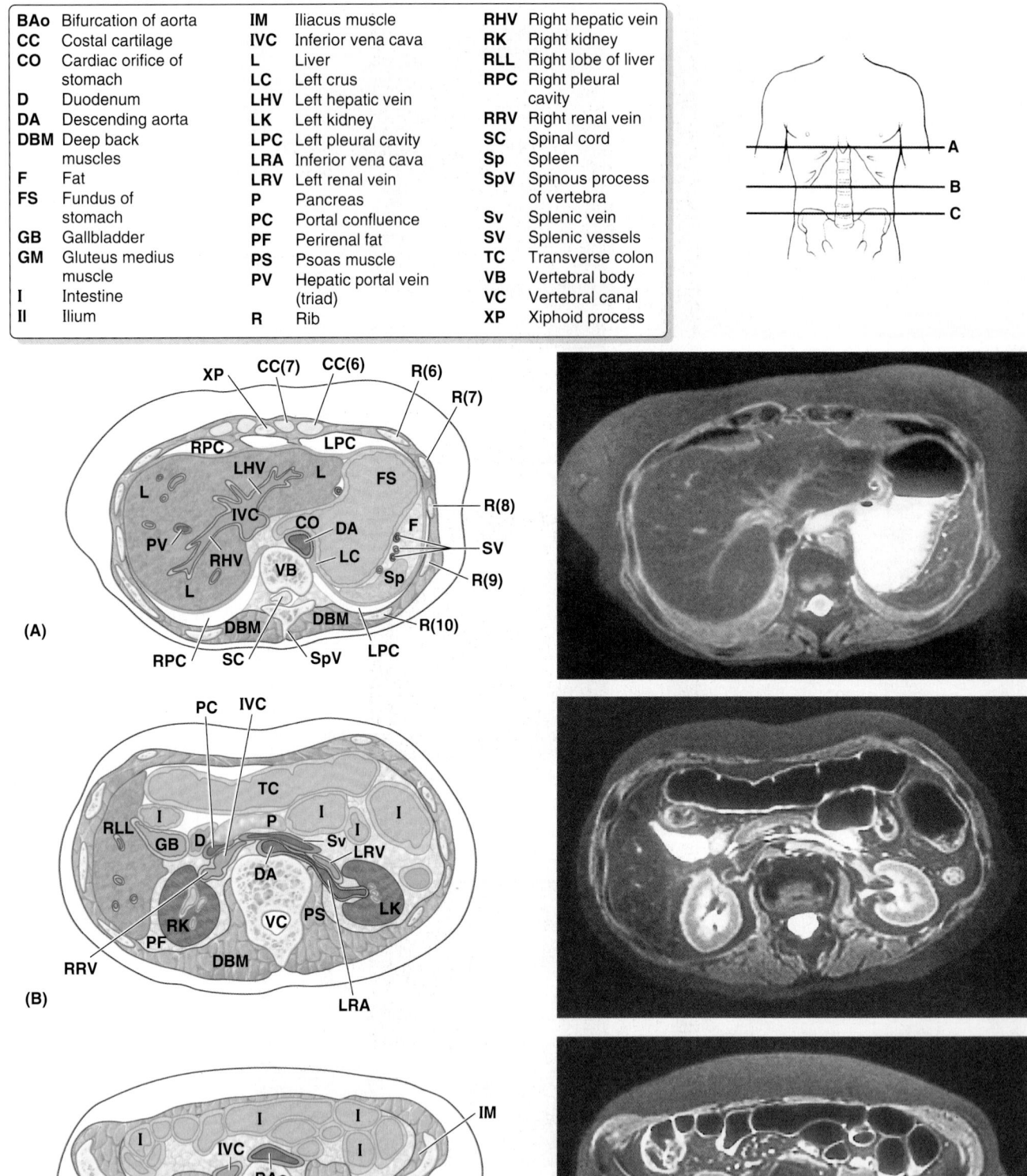

FIGURE 2.103. Transverse magnetic resonance images (MRIs) of abdomen. A. Level of T10 vertebra and esophageal hiatus. **B.** Level of L1–L2 vertebra and renal vessels and hilum. **C.** Level of L5 vertebra and bifurcation of aorta. (Courtesy of Dr. W. Kucharczyk, Professor of Medical Imaging, University of Toronto, and Clinical Director of Tri-Hospital Resonance Centre, Toronto, ON, Canada.)

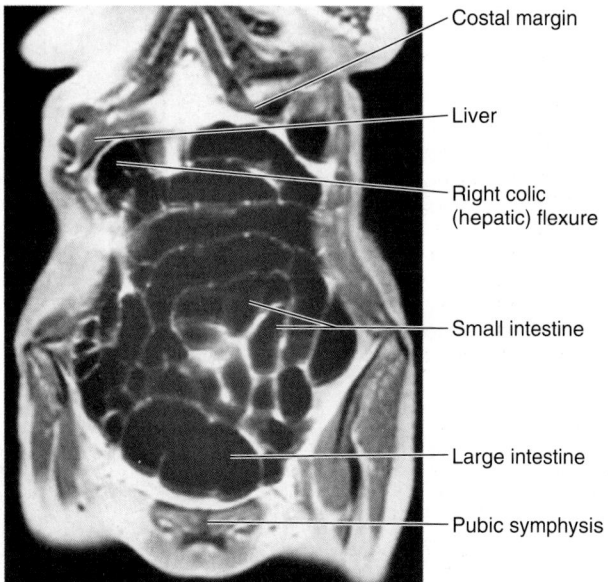

Costal margin
Liver
Right colic (hepatic) flexure
Small intestine
Large intestine
Pubic symphysis

(A) Anteroposterior view

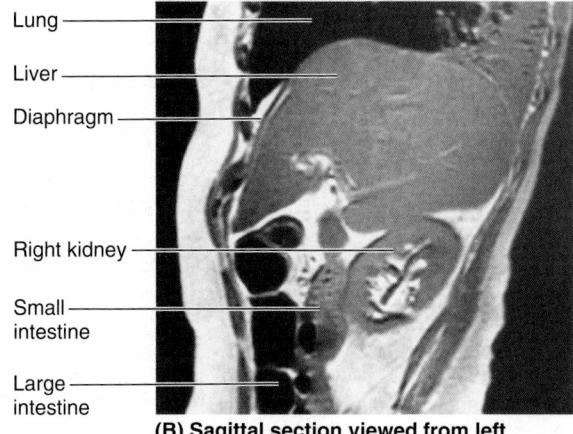

Lung
Liver
Diaphragm
Right kidney
Small intestine
Large intestine

(B) Sagittal section viewed from left

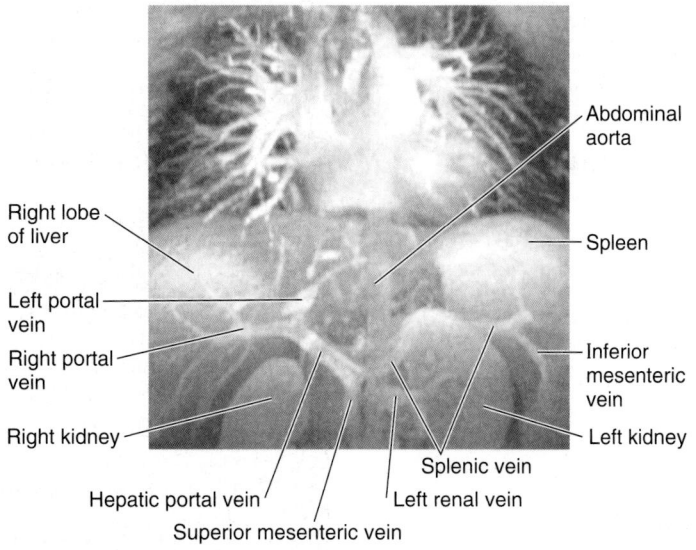

Abdominal aorta
Spleen
Right lobe of liver
Left portal vein
Right portal vein
Inferior mesenteric vein
Right kidney
Left kidney
Splenic vein
Hepatic portal vein
Left renal vein
Superior mesenteric vein

(C) Anteroposterior view

FIGURE 2.104. Magnetic resonance images (MRIs) and magnetic resonance (MR) angiogram of abdomen. A. Coronal MRI through viscera (almost all intestine) of anterior abdominal cavity. **B.** Sagittal MRI in right midclavicular line. **C.** Anteroposterior MR angiogram demonstrating great vessels of thorax and aorta and portal vein in abdomen.

3 Pelvis and Perineum

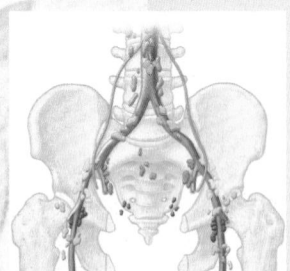

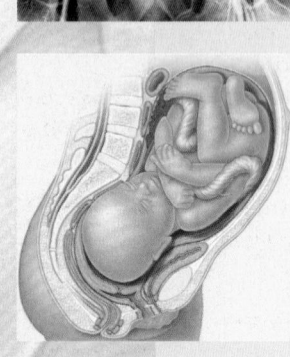

INTRODUCTION TO PELVIS AND PERINEUM

In common usage, the *pelvis* (L. basin) is the part of the trunk that is inferoposterior to the abdomen, and is the area of transition between the trunk and the lower limbs. The *pelvic cavity* is the inferiormost part of the abdominopelvic cavity. Anatomically, the pelvis is the part of the body surrounded by the *pelvic girdle* (bony pelvis), part of the appendicular skeleton of the lower limb (Fig. 3.1).

The pelvis is subdivided into greater and lesser pelves. The *greater pelvis* is surrounded by the superior pelvic girdle. The greater pelvis is occupied by inferior abdominal viscera, affording them protection similar to the way the superior abdominal viscera are protected by the inferior thoracic cage. The *lesser pelvis* is surrounded by the inferior pelvic girdle, which provides the skeletal framework for both the pelvic cavity and the perineum—compartments of the trunk separated by the musculofascial pelvic diaphragm. Externally, the pelvis is covered or overlapped by the inferior anterolateral abdominal wall anteriorly, the gluteal region of the lower limb posterolaterally, and the perineum inferiorly.

The term *perineum*[1] refers both to the area of the surface of the trunk between the thighs and the buttocks, extending from the coccyx to the pubis, and to the shallow compartment lying deep (superior) to this area but inferior to the pelvic diaphragm. The perineum includes the anus and external genitalia: the penis and scrotum of the male and the vulva of the female.

PELVIC GIRDLE

The **pelvic girdle** is a basin-shaped ring of bones that connects the vertebral column to the two femurs. The primary *functions of the pelvic girdle* are to:

- Bear the weight of the upper body when sitting and standing.
- Transfer that weight from the axial to the lower appendicular skeleton for standing and walking.
- Provide attachment for the powerful muscles of locomotion and posture and those of the abdominal wall, withstanding the forces generated by their actions.

Consequently, the pelvic girdle is strong and rigid, especially compared to the pectoral (shoulder) girdle. Other functions of the pelvic girdle are to:

- Contain and protect the pelvic viscera (inferior parts of the urinary tracts and the internal reproductive organs) and the inferior abdominal viscera (intestines), while permitting passage of their terminal parts (and, in females, a full-term fetus) via the perineum.
- Provide support for the abdominopelvic viscera and gravid (pregnant) uterus.
- Provide attachment for the erectile bodies of the external genitalia.
- Provide attachment for the muscles and membranes that assist the functions listed above by forming the pelvic floor and filling gaps that exist in or around it.

[1]The term *perineum* has been used in different ways, in different languages, and in different circumstances. In its most restricted sense, and in obstetrics, it has been used to refer to the area superficial to the perineal body, between the vulva or scrotum and the anus or to the perineal body itself. In an intermediate sense, it has included only the *perineal region,* a superficial (surface) area bounded by the thighs laterally, the mons pubis anteriorly, and the coccyx posteriorly. In its widest sense, as used in *Terminologia Anatomica* (the international anatomical terminology) and in this book, it refers to the region of the body that includes all structures of the anal and urogenital triangles, superficial and deep, extending as far superiorly as the inferior fascia of the pelvic diaphragm.

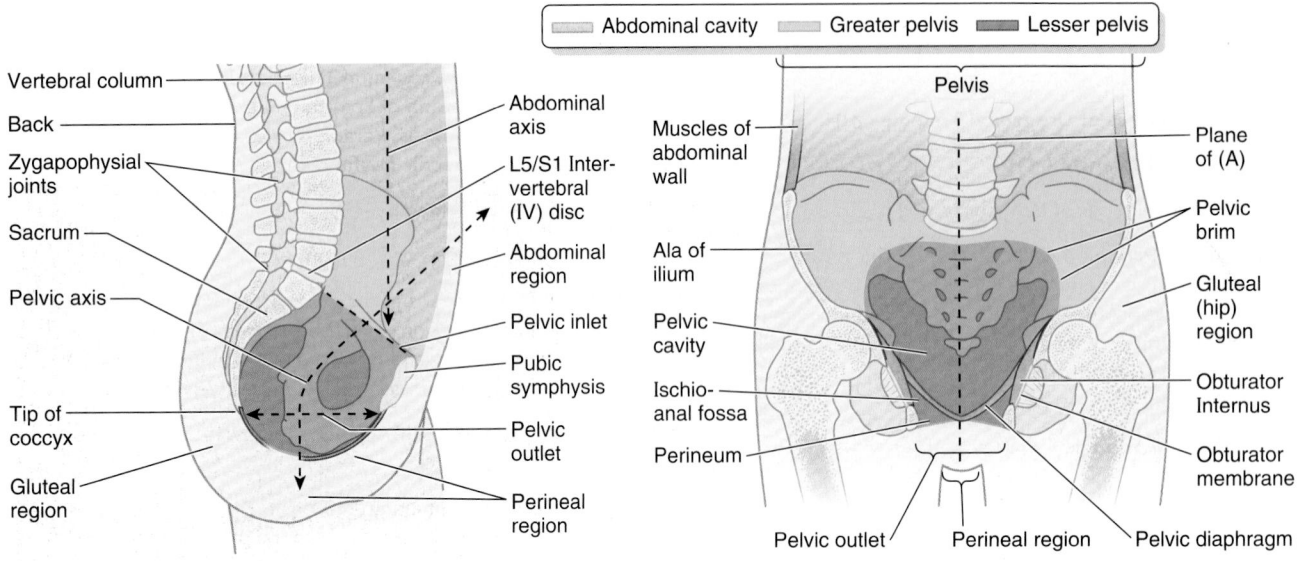

| Abdominal cavity | Greater pelvis | Lesser pelvis |

(A) Medial view of left half of bisected lower trunk

(B) Anterior view of posterior half of coronally-sectioned lower trunk

FIGURE 3.1. Pelvis and perineum. A and B. The pelvis (*green*) is the space within the pelvic girdle, overlapped externally by the abdominal and gluteal regions, perineum, and lower back. Thus the pelvis has no external surface area. The greater pelvis (*light green*) is pelvic by virtue of its bony boundaries, but is abdominal in terms of its contents. The lesser pelvis (*dark green*) provides the bony framework (skeleton) for the pelvic cavity and deep perineum.

Bones and Features of Pelvic Girdle

In the mature individual, the pelvic girdle is formed by three bones (Fig. 3.2A):

- Right and left *hip bones* (coxal or pelvic bones): large, irregularly shaped bones, each of which develops from the fusion of three bones, the *ilium, ischium,* and *pubis.*
- *Sacrum:* formed by the fusion of five, originally separate, sacral vertebrae.

The internal (medial or pelvic) aspects of the hip bones bound the pelvis, forming its lateral walls; these aspects of the bones are emphasized here. Their external aspects, primarily involved in providing attachment for the lower limb muscles, are discussed in Chapter 5. As they are part of the vertebral column, the sacrum and coccyx are discussed in detail in Chapter 4.

In infants and children, the hip bones consist of three separate bones that are united by a *triradiate cartilage* at the *acetabulum* (Fig. 3.2B), the cup-like depression in the lateral surface of the hip bone, which articulates with the head of the femur. After puberty, the ilium, ischium, and pubis fuse to form the hip bone. The two hip bones are joined anteriorly at the *pubic symphysis,* a secondary cartilaginous joint. The hip bones articulate posteriorly with the sacrum at the *sacro-iliac joints* to form the pelvic girdle.

The *ilium* is the superior, fan-shaped part of the hip bone (Fig. 3.2B & C). The *ala (wing) of the ilium* represents the spread of the fan, and the *body of the ilium,* the handle of the fan. On its external aspect, the body participates in formation of the acetabulum. The *iliac crest,* the rim of the fan, has a curve that follows the contour of the ala between the *anterior* and *posterior superior iliac spines.* The anteromedial concave surface of the ala forms the *iliac fossa.* Posteriorly, the **sacropelvic surface of the ilium** features an **auricular surface** and an **iliac tuberosity,** for synovial and syndesmotic articulation with the sacrum, respectively.

The *ischium* has a body and ramus (L. branch). The *body of the ischium* helps form the acetabulum and the *ramus of the ischium* forms part of the *obturator foramen.* The large postero-inferior protuberance of the ischium is the *ischial tuberosity.* The small pointed posteromedial projection near the junction of the ramus and body is the *ischial spine.* The concavity between the ischial spine and the ischial tuberosity is the *lesser sciatic notch.* The larger concavity, the *greater sciatic notch,* is superior to the ischial spine and is formed in part by the ilium.

The *pubis* is an angulated bone with a *superior ramus,* which helps form the acetabulum, and an *inferior ramus,* which helps form the obturator foramen. A thickening on the anterior part of the *body of the pubis* is the *pubic crest,* which ends laterally as a prominent swelling, the *pubic tubercle.* The lateral part of the superior pubic ramus has an oblique ridge, the *pecten pubis* (pectineal line of pubis).

The pelvis is divided into *greater* (*false*) and *lesser* (*true*) *pelves* by the oblique plane of the **pelvic inlet** (superior

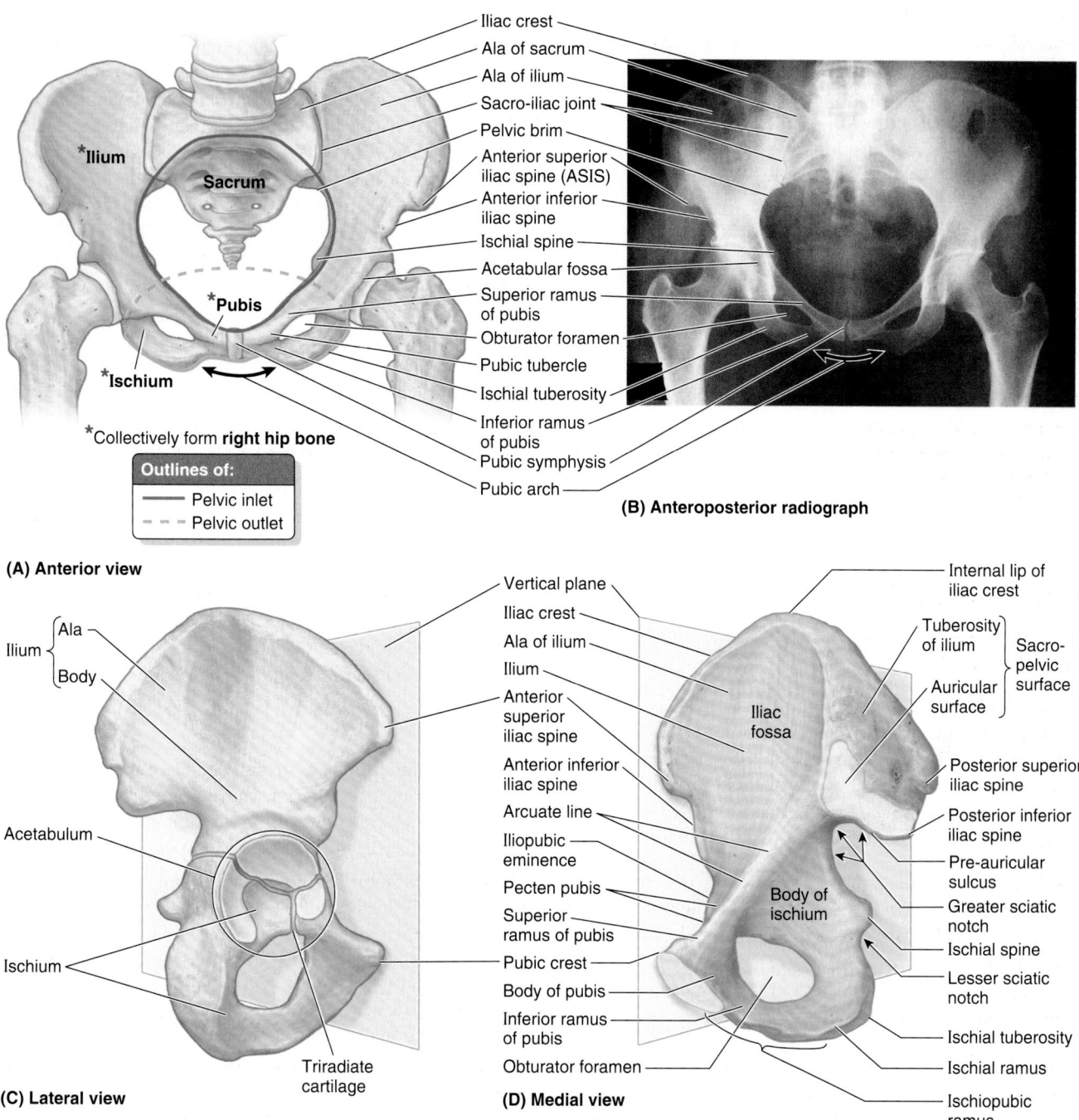

Iliac crest
Ala of sacrum
Ala of ilium
Sacro-iliac joint
Pelvic brim
Anterior superior iliac spine (ASIS)
Anterior inferior iliac spine
Ischial spine
Acetabular fossa
Superior ramus of pubis
Obturator foramen
Pubic tubercle
Ischial tuberosity
Inferior ramus of pubis
Pubic symphysis
Pubic arch

*Ilium
Sacrum
*Pubis
*Ischium
*Collectively form **right hip bone**

Outlines of:
— Pelvic inlet
- - - Pelvic outlet

(A) Anterior view

(B) Anteroposterior radiograph

Ilium { Ala / Body }
Acetabulum
Ischium
Triradiate cartilage

(C) Lateral view

Vertical plane
Iliac crest
Ala of ilium
Ilium
Anterior superior iliac spine
Anterior inferior iliac spine
Arcuate line
Iliopubic eminence
Pecten pubis
Superior ramus of pubis
Pubic crest
Body of pubis
Inferior ramus of pubis
Obturator foramen

Iliac fossa
Body of ischium

(D) Medial view

Internal lip of iliac crest
Tuberosity of ilium } Sacro-pelvic surface
Auricular surface
Posterior superior iliac spine
Posterior inferior iliac spine
Pre-auricular sulcus
Greater sciatic notch
Ischial spine
Lesser sciatic notch
Ischial tuberosity
Ischial ramus
Ischiopubic ramus

FIGURE 3.2. Pelvic girdle. A and **B.** Features of the pelvic girdle demonstrated anatomically (**A**) and radiographically (**B**). The pelvic girdle is formed by the two hip bones (of the inferior axial skeleton) anteriorly and laterally and the sacrum (of the axial skeleton) posteriorly. **C.** The hip bone is in the anatomical position when the anterior superior iliac spine (ASIS) and the anterior aspect of the pubis lie in the same vertical plane. The preadolescent hip bone is composed of three bones—ilium, ischium, and pubis—that meet in the cup-shaped acetabulum. Prior to their fusion, the bones are united by a triradiate cartilage along a Y-shaped line (*blue*). **D.** An adult's right hip bone in the anatomical position shows the bones when fused. (**B** courtesy of Dr. E. L. Lansdown, Professor of Medical Imaging, University of Toronto, Toronto, ON, Canada.)

pelvic aperture) (Figs. 3.1A and 3.2A). The bony edge (rim) surrounding and defining the pelvic inlet is the **pelvic brim,** formed by the:

- *Promontory* and *ala of the sacrum* (superior surface of its lateral part, adjacent to the body of the sacrum).
- A **right** and **left linea terminalis** (terminal line) together form a continuous oblique ridge consisting of the:
  - **Arcuate line** on the inner surface of the ilium.
  - *Pecten pubis* (pectineal line) and *pubic crest,* forming the superior border of the superior ramus and body of the pubis.

The **pubic arch** is formed by the **ischiopubic rami** (conjoined inferior rami of the pubis and ischium) of the two sides (Fig. 3.2A & C). These rami meet at the *pubic symphysis,* their inferior borders defining the **subpubic angle** (Fig. 3.3). The width of the subpubic angle is determined by the distance between the right and the left ischial tuberosities, which can be measured with the gloved fingers in the vagina during a pelvic examination.

The **pelvic outlet** (inferior pelvic aperture) is bounded by the (Figs. 3.1A and 3.2A):

- *Pubic arch* anteriorly.
- *Ischial tuberosities* laterally.
- Inferior margin of the *sacrotuberous ligament* (running between the coccyx and the ischial tuberosity) *posterolaterally.*
- *Tip of the coccyx* posteriorly.

The **greater pelvis** (false pelvis) is the part of the pelvis (Fig. 3.1):

- Superior to the pelvic inlet.
- Bounded by the iliac alae posterolaterally and the anterosuperior aspect of the S1 vertebra posteriorly.
- Occupied by abdominal viscera (e.g., the ileum and sigmoid colon).

The **lesser pelvis** (true pelvis) is the part of the pelvis:

- Between the *pelvic inlet* and the *pelvic outlet.*
- Bounded by the pelvic surfaces of the hip bones, sacrum, and coccyx.
- That includes the true *pelvic cavity* and the deep parts of the *perineum* (perineal compartment), specifically the ischio-anal fossae (Fig. 3.1B).
- That is of major obstetrical and gynecological significance.

The concave superior surface of the musculofascial *pelvic diaphragm* forms the floor of the true pelvic cavity, which is thus deepest centrally. The convex inferior surface of the pelvic diaphragm forms the roof of the perineum, which is therefore shallow centrally and deep peripherally. Its lateral parts (ischio-anal fossae) extending well up into the lesser pelvis. The terms *pelvis, lesser pelvis,* and *pelvic cavity* are commonly used incorrectly, as if they were synonymous terms.

Orientation of Pelvic Girdle

When a person is in the anatomical position, the right and left anterior superior iliac spines (ASISs) and the anterior aspect of the pubic symphysis lie in the same vertical plane (Fig. 3.2B & C). When a pelvic girdle in this position is viewed anteriorly (Fig. 3.2A), the tip of the coccyx appears close to the center of the pelvic inlet, and the pubic bones and pubic symphysis constitute more of a weight-bearing floor than an anterior wall. In the median view (Fig. 3.1A), the *sacral promontory* is located directly superior to the center of the pelvic outlet (site of the perineal body). Consequently, the curved axis of the pelvis intersects the axis of the abdominal cavity at an oblique angle.

The pelvic girdles of males and females differ in several respects (Fig. 3.3; Table 3.1). These sexual differences are related mainly to the heavier build and larger muscles of most men and to the adaptation of the pelvis (particularly the lesser pelvis) in women for parturition (childbearing). See the blue box "Variations in Male and Female Pelves," on p. 334.

Joints and Ligaments of Pelvic Girdle

The primary joints of the pelvic girdle are the *sacro-iliac joints* and the *pubic symphysis* (Fig. 3.4A). The sacro-iliac joints link the **axial skeleton** (skeleton of the trunk, composed of the vertebral column at this level) and the **inferior appendicular skeleton** (skeleton of the lower limb). The lumbosacral and sacrococcygeal joints, although joints of the axial skeleton, are directly related to the pelvic girdle. Strong ligaments support and strengthen these joints.

SACRO-ILIAC JOINTS

The **sacro-iliac joints** are strong, weight-bearing compound joints, consisting of an anterior synovial joint (between the ear-shaped *auricular surfaces* of the sacrum and ilium, covered with articular cartilage) and a posterior syndesmosis (between the *tuberosities* of these bones) (Fig. 3.4B). The auricular surfaces of this synovial joint have irregular but congruent elevations and depressions that interlock (Figs. 3.5A-C). The sacro-iliac joints differ from most synovial joints in that limited mobility is allowed, a consequence of their role in transmitting the weight of most of the body to the hip bones.

Weight is transferred from the axial skeleton to the ilia via the sacro-iliac ligaments (Fig. 3.4A), and then to the femurs during standing, and to the ischial tuberosities during sitting. As long as tight apposition is maintained between the articular surfaces, the sacro-iliac joints remain stable. Unlike a keystone at the top of an arch, the sacrum is actually suspended between the iliac bones and is firmly attached to them by posterior and interosseous sacro-iliac ligaments (Fig. 3.5A).

The thin **anterior sacro-iliac ligaments** are merely the anterior part of the fibrous capsule of the synovial part of the joint (Figs. 3.5A and 3.6). The abundant **interosseous sacro-iliac ligaments** (lying deep between the tuberosities of the sacrum and ilium and occupying an area of approximately 10 cm²) are the primary structures involved in transferring

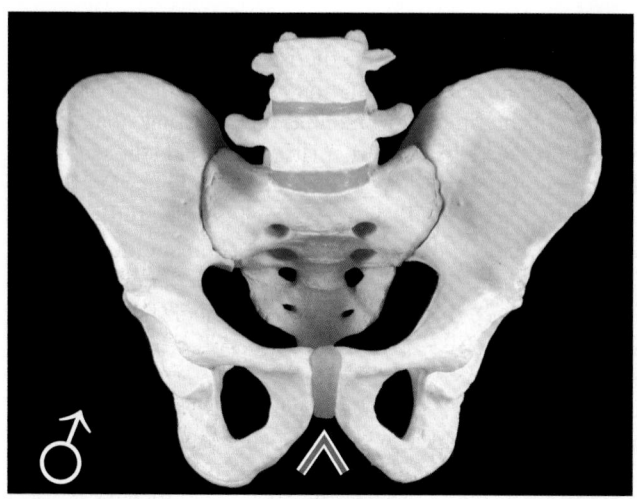
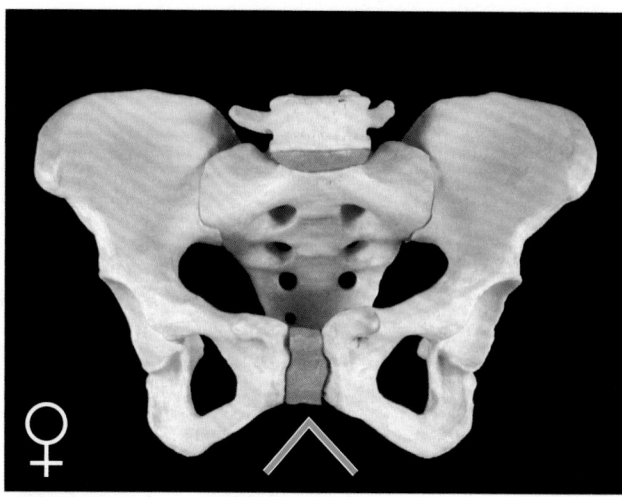

Antero-inferior Views

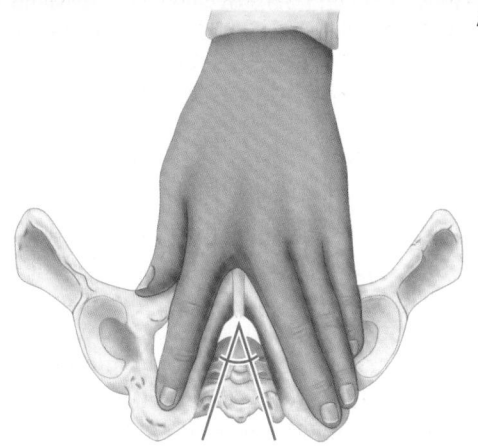
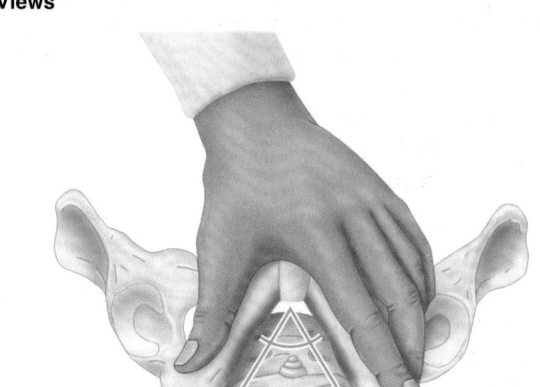

FIGURE 3.3. Pelvic girdles of male and female. Pubic arches, or subpubic angles typical for each gender (male = *red;* female = *green*) can be approximated by spreading index and middle finger (demonstrating narrow subpubic angle of male pelvis) or thumb and index finger (demonstrating wider subpubic angle of female pelvis).

TABLE 3.1. COMPARISON OF MALE AND FEMALE BONY PELVES

Bony Pelvis	Male (♂)	Female (♀)
General structure	Thick and heavy	Thin and light
Greater pelvis (false pelvis)	Deep	Shallow
Lesser pelvis (true pelvis)	Narrow and deep, tapering	Wide and shallow, cylindrical
Pelvic inlet (superior pelvic aperture)	Heart-shaped, narrow	Oval and rounded; wide
Pelvic outlet (inferior pelvic aperture)	Comparatively small	Comparatively large
Pubic arch and subpubic angle	Narrow (<70°)	Wide (>80°)
Obturator foramen	Round	Oval
Acetabulum	Large	Small
Greater sciatic notch	Narrow (~70°); inverted V	Almost 90°

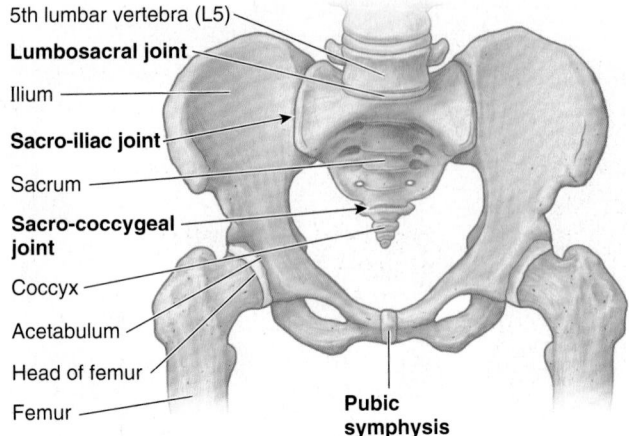

5th lumbar vertebra (L5)
Lumbosacral joint
Ilium
Sacro-iliac joint
Sacrum
Sacro-coccygeal joint
Coccyx
Acetabulum
Head of femur
Femur
Pubic symphysis

(A) Anterior view—joints of pelvic girdle

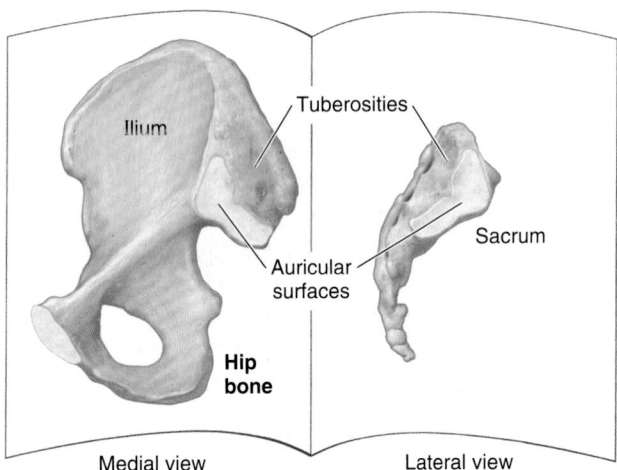

Ilium
Tuberosities
Sacrum
Auricular surfaces
Hip bone
Medial view
Lateral view

(B) Open book view of articulating surfaces of sacro-iliac joint

FIGURE 3.4. Joints of pelvic girdle. A. The sacro-iliac joints unite the axial and inferior appendicular skeletons. The lumbosacral and sacrococcygeal joints are joints of the axial skeleton directly related to the pelvic girdle. **B.** The auricular surfaces and tuberosities of the ilium and sacrum are demonstrated in an "opened book" view.

the weight of the upper body from the axial skeleton to the two ilia of the appendicular skeleton (Fig. 3.5A).

The **posterior sacro-iliac ligaments** are the posterior external continuation of the same mass of fibrous tissue (Figs. 3.5A and 3.6). Because the fibers of the interosseous and posterior sacro-iliac ligaments run obliquely upward and outward from the sacrum, the axial weight pushing down on the sacrum actually pulls the ilia inward (medially) so that they compress the sacrum between them, locking the irregular but congruent surfaces of the sacro-iliac joints together. The *iliolumbar ligaments* are accessory ligaments to this mechanism (Fig. 3.6).

Inferiorly, the posterior sacro-iliac ligaments are joined by fibers extending from the posterior margin of the ilium (between the posterior superior and posterior inferior iliac spines) and the base of the coccyx to form the massive *sacrotuberous ligament* (Fig. 3.6). This ligament passes from the posterior ilium and lateral sacrum and coccyx to the ischial tuberosity, transforming the sciatic notch of the hip bone into a

large sciatic foramen. The *sacrospinous ligament*, passing from lateral sacrum and coccyx to the ischial spine, further subdivides this foramen into *greater* and *lesser sciatic foramina.*

Most of the time, movement at the sacro-iliac joint is limited by interlocking of the articulating bones and the sacro-iliac ligaments to slight gliding and rotary movements (Fig. 3.5D). When landing after a high jump or when weightlifting in the standing position, exceptional force is transmitted through the bodies of the lumbar vertebrae to the superior end of the sacrum. Because this transfer of weight occurs anterior to the axis of the sacro-iliac joints, the superior end of the sacrum is pushed inferiorly and anteriorly. However, rotation of the superior sacrum is counterbalanced by the strong sacrotuberous and sacrospinous ligaments that anchor the inferior end of the sacrum to the ischium, preventing its superior and posterior rotation (Figs. 3.5D and 3.6). By allowing only slight upward movement of the inferior end of the sacrum relative to the hip bones, resilience is provided to the sacro-iliac region when the vertebral column sustains sudden increases in force or weight.

PUBIC SYMPHYSIS

The **pubic symphysis** consists of a fibrocartilaginous interpubic disc and surrounding ligaments uniting the bodies of the pubic bones in the median plane (Fig. 3.7). The **interpubic disc** is generally wider in women. The ligaments joining the bones are thickened at the superior and inferior margins of the symphysis, forming superior and inferior pubic ligaments. The **superior pubic ligament** connects the superior aspects of the pubic bodies and interpubic disc, extending as far laterally as the pubic tubercles. The **inferior** (arcuate) **pubic ligament** is a thick arch of fibers that connects the inferior aspects of the joint components, rounding off the *subpubic angle* as it forms the apex of the *pubic arch* (Fig. 3.3). The decussating, fibers of the tendinous attachments of the rectus abdominis and external oblique muscles also strengthen the pubic symphysis anteriorly (see Chapter 2).

LUMBOSACRAL JOINTS

L5 and S1 vertebrae articulate at the anterior *intervertebral (IV) joint* formed by the L5/S1 *IV disc* between their bodies (Fig. 3.4A) and at two posterior *zygapophysial joints* (facet joints) between the articular processes of these vertebrae (Fig. 3.1). The facets on the S1 vertebra face posteromedially, interlocking with the anterolaterally facing inferior articular facets of the L5 vertebra, preventing the lumbar vertebra from sliding anteriorly down the incline of the sacrum. These joints are further strengthened by fan-like **iliolumbar ligaments** radiating from the transverse processes of the L5 vertebra to the ilia (Fig. 3.6).

SACROCOCCYGEAL JOINT

The **sacrococcygeal joint** is a secondary cartilaginous joint (Fig. 3.4A) with an IV disc. Fibrocartilage and ligaments join the apex of the sacrum to the base of the coccyx. The **anterior** and **posterior sacrococcygeal ligaments** are long strands that reinforce the joint (Fig. 3.6).

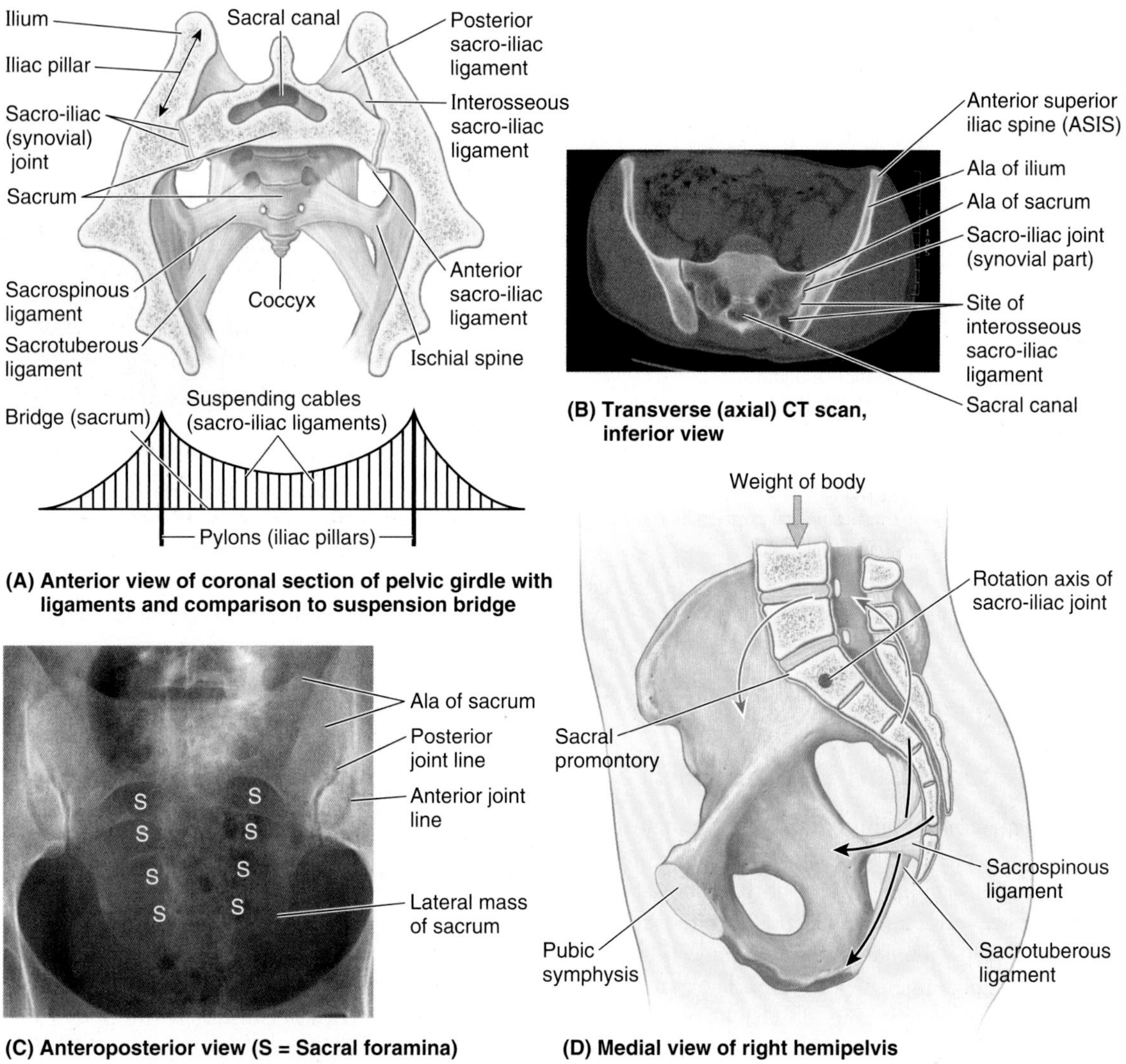

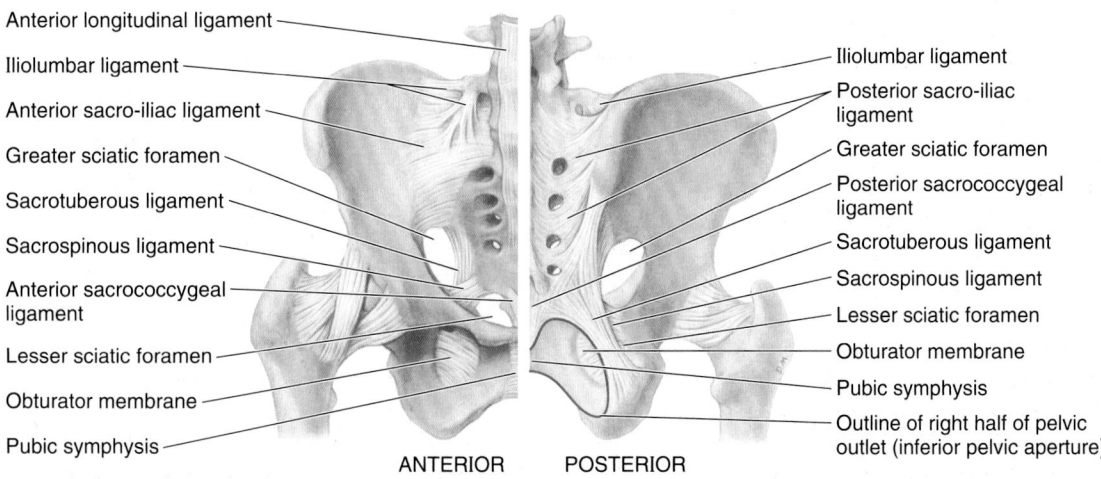

Ilium
Iliac pillar
Sacro-iliac (synovial) joint
Sacrum
Sacrospinous ligament
Sacrotuberous ligament
Sacral canal
Coccyx
Posterior sacro-iliac ligament
Interosseous sacro-iliac ligament
Anterior sacro-iliac ligament
Ischial spine

Bridge (sacrum)
Suspending cables (sacro-iliac ligaments)
Pylons (iliac pillars)

(A) Anterior view of coronal section of pelvic girdle with ligaments and comparison to suspension bridge

Anterior superior iliac spine (ASIS)
Ala of ilium
Ala of sacrum
Sacro-iliac joint (synovial part)
Site of interosseous sacro-iliac ligament
Sacral canal

(B) Transverse (axial) CT scan, inferior view

Ala of sacrum
Posterior joint line
Anterior joint line
Lateral mass of sacrum

(C) Anteroposterior view (S = Sacral foramina)

Weight of body
Rotation axis of sacro-iliac joint
Sacral promontory
Sacrospinous ligament
Sacrotuberous ligament
Pubic symphysis

(D) Medial view of right hemipelvis

FIGURE 3.5. Sacro-iliac joints and pubic symphysis with associated ligaments. A. The posterior half of a coronally sectioned pelvic girdle and its sacro-iliac joints are shown. The strong interosseous sacro-iliac ligaments lie deep (antero-inferior) to the posterior sacro-iliac ligaments and consist of shorter fibers connecting the tuberosity of the sacrum to the tuberosity of the ilium, suspending the sacrum from the ilia (left and right ilium) like the central portion of a suspension bridge is suspended from the pylons at each end. **B.** CT scan of the synovial and syndesmotic portions of the sacro-iliac joint. **C.** Because the articulating surfaces are irregular and slightly oblique, the anterior and posterior parts of the joint appear separately in an AP radiograph. **D.** The weight of the body is transmitted to the sacrum anterior to the axis of rotation at the sacro-iliac joint. The tendency for increased weight or force to rotate the upper sacrum anteriorly and inferiorly is resisted by the strong sacrotuberous and sacrospinous ligaments anchoring the inferior sacrum and coccyx to the ischium.

Anterior longitudinal ligament
Iliolumbar ligament
Anterior sacro-iliac ligament
Greater sciatic foramen
Sacrotuberous ligament
Sacrospinous ligament
Anterior sacrococcygeal ligament
Lesser sciatic foramen
Obturator membrane
Pubic symphysis

Iliolumbar ligament
Posterior sacro-iliac ligament
Greater sciatic foramen
Posterior sacrococcygeal ligament
Sacrotuberous ligament
Sacrospinous ligament
Lesser sciatic foramen
Obturator membrane
Pubic symphysis
Outline of right half of pelvic outlet (inferior pelvic aperture)

ANTERIOR   POSTERIOR

FIGURE 3.6. Ligaments of pelvic girdle. The ligaments of hip joint (shown but not labeled) are identified in Chapter 5 (Lower Limb).

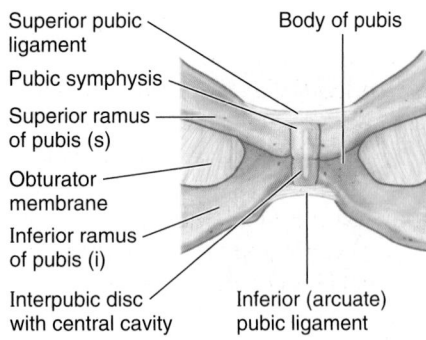

(A) Antero-inferior view of pubic bones and pubic symphysis

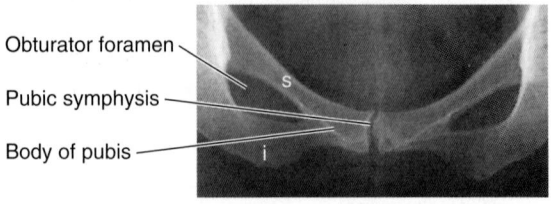

(B) Anteroposterior radiograph of pubic bones and symphysis

FIGURE 3.7. Pubic symphysis. A. The pubic symphysis is a secondary cartilaginous joint between the bodies of the pubic bones. **B.** Radiographic appearance of the pubic symphysis in the anatomical position. In this position, the bodies of the pubic bones are nearly horizontal, and the joint appears foreshortened.

PELVIC GIRDLE

Variations in Male and Female Pelves

Although anatomical differences between male and female pelves are usually clear cut, the pelvis of any person may have some features of the opposite sex. The pelvic types shown in Figure B3.1A and C are most common in males, B and A in white females, and B and C in black females, whereas D is uncommon in both sexes. The **gynecoid pelvis** is the normal female type (Fig. B3.1B); its pelvic inlet typically has a rounded oval shape and a wide transverse diameter. An **platypelloid** or markedly **android** (masculine or funnel-shaped) **pelvis** in a woman may present hazards to successful vaginal delivery of a fetus (Fig. B3.1A).

In *forensic medicine* (the application of medical and anatomical knowledge for the purposes of law), identification of human skeletal remains usually involves the diagnosis of sex. A prime focus of attention is the pelvic girdle because sexual differences usually are clearly visible. Even fragments of the pelvic girdle are useful in determining sex.

Pelvic Diameters (Conjugates)

The size of the lesser pelvis is particularly important in obstetrics because it is the bony canal through which the fetus passes during normal child birth. To determine the capacity of the female pelvis for childbearing, the diameters of the lesser pelvis are noted radiographically or manually during a pelvic examination. The minimum

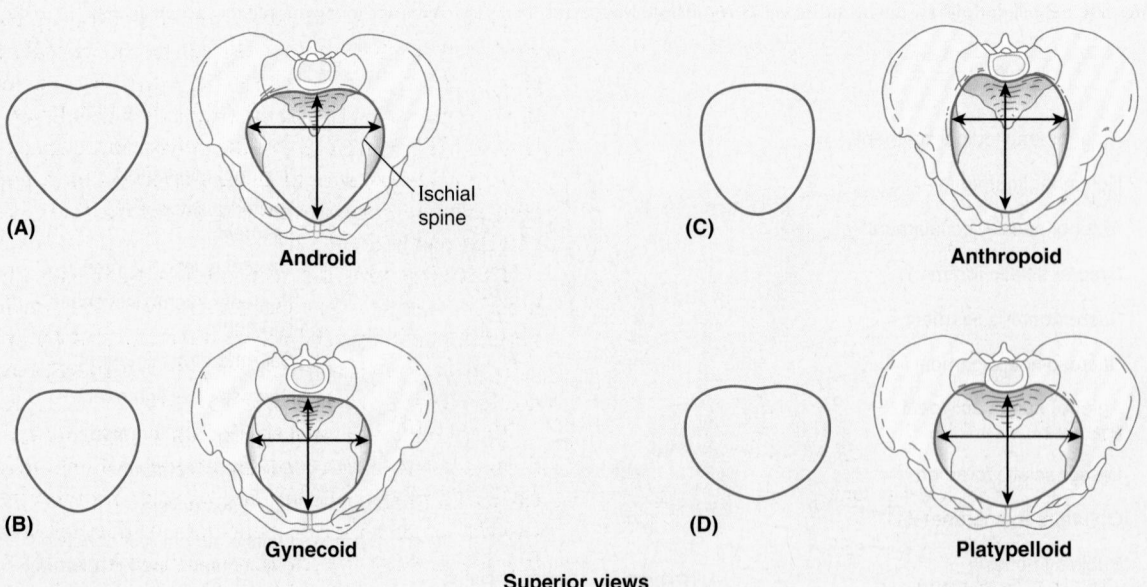

Superior views

FIGURE B3.1.

anteroposterior (AP) diameter of the lesser pelvis, the **true** (obstetrical) **conjugate** from the middle of the sacral promontory to the posterosuperior margin (closest point) of the pubic symphysis (Fig. B3.2A & B), is the narrowest *fixed distance* through which the baby's head must pass in a vaginal delivery. This distance, however, cannot be measured directly during a pelvic examination because of the presence of the

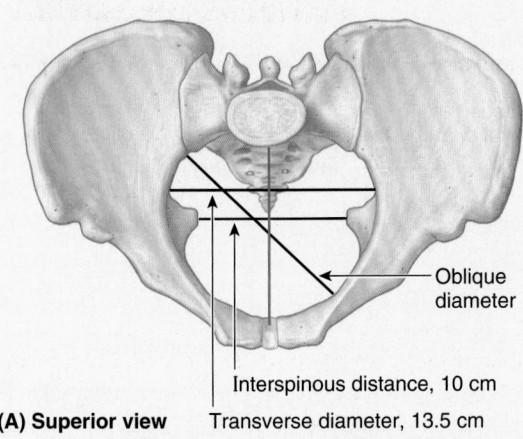

Oblique diameter

Interspinous distance, 10 cm

(A) Superior view   Transverse diameter, 13.5 cm

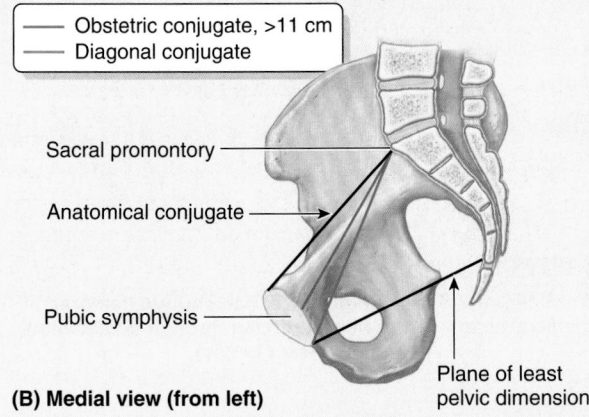

— Obstetric conjugate, >11 cm
— Diagonal conjugate

Sacral promontory —

Anatomical conjugate —

Pubic symphysis —

Plane of least pelvic dimension

(B) Medial view (from left)

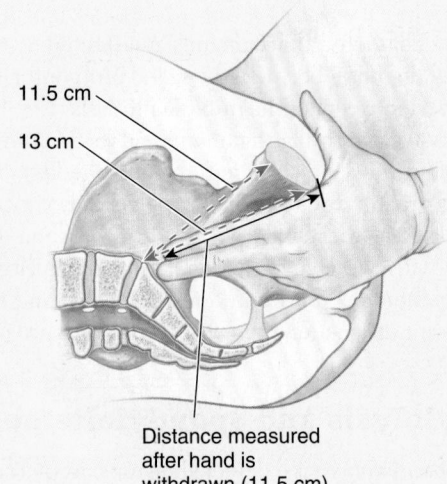

11.5 cm
13 cm

Distance measured after hand is withdrawn (11.5 cm)

(C) Medial view (from right)

FIGURE B3.2.

bladder. Consequently, the **diagonal conjugate** (Fig. B3.2B) is measured by palpating the sacral promontory with the tip of the *middle finger,* using the other hand to mark the level of the inferior margin of the pubic symphysis on the examining hand (Fig. B3.2C). After the examining hand is withdrawn, the distance between the tip of the *index finger* (1.5 cm shorter than the middle finger) and the marked level of the pubic symphysis is measured to estimate the true conjugate, which should be 11.0 cm or greater.

In all pelvic girdles, the ischial spines extend toward each other, and the **interspinous distance** between them is normally the narrowest part of the **pelvic canal** (the passageway through the pelvic inlet, lesser pelvis, and pelvic outlet) through which a baby's head must pass at birth (Fig. B3.2B), but it is not a fixed distance (see the blue box "Relaxation of Pelvic Ligaments and Increased Joint Mobility During Pregnancy," p. 336). During a pelvic examination, if the ischial tuberosities are far enough apart to permit three fingers to enter the vagina side by side, the subpubic angle is considered sufficiently wide to permit passage of an average fetal head at full term.

Pelvic Fractures

Anteroposterior compression of the pelvis occurs during crush accidents (as when a heavy object falls on the pelvis, Fig. B3.3A). This type of trauma commonly produces *fractures of the pubic rami*. When the pelvis is compressed laterally, the acetabula and ilia are squeezed toward each other and may be broken.

Fractures of the bony pelvic ring are almost always multiple fractures or a fracture combined with a joint dislocation. To illustrate this, try breaking a pretzel ring at just one point. Some pelvic fractures result from the tearing away of bone by the strong ligaments associated with the sacro-iliac joints. (These ligaments are shown in Figs 3.3 and 3.4A.)

Pelvic fractures can result from direct trauma to the pelvic bones, such as occurs during an automobile accident (Fig. B3.3A), or be caused by forces transmitted to these bones from the lower limbs during falls on the feet (Fig. B3.3B). Weak areas of the pelvis, where fractures often occur, are the pubic rami, the acetabula (or the area immediately surrounding them), the region of the sacro-iliac joints, and the alae of the ilium.

Pelvic fractures may cause injury to pelvic soft tissues, blood vessels, nerves, and organs. Fractures in the pubo-obturator area are relatively common and are often complicated because of their relationship to the urinary bladder and urethra, which may be ruptured or torn.

Falls on the feet or buttocks from a high ladder may drive the head of the femur through the acetabulum into the pelvic cavity, injuring pelvic viscera, nerves, and vessels. In individuals younger than 17 years of age, the acetabulum may fracture through the triradiate cartilage into its three developmental parts (Fig. 3.2C) or the bony acetabular margins may be torn away.

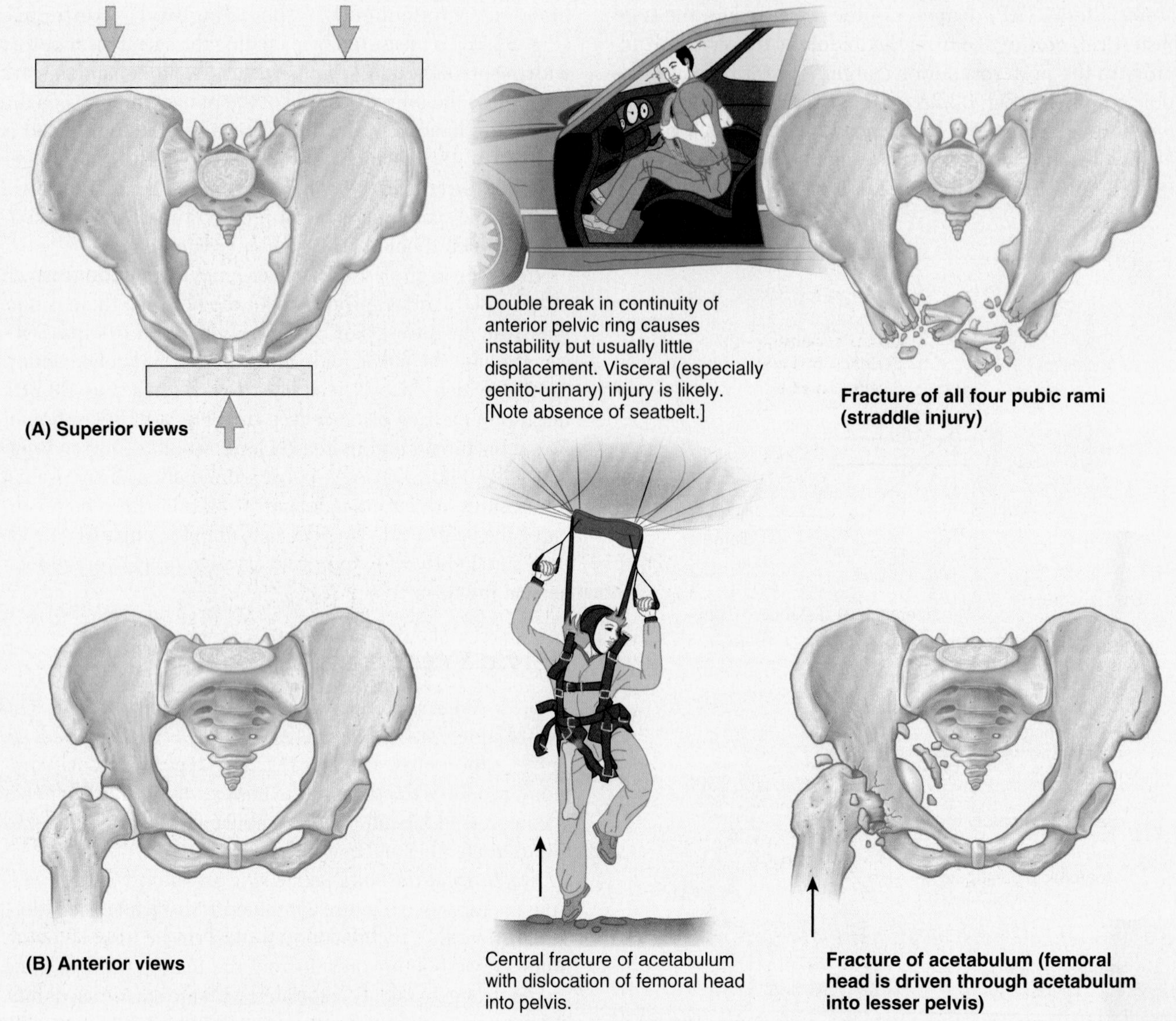

Double break in continuity of anterior pelvic ring causes instability but usually little displacement. Visceral (especially genito-urinary) injury is likely. [Note absence of seatbelt.]

Fracture of all four pubic rami (straddle injury)

(A) Superior views

(B) Anterior views

Central fracture of acetabulum with dislocation of femoral head into pelvis.

Fracture of acetabulum (femoral head is driven through acetabulum into lesser pelvis)

FIGURE B3.3.

Relaxation of Pelvic Ligaments and Increased Joint Mobility in Late Pregnancy

The larger cavity of the interpubic disc in females (Fig. 3.3) increases in size during pregnancy. This change increases the circumference of the lesser pelvis and contributes to increased flexibility of the pubic symphysis. Increased levels of sex hormones and the presence of the hormone *relaxin* cause the pelvic ligaments to relax during the latter half of pregnancy, allowing *increased movement at the pelvic joints*. Relaxation of the sacro-iliac joints and pubic symphysis permits as much as a 10–15% increase in diameters (mostly transverse, including the interspinous distance—Fig. B3.2A), facilitating passage of the fetus through the pelvic canal. The coccyx is also able to move posteriorly.

The one diameter that remains unaffected is the true (obstetrical) diameter between the sacral promontory and the posterosuperior aspect of the pubic symphysis (Fig. B3.2A & B). Relaxation of sacro-iliac ligaments causes the interlocking mechanism of the sacro-iliac joint to become less effective, permitting greater rotation of the pelvis and contributing to the lordotic ("swayback") posture often assumed during pregnancy with the change in the center of gravity. Relaxation of ligaments is not limited to the pelvis, and the possibility of joint dislocation increases during late pregnancy.

Spondylolysis and Spondylolisthesis

Spondylolysis is a defect allowing part of a *vertebral arch* (the posterior projection from the vertebral body that surrounds the spinal canal and bears the articular, transverse, and spinal processes) to be separated

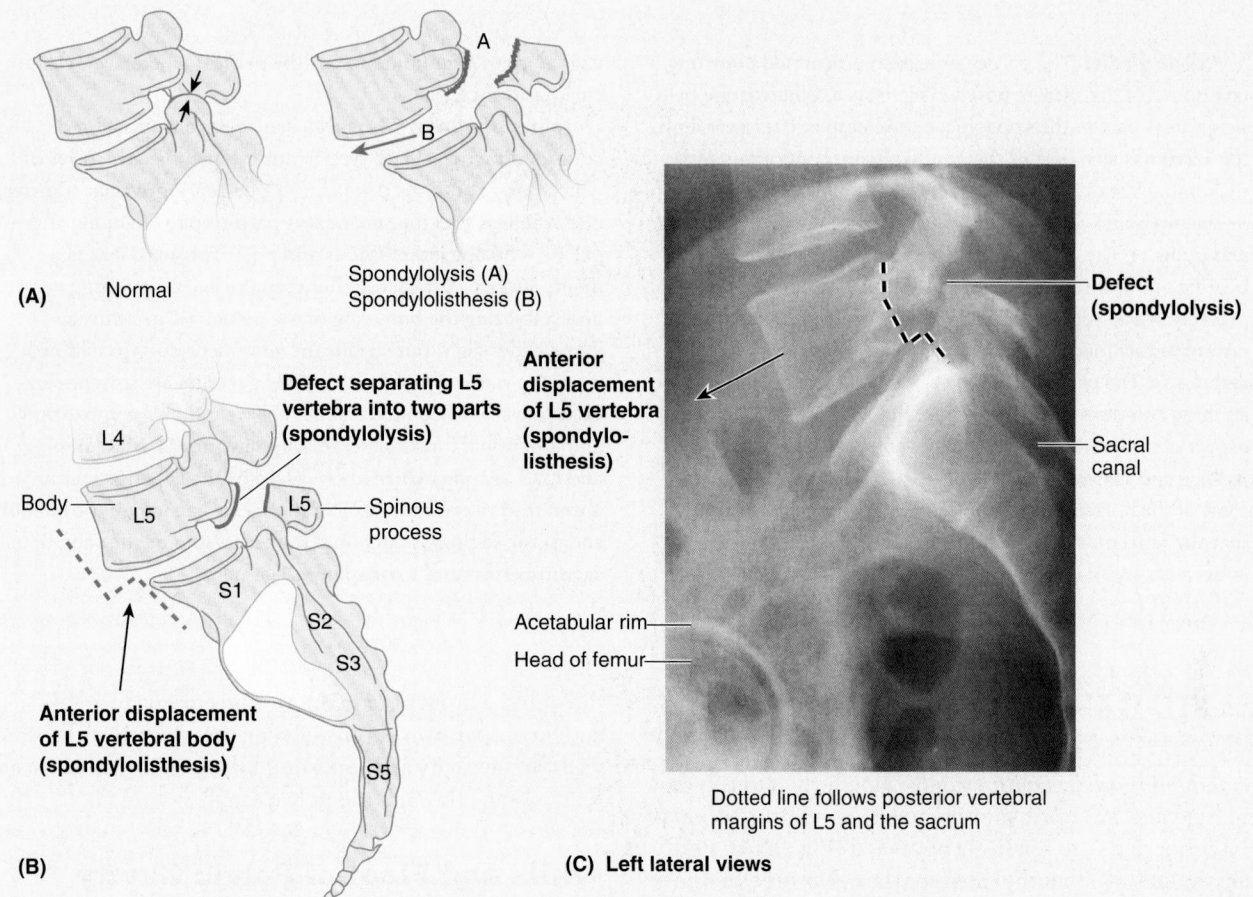

(A) Normal

Spondylolysis (A)
Spondylolisthesis (B)

Defect separating L5 vertebra into two parts (spondylolysis)

Anterior displacement of L5 vertebra (spondylo-listhesis)

L4

Body

L5

L5 — Spinous process

S1
S2
S3

S5

Anterior displacement of L5 vertebral body (spondylolisthesis)

(B)

Defect (spondylolysis)

Sacral canal

Acetabular rim

Head of femur

Dotted line follows posterior vertebral margins of L5 and the sacrum

(C) Left lateral views

FIGURE B3.4. **A.** Comparison of a normal L5 vertebra with intact articular processes that prevent spondylolisthesis and an L5 vertebra with spondylolysis (A) resulting in spondylolisthesis (B). **B.** Interpretive figure of radiograph (**C**) demonstrating spondylolysis and spondylolisthesis.

from its body. *Spondylolysis of vertebra L5* results in the separation of the vertebral body from the part of its vertebral arch bearing the inferior articular processes (Fig. B3.4A). The inferior articular processes of L5 normally interlock with the articular processes of the sacrum. When the defect is bilateral, the body of the L5 vertebrae may slide anteriorly on the sacrum (*spondylolisthesis*) so that it overlaps the sacral promontory (Fig. B3.4B–C). The intrusion of the L5 body into the pelvic inlet reduces the AP diameter of the pelvic inlet, which may interfere with parturition (child-

birth). It may also compress spinal nerves, causing low back or lower limb pain.

Obstetricians test for spondylolisthesis by running their fingers along the lumbar spinous processes. An abnormally prominent L5 process indicates that the anterior part of L5 vertebra and the vertebral column superior to it may have moved anteriorly relative to the sacrum and the vertebral arch of L5. Medical images, such as sagittal magnetic resonance imaging (MRI), are taken to confirm the diagnosis, and to measure the AP diameter of the pelvic inlet.

The Bottom Line

PELVIS AND PELVIC GIRDLE

Pelvis: The pelvis is the space enclosed by the pelvic girdle, which is subdivided into the greater pelvis (the inferior part of the abdominal cavity, which receives the protection of the alae of the ilia) and the lesser pelvis (the space inside the bony ring of pelvis inferior to the pelvic brim). ♦ The lesser pelvis provides the skeletal framework for both the pelvic cavity and

the perineum, which are separated by the musculofascial pelvic diaphragm. ♦ The term *perineum* refers both to the region that includes the anus and external genitalia and to a shallow compartment deep to that area. ♦ The inferior anterolateral abdominal wall, gluteal region, and perineum overlap the pelvis.

Pelvic girdle: The pelvic girdle is an articulated bony ring composed of the sacrum and two hip bones. Whereas the pelvic girdle is part of the appendicular skeleton of the lower limb, the sacrum is also part of the axial skeleton, continuous with the lumbar vertebrae superiorly and coccyx inferiorly.
♦ The hip bones are formed by the fusion of the ilium, ischium, and pubis. ♦ The primary functions of the pelvic girdle are bearing and transfer of weight; secondary functions include protection and support of abdominopelvic viscera and housing and attachment for structures of the genital and urinary systems. ♦ The pelvic girdle is in the anatomical position when its three anteriormost points (right and left ASISs and anterior aspect of pubic symphysis) lie in the same vertical plane.
♦ Male and female pelves are distinct. The characteristic features of the normal (gynecoid) female pelvis reflect the fact that the fetus must traverse the pelvic canal during childbirth.
♦ Because atypical female pelves may not be conducive to a vaginal birth, determination of the pelvic diameters is of clinical importance.

Joints of pelvis: The sacro-iliac joints are specialized compound synovial and syndesmotic joints, the structures of which reflect both the primary (weight-bearing/weight transfer and stability) and the secondary (parturition) functions of the pelvis. ♦ Strong interosseous and posterior sacro-iliac ligaments suspend the sacrum between the ilia, transferring weight and stabilizing the bony ring of the pelvis. ♦ The synovial joints allow slight but significant movement during childbirth, when the pubic symphysis and the ligaments are softened by hormones. ♦ To counterbalance the weight of the upper body and additional forces generated by activities such as jumping and load bearing, which are received by the superior sacrum anterior to the rotatory axis of the sacro-iliac joints, the inferior end of the sacrum is anchored to the ischium by the substantial sacrotuberous and sacrospinous ligaments.

PELVIC CAVITY

The *abdominopelvic cavity* extends superiorly into the thoracic cage and inferiorly into the pelvis, so that its superior and inferior parts are relatively protected (Fig. 3.8A). Perforating wounds in either the thorax or the pelvis may therefore involve the abdominopelvic cavity and its contents.

The funnel-shaped **pelvic cavity**—the space bounded peripherally by the bony, ligamentous, muscular pelvic walls and floor—is the inferoposterior part of the abdominopelvic cavity. The pelvic cavity is continuous with the abdominal cavity at the *pelvic inlet* but angulated posteriorly from it (Fig. 3.8A & C). Although continuous, the abdominal and pelvic cavities are described separately for descriptive purposes, facilitating the regional approach.

The pelvic cavity contains the terminal parts of the ureters, the urinary bladder, rectum, pelvic genital organs, blood vessels, lymphatics, and nerves. In addition to these distinctly pelvic viscera, it also contains what might be considered an overflow of abdominal viscera: loops of small intestine (mainly ileum) and, frequently, large intestine (appendix and transverse and/or sigmoid colon).

The pelvic cavity is limited inferiorly by the musculofascial *pelvic diaphragm,* which is suspended above (but descends centrally to the level of) the *pelvic outlet,* forming a bowl-like *pelvic floor.* The pelvic cavity is bounded posteriorly by the coccyx and inferiormost sacrum, with the superior part of the sacrum forming a roof over the posterior half of the cavity (Fig. 3.8A & B).

The bodies of the pubic bones, and the pubic symphysis uniting them, form an antero-inferior wall that is much shallower (shorter) than the posterosuperior wall and ceiling formed by sacrum and coccyx. Consequently, the **axis of the pelvis** (a line in the median plane defined by the center point of the pelvic cavity at every level) is curved, pivoting around the pubic symphysis (Fig.3.8A). The curving form of the axis and the disparity in depth between the anterior and posterior walls of the cavity are important factors in the mechanism of fetal passage through the pelvic canal.

Walls and Floor of Pelvic Cavity

The pelvic cavity has an antero-inferior wall, two lateral walls, a posterior wall, and a floor (Fig. 3.9A). The muscles forming the walls and floor of the pelvic cavity are demonstrated in Figure 3.10, and the proximal and distal attachments, innervation, and main actions of the muscles are described in Table 3.2.

ANTERO-INFERIOR PELVIC WALL

The antero-inferior pelvic wall (more of a weight-bearing floor than an anterior wall in the anatomical position) is formed primarily by the bodies and rami of the pubic bones and the pubic symphysis (Figs. 3.7 and 3.9B–D). It participates in bearing the weight of the urinary bladder.

LATERAL PELVIC WALLS

The lateral pelvic walls are formed by the right and left hip bones, each of which includes an obturator foramen closed by an **obturator membrane** (Figs. 3.8C & 3.9B). The fleshy attachments of the *obturator internus muscles* cover and thus pad most of the lateral pelvic walls (Figs. 3.9C and 3.10A). The fleshy fibers of each obturator internus converge posteriorly, become tendinous, and turn sharply laterally to pass from the lesser pelvis through the *lesser sciatic foramen* to attach to the greater trochanter of the femur. The medial surfaces of these muscles are covered by **obturator fascia,** thickened centrally as a *tendinous arch* that provides attachment for the pelvic diaphragm (Fig. 3.9D).

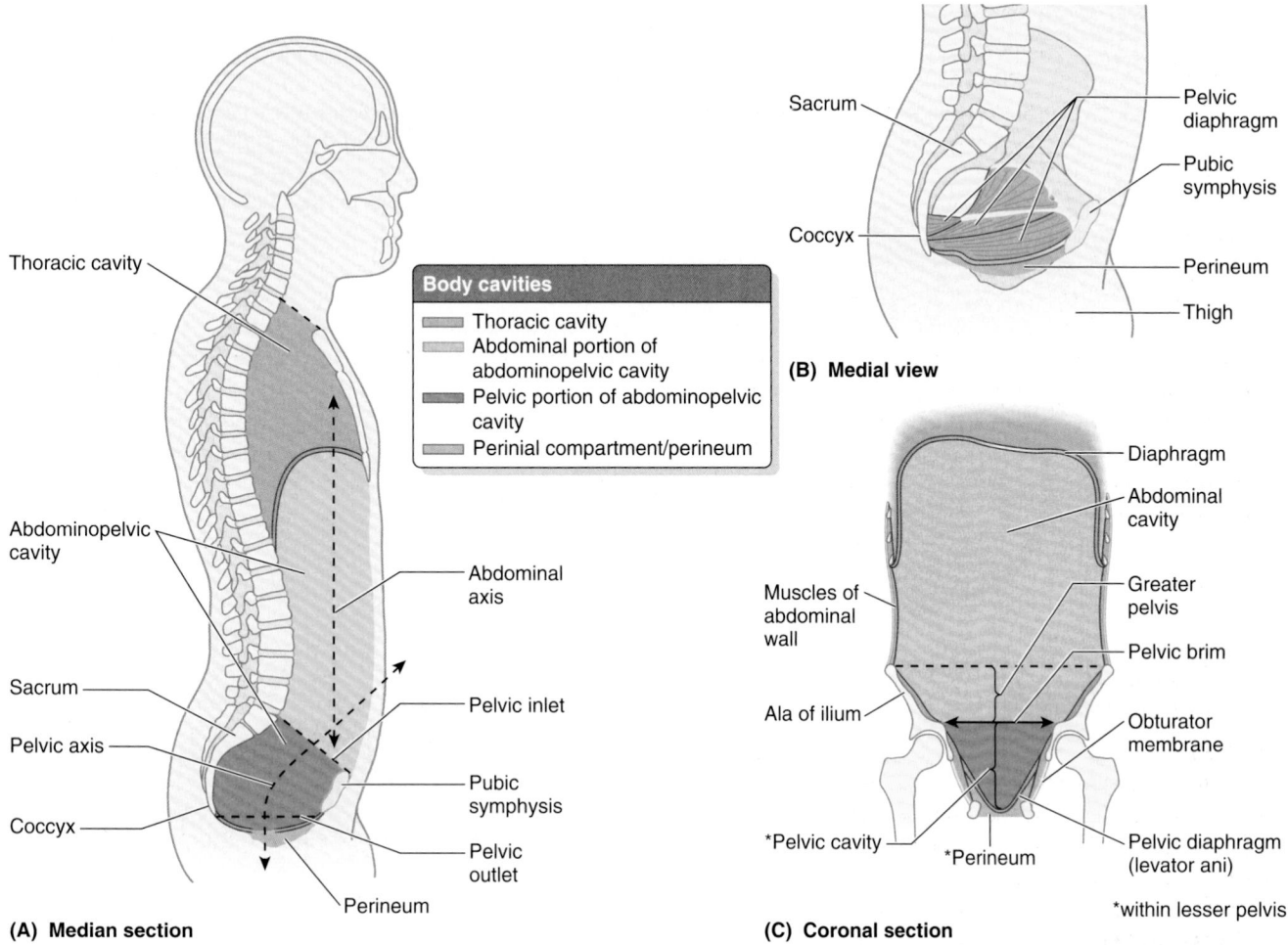

Thoracic cavity

Body cavities
- Thoracic cavity
- Abdominal portion of abdominopelvic cavity
- Pelvic portion of abdominopelvic cavity
- Perinial compartment/perineum

Abdominopelvic cavity

Sacrum

Pelvic axis

Coccyx

Abdominal axis

Pelvic inlet

Pubic symphysis

Pelvic outlet

Perineum

(A) Median section

Sacrum — Pelvic diaphragm — Pubic symphysis

Coccyx — Perineum — Thigh

(B) Medial view

Diaphragm — Abdominal cavity

Muscles of abdominal wall — Greater pelvis — Pelvic brim

Ala of ilium — Obturator membrane

*Pelvic cavity — *Perineum — Pelvic diaphragm (levator ani)

*within lesser pelvis

(C) Coronal section

FIGURE 3.8. Thoracic and abdominopelvic cavities. A and C. These sections of the trunk show the relationship of the thoracic and abdominopelvic cavities. Although the greater pelvis and pelvic cavity are actually continuous, they are demarcated by the plane of the pelvic inlet (defined by the pelvic brim). **B.** The pelvic diaphragm is a dynamic barrier separating the lesser pelvis and the perineum, forming the floor of the former and roof of the latter.

POSTERIOR WALL (POSTEROLATERAL WALL AND ROOF)

In the anatomical position, the posterior pelvic wall consists of a bony wall and roof in the midline (formed by the sacrum and coccyx) and musculoligamentous posterolateral walls, formed by the ligaments associated with the sacro-iliac joints and piriformis muscles (Fig. 3.9A–C). The ligaments include the anterior sacro-iliac, sacrospinous, and sacrotuberous ligaments.

The *piriformis muscles* arise from the superior sacrum, lateral to its pelvic foramina (Figs. 3.9A and 3.10A). The muscles pass laterally, leaving the lesser pelvis through the *greater sciatic foramen* to attach to the superior border of the greater trochanter of the femur (Fig. 3.10B). These muscles occupy much of the greater sciatic foramen, forming the posterolateral walls of the pelvic cavity (Fig. 3.9A). Immediately deep (anteromedial) to these muscles (often embedded in the fleshy fibers) are the nerves of the *sacral plexus* (Fig. 3.9D). A gap at the inferior border of the piriformis allows passage of neurovascular structures between the pelvis and the perineum and lower limb (gluteal region).

PELVIC FLOOR

The pelvic floor is formed by the bowl- or funnel-shaped **pelvic diaphragm,** which consists of the coccygeus and levator ani muscles and the fascias (L. *fasciae*) covering the superior and inferior aspects of these muscles (Figs. 3.9A, 3.10C, and 3.11; Table 3.2). The pelvic diaphragm lies within the lesser pelvis, separating the pelvic cavity from the perineum, for which it forms the roof.

The attachment of the diaphragm to the obturator fascia divides the obturator internus into a superior pelvic portion and an inferior perineal portion (Fig. 3.11B). Medial to the pelvic portions of the obturator internus muscles are the obturator nerves and vessels and other branches of the internal iliac vessels.

The **coccygeus muscles** arise from the lateral aspects of the inferior sacrum and coccyx, their fleshy fibers lying on and attaching to the deep surface of the sacrospinous ligament (Fig. 3.9B & C). The **levator ani** (a broad muscular sheet) is the larger and more important part of the pelvic floor. It is attached to the bodies of the pubic bones anteriorly, to the

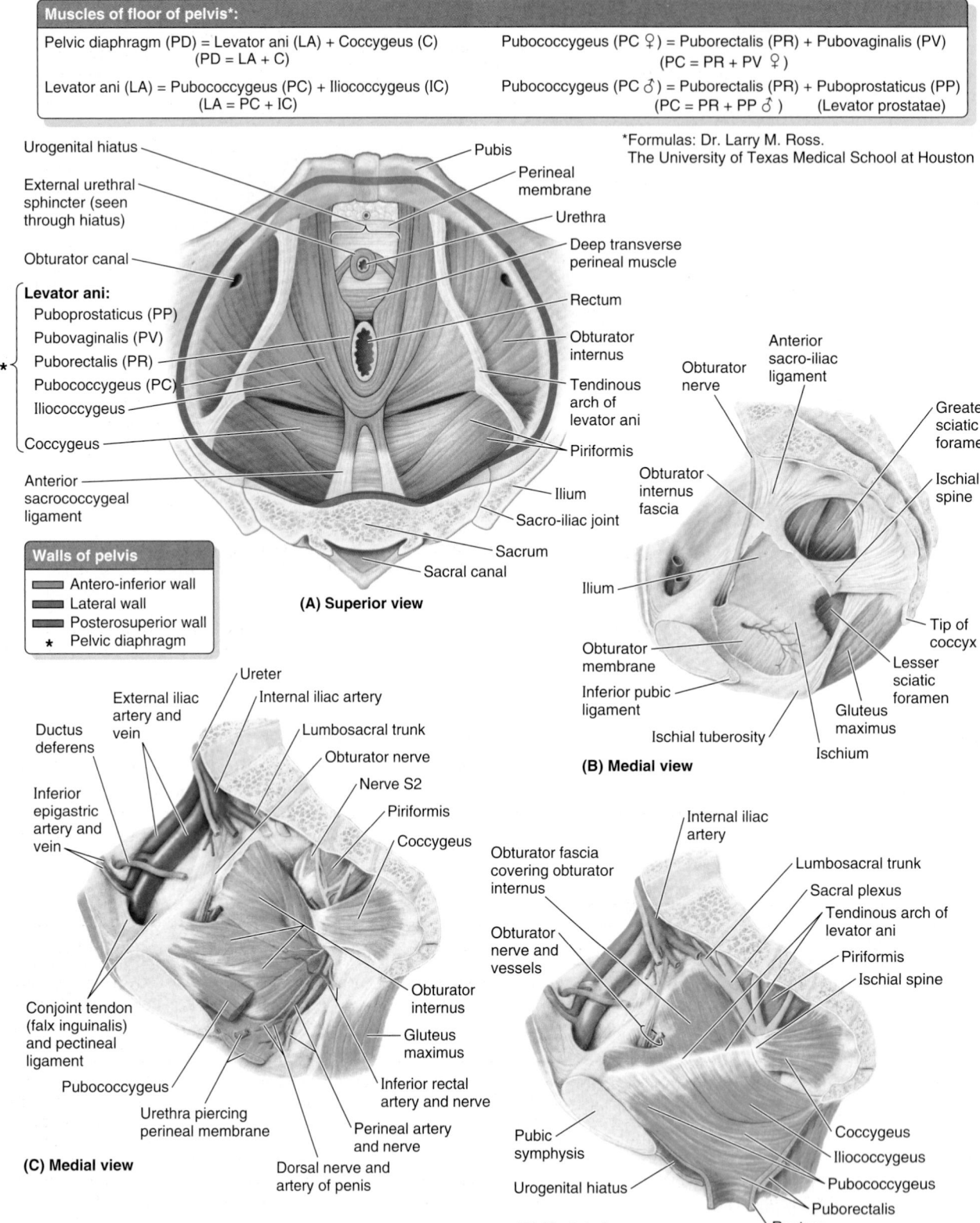

Muscles of floor of pelvis*:

Pelvic diaphragm (PD) = Levator ani (LA) + Coccygeus (C)
(PD = LA + C)

Levator ani (LA) = Pubococcygeus (PC) + Iliococcygeus (IC)
(LA = PC + IC)

Pubococcygeus (PC ♀) = Puborectalis (PR) + Pubovaginalis (PV)
(PC = PR + PV ♀)

Pubococcygeus (PC ♂) = Puborectalis (PR) + Puboprostaticus (PP)
(PC = PR + PP ♂)   (Levator prostatae)

*Formulas: Dr. Larry M. Ross.
The University of Texas Medical School at Houston

Urogenital hiatus
External urethral sphincter (seen through hiatus)
Obturator canal

Levator ani:
Puboprostaticus (PP)
Pubovaginalis (PV)
Puborectalis (PR)
Pubococcygeus (PC)
Iliococcygeus
Coccygeus

Anterior sacrococcygeal ligament

Pubis
Perineal membrane
Urethra
Deep transverse perineal muscle
Rectum
Obturator internus
Tendinous arch of levator ani
Piriformis
Ilium
Sacro-iliac joint
Sacrum
Sacral canal

Walls of pelvis
Antero-inferior wall
Lateral wall
Posterosuperior wall
* Pelvic diaphragm

(A) Superior view

Anterior sacro-iliac ligament
Obturator nerve
Obturator internus fascia
Greater sciatic foramen
Ischial spine
Tip of coccyx
Obturator membrane
Inferior pubic ligament
Ischial tuberosity
Gluteus maximus
Ischium
Lesser sciatic foramen
Ilium

(B) Medial view

Ureter
External iliac artery and vein
Internal iliac artery
Lumbosacral trunk
Obturator nerve
Nerve S2
Piriformis
Coccygeus
Ductus deferens
Inferior epigastric artery and vein
Conjoint tendon (falx inguinalis) and pectineal ligament
Pubococcygeus
Urethra piercing perineal membrane
Dorsal nerve and artery of penis
Obturator internus
Gluteus maximus
Inferior rectal artery and nerve
Perineal artery and nerve

(C) Medial view

Obturator fascia covering obturator internus
Obturator nerve and vessels
Internal iliac artery
Lumbosacral trunk
Sacral plexus
Tendinous arch of levator ani
Piriformis
Ischial spine
Pubic symphysis
Urogenital hiatus
Coccygeus
Iliococcygeus
Pubococcygeus
Puborectalis
Rectum

(D) Medial view

FIGURE 3.9. Floor and walls of pelvis. A. The floor of the pelvis is formed by the pelvic diaphragm, encircled by and suspended in part from the pubic symphysis and pubic bones anteriorly, the ilia laterally, and the sacrum and coccyx posteriorly. Parts **B** through **D** show the staged reconstruction of the parietal structures of the right hemipelvis. **B.** Posterolaterally, the coccyx and inferior part of the sacrum are attached to the ischial tuberosity by the sacrotuberous ligament and to the ischial spine by the sacrospinous ligament. The obturator membrane, composed of strong interlacing fibers, fills the obturator foramen. **C.** The muscles of the lesser pelvis are added. The obturator internus pads the lateral wall of the pelvis, its fibers converging to escape posteriorly through the lesser sciatic foramen (see part **B**). **D.** The levator ani is added, suspended from a thickening in the obturator fascia (the tendinous arch), which extends from the pubic body to the ischial spine.

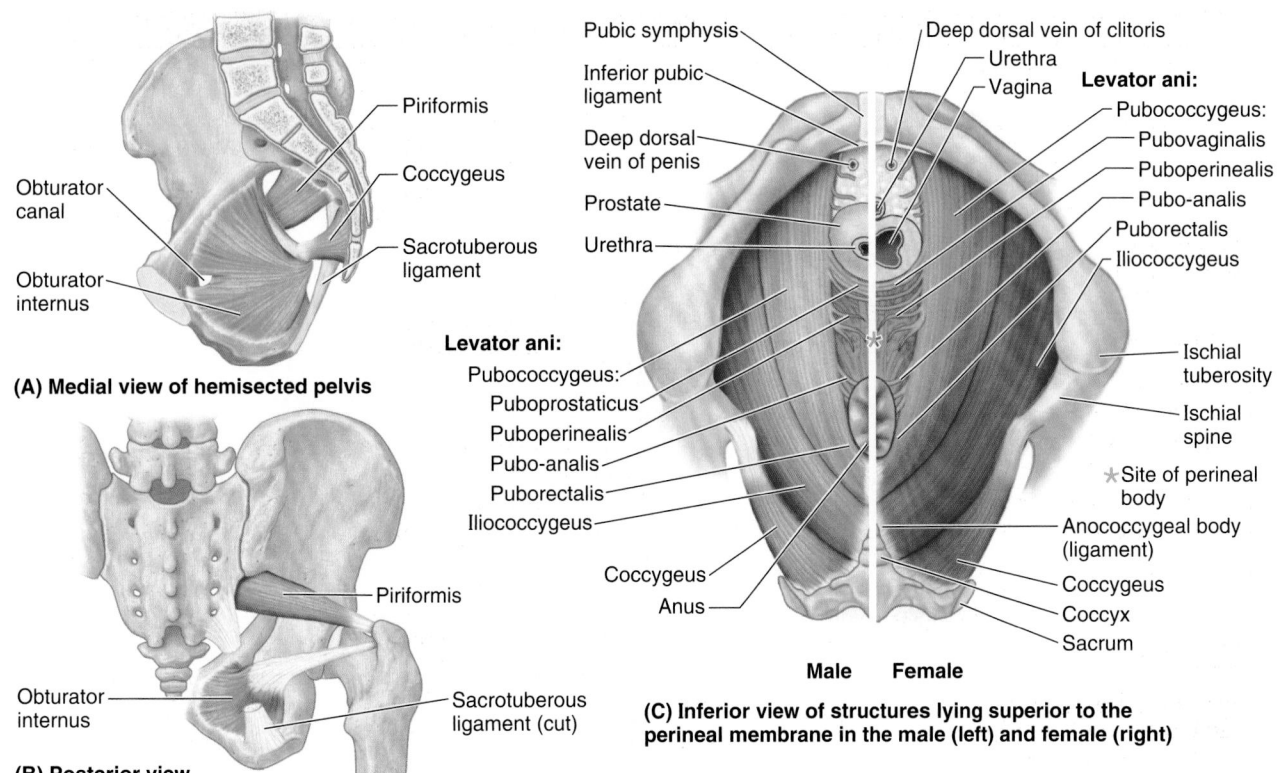

(A) Medial view of hemisected pelvis

(B) Posterior view

(C) Inferior view of structures lying superior to the perineal membrane in the male (left) and female (right)

FIGURE 3.10. Muscles of pelvic walls and floor. A and B. The obturator internis and piriformis are muscles that act on the lower limb but are also components of the pelvic walls. **C.** The muscles of the levator ani and the coccygeus comprise the pelvic diaphragm that forms the floor of the pelvic cavity. The fascia covering the inferior surface of the pelvic diaphragm forms the "roof" of the perineum.

ischial spines posteriorly, and to a thickening in the obturator fascia (the **tendinous arch of the levator ani**) between the two bony sites on each side.

The pelvic diaphragm thus stretches between the anterior, the lateral, and the posterior walls of the lesser pelvis, giving it the appearance of a hammock suspended from these attachments, closing much of the ring of the pelvic girdle. An anterior gap between the medial borders of the levator ani muscles of each side—the **urogenital hiatus**—gives passage to the urethra and, in females, the vagina (Fig. 3.9A).

TABLE 3.2. MUSCLES OF PELVIC WALLS AND FLOOR

Boundary	Muscle	Proximal Attachment	Distal Attachment	Innervation	Main Action
Lateral wall	**Obturator internus**	Pelvic surfaces of ilium and ischium; obturator membrane	Greater trochanter of femur	Nerve to obturator internus (L5, S1, S2)	Rotates thigh laterally; assists in holding head of femur in acetabulum
Postero-superior wall	**Piriformis**	Pelvic surface of S2–S4 segments; superior margin of greater sciatic notch and sacrotuberous ligament	Greater trochanter of femur	Anterior rami of S1 and S2	Rotates thigh laterally; abducts thigh; assists in holding head of femur in acetabulum
Floor	**Coccygeus** (ischiococcygeus)	Ischial spine	Inferior end of sacrum and coccyx	Branches of S4 and S5 spinal nerves	Forms small part of pelvic diaphragm that supports pelvic viscera; flexes coccyx
	Levator ani (pubo-rectalis, pubo-coccygeus, and iliococcygeus)	Body of pubis; tendinous arch of obturator fascia; ischial spine	Perineal body; coccyx; anococcygeal ligament; walls of prostate or vagina, rectum, and anal canal	Nerve to levator ani (branches of S4), inferior anal (rectal) nerve, and coccygeal plexus	Forms most of pelvic diaphragm that helps support pelvic viscera and resists increases in intra-abdominal pressure

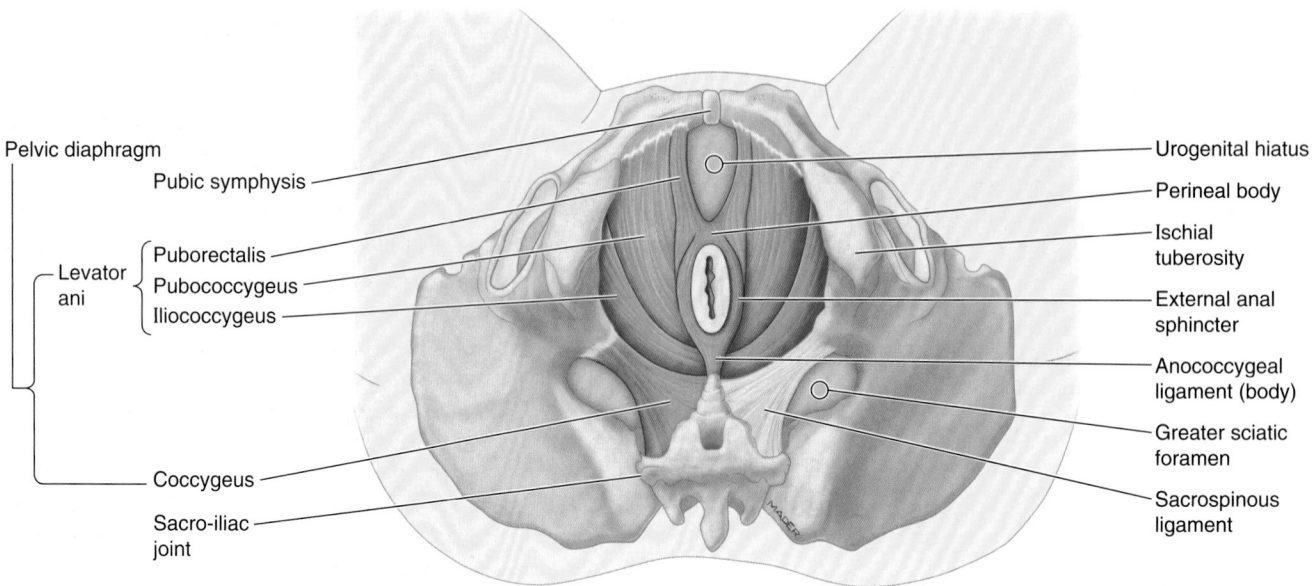

(A) Inferior view of perineum, lithotomy position

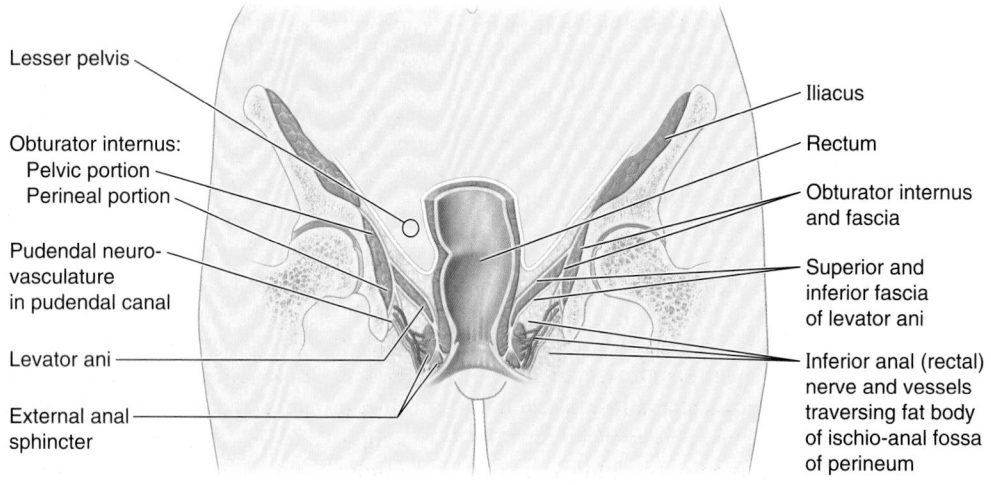

(B) Anterior view of schematic coronal section

FIGURE 3.11. Pelvic diaphragm and anorectum in situ. A and B. The components of the pelvic diaphragm (levator ani and coccygeus) form the floor of the pelvic cavity and the roof of the perineum. **B.** The basin-like nature for which the pelvis was named is evident in this coronal section. The fat-filled ischio-anal fossae of the perineum also lie within the bony ring of the lesser pelvis.

The levator ani consists of three parts, often poorly demarcated but designated according to attachments and fiber course (Figs. 3.9A & D, 3.10C, and 3.11):

- **Puborectalis:** the thicker, narrower, medial part of the levator ani, consisting of muscle fibers that are continuous between the posterior aspects of the bodies of the right and left pubic bones. It forms a U-shaped muscular sling (*puborectal sling*) that passes posterior to the anorectal junction (Fig. 3.12), bounding the urogenital hiatus. This part plays a major role in maintaining fecal continence.
- **Pubococcygeus:** the wider but thinner intermediate part of the levator ani, which arises lateral to the puborectalis

from the posterior aspect of the body of the pubis and anterior tendinous arch (Figs. 3.9A & D, 3.10C, and 3.11). It passes posteriorly in a nearly horizontal plane; its lateral fibers attach to the coccyx and its medial fibers merge with those of the contralateral muscle to form a fibrous raphe or tendinous plate, part of the **anococcygeal body** or ligament between the anus and the coccyx (often referred to clinically as the "levator plate").

Shorter muscular slips of the pubococcygeus extending medially and blending with the fascia around midline structures are named for the structure near their termination: *pubovaginalis* (females), *puboprostaticus* (males), *puboperinealis*, and *pubo-analis*.

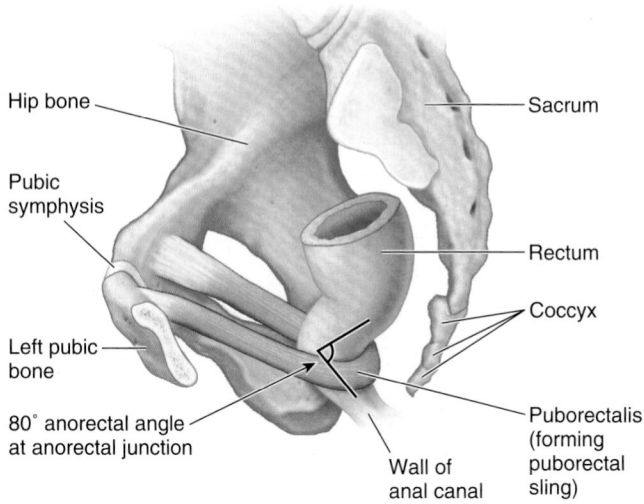

Hip bone

Pubic symphysis

Left pubic bone

80° anorectal angle at anorectal junction

Sacrum

Rectum

Coccyx

Puborectalis (forming puborectal sling)

Wall of anal canal

Medial view from left

FIGURE 3.12. Puborectalis muscle. Most of the left hip bone has been removed to demonstrate that this part of the levator ani is formed by continuous muscle fibers following a U-shaped course around the anorectal junction. The puborectalis thus forms a puborectal sling, the tonus of which is responsible for maintaining the anorectal angle (perineal flexure).

- **Iliococcygeus:** the posterolateral part of the levator ani, which arises from the posterior tendinous arch and ischial spine. It is thin and often poorly developed (appearing more aponeurotic than muscular), and also blends with the anococcygeal body posteriorly.

The levator ani forms a dynamic floor for supporting the abdominopelvic viscera. It is tonically contracted most of the time to support the abdominopelvic viscera, and to assist in maintaining urinary and fecal continence. It is actively contracted during activities such as forced expiration, coughing, sneezing, vomiting, and fixation of the trunk during strong movements of the upper limbs (e.g., when lifting heavy objects), primarily to increase support of the viscera during periods of increased intra-abdominal pressure, and perhaps secondarily to contribute to the increased pressure (to aid expulsion).

Penetrated centrally by the anal canal, the levator ani is funnel shaped, with the U-shaped puborectalis looping around the "funnel spout"; its tonic contraction bends the anorectum anteriorly. Active contraction of the (voluntary) puborectalis portion is important in maintaining fecal continence immediately after rectal filling or during peristalsis when the rectum is full and the involuntary sphincter muscle is inhibited (relaxed).

The levator ani must relax to allow urination and defecation. The increased intra-abdominal pressure for defecation is provided by contraction of the (thoracic) diaphragm and muscles of the anterolateral abdominal wall. Acting together, the parts of the levator ani elevate the pelvic floor after their relaxation and the consequent descent of the pelvic diaphragm that occurs during urination and defecation.

Peritoneum and Peritoneal Cavity of Pelvis

The parietal peritoneum lining the abdominal cavity continues inferiorly into the pelvic cavity, but does not reach the pelvic floor. Instead, it reflects onto the pelvic viscera, remaining separated from the pelvic floor by the pelvic viscera and the surrounding pelvic fascia (Table 3.3). Except for the ovaries and uterine tubes, the pelvic viscera are not completely ensheathed by the peritoneum, lying inferior to it for the main part. Only their superior and superolateral surfaces are covered with peritoneum. Only the uterine tubes (except for their ostia, which are open) are intraperitoneal and suspended by a mesentery. The ovaries, although suspended in the peritoneal cavity by a mesentery, are not covered with glistening peritoneum; instead a special, relatively dull epithelium of cuboidal cells (*germinal epithelium*) covers them.

A loose areolar (fatty) layer between the transversalis fascia and the parietal peritoneum of the inferior part of the anterior abdominal wall allows the bladder to expand between these layers as it becomes distended with urine. The region superior to the bladder (*1* in Table 3.3) is the only site where the parietal peritoneum is not firmly bound to the underlying structures. Consequently, the level at which the peritoneum reflects onto the superior surface of the bladder, creating the **supravesical fossa** (*2* in Table 3.3), is variable, depending on the fullness of the bladder. When the peritoneum reflects from the abdominopelvic wall onto the pelvic viscera and fascia, a series of folds and fossae is created (*2–7* in Table 3.3).

In the female, as the peritoneum at or near the midline reaches the posterior border of the roof of the bladder, it reflects onto the anterior aspect of the uterus at the isthmus of the uterus (see "Female Internal Genital Organs," p. 382); thus it is not related to the anterior vaginal fornix, which is subperitoneal in location. The peritoneum passes over the fundus of the uterus and descends the entire posterior aspect of the uterus onto the posterior vaginal wall before reflecting superiorly onto the anterior wall of the inferior rectum (rectal ampulla). The "pocket" thus formed between the uterus and the rectum is the **recto-uterine pouch** (cul-de-sac of Douglas) (*6* in Table 3.3C). The median recto-uterine pouch is often described as being the inferiormost extent of the peritoneal cavity in the female, but often its lateral extensions on each side of the rectum, the **pararectal fossae,** are deeper.

Prominent peritoneal ridges, the **recto-uterine folds,** formed by underlying fascial ligaments demarcate the lateral boundaries of the pararectal fossae (Table 3.3A). As the peritoneum passes up and over the uterus in the middle of the pelvic cavity, a double peritoneal fold, the *broad ligament of the uterus,* extends between the uterus and the lateral pelvic wall on each side, forming a partition that separates the paravesical fossae and pararectal fossae of each side. The uterine tubes, ovaries, ligaments of the ovaries, and round ligaments of the uterus are enclosed within the broad ligaments. Subdivisions of the broad ligament related to these structures will be discussed with the uterus later in this chapter. Recall that in

TABLE 3.3. PERITONEAL REFLECTIONS IN PELVIS[a]

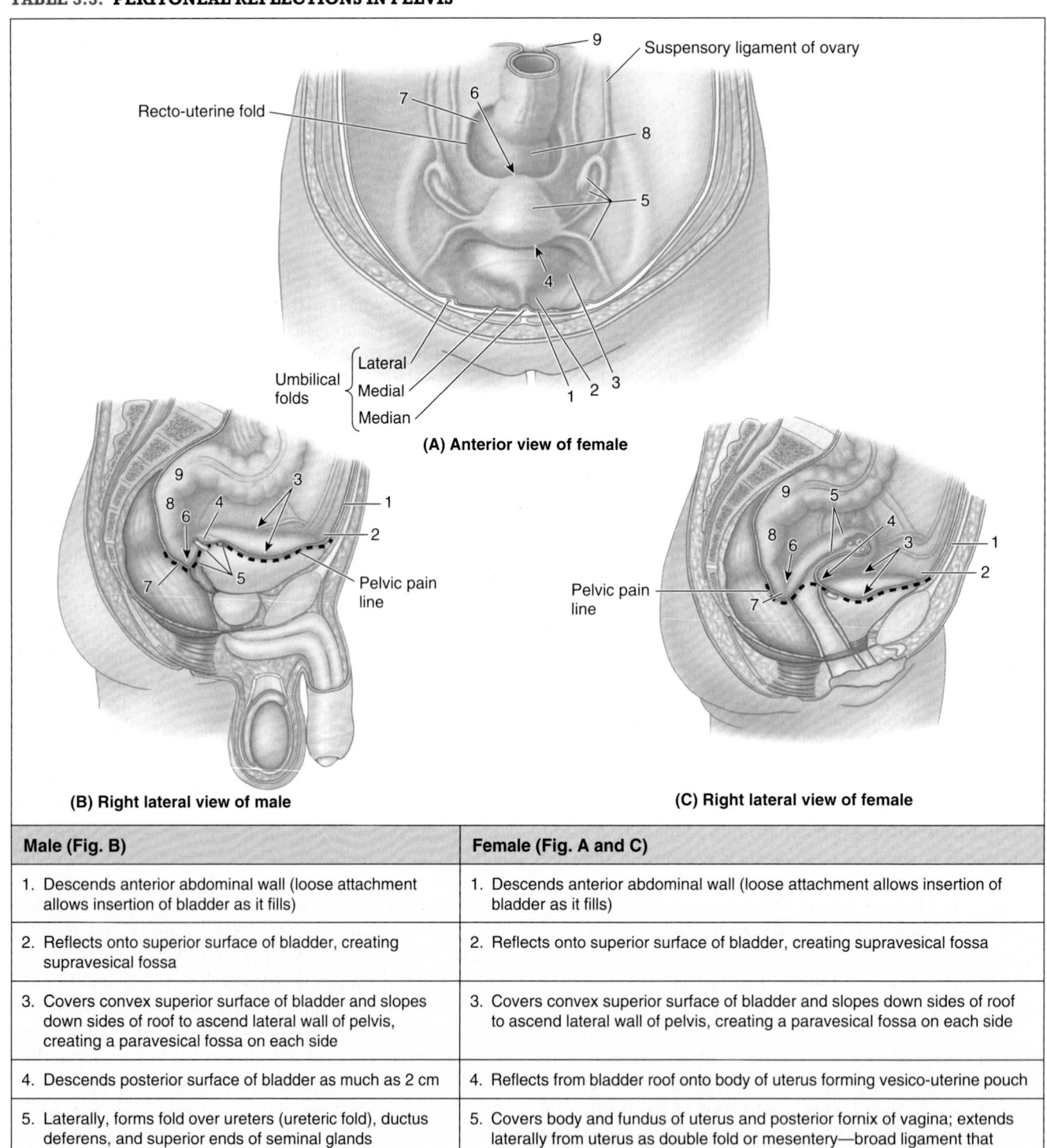

(A) Anterior view of female

(B) Right lateral view of male

(C) Right lateral view of female

Male (Fig. B)	Female (Fig. A and C)
1. Descends anterior abdominal wall (loose attachment allows insertion of bladder as it fills)	1. Descends anterior abdominal wall (loose attachment allows insertion of bladder as it fills)
2. Reflects onto superior surface of bladder, creating supravesical fossa	2. Reflects onto superior surface of bladder, creating supravesical fossa
3. Covers convex superior surface of bladder and slopes down sides of roof to ascend lateral wall of pelvis, creating a paravesical fossa on each side	3. Covers convex superior surface of bladder and slopes down sides of roof to ascend lateral wall of pelvis, creating a paravesical fossa on each side
4. Descends posterior surface of bladder as much as 2 cm	4. Reflects from bladder roof onto body of uterus forming vesico-uterine pouch
5. Laterally, forms fold over ureters (ureteric fold), ductus deferens, and superior ends of seminal glands	5. Covers body and fundus of uterus and posterior fornix of vagina; extends laterally from uterus as double fold or mesentery—broad ligament that engulfs uterine tubes and round ligaments of uterus and suspends ovaries
6. Reflects from bladder and seminal glands onto rectum, forming rectovesical pouch	6. Reflects from vagina onto rectum, forming recto-uterine pouch
7. Rectovesical pouch extends laterally and posteriorly to form a pararectal fossa on each side of rectum	7. Recto-uterine pouch extends laterally and posteriorly to form a pararectal fossa on each side of rectum
8. Ascends rectum; from inferior to superior, rectum is subperitoneal and then retroperitoneal	8. Ascends rectum; from inferior to superior, rectum is subperitoneal and then retroperitoneal
9. Engulfs sigmoid colon beginning at rectosigmoid junction	9. Engulfs sigmoid colon beginning at rectosigmoid junction

[a]Numbers refer to table figures.

females the pelvic peritoneal cavity communicates with the external environment via the uterine tubes, uterus, and vagina.

In males and in females who have had a *hysterectomy* (removal of the uterus), the central peritoneum descends a short distance (as much as 2 cm) down the posterior surface (base) of the bladder, and then reflects superiorly onto the anterior surface of the inferior rectum, forming the **rectovesical pouch.** The female recto-uterine pouch is normally deeper (extends farther caudally) than the male rectovesical pouch.

In the male, a gentle peritoneal fold or ridge, the **ureteric fold,** is formed as the peritoneum passes up and over the ureter and *ductus (vas) deferens* (secretory duct of the testis) on each side of the posterior bladder, separating the paravesical and pararectal fossae; in this regard, it is the male equivalent of the broad ligament. Posterior to the ureteric folds and lateral to the central rectovesical pouch, the peritoneum often descends far enough caudally to cover the superior ends or superior posterior surfaces of the *seminal glands (vesicles)* and *ampullae of the ductus deferens.* Except for these sites (and the testis in its tunica vaginalis, which is derived from peritoneum), the male reproductive organs are not in contact with the peritoneum.

In both sexes, the inferior third of the rectum is below the inferior limits of the peritoneum (i.e., it is subperitoneal), the middle third is covered with peritoneum only on its anterior surface, and the superior third is covered on both its anterior and its lateral surfaces. The rectosigmoid junction, near the pelvic brim, is intraperitoneal.

Pelvic Fascia

The **pelvic fascia** is connective tissue that occupies the space between the membranous peritoneum and the muscular pelvic walls and floor not occupied by the pelvic viscera. This "layer" is a continuation of the comparatively thin (except around kidneys) endoabdominal fascia that lies between the muscular abdominal walls and the peritoneum superiorly. Traditionally, the pelvic fascia has been described as having parietal and visceral components (Fig. 3.13).

MEMBRANOUS PELVIC FASCIA: PARIETAL AND VISCERAL

The **parietal pelvic fascia** is a membranous layer of variable thickness that lines the inner (deep or pelvic) aspect of the muscles forming the walls and floor of the pelvis—the obturator internus, piriformis, coccygeus, levator ani, and part of the urethral sphincter muscles. Specific parts of the parietal fascia are named for the muscle that is covered (e.g., obturator fascia). This layer is continuous superiorly with the transversalis and iliopsoas fascias.

The **visceral pelvic fascia** includes the membranous fascia that directly ensheathes the pelvic organs, forming the adventitial layer of each. The membranous parietal and visceral layers become continuous where the organs penetrate the pelvic floor (Figs. 3.13A & C and 3.14). Here the parietal fascia is thickened, forming the **tendinous arch of pelvic fascia,** a continuous

bilateral band running from the pubis to the sacrum along the pelvic floor adjacent to the viscera (Fig. 3.14A & B). The anteriormost part of this tendinous arch (**puboprostatic ligament** in males; **pubovesical ligament** in females) connects the prostate to the pubis in the male, or the fundus (base) of the bladder to the pubis in the female. The posteriormost part of the band runs as the **sacrogenital ligaments** from the sacrum around the side of the rectum to attach to the prostate in the male or the vagina in the female. In females, the lateral connection of the visceral fascia of the vagina with the tendinous arch of the pelvic fascia is the **paracolpium** (Fig. 3.13A). The paracolpia suspend the vagina between the tendinous arches, assisting the vagina in bearing the weight of the fundus of the bladder.

ENDOPELVIC FASCIA: LOOSE AND CONDENSED

Often the abundant connective tissue remaining between the parietal and visceral membranous layers is considered to be part of the visceral fascia, but sometimes it is labeled as parietal fascia. It is probably more realistic to consider this remaining fascia simply as extraperitoneal or *subperitoneal endopelvic fascia* (Fig. 3.13A & C), which lies adjacent to both the parietal and visceral membranous fascias. This fascia forms a connective tissue matrix or packing material for the pelvic viscera (Fig. 3.13B & D). It varies markedly in density and content. Some of it is an extremely *loose areolar (fatty) tissue,* relatively devoid of all but minor lymphatics and nutrient vessels. During dissection or surgery, the fingers can be pushed into this loose tissue with ease, creating actual spaces by blunt dissection, for example, between the pubis and bladder anteriorly and between the sacrum and rectum posteriorly. These *potential spaces,* normally consisting only of a layer of loose fatty tissue, are the **retropubic** (or *prevesical,* extended posterolaterally as *paravesical*) and **retrorectal** (or *presacral*) **spaces,** respectively. The presence of loose connective tissue here accommodates the expansion of the urinary bladder and rectal ampulla as they fill.

Although types of endopelvic fascia do not differ much in their gross appearance, other parts of the endopelvic fascia have a much more fibrous consistency, containing an abundance of collagen and elastic fibers, and a scattering of smooth muscle fibers. These parts are often described as "fascial condensations" or pelvic "ligaments." For example, during dissection, if you insert the fingers of one hand into the retropubic space and the fingers of the other hand into the presacral space and attempt to bring them together along the lateral pelvic wall, you will find that they do not meet or pass from one space to the other. They encounter the so-called **hypogastric sheath,** a thick band of condensed pelvic fascia. This fascial condensation is not merely a barrier separating the two potential spaces. It gives passage to essentially all the vessels and nerves passing from the lateral wall of the pelvis to the pelvic viscera, along with the ureters and, in the male, the ductus deferens.

As it extends medially from the lateral wall, the hypogastric sheath divides into three laminae (layers) that pass to or between the pelvic organs, conveying neurovascular structures and providing support. Because of the latter function,

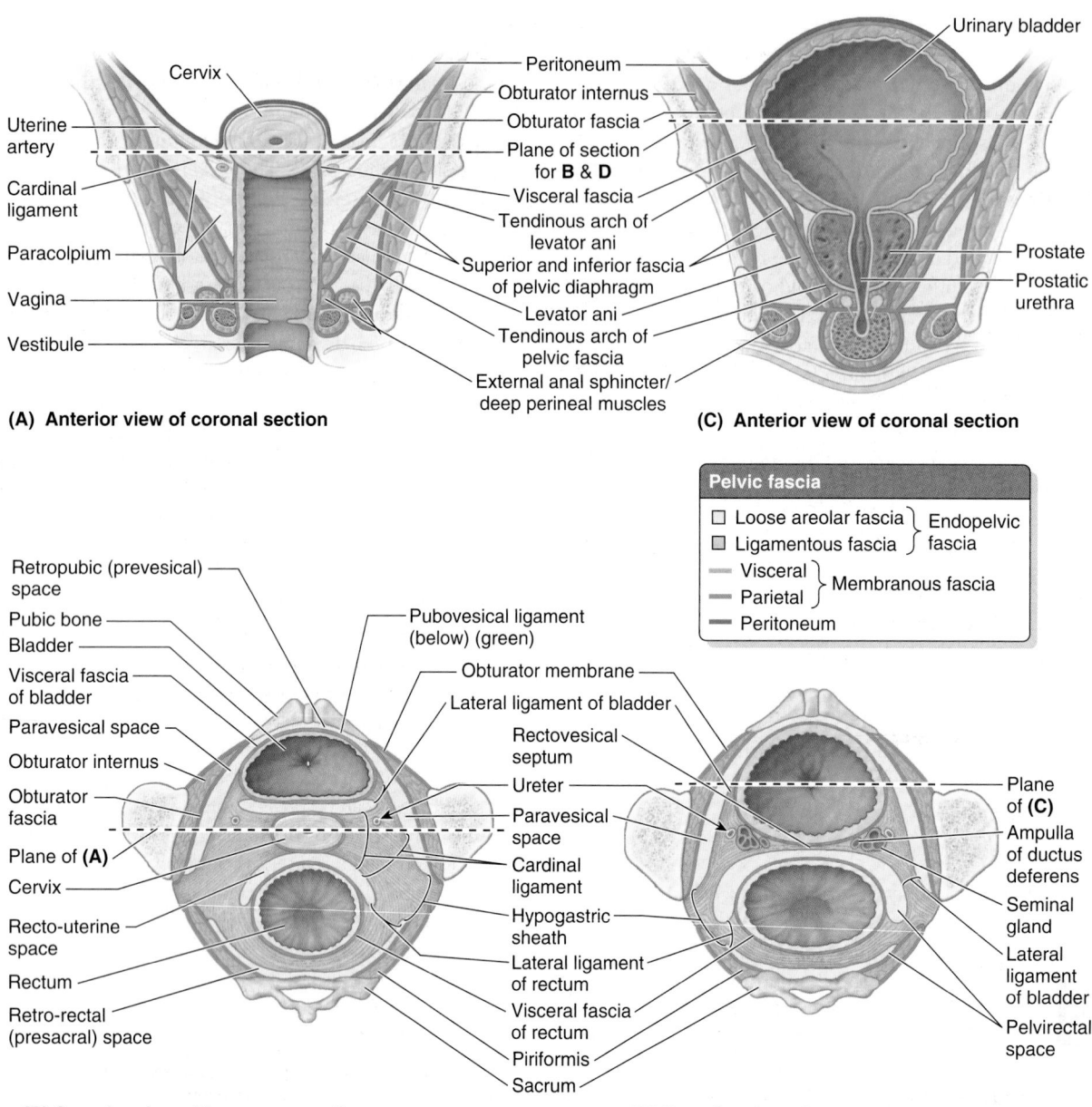

(A) Anterior view of coronal section

(C) Anterior view of coronal section

(B) Superior view of transverse section

(D) Superior view of transverse section

FIGURE 3.13. Pelvic fascia: endopelvic fascia and fascial ligaments. Coronal and transverse sections of female (**A** and **B**) and male (**C** and **D**) pelves demonstrating the parietal and visceral pelvic fascia and the endopelvic fascia between them, with its ligamentous and loose areolar components.

they are also referred to as ligaments. The anteriormost lamina, the **lateral ligament of the bladder,** passes to the bladder, conveying the superior vesical arteries and veins. The posteriormost lamina (*lateral rectal ligament*) passes to the rectum, conveying the middle rectal artery and vein.

In the male, the middle lamina forms a relatively thin fascial partition, the **rectovesical septum** (Fig. 3.13D), between the posterior surface of the bladder and the prostate anteriorly and the rectum posteriorly. In the female, the middle lamina is markedly more substantial than the other two, passing medially to the uterine cervix and vagina as the **cardinal ligament** (transverse cervical ligament) (Figs. 3.13B and 3.14A & B).

In its superiormost portion, at the base of the peritoneal broad ligament, the uterine artery runs medially toward the

cervix while the ureters pass immediately inferior to them. The ureters pass on each side of the cervix heading anteriorly toward the bladder. This relationship ("water passing under the bridge") is an especially important one for surgeons (see the blue box "Iatrogenic Injury of the Ureters" on p. 361). The cardinal ligament, and the way in which the uterus normally "rests" on top of the bladder, provide the main passive support for the uterus. The perineal muscles provide dynamic support for the uterus by contracting during moments of increased intra-abdominal pressure (sneezing, coughing, etc.). Passive and dynamic supports together resist the tendency for the uterus to fall or be pushed through the hollow tube formed by the vagina (uterine prolapse). The cardinal ligament has enough fibrous content to anchor wide loops of suture during surgical repairs.

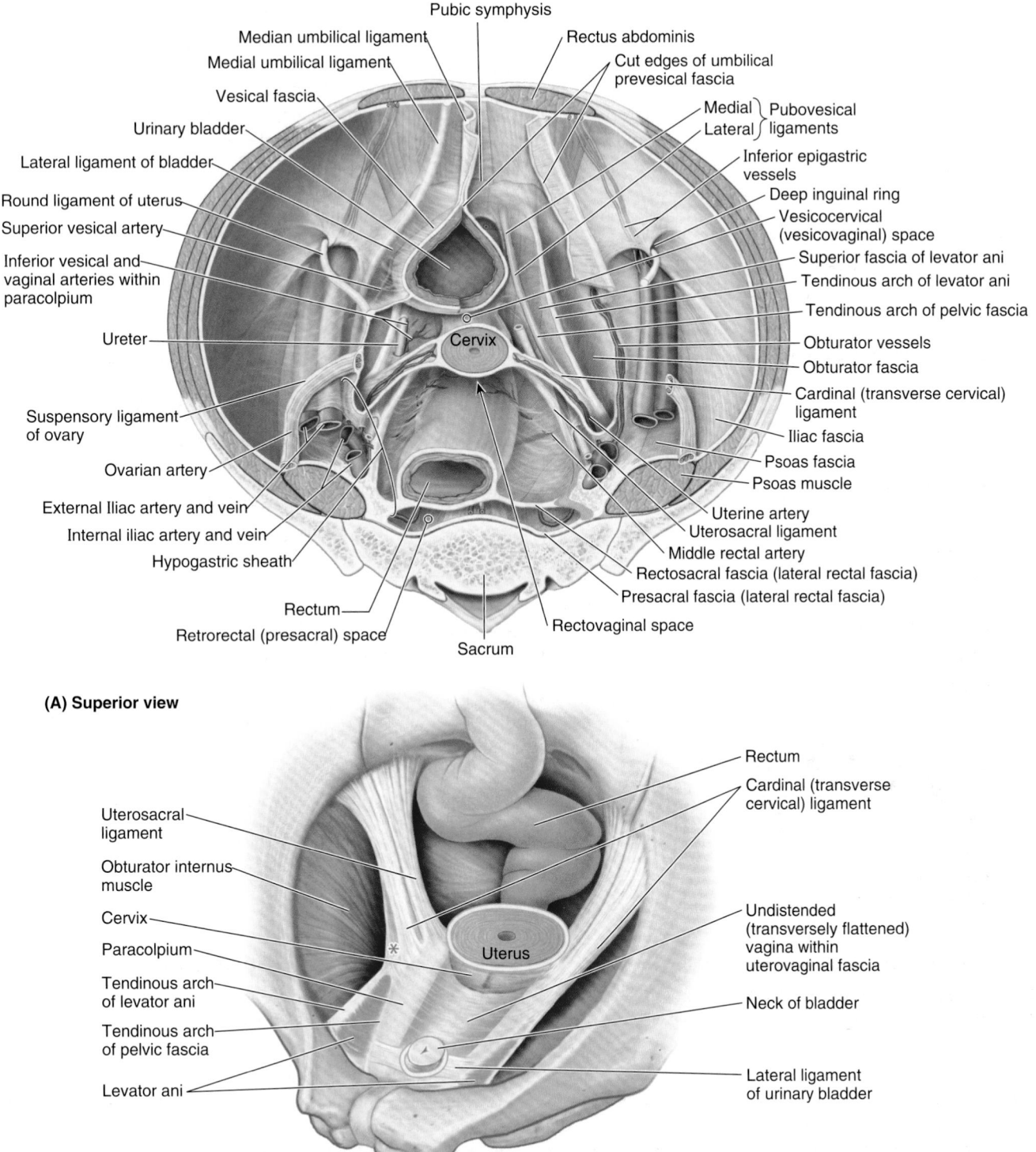

(A) Superior view

(B) Left anterolateral view

∗ Ischial spine

FIGURE 3.14. Pelvic fascial ligaments. A. Peritoneum and loose areolar endopelvic fascia have been removed to demonstrate the pelvic fascial ligaments found inferior to the peritoneum but superior to the female pelvic floor (pelvic diaphragm). The tendinous arch of the levator ani is a thickening of the obturator (parietal) fascia, providing the anterolateral attachment of the levator ani. The tendinous arch of the pelvic fascia (highlighted in *green*) is a thickening at the point of reflection of parietal membranous fascia onto the pelvic viscera, where it becomes visceral membranous fascia. **B.** Fascial ligaments supporting the vagina and cervix of the uterus. Since the posterior part of the urinary bladder rests on the anterior wall of the vagina, the paracolpium supports the vagina and contributes to the support of the bladder.

In addition to the ischio-anal fossae inferior to the pelvic diaphragm (i.e., in the perineum) (Fig. 3.13A & C), there is a surgically important potential **pelvirectal space** in the loose extraperitoneal connective tissue superior to the pelvic diaphragm (Fig. 3.13D). It is divided into anterior *recto-uterine* (female) or *rectovesical* (male) *spaces* and posterior **retrorectal** (presacral) **spaces** by the **rectosacral** (lateral rectal) **ligaments,** which are the posterior laminae of the hypogastric sheaths. These ligaments connect the rectum to the parietal pelvic fascia at the S2–S4 levels (Fig. 3.13B & D). The middle rectal arteries and rectal nerve plexuses are embedded in the lateral rectal ligaments.

PELVIC CAVITY

Injury to Pelvic Floor

During childbirth, the pelvic floor supports the fetal head while the cervix of the uterus is dilating to permit delivery of the fetus. The perineum, levator ani, and ligaments of the pelvic fascia may be injured during childbirth (Fig. B3.5A). The pubococcygeus and puborectalis, the main and most medial parts of the levator ani, are the muscles torn most often (Fig. B3.5B). These parts of the muscle are important because they encircle and support the urethra, vagina, and anal canal. Weakening of the levator ani and pelvic fascia (e.g., tearing of the paracolpium) from stretching or tearing during childbirth, may decrease support for the vagina, bladder, uterus, or rectum, or alter the position of the neck of the bladder and the urethra. These changes may cause *urinary stress incontinence,* characterized by dribbling of urine when intra-abdominal pressure is

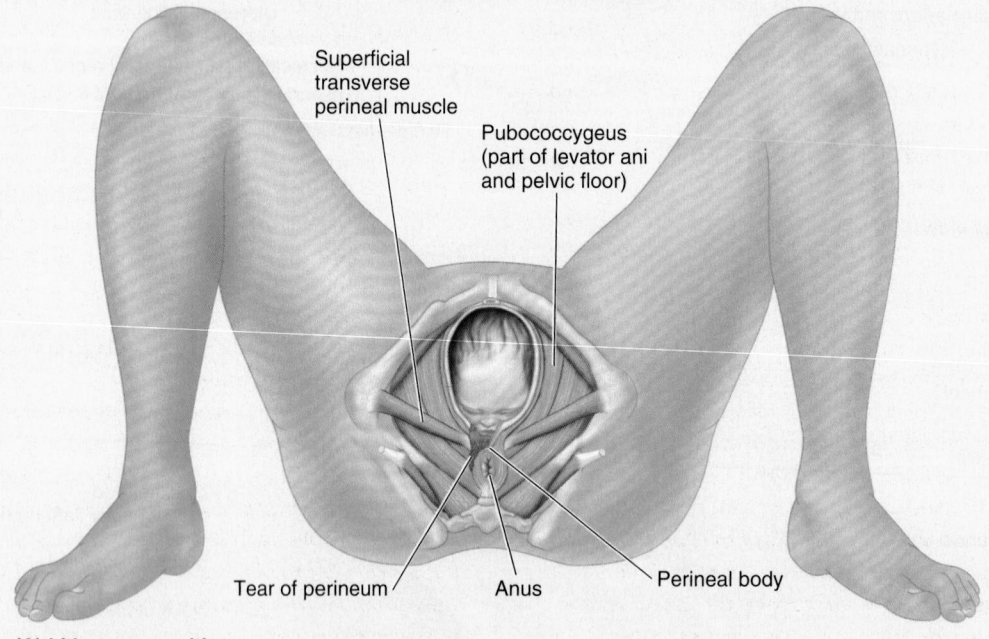

Superficial transverse perineal muscle

Pubococcygeus (part of levator ani and pelvic floor)

Tear of perineum   Anus   Perineal body

(A) Lithotomy position

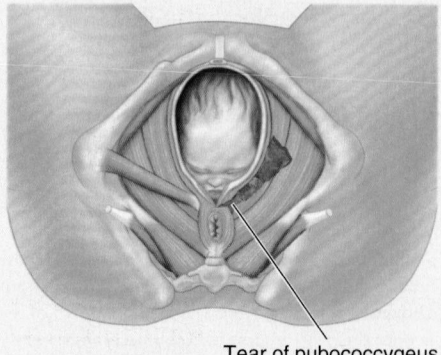

Tear of pubococcygeus

(B) Lithotomy position

FIGURE B3.5.

raised during coughing and lifting, for instance, or lead to the prolapse of one or more pelvic organs (see the blue box "Cystocele—Hernia of Bladder" on p. 373).

Prenatal "Relaxation" Training for Participatory Childbirth

 Parents wishing to participate actively in the birth of their baby may take *prenatal training* (e.g., Lamaze classes) that, among other things, attempts

to train women to learn how to relax voluntarily the muscles of the pelvic floor while simultaneously increasing intra-abdominal pressure through contraction of the diaphragm and anterolateral abdominal wall muscles. The aim of this method is to facilitate passage of the fetus through the birth canal, actively pushing ("bearing down") to aid the uterine contractions that expel the baby without providing resistance (and perhaps minimizing obstetrical tearing) caused by contraction of the pelvic muscles. Except when defecating or urinating, the natural reflex is to contract pelvic musculature in response to increased intra-abdominal pressure.

The Bottom Line

PELVIC CAVITY, PELVIC PERITONEUM, AND PELVIC FASCIA

Pelvic cavity: The pelvic cavity, between the pelvic inlet superiorly and the pelvic diaphragm inferiorly, contains the terminal parts of the urinary and alimentary systems, the internal genital organs, the associated vascular structures, and the nerves supplying both the pelvis and lower limbs. ♦ The pubic symphysis and bones of the lesser pelvis bound the cavity; they do so directly in the region of the midline anteriorly and posterosuperiorly. ♦ The lateral walls are padded by the obturator internus muscles. ♦ The sacrotuberous and sacrospinous ligaments form the greater sciatic foramen in the posterolateral walls. This foramen is filled by the structures that traverse it, including the piriformis muscle. ♦ The dynamic floor of the pelvic cavity is the hammock-like pelvic diaphragm, composed of the levator ani and coccygeus muscles. ♦ The levator ani is a tripartite, funnel-shaped muscular sheet formed by the puborectalis, pubococcygeus, and iliococcygeus muscles. ♦ In addition to the levator's general role of supporting abdomino-pelvic viscera as part of the pelvic diaphragm, the puborectalis is particularly involved in maintaining fecal continence. ♦ The ability of the musculofascial pelvic floor to relax and distend is critical to the functions of defecation and parturition.

Peritoneum: The peritoneum lining the abdominal cavity continues into the pelvic cavity, reflecting onto the superior aspects of most pelvic viscera (only the lengths of the uterine tubes, but not their free ends, are fully intraperitoneal and have a mesentery). In so doing, the peritoneum creates a number of folds and fossae. ♦ Because the peritoneum is not firmly bound to the suprapubic abdominal wall, the bladder is able to expand between the peritoneum and the anterior abdominal wall as it fills, elevating the supravesical fossae. ♦ The rectovesical pouch

and its lateral extensions, the pararectal fossae, are the inferiormost extents of the peritoneal cavity in males. ♦ In females, the uterus is located between the bladder and rectum, creating uterovesical and recto-uterine pouches. ♦ The lateral extensions of the peritoneal fold engulfing the uterine fundus form the broad ligament, a transverse duplication of peritoneum separating the paravesical and pararectal fossae. ♦ The recto-uterine fossa and its lateral extensions, the pararectal fossae, are the inferiormost extents of the peritoneal cavity in females.

Pelvic fascia: Membranous parietal pelvic fascia, continuous with the fascia lining the abdominal cavity, lines the pelvic walls and reflects onto the pelvic viscera as pelvic visceral fascia. ♦ The right and left lines of reflection are thickened into paramedian fascial bands extending from pubis to coccyx, the tendinous arches of the pelvic fascia. ♦ The subperitoneal space between the parietal and visceral pelvic fascias is occupied with fatty endopelvic fascia. This fascial matrix has loose areolar portions, occupying potential spaces, and condensed fibrous tissue, surrounding neurovascular structures in transit to the viscera while also tethering (supporting) the viscera. ♦ The two portions of endopelvic fascia are indistinct in appearance but have distinctly different textures. ♦ The primary fascial condensations form the hypogastric sheaths along the posterolateral pelvic walls. ♦ As these fascial sheaths extend toward the viscera, three laminae are formed, including the lateral ligament of the bladder anteriorly and the lateral rectal ligaments posteriorly. ♦ In females, the middle lamina is the cardinal ligament that passively supports the vagina and uterine cervix, while conveying their neurovasculature. ♦ In males, the middle lamina is the rectovesical septum.

NEUROVASCULAR STRUCTURES OF PELVIS

The major neurovascular structures of the pelvis lie extraperitoneally against the posterolateral walls. The somatic

nerves lie laterally (adjacent to the walls), with the vascular structures medial to them. Generally, the veins are lateral to the arteries (Fig. 3.15). Pelvic lymph nodes are mostly clustered around the pelvic veins, the lymphatic drainage often paralleling venous flow. In dissecting from the pelvic cavity toward the pelvic walls, the pelvic arteries are encountered

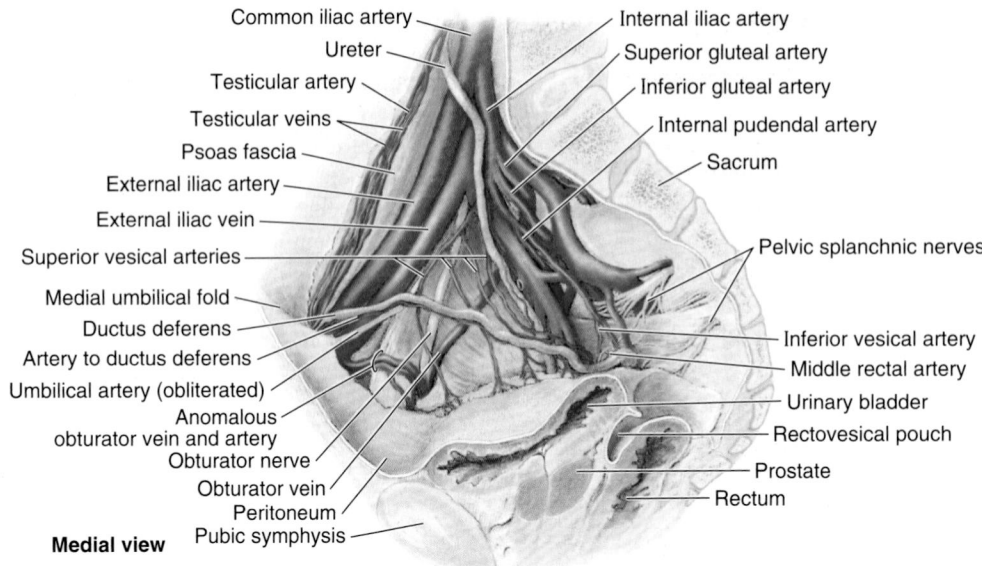

FIGURE 3.15. Neurovascular relationships of pelvis. The neurovascular structures of the male pelvis are shown. Generally, the pelvic veins lie between the pelvic arteries (which lie medially or internally), and the somatic nerves (which lie laterally or externally).

first, followed by the associated pelvic veins, and then the somatic nerves of the pelvis.

Pelvic Arteries

The pelvis is richly supplied with arteries, among which multiple anastomoses occur, providing an extensive collateral circulation. Information concerning the origin, course, distribution, and anastomoses of the arteries of the pelvis is provided in Figure 3.16 and Table 3.4. The following text provides additional information not provided in the table.

Six main arteries enter the lesser pelvis of females: the paired internal iliac and ovarian arteries, and the unpaired median sacral and superior rectal arteries. Since the testicular arteries do not enter the lesser pelvis, only four main arteries enter the lesser pelvis of males.

INTERNAL ILIAC ARTERY

The **internal iliac artery** is the principal artery of the pelvis, supplying most of the blood to the pelvic viscera and some to the musculoskeletal part of the pelvis; however, it also supplies branches to the gluteal region, medial thigh regions, and the perineum (Fig. 3.15).

Each internal iliac artery, approximately 4 cm long, begins as the *common iliac artery* and bifurcates into the internal and external iliac arteries at the level of the IV disc between the L5 and S1 vertebrae. The ureter crosses the common iliac artery or its terminal branches at or immediately distal to the bifurcation. The internal iliac artery is separated from the sacro-iliac joint by the internal iliac vein and the lumbosacral trunk. It descends posteromedially into the lesser pelvis, medial to the external iliac vein and obturator nerve, and lateral to the peritoneum.

Anterior Division of Internal Iliac Artery. Although variations are common, the internal iliac artery usually ends at the superior edge of the greater sciatic foramen by dividing into anterior and posterior divisions (trunks). The branches of the **anterior division of the internal iliac artery** are mainly visceral (i.e., they supply the bladder, rectum, and reproductive organs), but they also include parietal branches that pass to the thigh and buttocks (Fig. 3.17A & B). The arrangement of the visceral branches is variable.

Umbilical Artery. Before birth, the umbilical arteries are the main continuation of the internal iliac arteries, passing along the lateral pelvic wall and then ascending the anterior abdominal wall to and through the umbilical ring into the umbilical cord.

Prenatally, the **umbilical arteries** conduct oxygen- and nutrient-deficient blood to the placenta for replenishment. When the umbilical cord is cut, the distal parts of these vessels no longer function and become occluded distal to branches that pass to the bladder. The *occluded parts* form fibrous cords called the *medial umbilical ligaments* (Figs. 3.16 and 3.17A & B). The ligaments raise folds of peritoneum (*medial umbilical folds*) on the deep surface of the anterior abdominal wall (see Chapter 2).

Postnatally, the *patent parts of the umbilical arteries* run antero-inferiorly between the urinary bladder and the lateral wall of the pelvis.

Obturator Artery. The origin of the **obturator artery** is variable; usually it arises close to the origin of the umbilical artery, where it is crossed by the ureter. It runs antero-inferiorly on the obturator fascia on the lateral wall of the pelvis, and passes between the obturator nerve and vein (Figs. 3.16 and 3.17A & B).

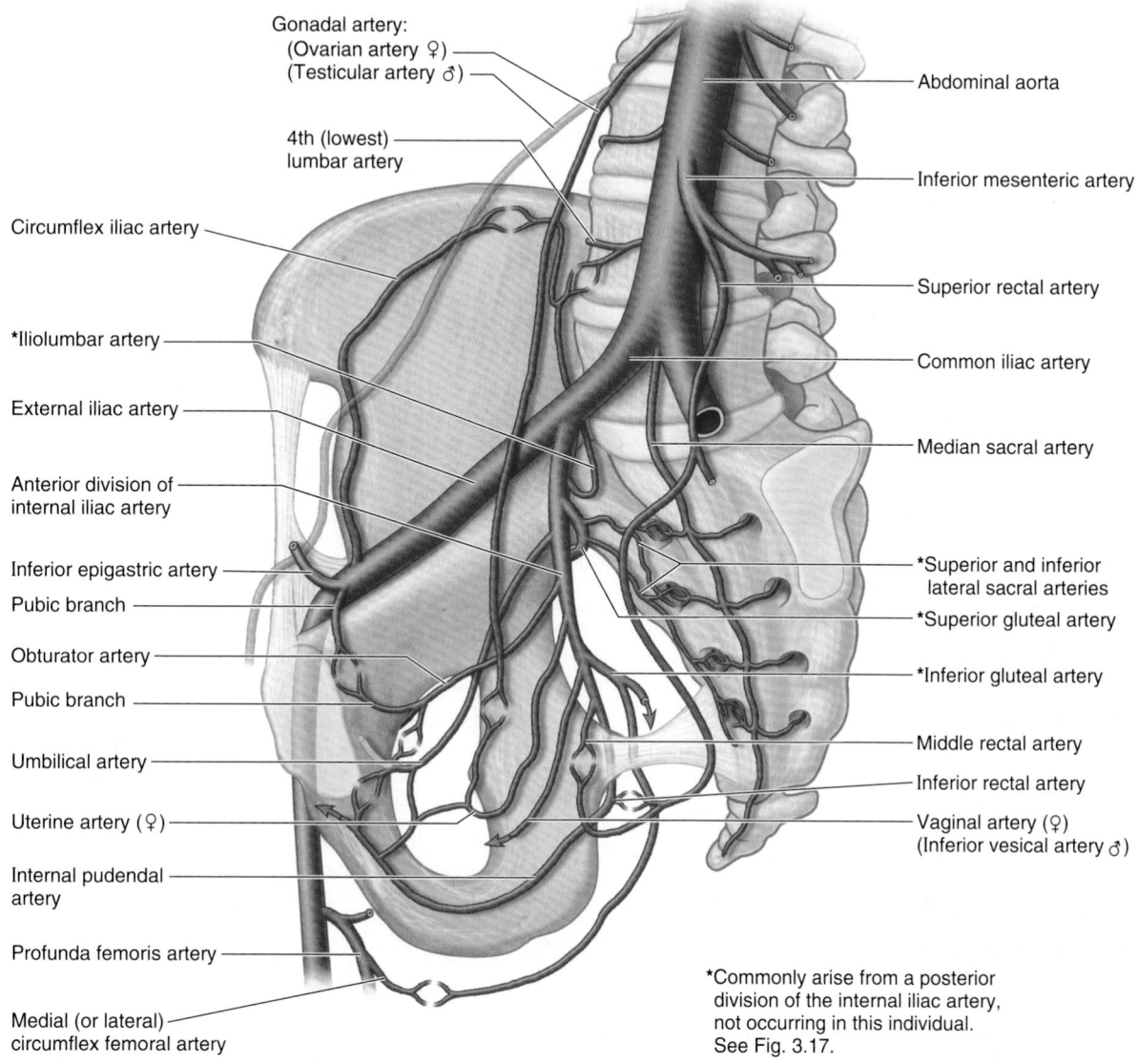

Gonadal artery:
(Ovarian artery ♀)
(Testicular artery ♂)

4th (lowest)
lumbar artery

Circumflex iliac artery

*Iliolumbar artery

External iliac artery

Anterior division of
internal iliac artery

Inferior epigastric artery

Pubic branch

Obturator artery

Pubic branch

Umbilical artery

Uterine artery (♀)

Internal pudendal
artery

Profunda femoris artery

Medial (or lateral)
circumflex femoral artery

Abdominal aorta

Inferior mesenteric artery

Superior rectal artery

Common iliac artery

Median sacral artery

*Superior and inferior
lateral sacral arteries

*Superior gluteal artery

*Inferior gluteal artery

Middle rectal artery

Inferior rectal artery

Vaginal artery (♀)
(Inferior vesical artery ♂)

*Commonly arise from a posterior
division of the internal iliac artery,
not occurring in this individual.
See Fig. 3.17.

FIGURE 3.16. Arteries and arterial anastomoses in pelvis. The origins, courses, and distribution of the arteries and the arterial anastomoses formed are described in Table 3.4.

Within the pelvis, the obturator artery gives off muscular branches, a nutrient artery to the ilium, and a pubic branch. The **pubic branch** arises just before the obturator artery leaves the pelvis. It ascends on the pelvic surface of the pubis to anastomose with its fellow of the opposite side, and the *pubic branch of the inferior epigastric artery*, a branch of the external iliac artery.

In a common variation (20%), an **aberrant** or **accessory obturator artery** arises from the inferior epigastric artery and descends into the pelvis along the usual route of the pubic branch (Figs. 3.15 and 3.16). Surgeons performing hernia repairs must keep this common variation in mind. The extrapelvic distribution of the obturator artery is described with the lower limb.

Inferior Vesical Artery. The **inferior vesical artery** occurs only in males (Figs. 3.16 and 3.17A), being replaced by the vaginal artery in females (Figs. 3.16 and 3.17B).

Uterine Artery. The **uterine artery** is an additional branch of the internal iliac artery in females, usually arising separately and directly from the internal iliac artery (Figs. 3.16 and 3.17B). It may arise from the umbilical artery. Developmentally, it is the homolog of the artery to the ductus deferens in the male. It descends on the lateral wall of the pelvis, anterior to the internal iliac artery, and passes medially to reach the junction of the uterus and vagina, where the cervix (neck) of the uterus protrudes into the superior vagina (Fig. 3.18A & B). As it passes medially, the uterine artery passes directly superior to the ureter. The relationship of ureter to artery is often remembered by the phrase *"water (urine) passes under the bridge (uterine artery)."* On reaching the side of the cervix, the uterine artery divides into a smaller descending **vaginal branch,** which supplies the cervix and vagina, and a larger **ascending branch,** which runs along the lateral margin of the uterus, supplying it. The ascending branch bifurcates into

TABLE 3.4. ARTERIES OF PELVIS

Artery	Origin	Course	Distribution	Anastomoses
Gonadal	Abdominal aorta	Descends retroperitoneally;		
Testicular (♂)		traverses inguinal canal and enters scrotum	Abdominal ureter, testis, and epididymis	Cremasteric artery, artery of ductus deferens
Ovarian (♀)		crosses pelvic brim, descends in suspensory ligament of ovary	Abdominal and/or pelvic ureter, ovary, and ampullary end of uterine tube	Uterine artery via tubal and ovarian branches
Superior rectal	Continuation of inferior mesenteric artery	Crosses left common iliac vessels and descends into pelvis between layers of sigmoid mesocolon	Superior part of rectum	Middle rectal artery; inferior rectal (internal pudendal) artery
Median sacral	Posterior aspect of abdominal aorta	Descends close to midline over L4 and L5 vertebrae, sacrum, and coccyx	Inferior lumbar vertebrae, sacrum, and coccyx	Lateral sacral artery (via medial sacral branches)
Internal iliac	Common iliac artery	Passes medially over pelvic brim and descends into pelvic cavity; often forms anterior and posterior divisions	Main blood supply to pelvic organs, gluteal muscles, and perineum	
Anterior division of internal iliac	Internal iliac artery	Passes anteriorly along lateral wall of pelvis, dividing into visceral, obturator, and internal pudendal arteries	Pelvic viscera, muscles of superior medial thigh, and perineum	
Umbilical	Anterior division of internal iliac artery	Runs a short pelvic course, gives off superior vesical arteries, then obliterates, becoming medial umbilical ligament	Superior aspect of urinary bladder and, in some males, ductus deferens (via superior vesical arteries and artery to ductus deferens)	(Occasionally the patent part of the umbilical artery)
Superior vesical	(Patent proximal umbilical artery)	Usually multiple; pass to superior aspect of urinary bladder	Superior aspect of urinary bladder; in some males, ductus deferens (via artery to ductus deferens)	Inferior vesical (♂); vaginal artery (♀)
Obturator		Runs antero-inferiorly on obturator fascia of lateral pelvic wall, exiting pelvis via obturator canal	Pelvic muscles, nutrient artery to ilium, head of femur, and muscles of medial compartment of thigh	Inferior epigastric (via pubic branch); umbilical artery
Inferior vesical (♂)		Passes subperitoneally in lateral ligament of bladder, giving rise to prostatic artery (♂) and occasionally the artery to the ductus deferens	Inferior aspect of male urinary bladder, pelvic part of ureter; prostate, and seminal glands; occasionally ductus deferens	Superior vesical artery
Artery to ductus deferens (♂)	(Superior or inferior vesical artery)	Runs subperitoneally to ductus deferens	Ductus deferens	Testicular artery; cremasteric artery
Prostatic branches (♂)	(Inferior vesical artery)	Descends on posterolateral aspects of prostate	Prostate and prostatic urethra	Deep perineal (internal pudendal)
Uterine (♀)		Runs anteromedially in base of broad ligament/superior cardinal ligament, gives rise to vaginal branch, then crosses ureter superiorly to reach lateral aspect of uterine cervix	Uterus, ligaments of uterus, medial parts of uterine tube and ovary, and superior vagina	Ovarian artery (via tubal and ovarian branches); vaginal artery

TABLE 3.4. ARTERIES OF PELVIS (Continued)

Artery	Origin	Course	Distribution	Anastomoses
Vaginal (♀)	(Uterine artery)	Divides into vaginal and inferior vesical branches, the former descending on the vagina, the latter passing to the urinary bladder	Vaginal branch: lower vagina, vestibular bulb, and adjacent rectum; inferior vesical branch: fundus of urinary bladder	Vaginal branch of uterine artery, superior vesical artery
Internal pudendal	Anterior division of internal iliac artery	Exits pelvis via infrapiriform part of greater sciatic foramen, enters perineum (ischio-anal fossa) via lesser sciatic foramen, passes via pudendal canal to UG triangle	Main artery of perineum, including muscles and skin of anal and urogenital triangles, erectile bodies	(Umbilical artery; prostatic branches of inferior vesical artery in males)
Middle rectal		Descends in pelvis to inferior part of rectum	Inferior part of rectum, seminal glands, prostate (vagina)	Superior and inferior rectal arteries
Inferior gluteal[a]		Exits pelvis via infrapiriform part of greater sciatic foramen	Pelvic diaphragm (coccygeus and levator ani), piriformis, quadratus femoris, superiormost hamstrings, gluteus maximus, and sciatic nerve	Profunda femoris artery (via medial and lateral circumflex femoral arteries)
Posterior division of internal iliac	Internal iliac artery	Passes posteriorly and gives rise to parietal branches	Pelvic wall and gluteal region	
Iliolumbar[b]	Posterior division of internal iliac artery	Ascends anterior to sacro-iliac joint and posterior to common iliac vessels and psoas major, dividing into iliac and lumbar branches	Psoas major, iliacus, and quadratus lumborum muscles; cauda equina in vertebral canal	Circumflex iliac artery and 4th (and lowest) lumbar artery
Lateral sacral (superior and inferior)		Runs on anteromedial aspect of piriformis to send branches into pelvic sacral foramina	Piriformis, structures in sacral canal, erector spinae, and overlying skin	Medial sacral arteries (from median sacral artery)
Superior gluteal		Passes between lumbosacral trunk and anterior ramus of S1 spinal nerve to exit pelvis via suprapiriform portion of greater sciatic foramen	Piriformis, all three gluteal muscles, and tensor fasciae latae	Lateral sacral, inferior gluteal, internal pudendal, deep circumflex femoral, lateral circumflex femoral

[a]May occur as terminal branch of posterior division of internal iliac artery.
[b]Often arises directly from internal iliac artery, proximal to divisions.

ovarian and **tubal branches,** which continue to supply the medial ends of the ovary and uterine tube, and anastomose with the ovarian and tubal branches of the ovarian artery.

Vaginal Artery. The **vaginal artery** is the homolog to the inferior vesical artery in males. It often arises from the initial part of the uterine artery instead of arising directly from the anterior division. The vaginal artery supplies numerous branches to the anterior and posterior surfaces of the vagina (Figs 3.16, 3.17B, and 3.18).

Middle Rectal Artery. The **middle rectal artery** may arise independently from the internal iliac artery, or it may arise in common with the inferior vesical artery or the internal pudendal artery (Figs 3.16 and 3.17).

Internal Pudendal Artery. The **internal pudendal artery,** larger in males than in females, passes inferolaterally, anterior to the piriformis muscle and sacral plexus. It

leaves the pelvis between the piriformis and coccygeus muscles by passing through the inferior part of the *greater sciatic foramen.* The internal pudendal artery then passes around the posterior aspect of the ischial spine or the sacrospinous ligament and enters the *ischio-anal fossa* through the *lesser sciatic foramen.*

The internal pudendal artery, along with the internal pudendal veins and branches of the pudendal nerve, passes through the pudendal canal in the lateral wall of the ischio-anal fossa (Fig. 3.11B). As it exits the pudendal canal, medial to the ischial tuberosity, the internal pudendal artery divides into its terminal branches, the *perineal artery* and *dorsal arteries of the penis or clitoris.*

Inferior Gluteal Artery. The **inferior gluteal artery** is the larger terminal branch of the anterior division of the internal iliac artery (Fig. 3.18A), but may arise from the posterior

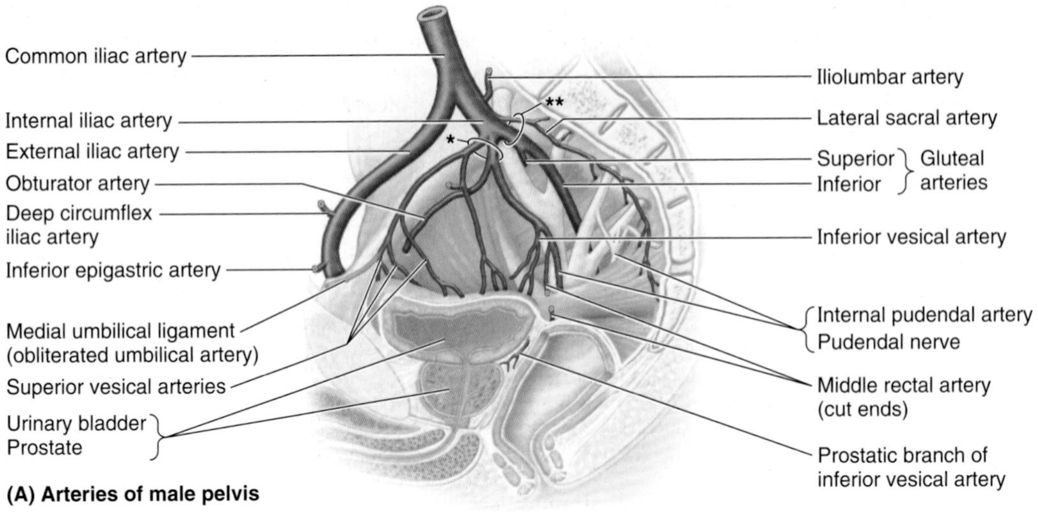

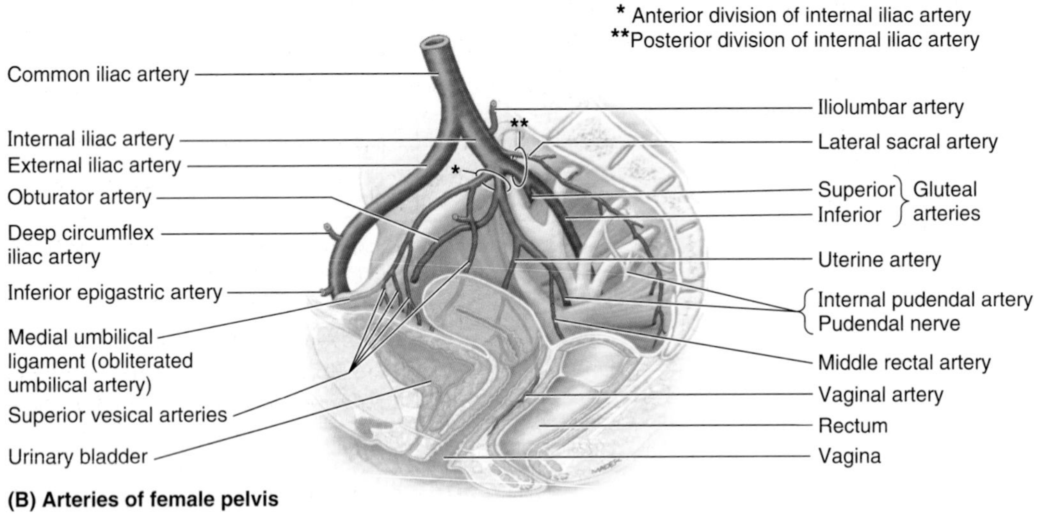

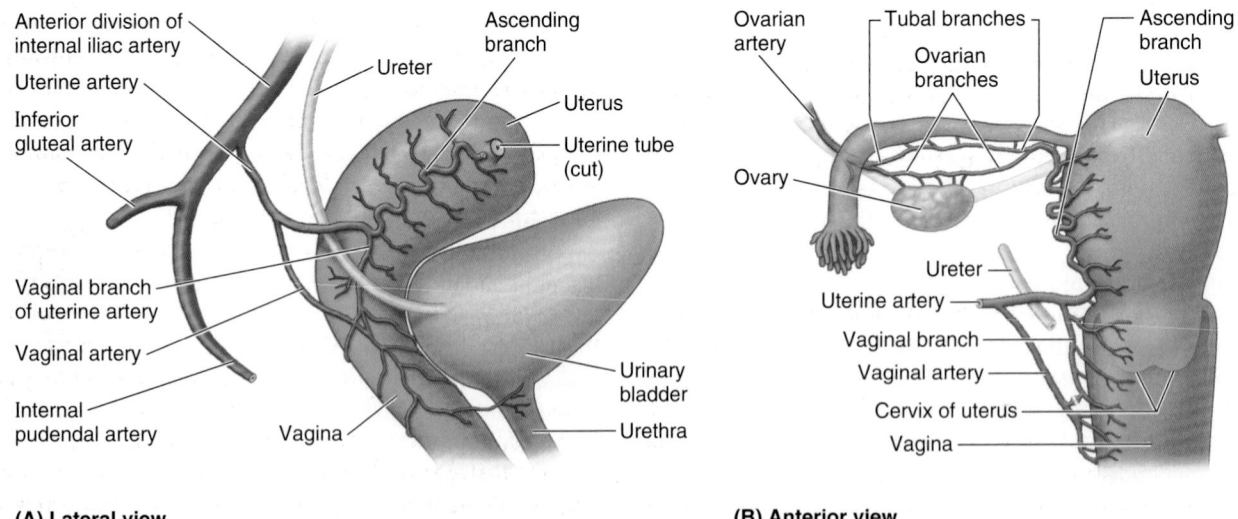

(A) Arteries of male pelvis

Common iliac artery
Internal iliac artery
External iliac artery
Obturator artery
Deep circumflex iliac artery
Inferior epigastric artery
Medial umbilical ligament (obliterated umbilical artery)
Superior vesical arteries
Urinary bladder }
Prostate }

Iliolumbar artery
Lateral sacral artery
Superior } Gluteal
Inferior } arteries
Inferior vesical artery
Internal pudendal artery
Pudendal nerve
Middle rectal artery (cut ends)
Prostatic branch of inferior vesical artery

* Anterior division of internal iliac artery
**Posterior division of internal iliac artery

(B) Arteries of female pelvis

Common iliac artery
Internal iliac artery
External iliac artery
Obturator artery
Deep circumflex iliac artery
Inferior epigastric artery
Medial umbilical ligament (obliterated umbilical artery)
Superior vesical arteries
Urinary bladder

Iliolumbar artery
Lateral sacral artery
Superior } Gluteal
Inferior } arteries
Uterine artery
Internal pudendal artery
Pudendal nerve
Middle rectal artery
Vaginal artery
Rectum
Vagina

FIGURE 3.17. **Arteries of pelvis.** Anterior divisions of the internal iliac arteries usually supply most of the blood to pelvic structures.

(A) Lateral view

Anterior division of internal iliac artery
Uterine artery
Inferior gluteal artery
Ureter
Ascending branch
Uterus
Uterine tube (cut)
Vaginal branch of uterine artery
Vaginal artery
Internal pudendal artery
Vagina
Urinary bladder
Urethra

(B) Anterior view

Ovarian artery
Tubal branches
Ovarian branches
Ascending branch
Uterus
Ovary
Ureter
Uterine artery
Vaginal branch
Vaginal artery
Cervix of uterus
Vagina

FIGURE 3.18. **Uterine and vaginal arteries. A.** The origin of the arteries from the anterior division of the internal iliac artery and distribution to the uterus and vagina are shown. **B.** The anastomoses between the ovarian and tubal branches of the ovarian and uterine arteries and between the vaginal branch of the uterine artery and the vaginal artery provide potential pathways of collateral circulation. These communications occur, and the ascending branch courses, between the layers of the broad ligament.

division (Fig. 3.17). It passes posteriorly between the sacral nerves (usually S2 and S3), and leaves the pelvis through the inferior part of the *greater sciatic foramen,* inferior to the piriformis muscle (Fig. 3.16). It supplies the muscles and skin of the buttock, and the posterior surface of the thigh.

Posterior Division of Internal Iliac Artery. When the internal iliac artery divides into anterior and posterior divisions, the posterior division typically gives rise to the following three parietal arteries (Fig. 3.17A & B):

- **Iliolumbar artery:** This artery runs superolaterally in a *recurrent fashion* (turning sharply backward relative to its source) to the iliac fossa. Within the fossa, the artery divides into an *iliac branch,* which supplies the iliacus muscle and ilium, and a *lumbar branch,* which supplies the psoas major and quadratus lumborum muscles.
- **Lateral sacral arteries:** Superior and inferior lateral sacral arteries may arise as independent branches or via a common trunk. The lateral sacral arteries pass medially and descend anterior to the sacral anterior rami, giving off *spinal branches,* which pass through the anterior sacral foramina, and supply the spinal meninges enclosing the roots of the sacral nerves. Some branches of these arteries pass from the sacral canal through the posterior sacral foramina and supply the erector spinae muscles of the back and the skin overlying the sacrum.
- **Superior gluteal artery:** The largest branch of the posterior division, the superior gluteal artery supplies the gluteal muscles in the buttocks.

OVARIAN ARTERY

The **ovarian artery** arises from the abdominal aorta inferior to the renal artery but considerably superior to the inferior mesenteric artery (Fig. 3.16). As it passes inferiorly, the ovarian artery adheres to the parietal peritoneum and runs anterior to the ureter on the posterior abdominal wall, usually giving branches to it. As the ovarian artery enters the lesser pelvis, it crosses the origin of the external iliac vessels. It then runs medially, dividing into an **ovarian branch** and a **tubal branch,** which supply the ovary and uterine tube, respectively (Fig. 3.18B). These branches anastomose with the corresponding branches of the uterine artery.

MEDIAN SACRAL ARTERY

The **median sacral artery** is a small unpaired artery that usually arises from the posterior surface of the abdominal aorta, just superior to its bifurcation, but it may arise from its anterior surface (Fig. 3.16). This vessel runs anterior to the bodies of the last one or two lumbar vertebrae, the sacrum, and the coccyx, its terminal branches participating in a series of anastomotic loops. Before the median sacral artery enters the lesser pelvis, it sometimes gives rise to a pair of *L5 arteries.*

As it descends over the sacrum, the median sacral artery gives off small parietal (lateral sacral) branches that anastomose with the lateral sacral arteries. It also gives rise to small visceral branches to the posterior part of the rectum, which anastomose with the superior and middle rectal arteries. The median sacral artery represents the caudal end of the embryonic dorsal aorta, which reduced in size as the tail-like caudal eminence of the embryo disappeared.

SUPERIOR RECTAL ARTERY

The **superior rectal artery** is the direct continuation of the inferior mesenteric artery (Fig. 3.16). It crosses the left common iliac vessels and descends in the sigmoid mesocolon to the lesser pelvis. At the level of the S3 vertebra, the superior rectal artery divides into two branches, which descend on each side of the rectum and supply it as far inferiorly as the internal anal sphincter.

Pelvic Veins

Pelvic venous plexuses are formed by the interjoining veins surrounding the pelvic viscera (Fig. 3.19B & C). These intercommunicating networks of veins are clinically important. The various plexuses within the lesser pelvis (rectal, vesical, prostatic, uterine, and vaginal) unite and are drained mainly by tributaries of the *internal iliac veins,* but some of them drain through the superior rectal vein into the inferior mesenteric vein of the hepatic portal system (Fig. 3.19A), or through lateral sacral veins into the *internal vertebral venous plexus* (see Chapter 4). Additional relatively minor paths of venous drainage from the lesser pelvis include the parietal **median sacral vein** and, in females, the ovarian veins.

The **internal iliac veins** form superior to the greater sciatic foramen and lie postero-inferior to the internal iliac arteries (Fig. 3.19A & B). Tributaries of the internal iliac veins are more variable than the branches of the internal iliac artery with which they share names, but roughly accompany them, draining the same territories that the arteries supply. However, there are no veins accompanying the umbilical arteries between the pelvis and the umbilicus, and the **iliolumbar veins** from the iliac fossae of the greater pelvis usually drain into the common iliac veins instead. The internal iliac veins merge with the external iliac veins to form the **common iliac veins,** which unite at the level of vertebra L4 or L5 to form the **inferior vena cava** (Fig. 3.19A).

The **superior gluteal veins,** the accompanying veins (L. *venae comitantes*) of the superior gluteal arteries of the gluteal region, are the largest tributaries of the internal iliac veins except during pregnancy, when the uterine veins become larger. Testicular veins traverse the greater pelvis as they pass from the deep inguinal ring toward their posterior abdominal terminations but do not usually drain pelvic structures.

The **lateral sacral veins** often appear disproportionately large in angiographs. They anastomose with the internal vertebral venous plexus (Chapter 4), providing an alternate collateral pathway to reach either the inferior or superior vena cava. It may also provide a pathway for metastasis of prostatic or ovarian cancer cells to vertebral or cranial sites.

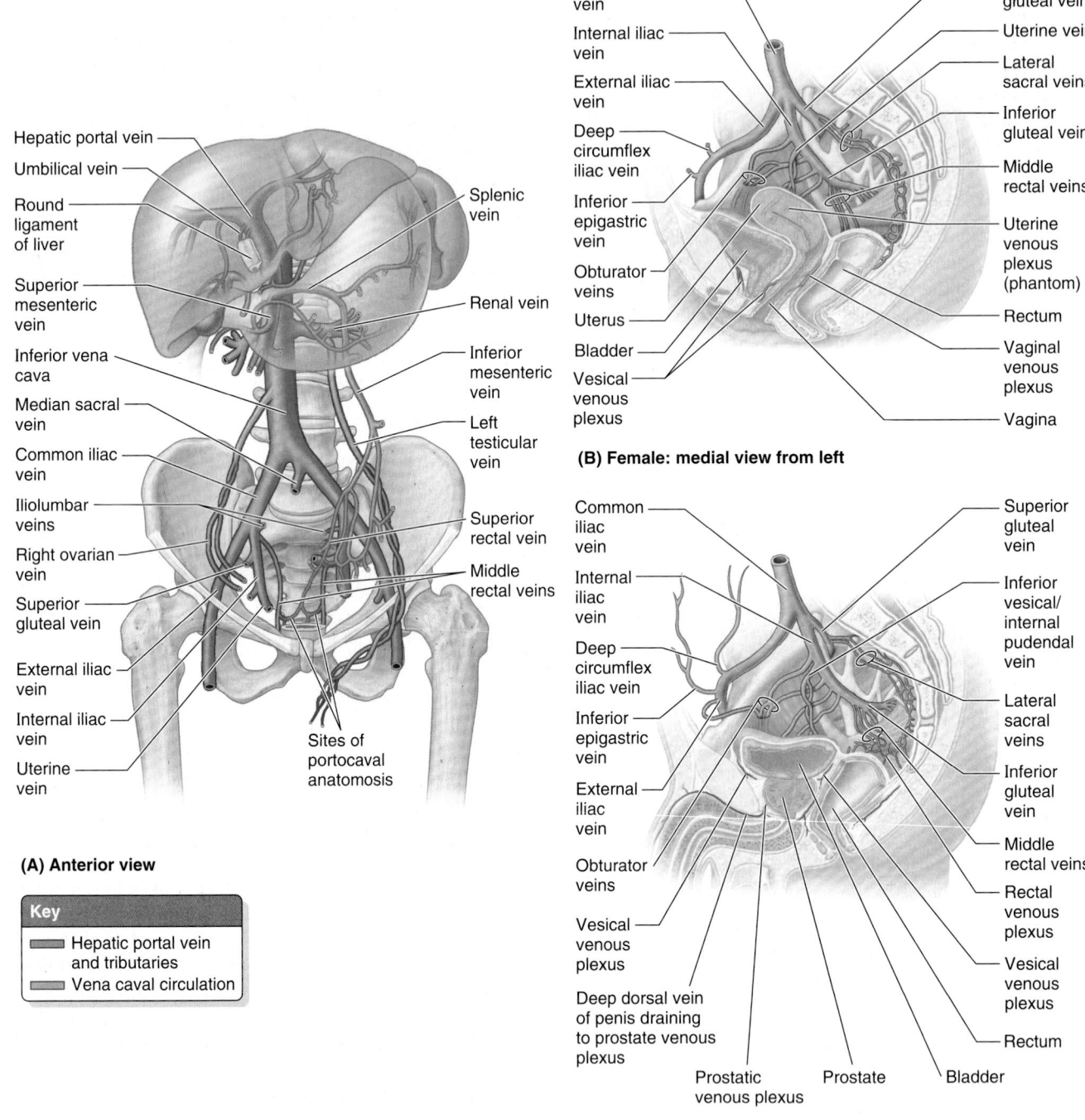

(A) Anterior view

Key
▬ Hepatic portal vein and tributaries
▬ Vena caval circulation

(B) Female: medial view from left

(C) Male: medial view from left

FIGURE 3.19. Pelvic veins. A. The female (*right*) and male (*left*) patterns of hepatic portal and systemic (vena caval) venous systems of the abdominopelvic cavity are shown. Venous drainage from pelvic organs flows mainly to the caval system via the internal iliac veins. The superior rectum normally drains into the hepatic portal system, although the superior rectal veins anastomose with the middle and inferior rectal veins, which are tributaries of the internal iliac veins. The female (**B**) and male (**C**) pelvic veins and venous plexuses are demonstrated.

Lymph Nodes of Pelvis

The lymph nodes receiving lymph drainage from pelvic organs are variable in number, size, and location. Dividing them into definite groups is often somewhat arbitrary. Four primary groups of nodes are located in or adjacent to the pelvis, named for the blood vessels with which they are associated (Fig. 3.20):

- **External iliac lymph nodes:** lie above the pelvic brim, along the external iliac vessels. They receive lymph mainly from the inguinal lymph nodes; however, they receive lymph

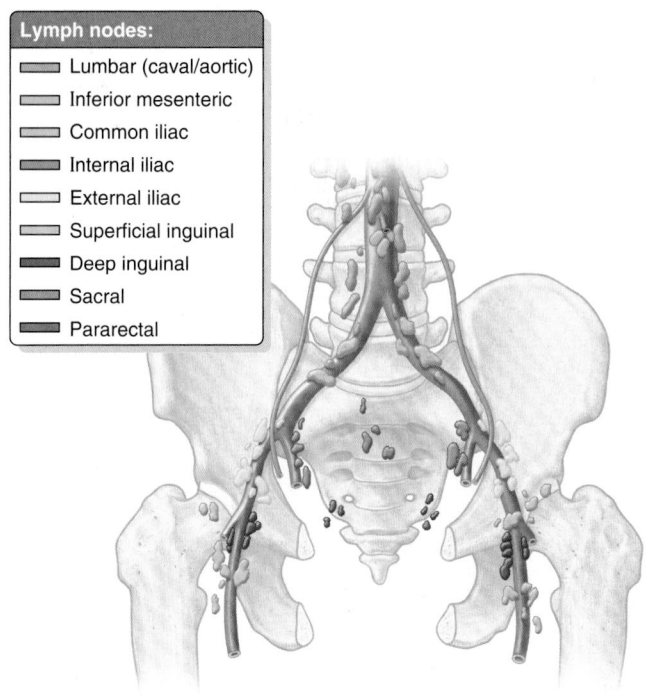

Lymph nodes:
- Lumbar (caval/aortic)
- Inferior mesenteric
- Common iliac
- Internal iliac
- External iliac
- Superficial inguinal
- Deep inguinal
- Sacral
- Pararectal

FIGURE 3.20. Lymph nodes of pelvis.

from pelvic viscera, especially the superior parts of the middle to anterior pelvic organs. Whereas most of the lymphatic drainage from the pelvis tends to parallel routes of venous drainage, the lymphatic drainage to the external iliac nodes does not. These nodes drain into the common iliac nodes.

- **Internal iliac lymph nodes:** clustered around the anterior and posterior divisions of the internal iliac artery and the origins of the gluteal arteries. They receive drainage from the inferior pelvic viscera, deep perineum, and gluteal region and drain into the common iliac nodes.
- **Sacral lymph nodes:** lie in the concavity of the sacrum, adjacent to the median sacral vessels. They receive lymph from postero-inferior pelvic viscera and drain either to internal or common iliac nodes.
- **Common iliac lymph nodes:** lie superior to the pelvic brim, along the common iliac blood vessels (Fig. 3.20), and receive drainage from the three main groups listed above. These nodes begin a common route for drainage from the pelvis that passes next to the lumbar (caval/aortic) nodes. Inconstant direct drainage to the common iliac nodes occurs from some pelvic organs (e.g., from the neck of the bladder and inferior vagina).

Additional minor groups of nodes (e.g., the **pararectal nodes**) occupy the connective tissue along the branches of the internal iliac vessels.

Both primary and minor groups of pelvic nodes are highly interconnected, so that many nodes can be removed without disturbing drainage. The interconnections also allow cancer to spread in virtually any direction, to any pelvic or abdominal viscus. While the lymphatic drainage tends to parallel the venous drainage (except for that to the external iliac nodes, where

proximity provides a rough guide), the pattern is not sufficiently predictable to allow the progress of metastatic cancer from pelvic organs to be anticipated or staged in a manner comparable to that of breast cancer progressing though the axillary nodes. Lymphatic drainage from the specific pelvic organs is described following the description of the pelvic viscera (p. 400).

Pelvic Nerves

The pelvis is innervated mainly by the **sacral** and **coccygeal spinal nerves** and the *pelvic part of the autonomic nervous system*. The piriformis and coccygeus muscles form a bed for the sacral and coccygeal nerve plexuses (Fig. 3.21). The anterior rami of the S2 and S3 nerves emerge between the digitations of these muscles.

OBTURATOR NERVE

The *obturator nerve* arises from the anterior rami of spinal nerves L2–L4 of the *lumbar plexus* in the abdomen (greater pelvis) and enters the lesser pelvis. It runs in the extraperitoneal fat along the lateral wall of the pelvis to the *obturator canal,* an opening in the obturator membrane that otherwise fills the obturator foramen. Here it divides into anterior and posterior parts, which leave the pelvis through this canal and supply the medial thigh muscles. No pelvic structures are supplied by the obturator nerve.

LUMBOSACRAL TRUNK

At or immediately superior to the pelvic brim, the descending part of the L4 nerve unites with the anterior ramus of the L5 nerve to form the thick, cord-like **lumbosacral trunk** (see Figs. 3.9D, 3.21, and 3.22). The trunk passes inferiorly, on the anterior surface of the ala of the sacrum, and joins the sacral plexus.

SACRAL PLEXUS

The plexus is demonstrated in Figure 3.22, and the segmental composition and distribution of the nerves derived from it are listed in Table 3.5. The following text provides additional information about the formation of the nerves and their courses.

The **sacral plexus** is located on the posterolateral wall of the lesser pelvis. The two main nerves arising from the sacral plexus, the *sciatic* and *pudendal nerves,* lie external to the parietal pelvic fascia. Most branches of the sacral plexus leave the pelvis through the *greater sciatic foramen.*

The *sciatic nerve* is the largest nerve in the body. It is formed as the large anterior rami of spinal nerves L4–S3 converge on the anterior surface of the piriformis (Figs. 3.21 and 3.22). As it is formed, the sciatic nerve passes through the greater sciatic foramen, usually inferior to the piriformis, to enter the gluteal region. It then descends along the posterior aspect of the thigh to supply the posterior aspect of the thigh and the entire leg and foot.

The **pudendal nerve** is the main nerve of the perineum and the chief sensory nerve of the external genitalia.

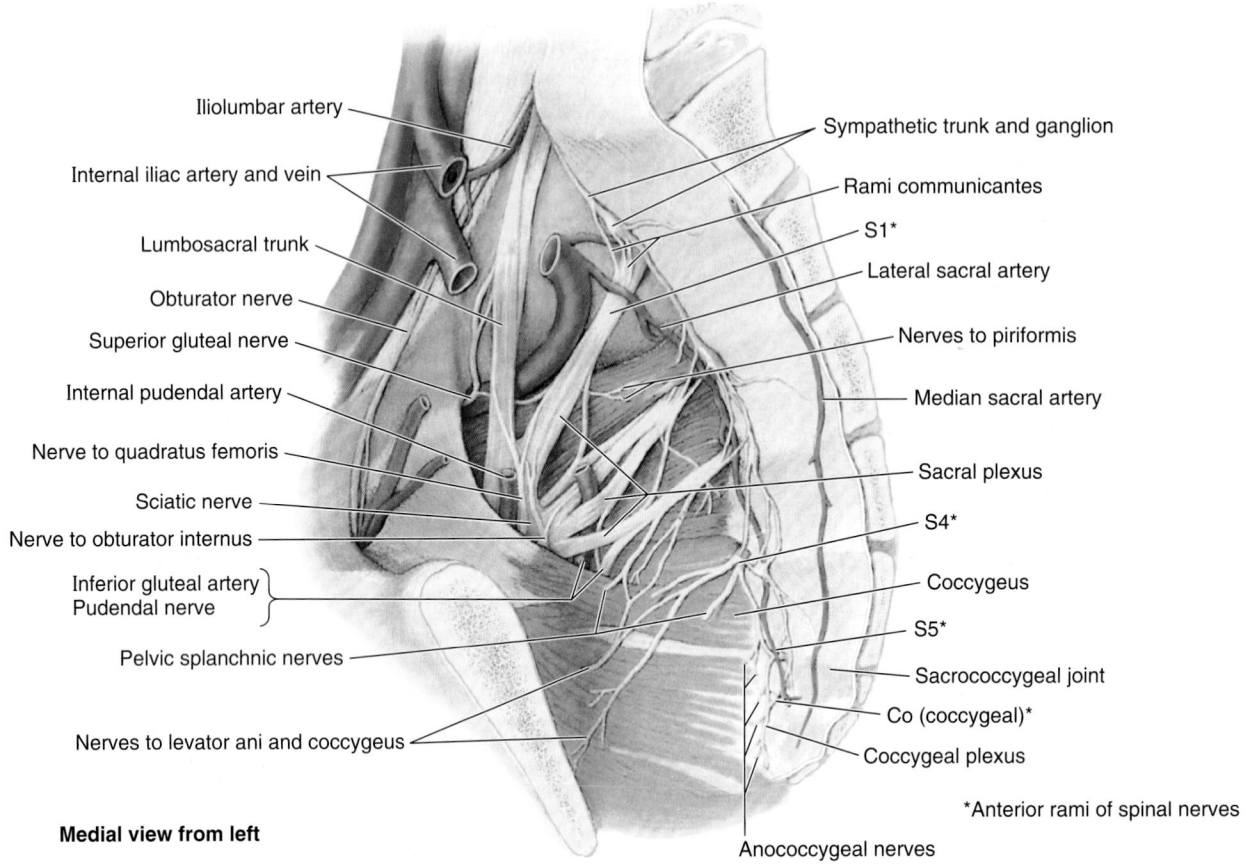

Iliolumbar artery

Internal iliac artery and vein

Lumbosacral trunk

Obturator nerve

Superior gluteal nerve

Internal pudendal artery

Nerve to quadratus femoris

Sciatic nerve

Nerve to obturator internus

Inferior gluteal artery
Pudendal nerve

Pelvic splanchnic nerves

Nerves to levator ani and coccygeus

Medial view from left

Sympathetic trunk and ganglion

Rami communicantes

S1*

Lateral sacral artery

Nerves to piriformis

Median sacral artery

Sacral plexus

S4*

Coccygeus

S5*

Sacrococcygeal joint

Co (coccygeal)*

Coccygeal plexus

*Anterior rami of spinal nerves

Anococcygeal nerves

FIGURE 3.21. Nerves and nerve plexuses of pelvis. Somatic nerves (sacral and coccygeal nerve plexuses) and the pelvic (sacral) part of the sympathetic trunk are shown. Although located in the pelvis, most of the nerves seen here are involved with the innervation of the lower limb rather than the pelvic structures.

Accompanied by the internal pudendal artery, it leaves the pelvis through the greater sciatic foramen between the piriformis and coccygeus muscles. It then hooks around the ischial spine and sacrospinous ligament and enters the perineum through the lesser sciatic foramen (Fig. 3.22).

The *superior gluteal nerve* leaves the pelvis through the greater sciatic foramen, superior to the piriformis to supply muscles in the gluteal region (Figs. 3.21 and 3.22).

The *inferior gluteal nerve* leaves the pelvis through the greater sciatic foramen (Fig. 3.22), inferior to the piriformis and superficial to the sciatic nerve, accompanying the inferior gluteal artery. Both break up into several branches that enter the deep surface of the overlying gluteus maximus muscle.

COCCYGEAL PLEXUS

The **coccygeal plexus** is a small network of nerve fibers formed by the anterior rami of S4 and S5 and the **coccygeal nerves** (Fig. 3.21). It lies on the pelvic surface of the coccygeus and supplies this muscle, part of the levator ani, and the sacrococcygeal joint. The **anococcygeal nerves** arising from this plexus pierce the coccygeus and anococcygeal ligament to supply a small area of skin between the tip of the coccyx and the anus.

PELVIC AUTONOMIC NERVES

Autonomic nerves enter the pelvic cavity via four routes (Fig. 3.23):

- *Sacral sympathetic trunks:* primarily provide sympathetic innervation to the lower limbs.
- *Peri-arterial plexuses:* postsynaptic, sympathetic, vasomotor fibers to superior rectal, ovarian, and internal iliac arteries and their derivative branches.
- *Hypogastric plexuses:* most important route by which sympathetic fibers are conveyed to the pelvic viscera.
- *Pelvic splanchnic nerves:* pathway for parasympathetic innervation of pelvic viscera and descending and sigmoid colon.

The **sacral sympathetic trunks** are the inferior continuation of the lumbar sympathetic trunks (Figs. 3.21 and 3.23). Each of the sacral trunks is diminished in size from that of the lumbar trunks and usually has four sympathetic ganglia. The sacral trunks descend on the pelvic surface of the sacrum just medial to the pelvic sacral foramina and converge to form the small median **ganglion impar** (coccygeal ganglion) anterior to the coccyx. The sacral sympathetic trunks descend posterior to the rectum in the extraperitoneal connective tissue and send communicating branches (gray rami communicantes) to each of the anterior rami of the sacral and coccygeal nerves.

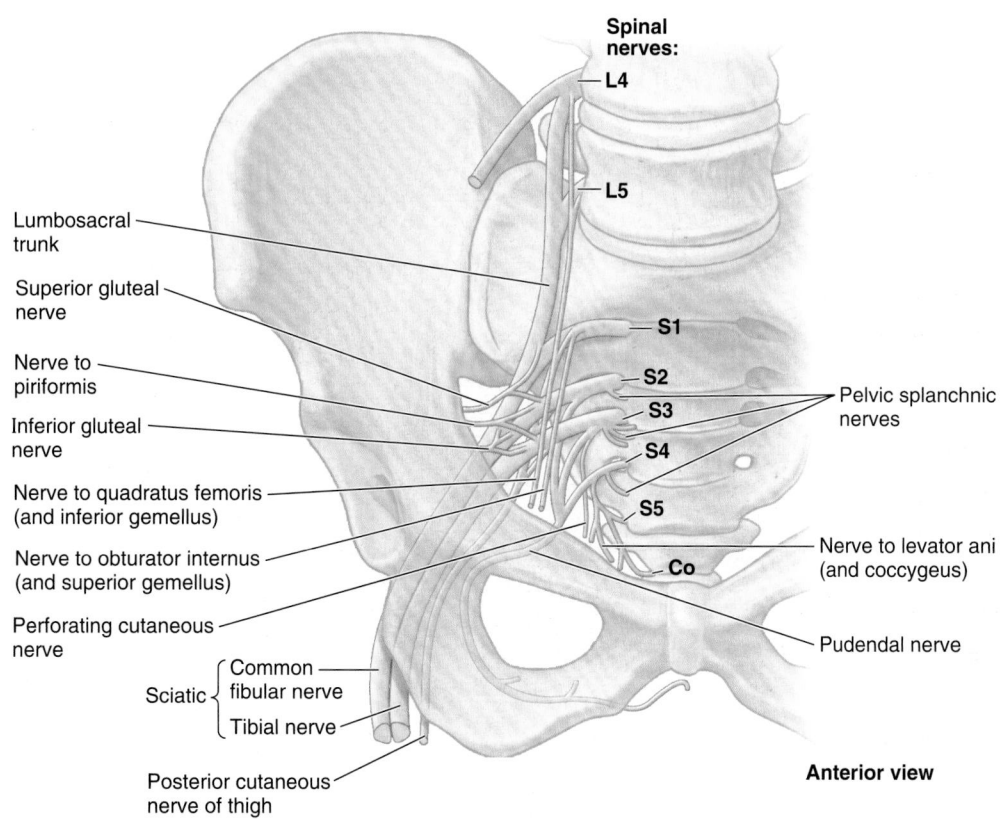

FIGURE 3.22. Somatic nerves of pelvis—the sacral plexus.

TABLE 3.5. SOMATIC NERVES OF PELVIS

Nerve	Origin	Distribution
Sciatic	L4, L5, S1, S2, S3	Articular branches to hip joint and muscular branches to flexors of knee in thigh and all muscles in leg and foot
Superior gluteal	L4, L5, S1	Gluteus medius and gluteus minimus muscles
Nerve to quadratus femoris (and inferior gemellus)	L4, L5, S1	Quadratus femoris and inferior gemellus muscles
Inferior gluteal	L5, S1, S2	Gluteus maximus
Nerve to obturator internus (and superior gemellus)	L5, S1, S2	Obturator internus and superior gemellus muscles
Nerve to piriformis	S1, S2	Piriformis muscle
Posterior cutaneous nerve of thigh	S2, S3	Cutaneous branches to buttocks and uppermost medial and posterior surfaces of thigh
Perforating cutaneous	S2, S3	Cutaneous branches to medial part of buttocks
Pudendal	S2, S3, S4	Structures in perineum: sensory to genitalia; muscular branches to perineal muscles, external urethral sphincter, and external anal sphincter
Pelvic splanchnic	S2, S3, S4	Pelvic viscera via inferior hypogastric and pelvic plexuses
Nerves to levator ani and coccygeus	S3, S4	Levator ani and coccygeus muscles

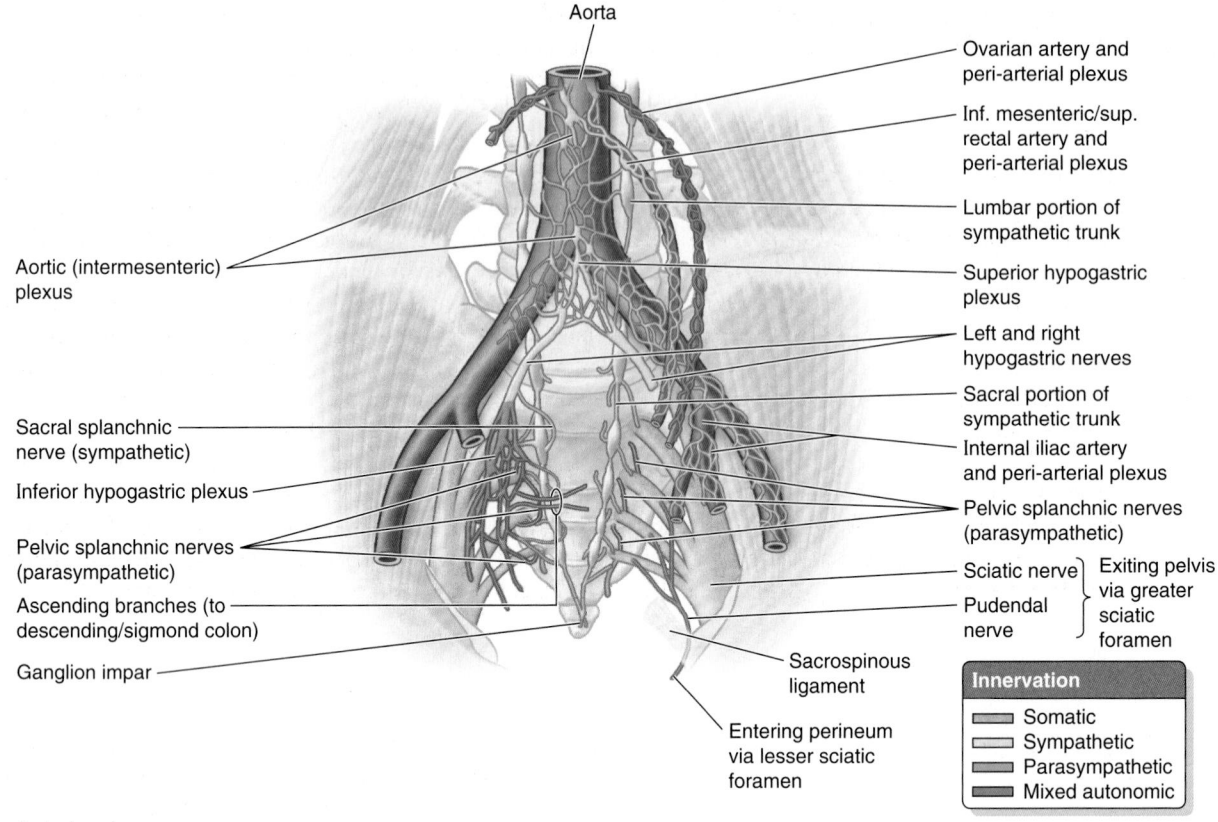

Anterior view

FIGURE 3.23. Autonomic nerves of pelvis. The superior hypogastric plexus is a continuation of the aortic plexus that divides into left and right hypogastric nerves as it enters the pelvis. The hypogastric and pelvic splanchnic nerves merge to form the inferior hypogastric plexuses, which thus consist of both sympathetic and parasympathetic fibers. Autonomic (sympathetic) fibers also enter the pelvis via the sympathetic trunks and peri-arterial plexuses.

They also send small branches to the median sacral artery and the inferior hypogastric plexus. The primary function of the sacral sympathetic trunks is to provide postsynaptic fibers to the sacral plexus for sympathetic (vasomotor, pilomotor, and sudomotor) innervation of the lower limb.

The *peri-arterial plexuses of the ovarian, superior rectal, and internal iliac arteries* are minor routes by which sympathetic fibers enter the pelvis. Their primary function is vasomotion of the arteries they accompany.

The **hypogastric plexuses** (superior and inferior) are networks of sympathetic and visceral afferent nerve fibers. The main part of the *superior hypogastric plexus* is a prolongation of the *intermesenteric plexus* (see Chapter 2), which lies inferior to the bifurcation of the aorta (Fig. 3.23). It carries fibers conveyed to and from the intermesenteric plexus by the L3 and L4 splanchnic nerves. The superior hypogastric plexus enters the pelvis, dividing into **right** and **left hypogastric nerves,** which descend on the anterior surface of the sacrum. These nerves descend lateral to the rectum within *hypogastric sheaths* and then spread in a fan-like fashion as they merge with the pelvic splanchnic nerves to form the right and left *inferior hypogastric plexuses.*

The **inferior hypogastric plexuses** thus contain both sympathetic and parasympathetic fibers as well as visceral afferent fibers, which continue through the lamina of the hypogastric sheath to the pelvic viscera, upon which they form sub-plexuses collectively referred to as the **pelvic plexuses.** In both sexes, sub-plexuses are associated with the lateral aspects of the rectum and inferolateral surfaces of the bladder. In addition, sub-plexuses in the male are also associated with the prostate and seminal glands. In females, sub-plexuses are also associated with the cervix of the uterus and the lateral fornices of the vagina.

Pelvic splanchnic nerves arise in the pelvis from the anterior rami of spinal nerves S2–S4 of the sacral plexus (Figs 3.21– 3.23). They convey presynaptic parasympathetic fibers derived from the S2–S4 spinal cord segments, which make up the sacral outflow of the parasympathetic (craniosacral) nervous system, and visceral afferent fibers from cell bodies in the spinal ganglia of the corresponding spinal nerves. The greatest contribution of these fibers is usually from the S3 nerve.

The hypogastric/pelvic system of plexuses, receiving sympathetic fibers via lumbar splanchnic nerves and parasympathetic fibers via pelvic splanchnic nerves, innervate the pelvic viscera. Although the sympathetic component largely produces vasomotion as elsewhere, here it also inhibits peristaltic contraction of the rectum and stimulates contraction of the internal genital organs during orgasm, producing ejaculation in the male.

Because the pelvis does not include a cutaneous area, pelvic sympathetic fibers do not produce pilomotion or vasomotion functions. The parasympathetic fibers distributed within the pelvis stimulate contraction of the rectum and bladder for defecation and urination, respectively. Parasympathetic fibers in the prostatic plexus penetrate the pelvic floor to reach the erectile bodies of the external genitalia, producing erection.

VISCERAL AFFERENT INNERVATION IN PELVIS

Visceral afferent fibers travel with autonomic nerve fibers, although the sensory impulses are conducted centrally, retrograde to the efferent impulses conveyed by the autonomic fibers. All visceral afferent fibers conducting reflexive sensation (information that does not reach consciousness) travel with parasympathetic fibers. Thus, in the case of the pelvis, they travel through the pelvic and inferior hypogastric plexuses and the pelvic splanchnic nerves to the spinal sensory ganglia of spinal nerves S2–S4.

The paths followed by visceral afferent fibers conducting pain from the pelvic viscera differ in terms of course and destination, depending on whether the viscus or part of the viscus from which the pain is emanating is located superior or inferior to the **pelvic pain line.** Except in the case of the alimentary canal, the pelvic pain line corresponds to the inferior limit of the peritoneum (see Table 3.3 figures B & C). Intraperitoneal abdominopelvic viscera, or aspects of visceral structures that are in contact with the peritoneum, are superior to the pain line; subperitoneal pelvic viscera or portions of viscera are inferior to the pain line. In the case of the alimentary tract (large intestine), the pain line does not correlate with the peritoneum; the pain line occurs in the middle of the sigmoid colon.

Visceral afferent fibers conducting pain impulses from abdominopelvic viscera superior to the pain line follow sympathetic fibers retrograde, ascending through hypogastric/aortic plexuses, abdominopelvic splanchnic nerves, lumbar sympathetic trunks, and white rami communicantes to reach cell bodies in the inferior thoracic/upper lumbar spinal ganglia. Afferent fibers conducting pain impulses from the viscera or portions of viscera inferior to the pain line follow the parasympathetic fibers retrograde through the pelvic and inferior hypogastric plexuses and pelvic splanchnic nerves to reach cell bodies in the spinal sensory ganglia of S2–S4.

NEUROVASCULAR STRUCTURES OF PELVIS

Iatrogenic Injury of Ureters

INJURY DURING LIGATION OF UTERINE ARTERY

The fact that the ureter passes immediately inferior to the uterine artery near the lateral part of the fornix of the vagina is clinically important. The ureter is in danger of being inadvertently clamped (crushed), ligated, or transected during a *hysterectomy* (excision of uterus) when the uterine artery is ligated and severed to remove the uterus. The point at which the uterine artery and ureter cross lies approximately 2 cm superior to the ischial spine.

INJURY DURING LIGATION OF OVARIAN ARTERY

The ureters are vulnerable to injury when the ovarian vessels are ligated during an *ovariectomy* (excision of ovary) because these structures are close to each other as they cross the pelvic brim.

Ligation of Internal Iliac Artery and Collateral Circulation in Pelvis

Occasionally the internal iliac artery becomes stenotic (the lumen becomes narrow) due to atherosclerotic cholesterol deposit (Fig. B3.6), or is surgically ligated to control pelvic hemorrhage. Because of the numerous anastomoses between the artery's branches and adjacent arteries (see Fig. 3.16; Table 3.4), ligation does not stop blood flow but it does reduce blood pressure, allowing *hemostasis* (arrest of bleeding) to occur. Examples of collateral pathways to the internal iliac artery include the following pairs of anastomosing arteries: lumbar and iliolumbar, median sacral and lateral sacral, superior rectal and middle rectal, and inferior gluteal and profunda femoris artery. Blood flow in the artery is maintained, although it may be reversed in the anastomotic branch. The collateral pathways may maintain the blood supply to the pelvic viscera, gluteal region, and genital organs.

Injury to Pelvic Nerves

During childbirth, the fetal head may compress the nerves of the mother's sacral plexus, producing pain in the lower limbs. The *obturator nerve* is vulnerable to injury during surgery (e.g., during removal of cancerous lymph nodes from the lateral pelvic wall). Injury to this nerve may cause painful spasms of the adductor muscles of the thigh and sensory deficits in the medial thigh region.

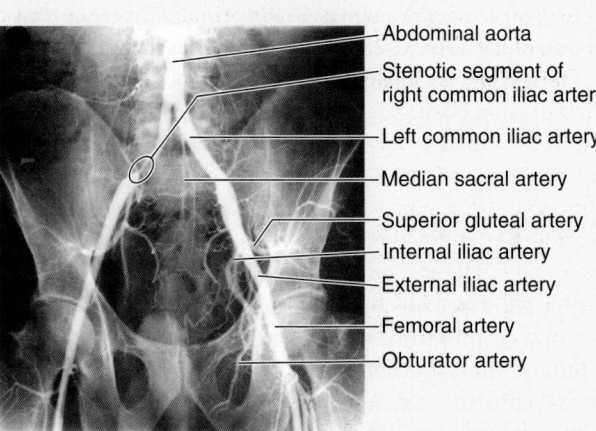

AP view

FIGURE B3.6. Iliac arteriogram. Radiopaque dye was injected into the abdominal aorta in the lumbar region. There is a site of narrowing (stenosis) of the right common iliac artery (*circled area*). (Courtesy of Dr. D. Sniderman, Associate Professor of Medical Imaging, University of Toronto, Toronto, ON, Canada.)

The Bottom Line

NEUROVASCULAR STRUCTURES OF PELVIS

Progressing from the pelvic cavity outward, as when dissecting the pelvis, the retroperitoneal hypogastric/pelvic autonomic nerve plexuses are encountered first (nearest the viscera), then pelvic arteries, pelvic veins, and finally the pelvic somatic nerves and sympathetic trunks, the latter two being adjacent to the pelvic walls.

Pelvic arteries: Multiple anastomosing arteries provide a collateral circulatory system that helps ensure an adequate blood supply to the greater and lesser pelves. Most arterial blood is delivered to the lesser pelvis by the internal iliac arteries, which commonly bifurcate into an anterior division (providing all the visceral branches) and a posterior division (usually exclusively parietal). ♦ Postnatally, the umbilical arteries become occluded distal to the origin of the superior vesical arteries and, in the male, the arteries to the ductus deferens. ♦ The inferior vesical (males) and vaginal arteries (females) supply the inferior bladder and pelvic urethra. The inferior vesical artery also supplies the prostate; the vaginal artery supplies the superior vagina. ♦ The uterine artery is exclusively female, but both sexes have middle rectal arteries.

Parietal branches of the anterior division of the internal iliac in both sexes include the obturator, inferior gluteal, and internal pudendal arteries, the main branches of which arise outside of the lesser pelvis. ♦ A clinically significant aberrant obturator artery arises from the inferior epigastric vessels in approximately 20% of the population. ♦ The iliolumbar, superior gluteal, and lateral sacral arteries are parietal branches of the posterior division of the internal iliac artery, distributed outside of the lesser pelvis. ♦ The iliolumbar artery is a major supplier to structures of the iliac fossae (greater pelvis). ♦ The gonadal arteries of both sexes descend into the greater pelvis from the abdominal aorta, but only the ovarian arteries enter the lesser pelvis.

Pelvic veins: The venous plexuses associated with and named for the various pelvic viscera intercommunicate with each other and the internal vertebral (epidural) venous plexuses of the vertebral canal. However, most venous blood exits the pelvis via the internal iliac veins.

Pelvic lymph drainage and nodes: Lymphatic drainage from the pelvis follows a pattern that mainly, but not completely, follows the pattern of venous drainage through variable minor and major groups of lymph nodes, the latter including the sacral and internal, external, and common iliac nodes. ♦ Aspects of the anterior to middle pelvic organs, approximately at the level of (and including) the roof of the undistended urinary bladder, drain to the external iliac nodes, independent of the venous drainage. ♦ The pelvic lymph nodes are highly interconnected, so that lymphatic drainage (and metastatic cancer cells) can pass in almost any direction, to any pelvic or abdominal organ.

Pelvic nerves: Somatic nerves within the pelvis form the sacral plexus, primarily concerned with innervation of the lower limbs and perineum. ♦ The pelvic portions of the sympathetic trunks are also primarily concerned with innervation of the lower limbs. ♦ Autonomic nerves are primarily brought to the pelvis via the superior hypogastric plexus (sympathetic fibers) and pelvic splanchnic nerves (parasympathetic fibers), the two merging to form the inferior hypogastric and pelvic plexuses. ♦ Sympathetic fibers to the pelvis produce vasomotion and contraction of internal genital organs during orgasm; they also inhibit rectal peristalsis. ♦ Pelvic parasympathetic fibers stimulate bladder and rectal emptying and extend to the erectile bodies of the external genitalia to produce erection. ♦ Visceral afferent fibers travel retrogradely along the autonomic nerve fibers. ♦ Visceral afferents conveying unconscious reflex sensation follow the course of the parasympathetic fibers to the spinal sensory ganglia of S2–S4, as do those transmitting pain sensations from the viscera inferior to the pelvic pain line (structures that do not contact the peritoneum plus the distal sigmoid colon and rectum). ♦ Visceral afferent fibers conducting pain from structures superior to the pelvic pain line (structures in contact with the peritoneum, except for the distal sigmoid colon and rectum) follow the sympathetic fibers retrogradely to inferior thoracic and superior lumbar spinal ganglia.

PELVIC VISCERA

The **pelvic viscera** include the distal parts of the urinary system and gastrointestinal tract, and the reproductive system. Although the sigmoid colon and parts of the small bowel extend into the pelvic cavity, they are abdominal rather than pelvic viscera. The bladder and rectum—true pelvic viscera—are inferior continuations of systems encountered in the abdomen. Except for features related to sharing of the male urethra by the urinary and reproductive tracts, and physical relationships to the respective reproductive organs, there are relatively few distinctions between the male and female pelvic urinary and gastrointestinal organs.

Urinary Organs

The pelvic urinary organs (Fig. 3.24A) are the:

- Pelvic portions of the ureters, which carry urine from the kidneys.
- Urinary bladder, which temporarily stores urine.
- Urethra, which conducts urine from the bladder to the exterior.

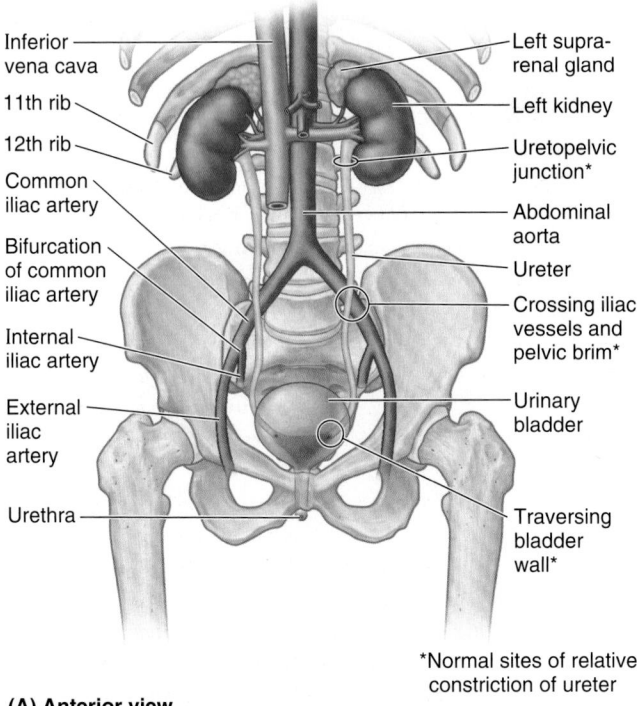

Inferior vena cava

11th rib

12th rib

Common iliac artery

Bifurcation of common iliac artery

Internal iliac artery

External iliac artery

Urethra

Left supra-renal gland

Left kidney

Uretopelvic junction*

Abdominal aorta

Ureter

Crossing iliac vessels and pelvic brim*

Urinary bladder

Traversing bladder wall*

*Normal sites of relative constriction of ureter

(A) Anterior view

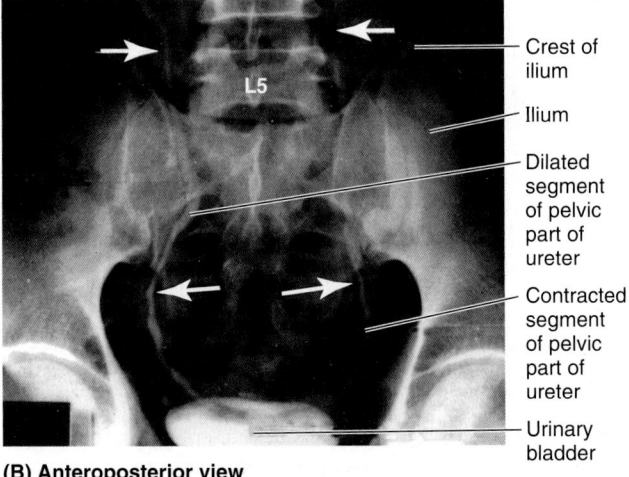

L5

Crest of ilium

Ilium

Dilated segment of pelvic part of ureter

Contracted segment of pelvic part of ureter

Urinary bladder

(B) Anteroposterior view

FIGURE 3.24. Genito-urinary viscera. A. The course and normal sites of relative constriction of the ureters. **B.** Normal intravenous urogram. The *arrows* indicate transient narrowing of the lumina of the ureters resulting from peristaltic contraction.

URETERS

The **ureters** are muscular tubes, 25–30 cm long, that connect the kidneys to the urinary bladder. The ureters are retroperitoneal; their superior abdominal portions are described in Chapter 2. As the ureters cross the bifurcation of the common iliac artery (or the beginning of the external iliac artery), they pass over the pelvic brim, thus leaving the abdomen and entering the lesser pelvis (Fig. 3.24A & B). The pelvic parts of the ureters run on the lateral walls of the pelvis, parallel to the anterior margin of the greater sciatic notch, between

the parietal pelvic peritoneum and the internal iliac arteries. Opposite the ischial spine, they curve anteromedially, superior to the levator ani, and enter the urinary bladder. The inferior ends of the ureters are surrounded by the *vesical venous plexus* (Fig. 3.19B & C).

The ureters pass obliquely through the muscular wall of the urinary bladder in an inferomedial direction, entering the outer surface of the bladder approximately 5 cm apart, but their internal openings into the lumen of the empty bladder are separated by only half that distance. This oblique passage through the bladder wall forms a one-way "flap valve," the internal pressure of the filling bladder causing the intramural passage to collapse. In addition, contractions of the bladder musculature act as a sphincter preventing the reflux of urine into the ureters when the bladder contracts, increasing internal pressure during micturition. Urine is transported down the ureters by means of peristaltic contractions, a few drops being transported at intervals of 12–20 sec (Fig. 3.24B).

In males, the only structure that passes between the ureter and the peritoneum is the *ductus deferens* (see Fig. 3.34); it crosses the ureter within the ureteric fold of peritoneum. The ureter lies posterolateral to the ductus deferens and enters the posterosuperior angle of the bladder, just superior to the seminal gland.

In females, the ureter passes medial to the origin of the uterine artery and continues to the level of the ischial spine, where it is crossed superiorly by the uterine artery (see the blue box "Iatrogenic Injury of the Ureters" on p. 361). It then passes close to the lateral part of the fornix of the vagina and enters the posterosuperior angle of the bladder.

Arterial Supply and Venous Drainage of Pelvic Portion of Ureters. The *arterial supply* to the pelvic parts of the ureters is variable, with ureteric branches extending from the common iliac, internal iliac, and ovarian arteries (Fig. 3.25; Table 3.4). The ureteric branches anastomose along the length of the ureter forming a continuous blood supply, although not necessarily effective collateral pathways. The most constant arteries supplying the terminal parts of the ureter in females are branches of the *uterine arteries*. The source of similar branches in males are the *inferior vesical arteries*. The blood supply of the ureters is a matter of great concern to surgeons operating in the region (see the blue box "Iatrogenic Compromise of Ureteric Blood Supply," p. 373).

The *venous drainage* from the pelvic parts of the ureters generally parallels the arterial supply, draining to veins with corresponding names. *Lymphatic vessels* pass primarily to common and internal iliac nodes (Fig. 3.20).

Innervation of Ureters. The nerves to the ureters derive from adjacent autonomic plexuses (renal, aortic, superior and inferior hypogastric—Fig. 3.26). The ureters are primarily superior to the pelvic pain line. Afferent (pain) fibers from the ureters follow sympathetic fibers in a retrograde direction to reach the spinal ganglia and spinal cord segments of T10–L2 or L3. *Ureteric pain* is usually referred to the ipsilateral lower quadrant of the abdomen, especially to the groin (inguinal region). See the blue box "Ureteric Calculi" on page 373.

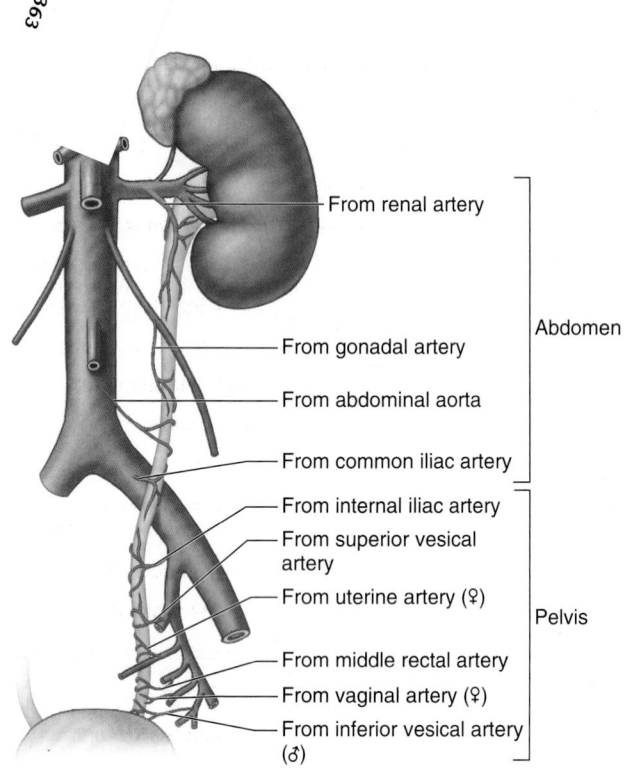

Anterior view of left side

FIGURE 3.25. Blood supply of ureter. Branches supplying the abdominal half of the ureter approach medially, while those supplying the pelvic half approach laterally. During surgery, the ureters are avoided and left undisturbed when possible. If necessary, traction of the ureters is applied gently and only toward the blood supply to avoid disruption of the small branches.

URINARY BLADDER

The **urinary bladder,** a hollow viscus with strong muscular walls, is characterized by its distensibility (Fig. 3.27A). The bladder is a temporary reservoir for urine, and varies in size, shape, position, and relationships according to its content, and the state of neighboring viscera. When empty, the adult urinary bladder is located in the lesser pelvis, lying partially superior to and partially posterior to the pubic bones (Fig. 3.27B). It is separated from these bones by the potential *retropubic space* (of Retzius) and lies mostly inferior to the peritoneum, resting on the pubic bones and pubic symphysis anteriorly and the prostate (males) or anterior wall of the vagina (females) posteriorly (Fig. 3.27A & B). The bladder is relatively free within the extraperitoneal subcutaneous fatty tissue, except for its neck, which is held firmly by the *lateral ligaments of bladder* and the *tendinous arch of the pelvic fascia*—especially its anterior component, the *puboprostatic ligament* in males and the *pubovesical ligament* in females (see also Fig. 3.14A). In females, since the posterior aspect of the bladder rests directly upon the anterior wall of the vagina, the lateral attachment of the vagina to the tendinous arch of the pelvic fascia, the *paracolpium,* is an indirect but important factor in supporting the urinary bladder (see Fig. 3.14B; DeLancey, 1992; Ashton-Miller and DeLancey, 2007).

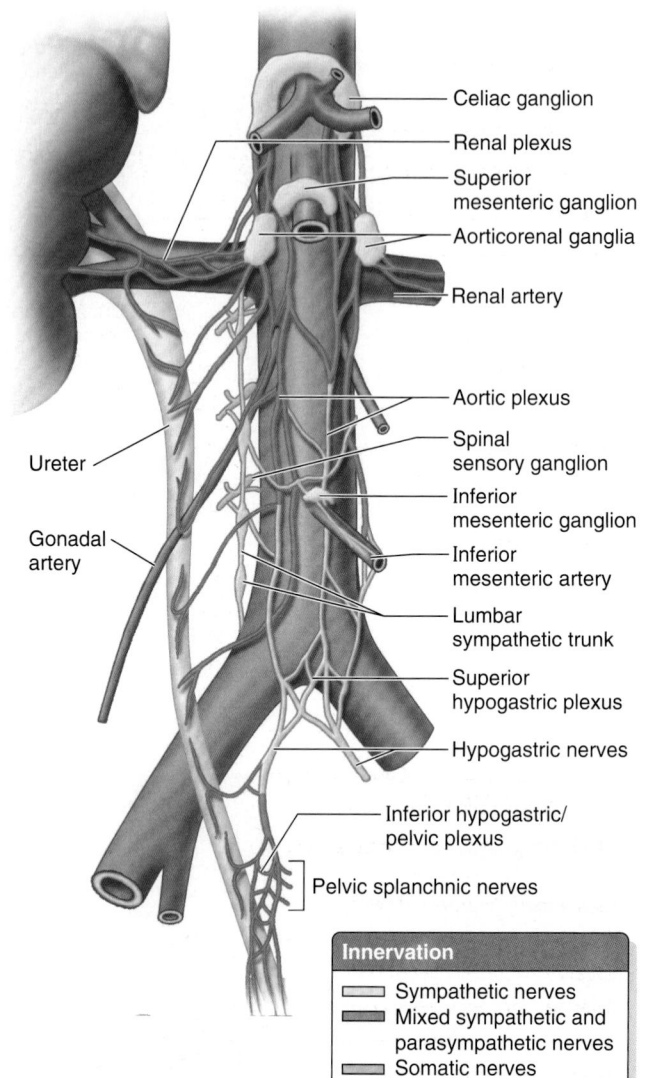

Anterior view

FIGURE 3.26. Innervation of ureters. Nerve fibers from the renal, aortic, and superior and inferior hypogastric plexuses extend to the ureter, carrying visceral afferent and sympathetic fibers to the T10–L2(3) spinal sensory ganglia and cord segments. Parasympathetic fibers, from the S2–S4 spinal cord segments, are distributed to the pelvic part of the ureter. Extrinsic ANS fibers are not essential for the initiation and propagation of ureteric peristalsis.

In infants and young children, the urinary bladder is in the abdomen even when empty (Fig. 3.28A). The bladder usually enters the greater pelvis by 6 years of age; however, it is not located entirely within the lesser pelvis until after puberty. An empty bladder in adults lies almost entirely in the lesser pelvis, its superior surface level with the superior margin of the pubic symphysis (Fig. 3.28B). As the bladder fills, it enters the greater pelvis as it ascends in the extraperitoneal fatty tissue of the anterior abdominal wall (Fig. 3.27A). In some individuals, a full bladder may ascend to the level of the umbilicus.

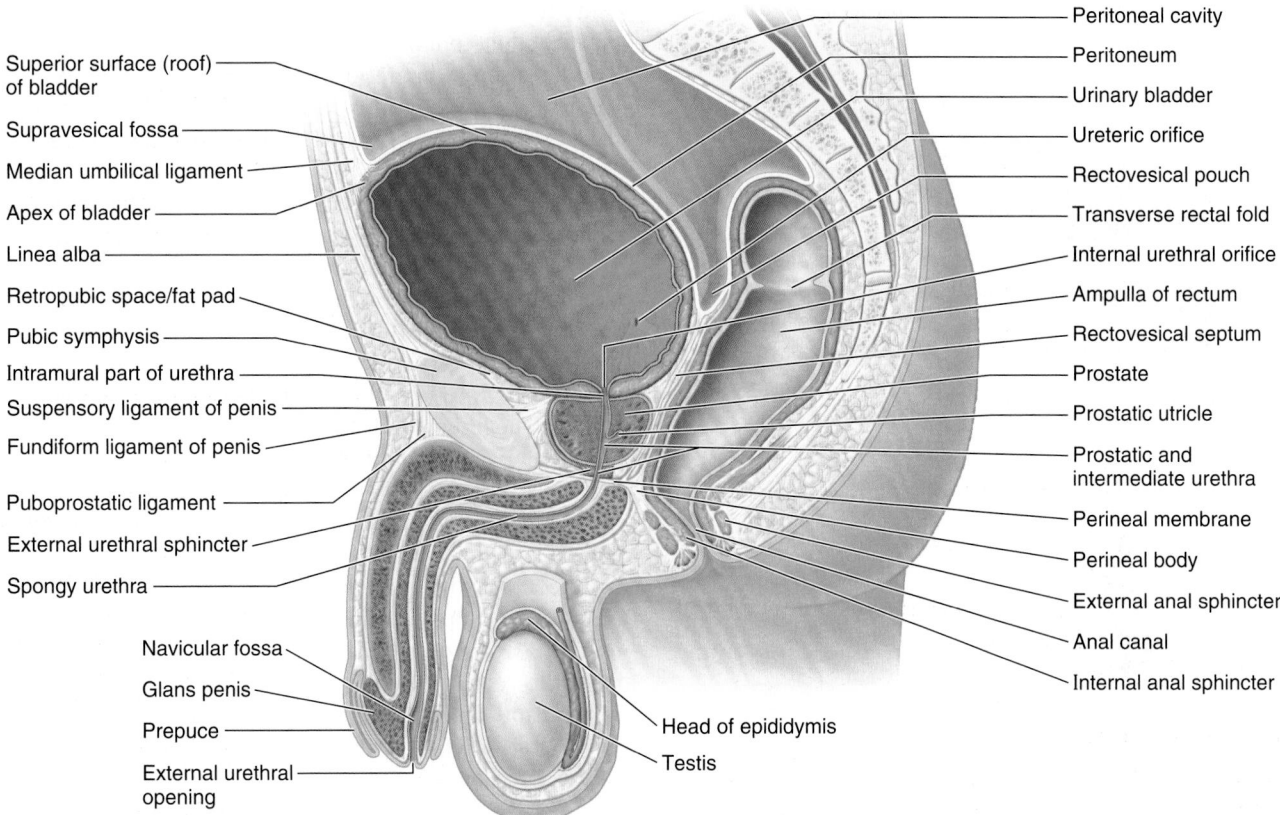

Peritoneal cavity
Peritoneum
Urinary bladder
Ureteric orifice
Rectovesical pouch
Transverse rectal fold
Internal urethral orifice
Ampulla of rectum
Rectovesical septum
Prostate
Prostatic utricle
Prostatic and intermediate urethra
Perineal membrane
Perineal body
External anal sphincter
Anal canal
Internal anal sphincter

Superior surface (roof) of bladder
Supravesical fossa
Median umbilical ligament
Apex of bladder
Linea alba
Retropubic space/fat pad
Pubic symphysis
Intramural part of urethra
Suspensory ligament of penis
Fundiform ligament of penis
Puboprostatic ligament
External urethral sphincter
Spongy urethra
Navicular fossa
Glans penis
Prepuce
External urethral opening
Head of epididymis
Testis

(A) Median section of male pelvis

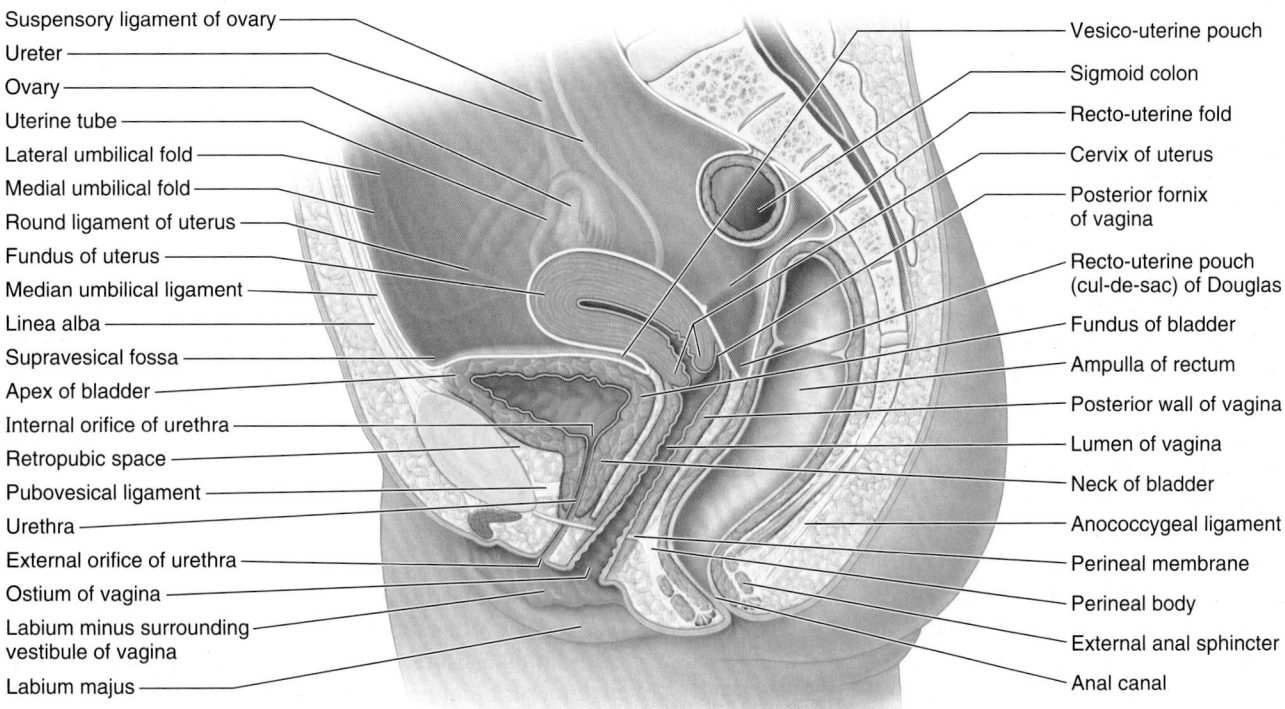

Suspensory ligament of ovary
Ureter
Ovary
Uterine tube
Lateral umbilical fold
Medial umbilical fold
Round ligament of uterus
Fundus of uterus
Median umbilical ligament
Linea alba
Supravesical fossa
Apex of bladder
Internal orifice of urethra
Retropubic space
Pubovesical ligament
Urethra
External orifice of urethra
Ostium of vagina
Labium minus surrounding vestibule of vagina
Labium majus

Vesico-uterine pouch
Sigmoid colon
Recto-uterine fold
Cervix of uterus
Posterior fornix of vagina
Recto-uterine pouch (cul-de-sac) of Douglas
Fundus of bladder
Ampulla of rectum
Posterior wall of vagina
Lumen of vagina
Neck of bladder
Anococcygeal ligament
Perineal membrane
Perineal body
External anal sphincter
Anal canal

(B) Median section of female pelvis

FIGURE 3.27. Viscera in hemisected male and female pelves. A. In this male pelvis, the urinary bladder is distended, as if full. Compare its relation to the anterior abdominal wall, pubic symphysis, and level of the supravesical fossa to that of the non-distended (empty) bladder in part **B. B.** In this female pelvis, the uterus was sectioned in its own median plane and is depicted as though it coincided with the median plane of the body, which is seldom the case. With the bladder empty, the normal disposition of the uterus shown here—bent on itself (anteflexed) at the junction of the body and cervix and tipped anteriorly (anteverted)—causes its weight to be borne mainly by the bladder. The urethra lies anterior and parallel to the lower half of the vagina.

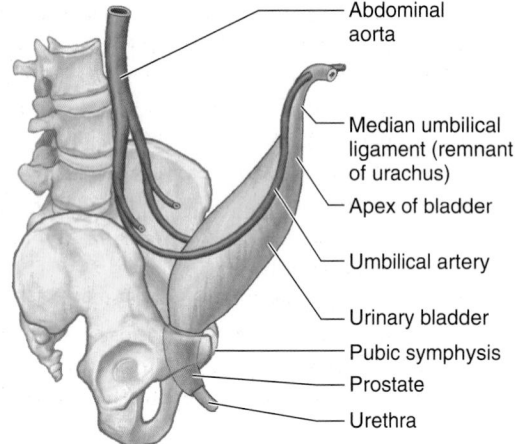

- Abdominal aorta
- Median umbilical ligament (remnant of urachus)
- Apex of bladder
- Umbilical artery
- Urinary bladder
- Pubic symphysis
- Prostate
- Urethra

(A) Right lateral view

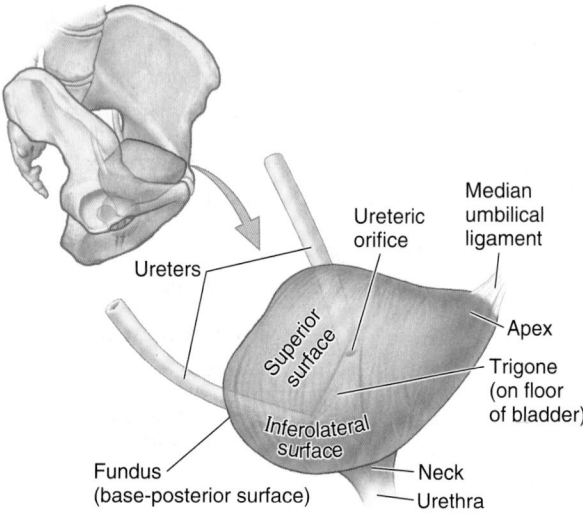

- Ureteric orifice
- Median umbilical ligament
- Ureters
- Superior surface
- Apex
- Trigone (on floor of bladder)
- Inferolateral surface
- Fundus (base-posterior surface)
- Neck
- Urethra

(B) Lateral view

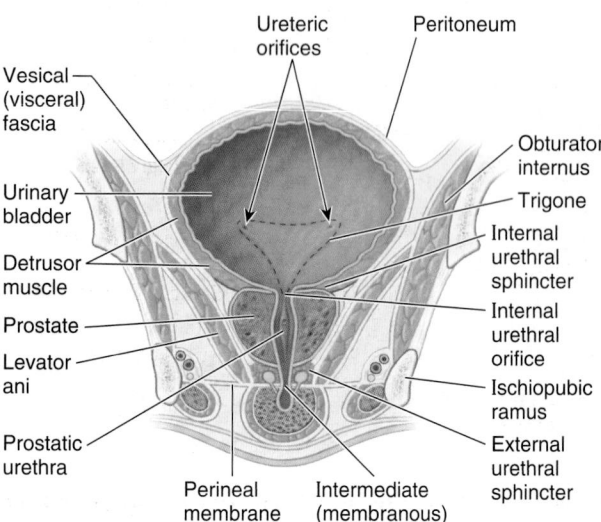

- Ureteric orifices
- Peritoneum
- Vesical (visceral) fascia
- Urinary bladder
- Detrusor muscle
- Prostate
- Levator ani
- Prostatic urethra
- Obturator internus
- Trigone
- Internal urethral sphincter
- Internal urethral orifice
- Ischiopubic ramus
- External urethral sphincter
- Perineal membrane
- Intermediate (membranous) part of urethra

(C) Anterior view

FIGURE 3.28. Urinary bladder and prostatic urethra. A. The bladder of an infant lies almost entirely in the abdominal cavity. **B.** Adult bladder and prostate demonstrating their pelvic location (inset) and the surfaces of the bladder. **C.** Coronal section of urinary bladder and prostate in the plane of the prostatic urethra.

At the end of micturition (urination), the bladder of a normal adult contains virtually no urine. When empty, the bladder is somewhat tetrahedral (Fig. 3.28B) and externally has an apex, body, fundus, and neck. The bladder's four surfaces (superior, two inferolateral, and posterior) are most apparent when viewing an empty, contracted bladder that has been removed from a cadaver, when the bladder appears rather boat shaped.

The **apex of the bladder** points toward the superior edge of the pubic symphysis when the bladder is empty. The **fundus of the bladder** is opposite the apex, formed by the somewhat convex posterior wall. The **body of the bladder** is the major portion of the bladder between the apex and the fundus. The fundus and *inferolateral surfaces* meet inferiorly at the **neck of the bladder**.

The **bladder bed** is formed by the structures that directly contact it. On each side, the pubic bones and fascia covering the levator ani and the superior obturator internus lie in contact with the inferolateral surfaces of the bladder (Fig. 3.28C). Only the superior surface is covered by peritoneum. Consequently, in males the fundus is separated from the rectum centrally by only the fascial rectovesical septum and laterally by the seminal glands and ampullae of the ductus deferentes (Fig. 3.27A). In females the fundus is directly related to the superior anterior wall of the vagina (Fig. 3.27B). The bladder is enveloped by a loose connective tissue visceral fascia.

The walls of the bladder are composed chiefly of the **detrusor muscle.** Toward the neck of the male bladder, the muscle fibers form the involuntary **internal urethral sphincter.** This sphincter contracts during ejaculation to prevent *retrograde ejaculation (ejaculatory reflux)* of semen into the bladder. Some fibers run radially and assist in opening the **internal urethral orifice.** In males, the muscle fibers in the neck of the bladder are continuous with the fibromuscular tissue of the prostate, whereas in females these fibers are continuous with muscle fibers in the wall of the urethra.

The **ureteric orifices** and the internal urethral orifice are at the angles of the **trigone of the bladder** (Fig. 3.28C). The ureteric orifices are encircled by loops of detrusor musculature that tighten when the bladder contracts to assist in preventing reflux of urine into the ureter. The **uvula of the bladder** is a slight elevation of the trigone; it is usually more prominent in older men owing to enlargement of the posterior lobe of the prostate (see Fig. 3.30A).

Arterial Supply and Venous Drainage of Bladder. The main arteries supplying the bladder are branches of the internal iliac arteries (see Table 3.4). The *superior vesical arteries* supply anterosuperior parts of the bladder. In males, the *inferior vesical arteries* supply the fundus and neck of the bladder. In females, the *vaginal arteries* replace the inferior vesical arteries and send small branches to posteroinferior parts of the bladder (see Fig. 3.17B). The obturator and inferior gluteal arteries also supply small branches to the bladder.

The veins draining blood from the bladder correspond to the arteries, and are tributaries of the internal iliac veins. In males, the *vesical venous plexus* is continuous with the *prostatic venous plexus* (see Fig. 3.19C), and the combined plexus complex envelops the fundus of the bladder and prostate, the seminal glands, the ductus deferentes, and the inferior ends of the ureters. It also receives blood from the deep dorsal vein of the penis, which drains into the prostatic venous plexus. The **vesical venous plexus** is the venous network that is most directly associated with the bladder itself. It mainly drains through the inferior vesical veins into the internal iliac veins; however, it may drain through the sacral veins into the *internal vertebral venous plexuses.* In females, the vesical venous plexus envelops the pelvic part of the urethra and the neck of the bladder, receives blood from the *dorsal vein of the clitoris,* and communicates with the *vaginal or uterovaginal venous plexus* (Fig. 3.19B).

Innervation of Bladder. *Sympathetic fibers* are conveyed from inferior thoracic and upper lumbar spinal cord levels to the vesical (pelvic) plexuses primarily through the hypogastric plexuses and nerves, whereas parasympathetic fibers from sacral spinal cord levels are conveyed by the pelvic splanchnic nerves and the inferior hypogastric plexus (Fig. 3.29). The *parasympathetic fibers* are motor to the detrusor muscle and inhibitory to the internal ure-

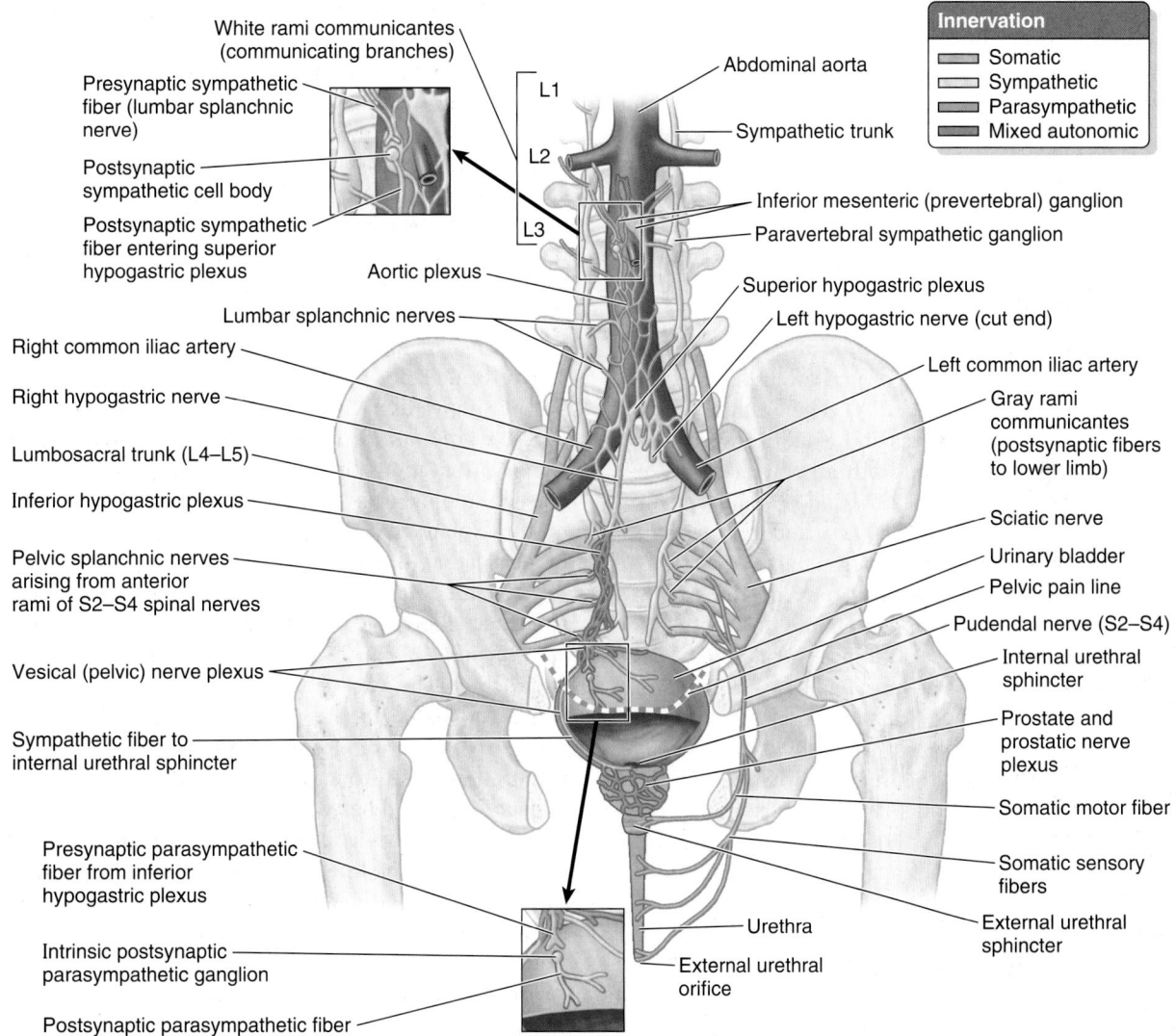

FIGURE 3.29. Innervation of bladder and urethra. Presynaptic sympathetic fibers from the T11–L2 or L3 spinal cord levels involved in innervation of the bladder, prostate, and proximal urethra pass via lumbar splanchnic nerves to the aortic/hypogastric system of plexuses, synapsing in the plexuses en route to the pelvic viscera. Presynaptic parasympathetic fibers to the bladder arise from neurons in the S2–S4 spinal cord segments and pass from the anterior rami of spinal nerves S2–S4 via the pelvic splanchnic nerves and inferior hypogastric and vesical (pelvic) plexuses to the bladder. They synapse with postsynaptic neurons located on or near the bladder wall. Visceral afferent fibers conveying reflex information and pain sensation from subperitoneal viscera (inferior to the pelvic pain line) follow parasympathetic fibers retrogradely to the S2–S4 spinal ganglia, whereas those conducting pain from the bladder roof (superior to the pelvic pain line) follow sympathetic fibers retrogradely to the T11–L2 or L3 spinal ganglia. The pelvic (sacral) sympathetic trunk primarily serves the lower limb. The somatic nerves shown here are distributed to the perineum.

of the male bladder. Hence, when visceral ㅁㅁㅁ re stimulated by stretching, the bladder contracts reflexively, the internal urethral sphincter relaxes (in males), and urine flows into the urethra. With toilet training, we learn to suppress this reflex when we do not wish to void. The sympathetic innervation that stimulates ejaculation simultaneously causes contraction of the internal urethral sphincter, to prevent reflux of semen into the bladder. A sympathetic response at moments other than ejaculation (e.g., self-consciousness when standing at the urinal in front of a waiting line) can cause the internal sphincter to contract, hampering the ability to urinate until parasympathetic inhibition of the sphincter occurs.

Sensory fibers from most of the bladder are visceral; reflex afferents follow the course of the parasympathetic fibers, as do those transmitting pain sensations (such as results from overdistension) from the inferior part of the bladder. The superior surface of the bladder is covered with peritoneum and is therefore superior to the *pelvic pain line;* thus pain fibers from the superior bladder follow the sympathetic fibers retrogradely to the inferior thoracic and upper lumbar spinal ganglia (T11–L2 or L3).

PROXIMAL (PELVIC) MALE URETHRA

The **male urethra** is a muscular tube (18–22 cm long) that conveys urine from the *internal urethral orifice* of the urinary bladder to the *external urethral orifice,* located at the tip of the glans penis in males (Fig. 3.27A). The urethra also provides an exit for semen (sperms and glandular secretions). For descriptive purposes, the urethra is divided into four parts, demonstrated in Figures 3.27A and 3.30 and described in Table 3.6. The distal *intermediate part* and *spongy urethra* will be described further with the perineum (p. 418).

The **intramural (preprostatic) part of the urethra** varies in diameter and length, depending on whether the bladder is filling (bladder neck is tonically contracted so the internal urethral orifice is small and high; the *filling internal urethral orifice*) or emptying (the neck is relaxed so the orifice is wide and low; the *emptying internal urethral orifice*). The most prominent feature of the **prostatic urethra** is the **urethral crest,** a median ridge between bilateral grooves, the **prostatic sinuses** (Fig. 3.30). The secretory **prostatic ducts** open into the prostatic sinuses. The **seminal colliculus** is a rounded eminence in the middle of the urethral crest with a slit-like orifice that opens into a small cul-de-sac, the **prostatic utricle.** The utricle is the vestigial remnant of the embryonic uterovaginal canal, the surrounding walls of which, in the female, constitute the primordium of the uterus and a part of the vagina (Moore et al., 2012). The *ejaculatory ducts* open into the prostatic urethra via minute, slit-like openings located adjacent to and occasionally just within the orifice of the prostatic utricle. Thus urinary and reproductive tracts merge at this point.

Arterial Supply and Venous Drainage of Proximal Male Urethra. The intramural and prostatic parts of the urethra are supplied by *prostatic branches* of the *inferior vesical* and *middle rectal arteries* (see Figs. 3.15–3.17A). The veins from the proximal two parts of the urethra drain into the *prostatic venous plexus* (see Fig. 3.19C).

Innervation of Proximal Male Urethra. The nerves are derived from the prostatic plexus (mixed sympathetic, parasympathetic, and visceral afferent fibers) (Fig. 3.29). The **prostatic plexus** is one of the pelvic plexuses (an inferior extension of the vesical plexus) arising as organ-specific extensions of the inferior hypogastric plexus.

FEMALE URETHRA

The **female urethra** (approximately 4 cm long and 6 mm in diameter) passes antero-inferiorly from the *internal urethral orifice* of the urinary bladder (Fig. 3.27B), posterior and then inferior to the pubic symphysis, to the *external urethral orifice.* The musculature surrounding the internal urethral orifice of the female bladder is not organized into an internal sphincter. The **female external urethral orifice** is located in the *vestibule of the vagina,* the cleft between the labia minora of the external genitalia, directly anterior to the *vaginal orifice.* The urethra lies anterior to the vagina (forming an elevation in the anterior vaginal wall) (Fig. 3.27C); its axis is parallel to that of the vagina. The urethra passes with the vagina through the pelvic diaphragm, external urethral sphincter, and perineal membrane.

Urethral glands are present, particularly in the superior part of the urethra. One group of glands on each side, the *paraurethral glands,* are homologs to the prostate. These glands have a common para-urethral duct, which opens (one on each side) near the external urethral orifice. The external urethral sphincter is located in the perineum and is discussed in that section (p. 408).

Arterial Supply and Venous Drainage of Female Urethra. Blood is supplied to the female urethra by the *internal pudendal* and *vaginal arteries* (see Figs. 3.16, 3.17B, and 3.18A). The veins follow the arteries and have similar names (see Fig. 3.19B).

Innervation of Female Urethra. The nerves to the urethra arise from the *vesical (nerve) plexus* and the *pudendal nerve.* The pattern is similar to that in the male (Fig. 3.29), given the absence of a prostatic plexus and an internal urethra sphincter. Visceral afferents from most of the urethra run in the pelvic splanchnic nerves, but the termination receives somatic afferents from the pudendal nerve. Both the visceral and the somatic afferent fibers extend from cell bodies in the S2–S4 spinal ganglia.

Rectum

The **rectum** is the pelvic part of the digestive tract and is continuous proximally with the sigmoid colon (Fig. 3.31) and distally with the anal canal. The **rectosigmoid junction** lies anterior to the S3 vertebra. At this point, the teniae coli of the sigmoid colon spread to form a continuous outer longitudinal

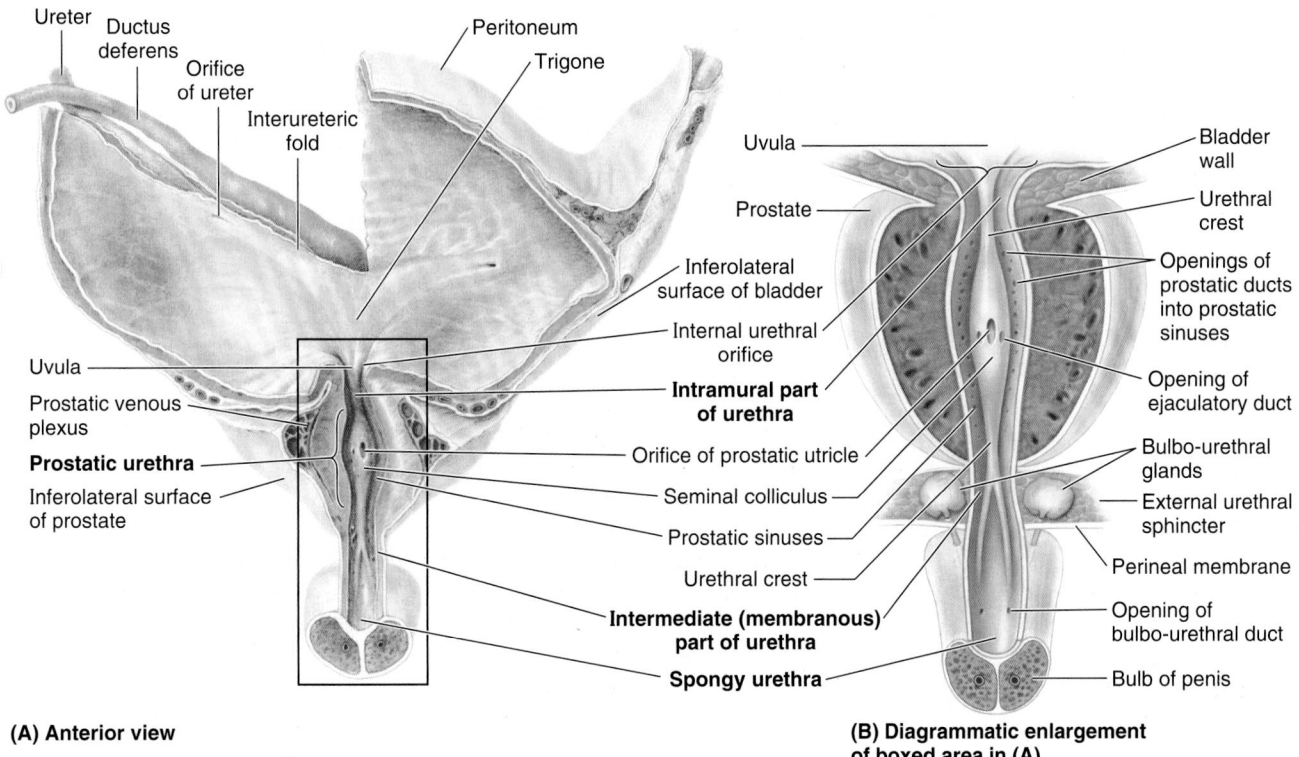

(A) Anterior view

**(B) Diagrammatic enlargement
of boxed area in (A)**

FIGURE 3.30. Interior of male bladder and urethra. A. The anterior parts of the bladder, prostate, and urethra are cut away. A portion of the posterior wall of the bladder has been removed to reveal the intramural part of ureter and the ductus deferens posterior to the bladder. The interureteric fold runs between the entrances of the ureters into the bladder lumen, demarcating the superior limit of the trigone of the bladder. The prominence of the posterior wall of the internal urethral orifice (at the tip of the leader line indicating this orifice), when exaggerated, becomes the uvula of the bladder. This small projection is produced by the middle lobe of the prostate. The prostatic fascia encloses the prostatic venous plexus. **B.** This enlarged detail of the boxed area in **A** demonstrates the bulbo-urethral glands embedded in the substance of the external urethral sphincter.

TABLE 3.6. PARTS OF MALE URETHRA

Part	Length[a]	Location/Disposition	Features
Intramural (preprostatic) part	0.5–1.5 cm	Extends almost vertically through neck of bladder	Surrounded by internal urethral sphincter; diameter and length vary, depending on whether bladder is filling or emptying
Prostatic urethra	3.0–4.0 cm	Descends through anterior prostate, forming a gentle, anteriorly concave curve; is bounded anteriorly by a vertical trough-like part (rhabdosphincter) of external urethral sphincter	Widest and most dilatable part; features urethral crest with seminal colliculus, flanked by prostatic sinuses into which prostatic ducts open; ejaculatory ducts open onto colliculus, hence urinary and reproductive tracts merge in this part
Intermediate (membranous) part	1.0–1.5 cm	Passes through deep perineal pouch, surrounded by circular fibers of external urethral sphincter; penetrates perineal membrane	Narrowest and least distensible part (except for external urethral orifice)
Spongy urethra	~15 cm	Courses through corpus spongiosum; initial widening occurs in bulb of penis; widens again distally as navicular fossa (in glans penis)	Longest and most mobile part; bulbo-urethral glands open into bulbous part; distally, urethral glands open into small urethral lacunae entering lumen of this part

[a]Lengths are provided for purposes of comparison—students should not memorize these lengths.

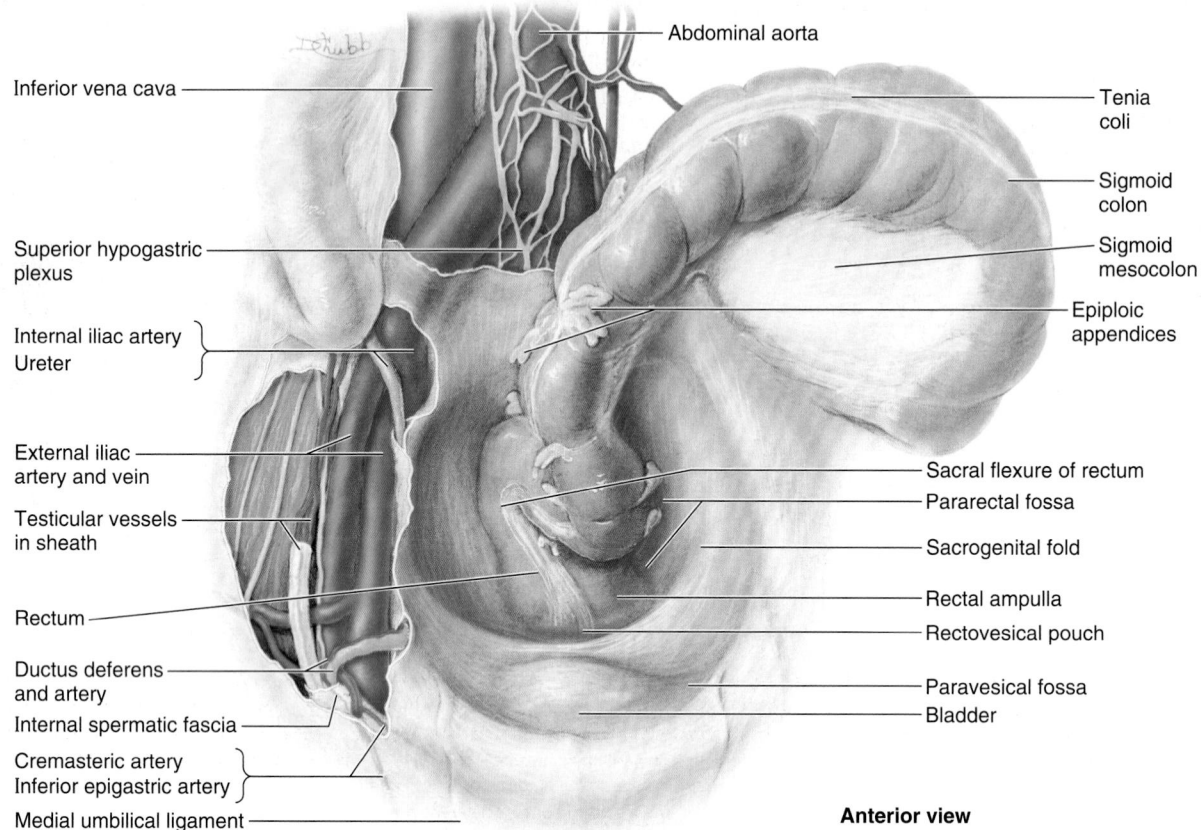

- Abdominal aorta
- Inferior vena cava
- Tenia coli
- Superior hypogastric plexus
- Sigmoid colon
- Sigmoid mesocolon
- Internal iliac artery
- Ureter
- Epiploic appendices
- External iliac artery and vein
- Testicular vessels in sheath
- Sacral flexure of rectum
- Pararectal fossa
- Sacrogenital fold
- Rectum
- Rectal ampulla
- Rectovesical pouch
- Ductus deferens and artery
- Internal spermatic fascia
- Paravesical fossa
- Cremasteric artery
- Inferior epigastric artery
- Bladder
- Medial umbilical ligament

Anterior view

FIGURE 3.31. Sigmoid colon entering lesser pelvis and becoming rectum. The sigmoid colon is intraperitoneal, suspended by the sigmoid mesocolon, but the rectum becomes retroperitoneal and then subperitoneal as it descends. The peritoneum has been removed superior to sacral promontory and right iliac fossa, revealing the superior hypogastric plexus lying in the bifurcation of the abdominal aorta and the internal iliac artery, ureter, and ductus deferens crossing the pelvic brim to enter the lesser pelvis.

layer of smooth muscle, and the fatty omental appendices are discontinued (see also Fig. 3.56).

Although its name is derived from the Latin term for "straight" (*rectus*), the term was coined during ancient studies on animals to describe the distal part of the colon; the human rectum is characterized by a number of flexures. The rectum follows the curve of the sacrum and coccyx, forming the **sacral flexure of the rectum.** The rectum ends antero-inferior to the tip of the coccyx, immediately before a sharp postero-inferior angle (the **anorectal flexure of the anal canal**) that occurs as the gut perforates the pelvic diaphragm (levator ani). The roughly 80° anorectal flexure is an important mechanism for fecal continence, being maintained during the resting state by the tonus of the puborectalis muscle, and by its active contraction during peristaltic contractions if defecation is not to occur. With the flexures of the rectosigmoid junction superiorly and the anorectal junction inferiorly, the rectum has an S shape when viewed laterally.

Three sharp **lateral flexures of the rectum** (**superior** and **inferior** on the left side, and **intermediate** on the right) are apparent when the rectum is viewed anteriorly (Fig. 3.32). The flexures are formed in relation to three internal infoldings (**transverse rectal folds**): two on the left and one on the right side. The folds overlie thickened parts of the circular muscle layer of the rectal wall. The dilated terminal part of the rectum, lying directly superior to and supported by the pelvic diaphragm (levator ani) and anococcygeal ligament, is the **ampulla of the rectum** (Figs. 3.27, 3.31, and 3.32). The ampulla receives and holds an accumulating fecal mass until it is expelled during defecation. The ability of the ampulla to relax to accommodate the initial and subsequent arrivals of fecal material is another essential element of maintaining fecal continence.

Peritoneum covers the anterior and lateral surfaces of the superior third of the rectum, only the anterior surface of the middle third, and no surface of the inferior third because it is subperitoneal (see Table 3.3). In males, the peritoneum reflects from the rectum to the posterior wall of the bladder, where it forms the floor of the *rectovesical pouch*. In females, the peritoneum reflects from the rectum to the posterior part of the fornix of the vagina, where it forms the floor of the *recto-uterine pouch*. In both sexes, lateral reflections of peritoneum from the superior third of the rectum form *pararectal fossae* (Fig. 3.31), which permit the rectum to distend as it fills with feces.

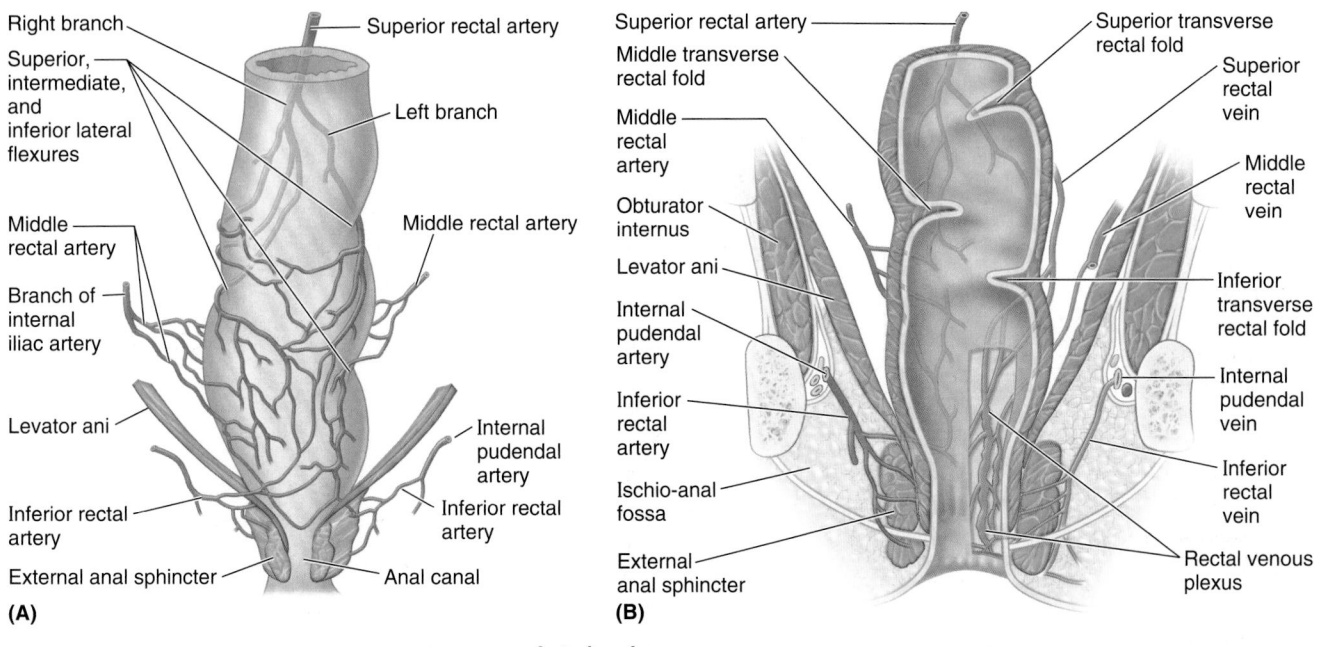

Anterior views

FIGURE 3.32. Arteries and veins of rectum and anal canal. A. Despite their name, the inferior rectal arteries, which are branches of the internal pudendal arteries, mainly supply the anal canal. The three sharp lateral flexures of the rectum reflect the way in which the lumen navigates the transverse rectal folds (shown in part **B**) on the internal surface. **B.** This coronal section of the rectum and anal canal demonstrates the arterial supply and venous drainage. The internal and external rectal venous plexuses are most directly related to the anal canal. The flexures and transverse rectal folds help support the weight of the feces.

The rectum lies posteriorly against the inferior three sacral vertebrae and the coccyx, anococcygeal ligament, median sacral vessels, and inferior ends of the sympathetic trunks and sacral plexuses. In males, the rectum is related anteriorly to the fundus of the urinary bladder, terminal parts of the ureters, ductus deferentes, seminal glands, and prostate (see Figs. 3.13D and 3.27A). The *rectovesical septum* lies between the fundus of the bladder and the ampulla of the rectum and is closely associated with the seminal glands and prostate. In females, the rectum is related anteriorly to the vagina and is separated from the posterior part of the fornix and the cervix by the *recto-uterine pouch* (see Figs. 3.13D and 3.27B). Inferior to this pouch, the weak rectovaginal septum separates the superior half of the posterior wall of the vagina from the rectum.

ARTERIAL SUPPLY AND VENOUS DRAINAGE OF RECTUM

The *superior rectal artery,* the continuation of the abdominal inferior mesenteric artery, supplies the proximal part of the rectum (Fig. 3.32). The right and left *middle rectal arteries,* which often arise from the anterior divisions of the internal iliac arteries in the pelvis, supply the middle and inferior parts of the rectum. The **inferior rectal arteries,** arising from the internal pudendal arteries in the perineum, supply the anorectal junction and anal canal. Anastomoses between the superior and inferior rectal arteries may provide potential collateral circulation, but anastomoses with the middle rectal arteries are sparse.

Blood from the rectum drains through the *superior, middle,* and *inferior rectal veins* (Fig. 3.32B). Anastomoses

occur between the portal and systemic veins in the wall of the anal canal. Because the superior rectal vein drains into the portal venous system and the middle and inferior rectal veins drain into the systemic system, these anastomoses are clinically important areas of portocaval anastomosis (see Fig. 2.75A). The submucosal rectal venous plexus surrounds the rectum, communicating with the vesical venous plexus in males and the uterovaginal venous plexus in females. The **rectal venous plexus** consists of two parts (Fig. 3.32B): the **internal rectal venous plexus** just deep to the mucosa of the anorectal junction and the subcutaneous **external rectal venous plexus** external to the muscular wall of the rectum. Although these plexuses bear the term *rectal,* they are primarily "anal" in terms of location, function, and clinical significance (see "Venous and Lymphatic Drainage of Anal Canal," p. 413).

INNERVATION OF RECTUM

The nerve supply to the rectum is from the sympathetic and parasympathetic systems (Fig. 3.33). The *sympathetic supply* is from the lumbar spinal cord, conveyed via lumbar splanchnic nerves and the hypogastric/pelvic plexuses, and through the peri-arterial plexus of the inferior mesenteric and superior rectal arteries. The *parasympathetic supply* is from the S2–S4 spinal cord level, passing via the pelvic splanchnic nerves and the left and right inferior hypogastric plexuses to the rectal (pelvic) plexus. Because the rectum is inferior (distal) to the pelvic pain line, all visceral afferent fibers follow the parasympathetic fibers retrogradely to the S2–S4 spinal sensory ganglia.

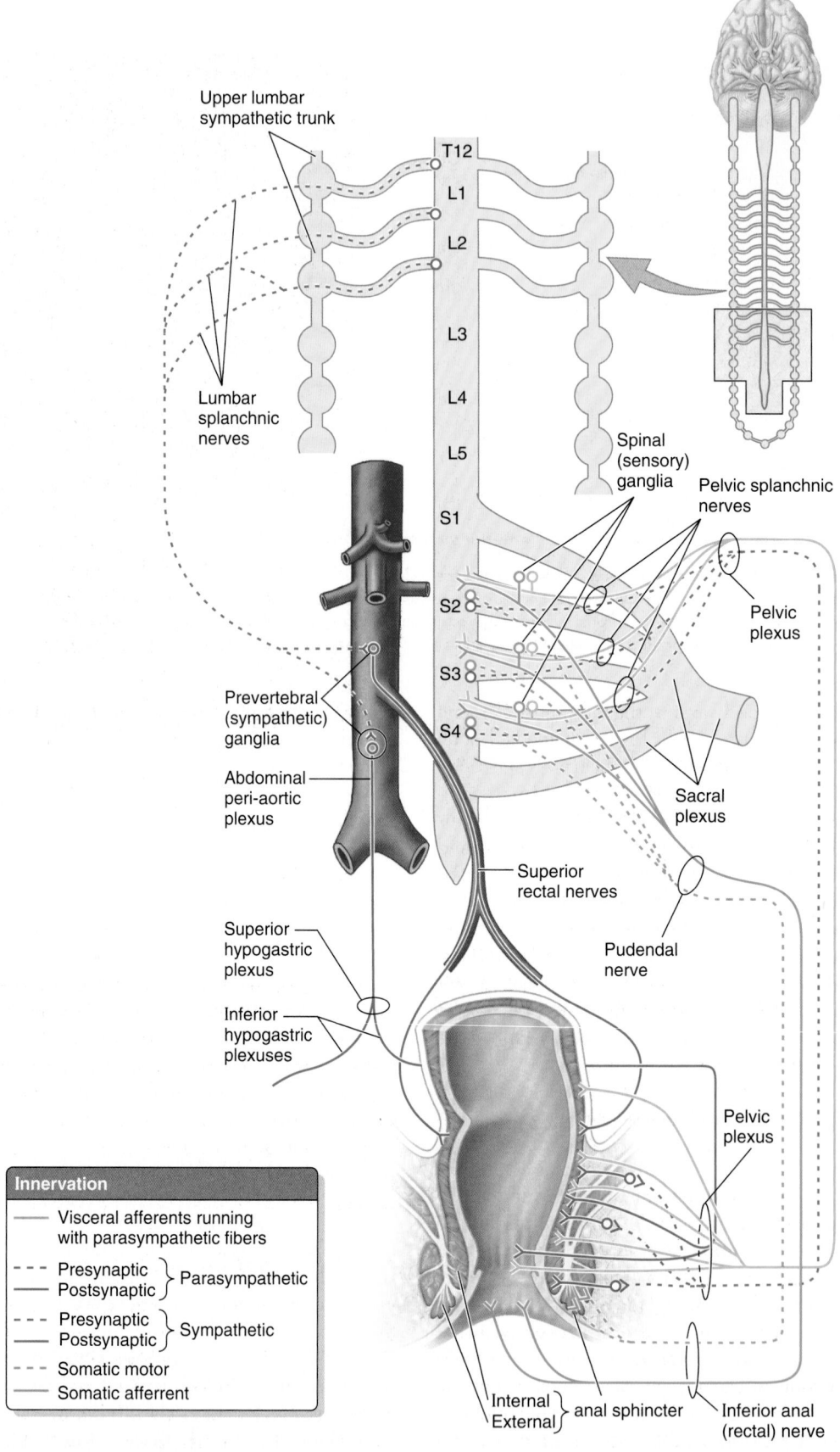

FIGURE 3.33. Innervation of rectum and anal canal. The lumbar and pelvic splanchnic nerves and hypogastric plexuses have been retracted laterally for clarity.

URINARY ORGANS AND RECTUM

Iatrogenic Compromise of Ureteric Blood Supply

The ureters may be injured during abdominal, retroperitoneal, pelvic, or gynecological operations as a result of inadvertently interrupting their blood supply. Identification of the ureters during their full course through the pelvis is an important preventive measure.

The longitudinal anastomoses between arterial branches to the ureter are usually adequate to maintain the blood supply along the length of the ureters, but occasionally they are not. Traction on the ureter during surgery may lead to delayed rupture of the ureter. The denuded ureteral segment becomes gangrenous and leaks, or ruptures 7–10 days after surgery. When traction is necessary, it is applied gently within a strictly limited range using padded, blunt retractors. It is useful to realize that although the blood supply to the abdominal segment of the ureter approaches from a medial direction, that of the pelvic segment approaches from a lateral direction (Fig. 3.25); the ureters should be retracted accordingly.

Ureteric Calculi

The ureters are expansile muscular tubes that dilate (along with the intrarenal collecting system—calices and renal pelvis) if obstructed (Fig. B3.7). Acute obstruction usually results from a *ureteric calculus* (L. pebble). The symptoms and severity depend on the location, type, and size of the calculus and on whether it is smooth or spiky. Although passage of small calculi (stones) usually causes little or no pain, larger ones produce severe pain. Stones that descend the length of the ureter cause pain described as migrating "from loin to groin" (from the lateral abdominal to inguinal regions).

The pain caused by a calculus is a *colicky pain,* which results from hyperperistalsis in the ureter, superior to the level of the obstruction. Ureteric calculi may cause complete or intermittent obstruction of urinary flow. The obstruction may occur anywhere along the ureter, but it occurs most often at the three sites where the ureters are normally relatively constricted (Fig. 3.24A): (1) at the junction of the ureters and renal pelves, (2) where they cross the external iliac artery and pelvic brim, and (3) during their passage through the wall of the urinary bladder (Fig. B3.7B).

The presence of calculi can often be confirmed by an abdominal radiograph, an intravenous urogram. Currently, computed tomography (CT) scanning is the preferred method. Ureteric calculi may be removed by open surgery, by endoscopy (*endourology*), or lithotripsy. *Lithotripsy* uses shock waves to break up a stone into small fragments that can be passed in the urine.

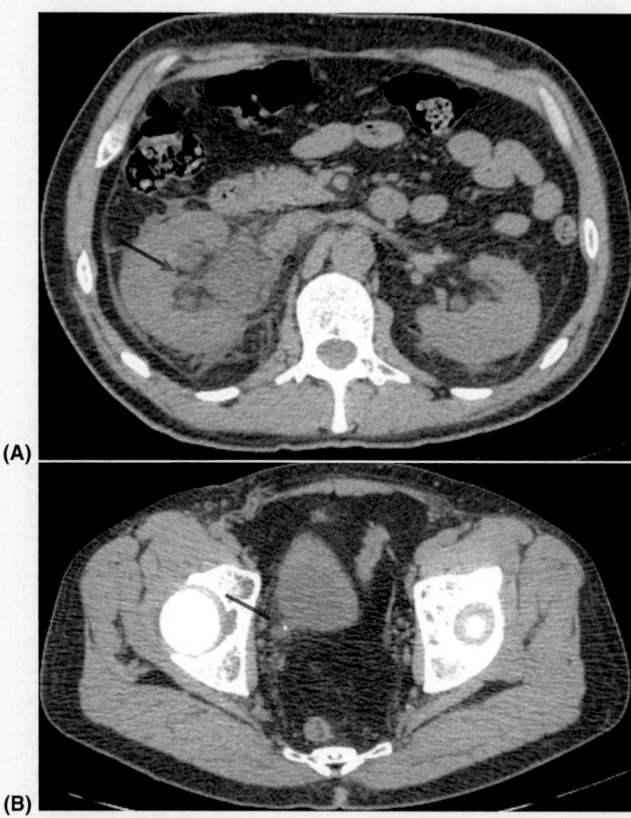

(A)

(B)
Transverse CT scan

FIGURE B3.7. Obstructing ureteric calculus. A. This image at the L1 vertebral level demonstrates an enlarged right kidney with a dilated intrarenal collecting system (*blue arrow*). **B.** In the lesser pelvis, a calcific density appears at the ureterovesical junction (*red arrow*) with dilation of the ureter.

Cystocele—Hernia of Bladder

Loss of bladder support in females by damage to the pelvic floor during childbirth (e.g., laceration of perineal muscles [see Fig. B3.5B] or a lesion of the nerves supplying them, or rupture of the fascial support of the vagina, the paracolpium [see Fig. 3.14B]) can result in collapse of the bladder onto the anterior vaginal wall. When intraabdominal pressure increases (as when "bearing down" during defecation), the anterior wall of the vagina may protrude through the vaginal orifice into the vestibule (Fig. B3.8).

Suprapubic Cystotomy

Although the superior surface of the empty bladder lies at the level of the superior margin of the pubic symphysis, as the bladder fills it extends superiorly above the symphysis into the loose areolar tissue between the parietal peritoneum and anterior abdominal wall (Fig. B3.9). The bladder then lies adjacent to this wall without the intervention of peritoneum. Consequently, the distended bladder may be punctured (*suprapubic cystotomy*) or approached surgically

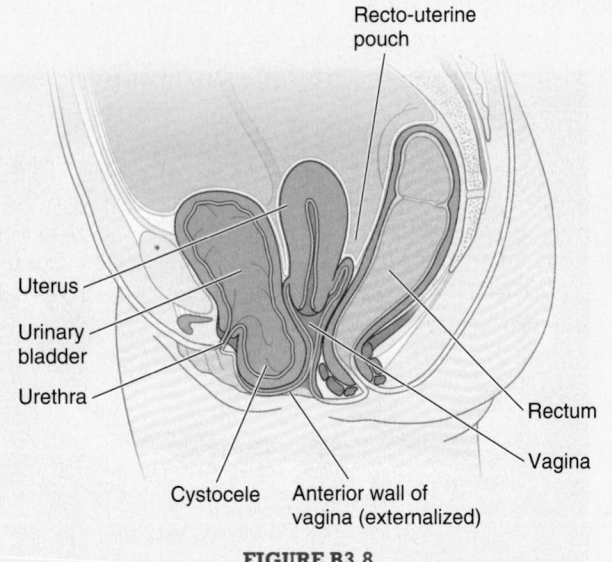

FIGURE B3.8.

superior to the pubic symphysis for the introduction of in-dwelling catheters or instruments without traversing the peritoneum and entering the peritoneal cavity. Urinary calculi, foreign bodies, and small tumors may also be removed from the bladder through a suprapubic extraperitoneal incision.

Rupture of Bladder

 Because of the superior position of the distended bladder, it may be ruptured by injuries to the inferior part of the anterior abdominal wall or by fractures

of the pelvis. The rupture may result in the escape of urine extraperitoneally or intraperitoneally. Rupture of the superior part of the bladder frequently tears the peritoneum, resulting in *extravasation (passage) of urine into the peritoneal cavity*. Posterior rupture of the bladder usually results in passage of urine extraperitoneally into the perineum.

Cystoscopy

The interior of the bladder and its three orifices can be examined with a *cystoscope*. During *transurethral resection of a tumor*, the instrument is passed into the bladder through the urethra (Fig. B3.10). Using a high-frequency electrical current, the tumor is removed in small fragments that are washed from the bladder with water.

Clinically Significant Differences Between Male and Female Urethrae

The female urethra is distensible because it contains considerable elastic tissue, as well as smooth muscle. It can be easily dilated without injury; consequently, the passage of catheters or cystoscopes is easier in females than in males. Infections of the urethra, and especially the bladder, are more common in women because the female urethra is short, more distensible, and is open to the exterior through the vestibule of the vagina.

Rectal Examination

Many structures related to the antero-inferior part of the rectum may be palpated through its walls (e.g., the prostate and seminal glands in males and the cervix in females). In both sexes, the pelvic surfaces of the sacrum and coccyx may be palpated. The ischial spines and tuberosities may also be palpated. Enlarged internal iliac lymph nodes, pathological thickening of the ureters, swellings in the ischio-anal fossae [e.g., ischio-anal abscesses (p. 416) and abnormal

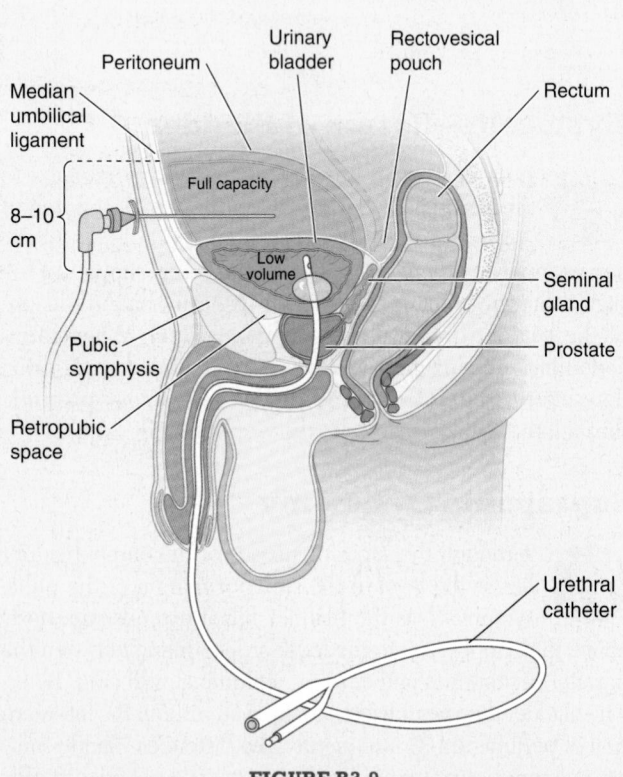

FIGURE B3.9.

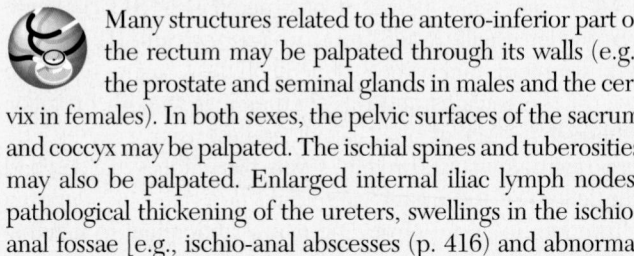

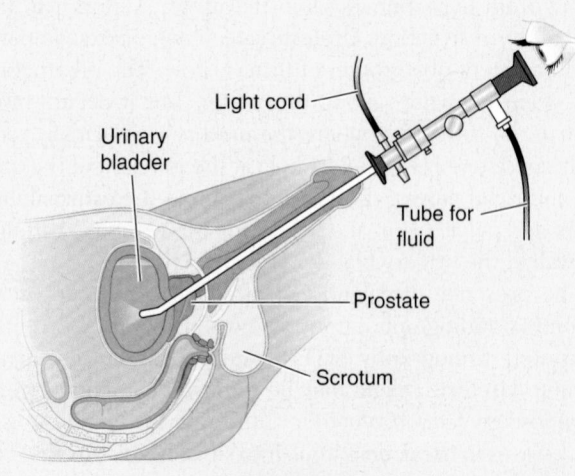

FIGURE B3.10.

contents in the rectovesical pouch in the male or the recto-uterine pouch in the female] may also be palpated. Tenderness of an inflamed appendix may also be detected rectally if it descends into the lesser pelvis (pararectal fossa).

The internal aspect of the rectum can be examined with a *proctoscope,* and biopsies of lesions may be taken through this instrument. During insertion of a *sigmoidoscope,* the curvatures of the rectum and its acute flexion at the recto-sigmoid junction have to be kept in mind so the patient does not undergo unnecessary discomfort. The operator must also know that the *transverse rectal folds,* which provide useful landmarks for the procedure, may temporarily impede passage of these instruments.

Resection of Rectum

When resecting the rectum in males (e.g., during cancer treatment), the plane of the rectovesical septum (a fascial septum extending superiorly from the perineal body) is located so that the prostate and urethra can be separated from the rectum. In this way, these organs are not damaged during the surgery.

The Bottom Line

PELVIC URINARY AND DIGESTIVE ORGANS

Ureters: The ureters carry urine from the renal pelves to the urinary bladder. ♦ The ureters descend subperitoneally into the pelvis, passing inferior to the ductus deferens of the male, or the uterine artery of the female; the latter relationship being of particular surgical importance. ♦ The ureters penetrate the bladder wall obliquely from its postero-inferior angle, creating a one-way valve. ♦ The pelvic portion of each ureter is served by the inferior vesical (male) or vaginal (female) artery, and the vesical venous plexus and internal iliac veins. ♦ Calculi, likely to become entrapped where the ureter crosses the pelvic brim or enters the bladder, produce severe groin pain.

Urinary bladder: The superior and inferior portions of the urinary bladder are quite distinct anatomically and functionally. ♦ The body of the bladder is highly distensible, embedded in loose extraperitoneal fat, and covered on its superior aspect with peritoneum, all of which allow expansion with filling. ♦ In contrast, the relatively indistensible neck of the bladder is anchored in place by pelvic ligaments and the floor of the bladder overlying it (which includes the trigone of the bladder), and remains relatively unchanged with filling. ♦ Most of the bladder body is served by superior vesical arteries and veins. ♦ The neck and adjacent inferior body are served by inferior vesical arteries and the vesical venous plexus. ♦ Sympathetic fibers from inferior thoracic and superior lumbar spinal cord segments maintain the tonus of the bladder neck and, in males during ejaculation, stimulate contraction of the internal urethral sphincter to prevent reflux of semen. ♦ Parasympathetic fibers conveyed by pelvic splanchnic nerves from the S2–S4 spinal cord segments inhibit the neck musculature and stimulate increased tonus of the detrusor muscle of the bladder walls for urination. ♦ Visceral afferent fibers conducting pain sensation from the roof of the bladder (superior to the pelvic pain line) follow the sympathetic fibers retrogradely to spinal sensory ganglia; the remaining visceral afferent fibers follow the parasympathetic fibers.

Urethra: *The male urethra* consists of four parts, two of which are the intramural and prostatic parts. ♦ The intramural part varies in length and caliber, depending on whether the bladder is filling or emptying. ♦ The prostatic urethra is distinguished both by its surroundings and the structures that open into it. It is surrounded by the prostate, the muscular anterior "lobe" that includes the trough-like superior extension of the external urethral sphincter anteriorly, and by the glandular lobes posteriorly. ♦ The prostatic ducts open into prostatic sinuses on each side of the urethral crest. ♦ The vestigial utricle is a relatively large opening in the center of the seminal colliculus, flanked by the tiny openings of the ejaculatory ducts. ♦ The reproductive and urinary tracts merge within the prostatic urethra.

The female urethra runs parallel to the vagina. It is firmly attached to and indents the anterior vaginal wall centrally and distally. ♦ Since it is not shared with the reproductive tract, an internal urethral sphincter is not required at the neck of the female bladder.

Rectum: The rectum accumulates and temporarily stores feces. ♦ The rectum begins at the rectosigmoid junction as the teniae of the sigmoid colon spread and unite into a continuous longitudinal layer of smooth muscle and the omental appendices cease. ♦ The rectum ends with the anorectal flexure as the gut penetrates the pelvic diaphragm, becoming the anal canal. ♦ Despite the Latin term *rectus* (straight), the rectum is concave anteriorly as the sacral flexure, and has three lateral flexures formed in relation to the internal transverse rectal folds. ♦ The rectum enlarges into the rectal ampulla directly above the pelvic floor. ♦ The superior, middle, and inferior parts of the rectum are, respectively, intraperitoneal, retroperitoneal, and subperitoneal. ♦ Collateral arterial circulation and a portocaval venous anastomosis result from anastomoses of the superior and middle rectal vessels. ♦ Sympathetic nerve fibers pass to the rectum (especially blood vessels and internal anal sphincter) from lumbar spinal cord segments via the hypogastric/pelvic plexuses, and the peri-arterial plexus of the superior rectal artery. ♦ Parasympathetic and visceral afferent fibers involve the middle sacral spinal cord segments and spinal ganglia.

Male Internal Genital Organs

The male internal genital organs include the testes, epididymides (singular = epididymis), ductus deferentes (singular = ductus deferens), seminal glands, ejaculatory ducts, prostate, and bulbo-urethral glands (Fig. 3.34). The testes and epididymides (described in Chapter 2) are considered internal genital organs on the basis of their developmental position and homology with the internal female ovaries. However, because of their external position postnatally and because in dissection they are encountered during the dissection of the inguinal region of the anterior abdominal wall, they were considered with the abdomen in Chapter 2.

DUCTUS DEFERENS

The **ductus deferens** (vas deferens) is the continuation of the *duct of the epididymis*. The ductus deferens:

- Has relatively thick muscular walls and a minute lumen, giving it a cord-like firmness.
- Begins in the tail of the epididymis, at the inferior pole of the testis (Fig. 2.21, p. 209).
- Ascends posterior to the testis, medial to the epididymis.
- Is the primary component of the spermatic cord.
- Penetrates the anterior abdominal wall via the inguinal canal.
- Crosses over the external iliac vessels and enters the pelvis.
- Passes along the lateral wall of the pelvis, where it lies external to the parietal peritoneum.
- Ends by joining the duct of the seminal gland to form the *ejaculatory duct.*

During the pelvic part of its course, the ductus deferens maintains direct contact with the peritoneum; no other structure intervenes between them. The ductus crosses superior to the

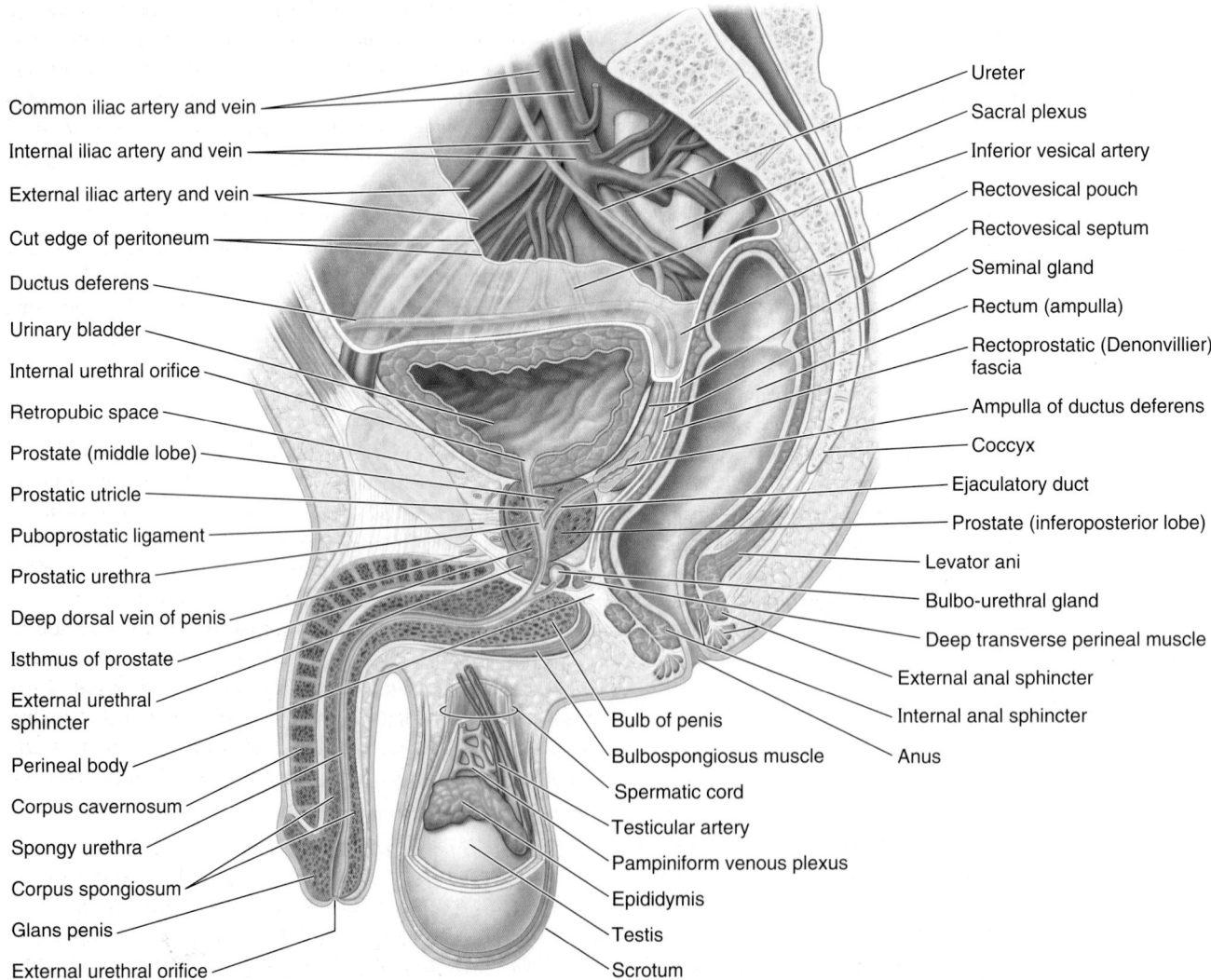

Common iliac artery and vein
Internal iliac artery and vein
External iliac artery and vein
Cut edge of peritoneum
Ductus deferens
Urinary bladder
Internal urethral orifice
Retropubic space
Prostate (middle lobe)
Prostatic utricle
Puboprostatic ligament
Prostatic urethra
Deep dorsal vein of penis
Isthmus of prostate
External urethral sphincter
Perineal body
Corpus cavernosum
Spongy urethra
Corpus spongiosum
Glans penis
External urethral orifice

Ureter
Sacral plexus
Inferior vesical artery
Rectovesical pouch
Rectovesical septum
Seminal gland
Rectum (ampulla)
Rectoprostatic (Denonvillier) fascia
Ampulla of ductus deferens
Coccyx
Ejaculatory duct
Prostate (inferoposterior lobe)
Levator ani
Bulbo-urethral gland
Deep transverse perineal muscle
External anal sphincter
Internal anal sphincter
Anus

Bulb of penis
Bulbospongiosus muscle
Spermatic cord
Testicular artery
Pampiniform venous plexus
Epididymis
Testis
Scrotum

Schematic medial section of male pelvis and penis, stepped dissection of scrotum and coverings of testis

FIGURE 3.34. Hemisected male pelvis and perineum (right half). The genital organs are demonstrated: testis, epididymis, ductus deferens, ejaculatory duct, and penis, with the accessory glandular structures (seminal gland, prostate, and bulbo-urethral gland). The spermatic cord connects the testis to the abdominal cavity, and the testis lies externally in a musculocutaneous pouch, the scrotum.

ureter near the posterolateral angle of the urinary bladder, running between the ureter and the peritoneum of the ureteric fold to reach the fundus of the bladder. The relationship of the ductus deferens to the ureter in the male is similar, although of lesser clinical importance, to that of the uterine artery to the ureter in the female. The developmental basis of this relationship is shown in Figure 3.35. Posterior to the bladder, the ductus deferens at first lies superior to the seminal gland, then descends medial to the ureter and the gland. Here the ductus deferens enlarges to form the **ampulla of the ductus deferens** before its termination (Fig. 3.36).

Arterial Supply and Venous Drainage of Ductus Deferens. The tiny *artery to the ductus deferens* usually arises from a superior (sometimes inferior) vesical artery (Figs. 3.16 and 3.34), and terminates by anastomosing with the testicular artery, posterior to the testis. Veins from most of the ductus drain into the testicular vein, including the distal pampiniform plexus. Its terminal portion drains into the vesicular/ prostatic venous plexus.

SEMINAL GLANDS

Each **seminal gland** (vesicle) is an elongated structure (approximately 5 cm long but sometimes much shorter) that lies between the fundus of the bladder and the rectum (Figs. 3.34 and 3.36). The seminal glands are obliquely placed superior to the prostate and do not store sperms (as the term "vesicle" implies). They secrete a thick alkaline fluid with fructose (an energy source for sperms), and a coagulating agent that mixes with the sperms as they pass into the ejaculatory ducts and urethra.

The superior ends of the seminal glands are covered with peritoneum and lie posterior to the ureters, where the peritoneum of the *rectovesical pouch* separates them

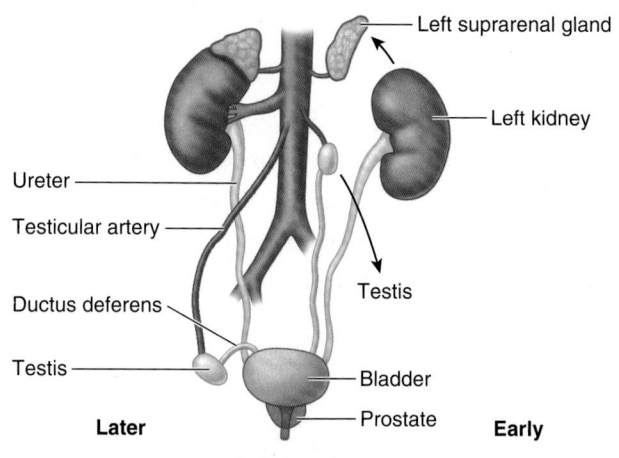

FIGURE 3.35. Structures crossing male ureter in abdomen and pelvis. During development, as the testis descends inferiorly and laterally from its original position (medial to the site of the kidneys on the posterior abdominal wall) to and then through the inguinal canal, the ureter is crossed by testicular vessels in the abdomen and by the ductus deferens in the pelvis. This relationship is retained throughout life.

from the rectum. The inferior ends of the seminal glands are closely related to the rectum and are separated from it only by the rectovesical septum (Fig. 3.34). The duct of the seminal gland joins the ductus deferens to form the *ejaculatory duct.*

Arterial Supply and Venous Drainage of Seminal Glands. The arteries to the seminal glands derive from the *inferior vesical* and *middle rectal arteries* (see Figs. 3.16 and 3.37; Table 3.4). The veins accompany the arteries and have similar names (Fig. 3.19C).

EJACULATORY DUCTS

The **ejaculatory ducts** are slender tubes that arise by the union of the ducts of the seminal glands with the ductus deferentes (Figs. 3.34, 3.36, and 3.37). The ejaculatory ducts (approximately 2.5 cm long) arise near the neck of the bladder, and run close together as they pass antero-inferiorly through the posterior part of the prostate and along the sides of the prostatic utricle. The ejaculatory ducts converge to open on the seminal colliculus by tiny, slit-like apertures on, or just within, the opening of the prostatic utricle (Fig. 3.30). Although the ejaculatory ducts traverse the glandular prostate, prostatic secretions do not join the seminal fluid until the ejaculatory ducts have terminated in the prostatic urethra.

Arterial Supply and Venous Drainage of Ejaculatory Ducts. The *arteries to the ductus deferens,* usually branches of the superior (but frequently inferior) vesical arteries, supply the ejaculatory ducts (Figs. 3.37). The veins join the *prostatic* and *vesical venous plexuses* (Fig. 3.19C).

PROSTATE

The **prostate** (approximately 3 cm long, 4 cm wide, and 2 cm in antero-posterior (AP) depth) is the largest accessory gland of the male reproductive system (Figs. 3.34, 3.36, and 3.37). The firm, walnut-size prostate surrounds the *prostatic urethra.* The glandular part makes up approximately two thirds of the prostate; the other third is fibromuscular.

The **fibrous capsule of the prostate** is dense and neurovascular, incorporating the prostatic plexuses of veins and nerves. All this is surrounded by the *visceral layer of the pelvic fascia,* forming a fibrous *prostatic sheath* that is thin anteriorly, continuous anterolaterally with the *puboprostatic ligaments,* and dense posteriorly where it blends with the *rectovesical septum.* The prostate has

- A base closely related to the neck of the bladder.
- An apex that is in contact with fascia on the superior aspect of the urethral sphincter and deep perineal muscles.
- A muscular anterior surface, featuring mostly transversely oriented muscle fibers forming a vertical, trough-like hemisphincter (rhabdosphincter), which is part of the urethral sphincter. The anterior surface is separated from the pubic symphysis by retroperitoneal fat in the retropubic space.

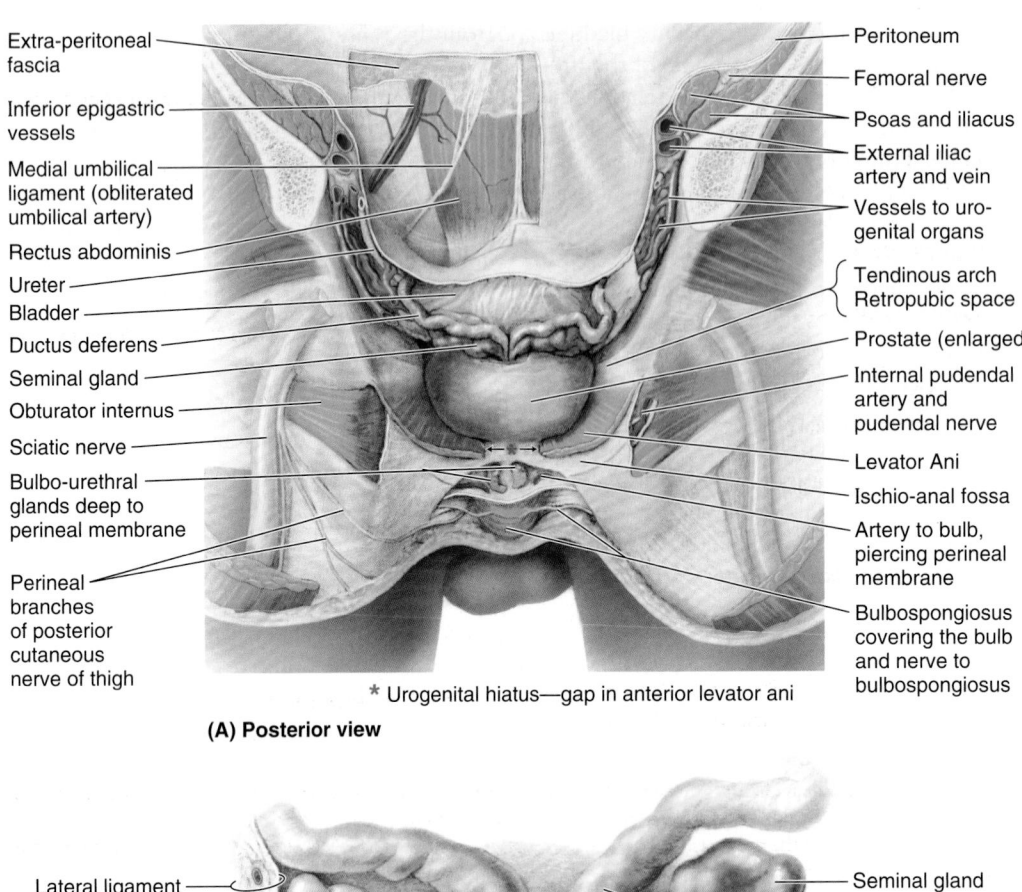

Extra-peritoneal fascia

Inferior epigastric vessels

Medial umbilical ligament (obliterated umbilical artery)

Rectus abdominis

Ureter

Bladder

Ductus deferens

Seminal gland

Obturator internus

Sciatic nerve

Bulbo-urethral glands deep to perineal membrane

Perineal branches of posterior cutaneous nerve of thigh

Peritoneum

Femoral nerve

Psoas and iliacus

External iliac artery and vein

Vessels to uro-genital organs

Tendinous arch
Retropubic space

Prostate (enlarged)

Internal pudendal artery and pudendal nerve

Levator Ani

Ischio-anal fossa

Artery to bulb, piercing perineal membrane

Bulbospongiosus covering the bulb and nerve to bulbospongiosus

* Urogenital hiatus—gap in anterior levator ani

(A) Posterior view

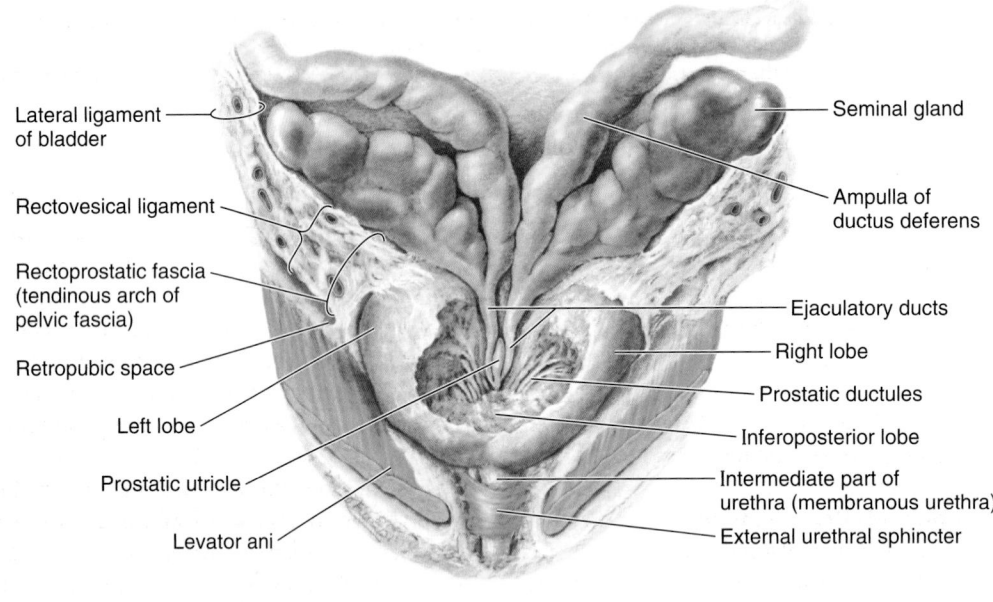

Lateral ligament of bladder

Rectovesical ligament

Rectoprostatic fascia (tendinous arch of pelvic fascia)

Retropubic space

Left lobe

Prostatic utricle

Levator ani

Seminal gland

Ampulla of ductus deferens

Ejaculatory ducts

Right lobe

Prostatic ductules

Inferoposterior lobe

Intermediate part of urethra (membranous urethra)

External urethral sphincter

(B) Posterior view

FIGURE 3.36. Posterior aspect of male pelvic viscera and posterior dissection of prostate. A. The posterior pelvic wall, rectum, and rectovesical septum have been removed. The umbilical ligaments, like the urinary bladder, are embedded in extraperitoneal or subperitoneal fascia (mostly removed in this dissection). **B.** The ejaculatory ducts are formed by the merger of the duct of the seminal gland and the ductus deferens. The vestigial prostatic utricle, usually seen as an invagination in an anterior view, appears in this posterior dissection as an evagination lying between the ejaculatory ducts.

- A posterior surface that is related to the ampulla of the rectum.
- Inferolateral surfaces that are related to the levator ani.

Although not clearly distinct anatomically, the following lobes of the prostate are traditionally described (Fig. 3.38A):

- The **isthmus of the prostate** (commissure of prostate; historically, the anterior "lobe") lies anterior to the ure-

thra. It is fibromuscular, the muscle fibers representing a superior continuation of the external urethral sphincter muscle to the neck of the bladder, and contains little, if any, glandular tissue.

- **Right** and **left lobes of the prostate,** separated anteriorly by the isthmus and posteriorly by a central, shallow longitudinal furrow, may each be subdivided for descriptive purposes into four indistinct lobules defined by their

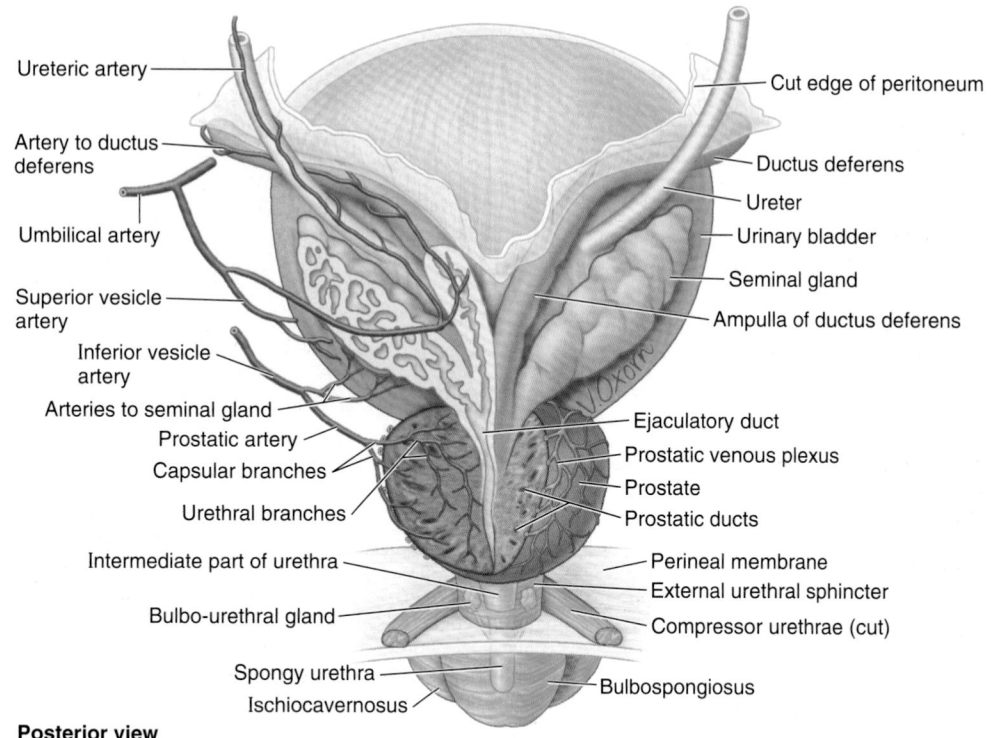

Ureteric artery

Artery to ductus deferens

Umbilical artery

Superior vesicle artery

Inferior vesicle artery

Arteries to seminal gland

Prostatic artery

Capsular branches

Urethral branches

Intermediate part of urethra

Bulbo-urethral gland

Spongy urethra

Ischiocavernosus

Cut edge of peritoneum

Ductus deferens

Ureter

Urinary bladder

Seminal gland

Ampulla of ductus deferens

Ejaculatory duct

Prostatic venous plexus

Prostate

Prostatic ducts

Perineal membrane

External urethral sphincter

Compressor urethrae (cut)

Bulbospongiosus

Posterior view

FIGURE 3.37. Pelvic part of ureters, urinary bladder, seminal glands, terminal parts of ductus deferens, and prostate. The left seminal gland and ampulla of the ductus deferens are dissected free and sliced open. Part of the prostate is also cut away to expose the ejaculatory duct. The perineal membrane lies between the external genitalia and the deep part of the perineum (anterior recess of ischio-anal fossa). It is pierced by the urethra, ducts of the bulbo-urethral glands, dorsal and deep arteries of the penis, cavernous nerves, and the dorsal nerve of the penis.

relationship to the urethra and ejaculatory ducts, and—although less apparent—by the arrangement of the ducts and connective tissue:

(1) an **inferoposterior** (lower posterior) **lobule** that lies posterior to the urethra and inferior to the ejaculatory ducts. This lobule constitutes the aspect of the prostate palpable by digital rectal examination.

(2) an **inferolateral** (lower lateral) **lobule** directly lateral to the urethra, forming the major part of the right or left lobe.

(3) a **superomedial lobule,** deep to the inferoposterior lobule, surrounding the ipsilateral ejaculatory duct.

(4) an **anteromedial lobule,** deep to the inferolateral lobule, directly lateral to the proximal prostatic urethra.

An embryonic *middle (median) lobe* gives rise to (3) and (4) above. This region tends to undergo hormone-induced hypertrophy in advanced age, forming a *middle lobule* that lies between the urethra and the ejaculatory ducts and is closely related to the neck of the bladder. Enlargement of the middle lobule is believed to be at least partially responsible for the formation of the *uvula* that may project into the internal urethral orifice (Fig. 3.30).

Some clinicians, especially urologists and sonographers, divide the prostate into peripheral and central (internal) zones (Fig. 3.38B). The central zone is comparable to the middle lobe.

The **prostatic ducts** (20–30) open chiefly into the *prostatic sinuses* that lie on either side of the seminal colliculus on the posterior wall of the prostatic urethra (Fig. 3.37). Prostatic fluid, a thin, milky fluid, provides approximately 20% of the volume of **semen** (a mixture of secretions produced by the testes, seminal glands, prostate, and bulbo-urethral glands that provides the vehicle by which sperms are transported) and plays a role in activating the sperms.

Arterial Supply and Venous Drainage of Prostate. The prostatic arteries are mainly branches of the internal iliac artery (see Table 3.4; Figs. 3.17A and 3.37), especially the *inferior vesical arteries,* but also the internal pudendal and middle rectal arteries. The veins join to form a plexus around the sides and base of the prostate (Figs. 3.19C and 3.37). This **prostatic venous plexus,** between the fibrous capsule of the prostate and the prostatic sheath, drains into the *internal iliac veins.* The prostatic venous plexus is continuous superiorly with the *vesical venous plexus* and communicates posteriorly with the *internal vertebral venous plexus.*

BULBO-URETHRAL GLANDS

The two pea-size **bulbo-urethral glands** (Cowper glands) lie posterolateral to the intermediate part of the urethra, largely embedded within the external urethral sphincter (Figs. 3.30B, 3.34, 3.36, and 3.37). The **ducts of the**

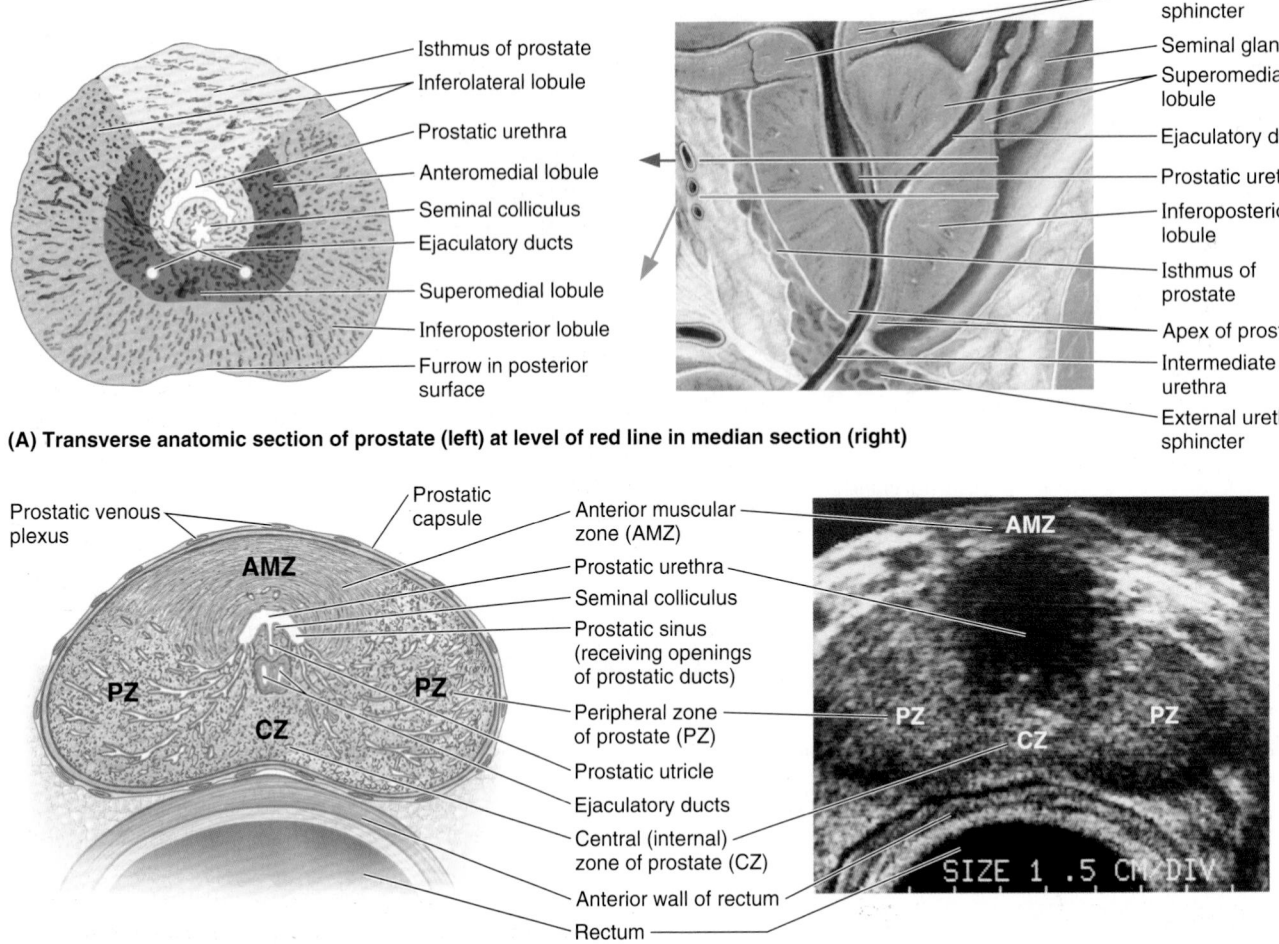

(A) Transverse anatomic section of prostate (left) at level of red line in median section (right)

(B) Graphic interpretation (left) of transverse ultrasound image (rt.) at level of green line in (A-rt).

FIGURE 3.38. Lobules and zones of prostate demonstrated by anatomical section and ultrasonographic imaging. A. Poorly demarcated lobules demonstrated in anatomic sections. **B.** The ultrasound (US) transducer was inserted into the rectum to scan the anteriorly located prostate. The ducts of the glands in the peripheral zone open into the prostatic sinuses, whereas the ducts of the glands in the central (internal) zone open into the prostatic sinuses and the seminal colliculus.

bulbo-urethral glands pass through the perineal membrane with the intermediate urethra, and open through minute apertures into the proximal part of the spongy urethra in the bulb of the penis. Their mucus-like secretion enters the urethra during sexual arousal.

INNERVATION OF INTERNAL GENITAL ORGANS OF MALE PELVIS

The ductus deferens, seminal glands, ejaculatory ducts, and prostate are richly innervated by sympathetic nerve fibers. *Presynaptic sympathetic fibers* originate from cell bodies in the intermediolateral cell column of the T12–L2 (or L3) spinal cord segments. They traverse the paravertebral ganglia of the sympathetic trunks to become components of lumbar (abdominopelvic) splanchnic nerves and the hypogastric and pelvic plexuses (see Fig. 3.29, p. 367).

Presynaptic parasympathetic fibers from S2 and S3 spinal cord segments traverse *pelvic splanchnic nerves*, which also join the inferior hypogastric/pelvic plexuses. Synapses with postsynaptic sympathetic and parasympathetic neurons occur within the plexuses, en route to or near the pelvic viscera. As part of an orgasm, the sympathetic system stimulates contraction of the internal urethral sphincter to prevent retrograde ejaculation. Simultaneously, it stimulates rapid peristaltic-like contractions of the ductus deferens, and the combined contraction of and secretion from the seminal glands and prostate that provide the vehicle (semen), and the expulsive force to discharge the sperms during ejaculation. The function of the parasympathetic innervation of the internal genital organs is unclear. However, parasympathetic fibers traversing the prostatic nerve plexus form the *cavernous nerves* that pass to the erectile bodies of the penis, which are responsible for producing penile erection.

MALE INTERNAL GENITAL ORGANS

Male Sterilization

The common method of sterilizing males is a *deferentectomy*, popularly called a *vasectomy*. During this procedure, part of the ductus deferens is ligated and/or excised through an incision in the superior part of the scrotum (Fig. B3.11). Hence, the subsequent ejaculated fluid from the seminal glands, prostate, and bulbo-urethral glands contains no sperms. The unexpelled sperms degenerate in the epididymis and the proximal part of the ductus deferens.

Reversal of a deferentectomy is successful in favorable cases (patients < 30 years of age and < 7 years postoperation) in most instances. The ends of the sectioned ductus deferentes are reattached under an operating microscope.

Abscesses in Seminal Glands

Localized collections of pus (abscesses) in the seminal glands may rupture, allowing pus to enter the peritoneal cavity. Seminal glands can be palpated during a rectal examination, especially if enlarged or full. They are palpated most easily when the bladder is moderately full. They can also be massaged to release their secretions for microscopic examination to detect *gonococci* (organisms that cause gonorrhea), for example.

Hypertrophy of Prostate

The prostate is of considerable medical interest because enlargement or *benign hypertrophy of the prostate (BHP)* is common after middle age, affecting virtually every male who lives long enough. An enlarged prostate projects into the urinary bladder and impedes urination by distorting the prostatic urethra. The middle lobule usually enlarges the most and obstructs the internal urethral orifice. The more the person strains, the more the valve-like prostatic mass occludes the urethra.

BHP is a common cause of urethral obstruction, leading to *nocturia* (need to void during the night), *dysuria* (difficulty and/or pain during urination), and *urgency* (sudden desire to void). BHP also increases the risk of bladder infections (cystitis) as well as kidney damage.

The prostate is examined for enlargement and tumors (focal masses or asymmetry) by *digital rectal examination* (Fig. B3.12). The palpability of the prostate depends on the fullness of the bladder. A full bladder offers resistance, holding the gland in place and making it more readily palpable. The malignant prostate feels hard and often irregular. In advanced stages, cancer cells metastasize both via lymphatic routes (initially to the internal iliac and sacral lymph nodes and later to distant nodes), and via venous routes (by way of the internal vertebral venous plexus, to the vertebrae and brain).

Because of the close relationship of the prostate to the prostatic urethra, obstructions may be relieved endoscopically. The instrument is inserted transurethrally through the external urethral orifice and spongy urethra into the prostatic urethra. All or part of the prostate, or just the hypertrophied part, is removed (*transurethral resection of the prostate; TURP*). In more serious cases, the entire prostate is removed along with the seminal glands, ejaculatory ducts, and terminal parts of the deferent ducts (*radical prostatectomy*).

TURP and improved open operative techniques attempt to preserve the nerves and blood vessels associated with the capsule of the prostate that pass to and from the penis, increasing the possibility for patients to retain sexual function after surgery, as well as restoring normal urinary control.

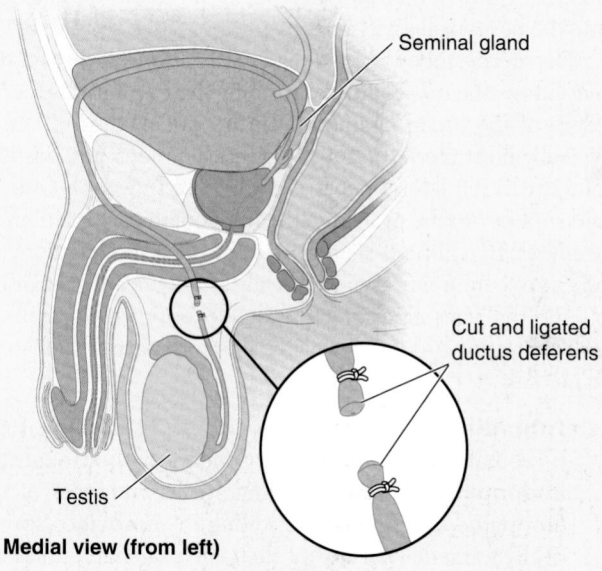

Medial view (from left)

FIGURE B3.11.

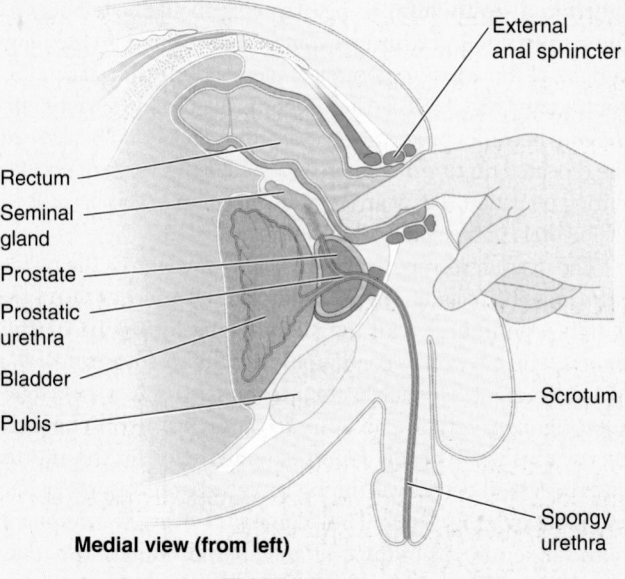

Medial view (from left)

FIGURE B3.12.

The Bottom Line

MALE INTERNAL GENITAL ORGANS

Ductus deferens: The cord-like ductus deferens is the primary component of the spermatic cord, conducting sperms from the epididymis to the ejaculatory duct. ♦ The distal portion of the ductus is superficial within the scrotum (and, therefore, easily accessible for deferentectomy or vasectomy) before it penetrates the anterior abdominal wall via the inguinal canal. ♦ The pelvic portion of the ductus lies immediately external to the peritoneum, with its terminal portion enlarging externally as its lumen becomes tortuous internally, forming the ampulla of the ductus deferens.

Seminal glands, ejaculatory ducts, and prostate: Obliquely placed seminal glands converge at the base of the bladder, where each of their ducts merges with the ipsilateral ductus deferens to form an ejaculatory duct. ♦ The two ejaculatory ducts immediately enter the posterior aspect of the prostate, running closely parallel through the gland to open on the seminal colliculus. ♦ Prostatic ducts open into prostatic sinuses, adjacent to the seminal colliculus. Thus, the major glandular secretions and sperms are delivered to the prostatic urethra. ♦ The seminal glands and prostate produce by far the greatest portion of the seminal fluid, indispensable for transport and delivery of sperms. ♦ These internal genital organs, located within the anterior male pelvis, receive blood from the inferior vesicle and middle rectal arteries, which drain into the continuous prostatic/vesicle venous plexus. ♦ Sympathetic fibers from lumbar levels stimulate the contraction and secretion resulting in ejaculation. ♦ The function of parasympathetic fibers from S2–S4 to the internal genital organs is unclear, but those traversing the prostatic nerve plexus to form the cavernous nerves produce erection.

Female Internal Genital Organs

The female internal genital organs include the ovaries, uterine tubes, uterus, and vagina.

OVARIES

The **ovaries** are almond-shaped and -sized female gonads in which the *oocytes* (female gametes or germ cells) develop. They are also endocrine glands that produce reproductive hormones. Each ovary is suspended by a short peritoneal fold or mesentery, the *mesovarium* (Fig. 3.39A). The mesovarium is a subdivision of a larger mesentery of the uterus, *the broad ligament.*

In prepubertal females, the connective tissue capsule (*tunica albuginea of the ovary*) comprising the surface of the ovary is covered by a smooth layer of **ovarian mesothelium** or **surface (germinal) epithelium,** a single layer of cuboidal cells that gives the surface a dull, grayish appearance, contrasting with the shiny surface of the adjacent peritoneal mesovarium with which it is continuous (Fig. 3.39B). After puberty, the ovarian surface epithelium becomes progressively scarred and distorted because of the repeated rupture of ovarian follicles and discharge of oocytes during ovulation. The scarring is less in women who have been taking oral contraceptives that inhibit ovulation.

The ovarian vessels, lymphatics, and nerves cross the pelvic brim, passing to and from the superolateral aspect of the ovary within a peritoneal fold, the **suspensory ligament of the ovary,** which becomes continuous with the mesovarium of the broad ligament. Medially within the mesovarium, a short *ligament of ovary* tethers the ovary to the uterus. Consequently the ovaries are typically found laterally between the uterus and the lateral pelvic wall during a manual or ultrasonic pelvic examination (Fig. 3.40). The ligament of ovary is a remnant of the superior part of the ovarian gubernaculum of the fetus (see Fig. 2.17B, p. 206). The ligament of the ovary connects the proximal (uterine) end of the ovary to the lateral angle of the uterus, just inferior to the entrance of the uterine tube (Fig. 3.39A). Because the ovary is suspended in the peritoneal cavity and its surface is not covered by peritoneum, the oocyte expelled at ovulation passes into the peritoneal cavity. However, its intraperitoneal life is short because it is normally trapped by the fimbriae of the infundibulum of the uterine tube and carried into the ampulla, where it may be fertilized.

UTERINE TUBES

The **uterine tubes** (formerly called oviducts or fallopian tubes) conduct the oocyte, discharged monthly from an ovary during child-bearing years, from the peri-ovarian peritoneal cavity to the uterine cavity. They also provide the usual site of fertilization. The tubes extend laterally from the *uterine horns* and open into the peritoneal cavity near the ovaries (Fig. 3.39A & B).

The uterine tubes (approximately 10 cm long) lie in a narrow mesentery, the **mesosalpinx,** forming the free anterosuperior edges of the broad ligaments. In the "ideal" disposition, as typically illustrated, the tubes extend symmetrically posterolaterally to the lateral pelvic walls, where they arch anterior and superior to the ovaries in the horizontally disposed broad ligament. In reality, as seen in an ultrasound examination, the tubes are commonly asymmetrically arranged with one or the other often lying superior and even posterior to the uterus.

The uterine tubes are divisible into four parts, from lateral to medial:

(1) **Infundibulum:** the funnel-shaped distal end of the tube that opens into the peritoneal cavity through the **abdominal ostium.** The finger-like processes of the fimbriated end of the infundibulum (**fimbriae**) spread over the medial surface of the ovary; one large **ovarian fimbria** is attached to the superior pole of the ovary.

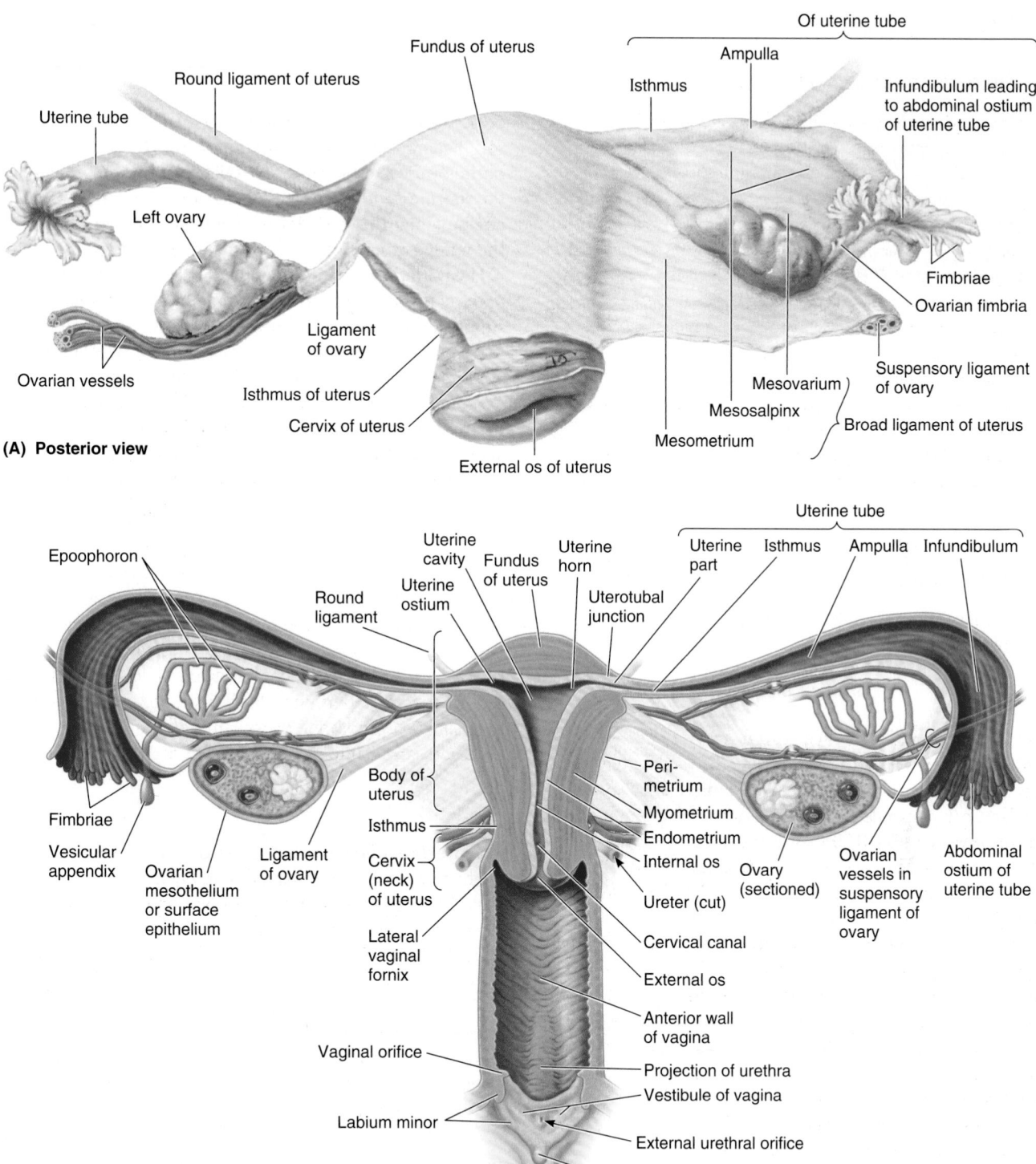

(A) Posterior view

(B) Posterior view

FIGURE 3.39. Internal female genital organs. A. Isolated dissection specimen consisting of the ovaries, uterine tubes, uterus, and related structures. The broad ligament is removed on the left side. **B.** This coronal section demonstrates the internal structure of the female genital organs. The epoophoron is a collection of rudimentary tubules in the mesosalpinx (mesentery of uterine tube). The epoophoron and vesicular appendage are vestiges of the embryonic mesonephros.

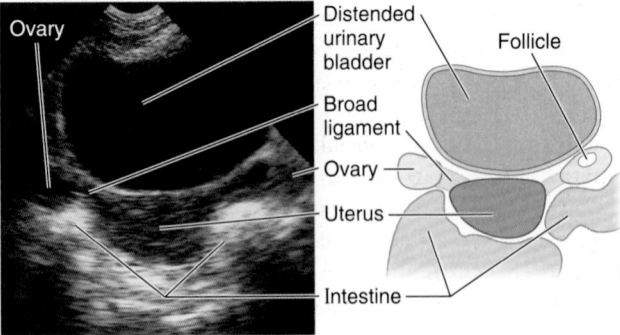

FIGURE 3.40. Ovaries and uterus as revealed by ultrasound scanning. The diagram on the right is a graphic interpretation of the image.

(2) **Ampulla:** the widest and longest part of the tube, which begins at the medial end of the infundibulum; fertilization of the oocyte usually occurs in the ampulla.

(3) **Isthmus:** the thick-walled part of the tube, which enters the uterine horn.

(4) **Uterine part:** the short intramural segment of the tube that passes through the wall of the uterus and opens via the uterine ostium into the uterine cavity at the uterine horn.

Arterial Supply and Venous Drainage of Ovaries and Uterine Tubes. The *ovarian arteries* arise from the abdominal aorta (see Fig. 3.16; Table 3.4) and descend along the posterior abdominal wall. At the pelvic brim, they cross over the external iliac vessels and enter the suspensory ligaments (Fig. 3.39A), approaching the lateral aspects of the ovaries and uterine tubes. The ascending branches of the *uterine arteries* (branches of the internal iliac arteries) course along the lateral aspects of the uterus to approach the medial aspects of the ovaries and tubes (Figs. 3.18B and 3.41). Both the ovarian and ascending uterine arteries terminate by bifurcating into *ovarian* and *tubal branches*, which supply the ovaries and tubes from opposite ends and anastomose with each other, providing a collateral circulation from abdominal and pelvic sources to both structures.

Veins draining the ovary form a vine-like **pampiniform plexus of veins** in the broad ligament near the ovary and uterine tube (Fig. 3.41). The veins of the plexus usually merge to form a singular **ovarian vein,** which leaves the lesser pelvis with the ovarian artery. The right ovarian vein ascends to enter the *inferior vena cava;* the left ovarian vein drains into the *left renal vein* (Fig. 3.19). The *tubal veins* drain into the *ovarian veins* and *uterine (uterovaginal) venous plexus* (Fig. 3.41).

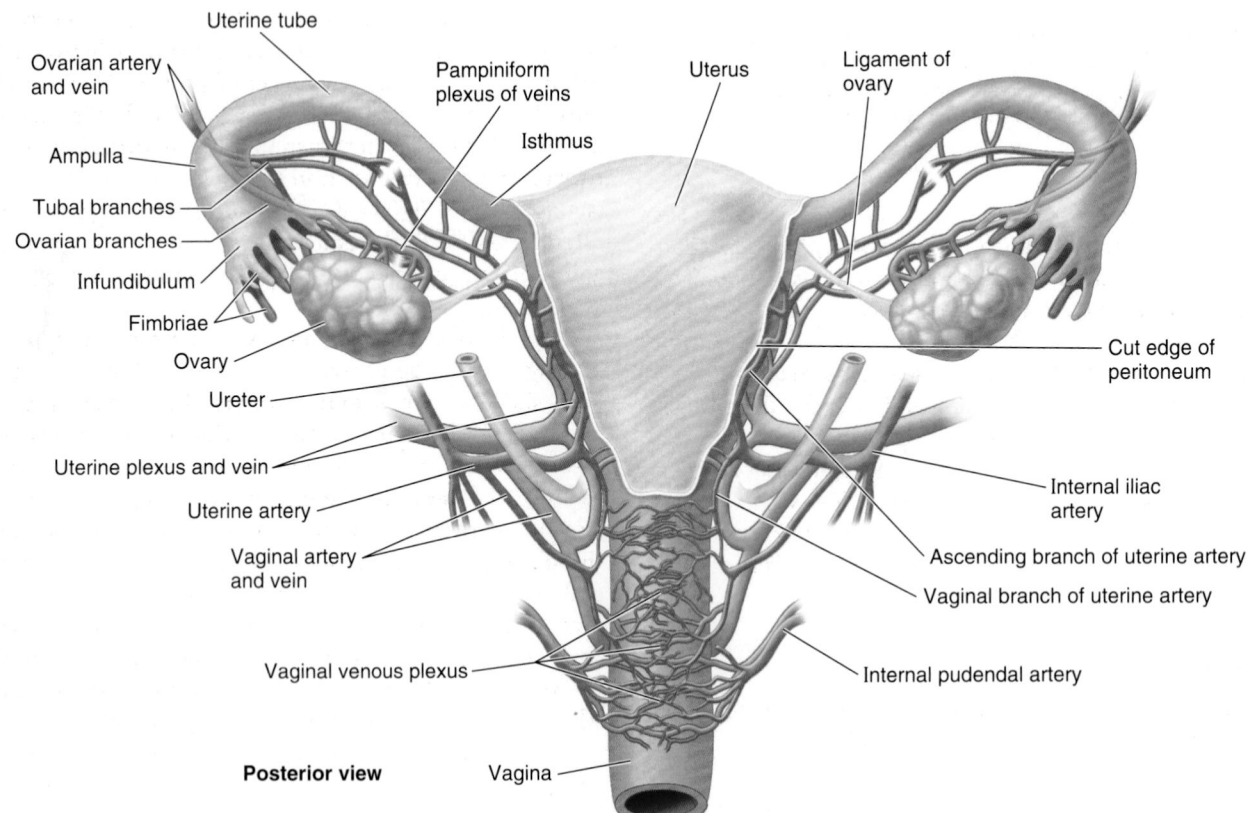

FIGURE 3.41. Blood supply and venous drainage of uterus, vagina, and ovaries. The broad ligament of the uterus is removed on each side of the uterus to show the anastomosing branches of the ovarian artery from the aorta and the uterine artery from the internal iliac artery supplying the ovary, uterine tube, and uterus. The veins follow a similar pattern, flowing retrograde to the arteries, but are more plexiform, including a pampiniform plexus related to the ovary and continuous uterine and vaginal plexuses (collectively, the uterovaginal plexus).

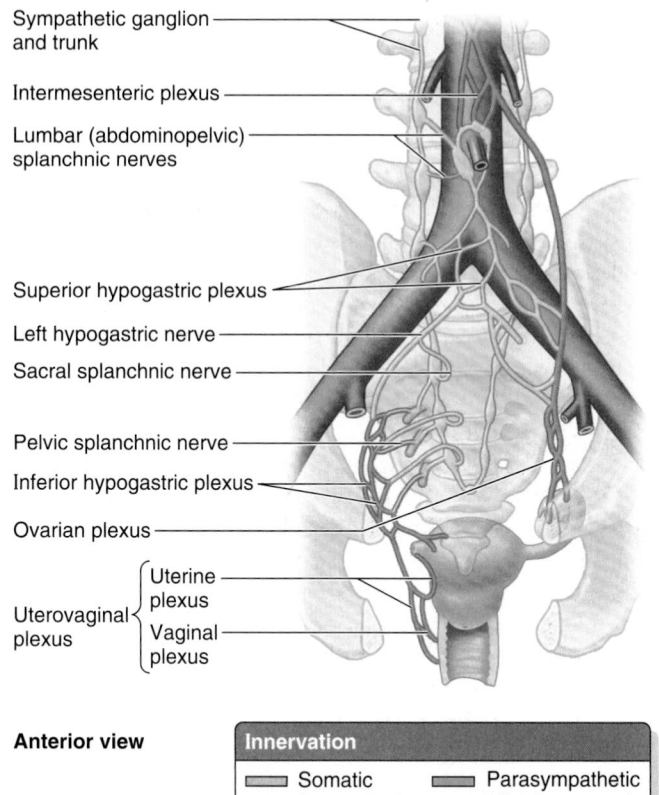

Sympathetic ganglion and trunk

Intermesenteric plexus

Lumbar (abdominopelvic) splanchnic nerves

Superior hypogastric plexus

Left hypogastric nerve

Sacral splanchnic nerve

Pelvic splanchnic nerve

Inferior hypogastric plexus

Ovarian plexus

Uterovaginal plexus { Uterine plexus / Vaginal plexus }

Anterior view

Innervation
- Somatic
- Sympathetic
- Parasympathetic
- Mixed autonomic

FIGURE 3.42. Nerve supply of ovaries and female internal genital organs. In addition to autonomic (visceral motor) fibers, these nerves convey visceral afferent fibers from these organs. The inferior part of the vagina is not depicted because it receives somatic innervation.

Innervation of Ovaries and Uterine Tubes. The nerve supply derives partly from the *ovarian plexus,* descending with the ovarian vessels, and partly from the *uterine (pelvic) plexus* (Fig. 3.42). The ovaries and uterine tubes are intraperitoneal and, therefore, are superior to the *pelvic pain line* (see Table 3.3). Thus, visceral afferent pain fibers ascend retrogradely with the descending sympathetic fibers of the ovarian plexus and lumbar splanchnic nerves to cell bodies in the T11–L1 spinal sensory ganglia. Visceral afferent reflex fibers follow parasympathetic fibers retrogradely through the uterine (pelvic) and inferior hypogastric plexuses and the pelvic splanchnic nerves to cell bodies in the S2–S4 spinal sensory ganglia.

UTERUS

The **uterus** (womb) is a thick-walled, pear-shaped, hollow muscular organ. The embryo and fetus develop in the uterus. Its muscular walls adapt to the growth of the fetus and then provide the power for its expulsion during childbirth. The non-gravid (non-pregnant) uterus usually lies in the lesser pelvis, with its body lying on the urinary bladder and its cervix between the urinary bladder and rectum (Fig. 3.43A).

The uterus is a very dynamic structure, the size and proportions of which change during the various changes of life (see the blue box "Lifetime Change in Normal Anatomy of Uterus," on p. 393).

The adult uterus is usually *anteverted* (tipped anterosuperiorly relative to the axis of the vagina) and *anteflexed* (flexed or bent anteriorly relative to the cervix, creating the *angle of flexion*) so that its mass lies over the bladder. Consequently, when the bladder is empty, the uterus typically lies in a nearly transverse plane (Figs. 3.43AB & 3.44A). The position of the uterus changes with the degree of fullness of the bladder (Fig. 3.44B) and rectum, and stage of pregnancy. Although its size varies considerably, the non-gravid uterus is approximately 7.5 cm long, 5 cm wide, and 2 cm thick and weighs approximately 90 g. The uterus is divisible into two main parts (Fig. 3.43B): the body and cervix.

The **body of the uterus,** forming the superior two thirds of the organ, includes the **fundus of the uterus,** the rounded part that lies superior to the uterine ostia (Fig. 3.39B). The body lies between the layers of the broad ligament and is freely movable (Fig. 3.39A). It has two surfaces: vesical (related to the bladder) and intestinal. The body is demarcated from the cervix by the **isthmus of the uterus,** a relatively constricted segment, approximately 1 cm long (Figs. 3.39A & B and 3.43B).

The **cervix of the uterus** is the cylindrical, relatively narrow inferior third of the uterus, approximately 2.5 cm long in an adult non-pregnant woman. For descriptive purposes, two parts are described: a **supravaginal part** between the isthmus and the vagina, and a **vaginal part,** which protrudes into the superiormost anterior vaginal wall (Fig. 3.43B). The rounded vaginal part surrounds the **external os of the uterus** and is surrounded in turn by a narrow recess, the *vaginal fornix* (Fig. 3.43C). The supravaginal part is separated from the bladder anteriorly by loose connective tissue and from the rectum posteriorly by the *recto-uterine pouch* (Fig. 3.43A).

The slit-like **uterine cavity** is approximately 6 cm in length from the external os to the wall of the fundus (Fig. 3.39B). The **uterine horns** (L. *cornua*) are the superolateral regions of the uterine cavity, where the uterine tubes enter. The uterine cavity continues inferiorly as the **cervical canal.** The fusiform canal extends from a narrowing inside the isthmus of the uterine body, the **anatomical internal os,** through the supravaginal and vaginal parts of the cervix, communicating with the lumen of the vagina through the external os. The uterine cavity (in particular, the cervical canal) and the lumen of the vagina together constitute the **birth canal,** through which the fetus passes at the end of gestation.

The wall of the body of the uterus consists of three coats, or layers:

- **Perimetrium**—the serosa or outer serous coat—consists of peritoneum supported by a thin layer of connective tissue.
- **Myometrium**—the middle coat of smooth muscle—becomes greatly distended (more extensive but much

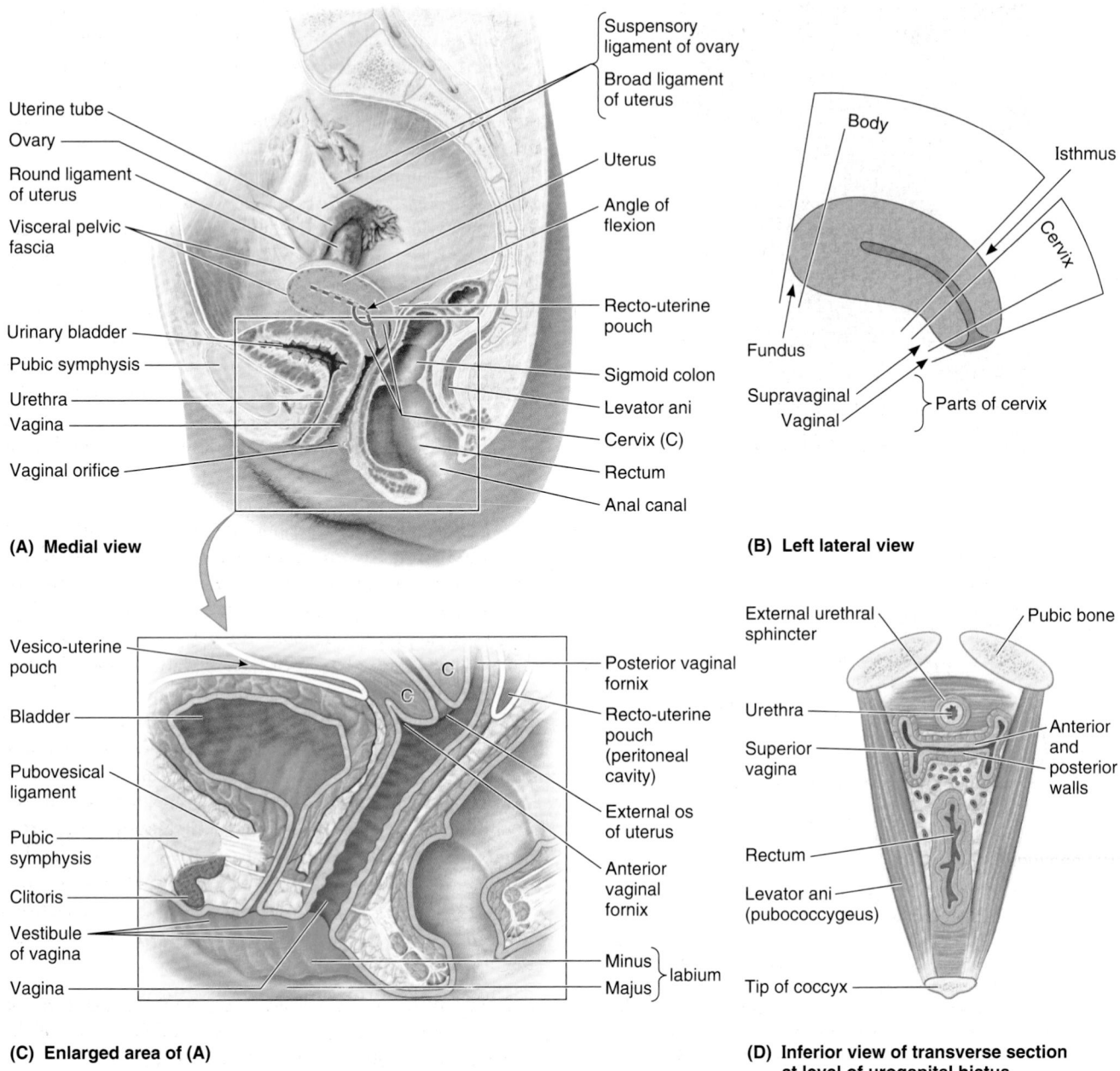

(A) Medial view

Suspensory ligament of ovary
Broad ligament of uterus
Uterine tube
Ovary
Round ligament of uterus
Visceral pelvic fascia
Uterus
Angle of flexion
Urinary bladder
Pubic symphysis
Urethra
Vagina
Vaginal orifice
Recto-uterine pouch
Sigmoid colon
Levator ani
Cervix (C)
Rectum
Anal canal

(B) Left lateral view

Body
Isthmus
Cervix
Fundus
Supravaginal
Vaginal
Parts of cervix

(C) Enlarged area of (A)

Vesico-uterine pouch
Bladder
Pubovesical ligament
Pubic symphysis
Clitoris
Vestibule of vagina
Vagina
Posterior vaginal fornix
Recto-uterine pouch (peritoneal cavity)
External os of uterus
Anterior vaginal fornix
Minus
Majus } labium

(D) Inferior view of transverse section at level of urogenital hiatus

External urethral sphincter
Pubic bone
Urethra
Superior vagina
Anterior and posterior walls
Rectum
Levator ani (pubococcygeus)
Tip of coccyx

FIGURE 3.43. Uterus and vagina. A and **B.** The disposition of the uterus is demonstrated in situ (**A**) and in isolation (**B**) in median sections. **A.** When the bladder is empty, the typical uterus is anteverted and anteflexed. **B.** The two main parts of the uterus, the body and cervix, are separated by the isthmus. Knowledge of further subdivisions of the main parts is especially important for such things as describing the location of tumors, sites of attachment of the placenta, and considering the consequences. **C** is an enlarged view of the outlined area in **A.** Note in (**A**) and (**C**) that the axes of the urethra and vagina are parallel, and the urethra is adherent to the anterior vaginal wall. Placing a gloved finger in the vagina can help direct the insertion of a catheter through the urethra into the bladder. **D.** A transverse section through the inferior female pelvic organs as they penetrate the pelvic floor through the urogenital hiatus (the gap between the right and the left sides of the levator ani) demonstrates the typical disposition of the non-distended lumina.

thinner) during pregnancy. The main branches of the blood vessels and nerves of the uterus are located in this coat. During childbirth, contraction of the myometrium is hormonally stimulated at intervals of decreasing length to dilate the cervical os and expel the fetus and placenta. During the menses, myometrial contractions may produce cramping.

• **Endometrium**—the inner mucous coat—is firmly adhered to the underlying myometrium. The endometrium is actively involved in the menstrual cycle, differing in structure with each stage of the cycle. If conception occurs, the blastocyst becomes implanted in this layer; if conception does not occur, the inner surface of this coat is shed during menstruation.

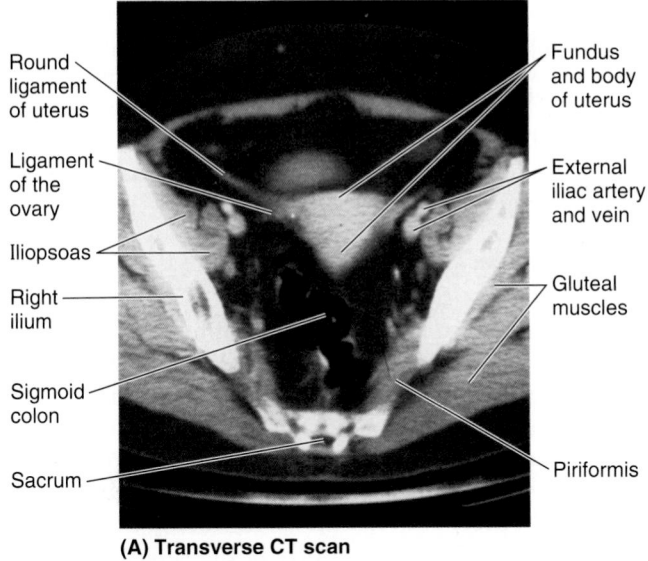

Round ligament of uterus

Ligament of the ovary

Iliopsoas

Right ilium

Sigmoid colon

Sacrum

Fundus and body of uterus

External iliac artery and vein

Gluteal muscles

Piriformis

(A) Transverse CT scan

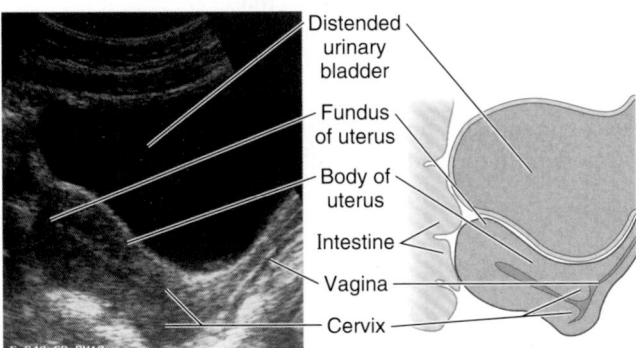

Distended urinary bladder

Fundus of uterus

Body of uterus

Intestine

Vagina

Cervix

(B) Longitudinal (median) ultrasound image

FIGURE 3.44. Imaging of female pelvic viscera. A. Because the uterus is nearly horizontally disposed when anteverted and anteflexed over the bladder, most of its body, including the fundus, appears in this transverse CT scan. (Courtesy of Dr. Donald R. Cahill, Department of Anatomy, Mayo Medical School, Rochester, MN.) **B.** Temporary retroversion and retroflexion result when a fully distended urinary bladder temporarily retroverts the uterus and decreases its angle of flexion. Compare with Figure 3.43A. (Courtesy of Dr. A. M. Arenson, Assistant Professor of Medical Imaging, University of Toronto, Toronto, ON, Canada.)

The amount of muscular tissue in the cervix is markedly less than in the body of the uterus. The cervix is mostly fibrous and is composed mainly of collagen with a small amount of smooth muscle and elastin.

Ligaments of Uterus. Externally, the **ligament of the ovary** attaches to the uterus postero-inferior to the uterotubal junction (Fig. 3.39A & B). The **round ligament of the uterus** (L. *ligamentum teres uteri*) attaches antero-inferiorly to this junction. These two ligaments are vestiges of the *ovarian gubernaculum*, related to the relocation of the gonad from its developmental position on the posterior abdominal wall (Fig. 2.17A).

The **broad ligament of the uterus** is a double layer of peritoneum (mesentery) that extends from the sides of the uterus to the lateral walls and floor of the pelvis (Fig. 3.39A). This ligament assists in keeping the uterus in position. The two layers of the broad ligament are continuous with each other at a free edge that surrounds the uterine tube. Laterally, the peritoneum of the broad ligament is prolonged superiorly over the vessels as the *suspensory ligament of the ovary*. Between the layers of the broad ligament on each side of the uterus, the *ligament of the ovary* lies posterosuperiorly and the *round ligament of the uterus* lies antero-inferiorly. The uterine tube lies in the antero-superior free border of the broad ligament, within a small mesentery called the *mesosalpinx*. Similarly, the ovary lies within a small mesentery called the *mesovarium* on the posterior aspect of the broad ligament. The largest part of the broad ligament, inferior to the mesosalpinx and mesovarium, which serves as a mesentery for the uterus itself, is the **mesometrium.**

The uterus is a dense structure located in the center of the pelvic cavity. The principal supports of the uterus holding it in this position are both passive and active or dynamic. *Dynamic support of the uterus* is provided by the pelvic diaphragm. Its tone during sitting and standing and active contraction during periods of increased intra-abdominal pressure (sneezing, coughing, etc.) is transmitted through the surrounding pelvic organs and the endopelvic fascia in which they are embedded. *Passive support of the uterus* is provided by its position—the way in which the normally *anteverted* and *anteflexed* uterus rests on top of the bladder. When intra-abdominal pressure is increased, the uterus is pressed against the bladder. The cervix is the least mobile part of the uterus because of the passive support provided by attached condensations of endopelvic fascia (ligaments), which may also contain smooth muscle (Figs. 3.13 and 3.14):

- *Cardinal (transverse cervical) ligaments* extend from the supravaginal cervix and lateral parts of the fornix of the vagina to the lateral walls of the pelvis (Fig. 3.14).
- *Uterosacral ligaments* pass superiorly and slightly posteriorly from the sides of the cervix to the middle of the sacrum; they are palpable during a rectal examination.

Together these passive and active supports keep the uterus centered in the pelvic cavity and resist the tendency for the uterus to fall or be pushed through the vagina (see the blue box "Disposition of Uterus and Uterine Prolapse," p. 392).

Relations of Uterus. Peritoneum covers the uterus anteriorly and superiorly, except for the cervix (Fig. 3.39A). The peritoneum is reflected anteriorly from the uterus onto the bladder and posteriorly over the posterior part of the fornix of the vagina to the rectum (Fig. 3.43A). Anteriorly, the uterine body is separated from the urinary bladder by the **vesico-uterine pouch,** where the peritoneum is reflected from the uterus onto the posterior margin of the superior surface of the bladder. Posteriorly, the uterine body and supravaginal part of the cervix are separated from the sigmoid colon by a layer of peritoneum, and the peritoneal cavity and from the rectum by the *recto-uterine pouch.* Laterally, the uterine artery crosses the ureter superiorly, near the cervix (Fig. 3.41).

Summary of the relations of the uterus (Fig. 3.45):

- *Anteriorly* (antero-inferiorly in its normal anteverted position): the vesico-uterine pouch and superior surface of the bladder; the supravaginal part of the cervix is related to the bladder and is separated from it by only fibrous connective tissue.
- *Posteriorly:* the recto-uterine pouch containing loops of small intestine and the anterior surface of rectum; only the visceral pelvic fascia uniting the rectum and uterus here resists increased intra-abdominal pressure.
- *Laterally:* the peritoneal broad ligament flanking the uterine body and the fascial cardinal ligaments on each side of the cervix and vagina; in the transition between the two ligaments, the ureters run anteriorly slightly superior to the lateral part of the vaginal fornix and inferior to the uterine arteries, usually approximately 2 cm lateral to the supravaginal part of the cervix (see Fig. 3.13A).

Arterial Supply and Venous Drainage of Uterus. The blood supply of the uterus derives mainly from the *uterine arteries,* with potential collateral supply from the ovarian arteries (Fig. 3.41). The *uterine veins* enter the broad ligaments with the arteries and form a **uterine venous plexus** on each side of the cervix. Veins from the uterine plexus drain into the internal iliac veins.

VAGINA

The **vagina,** a distensible musculomembranous tube (7–9 cm long), extends from the middle cervix of the uterus to the **vaginal orifice,** the opening at the inferior end of the vagina (Figs. 3.39B and 3.43A & C). The vaginal orifice, external urethral orifice, and ducts of the greater and lesser vestibular glands open into the **vestibule of the vagina,** the cleft between the labia minora. The vaginal part of the cervix lies anteriorly in the superior vagina. The vagina:

- serves as a canal for menstrual fluid,
- forms the inferior part of the birth canal,
- receives the penis and ejaculate during sexual intercourse, and
- communicates superiorly with the *cervical canal* and inferiorly with the vestibule of the vagina.

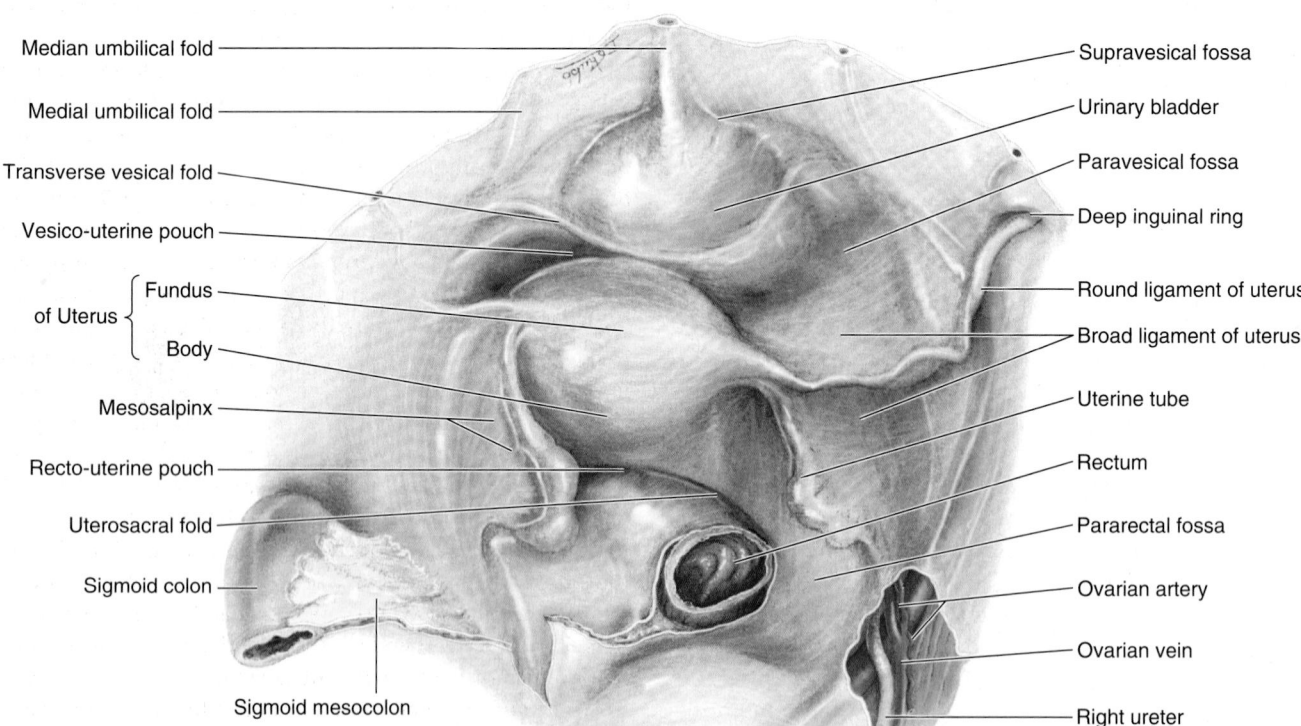

FIGURE 3.45. Relationship of the female pelvic viscera. The peritoneum is intact, lining the pelvic cavity and covering the superior aspect of the bladder, fundus and body of uterus, and much of the rectum. In this supine cadaver, the uterine tube and meso salpinx on each side are hanging down, obscuring the ovaries from view. The uterus is typically asymmetrically placed, as it is here. The round ligament of the uterus follows the same subperitoneal course as the ductus deferens of the male.

The vagina is usually collapsed. The orifice is usually collapsed toward the midline so that its lateral walls are in contact on each side of an anteroposterior slit. Superior to the orifice, however, the anterior and posterior walls are in contact on each side of a transverse potential cavity, H-shaped in cross section (Fig. 3.43D), except at its superior end where the cervix holds them apart. The vagina lies posterior to the urinary bladder and urethra, the latter projecting along the midline of its inferior anterior wall (Fig. 3.39B). The vagina lies anterior to the rectum, passing between the medial margins of the levator ani (puborectalis) muscles. The **vaginal fornix,** the recess around the cervix, has *anterior, posterior,* and *lateral parts* (Figs. 3.39A and 3.43C). The **posterior vaginal fornix** is the deepest part and is closely related to the recto-uterine pouch. Four muscles compress the vagina and act as sphincters: **pubovaginalis,** *external urethral sphincter,* **urethrovaginal sphincter,** and *bulbospongiosus* (Fig. 3.46).

The vagina is related (Fig. 3.27):

- anteriorly to the fundus of the urinary bladder and urethra;
- laterally to the levator ani, visceral pelvic fascia, and ureters; and
- posteriorly (from inferior to superior) to the anal canal, rectum, and recto-uterine pouch.

ARTERIAL SUPPLY AND VENOUS DRAINAGE OF VAGINA

The arteries supplying the superior part of the vagina derive from the *uterine arteries.* The arteries supplying the middle

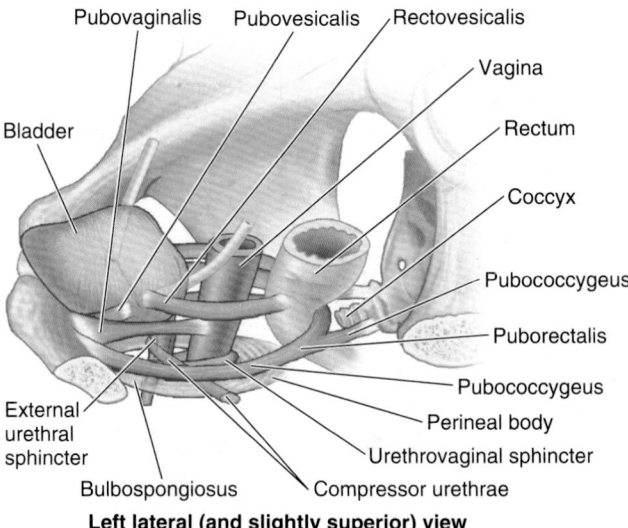

Left lateral (and slightly superior) view

FIGURE 3.46. Muscles compressing urethra and vagina. Muscles that compress the vagina and act as sphincters include the pubovaginalis, external urethral sphincter (especially its urethrovaginal sphincter part), and bulbospongiosus. The compressor urethrae and external urethral sphincter compress the urethra.

and inferior parts of the vagina derive from the *vaginal* and *internal pudendal arteries* (Figs. 3.18 and 3.41).

The vaginal veins form **vaginal venous plexuses** along the sides of the vagina and within the vaginal mucosa (Fig. 3.41). These veins are continuous with the *uterine venous plexus* as the **uterovaginal venous plexus,** and drain into the internal iliac veins through the uterine vein. This plexus also communicates with the vesical and rectal venous plexuses.

INNERVATION OF VAGINA AND UTERUS

Only the inferior one fifth to one quarter of the vagina is somatic in terms of innervation. Innervation of this part of the vagina is from the *deep perineal nerve,* a branch of the *pudendal nerve,* which conveys sympathetic and visceral afferent fibers but no parasympathetic fibers (Fig. 3.47). Only this somatically innervated part is sensitive to touch and temperature, even though the somatic and visceral afferent fibers have their cell bodies in the same (S2–S4) spinal ganglia.

Most of the vagina (superior three quarters to four fifths) is visceral in terms of its innervation. Nerves to this part of the vagina and to the uterus are derived from the **uterovaginal nerve plexus,** which travels with the uterine artery at the junction of the base of the (peritoneal) broad ligament and the superior part of the (fascial) transverse cervical ligament. The uterovaginal nerve plexus is one of the pelvic plexuses that extends to the pelvic viscera from the inferior hypogastric plexus. Sympathetic, parasympathetic, and visceral afferent fibers pass through this plexus.

Sympathetic innervation originates in the inferior thoracic spinal cord segments and passes through *lumbar splanchnic nerves* and the intermesenteric-hypogastric-pelvic series of plexuses. *Parasympathetic innervation* originates in the S2–S4 spinal cord segments and passes through the *pelvic splanchnic nerves* to the inferior hypogastric-uterovaginal plexus. The *visceral afferent innervation* of the superior (intraperitoneal; fundus and body) and inferior (subperitoneal; cervical) parts of the uterus and vagina differ in terms of course and destination. Visceral afferent fibers conducting pain impulses from the intraperitoneal uterine fundus and body (superior to the *pelvic pain line*) follow the sympathetic innervation retrograde to reach cell bodies in the inferior thoracic-superior lumbar spinal ganglia. Afferent fibers conducting pain impulses from the subperitoneal uterine cervix and vagina (inferior to the pelvic pain line) follow the parasympathetic fibers retrograde through the uterovaginal and inferior hypogastric plexuses and pelvic splanchnic nerves to reach cell bodies in the spinal sensory ganglia of S2–S4. The two different routes followed by visceral pain fibers is clinically significant in that it offers mothers a variety of types of anesthesia for childbirth (see the blue box "Anesthesia for Childbirth," on p. 397). All visceral afferent fibers from the uterus and vagina not concerned with pain (those conveying unconscious sensations) also follow the latter route.

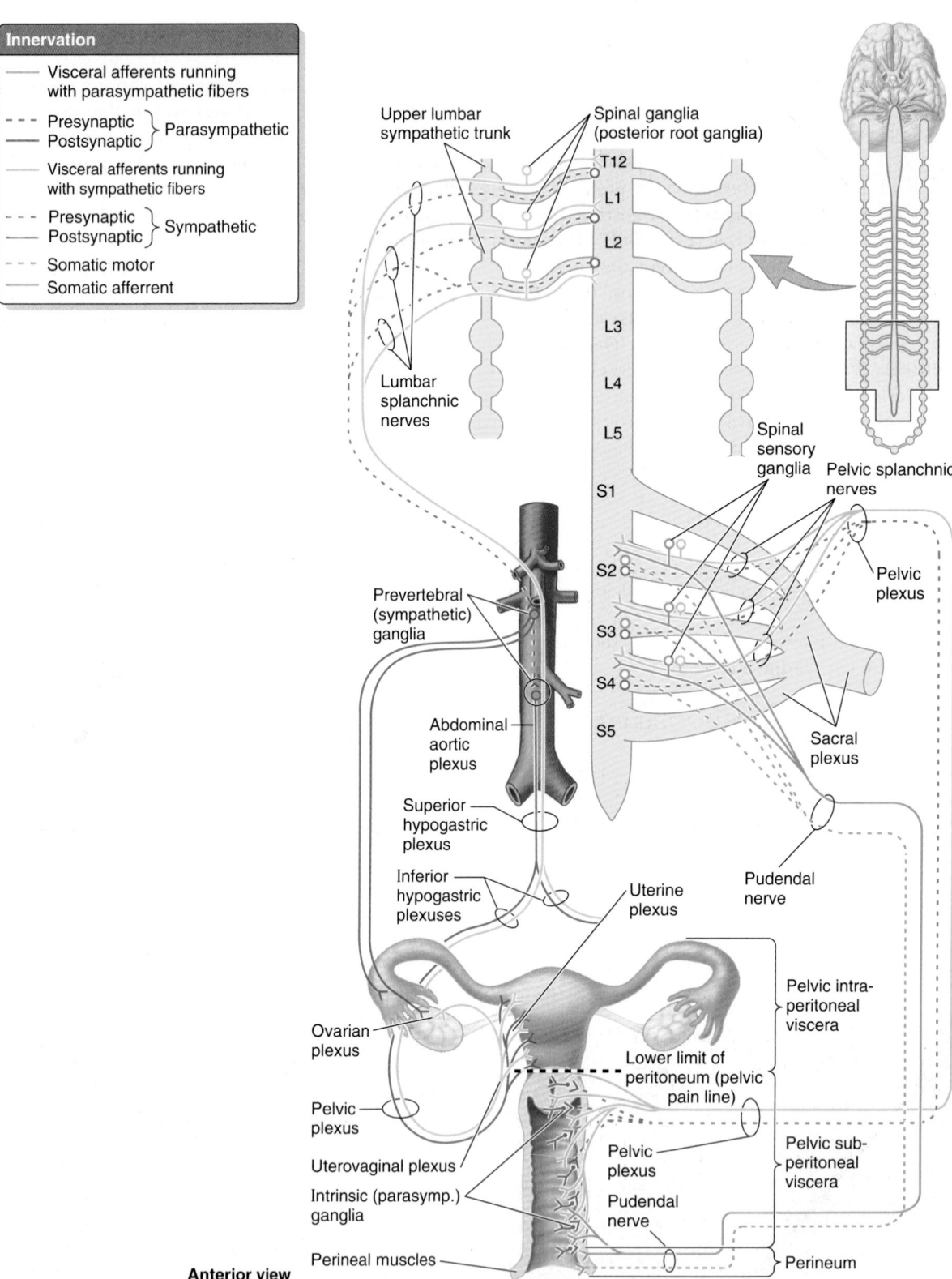

Innervation

— Visceral afferents running with parasympathetic fibers

- - - Presynaptic } Parasympathetic
—— Postsynaptic

— Visceral afferents running with sympathetic fibers

- - - Presynaptic } Sympathetic
—— Postsynaptic

- - - Somatic motor
—— Somatic afferent

Upper lumbar sympathetic trunk

Spinal ganglia (posterior root ganglia)

T12

L1

L2

L3

L4

L5

S1

S2

S3

S4

S5

Lumbar splanchnic nerves

Spinal sensory ganglia

Pelvic splanchnic nerves

Pelvic plexus

Sacral plexus

Prevertebral (sympathetic) ganglia

Abdominal aortic plexus

Superior hypogastric plexus

Inferior hypogastric plexuses

Uterine plexus

Pudendal nerve

Pelvic intra-peritoneal viscera

Lower limit of peritoneum (pelvic pain line)

Ovarian plexus

Pelvic plexus

Uterovaginal plexus

Intrinsic (parasymp.) ganglia

Perineal muscles

Pelvic plexus

Pudendal nerve

Pelvic sub-peritoneal viscera

Perineum

Anterior view

FIGURE 3.47. Innervation of female internal genital organs. Pelvic splanchnic nerves, arising from the S2–S4 anterior rami, supply parasympathetic motor fibers to the uterus and vagina (and vasodilator fibers to the erectile tissue of the clitoris and bulb of the vestibule; not shown). Presynaptic sympathetic fibers traverse the sympathetic trunk and pass through the lumbar splanchnic nerves to synapse in prevertebral ganglia with postsynaptic fibers; the latter fibers travel through the superior and inferior hypogastric plexuses to reach the pelvic viscera. Visceral afferent fibers conducting pain from intraperitoneal structures (such as the uterine body) travel with the sympathetic fibers to the T12–L2 spinal ganglia. Visceral afferent fibers conducting pain from subperitoneal structures, such as the cervix and vagina (i.e., the birth canal), travel with parasympathetic fibers to the S2–S4 spinal ganglia. Somatic sensation from the opening of the vagina also passes to the S2–S4 spinal ganglia via the pudendal nerve. In addition, muscular contractions of the uterus are hormonally induced.

FEMALE INTERNAL GENITAL ORGANS

Infections of Female Genital Tract

 Because the female genital tract communicates with the peritoneal cavity through the abdominal ostia of the uterine tubes, infections of the vagina, uterus, and tubes may result in *peritonitis*. Conversely, inflammation of a tube (*salpingitis*) may result from infections that spread from the peritoneal cavity. A major cause of infertility in women is blockage of the uterine tubes, often the result of salpingitis.

Patency of Uterine Tubes

HYSTEROSALPINGOGRAPHY

Patency of the uterine tubes may be determined by a radiographic procedure involving injection of a water-soluble radiopaque material or carbon dioxide gas into the uterus and tubes through the external os of the uterus (*hysterosalpingography*). The contrast medium travels through the uterine cavity and tubes (*arrowheads* in Fig. B3.13). Accumulation of radiopaque fluid or the appearance of gas bubbles in the pararectal fossae region of the peritoneal cavity indicates that the tubes are patent.

ENDOSCOPY

Patency of the uterine tubes can also be determined by *hysteroscopy*, examination of the interior of the tubes using a narrow endoscopic instrument (*hysteroscope*), which is introduced through the vagina and uterus.

Ligation of Uterine Tubes

Ligation of the uterine tubes is a surgical method of birth control. Oocytes discharged from the ovaries that enter the tubes of these patients degenerate

and are soon absorbed. Most surgical sterilizations are done by either abdominal tubal ligation or laparoscopic tubal ligation. *Open abdominal tubal ligation* is usually performed through a short suprapubic incision made at the pubic hairline (Fig. B3.14A) and involves interruption, often with removal of a segment of the tube, and tubal closure by suture ligation. *Laparoscopic tubal ligation* is done with a fiber optic laparoscope inserted through a small incision, usually near the umbilicus (Fig. B3.14B). In this procedure, tubal continuity is interrupted by applying cautery, rings, or clips.

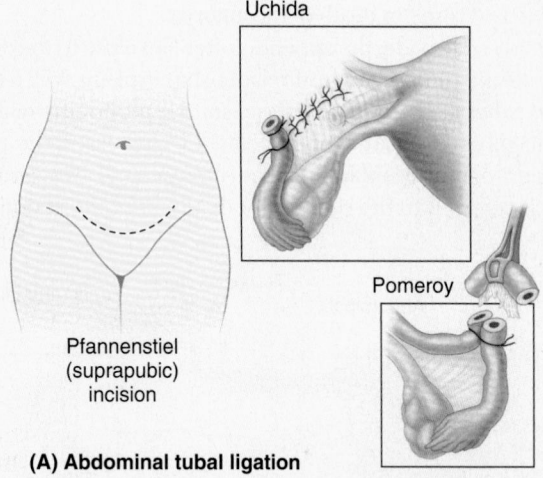

(A) Abdominal tubal ligation

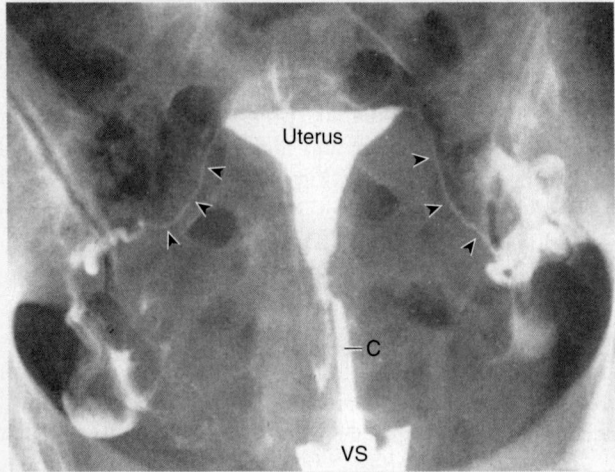

FIGURE B3.13. Hysterosalpingogram. Arrowheads, uterine tubes; c, catheter in the cervical canal, vs, vaginal speculum.

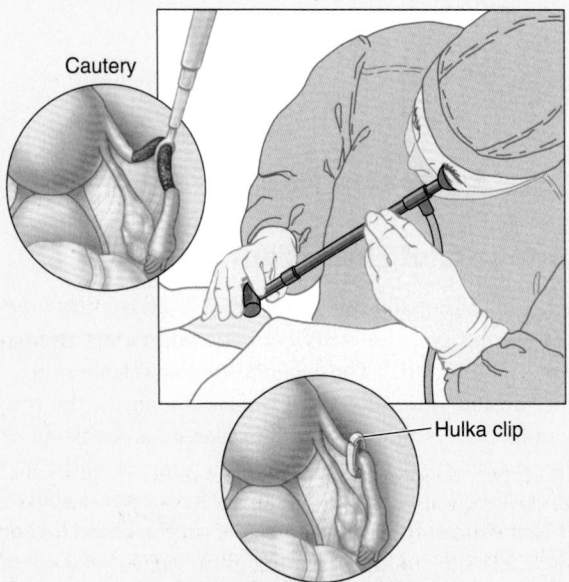

(B) Laparoscopic tubal ligation

FIGURE B3.14.

Ectopic Tubal Pregnancy

In some women, collections of pus may develop in a uterine tube (*pyosalpinx*) and the tube may be partly occluded by adhesions. In these cases, the morula (early embryo) may not be able to pass along the tube to the uterus, although sperms have obviously done so. When the blastocyst forms, it may implant in the mucosa of the uterine tube, producing an *ectopic tubal pregnancy*. Although implantation may occur in any part of the tube, the common site is in the ampulla (Fig. B3.15). Tubal pregnancy is the most common type of ectopic gestation; it occurs in approximately 1 of every 250 pregnancies in North America (Moore et al., 2012). If not diagnosed early, ectopic tubal pregnancies may result in rupture of the uterine tube, and severe hemorrhage into the abdominopelvic cavity during the first 8 weeks of gestation. Tubal rupture and hemorrhage constitute a threat to the mother's life and result in death of the embryo.

On the right side, the appendix often lies close to the ovary and uterine tube. This close relationship explains why a *ruptured tubal pregnancy* and the resulting peritonitis may be misdiagnosed as acute appendicitis. In both cases, the parietal peritoneum is inflamed in the same general area, and the pain is referred to the right lower quadrant of the abdomen.

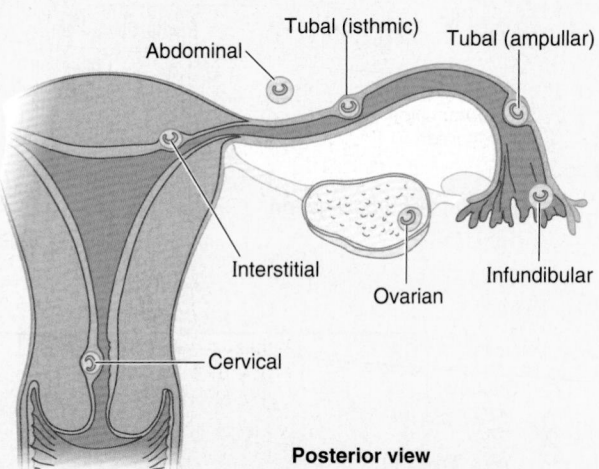

FIGURE B3.15. Sites of ectopic pregnancy.

Remnants of Embryonic Ducts

Occasionally the mesosalpinx between the uterine tube and the ovary contains embryonic remnants (Fig. 3.39B). The *epoophoron* forms from remnants of the mesonephric tubules of the *mesonephros,* the transitory embryonic kidney. There may also be a persistent *duct of the epoophoron* (duct of Gartner), a remnant of the mesonephric duct that forms the ductus deferens and ejaculatory duct in the male. It lies between layers of the broad ligament along each side of the uterus and/or vagina. A *vesicular appendage* is sometimes attached to the infundibulum of the uterine tube. It is the remains of the cranial end of the mesonephric duct that forms the *ductus epididymis*. Although

these vestigial structures are mostly of embryological and morphological interest, they occasionally accumulate fluid and form cysts (e.g., Gartner duct cysts).

Bicornate Uterus

Incomplete fusion of the embryonic paramesonephric ducts, from which the uterus is formed, results in a variety of congenital anomalies, ranging from formation of a unicornate uterus (receiving a uterine duct only from the right or left) to duplication in the form of a bicornate uterus (Fig. B3.16), doubled uterine cavites, or a completely doubled uterus (*uterus didelphys*).

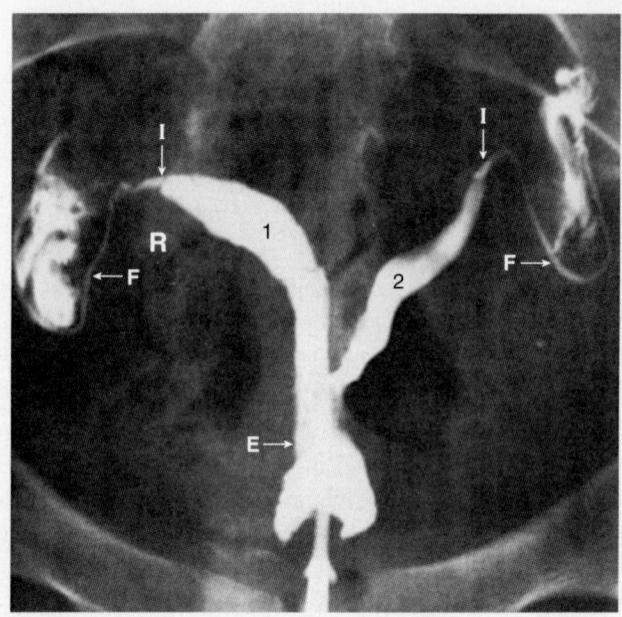

FIGURE B3.16. Bicornate uterus. *1* and *2*, uterine cavity; *E*, cervical canal; *F*, uterine tube; *I*, isthmus of tube, *R*, right side marker. (Reprinted with permission from Stuart GCE, Reid DF. Diagnostic studies. In: Copeland LJ, ed. Textbook of Gynecology, 2nd ed. Philadelphia: Saunders, 2000.)

Disposition of Uterus and Uterine Prolapse

When intra-abdominal pressure is increased, the normally anteverted and anteflexed uterus is pressed against the bladder (Fig. B3.17A). However, the uterus may assume other dispositions, including excessive anteflexion (Fig. B3.17B), anteflexion with retroversion (Fig. B3.17C), and retroflexion with retroversion (Fig. B3.17D). Instead of pressing the uterus against the bladder, increased intra-abdominal pressure tends to push the retroverted uterus, a solid mass positioned upright over the vagina (a flexible, hollow tube), into or even through the vagina (Fig. B3.17E–G). A retroverted uterus will not necessarily prolapse, but is more likely to do so. The situation is exacerbated in the presence of a disrupted perineal body or with atrophic ("relaxed") pelvic floor ligaments and muscles (see the blue box "Disruption of Perineal Body," p. 414).

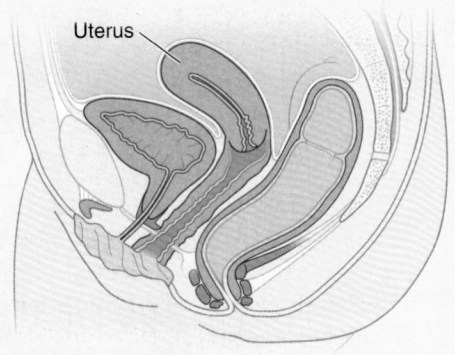

(A) Normal (anteverted, anteflexed)

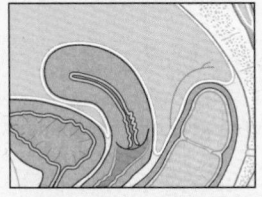

(B) Excessive anteflexion

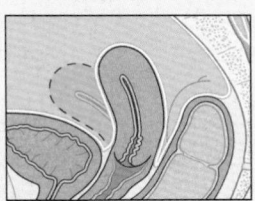

(C) Retroverted

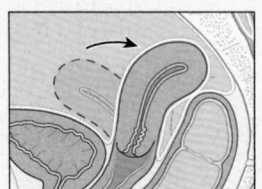

(D) Retroverted and retroflexed

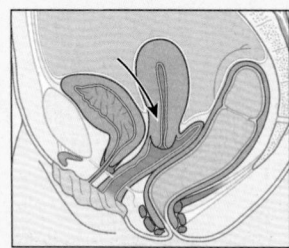

(E) First degree prolapse

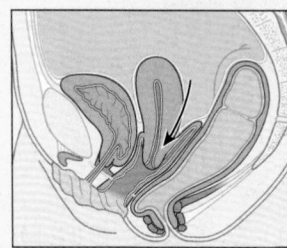

(F) Second degree prolapse

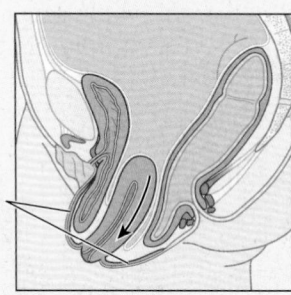

Everted vaginal wall

(G) Third degree prolapse

FIGURE B3.17.

Manual Examination of Uterus

 The size and disposition of the uterus may be examined by *bimanual palpation* (Fig. B3.18A). Two gloved fingers of the right hand are passed superiorly in the vagina, while the other hand is pressed inferoposteriorly on the pubic region of the anterior abdominal wall. The size and other characteristics of the uterus can be determined in this way (e.g., whether the uterus is in its normal anteverted position). When softening of the uterine isthmus occurs (*Hegar sign*), the cervix feels as though it were separated from the body. Softening of the isthmus is an early sign of pregnancy. The uterus can be further stabilized through rectovaginal examination, which is used if examination via the vagina alone does not yield clear findings (Fig. B3.18B).

Lifetime Changes in Anatomy of Uterus

The uterus is possibly the most dynamic structure in human anatomy (Fig. B3.19). At birth, the uterus is relatively large and has adult proportions (body to

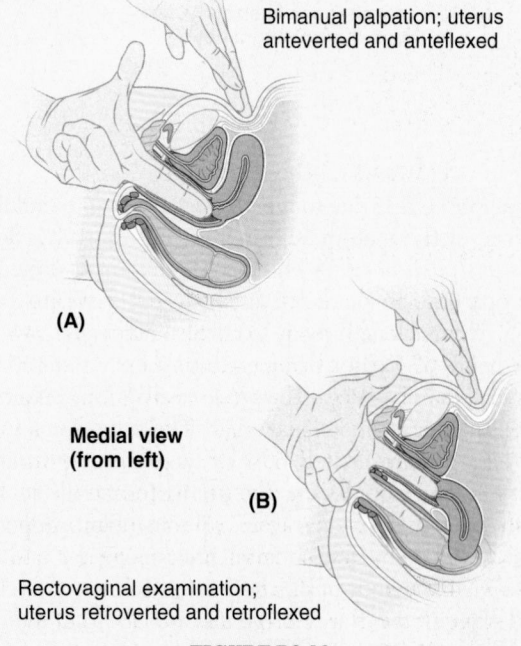

Bimanual palpation; uterus anteverted and anteflexed

(A)

Medial view (from left)

(B)

Rectovaginal examination; uterus retroverted and retroflexed

FIGURE B3.18.

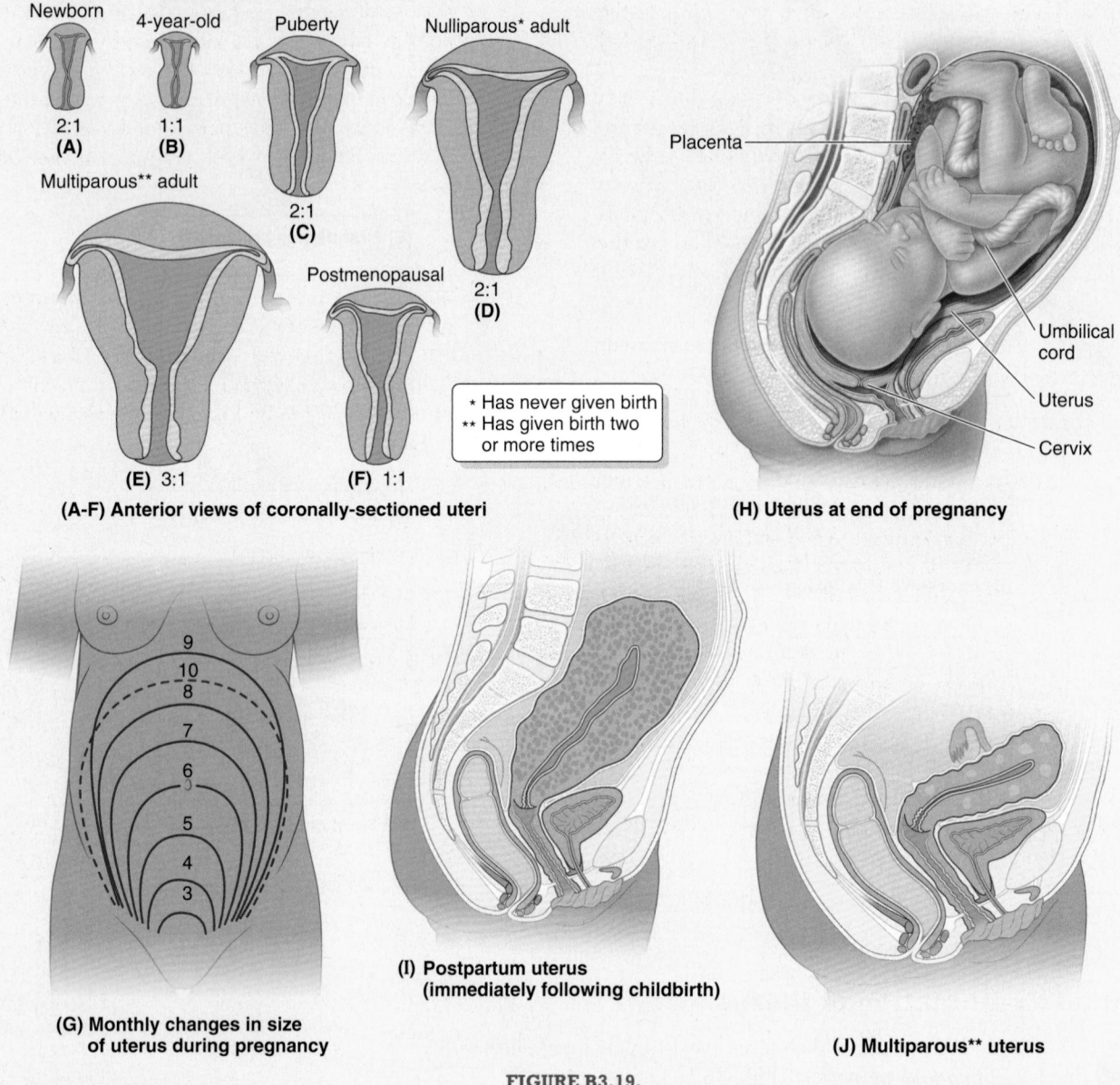

Newborn   4-year-old   Puberty   Nulliparous* adult

2:1   1:1
(A)   (B)

2:1
(C)

2:1
(D)

Multiparous** adult

Postmenopausal

* Has never given birth
** Has given birth two
   or more times

(E) 3:1   (F) 1:1

(A–F) Anterior views of coronally-sectioned uteri

Placenta

Umbilical
cord

Uterus

Cervix

(H) Uterus at end of pregnancy

9
10
8
7
6
5
4
3

(G) Monthly changes in size
of uterus during pregnancy

(I) Postpartum uterus
(immediately following childbirth)

(J) Multiparous** uterus

FIGURE B3.19.

cervical ratio = 2:1) due to the *prepartum* (before childbirth) influence of the maternal hormones (Fig. B3.19A). Several weeks *postpartum* (after childbirth), childhood dimensions and proportions are obtained: The body and cervix are approximately of equal length (body to cervical ratio = 1:1), with the cervix being of greater diameter (thickness) (Fig. B3.19B). Because of the small size of the pelvic cavity during infancy, the uterus is mainly an abdominal organ. The cervix remains relatively large (approximately 50% of total uterus) throughout childhood. During *puberty*, the uterus (especially its body) grows rapidly in size, once again assuming adult proportions (Fig. B3.19C). In the postpubertal, premenopausal, non-pregnant woman, the body of the uterus is pear shaped; the thick-walled superior two thirds of the uterus lies within the pelvic cavity (Fig. B3.19D). During this phase of life, the uterus

undergoes monthly changes in size, weight, and density in relation to the menstrual cycle.

Over the 9 months of pregnancy, the *gravid* uterus expands greatly to accommodate the fetus, becoming larger and increasingly thin walled (Fig. B3.19G). At the end of pregnancy (B3.19G, line 10), the fetus "drops," as the head becomes engaged in the lesser pelvis. The uterus becomes nearly membranous, with the fundus dropping below its highest level (achieved in the 9th month), at which time it extends superiorly to the costal margin, occupying most of the abdominopelvic cavity (Fig. B3.19H).

Immediately after delivery, the large uterus becomes thick walled and edematous (Fig. B3.19I), but its size reduces rapidly. The *multiparous* non-gravid uterus has a large and nodular body and usually extends into the lower abdominal cavity, often causing a slight protrusion of the inferior abdominal wall in lean women (Figs. B3.19E & J and 3.73B, p. 437).

During *menopause* (46–52 years of age), the uterus (again, especially the body) decreases in size. *Postmenopause,* the uterus is involuted and regresses to a markedly smaller size, once again assuming childhood proportions (Fig. B3.19F). All these stages represent normal anatomy for the particular age and reproductive status of the woman.

Cervical Cancer, Cervical Examination and Pap Smear

Until 1940, cervical cancer was the leading cause of death in North American women (Krebs, 2000). The decline in the incidence and number of women dying from cervical cancer is related to the accessibility of the cervix to direct visualization, and to cell and tissue study by means of Pap (Papanicolaou) smears, leading to the detection of premalignant cervical conditions (Copeland, 2000; Morris and Burke, 2000). The vagina can be distended with a *vaginal speculum* to enable inspection of the cervix (Fig. B3.20A & B). To prepare a Pap smear, a spatula is placed in the external os of the uterus (Fig. B3.20A). The spatula is rotated to scrape cellular material from the mucosa of the vaginal cervix (Fig. B3.20C), followed by insertion of a cytobrush into the cervical canal that is rotated to gather cellular material from the supravaginal cervical mucosa. The cellular material is then placed on glass slides for microscopic examination (Fig. B3.20D & E).

Because no peritoneum intervenes between the anterior cervix and the base of the bladder, cervical cancer may spread by contiguity to the bladder. It may also spread by lymphogenous (lymph borne) metastasis to external or internal iliac or sacral nodes. Hematogenous (blood borne) metastasis may occur via iliac veins or via the internal vertebral venous plexus.

Hysterectomy

Owing to the frequency of uterine and cervical cancer, hysterectomy, excision of the uterus (G. *hystera*), is a relatively common procedure. The uterus may be surgically approached and removed (A) through the anterior abdominal wall, or (B) through the vagina (Fig. B3.21).

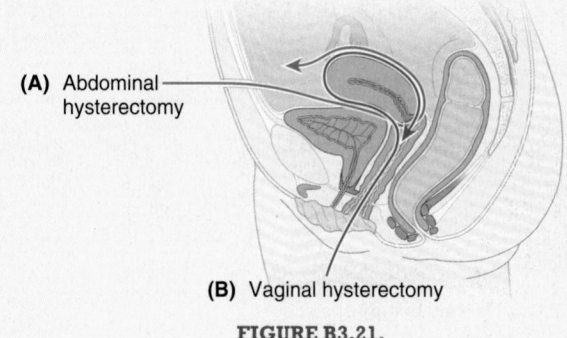

(A) Abdominal hysterectomy
(B) Vaginal hysterectomy

FIGURE B3.21.

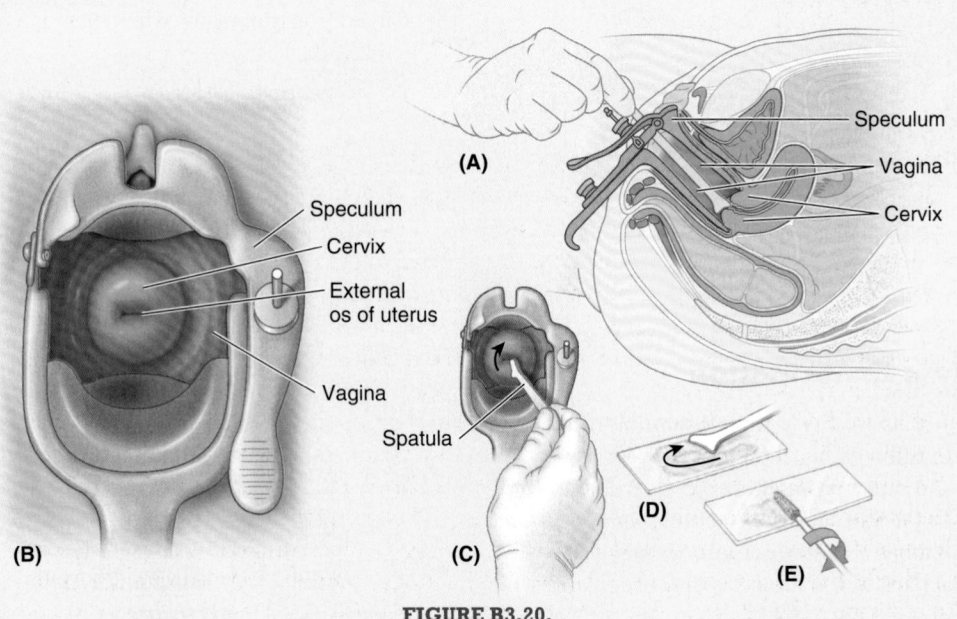

Speculum
Cervix
External os of uterus
Vagina
Speculum
Vagina
Cervix
Spatula
(A) (B) (C) (D) (E)

FIGURE B3.20.

Distension of Vagina

The vagina can be markedly distended, particularly in the region of the posterior part of the fornix. For example, distension of this part allows palpation of the sacral promontory during a pelvic examination (see the blue box "Pelvic Diameters (Conjugates)," p. 334). The distension also accommodates the erect penis during intercourse.

The vagina is especially distended by the fetus during parturition, particularly in an AP direction when the fetus's shoulders are delivered (Fig. B3.22). Lateral distension is limited by the ischial spines, which project posteromedially, and the sacrospinous ligaments extending from these spines to the lateral margins of the sacrum and coccyx. The birth canal is thus deep anteroposteriorly and narrow transversely at this point, causing the fetus's shoulders to rotate into the AP plane.

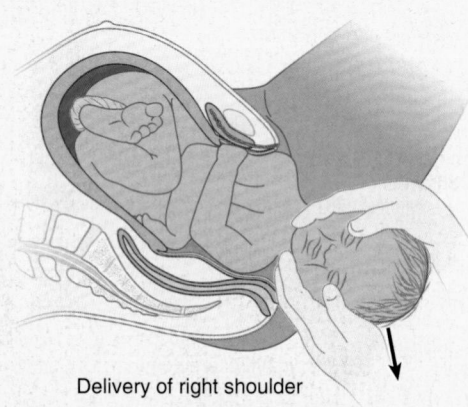

Delivery of right shoulder

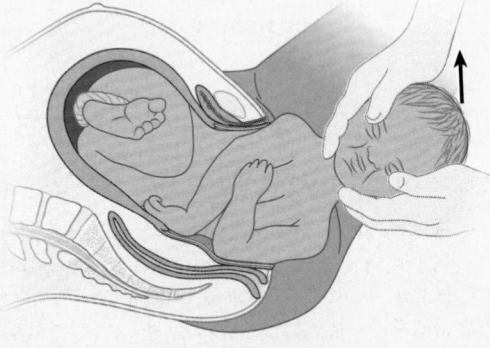

Delivery of left shoulder

FIGURE B3.22.

Digital Pelvic Examination

Because of its relatively thin, distensible walls and central location within the pelvis, the cervix, ischial spines, and sacral promontory can be palpated with the gloved digits in the vagina and/or rectum (*manual pelvic examination*). Pulsations of the uterine arteries may also be felt through the lateral parts of the fornix, as may irregularities of the ovaries, such as cysts (Fig. B3.23).

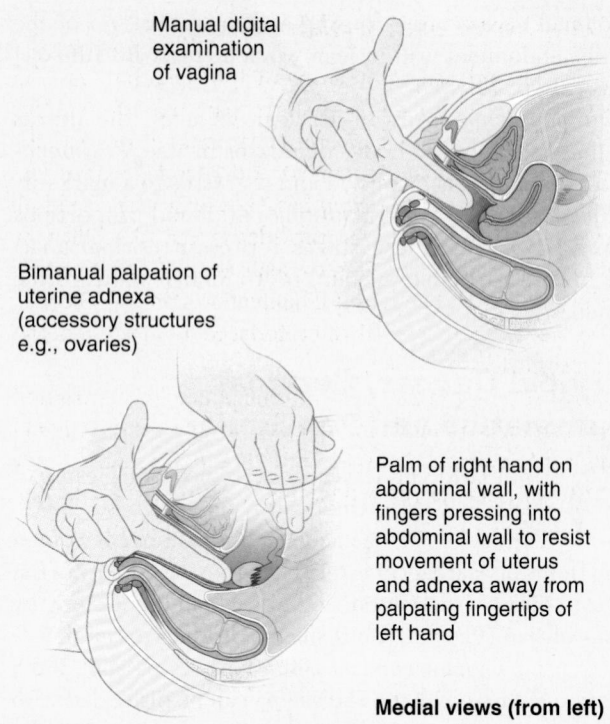

Manual digital examination of vagina

Bimanual palpation of uterine adnexa (accessory structures e.g., ovaries)

Palm of right hand on abdominal wall, with fingers pressing into abdominal wall to resist movement of uterus and adnexa away from palpating fingertips of left hand

Medial views (from left)

FIGURE B3.23.

Vaginal Fistulae

Because of the close relationship of the vagina to adjacent pelvic organs, obstetrical trauma during long and difficult labor may result in weaknesses, necrosis, or tears in the vaginal wall and sometimes beyond. These may form or subsequently develop into abnormal passages (*fistulas*) between the vaginal lumen and the lumina of the adjacent bladder, urethra, and rectum, or the perineum (Fig. B3.24). Urine enters the vagina from both *vesicovaginal* and *urethrovaginal fistulas*, but the flow is continuous from the former and occurs only during micturition from the latter. Fecal matter may be discharged from the vagina when there is a *rectovaginal fistula*.

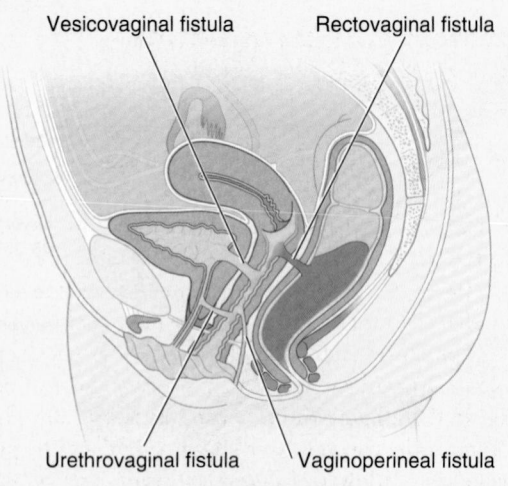

Vesicovaginal fistula   Rectovaginal fistula

Urethrovaginal fistula   Vaginoperineal fistula

Medial view (from left)

FIGURE B3.24.

Culdoscopy and Culdocentesis

An endoscopic instrument (*culdoscope*) can be inserted through the posterior part of the vaginal fornix to examine the ovaries or uterine tubes (e.g., for the presence of a tubal pregnancy). Although it involves less disruption of tissue, *culdoscopy* has been largely replaced by *laparoscopy,* which, however, provides greater flexibility for operative procedures and better visualization of pelvic organs (see the blue box "Laparoscopic Examination of Pelvic Viscera," p. 397). There is also less risk of bacterial contamination of the peritoneal cavity.

A pelvic abscess in the recto-uterine pouch can be drained through an incision made in the posterior part of the vaginal fornix (*culdocentesis*). Similarly, fluid in the peritoneal cavity (e.g., blood) can be aspirated by this technique (Fig. B3.25).

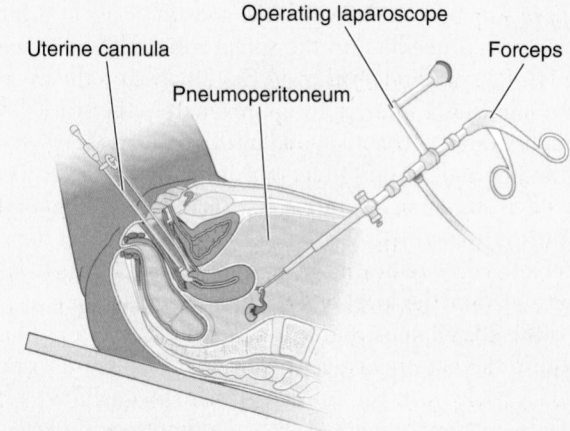

(A) Laparoscopy of pelvic viscera

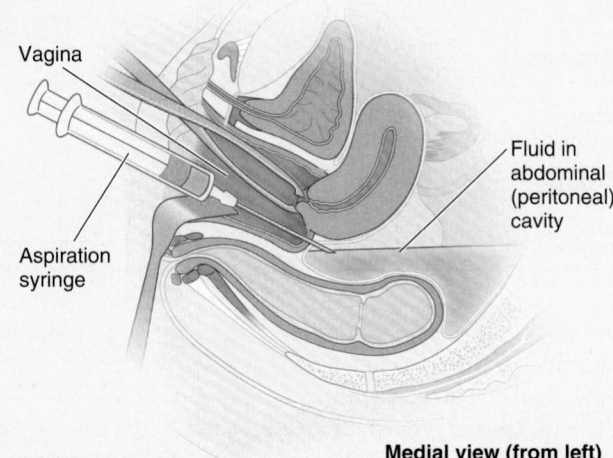

Medial view (from left)

FIGURE B3.25.

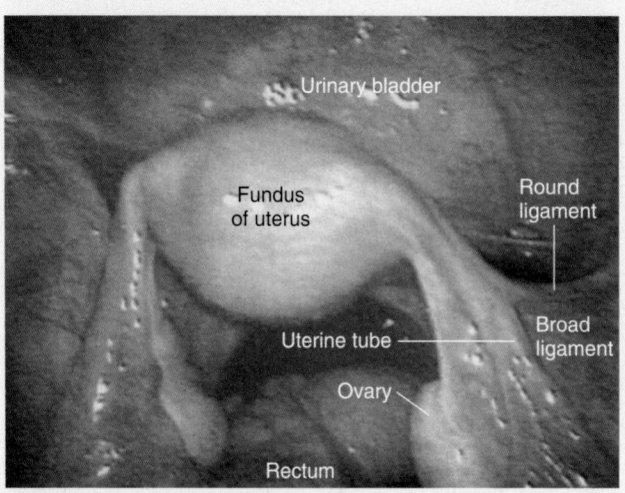

(B) Laparoscopic view of normal pelvis

FIGURE B3.26.

Laparoscopic Examination of Pelvic Viscera

Visual examination of the pelvic viscera is especially useful in diagnosing many conditions affecting the pelvic viscera, such as *ovarian cysts* and tumors, *endometriosis* (the presence of functioning endometrial tissue outside the uterus), and ectopic pregnancies. *Laparoscopy* involves inserting a *laparoscope* into the peritoneal cavity through a small (approximately 2 cm) incision below the umbilicus (Fig. B3.26). Insufflation of inert gas creates a pneumoperitoneum to provide space to visualize, and the pelvis is elevated so that gravity will pull the intestines into the abdomen. The uterus can be externally manipulated to facilitate visualization, or additional openings (ports) can be made to introduce other instruments for manipulation or to enable therapeutic procedures (e.g., ligation of the uterine tubes).

Anesthesia for Childbirth

Several options are available to women to reduce the pain and discomfort experienced during childbirth. *General anesthesia* has advantages for emergency procedures and for women who choose it over regional anesthesia. General anesthesia renders the mother unconscious; she is unaware of the labor and delivery. Clinicians monitor and regulate maternal respiration and both maternal and fetal cardiac function. Childbirth occurs passively under the control of maternal hormones with the assistance of an obstetrician. The mother is spared pain and discomfort but is unaware of the earliest moments of her baby's life.

Women who choose *regional anesthesia,* such as a spinal, pudendal nerve, or caudal epidural block, often wish to participate actively (e.g., using the Lamaze method) and be conscious of their uterine contractions to "bear down," or push, to assist the contractions and expel the fetus, yet do not wish to experience all the pain of labor.

Spinal anesthesia, in which the anesthetic agent is introduced with a needle into the spinal subarachnoid space at the L3–L4 vertebral level (A in Fig. B3.27), produces complete anesthesia inferior to approximately the waist level. The perineum, pelvic floor, and birth canal are anesthetized, and motor and sensory functions of the entire lower limbs, as well as sensation of uterine contractions, are temporarily eliminated. The mother is conscious, but she must depend on electronic monitoring of uterine contractions. If labor is extended or the level of anesthesia is inadequate, it may be difficult or impossible to re-administer the anesthesia. Because the anesthetic agent is heavier than cerebrospinal fluid, it remains in the inferior spinal subarachnoid space while the patient is inclined. The anesthetic agent circulates into the cerebral subarachnoid space in the cranial cavity when the patient lies flat following the delivery. A severe headache is a common sequel to spinal anesthesia.

A *pudendal nerve block* is a peripheral nerve block that provides local anesthesia over the S2–S4 dermatomes (the majority of the perineum) and the inferior quarter of the vagina (C in Fig. B3.27). It does not block pain from the superior birth canal (uterine cervix and superior vagina), so the mother is able to feel uterine contractions. It can be re-administered, but to do so may be disruptive and involve the use of a sharp instrument in close proximity to the infant's head. The anatomical basis of the administration of a pudendal block is provided in the blue box "Pudendal and Ilio-inguinal Nerve Blocks," p. 433.

The *caudal epidural block* is a popular choice for participatory childbirth (B in Fig. B3.27). It must be administered in advance of the actual delivery, which is not possible with a precipitous birth. The anesthetic agent is administered using an in-dwelling catheter in the sacral canal, enabling administration of more anesthetic agent for a deeper or more prolonged anesthesia, if necessary. Within the sacral canal, the

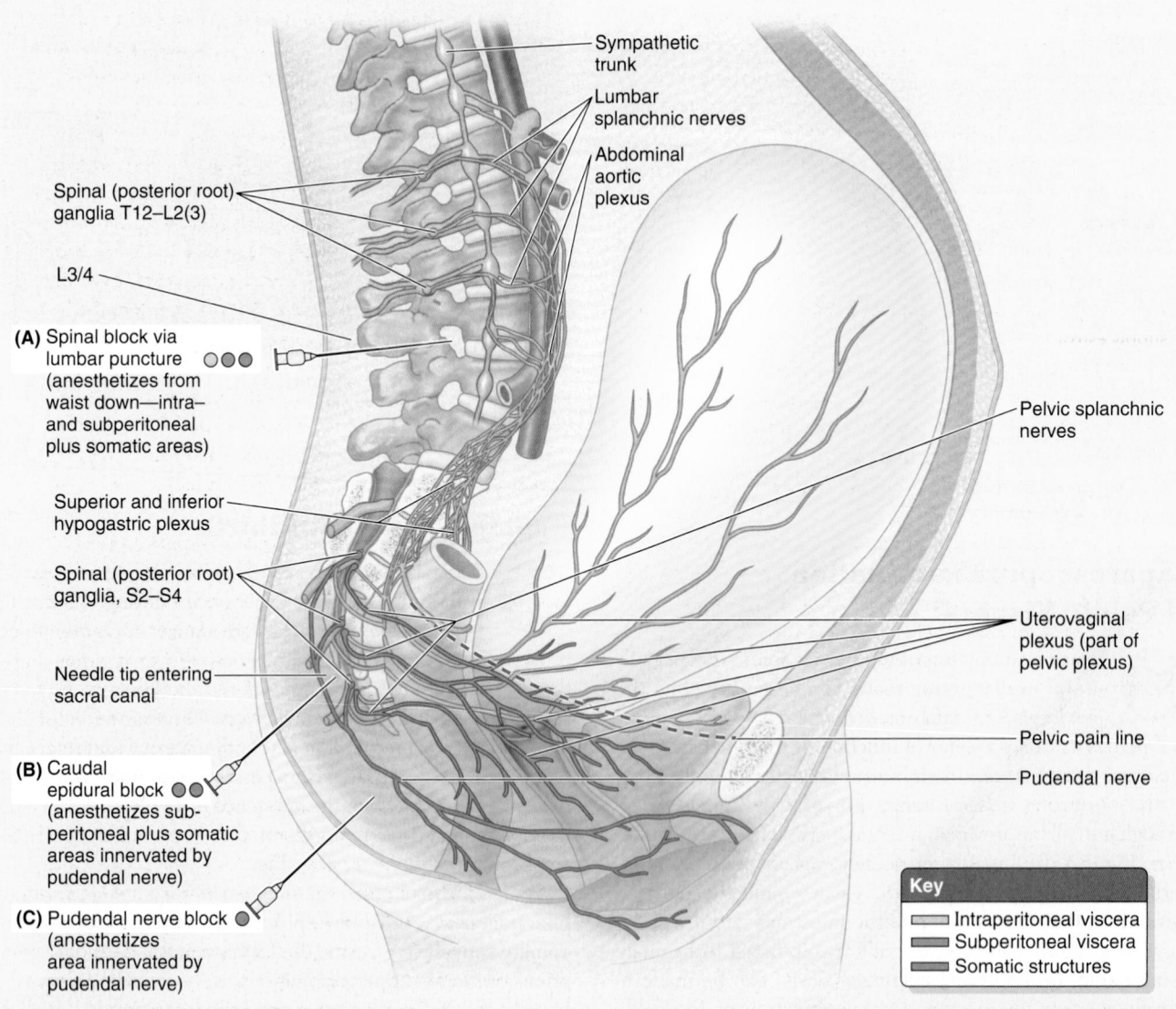

FIGURE B3.27.

anesthesia bathes the S2–S4 spinal nerve roots, including the pain fibers from the uterine cervix and superior vagina, and the afferent fibers from the pudendal nerve. Thus the entire birth canal, pelvic floor, and majority of the perineum are anesthetized, but the lower limbs are not usually affected. The pain fibers from the uterine body (superior to the *pelvic pain line*) ascend to the inferior thoracic-superior lumbar levels; these and the fibers superior to them are not affected by the anesthetic, so the mother is aware of her uterine contractions. With epidural anesthesia, no "spinal headache" occurs because the vertebral epidural space is not continuous with the cranial extradural (epidural) space (see Chapter 4).

The Bottom Line

FEMALE INTERNAL GENITAL ORGANS

Ovaries and uterine tubes: The ovaries are suspended by two peritoneal folds: the mesovarium from the posterosuperior aspect of the broad ligament, and the vascular suspensory ligament of the ovary from the lateral wall of the pelvis. ♦ They are attached to the uterus by the ligaments of the ovaries. ♦ The peritoneum ends at the ovary itself. It is replaced on the surface of the ovary with a duller, cuboidal epithelium.

The uterine tubes are the conduits and the site of fertilization for oocytes discharged into the peritoneal cavity. ♦ Coursing in a peritoneal fold (mesosalpinx) that makes up the superior margin of the broad ligament, each uterine tube has a fimbriated, funnel-like infundibulum, a wide ampulla, a narrow isthmus, and a short uterine part that traverses the uterine wall to enter the cavity.

The ovaries and uterine tubes receive a double (collateral) blood supply from the abdominal aorta via the ovarian arteries and from the internal iliac arteries via the uterine arteries. ♦ This collateral circulation allows the ovaries to be spared to supply estrogen when a hysterectomy necessitates ligation of the uterine arteries. ♦ Sympathetic and visceral afferent pain fibers travel with the ovarian vessels. ♦ Parasympathetic and visceral afferent reflex fibers traverse pelvic plexuses and pelvic splanchnic nerves.

Uterus: Shaped like an inverted pear, the uterus is the organ in which the blastocyst (early embryo) implants and develops into a mature embryo and then a fetus. ♦ Although its size and proportions change during the various phases of life, the non-gravid uterus consists of a body and cervix, demarcated by a relatively narrow isthmus. ♦ The uterus has a trilaminar wall consisting of (1) an inner vascular and secretory endometrium, which undergoes cyclical changes to prepare for implantation to occur and sheds with menstrual flow if it does not; (2) a hormonally stimulated intermediate smooth muscle myometrium, which dilates the cervical canal (exit) and expels the fetus during childbirth; and (3) visceral peritoneum (perimetrium), which covers most of the fundus and body (except for a bare area abutting the bladder) and continues bilaterally as the broad ligament (mesometrium).

The uterus is normally anteverted and anteflexed so that its weight is borne largely by the urinary bladder, although it also receives significant passive support from the cardinal ligaments and active support from the muscles of the pelvic floor. ♦ The uterine artery supplies the uterus and, during pregnancy, the placenta. ♦ The uterine veins drain to the uterovaginal venous plexus.

Vagina: The vagina is a musculomembranous passage connecting the uterine cavity to the exterior, allowing the entrance/insertion of the penis, ejaculate, tampons, or examining digits and the exit of a fetus or menstrual fluid. ♦ The vagina lies between and is closely related to the urethra anteriorly and rectum posteriorly, but is separated from the latter by the peritoneal recto-uterine pouch superiorly and the fascial rectovaginal septum inferiorly. The vagina is indented (invaginated) anterosuperiorly by the uterine cervix so that an encircling pocket or vaginal fornix is formed around it. ♦ Most of the vagina is located within the pelvis, receiving blood via pelvic branches of the internal iliac arteries (uterine and vaginal arteries) and draining directly into the uterovaginal venous plexus and, via deep (pelvic) routes, to the internal and external iliac and sacral lymph nodes. ♦ The inferiormost part of the vagina is located within the perineum, receiving blood from the internal pudendal artery and draining via superficial (perineal) routes into superficial inguinal nodes. ♦ The vagina is capable of remarkable distension, enabling manual examination (palpation) of pelvic landmarks and viscera (especially the ovaries) as well as of pathology (e.g., ovarian cysts).

Innervation of uterus and vagina: The inferiormost (perineal) portion of the vagina receives somatic innervation via the pudendal nerve (S2–S4) and is, therefore, sensitive to touch and temperature. ♦ The remainder of the vagina and uterus is pelvic and thus visceral in its location, receiving innervation from autonomic and visceral afferent fibers. ♦ All unconscious, reflex-type sensation travels retrogradely along the parasympathetic pathways to the S2–S4 spinal sensory ganglia, as does pain sensation arising in the subperitoneal uterus (primarily cervix) and vagina (inferior to the pelvic pain line)—that is, from the birth canal. ♦ However, pain sensation from the intraperitoneal uterus (superior to the pelvic pain line) travels retrogradely along the sympathetic pathway to the inferiormost thoracic and superior lumbar spinal ganglia. ♦ Epidural anesthesia may be administered to take advantage of the discrepancy in the pain pathways to facilitate participatory childbirth methods; uterine contractions are felt, but the birth canal is anesthetized.

Lymphatic Drainage of Pelvic Viscera

For the main part, the lymphatic vessels of the pelvis follow the venous system, following the tributaries of the internal iliac vein to the internal iliac nodes, directly or via the sacral lymph nodes (Fig. 3.48). However, structures located superiorly in the anterior portion of the pelvis drain to the external iliac nodes, a lymphatic pathway that does not parallel venous drainage. From both external and internal iliac nodes, lymph flows via common iliac and lumbar (caval/aortic) lymph nodes, draining via lumbar lymphatic trunks into the cisterna chyli.

LYMPHATIC DRAINAGE FROM URINARY SYSTEM

The superior portion of the pelvic part of the ureters drains primarily to external iliac nodes, while the inferior portion drains to the internal iliac nodes (Fig. 3.48A; Table 3.7). Lymphatic vessels from the superolateral aspects of the bladder pass to the *external iliac lymph nodes,* whereas those from the fundus and neck pass to the *internal iliac lymph nodes.* Some vessels from the neck of the bladder drain into the sacral or common iliac lymph nodes. Most lymphatic vessels from the female urethra and proximal part of the male urethra pass to

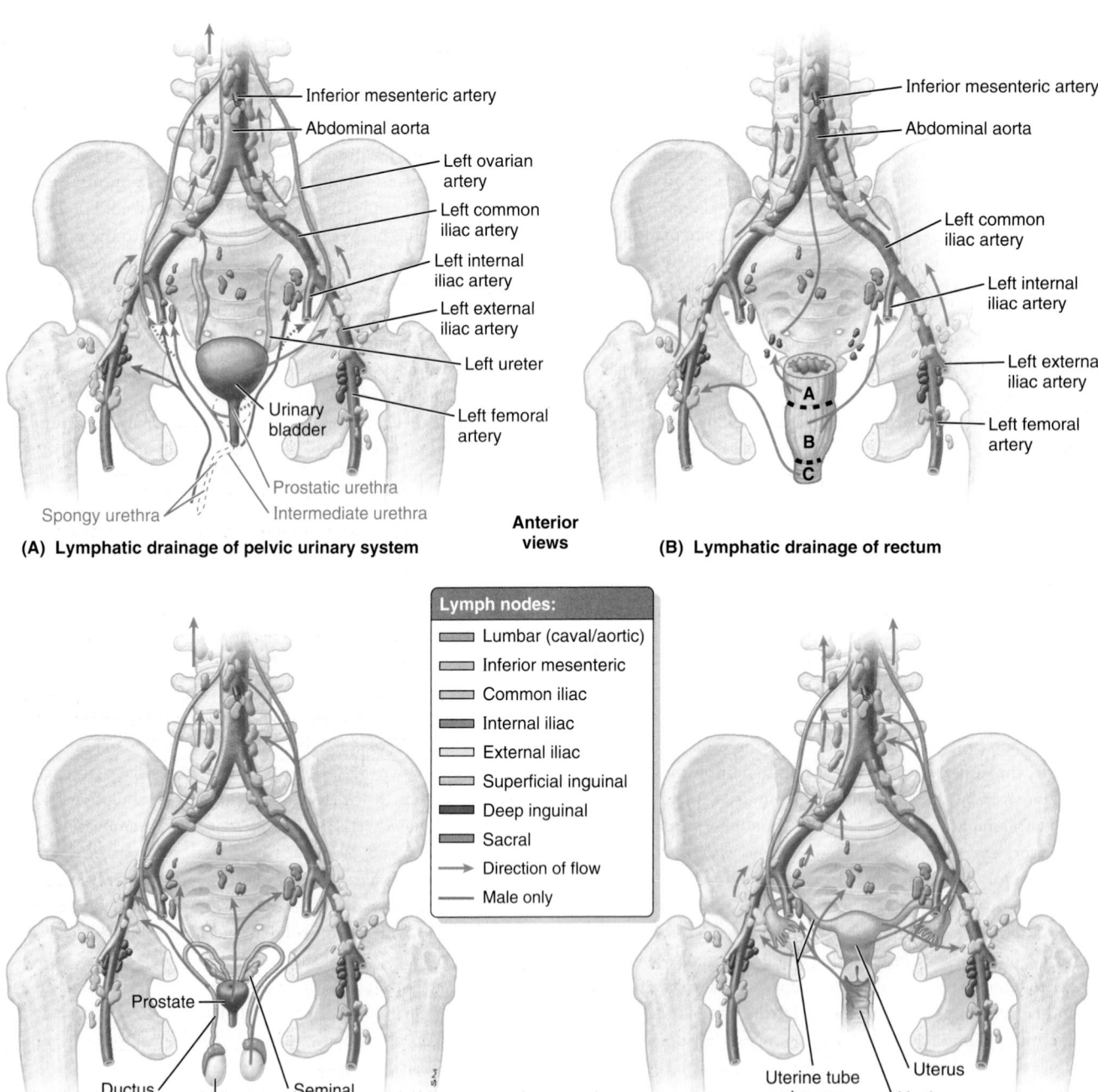

(A) Lymphatic drainage of pelvic urinary system

(B) Lymphatic drainage of rectum

Anterior views

Lymph nodes:
- Lumbar (caval/aortic)
- Inferior mesenteric
- Common iliac
- Internal iliac
- External iliac
- Superficial inguinal
- Deep inguinal
- Sacral
- Direction of flow
- Male only

(C) Lymphatic drainage of male internal genital organs

(D) Lymphatic drainage of female internal genital organs

FIGURE 3.48. Lymphatic drainage of pelvic viscera.

TABLE 3.7. LYMPHATIC DRAINAGE OF STRUCTURES OF PELVIS AND PERINEUM

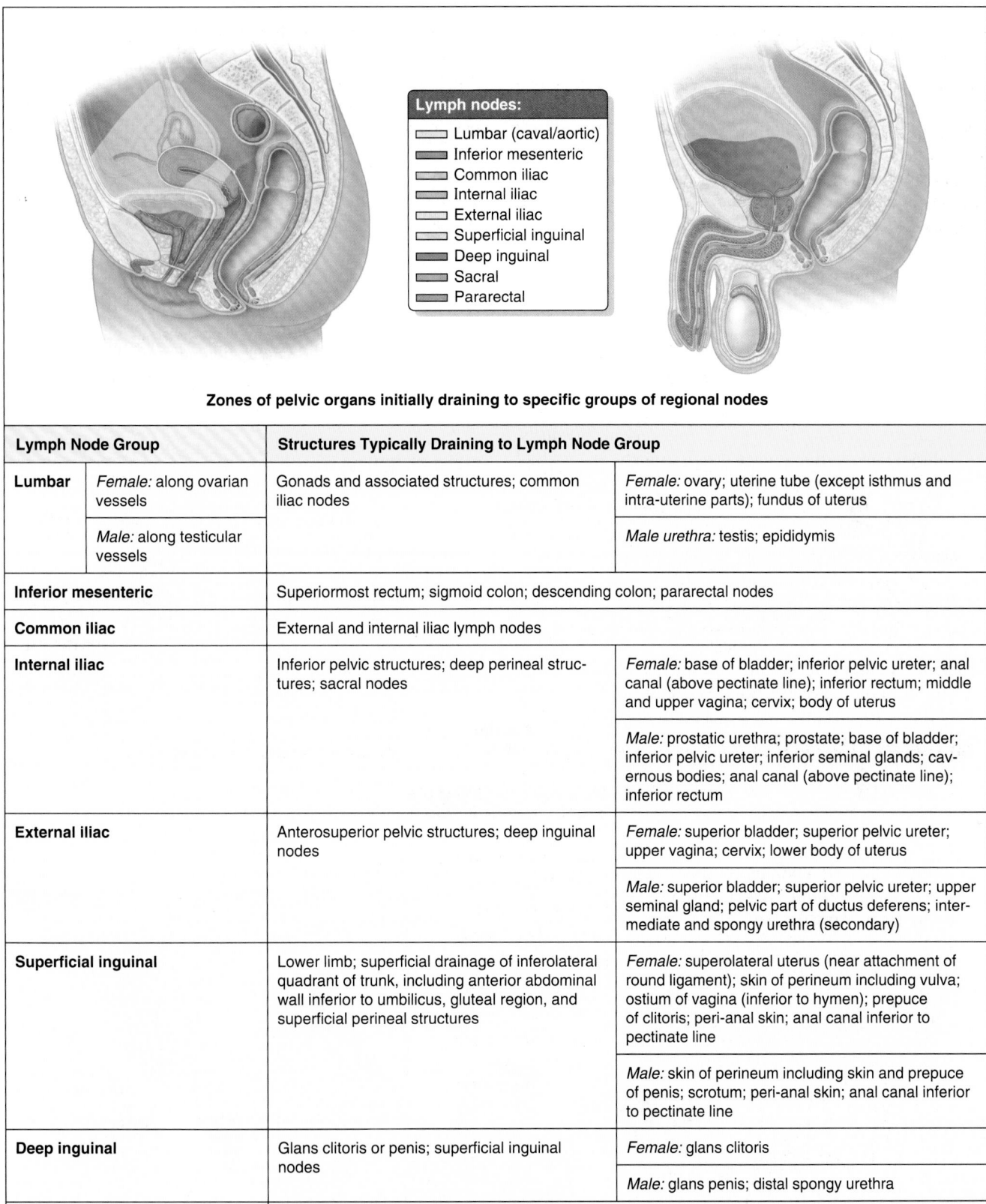

Zones of pelvic organs initially draining to specific groups of regional nodes

Lymph nodes:
- Lumbar (caval/aortic)
- Inferior mesenteric
- Common iliac
- Internal iliac
- External iliac
- Superficial inguinal
- Deep inguinal
- Sacral
- Pararectal

Lymph Node Group		Structures Typically Draining to Lymph Node Group		
Lumbar	*Female:* along ovarian vessels	Gonads and associated structures; common iliac nodes	*Female:* ovary; uterine tube (except isthmus and intra-uterine parts); fundus of uterus	
	Male: along testicular vessels		*Male urethra:* testis; epididymis	
Inferior mesenteric		Superiormost rectum; sigmoid colon; descending colon; pararectal nodes		
Common iliac		External and internal iliac lymph nodes		
Internal iliac		Inferior pelvic structures; deep perineal structures; sacral nodes	*Female:* base of bladder; inferior pelvic ureter; anal canal (above pectinate line); inferior rectum; middle and upper vagina; cervix; body of uterus	
			Male: prostatic urethra; prostate; base of bladder; inferior pelvic ureter; inferior seminal glands; cavernous bodies; anal canal (above pectinate line); inferior rectum	
External iliac		Anterosuperior pelvic structures; deep inguinal nodes	*Female:* superior bladder; superior pelvic ureter; upper vagina; cervix; lower body of uterus	
			Male: superior bladder; superior pelvic ureter; upper seminal gland; pelvic part of ductus deferens; intermediate and spongy urethra (secondary)	
Superficial inguinal		Lower limb; superficial drainage of inferolateral quadrant of trunk, including anterior abdominal wall inferior to umbilicus, gluteal region, and superficial perineal structures	*Female:* superolateral uterus (near attachment of round ligament); skin of perineum including vulva; ostium of vagina (inferior to hymen); prepuce of clitoris; peri-anal skin; anal canal inferior to pectinate line	
			Male: skin of perineum including skin and prepuce of penis; scrotum; peri-anal skin; anal canal inferior to pectinate line	
Deep inguinal		Glans clitoris or penis; superficial inguinal nodes	*Female:* glans clitoris	
			Male: glans penis; distal spongy urethra	
Sacral		Postero-inferior pelvic structures: inferior rectum; inferior vagina		
Pararectal		Superior rectum		

the *internal iliac lymph node.* However, a few vessels from the female urethra may also drain into the sacral nodes and, from the distal female urethra, to the *inguinal lymph nodes.*

LYMPHATIC DRAINAGE FROM RECTUM

Lymphatic vessels from the superior rectum pass to *inferior mesenteric lymph nodes,* many passing through **pararectal lymph nodes** (located directly on the muscle layer of the rectum) and/or sacral lymph nodes en route (Fig. 3.48B; Table 3.7). The inferior mesenteric nodes drain into the *lumbar (caval/aortic) lymph nodes.* Lymphatic vessels from the inferior half of the rectum drain directly to *sacral lymph nodes* or, especially from the distal ampulla, follow the middle rectal vessels to drain into the *internal iliac lymph nodes.*

LYMPHATIC DRAINAGE FROM
MALE PELVIC VISCERA

Lymphatic vessels from the ductus deferens, ejaculatory ducts, and inferior parts of the seminal glands drain to the *external iliac lymph nodes* (Fig. 3.48C; Table 3.7). The lymphatic vessels from the superior parts of the seminal glands and prostate terminate chiefly in the *internal iliac lymph nodes,* but some drainage from the latter may pass to the *sacral nodes.*

LYMPHATIC DRAINAGE FROM
FEMALE PELVIC VISCERA

Lymphatic vessels from the ovaries, joined by vessels from the uterine tubes and most from the fundus of the uterus, follow the ovarian veins as they ascend to the *right* and *left lumbar (caval/aortic) lymph nodes* (Fig. 3.48D; Table 3.7).

Lymphatic vessels from the uterus drain in many directions, coursing along the blood vessels that supply it as well as the ligaments attached to it:

- Most lymphatic vessels from the fundus and superior uterine body pass along the ovarian vessels to the *lumbar (caval/aortic) lymph nodes,* but some vessels from the fundus, particularly those near the entrance of the uterine

tubes and attachments of the round ligaments, run along the round ligament of the uterus to the *superficial inguinal lymph nodes.*
- Vessels from most of the uterine body and some from the cervix pass within the broad ligament to the *external iliac lymph nodes.*
- Vessels from the uterine cervix also pass along the uterine vessels, within the transverse cervical ligaments, to the *internal iliac lymph nodes,* and along uterosacral (sacrogenital) ligaments to the *sacral lymph nodes.*

Lymphatic vessels from the vagina drain from the parts of the vagina as follows:

- Superior part: to the internal and external iliac lymph nodes.
- Middle part: to the internal iliac lymph nodes.
- Inferior part: to the sacral and common iliac nodes (Fig. 3.48; Table 3.7).
- External orifice: to the superficial inguinal lymph nodes.

PERINEUM

The **perineum** refers to a shallow compartment of the body (*perineal compartment*) bounded by the pelvic outlet and separated from the pelvic cavity by the fascia covering the inferior aspect of the pelvic diaphragm, formed by the levator ani and coccygeus muscles (Fig. 3.49). In the anatomical position, the surface of the perineum—the **perineal region**—is the narrow region between the proximal parts of the thighs; however, when the lower limbs are abducted, it is a diamond-shaped area extending from the mons pubis anteriorly in females, the medial surfaces (insides) of the thighs laterally, and the gluteal folds and superior end of the intergluteal (natal) cleft posteriorly (Fig. 3.50).

The osseofibrous structures marking the boundaries of the perineum (perineal compartment) (Fig. 3.51A & B) are the:

- *Pubic symphysis,* anteriorly.
- **Ischiopubic rami** (combined *inferior pubic rami* and *ischial rami*), anterolaterally.

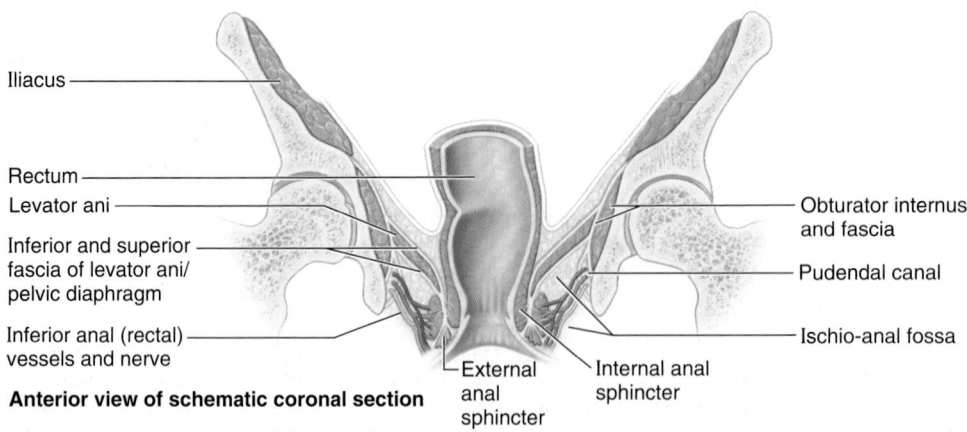

Iliacus

Rectum
Levator ani

Inferior and superior
fascia of levator ani/
pelvic diaphragm

Inferior anal (rectal)
vessels and nerve

Obturator internus
and fascia

Pudendal canal

Ischio-anal fossa

External anal sphincter

Internal anal sphincter

Anterior view of schematic coronal section

FIGURE 3.49. The inferior fascia of the pelvic diaphragm (levator ani) is the boundary separating pelvis from perineum.

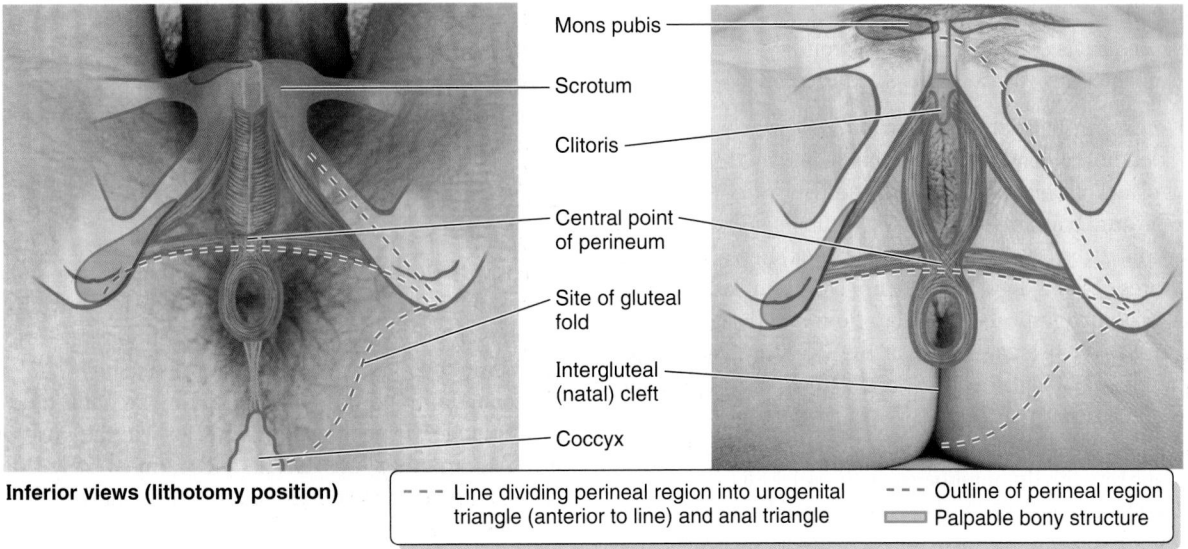

Inferior views (lithotomy position)

- - - Line dividing perineal region into urogenital   - - - Outline of perineal region
triangle (anterior to line) and anal triangle   ▭ Palpable bony structure

FIGURE 3.50. Male and female perineal regions. Boundaries and surface features of the perineal region with projections of the osseous boundaries and muscles of the superficial muscles of the perineum. The penis and some of the scrotum (part of the perineal region) are retracted anteriorly and thus not shown.

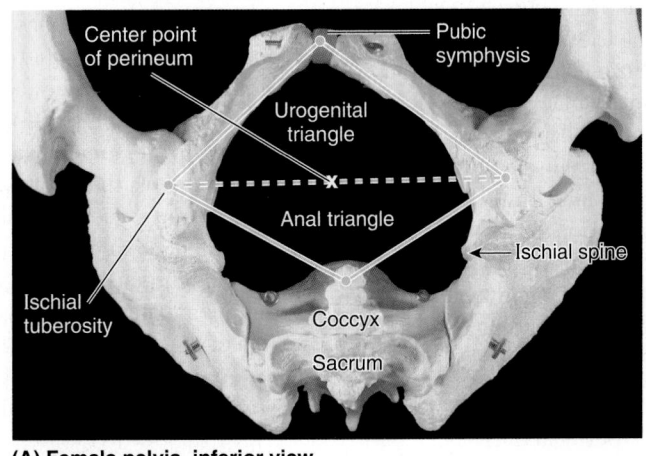

(A) Female pelvis, inferior view

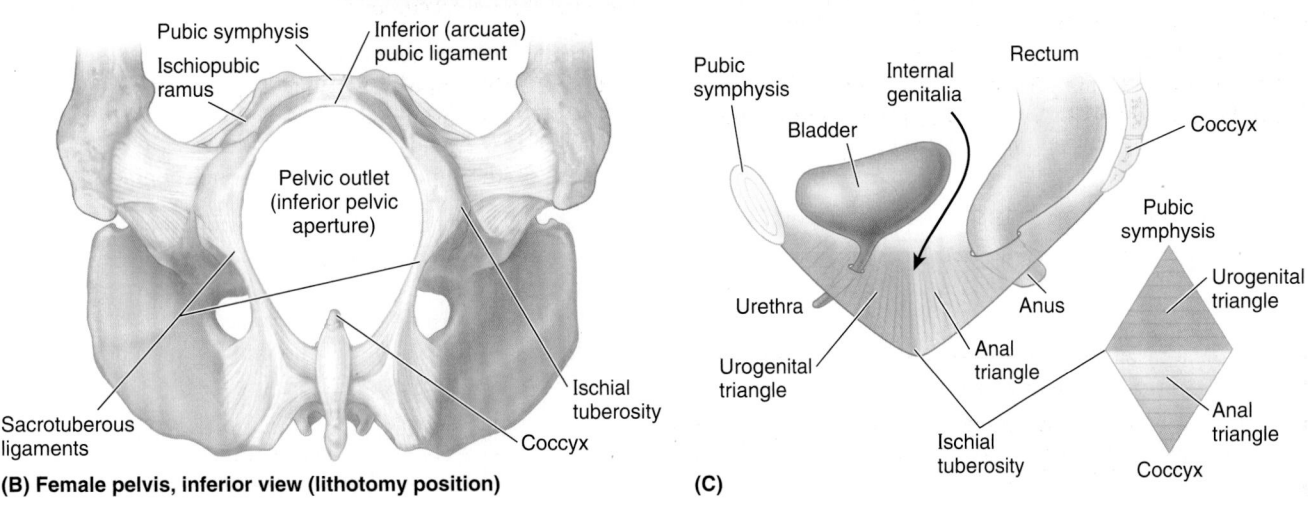

(B) Female pelvis, inferior view (lithotomy position)   **(C)**

FIGURE 3.51. Boundaries and disposition of perineum. A. Pelvic girdle demonstrating bony features bounding the perineum. The two triangles comprising the diamond-shaped perineum are superimposed. **B.** The osseofibrous structures bounding the pelvic outlet and perineum are identified. This view of the female pelvis is the one obstetricians visualize when the patient is on the examining table. **C.** The two triangles (urogenital and anal) that together comprise the perineum do not occupy the same plane. The plane between the bladder and rectum is occupied by internal genitalia and a septum formed during embryonic development as the urogenital sinus was partitioned into the urinary bladder and urethra anteriorly and the anorectum posteriorly.

- *Ischial tuberosities,* laterally.
- *Sacrotuberous ligaments,* posterolaterally.
- Inferiormost *sacrum* and *coccyx,* posteriorly.

A transverse line joining the anterior ends of the ischial tuberosities divides the diamond-shaped perineum into two triangles, the oblique planes of which intersect at the transverse line (Fig. 3.51B & C). The **anal triangle** lies posterior to this line. The anal canal and its orifice, the anus, constitute the major deep and superficial features of the triangle, lying centrally surrounded by ischioanal fat. The **urogenital (UG) triangle** is anterior to this line. In contrast to the open anal triangle, the UG triangle is "closed" by a thin sheet of tough, deep fascia, the **perineal membrane,** which stretches between the two sides of the pubic arch, covering the anterior part of the pelvic outlet (Fig. 3.52C). The perineal membrane thus fills the anterior gap in the pelvic diaphragm (the urogenital hiatus, Fig. 3.52A), but is perforated by the urethra in both sexes and by the vagina of the female. The membrane and the ischiopubic rami to which it attaches provide a foundation for the erectile bodies of the external genitalia—the penis and scrotum of males, and the pudendum or vulva of females—which are the superficial features of the triangle (Fig. 3.50).

The midpoint of the line joining the ischial tuberosities is the **central point of the perineum.** This is the location of the **perineal body,** which is an irregular mass, variable in size and consistency, and containing collagenous and elastic fibers, and both skeletal and smooth muscle (Fig. 3.52E). The perineal body lies deep to the skin, with relatively little overlying subcutaneous tissue, posterior to the vestibule of the vagina or bulb of the penis and anterior to the anus and anal canal. The perineal body is the site of convergence and interlacing of fibers of several muscles, including the:

- Bulbospongiosus.
- External anal sphincter.
- Superficial and deep transverse perineal muscles.
- Smooth and voluntary slips of muscle from the external urethral sphincter, levator ani, and muscular coats of the rectum.

Anteriorly, the perineal body blends with the posterior border of the perineal membrane and superiorly with the rectovesical or rectovaginal septum (Fig. 3.53A & B).

Fasciae and Pouches of Urogenital Triangle

PERINEAL FASCIAE[2]

The perineal fascia consists of superficial and deep layers. The **subcutaneous tissue of the perineum,** like that of the inferior anterior abdominal wall consists of a superficial *fatty layer* and a deep *membranous layer,* the (superficial) **perineal fascia** (Colles fascia).

[2]The terminology used in this section (in boldface) was recommended by the *Federative International Committee on Anatomical Terminology* (FICAT) in 1998; however, because many clinicians concerned with the perineum use eponyms, the authors have placed commonly used terms in parentheses so that the FICAT terminology will be understood by all readers.

In females, the **fatty layer of subcutaneous tissue of the perineum** makes up the substance of the labia majora and mons pubis, and is continuous anteriorly and superiorly with the fatty layer of subcutaneous tissue of the abdomen (Camper fascia) (Fig. 3.53A & C). In males, the fatty layer is greatly diminished in the urogenital triangle, being replaced altogether in the penis and scrotum with smooth (dartos) muscle. It is continuous between the penis or scrotum and thighs with the fatty layer of subcutaneous tissue of the abdomen (Fig. 3.53B & D). In both sexes, the fatty layer of subcutaneous tissue of the perineum is continuous posteriorly with the ischio-anal fat pad in the anal region (Fig. 3.53E).

The membranous **perineal fascia** does not extend into the anal triangle, being attached posteriorly to the posterior margin of the perineal membrane and perineal body (Fig. 3.53A & B). Laterally it is attached to the fascia lata (deep fascia) of the superiormost medial aspect of the thigh (Fig. 3.53C & E). Anteriorly in males, the perineal fascia is continuous with the **dartos fascia** of the penis and scrotum; however, on each side of and anterior to the scrotum, the perineal fascia becomes continuous with the membranous layer of subcutaneous tissue of the abdomen (Scarpa fascia) (Fig. 3.53B). In females, the perineal fascia passes superior to the fatty layer forming the labia majora and becomes continuous with the membranous layer of subcutaneous tissue of the abdomen (Fig. 3.53A & C).

The **deep perineal fascia** (investing or Gallaudet fascia) intimately invests the ischiocavernosus, bulbospongiosus, and superficial transverse perineal muscles (Fig. 3.53A & E). It is also attached laterally to the ischiopubic rami. Anteriorly it is fused to the suspensory ligament of the penis (see Fig. 3.63) and is continuous with the deep fascia covering the external oblique muscle of the abdomen and the rectus sheath. In females, the deep perineal fascia is fused with the suspensory ligament of the clitoris and, as in males, with the deep fascia of the abdomen.

SUPERFICIAL PERINEAL POUCH

The **superficial perineal pouch** (space or compartment) is a potential space between the perineal fascia and the perineal membrane, bounded laterally by the ischiopubic rami (Figs. 3.52D & E and 3.53).

In males, the superficial perineal pouch contains the:

- *Root* (bulb and crura) *of the penis* and associated muscles (*ischiocavernosus* and *bulbospongiosus*).
- Proximal (bulbous) part of the *spongy urethra.*
- *Superficial transverse perineal muscles.*
- *Deep perineal branches* of the internal pudendal vessels and pudendal nerves.

In females, the superficial perineal pouch contains the:

- *Clitoris* and associated muscle (ischiocavernosus).
- *Bulbs of the vestibule* and surrounding muscle (bulbospongiosus).
- *Greater vestibular glands.*
- *Superficial transverse perineal muscles.*
- Related vessels and nerves (*deep perineal branches* of the internal pudendal vessels and pudendal nerves).

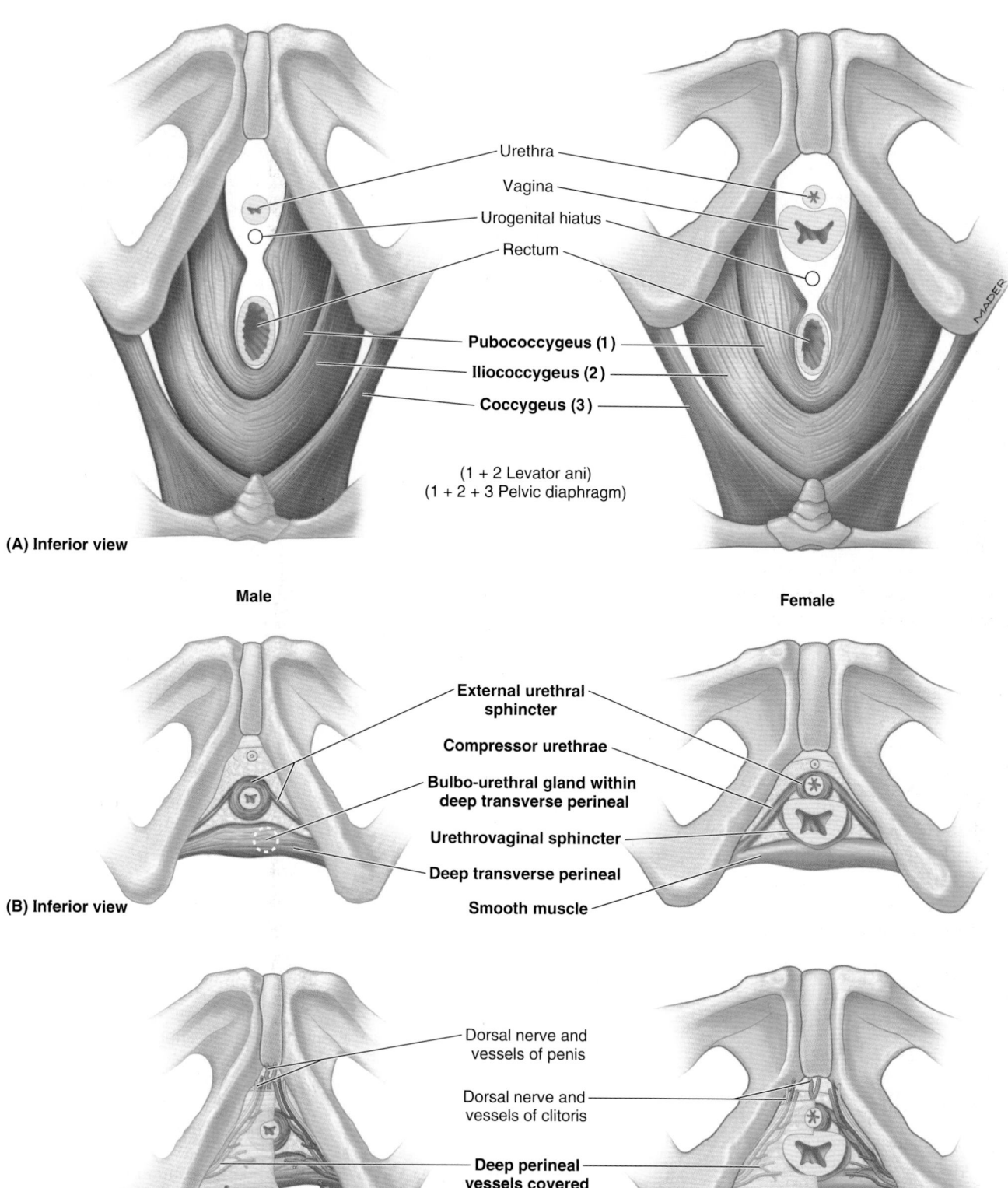

(A) Inferior view

Urethra
Vagina
Urogenital hiatus
Rectum
Pubococcygeus (1)
Iliococcygeus (2)
Coccygeus (3)

(1 + 2 Levator ani)
(1 + 2 + 3 Pelvic diaphragm)

Male

Female

External urethral sphincter
Compressor urethrae
Bulbo-urethral gland within deep transverse perineal
Urethrovaginal sphincter
Deep transverse perineal
Smooth muscle

(B) Inferior view

Dorsal nerve and vessels of penis
Dorsal nerve and vessels of clitoris
Deep perineal vessels covered with perineal membrane

(C) Inferior view

FIGURE 3.52. Layers of perineum of males and females. The layers are shown as being built up from deep (**A**) to superficial (**E**) layers. **A.** The pelvic outlet is almost closed by the pelvic diaphragm (levator ani and coccygeus muscles), forming the floor of the pelvic cavity and, as viewed here, the roof of the perineum. The urethra (and vagina in females) and rectum pass through the urogenital hiatus of the pelvic diaphragm. **B** and **C.** The external urethral sphincter and deep transverse perineal muscle span the region of the urogenital hiatus, which is closed inferiorly by the perineal membrane extending between the ischiopubic rami.

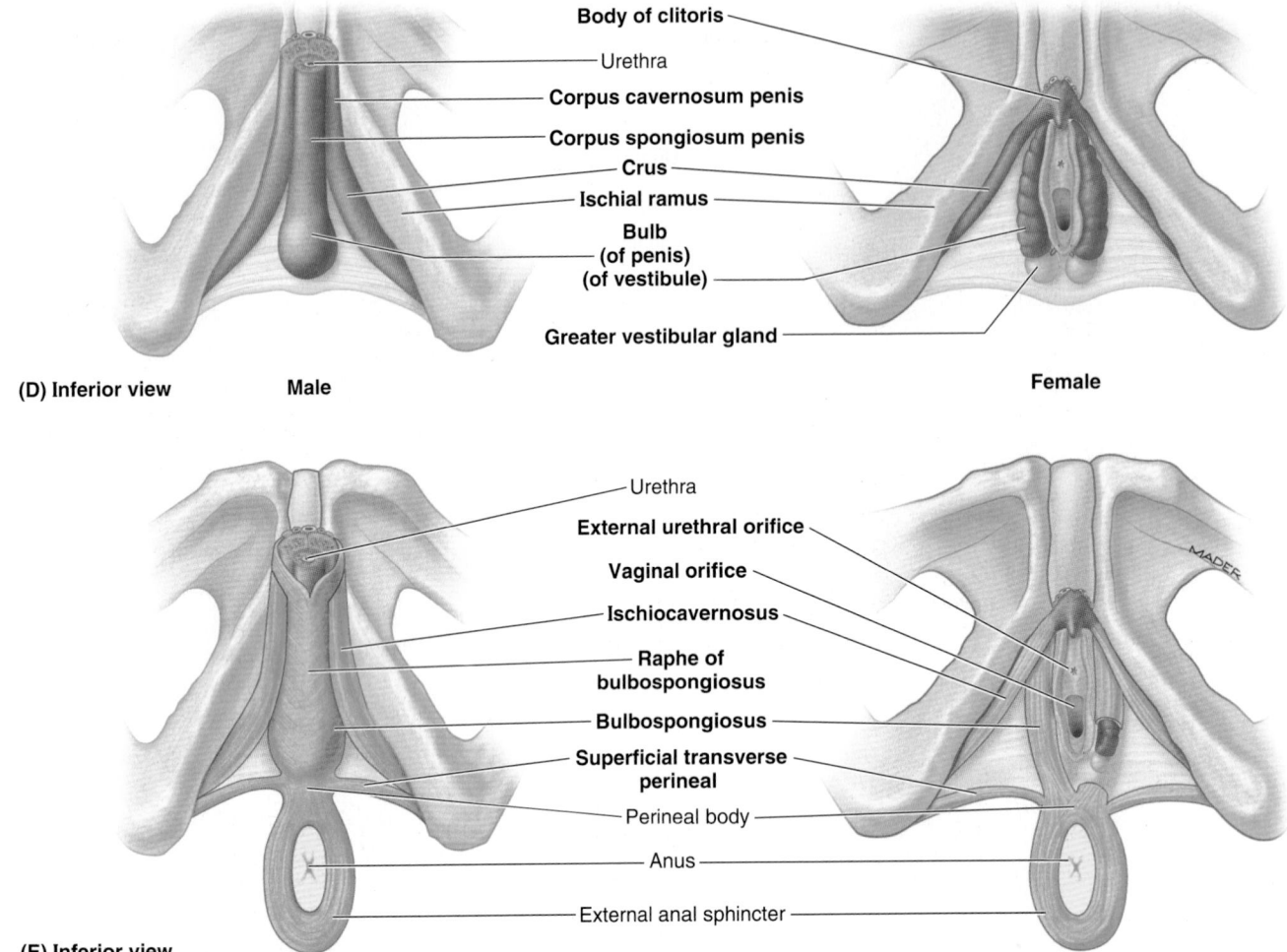

(D) Inferior view   **Male**

Body of clitoris
Urethra
Corpus cavernosum penis
Corpus spongiosum penis
Crus
Ischial ramus
Bulb
(of penis)
(of vestibule)
Greater vestibular gland

Female

(E) Inferior view

Urethra
External urethral orifice
Vaginal orifice
Ischiocavernosus
Raphe of bulbospongiosus
Bulbospongiosus
Superficial transverse perineal
Perineal body
Anus
External anal sphincter

FIGURE 3.52. (*Continued*) **D and E.** Inferior to the perineal membrane, the superficial perineal pouch (space) contains the erectile bodies and the muscles associated with them.

The structures of the superficial perineal pouch will be discussed in greater detail, specific to each sex, under "Male Perineum" and "Female Perineum," later in this chapter.

DEEP PERINEAL POUCH

The **deep perineal pouch** (space) is bounded inferiorly by the perineal membrane, superiorly by the inferior fascia of the pelvic diaphragm, and laterally by the inferior portion of the obturator fascia (covering the obturator internus muscle) (Fig. 3.53C & D). It includes the fat-filled anterior recesses of the ischio-anal fossae. The superior boundary in the region of the urogenital hiatus is indistinct.

In both sexes, the deep perineal pouch contains:

• Part of the urethra, centrally.
• The inferior part of the external urethral sphincter muscle, above the center of the perineal membrane, surrounding the urethra.
• Anterior extensions of the ischio-anal fat pads.

In males, the deep perineal pouch contains the:

• *Intermediate part of the urethra,* the narrowest part of the male urethra.
• *Deep transverse perineal muscles,* immediately superior to the perineal membrane (on its superior surface), running transversely along its posterior aspect.
• *Bulbo-urethral glands,* embedded within the deep perineal musculature.
• Dorsal neurovascular structures of the penis.

In females, the deep perineal pouch contains the:

• Proximal part of the *urethra.*
• A mass of smooth muscle in the place of deep transverse perineal muscles on the posterior edge of the perineal membrane, associated with the perineal body.
• Dorsal neurovasculature of the clitoris.

Past Concept of Deep Perineal Pouch and External Urethral Sphincter. Traditionally, a trilaminar, triangular *UG diaphragm* has been described as making up the deep

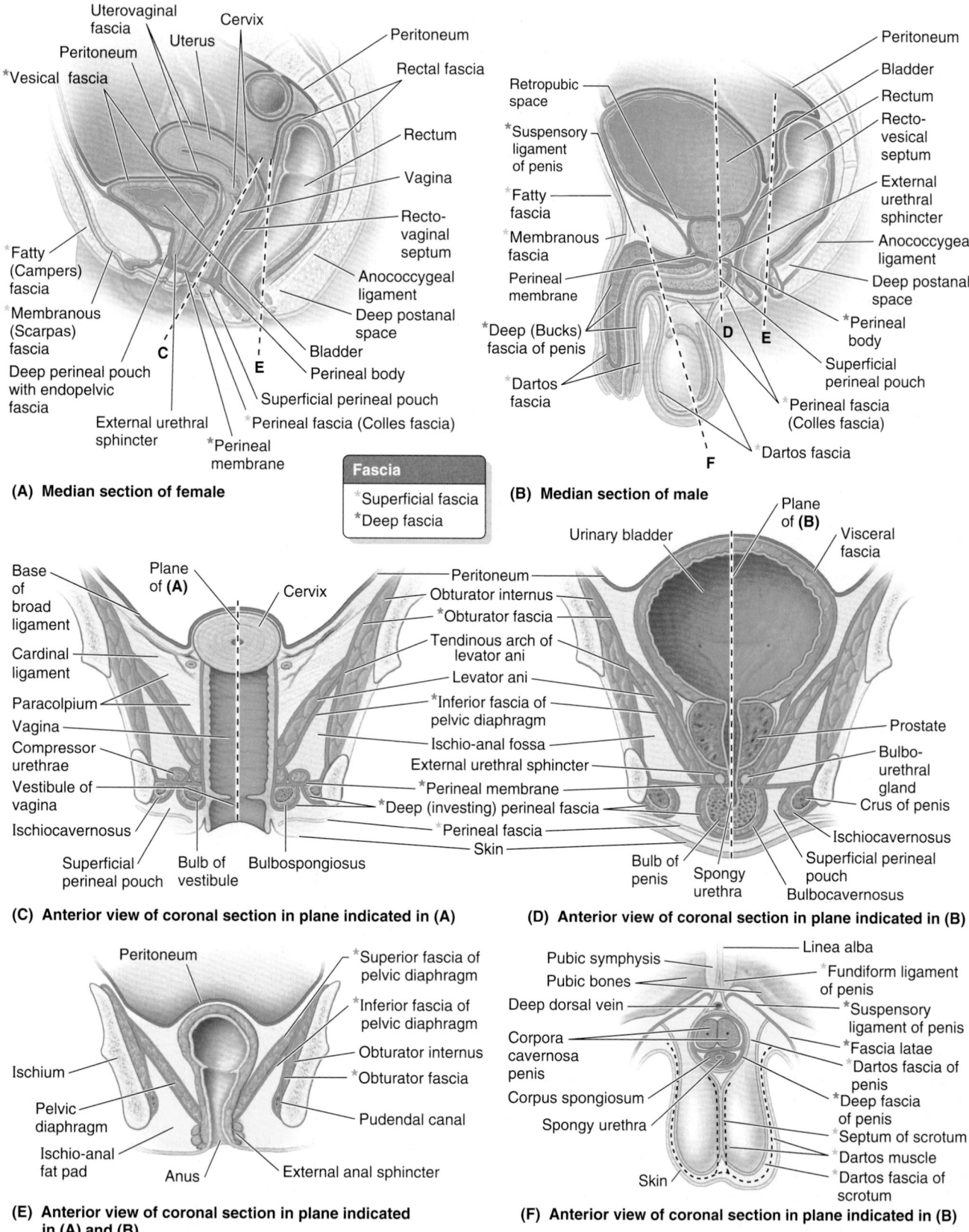

(A) Median section of female

Uterovaginal fascia
Cervix
Uterus
Peritoneum
*Vesical fascia
Peritoneum
Rectal fascia
Rectum
Vagina
Recto-vaginal septum
Anococcygeal ligament
Deep postanal space
Bladder
Perineal body
Superficial perineal pouch
*Perineal fascia (Colles fascia)
*Fatty (Campers) fascia
*Membranous (Scarpas) fascia
Deep perineal pouch with endopelvic fascia
External urethral sphincter
*Perineal membrane

(B) Median section of male

Retropubic space
*Suspensory ligament of penis
*Fatty fascia
*Membranous fascia
Perineal membrane
*Deep (Bucks) fascia of penis
*Dartos fascia
Peritoneum
Bladder
Rectum
Recto-vesical septum
External urethral sphincter
Anococcygeal ligament
Deep postanal space
*Perineal body
Superficial perineal pouch
*Perineal fascia (Colles fascia)
*Dartos fascia

Fascia
*Superficial fascia
*Deep fascia

(C) Anterior view of coronal section in plane indicated in (A)

Base of broad ligament
Plane of (A)
Cervix
Cardinal ligament
Paracolpium
Vagina
Compressor urethrae
Vestibule of vagina
Ischiocavernosus
Superficial perineal pouch
Bulb of vestibule
Bulbospongiosus
Peritoneum
Obturator internus
*Obturator fascia
Tendinous arch of levator ani
Levator ani
*Inferior fascia of pelvic diaphragm
Ischio-anal fossa
External urethral sphincter
*Perineal membrane
*Deep (investing) perineal fascia
*Perineal fascia
Skin

(D) Anterior view of coronal section in plane indicated in (B)

Plane of (B)
Urinary bladder
Visceral fascia
Prostate
Bulbo-urethral gland
Crus of penis
Ischiocavernosus
Superficial perineal pouch
Bulbocavernosus
Bulb of penis
Spongy urethra

(E) Anterior view of coronal section in plane indicated in (A) and (B)

Peritoneum
Ischium
Pelvic diaphragm
Ischio-anal fat pad
Anus
*Superior fascia of pelvic diaphragm
*Inferior fascia of pelvic diaphragm
Obturator internus
*Obturator fascia
Pudendal canal
External anal sphincter

(F) Anterior view of coronal section in plane indicated in (B)

Pubic symphysis
Pubic bones
Deep dorsal vein
Corpora cavernosa penis
Corpus spongiosum
Spongy urethra
Skin
Linea alba
*Fundiform ligament of penis
*Suspensory ligament of penis
*Fascia latae
*Dartos fascia of penis
*Deep fascia of penis
*Septum of scrotum
*Dartos muscle
*Dartos fascia of scrotum

FIGURE 3.53. Fasciae of perineum. A and **B.** Median sections, viewed from left, demonstrate the fasciae in the female (**A**) and male (**B**). The planes of the sections shown in parts **C–F** are indicated. **C.** This coronal section of the female urogenital triangle is in the plane of the vagina. Fibro-areolar components of the endopelvic fascia (cardinal ligament and paracolpium) are shown. **D.** This coronal section of the male urogenital triangle is in the plane of the prostatic urethra. **E.** This coronal section of the anal triangle is in the plane of the lower rectal and anal canals. **F.** This coronal section demonstrates the subcutaneous tissue of the proximal penis and scrotum. An enlarged view of the layers of the penis is provided in Figure 3.61C.

perineal pouch. Although the classical descriptions appear justified when viewing only the superficial aspect of the structures occupying the deep pouch (Fig. 3.54A), the long-held concept of a flat, essentially two-dimensional diaphragm is erroneous. According to this concept, a trilaminar "UG diaphragm" consisted of the perineal membrane (inferior fascia of the UG diaphragm) inferiorly, a superior fascia of the UG diaphragm superiorly, and deep perineal muscles in between. The deep pouch was the space between the two fascial membranes, occupied by what was perceived to be a flat muscular sheet consisting of a disc-like sphincter urethra anterior to or within an equally two-dimensional, transversely oriented deep transverse perineal muscle. In males, the bulbo-urethral glands were also considered occupants of the pouch. Only the descriptions of the perineal membrane and deep transverse perineal muscles of the male (with embedded glands)

appear to be supported by evidence, which includes medical imaging of live subjects (Myers et al., 1998a, 1998b). Many texts, atlases, and medical illustrations continue to feature the old model, and students are likely to encounter the outdated images and concepts in clinical training and practice, and need to be aware of the inaccuracies in this regard.

Current Concept of Deep Perineal Pouch and External Urethral Sphincter. In the female, the posterior edge of the perineal membrane is typically occupied by a mass of smooth muscle in the place of the deep transverse perineal muscles (Wendell-Smith, 1995). Immediately superior to the posterior half of the perineal membrane, the flat, sheet-like, deep transverse perineal muscle, when developed (typically only in males), offers dynamic support for the pelvic viscera. As described by Oelrich (1980), however, the urethral sphincter muscle is not a flat, planar structure,

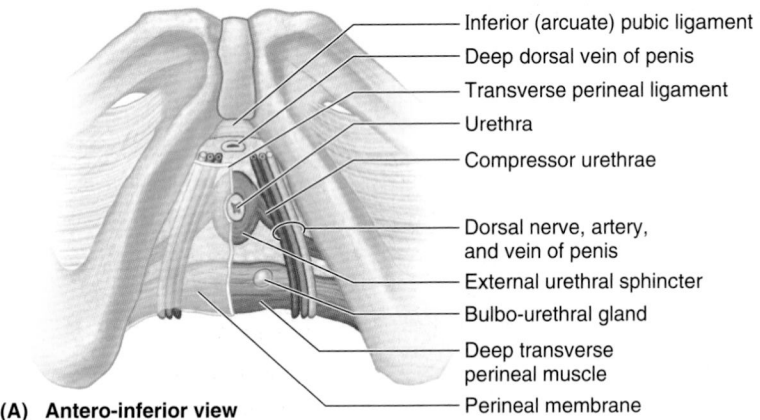

(A) Antero-inferior view

Inferior (arcuate) pubic ligament
Deep dorsal vein of penis
Transverse perineal ligament
Urethra
Compressor urethrae
Dorsal nerve, artery, and vein of penis
External urethral sphincter
Bulbo-urethral gland
Deep transverse perineal muscle
Perineal membrane

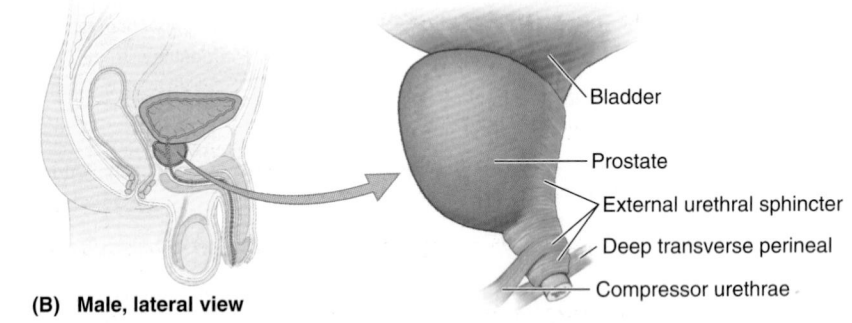

(B) Male, lateral view

Bladder
Prostate
External urethral sphincter
Deep transverse perineal
Compressor urethrae

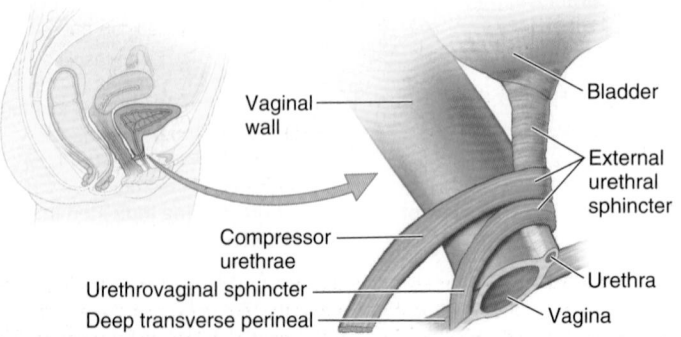

(C) Female, lateral view

Vaginal wall
Bladder
External urethral sphincter
Compressor urethrae
Urethrovaginal sphincter
Deep transverse perineal
Urethra
Vagina

FIGURE 3.54. Deep perineal pouch and male and female external urethral sphincters. A. The deep perineal pouch is seen through (left side) and after removal of the perineal membrane (right side). **B.** The trough-like fibers of the superior male external urethral sphincter ascend to the neck of the bladder as part of the isthmus of the prostate. The inferior sphincter includes cylindrical and loop-like portions (compressor urethrae). **C.** The female urethral sphincter complex is shown.

and the only "superior fascia" is the intrinsic fascia of the external urethral sphincter muscle. Contemporary views consider the *inferior fascia of the pelvic diaphragm* to be the superior boundary of the deep pouch (Fig. 3.53C–E). In both views, the strong *perineal membrane* is the inferior boundary (floor) of the deep pouch, separating it from the superficial pouch. The perineal membrane is indeed, with the perineal body, the final passive support of the pelvic viscera.

The male **external urethral sphincter** is more like a tube or trough than a disc. In the male only the inferior part of the muscle forms an encircling investment (a true sphincter) for the intermediate part of the urethra inferior to the prostate (Fig. 3.54B). Its larger, trough-like part extends vertically to the neck of the bladder as part of the *isthmus of the prostate*, displacing glandular tissue and investing the prostatic urethra anteriorly and anterolaterally only (see also Fig. 3.38). Apparently, the muscular primordium is established around the whole length of the urethra before development of the prostate. As the prostate develops from urethral glands, the posterior and posterolateral muscle atrophies, or is displaced by the prostate. Whether this part of the muscle compresses or dilates the prostatic urethra is a matter of some controversy.

The female **external urethral sphincter** is more properly a "urogenital sphincter" (Oelrich, 1983). Here, too, a part forms a true anular sphincter around the urethra (Fig. 3.54C), with several additional parts extending from it: a superior part, extending to the neck of the bladder; a subdivision described as extending inferolaterally to the ischial ramus on each side (the compressor urethrae muscle); and yet another band-like part, which encircles both the vagina and the urethra (urethrovaginal sphincter). In both the male and female, the musculature described is oriented perpendicular to the perineal membrane, rather than lying in a plane parallel to it.

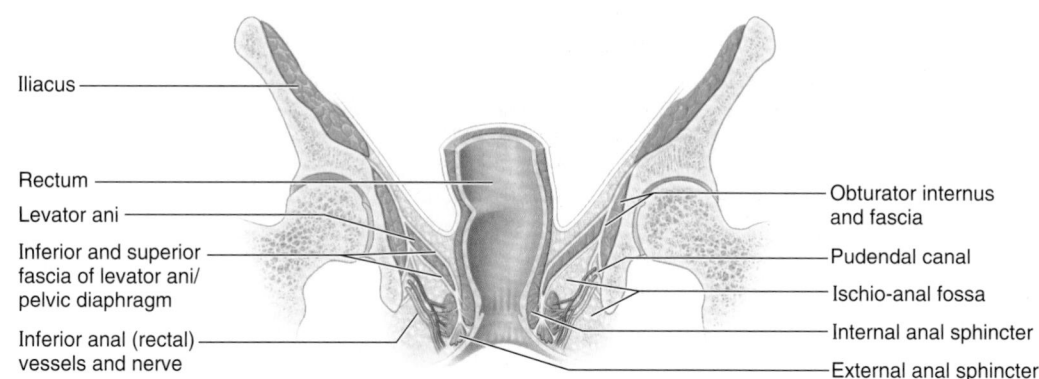

(A) Anterior view of schematic coronal section

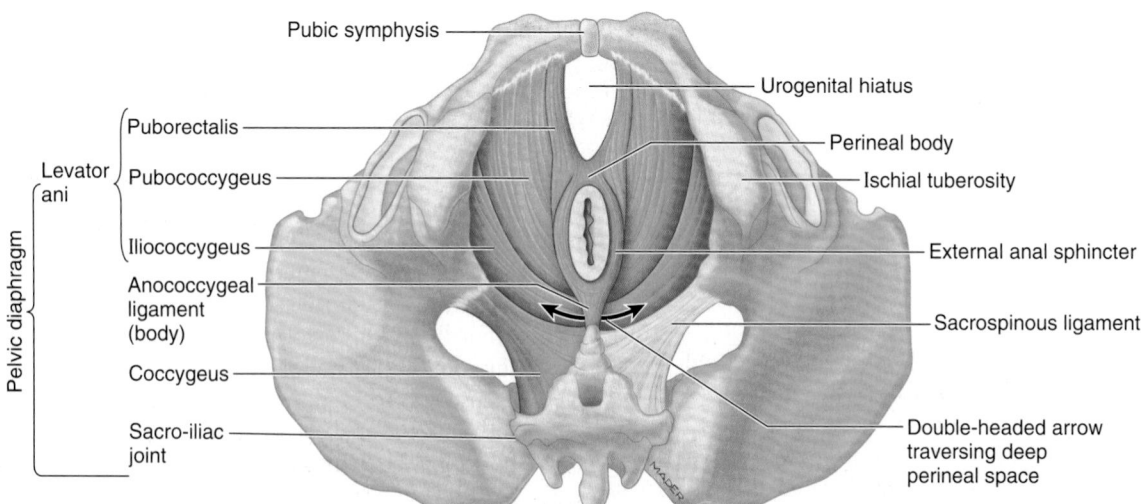

(B) Inferior view of perineum (lithotomy position)

FIGURE 3.55. Pelvic diaphragm and ischio-anal fossae. A. Coronal section of the pelvis in the plane of the rectum and anal canal, demonstrating lateral and medial walls and roof of the ischio-anal fossae. **B.** Fascia covering the inferior aspect of the pelvic diaphragm forms the roof of the ischio-anal fossae. The left sacrospinous ligament has been removed to reveal the coccygeus. Abscesses of the right or left ischio-anal fossa may extend to the contralateral fossa via the deep postanal space (*double-headed arrow*).

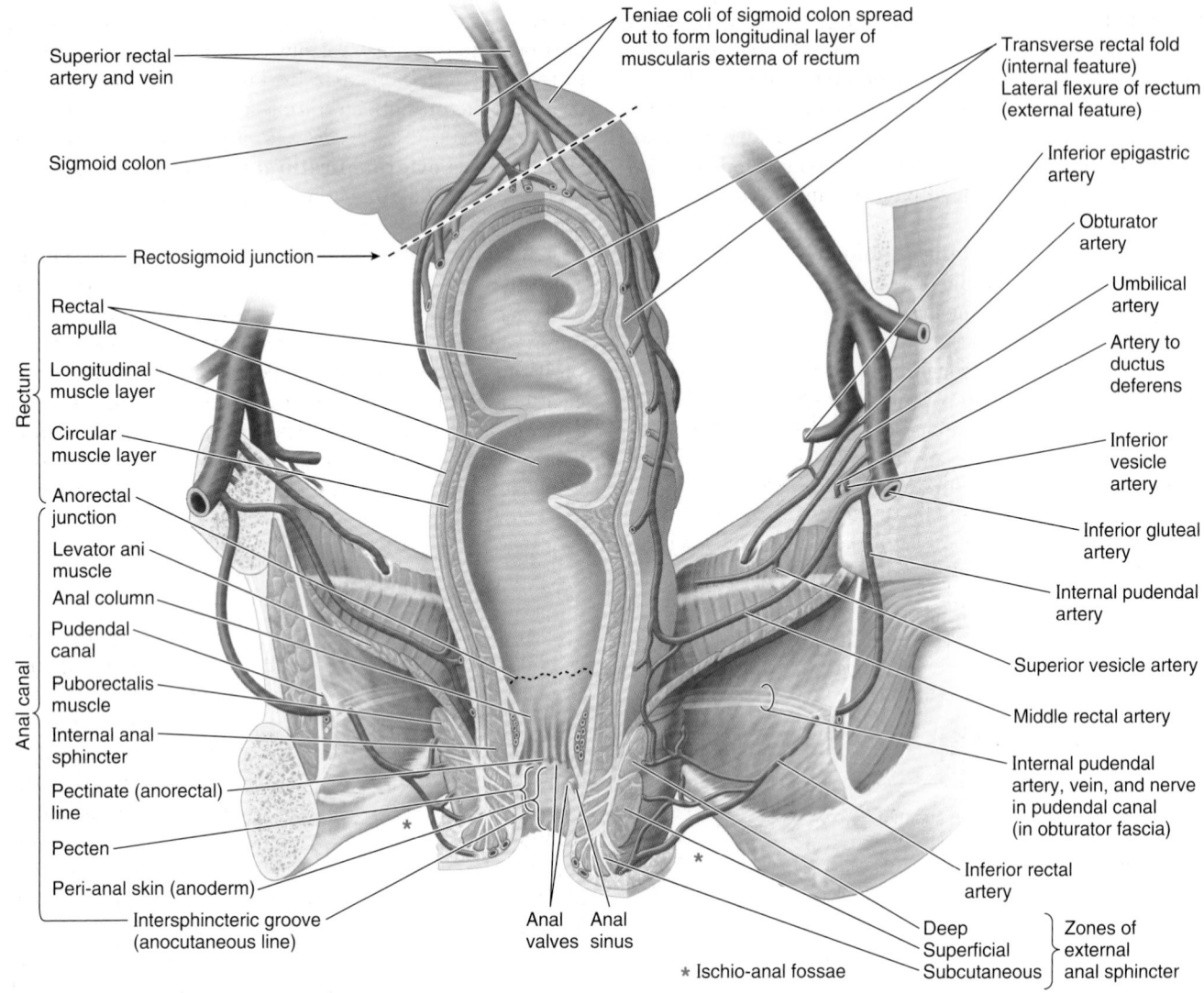

Superior rectal artery and vein

Teniae coli of sigmoid colon spread out to form longitudinal layer of muscularis externa of rectum

Transverse rectal fold (internal feature)
Lateral flexure of rectum (external feature)

Sigmoid colon

Inferior epigastric artery

Obturator artery

Rectosigmoid junction

Umbilical artery

Rectum
- Rectal ampulla
- Longitudinal muscle layer
- Circular muscle layer
- Anorectal junction

Artery to ductus deferens

Inferior vesicle artery

Anal canal
- Levator ani muscle
- Anal column
- Pudendal canal
- Puborectalis muscle
- Internal anal sphincter
- Pectinate (anorectal) line
- Pecten
- Peri-anal skin (anoderm)
- Intersphincteric groove (anocutaneous line)

Inferior gluteal artery

Internal pudendal artery

Superior vesicle artery

Middle rectal artery

Internal pudendal artery, vein, and nerve in pudendal canal (in obturator fascia)

Inferior rectal artery

Anal valves   Anal sinus

Deep } Zones of
Superficial } external
Subcutaneous } anal sphincter

* Ischio-anal fossae

Posterior view of anterior pelvis and perineum

FIGURE 3.56. Rectum and anal canal, levator ani, and ischio-anal fossa. The left posterolateral third of the rectum and anal canal have been removed to demonstrate the luminal features. The pudendal vessels and nerves are transmitted by the pudendal canal, a space within the obturator fascia that covers the medial surface of the obturator internus, lining the lateral wall of the ischio-anal fossa.

Features of Anal Triangle

ISCHIO-ANAL FOSSAE

The **ischio-anal fossae** (formerly called ischiorectal fossae) on each side of the anal canal are large fascia-lined, wedge-shaped spaces between the skin of the anal region and the pelvic diaphragm (Figs. 3.53D, 3.55A, and 3.56). The apex of each fossa lies superiorly where the levator ani muscle arises from the obturator fascia. The ischio-anal fossae, wide inferiorly and narrow superiorly, are filled with fat and loose connective tissue. The two ischio-anal fossae communicate by means of the *deep postanal space* over the *anococcygeal ligament* (body), a fibrous mass located between the anal canal and the tip of the coccyx (Figs. 3.53A & B and 3.55B).

Each ischio-anal fossa is bounded:

- Laterally by the ischium and overlapping inferior part of the obturator internus, covered with obturator fascia.
- Medially by the external anal sphincter, with a sloping superior medial wall or roof formed by the levator ani as it descends to blend with the sphincter; both structures surround the anal canal.
- Posteriorly by the sacrotuberous ligament and gluteus maximus.
- Anteriorly by the bodies of the pubic bones, inferior to the origin of the puborectalis. These parts of the fossae, extending into the UG triangle superior to the perineal membrane (and musculature on its superior surface), are known as the **anterior recesses of the ischio-anal fossae.**

Each ischio-anal fossa is filled with a **fat body of the ischio-anal fossa.** These fat bodies support the anal canal but they are readily displaced to permit descent and expansion of the anal canal during the passage of feces. The fat bodies are traversed by tough, fibrous bands, as well as by several neurovascular structures, including the inferior anal/rectal vessels and nerves and two other cutaneous nerves, the perforating branch of S2 and S3 and the perineal branch of S4 nerve.

PUDENDAL CANAL AND ITS NEUROVASCULAR BUNDLE

The **pudendal canal** (Alcock canal) is an essentially horizontal passageway within the obturator fascia that covers the medial aspect of the obturator internus and lines the lateral wall of the ischio-anal fossa (Figs. 3.55A and 3.56). The internal pudendal artery and vein, the pudendal nerve, and the nerve to the obturator internus enter this canal at the lesser sciatic notch, inferior to the ischial spine. The internal pudendal vessels and the pudendal nerve supply and drain blood from and innervate most of the perineum. As the artery and nerve enter the canal, they give rise to the **inferior rectal artery** and **nerve,** which pass medially to supply the external anal sphincter and the peri-anal skin (Figs. 3.56–3.58; Table 3.8). Toward the distal

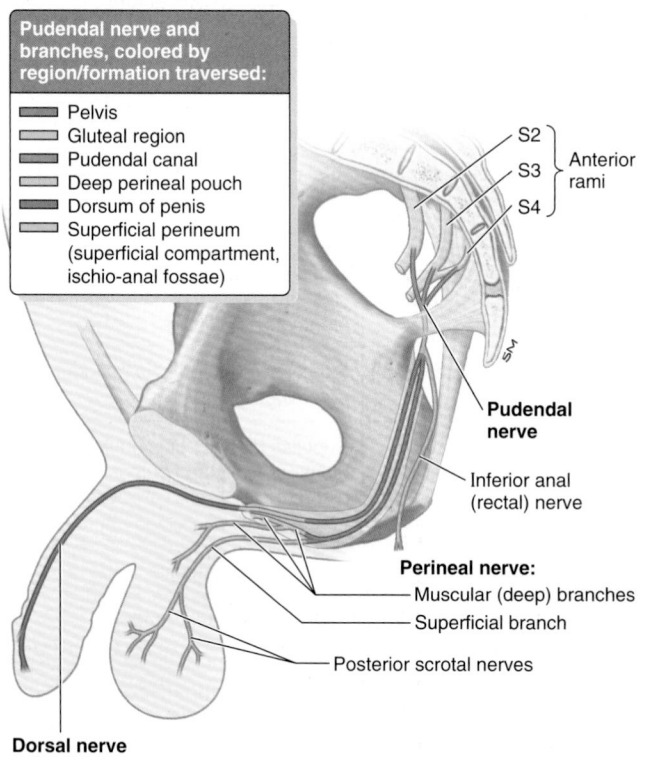

FIGURE 3.57. Distribution of pudendal nerve. The five regions traversed by the nerve are shown. The pudendal nerve supplies the skin, organs, and muscles of the perineum; therefore, it is concerned with micturition, defecation, erection, ejaculation, and, in the female, parturition. Although the pudendal nerve is shown here in the male, its distribution is similar in the female because the parts of the female perineum are homologs of those in the male.

(anterior) end of the pudendal canal, the artery and nerve both bifurcate, giving rise to the *perineal nerve* and *artery,* which are distributed mostly to the superficial pouch (inferior to the perineal membrane), and to the *dorsal artery* and *nerve of the penis or clitoris,* which run in the deep pouch (superior to the membrane). When the latter structures reach the dorsum of the penis or clitoris, the nerves run distally on the lateral side of the continuation of the internal pudendal artery as they both proceed to the glans penis or glans clitoris.

The perineal nerve has two branches: The **superficial perineal nerve** gives rise to posterior scrotal or labial (cutaneous) branches, and the **deep perineal nerve** supplies the muscles of the deep and superficial perineal pouches, the skin of the vestibule, and the mucosa of the inferiormost part of the vagina. The inferior rectal nerve communicates with the posterior scrotal or labial and perineal nerves. The **dorsal nerve of the penis** or **clitoris** is the primary sensory nerve serving the male or female organ, especially the sensitive glans at the distal end.

ANAL CANAL

The **anal canal** is the terminal part of the large intestine and of the entire digestive tract. It extends from the superior aspect of the pelvic diaphragm to the **anus** (Figs. 3.55B and 3.56). The canal (2.5–3.5 cm long) begins where the rectal ampulla narrows at the level of the U-shaped sling formed by the puborectalis muscle (Fig. 3.12). The anal canal ends at the anus, the external outlet of the alimentary tract. The anal canal, surrounded by internal and external anal sphincters, descends postero-inferiorly between the anococcygeal ligament and the perineal body. The canal is collapsed, except during passage of feces. Both sphincters must relax before defecation can occur.

The **internal anal sphincter** (Figs. 3.55A and 3.56) is an involuntary sphincter surrounding the superior two thirds of the anal canal. It is a thickening of the circular muscle layer. Its contraction (tonus) is stimulated and maintained by sympathetic fibers from the superior rectal (peri-arterial) and hypogastric plexuses. Its contraction is inhibited by parasympathetic fiber stimulation, both intrinsically in relation to peristalsis, and extrinsically by fibers conveyed by the pelvic splanchnic nerves. This sphincter is tonically contracted most of the time to prevent leakage of fluid or flatus; however, it relaxes (is inhibited) temporarily in response to distension of the rectal ampulla by feces or gas, requiring voluntary contraction of the puborectalis and external anal sphincter if defecation or flatulence is not to occur. The ampulla relaxes after initial distension (when peristalsis subsides) and tonus returns until the next peristalsis, or until a threshold level of distension occurs, at which point inhibition of the sphincter is continuous until distension is relieved.

The **external anal sphincter** is a large voluntary sphincter that forms a broad band on each side of the inferior two thirds of the anal canal (Figs. 3.52E, 3.55, and 3.56). This sphincter is attached anteriorly to the perineal body and posteriorly to the coccyx via the anococcygeal ligament. It blends superiorly with the puborectalis muscle.

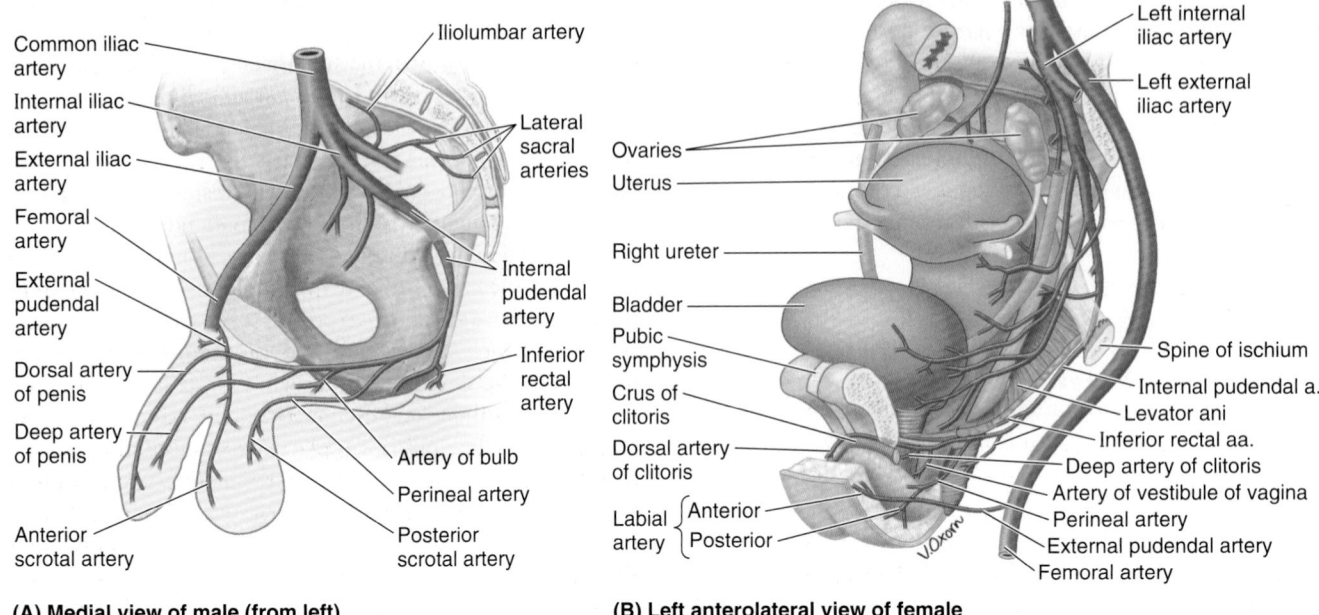

(A) Medial view of male (from left)   **(B) Left anterolateral view of female**

FIGURE 3.58. Arteries of perineum.

TABLE 3.8. ARTERIES OF PERINEUM

Artery	Origin	Course	Distribution in Perineum
Internal pudendal	Anterior division of internal iliac artery	Leaves pelvis through greater sciatic foramen; hooks around ischial spine to enter perineum via lesser sciatic foramen; enters pudendal canal	Primary artery of perineum and external genital organs
Inferior rectal	Internal pudendal artery	Arises at entrance to pudendal canal; traverses ischio-anal fossa to anal canal	Anal canal inferior to pectinate line; anal sphincters; peri-anal skin
Perineal		Arises within pudendal canal; passes to superficial pouch (space) on exit	Supplies superficial perineal muscles and scrotum of male/vestibule of female
Posterior scrotal (♂) or labial (♀)	Terminal branch of perineal artery	Runs in superficial fascia of posterior scrotum or labia majora	Skin of scrotum or labia majora and minora
Artery of bulb of penis (♂) or vestibule (♀)		Pierces perineal membrane to reach bulb of penis or vestibule of vagina	Supplies bulb of penis (including bulbar urethra) and bulbo-urethral gland (male) or bulb of vestibule and greater vestibular gland (female)
Deep artery of penis (♂) or clitoris (♀)	Terminal branch of internal pudendal artery	Pierces perineal membrane to enter crura of corpora cavernosa of penis or clitoris; branches run proximally and distally	Supplies most erectile tissue of corpora cavernosa of penis or clitoris via helicine arteries
Dorsal artery of penis (♂) or clitoris (♀)		Passes to deep pouch; pierces perineal membrane and traverses suspensory ligament of penis or clitoris to run on dorsum of penis or clitoris to glans	Deep perineal pouch; skin of penis; fascia of penis or clitoris; distal corpus spongiosum of penis, including spongy urethra; glans penis or clitoris
External pudendal, superficial, and deep branches	Femoral artery	Pass medially from thigh to reach anterior aspect of the urogenital triangle of perineum	Anterior aspect of scrotum and skin at root of penis of male; mons pubis and anterior aspect of labia of female

The external anal sphincter is described as having subcutaneous, superficial, and deep parts; these are zones rather than muscle bellies and are often indistinct. The external anal sphincter is supplied mainly by S4 through the inferior rectal nerve (Fig. 3.57), although its deep part also receives fibers from the nerve to the levator ani, in common with the puborectalis, with which it contracts in unison to maintain continence when the internal sphincter is relaxed (except during defecation).

Internally, the superior half of the mucous membrane of the anal canal is characterized by a series of longitudinal ridges called **anal columns** (Fig. 3.56). These columns contain the terminal branches of the superior rectal artery and vein. The **anorectal junction,** indicated by the superior ends of the anal columns, is where the rectum joins the anal canal. At this point, the wide rectal ampulla abruptly narrows as it traverses the pelvic diaphragm. The inferior ends of the anal columns are joined by **anal valves.** Superior to the valves are small recesses called **anal sinuses.** When compressed by feces, the anal sinuses exude mucus, which aids in evacuation of feces from the anal canal.

The inferior comb-shaped limit of the anal valves forms an irregular line, the **pectinate line,** that indicates the junction of the superior part of the anal canal (visceral; derived from the embryonic hindgut), and the inferior part (somatic; derived from the embryonic proctodeum).

The anal canal superior to the pectinate line differs from the part inferior to the pectinate line in its arterial supply, innervation, and venous and lymphatic drainage (Fig. 3.59). These differences result from the different embryological origins of the superior and inferior parts of the anal canal (Moore, Persaud, and Torchiam 2012).

Arterial Supply of Anal Canal. The *superior rectal artery* supplies the anal canal *superior to the pectinate line* (Figs. 3.32A and 3.59). The two *inferior rectal arteries* supply the anal canal *inferior to the pectinate line* as well as the surrounding muscles and peri-anal skin (Figs. 3.32, 3.58, and 3.59; Table 3.8). The *middle rectal arteries* assist with the blood supply to the anal canal by forming anastomoses with the superior and inferior rectal arteries.

Venous and Lymphatic Drainage of Anal Canal. The *internal rectal venous plexus* drains in both directions from the level of the pectinate line. *Superior to the pectinate line,* the internal rectal plexus drains chiefly into the *superior rectal vein* (a tributary of the inferior mesenteric vein) and the portal system (Figs. 3.32B and 3.59). *Inferior to the pectinate line,* the internal rectal plexus drains into the *inferior rectal veins* (tributaries of the caval venous system) around the margin of the external anal sphincter. The *middle rectal veins* (tributaries of the internal iliac veins) mainly drain the muscularis externa of the ampulla and form anastomoses with the superior and inferior rectal veins. In addition to the abundant venous anastomoses, the rectal plexuses receive multiple arteriovenous anastomoses (AVAs) from the superior and middle rectal arteries.

The normal submucosa of the anorectal junction is markedly thickened, and in section has the appearance of a cavernous (erectile) tissue, owing to the presence of the sacculated veins of the internal rectal venous plexus. The vascular submucosa is especially thickened in the left lateral, right anterolateral, and right posterolateral positions, forming anal cushions, or threshold pads, at the point of closure of the anal canal. Because these cushions contain plexuses of saccular veins capable of directly receiving arterial blood via multiple

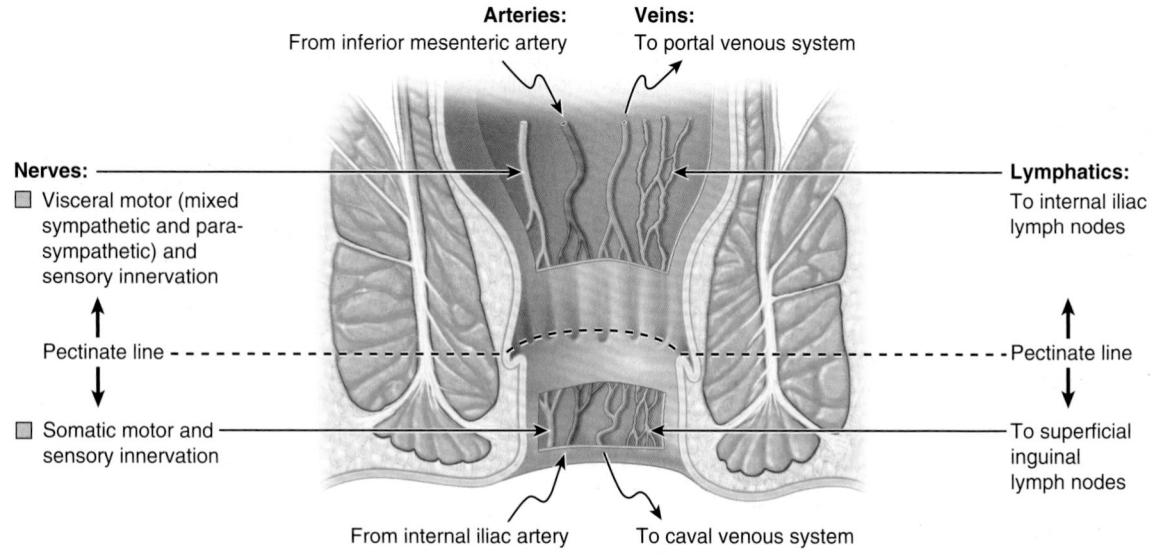

Separation of "visceral" and "parietal" at the pectinate line

FIGURE 3.59. **Transitions occurring at pectinate line.** Vessels and nerves superior to the pectinate line are visceral; those inferior to the pectinate line are parietal or somatic. This orientation reflects the embryological development of the anorectum.

AVAs, they are variably pliable and turgid, and form a sort of flutter valve that contributes to the normally water- and gas-tight closure of the anal canal.

Superior to the pectinate line, the lymphatic vessels drain deeply into the *internal iliac lymph nodes,* and through them into the common iliac and lumbar lymph nodes (Figs. 3.48B and 3.59; Table 3.7). *Inferior to the pectinate line,* the lymphatic vessels drain superficially into the **superficial inguinal lymph nodes,** as does most of the perineum.

Innervation of Anal Canal. The nerve supply to the anal canal *superior to the pectinate line* is visceral innervation from the *inferior hypogastric plexus,* involving sympathetic, parasympathetic, and visceral afferent fibers (Figs. 3.33 and 3.59). Sympathetic fibers maintain the tonus of the internal anal sphincter. Parasympathetic fibers inhibit the tonus of the internal sphincter and evoke peristaltic contraction for defecation. The superior part of the anal canal, like the rectum superior to it, is inferior to the *pelvic pain line* (see Table 3.3); all visceral afferents travel with the parasympathetic fibers to spinal sensory ganglia S2–S4. Superior to the pectinate line, the anal canal is sensitive only to stretching, which evokes sensations at both the conscious and unconscious (reflex) levels. For example, distension of the rectal ampulla inhibits (relaxes) the tonus of the internal sphincter.

The nerve supply of the anal canal *inferior to the pectinate line* is somatic innervation derived from the *inferior anal (rectal) nerves,* branches of the pudendal nerve. Therefore, this part of the anal canal is sensitive to pain, touch, and temperature. Somatic efferent fibers stimulate contraction of the voluntary external anal sphincter.

PERINEUM

Disruption of Perineal Body

The perineal body is an important structure, especially in women, because it is the final support of the pelvic viscera, linking muscles that extend across the pelvic outlet, like crossing beams supporting the overlying pelvic diaphragm. Stretching or tearing the attachments of perineal muscles from the perineal body can occur during childbirth, removing support from the pelvic floor. As a result, *prolapse of pelvic viscera,* including prolapse of the bladder (through the urethra) and prolapse of the uterus and/or vagina (through the vaginal orifice) may occur. (Various degrees of prolapse are illustrated in the blue box "Disposition of Uterus and Uterine Prolapse," p. 392).

The perineal body can also be disrupted by trauma, inflammatory disease, and infection, which can result in the formation of a *fistula* (abnormal canal) connected to the vestibule (see the blue box "Vaginal Fistulae," p. 396). Attenuation of the perineal body, associated with diastasis (separation) of the puborectalis and pubococcygeus parts of the levator ani, may also result in the formation of a *cystocele, rectocele,* and/or *enterocele,* hernial protrusions of part of the bladder, rectum, or rectovaginal pouch, respectively, into the vaginal wall (Fig. B3.28).

Episiotomy

During vaginal surgery and labor, an *episiotomy* (surgical incision of the perineum and inferoposterior vaginal wall) may be made to enlarge the vaginal orifice, with the intention of decreasing excessive traumatic tearing of the perineum and uncontrolled jagged tears of the perineal muscles. Episiotomies are still performed in a large portion of vaginal deliveries in the United States (Gabbe et al., 2007). It is generally agreed that episiotomy is indicated when

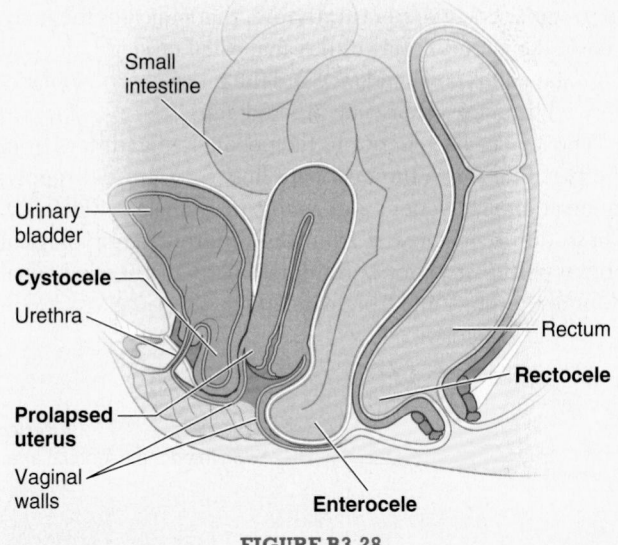

FIGURE B3.28.

descent of the fetus is arrested or protracted, when instrumentation is necessary (e.g., use of obstetrical forceps), or to expedite delivery when there are signs of fetal distress. However, routine prophylactic episiotomy is widely debated and declining in frequency.

The perineal body is the major structure incised during a *median episiotomy* (Fig. B3.29A & B). The rationale of the median incision is that the scar produced as the wound heals will not be greatly different from the fibrous tissue surrounding it. Also, because the incision extends only partially into this fibrous tissue, some surgeons believe that the incision is more likely to be self-limiting, resisting further tearing. However, when further tearing does occur, it is directed toward the anus, and sphincter damage or anovaginal fistulae are poten-

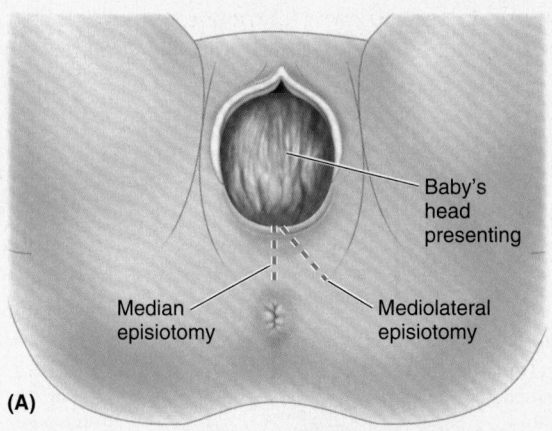

(A)

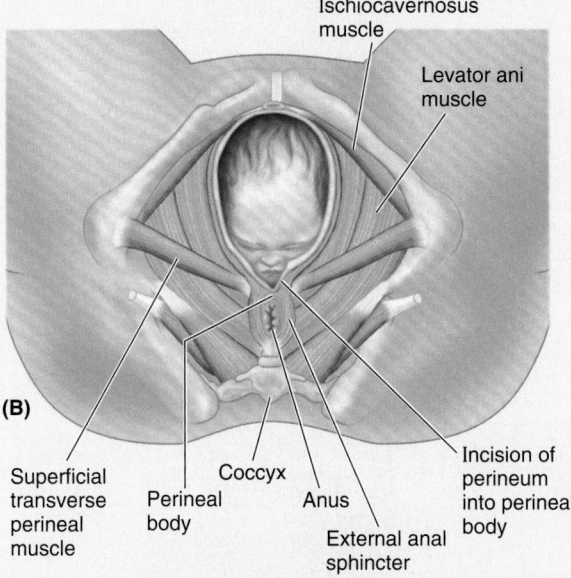

Ischiocavernosus muscle

Levator ani muscle

(B)

Superficial transverse perineal muscle

Perineal body

Coccyx

Anus

External anal sphincter

Incision of perineum into perineal body

Inferior view (lithotomy position)

FIGURE B3.29.

tial sequelae. Recent studies indicate median episiotomies are associated with an increased incidence of severe lacerations, associated in turn with an increased incidence of long-term incontinence, pelvic prolapse, and anovaginal fistulae.

Mediolateral episiotomies (Fig. B3.29A) appear to result in a lower incidence of severe laceration and are less likely to be associated with damage to the anal sphincters and canal. (*Note:* The clinical use of the term *mediolateral* is technically inappropriate here; it actually refers to an incision that is initially a median incision that then turns laterally as it proceeds posteriorly, circumventing the perineal body and directing further tearing away from the anus.)

Rupture of Urethra in Males and Extravasation of Urine

Fractures of the pelvic girdle, especially those resulting from separation of the pubic symphysis and puboprostatic ligaments, often cause a *rupture of the intermediate part of the urethra.* Rupture of this part of the urethra results in the *extravasation* (escape) *of urine and blood* into the deep perineal pouch (Fig. B3.30A); the fluid may pass superiorly through the urogenital hiatus and distribute extraperitoneally around the prostate and bladder.

The common site of *rupture of the spongy urethra* and *extravasation of urine* is in the bulb of the penis (Fig. B3.30B). This injury usually results from a forceful blow to the perineum (*straddle injury*), such as falling on a metal beam or, less commonly, from the incorrect passage (*false passage*) of a transurethral catheter or device that fails to negotiate the angle of the urethra in the bulb of the penis. Rupture of the corpus spongiosum and spongy urethra results in urine passing from it (extravasating) into the superficial perineal space. The attachments of the perineal fascia determine the direction of flow of the extravasated urine. Urine may pass into the loose connec-

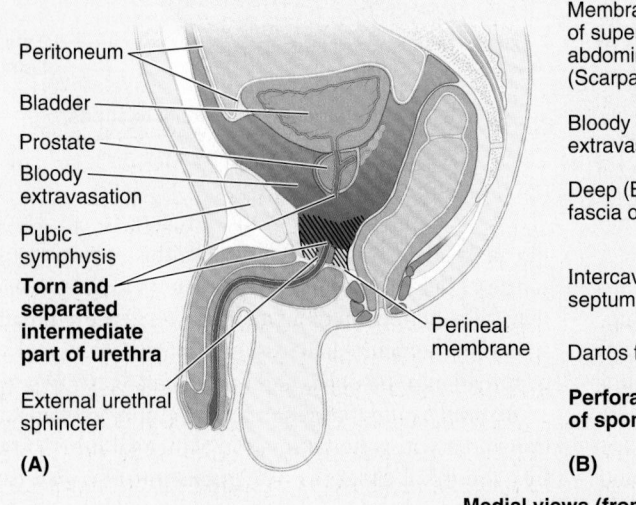

Peritoneum

Bladder

Prostate

Bloody extravasation

Pubic symphysis

Torn and separated intermediate part of urethra

External urethral sphincter

(A)

Perineal membrane

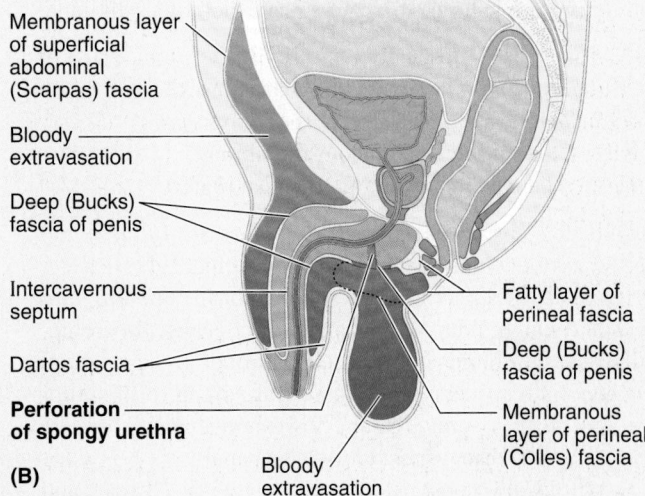

Membranous layer of superficial abdominal (Scarpas) fascia

Bloody extravasation

Deep (Bucks) fascia of penis

Intercavernous septum

Dartos fascia

Perforation of spongy urethra

Bloody extravasation

Fatty layer of perineal fascia

Deep (Bucks) fascia of penis

Membranous layer of perineal (Colles) fascia

(B)

Medial views (from left)

FIGURE B3.30.

tive tissue in the scrotum, around the penis, and superiorly, deep to the membranous layer of subcutaneous connective tissue of the inferior anterior abdominal wall.

The urine cannot pass far into the thighs because the membranous layer of superficial perineal fascia blends with the fascia lata, enveloping the thigh muscles, just distal to the inguinal ligament. In addition, urine cannot pass posteriorly into the anal triangle because the superficial and deep layers of perineal fascia are continuous with each other around the superficial perineal muscles and with the posterior edge of the perineal membrane between them. Rupture of a blood vessel into the superficial perineal pouch resulting from trauma would result in a similar containment of blood in the superficial perineal pouch.

Starvation and Rectal Prolapse

The fat bodies of the ischio-anal fossae are among the last reserves of fatty tissue to disappear with starvation. In the absence of the support provided by the ischio-anal fat, rectal prolapse is relatively common.

Pectinate Line—a Clinically Important Landmark

The *pectinate line* (also called the dentate or mucocutaneous line by some clinicians) is a particularly important landmark because it is visible and approximates the level of important anatomical changes related to the transition from visceral to parietal (Fig. 3.59), affecting such things as the types of tumors that occur, and the direction in which they metastasize.

Anal Fissures; Ischio-anal and Peri-anal Abscesses

The ischio-anal fossae are occasionally the sites of infection, which may result in the formation of *ischio-anal abscesses* (Fig. B3.31A). These collections of pus are painful. Infections may reach the ischio-anal fossae in several ways:

- After *cryptitis* (inflammation of anal sinuses).
- Extension from a pelvirectal abscess.
- After a tear in the anal mucous membrane.
- From a penetrating wound in the anal region.

Diagnostic signs of an ischio-anal abscess are fullness and tenderness between the anus and the ischial tuberosity. A peri-anal abscess may rupture spontaneously, opening into the anal canal, rectum, or peri-anal skin. Because the ischio-anal fossae communicate posteriorly through the *deep post-anal space*, an abscess in one fossa may spread to the other one, and form a semicircular "horseshoe-shaped" abscess around the posterior aspect of the anal canal.

In chronically constipated persons, the anal valves and mucosa may be torn by hard feces. An *anal fissure* (slit-like lesion) is usually located in the posterior midline, inferior to

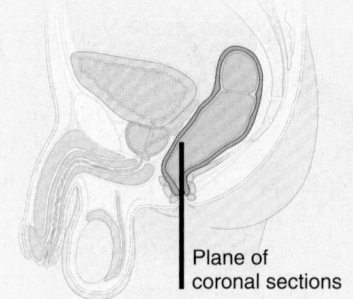

Plane of coronal sections

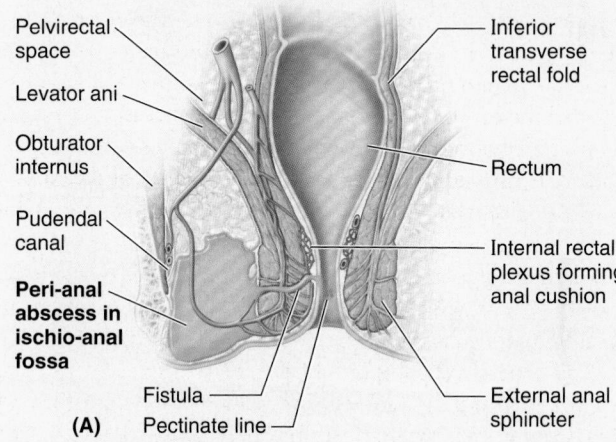

Pelvirectal space
Levator ani
Obturator internus
Pudendal canal
Peri-anal abscess in ischio-anal fossa
Fistula
Pectinate line
(A)
Inferior transverse rectal fold
Rectum
Internal rectal plexus forming anal cushion
External anal sphincter

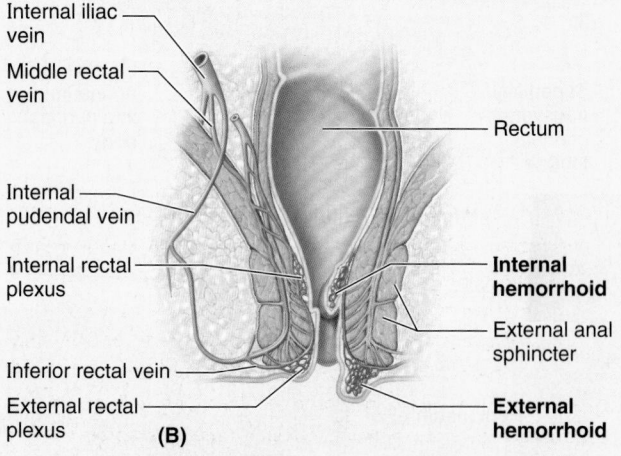

Internal iliac vein
Middle rectal vein
Internal pudendal vein
Internal rectal plexus
Inferior rectal vein
External rectal plexus
(B)
Rectum
Internal hemorrhoid
External anal sphincter
External hemorrhoid

Anterior views of coronal sections

FIGURE B3.31.

the anal valves. It is painful because this region is supplied by sensory fibers of the inferior rectal nerves. A *peri-anal abscess* may follow infection of an anal fissure, and the infection may spread to the ischio-anal fossae and form ischio-anal abscesses or spread into the pelvis and form a *pelvirectal abscess*.

An *anal fistula* may result from the spread of an anal infection and cryptitis (inflammation of an anal sinus). One end of this abnormal canal (fistula) opens into the anal canal, and the other end opens into an abscess in the ischio-anal fossa or into the peri-anal skin.

Hemorrhoids

Internal hemorrhoids (piles) are prolapses of rectal mucosa (more specifically, of the "anal cushions") containing the normally dilated veins of the *internal rectal venous plexus* (Fig. B3.31B). Internal hemorrhoids result from a breakdown of the muscularis mucosae, a smooth muscle layer deep to the mucosa. Internal hemorrhoids that prolapse into or through the anal canal are often compressed by the contracted sphincters, impeding blood flow. As a result, they tend to strangulate and ulcerate. Because of the presence of abundant arteriovenous anastomoses, bleeding from internal hemorrhoids is characteristically bright red. The current practice is to treat only prolapsed, ulcerated internal hemorrhoids. *External hemorrhoids* are thromboses (blood clots) in the veins of the *external rectal venous plexus* and are covered by skin. Predisposing factors for hemorrhoids include pregnancy, chronic constipation and prolonged toilet sitting and straining, and any disorder that impedes venous return, including increased intra-abdominal pressure.

The anastomoses between the superior, middle, and inferior rectal veins form clinically important communications between the portal and systemic venous systems (see Fig. 2.75A). The superior rectal vein drains into the inferior mesenteric vein, whereas the middle and inferior rectal veins drain through the systemic system into the inferior vena cava. Any abnormal increase in pressure in the valveless portal system or veins of the trunk may cause enlargement of the superior rectal veins, resulting in an increase in blood flow or stasis in the internal rectal venous plexus. In the *portal hypertension* that occurs in relation to *hepatic cirrhosis,* the portocaval anastomosis between the superior and the middle and inferior rectal veins, along with portocaval anastomoses elsewhere, may become varicose. It is important to note that the veins of the rectal plexuses *normally* appear varicose (dilated and tortuous), even in newborns, and that internal hemorrhoids occur most commonly in the absence of portal hypertension.

Regarding pain from and the treatment of hemorrhoids, it is important to note that the anal canal superior to the pectinate line is visceral; thus it is innervated by visceral afferent pain fibers, so that an incision or needle insertion in this region is painless. Internal hemorrhoids are not painful and can be treated without anesthesia. Inferior to the pectinate line, the anal canal is somatic, supplied by the inferior anal (rectal) nerves containing somatic sensory fibers. Therefore, it is sensitive to painful stimuli (e.g., to the prick of a hypodermic needle). External hemorrhoids can be painful but often resolve in a few days.

Anorectal Incontinence

Stretching of the pudendal nerve(s) during a traumatic childbirth can result in *pudendal nerve damage* and *anorectal incontinence*.

The Bottom Line

PERINEUM AND PERINEAL REGION

The perineum is the diamond-shaped compartment bounded peripherally by the osseofibrous pelvic outlet and deeply (superiorly) by the pelvic diaphragm. ♦ The surface area overlying this compartment is the perineal region. ♦ The urogenital (UG) triangle (anteriorly) and anal triangle (posteriorly) that together form this diamond-shaped area lie at angles to each other. ♦ The intersecting planes of the triangles define the transverse line (extending between ischial tuberosities) that is the base of each triangle. ♦ Centrally, the UG triangle is perforated by the urethra and, in females, by the vagina. ♦ The anal triangle is perforated by the anal canal. ♦ The perineal body is a musculofibrous mass between the UG and the anal perforating structures, at the center point of the perineum.

UG triangle: The subcutaneous tissue of the urogenital triangle includes a superficial fatty layer and a deeper membranous layer, the perineal fascia (Colles fascia), which are continuous with corresponding layers of the inferior anterior abdominal wall. ♦ In females, the fatty layer is thick within the mons pubis and labia majora, but in males it is replaced by smooth dartos muscle in the penis and scrotum. ♦ The perineal fascia is limited to the UG triangle, fusing with the deep fascia at the posterior border (base) of the triangle. ♦ In males, this layer extends into the penis and scrotum, where it is closely associated with the loose, mobile skin of these structures.
♦ The planar perineal membrane divides the urogenital triangle of the perineum into superficial and deep perineal pouches.
♦ The superficial perineal pouch is between the membranous layer of subcutaneous tissue of the perineum and the perineal membrane, and is bounded laterally by the ischiopubic rami.
♦ The deep perineal pouch is between the perineal membrane and the inferior fascia of the pelvic diaphragm, and is bounded laterally by the obturator fascia. ♦ The superficial perineal pouch contains the erectile bodies of the external genitalia and associated muscles, the superficial transverse perineal muscle, deep perineal nerves and vessels, and in females the greater vestibular glands. ♦ The deep pouch includes the fat-filled anterior recesses of the ischio-anal fossae (laterally), the deep perineal muscle and inferiormost part of the external urethral sphincter, the part of the urethra traversing the perineal membrane and inferiormost external urethral sphincter (the intermediate urethra of males), the dorsal nerves of the penis/clitoris, and in males the bulbo-urethral glands.

Anal triangle: The ischio-anal fossae are fascia-lined, wedge-shaped spaces occupied by ischio-anal fat bodies.

♦ The fat bodies provide supportive packing that can be compressed or pushed aside to permit the temporary descent and expansion of the anal canal or vagina for passage of feces or a fetus. ♦ The fat bodies are traversed by inferior anal/rectal neurovasculature. ♦ The pudendal canal is an important passageway in the lateral wall of the fossa, between layers of the obturator fascia, for neurovasculature passing to and from the UG triangle.

Anal canal: The anal canal is the terminal part of both the large intestine and the digestive tract, the anus being the external outlet. ♦ Closure (and thus fecal continence) is maintained by the coordinated action of the involuntary internal and voluntary external anal sphincters. ♦ The sympathetically stimulated

tonus of the internal sphincter maintains closure, except during filling of the rectal ampulla and when inhibited during a parasympathetically stimulated peristaltic contraction of the rectum. ♦ During these moments, closure is maintained (unless defecation is permitted) by voluntary contraction of the puborectalis and external anal sphincter. ♦ Internally, the pectinate line demarcates the transition from visceral to somatic neurovascular supply and drainage. ♦ The anal canal is surrounded by superficial and deep venous plexuses, the veins of which normally have a varicose appearance. ♦ Thromboses in the superficial plexus and mucosal prolapse, including portions of the deep plexus, constitute painful external and insensitive internal hemorrhoids, respectively.

Male Urogenital Triangle

The **male urogenital triangle** includes the external genitalia and perineal muscles. The **male external genitalia** include the distal urethra, scrotum, and penis.

DISTAL MALE URETHRA

The male urethra is subdivided into four parts: intramural (preprostatic), prostatic, intermediate, and spongy. The intramural and prostatic parts are described with the pelvis (earlier in this chapter). Details concerning all four parts of the male urethra are provided and compared in Table 3.6.

The **intermediate (membranous) part of the urethra** begins at the apex of the prostate and traverses the deep perineal pouch, surrounded by the external urethral sphincter. It then penetrates the perineal membrane, ending as the urethra enters the bulb of the penis (Fig. 3.60). Posterolateral to this part of the urethra are the small *bulbo-urethral glands* and their slender ducts, which open into the proximal part of the spongy urethra.

The **spongy urethra** begins at the distal end of the intermediate part of the urethra and ends at the **male external urethral orifice**, which is slightly narrower than any of the other parts of the urethra. The lumen of the spongy urethra is approximately 5 mm in diameter; however, it is expanded in the bulb of the penis to form the **intrabulbar fossa,** and in the glans penis to form the **navicular fossa.** On each side, the slender *ducts of the bulbo-urethral glands* open into the proximal part of the spongy urethra; the orifices of these ducts are extremely small. There are also many minute openings of the ducts of mucus-secreting **urethral glands** into the spongy urethra.

Arterial Supply of Distal Male Urethra. The arterial supply of the intermediate and spongy parts of the urethra is from branches of the *dorsal artery of the penis* (Figs. 3.52C and 3.58; Table 3.8).

Venous and Lymphatic Drainage of Distal Male Urethra. Veins accompany the arteries and have similar names. Lymphatic vessels from the intermediate part of the urethra drain mainly into the *internal iliac lymph nodes* (Table 3.7; see Fig. 3.65), whereas most vessels from the spongy urethra pass to the *deep inguinal lymph nodes*, but some lymph passes to the external iliac nodes.

Innervation of Distal Male Urethra. The innervation of the intermediate part of the urethra is the same as that of the prostatic part: autonomic (efferent) innervation via the *prostatic nerve plexus*, arising from the *inferior hypogastric plexus*. The *sympathetic innervation* is from the lumbar spinal cord levels via the lumbar splanchnic nerves, and the *parasympathetic innervation* is from the sacral levels via the *pelvic splanchnic nerves*. The visceral afferent fibers follow the parasympathetic fibers retrogradely to sacral spinal sensory ganglia. The dorsal nerve of the penis, a branch of the *pudendal nerve*, provides somatic innervation of the spongy part of the urethra (Fig. 3.57).

SCROTUM

The **scrotum** is a cutaneous fibromuscular sac for the testes and associated structures. It is situated postero-inferior to the penis and inferior to the pubic symphysis. The bilateral embryonic formation of the scrotum is indicated by the midline **scrotal raphe** (Fig. 3.61A & E), which is continuous on the ventral surface of the penis with the **penile raphe** and posteriorly along the median line of the perineum with the **perineal raphe**. Internally, deep to the scrotal raphe, the scrotum is divided into two compartments, one for each testis, by a prolongation of the dartos fascia, the **septum of the scrotum.** The testes and epididymides and their coverings are described with the abdomen (see Chapter 2).

Arterial Supply of Scrotum. **Anterior scrotal arteries,** terminal branches of the **external pudendal arteries** (from the femoral artery), supply the anterior aspect of the scrotum. **Posterior scrotal arteries,** terminal branches of the superficial perineal branches of the *internal pudendal arteries*, supply the posterior aspect (Fig. 3.58A; Table 3.8).

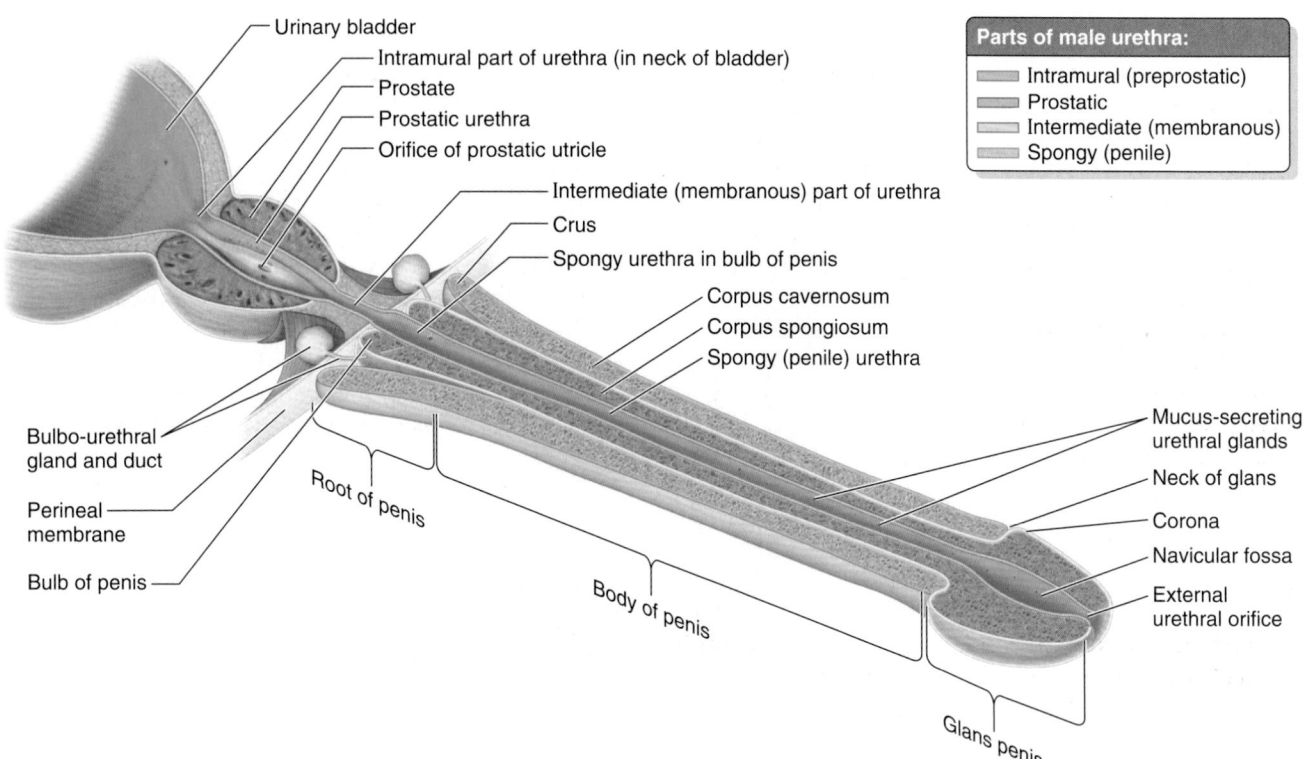

Parts of male urethra:
- Intramural (preprostatic)
- Prostatic
- Intermediate (membranous)
- Spongy (penile)

Urinary bladder
Intramural part of urethra (in neck of bladder)
Prostate
Prostatic urethra
Orifice of prostatic utricle
Intermediate (membranous) part of urethra
Crus
Spongy urethra in bulb of penis
Corpus cavernosum
Corpus spongiosum
Spongy (penile) urethra
Mucus-secreting urethral glands
Neck of glans
Corona
Navicular fossa
External urethral orifice
Bulbo-urethral gland and duct
Perineal membrane
Bulb of penis
Root of penis
Body of penis
Glans penis

FIGURE 3.60. Male urethra and associated structures. The urethra has four parts: the vesicular part (in the bladder neck), the prostatic urethra, the intermediate part (membranous urethra), and the spongy (cavernous) urethra. The ducts of the bulbo-urethral glands open into the proximal part of the spongy urethra. The urethra is not uniform in its caliber: The external urethral orifice and intermediate part are narrowest. Attempting to approach this "straight-line" position as much as possible facilitates passage of a catheter or other transurethral device.

The scrotum also receives branches from the cremasteric arteries (branches of the inferior epigastric arteries).

Venous and Lymphatic Drainage of Scrotum. The *scrotal veins* accompany the arteries, sharing the same names but draining primarily to the *external pudendal veins*. Lymphatic vessels from the scrotum carry lymph to the *superficial inguinal lymph nodes* (Table 3.6).

Innervation of Scrotum. The *anterior aspect of the scrotum* is supplied by derivatives of the *lumbar plexus:* **anterior scrotal nerves,** derived from the *ilio-inguinal nerve,* and the *genital branch of the genitofemoral nerve* (Table 3.10). The *posterior aspect of the scrotum* is supplied by derivatives of the sacral plexus: **posterior scrotal nerves,** branches of the *superficial perineal branches* of the *pudendal nerve,* and the *perineal branch of the posterior cutaneous nerve of thigh* (Figs. 3.57, 3.62A, and 3.64). Sympathetic fibers conveyed by these nerves assist in the thermoregulation of the testes, stimulating contraction of the smooth dartos muscle in response to cold or stimulating the scrotal sweat glands while inhibiting contraction of the dartos muscle in response to excessive warmth.

PENIS

The **penis** is the male copulatory organ and, by conveying the urethra, provides the common outlet for urine and semen (Figs. 3.60–3.62). The penis consists of a *root, body,* and *glans.* It is composed of three cylindrical cavernous bodies of erectile tissue: the paired **corpora cavernosa** dorsally and the single **corpus spongiosum** ventrally. In the anatomical position, the penis is erect; when the penis is flaccid, its dorsum is directed anteriorly. Each *cavernous body* has an outer fibrous covering or capsule, the **tunica albuginea** (Fig. 3.61C). Superficial to the outer covering is the **deep fascia of the penis** (Buck fascia), the continuation of the deep perineal fascia that forms a strong membranous covering for the corpora cavernosa and corpus spongiosum, binding them together (Fig. 3.61C & D). The corpus spongiosum contains the *spongy urethra.* The corpora cavernosa are fused with each other in the median plane, except posteriorly where they separate to form the **crura of the penis** (Figs. 3.60 and 3.62B). Internally, the cavernous tissue of the corpora is separated (usually incompletely) by the **septum penis** (Fig. 3.61C).

The **root of the penis,** the attached part, consists of the crura, bulb, and ischiocavernosus and bulbospongiosus muscles (Figs. 3.60 and 3.62A & B). The root is located in the superficial perineal pouch, between the perineal membrane superiorly and the deep perineal fascia inferiorly (see Fig. 3.53B & D). The **crura** and **bulb of the penis** consist of erectile tissue. Each crus is attached to the inferior part of the internal surface of the corresponding ischial ramus (see

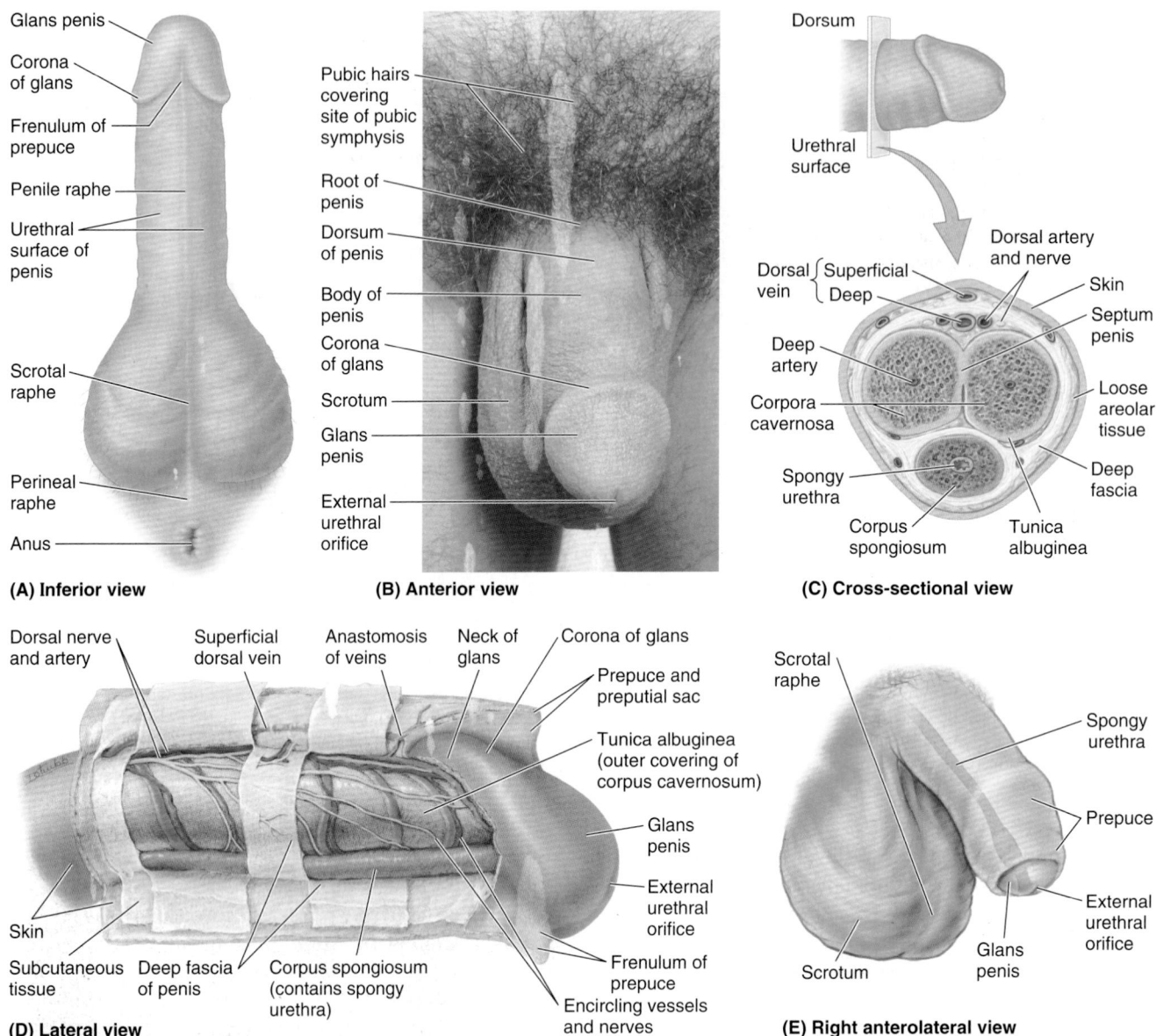

(A) Inferior view

Glans penis
Corona of glans
Frenulum of prepuce
Penile raphe
Urethral surface of penis
Scrotal raphe
Perineal raphe
Anus

(B) Anterior view

Pubic hairs covering site of pubic symphysis
Root of penis
Dorsum of penis
Body of penis
Corona of glans
Scrotum
Glans penis
External urethral orifice

(C) Cross-sectional view

Dorsum
Urethral surface
Dorsal vein { Superficial, Deep }
Deep artery
Corpora cavernosa
Spongy urethra
Dorsal artery and nerve
Skin
Septum penis
Loose areolar tissue
Deep fascia
Corpus spongiosum
Tunica albuginea

(D) Lateral view

Dorsal nerve and artery
Superficial dorsal vein
Anastomosis of veins
Neck of glans
Corona of glans
Prepuce and preputial sac
Tunica albuginea (outer covering of corpus cavernosum)
Glans penis
External urethral orifice
Frenulum of prepuce
Encircling vessels and nerves
Skin
Subcutaneous tissue
Deep fascia of penis
Corpus spongiosum (contains spongy urethra)

(E) Right anterolateral view

Scrotal raphe
Spongy urethra
Prepuce
External urethral orifice
Glans penis
Scrotum

FIGURE 3.61. Penis and scrotum. A. The urethral surface of the circumcised penis is shown. The spongy urethra is deep to the cutaneous penile raphe. The scrotum is divided into right and left halves by the cutaneous scrotal raphe, which is continuous with the penile and perineal raphes. **B.** The dorsum of the circumcised penis and the anterior surface of the scrotum are shown. The penis comprises a root, body, and glans. **C.** The penis contains three erectile masses: two corpora cavernosa and a corpus spongiosum (containing the spongy urethra). **D.** The skin of the penis extends distally as the prepuce, overlapping the neck and corona of the glans. **E.** An uncircumcised penis.

Fig. 3.52D), anterior to the ischial tuberosity. The enlarged posterior part of the bulb of the penis is penetrated superiorly by the urethra, continuing from its intermediate part (Figs. 3.60 and 3.62B).

The **body of the penis** is the free pendulous part that is suspended from the pubic symphysis. Except for a few fibers of the bulbospongiosus near the root of the penis and the ischiocavernosus that embrace the crura, the body of the penis has no muscles (Fig. 3.62).

The penis consists of thin skin, connective tissue, blood and lymphatic vessels, fascia, the corpora cavernosa, and corpus spongiosum containing the spongy urethra (Fig. 3.61C). Distally, the corpus spongiosum expands to form the conical

glans penis, or head of the penis (Figs. 3.61A, B, & D and 3.62B). The margin of the glans projects beyond the ends of the corpora cavernosa to form the **corona of the glans.** The corona overhangs an obliquely grooved constriction, the **neck of the glans,** which separates the glans from the body of the penis. The slit-like opening of the spongy urethra, the *external urethral orifice* (meatus), is near the tip of the glans penis.

The skin of the penis is thin, darkly pigmented relative to adjacent skin, and connected to the tunica albuginea by loose connective tissue. At the neck of the glans, the skin and fascia of the penis are prolonged as a double layer of skin, the **prepuce** (foreskin), which in uncircumcised males

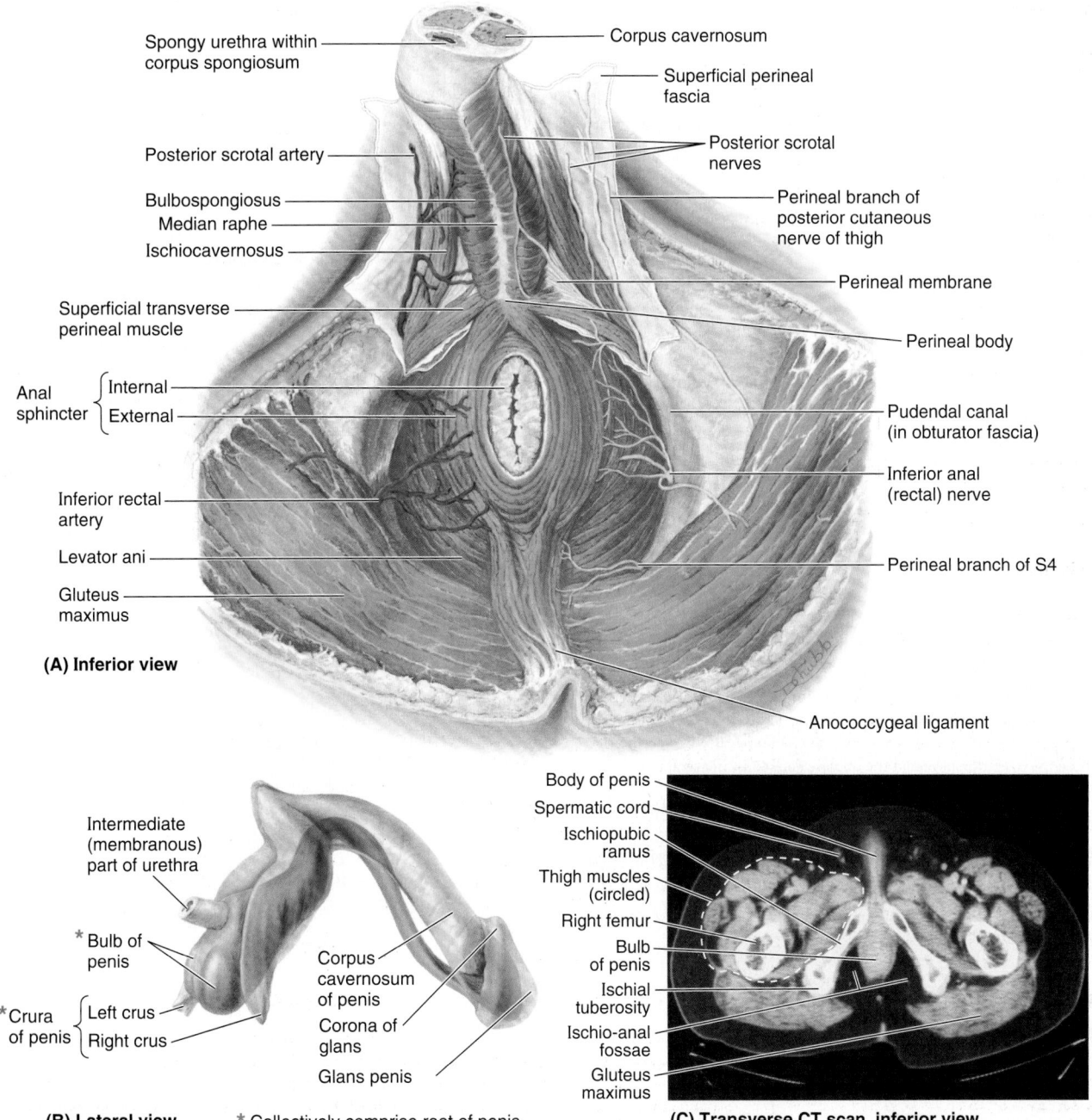

Spongy urethra within corpus spongiosum

Corpus cavernosum

Superficial perineal fascia

Posterior scrotal artery

Posterior scrotal nerves

Bulbospongiosus
Median raphe
Ischiocavernosus

Perineal branch of posterior cutaneous nerve of thigh

Superficial transverse perineal muscle

Perineal membrane

Perineal body

Anal sphincter { Internal — External — }

Pudendal canal (in obturator fascia)

Inferior anal (rectal) nerve

Inferior rectal artery

Levator ani

Perineal branch of S4

Gluteus maximus

(A) Inferior view

Anococcygeal ligament

Intermediate (membranous) part of urethra

* Bulb of penis

*Crura of penis { Left crus — Right crus }

Corpus cavernosum of penis

Corona of glans

Glans penis

(B) Lateral view   * Collectively comprise root of penis

Body of penis
Spermatic cord
Ischiopubic ramus
Thigh muscles (circled)
Right femur
Bulb of penis
Ischial tuberosity
Ischio-anal fossae
Gluteus maximus

(C) Transverse CT scan, inferior view

FIGURE 3.62. Male perineum and structure of penis. A. The anal canal is surrounded by the external anal sphincter, with an ischio-anal fossa on each side. The inferior anal (rectal) nerve branches from the pudendal nerve at the entrance to the pudendal canal and, with the perineal branch of S4, supplies the external anal sphincter. **B.** The corpus spongiosum has been separated from the corpora cavernosa. The natural flexures of the penis are preserved. The glans penis fits like a cap over the blunt ends of the corpora cavernosa. **C.** CT scan at level of superficial pouch of a male. (Courtesy of Dr. Donald R. Cahill, Department of Anatomy, Mayo Medical School, Rochester, MN.)

covers the glans penis to a variable extent (Fig 3.61E). The **frenulum of the prepuce** is a median fold that passes from the deep layer of the prepuce to the urethral surface of the glans (Fig. 3.61A & D).

The **suspensory ligament of the penis** is a condensation of deep fascia that arises from the anterior surface of the pubic symphysis (Fig. 3.63). The ligament passes inferiorly and splits to form a sling that is attached to the deep fascia

of the penis at the junction of its root and body. The fibers of the suspensory ligament are short and taut, anchoring the erectile bodies of the penis to the pubic symphysis.

The **fundiform ligament of the penis** is an irregular mass or condensation of collagen and elastic fibers of the subcutaneous tissue that descends in the midline from the linea alba anterior to the pubic symphysis (see Fig. 3.53F). The ligament splits to surround the penis and then unites and

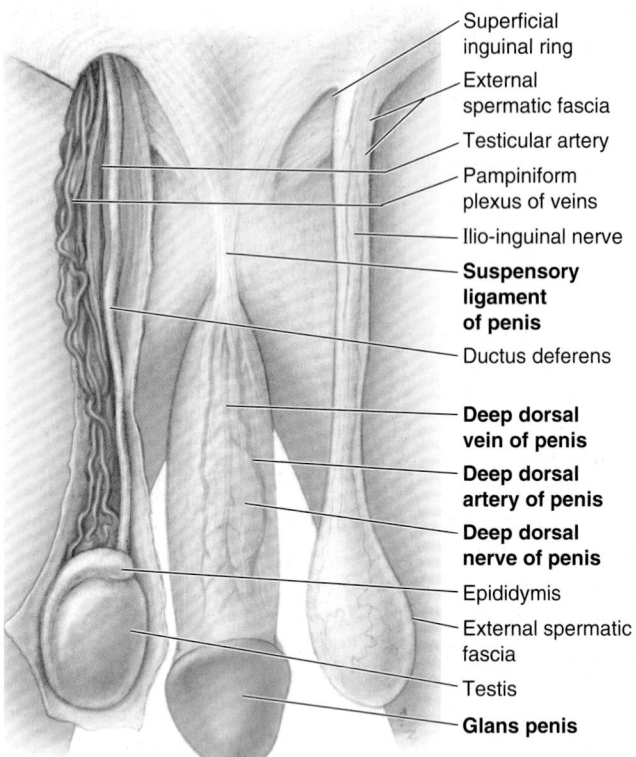

- Superficial inguinal ring
- External spermatic fascia
- Testicular artery
- Pampiniform plexus of veins
- Ilio-inguinal nerve
- **Suspensory ligament of penis**
- Ductus deferens
- **Deep dorsal vein of penis**
- **Deep dorsal artery of penis**
- **Deep dorsal nerve of penis**
- Epididymis
- External spermatic fascia
- Testis
- **Glans penis**

FIGURE 3.63. Vessels and nerves on dorsum of penis and contents of spermatic cord. The skin of the penis and scrotum has been removed. The superficial (dartos) fascia covering the penis has also been removed to expose the deep dorsal vein in the midline flanked by bilateral dorsal arteries and nerves.

blends inferiorly with the dartos fascia forming the scrotal septum. The fibers of the fundiform ligament are relatively long and loose and lie superficial (anterior) to the suspensory ligament.

Arterial Supply of Penis. The penis is supplied mainly by *branches of the internal pudendal arteries* (see Fig. 3.58A; Table 3.8).

- **Dorsal arteries of the penis** run on each side of the deep dorsal vein in the dorsal groove between the corpora cavernosa (Figs. 3.61C & D and 3.63), supplying the fibrous tissue around the corpora cavernosa, the corpus spongiosum and spongy urethra, and the penile skin.
- **Deep arteries of the penis** pierce the crura proximally and run distally near the center of the corpora cavernosa, supplying the erectile tissue in these structures (Figs. 3.58A and 3.61C).
- **Arteries of the bulb of the penis** supply the posterior (bulbous) part of the corpus spongiosum and the urethra within it as well as the bulbo-urethral gland (see Fig. 3.58A).

In addition, **superficial** and **deep branches of the external pudendal arteries** supply the penile skin, anastomosing with branches of the internal pudendal arteries.

The *deep arteries of the penis* are the main vessels supplying the cavernous spaces in the erectile tissue of the corpora cavernosa and are, therefore, involved in the erection of the penis. They give off numerous branches that open directly into the cavernous spaces. When the penis is flaccid, these arteries are coiled, restricting blood flow; they are called **helicine arteries of the penis** (G. *helix*, a coil).

Venous Drainage of Penis. Blood from the cavernous spaces is drained by a venous plexus that joins the **deep dorsal vein of the penis** in the deep fascia (Figs. 3.61C and 3.63). This vein passes between the laminae of the suspensory ligament of the penis, inferior to the inferior pubic ligament and anterior to the perineal membrane, to enter the pelvis, where it drains into the prostatic venous plexus. Blood from the skin and subcutaneous tissue of the penis drains into the **superficial dorsal vein(s),** which drain(s) into the *superficial external pudendal vein.* Some blood also passes to the internal pudendal vein.

Innervation of Penis. The nerves derive from the S2–S4 spinal cord segments and spinal ganglia, passing through the pelvic splanchnic and pudendal nerves, respectively (Fig. 3.64). Sensory and sympathetic innervation is provided

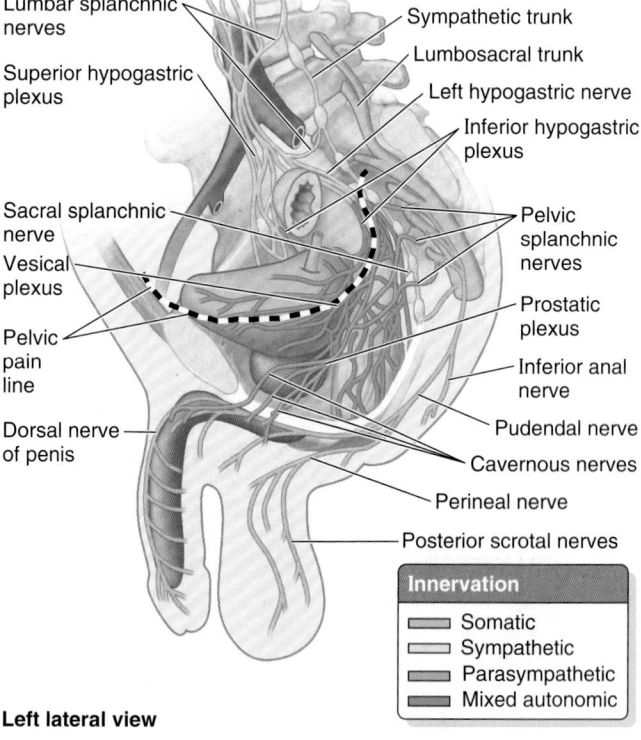

- Lumbar splanchnic nerves
- Superior hypogastric plexus
- Sacral splanchnic nerve
- Vesical plexus
- Pelvic pain line
- Dorsal nerve of penis
- Sympathetic trunk
- Lumbosacral trunk
- Left hypogastric nerve
- Inferior hypogastric plexus
- Pelvic splanchnic nerves
- Prostatic plexus
- Inferior anal nerve
- Pudendal nerve
- Cavernous nerves
- Perineal nerve
- Posterior scrotal nerves

Innervation
- Somatic
- Sympathetic
- Parasympathetic
- Mixed autonomic

Left lateral view

FIGURE 3.64. Nerves of perineum. The pudendal nerve conveys the majority of sensory, sympathetic, and somatic motor fibers to the perineum. Although originating from the same spinal cord segments from which the pudendal nerve is derived, the parasympathetic fibers of the cavernous nerves course independently of the pudendal nerve. With the exception of the cavernous nerves, parasympathetic fibers do not occur outside the head, neck, or cavities of the trunk. The cavernous nerves arise from the prostatic plexus of males and from the vesical plexus of females. They terminate on the arteriovenous anastomoses and helicine arteries of the erectile bodies, which, when stimulated, produce erection of the penis or engorgement of the clitoris and vestibular bulb in females.

primarily by the *dorsal nerve of the penis*, a terminal branch of the *pudendal nerve*, which arises in the pudendal canal and passes anteriorly into the deep perineal pouch. It then runs to the dorsum of the penis, where it runs lateral to the dorsal artery (Figs. 3.61C and 3.63). It supplies both the skin and glans penis. The penis is richly supplied with a variety of sensory nerve endings, especially the glans penis. Branches of the *ilio-inguinal nerve* supply the skin at the root of the penis. *Cavernous nerves*, conveying parasympathetic fibers independently from the prostatic nerve plexus, innervate the helicine arteries of the erectile tissue.

LYMPHATIC DRAINAGE OF MALE PERINEUM

Lymph from the skin of all parts of the perineum, including the hairless skin inferior to the pectinate line of the anorectum but excluding the glans penis, drains to the *superficial inguinal nodes* (Fig. 3.65).

Reflective of their abdominal origins, lymph from the testes follow a route, independent of the scrotal drainage, along the testicular veins to the intermesenteric portion of the *lumbar (caval/aortic)* and *pre-aortic lymph nodes.*

Lymphatic drainage from the intermediate and proximal parts of the urethra and cavernous bodies drain into the *internal iliac lymph nodes*, whereas most vessels from the distal spongy urethra and glans penis pass to the *deep inguinal nodes*, but some lymph passes to the external inguinal nodes.

PERINEAL MUSCLES OF MALE

The **superficial perineal muscles,** located in the superficial perineal pouch, include the superficial transverse peri-

neal, bulbospongiosus, and ischiocavernosus muscles (Figs. 3.62A and 3.66). Details about the attachments, innervation, and actions of these muscles are provided in Table 3.9.

The **superficial transverse perineal muscles** and the bulbospongiosus muscles join the external anal sphincter in attaching centrally to the perineal body. They cross the pelvic outlet like intersecting beams, supporting the perineal body to aid the pelvic diaphragm in supporting the pelvic viscera. Simultaneous contraction of the superficial perineal muscles (plus the deep transverse perineal muscle) during penile erection provides a firmer base for the penis.

The **bulbospongiosus muscles** form a constrictor that compresses the bulb of the penis and the corpus spongiosum, thereby aiding in emptying the spongy urethra of residual urine and/or semen. The anterior fibers of the bulbospongiosus, encircling the most proximal part of the body of the penis, also assist erection by increasing the pressure on the erectile tissue in the root of the penis (Fig. 3.62A). At the same time, they also compress the deep dorsal vein of the penis, impeding venous drainage of the cavernous spaces and helping promote enlargement and turgidity of the penis.

The **ischiocavernosus muscles** surround the crura in the root of the penis. They force blood from the cavernous spaces in the crura into the distal parts of the corpora cavernosa, which increases the turgidity (firm distension) of the penis during erection. Contraction of the ischiocavernosus muscles also compresses the tributaries of deep dorsal vein of the penis leaving the crus of the penis, thereby restricting venous outflow from the penis and helping maintain the erection.

Because of their function during erection and the activity of the bulbospongiosus subsequent to urination and ejaculation to expel the last drops of urine and semen, the perineal muscles are generally more developed in males than in females.

ERECTION, EMISSION, EJACULATION, AND REMISSION

When a male is stimulated erotically, arteriovenous anastomoses by which blood is normally able to bypass the "empty" potential spaces or sinuses of the corpora cavernosa are closed. The smooth muscle in the fibrous trabeculae and coiled helicine arteries relaxes (is inhibited) as a result of parasympathetic stimulation (S2–S4 through the cavernous nerves from the **prostatic nerve plexus**). Consequently, the *helicine arteries* straighten, enlarging their lumina and allowing blood to flow into and dilate the cavernous spaces in the corpora of the penis.

The bulbospongiosus and ischiocavernosus muscles compress veins egressing from the corpora cavernosa, impeding the return of venous blood. As a result, the corpora cavernosa and corpus spongiosum become engorged with blood near arterial pressure, causing the erectile bodies to become turgid (enlarged and rigid), and an **erection** occurs.

During **emission**, semen (sperms and other glandular secretions) is delivered to the prostatic urethra through the ejaculatory ducts after peristalsis of the ductus deferentes and seminal glands. Prostatic fluid is added to the seminal

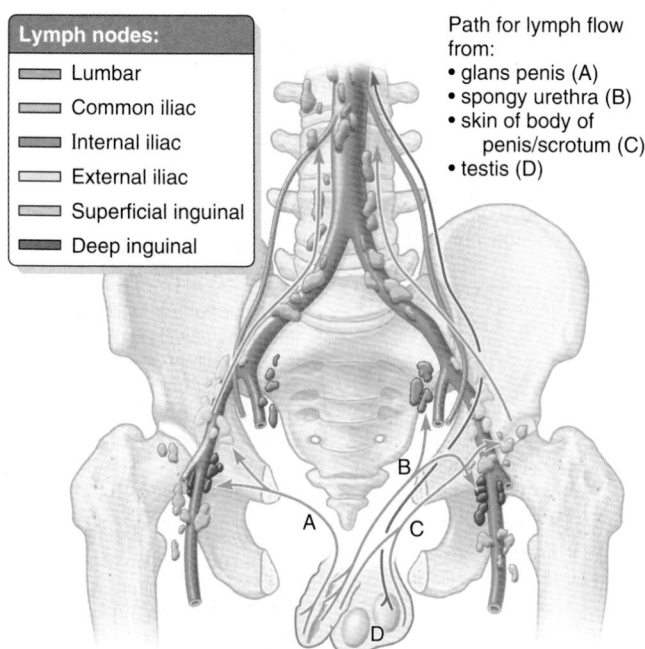

Lymph nodes:
- Lumbar
- Common iliac
- Internal iliac
- External iliac
- Superficial inguinal
- Deep inguinal

Path for lymph flow from:
- glans penis (A)
- spongy urethra (B)
- skin of body of penis/scrotum (C)
- testis (D)

FIGURE 3.65. Lymphatic drainage of male urogenital triangle—penis, spongy urethra, scrotum, and testis. The *arrows* indicate the direction of lymph flow to the lymph nodes.

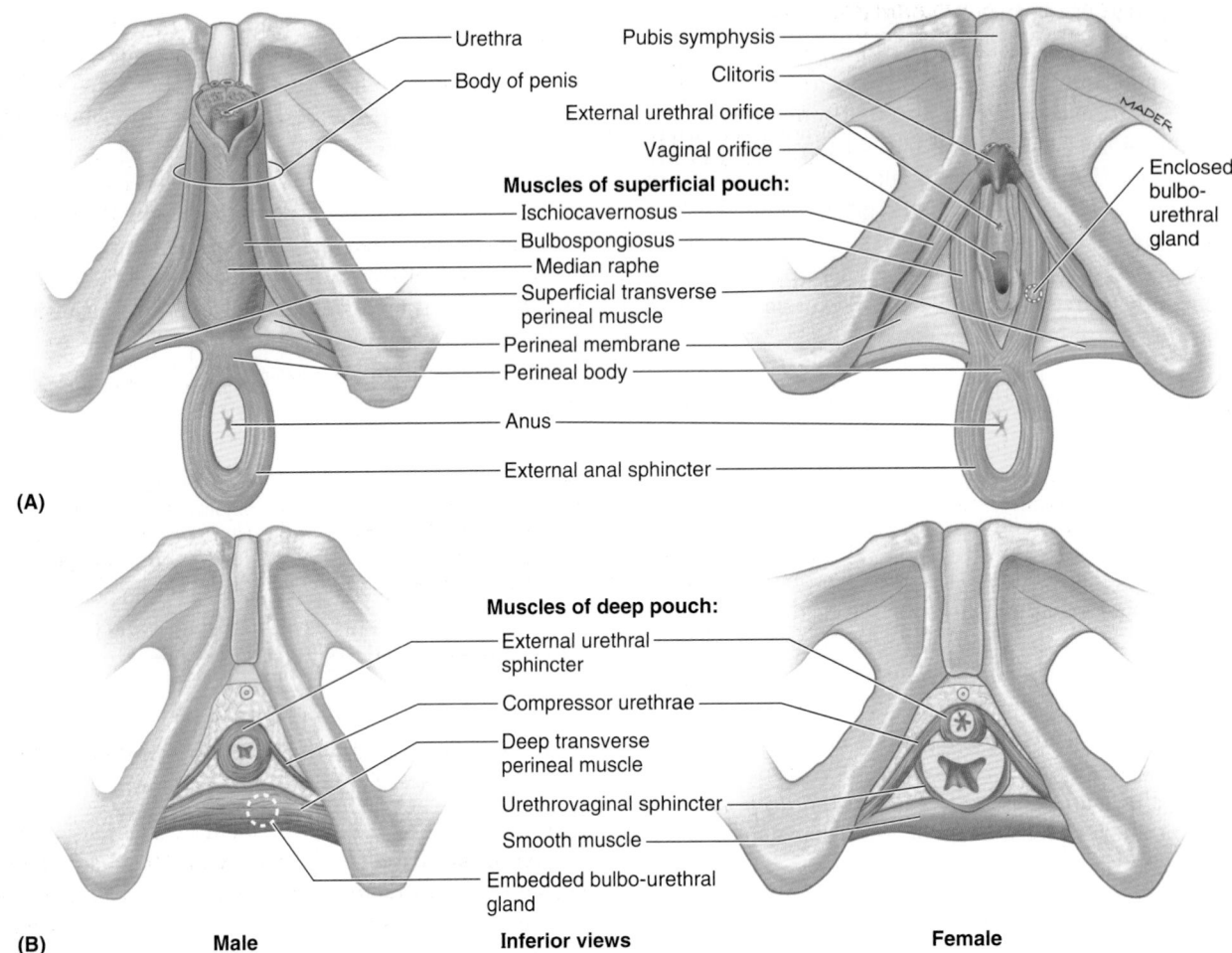

FIGURE 3.66. Muscles of perineum. A. Muscles of superficial perineal pouch. **B.** Muscles of deep perineal pouch.

TABLE 3.9. MUSCLES OF PERINEUM

Muscle	Origin	Course and Distribution	Innervation	Main Action
External anal sphincter	Skin and fascia surrounding anus; coccyx via anococcygeal ligament	Passes around lateral aspects of anal canal, insertion into perineal body	Inferior anal (rectal) nerve, a branch of pudendal nerve (S2–S4)	Constricts anal canal during peristalsis, resisting defecation; supports and fixes perineal body and pelvic floor
Bulbospongiosus	*Male:* median raphe on ventral surface of bulb of penis; perineal body	*Male:* surrounds lateral aspects of bulb of penis and most proximal part of body of penis, inserting into perineal membrane, dorsal aspect of corpus spongiosum and corpora cavernosa, and fascia of bulb of penis	Muscular (deep) branch of perineal nerve, a branch of pudendal nerve (S2–S4)	*Male:* supports and fixes perineal body/pelvic floor; compresses bulb of penis to expel last drops of urine/semen; assists erection by compressing outflow via deep perineal vein and by pushing blood from bulb into body of penis
	Female: perineal body	*Female:* passes on each side of lower vagina, enclosing bulb and greater vestibular gland; inserts into pubic arch and fascia of corpora cavernosa of clitoris		*Female:* supports and fixes perineal body/pelvic floor; "sphincter" of vagina; assists in erection of clitoris (and perhaps bulb of vestibule); compresses greater vestibular gland

TABLE 3.9. MUSCLES OF PERINEUM (Continued)

Muscle	Origin	Course and Distribution	Innervation	Main Action
Ischiocavernosus	Internal surface of ischiopubic ramus and ischial tuberosity	Embraces crus of penis or clitoris, inserting onto inferior and medial aspects of crus and to perineal membrane medial to crus	Muscular (deep) branch of perineal nerve, a branch of pudendal nerve (S2–S4)	Maintains erection of penis or clitoris by compressing outflow veins and pushing blood from the root of penis or clitoris into the body of penis or clitoris
Superficial transverse perineal muscle		Passes along inferior aspect of posterior border of perineal membrane to perineal body		Supports and fixes perineal body/pelvic floor to support abdominopelvic viscera and resist increased intra-abdominal pressure
Deep transverse perineal muscle		Passes along superior aspect of posterior border of perineal membrane to perineal body and external anal sphincter		
External urethral sphincter	(Compressor urethra portion only)	Surrounds urethra superior to perineal membrane; in males, it also ascends anterior aspect of prostate; in females, some fibers also enclose vagina (urethrovaginal sphincter)	Dorsal nerve of penis or clitoris, the terminal branch of the pudendal nerve (S2–S4)	Compresses urethra to maintain urinary continence; in females, urethrovaginal sphincter portion also compresses vagina

fluid as the smooth muscle in the prostate contracts. Emission is a sympathetic response (L1–L2 nerves). During **ejaculation,** semen is expelled from the urethra through the external urethral orifice.

Ejaculation results from:

- Closure of the internal urethral sphincter at the neck of the urinary bladder, *a sympathetic response* (L1–L2 nerves).
- Contraction of the urethral muscle, *a parasympathetic response* (S2–S4 nerves).

- Contraction of the bulbospongiosus muscles, from the pudendal nerves (S2–S4).

After ejaculation, the penis gradually returns to a flaccid state (**remission**), resulting from sympathetic stimulation, which causes constriction of the smooth muscle in the coiled helicine arteries. The bulbospongiosus and ischiocavernosus muscles relax, allowing more blood to be drained from the cavernous spaces in the corpora cavernosa into the deep dorsal vein.

MALE UROGENITAL TRIANGLE

Urethral Catheterization

Urethral catheterization is done to remove urine from a person who is unable to micturate. It is also performed to irrigate the bladder and to obtain an uncontaminated sample of urine. When inserting catheters and *urethral sounds* (slightly conical instruments for exploring and dilating a constricted urethra), the curves of the male urethra must be considered. Just distal to the perineal membrane, the spongy urethra is well covered inferiorly and posteriorly by erectile tissue of the bulb of the penis; however, a short segment of the intermediate part of the urethra is unprotected (Fig. B3.32). Because the urethral wall is thin and the angle that must be negotiated to enter the intermediate part of the spongy urethra, the wall is vulnerable to rupture during the insertion of urethral catheters and sounds. The intermediate part, the least distensible part, runs infero-

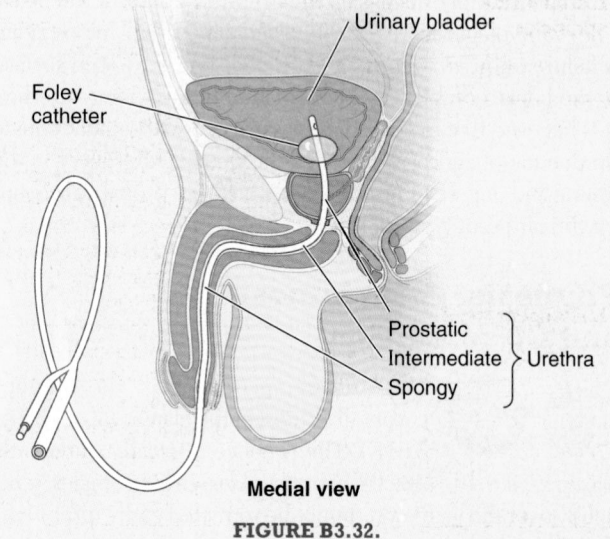

Medial view

FIGURE B3.32.

anteriorly as it passes through the external urethral sphincter. Proximally, the prostatic part takes a slight curve that is concave anteriorly as it traverses the prostate.

Urethral stricture may result from external trauma of the penis or infection of the urethra. Urethral sounds are used to dilate the constricted urethra in such cases. The spongy urethra will expand enough to permit passage of an instrument approximately 8 mm in diameter. The external urethral orifice is the narrowest and least distensible part of the urethra; hence, an instrument that passes through this opening normally passes through all other parts of the urethra.

Distension of Scrotum

The scrotum is easily distended. In persons with large indirect inguinal hernias, for example, the intestine may enter the scrotum, making it as large as a soccer ball. Similarly, inflammation of the testes (*orchitis*), associated with mumps, bleeding in the subcutaneous tissue, or chronic lymphatic obstruction (as occurs in the parasitic disease *elephantiasis*) may produce an enlarged scrotum.

Palpation of Testes

The soft, pliable skin of the scrotum makes it easy to palpate the testes and the structures related to them (e.g., the epididymis and ductus deferens). The left testis commonly lies at a more inferior level than does the right one.

Hypospadias

Hypospadias is a common congenital anomaly of the penis, occurring in 1 in 300 newborns. In the simplest and most common form, *glanular hypospadias,* the external urethral orifice is on the ventral aspect of the glans penis. In other infants, the defect is in the body of the penis (*penile hypospadias*) (Fig. B3.33A), or in the perineum (*penoscrotal* or *scrotal hypospadias*) (Fig. B3.33B). Hence, the external urethral orifice is on the urethral surface of the penis. The embryological basis of penile and penoscrotal hypospadias is failure of the *urogenital folds* to fuse on the ventral surface of the penis, completing the formation of the spongy urethra. It is believed that hypospadias is associated with an inadequate production of androgens by the fetal testes. Differences in the timing and degree of hormonal insufficiency probably account for the different types of hypospadias.

Phimosis, Paraphimosis, and Circumcision

In an uncircumcised penis, the prepuce covers all or most of the glans penis (see Fig. 3.61E). The prepuce is usually sufficiently elastic for it to be retracted over the glans. In some males, it fits tightly over the glans and cannot be retracted easily (*phimosis*) if at all. As there are modified sebaceous glands in the prepuce,

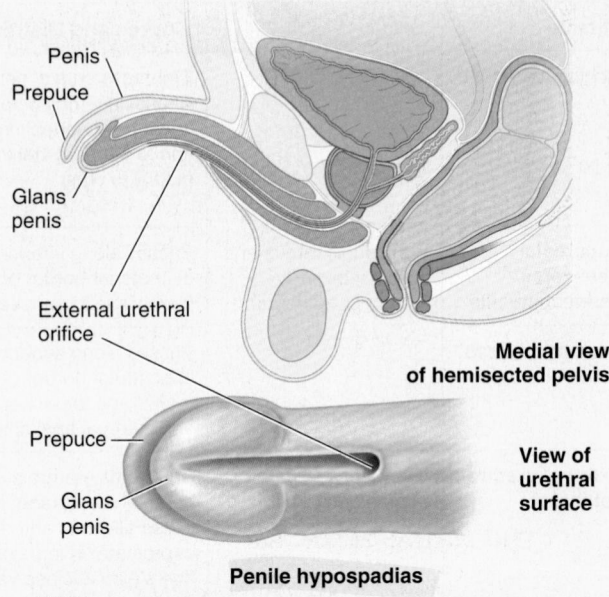

Penis
Prepuce
Glans penis
External urethral orifice

Medial view of hemisected pelvis

Prepuce
Glans penis

View of urethral surface

Penile hypospadias

(A)

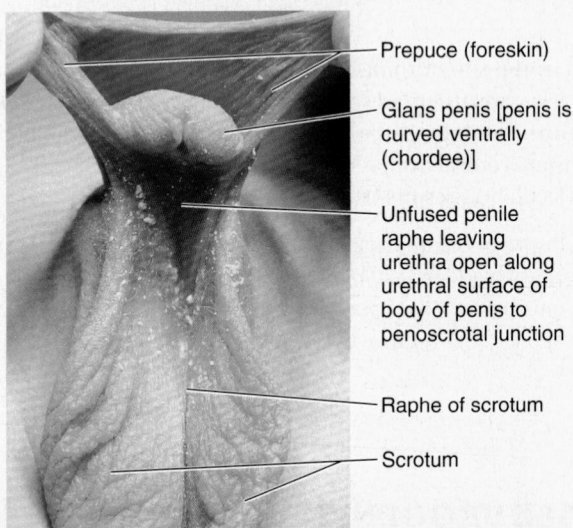

Prepuce (foreskin)

Glans penis [penis is curved ventrally (chordee)]

Unfused penile raphe leaving urethra open along urethral surface of body of penis to penoscrotal junction

Raphe of scrotum

Scrotum

(B) Antero-inferior view

FIGURE B3.33.

the oily secretions of cheesy consistency (*smegma*) from them accumulate in the **preputial sac,** located between the glans and prepuce, causing irritation.

In some males, retraction of the prepuce over the glans penis constricts the neck of the glans so much that there is interference with the drainage of blood and tissue fluid. In persons with this condition (*paraphimosis*), the glans may enlarge so much that the prepuce cannot be drawn over it. Circumcision is commonly performed in such cases.

Circumcision, surgical excision of the prepuce, is the most commonly performed minor surgical operation on male infants. Following circumcision, the glans penis is exposed (see Fig. 3.61B). Although it is a religious practice in Islam and Judaism, it is often done routinely for non-religious reasons (a preference usually explained in terms of tradition or hygiene) in North America. In adults, circumcision is usually performed when phimosis or paraphimosis occurs.

Impotence and Erectile Dysfunction

Inability to obtain an erection (*impotence*) may result from several causes. When a lesion of the prostatic plexus or cavernous nerves results in an inability to achieve an erection, a surgically implanted, semirigid or inflatable penile prosthesis may assume the role of the erectile bodies, providing the rigidity necessary to insert and move the penis within the vagina during intercourse.

Erectile dysfunction (ED) may occur in the absence of a nerve insult due to a variety of other causes. Central nervous system (hypothalamic) and endocrine (pituitary or testicular) disorders may result in reduced testosterone (male hormone) secretion. Nerve fibers may fail to stimulate erectile tissues, or blood vessels may be insufficiently responsive to autonomic stimulation. In many such cases erection can be achieved with the assistance of oral medications or injections that increase blood flow into the cavernous sinusoids by causing relaxation of smooth muscle.

The Bottom Line

MALE UROGENITAL TRIANGLE

Distal male urethra: The intermediate urethra is the shortest and narrowest part of the male urethra, the limit of its distension normally being the same as that of the external urethral orifice. ♦ It is encircled by voluntary muscle of the inferior part of the external urethral sphincter before perforating the perineal membrane. ♦ Immediately inferior to the membrane, the urethra enters the corpus spongiosum and becomes the spongy urethra, the longest part of the male urethra. ♦ The spongy urethra has expansions at each end, the intrabulbar and navicular fossae. ♦ The intermediate and spongy parts of the urethra are supplied and drained by the same dorsal (blood) vessels of the penis, but differ in terms of innervation and lymphatic drainage. The intermediate part follows visceral paths and the spongy part follows somatic paths.

Scrotum: The scrotum is a dynamic, fibromuscular cutaneous sac for the testes and epididymides. ♦ Its internal subdivision by a septum of dartos fascia is demarcated externally by a median scrotal raphe. ♦ The anterior aspect of the scrotum is served by anterior scrotal blood vessels and nerves, continuations of external pudendal blood vessels and branches of the lumbar nerve plexus. ♦ The posterior aspect of the scrotum is served by posterior scrotal blood vessels and nerves, continuations of internal pudendal blood vessels and branches of the sacral nerve plexus. ♦ Sympathetic innervation of smooth dartos muscle and sweat glands assists thermoregulation of the testes.

Penis: The penis is an organ of copulation and excretion of urine and semen. ♦ It is formed mainly of thin, mobile skin overlying three cylindrical bodies of erectile cavernous tissue, the paired corpora cavernosa, and a single corpus spongiosum containing the spongy urethra. ♦ The erectile bodies are bound together by deep fascia of the penis, except at the root where they separate into the crura and bulb of the penis. ♦ The crura attach to the ischiopubic rami, but all parts of the root are attached to the perineal membrane. ♦ At the junction of the root and body, the penis is attached to the pubic symphysis by the suspensory ligament of the penis. ♦ The ischiocavernosus muscles ensheath the crura, and the bulbospongiosus muscle ensheaths the bulb, its most anterior fibers encircling the most proximal part of the penile body and deep dorsal vessels. ♦ The glans penis is a distal expansion of the corpus spongiosum, which has the external urethral orifice at its tip and a projecting corona that overhangs the neck of the glans. ♦ Unless removed by circumcision, the neck is covered by the prepuce (foreskin).

Except for skin near its root, the penis is supplied mainly by branches of the internal pudendal arteries. ♦ The dorsal arteries supply most of the body and glans. ♦ The deep arteries supply the cavernous tissue. The terminal helicine arteries open to engorge the sinuses with blood at arterial pressure, causing penile erection. ♦ Superficial structures drain via the superficial dorsal vein to the external pudendal veins, whereas the erectile bodies drain via the deep dorsal vein to the prostatic venous plexus. ♦ Sensory and sympathetic innervation is provided mainly by the dorsal nerve of the penis, but the helicine arteries that produce erection are innervated by cavernous nerves, extensions of the prostatic nerve plexus.

Perineal muscles: In addition to their bony origins, the voluntary superficial and deep muscles of the perineum are also attached to the perineal membrane (by which they are separated) and the perineal body. ♦ In addition to the sphincteric functions of the external anal and urethral sphincters for maintaining fecal and urinary continence, the male perineal muscles function as a group to provide a base for the penis and support for the perineal body (which in turn supports the pelvic diaphragm). ♦ The ischiocavernosus and bulbospongiosus muscles both constrict venous outflow from the erectile bodies to assist erection, simultaneously pushing blood from the penile root into the penile body. ♦ In addition, the bulbospongiosus muscle constricts the bulb of the penis to express the final drops of urine or semen. ♦ Because of these multiple functions, the perineal muscles are generally relatively well developed in males. The perineal muscles are innervated by muscular branches of the pudendal nerve.

...e Urogenital Triangle

The **female urogenital triangle** includes the female external genitalia, perineal muscles, and anal canal.

FEMALE EXTERNAL GENITALIA

The **female external genitalia** (Fig. 3.67) include the mons pubis, labia majora (enclosing the pudendal cleft), labia minora (enclosing the vestibule of vagina), clitoris, bulbs of vestibule, and greater and lesser vestibular glands. The synonymous terms **vulva** and **pudendum** include all these parts; the term *pudendum* is commonly used clinically. The vulva serves:

- As sensory and erectile tissue for sexual arousal and intercourse.
- To direct the flow of urine.
- To prevent entry of foreign material into the urogenital tract.

Mons Pubis. The **mons pubis** is the rounded, fatty eminence anterior to the pubic symphysis, pubic tubercles, and superior pubic rami. The eminence is formed by a mass of fatty subcutaneous tissue. The amount of

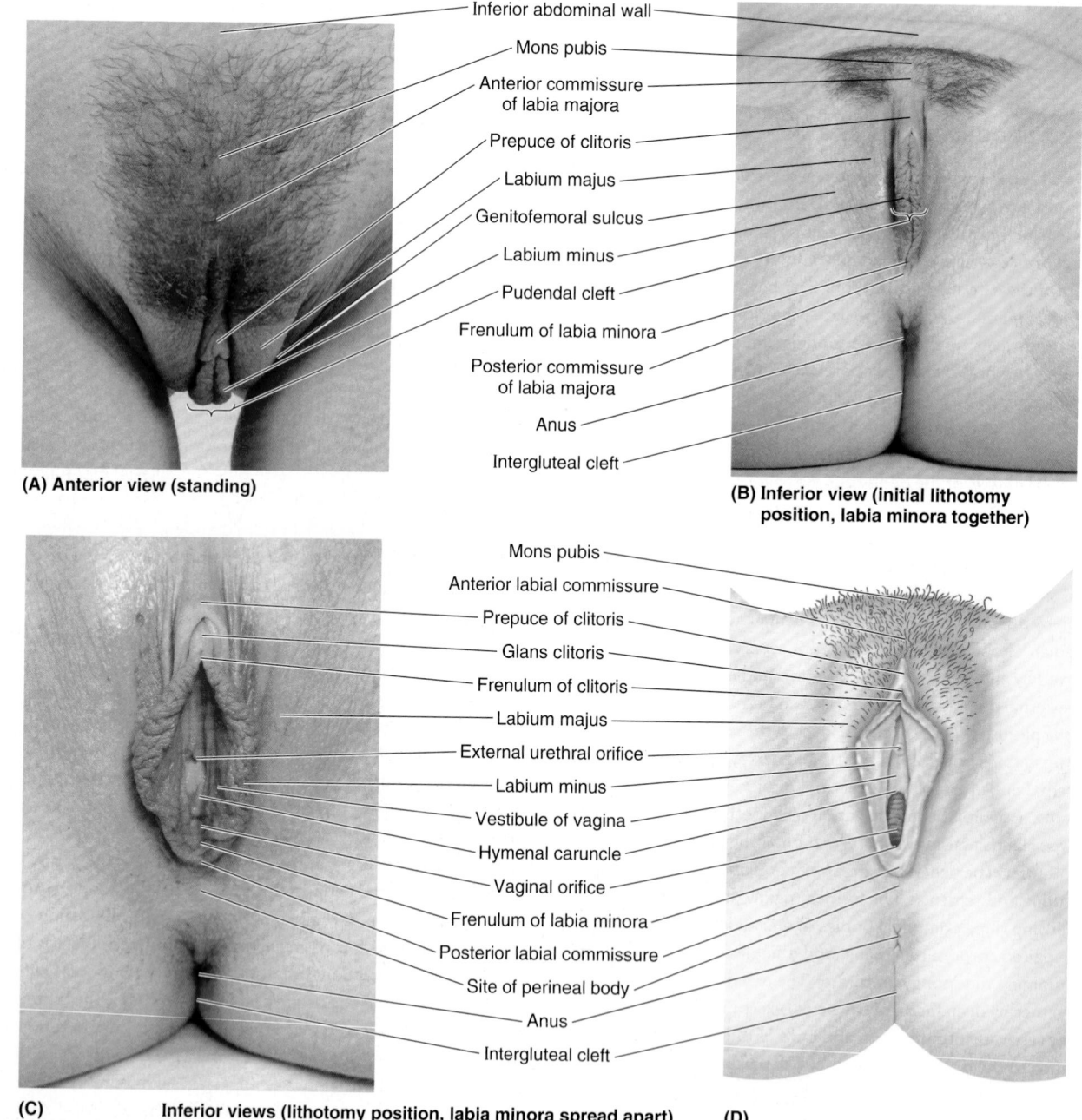

(A) Anterior view (standing)

Inferior abdominal wall
Mons pubis
Anterior commissure of labia majora
Prepuce of clitoris
Labium majus
Genitofemoral sulcus
Labium minus
Pudendal cleft
Frenulum of labia minora
Posterior commissure of labia majora
Anus
Intergluteal cleft

(B) Inferior view (initial lithotomy position, labia minora together)

Mons pubis
Anterior labial commissure
Prepuce of clitoris
Glans clitoris
Frenulum of clitoris
Labium majus
External urethral orifice
Labium minus
Vestibule of vagina
Hymenal caruncle
Vaginal orifice
Frenulum of labia minora
Posterior labial commissure
Site of perineal body
Anus
Intergluteal cleft

(C)　Inferior views (lithotomy position, labia minora spread apart)　**(D)**

FIGURE 3.67. Female external genitalia. A–C. Surface anatomy of vulva (pudendum) of vagina demonstrated in three positions. **D.** Illustration of vulva, similar to (**C**). Moisture typically keeps the labia minora passively apposed, keeping the vestibule of vagina closed (**B**) unless spread apart as in (**C**).

fat increases at puberty and decreases after menopause. The surface of the mons is continuous with the anterior abdominal wall. After puberty, the mons pubis is covered with coarse pubic hairs.

Labia Majora. The **labia majora** are prominent folds of skin that indirectly protect the clitoris and urethral and vaginal orifices (Fig. 3.67). Each labium majus is largely filled with a finger-like "digital process" of loose subcutaneous tissue containing smooth muscle, and the termination of the round ligament of the uterus (Fig. 3.68). The labium majus passes inferoposteriorly from the mons pubis toward the anus (Fig. 3.67D).

The labia majora lie on the sides of a central depression (a narrow slit when the thighs are adducted—Fig. 3.67A),

the **pudendal cleft,** within which are the labia minora and vestibule (Fig. 3.67C & D). The external aspects of the labia majora in the adult are covered with pigmented skin containing many sebaceous glands, and are covered with crisp pubic hairs. The internal aspects of the labia are smooth, pink, and hairless.

The labia majora are thicker anteriorly where they join to form the **anterior commissure.** Posteriorly, in nulliparous women (those never having borne children), they merge to form a ridge, the **posterior commissure,** which overlies the perineal body and is the posterior limit of the vulva. This commissure usually disappears after the first vaginal birth.

Labia Minora. The **labia minora** are rounded folds of fat-free, hairless skin. They are enclosed in the pudendal

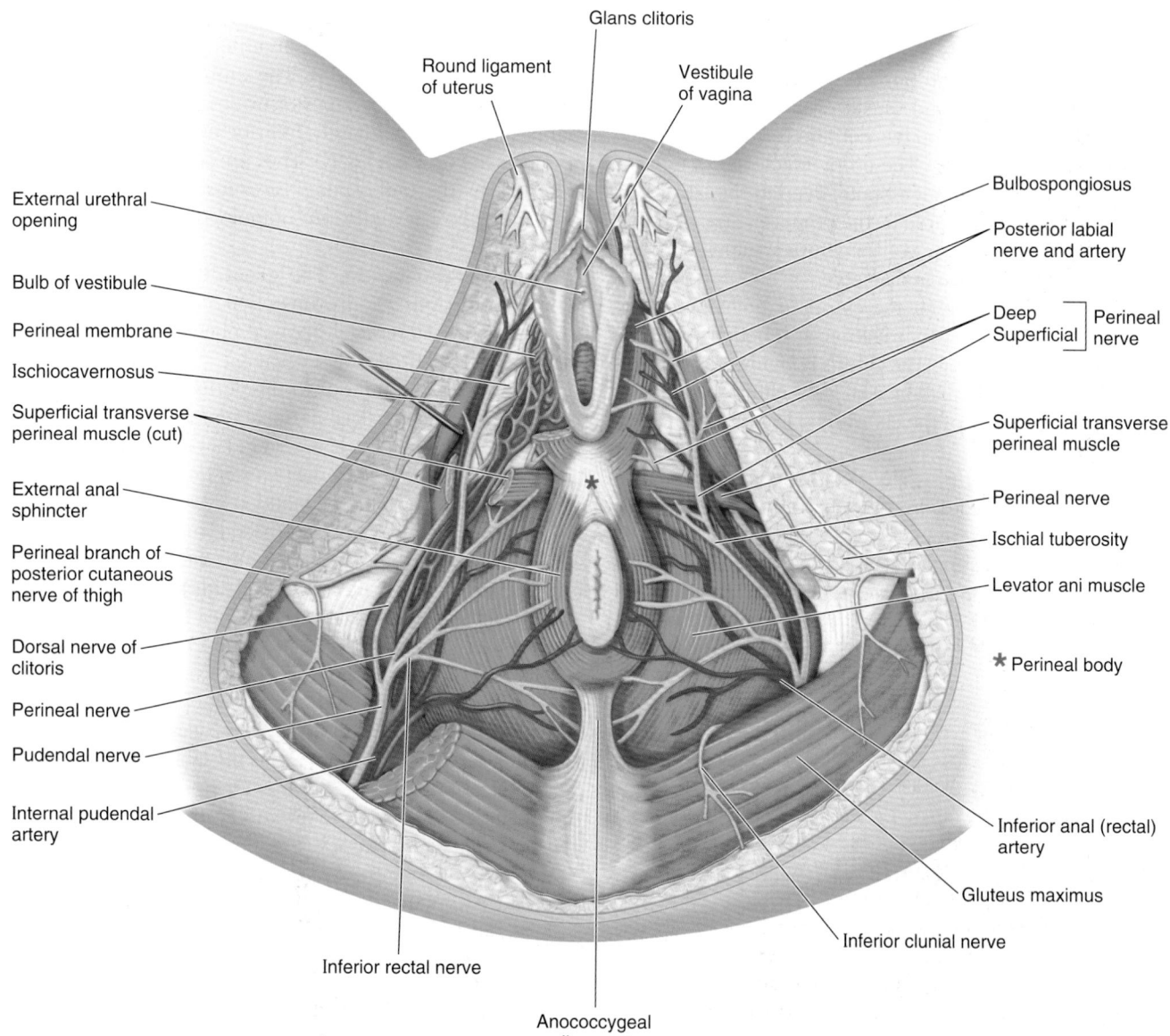

FIGURE 3.68. Female perineum. Skin, subcutaneous tissue (including perineal fascia and ischio-anal fat bodies), and the investing fascia of the muscles have been removed. On the right side, the bulbospongiosus muscle has been resected to reveal the bulb of the vestibule. Deeper dissection of the superficial pouch (right side) reveals the bulbs of the vestibule and the greater vestibular glands.

cleft and immediately surround and close over the *vestibule of vagina* into which both the external urethral and vaginal orifices open. They have a core of spongy connective tissue containing erectile tissue at their base and many small blood vessels. Anteriorly, the labia minora form two laminae. The medial laminae of each side unite as the **frenulum of the clitoris.** The lateral laminae unite anterior to (or often anterior and inferior to, thus overlapping and obscuring) the glans clitoris, forming the **prepuce** (foreskin) **of the clitoris.** In young women, especially virgins, the labia minora are connected posteriorly by a small transverse fold, the **frenulum of the labia minora** (fourchette). Although the internal surface of each labium minus consists of thin moist skin, it has the pink color typical of mucous membrane and contains many sebaceous glands and sensory nerve endings. (See the blue box "Female Circumcision," p. 432.)

Clitoris. The **clitoris** is an erectile organ located where the labia minora meet anteriorly (Figs. 3.67 and 3.68). The clitoris consists of a **root** and a small, cylindrical **body,** which are composed of two crura, two corpora cavernosa, and the **glans clitoris** (Fig. 3.69). The crura attach to the inferior pubic rami and perineal membrane, deep to the labia. The body of the clitoris is covered by the prepuce (Figs. 3.67 and 3.68). Together, the body and glans clitoris are approximately 2 cm in length and <1 cm in diameter.

In contrast to the penis, the clitoris is not functionally related to the urethra or to urination. It functions solely as an organ of sexual arousal. The clitoris is highly sensitive and enlarges on tactile stimulation. The glans clitoris is the most highly innervated part of the clitoris and is densely supplied with sensory endings.

Vestibule of Vagina. The **vestibule of the vagina** is the space surrounded by the labia minora into which the orifices of the urethra and vagina and the ducts of the greater and lesser vestibular glands open (Figs. 3.67C & D and 3.68). The *external urethral orifice* is located 2–3 cm postero-inferior to the glans clitoris and anterior to the vaginal orifice. On each side of the external urethral orifice are the openings of the ducts of the **para-urethral glands**. *Openings of the ducts of the greater vestibular glands* are located on the upper, medial aspects of the labia minora, in 5 and 7 o'clock positions relative to the vaginal orifice in the lithotomy position.

The size and appearance of the **vaginal orifice** vary with the condition of the **hymen,** a thin anular fold of mucus membrane, which partially or wholly occludes the vaginal orifice. After its rupture, only remnants of the hymen, **hymenal caruncles** (tags), are visible (Fig. 3.67C & D). These remnants demarcate the vagina from the vestibule. The hymen has no established physiological function. It is considered primarily a developmental vestige, but its condition (and that of the frenulum of the labia minora) often provides critical evidence in cases of child abuse and rape.

Bulbs of Vestibule. The **bulbs of the vestibule** are paired masses of elongated erectile tissue, approximately 3 cm in length (Fig. 3.68). The bulbs lie along the sides of the vaginal orifice, superior or deep to (not within) the labia minora, immediately inferior to the perineal membrane (see Fig. 3.52D & E). They are covered inferiorly and laterally by the bulbospongiosus muscles extending along their length. The bulbs are homologous with the bulb of the penis.

Vestibular Glands. The **greater vestibular glands** (Bartholin glands), approximately 0.5 cm in diameter, are located in the superficial perineal pouch. They lie on each side of the vestibule of the vagina, posterolateral to the vaginal orifice and inferior to the perineal membrane; thus they are in the superficial perineal pouch (Fig. 3.52D). The greater vestibular glands are round or oval and are partly overlapped posteriorly by the *bulbs of the vestibule.* Like the bulbs, they are partially surrounded by the bulbospongiosus muscles. The slender ducts of these glands pass deep to the bulbs of the vestibule and open into the vestibule on each side of the vaginal orifice. These glands secrete mucus into the vestibule during sexual arousal. (See the blue box "Infection of Greater Vestibular Glands," p. 433.)

The **lesser vestibular glands** are small glands on each side of the vestibule of the vagina that open into it between the urethral and vaginal orifices. These glands secrete mucus into the vestibule, which moistens the labia and vestibule.

Arterial Supply and Venous Drainage of Vulva. The abundant arterial supply to the vulva is from the *external* and *internal pudendal arteries* (Fig. 3.68-; see also Fig. 3.58B; Table 3.8). The *internal pudendal artery* supplies most of the skin, external genitalia, and perineal muscles. The labial arteries are branches of the internal pudendal artery, as are those of the clitoris.

The labial veins are tributaries of the *internal pudendal veins* and accompanying veins of the internal pudendal artery. Venous engorgement during the excitement phase of the sexual response causes an increase in the size and consistency of the clitoris and the bulbs of the vestibule of the vagina.

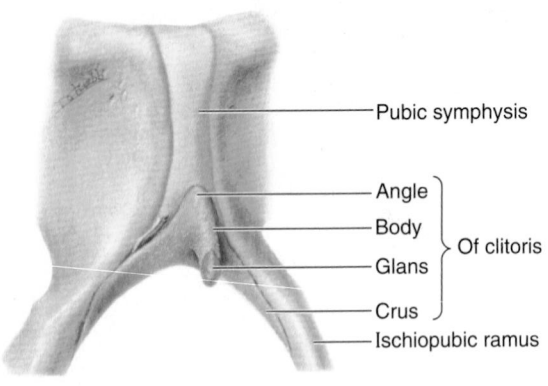

Pubic symphysis

Angle ⎤
Body ⎥
Glans ⎬ Of clitoris
Crus ⎦

Ischiopubic ramus

Anterior view

FIGURE 3.69. Clitoris. The surrounding soft tissues have been removed to reveal the parts of the clitoris.

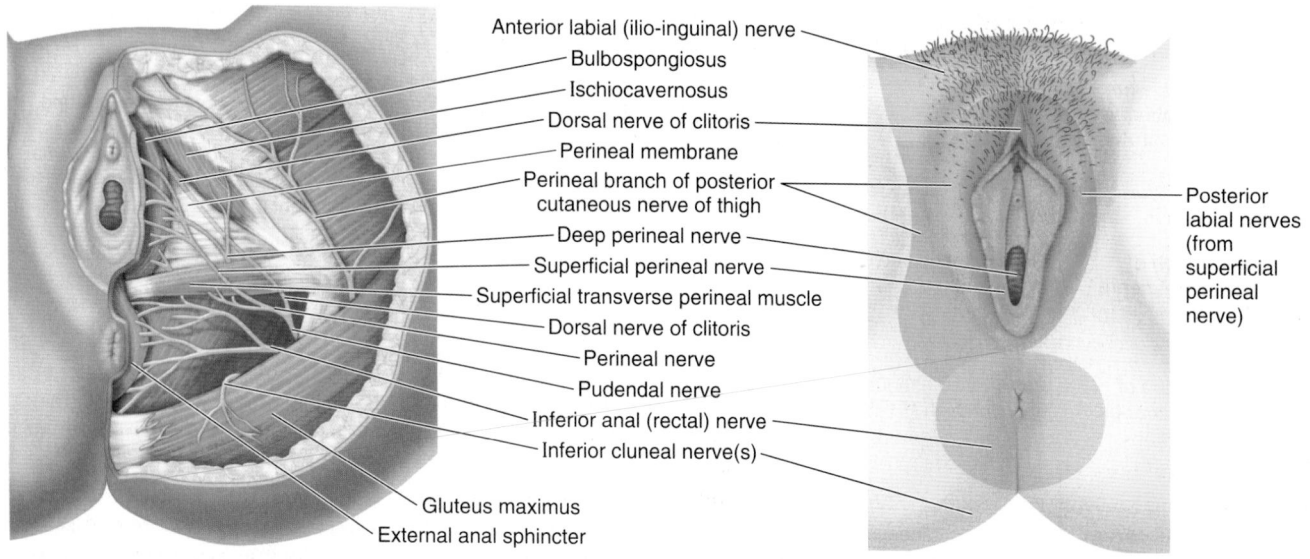

(A) Dissection of perineal nerves

Anterior labial (ilio-inguinal) nerve
Bulbospongiosus
Ischiocavernosus
Dorsal nerve of clitoris
Perineal membrane
Perineal branch of posterior cutaneous nerve of thigh
Deep perineal nerve
Superficial perineal nerve
Superficial transverse perineal muscle
Dorsal nerve of clitoris
Perineal nerve
Pudendal nerve
Inferior anal (rectal) nerve
Inferior cluneal nerve(s)
Gluteus maximus
External anal sphincter

Inferior views (lithotomy position)

Posterior labial nerves (from superficial perineal nerve)

(B) Distribution of perineal nerves

FIGURE 3.70. Nerves of female perineum. A. In this view, the skin, subcutaneous tissue, and ischio-anal fat bodies have been removed. Most of the area and most features of the perineum are innervated by branches of the pudendal nerve (S2–S4). **B.** Cutaneous zones of innervation.

Innervation of Vulva. The anterior aspect of the vulva (mons pubis, anterior labia) is supplied by derivatives of the lumbar plexus: the **anterior labial nerves,** derived from the *ilio-inguinal nerve,* and the *genital branch of the genitofemoral nerve.*

The posterior aspect of the vulva is supplied by derivatives of the sacral plexus: the *perineal branch of the posterior cutaneous nerve of the thigh* laterally, and the *pudendal nerve* centrally (Figs. 3.68 and 3.70). The latter is the primary nerve of the perineum. Its **posterior labial nerves** (terminal *superficial branches of the perineal nerve*) supply the labia. *Deep* and *muscular branches of the perineal nerve* supply the orifice of the vagina and superficial perineal muscles. The *dorsal nerve of the clitoris* supplies deep perineal muscles and sensation to the clitoris. (See the blue box "Pudendal and Ilio-inguinal Nerve Blocks," p. 433.)

The bulb of the vestibule and erectile bodies of the clitoris receive parasympathetic fibers via *cavernous nerves* from the uterovaginal nerve plexus. Parasympathetic stimulation produces increased vaginal secretion, erection of the clitoris, and engorgement of erectile tissue in the bulbs of the vestibule.

LYMPHATIC DRAINAGE OF FEMALE PERINEUM

The vulva contains a rich network of lymphatic vessels. Lymph from the skin of the perineum, including the anoderm inferior to the pectinate line of the anorectum and the inferiormost vagina, vaginal orifice, and vestibule, drains initially to the *superficial inguinal lymph nodes.* Lymph from the clitoris, vestibular bulb, and anterior labia minora drains to the *deep inguinal lymph nodes,* or directly to the *internal iliac lymph nodes,* and that from the urethra drains to the *internal iliac* or *sacral lymph nodes* (Fig. 3.71; Table 3.7).

PERINEAL MUSCLES OF FEMALE

The *superficial perineal muscles* include the *superficial transverse perineal,* **ischiocavernosus,** and *bulbospongiosus muscles* (Fig. 3.66A & B). Details of their attachments, innervation, and action are provided in Table 3.9. (See the blue boxes "Exercises for Increased Development of Female Perineal Muscles" and "Vaginismus," pp. 433 and 434.)

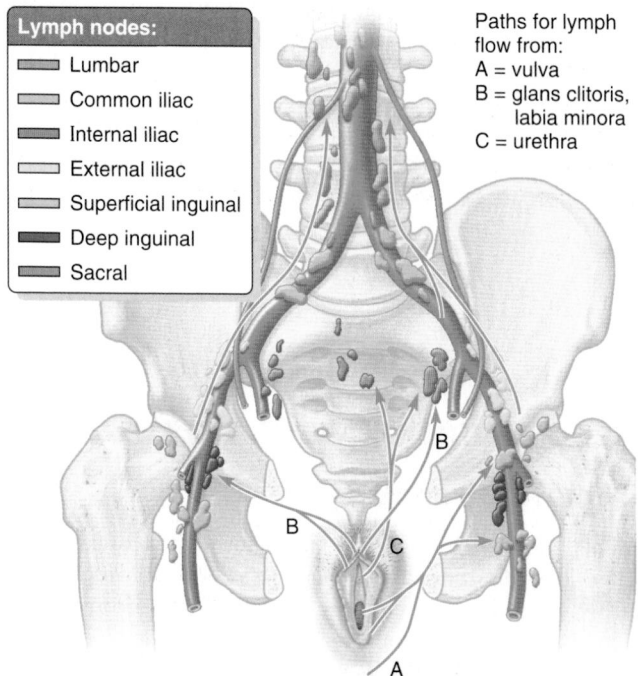

Lymph nodes:
Lumbar
Common iliac
Internal iliac
External iliac
Superficial inguinal
Deep inguinal
Sacral

Paths for lymph flow from:
A = vulva
B = glans clitoris, labia minora
C = urethra

FIGURE 3.71. Lymphatic drainage of vulva. The *arrows* indicate the direction of lymph flow to the lymph nodes.

TABLE 3.10. NERVES OF PERINEUM

Nerve	Origin	Course	Distribution
Anterior labial nerves (♀); **anterior scrotal nerves** (♂)	Terminal part of ilio-inguinal nerve (L1)	Arise as ilio-inguinal exits superficial inguinal ring; pass anteriorly and inferiorly	*In females,* sensory to mons pubis and anterior part of labium majus; *in males,* sensory to pubic region, skin of proximal penis, and anterior aspect of scrotum, and adjacent thigh
Genital branch of genitofemoral nerve	Genitofemoral nerve (L1 and L2)	Emerges through or near superficial inguinal ring	*In females,* sensory to anterior labia majora; *in males,* motor to cremaster muscle, sensory to anterior aspect of scrotum and adjacent thigh
Perineal branch of posterior cutaneous nerve of thigh	Posterior cutaneous nerve of thigh (S1–S3)	Arises deep to inferior border of gluteus maximus; passes medially over sacrotuberous ligament to parallel ischiopubic ramus	Sensory to lateral perineum (labia majora in ♀, scrotum in ♂), genitofemoral sulcus, and superiormost medial thigh; may overlap lateral parts of perineum supplied by pudendal nerve
Inferior clunial nerves	Posterior cutaneous nerve of thigh (S1–S3)	Arise deep to and emerge from inferior border of gluteus maximus, ascending in subcutaneous tissue	Skin of inferior and inferolateral gluteal region (buttocks)—gluteal fold and area superior to it
Pudendal nerve (S2–S4)	Sacral plexus (anterior rami of S2–S4)	Exits pelvis via infrapiriform part of greater sciatic foramen; passes posterior to sacrospinous ligament; enters perineum via lesser sciatic foramen, immediately dividing into branches as it enters pudendal canal	Motor to muscles of perineum and sensory to majority of perineal region via its branches, the inferior rectal and perineal nerves, and the dorsal nerve of clitoris or penis
Inferior anal (rectal) **nerve**	Pudendal nerve (S3–4)	Passes medially from area of ischial spine (entrance to pudendal canal), traversing ischio-anal fat body	External anal sphincter; participates in innervation of inferior and medial-most part of levator ani (puborectalis); sensory to anal canal inferior to pectinate line and circumanal skin
Perineal nerve	Pudendal nerve	Arises near entrance to pudendal canal, paralleling parent nerve to end of canal, then passes medially	Divides into superficial and deep branches, the posterior labial or scrotal nerve and the deep perineal nerve
Posterior labial nerves (♀), **posterior scrotal nerves** (♂)	Superficial terminal branch of perineal nerve	Arise in anterior (terminal) end of pudendal canal, passing medially and superficially	*In females,* labia minora and all but anterior labia majora; *in males,* posterior aspect of scrotum
Deep perineal nerve	Deep terminal branch of perineal nerve	Arise in anterior (terminal) end of pudendal canal, passing medially then deeply in superficial perineal pouch	Motor to muscles of superficial perineal pouch (ischiocavernosus, bulbospongiosus, and superficial perineal muscles); *in females,* sensory to vestibule of vagina and inferior part of vagina

FEMALE UROGENITAL TRIANGLE

Female Circumcision

 Although illegal and now being actively discouraged in most countries, *female circumcision* is widely practiced in some cultures. The operation performed during childhood removes the prepuce of the clitoris and commonly also removes part or all of the clitoris and labia minora. This disfiguring procedure is erroneously thought to inhibit sexual arousal and gratification.

Vulvar Trauma

 The highly vascular bulbs of the vestibule are susceptible to disruption of vessels as the result of trauma (e.g., athletic injuries such as jumping hurdles, sexual assault, and obstetrical injury). These injuries often result in *vulvar hematomas* (localized collection of blood) in the labia majora, for example.

Infection of Greater Vestibular Glands

The greater vestibular glands are usually not palpable, but are so when infected. Occlusion of the vestibular gland duct can predispose the individual to *infection of the greater vestibular gland*. The gland is the site or origin of most vulvar adenocarcinomas (cancers). *Bartholinitis*, inflammation of the greater vestibular (Bartholin) glands, may result from a number of pathogenic organisms. Infected glands may enlarge to a diameter of 4–5 cm and impinge on the wall of the rectum. Occlusion of the vestibular gland duct without infection can result in the accumulation of mucin (*Bartholin gland cyst*) (Fig. B3.34).

Pudendal and Ilio-inguinal Nerve Blocks

To relieve perineal pain during childbirth, *pudendal nerve block anesthesia* may be performed by injecting a local anesthetic agent into the tissues surrounding the pudendal nerve (Fig. B3.35). The injection is made where the pudendal nerve crosses the lateral aspect of the sacrospinous ligament, near its attachment to the ischial spine. The needle may be passed through the overlying skin (as illustrated) or, more commonly perhaps, through the vagina parallel to the palpating finger. Because the fetus's head is usually stationed within the lesser pelvis at this stage, it is important that the physician's finger is always positioned between the needle tip and the baby's head during the procedure.

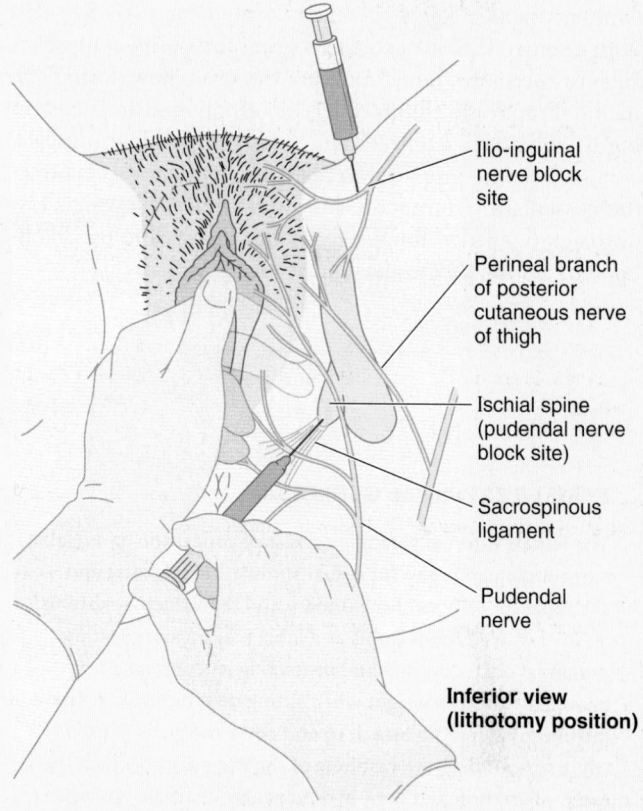

Ilio-inguinal nerve block site

Perineal branch of posterior cutaneous nerve of thigh

Ischial spine (pudendal nerve block site)

Sacrospinous ligament

Pudendal nerve

Inferior view (lithotomy position)

FIGURE B3.35.

To abolish sensation from the anterior part of the perineum, an *ilio-inguinal nerve block* is performed. When patients continue to complain of pain sensation after proper administration of a pudendal or pudendal and ilio-inguinal nerve blocks, it is usually the result of overlapping innervation by the perineal branch at the posterior cutaneous nerve of the thigh. Other types of anesthesia for childbirth are explained and compared in the blue box "Anesthesia for Childbirth," p. 397.

Exercises for Increased Development of Female Perineal Muscles

The superficial transverse perineal muscle, bulbospongiosus, and external anal sphincter, through their common attachment to the perineal body, form crossing beams over the pelvic outlet to support the perineal body, as in males. In the absence of the functional demands related to urination, penile erection, and ejaculation in males, the muscles are commonly relatively underdeveloped in women. However, when developed, they contribute to the support of the pelvic viscera, and help prevent urinary stress incontinence and postpartum prolapse of pelvic viscera. Therefore, many gynecologists as well as prepartum classes for participatory childbirth recommend that

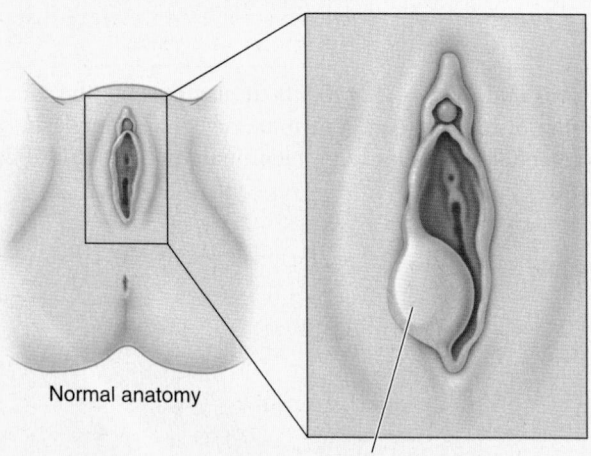

Normal anatomy

Bartholin gland cyst

FIGURE B3.34.

women practice *Kegel exercises* (named for J. H. Kegel, a 20th-century U.S. gynecologist) using the perineal muscles, such as successive interruption of the urine flow during urination. Prepartum childbirth classes emphasize that in learning to voluntarily contract and relax the perineal muscles, women become prepared to resist the tendency to contract the musculature during uterine contractions, allowing a less obstructed passage for the fetus and decreasing the likelihood of tearing the perineal muscles.

Vaginismus

The initial distension of the bulbospongiosus and transverse perineal muscles are thought to trigger the involuntary spasms of the perivaginal and levator ani muscles of *vaginismus,* an emotional (psychosomatic) gynecological disorder. Vaginismus is encountered clinically when pelvic examination is attempted. In mild forms, it causes *dyspareunia* (painful intercourse); in severe forms, it prevents vaginal entry.

The Bottom Line

FEMALE EXTERNAL GENITALIA

The female external genitalia consist of concentric folds (labia) surrounding an organ for sexual stimulation (clitoris) and the separate orifices of the urinary and reproductive systems. ♦ The fat-filled mons pubis and labia majora surround the pudendal cleft, covering and protecting its contents (e.g., bearing the body's weight when sitting on a bicycle). ♦ The fat-free labia minora attach to and cover the glans clitoris while surrounding the vestibule of the vagina, into which the external urethral and vaginal orifices and vestibular glands open. ♦ The erectile clitoris, consisting of a highly sensitive glans, short body, and crura that attach to the pubic rami and perineal membrane, functions solely as a tactile sensory organ. ♦ A hymen or its remnants, hymenal caruncles, demarcate the vagina from the vestibule and vaginal orifice. ♦ Immediately superior to the bases of the labia minora on each side of the vaginal orifice, the bulbs of the vestibule are paired masses of erectile tissue, homologous with the bulb of the penis. ♦ The internal pudendal vessels serve most of the vulva, with the external pudendal vessels serving a smaller anterior area. ♦ Except for the glans clitoris and associated structures (which

drain to deep inguinal and external iliac nodes), lymph from the perineum drains to the superficial inguinal nodes.

Innervation of the perineum is primarily from the pudendal nerve, with additional cutaneous innervation anteriorly from anterior labial nerves (ilio-inguinal and genitofemoral nerves), and laterally from the posterior cutaneous nerve of the thigh. ♦ Parasympathetic fibers, passing independently from the pelvis to the perineum as cavernous nerves, innervate the erectile tissues.

Female perineal muscles: Although homologous to the muscles of the male, the perineal muscles of the female are generally less well developed. ♦ In addition to the sphincteric functions of the external anal and urethral sphincters for maintaining fecal and urinary continence, the female perineal muscles are also capable of supporting the perineal body (which in turn supports the pelvic diaphragm). ♦ Learning to control perineal muscles through routine (Kegel) exercises may reduce the risk of obstetrical laceration of perineal muscles, and subsequent prolapse of pelvic viscera. ♦ The perineal muscles are innervated by muscular branches of the pudendal nerve.

SECTIONAL IMAGING OF PELVIS AND PERINEUM

Magnetic Resonance Imaging

MRI provides excellent evaluation of pelvic structures in any plane (Fig. 3.72) and permits outstanding delineation of the uterus and ovaries (Fig. 3.73). It also permits the identification of tumors (e.g., a myoma, or benign neoplasm) and congenital anomalies (e.g., bicornuate uterus—Fig. B3.16, p. 392).

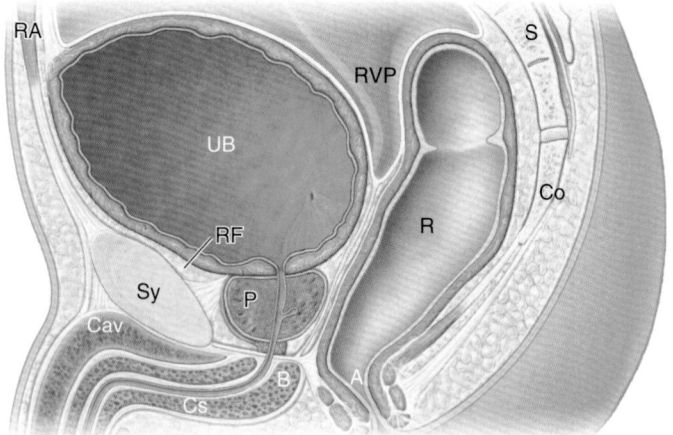

(A) Anatomical median section

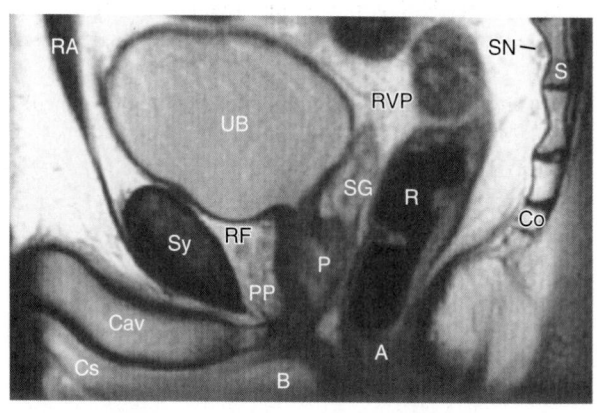

(B) Median MRI

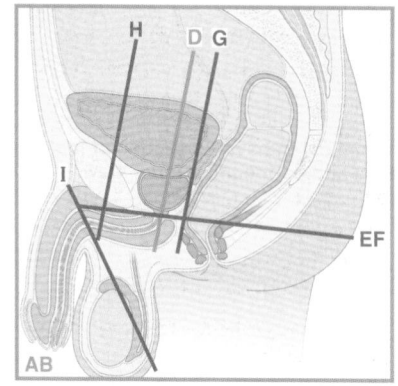

— Sections on this page
— Sections on next page

A	Anus	P	Prostate
B	Bulb of penis	PC	Peritoneal cavity
Co	Coccyx	PP	Prostatic venous plexus
Cav	Corpus cavernosum of penis	Ps	Psoas muscle
Cs	Corpus spongiosum of penis	R	Rectum
F	Femur	RA	Rectus abdominis
FN	Femoral nerve	RF	Retropubic fat
IL	Ilium	RVP	Rectovesical pouch
IM	Iliacus muscle	S	Sacrum
IR	Ischiopubic ramus	SG	Seminal gland
LA	Levator ani	SN	Sacral nerves
Oe	Obturator externus	Sy	Pubic symphysis
Oi	Obturator internus	UB	Urinary bladder

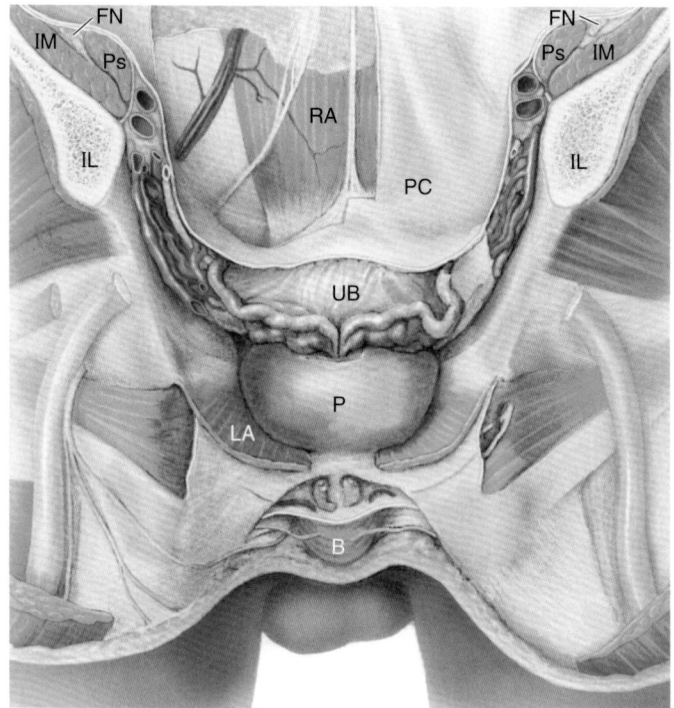

(C) Dissection, posterior view

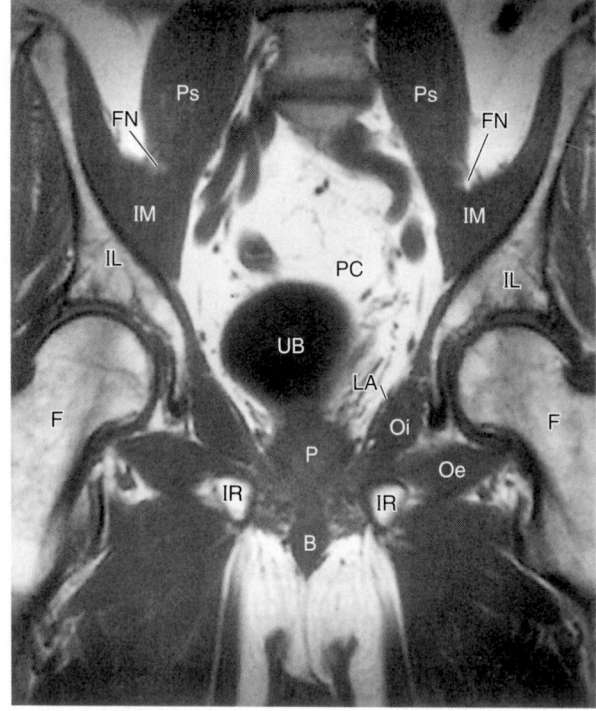

(D) Coronal MRI

FIGURE 3.72. MRI scans of male pelvis and perineum.

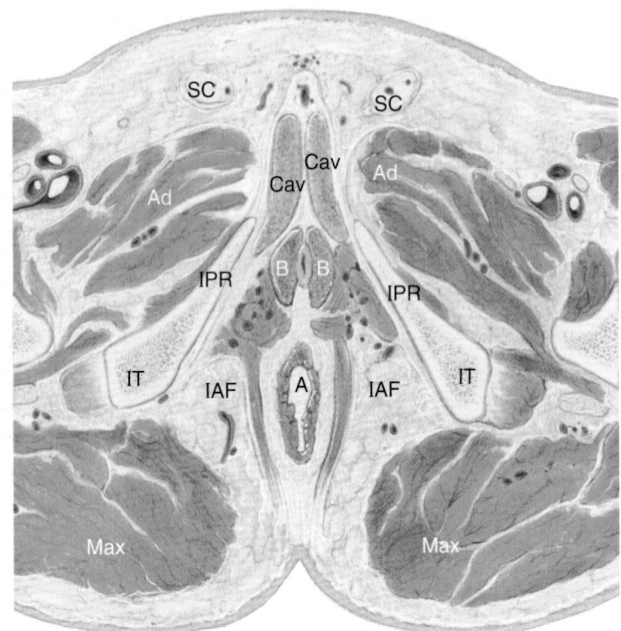

(E) Anatomical transverse section of male pelvis and perineum. Inferior view.

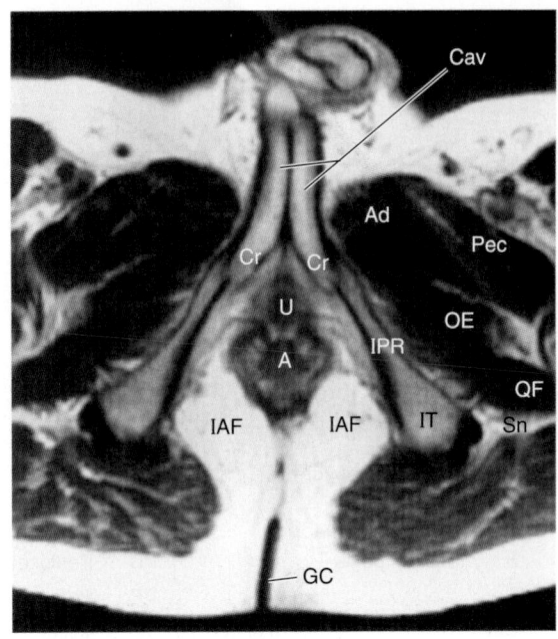

(F) Transverse (axial) MRI

A	Anal canal	In	Small intestine	Pec	Pectineus
Ad	Adductor muscles	IPR	Ischiopubic ramus	Pu	Pubic bone
B	Bulb of penis	IR	Inferior rectal nerve vessels	QF	Quadratus femoris
Cav	Corpus cavernosum	IT	Ischial tuberosity	R	Rectum
Cr	Crus penis	LA	Levator ani	SC	Spermatic cord
Cs	Corpus spongiosum	LS	Lumbosacral trunk	Sep	Scrotal septum
GC	Intergluteal cleft	Max	Gluteus maximus	Sig	Sigmoid colon
IAF	Ischio-anal fossa	OE	Obturator externus	Sk	Skin
IL	Ilium	OI	Obturator internus	Sn	Sciatic nerve
		P	Pampiniform plexus	SG	Seminal glands
				Sy	Pubic symphysis
				T	Testis
				U	Urethra

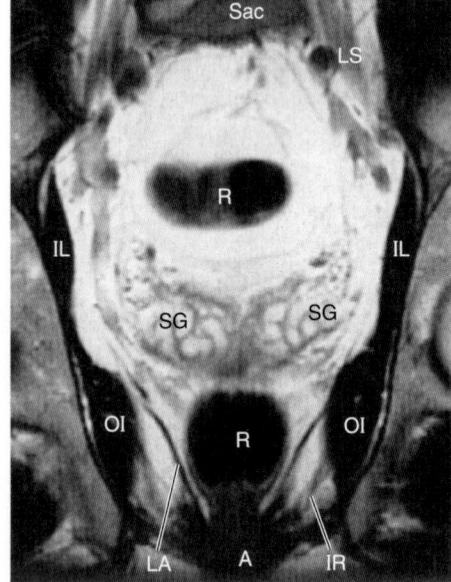

(G) Coronal MRI

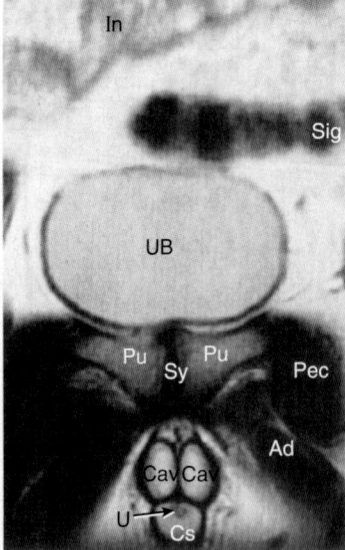

(H) Coronal MRI

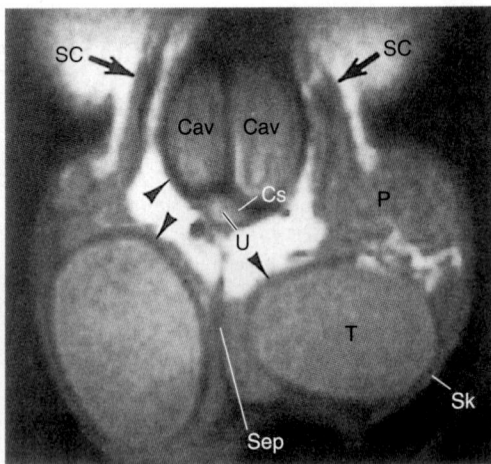

(I) Coronal section

FIGURE 3.72. *(Continued)*

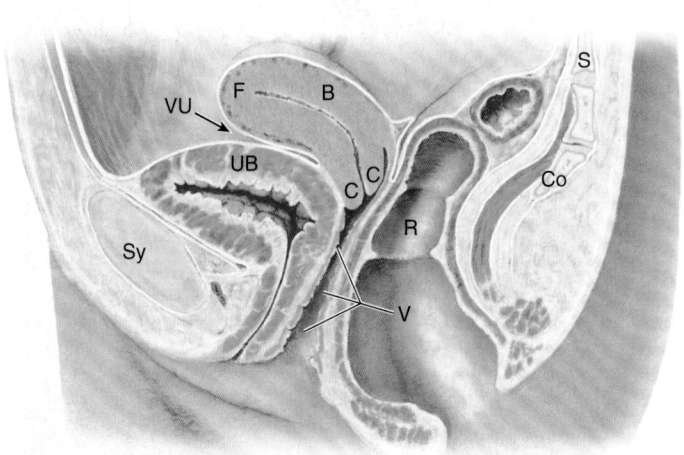

(A) Median anatomical section

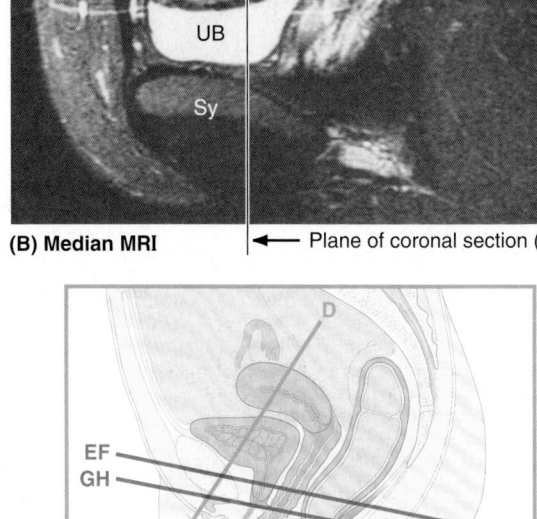

(B) Median MRI ← Plane of coronal section (below)

B	Body of uterus	R	Rectum
C	Cervix of uterus	RA	Rectus abdominis
CC	Cervical canal	RU	Recto-uterine pouch
Co	Coccyx	S	Sacrum
E	Endometrium	SB	Small intestine (bowel)
F	Fundus of uterus	Sy	Pubic symphysis
M	Myometrium	UB	Urinary bladder
Oe	Obturator externus	UC	Uterine cavity
Ov	Ovary	V	Vagina
PB	Pubic bone	VU	Vesico-uterine pouch

— Sections on this page
— Sections on next page

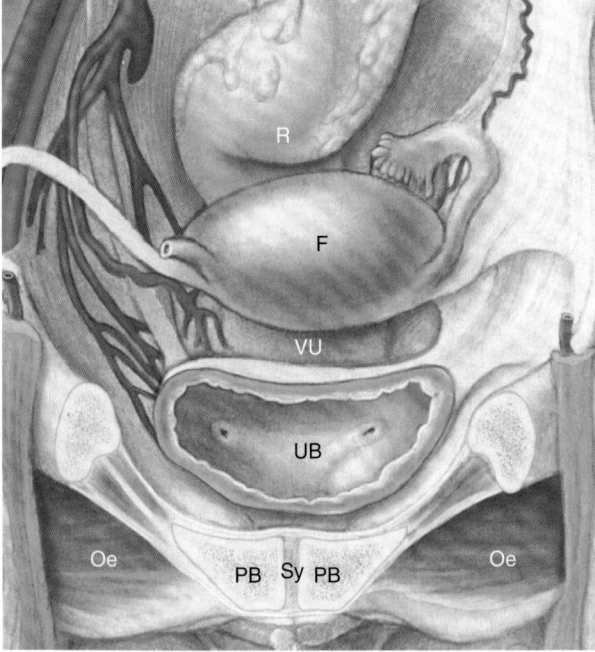

(C) Dissection, anterior view

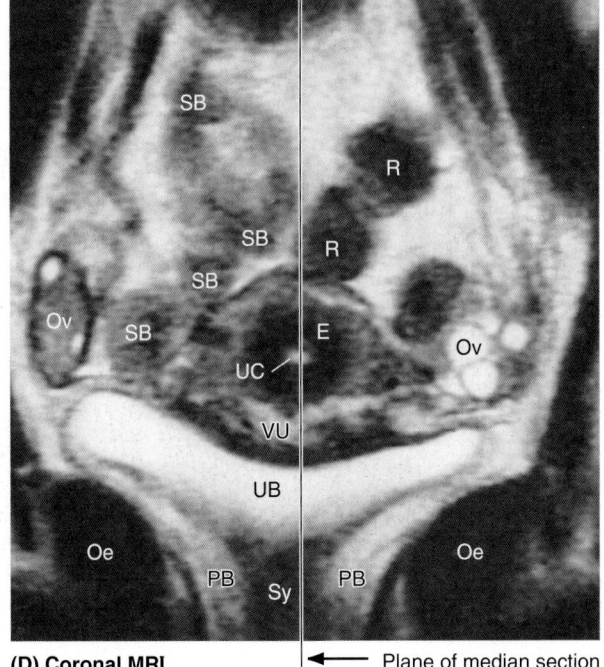

(D) Coronal MRI ← Plane of median section (above)

FIGURE 3.73. **MRI studies of female pelvis.** (**B** and **D** courtesy of Dr. Shirley McCarthy, Department of Diagnostic Radiology, Yale University and Yale-New Haven Hospital, New Haven, CT.)

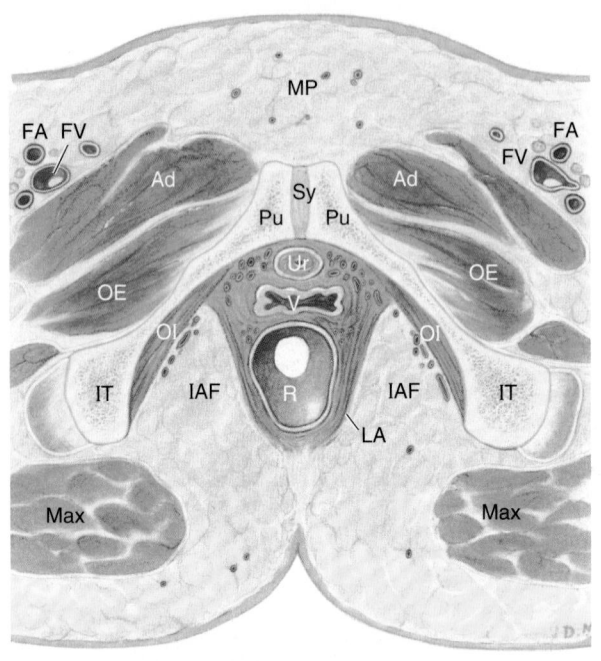

(E) Anatomical transverse section

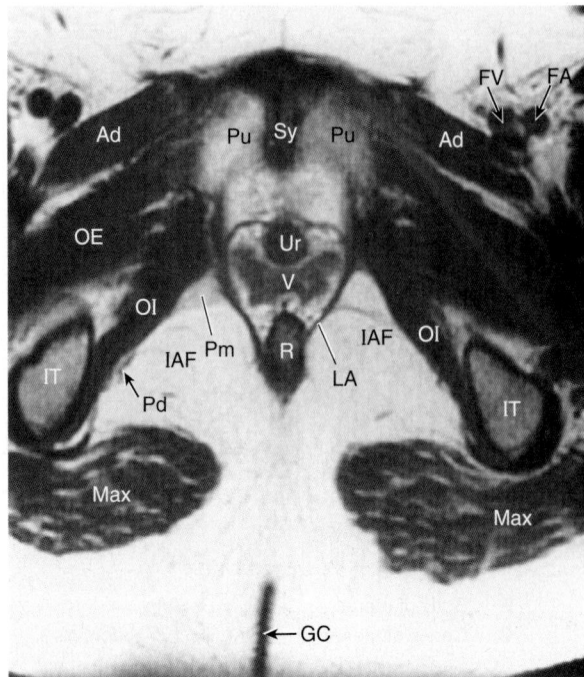

(F) Transverse MRI scan

AC	Anal canal	IT	Ischial tuberosity	Pm	Perineal membrane
Ad	Adductor muscles	LA	Levator ani	PR	Puborectalis
CC	Crus of clitoris	LM	Labium majus	Pu	Pubic bone
FA	Femoral artery	MP	Mons pubis	QF	Quadratus femoris
FV	Femoral vein	Max	Gluteus maximus	R	Rectum
GC	Gluteal cleft	OE	Obturator externus	Sy	Pubic symphysis
IAF	Ischio-anal fossa	OI	Obturator internus	Ur	Urethra
IPR	Ischiopubic ramus	Pec	Pectineus	V	Vagina
				Ve	Vestibule of vagina

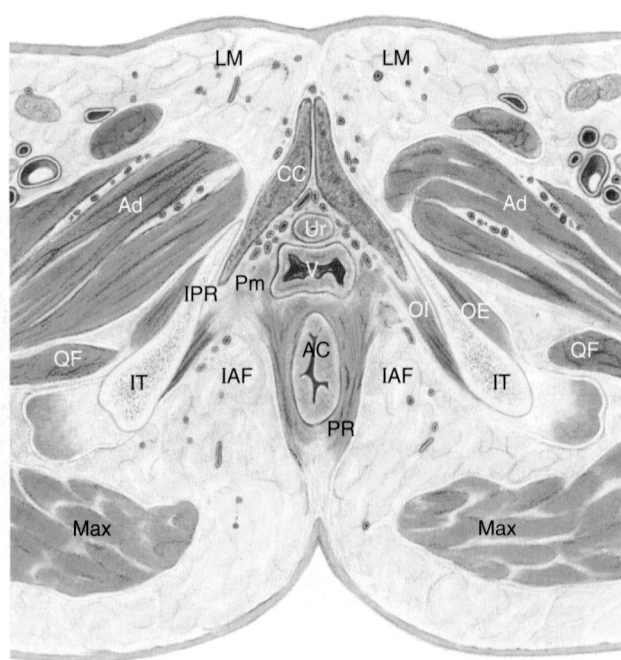

(G) Anatomical transverse section

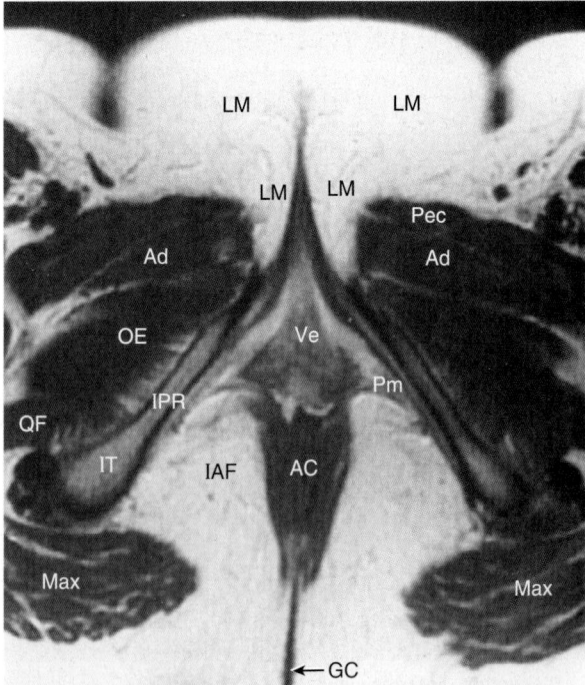

(H) Transverse MRI scan

FIGURE 3.73. *(Continued)*

4

Back

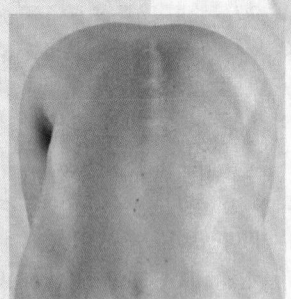

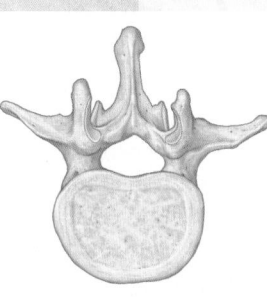

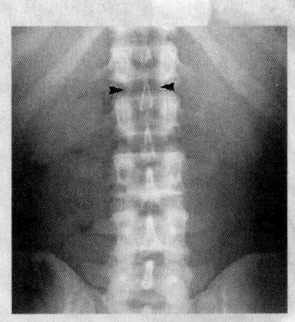

OVERVIEW OF BACK AND VERTEBRAL COLUMN

The **back** comprises the posterior aspect of the trunk, inferior to the neck and superior to the buttocks. It is the region
of the body to which the head, neck, and limbs are attached.
The back includes the:

- Skin and subcutaneous tissue.
- Muscles: a superficial layer, primarily concerned with
  positioning and moving the upper limbs, and deeper layers
  ("true back muscles"), specifically concerned with moving
  or maintaining the position of the axial skeleton (posture).
- Vertebral column: the vertebrae, intervertebral (IV) discs,
  and associated ligaments (Fig. 4.1).
- Ribs (in the thoracic region): particularly their posterior
  portions, medial to the angles of the ribs.
- Spinal cord and meninges (membranes that cover the spinal cord).
- Various segmental nerves and vessels.

Because of their close association with the trunk, the back
of the neck and the posterior and deep cervical muscles and
vertebrae are also described in this chapter. The scapulae,
although located in the back, are part of the appendicular
skeleton and are considered with the upper limb (Chapter 6).

Study of the soft tissues of the back is best preceded by examination of the *vertebrae* and the fibrocartilaginous *intervertebral
discs* that are interposed between the bodies of adjacent vertebrae. The vertebrae and IV discs collectively make up the **vertebral column** (spine), the skeleton of the neck and back that is
the main part of the *axial skeleton* (i.e., articulated bones of the
cranium, vertebral column, ribs, and sternum) (Fig. 4.1D). The
vertebral column extends from the cranium (skull) to the apex
of the coccyx. In adults, it is 72–75 cm long, of which approximately one quarter is formed by the IV discs that separate and
bind the vertebrae together. The vertebral column:

- Protects the spinal cord and spinal nerves.
- Supports the weight of the body superior to the level of
  the pelvis.
- Provides a partly rigid and flexible axis for the body and an
  extended base on which the head is placed and pivots.
- Plays an important role in posture and *locomotion* (the
  movement from one place to another).

VERTEBRAE

The vertebral column in an adult typically consists of 33
vertebrae arranged in five regions: 7 cervical, 12 thoracic,
5 lumbar, 5 sacral, and 4 coccygeal (Fig. 4.1A–D). Significant
motion occurs only between the 25 superior vertebrae. Of the
9 inferior vertebrae, the 5 sacral vertebrae are fused in adults to
form the *sacrum,* and after approximately age 30, the 4 coccygeal vertebrae fuse to form the *coccyx.* The **lumbosacral angle**
occurs at the junction of the long axes of the lumbar region of
the vertebral column and the sacrum (Fig. 4.1D). The vertebrae
gradually become larger as the vertebral column descends to
the sacrum and then become progressively smaller toward the
apex of the coccyx (Fig. 4.1A–D). The change in size is related
to the fact that successive vertebrae bear increasing amounts
of the body's weight as the column descends. The vertebrae
reach maximum size immediately superior to the sacrum, which
transfers the weight to the pelvic girdle at the sacro-iliac joints.

The vertebral column is flexible because it consists of
many relatively small bones, called **vertebrae** (singular =
vertebra), that are separated by resilient IV discs (Fig. 4.1D).
The 25 cervical, thoracic, lumbar, and first sacral vertebrae
also articulate at synovial *zygapophysial (facet) joints* (Fig.
4.2D), which facilitate and control the vertebral column's
flexibility. Although the movement between two adjacent
vertebrae is small, in aggregate the vertebrae and IV discs
uniting them form a remarkably flexible yet rigid column that
protects the spinal cord it surrounds.

Structure and Function of Vertebrae

Vertebrae vary in size and other characteristics from one
region of the vertebral column to another, and to a lesser
degree within each region; however, their basic structure is

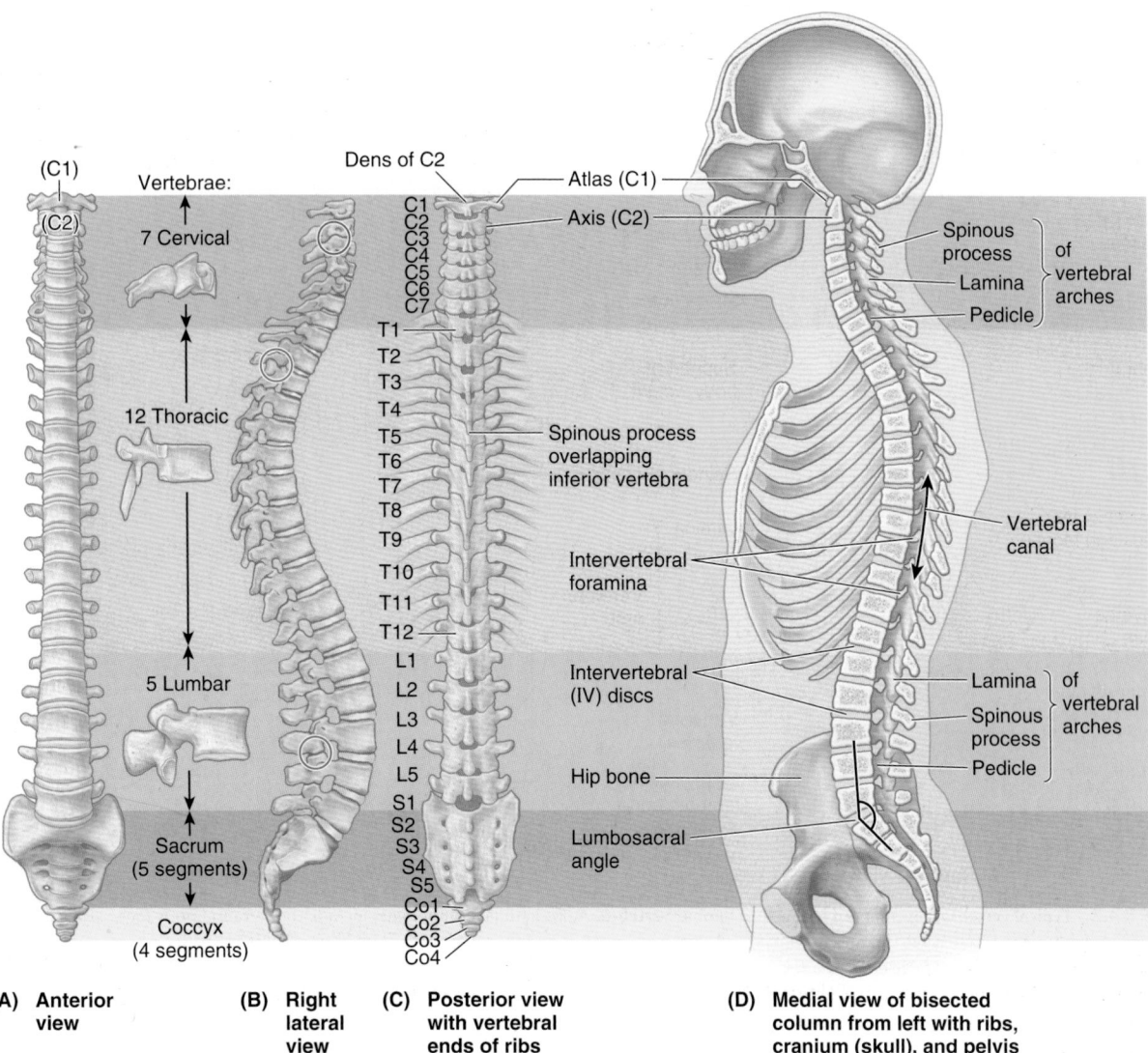

FIGURE 4.1. Vertebral column and its five regions. The isolated vertebrae between (**A**) and (**B**) are typical of each of the three mobile regions of the vertebral column. The continuous, weight-bearing column of vertebral bodies and IV discs increases in size as the column descends. Zygapophysial (facet) joints representative of each region are circled. The posterior view (**C**) includes the vertebral ends of ribs, representing the skeleton of the back. The bisected vertebral column in (**D**) demonstrates the vertebral canal. The intervertebral (IV) foramina (also seen in **B**) are openings in the lateral wall of the vertebral canal through which spinal nerves exit.

the same. A **typical vertebra** (Fig. 4.2) consists of a vertebral body, a vertebral arch, and seven processes.[1]

The **vertebral body** is the more massive, roughly cylindrical, anterior part of the bone that gives strength to the vertebral column and supports body weight. The size of the vertebral bodies increases as the column descends, most markedly from T4 inferiorly, as each bears progressively greater body weight.

The vertebral body consists of vascular, trabecular (spongy, cancellous) bone enclosed by a thin external layer

of compact bone (Fig. 4.3). The *trabecular bone* is a meshwork of mostly tall vertical trabeculae intersecting with short, horizontal trabeculae. The spaces between the trabeculae are occupied by red bone marrow that is among the most actively hematopoietic (blood-forming) tissues of the mature individual. One or more large foramina in the posterior surface of the vertebral body accommodate basivertebral veins that drain the marrow (see Fig. 4.26).

In life, most of the superior and inferior surfaces of the vertebral body are covered with discs of hyaline cartilage (vertebral "end plates"), which are remnants of the cartilaginous model from which the bone develops. In dried laboratory and museum skeletal specimens, this cartilage is absent, and the exposed bone appears spongy, except at the periphery where an **epiphysial rim** or ring of smooth bone, derived from an *anular epiphysis*, is fused to the body (Fig. 4.2B).

[1]In contemporary usage, the terms *vertebral body* and *centrum* and the terms *vertebral arch* and *neural arch* are often erroneously used as synonyms. Technically, however, in each case the former is a gross anatomy term applied to parts of the adult vertebrae, and the latter is an embryology term referring to parts of a developing vertebra ossifying from primary centers. The vertebral body includes the centrum and part of the neural arch; the vertebral arch is thus less extensive than the neural arch, and the centrum is less extensive than the vertebral body (O'Rahilly, 1986; Standring, 2008).

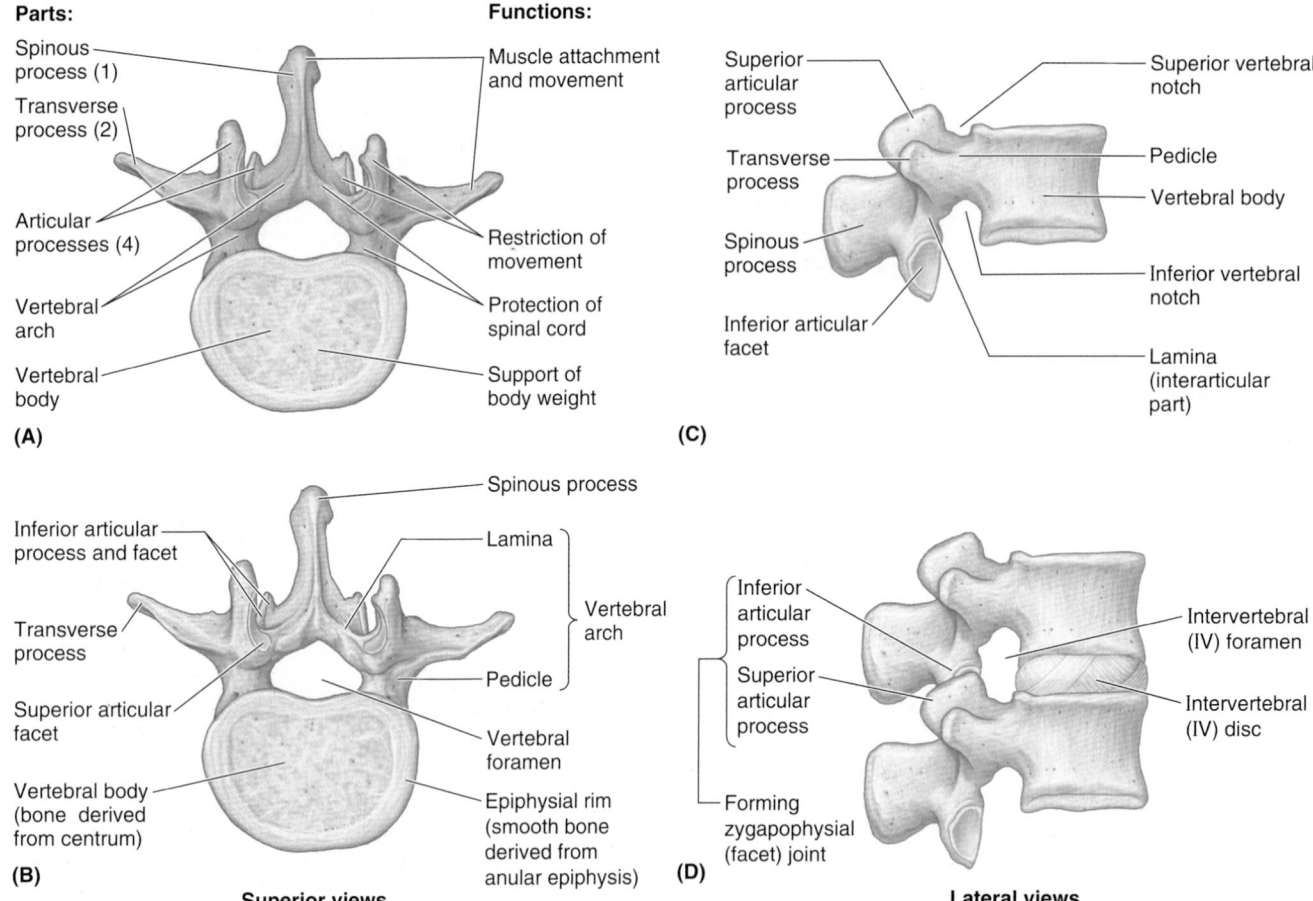

FIGURE 4.2. Typical vertebra, represented by the 2nd lumbar vertebra. A. Functional components include the vertebral body (*bone color*), a vertebral arch (*red*), and seven processes: three for muscle attachment and leverage (*blue*) and four that participate in synovial joints with adjacent vertebrae (*yellow*). **B and C.** Bony formations of vertebrae are demonstrated. The vertebral foramen is bounded by the vertebral arch and body. A small superior vertebral notch and a larger inferior vertebral notch flank the pedicle. **D.** The superior and inferior notches of adjacent vertebrae plus the IV disc that unites them form the IV foramen for passage of a spinal nerve and its accompanying vessels. Each articular process has an articular facet where contact occurs with the articular facets of adjacent vertebrae (**B–D**).

In addition to serving as growth zones, the anular epiphyses and their cartilaginous remnants provide some protection to the vertebral bodies and permit some diffusion of fluid between the IV disc and the blood vessels (capillaries) in the vertebral body (see Fig. 4.26). The superior and inferior epiphyses usually unite with the **centrum,** the primary ossification center for the central mass of the vertebral body (Fig. 4.2B), early in adult life (at approximately age 25) (see "Ossification of Vertebrae," p. 453).

The **vertebral arch** is posterior to the vertebral body and consists of two (right and left) pedicles and laminae (Fig. 4.2A & C). The **pedicles** are short, stout cylindrical processes that project posteriorly from the vertebral body to meet two broad, flat plates of bone, called **laminae,** which unite in the midline. The vertebral arch and the posterior surface of the vertebral body form the walls of the **vertebral foramen** (Fig. 4.2A & B). The succession of vertebral foramina in the articulated vertebral column forms the **vertebral canal** (spinal canal). The canal contains the spinal cord and the roots of the spinal nerves, along with the membranes (meninges), fat, and vessels that surround and serve them (Figs. 4.1D and 4.3). (See the blue box "Laminectomy," p. 457.)

The **vertebral notches** are indentations observed in lateral views of the vertebrae superior and inferior to each pedicle between the superior and inferior articular processes posteriorly and the corresponding projections of the body anteriorly (Fig. 4.2C). The **superior** and **inferior vertebral notches** of adjacent vertebrae and the IV discs connecting them form **intervertebral foramina** (Fig. 4.2D) through which the spinal nerves emerge from the vertebral column (see Fig. 4.27). Also, the spinal (posterior root) ganglia are located in these foramina.

Seven processes arise from the vertebral arch of a typical vertebra (Fig. 4.2A–C):

- One median **spinous process** projects posteriorly (and usually inferiorly, typically overlapping the vertebra below) from the vertebral arch at the junction of the laminae.
- Two **transverse processes** project posterolaterally from the junctions of the pedicles and laminae.
- Four **articular processes** (G. *zygapophyses*)—two **superior** and two **inferior**—also arise from the junctions of the pedicles and laminae, each bearing an **articular surface (facet).**

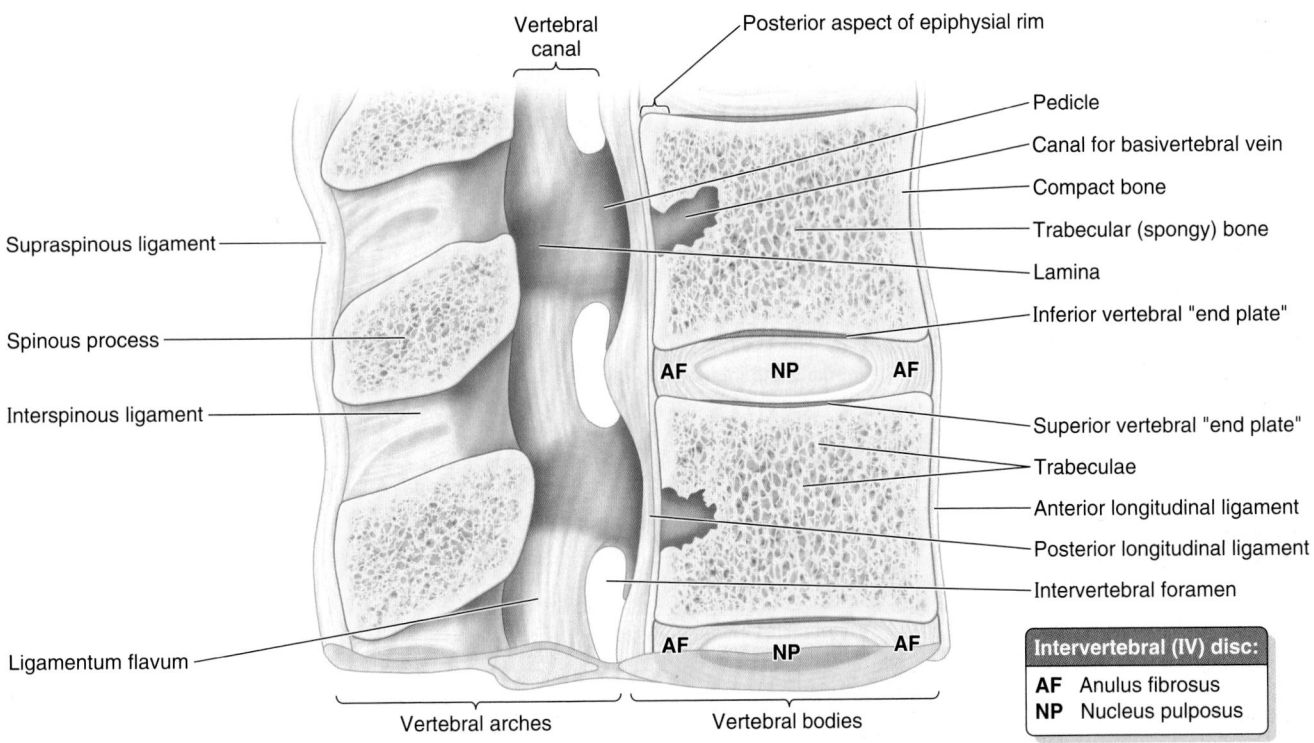

Vertebral canal

Posterior aspect of epiphysial rim

Pedicle

Canal for basivertebral vein

Compact bone

Trabecular (spongy) bone

Lamina

Inferior vertebral "end plate"

Supraspinous ligament

Spinous process

Interspinous ligament

AF   NP   AF

Superior vertebral "end plate"

Trabeculae

Anterior longitudinal ligament

Posterior longitudinal ligament

Intervertebral foramen

Ligamentum flavum

AF   NP   AF

Intervertebral (IV) disc:
AF Anulus fibrosus
NP Nucleus pulposus

Vertebral arches   Vertebral bodies

Medial view of left halves of two adjacent hemisected vertebrae and associated IV discs

FIGURE 4.3. Internal aspects of vertebral bodies and vertebral canal. The bodies consist largely of trabecular (spongy) bone—with tall, vertical supporting trabeculae linked by short horizontal trabeculae—covered by a relatively thin layer of compact bone. Hyaline cartilage "end plates" cover the superior and inferior surfaces of the bodies, surrounded by smooth bony epiphysial rims. The posterior longitudinal ligament, covering the posterior aspect of the bodies and linking the IV discs, forms the anterior wall of the vertebral canal. Lateral and posterior walls of the vertebral canal are formed by vertebral arches (pedicles and laminae) alternating with IV foramina and ligamenta flava.

The spinous and transverse processes provide attachment for deep back muscles and serve as levers, facilitating the muscles that fix or change the position of the vertebrae.

The articular processes are in apposition with corresponding processes of vertebrae adjacent (superior and inferior) to them, forming *zygapophysial (facet) joints* (Figs. 4.1B and 4.2D). Through their participation in these joints, these processes determine the types of movement permitted and restricted between the adjacent vertebrae of each region.

The articular processes also assist in keeping adjacent vertebrae aligned, particularly preventing one vertebra from slipping anteriorly on the vertebra below. Generally, the articular processes bear weight only temporarily, as when one rises from the flexed position, and unilaterally when the cervical vertebrae are laterally flexed to their limit. However, the inferior articular processes of the L5 vertebra bear weight even in the erect posture.

Regional Characteristics of Vertebrae

Each of the 33 vertebrae is unique; however, most of the vertebrae demonstrate characteristic features identifying them as belonging to one of the five regions of the vertebral column (e.g., vertebrae having foramina in their transverse processes are cervical vertebrae) (Fig. 4.4). In addition, certain individual vertebrae have distinguishing features; the C7 vertebra, for example, has the longest spinous process. It

forms a prominence under the skin at the back of the neck, especially when the neck is flexed (see Fig. 4.8A).

In each region, the **articular facets** are oriented on the articular processes of the vertebrae in a characteristic direction that determines the type of movement permitted between the adjacent vertebrae and, in aggregate, for the region. For example, the articular facets of thoracic vertebrae are nearly vertical, and together define an arc centered in the IV disc; this arrangement permits rotation and lateral flexion of the vertebral column in this region (Fig. 4.7). Regional variations in the size and shape of the vertebral canal accommodate the varying thickness of the spinal cord (Fig. 4.1D).

CERVICAL VERTEBRAE

Cervical vertebrae form the skeleton of the neck (Fig. 4.1). The smallest of the 24 movable vertebrae, the cervical vertebrae are located between the cranium and the thoracic vertebrae. Their smaller size reflects the fact that they bear less weight than do the larger inferior vertebrae. Although the cervical IV discs are thinner than those of inferior regions, they are relatively thick compared to the size of the vertebral bodies they connect. The relative thickness of the IV discs, the nearly horizontal orientation of the articular facets, and the small amount of surrounding body mass give the cervical region the greatest range and variety of movement of all the vertebral regions.

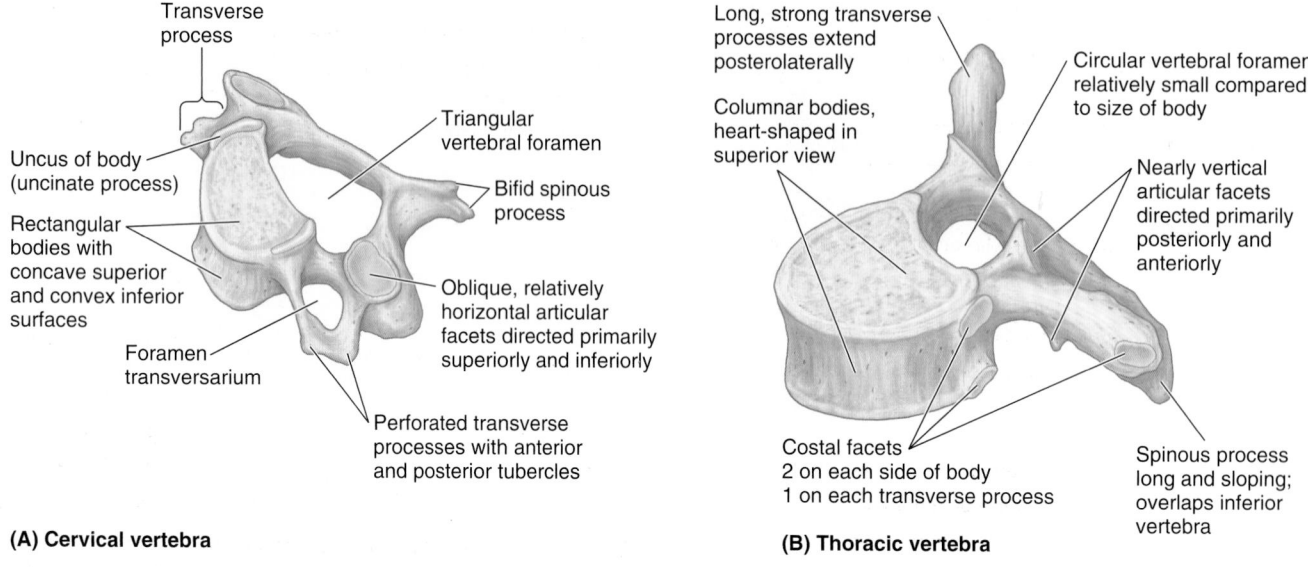

(A) Cervical vertebra

Transverse process

Uncus of body (uncinate process)

Rectangular bodies with concave superior and convex inferior surfaces

Foramen transversarium

Triangular vertebral foramen

Bifid spinous process

Oblique, relatively horizontal articular facets directed primarily superiorly and inferiorly

Perforated transverse processes with anterior and posterior tubercles

(B) Thoracic vertebra

Long, strong transverse processes extend posterolaterally

Columnar bodies, heart-shaped in superior view

Circular vertebral foramen, relatively small compared to size of body

Nearly vertical articular facets directed primarily posteriorly and anteriorly

Costal facets 2 on each side of body 1 on each transverse process

Spinous process long and sloping; overlaps inferior vertebra

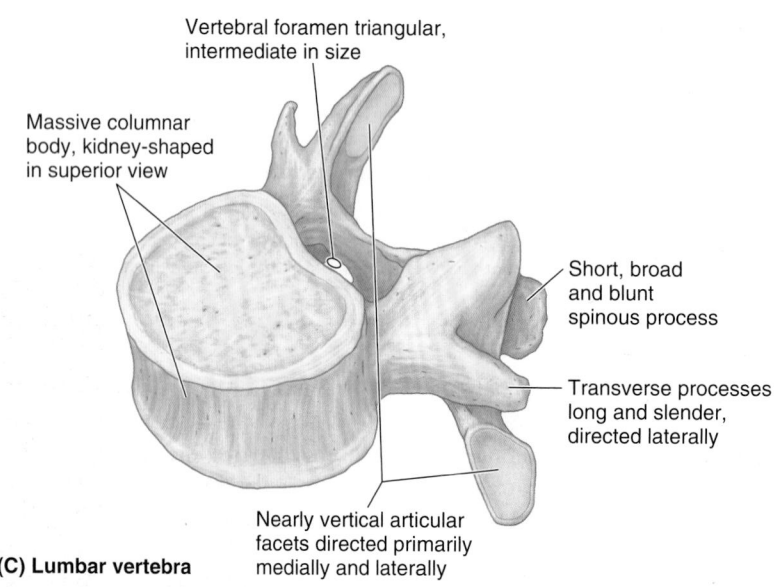

Vertebral foramen triangular, intermediate in size

Massive columnar body, kidney-shaped in superior view

Short, broad and blunt spinous process

Transverse processes long and slender, directed laterally

(C) Lumbar vertebra

Nearly vertical articular facets directed primarily medially and laterally

Left anterior superior oblique views of "typical" presacral vertebrae

FIGURE 4.4. Comparison of presacral vertebrae. As the vertebral column descends, bodies increase in size in relationship to increased weight-bearing. The size of the vertebral canal changes in relationship to the diameter of the spinal cord.

The distinctive features of cervical vertebrae are illustrated in Figures 4.4A and 4.5 and described in Table 4.1. The most distinctive feature of each cervical vertebra is the oval **foramen transversarium** (transverse foramen) in the *transverse process.* The vertebral arteries and their accompanying veins pass through the transverse foramina, except those in C7, which transmit only small accessory veins. Thus the foramina are smaller in C7 than those in other cervical vertebrae, and occasionally they are absent.

The *transverse processes of cervical vertebrae* end laterally in two projections: an **anterior tubercle** and a **posterior tubercle.** The tubercles provide attachment for a laterally

placed group of cervical muscles (levator scapulae and scalenes). The anterior rami of the cervical spinal nerves course initially on the transverse processes in **grooves for spinal nerves** between the tubercles (Fig. 4.5A & B). The anterior tubercles of vertebra C6 are called **carotid tubercles** (Fig. 4.5A) because the common carotid arteries may be compressed here, in the groove between the tubercle and body, to control bleeding from these vessels. Bleeding may continue because of the carotid's multiple anastomoses of distal branches with adjacent and contralateral branches, but at a slower rate.

Vertebrae C3–C7 are typical cervical vertebrae (Figs. 4.4A and 4.5A; Table 4.1). They have large vertebral foramina to

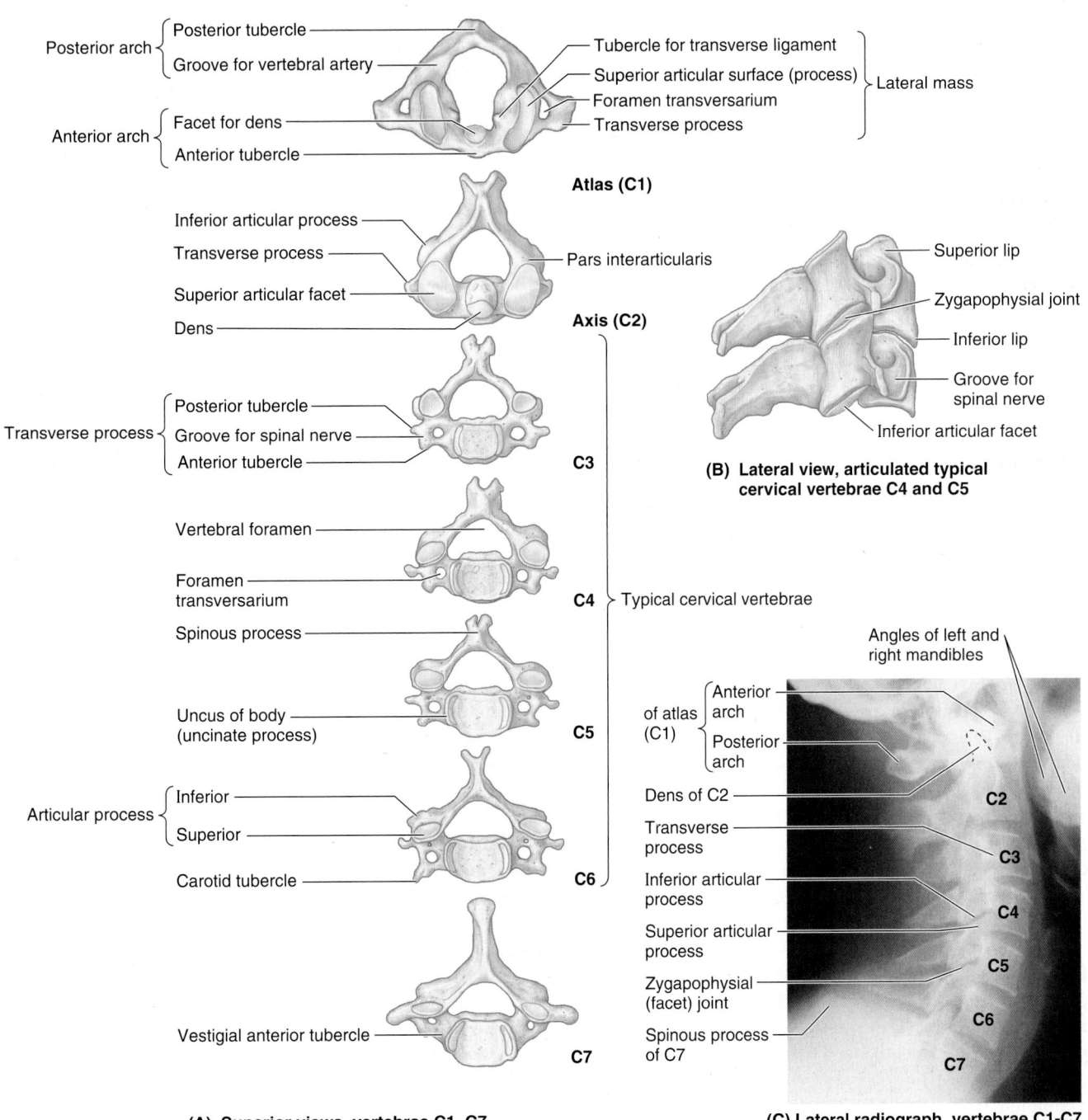

FIGURE 4.5. Cervical vertebrae. A. C1, C2, and C3 vertebrae are atypical. **B.** The superior and inferior surfaces of the bodies of the cervical vertebrae are reciprocally convex and concave. Combined with the oblique orientation of the articular facets, this facilitates flexion and extension as well as lateral flexion. **C.** The anterior arch of the atlas lies anterior to the continuous curved line formed by the anterior surfaces of the C2-C7 vertebral bodies. (Courtesy of Dr. J. Heslin, Toronto, Ontario, Canada.)

accommodate the *cervical enlargement of the spinal cord* as a consequence of this region's role in the innervation of the upper limbs. The superior borders of the transversely elongated bodies of the cervical vertebrae are elevated posteriorly and especially laterally, but they are depressed anteriorly, resembling somewhat a sculpted seat.

The inferior border of the body of the superiorly placed vertebra is reciprocally shaped. The adjacent cervical vertebrae articulate in a way that permits free flexion and extension and some lateral flexion but restricted rotation. The planar, nearly horizontal articular facets of the articular processes are also favorable for these movements. The elevated superolateral margin is the **uncus of the body** (uncinate process) (Fig. 4.4A).

The spinous processes of the C3–C6 vertebrae are short and usually bifid in white people, especially males, but usually

TABLE 4.1. CERVICAL VERTEBRAE[a]

Part	Characteristics
Vertebral body	Small and wider from side to side than anteroposteriorly; superior surface concave with uncus of body (uncinate process); inferior surface convex
Vertebral foramen	Large and triangular
Transverse processes	Foramina transversarii and anterior and posterior tubercles; vertebral arteries and accompanying venous and sympathetic plexuses pass through foramina transversarii of all cervical vertebrae except C7, which transmits only small accessory vertebral veins.
Articular processes	Superior facets directed superoposteriorly; inferior facets directed infero-anteriorly; obliquely placed facets are most nearly horizontal in this region
Spinous processes	Short (C3–C5) and bifid (C3–C6); process of C6 long, that of C7 is longer (thus C 7 is called "vertebra prominens")

[a]The C1, C2, and C7 vertebrae are atypical.

not as commonly in people of African descent or in females (Duray et al., 1999). C7 is a prominent vertebra that is characterized by a long spinous process. Because of this prominent process, C7 is called the **vertebra prominens.** Run your finger along the midline of the posterior aspect of your neck until you feel the prominent C7 spinous process. It is the most prominent spinous process in 70% of people (Fig. 4.8A).

The two superior-most cervical vertebrae are atypical. **Vertebra C1,** also called the **atlas,** is unique in that it has neither a body nor a spinous process (Figs. 4.5A and 4.6B). This ring-shaped bone has paired **lateral masses** that serve the place of a body by bearing the weight of the globe-like cranium in a manner similar to the way that Atlas of Greek mythology bore the weight of the world on his shoulders (Fig. 4.6E). The transverse processes of the atlas arise from the lateral masses, causing them to be more laterally placed than those of the inferior vertebrae. This feature makes the atlas the widest of the cervical vertebrae, thus providing increased leverage for attached muscles.

The kidney-shaped, concave **superior articular surfaces** of the lateral masses articulate with two large cranial protuberances, the **occipital condyles,** at the sides of the foramen magnum (Fig. 4.6A). **Anterior** and **posterior arches,** each of which bears a **tubercle** in the center of its external aspect, extend between the lateral masses, forming a complete ring (Fig. 4.6B). The posterior arch, which corresponds to the lamina of a typical vertebra, has a wide **groove for the vertebral artery** on its superior surface. The C1 nerve also runs in this groove.

Vertebra C2, also called the **axis,** is the strongest of the cervical vertebrae (Figs. 4.5A and 4.6C). C1, carrying the cranium, rotates on C2 (e.g., when a person turns the head to indicate "no"). The axis has two large, flat bearing surfaces, the **superior articular facets,** on which the atlas rotates. The distinguishing feature of C2 is the blunt tooth-like **dens** (odontoid process), which projects superiorly from its body. Both the dens (G. tooth) and the spinal cord inside its coverings (meninges) are encircled by the atlas. The dens lies anterior to the spinal cord and serves as the pivot about which the rotation of the head occurs.

The dens is held in position against the posterior aspect of the anterior arch of the atlas by the **transverse ligament of the atlas** (Fig. 4.6B). This ligament extends from one lateral mass of the atlas to the other, passing between the dens and spinal cord, forming the posterior wall of the "socket" that receives the dens. Thus it prevents posterior (horizontal) displacement of the dens and anterior displacement of the atlas. Either displacement would compromise the portion of the vertebral foramen of C1 that gives passage to the spinal cord. C2 has a large bifid spinous process (Fig. 4.6C & D) that can be felt deep in the *nuchal groove,* the superficial vertical groove at the back of the neck.

THORACIC VERTEBRAE

The **thoracic vertebrae** are in the upper back and provide attachment for the ribs (Fig. 4.1). Thus the primary characteristic features of thoracic vertebrae are the **costal facets** for articulation with ribs. The costal facets and other characteristic features of thoracic vertebrae are illustrated in Figures 4.4B and 4.7 and listed in Table 4.2.

The middle four thoracic vertebrae (T5–T8) demonstrate all the features typical of thoracic vertebrae. The articular processes extend vertically with paired, nearly coronally oriented articular facets that define an arc centered in the IV disc. This arc permits rotation and some lateral flexion of the vertebral column in this region. In fact, the greatest degree of rotation is permitted here (Fig. 4.7A). Attachment of the rib cage, combined with the vertical orientation of articular facets and overlapping spinous processes, limits flexion and extension as well as lateral flexion.

The T1–T4 vertebrae share some features of cervical vertebrae. T1 is atypical of thoracic vertebrae in that it has a long, almost horizontal spinous process that may be nearly as prominent as that of the vertebra prominens. T1 also has a complete costal facet on the superior edge of its body for the 1st rib and a demifacet on its inferior edge that contributes to the articular surface for the 2nd rib.

The T9–T12 vertebrae have some features of lumbar vertebrae (e.g., tubercles similar to the accessory processes).

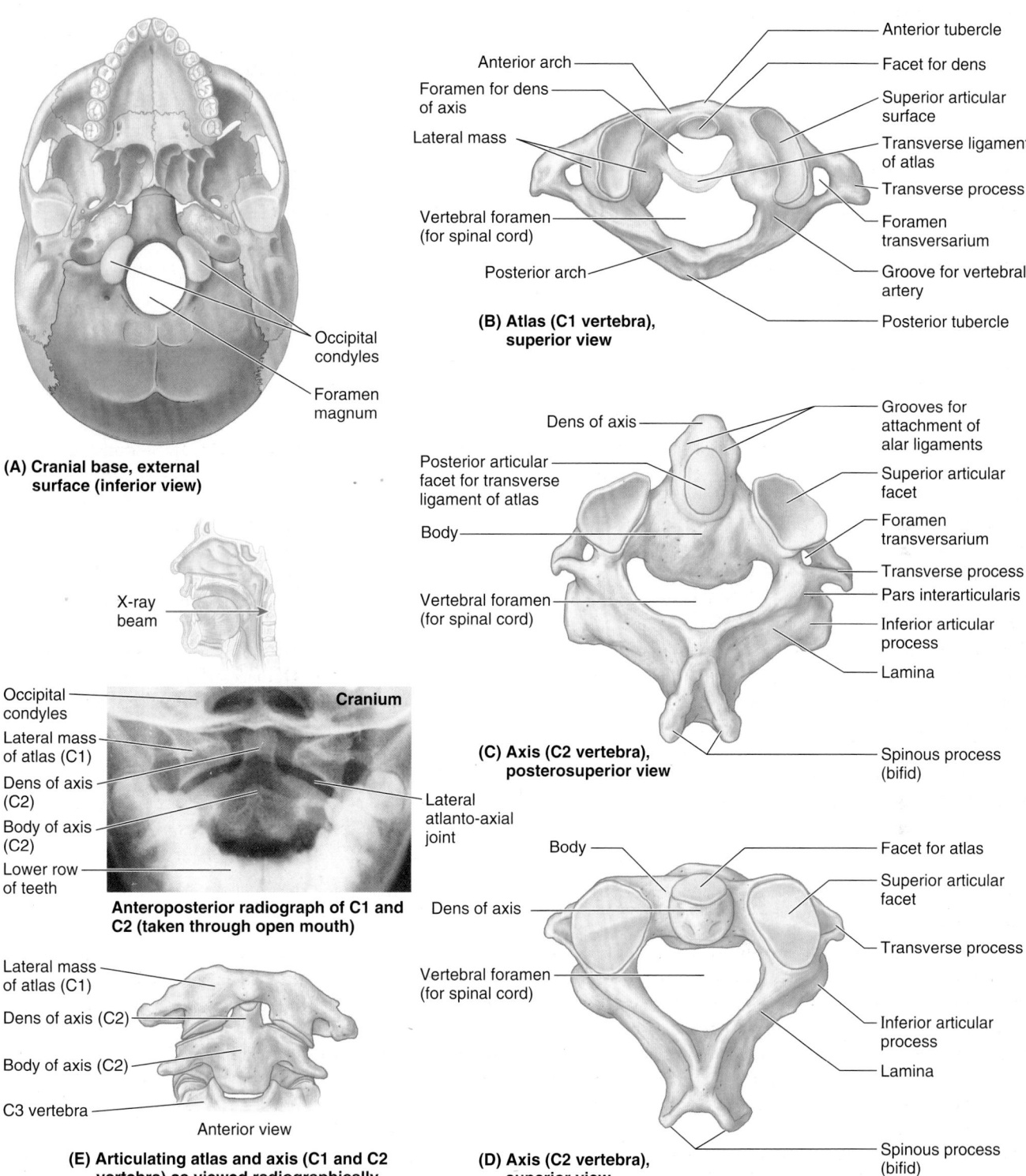

(A) Cranial base, external surface (inferior view)

Occipital condyles
Foramen magnum

(B) Atlas (C1 vertebra), superior view

Anterior tubercle
Facet for dens
Superior articular surface
Transverse ligament of atlas
Transverse process
Foramen transversarium
Groove for vertebral artery
Posterior tubercle
Anterior arch
Foramen for dens of axis
Lateral mass
Vertebral foramen (for spinal cord)
Posterior arch

X-ray beam

(C) Axis (C2 vertebra), posterosuperior view

Dens of axis
Posterior articular facet for transverse ligament of atlas
Body
Vertebral foramen (for spinal cord)
Grooves for attachment of alar ligaments
Superior articular facet
Foramen transversarium
Transverse process
Pars interarticularis
Inferior articular process
Lamina
Spinous process (bifid)

Occipital condyles
Lateral mass of atlas (C1)
Dens of axis (C2)
Body of axis (C2)
Lower row of teeth
Cranium
Lateral atlanto-axial joint

Anteroposterior radiograph of C1 and C2 (taken through open mouth)

Lateral mass of atlas (C1)
Dens of axis (C2)
Body of axis (C2)
C3 vertebra
Anterior view

(E) Articulating atlas and axis (C1 and C2 vertebra) as viewed radiographically

(D) Axis (C2 vertebra), superior view

Body
Dens of axis
Vertebral foramen (for spinal cord)
Facet for atlas
Superior articular facet
Transverse process
Inferior articular process
Lamina
Spinous process (bifid)

FIGURE 4.6. Cranial base and C1 and C2 vertebrae. A. The occipital condyles articulate with the superior articular facets of the atlas (vertebra C1). **B.** The atlas, on which the cranium rests, has neither a spinous process nor a body. It consists of two lateral masses connected by anterior and posterior arches. **C and D.** The tooth-like dens characterizes the axis (vertebra C2) and provides a pivot around which the atlas turns and carries the cranium. It articulates anteriorly with the anterior arch of the atlas ("Facet for dens of axis," in part **B**), and posteriorly with the transverse ligament of the atlas (see part **B**). **E.** Radiograph and articulated atlas and axis showing the dens projecting superiorly from the body of the axis between the lateral masses of the atlas. Since the atlas and axis lie posterior to the mandible (Fig. 4.5C), anteroposterior radiographs must be taken through the open mouth as indicated in the orientation figure.

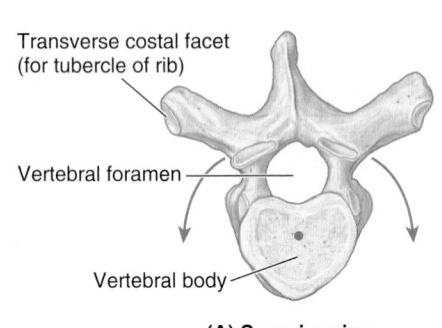

(A) Superior view

Transverse costal facet
(for tubercle of rib)

Vertebral foramen

Vertebral body

H	Head of rib
P	Pedicle of vertebra
TP	Transverse process of vertebra
SP	Spinous process (tip overlaps inferior IV disc and vertebra)
Arrows	Joints of heads of ribs

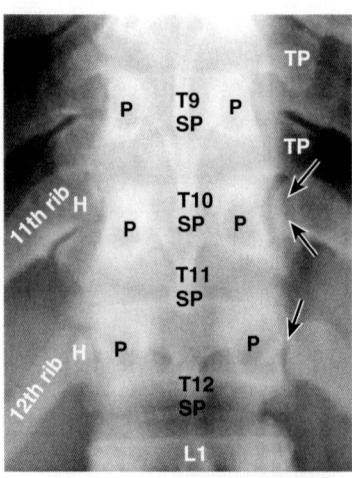

(C) AP projection radiograph, AP view

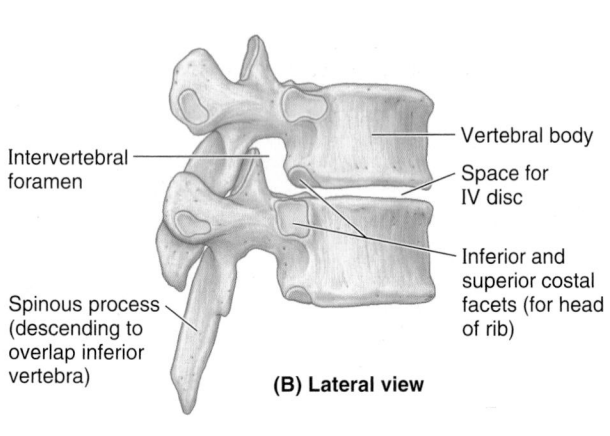

(B) Lateral view

Intervertebral foramen

Spinous process (descending to overlap inferior vertebra)

Vertebral body

Space for IV disc

Inferior and superior costal facets (for head of rib)

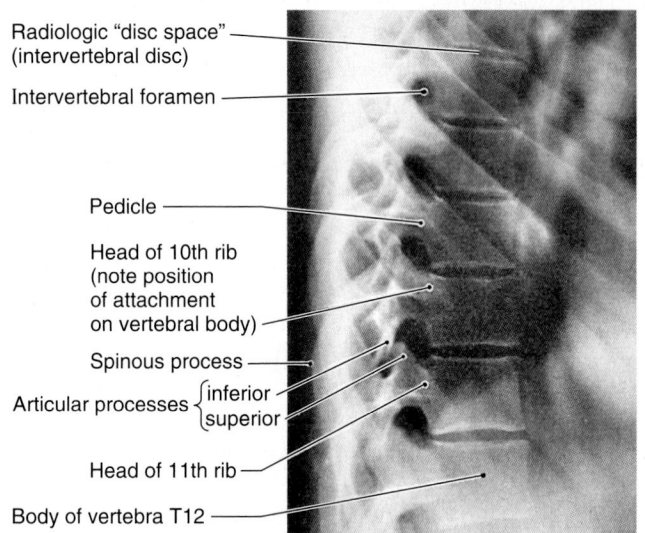

(D) Right lateral view

Radiologic "disc space" (intervertebral disc)

Intervertebral foramen

Pedicle

Head of 10th rib (note position of attachment on vertebral body)

Spinous process

Articular processes {inferior, superior}

Head of 11th rib

Body of vertebra T12

FIGURE 4.7. Thoracic vertebrae. Isolated (**A**) and articulated (**B**) typical thoracic vertebrae. (See also Fig. 4.4B.) In radiographs of the thoracic vertebrae, the articulating ribs obscure lateral features in anteroposterior views (**C**) and the vertebral arch components in lateral views (**D**). The uniformity of the vertebral bodies and radiographic "disc spaces" (caused by the radiolucency of the IV discs) are apparent.

TABLE 4.2. THORACIC VERTEBRAE

Part	Characteristics
Vertebral body	Heart shaped; one or two costal facets for articulation with head of rib
Vertebral foramen	Circular and smaller than those of cervical and lumbar vertebrae (admits the distal part of a medium-size index finger)
Transverse processes	Long and strong and extend posterolaterally; length diminishes from T1 to T12 (T1–T10 have facets for articulation with tubercle of rib)
Articular processes	Nearly vertical articular facets; superior facets directed posteriorly and slightly laterally; inferior facets directed anteriorly and slightly medially; planes of facets lie on an arc centered in the vertebral body
Spinous processes	Long; slope postero-inferiorly; tips extend to level of vertebral body below

Mammillary processes (small tubercles) also occur on vertebra T12. However, most of the transition in characteristics of vertebrae from the thoracic to the lumbar region occurs over the length of a single vertebra: vertebra T12. Generally, its superior half is thoracic in character, having costal facets and articular processes that permit primarily rotatory movement, whereas its inferior half is lumbar in character, devoid of costal facets and having articular processes that permit only flexion and extension. Consequently, vertebra T12 is subject to transitional stresses that cause it to be the most commonly fractured vertebra.

SURFACE ANATOMY OF CERVICAL AND THORACIC VERTEBRAE

Several of the *spinous processes* can usually be observed, especially when the back is flexed and the scapulae are protracted (Fig. 4.8A); most of them can be palpated—even in an obese patient—because fat does not normally accumulate in the midline.

The tip of the *C7 spinous process* is the most evident superficially. Often, when the patient stands erect, it is the only spinous process visible (Fig. 4.8B); hence the name *vertebra prominens*. The *spinous process of C2* can be felt deeply in the midline, inferior to the *external occipital protuberance*, a median projection located at the junction of the head and neck. C1 has no spinous process, and its small posterior tubercle is neither visible nor palpable.

The short bifid spinous processes of the C3–C5 vertebrae may be felt in the **nuchal groove** between the neck muscles, but they are not easy to palpate because the cervical lordosis, which is concave posteriorly, places them deep to the surface from which they are separated by the nuchal ligament. However, because it is considerably longer, the bifid *spinous process of C6 vertebra* is easily felt superior to the visible tip of the C7 process (vertebra prominens) when the neck is flexed.

When the neck and back are flexed, the *spinous processes of the upper thoracic vertebra* may also be seen. If the individual is especially lean, a continuous ridge appears linking their tips—the *supraspinous ligament* (Fig. 4.8C).

Although C7 is most commonly the most superior spinal process that is visible and readily palpable, the spinous process of T1 sometimes is more prominent. The spinous processes of the other thoracic vertebrae may be obvious in thin people, and in others can be identified by superior to inferior palpation beginning at the C7 spinous process. The tips of the *thoracic spinous processes* do not indicate the level of the corresponding vertebral bodies because they overlap (lie at the level of) the vertebra below (Figs. 4.1D and 4.7B & C).

When the back is not being flexed or the scapulae are not protracted, the tips of the thoracic spinous processes lie deep to a **median longitudinal furrow** (Fig. 4.8B & C). The tips of the spinous processes are normally in line with each other, even if the collective line wanders slightly from the midline. A sudden shift in the alignment of adjacent spinous processes may be the result of a unilateral dislocation of a zygapophysial joint;

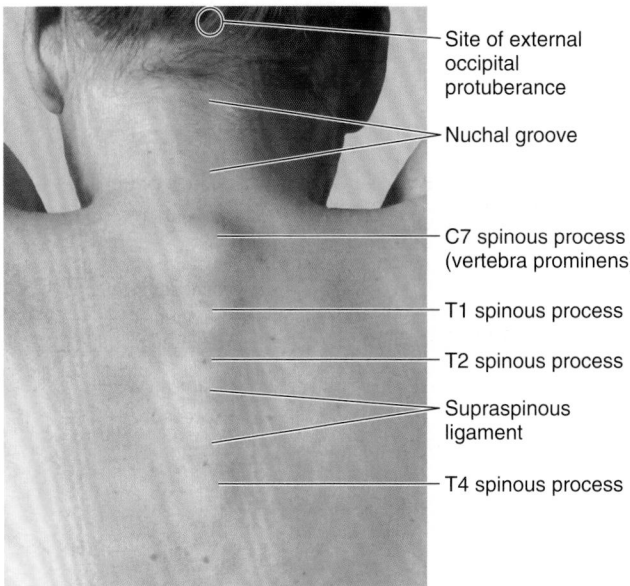

(A) Posterior view with neck and back flexed and scapulae protracted

- Site of external occipital protuberance
- Nuchal groove
- C7 spinous process (vertebra prominens)
- T1 spinous process
- T2 spinous process
- Supraspinous ligament
- T4 spinous process

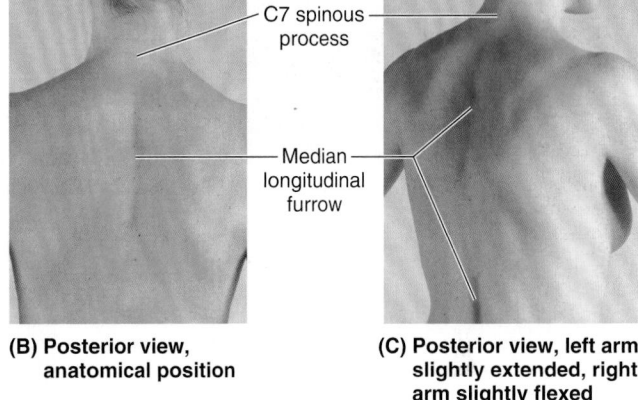

- C7 spinous process
- Median longitudinal furrow

(B) Posterior view, anatomical position

(C) Posterior view, left arm slightly extended, right arm slightly flexed

FIGURE 4.8. Surface anatomy of cervical and thoracic vertebrae. Except for the spinous process of the C7 vertebra (vertebra prominens), the visibility of the spinous processes depends on the abundance of subcutaneous tissue and the position of the back, neck, and upper limbs (especially protraction/retraction of scapulae). However, the spinous and thoracic transverse processes can usually be palpated in the mid- and paravertebral lines.

however, slight irregular misalignments may also result from a fracture of the spinous process. The short 12th rib, the lateral end of which can be palpated in the posterior axillary line, can be used to confirm identity of the T12 spinous process.

The *transverse processes of C1* may be felt laterally by deep palpation between the *mastoid processes* (prominences of the temporal bones posterior to the ears) and the angles of the jaws. The *carotid tubercle*, the anterior tubercle of the transverse process of C6 vertebra, may be large enough to be palpable; the *carotid artery* lies anterior to it. In most people, the transverse processes of thoracic vertebrae can be palpated on each side of the spinous processes in the thoracic region. In lean individuals, the ribs can be palpated from the tubercle to the angle, at least in the lower back (inferior to the scapula) (see Figs. 1.1 and 1.2A).

LUMBAR VERTEBRAE

Lumbar vertebrae are in the lower back between the thorax and sacrum (Fig. 4.1). Characteristic features of lumbar vertebrae are illustrated in Figures 4.4C and 4.9 and described in Table 4.3. Because the weight they support increases toward the inferior end of the vertebral column, lumbar vertebrae have massive bodies, accounting for much of the thickness of the lower trunk in the median plane. Their articular processes extend vertically, with articular facets sagittally oriented initially (beginning abruptly with the T12–L1

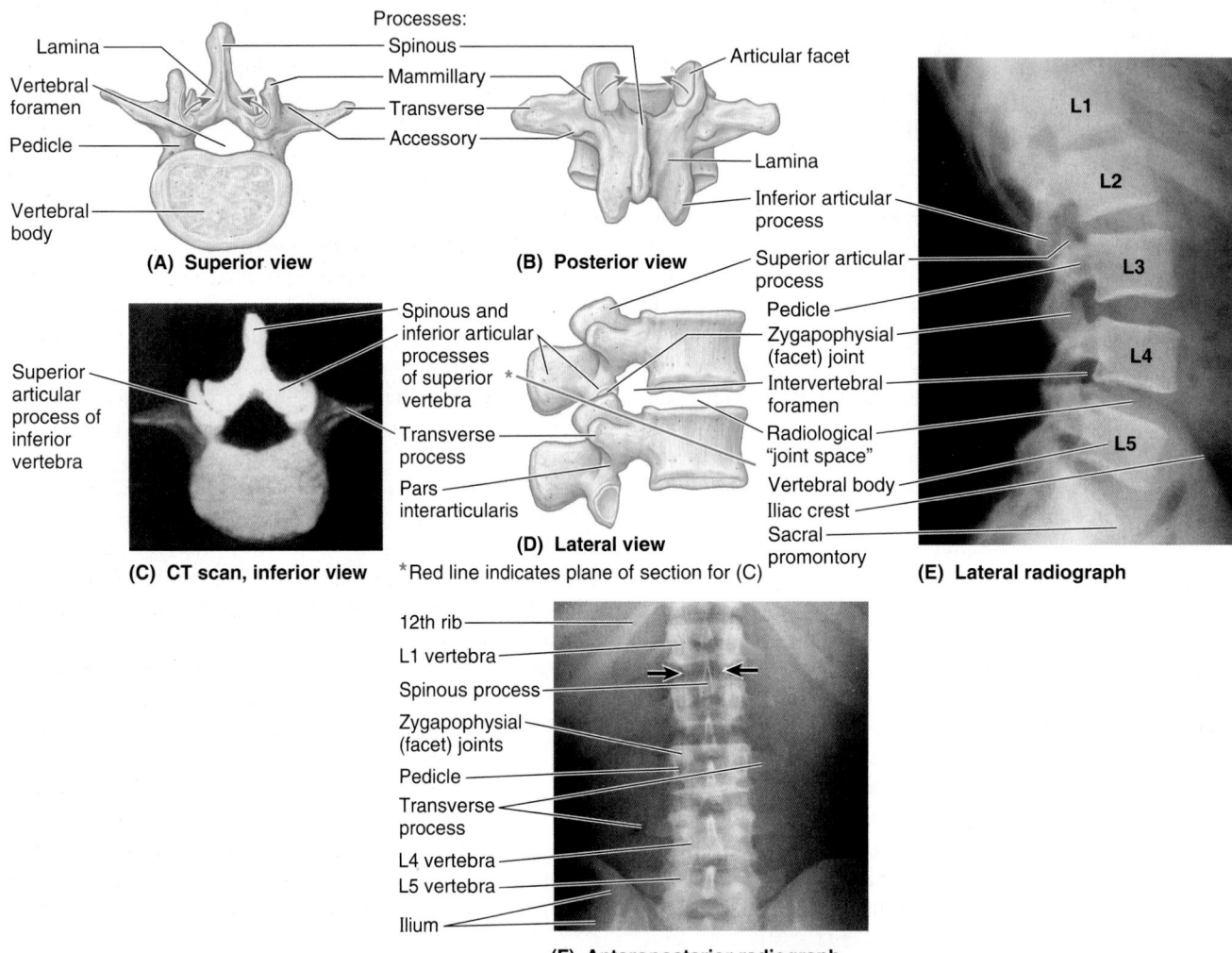

FIGURE 4.9. Lumbar vertebrae. Isolated (**A–C**) and articulated (**D–F**) typical lumbar vertebrae. In a lateral radiograph (**E**) the wedge shape of the lumbar vertebrae and especially the lumbar IV discs is evident. In anteroposterior radiographs (**F**) the vertebral canal is visible as a columnar shadow (between *arrowheads*); in lateral radiographs, the vertebral canal is primarily evident in the radiolucency of the IV foramina. (Courtesy of Dr. J. Heslin, Toronto, Ontario, Canada.)

TABLE 4.3. LUMBAR VERTEBRAE

Part	Characteristics
Vertebral body	Massive; kidney shaped when viewed superiorly
Vertebral foramen	Triangular; larger than in thoracic vertebrae and smaller than in cervical vertebrae
Transverse processes	Long and slender; accessory process on posterior surface of base of each process
Articular processes	Nearly vertical facets; superior facets directed posteromedially (or medially); inferior facets directed anterolaterally (or laterally); mammillary process on posterior surface of each superior articular process
Spinous processes	Short and sturdy; thick, broad, and hatchet shaped

joints), but becoming more coronally oriented as the column descends.

The L5–S1 facets are distinctly coronal in orientation. In the more sagittally oriented superior joints, the laterally facing facets of the inferior articular processes of the vertebra above are "gripped" by the medially facing facets of the superior processes of the vertebra below, in a manner that facilitates flexion and extension and allows lateral flexion, but prohibits rotation (Fig. 4.9A, B, D, & E).

The *transverse processes* project somewhat posterosuperiorly as well as laterally. On the posterior surface of the base of each transverse process is a small **accessory process,** which provides an attachment for the intertransversarii muscles. On the posterior surface of the superior articular processes are small tubercles, the **mammillary processes,** which give attachment to both the multifidus and intertransversarii muscles of the back.

Vertebra L5, distinguished by its massive body and transverse processes, is the largest of all movable vertebrae. It carries the weight of the whole upper body. The L5 body is markedly taller anteriorly; therefore, it is largely responsible for the lumbosacral angle between the long axis of the lumbar region of the vertebral column and that of the sacrum (Fig. 4.1D). Body weight is transmitted from L5 vertebra to the *base of the sacrum,* formed by the superior surface of S1 vertebra (Fig. 4.10A).

SACRUM

The wedged-shaped **sacrum** (L. sacred bone) is usually composed of five fused sacral vertebrae in adults (Fig. 4.10). It is located between the hip bones and forms the roof and posterosuperior wall of the posterior half of the pelvic cavity. The triangular shape of the sacrum results from the rapid decrease in the size of the inferior lateral masses of the sacral vertebrae during development. The inferior half of the sacrum is not weight-bearing; therefore, its bulk is diminished considerably. The sacrum provides strength and stability to the pelvis and transmits the weight of the body to the *pelvic girdle,* the bony ring formed by the hip bones and sacrum, to which the lower limbs are attached.

The **sacral canal** is the continuation of the vertebral canal in the sacrum (Fig. 4.10B & C). It contains the bundle of spinal nerve roots arising inferior to the L1 vertebra, known as the *cauda equina* (L. horse tail), that descend past the termination of the spinal cord. On the pelvic and posterior surfaces of the sacrum between its vertebral components are typically four pairs of **sacral foramina** for the exit of the posterior and anterior rami of the spinal nerves (Fig. 4.10A–D). The anterior (pelvic) sacral foramina are larger than the posterior (dorsal) ones.

The **base of the sacrum** is formed by the superior surface of the S1 vertebra (Fig. 4.10A). Its superior articular processes articulate with the inferior articular processes of the L5 vertebra. The anterior projecting edge of the body of the S1 vertebra is the **sacral promontory** (L. mountain ridge), an important obstetrical landmark (see Chapter 3).

The **apex of the sacrum,** its tapering inferior end, has an oval facet for articulation with the coccyx.

The sacrum supports the vertebral column and forms the posterior part of the bony pelvis. The sacrum is tilted so that it articulates with the L5 vertebra at the **lumbosacral angle** (Fig. 4.1D), which varies from 130° to 160°. The sacrum is often wider in proportion to length in the female than in the male, but the body of the S1 vertebra is usually larger in males.

The **pelvic surface of the sacrum** is smooth and concave (Fig. 4.10A). Four transverse lines on this surface of sacra from adults indicate where fusion of the sacral vertebrae occurred. During childhood, the individual sacral vertebrae are connected by hyaline cartilage and separated by IV discs. Fusion of the sacral vertebrae starts after age 20; however, most of the IV discs remain unossified up to or beyond middle life.

The **dorsal surface of the sacrum** is rough, convex, and marked by five prominent longitudinal ridges (Fig. 4.10B). The central ridge, the **median sacral crest,** represents the fused rudimentary spinous processes of the superior three or four sacral vertebra; S5 has no spinous process. The **intermediate sacral crests** represent the fused articular processes, and the **lateral sacral crests** are the tips of the transverse processes of the fused sacral vertebrae.

The clinically important features of the dorsal surface of the sacrum are the inverted U-shaped sacral hiatus and the sacral cornua (L. horns). The **sacral hiatus** results from the absence of the laminae and spinous process of S5 and sometimes S4. The sacral hiatus leads into the sacral canal. Its depth varies, depending on how much of the spinous process and laminae of S4 are present. The **sacral cornua,** representing the inferior articular processes of S5 vertebra, project inferiorly on each side of the sacral hiatus and are a helpful guide to its location.

The superior part of the **lateral surface of the sacrum** looks somewhat like an auricle (L. external ear); because of its shape, this area is called the **auricular surface** (Fig. 4.10B & C). It is the site of the synovial part of the sacro-iliac joint between the sacrum and ilium. During life, the auricular surface is covered with hyaline cartilage.

COCCYX

The **coccyx** (tail bone) is a small triangular bone that is usually formed by fusion of the four rudimentary coccygeal vertebrae, although in some people, there may be one less or one more (Fig. 4.10A–D). **Coccygeal vertebra 1 (Co1)** may remain separate from the fused group. The coccyx is the remnant of the skeleton of the embryonic tail-like caudal eminence, which is present in human embryos from the end of the 4th week until the beginning of the 8th week (Moore, Persaud and Torchia, 2012). The pelvic surface of the coccyx is concave and relatively smooth, and the posterior surface has rudimentary articular processes. Co1 is the largest and broadest of all the coccygeal vertebrae. Its short transverse processes are connected to the sacrum, and its rudimentary

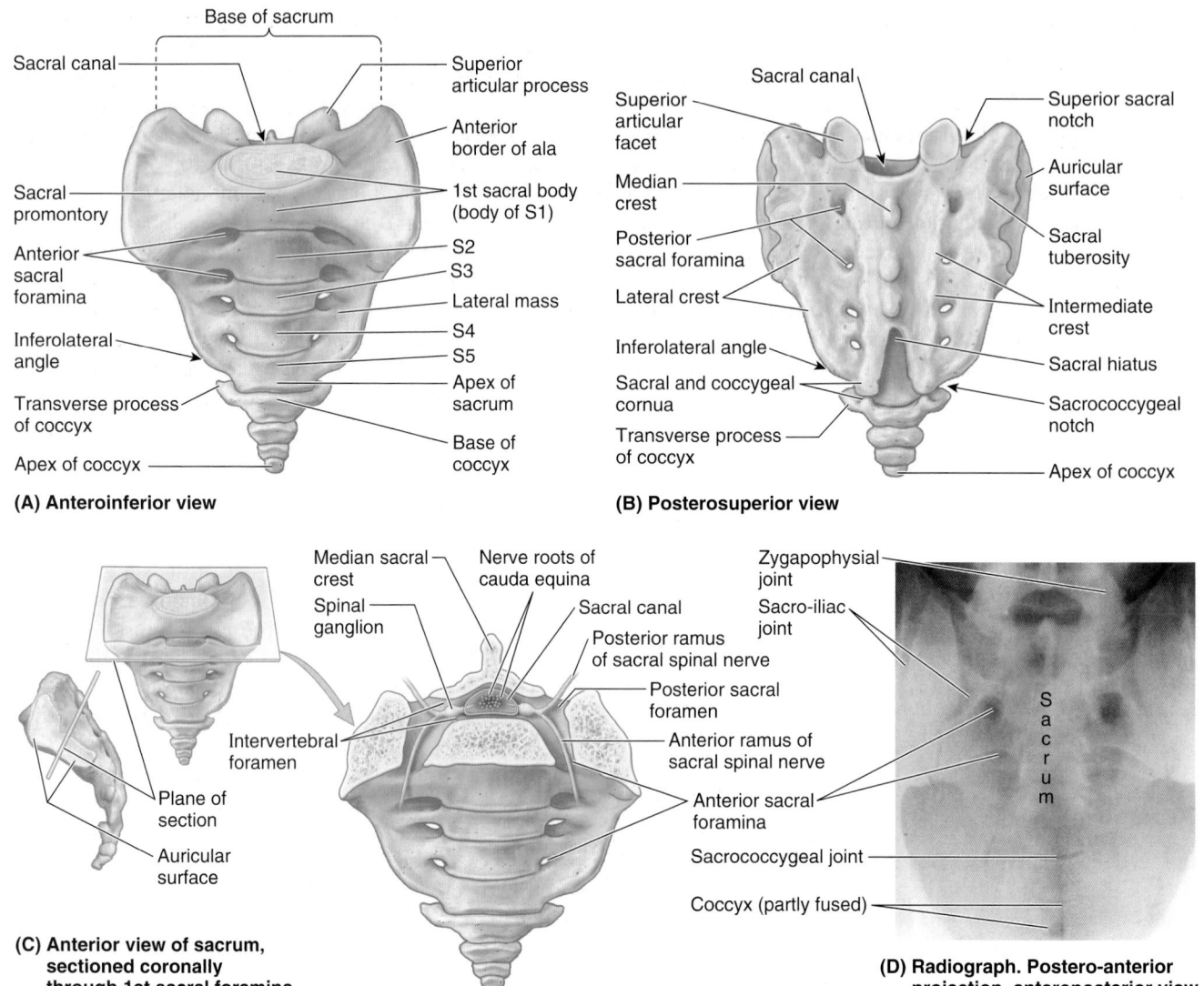

(A) Anteroinferior view

Base of sacrum

Sacral canal

Superior articular process

Anterior border of ala

Sacral promontory

1st sacral body (body of S1)

Anterior sacral foramina

S2
S3

Lateral mass

S4
S5

Inferolateral angle

Apex of sacrum

Transverse process of coccyx

Apex of coccyx

Base of coccyx

(B) Posterosuperior view

Superior articular facet

Sacral canal

Superior sacral notch

Median crest

Auricular surface

Posterior sacral foramina

Sacral tuberosity

Lateral crest

Intermediate crest

Inferolateral angle

Sacral hiatus

Sacral and coccygeal cornua

Sacrococcygeal notch

Transverse process of coccyx

Apex of coccyx

(C) Anterior view of sacrum, sectioned coronally through 1st sacral foramina

Median sacral crest

Nerve roots of cauda equina

Spinal ganglion

Sacral canal

Posterior ramus of sacral spinal nerve

Posterior sacral foramen

Intervertebral foramen

Anterior ramus of sacral spinal nerve

Plane of section

Anterior sacral foramina

Auricular surface

(D) Radiograph. Postero-anterior projection, anteroposterior view

Zygapophysial joint

Sacro-iliac joint

Sacrum

Anterior sacral foramina

Sacrococcygeal joint

Coccyx (partly fused)

FIGURE 4.10. Sacrum and coccyx. A. Base and pelvic surface of sacrum and coccyx. **B.** Dorsal surface of sacrum and coccyx. **C.** Lateral and anterior orientation drawings of the sacrum in its anatomical position demonstrate the essentially frontal plane and level at which the sacrum has been sectioned to reveal the sacral canal containing the cauda equina. Spinal ganglia lie within the IV foramina, as they do at superior vertebral levels. However, the sacral posterior and anterior rami of the spinal nerves exit via posterior and anterior (pelvic) sacral foramina, respectively. The lateral orientation drawing demonstrates the auricular surface that joins the ilium to form the synovial part of the sacro-iliac joint. In the anatomical position, the S1–S3 vertebrae lie in an essentially transverse plane, forming a roof for the posterior pelvic cavity. **D.** In anteroposterior radiographs the oblique plane of the auricular surfaces creates two lines indicating each sacro-iliac joint. The lateral line is the anterior aspect of the joint, the medial line is the posterior aspect.

articular processes form **coccygeal cornua,** which articulate with the sacral cornua. The last three coccygeal vertebrae often fuse during middle life, forming a beak-like coccyx; this accounts for its name (G. *coccyx,* cuckoo). With increasing age, Co1 often fuses with the sacrum, and the remaining coccygeal vertebrae usually fuse to form a single bone.

The coccyx does not participate with the other vertebrae in support of the body weight when standing; however, when sitting it may flex anteriorly somewhat, indicating that it is receiving some weight. The coccyx provides attachments for parts of the gluteus maximus and coccygeus muscles and the *anococcygeal ligament,* the median fibrous band of the pubococcygeus muscles (see Chapter 3).

SURFACE ANATOMY OF LUMBAR VERTEBRAE, SACRUM, AND COCCYX

The spinous processes of lumbar vertebrae are large and easy to observe when the trunk is flexed (Fig. 4.11A). They can also be palpated in the *posterior median furrow* (Fig. 4.11B & C). The *L2 spinous process* provides an estimate of the position of the inferior end of the spinal cord. A horizontal line joining the highest points of the iliac crests passes through the tip of the L4 spinous process and the L4–L5 IV disc. This is a useful landmark when performing a lumbar puncture to obtain a sample of cerebrospinal fluid (CSF) (see the blue box "Lumbar Spinal Puncture," p. 505).

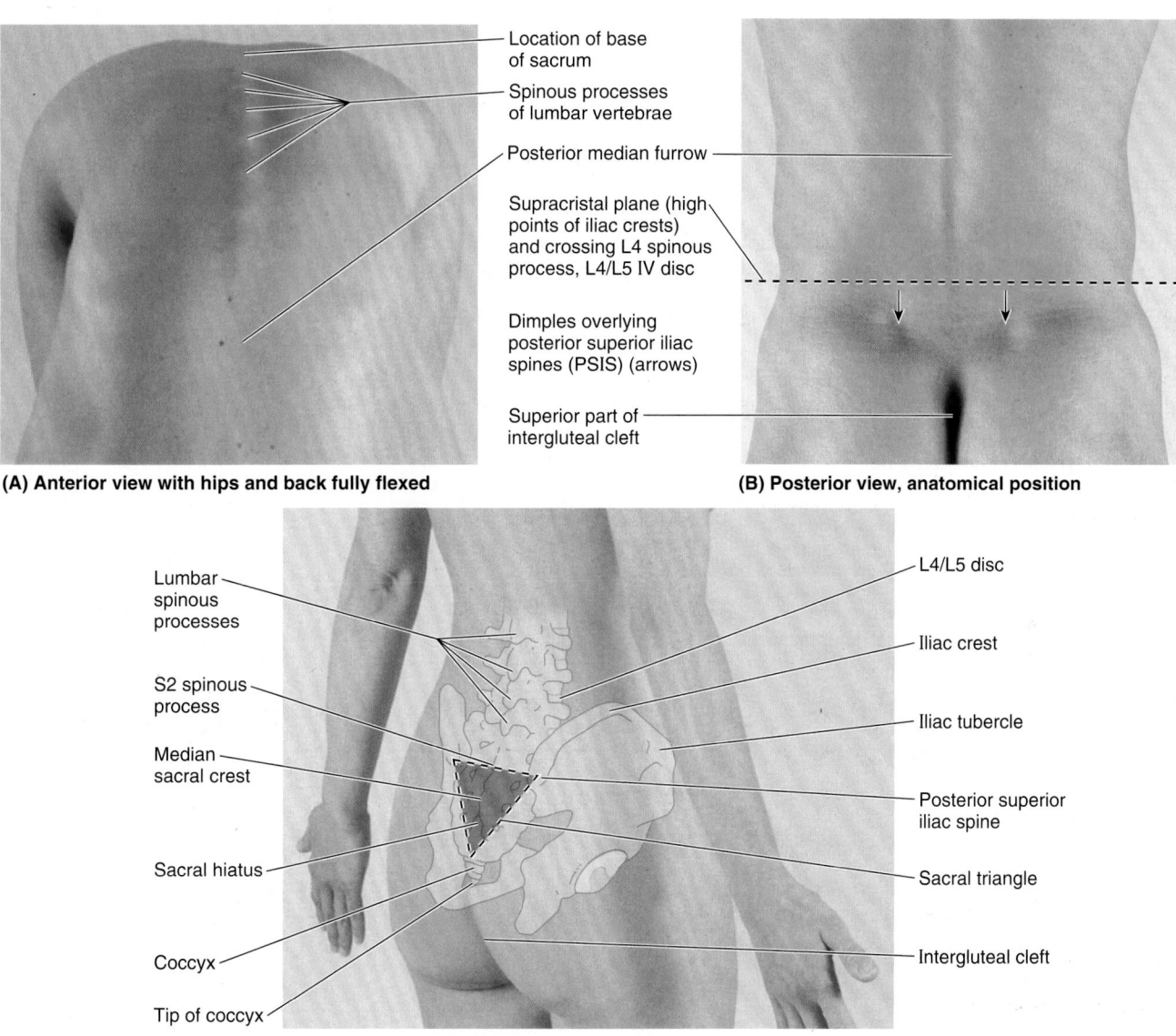

FIGURE 4.11. Surface anatomy of lumbar vertebrae, sacrum, and coccyx.

The *S2 spinous process* lies at the middle of a line drawn between the *posterior superior iliac spines,* indicated by the *skin dimples* (Fig. 4.11B). The dimples are formed by the attachment of skin and deep fascia to these spines. This level indicates the inferior extent of the subarachnoid space (lumbar cistern). The *median sacral crest* can be felt inferior to the L5 spinous process. The **sacral triangle** is formed by the lines joining the two posterior superior iliac spines and the superior part of the **intergluteal** (natal) **cleft** between the buttocks. The sacral triangle outlining the sacrum is a common area of pain resulting from low back sprains. The *sacral hiatus* can be palpated at the inferior end of the sacrum located in the superior part of the intergluteal cleft.

The transverse processes of thoracic and lumbar vertebrae are covered with thick muscles and may or may not be palpable. The *coccyx* can be palpated in the intergluteal cleft,

inferior to the apex of the sacral triangle. The **apex of the coccyx** can be palpated approximately 2.5 cm posterosuperior to the anus. Clinically, the coccyx is examined with a gloved finger in the anal canal.

Ossification of Vertebrae

Vertebrae begin to develop during the embryonic period as mesenchymal condensations around the notochord. Later, these mesenchymal bone models chondrify and cartilaginous vertebrae form. Typically, vertebrae begin to ossify toward the end of the embryonic period (8th week), with three **primary ossification centers** developing in each cartilaginous vertebra: an endochondral **centrum,** which will eventually constitute most of the body of the vertebra, and two perichondral centers, one in each half of the **neural arch** (Fig. 4.12A–J & M).

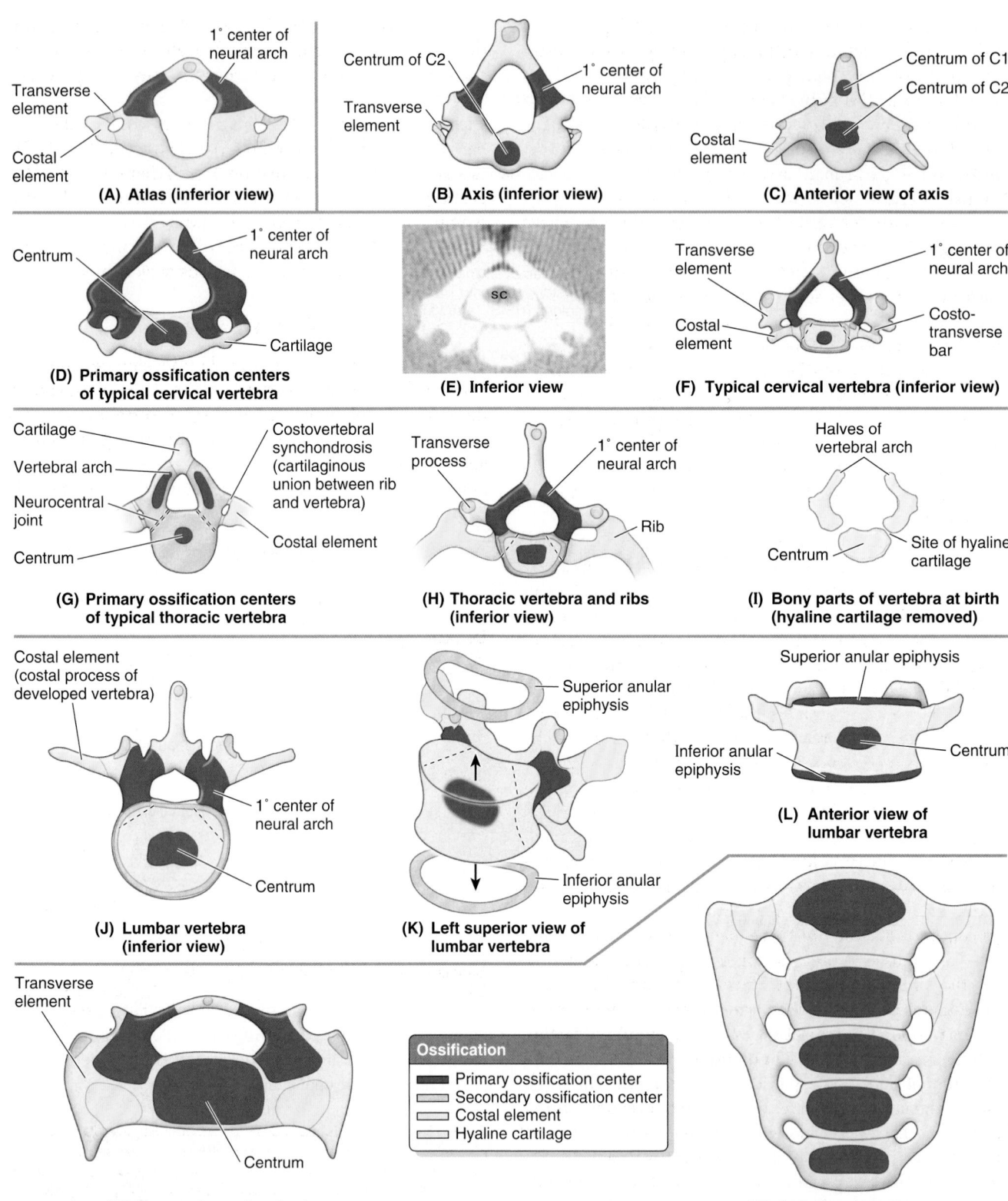

FIGURE 4.12. Ossification of vertebrae. A. Vertebra C1 (atlas) lacks a centrum. **B and C.** Vertebra C2 (axis) has two centra, one of which forms most of the dens. **D–F.** The development of "typical" cervical vertebrae is shown, including (**D**) the primary ossification centers within the hyaline cartilage, (**E**) a CT scan of the vertebra shown in part **D** (*SC*, spinal cord), and (**F**) the primary and secondary ossification centers. **G–I.** The development of thoracic vertebrae is shown, including (**G**) the three primary ossification centers in a cartilaginous vertebra of a 7-week-old embryo (observe the joints present at this stage), (**H**) the primary and secondary ossification centers (with ribs developed from costal elements), and (**I**) the bony parts of a thoracic vertebra after skeletonization (cartilage removed). **J–L.** The development of the lumbar vertebrae is shown, including (**J**) the primary and secondary ossification centers, (**K**) the anular epiphyses separated from the body, and (**L**) the anular epiphyses in place. **M and N.** The development of the sacrum is shown. Note that the ossification and fusion of sacral vertebrae may not be completed until age 35.

Ossification continues throughout the fetal period. At birth, typical vertebra and the superiormost sacral vertebrae consist of three bony parts united by hyaline cartilage. The inferior sacral vertebrae and all the coccygeal vertebrae are still entirely cartilaginous; they ossify during infancy. The halves of the neural arches articulate at **neurocentral joints,** which are primary cartilaginous joints (Fig. 4.12G). The halves of the neural/vertebral arch begin to fuse with each other posterior to the vertebral canal during the 1st year, beginning in the lumbar region and then in the thoracic and cervical regions. The neural arches begin fusing with the centra in the upper cervical region around the end of the 3rd year, but usually the process is not completed in the lower lumbar region until after the 6th year (Moore et al., 2012).

Five **secondary ossification centers** develop during puberty in each typical vertebra: one at the tip of the spinous process; one at the tip of each transverse process; and two anular epiphyses (ring epiphyses), one on the superior and one on the inferior edges of each vertebral body (i.e., around the margins of the superior and inferior surfaces of the vertebral body) (Fig. 4.12F & I–L).

The hyaline **anular epiphyses,** to which the IV discs attach, are sometimes referred to as *epiphysial growth plates* and form the zone from which the vertebral body grows in height. When growth ceases early in the adult period, the epiphyses usually unite with the vertebral body. This union results in the characteristic smooth raised margin, the **epiphysial rim,** around the edges of the superior and inferior surfaces of the body of the adult vertebra (Figs. 4.2B and 4.3). All secondary ossification centers have usually united with the vertebrae by age 25; however, the ages at which specific unions occur vary.

Exceptions to the typical pattern of ossification occur in vertebrae C1, C2, and C7 (Fig. 4.12A–C) and in the sacrum (Fig. 4.12M & N) and coccyx. In addition, at all levels, primordial "ribs" (**costal elements**) appear in association with the secondary ossification centers of the transverse processes (**transverse elements**). The costal elements normally develop into ribs only in the thoracic region; they become part of the transverse process or its equivalent at other levels.

In the cervical region, the costal element normally remains diminutive as part of the transverse process. *Foramina transversarii* develop as gaps between the two lateral ossification centers, medial to a linking **costotransverse bar,** which forms the lateral boundary of the foramina (Fig. 4.12A–F). Also, as a result of the cervical transverse processes being formed from the two developmental elements, the transverse processes of cervical vertebrae end laterally in an *anterior tubercle* (from the costal element) and a *posterior tubercle* (from the transverse element). The atypical morphology of vertebrae C1 and C2 is also established during development. The centrum of C1 becomes fused to that of C2 and loses its peripheral connection to the remainder of C1, thus forming the *dens* (Fig. 4.12C). Since these first two centra are fused and are now part of C2, no IV disc is formed between C1 and C2 to connect them. The part of the body that remains with C1 is represented by the *anterior arch* and *tubercle of C1.*

In the thoracic region, the costal elements separate from the developing vertebrae and elongate into ribs, and the transverse elements alone form the transverse processes (Fig. 4.12I).

All but the base of the transverse processes of the lumbar vertebrae develop from the costal element (Fig. 4.12J); this projecting bar of the mature bone is therefore called the **costal process.** The transverse elements of the lumbar vertebrae form the *mammillary processes.*

The ala and auricular surfaces of the sacrum are formed by the fusion of the transverse and costal elements.

Variations in Vertebrae

Most people have 33 vertebrae, but developmental errors may result in 32 or 34 vertebrae (Fig. 4.13). Estimates of the frequency of abnormal numbers of vertebrae superior to the sacrum (the normal number is 24) range between 5% and 12%. Variations in vertebrae are affected by race, gender, and developmental factors (genetic and environmental).

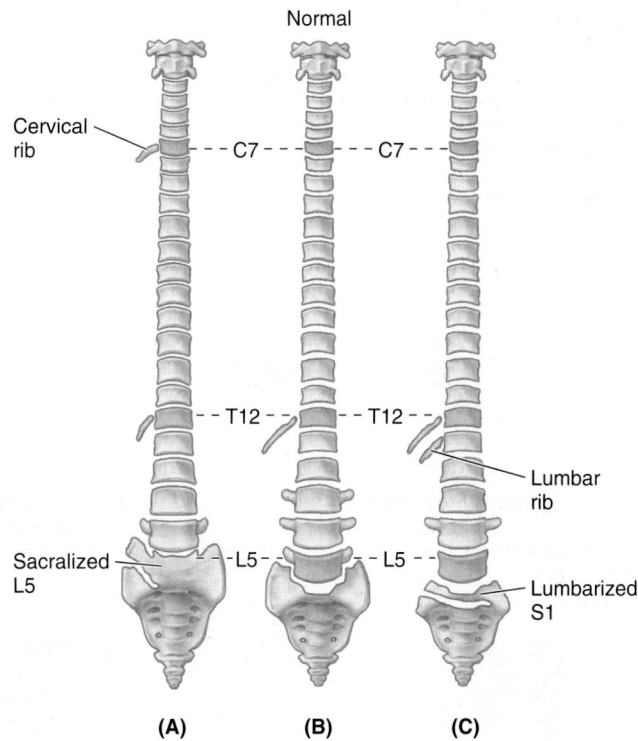

FIGURE 4.13. Variations in vertebrae and their relationship to ribs. A. A "cranial shift" is demonstrated, in which there are 13 ribs, including a cervical rib articulating with vertebra C7, and a diminished 12th rib articulating with vertebra T12. Vertebra L5 is shown partially incorporated into the sacrum, but such "sacralization" can also be complete. The lowest sacral segment (S5) is partially segmented. **B.** The common arrangement of the vertebrae and the position of the 1st and 12th ribs are shown. **C.** A "caudal shift" is shown, in which the 12th rib is increased in size, and there is a small lumbar rib. The transverse process of vertebra L4 is increased in size, whereas that of vertebra L5 is greatly reduced. The first sacral segment is shown partially separated from the rest of the sacrum, but such "lumbarization" can also be complete. The 1st coccygeal segment is incorporated into the sacrum—that is, it is "sacralized."

An increased number of vertebrae occurs more often in males, and a reduced number occurs more frequently in females. Some races show more variation in the number of vertebrae. Variations in the number of vertebrae may be clinically important. An increased length of the presacral region of the vertebral column increases the strain on the inferior part of the lumbar region of the column owing to the increased leverage. However, most numerical variations are detected incidentally during diagnostic medical imaging studies being performed for other reasons and during dissections and autopsies of persons with no history of back problems.

Caution is necessary, however, when describing an injury (e.g., when reporting the site of a vertebral fracture). When counting the vertebrae, begin at the base of the neck. The number of cervical vertebrae (seven) is remarkably constant (and not just in humans, but among vertebrates—even giraffes and snakes have seven cervical vertebrae!). When considering a numerical variation, the thoracic and lumbar regions must be considered together because people having more than five lumbar vertebrae often have a compensatory decrease in the number of thoracic vertebrae (O'Rahilly, 1986).

Variations in vertebrae also involve the relationship between the vertebrae and ribs, and the number of vertebrae that fuse to form the sacrum (Fig. 4.13). The relationship of presacral vertebrae to ribs and/or sacrum may occur higher (*cranial shift*) or lower (*caudal shift*) than normal. Note, however, that a C7 vertebra articulating with a rudimentary cervical rib(s) is still considered a cervical vertebra. The same is true for lumbar vertebrae and lumbar ribs. Likewise, an L5 vertebra fused to the sacrum is referred to as a "sacralized 5th lumbar vertebra" (see the blue box "Abnormal Fusion of Vertebrae," p. 462).

VERTEBRAE

Vertebral Body Osteoporosis

Vertebral body osteoporosis is a common metabolic bone disease that is often detected during routine radiographic studies. Osteoporosis results from a net demineralization of the bones caused by a disruption of the normal balance of calcium deposition and resorption. As a result, the quality of bone is reduced and atrophy of skeletal tissue occurs. Although osteoporosis affects the entire skeleton, the most affected areas are the neck of the femur, the bodies of vertebrae, the metacarpals (bones of the hand), and the radius. These bones become weakened and brittle, and are subject to fracture.

Radiographs taken during early to moderate osteoporosis demonstrate demineralization, which is evident as diminished radiodensity of the trabecular (spongy) bone of the vertebral bodies, causing the thinned cortical bone to appear relatively prominent (Fig. B4.1B). Osteoporosis especially affects the horizontal trabeculae of the trabecular bone of the vertebral body (see Fig. 4.3). Consequently, vertical striping may become apparent, reflecting the loss of the horizontal

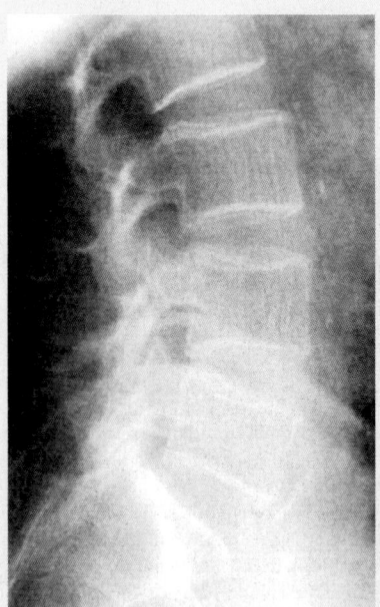 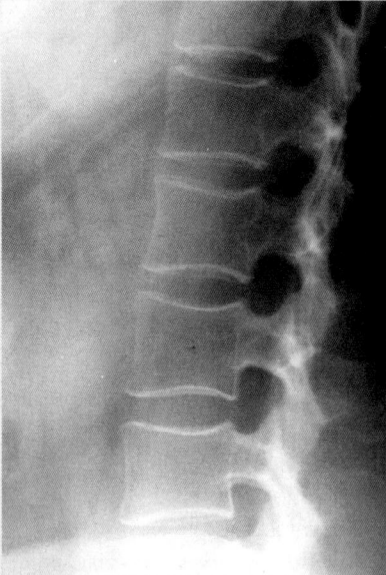 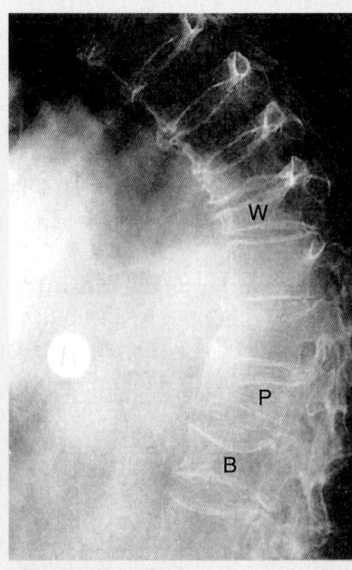

| (A) Right lateral view | (B) Left lateral view | (C) Left lateral view |

FIGURE B4.1. **Effects of osteoporosis on vertebral column. A.** Early to moderate osteoporosis, characterized by vertical striation in the vertebral bodies. **B.** Later, the striated pattern is lost as the continued loss of trabecular bone produces uniform radiolucency (less white, more "transparent"). In contrast, the cortical bone, while thinned, appears relatively prominent. **C.** Late osteoporosis in the thoracic region of the vertebral column demonstrates excessive thoracic kyphosis, a result of the collapse of the vertebral bodies, which have become wedge-shaped (*W*), planar (*P*), and biconcave (*B*).

supporting trabeculae and thickening of the vertical struts (Fig. B4.1A). Radiographs in later stages may reveal vertebral collapse (compression fractures) and increased thoracic kyphosis (Figs. 4.1C and B4.17B). Vertebral body osteoporosis occurs in all vertebrae but is most common in thoracic vertebrae and is an especially common finding in postmenopausal females.

Laminectomy

The surgical excision of one or more spinous processes and the adjacent supporting vertebral laminae in a particular region of the vertebral column is called a *laminectomy* (1 in Fig. B4.2A). The term is also commonly used to denote removal of most of the vertebral arch by transecting the pedicles (2 in Fig. B4.2A).

Laminectomies are performed surgically (or anatomically in the dissection laboratory) to gain access to the vertebral canal, providing posterior exposure of the spinal cord (if performed above the L2 level) and/or the roots of specific spinal nerves. Surgical laminectomy is often performed to relieve pressure on the spinal cord or nerve roots caused by a tumor, herniated IV disc, or bony hypertrophy (excess growth).

Dislocation of Cervical Vertebrae

Because of their more horizontally oriented articular facets, the cervical vertebrae are less tightly interlocked than other vertebrae. The cervical vertebrae, "stacked like coins," can be dislocated in neck injuries with less force than is required to fracture them (Fig. B4.3A–F). Because of the large vertebral canal in the cervical region, slight dislocation can occur here without damaging the spinal cord (Fig. B4.3B). Severe dislocations, or dislocations combined with fractures (fracture–dislocations) injure the spinal cord. If the dislocation does not result in "facet jumping" with locking of the displaced articular processes (Fig. B4.3F & G), the cervical vertebrae may self-reduce (slip back into place) so that a radiograph may not indicate that the cord has been injured. An MRI, however, may reveal the resulting soft tissue damage.

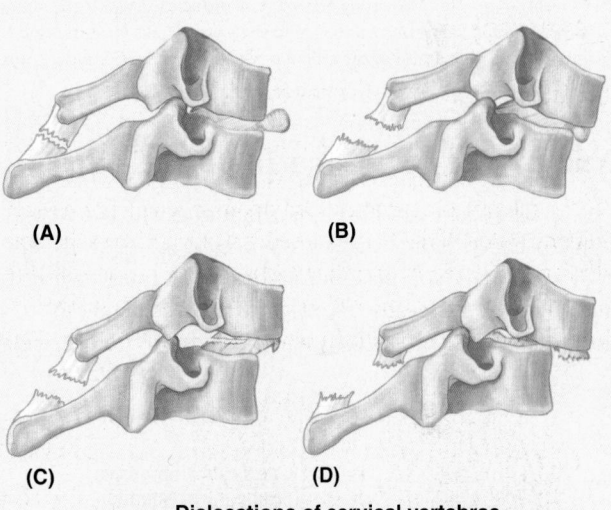

(A)   (B)

(C)   (D)

Dislocations of cervical vertebrae

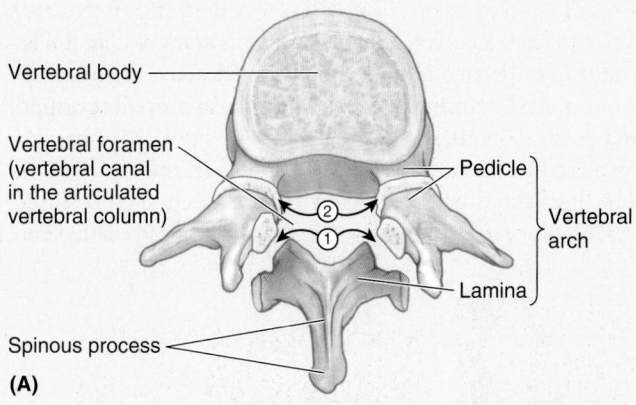

Vertebral body

Vertebral foramen
(vertebral canal
in the articulated
vertebral column)

Pedicle

Vertebral
arch

Lamina

Spinous process

(A)

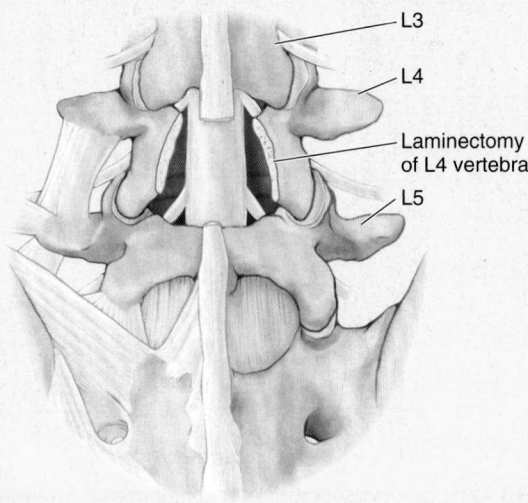

L3

L4

Laminectomy
of L4 vertebra

L5

(B)

FIGURE B4.2. A. Sites at which laminectomies are performed. **B.** Posterior view, postlaminectomy.

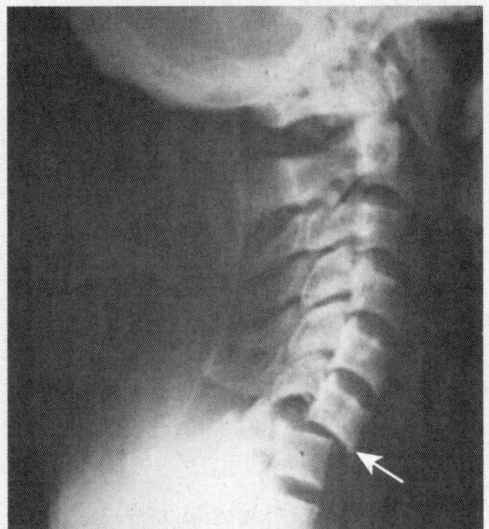

(E) Right lateral view

FIGURE B4.3 A-E. (*Continued on next page*)

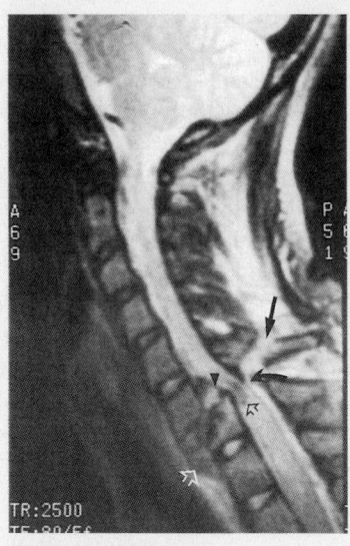
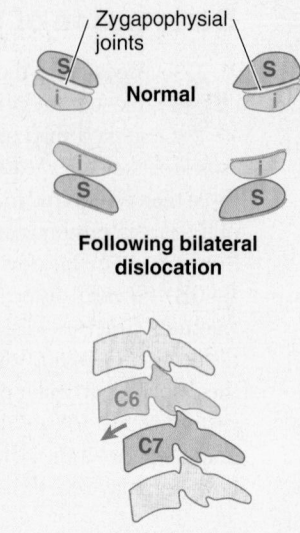
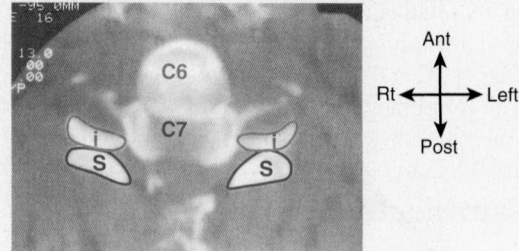

(F) Median MRI scan

Zygapophysial joints

Normal

Following bilateral dislocation

Bilateral facet dislocation (locked facets)

(G) Transverse CT scan – inferior view

i = inferior articular processes of C6
s = superior articular processes of C7

Normally, the superior articular processes are anterior to the inferior processes, with their flat articular facets in contact as a zygapophysial joint.

FIGURE B4.3. Dislocations of cervical vertebrae. Four stages of injury are shown: (**A**) stage I, flexion sprain; (**B**) stage II, anterior subluxation with 25% anterior translation; (**C**) stage III, 50% translation; and (**D**) stage IV, complete dislocation. **E.** This lateral view radiograph shows a stage III dislocation with 50% translation. **F.** This MRI study of a stage IV dislocation with cord injury reveals that the body of C7 is fractured (*open white arrowhead*). The ligamentum flavum is disrupted (*curved black arrow*), and the spinous process is avulsed (*straight black arrow*). **G.** This transverse CT scan (same individual shown in **F**) reveals the reversed position of the articular processes of the C6 and C7 vertebrae owing to "facet jumping."

Fracture and Dislocation of Atlas

The atlas (vertebra C1) is a bony ring, with two wedge-shaped lateral masses, connected by relatively thin anterior and posterior arches and a transverse ligament (Fig. 4.4A). Because the taller side of the lateral mass is directed laterally, vertical forces (as would result from striking the bottom of a pool in a diving accident) compressing the lateral masses between the occipital condyles and the axis drive them apart, fracturing one or both of the anterior or posterior arches (Fig. B4.4B).

If the force is sufficient, *rupture of the transverse ligament* that links them will also occur (Fig. B4.4C). The resulting *Jefferson* or *burst fracture* (Fig. B4.4C–E) in itself does not

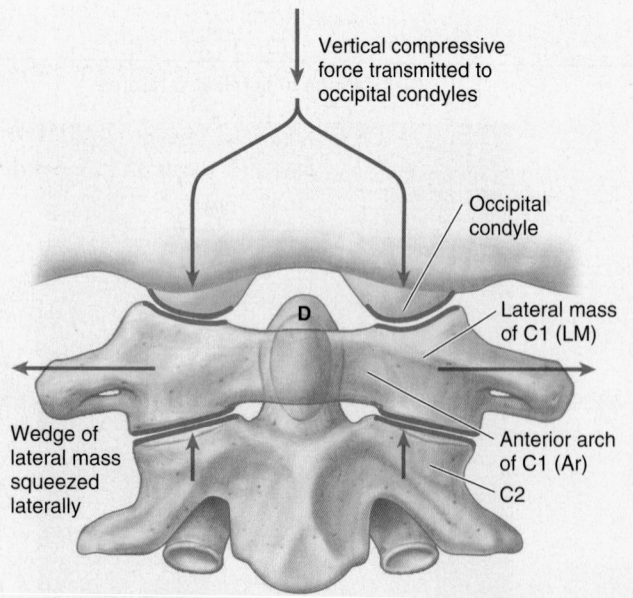
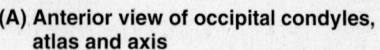
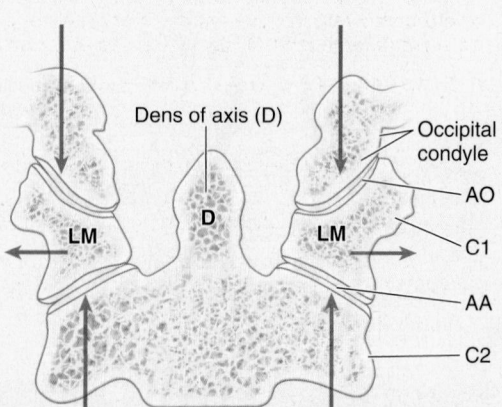

(A) Anterior view of occipital condyles, atlas and axis

(B) Anterior view of coronal section of craniovertebral [atlanto-occipital (AO) and lateral atlanto-axial (AA)] joints. Compare with radiograph in Fig 4.6E.

FIGURE B4.4 A-B. Jefferson fracture of atlas. (*Continued on next page*)

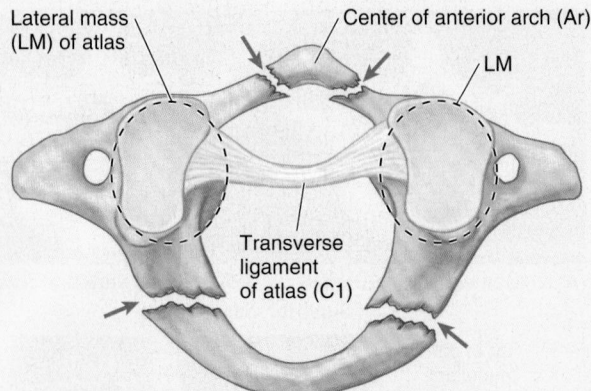

(C) Inferior view of Jefferson (burst) fracture of C1

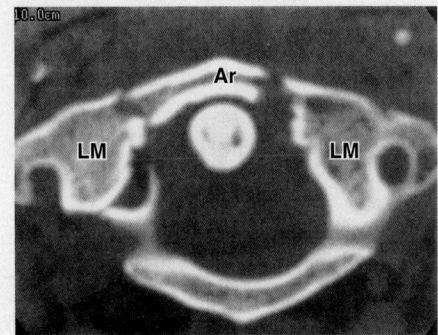

(D) Inferior view of CT scan of Jefferson fracture

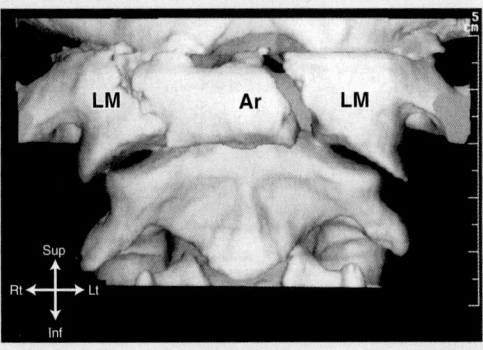

(E) Anterior view of reconstructed CT image of Jefferson fracture showing fragment of anterior arch (Ar) and outward (lateral) shift of lateral masses of C1 (LM)

FIGURE B4.4 C-E. Jefferson fracture of atlas.

necessarily result in spinal cord injury, because the dimensions of the bony ring actually increase. Spinal cord injury is more likely, however, if the transverse ligament has also been ruptured (see the blue box "Rupture of Transverse Ligament of Atlas" on p. 477) indicated radiographically by widely spread lateral masses.

Fracture and Dislocation of Axis

Fractures of the vertebral arch of the axis (vertebra C2) are one of the most common injuries of the cervical vertebrae (up to 40%) (Yochum and Rowe, 2004). Usually the fracture occurs in the bony column formed by the superior and inferior articular processes of the axis, the **pars interarticularis** (Fig. 4.5A). A fracture in this location, called a *traumatic spondylolysis of C2* (Fig. B4.5A, B, & D), usually occurs as a result of *hyperextension of the head on the neck*, rather than the combined hyperextension of the head and neck, which may result in *whiplash injury*.

Such hyperextension of the head was used to execute criminals by hanging, in which the knot was placed under the chin before the body suddenly dropped its length through the gallows floor (Fig. B4.5C); thus, this fracture has been called a *hangman's fracture*.

In more severe injuries, the body of the C2 vertebra is displaced anteriorly with respect to the body of the C3 vertebra. With or without such subluxation (incomplete dislocation) of

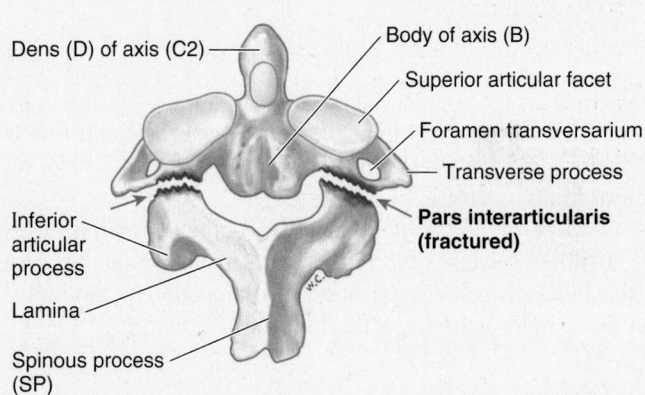

(A) Posterosuperior view

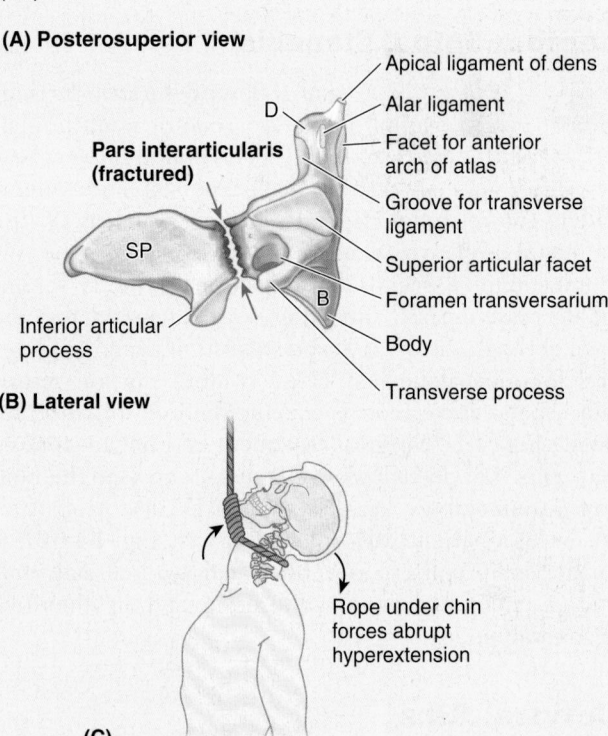

(B) Lateral view

(C)

FIGURE B4.5 A-C. Fracture and dislocation of axis. Posterosuperior (**A**) and lateral (**B**) views of a hangman's fracture of the C2 vertebra are shown (*arrows*). **C.** The position of the knot produces hyperextension during hanging (*arrows*). (*Continued on next page*)

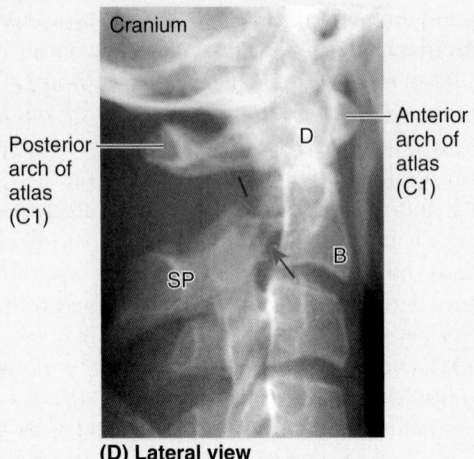

(D) Lateral view

FIGURE B4.5 D. **Fracture and dislocation of axis.** (*continued*) D. Right lateral view radiograph demonstrating a hangman's fracture (*arrow*) of C2.

the axis, injury of the spinal cord and/or of the brainstem is likely, sometimes resulting in *quadriplegia* (paralysis of all four limbs) or death.

Fractures of the dens are also common axis injuries (40–50%), which may result from a horizontal blow to the head or as a complication of *osteopenia* (pathological loss of bone mass) (see the blue box "Fracture of Dens," p. 476).

Lumbar Spinal Stenosis

Lumbar spinal stenosis describes a *stenotic* (narrow) vertebral foramen in one or more lumbar vertebrae (Fig. B4.6B). This condition may be a hereditary anomaly that can make a person more vulnerable to age-related degenerative changes such as IV disc bulging. Lumbar spinal nerves increase in size as the vertebral column descends, but paradoxically, the IV foramina decrease in size. Narrowing is usually maximal at the level of the IV discs. However, stenosis of a lumbar vertebral foramen alone may cause compression of one or more spinal nerve roots occupying the inferior vertebral canal (Fig. 4.1). Surgical treatment of lumbar stenosis may consist of decompressive laminectomy (see the blue box "Laminectomy," p. 457). When IV disc protrusion occurs in a patient with spinal stenosis (Fig. B4.6B), it further compromises a vertebral canal that is already limited, as does arthritic proliferation and ligamentous degeneration.

Cervical Ribs

A *cervical rib* is a relatively common anomaly. In 1–2% of people, the developmental costal element of C7, which normally becomes a small part of the transverse process that lies anterior to the

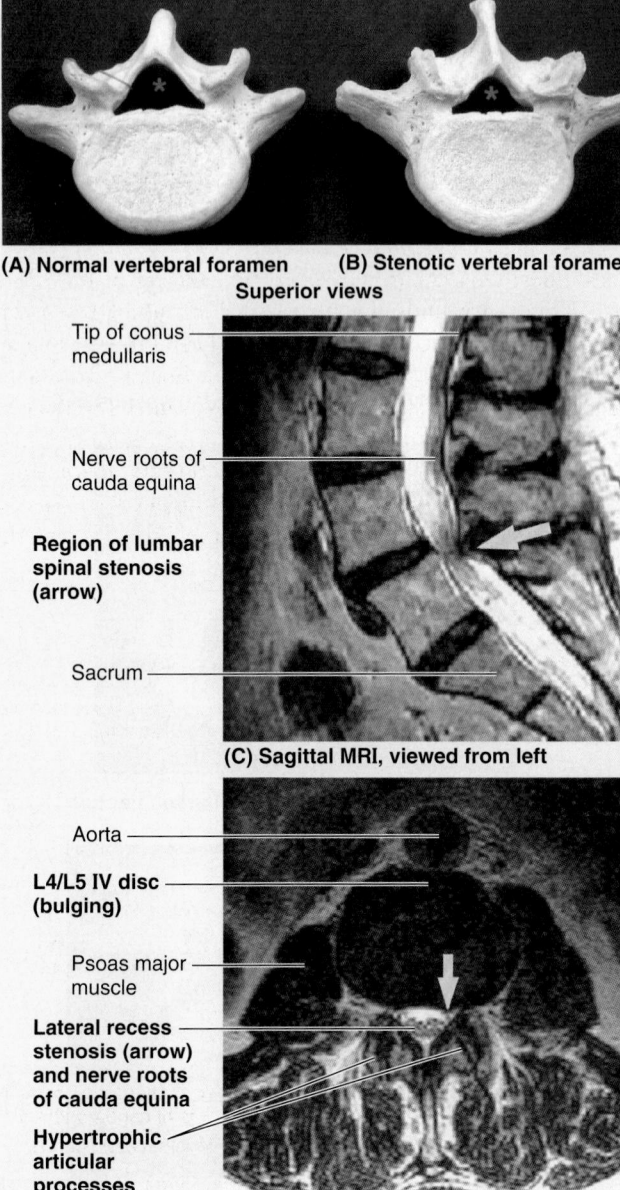

(A) Normal vertebral foramen   (B) Stenotic vertebral foramen
Superior views

Tip of conus medullaris

Nerve roots of cauda equina

Region of lumbar spinal stenosis (arrow)

Sacrum

(C) Sagittal MRI, viewed from left

Aorta

L4/L5 IV disc (bulging)

Psoas major muscle

Lateral recess stenosis (arrow) and nerve roots of cauda equina

Hypertrophic articular processes

(D) Transverse MRI, inferior view

FIGURE B4.6. **Lumbar spinal stenosis.** Normal (**A**) and stenotic (**B**) vertebral foramina are compared. The sagittal (**C**) and transverse (**D**) lumbar MRIs demonstrate a high-grade stenosis caused by hypertrophic articular processes and ligamenta flava and moderate peripheral bulging of the L4–L5 IV disc.

foramen transversarium (Fig. 4.5A), becomes abnormally enlarged. This structure may vary in size from a small protuberance to a complete rib that occurs bilaterally about 60% of the time.

The *supernumerary (extra) rib* or a fibrous connection extending from its tip to the first thoracic rib may elevate and place pressure on structures that emerge from the superior thoracic aperture, notably the subclavian artery or inferior trunk of the brachial plexus, and may cause *thoracic outlet syndrome.*

Caudal Epidural Anesthesia

In living persons, the sacral hiatus is closed by the membranous **sacrococcygeal ligament,** which is pierced by the *filum terminale* (a connective tissue strand extending from the tip of the spinal cord to the coccyx). Deep (superior) to the ligament, the epidural space of the sacral canal is filled with fatty connective tissue (Fig. B4.7A). In *caudal epidural anesthesia* or *caudal analgesia,* a local anesthetic agent is injected into the fat of the sacral canal that surrounds the proximal portions of the sacral nerves. This can be accomplished by several routes, including the sacral hiatus (Fig. B4.7B & C). Because the sacral hiatus is located between the sacral cornua and inferior to the S4 spinous process or median sacral crest, these palpable bony landmarks are important for locating the hiatus (Fig. B4.7A). The anesthetic solution spreads superiorly and extradurally, where it acts on the S2–Co1 spinal nerves of the cauda equina. The height to which the anesthetic ascends is controlled by the amount injected and the position of the patient. Sensation is lost inferior to the epidural block. Anesthetic agents can also be injected through the posterior sacral foramina into the sacral canal around the spinal nerve roots (*transsacral epidural anesthesia*) (Fig. B4.7B). Epidural anesthesia during childbirth is discussed in Chapter 3.

Injury of Coccyx

An abrupt fall onto the buttocks may cause a painful subperiosteal bruising or fracture of the coccyx, or a fracture—dislocation of the sacrococcygeal joint. Displacement is common, and surgical removal of the fractured bone may be required to relieve pain. An especially difficult childbirth occasionally injures the mother's coccyx. A troublesome syndrome, *coccygodynia* (or coccydynia), often follows coccygeal trauma; pain relief is commonly difficult.

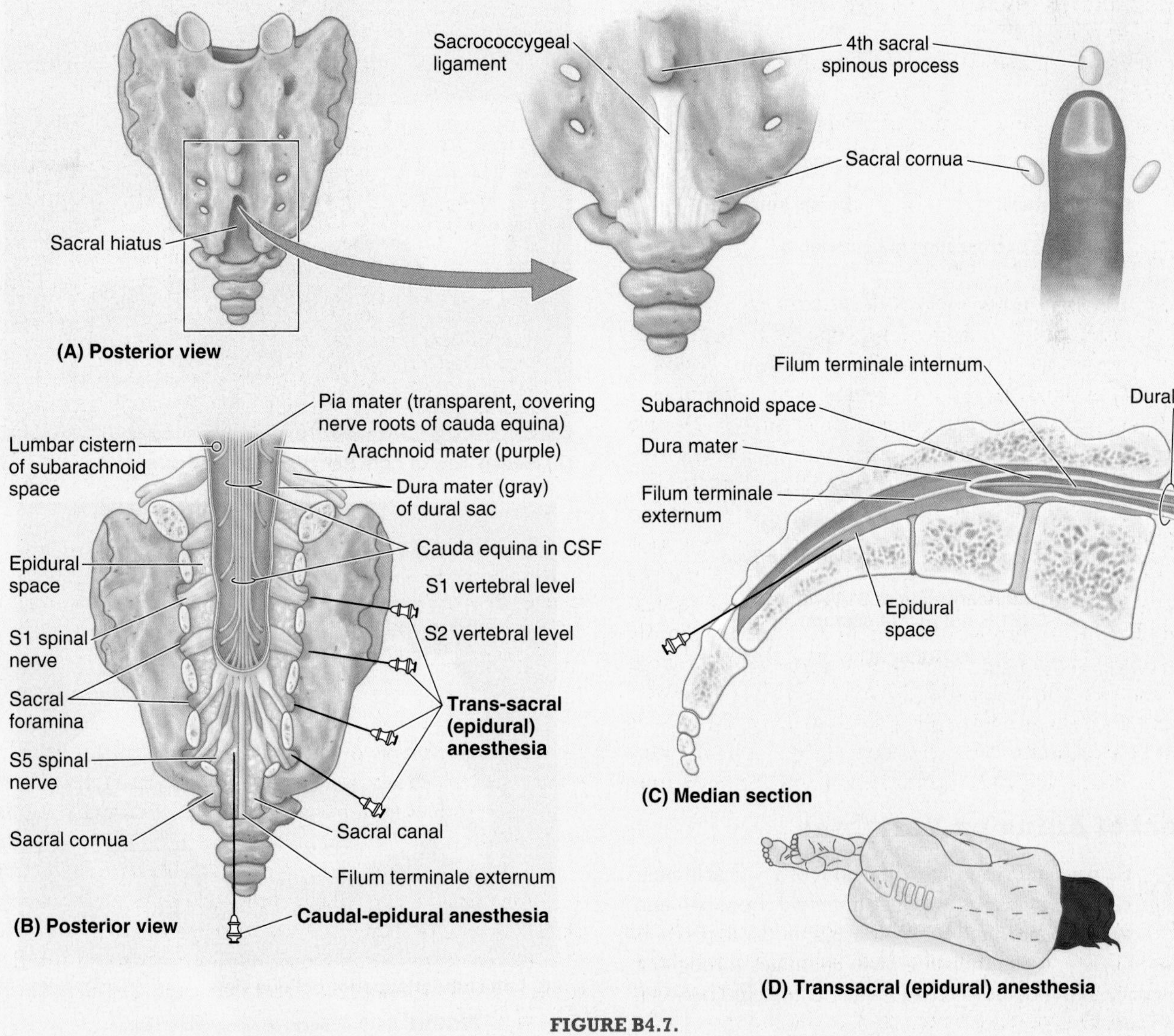

(A) Posterior view

Sacrococcygeal ligament · 4th sacral spinous process · Sacral cornua · Sacral hiatus

(B) Posterior view

Lumbar cistern of subarachnoid space · Pia mater (transparent, covering nerve roots of cauda equina) · Arachnoid mater (purple) · Dura mater (gray) of dural sac · Cauda equina in CSF · S1 vertebral level · S2 vertebral level · Epidural space · **Trans-sacral (epidural) anesthesia** · S1 spinal nerve · Sacral foramina · S5 spinal nerve · Sacral canal · Sacral cornua · Filum terminale externum · **Caudal-epidural anesthesia**

(C) Median section

Filum terminale internum · Subarachnoid space · Dura mater · Filum terminale externum · Dural sac · Epidural space

(D) Transsacral (epidural) anesthesia

FIGURE B4.7.

Abnormal Fusion of Vertebrae

In approximately 5% of people, L5 is partly or completely incorporated into the sacrum—conditions known as *hemisacralization* and *sacralization of the L5 vertebra* (Fig. B4.8A), respectively. In others, S1 is more or less separated from the sacrum and is partly or completely fused with L5 vertebra, which is called *lumbarization of the S1 vertebra* (Fig. B4.8B). When L5 is sacralized, the L5–S1 level is strong and the L4–L5 level degenerates, often producing painful symptoms.

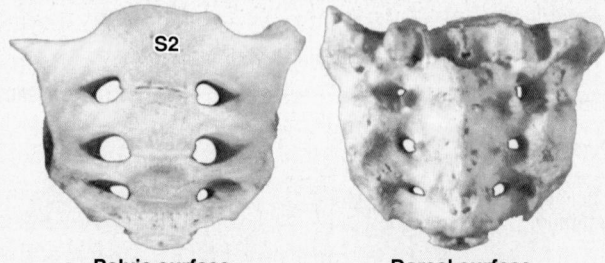

Pelvic surface   **Dorsal surface**

(A) Sacralization of L5 vertebra

Pelvic surface   **Dorsal surface**

(B) Lumbarization of S1 vertebra
(S1 is not part of sacrum)

FIGURE B4.8.

Effect of Aging on Vertebrae

Between birth and age 5, the body of a typical lumbar vertebra increases in height threefold (from 5–6 mm to 15–18 mm), and between ages 5 and 13, it increases another 45–50%. Longitudinal growth continues throughout adolescence, but the rate decreases and is completed between ages 18 and 25.

During middle and older age, there is an overall decrease in bone density and strength, particularly centrally within the vertebral body. Consequently, the articular surfaces gradually bow inward so that both the superior and inferior surfaces of the vertebrae become increasingly concave (Fig. B4.9A), and the IV discs become increasingly convex. The bone loss and consequent change in shape of the vertebral bodies may account in part for the slight loss in height that occurs with aging. The development of these concavities may cause an apparent narrowing of the intervertebral "space" on radiographs based on the distance between the margins of the vertebral bodies; however, this should not be interpreted as a loss of IV disc thickness.

Aging of the IV discs combined with the changing shape of the vertebrae results in an increase in compressive forces at the periphery of the vertebral bodies, where the discs attach. In response, *osteophytes* (bony spurs) commonly develop around the margins of the vertebral body (along the attach-

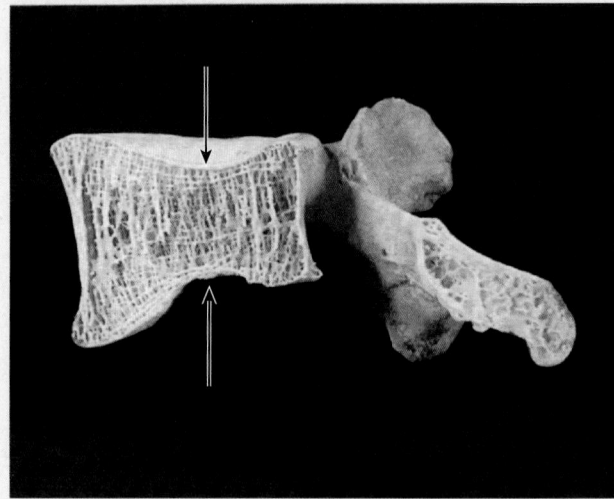

(A) Medial view of right half of lumbar vertebra

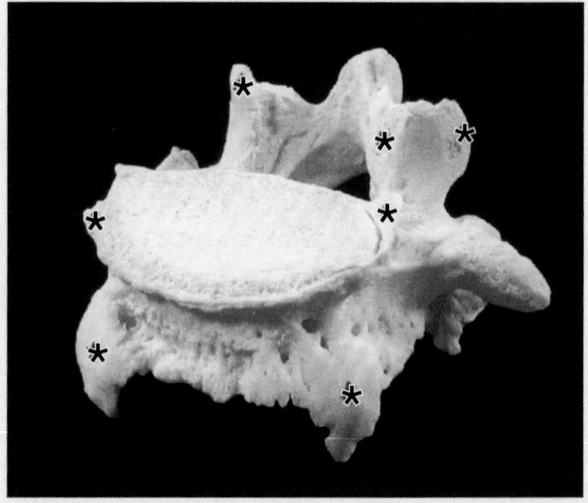

(B) Left anterior superior oblique view   *osteophytes

FIGURE B4.9. Effects of aging on vertebrae.

ments of the fibers of the outer part of the disc), especially anteriorly and posteriorly (Fig. B4.9B). Similarly, as altered mechanics place greater stresses on the zygapophysial joints, osteophytes develop along the attachments of the joint capsules and accessory ligaments, especially those of the superior articular process, whereas extensions of the articular cartilage develop around the articular facets of the inferior processes.

This bony or cartilaginous growth during advanced age has traditionally been viewed as a disease process (*spondylosis* in the case of the vertebral bodies, *osteoarthrosis* in the case of the zygapophysial joints), but it may be more realistic to view it as an expected morphological change with age, representing normal anatomy for a particular age range.

Correlation of these findings with pain is often difficult. Some people with these manifestations present with pain, others demonstrate the same age-related changes but have no pain, and still others exhibit little morphological change but complain of the same types of pain as those with evident change. In view of this and the typical occurrence of these findings, some clinicians have suggested that such age-related changes should not be considered pathological, but as the normal anatomy of aging (Bogduk, 2012).

Anomalies of Vertebrae

Sometimes the epiphysis of a transverse process fails to fuse. Therefore, caution must be exercised so that a persistent epiphysis is not mistaken for a vertebral fracture in a radiograph or computed tomographic (CT) scan.

A common birth defect of the vertebral column is *spina bifida occulta*, in which the neural arches of L5 and/or S1 fail to develop normally and fuse posterior to the vertebral canal. This bony defect, present in up to 24% of the population (Greer, 2009), usually occurs in the vertebral arch of L5 and/or S1. The defect is concealed by the overlying skin, but its location is often indicated by a tuft of hair. Most people with spina bifida occulta have no back problems. When examining a newborn, adjacent vertebrae should be palpated in sequence to be certain the vertebral arches are intact and continuous from the cervical to the sacral regions.

In severe types of spina bifida, *spina bifida cystica*, one or more vertebral arches may fail to develop completely. Spina bifida cystica is associated with herniation of the meninges (*meningocele*, a spina bifida associated with a meningeal cyst) and/or the spinal cord (*meningomyelocele*) (Fig. B4.10). Neurological symptoms are usually present in severe cases of meningomyelocele (e.g., paralysis of the limbs and disturbances in bladder and bowel control). Severe forms of spina bifida result from *neural tube defects*, such as the defective closure of the neural tube during the 4th week of embryonic development (Moore, et al., 2012).

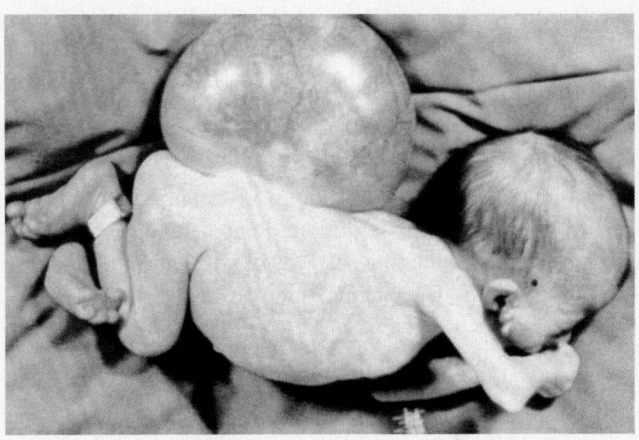

FIGURE B4.10. Infant with spina bifida cystica.

The Bottom Line

VERTEBRAE

Typical vertebrae: Vertebrae consist of vertebral bodies, which bear weight and increase in size proportionately, and vertebral arches, which collectively house and protect the spinal cord and the roots of the spinal nerves. ♦ Processes extending from the vertebral arch provide attachment and leverage for muscles, or direct movements between vertebrae.

Regional characteristics of vertebrae: The chief regional characteristics of vertebrae are ♦ foramina transversarii for cervical vertebrae, ♦ costal facets for thoracic vertebrae, ♦ the absence of foramina transversarii and costal facets for lumbar vertebrae, ♦ the fusion of adjacent sacral vertebrae, and ♦ the rudimentary nature of coccygeal vertebrae.

Ossification of vertebrae: Vertebrae typically ossify from three primary ossification centers within a cartilaginous model: a centrum that will form most of the body and a center in each half of the neural arch. ♦ Thus, by the time of birth, most vertebrae consist of three bony parts united by hyaline cartilage. ♦ Fusion occurs during the first 6 years in a centrifugal pattern from the lumbar region. ♦ During puberty, five secondary ossification centers appear: three related to the spinous and transverse processes and two anular epiphyses around the superior and inferior margins of the vertebral body. ♦ Costal elements formed in association with the ossification center of the transverse process usually form ribs only in the thoracic region. They form components of the transverse processes or their equivalents in other regions. ♦ Knowledge of the pattern of ossification of vertebrae allows understanding of the normal structure of typical and atypical vertebrae, as well as variations and malformations.

VERTEBRAL COLUMN

The **vertebral column** (spine) is an aggregate structure, normally made up of 33 vertebrae and the components that unite them into a single structural, functional entity—the "axis" of the axial skeleton. Because it provides the semirigid, central "core" about which movements of the trunk occur, "soft" or hollow structures that run a longitudinal course are subject to damage or kinking (e.g., the spinal cord, descending aorta, venae cavae, thoracic duct, and esophagus). Thus they lie in close proximity to the vertebral axis, where they receive its semirigid support and torsional stresses on them are minimized.

Joints of Vertebral Column

The joints of the vertebral column include the:

- Joints of the vertebral bodies.
- Joints of the vertebral arches.
- Craniovertebral (atlanto-axial and atlanto-occipital) joints.
- Costovertebral joints (see Chapter 1).
- Sacro-iliac joints (see Chapter 3).

JOINTS OF VERTEBRAL BODIES

The joints of the vertebral bodies are *symphyses (secondary cartilaginous joints)* designed for weight-bearing and strength. The articulating surfaces of adjacent vertebrae are connected by **IV discs** and ligaments (Fig. 4.14).

The IV discs provide strong attachments between the vertebral bodies, uniting them into a continuous semirigid column and forming the inferior half of the anterior border of the IV foramen. In aggregate, the discs account for 20–25%

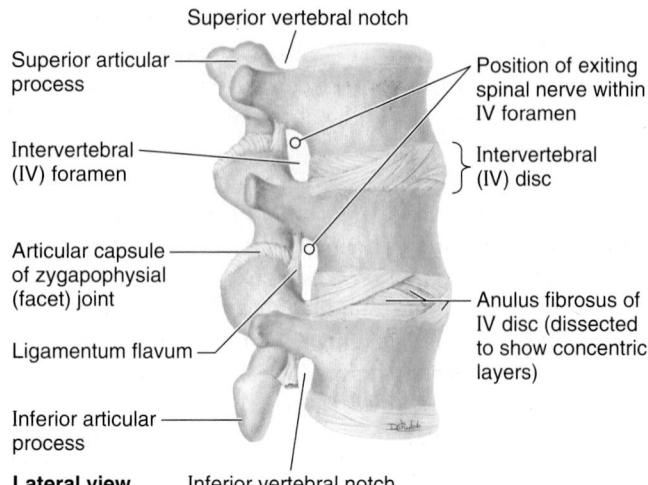

FIGURE 4.14. Lumbar vertebrae and IV discs. This view of the superior lumbar region shows the structure of the anuli fibrosi of the discs and the structures involved in formation of IV foramina. The disc forms the inferior half of the anterior boundary of an IV foramen (except in the cervical region). Thus herniation of the disc will not affect the spinal nerve exiting from the superior part of that foramen, bounded by bone.

of the length (height) of the vertebral column (Fig. 4.1). As well as permitting movement between adjacent vertebrae, their resilient deformability allows them to serve as shock absorbers. Each IV disc consists of an *anulus fibrosus*, an outer fibrous part, composed of concentric lamellae of fibrocartilage, and a gelatinous central mass, the *nucleus pulposus*.

The **anulus fibrosus** (L. *anus*, a ring) is a bulging fibrous ring consisting of concentric lamellae (layers) of fibrocartilage forming the circumference of the IV disc (Figs. 4.14 and

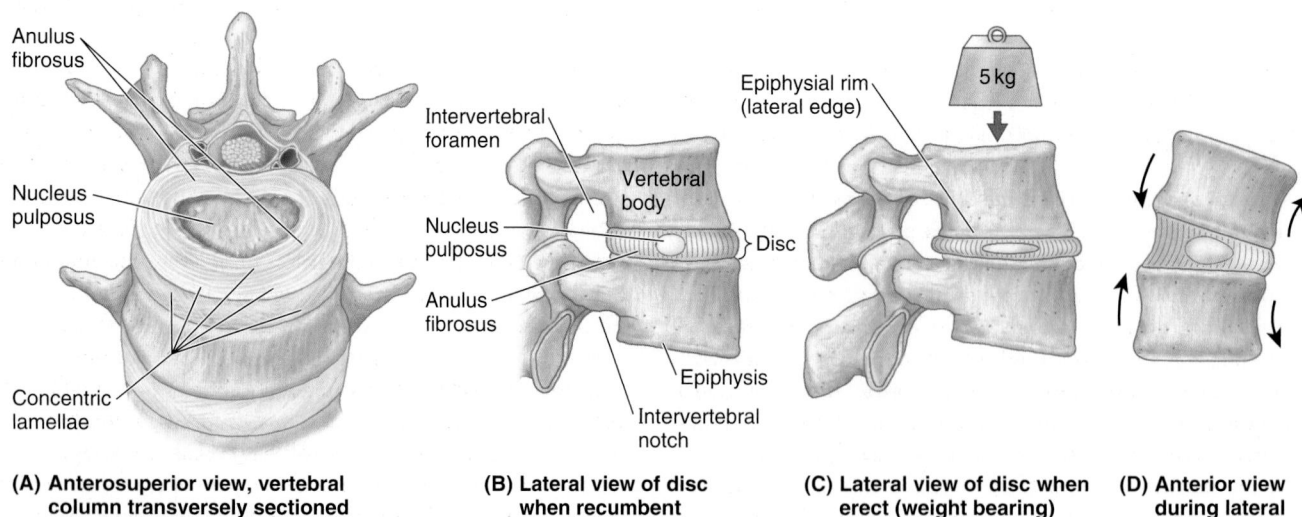

(A) Anterosuperior view, vertebral column transversely sectioned through IV disc

(B) Lateral view of disc when recumbent

(C) Lateral view of disc when erect (weight bearing)

(D) Anterior view during lateral flexion

FIGURE 4.15. Structure and function of IV discs. A. The disc consists of a nucleus pulposus and an anulus fibrosus. The superficial layers of the anulus have been cut and spread apart to show the direction of the fibers. Note that the combined thickness of the rings of the anulus is diminished posteriorly—that is, the anulus is thinner posteriorly. **B.** The fibrogelatinous nucleus pulposus occupies the center of the disc and acts as a cushion and shock-absorbing mechanism. **C.** The pulpy nucleus flattens and the anulus bulges when weight is applied, as occurs during standing and more so during lifting. **D.** During flexion and extension movements, the nucleus pulposus serves as a fulcrum. The anulus is simultaneously placed under compression on one side and tension on the other.

4.15A). The anuli insert into the smooth, rounded *epiphysial rims* on the articular surfaces of the vertebral bodies formed by the fused *anular epiphyses* (Figs. 4.2B and 4.15B & C). The fibers forming each lamella run obliquely from one vertebra to another, about 30 or more degrees from vertical. The fibers of adjacent lamellae cross each other obliquely in opposite directions at angles of more than 60 degrees (Fig. 4.14). This arrangement allows limited rotation between adjacent vertebrae, while providing a strong bond between them. The anulus is thinner posteriorly and may be incomplete posteriorly in the adult in the cervical region (Mercer and Bogduk, 1999). The anulus becomes decreasingly vascularized centrally, and only the outer third of the anulus receives sensory innervation.

The **nucleus pulposus** (L. *pulpa*, fleshy) is the core of the IV disc (Fig. 4.15A). At birth, these pulpy nuclei are about 88% water and are initially more cartilaginous than fibrous. Their semifluid nature is responsible for much of the flexibility and resilience of the IV disc and of the vertebral column as a whole.

Vertical forces deform the IV discs, which thus serve as shock absorbers. The nuclei become broader when compressed and thinner when tensed or stretched (as when hanging or suspended) (Fig. 4.15C). Compression and tension occur simultaneously in the same disc during anterior and lateral flexion and extension of the vertebral column (Fig. 4.15D). During these movements, as well as during rotation, the turgid nucleus acts as a semifluid fulcrum. Because the lamellae of the anulus fibrosus are thinner and less numerous posteriorly than they are anteriorly or laterally, the nucleus pulposus is not centered in the disc but is positioned between the center and posterior aspect of the disc (Fig. 4.15A). The nucleus pulposus is avascular; it receives its nourishment by diffusion from blood vessels at the periphery of the anulus fibrosus and vertebral body.

There is no IV disc between C1 and C2 vertebrae; the most inferior functional disc is between L5 and S1 vertebrae. The discs vary in thickness in different regions. The thickness of the discs increases as the vertebral column descends. However, their thickness relative to the size of the bodies they connect is most clearly related to the range of movement, and relative thickness is greatest in the cervical and lumbar regions. Their thickness is most uniform in the thoracic region. The discs are thicker anteriorly in the cervical and lumbar regions, their varying shapes producing the secondary curvatures of the vertebral column (see Fig. 4.1B).

Uncovertebral "joints" or **clefts** (of Luschka) commonly develop between the unci of the bodies of C3 or 4–C6 or 7 vertebrae and the beveled inferolateral surfaces of the vertebral bodies superior to them after 10 years of age (Fig. 4.16). The joints are at the lateral and posterolateral margins of the IV discs. The articulating surfaces of these joint-like structures are covered with cartilage moistened by fluid contained within an interposed potential space, or "capsule." They are considered synovial joints by some; others consider them to be degenerative spaces (clefts) in the discs occupied by extracellular fluid. The uncovertebral "joints" are frequent sites of bone spur formation in later years, which may cause neck pain.

The **anterior longitudinal ligament** is a strong, broad fibrous band that covers and connects the anterolateral

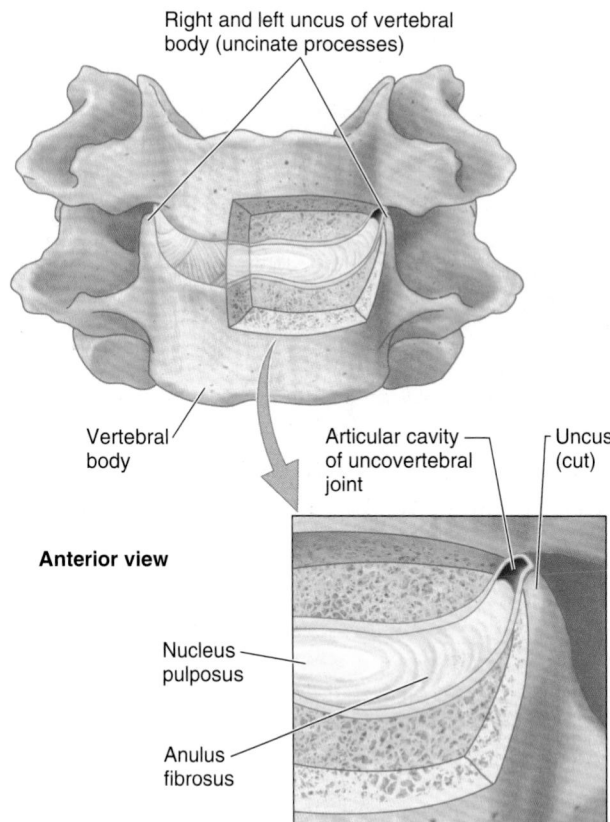

FIGURE 4.16. Uncovertebral joints. These small, synovial joint-like structures are between the unci of the bodies of the lower vertebrae and the beveled surfaces of the vertebral bodies superior to them. These joints are at the posterolateral margins of the IV discs.

aspects of the vertebral bodies and IV discs (Fig. 4.17). The ligament extends longitudinally from the pelvic surface of the sacrum to the anterior tubercle of vertebra C1 and the occipital bone anterior to the foramen magnum are the superiormost parts, the anterior atlanto-axial and atlanto-occipital ligaments. Although thickest on the anterior aspect of the vertebral bodies (illustrations often depict only this portion), the anterior longitudinal ligament also covers the lateral aspects of the bodies to the IV foramen. This ligament prevents hyperextension of the vertebral column, maintaining stability of the joints between the vertebral bodies. *The anterior longitudinal ligament is the only ligament that limits extension;* all other IV ligaments limit forms of flexion.

The **posterior longitudinal ligament** is a much narrower, somewhat weaker band than the anterior longitudinal ligament (Fig. 4.17; see also 4.18B). The posterior longitudinal ligament runs within the vertebral canal along the posterior aspect of the vertebral bodies. It is attached mainly to the IV discs and less so to the posterior aspects of the vertebral bodies from C2 to the sacrum, often bridging fat and vessels between the ligament and the bony surface. This ligament weakly resists hyperflexion of the vertebral column and helps prevent or redirect posterior herniation of the nucleus pulposus. It is well provided with nociceptive (pain) nerve endings.

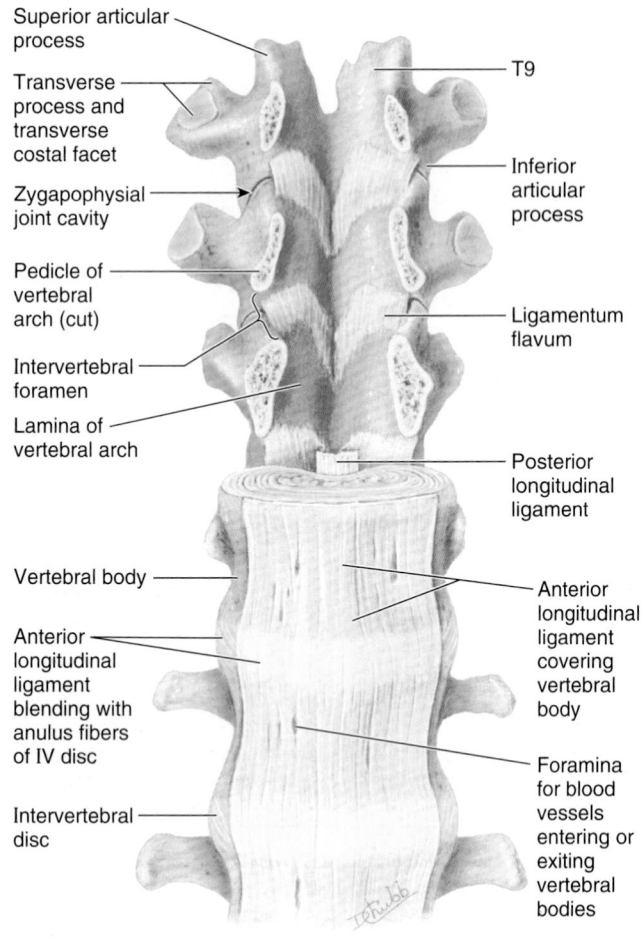

Superior articular process

Transverse process and transverse costal facet

Zygapophysial joint cavity

Pedicle of vertebral arch (cut)

Intervertebral foramen

Lamina of vertebral arch

Vertebral body

Anterior longitudinal ligament blending with anulus fibers of IV disc

Intervertebral disc

T9

Inferior articular process

Ligamentum flavum

Posterior longitudinal ligament

Anterior longitudinal ligament covering vertebral body

Foramina for blood vessels entering or exiting vertebral bodies

Anterior view

FIGURE 4.17. Relationship of ligaments to vertebrae and IV discs. The inferior thoracic (T9–T12) and superior lumbar (L1–L2) vertebrae, with associated discs and ligaments, are shown. The pedicles of the T9–T11 vertebrae have been sawn through and their bodies and intervening discs removed to provide an anterior view of the posterior wall of the vertebral canal. Between the adjacent left or right pedicles, the inferior and superior articular processes and the zygapophysial joints between them (from which joint capsules have been removed) and the lateralmost extent of the ligamenta flava form the posterior boundaries of IV foramina. The anterior longitudinal ligament is broad, whereas the posterior longitudinal ligament is narrow.

JOINTS OF VERTEBRAL ARCHES

The joints of the vertebral arches are the *zygapophysial joints (facet joints)*. These articulations are plane synovial joints between the superior and inferior articular processes (G. *zygapophyses*) of adjacent vertebrae (Figs. 4.15 and 4.17). Each joint is surrounded by a thin **joint capsule.** Those in the cervical region are especially thin and loose, reflecting the wide range of movement (Fig. 4.18). The joint capsule is attached to the margins of the articular surfaces of the articular processes of adjacent vertebrae. Accessory ligaments unite the laminae, transverse processes, and spinous processes and help stabilize the joints.

The zygapophysial joints permit gliding movements between the articular processes; the shape and disposition of the articu-

lar surfaces determine the types of movement possible. The range (amount) of movement is determined by the size of the IV disc relative to that of the vertebral body. In the cervical and lumbar regions, these joints bear some weight, sharing this function with the IV discs, particularly during lateral flexion.

The zygapophysial joints are innervated by articular branches that arise from the medial branches of the posterior rami of spinal nerves (Fig. 4.19). As these nerves pass postero-inferiorly, they lie in grooves on the posterior surfaces of the medial parts of the transverse processes. Each articular branch supplies two adjacent joints; therefore, each joint is supplied by two nerves.

ACCESSORY LIGAMENTS OF INTERVERTEBRAL JOINTS

The laminae of adjacent vertebral arches are joined by broad, pale yellow bands of elastic tissue called the **ligamenta flava** (L. *flavus*, yellow). These ligaments extend almost vertically from the lamina above to the lamina below, those of opposite sides meeting and blending in the midline (Figs. 4.14 and 4.17). The flaval ligaments bind the lamina of the adjoining vertebrae together, forming alternating sections of the posterior wall of the vertebral canal. The ligamenta flava are long, thin, and broad in the cervical region, thicker in the thoracic region, and thickest in the lumbar region. These ligaments resist separation of the vertebral lamina by limiting abrupt flexion of the vertebral column, and thereby prevent injury to the IV discs. The strong, elastic yellow ligaments help preserve the normal curvatures of the vertebral column and assist with straightening of the column after flexing.

Adjoining spinous processes are united by weak, often membranous interspinous ligaments and strong fibrous supraspinous ligaments (Fig. 4.18A & B). The thin **interspinous ligaments** connect adjoining spinous processes, attaching from the root to the apex of each process. The cord-like band forming the **supraspinous ligaments** connects the tips of the spinous processes from C7 to the sacrum and merge superiorly with the nuchal ligament at the back of the neck (Fr. *nuque*, back of neck) (Fig. 4.18A). Unlike the interspinous and supraspinous ligaments, the strong, broad **nuchal ligament** (L. *ligamentum nuchae*) is composed of thickened fibroelastic tissue, extending as a median band from the external occipital protuberance and posterior border of the foramen magnum to the spinous processes of the cervical vertebrae. Because of the shortness and depth of the C3–C5 spinous processes, the nuchal ligament provides attachment for muscles that attach to the spinous processes of vertebrae at other levels. The **intertransverse ligaments**, connecting adjacent transverse processes, consist of scattered fibers in the cervical region and fibrous cords in the thoracic region (Fig. 4.18B). In the lumbar region these ligaments are thin and membranous.

CRANIOVERTEBRAL JOINTS

There are two sets of craniovertebral joints, the *atlantooccipital joints*, formed between the atlas (C1 vertebra), and the occipital bone of the cranium, and the *atlanto-axial*

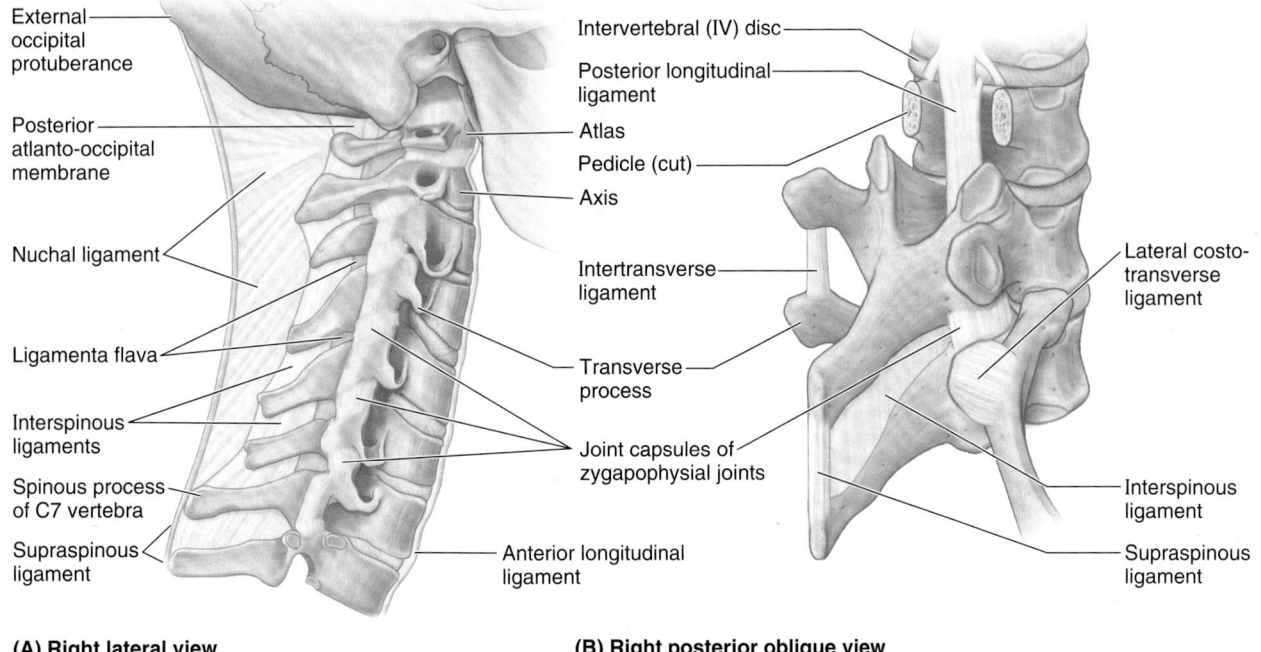

(A) Right lateral view

(B) Right posterior oblique view

FIGURE 4.18. **Joints and ligaments of vertebral column. A.** The ligaments in the cervical region are shown. Superior to the prominent spinous process of C7 (vertebra prominens), the spinous processes are deeply placed and attached to an overlying nuchal ligament. **B.** The ligaments in the thoracic region are shown. The pedicles of the superior two vertebrae have been sawn through and the vertebral arches removed to reveal the posterior longitudinal ligament. Intertransverse, supraspinous, and interspinous ligaments are demonstrated in association with the vertebrae with intact vertebral arches.

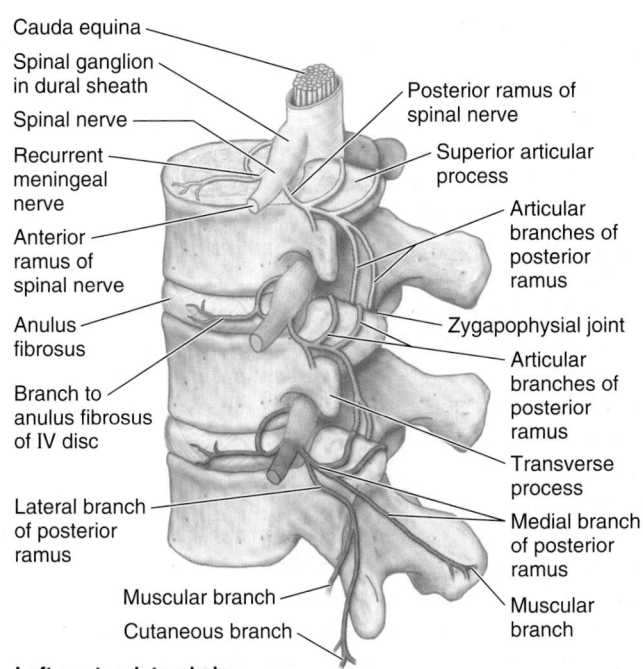

Left posterolateral view

FIGURE 4.19. **Innervation of zygapophysial joints.** The posterior rami arise from the spinal nerves outside the IV foramen and divide into medial and lateral branches. The medial branch gives rise to articular branches that are distributed to the zygapophysial joint at that level, and to the joint one level inferior to its exit. Thus, each zygapophysial joint receives articular rami from the medial branch of the posterior rami of two adjacent spinal nerves. The medial branches of both posterior rami have to be ablated to denervate a zygapophysial joint.

joints, formed between the atlas and axis (C2 vertebra) (Fig. 4.20). The Greek word *atlanto* refers to the atlas (C1 vertebra). The craniovertebral joints are synovial joints that have no IV discs. Their design gives a wider range of movement than in the rest of the vertebral column. The articulations involve the occipital condyles, atlas, and axis.

Atlanto-occipital Joints. The articulations are between the superior articular surfaces of the lateral masses of the atlas and the occipital condyles (Figs. 4.6A & B and 4.20A). These joints permit nodding of the head, such as the flexion and extension of the head occurring when indicating approval (the "yes" movement). These joints also permit sideways tilting of the head. The main movement is flexion, with a little lateral flexion and rotation. They are synovial joints of the condyloid type and have thin, loose joint capsules.

The cranium and C1 are also connected by **anterior** and **posterior atlanto-occipital membranes,** which extend from the anterior and posterior arches of C1 to the anterior and posterior margins of the foramen magnum (Figs. 4.20B and 4.21). The anterior membranes are composed of broad, densely woven fibers (especially centrally where they are continuous with the anterior longitudinal ligament). The posterior membranes are broad but relatively weak. The atlanto-occipital membranes help prevent excessive movement of the atlanto-occipital joints.

Atlanto-axial Joints. There are three atlanto-axial articulations (Fig. 4.20A–D): two (right and left) **lateral atlanto-axial joints** (between the inferior facets of the lateral masses of C1 and the superior facets of C2), and one

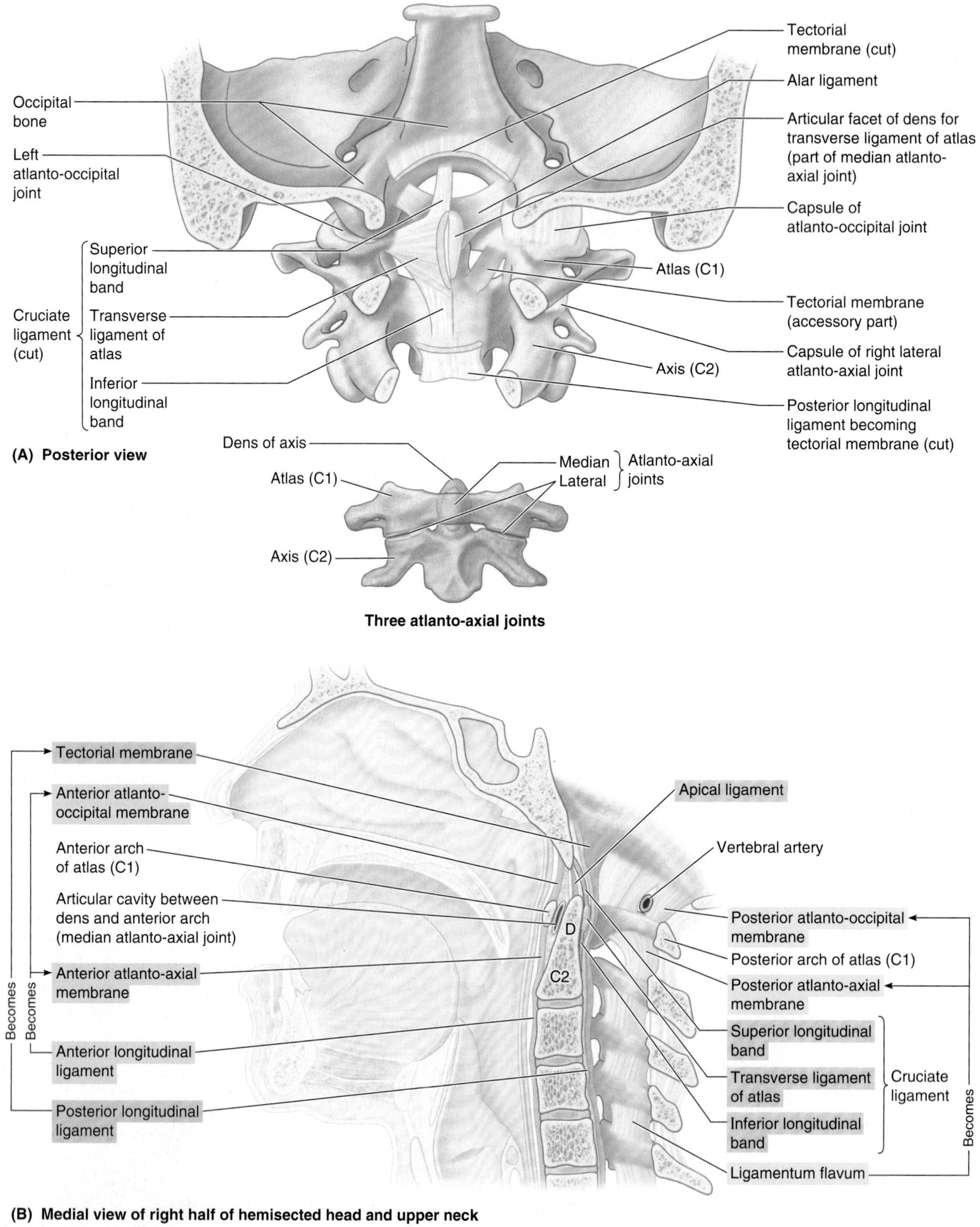

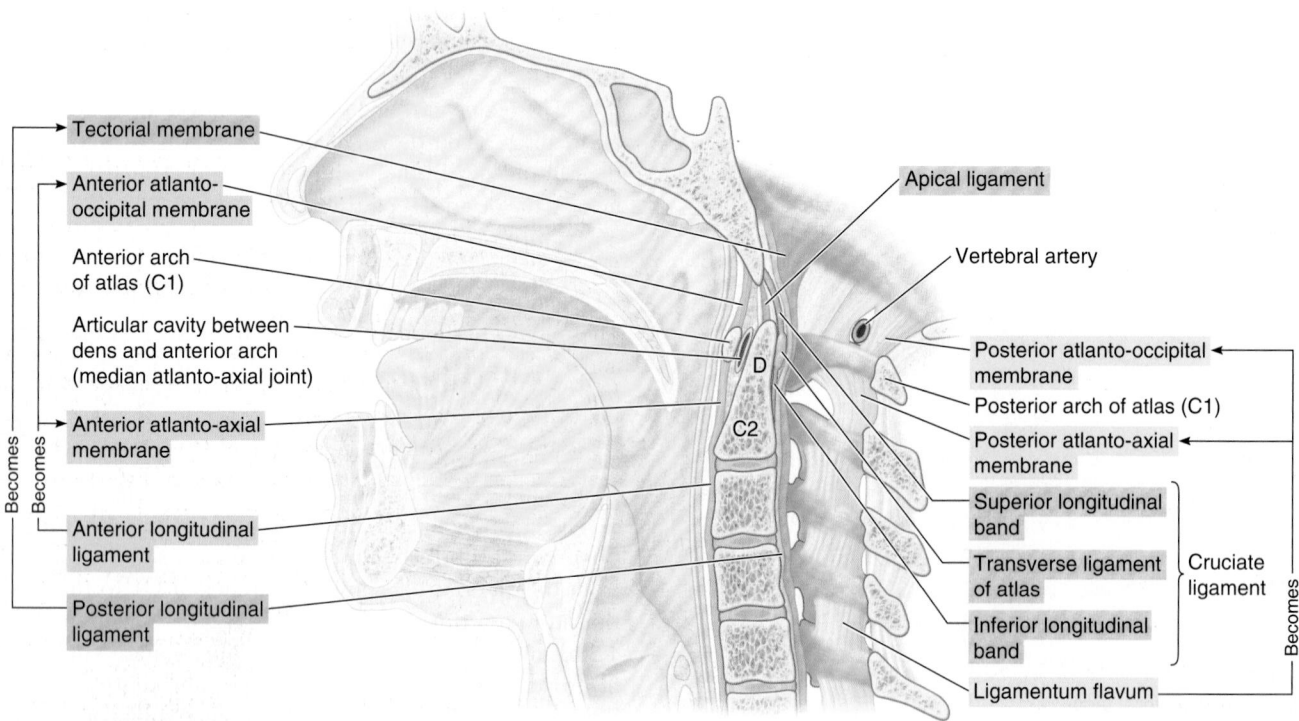

(A) Posterior view

Occipital bone

Left atlanto-occipital joint

Cruciate ligament (cut)
- Superior longitudinal band
- Transverse ligament of atlas
- Inferior longitudinal band

Tectorial membrane (cut)

Alar ligament

Articular facet of dens for transverse ligament of atlas (part of median atlanto-axial joint)

Capsule of atlanto-occipital joint

Atlas (C1)

Tectorial membrane (accessory part)

Capsule of right lateral atlanto-axial joint

Axis (C2)

Posterior longitudinal ligament becoming tectorial membrane (cut)

Dens of axis

Atlas (C1)

Median / Lateral } Atlanto-axial joints

Axis (C2)

Three atlanto-axial joints

Tectorial membrane

Anterior atlanto-occipital membrane

Anterior arch of atlas (C1)

Articular cavity between dens and anterior arch (median atlanto-axial joint)

Anterior atlanto-axial membrane

Anterior longitudinal ligament

Posterior longitudinal ligament

Becomes   Becomes

Apical ligament

Vertebral artery

Posterior atlanto-occipital membrane

Posterior arch of atlas (C1)

Posterior atlanto-axial membrane

Superior longitudinal band

Transverse ligament of atlas

Inferior longitudinal band

Ligamentum flavum

Cruciate ligament

Becomes

D

C2

(B) Medial view of right half of hemisected head and upper neck

FIGURE 4.20. Craniovertebral joints and ligaments. A. Ligaments of the atlanto-occipital and atlanto-axial joints. The tectorial membrane and the right side of the cruciate ligament of the atlas have been removed to show the attachment of the right alar ligament to the dens of C2 (axis). **B.** The hemisected craniovertebral region shows the median joints and membranous continuities of the ligamenta flava and longitudinal ligaments in the craniovertebral region. **C.** The articulated atlas and axis showing that the median atlanto-axial joint is formed as the anterior arch and transverse ligament of the atlas form a socket for the dens of the axis. **D.** During rotation of head, the cranium and atlas rotate as a unit around the pivot of the dens when the head is turned side to side (the "no" movement).

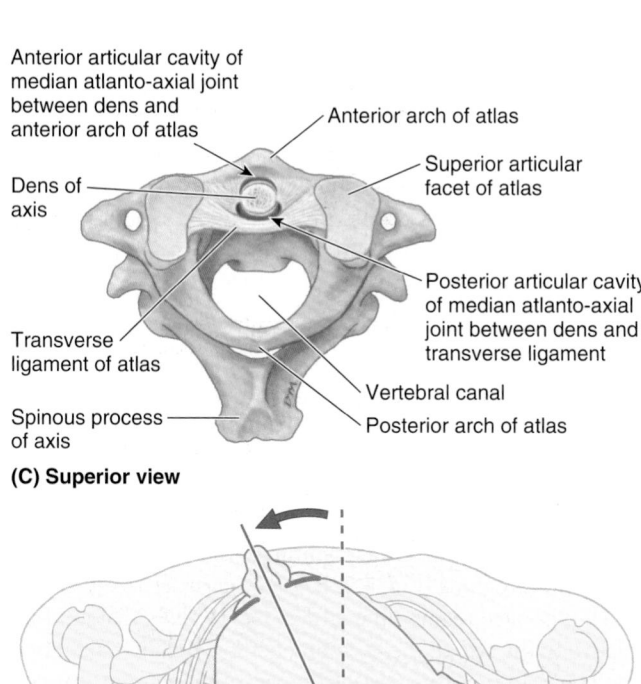

Anterior articular cavity of median atlanto-axial joint between dens and anterior arch of atlas

Anterior arch of atlas

Superior articular facet of atlas

Dens of axis

Posterior articular cavity of median atlanto-axial joint between dens and transverse ligament

Transverse ligament of atlas

Vertebral canal

Spinous process of axis

Posterior arch of atlas

(C) Superior view

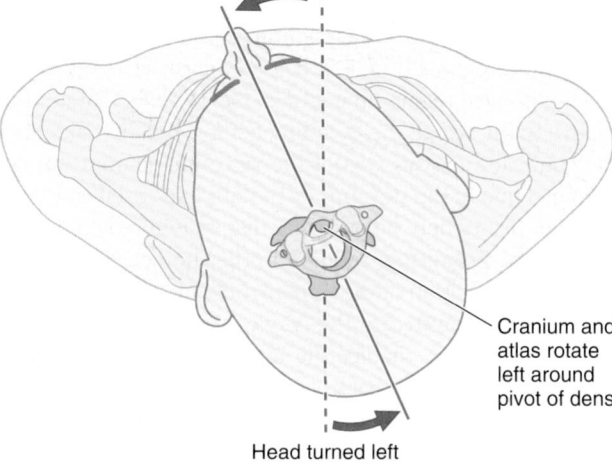

Cranium and atlas rotate left around pivot of dens

Head turned left

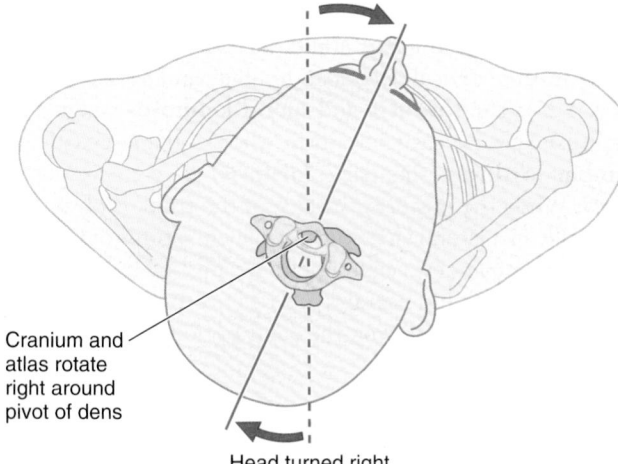

Cranium and atlas rotate right around pivot of dens

Head turned right

(D) Superior views

FIGURE 4.20. *(Continued)*

median atlanto-axial joint (between the dens of C2 and the anterior arch of the atlas). The lateral atlanto-axial joints are gliding-type synovial joints, whereas the median atlanto-axial joint is a pivot joint.

Movement at all three atlanto-axial joints permits the head to be turned from side to side (Fig. 4.20D), as occurs when rotating the head to indicate disapproval (the "no" movement). During this movement, the cranium and C1 rotate on

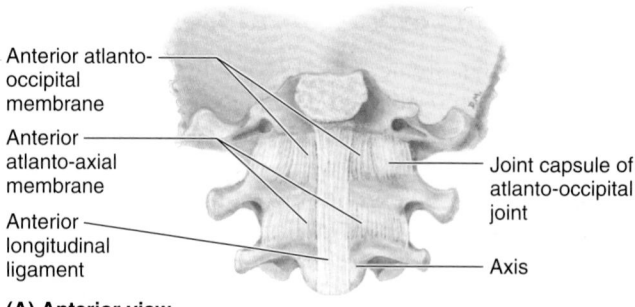

Anterior atlanto-occipital membrane

Anterior atlanto-axial membrane

Anterior longitudinal ligament

Joint capsule of atlanto-occipital joint

Axis

(A) Anterior view

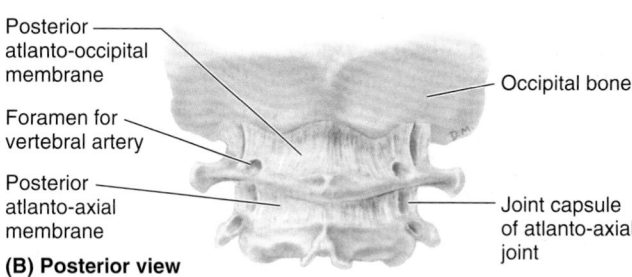

Posterior atlanto-occipital membrane

Foramen for vertebral artery

Posterior atlanto-axial membrane

Occipital bone

Joint capsule of atlanto-axial joint

(B) Posterior view

FIGURE 4.21. Membranes of craniovertebral joints. A. Only the thicker, most anterior part of the anterior longitudinal ligament is included here to demonstrate its superior continuation as the anterior atlanto-axial membrane and anterior atlanto-occipital membrane. Laterally, the membranes blend with the joint capsules of the lateral atlanto-axial and atlanto-occipital joints. **B.** The posterior atlanto-occipital and atlanto-axial membranes span the gaps between the posterior arch of the atlas (C1) and the occipital bone (posterior margin of the foramen magnum) superiorly, and the laminae of the axis (C2) inferiorly. The vertebral arteries penetrate the atlanto-occipital membrane before traversing the foramen magnum.

C2 as a unit. During rotation of the head, the dens of C2 is the axis or pivot that is held in a socket or collar formed anteriorly by the anterior arch of the atlas and posteriorly by the *transverse ligament of the atlas* (Fig. 4.20A–D); this strong band extends between the tubercles on the medial aspects of the lateral masses of C1 vertebrae.

Vertically oriented but much weaker superior and inferior *longitudinal bands* pass from the transverse ligament of the atlas to the occipital bone superiorly and to the body of C2 inferiorly. The **cruciate ligament of the atlas,** so named because of its resemblance to a cross, consists of the transverse ligament of the atlas plus the longitudinal bands (Fig. 4.20A).

The **alar ligaments** extend from the sides of the dens of the axis to the lateral margins of the foramen magnum. These short, rounded cords, approximately 0.5 cm in diameter, attach the cranium to the C1 vertebra and act as check ligaments in preventing excessive rotation at the joints.

The **tectorial membrane** (Fig. 4.20A & B) is the strong superior continuation of the posterior longitudinal ligament that broadens and passes posteriorly over the median atlanto-axial joint and its ligaments. It runs superiorly from the body of C2 through the foramen magnum to attach to the central part of the floor of the cranial cavity, formed by the internal surface of the occipital bone.

Movements of Vertebral Column

The range of movement of the vertebral column varies according to the region and the individual. Contortionists, who begin their training during early childhood, become capable of extraordinary movements. The normal range of movement possible in healthy young adults is typically reduced by 50% or more as they age.

The mobility of the vertebral column results primarily from the compressibility and elasticity of the IV discs. The vertebral column is capable of flexion, extension, lateral flexion and extension, and rotation (torsion) (Fig. 4.22). Bending of the vertebral column to the right or left from the neutral (erect) position is *lateral flexion;* returning to the erect posture from a position of lateral flexion is *lateral extension.*

The range of movement of the vertebral column is limited by the:

- Thickness, elasticity, and compressibility of the IV discs.
- Shape and orientation of the zygapophysial joints.
- Tension of the joint capsules of the zygapophysial joints.
- Resistance of the back muscles and ligaments (e.g., the ligamenta flava and the posterior longitudinal ligament).
- Attachment to the thoracic (rib) cage.
- Bulk of surrounding tissue.

Movements are not produced exclusively by the back muscles. They are assisted by gravity and the action of the anterolateral abdominal muscles. Movements between adjacent vertebrae occur at the resilient nuclei pulposi of the IV discs (serving as the axis of movement) and at the zygapophysial joints (Figs. 4.14 and 4.15).

The orientation of the latter joints permits some movements and restricts others. With the exception perhaps of C1–C2, movement never occurs at a single segment of the column. Although movements between adjacent vertebrae are relatively small, especially in the thoracic region, the summation of all the small movements produces a considerable range of movement of the vertebral column as a whole (e.g., when flexing to touch the floor; Fig. 4.22A). Movements of the vertebral column are freer in the cervical and lumbar regions than elsewhere. Flexion, extension, lateral flexion, and rotation of the neck are especially free because the:

- IV discs, although thin relative to most other discs, are thick relative to the size of the vertebral bodies at this level.
- Articular surfaces of the zygapophysial joints are relatively large and the joint planes are almost horizontal.
- Joint capsules of the zygapophysial joints are loose.
- Neck is relatively slender (with less surrounding soft tissue bulk compared with the trunk).

Flexion of the vertebral column is greatest in the cervical region. The sagittally oriented joint planes of the lumbar region are conducive to flexion and extension. Extension of the vertebral column is most marked in the lumbar region and is usually more extensive than flexion; however, the interlocking articular processes here prevent rotation (Fig. 4.9). The lumbar region, like the cervical region, has IV discs that are large relative to the size of the vertebral bodies. Lateral flexion of the vertebral column is greatest in the cervical and lumbar regions (Fig. 4.22B).

The thoracic region, in contrast, has IV discs that are thin relative to the size of the vertebral bodies. Relative stability is also conferred on this part of the vertebral column through its connection to the sternum by the ribs and costal cartilages. The joint planes here lie on an arc that is centered on the vertebral body, permitting rotation in the thoracic region (Fig. 4.22C). This rotation of the upper trunk, in combination with the rotation permitted in the cervical region and that at the atlanto-axial joints, enables the torsion of the axial skeleton that occurs as one looks back over the shoulder. However, flexion is limited in the thoracic region, including lateral flexion.

Curvatures of Vertebral Column

The vertebral column in adults has four *curvatures* that occur in the cervical, thoracic, lumbar, and sacral regions (Fig. 4.23). The **thoracic** and **sacral kyphoses** (singular = **kyphosis**) are concave anteriorly, whereas the **cervical** and **lumbar lordoses** (singular = **lordosis**) are concave posteriorly. When the posterior surface of the trunk is observed, especially in a lateral view, the normal curvatures of the vertebral column are especially apparent (Fig. 4.24).

The thoracic and sacral kyphoses are **primary curvatures** that develop during the fetal period in relationship

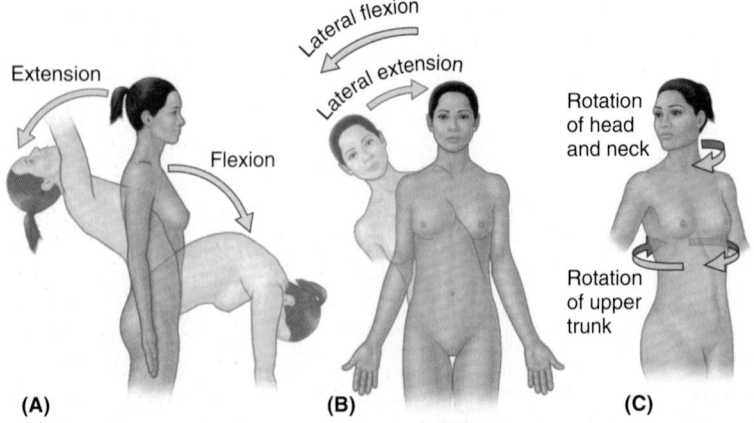

FIGURE 4.22. Movements of vertebral column. A. Flexion and extension, both in the median plane, are shown. Flexion and extension are occurring primarily in the cervical and lumbar regions. **B.** Lateral flexion (to the right or left in a frontal plane) is shown, also occurring mostly in the cervical and lumbar regions. **C.** Rotation around a longitudinal axis, which occurs primarily at the craniovertebral joints (augmented by the cervical region) and the thoracic region, is shown.

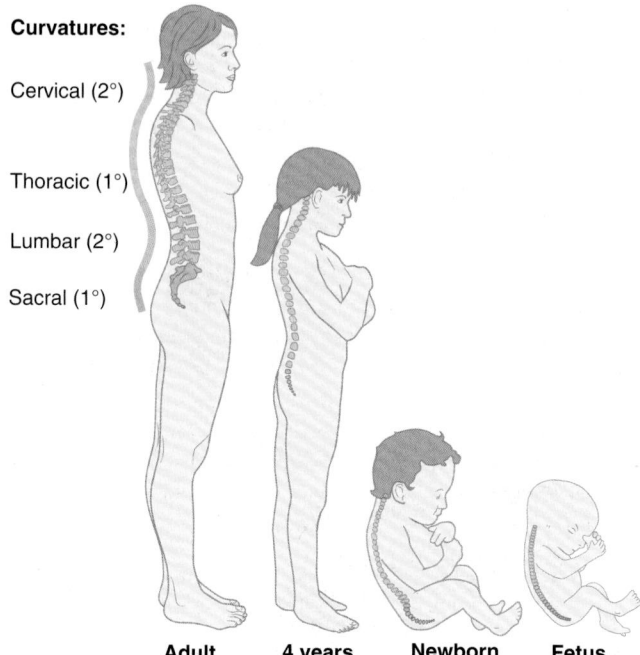

Curvatures:

Cervical (2°)

Thoracic (1°)

Lumbar (2°)

Sacral (1°)

Adult   4 years   Newborn   Fetus

FIGURE 4.23. Curvatures of vertebral column. The four curvatures of the adult vertebral column—cervical, thoracic, lumbar, and sacral—are contrasted with the C-shaped curvature of the column during fetal life, when only the primary (*1°*) curvatures exist. The secondary (*2°*) curvatures develop during infancy and childhood.

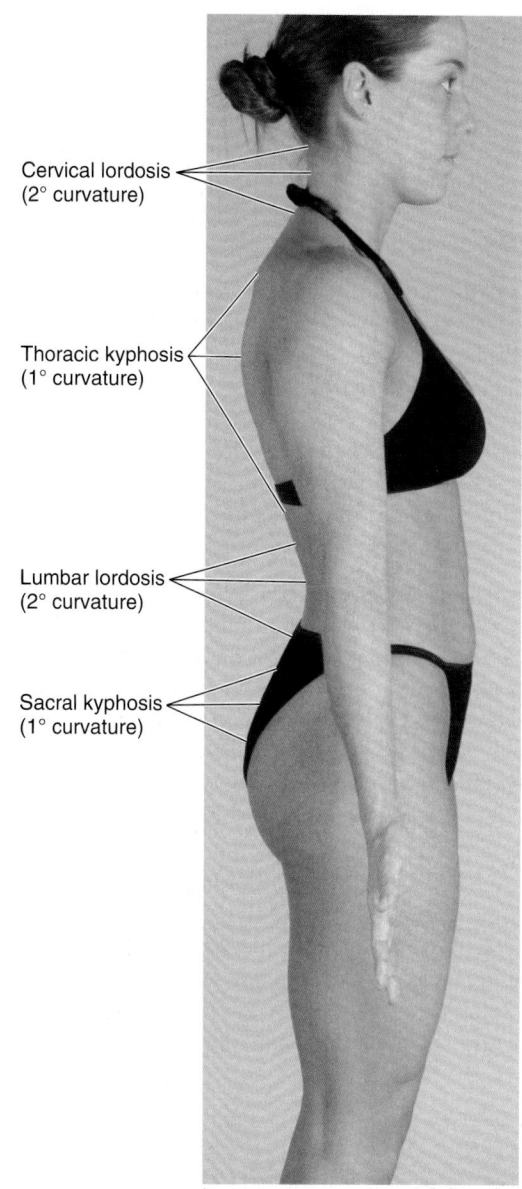

Cervical lordosis (2° curvature)

Thoracic kyphosis (1° curvature)

Lumbar lordosis (2° curvature)

Sacral kyphosis (1° curvature)

Lateral view

FIGURE 4.24. Surface anatomy of curvatures of vertebral column.

to the (flexed) fetal position (Moore et al., 2012). Compare the curvatures in Figure 4.23, noting that the primary curvatures are in the same direction as the main curvatures of the fetal vertebral column. The primary curvatures are retained throughout life as a consequence of differences in height between the anterior and posterior parts of the vertebrae.

The cervical and lumbar lordoses are **secondary curvatures** that result from extension from the flexed fetal position. They begin to appear during the late fetal period but do not become obvious until infancy. Secondary curvatures are maintained primarily by differences in thickness between the anterior and the posterior parts of the IV discs.

The *cervical lordosis* becomes fully evident when an infant begins to raise (extend) the head while prone and to hold the head erect while sitting. The *lumbar lordosis* becomes apparent when toddlers begin to assume the upright posture, standing and walking. This curvature, generally more pronounced in females, ends at the *lumbosacral angle* formed at the junction of L5 vertebra with the sacrum (Fig. 4.1D). The *sacral kyphosis* also differs in males and females, that of the female reduced so that the coccyx protrudes less into the pelvic outlet (see Chapter 3).

The curvatures of the vertebral column provide additional flexibility (shock-absorbing resilience), further augmenting that provided by the IV discs. When the load borne by the vertebral column is markedly increased (as by carrying a heavy backpack), both the IV discs and the flexible curvatures are compressed (i.e., the curvatures tend to increase).

While the flexibility provided by the IV discs is passive and limited primarily by the zygapophysial joints and longitudinal ligaments, that provided by the curvatures is actively (dynamically) resisted by the contraction of muscle groups antagonistic to the movement (e.g., the long extensors of the back resist excessive thoracic kyphosis, whereas the abdominal flexors resist excessive lumbar lordosis).

Carrying additional weight anterior to the body's normal gravitational axis (e.g., abnormally large breasts, a pendulous abdomen in men or during late pregnancy, or carrying a young child) also tends to increase these curvatures. The muscles that provide resistance to the increase in curvature often ache when the weight is borne for extended periods.

When sitting, especially in the absence of back support for long periods of time, one usually "cycles" between back flexion (slumping) and extension (sitting up straight) to minimize stiffness and fatigue. This allows alternation between the active support provided by the extensor muscles of the back and the passive resistance to flexion provided by ligaments.

Vasculature of Vertebral Column

Vertebrae are supplied by *periosteal* and *equatorial branches* of the major cervical and *segmental arteries* and their **spinal branches** (Fig. 4.25). Parent arteries of periosteal, equatorial, and spinal branches occur at all levels of the vertebral column, in close association with it, and include the following arteries (described in detail in other chapters):

- *Vertebral* and *ascending cervical arteries* in the neck (Chapter 8).
- The major *segmental arteries* of the trunk:
  - *Posterior intercostal arteries* in the thoracic region (Chapter 1).
  - *Subcostal* and *lumbar arteries* in the abdomen (Chapter 2).
  - *Iliolumbar* and *lateral and medial sacral arteries* in the pelvis (Chapter 3).

Periosteal and **equatorial branches** arise from these arteries as they cross the external (anterolateral) surfaces of the vertebrae. Spinal branches enter the IV foramina and divide. Smaller **anterior** and **posterior vertebral canal branches** pass to the vertebral body and vertebral arch, respectively, and give rise to ascending and descending branches that anastomose with the spinal canal branches of adjacent levels (Fig. 4.25). Anterior vertebral canal branches send **nutrient arteries** anteriorly into the vertebral bodies that supply most of the red marrow of the central vertebral body (Bogduk, 2012). The larger branches of the spinal branches continue as terminal *radicular or segmental medullary arteries* distributed to the posterior and anterior roots of the spinal nerves and their coverings and to the spinal cord, respectively (see "Vasculature of Spinal Cord and Spinal Nerve Roots," p. 501).

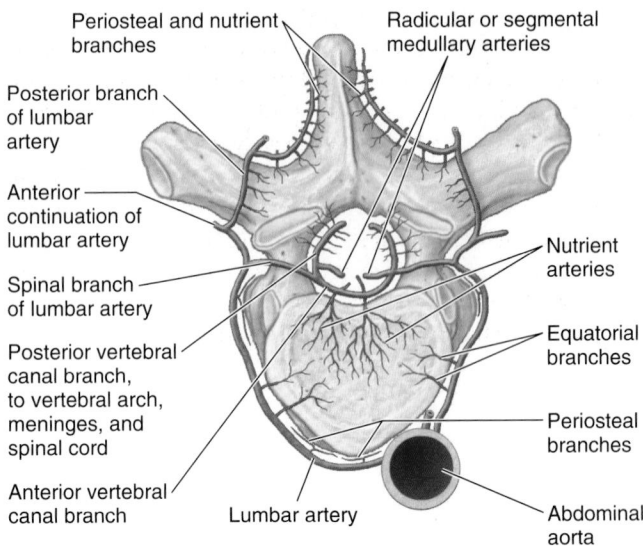

FIGURE 4.25. Blood supply of vertebrae. Typical vertebrae are supplied by segmental arteries—here lumbar arteries. In the thoracic and lumbar regions, each vertebra is encircled on three sides by paired intercostal or lumbar arteries that arise from the aorta. The segmental arteries supply equatorial branches to the vertebral body, and posterior branches supply the vertebral arch structures and the back muscles. Spinal branches enter the vertebral canal through the IV foramina to supply the bones, periosteum, ligaments, and meninges that bound the epidural space and radicular or segmental medullary arteries that supply nervous tissue (spinal nerve roots and spinal cord).

Spinal veins form venous plexuses along the vertebral column both inside and outside the vertebral canal. These plexuses are the **internal vertebral venous plexuses** (epidural venous plexuses) and **external vertebral venous plexuses,** respectively (Fig. 4.26). These plexuses communicate through the intervertebral foramina. Both plexuses are densest

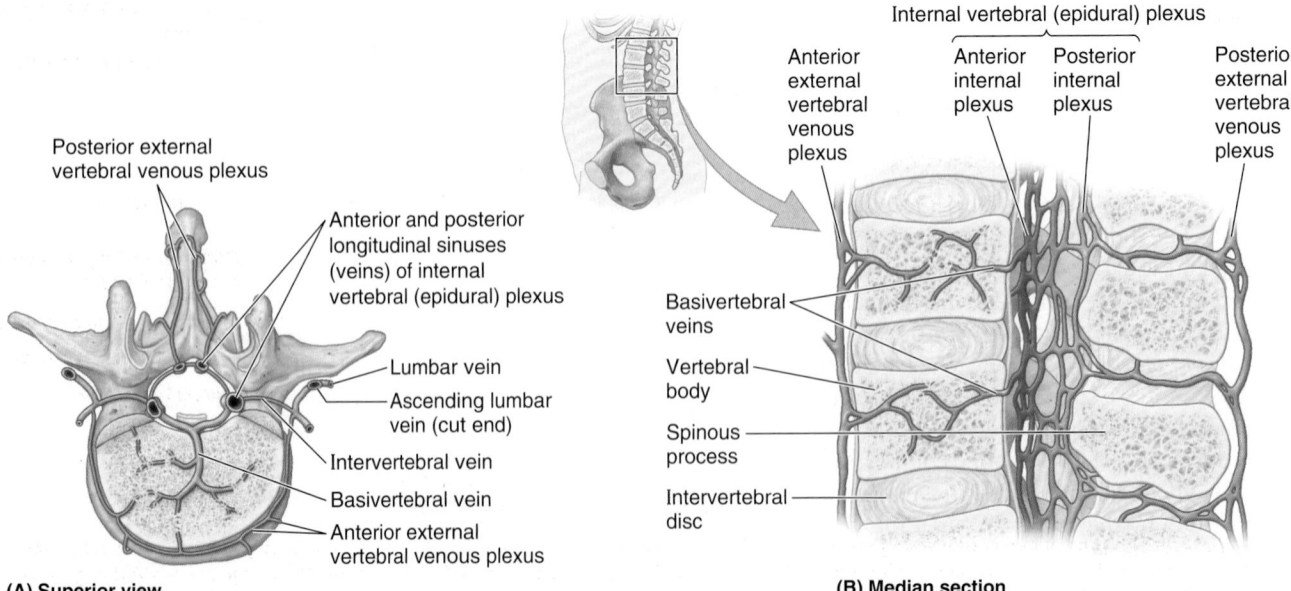

(A) Superior view

(B) Median section

FIGURE 4.26. Venous drainage of vertebral column. A. The venous drainage parallels the arterial supply and enters the external and internal vertebral venous plexuses. There is also anterolateral drainage from the external aspects of the vertebrae into segmental veins. **B.** The dense plexus of thin-walled vessels within the vertebral canal, the internal vertebral venous plexuses, consists of valveless anastomoses between anterior and posterior longitudinal venous sinuses.

anteriorly and posteriorly and relatively sparse laterally. The large, tortuous **basivertebral veins** form within the vertebral bodies. They emerge from foramina on the surfaces of the vertebral bodies (mostly the posterior aspect) and drain into the anterior external and especially the anterior internal vertebral venous plexuses, which may form large longitudinal sinuses. The **intervertebral veins** receive veins from the spinal cord and vertebral venous plexuses as they accompany the spinal nerves through the IV foramina to drain into the *vertebral veins* of the neck and *segmental (intercostal, lumbar,* and *sacral) veins* of the trunk (Figs. 4.26A and 4.27).

Nerves of Vertebral Column

Other than the zygapophysial joints (innervated by *articular branches of the medial branches of the posterior rami*, as described with these joints), the vertebral column is innervated by (**recurrent**) **meningeal branches of the spinal nerves** (Fig. 4.27). These small branches are the only branches to arise from the *mixed spinal nerve*, arising immediately after it is formed and before its division into anterior and posterior rami, or from the anterior ramus immediately after its formation.

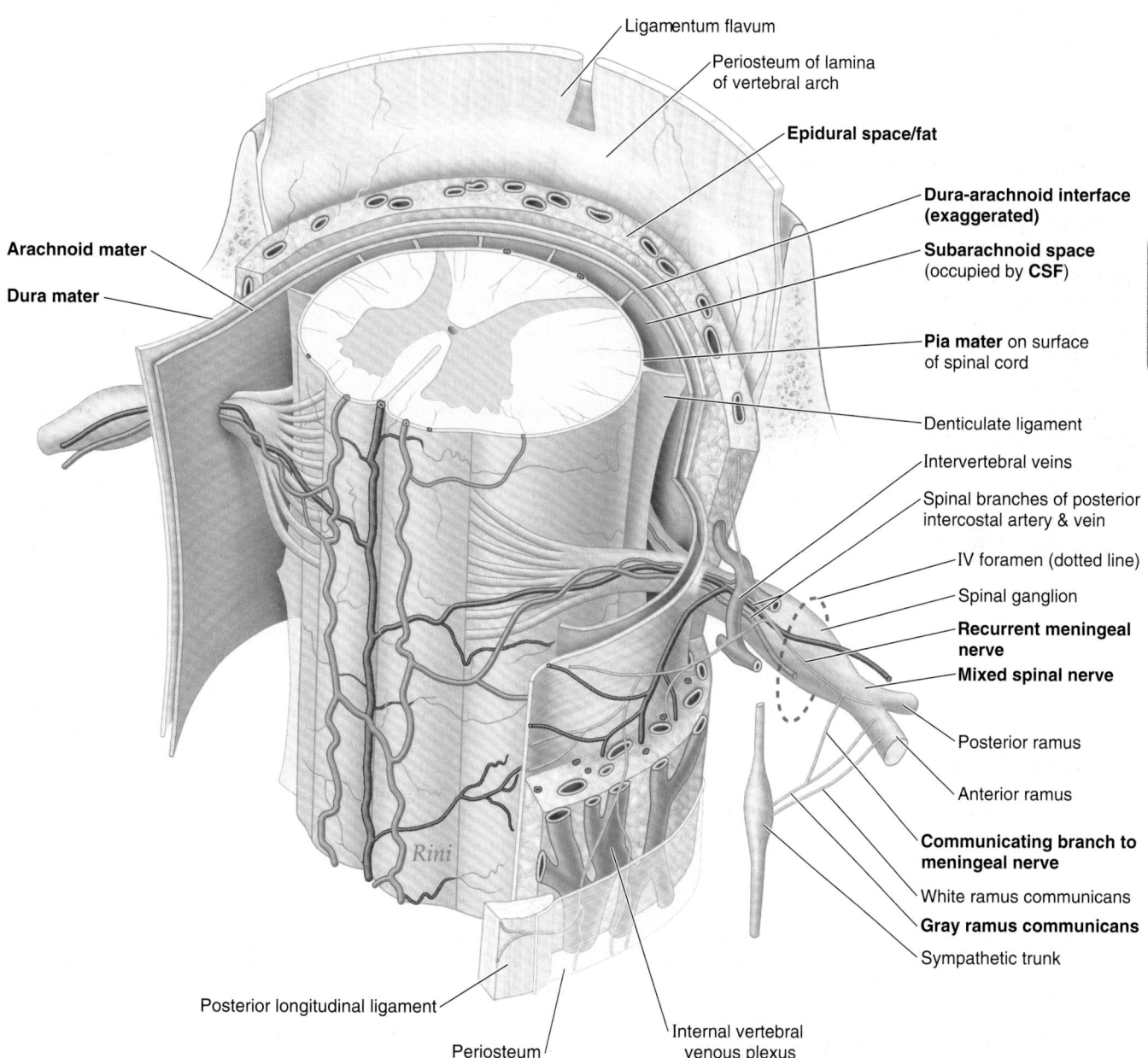

Ligamentum flavum
Periosteum of lamina of vertebral arch
Epidural space/fat
Dura-arachnoid interface (exaggerated)
Subarachnoid space (occupied by **CSF**)
Pia mater on surface of spinal cord
Denticulate ligament
Intervertebral veins
Spinal branches of posterior intercostal artery & vein
IV foramen (dotted line)
Spinal ganglion
Recurrent meningeal nerve
Mixed spinal nerve
Posterior ramus
Anterior ramus
Communicating branch to meningeal nerve
White ramus communicans
Gray ramus communicans
Sympathetic trunk
Internal vertebral venous plexus
Periosteum
Posterior longitudinal ligament
Arachnoid mater
Dura mater

FIGURE 4.27. **Innervation of periosteum and ligaments of vertebral column and of meninges.** Except for the zygapophysial joints and external elements of the vertebral arch, the fibroskeletal structures of the vertebral column (and the meninges) are supplied by the (recurrent) meningeal nerves. Although usually omitted from diagrams and illustrations of spinal nerves, these fine nerves are the first branches to arise from all 31 pairs of spinal nerves and are the nerves that initially convey localized pain sensation from the back produced by acute herniation of an IV disc or from sprains, contusions, fractures, or tumors of the vertebral column itself. (Based on Frick H, Kummer B, Putz R. *Wolf-Heidegger's atlas of human anatomy,* 4th ed. Basel: Karger AG, 1990:476.)

Two to four of these fine meningeal branches arise on each side at all vertebral levels. Close to their origin, the meningeal branches receive communicating branches from the nearby gray rami communicantes. As the spinal nerves exit the IV foramina, most of the meningeal branches run back through the foramina into the vertebral canal (hence the alternate term *recurrent*). However, some branches remain outside the canal and are distributed to the anterolateral aspect of the vertebral bodies and IV discs. They also supply the periosteum and especially the anuli fibrosi and anterior longitudinal ligament. Inside the vertebral canal, transverse, ascending, and descending branches distribute nerve fibers to the:

- Periosteum (covering the surface of the posterior vertebral bodies, pedicles, and laminae).
- Ligamenta flava.
- Anuli fibrosi of the posterior and posterolateral aspect of the IV discs.
- Posterior longitudinal ligament.
- Spinal dura mater.
- Blood vessels within the vertebral canal.

Nerve fibers to the periosteum, anuli fibrosi, and ligaments supply pain receptors. Those to the anuli fibrosi and ligaments also supply receptors for proprioception (the sense of one's position). Sympathetic fibers to the blood vessels stimulate vasoconstriction.

VERTEBRAL COLUMN

Aging of Intervertebral Discs

With advancing age, the nuclei pulposi dehydrate and lose elastin and proteoglycans while gaining collagen. As a result, the IV discs lose their turgor, becoming stiffer and more resistant to deformation. As the nucleus dehydrates, the two parts of the disc appear to merge as the distinction between them becomes increasingly diminished. With advancing age, the nucleus becomes dry and granular and may disappear altogether as a distinct formation. As these changes occur, the anulus fibrosis assumes an increasingly greater share of the vertical load and the stresses and strains that come with it. The lamellae of the anulus thicken and often develop fissures and cavities.

Although the margins of adjacent vertebral bodies may approach more closely as the superior and inferior surfaces of the body become shallow concavities (the most probable reason for slight loss of height with aging), it has been shown that the intervertebral discs *increase* in size with age. Not only do they become increasing convex but, between the ages of 20 and 70, their anteroposterior (AP) diameter increases about 10% in females and 2% in males, while thickness (height) increases centrally about 10% in both sexes. Overt or marked disc narrowing, especially when it is greater than that of more superiorly located discs, suggests pathology, not normal aging (Bogduk, 2012).

Herniation of Nucleus Pulposus

Herniation (protrusion) of the gelatinous nucleus pulposus into or through the anulus fibrosus is a well-recognized cause of lower back pain (LBP) and lower limb pain (Fig. B4.11A & C). However, there are many other causes of LBP; furthermore, herniations are often coincidental findings in asymptomatic individuals.

The IV discs in young persons are strong—usually so strong that the vertebrae often fracture during a fall before the discs rupture. Furthermore, the water content of their nuclei pulposi is high (approaching 90%), giving them great turgor (fullness). However, violent hyperflexion of the vertebral column may rupture an IV disc and fracture the adjacent vertebral bodies.

Flexion of the vertebral column produces compression anteriorly and stretching or tension posteriorly, squeezing the nucleus pulposus further posteriorly toward the thinnest part of the anulus fibrosus. If the anulus fibrosus has degenerated, the nucleus pulposus may herniate into the vertebral canal and compress the spinal cord (Fig. B4.11A & B) or the nerve roots of the cauda equina (Fig. B4.11A & C). A herniated IV disc is inappropriately called a "slipped disc" by some people.

Herniations of the nucleus pulposus usually extend posterolaterally, where the anulus fibrosus is relatively thin, and does not receive support from either the posterior or the anterior longitudinal ligaments. A posterolateral herniated IV disc is more likely to be symptomatic because of the proximity of the spinal nerve roots. The *localized back pain* of a herniated disc, which is usually *acute pain,* results from pressure on the longitudinal ligaments and periphery of the anulus fibrosus and from local inflammation caused by chemical irritation by substances from the ruptured nucleus pulposus. *Chronic pain* resulting from compression of the spinal nerve roots by the herniated disc is usually *referred pain,* perceived as coming from the area (dermatome) supplied by that nerve. Because the IV discs are largest in the lumbar and lumbosacral regions, where movements are consequently greater, posterolateral herniations of the nucleus pulposus are most common here (Fig. B4.11B).

Approximately 95% of *lumbar disc protrusions* occur at the L4–L5 or L5–S1 levels. The marked decrease in the radiographic intervertebral space (i.e., in disc height) that may occur as a result of acute herniation of a nucleus pulposus may also result in narrowing of the IV foramina, perhaps exacerbating the compression of the spinal nerve roots, especially if hypertrophy of the surrounding bone has also occurred. Because the nucleus becomes increasingly

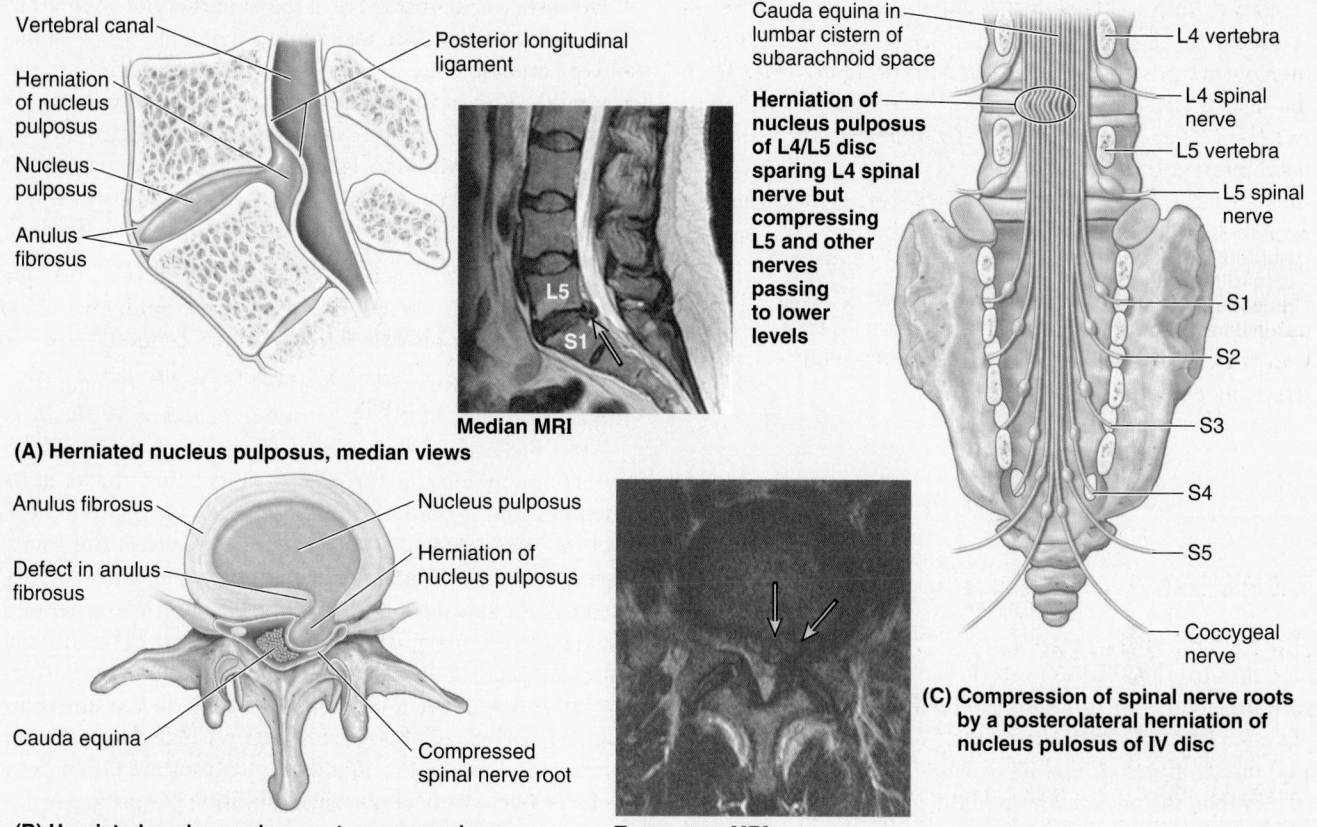

(A) Herniated nucleus pulposus, median views

Vertebral canal
Herniation of nucleus pulposus
Nucleus pulposus
Anulus fibrosus
Posterior longitudinal ligament
Median MRI

(B) Herniated nucleus pulposus, transverse views

Anulus fibrosus
Defect in anulus fibrosus
Cauda equina
Nucleus pulposus
Herniation of nucleus pulposus
Compressed spinal nerve root

Transverse MRI

Cauda equina in lumbar cistern of subarachnoid space
Herniation of nucleus pulposus of L4/L5 disc sparing L4 spinal nerve but compressing L5 and other nerves passing to lower levels
L4 vertebra
L4 spinal nerve
L5 vertebra
L5 spinal nerve
S1
S2
S3
S4
S5
Coccygeal nerve
(C) Compression of spinal nerve roots by a posterolateral herniation of nucleus pulosus of IV disc

FIGURE B4.11. **Herniation of nucleus pulposus. A.** Right half of hemisected lumbosacral joint and median MRI of lumbosacral region. **B.** Inferior views, transverse section and transverse MRI of herniated IV disc **C.** Posterior view, cauda equina. The arrows in the MRIs are indicating herniations.

dehydrated and fibrous, or even granular or solid with aging, a diagnosis of acute herniation in advanced years is regarded with suspicion. It is more likely that the nerve roots are being compressed by increased ossification of the IV foramen as they exit.

Acute middle and low back pain, may be caused by a mild posterolateral protrusion of a lumbar IV disc at the L5–S1 level that affects nociceptive (pain) endings in the region, such as those associated with the posterior longitudinal ligament. The clinical picture varies considerably, but pain of acute onset in the lower back is a common presenting symptom. Because muscle spasm is associated with low back pain, the lumbar region of the vertebral column becomes tense and increasingly cramped as relative ischemia occurs, causing painful movement.

Sciatica, pain in the lower back and hip radiating down the back of the thigh into the leg, is often caused by a herniated lumbar IV disc that compresses and compromises the L5 or S1 component of the sciatic nerve (Fig. 4.11C). The IV foramina in the lumbar region decrease in size and the lumbar nerves increase in size, which may explain why sciatica is so common. Bone spurs (*osteophytes*) developing around the zygapophysial joints or the posterolateral margins during aging may narrow the foramina even more, causing shooting pains down the lower limbs. Any maneuver that stretches the

sciatic nerve, such as flexing the thigh with the knee extended (*straight leg-raising test*), may produce or exacerbate (but in some individuals relieves) sciatic pain.

IV discs may also be damaged by violent rotation (e.g., during a golf swing) or flexing of the vertebral column. The general rule is that when an IV disc protrudes, it usually compresses the nerve root numbered one inferior to the herniated disc; for example, the L5 nerve is compressed by an L4–L5 IV disc herniation. Recall that in the thoracic and lumbar regions the IV disc forms the inferior half of the anterior border of the IV foramen and that the superior half is formed by the bone of the body of the superior vertebra (Fig. 4.14).

The spinal nerve roots descend to the IV foramen from which the spinal nerve formed by their merging will exit. The nerve that exits a given IV foramen passes through the superior bony half of the foramen and thus lies above and is not affected by a herniating disc at that level. However, the nerve roots passing to the IV foramen immediately and farther below pass directly across the area of herniation. Symptom-producing IV disc protrusions occur in the cervical region, almost as often as in the lumbar region.

Chronic or sudden forcible *hyperflexion of the cervical region,* as might occur during a head-on collision or during illegal head blocking in football (Fig. B4.12) for example,

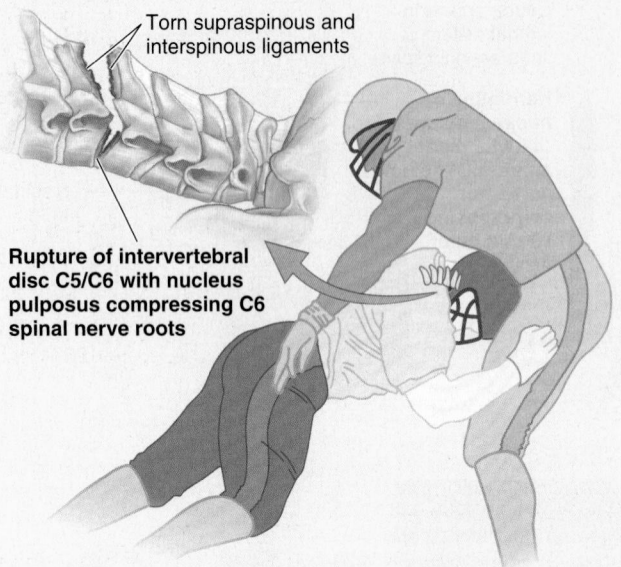

Torn supraspinous and interspinous ligaments

Rupture of intervertebral disc C5/C6 with nucleus pulposus compressing C6 spinal nerve roots

FIGURE B4.12.

may rupture the IV disc posteriorly without fracturing the vertebral body. In this region, the IV discs are centrally placed in the anterior border of the IV foramen, and a herniated disc compresses the nerve actually exiting at that level (rather than the level below as in the lumbar region).

However, recall that cervical spinal nerves exit superior to the vertebra of the same number, so the numerical relationship of herniating disc to nerve affected is the same (e.g., the cervical IV discs most commonly ruptured are those between C5–C6 and C6–C7, compressing spinal nerve roots C6 and C7, respectively). Cervical IV disc protrusions result in pain in the neck, shoulder, arm, and hand. Any sport or activity in which movement causes downward or twisting pressure on the neck or lower back may produce herniation of a nucleus pulposus.

Fracture of Dens of Axis

The transverse ligament of the atlas is stronger than the dens of the C2 vertebra. *Fractures of the dens* make up about 40% of fractures of the axis. The most common dens fracture occurs at its base—that is, at its junction with the body of the axis (Fig. B4.13A). Often these fractures are unstable (do not reunite) because the transverse ligament of the atlas becomes interposed between fragments (Crockard et al., 1993) and because the separated fragment (the dens) no longer has a blood supply, resulting in *avascular necrosis* (G., death). Almost as common are fractures of the vertebral body inferior to the base of the dens (Fig. B4.13B–E). This type of fracture heals more readily because the fragments retain their blood supply. Other dens fractures result from abnormal ossification patterns.

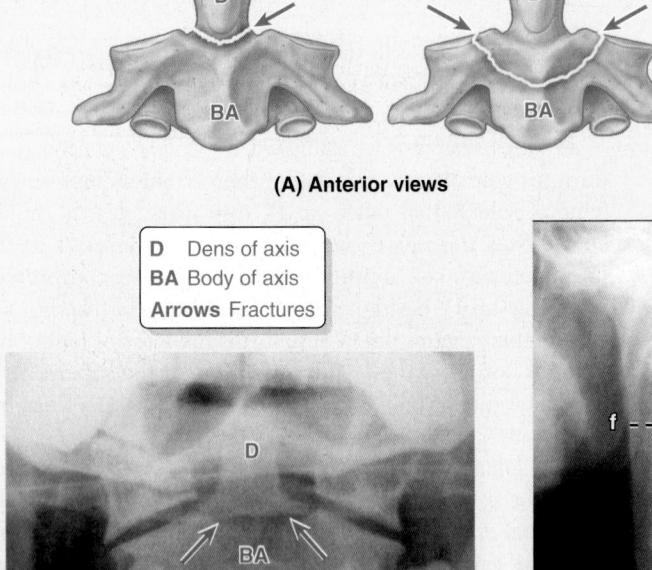

(A) Anterior views

D   Dens of axis
BA Body of axis
Arrows Fractures

(B) Open mouth radiograph (compare with Figure 4.6E)

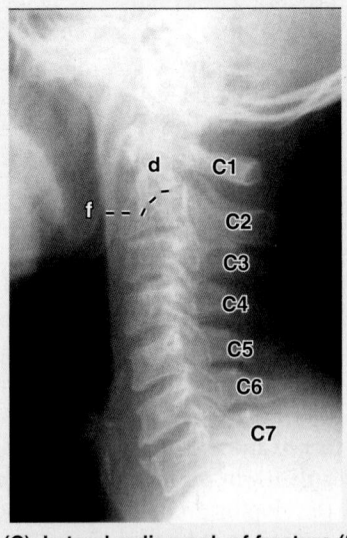

(C) Lateral radiograph of fracture (f) of base of dens (d)

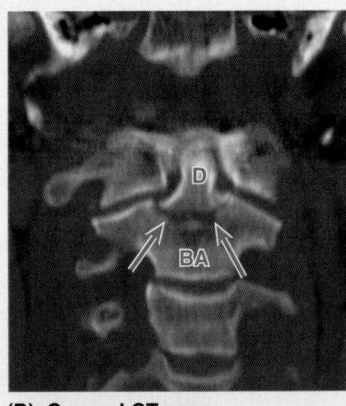

(D) Coronal CT scan

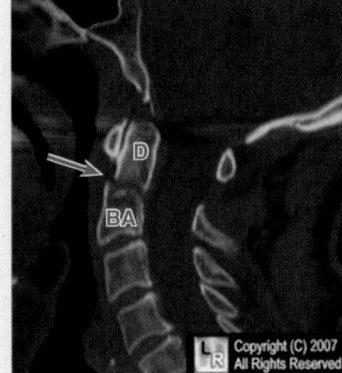

(E) Median CT scan

FIGURE B4.13. Fractures of dens of axis (C2).

Rupture of Transverse Ligament of Atlas

When the transverse ligament of the atlas ruptures, the dens of the axis is set free, resulting in *atlanto-axial subluxation*—incomplete dislocation of the median atlanto-axial joint (Fig. B4.14A). Pathological softening of the transverse and adjacent ligaments, usually resulting from disorders of connective tissue, may also cause atlanto-axial subluxation (Bogduk and Macintosh, 1984); 20% of people with Down syndrome exhibit laxity or agenesis of this ligament. Dislocation owing to transverse ligament rupture or agenesis is more likely to cause spinal cord compression than that resulting from fracture of the dens (Fig. B4.14B). In this fracture, the dens fragment is held in place against the anterior arch of the atlas by the transverse ligament, and the dens and atlas move as a unit.

In the absence of a competent ligament, the upper cervical region of the spinal cord may be compressed between the approximated posterior arch of the atlas and the dens (Fig. B4.14A), causing paralysis of all four limbs (*quadriplegia*), or into the medulla of the brainstem, resulting in death. *Steele's Rule of Thirds:* Approximately one third of the atlas ring is occupied by the dens, one third by the spinal cord, and the remaining third by the fluid-filled space and tissues surrounding the cord (Fig. B4.14C & D). This explains why some people with anterior displacement of the atlas may be relatively asymptomatic until a large degree of movement (greater than one third of the diameter of the atlas ring) occurs. Sometimes inflammation in the craniovertebral area may produce softening of the ligaments of the craniovertebral joints and cause dislocation of the atlanto-axial joints. Sudden movement of a patient from a bed to a chair, for example, may produce posterior displacement of the dens of the axis and injury to the spinal cord.

Rupture of Alar Ligaments

The alar ligaments are weaker than the transverse ligament of the atlas. Consequently, combined flexion and rotation of the head may tear one or both alar ligaments. *Rupture of an alar ligament* results in an increase of approximately 30% in the range of movement to the contralateral side (Dvorak et al., 1988).

Fractures and Dislocations of Vertebrae

Although the construction of the vertebral column permits a considerable amount of movement as well as support and protection, excessive or sudden violent movement or movement of a type not permitted in a specific region is likely to result in fractures, dislocations, and fracture–dislocations of the vertebral column.

Sudden forceful flexion, as occurs in automobile accidents or from a violent blow to the back of the head, commonly produces a *crush* or *compression fracture* of the body of one or more vertebrae. If violent anterior movement of the vertebra occurs in combination with compression, a vertebra may be displaced anteriorly on the vertebra inferior to it (e.g., dislocation of C6 or C7 vertebrae) (see the blue box "Dislocation of Cervical Vertebrae,"

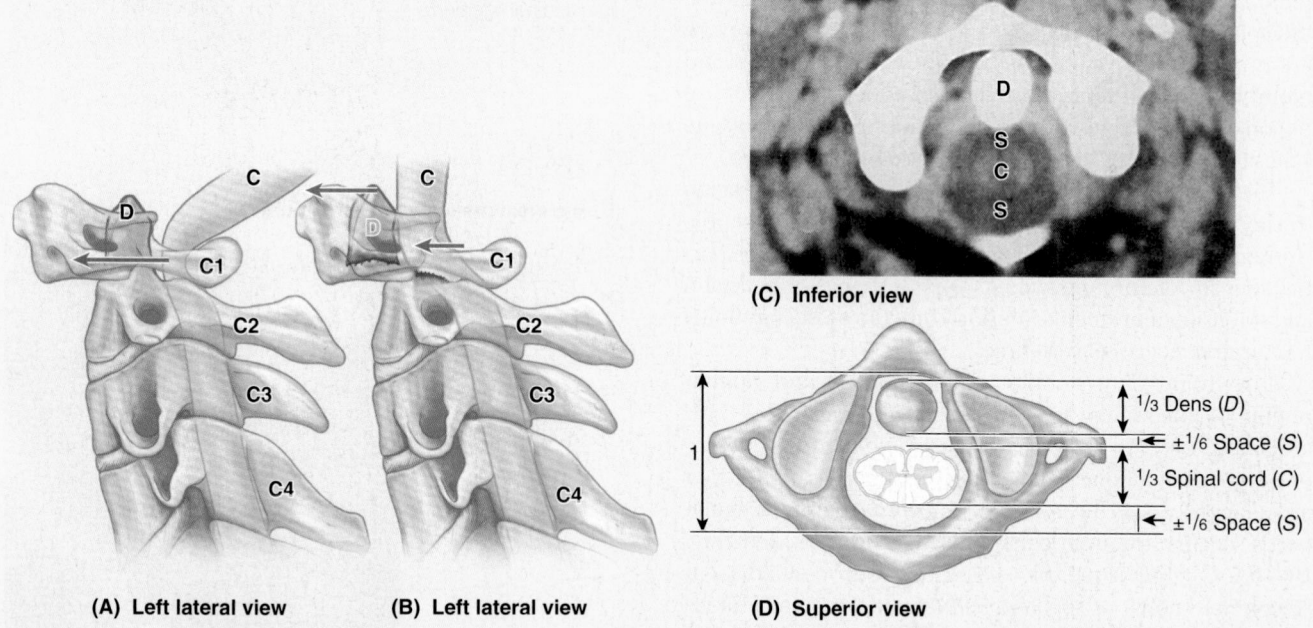

FIGURE B4.14. Rupture of transverse ligament of atlas. A. This left lateral view demonstrates that subluxation of the median atlanto-axial joint results from rupture of the transverse ligament. The atlas moves but the dens is fixed. *C, spinal cord; D, dens of axis.* **B.** This left lateral view of a fracture of the dens shows that the dens and atlas move together as a unit because the transverse ligament holds the dens to the anterior arch of the atlas. **C and D.** Inferior view of transverse CT scan and interpretive drawing showing a normal median atlanto-axial joint and demonstrating Steele's Rule of Thirds.

p. 457). Usually this displacement dislocates and fractures the articular facets between the two vertebrae and ruptures the interspinous ligaments. Irreparable injuries to the spinal cord accompany most severe flexion injuries of the vertebral column.

Sudden, forceful extension of the neck can also injure the vertebral column and spinal cord. Head butting or illegal face blocking in football may lead to a *hyperextension injury of the neck* (Fig. B4.15A). Such violent hyperextension is most likely to injure posterior parts of the vertebrae, fracturing by crush or compression of the vertebral arches and their processes. Fractures of cervical vertebrae may radiate pain to the back of the neck and scapular region because the same spinal sensory ganglia and spinal cord segments receiving pain impulses from the vertebrae are also involved in supplying neck muscles.

Severe hyperextension of the neck ("whiplash" injury) also occurs during rear-end motor vehicle collisions (Fig. B4.15B), especially when the head restraint is too low as illustrated. In these types of hyperextension injuries, the anterior longitudinal ligament is severely stretched and may be torn.

Hyperflexion injury of the vertebral column may also occur as the head "rebounds" after the hyperextension, snapping the head forward onto the thorax. "Facet jumping" or locking of the cervical vertebrae may occur because of dislocation of the vertebral arches (see the blue box "Dislocation of Cervical Vertebrae," p. 457). Severe hyperextension of the head on the upper neck may, in addition to producing a cervical spondylolysis or hangman's fracture (see the blue box "Fracture and Dislocation of Axis," p. 459), rupture the anterior longitudinal ligament and the adjacent anulus fibrosus of the C2–C3 IV disc. If this injury occurs, the cranium, C1, and the anterior portion (dens and body) of C2 are separated from the rest of the axial skeleton (Fig. B4.15C), and the spinal cord is usually severed. Persons with this severe injury seldom survive. Football, diving, falls from horses, and motor vehicle collisions cause most fractures of the cervical region of the vertebral column. Symptoms range from vague aches to progressive loss of motor and sensory functions.

The transition from the relatively inflexible thoracic region to the much more mobile lumbar region occurs abruptly. Consequently, vertebrae T11 and especially T12 (which participates in rotatory movements superiorly but only flexion and extension movements inferiorly) are the most commonly fractured noncervical vertebrae.

Dislocation of vertebrae in the thoracic and lumbar regions is uncommon because of the interlocking of their articular processes. However, when *spondylolysis*—fracture of the column of bones connecting the superior and inferior articular processes (the *pars interarticularis,* or interarticular part)—occurs, the interlocking mechanism is broken (Fig. B4.16A–C). Subsequently, dislocation between adjacent vertebrae, known as *spondylolisthesis,* may occur. Failure or fracture of the interarticular parts of the vertebral laminae of L5 (*spondylolysis of L5*) especially may result in spondylolisthesis of the L5 vertebral body relative to the sacrum (S1 vertebra) due to the downward tilt of the L5/S1 IV

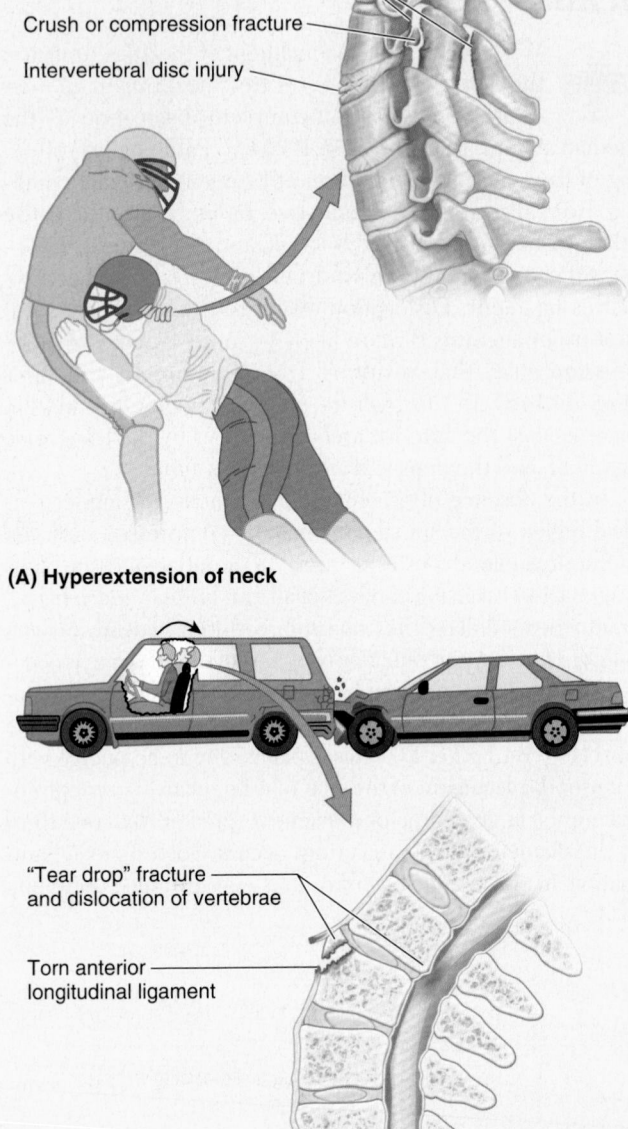

Fracture of posterior arch elements (pedicle, spine)

Crush or compression fracture

Intervertebral disc injury

(A) Hyperextension of neck

"Tear drop" fracture and dislocation of vertebrae

Torn anterior longitudinal ligament

(B) Hyperextension (whiplash) injury

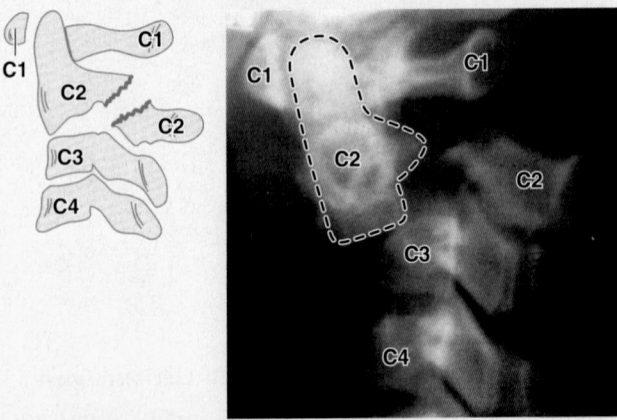

(C) Hangman's fracture with disruption of C2/C3 disc and anterior longitudinal ligament

FIGURE B4.15. Extension injuries of cervical vertebrae.

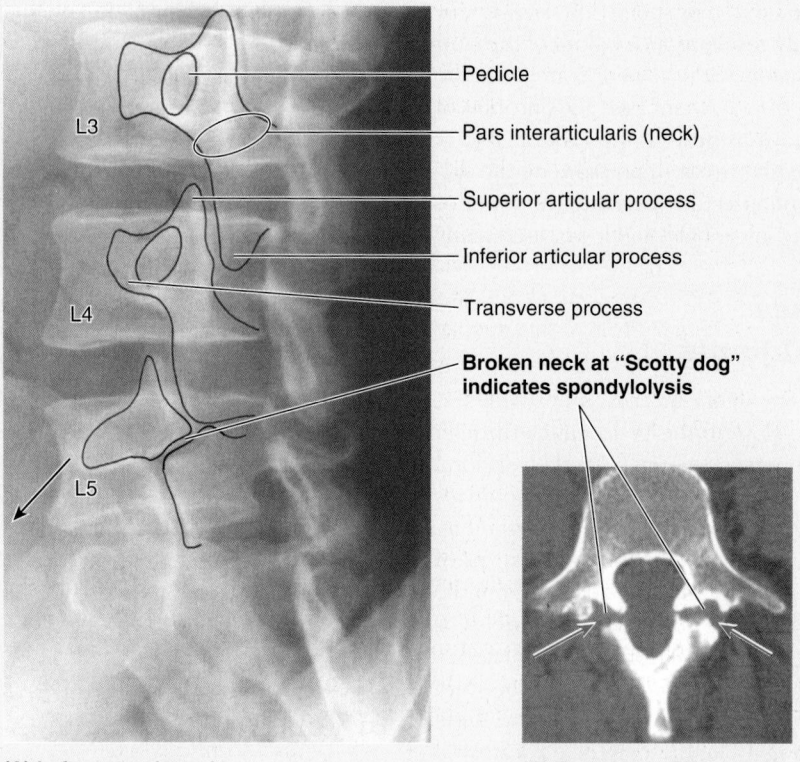

Pedicle

Pars interarticularis (neck)

Superior articular process

Inferior articular process

Transverse process

Broken neck at "Scotty dog" indicates spondylolysis

L3

L4

L5

(A) Left posterolateral oblique view

(B) Transverse CT

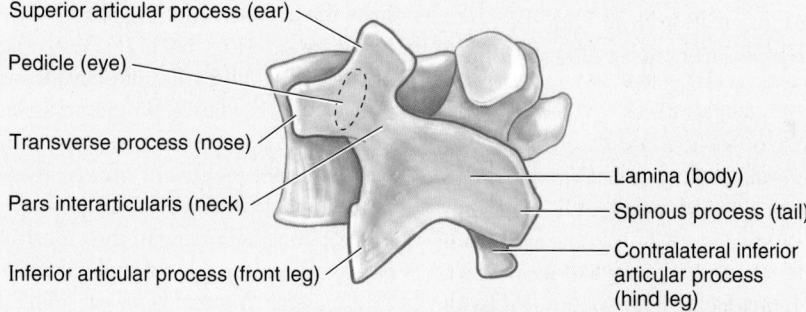

Superior articular process (ear)

Pedicle (eye)

Transverse process (nose)

Pars interarticularis (neck)

Inferior articular process (front leg)

Lamina (body)

Spinous process (tail)

Contralateral inferior articular process (hind leg)

(C) "Scotty dog sign" in posterolateral oblique view of lumbar vertebra

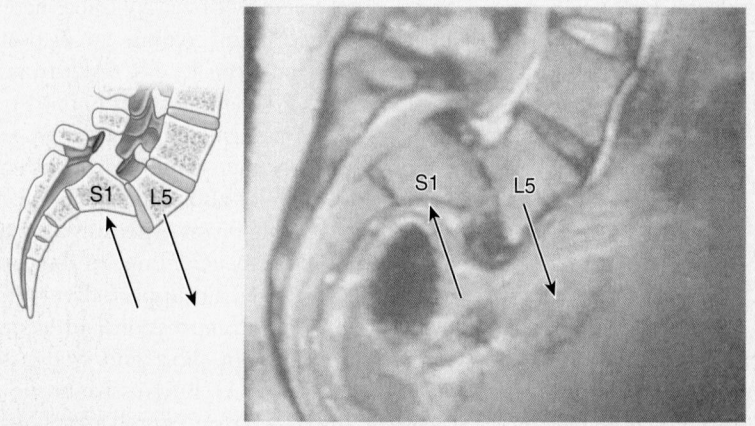

S1   L5

S1   L5

(D) Median MRI, spondylolisthesis secondary to spondylolysis of L5

FIGURE B4.16. Spondylolysis and spondylolisthesis.

joint (Fig. B4.16D). Most agree that spondylolysis of L5, or susceptibility to it, probably results from a failure of the centrum of L5 to unite adequately with the neural arches at the neurocentral joint during development (see "Ossification of Vertebrae," p. 453). Spondylolisthesis at the L5–S1 IV joint may (but does not necessarily) result in pressure on the spinal nerves of the cauda equina as they pass into the superior part of the sacrum, causing lower back and lower limb pain.

Injury and Disease of Zygapophysial Joints

The zygapophysial joints are of clinical interest because they are close to the IV foramina through which the spinal nerves emerge from the vertebral canal. When these joints are injured or develop osteophytes (*osteoarthritis*), the spinal nerves are often affected (see Fig. B4.9B). This causes pain along the distribution patterns of the *dermatomes* and spasm in the muscles derived from the associated *myotomes*. A myotome consists of all muscles or parts of muscles receiving innervation from one spinal nerve.

Denervation of lumbar zygapophysial joints is a procedure used for treatment of back pain caused by disease of these joints. The nerves are sectioned near the joints or are destroyed by radiofrequency *percutaneous rhizolysis* (G. *rhiza*, root + G. *lysis*, dissolution). The denervation is directed at the articular branches of two adjacent posterior rami of the spinal nerves because each joint receives innervation from both the nerve exiting at that level and the superjacent nerve (Fig. 4.19).

Back Pain

Back pain in general, and lower back pain in particular, is an immense health problem, second only to the common cold as a reason people visit their doctors. In terms of health factors causing lost work days, backache is second only to headache. The anatomical bases for the pain, especially the nerves initially involved in sensing and carrying pain from the vertebral column itself, are rarely described.

Five categories of structures receive innervation in the back and can be sources of pain:

- Fibroskeletal structures: periosteum, ligaments, and anuli fibrosi of IV discs.
- Meninges: coverings of the spinal cord.
- Synovial joints: capsules of the zygapophysial joints.
- Muscles: intrinsic muscles of the back.
- Nervous tissue: spinal nerves or nerve roots exiting the IV foramina.

Of these, the first two are innervated by (recurrent) meningeal branches of the spinal nerves and the next two are innervated by posterior rami (articular and muscular branches). Pain from nervous tissue—that is, caused by compression or irritation of spinal nerves or nerve roots—is typically *referred pain*, perceived as coming from the cutaneous or subcutaneous area (dermatome) supplied by that nerve

(see the blue box "Herniation of Nucleus Pulposus," p. 474), but it may be accompanied by localized pain.

Pain related to the meninges is relatively rare and is discussed later in this chapter; it is generally not considered to be a factor in back pain.

Localized *lower back pain* (*LBP*) (pain perceived as coming from the back) is generally muscular, joint, or fibroskeletal pain. *Muscular pain* is usually related to reflexive cramping (spasms) producing *ischemia*, often secondarily as a result of *guarding* (contraction of muscles in anticipation of pain). *Zygapophysial joint pain* is generally associated with aging (osteoarthritis) or disease (rheumatoid arthritis) of the joints. Pain from vertebral fractures and dislocations is no different than that from other bones and joints: The sharp pain following a fracture is mostly periosteal in origin, whereas pain from dislocations is ligamentous. The acute localized pain associated with an IV disc herniation undoubtedly emanates from the disrupted posterolateral anulus fibrosis and impingement on the posterior longitudinal ligament. Pain in all of these latter instances is conveyed initially by the meningeal branches of the spinal nerves.

Abnormal Curvatures of Vertebral Column

To detect an *abnormal curvature of the vertebral column*, have the individual stand in the anatomical position. Inspect the profile of the vertebral column from the person's side (Fig. B4.17A–C) and then from the posterior aspect (Fig. B4.17D). With the person bending over, observe the ability to flex directly forward and whether the back is level once the flexed position is assumed (Fig. B4.17E).

Abnormal curvatures in some people result from developmental anomalies; in others, the curvatures result from pathological processes. The most prevalent metabolic disease of bone occurring in the elderly, especially in women, is *osteoporosis* (atrophy of skeletal tissue).

Excessive thoracic kyphosis (clinically shortened to *kyphosis*, although this term actually applies to the normal curvature, and colloquially known as *humpback* or *hunchback*) is characterized by an abnormal increase in the thoracic curvature; the vertebral column curves posteriorly (Fig. B4.17B & F). This abnormality can result from erosion (due to osteoporosis) of the anterior part of one or more vertebrae. *Dowager's hump* is a colloquial name for excessive thoracic kyphosis in older women resulting from osteoporosis. However, this type of kyphosis also occurs in elderly men (Swartz, 2009).

Osteoporosis especially affects the horizontal trabeculae of the trabecular bone of the vertebral body (Fig. 4.3). The remaining, unsupported vertical trabeculae are less able to resist compression and sustain compression fractures, resulting in short and wedge-shaped thoracic vertebrae (Fig. B4.9A). Progressive erosion and collapse of vertebrae also result in an overall loss of height. The excessive kyphosis leads to an increase in the AP diameter of the thorax and a significant reduction in dynamic pulmonary capacity.

Excessive lumbar lordosis (clinically shortened to *lordosis*, although once again this term actually describes the normal curvature; colloquially, excessive lumbar lordosis is known as *hollow back* or *sway back*) is characterized by an anterior tilting of the pelvis (the upper sacrum is flexed or rotated antero-inferiorly—*nutation*), with increased extension of the lumbar vertebrae, producing an abnormal increase in the lumbar kyphosis (Fig. B4.17C).

This abnormal *extension deformity* is often associated with weakened trunk musculature, especially the anterolateral abdominal muscles. To compensate for alterations to their normal line of gravity, women develop a temporary excessive lumbar lordosis during late pregnancy. This lordotic curvature may cause lower back pain, but the discomfort normally disappears soon after childbirth.

Obesity in both sexes can also cause excessive lumbar lordosis and lower back pain because of the increased weight of the abdominal contents (e.g., "potbelly") anterior to the normal line of gravity. Loss of weight and exercise of the anterolateral abdominal muscles facilitate correction of this type of excessive lordosis.

Scoliosis (G., crookedness or curved back) is characterized by an *abnormal lateral curvature* that is accompanied by rotation of the vertebrae (Fig. B4.17D, E, & G). The spinous processes turn toward the cavity of the abnormal curvature, and when the individual bends over, the ribs rotate posteriorly (protrude) on the side of the increased convexity.

Deformities of the vertebral column, such as failure of half of a vertebra to develop (*hemivertebra*) are causes of *structural scoliosis*. Sometimes a structural scolioses is combined with excessive thoracic kyphosis—*kyphoscoliosis*—in which an abnormal AP diameter produces a severe restriction of the thorax and lung expansion (Swartz, 2009). Approximately 80% of all structural scolioses are idiopathic,

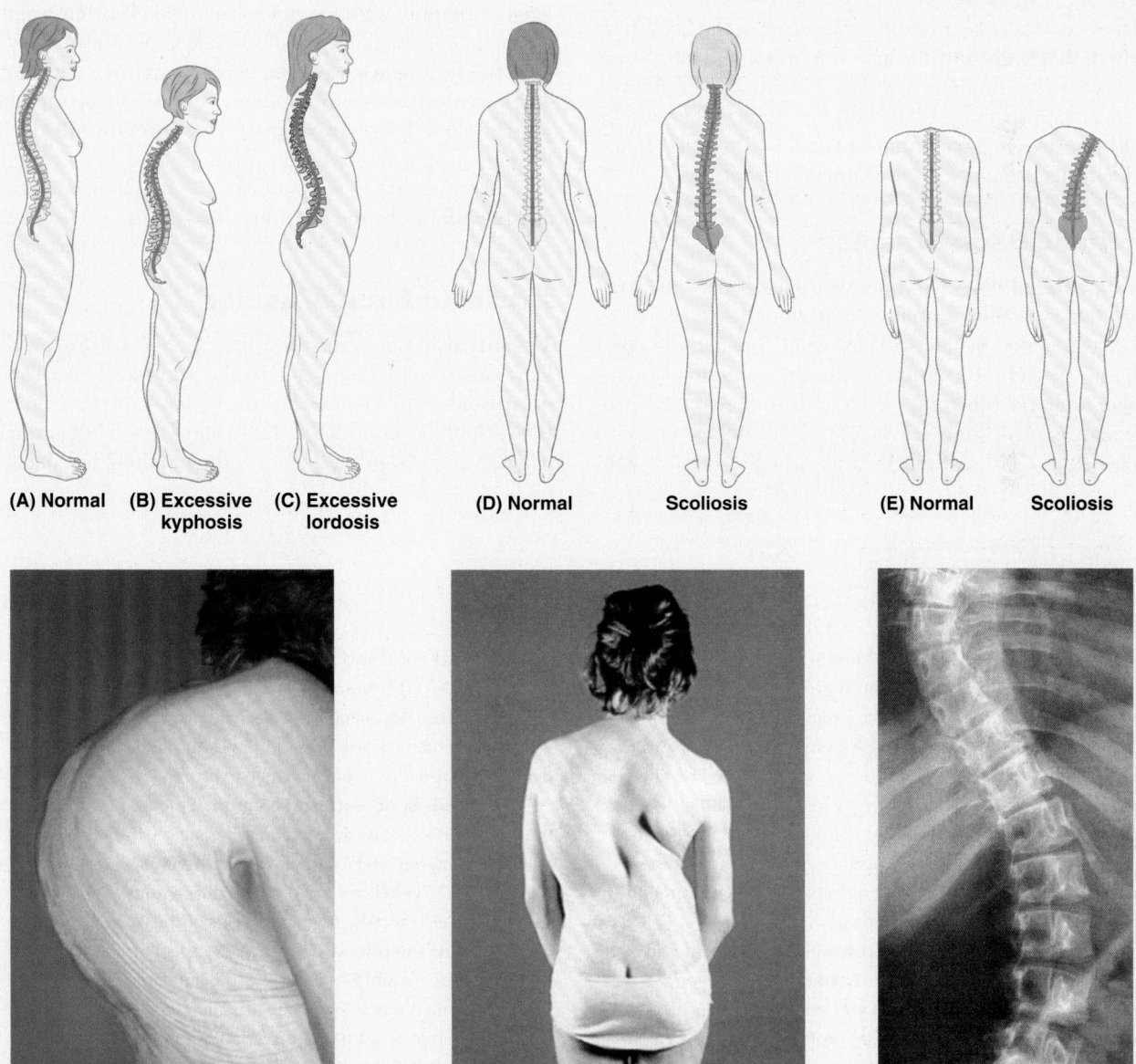

(A) Normal **(B) Excessive kyphosis** **(C) Excessive lordosis**   **(D) Normal**   **Scoliosis**   **(E) Normal**   **Scoliosis**

(F) Right lateral view, excessive kyphosis   **(G) Posterior views, thoracolumbar scoliosis**

FIGURE B4.17. Abnormal curvatures of vertebral column.

occurring without other associated health conditions or an identifiable cause. Idiopathic scoliosis first develops in girls between the ages of 10 and 14 and in boys between the ages of 12 and 15. It is most common and severe among females.

Problems extrinsic to a structurally normal vertebral column, such as asymmetrical weakness of the intrinsic back muscles (*myopathic scoliosis*), or a difference in the length of the lower limbs with a compensatory pelvic tilt, may lead to a *functional scoliosis*. When a person is standing, an obvious inclination or listing to one side may be a sign of scoliosis that is secondary to a herniated IV disc. *Habit scoliosis* is supposedly caused by habitual standing or sitting in an improper position. When the scoliosis is entirely postural, it disappears during maximum flexion of the vertebral column. Functional scolioses do not persist once the underlying problem has been effectively treated.

MUSCLES OF BACK

Most body weight lies anterior to the vertebral column, especially in obese people; consequently, the many strong muscles attached to the spinous and transverse processes of the vertebrae are necessary to support and move the column.

There are two major groups of muscles in the back. The **extrinsic back muscles** include *superficial* and *intermediate muscles* that produce and control limb and respiratory movements, respectively. The *intrinsic (deep) back muscles* include muscles that specifically act on the vertebral column, producing its movements and maintaining posture.

Extrinsic Back Muscles

The **superficial extrinsic back muscles** (trapezius, latissimus dorsi, levator scapulae, and rhomboids) are posterior axio-appendicular muscles that connect the axial skeleton (vertebral column) with the superior appendicular skeleton (pectoral girdle and humerus) and produce and control limb movements (Fig. 4.28A; see also Table 6.4, p. 700). Although located in the back region, for the most part these muscles receive their nerve supply from the anterior rami of cervical nerves and act on the upper limb. The trapezius receives its motor fibers from a cranial nerve, the spinal accessory nerve (CN XI).

The **intermediate extrinsic back muscles** (serratus posterior) are thin muscles, commonly designated as superficial respiratory muscles, but are more likely proprioceptive rather than motor in function (Vilensky et al., 2001). They are described with muscles of the thoracic wall (see Chapter 1). The **serratus posterior superior** lies deep to the rhomboid muscles, and the **serratus posterior inferior** lies deep to the latissimus dorsi. Both serratus muscles are innervated by intercostal nerves, the superior by the first four intercostals and the inferior by the last four.

Intrinsic Back Muscles

The **intrinsic back muscles** (*muscles of back proper, deep back muscles*) are innervated by the posterior rami of spinal nerves and act to maintain posture and control movements of the vertebral column (Figs. 4.28B and 4.29). These muscles, which extend from the pelvis to the cranium, are enclosed

The Bottom Line

VERTEBRAL COLUMN

Joints of vertebral column: Vertebrae are joined to form a semirigid column by IV discs and zygapophysial joints. ♦ The relative thickness of the discs determines the degree of mobility. ♦ The disposition of the zygapophysial joints controls the type of movement between adjacent vertebrae. ♦ The anterior longitudinal ligament resists hyperextension; all other ligaments resist forms of flexion. ♦ The atlanto-occipital joints enable the "yes" (nodding) movement of the head. ♦ The atlanto-axial joints enable the "no" (rotational) movement of the head. Alar ligaments limit rotation.

Movements of vertebral column: The cervical and lumbar regions are most mobile (and consequently most vulnerable to injury). ♦ Flexion and extension occur primarily in the cervical and lumbar regions. ♦ Rotation occurs in the cervical and thoracic regions.

Curvatures of vertebral column: Primary curvatures (thoracic and sacral kyphoses) are developmental; secondary curvatures (cervical and lumbar lordoses) are acquired in relation to the erect human posture. ♦ The curvatures provide shock-absorbing resilience and flexibility to the axial skeleton. ♦ Extensors of the back and abdominal flexors provide dynamic support to maintain the curvatures.

Vasculatures of vertebral column: Spinal branches of the major cervical and segmental arteries supply the vertebral column. ♦ Internal and external vertebral venous plexuses collect blood from the vertebrae and drain, in turn, into the vertebral veins of the neck and the segmental veins of the trunk.

Nerves of vertebral column: Zygapophysial joints are innervated by medial branches of adjacent posterior rami; (recurrent) meningeal branches of spinal nerves supply most bone (periosteum), IV discs, and ligaments as well as the meninges (coverings) of the spinal cord. ♦ These two (groups of) nerves convey all localized pain from the vertebral column.

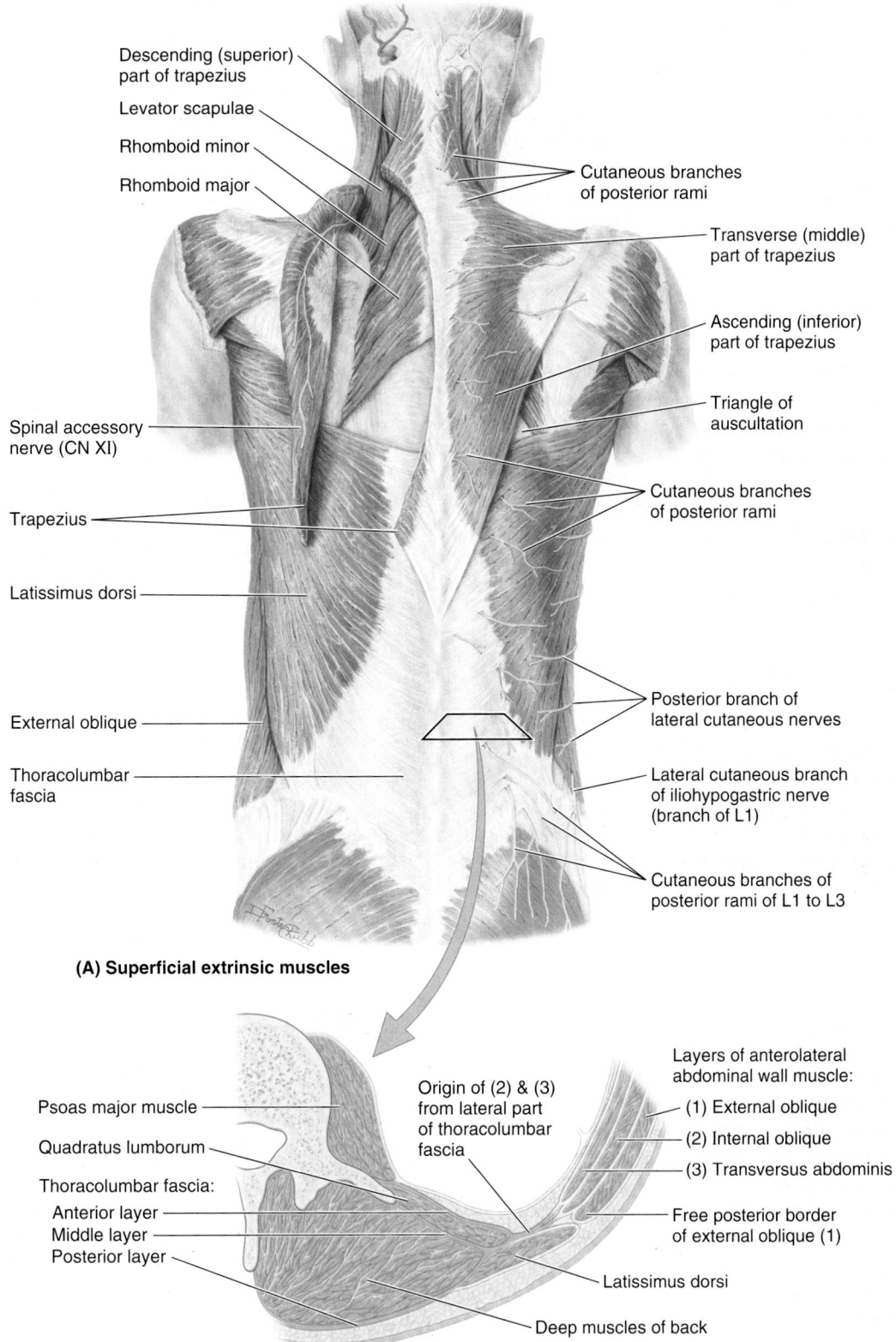

Descending (superior) part of trapezius

Levator scapulae

Rhomboid minor

Rhomboid major

Cutaneous branches of posterior rami

Transverse (middle) part of trapezius

Ascending (inferior) part of trapezius

Triangle of auscultation

Spinal accessory nerve (CN XI)

Cutaneous branches of posterior rami

Trapezius

Latissimus dorsi

External oblique

Posterior branch of lateral cutaneous nerves

Thoracolumbar fascia

Lateral cutaneous branch of iliohypogastric nerve (branch of L1)

Cutaneous branches of posterior rami of L1 to L3

(A) Superficial extrinsic muscles

Psoas major muscle

Quadratus lumborum

Origin of (2) & (3) from lateral part of thoracolumbar fascia

Layers of anterolateral abdominal wall muscle:

(1) External oblique

(2) Internal oblique

(3) Transversus abdominis

Thoracolumbar fascia:
Anterior layer
Middle layer
Posterior layer

Free posterior border of external oblique (1)

Latissimus dorsi

Deep muscles of back

(B) Inferior view of transverse section of posterolateral abdominal wall

FIGURE 4.28. Muscles of back. A. The superficial extrinsic muscles. The trapezius is reflected on the left to show the spinal accessory nerve (CN XI), coursing on its deep surface, and the levator scapulae and rhomboid muscles. **B.** This transverse section of part of the back shows the location of the intrinsic back muscles and the layers of fascia associated with them.

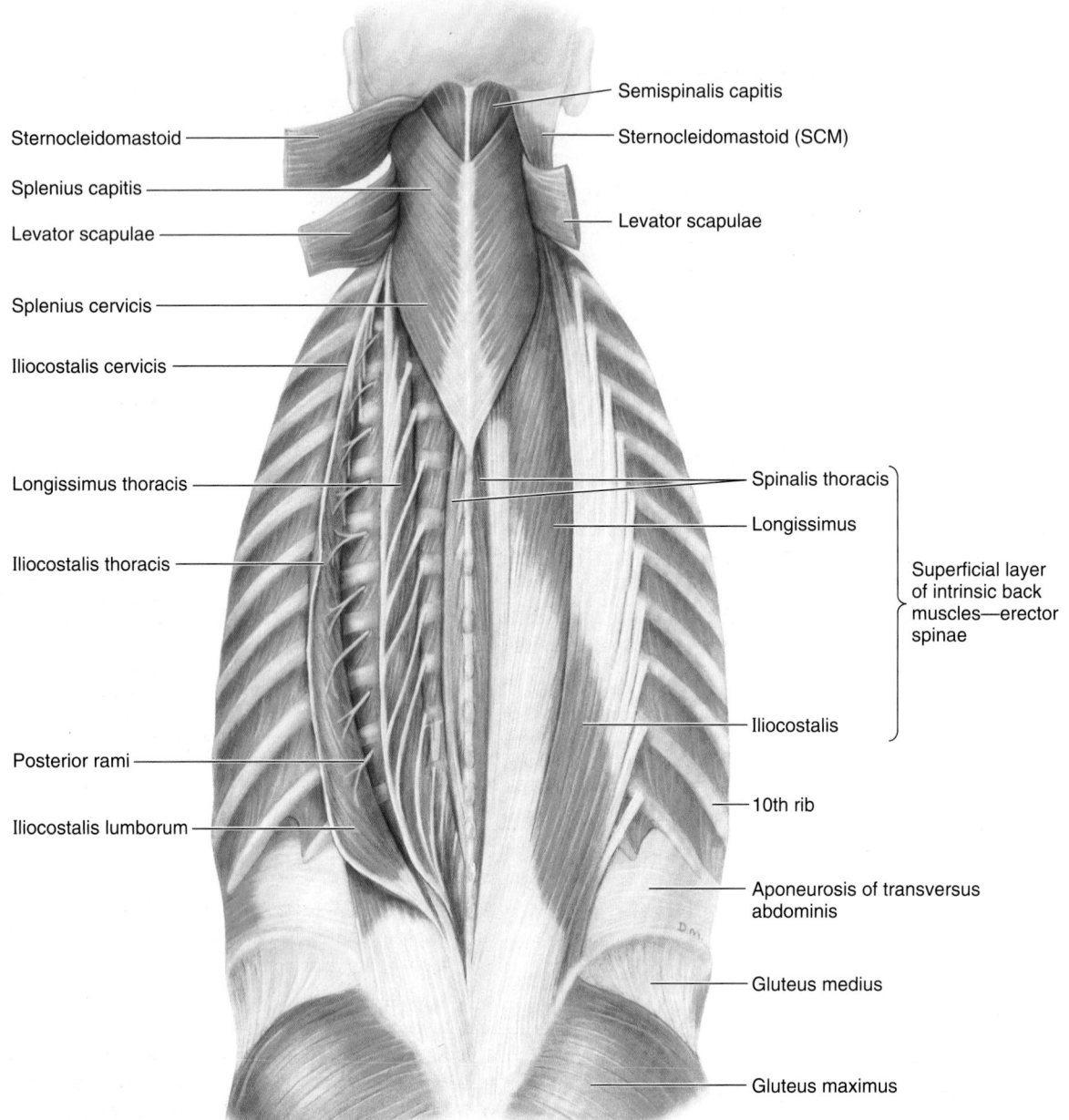

FIGURE 4.29. Superficial and intermediate layers of intrinsic back muscles: splenius and erector spinae. The sternocleidomastoid (SCM) and levator scapulae muscles are reflected to reveal the splenius capitis and splenius cervicis muscles. On the right side, the erector spinae is undisturbed (in situ) and shows the three columns of this massive muscle. On the left side, the spinalis muscle, the thinnest and most medial of the erector spinae columns, is displayed as a separate muscle by reflecting the longissimus and iliocostalis columns of the erector spinae. As they ascend, the direction of fibers is different in the three main groups of muscles: the superficial (splenius) muscles run from medial to lateral, the intermediate (erector spinae) muscles run mostly vertically, and the deep (transversospinalis) muscles run mainly from lateral to medial (see Fig. 4.32).

by *deep fascia* that attaches medially to the nuchal ligament, the tips of the spinous processes of the vertebrae, the supra-spinous ligament, and the median crest of the sacrum. The fascia attaches laterally to the cervical and lumbar transverse processes and the angles of the ribs. The thoracic and lumbar parts of the deep fascia constitute the *thoracolumbar fascia*.

It extends laterally from the spinous processes and forms a thin covering over the intrinsic back muscles in the thoracic region and a strong thick covering for muscles in the lumbar region. The intrinsic back muscles are grouped into super-ficial, intermediate, and deep layers according to their rela-tionship to the surface.

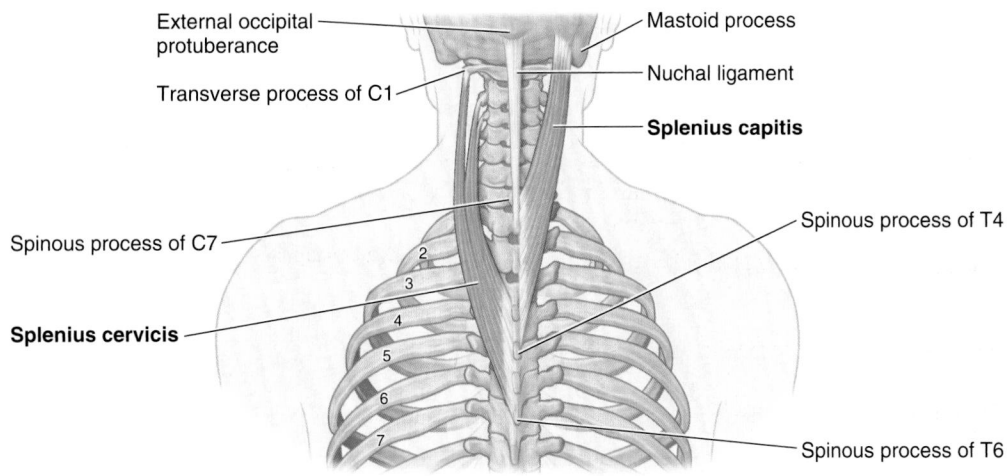

External occipital protuberance
Transverse process of C1
Spinous process of C7
Splenius cervicis
Mastoid process
Nuchal ligament
Splenius capitis
Spinous process of T4
Spinous process of T6

FIGURE 4.30. Superficial layer of intrinsic back muscles (splenius muscles).

TABLE 4.4. SUPERFICIAL LAYER OF INTRINSIC BACK MUSCLES

Muscle	Proximal Attachment	Distal Attachment	Nerve Supply	Main Action(s)
Splenius	Nuchal ligament and spinous processes of C7–T6 vertebrae	*Splenius capitis:* fibers run superolaterally to mastoid process of temporal bone and lateral third of superior nuchal line of occipital bone *Splenius cervicis:* tubercles of transverse processes of C1–C3 or C4 vertebrae	Posterior rami of spinal nerves	*Acting alone:* laterally flex neck and rotate head to side of active muscles *Acting together:* extend head and neck

SUPERFICIAL LAYER

The **splenius muscles** (L. *musculi splenii*) are thick and flat and lie on the lateral and posterior aspects of the neck, covering the vertical muscles somewhat like a bandage, which explains their name (L. *splenion,* bandage) (Figs. 4.29 and 4.30). The splenius muscles arise from the midline and extend superolaterally to the cervical vertebrae (**splenius cervicis**) and cranium (**splenius capitis**). The splenius muscles cover and hold the deep neck muscles in position. The superficial layer of intrinsic muscles is illustrated in Figure 4.30, and information on their attachments, nerve supply, and actions is provided in Table 4.4.

INTERMEDIATE LAYER

The massive **erector spinae muscles** lie in a "groove" on each side of the vertebral column between the spinous processes centrally and the angles of the ribs laterally (Fig. 4.29). The erector spinae are the *chief extensors of the vertebral column* and are divided into three columns: The **iliocostalis** forms the lateral column, the **longissimus** forms the intermediate column, and the **spinalis** forms the medial column. Each column is divided regionally into three parts according to the superior attachments (e.g., iliocostalis lumborum, iliocostalis thoracis, and iliocostalis cervicis). The common

origin of the three erector spinae columns is through a broad tendon that attaches inferiorly to the posterior part of the iliac crest, the posterior aspect of the sacrum, the sacro-iliac ligaments, and the sacral and inferior lumbar spinous processes.

The erector spinae are often referred to as the "long muscles" of the back. In general, they are dynamic (motion-producing) muscles, acting bilaterally to extend (straighten) the flexed trunk. The muscles of the intermediate layer of intrinsic muscles are illustrated in isolation in Figure 4.31, and information on their attachments, nerve supply, and actions is provided in Table 4.5.

DEEP LAYER

Deep to the erector spinae is an obliquely disposed group of much shorter muscles, the **transversospinalis muscle group** consisting of the semispinalis, multifidus, and rotatores. These muscles originate from transverse processes of vertebrae and pass to spinous processes of more superior vertebrae. They occupy the "gutter" between the transverse and the spinous processes and are attached to these processes, the laminae between them, and the ligaments linking them together (Fig. 4.32).

(text continues on p. 488)

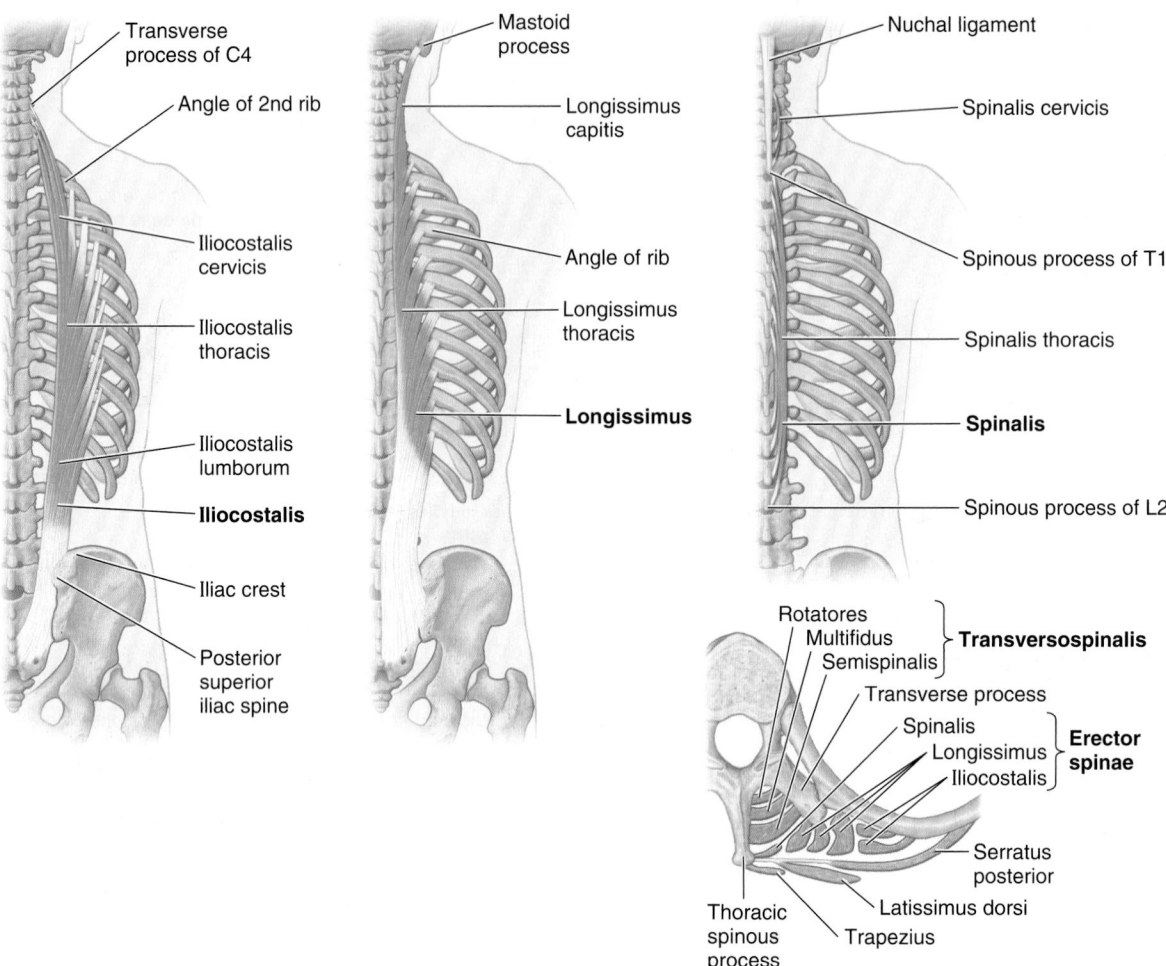

FIGURE 4.31. Intermediate layer of intrinsic back muscles (erector spinae muscles).

TABLE 4.5. INTERMEDIATE LAYER OF INTRINSIC BACK MUSCLES

Muscle	Proximal Attachment	Distal Attachment	Nerve Supply	Main Action(s)
Erector spinae **Iliocostalis** **Longissimus** **Spinalis**	Arises by a broad tendon from posterior part of iliac crest, posterior surface of sacrum, sacro-iliac ligaments, sacral and inferior lumbar spinous processes, and supraspinous ligament	*Iliocostalis:* lumborum, thoracis, cervicis; fibers run superiorly to angles of lower ribs and cervical transverse processes *Longissimus:* thoracis, cervicis, capitis; fibers run superiorly to ribs between tubercles and angles to transverse processes in thoracic and cervical regions, and to mastoid process of temporal bone *Spinalis:* thoracis, cervicis, capitis; fibers run superiorly to spinous processes in the upper thoracic region and to cranium	Posterior rami of spinal nerves	*Acting bilaterally:* extend vertebral column and head; as back is flexed, control movement via eccentric contraction *Acting unilaterally:* laterally flex vertebral column

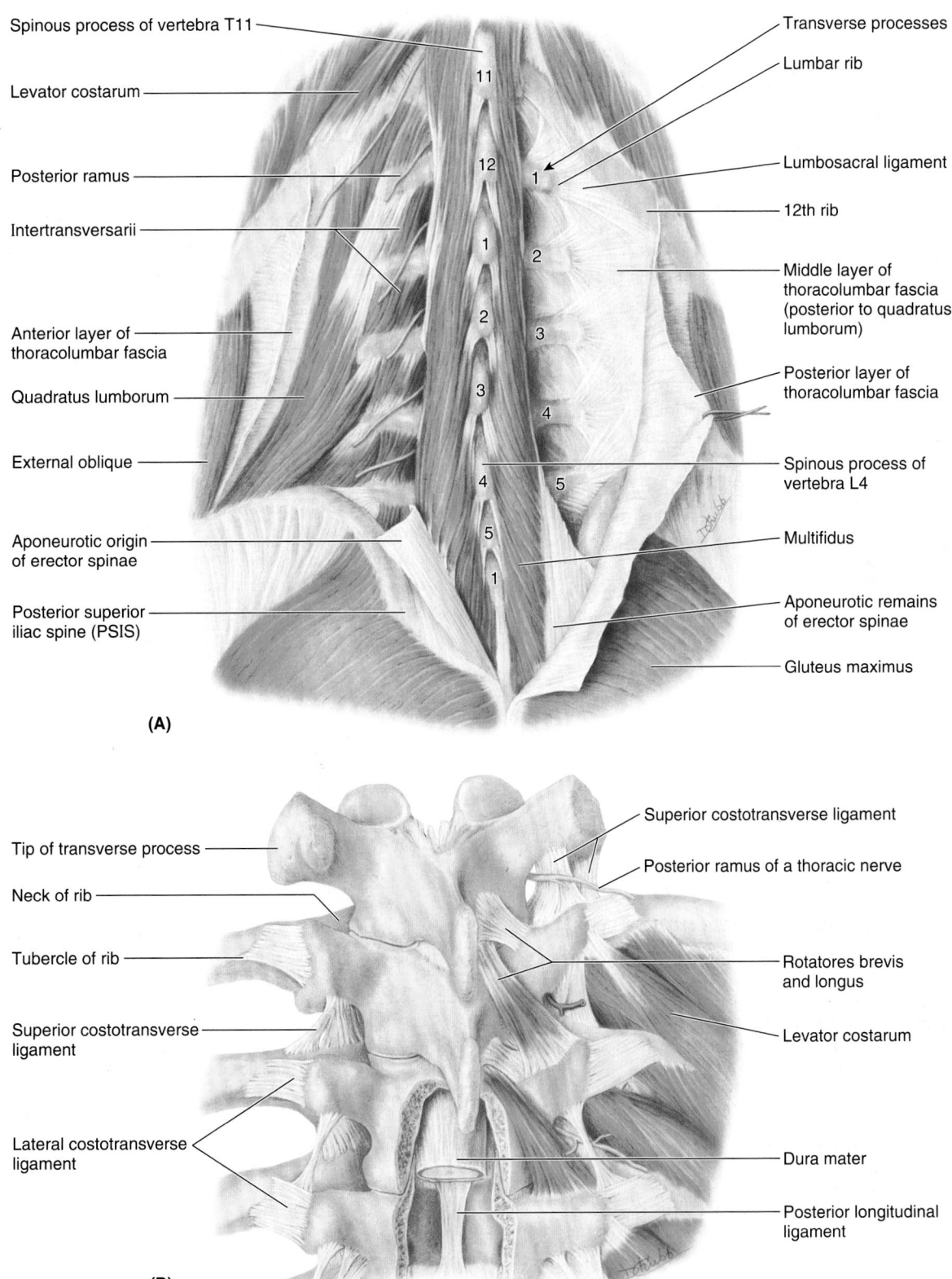

Spinous process of vertebra T11

Levator costarum

Posterior ramus

Intertransversarii

Anterior layer of thoracolumbar fascia

Quadratus lumborum

External oblique

Aponeurotic origin of erector spinae

Posterior superior iliac spine (PSIS)

(A)

Transverse processes

Lumbar rib

Lumbosacral ligament

12th rib

Middle layer of thoracolumbar fascia (posterior to quadratus lumborum)

Posterior layer of thoracolumbar fascia

Spinous process of vertebra L4

Multifidus

Aponeurotic remains of erector spinae

Gluteus maximus

Tip of transverse process

Neck of rib

Tubercle of rib

Superior costotransverse ligament

Lateral costotransverse ligament

(B)

Superior costotransverse ligament

Posterior ramus of a thoracic nerve

Rotatores brevis and longus

Levator costarum

Dura mater

Posterior longitudinal ligament

FIGURE 4.32. Deep layer of intrinsic back muscles (transversospinalis muscles).

The **semispinalis** is the superficial member of the group. As its name indicates, it arises from approximately half of the vertebral column. It is divided into three parts according to the superior attachments (Table 4.6): **semispinalis capitis, semispinalis thoracis,** and **semispinalis cervicis.** Semispinalis capitis forms the longitudinal bulge in the back of the neck near the median plane (Fig. 4.33A).

The **multifidus** is the middle layer of the group and consists of short, triangular muscular bundles that are thickest in the lumbar region (Fig. 4.33B).

The **rotatores,** or rotator muscles, are the deepest of the three layers of transversospinal muscles and are best developed in the thoracic region. The transversospinalis group of the deep layer of intrinsic back muscles are illustrated separately in Figure 4.33, and details concerning their attachments, innervation, and action are provided in Table 4.6.

The **interspinales, intertransversarii,** and **levatores costarum** are minor deep back muscles that are relatively sparse in the thoracic region. The interspinales and intertransversarii muscles connect spinous and transverse processes, respectively. The elevators of the ribs represent the posterior intertransversarii muscles of the neck. Details concerning the attachments, nerve supply, and actions of the minor muscles of the deep layer of intrinsic muscles are provided in Table 4.6.

PRINCIPAL MUSCLES PRODUCING MOVEMENTS OF INTERVERTEBRAL JOINTS

The principal muscles producing movements of the cervical, thoracic, and lumbar IV joints are illustrated in Figures 4.34 and 4.35, with details summarized in Tables 4.7 and 4.8. Many of the muscles acting on the cervical vertebrae are discussed in greater detail in Chapter 8(Neck). The back muscles are relatively inactive in the stand-easy position, but they (especially the shorter deep layer of intrinsic muscles) act as static postural muscles (fixators, or steadiers) of the vertebral column, maintaining tension and stability as required for the erect posture.

Note in Table 4.8 that all movements of the IV joints (i.e., all movements of the vertebral column) except pure extension involve or are solely produced by the *concentric contraction* of abdominal muscles. However, bear in mind that in these as in all movements, the *eccentric contraction* (controlled relaxation) of the antagonist muscles is vital to smooth, controlled movement (see "Muscle Tissue and the Muscular System" in the Introduction, p. 29). Thus it is actually the interaction of anterior (abdominal) and posterior (back) muscles (as well as the contralateral pairs of each) that provides the stability and produces motion of the axial skeleton, much like guy (guide) wires support a pole. Often chronic back strain (such as that caused

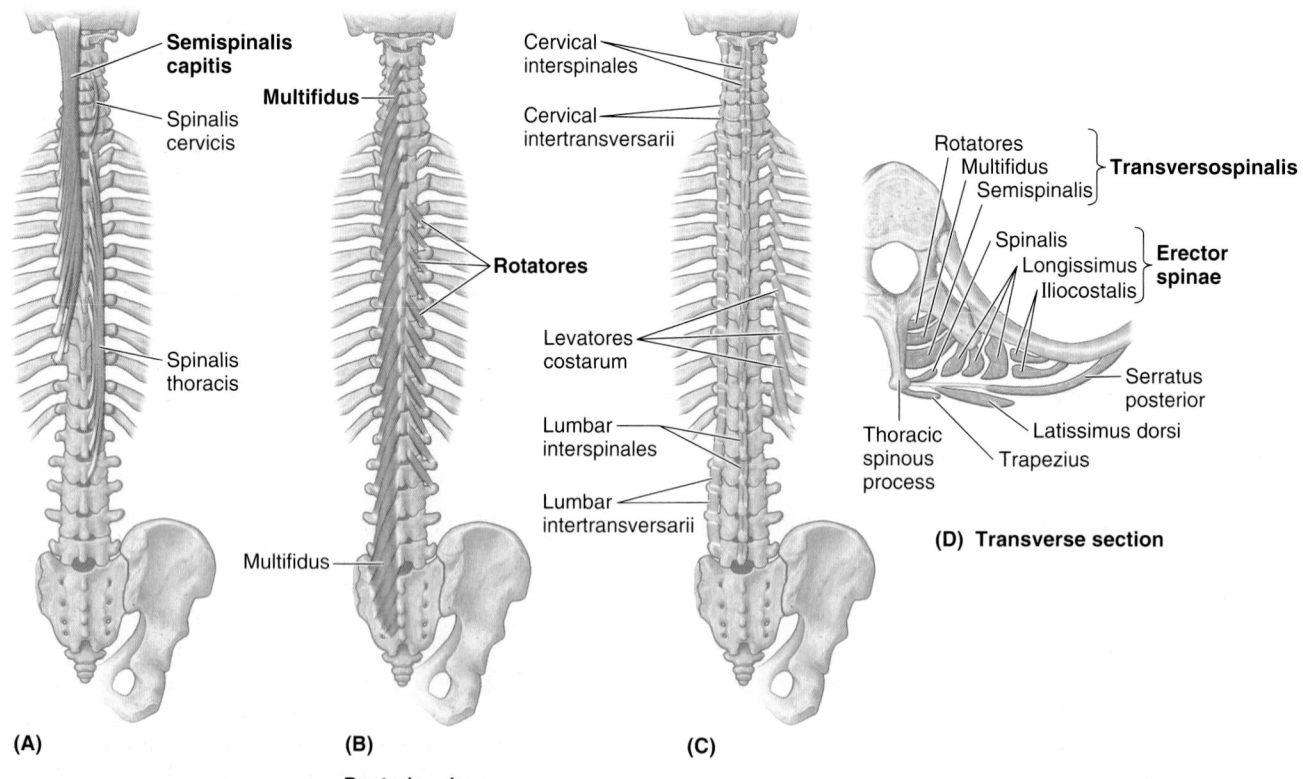

(A)   **(B)**   **(C)**

Posterior views

FIGURE 4.33. **Muscles of deep layer of intrinsic back muscles. A.** The transversospinalis muscle group is deep to the erector spinae (see **D**). The short lumbar rib is articulating with the transverse process of L1 vertebra. This common variation does not usually cause a problem; however, those unfamiliar with its possible presence may think it is a fractured transverse process. **B.** Deeper dissection showing the rotatores. The levatores costarum muscles represent the intertransversarii muscles in the thoracic region.

TABLE 4.6. DEEP LAYERS OF INTRINSIC BACK MUSCLES

Muscle	Proximal Attachment	Distal Attachment	Nerve Supply	Main Action(s)
Deep layer				

Transversospinalis
  Semispinalis
  Multifidus
  Rotatores (brevis and longus)
	Transverse processes
	
	Semispinalis: arises from transverse processes of C4–T12 vertebrae
	
	Multifidus: arises from posterior sacrum, posterior superior iliac spine of ilium, aponeurosis of erector spinae, sacro-iliac ligaments, mammillary processes of lumbar vertebrae, transverse processes of T1–T3, articular processes of C4–C7
	
	Rotatores: arise from transverse processes of vertebrae; best developed in thoracic region
		Spinous processes of more superior vertebrae
		
		Semispinalis: thoracis, cervicis, capitis; fibers run superomedially to occipital bone and spinous processes in thoracic and cervical regions, spanning 4–6 segments
		
		Multifidus: thickest in lumbar region; fibers pass obliquely superomedially to entire length of spinous processes, located 2–4 segments superior to proximal attachment
		
		Rotatores: fibers pass superomedially to attach to junction of lamina and transverse process or spinous process of vertebra immediately (brevis) or 2 segments (longus) superior to vertebra of attachment
			Posterior rami of spinal nerves[a]	Extension
				
				Semispinalis: extends head and thoracic and cervical regions of vertebral column and rotates them contralaterally
				
				Multifidus: stabilizes vertebrae during local movements of vertebral column
				
				Rotatores: stabilize vertebrae and assist with local extension and rotatory movements of vertebral column; may function as organs of proprioception

Minor deep layer				
Interspinales	Superior surfaces of spinous processes of cervical and lumbar vertebrae	Inferior surfaces of spinous processes of vertebra superior to vertebra of proximal attachment	Posterior rami of spinal nerves	Aid in extension and rotation of vertebral column
Intertransversarii	Transverse processes of cervical and lumbar vertebrae	Transverse processes of adjacent vertebrae	Posterior and anterior rami of spinal nerves[a]	Aid in lateral flexion of vertebral column; acting bilaterally, stabilize vertebral column
Levatores costarum	Tips of transverse processes of C7 and T1–T11 vertebrae	Pass inferolaterally and insert on rib between tubercle and angle	Posterior rami of C8–T11 spinal nerves	Elevate ribs, assisting respiration; assist with lateral flexion of vertebral column

[a]Most back muscles are innervated by posterior rami of spinal nerves, but a few are innervated by anterior rami. Intertransversarii of the cervical region are supplied by anterior rami.

by excessive lumbar lordosis; B4.17C) results from imbalance in this support (lack of tonus of abdominal muscles in the case of lordosis). Exercise or elimination of excessive, unevenly distributed weight may be required to restore balance.

Smaller muscles generally have higher densities of **muscle spindles** (sensors of proprioception that are interdigitated among the muscle's fibers) than do large muscles. It was assumed that the higher concentration of spindles occurred because small muscles produce the most precise movements, such as fine postural movements or manipulation and, therefore, require more proprioceptive feedback.

The movements described for small muscles are deduced from the location of their attachments and the direction of the muscle fibers and from activity measured by electromyography as movements are performed. Muscles such as the rotatores, however, are so small and are placed in positions of such relatively poor mechanical advantage that their ability to produce the movements described is somewhat questionable. Furthermore, such small muscles are often redundant to other larger muscles that have superior mechanical advantage. Hence, it has been proposed (Buxton and Peck, 1989) that the smaller muscles of small–large muscle pairs function more as "kinesiological monitors," or organs of proprioception and that the larger muscles are the producers of motion.

(text continues on p. 492)

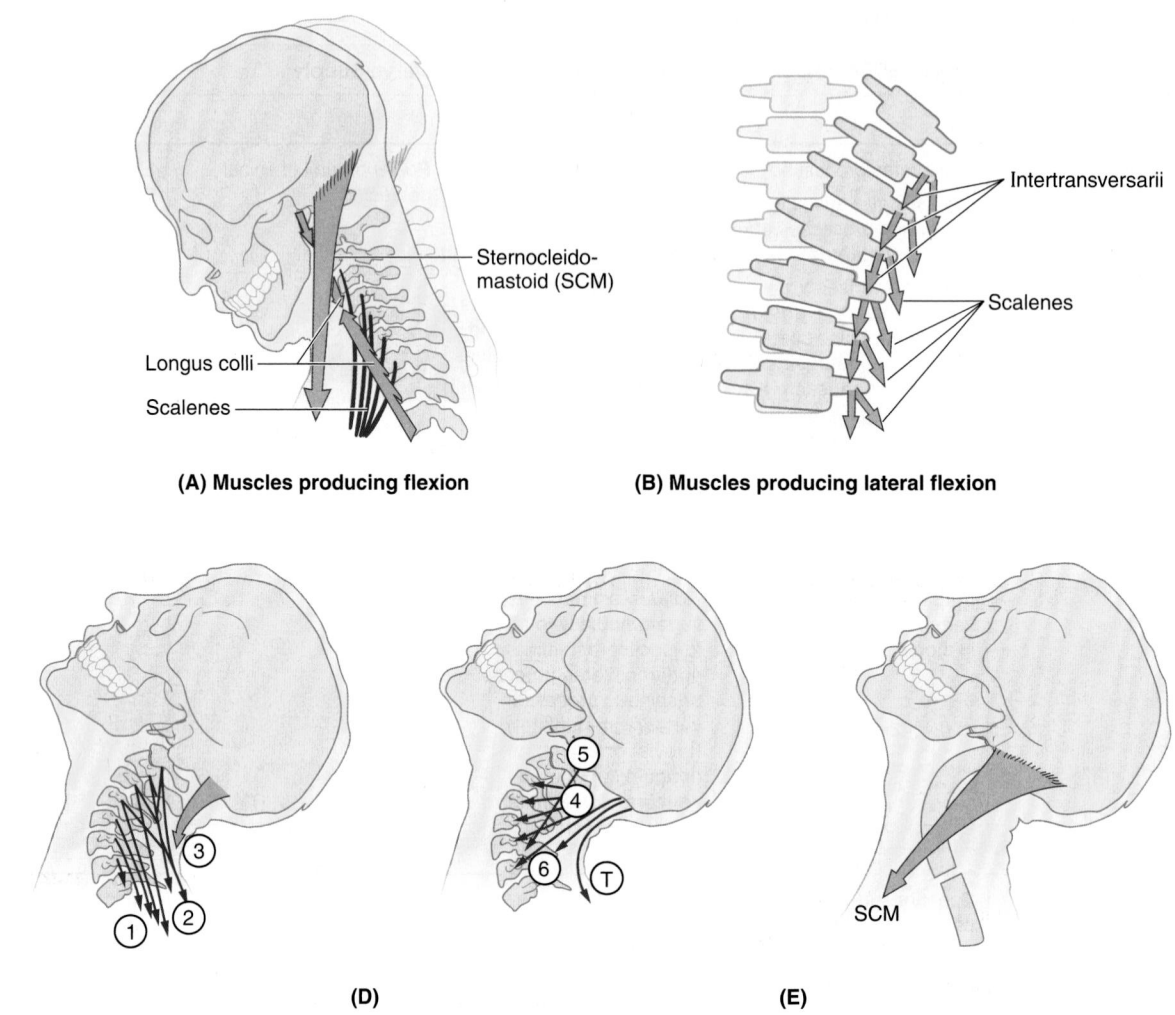

(A) Muscles producing flexion

(B) Muscles producing lateral flexion

(C)

(D)

(E)

B–D = Muscles Producing Extension

FIGURE 4.34. Principal muscles producing movements of cervical intervertebral joints.

TABLE 4.7. PRINCIPAL MUSCLES PRODUCING MOVEMENT OF CERVICAL INTERVERTEBRAL JOINTS

Flexion	Extension	Lateral Bending	Rotation (not shown)

Bilateral action of
  Longus coli
  Scalene
  Sternocleidomastoid
	Deep neck muscles
	  1, semispinalis cervicis and iliocostalis cervicis
	  2, splenius cervicis and levator scapulae
	  3, splenius capitis
	  4, multifidus
	  5, longissimus capitis
	  6, semispinalis capitis
	  T, trapezius
		Unilateral action of
		  Iliocostalis cervicis
		  Longissimus capitis and cervicis
		  Splenius capitis and cervicis
		  Intertransversarii and scalenes
			Unilateral action of
			  Rotatores
			  Semispinalis capitis and cervicis
			  Multifidus
			  Splenius cervicis

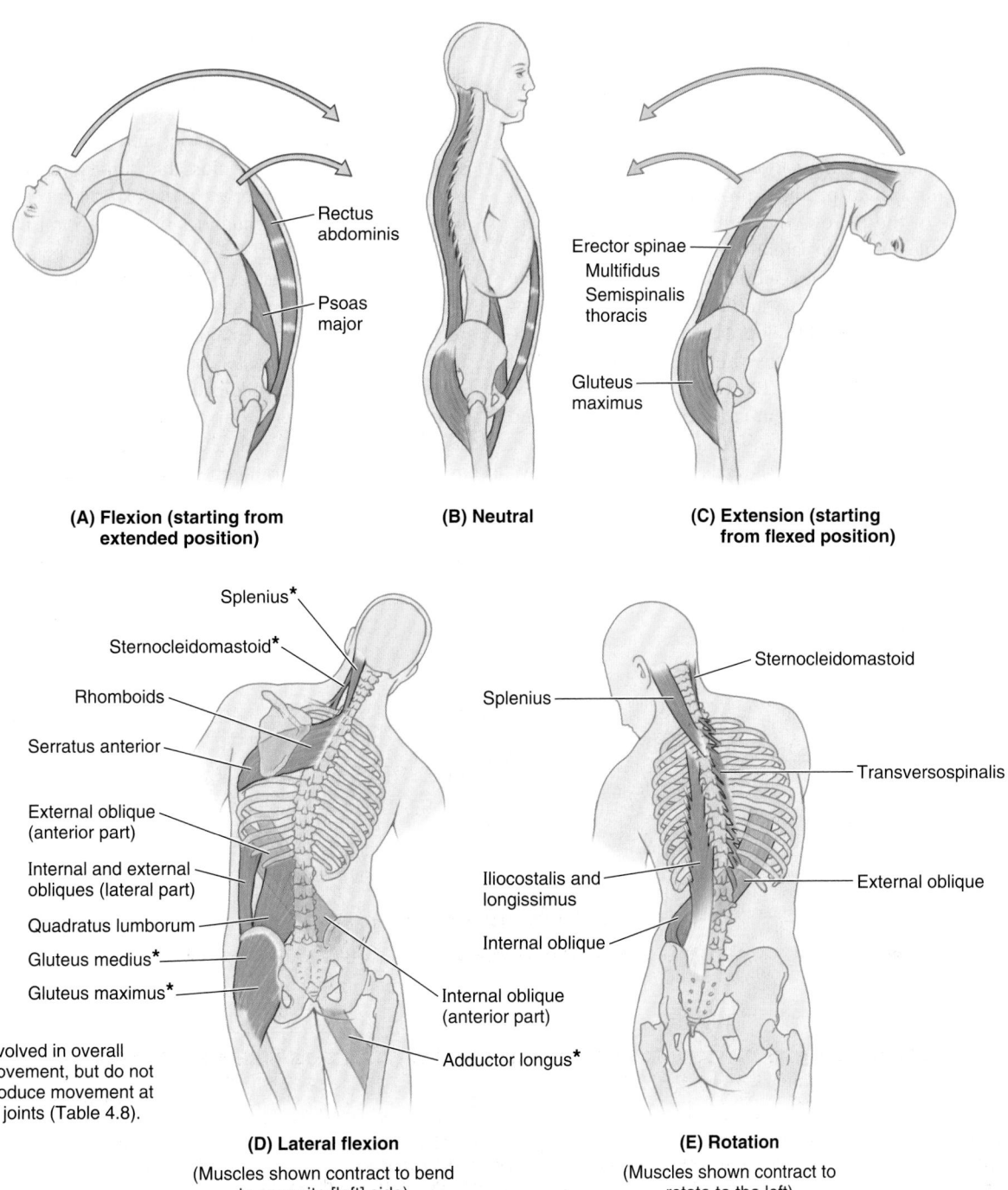

FIGURE 4.35. Principal muscles producing movements of thoracic and lumbar intervertebral joints.

TABLE 4.8. PRINCIPAL MUSCLES PRODUCING MOVEMENTS OF THORACIC AND LUMBAR INTERVERTEBRAL (IV) JOINTS

Flexion	Extension	Lateral Bending	Rotation
Bilateral action of Rectus abdominis Psoas major Gravity	Bilateral action of Erector spinae Multifidus Semispinalis thoracis	Unilateral action of Iliocostalis thoracis and lumborum Longissimus thoracis Multifidus External and internal oblique Quadratus lumborum Rhomboids Serratus anterior	Unilateral action of Rotatores Multifidus Iliocostalis Longissimus External oblique acting synchronously with opposite internal oblique Splenius thoracis

Surface Anatomy of Back Muscles

The *posterior median furrow* overlies the tips of the spinous processes of the vertebrae (Fig. 4.36). The furrow is continuous superiorly with the *nuchal groove* in the neck and is deepest in the lower thoracic and upper lumbar regions.

The *erector spinae* produce prominent vertical bulges on each side of the furrow. In the lumbar region, they are readily palpable, and their lateral borders coincide with the angles of the ribs and are indicated by shallow grooves in the skin. When the individual is standing, the lumbar spinous processes may be indicated by depressions in the skin. These processes usually become visible when the vertebral column is flexed (see Figs. 4.8A and 4.11A & C). The median furrow ends in the flattened triangular area covering the sacrum and is replaced inferiorly by the *intergluteal cleft*.

When the upper limbs are elevated, the scapulae move laterally on the thoracic wall, making the *rhomboid* and *teres major muscles* visible. The superficially located *trapezius* and *latissimus dorsi muscles* connecting the upper limbs to the vertebral column are also clearly visible (Fig. 4.36).

Suboccipital and Deep Neck Muscles

Often misrepresented as a surface region, the **suboccipital region** is a muscle "compartment" deep to the superior part of the posterior cervical region, underlying the trapezius, sternocleidomastoid, splenius, and semispinalis muscles. It is a pyramidal space inferior to the external occipital prominence of the head that includes the posterior aspects of vertebrae C1 and C2 (Fig. 4.37 orientation figure).

The four small muscles of the suboccipital region lie deep (anterior) to the semispinalis capitis muscles and consist of two rectus capitis posterior (major and minor) and two obliquus muscles. All four muscles are innervated by the *posterior ramus of C1*, the **suboccipital nerve.** The nerve emerges as the vertebral artery courses deeply between the occipital bone and the atlas (vertebra C1) within the **suboccipital triangle.** Details concerning the boundaries and contents of this triangle and the attachments of the *suboccipital muscles* are illustrated in Figure 4.37 and described in Table 4.9.

Note that the **obliquus capitis inferior** is the only "capitis" muscle that has no attachment to the cranium (skull).

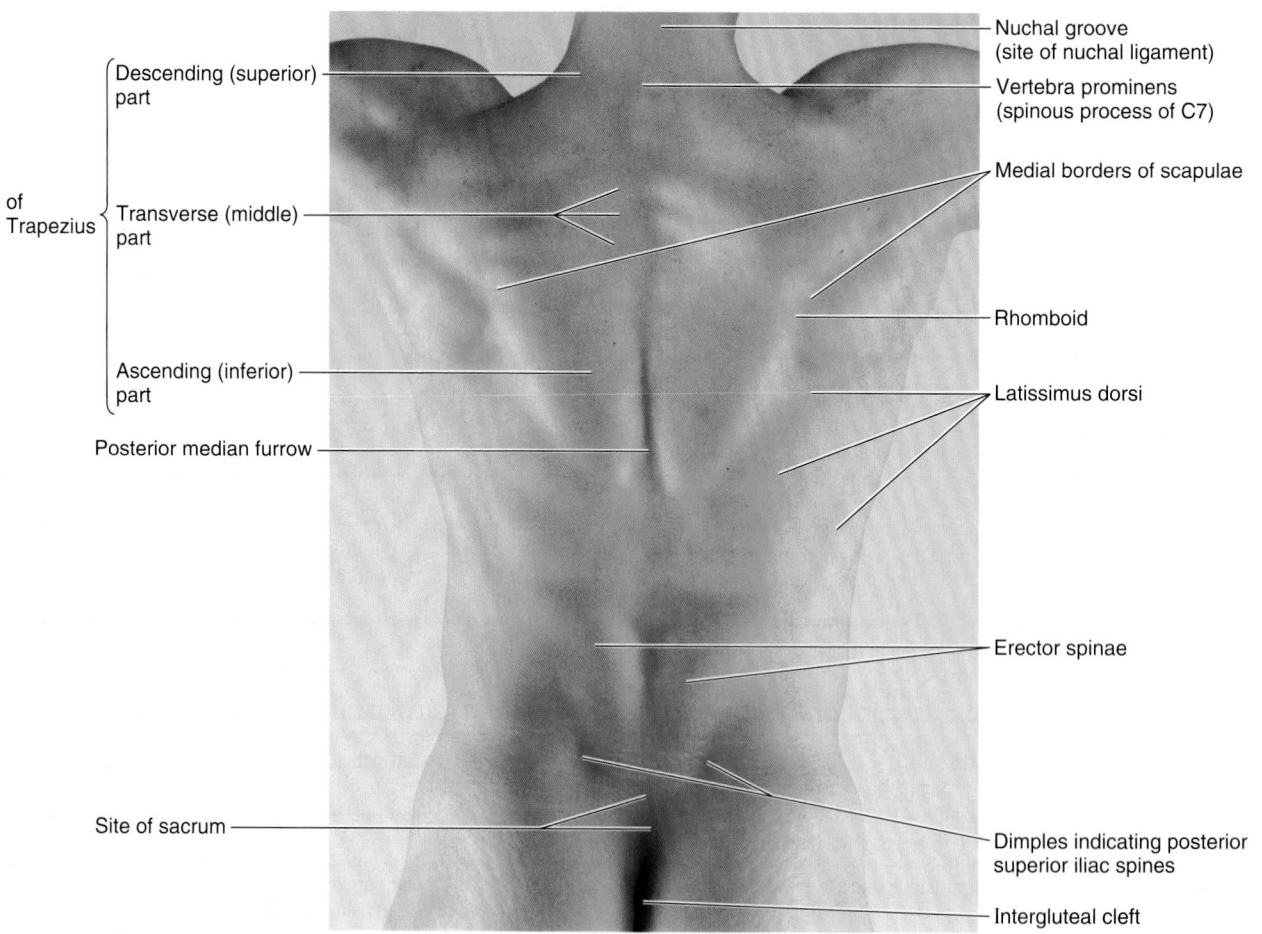

Posterior view

FIGURE 4.36. Surface anatomy of muscles of back.

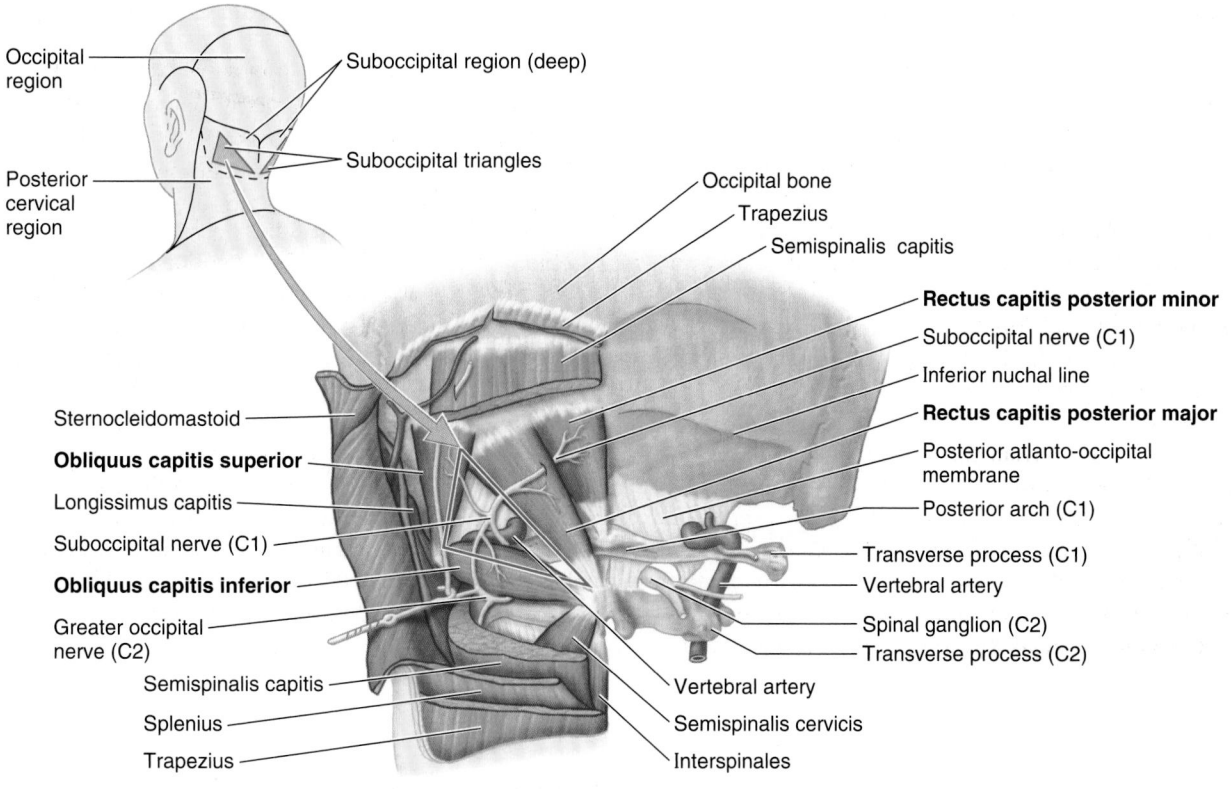

FIGURE 4.37. Suboccipital muscles and suboccipital triangle.

TABLE 4.9. SUBOCCIPITAL MUSCLES AND SUBOCCIPITAL TRIANGLE

Suboccipital Muscles		
Muscle	**Origin**	**Insertion**
Rectus capitis posterior major	Spinous process of vertebra C2	Lateral part of inferior nuchal line of occipital bone
Rectus capitis posterior minor	Posterior tubercle of posterior arch of vertebra C1 (atlas)	Medial part of inferior nuchal line of occipital bone
Obliquus capitis inferior	Posterior tubercle of posterior arch of vertebra C2 (axis)	Transverse process of vertebra C1 (atlas)
Obliquus capitis superior	Transverse process of vertebra C1	Occipital bone between superior and inferior nuchal lines
Suboccipital Triangle		
Aspect of Triangle	**Structures**	
Superomedial boundary	Rectus capitis posterior major	
Superolateral boundary	Obliquus capitis superior	
Inferolateral boundary	Obliquus capitis inferior	
Floor	Posterior atlanto-occipital membrane and posterior arch of vertebra C1 (atlas)	
Roof	Semispinalis capitis	
Contents	Vertebral artery and suboccipital nerve	

TABLE 4.10.   PRINCIPAL MUSCLES PRODUCING MOVEMENT OF ATLANTO-OCCIPITAL JOINTS

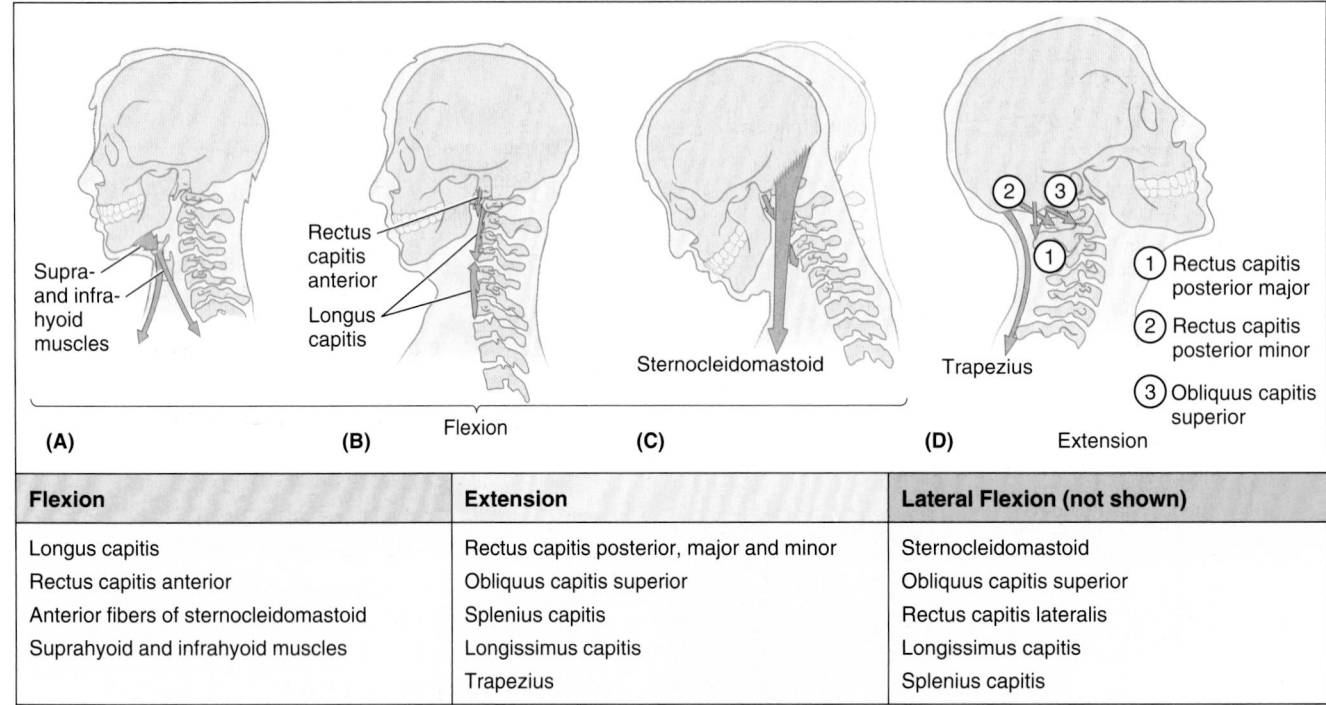

Flexion	Extension	Lateral Flexion (not shown)
Longus capitis	Rectus capitis posterior, major and minor	Sternocleidomastoid
Rectus capitis anterior	Obliquus capitis superior	Obliquus capitis superior
Anterior fibers of sternocleidomastoid	Splenius capitis	Rectus capitis lateralis
Suprahyoid and infrahyoid muscles	Longissimus capitis	Longissimus capitis
	Trapezius	Splenius capitis

These muscles are mainly postural muscles, but actions are typically described for each muscle in terms of producing movement of the head.

The suboccipital muscles act on the head directly or indirectly (explaining the inclusion of *capitis* in their names) by extending it on vertebra C1 and rotating it on vertebrae C1 and C2. However, recall the discussion of the small member of the small–large muscle pair functioning as a kinesiological monitor for the sense of proprioception (p. 489).

The principal muscles producing movements of the craniovertebral joints are summarized in Tables 4.10 and 4.11, and the nerves of the posterior cervical region, including the suboccipital region/triangles, are illustrated in Figure 4.38 and summarized in Table 4.12.

TABLE 4.11.   PRINCIPAL MUSCLES PRODUCING MOVE-MENT OF ATLANTO-AXIAL JOINTS[a]

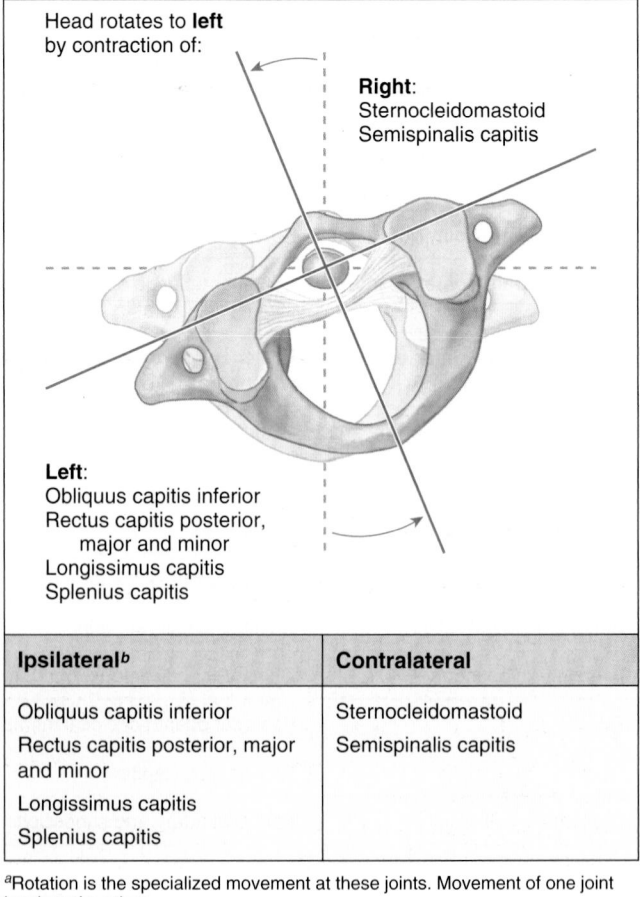

Ipsilateral[b]	Contralateral
Obliquus capitis inferior	Sternocleidomastoid
Rectus capitis posterior, major and minor	Semispinalis capitis
Longissimus capitis	
Splenius capitis	

[a]Rotation is the specialized movement at these joints. Movement of one joint involves the other.

[b]Same side to which head is rotated.

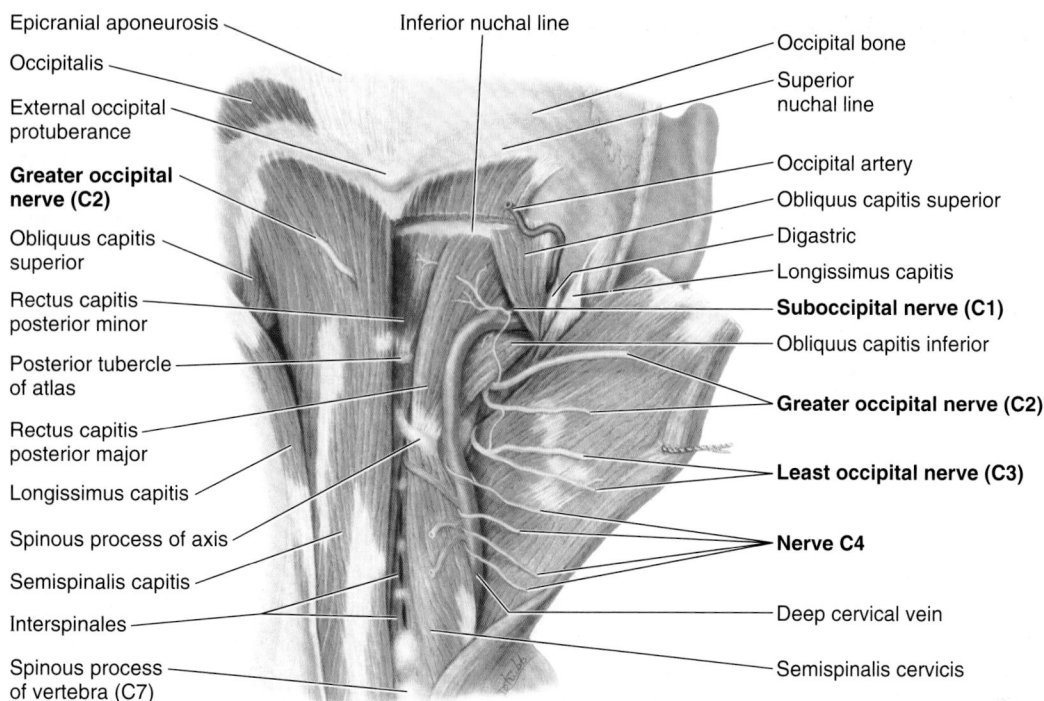

Epicranial aponeurosis
Occipitalis
External occipital protuberance
Greater occipital nerve (C2)
Obliquus capitis superior
Rectus capitis posterior minor
Posterior tubercle of atlas
Rectus capitis posterior major
Longissimus capitis
Spinous process of axis
Semispinalis capitis
Interspinales
Spinous process of vertebra (C7)

Inferior nuchal line

Occipital bone
Superior nuchal line
Occipital artery
Obliquus capitis superior
Digastric
Longissimus capitis
Suboccipital nerve (C1)
Obliquus capitis inferior
Greater occipital nerve (C2)
Least occipital nerve (C3)
Nerve C4
Deep cervical vein
Semispinalis cervicis

FIGURE 4.38. Nerves of posterior cervical region, including suboccipital region/triangles.

TABLE 4.12. NERVES OF POSTERIOR CERVICAL REGION, INCLUDING SUBOCCIPITAL REGION/TRIANGLES

Nerve	Origin	Course	Distribution
Suboccipital	Posterior ramus of spinal nerve C1	Runs between cranium and C1 vertebra to reach suboccipital triangle	Muscles of suboccipital triangle
Greater occipital	Posterior ramus of spinal nerve C2	Emerges inferior to obliquus capitis inferior and ascends to posterior scalp	Skin over neck and occipital bone
Lesser occipital	Anterior rami of spinal nerves C2–C3	Passes directly to skin	Skin of superior posterolateral neck and scalp posterior to external ear
Posterior rami, nerves C3–C7	Posterior rami of spinal nerves C3–C7	Pass segmentally to muscles and skin	Intrinsic muscles of back and overlying skin (adjacent to vertebral column)

MUSCLES OF BACK

Back Strains, Sprains, and Spasms

Adequate warmup and stretching, and exercises to increase the tonus of the "core muscles" (muscles of the anterolateral abdominal wall—especially the transversus abdominis—determined to play a role in lumbar stabilization) prevent many back strains and sprains, common causes of lower back pain.

Back sprain is an injury in which only ligamentous tissue, or the attachment of ligament to bone, is involved, without dislocation or fracture. It results from excessively strong contractions related to movements of the vertebral column, such as excessive extension or rotation.

Back strain is a common injury in people who participate in sports; it results from overly strong muscular contraction. The strain involves some degree of stretching or microscopic tearing of muscle fibers. The muscles usually involved are those producing movements of the lumbar IV joints, especially the erector spinae. If the weight is not properly balanced on the vertebral column, strain is exerted on the muscles.

Using the back as a lever when lifting puts an enormous strain on the vertebral column and its ligaments and muscles. Strains can be minimized if the lifter crouches, holds the back as straight as possible, and uses the muscles of the buttocks and lower limbs to assist with the lifting.

As a protective mechanism, the back muscles go into *spasm* after an injury or in response to inflammation (e.g., of ligaments). A spasm is a sudden involuntary contraction of one or more muscle groups. Spasms are attended by cramps, pain, and interference with function, producing involuntary movement and distortion.

Reduced Blood Supply to the Brainstem

 The winding course of the vertebral arteries through the foramina transversarii of the transverse processes of the cervical vertebrae and through the suboccipital triangles becomes clinically significant when blood flow through these arteries is reduced, as occurs with *arteriosclerosis* (hardening of arteries). Under these conditions, prolonged turning of the head, as occurs when backing up a motor vehicle, may cause light-headedness, dizziness, and other symptoms from the interference with the blood supply to the brainstem.

The Bottom Line

MUSCLES OF BACK

The superficial extrinsic back muscles are axio-appendicular muscles that serve the upper limb. ♦ Except for the trapezius—innervated by CN XI—the extrinsic back muscles are innervated by the anterior rami of spinal nerves. ♦ The deep intrinsic back muscles connect elements of the axial skeleton, are mostly innervated by posterior rami of spinal nerves, and are arranged in three layers: superficial (splenius muscles), intermediate (erector spinae), and deep (transversospinalis muscles). ♦ The intrinsic muscles provide primarily extension and proprioception for posture, and work synergistically with the muscles of the anterolateral abdominal wall to stabilize and produce movements of the trunk. ♦ Suboccipital muscles extend between vertebrae C1 (atlas) and C2 (axis) and the occipital bone and produce—and/or provide proprioceptive information concerning—movements at the craniovertebral joints.

CONTENTS OF VERTEBRAL CANAL

The spinal cord, spinal nerve roots, spinal meninges, and the neurovascular structures that supply them are located in the vertebral canal (Fig. 4.27).

Spinal Cord

The **spinal cord** is the major reflex center and conduction pathway between the body and brain. This cylindrical structure, slightly flattened anteriorly and posteriorly, is protected by the vertebrae, their associated ligaments and muscles, the spinal meninges, and the cerebrospinal fluid (CSF).

The spinal cord begins as a continuation of the **medulla oblongata** (often called the *medulla*), the caudal part of the brainstem (see Fig. 7.36, p. 879). In adults, the spinal cord is 42–45 cm long and extends from the foramen magnum in the occipital bone to the level of the L1 or L2 vertebra (Fig. 4.39). However, its tapering inferior end, the **conus medullaris,** may terminate as high as T12 vertebra or as low as L3 vertebra. Thus the spinal cord occupies only the superior two thirds of the vertebral canal.

The spinal cord is enlarged in two regions in relationship to innervation of the limbs. The **cervical enlargement** extends from C4 through T1 segments of the spinal cord, and most of the anterior rami of the spinal nerves arising from it form the *brachial plexus of nerves* that innervates the upper limbs. The **lumbosacral enlargement** extends from T11 through S1 segments of the spinal cord, inferior to which the cord continues to diminish as the conus medullaris. The anterior rami of the spinal nerves arising from this enlargement make up the *lumbar* and *sacral plexuses of nerves* that innervate the lower limbs.

Spinal Nerves and Nerve Roots

The formation and composition of *spinal nerves* and *nerve roots* are discussed in the Introduction (p. 50). Readers are urged to read this information now if they have not done so previously. The portion of the spinal cord giving rise to the rootlets and roots that ultimately form one bilateral pair of spinal nerves is designated a **spinal cord segment,** the identity of which is the same as the spinal nerves arising from it.

Cervical spinal nerves (except C8) bear the same alphanumeric designation as the vertebrae forming the inferior margin of the IV foramina through which the nerve exits the vertebral canal. The more inferior spinal (T1 through Co1) nerves bear the same alphanumeric designation as the vertebrae forming the superior margin of their exit (Table 4.13). first cervical nerves lack posterior roots in 50% of people, and the coccygeal nerve may be absent.

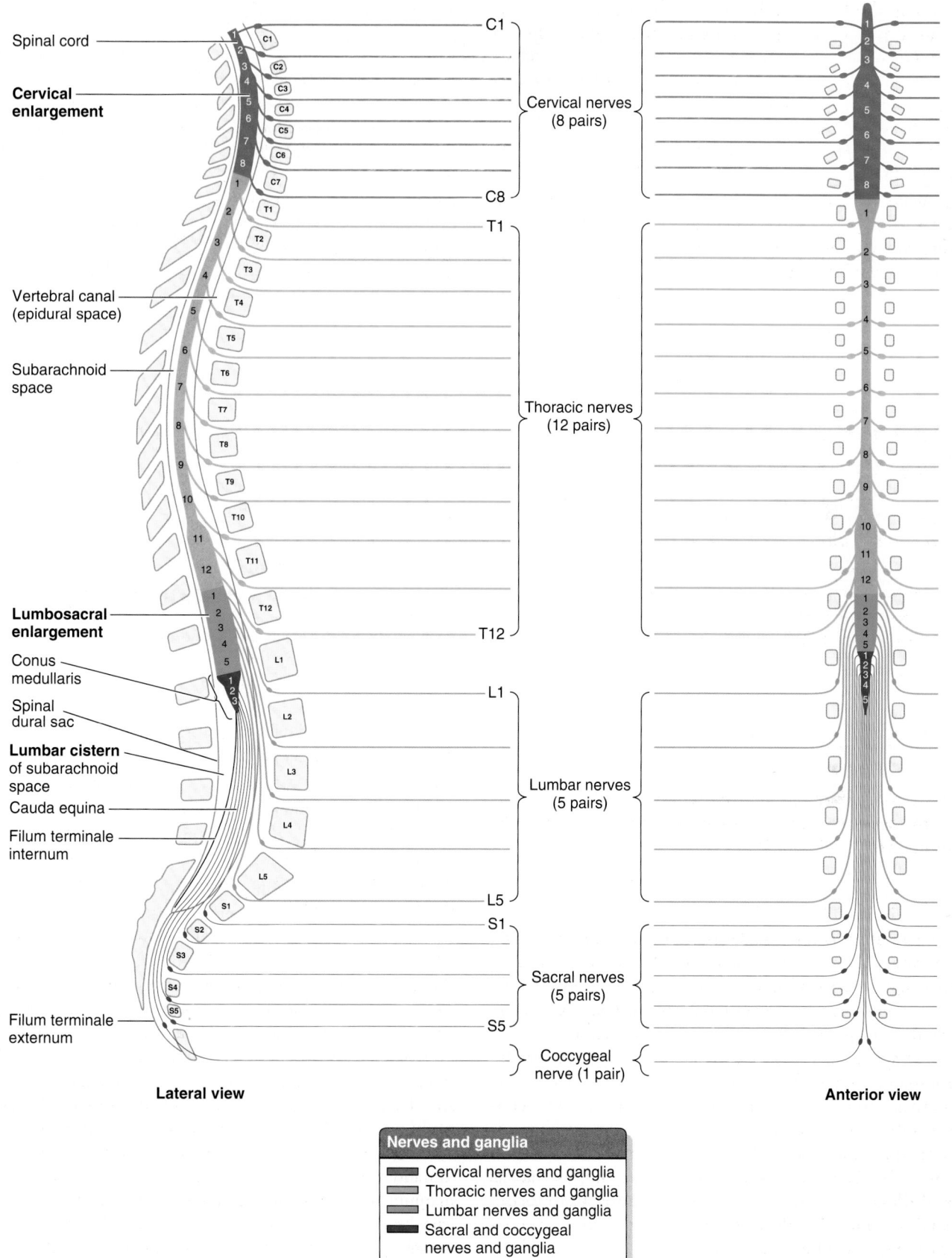

Spinal cord

Cervical enlargement

Vertebral canal (epidural space)

Subarachnoid space

Lumbosacral enlargement

Conus medullaris

Spinal dural sac

Lumbar cistern of subarachnoid space

Cauda equina

Filum terminale internum

Filum terminale externum

C1
C8
T1
T12
L1
L5
S1
S5

Cervical nerves (8 pairs)

Thoracic nerves (12 pairs)

Lumbar nerves (5 pairs)

Sacral nerves (5 pairs)

Coccygeal nerve (1 pair)

Lateral view

Anterior view

Nerves and ganglia

- Cervical nerves and ganglia
- Thoracic nerves and ganglia
- Lumbar nerves and ganglia
- Sacral and coccygeal nerves and ganglia

FIGURE 4.39. Vertebral column, spinal cord, spinal ganglia, and spinal nerves. Lateral and anterior views illustrating the relation of the spinal cord segments (the numbered segments) and spinal nerves to the adult vertebral column.

TABLE 4.13. NUMBERING OF SPINAL NERVES AND VERTEBRAE

Segmental Level	Number of Nerves	Level of Exit from Vertebral Column
Cervical	8 (C1–C8)	Nerve C1[a] (suboccipital nerve) passes superior to arch of vertebra C1 Nerves C2–C7 pass through IV foramina superior to the corresponding vertebrae Nerve C8 passes through the IV foramen between vertebra C7 and T1
Thoracic	12 (T1–T12)	Nerves T1–L5 pass through IV foramina inferior to the corresponding vertebrae
Lumbar	5 (L1–L5)	
Sacral	5 (S1–S5)	Nerves S1–S4 branch into anterior and posterior rami within the sacrum, with the respective rami passing through the anterior and posterior sacral foramina
Coccygeal[a]	1 (Co1)	5th sacral and coccygeal nerves pass through the sacral hiatus

[a]The first cervical nerves lack posterior roots in 50% of people, and the coccygeal nerves may be absent.

(Modified from *Barr's The Human Nervous System*.)

In embryos, the spinal cord occupies the whole length of the vertebral canal (Fig. 4.23); thus cord segments lie approximately at the vertebral level of the same number, the spinal nerves passing laterally to exit the corresponding IV foramen. By the end of the embryonic period (8th week), the tail-like caudal eminence has disappeared, and the number of coccygeal vertebrae is reduced from six to four segments. The spinal cord in the vertebral canal of the coccyx atrophies.

During the fetal period, the vertebral column grows faster than the spinal cord; as a result, the cord "ascends" relative to the vertebral canal. At birth, the tip of the conus medullaris is at the L4–L5 level. Thus, in postnatal life, the spinal cord is shorter than the vertebral column; consequently, there is a progressive obliquity of the spinal nerve roots (Figs. 4.39 and 4.40). Because the distance between the origin of a nerve's roots from the spinal cord and the nerve's exit from the vertebral canal increases as the inferior end of the vertebral column is approached, the length of the nerve roots also increases progressively.

The lumbar and sacral nerve roots are therefore the longest, extending far beyond the termination of the adult spinal cord at approximately the L2 level to reach the remaining lumbar, sacral, and coccygeal IV foramina. This loose bundle of spinal nerve roots, arising from the lumbosacral enlargement and the conus medullaris and coursing within the *lumbar cistern* of CSF caudal to the termination of the spinal cord, resembles a horse's tail, hence its name—the **cauda equina** (L. horse tail).

Arising from the tip of the conus medullaris, the filum terminale descends among the spinal nerve roots in the cauda equina. The **filum terminale** is the vestigial remnant of the caudal part of the spinal cord that was in the tail-like caudal eminence of the embryo. Its proximal end (the **filum terminale internum,** or *pial part of the terminal filum*) consists of vestiges of neural tissue, connective tissue, and neuroglial tissue covered by pia mater. The filum terminale perforates the inferior end of the dural sac, gaining a layer of dura and continuing through the sacral hiatus as the **filum terminale**

externum (or *dural part of the terminal filum*, also known as the *coccygeal ligament*) to attach to the dorsum of the coccyx. The filum terminale is an anchor for the inferior end of the spinal cord and spinal meninges (Fig. 4.44).

Spinal Meninges and Cerebrospinal Fluid (CSF)

Collectively, the spinal dura mater, arachnoid mater, and pia mater surrounding the spinal cord constitute the **spinal meninges** (Figs. 4.41 and 4.42; Table 4.14). These membranes surround, support, and protect the spinal cord and spinal nerve roots, including those of the cauda equina, and contain the CSF in which these structures are suspended.

SPINAL DURA MATER

The **spinal dura mater** (or simply, spinal dura), composed mainly of tough fibrous tissue with some elastic fibers, is the outermost covering membrane of the spinal cord. The spinal dura is separated from the periosteum-covered bone and the ligaments that form the walls of the vertebral canal by the **epidural space.** This space is occupied by the internal vertebral venous plexus embedded in a fatty matrix (**epidural fat**). The epidural space runs the length of the vertebral canal, terminating superiorly at the foramen magnum and laterally at the IV foramina, as the spinal dura adheres to the periosteum surrounding each opening, and inferiorly, as the sacral hiatus is sealed by the sacrococcygeal ligament.

The spinal dura forms the **spinal dural sac,** a long tubular sheath within the vertebral canal (Figs. 4.39 and 4.40). This sac adheres to the margin of the foramen magnum of the cranium, where it is continuous with the cranial dura mater. The sac is anchored inferiorly to the coccyx by the *filum terminale*. The spinal dural sac is evaginated by each pair of posterior and anterior roots as they extend laterally toward their exit from the vertebral canal (Fig. 4.43). Thus tapering lateral extensions of the spinal dura surround each pair of

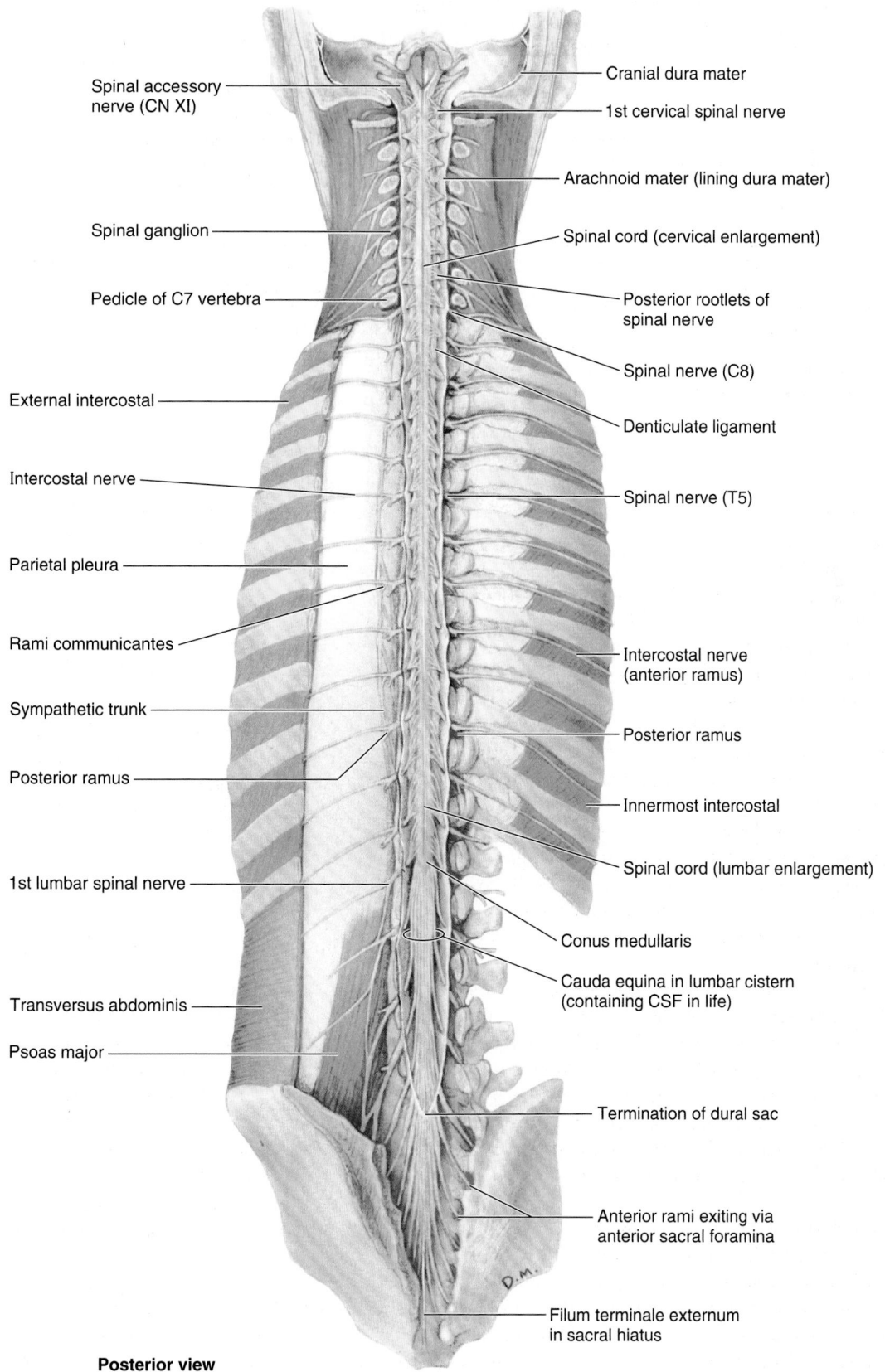

Spinal accessory nerve (CN XI)

Spinal ganglion

Pedicle of C7 vertebra

External intercostal

Intercostal nerve

Parietal pleura

Rami communicantes

Sympathetic trunk

Posterior ramus

1st lumbar spinal nerve

Transversus abdominis

Psoas major

Cranial dura mater

1st cervical spinal nerve

Arachnoid mater (lining dura mater)

Spinal cord (cervical enlargement)

Posterior rootlets of spinal nerve

Spinal nerve (C8)

Denticulate ligament

Spinal nerve (T5)

Intercostal nerve (anterior ramus)

Posterior ramus

Innermost intercostal

Spinal cord (lumbar enlargement)

Conus medullaris

Cauda equina in lumbar cistern (containing CSF in life)

Termination of dural sac

Anterior rami exiting via anterior sacral foramina

Filum terminale externum in sacral hiatus

Posterior view

FIGURE 4.40. Spinal cord in situ. The vertebral arches and the posterior aspect of the sacrum have been removed to expose the spinal cord in the vertebral canal. The spinal dural sac has also been opened to reveal the spinal cord and posterior nerve roots, the termination of the spinal cord between the L1 and the L2 vertebral level, and the termination of the spinal dural sac at the S2 segment.

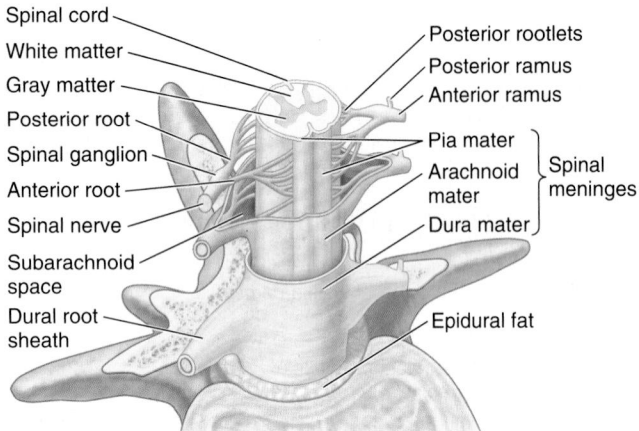

FIGURE 4.41. Spinal cord, spinal nerves, and spinal meninges. Three membranes (the spinal meninges) cover the spinal cord: dura mater, arachnoid mater, and pia mater. As the spinal nerve roots extend toward an IV foramen, they are surrounded by a dural root sheath (sleeve) that is continuous distally with the epineurium of the spinal nerve.

posterior and anterior nerve roots as **dural root sheaths,** or sleeves (Figs. 4.41 and 4.44). Distal to the spinal ganglia, these sheaths blend with the **epineurium** (outer connective tissue covering of spinal nerves) that adheres to the periosteum lining the IV foramina.

Innervation of Dura Mater. Nerve fibers are distributed to the spinal dura by the *(recurrent) meningeal nerves* (see Fig. 4.27). The function of these afferent and sympathetic fibers is unclear, although it is known that the afferent fibers supply pain receptors that are involved in the referred pain characteristic of spinal disorders and become irritated when there is inflammation of the meninges *(meningitis).*

SPINAL ARACHNOID MATER

The **spinal arachnoid mater** is a delicate, avascular membrane composed of fibrous and elastic tissue that lines the spinal dural sac and its dural root sheaths. It encloses the CSF-filled subarachnoid space containing the spinal cord,

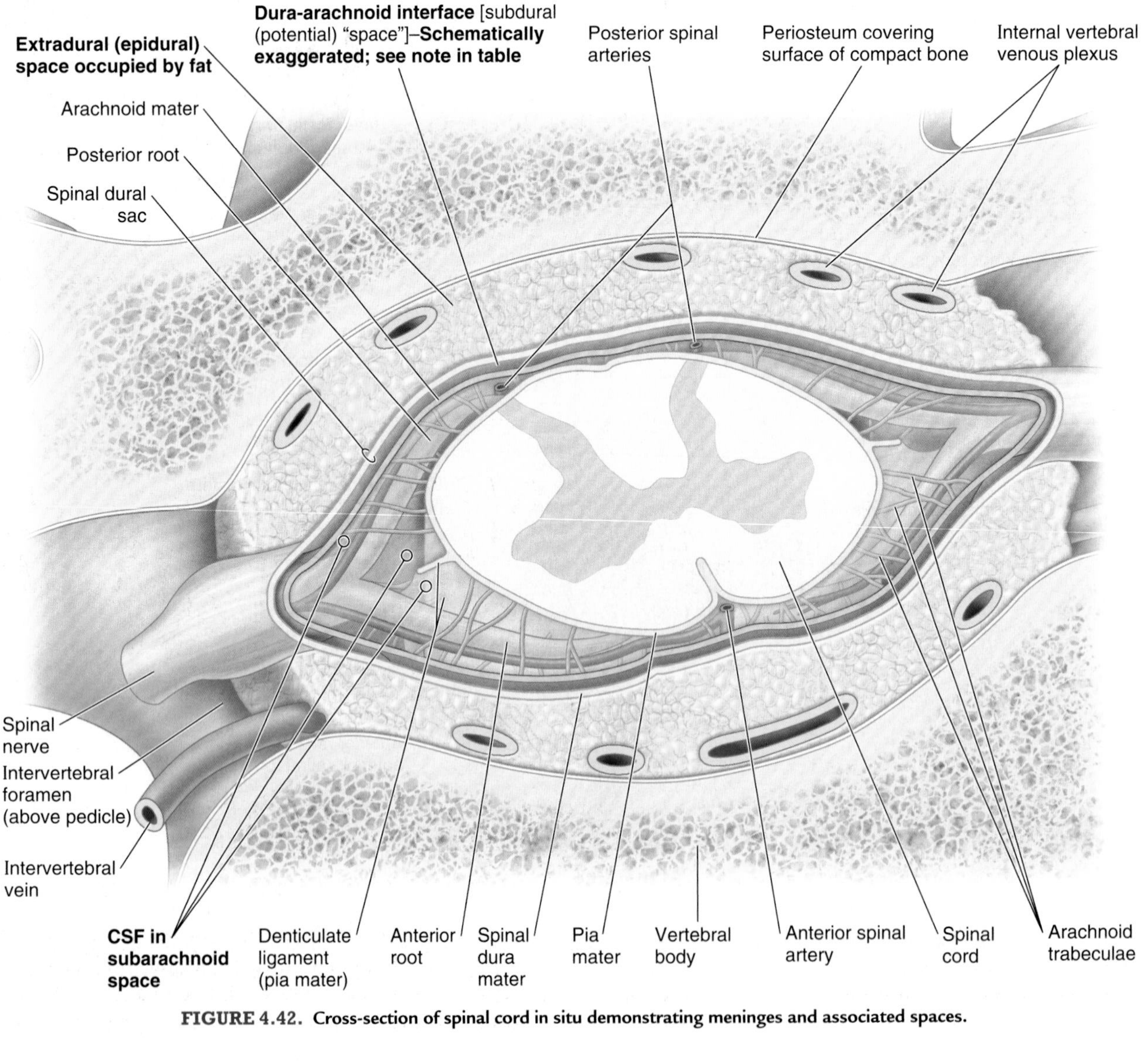

FIGURE 4.42. Cross-section of spinal cord in situ demonstrating meninges and associated spaces.

TABLE 4.14. SPACES ASSOCIATED WITH SPINAL MENINGES[a]

Space	Location	Contents
Epidural	Space between periosteum lining bony wall of vertebral canal and spinal dura mater	Fat (loose connective tissue); internal vertebral venous plexuses; inferior to L2 vertebra, ensheathed roots of spinal nerves
Subarachnoid (leptomeningeal)	Naturally occurring space between arachnoid mater and pia mater	CSF; radicular, segmental, medullary, and spinal arteries; veins; arachnoid trabeculae

[a]Although it is common to refer to a "subdural space," there is no naturally occurring space at the arachnoid–dura junction (Haines, 2006).

spinal nerve roots, and spinal ganglia (Figs. 4.41, 4.42, and 4.43).

The spinal arachnoid is not attached to the spinal dura but is held against its inner surface by the pressure of the CSF. In a lumbar spinal puncture, the needle traverses the spinal dura and arachnoid simultaneously. Their apposition is the **dura–arachnoid interface** (Fig. 4.42), often erroneously referred to as the "subdural space." No actual space occurs naturally at this site; it is, rather, a weak cell layer (Haines, 2006). Bleeding into this layer creates a pathological space at the dura–arachnoid junction in which a *subdural hematoma* is formed. In the cadaver, because of the absence of CSF, the spinal arachnoid falls away from the inner surface of the dura and lies loosely on the spinal cord.

The spinal arachnoid is separated from the pia mater on the surface of the spinal cord by the *subarachnoid space* containing CSF. Delicate strands of connective tissue, the **arachnoid trabeculae,** span the subarachnoid space connecting the spinal arachnoid and pia.

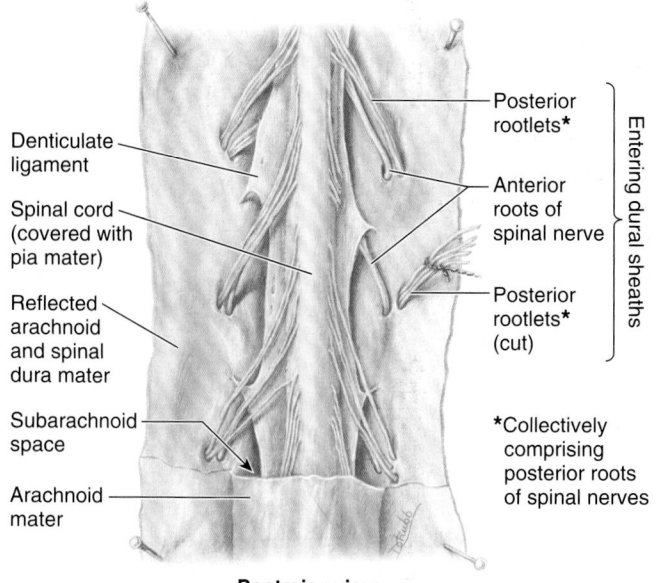

Denticulate ligament

Spinal cord (covered with pia mater)

Reflected arachnoid and spinal dura mater

Subarachnoid space

Arachnoid mater

Posterior rootlets*

Anterior roots of spinal nerve

Posterior rootlets* (cut)

Entering dural sheaths

*Collectively comprising posterior roots of spinal nerves

Posterior view

FIGURE 4.43. Spinal cord within its meninges. The spinal dura and arachnoid mater have been split and pinned flat to expose the spinal cord and denticulate ligaments between posterior and anterior spinal nerve roots.

SPINAL PIA MATER

The **spinal pia mater,** the innermost covering membrane of the spinal cord, is thin and transparent, and closely follows all the surface features of the spinal cord (Haines, 2006). The spinal pia also directly covers the roots of the spinal nerves and the spinal blood vessels. Inferior to the conus medullaris, the spinal pia continues as the filum terminale (Fig. 4.39).

The spinal cord is suspended in the dural sac by the *filum terminale* and the right and left **denticulate ligaments** (L. *denticulus,* small tooth), which run longitudinally along each side of the spinal cord (Figs. 4.43–4.45). The denticulate ligaments consist of a fibrous sheet of pia extending midway between the posterior and anterior nerve roots from the lateral surfaces of the spinal cord. The 20–22 sawtooth-like processes attach to the inner surface of the arachnoid-lined dural sac. The most superior process of the right and left denticulate ligaments attaches to the cranial dura immediately superior to the foramen magnum, and the inferior process extends from the conus medullaris, passing between the T12 and the L1 nerve roots.

SUBARACHNOID SPACE

The **subarachnoid space** is located between the arachnoid and pia mater and is filled with CSF (Figs. 4.41–4.43). The enlargement of the subarachnoid space in the dural sac, caudal to the conus medullaris and containing CSF and the cauda equina, is the **lumbar cistern** (Figs. 4.39 and 4.40). It extends from the L2 vertebra to the second segment of the sacrum. Dural root sheaths, enclosing spinal nerve roots in extensions of the subarachnoid space, protrude from the sides of the lumbar cistern (Fig. 4.44A & B).

Vasculature of Spinal Cord and Spinal Nerve Roots

ARTERIES OF SPINAL CORD AND NERVE ROOTS

The arteries supplying the spinal cord are branches of the vertebral, ascending cervical, deep cervical, intercostal, lumbar, and lateral sacral arteries (Figs. 4.46 and 4.47). Three longitudinal arteries supply the spinal cord: an *anterior*

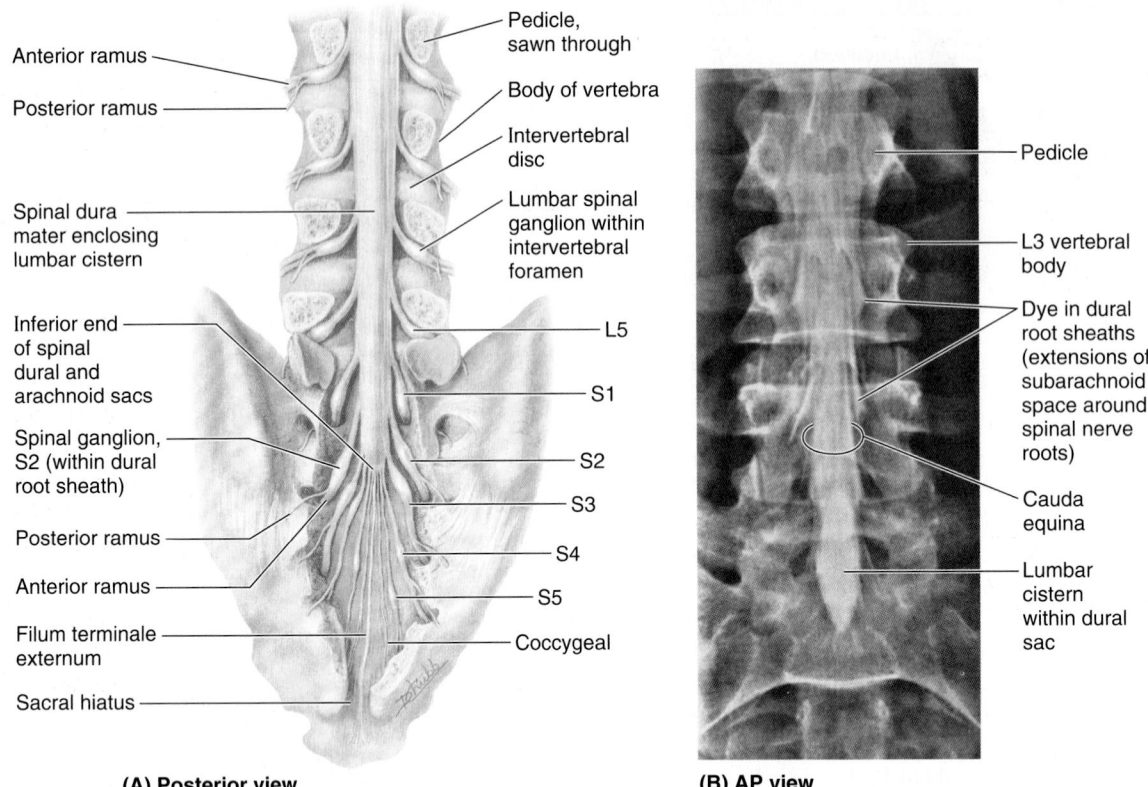

Anterior ramus

Posterior ramus

Spinal dura mater enclosing lumbar cistern

Inferior end of spinal dural and arachnoid sacs

Spinal ganglion, S2 (within dural root sheath)

Posterior ramus

Anterior ramus

Filum terminale externum

Sacral hiatus

Pedicle, sawn through

Body of vertebra

Intervertebral disc

Lumbar spinal ganglion within intervertebral foramen

L5

S1

S2

S3

S4

S5

Coccygeal

(A) Posterior view

Pedicle

L3 vertebral body

Dye in dural root sheaths (extensions of subarachnoid space around spinal nerve roots)

Cauda equina

Lumbar cistern within dural sac

(B) AP view

FIGURE 4.44. Inferior end of spinal dural sac. A. A laminectomy has been performed (i.e., the vertebral arches of the lumbar and sacral vertebrae have been removed) to show the inferior end of the dural sac, which encloses the lumbar cistern containing CSF and the cauda equina. The lumbar spinal ganglia lie within the IV foramina, but the sacral spinal ganglia (S1–S5) are in the sacral canal. In the lumbar region, the nerves exiting the IV foramina pass superior to the IV discs at that level; thus herniation of the nucleus pulposus tends to impinge on nerves passing to lower levels. **B.** Myelogram of lumbar region. Contrast medium was injected into the lumbar cistern. The lateral projections indicate extensions of the subarachnoid space into the dural root sheaths around the spinal nerve roots.

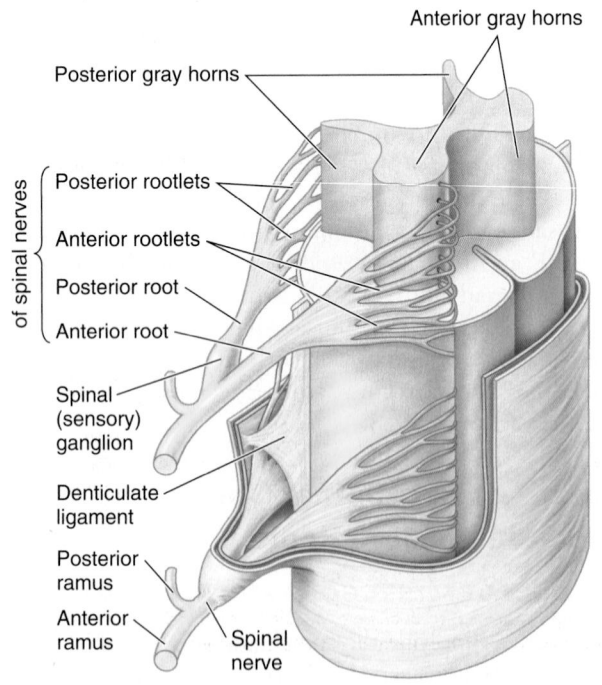

Anterior gray horns

Posterior gray horns

Posterior rootlets

Anterior rootlets

Posterior root

Anterior root

of spinal nerves

Spinal (sensory) ganglion

Denticulate ligament

Posterior ramus

Anterior ramus

Spinal nerve

FIGURE 4.45. Spinal cord, anterior and posterior nerve rootlets and roots, spinal ganglia, spinal nerves, and meninges.

spinal artery and paired *posterior spinal arteries.* These arteries run longitudinally from the medulla of the brainstem to the conus medullaris of the spinal cord.

The **anterior spinal artery,** formed by the union of branches of the vertebral arteries, runs inferiorly in the anterior median fissure. **Sulcal arteries** arise from the anterior spinal artery and enter the spinal cord through this fissure (Fig. 4.47B). The sulcal arteries supply approximately two thirds of the cross-sectional area of the spinal cord (Standring, 2008).

Each **posterior spinal artery** is a branch of either the *vertebral artery* or the *posteroinferior cerebellar artery* (Figs. 4.46B and 4.47). The posterior spinal arteries commonly form anastomosing channels in the pia mater.

By themselves, the anterior and posterior spinal arteries can supply only the short superior part of the spinal cord. The circulation to much of the spinal cord depends on segmental medullary and radicular arteries running along the spinal nerve roots. The **anterior** and **posterior segmental medullary arteries** are derived from spinal branches of the ascending cervical, deep cervical, vertebral, posterior intercostal, and lumbar arteries. The segmental medullary arteries occur mainly in association with the cervical and lumbosacral

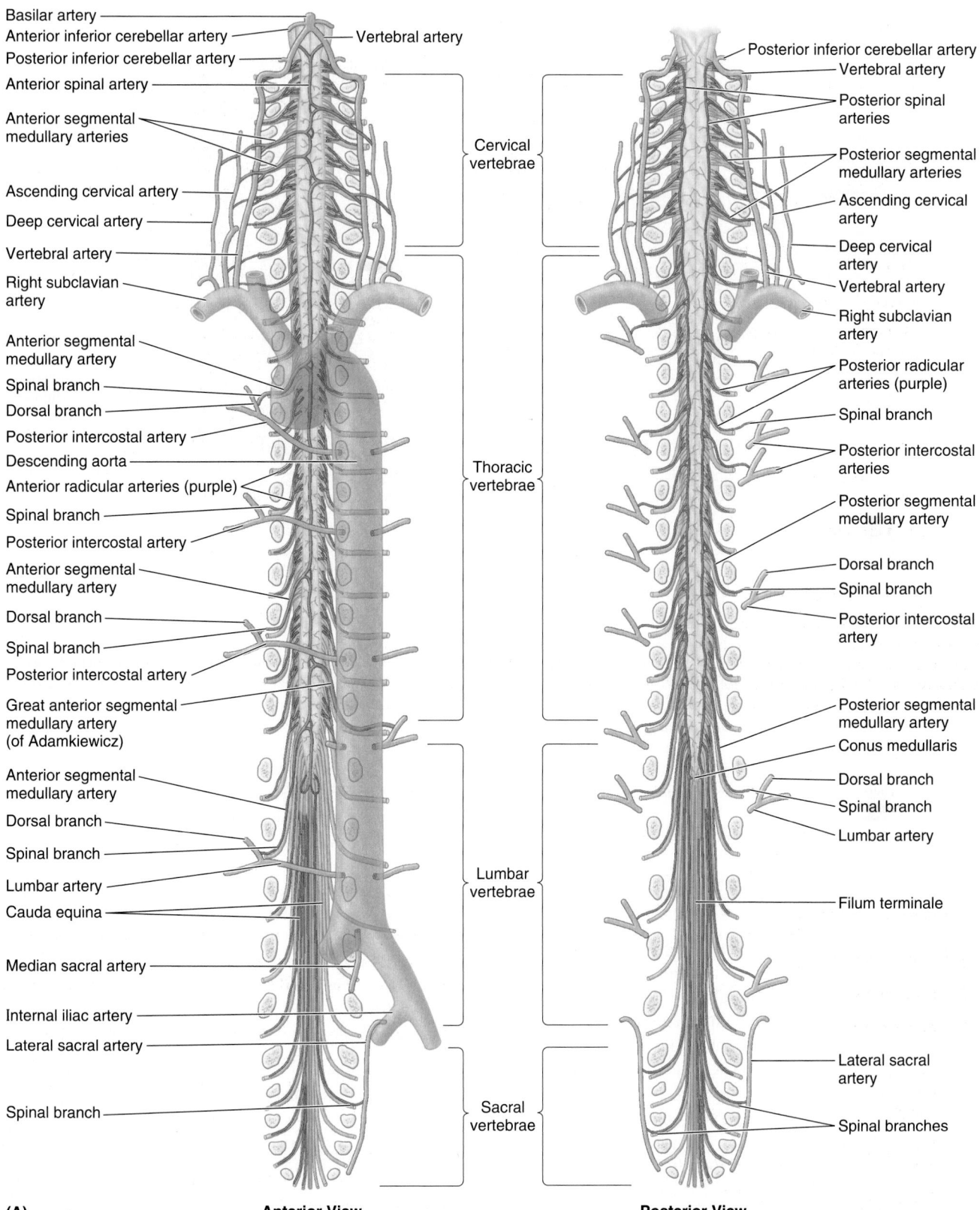

Basilar artery
Anterior inferior cerebellar artery
Posterior inferior cerebellar artery
Anterior spinal artery
Anterior segmental medullary arteries
Ascending cervical artery
Deep cervical artery
Vertebral artery
Right subclavian artery
Anterior segmental medullary artery
Spinal branch
Dorsal branch
Posterior intercostal artery
Descending aorta
Anterior radicular arteries (purple)
Spinal branch
Posterior intercostal artery
Anterior segmental medullary artery
Dorsal branch
Spinal branch
Posterior intercostal artery
Great anterior segmental medullary artery (of Adamkiewicz)
Anterior segmental medullary artery
Dorsal branch
Spinal branch
Lumbar artery
Cauda equina
Median sacral artery
Internal iliac artery
Lateral sacral artery
Spinal branch

Vertebral artery

Cervical vertebrae
Thoracic vertebrae
Lumbar vertebrae
Sacral vertebrae

Posterior inferior cerebellar artery
Vertebral artery
Posterior spinal arteries
Posterior segmental medullary arteries
Ascending cervical artery
Deep cervical artery
Vertebral artery
Right subclavian artery
Posterior radicular arteries (purple)
Spinal branch
Posterior intercostal arteries
Posterior segmental medullary artery
Dorsal branch
Spinal branch
Posterior intercostal artery
Posterior segmental medullary artery
Conus medullaris
Dorsal branch
Spinal branch
Lumbar artery
Filum terminale
Lateral sacral artery
Spinal branches

(A)   **Anterior View**   **Posterior View**

FIGURE 4.46. Arterial supply of spinal cord. A and B. Three longitudinal arteries supply the spinal cord: an anterior spinal artery and two posterior spinal arteries. Radicular arteries are shown at only the cervical and thoracic levels, but they also occur at the lumbar and sacral levels.

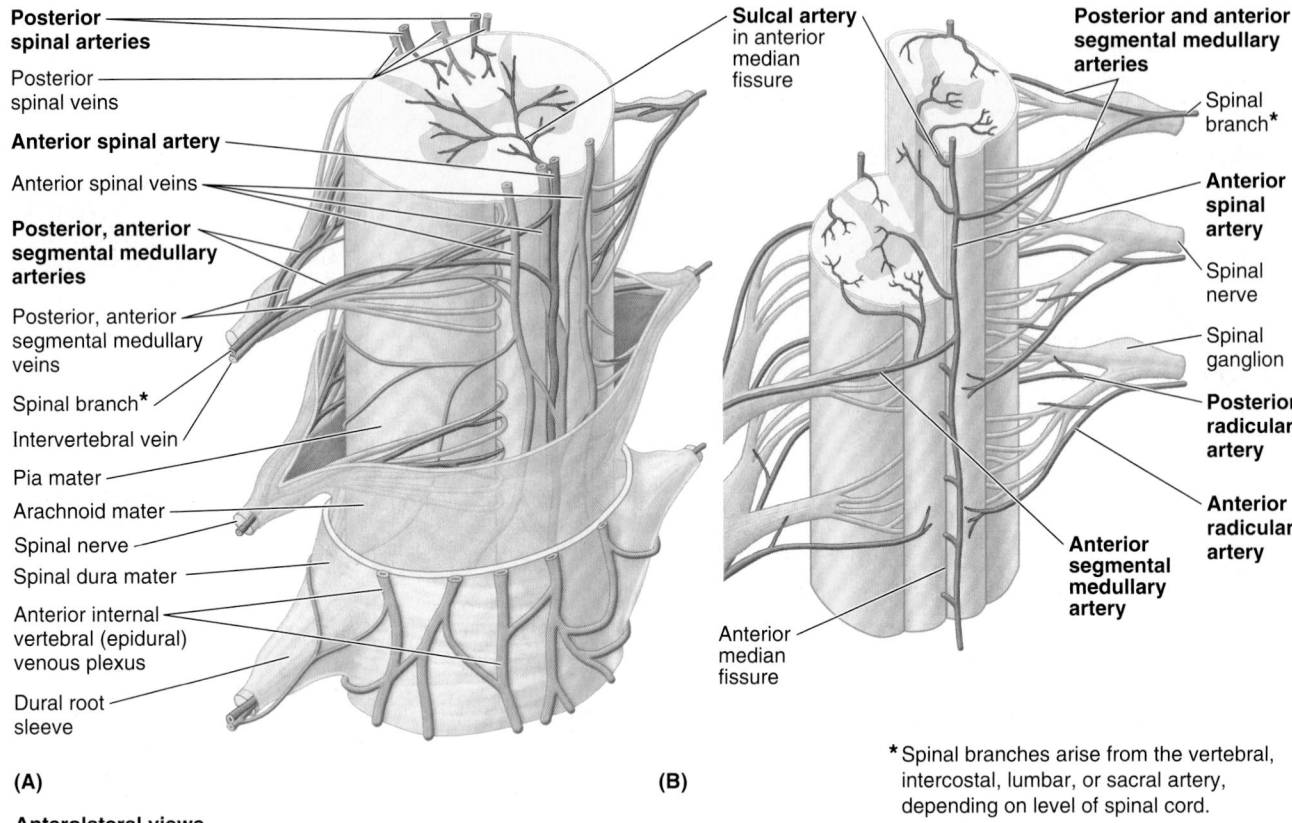

Posterior spinal arteries

Posterior spinal veins

Anterior spinal artery

Anterior spinal veins

Posterior, anterior segmental medullary arteries

Posterior, anterior segmental medullary veins

Spinal branch*

Intervertebral vein

Pia mater

Arachnoid mater

Spinal nerve

Spinal dura mater

Anterior internal vertebral (epidural) venous plexus

Dural root sleeve

(A)

Anterolateral views

Sulcal artery in anterior median fissure

Posterior and anterior segmental medullary arteries

Spinal branch*

Anterior spinal artery

Spinal nerve

Spinal ganglion

Posterior radicular artery

Anterior radicular artery

Anterior segmental medullary artery

Anterior median fissure

(B)

*Spinal branches arise from the vertebral, intercostal, lumbar, or sacral artery, depending on level of spinal cord.

Most proximal spinal nerves and roots are accompanied by **radicular arteries**, which do not reach the posterior or anterior spinal arteries. **Segmental medullary arteries** occur irregularly *in the place of* radicular arteries—they are really just larger vessels that make it all the way to the spinal arteries.

FIGURE 4.47. Arterial supply and venous drainage of spinal cord and spinal nerve roots. A. The veins that drain the spinal cord, as well as internal vertebral venous plexuses, drain into the intervertebral veins, which in turn drain into segmental veins. **B.** The pattern of the arterial supply of the spinal cord is from three longitudinal arteries: one anterior lying in the anteromedian position and the other two lying posterolaterally. These vessels are reinforced by medullary branches derived from the segmental arteries. The sulcal arteries are small branches of the anterior spinal artery coursing in the anterior median fissure.

enlargements, regions where the need for a good blood supply is greatest. They enter the vertebral canal through the IV foramina.

The **great anterior segmental medullary artery** (of Adamkiewicz), which is on the left side in about 65% of people, reinforces the circulation to two thirds of the spinal cord, including the lumbosacral enlargement (Figs. 4.39 and 4.46A). The great artery, much larger than the other segmental medullary arteries, usually arises via a spinal branch from an inferior intercostal or upper lumbar artery and enters the vertebral canal through the IV foramen at the lower thoracic or upper lumbar level.

The posterior and anterior roots of the spinal nerves and their coverings are supplied by **posterior** and **anterior radicular arteries** (L. *radix*, root), which run along the nerve roots (Figs. 4.46 and 4.47). The radicular arteries do not reach the posterior, anterior, or spinal arteries. Segmental medullary arteries replace the radicular arteries at the irregular levels at which they occur. Most radicular arteries are small and supply only the nerve roots; however,

some of them may assist with the supply of superficial parts of the gray matter in the posterior and anterior horns of the spinal cord.

VEINS OF SPINAL CORD

In general, the veins of the spinal cord have a distribution similar to that of the spinal arteries. There are usually *three* **anterior** and *three* **posterior spinal veins** (Fig. 4.47A). The spinal veins are arranged longitudinally, communicate freely with each other, and are drained by up to 12 **anterior** and **posterior medullary** and **radicular veins.** The veins of the spinal cord join the internal vertebral (epidural) venous plexuses in the epidural space (see Fig. 4.27). The *internal vertebral venous plexuses* pass superiorly through the foramen magnum to communicate with dural sinuses and vertebral veins in the cranium. The internal vertebral plexuses also communicate with the external vertebral venous plexuses on the external surface of the vertebrae.

CONTENTS OF VERTEBRAL CANAL

Compression of Lumbar Spinal Nerve Roots

The lumbar spinal nerves increase in size from superior to inferior, whereas the IV foramina decrease in diameter. Consequently, the L5 spinal nerve roots are the thickest and their foramina, the narrowest. This increases the chance that these nerve roots will be compressed if *osteophytes* (bony spurs) develop (see Fig. B4.9B), or herniation of an IV disc occurs.

Myelography

Myelography is a radiopaque contrast procedure that allows visualization of the spinal cord and spinal nerve roots (Fig. 4.44B). In this procedure, CSF is withdrawn by lumbar puncture and replaced with a contrast material injected into the spinal subarachnoid space. This technique shows the extent of the subarachnoid space and its extensions around the spinal nerve roots within the dural root sheaths. High-resolution MRI has largely supplanted myelography.

Development of Meninges and Subarachnoid Space

Together, the arachnoid and pia mater form the **leptomeninges** (G. slender membranes). They develop as a single layer from the mesenchyme surrounding the embryonic spinal cord. Fluid-filled spaces form within this layer and coalesce to produce the subarachnoid space (Moore, et al., 2012). The origin of both pia and arachnoid from a single membrane is reflected by the numerous arachnoid trabeculae passing between them (Fig. 4.42). In adults, the arachnoid is thick enough to be manipulated with forceps. The delicate pia mater gives a shiny appearance to the surface of the spinal cord but is barely visible to the unaided eye as a distinct layer.

Lumbar Spinal Puncture

Lumbar puncture (LP, spinal tap), the withdrawal of CSF from the lumbar cistern, is an important diagnostic tool for evaluating a variety of central nervous system (CNS) disorders. *Meningitis* and diseases of the CNS may alter the cells in the CSF or change the concentration of its chemical constituents. Examination of CSF can also determine if blood is present.

LP is performed with the patient lying on the side with the back and hips flexed (knee-chest position, Fig. B4.18). Flexion of the vertebral column facilitates insertion of the needle by spreading apart the vertebral laminae and spinous processes, stretching the ligamenta flava.

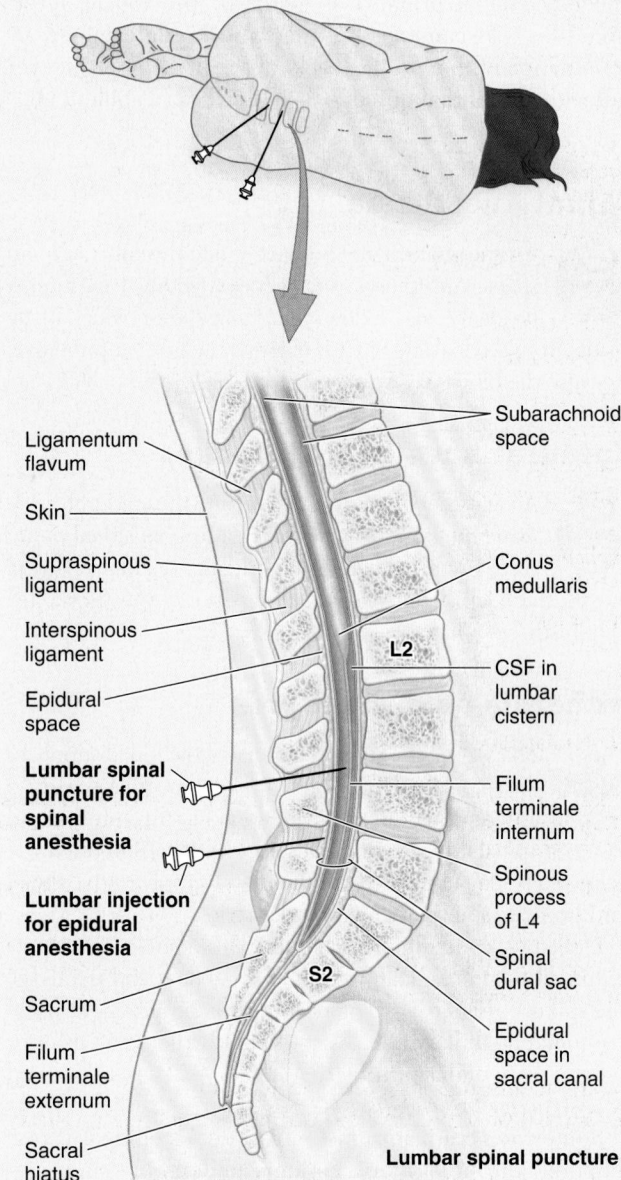

Lumbar spinal puncture

FIGURE B4.18.

The skin covering the lower lumbar vertebrae is anesthetized, and a *lumbar puncture needle*, fitted with a *stylet*, is inserted in the midline between the spinous processes of the L3 and L4 (or L4 and L5) vertebrae. Recall that a plane transecting the highest points of the iliac crests—the *supracristal plane*—usually passes through the L4 spinous process. At these levels, there is no danger of damaging the spinal cord.

After passing 4–6 cm in adults (more in obese persons), the needle "pops" through the ligamentun flavum, then punctures the dura and arachnoid and enters the lumbar cistern. When the stylet is removed, CSF escapes at the rate of approximately one drop per second. If subarachnoid pressure is high, CSF flows out or escapes as a jet.

Lumbar puncture is not performed in the presence of increased intracranial pressure (within the cranial cavity). The intracranial pressure is generally previously determined by CT scanning, but may also be determined by examination of the fundus (back) of the interior of the eyeball with an ophthalmoscope (see blue box "Papilledema," p. 911).

Spinal Anesthesia

 An anesthetic agent is injected into the subarachnoid space. Anesthesia usually occurs within 1 minute. A headache may follow *spinal anesthesia,* which likely results from the leakage of CSF through the lumbar puncture. (See also the blue box "Anesthesia for Childbirth," p. 397.)

Epidural Anesthesia (Blocks)

 An anesthetic agent is injected into the epidural space using the position described for lumbar spinal puncture, or through the sacral hiatus (caudal epidural anesthesia/block). (See also the blue box "Anesthesia for Childbirth," p. 397.)

Ischemia of Spinal Cord

 The segmental reinforcements of the blood supply to the spinal cord from the segmental medullary arteries are important in supplying blood to the anterior and posterior spinal arteries. Fractures, dislocations, and fracture–dislocations may interfere with the blood supply to the spinal cord from the spinal and medullary arteries.

Deficient blood supply (ischemia) of the spinal cord affects its function and can lead to muscle weakness and paralysis. The spinal cord may also suffer circulatory impairment if the segmental medullary arteries, particularly the great anterior segmental medullary artery (of Adamkiewicz), are narrowed by *obstructive arterial disease.*

Sometimes the aorta is purposely occluded (cross clamped) during surgery. Patients undergoing such surgeries, and those with ruptured aneurysms of the aorta or occlusion of the great anterior segmental medullary artery, may lose all sensation and voluntary movement inferior to the level of impaired blood supply to the spinal cord (*paraplegia*) secondary to death of neurons in the part of the spinal cord supplied by the anterior spinal artery (Figs. 4.46 and 4.47).

Neurons with cell bodies distant from the site of ischemia of the spinal cord will also die, secondary to the degeneration of axons traversing the site. The likelihood of *iatrogenic paraplegia* depends on such factors as the age of the patient,

the extent of the disease, and the length of time the aorta is cross clamped.

When systemic blood pressure drops severely for 3–6 min, blood flow from the segmental medullary arteries to the anterior spinal artery supplying the midthoracic region of the spinal cord may be reduced or stopped. These people may also lose sensation and voluntary movement in the areas supplied by the affected level of the spinal cord.

Spinal Cord Injuries

 The vertebral canal varies considerably in size and shape from level to level, particularly in the cervical and lumbar regions. A narrow vertebral canal in the cervical region, into which the spinal cord fits tightly, is potentially dangerous because a minor fracture and/or dislocation of a cervical vertebra may damage the spinal cord. The protrusion of a cervical IV disc into the vertebral canal after a neck injury may cause *spinal cord shock* associated with paralysis inferior to the site of the lesion.

In some people, no fracture or dislocation of cervical vertebrae can be found. If the individual dies and an autopsy is performed, a softening of the spinal cord may be detected at the site of the cervical disc protrusion. Encroachment of the vertebral canal by a protruding IV disc, by swollen ligamenta flava, or resulting from *osteoarthritis of the zygapophysial joints* may exert pressure on one or more of the spinal nerve roots of the cauda equina. Pressure may produce sensory and motor symptoms in the area of distribution of the involved spinal nerve. This group of bone and joint abnormalities, called *lumbar spondylosis* (degenerative joint disease), also causes localized pain and stiffness.

Transection of the spinal cord results in loss of all sensation and voluntary movement inferior to the lesion. Transection between the following levels will result in the indicated effects:

- C1–C3: no function below head level; a ventilator is required to maintain respiration.
- C4–C5: *quadriplegia* (no function of upper and lower limbs); respiration occurs.
- C6–C8: loss of lower limb function combined with a loss of hand and a variable amount of upper limb function; the individual may be able to self-feed or propel a wheelchair.
- T1–T9 *paraplegia* (paralysis of both lower limbs); the amount of trunk control varies with the height of the lesion.
- T10–L1: some thigh muscle function, which may allow walking with long leg braces.
- L2–L3: retention of most leg muscle function; short leg braces may be required for walking.

The Bottom Line

CONTENTS OF VERTEBRAL CANAL

The spinal cord, spinal nerve roots, CSF, and meninges that surround them are the main contents of the vertebral canal.

Spinal cord: In adults, the spinal cord occupies only the superior two thirds of the vertebral canal and has two (cervical and lumbosacral) enlargements related to innervation of the limbs. ♦ The inferior, tapering end of the spinal cord, the conus medullaris, ends at the level of the L1 or L2 vertebra. ♦ However, the filum terminale and spinal nerve roots from the lumbosacral part of the spinal cord that form the cauda equina continue inferiorly within the lumbar cistern containing CSF.

Spinal meninges and CSF: Nerve tissues and neurovascular structures of the vertebral canal are suspended in CSF contained within the dural sac and dural root sheaths. ♦ The fluid-filled subarachnoid space is lined with pia and arachnoid mater, which are continuous membranes (leptomeninges). ♦ Because the spinal cord does not extend into the lumbar cistern (the inferior part of the subarachnoid space), it is an ideal site for sampling CSF or for injection of anesthetic agents.

Vasculature of spinal cord and spinal nerve roots: Longitudinal spinal arteries supplying the spinal cord are reinforced by asymmetric segmental medullary arteries occurring at irregular levels (mostly in association with the cervical and lumbar enlargements) that also supply the spinal nerve roots at those levels ♦ At levels and on the sides where segmental medullary arteries do not occur, radicular arteries supply the nerve roots. ♦ The veins draining the spinal cord have a distribution and drainage generally reflective of the spinal arteries, although there are normally three longitudinal spinal veins both anteriorly and posteriorly.

thePoint   **Board-review questions, case studies, and additional resources are available at thePoint.lww.com.**

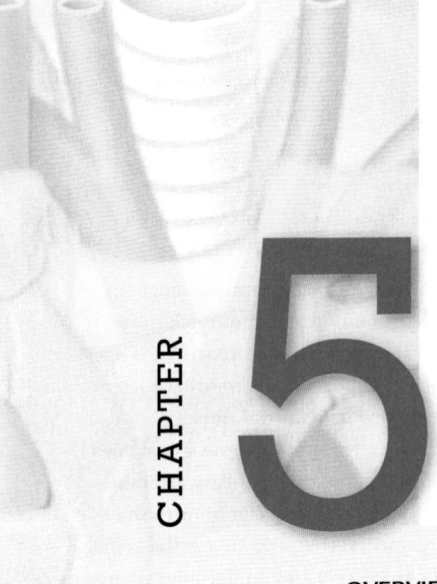

OVERVIEW OF LOWER LIMB

The **lower limbs** (extremities) are extensions from the trunk specialized to support body weight, for *locomotion* (the ability to move from one place to another), and to maintain balance.

The lower limbs have six major regions (Fig. 5.1):

1. The **gluteal region** (G. *gloutos*, buttocks) is the transitional region between the trunk and free lower limbs. It

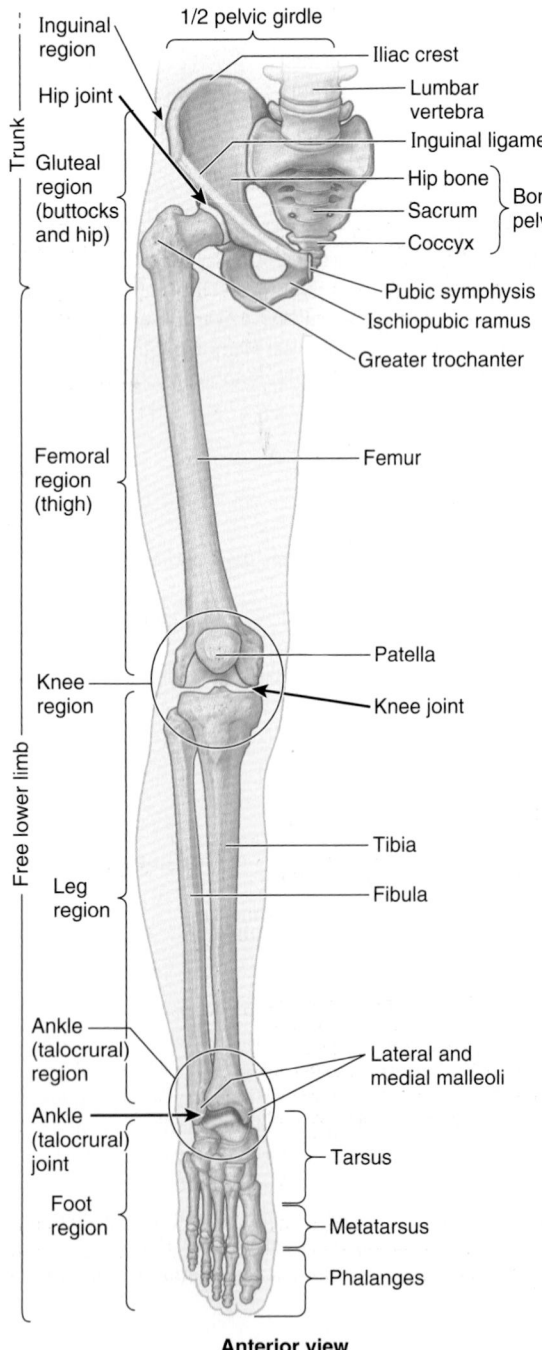

Anterior view

FIGURE 5.1. Regions and bones of lower limb.

includes two parts of the lower limb: the rounded, prominent posterior region, the **buttocks** (L. *nates, clunes*), and the lateral, usually less prominent **hip region** (L. *regio coxae*), which overlies the hip joint and greater trochanter of the femur. The "width of the hips" in common terminology is a reference to one's transverse dimensions at the level of the greater trochanters. The gluteal region is bounded superiorly by the iliac crest, medially by the *intergluteal cleft (natal cleft)*, and inferiorly by the skin fold (groove) underlying the buttocks, the **gluteal fold** (L. *sulcus glutealis*). The gluteal muscles, overlying the pelvic girdle, constitute the bulk of this region.

2. The **femoral region** (thigh) is the region of the free lower limb that lies between the gluteal, abdominal, and perineal regions proximally and the knee region distally. It includes most of the *femur* (thigh bone). The transition from trunk to free lower limb occurs abruptly in the *inguinal region* or *groin*. Here the boundary between the abdominal and perineal regions and the femoral region is demarcated by the *inguinal ligament* anteriorly and the ischiopubic ramus of the *hip bone* (part of the *pelvic girdle* or skeleton of the pelvis) medially. Posteriorly, the gluteal fold separates the gluteal and femoral regions (see Fig. 5.46A).

3. The **knee region** (L. *regio genus*) includes the prominences (condyles) of the distal femur and proximal tibia, the head of the fibula, and the *patella* (knee cap, which lies anterior to the distal end of the femur), as well as the joints between these bony structures. The **posterior region of the knee** (L. *poples*) includes a well-defined, fat-filled hollow, transmitting neurovascular structures, called the *popliteal fossa*.

4. The **leg region** (L. *regio cruris*) is the part that lies between the knee and the narrow, distal part of the leg. It includes most of the *tibia* (shin bone) and *fibula* (calf bone). The **leg** (L., *crus*) connects the knee and foot. Often laypersons refer incorrectly to the entire lower limb as "the leg."

5. The *ankle* (L. *tarsus*) or **talocrural region** (L. *regio talocruralis*) includes the medial and lateral prominences (*malleoli*) that flank the ankle (talocrural) joint.

6. The *foot* (L. *pes*) or **foot region** (L. *regio pedis*) is the distal part of the lower limb containing the *tarsus, metatarsus*, and *phalanges* (toe bones). The **toes** are the **digits of the foot.** The **great toe** (L. *hallux*), like the thumb, has only two *phalanges* (digital bones); the other digits have three.

DEVELOPMENT OF LOWER LIMB

Development of the lower limb is illustrated, explained, and contrasted with that of the upper limb in Figure 5.2. Initially, the development of the lower limb is similar to that of the upper limb, although occurring about a week later. During the 5th week, **lower limb buds** bulge from the lateral aspect of the L2–S2 segments of the trunk (a broader base than

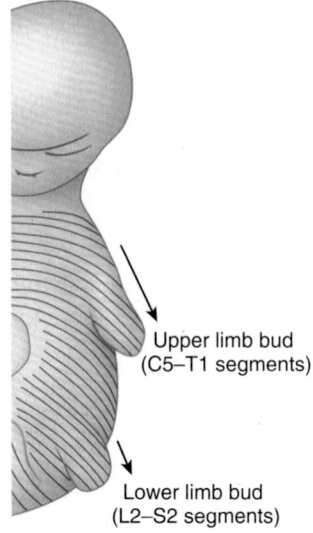

Upper limb bud
(C5–T1 segments)

Lower limb bud
(L2–S2 segments)

(A)

During early development, the trunk is divided into segments (metameres) that correspond to and receive innervation from the corresponding spinal cord segments. During the 4th week of development, the upper limb buds appear as elevations of the C5–T1 segments of the anterolateral body wall. Following the cranial to caudal pattern of development common to other systems, the lower limb buds appear about a week later (5th week). The lower limb buds grow laterally from broader bases formed by the L2–S2 segments.

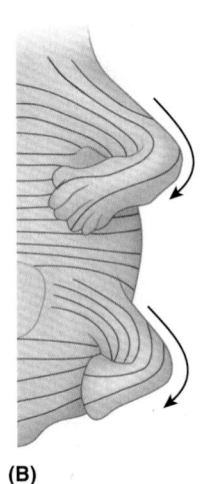

(B)

The distal ends of the limb buds flatten into paddle-like hand and foot plates that are elongated in the craniocaudal axis. Initially, both the thumb and the great toe are on the cranial sides of the developing hand and foot, directed superiorly, with the palms and soles directed anteriorly. Flexures occur where gaps develop between the precursors of the long bones [see (E)]. At first, the limbs bend anteriorly, so that the elbow and knee are directed laterally, causing the palm and sole to be directed medially (toward the trunk).

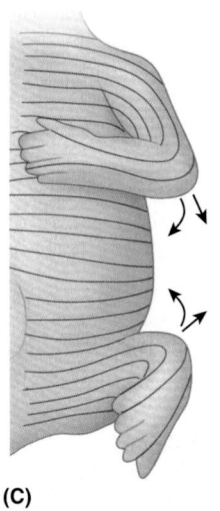

(C)

By the end of the 7th week, the proximal parts of the upper and lower limbs undergo a 90° torsion around their long axes, but in opposite directions, so that the elbow becomes directed caudally and the knee cranially.

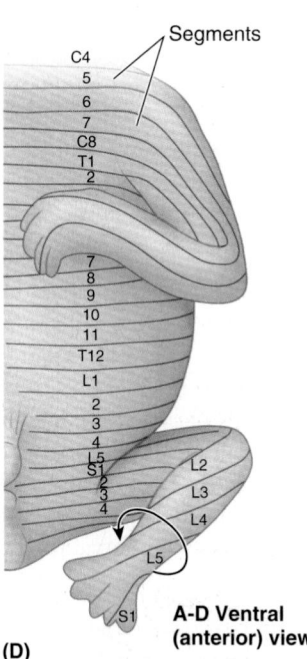

**A-D Ventral
(anterior) views**

(D)

In the lower limb, the torsion of the proximal limb is accompanied by a permanent pronation (twisting) of the leg, so that the foot becomes oriented with the great toe on the medial side.

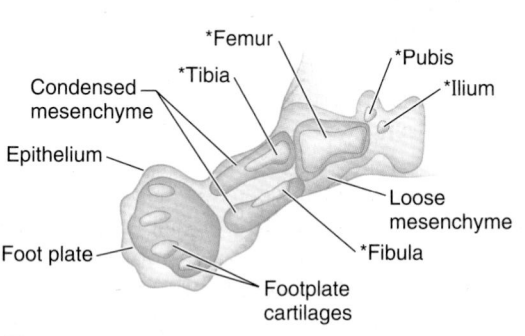

(E)

As the limb buds elongate, the loose mesenchyme within them condenses centrally, and cartilaginous models of the limb bones appear. The distal ends of the limb buds flatten into paddle-like plates (hand plates and foot plates) elongated in the craniocaudal axis. Gaps develop between precursors of the long bones where flexures (future elbow and knee joints) will occur.

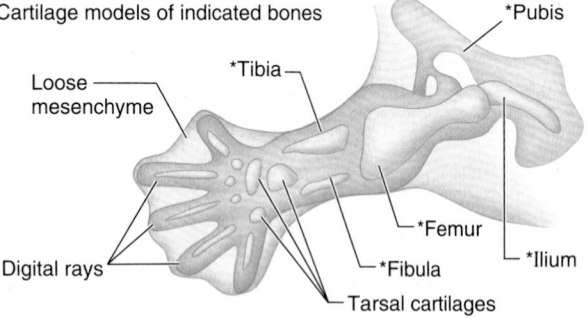

(F)

During the 7th week, digital rays, the first indication of future digits, appear. The thinner tissue between the digital rays undergoes apoptosis (programmed cell death), causing notches to develop, so that the rays soon appear as webbed fingers and toes. As the tissue breakdown progresses, separate digits are formed by the end of the eigth week. (Moore et al., 2012)

FIGURE 5.2. Development of lower limbs. A–D. The upper and lower limbs develop from limb buds that arise from the lateral body wall during the 4th and 5th weeks, respectively. They then elongate, develop flexures, and rotate in opposite directions. Segmental innervation is maintained, the dermatomal pattern reflecting the elongation and spiraling of the limb. **E and F.** Future bones develop from cartilage models, demonstrated at the end of the 6th week (**E**) and beginning of the 7th week (**F**).

for the upper limbs) (Fig. 5.2A). Both limbs initially extend from the trunk with their developing thumbs and great toes directed superiorly and the palms and soles directed anteriorly. Both limbs then undergo torsion around their long axes, but in opposite directions (Fig. 5.2B–D).

The medial rotation and permanent pronation of the lower limb explain how:

- the knee, unlike the joints superior to it, extends anteriorly and flexes posteriorly, as do the joints inferior to the knee (e.g., interphalangeal joints of the toes);
- the foot becomes oriented with the great toe on the medial side (Fig. 5.2D), whereas the hand (in the anatomical position) becomes oriented with the thumb on the lateral side; and
- the "barber-pole" pattern of the segmental innervation of the skin (dermatomes) of the lower limb develops (see "Cutaneous Innervation of Lower Limb," p. 536).

The torsion and twisting of the lower limb is still in progress at birth (note that babies' feet tend to meet sole to sole when they are brought together, like clapping). Completion of the process coincides with the mastering of walking skills.

BONES OF LOWER LIMB

The skeleton of the lower limb (inferior appendicular skeleton) may be divided into two functional components: the pelvic girdle and the bones of the free lower limb (Fig. 5.1). The **pelvic girdle** (bony pelvis) is a bony ring composed of the sacrum and right and left hip bones joined anteriorly at the pubic symphysis.

The pelvic girdle attaches the free lower limb to the axial skeleton, the sacrum being common to the axial skeleton and the pelvic girdle. The pelvic girdle also makes up the skeleton of the lower part of the trunk. Its protective and supportive functions serve the abdomen, pelvis, and perineum as well as the lower limbs. The *bones of the free lower limb* are contained within and specifically serve that part of the limb.

Arrangement of Lower Limb Bones

Body weight is transferred from the vertebral column through the *sacro-iliac joints* to the pelvic girdle and from the pelvic girdle through the hip joints to the femurs (L. *femora*) (Fig. 5.3A).

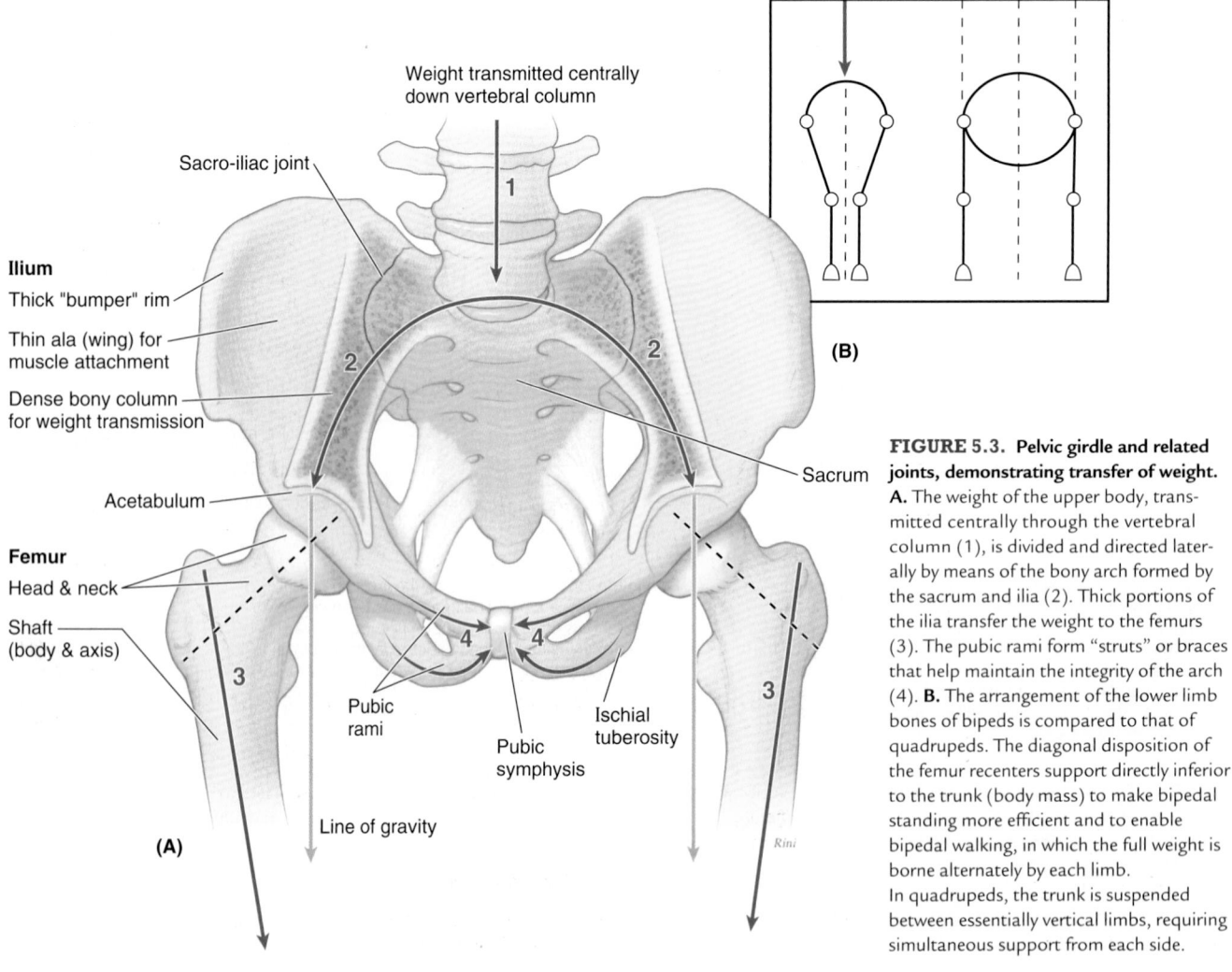

FIGURE 5.3. **Pelvic girdle and related joints, demonstrating transfer of weight.** **A.** The weight of the upper body, transmitted centrally through the vertebral column (1), is divided and directed laterally by means of the bony arch formed by the sacrum and ilia (2). Thick portions of the ilia transfer the weight to the femurs (3). The pubic rami form "struts" or braces that help maintain the integrity of the arch (4). **B.** The arrangement of the lower limb bones of bipeds is compared to that of quadrupeds. The diagonal disposition of the femur recenters support directly inferior to the trunk (body mass) to make bipedal standing more efficient and to enable bipedal walking, in which the full weight is borne alternately by each limb.

In quadrupeds, the trunk is suspended between essentially vertical limbs, requiring simultaneous support from each side.

To support the erect bipedal posture better, the femurs are oblique (directed inferomedially) within the thighs so that when standing the knees are adjacent and placed directly inferior to the trunk, returning the center of gravity to the vertical lines of the supporting legs and feet (Figs. 5.1, 5.3, and 5.4). Compare this oblique position of the femurs with that of quadrupeds, in whom the femurs are vertical and the knees are apart, with the trunk mass suspended between the limbs (Fig. 5.3B).

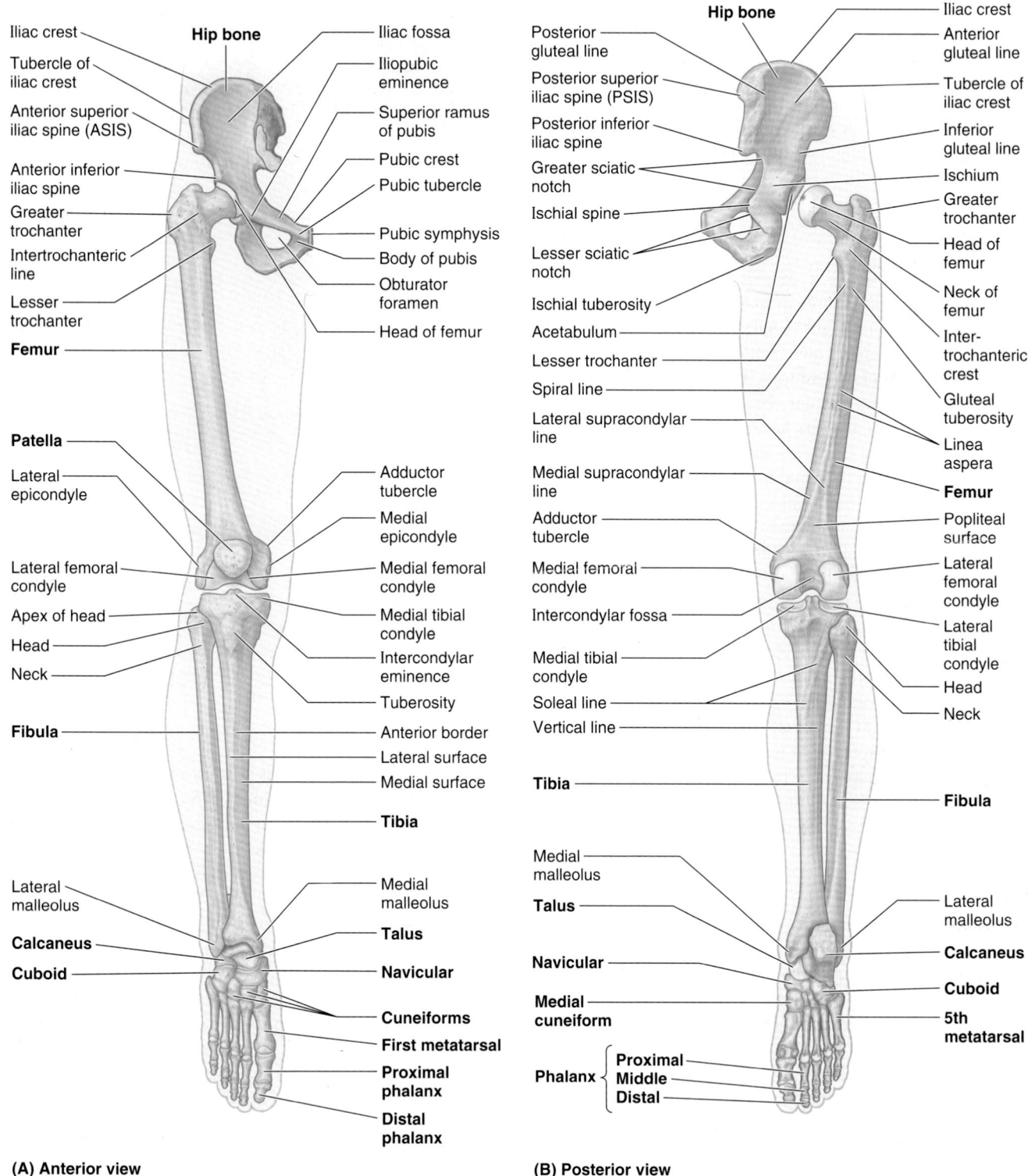

(A) Anterior view

(B) Posterior view

FIGURE 5.4. Bones of lower limb. A and B. Individual bones and bony formations are identified. The foot is in full plantarflexion. The hip joint is disarticulated (**B**) to demonstrate the acetabulum of the hip bone, which receives the head of the femur.

The femurs of human females are slightly more oblique than those of males, reflecting the greater width of their pelves. At the knees, the distal end of each femur articulates with the patella and tibia of the corresponding leg. Weight is transferred from the knee joint to the ankle joint by the tibia. The fibula does not articulate with the femur and does not bear or transfer weight, but it provides for muscle attachment and contributes to the formation of the ankle joint.

At the ankle, the weight borne by the tibia is transferred to the *talus* (Fig. 5.4). The talus is the keystone of a longitudinal arch formed by the tarsal and metatarsal bones of each foot that distributes the weight evenly between the heel and forefoot when standing, creating a flexible but stable bony platform to support the body.

Hip Bone

The mature **hip bone** (L. *os coxae*) is the large, flat pelvic bone formed by the fusion of three primary bones—*ilium, ischium, and pubis*—at the end of the teenage years. Each of the three bones is formed from its own primary center of ossification; five secondary centers of ossification appear later.

At birth, the three primary bones are joined by hyaline cartilage; in children, they are incompletely ossified (Fig. 5.5). At puberty, the three bones are still separated by a Y-shaped **triradiate cartilage** centered in the acetabulum, although the two parts of the ischiopubic rami fuse by the 9th year (Fig. 5.5B). The bones begin to fuse between 15 and 17 years of age; fusion is complete between 20 and 25 years of age. Little or no trace of the lines of fusion of the primary bones is visible in older adults (Fig. 5.6). Although the bony components are rigidly fused, their names are still used in adults to describe the three parts of the hip bone.

Because much of the medial aspect of the hip bones/bony pelvis is primarily concerned with pelvic and perineal structures and functions (Chapter 3) or their union with the vertebral column (Chapter 4), it is described more thoroughly in those chapters. Aspects of the hip bones concerned with lower limb structures and functions, mainly involving their lateral aspects, are described in this chapter.

ILIUM

The **ilium** forms the largest part of the hip bone and contributes the superior part of the acetabulum (Fig. 5.5B). The ilium has thick medial portions (columns) for weight bearing and thin, wing-like, posterolateral portions, the **alae** (L. wings), that provide broad surfaces for the fleshy attachment of muscles (Fig. 5.3).

The **body of the ilium** joins the pubis and ischium to form the acetabulum. Anteriorly, the ilium has stout **anterior superior** and **anterior inferior iliac spines** that provide attachment for ligaments and tendons of lower limb muscles (Fig. 5.6).

Beginning at the anterior superior iliac spine (ASIS), the long curved and thickened superior border of the ala of the

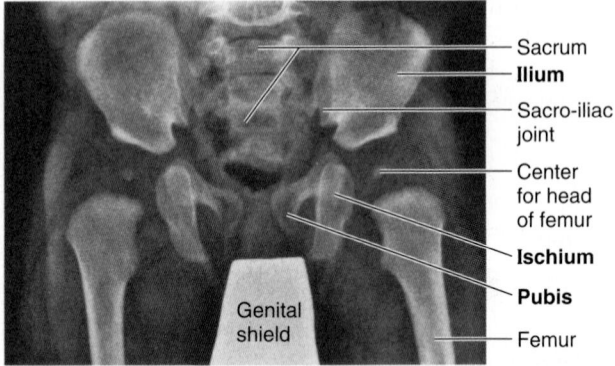

(A) Anterior view

Sacrum
Ilium
Sacro-iliac joint
Center for head of femur
Ischium
Pubis
Genital shield
Femur

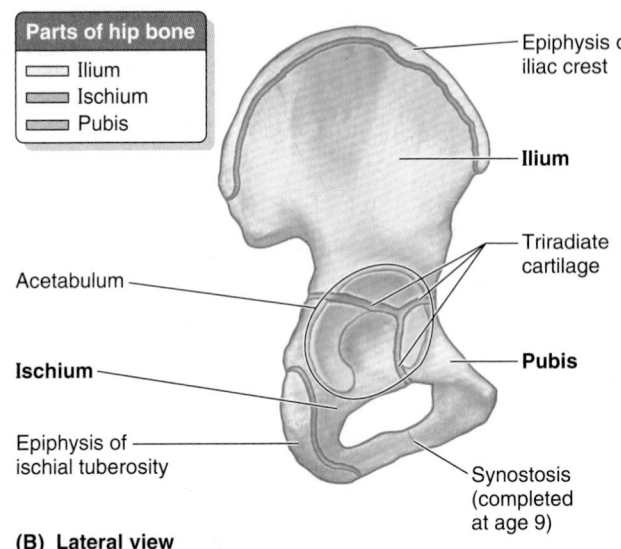

Parts of hip bone
☐ Ilium
▨ Ischium
▨ Pubis

Epiphysis of iliac crest
Ilium
Triradiate cartilage
Acetabulum
Ischium
Pubis
Epiphysis of ischial tuberosity
Synostosis (completed at age 9)

(B) Lateral view

FIGURE 5.5. Parts of hip bones. A. An anteroposterior radiograph of an infant's hips shows the three parts of the incompletely ossified hip bones (ilium, ischium, and pubis). **B.** The right hip bone of a 13-year-old demonstrating the Y-shaped triradiate cartilage.

ilium, the **iliac crest,** extends posteriorly, terminating at the **posterior superior iliac spine** (PSIS). The crest serves as a protective "bumper" and is an important site of aponeurotic attachment for thin, sheet-like muscles and deep fascia. A prominence on the external lip of the crest, the **tubercle of the iliac crest** (iliac tubercle), lies 5–6 cm posterior to the ASIS. The **posterior inferior iliac spine** marks the superior end of the *greater sciatic notch.*

The lateral surface of the ala of the ilium has three rough curved lines—the posterior, anterior, and inferior **gluteal lines**—that demarcate the proximal attachments of the three large gluteal muscles (pl., *glutei*). Medially, each ala has a large, smooth depression, the **iliac fossa** (Fig. 5.6B), that provides proximal attachment for the iliacus muscle. The bone forming the superior part of this fossa may become thin and translucent, especially in older women with osteoporosis.

Posteriorly, the medial aspect of the ilium has a rough, ear-shaped articular area called the **auricular surface** (L. *auricula*, a little ear), and an even rougher **iliac tuberosity**

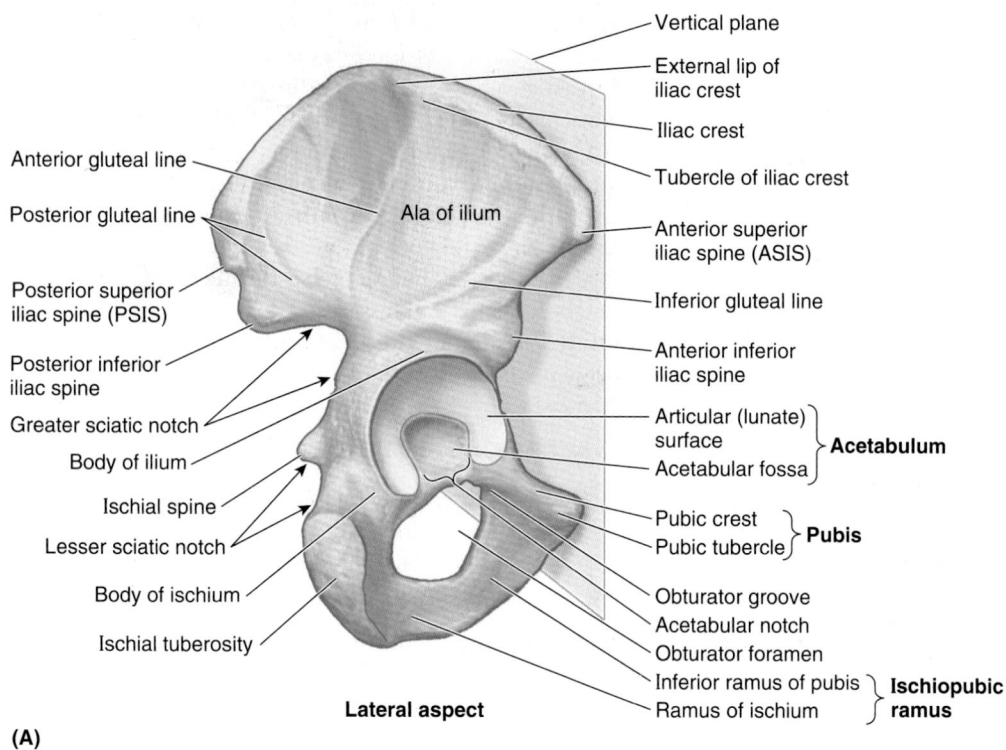

Vertical plane
External lip of iliac crest
Iliac crest
Tubercle of iliac crest
Anterior superior iliac spine (ASIS)
Inferior gluteal line
Anterior inferior iliac spine
Articular (lunate) surface ⎫
Acetabular fossa ⎭ **Acetabulum**
Pubic crest ⎫ **Pubis**
Pubic tubercle ⎭
Obturator groove
Acetabular notch
Obturator foramen
Inferior ramus of pubis ⎫ **Ischiopubic**
Ramus of ischium ⎭ **ramus**

Anterior gluteal line
Posterior gluteal line
Ala of ilium
Posterior superior iliac spine (PSIS)
Posterior inferior iliac spine
Greater sciatic notch
Body of ilium
Ischial spine
Lesser sciatic notch
Body of ischium
Ischial tuberosity

Lateral aspect

(A)

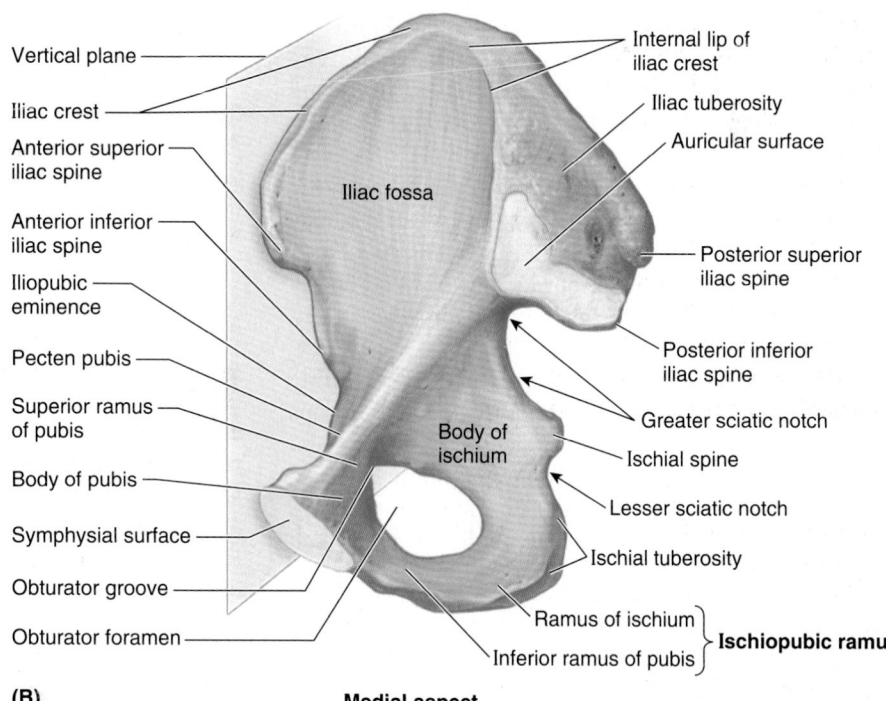

Vertical plane
Iliac crest
Anterior superior iliac spine
Anterior inferior iliac spine
Iliopubic eminence
Pecten pubis
Superior ramus of pubis
Body of pubis
Symphysial surface
Obturator groove
Obturator foramen

Internal lip of iliac crest
Iliac tuberosity
Auricular surface
Iliac fossa
Posterior superior iliac spine
Posterior inferior iliac spine
Greater sciatic notch
Body of ischium
Ischial spine
Lesser sciatic notch
Ischial tuberosity
Ramus of ischium ⎫ **Ischiopubic ramus**
Inferior ramus of pubis ⎭

(B)   **Medial aspect**

FIGURE 5.6. Right hip bone of adult in anatomical position. In this position, the anterior superior iliac spine (ASIS) and the anterior aspect of the pubis lie in the same coronal plane (blue). **A.** The large hip bone is constricted in the middle and expanded at its superior and inferior ends. **B.** The symphysial surface of the pubis articulates with the corresponding surface of the contralateral hip bone. The auricular surface of the ilium articulates with a corresponding surface of the sacrum to form the sacro-iliac joint.

superior to it for synovial and syndesmotic articulation with the reciprocal surfaces of the sacrum at the sacro-iliac joint.

ISCHIUM

The **ischium** forms the postero-inferior part of the hip bone. The superior part of the **body of the ischium** fuses with the pubis and ilium, forming the postero-inferior aspect of the acetabulum. The **ramus of the ischium** joins the *inferior ramus of the pubis* to form a bar of bone, the **ischiopubic ramus** (Fig. 5.6A), which constitutes the inferomedial boundary of the *obturator foramen.* The posterior border of the ischium forms the inferior margin of a deep indentation called the **greater sciatic notch.** The large, triangular **ischial spine** at the inferior margin of this notch provides ligamentous attachment. This sharp demarcation separates the greater sciatic notch from a more inferior, smaller, rounded, and smooth-surfaced indentation, the **lesser sciatic notch.** The lesser sciatic notch serves as a trochlea or pulley for a muscle that emerges from the bony pelvis. The rough bony projection at the junction of the inferior end of the body of the ischium and its ramus is the large **ischial tuberosity.** The body's weight rests on this tuberosity when sitting, and it provides the proximal, tendinous attachment of posterior thigh muscles.

PUBIS

The **pubis** forms the anteromedial part of the hip bone, contributing the anterior part of the acetabulum, and provides proximal attachment for muscles of the medial thigh. The pubis is divided into a flattened medially placed **body** and **superior** and **inferior rami** that project laterally from the body (Fig. 5.6).

Medially, the **symphysial surface** of the body of the pubis articulates with the corresponding surface of the body of the contralateral pubis by means of the *pubic symphysis* (Fig. 5.3A). The anterosuperior border of the united bodies and symphysis forms the **pubic crest,** which provides attachment for abdominal muscles.

Small projections at the lateral ends of this crest, the **pubic tubercles,** are important landmarks of the inguinal regions (Fig. 5.6). The tubercles provide attachment for the main part of the inguinal ligament and thereby indirect muscle attachment. The posterior margin of the superior ramus of the pubis has a sharp raised edge, the **pecten pubis,** which forms part of the pelvic brim (see Chapter 3).

OBTURATOR FORAMEN

The **obturator foramen** is a large oval or irregularly triangular opening in the hip bone. It is bounded by the pubis and ischium and their rami. Except for a small passageway for the obturator nerve and vessels (the *obturator canal*), the obturator foramen is closed by the thin, strong *obturator membrane.* The presence of the foramen minimizes bony mass (weight) while its closure by the obturator membrane still provides extensive surface area on both sides for fleshy muscle attachment.

ACETABULUM

The **acetabulum** (L., shallow vinegar cup) is the large cup-shaped cavity or socket on the lateral aspect of the hip bone that articulates with the head of the femur to form the hip joint (Fig. 5.6A). All three primary bones forming the hip bone contribute to the formation of the acetabulum (Fig. 5.5).

The margin of the acetabulum is incomplete inferiorly at the **acetabular notch,** which makes the fossa resemble a cup with a piece of its lip missing (Fig. 5.6A). The rough depression in the floor of the acetabulum extending superiorly from the acetabular notch is the **acetabular fossa.** The acetabular notch and fossa also create a deficit in the smooth **lunate surface of the acetabulum,** the articular surface receiving the head of the femur.

ANATOMICAL POSITION OF HIP BONE

Surfaces, borders, and relationships of the hip bone are described assuming that the body is in the *anatomical position.* To place an isolated hip bone or bony pelvis in this position, situate it so that the:

- ASIS and the anterosuperior aspect of the pubis lie in the same coronal plane.
- Symphysial surface of the pubis is vertical, parallel to the median plane (Fig. 5.6).

In the anatomical position, the:

- Acetabulum faces inferolaterally, with the acetabular notch directed inferiorly.
- Obturator foramen lies inferomedial to the acetabulum.
- Internal aspect of the body of the pubis faces almost directly superiorly. (It essentially forms a floor on which the urinary bladder rests.)
- Superior pelvic aperture (pelvic inlet) is more vertical than horizontal; in the anteroposterior (AP) view, the tip of the coccyx appears near its center (Fig. 5.3).

Femur

The **femur** is the longest and heaviest bone in the body. It transmits body weight from the hip bone to the tibia when a person is standing (Fig. 5.4). Its length is approximately a quarter of the person's height. The femur consists of a **shaft** (body) and two ends, superior or proximal and inferior or distal (Fig. 5.7).

The superior (proximal) end of the femur consists of a head, neck, and two trochanters (greater and lesser). The round **head of the femur** makes up two thirds of a sphere that is covered with articular cartilage, except for a medially placed depression or pit, the **fovea for the ligament of the head.** In early life, the ligament gives passage to an artery supplying the epiphysis of the head. The **neck of the femur** is trapezoidal, with its narrow end supporting the head and its broader base being continuous with the shaft. Its average diameter is three quarters that of the femoral head.

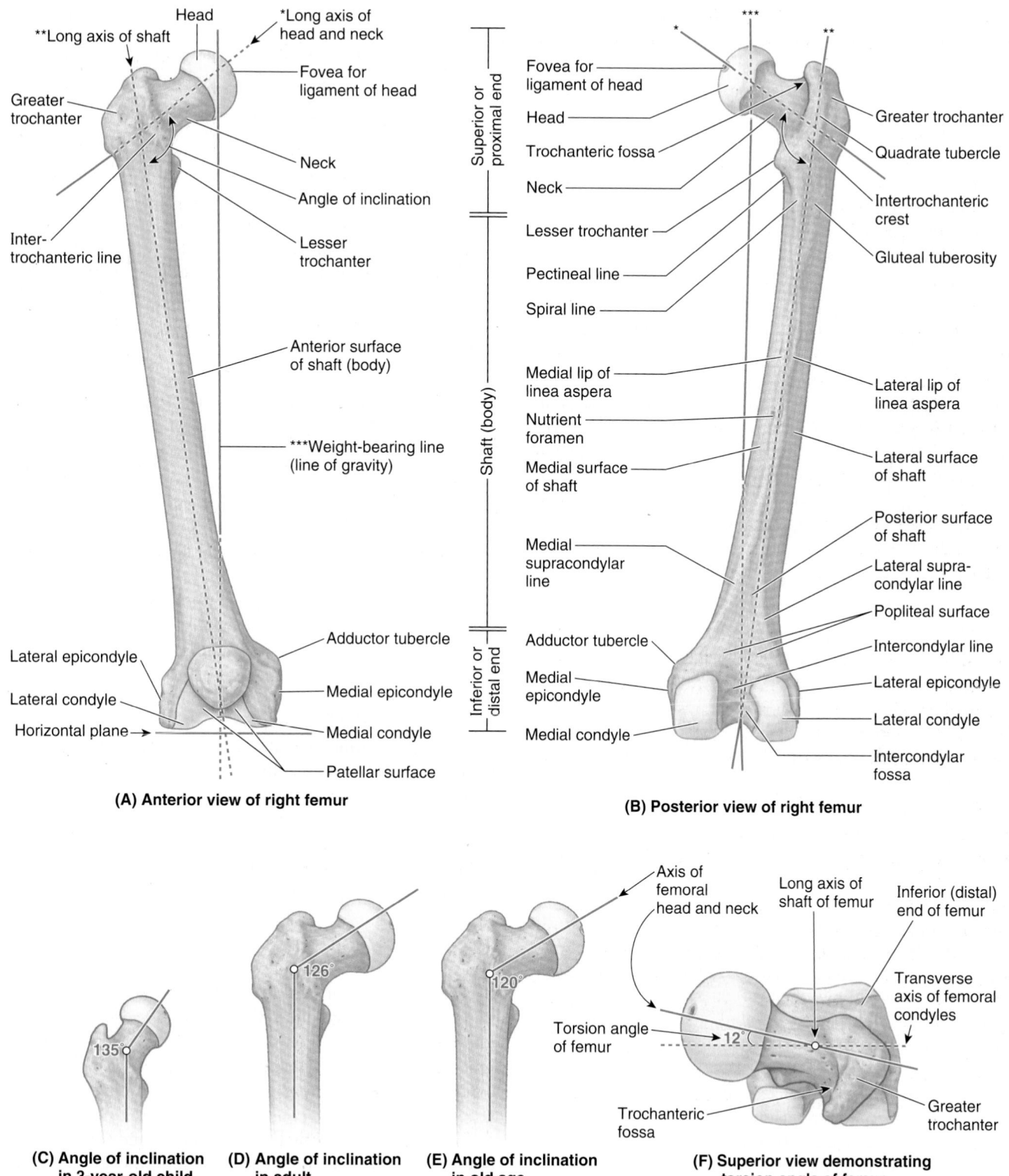

(A) Anterior view of right femur

(B) Posterior view of right femur

(C) Angle of inclination in 3-year-old child

(D) Angle of inclination in adult

(E) Angle of inclination in old age

(F) Superior view demonstrating torsion angle of femur

FIGURE 5.7. Right femur. A and B. The bony features of an adult femur are shown. Functionally and morphologically, the bone consists of highly modified superior and inferior ends and an intervening cylindrical shaft. **A–E.** The femur is "bent" so that the long axis of the head and neck lies at an angle (angle of inclination) to that of the shaft. When the massive femoral condyles rest on a horizontal surface, the femur assumes its oblique anatomical position in which the center of the round femoral head lies directly superior to the intercondylar fossa. **C–E.** The angle of inclination decreases (becomes more acute) with age, resulting in greater stress at a time when bone mass is reduced. When the femur is viewed along the long axis of the femoral shaft, so that the proximal end is superimposed over the distal end (**F**), it can be seen that the axis of the head and neck of the femur forms a 12° angle with the transverse axis of the femoral condyles (angle of torsion).

The proximal femur is "bent" (L-shaped) so that the long axis of the head and neck projects superomedially at an angle to that of the obliquely oriented shaft (Fig. 5.7A & B). This obtuse **angle of inclination** is greatest (most nearly straight) at birth and gradually diminishes (becomes more acute) until the adult angle is reached (115–140°, averaging 126°) (Fig. 5.7C–E).

The angle of inclination is less in females because of the increased width between the acetabula (a consequence of a wider lesser pelvis) and the greater obliquity of the femoral shaft. The angle of inclination allows greater mobility of the femur at the hip joint because it places the head and neck more perpendicular to the acetabulum in the neutral position. The abductors and rotators of the thigh attach mainly to the apex of the angle (the *greater trochanter*) so they are pulling on a lever (the short limb of the L) that is directed more laterally than vertically. This provides increased leverage for the abductors and rotators of the thigh, and allows the considerable mass of the abductors of the thigh to be placed superior to the femur (in the gluteal region) instead of lateral to it, freeing the lateral aspect of the femoral shaft to provide an increased area for the fleshy attachment of the extensors of the knee.

The angle of inclination also allows the obliquity of the femur within the thigh, which permits the knees to be adjacent and inferior to the trunk, as explained previously. All of this is advantageous for bipedal walking; however, it imposes considerable strain on the neck of the femur. Consequently, fractures of the femoral neck can occur in older people as a result of a slight stumble if the neck has been weakened by osteoporosis (pathologic reduction of bone mass).

The torsion of the proximal lower limb (femur) that occurred during development does not conclude with the long axis of the superior end of the femur (head and neck) parallel to the transverse axis of the inferior end (femoral condyles). When the femur is viewed superiorly (so that one is looking along the long axis of the shaft), it is apparent that the two axes lie at an angle (the **torsion angle,** or **angle of declination**), the mean of which is 7° in males and 12° in females. The torsion angle, combined with the angle of inclination, allows rotatory movements of the femoral head within the obliquely placed acetabulum to convert into flexion and extension, abduction and adduction, and rotational movements of the thigh.

Where the femoral neck and shaft join there are two large, blunt elevations called trochanters (Fig. 5.7A, B, & F). The abrupt, conical and rounded **lesser trochanter** (G., a runner) extends medially from the posteromedial part of the junction of the neck and shaft to give tendinous attachment to the primary flexor of the thigh (the iliopsoas).

The **greater trochanter** is a large, laterally placed bony mass that projects superiorly and posteriorly where the neck joins the femoral shaft, providing attachment and leverage for abductors and rotators of the thigh. The site where the neck and shaft join is indicated by the **intertrochanteric line,** a roughened ridge formed by the attachment of a powerful ligament (iliofemoral ligament). The intertrochanteric line runs from the greater trochanter and winds around the lesser trochanter to continue posteriorly and inferiorly as a less distinct ridge, the **spiral line.**

A similar but smoother and more prominent ridge, the **intertrochanteric crest,** joins the trochanters posteriorly. The rounded elevation on the crest is the **quadrate tubercle.** In anterior and posterior views (Fig. 5.7A & B), the greater trochanter is in line with the femoral shaft. In posterior and superior views (Fig. 5.7B & F), it overhangs a deep depression medially, the **trochanteric fossa.**

The **shaft of the femur** is slightly bowed (convex) anteriorly. This convexity may increase markedly, proceeding laterally as well as anteriorly, if the shaft is weakened by a loss of calcium, as occurs in *rickets* (a disease attributable to vitamin D deficiency). Most of the shaft is smoothly rounded, providing fleshy origin to extensors of the knee, except posteriorly where a broad, rough line, the **linea aspera,** provides aponeurotic attachment for adductors of the thigh. This vertical ridge is especially prominent in the middle third of the femoral shaft, where it has **medial** and **lateral lips** (margins). Superiorly, the lateral lip blends with the broad, rough **gluteal tuberosity,** and the medial lip continues as a narrow, rough spiral line.

The *spiral line* extends toward the lesser trochanter but then passes to the anterior surface of the femur, where it is continuous with the intertrochanteric line. A prominent intermediate ridge, the **pectineal line,** extends from the central part of the linea aspera to the base of the lesser trochanter. Inferiorly, the linea aspera divides into medial and lateral **supracondylar lines,** which lead to the medial and lateral femoral condyles (Fig. 5.7B).

The **medial** and **lateral femoral condyles** make up nearly the entire inferior (distal) end of the femur. The two condyles are on the same horizontal level when the bone is in its anatomical position, so that if an isolated femur is placed upright with both condyles contacting the floor or tabletop, the femoral shaft will assume the same oblique position it occupies in the living body (about 9° from vertical in males and slightly greater in females).

The femoral condyles articulate with menisci (crescentic plates of cartilage) and tibial condyles to form the knee joint (Fig. 5.4). The menisci and tibial condyles glide as a unit across the inferior and posterior aspects of the femoral condyles during flexion and extension. The convexity of the articular surface of the condyles increases as it descends the anterior surface, covering the inferior end, and then ascends posteriorly. The condyles are separated posteriorly and inferiorly by an **intercondylar fossa** but merge anteriorly, forming a shallow longitudinal depression, the **patellar surface** (Fig. 5.7), which articulates with the patella. The lateral surface of the lateral condyle has a central projection called the *lateral epicondyle.* The medial surface of the medial condyle has a larger and more prominent *medial epicondyle,* superior to which another elevation, the **adductor tubercle,** forms in relation to a tendon attachment. The epicondyles provide proximal attachment for the medial and lateral collateral ligaments of the knee joint.

SURFACE ANATOMY OF PELVIC GIRDLE AND FEMUR

Bony landmarks are helpful during physical examinations and surgery because they can be used to evaluate normal development, detect and assess fractures and dislocations, and locate the sites of structures such as nerves and blood vessels.

When your hands are on your hips, they rest on your *iliac crests* (Fig. 5.8A). The anterior third of the crests is easily palpated because the crests are subcutaneous (Fig. 5.8C & D). The posterior two thirds of the crests are more difficult to palpate because they are usually covered with fat. The iliac crest ends anteriorly at the rounded *ASIS* (anterior superior iliac spine), which is easy to palpate by tracing the iliac crest antero-inferiorly. The ASIS is often visible in thin individuals. In obese people these spines are covered with fat and may be difficult to locate; however, they are easier to palpate

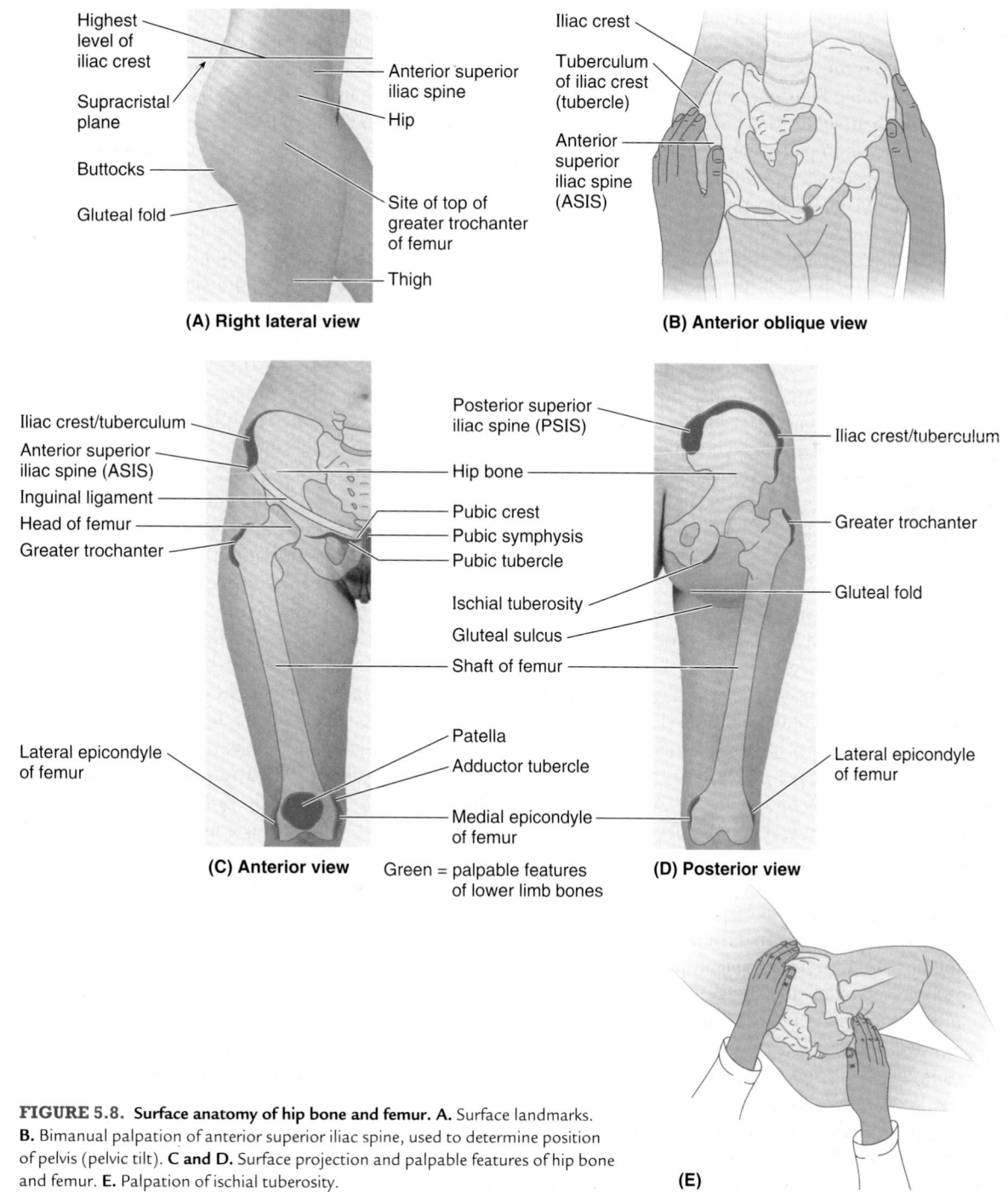

FIGURE 5.8. **Surface anatomy of hip bone and femur. A.** Surface landmarks. **B.** Bimanual palpation of anterior superior iliac spine, used to determine position of pelvis (pelvic tilt). **C and D.** Surface projection and palpable features of hip bone and femur. **E.** Palpation of ischial tuberosity.

when the person is sitting and the muscles attached to them are relaxed.

The *iliac tubercle,* 5–6 cm posterior to the ASIS, marks the widest point of the iliac crest. To palpate this tubercle, place your thumb on the ASIS and move your fingers posteriorly along the external lip of the iliac crest (Fig. 5.8B). The iliac tubercle lies at the level of the spinous process of the L5 vertebra.

Approximately a hand's width inferior to the umbilicus, the *bodies* and *superior rami of the pubic bones* may be palpated (Fig. 5.8C). The *pubic tubercle* can be palpated about 2 cm from the pubic symphysis at the anterior extremity of the *pubic crest.* The iliac crest ends posteriorly at the sharp *PSIS* (posterior superior iliac spine) (Fig. 5.8D), which may be difficult to palpate; however, its position is easy to locate because it lies at the bottom of a skin dimple, approximately 4 cm lateral to the midline. The dimple exists because the skin and underlying fascia attach to the PSIS. The skin dimples are useful landmarks when palpating the area of the sacro-iliac joints in search of edema (swelling) or local tenderness. These dimples also indicate the termination of the iliac crests from which bone marrow and pieces of bone for grafts can be obtained (e.g., to repair a fractured tibia).

The *ischial tuberosity* is easily palpated in the inferior part of the buttocks when the thigh is flexed (Fig. 5.8E). The buttocks covers and obscures the tuberosity when the thigh is extended (Fig. 5.8D). The *gluteal fold* coincides with the inferior border of the gluteus maximus and indicates the separation of the buttocks from the thigh.

The center of the *femoral head* can be palpated deep to a point approximately a thumb's breadth inferior to the mid-point of the inguinal ligament (Fig. 5.8C). The *shaft of the femur* is covered with muscles and is not usually palpable. Only the superior and inferior ends of the femur are palpable. The laterally placed *greater trochanter* projects superior to the junction of the shaft with the femoral neck and can be palpated on the lateral side of the thigh approximately 10 cm inferior to the iliac crest (Fig. 5.8B).

The *greater trochanter* forms a prominence anterior to the hollow on the lateral side of the buttocks (Fig. 5.8D). The prominences of the greater trochanters are normally responsible for the width of the adult pelvis. The posterior edge of the greater trochanter is relatively uncovered and most easily palpated when the limb is not weight-bearing. The anterior and lateral parts of the trochanter are not easy to palpate because they are covered by fascia and muscle. Because it lies close to the skin, the greater trochanter causes discomfort when you lie on your side on a hard surface. In the anatomical position, a line joining the tips of the greater trochanters normally passes through the pubic tubercles and the center of the femoral heads. The *lesser trochanter* is indistinctly palpable superior to the lateral end of the gluteal fold.

The *femoral condyles* are subcutaneous and easily palpated when the knee is flexed or extended (Fig. 5.8C & D). At the center of the lateral aspect of each condyle is a prominent *epicondyle* that is easily palpable. The patellar surface of the femur is where the *patella* slides during flexion and extension

of the leg at the knee joint. The lateral and medial margins of the patellar surface can be palpated when the leg is flexed. The *adductor tubercle,* a small prominence of bone, may be felt at the superior part of the medial femoral condyle by pushing your thumb inferiorly along the medial side of the thigh until it encounters the tubercle.

Tibia and Fibula

The tibia and fibula are the bones of the leg (Figs. 5.4 and 5.9). The *tibia* articulates with the condyles of the femur superiorly and the talus inferiorly, and in so doing transmits the body's weight. The *fibula* mainly functions as an attachment for muscles, but it is also important for the stability of the ankle joint. The shafts of the tibia and fibula are connected by a dense **interosseous membrane** composed of strong oblique fibers descending from the tibia to the fibula.

TIBIA

Located on the anteromedial side of the leg, nearly parallel to the fibula, the **tibia** (shin bone) is the second largest bone in the body. It flares outward at both ends to provide an increased area for articulation and weight transfer. The superior (proximal) end widens to form **medial** and **lateral condyles** that overhang the shaft medially, laterally, and posteriorly, forming a relatively flat **superior articular surface,** or *tibial plateau.* This plateau consists of two smooth articular surfaces (the medial one slightly concave and the lateral one slightly convex) that articulate with the large condyles of the femur. The articular surfaces are separated by an **intercondylar eminence** formed by two **intercondylar tubercles** (medial and lateral) flanked by relatively rough **anterior** and **posterior intercondylar areas.**

The tubercles fit into the **intercondylar fossa** between femoral condyles (Fig. 5.7B). The intercondylar tubercles and areas provide attachment for the menisci and principal ligaments of the knee, which hold the femur and tibia together, maintaining contact between their articular surfaces.

The anterolateral aspect of the lateral tibial condyle bears an **anterolateral tibial tubercle** (Gerdy tubercle) inferior to the articular surface (Fig. 5.9A), which provides the distal attachment for a dense thickening of the fascia covering the lateral thigh, adding stability to the knee joint. The lateral condyle also bears a **fibular articular facet** posterolaterally on its inferior aspect for the head of the fibula.

Unlike that of the femur, the **shaft of the tibia** is truly vertical within the leg. It is somewhat triangular in cross-section, having three surfaces and borders: medial, lateral/interosseous, and posterior.

The **anterior border of the tibia** is the most prominent border. It and the adjacent **medial surface** are subcutaneous throughout their lengths and are commonly known as the "shin". Their periosteal covering and overlying skin are vulnerable to bruising. At the superior end of the anterior border, a broad, oblong **tibial tuberosity** provides distal

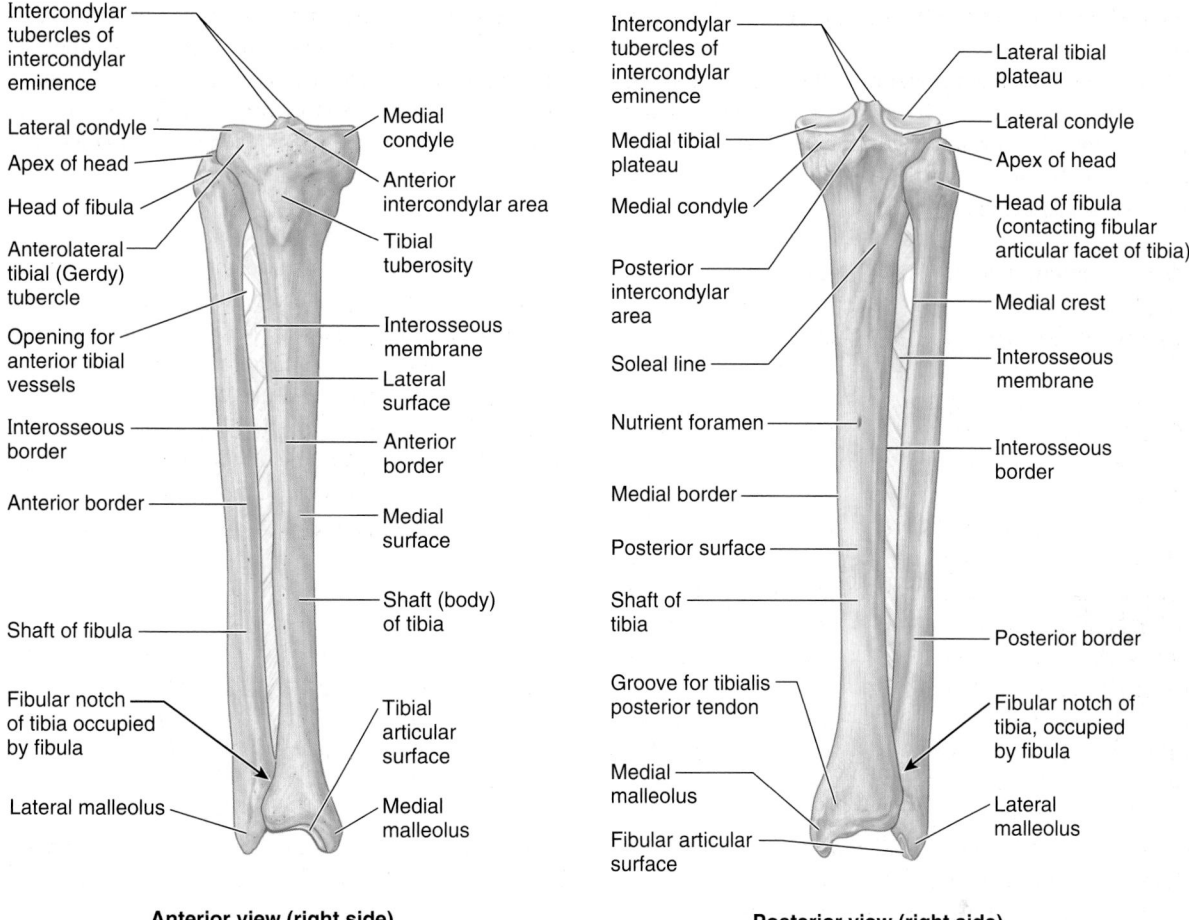

Anterior view (right side)   **Posterior view (right side)**

FIGURE 5.9. Right tibia and fibula. Tibiofibular syndesmoses, including the dense interosseous membrane, tightly connect the tibia and fibula. The interosseous membrane also provides additional surface area for muscular attachment. The anterior tibial vessels traverse the opening in the membrane to enter the anterior compartment of the leg.

attachment for the *patellar ligament,* which stretches between the inferior margin of the patella and the tibial tuberosity.

The tibial shaft is thinnest at the junction of its middle and distal thirds. The distal end of the tibia is smaller than the proximal end, flaring only medially; the medial expansion extends inferior to the rest of the shaft as the **medial malleolus.** The inferior surface of the shaft and the lateral surface of the medial malleolus articulate with the talus and are covered with articular cartilage (Fig. 5.4).

The **interosseous border** of the tibia is sharp where it gives attachment to the *interosseous membrane* that unites the two leg bones (Fig. 5.9). Inferiorly, the sharp border is replaced by a groove, the **fibular notch,** that accommodates and provides fibrous attachment to the distal end of the fibula.

On the posterior surface of the proximal part of the tibial shaft is a rough diagonal ridge, called the **soleal line,** which runs inferomedially to the medial border. This line is formed in relationship to the aponeurotic origin of the soleus muscle approximately one third of the way down the shaft. Immediately distal to the soleal line is an obliquely directed vascular groove, which leads to a large **nutrient foramen** for passage of the main artery supplying the proximal end of the bone

and its marrow. From it, the nutrient canal runs inferiorly in the tibia before it opens into the medullary (marrow) cavity.

FIBULA

The slender **fibula** lies posterolateral to the tibia and is firmly attached to it by the *tibiofibular syndesmosis,* which includes the *interosseous membrane* (Fig. 5.9). The fibula has no function in weight-bearing. It serves mainly for muscle attachment, providing distal attachment (insertion) for one muscle and proximal attachment (origin) for eight muscles. The fibers of the tibiofibular syndesmosis are arranged to resist the resulting net downward pull on the fibula.

The distal end enlarges and is prolonged laterally and inferiorly as the **lateral malleolus.** The malleoli form the outer walls of a rectangular socket (*mortise*), which is the superior component of the ankle joint (Fig. 5.4A), and provide attachment for the ligaments that stabilize the joint. The lateral malleolus is more prominent and posterior than the medial malleolus and extends approximately 1 cm more distally.

The proximal end of the fibula consists of an enlarged **head** superior to a small **neck** (Fig. 5.9). The head has a pointed **apex.** The head of the fibula articulates with the fibular facet on the posterolateral, inferior aspect of the lateral tibial condyle. The **shaft of the fibula** is twisted and marked by the sites of muscular attachments. Like the shaft of the tibia, it is triangular in cross-section, having three borders (anterior, interosseous, and posterior) and three surfaces (medial, posterior, and lateral).

SURFACE ANATOMY OF TIBIA AND FIBULA

The *tibial tuberosity,* an oval elevation on the anterior surface of the tibia, is easily palpated approximately 5 cm distal to the apex of the patella (Fig. 5.10A). The subcutaneous, flat *anteromedial surface of the tibia* is also easy to palpate. The skin covering this surface is freely movable. The *tibial condyles* can be palpated anteriorly at the sides of the patellar ligament, especially when the knee is flexed.

The *head of the fibula* is prominent at the level of the superior part of the tibial tuberosity because the knob-like head is subcutaneous at the posterolateral aspect of the knee. The *neck of the fibula* can be palpated just distal to the lateral side of the fibular head. Doing so may evoke a mildly unpleasant sensation because of the presence of the nerve passing there.

The *medial malleolus,* the prominence on the medial side of the ankle, is also subcutaneous and prominent. Note that its inferior end is blunt and does not extend as far distally as the lateral malleolus. The medial malleolus lies approximately 1.25 cm proximal to the level of the tip of the lateral malleolus (Figs. 5.10A & B).

Only the distal quarter of the shaft of the fibula is palpable. Palpate your *lateral malleolus,* noting that it is subcutaneous and that its inferior end is sharp. Note that the tip of the lateral malleolus extends farther distally and more posteriorly than does the tip of the medial malleolus.

Bones of Foot

The bones of the foot include the *tarsus, metatarsus,* and *phalanges.* There are 7 tarsal bones, 5 metatarsal bones, and 14 phalanges (Figs. 5.1, 5.4, and 5.11). Although knowledge of the characteristics of individual bones is necessary for an understanding of the structure of the foot, it is important to study the skeleton of the foot as a whole and to identify its principal bony landmarks in the living foot (see "Surface Anatomy of Bones of Foot," p. 524, and Surface Anatomy of Ankle Region and Foot," p. 622).

TARSUS

The **tarsus** (posterior or proximal foot; *hindfoot + midfoot—*Fig 5.11C) consists of seven bones (Fig. 5.11A & B): talus, calcaneus, cuboid, navicular, and three cuneiforms. Only one bone, the talus, articulates with the leg bones.

The **talus** (L., ankle bone) has a body, neck, and head (Fig. 5.11D). The superior surface, or **trochlea of the talus,** is gripped by the two malleoli (Fig. 5.4) and receives the weight of the body from the tibia. The talus transmits that weight in turn, dividing it between the *calcaneus,* on which the **body of talus** rests, and the forefoot, via an osseoligamentous

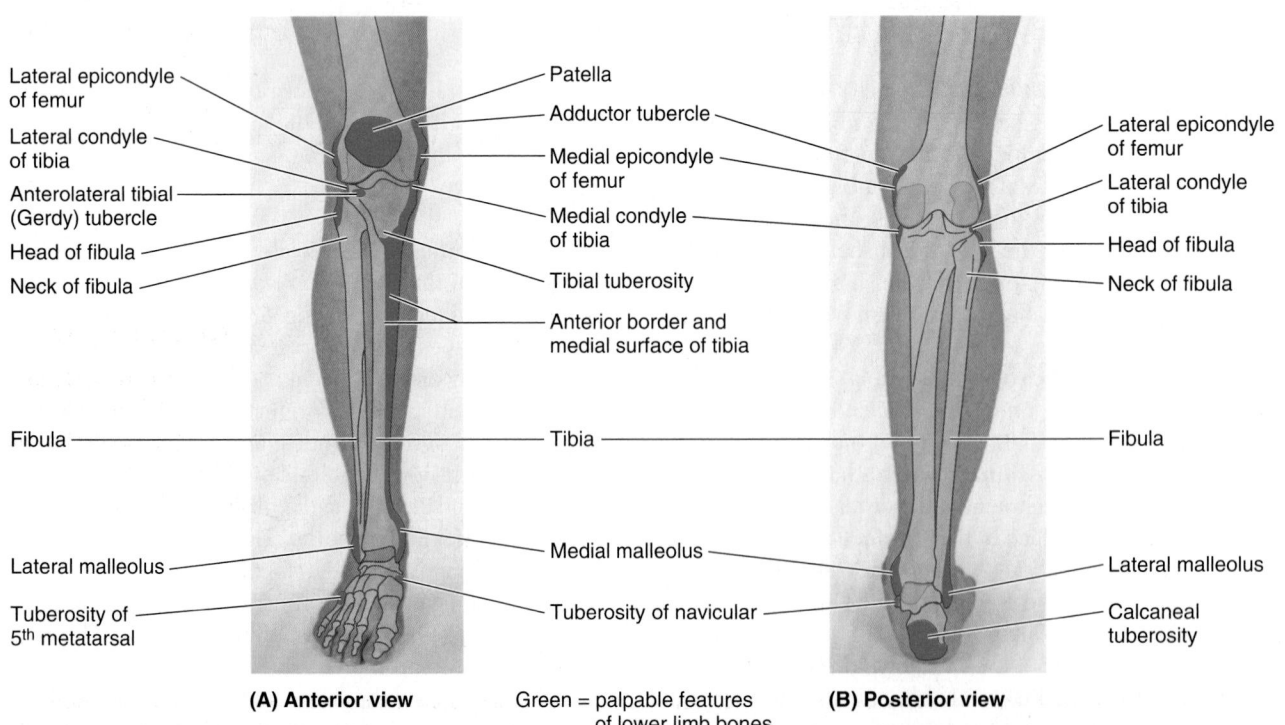

Lateral epicondyle of femur
Lateral condyle of tibia
Anterolateral tibial (Gerdy) tubercle
Head of fibula
Neck of fibula
Fibula
Lateral malleolus
Tuberosity of 5th metatarsal

Patella
Adductor tubercle
Medial epicondyle of femur
Medial condyle of tibia
Tibial tuberosity
Anterior border and medial surface of tibia
Tibia
Medial malleolus
Tuberosity of navicular

Lateral epicondyle of femur
Lateral condyle of tibia
Head of fibula
Neck of fibula
Fibula
Lateral malleolus
Calcaneal tuberosity

(A) Anterior view   Green = palpable features of lower limb bones   **(B) Posterior view**

FIGURE 5.10. Surface projection and palpable features of bones of leg, ankle, and heel.

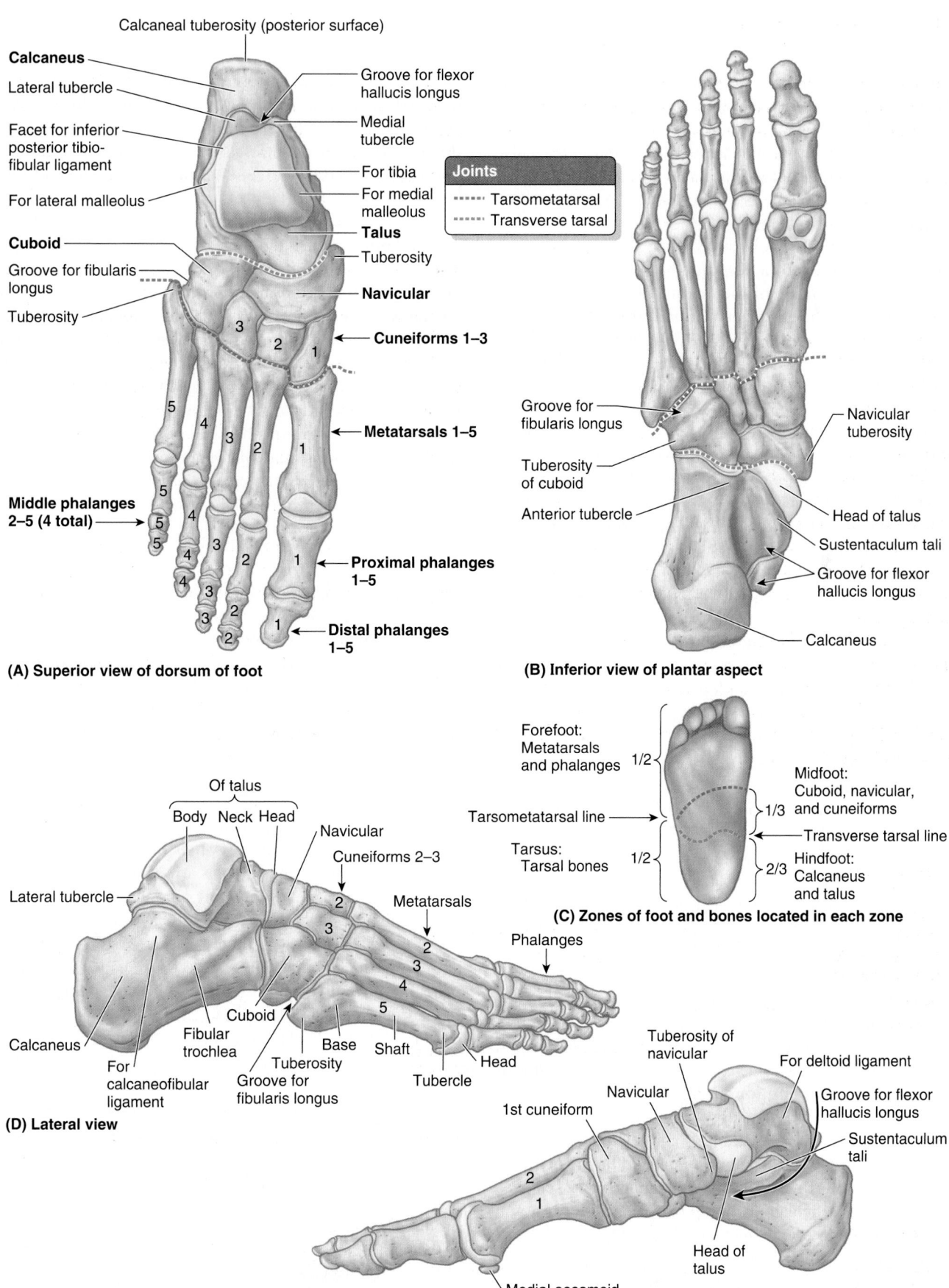

Calcaneal tuberosity (posterior surface)

Calcaneus
Lateral tubercle
Facet for inferior posterior tibio-fibular ligament
For lateral malleolus
Cuboid
Groove for fibularis longus
Tuberosity

Groove for flexor hallucis longus
Medial tubercle
For tibia
For medial malleolus
Talus
Tuberosity
Navicular

3 2 1 ← **Cuneiforms 1–3**

5 4 3 2 1 ← **Metatarsals 1–5**

Joints
----- Tarsometatarsal
----- Transverse tarsal

5 5 4 5 5 4 4 3 3 3 2 2 1 ← **Proximal phalanges 1–5**

Middle phalanges 2–5 (4 total) →

3 3 2 2 1 ← **Distal phalanges 1–5**

(A) Superior view of dorsum of foot

Groove for fibularis longus
Tuberosity of cuboid
Anterior tubercle

Navicular tuberosity
Head of talus
Sustentaculum tali
Groove for flexor hallucis longus
Calcaneus

(B) Inferior view of plantar aspect

Forefoot: Metatarsals and phalanges 1/2
Tarsometatarsal line
Tarsus: Tarsal bones 1/2

Midfoot: Cuboid, navicular, and cuneiforms 1/3
Transverse tarsal line
Hindfoot: Calcaneus and talus 2/3

(C) Zones of foot and bones located in each zone

Of talus
Body Neck Head
Navicular
Cuneiforms 2–3
2
3
Metatarsals
2
3
4
5
Phalanges

Lateral tubercle
Calcaneus
For calcaneofibular ligament
Fibular trochlea
Cuboid
Tuberosity
Groove for fibularis longus
Base Shaft
Tubercle
Head

(D) Lateral view

Tuberosity of navicular
For deltoid ligament
Navicular
Groove for flexor hallucis longus
Sustentaculum tali
1st cuneiform
2
1
Head of talus
Medial sesamoid

(E) Medial view

FIGURE 5.11. Bones of right foot. A, B, D, and E. Bones of foot in four views, demonstrating articular surfaces and major prominences and grooves.
C. The seven bones of the tarsus make up the posterior half of the foot. The talus and calcaneus occupy the posterior two thirds of the tarsus, or the hindfoot, and the cuboid, navicular, and medial, lateral, and intermediate cuneiforms occupy the anterior third, or midfoot. The metatarsus connects the tarsus posteriorly with the phalanges anteriorly. Together, the metatarsus and phalanges make up the anterior half of the foot (forefoot).

"hammock" that receives the rounded and anteromedially directed **head of talus.** The hammock (spring ligament) is suspended across a gap between a shelf-like medial projection of the calcaneus (sustentaculum tali) and the navicular bone, which lies anteriorly (Fig. 5.11B & E).

The talus is the only tarsal bone that has no muscular or tendinous attachments. Most of its surface is covered with articular cartilage. The talar body bears the trochlea superiorly and narrows into a *posterior process* that features a **groove for the tendon of the flexor hallucis longus** (Fig. 5.11E), flanked by a prominent **lateral tubercle** and a less prominent **medial tubercle** (Fig. 5.11A & D).

The **calcaneus** (L., heel bone) is the largest and strongest bone in the foot (Fig. 5.11). When standing, the calcaneus transmits the majority of the body's weight from the talus to the ground. The anterior two thirds of the calcaneus's superior surface articulates with the talus and its anterior surface articulates with the cuboid.

The lateral surface of the calcaneus has an oblique ridge (Fig. 5.11D), the **fibular trochlea,** that lies between the tendons of the fibularis longus and brevis. This trochlea anchors a tendon pulley for the evertors of the foot (muscles that move the sole of the foot away from the median plane). The **sustentaculum tali** (L., talar shelf), the shelf-like support of the head of the talus, projects from the superior border of the medial surface of the calcaneus (Fig. 5.11B & E). The posterior part of the calcaneus has a massive, weight-bearing prominence, the **calcaneal tuberosity** (L. *tuber calcanei*), which has **medial, lateral,** and **anterior tubercles.** Only the medial tubercle contacts the ground during standing.

The **navicular** (L., little ship) is a flattened, boat-shaped bone located between the head of the talus posteriorly and the three cuneiforms anteriorly (Fig. 5.11). The medial surface of the navicular projects inferiorly to form the **navicular tuberosity,** an important site for tendon attachment because the medial border of the foot does not rest on the ground, as does the lateral border. Instead, it forms a *longitudinal arch of the foot,* which must be supported centrally. If this tuberosity is too prominent, it may press against the medial part of the shoe and cause foot pain.

The **cuboid,** approximately cubical in shape, is the most lateral bone in the distal row of the tarsus (Fig. 5.11A & D). Anterior to the **tuberosity of the cuboid** on the lateral and inferior surfaces of the bone is a **groove for the tendon of the fibularis (peroneus) longus** muscle.

The three cuneiform bones (Fig. 5.11A, D, & E) are the medial (1st), intermediate (2nd), and lateral (3rd). The **medial cuneiform** is the largest bone, and the **intermediate cuneiform** is the smallest. Each cuneiform (L. *cuneus,* wedge shaped) articulates with the navicular posteriorly and the base of its appropriate metatarsal anteriorly. The **lateral cuneiform** also articulates with the cuboid.

METATARSUS

The **metatarsus** (anterior or distal foot, forefoot—Fig. 5.11C) consists of five metatarsals that are numbered from the medial side of the foot (Fig. 5.11A). In the articulated skeleton of the foot (Figs. 5.1, 5.4, and 5.11), the tarsometatarsal joints form an oblique **tarsometatarsal line** joining the midpoints of the medial and shorter lateral borders of the foot. Thus, the metatarsals and phalanges are located in the anterior half (forefoot) and the tarsals are in the posterior half (hindfoot) (Fig. 5.11A & C).

The **1st metatarsal** is shorter and stouter than the others. The **2nd metatarsal** is the longest. Each metatarsal has a base proximally, a shaft, and a head distally (Fig. 5.11C). The base of each metatarsal is the larger, proximal end. The bases of the metatarsals articulate with the cuneiform and cuboid bones, and the heads articulate with the proximal phalanges. The bases of the 1st and 5th metatarsals have large tuberosities that provide for tendon attachment; the **tuberosity of the 5th metatarsal** projects laterally over the cuboid (Fig. 5.11A & D). On the plantar surface of the head of the 1st metatarsal are prominent **medial** and **lateral sesamoid bones** (Fig. 5.11E); they are embedded in the tendons passing along the plantar surface (see the following section on Surface Anatomy).

PHALANGES

The 14 **phalanges** of the lower limb are as follows: the 1st digit (great toe) has 2 phalanges (proximal and distal); the other four digits have 3 phalanges each: proximal, middle, and distal (Fig. 5.11A & D). Each **phalanx** has a **base** (proximally), a **shaft,** and a **head** (distally). The phalanges of the 1st digit are short, broad, and strong. The middle and distal phalanges of the 5th digit may be fused in elderly people.

Surface Anatomy of Bones of Foot

The *head of the talus* is palpable anteromedial to the proximal part of the lateral malleolus when the foot is inverted, and anterior to the medial malleolus when the foot is everted (Fig. 5.12A). Eversion of the foot makes the talar head more prominent as it moves away from the navicular. The head of the talus occupies the space between the sustentaculum tali and the navicular tuberosity. If the talar head is difficult to palpate, draw a line from the tip of the medial malleolus to the navicular tuberosity; the head of the talus lies deep to the center of this line. When the foot is plantarflexed, the superior surface of the body of the talus can be palpated on the anterior aspect of the ankle, anterior to the inferior end of the tibia.

The weight-bearing *medial tubercle of the calcaneus* on the plantar surface of the foot is broad and large (Fig. 5.12D), but often it is not palpable because of the overlying skin and subcutaneous tissue. The *sustentaculum tali* is the only part of the medial aspect of the calcaneus that may be palpated as a small prominence approximately a finger's breadth distal to the tip of the medial malleolus (Fig. 5.12B). The entire lateral surface of the calcaneus is subcutaneous. The *fibular trochlea,* a small lateral extension of the calcaneus, may be detectable as a small tubercle on the lateral aspect of the calcaneus, antero-inferior to the tip of the lateral malleolus (Fig 5.12C).

Usually, palpation of bony prominences on the plantar surface of the foot is difficult because of the thick skin, fascia, and

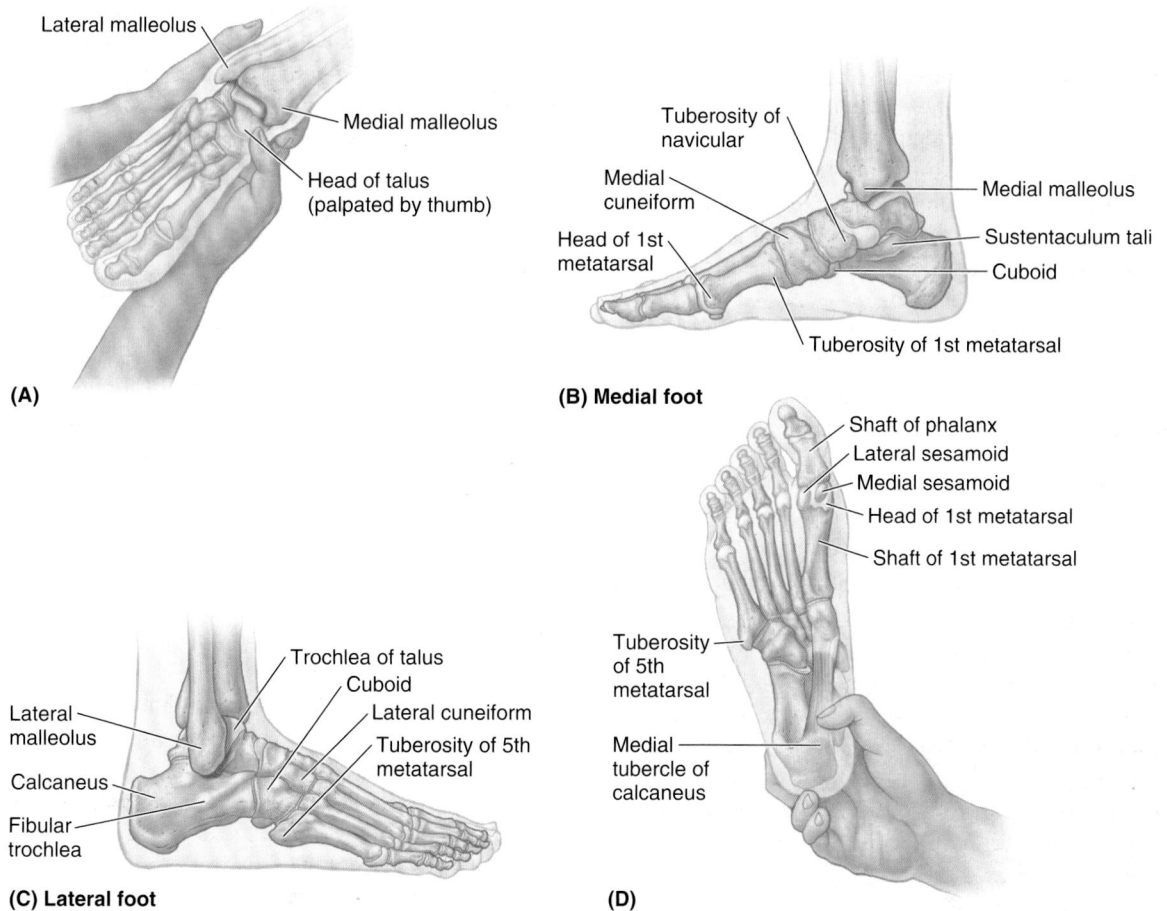

FIGURE 5.12. Surface projection and palpation of bony prominences of foot.

pads of fat. The medial and lateral *sesamoid bones* inferior to the head of the 1st metatarsal can be felt to slide when the great toe is moved passively. The *heads of the metatarsals* can be palpated by placing the thumb on their plantar surfaces and the index finger on their dorsal surfaces. If *callosities* (calluses), thickenings of the keratin layer of the epidermis, are present, the metatarsal heads are difficult to palpate.

The *tuberosity of the 5th metatarsal* forms a prominent landmark on the lateral aspect of the foot (Fig. 5.12C & D), that can easily be palpated at the midpoint of the lateral border of the foot. The *shafts of the metatarsals and phalanges*

can be felt on the dorsum of the foot between the extensor tendons.

The *cuboid* can be felt on the lateral aspect of the foot, posterior to the base of the 5th metatarsal. The *medial cuneiform* can be palpated between the tuberosity of the navicular and the base of the 1st metatarsal (Fig. 5.12B). The *head of the 1st metatarsal* forms a prominence on the medial aspect of the foot. The *tuberosity of the navicular* is easily seen and palpated on the medial aspect of the foot (Fig. 5.12B), inferoanterior to the tip of the medial malleolus. The cuboid and cuneiforms are difficult to identify individually by palpation.

BONES OF LOWER LIMB

Lower Limb Injuries

Knee, leg, and foot injuries are the most common lower limb injuries. Injuries to the hips make up less than 3% of lower limb injuries. In general, most injuries result from acute trauma during contact sports such as hockey and football and from overuse during endurance sports such as marathon races.

Adolescents are most vulnerable to these injuries because of the demands of sports on their maturing musculoskeletal systems. The cartilaginous models of the bones in the developing lower limbs are transformed into bone by endochondral ossification (Fig. 5.2E & F). Because the process is not completed until early adulthood, cartilaginous epiphysial plates still exist during the teenage years, when physical activity often peaks and involvement in competitive sports is most common.

The **epiphysial plates** are discs of hyaline cartilage between the metaphysis and epiphysis of a mature long bone that permit the bone to grow longer. During growth spurts, bones actually grow faster than the attached muscle. The combined stress on the epiphysial plates resulting from physical activity and rapid growth may result in irritation and injury of the plates and developing bone (*osteochondrosis*).

Injuries of Hip Bone

Fractures of the hip bone are referred to as *pelvic fractures* (see the blue box "Pelvic Fractures" in Chapter 3, p. 335). The term *hip fracture* is most commonly applied (unfortunately) to fractures of the femoral head, neck, or trochanters.

Avulsion fractures of the hip bone may occur during sports that require sudden acceleration or deceleration forces, such as sprinting or kicking in football, soccer, hurdle jumping, basketball, and martial arts (Fig. B5.1). A small part of bone

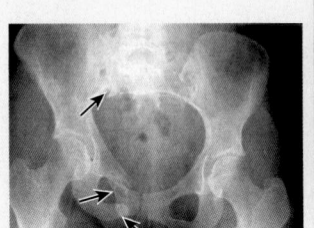

(A) Pelvic fracture (radiograph)

(B) Hip fracture (fracture of neck of left femur) -- MRI

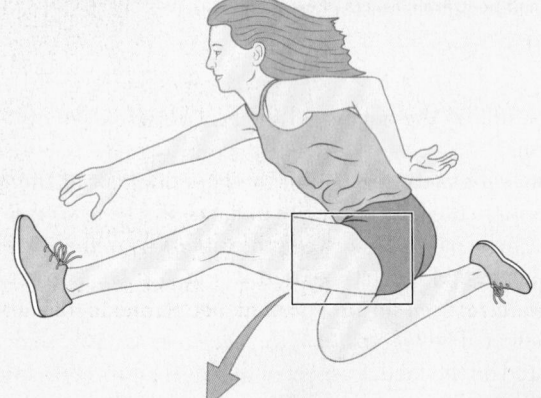

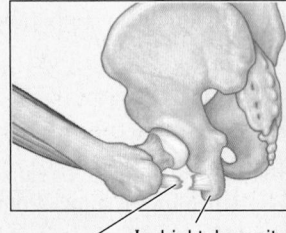

Ischial tuberosity

Hamstrings tendon (torn and avulsed from tuberosity)

(C) Tendon avulsion

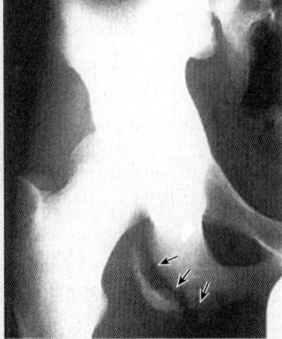

(D) Avulsion fracture of pelvis in an adolescent competitive athlete

FIGURE B5.1.

with a piece of a tendon or ligament attached is "avulsed" (torn away). These fractures occur at **apophyses** (bony projections that lack secondary ossification centers). Avulsion fractures occur where muscles are attached: anterior superior and inferior iliac spines, ischial tuberosities, and ischiopubic rami.

Coxa Vara and Coxa Valga

The angle of inclination between the long axis of the femoral neck and the femoral shaft (Fig. 5.7C–E) varies with age, sex, and development of the femur (e.g., a congenital defect in the ossification of the femoral neck). It may also change with any pathological process that weakens the neck of the femur (e.g., rickets). When the angle of inclination is decreased, the condition is *coxa vara* (Fig. B5.2A); when it is increased, it is *coxa valga* (Fig. B5.2B). The term "vara" or "varus" is a Latin adjective describing any bone or joint in a limb that is deformed so that the distal element (the shaft of the femur relative to the femoral neck in this case) deviates toward the midline. Conversely, the term "valga" or "valgus" describes a bone or joint in a limb that is deformed so that the distal element deviates away from the midline. Coxa vara causes a mild shortening of the lower limb and limits passive abduction of the hip.

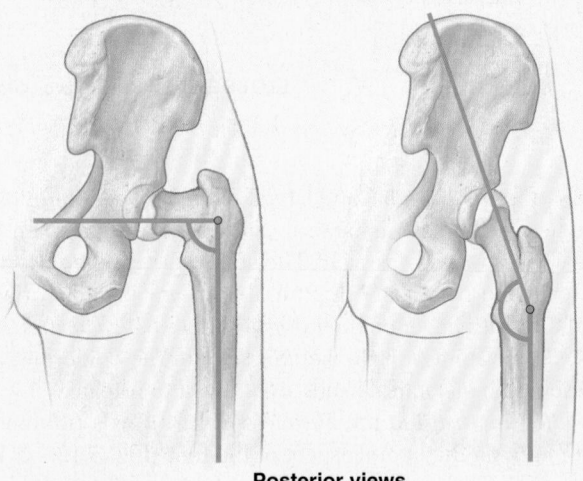

Posterior views

(A) Coxa vara (decreased angle of inclination)

(B) Coxa valga (increased angle of inclination)

FIGURE B5.2.

Dislocated Epiphysis of Femoral Head

In older children and adolescents (10–17 years of age), the epiphysis of the femoral head may slip away from the femoral neck because of a weakened epiphysial plate. This injury may be caused by acute trauma or repetitive microtraumas that place increased shearing stress on the epiphysis, especially with abduction and lateral rotation of the thigh. The epiphysis often dislocates (slips) slowly resulting in a progressive *coxa vara*. The common initial symptom of the injury is hip discomfort that may be referred to the knee. Radiographic examination of the superior end of the femur is usually required to confirm a diagnosis of a dislocated epiphysis of the head of the femur.

Femoral Fractures

Despite its large size and strength, the femur is commonly fractured. The type of fracture sustained is frequently age- and even sex-related. The neck of the femur is most frequently fractured because it is the narrowest and weakest part of the bone and it lies at a marked angle to the line of weight-bearing (pull of gravity). It becomes increasingly vulnerable with age, especially in females, secondary to *osteoporosis*.

Fractures of the proximal femur occur at several locations; two examples are transcervical (middle of neck) and intertrochanteric (Fig. B5.3). These fractures usually occur as a result of indirect trauma (stumbling or stepping down hard, as off a curb or step). Because of the angle of inclination, these fractures are inherently unstable and *impaction* (overriding of fragments resulting in foreshortening of the limb) occurs. Muscle spasm also contributes to the shortening of the limb.

Intracapsular fractures (occurring within the hip joint capsule) are complicated by degeneration of the femoral head, owing to vascular trauma (see the blue boxes "Fractures of the Femoral Neck" on p. 659 and "Surgical Hip Replacement" on p. 660).

Fractures of the greater trochanter and femoral shaft usually result from direct trauma (direct blows sustained by the bone resulting from falls or being hit) and are most common during the more active years. They frequently occur during motor vehicle accidents and sports, such as skiing and climbing. In some cases, a *spiral fracture* of the femoral shaft occurs, resulting in foreshortening as the fragments override, or the fracture may be comminuted (broken into several pieces), with the fragments displaced in various directions as a result of muscle pull and depending on the level of the fracture. Union of this serious type of fracture may take up to a year.

Fractures of the inferior or distal femur may be complicated by separation of the condyles, resulting in misalignment of the articular surfaces of the knee joint, or by hemorrhage from the large popliteal artery that runs directly on the posterior surface of the bone. This fracture compromises the blood supply to the leg (an occurrence that should always be considered in knee fractures or dislocations).

Tibial Fractures

The tibial shaft is narrowest at the junction of its middle and inferior thirds, which is the most frequent site of fracture. Unfortunately, this area of the bone also has the poorest blood supply. Because its anterior surface is subcutaneous, the tibial shaft is the

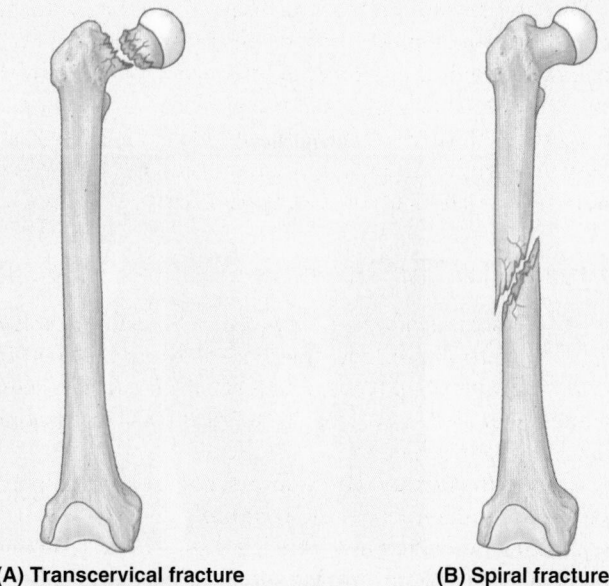

(A) Transcervical fracture of femoral neck

(B) Spiral fracture

A - C Anterior views

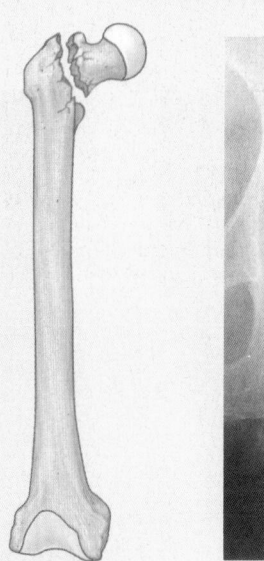

(C) Intertrochanteric fracture

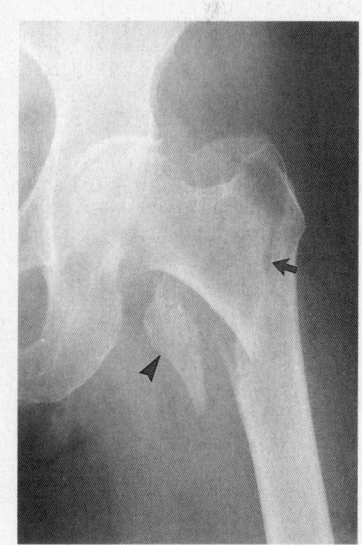

(D) Comminuted intertrochanteric fracture of left proximal femur

FIGURE B5.3.

most common site for a *compound fracture* (Fig. B5.4A). Compound tibial fractures may also result from direct trauma (e.g., a "bumper fracture" caused when a car bumper strikes the leg). Fracture of the tibia through the nutrient canal predisposes the patient to non-union of the bone fragments resulting from damage to the nutrient artery.

Transverse march (stress) fractures of the inferior third of the tibia (Fig. B5.4B) are common in people who take long hikes before they are conditioned for this activity. The strain may fracture the anterior cortex of the tibia. Indirect violence applied to the tibial shaft when the bone turns with the foot fixed during a fall may produce a fracture (e.g., when a person is tackled in football).

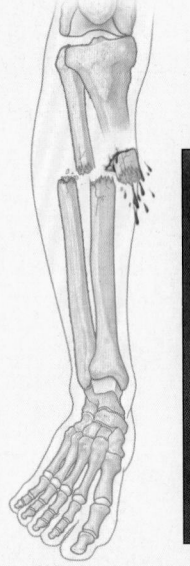

Lateral view | Anterior view

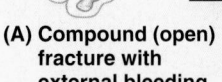

(A) Compound (open) fracture with external bleeding

(B) March (stress) fracture of tibia (arrows), most apparent in the MRI study at the right

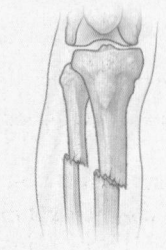

(C) Diagonal fracture with shortening

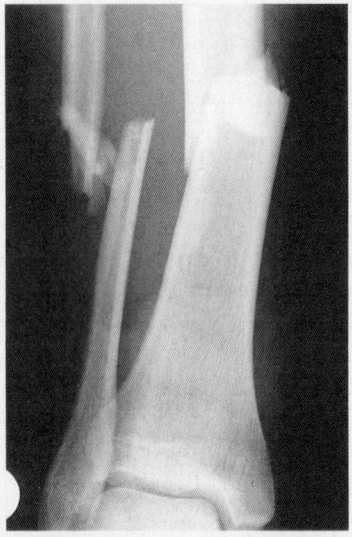

(E) Transverse "boot top" fracture with shortening due to overriding of fracture fragments

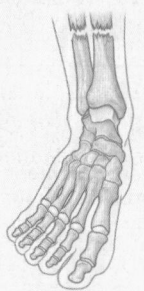

(D) Transverse "boot top" fracture

A - E Anterior views

FIGURE B5.4.

Fractures Involving Epiphysial Plates

The primary ossification center for the superior end of the tibia appears shortly after birth and joins the shaft of the tibia during adolescence (usually 16–18 years of age). Tibial fractures in children are more serious if they involve the epiphysial plates because continued normal growth of the bone may be jeopardized. The tibial tuberosity usually forms by inferior bone growth from the superior epiphysial center at approximately 10 years of age, but a separate center for the tibial tuberosity may appear at approximately 12 years of age. Disruption of the epiphysial plate at the tibial tuberosity may cause inflammation of the tuberosity and chronic recurring pain during adolescence (*Osgood-Schlatter disease*), especially in young athletes (Fig. B5.5).

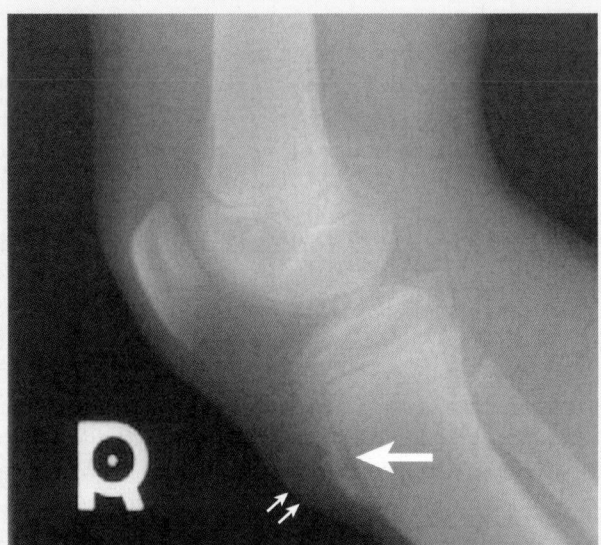

Osgood-Schlatter Disease

Prominence of tibial tuberosity (*double arrow*) elongated and fragmented (*single arrow*), with overlying soft tissue swelling

FIGURE B5.5.

Fibular Fractures

Fibular fractures commonly occur 2–6 cm proximal to the distal end of the lateral malleolus and are often associated with *fracture–dislocations of the ankle joint*, which are combined with tibial fractures (Fig. B5.6B). When a person slips and the foot is forced into an excessively inverted position, the ankle ligaments tear, forcibly tilting the talus against the lateral malleolus, and may shear it off (Fig. B5.6C).

Fractures of the lateral and medial malleoli are relatively common in soccer and basketball players. Fibular fractures can be painful owing to disrupted muscle attachments. Walking is compromised because of the bone's role in ankle stability.

In addition, severe torsion during skiing may produce a *diagonal fracture* (Fig. B5.4C) of the tibial shaft at the junction of the middle and inferior thirds, as well as a *fracture of the fibula*. Diagonal fractures are often associated with limb shortening caused by overriding of the fractured ends. Frequently during skiing, a fracture results from a high-speed forward fall, which angles the leg over the rigid ski boot, producing a "boot-top fracture" (Fig. B5.4D & E).

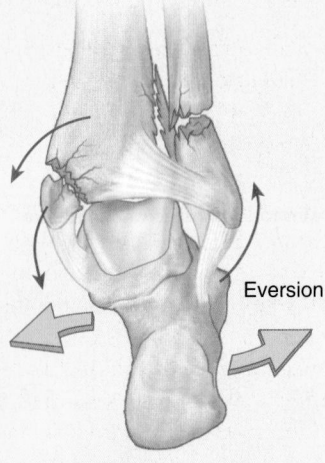

(A) Tibial and fibular fractures

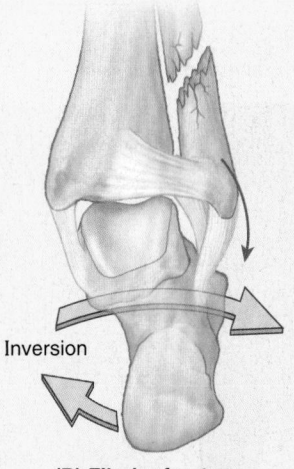

Inversion

(B) Fibular fracture with excessive inversion of foot

Posterior views

FIGURE B5.6.

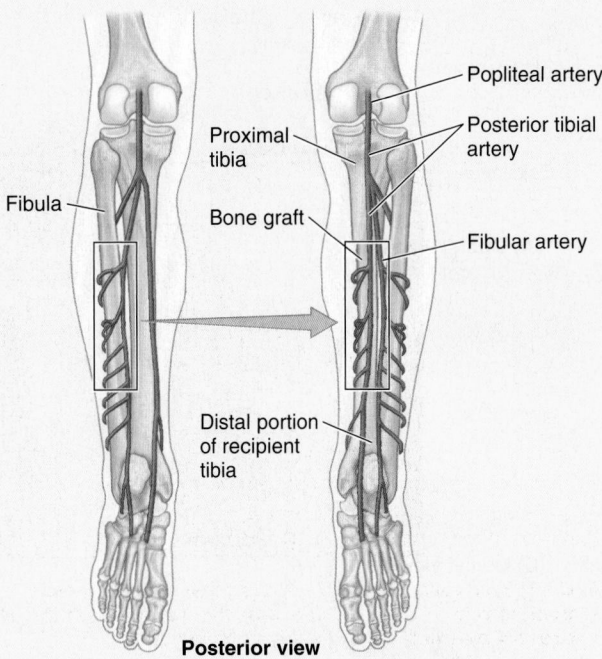

Posterior view

Section of left fibula grafted into right tibia

FIGURE B5.7.

which it is now attached. Healing proceeds as if a fracture had occurred at each of its ends.

Awareness of the location of the nutrient foramen in the fibula is important when performing free vascularized fibular transfers. Because the nutrient foramen is located in the middle third of the fibula in most cases (Fig. 5.9, posterior view), this segment of the bone is used for transplanting when the graft must include a blood supply to the medullary cavity as well as to the compact bone of the surface (via the periosteum).

Because of its extensive subcutaneous location, the anterior tibia is accessible for obtaining pieces of bone for grafting in children; it is also used as a site for *intramedullary infusion* in dehydrated or shocked children.

Bone Grafts

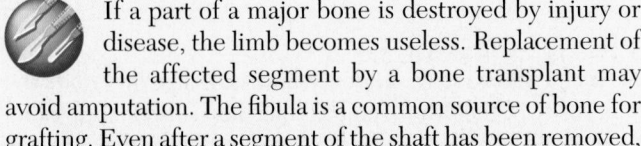 If a part of a major bone is destroyed by injury or disease, the limb becomes useless. Replacement of the affected segment by a bone transplant may avoid amputation. The fibula is a common source of bone for grafting. Even after a segment of the shaft has been removed, walking, running, and jumping can be normal.

Free vascularized fibulas have been used to restore skeletal integrity to upper and lower limbs in which congenital bone defects exist and to replace segments of bone after trauma or excision of a malignant tumor (Fig. B5.7). The remaining parts of the fibula usually do not regenerate because the periosteum and nutrient artery are generally removed with the piece of bone, so that the graft will remain alive and grow when transplanted to another site. Secured in its new site, the fibular segment restores the blood supply of the bone to

Calcaneal Fractures

A hard fall onto the heel, from a ladder for example, may fracture the calcaneus into several pieces, producing a *comminuted fracture* (Fig. B5.8A). A calcaneal fracture is usually disabling because it disrupts the subtalar (talocalcaneal) joint, where the talus articulates with the calcaneus.

Fractures of Talar Neck

Fractures of the talar neck (Fig. B5.8B) may occur during severe dorsiflexion of the ankle (e.g., when a person is pressing extremely hard on the brake pedal of a vehicle during a head-on collision). In some cases, the body of the talus dislocates posteriorly.

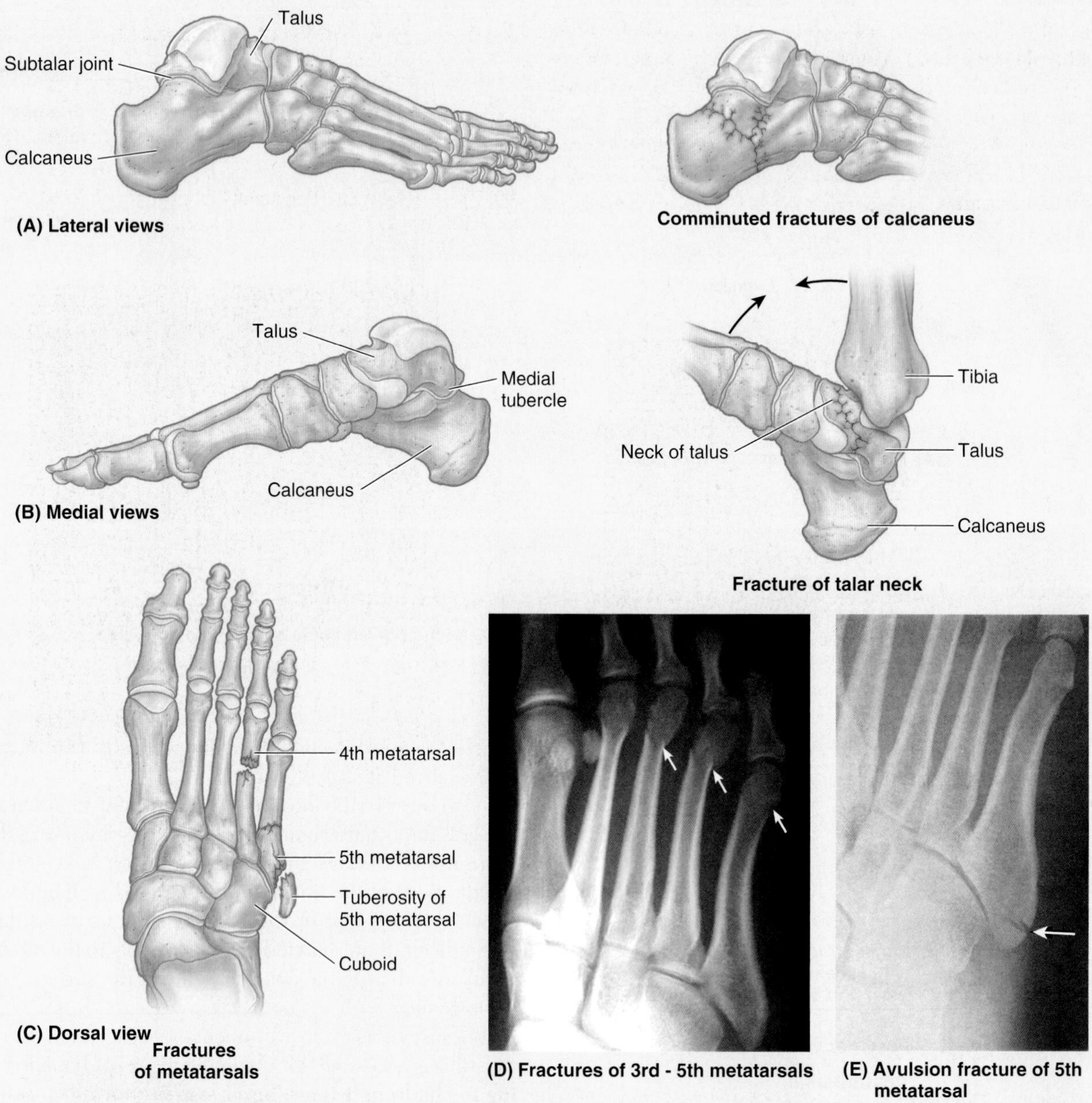

(A) Lateral views

Subtalar joint

Talus

Calcaneus

Comminuted fractures of calcaneus

Talus

Medial tubercle

Calcaneus

(B) Medial views

Tibia

Neck of talus

Talus

Calcaneus

Fracture of talar neck

4th metatarsal

5th metatarsal

Tuberosity of 5th metatarsal

Cuboid

(C) Dorsal view
Fractures of metatarsals

(D) Fractures of 3rd - 5th metatarsals

(E) Avulsion fracture of 5th metatarsal

FIGURE B5.8.

Fractures of Metatarsals

Metatarsal fractures occur when a heavy object falls on the foot, for example, or when it is run over by a heavy object such as a metal wheel (Fig. B5.8C & D). Metatarsal fractures are also common in dancers, especially female ballet dancers who use the demi-pointe technique. The *dancer's fracture* usually occurs when the dancer loses balance, putting the full body weight on the metatarsal and fracturing the bone. *Fatigue fractures of the metatarsals* may result from prolonged walking. These fractures, usually transverse, result from repeated stress on the metatarsals.

When the foot is suddenly and violently inverted, the tuberosity of the 5th metatarsal may be avulsed (torn away) by the tendon of the fibularis brevis muscle. *An avulsion fracture of the tuberosity of the 5th metatarsal* (Fig. B5.8C & E) is common in basketball and tennis players. This injury produces pain and edema at the base of the 5th metatarsal and may be associated with a severe ankle sprain.

Os Trigonum

During ossification of the talus, the secondary ossification center, which becomes the lateral tubercle of the talus, occasionally fails to unite with the body of the talus. This failure may be caused by applied stress (forceful plantarflexion) during the early teens. Occasionally, a partly or even fully ossified center may fracture and progress to non-union. Either event may result in a bone (accessory ossicle) known as an **os trigonum,** which occurs in 14–25% of adults, more commonly bilaterally (Fig. B5.9). It has an increased prevalence among soccer players and ballet dancers.

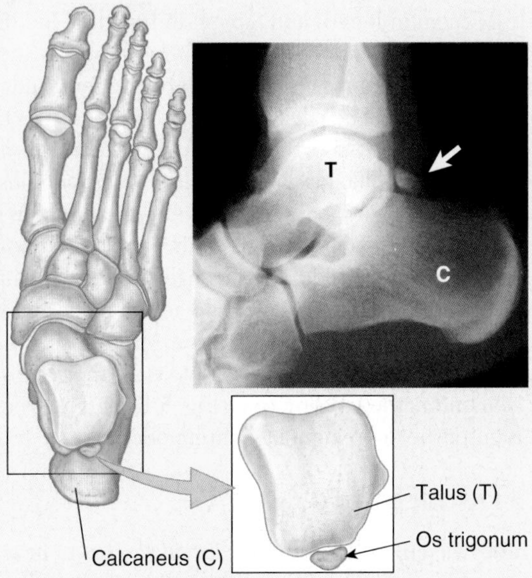

Fracture of Sesamoid Bones

The sesamoid bones of the great toe (Fig. 5.12D) in the tendon of the flexor hallucis longus bear the weight of the body, especially during the latter part of the stance phase of walking. The sesamoids develop before birth and begin to ossify during late childhood. *Fracture of the sesamoid bones* may result from a crushing injury (Fig B5.10).

Talus (T)
Os trigonum

Calcaneus (C)

FIGURE B5.9.

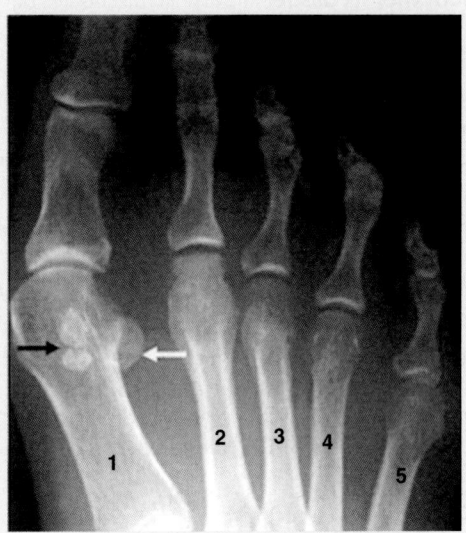

Black arrow: Fractured sesamoid bone
White arrow: Normal sesamoid bone
Metatarsals 1-5

FIGURE B5.10.

The Bottom Line

BONES OF LOWER LIMB

Hip bone: Formed by the union of three primary bones (ilium, ischium, and pubis), the hip bones are joined to the sacrum posteriorly and to each other anteriorly (at the pubic symphysis) to form the pelvic girdle. ♦ Each hip bone is specialized to receive half the weight of the upper body when standing and all of it periodically during walking. ♦ Thick parts of the bone transfer weight to the femur. ♦ Thin parts of the bone provide a broad surface for attachment of powerful muscles that move the femur. ♦ The pelvic girdle encircles and protects the pelvic viscera, particularly the reproductive organs.

Femur: Through development, our largest bone, the femur, has developed a bend (angle of inclination) and has twisted (medial rotation and torsion so that the knee and all joints inferior to it flex posteriorly) to accommodate our erect posture and to enable bipedal walking and running. ♦ The angle of inclination and attachment of the abductors and rotators to the greater trochanter allow increased leverage, superior placement of the abductors, and oblique orientation of the

femur in the thigh. ♦ Combined with the torsion angle, oblique rotatory movements at the hip joint are converted into movements of flexion–extension and abduction–adduction (in the sagittal and coronal planes, respectively) as well as of rotation.

Tibia and fibula: Our second largest bone, the tibia, is a ver-tical column bearing the weight of all superior to it. ♦ The slender fibula does not bear weight but, along with the interosseous membrane that binds it to the tibia, is accessory to the tibia in providing an additional surface area for fleshy muscle attachment and in forming the socket of the ankle joint. ♦ Through development, the two bones have become permanently pronated to provide for a stable stance and facilitate locomotion.

Bones of foot: The many bones of the foot form a functional unit that allows weight to be distributed to a wide platform to maintain balance when standing, enable conformation and adjustment to terrain variations, and perform shock absorption. ♦ They also transfer weight from the heel to the forefoot as required in walking and running.

FASCIA, VEINS, LYMPHATICS, EFFERENT VESSELS, AND CUTANEOUS NERVES OF LOWER LIMB

Subcutaneous Tissue and Fascia

The **subcutaneous tissue** (superficial fascia) lies deep to the skin (Fig. 5.13) and consists of loose connective tissue that contains a variable amount of fat, cutaneous nerves, superficial veins (great and small saphenous veins and their tributaries), lymphatic vessels, and lymph nodes.

The subcutaneous tissue of the hip and thigh is continuous with that of the inferior part of the anterolateral abdominal wall and buttocks. At the knee, the subcutaneous tissue loses its fat and blends with the deep fascia, but fat is again present distal to the knee in the subcutaneous tissue of the leg.

The **deep fascia of the lower limb** is especially strong, investing the limb like an elastic stocking (Fig. 5.13A & B). This fascia limits outward expansion of contracting muscles, making muscular contraction more efficient in compressing veins to push blood toward the heart.

FASCIA LATA

The deep fascia of the thigh is called **fascia lata** (L. *lata*, broad). Superiorly, the fascia lata attaches to and is continuous with:

- The inguinal ligament, pubic arch, body of pubis, and pubic tubercle anteriorly.
- The membranous layer of subcutaneous tissue (Scarpa fascia) of the inferior abdominal wall also attaches to the fascia lata approximately a finger's breadth inferior to the inguinal ligament.
- The iliac crest laterally and posteriorly.
- The sacrum, coccyx, sacrotuberous ligament, and ischial tuberosity/ischiopubic ramus posteriorly and medially.

Inferiorly, the fascia lata attaches to and is continuous with:

- Exposed parts of bones around the knee.
- The deep fascia of the leg inferior to the knee.

The fascia lata is substantial because it encloses the large thigh muscles, especially laterally, where it is thickened and strengthened by additional reinforcing longitudinal fibers to form the **iliotibial tract** (Fig. 5.13B). This broad band of fibers is the shared aponeurosis of the tensor fasciae latae and gluteus maximus muscles. The iliotibial tract extends from the iliac tubercle to the **anterolateral tubercle of the tibia** (*Gerdy tubercle*).

The thigh muscles are separated into three compartments—anterior, medial, and posterior. The walls of these compartments are formed by the fascia lata and three fascial intermuscular septa that arise from its deep aspect and attach to the linea aspera of the femur (Fig. 5.13D). The **lateral** intermuscular septum is especially strong; the other two septa are relatively weak. The lateral intermuscular septum extends deeply from the iliotibial tract to the lateral lip of the linea aspera and lateral supracondylar line of the femur. This septum offers an *internervous plane* (plane between nerves) to surgeons needing wide exposure of the femur.

The **saphenous opening** in the fascia lata (Fig. 5.13A) is a gap or hiatus in the fascia lata inferior to the medial part of the inguinal ligament, approximately 4 cm inferolateral to the pubic tubercle. The saphenous opening is usually approximately 3.75 cm in length and 2.5 cm in breadth, and its long axis is vertical. Its medial margin is smooth but its superior, lateral, and inferior margins form a sharp crescentic edge, the **falciform margin.** The falciform margin is joined at its medial margin by fibrofatty tissue, the **cribriform fascia** (L. *cribrum*, a sieve). This sieve-like fascia is a localized membranous layer of subcutaneous tissue that spreads over the saphenous opening, closing it. The connective tissue is pierced by numerous openings (thus its name) for the passage of efferent lymphatic vessels from the superficial inguinal lymph nodes, and by the great saphenous vein and its tributaries. After passing through the saphenous opening and cribriform fascia, the *great saphenous vein* enters the femoral vein (Fig. 5.13A). The lymphatic vessels enter the deep inguinal lymph nodes.

DEEP FASCIA OF LEG

The **deep fascia of the leg,** or *crural fascia* (L. *crus*, leg), attaches to the anterior and medial borders of the tibia, where it is continuous with its periosteum. The deep fascia of the leg is thick in the proximal part of the anterior aspect of the leg, where it forms part of the proximal attachments of the underlying muscles. Although thinner distally, the deep fascia of the leg forms thickened bands both superior and anterior to the ankle joint, the **extensor retinacula** (Fig. 5.13A).

Anterior and **posterior intermuscular septa** pass from the deep surface of the lateral deep fascia of the leg and attach to the corresponding margins of the fibula. The *interosseous membrane* and intermuscular septa divide the leg into three compartments: anterior (dorsiflexor), lateral (fibular), and posterior (plantarflexor) (Fig. 5.13C). The posterior compartment is further subdivided by the **transverse intermuscular septum,** separating superficial and deep plantarflexor muscles.

Venous Drainage of Lower Limb

The lower limb has superficial and deep veins: the superficial veins are in the subcutaneous tissue and run independent from named arteries; the deep veins are deep to (beneath) the deep fascia and accompany all major arteries. Superficial and deep veins have valves, which are more numerous in deep veins.

SUPERFICIAL VEINS OF LOWER LIMB

The two major superficial veins in the lower limb are the *great and small saphenous veins* (Fig. 5.14A & B). Most of their tributaries are unnamed.

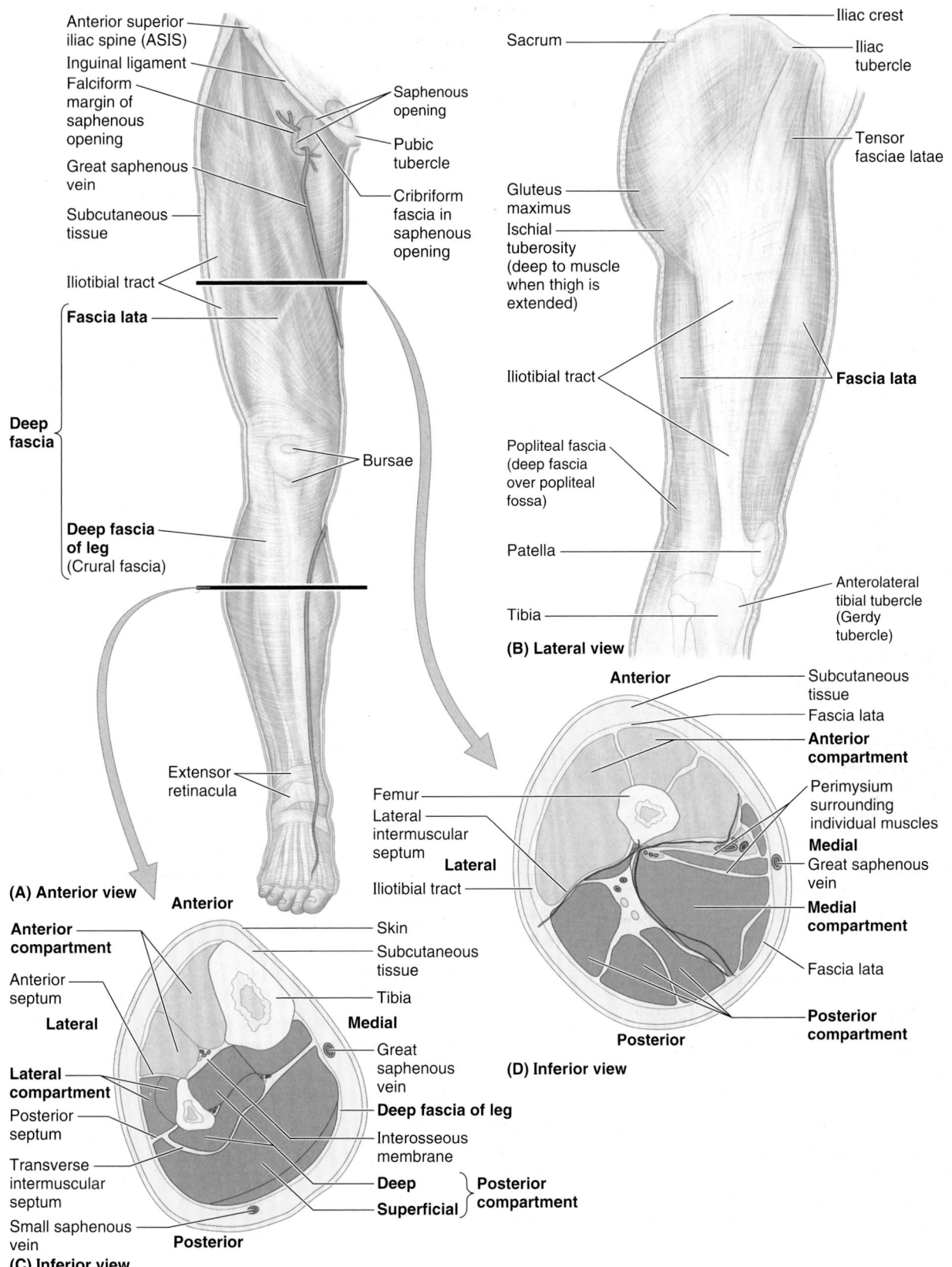

FIGURE 5.13. Fascia, intermuscular septa, and fascial compartments of lower limb. A. The anterior skin and subcutaneous tissue have been removed to reveal the deep fascia. **B.** The fascia lata is reinforced laterally by longitudinal fibers of the iliotibial tract, the common aponeurotic tendon of the gluteus maximus and tensor fasciae latae. **C and D.** The fascial compartments of the thigh and leg, containing muscles sharing common functions and innervation, are demonstrated in transverse sections.

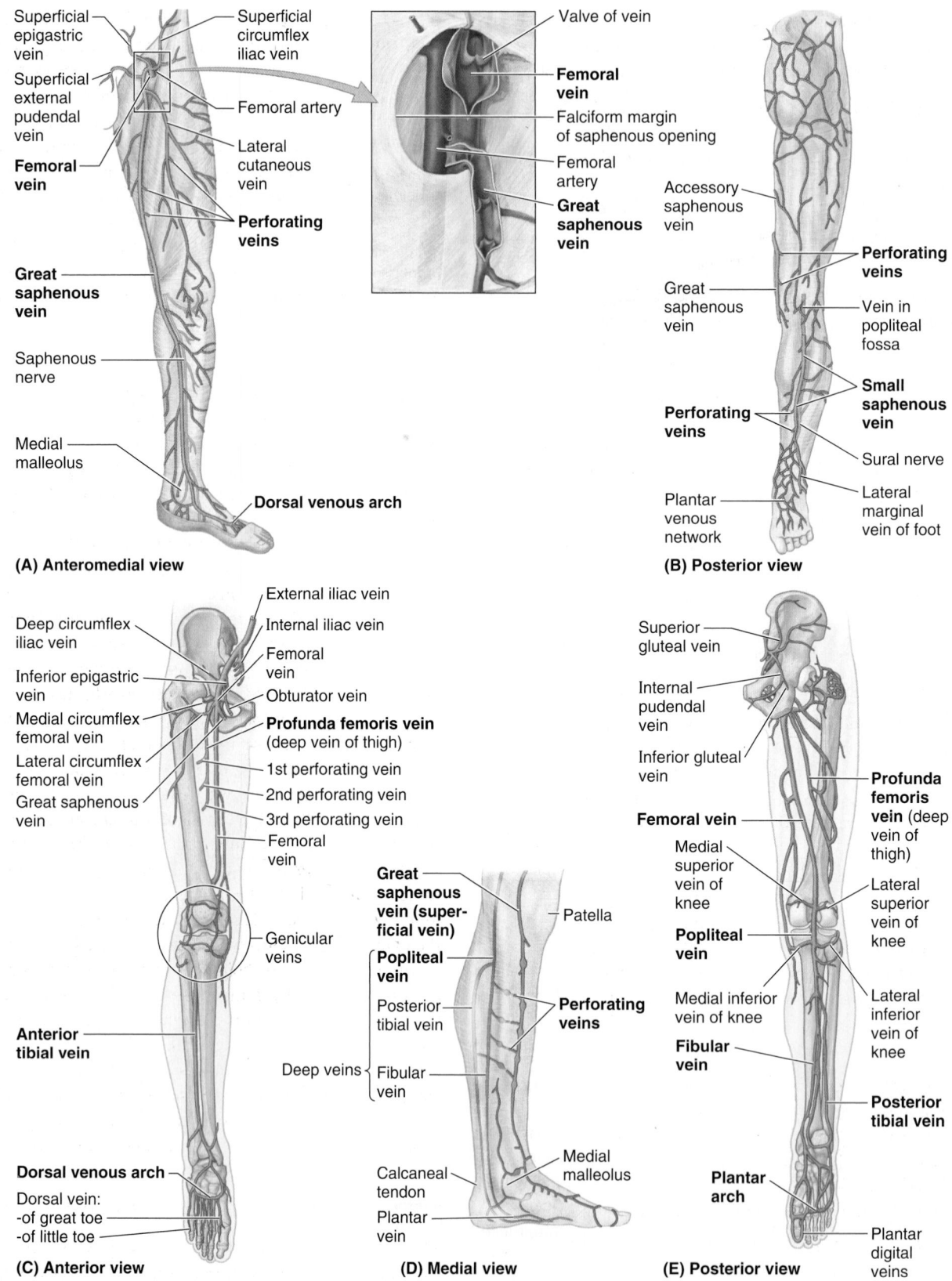

FIGURE 5.14. Veins of lower limb. The veins are divided into superficial (**A and B**) and deep (**C and E**) groups. The superficial veins, usually unaccompanied, course within the subcutaneous tissue; the deep veins are internal to the deep fascia and usually accompany arteries. **A, inset.** The proximal ends of the femoral and great saphenous veins are opened and spread apart to show the valves. Although depicted as single veins in parts **C** and **E,** the deep veins usually occur as duplicate or multiple accompanying veins. **D.** Multiple perforating veins pierce the deep fascia to shunt blood from the superficial veins to the deep veins.

The **great saphenous vein** is formed by the union of the *dorsal vein of the great toe* and the *dorsal venous arch of the foot* (Figs. 5.14A & 5.15A). The great saphenous vein:

- Ascends anterior to the medial malleolus.
- Passes posterior to the medial condyle of the femur (about a hand's breadth posterior to the medial border of the patella).
- Anastomoses freely with the small saphenous vein.
- Traverses the saphenous opening in the fascia lata.
- Empties into the femoral vein.

The great saphenous vein has 10–12 valves, which are more numerous in the leg than in the thigh (Fig. 5.14A & D). These valves are usually located just inferior to the perforating veins. The perforating veins also have valves.

Venous valves are cusps (flaps) of endothelium with cup-like *valvular sinuses* that fill from above. When they are full, the valve cusps occlude the lumen of the vein, thereby preventing reflux of blood distally, making flow unidirectional. The valvular mechanism also breaks the column of blood in the saphenous vein into shorter segments, reducing back pressure. Both effects make it easier for the *musculovenous pump* (discussed in the Introduction) to overcome the force of gravity to return the blood to the heart.

As it ascends in the leg and thigh, the great saphenous vein receives numerous tributaries and communicates in several locations with the small saphenous vein. Tributaries from the medial and posterior aspects of the thigh frequently unite to form an **accessory saphenous vein** (Fig. 5.14B). When present, this vein becomes the main communication between the great and small saphenous veins.

Also, fairly large vessels, the **lateral** and **anterior cutaneous veins**, arise from networks of veins in the inferior part of the thigh and enter the great saphenous vein superiorly, just before it enters the femoral vein. Near its termination, the great saphenous vein also receives the superficial circumflex iliac, superficial epigastric, and external pudendal veins (Fig. 5.14A).

The **small saphenous vein** arises on the lateral side of the foot from the union of the *dorsal vein of the little toe* with the *dorsal venous arch* (Fig. 5.14B & C). The small saphenous vein:

- Ascends posterior to the lateral malleolus as a continuation of the lateral marginal vein.
- Passes along the lateral border of the calcaneal tendon.
- Inclines to the midline of the fibula and penetrates the deep fascia.
- Ascends between the heads of the gastrocnemius muscle.
- Empties into the popliteal vein in the popliteal fossa.

Although many tributaries are received by the saphenous veins, their diameters remain remarkably uniform as they ascend the limb. This is possible because the blood received by the saphenous veins is continuously shunted from these superficial veins in the subcutaneous tissue to the deep veins internal to the deep fascia by means of many perforating veins.

The **perforating veins** penetrate the deep fascia close to their origin from the superficial veins and contain valves that allow blood to flow only from the superficial veins to the deep veins (Fig. 5.14A & D). The perforating veins pass through the deep fascia at an oblique angle so that when muscles contract and the pressure increases inside the deep fascia, the perforating veins are compressed. Compression of these veins also prevents blood from flowing from the deep to the superficial veins. This pattern of venous blood flow—from superficial to deep—is important for proper venous return from the lower limb because it enables muscular contractions to propel blood toward the heart against gravity (**musculovenous pump**—see Fig. I.25 in Introduction).

DEEP VEINS OF LOWER LIMB

Deep veins accompany all the major arteries and their branches. Instead of occurring as a single vein in the limbs (although they are frequently illustrated as one and are often referred to as a single vein), the **accompanying veins** (L. *venae comitantes*) usually occur as paired, frequently interconnecting veins that flank the artery they accompany (Fig. 5.14C & E). *They are contained within a vascular sheath with the artery,* whose pulsations also help compress and move blood in the veins.

Although the *dorsal venous arch* drains primarily via the saphenous veins, perforating veins penetrate the deep fascia, forming and continually supplying an **anterior tibial vein** in the anterior leg. **Medial** and **lateral plantar veins** from the plantar aspect of the foot form the **posterior tibial** and **fibular veins** posterior to the medial and lateral malleoli (Fig. 5.14C–E). All three deep veins from the leg flow into the popliteal vein posterior to the knee, which becomes the femoral vein in the thigh. Veins accompanying the perforating arteries of the profunda femoris vein drain blood from the thigh muscles and terminate in the *profunda femoris vein* (deep vein of thigh), which joins the terminal portion of the femoral vein (Fig 5.14C & E). The femoral vein passes deep to the inguinal ligament to become the external iliac vein.

Because of the effect of gravity, blood flow is slower when a person stands quietly. During exercise, blood received by the deep veins from the superficial veins is propelled by muscular contraction to the femoral and then the external iliac veins. Flow in the reverse direction is prevented if the valves are competent. The deep veins are more variable and anastomose much more frequently than the arteries they accompany. Both superficial and deep veins can be ligated if necessary.

Lymphatic Drainage of Lower Limb

The lower limb has superficial and deep lymphatic vessels. The **superficial lymphatic vessels** converge on and accompany the saphenous veins and their tributaries (Fig. 5.15A). The lymphatic vessels accompanying the great saphenous vein end in the vertical group of **superficial inguinal lymph nodes.** Most lymph from these nodes passes directly to the *external iliac lymph nodes,* located along the external iliac

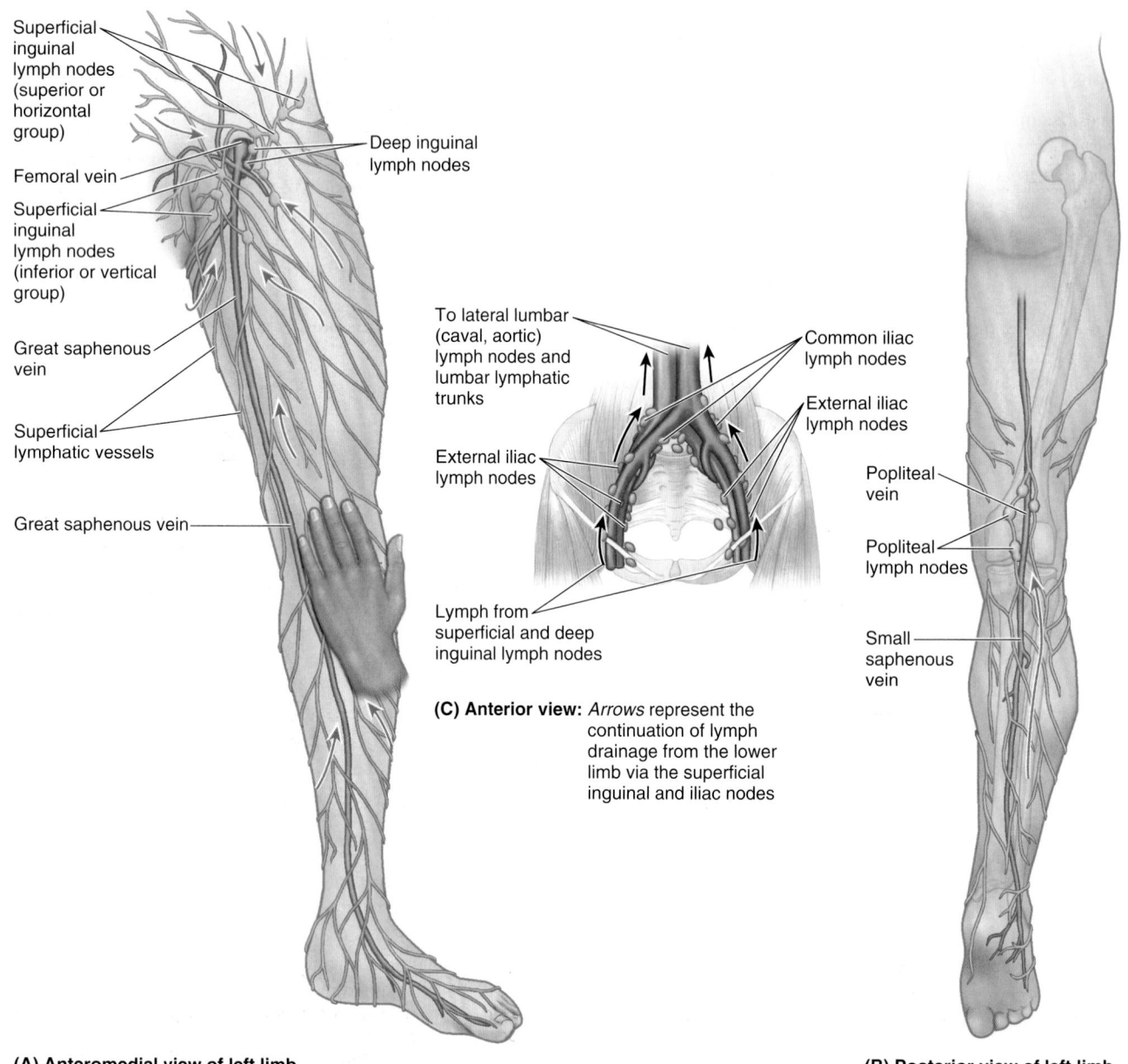

(A) Anteromedial view of left limb   **(B) Posterior view of left limb**

FIGURE 5.15. Superficial veins and lymphatics of lower limb. A. The superficial lymphatic vessels converge toward and accompany the great saphenous vein, draining into the inferior (vertical) group of superficial inguinal lymph nodes. The great saphenous vein passes anterior to the medial malleolus, approximately a hand's breadth posterior to the patella. **B.** Superficial lymphatic vessels of the lateral foot and posterolateral leg accompany the small saphenous vein and drain initially into the popliteal lymph nodes. The efferent vessels from these nodes join other deep lymphatics, which accompany the femoral vessels to drain into the deep inguinal lymph nodes. **C.** Lymph from the superficial and deep inguinal lymph nodes traverses the external and common iliac nodes before entering the lateral lumbar (aortic) lymph nodes and the lumbar lymphatic trunk.

vein. Some also passes to the **deep inguinal lymph nodes,** located under the deep fascia on the medial aspect of the femoral vein. The lymphatic vessels accompanying the small saphenous vein enter the **popliteal lymph nodes,** which surround the popliteal vein in the fat of the popliteal fossa (Fig. 5.15B).

Deep lymphatic vessels from the leg accompany deep veins and also enter the popliteal lymph nodes. Most lymph from these nodes ascends through deep lymphatic vessels to the *deep inguinal lymph nodes.* Lymph from the deep nodes

passes to the external and common iliac lymph nodes and then enters the *lumbar lymphatic trunks* (Fig. 5.15C).

Cutaneous Innervation of Lower Limb

Cutaneous nerves in the subcutaneous tissue supply the skin of the lower limb (Fig. 5.16; Table 5.1). These nerves, except for some proximal unisegmental nerves arising from the T12 or L1 spinal nerves, are branches of the *lumbar* and *sacral plexuses.* The areas of skin supplied

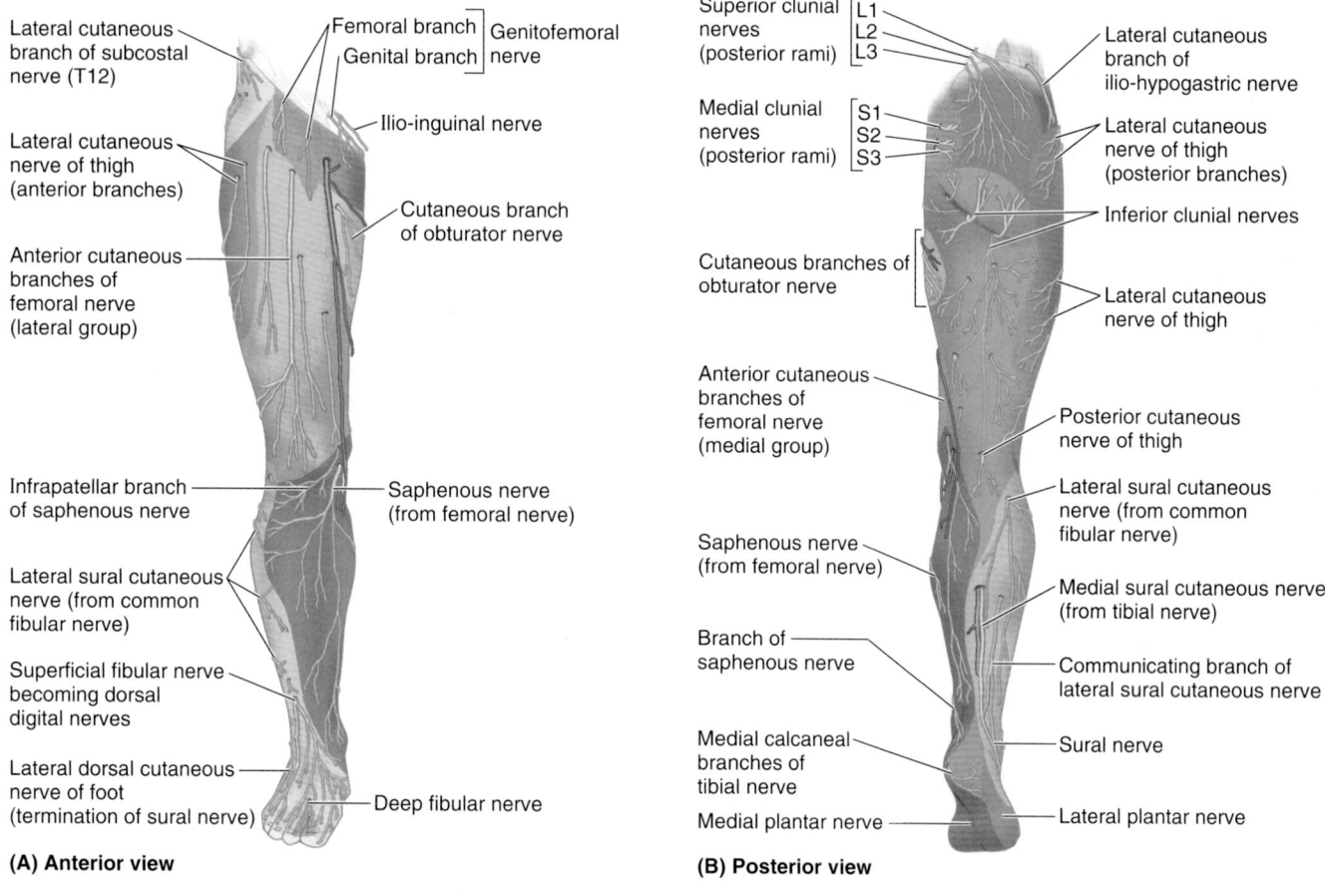

FIGURE 5.16. Cutaneous nerves of lower limb.

TABLE 5.1. CUTANEOUS NERVES OF LOWER LIMB

Nerve	Origin (Contributing Spinal Nerves)	Course	Distribution in Lower Limb
Subcostal	T12 anterior ramus	Courses along inferior border of 12th rib; lateral cutaneous branch descends over iliac crest	Lateral cutaneous branch supplies skin of hip region inferior to anterior part of iliac crest and anterior to greater trochanter
Iliohypogastric	Lumbar plexus (L1; occasionally T12)	Parallels iliac crest; divides into lateral and anterior cutaneous branches	Lateral cutaneous branch supplies superolateral quadrant of buttocks
Ilio-inguinal	Lumbar plexus (L1; occasionally T12)	Passes through inguinal canal; divides into femoral and scrotal or labial branches	Femoral branch supplies skin over medial femoral triangle
Genitofemoral	Lumbar plexus (L1–L2)	Descends anterior surface of psoas major; divides into genital and femoral branches	Femoral branch supplies skin over lateral part of femoral triangle; genital branch supplies anterior scrotum or labia majora
Lateral cutaneous nerve of thigh	Lumbar plexus (L2–L3)	Passes deep to inguinal ligament, 2–3 cm medial to anterior superior iliac spine	Supplies skin on anterior and lateral aspects of thigh
Anterior cutaneous branches	Lumbar plexus via femoral nerve (L2–L4)	Arise in femoral triangle; pierce fascia lata along path of sartorius muscle	Supply skin of anterior and medial aspects of thigh
Cutaneous branch of obturator nerve	Lumbar plexus via obturator nerve, anterior branch (L2–L4)	Following its descent between adductors longus and brevis, anterior division of obturator nerve pierces fascia lata to reach skin of thigh	Skin of middle part of medial thigh

TABLE 5.1. CUTANEOUS NERVES OF LOWER LIMB (Continued)

Nerve	Origin (Contributing Spinal Nerves)	Course	Distribution in Lower Limb
Posterior cutaneous nerve of thigh	Sacral plexus (S1–S3)	Enters gluteal region via infrapiriform portion of greater sciatic foramen deep to gluteus maximus; then descends deep to fascia lata	Terminal branches pierce fascia lata to supply skin of posterior thigh and popliteal fossa
Saphenous nerve	Lumbar plexus via femoral nerve (L3–L4)	Traverses adductor canal but does not pass through adductor hiatus; crossing medial side of knee deep to sartorius tendon	Skin on medial side of leg and foot
Superficial fibular nerve	Common fibular nerve (L4–S1)	Courses through lateral compartment of leg; after supplying fibular muscles, perforates deep fascia of leg	Skin of anterolateral leg and dorsum of foot, excluding web between great and 2nd toes
Deep fibular nerve	Common fibular nerve (L5)	After supplying muscles on dorsum of foot, pierces deep fascia superior to heads of 1st and 2nd metatarsals	Skin of web between great and 2nd toes
Sural nerve	Tibial and common fibular nerves (S1–S2)	Medial sural cutaneous branch of tibial nerve and lateral sural cutaneous branch of fibular nerve merge at varying levels on posterior leg	Skin of posterolateral leg and lateral margin of foot
Medial plantar nerve	Tibial nerve (L4–L5)	Passes between first and second layers of plantar muscles; then between medial and middle muscles of first layer	Skin of medial side of sole, and plantar aspect, sides, and nail beds of medial 3½ toes
Lateral plantar nerve	Tibial nerve (S1–S2)	Passes between first and second layers of plantar muscles; then between middle and lateral muscles of first layer	Skin of lateral sole, and plantar aspect, sides, and nail beds of lateral 1½ toes
Calcaneal nerves	Tibial and sural nerves (S1–S2)	Lateral and medial branches of tibial and sural nerves, respectively, over calcaneal tuberosity	Skin of heel
Superior clunial nerves	L1–L3 posterior rami	Penetrate thoracodorsal fascia; course laterally and inferiorly in subcutaneous tissue	Skin overlying superior and central parts of buttocks
Medial clunial nerves	S1–S3 posterior rami	Emerge from dorsal sacral foramina; directly enter overlying subcutaneous tissue	Skin of medial buttocks and intergluteal cleft
Inferior clunial nerves	Posterior cutaneous nerve of thigh (S2–S3)	Arise deep to gluteus maximus; emerge from beneath inferior border of muscle	Skin of inferior buttocks (overlying gluteal fold)

by the individual spinal nerves, including those contributing to the plexuses, are called *dermatomes.* The dermatomal (segmental) pattern of skin innervation is retained throughout life but is distorted by limb lengthening and the torsion of the limb that occurs during development (Figs. 5.2 and 5.17).

Although simplified into distinct zones in dermatome maps, adjacent dermatomes overlap, except at the **axial line,** the line of junction of dermatomes supplied from discontinuous spinal levels. The cutaneous nerves of the lower limb are illustrated in Figure 5.16 and their origin (including contributing spinal nerves), course, and distribution are listed in Table 5.1.

Motor Innervation of Lower Limb

Somatic motor (general somatic efferent) fibers traveling in the same mixed peripheral nerves that convey sensory fibers to the cutaneous nerves transmit impulses to the muscles of the lower limb. The unilateral embryological muscle mass receiving innervation from a single spinal cord segment or spinal nerve comprise a *myotome.* Lower limb muscles usually receive motor fibers from several spinal cord segments or nerves. Thus, most muscles are composed of more than one myotome, and most often multiple spinal cord segments are involved in producing the movement of the lower limb (Fig. 5.18).

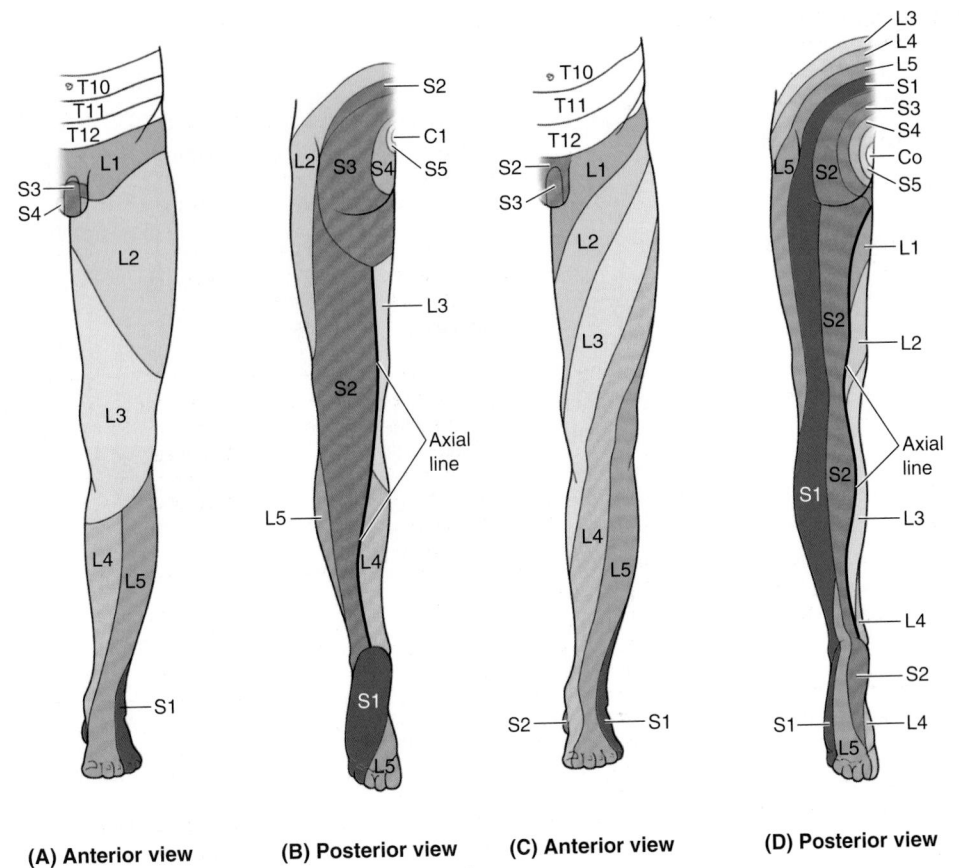

(A) Anterior view **(B) Posterior view** **(C) Anterior view** **(D) Posterior view**

FIGURE 5.17. Dermatomes of lower limb. The dermatomal or segmental pattern of distribution of sensory nerve fibers persists despite the merging of spinal nerves in plexus formation during development. Two different dermatome maps are commonly used. **A and B.** The dermatome pattern of the lower limb according to Foerster (1933) is preferred by many because of its correlation with clinical findings. **C and D.** The dermatome pattern of the lower limb according to Keegan and Garrett (1948) is preferred by others for its aesthetic uniformity and obvious correlation with development. Although depicted as distinct zones, adjacent dermatomes overlap considerably, except along the axial line.

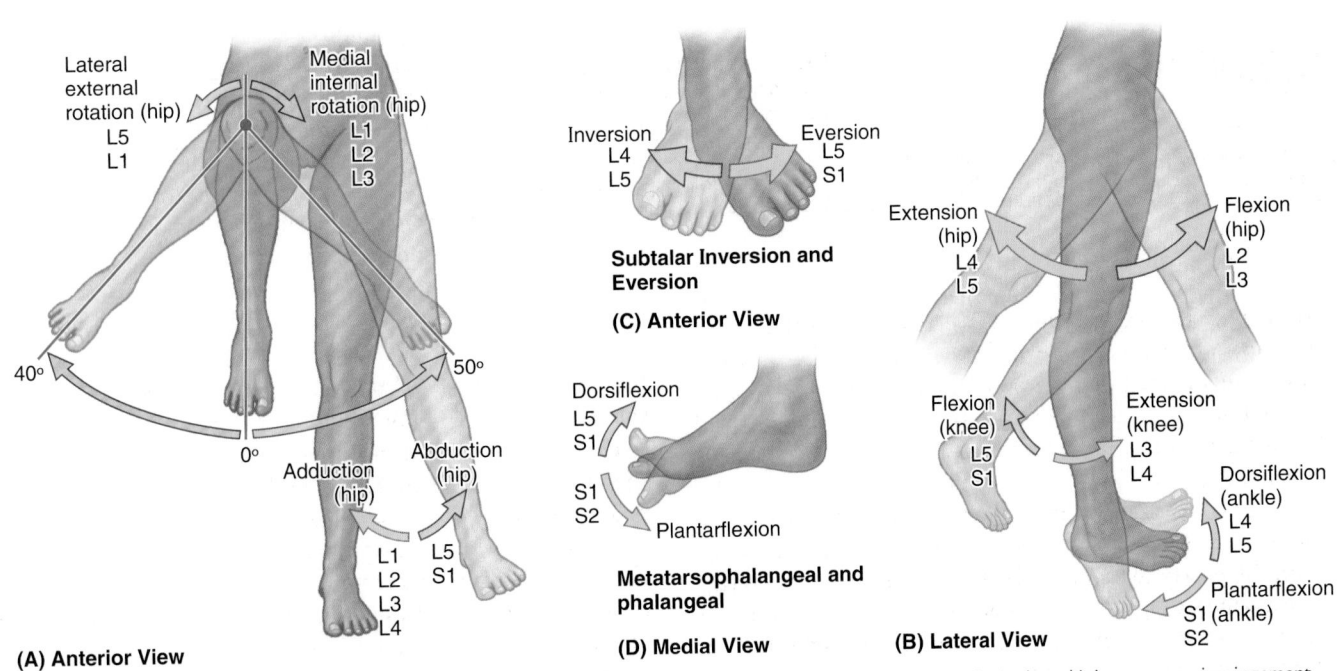

(A) Anterior View **(D) Medial View** **(B) Lateral View**

FIGURE 5.18. Myotomes: segmental innervation of muscle groups and movements of lower limb. The level of spinal cord injury or nerve impingement may be determined by the strength and ability to perform particular movements.

FASCIA, VEINS, LYMPHATICS, AND CUTANEOUS NERVES OF LOWER LIMB

Compartment Syndromes and Fasciotomy

 The fascial compartments of the lower limbs are generally closed spaces, ending proximally and distally at the joints. Trauma to muscles and/or vessels in the compartments from burns, sustained intense use of muscles, or blunt trauma may produce hemorrhage, edema, and inflammation of the muscles. Because the septa and deep fascia of the leg forming the boundaries of the leg compartments are strong, the increased volume consequent to any of these processes increases intracompartmental pressure.

The pressure may reach levels high enough to compress structures significantly in the compartment(s) concerned. The small vessels of muscles and nerves *(vasa nervorum)* are particularly vulnerable to compression. Structures distal to the compressed area may become ischemic and permanently injured (e.g., loss of motor function in muscles whose blood supply and/or innervation is affected). Increased pressure in a confined anatomical space adversely affects the circulation and threatens the function and viability of tissue within or distally, constituting *compartment syndromes.*

Loss of distal leg pulses is an obvious sign of arterial compression, as is lowering of the temperature of tissues distal to the compression. A *fasciotomy* (incision of overlying fascia or a septum) may be performed to relieve the pressure in the compartment(s) concerned.

Varicose Veins, Thrombosis, and Thrombophlebitis

 Frequently, the great saphenous vein and its tributaries become *varicose* (dilated so that the cusps of their valves do not close). *Varicose veins* are common in the posteromedial parts of the lower limb and may cause discomfort (Fig. B5.11A). In a healthy vein, the valves allow blood to flow toward the heart (B) while keeping blood from flowing away from the heart (C). Valves in varicose veins (D) are incompetent due to dilation or rotation and no longer function properly. As a result, blood flows inferiorly in the veins, producing varicose veins.

Deep venous thrombosis (DVT) of one or more of the deep veins of the lower limb is characterized by swelling, warmth, and *erythema* (inflammation and infection). *Venous stasis* (stagnation) is an important cause of thrombus formation. Venous stasis can be caused by:

- Incompetent, loose fascia that fails to resist muscle expansion, diminishing the effectiveness of the musculovenous pump. (Fig I.26, p. 42)

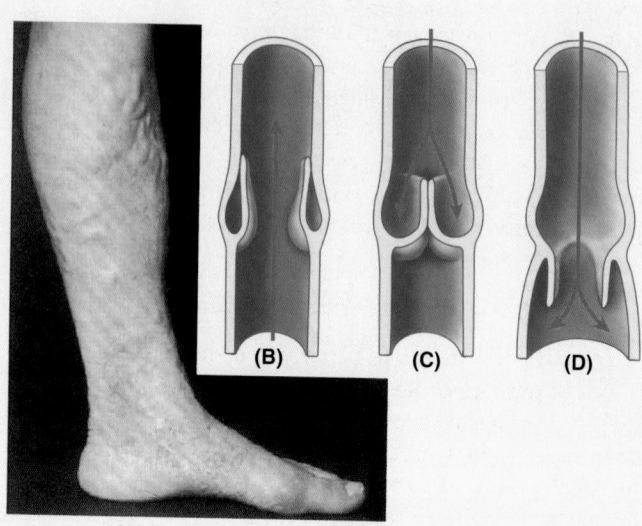

(A) Varicose veins

FIGURE B5.11.

- External pressure on the veins from bedding during a prolonged hospital stay or from a tight cast or bandage.
- Muscular inactivity (e.g., during an overseas aircraft flight).

DVT with inflammation around the involved veins *(thrombophlebitis)* may develop. A large thrombus that breaks free from a lower limb vein may travel to a lung, forming a *pulmonary thromboembolism* (obstruction of a pulmonary artery). A large embolus may obstruct a main pulmonary artery and may cause death.

Saphenous Vein Grafts

 The great saphenous vein is sometimes used for coronary arterial bypasses because (1) it is readily accessible, (2) a sufficient distance occurs between the tributaries and the perforating veins so that usable lengths can be harvested, and (3) its wall contains a higher percentage of muscular and elastic fibers than do other superficial veins.

Saphenous vein grafts are used to bypass obstructions in blood vessels (e.g., in an intracoronary thrombus). When part of the great saphenous vein is removed for a bypass, the vein is inverted so that the valves do not obstruct blood flow in the venous graft. Because there are so many other leg veins, removal of the great saphenous vein rarely produces a significant problem in the lower limb or seriously affects circulation, provided the deep veins are intact. In fact, removal of this vein may facilitate the superficial to deep drainage pattern to take advantage of the musculovenous pump.

Saphenous Cutdown and Saphenous Nerve Injury

 Even when it is not visible in infants, in obese people, or in patients in shock whose veins are collapsed, the great saphenous vein

can always be located by making a skin incision anterior to the medial malleolus (Fig. 5.14A). This procedure, called a *saphenous cutdown,* is used to insert a cannula for prolonged administration of blood, plasma expanders, electrolytes, or drugs.

The saphenous nerve accompanies the great saphenous vein anterior to the medial malleolus. Should this nerve be cut during a saphenous cutdown or caught by a ligature during closure of a surgical wound, the patient may complain of pain or numbness along the medial border of the foot.

Enlarged Inguinal Lymph Nodes

Lymph nodes enlarge when diseased. *Abrasions* and minor sepsis, caused by pathogenic microorganisms or their toxins in the blood or other tissues, may produce moderate enlargement of the superficial inguinal lymph nodes *(lymphadenopathy)* in otherwise healthy people. Because these enlarged nodes are located in subcutaneous tissue, they are usually easy to palpate.

When inguinal lymph nodes are enlarged, their entire field of drainage—the trunk inferior to the umbilicus, including the perineum, as well as the entire lower limb—should be examined to determine the cause of their enlargement. In female patients, the relatively remote possibility of metastasis of cancer from the uterus should also be considered because some lymphatic drainage from the uterine fundus may flow along lymphatics accompanying the round ligament of the uterus through the inguinal canal to reach the superficial inguinal lymph nodes. All palpable lymph nodes should also be examined.

Regional Nerve Blocks of Lower Limbs

Interruption of the conduction of impulses in peripheral nerves (nerve block) may be achieved by making perineural injections of anesthetics close to the nerves whose conductivity is to be blocked.

The femoral nerve (L2–L4) can be blocked 2 cm inferior to the inguinal ligament, approximately a finger's breadth lateral to the femoral artery. *Paresthesia* (tingling, burning, tickling) radiates to the knee and over the medial side of the leg if the saphenous nerve (terminal branch of femoral) is affected.

Abnormalities of Sensory Function

In most instances, a peripheral nerve sensitizing an area of skin represents more than one segment of the spinal cord. Therefore, to interpret abnormalities of peripheral sensory function, peripheral nerve distribution of the major cutaneous nerves must be interpreted as anatomically different from dermatome distribution of the spinal cord segments (Fig. 5.17). Neighboring dermatomes may overlap.

Pain sensation is tested by using a sharp object and asking the patient if pain is felt. If there is no sensation, the spinal cord segment(s) involved can be determined.

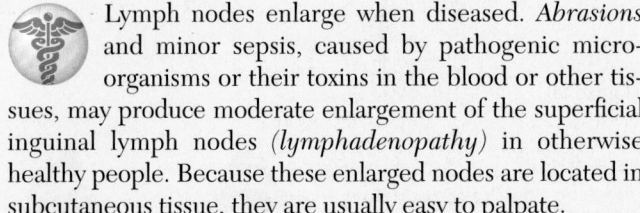

The Bottom Line

FASCIA, EFFERENT VESSELS, AND CUTANEOUS NERVES OF LOWER LIMB

Fascia: The lower limb is invested by subcutaneous tissue and deep fascia. ♦ The former insulates, stores fat, and provides passage for cutaneous nerves and superficial vessels (lymphatics and veins). ♦ The deep fascia of the thigh (fascia lata) and leg (crural fascia) (1) surround the thigh and leg, respectively, limiting outward bulging of muscles and facilitating venous return in deep veins; (2) separate muscles with similar functions and innervation into compartments; and (3) surround individual muscles, allowing them to act independently. ♦ Modifications of the deep fascia include openings that allow the passage of neurovascular structures (e.g., the saphenous opening) and thickenings that retain tendons close to the joints they act on (retinacula).

Veins: The veins of the lower limb include both superficial (in the subcutaneous tissue) and deep (internal to the deep fascia) veins. ♦ The superficial great and small saphenous veins mainly drain the integument or skin and, via many perforating veins, continuously shunt blood to the deep veins accompanying the arteries. ♦ Deep veins are subject to muscle compression (musculovenous pump) to aid venous return. ♦ All lower limb veins have valves to overcome the effects of gravity.

Lymphatic vessels: Most lymph from the lower limb drains via lymphatics that follow the superficial veins (e.g., the saphenous veins) to the superficial inguinal nodes. ♦ Some lymphatics follow deep veins to deep inguinal nodes. Lymph drainage from the lower limb then passes deep to the external and common iliac nodes of the trunk.

Cutaneous nerves: The cutaneous innervation of the lower limb reflects both the original segmental innervation of the skin via separate spinal nerves in its dermatomal pattern, and the result of plexus formation in the distribution of multisegmental peripheral nerves. ♦ Most innervation of the thigh is supplied by lateral and posterior cutaneous nerves of the thigh and anterior cutaneous branches of the femoral nerve, the names of which describe their distribution. The latter branches also supply most of the medial aspect of the thigh. ♦ The innervation of the leg and dorsum of the foot is supplied by saphenous (anteromedial leg), sural (posterolateral leg), and fibular nerves (anterolateral leg and dorsum of foot). ♦ The plantar aspect (sole) of the foot is supplied by calcaneal branches of the tibial and sural nerves (heel region) and the medial and lateral plantar nerves; the areas of distribution of the latter are demarcated by a line bisecting the 4th toe.

POSTURE AND GAIT

The lower limbs function primarily in standing and walking. Typically the actions of lower limb muscles are described as if the muscle were acting in isolation, which rarely occurs.

In this book, including the comments in the tables, the role of each muscle (or of the functional group of which it is a member) is described in typical activities, especially standing and walking. It is important to be familiar with lower limb movements and concentric and eccentric contractions of muscles, as described in the Introduction (p. 29), and to have a basic understanding of the processes of standing and walking.

Standing at Ease

When a person is standing at ease with the feet slightly apart and rotated laterally so the toes point outward, only a few of the back and lower limb muscles are active (Fig. 5.19). The mechanical arrangement of the joints and muscles are such that a minimum of muscular activity is required to keep from falling. In the stand-easy position, the hip and knee joints are extended and are in their most stable positions (maximal contact of articular surfaces for weight transfer, with supporting ligaments taut).

The ankle joint is less stable than the hip and knee joints, and the line of gravity falls between the two limbs, just anterior to the axis of rotation of the ankle joints. Consequently, a tendency to fall forward (*forward sway*) must be countered periodically by bilateral contraction of the calf muscles (plantarflexion). The spread or splay of the feet increases lateral stability. However, when *lateral sway* occurs, it is countered by the hip abductors (acting through the iliotibial tract). The fibular collateral ligament of the knee joint and the evertor muscles of one side act with the thigh adductors, tibial collateral ligament, and invertor muscles of the contralateral side.

Walking: The Gait Cycle

Locomotion is a complex function. The movements of the lower limbs during walking on a level surface may be divided into alternating swing and stance phases, illustrated in Figure 5.20 and described in Table 5.2. The **gait cycle** consists of one cycle of swing and stance by one limb. The **stance phase** begins with a **heel strike** (Fig. 5.20A), when the heel strikes the ground and begins to assume the body's full weight (loading response), and ends with a *push off* by the forefoot (Fig 5.20G)—a result of plantarflexion. (See the blue box "Absence of Plantarflexion," p. 607).

The **swing phase** begins after push off when the toes leave the ground and ends when the heel strikes the ground. The swing phase occupies approximately 40% of the walking cycle and the stance phase, 60%. The stance phase of walking is longer than the swing phase because

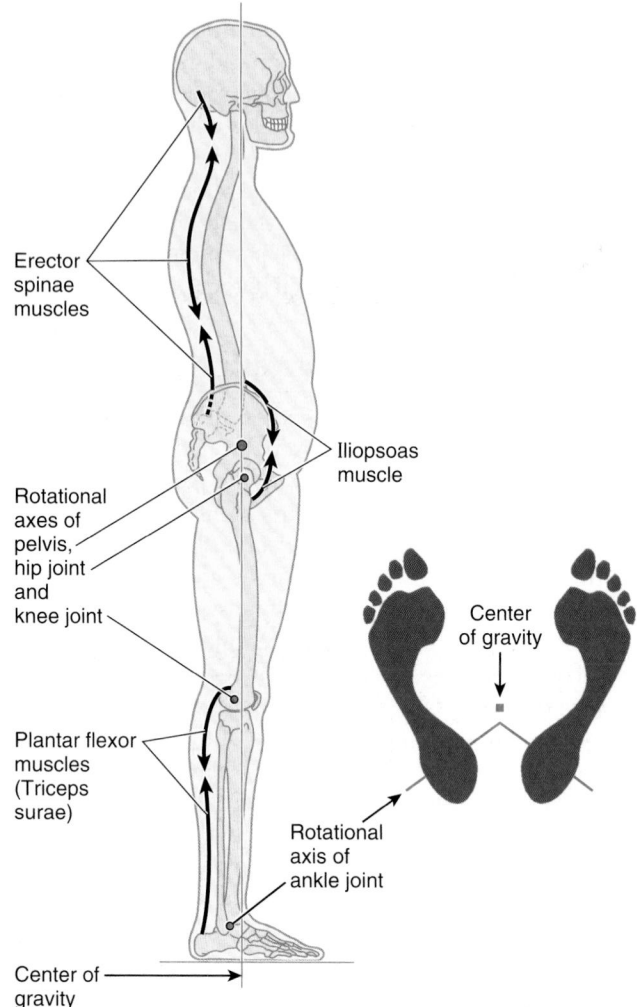

(A) Lateral view   **(B) Inferior view**

FIGURE 5.19. Relaxed standing. A. The relationship of the line of gravity to the transverse rotational axes of the pelvis and lower limb in the relaxed standing (stand-easy) position is demonstrated. Only minor postural adjustments, mainly by the extensors of the back and the plantarflexors of the ankle, are necessary to maintain this position because the ligaments of the hip and knee are being tightly stretched to provide passive support. **B.** A bipedal platform is formed by the feet during relaxed standing. The weight of the body is symmetrically distributed around the center of gravity, which falls in the posterior third of a median plane between the slightly parted and laterally rotated feet, anterior to the rotational axes of the ankle joints.

it begins and ends with relatively short periods (each 10% of the cycle) of double support (both feet are contacting the ground) as the weight is transferred from one side to the other, with a more extended period of single support (only one foot on the ground bearing all body weight) in between as the contralateral limb swings forward. In **running,** there is no period of double support; consequently, the time and percentage of the gait cycle represented by the stance phase are reduced.

Walking is a remarkably efficient activity, taking advantage of gravity and momentum so that a minimum

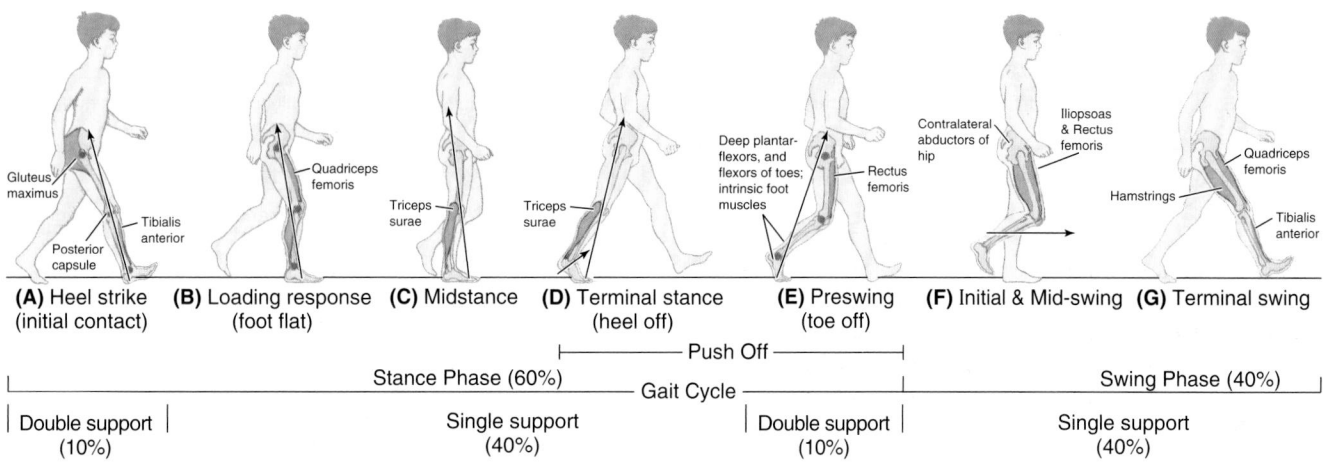

(A) Heel strike (initial contact) **(B)** Loading response (foot flat) **(C)** Midstance **(D)** Terminal stance (heel off) **(E)** Preswing (toe off) **(F)** Initial & Mid-swing **(G)** Terminal swing

Push Off

Stance Phase (60%) — Gait Cycle — Swing Phase (40%)

| Double support (10%) | Single support (40%) | Double support (10%) | Single support (40%) |

FIGURE 5.20. Gait cycle. The activity of one limb between two repeated events of walking. Eight phases are typically described, two of which have been combined in (**F**) for simplification.

TABLE 5.2. MUSCLE ACTION DURING GAIT CYCLE

	Phase of Gait	Mechanical Goals	Active Muscle Groups	Examples
STANCE PHASE	**Heel strike** (initial contact)	**Lower forefoot to ground**	Ankle dorsiflexors (eccentric contraction)	**Tibialis anterior**
		Continue deceleration (reverse forward swing)	Hip extensors	Gluteus maximus
		Preserve longitudinal arch of foot	Intrinsic muscles of foot	Flexor digitorum brevis
			Long tendons of foot	Tibialis anterior
	Loading response (flat foot)	**Accept weight**	Knee extensors	**Quadriceps**
		Decelerate mass (slow dorsiflexion)	Ankle plantarflexors	Triceps surae (soleus and gastrocnemius)
		Stabilize pelvis	Hip abductors	**Gluteus medius and minimus; tensor fasciae latae**
		Preserve longitudinal arch of foot	Intrinsic muscles of foot	Flexor digitorum brevis
			Long tendons of foot	Tibialis posterior; long flexors of digits
	Midstance	Stabilize knee	Knee extensors	Quadriceps
		Control dorsiflexion (preserve momentum)	Ankle plantarflexors (eccentric contraction)	Triceps surae (soleus and gastrocnemius)
		Stabilize pelvis	Hip abductors	**Gluteus medius and minimus; tensor fasciae latae**
		Preserve longitudinal arch of foot	Intrinsic muscles of foot	Flexor digitorum brevis
			Long tendons of foot	Tibialis posterior; long flexors of digits
	Terminal stance (heel off)	**Accelerate mass**	Ankle plantarflexors (concentric contraction)	**Triceps surae (soleus and gastrocnemius)**
		Stabilize pelvis	Hip abductors	**Gluteus medius and minimus; tensor fasciae latae**
		Preserve arches of foot; fix forefoot	Intrinsic muscles of foot	Adductor hallucis
			Long tendons of foot	Tibialis posterior; long flexors of digits

TABLE 5.2. MUSCLE ACTION DURING GAIT CYCLE (Continued)

	Phase of Gait	Mechanical Goals	Active Muscle Groups	Examples
STANCE PHASE (cont'd.)	**Presswing** (toe off)	Accelerate mass	Long flexors of digits	Flexor hallucis longus; flexor digitorum longus
		Preserve arches of foot; fix forefoot	Intrinsic muscles of foot	Adductor hallucis
			Long tendons of foot	Tibialis posterior; long flexors of digits
		Decelerate thigh; prepare for swing	Flexor of hip (eccentric contraction)	Iliopsoas; rectus femoris
SWING PHASE	**Initial swing**	**Accelerate thigh; vary cadence**	**Flexor of hip (concentric contraction)**	**Iliopsoas; rectus femoris**
		Clear foot	**Ankle dorsiflexors**	**Tibialis anterior**
	Midswing	**Clear foot**	**Ankle dorsiflexors**	**Tibialis anterior**
	Terminal swing	Decelerate thigh	Hip extensors (eccentric contraction)	Gluteus maximus; hamstrings
		Decelerate leg	Knee flexors (eccentric contraction)	Hamstrings
		Position foot	**Ankle dorsiflexors**	Tibialis anterior
		Extend knee to place foot (control stride); prepare for contact	**Knee extensors**	Quadriceps

of physical exertion is required. Most energy is used (1) in the eccentric contraction of the dorsiflexors during the beginning (**loading response**) phase of stance (Fig. 5.20B) as the heel is lowered to the ground following heel strike, and (2) especially at the end of stance (**terminal stance;** Fig. 5.20D) as the plantarflexors concentrically contract, pushing the forefoot (metatarsals and phalanges) down to produce push off, thus providing most of the propulsive force.

During the last part of the stance phase (**push off** or *toe off*; Fig. 5.20E), the toes flex to grip the ground and augment the push off initiated from the ball of the foot (sole underlying the heads of the medial two metatarsals). The long flexors and intrinsic muscles of the foot stabilize the forefoot and toes so that the effect of plantarflexion at the ankle and flexion of the toes is maximized.

The swing phase also involves flexion of the hip so that the free limb accelerates faster than the forward movement of the body. During **initial swing** (Fig. 5.20F), the knee flexes almost simultaneously, owing to momentum (without expenditure of energy), followed by dorsiflexion (lifting the forefoot up) at the ankle joint. The latter two movements have the effect of shortening the free limb so that it will clear the ground as it swings forward. By **midswing,** knee extension is added to the flexion and momentum of the thigh to realize anterior swing fully.

The extensors of the hip and flexors of the knee contract eccentrically at the end of swing phase (**terminal swing;** Fig. 5.20G) to decelerate the forward movement, while extensors of the knee (quadriceps) contract as necessary to extend the leg for the desired length of stride and to position the foot (present the heel) for heel strike.

Contraction of the knee extensors is maintained through the heel strike into the loading phase to absorb shock and keep the knee from buckling until it reaches full extension. Because the unsupported side of the hip tends to drop during the swing phase (which would negate the effect of limb shortening), abductor muscles on the supported side contract strongly during the single support part of the stance phase (Fig. 5.20F & G), pulling on the fixed femur to resist the tilting and keep the pelvis level. The same muscles also rotate (advance) the contralateral side of the pelvis forward, concurrent with the swing of its free limb.

Of course, all these actions alternate from side to side with each step. The extensors of the hip normally make only minor contributions to level walking. Primarily, the hip is passively extended by momentum during stance, except when accelerating or walking fast, and becomes increasingly active with increase in slope (steepness) during walking uphill or up stairs. Concentric hip flexion and knee extension are used during the swing phase of level walking and so are not weight-bearing actions; however, they are affected by body weight when their eccentric contraction is necessary for deceleration or walking downhill or down stairs.

Stabilization and resilience are important during locomotion. The invertors and evertors of the foot are principal stabilizers of the foot during the stance phase. Their long tendons, plus those of the flexors of the digits, also help support the arches of the foot during the stance phase, assisting the intrinsic muscles of the sole.

ANTERIOR AND MEDIAL REGIONS OF THIGH

Organization of Proximal Lower Limb

During evolution, the development of a prominent gluteal region is closely associated with the assumption of bipedalism and an erect posture. The prominent gluteal region is unique to humans. Modification of the shape of the femur necessary for bipedal walking and running (specifically the "bending" of the bone, creating the angle of inclination and the trochanters, p. 518) allows the superior placement of the abductors of the thigh into the gluteal region.

The remaining thigh muscles are organized into three compartments by intermuscular septa that pass deeply between the muscle groups from the inner surface of the fascia lata to the linea aspera of the femur (see Fig. 5.13D). The compartments are *anterior* or *extensor, medial* or *adductor,* and *posterior* or *flexor,* so named on the basis of their location or action at the knee joint. Generally, the anterior group is innervated by the femoral nerve, the medial group by the obturator nerve, and the posterior group by the tibial portion of the sciatic nerve. Although the compartments vary in absolute and relative size depending on the level, the anterior compartment is the largest overall and includes the femur.

To facilitate continuity and follow an approach commonly used in dissection courses, the anterior and medial compartments of the thigh are addressed initially, followed by continuous examination of the posterior aspect of the proximal limb: gluteal region and posterior thigh. This approach is then continued by consideration of the popliteal fossa and leg.

Anterior Thigh Muscles

The large **anterior compartment of the thigh** contains the **anterior thigh muscles,** the *flexors of the hip* (Fig. 5.21A–D) and *extensors of the knee* (Fig. 5.21E–I). For attachments, nerve supply, and main actions of these muscles, see Tables 5.3.I and 5.3.II. The anterior thigh muscles include the pectineus, iliopsoas, sartorius, and quadriceps femoris.[1]

The major muscles of the anterior compartment tend to atrophy (diminish) rapidly with disease, and physical therapy is often necessary to restore strength, tone, and symmetry with the opposite limb after immobilization of the thigh or leg.

PECTINEUS

The **pectineus** is a flat quadrangular muscle located in the anterior part of the superomedial aspect of the thigh (Fig. 5.21A & B; Table 5.3.I). It often appears to be composed of two layers, superficial and deep, and these are generally

[1]Because of its anterior position, the tensor fasciae latae is often studied with the anterior thigh muscles for convenience (i.e., when the cadaver is supine); however, it is actually part of the gluteal group and will be included with that group in this book.

innervated by two different nerves. Because of the dual nerve supply and the muscle's actions (the pectineus adducts and flexes the thigh and assists in medial rotation of the thigh), it is actually a transitional muscle between the anterior and medial compartments.

obturator nerve & femoral nerve.

ILIOPSOAS

The **iliopsoas,** the chief flexor of the thigh, is the most powerful of the hip flexors with the longest range. Although it is one of the body's most powerful muscles, it is relatively hidden, with most of its mass located in the posterior wall of the abdomen and greater pelvis. Its broad lateral part, the **iliacus,** and its long medial part, the **psoas major,** arise from the iliac fossa and lumbar vertebrae, respectively (Fig. 5.21C; Table 5.3.I). Thus, it is the only muscle attached to the vertebral column, pelvis, and femur. It is in a unique position not only to produce movement but to stabilize (fixate). However, it can also perpetuate and even contribute to deformity and disability when it is malformed (especially if it is shortened), dysfunctional, or diseased.

Concentric contraction of the iliopsoas typically moves the free limb, producing flexion at the hip to lift the limb and initiate its forward swing during walking (i.e., during the pre-swing and initial swing phases) as the opposite limb accepts weight (Fig. 5.20E & F), or to elevate the limb during climbing. However, it is also capable of moving the trunk. Bilateral contraction of the iliopsoas muscles initiates flexion of the trunk at the hip on the fixed thigh—as when (incorrectly) doing sit-ups—and decreases the lumbar lordosis (curvature) of the vertebral column. It is active during walking downhill, its eccentric contraction resisting acceleration.

The iliopsoas is also a postural muscle, active during standing in maintaining normal lumbar lordosis (and indirectly the compensatory thoracic kyphosis; see Chapter 4) and resisting hyperextension of the hip joint (see Fig. 5.19).

SARTORIUS

The **sartorius,** the "tailor's muscle" (L. *sartus,* patched or repaired), is long and ribbon-like. It passes lateral to medial across the superoanterior part of the thigh (Fig. 5.21D; Table 5.3.I). The sartorius lies superficially in the anterior compartment, within its own relatively distinct fascial sheath. It descends inferiorly as far as the medial side of the knee.

The sartorius, the longest muscle in the body, acts across two joints. It flexes the hip joint and participates in flexion of the knee joint. It also weakly abducts the thigh and laterally rotates it. The actions of both sartorius muscles bring the lower limbs into the cross-legged sitting position. None of the actions of the sartorius is strong; therefore, it is mainly a synergist, acting with other thigh muscles that produce these movements.

QUADRICEPS FEMORIS

The **quadriceps femoris** (L., four-headed femoral muscle) forms the main bulk of the anterior thigh muscles, and

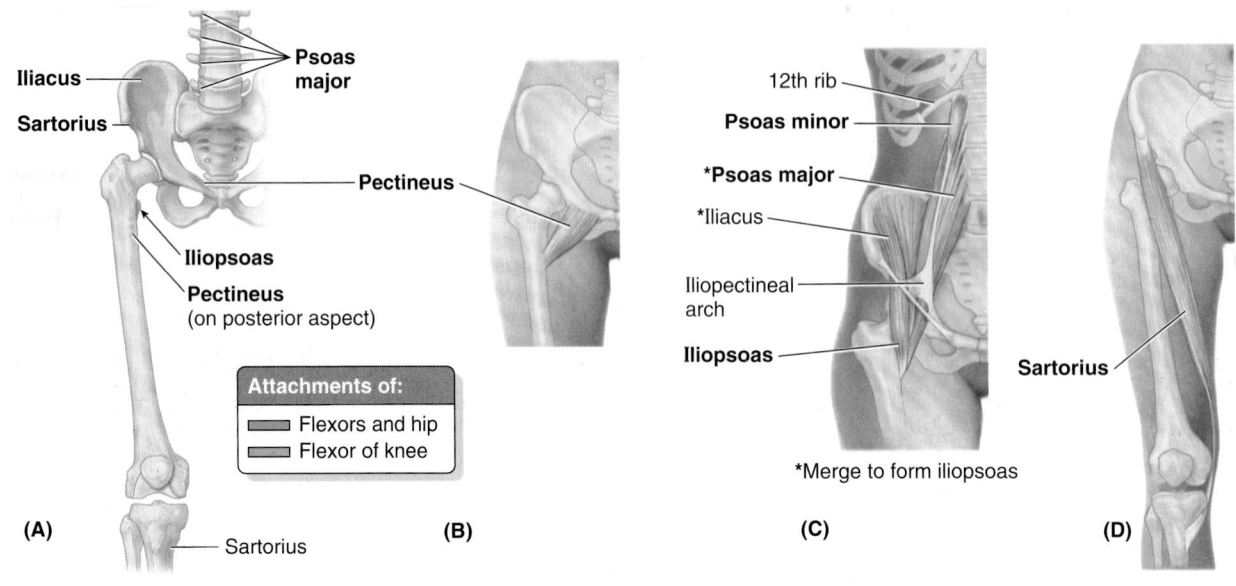

Anterior views

FIGURE 5.21. Muscles of anterior thigh: flexors of hip joint.

TABLE 5.3.I. MUSCLES OF ANTERIOR THIGH: FLEXORS OF HIP JOINT

Muscle	Proximal Attachment[a]	Distal Attachment	Innervation[b]	Main Action(s)
Pectineus (Fig. 5.21A & B)	Superior ramus of pubis	Pectineal line of femur, just inferior to lesser trochanter	Femoral nerve (**L2**, L3); may receive a branch from obturator nerve	Adducts and flexes thigh; assists with medial rotation of thigh
Iliopsoas (Fig. 5.21A & C) **Psoas major**	Sides of T12–L5 vertebrae and discs between them; transverse processes of all lumbar vertebrae	Lesser trochanter of femur	Anterior rami of lumbar nerves (**L1, L2**, L3)	Act conjointly in flexing thigh at hip joint and in stabilizing this joint[c]
Psoas minor	Sides of T12–L1 vertebrae and intervertebral discs	Pectineal line, iliopectineal eminence via iliopectineal arch	Anterior rami of lumbar nerves (L1, L2)	
Iliacus	Iliac crest, iliac fossa, ala of sacrum, and anterior sacro-iliac ligaments	Tendon of psoas major, lesser trochanter, and femur distal to it	Femoral nerve (**L2**, L3)	
Sartorius (Fig. 5.21A & D)	Anterior superior iliac spine and superior part of notch inferior to it	Superior part of medial surface of tibia	Femoral nerve (L2, L3)	Flexes, abducts, and laterally rotates thigh at hip joint; flexes leg at knee joint, (medially rotating leg when knee is flexed)[d]

[a] The Latin word *insertio* means *attachment*. The terms insertion and origin (L. *origo*) have not been used here (or elsewhere) since they change with function.

[b] The spinal cord segmental innervation is indicated (e.g., "L1, L2, L3" means that the nerves supplying the psoas major are derived from the first three lumbar segments of the spinal cord). Numbers in boldface (**L1, L2**) indicate the main segmental innervation. Damage to one or more of the listed spinal cord segments, or to the motor nerve roots arising from them, results in paralysis of the muscles concerned.

[c] The psoas major is also a postural muscle that helps control the deviation of the trunk and is active during standing.

[d] The four actions of the sartorius (L. *sartor,* tailor) produce the once common cross-legged sitting position used by tailors, hence the name.

collectively constitutes the largest and one of the most powerful muscles in the body. It covers almost all the anterior aspect and sides of the femur (Fig. 5-21E–I). The quadriceps femoris (usually shortened to quadriceps) consists of four parts: (1) rectus femoris, (2) vastus lateralis, (3) vastus intermedius, and (4) vastus medialis. Collectively, the quadriceps is a two-joint muscle capable of producing action at both the hip and knee.

The quadriceps is the great extensor of the leg. Concentric contraction of the quadriceps to extend the knee against gravity is important during rising from sitting or squatting, climbing and walking up stairs, and for acceleration and projection (running and jumping) when it is lifting or moving the body's weight. Consequently, it may be three times stronger than its antagonistic muscle group, the hamstrings.

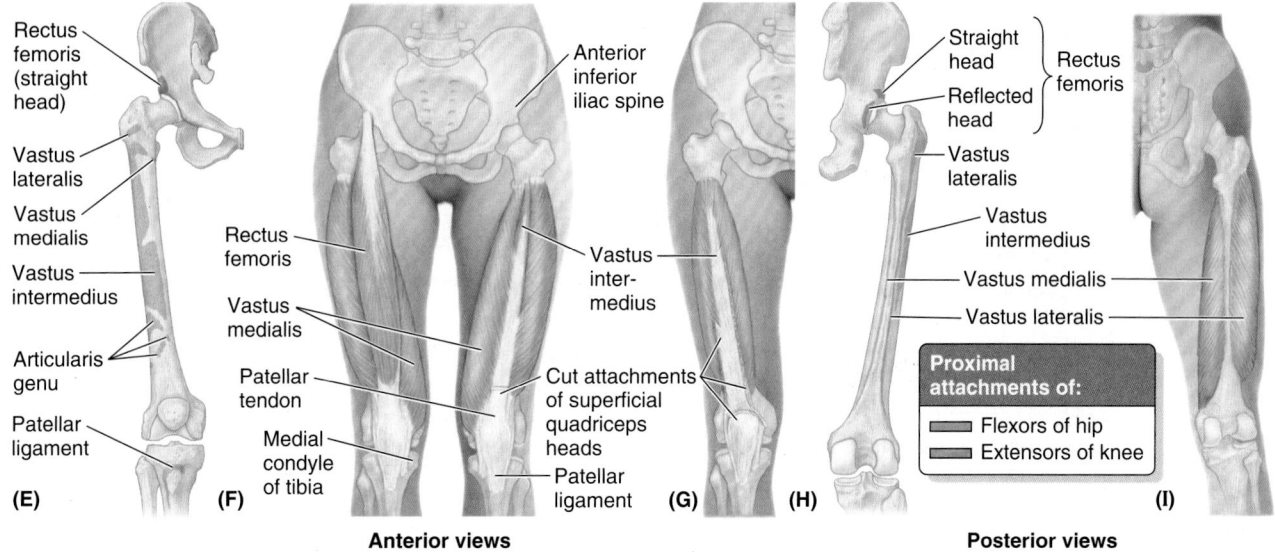

(E)   (F)   Medial condyle of tibia   (G)   (H)   (I)

Anterior views   **Posterior views**

FIGURE 5.21. Muscles of anterior thigh *(continued)*: extensors of knee.

TABLE 5.3.II. **MUSCLES OF ANTERIOR THIGH: EXTENSORS OF KNEE**

Muscle	Proximal Attachment	Distal Attachment	Innervation[a]	Main Action
Quadriceps femoris (Fig. 5.21E–H)				
Rectus femoris	Anterior inferior iliac spine and ilium superior to acetabulum	Via common tendinous (quadriceps tendon) and independent attachments to base of patella; indirectly via patellar ligament to tibial tuberosity; medial and lateral vasti also attach to tibia and patella via aponeuroses (medial and lateral patellar retinacula)	Femoral nerve (L2, **L3, L4**)	Extend leg at knee joint; rectus femoris also steadies hip joint and helps iliopsoas flex thigh
Vastus lateralis	Greater trochanter and lateral lip of linea aspera of femur			
Vastus medialis	Intertrochanteric line and medial lip of linea aspera of femur			
Vastus intermedius	Anterior and lateral surfaces of shaft of femur			

[a] The spinal cord segmental innervation is indicated (e.g., "L1, L2, L3" means that the nerves supplying the quadriceps femoris are derived from the first three lumbar segments of the spinal cord). Numbers in boldface (**L3, L4**) indicate the main segmental innervation. Damage to one or more of the listed spinal cord segments or to the motor nerve roots arising from them results in paralysis of the muscles concerned.

In level walking, the quadriceps muscles become active during the termination of the swing phase, preparing the knee to accept weight (Fig. 5.20G; Table 5.2). The quadriceps is primarily responsible for absorbing the jarring shock of heel strike, and its activity continues as the weight is assumed during the early stance phase (loading response). It also functions as a fixator during bent-knee sports, such as skiing and tennis, and contracts eccentrically during downhill walking and descending stairs.

The tendons of the four parts of the quadriceps unite in the distal portion of the thigh to form a single, strong, broad **quadriceps tendon** (Fig.5.21F). The **patellar ligament** (L. *ligamentum patellae*), attached to the tibial tuberosity, is the continuation of the quadriceps tendon in which the patella is embedded. The *patella* is thus the largest sesamoid bone in the body.

The medial and lateral vasti muscles also attach independently to the patella and form aponeuroses, the **medial** and **lateral patellar retinacula,** which reinforce the joint capsule of the knee joint on each side of the patella en route to attachment to the anterior border of the tibial plateau. The retinacula also play a role in keeping the patella aligned over the patellar surface of the femur.

The **patella** provides a bony surface that is able to withstand the compression placed on the quadriceps tendon during kneeling and the friction occurring when the knee is flexed and extended during running. The patella also provides additional leverage for the quadriceps in placing the tendon more anteriorly, farther from the joint's axis, causing it to approach the tibia from a position of greater mechanical advantage. The inferiorly directed apex of the patella indicates the level of the joint plane of the knee when the leg is extended and the patellar ligament is taut (Fig. 5.22C).

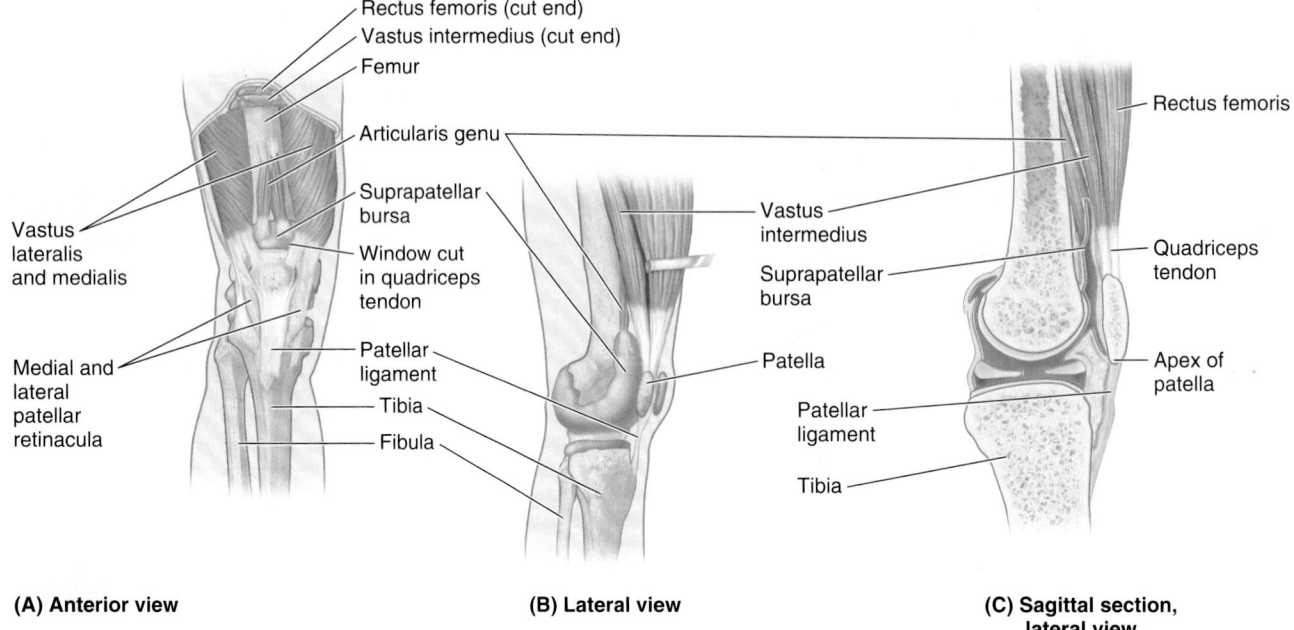

(A) Anterior view   **(B) Lateral view**   **(C) Sagittal section, lateral view**

FIGURE 5.22. Suprapatellar bursa and articularis genu. In **A** and **B,** the suprapatellar bursa, normally a potential space extending between the quadriceps and the femur (exaggerated for schematic purposes in **C**), is depicted as if injected with latex.

Testing the quadriceps[2] is performed with the person in the supine position with the knee partly flexed. The person extends the knee against resistance. During the test, contraction of the rectus femoris should be observable and palpable if the muscle is acting normally, indicating that its nerve supply is intact.

Rectus Femoris. The **rectus femoris** received its name because it runs straight down the thigh (L. *rectus*, straight). Because of its attachments to the hip bone and tibia via the patellar ligament (Fig. 5.21E & F), *it crosses two joints;* hence it is capable of flexing the thigh at the hip joint and extending the leg at the knee joint. The rectus femoris is the only part of the quadriceps that crosses the hip joint, and as a hip flexor it acts with and like the iliopsoas during the preswing and initial swing phases of walking (Fig. 5.20F; Table 5.2).

The ability of the rectus femoris to extend the knee is compromised during hip flexion, but it does contribute to the extension force during the toe off phase of walking, when the thigh is extended. It is particularly efficient in movements combining knee extension and hip flexion from a position of hip hyperextension and knee flexion, as in the preparatory position for kicking a soccer ball. The rectus femoris is susceptible to injury and avulsion from the anterior inferior iliac spine during kicking, hence the name "kicking muscle." A loss of function of the rectus femoris may reduce thigh flexion strength by as much as 17%.

Vastus Muscles. The names of the three large **vastus muscles** (pl., vasti) indicate their position around the femoral shaft (Fig. 5.21E–I; Table 5.3.II):

- **Vastus lateralis,** the largest component of the quadriceps, lies on the lateral side of the thigh.
- **Vastus medialis** covers the medial side of the thigh.
- **Vastus intermedius** lies deep to the rectus femoris, between the vastus medialis and vastus lateralis.

It is difficult to isolate the function of the three vastus muscles.

The small, flat **articularis genu** (articular muscle of the knee), a derivative of the vastus intermedius, usually consists of a variable number of muscular slips that attach superiorly to the inferior part of the anterior aspect of the femur, and inferiorly to the synovial membrane of the knee joint and the wall of the *suprapatellar bursa* (Figs. 5.21E and 5.22). The articularis genu muscle pulls the synovial membrane superiorly during extension of the leg, thereby preventing folds of the membrane from being compressed between the femur and the patella within the knee joint.

Medial Thigh Muscles

The muscles of the medial compartment of the thigh comprise the **adductor group,** consisting of the adductor longus, adductor brevis, adductor magnus, gracilis, and obturator externus (Fig. 5.23). In general, they attach proximally to the antero-inferior external surface of the bony pelvis (pubic bone, ischiopubic ramus, and ischial tuberosity), and adjacent obturator membrane, and distally to the linea aspera of the femur (Fig. 5.23A; Table 5.4)

All adductor muscles, except the "hamstring part" of the adductor magnus and part of the pectineus are supplied by the *obturator nerve* (L2–L4). The hamstring part of the adductor magnus is supplied by the tibial part of the sciatic

---

[2]There are entire texts dedicated to the testing of muscles. We are providing only a few important examples useful to primary care health professionals.

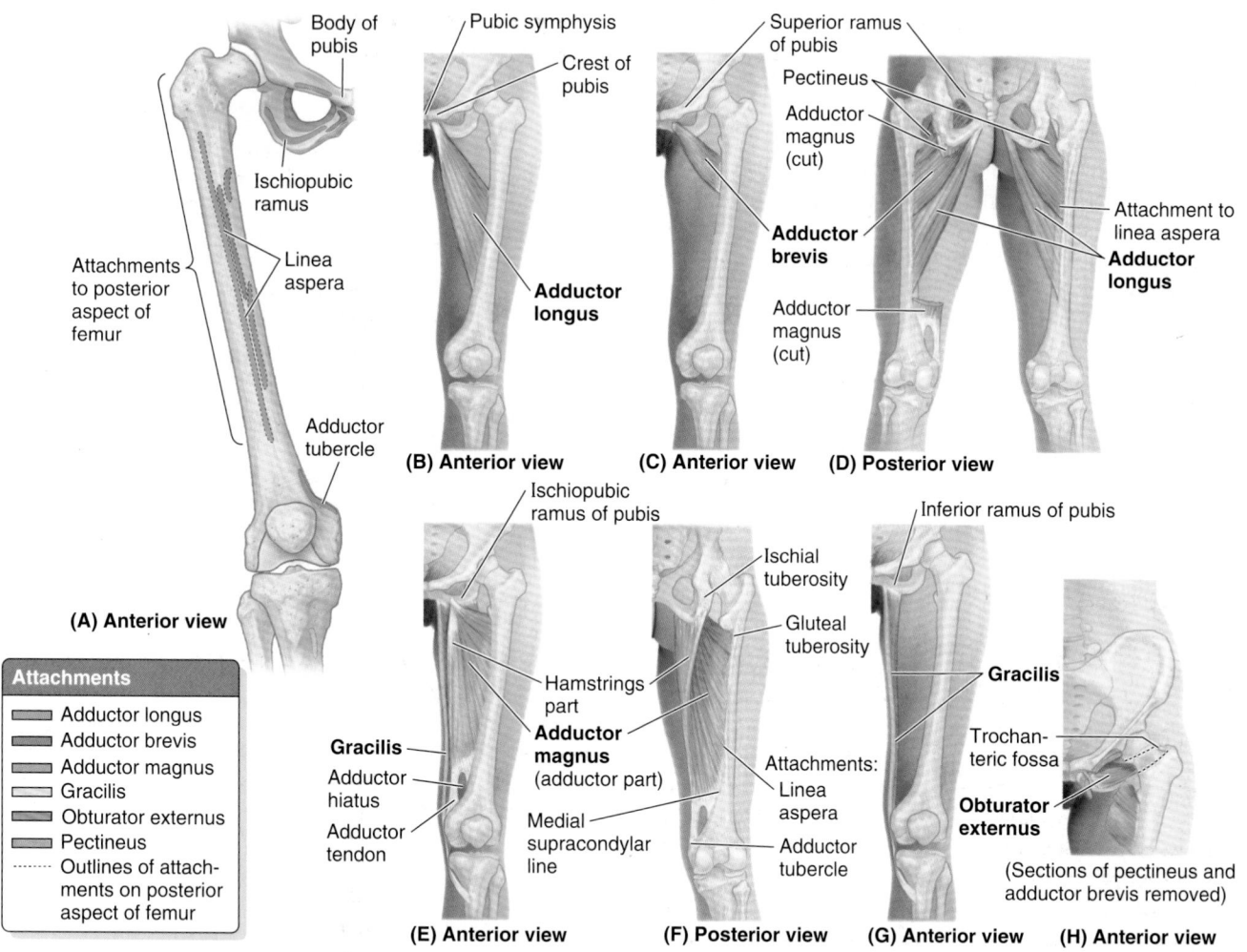

FIGURE 5.23. Muscles of medial thigh: adductors of thigh.

TABLE 5.4. MUSCLES OF MEDIAL THIGH: ADDUCTORS OF THIGH

Muscle[a]	Proximal Attachment	Distal Attachment	Innervation[b]	Main Action
Adductor longus (Fig. 5.23E & G)	Body of pubis inferior to pubic crest	Middle third of linea aspera of femur	Obturator nerve, branch of, anterior division (L2, **L3**, L4)	Adducts thigh
Adductor brevis (Fig. 5.23F & G)	Body and inferior ramus of pubis	Pectineal line and proximal part of linea aspera of femur		Adducts thigh; to some extent flexes it
Adductor magnus (Fig. 5.23C, D, & G)	Adductor part: inferior ramus of pubis, ramus of ischium Hamstrings part: ischial tuberosity	Adductor part: gluteal tuberosity, linea aspera, medial supracondylar line Hamstring part: adductor tubercle of femur	Adductor part: obturator nerve (L2, **L3**, L4), branches of posterior division Hamstring part: tibial part of sciatic nerve (**L4**)	Adducts thigh Adductor part: flexes thigh Hamstrings part: extends thigh
Gracilis (Fig. 5.23H)	Body and inferior ramus of pubis	Superior part of medial surface of tibia.	Obturator nerve (**L2**, L3)	Adducts thigh; flexes leg; helps rotate leg medially
Obturator externus	Margins of obturator foramen and obturator membrane	Trochanteric fossa of femur	Obturator nerve (L3, **L4**)	Laterally rotates thigh; steadies head of femur in acetabulum

[a]Collectively, the five muscles listed are the adductors of the thigh, but their actions are more complex (e.g., they act as flexors of the hip joint during flexion of the knee joint and are active during walking).

[b]The spinal cord segmental innervation is indicated (e.g., "L2, L3, L4" means that the nerves supplying the adductor longus are derived from the second to fourth lumbar segments of the spinal cord). Numbers in boldface (**L3**) indicate the main segmental innervation. Damage to one or more of the listed spinal cord segments or to the motor nerve roots arising from them results in paralysis of the muscles concerned.

nerve (L4). The details of their attachments, nerve supply, and actions of the muscles are provided in Table 5.4.

ADDUCTOR LONGUS

The **adductor longus** is a large, fan-shaped muscle and is the most anteriorly placed of the adductor group. The triangular long adductor arises by a strong tendon from the anterior aspect of the body of the pubis, just inferior to the pubic tubercle (apex of triangle), and expands to attach to the linea aspera of the femur (base of triangle) (Fig. 5.23A & B); in so doing it covers the anterior aspects of the adductor brevis and the middle of the adductor magnus.

ADDUCTOR BREVIS

The **adductor brevis,** the short adductor, lies deep to the pectineus and adductor longus, where it arises from the body and inferior ramus of the pubis. It widens as it passes distally to attach to the superior part of the linea aspera (Fig. 5.23A, C, & D).

As the obturator nerve emerges from the obturator canal to enter the medial compartment of the thigh, it splits into an anterior and a posterior division. The two divisions pass anterior and posterior to the adductor brevis. This unique relationship is useful in identifying the muscle during dissection and in anatomical cross-sections.

ADDUCTOR MAGNUS

The **adductor magnus** is the largest, most powerful, and most posterior muscle in the adductor group. This adductor is a composite, triangular muscle with a thick, medial margin that has an *adductor part* and a *hamstring part*. The two parts differ in their attachments, nerve supply, and main actions (Table 5.4).

The adductor part fans out widely for aponeurotic distal attachment along the entire length of the linea aspera of the femur, extending inferiorly onto the medial supracondylar ridge (Fig. 5.23A, E, & F). The hamstring part has a tendinous distal attachment to the adductor tubercle.

GRACILIS

The **gracilis** (L., slender) is a long, strap-like muscle and is the most medial muscle of the thigh. It is the most superficial of the adductor group and the weakest member. It is the only one of the group to cross the knee joint as well as the hip joint. The gracilis joins with two other two-joint muscles from the other two compartments (the *sartorius* and *semitendinosus muscles*) (Fig. 5.24). Thus, the three muscles are innervated by three different nerves. They have a common tendinous insertion, the **pes anserinus** (L., goose's foot), into the superior part of the medial surface of the tibia.

The gracilis is a synergist in adducting the thigh, flexing the knee, and rotating the leg medially when the knee is flexed. It acts with the other two "pes anserinus" muscles to

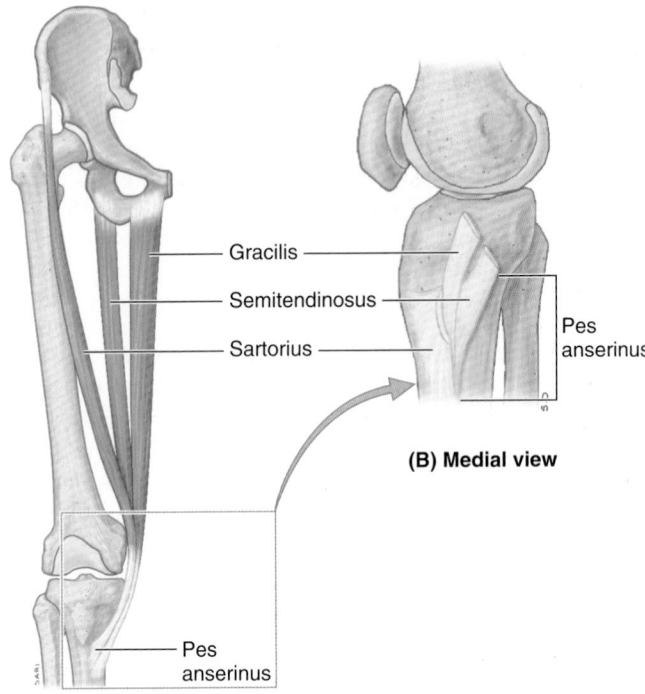

(A) Anterior view

(B) Medial view

FIGURE 5.24. Pes anserinus. A. Contributing muscles. **B.** Converging tendons forming pes anserinus.

add stability to the medial aspect of the extended knee, much as the gluteus maximus and tensor fasciae latae do via the iliotibial tract on the lateral side.

OBTURATOR EXTERNUS

The **obturator externus** is a flat, relatively small, fan-shaped muscle that is deeply placed in the superomedial part of the thigh. It extends from the external surface of the obturator membrane and surrounding bone of the pelvis to the posterior aspect of the greater trochanter, passing directly under the acetabulum and neck of the femur (Fig 5.23H).

ACTIONS OF ADDUCTOR MUSCLE GROUP

From the anatomical position, the main action of the adductor group is to pull the thigh medially, toward or past the median plane. Three adductors (longus, brevis, and magnus) are used in all movements in which the thighs are adducted (e.g., pressed together when riding a horse).

They are also used to stabilize the stance when standing on both feet, to correct a lateral sway of the trunk, or when there is a side-to-side shift of the surface on which one is standing (rocking a boat, standing on a balance board). These muscles are also used in kicking with the medial side of the foot in soccer and in swimming. Finally, they contribute to flexion of the extended thigh and extension of the flexed thigh when running or against resistance.

The adductors as a group constitute a large muscle mass. Although they are important in many activities, it has been shown that a reduction of as much as 70% in their function will result in only a slight to moderate impairment of hip function (Markhede and Stener, 1981).

Testing of the medial thigh muscles is performed while the person is lying supine with the knee straight. The individual adducts the thigh against resistance, and if the adductors are normal, the proximal ends of the gracilis and adductor longus can easily be palpated.

ADDUCTOR HIATUS

The **adductor hiatus** is an opening or aperture between the aponeurotic distal attachment of the adductor part of the adductor magnus and the tendinous distal attachment of the hamstring part (Fig. 5.23E). The adductor hiatus transmits the femoral artery and vein from the adductor canal in the thigh to the popliteal fossa posterior to the knee. The opening is located just lateral and superior to the adductor tubercle of the femur.

Neurovascular Structures and Relationships in Anteromedial Thigh

FEMORAL TRIANGLE

The **femoral triangle,** a subfascial formation, is a triangular landmark useful in dissection and in understanding relationships in the groin (Fig. 5.25A & B). In living people, it appears as a triangular depression inferior to the inguinal ligament when the thigh is flexed, abducted, and laterally rotated (Fig. 5.25A). The femoral triangle is bounded (Fig 5.25B):

- Superiorly by the *inguinal ligament* (thickened inferior margin of external oblique aponeurosis) that forms the *base of the femoral triangle.*
- Medially by the lateral border of the adductor longus.
- Laterally by the sartorius; the *apex* of the femoral triangle is where the medial border of the sartorius crosses the lateral border of the adductor longus.

The muscular *floor of the femoral triangle* is formed by the iliopsoas laterally and the pectineus medially. The *roof of the femoral triangle* is formed by the fascia lata and cribriform fascia, subcutaneous tissue, and skin.

The inguinal ligament actually serves as a flexor retinaculum, retaining structures that pass anterior to the hip joint against the joint during flexion of the thigh. Deep to the inguinal ligament, the **retro-inguinal space** (created as the inguinal ligament spans the gap between the two bony prominences to which it is attached, the ASIS and pubic tubercle) is an important passageway connecting the trunk/abdominopelvic cavity to the lower limb (Fig 5.26A & B).

The retro-inguinal space is divided into two compartments (L. *lacunae*) by a thickening of the iliopsoas fascia, the **iliopectineal arch,** which passes between the deep surface of the inguinal ligament and the *iliopubic eminence* (see Fig. 5.6B). Lateral to the iliopectineal arch is the **muscular compartment of the retro-inguinal space,** through which the iliopsoas muscle and femoral nerve pass from the greater

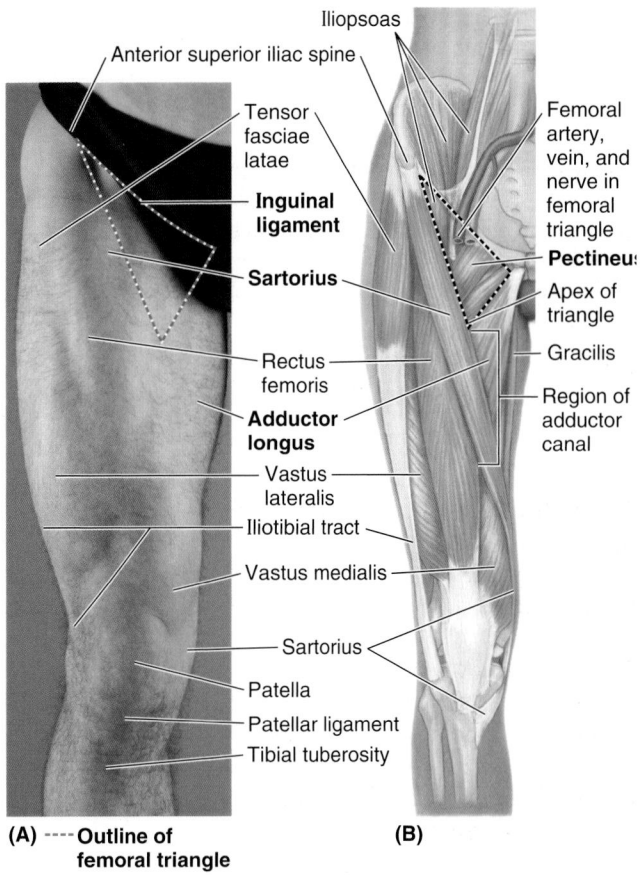

(A) ----**Outline of femoral triangle**   **(B)**

Anterior views

FIGURE 5.25. **Surface anatomy of femoral triangle. A.** Surface anatomy. **B.** Underlying structures.

pelvis into the anterior thigh (Fig. 5.26A & B). Medial to the arch, the **vascular compartment of the retro-inguinal space** allows passage of the major vascular structures (veins, artery, and lymphatics) between the greater pelvis and the femoral triangle of the anterior thigh. As they enter the femoral triangle, the names of the vessels change from *external iliac* to *femoral.*

The contents of the femoral triangle, from lateral to medial, are (Figs. 5.26B and 5.27A & B) the:

- Femoral nerve and its (terminal) branches.
- Femoral sheath and its contents:
  - Femoral artery and several of its branches.
  - Femoral vein and its proximal tributaries (e.g., the great saphenous and profunda femoris veins).
  - Deep inguinal lymph nodes and associated lymphatic vessels.

The femoral triangle is bisected by the femoral artery and vein, which pass to and from the adductor canal inferiorly at the triangle's apex (Fig. 5.27A). The **adductor canal** is an intermuscular passageway deep to the sartorius by which the major neurovascular bundle of the thigh traverses the middle third of the thigh (Figs. 5.27B and 5.30).

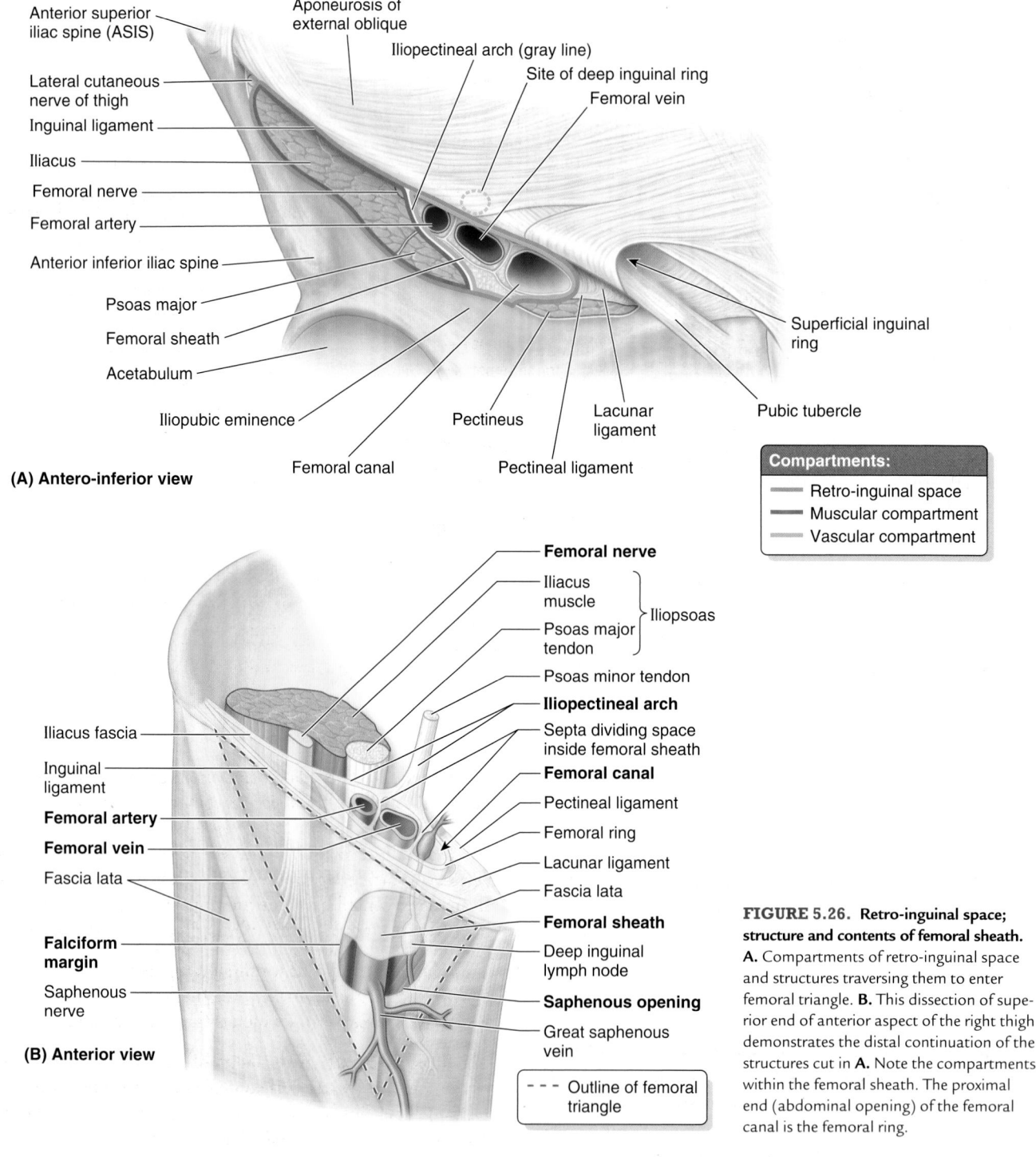

(A) Antero-inferior view

Anterior superior iliac spine (ASIS)
Lateral cutaneous nerve of thigh
Inguinal ligament
Iliacus
Femoral nerve
Femoral artery
Anterior inferior iliac spine
Psoas major
Femoral sheath
Acetabulum
Iliopubic eminence
Femoral canal
Pectineus
Pectineal ligament
Lacunar ligament
Pubic tubercle
Superficial inguinal ring
Femoral vein
Site of deep inguinal ring
Iliopectineal arch (gray line)
Aponeurosis of external oblique

Compartments:
Retro-inguinal space
Muscular compartment
Vascular compartment

(B) Anterior view

Iliacus fascia
Inguinal ligament
Femoral artery
Femoral vein
Fascia lata
Falciform margin
Saphenous nerve

Femoral nerve
Iliacus muscle
Psoas major tendon
}Iliopsoas
Psoas minor tendon
Iliopectineal arch
Septa dividing space inside femoral sheath
Femoral canal
Pectineal ligament
Femoral ring
Lacunar ligament
Fascia lata
Femoral sheath
Deep inguinal lymph node
Saphenous opening
Great saphenous vein

- - - Outline of femoral triangle

FIGURE 5.26. Retro-inguinal space; structure and contents of femoral sheath. A. Compartments of retro-inguinal space and structures traversing them to enter femoral triangle. **B.** This dissection of superior end of anterior aspect of the right thigh demonstrates the distal continuation of the structures cut in **A.** Note the compartments within the femoral sheath. The proximal end (abdominal opening) of the femoral canal is the femoral ring.

FEMORAL NERVE

The **femoral nerve** (L2–L4) is the largest branch of the lumbar plexus. The nerve originates in the abdomen within the psoas major and descends posterolaterally through the pelvis to approximately the midpoint of the inguinal ligament (Figs. 5.26B and 5.27A). It then passes deep to this ligament and enters the femoral triangle, lateral to the femoral vessels.

After entering the femoral triangle, the femoral nerve divides into several branches to the anterior thigh muscles. It also sends articular branches to the hip and knee joints and provides several cutaneous branches to the anteromedial side of the thigh (Table 5.1, p. 537).

The terminal cutaneous branch of the femoral nerve, the **saphenous nerve,** descends through the femoral triangle, lateral to the femoral sheath containing the femoral

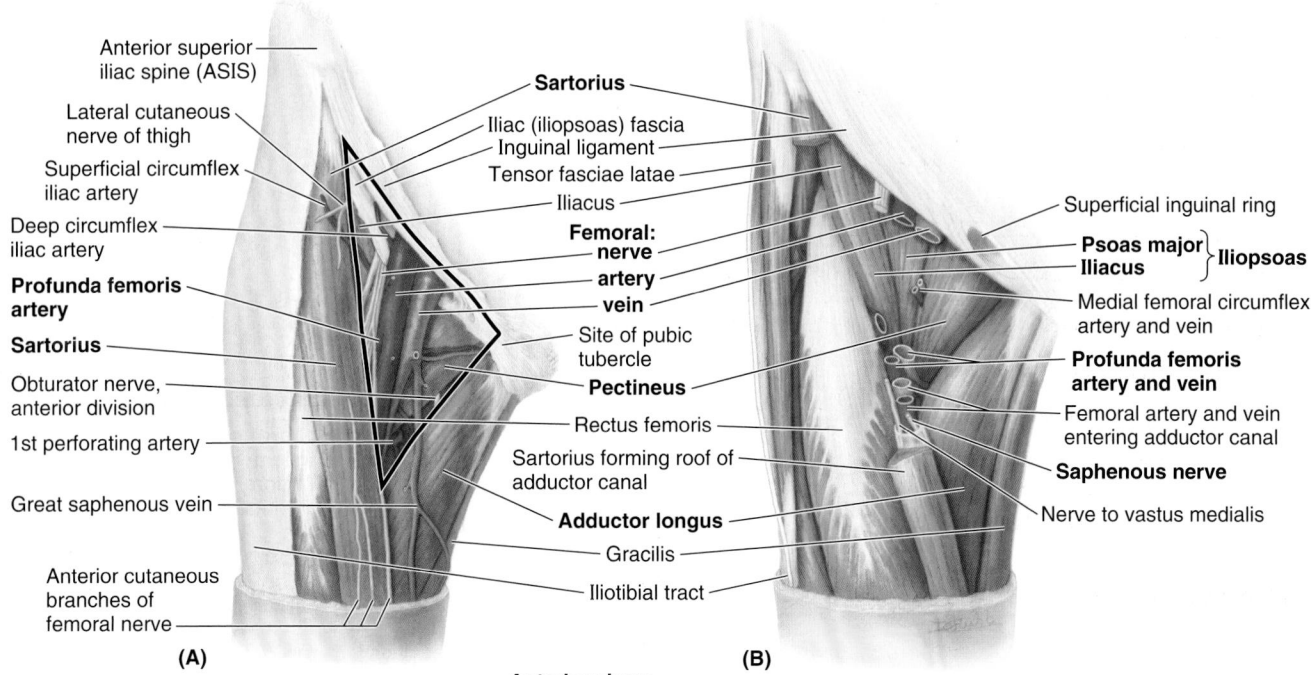

FIGURE 5.27. **Structures of femoral triangle. A.** The boundaries and contents of femoral triangle. The triangle is bound by the inguinal ligament superiorly, the adductor longus medially, and the sartorius laterally. The femoral nerve and vessels enter the base of the triangle superiorly and exit from its apex inferiorly. **B.** In this deeper dissection, sections have been removed from the sartorius and femoral vessels and nerve. Observe the muscles forming the floor of the femoral triangle: the iliopsoas laterally and the pectineus medially. Of the neurovascular structures at the apex of the femoral triangle, the two anterior vessels (femoral artery and vein) and the two nerves enter the adductor canal (anterior to adductor longus), and the two posterior vessels (profunda femoris artery and vein) pass deep (posterior) to the adductor longus.

vessels (Figs. 5.26B and 5.27B; Table 5.1 p. 538). The saphenous nerve accompanies the femoral artery and vein through the adductor canal and becomes superficial by passing between the sartorius and gracilis when the femoral vessels traverse the adductor hiatus at the distal end of the canal. It runs antero-inferiorly to supply the skin and fascia on the anteromedial aspects of the knee, leg, and foot.

FEMORAL SHEATH

The **femoral sheath** is a funnel-shaped fascial tube of varying length (usually 3–4 cm) that passes deep to the inguinal ligament, lining the vascular compartment of the retro-inguinal space (Fig. 5.28). It terminates inferiorly by blending with the adventitia of the femoral vessels. The sheath encloses proximal parts of the femoral vessels and creates the femoral canal medial to them (Figs. 5.26B and 5.28B).

The femoral sheath is formed by an inferior prolongation of transversalis and iliopsoas fascia from the abdomen. The femoral sheath does not enclose the femoral nerve because it passes through the muscular compartment. When a long femoral sheath occurs (extends farther distally), its medial wall is pierced by the great saphenous vein and lymphatic vessels (Fig. 5.28).

The femoral sheath allows the femoral artery and vein to glide deep to the inguinal ligament during movements of the hip joint.

The femoral sheath lining the vascular compartment is subdivided internally into three smaller compartments by vertical septa of extraperitoneal connective tissue that extend from the abdomen along the femoral vessels (Figs. 5.26B and 5.28B). The compartments of the femoral sheath are the:

- *Lateral compartment* for the femoral artery.
- *Intermediate compartment* for the femoral vein.
- *Medial compartment,* which is the femoral canal.

The **femoral canal** is the smallest of the three compartments of the femoral sheath. It is conical and short (approximately 1.25 cm) and lies between the medial edge of the femoral sheath and the femoral vein. The femoral canal:

- Extends distally to the level of the proximal edge of the saphenous opening.
- Allows the femoral vein to expand when venous return from the lower limb is increased, or when increased intra-abdominal pressure causes a temporary stasis (blockage) in the vein (as during a *Valsalva maneuver,* i.e., taking a breath and holding it, often while bearing down).
- Contains loose connective tissue, fat, a few lymphatic vessels, and sometimes a deep inguinal lymph node (lacunar lymph node).

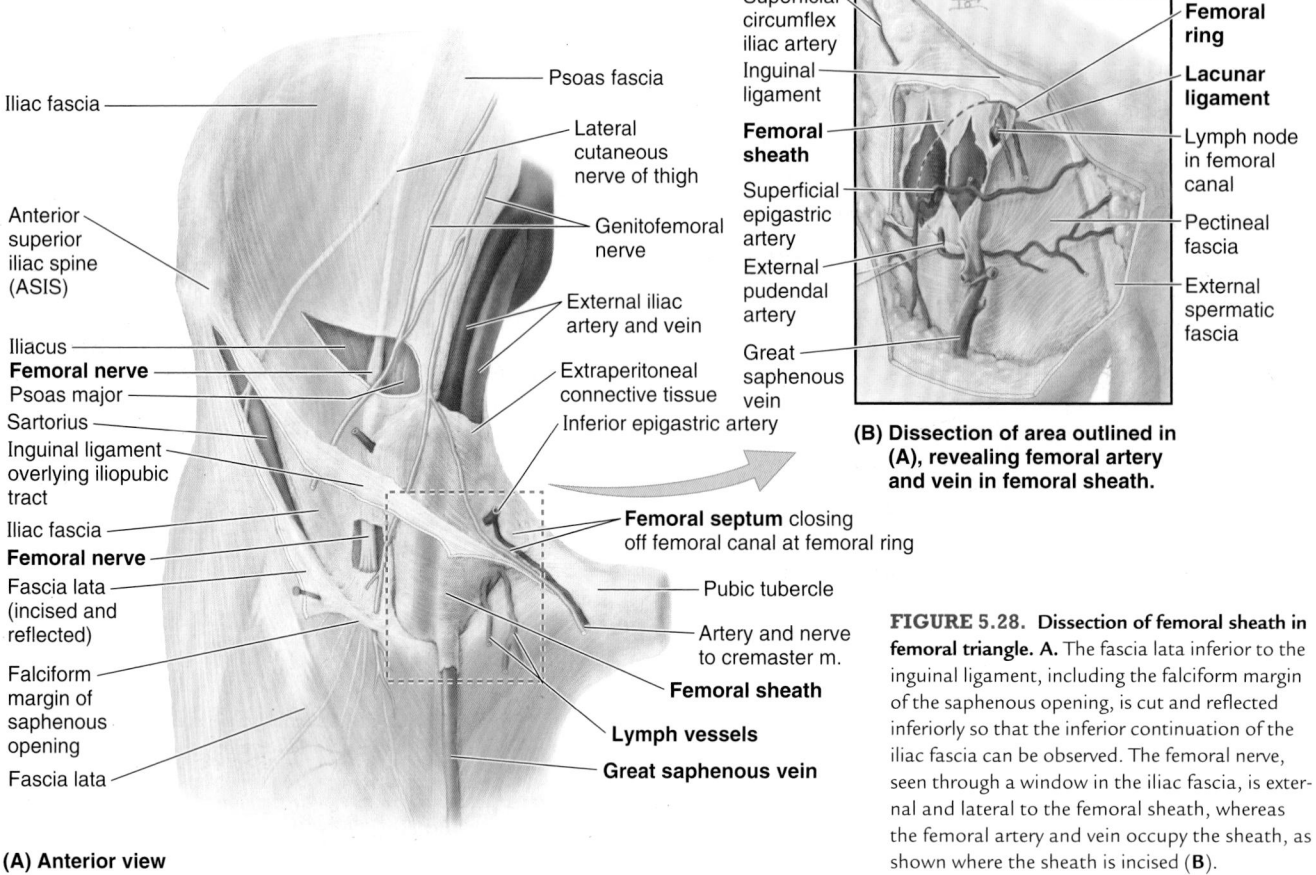

(A) Anterior view

Labels (A):
Iliac fascia — Psoas fascia — Lateral cutaneous nerve of thigh — Genitofemoral nerve — Anterior superior iliac spine (ASIS) — External iliac artery and vein — Extraperitoneal connective tissue — Inferior epigastric artery — Iliacus — **Femoral nerve** — Psoas major — Sartorius — Inguinal ligament overlying iliopubic tract — Iliac fascia — **Femoral nerve** — Fascia lata (incised and reflected) — Falciform margin of saphenous opening — Fascia lata — **Femoral septum** closing off femoral canal at femoral ring — Pubic tubercle — Artery and nerve to cremaster m. — **Femoral sheath** — **Lymph vessels** — **Great saphenous vein**

Labels (B):
Superficial circumflex iliac artery — Inguinal ligament — **Femoral sheath** — Superficial epigastric artery — External pudendal artery — Great saphenous vein — **Femoral ring** — **Lacunar ligament** — Lymph node in femoral canal — Pectineal fascia — External spermatic fascia

(B) Dissection of area outlined in (A), revealing femoral artery and vein in femoral sheath.

FIGURE 5.28. Dissection of femoral sheath in femoral triangle. A. The fascia lata inferior to the inguinal ligament, including the falciform margin of the saphenous opening, is cut and reflected inferiorly so that the inferior continuation of the iliac fascia can be observed. The femoral nerve, seen through a window in the iliac fascia, is external and lateral to the femoral sheath, whereas the femoral artery and vein occupy the sheath, as shown where the sheath is incised (**B**).

The *base of the femoral canal* is the oval **femoral ring** formed by the small (approximately 1 cm wide) proximal opening at its abdominal end. This opening is closed by extraperitoneal fatty tissue that forms the transversely oriented **femoral septum** (Fig. 5.28A). The abdominal surface of the septum is covered by parietal peritoneum. The femoral septum is pierced by lymphatic vessels connecting the inguinal and external iliac lymph nodes.

The *boundaries of the femoral ring* are (Fig. 5.26B):

- *Laterally,* the vertical septum between the femoral canal and femoral vein.
- *Posteriorly,* the superior ramus of the pubis covered by the pectineus muscle and its fascia.
- *Medially,* the lacunar ligament.
- *Anteriorly,* the medial part of the inguinal ligament.

FEMORAL ARTERY

Details concerning the origin, course, and distribution of the arteries of the thigh are illustrated in Figure 5.29 and described in Table 5.5.

The **femoral artery,** the continuation of the external iliac artery distal to the inguinal ligament, is the primary artery of the lower limb (Figs. 5.26–29; Table 5.5). It enters the *femoral triangle* deep to the midpoint of the inguinal ligament (midway between the ASIS and the pubic tubercle),

lateral to the femoral vein (Fig. 5.30A). The pulsations of the femoral artery are palpable within the triangle because of its relatively superficial position deep (posterior) to the fascia lata. The artery lies and descends on the adjacent borders of the iliopsoas and pectineus muscles that form the floor of the triangle. The superficial epigastric artery, superficial (and sometimes the deep) circumflex iliac arteries, and the superficial and deep external pudendal arteries arise from the anterior aspect of the proximal part of the femoral artery.

The **profunda femoris artery** (deep artery of thigh) is the largest branch of the femoral artery and the chief artery to the thigh (Fig 5.29). It arises from the lateral or posterior side of the femoral artery in the femoral triangle. In the middle third of the thigh, where it is separated from the femoral artery and vein by the adductor longus (Figs. 5.27B and 5.30B), it gives off 3–4 *perforating arteries* that wrap around the posterior aspect of the femur (Fig. 5.29; Table 5.5). The perforating arteries supply muscles of all three fascial compartments (adductor magnus, hamstrings, and vastus lateralis).

The **circumflex femoral arteries** encircle the uppermost shaft of the femur and anastomose with each other and other arteries, supplying the thigh muscles and the superior (proximal) end of the femur. The **medial circumflex femoral artery** is especially important because it supplies

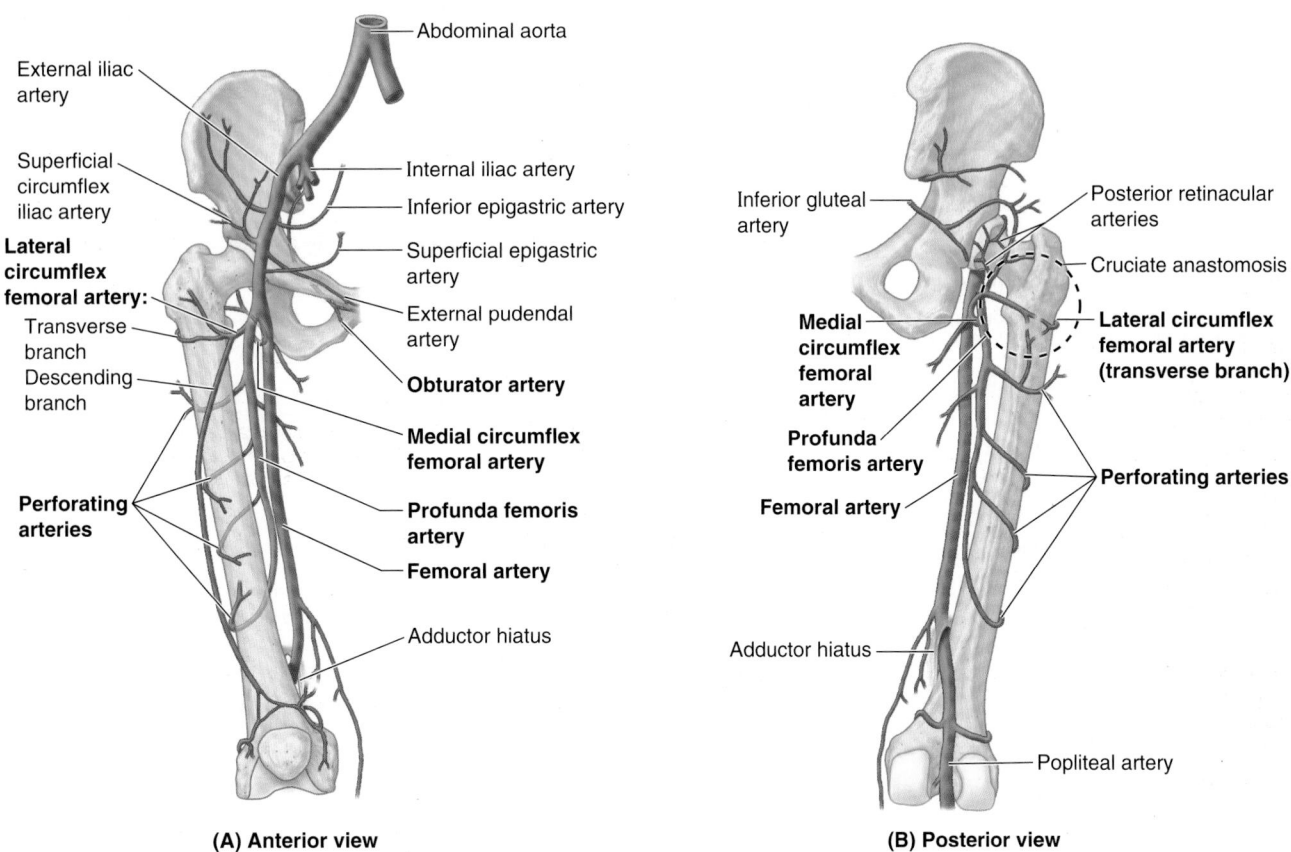

FIGURE 5.29. Arteries of anterior and medial thigh.

TABLE 5.5. ARTERIES OF ANTERIOR AND MEDIAL THIGH

Artery	Origin	Course	Distribution
Femoral	Continuation of external iliac artery distal to inguinal ligament	Descends through femoral triangle bisecting it; then courses through adductor canal; terminates as it traverses adductor hiatus, where its name becomes popliteal artery	Branches supply anterior and anteromedial aspects of thigh
Profunda femoris artery (deep artery of thigh)	Femoral artery 1–5 cm inferior to inguinal ligament	Passes deeply between pectineus and adductor longus; descending posterior to latter on medial side of femur	Three to four perforating arteries pass through adductor magnus muscle, winding around femur to supply muscles in medial, posterior, and lateral part of anterior compartments
Medial circumflex femoral	Profunda femoris artery; may arise from femoral artery	Passes medially and posteriorly between pectineus and iliopsoas; enters gluteal region and gives rise to posterior retinacular arteries; then terminates by dividing into transverse and ascending branches	Supplies most of blood to head and neck of femur; transverse branch takes part in cruciate anastomosis of thigh; ascending branch joins inferior gluteal artery
Lateral circumflex femoral		Passes laterally deep to sartorius and rectus femoris, dividing into ascending, transverse, and descending arteries	Ascending branch supplies anterior part of gluteal region; transverse branch winds around femur; descending branch joins genicular peri-articular anastomosis
Obturator	Internal iliac artery or (in ~20%) as an accessory or replaced obturator artery from the inferior epigastric artery	Passes through obturator foramen; enters medial compartment of thigh and divides into anterior and posterior branches, which pass on respective sides of adductor brevis	Anterior branch supplies obturator externus, pectineus, adductors of thigh, and gracilis; posterior branch supplies muscles attached to ischial tuberosity

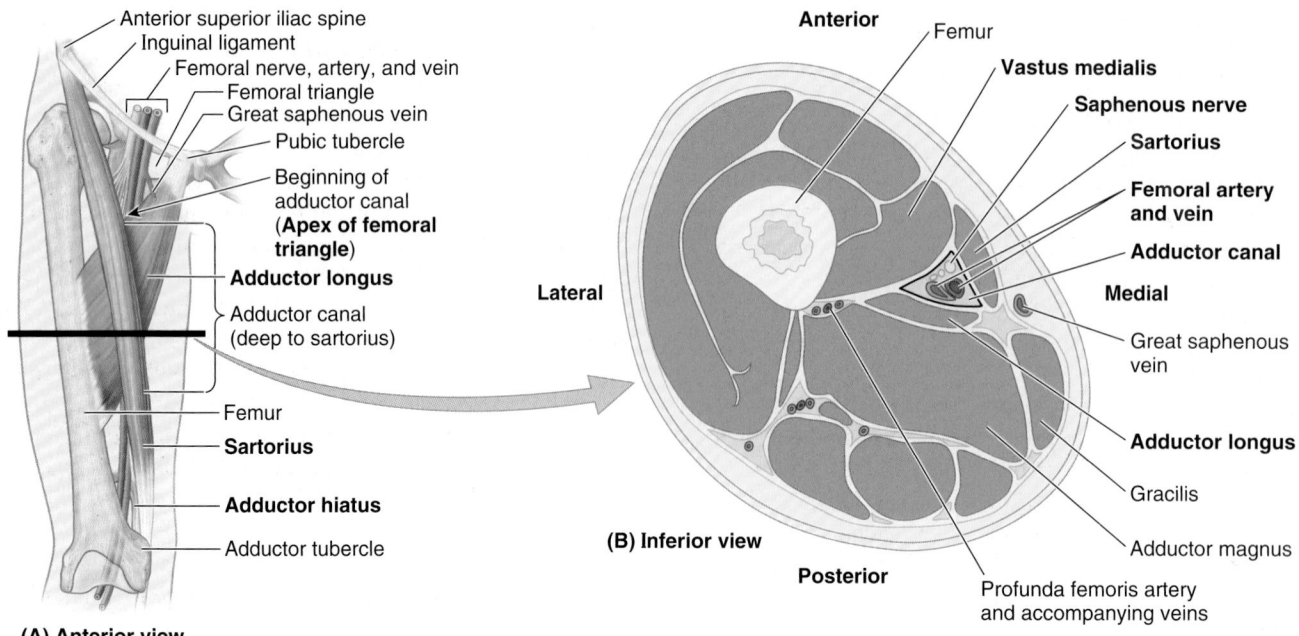

(A) Anterior view

FIGURE 5.30. **Adductor canal in medial part of middle third of thigh. A.** Orientation drawing showing the adductor canal and the level of the section shown in **B. B.** This transverse section of the thigh shows the muscles bounding the adductor canal and its neurovascular contents.

most of the blood to the head and neck of the femur via its branches, the **posterior retinacular arteries.** The retinacular arteries are often torn when the femoral neck is fractured or the hip joint is dislocated. The **lateral circumflex femoral artery,** less able to supply the femoral head and neck as it passes laterally across the thickest part of the joint capsule of the hip joint, mainly supplies muscles on the lateral side of the thigh.

Obturator Artery. The **obturator artery** helps the profunda femoris artery supply the adductor muscles via anterior and posterior branches, which anastomose. The posterior branch gives off an acetabular branch that supplies the head of the femur.

FEMORAL VEIN

The **femoral vein** is the continuation of the popliteal vein proximal to the adductor hiatus. As it ascends through the adductor canal, the femoral vein lies posterolateral and then posterior to the femoral artery (Figs. 5.26B and 5.27A & B). The femoral vein enters the femoral sheath lateral to the femoral canal and ends posterior to the inguinal ligament, where it becomes the external iliac vein.

In the inferior part of the femoral triangle, the femoral vein receives the profunda femoris vein, the great saphenous vein, and other tributaries. The **profunda femoris vein** (deep vein of thigh), formed by the union of three or four perforating veins, enters the femoral vein approximately 8 cm inferior to the inguinal ligament and approximately 5 cm inferior to the termination of the great saphenous vein.

ADDUCTOR CANAL

The **adductor canal** (subsartorial canal; Hunter canal) is a long (approximately 15 cm), narrow passageway in the middle third of the thigh. It extends from the apex of the femoral triangle, where the sartorius crosses over the adductor longus, to the *adductor hiatus* in the tendon of the adductor magnus (Fig. 5.30A).

The adductor canal provides an *intermuscular passage* for the femoral artery and vein, the saphenous nerve, and the slightly larger nerve to vastus medialis, delivering the femoral vessels to the popliteal fossa where they become popliteal vessels.

The *adductor canal is bounded* (Fig. 5.30B):

- Anteriorly and laterally by the vastus medialis.
- Posteriorly by the adductors longus and magnus.
- Medially by the sartorius, which overlies the groove between the above muscles, forming the roof of the canal.

In the inferior third to half of the canal, a tough subsartorial or vastoadductor fascia spans between the adductor longus and the vastus medialis muscles, forming the anterior wall of the canal deep to the sartorius. Because this fascia has a distinct superior margin, novices dissecting in this area commonly assume when they see the femoral vessels pass deep to this fascia that they are traversing the adductor hiatus. The *adductor hiatus,* however, is located at a more inferior level, just proximal to the medial supracondylar ridge. This hiatus is a gap between the aponeurotic adductor and the tendinous hamstrings attachments of the adductor magnus (Fig. 5.23E).

Surface Anatomy of Anterior and Medial Regions of Thigh

In fairly muscular individuals, some of the bulky anterior thigh muscles can be observed. The prominent muscles are the *quadriceps* and *sartorius*, whereas laterally the tensor fasciae latae is palpable as is the *iliotibial tract* to which this muscle attaches (Fig. 5.31A).

Three of the four parts of the *quadriceps* are visible or can be approximated (Fig. 5.31A & B). The fourth part (vastus intermedius) is deep and almost hidden by the other muscles and cannot be palpated.

The *rectus femoris* may be easily observed as a ridge passing down the thigh when the lower limb is raised from the floor while sitting. Observe the large bulges formed by the vastus lateralis and medialis at the knee (Fig. 5.31B). The *patellar ligament* is easily observed, especially in thin people, as a thick band running from the patella to the tibial tuberosity. You can also palpate the *infrapatellar fat pads*, the masses of loose fatty tissue on each side of the patellar ligament.

On the medial aspect of the inferior part of the thigh, the *gracilis* and *sartorius* muscles form a well-marked prominence, which is separated by a depression from the large bulge formed by the *vastus medialis* (Fig. 5.31A & B). Deep in this depressed area, the *large tendon of the adductor magnus* can be palpated as it passes to its attachment to the adductor tubercle of the femur.

Measurements of the lower limb are taken to detect shortening (e.g., resulting from a femoral fracture). To make such measurements, compare the affected limb with the corresponding limb. Real limb shortening is detected by comparing the measurements from the ASIS to the distal tip of the medial malleolus on both sides.

To determine if the shortening is in the thigh, the measurement is taken from the top of the ASIS to the distal edge of the lateral femoral condyle on both sides. Keep in mind that small differences between the two sides—such as a difference of 1.25 cm in total length of the limb—may be normal.

The proximal two thirds of a line drawn from the midpoint of the inguinal ligament to the *adductor tubercle* when the thigh is flexed, abducted, and rotated laterally represents the course of the *femoral artery* (Fig. 5.30A). The proximal third of the line represents this artery as it passes through the

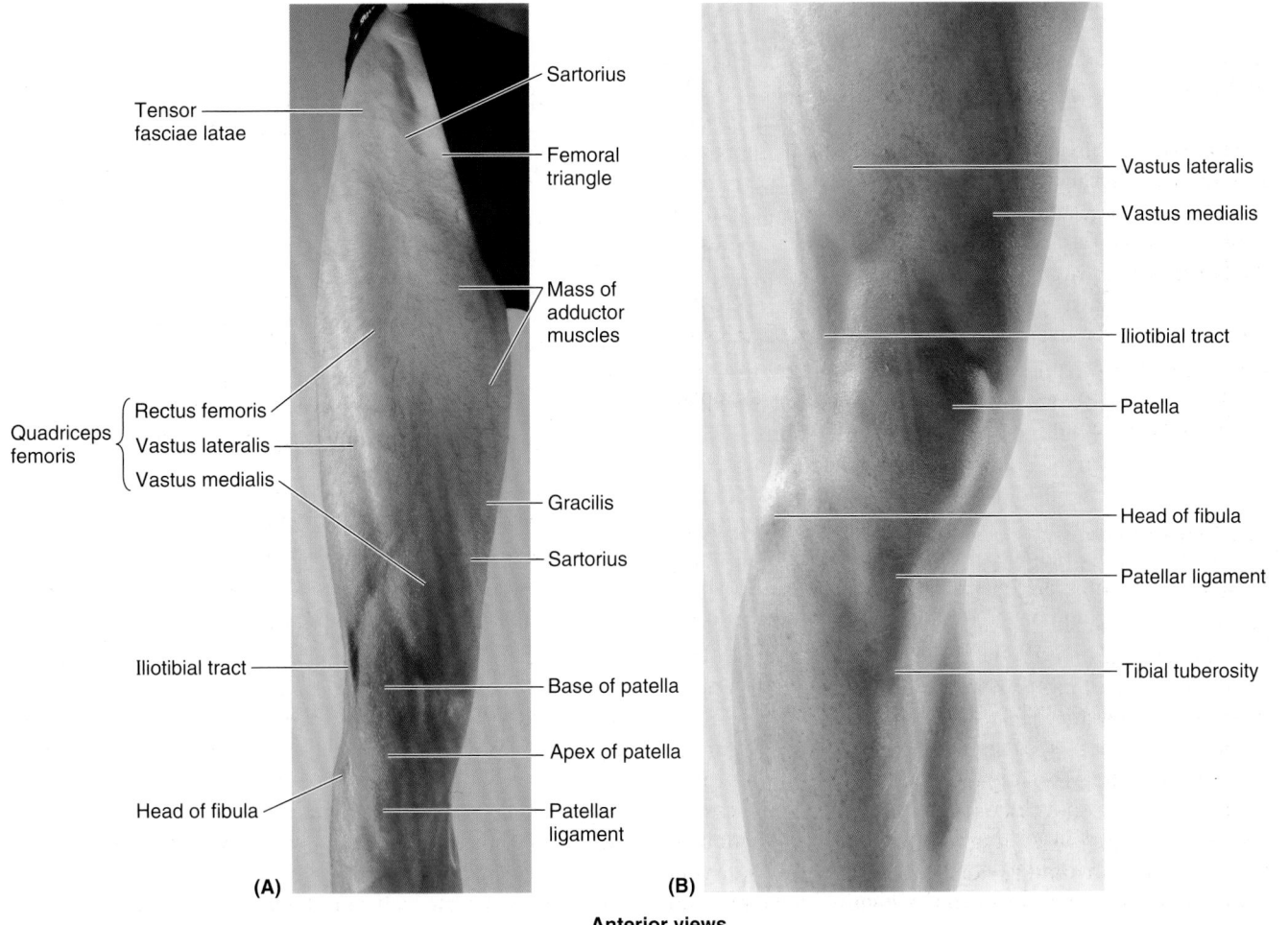

(A)   **(B)**

Anterior views

FIGURE 5.31. Surface anatomy of anterior and medial thigh.

femoral triangle, whereas the middle third represents the artery while it is in the *adductor canal.* Approximately 3.75 cm along this line distal to the inguinal ligament, the profunda femoris artery arises from the femoral artery.

The *femoral vein* is (Figs. 5.26B & 5.27A):

• Medial to the femoral artery at the base of the femoral triangle (indicated by inguinal ligament).
• Posterior to the femoral artery at the apex of the femoral triangle.
• Posterolateral to the artery in the adductor canal.

The *femoral triangle,* in the supero-anterior aspect of the thigh, is not a prominent surface feature in most people. When some people sit cross-legged, the sartorius and adductor longus stand out, delineating the femoral triangle. The surface anatomy of the femoral triangle is clinically important because of its contents (Fig. 5.26B).

The *femoral artery* can be felt pulsating just inferior to the midinguinal point. When you palpate the femoral pulse, the *femoral vein* is just medial, the femoral nerve is a finger's breadth lateral, and the *femoral head* is just posterior. The femoral artery runs a 5-cm superficial course through the femoral triangle before it is covered by the sartorius in the adductor canal.

The *great saphenous vein* enters the thigh posterior to the medial femoral condyle and passes superiorly along a line from the adductor tubercle to the *saphenous opening.* The central point of this opening, where the great saphenous vein enters the femoral vein, is located 3.75 cm inferior and 3.75 cm lateral to the pubic tubercle (Fig. 5.27A).

ANTERIOR AND MEDIAL REGIONS OF THIGH

Hip and Thigh Contusions

Sports broadcasters and trainers refer to a "hip pointer," which is a *contusion of the iliac crest* that usually occurs at its anterior part (e.g., where the sartorius attaches to the ASIS). This is one of the most common injuries to the hip region, usually occurring in association with collision sports, such as the various forms of football, ice hockey, and volleyball.

Contusions cause bleeding from ruptured capillaries and infiltration of blood into the muscles, tendons, and other soft tissues. The term *hip pointer* may also refer to avulsion of bony muscle attachments, for example, of the sartorius or rectus femoris to the anterior superior and inferior iliac spines, respectively, of the hamstrings from the ischium. However, these injuries should be called *avulsion fractures.*

Another term commonly used is "charley horse," which may refer either to the cramping of an individual thigh muscle because of ischemia or to contusion and rupture of blood vessels sufficient enough to form a *hematoma.* The injury is usually the consequence of tearing of fibers of the rectus femoris; sometimes the quadriceps tendon is also partially torn. The most common site of a thigh hematoma is in the quadriceps. A charley horse is associated with localized pain and/or muscle stiffness and commonly follows direct trauma (e.g., a stick slash in hockey or a tackle in football).

Psoas Abscess

The psoas major muscle arises in the abdomen from the intervertebral discs, the sides of the T12–L5 vertebrae, and their transverse processes (see Fig. B2.35). The medial arcuate ligament of the diaphragm arches obliquely over the proximal part of the psoas major. The transversalis fascia on the internal abdominal wall is continuous with the psoas fascia, where it forms a fascial covering for the psoas major that accompanies the muscle into the anterior region of the thigh.

There is a resurgence of tuberculosis (TB) in Africa, Asia, and elsewhere. A retroperitoneal *pyogenic infection* (pusforming) in the abdomen or greater pelvis, characteristically occurring in association with TB of the vertebral column, or secondary to regional enteritis of the ileum (*Crohn disease*), may result in the formation of a *psoas abscess.* When the abscess passes between the psoas and its fascia to the inguinal and proximal thigh regions, severe pain may be referred to the hip, thigh, or knee joint. A psoas abscess should always be considered when edema occurs in the proximal part of the thigh. Such an abscess may be palpated or observed in the inguinal region, just inferior or superior to the inguinal ligament, and may be mistaken for an indirect inguinal hernia or a femoral hernia, an enlargement of the inguinal lymph nodes, or a saphenous varix. The lateral border of the psoas is commonly visible in radiographs of the abdomen; an obscured psoas shadow may be an indication of abdominal pathology.

Paralysis of Quadriceps

A person with *paralyzed quadriceps muscles* cannot extend the leg against resistance and usually presses on the distal end of the thigh during walking to prevent inadvertent flexion of the knee joint.

Weakness of the vastus medialis or vastus lateralis, resulting from arthritis or trauma to the knee joint, can result in abnormal patellar movement and loss of joint stability.

Chondromalacia Patellae

Chondromalacia patellae (runner's knee) is a common knee injury for marathon runners. Such overstressing of the knee region can also occur in running sports such as basketball. The soreness and aching around or deep to the patella results from *quadriceps imbalance*. Chondromalacia patellae may also result from a blow to the patella or extreme flexion of the knee (e.g., during squatting when power lifting).

Patellar Fractures

A direct blow to the patella may fracture it into two or more fragments (Fig B5.12). *Transverse patellar fractures* may result from a blow to the knee or sudden contraction of the quadriceps (e.g., when one slips and attempts to prevent a backward fall). The proximal fragment is pulled superiorly with the quadriceps tendon, and the distal fragment remains with the patellar ligament.

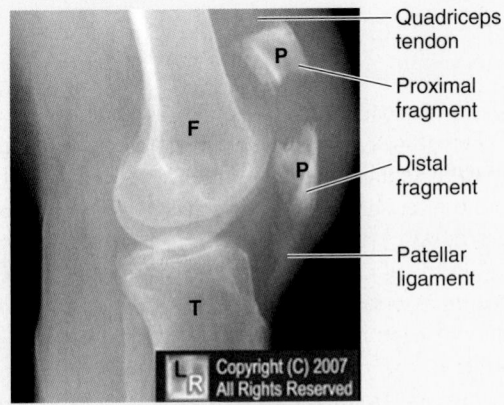

Femur (F), Patella (P), Tibia (T)

FIGURE B5.12.

Abnormal Ossification of Patella

The patella is cartilaginous at birth. It ossifies during the 3rd–6th years, frequently from more than one ossification center. Although these centers usually coalesce and form a single bone, they may remain separate on one or both sides, giving rise to a *bipartite* or *tripartite* *patella* (Fig. B5.13). An unwary observer might interpret this

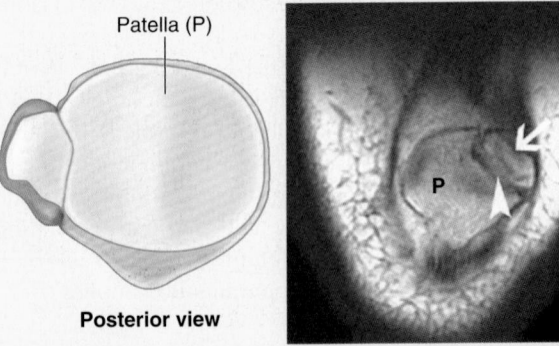

Patella (P)

Posterior view

(A) Bipartite patella

(B) Coronal MRI image of bipartite patella (arrow = bipartate segment; arrowhead = false "fracture" line)

FIGURE B5.13.

condition on a radiograph or CT as a patellar fracture. *Ossification abnormalities* are nearly always bilateral; therefore, diagnostic images should be examined from both sides. If the defects are bilateral, the defects are likely ossification abnormalities.

Patellar Tendon Reflex

Tapping the patellar ligament with a reflex hammer (Fig. B5.14) normally elicits the *patellar tendon reflex* ("knee jerk"). This myotatic (deep tendon) reflex is routinely tested during a physical examination by having the person sit with the legs dangling. A firm strike on the ligament with a reflex hammer usually causes the leg to extend. If the reflex is normal, a hand on the person's quadriceps should feel the muscle contract. This tendon reflex tests the integrity of the femoral nerve and the L2–L4 spinal cord segments.

Tapping the ligament activates muscle spindles in the quadriceps. Afferent impulses from the spindles travel in the femoral nerve to the L2–L4 segments of the spinal cord. From here, efferent impulses are transmitted via motor fibers in the femoral nerve to the quadriceps, resulting in a jerk-like contraction of the muscle and extension of the leg at the knee joint.

Diminution or absence of the patellar tendon reflex may result from any lesion that interrupts the innervation of the quadriceps (e.g., peripheral nerve disease).

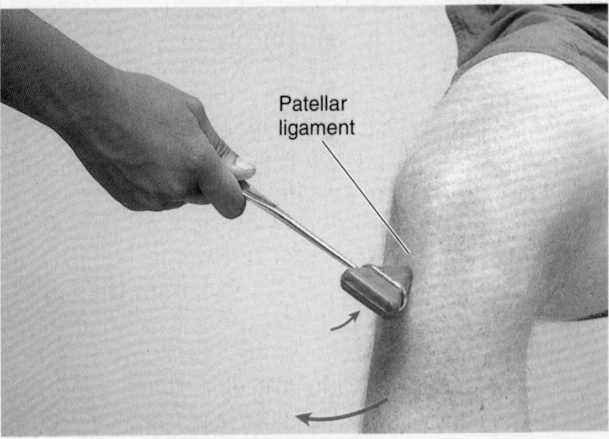

Patellar ligament

FIGURE B5.14.

Transplantation of Gracilis

Because the gracilis is a relatively weak member of the adductor group of muscles, it can be removed without noticeable loss of its actions on the leg. Surgeons often transplant the gracilis, or part of it, with its nerve and blood vessels to replace a damaged muscle in the hand, for example. Once the muscle is transplanted, it soon produces good digital flexion and extension.

Freed from its distal attachment, the muscle has also been relocated and repositioned to create a replacement for a nonfunctional external anal sphincter.

Groin Pull

Sports broadcasters refer to a "pulled groin" or "groin injury." These terms mean that a strain, stretching, and probably some tearing of the proximal attachments of the anteromedial thigh muscles have occurred. The injury usually involves the flexor and adductor thigh muscles. The proximal attachments of these muscles are in the inguinal region (groin), the junction of the thigh and trunk.

Groin pulls usually occur in sports that require quick starts (e.g., sprinting and base stealing in baseball), or extreme stretching (e.g., gymnastics).

Injury to Adductor Longus

Muscle strains of the adductor longus may occur in horseback riders and produce pain (rider's strain). Ossification sometimes occurs in the tendons of these muscles because the horseback riders actively adduct their thighs to keep from falling from their animals. The ossified tendons are sometimes wrongly called "riders' bones."

Palpation, Compression, and Cannulation of Femoral Artery

The initial part of the femoral artery, proximal to the branching of the profunda femoris artery, is superficial in position, making it especially accessible and useful for a number of clinical procedures. Some vascular surgeons refer to this part of the femoral artery as the *common femoral artery* and to its continuation distally as the *superficial femoral artery*. This terminology is not recommended by the Federative International Committee on Anatomical Terminology because it is a deep artery. The term is not used in this book because it may cause misunderstanding.

With the person lying in the supine position, the *femoral pulse* may be palpated midway between the ASIS and the pubic symphysis (Fig. B5.15A & B). By placing the tip of the little finger (of the right hand when dealing with the right side) on the ASIS and the tip of the thumb on the pubic tubercle, the femoral pulse can be palpated with the midpalm just inferior to the midpoint of the inguinal ligament by pressing firmly. Normally the pulse is strong; however, if the common or external iliac arteries are partially occluded, the pulse may be diminished.

Compression of the femoral artery may also be accomplished at this site by pressing directly posteriorly against the superior pubic ramus, psoas major, and femoral head (Fig. B5.15C). Compression at this point will reduce blood flow through the femoral artery and its branches, such as the profunda femoris artery.

The femoral artery may be cannulated just inferior to the midpoint of the inguinal ligament. In *left cardial (cardiac) angiography*, a long, slender catheter is inserted into the artery and passed up the external iliac artery, common iliac artery, and aorta to the left ventricle of the heart. This same approach is used to visualize the coronary arteries in *coronary arteriography*.

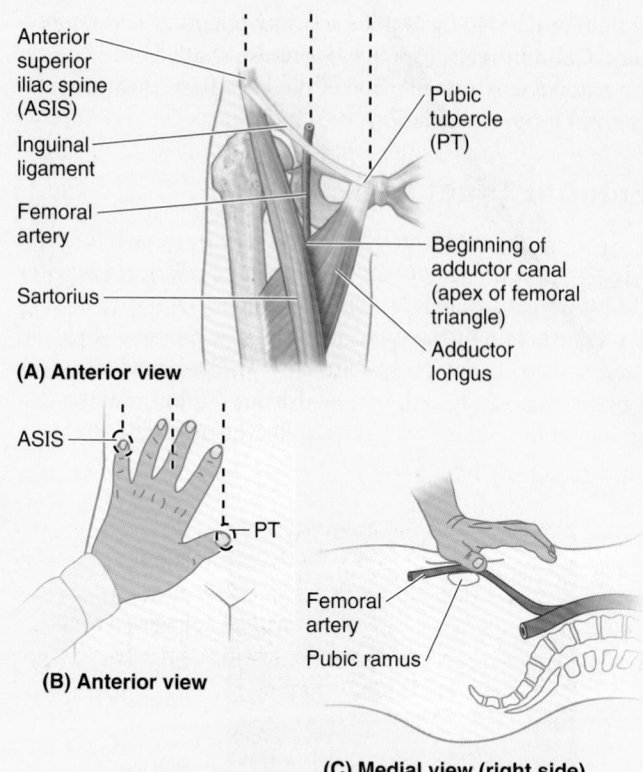

(A) Anterior view

(B) Anterior view

(C) Medial view (right side)

FIGURE B5.15.

Blood may also be taken from the femoral artery for *blood gas analysis* (the determination of oxygen and carbon dioxide concentrations and pressures with the pH of the blood by laboratory tests).

Laceration of Femoral Artery

The superficial position of the femoral artery in the femoral triangle makes it vulnerable to traumatic injury (see Fig. 5.26B), especially laceration. Commonly, both the femoral artery and vein are lacerated in anterior thigh wounds because they lie close together. In some cases, an *arteriovenous shunt* occurs as a result of communication between the injured vessels.

When it is necessary to ligate the femoral artery, anastomosis of branches of the femoral artery with other arteries that cross the hip joint may supply blood to the lower limb. The *cruciate anastomosis* is a four-way common meeting of the medial and lateral circumflex femoral arteries with the inferior gluteal artery superiorly, and the first perforating artery inferiorly, posterior to the femur (Fig. 5.29; Table 5.5), occurring less often than its frequent mention implies.

Potentially Lethal Misnomer

Clinical staff, some vascular laboratories, and even some text and reference books use the term "superficial femoral vein" when referring to the femoral vein before it is joined by the accompanying veins of the profunda femoris artery (profunda femoris veins). Some primary care physicians may not have been taught and/or may not realize that the so-called superficial femoral vein is actually a deep vein, and that acute thrombosis of this vessel is potentially life threatening. The adjective *superficial* should not be used because it implies that this vein is a superficial vein. Most *pulmonary emboli* originate in deep veins, not in superficial veins. The risk of embolism can be greatly reduced by anticoagulant treatment. The use of imprecise language here creates the possibility that an acute thrombosis of this truly deep vessel could be overlooked as an acute clinical issue, and a life-threatening situation created.

Saphenous Varix

A localized dilation of the terminal part of the great saphenous vein, called a *saphenous varix* (L. dilated vein), may cause edema in the femoral triangle. A saphenous varix may be confused with other groin swellings, such as a psoas abscess; however, a varix should be considered when varicose veins are present in other parts of the lower limb.

Location of Femoral Vein

The femoral vein is not usually palpable but its position can be located inferior to the inguinal ligament by feeling the pulsations of the femoral artery, which is immediately lateral to the vein. In thin people, the femoral vein may be close to the surface and may be mistaken for the great saphenous vein. It is important therefore to know that the femoral vein has no tributaries at this level, except for the great saphenous vein that joins it approximately 3 cm inferior to the inguinal ligament. In *varicose*

vein operations, it is obviously important to identify the great saphenous vein correctly, and not tie off the femoral vein by mistake.

Cannulation of Femoral Vein

To secure blood samples and take pressure recordings from the chambers of the right side of the heart and/or from the pulmonary artery and to perform *right cardiac angiography*, a long, slender catheter is inserted into the femoral vein as it passes through the femoral triangle. Under fluoroscopic control, the catheter is passed superiorly through the external and common iliac veins into the inferior vena cava and right atrium of the heart. *Femoral venous puncture* may also be used for the administration of fluids.

Femoral Hernias

The femoral ring is a weak area in the anterior abdominal wall that normally is of a size sufficient to admit the tip of the little finger (Fig. B5.16). The femoral ring is the usual originating site of a *femoral hernia*, a protrusion of abdominal viscera (often a loop of small intestine) through the femoral ring into the femoral canal. A femoral hernia appears as a mass, often tender, in the femoral triangle, inferolateral to the pubic tubercle.

The hernia is bounded by the femoral vein laterally and the lacunar ligament medially. The hernial sac compresses the contents of the femoral canal (loose connective tissue, fat, and lymphatics) and distends the wall of the canal. Initially, the hernia is small because it is contained within the canal, but it can enlarge by passing inferiorly through the saphenous opening into the subcutaneous tissue of the thigh.

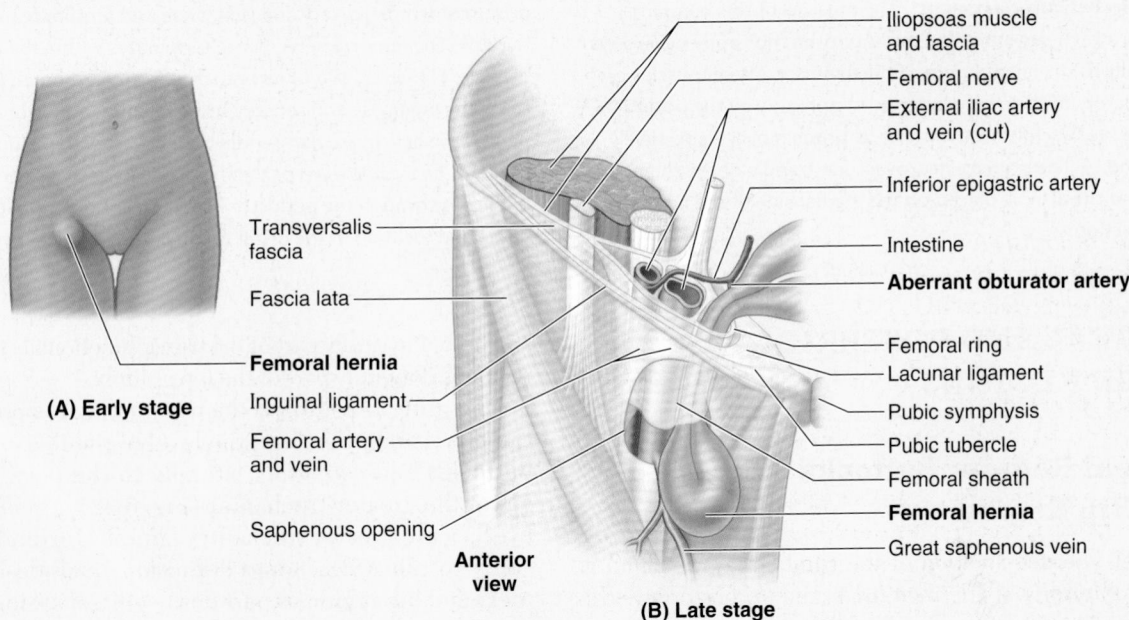

(A) Early stage

Transversalis fascia
Fascia lata
Femoral hernia
Inguinal ligament
Femoral artery and vein
Saphenous opening

Anterior view

(B) Late stage

Iliopsoas muscle and fascia
Femoral nerve
External iliac artery and vein (cut)
Inferior epigastric artery
Intestine
Aberrant obturator artery
Femoral ring
Lacunar ligament
Pubic symphysis
Pubic tubercle
Femoral sheath
Femoral hernia
Great saphenous vein

FIGURE B5.16.

Femoral hernias are more common in females because of their wider pelves. *Strangulation of a femoral hernia* may occur because of the sharp, rigid boundaries of the femoral ring, particularly the concave margin of the lacunar ligament. Strangulation of a femoral hernia interferes with the blood supply to the herniated intestine. This vascular impairment may result in *necrosis* (death of the tissues).

Replaced or Accessory Obturator Artery

An enlarged pubic branch of the inferior epigastric artery either takes the place of the obturator artery (**replaced obturator artery**), or joins it as an **accessory obturator artery,** in approximately 20% of people (Fig. B5.17). This artery runs close to or across the femoral ring to reach the obturator foramen and could be closely related to the neck of a femoral hernia. Consequently, this artery could be involved in a *strangulated femoral hernia.* Surgeons placing staples during endoscopic repair of both

inguinal and femoral hernias must also be vigilant concerning the possible presence of this common arterial variant.

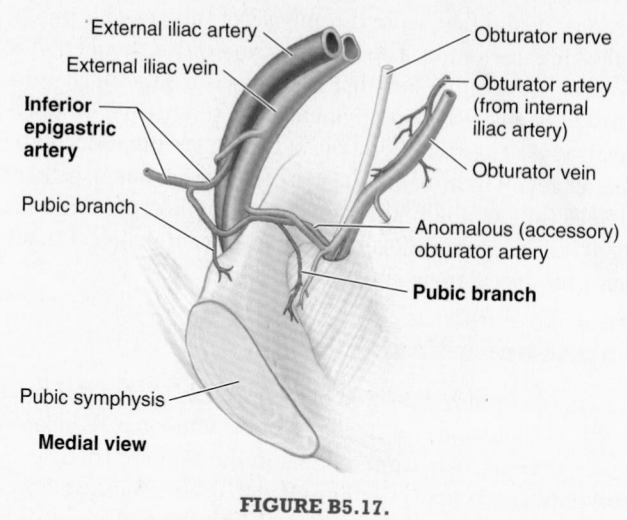

FIGURE B5.17.

The Bottom Line

ANTERIOR AND MEDIAL COMPARTMENTS OF THIGH

Anterior compartment: This large anterior compartment includes the flexors of the hip and extensors of the knee, with most muscles innervated by the femoral nerve. ♦ The quadriceps femoris accounts for most of the mass of this compartment. It surrounds the femur on three sides and has a common tendon of attachment to the tibia, which includes the patella as a sesamoid bone. ♦ Major muscles of this compartment atrophy rapidly with disease or disuse, requiring physical therapy to retain or restore function.

Medial compartment: The muscles of this compartment attach proximally to the antero-inferior bony pelvis and distally to the linea aspera of the femur. ♦ These muscles are adductors of the thigh, innervated primarily by the obturator nerve. Use of these muscles as prime movers is relatively limited. ♦ The primary neurovascular bundle of the thigh, like that of the arm, is placed on the medial side of the limb for protection.

Neurovascular structures and relationships in antero-medial thigh: In the upper third of the thigh, the neurovascular bundle is most superficial as it enters deep to the inguinal ligament. This relatively superficial position is important for clinical procedures. ♦ Although they are essentially adjacent, the femoral nerve traverses the muscular lacunae of the subinguinal space, whereas the femoral vessels traverse the vascular lacunae within the femoral sheath. ♦ The femoral vessels bisect the femoral triangle, where the primary vessels of the thigh, the profunda femoris artery and vein, arise and terminate, respectively. ♦ The femoral nerve per se terminates within the femoral triangle. However, two of its branches, a motor branch (nerve to vastus medialis) and sensory branch (saphenous nerve), are part of the neurovascular bundle that traverses the adductor canal in the middle third of the thigh. ♦ The vascular structures then pass through the adductor hiatus, becoming popliteal in name and location in the distal thigh/posterior knee region.

GLUTEAL AND POSTERIOR THIGH REGIONS

Gluteal Region: Buttocks and Hip Region

Although the demarcation of the trunk and lower limb is abrupt anteriorly at the inguinal ligament, posteriorly the gluteal region is a large transitional zone between the trunk

and limb. Physically part of the trunk, functionally the gluteal region is definitely part of the lower limb.

The **gluteal region** is the prominent area posterior to the pelvis and inferior to the level of the iliac crests (the buttocks) and extending laterally to the posterior margin of the greater trochanter (Fig. 5.32). The *hip region* overlies the greater trochanter laterally, extending anteriorly to the ASIS. Some definitions include both buttocks and hip region as part of the gluteal region, but the two parts are commonly distinguished. The *intergluteal*

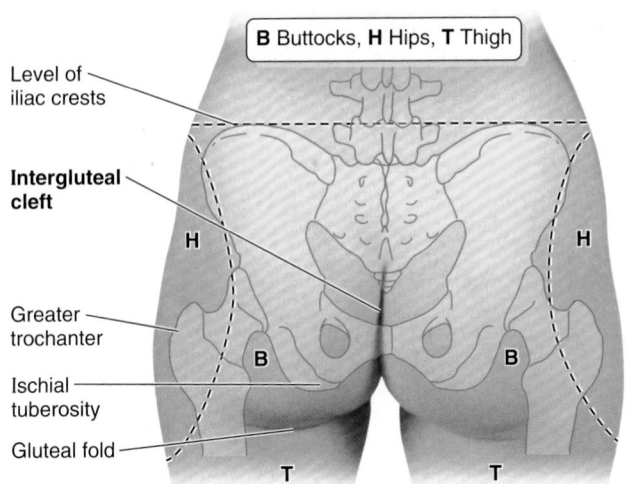

FIGURE 5.32. Gluteal region, comprising buttocks and hip region.

ligament extends across the sciatic notch of the hip bone, converting the notch into a foramen that is further subdivided by the **sacrospinous ligament** and *ischial spine*, creating the greater and lesser sciatic foramina. The **greater sciatic foramen** is the passageway for structures entering or leaving the pelvis (e.g., sciatic nerve), whereas the **lesser sciatic foramen** is the passageway for structures entering or leaving the perineum (e.g., pudendal nerve).

It is helpful to think of the greater sciatic foramen as the "door" through which all lower limb arteries and nerves leave the pelvis and enter the gluteal region. The *piriformis muscle* (Fig. 5.34D–G; Table 5.6) also enters the gluteal region through the greater sciatic foramen and fills most of it.

Muscles of Gluteal Region

The **muscles of the gluteal region** (Fig. 5.35) share a common compartment, but are organized into two layers, superficial and deep:

- The *superficial layer of muscles of the gluteal region* consists of the three large overlapping glutei (maximus, medius, and minimus) and the tensor fasciae latae (Figs. 5.34A, C–E, & J and 5.35). These muscles all have proximal attachments to the posterolateral (external) surface and margins of the ala of the ilium, and are mainly extensors, abductors, and medial rotators of the thigh.
- The *deep layer of muscles of the gluteal region* consists of smaller muscles (piriformis, obturator internus, superior and inferior gemelli, and quadratus femoris) covered by the inferior half of the gluteus maximus (Figs. 5.34F–I and 5.35).

cleft (natal cleft) is the groove that separates the buttocks from each other. The **gluteal muscles** (gluteus maximus, medius, and minimus and tensor fasciae latae) form the bulk of the region. The *gluteal fold* demarcates the inferior boundary of the buttocks and the superior boundary of the thigh.

GLUTEAL LIGAMENTS

The parts of the bony pelvis—hip bones, sacrum, and coccyx—are bound together by dense ligaments (Fig. 5.33). The **posterior sacro-iliac ligament** is continuous inferiorly with the sacrotuberous ligament. The **sacrotuberous**

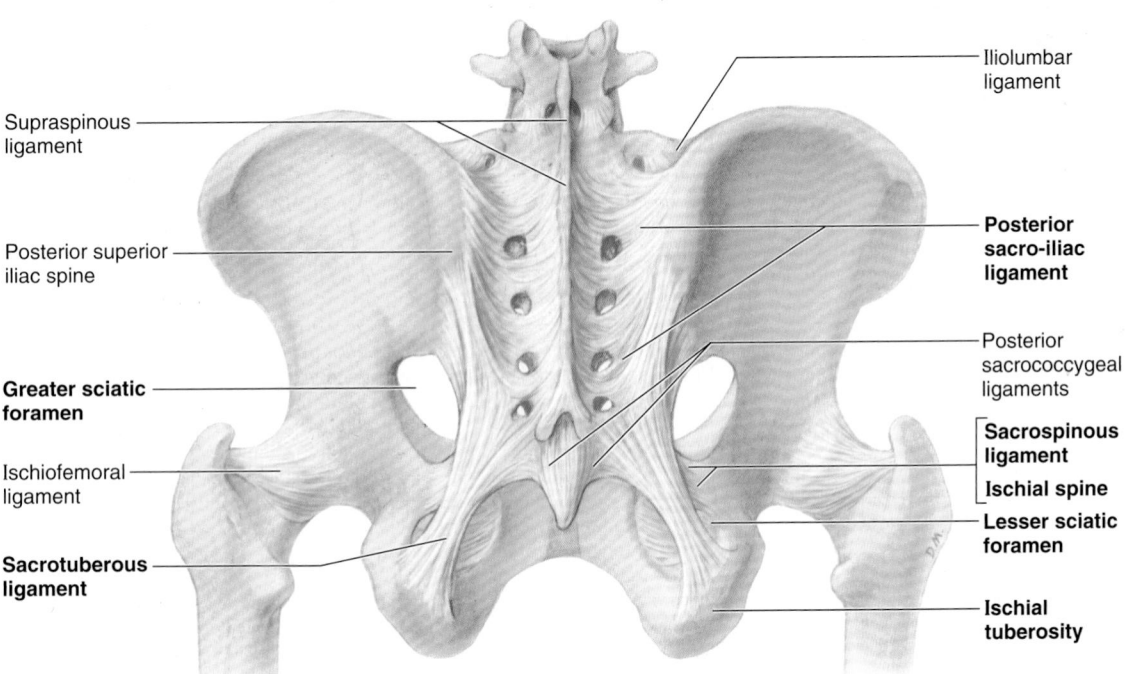

FIGURE 5.33. Ligaments of pelvic girdle. The sacrotuberous and sacrospinous ligaments convert the greater and lesser sciatic notches into foramina.

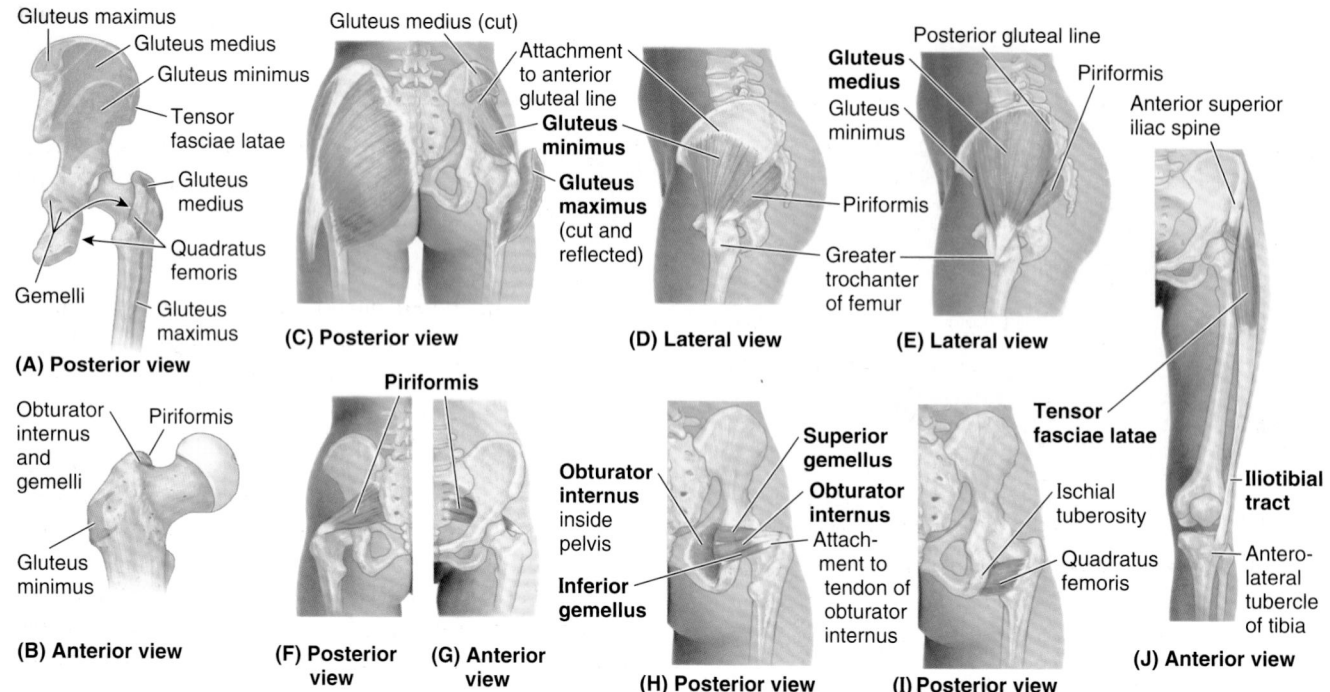

FIGURE 5.34. Muscles of gluteal region: abductors and lateral rotators.

TABLE 5.6. MUSCLES OF GLUTEAL REGION: ABDUCTORS AND ROTATORS OF THIGH

Muscle	Proximal Attachment	Distal Attachment	Innervation[a]	Main Action
Gluteus maximus (Fig. 5.34A & C)	Ilium posterior to posterior gluteal line; dorsal surface of sacrum and coccyx; sacrotuberous ligament	Most fibers end in iliotibial tract, which inserts into lateral condyle of tibia; some fibers insert on gluteal tuberosity	Inferior gluteal nerve (L5, **S1, S2**)	Extends thigh (especially from flexed position) and assists in its lateral rotation; steadies thigh and assists in rising from sitting position
Gluteus medius (Fig. 5.34A, C, & E)	External surface of ilium between anterior and posterior gluteal lines	Lateral surface of greater trochanter of femur	Superior gluteal nerve (**L5**, S1)	Abduct and medially rotate thigh; keep pelvis level when ipsilateral limb is weight-bearing and advance opposite (unsupported) side during its swing phase
Gluteus minimus (Fig. 5.34A–D)	External surface of ilium between anterior and inferior gluteal lines	Anterior surface of greater trochanter of femur		
Tensor fasciae latae (Fig. 5.34J)	Anterior superior iliac spine; anterior part of iliac crest	Iliotibial tract, which attaches to lateral condyle of tibia		
Piriformis (Fig. 5.34F & G)	Anterior surface of sacrum; sacrotuberous ligament	Superior border of greater trochanter of femur	Branches of anterior rami of **S1**, S2	Laterally rotate extended thigh and abduct flexed thigh; steady femoral head in acetabulum
Obturator internus (Fig. 5.34H)	Pelvic surface of obturator membrane and surrounding bones	Medial surface of greater trochanter (trochanteric fossa) of femur[b]	Nerve to obturator internus (L5, **S1**)	
Superior and inferior gemelli (Fig. 5.34H)	Superior: ischial spine Inferior: ischial tuberosity	Medial surface of greater trochanter (trochanteric fossa) of femur[b]	Superior gemellus: same nerve supply as obturator internus Inferior gemellus: same nerve supply as quadratus femoris	
Quadratus femoris (Fig. 5.34I)	Lateral border of ischial tuberosity	Quadrate tubercle on intertrochanteric crest of femur and area inferior to it	Nerve to quadratus femoris (L5, S1)	Laterally rotates thigh[c]; steadies femoral head in acetabulum

[a] The spinal cord segmental innervation is indicated (e.g., "S1, S2" means that the nerves supplying the piriformis are derived from the first two sacral segments of the spinal cord). Numbers in boldface (**S1**) indicate the main segmental innervation. Damage to one or more of the listed spinal cord segments, or to the motor nerve roots arising from them, results in paralysis of the muscles concerned.

[b] The gemelli muscles blend with and share the tendon of the obturator internus as it attaches to the greater trochanter of the femur, collectively forming the triceps coxae.

[c] There are six lateral rotators of the thigh: piriformis, obturator internus, superior and inferior gemelli, quadratus femoris, and obturator externus. These muscles also stabilize the hip joint.

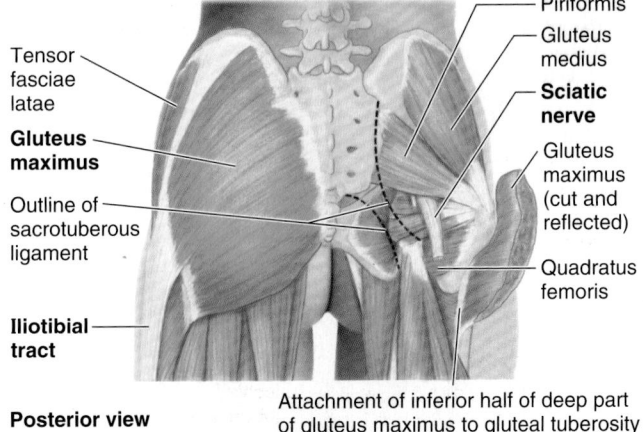

Posterior view

FIGURE 5.35. Muscles of gluteal region: superficial and deep dissections.

These muscles all have distal attachments on or adjacent to the intertrochanteric crest of the femur. These muscles are lateral rotators of the thigh, but they also stabilize the hip joint, working with the strong ligaments of the hip joint to steady the femoral head in the acetabulum.

The attachments of the muscles of the gluteal region are illustrated in Figure 5.34A–J, and their innervation and main actions are described in Table 5.6.

GLUTEUS MAXIMUS

The **gluteus maximus** is the most superficial gluteal muscle (Figs. 5.34C and 5.35). It is the largest, heaviest, and most coarsely fibered muscle of the body. The gluteus maximus covers all of the other gluteal muscles, except for the anterosuperior third of the gluteus medius.

The *ischial tuberosity* can be felt on deep palpation through the inferior part of the muscle, just superior to the medial part of the gluteal fold (see Fig. 5.32). When the thigh is flexed, the inferior border of the gluteus maximus moves superiorly, leaving the ischial tuberosity subcutaneous. You do not sit on your gluteus maximus; you sit on the fatty fibrous tissue and the ischial bursa that lie between the ischial tuberosity and skin.

The gluteus maximus slopes inferolaterally at a 45° angle from the pelvis to the buttocks. The fibers of the superior and larger part of the gluteus maximus and superficial fibers of its inferior part insert into the *iliotibial tract* and indirectly, via the lateral intermuscular septum, into the linea aspera of the femur (Fig. 5.36A & B). Some deep fibers of the inferior part of the muscle (roughly the deep anterior and inferior quarter) attach to the *gluteal tuberosity* of the femur.

The inferior gluteal nerve and vessels enter the deep surface of the gluteus maximus at its center. It is supplied by both the inferior and superior gluteal arteries. In the superior part of its course, the *sciatic nerve* passes deep to the gluteus maximus (Fig. 5.35).

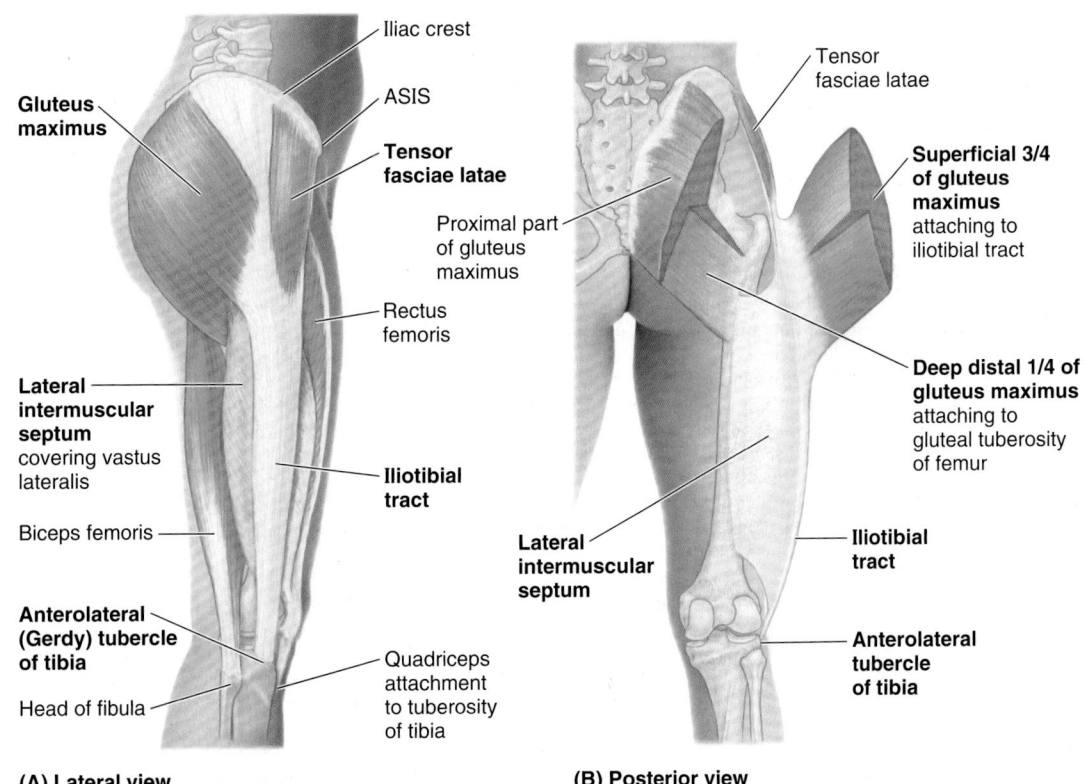

(A) Lateral view   **(B) Posterior view**

FIGURE 5.36. Gluteus maximus and tensor fasciae latae. Shown are superficial (**A**) and deep (**B**) views of the lateral musculofibrous complex formed by the tensor fasciae latae and gluteus maximus muscles and their shared aponeurotic tendon, the iliotibial tract. The iliotibial tract is continuous posteriorly and deeply with the dense lateral intermuscular septum.

The main actions of the gluteus maximus are extension and lateral rotation of the thigh. When the distal attachment of the gluteus maximus is fixed, the muscle extends the trunk on the lower limb. Although it is the strongest extensor of the hip, it acts mostly when force is necessary (rapid movement or movement against resistance). The gluteus maximus functions primarily between the flexed and standing (straight) positions of the thigh, as when rising from the sitting position, straightening from the bending position, walking uphill and up stairs, and running. It is used only briefly during casual walking and usually not at all when standing motionless.

Paralysis of the gluteus maximus does not seriously affect walking on level ground. Verify this by placing your hand on your buttocks when walking slowly. The gluteus maximus contracts only briefly during the earliest part of the stance phase (from heel strike to when the foot is flat on the ground, to resist further flexion as weight is assumed by the partially flexed limb) (Fig. 5.20A and Table 5.2). If you climb stairs and put your hands on your buttocks, you will feel the gluteus maximus contract strongly.

Because the iliotibial tract crosses the knee and attaches to the anterolateral tubercle of the tibia (Gerdy) (Figs. 5.34J and 5.36A & B), the gluteus maximus and tensor fasciae latae together are also able to assist in making the extended knee stable, but they are not usually called on to do so during normal standing. Because the iliotibial tract attaches to the femur via the lateral intermuscular septum, it does not have the freedom necessary to produce motion at the knee.

Testing the gluteus maximus is performed when the person is prone with the lower limb straight. The person tightens the buttocks and extends the hip joint as the examiner observes and palpates the gluteus maximus.

Gluteal Bursae. Gluteal bursae (L., purses) separate the gluteus maximus from adjacent structures (Fig. 5.37). Bursae are membranous sacs lined by a synovial membrane containing a capillary layer of slippery fluid resembling egg white. Bursae are located in areas subject to friction (e.g., where the iliotibial tract crosses the greater trochanter). The purpose of bursae is to reduce friction and permit free movement. Usually three bursae are associated with the gluteus maximus:

1. The **trochanteric bursa** separates superior fibers of the gluteus maximus from the greater trochanter. This bursa is commonly the largest of the bursae formed in relation to bony prominences and is present at birth. Other bursae appear to form as a result of postnatal movement.
2. The **ischial bursa** separates the inferior part of the gluteus maximus from the ischial tuberosity; it is often absent.
3. The **gluteofemoral bursa** separates the iliotibial tract from the superior part of the proximal attachment of the vastus lateralis.

See the blue boxes "Trochanteric Bursitis" and "Ischial Bursitis" on p. 581.

GLUTEUS MEDIUS AND GLUTEUS MINIMUS

The smaller gluteal muscles, **gluteus medius** and **gluteus minimus,** are fan shaped, and their fibers converge in

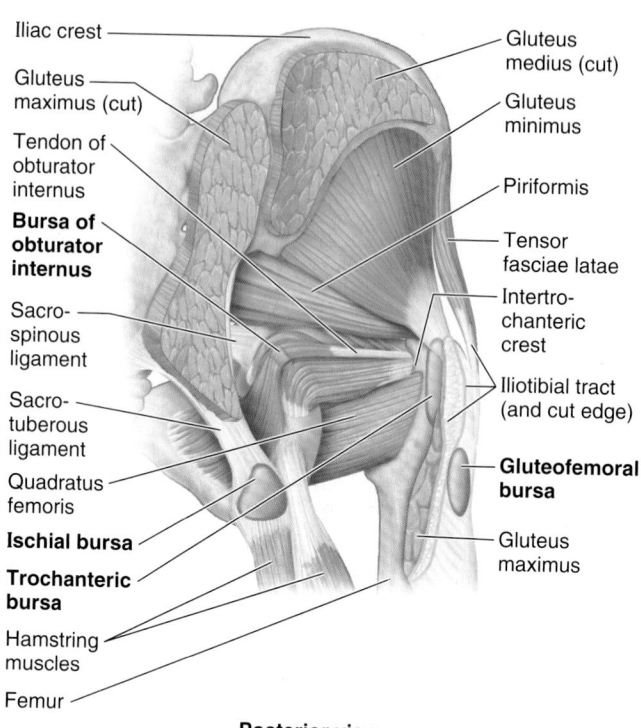

Posterior view

FIGURE 5.37. Gluteal muscles and bursae. Three bursae (trochanteric, gluteofemoral, and ischial) usually separate the gluteus maximus from underlying bony prominences. The bursa of the obturator internus underlies the tendon of the obturator internus.

the same manner toward essentially the same target (Figs. 5.34C–E, 5.35, 5.37, and 5.38). They share the same actions and nerve supply (Table 5.6) and are supplied by the same blood vessel, the superior gluteal artery. The gluteus minimus and most of the gluteus medius lie deep to the gluteus maximus on the external surface of the ilium. The gluteus medius and minimus abduct or stabilize the thigh and rotate it medially (Fig. 5.20F and 5.39; Table 5.2).

Testing the gluteus medius and minimus is performed while the person is side-lying with the test limb uppermost and the lowermost limb flexed at the hip and knee for stability. The person abducts the thigh without flexion or rotation against straight downward resistance. The gluteus medius can be palpated inferior to the iliac crest, posterior to the tensor fasciae latae, which is also contracting during abduction of the thigh.

TENSOR FASCIAE LATAE

The **tensor fasciae latae** is a fusiform muscle approximately 15 cm long that is enclosed between two layers of fascia lata (Figs. 5.34C & J, 5.36, and 5.37). Its attachments, innervation, and action are provided in Table 5.6.

The tensor fasciae latae and the superficial and anterior part of the gluteus maximus share a common distal attachment to the *anterolateral tubercle of the tibia* via the *iliotibial tract*, which acts as a long aponeurosis for the muscles.

(text continues on p. 569)

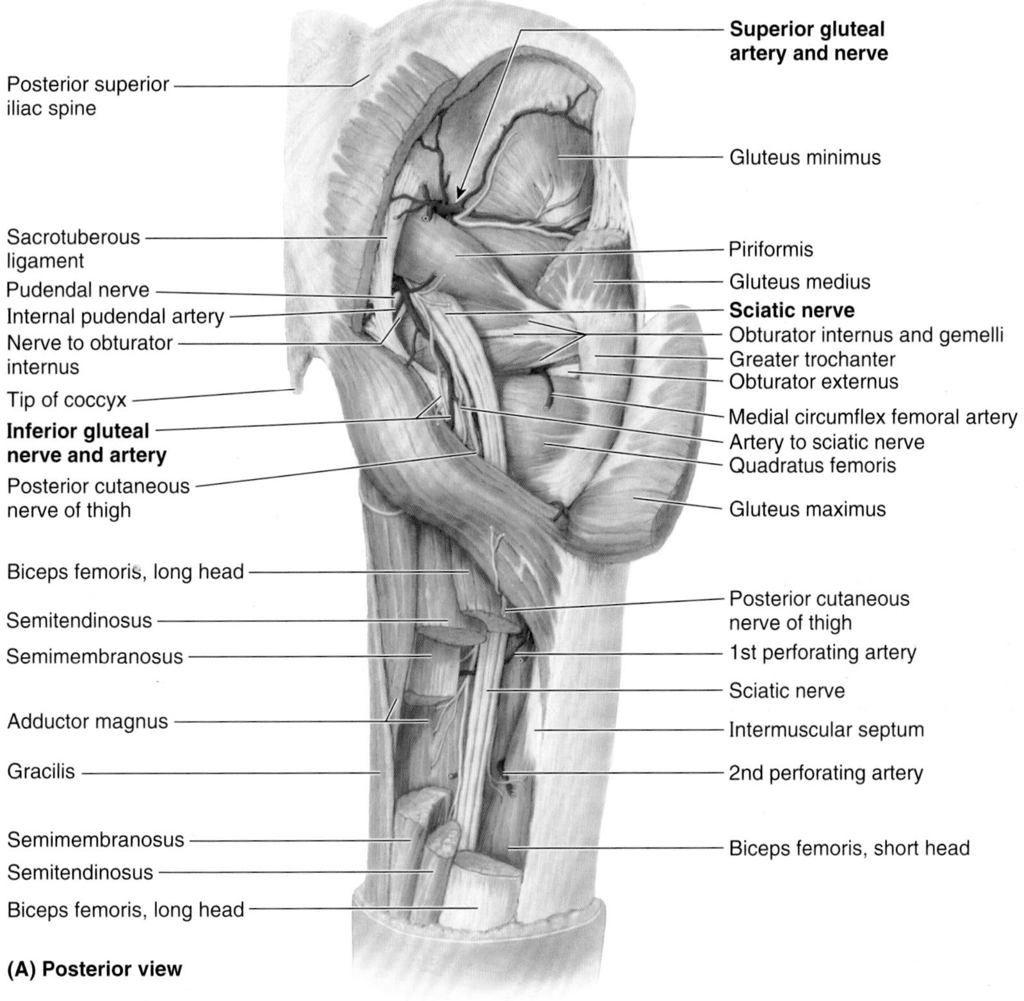

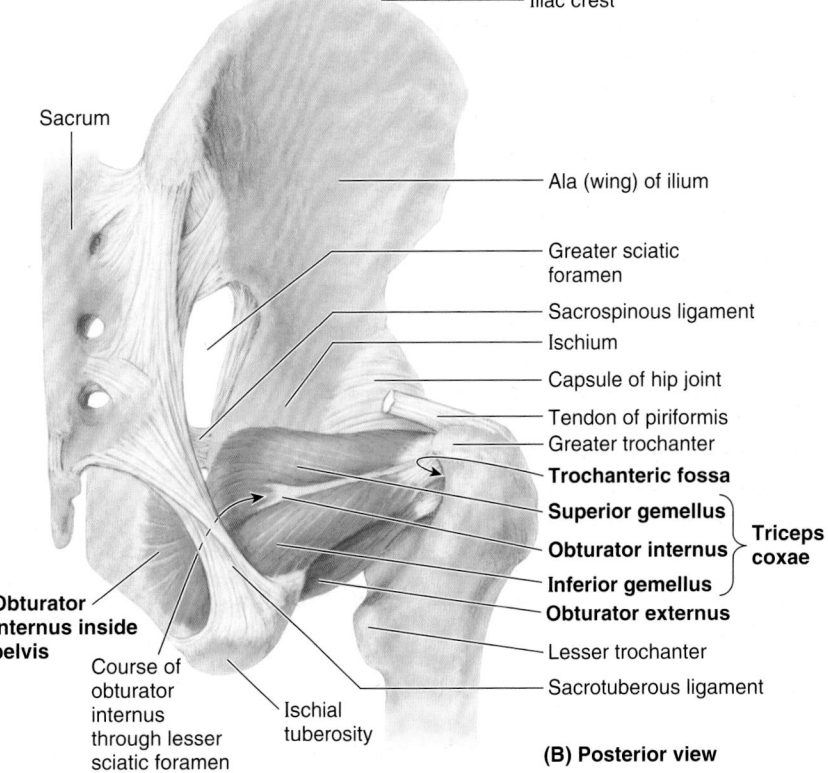

Superior gluteal artery and nerve

Posterior superior iliac spine

Gluteus minimus

Sacrotuberous ligament

Piriformis

Pudendal nerve

Gluteus medius

Internal pudendal artery

Sciatic nerve

Nerve to obturator internus

Obturator internus and gemelli
Greater trochanter
Obturator externus

Tip of coccyx

Inferior gluteal nerve and artery

Medial circumflex femoral artery
Artery to sciatic nerve
Quadratus femoris

Posterior cutaneous nerve of thigh

Gluteus maximus

Biceps femoris, long head

Semitendinosus

Posterior cutaneous nerve of thigh

Semimembranosus

1st perforating artery

Sciatic nerve

Adductor magnus

Intermuscular septum

Gracilis

2nd perforating artery

Semimembranosus

Biceps femoris, short head

Semitendinosus

Biceps femoris, long head

(A) Posterior view

Iliac crest

Sacrum

Ala (wing) of ilium

Greater sciatic foramen

Sacrospinous ligament

Ischium

Capsule of hip joint

Tendon of piriformis
Greater trochanter

Trochanteric fossa

Superior gemellus

Obturator internus } **Triceps coxae**

Inferior gemellus

Obturator externus

Lesser trochanter

Sacrotuberous ligament

Obturator internus inside pelvis

Course of obturator internus through lesser sciatic foramen

Ischial tuberosity

(B) Posterior view

FIGURE 5.38. Dissection of gluteal region and posterior thigh. A. Most of the gluteus maximus and medius are removed, and segments of the hamstrings are excised, to reveal the neurovascular structures of the gluteal region and proximal posterior thigh. The sciatic nerve runs deep (anterior) to and is protected by the overlying gluteus maximus initially and then the biceps femoris. **B.** This dissection shows some of the lateral rotators of the thigh. The components of the triceps coxae share a common attachment into the trochanteric fossa adjacent to that of the obturator externus.

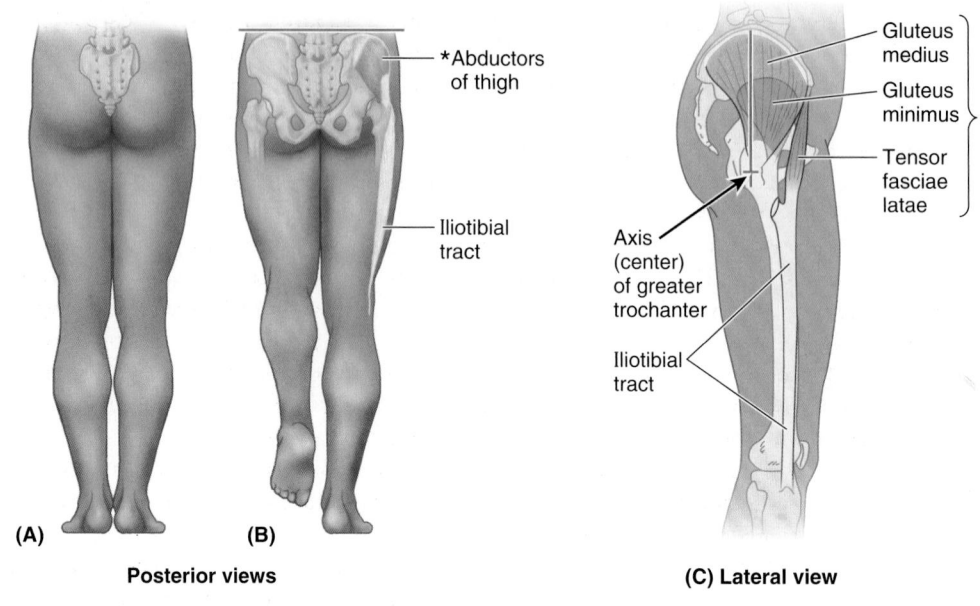

Posterior views

(C) Lateral view

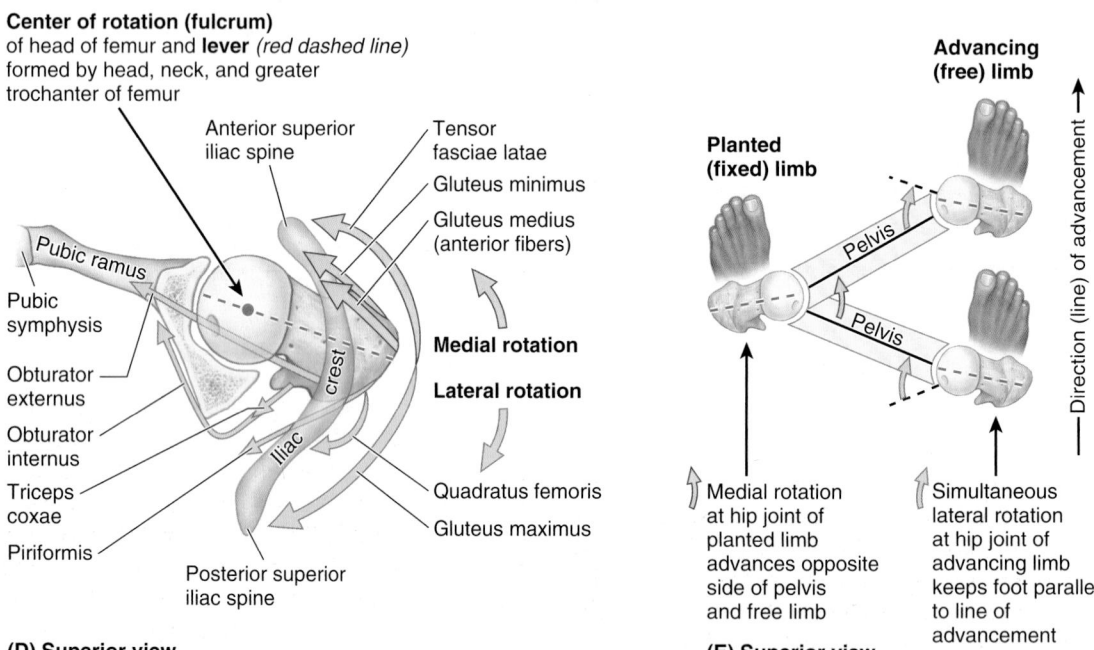

(D) Superior view

(E) Superior view

FIGURE 5.39. Action of abductors/medial rotators of thigh in walking. A and B. The role of the abductors (gluteus medius and minimus, tensor fasciae latae) is demonstrated. When the weight is on both feet (**A**), the pelvis is evenly supported and does not sag. When the weight is borne by one limb (**B**), the muscles on the supported side fix the pelvis so that it does not sag to the unsupported side. Keeping the pelvis level enables the non-weight-bearing limb to clear the ground as it is brought forward during the swing phase. **C and D.** The role of the rotators of the thigh is demonstrated in lateral (**C**) and superior (**D**) views. Note that most abductors—the tensor fasciae latae, gluteus minimus, and most (the anterior fibers) of the gluteus medius—lie anterior to the lever provided by the axis of the head, neck, and greater trochanter of the femur to rotate the thigh around the vertical axis traversing the femoral head. The superior view of the right hip joint (**D**) includes the superior pubic ramus, acetabulum, and iliac crest; the inferior part of the ilium has been removed to reveal the head and neck of the femur. The lines of pull of the rotators of the hip are indicated by arrows, demonstrating the antagonistic relationship resulting from their positions relative to the lever and the center of rotation (fulcrum). The medial rotators pull the greater trochanter anteriorly and the lateral rotators pull the trochanter posteriorly, resulting in rotation of the thigh around the vertical axis. Note that all of these muscles also pull the head and neck of the femur medially into the acetabulum, strengthening the joint. In walking (**E**), the same muscles that act unilaterally during the stance phase (planted limb) to keep the pelvis level via abduction can simultaneously produce medial rotation at the hip joint, advancing the opposite unsupported side of the pelvis (augmenting advancement of the free limb). The lateral rotators of the advancing (free) limb act during the swing phase to keep the foot parallel to the direction (line) of advancement.

However, unlike the gluteus maximus, the tensor fasciae latae is served by the superior gluteal neurovascular bundle. Despite its gluteal innervation and shared attachment, this tensor is primarily a flexor of the thigh because of its anterior location; however, it generally does not act independently.

To produce flexion, the tensor fasciae latae acts in concert with the iliopsoas and rectus femoris. When the iliopsoas is paralyzed, the tensor fasciae latae undergoes hypertrophy in an attempt to compensate for the paralysis. It also works in conjunction with other abductor/medial rotator muscles (gluteus medius and minimus) (Fig. 5.39). It lies too far anteriorly to be a strong abductor and thus probably contributes primarily as a synergist or fixator.

The tensor fasciae latae tenses the fascia lata and iliotibial tract. Because the iliotibial tract is attached to the femur via the lateral intermuscular septum, the tensor produces little if any movement of the leg (Fig. 5.36B). However, when the knee is fully extended, it contributes to (increases) the extending force, adding stability, and plays a role in supporting the femur on the tibia when standing if lateral sway occurs. When the knee is flexed by other muscles, the tensor fasciae latae can synergistically augment flexion and lateral rotation of the leg.

The abductors/medial rotators of the hip joint play an essential role during locomotion, advancing and preventing the sagging of the unsupported side of the pelvis during walking, as illustrated and explained in Figure 5.39. The supportive and action-producing functions of the abductors/medial rotators depend on normal:

- Muscular activity and innervation from the superior gluteal nerve.
- Articulation of the hip joint components.
- Strength and angulation of the femoral neck.

PIRIFORMIS

The pear-shaped **piriformis** (L. *pirum*, a pear) is located partly on the posterior wall of the lesser pelvis, and partly posterior to the hip joint (Figs. 5.34F & G, 5.35, and 5.37; Table 5.6). The piriformis leaves the pelvis through the *greater sciatic foramen*, almost filling it, to reach its attachment to the superior border of the *greater trochanter*.

Because of its key position in the buttocks, the *piriformis is the landmark of the gluteal region*. The piriformis provides the key to understanding relationships in the gluteal region because it determines the names of the blood vessels and nerves (Fig. 5.38A):

- The superior gluteal vessels and nerve emerge superior to it.
- The inferior gluteal vessels and nerve emerge inferior to it.

See the blue box "Injury to Sciatic Nerve," p. 582.

OBTURATOR INTERNUS AND GEMELLI

The **obturator internus** and the **superior** and **inferior gemelli** (L. *geminus*, small twin) form a tricipital (three-headed) muscle, the **triceps coxae** (triceps of the hip), which occupies the gap between the piriformis and the quadratus femoris (Figs. 5.34H and 5.38A & B). The common tendon of these muscles lies horizontally in the buttocks as it passes to the greater trochanter of the femur.

The attachments, action, and innervation are described in Table 5.6. The *obturator internus* is located partly in the pelvis, where it covers most of the lateral wall of the lesser pelvis (Fig. 5.38B). It leaves the pelvis through the *lesser sciatic foramen*, makes a right-angle turn (Figs. 5.38B and 5.39D), becomes tendinous, and receives the distal attachments of the gemelli before attaching to the medial surface of the greater trochanter (trochanteric fossa) of the femur.

The small *gemelli* are narrow, triangular extrapelvic reinforcements of the obturator internus. Although the inferior gemellus receives separate innervation from the nerve to the quadratus femoris, it is more realistic to consider these three muscles as a unit (i.e., the *triceps coxae*) because they are incapable of independent action.

The **bursa of the obturator internus** allows free movement of the muscle over the posterior border of the ischium, where the border forms the lesser sciatic notch and the trochlea over which the tendon glides as it turns (Fig. 5.37).

QUADRATUS FEMORIS

The **quadratus femoris,** a short, flat quadrangular muscle, is located inferior to the obturator internus and gemelli (Figs. 5.34I, 5.35, 5.37, and 5.38A). True to its name, the quadratus femoris is a rectangular muscle that is a strong lateral rotator of the thigh.

OBTURATOR EXTERNUS

Based on its location (posterior to the pectineus and the superior ends of the adductor muscles), and its innervation (*obturator nerve*), the obturator externus was described earlier in this chapter (p. 550) with the medial thigh muscles (Fig. 5.23H; Table 5.4). However, it functions as a lateral rotator of the thigh, and its distal attachment is visible only during dissection of the gluteal region (Fig. 5.38B) or hip joint. That is why it is mentioned again in this context.

The belly of the obturator externus lies deep in the proximal thigh, with its tendon passing inferior to the neck of the femur and deep to the quadratus femoris, on the way to its attachment to the *trochanteric fossa of the femur* (Figs. 5.38A and 5.39D). The obturator externus, with other short muscles around the hip joint, stabilizes the head of the femur in the acetabulum. It is most effective as a lateral rotator of the thigh when the hip joint is flexed.

Posterior Thigh Region

The **posterior thigh muscles** and their attachments are illustrated in Figure 5.40, and their attachments, innervation, and actions are described in Table 5.7.

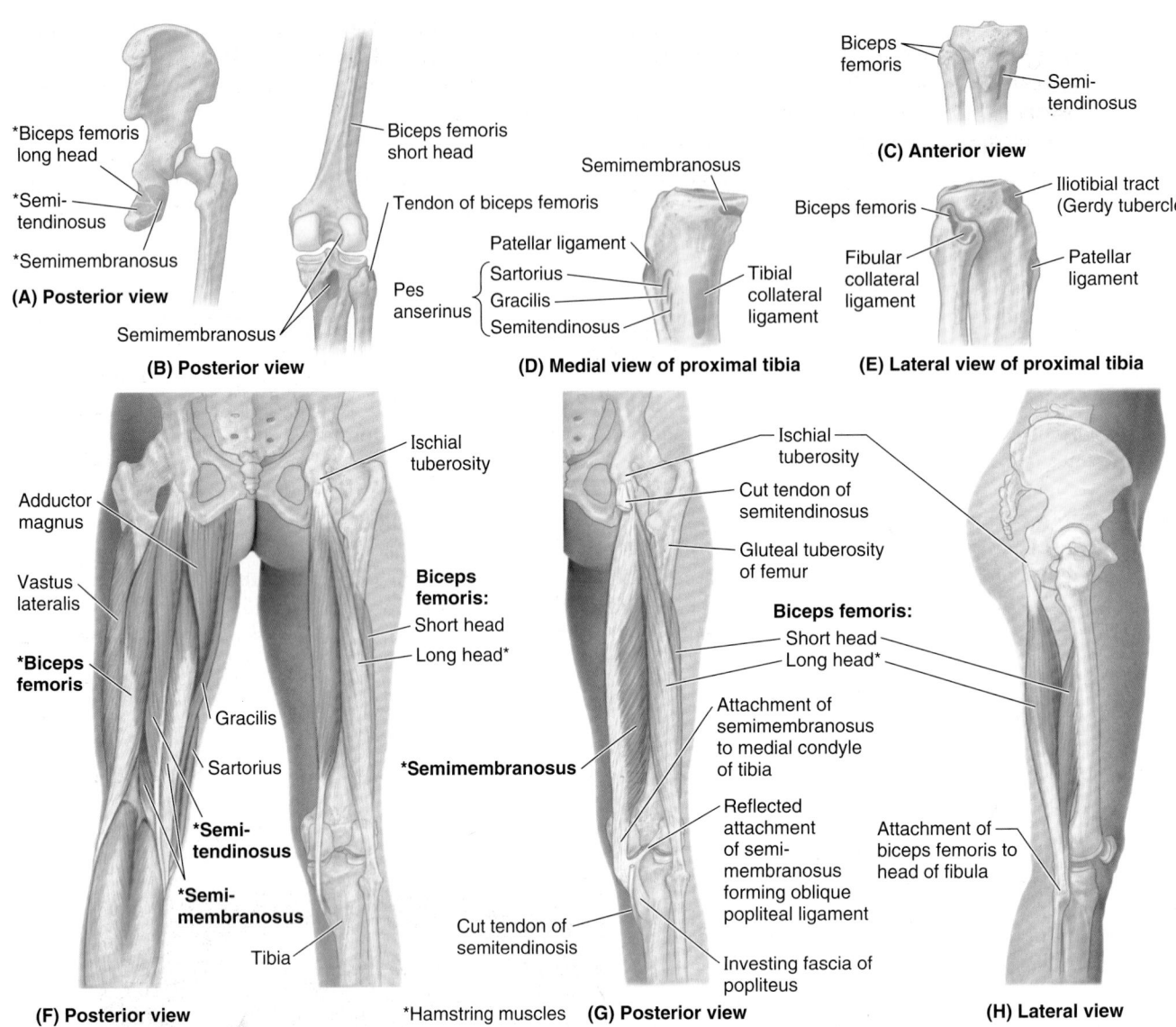

FIGURE 5.40. Muscles of posterior thigh: extensors of hip and flexors of knee.

TABLE 5.7. MUSCLES OF POSTERIOR THIGH: EXTENSORS OF HIP AND FLEXORS OF KNEE

Muscle[a]	Proximal Attachment	Distal Attachment	Innervation[b]	Main Action
Semitendinosus	Ischial tuberosity	Medial surface of superior part of tibia	Tibial division of sciatic nerve part of tibia (**L5, S1**, S2)	Extend thigh; flex leg and rotate it medially when knee is flexed; when thigh and leg are flexed, these muscles can extend trunk
Semimembranosus		Posterior part of medial condyle of tibia; reflected attachment forms oblique popliteal ligament (to lateral femoral condyle)		
Biceps femoris	Long head: ischial tuberosity Short head: linea aspera and lateral supracondylar line of femur	Lateral side of head of fibula; tendon is split at this site by fibular collateral ligament of knee	Long head: tibial division of sciatic nerve (L5, **S1**, S2) Short head: common fibular division of sciatic nerve (L5, **S1**, S2)	Flexes leg and rotates it laterally when knee is flexed; extends thigh (e.g., accelerating mass during first step of gait).

[a]Collectively these three muscles are known as hamstrings.

[b]The spinal cord segmental innervation is indicated (e.g., "L5, S1, S2" means that the nerves supplying the semitendinosus are derived from the fifth lumbar segment and first two sacral segments of the spinal cord). Numbers in boldface (**L5, S1**) indicate the main segmental innervation. Damage to one or more of the listed spinal cord segments or to the motor nerve roots arising from them results in paralysis of the muscles concerned.

Three of the four muscles in the posterior aspect of the thigh are *hamstrings*. The **hamstring muscles** (Figs. 5.40A–D and 5.41B) are: (1) *semitendinosus*, (2) *semimembranosus*, and (3) *biceps femoris* (long head). The hamstring muscles ("hamstrings" for short) share common features:

- Proximal attachment to the ischial tuberosity deep to the gluteus maximus (Fig. 5.40A , F–H).
- Distal attachment to the bones of the leg (Fig. 5.40B–E).

- Thus they span and act on two joints, producing extension at the hip joint and flexion at the knee joint.
- Innervation by the tibial division of the sciatic nerve (Fig. 5.41A).

The long head of the biceps femoris meets all these conditions, but the short head of the biceps, the fourth muscle of the posterior compartment, fails to meet any of them.

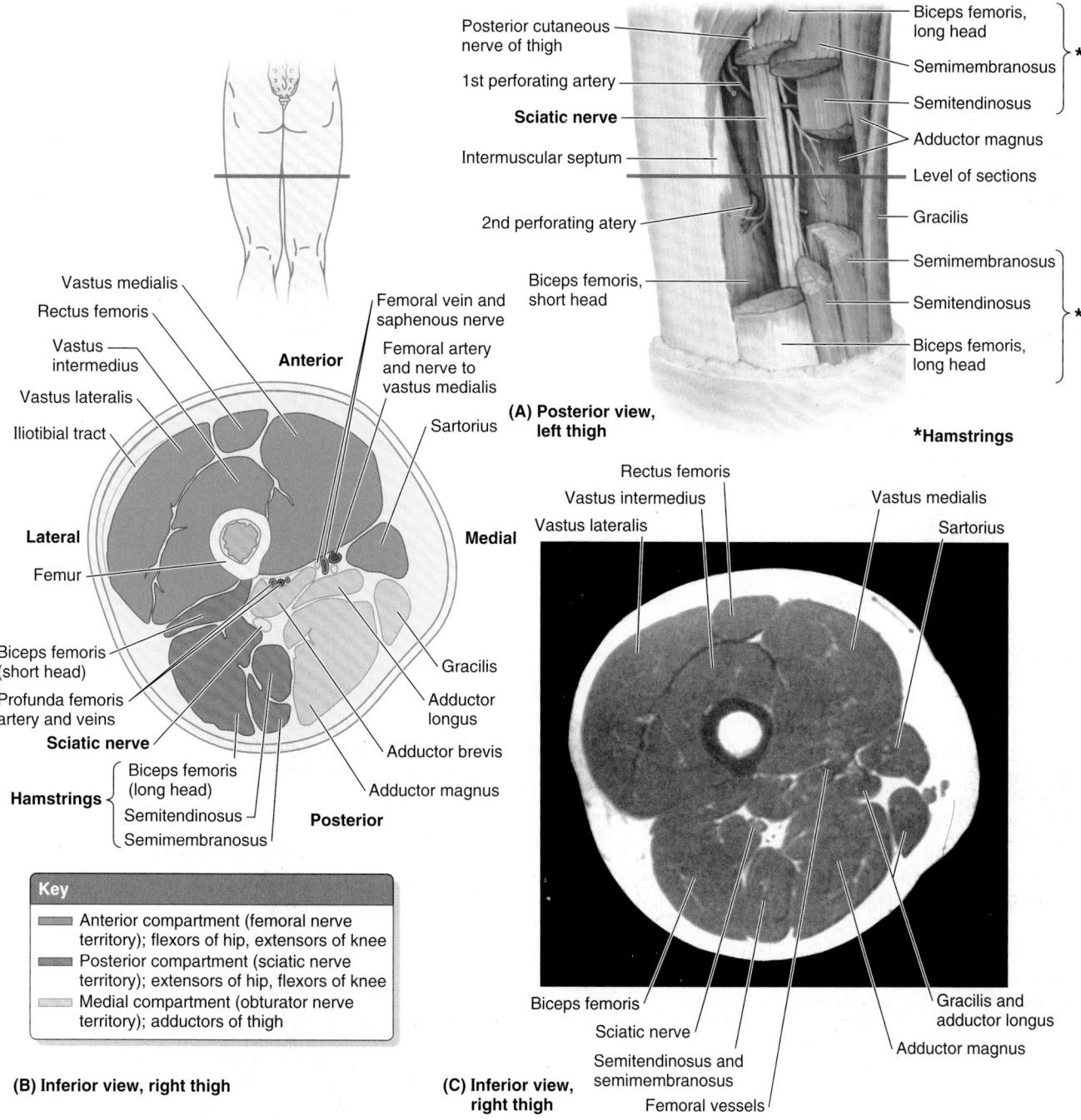

FIGURE 5.41. Muscles and fascial compartments of thigh. A. Segments of the hamstring muscles are excised to reveal the sciatic nerve. The level of the sections shown in parts **B** and **C** is indicated. **B.** An anatomical transverse section through the middle thigh, 10–15 cm inferior to the inguinal ligament. The three compartments of the thigh are demonstrated in different shades of color. Note that each has its own nerve supply and functional group(s) of muscles. **C.** Transverse MRI of the right thigh corresponding to **B.** (Courtesy of Dr. W. Kucharczyk, Chair of Medical Imaging, Faculty of Medicine, University of Toronto and Clinical Director of the Tri-Hospital Resonance Centre, Toronto, Ontario, Canada.)

The hamstrings received their name because it is common to tie hams (pork thighs) up for curing and/or smoking using the long tendons of these muscles (Fig. 5.40F). This also explains the expression "hamstringing the enemy" by slashing these tendons lateral and medial to the knees.

The two actions of the hamstrings cannot be performed maximally at the same time: full flexion of the knee requires so much shortening of the hamstrings that they cannot provide the additional contraction that would be necessary for simultaneous full extension of the thigh; similarly, full extension of the hip shortens the hamstrings so they cannot further contract to act fully on the knee. When the thighs and legs are fixed, the hamstrings can help extend the trunk at the hip joint.

The hamstrings are active in thigh extension under all situations except full flexion of the knee, including maintenance of the relaxed standing posture (standing at ease). A person with paralyzed hamstrings tends to fall forward because the gluteus maximus muscles cannot maintain the necessary muscle tone to stand straight.

The hamstrings are the hip extensors involved in walking on flat ground, when the gluteus maximus demonstrates minimal activity. However, rather than producing either hip extension or knee flexion per se during normal walking, the hamstrings demonstrate most activity when they are eccentrically contracting, resisting (decelerating) hip flexion and knee extension during terminal swing (between midswing and heel strike) (see Fig. 5.20G; Table 5.2).

The length of the hamstrings varies, but this is usually a matter of conditioning. In some people, they are not long enough to allow them to touch their toes when the knees are extended. Routine stretch exercise can lengthen these muscles and tendons.

To test the hamstrings, the person flexes their leg against resistance. Normally, these muscles—especially their tendons on each side of the popliteal fossa—should be prominent as they bend the knee (see Fig. 5.48C).

SEMITENDINOSUS

The **semitendinosus,** as its name indicates, is half tendinous (Fig. 5.40F). It has a fusiform belly that is usually interrupted by a tendinous intersection and a long, cord-like tendon that begins approximately two thirds of the way down the thigh. Distally, the tendon attaches to the medial surface of the superior part of the tibia as part of the *pes anserinus* formation in conjunction with the tendinous insertions of the sartorius and gracilis (Fig. 5.40D).

SEMIMEMBRANOSUS

The **semimembranosus** is a broad muscle that is also aptly named because of the flattened membranous form of its proximal attachment to the ischial tuberosity (Fig. 5.40G; Table 5.7). The tendon of the semimembranosus forms around the middle of the thigh and descends to the posterior part of the medial condyle of the tibia.

The semimembranous tendon divides distally into three parts: (1) a direct attachment to the posterior aspect of the medial tibial condyle, (2) a part that blends with the popliteal fascia, and (3) a reflected part that reinforces the intercondylar part of the joint capsule of the knee as the **oblique popliteal ligament** (Fig. 5.40G; see also Fig. 5.58B).

When the knee is flexed to 90°, the tendons of the medial hamstrings or "semi-" muscles (semitendinosus and semimembranosus) pass to the medial side of the tibia. In this position, contraction of the medial hamstrings (and of synergists including the gracilis, sartorius, and popliteus) produces a limited amount (about 10°) of medial rotation of the tibia at the knee. The two medial hamstrings are not as active as the lateral hamstring, the biceps femoris, which is the "workhorse" of extension at the hip (Hamill and Knutzen, 2008).

BICEPS FEMORIS

The fusiform **biceps femoris,** as its name indicates, has two heads: a *long head* and a *short head* (Fig. 5.40F–H). In the inferior part of the thigh, the long head becomes tendinous and is joined by the short head. The rounded common tendon of these heads attaches to the head of the fibula and can easily be seen and felt as it passes the knee, especially when the knee is flexed against resistance.

The **long head of the biceps femoris** crosses and provides protection for the sciatic nerve after it descends from the gluteal region into the posterior aspect of the thigh (Figs. 5.38A and 5.41A–C). When the sciatic nerve divides into its terminal branches, the lateral branch (common fibular nerve) continues this relationship, running with the biceps tendon.

The **short head of the biceps femoris** arises from the lateral lip of the inferior third of the linea aspera and supracondylar ridge of the femur (Fig. 5.40B & H). Whereas the hamstrings have a common nerve supply from the tibial division of the sciatic nerve, *the short head of the biceps is innervated by the fibular division* (Table 5.7). Because each of the two heads of the biceps femoris has a different nerve supply, a wound in the posterior thigh with nerve injury may paralyze one head and not the other.

When the knee is flexed to 90°, the tendons of the lateral hamstring (biceps), as well as the iliotibial tract, pass to the lateral side of the tibia. In this position, contraction of the biceps and tensor fasciae latae produces about 40° lateral rotation of the tibia at the knee. Rotation of the flexed knee is especially important in snow skiing.

Neurovascular Structures of Gluteal and Posterior Thigh Regions

Several important nerves arise from the *sacral plexus* and either supply the gluteal region (e.g., superior and inferior gluteal nerves) or pass through it to supply the perineum and thigh (e.g., the pudendal and sciatic nerves, respectively). Figure 5.42 depicts the nerves of the gluteal region and posterior thigh, and Table 5.8 describes their origin, course, and distribution.

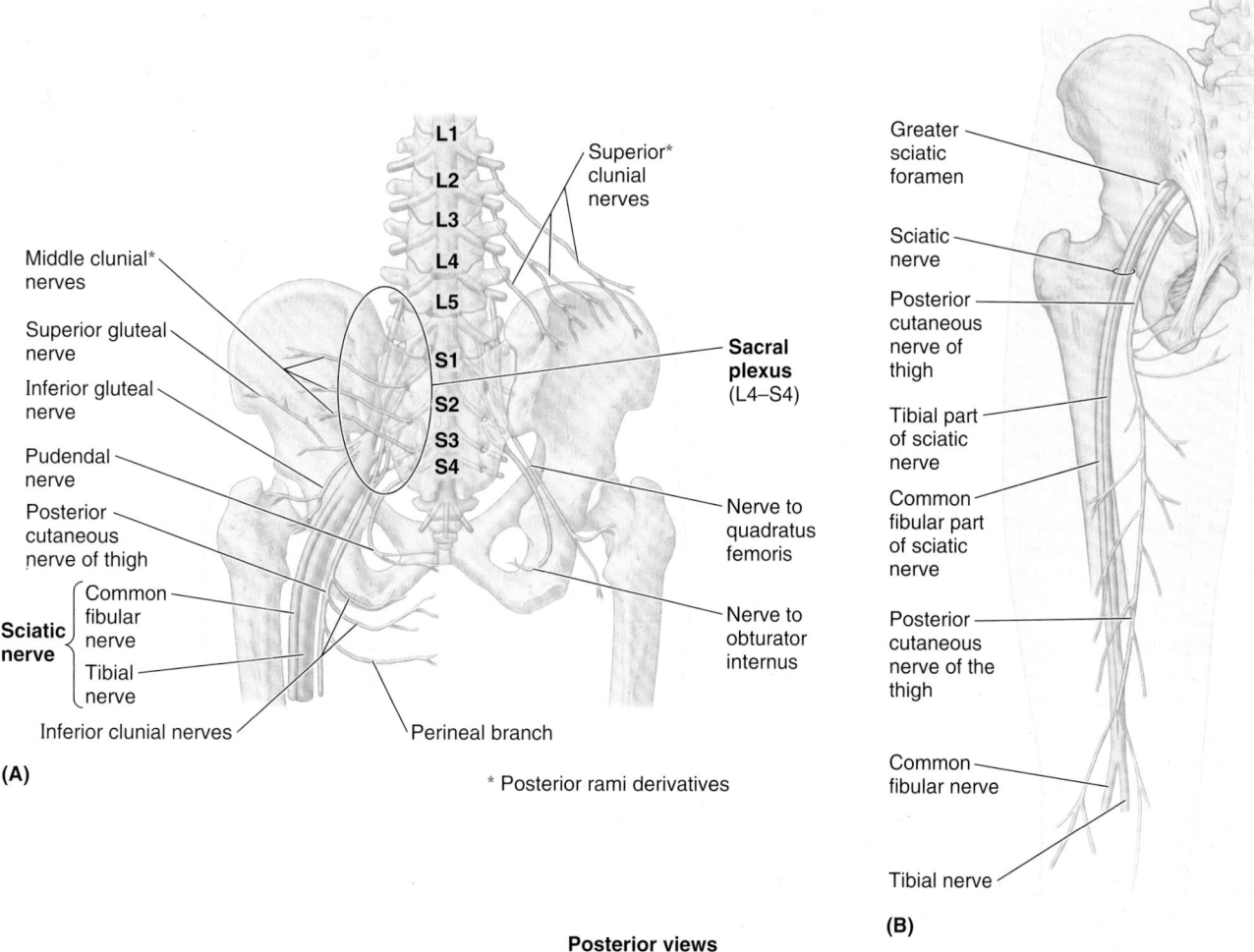

FIGURE 5.42. Nerves of the gluteal and posterior thigh regions.

TABLE 5.8. NERVES OF GLUTEAL AND POSTERIOR THIGH REGIONS

Nerve	Origin	Course	Distribution
Clunial			
Superior	As lateral cutaneous branches of posterior rami of L1–L3 spinal nerves	Pass inferolaterally across iliac crest	Supply skin of superior buttocks as far as tubercle of iliac crest
Middle	As lateral cutaneous branches of posterior rami of S1–S3 spinal nerves	Exit through posterior sacral foramina and pass laterally to gluteal region	Supply skin over sacrum and adjacent area of buttocks
Inferior	Posterior cutaneous nerve of thigh (anterior rami of S2–S3 spinal nerves)	Emerges from inferior border of gluteus maximus and ascends superficial to it	Supplies skin of inferior half of buttocks as far as greater trochanter

TABLE 5.8. NERVES OF GLUTEAL AND POSTERIOR THIGH REGIONS (Continued)

Nerve	Origin	Course	Distribution
Sciatic	Sacral plexus (anterior and posterior divisions of anterior rami of L4–S3 spinal nerves)	Enters gluteal region via greater sciatic foramen inferior to piriformis and deep to gluteus maximus; descends in posterior thigh deep to biceps femoris; bifurcates into tibial and common fibular nerves at apex of popliteal fossa	Supplies no muscles in gluteal region; supplies all muscles of posterior compartment of thigh (tibial division supplies all but short head of biceps, which is supplied by common fibular division)
Posterior cutaneous nerve of thigh	Sacral plexus (anterior and posterior divisions of anterior rami of S1–S3 spinal nerves)	Enters gluteal region via greater sciatic foramen inferior to piriformis and deep to gluteus maximus, emerging from inferior border of latter; descends in posterior thigh deep to fascia lata	Supplies skin of inferior half of buttocks (through inferior clunial nerves), skin over posterior thigh and popliteal fossa, and skin of lateral perineum and upper medial thigh (via its perineal branch)
Superior gluteal	Sacral plexus (posterior divisions of anterior rami of L4–S1 spinal nerves)	Enters gluteal region via greater sciatic foramen superior to piriformis; courses laterally between gluteus medius and minimus as far as tensor fasciae latae	Innervates gluteus medius, gluteus minimus, and tensor fasciae latae muscles
Inferior gluteal	Sacral plexus (posterior divisions of anterior rami of L5–S2 spinal nerves)	Enters gluteal region via greater sciatic foramen inferior to piriformis and deep to inferior part of gluteus maximus, dividing into several branches	Supplies gluteus maximus
Nerve to quadratus femoris	Sacral plexus (anterior divisions of anterior rami of L4–S1 spinal nerves)	Enters gluteal region via greater sciatic foramen inferior to piriformis, deep (anterior) to sciatic nerve	Innervates hip joint, inferior gemellus, and quadratus femoris
Pudendal	Sacral plexus (anterior divisions of anterior rami of S2–S4 spinal nerves)	Exits pelvis via greater sciatic foramen inferior to piriformis; descends posterior to sacrospinous ligament; enters perineum through lesser sciatic foramen	Supplies no structures in gluteal region or posterior thigh (principal nerve to perineum)
Nerve to obturator internus	Sacral plexus (posterior divisions of anterior rami of L5–S2 spinal nerves)	Exits pelvis via greater sciatic foramen inferior to piriformis; descends posterior to sacrospinous ligament; enters perineum through lesser sciatic foramen	Supplies superior gemellus and obturator internus

CLUNIAL NERVES

The skin of the gluteal region is richly innervated by **superior, middle,** and **inferior clunial nerves** (L. *clunes*, buttocks). These superficial nerves supply the skin over the iliac crest, between the posterior superior iliac spines, and over the iliac tubercles. Consequently, these nerves are vulnerable to injury when bone is taken from the ilium for grafting.

DEEP GLUTEAL NERVES

The *deep gluteal nerves* are the superior and inferior gluteal nerves, sciatic nerve, nerve to quadratus femoris, posterior cutaneous nerve of the thigh, nerve to obturator internus, and pudendal nerve (Figs. 5.38A and 5.42; Table 5.8). All of these nerves are branches of the sacral plexus and leave the pelvis through the greater sciatic foramen. Except for the superior gluteal nerve, they all emerge inferior to the piriformis.

Superior Gluteal Nerve. The **superior gluteal nerve** runs laterally between the gluteus medius and minimus with the deep branch of the superior gluteal artery. It divides into a superior branch that supplies the gluteus medius and an inferior branch that continues to pass between the gluteus medius and the gluteus minimus to supply both muscles and the tensor fasciae latae.

See the blue box "Injury to Superior Gluteal Nerve" on p. 581.

Inferior Gluteal Nerve. The **inferior gluteal nerve** leaves the pelvis through the greater sciatic foramen, inferior to the piriformis and superficial to the sciatic nerve, accompanied by multiple branches of the inferior gluteal artery and vein. The inferior gluteal nerve also divides into several branches, which provide motor innervation to the overlying gluteus maximus.

Sciatic Nerve. The **sciatic nerve** is the largest nerve in the body and is the continuation of the main part of the *sacral plexus*. The branches (rami) converge at the inferior border of the piriformis to form the sciatic nerve, a thick, flattened band approximately 2 cm wide. The sciatic nerve is the most lateral structure emerging through the greater sciatic foramen inferior to the piriformis.

Medial to the sciatic nerve are the inferior gluteal nerve and vessels, the internal pudendal vessels, and the pudendal nerve. The sciatic nerve runs inferolaterally under cover of the gluteus maximus, midway between the greater trochanter and ischial tuberosity. The nerve rests on the ischium and then passes posterior to the obturator internus, quadratus femoris, and adductor magnus muscles. The sciatic nerve is so large that it receives a named branch of the inferior gluteal artery, the **artery to the sciatic nerve.**

The sciatic nerve supplies no structures in the gluteal region. It supplies the posterior thigh muscles, all leg and foot muscles, and the skin of most of the leg and foot. It also supplies the articular branches to all joints of the lower limb. The sciatic nerve is really two nerves, the *tibial nerve*, derived from anterior (preaxial) divisions of the anterior rami, and the *common fibular nerve*, derived from posterior (postaxial) divisions of the anterior rami, which are loosely bound together in the same connective tissue sheath (Figs. 5.42 and 5.43A).

The tibial and common fibular nerves usually separate in the distal thigh (Fig. 5.42B); however, in approximately 12% of people, the nerves separate as they leave the pelvis (Fig. 5.43A). In these cases, the tibial nerve passes inferior to the piriformis, and the common fibular nerve pierces this muscle or passes superior to it (Fig. 5.43B & C).

Nerve to Quadratus Femoris. The **nerve to the quadratus femoris** leaves the pelvis anterior to the sciatic nerve and obturator internus, and passes over the posterior surface of the hip joint (Fig. 5.42). It supplies an articular branch to this joint and innervates the inferior gemellus and quadratus femoris muscles.

Posterior Cutaneous Nerve of Thigh. The **posterior cutaneous nerve of the thigh** supplies more skin than any other cutaneous nerve (Fig. 5.42). Its fibers from the anterior divisions of S2 and S3 supply the skin of the perineum via its *perineal branch*. Some of the fibers from the posterior divisions of the anterior rami of S1 and S2 supply the skin of the inferior part of the buttocks (via the *inferior clunial nerves*). Other fibers continue inferiorly in branches that supply the skin of the posterior thigh and proximal part of the leg. Unlike most nerves bearing the name *cutaneous*, the main part of this nerve lies deep to the deep fascia (fascia lata), with only its terminal branches penetrating the subcutaneous tissue for distribution to the skin.

Pudendal Nerve. The *pudendal nerve* is the most medial structure to exit the pelvis through the greater sciatic foramen. It descends inferior to the piriformis, posterolateral to the sacrospinous ligament, and enters the perineum through the lesser sciatic foramen to supply structures in this region. The pudendal nerve supplies no structures in the gluteal region or posterior thigh, and is discussed in detail in Chapter 3.

Nerve to Obturator Internus. The **nerve to the obturator internus** arises from the anterior divisions of the anterior rami of the L5–S2 nerves and parallels the course of the pudendal nerve (Fig. 5.42A). As it passes around the base of the ischial spine, the nerve supplies the superior gemellus. After entering the perineum via the lesser sciatic foramen, the nerve supplies the obturator internus muscle.

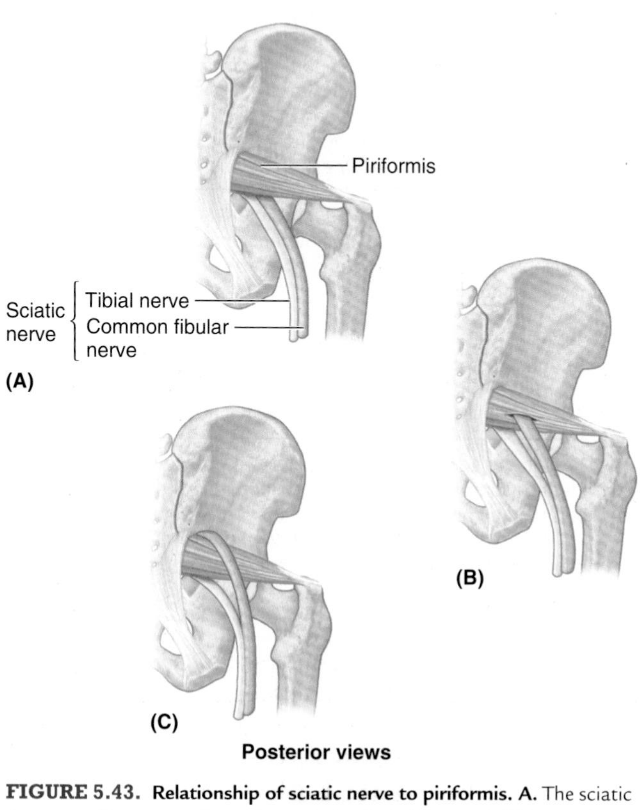

ARTERIES OF GLUTEAL AND POSTERIOR THIGH REGIONS

The **arteries of the gluteal region** arise, directly or indirectly, from the *internal iliac arteries*, but the patterns of origin of the arteries are variable (Figs. 5.38A and 5.44; Table 5.9). The major branches of the internal iliac artery that supply or traverse the gluteal region are the (1) superior gluteal artery, (2) inferior gluteal artery, and (3) internal pudendal artery. The posterior compartment of the thigh has no major artery exclusive to the compartment; it receives blood from multiple sources: inferior gluteal, medial circumflex femoral, perforating, and popliteal arteries.

Superior Gluteal Artery. The **superior gluteal artery** is the largest branch of the internal iliac artery and passes posteriorly between the lumbosacral trunk and the S1 nerve. This artery leaves the pelvis through the greater sciatic foramen, superior to the piriformis, and divides immediately into superficial and deep branches. The *superficial branch* supplies the gluteus maximus and skin over the proximal attachment of this muscle. The *deep branch* supplies the gluteus medius, gluteus minimus, and tensor fasciae latae. The superior gluteal artery anastomoses with the inferior gluteal and medial circumflex femoral arteries.

Inferior Gluteal Artery. The **inferior gluteal artery** arises from the internal iliac artery and passes posteriorly through the parietal pelvic fascia, between the S1 and the

(A)

Piriformis

Sciatic nerve { Tibial nerve
Common fibular nerve

(B)

(C)

Posterior views

FIGURE 5.43. Relationship of sciatic nerve to piriformis. A. The sciatic nerve usually emerges from the greater sciatic foramen inferior to the piriformis. **B.** In 12.2% of 640 limbs studied by Dr. J. C. B. Grant, the sciatic nerve divided before exiting the greater sciatic foramen; the common fibular division (*yellow*) passed through the piriformis. **C.** In 0.5% of cases, the common fibular division passed superior to the muscle, where it is especially vulnerable to injury during intragluteal injections.

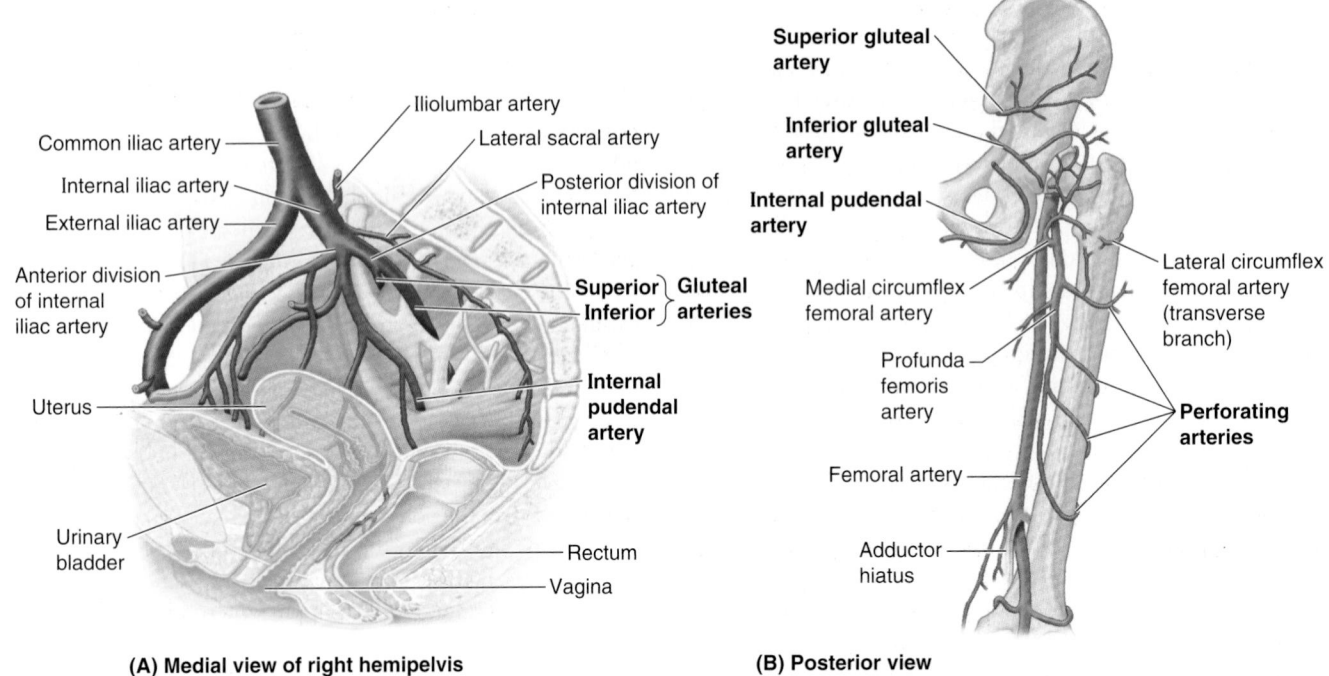

FIGURE 5.44. Arteries of the gluteal and posterior thigh regions.

TABLE 5.9. ARTERIES OF GLUTEAL AND POSTERIOR THIGH REGIONS

Artery[a]	Course	Distribution
Superior gluteal	Enters gluteal region through greater sciatic foramen superior to piriformis; divides into superficial and deep branches; anastomoses with inferior gluteal and medial circumflex femoral arteries (not shown in Fig. 5.44)	Superficial branch: supplies gluteus maximus Deep branch: runs between gluteus medius and minimus and supplies them and tensor fasciae latae
Inferior gluteal	Enters gluteal region through greater sciatic foramen inferior to piriformis; descends on medial side of sciatic nerve; anastomoses with superior gluteal artery and participates in cruciate anastomosis of thigh, involving first perforating artery of profunda femoris and medial and lateral circumflex arteries (not shown in Fig. 5.44)	Supplies gluteus maximus, obturator internus, quadratus femoris, and superior parts of hamstrings
Internal pudendal	Enters gluteal region through greater sciatic foramen; descends posterior to ischial spine; enters perineum through lesser sciatic foramen	Supplies external genitalia and muscles in perineal region; does not supply gluteal region
Perforating	Enters posterior compartment by perforating aponeurotic portion of adductor magnus attachment and medial intermuscular septum; after providing muscular branches to hamstrings, continues to anterior compartment by piercing lateral intermuscular septum	Supplies majority (central portions) of hamstring muscles, then continues to supply vastus lateralis in anterior compartment

[a]All of these arteries arise from the internal iliac artery (see Fig. 5.29 for an anterior view).

S2 (or S2 and S3) nerves. The inferior gluteal artery leaves the pelvis through the greater sciatic foramen, inferior to the piriformis. It enters the gluteal region deep to the gluteus maximus and descends medial to the sciatic nerve.

The inferior gluteal artery supplies the gluteus maximus, obturator internus, quadratus femoris, and superior parts of the hamstrings. It anastomoses with the superior gluteal artery and frequently, but not always, participates in the *cruciate anastomosis of the thigh*, involving the first perforating arteries of the profunda femoris artery and the medial and lateral circumflex

femoral arteries (Table 5.5). These vessels all participate in supplying the structures of the proximal posterior thigh.

Before birth, the inferior gluteal artery is the major artery of the posterior compartment, traversing its length and becoming continuous with the popliteal artery. This part of the artery diminishes, however, persisting postnatally as the *artery to the sciatic nerve*.

Internal Pudendal Artery. The **internal pudendal artery** arises from the internal iliac artery and lies anterior to the inferior gluteal artery. Its course parallels that of the pudendal nerve, entering the gluteal region through the greater sciatic foramen inferior to the piriformis. The internal pudendal artery leaves the gluteal region immediately by crossing the ischial spine/sacrospinous ligament and enters the perineum through the lesser sciatic foramen. Like the pudendal nerve, it supplies the skin, external genitalia, and muscles in the perineal region. It does not supply any structures in the gluteal region or posterior thigh.

Perforating Arteries. There are usually four **perforating arteries** of the profunda femoris artery, three arising in the anterior compartment and the fourth being the terminal branch of the profunda femoris artery itself (Fig. 5.44; Table 5.9). The perforating arteries are large vessels, unusual in the limbs for their transverse, intercompartmental course.

Surgeons operating in the posterior compartment are careful to identify them to avoid inadvertent injury. They perforate the aponeurotic portion of the distal attachment of the adductor magnus to enter the posterior compartment. Within the posterior compartment, they typically give rise to muscular branches to the hamstrings and anastomotic branches that ascend or descend to unite with those arising superiorly or inferiorly from the other perforating arteries or the inferior gluteal and popliteal artery.

A continuous anastomotic arterial chain thus extends from the gluteal to popliteal regions, which gives rise to additional branches to muscles and to the sciatic nerve. After giving off their posterior compartment branches, the perforating arteries pierce the lateral intermuscular septum to enter the anterior compartment, where they supply the vastus lateralis muscle.

VEINS OF GLUTEAL AND POSTERIOR THIGH REGIONS

The **gluteal veins** are tributaries of the internal iliac veins that drain blood from the gluteal region. The **superior and inferior gluteal veins** accompany the corresponding arteries through the greater sciatic foramen, superior and inferior to the piriformis, respectively (Fig. 5.45A). They communicate with tributaries of the femoral vein, thereby providing alternative routes for the return of blood from the lower limb (e.g., if the femoral vein is occluded or has to be ligated).

The **internal pudendal veins** accompany the internal pudendal arteries and join to form a single vein that enters the internal iliac vein. These veins drain blood from the external genitalia or pudendum (L. *pudere,* to be ashamed). *Perforating veins* accompany the arteries of the same name to drain blood from the posterior compartment of the thigh into the *profunda femoris vein*. The perforating veins, like the arteries, usually communicate inferiorly with the popliteal vein and superiorly with the inferior gluteal vein.

LYMPHATIC DRAINAGE OF GLUTEAL AND THIGH REGIONS

Lymph from the deep tissues of the buttocks follows the gluteal vessels to the **superior** and **inferior gluteal lymph nodes,** and from them to the *internal, external,* and *common*

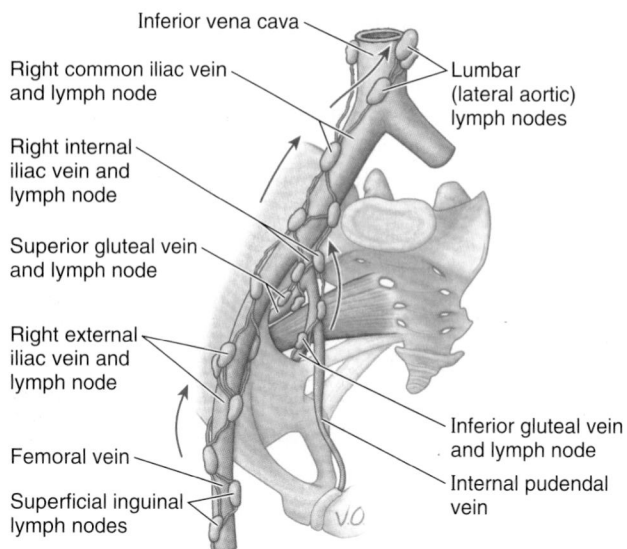

(A) Deep lymphatic drainage from gluteal and thigh regions

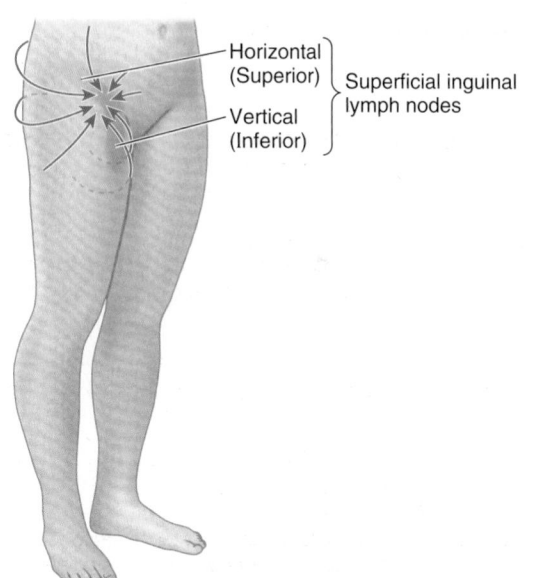

(B) Superficial lymphatic drainage from gluteal region and thigh

FIGURE 5.45. Lymphatic drainage of gluteal and thigh regions.
A. Lymph from the deep tissues of the gluteal region enters the pelvis along the gluteal veins, draining to the superior and inferior gluteal lymph nodes; from them, it passes to the iliac and lateral lumbar (caval/aortic) lymph nodes. **B.** Lymph from superficial tissues of the gluteal region passes initially to the superficial inguinal nodes, which also receive lymph from the thigh. Lymph from all the superficial inguinal nodes passes via efferent lymph vessels to the external and common iliac, and right and left lumbar (caval/aortic) lymph nodes, draining via lumbar lymphatic trunks to the cisterna chyli and thoracic duct.

iliac lymph nodes (Fig. 5.45A) and from them to the *lateral lumbar (aortic/caval) lymph nodes.*

Lymph from the superficial tissues of the gluteal region enters the *superficial inguinal lymph nodes,* which also receive lymph from the thigh (Fig. 5.45A & B). All the superficial inguinal nodes send efferent lymphatic vessels to the external iliac lymph nodes.

In terms of the vascular supply to the lower limb as a whole, the majority of the arterial blood coming to the limb and most of the venous blood and lymph exiting from it pass along the more protected anteromedial aspect of the limb.

Flexor muscles are generally better protected than are extensor muscles, the latter being exposed and therefore vulnerable in the flexed, defensive (fetal) position (vertebral column and limbs flexed).

Surface Anatomy of Gluteal and Posterior Thigh Regions

The skin of the gluteal region is usually thick and coarse, especially in men, whereas the skin of the thigh is relatively thin and loosely attached to the underlying subcutaneous tissue. A line joining the *highest points of the iliac crests* (Fig. 5.46A) crosses the L4–L5 intervertebral (IV) disc and is a useful landmark when a *lumbar spinal puncture* is performed (see Chapter 4), indicating the middle of the *lumbar cistern.*

The *intergluteal cleft*, beginning inferior to the apex of the sacrum, is the deep groove between the buttocks. It extends as far superiorly as the S3 or S4 segment. The *coccyx* is palpable in the superior part of the intergluteal cleft.

The *posterior superior iliac spines (PSIS)* are located at the posterior extremities of the iliac crests, and may be difficult to palpate; however, their position can always be located at the bottom of the permanent skin dimples approximately 3.75

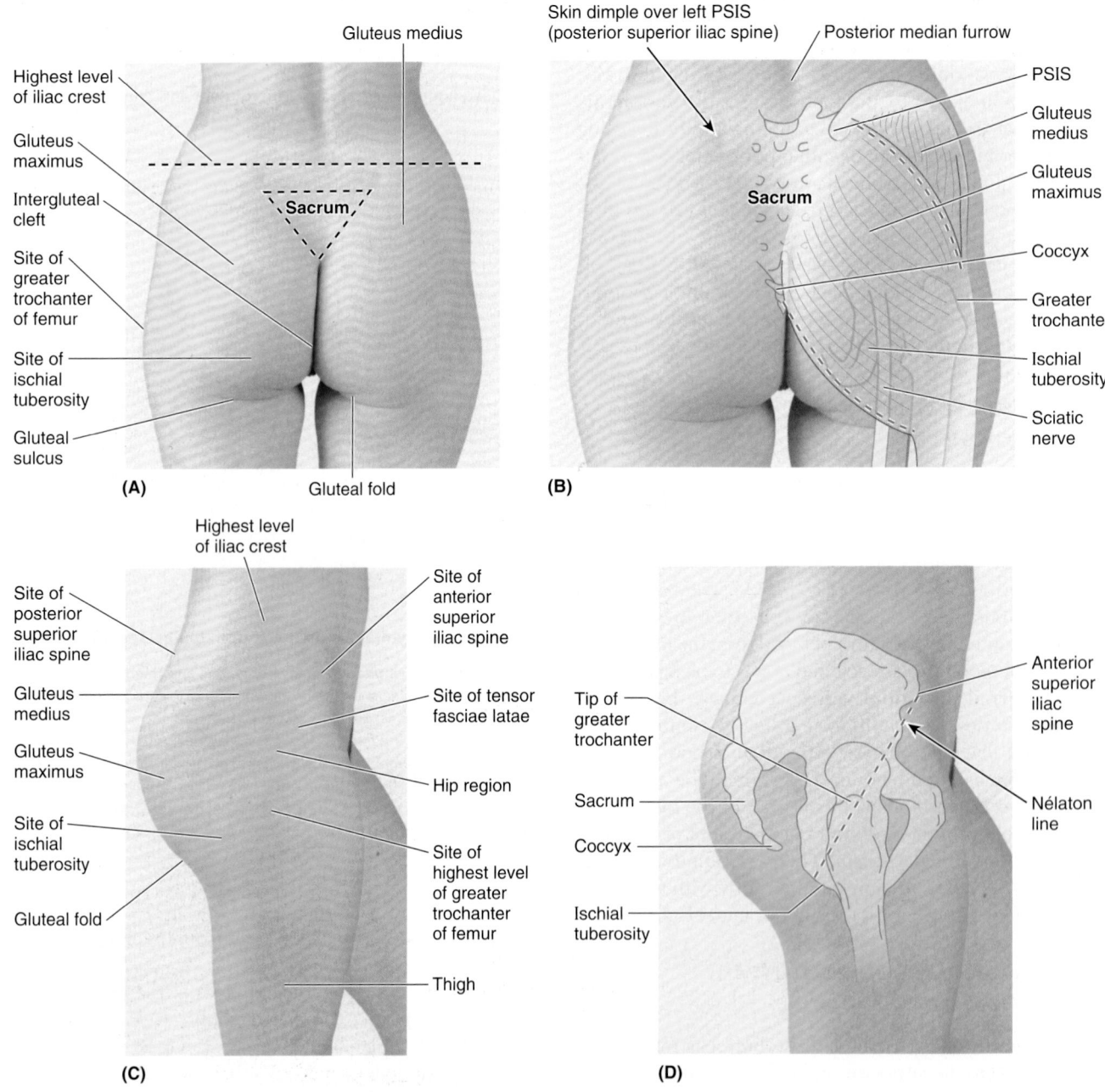

FIGURE 5.46. Surface anatomy of the gluteal region.

cm from midline (Fig. 5.46B). A line joining these dimples, often more visible in women than in men, passes through the S2 spinous process, indicating the level of the lowest limit of the *dural sac*, the middle of the sacro-iliac joints, and the bifurcation of the common iliac arteries (see Fig. 4.39).

The location of only two of the gluteal muscles can be observed. The *gluteus maximus* covering most structures in the gluteal region can be felt to contract when straightening up from bending over. The inferior edge of this large muscle is located just superior to the *gluteal fold*, which contains a variable amount of subcutaneous fat (Fig. 5.46A & C). The gluteal fold disappears when the hip joint is flexed. The degree of prominence of the gluteal fold changes in certain abnormal conditions, such as *atrophy of the gluteus maximus*. An imaginary line drawn from the coccyx to the ischial tuberosity indicates the inferior edge of the gluteus maximus (Fig. 5.46B). Another line drawn from the PSIS to a point slightly superior to the greater trochanter indicates the superior edge of this muscle.

The **gluteal sulcus,** the skin crease inferior to the gluteal fold, delineates the buttocks from the posterior aspect of the thigh (Fig. 5.46A & B). When the thigh is extended as in the figures, the *ischial tuberosity* is covered by the inferior part of the gluteus maximus; however, the tuberosity is easy to palpate when the thigh is flexed because

the gluteus maximus slips superiorly off the tuberosity, which is then subcutaneous. Feel the ischial tuberosity as you bend to sit.

The superior part of the *gluteus medius* can be palpated between the superior part of the gluteus maximus and the iliac crest (Figs. 5.46B and 5.47A & B). The gluteus medius of one buttocks can be felt when all the body weight shifts onto the ipsilateral limb (the one on the same side).

The *greater trochanter,* the most lateral bony point in the gluteal region, may be felt on the lateral aspect of the hip, especially its inferior part (Fig. 5.46A–C). It is easier to palpate when you passively abduct your lower limb to relax the gluteus medius and minimus. The top of the greater trochanter lies approximately a hand's breadth inferior to the *tubercle of the iliac crest* (Fig. 5.47).

The prominence of the trochanter increases when a dislocated hip causes atrophy of the gluteal muscles and displacement of the trochanter. A line drawn from the ASIS to the ischial tuberosity (*Nélaton line*), passing over the lateral aspect of the hip region, normally passes over or near the top of the greater trochanter (Fig. 5.46D). The trochanter can be felt superior to this line in a person with a dislocated hip or a fractured femoral neck. The *lesser trochanter* is palpable with difficulty from the posterior aspect when the thigh is extended and rotated medially.

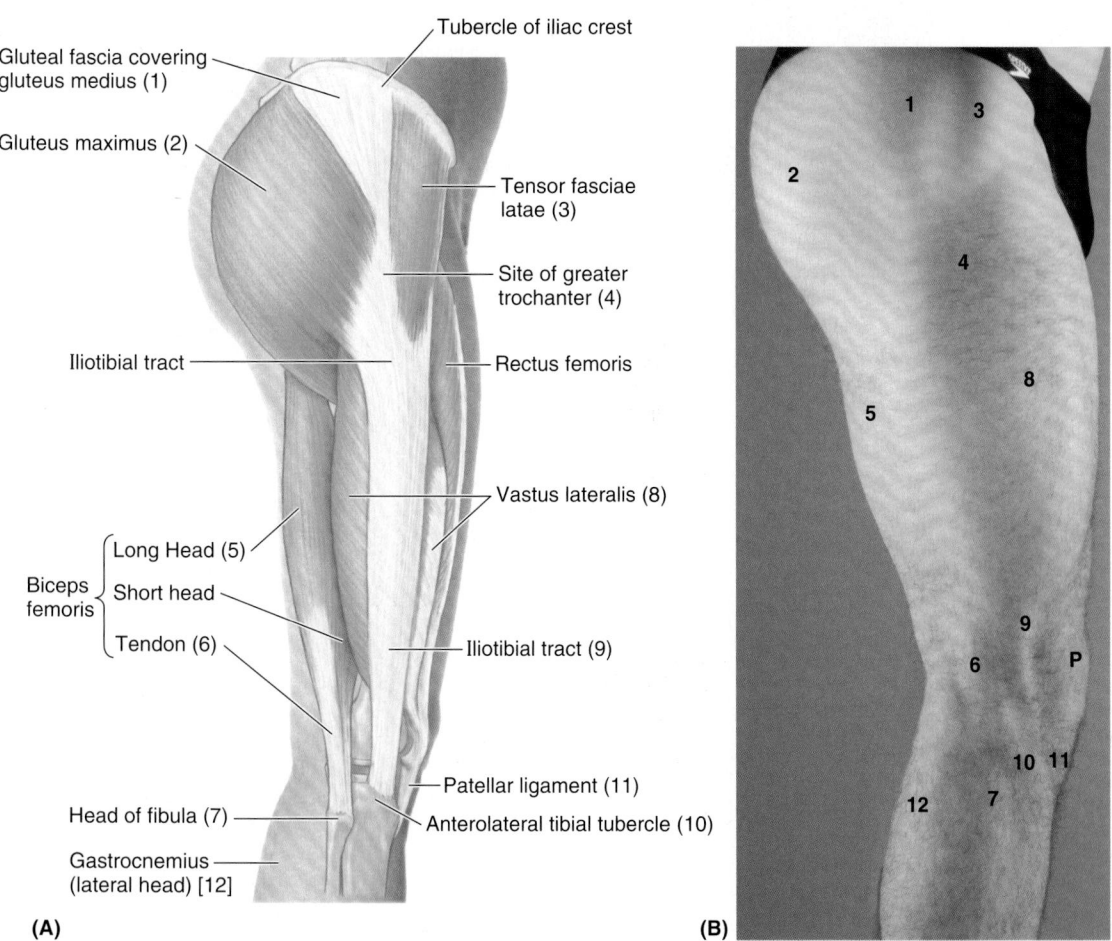

Tubercle of iliac crest

Gluteal fascia covering gluteus medius (1)

Gluteus maximus (2)

Tensor fasciae latae (3)

Site of greater trochanter (4)

Iliotibial tract

Rectus femoris

Long Head (5)

Biceps femoris — Short head

Tendon (6)

Vastus lateralis (8)

Iliotibial tract (9)

Head of fibula (7)

Patellar ligament (11)

Anterolateral tibial tubercle (10)

Gastrocnemius (lateral head) [12]

(A)

(B)

FIGURE 5.47. Surface anatomy of the hip region and lateral thigh.

The surface marking of the superior border of the *piriformis* is indicated by a line joining the skin dimple formed by the posterior superior iliac spine to the superior border of the greater trochanter of the femur (Fig. 5.48A).

The *sciatic nerve*, the most important structure inferior to the piriformis, is represented by a line that extends from a point midway between the greater trochanter and ischial tuberosity (Fig. 5.48A) down the middle of the posterior aspect of the thigh (Fig. 5.48B). The level of the bifurcation of the sciatic nerve into the tibial and common fibular nerves varies. The separation usually occurs between the middle and inferior third of the thigh. Less commonly, the division of the sciatic nerve occurs as it

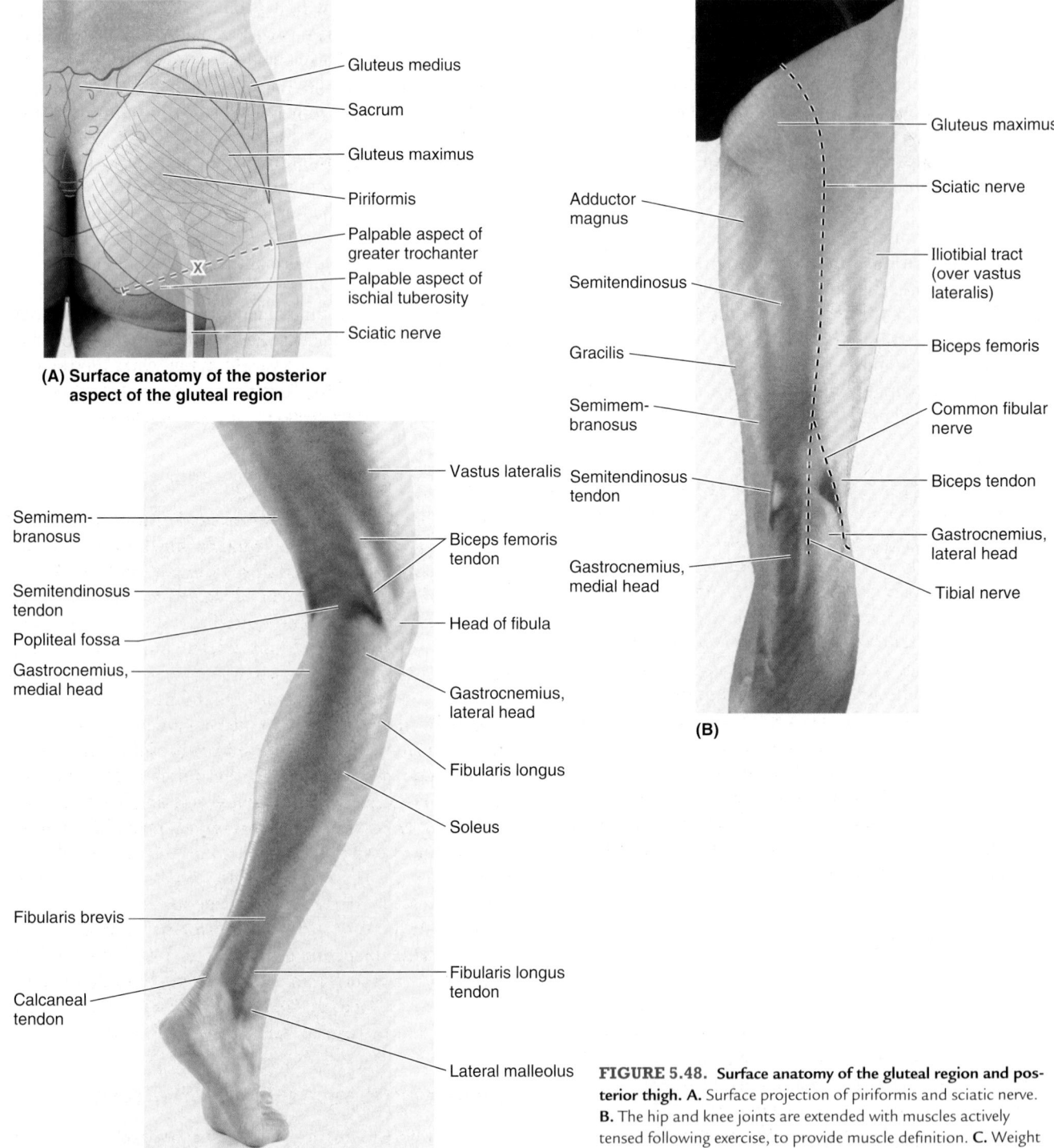

(A) Surface anatomy of the posterior aspect of the gluteal region

- Gluteus medius
- Sacrum
- Gluteus maximus
- Piriformis
- Palpable aspect of greater trochanter
- Palpable aspect of ischial tuberosity
- Sciatic nerve

(B)

- Gluteus maximus
- Sciatic nerve
- Iliotibial tract (over vastus lateralis)
- Biceps femoris
- Common fibular nerve
- Biceps tendon
- Gastrocnemius, lateral head
- Tibial nerve

Adductor magnus
Semitendinosus
Gracilis
Semimembranosus
Semitendinosus tendon
Gastrocnemius, medial head

(C)

Semimembranosus
Semitendinosus tendon
Popliteal fossa
Gastrocnemius, medial head
Fibularis brevis
Calcaneal tendon

- Vastus lateralis
- Biceps femoris tendon
- Head of fibula
- Gastrocnemius, lateral head
- Fibularis longus
- Soleus
- Fibularis longus tendon
- Lateral malleolus

FIGURE 5.48. Surface anatomy of the gluteal region and posterior thigh. A. Surface projection of piriformis and sciatic nerve. **B.** The hip and knee joints are extended with muscles actively tensed following exercise, to provide muscle definition. **C.** Weight is being borne by the right limb while the hip, knee, and metatarsophalangeal joints are in a flexed position.

passes through the greater sciatic foramen. The sciatic nerve stretches when the thigh is flexed and the knee is extended, and it relaxes when the thigh is extended and the knee is flexed.

The *tibial nerve* bisects the popliteal fossa. The *common fibular nerve* follows the tendon of the biceps femoris.

The *hamstrings* can be felt as a group as they arise from the ischial tuberosity and extend along the lateral and posterior aspects of the thigh (Fig. 5.48B & C). The *iliotibial tract*, the fibrous band that reinforces the fascia lata laterally, can be observed on the lateral aspect of the thigh as it passes to the *lateral tibial condyle* (Fig. 5.47A & B).

While sitting down with your lower limb extended, raise your heel off the floor and feel the anterior border of the iliotibial tract passing a finger's breadth posterior to the lateral border of the patella. Note that the iliotibial tract is promi-

nent and taut when the heel is raised and indistinct when the heel is lowered.

The *tendons of the hamstrings* can be observed and palpated at the borders of the *popliteal fossa*, the depression between the tendons at the back of the flexed knee (Fig. 5.48B & C). The *biceps femoris tendon* is on the lateral side of the fossa. The most lateral tendon on the medial side when the knee is flexed against resistance is the *semimembranosus tendon*.

While sitting on a chair with your knee flexed, press your heel against the leg of the chair and feel your biceps femoris tendon laterally and trace it to the head of the fibula. Also feel the narrow and more prominent *semitendinosus tendon* medially, which pulls away from the semimembranosus tendon that attaches to the superomedial part of the tibia.

See the blue box "Hamstring Injuries" below.

GLUTEAL AND POSTERIOR THIGH REGIONS

Trochanteric Bursitis

Trochanteric bursitis, inflammation of the trochanteric bursa (see Fig. 5.37), may result from repetitive actions such as climbing stairs while carrying heavy objects, or running on a steeply elevated treadmill. These movements involve the gluteus maximus and move the superior tendinous fibers repeatedly back and forth over the bursae of the greater trochanter. Trochanteric bursitis causes deep diffuse pain in the lateral thigh region.

This type of *friction bursitis* is characterized by point tenderness over the great trochanter; however, the pain radiates along the iliotibial tract that extends from the iliac tubercle to the tibia (see Figs. 5.36 and 5.39C). This thickening of the fascia lata receives tendinous reinforcements from the tensor fasciae latae and gluteus maximus muscles. The pain from an inflamed trochanteric bursa, usually localized just posterior to the greater trochanter, is generally elicited by manually resisting abduction and lateral rotation of the thigh while the person is lying on the unaffected side.

Ischial Bursitis

Recurrent microtrauma resulting from repeated stress (e.g., as from cycling, rowing, or other activities involving repetitive hip extension while seated) may overwhelm the ability of the ischial bursa (see Fig. 5.37) to dissipate applied stress. The recurrent trauma results in inflammation of the bursa (*ischial bursitis*).

Ischial bursitis is a *friction bursitis* resulting from excessive friction between the ischial bursae and the ischial tuberosities. Localized pain occurs over the bursa, and the pain increases

with movement of the gluteus maximus. Calcification may occur in the bursa with chronic bursitis. Because the ischial tuberosities bear the body's weight during sitting, these pressure points may lead to *pressure sores* in debilitated people, particularly paraplegic persons with poor nursing care.

Hamstring Injuries

Hamstring strains (pulled and/or torn hamstrings) are common in individuals who run and/or kick hard (e.g., in running, jumping, and quick-start sports such as baseball, basketball, football, and soccer). The violent muscular exertion required to excel in these sports may avulse (tear) part of the proximal tendinous attachments of the hamstrings to the ischial tuberosity. Hamstring strains are twice as common as quadriceps strains.

Usually thigh strains are accompanied by contusion (bruising) and tearing of muscle fibers, resulting in rupture of the blood vessels supplying the muscles. The resultant *hematoma* is contained by the dense stocking-like fascia lata.

Tearing of hamstring fibers is often so painful when the athlete moves or stretches the leg that the person falls and writhes in pain. These injuries often result from inadequate warming up before practice or competition.

Avulsion (tearing away) of the ischial tuberosity at the proximal attachment of the biceps femoris and semitendinosus may result from forcible flexion of the hip with the knee extended (e.g., kicking a football). (See Fig. B5.1 and the blue box "Injuries of Hip Bone" on p. 526.)

Injury to Superior Gluteal Nerve

Injury to this nerve results in a characteristic motor loss, resulting in a disabling *gluteus medius limp,* to compensate for weakened abduction of the thigh by the gluteus medius and minimus, and/or a *gluteal gait,* a compensatory list of the body to the weakened gluteal side. This compensation places the center of gravity over the supporting lower limb. Medial rotation of the thigh is also severely impaired. When a standing person is asked to lift one foot off the ground and stand on one foot, the gluteus medius and minimus normally contract as soon as the contralateral foot leaves the floor, preventing tipping of the pelvis to the unsupported side (Fig. B5.18A & B).

When a person who has suffered a lesion of the superior gluteal nerve is asked to stand on one leg, the pelvis on the unsupported side descends (Fig. B5.18C), indicating that the gluteus medius and minimus on the supported side are weak or non-functional. This sign is referred to clinically as a *positive Trendelenburg test.* Other causes of this sign include *fracture of the greater trochanter* (the distal attachment of gluteus medius) and *dislocation of the hip joint.*

When the pelvis descends on the unsupported side, the lower limb becomes, in effect, too long and does not clear the ground when the foot is brought forward in the swing phase of walking. To compensate, the individual leans away from the unsupported side, raising the pelvis to allow adequate room for the foot to clear the ground as it swings forward. This results in a characteristic "waddling" or *gluteal gait.*

Other ways to compensate is to lift the foot higher as it is brought forward, resulting in the so-called *steppage gait,* or to swing the foot outward (laterally), the so-called *swing-out gait.* These same gaits are adopted to compensate for the *footdrop* that results from common fibular nerve paralysis. (See these abnormal gaits illustrated in Fig. B5.20 of the blue box "Injury to Common Fibular Nerve and Footdrop," p. 605.)

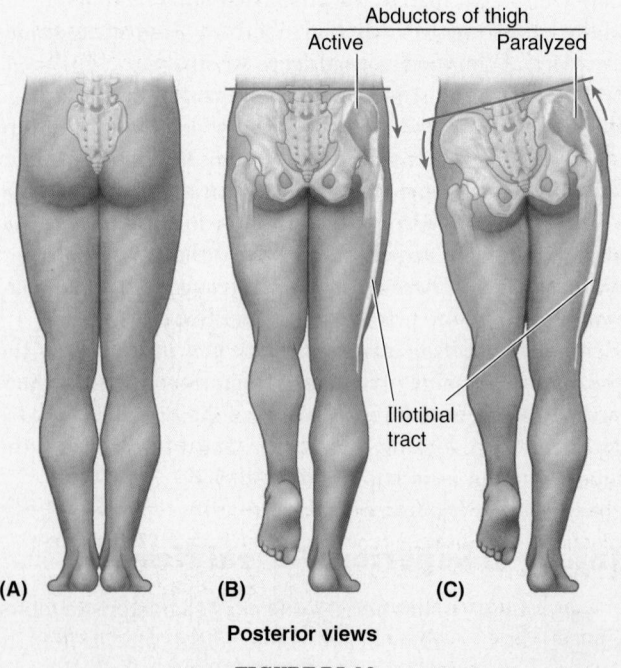

Abductors of thigh
Active   Paralyzed

Iliotibial tract

(A)   (B)   (C)

Posterior views

FIGURE B5.18.

Anesthetic Block of Sciatic Nerve

Sensation conveyed by the sciatic nerve can be blocked by the injection of an anesthetic agent a few centimeters inferior to the midpoint of the line joining the posterior superior iliac spine (PSIS) and the superior border of the greater trochanter. *Paresthesia* (nonpainful anesthesia) radiates to the foot because of anesthesia of the plantar nerves, which are terminal branches of the tibial nerve derived from the sciatic nerve.

Injury to Sciatic Nerve

A pain in the buttocks may result from compression of the sciatic nerve by the piriformis (*piriformis syndrome*). Individuals involved in sports that require excessive use of the gluteal muscles (e.g., ice skaters, cyclists, and rock climbers) and women are more likely to develop this syndrome. In approximately 50% of cases, the histories indicate trauma to the buttocks associated with hypertrophy (increase in bulk) and *spasm of the piriformis.* In the approximately 12% of people in whom the common fibular division of the sciatic nerve passes through the piriformis (Fig. 5.43B), this muscle may compress the nerve.

Complete section of the sciatic nerve is uncommon; however, when it occurs, the leg is useless because extension of the hip is impaired, as is flexion of the leg. All ankle and foot movements are also lost.

Incomplete section of the sciatic nerve (e.g., from stab wounds) may also involve the inferior gluteal and/or the posterior femoral cutaneous nerves. Recovery from a lesion of the sciatic nerve is slow and usually incomplete.

With respect to the sciatic nerve, the buttocks has a *side of safety* (its lateral side) and a *side of danger* (its medial side). Wounds or surgery on the medial side of the buttocks may injure the sciatic nerve and its branches to the hamstrings (semitendinosus, semimembranosus, and biceps femoris) on the posterior aspect of the thigh. Paralysis of these muscles results in impairment of thigh extension and leg flexion.

Intragluteal Injections

The gluteal region (buttocks) is a common site for intramuscular (IM) injection of drugs. *Gluteal IM injections* penetrate the skin, fascia, and muscles. The gluteal region is a common injection site because the muscles are thick and large; consequently, they provide a substantial volume for absorption of injected substances by intramuscular veins. It is important to be aware of the extent of the gluteal region and the safe region for giving injections. Some people restrict the area of the buttocks to the most prominent part. This misunderstanding may be dangerous because the sciatic nerve lies deep to this area (Fig. B5.19A).

Injections into the buttocks are safe only in the superolateral quadrant of the buttocks or superior to a line extending from the PSIS to the superior border of the greater trochanter (approximating the superior border of the gluteus maximus).

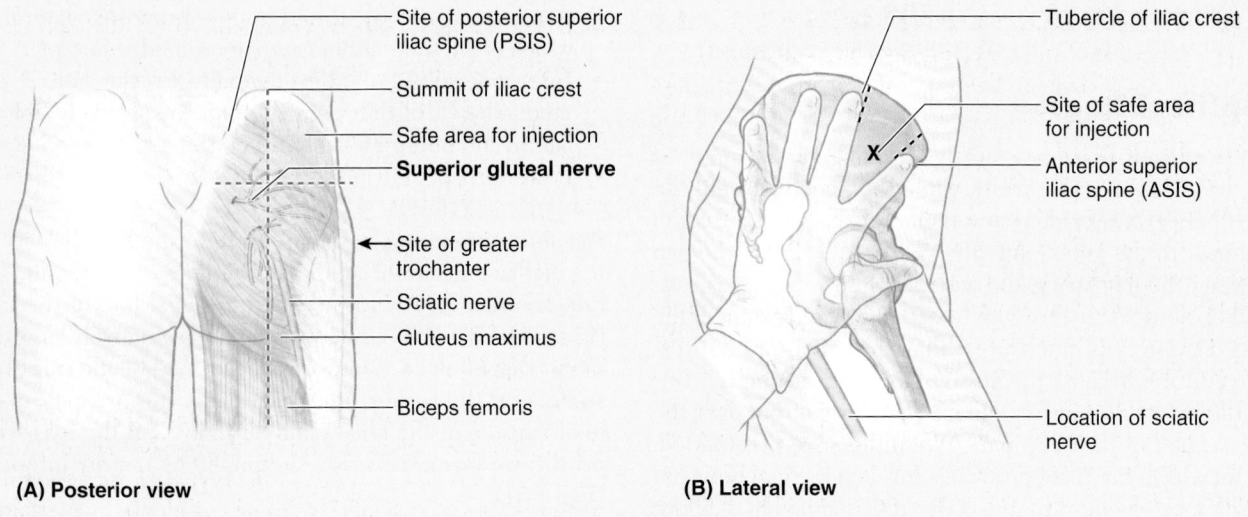

Site of posterior superior
iliac spine (PSIS)

Summit of iliac crest

Safe area for injection

Superior gluteal nerve

Site of greater
trochanter

Sciatic nerve

Gluteus maximus

Biceps femoris

(A) Posterior view

Tubercle of iliac crest

Site of safe area
for injection

Anterior superior
iliac spine (ASIS)

Location of sciatic
nerve

(B) Lateral view

FIGURE B5.19.

IM injections can also be given safely into the anterolateral part of the thigh, where the needle enters the tensor fasciae latae (Fig. 5.47A) as it extends distally from the iliac crest and ASIS. The index finger is placed on the ASIS and the fingers are spread posteriorly along the iliac crest until the tubercle of the crest is felt by the middle finger (Fig. B5.19B). An IM injection can be made safely in the triangular area between the fingers (just anterior to the proximal joint of the middle finger) because it is superior to the sciatic nerve. Complications of improper technique include nerve injury, hematoma, and abscess formation.

The Bottom Line

GLUTEAL AND POSTERIOR THIGH REGIONS

Gluteal region: The femur is bent at the angle of inclination, creating a relatively transverse lever formed by the proximal femur. ♦ This enables superior placement of the abductors of the thigh, and provides mechanical advantage for the deeper medial and lateral rotators of the thigh, critical for bipedal locomotion. ♦ Despite their designations, the abductors/medial rotators (the superficial gluteal muscles) are most active during the stance phase, when they simultaneously elevate and advance the contralateral unsupported side of the pelvis during ambulation. ♦ The lateral rotators (deep gluteal muscles) of the unsupported side rotate the free limb during the swing phase so that the foot remains parallel to the line of advancement.

Posterior femoral region: Although they have only about two thirds the strength of the gluteus maximus, the hamstrings are the main extensors of the hip used in normal walking. ♦ The hamstrings are two-joint muscles, and their concentric contraction produces either extension of the hip or flexion of the knee. ♦ However, in walking, the hamstrings are most active in eccentrically contracting to decelerate hip flexion and knee extension during terminal swing. ♦ The hamstrings also rotate the flexed knee. ♦ If resistance to hip extension is increased or more vigorous extension is required, the gluteus maximus is called into action.

Neurovascular structures of gluteal and posterior femoral regions: Because it overlies the major doorway (greater sciatic foramen) by which derivatives of the sacral plexus exit the bony pelvis, the gluteal region includes a disproportionate number of nerves of all sizes, both motor and sensory. ♦ Fortunately, most nerves are in the inferomedial quadrant; thus properly administered IM injections avoid these structures. ♦ Because the sciatic nerve includes fibers from the L4–S3 spinal nerves, it is affected by the most common nerve compression syndromes (e.g., radiculopathies [disorders] of the L4 and L5 spinal nerve roots; see Chapter 4). ♦ Even though occurring outside the lower limb per se, these syndromes result in sciatica—pain that radiates down the lower limb along the course of the sciatic nerve and its terminal branches. ♦ Pain experienced in the lower limb may not necessarily arise from a problem in the limb! ♦ Arteries and veins serving the gluteal region and the proximal part of the posterior compartment of the thigh are branches and tributaries of the internal iliac artery and vein that pass to and from the region via the greater sciatic foramen. ♦ All but the superior gluteal vessels exit the greater sciatic foramen inferior to the piriformis muscle. ♦ Although the pudendal vessels follow the same route, they traverse the gluteal region only briefly en route to and from the perineum via the lesser sciatic foramen. ♦ The posterior compartment of the thigh does not have a major artery coursing through it with primary responsibility for the compartment. Rather, branches from several arteries in other compartments supply it.

POPLITEAL FOSSA AND LEG

Popliteal Region

The **popliteal fossa** is a mostly fat-filled compartment of the lower limb. Superficially, when the knee is flexed, the popliteal fossa is evident as a diamond-shaped depression posterior to the knee joint (Fig. 5.49). The size of the gap between the hamstring and gastrocnemius muscles is misleading, however, in terms of the actual size and extent of the fossa. Deeply, it is much larger than the superficial depression indicates because the heads of the gastrocnemius forming the inferior boundary superficially form a roof over the inferior half of the deep part. When the knee is extended, the fat within the fossa protrudes through the gap between muscles, producing a rounded elevation flanked by shallow, longitudinal grooves overlying the hamstring tendons. In dissection, if the heads of the gastrocnemius are separated and retracted (Fig. 5.50), a much larger space is revealed.

Superficially, the popliteal fossa is bounded:

- Superolaterally by the biceps femoris (superolateral border).

- Superomedially by the semimembranosus, lateral to which is the semitendinosus (superomedial border).
- Inferolaterally and inferomedially by the lateral and medial heads of the gastrocnemius, respectively (inferolateral and inferomedial borders).
- Posteriorly by skin and *popliteal fascia* (roof).

Deeply, the superior boundaries are formed by the diverging medial and lateral supracondylar lines of the femur. The inferior boundary is formed by the soleal line of the tibia (Fig. 5.4B). These boundaries surround a relatively large diamond-shaped floor (anterior wall), formed by the **popliteal surface of the femur** superiorly, the posterior aspect of the joint capsule of the knee joint centrally, and the **investing popliteus fascia** covering the popliteus muscle inferiorly (Fig. 5.51).

The *contents of the popliteal fossa* (Figs. 5.49B, 5.50, and 5.51) include the:

- Termination of the small saphenous vein.
- Popliteal arteries and veins and their branches and tributaries.
- Tibial and common fibular nerves.

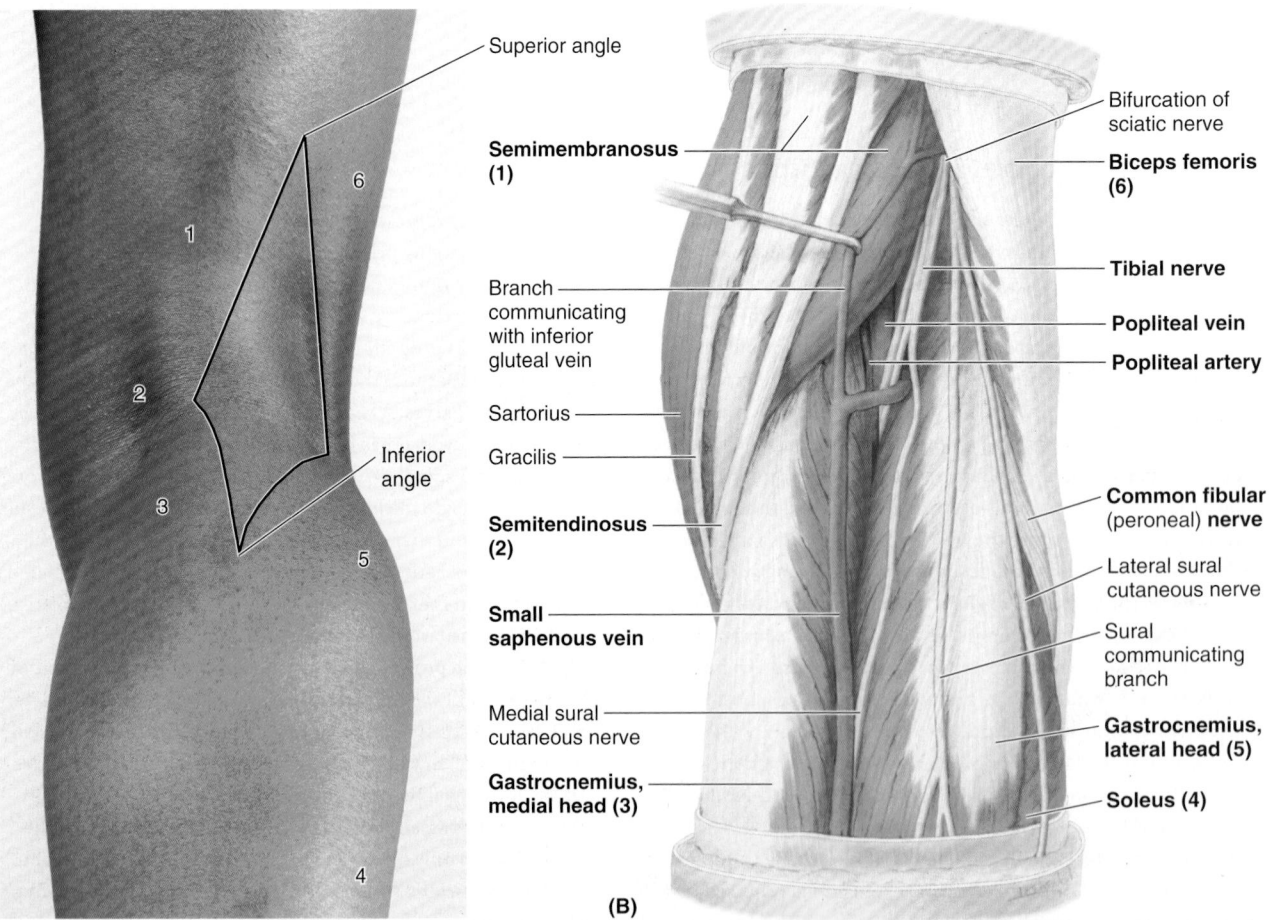

(A)

(B)

FIGURE 5.49. Superficial popliteal region. A. *Numbers* on the surface anatomy refer to structures identified in **B.** The diamond-shaped gap in the roof of the popliteal fossa, formed by the overlying muscles, is outlined. **B.** Superficial dissection of the popliteal region showing the muscles that cover most of the popliteal fossa.

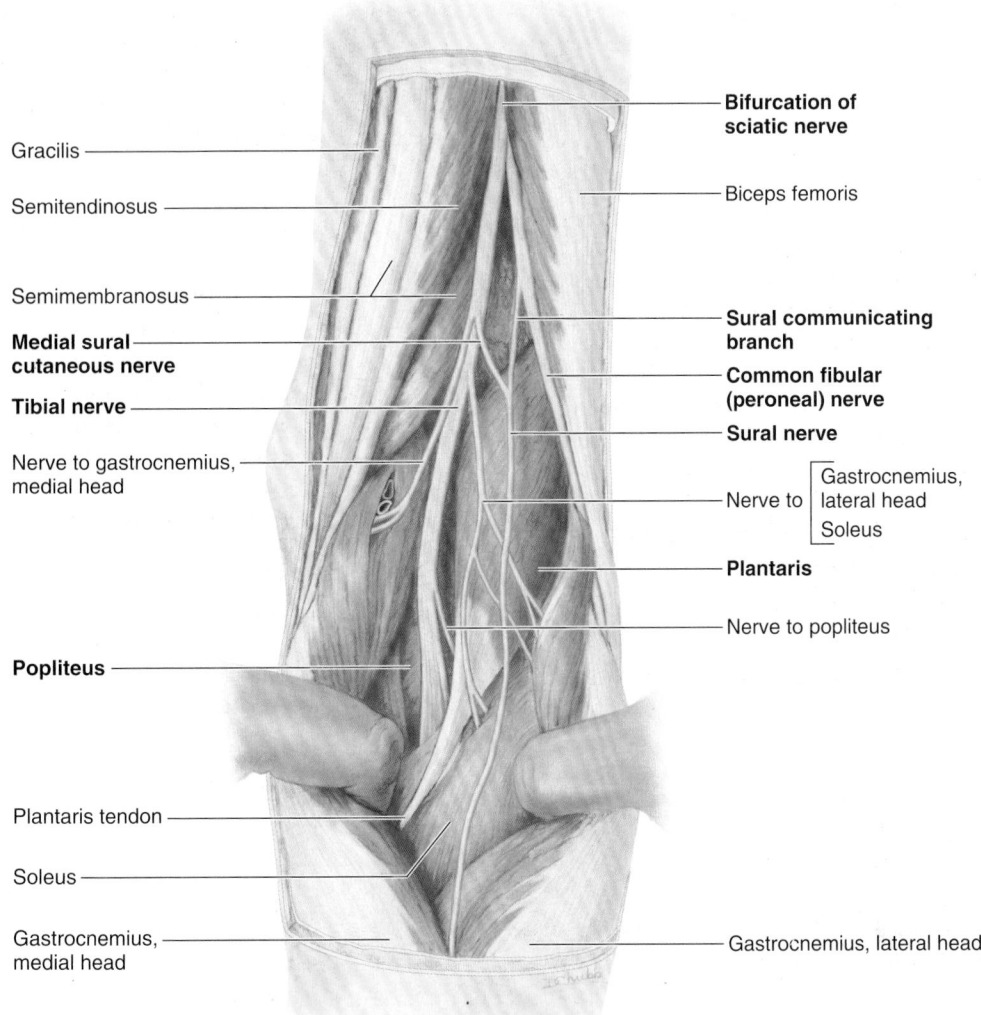

Posterior view

FIGURE 5.50. Exposure of popliteal fossa and nerves in it. The two heads of the gastrocnemius muscle have been separated and are being retracted. The sciatic nerve separates into its components at the apex of the popliteal fossa (or higher; Fig. 5.43B). The common fibular nerve courses along the medial border of the biceps femoris. All the motor branches arising from the tibial nerve, except one, arise from the lateral side; consequently, in surgery it is safer to dissect on the medial side. The level at which the medial sural cutaneous nerve and sural communicating branch merge to form the sural nerve—occurring high here—is quite variable; it may even occur at the level of the ankle.

- Posterior cutaneous nerve of thigh (see Fig. 5.42B).
- Popliteal lymph nodes and lymphatic vessels (see Fig. 5.15B).

FASCIA OF POPLITEAL FOSSA

The *subcutaneous tissue* (superficial fascia) overlying the popliteal fossa contains the small saphenous vein (Fig. 5.14B—unless it has penetrated the deep fascia of the leg at a more inferior level) and three cutaneous nerves: the terminal branch(es) of the *posterior cutaneous nerve of the thigh* and the *medial* and *lateral sural cutaneous nerves* (Fig. 5.49B).

The **popliteal fascia** is a strong sheet of deep fascia, continuous superiorly with the *fascia lata* and inferiorly with the *deep fascia of the leg* (Fig. 5.13B). The popliteal fascia forms a protective covering for neurovascular structures passing from the thigh through the popliteal fossa to the leg, and a relatively loose but functional retaining "retinaculum" (retaining band) for the hamstring tendons. Often the fascia is pierced by the small saphenous vein.

When the leg extends, the fat within the fossa is relatively compressed as the popliteal fascia becomes taut, and the semimembranosus muscle moves laterally, providing further protection to the contents of the fossa.

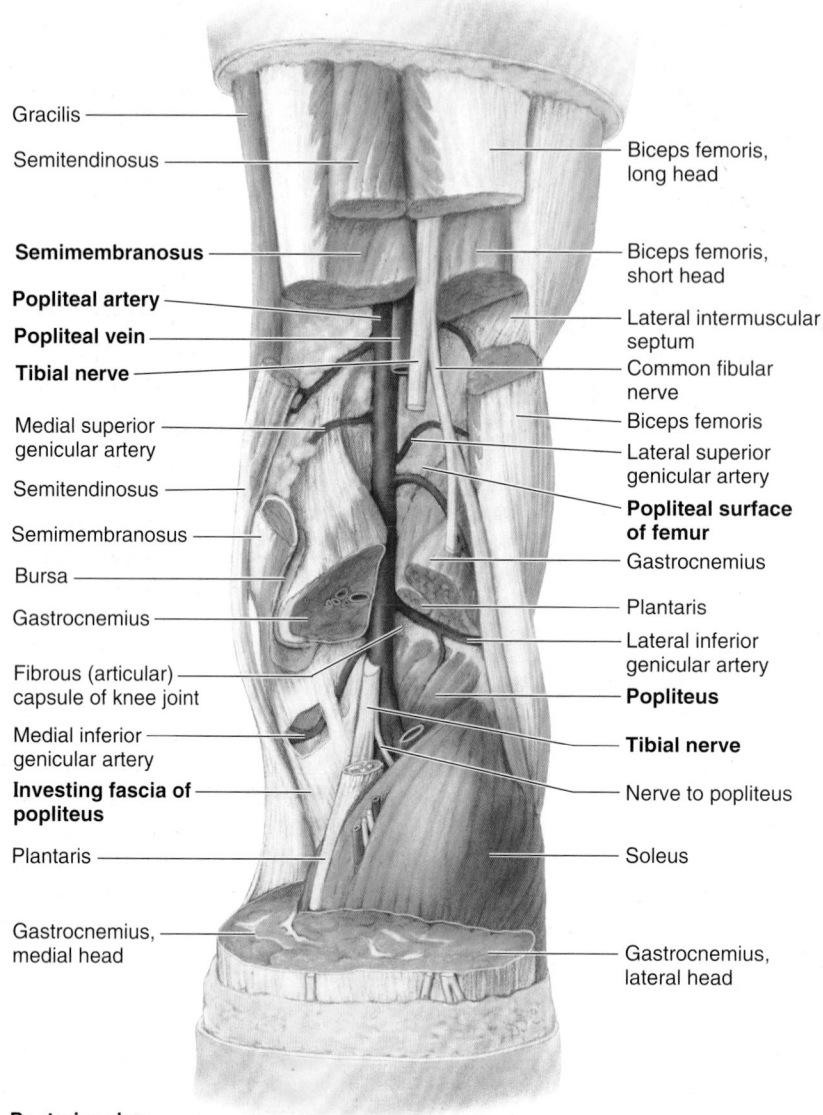

Gracilis

Semitendinosus

Semimembranosus

Popliteal artery

Popliteal vein

Tibial nerve

Medial superior
genicular artery

Semitendinosus

Semimembranosus

Bursa

Gastrocnemius

Fibrous (articular)
capsule of knee joint

Medial inferior
genicular artery

**Investing fascia of
popliteus**

Plantaris

Gastrocnemius,
medial head

Biceps femoris,
long head

Biceps femoris,
short head

Lateral intermuscular
septum

Common fibular
nerve

Biceps femoris

Lateral superior
genicular artery

**Popliteal surface
of femur**

Gastrocnemius

Plantaris

Lateral inferior
genicular artery

Popliteus

Tibial nerve

Nerve to popliteus

Soleus

Gastrocnemius,
lateral head

Posterior view

FIGURE 5.51. Deep dissection of popliteal fossa. The popliteal artery runs on the floor of the fossa, formed by the popliteal surface of the femur, the joint capsule of the knee, and the investing fascia of the popliteus.

The contents, most important the popliteal artery and lymph nodes, are most easily palpated with the knee semi-flexed. Because of the deep fascial roof and osseofibrous floor, the fossa is a relatively confined space. Many disorders produce swelling of the fossa, making knee extension painful. (See the blue boxes "Popliteal Abscess and Tumor" and "Popliteal Aneurysm and Hemorrhage" on p. 604, and "Popliteal Cysts" on p. 665.)

NEUROVASCULAR STRUCTURES
AND RELATIONSHIPS IN POPLITEAL FOSSA

All important neurovascular structures that pass from the thigh to the leg do so by traversing the popliteal fossa. Progressing from superficial to deep (posterior to anterior) within the fossa, as in dissection, the nerves are encountered first, then the veins. The arteries lie deepest, directly on the popliteal surface of the femur, joint capsule, and *investing fascia of the popliteus* forming the floor of the fossa (Fig. 5.51).

Nerves in Popliteal Fossa. The *sciatic nerve* usually ends at the superior angle of the popliteal fossa by dividing into the tibial and common fibular nerves (Figs. 5.49B, 5.50, and 5.51).

The **tibial nerve** is the medial, larger terminal branch of the sciatic nerve derived from anterior (preaxial) divisions of the anterior rami of the L4–S3 spinal nerves. *The tibial nerve is the most superficial of the three main central components of the popliteal fossa* (i.e., nerve, vein, and artery); however, it is still in a deep and protected position. The tibial nerve bisects the fossa as it passes from its superior to its inferior angle.

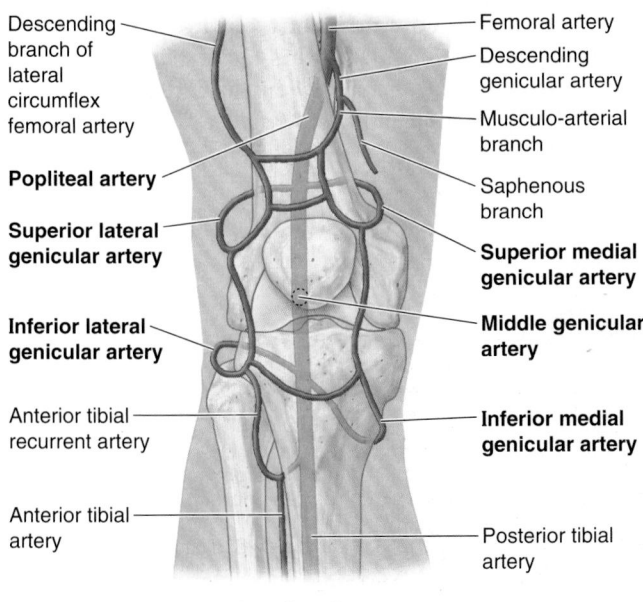

Descending branch of lateral circumflex femoral artery

Popliteal artery

Superior lateral genicular artery

Inferior lateral genicular artery

Anterior tibial recurrent artery

Anterior tibial artery

Femoral artery

Descending genicular artery

Musculo-arterial branch

Saphenous branch

Superior medial genicular artery

Middle genicular artery

Inferior medial genicular artery

Posterior tibial artery

Anterior view

FIGURE 5.52. Genicular anastomosis. The many arteries making up the peri-articular anastomosis around the knee provide an important collateral circulation for bypassing the popliteal artery when the knee joint has been maintained too long in a fully flexed position or when the vessels are narrowed or occluded.

While in the fossa, the tibial nerve gives branches to the soleus, gastrocnemius, plantaris, and popliteus muscles. The **medial sural cutaneous nerve** is also derived from the tibial nerve in the popliteal fossa. It is joined by the **sural communicating branch of the common fibular nerve** at a highly variable level to form the *sural nerve.* This nerve supplies the lateral side of the leg and ankle.

The **common fibular** (peroneal) **nerve** is the lateral, smaller terminal branch of the sciatic nerve derived from posterior (postaxial) divisions of the anterior rami of the L4–S2 spinal nerves. The common fibular nerve begins at the superior angle of the popliteal fossa and follows closely the medial border of the biceps femoris and its tendon along the superolateral boundary of the fossa. The nerve leaves the fossa by passing superficial to the lateral head of the gastrocnemius and then passes over the posterior aspect of the head of the fibula. The *common fibular nerve winds around the neck of the fibula* and divides into its terminal branches.

The most inferior branches of the *posterior cutaneous nerve of the thigh* supply the skin that overlies the popliteal fossa (see Fig 5.42B). The nerve traverses most of the length of the posterior compartment of the thigh deep to the fascia lata; only its terminal branches enter the subcutaneous tissue as cutaneous nerves.

Blood Vessels in Popliteal Fossa. The **popliteal artery,** the continuation of the femoral artery (Figs. 5.51 and 5.52), begins when the latter passes through the adductor hiatus (see Fig. 5.30A). The popliteal artery passes inferolaterally through the fossa and ends at the inferior border of the popliteus by dividing into the *anterior* and *posterior*

tibial arteries. The deepest (most anterior) structure in the fossa, the popliteal artery, runs in close proximity to the joint capsule of the knee as it spans the intercondylar fossa.

Five *genicular branches of the popliteal artery* supply the capsule and ligaments of the knee joint. The genicular arteries are the **superior lateral, superior medial, middle, inferior lateral,** and **inferior medial genicular arteries** (Fig. 5.52). They participate in the formation of the peri-articular **genicular anastomosis,** a network of vessels surrounding the knee that provides collateral circulation capable of maintaining blood supply to the leg during full knee flexion, which may kink the popliteal artery. Other contributors to this important genicular anastomosis are the:

- **Descending genicular artery,** a branch of the femoral artery, superomedially.
- **Descending branch of the lateral femoral circumflex artery,** superolaterally.
- **Anterior tibial recurrent artery,** a branch of the anterior tibial artery, inferolaterally.

Muscular branches of the popliteal artery supply the hamstring, gastrocnemius, soleus, and plantaris muscles. The superior muscular branches of the popliteal artery have clinically important anastomoses with the terminal part of the profunda femoris and gluteal arteries.

The **popliteal vein** begins at the distal border of the popliteus as a continuation of the *posterior tibial vein* (Fig. 5.51). Throughout its course, the vein lies close to the popliteal artery, lying superficial to it and in the same fibrous sheath. The popliteal vein is initially posteromedial to the artery and lateral to the tibial nerve. More superiorly, the popliteal vein lies posterior to the artery, between this vessel and the overlying tibial nerve. Superiorly, the popliteal vein, which has several valves, becomes the *femoral vein* as it traverses the adductor hiatus. The *small saphenous vein* passes from the posterior aspect of the lateral malleolus to the popliteal fossa, where it pierces the deep popliteal fascia and enters the popliteal vein.

Lymph Nodes in Popliteal Fossa. The **superficial popliteal lymph nodes** are usually small and lie in the subcutaneous tissue. A lymph node lies at the termination of the small saphenous vein and receives lymph from the lymphatic vessels that accompany this vein (see Fig. 5.15). The **deep popliteal lymph nodes** surround the vessels and receive lymph from the joint capsule of the knee and the lymphatic vessels that accompany the deep veins of the leg. The lymphatic vessels from the popliteal lymph nodes follow the femoral vessels to the *deep inguinal lymph nodes.*

Anterior Compartment of Leg

ORGANIZATION OF LEG

The bones of the leg (*tibia* and *fibula*) that connect the knee and ankle, and the three fascial compartments (*anterior, lateral,* and *posterior compartments of the leg*), formed by the *anterior* and *posterior intermuscular septa*, the *interosseous membrane*, and the two leg bones to which they attach, were discussed at the beginning of this chapter and are illustrated in cross-section in Figure 5.53. The muscles of each compartment share common functions and innervations.

The **anterior compartment of the leg,** or *dorsiflexor (extensor) compartment*, is located anterior to the *interosseous membrane*, between the lateral surface of the shaft of the tibia and the medial surface of the shaft of the fibula.

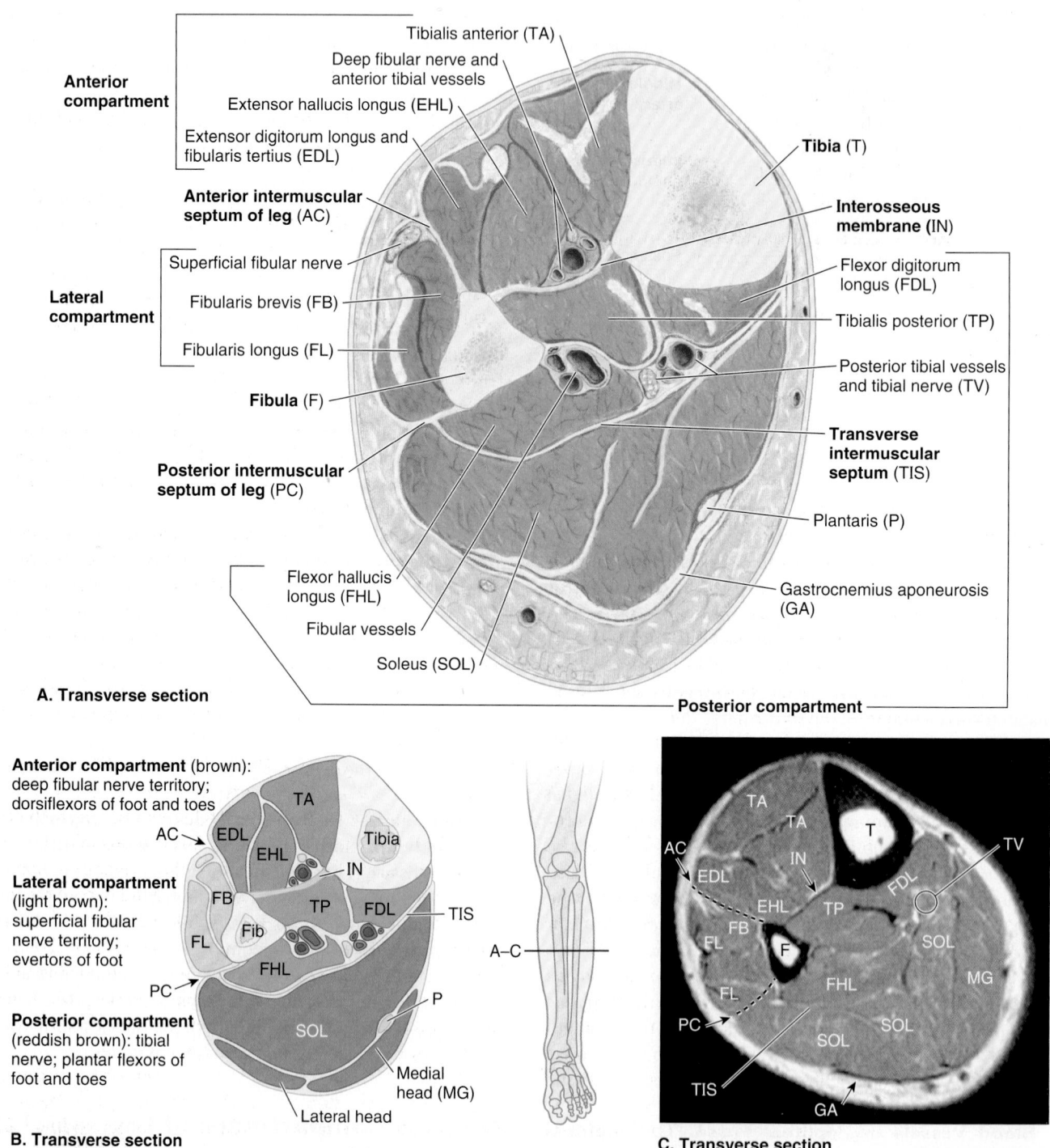

A. Transverse section

B. Transverse section

Anterior compartment (brown):
deep fibular nerve territory;
dorsiflexors of foot and toes

Lateral compartment
(light brown):
superficial fibular
nerve territory;
evertors of foot

Posterior compartment
(reddish brown): tibial
nerve; plantar flexors of
foot and toes

C. Transverse section

FIGURE 5.53. Compartments of leg at midcalf level in transverse anatomical section. A. The anterior (dorsiflexor or extensor) compartment contains four muscles (the fibularis tertius lies inferior to the level of this section). The lateral (fibular) compartment contains two evertor muscles. The posterior (plantarflexor or flexor) compartment, containing seven muscles, is subdivided by an intracompartmental transverse intermuscular septum into a superficial group of three (two of which are commonly tendinous/aponeurotic at this level) and a deep group of four. The popliteus (part of the deep group) lies superior to the level of this section. **B.** Overview of compartments of leg. **C.** MRI of the leg. Abbreviations are defined in the labels for parts **A** and **B.**

The anterior compartment is bounded anteriorly by the *deep fascia of the leg* and skin. The deep fascia overlying the anterior compartment is dense superiorly, providing part of the proximal attachment of the muscle immediately deep to it. With unyielding structures on three sides (the two bones and the interosseous membrane) and a dense fascia on the remaining side, the relatively small anterior compartment is especially confined and therefore most susceptible to *compartment syndromes* (see the blue box "Containment and Spread of Compartmental Infections in the Leg," p. 605).

Inferiorly, two band-like thickenings of the fascia form **retinacula** that bind the tendons of the anterior compartment muscles before and after they cross the ankle joint, preventing them from bowstringing anteriorly during dorsiflexion of the joint (Fig. 5.54):

1. The **superior extensor retinaculum** is a strong, broad band of deep fascia, passing from the fibula to the tibia, proximal to the malleoli.
2. The **inferior extensor retinaculum,** a Y-shaped band of deep fascia, attaches laterally to the anterosuperior surface of the calcaneus. It forms a strong loop around the tendons of the fibularis tertius and the extensor digitorum longus muscles.

MUSCLES OF ANTERIOR COMPARTMENT OF LEG

The four muscles in the anterior compartment of the leg are the *tibialis anterior, extensor digitorum longus, extensor hallucis longus,* and *fibularis tertius* (Figs. 5.53A & B and 5.55; Table 5.10). These muscles pass and insert anterior to the transversely oriented axis of the ankle (talocrural) joint and, therefore, are **dorsiflexors of the ankle joint,** elevating the forefoot and depressing the heel. The long extensors also pass along and attach to the dorsal aspect of the digits and are thus extensors (elevators) of the toes.

Although it is a relatively weak and short movement—only about a quarter the strength of plantarflexion (Soderberg, 1986), with a range of about 20° from neutral—dorsiflexion is actively used in the swing phase of walking, when concentric contraction keeps the forefoot elevated to clear the ground as the free limb swings forward (see Fig. 5.20F & G and Table 5.2). Immediately after, in the stance phase, eccentric contraction of the tibialis anterior controls the lowering of the forefoot to the floor following heel strike (see Fig. 5.20A and Table 5.2). The latter is important to a smooth gait and is important to deceleration (braking) relative to running and walking downhill. During standing, the dorsiflexors reflexively pull the leg (and thus the center of gravity) anteriorly on the fixed foot when the body starts to lean (the center of gravity begins to shift too far) posteriorly. When descending a slope, especially if the surface is loose (sand, gravel, or snow), dorsiflexion is used to "dig in" one's heels.

Tibialis Anterior. The **tibialis anterior** (**TA**), the most medial and superficial dorsiflexor, is a slender muscle that lies against the lateral surface of the tibia (Figs. 5.53 and 5.56). The long tendon of TA begins halfway down the leg and descends along the anterior surface of the tibia. Its tendon passes within its own synovial sheath deep to the superior and inferior extensor retinacula (Fig. 5.54) to its attachment on the medial side of the foot. In so doing, its tendon is located farthest from the axis of the ankle joint, giving it the most mechanical advantage and making it the strongest dorsiflexor. Although antagonists at the ankle joint, TA and the tibialis posterior (in the posterior compartment) both cross the subtalar and transverse tarsal joints to attach to the medial border of the foot. Thus, they act synergistically to invert the foot.

To test the TA, the person is asked to stand on their heels or dorsiflex the foot against resistance; if normal, its tendon can be seen and palpated.

Extensor Digitorum Longus. The **extensor digitorum longus** (**EDL**) is the most lateral of the anterior leg muscles (Figs. 5.53–5.56). A small part of the proximal attachment of the muscle is to the lateral tibial condyle; however, most of it attaches to the medial surface of the fibula and the superior part of the anterior surface of the interosseous membrane (Fig. 5.55A; Table 5.10).

The EDL becomes tendinous superior to the ankle, forming four tendons that attach to the phalanges of the lateral four toes. A common synovial sheath surrounds the four tendons of the EDL (plus that of the fibularis tertius) as they diverge on the dorsum of the foot and pass to their distal attachments (Fig. 5.54B).

Each tendon of EDL forms a membranous *extensor expansion* (dorsal aponeurosis) over the dorsum of the proximal phalanx of the toe, which divides into two lateral bands and one central band (Fig. 5.54A). The central band inserts into the base of the middle phalanx, and the lateral slips converge to insert into the base of the distal phalanx.

To test the EDL, the lateral four toes are dorsiflexed against resistance; if acting normally, the tendons can be seen and palpated.

Fibularis Tertius. The **fibularis tertius** (**FT**) is a separated part of EDL, which shares its synovial sheath (Figs. 5.54 and 5.56). Proximally, the attachments and fleshy parts of the EDL and FT are continuous; however, distally the FT tendon is separate and attaches to the 5th metatarsal, not to a phalanx (Fig. 5.55F; Table 5.10). Although FT contributes (weakly) to dorsiflexion, it also acts at the subtalar and transverse tarsal joints, contributing to eversion (pronation) of the foot. It may play a special proprioceptive role in sensing sudden inversion and then contracting reflexively to protect the anterior tibiofibular ligament, the most commonly sprained ligament of the body. FT is not always present.

Extensor Hallucis Longus. The **extensor hallucis longus** (**EHL**) is a thin muscle that lies deeply between the TA and EDL at its superior attachment to the middle half of the fibula and interosseous membrane (Fig. 5.55E; Table 5.10). EHL rises to the surface in the distal third of the leg, passing deep to the extensor retinacula (Figs. 5.54 and 5.56). It courses distally along the crest of the dorsum of the foot to the great toe.

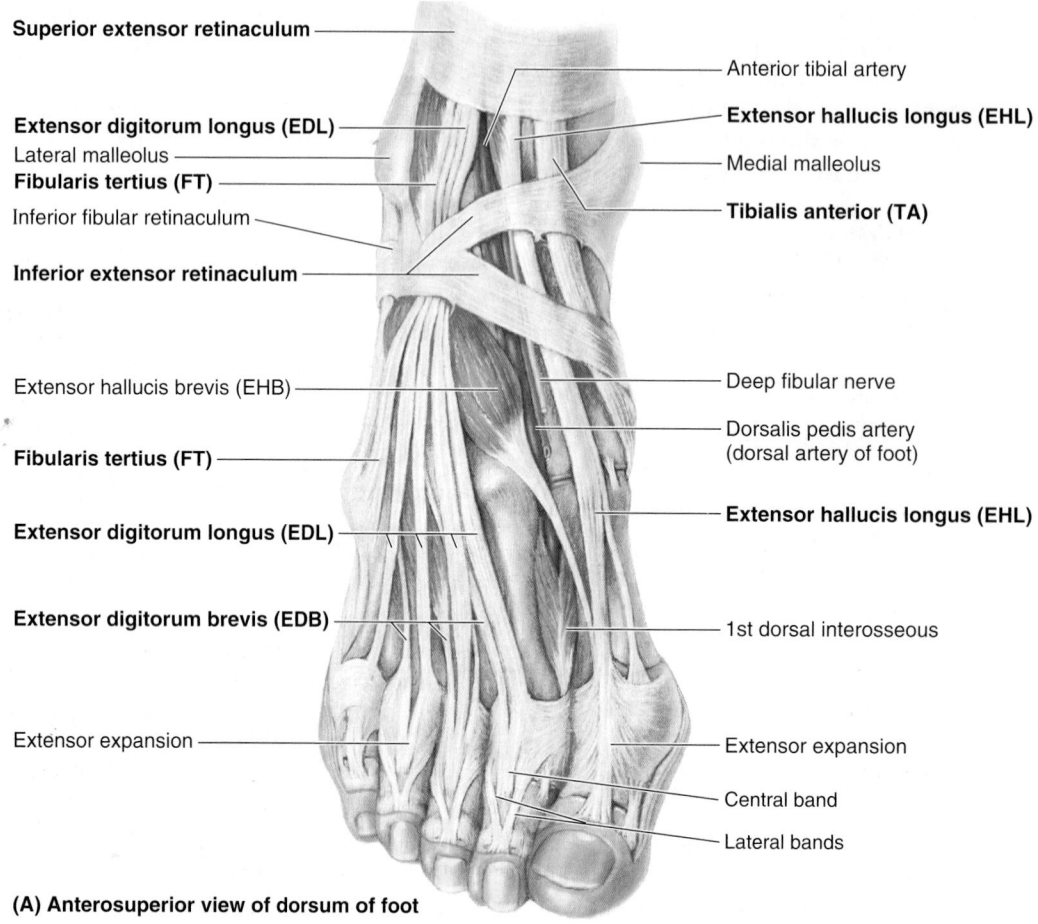

Superior extensor retinaculum

Anterior tibial artery

Extensor digitorum longus (EDL)

Extensor hallucis longus (EHL)

Lateral malleolus

Medial malleolus

Fibularis tertius (FT)

Inferior fibular retinaculum

Tibialis anterior (TA)

Inferior extensor retinaculum

Extensor hallucis brevis (EHB)

Deep fibular nerve

Dorsalis pedis artery
(dorsal artery of foot)

Fibularis tertius (FT)

Extensor hallucis longus (EHL)

Extensor digitorum longus (EDL)

Extensor digitorum brevis (EDB)

1st dorsal interosseous

Extensor expansion

Extensor expansion

Central band

Lateral bands

(A) Anterosuperior view of dorsum of foot

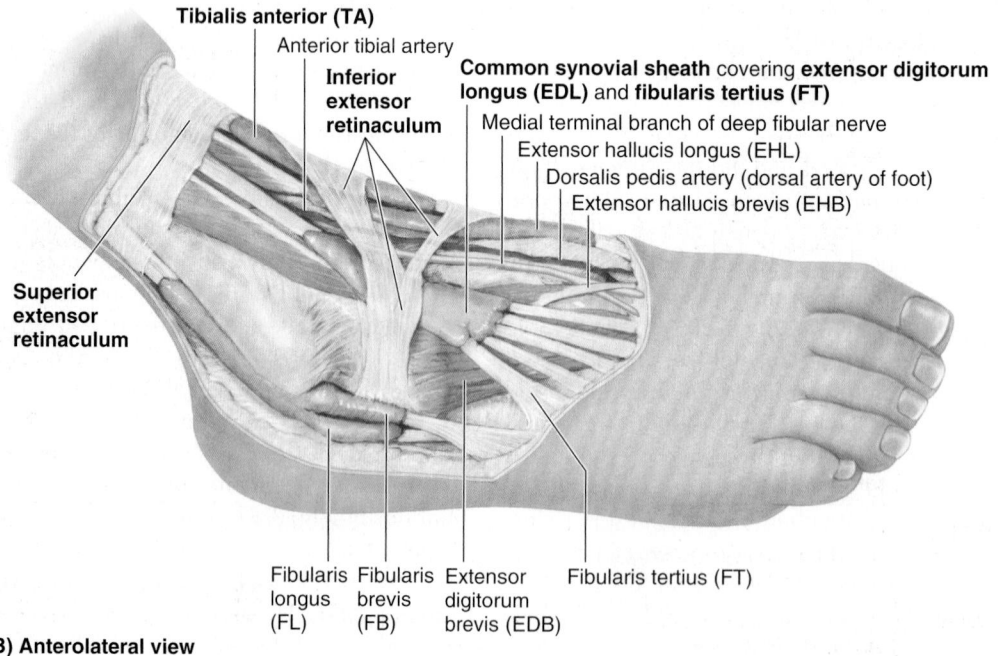

Tibialis anterior (TA)

Anterior tibial artery

**Inferior
extensor
retinaculum**

Common synovial sheath covering **extensor digitorum
longus (EDL)** and **fibularis tertius (FT)**

Medial terminal branch of deep fibular nerve

Extensor hallucis longus (EHL)

Dorsalis pedis artery (dorsal artery of foot)

Extensor hallucis brevis (EHB)

**Superior
extensor
retinaculum**

Fibularis
longus
(FL)

Fibularis
brevis
(FB)

Extensor
digitorum
brevis (EDB)

Fibularis tertius (FT)

(B) Anterolateral view

FIGURE 5.54. Dissections of foot. These dissections demonstrate the continuation of the anterior and lateral leg muscles into the foot. The thinner portions of the deep fascia of the leg have been removed, leaving the thicker portions that make up the extensor and fibular retinacula, which retain the tendons as they cross the ankle. **A.** The vessels and nerves are cut short. At the ankle, the vessels and the deep fibular nerve lie midway between the malleoli and between the tendons of the long dorsiflexors of the toes. **B.** Synovial sheaths surround the tendons as they pass beneath the retinacula of the ankle.

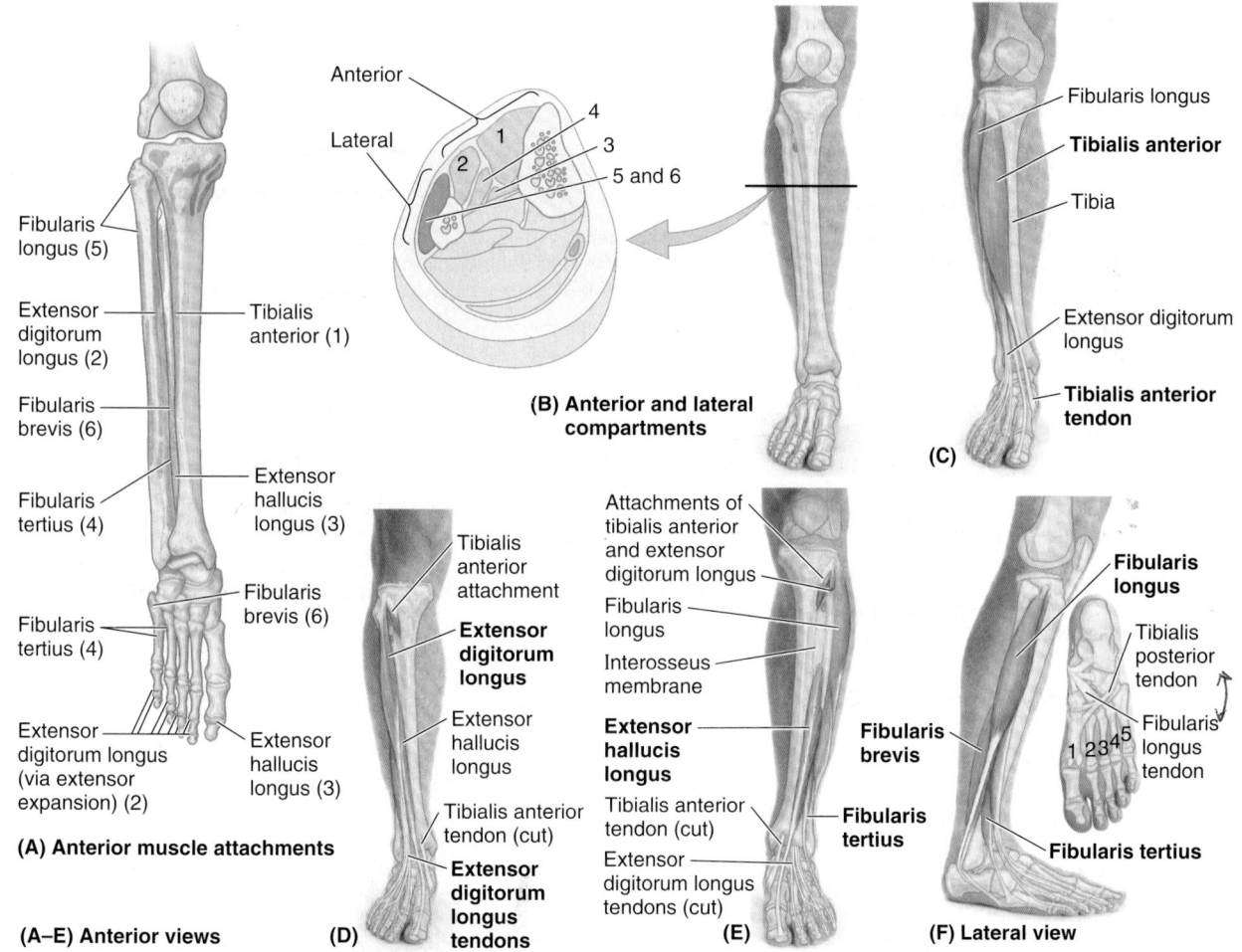

FIGURE 5.55. Muscles of anterior and lateral compartments of leg.

TABLE 5.10. MUSCLES OF ANTERIOR AND LATERAL COMPARTMENTS OF LEG

Muscle[a]	Proximal Attachment	Distal Attachment	Innervation[b]	Main Action
Anterior compartment **Tibialis anterior** (1)	Lateral condyle and superior half of lateral surface of tibia and interosseous membrane	Medial and inferior surfaces of medial cuneiform and base of 1st metatarsal	Deep fibular nerve (**L4**, L5)	Dorsiflexes ankle and inverts foot
Extensor digitorum longus (2)	Lateral condyle of tibia and superior three quarters of medial surface of fibula and interosseous membrane	Middle and distal phalanges of lateral four digits		Extends lateral four digits and dorsiflexes ankle
Extensor hallucis longus (3)	Middle part of anterior surface of fibula and interosseous membrane	Dorsal aspect of base of distal phalanx of great toe (hallux)		Extends great toe and dorsiflexes ankle
Fibularis tertius (4)	Inferior third of anterior surface of fibula and interosseous membrane	Dorsum of base of 5th metatarsal		Dorsiflexes ankle and aids in eversion of foot
Lateral compartment **Fibularis longus** (5)	Head and superior two thirds of lateral surface of fibula	Base of 1st metatarsal and medial cuneiform	Superficial fibular nerve (**L5**, **S1**, S2)	Everts foot and weakly plantarflexes ankle
Fibularis brevis (6)	Inferior two thirds of lateral surface of fibula	Dorsal surface of tuberosity on lateral side of base of 5th metatarsal		

[a] Numbers refer to Figure 5.55A & B.

[b] The spinal cord segmental innervation is indicated (e.g., "L4, L5" means that the nerves supplying the tibialis anterior are derived from the fourth and fifth lumbar segments of the spinal cord). Numbers in boldface (**L4**) indicate the main segmental innervation. Damage to one or more of the listed spinal cord segments or to the motor nerve roots arising from them results in paralysis of the muscles concerned.

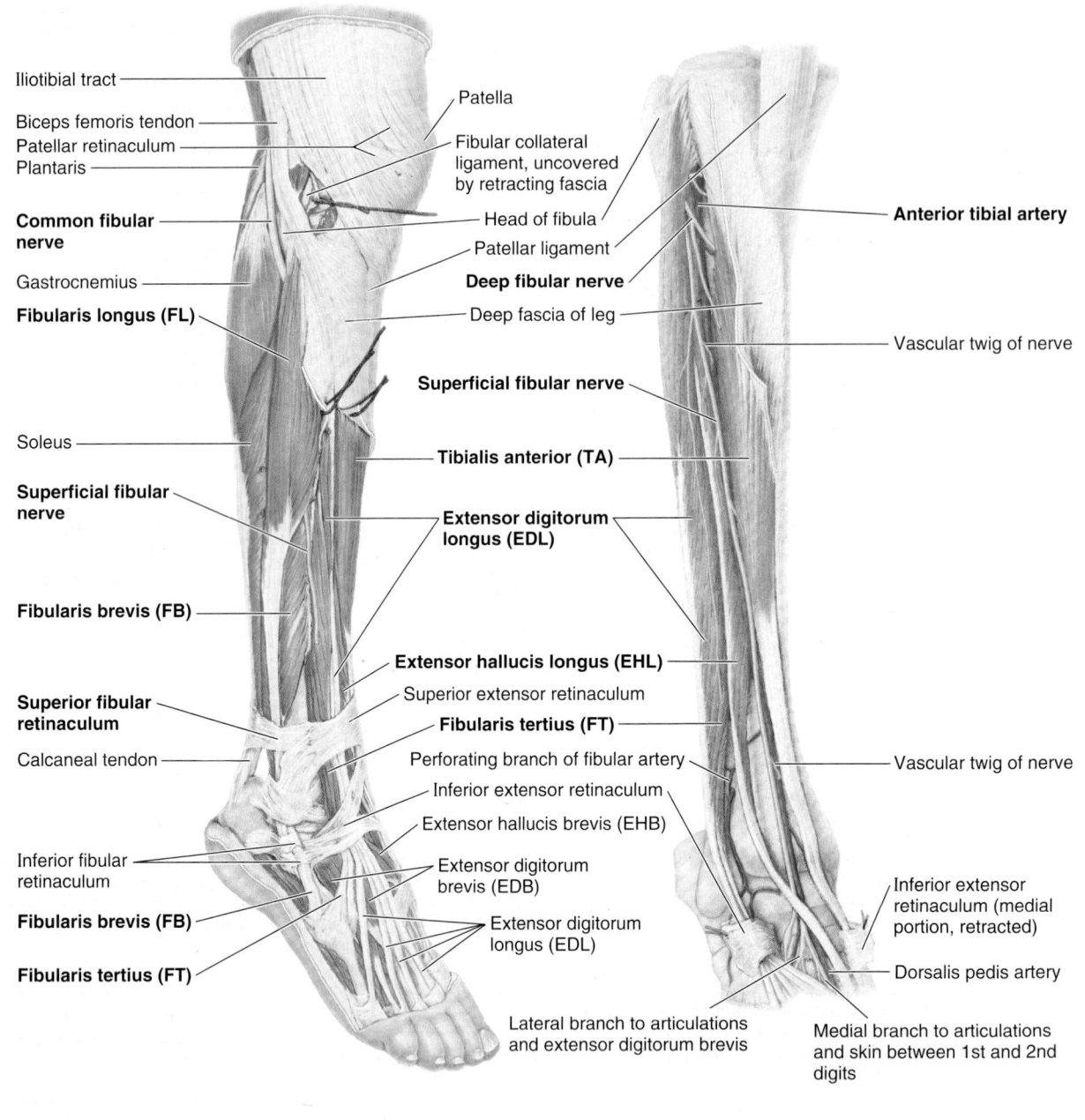

(A) Anterolateral view   **(B) Anterior oblique view**

FIGURE 5.56. Dissections of anterior and lateral compartments of leg. A. This dissection shows the muscles of the anterolateral leg and dorsum of the foot. The common fibular nerve, coursing subcutaneously across the lateral aspect of the head and neck of the fibula, is the most commonly injured peripheral nerve. **B.** In this deeper dissection of the anterior compartment, the muscles and inferior extensor retinaculum are retracted to display the arteries and nerves.

To test the EHL, the great toe is dorsiflexed against resistance; if acting normally, its entire tendon can be seen and palpated.

NERVE OF ANTERIOR COMPARTMENT OF LEG

The **deep fibular** (peroneal) **nerve** is the nerve of the anterior compartment (Figs. 5.53A, 5.56B, and 5.57; Table 5.11). It is one of the two terminal branches of the common fibular

nerve, arising between the fibularis longus muscle and the neck of the fibula. After its entry into the anterior compartment, the deep fibular nerve accompanies the anterior tibial artery, first between the TA and EDL and then between the TA and EHL. The deep fibular nerve then exits the compartment, continuing across the ankle joint to supply intrinsic muscles (extensors digitorum and hallucis brevis), and a small area of the skin of the foot. A lesion of this nerve results in an inability to dorsiflex the ankle (footdrop).

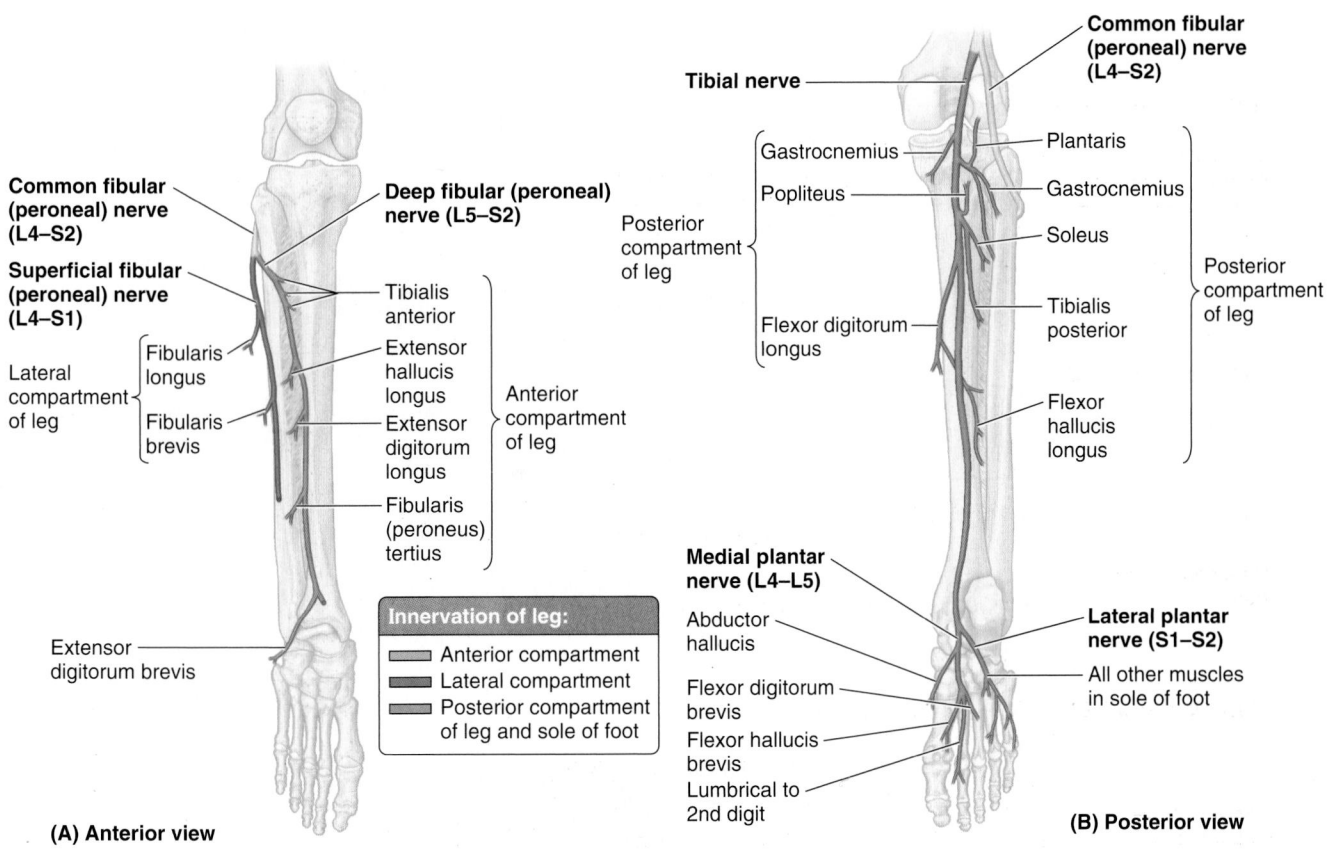

Common fibular (peroneal) nerve (L4–S2)

Common fibular (peroneal) nerve (L4–S2)

Superficial fibular (peroneal) nerve (L4–S1)

Deep fibular (peroneal) nerve (L5–S2)

Lateral compartment of leg
- Fibularis longus
- Fibularis brevis

- Tibialis anterior
- Extensor hallucis longus
- Extensor digitorum longus
- Fibularis (peroneus) tertius

Anterior compartment of leg

Extensor digitorum brevis

Tibial nerve

Posterior compartment of leg
- Gastrocnemius
- Popliteus
- Flexor digitorum longus

- Plantaris
- Gastrocnemius
- Soleus
- Tibialis posterior
- Flexor hallucis longus

Posterior compartment of leg

Medial plantar nerve (L4–L5)
- Abductor hallucis
- Flexor digitorum brevis
- Flexor hallucis brevis
- Lumbrical to 2nd digit

Lateral plantar nerve (S1–S2)
- All other muscles in sole of foot

Innervation of leg:
- Anterior compartment
- Lateral compartment
- Posterior compartment of leg and sole of foot

(A) Anterior view

(B) Posterior view

FIGURE 5.57. Nerves of leg.

TABLE 5.11. NERVES OF LEG

Nerve	Origin	Course	Distribution in Leg
Saphenous	Femoral nerve	Descends with femoral vessels through femoral triangle and adductor canal and then descends with great saphenous vein	Supplies skin on medial side of ankle and foot
Sural	Usually arises from branches of both tibial and common fibular nerves	Descends between heads of gastrocnemius and becomes superficial at middle of leg; descends with small saphenous vein and passes inferior to lateral malleolus to lateral side of foot	Supplies skin on posterior and lateral aspects of leg and lateral side of foot
Tibial	Sciatic nerve	Forms as sciatic bifurcates at apex of popliteal fossa; descends through popliteal fossa and lies on popliteus; runs inferiorly on tibialis posterior with posterior tibial vessels; terminates beneath flexor retinaculum by dividing into medial and lateral plantar nerves	Supplies posterior muscles of leg and knee joint
Common fibular (peroneal)	Sciatic nerve	Forms as sciatic bifurcates at apex of popliteal fossa and follows medial border of biceps femoris and its tendon; passes over posterior aspect of head of fibula and then winds around neck of fibula deep to fibularis longus, where it divides into deep and superficial fibular nerves	Supplies skin on lateral part of posterior aspect of leg via the lateral sural cutaneous nerve; also supplies knee joint via its articular branch
Superficial fibular (peroneal)	Common fibular nerve	Arises between fibularis longus and neck of fibula and descends in lateral compartment of leg; pierces deep fascia at distal third of leg to become subcutaneous	Supplies fibularis longus and brevis and skin on distal third of anterior surface of leg and dorsum of foot
Deep fibular (peroneal)	Common fibular nerve	Arises between fibularis longus and neck of fibula; passes through extensor digitorum longus and descends on interosseous membrane; crosses distal end of tibia and enters dorsum of foot	Supplies anterior muscles of leg, dorsum of foot, and skin of first interdigital cleft; sends articular branches to joints it crosses

ARTERY IN ANTERIOR COMPARTMENT OF LEG

The **anterior tibial artery** supplies structures in the anterior compartment (Figs. 5.53A, 5.58B, and 5.59; Table 5.12). The smaller terminal branch of the popliteal artery, the anterior tibial artery, begins at the inferior border of the popliteus muscle (i.e., as the popliteal artery passes deep to the tendinous arch of the soleus). The artery immediately passes anteriorly through a gap in the superior part of the interosseous membrane to descend on the anterior surface of this membrane between the TA and EDL muscles. At the ankle joint, midway between the malleoli, the anterior tibial artery changes names, becoming the *dorsalis pedis artery* (dorsal artery of the foot).

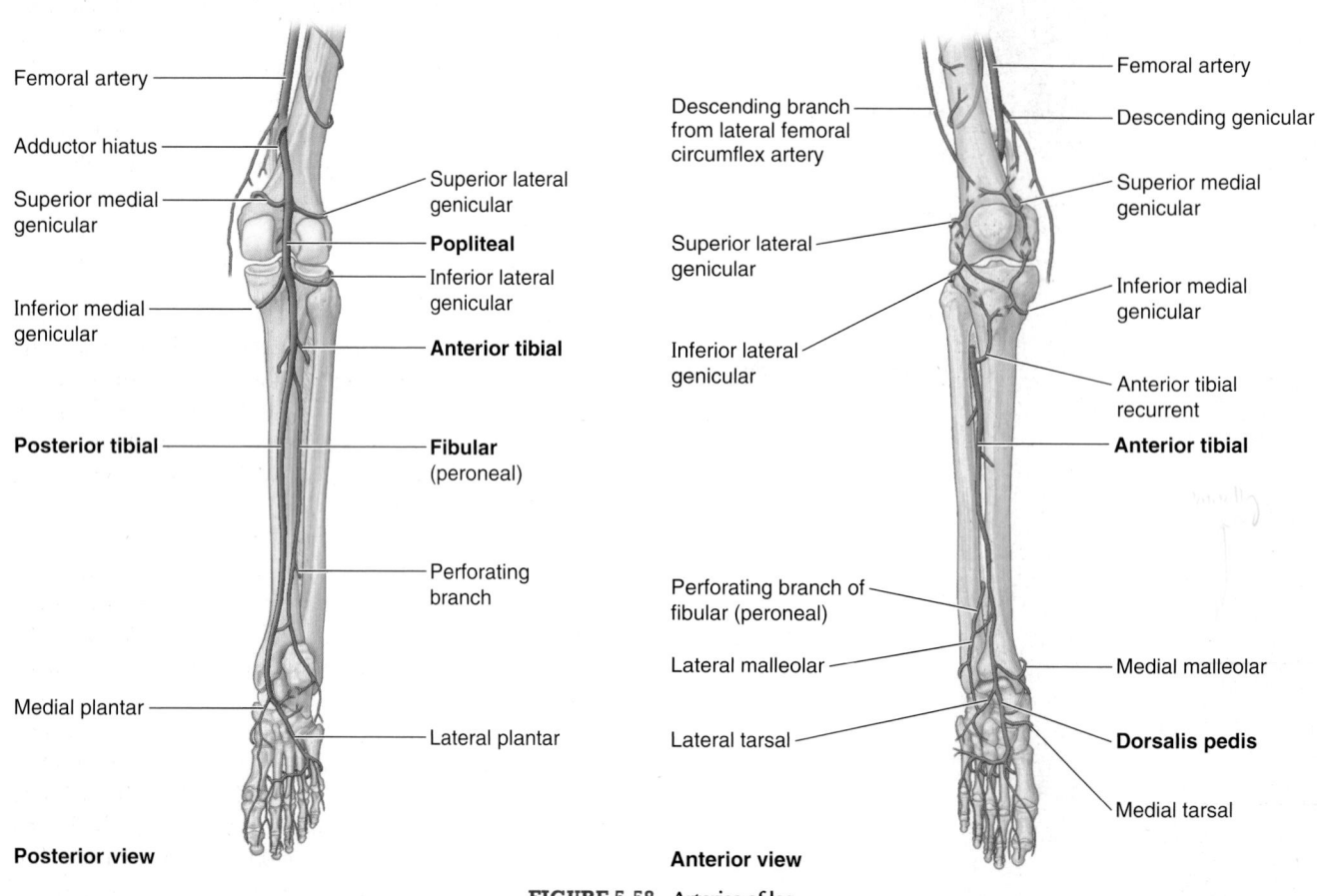

FIGURE 5.58. Arteries of leg.

TABLE 5.12. ARTERIES OF LEG

Artery	Origin	Course	Distribution in Leg
Popliteal	Continuation of femoral artery at adductor hiatus in adductor magnus	Passes through popliteal fossa to leg; ends at lower border of popliteus muscle by dividing into anterior and posterior tibial arteries	Superior, middle, and inferior genicular arteries to both lateral and medial aspects of knee
Anterior tibial	Popliteal	Passes between tibia and fibula into anterior compartment through gap in superior part of interosseous membrane and descends this membrane between tibialis anterior and extensor digitorum longus	Anterior compartment of leg
Dorsalis pedis (Dorsal artery of foot)	Continuation of anterior tibial artery distal to inferior extensor retinaculum	Descends anteromedially to first interosseous space and divides into plantar and arcuate arteries	Muscles on dorsum of foot; pierces first dorsal interosseous muscles as deep plantar artery to contribute to formation of plantar arch
Posterior tibial	Popliteal	Passes through posterior compartment of leg and terminates distal to flexor retinaculum by dividing into medial and lateral plantar arteries	Posterior and lateral compartments of leg; circumflex fibular branch joins anastomoses around knee; nutrient artery passes to tibia
Fibular	Posterior tibial	Descends in posterior compartment adjacent to posterior intermuscular septum	Posterior compartment of leg: perforating branches supply lateral compartment of leg

Lateral Compartment of Leg

The **lateral compartment of the leg,** or *evertor compartment,* is the smallest (narrowest) of the leg compartments. It is bounded by the lateral surface of the fibula, the anterior and posterior intermuscular septa, and the deep fascia of the leg (Figs. 5.53A & B and 5.55F; Table 5.10). The lateral compartment ends inferiorly at the **superior fibular retinaculum,** which spans between the distal tip of the fibula and the calcaneus (Fig. 5.56A). Here the tendons of the two muscles of the lateral compartment (fibularis longus and brevis) enter a *common synovial sheath* to accommodate their passage between the superior fibular retinaculum and the lateral malleolus, using the latter as a trochlea as they cross the ankle joint.

MUSCLES IN LATERAL COMPARTMENT OF LEG

The lateral compartment contains the *fibularis longus and brevis muscles.* These muscles have their fleshy bellies in the lateral compartment but are tendinous as they exit the compartment within the common synovial sheath deep to the superior fibular retinaculum. Both muscles are **evertors of the foot,** elevating the lateral margin of the foot. Developmentally, the fibularis muscles are postaxial muscles, receiving innervation from the posterior divisions of the spinal nerves, which contribute to the sciatic nerve. However, because the fibularis longus and brevis pass posterior to the transverse axis of the ankle (talocrural) joint, they contribute to plantarflexion at the ankle—unlike the postaxial muscles of the anterior compartment (including the fibularis tertius), which are dorsiflexors.

As evertors, the fibularis muscles act at the subtalar and transverse tarsal joints. From the neutral position, only a few degrees of eversion are possible. In practice, the primary function of the evertors of the foot is not to elevate the lateral margin of the foot (the common description of eversion) but to depress or fix the medial margin of the foot in support of the toe off phase of walking and, especially, running and to resist inadvertent or excessive inversion of the foot (the position in which the ankle is most vulnerable to injury). When standing (and particularly when balancing on one foot), the fibularis muscles contract to resist medial sway (to recenter a line of gravity, which has shifted medially) by pulling laterally on the leg while depressing the medial margin of the foot.

To test the fibularis longus and brevis, the foot is everted strongly against resistance; if acting normally, the muscle tendons can be seen and palpated inferior to the lateral malleolus.

Fibularis Longus. The **fibularis longus (FL)** is the longer and more superficial of the two fibularis muscles, arising much more superiorly on the shaft of the fibula (Figs. 5.53, 5.55F, and 5.56A; Table 5.10). The narrow FL extends from the head of the fibula to the sole of the foot. Its tendon can be palpated and observed proximal and posterior to the lateral malleolus. Distal to the superior fibular retinaculum, the common sheath shared by the fibular muscles splits to extend through separate compartments deep to the **inferior fibular retinaculum** (Figs. 5.54A and 5.56). The FL passes through

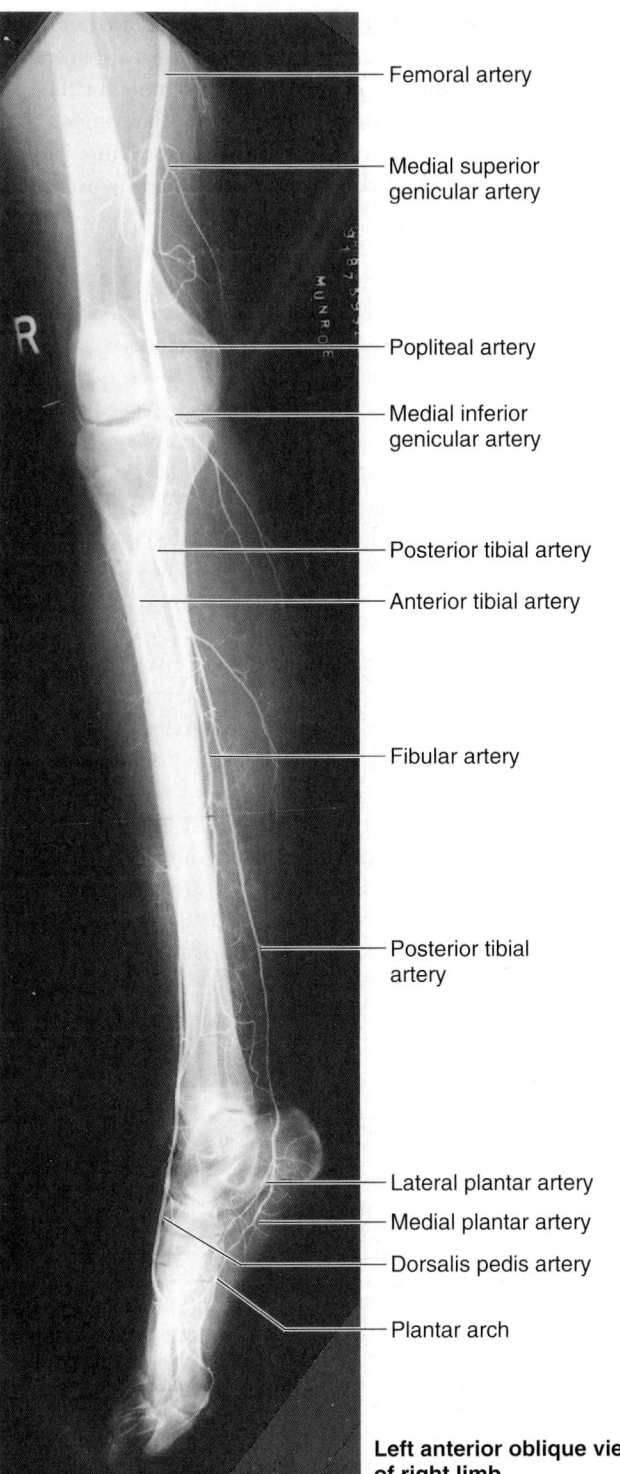

— Femoral artery

— Medial superior genicular artery

— Popliteal artery

— Medial inferior genicular artery

— Posterior tibial artery

— Anterior tibial artery

— Fibular artery

— Posterior tibial artery

— Lateral plantar artery

— Medial plantar artery

— Dorsalis pedis artery

— Plantar arch

Left anterior oblique view of right limb

FIGURE 5.59. Popliteal arteriogram. The popliteal artery begins at the site of the adductor hiatus (where it may be compressed), and then lies successively on the distal end of the femur, joint capsule of the knee joint, and popliteus muscle (not visible) before dividing into the anterior and posterior tibial arteries at the inferior angle of the popliteus fossa. Here it is subject to entrapment as it passes beneath the tendinous arch of the soleus muscle. (Courtesy of Dr. K. Sniderman, Associate Professor of Medical Imaging, University of Toronto, Toronto, Ontario, Canada.)

the inferior compartment—inferior to the *fibular trochlea* on the calcaneus—and enters a groove on the antero-inferior aspect of the cuboid bone (see Fig. 5.11D). It then crosses the sole of the foot, running obliquely and distally to reach its attachment to the *1st metatarsal and medial cuneiform bones* (see Fig. 5.11B). When a person stands on one foot, the FL helps steady the leg on the foot.

Fibularis Brevis. The **fibularis brevis** (**FB**) is a fusiform muscle that lies deep to the FL, and, true to its name, the FB is shorter than its partner in the lateral compartment (Figs. 5.53, 5.55F, and 5.56A; Table 5.10). Its broad tendon grooves the posterior aspect of the lateral malleolus and can be palpated inferior to it. The narrower tendon of the FL lies posterior to the tendon of the FB and does not contact the lateral malleolus. The tendon of the FB traverses the superior compartment of the inferior fibular retinaculum, passing superior to the fibular trochlea of the calcaneus; it can be easily traced to its distal attachment to the *base of the 5th metatarsal* (Fig. 5.11D). The tendon of the *fibularis tertius*, a slip of muscle from the extensor digitorum longus, often merges with the tendon of the FB (Fig. 5.56A). Occasionally, however, the fibularis tertius passes anteriorly to attach directly to the proximal phalanx of the 5th digit.

NERVES IN LATERAL COMPARTMENT OF LEG

The **superficial fibular** (peroneal) **nerve,** a terminal branch of the common fibular nerve, is the nerve of the lateral compartment (Figs. 5.43A, 5.56A, and 5.57A; Table 5.11). After supplying the FL and FB, the superficial fibular nerve continues as a cutaneous nerve, supplying the skin on the distal part of the anterior surface of the leg and nearly all the dorsum of the foot.

BLOOD VESSELS IN LATERAL COMPARTMENT OF LEG

The lateral compartment does not have an artery coursing through it. Instead, *perforating branches* and accompanying veins supply blood to and drain blood from the compartment. Proximally, **perforating branches of the anterior tibial artery** penetrate the anterior intermuscular septum. Inferiorly, **perforating branches of the fibular artery** penetrate the posterior intermuscular septum, along with their accompanying veins (L. *venae comitantes*) (Figs. 5.58 and 5.59; Table 5.12).

Posterior Compartment of Leg

The **posterior compartment of the leg** (plantarflexor compartment) is the largest of the three leg compartments (Fig. 5.53A). The posterior compartment and the muscles within it are divided into superficial and deep subcompartments/muscle groups by the *transverse intermuscular septum.* The tibial nerve and posterior tibial and fibular vessels supply both parts of the posterior compartment but run in the deep subcompartment deep (anterior) to the transverse intermuscular septum.

The larger **superficial subcompartment** is the least confined compartmental area. The smaller **deep subcompartment,** like the anterior compartment, is bounded by the two leg bones and the *interosseous membrane* that binds them together, plus the transverse intermuscular septum. Therefore, the deep subcompartment is quite tightly confined. Because the nerve and blood vessels supplying the entire posterior compartment and the sole of the foot pass through the deep subcompartment, when swelling occurs it leads to a compartment syndrome that has serious consequences, such as muscular necrosis (tissue death) and paralysis.

Inferiorly, the deep subcompartment tapers as the muscles it contains become tendinous. The transverse intermuscular septum ends as reinforcing transverse fibers that extend between the tip of the medial malleolus and the calcaneus to form the **flexor retinaculum** (see Fig. 5.61). The retinaculum is subdivided deeply, forming separate compartments for each tendon of the deep muscle group, as well as for the tibial nerve and posterior tibial artery as they bend around the medial malleolus.

Muscles of the posterior compartment produce *plantarflexion* at the ankle, *inversion* at the subtalar and transverse tarsal joints, and *flexion* of the toes. **Plantarflexion** is a powerful movement (four times stronger than dorsiflexion) produced over a relatively long range (approximately 50° from neutral) by muscles that pass posterior to the transverse axis of the ankle joint. Plantarflexion develops thrust, applied primarily at the ball of the foot, that is used to propel the body forward and upward, and is the major component of the forces generated during the push-off (heel off and toe off) parts of the stance phase of walking and running (see Fig. 5.20D & E; Table 5.2).

SUPERFICIAL MUSCLE GROUP IN POSTERIOR COMPARTMENT

The *superficial group of calf muscles* (muscles forming prominence of "calf" of posterior leg) includes the *gastrocnemius, soleus,* and *plantaris.* Details concerning their attachments, innervation, and actions are provided in Figure 5.60A–E and Table 5.13.I). The gastrocnemius and soleus share a common tendon, the *calcaneal tendon,* which attaches to the calcaneus. Collectively these two muscles make up the three-headed **triceps surae** (L. *sura,* calf) (Figs. 5.60 and 5.61A). This powerful muscular mass tugs on the lever provided by the calcaneal tuberosity, elevating the heel and thus depressing the forefoot, generating as much as 93% of the plantarflexion force.

The large size of the gastrocnemius and soleus muscles is a human characteristic that is directly related to our upright stance. These muscles are strong and heavy because they lift, propel, and accelerate the weight of the body when walking, running, jumping, or standing on the toes.

The **calcaneal tendon** (L. *tendo calcaneus,* Achilles tendon) is the most powerful (thickest and strongest) tendon in the body. Approximately 15 cm in length, it is a continuation

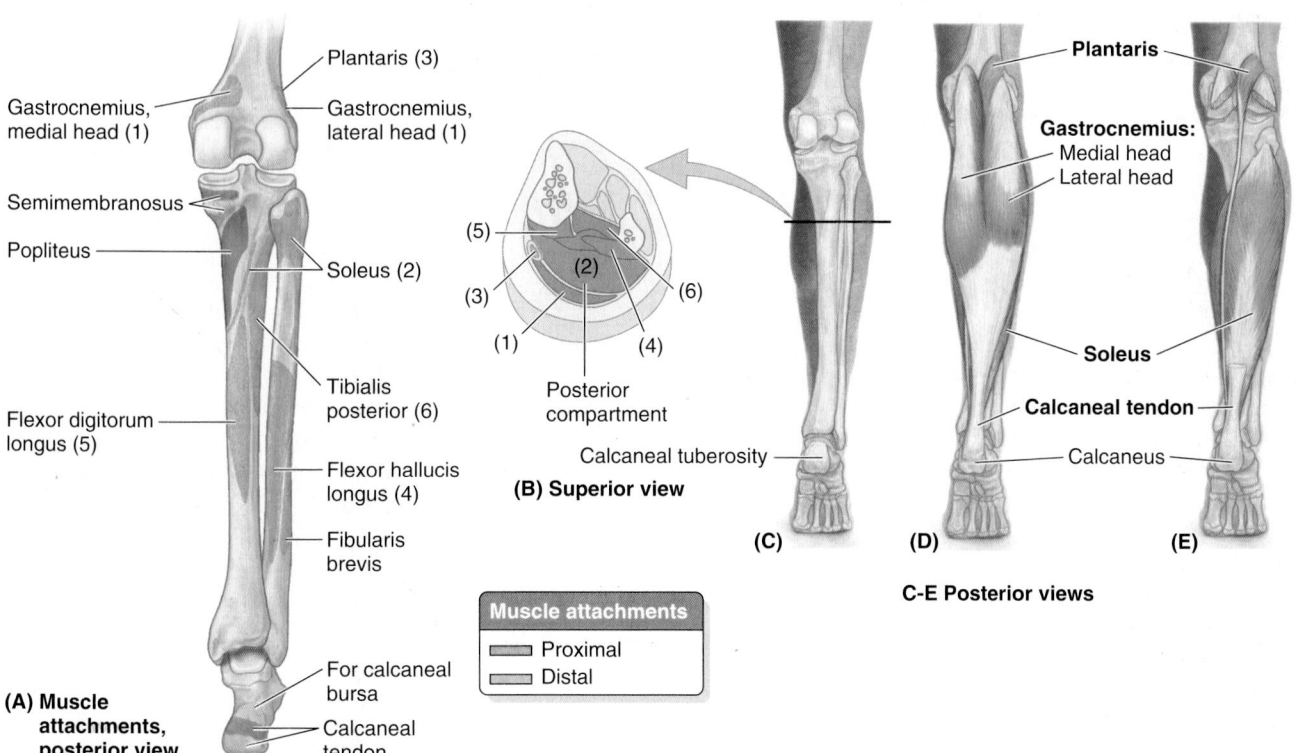

FIGURE 5.60A–E. Superficial (calf) muscles of posterior compartment of leg.

TABLE 5.13.I. SUPERFICIAL MUSCLES OF POSTERIOR COMPARTMENT OF LEG

Muscle[a]	Proximal Attachment	Distal Attachment	Innervation[b]	Main Action
Gastrocnemius (1)	Lateral head: lateral aspect of lateral condyle of femur	Posterior surface of calcaneus via calcaneal tendon	Tibial nerve (S1, S2)	Plantarflexes ankle when knee is extended; raises heel during walking; flexes leg at knee joint
	Medial head: popliteal surface of femur; superior to medial condyle			
Soleus (2)	Posterior aspect of head and superior quarter of posterior surface of fibula; soleal line and middle third of medial border of tibia; and tendinous arch extending between the bony attachments			Plantarflexes ankle independent of position of knee; steadies leg on foot
Plantaris (3)	Inferior end of lateral supracondylar line of femur; oblique popliteal ligament			Weakly assists gastrocnemius in plantarflexing ankle

[a] Numbers refer to Figure 5.60A.

[b] The spinal cord segmental innervation is indicated (e.g., "S1, S2" means that the nerves supplying these muscles are derived from the first and second sacral segments of the spinal cord). Damage to one or more of the listed spinal cord segments or to the motor nerve roots arising from them results in paralysis of the muscles concerned.

of the flat aponeurosis formed halfway down the calf where the bellies of the gastrocnemius terminate (Figs. 5.60D & E and 5.61). Proximally, the aponeurosis receives fleshy fibers of the soleus directly on its deep surface but distally the soleus fibers become tendinous. The tendon thus becomes thicker (deeper) but narrower as it descends until it becomes essentially round in cross-section superior to the calcaneus. It then widens as it inserts on the posterior surface of the *calcaneal tuberosity*. The calcaneal tendon typically spirals a quarter turn (90°) during its descent, so that the gastrocnemius fibers attach laterally and the soleal fibers attach

medially. This arrangement is thought to be significant to the tendon's elastic ability to absorb energy (shock) and recoil, releasing the energy as part of the propulsive force it exerts. Although they share a common tendon, the two muscles of the triceps surae are capable of acting alone, and often do so: "You stroll with the soleus but win the long jump with the gastrocnemius."

To test the triceps surae, the foot is plantarflexed against resistance (e.g., by "standing on the toes," in which case body weight [gravity] provides resistance). If normal, the calcaneal tendon and triceps surae can be observed and palpated.

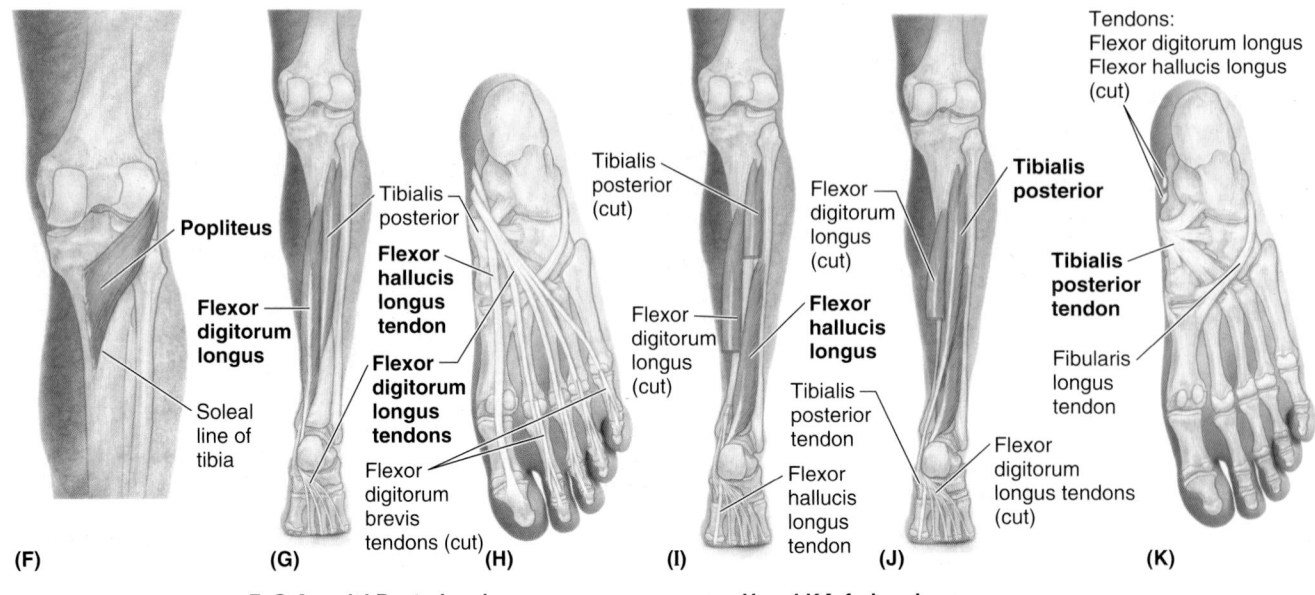

F, G, I, and J Posterior views   H and K Inferior views

FIGURE 5.60F-K. Deep muscles of posterior compartment of leg.

TABLE 5.13.II. DEEP MUSCLES OF POSTERIOR COMPARTMENT OF LEG

Muscle[a]	Proximal Attachment	Distal Attachment	Innervation[b]	Main Action
Popliteus	Lateral surface of lateral condyle of femur and lateral meniscus	Posterior surface of tibia, superior to soleal line	Tibial nerve (L4, L5, S1)	Weakly flexes knee and unlocks it by rotating femur 5° on fixed tibia; medially rotates tibia of unplanted limb
Flexor hallucis longus (4)	Inferior two thirds of posterior surface of fibula; inferior part of interosseous membrane	Base of distal phalanx of great toe (hallux)	Tibial nerve (S2, S3)	Flexes great toe at all joints; weakly plantarflexes ankle; supports medial longitudinal arch of foot
Flexor digitorum longus (5)	Medial part of posterior surface of tibia inferior to soleal line; by a broad tendon to fibula	Bases of distal phalanges of lateral four digits		Flexes lateral four digits; plantarflexes ankle; supports longitudinal arches of foot
Tibialis posterior (6)	Interosseous membrane; posterior surface of tibia inferior to soleal line; posterior surface of fibula	Tuberosity of navicular, cuneiform, cuboid, and sustentaculum tali of calcaneus; bases of 2nd, 3rd, and 4th metatarsals	Tibial nerve (L4, L5)	Plantarflexes ankle; inverts foot

[a] Numbers refer to Figure 5.60A.

[b] The spinal cord segmental innervation is indicated (e.g., "S2, S3" means that the nerves supplying the flexor hallucis longus are derived from the second and third sacral segments of the spinal cord). Damage to one or more of the listed spinal cord segments or to the motor nerve roots arising from them results in paralysis of the muscles concerned.

A *subcutaneous calcaneal bursa,* located between the skin and the calcaneal tendon, allows the skin to move over the taut tendon. A deep *bursa of the calcaneal tendon* (retrocalcaneal bursa), located between the tendon and the calcaneus, allows the tendon to glide over the bone.

Gastrocnemius. The **gastrocnemius** is the most superficial muscle in the posterior compartment and forms the proximal, most prominent part of the *calf* (Figs. 5.60D and 5.61A; Table 5.13.I). It is a fusiform, two-headed, two-joint muscle with the medial head slightly larger and extending more distally than its lateral partner. The heads come together

at the inferior margin of the popliteal fossa, where they form the inferolateral and inferomedial boundaries of this fossa. Because its fibers are largely of the white, fast-twitch (type 2) variety, contractions of the gastrocnemius produce rapid movements during running and jumping. It is recruited into action only intermittently during symmetrical standing.

The gastrocnemius crosses and is capable of acting on both the knee and the ankle joints; however, it cannot exert its full power on both joints at the same time. It functions most effectively when the knee is extended (and is maximally activated when knee extension is combined with dorsiflexion,

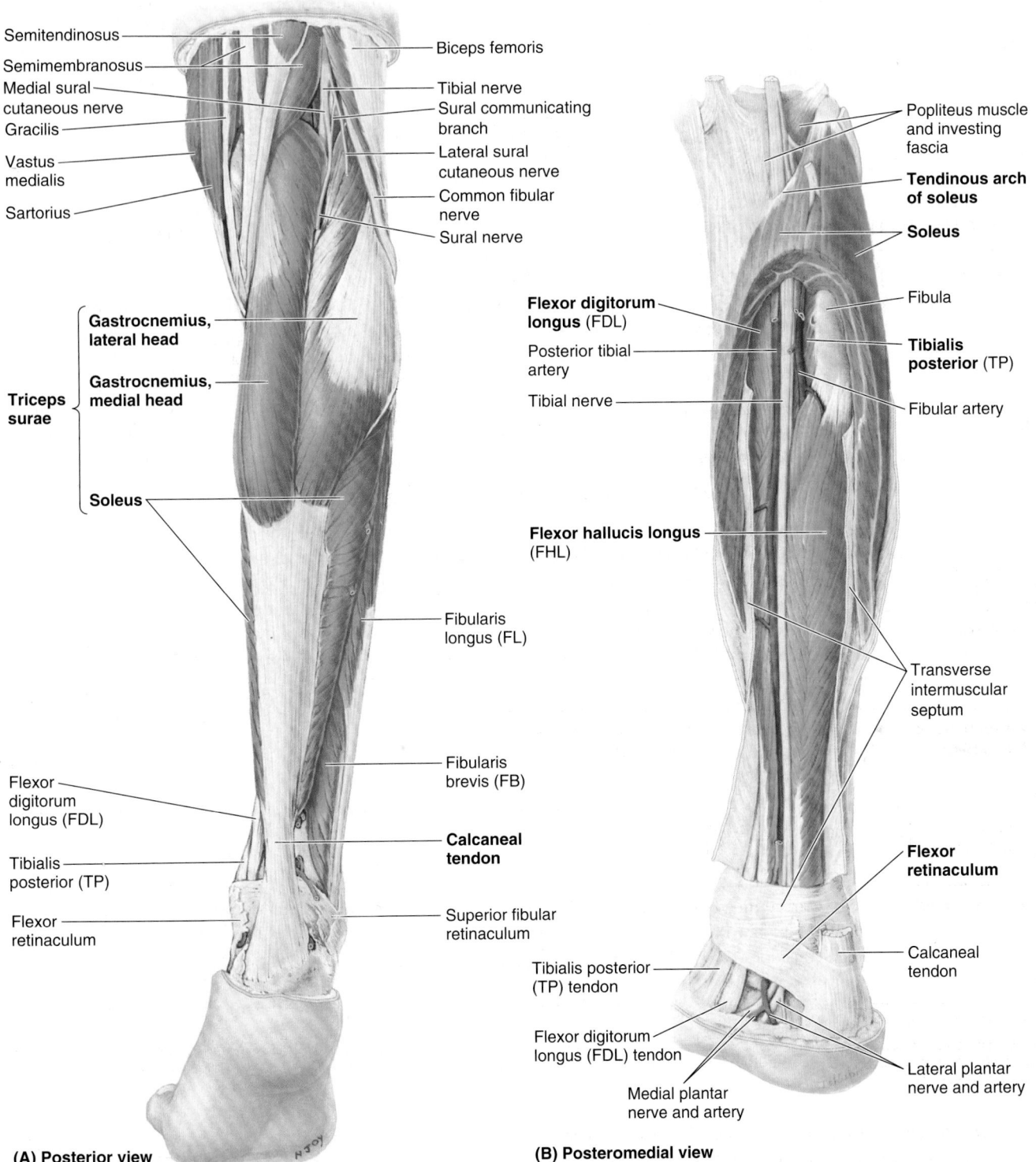

Semitendinosus

Semimembranosus

Medial sural
cutaneous nerve

Gracilis

Vastus
medialis

Sartorius

Biceps femoris

Tibial nerve

Sural communicating
branch

Lateral sural
cutaneous nerve

Common fibular
nerve

Sural nerve

**Triceps
surae**

**Gastrocnemius,
lateral head**

**Gastrocnemius,
medial head**

Soleus

Flexor
digitorum
longus (FDL)

Tibialis
posterior (TP)

Flexor
retinaculum

Fibularis
longus (FL)

Fibularis
brevis (FB)

**Calcaneal
tendon**

Superior fibular
retinaculum

(A) Posterior view

Popliteus muscle
and investing
fascia

**Tendinous arch
of soleus**

Soleus

Fibula

**Tibialis
posterior (TP)**

Fibular artery

**Flexor digitorum
longus (FDL)**

Posterior tibial
artery

Tibial nerve

**Flexor hallucis longus
(FHL)**

Transverse
intermuscular
septum

**Flexor
retinaculum**

Calcaneal
tendon

Tibialis posterior
(TP) tendon

Flexor digitorum
longus (FDL) tendon

Medial plantar
nerve and artery

Lateral plantar
nerve and artery

(B) Posteromedial view

FIGURE 5.61. Dissections of posterior aspect of leg. A. Superficial dissection. Except for the retinacula in the ankle region, the deep fascia has been removed to reveal the nerves and muscles. The three heads of the triceps surae muscle attach distally to the calcaneus via the spiraling fibers of the calcaneal tendon. **B.** Deep dissection. The gastrocnemius and most of the soleus are removed, leaving only a horseshoe-shaped section of the soleus close to its proximal attachments and the distal part of the calcaneal tendon. The transverse intermuscular septum has been split to reveal the deep muscles, vessels, and nerves.

as in the sprint start). It is incapable of producing plantarflexion when the knee is fully flexed.

Soleus. The **soleus** is located deep to the gastrocnemius and is the "workhorse" of plantarflexion (Figs. 5.60E and 5.61A & B; Table 5.13.I). It is a large muscle, flatter than the gastrocnemius, that is named for its resemblance to a sole—the flat fish that reclines on its side on the sea floor. The soleus has a continuous proximal attachment in the shape of an inverted U to the posterior aspects of the fibula and tibia, and a tendinous arch between them, the **tendinous arch of soleus** (L. *arcus tendineus soleus*) (Figs. 5.60A and 5.61B). The popliteal artery and tibial nerve exit the popliteal fossa by passing through this arch, the popliteal artery simultaneously bifurcating into its terminal branches, the anterior and posterior tibial arteries.

The soleus can be palpated on each side of the gastrocnemius when the individual is standing on their tiptoes. The soleus may act with the gastrocnemius in plantarflexing the ankle joint; it cannot act on the knee joint and acts alone when the knee is flexed. Soleus has many parts, each with fiber bundles coursing in different direction.

When the foot is planted, the soleus pulls posteriorly on the bones of the leg. This is important to standing because the line of gravity passes anterior to the leg's bony axis. The soleus is thus an antigravity muscle (the predominant plantarflexor for standing and strolling), which contracts antagonistically but cooperatively (alternately) with the dorsiflexor muscles of the leg to maintain balance. Composed largely of red, fatigue-resistant, slow-twitch (type 1) muscle fibers, it is a strong but relatively slow plantarflexor of the ankle joint, capable of sustained contraction. Electromyography (EMG) studies show that during symmetrical standing, the soleus is continuously active.

Plantaris. The **plantaris** is a small muscle with a short belly and a long tendon (Figs. 5.50, 5.53A, and 5.60A & E; Table 5.13.I). This vestigial muscle is absent in 5–10% of people and is highly variable in size and form when present (most commonly a tapering slip about the size of the small finger). It acts with the gastrocnemius but is insignificant as either a flexor of the knee or a plantarflexor of the ankle.

The plantaris has been considered to be an organ of proprioception for the larger plantarflexors, as it has a high density of muscle spindles (receptors for proprioception). Its long, slender tendon is easily mistaken for a nerve (and hence dubbed by some as the "freshman's nerve").

The plantaris tendon runs distally between the gastrocnemius and soleus (Figs. 5.53A and 5.60B), and occasionally suddenly ruptures with a painful *pop* during activities such as racquet sports. Because of its minor role, the plantaris tendon can be removed for grafting (e.g., during reconstructive surgery of the tendons of the hand) without causing disability.

DEEP MUSCLE GROUP IN POSTERIOR COMPARTMENT

Four muscles make up the deep group in the posterior compartment of the leg (Figs. 5.53, 5.61B, 5.62–5.64; Table 5.13.II): *popliteus, flexor digitorum longus, flexor hallucis longus,* and

tibialis posterior. The popliteus acts on the knee joint, whereas the other muscles plantarflex the ankle with two continuing on to flex the toes. However, because of their smaller size and the close proximity of their tendons to the axis of the ankle joint, the "non-triceps" plantarflexors collectively produce only about 7% of the total force of plantarflexion, and in this the fibularis longus and brevis are most significant. When the calcaneal tendon is ruptured, these muscles cannot generate the power necessary to lift the body's weight (i.e., to stand on the toes).

The two muscles of the posterior compartment that pass to the toes are crisscrossed—that is, the muscle attaching medially to the great toe (flexor hallucis longus) arises laterally (from the fibula) in the deep subcompartment, and the muscle attaching to the lateral four toes (flexor digitorum longus) arises medially (from the tibia) (Fig. 5.40). Their tendons cross in the sole of the foot.

Popliteus. The **popliteus** is a thin, triangular muscle that forms the inferior part of the floor of the popliteal fossa (Figs. 5.50, 5.51, 5.60A & F, and 5.62; Table 5.13.II). Proximally, its tendinous attachment to the lateral aspect of the lateral femoral condyle and its broader attachment to the lateral meniscus occur between the fibrous layer and the synovial membrane of the joint capsule of the knee. The apex of its fleshy belly emerges from the joint capsule of the knee joint. It has a fleshy distal attachment to the tibia that is covered by investing fascia reinforced by a fibrous expansion from the semimembranosus muscle (**investing fascia of popliteus**— Fig. 5.62).

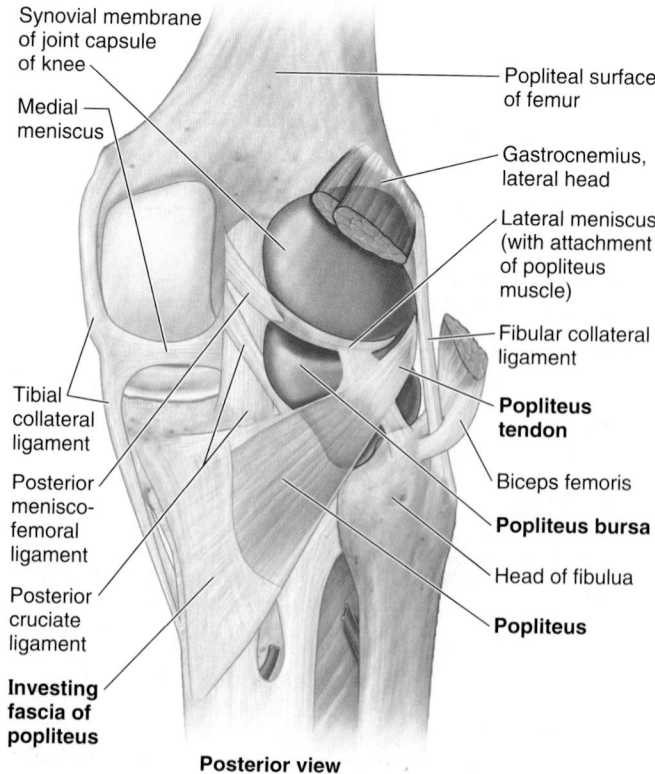

Synovial membrane of joint capsule of knee
Medial meniscus
Tibial collateral ligament
Posterior menisco-femoral ligament
Posterior cruciate ligament
Investing fascia of popliteus
Popliteal surface of femur
Gastrocnemius, lateral head
Lateral meniscus (with attachment of popliteus muscle)
Fibular collateral ligament
Popliteus tendon
Biceps femoris
Popliteus bursa
Head of fibulua
Popliteus

Posterior view

FIGURE 5.62. Deep dissection of popliteal fossa and posterior knee joint.

The popliteus is insignificant as a flexor of the knee joint per se; but during flexion at the knee, it assists in pulling the lateral meniscus of the knee joint posteriorly, a movement otherwise produced passively by compression (as it is for the medial meniscus). When a person is standing with the knee partly flexed, the popliteus contracts to assist the posterior cruciate ligament (PCL) in preventing anterior displacement of the femur on the inclined tibial plateau (see Fig. 5.89C).

The **popliteus bursa** lies deep to the popliteus tendon (Fig. 5.62). When standing with the knees locked in the fully extended position, the popliteus acts to rotate the femur laterally 5° on the tibial plateaus, releasing the knee from its close-packed or locked position so that flexion can occur. When the foot is off the ground and the knee is flexed, the popliteus can aid the medial hamstrings (the "semimuscles") in rotating the tibia medially beneath the femoral condyles.

Flexor Hallucis Longus. The **flexor hallucis longus** (FHL) is a powerful flexor of all of the joints of the great toe (Fig. 5.63A). Immediately after the triceps surae has delivered the thrust of plantarflexion to the *ball of the foot* (the prominence of the sole underlying the heads of the 1st and 2nd metatarsals), the FHL delivers a final thrust via flexion of the great toe for the preswing phase (toe off) of the gait cycle (see Fig. 5.20E; Table 5.2). When barefoot, this thrust is delivered by the great toe; but with soled shoes on, it becomes part of the thrust of plantarflexion delivered by the forefoot.

The tendon of the FHL passes posterior to the distal end of the tibia and occupies a shallow groove on the posterior surface of the talus, which is continuous with the groove on the

plantar surface of the sustentaculum tali (Figs. 5.60H–K and 5.63A & B; Table 5.13.II). The tendon then crosses deep to the tendon of the flexor digitorum longus in the sole of the foot. As it passes to the distal phalanx of the great toe, the FHL tendon runs between two *sesamoid bones* in the tendons of the flexor hallucis brevis (Fig. 5.63B). These bones protect the tendon from the pressure of the head of the 1st metatarsal bone.

To test the FHL, the distal phalanx of the great toe is flexed against resistance; if normal, the tendon can be seen and palpated on the plantar aspect of the great toe as it crosses the joints of the toe.

Flexor Digitorum Longus. The **flexor digitorum longus** (FDL) is smaller than the FHL, even though it moves four digits (Figs. 5.60G–K, 5.61B, and 5.63A & B; Table 5.13. II). It passes diagonally into the sole of the foot, superficial to the tendon of the FHL. However, its direction of pull is realigned by the *quadratus plantae muscle,* which is attached to the posterolateral aspect of the FDL tendon as it divides into four tendons (Figs. 5.60H and 5.63B), which in turn pass to the distal phalanges of the lateral four digits.

To test the FDL, the distal phalanges of the lateral four toes are flexed against resistance; if they are acting normally, the tendons of the toes can be seen and palpated.

Tibialis Posterior. The **tibialis posterior** (TP), the deepest (most anterior) muscle in the posterior compartment, lies between the FDL and the FHL in the same plane as the tibia and fibula within the deep subcompartment (Figs. 5.60J & K, 5.61B, and 5.63A & B; Table 5.13.II). Distally, the TP attaches primarily to the navicular bone (in close proximity to

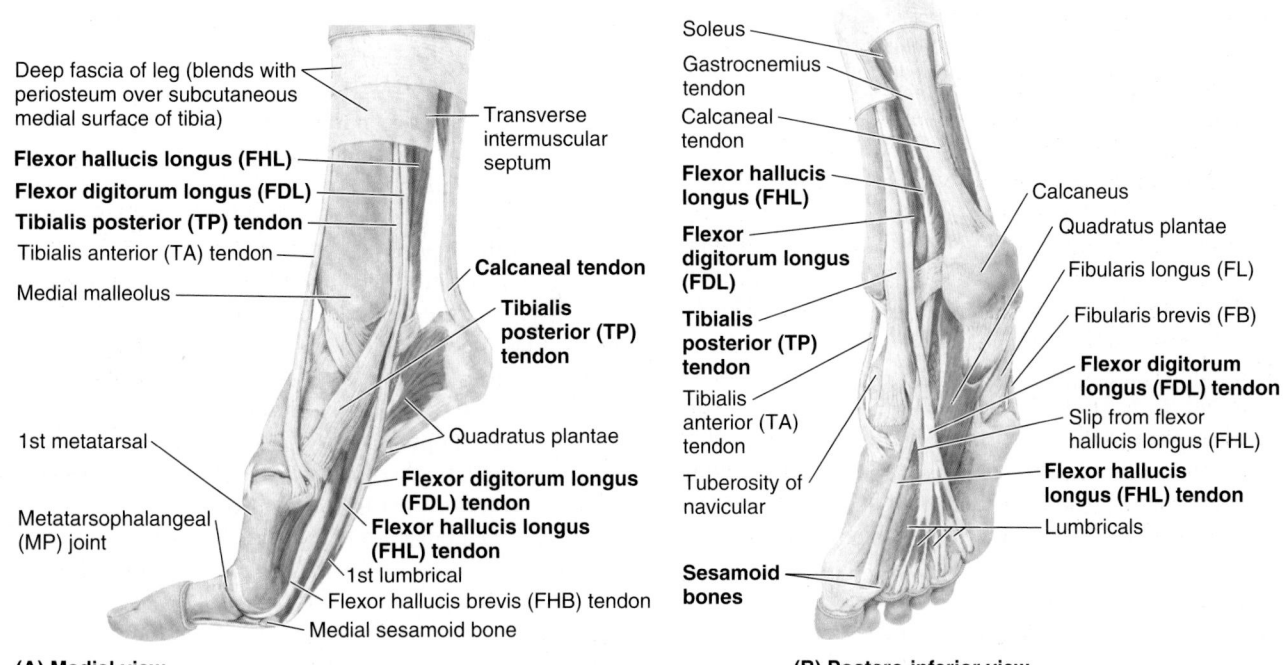

FIGURE 5.63. Dissection demonstrating continuation of plantarflexor tendons. A. The foot is raised as in the push-off phase of walking, demonstrating the position of the plantarflexor tendons as they cross the ankle. Observe the sesamoid bone acting as a "foot stool" for the 1st metatarsal, giving it extra height and protecting the flexor hallucis longus tendon. **B.** This view demonstrates the disposition of the deep plantarflexor tendons in the sole of the foot.

the high point of the medial longitudinal arch of the foot) but has attachments to other tarsal and metatarsal bones.

The TP is traditionally described as an invertor of the foot. Indeed, when the foot is off the ground, it can act synergistically with the TA to invert the foot, their otherwise antagonistic functions canceling each other out. However, the primary role of the TP is to support or maintain (fix) the medial longitudinal arch during weight-bearing; consequently, the muscle contracts statically throughout the stance phase of gait (see Fig. 5.20A–E; Table 5.2; see also Fig. 5.103C & E). In so doing, it acts independently of the TA because, once the foot is flat on the ground after heel strike, that muscle is relaxed during the stance phase (the dorsiflexion that occurs as the body passes over the planted foot is passive), unless braking requires its eccentric contraction.

While standing (especially on one foot), however, the two muscles may cooperate to depress the lateral side of the foot and pull medially on the leg as needed to counteract lateral leaning for balance.

To test the TA, the foot is inverted against resistance with the foot in slight plantarflexion; if normal, the tendon can be seen and palpated posterior to the medial malleolus.

NERVES IN POSTERIOR COMPARTMENT

The *tibial nerve* (L4, L5, and S1–S3) is the larger of the two terminal branches of the *sciatic nerve* (Fig. 5.57B; Table 5.11). It runs vertically through the popliteal fossa with the popliteal artery, passing between the heads of the gastrocnemius, the two structures exiting the fossa by passing deep to the tendinous arch of the soleus (Fig. 5.61B).

The tibial nerve supplies all muscles in the posterior compartment of the leg (Figs. 5.53A and 5.61B; Table 5.11). At the ankle, the nerve lies between the tendons of the FHL and the FDL. Postero-inferior to the medial malleolus, the tibial nerve divides into the medial and lateral plantar nerves. A branch of the tibial nerve, the *medial sural cutaneous nerve*, is usually joined by the *sural communicating branch of the common fibular nerve* to form the *sural nerve* (see Figs. 5.49B, 5.50, and 5.57). This nerve supplies the skin of the lateral and posterior part of the inferior third of the leg and the lateral side of the foot. Articular branches of the tibial nerve supply the knee joint, and medial calcaneal branches supply the skin of the heel.

ARTERIES IN POSTERIOR COMPARTMENT

The **posterior tibial artery,** the larger and more direct terminal branch of the *popliteal artery*, provides the blood supply to the posterior compartment of the leg and to the foot (Figs. 5.53A, 5.58, 5.61B, and 5.64; Table 5.12). It begins at the distal border of the popliteus, as the popliteal artery passes deep to the tendinous arch of the soleus and simultaneously bifurcates into its terminal branches. Close to its origin, the posterior tibial artery gives rise to its largest branch, the *fibular artery*, which runs lateral and parallel to it, also within the deep subcompartment.

During its descent, the posterior tibial artery is accompanied by the tibial nerve and veins. The artery runs posterior to

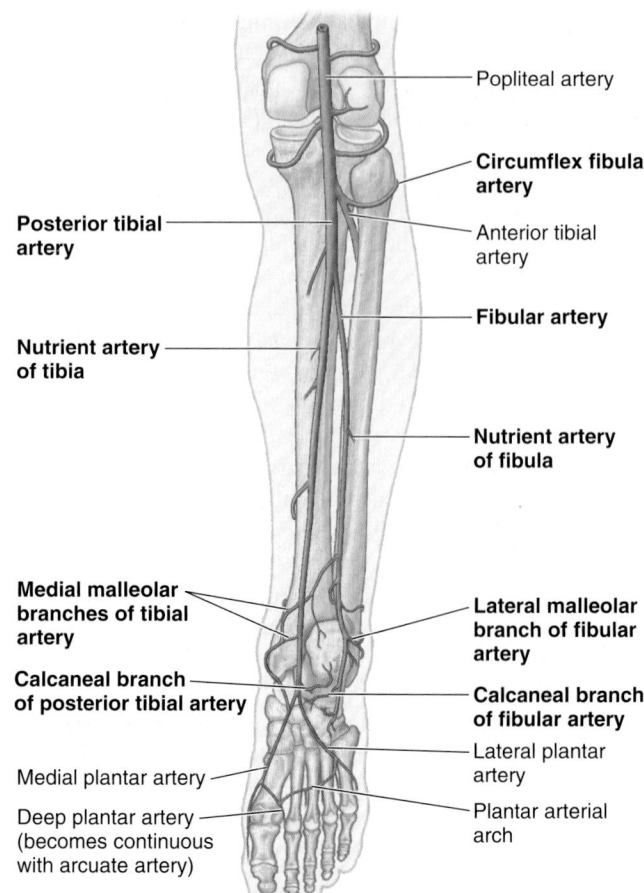

Posterior view with foot plantar flexed

FIGURE 5.64. Arteries of knee, posterior leg, and sole of foot.

the medial malleolus, from which it is separated by the tendons of the TP and FDL (Fig. 5.61B). Inferior to the medial malleolus, it runs between the tendons of the FHL and FDL. Deep to the flexor retinaculum and the origin of the abductor hallucis, the posterior tibial artery divides into *medial and lateral plantar arteries*, the arteries of the sole of the foot.

The **fibular** (peroneal) **artery,** the largest and *most important branch of the tibial artery*, arises inferior to the distal border of the popliteus and the tendinous arch of the soleus (Figs. 5.58A, 5.61B, and 5.64; Table 5.12). It descends obliquely toward the fibula and passes along its medial side, usually within the FHL. The fibular artery gives muscular branches to the popliteus and other muscles in both the posterior and the lateral compartments of the leg. It also gives rise to the **nutrient artery of the fibula** (Fig. 5.64).

Distally, the fibular artery gives rise to a perforating branch and terminal lateral malleolar and calcaneal branches. The perforating branch pierces the *interosseous membrane* and passes to the dorsum of the foot, where it anastomoses with the arcuate artery. The *lateral calcaneal branches* supply the heel, and the *lateral malleolar branch* joins other malleolar branches to form a peri-articular *arterial anastomosis of the ankle*.

The **circumflex fibular artery** arises from the origin of the anterior or posterior tibial artery at the knee and passes laterally over the neck of the fibula to the anastomoses around the knee.

The **nutrient artery of tibia,** the largest nutrient artery in the body, arises from the origin of the anterior or posterior tibial artery. It pierces the tibialis posterior, to which it supplies branches, and enters the nutrient foramen in the proximal third of the posterior surface of the tibia (see Fig. 5.9).

Surface Anatomy of Leg

The *tibial tuberosity* is an easily palpable elevation on the anterior aspect of the proximal part of the tibia, approximately 5 cm distal to the apex of the patella (Fig. 5.65A & B). This oval elevation indicates the level of the head of the fibula and the bifurcation of the popliteal artery into the anterior and posterior tibial arteries.

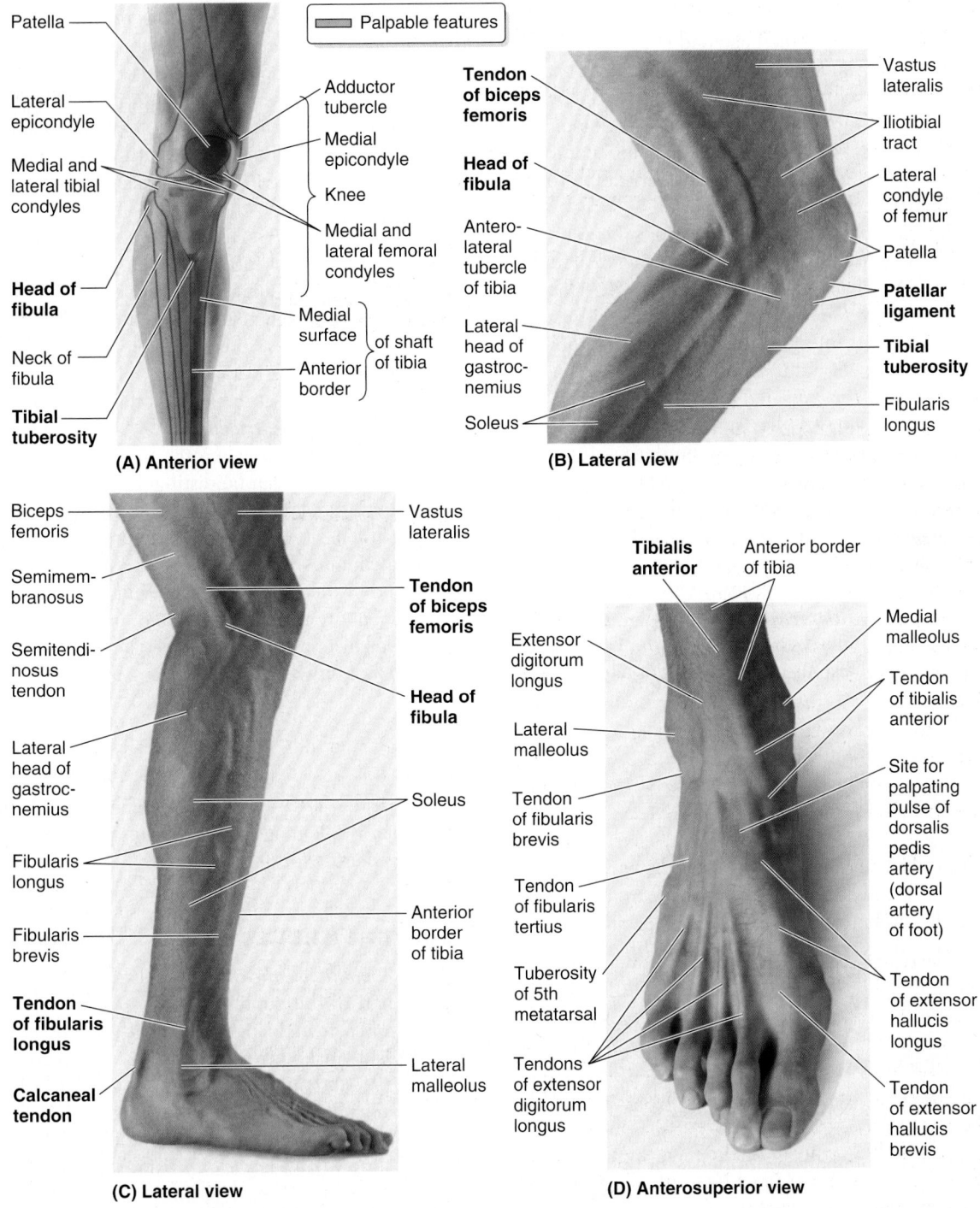

FIGURE 5.65. Surface anatomy of the leg. A. Standing at ease. **B and C.** Knee is flexed while weight-bearing. **D.** Extensors and flexors of toes are being contracted simultaneously, demonstrating extensor tendons without elevating toes from ground.

The *patellar ligament* may be felt as it extends from the inferior border of the apex of the patella. It is most easily felt when the knee is extended. When the knee flexes to a right angle, a depression may be felt on each side of the patellar ligament. The joint cavity is superficial in these depressions.

The *head of the fibula* is subcutaneous and may be palpated at the posterolateral aspect of the knee, at the level of the tibial tuberosity (Fig. 5.65B & C). The *neck of the fibula* can be palpated just distal to the head.

The *tendon of the biceps femoris* may be traced by palpating its distal attachment to the lateral side of the head of the fibula. This tendon and the head and neck of the fibula guide the examining finger to the *common fibular nerve* (Fig. 5.61A). The nerve is indicated by a line along the biceps femoris tendon, posterior to the head of the fibula, and around the lateral aspect of the fibular neck to its anterior aspect, just distal to the fibular head. Here it can be rolled against the fibular neck with the fingertips.

The *anterior border of the tibia* is sharp, subcutaneous, and easily followed distally by palpation from the tibial tuberosity to the medial malleolus (Fig. 5.65A–D). The *medial surface of the shaft of the tibia* is also subcutaneous, except at its proximal end. Its inferior third is crossed obliquely by the great saphenous vein as it passes proximally to the medial aspect of the knee.

The *tibialis anterior (TA)* lies superficially and is easily palpable just lateral to the anterior border of the tibia (Fig. 5.65D). As the foot is inverted and dorsiflexed, the large *tendon of the TA* can be seen and palpated as it runs distally and slightly medially over the anterior surface of the ankle joint to the medial side of the foot. If the 1st digit is dorsiflexed, the *tendon of the EHL* can be palpated just lateral to the tendon of TA. The *tendon of the EHB* may also be visible.

As the toes are extended, the *tendons of the EDL* can be palpated lateral to the extensor hallucis longus and followed to the four lateral digits. The *tendon of the FT* may be palpable lateral to the tendons of the EDL, especially when the foot is dorsiflexed and everted.

The *shaft of the fibula* is subcutaneous only in its distal part, proximal to the lateral malleolus; this is the common site of fractures. The *medial* and *lateral malleoli* are subcutaneous and prominent. Palpate them, noting that the tip of the lateral malleolus extends farther distally and posteriorly than the medial malleolus.

The *fibularis longus (FL)* is subcutaneous throughout its course (Fig. 5.65C). The tendons of this muscle and the *fibularis brevis (FB)* are palpable when the foot is everted as they pass around the posterior aspect of the lateral malleolus. These tendons may be followed anteriorly along the lateral side of the foot. The tendon of the FL runs as far anteriorly as the cuboid and then disappears by turning into the sole of the foot. The *tendon of the FB* may be traced to its attachment to the base of the 5th metatarsal.

The *calcaneal tendon* can be easily followed to its attachment to the calcaneal tuberosity, the posterior part of the calcaneus. The ankle joint is fairly superficial in the depression on each side of the calcaneal tendon. The *heads of the gastrocnemius* are easily recognizable in the superior part of the calf of the leg (Fig. 5.65B & C). The *soleus* can be palpated deep to and at the sides of the superior part of the calcaneal tendon. The *triceps surae* (soleus and gastrocnemius) is easy to palpate when the individual is standing on the toes. The soleus can be distinguished from the gastrocnemius during squatting (flexing the knees while standing on toes) because flexion of the knee to approximately 90° makes the gastrocnemius flaccid; plantarflexion in this position is maintained by the soleus. The deep muscles of the posterior compartment are not easily palpated, but their tendons can be observed just posterior to the medial malleolus, especially when the foot is inverted and the toes are flexed.

POPLITEAL FOSSA AND LEG

Popliteal Abscess and Tumor

 Because the deep popliteal fascia is strong and limits expansion, pain from an abscess or tumor in the popliteal fossa is usually severe. *Popliteal abscesses* tend to spread superiorly and inferiorly because of the toughness of the popliteal fascia.

Popliteal Pulse

 Because the popliteal artery is deep, it may be difficult to feel the *popliteal pulse*. Palpation of this pulse is commonly performed with the person in the prone position with the knee flexed to relax the popliteal fascia and hamstrings. The pulsations are best felt in the inferior part of the fossa where the popliteal artery is related to the tibia. Weakening or loss of the popliteal pulse is a sign of a femoral artery obstruction.

Popliteal Aneurysm and Hemorrhage

 A *popliteal aneurysm* (abnormal dilation of all or part of the popliteal artery) usually causes edema and pain in the popliteal fossa. A popliteal aneurysm may be distinguished from other masses by palpable pulsations (*thrills*) and abnormal arterial sounds (*bruits*) detectable with a stethoscope. Because the artery lies deep to the tibial nerve, an aneurysm may stretch the nerve or compress its blood supply (*vasa vasorum*). Pain from such nerve compression is usually referred, in this case to the skin overlying the medial aspect of the calf, ankle, or foot.

Because the popliteal artery is closely applied to the popliteal surface of the femur and the joint capsule (Fig. 5.59), fractures of the distal femur or dislocations of the knee may rupture the artery, resulting in hemorrhage. Furthermore, because of their proximity and confinement within the fossa, an injury of the artery and vein may result in an *arteriovenous fistula* (communication between an artery and a vein). Failure to recognize these occurrences and to act promptly may result in the loss of the leg and foot.

If the femoral artery must be ligated, blood can bypass the occlusion through the genicular anastomosis and reach the popliteal artery distal to the ligation (see Fig. 5.52).

Injury to Tibial Nerve

Injury to the tibial nerve is uncommon because of its deep and protected position in the popliteal fossa; however, the nerve may be injured by deep lacerations in the fossa. *Posterior dislocation of the knee joint* may also damage the tibial nerve. *Severance of the tibial nerve* produces paralysis of the flexor muscles in the leg and the intrinsic muscles in the sole of the foot. People with a tibial nerve injury are unable to plantarflex their ankle or flex their toes. Loss of sensation also occurs on the sole of the foot.

Containment and Spread of Compartmental Infections in the Leg

The fascial compartments of the lower limbs are generally closed spaces, ending proximally and distally at the joints. Because the septa and deep fascia of the leg forming the boundaries of the leg compartments are strong, the increased volume consequent to infection with *suppuration* (formation of pus) increases intracompartmental pressure. Inflammations within the anterior and posterior compartments of the leg spread chiefly in a distal direction; however, a purulent (pus-forming) infection in the lateral compartment of the leg can ascend proximally into the popliteal fossa, presumably along the course of the fibular nerve. *Fasciotomy* (incision of fascia) may be necessary to relieve pressure and debride (scrape away) pockets of infection.

Tibialis Anterior Strain (Shin Splints)

Shin splints—edema and pain in the area of the distal two thirds of the tibia—result from repetitive microtrauma of the tibialis anterior (TA; Fig. 5.56A) which causes small tears in the periosteum covering the shaft of the tibia and/or of fleshy attachments to the overlying deep fascia of the leg. Shin splints are a mild form of the anterior compartment syndrome. Shin splints commonly occur during traumatic injury or athletic overexertion of muscles in the anterior compartment, especially TA, by untrained persons.

Often persons who lead sedentary lives develop shin splints when they participate in long-distance walks.

Shin splints also occur in trained runners who do not warm up and cool down sufficiently. Muscles in the anterior compartment swell from sudden overuse, and the edema and muscle–tendon inflammation reduce the blood flow to the muscles. The swollen muscles are painful and tender to pressure.

Fibularis Muscles and Evolution of the Human Foot

Whereas the feet of anthropoids (higher primates) are inverted so that they walk on the outer border of the foot, the feet of humans are relatively everted (pronated) so that the soles lie more fully on the ground. This pronation is the result, at least in part, of the medial migration of the distal attachment of the fibularis longus across the sole of the foot (Fig. 5.60K), and the development of a fibularis tertius that is attached to the base of the 5th metatarsal. These features are unique to the human foot.

Injury to Common Fibular Nerve and Footdrop

Because of its superficial position, *the common fibular is the nerve most often injured in the lower limb*, mainly because it winds subcutaneously around the fibular neck, leaving it vulnerable to direct trauma (Fig. 5.57A). This nerve may also be severed during fracture of the fibular neck, or severely stretched when the knee joint is injured or dislocated. *Severance of the common fibular nerve* results in flaccid paralysis of all muscles in the anterior and lateral compartments of the leg (dorsiflexors of ankle and evertors of foot). The loss of dorsiflexion of the ankle causes *footdrop*, which is further exacerbated by unopposed inversion of the foot. This has the effect of making the limb "too long": The toes do not clear the ground during the swing phase of walking (Fig. B5.20A).

There are several other conditions that may result in a lower limb that is "too long" functionally, for example, pelvic tilt (see Fig. B5.18C) and spastic paralysis or contraction of the soleus. There are at least three means of compensating for this problem:

1. A *waddling gait*, in which the individual leans to the side opposite the long limb, "hiking" the hip (Fig. B5.20B).
2. A *swing-out gait*, in which the long limb is swung out laterally (abducted) to allow the toes to clear the ground (Fig. B5.20C).
3. A high-stepping *steppage gait*, in which extra flexion is employed at the hip and knee to raise the foot as high as necessary to keep the toes from hitting the ground (Fig. B5.20D).

Because the dropped foot makes it difficult to make the heel strike the ground first as in a normal gait, a steppage gait is commonly employed in the case of flaccid paralysis. Sometimes an extra "kick" is added as the free limb swings

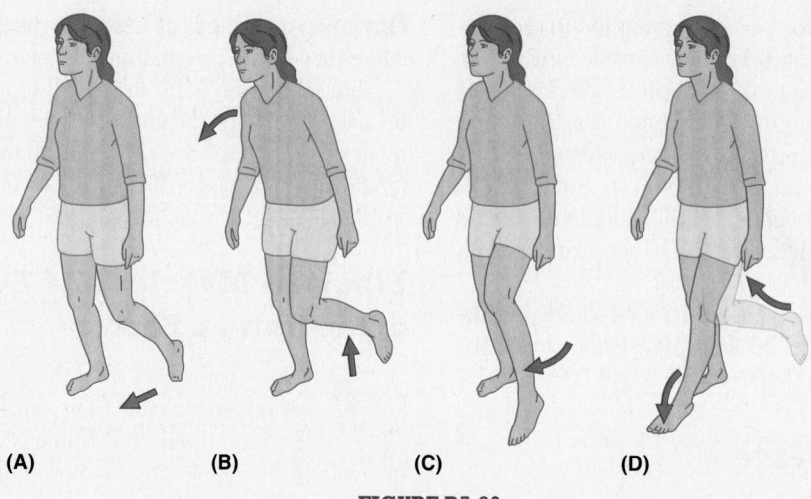

(A)   (B)   (C)   (D)

FIGURE B5.20.

forward in an attempt to flip the forefoot upward just before setting the foot down.

The braking action normally produced by eccentric contraction of the dorsiflexors is also lost in flaccid paralysis footdrop. Therefore, the foot is not lowered to the ground in a controlled manner after heel strike; instead the foot slaps the ground suddenly, producing a distinctive "*clop*," and greatly increasing the shock both received by the forefoot and transmitted up the tibia to the knee. Individuals with a common fibular nerve injury may also experience a variable loss of sensation on the anterolateral aspect of the leg and the dorsum of the foot.

Deep Fibular Nerve Entrapment

Excessive use of muscles supplied by the deep fibular nerve (e.g., during skiing, running, and dancing) may result in muscle injury and edema in the anterior compartment. This entrapment may cause compression of the deep fibular nerve and pain in the anterior compartment.

Compression of the nerve by tight-fitting ski boots, for example, may occur where the nerve passes deep to the inferior extensor retinaculum and the extensor hallucis brevis (see Fig 5.54A). Pain occurs in the dorsum of the foot and usually radiates to the web space between the 1st and 2nd toes. Because ski boots are a common cause of this type of nerve entrapment, this condition has been called the "ski boot syndrome"; however, the syndrome also occurs in soccer players and runners and can also result from tight shoes.

Superficial Fibular Nerve Entrapment

Chronic ankle sprains may produce recurrent stretching of the superficial fibular nerve, which may cause pain along the lateral side of the leg and the dorsum of the ankle and foot. Numbness and *paresthesia* (tickling or tingling) may be present and increase with activity.

Fabella in Gastrocnemius

Close to its proximal attachment, the lateral head of the gastrocnemius may contain a sesamoid bone, the **fabella** (L., bean), which articulates with the lateral femoral condyle. The fabella is visible in lateral radiographs of the knee in 3–5% of people (Fig. B5.21).

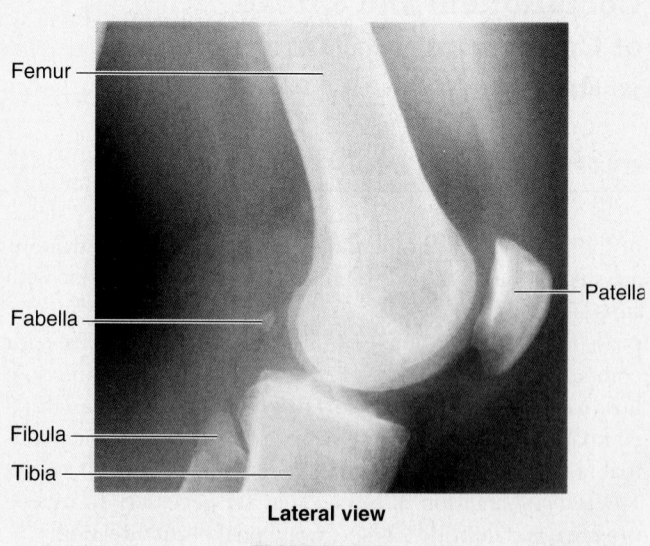

Femur
Fabella
Fibula
Tibia
Patella

Lateral view

FIGURE B5.21.

Calcaneal Tendinitis

Inflammation of the calcaneal tendon constitutes 9–18% of running injuries. Microscopic tears of collagen fibers in the tendon, particularly just superior to its attachment to the calcaneus, result in *tendinitis*, which causes pain during walking, especially when wearing rigid-soled shoes. Calcaneal tendinitis often occurs during repetitive activities, especially in individuals who take up running after prolonged inactivity, or suddenly increase the intensity of their training, but it may also result from poor footwear or training surfaces.

Ruptured Calcaneal Tendon

Rupture of the calcaneal tendon is often sustained by poorly conditioned people with a history of *calcaneal tendinitis*. The injury is typically experienced as an audible snap during a forceful push off (plantarflexion with the knee extended) followed immediately by sudden calf pain and sudden dorsiflexion of the plantarflexed foot. In a completely ruptured tendon, a gap is palpable, usually 1–5 cm proximal to the calcaneal attachment. The muscles affected are the gastrocnemius, soleus, and plantaris.

Calcaneal tendon rupture is probably the most severe acute muscular problem of the leg. Individuals with this injury cannot plantarflex against resistance (cannot raise the heel from the ground or balance on the affected side), and passive dorsiflexion (usually limited to 20° from neutral) is excessive.

Ambulation (walking) is possible only when the limb is laterally (externally) rotated, rolling over the transversely placed foot during the stance phase without push off. Bruising appears in the malleolar region, and a lump usually appears in the calf owing to shortening of the triceps surae. In older or non-athletic people, non-surgical repairs are often adequate, but surgical intervention is usually advised for those with active lifestyles, such as tennis players.

Calcaneal Tendon Reflex

The ankle jerk reflex, or triceps surae reflex, is a *calcaneal tendon reflex*. It is a myotatic reflex elicited while the person's legs are dangling over the side of the examining table. The calcaneal tendon is struck briskly with a reflex hammer just proximal to the calcaneus (Fig. B5.22). The normal result is plantarflexion of the ankle joint. *The calcaneal tendon reflex tests the S1 and S2 nerve roots.* If the S1 nerve root is injured or compressed, the ankle reflex is virtually absent.

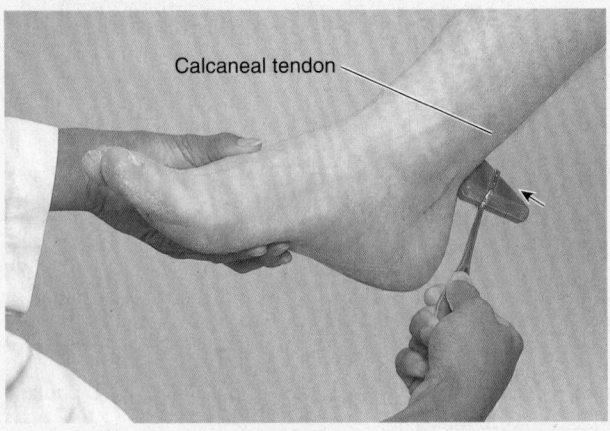

Calcaneal tendon

FIGURE B5.22.

Absence of Plantarflexion

If the muscles of the calf are paralyzed, the calcaneal tendon is ruptured, or normal push off is painful, a much less effective and efficient push off (from the midfoot) can still be accomplished by the actions of the gluteus maximus and hamstrings in extending the thigh at the hip joint and the quadriceps in extending the knee. Because push off from the forefoot is not possible (in fact, the ankle will be passively dorsiflexed as the body's weight moves anterior to the foot), those attempting to walk in the absence of plantarflexion often rotate the foot as far laterally (externally) as possible during the stance phase to disable passive dorsiflexion and allow a more effective push off through hip and knee extension exerted at the midfoot.

Gastrocnemius Strain

Gastrocnemius strain (tennis leg) is a painful acute injury resulting from partial tearing of the medial belly of the gastrocnemius at or near its musculotendinous junction, often seen in individuals older than 40 years of age. It is caused by overstretching the muscle by concomitant full extension of the knee and dorsiflexion of the ankle joint. Usually, an abrupt onset of stabbing pain is followed by edema and spasm of the gastrocnemius.

Calcaneal Bursitis

Calcaneal bursitis (retro-achilles bursitis) results from *inflammation of the deep bursa of the calcaneal tendon,* located between the calcaneal tendon and the superior part of the posterior surface of the calcaneus (Fig. B5.23). Calcaneal bursitis causes pain posterior to the heel, and occurs commonly during long-distance running, basketball, and tennis. It is caused by excessive friction on the bursa as the tendon continuously slides over it.

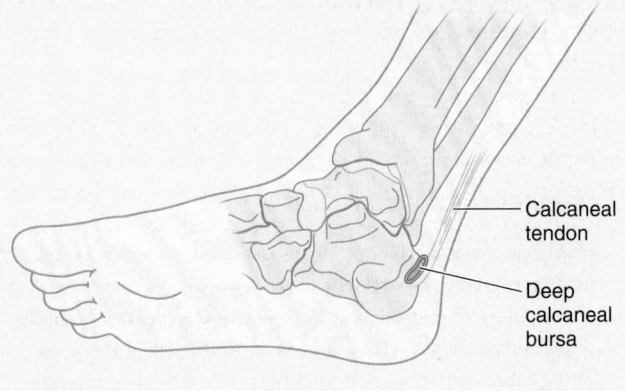

Calcaneal tendon

Deep calcaneal bursa

FIGURE B5.23.

Venous Return From Leg

A venous plexus deep to the triceps surae is involved in the return of blood from the leg. When a person is standing, the venous return from the leg depends largely on the muscular activity of the triceps surae (see "Venous Drainage of Lower Limb," p. 532). Contraction of the calf muscles pumps blood superiorly in the deep veins. The *musculovenous pump* is improved by the deep fascia that invests the muscles like an elastic stocking (p. 535).

Accessory Soleus

An accessory soleus is present in approximately 3% of people (Fig. B5.24). The accessory muscle usually appears as a distal belly medial to the calcaneal tendon. Clinically, an accessory soleus may be associated with pain and edema (swelling) during prolonged exercise.

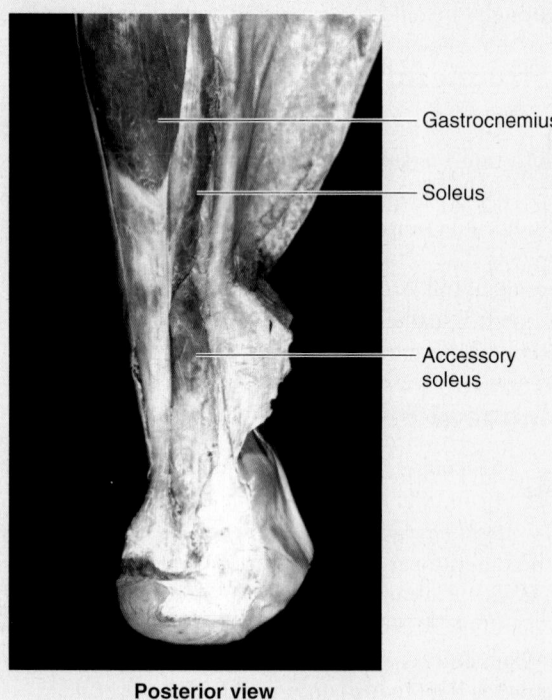

Gastrocnemius

Soleus

Accessory soleus

Posterior view

FIGURE B5.24.

Posterior Tibial Pulse

The *posterior tibial pulse* can usually be palpated between the posterior surface of the medial malleolus and the medial border of the calcaneal tendon (Fig. B5.25). Because the posterior tibial artery passes deep to the flexor retinaculum, it is important when palpating this pulse to have the person invert the foot to relax the retinaculum. Failure to do so may lead to the erroneous conclusion that a pulse is absent.

Both arteries are examined simultaneously for equality of force. Palpation of the posterior tibial pulses is essential for examining patients with *occlusive peripheral arterial disease.* Although posterior tibial pulses are absent in approximately 15% of normal young people, absence of posterior tibial pulses is a sign of occlusive peripheral arterial disease in people older than 60 years. For example, *intermittent claudication,* characterized by leg pain and cramps, develops during walking and disappears after rest. These conditions result from ischemia of the leg muscles caused by narrowing or occlusion of the leg arteries.

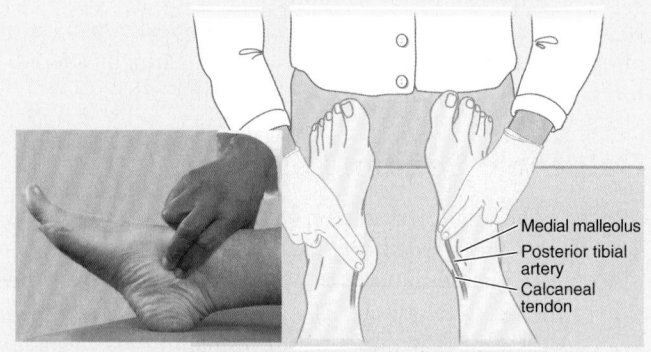

Medial malleolus
Posterior tibial artery
Calcaneal tendon

FIGURE B5.25.

The Bottom Line

POPLITEAL FOSSA AND LEG

Popliteal fossa: The popliteal fossa is a fat-filled and relatively confined compartment posterior to the knee that is traversed by all neurovascular structures passing between the thigh and the leg. ♦ The sciatic nerve bifurcates at the apex of the fossa, with the common fibular nerve passing laterally along the biceps tendon. ♦ The tibial nerve, popliteal vein, and popliteal artery bisect the fossa—in that order, from superficial (posterior) to deep (anterior). ♦ Genicular branches of the popliteal artery form a peri-articular genicular anastomosis around the knee, providing collateral circulation to maintain blood flow in all positions of the knee.

Anterior compartment of leg: The anterior compartment, confined by mostly unyielding bones and membranes, is susceptible to compartment syndromes. ♦ The contained muscles are ankle dorsiflexors/toe extensors that are active in walking as they (1) concentrically contract to raise the forefoot to clear the ground during the swing phase of the gait cycle and (2) eccentrically contract to lower the forefoot to the ground after the heel strike of the stance phase. ♦ The deep fibular nerve and anterior tibial artery course within and supply the anterior compartment. ♦ Injury of the common or deep fibular nerve results in **footdrop.**

Lateral compartment of leg: The small lateral compartment contains the primary evertors of the foot and the superficial fibular nerve that supplies them. ♦ Because no artery courses within this compartment, perforating branches from the anterior tibial and fibular arteries (and their accompanying veins) penetrate the intermuscular septa to supply (and drain) blood. ♦ Eversion is used to support/depress the medial foot during the toe off of the stance phase, and to resist inadvertent inversion, preventing injury.

Posterior compartment of leg: The posterior or plantar-flexor compartment is subdivided by the transverse intermuscular septum into superficial and deep subcompartments. ♦ In the superficial subcompartment, the gastrocnemius and soleus muscles (triceps surae) share a common tendon (the calcaneal tendon, the body's strongest tendon). ♦ The triceps surae provides the power of plantarflexion that propels the body in walking, and plays a major role in running and jumping via push off. ♦ The deep muscles in the posterior compartment augment the plantar flexor action through flexion of the digits and support of the longitudinal arches of the foot. ♦ The contents of the posterior compartment are supplied by the tibial nerve and two arteries, the (medial) posterior tibial and fibular arteries. ♦ All three structures (tibial nerve and two arteries) course within the confined deep subcompartment, where swelling may have profound consequences for the entire posterior compartment, the distal lateral compartment, and the foot.

FOOT

The clinical importance of the *foot* is indicated by the considerable amount of time primary care physicians devote to foot problems. *Podiatry* is the specialized field that deals with the study and care of the feet.

The *ankle* refers to the narrowest and malleolar parts of the distal leg, proximal to the dorsum and heel of the foot, including the ankle joint. The foot, distal to the ankle, provides a platform for supporting the body when standing and has an important role in locomotion.

The *skeleton of the foot* consists of 7 *tarsals*, 5 *metatarsals*, and 14 *phalanges* (Fig. 5.66). The foot and its bones may be considered in terms of three anatomical and functional zones (see Fig. 5.11C, p. 523):

- The **hindfoot:** talus and calcaneus.
- The **midfoot:** navicular, cuboid, and cuneiforms.
- The **forefoot:** metatarsals and phalanges.

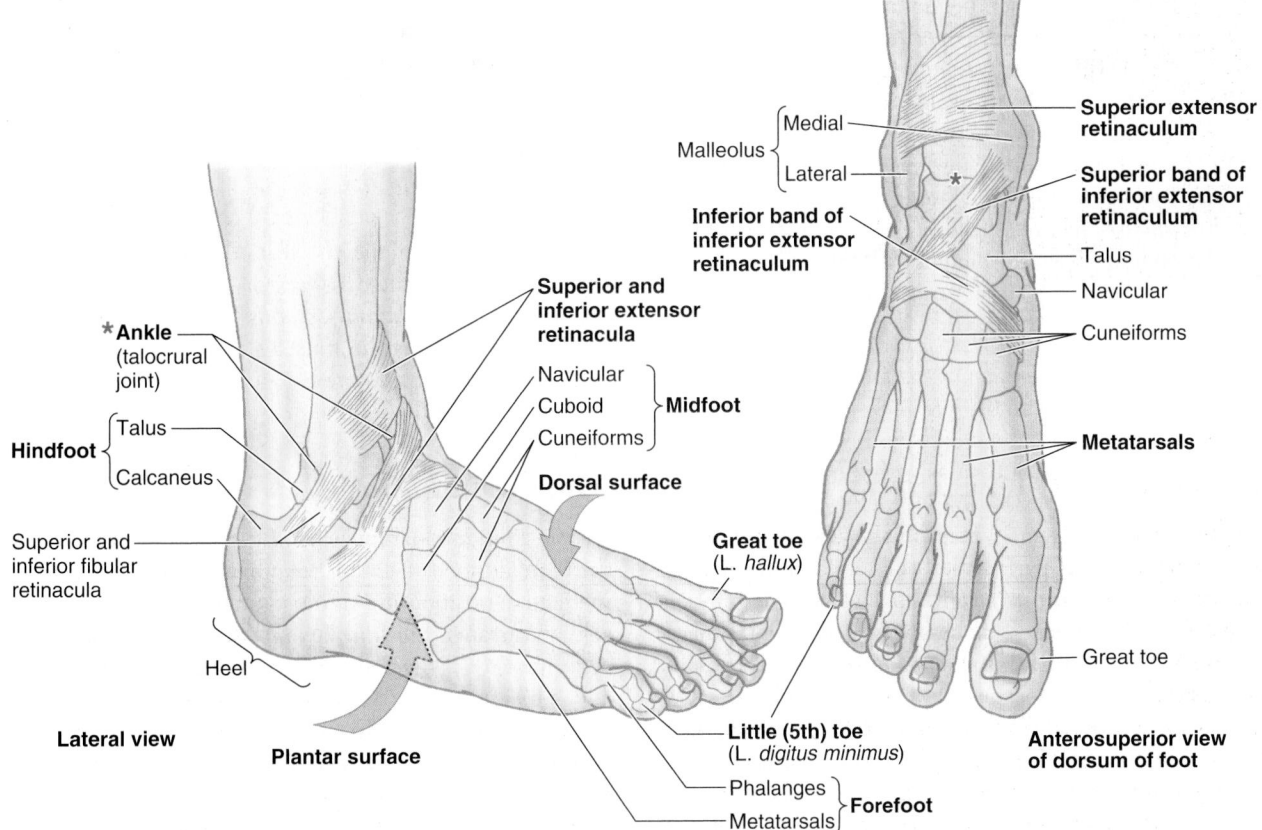

FIGURE 5.66. Surfaces, parts, bones, and retinacula of ankle and foot. The disposition of the bones of the foot and the superior and inferior extensor and fibular retinacula relative to surface features are demonstrated.

The part/region of the foot contacting the floor or ground is the **sole** (*plantar region*, L. *regio plantaris*). The part directed superiorly is the *dorsum of the foot* (L. *dorsum pedis*), or **dorsal region of the foot** (L. *regio dorsalis pedis*). The sole of the foot underlying the calcaneus is the *heel* or **heel region** (L. *regio calcanea*), and the sole underlying the heads of the medial two metatarsals is the **ball of the foot.** The **great toe** (L. *hallux*) is also the **1st toe** (digit of foot, L. *digitus primus*), and the **little toe** (L. *digitus minimus*) is also the **5th toe.**

Skin and Fascia of Foot

Marked variations occur in the thickness (strength) and texture of skin, subcutaneous tissue (superficial fascia), and deep fascia in relationship to weight-bearing and distribution, ground contact (grip, abrasion), and the need for containment or compartmentalization.

SKIN AND SUBCUTANEOUS TISSUE

The *skin of the dorsum of the foot* is much thinner and less sensitive than skin on most of the sole. The *subcutaneous tissue* is loose deep to the dorsal skin; therefore, *edema* (swelling) is most marked over this surface, especially anterior to and around the medial malleolus. The skin over the major weight-bearing areas of the sole—the heel, lateral margin, and ball of the foot—is thick. The subcutaneous tissue in the sole is more fibrous than in other areas of the foot.

Fibrous septa—highly developed *skin ligaments* (L., retinacula cutis)—divide this tissue into fat-filled areas, making it a shock-absorbing pad, especially over the heel. The skin ligaments also anchor the skin to the underlying deep fascia (plantar aponeurosis), improving the "grip" of the sole. The skin of the sole is hairless and sweat glands are numerous; the entire sole is sensitive ("ticklish"), especially the thinnerskinned area underlying the arch of the foot.

DEEP FASCIA OF FOOT

The *deep fascia of the dorsum of the foot* is thin where it is continuous proximally with the *inferior extensor retinaculum* (Fig. 5.67A). Over the lateral and posterior aspects of the foot, the deep fascia is continuous with the **plantar fascia,** the deep fascia of the sole (Fig. 5.67B & C). The plantar fascia has a thick central part and weaker medial and lateral parts.

The thick, central part of the plantar fascia forms the strong *plantar aponeurosis,* longitudinally arranged bundles of dense fibrous connective tissue investing the central plantar muscles. It resembles the palmar aponeurosis of the palm of the hand but is tougher, denser, and elongated.

The plantar fascia holds the parts of the foot together, helps protect the sole from injury, and helps to support the longitudinal arches of the foot.

The **plantar aponeurosis** arises posteriorly from the calcaneus and functions like a superficial ligament. Distally, the longitudinal bundles of collagen fibers of the aponeurosis divide into five bands that become continuous with the **fibrous digital sheaths** that enclose the flexor tendons that pass to the toes. At the anterior end of the sole, inferior to the heads of the metatarsals, the aponeurosis is reinforced by transverse fibers forming the **superficial transverse metatarsal ligament.**

In the midfoot and forefoot, vertical *intermuscular septa* extend deeply (superiorly) from the margins of the plantar aponeurosis toward the 1st and 5th metatarsals, forming the three *compartments of the sole* (Fig. 5.67C):

1. The **medial compartment of the sole** is covered superficially by thinner *medial plantar fascia.* It contains the abductor hallucis, flexor hallucis brevis, the tendon of the flexor hallucis longus, and the medial plantar nerve and vessels.
2. The **central compartment of the sole** is covered superficially by the dense *plantar aponeurosis.* It contains the flexor digitorum brevis, the tendons of the flexor hallucis longus and flexor digitorum longus, plus the muscles associated with the latter, the quadratus plantae and lumbricals, and the adductor hallucis. The lateral plantar nerve and vessels are also located here.
3. The **lateral compartment of the sole** is covered superficially by the thinner *lateral plantar fascia* and contains the abductor and flexor digiti minimi brevis.

In the forefoot only, a fourth compartment, the **interosseous compartment of the foot,** is surrounded by the plantar and dorsal interosseous fascias. It contains the metatarsals, the dorsal and plantar interosseous muscles, and the deep plantar and metatarsal vessels. Whereas the plantar interossei and plantar metatarsal vessels are distinctly plantar in position, the remaining structures of the compartment are located intermediate between the plantar and dorsal aspects of the foot.

A fifth compartment, the **dorsal compartment of the foot,** lies between the dorsal fascia of the foot and the tarsal bones and the dorsal interosseous fascia of the midfoot and forefoot. It contains the muscles (extensors hallucis brevis and extensor digitorum brevis) and neurovascular structures of the dorsum of the foot.

Muscles of Foot

Of the 20 individual muscles of the foot, 14 are located on the plantar aspect, 2 are on the dorsal aspect, and 4 are intermediate in position. From the plantar aspect, muscles of the sole are arranged in four layers within four compartments. The muscles of the foot are illustrated in Figures 5.68A–J and 5.69; their attachments, innervation, and actions are described in Table 5.14.

Despite their compartmental and layered arrangement, the plantar muscles function primarily as a group during the support phase of stance, maintaining the arches of the foot (see Fig 5.20B–E; Table 5.2). They basically resist forces that tend to reduce the longitudinal arch as weight is received at

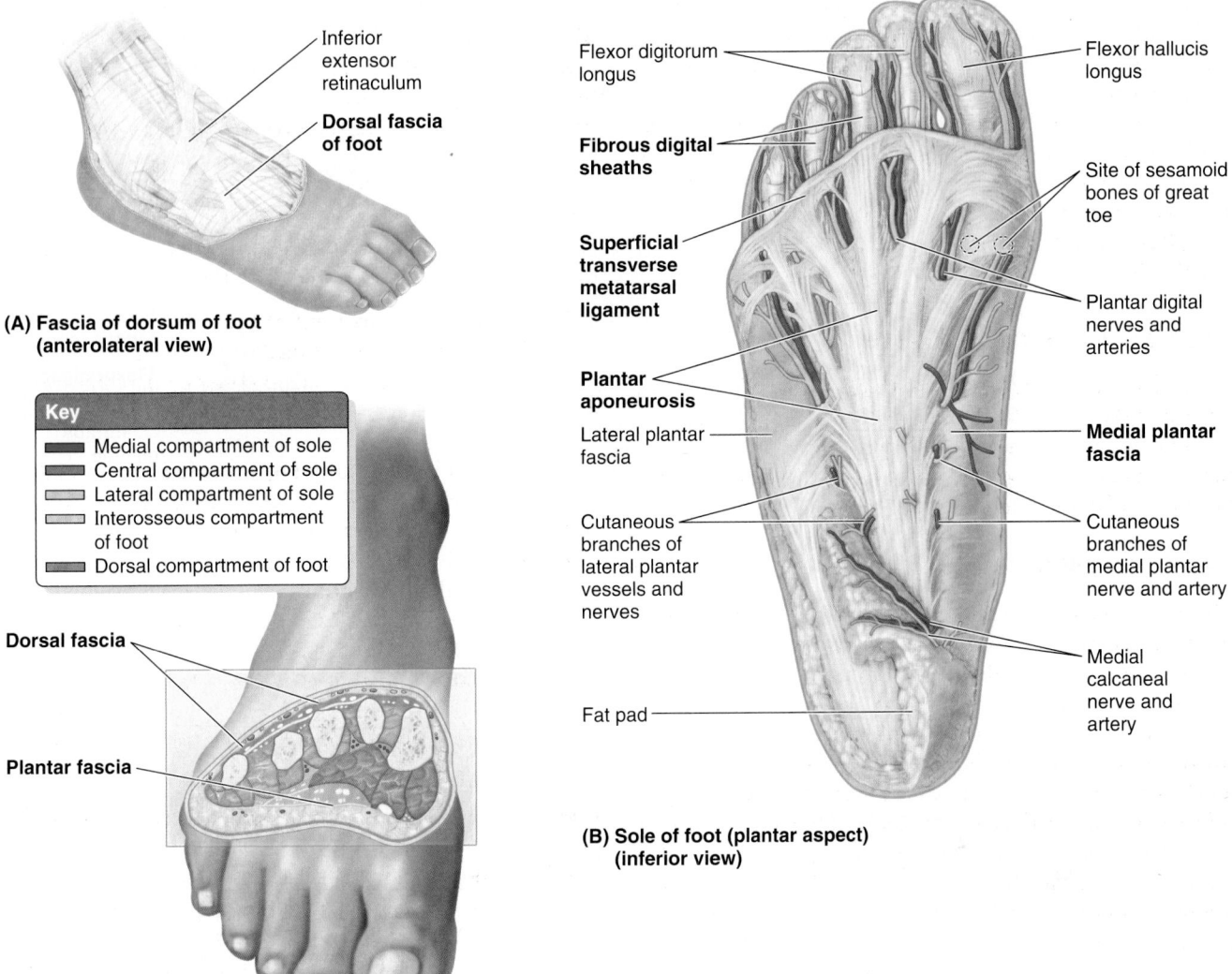

(A) Fascia of dorsum of foot (anterolateral view)

Inferior extensor retinaculum

Dorsal fascia of foot

Key
- Medial compartment of sole
- Central compartment of sole
- Lateral compartment of sole
- Interosseous compartment of foot
- Dorsal compartment of foot

Dorsal fascia

Plantar fascia

(C) Transverse section of foot compartments (anterior view)

Flexor digitorum longus

Fibrous digital sheaths

Superficial transverse metatarsal ligament

Plantar aponeurosis

Lateral plantar fascia

Cutaneous branches of lateral plantar vessels and nerves

Fat pad

Flexor hallucis longus

Site of sesamoid bones of great toe

Plantar digital nerves and arteries

Medial plantar fascia

Cutaneous branches of medial plantar nerve and artery

Medial calcaneal nerve and artery

(B) Sole of foot (plantar aspect) (inferior view)

FIGURE 5.67. Fascia and compartments of foot. A. The skin and subcutaneous tissue have been removed to demonstrate the deep fascia of the leg and dorsum of the foot. **B.** The deep plantar fascia consists of the thick plantar aponeurosis and the thinner medial and lateral plantar fascia. Thinner parts of the plantar fascia have been removed, revealing the plantar digital vessels and nerves. **C.** The bones and muscles of the foot are surrounded by the deep dorsal and plantar fascia. A large central and smaller medial and lateral compartments of the sole are created by intermuscular septa that extend deeply from the plantar aponeurosis.

the heel (posterior end of the arch) and then transferred to the ball of the foot and great toe (anterior end of the arch).

The muscles become most active in the later portion of the movement to stabilize the foot for propulsion (push off), a time when forces also tend to flatten the foot's transverse arch. Concurrently, they are also able to refine further the efforts of the long muscles, producing supination and pronation in enabling the platform of the foot to adjust to uneven ground.

The muscles of the foot are of little importance individually because fine control of the individual toes is not important to most people. Rather than producing actual movement, they are most active in fixing the foot or in increasing the pressure applied against the ground by various aspects of the sole or toes to maintain balance.

Although the adductor hallucis resembles a similar muscle of the palm that adducts the thumb, despite its name the adductor hallucis is probably most active during the push-off phase of stance in pulling the lateral four metatarsals toward the great toe, fixing the transverse arch of the foot, and resisting forces that would spread the metatarsal heads as weight and force are applied to the forefoot (Table 5.2).

In Table 5.14, note that the:

- **P**lantar interossei **AD**duct (**PAD**) and arise from a single metatarsal as unipennate muscles.
- **D**orsal interossei **AB**duct (**DAB**) and arise from two metatarsals as bipennate muscles.

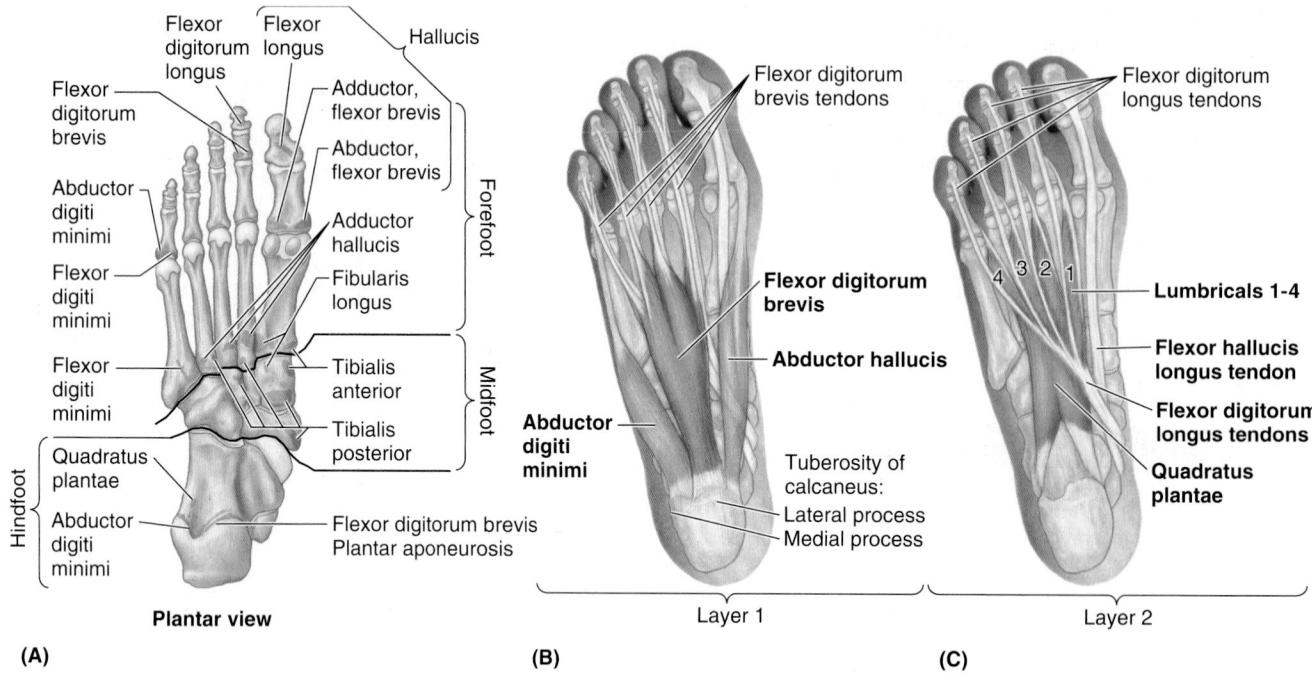

FIGURE 5.68A–C. Muscles of foot: 1st and 2nd layers of sole.

TABLE 5.14.I. MUSCLES OF FOOT: 1ST AND 2ND LAYERS OF SOLE

Muscle	Proximal Attachment	Distal Attachment	Innervation[a]	Main Action[b]
1st layer				
Abductor hallucis	Medial tubercle of tuberosity of calcaneus; flexor retinaculum; plantar aponeurosis	Medial side of base of proximal phalanx of 1st digit	Medial plantar nerve (S2, S3)	Abducts and flexes 1st digit (great toe, hallux)
Flexor digitorum brevis	Medial tubercle of tuberosity of calcaneus; plantar aponeurosis; intermuscular septa	Both sides of middle phalanges of lateral four digits	Medial plantar nerve (S2, S3)	Flexes lateral four digits
Abductor digiti minimi	Medial and lateral tubercles of tuberosity of calcaneus; plantar aponeurosis; intermuscular septa	Lateral side of base of proximal phalanx of 5th digit	Lateral plantar nerve (S2, S3)	Abducts and flexes little toe (5th digit)
2nd layer				
Quadratus plantae	Medial surface and lateral margin of plantar surface of calcaneus	Posterolateral margin of tendon of flexor digitorum longus	Lateral plantar nerve (S2, S3)	Assists flexor digitorum longus in flexing lateral four digits
Lumbricals	Tendons of flexor digitorum longus	Medial aspect of expansion over lateral four digits	Medial one: medial plantar nerve (S2, S3)	Flex proximal phalanges, extend middle and distal phalanges of lateral four digits
			Lateral three: lateral plantar nerve (S2, S3)	

[a] The spinal cord segmental innervation is indicated (e.g., "S2, S3" means that the nerves supplying the abductor hallucis are derived from the second and third sacral segments of the spinal cord). Damage to one or more of the listed spinal cord segments or to the motor nerve roots arising from them results in paralysis of the muscles concerned.

[b] Despite individual actions, the primary function of the intrinsic muscles of the sole of the foot is to resist flattening or maintain the arch of the foot.

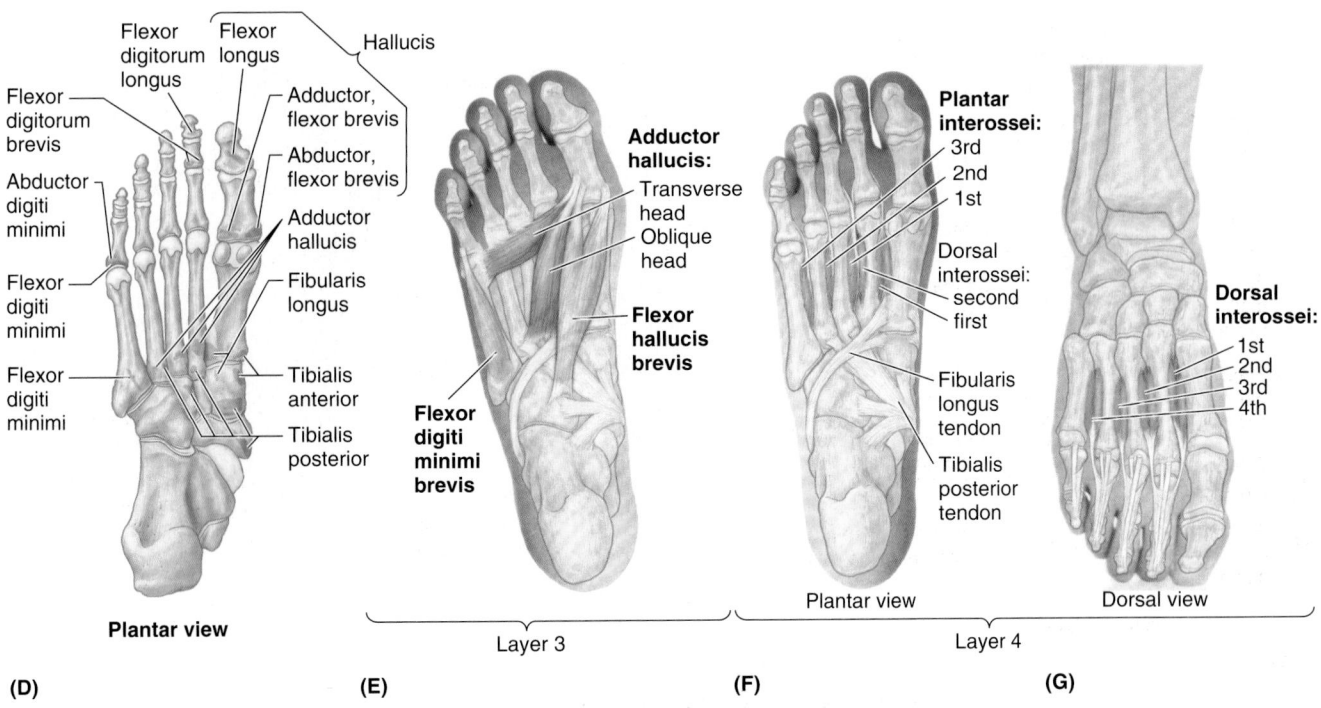

FIGURE 5.68D–G. Muscles of foot: 3rd and 4th layers of sole.

TABLE 5.14.II. MUSCLES OF FOOT: 3RD AND 4TH LAYERS OF SOLE

Muscle	Proximal Attachment	Distal Attachment	Innervation[a]	Main Action[b]
3rd layer				
Flexor hallucis brevis	Plantar surfaces of cuboid and lateral cuneiforms	Both sides of base of proximal phalanx of 1st digit	Medial plantar nerve (S2, S3)	Flexes proximal phalanx of 1st digit
Adductor hallucis	Oblique head: bases of metatarsals 2–4	Tendons of both heads attach to lateral side of base of proximal phalanx of 1st digit	Deep branch of lateral plantar nerve (S2, S3)	Traditionally said to adduct 1st digit; assists in transverse arch of foot by metatarsals medially
	Transverse head: plantar ligaments of metatarsophalangeal (MTP) joints			
Flexor digit minimi brevis	Base of 5th metatarsal	Base of proximal phalanx of 5th digit	Superficial branch of lateral plantar nerve (S2, S3)	Flexes proximal phalanx of 5th digit, thereby assisting with its flexion
4th layer				
Plantar interossei (three muscles)	Plantar aspect of medial sides of shafts of metatarsals 3–5	Medial sides of bases of phalanges of 3rd–5th digits	Lateral plantar nerve (S2, S3)	Adduct digits 3–5 and flex metatarsophalangeal joints
Dorsal interossei (four muscles)	Adjacent sides of shafts of metatarsals 1–5	1st: medial side of proximal phalanx of 2nd digit;		Abduct digits 2–4 and flex metatarsophalangeal joints
		2nd–4th: lateral sides of 2nd–4th digits		

[a]The spinal cord segmental innervation is indicated (e.g., "S2, S3" means that the nerves supplying the flexor hallucis brevis are derived from the second and third sacral segments of the spinal cord). Damage to one or more of the listed spinal cord segments or to the motor nerve roots arising from them results in paralysis of the muscles concerned.

[b]Despite individual actions, the primary function of the intrinsic muscles of the sole of the foot is to resist flattening or maintain the arch of the foot.

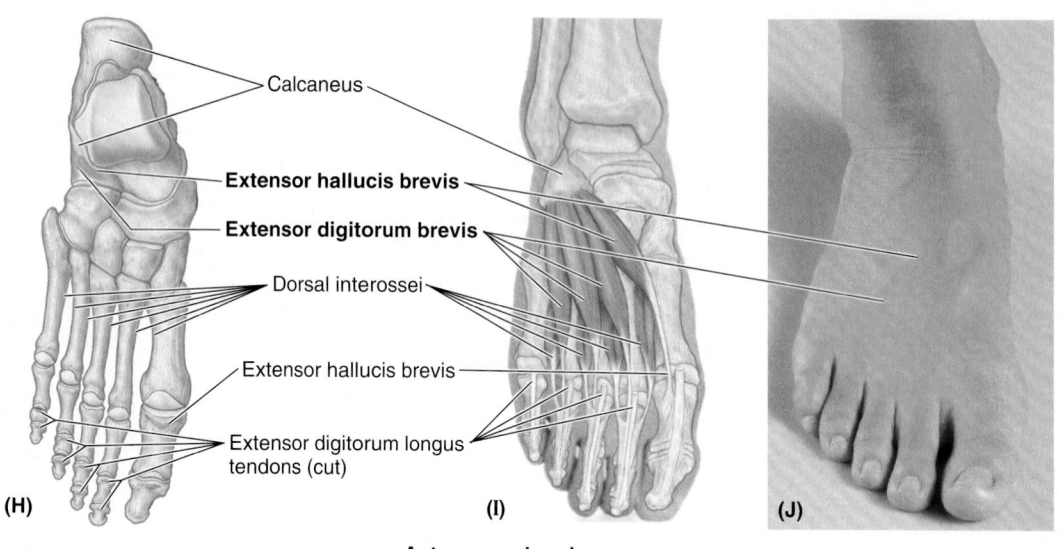

Anterosuperior views

FIGURE 5.68H–J. Muscles of foot: dorsum of foot.

TABLE 5.14.III. MUSCLES OF FOOT: DORSUM OF FOOT

Muscle	Proximal Attachment	Distal Attachment	Innervation[a]	Main Action
Extensor digitorum brevis	Calcaneus (floor of tarsal sinus); interosseous talocalcaneal ligament; stem of inferior extensor retinaculum	Long extensor tendons of four medial digits (toes 2–4)	Deep fibular nerve (L5 or S1, or both)	Aids the extensor digitorum longus in extending the four medial toes at the metatarsophalangeal and interphalangeal joints
Extensor hallucis brevis	In common with extensor digitorum brevis (above)	Dorsal aspect of base of proximal phalanx of great toe (digit 1)		Aids the extensor hallucis longus in extending the great toe at the metatarsophalangeal joint

[a]The spinal cord segmental innervation is indicated (e.g., "L5 or S1" means that the nerve supplying the extensor digitorum brevis is derived from either the fifth lumbar segment or first sacral segment of the spinal cord). Damage to one or more of the listed spinal cord segments or to the motor nerve roots arising from them results in paralysis of the muscles concerned.

There are two neurovascular planes between the muscle layers of the sole of the foot (Figs. 5.69 and 5.70B): (1) a superficial one between the 1st and the 2nd muscular layers and (2) a deep one between the 3rd and the 4th muscular layers. The *tibial nerve* divides posterior to the medial malleolus into the *medial* and *lateral plantar nerves* (Figs. 5.61B, 5.71, and 5.72; Table 5.15). These nerves supply the intrinsic muscles of the plantar aspect of the foot.

The medial plantar nerve courses within the medial compartment of the sole between the 1st and 2nd muscle layers. Initially, the lateral plantar nerve (and artery) run laterally between the muscles of the 1st and 2nd layers of plantar muscles (Figs. 5.69C and 5.70B). Their deep branches then pass medially between the muscles of the 3rd and 4th layers (Fig. 5.70B).

Two closely connected muscles on the dorsum of the foot are the **extensor digitorum brevis (EDB)** and **extensor hallucis brevis (EHB)** (Figs. 5.54A & B and 5.56A). The EHB is actually part of the EDB. These thin, broad muscles form a fleshy mass on the lateral part of the dorsum of the foot, anterior to the lateral malleolus. Its small fleshy belly may be felt when the toes are extended.

Neurovascular Structures and Relationships in Foot

NERVES OF FOOT

The *cutaneous innervation of the foot* is supplied (Fig. 5.72; Table 5.15):

- Medially by the *saphenous nerve*, which extends distally to the head of 1st metatarsal.
- Superiorly (dorsum of foot) by the *superficial* (primarily) and *deep fibular nerves.*
- Inferiorly (sole of foot) by the *medial and lateral plantar nerves;* the common border of their distribution extends along the 4th metacarpal and toe or digit. (This is similar to the pattern of innervation of the palm of the hand.)
- Laterally by the *sural nerve,* including part of the heel.
- Posteriorly (heel) by *medial and lateral calcaneal branches* of the tibial and sural nerves, respectively.

(text continues on p. 617)

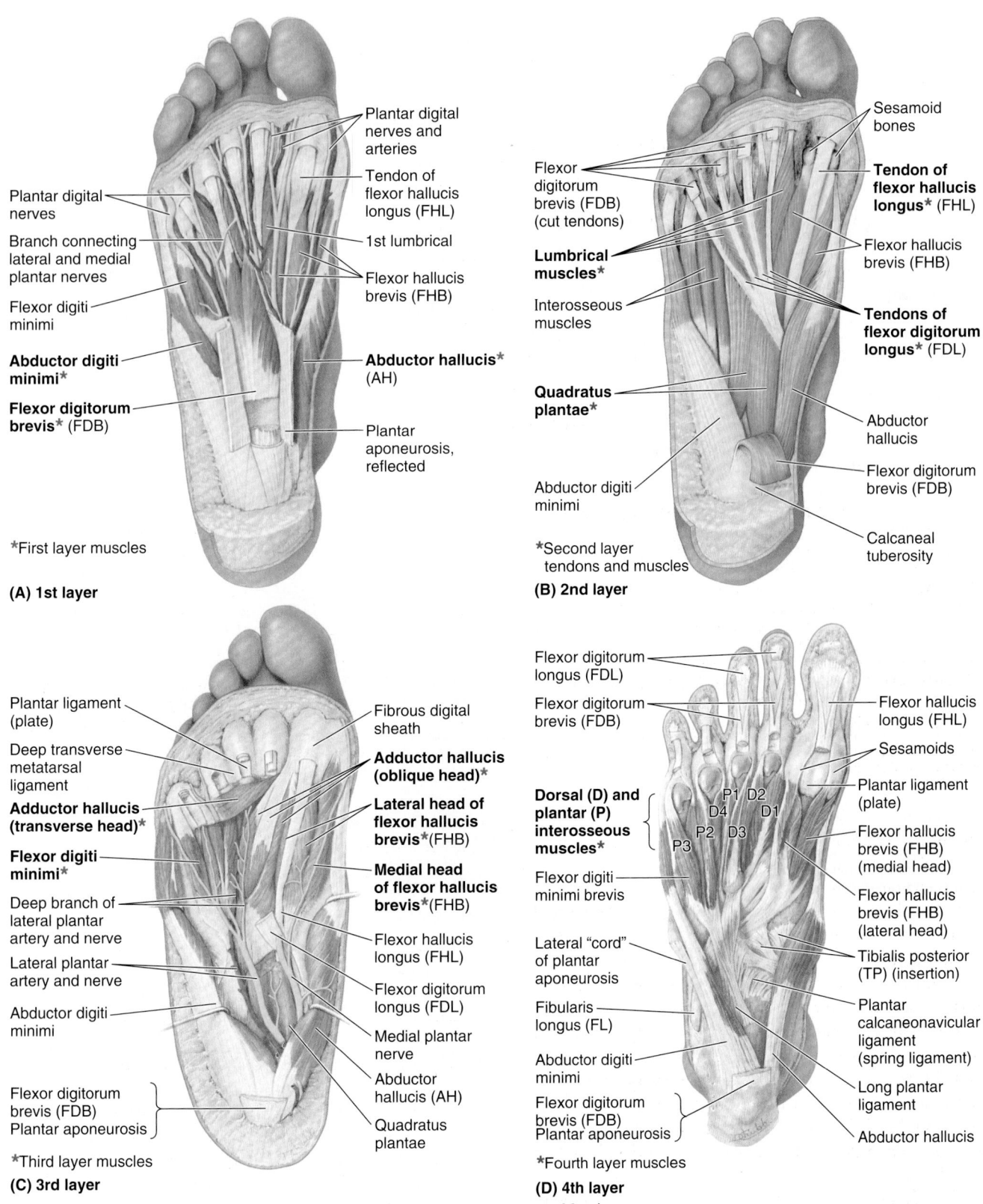

(A) 1st layer

Plantar digital nerves and arteries

Tendon of flexor hallucis longus (FHL)

1st lumbrical

Flexor hallucis brevis (FHB)

Abductor hallucis* (AH)

Plantar aponeurosis, reflected

Plantar digital nerves

Branch connecting lateral and medial plantar nerves

Flexor digiti minimi

Abductor digiti minimi*

Flexor digitorum brevis* (FDB)

*First layer muscles

(B) 2nd layer

Flexor digitorum brevis (FDB) (cut tendons)

Lumbrical muscles*

Interosseous muscles

Quadratus plantae*

Abductor digiti minimi

Sesamoid bones

Tendon of flexor hallucis longus* (FHL)

Flexor hallucis brevis (FHB)

Tendons of flexor digitorum longus* (FDL)

Abductor hallucis

Flexor digitorum brevis (FDB)

Calcaneal tuberosity

*Second layer tendons and muscles

(C) 3rd layer

Plantar ligament (plate)

Deep transverse metatarsal ligament

Adductor hallucis (transverse head)*

Flexor digiti minimi*

Deep branch of lateral plantar artery and nerve

Lateral plantar artery and nerve

Abductor digiti minimi

Flexor digitorum brevis (FDB) Plantar aponeurosis

Fibrous digital sheath

Adductor hallucis (oblique head)*

Lateral head of flexor hallucis brevis* (FHB)

Medial head of flexor hallucis brevis* (FHB)

Flexor hallucis longus (FHL)

Flexor digitorum longus (FDL)

Medial plantar nerve

Abductor hallucis (AH)

Quadratus plantae

*Third layer muscles

(D) 4th layer

Flexor digitorum longus (FDL)

Flexor digitorum brevis (FDB)

Dorsal (D) and plantar (P) interosseous muscles*

Flexor digiti minimi brevis

Lateral "cord" of plantar aponeurosis

Fibularis longus (FL)

Abductor digiti minimi

Flexor digitorum brevis (FDB) Plantar aponeurosis

P1 D2
D4   D1
P2 D3
P3

Flexor hallucis longus (FHL)

Sesamoids

Plantar ligament (plate)

Flexor hallucis brevis (FHB) (medial head)

Flexor hallucis brevis (FHB) (lateral head)

Tibialis posterior (TP) (insertion)

Plantar calcaneonavicular ligament (spring ligament)

Long plantar ligament

Abductor hallucis

*Fourth layer muscles

Inferior views (plantar aspect of foot)

FIGURE 5.69. Layers of plantar muscles. A. The 1st layer consists of the abductors of the large and small toes and the short flexor of the toes. **B.** The 2nd layer consists of the long flexor tendons and associated muscles: four lumbricals and the quadratus plantae. **C.** The 3rd layer consists of the flexor of the little toe and the flexor and adductor of the great toe. Also demonstrated are the neurovascular structures that course in a plane between the 1st and 2nd layers. **D.** The 4th layer consists of the dorsal and plantar interosseous muscles.

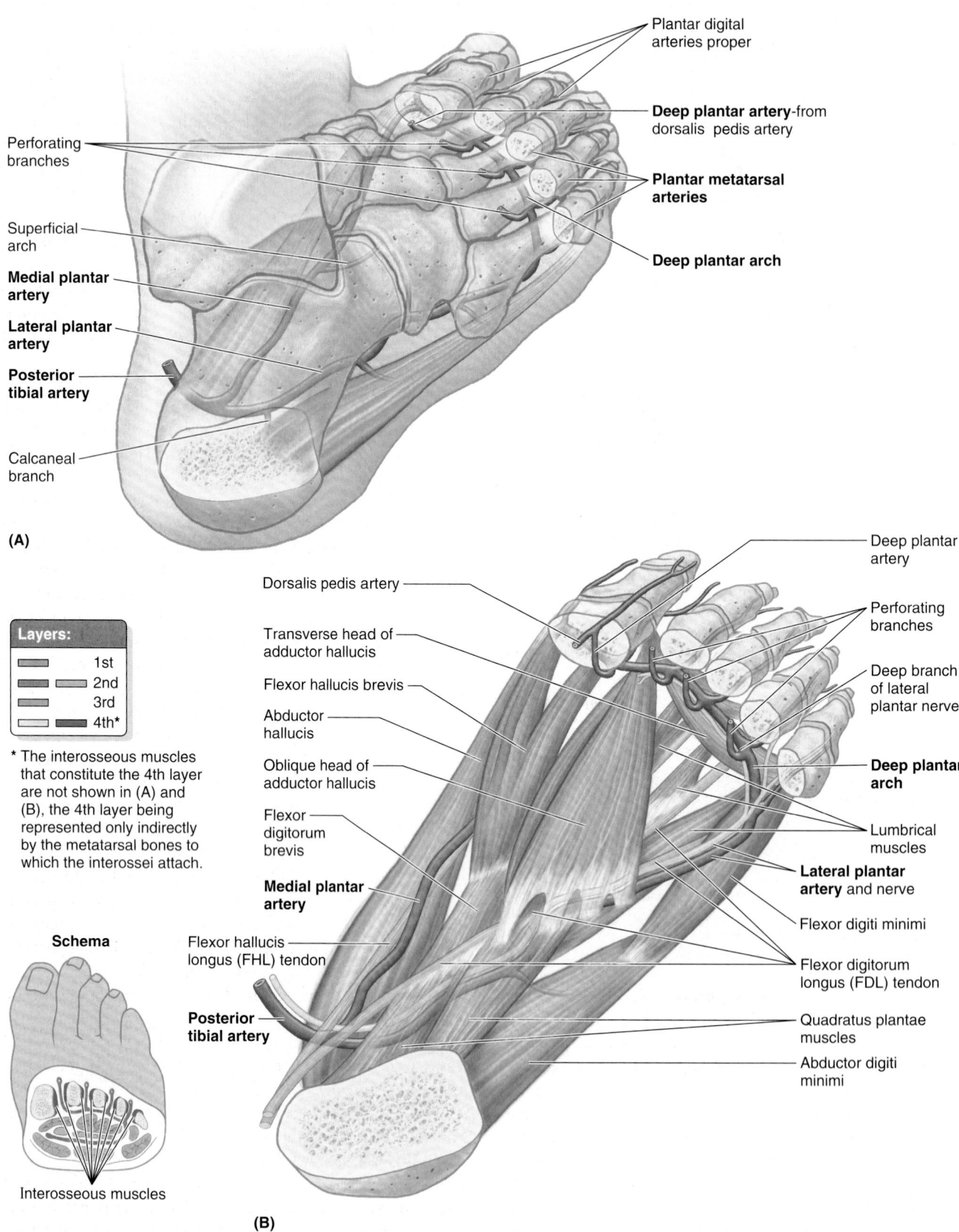

Plantar digital
arteries proper

Deep plantar artery-from
dorsalis pedis artery

**Plantar metatarsal
arteries**

Deep plantar arch

Perforating
branches

Superficial
arch

**Medial plantar
artery**

**Lateral plantar
artery**

**Posterior
tibial artery**

Calcaneal
branch

(A)

Deep plantar
artery

Dorsalis pedis artery

Perforating
branches

Transverse head of
adductor hallucis

Deep branch
of lateral
plantar nerve

Flexor hallucis brevis

Abductor
hallucis

**Deep plantar
arch**

Oblique head of
adductor hallucis

Flexor
digitorum
brevis

Lumbrical
muscles

**Medial plantar
artery**

**Lateral plantar
artery** and nerve

Flexor hallucis
longus (FHL) tendon

Flexor digiti minimi

**Posterior
tibial artery**

Flexor digitorum
longus (FDL) tendon

Quadratus plantae
muscles

Abductor digiti
minimi

Layers:

	1st
	2nd
	3rd
	4th*

* The interosseous muscles
that constitute the 4th layer
are not shown in (A) and
(B), the 4th layer being
represented only indirectly
by the metatarsal bones to
which the interossei attach.

Schema

Interosseous muscles

(B)

FIGURE 5.70. Arteries and muscle layers of foot. A and B. The posterior tibial artery terminates as it enters the foot by dividing into the medial and lateral plantar arteries. Observe the distal anastomoses of these vessels with the deep plantar artery from the dorsal artery of the foot and the perforating branches to the arcuate artery on the dorsum of the foot (see Fig. 5.73). Note that the plantar arteries enter and run in the plane between the 1st and the 2nd layers, with the lateral plantar artery passing from medial to lateral. The deep branches of the artery then pass from lateral to medial between the 3rd and the 4th layers.

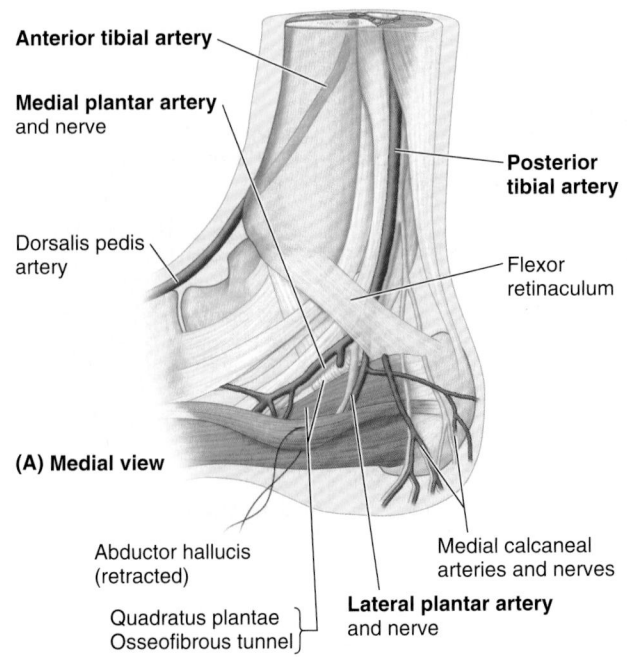

Anterior tibial artery

Medial plantar artery
and nerve

Dorsalis pedis
artery

**Posterior
tibial artery**

Flexor
retinaculum

(A) Medial view

Abductor hallucis
(retracted)

Quadratus plantae
Osseofibrous tunnel

Medial calcaneal
arteries and nerves

Lateral plantar artery
and nerve

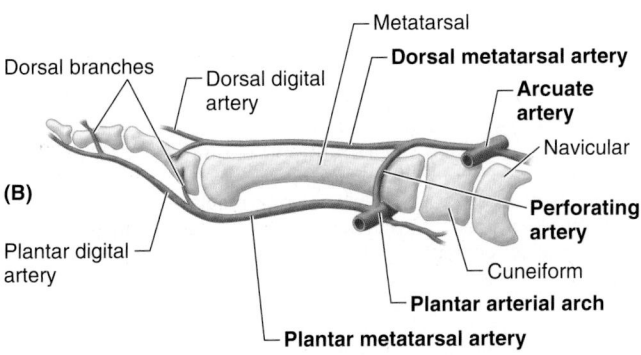

Dorsal branches

Metatarsal

Dorsal metatarsal artery

Dorsal digital
artery

**Arcuate
artery**

Navicular

(B)

**Perforating
artery**

Plantar digital
artery

Cuneiform

Plantar arterial arch

Plantar metatarsal artery

FIGURE 5.71. Arteries of foot: branching and communicating. **A.** Branching of the parent neurovascular structures that give rise to plantar vessels and nerves. **B.** The arteries of the midfoot and forefoot resemble those of the hand in that (1) arches on the two aspects give rise to metatarsal (metacarpal) arteries, which in turn give rise to digital arteries; (2) the dorsal arteries are exhausted before reaching the distal ends of the toes or digits, so the plantar (palmar) digital arteries send branches dorsally to supply the distal dorsal aspects of the digits, including the nail beds; and (3) perforating branches extend between the metatarsals (metacarpals) forming anastomoses between the arches of each side.

Saphenous Nerve. The *saphenous nerve* is the longest and most widely distributed cutaneous branch of the femoral nerve; it is the only branch to extend beyond the knee (Fig. 5.72A; Table 5.15; see also Fig. 5.74B). In addition to supplying the skin and fascia on the anteromedial aspect of the leg, the saphenous nerve passes anterior to the medial malleolus to the dorsum of the foot, where it supplies articular branches to the ankle joint and continues to supply skin along the medial side of the foot as far anteriorly as the head of the 1st metatarsal.

Superficial and Deep Fibular Nerves. After coursing between and supplying the fibular muscles in the lateral compartment of the leg, the *superficial fibular nerve* emerges as a cutaneous nerve about two thirds of the way down the leg. It then supplies the skin on the anterolateral aspect of the

leg and divides into the **medial** and **intermediate dorsal cutaneous nerves,** which continue across the ankle to supply most of the skin on the dorsum of the foot. Its terminal branches are the dorsal digital nerves (common and proper) that supply the skin of the proximal aspect of the medial half of the great toe and that of the lateral three and a half digits.

After supplying the muscles of the anterior compartment of the leg, the *deep fibular nerve* passes deep to the extensor retinaculum and supplies the intrinsic muscles on the dorsum of the foot (extensors digitorum and hallucis longus) and the tarsal and tarsometatarsal joints. When it finally emerges as a cutaneous nerve, it is so far distal in the foot that only a small area of skin remains available for innervation: the web of skin between and contiguous sides of the 1st and 2nd toes. It innervates this area as the **1st common dorsal** (and then **proper dorsal**) **digital nerve(s).**

Medial Plantar Nerve. The **medial plantar nerve,** the larger and more anterior of the two terminal branches of the tibial nerve, arises deep to the flexor retinaculum. It enters the sole of the foot by passing deep to the abductor hallucis (AH) (Figs. 5.69C and 5.71A). It then runs anteriorly between the AH muscle and the flexor digitorum brevis (FDB), supplying both with motor branches on the lateral side of the medial plantar artery (Fig. 5.69A & C). After sending motor branches to the flexor hallucis brevis (FHB) and 1st lumbrical muscle, the medial plantar nerve terminates near the bases of the metatarsals by dividing into three sensory branches (*common plantar digital nerves*). These branches supply the skin of the medial three and a half digits (including the dorsal skin and nail beds of their distal phalanges), and the skin of the sole proximal to them. Compared to the other terminal branch of the tibial nerve, the medial plantar nerve supplies more skin area but fewer muscles. Its distribution to both skin and muscles of the foot is comparable to that of the median nerve in the hand.

Lateral Plantar Nerve. The **lateral plantar nerve,** the smaller and more posterior of the two terminal branches of the tibial nerve, also courses deep to the AH (Fig. 5.71A), but runs anterolaterally between the 1st and 2nd layers of plantar muscles, on the medial side of the lateral plantar artery (Fig. 5.69C). The lateral plantar nerve terminates as it reaches the lateral compartment, dividing into superficial and deep branches (Fig. 5.72B; Table 5.15).

The *superficial branch* divides, in turn, into two **plantar digital nerves** (one common and one proper) that supply the skin of the plantar aspects of the lateral one and a half digits, the dorsal skin and nail beds of their distal phalanges, and skin of the sole proximal to them. The *deep branch of the lateral plantar nerve* courses with the plantar arterial arch between the 3rd and the 4th muscle layers.

The superficial and deep branches of the lateral plantar nerve supply all muscles of the sole not supplied by the medial plantar nerve. Compared to the medial plantar nerve, the lateral plantar nerve supplies less skin area but more individual muscles. Its distribution to both skin and muscles of the foot is comparable to that of the ulnar nerve in the hand (Chapter 6). The medial and lateral plantar nerves also provide innervation to the plantar aspects of all the joints of the foot.

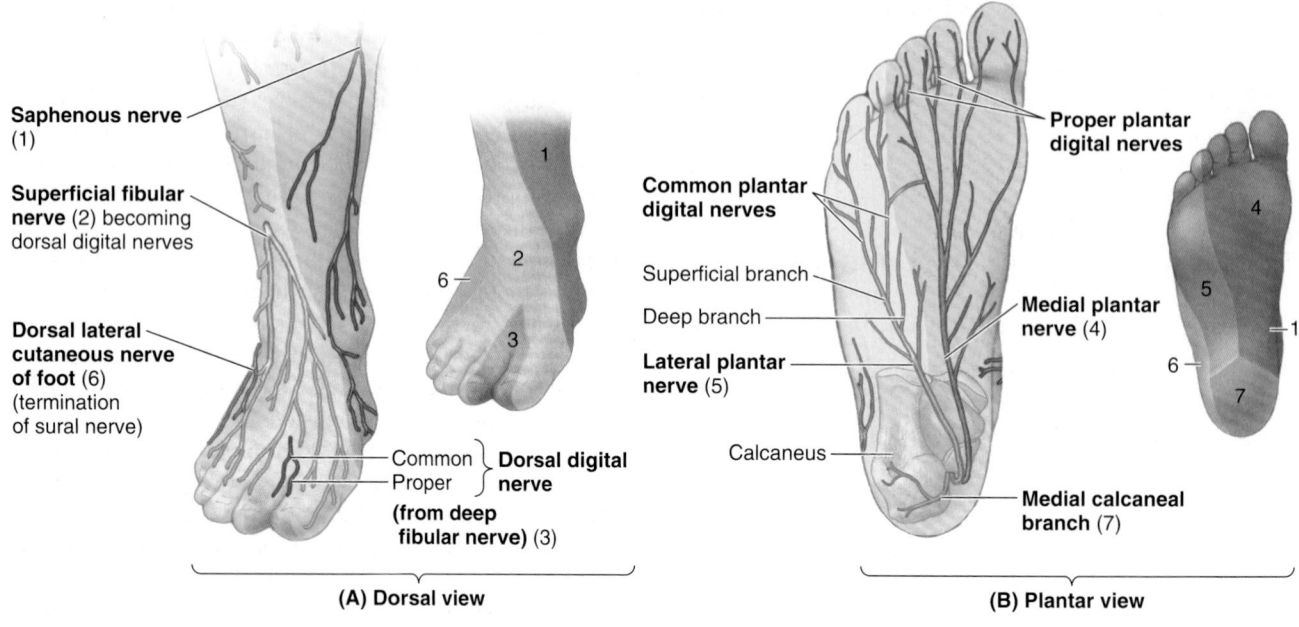

FIGURE 5.72. Nerves of foot.

TABLE 5.15. NERVES OF FOOT

Nerve[a]	Origin	Course	Distribution in Foot
Saphenous (1)	Femoral nerve	Arises in femoral triangle and descends through thigh and leg; accompanies great saphenous vein anterior to medial malleolus; ends on medial side of foot	Supplies skin on medial side of foot as far anteriorly as head of 1st metatarsal
Superficial fibular (2)	Common fibular nerve	Pierces deep fascia in distal third of leg to become cutaneous; then sends branches to foot and digits	Supplies skin on dorsum of foot and all digits, except lateral side of 5th and adjoining sides of the 1st and 2nd digits
Deep fibular (3)		Passes deep to extensor retinaculum to enter dorsum of foot	Supplies extensor digitorum brevis and skin on contiguous sides of 1st and 2nd digits
Medial plantar (4)	Larger terminal branch of tibial nerve	Passes distally in foot between abductor hallucis and flexor digitorum brevis; divides into muscular and cutaneous branches	Supplies skin of medial side of sole of foot and sides of first three digits; also supplies abductor hallucis, flexor digitorum brevis, flexor hallucis brevis, and first lumbrical
Lateral plantar (5)	Smaller terminal branch of tibial nerve	Passes laterally in foot between quadratus plantae and flexor digitorum brevis muscles; divides into superficial and deep branches	Supplies quadratus plantae, abductor digiti minimi, digiti minimi brevis; deep branch supplies plantar and dorsal interossei, lateral three lumbricals, and adductor hallucis; supplies skin on sole lateral to a line splitting 4th digit
Sural (6)	Usually arises from branches of both tibial and common fibular nerves	Passes inferior to the lateral malleolus to lateral side of foot	Lateral aspect of hindfoot and midfoot
Calcaneal branches (7)	Tibial and sural nerves	Pass from distal part of the posterior aspect of leg to skin on heel	Skin of heel

[a] Numbers refer to Figure 5.72.

Sural Nerve. The **sural nerve** is formed by union of the *medial sural cutaneous nerve* (from the tibial nerve) and *sural communicating branch of the common fibular nerve,* respectively (see Fig. 5.57B; Table 5.11). The level of junction of these branches is variable; it may be high (in the popliteal fossa), or low (proximal to heel). Sometimes the branches do not join and, therefore, no sural nerve is formed. In these people, the skin normally innervated by the sural nerve is supplied by the medial and lateral sural cutaneous branches. The sural nerve accompanies the small saphenous vein and enters the foot posterior to the lateral malleolus to supply the ankle joint and skin along the lateral margin of the foot (Fig. 5.72A; Table 5.15).

ARTERIES OF FOOT

The arteries of the foot are terminal branches of the anterior and posterior tibial arteries (Figs. 5.71A and 5.73), respectively: the dorsalis pedis and plantar arteries.

Dorsalis Pedis Artery. Often a major source of blood supply to the forefoot (e.g., during extended periods of standing), the **dorsalis pedis artery** (dorsal artery of foot) is *the direct continuation of the anterior tibial artery.* The dorsalis pedis artery begins midway between the malleoli and runs anteromedially, deep to the inferior extensor retinaculum between the extensor hallucis longus and the extensor digitorum longus tendons on the dorsum of the foot.

The dorsalis pedis artery passes to the first interosseous space, where it divides into the *1st dorsal metatarsal artery* and a **deep plantar artery.** The latter passes deeply between the heads of the first dorsal interosseous muscle to enter the sole of the foot, where it joins the lateral plantar artery to form the *deep plantar arch.* The course and destination of the dorsal artery and its major continuation, the deep plantar artery, are comparable to the radial artery of the hand, which completes a deep arterial arch in the palm.

The **lateral tarsal artery,** a branch of the dorsalis pedis artery, runs laterally in an arched course beneath the EDB to supply this muscle and the underlying tarsals and joints. It anastomoses with other branches, such as the arcuate artery.

The **1st dorsal metatarsal artery** divides into branches that supply both sides of the great toe and the medial side of the 2nd toe.

The **arcuate artery** runs laterally across the bases of the lateral four metatarsals, deep to the extensor tendons, to reach the lateral aspect of the forefoot, where it may anastomose with the lateral tarsal artery to form an arterial loop. The arcuate artery gives rise to the **2nd, 3rd,** and

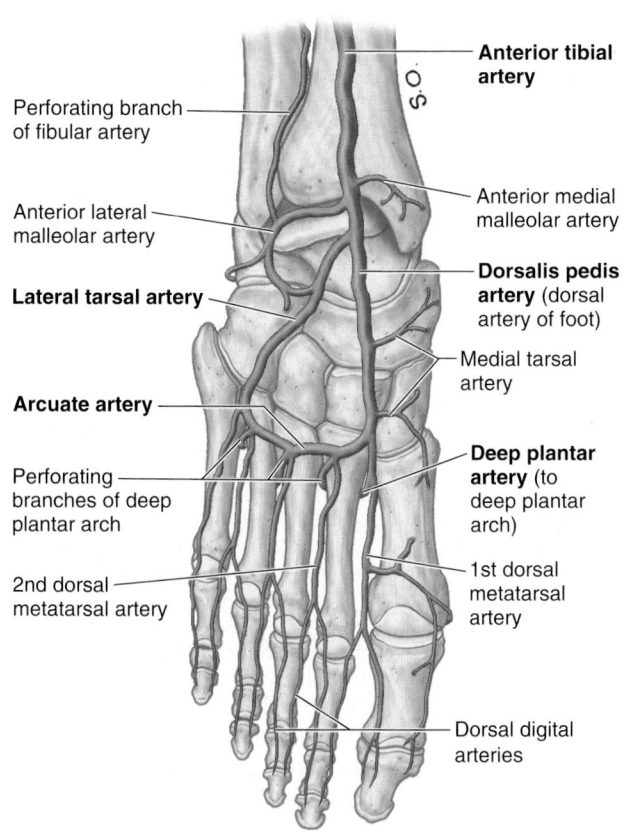

(A) Dorsum of foot

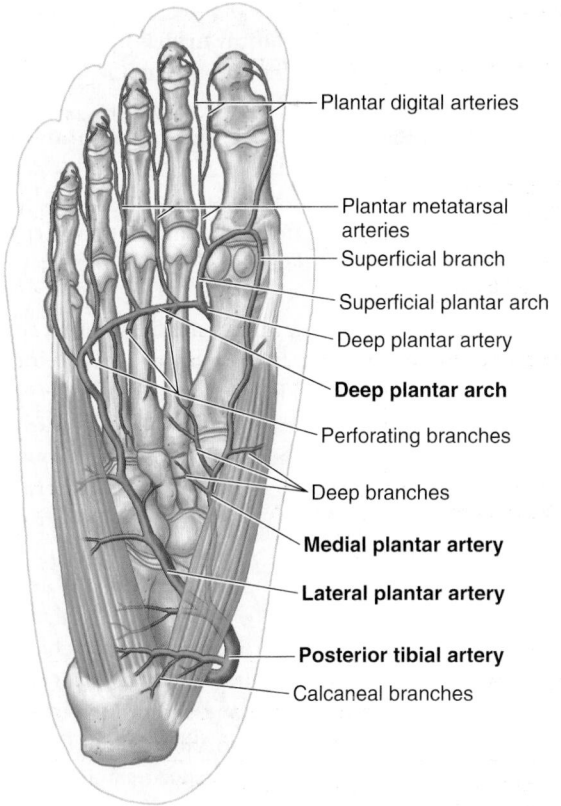

(B) Plantar aspect of foot

FIGURE 5.73. Arteries of foot: overview. A. The anterior tibial artery becomes the dorsalis pedis artery when it crosses the talocrural joint. **B.** The medial and lateral plantar arteries are terminal branches of the posterior tibial artery. The deep plantar artery and perforating branches of the deep plantar arch provide anastomoses between the dorsal and the plantar arteries.

4th dorsal metatarsal arteries. These vessels run distally to the clefts of the toes and are connected to the plantar arch and the plantar metatarsal arteries by *perforating branches* (Figs. 5.70A & B, 5.71B, and 5.73A & B). Distally, each dorsal metatarsal artery divides into two **dorsal digital arteries** for the dorsal aspect of the sides of adjoining toes (Fig. 5.73A); however, these arteries generally end proximal to the distal interphalangeal joint (Fig. 5.71B), and are replaced by or receive replenishment from dorsal branches of the plantar digital arteries.

ARTERIES OF SOLE OF FOOT

The sole of the foot has a prolific blood supply from the posterior tibial artery, which divides deep to the flexor retinaculum (Figs. 5.69A, 5.71A, and 5.73B). The terminal branches pass deep to the *abductor hallucis* (AH) as the medial and lateral plantar arteries, which accompany the similarly named nerves.

Medial Plantar Artery. The **medial plantar artery** is the smaller terminal branch of the posterior tibial artery. It gives rise to a deep branch (or branches) that supplies mainly muscles of the great toe. The larger superficial branch of the medial plantar artery supplies the skin on the medial side of the sole and has digital branches that accompany digital branches of the medial plantar nerve, the more lateral of which anastomose with medial plantar metatarsal arteries. Occasionally, a **superficial plantar arch** is formed when the superficial branch anastomoses with the lateral plantar artery or the deep plantar arch (Fig. 5.73B).

Lateral Plantar Artery. The **lateral plantar artery,** much larger than the medial plantar artery, arises with and accompanies the nerve of the same name (Figs. 5.69C, 5.70B, 5.71A, and 5.73B). It runs laterally and anteriorly, at first deep to the AH and then between the FDB and quadratus plantae.

The lateral plantar artery arches medially across the foot with the deep branch of the lateral plantar nerve to form the **deep plantar arch,** which is completed by union with the *deep plantar artery,* a branch of the dorsalis pedis artery. As it crosses the foot, the deep plantar arch gives rise to four **plantar metatarsal arteries;** three **perforating branches;** and many branches to the skin, fascia, and muscles in the sole. The plantar metatarsal arteries divide near the base of the proximal phalanges to form the **plantar digital arteries,** supplying adjacent digits (toes); the more medial metatarsal arteries are joined by superficial digital branches of the medial plantar artery. The plantar digital arteries typically provide most of the blood reaching the distal toes, including the nail bed, via perforating and dorsal branches (Fig. 5.71B and 5.73)—an arrangement that also occurs in the fingers.

VENOUS DRAINAGE OF FOOT

As in the rest of the lower limb, there are both superficial and deep veins in the foot. The *deep veins* take the form of interanastomosing paired veins accompanying all arteries internal to the deep fascia (Fig. 5.74A). The *superficial veins* are subcutaneous and unaccompanied by arteries (Fig. 5.74B). Unlike the leg and thigh, however, the venous drainage of the foot is primarily to the major superficial veins, both from the deep accompanying veins and other smaller superficial veins.

Perforating veins begin the one-way shunting of blood from superficial to deep veins, a pattern essential to operation of the *musculovenous pump,* proximal to the ankle joint. Most blood is drained from the foot through the superficial veins.

Dorsal digital veins continue proximally as **dorsal metatarsal veins,** which also receive branches from **plantar digital veins.** These veins drain to the **dorsal venous arch of the foot,** proximal to which a **dorsal venous network** covers the remainder of the dorsum of the foot. Both the arch and the network are located in the subcutaneous tissue.

For the main part, superficial veins from a **plantar venous network** either drain around the medial border of the foot to converge with the medial part of the dorsal venous arch and network to form a **medial marginal vein,** which becomes the *great saphenous vein,* or drain around the lateral margin to converge with the lateral part of the dorsal venous arch and network to form the **lateral marginal vein,** which becomes the *small saphenous vein.*

Perforating veins from the great and small saphenous veins then continuously shunt blood deeply as they ascend to take advantage of the musculovenous pump.

LYMPHATIC DRAINAGE OF FOOT

The lymphatics of the foot begin in subcutaneous plexuses. The collecting vessels consist of superficial and deep lymphatic vessels that follow the superficial veins and major vascular bundles, respectively.

Superficial lymphatic vessels are most numerous in the sole of the foot (Fig. 5.75). The *medial superficial lymphatic vessels,* larger and more numerous than the lateral ones, drain the medial side of the dorsum and sole of the foot (Fig. 5.75A). These vessels converge on the *great saphenous vein* and accompany it to the vertical group of *superficial inguinal lymph nodes,* located along the vein's termination, and then to the *deep inguinal lymph nodes* along the proximal femoral vein (see Fig. 5.45B). The *lateral superficial lymphatic vessels* drain the lateral side of the dorsum and sole of the foot. Most of these vessels pass posterior to the lateral malleolus and accompany the small saphenous vein to the popliteal fossa, where they enter the *popliteal lymph nodes* (Fig. 7.75B).

The *deep lymphatic vessels* from the foot follow the main blood vessels: fibular, anterior and posterior tibial, popliteal, and femoral veins. The deep vessels from the foot also drain into the popliteal lymph nodes. Lymphatic vessels from them follow the femoral vessels, carrying lymph to the deep inguinal lymph nodes. From the deep inguinal nodes, all lymph from the lower limb passes deep to the inguinal ligament to the *iliac lymph nodes* (see Fig. 5.45A).

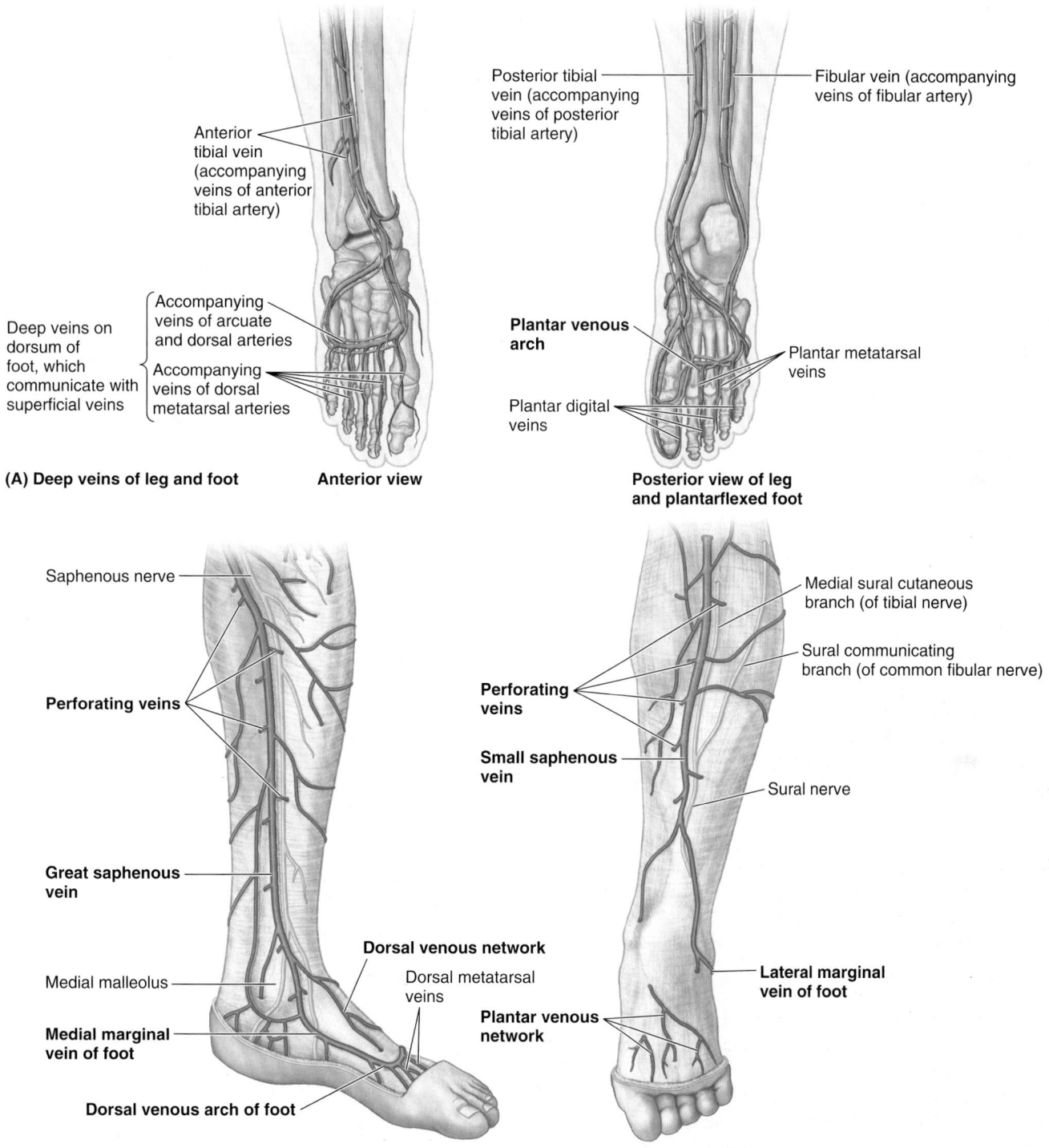

Anterior tibial vein (accompanying veins of anterior tibial artery)

Deep veins on dorsum of foot, which communicate with superficial veins
{ Accompanying veins of arcuate and dorsal arteries
Accompanying veins of dorsal metatarsal arteries }

(A) Deep veins of leg and foot   **Anterior view**

Posterior tibial vein (accompanying veins of posterior tibial artery)

Fibular vein (accompanying veins of fibular artery)

Plantar venous arch

Plantar metatarsal veins

Plantar digital veins

Posterior view of leg and plantarflexed foot

Saphenous nerve

Perforating veins

Great saphenous vein

Medial malleolus

Medial marginal vein of foot

Dorsal venous network

Dorsal metatarsal veins

Dorsal venous arch of foot

(B) Superficial veins of leg and foot   **Medial view**

Medial sural cutaneous branch (of tibial nerve)

Sural communicating branch (of common fibular nerve)

Perforating veins

Small saphenous vein

Sural nerve

Lateral marginal vein of foot

Plantar venous network

Posterior view of leg and plantarflexed foot

FIGURE 5.74. Veins of leg and foot. A. The deep veins accompany the arteries and their branches; they anastomose frequently and have numerous valves. **B.** The main superficial veins drain into the deep veins as they ascend the limb by means of perforating veins so that muscular compression can propel blood toward the heart against the pull of gravity. The distal great saphenous vein is accompanied by the saphenous nerve, and the small saphenous vein is accompanied by the sural nerve and its medial root (medial sural cutaneous nerve).

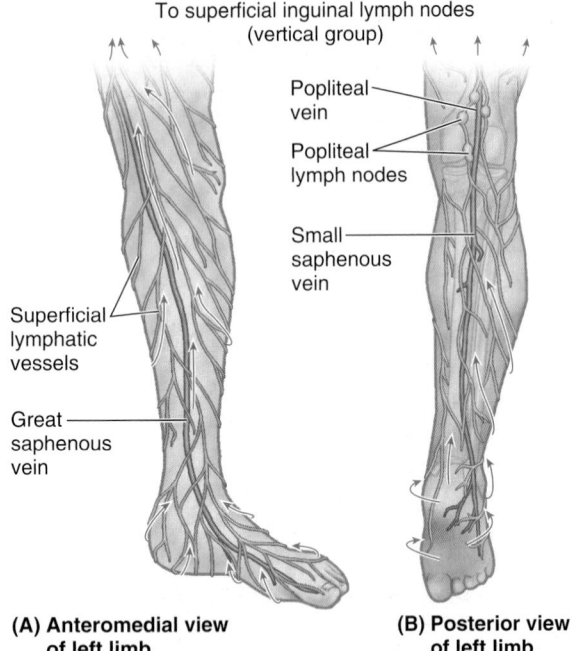

(A) Anteromedial view of left limb

To superficial inguinal lymph nodes (vertical group)

Popliteal vein

Popliteal lymph nodes

Small saphenous vein

Superficial lymphatic vessels

Great saphenous vein

(B) Posterior view of left limb

FIGURE 5.75. Lymphatic drainage of foot. Lymphatic drainage from the sole drains dorsally and proximally. **A.** Superficial lymphatic vessels from the medial foot drain are joined by those from the anteromedial leg in draining to the superficial inguinal lymph nodes via lymphatics that accompany the great saphenous vein. **B.** Superficial lymphatic vessels from the lateral foot join those from the posterolateral leg, converging to vessels accompanying the small saphenous vein and draining into the popliteal lymph nodes.

Surface Anatomy of Ankle and Foot Regions

The tendons in the ankle region can be identified satisfactorily only when their muscles are acting. If the foot is actively inverted, the *tendon of the tibialis posterior* may be palpated as it passes posterior and distal to the medial malleolus, then superior to the sustentaculum tali, to reach its attachment to the tuberosity of the navicular (Fig. 5.76A–C). Hence the tibialis posterior tendon is the guide to the navicular. The tendon of the tibialis posterior also indicates the site for palpating the *posterior tibial pulse* (halfway between the medial malleolus and the calcaneal tendon—Fig. B5.25).

The tendons of the fibularis longus and brevis may be followed distally, posterior and inferior to the lateral malleolus, and then anteriorly along the lateral aspect of the foot (Fig. 5.76D & E). The *fibularis longus tendon* can be palpated as far as the cuboid, and then it disappears as it turns into the sole. The *fibularis brevis tendon* can easily be traced to its attachment to the dorsal surface of the tuberosity on the base of the 5th metatarsal. This tuberosity is located at the middle of the lateral border of the foot. With toes actively extended, the small fleshy *belly of the* **extensor digitorum brevis** may be seen and palpated anterior to the lateral malleolus. Its position should be observed and palpated so that it may not be mistaken subsequently for an abnormal edema (swelling).

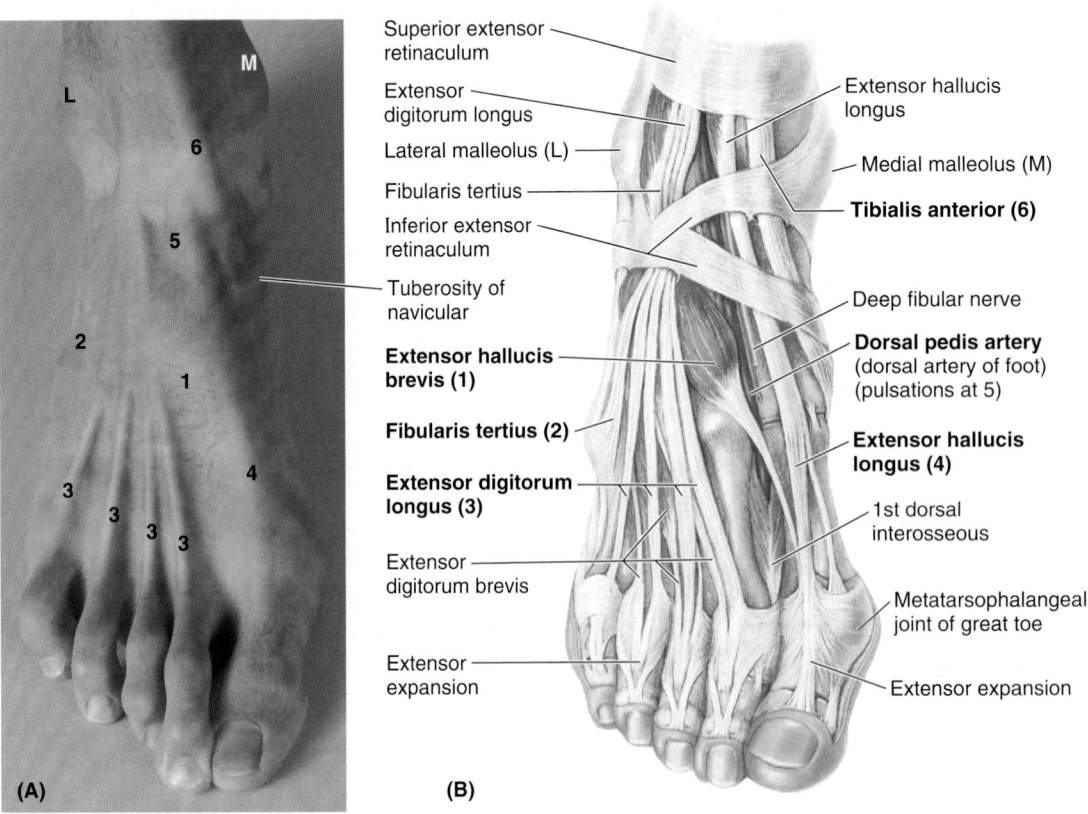

(A)

Superior extensor retinaculum

Extensor digitorum longus

Lateral malleolus (L)

Fibularis tertius

Inferior extensor retinaculum

Tuberosity of navicular

Extensor hallucis brevis (1)

Fibularis tertius (2)

Extensor digitorum longus (3)

Extensor digitorum brevis

Extensor expansion

Extensor hallucis longus

Medial malleolus (M)

Tibialis anterior (6)

Deep fibular nerve

Dorsal pedis artery (dorsal artery of foot) (pulsations at 5)

Extensor hallucis longus (4)

1st dorsal interosseous

Metatarsophalangeal joint of great toe

Extensor expansion

(B)

FIGURE 5.76. Surface anatomy of the foot. A. Visible features. **B.** Underlying structures.

Key for (C):

1 Abductor hallucis
2 Ball of foot
3 Calcaneal tendon
4 Extensor hallucis longus tendon
5 Medial malleolus
6 Medial longitudinal arch of foot
7 Navicular tuberosity
8 Sustentaculum tali
9 Tibialis anterior tendon
10 Tibialis posterior tendon
11 Head of 1st metatarsal

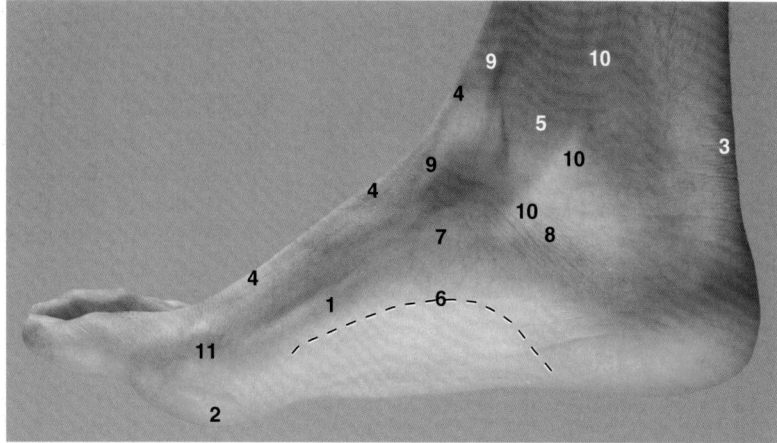

(C) Medial view

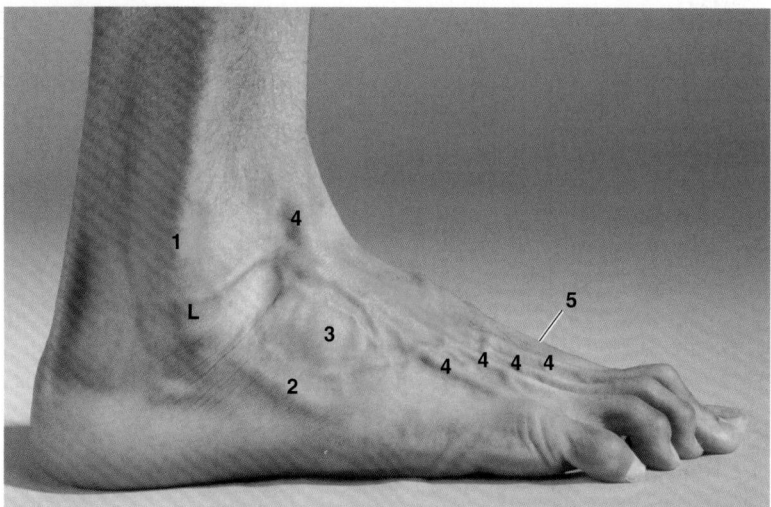

(D) Lateral view

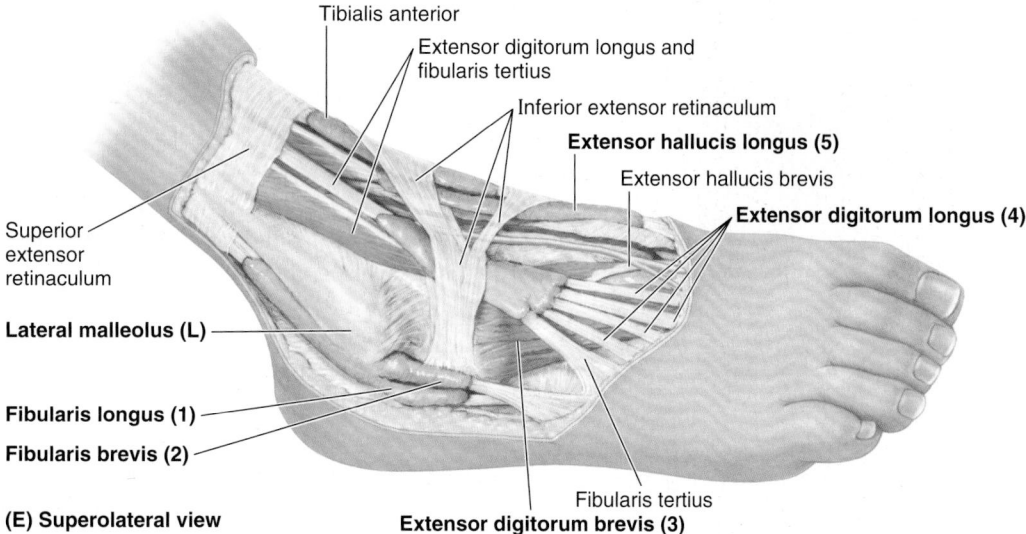

(E) Superolateral view

FIGURE 5.76. Surface anatomy of the foot (*continued*). **C and D.** Visible features. **E.** Underlying structures. Numbers in parentheses in (**E**) refer to structures identified in (**D**).

The tendons on the anterior aspect of the ankle (from medial to lateral side) are easily palpated when the foot is dorsiflexed (Fig. 5.76A–C):

- The large *tendon of the tibialis anterior* leaves the cover of the superior extensor tendon, from which level the tendon is invested by a continuous synovial sheath; the tendon may be traced to its attachment to the 1st cuneiform and the base of the 1st metatarsal.

- The *tendon of the extensor hallucis longus*, obvious when the great toe is extended against resistance, may be followed to its attachment to the base of the distal phalanx of the great toe.
- The *tendons of the extensor digitorum longus* may be followed easily to their attachments to the lateral four toes.
- The *tendon of the fibularis tertius* may also be traced to its attachment to the base of the 5th metatarsal. This muscle is of minor importance and may be absent.

FOOT

Plantar Fasciitis

Inflammation of the plantar fascia—*plantar fasciitis*—is often caused by an overuse mechanism. It may result from running and high-impact aerobics, especially when inappropriate footwear is worn. Plantar fasciitis is the most common hindfoot problem in runners. It causes pain on the plantar surface of the foot and heel. The pain is often most severe after sitting, and when beginning to walk in the morning. It usually dissipates after 5–10 minutes of activity and often recurs again following rest.

Point tenderness is located at the proximal attachment of the aponeurosis to the medial tubercle of the calcaneus and on the medial surface of this bone. The pain increases with passive extension of the great toe and may be further exacerbated by dorsiflexion of the ankle and/or weight-bearing.

If a *calcaneal spur* (abnormal bony process) protrudes from the medial tubercle, plantar fasciitis is likely to cause pain on the medial side of the foot when walking (Fig. B5.26). Usually a bursa develops at the end of the spur that may also become inflamed and tender.

Infections of Foot

Foot infections are common, especially in seasons, climates, and cultures where shoes are less commonly worn. A neglected puncture wound may lead to an extensive deep infection, resulting in swelling, pain, and fever.

Deep infections of the foot often localize within the compartments between the muscular layers (Fig. 5.70B). A well-established infection in one of the enclosed fascial or muscular spaces usually requires surgical incision and drainage. When possible, the incision is made on the medial side of the foot, passing superior to the abductor hallucis to allow visualization of critical neurovascular structures, while avoiding production of a painful scar in a weight-bearing area.

Contusion of Extensor Digitorum Brevis

Functionally, the EDB and EHB muscles are relatively unimportant. Clinically, knowing the location of the belly of the EDB is important for distinguishing it from abnormal edema. Contusion and tearing of muscle fibers and associated blood vessels result in a *hematoma* (clotted extravasated blood), producing edema anteromedial to the lateral malleolus. Most people who have not seen this inflamed muscle assume they have a severely sprained ankle.

Sural Nerve Grafts

Pieces of the sural nerve are often used for nerve grafts in procedures such as repairing nerve defects resulting from wounds. The surgeon is usually able to locate this nerve in relation to the small saphenous vein (Fig. 5.74B). Because of the variations in the level of formation of the sural nerve, the surgeon may have to make incisions in both legs, and then select the better specimen.

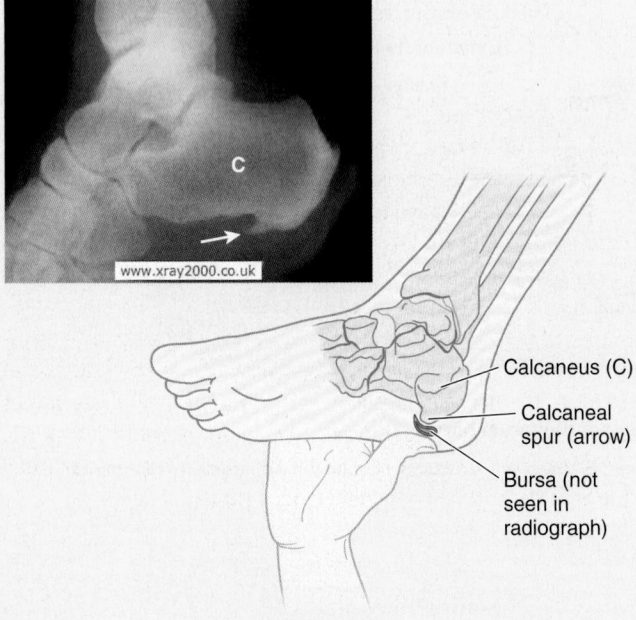

Calcaneus (C)
Calcaneal spur (arrow)
Bursa (not seen in radiograph)

FIGURE B5.26.

Anesthetic Block of Superficial Fibular Nerve

After the superficial fibular nerve pierces the deep fascia to become a cutaneous nerve, it divides into medial and intermediate cutaneous nerves (Fig. 5.72A). In thin people, these branches can often be seen or felt as ridges under the skin when the foot is plantarflexed. Injections of an anesthetic agent around these branches in the ankle region, anterior to the palpable portion of the fibula, anesthetizes the skin on the dorsum of the foot (except the web between and adjacent surfaces of the 1st and 2nd toes) more broadly and effectively than more local injections on the dorsum of the foot for superficial surgery.

Plantar Reflex

The **plantar reflex** (L4, L5, S1, and S2 nerve roots) is a myotatic (deep tendon) reflex that is routinely tested during neurologic examinations. The lateral aspect of the sole of the foot is stroked with a blunt object, such as a tongue depressor, beginning at the heel and crossing to the base of the great toe. The motion is firm and continuous but neither painful nor ticklish. Flexion of the toes is a normal response. Slight fanning of the lateral four toes and dorsiflexion of the great toe is an abnormal response (*Babinski sign*), indicating brain injury or cerebral disease, except in infants. Because the corticospinal tracts are not fully developed in newborns, a Babinski sign is usually elicited and may be present until children are 4 years of age (except in infants with a brain injury or cerebral disease).

Medial Plantar Nerve Entrapment

Compressive irritation of the medial plantar nerve as it passes deep to the flexor retinaculum, or curves deep to the abductor hallucis, may cause aching, burning, numbness, and tingling (paresthesia) on the medial side of the sole of the foot and in the region of the navicular tuberosity. *Medial plantar nerve compression* may occur during repetitive eversion of the foot (e.g., during gymnastics and running). Because of its frequency in runners, these symptoms have been called "jogger's foot."

Palpation of Dorsalis Pedis Pulse

The **dorsalis pedis artery pulse** is evaluated during a physical examination of the peripheral vascular system. Dorsalis pedis pulses may be palpated with the feet slightly dorsiflexed. The pulses are usually easy to palpate because these dorsal arteries are subcutaneous and pass along a line from the extensor retinaculum to a point just lateral to the EHL tendons (Swartz, 2009) (Fig. B5.27). A diminished or absent dorsalis pedis pulse usually suggests vascular insufficiency resulting from arterial disease. The five *P signs of acute arterial occlusion* are **p**ain, **p**allor, **p**aresthesia, **p**aralysis, and **p**ulselessness. Some healthy adults (and even children) have *congenitally non-palpable dorsalis pedis pulses;* the variation is usually bilateral. In these cases, the dorsalis pedis artery is replaced by an enlarged perforating fibular artery.

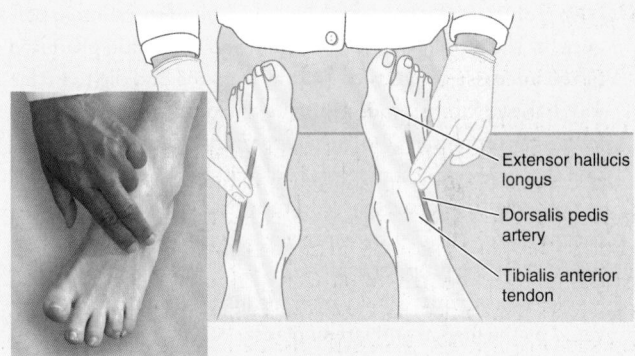

Extensor hallucis longus

Dorsalis pedis artery

Tibialis anterior tendon

FIGURE B5.27.

Hemorrhaging Wounds of Sole of Foot

Puncture wounds of the sole of the foot involving the deep plantar arch and its branches usually result in severe bleeding, typically from both ends of the cut artery because of the abundant anastomoses. Ligation of the deep arch is difficult because of its depth and the structures that surround it.

Lymphadenopathy

Infections of the foot may spread proximally, causing enlargement of the popliteal and inguinal lymph nodes (*lymphadenopathy*). Infections on the lateral side of the foot initially produce enlargement of popliteal lymph nodes (*popliteal lymphadenopathy*); later, the inguinal lymph nodes may enlarge.

Inguinal lymphadenopathy without popliteal lymphadenopathy can result from infection of the medial side of the foot, leg, or thigh; however, enlargement of these nodes can also result from an infection or tumor in the vulva, penis, scrotum, perineum, and gluteal region, and from terminal parts of the urethra, anal canal, and vagina.

The Bottom Line

FOOT

Muscles of foot: The intrinsic muscles of the plantar surface of the foot are arranged in four layers and divided into four fascial compartments. ♦ A tough plantar aponeurosis overlies the central compartment, passively contributing to arch maintenance and, along with firmly bound fat, protecting the vessels and nerves from compression. ♦ There is similarity to the arrangement of muscles in the palm of the hand, but the muscles of the foot generally respond as a group rather than individually, acting to maintain the longitudinal arch of the foot or push a portion of it harder against the ground to maintain balance. ♦ The movements of abduction and adduction produced by the interossei are toward or away from the 2nd digit. ♦ The foot has two intrinsic muscles on its dorsum that augment the long extensor muscles. ♦ The plantar intrinsic muscles function throughout the stance phase of gait, from heel strike to toe off, resisting forces that tend to spread the arches of the foot. ♦ These muscles are especially active in fixing the medial forefoot for the propulsive push off.

Nerves of foot: The plantar intrinsic muscles are innervated by the medial and lateral plantar nerves, whereas the dorsal muscles are innervated by the deep fibular nerve. ♦ Most of the dorsum of the foot receives cutaneous innervation from the superficial fibular nerve, the exception being the skin of the web between and the adjacent sides of the 1st and 2nd toes. The latter receives innervation from the deep fibular nerve after it supplies the muscles on the dorsum of the foot. ♦ The skin of the medial and lateral sides of the foot is innervated by the saphenous and sural nerves, respectively. ♦ The plantar aspect of the foot receives innervation from the larger medial and smaller lateral plantar nerves. ♦ The medial plantar nerve supplies more skin (the plantar aspect of the medial three and half toes and adjacent sole) but fewer muscles (the

medial hallux and 1st lumbrical muscles only) than the lateral plantar nerve. ♦ The lateral planar nerve supplies the remaining muscles and skin of the plantar aspect. ♦ The distribution of the medial and lateral plantar nerves is comparable to that of the median and ulnar nerves in the palm.

Arteries of foot: The dorsal and plantar arteries of the foot are terminal branches of the anterior and posterior tibial arteries, respectively. ♦ The dorsalis pedis artery supplies all of the dorsum of the foot and, via the arcuate artery, the proximal dorsal aspect of the toes. It also contributes to formation of the deep plantar arch via its terminal deep plantar artery. ♦ The smaller medial and larger lateral plantar arteries supply the plantar aspect of the foot, the latter running in vascular planes between the 1st and 2nd layers and then, as the plantar arch, the 3rd and 4th layers of the intrinsic muscles. ♦ Anastomoses between the dorsalis pedis and plantar arteries are abundant and important for the health of the foot. ♦ Except for the scarcity of a superficial plantar arch, the arterial pattern of the foot is similar to that of the hand.

Efferent vessels of foot: Venous drainage of the foot primarily follows a superficial route, draining to the dorsum of the foot and then medially via the great saphenous vein or laterally via the small saphenous veins. ♦ From these veins, blood is shunted by perforating veins to the deep veins of the leg and thigh that participate in the musculovenous pump. ♦ The lymphatics carrying lymph from the foot drain toward and then along the superficial veins draining the foot. ♦ Lymph from the medial foot follows the great saphenous vein and drains directly to superficial inguinal lymph nodes. ♦ Lymph from the lateral foot follows the small saphenous vein and drains initially to the popliteal lymph nodes and then by deep lymphatic vessels to the deep inguinal nodes.

JOINTS OF LOWER LIMB

The joints of the lower limb include the articulations of the pelvic girdle—lumbosacral joints, sacro-iliac joints, and pubic symphysis, which are discussed in Chapter 3. The remaining joints of the lower limb are the hip joints, knee joints, tibiofibular joints, ankle joints, and foot joints (Fig. 5.77).

Hip Joint

The **hip joint** forms the connection between the lower limb and the pelvic girdle (Fig. 5.77A). It is a strong and stable *multiaxial ball and socket type of synovial joint*. The head of the femur is the ball and the acetabulum is the socket (Fig. 5.78). The hip joint is designed for stability over a wide range of movement. Next to the glenohumeral (shoulder) joint, it is the most movable of all joints. During

standing, the entire weight of the upper body is transmitted through the hip bones to the heads and necks of the femora.

ARTICULAR SURFACES OF HIP JOINT

The round head of the femur articulates with the cup-like acetabulum of the hip bone (Figs. 5.77–5.80). The *head of the femur* forms approximately two thirds of a sphere. Except for the depression or *fovea for the ligament of the femoral head*, all of the head is covered with articular cartilage, which is thickest over weight-bearing areas.

The *acetabulum*, a hemispherical hollow on the lateral aspect of the hip bone, is formed by the fusion of three bony parts (see Fig. 5.5). The heavy, prominent **acetabular rim** of the acetabulum consists of a semilunar articular part covered with articular cartilage, the *lunate surface of the acetabulum*

(text continues on p. 629)

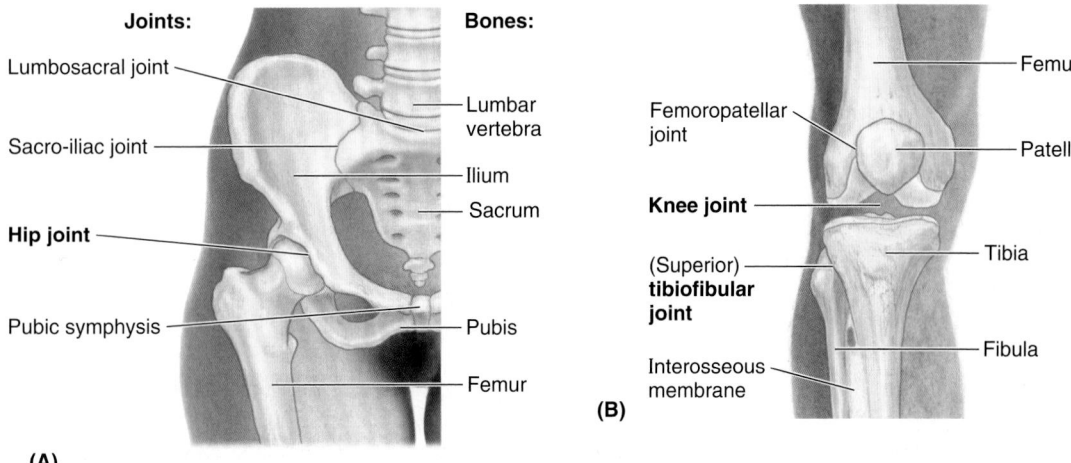

Joints:

Lumbosacral joint

Sacro-iliac joint

Hip joint

Pubic symphysis

Bones:

Lumbar vertebra

Ilium

Sacrum

Pubis

Femur

(A)

Femur

Femoropatellar joint

Patella

Knee joint

Tibia

(Superior) **tibiofibular joint**

Fibula

Interosseous membrane

(B)

Fibula

Tibia

Tibiofibular syndesmosis

Ankle joint

Transverse tarsal

Intertarsal

Tarsometatarsal

Joints

Metatarsophalangeal

Interphalangeal

(C)

Anterior views

FIGURE 5.77. Joints of lower limb. The lower limb joints are (**A**) those of the pelvic girdle connecting the free lower limb to the vertebral column, (**B**) the knee and tibiofibular joint, and (**C**) tibiofibular syndesmosis, ankle joint, and the many joints of the foot.

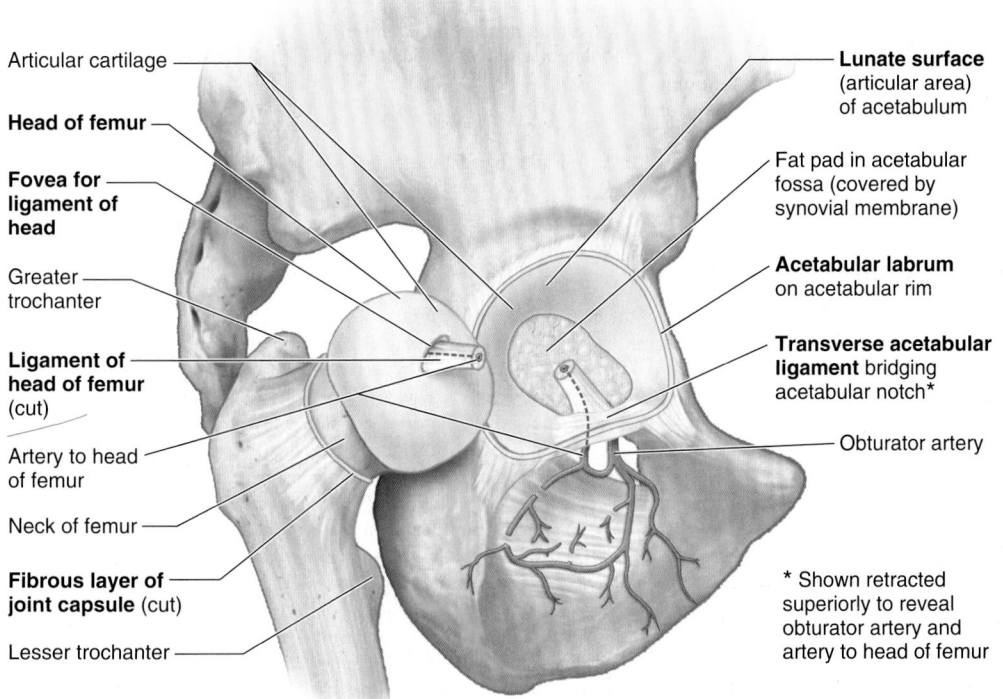

Articular cartilage

Head of femur

Fovea for ligament of head

Greater trochanter

Ligament of head of femur (cut)

Artery to head of femur

Neck of femur

Fibrous layer of joint capsule (cut)

Lesser trochanter

Lunate surface (articular area) of acetabulum

Fat pad in acetabular fossa (covered by synovial membrane)

Acetabular labrum on acetabular rim

Transverse acetabular ligament bridging acetabular notch*

Obturator artery

* Shown retracted superiorly to reveal obturator artery and artery to head of femur

Lateral view

FIGURE 5.78. Hip joint. The joint was disarticulated by cutting the ligament of the head of the femur and retracting the head from the acetabulum. The transverse acetabular ligament is retracted superiorly to show the obturator canal, which transmits the obturator nerve and vessels passing from the pelvic cavity to the medial thigh.

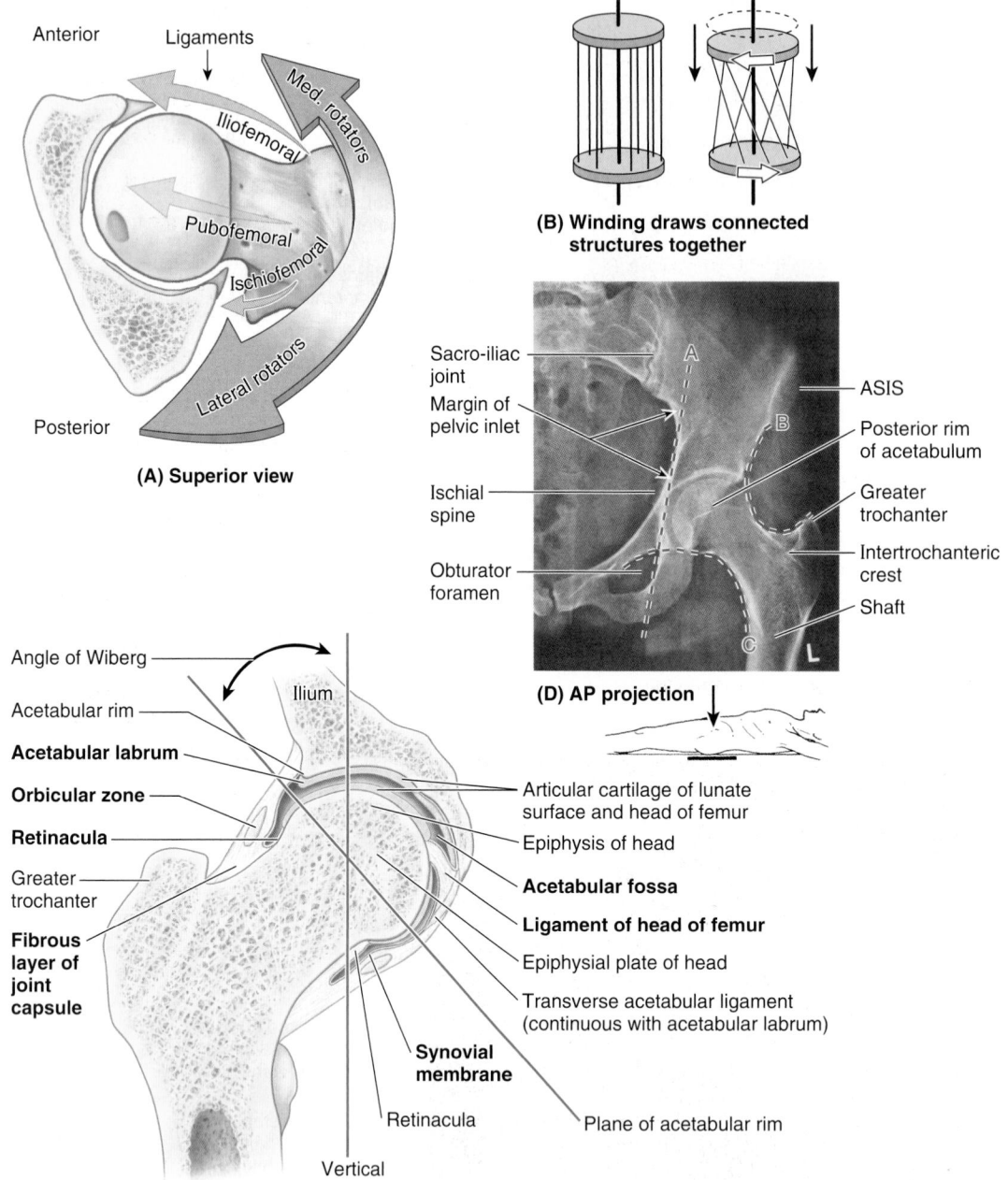

(A) Superior view

(B) Winding draws connected structures together

(D) AP projection

(C) Anterior view of coronal section

FIGURE 5.79. Factors increasing stability of hip joint. A. This superior view of the hip joint demonstrates the medial and reciprocal pull of the peri-articular muscles (medial and lateral rotators; *reddish brown arrows*) and intrinsic ligaments of the hip joint (*gray arrows*) on the femur. Relative strengths are indicated by arrow width: Anteriorly, the muscles are less abundant but the ligaments are robust; posteriorly, the muscles predominate. **B.** Parallel fibers linking two discs resemble those making up the tube-like fibrous layer of the hip joint capsule. When one disc (the femur) rotates relative to the other (the acetabulum), the fibers become increasingly oblique and draw the two discs together. Similarly, extension of the hip joint winds (increases the obliquity of) the fibers of the fibrous layer, pulling the head and neck of the femur tightly into the acetabulum, increasing the stability of the joint. Flexion unwinds the fibers of the capsule. **C.** In this coronal section of hip joint, the acetabular labrum and transverse acetabular ligament, spanning the acetabular notch (and included in the plane of section here), extend the acetabular rim so that a complete socket is formed. Thus the acetabular complex engulfs the head of the femur. The epiphysis of the femoral head is entirely within the joint capsule. The thick weight-bearing bone of the ilium normally lies directly superior to the head of the femur for efficient transfer of weight to the femur (Fig. 5.3). The angle of Wiberg (see text) is used radiographically to determine the degree to which the acetabulum overhangs the head of the femur. **D.** Several different lines and curvatures are used in the detection of hip abnormalities (dislocations, fractures, or slipped epiphyses). The Kohler line (*red A*) is normally tangential to the pelvic inlet and the obturator foramen. The acetabular fossa should lie lateral to this line. A fossa that crosses the line suggests an acetabular fracture with inward displacement. The Shenton line (*red B*) and the iliofemoral line (*red C*) should appear in a normal AP radiograph as smooth, continuous lines that are bilaterally symmetrical. The Shenton line is a radiographic indication of the angle of inclination (*ASIS* = anterior superior iliac spine).

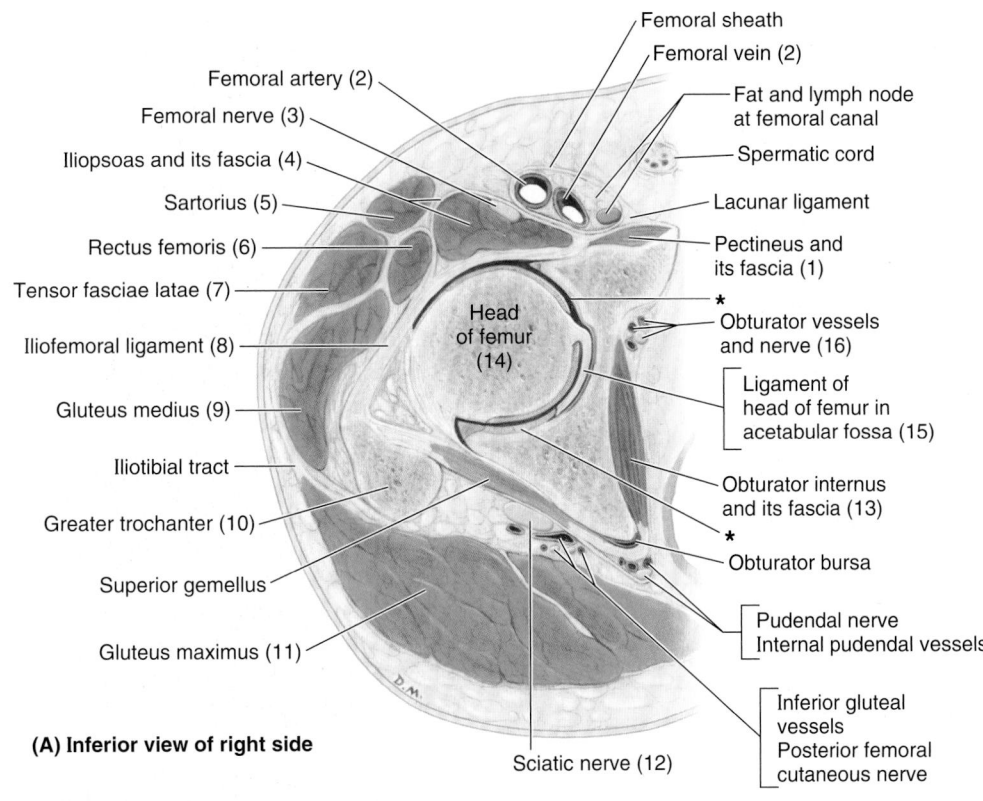

Femoral sheath
Femoral vein (2)
Femoral artery (2)
Femoral nerve (3)
Iliopsoas and its fascia (4)
Sartorius (5)
Rectus femoris (6)
Tensor fasciae latae (7)
Iliofemoral ligament (8)
Gluteus medius (9)
Iliotibial tract
Greater trochanter (10)
Superior gemellus
Gluteus maximus (11)

Fat and lymph node at femoral canal
Spermatic cord
Lacunar ligament
Pectineus and its fascia (1)
★
Obturator vessels and nerve (16)
Ligament of head of femur in acetabular fossa (15)
Obturator internus and its fascia (13)
★
Obturator bursa
Pudendal nerve
Internal pudendal vessels
Inferior gluteal vessels
Posterior femoral cutaneous nerve

Head of femur (14)

(A) Inferior view of right side

Sciatic nerve (12)

★ Lunate surface of acetabulum

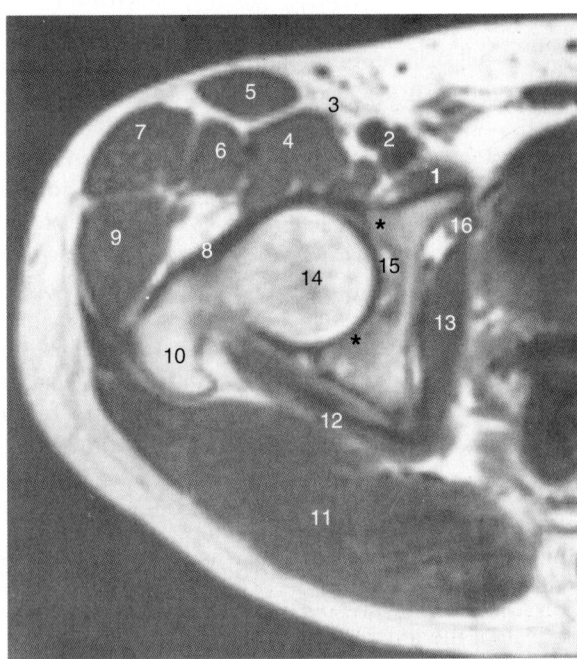

(B) Inferior view of right side

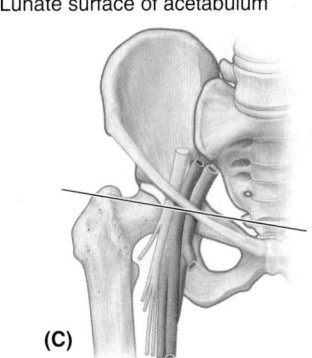

(C)

FIGURE 5.80. Sectional and radiographic anatomy of gluteal region and proximal anterior thigh at level of hip joint. A and B. A descriptive drawing and transverse (axial MRI) study of an anatomical section of the thigh are shown. Numbers in parentheses in (**A**) refer to structures identified in (**B**). **C.** The orientation drawing shows the level of the section.

(Figs. 5.78–5.80). The acetabular rim and lunate surface form approximately three quarters of a circle; the missing inferior segment of the circle is the *acetabular notch*.

The lip-shaped **acetabular labrum** (L. *labrum*, lip) is a fibrocartilaginous rim attached to the margin of the acetabulum, increasing the acetabular articular area by nearly 10%. The **transverse acetabular ligament,** a continuation of the

acetabular labrum, bridges the acetabular notch (Figs. 5.78 and 5.79C). As a result of the height of the rim and labrum, more than half of the femoral head fits within the acetabulum (Figs. 5.79C and 5.80). Thus during dissection, the femoral head must be cut from the acetabular rim to enable disarticulation of the joint. Centrally a deep non-articular part, called the *acetabular fossa*, is formed mainly by the ischium

(Figs. 5.78, 5.79C, and 5.80). This fossa is thin walled (often translucent) and continuous inferiorly with the acetabular notch.

The articular surfaces of the acetabulum and femoral head are most congruent when the hip is flexed 90°, abducted 5°, and rotated laterally 10° (the position in which the axis of the acetabulum and the axis of the femoral head and neck are aligned), which is the quadruped position!

In other words, in assuming the upright position, a relatively small degree of joint stability was sacrificed to maximize weight-bearing when erect. Even so, the hip joint is our most stable joint, owing also to its complete ball and socket construction (depth of socket), the strength of its joint capsule, and the attachments of muscles crossing the joint, many of which are located at some distance from the center of movement (Palastanga et al., 2011).

JOINT CAPSULE OF HIP JOINT

The hip joints are enclosed within strong *joint capsules,* formed of a loose external *fibrous layer* (fibrous capsule) and an internal *synovial membrane* (Fig. 5.79C). Proximally, the fibrous layer attaches to the acetabulum, just peripheral to the acetabular rim to which the labrum is attached, and to the transverse acetabular ligament (Figs. 5.79C and 5.81A, C, & *D*). Distally, the fibrous layer attaches to the femoral neck only anteriorly at the *intertrochanteric line* and root of the greater trochanter (Fig. 5.81B). Posteriorly, the fibrous layer crosses the neck proximal to the *intertrochanteric crest* but is not attached to it.

Most fibers of the fibrous layer of the capsule take a spiral course from the hip bone to the intertrochanteric line of the femur, but some deep fibers pass circularly around the neck, forming the **orbicular zone** (Fig. 5.79C and 5.81D). Thick parts of the fibrous layer form the **ligaments of the hip joint,** which pass in a spiral fashion from the pelvis to the femur (Fig. 5.81A, C, & *D*). Extension winds the spiraling ligaments and fibers more tightly, constricting the capsule and drawing the femoral head tightly into the acetabulum (Fig. 5.79B). The tightened fibrous layer increases the stability of the joint, but restricts extension of the joint to 10–20° beyond the vertical position. Flexion increasingly unwinds the spiraling ligaments and fibers. This permits considerable flexion of the hip joint with increasing mobility.

Of the three intrinsic ligaments of the joint capsule below, it is the first one that reinforces and strengthens the joint:

- *Anteriorly and superiorly* is the strong, Y-shaped **iliofemoral ligament,** which attaches to the anterior inferior iliac spine and the acetabular rim proximally and the intertrochanteric line distally (Fig. 5.81A & C). Said to be the body's strongest ligament, the iliofemoral ligament specifically prevents hyperextension of the hip joint during standing by screwing the femoral head into the acetabulum via the mechanism described above.

- *Anteriorly and inferiorly* is the **pubofemoral ligament,** which arises from the obturator crest of the pubic bone and passes laterally and inferiorly to merge with the fibrous layer of the joint capsule (Fig. 5.81A). This ligament blends with the medial part of the iliofemoral ligament, and tightens during both extension and abduction of the hip joint. The pubofemoral ligament prevents overabduction of the hip joint.

- *Posteriorly* is the **ischiofemoral ligament,** which arises from the ischial part of the acetabular rim (Fig. 5.81D). The weakest of the three ligaments, it spirals superolaterally to the femoral neck, medial to the base of the greater trochanter.

The relative size, strengths, and positions of the three ligaments of the hip joint are shown in Figure 5.79A. The ligaments and peri-articular muscles (the medial and lateral rotators of the thigh) play a vital role in maintaining the structural integrity of the joint.

Both muscles and ligaments pull the femoral head medially into the acetabulum, and they are reciprocally balanced when doing so. The medial flexors, located anteriorly, are fewer, weaker, and less mechanically advantaged, whereas the anterior ligaments are strongest. Conversely, the ligaments are weaker posteriorly where the medial rotators are abundant, stronger, and more mechanically advantaged.

In all synovial joints, a synovial membrane lines the internal surfaces of the fibrous layer, as well as any intracapsular bony surfaces not lined with articular cartilage. Thus in the hip joint, where the fibrous layer attaches to the femur distant from the articular cartilage covering the femoral head, the **synovial membrane of the hip joint** reflects proximally along the femoral neck to the edge of the femoral head. Longitudinal synovial folds (*retinacula*) occur in the synovial membrane covering the femoral neck (Fig. 5.79C). Subsynovial **retinacular arteries** (branches of the medial, and a few of the lateral, circumflex femoral artery) that supply the femoral head and neck course within the synovial folds (Fig. 5.82).

The **ligament of the head of the femur** (Figs. 5.78, 5.79C, 5.80, and 5.82), primarily a synovial fold conducting a blood vessel, is weak and of little importance in strengthening the hip joint. Its wide end attaches to the margins of the acetabular notch and the transverse acetabular ligament; its narrow end attaches to the *fovea for the ligament of the head.*

Usually, the ligament contains a small artery to the head of the femur. A *fat pad* in the acetabular fossa fills the part of the acetabular fossa that is not occupied by the ligament of the femoral head (Fig. 5.78). Both the ligament and the fat pad are covered with synovial membrane. The malleable nature of the fat pad permits it to change shape to accommodate the variations in the congruity of the femoral head and acetabulum, as well as changes in the position of the ligament of the head during joint movements. A *synovial protrusion* beyond the free margin of the joint capsule onto the posterior aspect of the femoral neck forms a bursa for the obturator externus tendon (Fig. 5.81D).

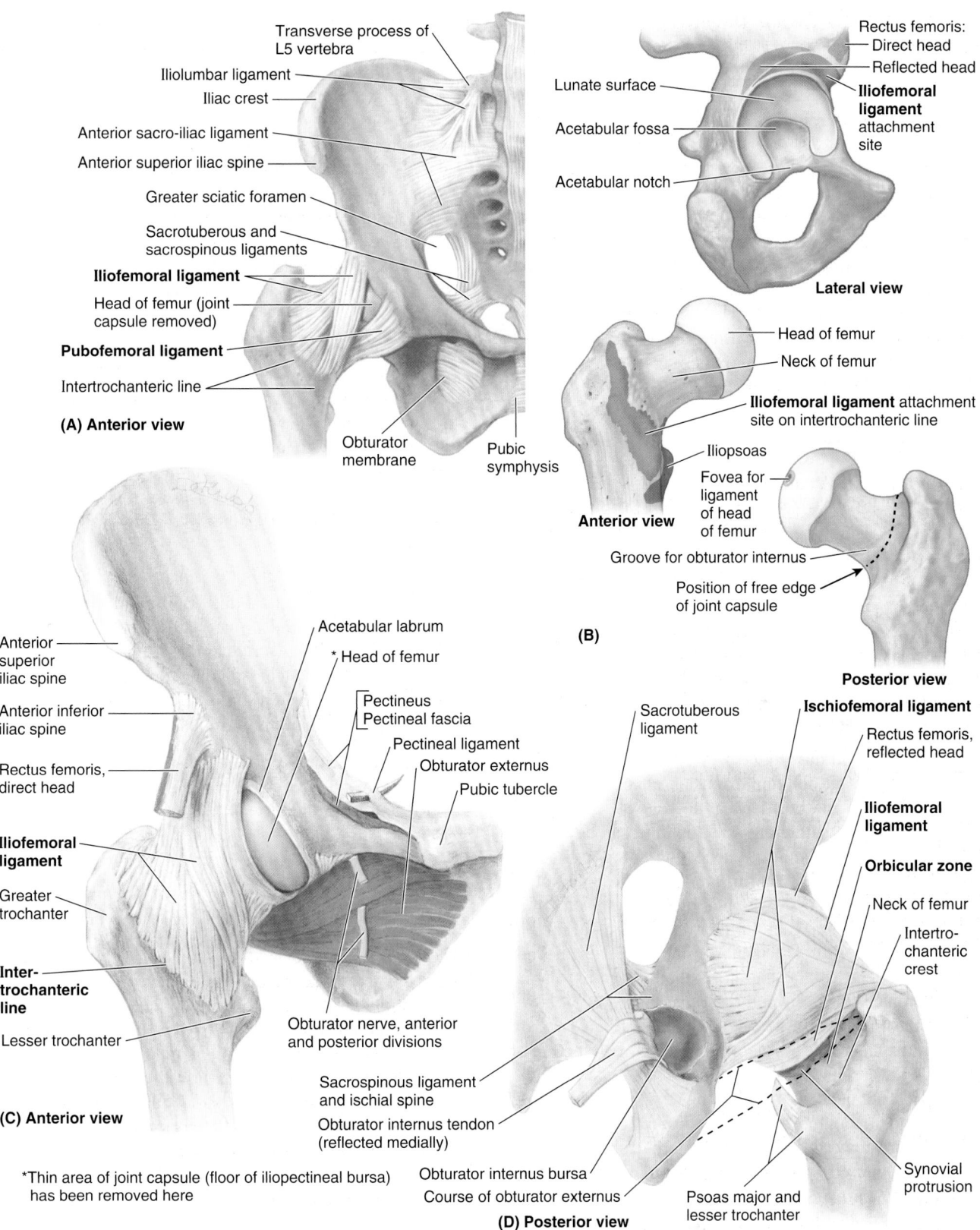

(A) Anterior view

Transverse process of L5 vertebra
Iliolumbar ligament
Iliac crest
Anterior sacro-iliac ligament
Anterior superior iliac spine
Greater sciatic foramen
Sacrotuberous and sacrospinous ligaments
Iliofemoral ligament
Head of femur (joint capsule removed)
Pubofemoral ligament
Intertrochanteric line
Obturator membrane
Pubic symphysis

Rectus femoris:
Direct head
Reflected head
Iliofemoral ligament attachment site
Lunate surface
Acetabular fossa
Acetabular notch

Lateral view

Head of femur
Neck of femur
Iliofemoral ligament attachment site on intertrochanteric line
Iliopsoas
Fovea for ligament of head of femur
Groove for obturator internus
Position of free edge of joint capsule

Anterior view

(B)

Posterior view

(C) Anterior view

Anterior superior iliac spine
Anterior inferior iliac spine
Rectus femoris, direct head
Iliofemoral ligament
Greater trochanter
Inter-trochanteric line
Lesser trochanter

Acetabular labrum
*Head of femur
Pectineus
Pectineal fascia
Pectineal ligament
Obturator externus
Pubic tubercle
Obturator nerve, anterior and posterior divisions
Sacrospinous ligament and ischial spine
Obturator internus tendon (reflected medially)

*Thin area of joint capsule (floor of iliopectineal bursa) has been removed here

Sacrotuberous ligament
Ischiofemoral ligament
Rectus femoris, reflected head
Iliofemoral ligament
Orbicular zone
Neck of femur
Intertrochanteric crest
Synovial protrusion
Psoas major and lesser trochanter
Course of obturator externus
Obturator internus bursa

(D) Posterior view

FIGURE 5.81. Ligaments of pelvis and hip joint. A. Weight transfer from the vertebral column to the pelvic girdle is a function of the sacro-iliac ligaments. Weight transfer at the hip joint is accomplished primarily by the disposition of the bones, with the ligaments limiting the range of movement and adding stability. **B.** Articulating surfaces of hip joint and sites of attachment and tendinous relationships of iliofemoral ligaments and joint capsule. **C.** Iliofemoral ligament. **D.** The ischiofemoral ligament. Because the joint capsule does not attach to the posterior aspect of the femur, the synovial membrane protrudes from the joint capsule, forming the obturator externus bursa to facilitate movement of the tendon of the obturator externus (shown in part **C**) over the bone.

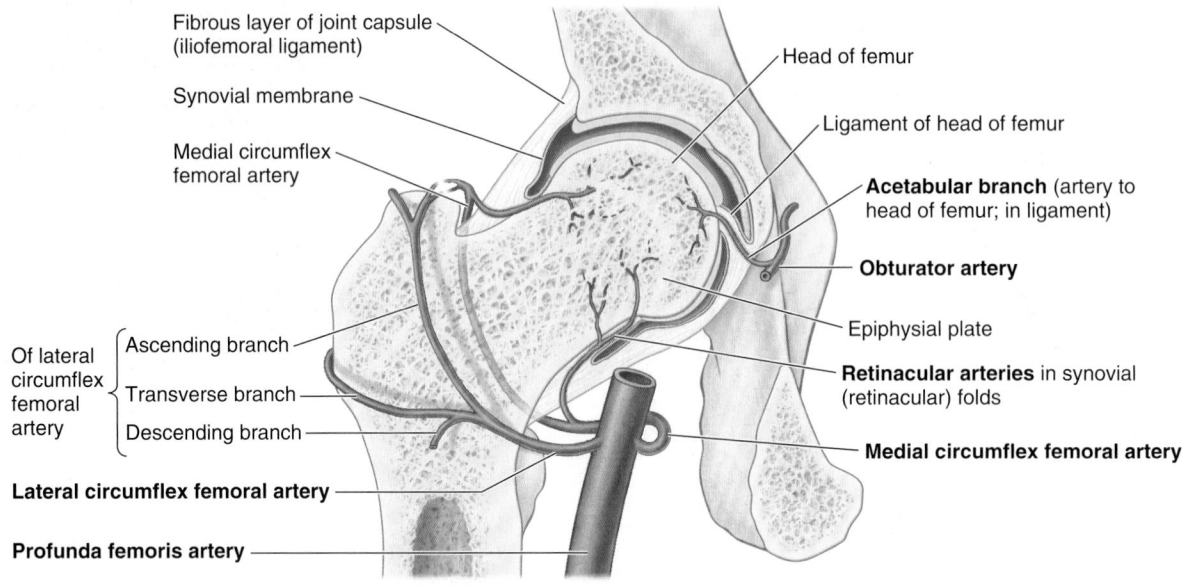

Anterior view of coronally sectioned hip joint

FIGURE 5.82. Blood supply of head and neck of femur. Branches of the medial and lateral circumflex femoral arteries, branches of the profunda femoris artery, and the artery to the femoral head (a branch of the obturator artery) supply the head and neck of the femur. In the adult, the medial circumflex femoral artery is the most important source of blood to the femoral head and adjacent (proximal) neck.

MOVEMENTS OF HIP JOINT

Hip movements are flexion–extension, abduction–adduction, medial–lateral rotation, and circumduction (Fig. 5.83). Movements of the trunk at the hip joints are also important, such as those occurring when a person lifts their trunk from the supine position during sit-ups, or keeps the pelvis level when one foot is off the ground.

The degree of flexion and extension possible at the hip joint depends on the position of the knee. If the knee is flexed, relaxing the hamstrings, the hip joint can be actively flexed until the thigh almost reaches the anterior abdominal wall and can reach it via further passive flexion. Not all of this movement occurs at the hip joint; some results from flexion of the vertebral column. During extension of the hip joint, the fibrous layer of the joint capsule, especially the iliofemoral ligament, is taut; therefore, the hip can usually be extended only slightly beyond the vertical except by movement of the bony pelvis (flexion of lumbar vertebrae).

From the anatomical position, the range of abduction of the hip joint is usually somewhat greater than for adduction. About 60° of abduction is possible when the thigh is extended at the hip joint, and more when it is flexed. Lateral rotation is much more powerful than medial rotation.

The main muscles producing movements of the hip joint are listed in Figure 5.83B. Note that:

1. The *iliopsoas* is the strongest flexor of the hip.
2. In addition to its function as an adductor, the *adductor magnus* also serves as a flexor (anterior or aponeurotic part) and an extensor (posterior or hamstrings part).
3. Several muscles participate in both flexion and adduction (*pectineus* and *gracilis* as well all three "adductor" muscles).

4. In addition to serving as abductors, the anterior portions of the *gluteus medius* and *minimus* are also medial rotators.
5. The *gluteus maximus* serves as the primary extensor from the flexed to the straight (standing) position, and from this point posteriorly, extension is achieved primarily by the hamstrings. The gluteus maximus is also a lateral rotator.

BLOOD SUPPLY OF HIP JOINT

Arteries supplying the hip joint (Fig. 5.82) include the following:

- The *medial* and *lateral circumflex femoral arteries,* which are usually branches of the *profunda femoris artery* but occasionally they arise as branches of the *femoral artery.*
- The **artery to the head of the femur,** which is a branch of the obturator artery of variable size; it traverses the ligament of the head.

The main blood supply of the hip joint is from the *retinacular arteries* arising as branches of the circumflex femoral arteries. Retinacular arteries arising from the medial circumflex femoral artery are most abundant, bringing more blood to the head and neck of the femur because they are able to pass beneath the unattached posterior border of the joint capsule. Retinacular arteries arising from the lateral circumflex femoral must penetrate the thick iliofemoral ligament and are smaller and fewer.

NERVE SUPPLY OF HIP JOINT

Hilton's law states that the nerves supplying the muscles extending directly across and acting at a given joint also innervate the joint. Articular rami arise from the intramuscular rami

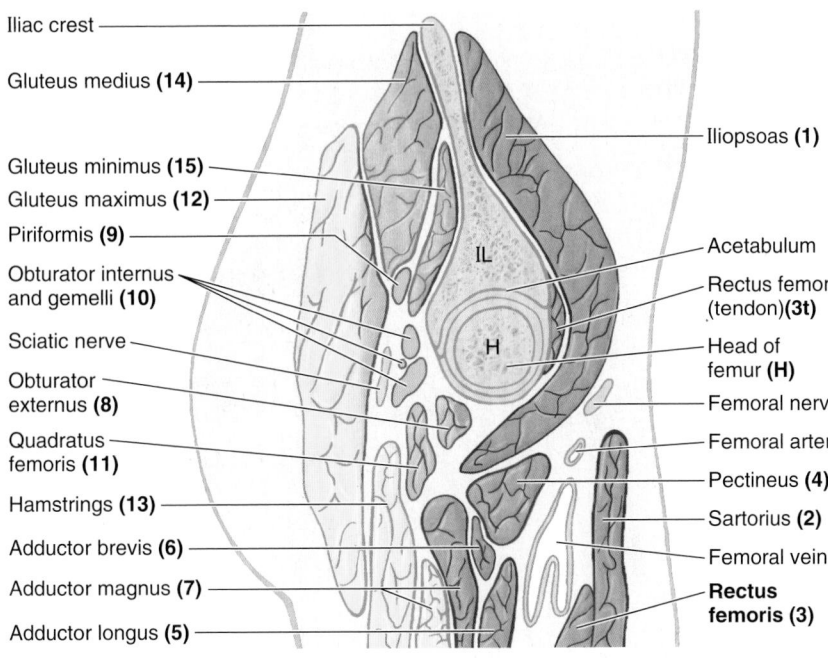

Iliac crest
Gluteus medius (14)
Gluteus minimus (15)
Gluteus maximus (12)
Piriformis (9)
Obturator internus and gemelli (10)
Sciatic nerve
Obturator externus (8)
Quadratus femoris (11)
Hamstrings (13)
Adductor brevis (6)
Adductor magnus (7)
Adductor longus (5)

Iliopsoas (1)
Acetabulum
Rectus femoris (tendon)(3t)
Head of femur (H)
Femoral nerve
Femoral artery
Pectineus (4)
Sartorius (2)
Femoral vein
Rectus femoris (3)

IL
H

(A) Lateral view of sagittal section through femoral head

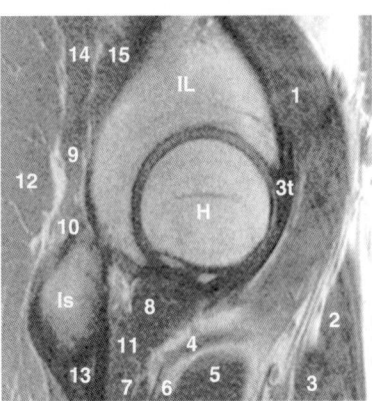

(B) MRI of hip joint, lateral view (Is, Ischium)

Functional groups of muscles acting at hip joint

Flexors
Iliopsoas (1)
Sartorius (2)
Tensor fasciae latae
Rectus femoris (3) tendon (3t)
Pectineus (4)
Adductor longus (5)
Adductor brevis (6)
Adductor magnus—anterior part (7)
Gracilis

Adductors
Pectineus (4)
Adductor longus (5)
Adductor brevis (6)
Adductor magnus (7)
Obturator externus (8)
Gracilis

Lateral rotators
Obturator externus (8)
Piriformis (9)
Obturator internus (10)
Gemelli (10)
Quadratus femoris (11)
Gluteus maximus (12)

Extensors
Gluteus maximus (12)
Hamstrings: (13)
Semitendinosus
Semimembranosus
Long head, biceps femoris
Adductor magnus—posterior part

Abductors
Gluteus medius (14)
Gluteus minimus (15)
Tensor fasciae latae

Medial rotators
Gluteus medius (14) } Anterior parts
Gluteus minimus (15) }
Tensor fasciae latae

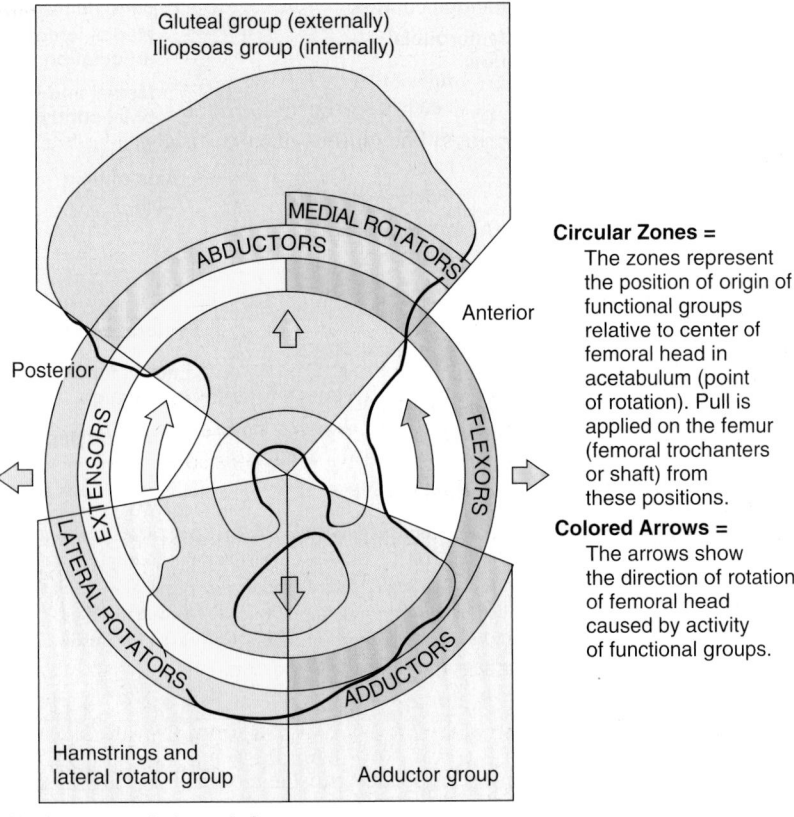

Gluteal group (externally)
Iliopsoas group (internally)

MEDIAL ROTATORS
ABDUCTORS
Anterior
Posterior
EXTENSORS
FLEXORS
LATERAL ROTATORS
ADDUCTORS

Hamstrings and lateral rotator group

Adductor group

Circular Zones =
The zones represent the position of origin of functional groups relative to center of femoral head in acetabulum (point of rotation). Pull is applied on the femur (femoral trochanters or shaft) from these positions.

Colored Arrows =
The arrows show the direction of rotation of femoral head caused by activity of functional groups.

(C) Diagrammatic lateral view

FIGURE 5.83. Relations of hip joint and muscles producing movements of joint. A. Sagittal section of the hip joint showing the muscles, vessels, and nerves related to it. The muscles are color coded to indicate their function(s). Applying Hilton's law, it is possible to deduce the innervation of the hip joint by knowing which muscles directly cross and act on the joint and their nerve supply. **B.** The relative positions of the muscles producing movements of the hip joint and the direction of the movement are demonstrated.

of the muscular branches and directly from named nerves. A knowledge of the nerve supply of the muscles and their relationship to the joints can allow one to deduce the nerve supply of many joints. Possible deductions regarding the hip joint and its muscular relationships include (Fig. 5.83):

- Flexors innervated by the femoral nerve pass anterior to the hip joint; the anterior aspect of the hip joint is innervated by the femoral nerve (directly and via articular rami of the muscular branches to the pectineus and rectus femoris).
- Lateral rotators pass inferior and posterior to the hip joint; the inferior aspect of the joint is innervated by the obturator nerve (directly and via articular rami of the muscular branch to the obturator externus), and the posterior aspect is innervated by the nerve to the quadratus femoris.
- Abductors innervated by the superior gluteal nerve pass superior to the hip joint; the superior aspect of the joint is innervated by the superior gluteal nerve.

Pain perceived as coming from the hip joint may be misleading because pain can be referred from the vertebral column.

Knee Joint

The **knee joint** is the largest and most superficial joint. It is primarily a hinge type of synovial joint, allowing flexion and extension; however, the hinge movements are combined with gliding and rolling, and with rotation about a vertical axis. Although the knee joint is well constructed, its function is commonly impaired when it is hyperextended (e.g., in body contact sports, such as ice hockey and football).

ARTICULATIONS, ARTICULAR SURFACES, AND STABILITY OF KNEE JOINT

Relevant anatomical details of the involved bones, including their articulating surfaces, were discussed in "Bones of Lower Limb," p. 512. The articular surfaces of the knee joint are characterized by their large size and their complicated and incongruent shapes. The knee joint consists of three articulations (Figs. 5.84 and 5.85):

- Two **femorotibial articulations** (**lateral** and **medial**) between the lateral and the medial femoral and tibial condyles.
- One intermediate **femoropatellar articulation** between the patella and the femur.

The fibula is not involved in the knee joint.

The knee joint is relatively weak mechanically because of the incongruence of its articular surfaces, which has been compared to two balls sitting on a warped tabletop. The stability of the knee joint depends on (1) the strength and actions of the surrounding muscles and their tendons, and (2) the ligaments that connect the femur and tibia. Of these supports, the muscles are most important; therefore, many sport injuries are preventable through appropriate conditioning and training.

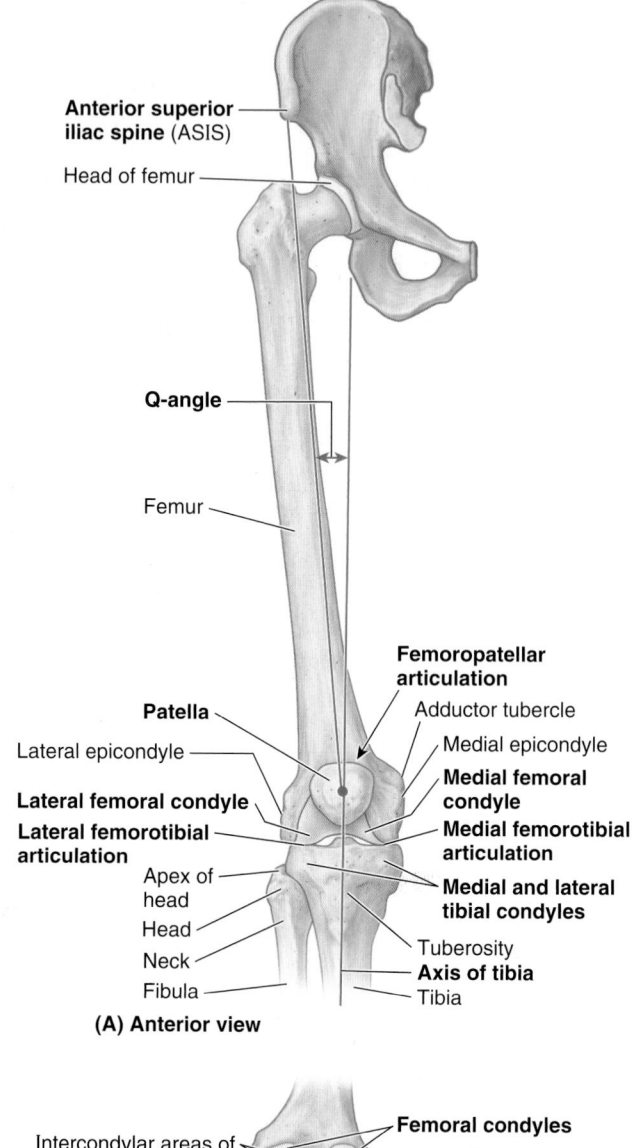

FIGURE 5.84. Bones of knee joint. A The bones articulating at the knee joint are shown. The hip bone and proximal femur are included to demonstrate the Q-angle, determined during physical examination to indicate alignment of the femur and tibia and to evaluate valgus or varus stress at the knee. **B.** The bones and bony features of the posterior aspect of the knee joint and knee are shown.

The most important muscle in stabilizing the knee joint is the large *quadriceps femoris*, particularly the inferior fibers of the vastus medialis and lateralis (Fig. 5.86A). The knee joint functions surprisingly well after a ligament strain if the quadriceps is well conditioned.

The erect, extended position is the most stable position of the knee joint. In this position, the articular surfaces are most

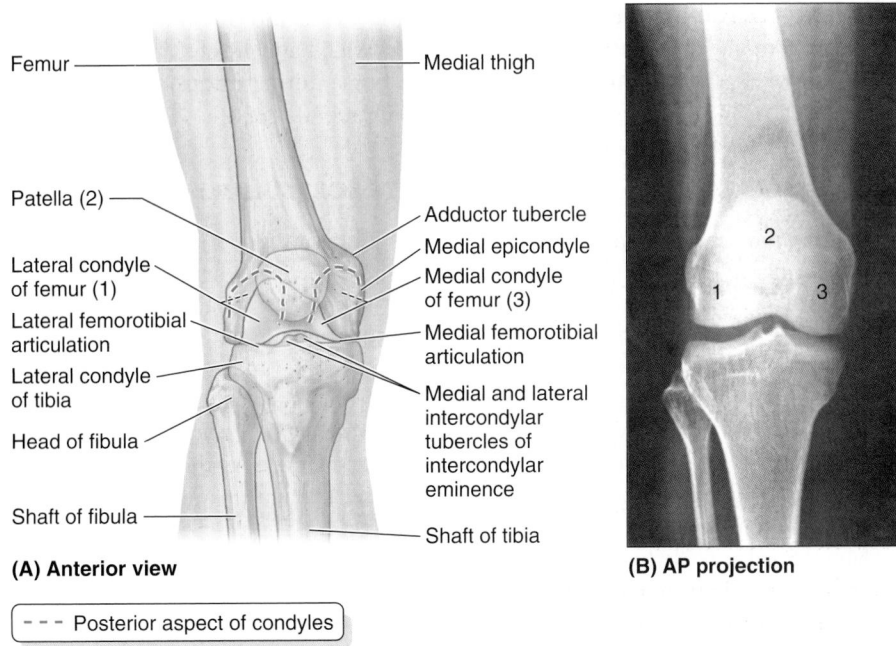

(A) Anterior view

Femur
Medial thigh
Patella (2)
Adductor tubercle
Medial epicondyle
Lateral condyle of femur (1)
Medial condyle of femur (3)
Lateral femorotibial articulation
Lateral condyle of tibia
Medial femorotibial articulation
Head of fibula
Medial and lateral intercondylar tubercles of intercondylar eminence
Shaft of fibula
Shaft of tibia

(B) AP projection

--- Posterior aspect of condyles

FIGURE 5.85. Radiography of knee joint. A and B. The orientation drawing depicts the structures visible in the AP radiograph of the right knee joint.

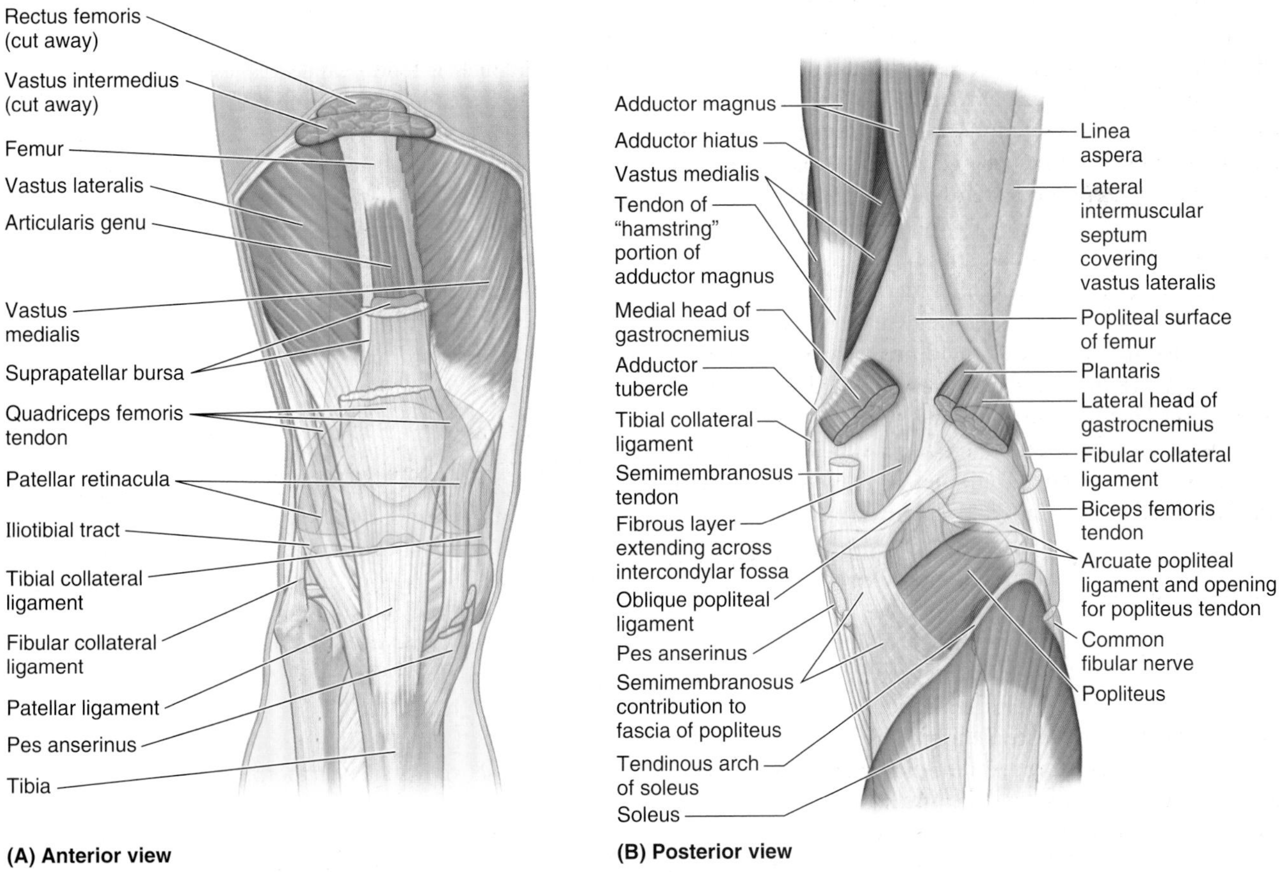

(A) Anterior view

Rectus femoris (cut away)
Vastus intermedius (cut away)
Femur
Vastus lateralis
Articularis genu
Vastus medialis
Suprapatellar bursa
Quadriceps femoris tendon
Patellar retinacula
Iliotibial tract
Tibial collateral ligament
Fibular collateral ligament
Patellar ligament
Pes anserinus
Tibia

(B) Posterior view

Adductor magnus
Adductor hiatus
Vastus medialis
Tendon of "hamstring" portion of adductor magnus
Medial head of gastrocnemius
Adductor tubercle
Tibial collateral ligament
Semimembranosus tendon
Fibrous layer extending across intercondylar fossa
Oblique popliteal ligament
Pes anserinus
Semimembranosus contribution to fascia of popliteus
Tendinous arch of soleus
Soleus
Linea aspera
Lateral intermuscular septum covering vastus lateralis
Popliteal surface of femur
Plantaris
Lateral head of gastrocnemius
Fibular collateral ligament
Biceps femoris tendon
Arcuate popliteal ligament and opening for popliteus tendon
Common fibular nerve
Popliteus

FIGURE 5.86. External aspect of joint capsule of knee. The fibrous layer of the joint capsule is relatively thin in some places and thickened in others to form reinforcing intrinsic (capsular) ligaments. **A.** Modifications of the anterior aspect and sides of the fibrous layer include the patellar retinacula, which attach to the sides of the quadriceps tendon, patella, and patellar ligament, and incorporation of the iliotibial tract (laterally) and the medial collateral ligament (medially). **B.** The hamstring and gastrocnemius muscles and the posterior intermuscular septum have been cut and removed to expose the adductor magnus, lateral intermuscular septum, and the floor of the popliteal fossa. Posterior modifications of the fibrous layer include the oblique and arcuate popliteal ligaments, and a perforation inferior to the arcuate popliteal ligament to allow passage of the popliteus tendon.

congruent (contact is minimized in all other positions); the primary ligaments of the joint (collateral and cruciate ligaments) are taut, and the many tendons surrounding the joint provide a splinting effect.

JOINT CAPSULE OF KNEE JOINT

The **joint capsule of the knee joint** is typical in consisting of an external *fibrous layer of the capsule* (fibrous capsule) and an internal *synovial membrane* that lines all internal surfaces of the articular cavity not covered with articular cartilage (5.87B). The fibrous layer has a few thickened parts that make up intrinsic ligaments, but for the main part, it is thin and is actually incomplete in some areas (5.86B). The fibrous layer attaches to the femur superiorly, just proximal to the articular margins of the condyles. Posteriorly, the fibrous layer encloses the condyles and the *intercondylar fossa*. The fibrous layer has an opening or gap posterior to the lateral tibial condyle to allow the tendon of the popliteus to pass out of the joint capsule to attach to the tibia (Fig. 5.87B). Inferiorly, the fibrous layer attaches to the margin of the superior articular surface (tibial plateau) of the tibia, except where the tendon of the popliteus crosses the bone (Figs. 5.86A & B and 5.87B). The quadriceps tendon, patella, and patellar ligament replace the fibrous layer anteriorly—that is, the fibrous layer is continuous with the lateral and medial margins of these structures, and there is no separate fibrous layer in the region of these structures (Figs. 5.86A and 5.87B).

The extensive *synovial membrane of the capsule* lines all surfaces bounding the articular cavity (the space containing synovial fluid) not covered by articular cartilage (Fig. 5.87A & B). Thus, it attaches to the periphery of the articular cartilage covering the femoral and tibial condyles; the posterior surface of the patella; and the edges of the *menisci*, the fibrocartilaginous discs between the tibial and femoral articular surfaces. The synovial membrane lines the internal surface of the fibrous layer laterally and medially, but centrally it becomes separated from the fibrous layer.

From the posterior aspect of the joint, the synovial membrane reflects anteriorly into the intercondylar region, covering the cruciate ligaments and the **infrapatellar fat pad,** so that they are excluded from the articular cavity. This creates a median **infrapatellar synovial fold,** a vertical fold of synovial membrane that approaches the posterior aspect of the patella, occupying all but the most anterior part of the intercondylar region. Thus, it almost subdivides the articular cavity into right and left femorotibial articular cavities; indeed, this is how arthroscopic surgeons consider the articular cavity. Fat-filled **lateral** and **medial alar folds** cover the inner surface of fat pads that occupy the space on each side of the patellar ligament internal to the fibrous layer.

Superior to the patella, the knee joint cavity extends deep to the vastus intermedius as the *suprapatellar bursa* (Figs. 5.86A and 5.88A & B). The synovial membrane of the joint capsule is continuous with the synovial lining of this bursa. This large bursa usually extends approximately 5 cm superior to the patella; however, it may extend halfway up the anterior aspect of the femur. Muscle slips deep

to the vastus intermedius form the *articular muscle of the knee,* which attaches to the synovial membrane and retracts the bursa during extension of the knee (see Figs. 5.22 and 5.86A).

EXTRACAPSULAR LIGAMENTS OF KNEE JOINT

The joint capsule is strengthened by five extracapsular or capsular (intrinsic) ligaments: patellar ligament, fibular collateral ligament, tibial collateral ligament, oblique popliteal ligament, and arcuate popliteal ligament (Fig. 5.86A & B). They are sometimes called *external ligaments* to differentiate them from internal ligaments, such as the cruciate ligaments.

The *patellar ligament,* the distal part of the quadriceps femoris tendon, is a strong, thick fibrous band passing from the apex and adjoining margins of the patella to the tibial tuberosity (Fig. 5.86A). The patellar ligament is the anterior ligament of the knee joint. Laterally, it receives the *medial* and *lateral patellar retinacula,* aponeurotic expansions of the vastus medialis and lateralis and overlying deep fascia. The retinacula make up the joint capsule of the knee on each side of the patella (Figs. 5.86A and 5.87B) and play an important role in maintaining alignment of the patella relative to the patellar articular surface of the femur. The oblique placement of the femur and/or line of pull of the quadriceps femoris muscle relative to the axis of the patellar tendon and tibia, assessed clinically as the *Q-angle,* favors lateral displacement of the patella (Fig. 5.84).

The *collateral ligaments of the knee* are taut when the knee is fully extended, contributing to stability while standing (Fig. 5.88A & D). As flexion proceeds, they become increasingly slack, permitting and limiting (serving as check ligaments for) rotation at the knee.

The **fibular collateral ligament** (**FCL;** lateral collateral ligament), a cord-like extracapsular ligament, is strong. It extends inferiorly from the lateral epicondyle of the femur to the lateral surface of the fibular head (Fig. 5.88A & C). The tendon of the popliteus passes deep to the FCL, separating it from the lateral meniscus. The tendon of the biceps femoris is split into two parts by the FCL (Fig. 5.88A).

The **tibial collateral ligament** (**TCL;** medial collateral ligament) is a strong, flat, intrinsic (capsular) band that extends from the medial epicondyle of the femur to the medial condyle and the superior part of the medial surface of the tibia (Fig. 5.88D & E). At its midpoint, the deep fibers of the TCL are firmly attached to the medial meniscus. The TCL, weaker than the FCL, is more often damaged. As a result, the TCL and medial meniscus are commonly torn during contact sports such as football and ice hockey.

The **oblique popliteal ligament** is a recurrent expansion of the tendon of the semimembranosus that reinforces the joint capsule posteriorly as it spans the intracondylar fossa (Fig. 5.86B). The ligament arises posterior to the medial tibial condyle and passes superolaterally toward the lateral femoral condyle, blending with the central part of the posterior aspect of the joint capsule.

The **arcuate popliteal ligament** also strengthens the joint capsule posterolaterally. It arises from the posterior

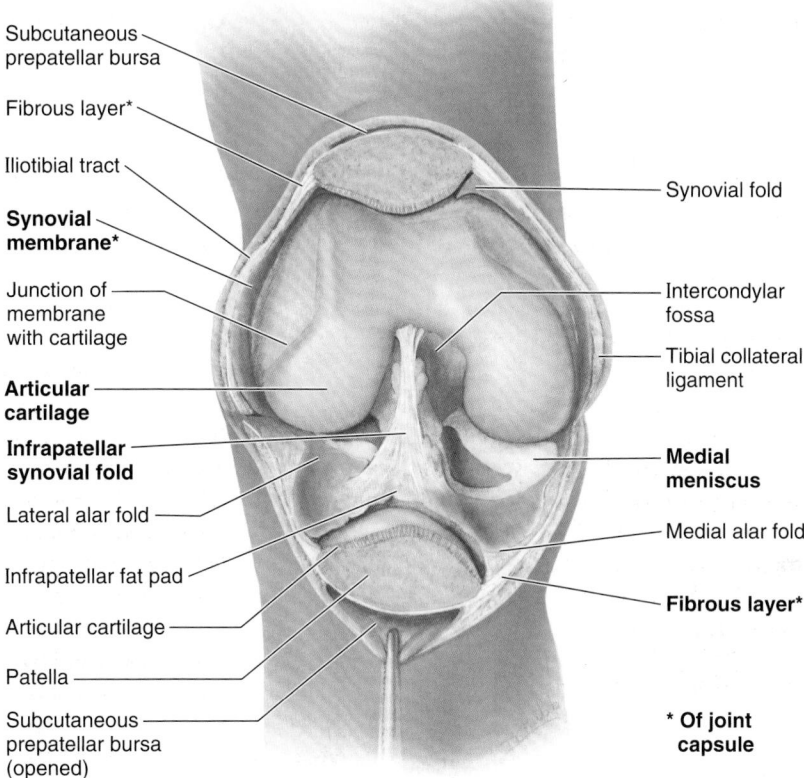

Subcutaneous prepatellar bursa

Fibrous layer*

Iliotibial tract

Synovial membrane*

Junction of membrane with cartilage

Articular cartilage

Infrapatellar synovial fold

Lateral alar fold

Infrapatellar fat pad

Articular cartilage

Patella

Subcutaneous prepatellar bursa (opened)

Synovial fold

Intercondylar fossa

Tibial collateral ligament

Medial meniscus

Medial alar fold

Fibrous layer*

* **Of joint capsule**

(A) Anterior view of flexed knee

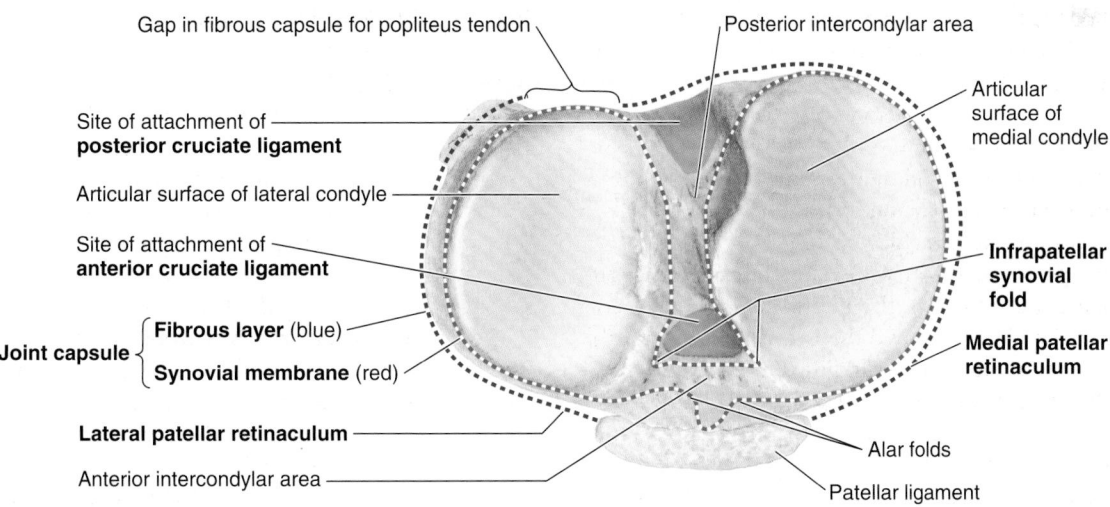

Gap in fibrous capsule for popliteus tendon

Posterior intercondylar area

Site of attachment of **posterior cruciate ligament**

Articular surface of lateral condyle

Site of attachment of **anterior cruciate ligament**

Joint capsule { **Fibrous layer** (blue)   **Synovial membrane** (red) }

Lateral patellar retinaculum

Anterior intercondylar area

Articular surface of medial condyle

Infrapatellar synovial fold

Medial patellar retinaculum

Alar folds

Patellar ligament

(B) Superior view of superior articular surface of tibia (tibial plateau)

FIGURE 5.87. Internal aspect of joint capsule of knee: layers, articular cavity, and articular surfaces. A. The joint capsule was incised transversely, the patella was sawn through, and then the knee was flexed, opening the articular cavity. The infrapatellar fold of synovial membrane encloses the cruciate ligaments, excluding them from the joint cavity. All internal surfaces not covered with or made of articular cartilage (*blue*, or *gray* in the case of the menisci) are lined with synovial membrane (mostly *purple*, but transparent and colorless where it is covering non-articular surfaces of the femur). **B.** The attachments of the fibrous layer and synovial membrane to the tibia are shown. Note that although they are adjacent on each side, they part company centrally to accommodate intercondylar and infrapatellar structures that are intracapsular (inside the fibrous layer) but extra-articular (excluded from the articular cavity by synovial membrane).

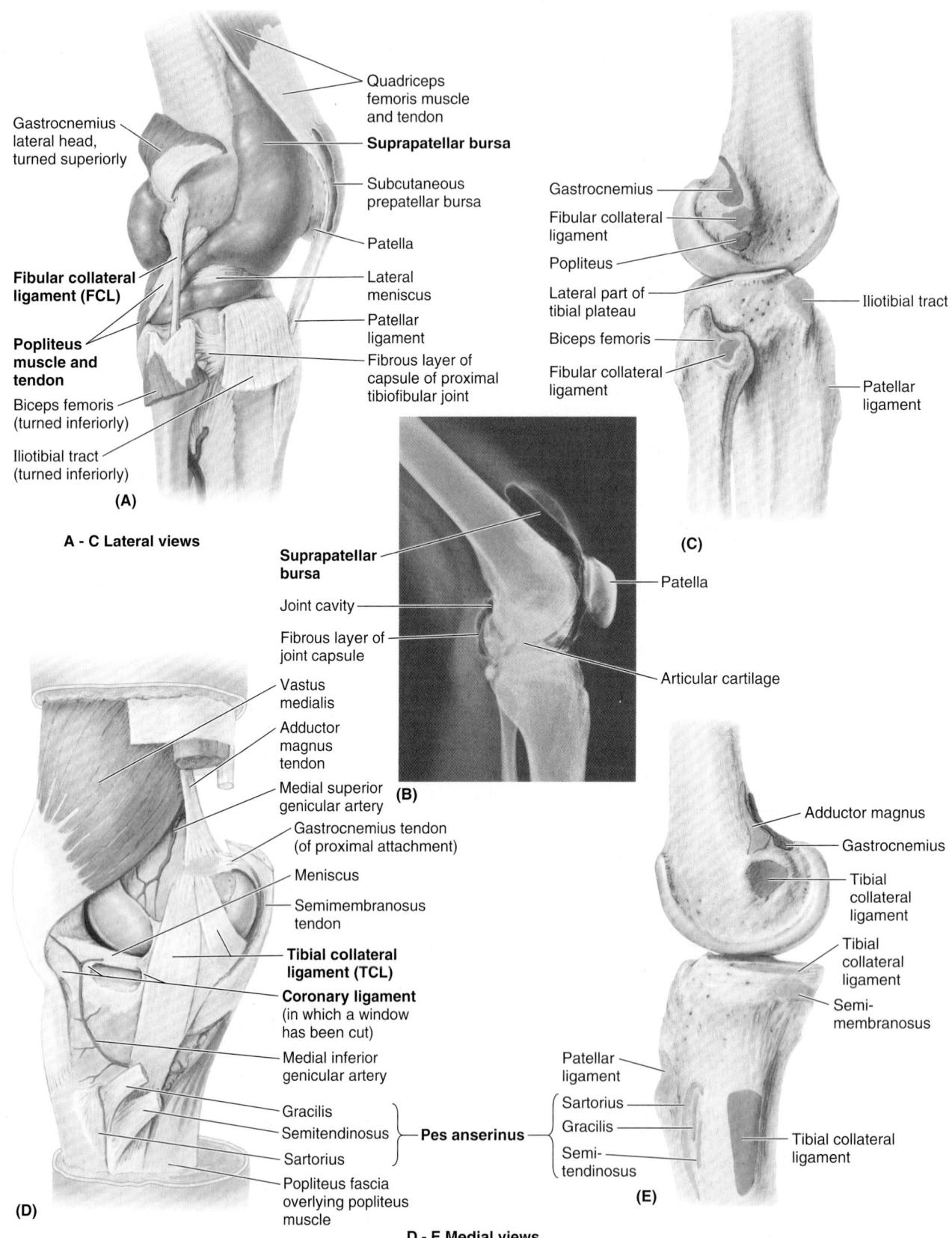

Quadriceps
femoris muscle
and tendon

Suprapatellar bursa

Subcutaneous
prepatellar bursa

Patella

Lateral
meniscus

Patellar
ligament

Fibrous layer of
capsule of proximal
tibiofibular joint

Gastrocnemius
lateral head,
turned superiorly

**Fibular collateral
ligament (FCL)**

**Popliteus
muscle and
tendon**

Biceps femoris
(turned inferiorly)

Iliotibial tract
(turned inferiorly)

(A)

A - C Lateral views

Gastrocnemius

Fibular collateral
ligament

Popliteus

Lateral part of
tibial plateau

Biceps femoris

Fibular collateral
ligament

Iliotibial tract

Patellar
ligament

(C)

**Suprapatellar
bursa**

Joint cavity

Fibrous layer of
joint capsule

Patella

Articular cartilage

(B)

Vastus
medialis

Adductor
magnus
tendon

Medial superior
genicular artery

Gastrocnemius tendon
(of proximal attachment)

Meniscus

Semimembranosus
tendon

**Tibial collateral
ligament (TCL)**

Coronary ligament
(in which a window
has been cut)

Medial inferior
genicular artery

Gracilis
Semitendinosus } **Pes anserinus**
Sartorius

Popliteus fascia
overlying popliteus
muscle

(D)

Adductor magnus

Gastrocnemius

Tibial
collateral
ligament

Tibial
collateral
ligament

Semi-
membranosus

Patellar
ligament

{ Sartorius
Gracilis
Semi-
tendinosus }

Pes anserinus

Tibial collateral
ligament

(E)

D - E Medial views

FIGURE 5.88. Collateral ligaments and bursae of knee joint. A. Fibular collateral ligament. Purple latex was injected to demonstrate the extensive and complex articular cavity. The cavity/synovial membrane extends superiorly deep to the quadriceps, forming the suprapatellar bursa. **B.** Arthrogram, knee joint slightly flexed. The suprapatellar bursa is inflated with CO_2. **C.** The attachment sites of the FCL (*green*) and related muscles (*red,* proximal; *blue,* distal). **D.** Tibial collateral ligament (isolated from the fibrous layer of the joint capsule, of which it is a part). **E.** The attachment sites of the TCL and related muscles.

aspect of the fibular head, passes superomedially over the tendon of the popliteus, and spreads over the posterior surface of the knee joint. Its development appears to be inversely related to the presence and size of a *fabella* in the proximal attachment of the lateral head of gastrocnemius (see blue box "Fabella in Gastrocnemius;" Fig. B5.21 on p. 606). Both structures are thought to contribute to posterolateral stability of the knee.

INTRA-ARTICULAR LIGAMENTS OF KNEE JOINT

The intra-articular ligaments within the knee joint consist of the cruciate ligaments and menisci. The tendon of the popliteus is also intra-articular during part of its course.

The **cruciate ligaments** (L. *crux*, a cross) crisscross within the joint capsule of the joint but outside the synovial cavity (Figs. 5.89 and 5.90). The cruciate ligaments are located in

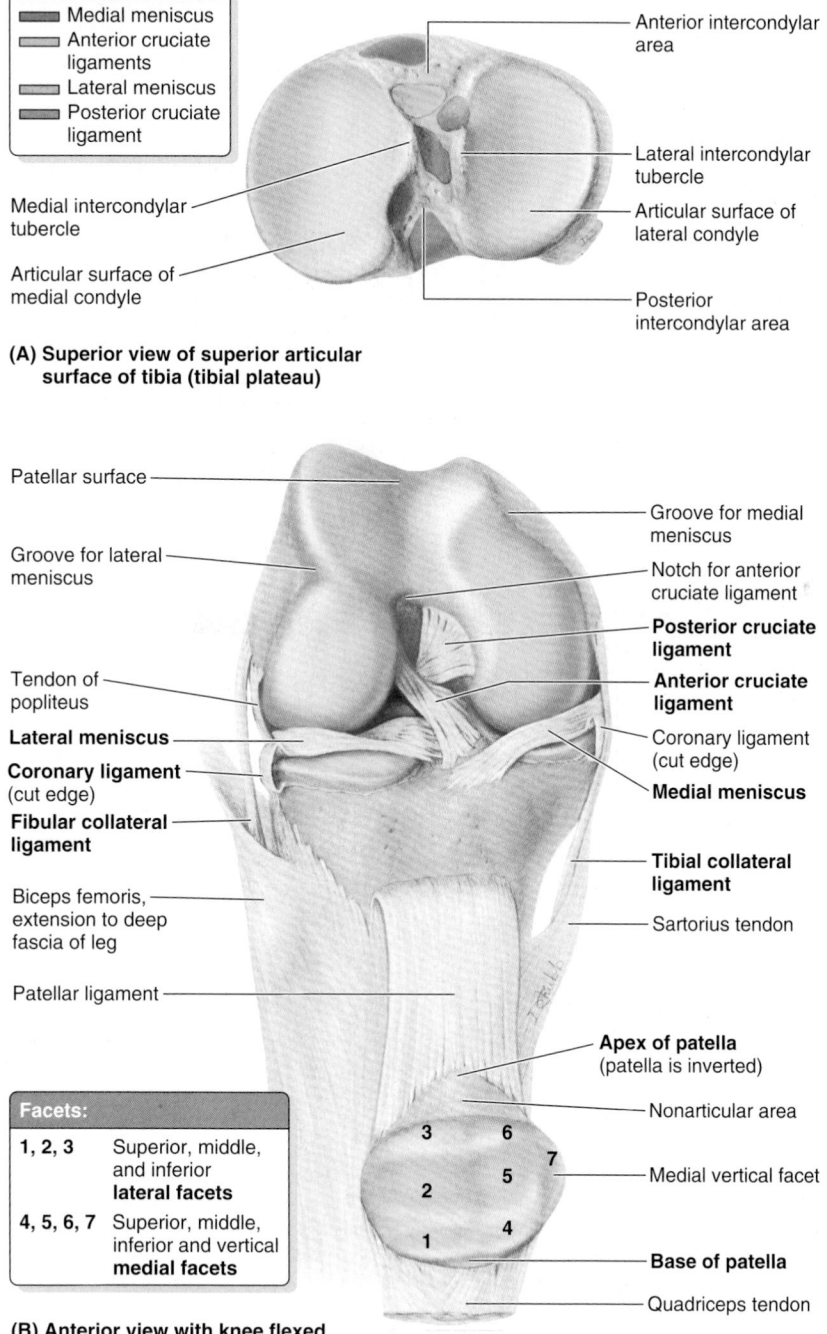

Attachment of:
- ▬ Medial meniscus
- ▬ Anterior cruciate ligaments
- ▬ Lateral meniscus
- ▬ Posterior cruciate ligament

Anterior intercondylar area

Medial intercondylar tubercle

Lateral intercondylar tubercle

Articular surface of lateral condyle

Articular surface of medial condyle

Posterior intercondylar area

(A) Superior view of superior articular surface of tibia (tibial plateau)

Patellar surface

Groove for medial meniscus

Groove for lateral meniscus

Notch for anterior cruciate ligament

Posterior cruciate ligament

Anterior cruciate ligament

Tendon of popliteus

Coronary ligament (cut edge)

Lateral meniscus

Medial meniscus

Coronary ligament (cut edge)

Fibular collateral ligament

Tibial collateral ligament

Biceps femoris, extension to deep fascia of leg

Sartorius tendon

Patellar ligament

Apex of patella (patella is inverted)

Nonarticular area

Medial vertical facet

Base of patella

Quadriceps tendon

Facets:

1, 2, 3	Superior, middle, and inferior **lateral facets**
4, 5, 6, 7	Superior, middle, inferior and vertical **medial facets**

3 6
7
2 5
1 4

(B) Anterior view with knee flexed

FIGURE 5.89. Cruciate ligaments of knee joint. A. Superior aspect of the superior articular surface of the tibia (tibial plateau), showing the medial and lateral condyles (articular surfaces) and the intercondylar eminence between them. The sites of attachment of the cruciate ligaments are colored *green;* those of the medial meniscus, *purple;* and those of the lateral meniscus, *orange.* **B.** The quadriceps tendon has been severed and the patella (within the tendon and its continuation, the patellar ligament) has been reflected inferiorly. The knee is flexed to demonstrate the cruciate ligaments.

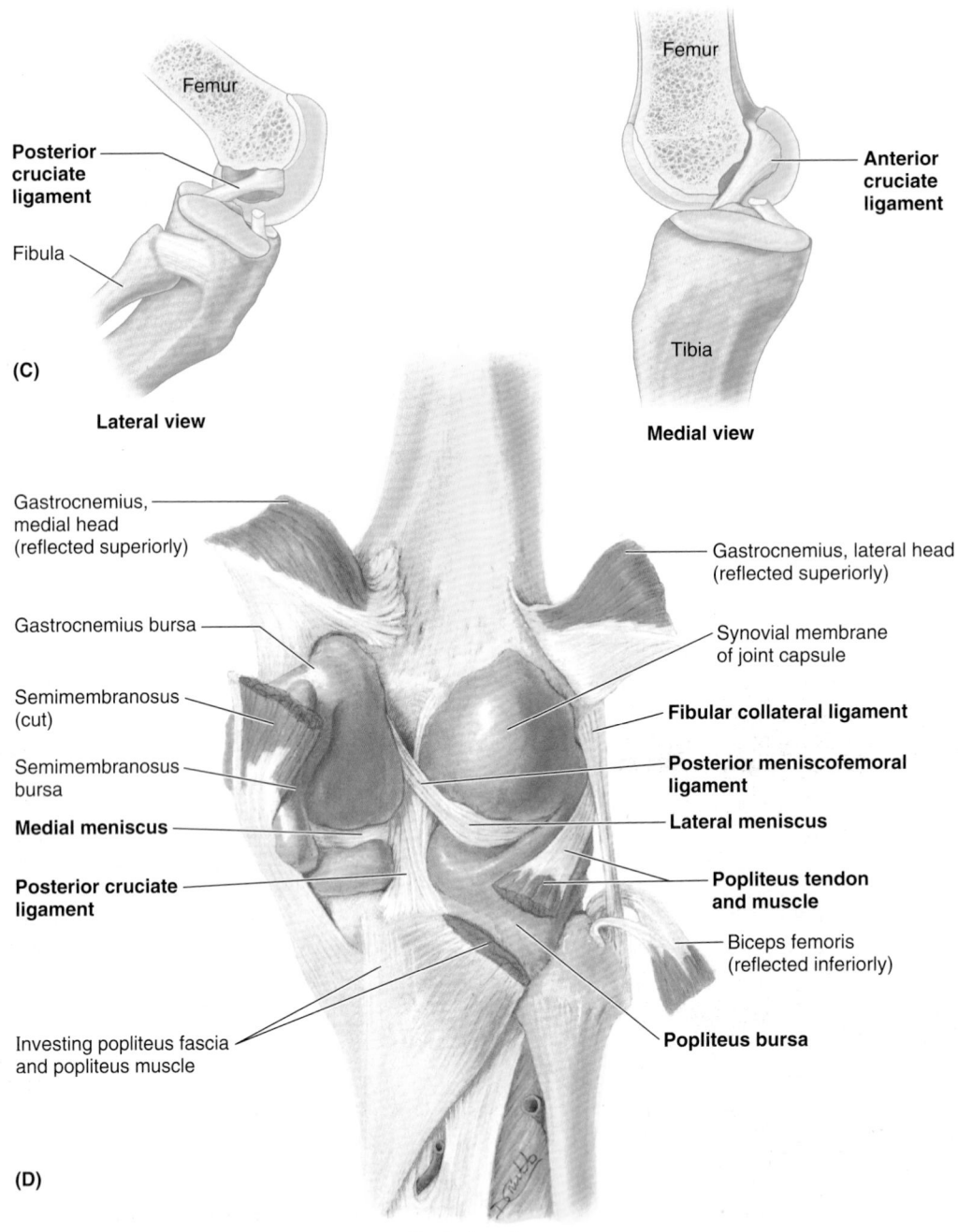

(C)

Lateral view

Posterior
cruciate
ligament

Fibula

Femur

Femur

Anterior
cruciate
ligament

Tibia

Medial view

Gastrocnemius,
medial head
(reflected superiorly)

Gastrocnemius bursa

Semimembranosus
(cut)

Semimembranosus
bursa

Medial meniscus

**Posterior cruciate
ligament**

Investing popliteus fascia
and popliteus muscle

Gastrocnemius, lateral head
(reflected superiorly)

Synovial membrane
of joint capsule

Fibular collateral ligament

**Posterior meniscofemoral
ligament**

Lateral meniscus

**Popliteus tendon
and muscle**

Biceps femoris
(reflected inferiorly)

Popliteus bursa

(D)

Posterior view

FIGURE 5.89. Cruciate ligaments of knee joint (*continued*). **C.** In these lateral and medial views, the femur has been sectioned longitudinally and the near half has been removed with the proximal part of the corresponding cruciate ligament. The lateral view demonstrates how the posterior cruciate ligament resists anterior displacement of the femur on the tibial plateau. The medial view demonstrates how the anterior cruciate ligament resists posterior displacement of the femur on the tibial plateau. **D.** Both heads of the gastrocnemius are reflected superiorly, and the biceps femoris is reflected inferiorly. The articular cavity has been inflated with purple latex to demonstrate its continuity with the various bursae and the reflections and attachments of the complex synovial membrane.

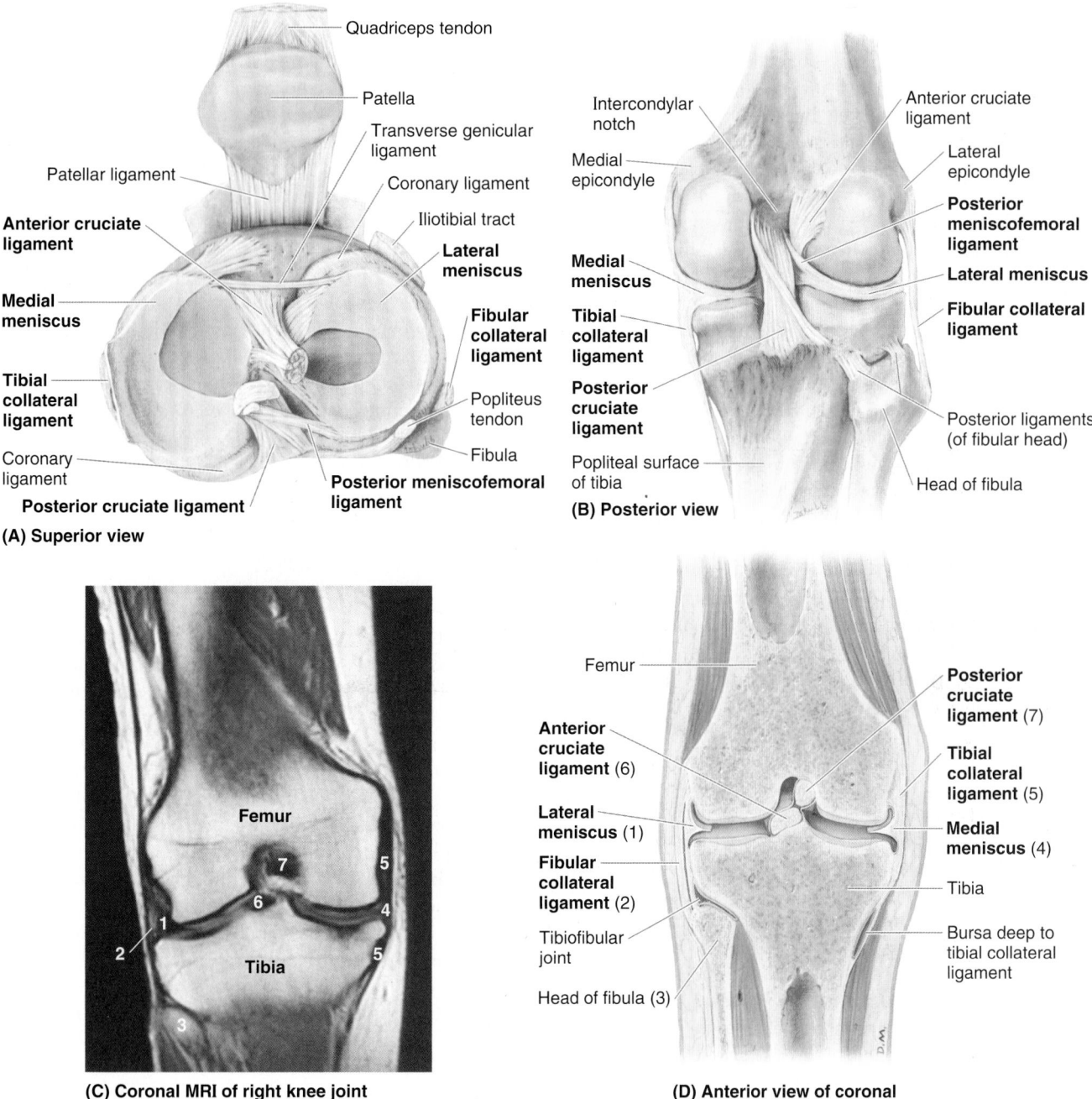

(A) Superior view

(B) Posterior view

(C) Coronal MRI of right knee joint

(D) Anterior view of coronal section of right knee joint

FIGURE 5.90. Menisci of knee joint. A. The quadriceps tendon is cut, and the patella and patellar ligament are reflected inferiorly and anteriorly. The menisci, their attachments to the intercondylar area of the tibia, and the tibial attachments of the cruciate ligaments are shown. **B.** The band-like tibial collateral ligament is attached to the medial meniscus. The cord-like fibular collateral ligament is separated from the lateral meniscus. The posterior meniscofemoral ligament attaches the lateral meniscus to the medial femoral condyle. **C and D.** The numbers on the MRI study refer to the structures labeled in the corresponding anatomical coronal section. (Part **C** courtesy of Dr. W. Kucharczyk, Professor and Neuroradiologist Senior Scientist, Department of Medical Resonance Imaging, University Health Network, Toronto, Ontario, Canada.)

the center of the joint and cross each other obliquely, like the letter X. During medial rotation of the tibia on the femur, the cruciate ligaments wind around each other; thus the amount of medial rotation possible is limited to about 10°. Because they become unwound during lateral rotation, nearly 60° of lateral rotation is possible when the knee is flexed approximately 90°, the movement being ultimately limited by the TCL. The chiasm (crossing) of the cruciate ligaments serves as the pivot for rotatory movements at the knee. Because of their oblique orientation, in every position one cruciate ligament, or parts of one or both ligaments, is tense. It is the cruciate ligaments that maintain contact with the femoral and tibial articular surfaces during flexion of the knee (Fig. 5.89C).

The **anterior cruciate ligament** (**ACL**), the weaker of the two cruciate ligaments, arises from the anterior intercondylar area of the tibia, just posterior to the attachment of the medial meniscus (Fig. 5.89A & B). The ACL has a relatively poor blood supply. It extends superiorly, posteriorly, and laterally to attach to the posterior part of the medial side of the lateral condyle of the femur (Fig. 5.89C). It limits posterior rolling (turning and traveling) of the femoral condyles on the tibial plateau during flexion, converting it to spin (turning in place). It also prevents posterior displacement of the femur on the tibia and hyperextension of the knee joint. When the joint is flexed at a right angle, the tibia cannot be pulled anteriorly (like pulling out a drawer) because it is held by the ACL.

The **posterior cruciate ligament** (**PCL**), the stronger of the two cruciate ligaments, arises from the posterior intercondylar area of the tibia (Fig. 5.89A & D). The PCL passes superiorly and anteriorly on the medial side of the ACL to attach to the anterior part of the lateral surface of the medial condyle of the femur (Fig. 5.89B & C). The PCL limits anterior rolling of the femur on the tibial plateau during extension, converting it to spin. It also prevents anterior displacement of the femur on the tibia or posterior displacement of the tibia on the femur and helps prevent hyperflexion of the knee joint. In the weight-bearing flexed knee, the PCL is the main stabilizing factor for the femur (e.g., when walking downhill).

The **menisci of the knee joint** are crescentic plates ("wafers") of fibrocartilage on the articular surface of the tibia that deepen the surface and play a role in shock absorption (Figs. 5.89 and 5.90). The menisci (G. *meniskos,* crescent) are thicker at their external margins and taper to thin, unattached edges in the interior of the joint. Wedge shaped in transverse section, the menisci are firmly attached at their ends to the *intercondylar area of the tibia* (Fig. 5.89A). Their external margins attach to the joint capsule of the knee. The **coronary ligaments** are portions of the joint capsule extending between the margins of the menisci and most of the periphery of the tibial condyles (Figs. 5.89B and 5.90A). A slender fibrous band, the **transverse ligament of the knee,** joins the anterior edges of the menisci, crossing the anterior intercondylar area (Fig. 5.9A) and tethering the menisci to each other during knee movements.

The **medial meniscus** is C shaped, broader posteriorly than anteriorly (Fig. 5.90A). Its anterior end (horn) is attached to the anterior intercondylar area of the tibia,

anterior to the attachment of the ACL (Figs. 5.89A & B and 5.90A). Its posterior end is attached to the posterior intercondylar area, anterior to the attachment of the PCL. The medial meniscus firmly adheres to the deep surface of the TCL (Figs. 5.88D and 5.90A–D). Because of its widespread attachments laterally to the tibial intercondylar area and medially to the TCL, the medial meniscus is less mobile on the tibial plateau than is the lateral meniscus.

The lateral meniscus is nearly circular, smaller, and more freely movable than the medial meniscus (Fig. 5.90A). The tendon of the popliteus has two parts proximally. One part attaches to the lateral epicondyle of the femur and passes between the lateral meniscus and inferior part of the lateral epicondylar surface of the femur (on the tendon's medial aspect) and the FCL that overlies its lateral aspect (Figs. 5.88A and 5.89B & D). The other, more medial part of the popliteal tendon attaches to the posterior limb of the lateral meniscus. A strong tendinous slip, the posterior meniscofemoral ligament, joins the lateral meniscus to the PCL and the medial femoral condyle (Figs. 5.89D and 5.90A & B).

MOVEMENTS OF KNEE JOINT

Flexion and extension are the main knee movements; some rotation occurs when the knee is flexed. The main movements of the knee joint are illustrated in Figure 5.91, and the muscles producing them and relevant details are provided in Table 5.16.

When the knee is fully extended with the foot on the ground, the knee passively "locks" because of medial rotation of the femoral condyles on the tibial plateau (the "screw-home mechanism"). This position makes the lower limb a solid column and more adapted for weight-bearing. When the knee is "locked," the thigh and leg muscles can relax briefly without making the knee joint too unstable. To unlock the knee, the popliteus contracts, rotating the femur laterally about 5° on the tibial plateau so that flexion of the knee can occur.

Movements of Menisci. Although the rolling movement of the femoral condyles during flexion and extension is limited (converted to spin) by the cruciate ligaments, some rolling does occur, and the point of contact between the femur and the tibia moves posteriorly with flexion and returns anteriorly with extension. Furthermore, during rotation of the knee, one femoral condyle moves anteriorly on the corresponding tibial condyle while the other femoral condyle moves posteriorly, rotating about the cruciate ligaments. The menisci must be able to migrate on the tibial plateau as the points of contact between femur and tibia change.

BLOOD SUPPLY OF KNEE JOINT

The arteries supplying the knee joint are the 10 vessels that form the peri-articular **genicular anastomoses** around the knee: the genicular branches of the femoral, popliteal, and anterior and posterior recurrent branches of the anterior tibial recurrent and circumflex fibular arteries (Figs. 5.92 and 5.93B). The middle genicular branches of the popliteal artery

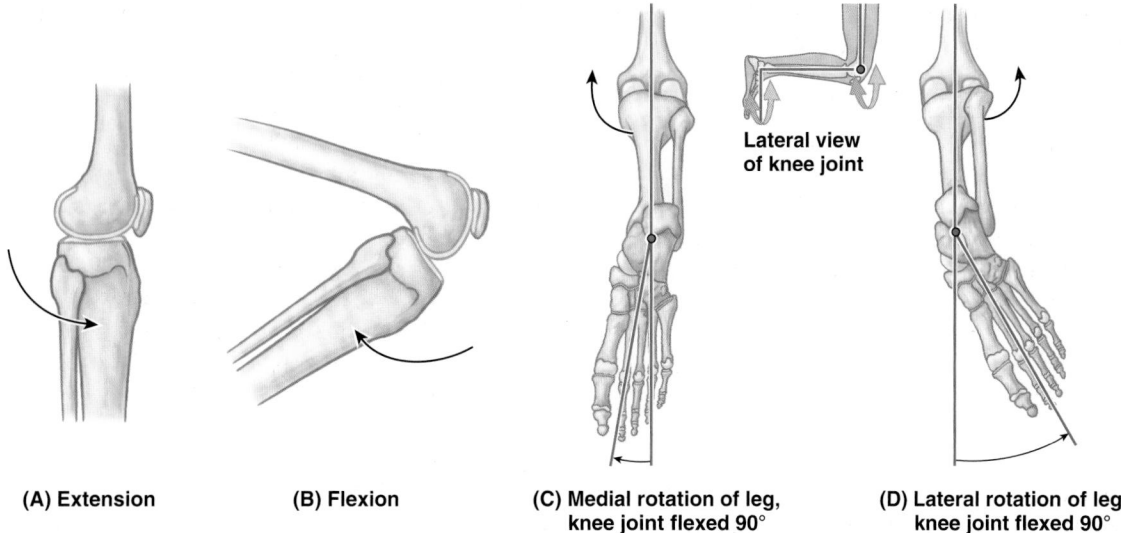

Lateral view of knee joint

(A) Extension   (B) Flexion   (C) Medial rotation of leg, knee joint flexed 90°   (D) Lateral rotation of leg, knee joint flexed 90°

FIGURE 5.91. Movements of knee joint.

TABLE 5.16. MOVEMENTS OF KNEE JOINT AND MUSCLES PRODUCING THEM

Movement	Degrees Possible	Muscles Producing Movement		Factors Limiting (Checking) Movement	Comments
		Primary	Secondary		
Extension	Normal knees extend to 0° (straight alignment of axes of tibia and femur)[a]	Quadriceps femoris	Weakly: tensor fasciae latae	Anterior edge of lateral meniscus contacts shallow groove between tibial and patellar surfaces of femoral condyles; anterior cruciate ligament contacts groove in intercondylar fossa	Ability of quadriceps to produce extension is most effective when hip joint is extended; flexion diminishes its efficiency
Flexion	120° (hip extended); 140° (hip flexed); 160° passively	Hamstrings (semitendinosus, semimembranosus, long head of biceps); short head of biceps	Gracilis, sartorius, gastrocnemius, popliteus	Calf of leg contacts thigh; length of hamstrings is also a factor—more knee flexion is possible when hip joint is flexed; cannot fully flex knee when hip is extended	Normally, role of gastrocnemius is minimal, but in presence of a supracondylar fracture, it rotates (flexes) distal fragment of femur
Medial rotation	10° with knee flexed; 5° with knee extended	Semitendinosus and semimembranosus when knee is flexed; popliteus when non-bearing knee is extended	Gracilis, sartorius	Collateral ligaments, loose during flexion without rotation, become taut at limits of rotation	When extended knee is bearing weight, action of popliteus laterally rotates femur; when not bearing weight, popliteus medially rotates patella
Lateral rotation	30°	Biceps femoris when knee is flexed		Collateral ligaments become taut; anterior cruciate ligament becomes wound around posterior cruciate ligament	At end of rotation, with no opposition, tensor fasciae latae can assist in maintaining position

[a]Straight alignment of axis of tibia with axis of femur is 0°; normal range extends to −3° (3° of hyperextension).

penetrate the fibrous layer of the joint capsule and supply the cruciate ligaments, synovial membrane, and peripheral margins of the menisci.

INNERVATION OF KNEE JOINT

Reflecting Hilton's law, the nerves supplying the muscles crossing (acting on) the knee joint also supply the joint (Fig. 5.93D); thus articular branches from the femoral (the

branches to the vasti), tibial, and common fibular nerves supply its anterior, posterior, and lateral aspects, respectively. In addition, however, the obturator and saphenous (cutaneous) nerves supply articular branches to its medial aspect.

BURSAE AROUND KNEE JOINT

There are at least 12 bursae around the knee joint because most tendons run parallel to the bones and pull lengthwise

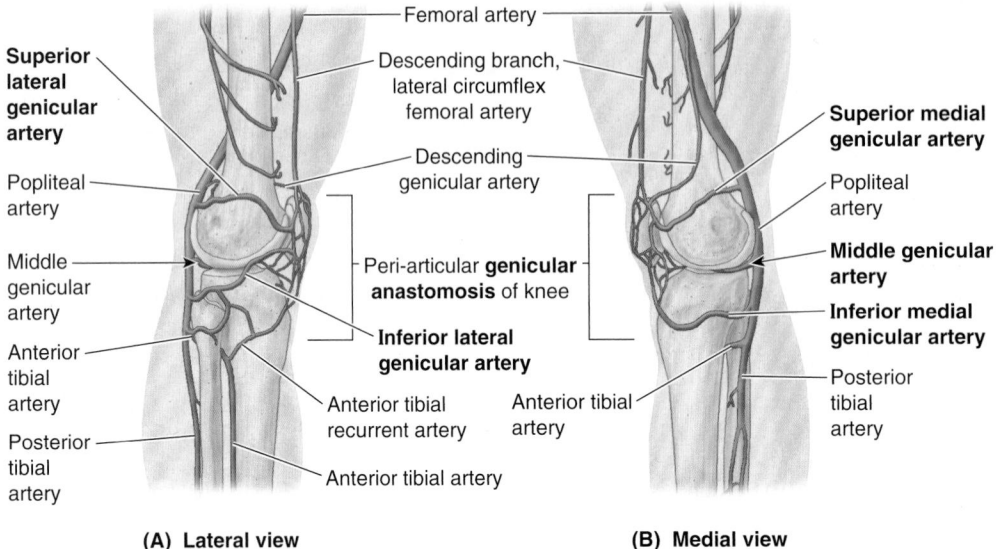

Superior lateral genicular artery

Femoral artery

Descending branch, lateral circumflex femoral artery

Descending genicular artery

Superior medial genicular artery

Popliteal artery

Popliteal artery

Middle genicular artery

Peri-articular **genicular anastomosis** of knee

Middle genicular artery

Anterior tibial artery

Inferior lateral genicular artery

Inferior medial genicular artery

Posterior tibial artery

Anterior tibial recurrent artery

Anterior tibial artery

Posterior tibial artery

Anterior tibial artery

(A) Lateral view

(B) Medial view

FIGURE 5.92. Arterial anastomoses around knee. In addition to providing collateral circulation, the genicular arteries of the genicular anastomosis supply blood to the structures surrounding the joint as well as to the joint itself (e.g., its joint or articular capsule). Compare these views with the anterior view in Figure 5.93B.

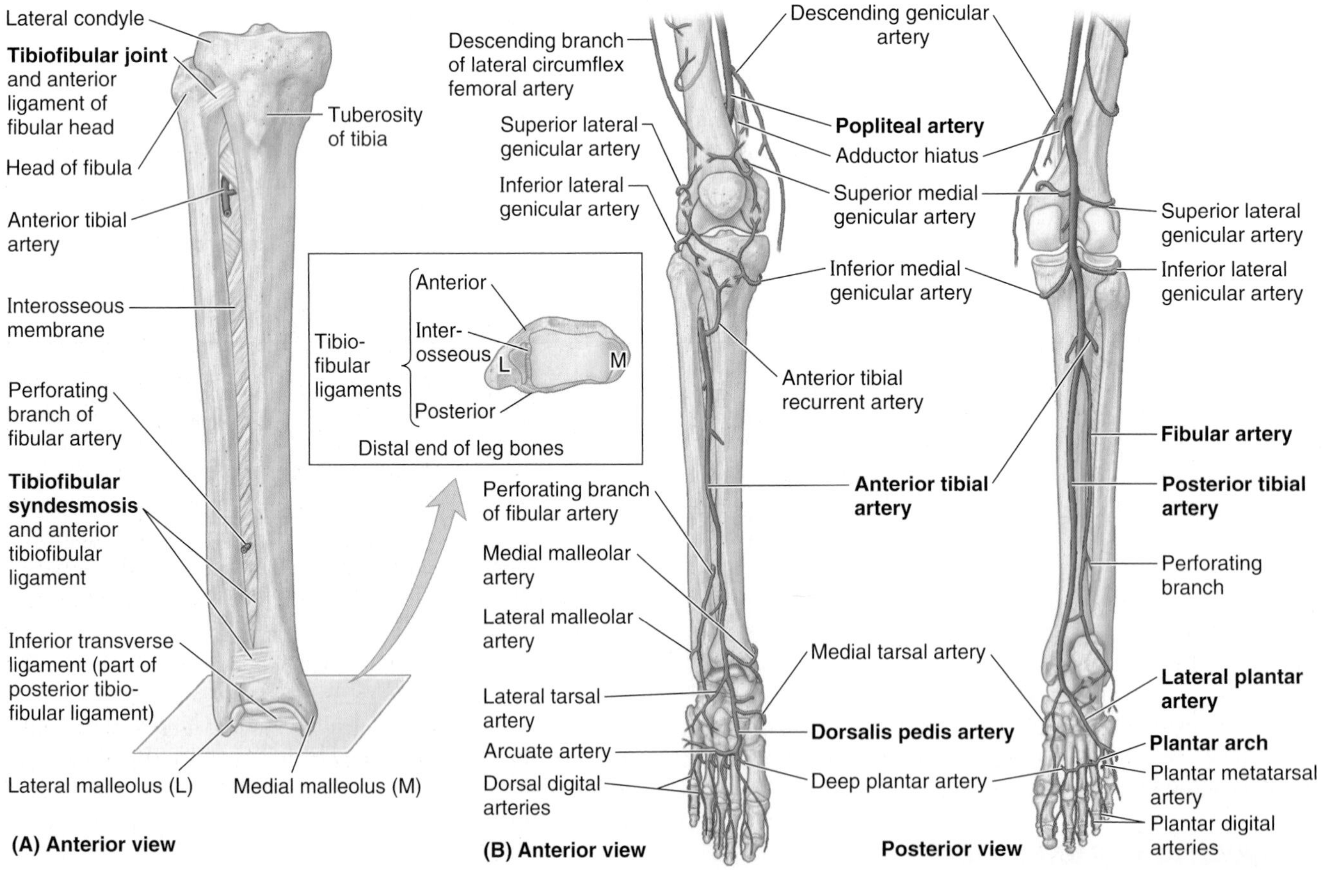

Lateral condyle

Tibiofibular joint and anterior ligament of fibular head

Head of fibula

Tuberosity of tibia

Anterior tibial artery

Interosseous membrane

Perforating branch of fibular artery

Tibiofibular syndesmosis and anterior tibiofibular ligament

Inferior transverse ligament (part of posterior tibio-fibular ligament)

Lateral malleolus (L)   Medial malleolus (M)

Tibio-fibular ligaments

Anterior

Inter-osseous   L   M

Posterior

Distal end of leg bones

Descending branch of lateral circumflex femoral artery

Superior lateral genicular artery

Inferior lateral genicular artery

Perforating branch of fibular artery

Medial malleolar artery

Lateral malleolar artery

Lateral tarsal artery

Arcuate artery

Dorsal digital arteries

Descending genicular artery

Popliteal artery

Adductor hiatus

Superior medial genicular artery

Inferior medial genicular artery

Anterior tibial recurrent artery

Anterior tibial artery

Medial tarsal artery

Dorsalis pedis artery

Deep plantar artery

Superior lateral genicular artery

Inferior lateral genicular artery

Fibular artery

Posterior tibial artery

Perforating branch

Lateral plantar artery

Plantar arch

Plantar metatarsal artery

Plantar digital arteries

(A) Anterior view

(B) Anterior view

Posterior view

FIGURE 5.93. Joints and neurovascular structures of leg and foot. A. The tibiofibular articulations include the synovial tibiofibular joint and the tibiofib-ular syndesmosis; the latter is made up of the interosseous membrane of the leg and the anterior and posterior tibiofibular ligaments. The oblique direction of the fibers of the interosseous membrane, primarily extending inferolaterally from the tibia, allows slight upward movement of the fibula but resists down-ward pull on it. **B.** The arterial supply of the joints of the leg and foot is demonstrated. Peri-articular anastomoses surround the knee and ankle.

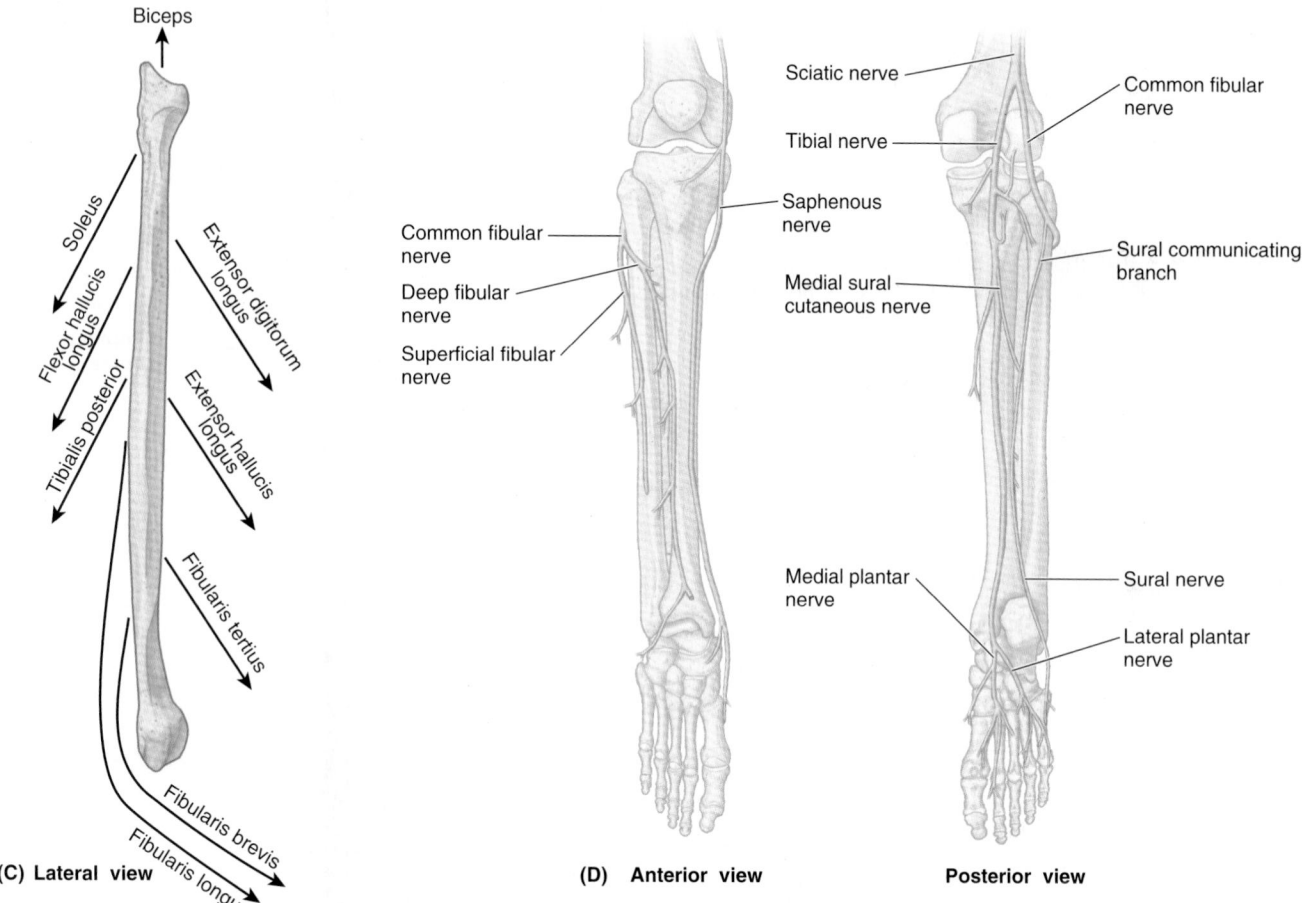

FIGURE 5.93. Joints and neurovascular structures of leg and foot (*continued*). **C.** Of the nine muscles attached to the fibula, all except one exert a downward pull on the fibula. **D.** The nerve supply of the leg and foot is demonstrated. Starting with the knee and progressing distally in the limb, cutaneous nerves become increasingly involved in providing innervation to joints, taking over completely in the distal foot and toes.

across the joint during knee movements. The main bursae of the knee are illustrated in Figure 5.94 and described in Table 5.17.

The **subcutaneous prepatellar** and **infrapatellar bursae** are located at the convex surface of the joint, allowing the skin to be able to move freely during movements of the knee (see Figs. 5.87A and 5.88A).

Four bursae communicate with the synovial cavity of the knee joint: suprapatellar bursa, popliteus bursa (deep to the distal quadriceps), anserine bursa (deep to the tendinous distal attachments of the sartorius, gracilis, and semitendinosus), and gastrocnemius bursa (Figs. 5.88A and 5.89D). The large **suprapatellar bursa** (Figs. 5.86A and 5.88A) is especially important because an infection in it may spread to the knee joint cavity. Although it develops separately from the knee joint, the bursa becomes continuous with it.

Tibiofibular Joints

The tibia and fibula are connected by two joints: the *tibiofibular joint* and the *tibiofibular syndesmosis* (inferior tibiofibular) *joint*. In addition, an *interosseous membrane* joins

the shafts of the two bones (Fig. 5.93A). The fibers of the interosseous membrane and all ligaments of both tibiofibular articulations run inferiorly from the tibia to the fibula. Thus the membrane and ligaments strongly resist the downward pull placed on the fibula by eight of the nine muscles attached to it (Fig. 5.93C). However, they allow slight upward movement of the fibula that occurs when the wide (posterior) end of the trochlea of the talus is wedged between the malleoli during dorsiflexion at the ankle. Movement at the superior tibiofibular joint is impossible without movement at the inferior tibiofibular syndesmosis.

The anterior tibial vessels pass through a hiatus at the superior end of the interosseous membrane (Fig. 5.93A & B). At the inferior end of the membrane is a smaller hiatus through which the perforating branch of the fibular artery passes.

TIBIOFIBULAR JOINT

The **tibiofibular joint** (superior tibiofibular joint) is a plane type of synovial joint between the flat facet on the fibular head and a similar articular facet located posterolaterally

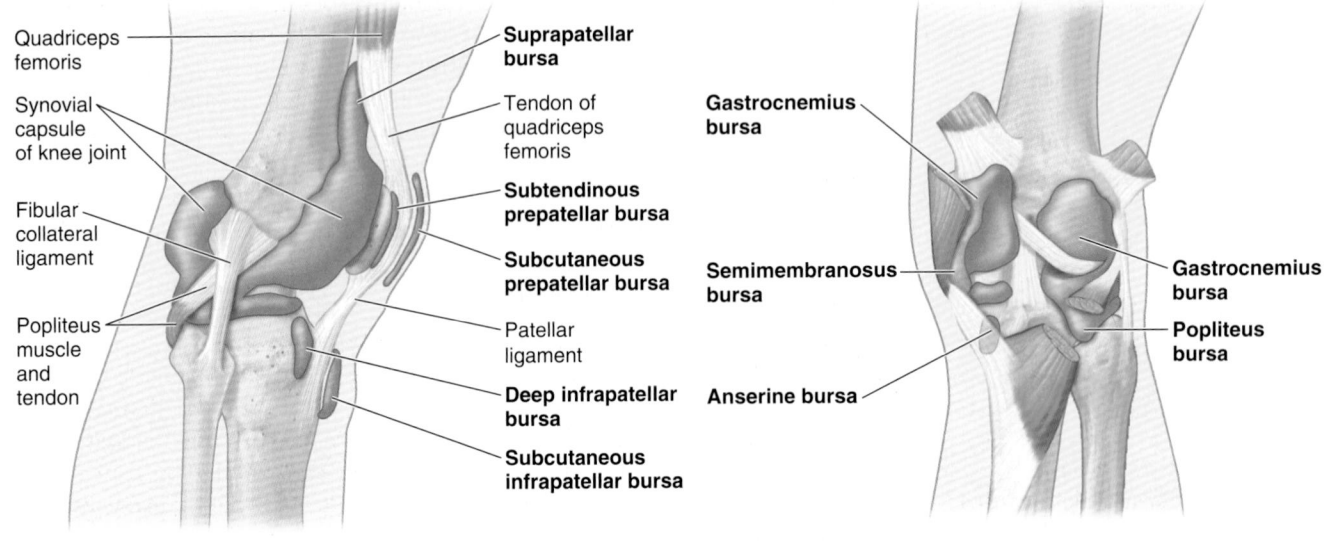

(A) Lateral view   **(B) Posterior view**

FIGURE 5.94. Bursae around knee joint and proximal leg.

TABLE 5.17. **BURSAE AROUND KNEE JOINT**

Bursae	Locations	Comments
Suprapatellar	Between femur and tendon of quadriceps femoris	Held in position by articular muscle of knee; communicates freely with (superior extension of) synovial cavity of knee joint
Popliteus	Between tendon of popliteus and lateral condyle of tibia	Opens into synovial cavity of knee joint inferior to lateral meniscus
Anserine	Separates tendons of sartorius, gracilis, and semi-tendinosus from tibia and tibial collateral ligament	Area where tendons of these muscles attach to tibia; resembles a goose's foot (L. *pes anserinus*)
Gastrocnemius	Deep to proximal attachment of tendon of medial head of gastrocnemius	An extension of synovial cavity of knee joint
Semimembranosus	Between medial head of gastrocnemius and semi-membranosus tendon	Related to distal attachment of semimembranosus
Subcutaneous pre-patellar	Between skin and anterior surface of patella	Allows free movement of skin over patella during movements of leg
Subcutaneous infrapatellar	Between skin and tibial tuberosity	Helps knee withstand pressure when kneeling
Deep infrapatellar	Between patellar ligament and anterior surface of tibia	Separated from knee joint by infrapatellar fat pad

on the lateral tibial condyle (Figs. 5.90B & D and 5.93A). A tense joint capsule surrounds the joint and attaches to the margins of the articular surfaces of the fibula and tibia. The joint capsule is strengthened by *anterior* and *posterior ligaments of the fibular head*, which pass superomedially from the fibular head to the lateral tibial condyle (Fig. 5.90B). The joint is crossed posteriorly by the tendon of the popliteus. A pouch of synovial membrane from the knee joint, the *popliteus bursa* (Fig. 5.94; Table 5.17), passes between the tendon of the popliteus and the lateral condyle of the tibia. About

20% of the time, the bursa also communicates with the synovial cavity of the tibiofibular joint, enabling transmigration of inflammatory processes between the two joints.

Movement. Slight movement of the joint occurs during dorsiflexion of the foot as a result of wedging of the trochlea of the talus between the malleoli (see "Articular Surfaces of Ankle Joint," p. 647).

Blood Supply. The arteries of the superior tibiofibular joint are from the inferior lateral genicular and anterior tibial recurrent arteries (Figs. 5.92A and 5.93B).

Nerve Supply. The nerves of the tibiofibular joint are from the common fibular nerve and the nerve to the popliteus (Fig. 5.93D).

TIBIOFIBULAR SYNDESMOSIS

The **tibiofibular syndesmosis** is a compound fibrous joint. It is the fibrous union of the tibia and fibula by means of the *interosseous membrane* (uniting the shafts) and the *anterior, interosseous,* and *posterior tibiofibular ligaments* (the latter making up the **inferior tibiofibular joint,** uniting the distal ends of the bones). The integrity of the inferior tibiofibular joint is essential for the stability of the ankle joint because it keeps the lateral malleolus firmly against the lateral surface of the talus.

Articular Surfaces and Ligaments. The rough, triangular articular area on the medial surface of the inferior end of the fibula articulates with a facet on the inferior end of the tibia (Fig. 5.93A). The strong deep **interosseous tibiofibular ligament,** continuous superiorly with the interosseous membrane, forms the principal connection between the tibia and the fibula. The joint is also strengthened anteriorly and posteriorly by the strong external **anterior** and **posterior tibiofibular ligaments.** The distal deep continuation of the posterior tibiofibular ligament, the **inferior transverse (tibiofibular) ligament,** forms a strong connection between the distal ends of the tibia (medial malleolus) and the fibula (lateral malleolus). It contacts the talus and forms the posterior "wall" of a square socket (with three deep walls, and a shallow or open anterior wall), the **malleolar mortise,** for the trochlea of the talus. The lateral and medial walls of the mortise are formed by the respective malleoli (Fig. 5.95).

Movement. Slight movement of the joint occurs to accommodate wedging of the wide portion of the trochlea of the talus between the malleoli during dorsiflexion of the foot.

Blood Supply. The arteries are from the perforating branch of the fibular artery and from medial malleolar branches of the anterior and posterior tibial arteries (Fig. 5.93B).

Nerve Supply. The nerves to the syndesmosis are from the deep fibular, tibial, and saphenous nerves (Fig. 5.93D).

Ankle Joint

The **ankle joint** (talocrural articulation) is a hinge-type synovial joint. It is located between the distal ends of the tibia and the fibula and the superior part of the talus (Figs. 5.95 and 5.96). The ankle joint can be felt between the tendons on the anterior surface of the ankle as a slight depression, approximately 1 cm proximal to the tip of the medial malleolus.

ARTICULAR SURFACES OF ANKLE JOINT

The distal ends of the tibia and fibula (along with the inferior transverse part of the posterior tibiofibular ligament) (Fig. 5.93A) form a *malleolar mortise* into which the pulley-shaped *trochlea of the talus* fits (Figs. 5.95B and 5.96). The trochlea (L., pulley) is the rounded superior articular surface of the talus (see Fig. 5.99C). The medial surface of the lateral malleolus articulates with the lateral surface of the talus. The tibia articulates with the talus in two places:

1. Its inferior surface forms the roof of the malleolar mortise, transferring the body's weight to the talus.
2. Its medial malleolus articulates with the medial surface of the talus.

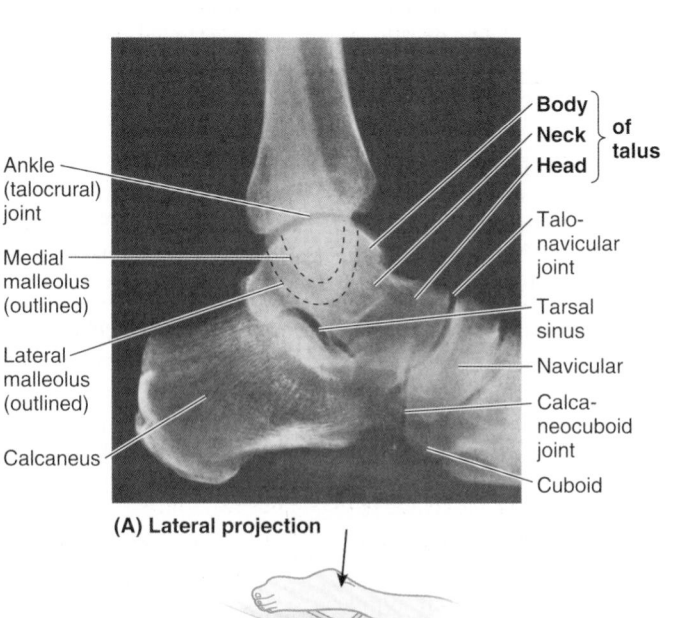

(A) Lateral projection

Labels: Ankle (talocrural) joint; Medial malleolus (outlined); Lateral malleolus (outlined); Calcaneus; Body, Neck, Head of talus; Talo-navicular joint; Tarsal sinus; Navicular; Calcaneocuboid joint; Cuboid

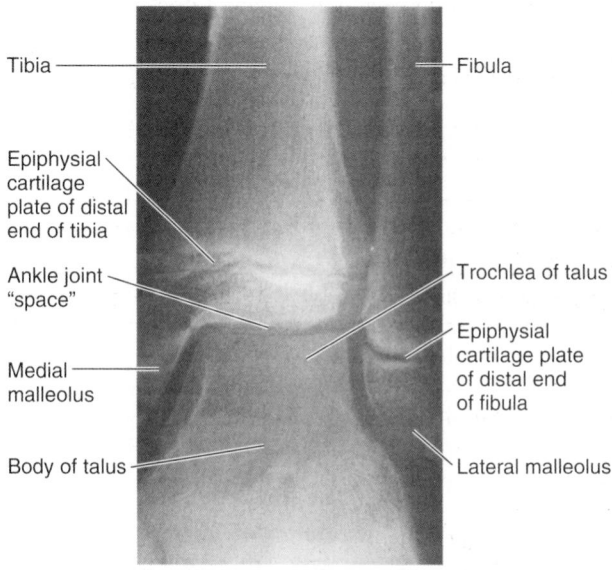

(B) Posteroanterior view (slightly oblique)

Labels: Tibia; Epiphysial cartilage plate of distal end of tibia; Ankle joint "space"; Medial malleolus; Body of talus; Fibula; Trochlea of talus; Epiphysial cartilage plate of distal end of fibula; Lateral malleolus

FIGURE 5.95. Ankle joint demonstrated radiographically. A. Left ankle (courtesy of Dr. P. Bobechko and Dr. E. Becker, Department of Medical Imaging, University of Toronto, Toronto, Ontario, Canada.) **B.** Ankle joint of 14-year-old boy. Epiphysial cartilage plates are evident at this age.

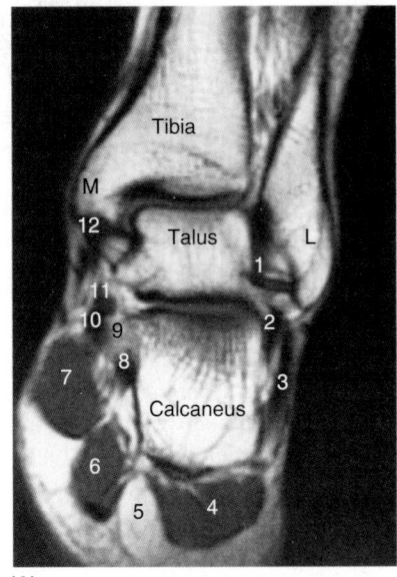

(A)

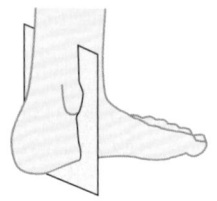

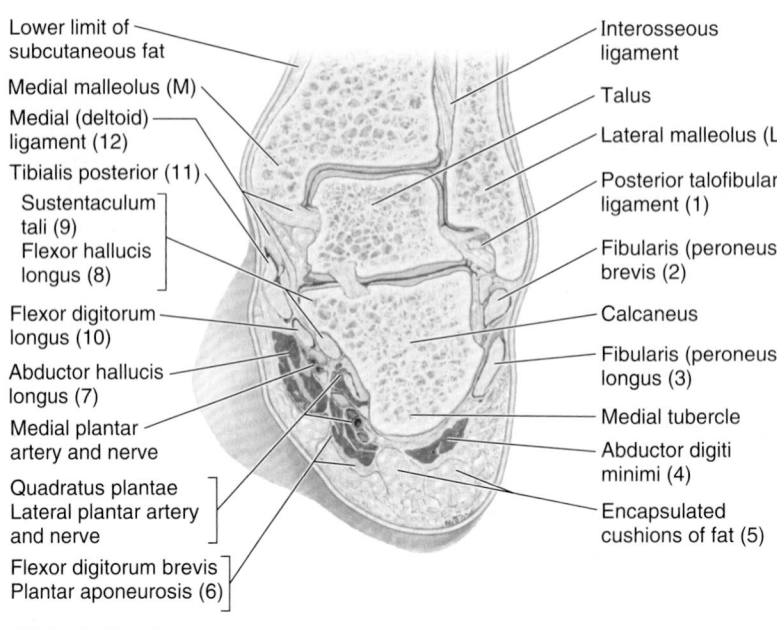

Lower limit of subcutaneous fat

Medial malleolus (M)

Medial (deltoid) ligament (12)

Tibialis posterior (11)

Sustentaculum tali (9)

Flexor hallucis longus (8)

Flexor digitorum longus (10)

Abductor hallucis longus (7)

Medial plantar artery and nerve

Quadratus plantae Lateral plantar artery and nerve

Flexor digitorum brevis Plantar aponeurosis (6)

Interosseous ligament

Talus

Lateral malleolus (L)

Posterior talofibular ligament (1)

Fibularis (peroneus) brevis (2)

Calcaneus

Fibularis (peroneus) longus (3)

Medial tubercle

Abductor digiti minimi (4)

Encapsulated cushions of fat (5)

(B) Posterior view

FIGURE 5.96. Sectional anatomy of ankle region. A and B. The orientation drawing depicts the structures visible in the MRI of the ankle. (Courtesy of Dr. W. Kucharczyk, Professor and Neuroradiologist Senior Scientist, Department of Medical Resonance Imaging, University Health Network, Toronto, Ontario, Canada.)

The malleoli grip the talus tightly as it rocks in the mortise during movements of the joint. The grip of the malleoli on the trochlea is strongest during dorsiflexion of the foot (as when "digging in one's heels" when descending a steep slope or during tug-of-war) because this movement forces the wider, anterior part of the trochlea posteriorly between the malleoli, spreading the tibia and fibula slightly apart. This spreading is limited especially by the strong interosseous tibiofibular ligament as well as the anterior and posterior tibiofibular ligaments that unite the tibia and fibula (Figs. 5.96 and 5.97).

The *interosseous ligament* is deeply placed between the nearly congruent surfaces of the tibia and fibula; although demonstrated in the inset for Figure 5.93A, the ligament can actually be observed only by rupturing it or in a cross-section.

The ankle joint is relatively unstable during plantarflexion because the trochlea is narrower posteriorly and, therefore, lies relatively loosely within the mortise. It is during plantarflexion that most injuries of the ankle occur (usually as a result of sudden, unexpected—and therefore inadequately resisted—inversion of the foot).

JOINT CAPSULE OF ANKLE JOINT

The **joint capsule of the ankle joint** is thin anteriorly and posteriorly but is supported on each side by strong *lateral and medial* (collateral) *ligaments* (Figs. 5.97 and 5.98; thin

areas of the capsule have been removed in Fig. 5.97, leaving only the reinforced parts—the ligaments—and a synovial fold). Its fibrous layer is attached superiorly to the borders of the articular surfaces of the tibia and the malleoli and inferiorly to the talus. The synovial membrane is loose and lines the fibrous layer of the capsule. The synovial cavity often extends superiorly between the tibia and the fibula as far as the interosseous tibiofibular ligament.

LIGAMENTS OF ANKLE JOINT

The ankle joint is reinforced laterally by the **lateral ligament of the ankle,** a compound structure consisting of three completely separate ligaments (Fig. 5.97A & B):

1. **Anterior talofibular ligament,** a flat, weak band that extends anteromedially from the lateral malleolus to the neck of the talus.
2. **Posterior talofibular ligament,** a thick, fairly strong band that runs horizontally medially and slightly posteriorly from the malleolar fossa to the lateral tubercle of the talus.
3. **Calcaneofibular ligament,** a round cord that passes postero-inferiorly from the tip of the lateral malleolus to the lateral surface of the calcaneus.

The joint capsule is reinforced medially by the large, strong **medial ligament of the ankle** (deltoid ligament) that attaches proximally to the medial malleolus (Fig. 5.98). The

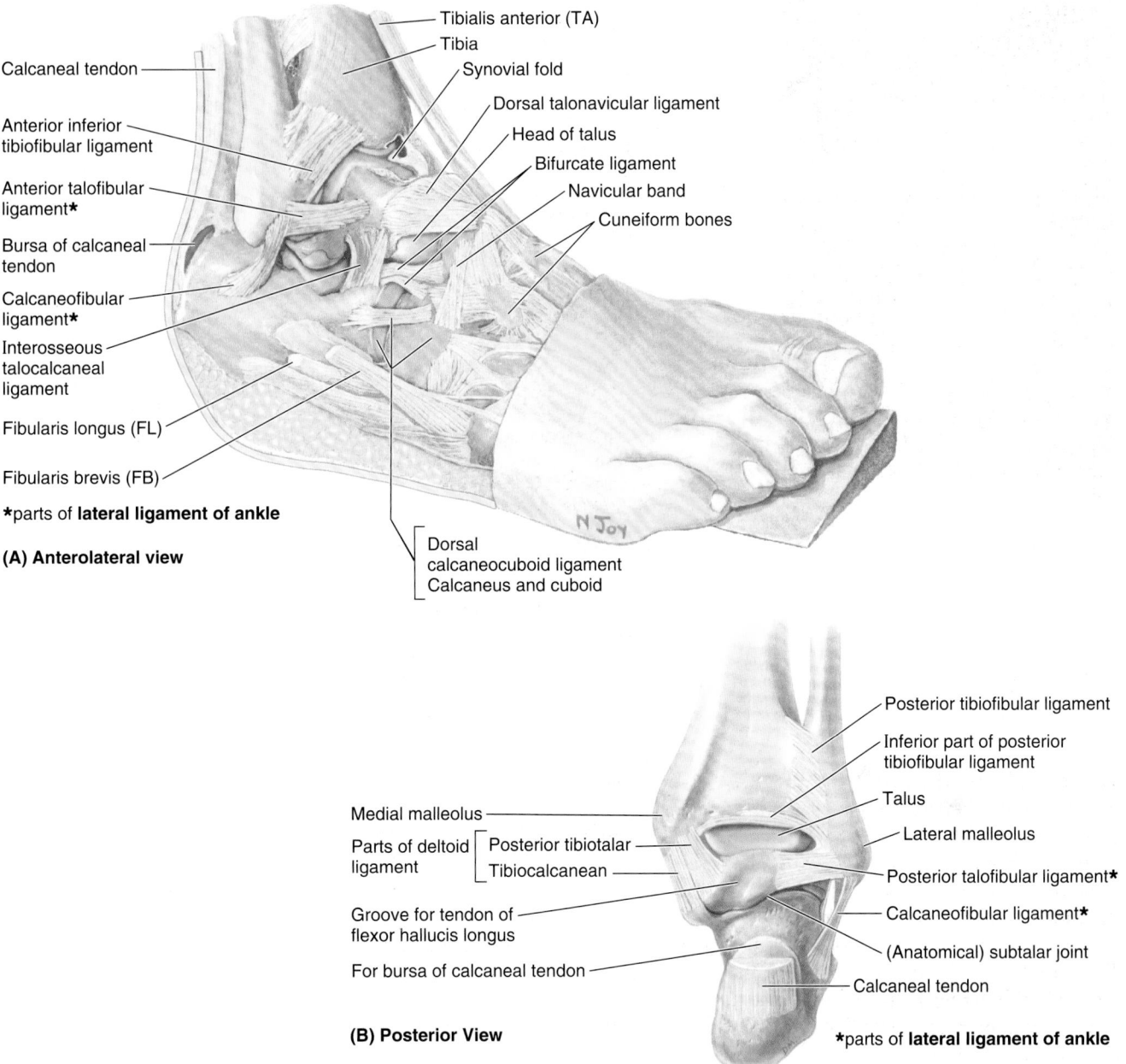

Tibialis anterior (TA)
Tibia
Synovial fold
Dorsal talonavicular ligament
Head of talus
Bifurcate ligament
Navicular band
Cuneiform bones

Calcaneal tendon
Anterior inferior tibiofibular ligament
Anterior talofibular ligament★
Bursa of calcaneal tendon
Calcaneofibular ligament★
Interosseous talocalcaneal ligament
Fibularis longus (FL)
Fibularis brevis (FB)

★parts of **lateral ligament of ankle**

(A) Anterolateral view

Dorsal calcaneocuboid ligament
Calcaneus and cuboid

Posterior tibiofibular ligament
Inferior part of posterior tibiofibular ligament
Talus
Lateral malleolus
Posterior talofibular ligament★
Calcaneofibular ligament★
(Anatomical) subtalar joint
Calcaneal tendon

Medial malleolus
Parts of deltoid ligament { Posterior tibiotalar — Tibiocalcanean — }
Groove for tendon of flexor hallucis longus
For bursa of calcaneal tendon

(B) Posterior View

★parts of **lateral ligament of ankle**

FIGURE 5.97. Dissection of ankle joint and joints of inversion and eversion. In (**A**), the foot has been inverted (by placing a wedge under the foot) to demonstrate the articular surfaces and make the lateral ligaments taut.

medial ligament fans out from the malleolus, attaching distally to the talus, calcaneus, and navicular via four adjacent and continuous parts: the **tibionavicular part,** the **tibiocalcaneal part,** and the **anterior** and **posterior tibiotalar parts.** The medial ligament stabilizes the ankle joint during eversion and prevents subluxation (partial dislocation) of the joint.

MOVEMENTS OF ANKLE JOINT

The main movements of the ankle joint are dorsiflexion and plantarflexion of the foot, which occur around a transverse axis passing through the talus (Fig. 5.99B). Because the narrow end of the trochlea of the talus lies loosely between the

malleoli when the foot is plantarflexed, some "wobble" (small amounts of abduction, adduction, inversion, and eversion) is possible in this unstable position.

- *Dorsiflexion of the ankle* is produced by the muscles in the anterior compartment of the leg (see Table 5.10). Dorsiflexion is usually limited by the passive resistance of the triceps surae to stretching and by tension in the medial and lateral ligaments.
- *Plantarflexion of the ankle* is produced by the muscles in the posterior compartment of the leg (see Table 5.13). In toe dancing by ballet dancers, for example, the dorsum of the foot is in line with the anterior surface of the leg.

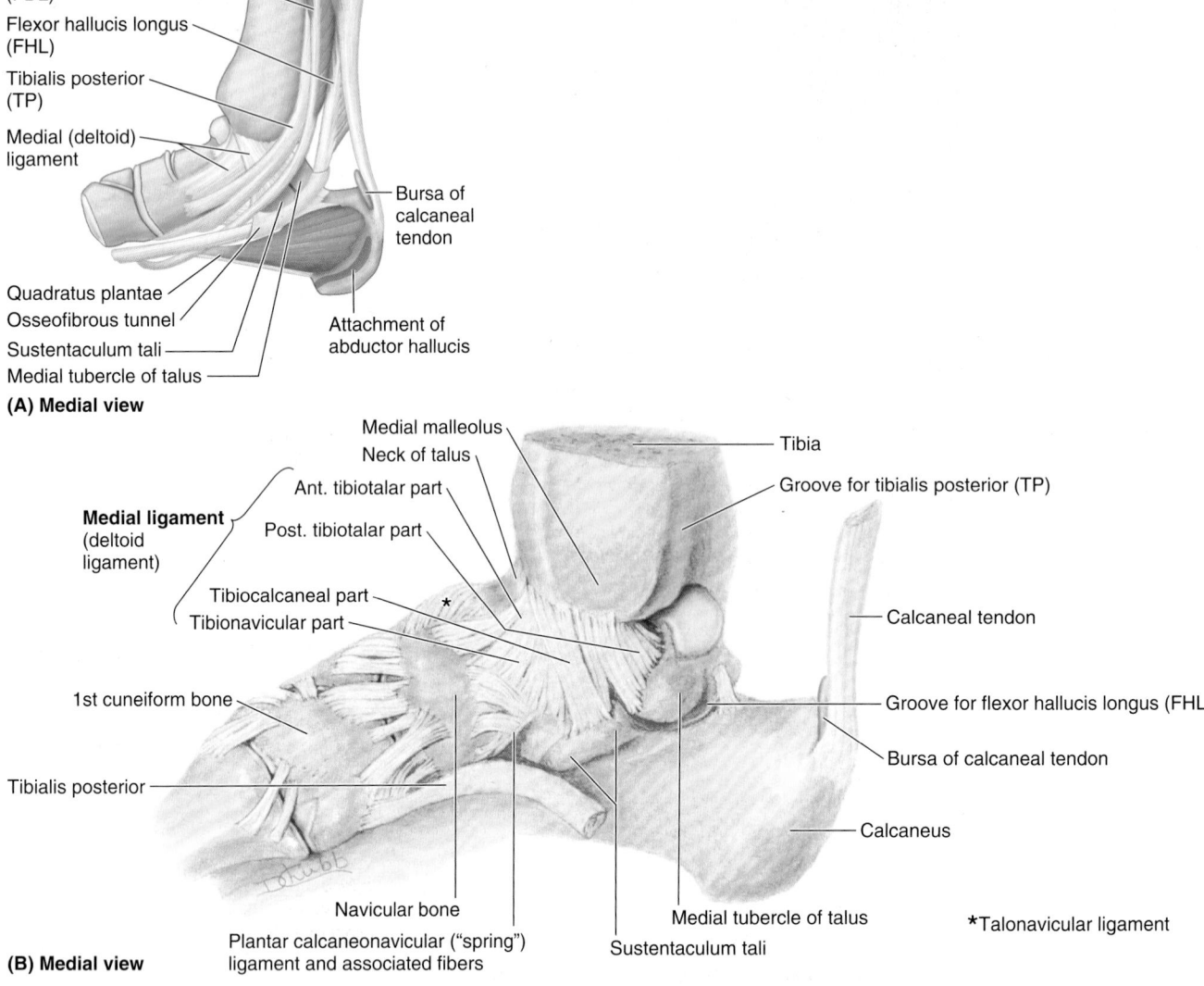

Calcaneal tendon
Flexor digitorum longus (FDL)
Flexor hallucis longus (FHL)
Tibialis posterior (TP)
Medial (deltoid) ligament
Bursa of calcaneal tendon
Quadratus plantae
Osseofibrous tunnel
Sustentaculum tali
Medial tubercle of talus
Attachment of abductor hallucis
(A) Medial view

Medial malleolus
Neck of talus
Ant. tibiotalar part
Medial ligament (deltoid ligament)
Post. tibiotalar part
Tibiocalcaneal part
Tibionavicular part
Tibia
Groove for tibialis posterior (TP)
Calcaneal tendon
Groove for flexor hallucis longus (FHL)
Bursa of calcaneal tendon
Calcaneus
1st cuneiform bone
Tibialis posterior
Navicular bone
Plantar calcaneonavicular ("spring") ligament and associated fibers
Sustentaculum tali
Medial tubercle of talus
*Talonavicular ligament
(B) Medial view

FIGURE 5.98. Tendons and ligaments on medial aspect of ankle and foot. A. The relationships of the flexor tendons to the medial malleolus and sustentaculum tali are shown as they descend the posterolateral aspect of the ankle region and enter the foot. Except for the part tethering the flexor hallucis longus tendon, the flexor retinaculum has been removed. **B.** The four parts of the medial (deltoid) ligament of the ankle are demonstrated in this dissection.

BLOOD SUPPLY OF ANKLE JOINT

The arteries are derived from malleolar branches of the fibular and anterior and posterior tibial arteries (Fig. 5.93B).

NERVE SUPPLY OF ANKLE JOINT

The nerves are derived from the tibial nerve and the deep fibular nerve, a division of the common fibular nerve (Fig. 5.93D).

Foot Joints

The many joints of the foot involve the tarsals, metatarsals, and phalanges (Fig. 5.99; Table 5.18). The important intertarsal joints are the *subtalar (talocalcaneal) joint* and

the *transverse tarsal joint (calcaneocuboid* and *talonavicular joints).* Inversion and eversion of the foot are the main movements involving these joints. The other intertarsal joints (e.g., *intercuneiform joints)* and the *tarsometatarsal* and *intermetatarsal joints* are relatively small and are so tightly joined by ligaments that only slight movement occurs between them. In the foot, flexion and extension occur in the forefoot at the metatarsophalangeal and interphalangeal joints (Fig. 5.100A & B; Table 5.19). Inversion is augmented by flexion of the toes (especially the great and 2nd toes), and eversion by their extension (especially of the lateral toes). All bones of the foot proximal to the metatarsophalangeal joints are united by dorsal and plantar ligaments. The bones of the metatarsophalangeal and interphalangeal joints are united by lateral and medial collateral ligaments.

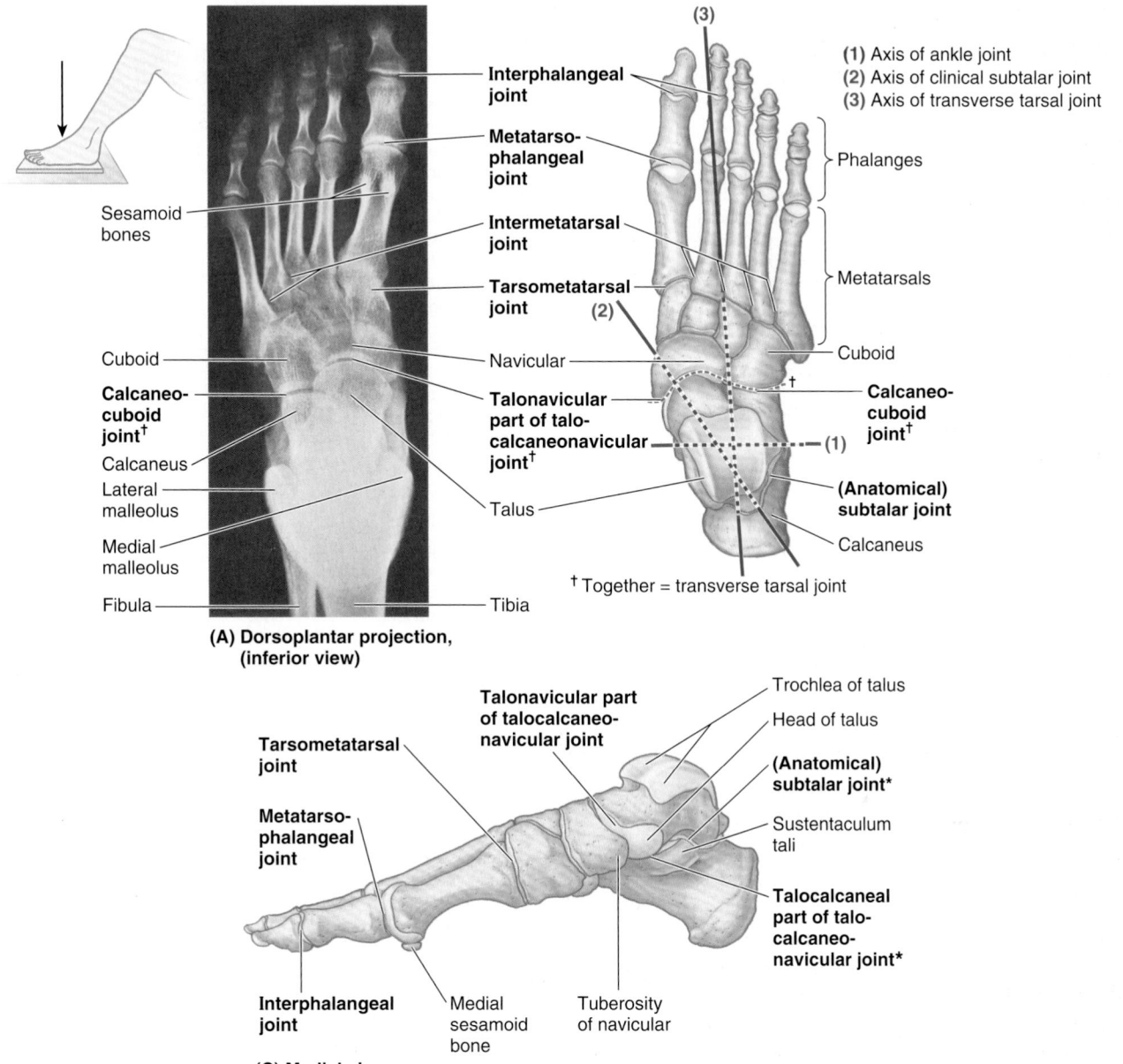

(1) Axis of ankle joint
(2) Axis of clinical subtalar joint
(3) Axis of transverse tarsal joint

Interphalangeal joint

Metatarso-phalangeal joint

Intermetatarsal joint

Tarsometatarsal joint

Navicular

Talonavicular part of talo-calcaneonavicular joint†

Phalanges

Metatarsals

Cuboid

Calcaneo-cuboid joint†

(Anatomical) subtalar joint

Calcaneus

Sesamoid bones

Cuboid

Calcaneo-cuboid joint†

Calcaneus

Lateral malleolus

Medial malleolus

Fibula

Talus

Tibia

† Together = transverse tarsal joint

(A) Dorsoplantar projection, (inferior view)

Talonavicular part of talocalcaneo-navicular joint

Tarsometatarsal joint

Metatarso-phalangeal joint

Interphalangeal joint

Medial sesamoid bone

Tuberosity of navicular

Trochlea of talus

Head of talus

(Anatomical) subtalar joint*

Sustentaculum tali

Talocalcaneal part of talo-calcaneo-navicular joint*

(C) Medial view

*Together = surgical subtalar joint

FIGURE 5.99. Joints of foot.

The **subtalar joint** occurs where the talus rests on and articulates with the calcaneus. The *anatomical subtalar joint* is a single synovial joint between the slightly concave posterior calcaneal articular surface of the talus and the convex posterior articular facet of the calcaneus (Figs. 5.96B and 5.97B). The joint capsule is weak but is supported by medial, lateral, posterior, and interosseous *talocalcaneal ligaments* (Figs. 5.96B and 5.97A). The **interosseous talocalcaneal ligament** lies within the *tarsal sinus*, which separates the subtalar and talocalcaneonavicular joints, and is especially strong. Orthopaedic surgeons use the term *subtalar joint* for the compound functional joint consisting of the anatomical subtalar joint plus the **talocalcaneal part of the talocalcaneonavicular joint.** The two separate

elements of the *clinical subtalar joint* straddle the talocalcaneal interosseous ligament. Structurally, the anatomical definition is logical because the anatomical subtalar joint is a discrete joint, having its own joint capsule and articular cavity. Functionally, however, the clinical definition is logical because the two parts of the compound joint function as a unit; it is impossible for them to function independently. The subtalar joint (by either definition) is where the majority of inversion and eversion occurs, around an axis that is oblique.

The **transverse tarsal joint** is a compound joint formed by two separate joints aligned transversely: the **talonavicular part of the talocalcaneonavicular joint**

(text continues on p. 654)

TABLE 5.18. JOINTS OF FOOT

Joint	Type	Articulating Surfaces	Joint Capsule	Ligaments	Movements	Blood Supply	Nerve Supply
Subtalar (talocalcaneal, anatomical sub-talar joint)	Plane synovial joint	Inferior surface of body of talus (posterior calcaneal articular facet) articulates with superior surface (posterior talar articular surface) of calcaneus	Fibrous layer of joint capsule is attached to margins of articular surfaces	Medial, lateral, and posterior talocalcaneal ligaments support capsule; interosseous talocalcaneal ligament binds bones together	Inversion and eversion of foot	Posterior tibial and fibular arteries	Plantar aspect: medial or lateral plantar nerve

Dorsal aspect: deep fibular nerve |
Talocalcaneo-navicular	Synovial joint; talonavicular part is ball and socket type	Head of talus articulates with calcaneus and navicular bones	Joint capsule incompletely encloses joint	Plantar calcaneonavicular (spring) ligament supports head of talus	Gliding and rotatory movements possible	Anterior tibial artery via lateral tarsal artery, a branch of dorsalis pedis artery (dorsal artery of foot)	
Calcaneocuboid	Plane synovial joint	Anterior end of calcaneus articulates with posterior surface of cuboid	Fibrous capsule encloses joint	Dorsal calcaneocuboid ligament, plantar calcaneocuboid, and long plantar ligaments support joint capsule	Inversion and eversion of foot; circumduction		
Cuneonavicular joint		Anterior navicular articulates with posterior surfaces of cuneiforms	Common capsule encloses joints	Dorsal and plantar cuneonavicular ligaments	Little movement occurs		
Tarsometatarsal		Anterior tarsal bones articulate with bases of metatarsal bones	Separate joint capsules enclose each joint	Dorsal, plantar, and interosseous tarsometatarsal ligaments bind bones together	Gliding or sliding		Deep fibular; medial and lateral plantar nerves; sural nerve
Intermetatarsal	Plane synovial joint	Bases of metatarsal bones articulate with each other		Dorsal, plantar, and interosseous intermetatarsal ligaments bind lateral four metatarsal bones together	Little individual movement occurs	Lateral metatarsal artery (a branch of dorsalis pedis artery)	
Metatarso-phalangeal	Condyloid synovial joint	Heads of metatarsal bones articulate with bases of proximal phalanges	Separate joint capsules enclose each joint	Collateral ligaments support capsule on each side; plantar ligament supports plantar part of capsule	Flexion, extension, and some abduction, adduction, and circumduction		Digital nerves
Interphalangeal	Hinge synovial joint	Head of one phalanx articulates with base of one distal to it		Collateral and plantar ligaments support joints	Flexion and extension	Digital branches of plantar arch	

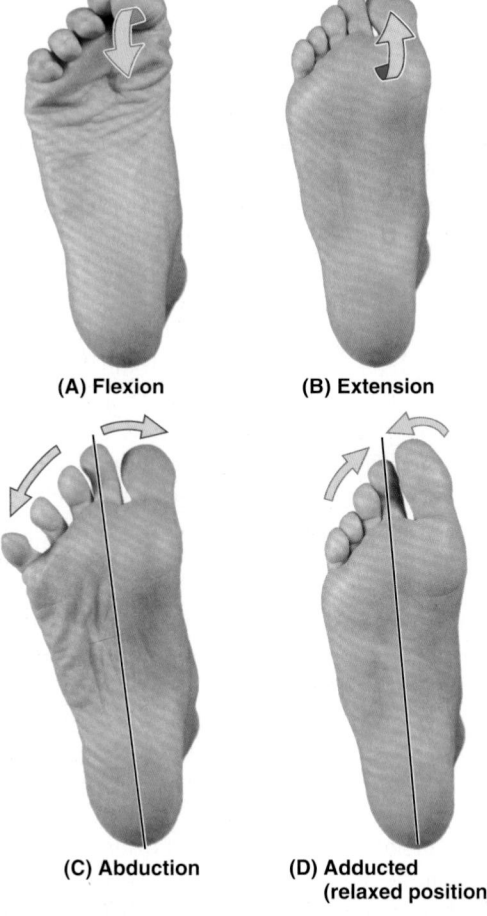

(A) Flexion

(B) Extension

(C) Abduction

**(D) Adducted
(relaxed position)**

FIGURE 5.100. Movements of joints of forefoot.

TABLE 5.19. **MOVEMENTS OF JOINTS OF FOREFOOT AND MUSCLES PRODUCING THEM**

Movement (letters refer to Fig. 5.100)	Muscles[a]
Metatarsophalangeal joints	
Flexion (*A*)	**Flexor digitorum brevis** **Lumbricals** **Interossei** **Flexor hallucis brevis** **Flexor hallucis longus** Flexor digit minimi brevis Flexor digitorum longus
Extension (*B*)	**Extensor hallucis longus** **Extensor digitorum longus** **Extensor digitorum brevis**
Abduction (*C*)	**Abductor hallucis** **Abductor digiti minimi** **Dorsal interossei**
Adduction (*D*)	**Adductor hallucis** **Plantar interossei**
Interphalangeal joints	
Flexion (fig. *A*)	**Flexor hallucis longus** **Flexor digitorum longus** **Flexor digitorum brevis** Quadratus plantae
Extension (fig. *B*)	**Extensor hallucis longus** **Extensor digitorum longus** **Extensor digitorum brevis**

[a] Muscles in boldface are chiefly responsible for the movement; the other muscles assist them.

and the **calcaneocuboid joint** (Fig. 5.99B & C). At this joint, the midfoot and forefoot rotate as a unit on the hindfoot around a longitudinal (AP) axis, augmenting the inversion and eversion movements occurring at the clinical subtalar joint. Transection across the transverse tarsal joint is a standard method for *surgical amputation of the foot.*

MAJOR LIGAMENTS OF FOOT

The major ligaments of the plantar aspect of the foot (Fig. 5.101) are the:

- **Plantar calcaneonavicular ligament** (spring ligament), which extends across and fills a wedge-shaped gap between the sustentaculum tali and the inferior margin of the posterior articular surface of the navicular (Fig. 5.101A & B). The spring ligament supports the head of the talus and plays important roles in the transfer of weight from the talus and in maintaining the longitudinal arch of the foot, of which it is the keystone (superiormost element).
- **Long plantar ligament,** which passes from the plantar surface of the calcaneus to the groove on the cuboid. Some of its fibers extend to the bases of the metatarsals, thereby forming a tunnel for the tendon of the fibularis longus (Fig. 5.101A). The long plantar ligament is important in maintaining the longitudinal arch of the foot.
- **Plantar calcaneocuboid ligament** (short plantar ligament), which is located on a plane between the plantar calcaneonavicular and the long plantar ligaments (Fig. 5.101B). It extends from the anterior aspect of the inferior surface of the calcaneus to the inferior surface of the cuboid. It is also involved in maintaining the longitudinal arch of the foot.

ARCHES OF FOOT

If the feet were more rigid structures, each impact with the ground would generate extremely large forces of short duration (shocks) that would be propagated through the skeletal system. Because the foot is composed of numerous bones connected by ligaments, it has considerable flexibility that allows it to deform with each ground contact, thereby absorbing much of the shock. Furthermore, the tarsal and metatarsal bones are arranged in longitudinal and transverse arches passively supported and actively restrained by flexible tendons that add to the weight-bearing capabilities and resiliency of the foot. Thus, much smaller forces of longer duration are transmitted through the skeletal system.

The arches distribute weight over the pedal platform (foot), acting not only as shock absorbers but also as springboards for propelling it during walking, running, and jumping. The resilient arches add to the foot's ability to adapt to changes in surface contour. The weight of the body is transmitted to the talus from the tibia. Then it is transmitted posteriorly to the calcaneus and anteriorly to the "ball of the foot" (the sesamoids of the 1st metatarsal and the head of the 2nd metatarsal), and that weight/pressure is shared laterally with the heads of the 3rd–5th metatarsals as necessary for balance and comfort (Fig. 5.102). Between these weight-bearing points are the relatively elastic arches of the foot, which become slightly flattened by body weight during standing. They normally resume their curvature (recoil) when body weight is removed.

The **longitudinal arch of the foot** is composed of medial and lateral parts (Fig. 5.103). Functionally, both parts act as a unit with the transverse arch of the foot, spreading the weight in all directions. The **medial longitudinal arch** is higher and more important than the lateral longitudinal arch (Fig. 5.103A & B). The medial longitudinal arch is composed of the calcaneus, talus, navicular, three cuneiforms, and three metatarsals. *The talar head is the keystone of the medial longitudinal arch.* The tibialis anterior and posterior, via their tendinous attachments, help support the medial longitudinal arch. The fibularis longus tendon, passing from lateral to medial, also helps support this arch (Fig. 5.103C & E). The **lateral longitudinal arch** is much flatter than the medial part of the arch and rests on the ground during standing (Fig. 5.103B & D). It is made up of the calcaneus, cuboid, and lateral two metatarsals.

The **transverse arch of the foot** runs from side to side (Fig. 5.103C). It is formed by the cuboid, cuneiforms, and bases of the metatarsals. The medial and lateral parts of the longitudinal arch serve as pillars for the transverse arch. The tendons of the fibularis longus and tibialis posterior, crossing under the sole of the foot like a stirrup (Fig. 5.103C), help maintain the curvature of the transverse arch. The integrity of the bony arches of the foot is maintained by both passive factors and dynamic supports (Fig. 5.103E).

Passive factors involved in forming and maintaining the arches of the foot include:

- The shape of the united bones (both arches, but especially the transverse arch).
- Four successive layers of fibrous tissue that bowstring the longitudinal arch (superficial to deep):
  1. Plantar aponeurosis.
  2. Long plantar ligament.
  3. Plantar calcaneocuboid (short plantar) ligament.
  4. Plantar calcaneonavicular (spring) ligament.

Dynamic supports involved in maintaining the arches of the foot include:

- Active (reflexive) bracing action of intrinsic muscles of foot (longitudinal arch).
- Active and tonic contraction of muscles with long tendons extending into foot:
  - Flexors hallucis and digitorum longus for the longitudinal arch.
  - Fibularis longus and tibialis posterior for the transverse arch.

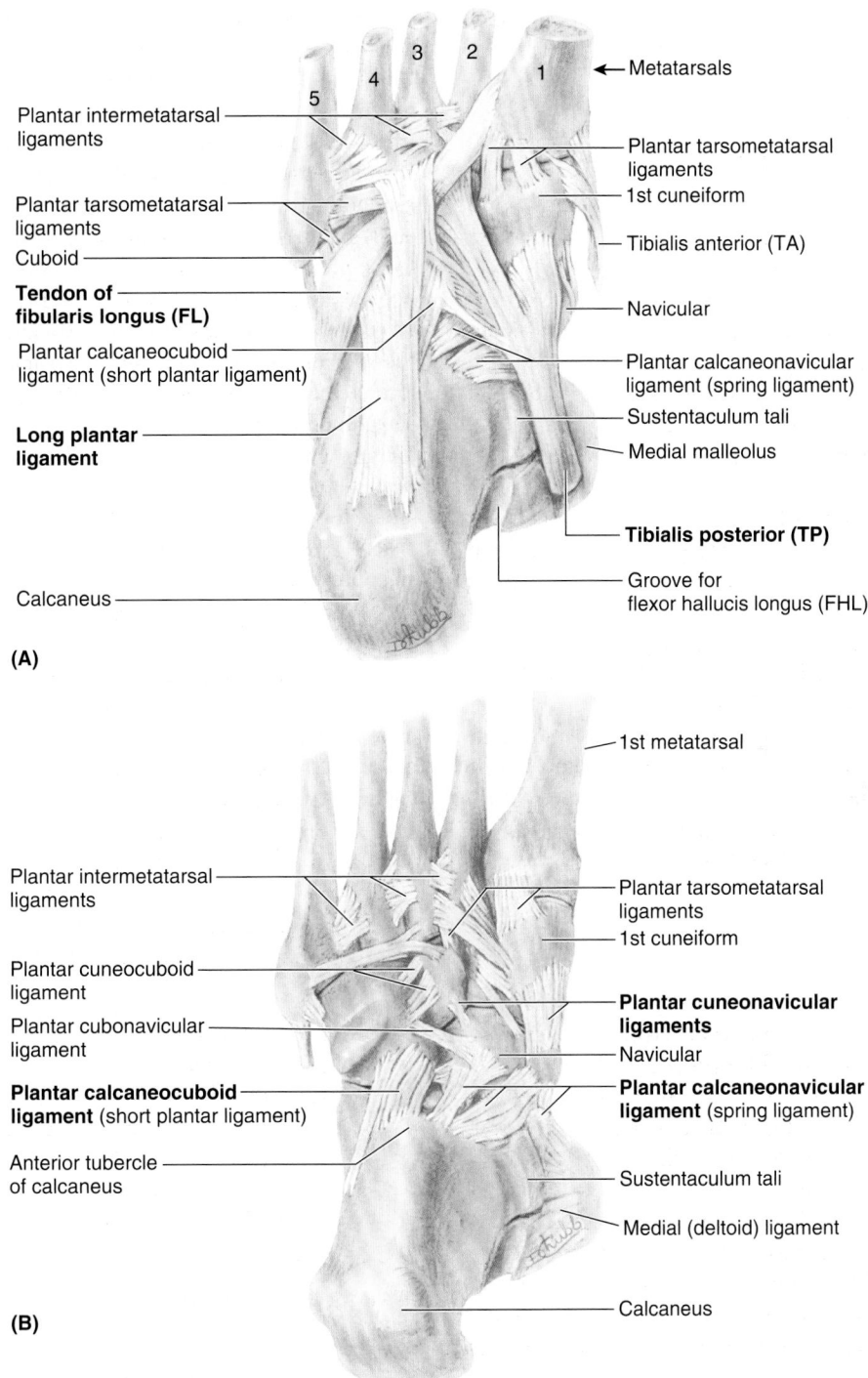

Plantar views

FIGURE 5.101. Plantar ligaments. A and B. Sequential stages of a deep dissection of the sole of the right foot showing the attachments of the ligaments and the tendons of the long evertor and invertor muscles.

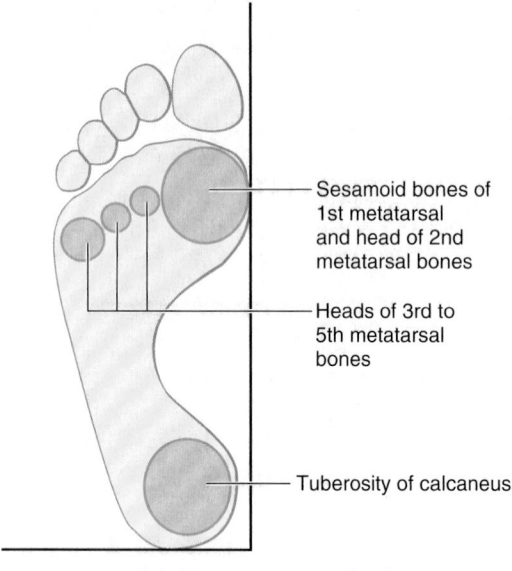

Plantar view

FIGURE 5.102. Weight-bearing areas of foot. Body weight is divided approximately equally between the hindfoot (calcaneus) and the forefoot (heads of the metatarsals). The forefoot has five points of contact with the ground: a large medial one that includes the two sesamoid bones associated with the head of the 1st metatarsal and the heads of the lateral four metatarsals. The 1st metatarsal supports the major share of the load, with the lateral forefoot providing balance.

Of these factors, the plantar ligaments and the plantar aponeurosis bear the greatest stress and are most important in maintaining the arches of the foot.

Surface Anatomy of Joints of Knee, Ankle, and Foot

The ***knee region*** is between the thigh and the leg (Fig. 5.104A). Superior to it are the large bulges formed by the *vasta lateralis* and *medialis*. Superolateral to the knee is the *iliotibial tract*, which can be followed inferiorly to the *anterolateral (Gerdy) tubercle of the tibia*. The *patella*, easily palpated and moveable from-side-to side during extension, lies anterior to the *femoral condyles* (palpable to each side of the middle of the patella). Extending from the *apex of the patella*, the *patellar ligament* is easily visible, especially in thin people, as a thick band attached to the prominent *tibial tuberosity*. The *plane of the knee joint*, between femoral condyles and tibial plateau, may be palpated on each side of the junction of patellar apex and ligament when the knee is extended. Laterally, the *head of the fibula* is readily located by following the *tendon of the biceps femoris* inferiorly. This tendon is particularly prominent when the knee is partially flexed (Fig. 5.104B). The *fibular collateral ligament* may be palpated as a cord-like structure superior to the fibular head and anterior to biceps tendon, when the knee is fully flexed.

The prominences of the *lateral and medial malleoli* provide an approximation of the *axis of the ankle joint* (Fig. 5.104C–E). When the ankle is plantarflexed, the *anterior* border of the distal end of the tibia is palpable proximal to the malleoli, providing an indication of the joint plane of the ankle joint. The *sustentaculum tali*, approximately 2 cm distal to the tip of the medial malleolus, is best felt by palpating it from below where it is somewhat obscured by the *tendon of the flexor digitorum longus*, which crosses it. On the lateral side, when the foot is inverted, the

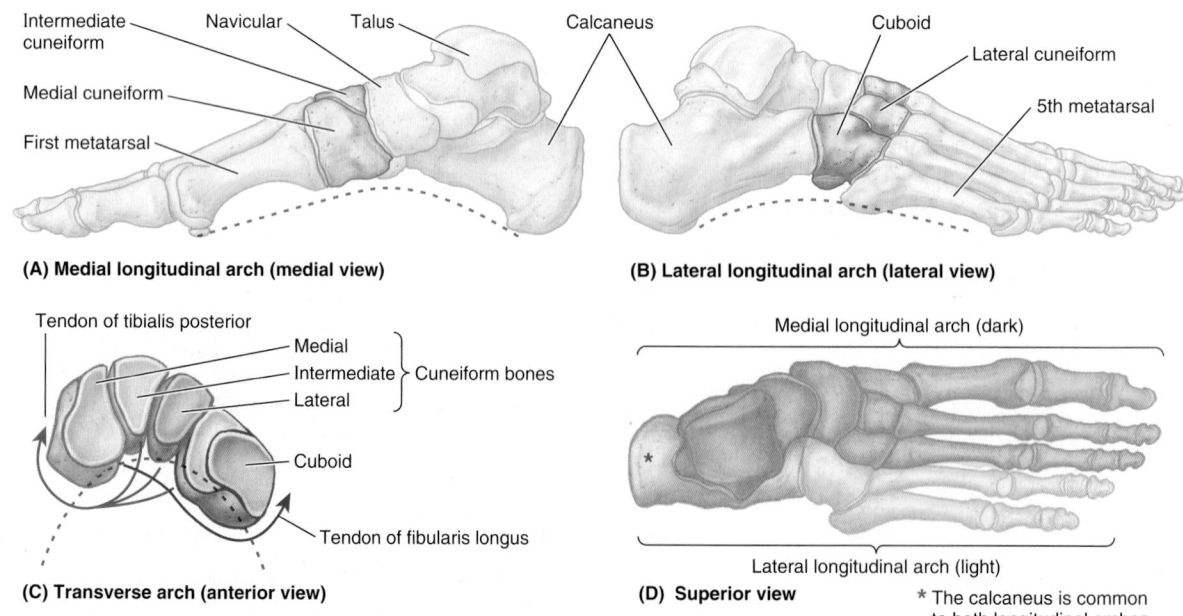

(A) Medial longitudinal arch (medial view)

(B) Lateral longitudinal arch (lateral view)

(C) Transverse arch (anterior view)

(D) Superior view

* The calcaneus is common to both longitudinal arches

FIGURE 5.103. Arches of foot. A and B. The medial longitudinal arch is higher than the lateral longitudinal arch, which may contact the ground when standing erect. **C.** The transverse arch is demonstrated at the level of the cuneiforms, receiving stirrup-like support from a major invertor (tibialis posterior) and evertor (fibularis longus). **D.** The components of the medial (*dark gray*) and lateral (*light gray*) longitudinal arches are indicated. The calcaneus (*medium gray*) is common to both. The medial arch is primarily weight-bearing, whereas the lateral arch provides balance.

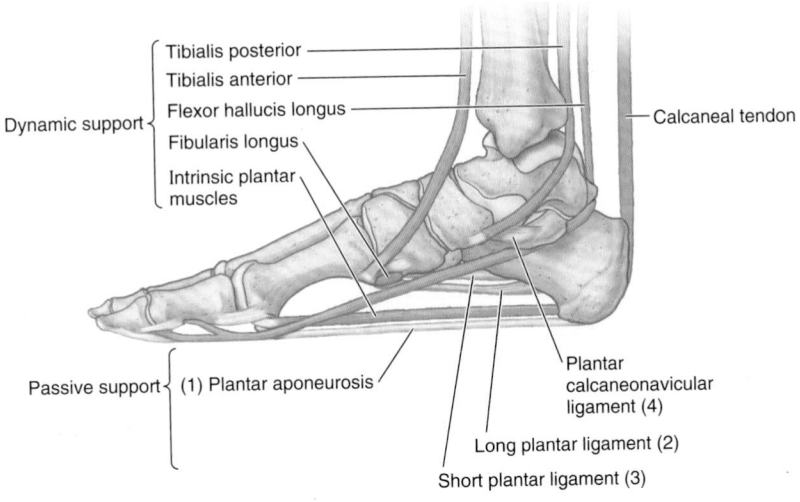

(E) Medial longitudinal arch (medial view)

FIGURE 5.103. **Arches of foot** (*continued*). **E.** The active (*red lines*) and passive (*gray*) supports of the longitudinal arches are represented. There are four layers of passive support (*1–4*).

lateral margin of the ***anterior surface of the calcaneus*** is uncovered and palpable. This indicates the site of the *calcaneocuboid joint*. When the foot is plantarflexed, the *head of the talus* is exposed. Palpate it dorsal to where the anterior surface of the calcaneus is felt. The *calcaneal tendon* at the posterior aspect of the ankle is easily palpated and traced to its attachment to the *calcaneal tuberosity*. In the depression on each side of the tendon, the ankle joint is superficial. When the joint is overfilled with fluid, these depressions may be obliterated. The *transverse tarsal joint* is indicated by a line from the posterior aspect of the

tuberosity of the navicular to a point halfway between the lateral malleolus and the tuberosity of the 5th metatarsal.

The ***metatarsophalangeal joint of the great toe*** lies distal to the knuckle formed by the head of the 1st metatarsal. *Gout*, a metabolic disorder, commonly causes edema and tenderness of this joint, as does *osteoarthritis* (degenerative joint disease). Severe pain in the 1st metatarsophalangeal joint is called *podagra* (from G. *pous* + G. *agra*, a seizure). Often the 1st metatarsophalangeal joint is the first one affected by arthritis.

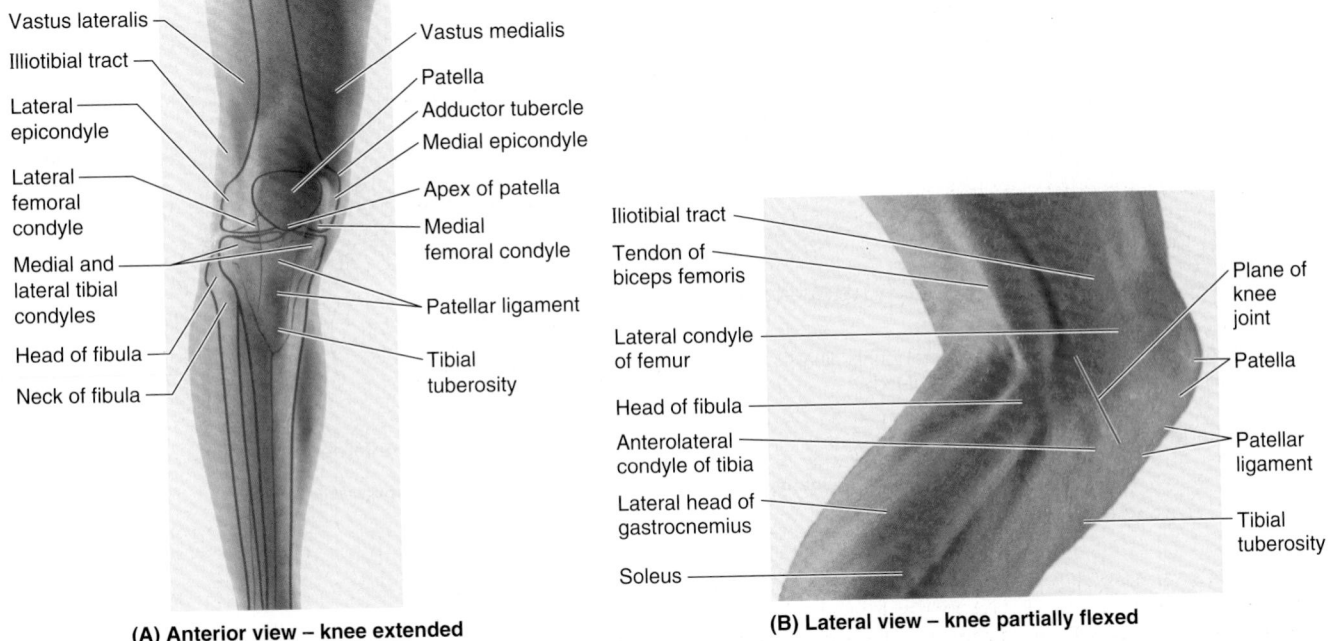

(A) Anterior view – knee extended

(B) Lateral view – knee partially flexed

FIGURE 5.104. **Surface anatomy of the joints of the knee, leg, ankle, and foot.**

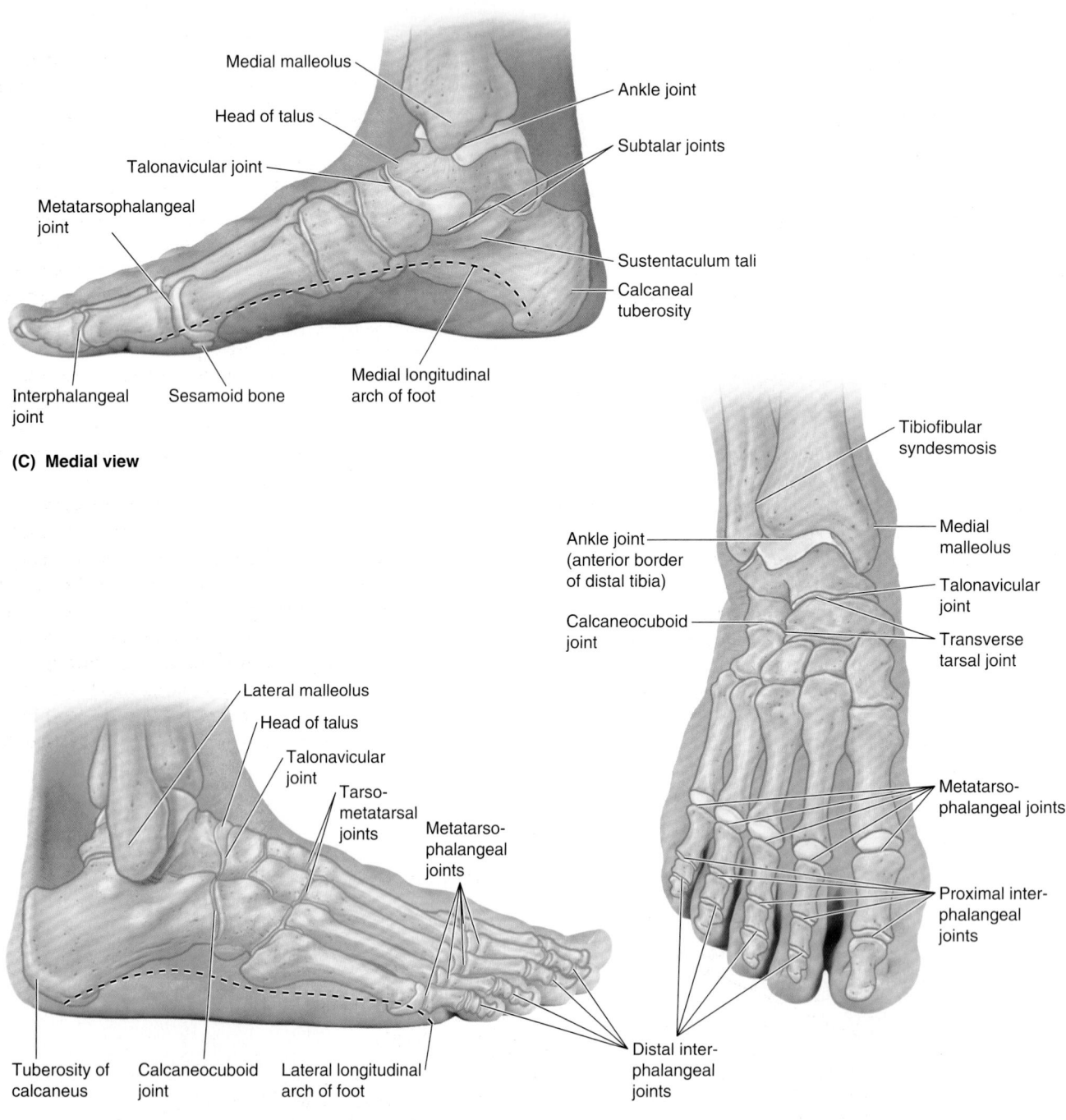

(C) Medial view

(D) Lateral view

(E) Anterosuperior view

FIGURE 5.104. Surface anatomy of the joints of the knee, leg, ankle, and foot (*continued*).

JOINTS OF LOWER LIMB

Bipedalism and Congruity of Articular Surfaces of Hip Joint

The acetabulum is directed inferiorly, laterally, and anteriorly in humans. The weight-bearing iliac portion of the acetabular rim overlies the femoral head, which is important for transfer of weight to the femur in the erect (standing/walking) position (Figs. 5.77A and 5.79C).

Consequently, of the positions commonly assumed by humans, the hip joint is mechanically most stable when a person is bearing weight, as when lifting a heavy object, for example. Decreases in the degree to which the ilium overlies the femoral head (detectable radiographically as the *angle of Wiberg*—Fig. 5.79C & D) may indicate joint instability.

Because of the anterior direction the axis of the acetabulum and the posterior direction of the axis of the femoral head and neck as it extends laterally (owing to the torsion angle—discussed earlier on page. 518), there is an angle of 30–40° between their axes (Fig. B5.28). Consequently, the articular surfaces of the head and acetabulum are not fully congruent in the erect (bipedal) posture. The anterior part of the femoral head is "exposed" and articulates mostly with the joint capsule (Figs. 5.79C, 5.80, 5.81A & C, and 5.84). Nonetheless, rarely is >40% of the available articular surface of the femoral head in contact with the surface of the acetabulum in any position.

Relative to other joints and in view of the large size of the hip joint, this is extensive contact, contributing considerably to the joint's great stability.

Fractures of Femoral Neck

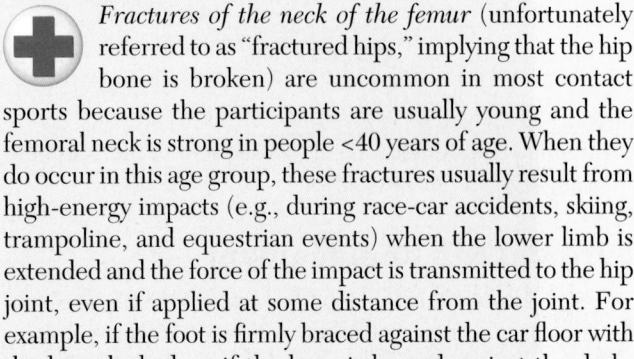

Fractures of the neck of the femur (unfortunately referred to as "fractured hips," implying that the hip bone is broken) are uncommon in most contact sports because the participants are usually young and the femoral neck is strong in people <40 years of age. When they do occur in this age group, these fractures usually result from high-energy impacts (e.g., during race-car accidents, skiing, trampoline, and equestrian events) when the lower limb is extended and the force of the impact is transmitted to the hip joint, even if applied at some distance from the joint. For example, if the foot is firmly braced against the car floor with the knee locked, or if the knee is braced against the dashboard during a head-on collision, the force of the impact may be transmitted superiorly and produce a femoral neck fracture. Femoral neck fractures are especially common in individuals >60 years, especially in women, because their femoral necks are more often weak and brittle, as a result of *osteoporosis* (Fig. B5.29). Fractures of the femoral neck are often intracapsular, and realignment of the neck fragments requires internal skeletal fixation.

Fractures of the femoral neck cause lateral rotation of the lower limb. Fractures of the femoral neck often disrupt the blood supply to the head of the femur. Most of the blood to the head and neck of the femur is supplied by the medial circumflex femoral artery (Fig. 5.82). The retinacular arteries arising from this artery are often torn when the femoral neck is fractured or the hip joint is dislocated. Following some femoral neck fractures, the artery to the ligament of the femoral head may be the only remaining source of blood to the proximal fragment. This artery is frequently inadequate for

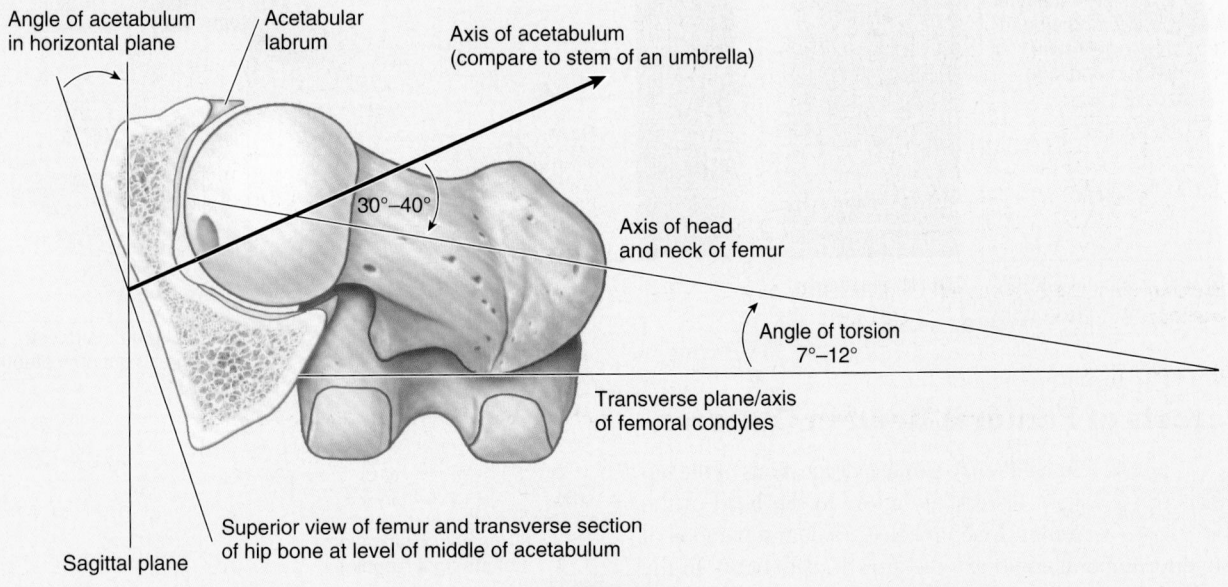

Angle of acetabulum in horizontal plane

Acetabular labrum

Axis of acetabulum (compare to stem of an umbrella)

30°–40°

Axis of head and neck of femur

Angle of torsion 7°–12°

Transverse plane/axis of femoral condyles

Sagittal plane

Superior view of femur and transverse section of hip bone at level of middle of acetabulum

FIGURE B5.28.

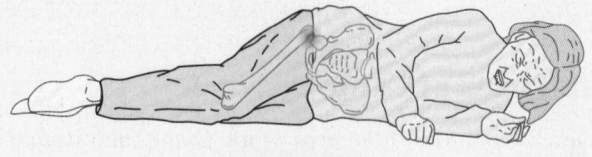

FIGURE B5.29.

maintaining the femoral head; consequently, the fragment may undergo *aseptic vascular necrosis* (tissue death).

Surgical Hip Replacement

Although the hip joint is strong and stable, it is subject to severe traumatic injury and degenerative disease. *Osteoarthritis of the hip joint*, characterized by pain, edema, limitation of motion, and erosion of articular cartilage, is a common cause of disability (Fig. B5.30A). During hip replacement, a metal prosthesis anchored to the person's femur by bone cement replaces the femoral head and neck (Fig. B5.30B). A plastic socket cemented to the hip bone replaces the acetabulum.

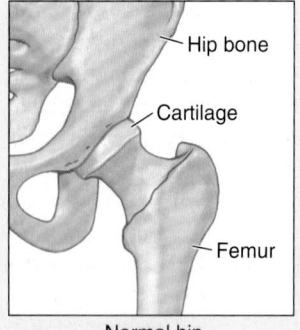

- Hip bone
- Cartilage
- Femur

Normal hip

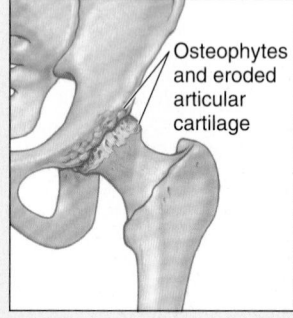

Osteophytes and eroded articular cartilage

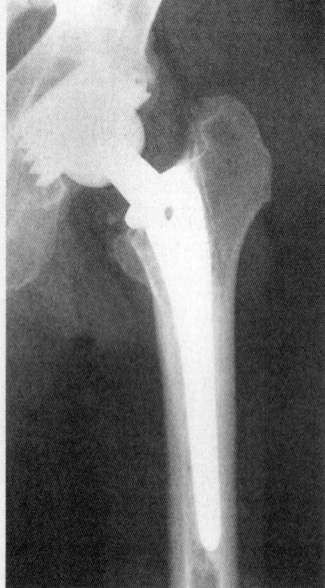

(A) Hip with moderate arthritis   **(B)** Hip prosthesis

FIGURE B5.30.

Necrosis of Femoral Head in Children

  In children, traumatic dislocations of the hip joint disrupt the artery to the head of the femur. Fractures that result in separation of the superior femoral epiphysis (the growth plate between the femoral head and neck) are also likely to result in an inadequate

blood supply to the femoral head and in *post-traumatic avascular necrosis of the head of the femur.* As a result, incongruity of the joint surfaces develops, and growth at the epiphysis is retarded. Such conditions, most common in children 3–9 years of age, produce hip pain that may radiate to the knee.

Dislocation of Hip Joint

Congenital dislocation of the hip joint is common, occurring in approximately 1.5 per 1000 neonates; it is bilateral in approximately half the cases. Girls are affected at least eight times more often than boys (Salter, 1999). Dislocation occurs when the femoral head is not properly located in the acetabulum. Inability to abduct the thigh is characteristic of congenital dislocation. In addition, the affected limb appears (and functions as if it is) shorter because the dislocated femoral head is more superior than on the normal side, resulting in a positive *Trendelenburg sign* (hip appears to drop on one side during walking). Approximately 25% of all cases of arthritis of the hip in adults are the direct result of residual defects from congenital dislocation of the hip.

Acquired dislocation of the hip joint is uncommon because this articulation is so strong and stable. Nevertheless, dislocation may occur during an automobile accident when the hip is flexed, adducted, and medially rotated, the usual position of the lower limb when a person is riding in a car.

Posterior dislocations of the hip joint are most common. A head-on collision that causes the knee to strike the dashboard

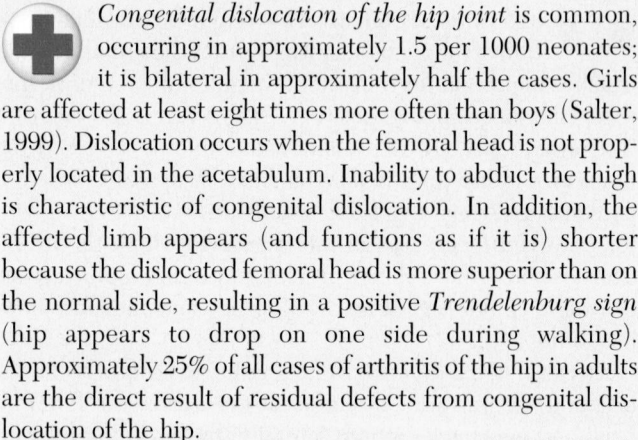

(A)

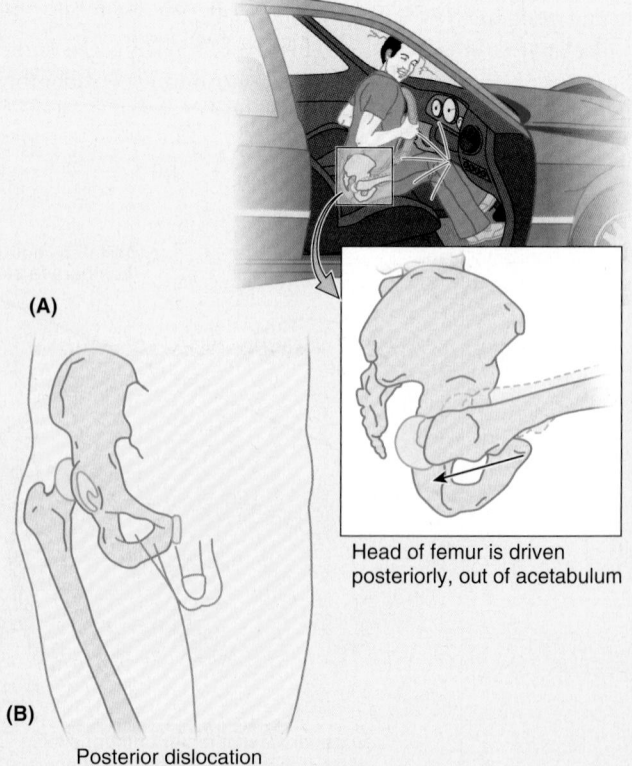

Head of femur is driven posteriorly, out of acetabulum

(B)

Posterior dislocation of the right hip joint

FIGURE B5.31.

may dislocate the hip when the femoral head is forced out of the acetabulum (Fig. B5.31A). The joint capsule ruptures inferiorly and posteriorly, allowing the femoral head to pass through the tear in the capsule, and over the posterior margin of the acetabulum onto the lateral surface of the ilium, shortening and medial rotating the limb (Fig. B5.31B).

Because of the close relationship of the *sciatic nerve* to the hip joint (Fig. 5.80A), it may be injured (stretched and/or compressed) during posterior dislocations or fracture–dislocations of the hip joint. This kind of injury may result in paralysis of the hamstrings and muscles distal to the knee supplied by the sciatic nerve. Sensory changes may also occur in the skin over the posterolateral aspects of the leg and over much of the foot because of injury to sensory branches of the sciatic nerve.

Anterior dislocation of the hip joint results from a violent injury that forces the hip into extension, abduction, and lateral rotation (e.g., catching a ski tip when snow skiing). In these cases, the femoral head is inferior to the acetabulum. Often, the acetabular margin fractures, producing a *fracture–dislocation of the hip joint.* When the femoral head dislocates, it usually carries the acetabular bone fragment and acetabular labrum with it. These injuries also occur with posterior dislocations.

Genu Valgum and Genu Varum

 The femur is placed diagonally within the thigh, whereas the tibia is almost vertical within the leg, creating an angle at the knee between the long axes of the bones (Fig. B5.32A). The angle between the two bones, referred to clinically as the **Q-angle,** is assessed by drawing a line from the ASIS to the middle of the patella and extrapolating a second (vertical) line passing through the middle of the patella and tibial tuberosity (Fig. 5.84). The Q-angle is typically greater in adult females, owing to their wider pelves. When normal, the angle of the femur within the thigh places the middle of the knee joint directly inferior to the head of the femur when standing, centering the weight-bearing line in the intercondylar region of the knee (Fig. B5.32A).

A medial angulation of the leg in relation to the thigh, in which the femur is abnormally vertical and the Q-angle is small, is a deformity called *genu varum* (bowleg) that causes unequal weight-bearing: The line of weight-bearing falls medial to the center of the knee (Fig. B5.32B). Excess pressure is placed on the medial aspect of the knee joint, which results in *arthrosis* (destruction of knee cartilages), and the fibular collateral ligament is overstressed (Fig. B.32D). A lateral angulation of the leg (large Q-angle, >17°) in relation to the thigh (exaggeration of the knee angle) is called *genu valgum* (knock-knee) (Fig. B5.32C). Because of the exaggerated knee angle in genu valgum, the weight-bearing line falls lateral to the center of the knee. Consequently, the tibial collateral ligament is overstretched, and there is excess stress on the lateral meniscus and cartilages of the lateral femoral and tibial condyles. The patella, normally pulled laterally by the tendon of the vastus lateralis, is pulled even farther laterally when the leg is extended in the presence of genu valgum so that its articulation with the femur is abnormal.

Children commonly appear bowlegged for 1–2 years after starting to walk, and knock-knees are frequently observed in children 2–4 years of age. Persistence of these abnormal knee angles in late childhood usually means congenital deformities exist that may require correction. Any irregularity of a joint eventually leads to wear and tear (arthrosis) of the articular cartilages and degenerative joint changes (*osteoarthritis [arthrosis]*).

Patellar Dislocation

When the patella is dislocated, it nearly always dislocates laterally. *Patellar dislocation* is more common in women, presumably because of their greater

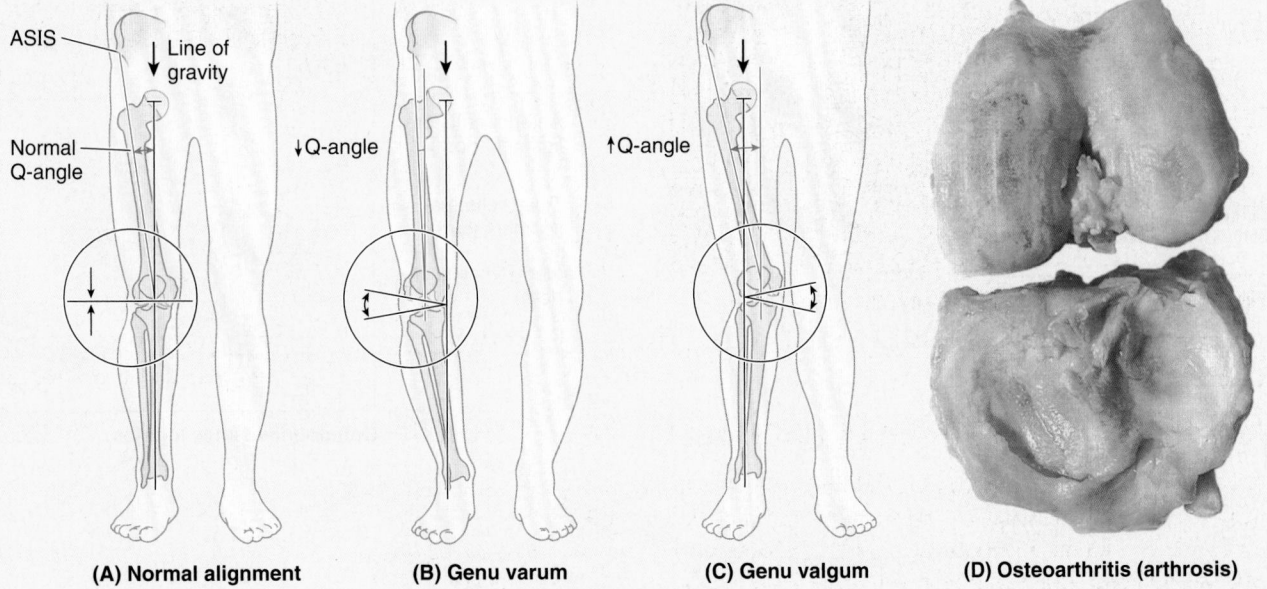

| (A) Normal alignment | (B) Genu varum | (C) Genu valgum | (D) Osteoarthritis (arthrosis) |

FIGURE B5.32.

Q-angle, which, in addition to representing the oblique placement of the femur relative to the tibia, represents the angle of pull of the quadriceps relative to the axis of the patella and tibia (the term *Q-angle* was actually coined in reference to the angle of pull of the quadriceps). The tendency toward lateral dislocation is normally counterbalanced by the medial, more horizontal pull of the powerful vastus medialis. In addition, the more anterior projection of the lateral femoral condyle and deeper slope for the larger lateral patellar facet provide a mechanical deterrent to lateral dislocation. An imbalance of the lateral pull and the mechanisms resisting it result in abnormal tracking of the patella within the patellar groove and chronic patellar pain, even if actual dislocation does not occur.

Patellofemoral Syndrome

 Pain deep to the patella often results from excessive running, especially downhill; hence, this type of pain is often called "runner's knee." The pain results from repetitive microtrauma caused by abnormal tracking of the patella relative to the patellar surface of the femur, a condition known as the *patellofemoral syndrome.* This syndrome may also result from a direct blow to the patella and from *osteoarthritis of the patellofemoral compartment* (degenerative wear and tear of articular cartilages). In some cases, strengthening of the vastus medialis corrects *patellofemoral dysfunction.* This muscle tends to prevent lateral dislocation of the patella resulting from the Q-angle because the vastus medialis attaches to and pulls on the medial border of the patella. Hence, weakness of the vastus medialis predisposes

the individual to the patellofemoral dysfunction and patellar dislocation.

Knee Joint Injuries

Knee joint injuries are common because the knee is a low-placed, mobile, weight-bearing joint, serving as a fulcrum between two long levers (thigh and leg). Its stability depends almost entirely on its associated ligaments and surrounding muscles.

The knee joint is essential for everyday activities such as standing, walking, and climbing stairs. It is also a main joint for sports that involve running, jumping, kicking, and changing directions. To perform these activities, the knee joint must be mobile; however, this mobility makes it susceptible to injuries.

The most common knee injury in contact sports is *ligament sprain,* which occurs when the foot is fixed in the ground (Fig. B5.33A). If a force is applied against the knee when the foot cannot move, ligament injuries are likely to occur. The tibial and fibular collateral ligaments (TCL and FCL) are tightly stretched when the leg is extended, normally preventing disruption of the sides of the knee joint.

The firm attachment of the TCL to the medial meniscus is of considerable clinical significance because tearing of this ligament frequently results in concomitant tearing of the medial meniscus. The injury is frequently caused by a blow to the lateral side of the extended knee, or excessive lateral twisting of the flexed knee that disrupts the TCL and concomitantly tears and/or detaches the medial meniscus from the joint capsule (Fig. B5.33A). This injury is common in

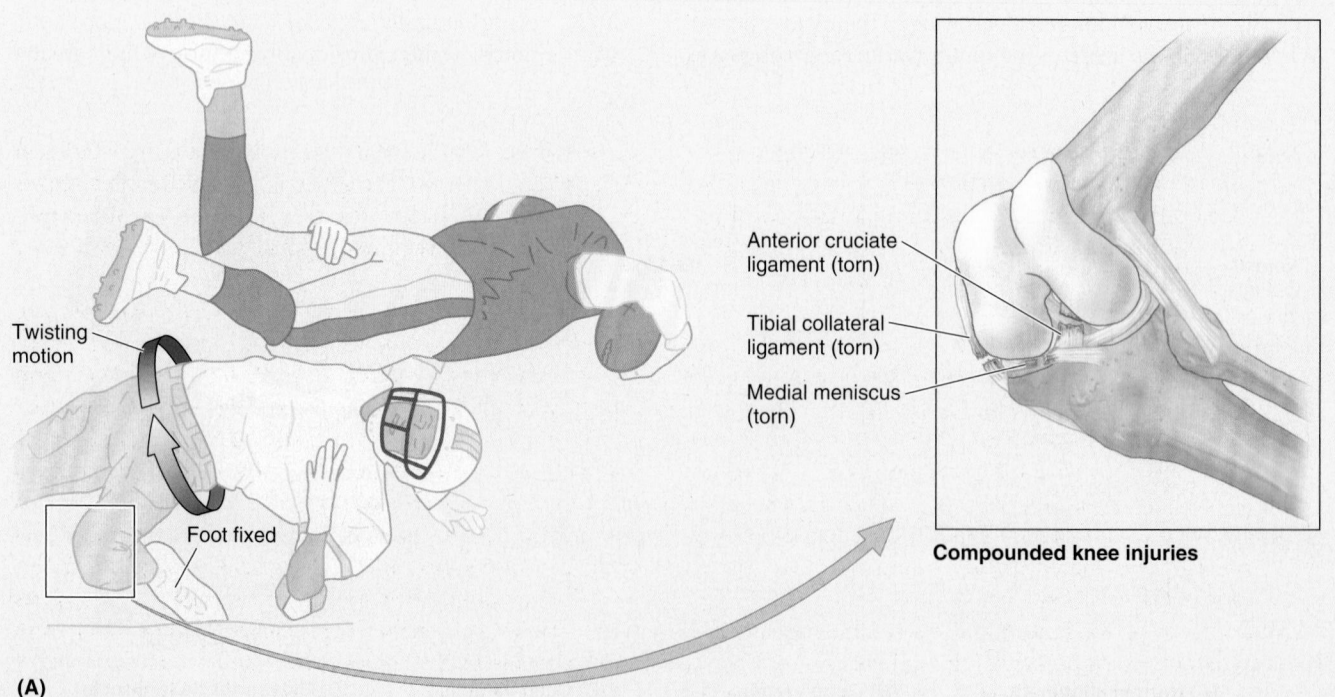

Twisting motion

Foot fixed

Anterior cruciate ligament (torn)

Tibial collateral ligament (torn)

Medial meniscus (torn)

Compounded knee injuries

(A)

FIGURE B5.33.

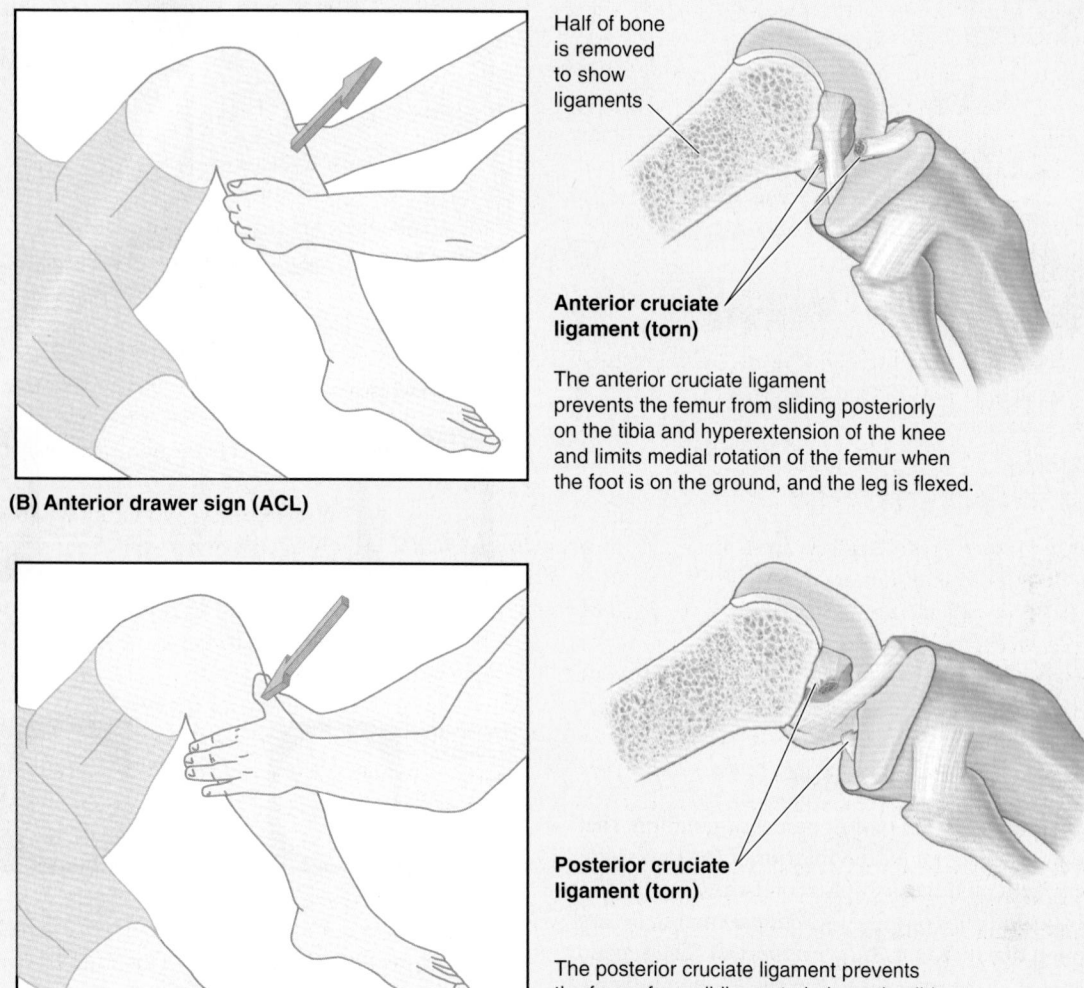

(B) Anterior drawer sign (ACL)

Half of bone is removed to show ligaments

Anterior cruciate ligament (torn)

The anterior cruciate ligament prevents the femur from sliding posteriorly on the tibia and hyperextension of the knee and limits medial rotation of the femur when the foot is on the ground, and the leg is flexed.

(C) Posterior drawer sign (PCL)

Posterior cruciate ligament (torn)

The posterior cruciate ligament prevents the femur from sliding anteriorly on the tibia, particularly when the knee is flexed.

FIGURE B5.33. (continued).

athletes who twist their flexed knees while running (e.g., in basketball, the various forms of football, and volleyball). The ACL, which serves as a pivot for rotatory movements of the knee, and is taut during flexion, may also tear subsequent to the rupture of the TCL, creating an "unhappy triad" of knee injuries.

Hyperextension and severe force directed anteriorly against the femur with the knee semiflexed (e.g., a cross-body block in football) may tear the ACL. *ACL rupture* is also a common knee injury in skiing accidents. This injury causes the free tibia to slide anteriorly under the fixed femur, known as the *anterior drawer sign* (Fig. B5.33B); it is tested clinically via the *Lachman test*. The ACL may tear away from the femur or tibia; however, tears commonly occur in the midportion of the ligament.

Although strong, *PCL ruptures* may occur when a player lands on the tibial tuberosity with the knee flexed (e.g., when knocked to the floor in basketball). PCL ruptures usually occur in conjunction with tibial or fibular ligament tears.

These injuries can also occur in head-on collisions when seat belts are not worn and the proximal end of the tibia strikes the dashboard. PCL ruptures allow the free tibia to slide posteriorly under the fixed femur, known as *the posterior drawer sign* (Fig. B5.33C).

Meniscal tears usually involve the medial meniscus. The lateral meniscus does not usually tear because of its mobility. *Pain on lateral rotation of the tibia* on the femur indicates injury of the lateral meniscus (Fig. B5.34A), whereas *pain on medial rotation of the tibia* on the femur indicates injury of the medial meniscus (Fig. B5.34B). Most meniscal tears occur in conjunction with TCL or ACL tears. Peripheral meniscal tears can often be repaired, or they may heal on their own because of the generous blood supply to this area. If tears do not heal or cannot repaired, the meniscus is removed (e.g., by arthroscopic surgery). Knee joints from which a meniscus has been removed suffer no loss of mobility; however, the knee may be less stable and the tibial plateaus often undergoes inflammatory reactions.

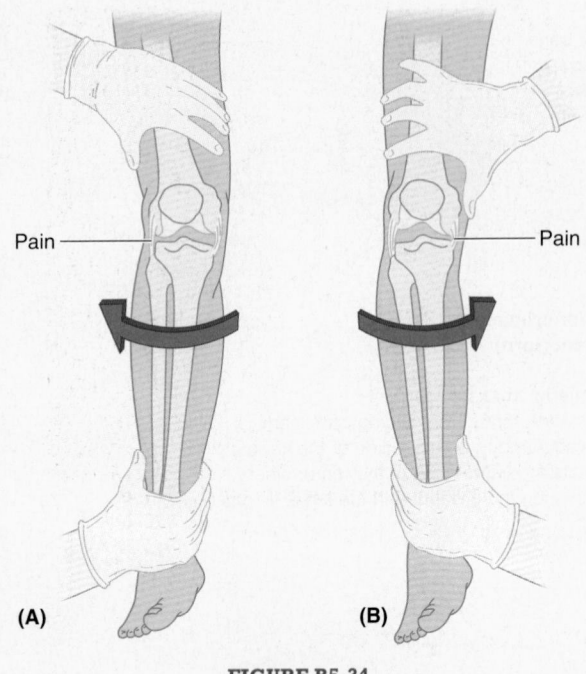

(A)   **(B)**

FIGURE B5.34.

Arthroscopy of Knee Joint

Arthroscopy is an endoscopic examination that allows visualization of the interior of the knee joint cavity with minimal disruption of tissue (Fig. B5.35). The arthroscope and one (or more) additional cannula(e) are inserted through tiny incisions, known as *portals*. The second cannula is for passage of specialized tools (e.g., manipulative probes or forceps) or equipment for trimming, shaping, or removing damaged tissue. This technique allows removal of torn menisci, loose bodies in the joint (such as bone chips), and *débridement* (the excision of devitalized articular cartilaginous material) in advanced cases of arthritis. Ligament repair or replacement may also be performed using an arthroscope. Although general anesthesia is usually preferable, knee arthroscopy can be performed using local or regional anesthesia. During arthroscopy, the articular cavity of the knee must be treated essentially as two separate (medial and lateral) femorotibial articulations, owing to the imposition of the synovial fold around the cruciate ligaments.

Aspiration of Knee Joint

Fractures of the distal end of the femur, or lacerations of the anterior thigh, may involve the suprapatellar bursa and result in infection of the knee joint. When the knee joint is infected and inflamed, the amount of synovial fluid may increase. *Joint effusions,* the escape of fluid from blood or lymphatic vessels, results in increased amounts of fluid in the joint cavity. Because the suprapatellar bursa communicates freely with the synovial cavity of the knee joint, fullness of the thigh in the region of

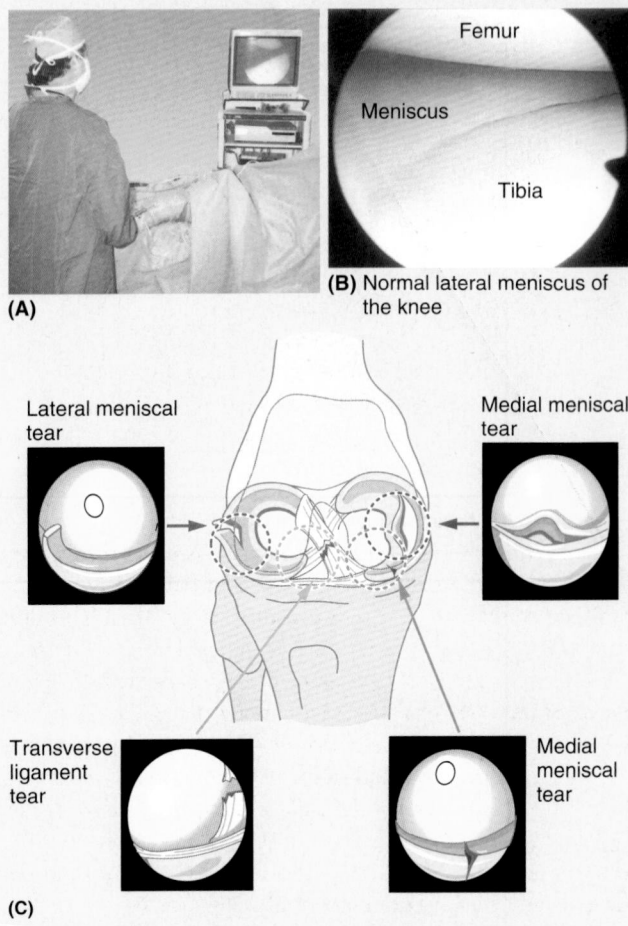

(A)

(B) Normal lateral meniscus of the knee

Lateral meniscal tear

Medial meniscal tear

Transverse ligament tear

Medial meniscal tear

(C)

FIGURE B5.35.

the suprapatellar bursa may indicate increased synovial fluid. This bursa can be aspirated to remove the fluid for examination. Direct *aspiration of the knee joint* is usually performed with the patient sitting on a table with the knee flexed. The joint is approached laterally, using three bony points as landmarks for needle insertion: the anterolateral tibial (Gerdy) tubercle, the lateral epicondyle of the femur, and the apex of the patella. In addition to being the route for aspiration of serous and sanguineous (bloody) fluid, this triangular area also lends itself to drug injection for treating pathology of the knee joint.

Bursitis in Knee Region

Prepatellar bursitis is caused by friction between the skin and the patella; however, the bursa may also be injured by compressive forces resulting from a direct blow or from falling on the flexed knee. If the inflammation is chronic, the bursa becomes distended with fluid and forms a swelling anterior to the knee. This condition has been called "housemaid's knee" (Fig. B5.36); however, other people who work on their knees without knee pads, such as hardwood floor and rug installers, may also develop prepatellar bursitis.

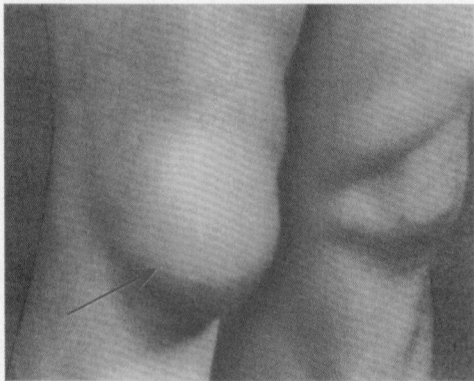

Prepatellar bursitis (arrow)

FIGURE B5.36.

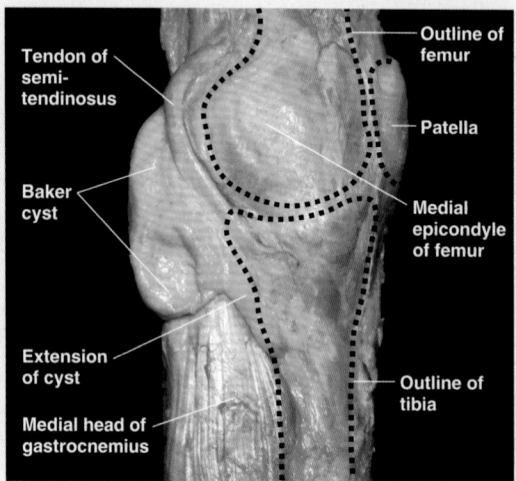

Medial view of dissection of left knee

FIGURE B5.37.

Subcutaneous infrapatellar bursitis is caused by excessive friction between the skin and the tibial tuberosity; the edema occurs over the proximal end of the tibia. This condition was formerly called "clergyman's knee" because of frequent genuflecting (L. *genu,* knee); however, it occurs more commonly in roofers and floor tilers if they do not wear knee pads. *Deep infrapatellar bursitis* results in edema between the patellar ligament and the tibia, superior to the tibial tuberosity. The inflammation is usually caused by overuse and subsequent friction between the patellar tendon and the structures posterior to it, the infrapatellar fat pad and tibia (Anderson et al., 2000). Enlargement of the deep infrapatellar bursa obliterates the dimples normally occurring on each side of the patellar ligament when the leg is extended (see Fig. 5.104A).

Abrasions or penetrating wounds may result in *suprapatellar bursitis,* an infection caused by bacteria entering the suprapatellar bursa from the torn skin (see Fig. 5.94A). The infection may spread to the cavity of the knee joint, causing localized redness and enlarged popliteal and inguinal lymph nodes.

Popliteal Cysts

Popliteal cysts (Baker cysts) are abnormal fluid-filled sacs of synovial membrane in the region of the popliteal fossa. A popliteal cyst is almost always a complication of chronic knee joint effusion. The cyst may be a herniation of the gastrocnemius or semimembranosus bursa through the fibrous layer of the joint capsule into the popliteal fossa, communicating with the synovial cavity of the knee joint by a narrow stalk (Fig. B5.37). Synovial fluid may also escape from the knee joint (*synovial effusion*) or a bursa around the knee and collect in the popliteal fossa. Here it forms a new synovial-lined sac, or popliteal cyst. Popliteal cysts are common in children but seldom cause symptoms. In adults, popliteal cysts can be large, extending as far as the midcalf, and may interfere with knee movements.

Knee Replacement

If a person's knee is diseased, resulting from osteoarthritis, for example, an artificial knee joint may be inserted (*total knee replacement arthroplasty*) (Fig. B5.38). The artificial knee joint consists of plastic and metal components that are cemented to the femoral and tibial bone ends after removal of the defective areas. The combination of metal and plastic mimics the smoothness of cartilage on cartilage and produces good results in "low-demand" people who have a relatively sedentary life. In "high-demand" people who are active in sports, the bone–cement junctions may break down, and the artificial knee components may loosen; however, improvements in bioengineering and surgical technique have provided better results.

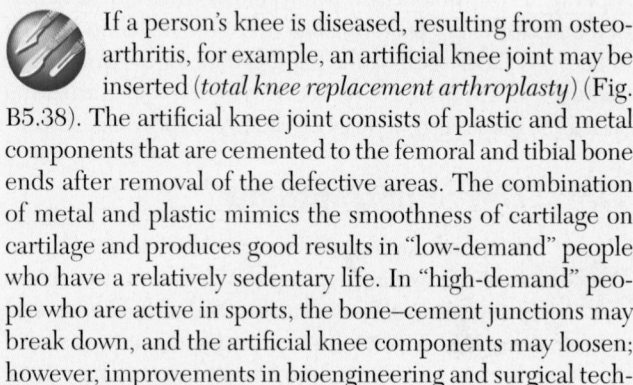

FIGURE B5.38.

Ankle Injuries

The ankle is the most frequently injured major joint in the body. *Ankle sprains* (torn fibers of ligaments) are most common. A sprained ankle is nearly always an *inversion injury,* involving twisting of the weight-bearing

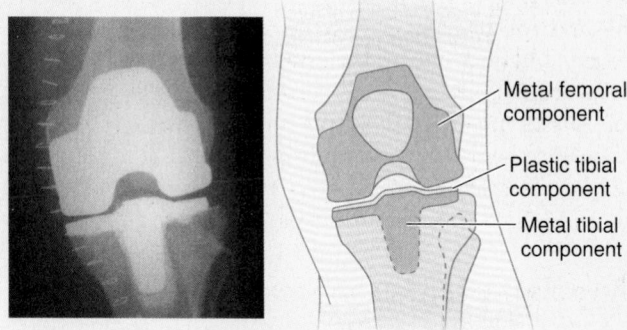

plantarflexed foot. The person steps on an uneven surface and the foot is forcibly inverted. *Lateral ligament sprains* occur in running and jumping sports, particularly basketball (70–80% of players have had at least one sprained ankle). The lateral ligament is injured because it is much weaker than the medial ligament, and is the ligament that resists inversion at the talocrural joint. The *anterior talofibular ligament*—part of the lateral ligament—is most vulnerable and most commonly torn during ankle sprains, either partially or completely, resulting in instability of the ankle joint (Fig. B5.39). The *calcaneofibular ligament* may also be torn. In severe sprains, the lateral malleolus of the fibula may also be fractured. *Shearing injuries fracture the lateral malleolus* at or superior to the ankle joint. *Avulsion fractures* break the malleolus inferior to the ankle joint; a fragment of bone is pulled off by the attached ligament(s).

A *Pott fracture–dislocation of the ankle* occurs when the foot is forcibly everted (Fig. B5.40). This action pulls on the extremely strong medial ligament, often tearing off the medial malleolus. The talus then moves laterally, shearing off the lateral malleolus or, more commonly, breaking the fibula superior to the tibiofibular syndesmosis. If the tibia is carried anteriorly, the posterior margin of the distal end of the tibia is also sheared off by the talus, producing a "trimalleolar fracture." In applying this term to this injury, the entire distal end of the tibia is erroneously considered to be a "malleolus."

Tibial Nerve Entrapment

The tibial nerve leaves the posterior compartment of the leg by passing deep to the flexor retinaculum in the interval between the medial malleolus and the calcaneus (Fig. 5.71A). Entrapment and compression of the tibial nerve (*tarsal tunnel syndrome*) occurs when there

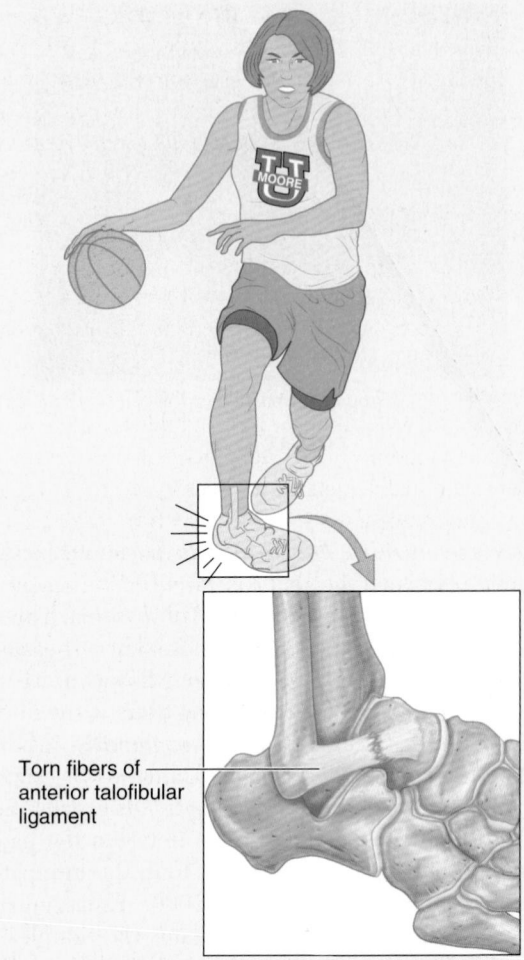

Torn fibers of anterior talofibular ligament

FIGURE B5.39.

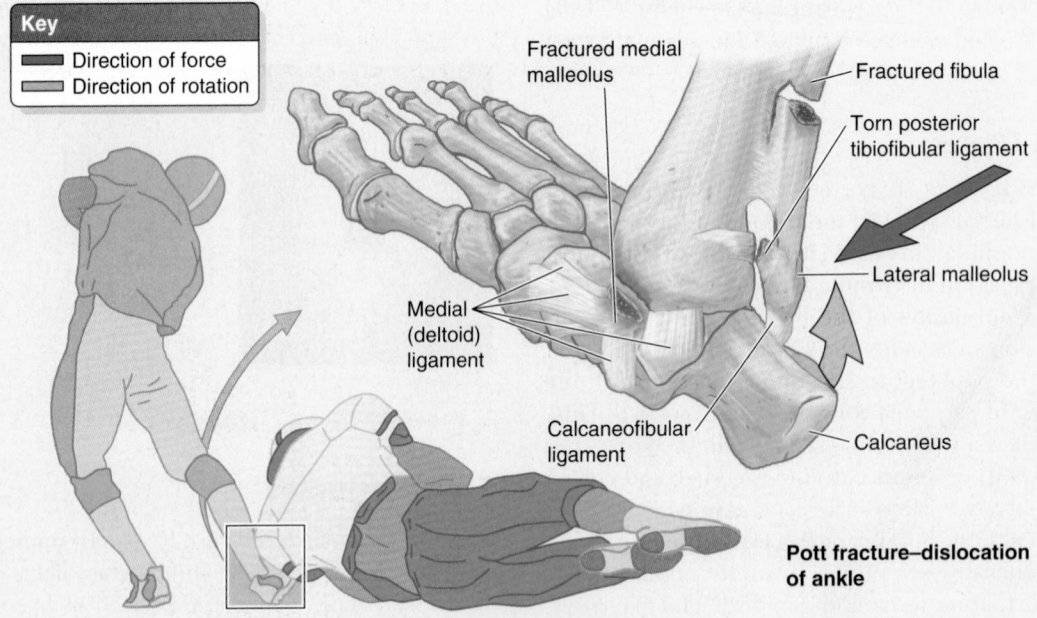

Key
- Direction of force
- Direction of rotation

Fractured medial malleolus

Fractured fibula

Torn posterior tibiofibular ligament

Lateral malleolus

Medial (deltoid) ligament

Calcaneofibular ligament

Calcaneus

Pott fracture–dislocation of ankle

FIGURE B5.40.

is edema and tightness in the ankle involving the synovial sheaths of the tendons of muscles in the posterior compartment of the leg. The area involved is from the medial malleolus to the calcaneus, and the heel pain results from compression of the tibial nerve by the flexor retinaculum.

Hallux Valgus

Hallux valgus is a foot deformity caused by pressure from footwear and degenerative joint disease; it is characterized by lateral deviation of the great toe (Fig. B5.41). The *L* in va*l*gus indicates *lateral deviation*. In some people, the painful deviation is so large that the great toe overlaps the 2nd toe (Fig. B5.41A), and there is a decrease in the medial longitudinal arch. Such deviation occurs especially in females, and its frequency increases with age. These individuals cannot move their 1st digit away from their 2nd digit because the sesamoids under the head of the 1st metatarsal are usually displaced, and lie in the space between the heads of the 1st and 2nd metatarsals (Fig. B5.41B). The 1st metatarsal shifts medially and the sesamoids shift laterally. Often the surrounding tissues swell and the resultant pressure and friction against the shoe cause a subcutaneous bursa to form; when tender and inflamed, the bursa is called a *bunion* (Fig. B5.41A). Often hard *corns* (inflamed areas of thick skin) also form over the proximal interphalangeal joints, especially of the little toe.

Hammer Toe

Hammer toe is a foot deformity in which the proximal phalanx is permanently and markedly dorsiflexed (hyperextended) at the metatarsophalangeal joint, and the middle phalanx strongly plantarflexed at the proximal interphalangeal joint. The distal phalanx of the digit is often also hyperextended. This gives the digit (usually the 2nd) a hammer-like appearance (Fig. B5.42A). This deformity of one or more toes may result from weakness of the lumbrical and interosseous muscles, which flex the metatarsophalangeal joints and extend

the interphalangeal joints. A *callosity* or *callus*, hard thickening of the keratin layer of the skin, often develops where the dorsal surface of the toe repeatedly rubs on the shoe.

Claw Toes

Claw toes are characterized by hyperextension of the metatarsophalangeal joints and flexion of the distal interphalangeal joints (Fig. B5.42B). Usually, the lateral four toes are involved. Callosities develop on the dorsal surfaces of the toes because of pressure of the shoe. They may also form on the plantar surfaces of the metatarsal heads and the toe tips because they bear extra weight when claw toes are present.

Pes Planus (Flatfeet)

The flat appearance of the sole of the foot before age 3 is normal; it results from the thick subcutaneous fat pad in the sole. As children get older, the fat is lost, and a normal medial longitudinal arch becomes visible (Fig. B5.42C). Flatfeet can either be *flexible* (flat, lacking a medial arch, when weight-bearing but normal in appearance when not bearing weight [Fig. B5.42D]), or *rigid* (flat even when not bearing weight). The more common *flexible flatfeet* result from loose or degenerated intrinsic ligaments (inadequate passive arch support). Flexible flatfeet is common in childhood but usually resolves with age as the ligaments grow and mature. The condition occasionally persists into adulthood and may or may not be symptomatic.

Rigid flatfeet with a history that goes back to childhood are likely to result from a bone deformity (such as a fusion of adjacent tarsal bones). *Acquired flatfeet* ("fallen arches") are likely to be secondary to dysfunction of the tibialis posterior (dynamic arch support) owing to trauma, degeneration with age, or denervation. In the absence of normal passive or dynamic support, the plantar calcaneonavicular ligament fails to support the head of the talus. Consequently, the head of the talus displaces inferomedially and becomes prominent

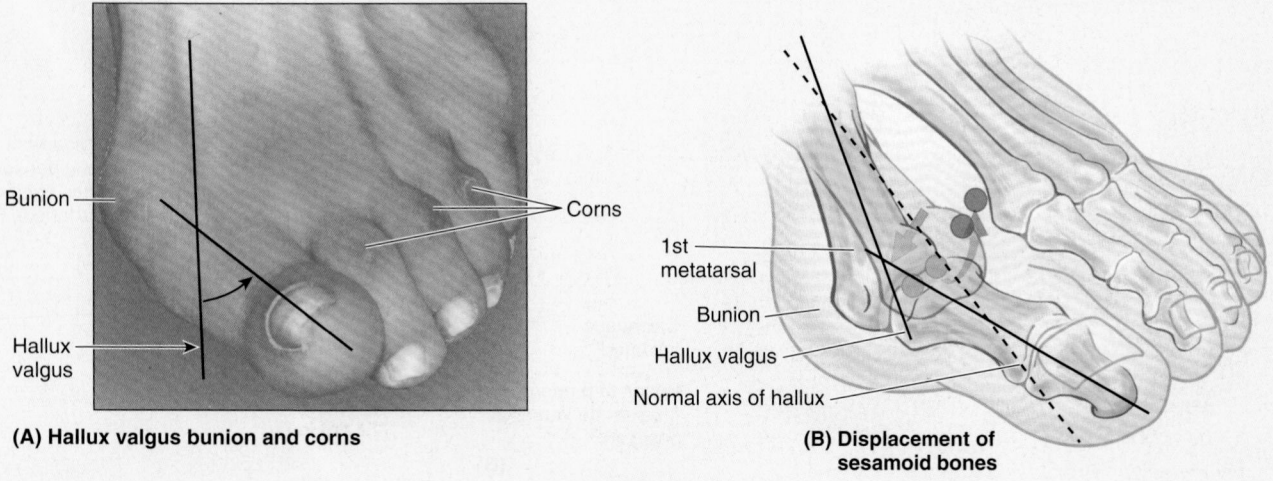

(A) Hallux valgus bunion and corns

(B) Displacement of sesamoid bones

FIGURE B5.41.

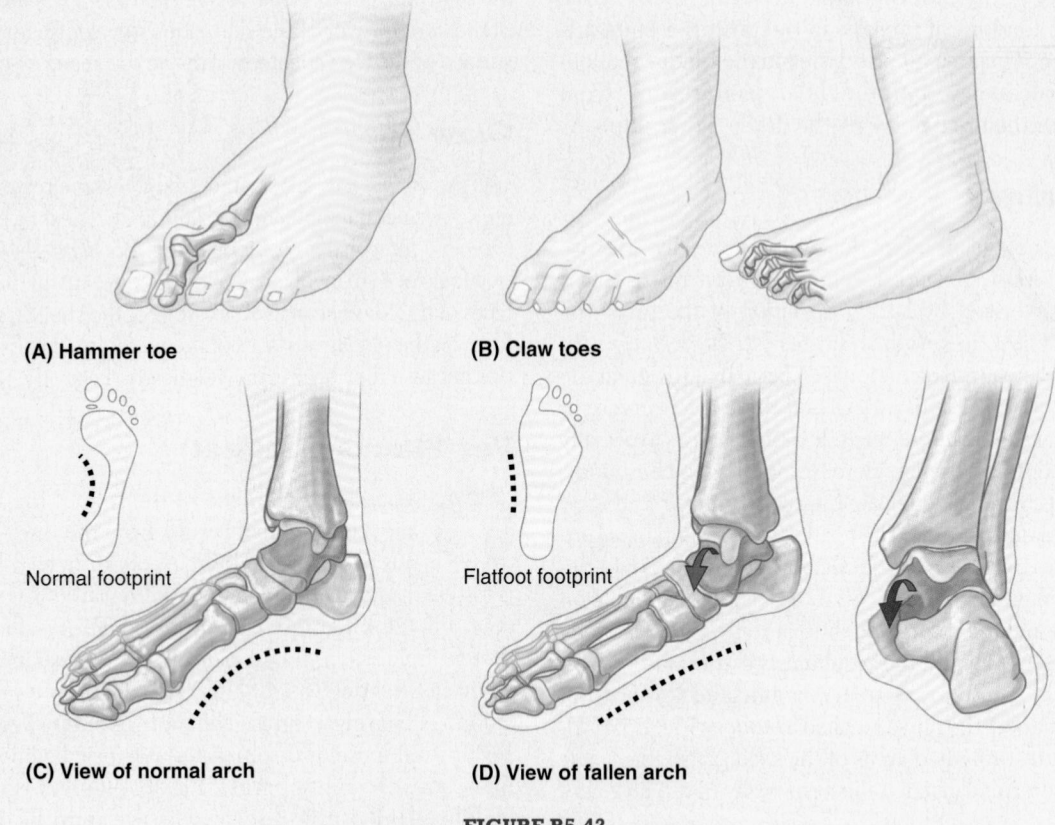

(A) Hammer toe

(B) Claw toes

Normal footprint

Flatfoot footprint

(C) View of normal arch

(D) View of fallen arch

FIGURE B5.42.

(Fig. B5.42D, red arrows). As a result, some flattening of the medial part of the longitudinal arch occurs, along with lateral deviation of the forefoot. Flatfeet are common in older people, particularly if they undertake much unaccustomed standing or gain weight rapidly, adding stress on the muscles and increasing the strain on the ligaments supporting the arches.

Clubfoot (Talipes equinovarus)

 Clubfoot refers to a foot that is twisted out of position. Of the several types, all are *congenital* (present at birth). *Talipes equinovarus,* the common type

(2 per 1000 neonates), involves the subtalar joint; boys are affected twice as often as girls. The foot is inverted, the ankle is plantarflexed, and the forefoot is adducted (turned toward the midline in an abnormal manner) (Fig. B5.43A). The foot assumes the position of a horse's hoof, hence the prefix "equino" (L. *equinus,* horse). In half of those affected, both feet are malformed. A person with an uncorrected clubfoot cannot put the heel and sole flat and must bear the weight on the lateral surface of the forefoot. Consequently, walking is painful. The main abnormality is shortness and tightness of the muscles, tendons, ligaments, and joint capsules on the medial side and posterior aspect of the foot and ankle (Fig. B5.43B).

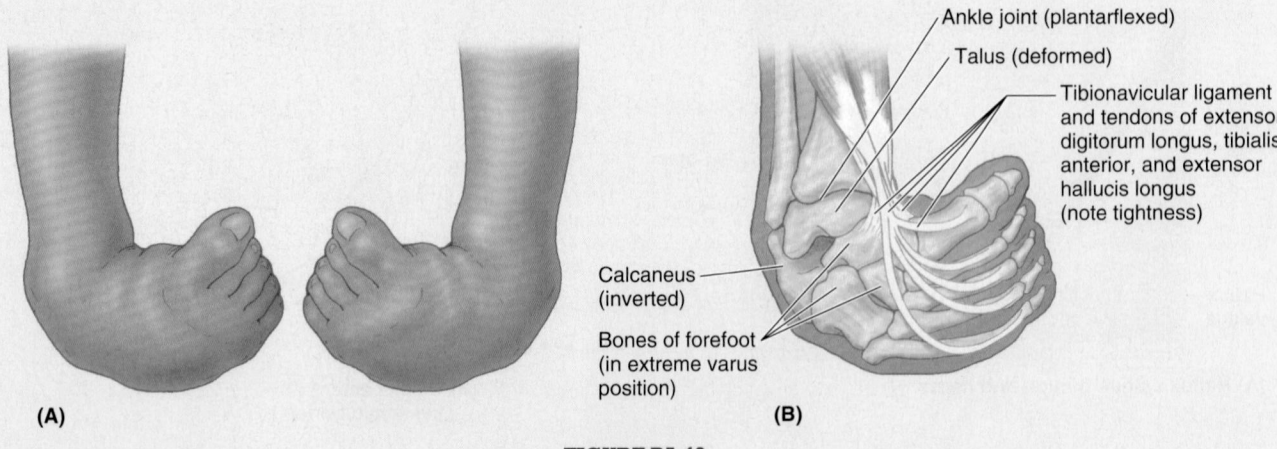

Ankle joint (plantarflexed)

Talus (deformed)

Tibionavicular ligament and tendons of extensor digitorum longus, tibialis anterior, and extensor hallucis longus (note tightness)

Calcaneus (inverted)

Bones of forefoot (in extreme varus position)

(A)

(B)

FIGURE B5.43.

The Bottom Line

JOINTS OF LOWER LIMB

Hip joint: The hip joint is the strongest and most stable joint. ♦ Its stability results from (1) the mechanical strength of its ball and (deep) socket construction, allowing extensive articular sur-face contact; (2) its strong joint capsule; and (3) its many surrounding muscles. ♦ However, it remains vulnerable, especially in older age, because of the angle of the femoral neck (inclination) and close association of the blood supply of the femoral head to the neck. Thus fractures result in avascular necrosis of the femoral head. ♦ Major movements of the hip joint include flexion and extension, possible over a wide range; medial and lateral rotation with abduction are part of every step of normal, bipedal walking.

Knee joint: The knee is a hinge joint with a wide range of motion (primarily flexion and extension, with rotation increasingly possible with flexion). ♦ It is our most vulnerable joint, owing to its incongruous articular surfaces and the mechanical disadvantage resulting from bearing the body's weight plus momentum while serving as a fulcrum between two long levers. ♦ Compensation is attempted by several features, including (1) strong intrinsic, extracapsular, and intracapsular ligaments; (2) splinting by many surrounding tendons (including the iliotibial tract); and (3) menisci that fill the spatial void, providing mobile articular surfaces. ♦ Of particular clinical importance are (1) collateral ligaments that are taut during (and limit) extension and are relaxed during flexion, allowing rotation for which they serve as check ligaments; (2) cruciate ligaments that maintain the joint during flexion, providing the pivot for rotation; and (3) the medial meniscus that is attached to the tibial collateral ligament, and is frequently injured because of this attachment.

Tibiofibular joints: The tibiofibular joints include a proximal synovial joint, an interosseous membrane, and a distal tibiofibular syndesmosis, consisting of anterior, interosseous, and posterior tibiofibular ligaments. ♦ Together these joints make up a compensatory system that allows a slight upward movement of the fibula owing to forced transverse expansion of the malleolar mortise (deep square socket) during maximal dorsiflexion of the ankle. ♦ All fibrous tibiofibular connections run downward from tibia to fibula, allowing this slight upward movement while strongly resisting the downward pull applied to the fibula by the contraction of eight of the nine muscles attached to it.

Ankle joint: The ankle (talocrural) joint is composed of a superior mortise, formed by the weight-bearing inferior surface of the tibia and the two malleoli, which receive the trochlea of the talus. ♦ The ankle joint is maintained medially by a strong, medial (deltoid) ligament, and a much weaker lateral ligament. ♦ The lateral ligament (specifically its anterior talofibular ligament component) is the most frequently injured ligament of the body. ♦ Injury occurs primarily by inadvertent inversion of the plantarflexed, weight-bearing foot. ♦ About 70° of dorsiflexion and plantarflexion is possible at the ankle joint, in addition to which small amounts of wobble occur in the less stable plantarflexed position.

Joints of foot: Functionally, there are three compound joints in the foot: (1) the clinical subtalar joint between the talus and the calcaneus, where inversion and eversion occur about an oblique axis; (2) the transverse tarsal joint, where the midfoot and forefoot rotate as a unit on the hindfoot around a longitudinal axis, augmenting inversion and eversion; and (3) the remaining joints of the foot, which allow the pedal platform (foot) to form dynamic longitudinal and transverse arches. ♦ The arches provide the resilience necessary for walking, running, and jumping, and are maintained by four layers of passive, fibrous support, plus the dynamic support provided by the intrinsic muscles of the foot, and the long fibular, tibial, and flexor tendons.

thePoint **Board-review questions, case studies, and additional resources are available at thePoint.lww.com.**

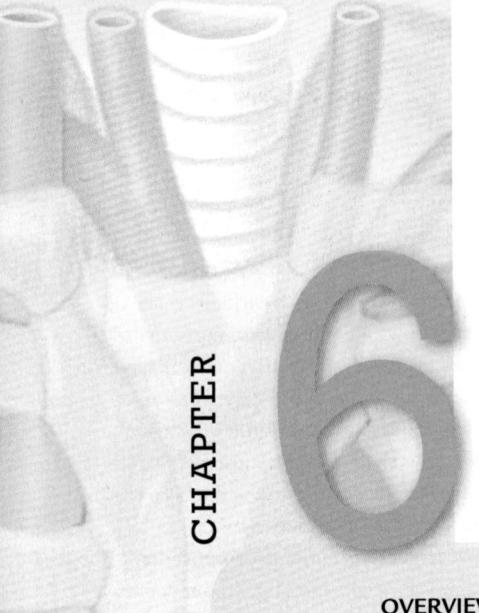

CHAPTER

6 Upper Limb

OVERVIEW OF UPPER LIMB

The upper limb is characterized by its mobility and ability to grasp, strike, and conduct fine motor skills (*manipulation*). These characteristics are especially marked in the hand when performing manual activities, such as buttoning a shirt.

Synchronized interplay occurs between the joints of the upper limb to coordinate the intervening segments to perform smooth, efficient motion at the most workable distance or position required for a specific task. Efficiency of hand function results in large part from the ability to place it in the proper position by movements at the scapulothoracic, glenohumeral, elbow, radio-ulnar, and wrist joints.

The upper limb consists of four major segments, which are further subdivided into regions for precise description (Figs. 6.1 and 6.2):

1. **Shoulder:** proximal segment of the limb that overlaps parts of the trunk (thorax and back) and lower lateral neck.

It includes the **pectoral, scapular,** and **deltoid regions** of the upper limb, and the lateral part (greater supraclavicular fossa) of the lateral cervical region. It overlies half of the pectoral girdle. The **pectoral girdle (shoulder girdle)** is a bony ring, incomplete posteriorly, formed by the *scapulae* and *clavicles,* and completed anteriorly by the *manubrium of the sternum* (part of the axial skeleton).

2. **Arm** (L. *brachium*): first segment of the free upper limb (more mobile part of the upper limb independent of the trunk) and the longest segment of the limb. It extends between and connects the shoulder and the elbow, and consists of **anterior** and **posterior regions of the arm,** centered around the humerus.

3. **Forearm** (L. *antebrachium*): second longest segment of the limb. It extends between and connects the elbow and wrist and includes **anterior** and **posterior regions of the forearm** overlying the radius and ulna.

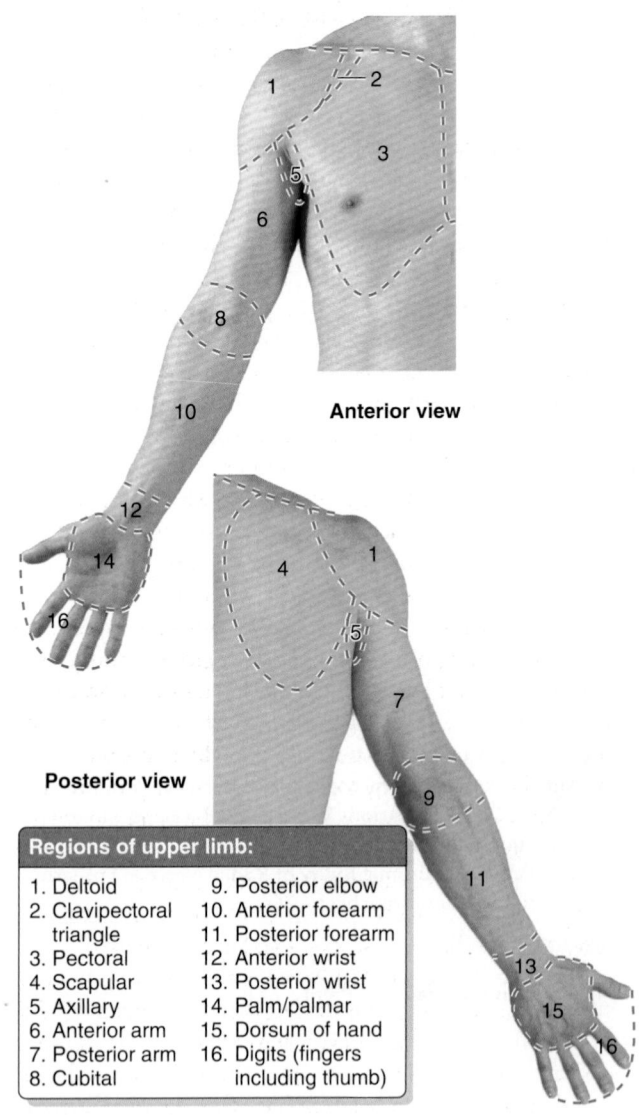

Regions of upper limb:

1. Deltoid	9. Posterior elbow
2. Clavipectoral triangle	10. Anterior forearm
3. Pectoral	11. Posterior forearm
4. Scapular	12. Anterior wrist
5. Axillary	13. Posterior wrist
6. Anterior arm	14. Palm/palmar
7. Posterior arm	15. Dorsum of hand
8. Cubital	16. Digits (fingers including thumb)

FIGURE 6.1. Segments and bones of upper limb. The joints divide the superior appendicular skeleton, and thus the limb itself, into four main segments: shoulder, arm, forearm, and hand.

FIGURE 6.2. Regions of upper limb. For exact description, the upper limb is divided into regions based on the external features (surface anatomy) of the underlying muscular formations, bones, and joints.

4. **Hand** (L. *manus*): part of the upper limb distal to the forearm that is formed around the carpus, metacarpus, and phalanges. It is composed of the **wrist, palm, dorsum of hand,** and **digits** (fingers, including an opposable thumb), and is richly supplied with sensory endings for touch, pain, and temperature.

COMPARISON OF UPPER AND LOWER LIMBS

Developing in a similar fashion, the upper and lower limbs share many common features (see Chapter 5). However, they are sufficiently distinct in structure to enable markedly different functions and abilities. Because the upper limb is not usually involved in weight bearing or motility, its stability has been sacrificed to gain mobility. The upper limb still possesses remarkable strength; and because of the hand's ability to conform to a paddle or assume a gripping or platform configuration, it may assume a role in motility in certain circumstances.

Both the upper and the lower limbs are connected to the **axial skeleton** (cranium, vertebral column, and associated thoracic cage) via the bony pectoral and pelvic girdles, respectively. The *pelvic girdle* consists of the two hip bones connected to the sacrum (see Chapter 5). The *pectoral girdle* consists of the scapulae and clavicles, connected to the manubrium of the sternum. Both girdles possess a large flat bone located posteriorly, which provides for attachment of proximal muscles, and connects with its contralateral partner anteriorly via small bony braces, the pubic rami and clavicles. However, the flat iliac bones of the pelvic girdle are also connected posteriorly through their primary attachment to the sacrum via the essentially rigid, weight-transferring sacro-iliac joints. This posterior connection to the axial skeleton places the lower limbs inferior to the trunk, enabling them to be supportive as they function primarily in relation to the line of gravity. Furthermore, because the two sides are connected both anteriorly and posteriorly, the pelvic girdle forms a complete rigid ring that limits mobility, making the movements of one limb markedly affect the movements of the other. The pectoral girdle, however, is connected to the trunk only anteriorly, via the sternum, by flexible joints with 3 degrees of freedom. It is an incomplete ring because the scapulae are not connected with each other posteriorly. Thus, the motion of one upper limb is independent of the other, and the limbs are able to operate effectively anterior to the body, at a distance and level that enable precise eye–hand coordination.

In both the upper and the lower limbs, the long bone of the most proximal segment is the largest and is unpaired. The long bones increase progressively in number but decrease in size in the more distal segments of the limb. The second most proximal segment of both limbs (i.e., the leg and forearm) has two parallel bones, although only in the forearm do both articulate with the bone of the proximal segment, and only in the leg do both articulate directly with the distal segment.

Although the paired bones of both the leg and forearm flex and extend as a unit, only those of the upper limb are able to move (supinate and pronate) relative to each other; the bones of the leg are fixed in the pronated position.

The wrist and ankle have a similar number of short bones (eight and seven, respectively). Both groups of short bones interrupt a series of long bones that resumes distally with several sets of long bones of similar lengths, with a similar number of joints of essentially the same type. The digits of the upper limb (fingers including the thumb) are the most mobile parts of either limb. However, all other parts of the upper limb are more mobile than the comparable parts of the lower limb.

BONES OF UPPER LIMB

The pectoral girdle and bones of the free part of the upper limb form the **superior appendicular skeleton** (Fig. 6.3); the pelvic girdle and bones of the free part of the lower limb form the **inferior appendicular skeleton.** The superior appendicular skeleton articulates with the axial skeleton only at the *sternoclavicular joint,* allowing great mobility. The clavicles and scapulae of the pectoral girdle are supported, stabilized, and moved by **axio-appendicular muscles** that attach to the relatively fixed ribs, sternum, and vertebrae of the *axial skeleton.*

Clavicle

The **clavicle** (collar bone) connects the upper limb to the trunk (Figs. 6.3 and 6.4). The **shaft of the clavicle** has a double curve in a horizontal plane. Its medial half is convex anteriorly, and its **sternal end** is enlarged and triangular where it articulates with the *manubrium of the sternum* at the *sternoclavicular* (SC) *joint.* Its lateral half is concave anteriorly, and its **acromial end** is flat where it articulates with the **acromion** of the scapula at the *acromioclavicular* (AC) *joint* (Figs. 6.3B and 6.4). The medial two thirds of the shaft of the clavicle are convex anteriorly, whereas the lateral third is flattened and concave anteriorly. These curvatures increase the resilience of the clavicle, and give it the appearance of an elongated capital S.

The clavicle:

- Serves as a moveable, crane-like strut (rigid support) from which the scapula and free limb are suspended, keeping them away from the trunk so that the limb has maximum freedom of motion. The strut is movable and allows the scapula to move on the thoracic wall at the "*scapulothoracic joint,*"[1] increasing the range of motion of the limb.

[1]The **scapulothoracic joint** is a physiological "joint," in which movement occurs between musculoskeletal structures (between the scapula and associated muscles and the thoracic wall), rather than an anatomical joint, in which movement occurs between directly articulating skeletal elements. The scapulothoracic joint is where the scapular movements of elevation–depression, protraction–retraction, and rotation occur.

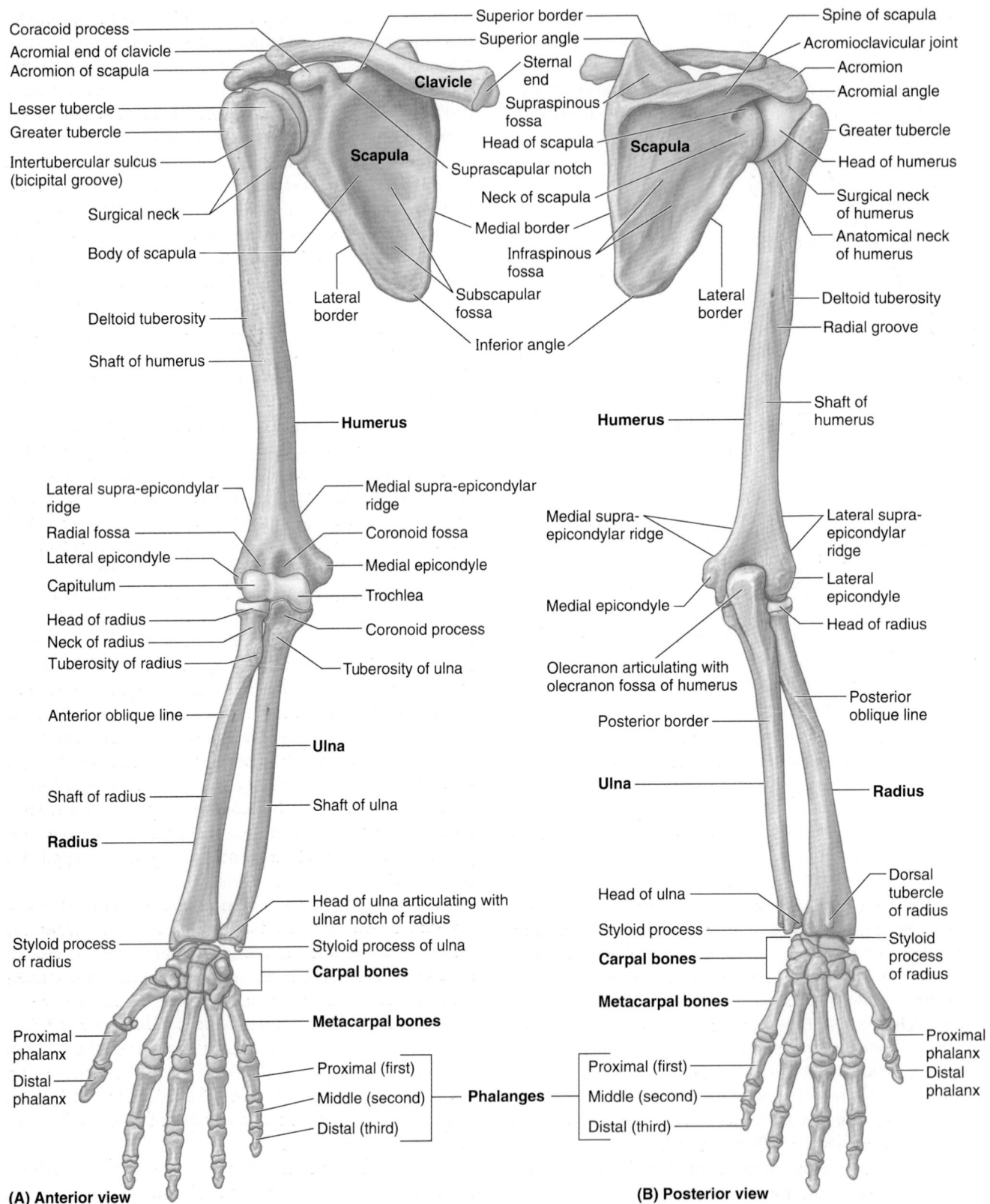

Coracoid process
Acromial end of clavicle
Acromion of scapula
Lesser tubercle
Greater tubercle
Intertubercular sulcus
(bicipital groove)
Surgical neck
Body of scapula
Deltoid tuberosity
Shaft of humerus

Humerus

Lateral supra-epicondylar
ridge
Radial fossa
Lateral epicondyle
Capitulum
Head of radius
Neck of radius
Tuberosity of radius
Anterior oblique line

Ulna

Shaft of radius

Radius

Styloid process
of radius

Proximal
phalanx
Distal
phalanx

(A) Anterior view

Superior border
Superior angle
Sternal
end
Supraspinous
fossa
Head of scapula

Clavicle

Scapula

Suprascapular notch
Neck of scapula
Medial border
Infraspinous
fossa
Subscapular
fossa
Inferior angle

Lateral
border

Medial supra-epicondylar
ridge
Coronoid fossa
Medial epicondyle
Trochlea
Coronoid process
Tuberosity of ulna

Shaft of ulna

Head of ulna articulating with
ulnar notch of radius
Styloid process of ulna

Carpal bones

Metacarpal bones

Proximal (first)
Middle (second)   **Phalanges**
Distal (third)

Spine of scapula
Acromioclavicular joint
Acromion
Acromial angle
Greater tubercle
Head of humerus
Surgical neck
of humerus
Anatomical neck
of humerus
Deltoid tuberosity
Radial groove

Scapula

Lateral
border

Shaft of
humerus

Humerus

Medial supra-
epicondylar ridge
Medial epicondyle
Olecranon articulating with
olecranon fossa of humerus
Posterior border

Ulna

Lateral supra-
epicondylar
ridge
Lateral
epicondyle
Head of radius
Posterior
oblique line

Radius

Dorsal
tubercle
of radius
Styloid
process
of radius

Head of ulna
Styloid process

Carpal bones

Metacarpal bones

Proximal (first)
Middle (second)
Distal (third)

Proximal
phalanx
Distal
phalanx

(B) Posterior view

FIGURE 6.3. Bones of upper limb.

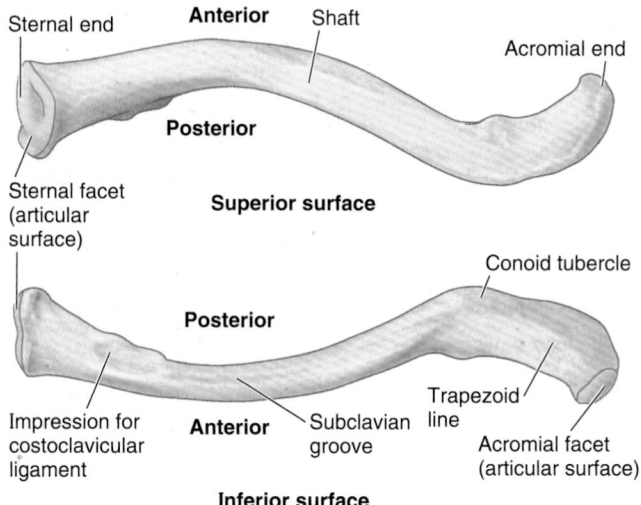

FIGURE 6.4. **Right clavicle.** Prominent features of the superior and inferior surfaces of the clavicle. The bone acts as a mobile strut (supporting brace) connecting the upper limb to the trunk; its length allows the limb to pivot around the trunk.

Fixing the strut in position, especially after its elevation, enables elevation of the ribs for deep inspiration.

- Forms one of the bony boundaries of the *cervico-axillary canal* (passageway between the neck and arm), affording protection to the neurovascular bundle supplying the upper limb.
- Transmits shocks (traumatic impacts) from the upper limb to the axial skeleton.

Although designated as a long bone, the clavicle has no medullary (marrow) cavity. It consists of spongy (trabecular) bone with a shell of compact bone.

The **superior surface of the clavicle,** lying just deep to the skin and platysma (G. flat plate) muscle in the subcutaneous tissue, is smooth.

The **inferior surface of the clavicle** is rough because strong ligaments bind it to the 1st rib near its sternal end and suspend the scapula from its acromial end. The **conoid tubercle,** near the acromial end of the clavicle (Fig. 6.4), gives attachment to the *conoid ligament,* the medial part of the *coracoclavicular ligament* by which the remainder of the upper limb is passively suspended from the clavicle. Also, near the acromial end of the clavicle is the **trapezoid line,** to which the *trapezoid ligament* attaches; it is the lateral part of the coracoclavicular ligament.

The **subclavian groove** (groove for the subclavius) in the medial third of the shaft of the clavicle is the site of attachment of the subclavius muscle. More medially is the **impression for the costoclavicular ligament,** a rough, often depressed, oval area that gives attachment to the ligament binding the 1st rib (L. *costa*) to the clavicle, limiting elevation of the shoulder.

Scapula

The **scapula** (shoulder blade) is a triangular flat bone that lies on the posterolateral aspect of the thorax, overlying

the 2nd–7th ribs (see Fig. I.11, p. 20). The convex **posterior surface** of the scapula is unevenly divided by a thick projecting ridge of bone, the **spine of the scapula,** into a small **supraspinous fossa** and a much larger **infraspinous fossa** (Fig. 6.5A). The concave **costal surface** of most of the scapula forms a large **subscapular fossa.** The broad bony surfaces of the three fossae provide attachments for fleshy muscles. The triangular **body of the scapula** is thin and translucent superior and inferior to the spine of the scapula; although its borders, especially the lateral one, are somewhat thicker. The spine continues laterally as the flat, expanded **acromion** (G. *akros,* point), which forms the subcutaneous point of the shoulder and articulates with the acromial end of the clavicle. The **deltoid tubercle** of the scapular spine is the prominence indicating the medial point of attachment of the deltoid. The spine and acromion serve as levers for the attached muscles, particularly the trapezius.

Because the acromion is a lateral extension of the scapula, the AC joint is placed lateral to the mass of the scapula and its attached muscles (Fig. 6.5C). The *glenohumeral* (shoulder) *joint* on which these muscles operate is almost directly inferior to the AC joint; thus the scapular mass is balanced with that of the free limb, and the suspending structure (coracoclavicular ligament) lies between the two masses.

Superolaterally, the lateral surface of the scapula has a **glenoid cavity** (G. socket), which receives and articulates with the head of the humerus at the glenohumeral joint (Fig. 6.5A & C). The glenoid cavity is a shallow, concave, oval fossa (L. *fossa ovalis*), directed anterolaterally and slightly superiorly—that is considerably smaller than the ball (head of the humerus) for which it serves as a socket. The beaklike **coracoid process** (G. *korak-odés,* like a crow's beak) is superior to the glenoid cavity, and projects anterolaterally. This process also resembles in size, shape, and direction a bent finger pointing to the shoulder, the knuckle of which provides the inferior attachment for the passively supporting coracoclavicular ligament.

The scapula has medial, lateral, and superior borders and superior, lateral, and inferior angles (Fig. 6.5B). When the scapular body is in the anatomical position, the thin **medial border of the scapula** runs parallel to and approximately 5 cm lateral to the spinous processes of the thoracic vertebrae; hence it is often called the *vertebral border* (Fig. 6.5B). From the inferior angle, the **lateral border of the scapula** runs superolaterally toward the apex of the axilla; hence it is often called the *axillary border.* The lateral border is made up of a thick bar of bone that prevents buckling of this stress-bearing region of the scapula.

The lateral border terminates in the truncated **lateral angle of the scapula,** the thickest part of the bone that bears the broadened **head of the scapula** (Fig. 6.5A & B). The glenoid cavity is the primary feature of the head. The shallow constriction between the head and body defines the **neck** of the scapula. The **superior border of the scapula** is marked near the junction of its medial two thirds and lateral third by the **suprascapular notch,** which is located where the superior border joins the base of the coracoid process.

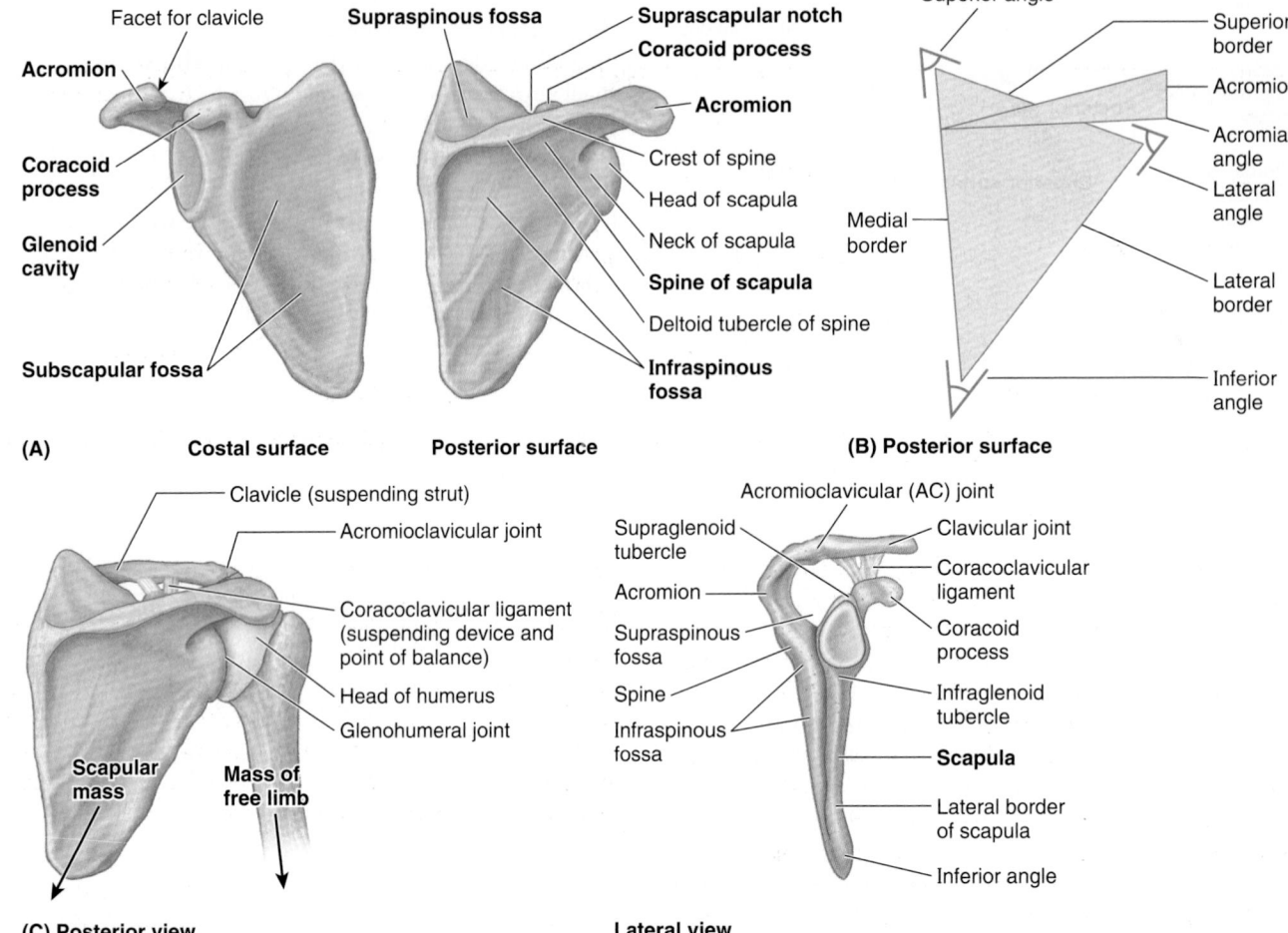

FIGURE 6.5. Right scapula. A. The bony features of the costal and posterior surfaces of the scapula. **B.** The borders and angles of the scapula. **C.** The scapula is suspended from the clavicle by the coracoclavicular ligament, at which a balance is achieved among the weight of the scapula and its attached muscles plus the muscular activity medially and the weight of the free limb laterally.

The superior border is the thinnest and shortest of the three borders.

The scapula is capable of considerable movement on the thoracic wall at the physiological *scapulothoracic joint*, providing the base from which the upper limb operates. These movements, enabling the arm to move freely, are discussed later in this chapter with the muscles that move the scapula.

Humerus

The **humerus** (arm bone), the largest bone in the upper limb, articulates with the scapula at the glenohumeral joint, and the radius and ulna at the elbow joint (Figs. 6.1, 6.3, and 6.5C). The proximal end of the humerus has a head, surgical and anatomical necks, and greater and lesser tubercles. The spherical **head of the humerus** articulates with the glenoid cavity of the scapula. The **anatomical neck of the humerus** is formed by the groove circumscribing the head and separating it from the greater and lesser tubercles. It indicates the line of attachment of the glenohumeral joint capsule. The **surgical neck of the humerus,** a common site of fracture, is the narrow part distal to the head and tubercles (Fig. 6.3B).

The junction of the head and neck with the shaft of the humerus is indicated by the greater and lesser tubercles, which provide attachment and leverage to some scapulohumeral muscles (Fig. 6.3A & B). The **greater tubercle** is at the lateral margin of the humerus, whereas the **lesser tubercle** projects anteriorly from the bone. The **intertubercular sulcus (bicipital groove)** separates the tubercles, and provides protected passage for the slender tendon of the long head of the biceps muscle.

The **shaft of the humerus** has two prominent features: the **deltoid tuberosity** laterally, for attachment of the deltoid muscle, and the oblique **radial groove (groove for radial nerve,** spiral groove) posteriorly, in which the radial nerve and profunda brachii artery lie as they pass anterior to the long head and between the medial and the lateral heads of the triceps brachii muscle. The inferior end of the humeral shaft widens as the sharp medial and lateral **supra-epicondylar** (supracondylar) **ridges** form, and then end distally in the especially prominent **medial epicondyle** and the **lateral epicondyle,** providing for muscle attachment.

The distal end of the humerus—including the trochlea, capitulum, olecranon, coronoid, and radial fossae—makes

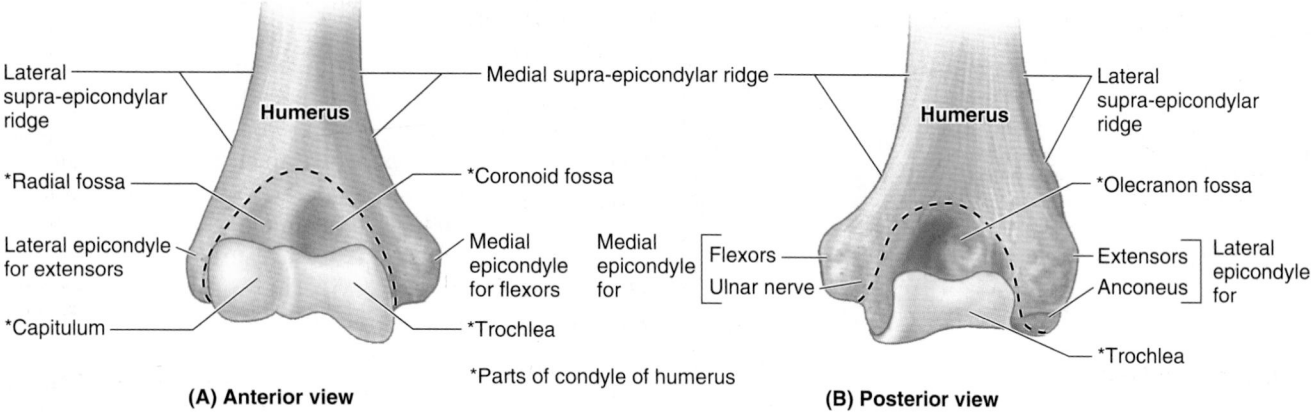

FIGURE 6.6. Distal end of right humerus. A and B. The condyle (the boundaries of which are indicated by the *dashed line*) consists of the capitulum; the trochlea; and the radial, coronoid, and olecranon fossae.

up the **condyle of the humerus** (Fig. 6.6). The condyle has two articular surfaces: a lateral **capitulum** (L. little head) for articulation with the head of the radius, and a medial, spool-shaped or pulley-like **trochlea** (L. pulley) for articulation with the proximal end (trochlear notch) of the ulna. Two hollows, or fossae, occur back to back superior to the trochlea, making the condyle quite thin between the epicondyles. Anteriorly, the **coronoid fossa** receives the coronoid process of the ulna during full flexion of the elbow. Posteriorly, the **olecranon fossa** accommodates the olecranon of the ulna during full extension of the elbow. Superior to the capitulum anteriorly, a shallower **radial fossa** accommodates the edge of the head of the radius when the forearm is fully flexed.

Bones of Forearm

The two forearm bones serve together to form the second unit of an articulated mobile strut (the first unit being the humerus), with a mobile base formed by the shoulder, that positions the hand. However, because this unit is formed by two parallel bones, one of which (the radius) can pivot about the other (the ulna), supination and pronation are possible. This makes it possible to rotate the hand when the elbow is flexed.

ULNA

The **ulna** is the stabilizing bone of the forearm and is the medial and longer of the two forearm bones (Figs. 6.7 and 6.8). Its more massive proximal end is specialized for articulation

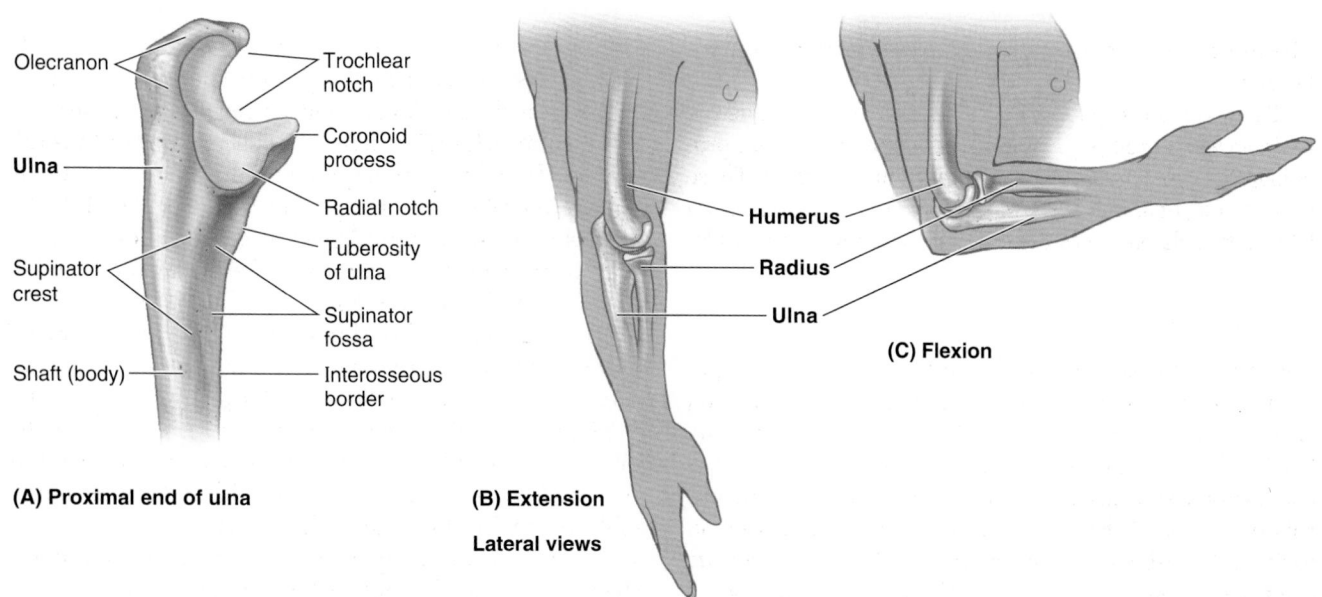

FIGURE 6.7. Bones of right elbow region. A. The proximal part of the ulna. **B.** The bones of the elbow region, demonstrating the relationship of the distal humerus and proximal ulna and radius during extension of the elbow joint. **C.** The relationship of the humerus and forearm bones during flexion of the elbow joint.

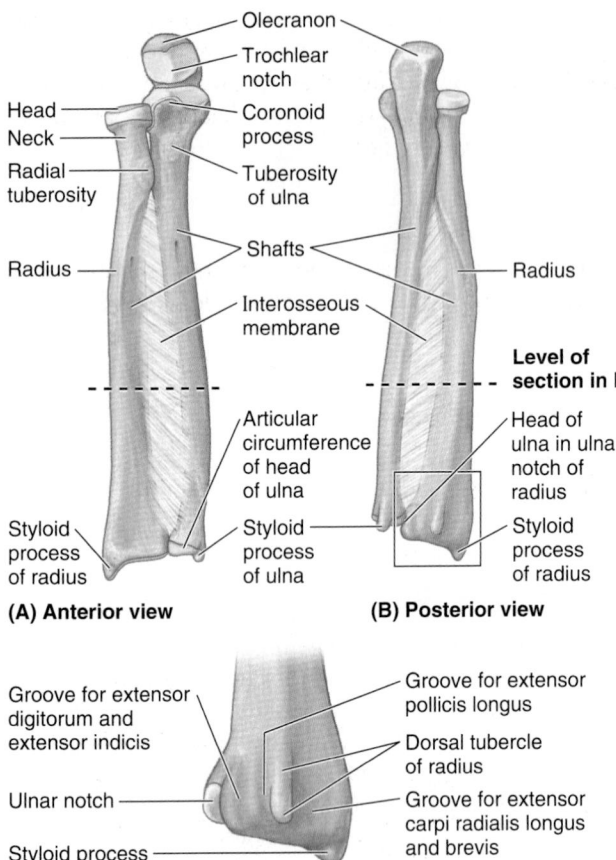

(A) Anterior view

(B) Posterior view

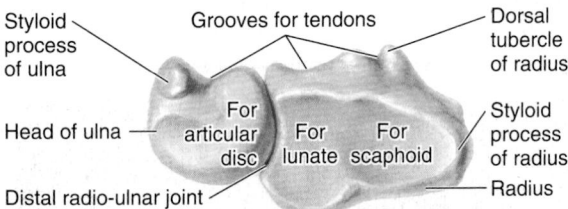

(C) Posterior view of distal end of radius

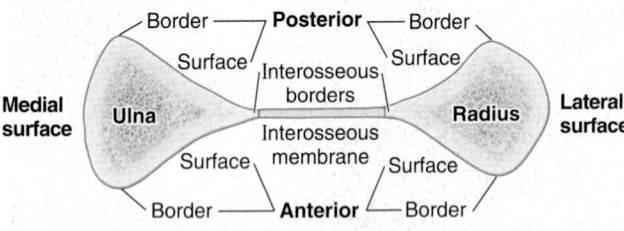

(D) Inferior view of distal ends of ulna and radius

(E) Inferior view of cross section through shafts of the ulna and radius and interosseous membrane

FIGURE 6.8. Right radius and ulna. A and B. The radius and ulna are shown in the articulated position, connected by the interosseous membrane. **C and D.** The features of the distal ends of the forearm bones. **E.** In cross section, the shafts of the radius and ulna appear almost as mirror images of one another for much of the middle and distal thirds of their lengths.

with the humerus proximally, and the head of the radius laterally. For articulation with the humerus, the ulna has two prominent projections: (1) the **olecranon,** which projects proximally from its posterior aspect (forming the point of the elbow), and serves as a short lever for extension of the elbow, and (2) the **coronoid process,** which projects anteriorly.

The olecranon and coronoid processes form the walls of the **trochlear notch,** which in profile resembles the jaws of a crescent wrench as it "grips" (articulates with) the trochlea of the humerus (Fig. 6.7B & C). The articulation between the ulna and humerus primarily allows only flexion and extension of the elbow joint, although a small amount of abduction–adduction occurs during pronation and supination of the forearm. Inferior to the coronoid process is the **tuberosity of the ulna** for attachment of the tendon of the brachialis muscle (Fig. 6.7A and 6.8A & B).

On the lateral side of the coronoid process is a smooth, rounded concavity, the **radial notch,** which receives the broad periphery of the head of the radius. Inferior to the radial notch on the lateral surface of the ulnar shaft is a prominent ridge, the **supinator crest.** Between it and the distal part of the coronoid process is a concavity, the **supinator fossa.** The deep part of the supinator muscle attaches to the supinator crest and fossa (6.7A).

The **shaft of the ulna** is thick and cylindrical proximally, but it tapers, diminishing in diameter, as it continues distally (Fig. 6.8A). At the narrow distal end of the ulna is a small but abrupt enlargement, the disc-like **head of the ulna** with a small, conical **ulnar styloid process.** The ulna does not reach—and therefore does not participate in—the wrist (radiocarpal) joint (Fig. 6.8).

RADIUS

The **radius** is the lateral and shorter of the two forearm bones. Its proximal end includes a short head, neck, and medially directed tuberosity (Fig. 6.8A). Proximally, the smooth superior aspect of the discoid **head of the radius** is concave for articulation with the capitulum of the humerus during flexion and extension of the elbow joint. The head also articulates peripherally with the radial notch of the ulna; thus the head is covered with articular cartilage.

The **neck of the radius** is a constriction distal to the head. The oval **radial tuberosity** is distal to the medial part of the neck, and demarcates the proximal end (head and neck) of the radius from the shaft.

The **shaft of the radius,** in contrast to that of the ulna, gradually enlarges as it passes distally. The distal end of the radius is essentially four sided when sectioned transversely. Its medial aspect forms a concavity, the **ulnar notch** (Fig. 6.8C & D), which accommodates the head of the ulna. Its lateral aspect becomes increasingly ridge-like, terminating distally in the **radial styloid process.**

Projecting dorsally, the **dorsal tubercle of the radius** lies between otherwise shallow grooves for the passage of the tendons of forearm muscles. The radial styloid process is larger than the ulnar styloid process, and extends farther distally (Fig. 6.8A & B). This relationship is of clinical importance when the ulna and/or the radius is fractured.

Most of the length of the shafts of the radius and ulna are essentially triangular in cross section, with a rounded, superficially directed base and an acute, deeply directed apex (Fig. 6.8A & E). The apex is formed by a section of the sharp **interosseous border of the radius or ulna** that connects to the thin, fibrous **interosseous membrane of the forearm** (Fig. 6.8A, B, & E). The majority of the fibers

of the interosseous membrane run an oblique course, passing inferiorly from the radius as they extend medially to the ulna (Fig. 6.8A & B). Thus, they are positioned to transmit forces received by the radius (via the hands) to the ulna for transmission to the humerus.

Bones of Hand

The **wrist**, or **carpus**, is composed of eight **carpal bones**, arranged in proximal and distal rows of four (Fig. 6.9A–C).

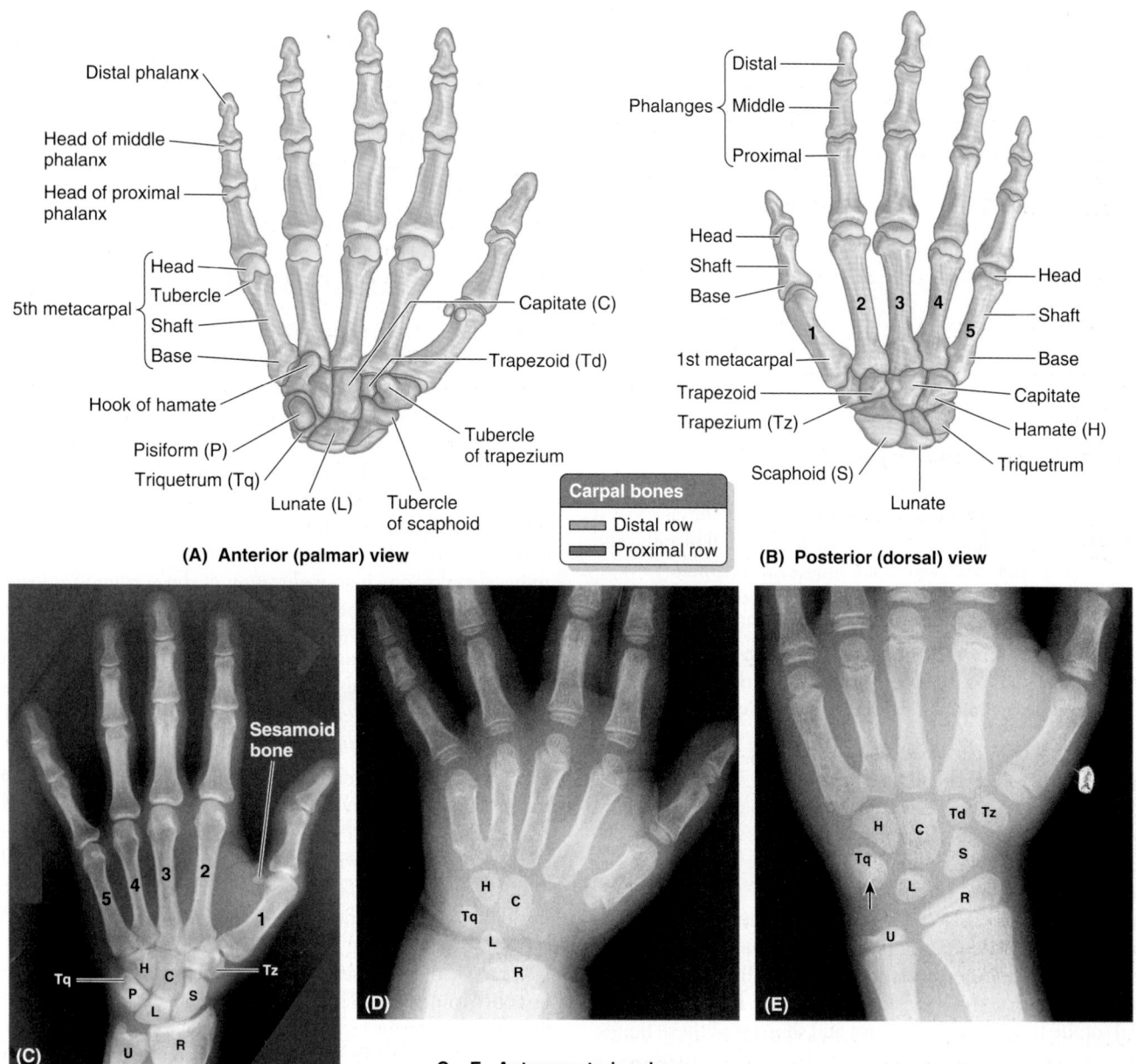

FIGURE 6.9. Bones of right hand. A–C. The skeleton of the hand consists of three segments: the carpals of the wrist (subdivided into proximal and distal rows), the metacarpals of the palm, and the phalanges of the fingers or digits. *U,* ulna; *R,* radius. **D.** The distal end of the forearm and hand of a 2.5-year-old child. Ossification centers of only four carpal bones are visible. Observe the distal radial epiphysis (*R*). **E.** The distal end of the forearm and hand of an 11-year-old child. Ossification centers of all carpal bones are visible. The *arrow* indicates the pisiform lying on the anterior surface of the triquetrum. The distal epiphysis of the ulna has ossified, but all the epiphyseal plates (lines) remain open (i.e., they are still unossified). (Parts C and D courtesy of Dr. D. Armstrong, Associate Professor of Medical Imaging, University of Toronto, Toronto, Ontario, Canada.).

These small bones give flexibility to the wrist. The carpus is markedly convex from side to side posteriorly, and concave anteriorly. Augmenting movement at the wrist joint, the two rows of carpal bones glide on each other; in addition, each bone glides on those adjacent to it.

From lateral to medial, the four carpal bones in the proximal row (purple in Fig. 6.9A & B) are the:

- **Scaphoid** (G. *skaphé,* skiff, boat): a boat-shaped bone that articulates proximally with the radius, and has a prominent **scaphoid tubercle;** it is the largest bone in the proximal row of carpals.
- **Lunate** (L. *luna,* moon): a moon-shaped bone between the scaphoid and the triquetral bones; it articulates proximally with the radius and is broader anteriorly than posteriorly.
- **Triquetrum** (L. *triquetrus,* three-cornered): a pyramidal bone on the medial side of the carpus; it articulates proximally with the articular disc of the distal radio-ulnar joint.
- **Pisiform** (L. *pisum,* pea), a small, pea-shaped bone that lies on the palmar surface of the triquetrum.

From lateral to medial, the four carpal bones in the distal row (green in Fig. 6.9A & B) are the:

- **Trapezium** (G. *trapeze,* table): a four-sided bone on the lateral side of the carpus; it articulates with the 1st and 2nd metacarpals, scaphoid, and trapezoid bones.
- **Trapezoid:** a wedge-shaped bone that resembles the trapezium; it articulates with the 2nd metacarpal, trapezium, capitate, and scaphoid bones.
- **Capitate** (L. *caput,* head): a head-shaped bone with a rounded extremity is the largest bone in the carpus; it articulates primarily with the 3rd metacarpal distally, and with the trapezoid, scaphoid, lunate, and hamate.
- **Hamate** (L. *hamulus,* a little hook): a wedge-shaped bone on the medial side of the hand; it articulates with the 4th and 5th metacarpal, capitate, and triquetral bones; it has a distinctive hooked process, the **hook of the hamate,** that extends anteriorly.

The proximal surfaces of the distal row of carpal bones articulate with the proximal row of carpal bones, and their distal surfaces articulate with the metacarpals.

The **metacarpus** forms the skeleton of the palm of the hand between the carpus and the phalanges. It is composed of five **metacarpal bones** (metacarpals). Each metacarpal consists of a base, shaft, and head. The proximal **bases of the metacarpals** articulate with the carpal bones, and the distal **heads of the metacarpals** articulate with the proximal phalanges, and form the knuckles of the hand. The 1st metacarpal (of the thumb) is the thickest and shortest of these bones. The 3rd metacarpal is distinguished by a **styloid process** on the lateral side of its base (Fig. 6.10).

Each digit (finger) has three **phalanges** except for the first (the thumb), which has only two; however, the phalanges of the first digit are stouter than those in the other fingers. Each phalanx has a **base** proximally, a **shaft** (body), and a **head** distally (Fig. 6.9). The proximal phalanges are the largest, the middle ones are intermediate in size, and the distal ones are the smallest. The shafts of the phalanges taper distally. The terminal phalanges are flattened and expanded at their distal ends, which underlie the nail beds.

OSSIFICATION OF BONES OF HAND

Radiographs of the wrist and hand are commonly used to assess skeletal age. For clinical studies, the radiographs are compared with a series of standards in a radiographic atlas of skeletal development to determine the skeletal age. Ossification centers are usually obvious during the 1st year; however, they may appear before birth. Each carpal bone usually ossifies from one center postnatally (Fig. 6.9D). The centers for the capitate and hamate appear first.

The shaft of each metacarpal begins to ossify during fetal life, and ossification centers appear postnatally in the heads of the four medial metacarpals and in the base of the 1st metacarpal. By age 11, ossification centers of all carpal bones are visible (Fig. 6.9E).

Surface Anatomy of Upper Limb Bones

Most bones of the upper limb offer a palpable segment or surface (notable exceptions being the lunate and trapezoid), enabling the skilled examiner to discern abnormalities owing to trauma (fracture or dislocation) or malformation (Fig. 6.10).

The *clavicle* is subcutaneous and can be easily palpated throughout its length. Its sternal end projects superior to the manubrium (Fig. 6.10). Between the elevated sternal ends of the clavicles is the **jugular notch** (suprasternal notch). The acromial end of the clavicle often rises higher than the acromion, forming a palpable elevation at the *acromioclavicular (AC) joint.* The acromial end can be palpated 2–3 cm medial to the lateral border of the acromion, particularly when the arm is alternately flexed and extended. Either or both ends of the clavicle may be prominent; when present, this condition is usually bilateral.

Note the elasticity of the skin over the clavicle and how easily it can be pinched into a mobile fold. This property of the skin is useful when ligating (tying a knot around) the third part of the subclavian artery: The skin lying superior to the clavicle is pulled down onto the clavicle and then incised; after the incision is made, the skin is allowed to return to its position superior to the clavicle, where it overlies the artery (thus not endangering it during the incision).

As the clavicle passes laterally, its medial part can be felt to be convex anteriorly. The large vessels and nerves to the upper limb pass posterior to this convexity. The flattened acromial end of the clavicle does not reach the point of the shoulder, formed by the lateral tip of the acromion of the scapula.

The *acromion* of the scapula is easily felt and often visible, especially when the deltoid contracts against resistance. The superior surface of the acromion is subcutaneous and may be traced medially to the AC joint. The lateral and posterior borders of the acromion meet to form the **acromial**

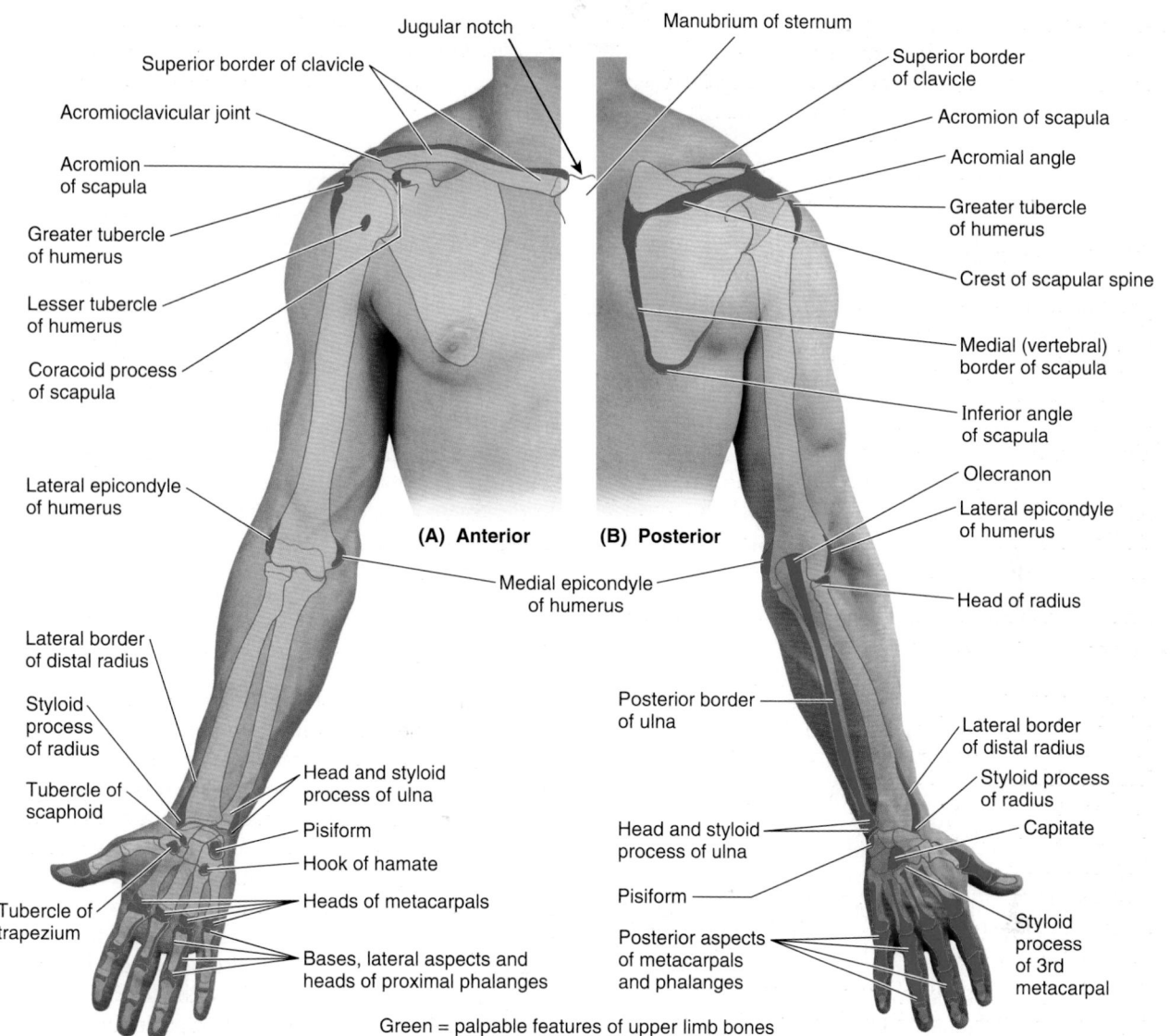

Jugular notch

Manubrium of sternum

Superior border of clavicle

Acromioclavicular joint

Acromion of scapula

Greater tubercle of humerus

Lesser tubercle of humerus

Coracoid process of scapula

Lateral epicondyle of humerus

(A) Anterior

(B) Posterior

Superior border of clavicle

Acromion of scapula

Acromial angle

Greater tubercle of humerus

Crest of scapular spine

Medial (vertebral) border of scapula

Inferior angle of scapula

Olecranon

Lateral epicondyle of humerus

Head of radius

Medial epicondyle of humerus

Lateral border of distal radius

Styloid process of radius

Tubercle of scaphoid

Head and styloid process of ulna

Pisiform

Hook of hamate

Heads of metacarpals

Posterior border of ulna

Lateral border of distal radius

Styloid process of radius

Capitate

Tubercle of trapezium

Head and styloid process of ulna

Pisiform

Styloid process of 3rd metacarpal

Bases, lateral aspects and heads of proximal phalanges

Posterior aspects of metacarpals and phalanges

Green = palpable features of upper limb bones

FIGURE 6.10. Surface anatomy of bones of upper limb.

angle (Fig. 6.10B). The humerus in the glenoid cavity and the *deltoid muscle* form the rounded curve of the shoulder. The **crest of the scapular spine** is subcutaneous throughout and easily palpated.

When the upper limb is in the anatomical position, the:

- *Superior angle of the scapula* lies at the level of the T2 vertebra.
- Medial end of the root of the scapular spine is opposite the spinous process of the T3 vertebra.
- *Inferior angle of the scapula* lies at the level of the T7 vertebra, near the inferior border of the 7th rib and 7th intercostal space.

The *medial border of the scapula* is palpable inferior to the root of the spine of the scapula as it crosses the 3rd–7th ribs. The lateral border of the scapula is not easily palpated because it is covered by the teres major and minor muscles. When the upper limb is abducted and the hand is placed on

the back of the head, the scapula is rotated, elevating the glenoid cavity such that the medial border of the scapula parallels the 6th rib and thus can be used to estimate its position and, deep to the rib, the oblique fissure of the lung. The inferior angle of the scapula is easily felt and is often visible. It is grasped when testing movements of the glenohumeral joint to immobilize the scapula. The *coracoid process of the scapula* can be felt by palpating deeply at the lateral side of the *clavipectoral (deltopectoral) triangle* (Fig. 6.11).

The *head of the humerus* is surrounded by muscles, except inferiorly; consequently, it can be palpated only by pushing the fingers well up into the *axillary fossa* (armpit). The arm should not be fully abducted, otherwise the fascia in the axilla will be tense and impede palpation of the humeral head. When the arm is moved and the scapula is fixed (held in place), the head of the humerus can be palpated.

The *greater tubercle of the humerus* may be felt with the person's arm by the side on deep palpation through the deltoid,

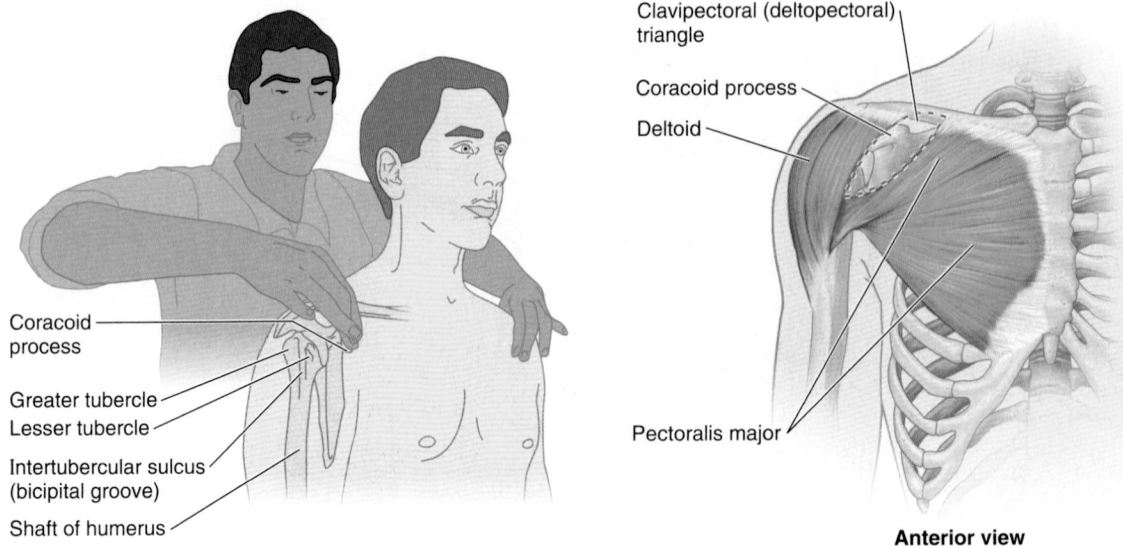

Clavipectoral (deltopectoral) triangle

Coracoid process

Deltoid

Pectoralis major

Coracoid process

Greater tubercle
Lesser tubercle

Intertubercular sulcus (bicipital groove)

Shaft of humerus

Anterior view

FIGURE 6.11. **Palpation of coracoid process of scapula.**

inferior to the lateral border of the acromion. In this position, the greater tubercle is the most lateral bony point of the shoulder and, along with the deltoid, gives the shoulder its rounded contour. When the arm is abducted, the greater tubercle is pulled beneath the acromion and is no longer palpable.

The *lesser tubercle of the humerus* may be felt with difficulty by deep palpation through the deltoid on the anterior aspect of the arm, approximately 1 cm lateral and slightly inferior to the tip of the coracoid process. Rotation of the arm facilitates palpation of this tubercle. The location of the *intertubercular sulcus or bicipital groove*, between the greater and the lesser tubercles, is identifiable during flexion and extension of the elbow joint by palpating in an upward direction along the tendon of the long head of the biceps brachii as it moves through the intertubercular groove.

The *shaft of the humerus* may be felt with varying distinctness through the muscles surrounding it. No part of the proximal part of the humeral shaft is subcutaneous.

The medial and lateral epicondyles of the humerus are subcutaneous and easily palpated on the medial and lateral aspects of the elbow region. The knob-like *medial epicondyle,* projecting posteromedially, is more prominent than the lateral epicondyle.

When the elbow joint is partially flexed, the *lateral epicondyle* is visible. When the elbow joint is fully extended, the lateral epicondyle can be palpated but not seen deep to a depression on the posterolateral aspect of the elbow.

The *olecranon* of the ulna can be easily palpated (Fig. 6.12). When the elbow joint is extended, observe that the tip of the olecranon and the humeral epicondyles lie in a straight line (Fig. 6.12A & B). When the elbow is flexed, the olecranon descends until its tip forms the apex of an approximately equilateral triangle, of which the epicondyles form the angles at its base (Fig. 6.12C). These normal relationships are important in the diagnosis of certain elbow injuries (e.g., dislocation of the elbow joint).

The *posterior border of the ulna,* palpable throughout the length of the forearm, demarcates the posteromedial boundary between the flexor–pronator and the extensor–supinator compartments of the forearm. The *head of the ulna* forms a large, rounded subcutaneous prominence that can be easily seen and palpated on the medial side of the dorsal aspect of the wrist, especially when the hand is pronated. The pointed subcutaneous *ulnar styloid process* may be felt slightly distal to the rounded ulnar head when the hand is supinated.

The *head of the radius* can be palpated and felt to rotate in the depression on the posterolateral aspect of the extended elbow joint, just distal to the lateral epicondyle of the humerus. The radial head can also be palpated as it rotates during pronation and supination of the forearm. The *ulnar nerve* feels like a thick cord where it passes posterior to the medial epicondyle of the humerus; pressing the nerve here evokes an unpleasant "funny bone" sensation.

The *radial styloid process* can be easily palpated in the *anatomical snuff box* on the lateral side of the wrist (see Fig. 6.65A); it is larger and approximately 1 cm more distal than the ulnar styloid process. The radial styloid process is easiest to palpate when the thumb is abducted. It is overlaid by the tendons of the thumb muscles. Because the radial styloid process extends more distally than the ulnar styloid process, more ulnar deviation than radial deviation of the wrist is possible.

The relationship of the radial and ulnar styloid processes is important in the diagnosis of certain wrist injuries (e.g., Colles fracture). Proximal to the radial styloid process, the anterior, lateral, and posterior surfaces of the radius are palpable for several centimeters. The *dorsal tubercle of radius* is easily felt around the middle of the dorsal aspect of the distal end of the radius. The dorsal tubercle acts as a pulley for the long extensor tendon of the thumb, which passes medial to it.

The *pisiform* can be felt on the anterior aspect of the medial border of the wrist and can be moved from side to side when the hand is relaxed. The *hook of the hamate* can

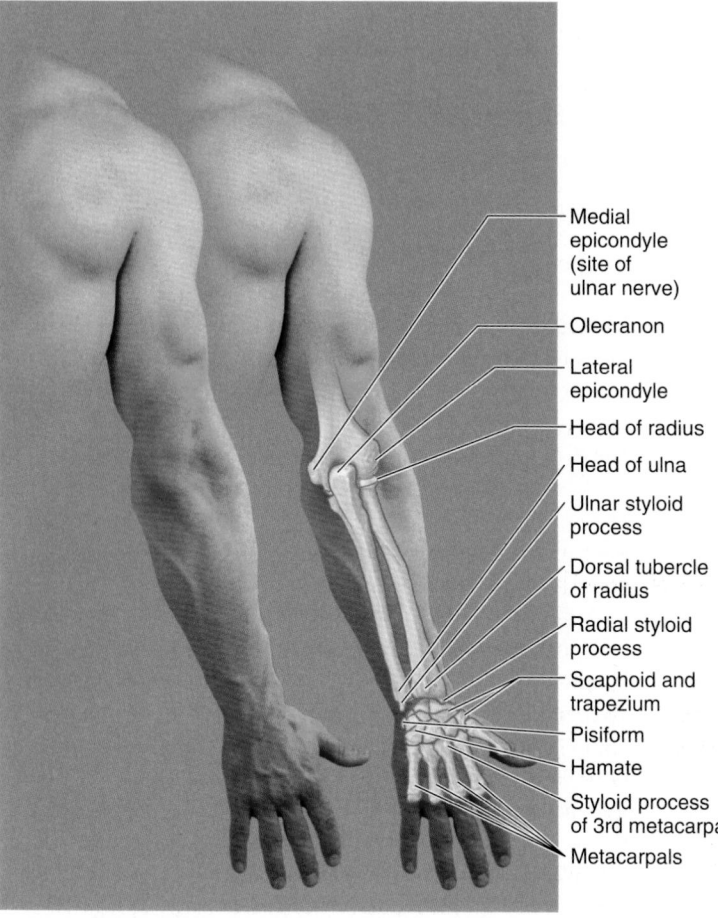

(A) Posterior view

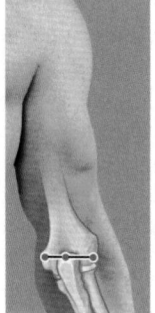

Posterior view

(B) Extension: epicondyles and olecranon aligned during extension

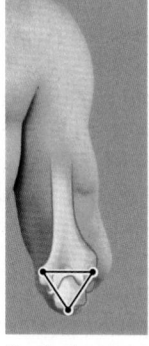

Posterior view   **Lateral view**

(C) Flexion: epicondyles form triangle and align vertically with olecranon during flexion

Labels (figure A): Medial epicondyle (site of ulnar nerve); Olecranon; Lateral epicondyle; Head of radius; Head of ulna; Ulnar styloid process; Dorsal tubercle of radius; Radial styloid process; Scaphoid and trapezium; Pisiform; Hamate; Styloid process of 3rd metacarpal; Metacarpals

FIGURE 6.12. Surface anatomy of bones and bony formations of elbow region.

be palpated on deep pressure over the medial side of the palm, approximately 2 cm distal and lateral to the pisiform. The *tubercles of the scaphoid and trapezium* can be palpated at the base and medial aspect of the *thenar eminence* (ball of thumb) when the hand is extended.

The *metacarpals*, although overlain by the long extensor tendons of the digits, can be palpated on the dorsum of the hand. The heads of these bones form the knuckles of the fist; the 3rd metacarpal head is most prominent. The *styloid process of the 3rd metacarpal* can be palpated approximately 3.5 cm from the

dorsal tubercle of radius. The dorsal aspects of the phalanges can also be easily palpated. The knuckles of the fingers are formed by the heads of the proximal and middle phalanges.

When measuring the upper limb, or segments of it, for comparison with the contralateral limb, or with standards for normal limb growth or size, the *acromial angle (Fig. 6.10B), lateral epicondyle of the humerus, styloid process of the radius,* and *tip of the third digit* are most commonly used as measuring points, with the limb relaxed (dangling), but with palms directed anteriorly.

BONES OF UPPER LIMB

Upper Limb Injuries

 Because the disabling effects of an injury to an upper limb, particularly the hand, are far out of proportion to the extent of the injury, a sound understanding of the structure and function of the upper limb is of the highest importance. Knowledge of its structure without an understanding of its functions is almost useless clinically

because the aim of treating an injured limb is to preserve or restore its functions.

Variations of Clavicle

The clavicle varies more in shape than most other long bones. Occasionally, the clavicle is pierced by a branch of the supraclavicular nerve. The clavicle is thicker and more curved in manual workers, and the sites of muscular attachments are more marked.

Fracture of Clavicle

The clavicle is one of the most frequently fractured bones. *Clavicular fractures* are especially common in children, and are often caused by an indirect force transmitted from an outstretched hand through the bones of the forearm and arm to the shoulder during a fall. A fracture may also result from a fall directly on the shoulder. The weakest part of the clavicle is the junction of its middle and lateral thirds.

After fracture of the clavicle, the sternocleidomastoid muscle elevates the medial fragment of bone (Fig. B6.1). Because of the subcutaneous position of the clavicle, the end of the superiorly directed fragment is prominent—readily palpable and/or apparent. The trapezius muscle is unable to hold the lateral fragment up owing to the weight of the upper limb; thus, the shoulder drops. The strong coracoclavicular ligament usually prevents dislocation of the acromioclavicular (AC) joint. People with fractured clavicles support the sagging limb with the other limb. In addition to being depressed, the lateral fragment of the clavicle may be pulled medially by the adductor muscles of the arm, such as the pectoralis major. Overriding of the bone fragments shortens the clavicle.

The slender clavicles of neonates may be fractured during delivery if they have broad shoulders; however, the bones usually heal quickly. A fracture of the clavicle is often incomplete in younger children—that is, it is a *greenstick fracture*, in which one side of a bone is broken and the other is bent. This fracture was so named because the parts of the bone do not separate; the bone resembles a tree branch (greenstick) that has been sharply bent but not disconnected.

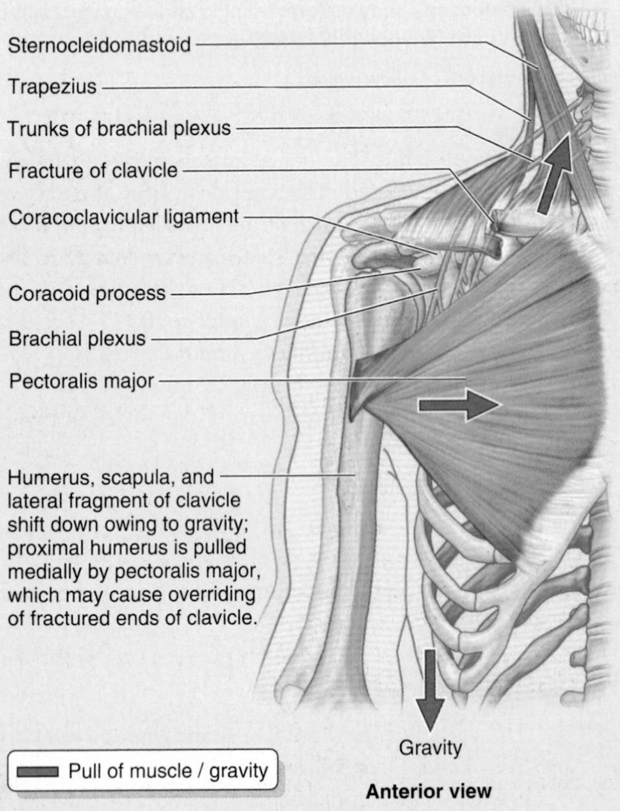

Ossification of Clavicle

The clavicle is the first long bone to ossify (via *intramembranous ossification*), beginning during the 5th and 6th embryonic weeks from medial and lateral primary ossification centers that are close together in the shaft of the clavicle. The ends of the clavicle later pass through a cartilaginous phase (*endochondral ossification*); the cartilages form growth zones similar to those of other long bones. A secondary ossification center appears at the sternal end, and forms a scale-like epiphysis that begins to fuse with the shaft (diaphysis) between 18 and 25 years of age, and is completely fused to it between 25 and 31 years of age. This is the last of the epiphyses of long bones to fuse. An even smaller scale-like epiphysis may be present at the acromial end of the clavicle; it must not be mistaken for a fracture.

Sometimes fusion of the two ossification centers of the clavicle fails to occur; as a result, a bony defect forms between the lateral and medial thirds of the clavicle. Awareness of this possible congenital defect should prevent diagnosis of a fracture in an otherwise normal clavicle. When doubt exists, both clavicles are radiographed because this defect is usually bilateral (Ger et al., 1996).

Fracture of Scapula

Fracture of the scapula is usually the result of severe trauma, as occurs in pedestrian–vehicle accidents. Usually there are also fractured ribs. Most fractures require little treatment because the scapula is covered on both sides by muscles. Most fractures involve the protruding subcutaneous acromion.

Fractures of Humerus

Most injuries of the proximal end of the humerus are *fractures of the surgical neck*. These injuries are especially common in elderly people with *osteoporosis*, whose demineralized bones are brittle. Humeral fractures often result in one fragment being driven into the spongy bone of the other fragment (*impacted fracture*). The injuries usually result from a minor fall on the hand, with the force being transmitted up the forearm bones of the extended limb. Because of impaction of the fragments, the fracture site is sometimes stable and the person is able to move the arm passively with little pain.

An *avulsion fracture of the greater tubercle of the humerus* is seen most commonly in middle-aged and elderly people (Fig. B6.2A). A small part of the tubercle is "avulsed" (torn away). The fracture usually results from a fall on the acromion, the point of the shoulder. In younger people, an avulsion fracture of the greater tubercle usually results from a fall on the hand when the arm is abducted. Muscles (especially the subscapularis) that remain attached to the humerus pull the limb into medial rotation.

Sternocleidomastoid
Trapezius
Trunks of brachial plexus
Fracture of clavicle
Coracoclavicular ligament
Coracoid process
Brachial plexus
Pectoralis major

Humerus, scapula, and lateral fragment of clavicle shift down owing to gravity; proximal humerus is pulled medially by pectoralis major, which may cause overriding of fractured ends of clavicle.

Gravity

█ Pull of muscle / gravity

Anterior view

FIGURE B6.1. Fracture of clavicle.

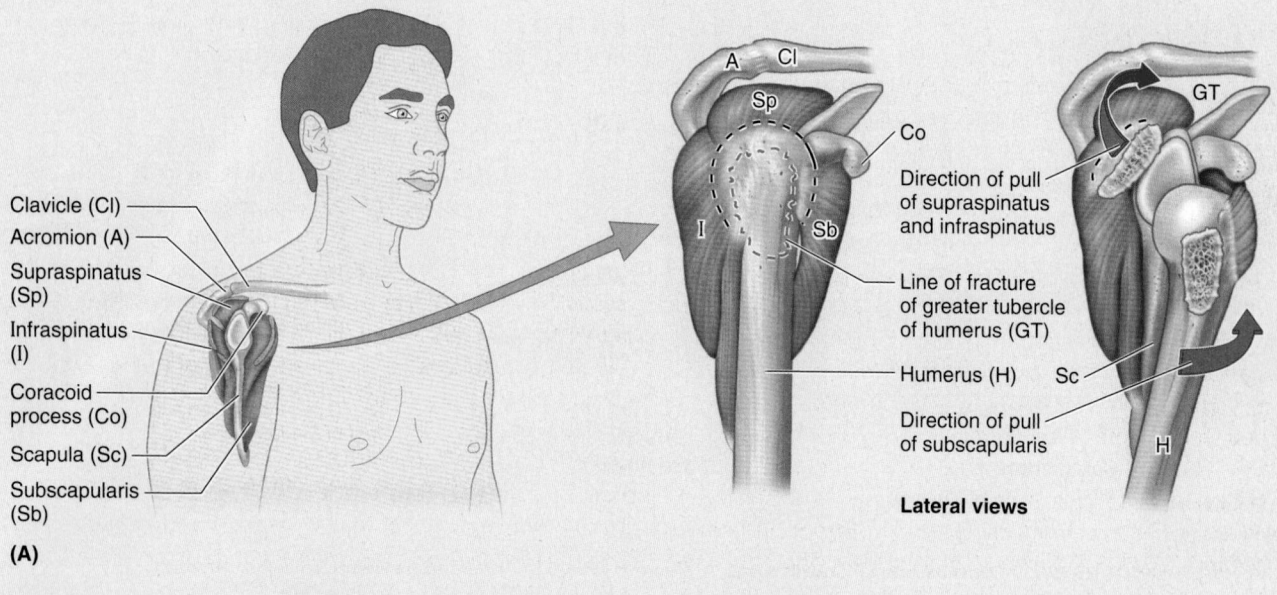

Clavicle (Cl)
Acromion (A)
Supraspinatus (Sp)
Infraspinatus (I)
Coracoid process (Co)
Scapula (Sc)
Subscapularis (Sb)

(A)

A   Cl
Sp
Co
I   Sb
Direction of pull of supraspinatus and infraspinatus
Line of fracture of greater tubercle of humerus (GT)
Humerus (H)   Sc
Direction of pull of subscapularis
GT
H

Lateral views

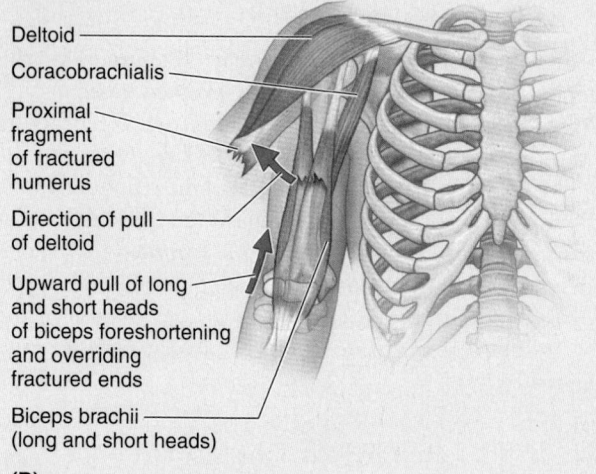

Deltoid
Coracobrachialis
Proximal fragment of fractured humerus
Direction of pull of deltoid
Upward pull of long and short heads of biceps foreshortening and overriding fractured ends
Biceps brachii (long and short heads)

(B)

FIGURE B6.2. Humeral fractures. A. An avulsion fracture of the greater tubercle of the humerus. **B.** A transverse fracture of humeral body.

A *transverse fracture of the shaft of the humerus* frequently results from a direct blow to the arm. The pull of the deltoid muscle carries the proximal fragment laterally (Fig. B6.2B). Indirect injury resulting from a fall on the outstretched hand may produce a *spiral fracture of the humeral shaft*. Overriding of the oblique ends of the fractured bone may result in foreshortening. Because the humerus is surrounded by muscles and has a well-developed periosteum, the bone fragments usually unite well.

An *intercondylar fracture of the humerus* results from a severe fall on the flexed elbow. The olecranon of the ulna is driven like a wedge between the medial and lateral parts of the condyle of the humerus, separating one or both parts from the humeral shaft.

The following parts of the humerus are in direct contact with the indicated nerves:

- Surgical neck: axillary nerve.
- Radial groove: radial nerve.
- Distal end of humerus: median nerve.
- Medial epicondyle: ulnar nerve.

These nerves may be injured when the associated part of the humerus is fractured. These injuries are discussed later in this chapter.

Fractures of Radius and Ulna

Fractures of both the radius and the ulna are usually the result of severe injury. A direct injury usually produces transverse fractures at the same level, usually in the middle third of the bones. Isolated fractures of the radius or ulna also occur. Because the shafts of these bones are firmly bound together by the interosseous membrane, a fracture of one bone is likely to be associated with dislocation of the nearest joint.

Fracture of the distal end of the radius is a common fracture in adults > 50 years of age, and occurs more frequently in women because their bones are more commonly weakened by *osteoporosis*. A complete transverse fracture of the distal 2 cm of the radius, called a *Colles fracture*, is the most common fracture of the forearm (Fig. B6.3). The distal fragment is displaced dorsally and is often *comminuted* (bro-

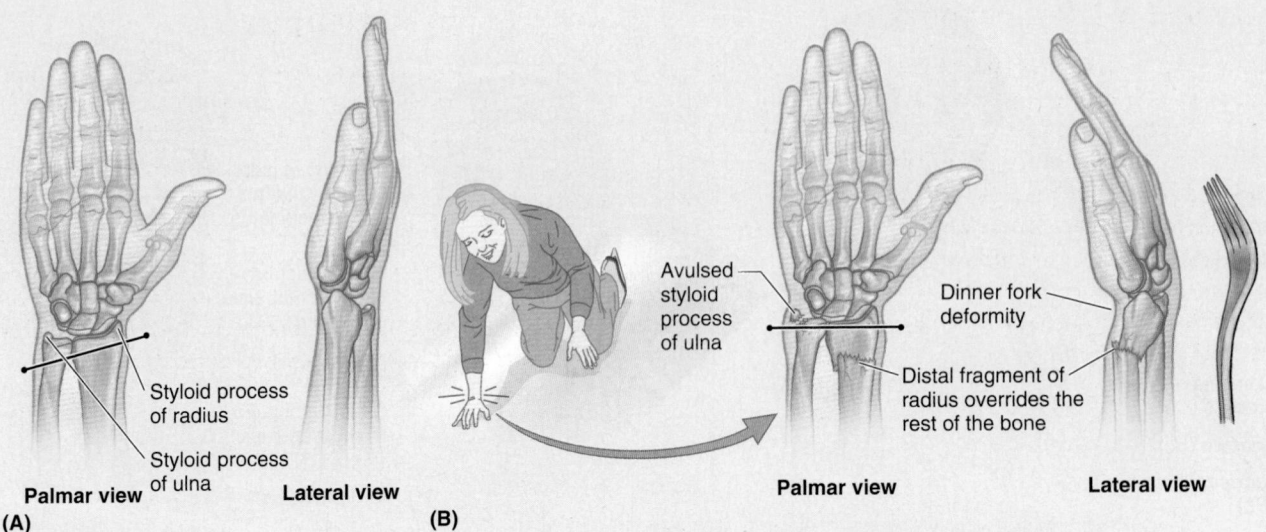

Styloid process of radius

Styloid process of ulna

Palmar view
(A)

Lateral view

(B)

Avulsed styloid process of ulna

Dinner fork deformity

Distal fragment of radius overrides the rest of the bone

Palmar view

Lateral view

FIGURE B6.3. Distal fracture of forearm bones. A. A normal wrist. **B.** A Colles fracture with a dinner fork deformity.

ken into pieces). The fracture results from forced extension of the hand, usually as the result of trying to ease a fall by outstretching the upper limb.

Often the ulnar styloid process is *avulsed* (broken off). Normally the radial styloid process projects farther distally than the ulnar styloid (Fig. B6.3A); consequently, when a Colles fracture occurs, this relationship is reversed because of shortening of the radius (Fig. B6.3B). This fracture is often referred to as a *dinner fork deformity* because a posterior angulation occurs in the forearm just proximal to the wrist and the normal anterior curvature of the relaxed hand. The posterior bending is produced by the posterior displacement and tilt of the distal fragment of the radius.

The typical history of a person with a Colles fracture includes slipping or tripping and, in an attempt to break the fall, landing on the outstretched limb with the forearm and hand pronated. Because of the rich blood supply to the distal end of the radius, bony union is usually good.

When the distal end of the radius fractures in children, the fracture line may extend through the distal epiphysial plate. *Epiphysial plate injuries* are common in older children because of their frequent falls in which the forces are transmitted from the hand to the radius and ulna. The healing process may result in malalignment of the epiphysial plate and disturbance of radial growth.

Fracture of Scaphoid

The scaphoid is the most frequently fractured carpal bone. It often results from a fall on the palm when the hand is abducted, the fracture occurring across the narrow part of the scaphoid (Fig. B6.4). Pain occurs primarily on the lateral side of the wrist, especially during dorsiflexion and abduction of the hand. Initial

radiographs of the wrist may not reveal a fracture; often this injury is (mis)diagnosed as a *severely sprained wrist*.

Radiographs taken 10–14 days later reveal a fracture because bone resorption has occurred there. Owing to the poor blood supply to the proximal part of the scaphoid, union of the fractured parts may take at least 3 months. *Avascular necrosis of the proximal fragment of the scaphoid* (pathological death of bone, resulting from inadequate blood supply) may occur, and produce *degenerative joint disease of the wrist*. In some cases, it is necessary to fuse the carpals surgically (*arthrodesis*).

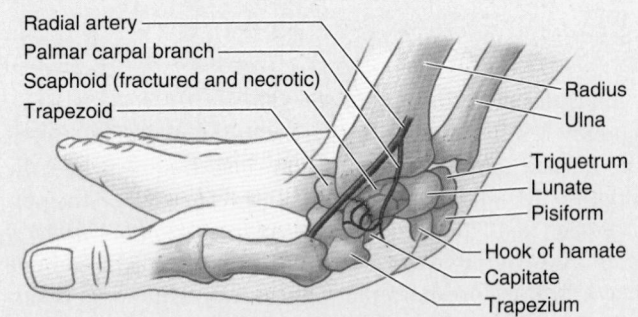

Radial artery
Palmar carpal branch
Scaphoid (fractured and necrotic)
Trapezoid

Radius
Ulna

Triquetrum
Lunate
Pisiform

Hook of hamate
Capitate
Trapezium

FIGURE B6.4. Fracture of scaphoid.

Fracture of Hamate

Fracture of the hamate may result in non-union of the fractured bony parts because of the traction produced by the attached muscles. Because the ulnar nerve is close to the hook of the hamate, the nerve may be injured by this fracture, causing decreased grip strength of the hand. The ulnar artery may also be damaged when the hamate is fractured.

Fracture of Metacarpals

The metacarpals (except the 1st) are closely bound together; hence isolated fractures tend to be stable. Furthermore, these bones have a good blood supply, and fractures usually heal rapidly. *Severe crushing injuries of the hand* may produce multiple metacarpal fractures, resulting in instability of the hand. Fracture of the 5th metacarpal, often referred to as a *boxer's fracture,* occurs when an unskilled person punches someone with a closed and abducted fist. The head of the bone rotates over the distal end of the shaft, producing a flexion deformity.

Fracture of Phalanges

Crushing injuries of the distal phalanges are common (e.g., when a finger is caught in a car door). Because of the highly developed sensation in the fingers, these injuries are extremely painful. A *fracture of a distal phalanx* is usually comminuted, and a painful hematoma (local collection of blood) soon develops. *Fractures of the proximal and middle phalanges* are usually the result of crushing or hyperextension injuries. Because of the close relationship of phalangeal fractures to the flexor tendons, the bone fragments must be carefully realigned to restore normal function of the fingers.

The Bottom Line

BONES OF UPPER LIMB

Comparison of upper and lower limbs: The development and structure of the upper and lower limbs have much in common; however, the upper limb has become a mobile organ that allows humans not only to respond to their environment, but to manipulate and control it to a large degree. ♦ The upper limb is composed of four increasingly mobile segments: The proximal three (shoulder, arm, and forearm) serve primarily to position the fourth segment (hand), which is used for grasping, manipulation, and touch. ♦ Four characteristics allow the independent operation of the upper limbs, allowing the hands to be precisely positioned and enabling accurate eye–hand coordination: (1) the upper limbs are not involved in weight bearing or ambulation, (2) the pectoral girdle is attached to the axial skeleton only anteriorly via a very mobile joint, (3) paired bones of the forearm can be moved relative to each other, and (4) the hands have long, mobile fingers and an opposable thumb.

Clavicle: The subcutaneously-located clavicle connects the upper limb (superior appendicular skeleton) to the trunk (axial skeleton). ♦ The clavicle serves as a movable crane-like strut (extended support) from which the scapula and free limb are suspended at a distance from the trunk that enables freedom of motion. ♦ Shocks received by the upper limb (especially the shoulder) are transmitted through the clavicle, resulting in a fracture that most commonly occurs between its middle and lateral thirds. ♦ The clavicle is the first long bone to ossify and the last to be fully formed.

Scapula: The scapula forms the mobile base from which the free upper limb acts. ♦ This triangular flat bone is curved to conform to the thoracic wall, and provides large surface areas and edges for attachment of muscles. ♦ These muscles (1) move the scapula on the thoracic wall at the physiological scapulothoracic joint, and (2) extend to the proximal humerus maintaining the integrity of—and producing motion at—the glenohumeral joint. ♦ The spine of the scapula and acromion serve as levers; the acromion enables the scapula and attached muscles to be located medially against the trunk with the acromioclavicular (AC) and glenohumeral joints, thereby allowing movement lateral to the

trunk. ♦ The coracoid process of the scapula is the site of attachment for the coracoclavicular ligament, which passively supports the upper limb, and a site for muscular (tendon) attachment.

Humerus: The long, strong humerus is a mobile strut—the first in a series of two—used to position the hand at a height (level) and distance from the trunk to maximize its efficiency. ♦ The spherical head of the humerus enables a great range of motion on the mobile scapular base; the trochlea and capitulum at its distal end facilitate the hinge movements of the elbow and, at the same time, the pivoting of the radius. ♦ The long shaft of the humerus enables reaching, and makes it an effective lever for power in lifting, as well as providing surface area for attachment of muscles that act primarily at the elbow. ♦ Added surface area for attachment of flexors and extensors of the wrist is provided by the epicondyles, the medial and lateral extensions of the distal end of the humerus.

Ulna and radius: The ulna and radius together make up the second unit of a two-unit articulated strut (the first unit being the humerus), projecting from a mobile base (shoulder) that serves to position the hand. ♦ Because the forearm unit is formed by two parallel bones, and the radius is able to pivot about the ulna, supination and pronation of the hand are possible during elbow flexion. ♦ Proximally, the larger medial ulna forms the primary articulation with the humerus, whereas distally, the shorter lateral radius forms the primary articulation with the hand via the wrist. ♦ Because the ulna does not reach the wrist, forces received by the hand are transmitted from the radius to the ulna via the interosseous membrane.

Hand: Each segment of the upper limb increases the functionality of the end unit, the hand. ♦ Located on the free end of a two-unit articulated strut (arm and forearm) projecting from a mobile base (shoulder), the hand can be positioned over a wide range relative to the trunk. ♦ The hand's connection to the flexible strut via the multiple small bones of the wrist, combined with the pivoting of the forearm, greatly increases its ability to be placed in a particular position with the digits able to flex (push or grip) in the necessary direction. ♦ The carpal bones are organized into two rows of four bones each and, as a group, articulate with the radius proximally and the metacarpals distally. ♦ The highly flexible, elongated digits—extending from a semirigid base (the palm)—enable the ability to grip, manipulate, or perform complex tasks involving multiple and simultaneous individual motions (e.g., when typing or playing a piano).

Surface anatomy: The upper limb presents multiple palpable bony features that are useful (1) when diagnosing fractures, dislocations, or malformations; (2) for approximating the position of deeper structures; and (3) for precisely describing the location of incisions and sites for therapeutic puncture, or areas of pathology or injury.

FASCIA, EFFERENT VESSELS, CUTANEOUS INNERVATION, AND MYOTOMES OF UPPER LIMB

Fascia of Upper Limb

Deep to the skin is (1) **subcutaneous tissue** (superficial fascia) containing fat, and (2) **deep fascia** compartmentalizing and investing the muscles (Fig. 6.13). If no structure (muscle, tendon, or bursa, for example) intervenes between the skin and bone, the deep fascia is usually attached to bone.

The fascia of the pectoral region is attached to the clavicle and sternum. The **pectoral fascia** invests the pectoralis major, and is continuous inferiorly with the fascia of the anterior abdominal wall. The pectoral fascia leaves the lateral border of the pectoralis major and becomes the **axillary fascia,** which forms the floor of the axilla (compartment deep to armpit). Deep to the pectoral fascia and pectoralis major, another fascial layer, the **clavipectoral fascia,** descends from the clavicle, enclosing the subclavius and then pectoralis minor, becoming continuous inferiorly with the axillary fascia.

The part of the clavipectoral fascia between the pectoralis minor and subclavius, the **costocoracoid membrane,** is pierced by the lateral pectoral nerve, which primarily supplies the pectoralis major. The part of the clavipectoral fascia inferior to the pectoralis minor, the **suspensory ligament of the axilla,** supports the axillary fascia, and pulls it and the overlying skin upward during abduction of the arm, forming the **axillary fossa** (armpit).

The scapulohumeral muscles that cover the scapula, and form the bulk of the shoulder, are also ensheathed by deep fascia. The **deltoid fascia** descends over the superficial surface of the deltoid from the clavicle, acromion, and scapular spine. From the deep surface of the deltoid fascia, numerous septa penetrate between the fascicles (bundles) of the muscle. Inferiorly, the deltoid fascia is continuous with the pectoral fascia anteriorly, and the dense infraspinous fascia posteriorly. The muscles that cover the anterior and posterior surfaces of the scapula are covered superficially with deep fascia, which is attached to the margins of the scapula and posteriorly to the spine of the scapula.

This arrangement creates osseofibrous *subscapular, supraspinous,* and *infraspinous compartments;* the muscles

in each compartment attach to (originate from) the deep surface of the overlying fascia in part, allowing the muscles to have greater bulk (mass) than would be the case if only bony attachments occurred. The **supraspinous** and **infraspinous fascia** overlying the supraspinatus and infraspinatus muscles, respectively, on the posterior aspect of the scapula are so dense and opaque that they must be removed during dissection to view the muscles.

The **brachial fascia,** a sheath of deep fascia, encloses the arm like a snug sleeve deep to the skin and subcutaneous tissue (Figs. 6.13A and 6.14A & B). It is continuous superiorly with the deltoid, pectoral, axillary, and infraspinous fascias. The brachial fascia is attached inferiorly to the epicondyles of the humerus and the olecranon of the ulna. This fascia is continuous with the **antebrachial fascia,** the deep fascia of the forearm. Two intermuscular septa—the **medial** and **lateral intermuscular septa**—extend from the deep surface of the brachial fascia to the central shaft and medial and lateral supra-epicondylar ridges of the humerus (Fig. 6.14B). These intermuscular septa divide the arm into **anterior (flexor)** and **posterior (extensor) fascial compartments,** each of which contains muscles serving similar functions and sharing common innervation. The fascial compartments of the upper limb are important clinically because they also contain and direct the spread of infection or hemorrhage in the limb.

In the forearm, similar fascial compartments are surrounded by the *antebrachial fascia* and are separated by the **interosseous membrane** connecting the radius and ulna (Fig. 6.14C). The antebrachial fascia thickens posteriorly over the distal ends of the radius and ulna to form a transverse band, the **extensor retinaculum,** which retains the extensor tendons in position (Fig. 6.14D).

The antebrachial fascia also forms an anterior thickening, which is continuous with the extensor retinaculum but is officially unnamed; some authors identify it as the *palmar carpal ligament.* Immediately distal and at a deeper level to the latter, the antebrachial fascia is also continued as the **flexor retinaculum (transverse carpal ligament).**[2] This fibrous band extends between the anterior prominences of the outer

---

[2]It is awkward that the structure officially identified as the flexor retinaculum does not correspond in position and structure to the extensor retinaculum when there is another structure (the palmar carpal ligament, currently unrecognized by *Terminologia Anatomica*) that does. The clinical community has proposed and widely adopted the use of the more structurally based term *transverse carpal ligament* to replace the term *flexor retinaculum.*

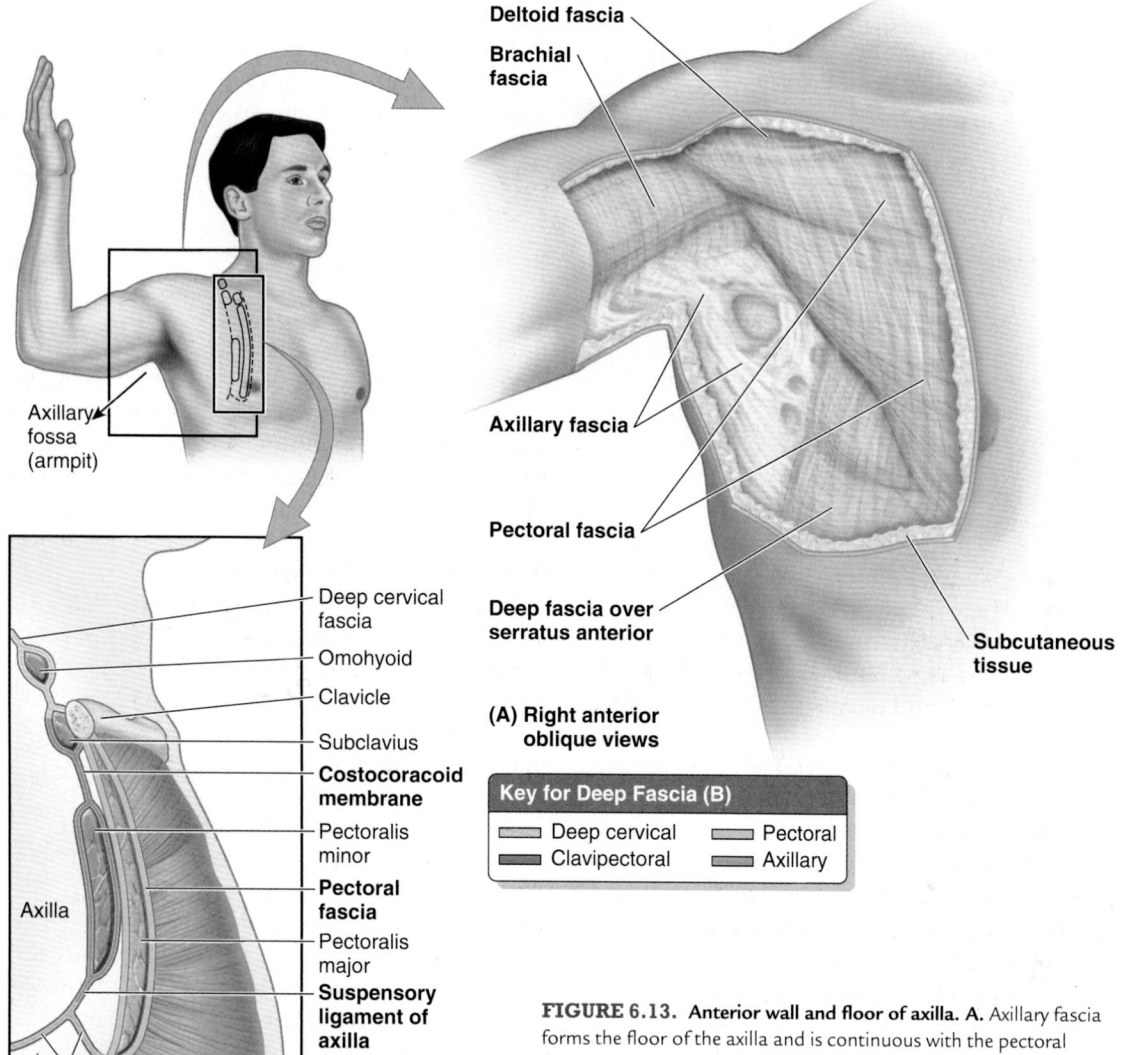

Deltoid fascia

Brachial fascia

Axillary fossa (armpit)

Axillary fascia

Pectoral fascia

Deep fascia over serratus anterior

Subcutaneous tissue

(A) Right anterior oblique views

Deep cervical fascia

Omohyoid

Clavicle

Subclavius

Costocoracoid membrane

Pectoralis minor

Pectoral fascia

Pectoralis major

Suspensory ligament of axilla

Axillary fascia

Axillary fossa

Axilla

Key for Deep Fascia (B)

Deep cervical	Pectoral
Clavipectoral	Axillary

(B) Lateral view of sagittal section

FIGURE 6.13. Anterior wall and floor of axilla. A. Axillary fascia forms the floor of the axilla and is continuous with the pectoral fascia. **B.** The pectoral fascia surrounds the pectoralis major, forming the anterior layer of the anterior axillary wall. The clavipectoral fascia extends between the coracoid process of the scapula, the clavicle, and the axillary fascia.

carpal bones, and converts the anterior concavity of the carpus into a *carpal tunnel,* through which the flexor tendons and median nerve pass.

The *deep fascia of the upper limb* continues beyond the extensor and flexor retinacula as the *palmar fascia.* The central part of the palmar fascia, the *palmar aponeurosis,* is thick, tendinous, and triangular and overlies the central compartment of the palm. Its *apex,* located proximally, is continuous with the *tendon of the palmaris longus* (when it is present) (Fig. 6.14A). The aponeurosis forms four distinct thickenings that radiate to the bases of the fingers and become continuous with the fibrous tendon sheaths of the digits. The bands are traversed distally by the **superficial transverse metacarpal ligament,** which forms the base of the palmar aponeurosis. Innumerable minute, strong *skin ligaments* (L. *retinacula cutis*) extend from the palmar aponeurosis to the skin (see

the Introduction; Fig. I.8B, p. 17). These ligaments hold the palmar skin close to the aponeurosis, allowing little sliding movement of the skin.

Venous Drainage of Upper Limb

SUPERFICIAL VEINS OF UPPER LIMB

The main superficial veins of the upper limb, the cephalic and basilic veins, originate in the subcutaneous tissue on the dorsum of the hand from the **dorsal venous network** (Fig. 6.15A). **Perforating veins** form communications between the superficial and deep veins (Fig. 6.15B). Like the dermatomal pattern, the logic for naming the main superficial veins of the upper limb cephalic (toward the head) and basilic (toward the base) becomes apparent when the limb is placed in its initial embryonic position.

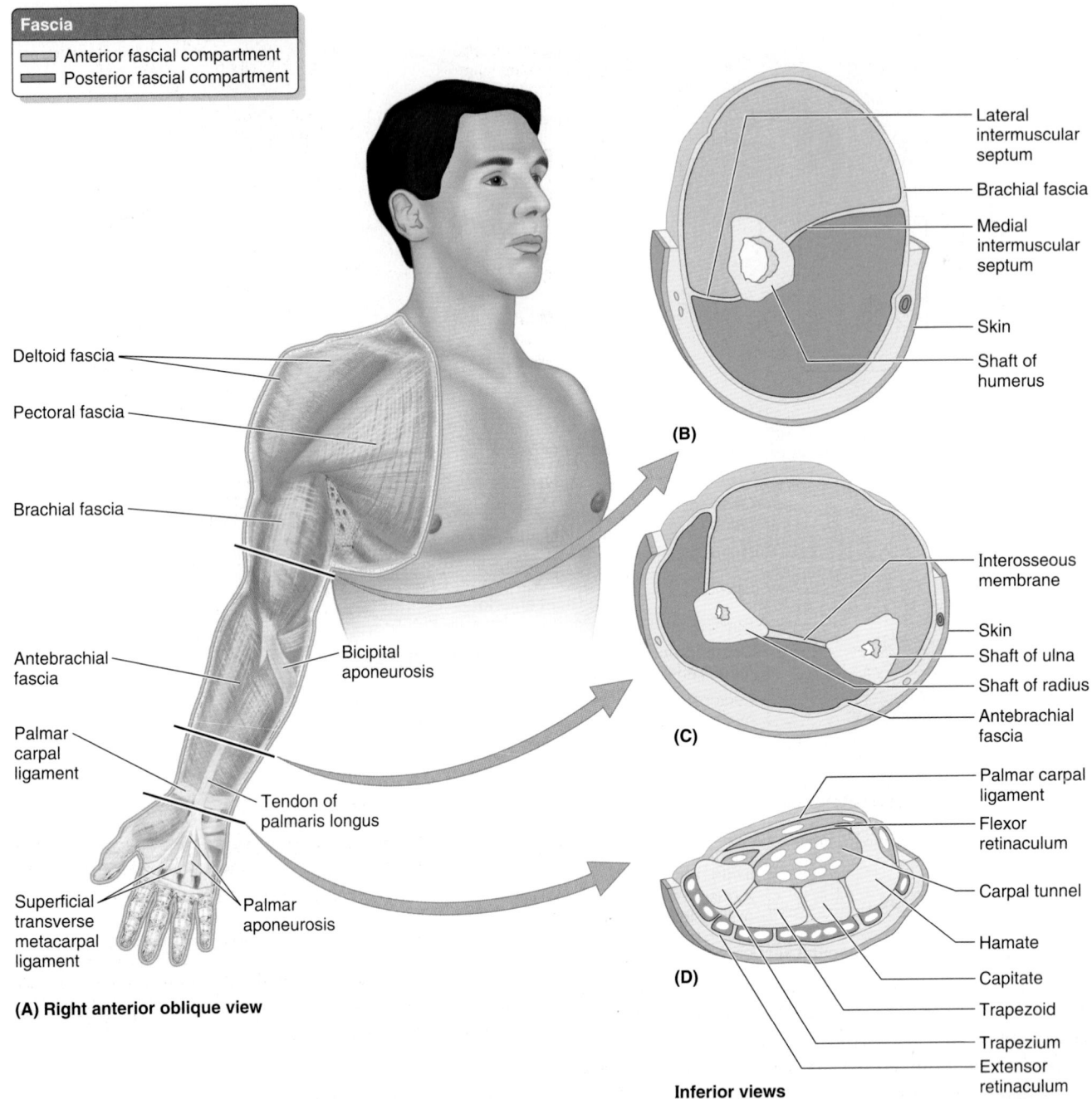

Fascia
- Anterior fascial compartment
- Posterior fascial compartment

Deltoid fascia

Pectoral fascia

Brachial fascia

Antebrachial fascia

Bicipital aponeurosis

Palmar carpal ligament

Tendon of palmaris longus

Superficial transverse metacarpal ligament

Palmar aponeurosis

(A) Right anterior oblique view

Lateral intermuscular septum

Brachial fascia

Medial intermuscular septum

Skin

Shaft of humerus

(B)

Interosseous membrane

Skin

Shaft of ulna

Shaft of radius

Antebrachial fascia

(C)

Palmar carpal ligament

Flexor retinaculum

Carpal tunnel

Hamate

Capitate

Trapezoid

Trapezium

Extensor retinaculum

(D)

Inferior views

FIGURE 6.14. **Fascia and compartments of upper limb. A.** Brachial and antebrachial fascia surround the structures of the free upper limb. **B.** The intermuscular septa and humerus divide the space inside the brachial fascia into anterior and posterior compartments, each of which contains muscles serving similar functions and the nerves and vessels supplying them. **C.** The interosseous membrane and the radius and ulna similarly separate the space inside the antebrachial fascia into anterior and posterior compartments. **D.** The deep fascia of the forearm thickens to form the extensor retinaculum posteriorly, and a corresponding thickening anteriorly (palmar carpal ligament). At a deeper level, the flexor retinaculum extends between the anterior prominences of the outer carpal bones, converting the anterior concavity of the carpus into an osseofibrous carpal tunnel.

The **cephalic vein** (G. *kephalé*, head) ascends in the subcutaneous tissue from the lateral aspect of the dorsal venous network, proceeding along the lateral border of the wrist and the anterolateral surface of the proximal forearm and arm; it is often visible through the skin. Anterior to the elbow, the cephalic vein communicates with the **median cubital vein,** which passes obliquely across the anterior aspect of the elbow in the cubital fossa (a depression in front of the elbow), and joins the basilic vein. The cephalic vein courses superiorly between the deltoid and pectoralis major muscles along the deltopectoral groove, then enters the *clavipectoral triangle* (Figs. 6.2 and 6.15B).

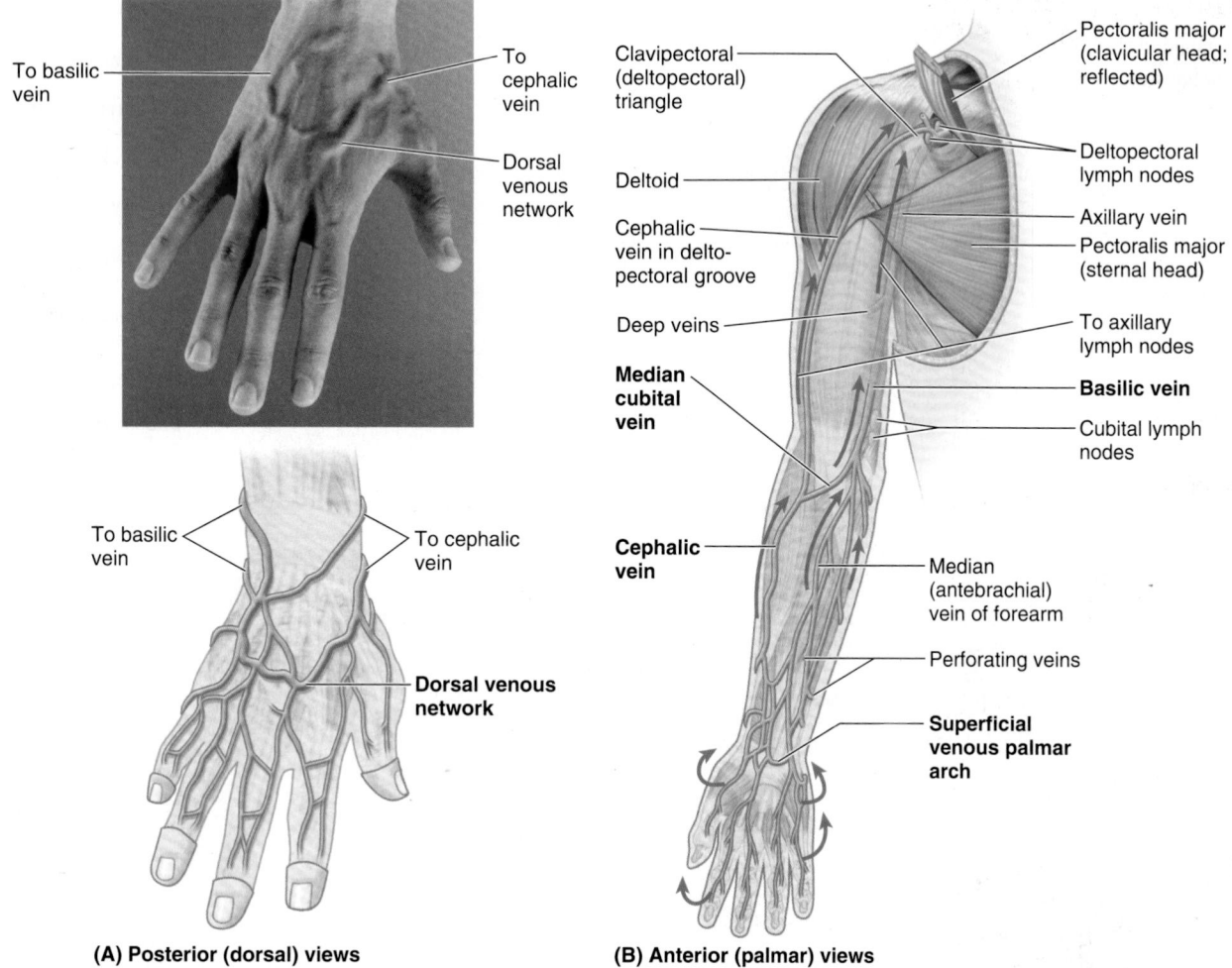

To basilic vein

To cephalic vein

Dorsal venous network

To basilic vein

To cephalic vein

Dorsal venous network

(A) Posterior (dorsal) views

Clavipectoral (deltopectoral) triangle

Deltoid

Cephalic vein in delto-pectoral groove

Deep veins

Median cubital vein

Cephalic vein

Pectoralis major (clavicular head; reflected)

Deltopectoral lymph nodes

Axillary vein

Pectoralis major (sternal head)

To axillary lymph nodes

Basilic vein

Cubital lymph nodes

Median (antebrachial) vein of forearm

Perforating veins

Superficial venous palmar arch

(B) Anterior (palmar) views

FIGURE 6.15. Superficial veins and lymph nodes of upper limb. A. The digital veins and dorsal venous network on the dorsum of the hand. **B.** Basilic and cephalic veins. *Arrows* indicate the flow of lymph within lymphatic vessels, which converge toward the vein and drain into the cubital and axillary lymph nodes.

It then pierces the costocoracoid membrane and part of the clavipectoral fascia, joining the terminal part of the axillary vein.

The **basilic vein** ascends in the subcutaneous tissue from the medial end of the dorsal venous network along the medial side of the forearm and the inferior part of the arm; it is often visible through the skin. It then passes deeply near the junction of the middle and inferior thirds of the arm, piercing the brachial fascia and running superiorly parallel to the brachial artery and the medial cutaneous nerve of the forearm to the axilla, where it merges with the accompanying veins (L. *venae comitantes*) of the axillary artery to form the axillary vein.

The **median antebrachial vein (median vein of the forearm)** is highly variable. It begins at the base of the dorsum of the thumb, curves around the lateral side of the wrist, and ascends in the middle of the anterior aspect of the forearm between the cephalic and the basilic veins. The median

antebrachial vein sometimes divides into a *median basilic vein*, which joins the basilic vein, and a *median cephalic vein*, which joins the cephalic vein.

DEEP VEINS OF UPPER LIMB

Deep veins lie internal to the deep fascia, and—in contrast to the superficial veins—usually occur as paired (continually interanastomosing) *accompanying veins* that travel with, and bear the same name as, the major arteries of the limb (Fig. 6.16).

Lymphatic Drainage of Upper Limb

Superficial lymphatic vessels arise from *lymphatic plexuses* in the skin of the fingers, palm, and dorsum of the hand and ascend mostly with the superficial veins, such as the cephalic and basilic veins (Fig. 6.17). Some vessels accompanying the basilic vein enter the **cubital lymph nodes,** located

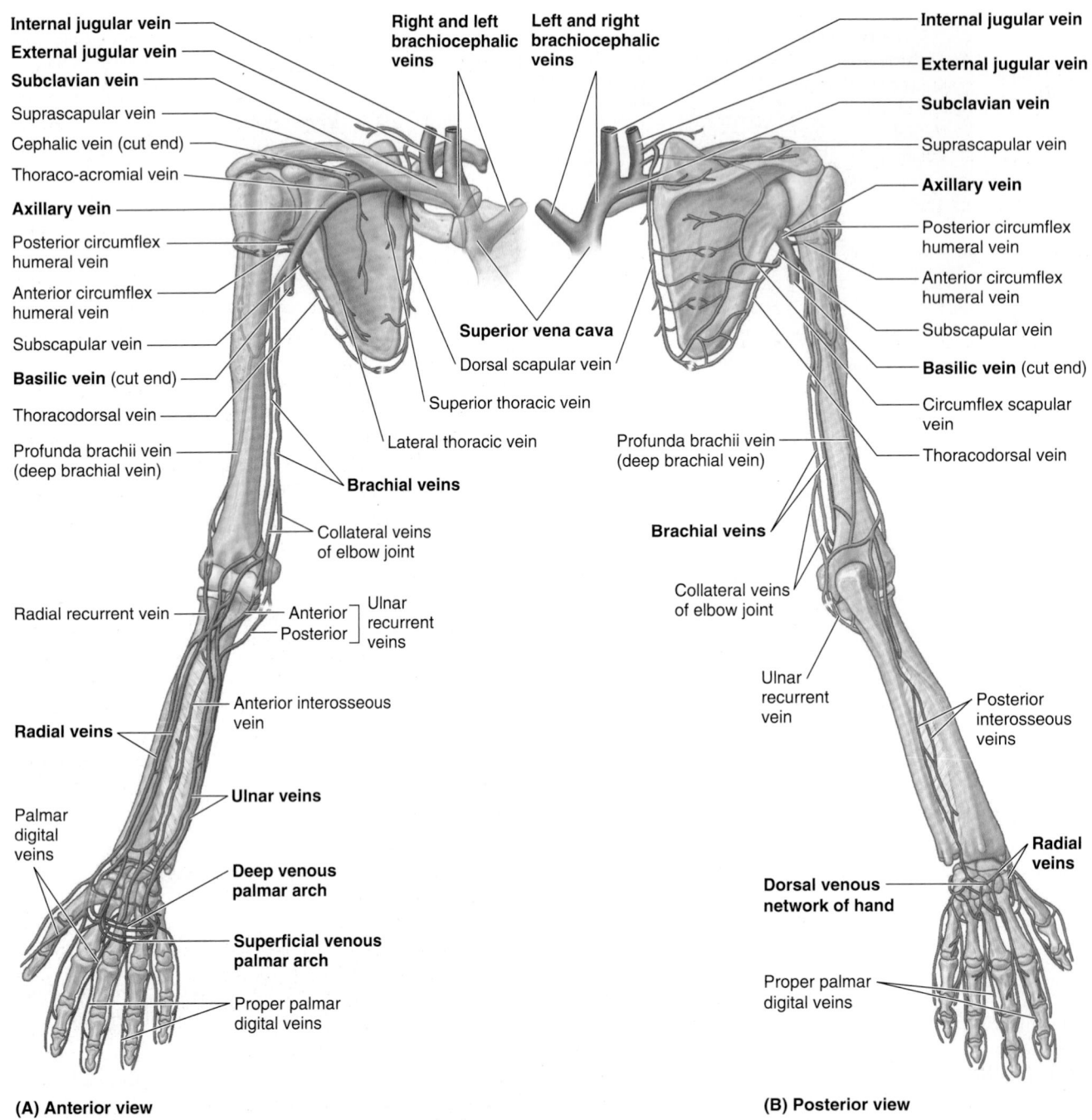

Internal jugular vein

External jugular vein

Subclavian vein

Suprascapular vein

Cephalic vein (cut end)

Thoraco-acromial vein

Axillary vein

Posterior circumflex humeral vein

Anterior circumflex humeral vein

Subscapular vein

Basilic vein (cut end)

Thoracodorsal vein

Profunda brachii vein (deep brachial vein)

Right and left brachiocephalic veins

Left and right brachiocephalic veins

Superior vena cava

Dorsal scapular vein

Superior thoracic vein

Lateral thoracic vein

Brachial veins

Collateral veins of elbow joint

Radial recurrent vein

Anterior ⎤ Ulnar
Posterior ⎦ recurrent veins

Anterior interosseous vein

Radial veins

Palmar digital veins

Ulnar veins

Deep venous palmar arch

Superficial venous palmar arch

Proper palmar digital veins

(A) Anterior view

Internal jugular vein

External jugular vein

Subclavian vein

Suprascapular vein

Axillary vein

Posterior circumflex humeral vein

Anterior circumflex humeral vein

Subscapular vein

Basilic vein (cut end)

Circumflex scapular vein

Thoracodorsal vein

Profunda brachii vein (deep brachial vein)

Brachial veins

Collateral veins of elbow joint

Ulnar recurrent vein

Posterior interosseous veins

Radial veins

Dorsal venous network of hand

Proper palmar digital veins

(B) Posterior view

FIGURE 6.16. Deep veins of upper limb. The deep veins bear the same name as the arteries they accompany.

proximal to the medial epicondyle and medial to the basilic vein. Efferent vessels from these lymph nodes ascend in the arm and terminate in the **humeral (lateral) axillary lymph nodes** (see Chapter 1).

Most superficial lymphatic vessels accompanying the cephalic vein cross the proximal part of the arm and the anterior aspect of the shoulder to enter the **apical axillary lymph nodes;** however, some vessels previously enter the more superficial **deltopectoral lymph nodes.**

Deep lymphatic vessels, less numerous than superficial vessels, accompany the major deep veins in the upper limb (radial, ulnar, and brachial--Fig. 6.16) and terminate in the humeral axillary lymph nodes. They drain lymph from the joint capsules, periosteum, tendons, nerves, and muscles and ascend with the deep veins; a few deep lymph nodes may occur along their course. The axillary lymph nodes are drained by the *subclavian lymphatic trunk;* both are discussed in greater detail with the axilla, later in this chapter.

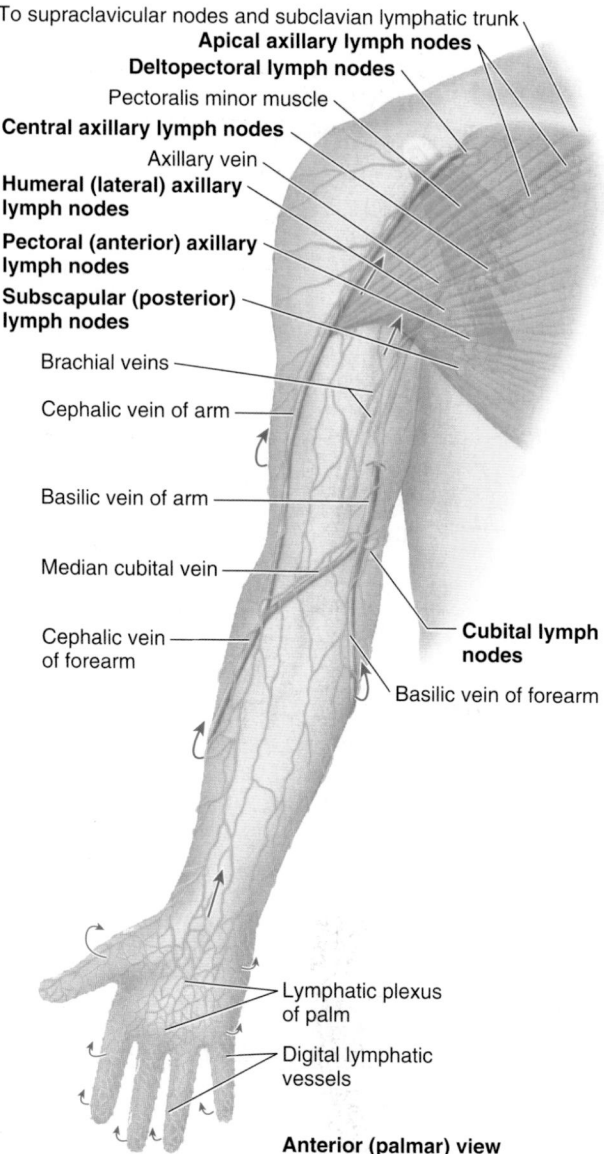

To supraclavicular nodes and subclavian lymphatic trunk
Apical axillary lymph nodes
Deltopectoral lymph nodes
Pectoralis minor muscle
Central axillary lymph nodes
Axillary vein
**Humeral (lateral) axillary
lymph nodes**
**Pectoral (anterior) axillary
lymph nodes**
**Subscapular (posterior)
lymph nodes**
Brachial veins
Cephalic vein of arm
Basilic vein of arm
Median cubital vein
Cephalic vein
of forearm
**Cubital lymph
nodes**
Basilic vein of forearm
Lymphatic plexus
of palm
Digital lymphatic
vessels
Anterior (palmar) view

FIGURE 6.17. Lymphatic drainage of upper limb. Superficial lymphatic vessels originate from the digital lymphatic vessels of the digits and lymphatic plexus of the palm. Most drainage from the palm passes to the dorsum of the hand (*small arrows*).

Cutaneous Innervation of Upper Limb

The cutaneous nerves of the upper limb follow a general pattern that is easy to understand if it is noted that developmentally the limbs grow as lateral protrusions of the trunk, with the 1st digit (thumb or great toe) located on the cranial side (thumb is directed superiorly). Thus the lateral aspect of the upper limb is innervated by more cranial spinal cord segments or nerves than the medial aspect.

There are two *dermatome maps* in common use. One has gained popular acceptance because of its more intuitive aesthetic qualities, corresponding to concepts of limb development (Keegan and Garrett, 1948); the other is based on clinical findings and is generally preferred by neurologists (Foerster, 1933). Both maps are approximations, delineating dermatomes as distinct zones when actually there is much overlap between adjacent dermatomes and much variation (even from side to side in the same individual). In both schemes, observe the progression of the segmental innervation of the various cutaneous areas around the limb when it is placed in its "initial embryonic position" (abducted with thumb directed superiorly) (Fig. 6.18; Table 6.1).

Most cutaneous nerves of the upper limb are derived from the *brachial plexus*, a major nerve network formed by the anterior rami of the C5–T1 spinal nerves (see "Brachial Plexus" on p. 721). The nerves to the shoulder, however, are derived from the *cervical plexus*, a nerve network consisting of a series of nerve loops formed between adjacent anterior rami of the first four cervical nerves. The cervical plexus lies deep to the sternocleidomastoid muscle on the anterolateral aspect of the neck.

The cutaneous nerves of the arm and forearm[3] are illustrated in Figure 6.19, and their contributing spinal nerves, source, and course and distribution are provided in Table 6.2

Note that there are lateral, medial, and posterior (but no anterior) cutaneous nerves of the arm and forearm; as discussed later in this chapter, this pattern corresponds to that of the cords of the brachial plexus.

Motor Innervation (Myotomes) of Upper Limb

Somatic motor (general somatic efferent) fibers traveling in the same mixed peripheral nerves that convey sensory fibers to the cutaneous nerves transmit impulses to the voluntary muscles of the upper limb. The unilateral embryological muscle mass (and derived muscle) receiving innervation from a single spinal cord segment or spinal nerve constitutes a *myotome*. Upper limb muscles usually receive motor fibers from several spinal cord segments or nerves. Thus most muscles are made up of more than one myotome, and multiple spinal cord segments are usually involved in producing the movement of the upper limb (Fig. 6.20). The intrinsic muscles of the hand constitute a single myotome (T1).

(text continues on p. 697)

[3]The preferred English-equivalent terms listed by *Terminologia Anatomica* (*TA*) are used here. Official alternate *TA* terms replace *of the arm* with *brachial*, and *of the forearm* with *antebrachial*.

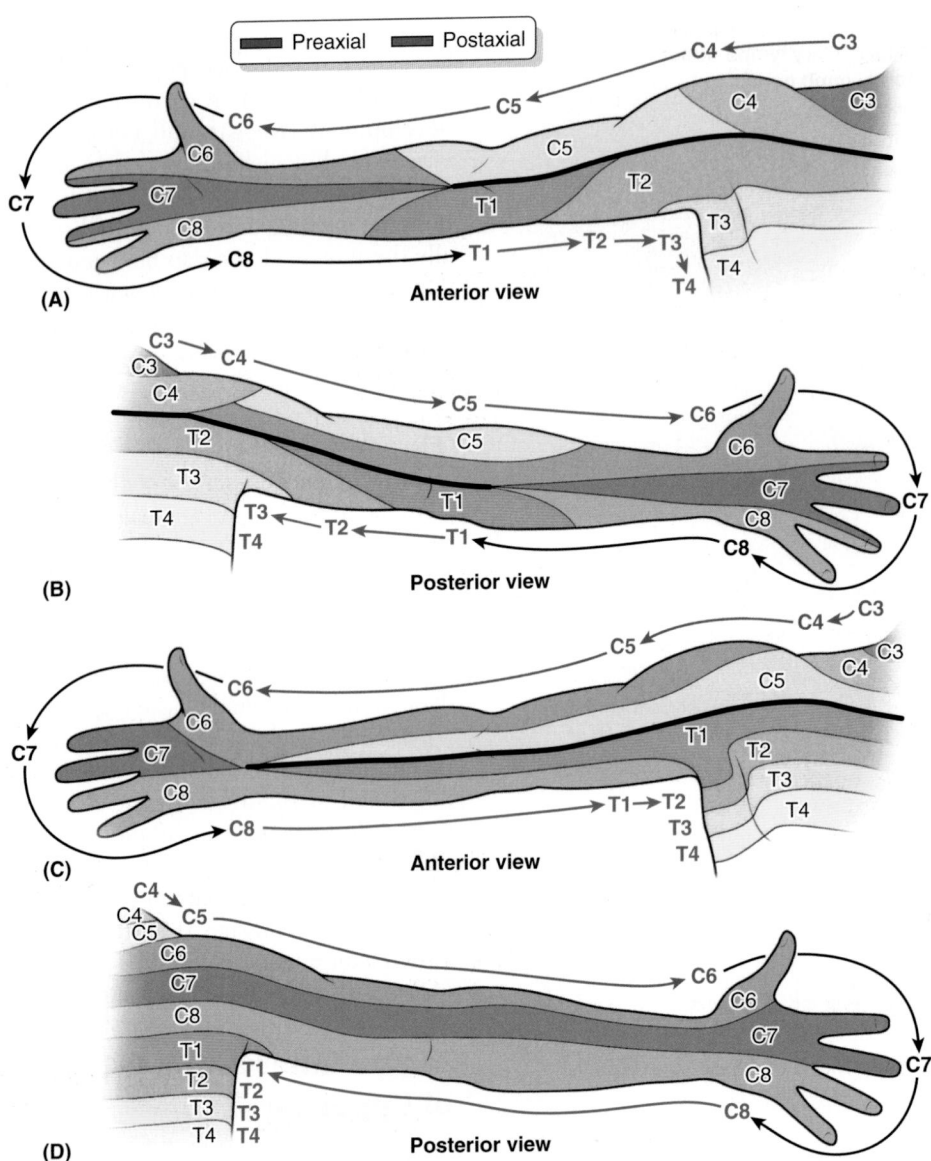

FIGURE 6.18. Segmental (dermatomal) and peripheral (cutaneous nerve) innervation of upper limb. A and B. The pattern of segmental (dermatomal) innervation of the upper limb proposed by Foerster (1933) depicts innervation of the medial aspect of the limb by upper thoracic (T1–T3) spinal cord segments, consistent with the experience of heart pain (angina pectoris) referred to that area. **C and D.** The pattern of segmental innervation proposed by Keegan and Garrett (1948) has gained popular acceptance, perhaps because of the regular progression of its stripes and correlation with developmental concepts. In both patterns, the dermatomes progress sequentially around the periphery of the outstretched limb (with the thumb directed superiorly), providing a way to approximate the segmental innervation.

TABLE 6.1. DERMATOMES OF UPPER LIMB

Spinal Segment/Nerve(s)	Description of Dermatome(s)
C3, C4	Region at base of neck, extending laterally over shoulder
C5	Lateral aspect of arm (i.e., superior aspect of abducted arm)
C6	Lateral forearm and thumb
C7	Middle and ring fingers (or middle three fingers) and center of posterior aspect of forearm
C8	Little finger, medial side of hand and forearm (i.e., inferior aspect of abducted arm)
T1	Medial aspect of forearm and inferior arm
T2	Medial aspect of superior arm and skin of axilla[a]

[a]Not indicated on the Keegan and Garrett (1948) dermatome map. However, pain experienced during a heart attack, considered to be mediated by T1 and T2, is commonly described as "radiating down the medial side of the left arm."

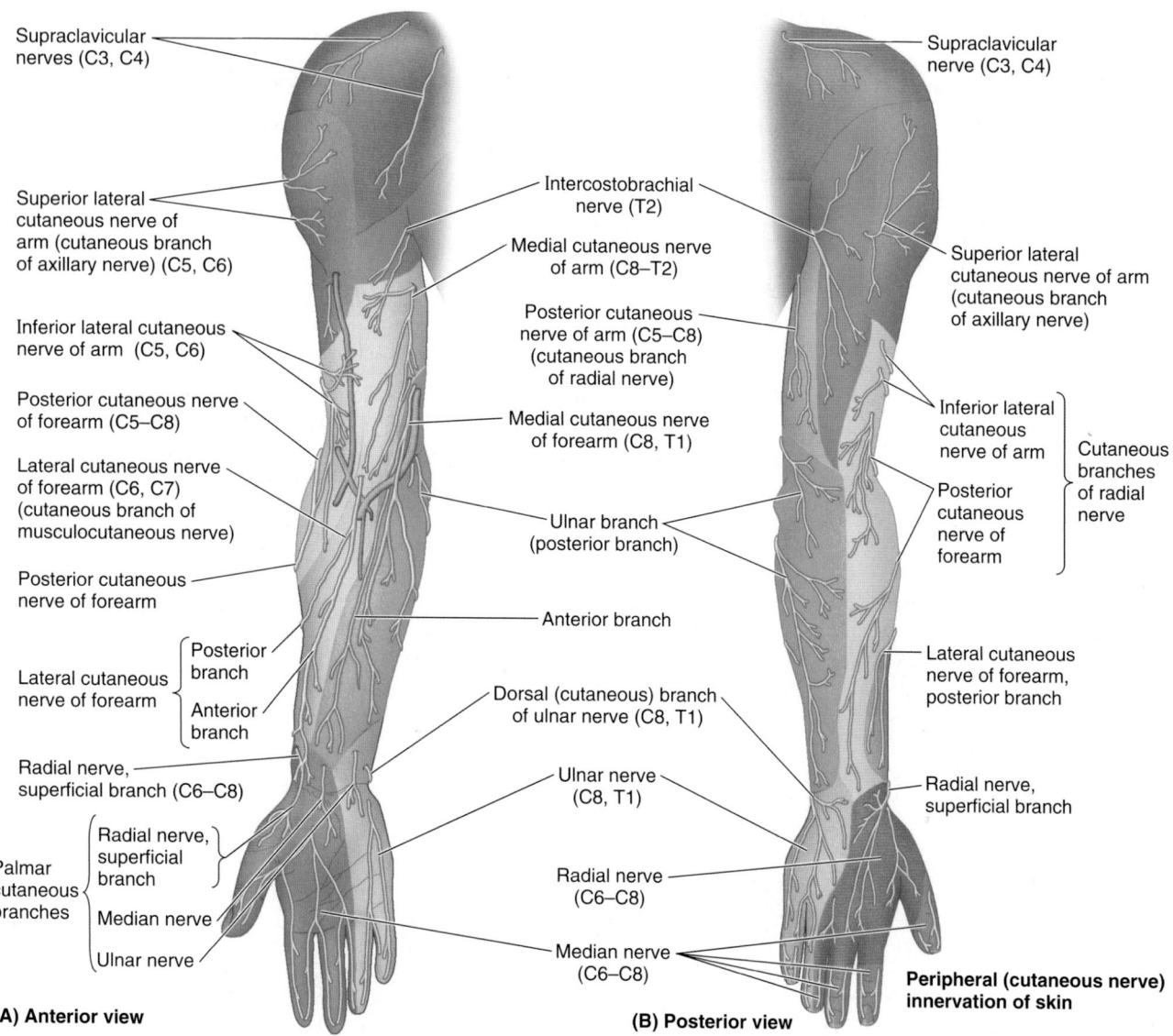

Supraclavicular nerves (C3, C4)

Superior lateral cutaneous nerve of arm (cutaneous branch of axillary nerve) (C5, C6)

Inferior lateral cutaneous nerve of arm (C5, C6)

Posterior cutaneous nerve of forearm (C5–C8)

Lateral cutaneous nerve of forearm (C6, C7) (cutaneous branch of musculocutaneous nerve)

Posterior cutaneous nerve of forearm

Lateral cutaneous nerve of forearm { Posterior branch / Anterior branch

Radial nerve, superficial branch (C6–C8)

Palmar cutaneous branches { Radial nerve, superficial branch / Median nerve / Ulnar nerve

(A) Anterior view

Intercostobrachial nerve (T2)

Medial cutaneous nerve of arm (C8–T2)

Posterior cutaneous nerve of arm (C5–C8) (cutaneous branch of radial nerve)

Medial cutaneous nerve of forearm (C8, T1)

Ulnar branch (posterior branch)

Anterior branch

Dorsal (cutaneous) branch of ulnar nerve (C8, T1)

Ulnar nerve (C8, T1)

Radial nerve (C6–C8)

Median nerve (C6–C8)

(B) Posterior view

Supraclavicular nerve (C3, C4)

Superior lateral cutaneous nerve of arm (cutaneous branch of axillary nerve)

Inferior lateral cutaneous nerve of arm

Posterior cutaneous nerve of forearm

} Cutaneous branches of radial nerve

Lateral cutaneous nerve of forearm, posterior branch

Radial nerve, superficial branch

Peripheral (cutaneous nerve) innervation of skin

FIGURE 6.19. Distribution of peripheral (named) cutaneous nerves in upper limb. Most of the nerves are branches of nerve plexuses and, therefore, contain fibers from more than one spinal nerve or spinal cord segment.

TABLE 6.2. CUTANEOUS NERVES OF UPPER LIMB

Cutaneous Nerve	Contributing Spinal Nerves	Source	Course and Distribution
Supraclavicular nerves	C3, C4	Cervical plexus	Pass anterior to clavicle, immediately deep to platysma, and supply skin over clavicle and supero-lateral aspect of pectoralis major
Superior lateral cutaneous nerve of arm	C5, C6	Terminal branch of axillary nerve	Emerges from beneath posterior margin of deltoid and supplies skin over lower part of this muscle and on lateral side of midarm
Inferior lateral cutaneous nerve of arm	C5, C6	Radial nerve (or posterior cutaneous nerve of arm)	Perforates lateral head of triceps, passing close to cephalic vein to supply skin over inferolateral aspect of arm
Posterior cutaneous nerve of arm	C5–C8	Radial nerve (in axilla)	Crosses posterior to and communicates with inter-costobrachial nerve and supplies skin on posterior arm as far as olecranon

TABLE 6.2. CUTANEOUS NERVES OF UPPER LIMB (Continued)

Cutaneous Nerve	Contributing Spinal Nerves	Source	Course and Distribution
Posterior cutaneous nerve of forearm	C5–C8	Radial nerve (with inferior lateral cutaneous nerve of arm)	Perforates lateral head of triceps, descends laterally in arm, then runs along and supplies posterior forearm to wrist.
Lateral cutaneous nerve of forearm	C6–C7	Musculocutaneous nerve (terminal branch)	Emerges lateral to biceps tendon deep to cephalic vein, supplying skin of anterolateral forearm to wrist
Medial cutaneous nerve of forearm	C8, T1	Medial cord of brachial plexus (in axilla)	Descends medial to brachial artery, pierces deep fascia with basilic vein in midarm, dividing into anterior and posterior branches that enter forearm and supply skin of anteromedial aspect to wrist
Medial cutaneous nerve of arm	C8–T2	Medial cord of brachial plexus (in axilla)	Communicates with intercostobrachial nerve, continuing to supply skin of medial aspect of distal arm
Intercostobrachial nerve	T2	Second intercostal nerve (as its lateral cutaneous branch)	Extends laterally, communicating with posterior and medial cutaneous nerves of arm, supplying skin of axilla and medial aspect of proximal arm

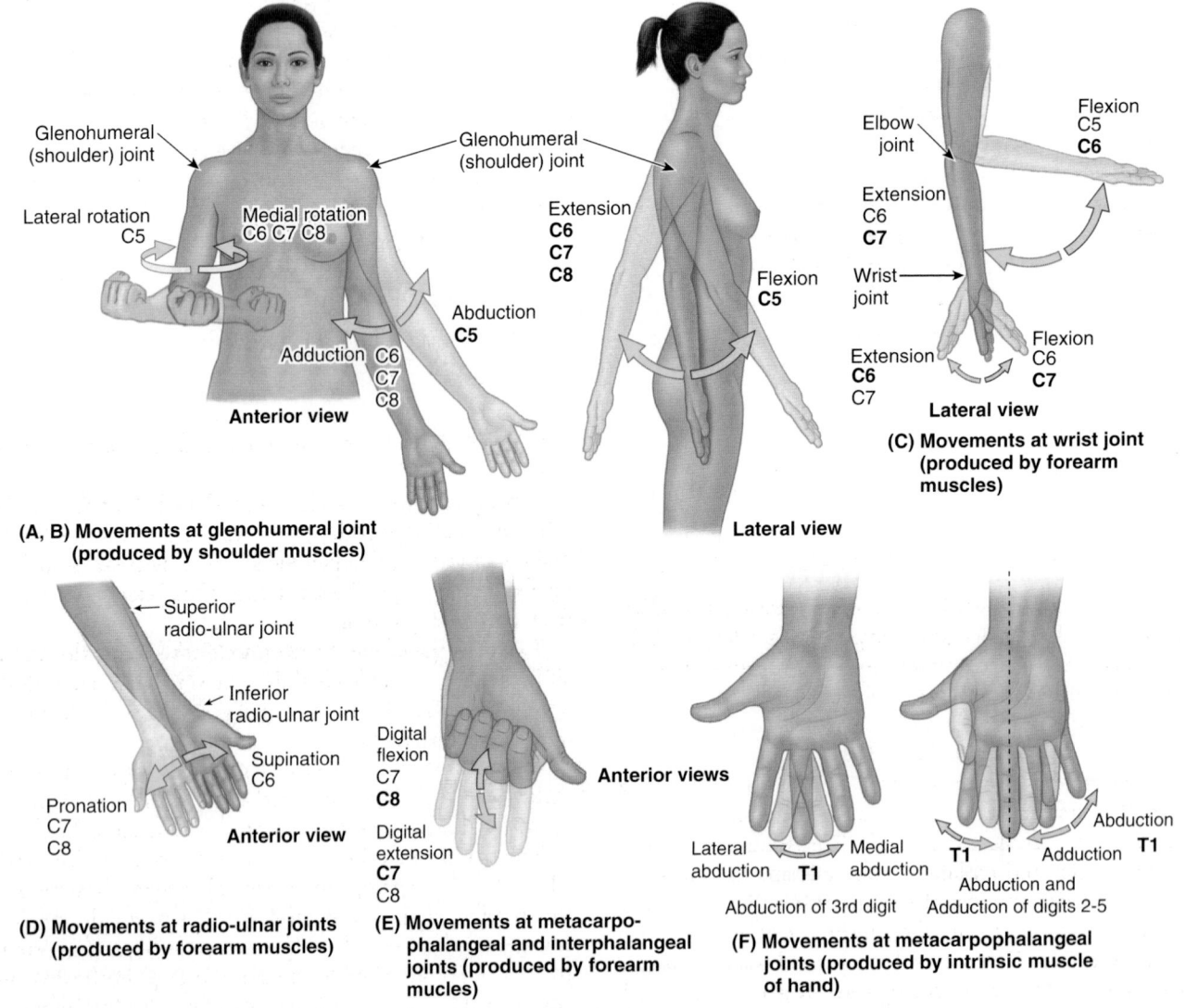

(A, B) Movements at glenohumeral joint (produced by shoulder muscles)

Lateral view

(C) Movements at wrist joint (produced by forearm muscles)

(D) Movements at radio-ulnar joints (produced by forearm muscles)

(E) Movements at metacarpophalangeal and interphalangeal joints (produced by forearm mucles)

(F) Movements at metacarpophalangeal joints (produced by intrinsic muscle of hand)

FIGURE 6.20. Segmental innervation of movements of the upper limb. A–F. Most movements involve portions of multiple myotomes; however, the intrinsic muscles of the hand involve a single myotome (T1).

The Bottom Line

FASCIA, EFFERENT VESSELS, CUTANEOUS INNERVATION, AND MYOTOMES OF UPPER LIMB

Fascia: The firm deep fascia of the upper limb surrounds and contains the structures of the upper limb as an expansion-limiting membrane deep to the skin and subcutaneous tissue. ♦ The deep surface of the fascia, which occasionally serves to extend the surface area available for muscular origin, is attached directly or via intermuscular septa to the enclosed bones. ♦ The deep fascia thus forms fascial compartments containing individual muscles or muscle groups of similar function and innervation. ♦ The compartments also contain or direct the spread of infection or hemorrhage.

Superficial veins: The cephalic vein courses along the cranial (cephalic) margin of the limb, while the basilic vein courses along the caudal (basic) margin of the limb. ♦ Both veins come from the dorsal venous network on the dorsum of the hand, and terminate by draining into the beginning (basilic vein) and end (cephalic vein) of the axillary vein.

Deep veins: Deep veins in the limbs usually take the form of paired accompanying veins, bearing the same name as the artery they accompany.

Lymphatic vessels: The superficial lymphatic vessels generally converge on and follow the superficial veins, and the deep lymphatics follow the deep veins. ♦ The lymph collected from the upper limb by both superficial and deep lymphatics drains into the axillary lymph nodes.

Dermatomes: As a consequence of plexus formation, two patterns of cutaneous innervation occur in the upper limb: (1) segmental innervation (dermatomes) by spinal nerves and (2) innervation by multisegmental peripheral (named) nerves. The former pattern is easiest to visualize if the limb is placed in its initial embryonic position (abducted with the thumb directed superiorly). ♦ The segments then progress in descending order around the limb (starting with C4 dermatome at the root of the neck, proceeding laterally or distally along the superior surface and then medially or proximally along the inferior surface, as the T2 dermatome continues onto the thoracic wall).

Cutaneous innervation: Like the brachial plexus, which forms posterior, lateral, and medial (but no anterior) cords, the arm and forearm have posterior, lateral, and medial (but no anterior) cutaneous nerves. ♦ The medial cutaneous nerves are branches of the medial cord of the brachial plexus. ♦ The posterior cutaneous nerves are branches of the radial nerve. ♦ Each of the lateral cutaneous nerves arise from a separate source (axillary, radial, and musculocutaneous nerves).

Myotomes: Most upper limb muscles include components of more than one myotome and thus receive motor fibers from several spinal cord segments or spinal nerves. ♦ Thus multiple spinal cord segments are involved in producing the movements of the upper limb. ♦ The intrinsic muscles of the hand constitute a single myotome (T1).

PECTORAL AND SCAPULAR REGIONS

Anterior Axio-appendicular Muscles

Four anterior axio-appendicular muscles (thoraco-appendicular or pectoral muscles) move the pectoral girdle: pectoralis major, pectoralis minor, subclavius, and serratus anterior. These muscles and their attachments are illustrated in Figure 6.21, and their attachments, nerve supply, and main actions are summarized in Table 6.3.

The **pectoralis major** is a large, fan-shaped muscle that covers the superior part of the thorax (Fig. 6.21A). It has *clavicular and sternocostal heads*. The sternocostal head is much larger, and its lateral border forms the muscular mass that makes up most of the anterior wall of the axilla. Its inferior border forms the *anterior axillary fold* (see "Axilla" on p. 713). The pectoralis major and adjacent deltoid muscles form the narrow **deltopectoral groove,** in which the cephalic vein runs (Fig. 6.15B); however, the muscles diverge slightly from each other superiorly and, along with the clavicle, form the *clavipectoral (deltopectoral) triangle* (Figs. 6.2 and 6.15B).

Producing powerful adduction and medial rotation of the arm when acting together, the two parts of the pectoralis major can also act independently: the clavicular head flexing the humerus, and the sternocostal head extending it back from the flexed position.

To test the clavicular head of pectoralis major, the arm is abducted 90°; the individual then moves the arm anteriorly against resistance. If acting normally, the clavicular head can be seen and palpated.

To test the sternocostal head of pectoralis major, the arm is abducted 60° and then adducted against resistance. If acting normally, the sternocostal head can be seen and palpated.

The **pectoralis minor** lies in the anterior wall of the axilla where it is almost completely covered by the much larger pectoralis major (Figs. 6.21B and 6.22). The pectoralis minor is triangular in shape: Its base (proximal attachment) is formed by fleshy slips attached to the anterior ends of the 3rd–5th ribs near their costal cartilages; its apex (distal attachment) is on the coracoid process of the scapula. Variations in the costal attachments of the muscle are common.

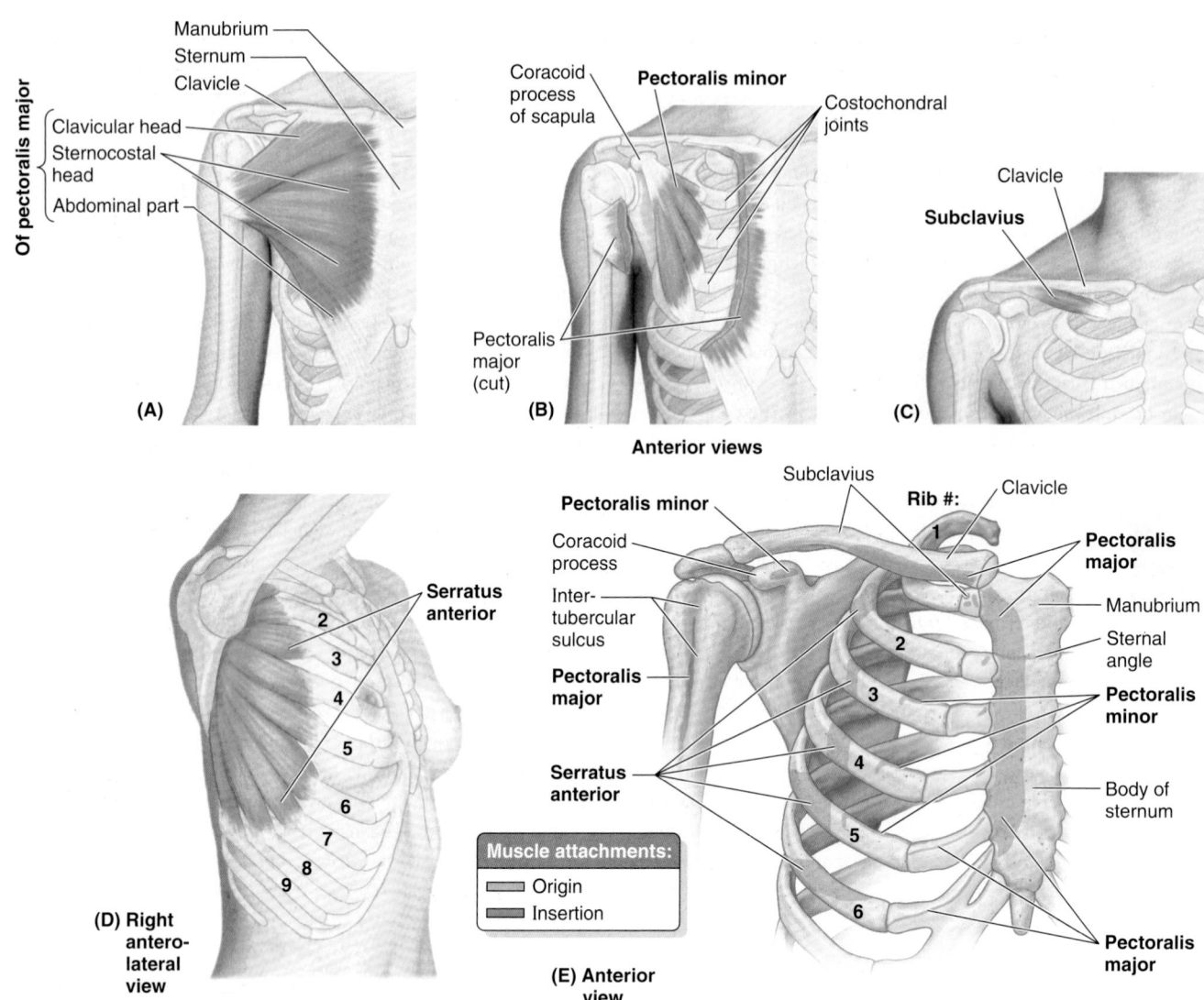

FIGURE 6.21. Anterior axio-appendicular muscles.

TABLE 6.3. ANTERIOR AXIO-APPENDICULAR MUSCLES

Muscle	Proximal Attachment	Distal Attachment	Innervation[a]	Main Action
Pectoralis major	Clavicular head: anterior surface of medial half of clavicle Sternocostal head: anterior surface of sternum, superior six costal cartilages, aponeurosis of external oblique muscle	Lateral lip of intertubercular sulcus of humerus	Lateral and medial pectoral nerves; clavicular head (C5, **C6**), sternocostal head (**C7, C8**, T1)	Adducts and medially rotates humerus; draws scapula anteriorly and inferiorly Acting alone, clavicular head flexes humerus and sternocostal head extends it from the flexed position
Pectoralis minor	3rd–5th ribs near their costal cartilages	Medial border and superior surface of coracoid process of scapula	Medial pectoral nerve (C8, T1)	Stabilizes scapula by drawing it inferiorly and anteriorly against thoracic wall
Subclavius	Junction of 1st rib and its costal cartilage	Inferior surface of middle third of clavicle	Nerve to subclavius (**C5**, C6)	Anchors and depresses clavicle
Serratus anterior	External surfaces of lateral parts of 1st–8th ribs	Anterior surface of medial border of scapula	Long thoracic nerve (C5, **C6, C7**)	Protracts scapula and holds it against thoracic wall; rotates scapula

[a]The spinal cord segmental innervation is indicated (e.g., "**C5**, C6" means that the nerves supplying the subclavius are derived from the fifth and sixth cervical segments of the spinal cord). Numbers in boldface (**C5**) indicate the main segmental innervation. Damage to one or more of the listed spinal cord segments or to the motor nerve roots arising from them results in paralysis of the muscles concerned.

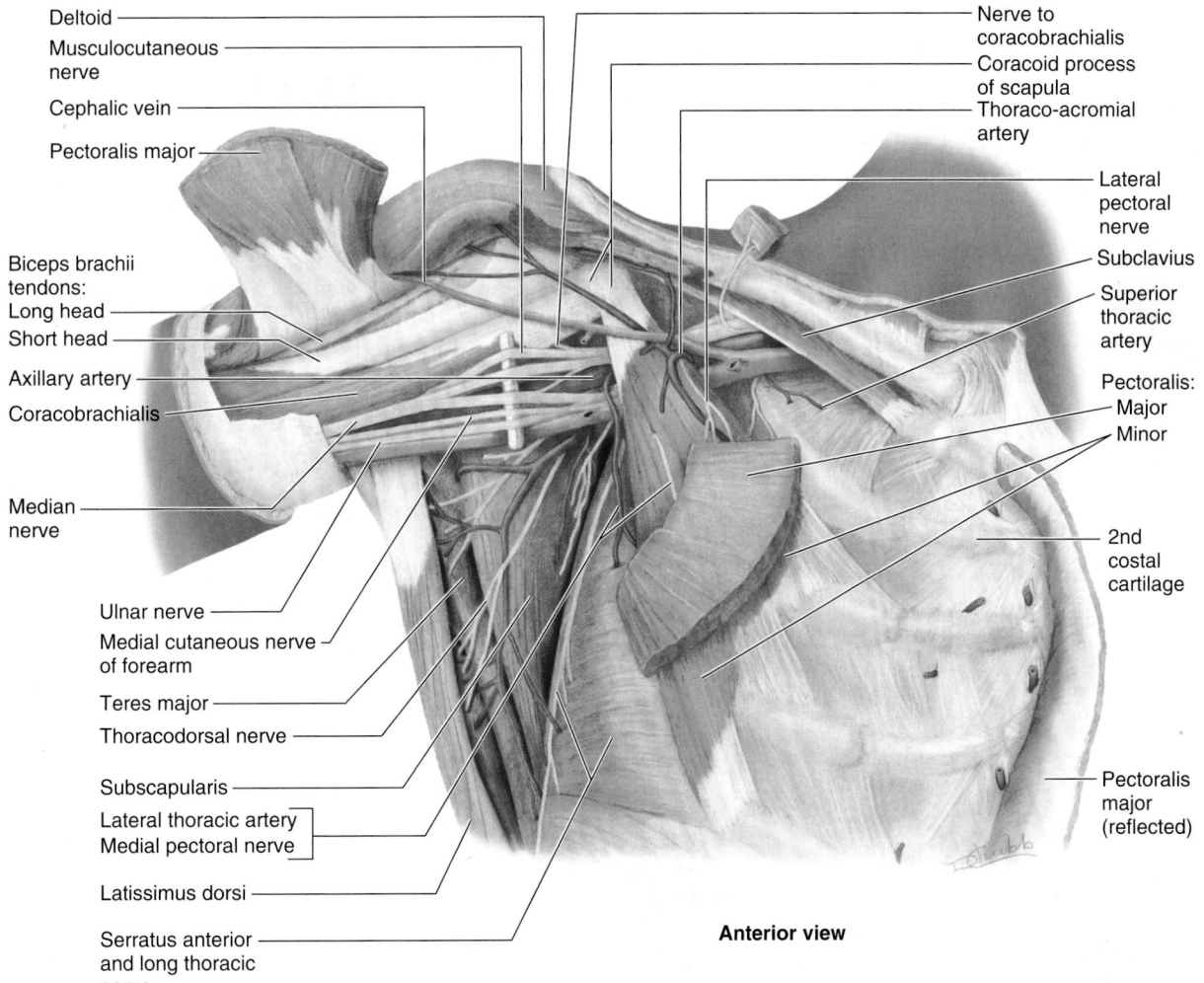

Deltoid
Musculocutaneous nerve
Cephalic vein
Pectoralis major
Biceps brachii tendons:
Long head
Short head
Axillary artery
Coracobrachialis
Median nerve
Ulnar nerve
Medial cutaneous nerve of forearm
Teres major
Thoracodorsal nerve
Subscapularis
Lateral thoracic artery
Medial pectoral nerve
Latissimus dorsi
Serratus anterior and long thoracic nerve

Nerve to coracobrachialis
Coracoid process of scapula
Thoraco-acromial artery
Lateral pectoral nerve
Subclavius
Superior thoracic artery
Pectoralis:
Major
Minor
2nd costal cartilage
Pectoralis major (reflected)

Anterior view

FIGURE 6.22. Axio-appendicular muscles contributing to walls of axilla. Of the anterior axio-appendicular muscles forming the anterior wall, only portions of the pectoralis major (attaching ends, a central part overlying the pectoralis minor, and a cube of muscle reflected superior to the clavicle), the pectoralis minor, and the subclavius remain. All the clavipectoral fascia and axillary fat have been removed, as has the axillary sheath surrounding the neurovascular bundle. This enables observation of the medial wall of the axilla, formed by the serratus anterior overlying the lateral thoracic wall, and of the latissimus dorsi contributing to the posterior wall.

The pectoralis minor stabilizes the scapula and is used when stretching the upper limb forward to touch an object that is just out of reach. It also assists in elevating the ribs for deep inspiration when the pectoral girdle is fixed or elevated. The pectoralis minor is a useful anatomical and surgical landmark for structures in the axilla (e.g., the axillary artery). With the coracoid process, the pectoralis minor forms a "bridge" under which vessels and nerves must pass to the arm.

The **subclavius** lies almost horizontally when the arm is in the anatomical position (Figs. 6.21C and 6.22). This small, round muscle is located inferior to the clavicle and affords some protection to the subclavian vessels and the superior trunk of the brachial plexus if the clavicle fractures. The subclavius anchors and depresses the clavicle, stabilizing it during movements of the upper limb. It also helps resist the tendency for the clavicle to dislocate at the sternoclavicular (SC) joint—for example, when pulling hard during a tug-of-war game.

The **serratus anterior** overlies the lateral part of the thorax and forms the medial wall of the axilla (Fig. 6.21D). This broad sheet of thick muscle was named because of the sawtoothed appearance of its fleshy slips or digitations (L. *serratus*, a saw). The muscular slips pass posteriorly and then medially to attach to the whole length of the anterior surface of the medial border of the scapula, including its inferior angle. The serratus anterior is one of the most powerful muscles of the pectoral girdle. It is a strong protractor of the scapula and is used when punching or reaching anteriorly (sometimes called the "boxer's muscle").

The strong inferior part of the serratus anterior rotates the scapula, elevating its glenoid cavity so the arm can be raised above the shoulder. It also anchors the scapula, keeping it closely applied to the thoracic wall, enabling other muscles to use it as a fixed bone for movements of the humerus. The serratus anterior holds the scapula against the thoracic wall when doing push-ups, or when pushing against resistance (e.g., pushing a car).

To test the serratus anterior (or the function of the long thoracic nerve that supplies it), the hand of the outstretched limb is pushed against a wall. If the muscle is acting normally, several digitations of the muscle can be seen and palpated.

Posterior Axio-appendicular and Scapulohumeral Muscles

The **posterior axio-appendicular muscles** (superficial and intermediate groups of extrinsic back muscles) attach the superior appendicular skeleton (of the upper limb) to the axial skeleton (in the trunk). *The intrinsic back muscles, which maintain posture and control movements of the vertebral column, are described on p. 485.* The posterior shoulder muscles are divided into three groups (Table 6.4):

- *Superficial posterior axio-appendicular (extrinsic shoulder) muscles:* trapezius and latissimus dorsi.
- *Deep posterior axio-appendicular (extrinsic shoulder) muscles:* levator scapulae and rhomboids.
- *Scapulohumeral (intrinsic shoulder) muscles:* deltoid, teres major, and the four rotator cuff muscles (supraspinatus, infraspinatus, teres minor, and subscapularis).

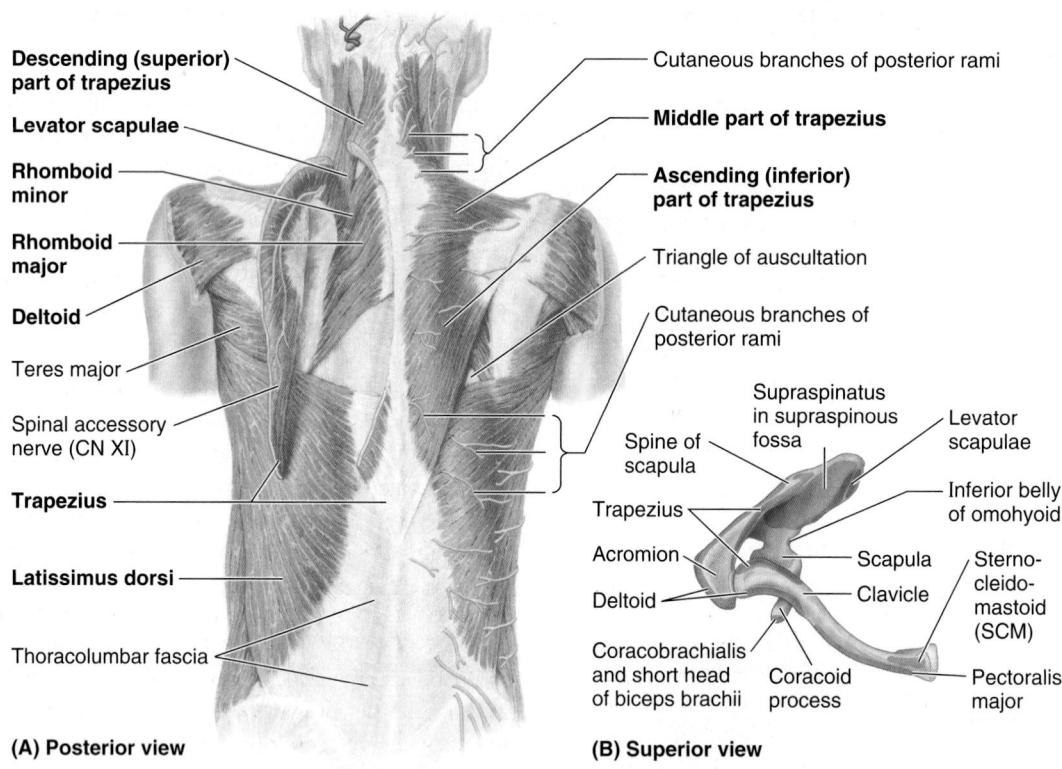

(A) Posterior view

Descending (superior) part of trapezius
Levator scapulae
Rhomboid minor
Rhomboid major
Deltoid
Teres major
Spinal accessory nerve (CN XI)
Trapezius
Latissimus dorsi
Thoracolumbar fascia

Cutaneous branches of posterior rami
Middle part of trapezius
Ascending (inferior) part of trapezius
Triangle of auscultation
Cutaneous branches of posterior rami

(B) Superior view

Spine of scapula
Supraspinatus in supraspinous fossa
Levator scapulae
Trapezius
Inferior belly of omohyoid
Acromion
Scapula
Clavicle
Sterno-cleido-mastoid (SCM)
Deltoid
Coracobrachialis and short head of biceps brachii
Coracoid process
Pectoralis major

FIGURE 6.23. Posterior axio-appendicular muscles.

TABLE 6.4. POSTERIOR AXIO-APPENDICULAR MUSCLES

Muscle	Proximal Attachment	Distal Attachment	Innervation[a]	Muscle Action
Superficial posterior axio-appendicular (extrinsic shoulder) muscles				
Trapezius	Medial third of superior nuchal line; external occipital protuberance; nuchal ligament; spinous processes of C7–T12 vertebrae	Lateral third of clavicle; acromion and spine of scapula	Spinal accessory nerve (CN XI) (motor fibers) and C3, C4 spinal nerves (pain and proprioceptive fibers)	Descending part elevates; ascending part depresses; and middle part (or all parts together) retracts scapula; descending and ascending parts act together to rotate glenoid cavity superiorly
Latissimus dorsi	Spinous processes of inferior 6 thoracic vertebrae, thoraco-lumbar fascia, iliac crest, and inferior 3 or 4 ribs	Floor of intertubercular sulcus of humerus	Thoracodorsal nerve (**C6, C7**, C8)	Extends, adducts, and medially rotates humerus; raises body toward arms during climbing

(continued)

TABLE 6.4. POSTERIOR AXIO-APPENDICULAR MUSCLES (Continued)

Muscle	Proximal Attachment	Distal Attachment	Innervation[a]	Muscle Action
Deep posterior axio-appendicular (extrinsic shoulder) muscles				
Levator scapulae	Posterior tubercles of transverse processes of C1–C4 vertebrae	Medial border of scapula superior to root of scapular spine	Dorsal scapular (C4, C5) and cervical (C3, C4) nerves	Elevates scapula and rotates its glenoid cavity inferiorly by rotating scapula
Rhomboid minor and major	Minor: nuchal ligament; spinous processes of C7 and T1 vertebrae Major: spinous processes of T2–T5 vertebrae	Minor: smooth triangular area at medial end of scapular spine Major: medial border of scapula from level of spine to inferior angle	Dorsal scapular nerve (C4, **C5**)	Retract scapula and rotate its glenoid cavity inferiorly; fix scapula to thoracic wall

[a]The spinal cord segmental innervation is indicated (e.g., "C4, **C5**" means that the nerves supplying the rhomboids are derived from the fourth and fifth cervical segments of the spinal cord). Numbers in boldface (**C5**) indicate the main segmental innervation. Damage to one or more of the listed spinal cord segments or to the motor nerve roots arising from them results in paralysis of the muscles concerned.

SUPERFICIAL POSTERIOR AXIO-APPENDICULAR (EXTRINSIC SHOULDER) MUSCLES

The superficial axio-appendicular muscles are the trapezius and latissimus dorsi. These muscles are illustrated in Figure 6.23, and their attachments, nerve supply, and main actions are listed in Table 6.4.

Trapezius. The **trapezius** provides a direct attachment of the pectoral girdle to the trunk. This large, triangular muscle covers the posterior aspect of the neck and the superior half of the trunk (Fig. 6.24). It was given its name because the muscles of the two sides form a *trapezium* (G. irregular four-sided figure). The trapezius attaches the pectoral girdle to the cranium and vertebral column, and assists in suspending the upper limb. The fibers of the trapezius are divided into three parts, which have different actions at the physiological scapulothoracic joint between the scapula and thoracic wall (Fig. 6.25; Table 6.5):

- Descending (superior) fibers elevate the scapula (e.g., when squaring the shoulders).
- Middle fibers retract the scapula (i.e., pull it posteriorly).
- Ascending (inferior) fibers depress the scapula and lower the shoulder.

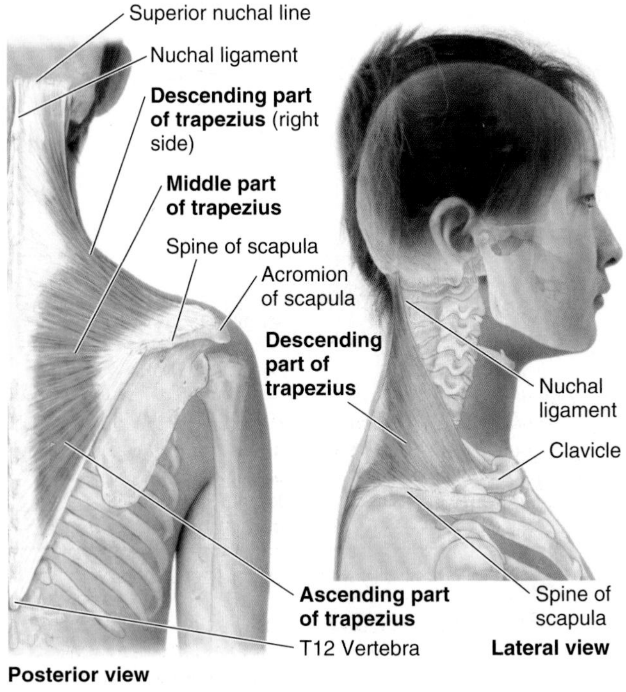

Descending and ascending trapezius fibers act together in rotating the scapula on the thoracic wall in different directions, twisting it like a wing nut. The trapezius also braces the shoulders by pulling the scapulae posteriorly and superiorly, fixing them in position on the thoracic wall with tonic contraction; consequently, weakness of the trapezius causes drooping of the shoulders.

To test the trapezius (or the function of the spinal accessory nerve [CN XI] that supplies it), the shoulder is shrugged against resistance (the person attempts to raise the shoulders as the examiner presses down on them). If the muscle is acting normally, the superior border of the muscle can be easily seen and palpated.

Latissimus Dorsi. The name **latissimus dorsi** (L. widest of back) was well chosen because this muscle covers a wide area of the back (Figs. 6.23 and 6.26; Table 6.4). This large, fan-shaped muscle passes from the trunk to the humerus, and acts directly on the glenohumeral joint and indirectly on the pectoral girdle (scapulothoracic joint). The latissimus dorsi extends, retracts, and rotates the humerus medially (e.g., when folding your arms behind your back, or scratching the skin over the opposite scapula).

In combination with the pectoralis major, the latissimus dorsi is a powerful adductor of the humerus, and plays a major role in downward rotation of the scapula in association with this movement (Fig. 6.25, Table 6.5). It is also useful in restoring the upper limb from abduction superior to the shoulder; hence the latissimus dorsi is important in climbing. In conjunction with the pectoralis major, the latissimus dorsi raises the trunk to the arm, which occurs when performing

Superior nuchal line
Nuchal ligament
Descending part of trapezius (right side)
Middle part of trapezius
Spine of scapula
Acromion of scapula
Descending part of trapezius
Nuchal ligament
Clavicle
Ascending part of trapezius
T12 Vertebra
Spine of scapula
Lateral view
Posterior view

FIGURE 6.24. Trapezius.

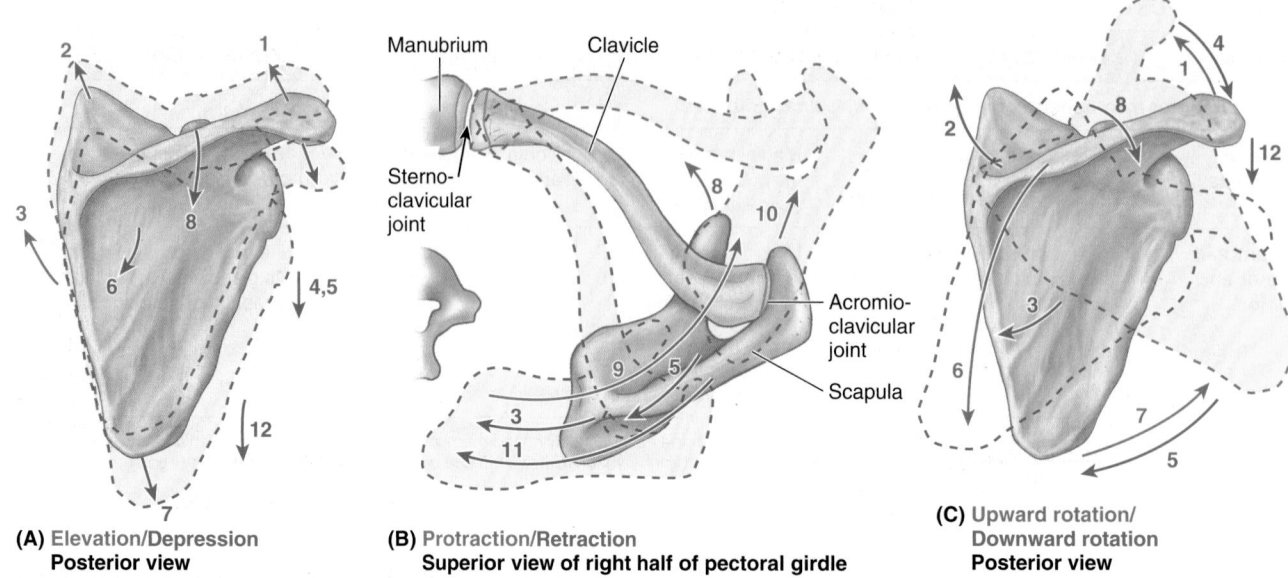

FIGURE 6.25. Movements of scapula and muscles producing them. *Arrows* indicate the direction of pull; the muscles (and gravity) producing each movement are identified by *numbers,* which are listed in Table 6.5.

TABLE 6.5. MOVEMENTS OF SCAPULA

Movement of Scapula	Muscles Producing Movement[a]	Nerve to Muscles	Range of Movement (Angular Rotation; Linear Displacement)
Elevation	**Trapezius, descending part** (1) Levator scapulae (2) Rhomboids (3)	Spinal accessory (CN XI) } Dorsal scapular	10–12 cm
Depression	**Gravity** (12) Pectoralis major, inferior sternocostal head (4) Latissimus dorsi (5) Trapezius, ascending part (6) Serratus anterior, inferior part (7) Pectoralis minor (8)	Pectoral nerves Thoracodorsal Spinal accessory (CN XI) Long thoracic Medial pectoral	
Protraction	**Serratus anterior** (9) Pectoralis major (10) Pectoralis minor (8)	Long thoracic Pectoral nerves Medial pectoral	40–45°; 15 cm
Retraction	**Trapezius, middle part** (11) Rhomboids (3) Latissimus dorsi (5)	Spinal accessory (CN XI) Dorsal scapular Thoracodorsal	
Upward rotation[b]	**Trapezius, descending part** (1) Trapezius, ascending part (6) **Serratus anterior, inferior part** (7)	} Spinal accessory (CN XI) Long thoracic	
Downward rotation[c]	Gravity (12) Levator scapulae (2) Rhomboids (3) **Latissimus dorsi** (5) Pectoralis minor (8) Pectoralis major, inferior sternocostal head (4)	} Dorsal scapular Thoracodorsal Medial pectoral Medial and lateral pectoral nerves	60°; inferior angle: 10–12 cm, superior angle: 5–6 cm

[a]Boldface indicates prime or essential mover(s). Numbers refer to Figure 6.25.

[b]The glenoid cavity moves superiorly, as in abduction of the arm.

[c]The glenoid cavity moves inferiorly, as in adduction of the arm.

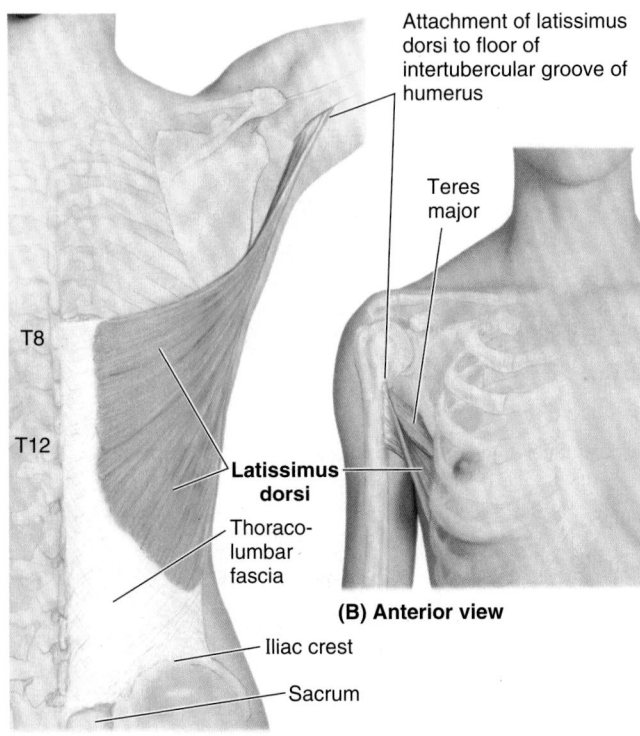

(B) Anterior view

(A) Posterior view

FIGURE 6.26. Latissimus dorsi. **A.** Proximal attachments. **B.** Distal attachment. See Table 6.4 for details.

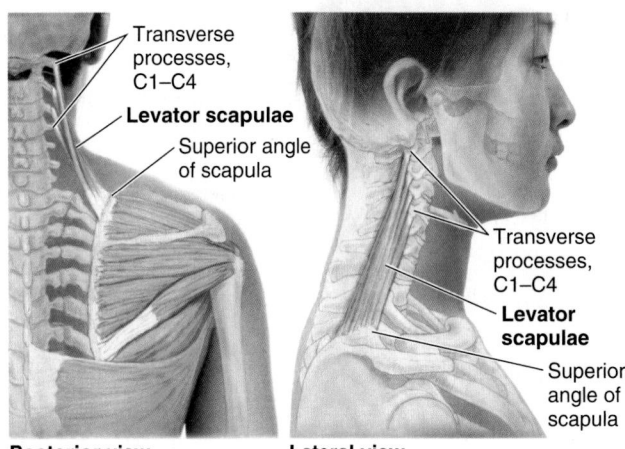

Posterior view   **Lateral view**

FIGURE 6.27. Levator scapulae.

chin-ups (hoisting oneself so the chin touches an overhead bar) or climbing a tree, for example. These movements are also used when chopping wood, paddling a canoe, and swimming (particularly during the crawl stroke).

To test the latissimus dorsi (or the function of the thoracodorsal nerve that supplies it), the arm is abducted 90° and then adducted against resistance provided by the examiner. If the muscle is normal, the anterior border of the muscle can be seen and easily palpated in the posterior axillary fold (see "Axilla" on p. 713).

DEEP POSTERIOR AXIO-APPENDICULAR (EXTRINSIC SHOULDER) MUSCLES

The **deep posterior axio-appendicular** (axio-scapular or thoraco-appendicular) **muscles** are the levator scapulae and rhomboids. These muscles provide direct attachment of the appendicular skeleton to the axial skeleton. The attachments, nerve supply, and main actions are given in Table 6.4, p.701.

Levator Scapulae. The superior third of the strap-like **levator scapulae** lies deep to the sternocleidomastoid; the inferior third is deep to the trapezius. From the transverse processes of the upper cervical vertebrae, the fibers of the levator of the scapula pass inferiorly to the superomedial border of the scapula (Figs. 6.23 and 6.27, Table 6.4). True to its name, the levator scapulae acts with the descending part of the trapezius to elevate the scapula, or fix it (resists forces that would depress it, as when carrying a load) (Fig. 6.25, Table 6.5).

With the rhomboids and pectoralis minor, the levator scapulae rotates the scapula, depressing the glenoid cavity (tilting it inferiorly by rotating the scapula). Acting bilaterally (also with the trapezius), the levators extend the neck; acting unilaterally, the muscle may contribute to lateral flexion of the neck (toward the side of the active muscle).

Rhomboids. The **rhomboids** (major and minor), which are not always clearly separated from each other, have a rhomboid appearance—that is, they form an oblique equilateral parallelogram (Figs. 6.23 and 6.28, Table 6.4). The rhomboids lie deep to the trapezius, and form broad parallel bands that pass inferolaterally from the vertebrae to the medial border of the scapula. The thin, flat **rhomboid major** is approximately two times wider than the thicker **rhomboid minor** lying superior to it.

The rhomboids retract and rotate the scapula, depressing its glenoid cavity (Table 6.5). They also assist the serratus anterior in holding the scapula against the thoracic

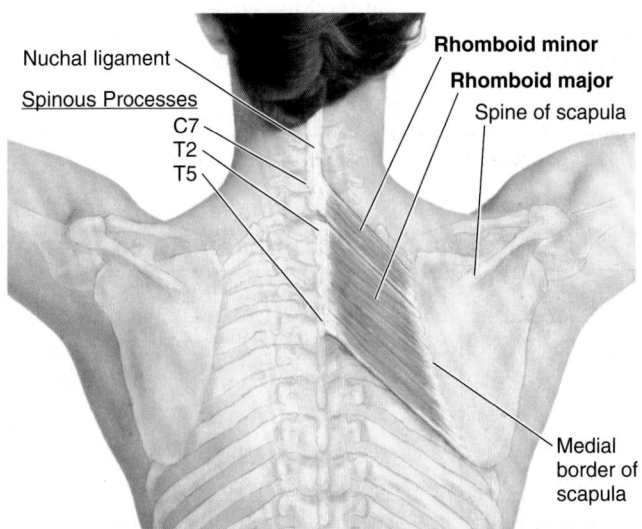

Posterior view

FIGURE 6.28. Rhomboids.

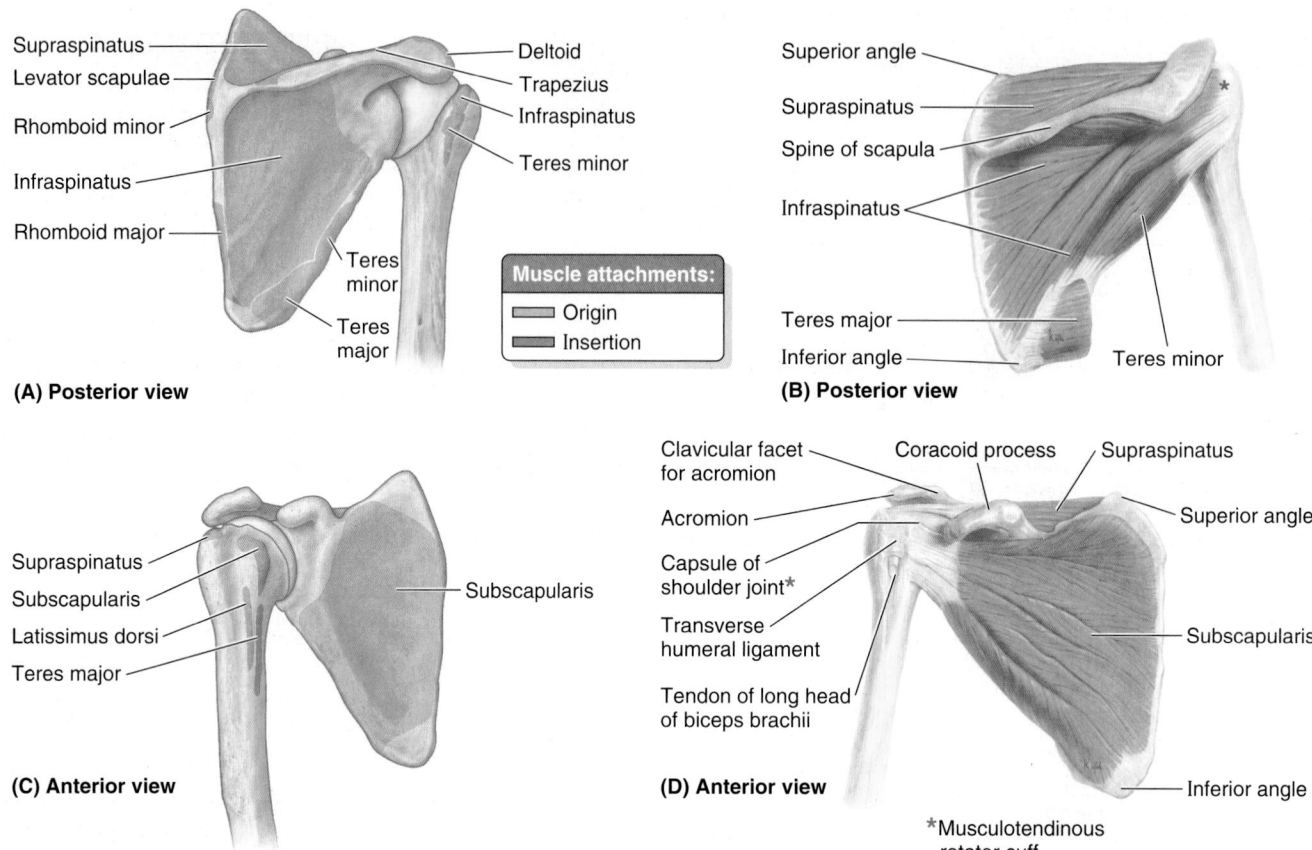

FIGURE 6.29. Scapulohumeral muscles. A–D. These muscles pass from the scapula to the humerus and act on the glenohumeral joint. Not included here is the deltoid, featured in Figure 6.30.

wall, and fixing the scapula during movements of the upper limb. The rhomboids are used when forcibly lowering the raised upper limbs (e.g., when driving a stake with a sledge hammer).

To test the rhomboids (or the function of the dorsal scapular nerve that supplies them), the individual places his or her hands posteriorly on the hips, and pushes the elbows posteriorly against resistance provided by the examiner. If the rhomboids are acting normally, they can be palpated along the medial borders of the scapulae; because they lie deep to the trapezius, they are unlikely to be visible during testing.

Scapulohumeral (Intrinsic Shoulder) Muscles

The **six scapulohumeral muscles** (deltoid, teres major, supraspinatus, infraspinatus, subscapularis, and teres minor) are relatively short muscles that pass from the scapula to the humerus and act on the glenohumeral joint. These muscles are illustrated in Figures 6.23 and 6.29, and their attachments, nerve supply, and main actions are summarized in Table 6.6.

Deltoid. The **deltoid** is a thick, powerful, coarse-textured muscle covering the shoulder and forming its rounded contour (Figs. 6.23 and 6.30; Table 6.6). As its name indicates,

TABLE 6.6. SCAPULOHUMERAL (INTRINSIC SHOULDER) MUSCLES

Muscle	Proximal Attachment	Distal Attachment	Innervation[a]	Muscle Action
Deltoid	Lateral third of clavicle; acromion and spine of scapula	Deltoid tuberosity of humerus	Axillary nerve (**C5**, C6)	Clavicular (anterior) part: flexes and medially rotates arm Acromial (middle) part: abducts arm Spinal (posterior) part: extends and laterally rotates arm
Supraspinatus[b]	Supraspinous fossa of scapula	Superior facet of greater tubercle of humerus	Suprascapular nerve (C4, **C5**, C6)	Initiates and assists deltoid in abduction of arm and acts with rotator cuff muscles[b]

TABLE 6.6. SCAPULOHUMERAL (INTRINSIC SHOULDER) MUSCLES (Continued)

Muscle	Proximal Attachment	Distal Attachment	Innervation[a]	Muscle Action
Infraspinatus[b]	Infraspinous fossa of scapula	Middle facet of greater tubercle of humerus	Suprascapular nerve (**C5**, C6)	Laterally rotates arm; and acts with rotator cuff muscles[b]
Teres minor[b]	Middle part of lateral border of scapula	Inferior facet of greater tubercle of humerus	Axillary nerve (**C5**, C6)	Laterally rotates arm; and acts with rotator cuff muscles[b]
Teres major	Posterior surface of inferior angle of scapula	Medial lip of inter-tubercular sulcus of humerus	Lower subscapular nerve (C5, **C6**)	Adducts and medially rotates arm
Subscapularis[b]	Subscapular fossa (most of anterior surface of scapula)	Lesser tubercle of humerus	Upper and lower subscapular nerves (C5, **C6**, C7)	Medially rotates arm; as part of rotator cuff, helps hold head of humerus in glenoid cavity

[a]The spinal cord segmental innervation is indicated (e.g., "**C5**, C6" means that the nerves supplying the deltoid are derived from the fifth and sixth cervical segments of the spinal cord). Numbers in boldface (**C5**) indicate the main segmental innervation. Damage to one or more of the listed spinal cord segments or to the motor nerve roots arising from them results in paralysis of the muscles concerned.

[b]Collectively, the supraspinatus, infraspinatus, teres minor, and subscapularis muscles are referred to as the rotator cuff, or SITS, muscles. Their primary function during all movements of the glenohumeral (shoulder) joint is to hold the humeral head in the glenoid cavity of the scapula.

the deltoid is shaped like the inverted Greek letter delta (Δ). The muscle is divided into unipennate anterior and posterior parts and a multipennate middle part (see Introduction, p. 31). The parts of the deltoid can act separately or as a whole. When all three parts of the deltoid contract simultaneously, the arm is abducted. The anterior and posterior parts act like guy ropes to steady the arm as it is abducted.

To initiate movement during the first 15° of abduction, the deltoid is assisted by the supraspinatus (Fig. 6.29B). When the arm is fully adducted, the line of pull of the deltoid coincides with the axis of the humerus; thus, it pulls directly upward on the bone and cannot initiate or produce abduction. It is, however, able to act as a shunt muscle, resisting inferior displacement of the head of the humerus from the

glenoid cavity, as when lifting and carrying suitcases. From the fully adducted position, abduction must be initiated by the supraspinatus, or by leaning to the side, allowing gravity to initiate the movement. The deltoid becomes fully effective as an abductor after the initial 15° of abduction.

The anterior and posterior parts of the deltoids are used to swing the limbs during walking. The anterior part assists the pectoralis major in flexing the arm, and the posterior part assists the latissimus dorsi in extending the arm. The deltoid also helps stabilize the glenohumeral joint and hold the head of the humerus in the glenoid cavity during movements of the upper limb.

To test the deltoid (or the function of the axillary nerve that supplies it), the arm is abducted, starting from approximately 15°, against resistance (Fig. 6.31). If acting normally, the deltoid can easily be seen and palpated. The influence of gravity is avoided when the person is supine.

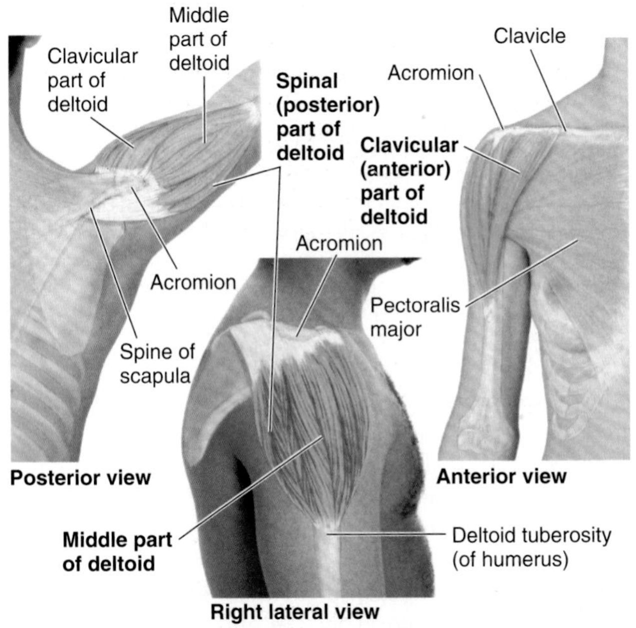

Middle part of deltoid

Clavicular part of deltoid

Acromion

Spine of scapula

Posterior view

Spinal (posterior) part of deltoid

Clavicular (anterior) part of deltoid

Acromion

Clavicle

Acromion

Pectoralis major

Anterior view

Middle part of deltoid

Deltoid tuberosity (of humerus)

Right lateral view

FIGURE 6.30. Deltoid.

FIGURE 6.31. Testing deltoid muscle. The examiner resists the patient's abduction of the limb by the deltoid. If the deltoid is acting normally, contraction of the middle part of the muscle can be palpated.

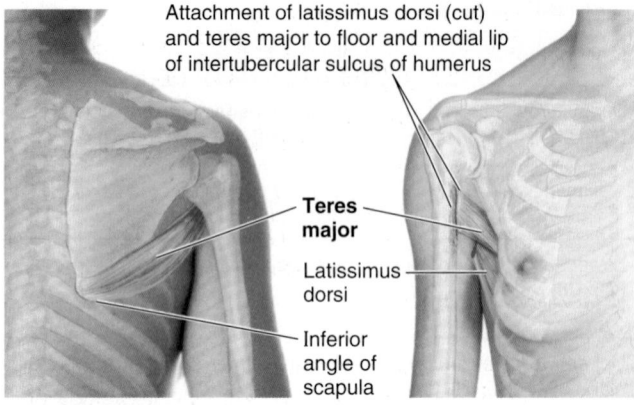

Attachment of latissimus dorsi (cut) and teres major to floor and medial lip of intertubercular sulcus of humerus

Teres major

Latissimus dorsi

Inferior angle of scapula

FIGURE 6.32. Teres major.

Teres Major. The **teres major** (L. *teres*, round) is a thick, rounded muscle passing laterally from the inferolateral third of the scapula (Figs. 6.23, 6.29A & B, and 6.32; Table 6.6). The inferior border of the teres major forms the inferior border of the lateral part of the posterior wall of the axilla. The teres major adducts and medially rotates the arm. It can also help extend it from the flexed position, and is an important stabilizer of the humeral head in the glenoid cavity—that is, it steadies the head in its socket.

To test the teres major (or the lower subscapular nerve that supplies it), the abducted arm is adducted against resistance. If acting normally, the muscle can be easily seen and palpated in the posterior axillary fold (Fig. 6.34).

ROTATOR CUFF MUSCLES

Four of the scapulohumeral muscles (intrinsic shoulder muscles)—**s**upraspinatus, **i**nfraspinatus, **t**eres minor, and subscapularis the *SITS muscles*)—are called **rotator cuff muscles** because they form a *musculotendinous rotator cuff* around the glenohumeral joint (Figs. 6.29B & D and 6.33). All except the supraspinatus are rotators of the humerus; the supraspinatus, besides being part of the rotator cuff, initiates and assists the deltoid in the first 15° of abduction of the arm.

The tendons of the SITS muscles blend with and reinforce the fibrous layer of the joint capsule of the glenohumeral joint, thus forming the rotator cuff that protects the joint and gives it stability. The tonic contraction of the contributing muscles holds the relatively large head of the humerus in the small, shallow glenoid cavity of the scapula during arm movements. The rotator cuff (SITS) muscles and their attachments are illustrated in Figure 6.29, and their attachments, nerve supply, and main actions are listed in Table 6.6.

Supraspinatus. The **supraspinatus** occupies the supraspinous fossa of the scapula (Figs. 6.5A, 6.29A & B and 6.33A). A bursa separates it from the lateral quarter of the fossa. (See "Deltoid" on p. 704 for a discussion of this muscle's cooperative action in abducting the upper limb.)

To test the supraspinatus, abduction of the arm is attempted from the fully adducted position against resistance, while the muscle is palpated superior to the spine of the scapula.

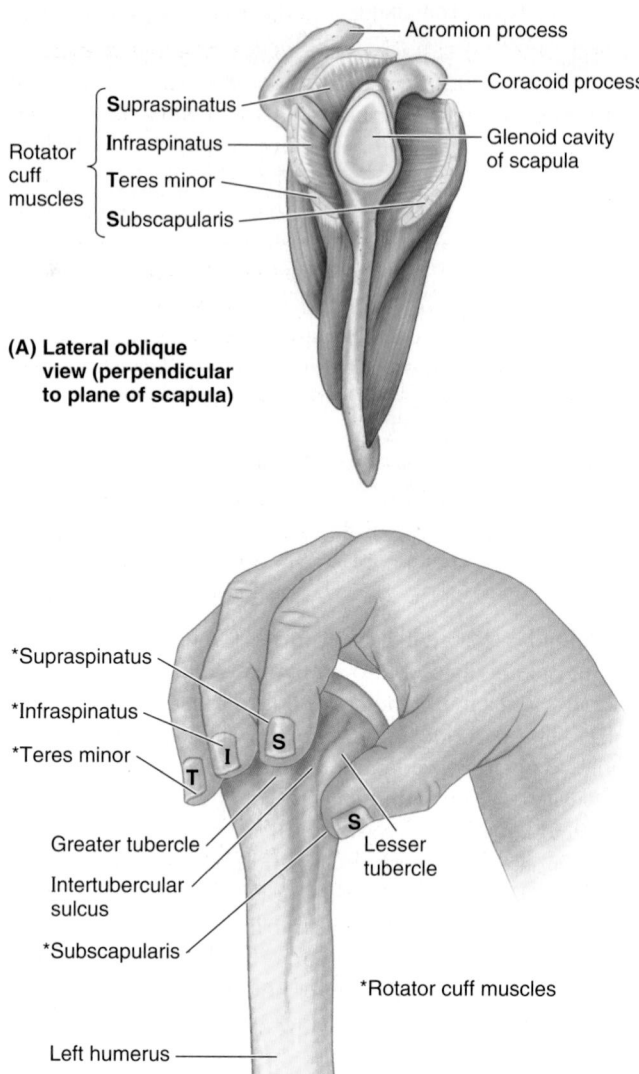

Acromion process

Coracoid process

Rotator cuff muscles
- **S**upraspinatus
- **I**nfraspinatus
- **T**eres minor
- **S**ubscapularis

Glenoid cavity of scapula

(A) Lateral oblique view (perpendicular to plane of scapula)

*Supraspinatus

*Infraspinatus

*Teres minor

Greater tubercle

Intertubercular sulcus

*Subscapularis

Left humerus

Lesser tubercle

*Rotator cuff muscles

(B) Right anterolateral view

FIGURE 6.33. Disposition of rotator cuff muscles. A. Coming from opposite sides and three separate fossae on the scapulae, the four rotator cuff (SITS) muscles pass laterally to engulf the head of the humerus. **B.** The primary combined function of the four SITS muscles is to "grasp" and pull the relatively large head of the humerus medially, holding it against the smaller, shallow glenoid cavity of the scapula. The tendons of the muscles (represented by three fingers and the thumb) blend with the fibrous layer of the capsule of the shoulder joint to form a musculotendinous rotator cuff, which reinforces the capsule on three sides (anteriorly, superiorly, and posteriorly) as it provides active support for the joint.

Infraspinatus. The **infraspinatus** occupies the medial three quarters of the infraspinous fossa (Fig. 6.5A) and is partly covered by the deltoid and trapezius. In addition to helping stabilize the glenohumeral joint, the infraspinatus is a powerful lateral rotator of the humerus.

To test the infraspinatus, the person flexes the elbow and adducts the arm. The arm is then laterally rotated against resistance. If acting normally, the muscle can be palpated inferior to the scapular spine. *To test the function of the*

suprascapular nerve, which supplies the supraspinatus and infraspinatus, both muscles must be tested as described.

Teres Minor. The **teres minor** is a narrow, elongate muscle that is completely hidden by the deltoid, and is often not clearly delineated from the infraspinatus. The teres minor works with the infraspinatus to rotate the arm laterally and assist in its adduction. The teres minor is most clearly distinguished from the infraspinatus by its nerve supply. The teres minor is supplied by the axillary nerve, whereas the infraspinatus is supplied by the suprascapular nerve (Table 6.6).

Subscapularis. The **subscapularis** is a thick, triangular muscle that lies on the costal surface of the scapula, and forms part of the posterior wall of the axilla (Figs. 6.29C & D and 6.33A). It crosses the anterior aspect of the scapulohumeral joint on its way to the humerus. The subscapularis is the primary medial rotator of the arm and also adducts it. It joins the other rotator cuff muscles in holding the head of the humerus in the glenoid cavity during all movements of the glenohumeral joint (i.e., it helps stabilize this joint during movements of the elbow, wrist, and hand).

Surface Anatomy of Pectoral, Scapular, and Deltoid Regions

The *clavicle* is the boundary demarcating the root of the neck from the thorax. It also indicates the "divide" between the deep cervical and axillary "lymph sheds" (like a mountain range dividing watershed areas): Lymph from structures superior to the clavicles drain via the deep cervical nodes, and lymph from structures inferior to the clavicles, as far inferiorly as the umbilicus, drain via the axillary lymph nodes.

The **infraclavicular fossa** is the depressed area just inferior to the lateral part of the clavicle (Fig. 6.34). This depression overlies the **clavipectoral (deltopectoral) triangle**—bounded by the clavicle superiorly, the pectoralis major medially, and the deltoid laterally—which may be evident in the fossa in lean individuals. The *cephalic vein,* ascending from the upper limb, enters the clavipectoral triangle and pierces the clavipectoral fascia to enter the *axillary vein.* The *coracoid process* of the scapula is not subcutaneous; it is covered by the anterior border of the deltoid; however, the tip of the process can be felt on deep palpation on the lateral aspect of the clavipectoral triangle. The coracoid process is used as a bony landmark when performing a brachial plexus block, and its position is of importance in diagnosing shoulder dislocations.

While lifting a weight, palpate the anterior sloping border of the *trapezius,* and where its superior fibers attach to the lateral third of the clavicle. When the arm is abducted and then adducted against resistance, the *sternocostal part of the pectoralis major* can be seen and palpated. If the *anterior axillary fold* bounding the axilla is grasped between the fingers and thumb, the inferior border of the *sternocostal head of the pectoralis major* can be felt. Several digitations of the *serratus anterior* are visible inferior to the anterior axillary fold. The *posterior axillary fold* is composed of skin and muscular tissue (latissimus dorsi and teres major) bounding the axilla posteriorly.

The lateral border of the acromion may be followed posteriorly with the fingers until it ends at the *acromial angle* (Fig. 6.35A). Clinically, the *length of the arm* is measured from the acromial angle to the lateral condyle of the humerus. The *spine of the scapula* is subcutaneous

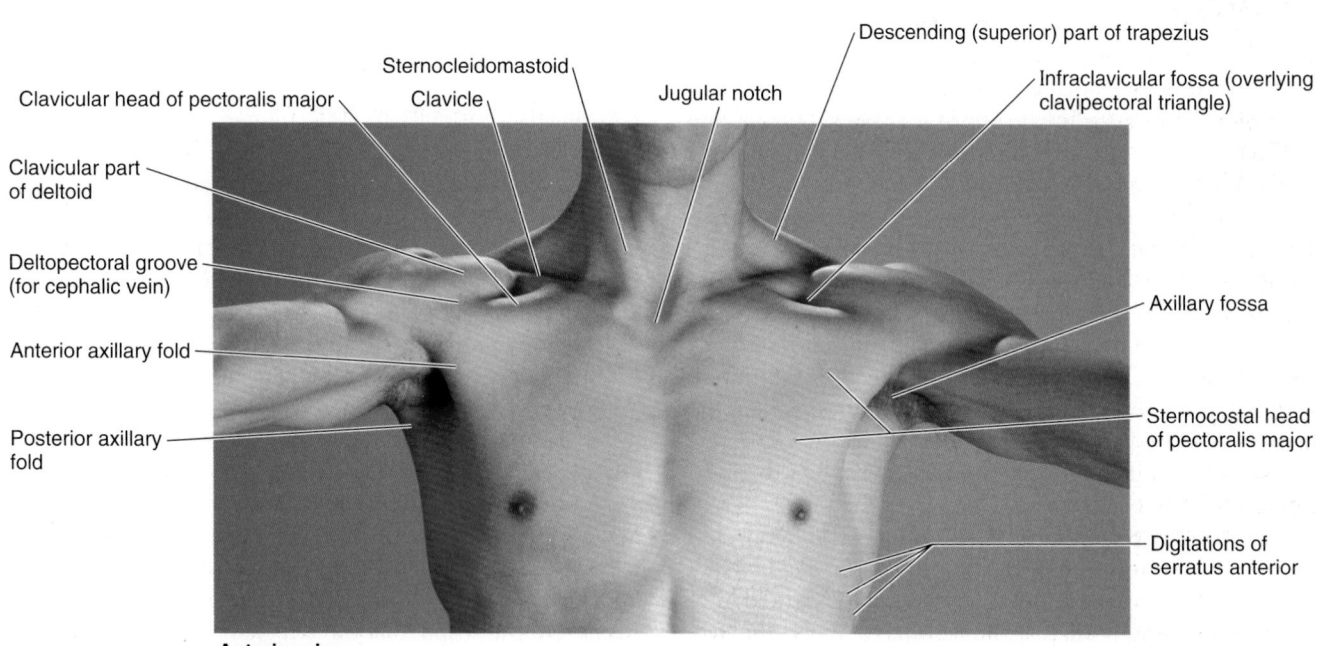

Anterior view

FIGURE 6.34. Surface anatomy of pectoral and deltoid regions.

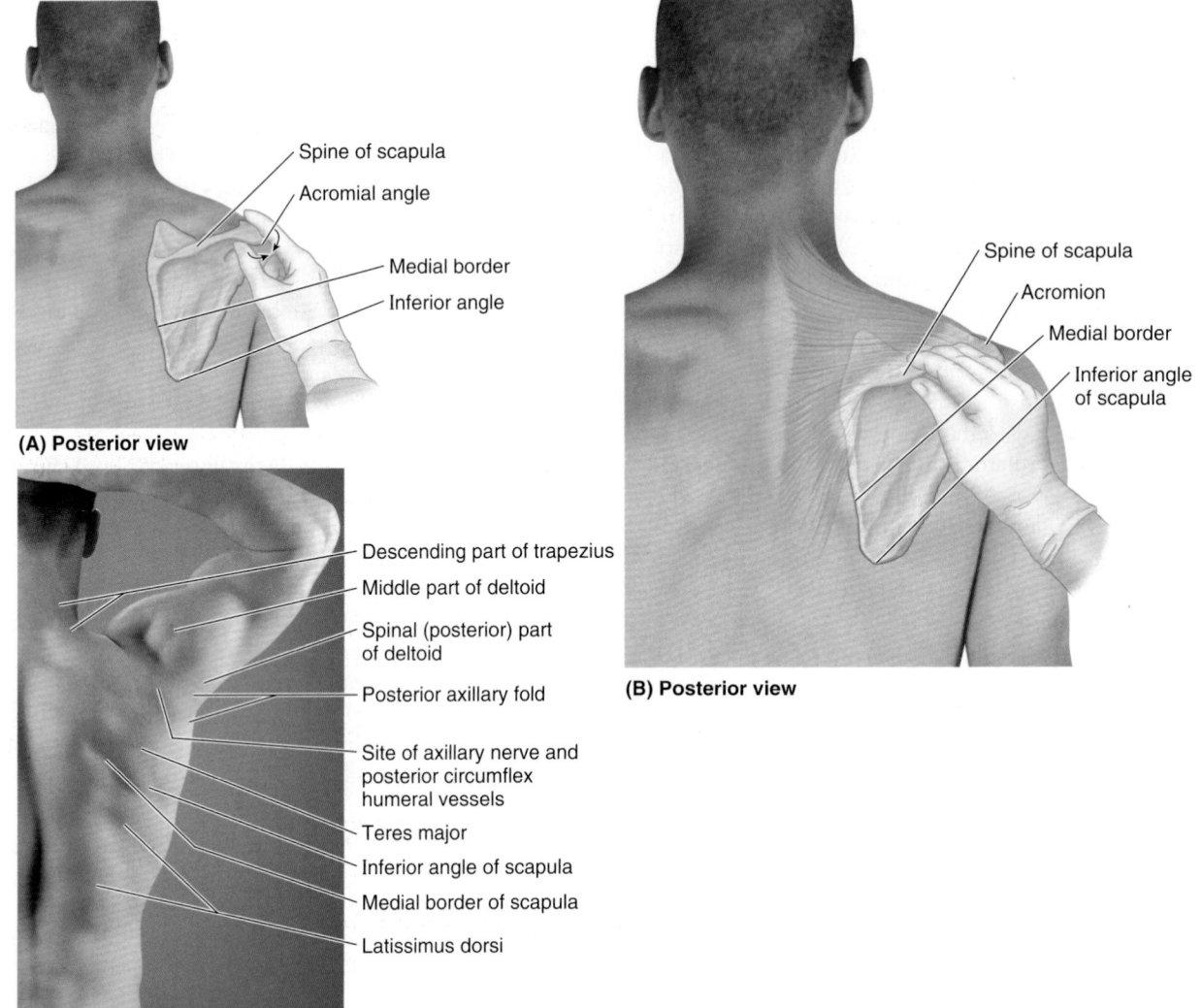

(A) Posterior view

- Spine of scapula
- Acromial angle
- Medial border
- Inferior angle

(B) Posterior view

- Spine of scapula
- Acromion
- Medial border
- Inferior angle of scapula

(C) Posterior view

- Descending part of trapezius
- Middle part of deltoid
- Spinal (posterior) part of deltoid
- Posterior axillary fold
- Site of axillary nerve and posterior circumflex humeral vessels
- Teres major
- Inferior angle of scapula
- Medial border of scapula
- Latissimus dorsi

FIGURE 6.35. Surface anatomy of scapula and scapular region.

throughout, and is easily palpated as it extends medially and slightly inferiorly from the acromion (Fig. 6.35B). The *root of the scapular spine* (medial end) is located opposite the tip of the T3 spinous process when the arm is adducted. The *medial border of the scapula* may be palpated inferior to the root of the spine as it crosses ribs 3–7 (Fig. 6.35C). It may be visible in some people, especially thin people. The *inferior angle of the scapula* is easily palpated and is usually visible. Grasp the inferior scapular angle with the thumb and fingers and move the scapula up and down. When the arm is adducted, the inferior scapular angle is opposite the tip of the T7 spinous process and lies over the 7th rib or intercostal space.

The **greater tubercle of the humerus** is the most lateral bony point in the shoulder when the arm is adducted, and may be felt on deep palpation through the deltoid inferior to the lateral border of the acromion. When the arm is abducted, observe that the greater tubercle disappears beneath the acromion and is no longer palpable. The **deltoid**

covering the proximal part of the humerus forms the rounded muscular contour of the shoulder. The borders and parts of the deltoid are usually visible when the arm is abducted against resistance (Fig. 6.36). Loss of the rounded muscular appearance of the shoulder, and the appearance of a surface depression distal to the acromion, are characteristic of a *dislocated shoulder*, or dislocation of the glenohumeral joint. The depression results from displacement of the humeral head. The **teres major** is prominent when the abducted arm is adducted and medially rotated against resistance (as when a gymnast stabilizes or fixes the shoulder joint during an iron cross maneuver on the rings).

When the upper limbs are abducted, the scapulae move laterally on the thoracic wall, enabling the **rhomboids** to be palpated. Because they are deep to the *trapezius*, the rhomboids are not always visible. If the rhomboids of one side are paralyzed, the scapula on the affected side remains farther from the midline than on the normal side because the paralyzed muscles are unable to retract it.

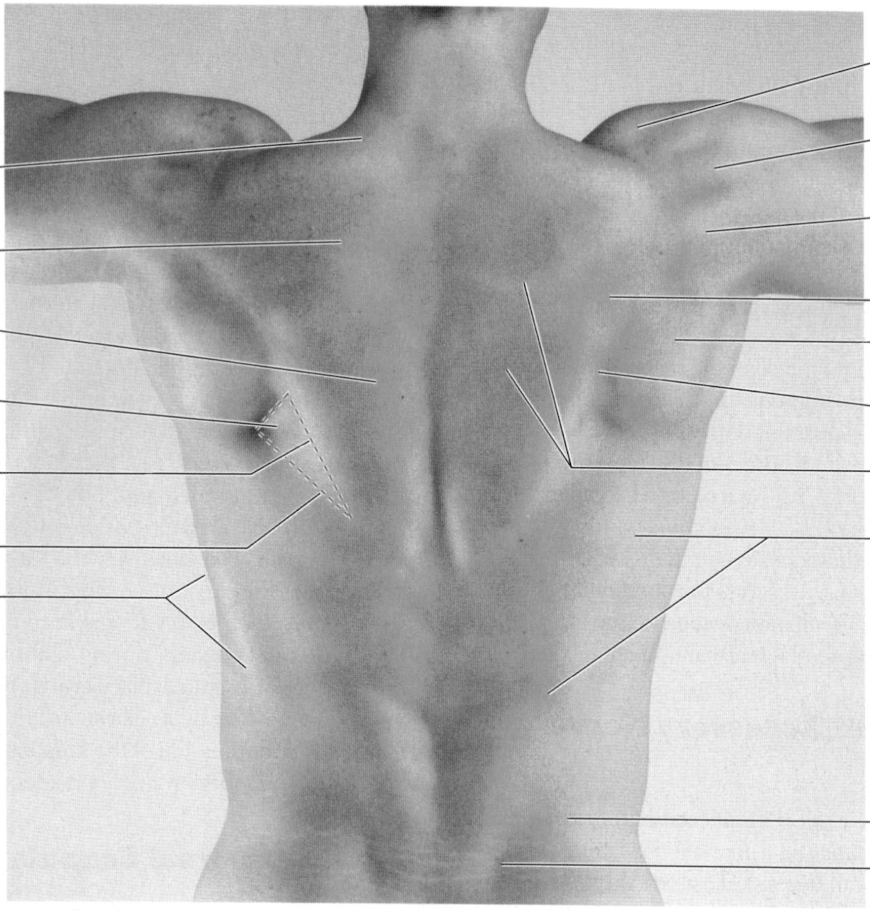

Trapezius {
Descending part
Middle part
Ascending part
}

Triangle of auscultation

Lateral border of trapezius

{ Superior border

Lateral border }

of latissiumus dorsi

Clavicular part
Middle part
Spinal part
} of deltoid

Spine of scapula

Teres major

Medial border of scapula

Rhomboids (deep to trapezius)

Latissimus dorsi

Iliac crest

Site of posterior superior iliac spine

Posterior view

FIGURE 6.36. Surface anatomy of posterior axio-appendicular and scapulohumeral muscles.

PECTORAL, SCAPULAR, AND DELTOID REGIONS

Absence of Pectoral Muscles

Absence of part of the pectoralis major, usually its sternocostal part, is uncommon, but when it occurs, no disability usually results. However, the anterior axillary fold, formed by the skin and fascia overlying the inferior border of the pectoralis major, is absent on the affected side, and the nipple is more inferior than usual. In *Poland syndrome*, both the pectoralis major and minor are absent; breast hypoplasia and absence of two to four rib segments are also seen.

Paralysis of Serratus Anterior

When the serratus anterior is paralyzed owing to *injury to the long thoracic nerve* (Fig. 6.22), the medial border of the scapula moves laterally and posteriorly away from the thoracic wall, giving the scapula the appearance of a wing, especially when the person leans on a hand or presses the upper limb against a wall. When the arm is raised, the medial border and inferior angle of the scapula pull markedly away from the posterior thoracic wall, a deformation known as a *winged scapula* (Fig. B6.5). In addition, the upper limb may not be able to be abducted above the horizontal position because the serratus anterior is unable to rotate the glenoid

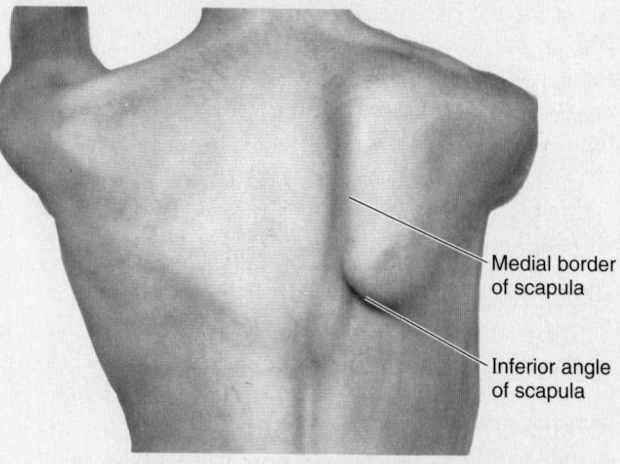

Medial border of scapula

Inferior angle of scapula

FIGURE B6.5. Right winged scapula

cavity superiorly to allow complete abduction of the limb. Remember, the trapezius also helps raise the arm above the horizontal. Although protected when the limbs are at one's sides, the long thoracic nerve is exceptional in that it courses on the superficial aspect of the serratus anterior, which it supplies. Thus when the limbs are elevated, as in a knife fight, the nerve is especially vulnerable. Weapons, including bullets directed toward the thorax, are a common source of injury.

Triangle of Auscultation

Near the inferior angle of the scapula is a small triangular gap in the musculature. The superior horizontal border of the latissimus dorsi, the medial border of the scapula, and the inferolateral border of the trapezius form a *triangle of auscultation* (Figs. 6.23 and 6.36). This gap in the thick back musculature is a good place to examine posterior segments of the lungs with a stethoscope. When the scapulae are drawn anteriorly by folding the arms across the chest and the trunk is flexed, the triangle of auscultation enlarges and parts of the 6th and 7th ribs and 6th intercostal space are subcutaneous.

Injury of Spinal Accessory Nerve (CN XI)

The primary clinical manifestation of *spinal accessory nerve palsy* is a marked ipsilateral weakness when the shoulders are elevated (shrugged) against

resistance. Injury of the spinal accessory nerve is discussed in greater detail in Chapters 8 and 9.

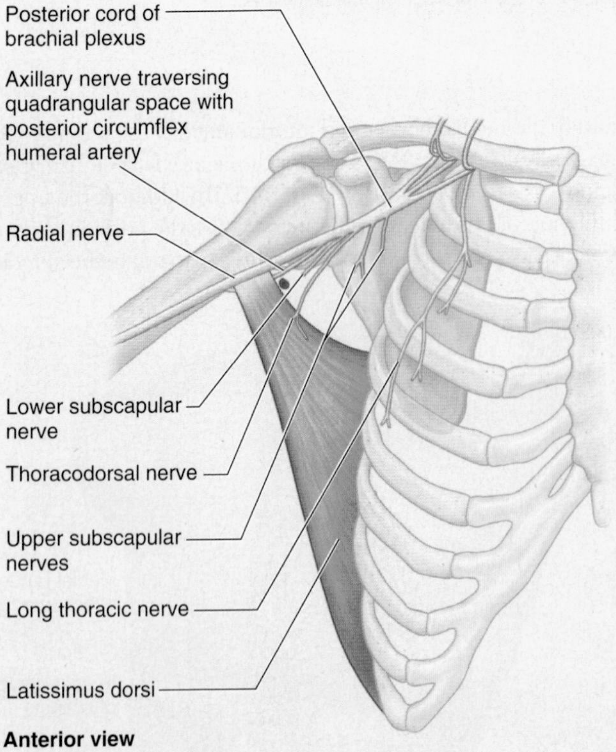

Injury of Thoracodorsal Nerve

Surgery in the inferior part of the axilla puts the thoracodorsal nerve (C6–C8), supplying the latissimus dorsi, at risk of injury. This nerve passes inferiorly along the posterior wall of the axilla, and enters the medial surface of the latissimus dorsi close to where it becomes tendinous (Fig. B6.6). The nerve is also vulnerable to injury during mastectomies when the axillary tail of the breast is removed (see p. 105). The nerve is also vulnerable during surgery on scapular lymph nodes because its terminal part lies anterior to them and the subscapular artery (Fig. B6.7).

The latissimus dorsi and the inferior part of the pectoralis major form an anteroposterior muscular sling between the trunk and the arm; however, the latissimus dorsi forms the more powerful part of the sling. With *paralysis of the latissimus dorsi,* the person is unable to raise the trunk with the upper limbs, as occurs during climbing. Furthermore, the person cannot use an axillary crutch because the shoulder is pushed superiorly by it. These are the primary activities for which active depression of the scapula is required; the passive depression provided by gravity is adequate for most activities.

Injury to Dorsal Scapular Nerve

Injury to the dorsal scapular nerve, the nerve to the rhomboids, affects the actions of these muscles. If the rhomboids on one side are paralyzed, the scapula on the affected side is located farther from the midline than that on the normal side.

Injury to Axillary Nerve

The deltoid atrophies when the axillary nerve (C5 and C6) is severely damaged. Because it passes inferior to the humeral head and winds around the surgical neck of the humerus (Fig. B6.8A), the axillary nerve is usually injured during fracture of this part of the humerus. It may also be damaged during dislocation of the glenohumeral joint, and by compression from the incorrect use of crutches. As the deltoid atrophies, the rounded contour of the shoulder is flattened compared to the uninjured side. This gives the shoulder a flattened appearance and produces a slight hollow inferior to the acromion. In addition to atrophy of the deltoid, a loss of sensation may occur over the lateral side of the proximal part of the arm, the area supplied by the *superior lateral cutaneous nerve of the arm,* the cutaneous branch of the axillary nerve (*red* in Fig. B6.8B).

The deltoid is a common site for the intramuscular injection of drugs. The axillary nerve runs transversely under cover of the deltoid at the level of the surgical neck of the humerus (Fig. B6.8A). Awareness of its location also avoids injury to it during surgical approaches to the shoulder.

Posterior cord of brachial plexus

Axillary nerve traversing quadrangular space with posterior circumflex humeral artery

Radial nerve

Lower subscapular nerve

Thoracodorsal nerve

Upper subscapular nerves

Long thoracic nerve

Latissimus dorsi

Anterior view

FIGURE B6.6. Branches of posterior cord of brachial plexus, including thoracodorsal nerve.

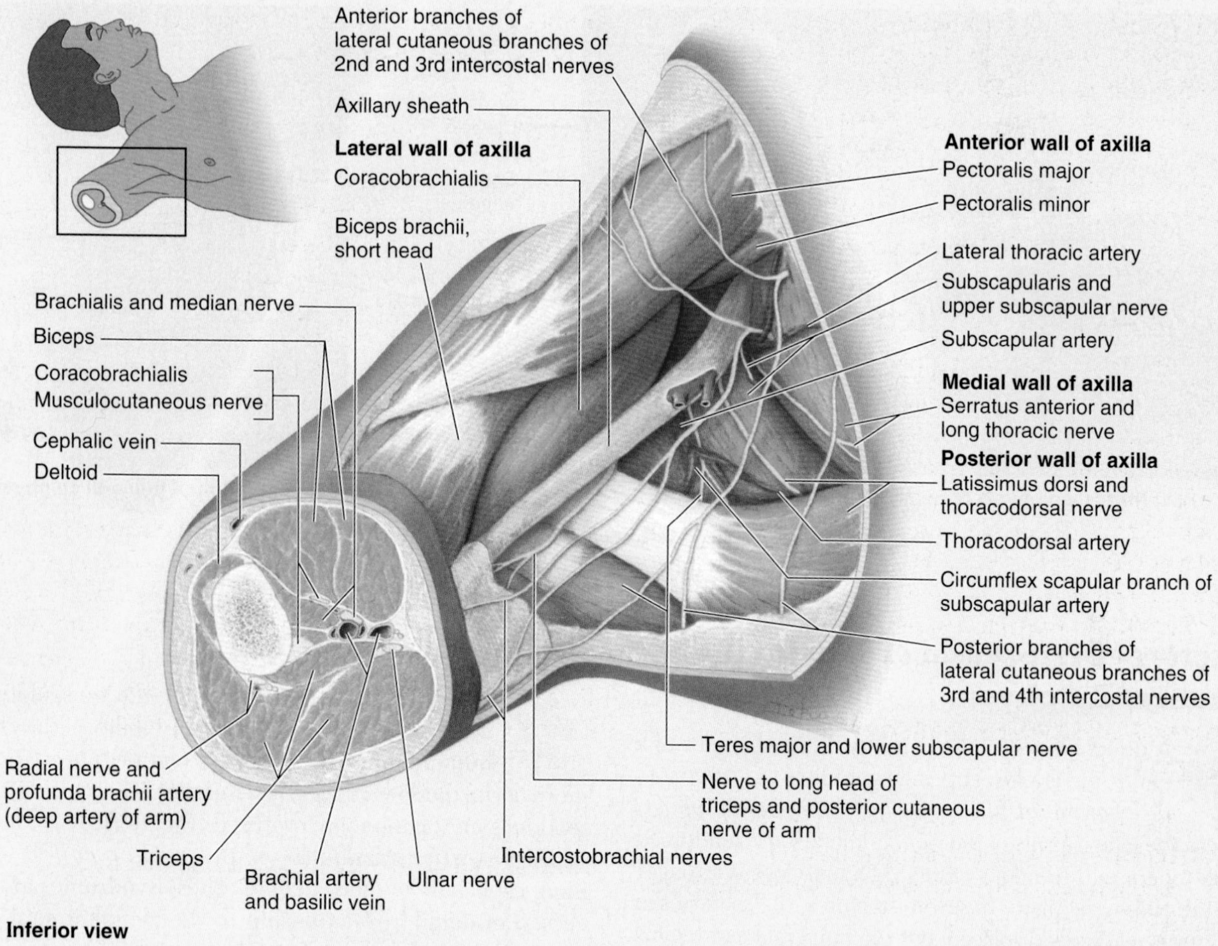

Anterior branches of
lateral cutaneous branches of
2nd and 3rd intercostal nerves

Axillary sheath

Lateral wall of axilla
Coracobrachialis

Biceps brachii,
short head

Brachialis and median nerve

Biceps

Coracobrachialis
Musculocutaneous nerve

Cephalic vein

Deltoid

Radial nerve and
profunda brachii artery
(deep artery of arm)

Triceps

Brachial artery
and basilic vein

Ulnar nerve

Intercostobrachial nerves

Inferior view

Anterior wall of axilla
Pectoralis major

Pectoralis minor

Lateral thoracic artery
Subscapularis and
upper subscapular nerve

Subscapular artery

Medial wall of axilla
Serratus anterior and
long thoracic nerve

Posterior wall of axilla
Latissimus dorsi and
thoracodorsal nerve

Thoracodorsal artery

Circumflex scapular branch of
subscapular artery

Posterior branches of
lateral cutaneous branches of
3rd and 4th intercostal nerves

Teres major and lower subscapular nerve

Nerve to long head of
triceps and posterior cutaneous
nerve of arm

FIGURE B6.7. Nerves closely related to walls of axilla.

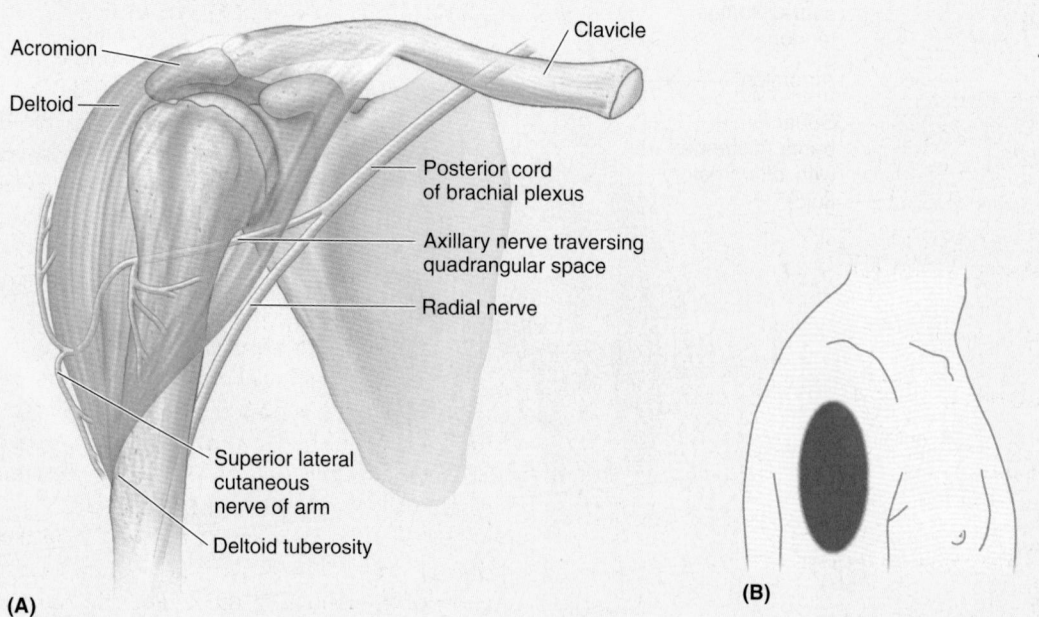

Acromion

Deltoid

Clavicle

Posterior cord
of brachial plexus

Axillary nerve traversing
quadrangular space

Radial nerve

Superior lateral
cutaneous
nerve of arm

Deltoid tuberosity

(A)

(B)

FIGURE B6.8. **A.** Normal course of axillary nerve. **B.** Area of anesthesia (*red*) following injury to axillary nerve.

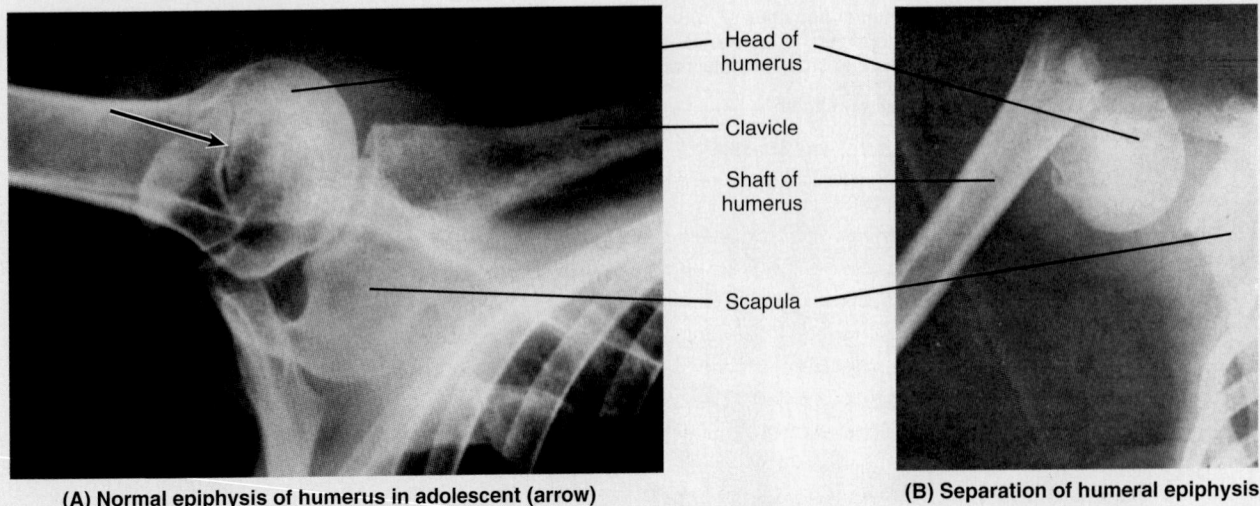

Head of
humerus

Clavicle

Shaft of
humerus

Scapula

(A) Normal epiphysis of humerus in adolescent (arrow)   **(B) Separation of humeral epiphysis**

FIGURE B6.9. Fracture-dislocation of proximal humeral epiphysis.

Fracture–Dislocation of Proximal Humeral Epiphysis

A direct blow or indirect injury of the shoulder of a child or adolescent may produce a *fracture–dislocation of the proximal humeral epiphysis* because the joint capsule of the glenohumeral joint, reinforced by the rotator cuff (tendons of the SITS muscles), is stronger than the epiphysial plate. In severe fractures, the shaft of the humerus is markedly displaced, but the humeral head retains its normal relationship with the glenoid cavity of the scapula (Fig. B6.9B).

Rotator Cuff Injuries

Injury or disease may damage the musculotendinous rotator cuff, producing instability of the glenohumeral joint. Trauma may tear or rupture one or more of the tendons of the SITS muscles. The supraspinatus tendon is most commonly ruptured (Fig. B6.10).

Degenerative tendonitis of the rotator cuff is common, especially in older people. These syndromes are discussed in detail in relationship to the glenohumeral joint (p. 814).

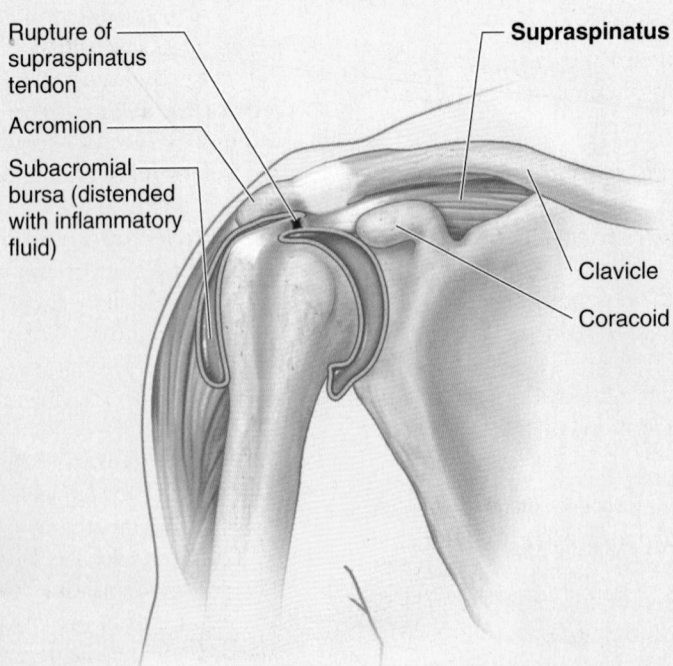

Rupture of
supraspinatus
tendon

Acromion

Subacromial
bursa (distended
with inflammatory
fluid)

Supraspinatus

Clavicle

Coracoid

FIGURE B6.10. Rotator cuff injury.

The Bottom Line

MUSCLES OF PROXIMAL UPPER LIMB

In terms of their attachments, the muscles of the proximal upper limb are axio-appendicular or scapulothoracic.

Axio-appendicular muscles: The axio-appendicular muscles serve to position the base from which the upper limb will be extended and function relative to the trunk. ♦ These muscles consist of anterior, superficial posterior, and deep posterior groups. ♦ The groups work antagonistically to elevate–depress and protract–retract the entire scapula, or rotate it to elevate or depress the glenoid cavity and glenohumeral joint (Table 6.5). ♦ These movements extend the functional range of movements that occur at the glenohumeral joint. ♦ All of these movements involve both the clavicle and the scapula; the limits to all movements of the latter are imposed by the former, which provides the only attachment to the axial skeleton. ♦ Most of these movements involve the cooperation of a number of muscles with different innervations. Therefore, single nerve injuries typically weaken, but do not eliminate, most movements. ♦ Notable exceptions are upward rotation of the lateral angle of the scapula (superior trapezius/spinal accessory nerve only), and lateral rotation of the inferior angle of the scapula (inferior serratus anterior/long thoracic nerve only).

Scapulohumeral muscles: The scapulohumeral muscles (deltoid, teres major, and SITS muscles), along with certain axioappendicular muscles, act in opposing groups to position the proximal strut of the upper limb (the humerus), producing abduction–adduction, flexion–extension, medial–lateral rotation, and circumduction of the arm. ♦ This establishes the height, distance from the trunk, and direction from which the forearm and hand will operate. ♦ Essentially all movements produced by the scapulohumeral muscles at the glenohumeral joint are accompanied by movements produced by axio-appendicular muscles at the sternoclavicular and scapulothoracic joints, especially beyond the initial stages of the movement. ♦ A skilled examiner, knowledgeable in anatomy, can manually fix or position the limb to isolate and test distinctive portions of specific upper limb movements. ♦ The SITS muscles contribute to the formation of the rotator cuff, which both rotates the humeral head (abducting and medially and laterally rotating the humerus), and holds it firmly against the shallow socket of the glenoid cavity, increasing the integrity of the glenohumeral joint capsule.

AXILLA

The **axilla** is the pyramidal space inferior to the glenohumeral joint and superior to the axillary fascia at the junction of the arm and thorax (Fig. 6.37). The axilla provides a passageway, or "distribution center," usually protected by the adducted upper limb, for the neurovascular structures that serve the upper limb. From this distribution center, neurovascular structures pass

- Superiorly via the *cervico-axillary canal* to (or from) the root of the neck (Fig. 6.37A).
- Anteriorly via the *clavipectoral triangle* to the pectoral region (Fig. 6.37D).
- Inferiorly and laterally into the limb itself.
- Posteriorly via the *quadrangular space* to the scapular region.
- Inferiorly and medially along the thoracic wall to the inferiorly placed axio-appendicular muscles (serratus anterior and latissimus dorsi).

The shape and size of the axilla varies, depending on the position of the arm; it almost disappears when the arm is fully abducted—a position in which its contents are vulnerable. A "tickle" reflex causes most people to rapidly resume the protected position when invasion threatens.

The axilla has an apex, a base, and four walls (three of which are muscular):

- The *apex of axilla* is the **cervico-axillary canal,** the passageway between the neck and axilla, bounded by the 1st rib, clavicle, and superior edge of the scapula. The arteries, veins, lymphatics, and nerves traverse this superior opening of the axilla to pass to or from the arm (Fig. 6.37A).
- The *base of axilla* is formed by the concave skin, subcutaneous tissue, and axillary (deep) fascia extending from the arm to the thoracic wall (approximately the 4th rib level), forming the **axillary fossa** (armpit). The base of the axilla and axillary fossa are bounded by the anterior and posterior axillary folds, the thoracic wall, and the medial aspect of the arm (Fig. 6.37C).
- The *anterior wall of axilla* has two layers, formed by the pectoralis major and pectoralis minor and the pectoral and clavicopectoral fascia associated with them (Figs. 6.13B and 6.37B & C). The **anterior axillary fold** is the inferiormost part of the anterior wall that may be grasped between the fingers; it is formed by the pectoralis major, as it bridges from thoracic wall to humerus, and the overlying integument (Fig. 6.37C & D).
- The *posterior wall of axilla* is formed chiefly by the scapula and subscapularis on its anterior surface and inferiorly by the teres major and latissimus dorsi (Fig. 6.37B & C). The **posterior axillary fold** is the inferiormost part of the posterior wall that may be grasped. It extends farther inferiorly than the anterior wall and is formed by latissimus dorsi, teres major, and overlying integument.

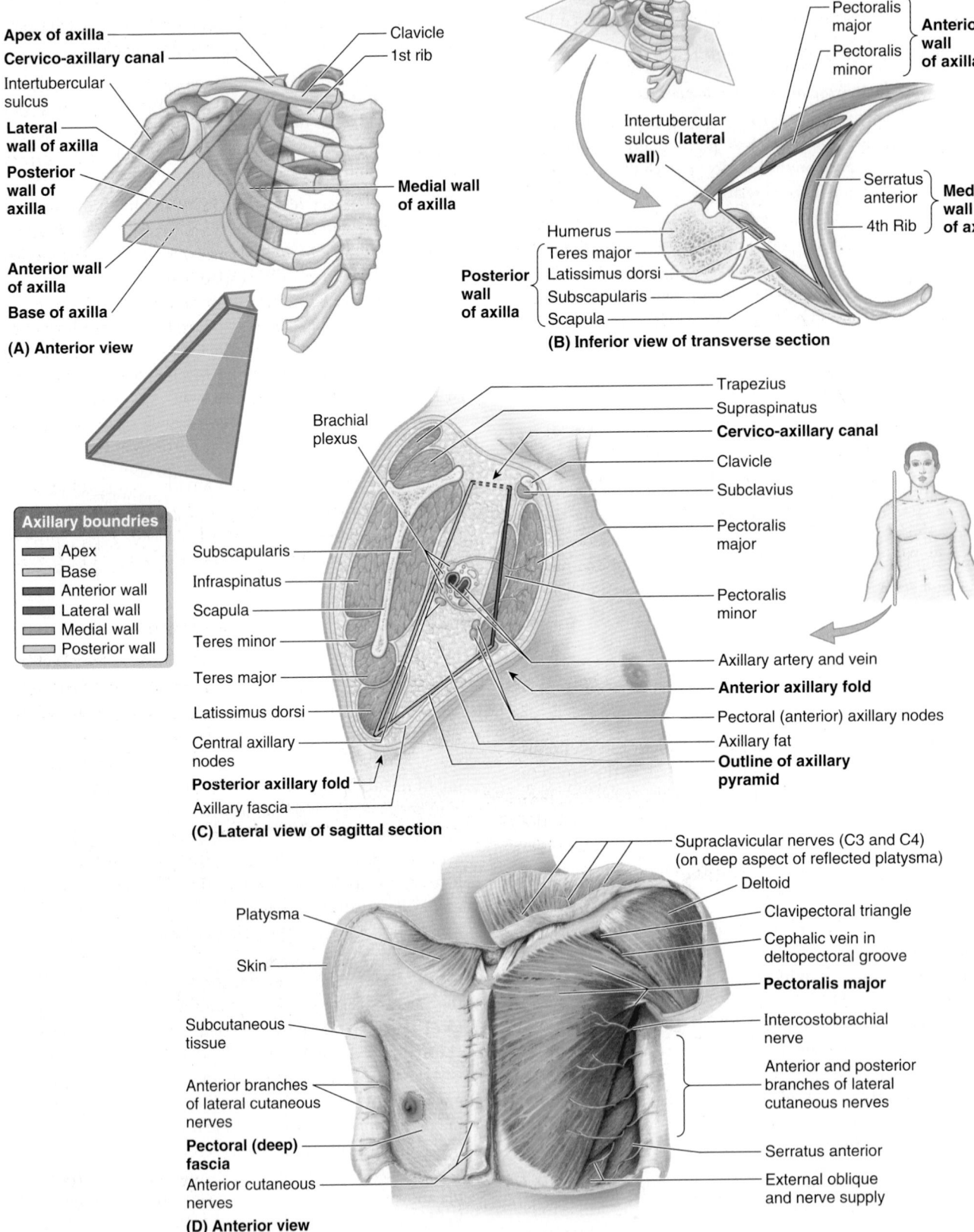

FIGURE 6.37. Location, boundaries, and contents of axilla. A. The axilla is a space inferior to the glenohumeral joint and superior to the skin of the axillary fossa at the junction of the arm and thorax. **B.** Note the axilla's three muscular walls. The small, lateral bony wall of the axilla is the intertubercular sulcus of the humerus. **C.** The contents of the axilla and the scapular and pectoral muscles forming its posterior and anterior walls, respectively. The inferior border of the pectoralis major forms the anterior axillary fold, and the latissimus dorsi and teres major form the posterior axillary fold. **D.** Superficial dissection of the pectoral region. Note that the subcutaneous platysma muscle is cut short on the right side. The severed muscle is reflected superiorly on the left side, together with the supraclavicular nerves, so that the clavicular attachments of the pectoralis major and deltoid can be observed.

- The *medial wall of axilla* is formed by the thoracic wall (1st–4th ribs and intercostal muscles) and the overlying serratus anterior (Fig. 6.37A & B).
- The *lateral wall of axilla* is a narrow bony wall formed by the *intertubercular sulcus* in the humerus.

The axilla contains axillary blood vessels (axillary artery and its branches, axillary vein and its tributaries), lymphatic vessels, and groups of *axillary lymph nodes,* all embedded in a matrix of *axillary fat* (Fig. 6.37C). The axilla also contains large nerves that make up the cords and branches of the *brachial plexus,* a network of interjoining nerves that pass from the neck to the upper limb (Fig. 6.38B). Proximally, these neurovascular structures are ensheathed in a sleeve-like extension of the cervical fascia, the **axillary sheath** (Fig. 6.38A).

Axillary Artery

The **axillary artery** begins at the lateral border of the 1st rib as the continuation of the *subclavian artery,* and ends at the inferior border of the teres major (Fig. 6.39). It passes posterior to the pectoralis minor into the arm, and becomes the *brachial artery* when it passes the inferior border of the teres major, at which point it usually has reached the humerus

(Fig. 6.39). For descriptive purposes, the axillary artery is divided into three parts by the pectoralis minor (the part number also indicates the number of its branches):

1. The **first part of the axillary artery** is located between the lateral border of the 1st rib and the medial border of the pectoralis minor; it is enclosed in the *axillary sheath* and has one branch—the *superior thoracic artery* (Figs. 6.38B & 6.39A; Table 6.7).
2. The **second part of the axillary artery** lies posterior to pectoralis minor and has two branches—the *thoracoacromial* and *lateral thoracic arteries*—which pass medial and lateral to the muscle, respectively.
3. The **third part of the axillary artery** extends from the lateral border of pectoralis minor to the inferior border of teres major; it has three branches. The *subscapular artery* is the largest branch of the axillary artery. Opposite the origin of this artery, the *anterior circumflex humeral and posterior circumflex humeral arteries* arise, sometimes by means of a common trunk.

The branches of the axillary artery are illustrated in Fig. 6.39, and their origin and course are described in Table 6.7.

The **superior thoracic artery** is a small, highly variable vessel that arises just inferior to the subclavius (Fig. 6.39A).

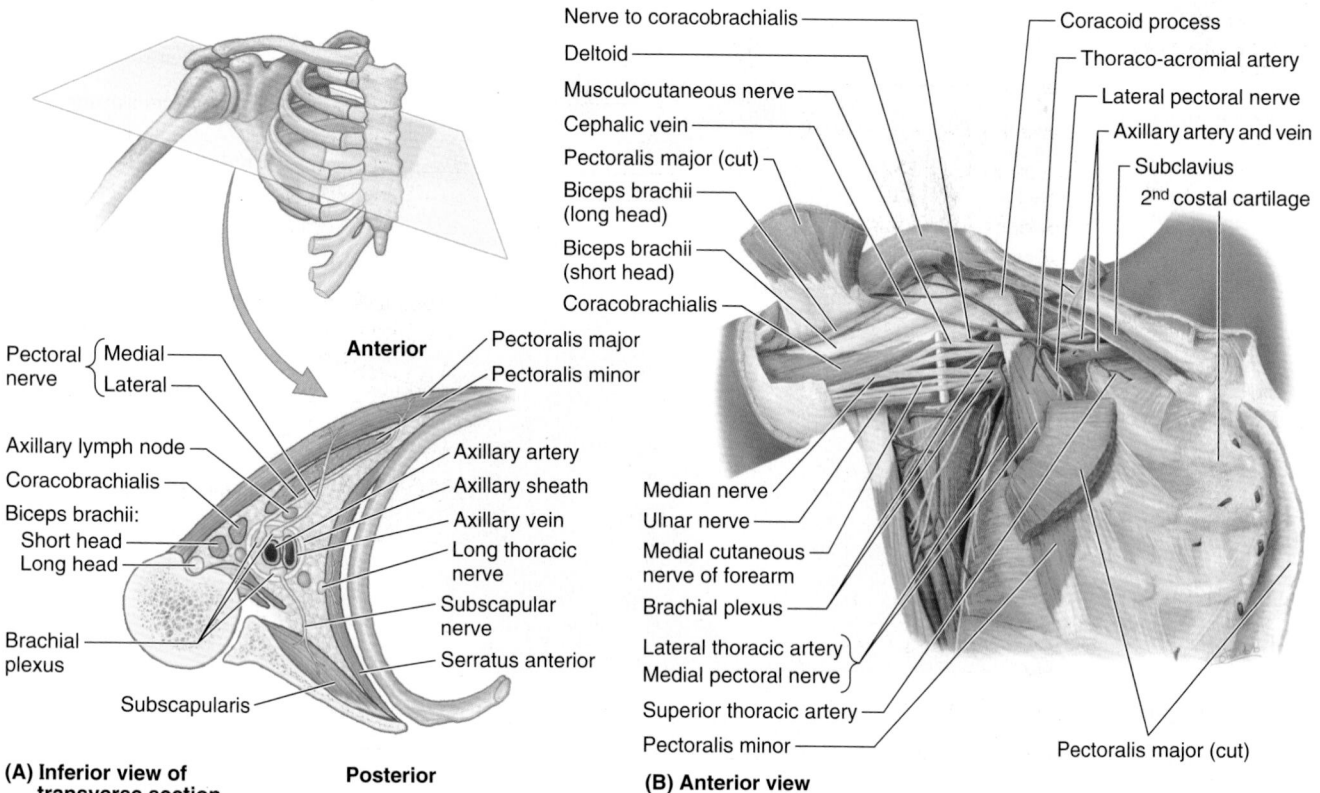

(A) Inferior view of transverse section

Anterior

Posterior

(B) Anterior view

FIGURE 6.38. Contents of axilla. A. Note the axillary sheath enclosing the axillary artery and vein and the three cords of the brachial plexus. The innervation of the muscular walls of the axilla is also shown. The tendon of biceps brachii slides within the intertubercular sulcus **B.** Dissection in which most of the pectoralis major has been removed and the clavipectoral fascia, axillary fat, and axillary sheath have been completely removed. The brachial plexus of nerves surrounds the axillary artery on its lateral and medial aspects (appearing here to be its superior and inferior aspects because the limb is abducted) and on its posterior aspect (not visible from this view). Figure 6.22 on p. 699 is an enlarged view of part B.

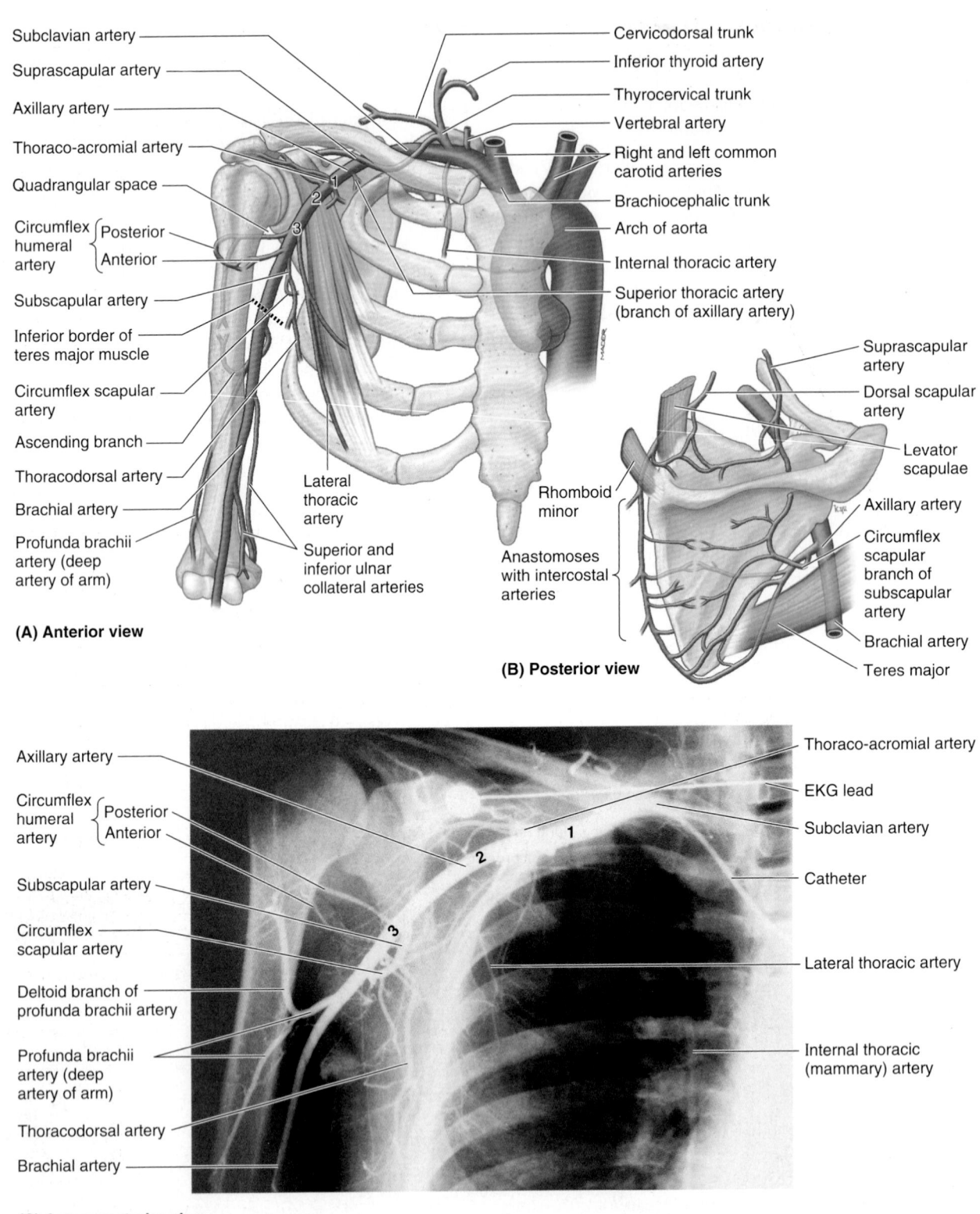

Subclavian artery

Suprascapular artery

Axillary artery

Thoraco-acromial artery

Quadrangular space

Circumflex { Posterior
humeral
artery { Anterior

Subscapular artery

Inferior border of
teres major muscle

Circumflex scapular
artery

Ascending branch

Thoracodorsal artery

Brachial artery

Profunda brachii
artery (deep
artery of arm)

Cervicodorsal trunk

Inferior thyroid artery

Thyrocervical trunk

Vertebral artery

Right and left common
carotid arteries

Brachiocephalic trunk

Arch of aorta

Internal thoracic artery

Superior thoracic artery
(branch of axillary artery)

Lateral
thoracic
artery

Rhomboid
minor

Anastomoses
with intercostal
arteries

Superior and
inferior ulnar
collateral arteries

(A) Anterior view

Suprascapular
artery

Dorsal scapular
artery

Levator
scapulae

Axillary artery

Circumflex
scapular
branch of
subscapular
artery

Brachial artery

Teres major

(B) Posterior view

Axillary artery

Circumflex { Posterior
humeral
artery { Anterior

Subscapular artery

Circumflex
scapular artery

Deltoid branch of
profunda brachii artery

Profunda brachii
artery (deep
artery of arm)

Thoracodorsal artery

Brachial artery

Thoraco-acromial artery

EKG lead

Subclavian artery

Catheter

Lateral thoracic artery

Internal thoracic
(mammary) artery

(C) Anteroposterior view

1: First part of the axillary artery is located between the lateral border of the 1st rib and the medial border of pectoralis minor.
2: Second part of the axillary artery lies posterior to pectoralis minor.
3: Third part of the axillary artery extends from the lateral border of pectoralis minor to the inferior border of teres major, where
   it becomes the brachial artery.

FIGURE 6.39. Arteries of proximal upper limb.

TABLE 6.7. ARTERIES OF PROXIMAL UPPER LIMB (SHOULDER REGION AND ARM)

Artery	Origin		Course
Internal thoracic	Inferior surface of the first part	Subclavian artery	Descends, inclining anteromedially, posterior to sternal end of clavicle and first costal cartilage; enters thorax to descend in parasternal plane; gives rise to perforating branches, anterior intercostal, musculophrenic, and superior epigastric arteries
Thyrocervical trunk	Anterior surface of first part		Ascends as a short, stout trunk, giving rise to four branches: suprascapular, ascending cervical, inferior thyroid arteries, and the cervicodorsal trunk
Suprascapular	Thyrocervical (or as direct branch of subclavian artery)		Passes inferolaterally crossing anterior scalene muscle, phrenic nerve, subclavian artery, and brachial plexus running laterally posterior and parallel to clavicle; next it passes over transverse scapular ligament to supraspinous fossa; then lateral to scapular spine (deep to acromion) to infraspinous fossa on posterior surface of scapula
Superior thoracic	First part (as only branch)	Axillary artery	Runs anteromedially along superior border of pectoralis minor; then passes between it and pectoralis major to thoracic wall; helps supply 1st and 2nd intercostal spaces and superior part of serratus anterior
Thoraco-acromial	Second part (first branch)		Curls around superomedial border of pectoralis minor; pierces costocoracoid membrane (clavipectoral fascia); divides into four branches: pectoral, deltoid, acromial, and clavicular
Lateral thoracic	Second part (second branch)		Descends along axillary border of pectoralis minor; follows it onto thoracic wall, supplying lateral aspect of breast
Circumflex humeral (anterior and posterior)	Third part (sometimes via a common trunk)		Encircle surgical neck of humerus, anastomosing with each other laterally; larger posterior branch traverses quadrangular space
Subscapular	Third part (largest branch of any part)		Descends from level of inferior border of subscapularis along lateral border of scapula, dividing within 2–3 cm into terminal branches, the circumflex scapular and thoracodorsal arteries
Circumflex scapular	Subscapular artery		Curves around lateral border of scapula to enter infraspinous fossa, anastomosing with suprascapular artery
Thoracodorsal			Continues course of subscapular artery, descending with thoracodorsal nerve to enter apex of latissimus dorsi
Profunda brachii (deep artery of arm)	Near its origin	Brachial artery	Accompanies radial nerve along radial groove of humerus, supplying posterior compartment of arm and participating in periarticular arterial anastomosis around elbow joint
Superior ulnar collateral	Near middle of arm		Accompanies ulnar nerve to posterior aspect of elbow; anastomoses with posterior ulnar recurrent artery
Inferior ulnar collateral	Superior to medial epicondyle of humerus		Passes anterior to medial epicondyle of humerus to anastomose with anterior ulnar collateral artery

It commonly runs inferomedially posterior to the axillary vein and supplies the subclavius, muscles in the 1st and 2nd intercostal spaces, superior slips of the serratus anterior, and overlying pectoral muscles. It anastomoses with the intercostal and/or internal thoracic arteries.

The **thoraco-acromial artery,** a short wide trunk, pierces the costocoracoid membrane and divides into four branches (acromial, deltoid, pectoral, and clavicular), deep to the clavicular head of the pectoralis major (Fig. 6.40).

The **lateral thoracic artery** has a variable origin. It usually arises as the second branch of the second part of the axillary artery and descends along the lateral border of the pectoralis minor, following it onto the thoracic wall (Fig. 6.38B and 6.39A); however, it may arise instead from the thoraco-acromial, suprascapular, or subscapular arteries. The lateral thoracic artery supplies the pectoral, serratus anterior, and intercostal muscles, the axillary lymph nodes, and the lateral aspect of the breast.

The **subscapular artery,** the branch of the axillary artery with the greatest diameter but shortest length descends along the lateral border of the subscapularis on the posterior axillary wall. It soon terminates by dividing into the circumflex scapular and thoracodorsal arteries.

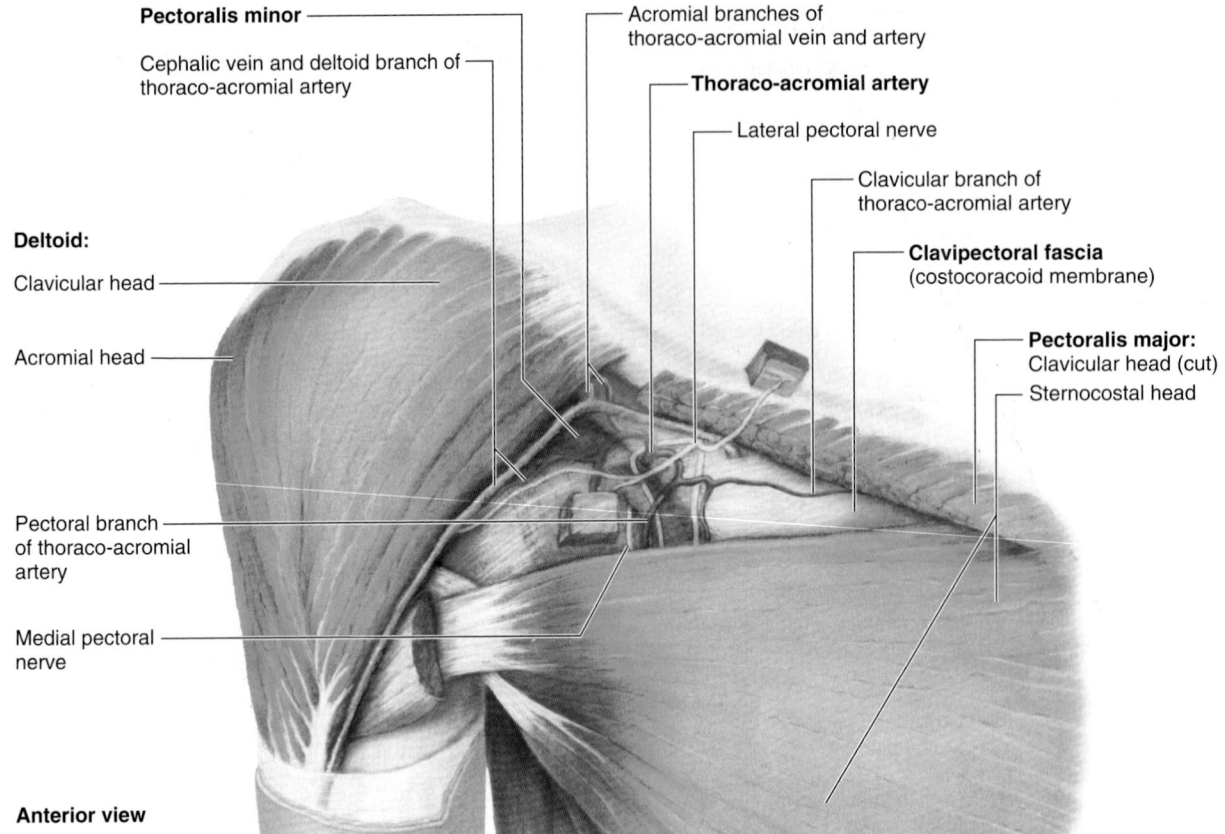

Pectoralis minor

Cephalic vein and deltoid branch of thoraco-acromial artery

Acromial branches of thoraco-acromial vein and artery

Thoraco-acromial artery

Lateral pectoral nerve

Clavicular branch of thoraco-acromial artery

Clavipectoral fascia (costocoracoid membrane)

Deltoid:

Clavicular head

Acromial head

Pectoralis major: Clavicular head (cut) Sternocostal head

Pectoral branch of thoraco-acromial artery

Medial pectoral nerve

Anterior view

FIGURE 6.40. Anterior wall of axilla. The clavicular head of the pectoralis major is excised except for its clavicular and humeral attaching ends and two cubes, which remain to identify its nerves.

The **circumflex scapular artery,** often the larger terminal branch of the subscapular artery, curves posteriorly around the lateral border of the scapula, passing posteriorly between the subscapularis and the teres major to supply muscles on the dorsum of the scapula (Fig. 6.39B). It participates in the anastomoses around the scapula.

The **thoracodorsal artery** continues the general course of the subscapular artery to the inferior angle of the scapula and supplies adjacent muscles, principally the latissimus dorsi (Fig. 6.39A & C). It also participates in the arterial anastomoses around the scapula.

The *circumflex humeral arteries* encircle the surgical neck of the humerus, anastomosing with each other. The smaller **anterior circumflex humeral artery** passes laterally, deep to the coracobrachialis and biceps brachii. It gives off an ascending branch that supplies the shoulder. The larger **posterior circumflex humeral artery** passes medially through the posterior wall of the axilla via the **quadrangular space** with the axillary nerve to supply the glenohumeral joint and surrounding muscles (e.g., the deltoid, teres major and minor, and long head of the triceps) (Fig. 6.39A & C; Table 6.7).

Axillary Vein

The **axillary vein** lies initially (distally) on the anteromedial side of the axillary artery, with its terminal part antero-inferior

to the artery (Fig. 6.41). This large vein is formed by the union of the *brachial vein* (the accompanying veins of the brachial artery) and the *basilic vein* at the inferior border of the teres major.

The axillary vein has three parts, which correspond to the three parts of the axillary artery. Thus the initial, distal end is the third part, whereas the terminal, proximal end is the first part. The axillary vein (first part) ends at the lateral border of the 1st rib, where it becomes the **subclavian vein.** The veins of the axilla are more abundant than the arteries, are highly variable, and frequently anastomose. The axillary vein receives tributaries that generally correspond to branches of the axillary artery with a few major exceptions:

- The veins corresponding to the branches of the thoracoacromial artery do not merge to enter by a common tributary; some enter independently into the axillary vein, but others empty into the cephalic vein, which then enters the axillary vein superior to the pectoralis minor, close to its transition into the subclavian vein.
- The axillary vein receives, directly or indirectly, the *thoraco-epigastric vein(s),* which is(are) formed by the anastomoses of superficial veins from the inguinal region with tributaries of the axillary vein (usually the lateral thoracic vein). These veins constitute a collateral route that

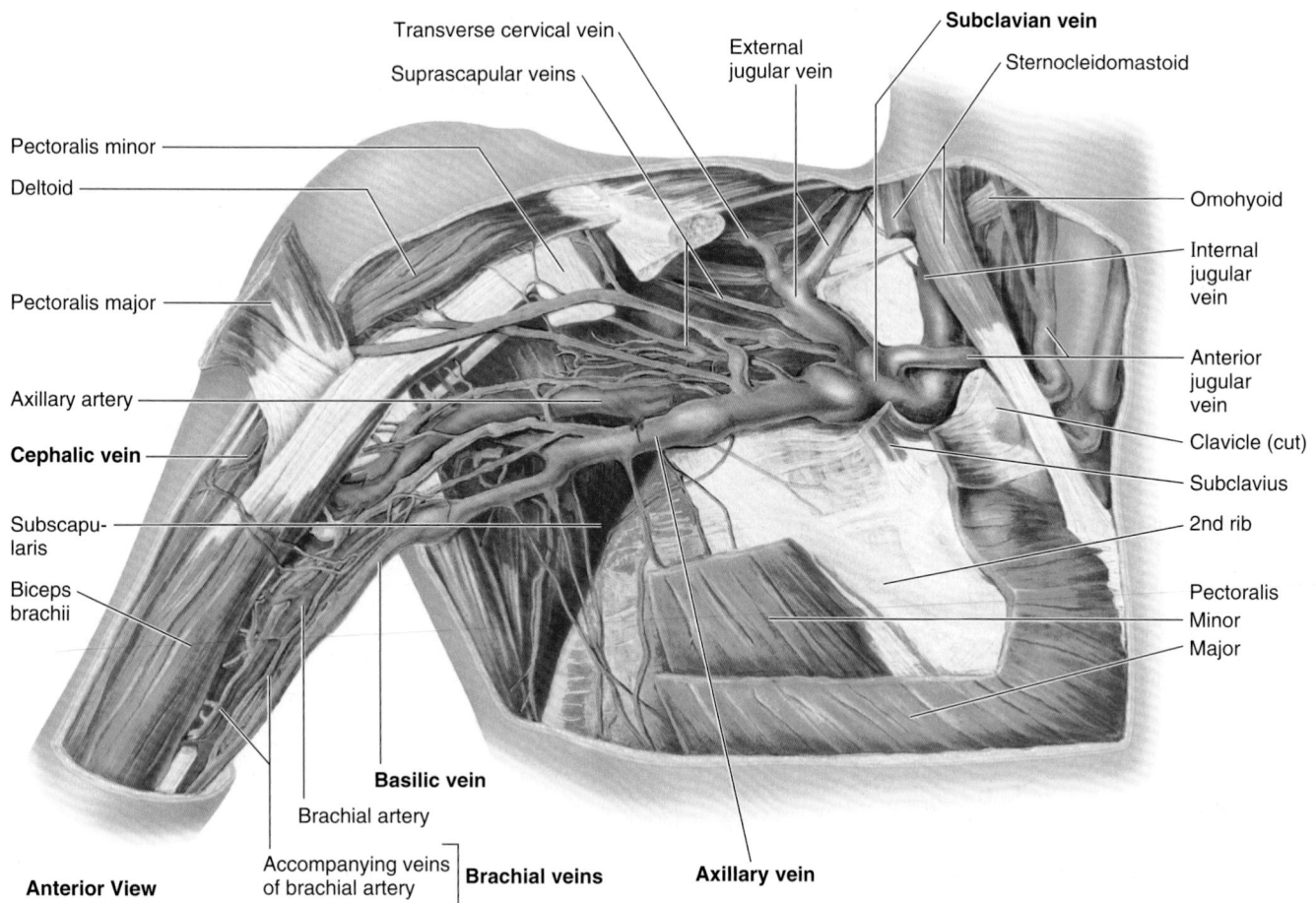

Transverse cervical vein
Suprascapular veins
External jugular vein
Subclavian vein
Sternocleidomastoid
Pectoralis minor
Deltoid
Omohyoid
Internal jugular vein
Pectoralis major
Anterior jugular vein
Axillary artery
Clavicle (cut)
Cephalic vein
Subclavius
2nd rib
Subscapu-laris
Pectoralis
Minor
Major
Biceps brachii
Basilic vein
Brachial artery
Accompanying veins of brachial artery | **Brachial veins**
Axillary vein
Anterior View

FIGURE 6.41. **Veins of axilla.** The basilic vein parallels the brachial artery to the axilla, where it merges with the accompanying veins (L. *venae comitantes*) of the axillary artery to form the axillary vein. The large number of smaller, highly variable veins in the axilla are also tributaries of the axillary vein.

enables venous return in the presence of obstruction of the inferior vena cava (see the blue box "Collateral Routes for Abdominopelvic Venous Blood" on p. 315).

Axillary Lymph Nodes

The fibrofatty connective tissue of the axilla (axillary fat) contains many lymph nodes. The axillary lymph nodes are arranged in five principal groups: pectoral, subscapular, humeral, central, and apical. The groups are arranged in a manner that reflects the pyramidal shape of the axilla (Fig. 6.37A). Three groups of axillary nodes are related to the triangular base, one group at each corner of the pyramid (Fig. 6.42A & C).

The **pectoral (anterior) nodes** consist of three to five nodes that lie along the medial wall of the axilla, around the lateral thoracic vein and the inferior border of the pectoralis minor. The pectoral nodes receive lymph mainly from the anterior thoracic wall, including most of the breast (especially the superolateral [upper outer] quadrant and subareolar plexus; see Chapter 1).

The **subscapular (posterior) nodes** consist of six or seven nodes that lie along the posterior axillary fold and subscapular blood vessels. These nodes receive lymph from the posterior aspect of the thoracic wall and scapular region.

The **humeral (lateral) nodes** consist of four to six nodes that lie along the lateral wall of the axilla, medial and posterior to the axillary vein. These nodes receive nearly all the lymph from the upper limb, except that carried by the lymphatic vessels accompanying the cephalic vein, which primarily drain directly to the apical axillary and infraclavicular nodes.

Efferent lymphatic vessels from these three groups pass to the **central nodes.** There are three or four of these large nodes situated deep to the pectoralis minor near the base of the axilla, in association with the second part of the axillary artery. Efferent vessels from the central nodes pass to the **apical nodes,** which are located at the apex of the axilla along the medial side of the axillary vein and the first part of the axillary artery.

The apical nodes receive lymph from all other groups of axillary nodes as well as from lymphatics accompanying the proximal cephalic vein. Efferent vessels from the apical group traverse the *cervico-axillary canal.*

These efferent vessels ultimately unite to form the **sub-clavian lymphatic trunk,** although some vessels may drain

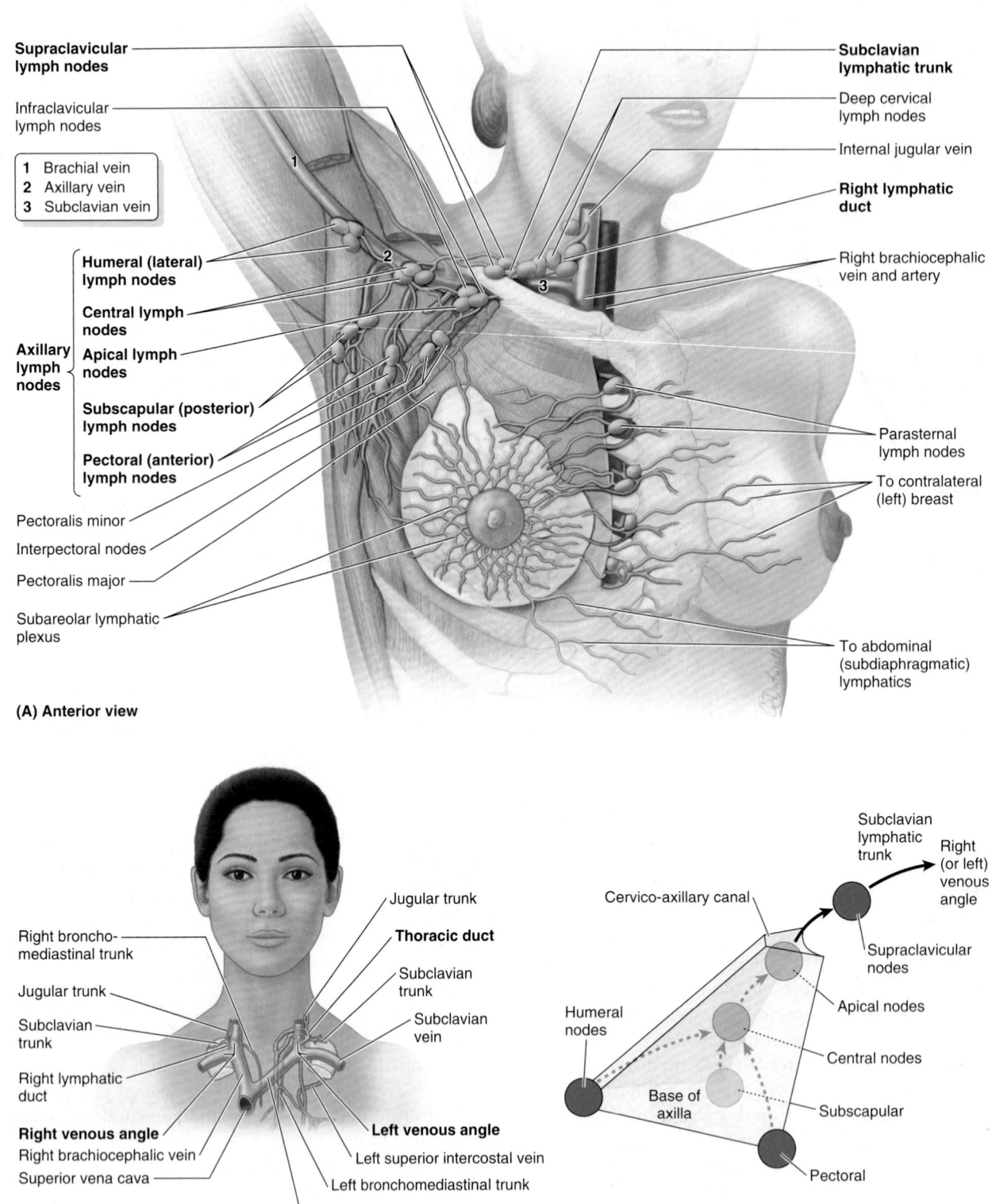

Supraclavicular lymph nodes

Infraclavicular lymph nodes

1 Brachial vein
2 Axillary vein
3 Subclavian vein

Humeral (lateral) lymph nodes

Central lymph nodes

Apical lymph nodes

Axillary lymph nodes

Subscapular (posterior) lymph nodes

Pectoral (anterior) lymph nodes

Pectoralis minor

Interpectoral nodes

Pectoralis major

Subareolar lymphatic plexus

(A) Anterior view

Subclavian lymphatic trunk

Deep cervical lymph nodes

Internal jugular vein

Right lymphatic duct

Right brachiocephalic vein and artery

Parasternal lymph nodes

To contralateral (left) breast

To abdominal (subdiaphragmatic) lymphatics

Right broncho-mediastinal trunk

Jugular trunk

Subclavian trunk

Right lymphatic duct

Right venous angle

Right brachiocephalic vein

Superior vena cava

Jugular trunk

Thoracic duct

Subclavian trunk

Subclavian vein

Left venous angle

Left superior intercostal vein

Left bronchomediastinal trunk

Left brachiocephalic vein

(B) Anterior view

Subclavian lymphatic trunk

Right (or left) venous angle

Cervico-axillary canal

Supraclavicular nodes

Apical nodes

Humeral nodes

Central nodes

Base of axilla

Subscapular

Pectoral

(C) Axillary lymph nodes

FIGURE 6.42. Axillary lymph nodes and lymphatic drainage of right upper limb and breast. A. Of the five groups of axillary lymph nodes, most lymphatic vessels from the upper limb terminate in the humeral (lateral) and central lymph nodes, but those accompanying the upper part of the cephalic vein terminate in the apical lymph nodes. The lymphatics of the breast are discussed in Chapter 1 (pp. 99–101). **B.** Lymph passing through the axillary nodes enters efferent lymphatic vessels that form the subclavian lymphatic trunk, which usually empties into the junctions of the internal jugular and subclavian veins (the venous angles). Occasionally, on the right side, this trunk merges with the jugular lymphatic and/or bronchomediastinal trunks to form a short right lymphatic duct; usually on the left side, it enters the termination of the thoracic duct. **C.** The positions of the five groups of axillary nodes, relative to each other and the pyramidal axilla. The typical pattern of drainage is indicated.

en route through the **clavicular (infraclavicular and supraclavicular) nodes.** Once formed, the subclavian trunk may be joined by the jugular and bronchomediastinal trunks on the right side to form the **right lymphatic duct,** or it may enter the right venous angle independently. On the left side, the subclavian trunk commonly joins the **thoracic duct** (Fig. 6.42A & B).

Brachial Plexus

Most nerves in the upper limb arise from the **brachial plexus,** a *major nerve network* (Figs. 6.38B and 6.43) supplying the upper limb; it begins in the neck and extends into the axilla. Almost all branches of the plexus arise in the axilla (after the plexus has crossed the 1st rib). The brachial plexus is formed by the union of the anterior rami of the last four cervical (C5–C8) and the first thoracic (T1) nerves, which constitute the **roots of the brachial plexus** (Figs. 6.43 and 6.44; Table 6.8).

The roots of the plexus usually pass through the gap between the anterior and the middle scalene (L. *scalenus anterior* and *medius*) muscles with the subclavian artery (Fig. 6.45). The sympathetic fibers carried by each root of the plexus are received from the gray rami of the middle and inferior cervical ganglia as the roots pass between the scalene muscles.

In the inferior part of the neck, the roots of the brachial plexus unite to form three trunks (Figs. 6.43–6.46A; Table 6.8):

1. A **superior trunk,** from the union of the C5 and C6 roots.
2. A **middle trunk,** which is a continuation of the C7 root.
3. An **inferior trunk,** from the union of the C8 and T1 roots.

Each trunk of the brachial plexus divides into anterior and posterior divisions as the plexus passes through the **cervico-axillary canal** posterior to the clavicle (Fig. 6.43).

(text continues on p. 724)

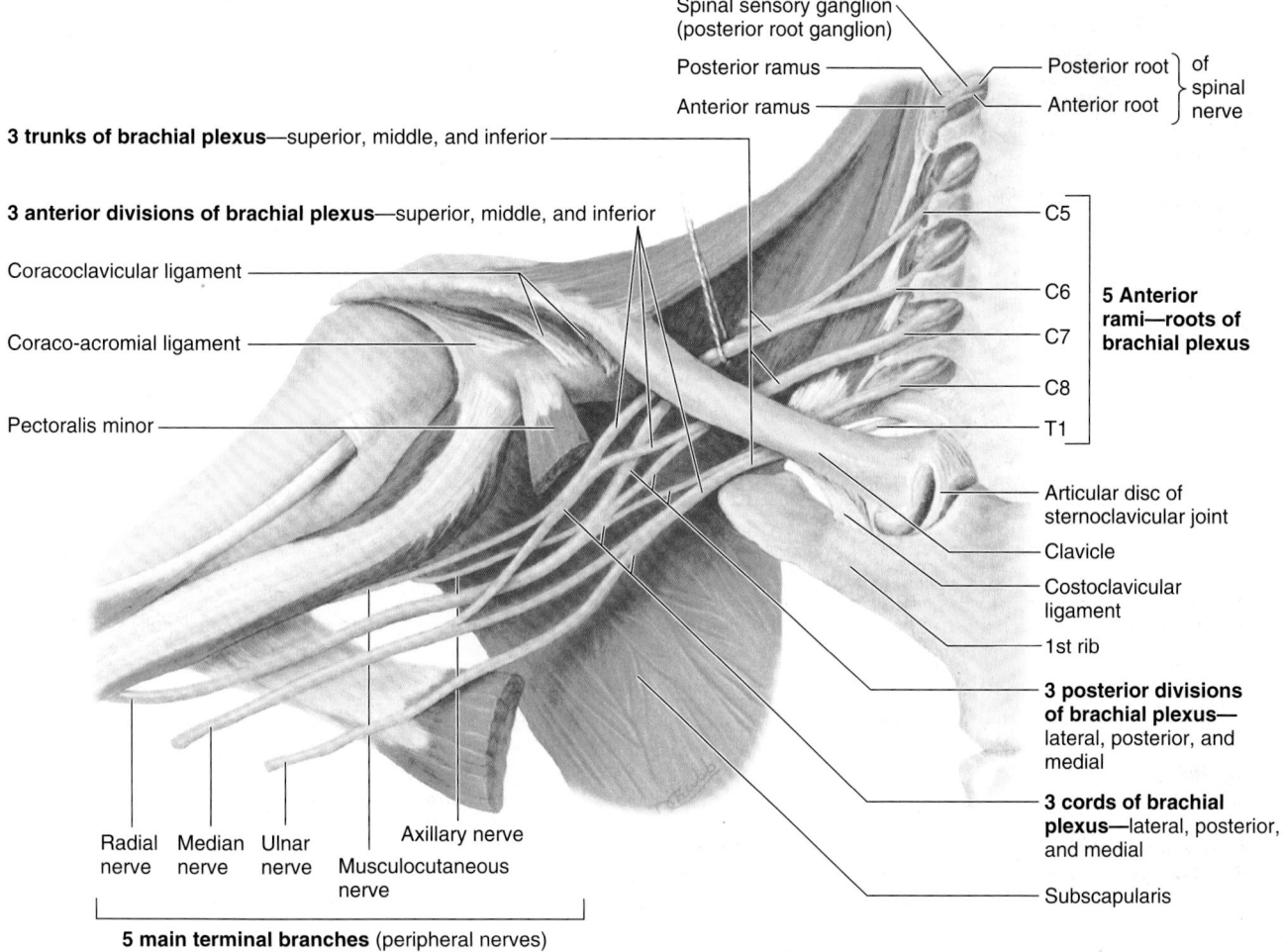

FIGURE 6.43. Formation of brachial plexus. This large nerve network extends from the neck to the upper limb via the cervico-axillary canal (bound by the clavicle, 1st rib, and superior scapula) to provide innervation to the upper limb and shoulder region. The brachial plexus is typically formed by the anterior rami of the C5–C8 nerves and the greater part of the anterior ramus of the T1 nerve (the *roots* of the brachial plexus). Observe the merging and continuation of certain roots of the plexus to three *trunks,* the separation of each trunk into anterior and posterior *divisions,* the union of the divisions to form three *cords,* and the derivation of the main terminal branches (*peripheral nerves*) from the cords as the products of plexus formation.

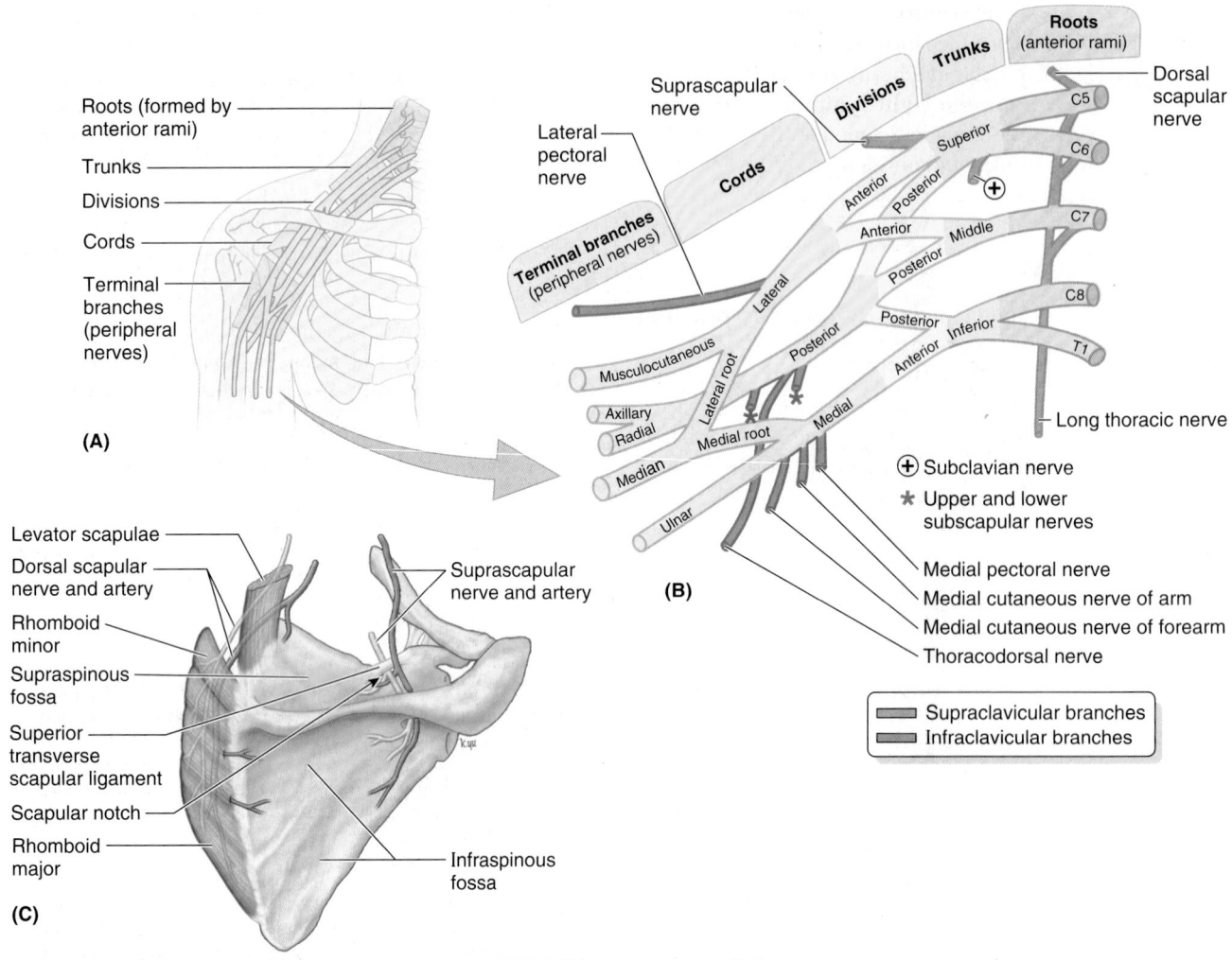

FIGURE 6.44. Nerves of upper limb.

TABLE 6.8. BRACHIAL PLEXUS AND NERVES OF UPPER LIMB

Nerve	Origin[a]	Course	Structures Innervated
Supraclavicular branches			
Dorsal scapular	Posterior aspect of anterior ramus of **C5** with a frequent contribution from C4	Pierces middle scalene; descends deep to levator scapulae and rhomboids	Rhomboids; occasionally supplies levator scapulae
Long thoracic	Posterior aspect of anterior rami of **C5, C6,** C7	Passes through cervico-axillary canal (Fig. 6.14), descending posterior to C8 and T1 roots of plexus (anterior rami); runs inferiorly on superficial surface of serratus anterior	Serratus anterior
Suprascapular	Superior trunk, receiving fibers from **C5,** C6 and often C4	Passes laterally across lateral cervical region (posterior triangle of neck), superior to brachial plexus; then through scapular notch inferior to superior transverse scapular ligament	Supraspinatus and infraspinatus muscles; glenohumeral (shoulder) joint
Subclavian nerve (nerve to subclavius)	Superior trunk, receiving fibers from C5, **C6** and often C4 (Fig. 6.44B)	Descends posterior to clavicle and anterior to brachial plexus and subclavian artery (Fig. 6.29); often giving an *accessory root to phrenic nerve*	Subclavius and sternoclavicular joint (accessory phrenic root innervates diaphragm)

TABLE 6.8. BRACHIAL PLEXUS AND NERVES OF UPPER LIMB (Continued)

Nerve	Origin[a]	Course	Structures Innervated
Infraclavicular branches			
Lateral pectoral	Side branch of lateral cord, receiving fibers from C5, **C6**, C7	Pierces costocoracoid membrane to reach deep surface of pectoral muscles; a *communicating branch to the medial pectoral nerve* passes anterior to axillary artery and vein	Primarily pectoralis major; but some lateral pectoral nerve fibers pass to pectoralis minor via branch to medial pectoral nerve (Fig. 6.46A)
Musculocutaneous	Terminal branch of lateral cord, receiving fibers from C5–C7	Exits axilla by piercing coracobrachialis (Fig. 6.43); descends between biceps brachii and brachialis (Figs. 6.47B and 6.48), supplying both; continues as *lateral cutaneous nerve of forearm*	Muscles of anterior compartment of arm (coracobrachialis, biceps brachii and brachialis) (Fig. 6.46B); skin of lateral aspect of forearm
Median	**Lateral root of median nerve** is a terminal branch of lateral cord (C6, C7) **Medial root of median nerve** is a terminal branch of medial cord (C8, T1)	Lateral and medial roots merge to form median nerve lateral to axillary artery; descends through arm adjacent to brachial artery, with nerve gradually crossing anterior to artery to lie medial to artery in cubital fossa (see Fig. 6.53, p. 738)	Muscles of anterior forearm compartment (except for flexor carpi ulnaris and ulnar half of flexor digitorum profundus), five intrinsic muscles in thenar half of palm and palmar skin (Fig. 6.46B)
Medial pectoral		Passes between axillary artery and vein; then pierces pectoralis minor and enters deep surface of pectoralis major; although it is called *medial* for its origin from medial cord, it lies lateral to lateral pectoral nerve	Pectoralis minor and sternocostal part of pectoralis major
Medial cutaneous nerve of arm	Side branches of medial cord, receiving fibers from C8, T1	Smallest nerve of plexus; runs along medial side of axillary and brachial veins; communicates with *intercostobrachial nerve*	Skin of medial side of arm, as far distal as medial epicondyle of humerus and olecranon of ulna
Median cutaneous nerve of forearm		Initially runs with ulnar nerve (with which it may be confused) but pierces deep fascia with basilic vein and enters subcutaneous tissue, dividing into anterior and posterior branches	Skin of medial side of forearm, as far distal as wrist
Ulnar	Larger terminal branch of medial cord, receiving fibers from C8, T1 and often C7	Descends medial arm; passes posterior to medial epicondyle of humerus; then descends ulnar aspect of forearm to hand (Figs. 6.46C and 6.47A)	Flexor carpi ulnaris and ulnar half of flexor digitorum profundus (forearm); most intrinsic muscles of hand; skin of hand medial to axial line of digit 4
Upper subscapular	Side branch of posterior cord, receiving fibers from **C5**	Passes posteriorly, entering subscapularis directly	Superior portion of subscapularis
Lower subscapular	Side branch of posterior cord, receiving fibers from **C6**	Passes inferolaterally, deep to subscapular artery and vein	Inferior portion of subscapularis and teres major
Thoracodorsal	Side branch of posterior cord, receiving fibers from C6, **C7**, C8	Arises between upper and lower subscapular nerves and runs inferolaterally along posterior axillary wall to apical part of latissimus dorsi	Latissimus dorsi
Axillary	Terminal branch of posterior cord, receiving fibers from **C5**, C6	Exits axillary fossa posteriorly, passing through quadrangular space[b] with posterior circumflex humeral artery (Fig. 6.48); gives rise to *superior lateral brachial cutaneous nerve;* then winds around surgical neck of humerus deep to deltoid (Fig. 6.46D)	Glenohumeral (shoulder) joint; teres minor and deltoid muscles (Fig. 6.46D); skin of superolateral arm (over inferior part of deltoid)
Radial	Larger terminal branch of posterior cord (largest branch of plexus), receiving fibers from C5–T1	Exits axillary fossa posterior to axillary artery; passes posterior to humerus in radial groove with deep brachial artery, between lateral and medial heads of triceps; perforates lateral intermuscular septum; enters cubital fossa, dividing into *superficial* (cutaneous) and *deep* (motor) *radial nerves* (Fig. 6.46D)	All muscles of posterior compartments of arm and forearm (Fig. 6.46D); skin of posterior and inferolateral arm, posterior forearm, and dorsum of hand lateral to axial line of digit 4

[a]Boldface (**C5**) indicates primary component of the nerve.

[b]Bounded superiorly by the subscapularis, head of humerus, and teres minor; inferiorly by the teres major; medially by the long head of the triceps; and laterally by the coracobrachialis and surgical neck of the humerus (Fig. 6.48).

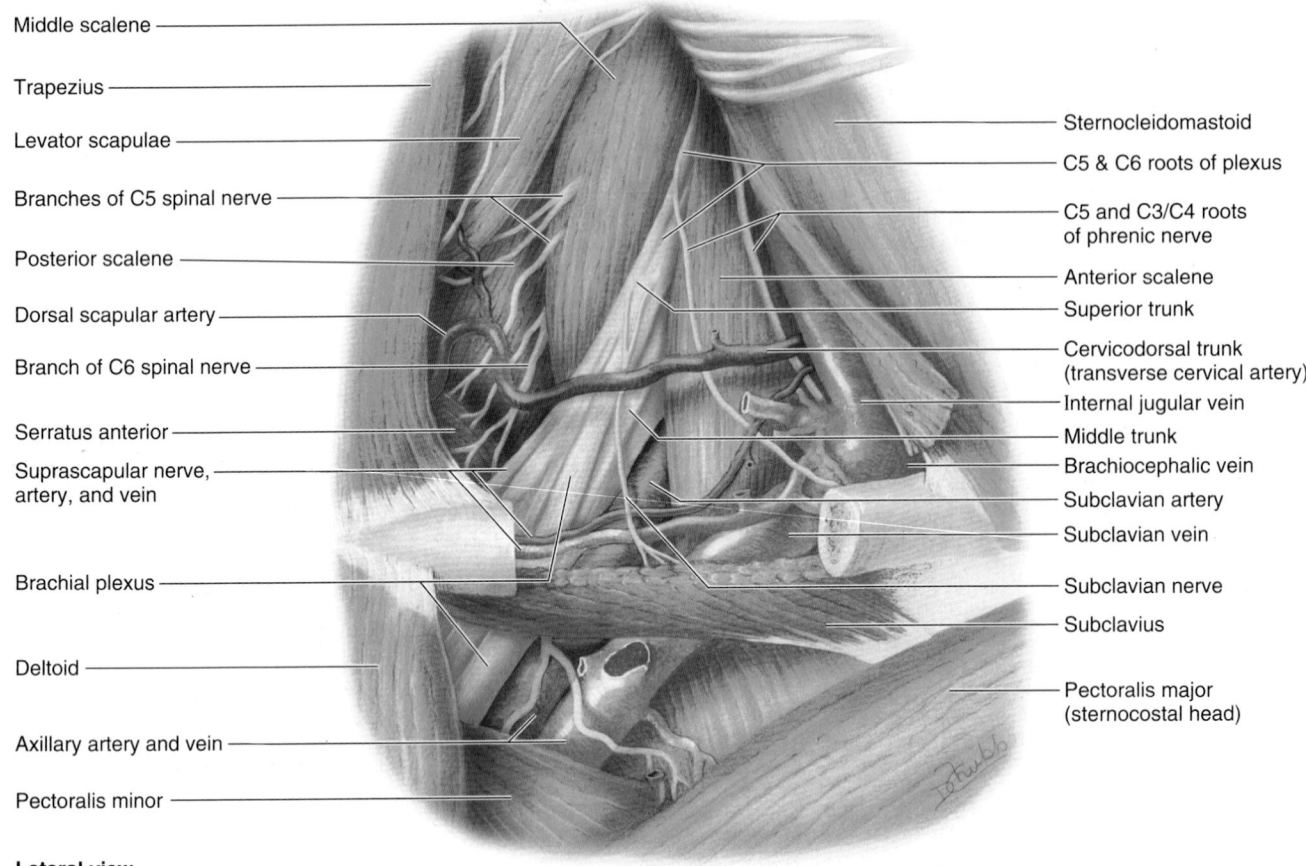

Middle scalene

Trapezius

Levator scapulae

Branches of C5 spinal nerve

Posterior scalene

Dorsal scapular artery

Branch of C6 spinal nerve

Serratus anterior

Suprascapular nerve, artery, and vein

Brachial plexus

Deltoid

Axillary artery and vein

Pectoralis minor

Sternocleidomastoid

C5 & C6 roots of plexus

C5 and C3/C4 roots of phrenic nerve

Anterior scalene

Superior trunk

Cervicodorsal trunk (transverse cervical artery)

Internal jugular vein

Middle trunk

Brachiocephalic vein

Subclavian artery

Subclavian vein

Subclavian nerve

Subclavius

Pectoralis major (sternocostal head)

Lateral view

FIGURE 6.45. Dissection of right lateral cervical region (posterior triangle). The brachial plexus and subclavian vessels have been dissected. The anterior rami of spinal nerves C5–C8 (plus T1, concealed here by the third part of the subclavian artery) constitute the roots of the brachial plexus. Merging and subsequent splitting of the nerve fibers conveyed by the roots form the trunks and divisions at the level shown. The subclavian artery emerges between the middle and the anterior scalene muscles with the roots of the plexus.

Anterior divisions of the trunks supply *anterior (flexor) compartments* of the upper limb, and **posterior divisions of the trunks** supply *posterior (extensor) compartments.*

The divisions of the trunks form three cords of the brachial plexus (Figs. 6.43, 6.44, and 6.46, Table 6.8):

1. Anterior divisions of the superior and middle trunks unite to form the **lateral cord.**
2. Anterior division of the inferior trunk continues as the **medial cord.**
3. Posterior divisions of all three trunks unite to form the **posterior cord.**

The cords bear the relationship to the second part of the axillary artery that is indicated by their names. For example, the lateral cord is lateral to the axillary artery, although it may appear to lie superior to the artery because it is most easily seen when the limb is abducted.

The products of plexus formation are multisegmental, peripheral (named) nerves. The brachial plexus is divided into **supraclavicular** and **infraclavicular parts** by the clavicle (Fig. 6.44B; Table 6.8). Four *branches of the*

supraclavicular part of the plexus arise from the roots (anterior rami) and trunks of the brachial plexus (dorsal scapular nerve, long thoracic nerve, nerve to subclavius, and suprascapular nerve), and are approachable through the neck. In addition, officially unnamed *muscular branches* arise from all five roots of the plexus (anterior rami C5–T1), which supply the scaleni and longus colli muscles. The C5 root of the *phrenic nerve* (considered a branch of the *cervical plexus*) arises from the C5 plexus root, joining the C3–C4 components of the nerve on the anterior surface of the anterior scalene muscle (Fig. 6.45). *Branches of the infraclavicular part of the plexus* arise from the cords of the brachial plexus and are approachable through the axilla (Figs. 6.44B and 6.46). Counting side and terminal branches, three branches arise from the lateral cord, whereas the medial and posterior cords each give rise to five branches (counting the roots of the median nerve as individual branches). The branches of the supraclavicular and infraclavicular parts of the brachial plexus are illustrated in Figs. 6.44B and 6.46 and listed in Table 6.8, along with the origin, course, and distribution of each branch.

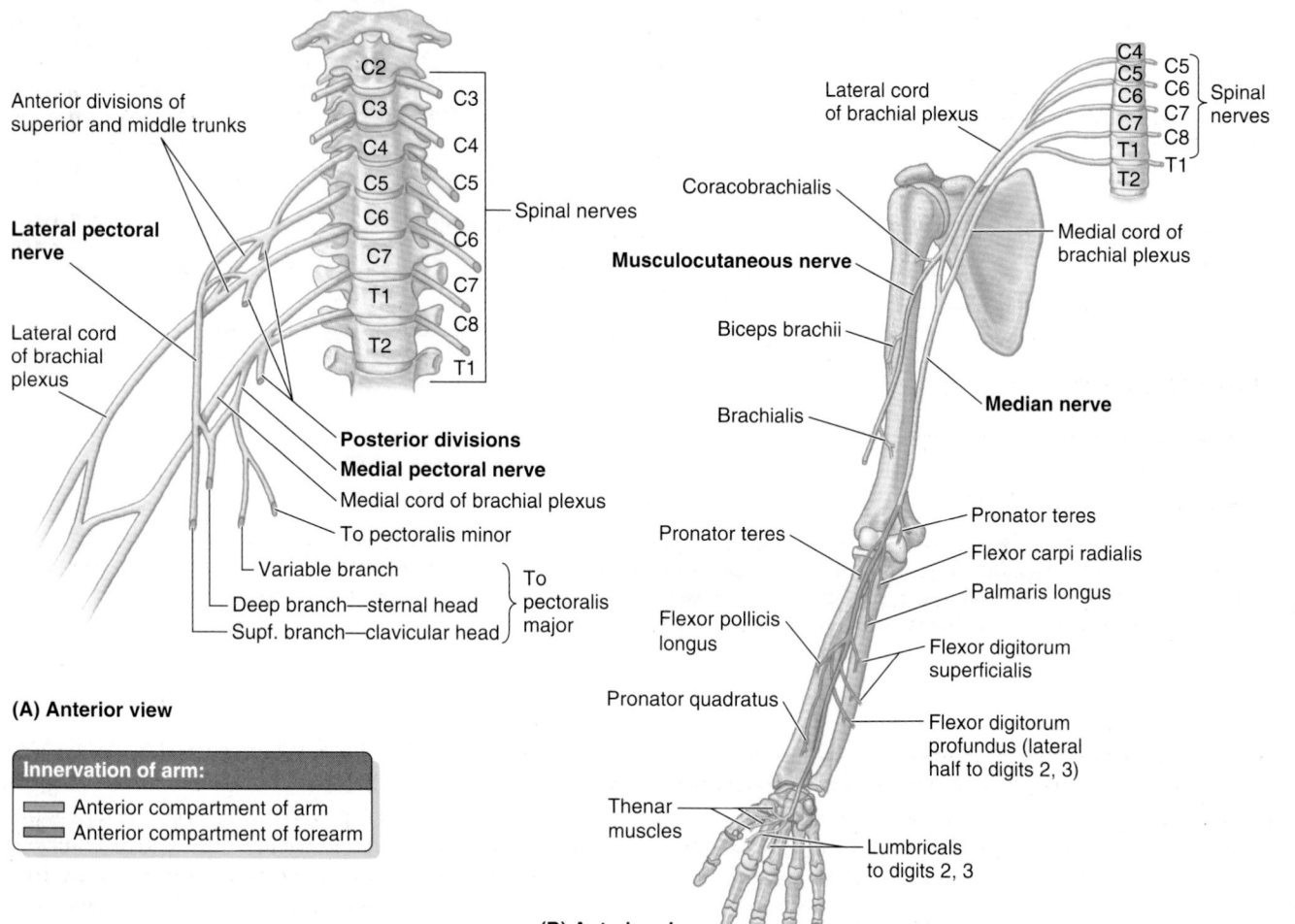

(A) Anterior view

Anterior divisions of superior and middle trunks

Lateral pectoral nerve

Lateral cord of brachial plexus

C2
C3
C4
C5
C6
C7
T1
T2

C3
C4
C5
C6
C7
C8
T1

Spinal nerves

Posterior divisions
Medial pectoral nerve
Medial cord of brachial plexus
To pectoralis minor
Variable branch
Deep branch—sternal head
Supf. branch—clavicular head

To pectoralis major

Innervation of arm:
Anterior compartment of arm
Anterior compartment of forearm

Lateral cord of brachial plexus

C4
C5
C6
C7
T1
T2

C5
C6
C7
C8
T1

Spinal nerves

Coracobrachialis

Musculocutaneous nerve

Medial cord of brachial plexus

Biceps brachii

Median nerve

Brachialis

Pronator teres

Flexor pollicis longus

Pronator quadratus

Thenar muscles

Pronator teres
Flexor carpi radialis
Palmaris longus
Flexor digitorum superficialis
Flexor digitorum profundus (lateral half to digits 2, 3)

Lumbricals to digits 2, 3

(B) Anterior view

FIGURE 6.46. Motor branches derived from cords of brachial plexus. A. The medial and lateral pectoral nerves arise from the medial and lateral cords of the brachial plexus, respectively (or from the anterior divisions of the trunks that form them, as shown here for the lateral pectoral nerve). **B.** The courses of the median and musculocutaneous nerves, and the typical pattern of branching of their motor branches are shown.

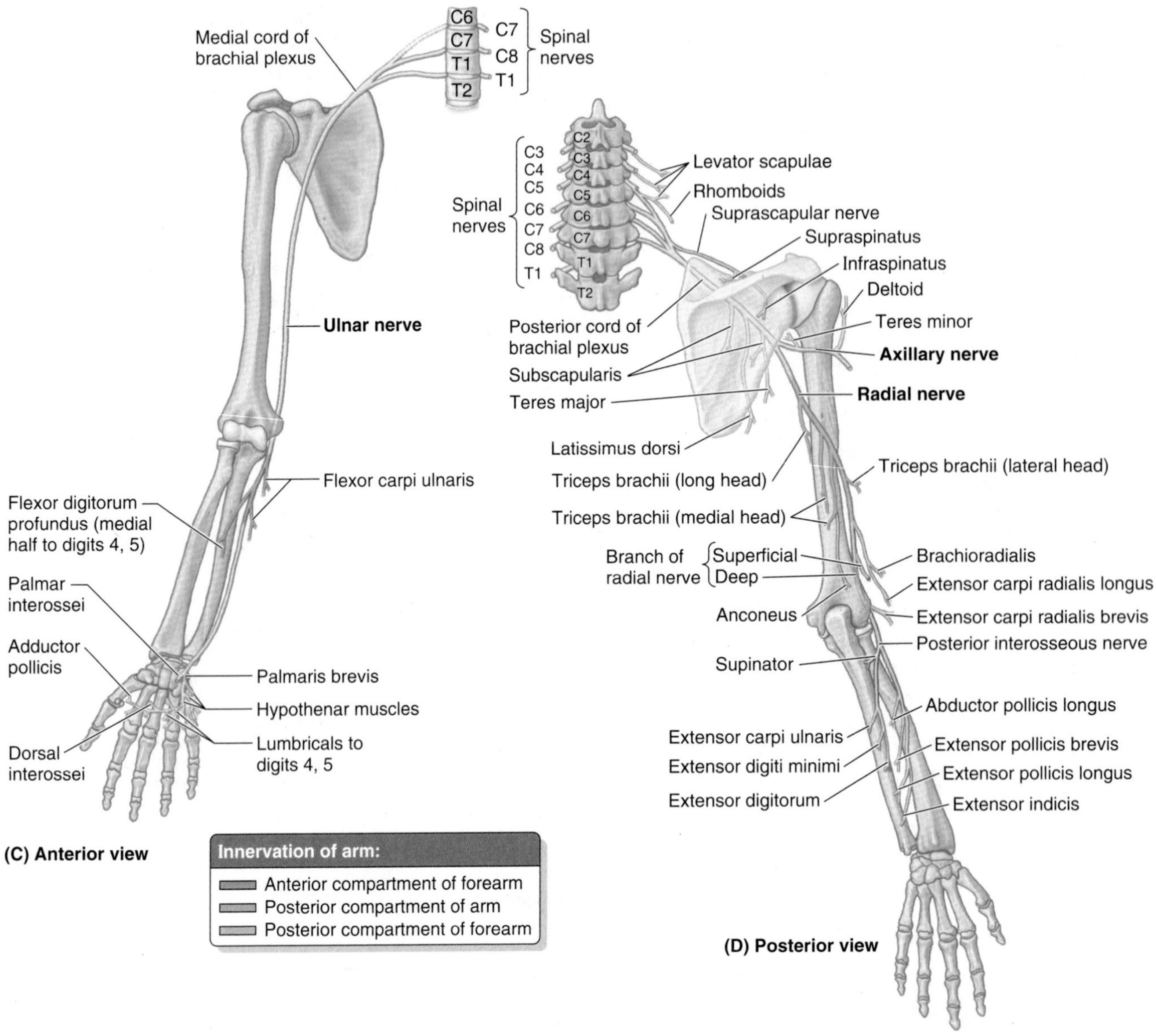

Medial cord of
brachial plexus

C6
C7
T1
T2

C7
C8
T1

Spinal
nerves

C3
C4
C5
C6
C7
C8
T1

Spinal
nerves

C2
C3
C4
C5
C6
C7
T1
T2

Levator scapulae

Rhomboids

Suprascapular nerve

Supraspinatus

Infraspinatus

Deltoid

Ulnar nerve

Teres minor

Axillary nerve

Posterior cord of
brachial plexus

Radial nerve

Subscapularis

Teres major

Latissimus dorsi

Triceps brachii (long head)

Triceps brachii (lateral head)

Triceps brachii (medial head)

Flexor carpi ulnaris

Branch of
radial nerve

Superficial
Deep

Brachioradialis

Extensor carpi radialis longus

Flexor digitorum
profundus (medial
half to digits 4, 5)

Anconeus

Extensor carpi radialis brevis

Posterior interosseous nerve

Palmar
interossei

Supinator

Adductor
pollicis

Abductor pollicis longus

Palmaris brevis

Extensor carpi ulnaris

Extensor pollicis brevis

Hypothenar muscles

Extensor digiti minimi

Extensor pollicis longus

Dorsal
interossei

Lumbricals to
digits 4, 5

Extensor digitorum

Extensor indicis

(C) Anterior view

Innervation of arm:

Anterior compartment of forearm
Posterior compartment of arm
Posterior compartment of forearm

(D) Posterior view

FIGURE 6.46. (*Continued*) **Motor branches derived from cords of brachial plexus. C.** The course of the ulnar nerve and the typical pattern of branching of its motor branches. **D.** The courses of the axillary and radial nerves and the typical pattern of branching of their motor branches. The posterior interosseous nerve is the continuation of the deep branch of the radial nerve, shown here bifurcating into two branches to supply all the muscles with fleshy bellies located entirely in the posterior compartment of the forearm. The dorsum of the hand has no fleshy muscle fibers; therefore, no motor nerves are distributed there.

AXILLA

Arterial Anastomoses Around Scapula

Many arterial anastomoses occur around the scapula. Several vessels join to form networks on the anterior and posterior surfaces of the scapula: the dorsal scapular, suprascapular, and (via the circumflex scapular) subscapular arteries (Fig. B6.11).

The importance of the *collateral circulation* made possible by these anastomoses becomes apparent when ligation

of a lacerated subclavian or axillary artery is necessary. For example, the axillary artery may have to be ligated between the 1st rib and subscapular artery; in other cases, *vascular stenosis* of the axillary artery may result from an atherosclerotic lesion that causes reduced blood flow. In either case, the direction of blood flow in the subscapular artery is reversed, enabling blood to reach the third part of the axillary artery.

Note that the subscapular artery receives blood through several anastomoses with the suprascapular artery, dorsal scapular artery, and intercostal arteries. *Slow occlusion* of the axillary artery (e.g., resulting from disease or trauma) often enables sufficient collateral circulation to develop, preventing ischemia (loss of blood supply). *Sudden occlusion* usually does not allow sufficient time for adequate collateral circulation to develop; as a result, there is an inadequate supply of blood to the arm, fore-arm, and hand. While potential collateral pathways (peri-articular

anastomoses) exist around the shoulder joint proximally, and the elbow joint distally, *surgical ligation of the axillary artery between the origins of the subscapular artery and the profunda brachii artery* will cut off the blood supply to the arm because the collateral circulation is inadequate.

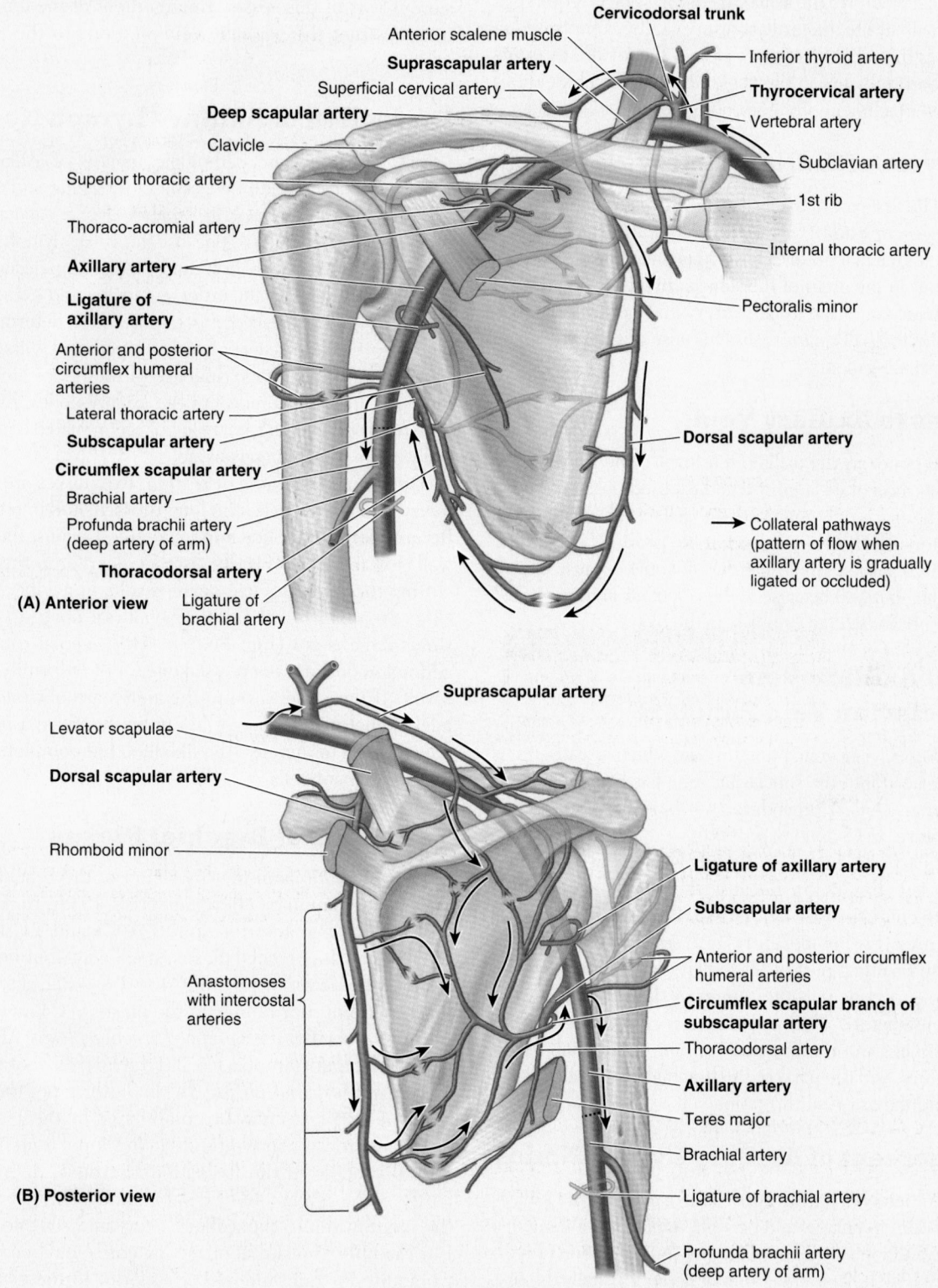

FIGURE B6.11.

Compression of Axillary Artery

The axillary artery can be palpated in the inferior part of the lateral wall of the axilla. Compression of the third part of this artery against the humerus may be necessary when profuse bleeding occurs (e.g., resulting from a stab or bullet wound in the axilla). If compression is required at a more proximal site, the axillary artery can be compressed at its origin (as the subclavian artery crosses the 1st rib) by exerting downward pressure in the angle between the clavicle and the inferior attachment of the sternocleidomastoid muscle.

Aneurysm of Axillary Artery

The first part of the axillary artery may enlarge (*aneurysm of axillary artery*) and compress the trunks of the brachial plexus, causing pain and anesthesia (loss of sensation) in the areas of the skin supplied by the affected nerves. Aneurysm of the axillary artery may occur in baseball pitchers and football quarterbacks because of their rapid and forceful arm movements.

Injuries to Axillary Vein

Wounds in the axilla often involve the axillary vein because of its large size and exposed position. When the arm is fully abducted, the axillary vein overlaps the axillary artery anteriorly. A wound in the proximal part of the axillary vein is particularly dangerous, not only because of profuse bleeding but also because of the risk of air entering it and producing *air emboli* (air bubbles) in the blood.

Role of Axillary Vein in Subclavian Vein Puncture

Subclavian vein puncture, in which a catheter is placed into the subclavian vein, has become a common clinical procedure (see blue box "Subclavian Vein Puncture" in Chapter 8, p. 1008).

The axillary vein becomes the subclavian vein as the first rib is crossed (Fig. 6.45). Because the needle is advanced medially to enter the vein as it crosses the rib, the vein actually punctured (the point of entry) in a "subclavian vein puncture" is the terminal part of the axillary vein. However, the needle tip proceeds into the lumen of the subclavian vein almost immediately. Thus it is clinically significant that the axillary vein lies anterior and inferior (i.e., superficial) to the axillary artery, and the parts of the brachial plexus that begin to surround the artery at this point.

Enlargement of Axillary Lymph Nodes

An infection in the upper limb can cause the axillary nodes to enlarge and become tender and inflamed, a condition called *lymphangitis* (inflammation of lymphatic vessels). The humeral group of nodes is usually the first to be involved.

Lymphangitis is characterized by the development of warm, red, tender streaks in the skin of the limb. Infections in the pectoral region and breast, including the superior part of the abdomen, can also produce enlargement of axillary nodes. In metastatic cancer of the apical group, the nodes often adhere to the axillary vein, which may necessitate excision of part of this vessel. Enlargement of the apical nodes may obstruct the cephalic vein superior to the pectoralis minor.

Dissection of Axillary Lymph Nodes

Excision and pathologic analysis of axillary lymph nodes are often necessary for *staging* and determining the appropriate treatment of a cancer, such as breast cancer (see p. 104). Because the axillary lymph nodes are arranged and receive lymph (and therefore metastatic breast cancer cells) in a specific order, removing and examining the lymph nodes in that order is important in determining the degree to which the cancer has developed, and is likely to have metastasized. Lymphatic drainage of the upper limb may be impeded after the removal of the axillary nodes, resulting in *lymphedema*, swelling as a result of accumulated lymph, especially in the subcutaneous tissue.

During *axillary node dissection,* two nerves are at risk of injury. During surgery, the long thoracic nerve to the serratus anterior is identified and maintained against the thoracic wall (Fig. B6.7, p. 711). As discussed earlier in this chapter, cutting the long thoracic nerve results in a winged scapula (Fig. B6.5 p. 709). If the thoracodorsal nerve to the latissimus dorsi is cut (Fig. B6.6, p. 710), medial rotation and adduction of the arm are weakened, but deformity does not result. If the nodes around this nerve are obviously malignant, sometimes the nerve has to be sacrificed as the nodes are resected to increase the likelihood of complete removal of all malignant cells.

Variations of Brachial Plexus

Variations in the formation of the brachial plexus are common (Bergman et al., 1988). In addition to the five anterior rami (C5–C8 and T1) that form the roots of the brachial plexus, small contributions may be made by the anterior rami of C4 or T2. When the superiormost root (anterior ramus) of the plexus is C4 and the inferiormost root is C8, it is a *prefixed brachial plexus.* Alternately, when the superior root is C6 and the inferior root is T2, it is a *postfixed brachial plexus.* In the latter type, the inferior trunk of the plexus may be compressed by the 1st rib, producing neurovascular symptoms in the upper limb. Variations may also occur in the formation of trunks, divisions, and cords; in the origin and/or combination of branches; and in the relationship to the axillary artery and scalene muscles. For example, the lateral or medial cords may receive fibers from anterior rami inferior or superior to the usual levels, respectively.

In some individuals, trunk divisions or cord formations may be absent in one or other parts of the plexus; however, the makeup of the terminal branches is unchanged. Because each peripheral nerve is a collection of nerve fibers bound together by connective tissue, it is understandable that the median nerve, for instance, may have two medial roots instead of one (i.e., the nerve fibers are simply grouped differently). This results from the fibers of the medial cord of the brachial plexus dividing into three branches, two forming the median nerve and the third forming the ulnar nerve. Sometimes it may be more confusing when the two medial roots are completely separate; however, understand that although the median nerve may have two medial roots the components of the nerve are the same (i.e., the impulses arise from the same place and reach the same destination whether they go through one or two roots).

Brachial Plexus Injuries

Injuries to the brachial plexus affect movements and cutaneous sensations in the upper limb. Disease, stretching, and wounds in the lateral cervical region (posterior triangle) of the neck (see Chapter 8), or in the axilla may produce brachial plexus injuries. Signs and symptoms depend on the part of the plexus involved. Injuries to the bra-

chial plexus result in *paralysis* and *anesthesia.* Testing the person's ability to perform movements assesses the degree of paralysis. In *complete paralysis,* no movement is detectable. In *incomplete paralysis,* not all muscles are paralyzed; therefore, the person can move, but the movements are weak compared with those on the normal side. Determining the ability of the person to feel pain (e.g., from a pinprick of the skin) tests the degree of anesthesia.

Injuries to superior parts of the brachial plexus (C5 and C6) usually result from an excessive increase in the angle between the neck and shoulder. These injuries can occur in a person who is thrown from a motorcycle or a horse, and lands on the shoulder in a way that widely separates the neck and shoulder (Fig. B6.12A). When thrown, the person's shoulder often hits something (e.g., a tree or the ground) and stops, but the head and trunk continue to move. This stretches or ruptures superior parts of the brachial plexus or *avulses* (tears) the roots of the plexus from the spinal cord.

Injury to the superior trunk of the plexus is apparent by the characteristic position of the limb ("waiter's tip position"), in which the limb hangs by the side in medial rotation (Fig. B6.12B; arrow). *Upper brachial plexus injuries* can also occur in a neonate when excessive stretching of the neck occurs during delivery (Fig. B6.12C).

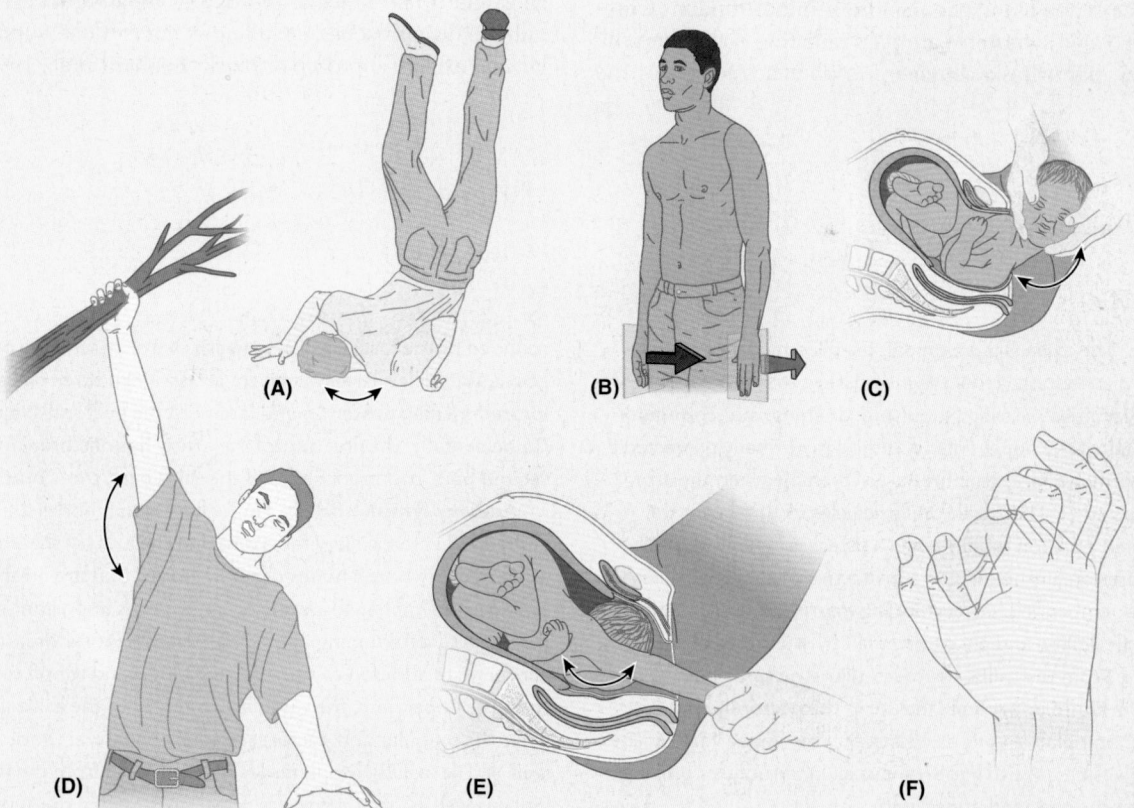

FIGURE B6.12. Injuries to brachial plexus. A. Note the excessive increase in the angle between the head and left shoulder. **B.** The waiter's tip position (left upper limb). **C.** Observe the excessive increase in the angle between the head and the left shoulder during this delivery. **D and E.** Excessive increases in the angle between the trunk and the right upper limb. **F.** A claw hand (person is attempting to assume lightly shaded "fist" position).

As a result of *injuries to the superior parts of the brachial plexus* (*Erb-Duchenne palsy*), paralysis of the muscles of the shoulder and arm supplied by the C5 and C6 spinal nerves occurs: deltoid, biceps, and brachialis. The usual clinical appearance is an upper limb with an adducted shoulder, medially rotated arm, and extended elbow. The lateral aspect of the forearm also experiences some loss of sensation. Chronic microtrauma to the superior trunk of the brachial plexus from carrying a heavy backpack can produce motor and sensory deficits in the distribution of the musculocutaneous and radial nerves. A *superior brachial plexus injury* may produce muscle spasms and severe disability in hikers (*backpacker's palsy*) who carry heavy backpacks for long periods.

Acute brachial plexus neuritis (*brachial plexus neuropathy*) is a neurologic disorder of unknown cause that is characterized by the sudden onset of severe pain, usually around the shoulder. Typically, the pain begins at night and is followed by muscle weakness and sometimes muscular atrophy (*neurologic amyotrophy*). Inflammation of the brachial plexus (*brachial neuritis*) is often preceded by some event (e.g., upper respiratory infection, vaccination, or non-specific trauma). The nerve fibers involved are usually derived from the superior trunk of the brachial plexus.

Compression of cords of the brachial plexus may result from prolonged hyperabduction of the arm during performance of manual tasks over the head, such as painting a ceiling. The cords are impinged or compressed between the coracoid process of the scapula and the pectoralis minor tendon. Common neurologic symptoms are pain radiating down the arm, numbness, paresthesia (tingling), erythema (redness of the skin caused by capillary dilation), and weakness of the hands. Compression of the axillary artery and vein causes ischemia of the upper limb and distension of the superficial veins. These signs and symptoms of *hyperabduction syndrome* result from compression of the axillary vessels and nerves.

Injuries to inferior parts of the brachial plexus (*Klumpke paralysis*) are much less common. Inferior brachial plexus injuries may occur when the upper limb is suddenly pulled superiorly—for example, when a person grasps something to break a fall (Fig. B6.12D), or a baby's upper limb is pulled excessively during delivery (Fig. B6.12E). These events injure the inferior trunk of the brachial plexus (C8 and T1), and may avulse the roots of the spinal nerves from the spinal cord. The short muscles of the hand are affected, and a *claw hand* results (Fig. B6.12F).

Brachial Plexus Block

Injection of an anesthetic solution into or immediately surrounding the axillary sheath interrupts conduction of impulses of peripheral nerves, and produces anesthesia of the structures supplied by the branches of the cords of the plexus (Fig. 6.38A). Sensation is blocked in all deep structures of the upper limb, and the skin distal to the middle of the arm. Combined with an occlusive tourniquet technique to retain the anesthetic agent, this procedure enables surgeons to operate on the upper limb without using a general anesthetic. The brachial plexus can be anesthetized using a number of approaches, including an interscalene, supraclavicular, and axillary approach or block (Leonard et al., 1999).

The Bottom Line

AXILLA

Axilla: The axilla is a pyramidal, fat-filled fascial compartment (distribution center) giving passage to or housing the major "utilities" serving (supplying, draining, and communicating with) the upper limb. ♦ Although normally protected by the arm, axillary structures are vulnerable when the arm is abducted. ♦ The "tickle" reflex causes us to recover the protected position rapidly when a threat is perceived. ♦ The structures are ensheathed in a protective wrapping (axillary sheath), embedded in a cushioning matrix (axillary fat) that allows flexibility, and are surrounded by musculoskeletal walls. ♦ From the axilla, neurovascular structures pass to and from the entire upper limb, including the pectoral, scapular, and subscapular regions as well as the free upper limb. ♦ The axilla gives passage to important vascular structures passing between the neck and upper limb.

Axillary vein and artery: The axillary vein lies anterior and slightly inferior to the axillary artery, both being surrounded by the fascial axillary sheath. ♦ For descriptive purposes, the axillary artery and vein are assigned three parts located medial, posterior, and lateral to the pectoralis minor. Coincidentally, the first part of the artery has one branch; the second part, two branches; and the third part, three branches.

Axillary lymph nodes: The axillary lymph nodes are embedded in the axillary fat external to the axillary sheath. ♦ The axillary lymph nodes occur in groups that are arranged and receive lymph in a specific order, which is important in staging and determining appropriate treatment for breast cancer. ♦ In addition to transporting blood and lymph to and from the upper limb, the vascular structures of the axilla also serve the scapular and pectoral regions and lateral thoracic wall. ♦ The axillary lymph nodes receive lymph from the upper limb, as well as from the entire upper quadrant of the superficial body wall, from the level of the clavicles to the umbilicus including most from the breast.

Brachial Plexus: The brachial plexus is an organized intermingling of the nerve fibers of the five adjacent anterior rami (C5–T1, the roots of the plexus) innervating the upper limb. ♦ Although their segmental identity is lost in forming the plexus, the original segmental distribution to skin (dermatomes) and muscles (myotomes) remains, exhibiting a cranial to caudal distribution for the skin (see "Cutaneous Innervation of Upper Limb" on p. 693), and a proximal to distal distribution for the muscles. For example, C5 and C6 fibers primarily innervate muscles that act at the shoulder or flex the elbow; C7 and C8 fibers innervate muscles that extend the elbow or are part of the forearm; and T1 fibers innervate the intrinsic muscles of the hand. ♦ Formation of the brachial plexus initially involves merging of the superior and inferior pairs of roots, resulting in three trunks that each divide into anterior and posterior divisions. ♦ The fibers traversing anterior divisions innervate flexors and pronators of the anterior compartments of the limb, whereas the fibers traversing posterior divisions innervate extensors and supinators of the posterior compartments of the limb. ♦ The six divisions merge to form three cords that surround the axillary artery. ♦ Two of the three cords give rise in turn to 5 nerves, and the third (lateral cord) gives rise to 3 nerves. ♦ In addition to the nerves arising from the cords, 10 more nerves arise from the various parts of the plexus. ♦ Most nerves arising from the plexus contain fibers from two or more adjacent anterior rami.

ARM

The **arm** extends from the shoulder to the elbow. Two types of movement occur between the arm and forearm at the elbow joint: flexion–extension and pronation–supination. The muscles performing these movements are clearly divided into anterior and posterior groups, separated by the humerus and medial and lateral intermuscular septae (Fig. 6.47). The chief action of both groups is at the elbow joint, but some muscles also act at the glenohumeral joint. The superior part of the humerus provides attachments for tendons of the shoulder muscles.

Muscles of Arm

Of the four major arm muscles, three flexors (biceps brachii, brachialis, and coracobrachialis) are in the anterior (flexor) compartment, supplied by the musculocutaneous nerve, and one extensor (triceps brachii) is in the posterior compartment, supplied by the radial nerve (Figs. 6.48 and 6.49B–D & F; Table 6.9). A distally placed assistant to the triceps, the anconeus, also lies within the posterior compartment (6.49G). The flexor muscles of the anterior compartment are almost twice as strong as the extensors in all positions; consequently, we are better pullers than pushers. It should be noted, however, that the extensors of the elbow are particularly important for raising oneself out of a chair, and for wheelchair activity. Therefore, conditioning of the triceps is of particular importance in elderly or disabled persons.

The arm muscles and their attachments are illustrated in Figure 6.49 and their attachments, innervation, and actions are described in Table 6.9.

BICEPS BRACHII

As the term **biceps brachii** indicates, the proximal attachment of this fusiform muscle usually has two heads (*bi*, two + L. *caput*, head). The two heads of the biceps arise proximally by tendinous attachments to processes of the scapula, their fleshy bellies uniting just distal to the middle of the arm (Fig. 6.49B).

Approximately 10% of people have a third head to the biceps. When present, the third head extends from the superomedial part of the brachialis (with which it is blended), usually lying posterior to the brachial artery. In either case, a single **biceps tendon** forms distally and attaches primarily to the radius.

Although the biceps is located in the anterior compartment of the arm, it has no attachment to the humerus (Figs. 6.47B & C and 6.49A & B). The biceps is a "three-joint muscle," crossing and capable of effecting movement at the glenohumeral, elbow, and radio-ulnar joints, although it primarily acts at the latter two. Its action and effectiveness are markedly affected by the position of the elbow and forearm. When the elbow is extended, the biceps is a simple flexor of the forearm; however, as elbow flexion approaches 90° and more power is needed against resistance, the biceps is capable of two powerful movements, depending on the position of the forearm. When the elbow is flexed close to 90° and the forearm is supinated, the biceps is most efficient in producing flexion. Alternately, when the forearm is pronated, the biceps is the primary (most powerful) supinator of the forearm. For example, it is used when right-handed people drive a screw into hard wood, and when inserting a corkscrew and pulling the cork from a wine bottle. The biceps barely operates as a flexor when the forearm is pronated, even against resistance. In the semiprone position, it is active only against resistance (Hamill and Knutzen, 2008).

Arising from the supraglenoid tubercle of the scapula, and crossing the head of the humerus within the cavity of the glenohumeral joint, the rounded tendon of the long head of the biceps continues to be surrounded by synovial membrane as it descends in the intertubercular sulcus of the humerus. A broad band, the **transverse humeral ligament,** passes from the lesser to the greater tubercle of the humerus and converts the intertubercular groove into a canal (Fig. 6.49B). The ligament holds the tendon of the long head of the biceps in the groove.

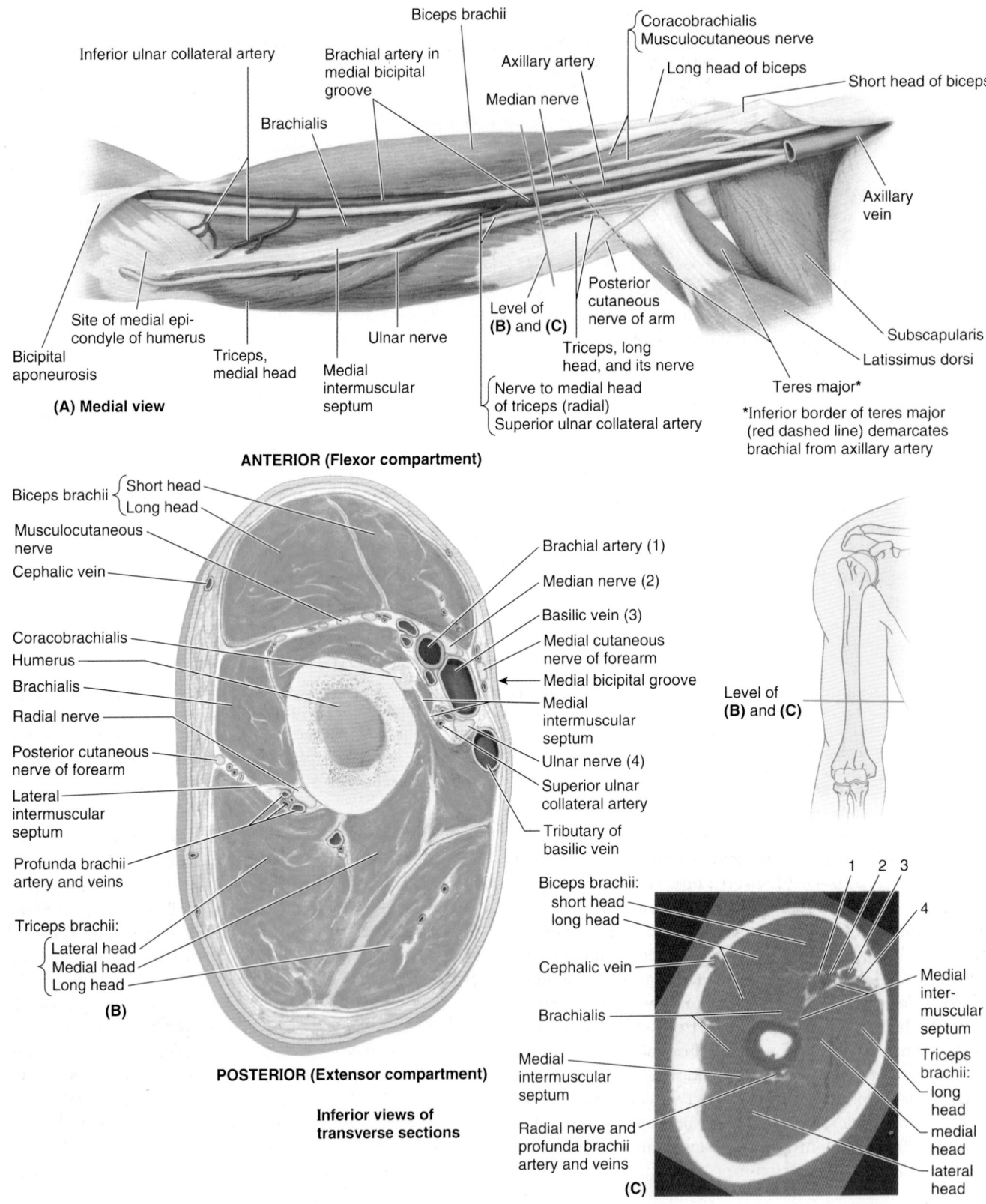

FIGURE 6.47. Muscles, neurovascular structures, and compartments of arm. A. In this dissection of the right arm, the veins have been removed, except for the proximal part of the axillary vein. Note the courses of the musculocutaneous, median, and ulnar nerves and the brachial artery along the medial (protected) aspect of the arm. Their courses generally parallel the medial intermuscular septum that separates the anterior and posterior compartments in the distal two thirds of the arm. **B.** In this transverse section of the right arm, the three heads of the triceps and the radial nerve and its companion vessels (in contact with the humerus) lie in the posterior compartment. **C.** This transverse MRI demonstrates the features shown in part **B**; the numbered structures are identified in part **B**. (Courtesy of Dr. W. Kucharczyk, Professor and Neuroradiologist Senior Scientist, Department of Medical Resonance Imaging, University Health Network, Toronto, Ontario, Canada.)

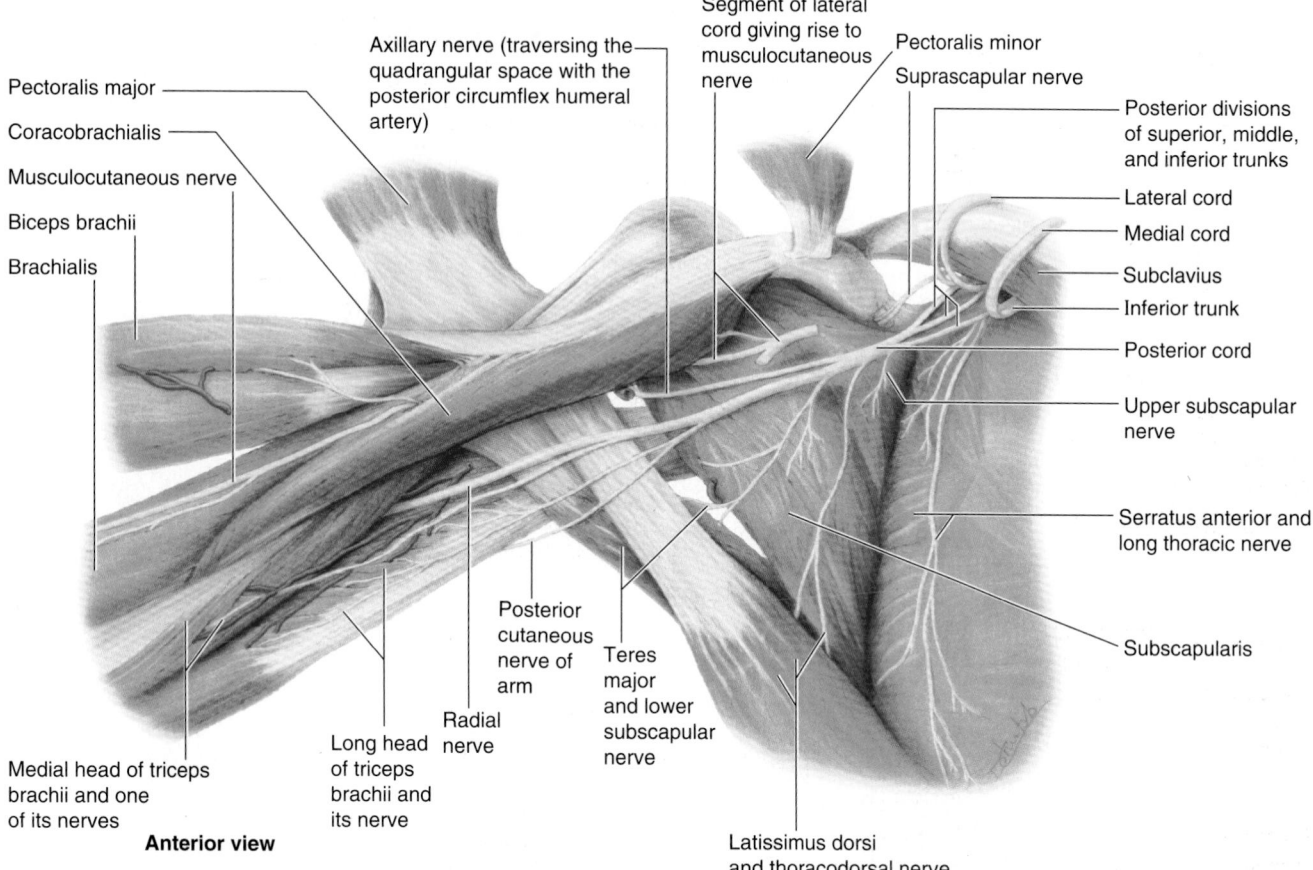

Pectoralis major

Coracobrachialis

Musculocutaneous nerve

Biceps brachii

Brachialis

Axillary nerve (traversing the quadrangular space with the posterior circumflex humeral artery)

Segment of lateral cord giving rise to musculocutaneous nerve

Pectoralis minor

Suprascapular nerve

Posterior divisions of superior, middle, and inferior trunks

Lateral cord

Medial cord

Subclavius

Inferior trunk

Posterior cord

Upper subscapular nerve

Serratus anterior and long thoracic nerve

Subscapularis

Posterior cutaneous nerve of arm

Teres major and lower subscapular nerve

Radial nerve

Long head of triceps brachii and its nerve

Medial head of triceps brachii and one of its nerves

Anterior view

Latissimus dorsi and thoracodorsal nerve

FIGURE 6.48. Nerves supplying medial and posterior walls of axilla, and muscles of arm. The pectoralis major and minor muscles are reflected supero-laterally, and the lateral and medial cords of the brachial plexus are reflected superomedially. All major vessels and the nerves arising from the medial and lateral cords of the brachial plexus (except for the musculocutaneous nerve arising from a segment of the lateral cord) are removed. The posterior cord, formed by the merging of the posterior divisions of all three trunks of the brachial plexus, is demonstrated. It gives rise to five peripheral nerves, four of which supply the muscles of the posterior wall of the axilla and posterior compartments of the upper limb.

Distally, the major attachment of the biceps is to the radial tuberosity via the biceps tendon. However, a triangular membranous band, the **bicipital aponeurosis,** runs from the biceps tendon across the cubital fossa, and merges with the antebrachial (deep) fascia covering the flexor muscles in the medial side of the forearm. It attaches indirectly by means of the fascia to the subcutaneous border of the ulna. The proximal part of the aponeurosis can be easily felt where it passes obliquely over the brachial artery and median nerve (Figs. 6.47A and 6.52A). The aponeurosis affords protection for these and other structures in the cubital fossa. It also helps lessen the pressure of the biceps tendon on the radial tuberosity during pronation and supination of the forearm.

To test the biceps brachii, the elbow joint is flexed against resistance when the forearm is supinated. If acting normally, the muscle forms a prominent bulge on the anterior aspect of the arm that is easily palpated.

BRACHIALIS

The **brachialis** is a flattened fusiform muscle that lies posterior (deep) to the biceps. Its distal attachment covers the

anterior part of the elbow joint (Figs. 6.47, 6.48, and 6.49D; Table 6.9). The brachialis is the main flexor of the forearm. It is the only pure flexor, producing the greatest amount of flexion force. It flexes the forearm in all positions, not being affected by pronation and supination; during both slow and quick movements; and in the presence or absence of resistance. When the forearm is extended slowly, the brachialis steadies the movement by slowly relaxing—that is, eccentric contraction (you use it to pick up and put down a teacup carefully, for example). The brachialis always contracts when the elbow is flexed and is primarily responsible for sustaining the flexed position. Because of its important and almost constant role, it is regarded as the workhorse of the elbow flexors.

To test the brachialis, the forearm is semipronated and flexed against resistance. If acting normally, the contracted muscle can be seen and palpated.

CORACOBRACHIALIS

The **coracobrachialis** is an elongated muscle in the supero-medial part of the arm. It is a useful landmark for locating other structures in the arm (Figs. 6.47, 6.48, and 6.49C;

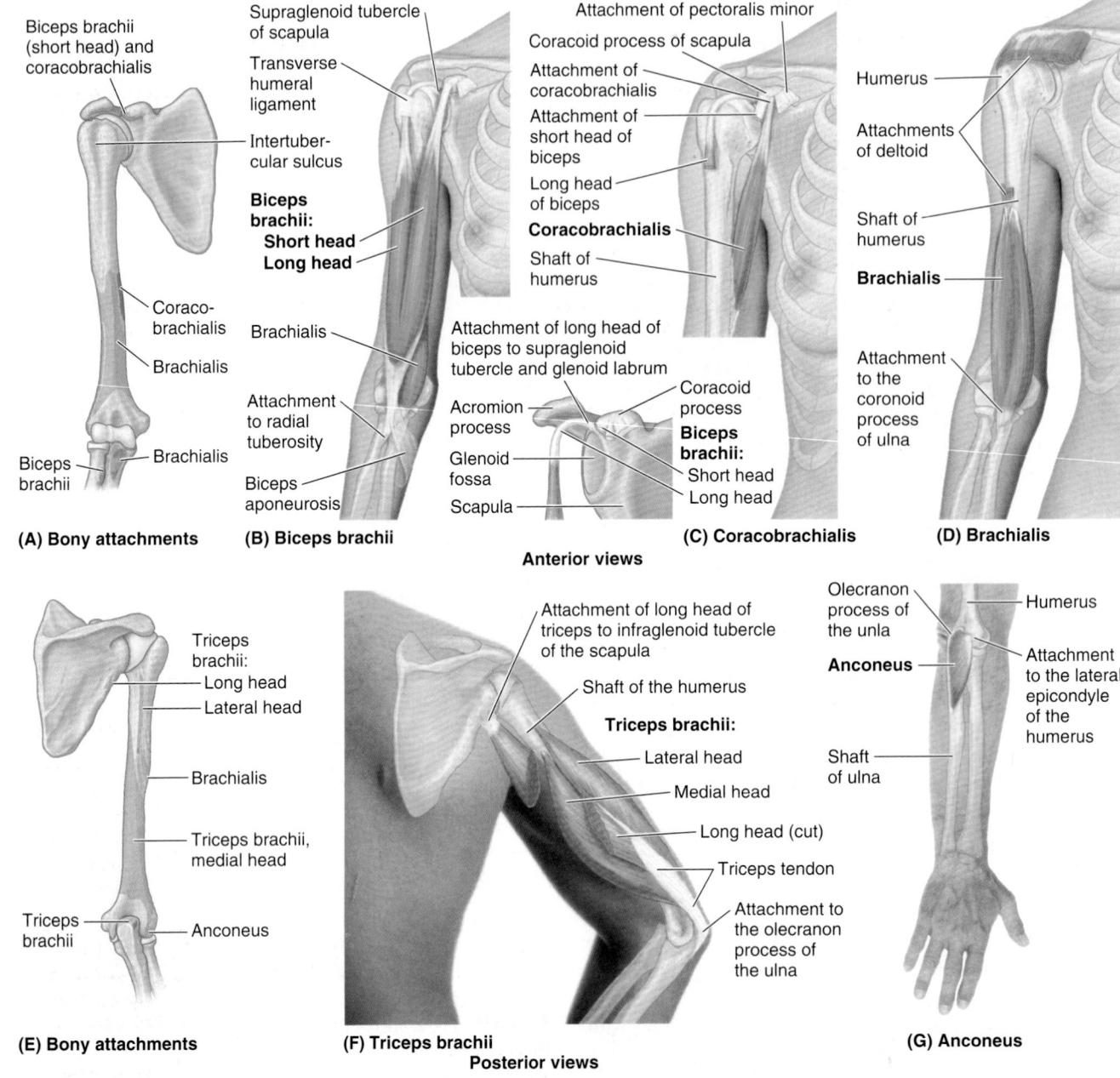

(A) Bony attachments   **(B) Biceps brachii**   **(C) Coracobrachialis**   **(D) Brachialis**

Anterior views

(E) Bony attachments   **(F) Triceps brachii**   **(G) Anconeus**

Posterior views

FIGURE 6.49. Muscles of arm.

TABLE 6.9. MUSCLES OF ARM

Muscle	Proximal Attachment	Distal Attachment	Innervation[a]	Muscle Action
Biceps brachii	*Short head*: tip of coracoid process of scapula *Long head*: supraglenoid tubercle of scapula	Tuberosity of radius and fascia of forearm via bicipital aponeurosis	Musculocutaneous nerve (C5, **C6**, C7)	Supinates forearm and, when it is supine. flexes forearm; short head resists dislocation of shoulder
Coracobrachialis	Tip of coracoid process of scapula	Middle third of medial surface of humerus		Helps flex and adduct arm; resists dislocation of shoulder
Brachialis	Distal half of anterior surface of humerus	Coronoid process and tuberosity ulna	Musculocutaneous nerve[b] (C5, C6) and radial nerve (C5, C7)	Flexes forearm in all positions

(continued)

TABLE 6.9. MUSCLES OF ARM (Continued)

Muscle	Proximal Attachment	Distal Attachment	Innervation[a]	Muscle Action
Triceps brachii	*Long head*: infraglenoid tubercle of scapula *Lateral head*: posterior surface of humerus, superior to radial groove *Medial head*: posterior surface of humerus, inferior to radial groove	Proximal end of olecranon of ulna and fascia of forearm	Radial nerve (C6, **C7, C8**)	Chief extensor of forearm; long head resists dislocation of humerus; especially important during adduction
Anconeus	Lateral epicondyle of humerus	Lateral surface of olecranon and superior part of posterior surface of ulna	Radial nerve (C7, C8, T1)	Assists triceps in extending forearm; stabilizes elbow joint; may abduct ulna during pronation

[a]The spinal cord segmental innervation is indicated (e.g., "C5, **C6**, C7" means that the nerves supplying the biceps brachii are derived from the fifth and sixth cervical segments of the spinal cord). Numbers in boldface (**C6**) indicate the main segmental innervation. Damage to one or more of the listed spinal cord segments or to the motor nerve roots arising from them results in paralysis of the muscles concerned.

[b]Some of the lateral part of the brachialis is innervated by a branch of the radial nerve.

Table 6.9). For example, the musculocutaneous nerve pierces it, and the distal part of its attachment indicates the location of the nutrient foramen of the humerus. The coracobrachialis helps flex and adduct the arm and stabilize the glenohumeral joint. With the deltoid and long head of the triceps, it serves as a *shunt muscle*, resisting downward dislocation of the head of the humerus, as when carrying a heavy suitcase. The median nerve and/or the brachial artery may run deep to the coracobrachialis and be compressed by it.

TRICEPS BRACHII

The **triceps brachii** is a large fusiform muscle in the posterior compartment of the arm (Figs. 6.47, 6.48, 6.49F, and 6.50; Table 6.9). As indicated by its name, *the triceps has three heads*: long, lateral, and medial. The triceps is the main extensor of the forearm.

Because its **long head** crosses the glenohumeral joint, the triceps helps stabilize the adducted glenohumeral joint by serving as a shunt muscle, resisting inferior displacement of the head of the humerus. The long head also aids in extension and adduction of the arm, but it is the least active head.

The **medial head** is the workhorse of forearm extension, active at all speeds and in the presence or absence of resistance.

The **lateral head** is the strongest but is it recruited into activity primarily against resistance (Hamill and Knutzen, 2008). Pronation and supination of the forearm do not affect triceps operation. Just proximal to the distal attachment of the triceps is a friction-reducing *subtendinous olecranon bursa*, between the triceps tendon and the olecranon.

To *test the triceps* (or to determine the level of a radial nerve lesion), the arm is abducted 90° and then the flexed forearm is extended against resistance provided by the examiner. If acting normally, the triceps can be seen and palpated. Its strength should be comparable with the contralateral

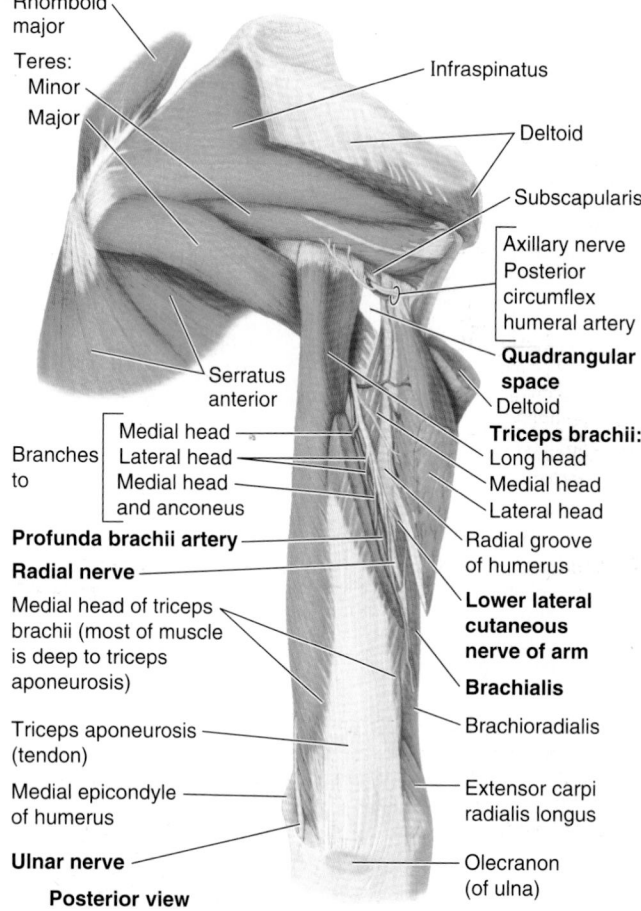

FIGURE 6.50. Muscles of scapular region and posterior region of arm. The lateral head of the triceps brachii is divided and displaced to show the structures traversing the *quadrangular space* and the radial nerve and the profunda brachii artery. The exposed bone of the radial groove, which is devoid of muscular attachment, separates the humeral attachments of the lateral and medial heads of the triceps. (Bony attachments are illustrated in Figure 6.49E.)

Figure labels:
Rhomboid major
Teres: Minor, Major
Serratus anterior
Branches to: Medial head, Lateral head, Medial head and anconeus
Profunda brachii artery
Radial nerve
Medial head of triceps brachii (most of muscle is deep to triceps aponeurosis)
Triceps aponeurosis (tendon)
Medial epicondyle of humerus
Ulnar nerve
Posterior view

Infraspinatus
Deltoid
Subscapularis
Axillary nerve
Posterior circumflex humeral artery
Quadrangular space
Deltoid
Triceps brachii: Long head, Medial head, Lateral head
Radial groove of humerus
Lower lateral cutaneous nerve of arm
Brachialis
Brachioradialis
Extensor carpi radialis longus
Olecranon (of ulna)

muscle, given consideration for lateral dominance (right or left handedness).

ANCONEUS

The **anconeus** is a small, triangular muscle on the posterolateral aspect of the elbow; it is usually partially blended with the triceps (Fig. 6.49G; Table 6.9). The anconeus helps the triceps extend the forearm and tenses the capsule of the elbow joint, preventing its being pinched during extension. It is also said to abduct the ulna during pronation of the forearm.

Brachial Artery

The **brachial artery** provides the main arterial supply to the arm and is the continuation of the axillary artery (Fig. 6.51). It begins at the inferior border of the teres major (Figs. 6.47A and 6.51), and ends in the cubital fossa opposite the neck of the radius where, under cover of the bicipital aponeurosis, it divides into the radial and ulnar arteries (Figs. 6.51 and 6.52).

The brachial artery, relatively superficial and palpable throughout its course, lies anterior to the triceps and brachialis. At first it lies medial to the humerus where its pulsations are palpable in the **medial bicipital groove** (Fig. 6.47A & B).

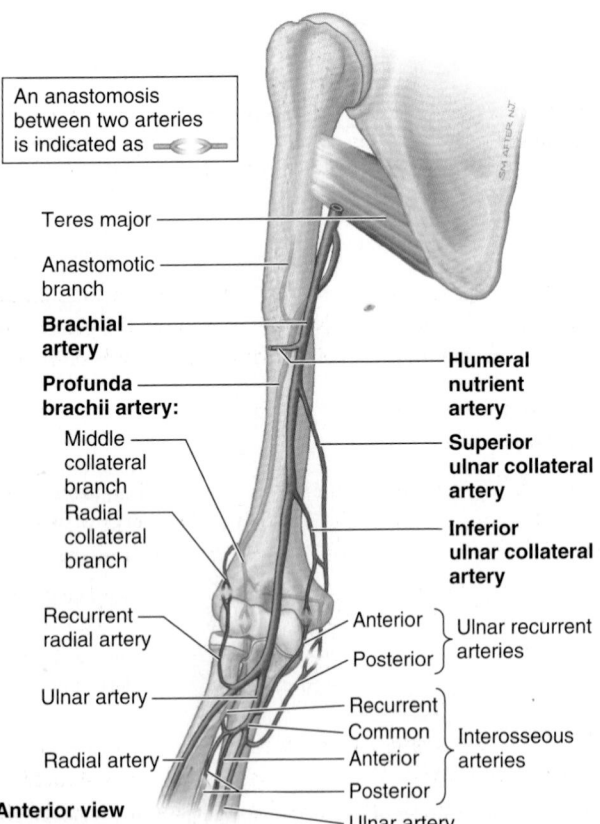

An anastomosis between two arteries is indicated as ⤜⤛

- Teres major
- Anastomotic branch
- **Brachial artery**
- **Profunda brachii artery:**
  - Middle collateral branch
  - Radial collateral branch
- Recurrent radial artery
- Ulnar artery
- Radial artery
- **Anterior view**

- **Humeral nutrient artery**
- **Superior ulnar collateral artery**
- **Inferior ulnar collateral artery**
- Anterior / Posterior } Ulnar recurrent arteries
- Recurrent / Common / Anterior / Posterior } Interosseous arteries
- Ulnar artery

FIGURE 6.51. Arterial supply of arm and proximal forearm. Functionally and clinically important peri-articular arterial anastomoses surround the elbow. The resulting collateral circulation allows blood to reach the forearm when flexion of the elbow compromises flow through the terminal part of the brachial artery.

It then passes anterior to the medial supra-epicondylar ridge and trochlea of the humerus (Figs. 6.51 and 6.53).

As it passes inferolaterally, the brachial artery accompanies the median nerve, which crosses anterior to the artery (Figs. 6.47A and 6.53). During its course through the arm, the brachial artery gives rise to many unnamed *muscular branches,* and the **humeral nutrient artery** (Fig. 6.51), which arise from its lateral aspect. The unnamed muscular branches are often omitted from illustrations, but they are evident during dissection.

The main named **branches of the brachial artery** arising from its medial aspect are the *profunda brachii artery (deep artery of the arm)* and the *superior* and *inferior ulnar collateral arteries.* The collateral arteries help form the **peri-articular arterial anastomoses of the elbow region** (Fig. 6.51). Other arteries involved are recurrent branches, sometimes double, from the radial, ulnar, and interosseous arteries, which run superiorly anterior and posterior to the elbow joint. These arteries anastomose with descending articular branches of the deep artery of the arm and the ulnar collateral arteries.

PROFUNDA BRACHII ARTERY

The **profunda brachii artery** (deep artery of the arm) is the largest branch of the brachial artery and has the most superior origin. The profunda brachii accompanies the radial nerve along the radial groove as it passes posteriorly around the shaft of the humerus (Figs. 6.50 and 6.53). The profunda brachii terminates by dividing into **middle** and **radial collateral arteries,** which participate in the peri-articular arterial anastomoses around the elbow (Fig. 6.51).

HUMERAL NUTRIENT ARTERY

The main **humeral nutrient artery** arises from the brachial artery around the middle of the arm, and enters the *nutrient canal* on the anteromedial surface of the humerus (Fig. 6.51). The artery runs distally in the canal toward the elbow. Other smaller humeral nutrient arteries also occur.

SUPERIOR ULNAR COLLATERAL ARTERY

The **superior ulnar collateral artery** arises from the medial aspect of the brachial artery near the middle of the arm, and accompanies the ulnar nerve posterior to the medial epicondyle of the humerus (Figs. 6.47A and 6.51). Here it anastomoses with the posterior ulnar recurrent artery and the inferior ulnar collateral artery, participating in the peri-articular arterial anastomoses of the elbow.

INFERIOR ULNAR COLLATERAL ARTERY

The **inferior ulnar collateral artery** arises from the brachial artery approximately 5 cm proximal to the elbow crease (Figs. 6.47A, 6.51, and 6.52B). It then passes inferomedially anterior to the medial epicondyle of the humerus, and joins the peri-articular arterial anastomoses of the elbow region by anastomosing with the anterior ulnar recurrent artery.

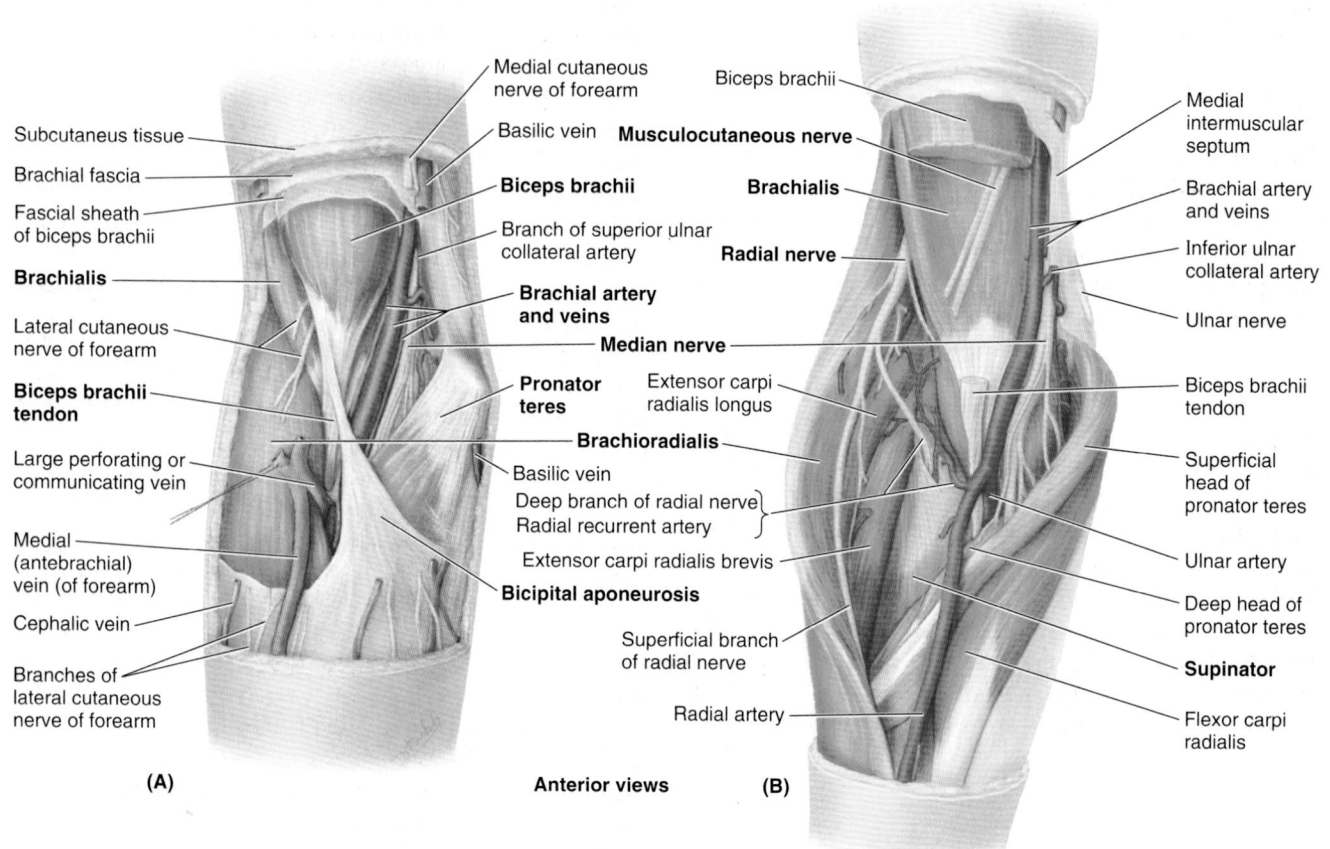

FIGURE 6.52. Dissections of cubital fossa. A. Superficial dissection. **B.** In this deep dissection, part of the biceps is excised and the cubital fossa is opened widely by retracting the forearm extensor muscles laterally and the flexor muscles medially. The radial nerve, which has just left the posterior compartment of the arm by piercing the lateral intermuscular septum, emerges between the brachialis and the brachioradialis and divides into a superficial (sensory) and a deep (motor) branch (details shown in Figure 6.57A & B).

Veins of Arm

Two sets of **veins of the arm,** superficial and deep, anastomose freely with each other. The superficial veins are in the subcutaneous tissue, and the deep veins accompany the arteries. Both sets of veins have valves, but they are more numerous in the deep veins than in the superficial veins.

SUPERFICIAL VEINS

The two main **superficial veins of the arm,** the **cephalic** and **basilic veins** (Figs. 6.47B & C and 6.52A), are described in "Superficial Veins of Upper Limb" on p. 689.

DEEP VEINS

Paired deep veins, collectively constituting the **brachial vein,** accompany the brachial artery (Fig. 6.52A). Their frequent connections encompass the artery, forming an anastomotic network within a common vascular sheath. The pulsations of the brachial artery help move the blood through this venous network.

The brachial vein begins at the elbow by union of the *accompanying veins of the ulnar and radial arteries,* and

ends by merging with the basilic vein to form the *axillary vein* (Figs. 6.16 and 6.41). Not uncommonly, the deep veins join to form one brachial vein during part of their course.

Nerves of Arm

Four main nerves pass through the arm: median, ulnar, musculocutaneous, and radial (Fig. 6.53). Their origins from the brachial plexus, courses in the upper limb, and the structures innervated by them are summarized in Table 6.8. The median and ulnar nerves supply no branches to the arm.

MUSCULOCUTANEOUS NERVE

The **musculocutaneous nerve** begins opposite the inferior border of the pectoralis minor, pierces the coracobrachialis, and continues distally between the biceps and the brachialis (Fig. 6.52B). After supplying all three muscles of the anterior compartment of the arm, the musculocutaneous nerve emerges lateral to the biceps as the *lateral cutaneous nerve of the forearm* (Fig. 6.53). It becomes truly subcutaneous when it pierces the deep fascia proximal to the cubital fossa to course initially

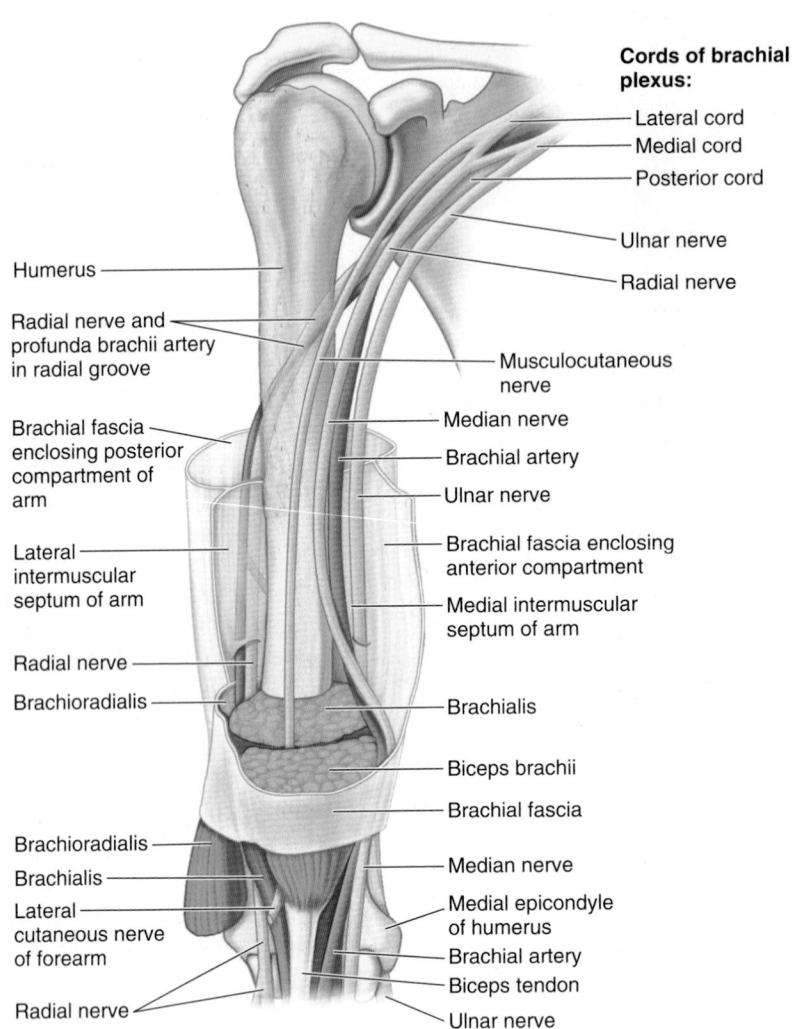

Cords of brachial plexus:
— Lateral cord
— Medial cord
— Posterior cord

— Ulnar nerve
— Radial nerve

Humerus —

Radial nerve and profunda brachii artery in radial groove

— Musculocutaneous nerve
— Median nerve

Brachial fascia enclosing posterior compartment of arm

— Brachial artery
— Ulnar nerve

Lateral intermuscular septum of arm

— Brachial fascia enclosing anterior compartment
— Medial intermuscular septum of arm

Radial nerve —
Brachioradialis —

— Brachialis

— Biceps brachii

— Brachial fascia

Brachioradialis —
Brachialis —
Lateral cutaneous nerve of forearm —

— Median nerve
— Medial epicondyle of humerus
— Brachial artery
— Biceps tendon

Radial nerve —

— Ulnar nerve

FIGURE 6.53. Relationship of arteries and nerves of arm to humerus and compartments of arm. The radial nerve and accompanying profunda brachii artery wind posteriorly around, and directly on the surface of, the humerus in the radial groove. The radial nerve and radial collateral artery then pierce the lateral intermuscular septum to enter the anterior compartment. The ulnar nerve pierces the medial intermuscular septum to enter the posterior compartment and then lies in the groove for the ulnar nerve on the posterior aspect of the medial epicondyle of the humerus. The median nerve and brachial artery descend in the arm to the medial side of the cubital fossa, where it is well protected and rarely injured. (Details are shown in Figures 6.50 and 6.57A & B.)

with the cephalic vein in the subcutaneous tissue (Fig. 6.52A). After crossing the anterior aspect of the elbow, it continues to supply the skin of the lateral aspect of the forearm.

RADIAL NERVE

The **radial nerve** in the arm supplies all the muscles in the posterior compartment of the arm (and forearm). The radial nerve enters the arm posterior to the brachial artery, medial to the humerus, and anterior to the long head of the triceps, where it gives branches to the long and medial heads of the triceps (Fig. 6.48). The radial nerve then descends inferolaterally with the profunda brachii artery and passes around the humeral shaft in the radial groove (Figs. 6.47B, 6.50, and 6.53). The branch of the radial nerve to the lateral head of the triceps arises within the radial groove. When it reaches the lateral border of the humerus, the radial nerve

pierces the lateral intermuscular septum, and continues inferiorly in the anterior compartment of the arm between the brachialis and the brachioradialis to the level of the lateral epicondyle of the humerus (Fig. 6.52B).

Anterior to the lateral epicondyle, the radial nerve divides into deep and superficial branches.

- The **deep branch of the radial nerve** is entirely muscular and articular in its distribution.
- The **superficial branch of the radial nerve** is entirely cutaneous in its distribution, supplying sensation to the dorsum of the hand and fingers.

MEDIAN NERVE

The *median nerve in the arm* runs distally in the arm on the lateral side of the brachial artery until it reaches the middle of the arm, where it crosses to the medial side and contacts the brachialis (Fig. 6.53). The median nerve then descends into the cubital fossa, where it lies deep to the bicipital aponeurosis and median cubital vein (Fig. 6.52). *The median nerve has no branches in the axilla or arm*, but it does supply articular branches to the elbow joint.

ULNAR NERVE

The *ulnar nerve in the arm* passes distally from the axilla anterior to the insertion of the teres major and to the long head of the triceps, on the medial side of the brachial artery (Fig. 6.47). Around the middle of the arm, it pierces the medial intermuscular septum with the superior ulnar collateral artery and descends between the septum and the medial head of the triceps (Fig. 6.53). The ulnar nerve passes posterior to the medial epicondyle and medial to the olecranon to enter the forearm (Fig. 6.46C). Posterior to the medial epicondyle, where the ulnar nerve is referred to in lay terms as the "funny bone." The ulnar nerve is superficial, easily palpable, and vulnerable to injury. Like the median nerve, the ulnar nerve has no branches in the arm, but it also supplies articular branches to the elbow joint.

Cubital Fossa

The **cubital fossa** is seen superficially as a depression on the anterior aspect of the elbow (Fig. 6.55A). Deeply, it is a space filled with a variable amount of fat anterior to the most distal part of the humerus and the elbow joint. The three *boundaries of the triangular cubital fossa* are as follows (Fig. 6.52):

1. Superiorly, an imaginary line connecting the *medial and lateral epicondyles.*
2. Medially, the mass of flexor muscles of the forearm arising from the common flexor attachment on the medial epicondyle; most specifically, the *pronator teres.*
3. Laterally, the mass of extensor muscles of the forearm arising from the lateral epicondyle and supra-epicondylar ridge; most specifically, the *brachioradialis.*

The *floor of the cubital fossa* is formed by the brachialis and supinator muscles of the arm and forearm, respectively. The *roof of the cubital fossa* is formed by the continuity of brachial and antebrachial (deep) fascia reinforced by the bicipital aponeurosis (Figs. 6.52 and 6.58), subcutaneous tissue, and skin. The *contents of the cubital fossa* are the (Figs. 6.52 and 6.57A):

- Terminal part of the brachial artery and the commencement of its terminal branches, the radial and ulnar arteries. The brachial artery lies between the biceps tendon and the median nerve.
- (Deep) accompanying veins of the arteries.
- Biceps brachii tendon.
- Median nerve.
- Radial nerve, deep between the muscles forming lateral boundary of the fossa (the brachioradialis, in particular)

and the brachialis, dividing into its superficial and deep branches. The muscles must be retracted to expose the nerve.

Superficially, in the subcutaneous tissue overlying the cubital fossa are the *median cubital vein,* lying anterior to the brachial artery, and the *medial and lateral cutaneous nerves of the forearm,* related to the basilic and cephalic veins (Fig. 6.55).

Surface Anatomy of Arm and Cubital Fossa

The borders of the *deltoid* are visible when the arm is abducted against resistance. The *distal attachment of the deltoid* can be palpated on the lateral surface of the humerus (Fig. 6.54A).

The *long, lateral, and medial heads of the triceps brachii* forms bulges on the posterior aspect of the arm and are identifiable when the forearm is extended from the flexed position against resistance. The *olecranon,* to which the triceps tendon attaches distally, is easily palpated. It is separated from the skin by only the *olecranon bursa,* which accounts for the mobility of the overlying skin. The *triceps tendon* is easily felt as it descends along the posterior aspect of the arm to the olecranon. The fingers can be pressed inward on each side of the tendon, where the elbow joint is superficial. An abnormal collection of fluid in the elbow joint or in the *subtendinous bursa of the triceps brachii* is palpable at these sites; the bursa lies deep to the triceps tendon (see Figs. 6.97 and 6.101).

The *biceps brachii* forms a bulge on the anterior aspect of the arm; its belly becomes more prominent when the elbow is flexed and supinated against resistance (Fig. 6.54B). The *biceps brachii tendon* can be palpated in the cubital fossa, immediately lateral to the midline, especially when the elbow is flexed against resistance. The proximal part of the *bicipital aponeurosis* can be palpated where it passes obliquely over the brachial artery and median nerve. Medial and lateral *bicipital grooves* separate the bulges formed by the biceps and triceps and indicate the location of the medial and lateral intermuscular septa (Fig. 6.54C). The cephalic vein runs superiorly in the lateral bicipital groove, and the basilic vein ascends in the medial bicipital groove. Deep to the latter is the main neurovascular bundle of the limb.

No part of the shaft of the humerus is subcutaneous; however, it can be palpated with varying distinctness through the muscles surrounding it, especially in many elderly people.

The *head of the humerus* is surrounded by muscles on all sides, except inferiorly; thus it can be palpated by pushing the fingers well up into the axilla. The arm should be close to the side so the axillary fascia is loose. The humeral head can be palpated when the arm is moved while the inferior angle of the scapula is held in place.

The *brachial artery* may be felt pulsating deep to the medial border of the biceps. The medial and lateral *epicondyles of the humerus* are subcutaneous and can be

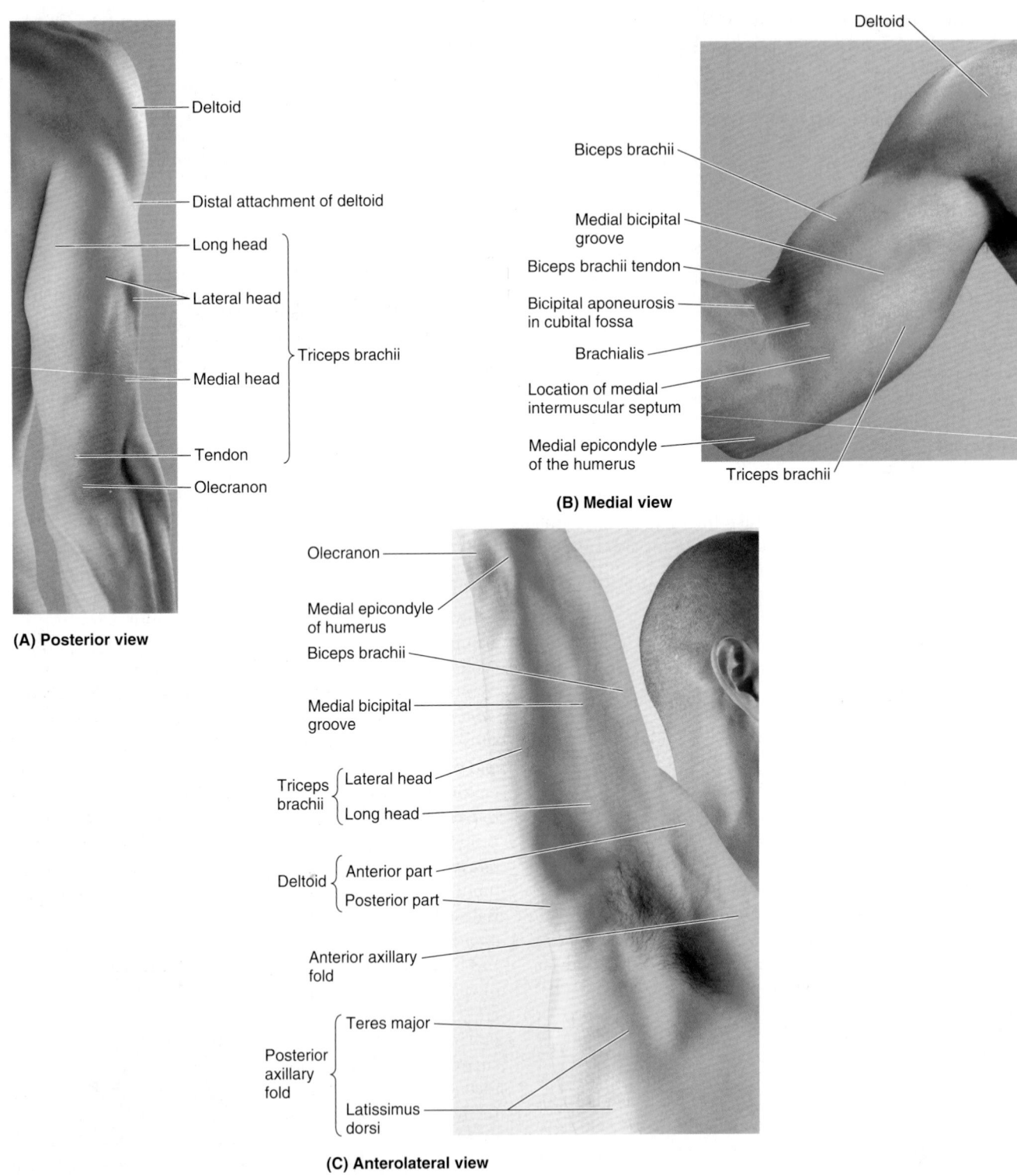

(A) Posterior view

Deltoid

Distal attachment of deltoid

Long head

Lateral head

Triceps brachii

Medial head

Tendon

Olecranon

(B) Medial view

Deltoid

Biceps brachii

Medial bicipital groove

Biceps brachii tendon

Bicipital aponeurosis in cubital fossa

Brachialis

Location of medial intermuscular septum

Medial epicondyle of the humerus

Triceps brachii

(C) Anterolateral view

Olecranon

Medial epicondyle of humerus

Biceps brachii

Medial bicipital groove

Triceps brachii { Lateral head / Long head }

Deltoid { Anterior part / Posterior part }

Anterior axillary fold

Posterior axillary fold { Teres major / Latissimus dorsi }

FIGURE 6.54. Surface anatomy of arm.

easily palpated at the medial and lateral aspects of the elbow. The medial epicondyle is more prominent.

In the cubital fossa, the *cephalic and basilic veins* in the subcutaneous tissue are clearly visible when a tourniquet is applied to the arm, as is the *median cubital vein*. This vein crosses the bicipital aponeurosis as it runs superomedially connecting the cephalic to the basilic vein (Fig. 6.55).

If the thumb is pressed into the cubital fossa, the muscular masses of the *long flexors of the forearm* will be felt forming the medial border, the pronator teres most directly.

The *lateral group of forearm extensors* (a soft mass that can be grasped separately), the brachioradialis (most medial) and the long and short extensors of the wrist, can be grasped between the fossa and the lateral epicondyle.

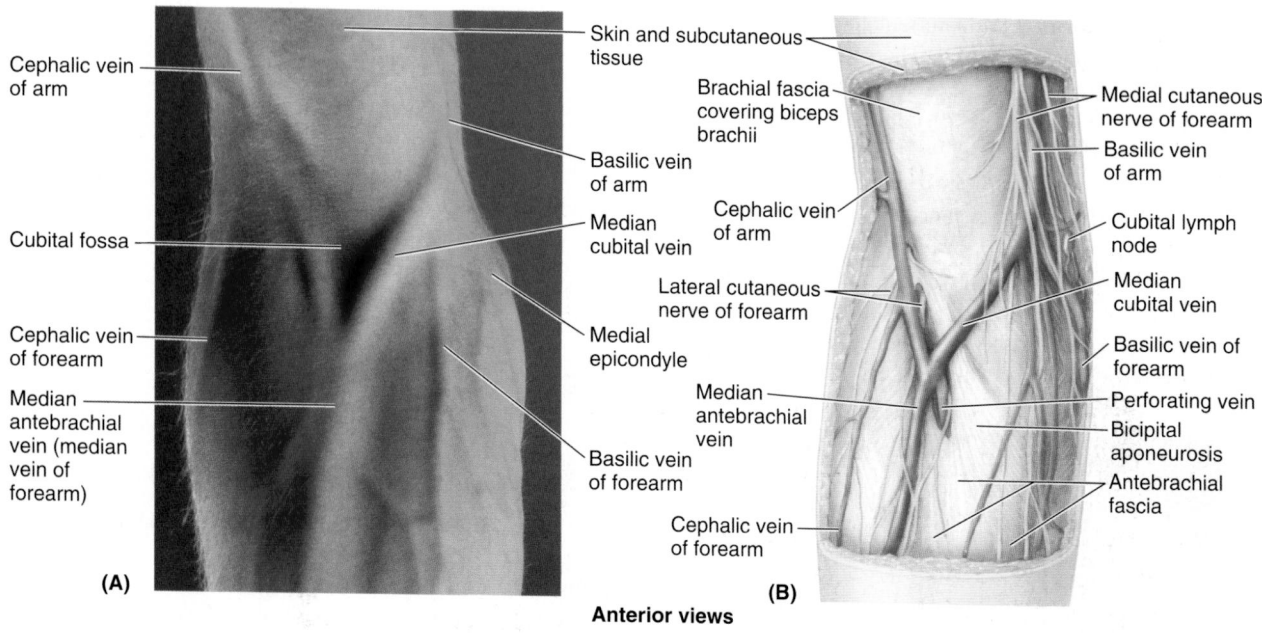

Cephalic vein of arm

Cubital fossa

Cephalic vein of forearm

Median antebrachial vein (median vein of forearm)

(A)

Skin and subcutaneous tissue

Brachial fascia covering biceps brachii

Cephalic vein of arm

Lateral cutaneous nerve of forearm

Median antebrachial vein

Cephalic vein of forearm

Basilic vein of arm

Median cubital vein

Medial epicondyle

Basilic vein of forearm

Medial cutaneous nerve of forearm

Basilic vein of arm

Cubital lymph node

Median cubital vein

Basilic vein of forearm

Perforating vein

Bicipital aponeurosis

Antebrachial fascia

(B)

Anterior views

FIGURE 6.55. Surface anatomy of cubital fossa.

ARM AND CUBITAL FOSSA

Bicipital Myotatic Reflex

The *biceps reflex* is one of several deep-tendon reflexes that are routinely tested during physical examinations. The relaxed limb is passively pronated and partially extended at the elbow. The examiner's thumb is firmly placed on the biceps tendon, and the reflex hammer is briskly tapped at the base of the nail bed of the examiner's thumb (Fig. B6.13). A normal (positive) response is an involuntary contraction of the biceps, felt as a momentarily tensed tendon, usually with a brief jerk-like flexion of the elbow. A positive response confirms the integrity of the musculocutaneous nerve and the C5 and C6 spinal cord segments. Excessive, diminished, or prolonged (hung) responses may indicate central or peripheral nervous system disease, or metabolic disorders (e.g., thyroid disease).

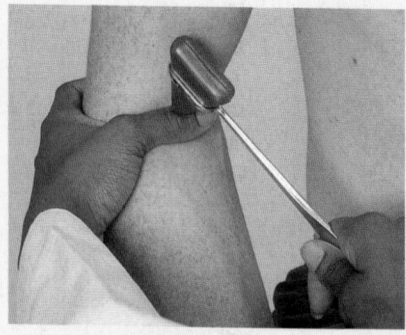

FIGURE B6.13. Method of eliciting biceps reflex.

Biceps Tendinitis

The tendon of the long head of the biceps is enclosed by a synovial sheath, and moves back and forth in the intertubercular sulcus (bicipital groove) of the humerus (Fig. 6.49B). Wear and tear of this mechanism can cause shoulder pain. Inflammation of the tendon (*biceps tendinitis*), usually the result of repetitive microtrauma, is common in sports involving throwing (e.g., baseball and cricket), and use of a racquet (e.g., tennis). A tight, narrow, and/or rough intertubercular sulcus may irritate and inflame the tendon, producing tenderness and *crepitus* (crackling sound).

Dislocation of Tendon of Long Head of Biceps Brachii

The tendon of the long head of the biceps can be partially or completely dislocated from the intertubercular sulcus in the humerus. This painful condition may occur in young persons during traumatic separation of the proximal epiphysis of the humerus. The injury also occurs in older persons with a history of biceps tendinitis. Usually a sensation of popping or catching is felt during arm rotation.

Rupture of Tendon of Long Head of Biceps Brachii

Rupture of the tendon usually results from wear and tear of an inflamed tendon as it moves back and forth in the intertubercular sulcus of the humerus. This injury usually occurs in individuals > 35 years of age. Typically, the tendon is torn from its attachment to the supraglenoid tubercle of the scapula (Fig. 6.5D). The rupture is commonly dramatic and is associated with a snap or pop. The detached muscle belly forms a ball near the center of the distal part of the anterior aspect of the arm (Popeye deformity) (Fig. B6.14). *Rupture of the biceps tendon* may result from forceful flexion of the arm against excessive resistance, as occurs in weight lifters (Anderson et al., 2000). However, the tendon ruptures more often as the result of prolonged tendinitis that weakens it. The rupture results from repetitive overhead motions, such as occurs in swimmers and baseball pitchers, that tear the weakened tendon in the intertubercular sulcus.

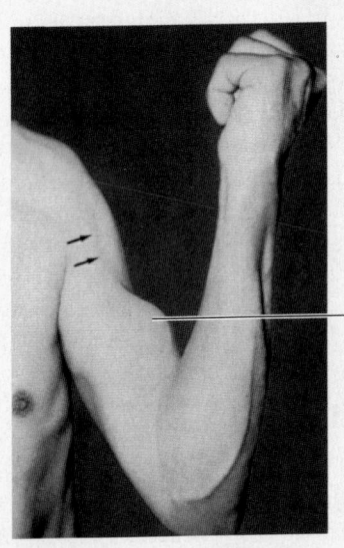

FIGURE B6.14. Rupture of biceps tendon (*arrows*).

Interruption of Blood Flow in Brachial Artery

Stopping bleeding through manual or surgical control of blood flow is called *hemostasis*. The best place to compress the brachial artery to control hemorrhage is medial to the humerus near the middle of the arm (Fig. B6.15). Because the arterial anastomoses around the elbow provide a functionally and surgically important collateral circulation, the brachial artery may be clamped distal to the origin of the deep artery of the arm without producing tissue damage (Fig. 6.51). The anatomical basis for this procedure is that the ulnar and radial arteries will still receive sufficient blood through the anastomoses around the elbow.

Although collateral pathways confer some protection against gradual temporary and partial occlusion, sudden complete occlusion or laceration of the brachial artery creates a surgical emergency because paralysis of muscles results from *ischemia of the elbow and forearm* within a few hours.

Muscles and nerves can tolerate up to 6 hours of ischemia (Salter, 1999). After this, fibrous scar tissue replaces necrotic tissue and causes the involved muscles to shorten permanently, producing a flexion deformity, the *ischemic compartment syndrome* (Volkmann or ischemic contracture). Flexion of the

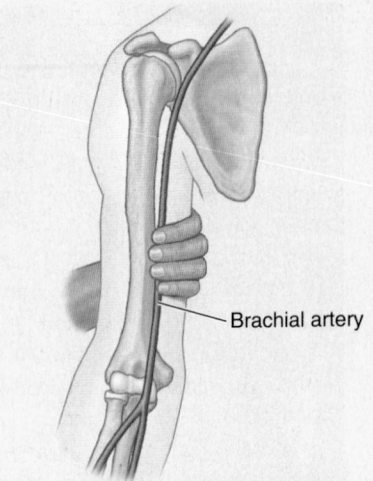

FIGURE B6.15. Compression of brachial artery.

fingers and sometimes the wrist results in loss of hand power as a result of irreversible necrosis of the forearm flexor muscles.

Fracture of Humeral Shaft

A *midhumeral fracture* may injure the radial nerve in the radial groove in the humeral shaft. When this nerve is damaged, the fracture is not likely to paralyze the triceps because of the high origin of the nerves to two of its three heads. A fracture of the distal part of the humerus, near the supra-epicondylar ridges, is called a *supra-epicondylar fracture* (Fig. B6.16). The distal bone fragment may be displaced anteriorly or posteriorly. The actions of the brachialis and triceps tend to pull the distal fragment over the proximal fragment, shortening the limb. Any of the nerves or branches of the brachial vessels related to the humerus may be injured by a displaced bone fragment.

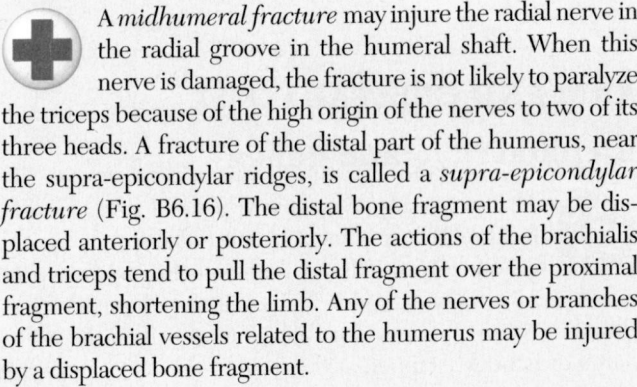
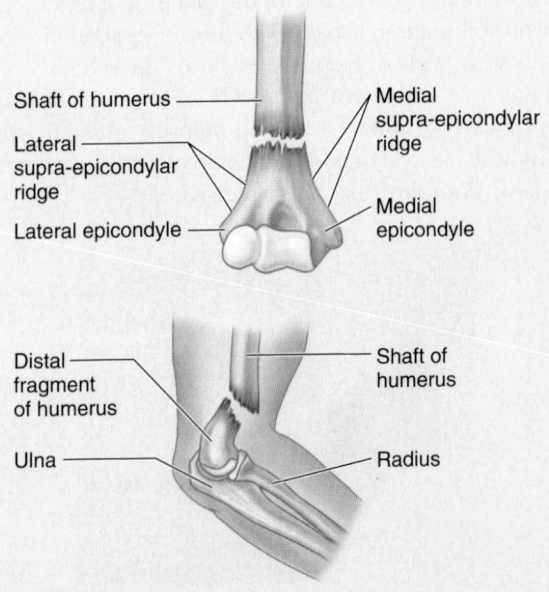

FIGURE B6.16. Supra-epicondylar fracture.

Injury to Musculocutaneous Nerve

Injury to the musculocutaneous nerve in the axilla (uncommon in this protected position) is typically inflicted by a weapon such as a knife. A musculocutaneous nerve injury results in *paralysis of the coracobrachialis, biceps, and brachialis.* Weak flexion may occur at the glenohumeral (shoulder) joint owing to the injury of the musculocutaneous nerve affecting the long head of the biceps brachii and the coracobrachialis. Consequently, flexion of the elbow joint and supination of the forearm are greatly weakened, but not lost. Weak flexion and supination are still possible, produced by the brachioradialis and supinator, respectively, both of which are supplied by the radial nerve. Loss of sensation may occur on the lateral surface of the forearm supplied by the lateral antebrachial cutaneous nerve, the continuation of the musculocutaneous nerve (Fig. 6.53)*.

Injury to Radial Nerve in Arm

Injury to the radial nerve superior to the origin of its branches to the triceps brachii results in *paralysis of the triceps, brachioradialis, supinator, and extensor muscles of the wrist and fingers.* Loss of sensation in areas of skin supplied by this nerve also occurs.

When the nerve is injured in the radial groove, the triceps is usually not completely paralyzed but only weakened because only the medial head is affected; however, the muscles in the posterior compartment of the forearm that are supplied by more distal branches of the nerve are paralyzed. The characteristic clinical sign of radial nerve injury is *wrist-drop*—inability to extend the wrist and the fingers at

the metacarpophalangeal joints (Fig. B6.17A). Instead, the relaxed wrist assumes a partly flexed position owing to unopposed tonus of flexor muscles and gravity (Fig. B6.17B).

Venipuncture in Cubital Fossa

The cubital fossa is the common site for sampling and transfusion of blood and intravenous injections because of the prominence and accessibility of veins. When the most common pattern of superficial veins is present, the median cubital vein is selected (Fig. 6.55). This vein lies directly on the deep fascia, running diagonally from the cephalic vein of the forearm to the basilic vein of the arm. It crosses the bicipital aponeurosis, which separates it from the underlying brachial artery and median nerve and provides some protection to the latter. Historically, during the days of bloodletting, the bicipital aponeurosis was known as the *grace Deux* (Fr., grace of God) *tendon,* by the grace of which arterial hemorrhage was usually avoided. A tourniquet is placed around the midarm to distend the veins in the cubital fossa. Once the vein is punctured, the tourniquet is removed so that when the needle is removed the vein will not bleed extensively. The median cubital vein is also a site for the introduction of cardiac catheters to secure blood samples from the great vessels and chambers of the heart. These veins may also be used for *coronary angiography* (see p. 154).

Variation of Veins in Cubital Fossa

The pattern of veins in the cubital fossa varies greatly. In approximately 20% of people, a **median antebrachial vein** (median vein of the forearm)

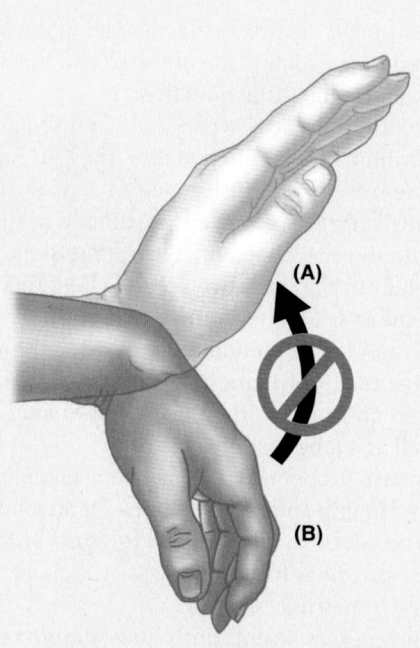

FIGURE B6.17. Wrist-drop.

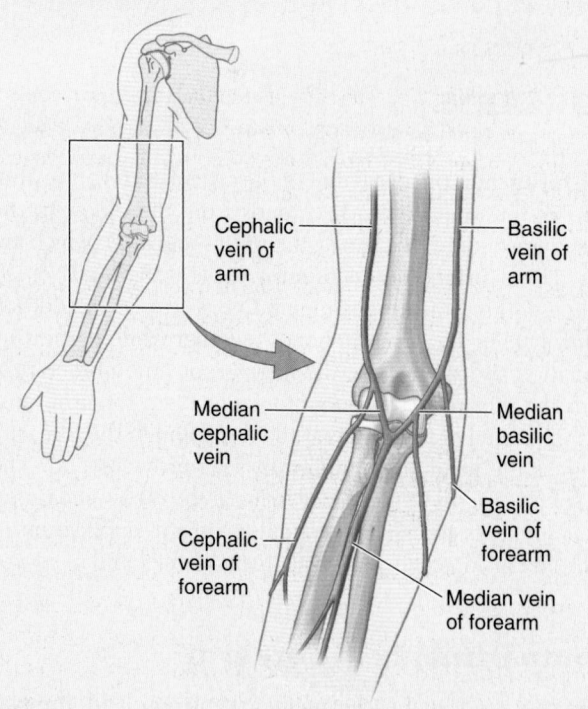

FIGURE B6.18.

divides into a **median basilic vein,** which joins the basilic vein of the arm. and a **median cephalic vein,** that joins the cephalic vein of the arm (Fig. B6.18). In these cases, a clear *M* formation is produced by the cubital veins. It is important to observe and remember that either the median cubital vein or the median basilic vein, whichever pattern is present,

crosses superficial to the brachial artery, from which it is separated by the bicipital aponeurosis. These veins are good sites for drawing blood but are not ideal for injecting an irritating drug because of the danger of injecting it into the brachial artery. In obese people, a considerable amount of fatty tissue may overlie the vein.

The Bottom Line

ARM AND CUBITAL FOSSA

Arm: The arm forms a column with the humerus at its center. ♦ The humerus, along with intermuscular septa in its distal two thirds, divides the arm lengthwise (or more specifically, the space inside the brachial fascia) into anterior or flexor and posterior or extensor compartments.

The anterior compartment contains three flexor muscles supplied by the musculocutaneous nerve. ♦ The coracobrachialis acts (weakly) at the shoulder, and the biceps and brachialis act at the elbow. ♦ The biceps is also the primary supinator of the forearm (when the elbow is flexed). ♦ The brachialis is the primary flexor of the forearm.

The posterior compartment contains a three-headed extensor muscle, the triceps, which is supplied by the radial nerve. ♦ One of the heads (the long head) acts at the shoulder, but mostly the heads work together to extend the elbow.

Both compartments of the arm are supplied by the brachial artery, the posterior compartment primarily via its major branch, the profunda brachii artery. ♦ The primary

neurovascular bundle is located on the medial side of the limb; thus it is usually protected by the limb it serves.

Cubital fossa: The triangular cubital fossa is bound by a line connecting the medial and lateral epicondyles of the humerus, and the pronator teres and brachioradialis muscles arising, respectively, from the epicondyles. ♦ The brachialis and supinator form the floor. ♦ The biceps tendon descends into the triangle to insert on the radial tuberosity. ♦ Medial to the tendon are the median nerve and terminal part of the brachial artery. ♦ Lateral to the tendon is the lateral cutaneous nerve of the forearm superficially, and—at a deeper level—the terminal part of the radial nerve. ♦ In the subcutaneous tissue, most commonly a median cubital veins runs obliquely across the fossa, connecting the cephalic vein of the forearm and basilic vein of the arm, providing an advantageous site for venipuncture. ♦ In about one fifth of the population, a median antebrachial vein bifurcates into median cephalic and median basilic veins, which replace the diagonal median cubital vein.

FOREARM

The **forearm** is the distal unit of the articulated strut (extension) of the upper limb. It extends from the elbow to the wrist and contains two bones, the *radius* and *ulna*, which are joined by an **interosseous membrane** (Fig. 6.56A, B, & D). Although thin, this fibrous membrane is strong. In addition to firmly tying the forearm bones together while permitting pronation and supination, the interosseous membrane provides the proximal attachment for some deep forearm muscles. The head of the ulna is at the distal end of the forearm, whereas the head of the radius is at its proximal end. The role of forearm movement, occurring at the elbow and radioulnar joints, is to assist the shoulder in the application of force and in controlling the placement of the hand in space.

Compartments of Forearm

As in the arm, the muscles of similar purpose and innervation are grouped within the same fascial compartments in the

forearm. Although the proximal boundary of the forearm per se is defined by the joint plane of the elbow, functionally the forearm includes the distal humerus.

For the distal forearm, wrist, and hand to have minimal bulk to maximize their functionality, they are operated by "remote control" by extrinsic muscles having their bulky, fleshy, contractile parts located proximally in the forearm, distant from the site of action. Their long, slender tendons extend distally to the operative site, like long ropes reaching to distant pulleys. Furthermore, because the structures on which the muscles and tendons act (wrist and fingers) have an extensive range of motion, a long range of contraction is needed, requiring that the muscles have long contractile parts as well as a long tendon(s).

The forearm proper is not, in fact, long enough to provide the required length and sufficient area for attachment proximally, so the proximal attachments (origins) of the muscles must occur proximal to the elbow—in the arm—and provided by the humerus.

Generally, flexors lie anteriorly and extensors posteriorly; however, the anterior and posterior aspects of the distal

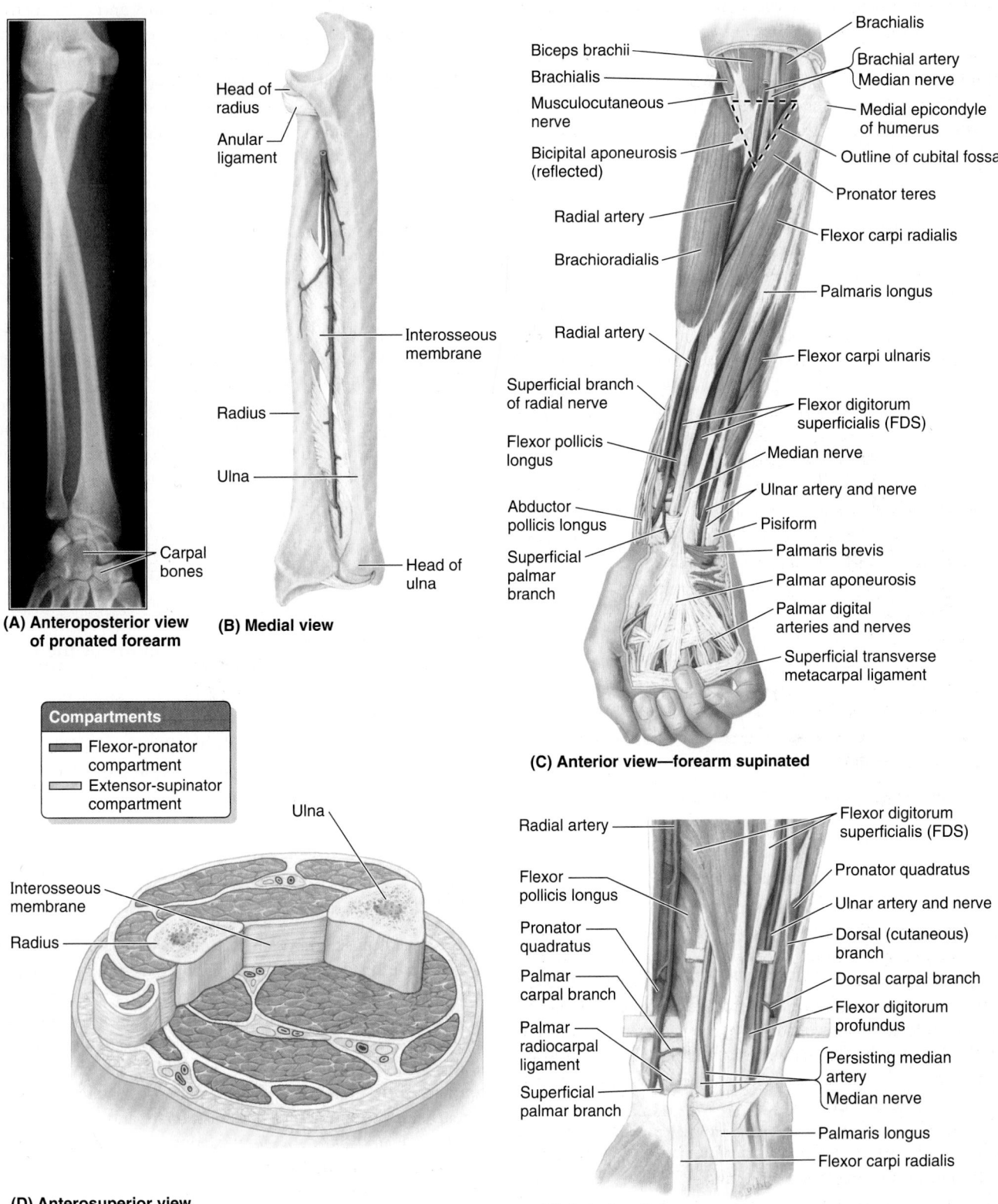

(A) Anteroposterior view of pronated forearm

Head of radius
Anular ligament
Interosseous membrane
Radius
Ulna
Head of ulna
Carpal bones

(B) Medial view

Biceps brachii
Brachialis
Musculocutaneous nerve
Bicipital aponeurosis (reflected)
Radial artery
Brachioradialis
Radial artery
Superficial branch of radial nerve
Flexor pollicis longus
Abductor pollicis longus
Superficial palmar branch

Brachialis
Brachial artery
Median nerve
Medial epicondyle of humerus
Outline of cubital fossa
Pronator teres
Flexor carpi radialis
Palmaris longus
Flexor carpi ulnaris
Flexor digitorum superficialis (FDS)
Median nerve
Ulnar artery and nerve
Pisiform
Palmaris brevis
Palmar aponeurosis
Palmar digital arteries and nerves
Superficial transverse metacarpal ligament

(C) Anterior view—forearm supinated

Compartments
▭ Flexor-pronator compartment
▭ Extensor-supinator compartment

Ulna
Interosseous membrane
Radius

(D) Anterosuperior view

Radial artery
Flexor pollicis longus
Pronator quadratus
Palmar carpal branch
Palmar radiocarpal ligament
Superficial palmar branch

Flexor digitorum superficialis (FDS)
Pronator quadratus
Ulnar artery and nerve
Dorsal (cutaneous) branch
Dorsal carpal branch
Flexor digitorum profundus
Persisting median artery
Median nerve
Palmaris longus
Flexor carpi radialis

(E) Anterior view

FIGURE 6.56. Bones, muscles, and flexor–pronator compartment of forearm. A. Anteroposterior (AP) radiograph of the forearm in pronation. (Courtesy of Dr. J. Heslin, Toronto, Ontario, Canada.) **B.** Bones of the forearm and radio-ulnar ligaments. **C.** Dissection showing the superficial muscles of the forearm and the palmar aponeurosis. **D.** Stepped transverse section demonstrating the compartments of the forearm. **E.** The flexor digitorum superficialis (FDS) and related structures. The ulnar artery emerges from its oblique course posterior to the FDS to meet and accompany the ulnar nerve.

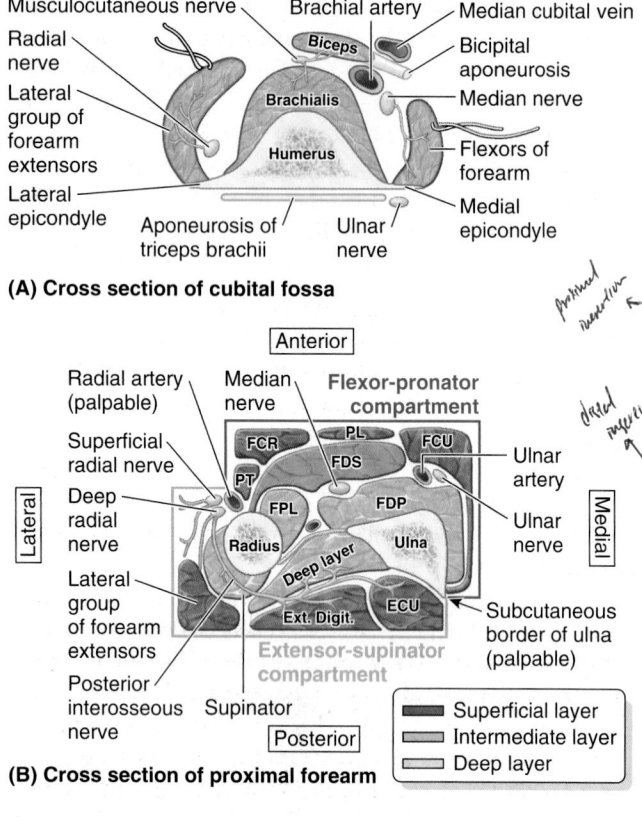

(A) Cross section of cubital fossa

(B) Cross section of proximal forearm

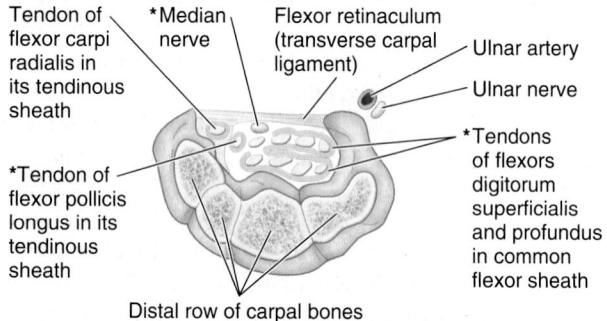

(C) Cross section of carpus (wrist)

Inferior views of transverse cross sections, right upper limb

FIGURE 6.57. **Cross sections demonstrating relationships at cubital fossa, proximal forearm, and wrist. A.** At the level of the cubital fossa, the flexors and extensor of the elbow occupy the anterior and posterior aspects of the humerus. Lateral and medial extensions (epicondyles and supra-epicondylar ridges) of the humerus provide proximal attachment (origin) for the forearm flexors and extensors. **B.** Consequently, in the proximal forearm, the "anterior" flexor–pronator compartment actually lies anteromedially, and the "posterior" extensor–supinator compartment lies posterolaterally. The radial artery (laterally) and the sharp, subcutaneous posterior border of the ulna (medially) are palpable features separating the anterior and posterior compartments. No motor nerves cross either demarcation, making them useful for surgical approaches. *Ext. digit.*, extensor digitorum; *ECU*, extensor carpi ulnaris; *FCR*, flexor carpi radialis; *FCU*, flexor carpi ulnaris; *FDP*, flexor digitorum profundus; *FDS*, flexor digitorum superficialis; *FPL*, flexor pollicis longus; *PL*, palmaris longus; *PT*, pronator teres. **C.** At the level of the wrist, nine tendons from three muscles (and one nerve) of the anterior compartment of the forearm traverse the carpal tunnel; eight of the tendons share a common synovial flexor sheath.

humerus are occupied by the chief flexors and extensors of the elbow (Fig. 6.57A). To provide the required attachment sites for the flexors and extensors of the wrist and fingers, medial and lateral extensions (epicondyles and supra-epicondylar ridges) have developed from the distal humerus.

The medial epicondyle and supra-epicondylar ridge provide attachment for the forearm flexors, and the lateral formations provide attachment for the forearm extensors. Thus, rather than lying strictly anteriorly and posteriorly, the proximal parts of the "anterior" (flexor–pronator) compartment of the forearm lie anteromedially, and the "posterior" (extensor–supinator) compartment lies posterolaterally (Figs. 6.56D, 6.57B, and 6.61C).

Spiraling gradually over the length of the forearm, the compartments become truly anterior and posterior in position in the distal forearm and wrist. These fascial compartments, containing the muscles in functional groups, are demarcated by the subcutaneous border of the ulna posteriorly (in the proximal forearm) and then medially (distal forearm) and by the radial artery anteriorly and then laterally. These structures are palpable (the artery by its pulsations) throughout the forearm. Because neither boundary is crossed by motor nerves, they also provide sites for surgical incision.

The **flexors and pronators of the forearm** are in the anterior compartment and are served mainly by the *median nerve;* the one and a half exceptions are innervated by the *ulnar nerve.* The **extensors and supinators of the forearm** are in the posterior compartment and are all served by the *radial nerve* (directly or by its deep branch).

The fascial compartments of the limbs generally end at the joints; therefore, fluids and infections in compartments are usually contained and cannot readily spread to other compartments. The anterior compartment is exceptional in this regard because it communicates with the central compartment of the palm through the carpal tunnel (Fig. 6.57C).

Muscles of Forearm

There are 17 muscles crossing the elbow joint, some of which act on the elbow joint exclusively, whereas others act at the wrist and fingers.

In the proximal part of the forearm, the muscles form fleshy masses extending inferiorly from the medial and lateral epicondyles of the humerus (Figs. 6.56C and 6.57A). The tendons of these muscles pass through the distal part of the forearm and continue into the wrist, hand, and fingers (Figs. 6.56C & E and 6.57C). The flexor muscles of the anterior compartment have approximately twice the bulk and strength of the extensor muscles of the posterior compartment.

FLEXOR–PRONATOR MUSCLES OF FOREARM

The **flexor muscles of the forearm** are in the **anterior (flexor–pronator) compartment of the forearm** and are separated from the *extensor muscles of the forearm* by the radius and ulna (Fig. 6.57B) and, in the distal two thirds of the forearm, by the interosseous membrane that connects them (Fig. 6.56B & D).

The tendons of most flexor muscles are located on the anterior surface of the wrist and are held in place by the **palmar carpal ligament** and the *flexor retinaculum (transverse carpal ligament)*, thickenings of the antebrachial fascia (Figs. 6.56C and 6.58).

The flexor/pronator muscles are arranged in three layers or groups (Fig. 6.59; Table 6.10):

1. A **superficial layer or group** of four muscles (pronator teres, flexor carpi radialis, palmaris longus, and flexor carpi ulnaris). These muscles are all attached proximally by a *common flexor tendon* to the medial epicondyle of the humerus, the *common flexor attachment.*

2. An **intermediate layer,** consisting of one muscle (flexor digitorum superficialis).

3. A **deep layer or group** of three muscles (flexor digitorum profundus, flexor pollicis longus, and pronator quadratus).

The five superficial and intermediate muscles cross the elbow joint; the three deep muscles do not. With the exception of the *pronator quadratus,* the more distally placed a muscle's distal attachment lies, the more distally and deeply placed is its proximal attachment.

All muscles in the anterior (flexor-pronator) compartment of the forearm are supplied by the median and/or ulnar nerves (most by the median; only one and a half exceptions are supplied by the ulnar).

Functionally, the brachioradialis is a flexor of the forearm, but it is located in the posterior (posterolateral) or extensor compartment and is thus supplied by the radial nerve. Therefore, the brachioradialis is a major exception to the rule that (1) the radial nerve supplies only extensor muscles and (2) that all flexors lie in the anterior (flexor) compartment.

The **long flexors of the digits** (flexor digitorum superficialis and flexor digitorum profundus) also flex the metacarpophalangeal and wrist joints. The flexor digitorum profundus flexes the fingers in slow action; this action is reinforced by the flexor digitorum superficialis when speed and flexion against resistance are required. When the wrist is flexed at the same time that the metacarpophalangeal and interphalangeal joints are flexed, the long flexor muscles of the fingers are operating over a shortened distance between attachments, and the action resulting from their contraction is consequently weaker. Extending the wrist increases their operating distance, and thus their contraction is more efficient in producing a strong grip (Fig. 6.73A).

Tendons of the long flexors of the digits pass through the distal part of the forearm, wrist, and palm and continue to the medial four fingers. The flexor digitorum superficialis flexes the middle phalanges, and the flexor digitorum profundus flexes the middle and distal phalanges.

The muscles of the anterior compartment of the forearm are illustrated in Figure 6.59 and their attachments, innervation, and main actions are listed by layers in Table 6.10. The following discussion provides additional details, beginning with the muscles of the superficial and intermediate layers.

Pronator Teres. The **pronator teres,** a fusiform muscle, is the most lateral of the superficial forearm flexors. Its lateral border forms the medial boundary of the cubital fossa.

To test the pronator teres, the person's forearm is flexed at the elbow and pronated from the supine position against resistance provided by the examiner. If acting normally, the muscle is prominent and can be palpated at the medial margin of the cubital fossa.

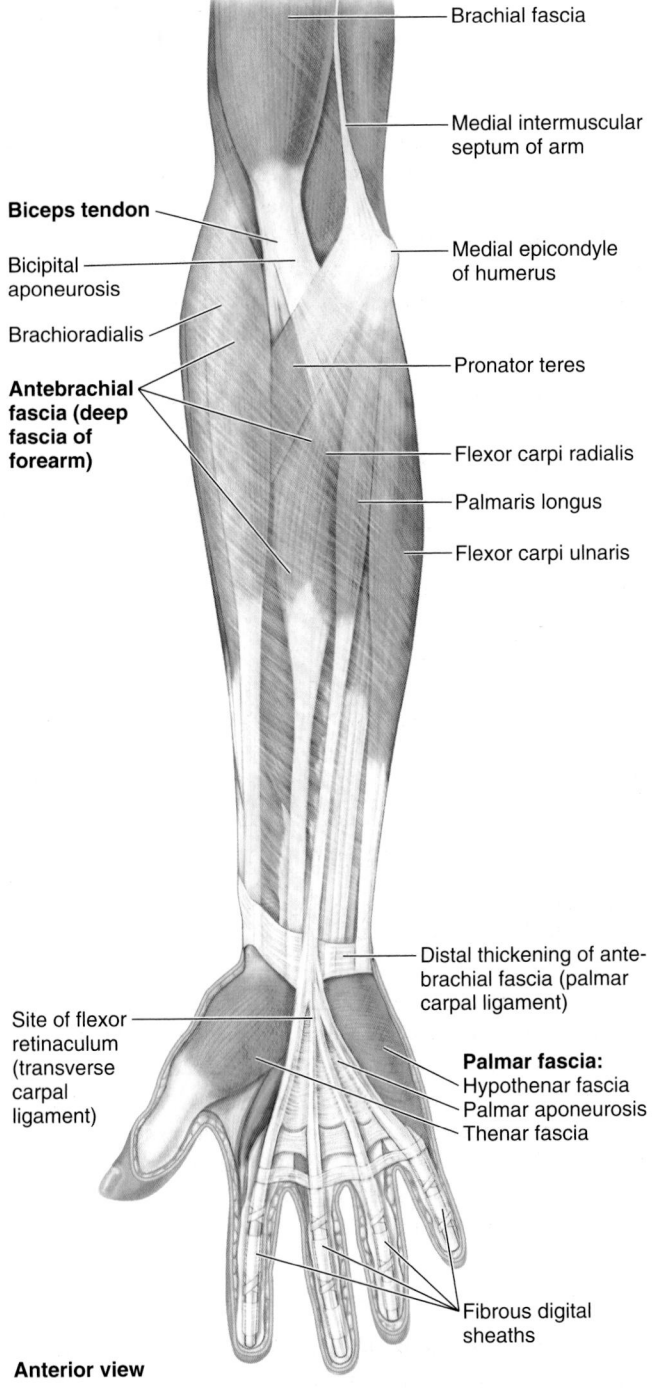

Brachial fascia

Medial intermuscular septum of arm

Biceps tendon

Bicipital aponeurosis

Brachioradialis

Medial epicondyle of humerus

Antebrachial fascia (deep fascia of forearm)

Pronator teres

Flexor carpi radialis

Palmaris longus

Flexor carpi ulnaris

Distal thickening of antebrachial fascia (palmar carpal ligament)

Site of flexor retinaculum (transverse carpal ligament)

Palmar fascia:
Hypothenar fascia
Palmar aponeurosis
Thenar fascia

Fibrous digital sheaths

Anterior view

FIGURE 6.58. Fascia of distal upper limb and superficial muscles of forearm.

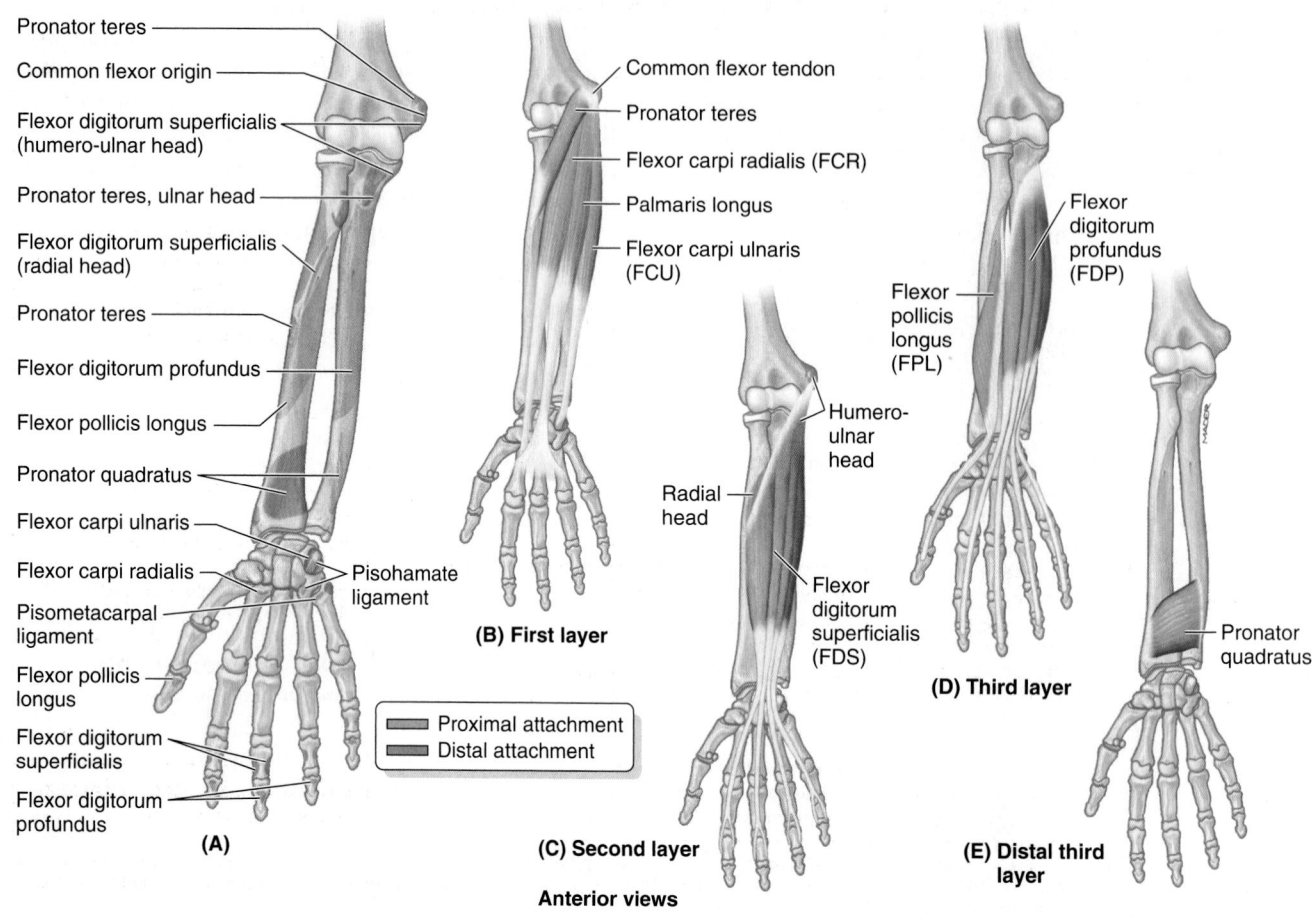

FIGURE 6.59. Flexor muscles of forearm.

TABLE 6.10. MUSCLES OF ANTERIOR COMPARTMENT OF FOREARM

Muscle	Proximal Attachment	Distal Attachment	Innervation[a]	Main Action
Superficial (first) layer				
Pronator teres				
Ulnar head	Coronoid process	Middle of convexity of lateral surface of radius		Pronates and flexes forearm (at elbow)
Humeral head			Median nerve (C6, **C7**)	
Flexor carpi radialis (FCR)		Base of 2nd metacarpal		Flexes and abducts hand (at wrist)
Palmaris longus	Medial epicondyle of humerus (common flexor origin)	Distal half of flexor reti-naculum and apex of palmar aponeurosis	Median nerve (C7, C8)	Flexes hand (at wrist) and tenses palmar aponeurosis
Flexor carpi ulnaris (FCU)				
Humeral head		Pisiform, hook of hamate, 5th metacarpal	Ulnar nerve (C7, **C8**)	Flexes and adducts hand (at wrist)
Ulnar head	Olecranon and posterior border of ulna (via aponeurosis)			

(continued)

TABLE 6.10. MUSCLES OF ANTERIOR COMPARTMENT OF FOREARM (Continued)

Muscle	Proximal Attachment	Distal Attachment	Innervation[a]	Main Action
Intermediate (second) layer				
Flexor digitorum superficialis (FDS)				
Humero-ulnar head	Medial epicondyle (common flexor origin and coronoid process)	Shafts of middle phalanges of medial four digits	Median nerve (C7, C8, T1)	Flexes middle phalanges at proximal interphalangeal joints of middle four digits; acting more strongly, it also flexes proximal phalanges at metacarpophalangeal joints
Radial head	Superior half of anterior border			
Deep (third) layer				
Flexor digitorum profundus (FDP)				
Medial part	Proximal three quarters of medial and anterior surfaces of ulna and interosseous membrane	Bases of distal phalanges of 4th and 5th digits	Ulnar nerve (C8, **T1**)	Flexes distal phalanges 4 and 5 at distal interphalangeal joints
Lateral part		Bases of distal phalanges of 2nd and 3rd digits		Flexes distal phalanges 2 and 3 at distal interphalangeal joints
Flexor pollicis longus (FPL)	Anterior surface of radius and adjacent interosseous membrane	Base of distal phalanx of thumb	Anterior interosseous nerve, from median nerve (**C8**, T1)	Flexes phalanges of 1st digit (thumb)
Pronator quadratus	Distal quarter of anterior surface of ulna	Distal quarter of anterior surface of radius		Pronates forearm; deep fibers bind radius and ulna together

[a]The spinal cord segmental innervation is indicated (e.g., "C6, **C7**" means that the nerves supplying the pronator teres are derived from the sixth and seventh cervical segments of the spinal cord). Numbers in boldface (**C7**) indicate the main segmental innervation. Damage to one or more of the listed spinal cord segments or to the motor nerve roots arising from them results in paralysis of the muscles concerned.

Flexor Carpi Radialis. The **flexor carpi radialis (FCR)** is a long fusiform muscle located medial to the pronator teres. In the middle of the forearm, its fleshy belly is replaced by a long, flattened tendon that becomes cord-like as it approaches the wrist. The FCR produces flexion (when acting with the flexor carpi ulnaris) and abduction of the wrist (when acting with the extensors carpi radialis longus and brevis). When acting alone, the FCR produces a combination of flexion and abduction simultaneously at the wrist so that the hand moves anterolaterally.

To reach its distal attachment, the FCR tendon passes through a canal in the lateral part of the flexor retinaculum, and through a vertical groove in the trapezium in its own synovial **tendinous sheath of the flexor carpi radialis** (Fig. 6.57C). The FCR tendon is a good guide to the radial artery, which lies just lateral to it (Fig. 6.56C).

To test the flexor carpi radialis, the person is asked to flex their wrist against resistance. If acting normally, its tendon can be easily seen and palpated.

Palmaris Longus. The **palmaris longus,** a small fusiform muscle, is absent on one or both sides (usually the left) in approximately 14% of people, but its actions are not missed. It has a short belly and a long, cord-like tendon that passes superficial to the flexor retinaculum and attaches to it and the apex of the palmar aponeurosis (Figs. 6.56C and 6.58). The palmaris longus tendon is a useful guide to the median nerve at the wrist. The tendon lies deep and slightly medial to this nerve before it passes deep to the flexor retinaculum.

To test the palmaris longus, the wrist is flexed and the pads of the little finger and thumb are tightly pinched together. If present and acting normally, the tendon can be easily seen and palpated.

Flexor Carpi Ulnaris. The **flexor carpi ulnaris (FCU)** is the most medial of the superficial flexor muscles. The FCU simultaneously flexes and adducts the hand at the wrist if acting alone. It flexes the wrist when it acts with the FCR and adducts it when acting with the extensor carpi ulnaris. The ulnar nerve enters the forearm by passing between the humeral and the ulnar heads of its proximal attachment (Fig. 6.56C). This muscle is exceptional among muscles of the anterior compartment, being fully innervated by the ulnar nerve. The tendon of the FCU is a guide to the ulnar nerve and artery, which are on its lateral side at the wrist (Fig. 6.56C & E).

To test the flexor carpi ulnaris, the person puts the posterior aspect of the forearm and hand on a flat table, and is then asked to flex the wrist against resistance while the examiner palpates the muscle and its tendon.

Flexor Digitorum Superficialis. The **flexor digitorum superficialis (FDS)** is sometimes considered one of the superficial muscles of the forearm, which attach to the common flexor origin and therefore cross the elbow (Table 6.10). When considered this way, it is the largest superficial muscle in the forearm. However, the FDS actually forms an intermediate layer between the superficial and the deep groups of forearm muscles (Figs. 6.56C and 6.57B). The median nerve and ulnar artery enter the forearm by passing between its humero-ulnar and radial heads (Fig. 6.59A & C). Near the wrist, the FDS gives rise to four tendons, which pass deep to the flexor retinaculum through the carpal tunnel to the fingers. The four tendons are enclosed (along with the four tendons of the flexor digitorum profundus) in a synovial *common flexor sheath* (Fig. 6.57C). The FDS flexes the middle phalanges of the medial four fingers at the proximal interphalangeal joints. In continued action, the FDS also flexes the proximal phalanges at the metacarpophalangeal joints and the wrist joint. The FDS is capable of flexing each finger it serves independently.

To test the flexor digitorum superficialis, one finger is flexed at the proximal interphalangeal joint against resistance and the other three fingers are held in an extended position to inactivate the flexor digitorum profundus.

The fascial plane between the intermediate and deep layers of muscles makes up the primary neurovascular plane of the anterior (flexor–pronator) compartment; the main neurovascular bundles exclusive to this compartment course within it. The following three muscles form the deep layer of forearm flexor muscles.

Flexor Digitorum Profundus. The **flexor digitorum profundus (FDP)** is the only muscle that can flex the distal interphalangeal joints of the fingers (Fig. 6.59A & E). This thick muscle "clothes" the anterior aspect of the ulna. The FDP flexes the distal phalanges of the medial four fingers after the FDS has flexed their middle phalanges (i.e., it curls the fingers and assists with flexion of the hand, making a fist). Each tendon is capable of flexing two interphalangeal joints, the metacarpophalangeal joint and the wrist joint. The FDP divides into four parts, which end in four tendons that pass posterior to the FDS tendons and the flexor retinaculum within the common flexor sheath (Fig. 6.57C). The part of the muscle going to the index finger usually separates from the rest of the muscle relatively early in the distal part of the forearm and is capable of independent contraction. Each tendon enters the fibrous sheath of its digit, posterior to the FDS tendons. Unlike the FDS, the FDP can flex only the index finger independently; thus the fingers can be independently flexed at the proximal but not the distal interphalangeal joints.

To test the flexor digitorum profundus, the proximal interphalangeal joint is held in the extended position while the person attempts to flex the distal interphalangeal joint. The integrity of the median nerve in the proximal forearm can be tested by performing this test using the index finger, and that of the ulnar nerve can be assessed by using the little finger.

Flexor Pollicis Longus. The **flexor pollicis longus (FPL),** the long flexor of the thumb (L. *pollex,* thumb), lies lateral to the FDP, where it clothes the anterior aspect of the radius distal to the attachment of the supinator (Figs. 6.56C & E and 6.59A & D; Table 6.10). The flat FPL tendon passes deep to the flexor retinaculum, enveloped in its own synovial **tendinous sheath of the flexor pollicis longus** on the lateral side of the common flexor sheath (Fig. 6.57C). The FPL primarily flexes the distal phalanx of the thumb at the interphalangeal joint and, secondarily, the proximal phalanx and 1st metacarpal at the metacarpophalangeal and carpometacarpal joints, respectively. The FPL is the only muscle that flexes the interphalangeal joint of the thumb. It also may assist in flexion of the wrist joint.

To test the flexor pollicis longus, the proximal phalanx of the thumb is held and the distal phalanx is flexed against resistance.

Pronator Quadratus. The **pronator quadratus (PQ),** as its name indicates, is quadrangular and pronates the forearm (Fig. 6.59E). It cannot be palpated or observed, except in dissections, because it is the deepest muscle in the anterior aspect of the forearm. Sometimes it is considered to constitute a fourth muscle layer. The PQ clothes the distal fourth of the radius and ulna and the interosseous membrane between them (Fig. 6.59A & E; Table 6.10). The PQ is the only muscle that attaches only to the ulna at one end and only to the radius at the other end.

The PQ is the prime mover for pronation. The muscle initiates pronation, and is assisted by the PT when more speed and power are needed. The pronator quadratus also helps the interosseous membrane hold the radius and ulna together, particularly when upward thrusts are transmitted through the wrist (e.g., during a fall on the hand).

EXTENSOR MUSCLES OF FOREARM

The muscles of the posterior compartment of the forearm are illustrated in Figure 6.60 and their attachments, innervation, and main actions of the are provided by layer in Table 6.11. The following discussion provides additional details.

The **extensor muscles** are in the **posterior (extensor–supinator) compartment of the forearm,** and all are innervated by branches of the radial nerve (Fig. 6.57B). These muscles can be organized physiologically into three functional groups:

1. Muscles that extend and abduct or adduct the hand at the wrist joint (extensor carpi radialis longus, extensor carpi radialis brevis, and extensor carpi ulnaris).
2. Muscles that extend the medial four fingers (extensor digitorum, extensor indicis, and extensor digiti minimi).
3. Muscles that extend or abduct the thumb (abductor pollicis longus, extensor pollicis brevis, and extensor pollicis longus).

The extensor tendons are held in place in the wrist region by the *extensor retinaculum,* which prevents bowstringing of the tendons when the hand is extended at the wrist joint. As the tendons pass over the dorsum of the wrist, they are provided with **synovial tendon sheaths** that reduce friction for the extensor tendons as they traverse the osseofibrous tunnels formed by the attachment of the extensor retinaculum to

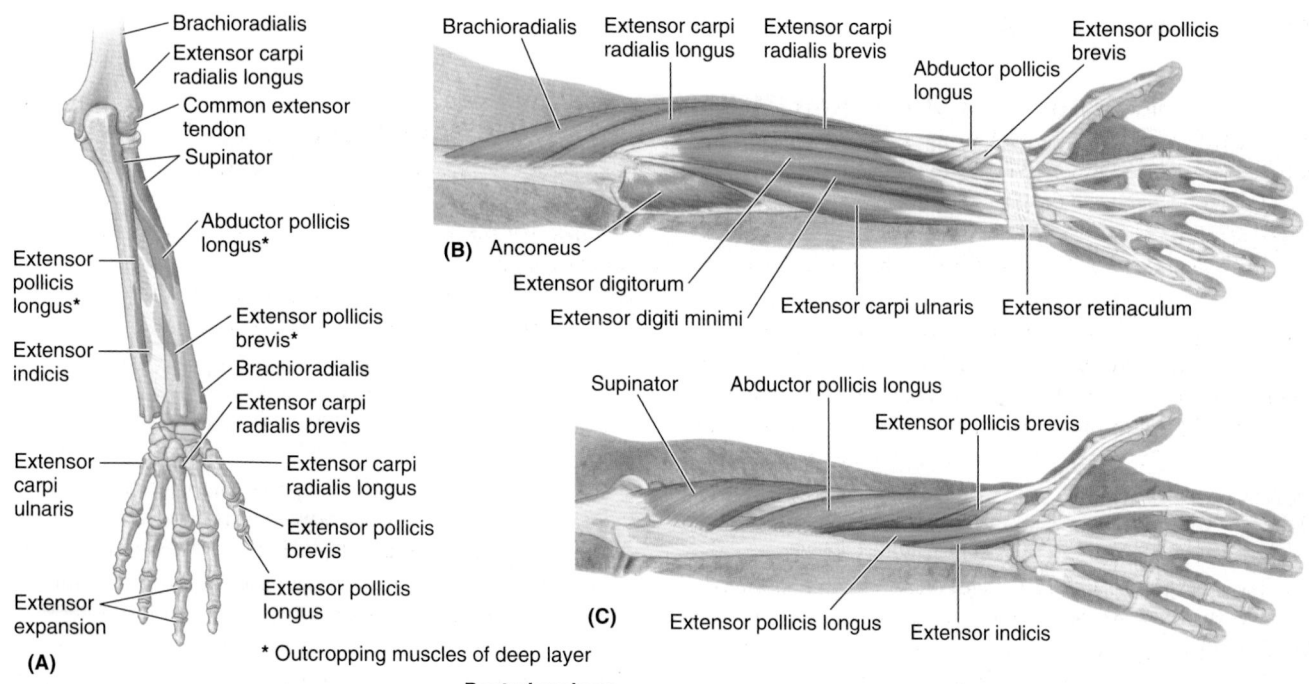

FIGURE 6.60. Extensor muscles of forearm.

Posterior views

(A)

* Outcropping muscles of deep layer

Labels in figure (A):
Brachioradialis
Extensor carpi radialis longus
Common extensor tendon
Supinator
Abductor pollicis longus*
Extensor pollicis longus*
Extensor indicis
Extensor pollicis brevis*
Brachioradialis
Extensor carpi radialis brevis
Extensor carpi radialis longus
Extensor pollicis brevis
Extensor pollicis longus
Extensor carpi ulnaris
Extensor expansion

Labels in figure (B):
Brachioradialis
Extensor carpi radialis longus
Extensor carpi radialis brevis
Abductor pollicis longus
Extensor pollicis brevis
Anconeus
Extensor digitorum
Extensor digiti minimi
Extensor carpi ulnaris
Extensor retinaculum

Labels in figure (C):
Supinator
Abductor pollicis longus
Extensor pollicis brevis
Extensor pollicis longus
Extensor indicis

TABLE 6.11. MUSCLES OF POSTERIOR COMPARTMENT OF FOREARM

Muscle	Proximal Attachment	Distal Attachment	Innervation[a]	Main Action
Superficial layer				
Brachioradialis	Proximal two thirds of supra-epicondylar ridge of humerus	Lateral surface of distal end of radius proximal to styloid process	Radial nerve (C5, **C6**, C7)	Relatively weak flexion of forearm; maximal when forearm is in mid-pronated position
Extensor carpi radialis longus (ECRL)	Lateral supra-epicondylar ridge of humerus	Dorsal aspect of base of 2nd metacarpal	Radial nerve (C6, C7)	Extend and abduct hand at the wrist joint; ECRL active during fist clenching
Extensor carpi radialis brevis (ECRB)	Lateral epicondyle of humerus (common extensor origin)	Dorsal aspect of base of 3rd metacarpal	Deep branch of radial nerve (**C7**, C8)	
Extensor digitorum		Extensor expansions of medial four digits		Extends medial four digits primarily at metacarpophalangeal joints, secondarily at interphalangeal joints
Extensor digiti minimi (EDM)		Extensor expansion of 5th digit		Extends 5th digit primarily at metacarpophalangeal joint, secondarily at interphalangeal joint
Extensor carpi ulnaris (ECU)	Lateral epicondyle of humerus; posterior border of ulna via a shared aponeurosis	Dorsal aspect of base of 5th metacarpal		Extends and adducts hand at wrist joint (also active during fist clenching)
Deep layer				
Supinator	Lateral epicondyle of humerus; radial collateral and anular ligaments; supinator fossa; crest of ulna	Lateral, posterior, and anterior surfaces of proximal third of radius	Deep branch of radial nerve (C7, **C8**)	Supinates forearm; rotates radius to turn palm anteriorly or superiorly (if elbow is flexed)
Extensor indicis	Posterior surface of distal third of ulna and interosseous membrane	Extensor expansion of 2nd digit	Posterior interosseous nerve (**C7**, C8), continuation of deep branch of radial nerve	Extends 2nd digit (enabling its independent extension); helps extend hand at wrist

TABLE 6.11. MUSCLES OF POSTERIOR COMPARTMENT OF FOREARM (Continued)

Muscle	Proximal Attachment	Distal Attachment	Innervation[a]	Main Action
Outcropping muscles of deep layer				
Abductor pollicis longus (APL)	Posterior surface of proximal halves of ulna, radius, and interosseous membrane	Base of 1st metacarpal	Posterior interosseous nerve (C7, **C8**), continuation of deep branch of radial nerve	Abducts thumb and extends it at carpometacarpal joint
Extensor pollicis longus (EPL)	Posterior surface of middle third of ulna and interosseous membrane	Dorsal aspect of base of distal phalanx of thumb		Extends distal phalanx of thumb at interphalangeal joint; extends metacarpophalangeal and carpometacarpal joints
Extensor pollicis brevis (EPB)	Posterior surface of distal third of radius and interosseous membrane	Dorsal aspect of base of proximal phalanx of thumb		Extends proximal phalanx of thumb at metacarpophalangeal joint; extends carpometacarpal joint

[a]The spinal cord segmental innervation is indicated (e.g., "**C7**, C8" means that the nerves supplying the extensor carpi radialis brevis are derived from the seventh and eighth cervical segments of the spinal cord). Numbers in boldface (**C7**) indicate the main segmental innervation. Damage to one or more of the listed spinal cord segments or to the motor nerve roots arising from them results in paralysis of the muscles concerned.

the distal radius and ulna (Fig. 6.61). The extensor muscles of the forearm are organized anatomically into superficial and deep layers (Fig. 6.57B).

Four of the *superficial extensors* (extensor carpi radialis brevis, extensor digitorum, extensor digiti minimi, and extensor carpi ulnaris) are attached proximally by a *common extensor tendon* to the lateral epicondyle (Figs. 6.60A and 6.61A & B; Table 6.11). The proximal attachment of the other two muscles in the superficial group (brachioradialis and extensor carpi radialis longus) is to the lateral supra-epicondylar ridge of the humerus and adjacent lateral intermuscular septum (Fig. 6.60A & B). The four flat tendons of the extensor digitorum pass deep to the extensor retinaculum to the medial four fingers (Fig. 6.62). The common tendons of the index and little fingers are joined on their medial sides near the knuckles by the respective tendons of the extensor indicis and extensor digiti minimi (extensors of the index and little fingers, respectively).

Brachioradialis. The **brachioradialis,** a fusiform muscle, lies superficially on the anterolateral surface of the forearm (Figs. 6.58 and 6.61A). It forms the lateral border of the cubital fossa (Fig. 6.56C). As mentioned previously, the brachioradialis is exceptional among muscles of the posterior (extensor) compartment in that it has rotated to the anterior aspect of the humerus and thus flexes the forearm at the elbow. It is especially active during quick movements or in the presence of resistance during flexion of the forearm (e.g., when a weight is lifted), acting as a shunt muscle resisting subluxation of the head of the radius. The brachioradialis and the supinator are the only muscles of the compartment that do not cross and therefore are incapable of acting at the wrist. As it descends, the brachioradialis overlies the radial nerve and artery where they lie together on the supinator, pronator teres tendon, FDS, and FPL. The distal part of the tendon is covered by the abductors pollicis longus and brevis as they pass to the thumb (Fig. 6.61B).

To test the brachioradialis, the elbow joint is flexed against resistance with the forearm in the midprone position. If the brachioradialis is acting normally, the muscle can be seen and palpated.

Extensor Carpi Radialis Longus. The **extensor carpi radialis longus (ECRL),** a fusiform muscle, is partly overlapped by the brachioradialis, with which it often blends (Fig. 6.61). As it passes distally, posterior to the brachioradialis, its tendon is crossed by the abductor pollicis brevis and extensor pollicis brevis. The ECRL is indispensable when clenching the fist.

To test the extensor carpi radialis longus, the wrist is extended and abducted with the forearm pronated. If acting normally, the muscle can be palpated inferoposterior to the lateral side of the elbow. Its tendon can be palpated proximal to the wrist.

Extensor Carpi Radialis Brevis. The **extensor carpi radialis brevis (ECRB),** as its name indicates, is a shorter muscle than the ECRL because it arises distally in the limb, yet it attaches adjacent to the ECRL in the hand (but to the base of the 3rd metacarpal rather than the 2nd). As it passes distally, it is covered by the ECRL. The ECRB and ECRL pass under the extensor retinaculum together within the **tendinous sheath of the extensor carpi radiales** (Fig. 6.62). The two muscles act together to various degrees, usually as synergists to other muscles. When the two muscles act by themselves, they abduct the hand as they extend it. Acting with the extensor carpi ulnaris, they extend the hand (the brevis is more involved in this action); acting with the FCR, they produce pure abduction. Their synergistic action with the extensor carpi ulnaris is important in steadying the wrist during tight flexion of the medial four digits (clenching a fist), a function in which the longus is more active.

Extensor Digitorum. The **extensor digitorum,** the principal extensor of the medial four digits, occupies much of the posterior surface of the forearm (Figs. 6.60 and 6.61A). Proximally, its four tendons join the tendon of the extensor

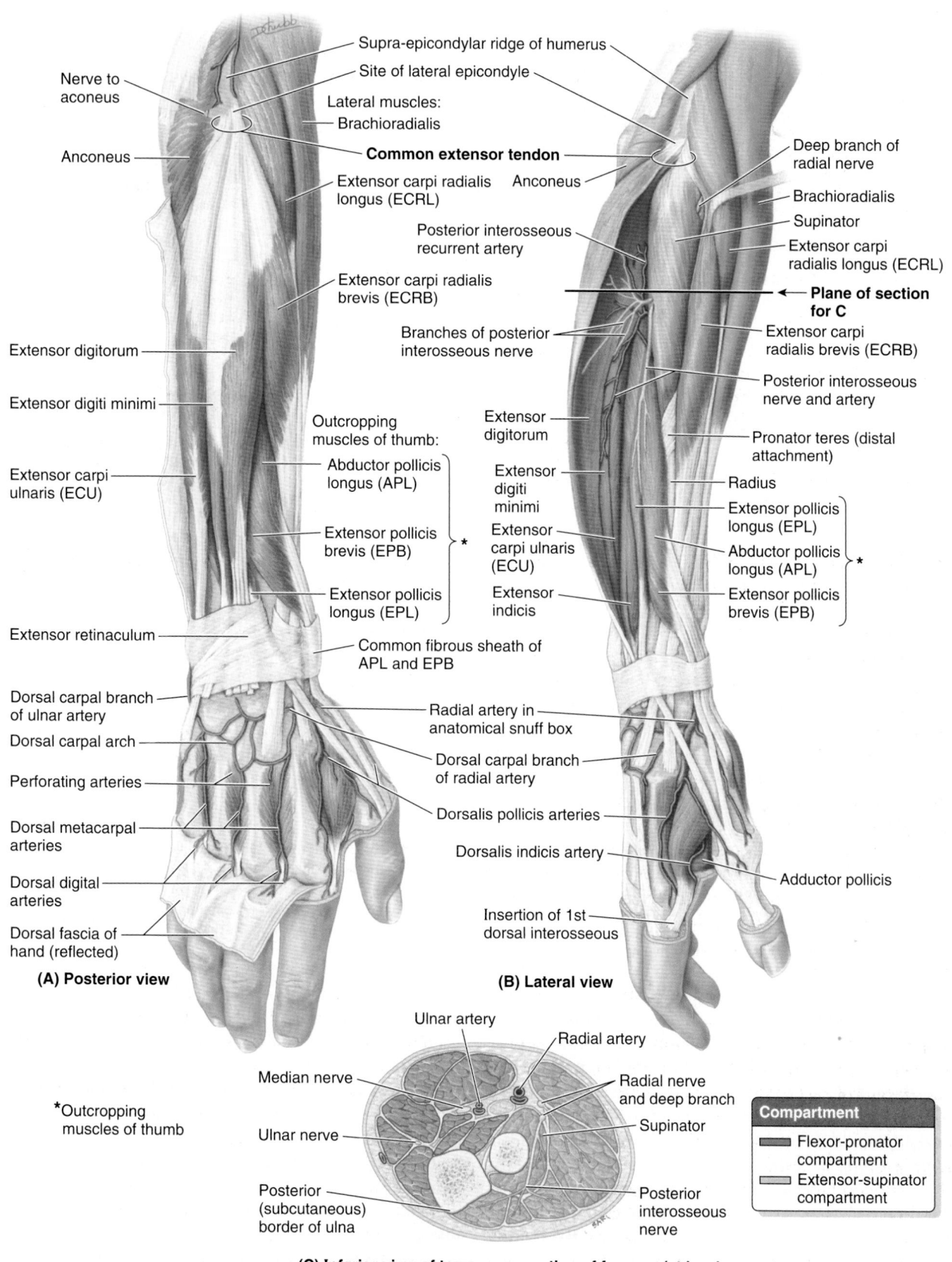

Nerve to aconeus

Anconeus

Supra-epicondylar ridge of humerus
Site of lateral epicondyle
Lateral muscles:
Brachioradialis
Common extensor tendon
Extensor carpi radialis longus (ECRL)
Posterior interosseous recurrent artery

Extensor carpi radialis brevis (ECRB)

Extensor digitorum

Extensor digiti minimi

Extensor carpi ulnaris (ECU)

Outcropping muscles of thumb:
Abductor pollicis longus (APL)
Extensor pollicis brevis (EPB)
Extensor pollicis longus (EPL)

*

Extensor retinaculum

Dorsal carpal branch of ulnar artery

Dorsal carpal arch

Perforating arteries

Dorsal metacarpal arteries

Dorsal digital arteries

Dorsal fascia of hand (reflected)

Common fibrous sheath of APL and EPB

Radial artery in anatomical snuff box

Dorsal carpal branch of radial artery

Dorsalis pollicis arteries

(A) Posterior view

Anconeus

Deep branch of radial nerve
Brachioradialis
Supinator
Extensor carpi radialis longus (ECRL)
← **Plane of section for C**
Extensor carpi radialis brevis (ECRB)
Posterior interosseous nerve and artery

Branches of posterior interosseous nerve

Extensor digitorum

Extensor digiti minimi

Extensor carpi ulnaris (ECU)

Extensor indicis

Pronator teres (distal attachment)
Radius
Extensor pollicis longus (EPL)
Abductor pollicis longus (APL)
Extensor pollicis brevis (EPB)

*

Dorsalis indicis artery

Adductor pollicis

Insertion of 1st dorsal interosseous

(B) Lateral view

*Outcropping muscles of thumb

Ulnar artery
Radial artery

Median nerve

Radial nerve and deep branch
Supinator

Ulnar nerve

Posterior (subcutaneous) border of ulna

Posterior interosseous nerve

Compartment		
▬	Flexor-pronator compartment	
▭	Extensor-supinator compartment	

(C) Inferior view of transverse section of forearm (at level indicated in part B)

FIGURE 6.61. **Extensor-supinator compartment of right forearm. A.** The superficial layer of extensor muscles. The distal extensor tendons have been removed from the dorsum of the hand without disturbing the arteries because they lie on the skeletal plane. The fascia on the posterior aspect of the distal-most forearm is thickened to form the extensor retinaculum, which is anchored on its deep aspect to the radius and ulna. **B.** The deep layer of extensor muscles is shown. Three outcropping muscles of the thumb (*star*) emerge from between the extensor carpi radialis brevis and the extensor digitorum: abductor pollicis longus, extensor pollicis brevis, and extensor pollicis longus. The furrow from which the three muscles emerge has been opened proximally to the lateral epicondyle, exposing the supinator muscle. **C.** This transverse section of the forearm shows the superficial and deep layers of muscles in the posterior compartment (*pink*), supplied by the radial nerve, and the anterior compartment (*gold*), supplied by the ulnar and median nerves.

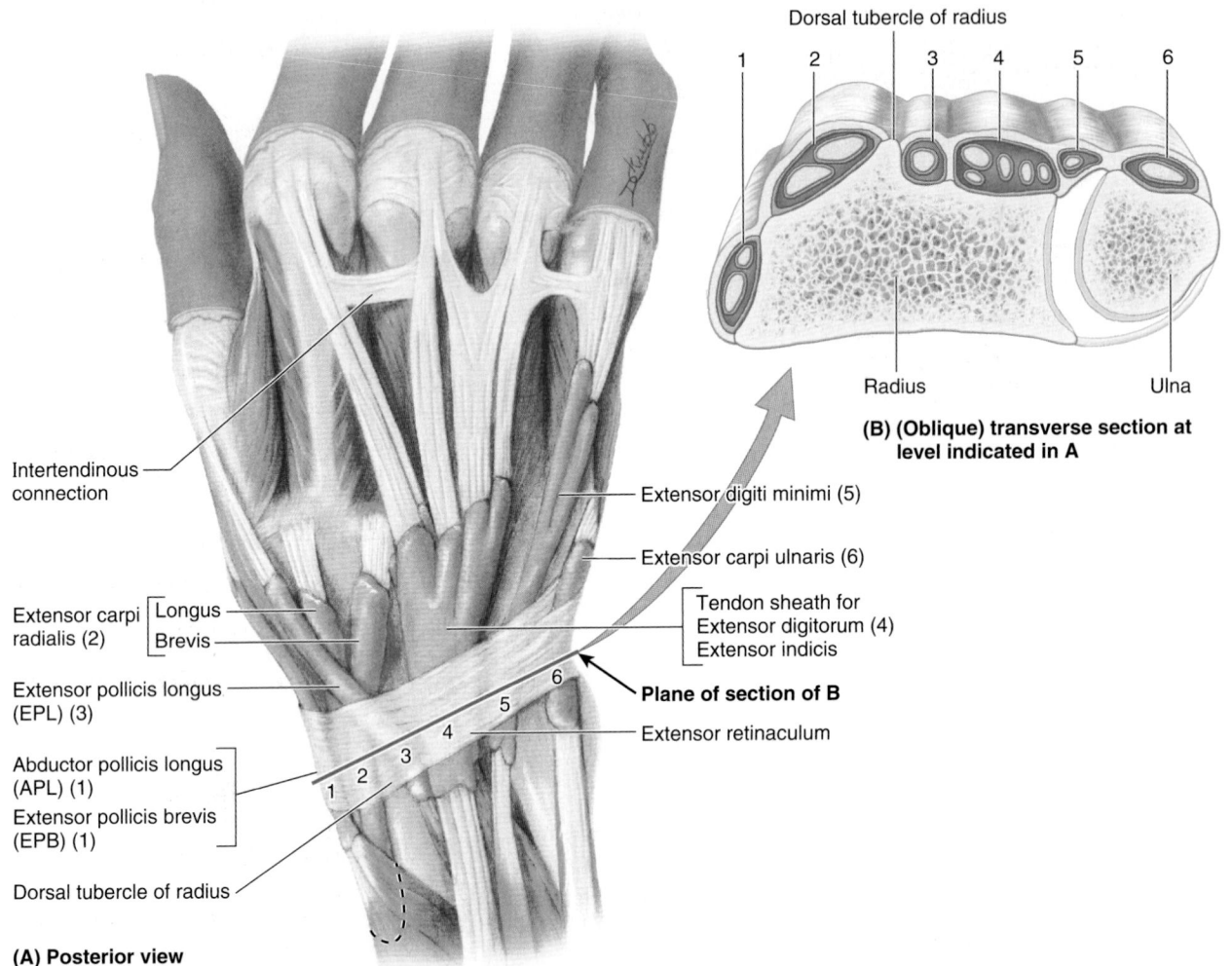

Dorsal tubercle of radius

**(B) (Oblique) transverse section at
level indicated in A**

Radius   Ulna

Intertendinous
connection

Extensor carpi [Longus
radialis (2)   [Brevis

Extensor pollicis longus
(EPL) (3)

Abductor pollicis longus
(APL) (1)

Extensor pollicis brevis
(EPB) (1)

Dorsal tubercle of radius

(A) Posterior view

Extensor digiti minimi (5)

Extensor carpi ulnaris (6)

Tendon sheath for
Extensor digitorum (4)
Extensor indicis

Plane of section of B

Extensor retinaculum

FIGURE 6.62. Synovial sheaths and tendons on distal forearm and dorsum of hand. A. Observe that the six synovial tendon sheaths (*purple*) occupy six osseofibrous tunnels formed by attachments of the extensor retinaculum to the ulna and especially the radius, which give passage to 12 tendons of nine extensor muscles. The tendon of the extensor digitorum to the little finger is shared between the ring finger and continues to the little finger via an intertendinous connection, then receives additional fibers from the tendon of the extensor digiti minimi. Such variations are common. Numbers refer to the labeled osseofibrous tunnels shown in part **B. B.** This slightly oblique transverse section of the distal end of the forearm shows the extensor tendons traversing the six osseofibrous tunnels deep to the extensor retinaculum.

indicis to pass deep to the extensor retinaculum through the **tendinous sheath of the extensor digitorum and extensor indicis** (common extensor synovial sheath) (Fig. 6.62A & B). On the dorsum of the hand, the tendons spread out as they run toward the digits. Adjacent tendons are linked proximal to the knuckles (metacarpophalangeal joints) by three oblique *intertendinous connections* that restrict independent extension of the four medial digits (especially the ring finger). Consequently, normally none of these digits can remain fully flexed as the other ones are fully extended. Commonly, the fourth tendon is fused initially with the tendon to the ring finger and reaches the little finger by an intertendinous connection.

On the distal ends of the metacarpals and along the phalanges of the four medial digits, the four tendons flatten to form **extensor expansions** (Fig. 6.63). Each extensor digital expansion (dorsal expansion or hood) is a triangular, tendinous aponeurosis that wraps around the dorsum and sides of

a head of the metacarpal and proximal phalanx. The visor-like "hood" formed by the extensor expansion over the head of the metacarpal, holding the extensor tendon in the middle of the digit, is anchored on each side to the **palmar ligament** (a reinforced portion of the fibrous layer of the joint capsule of the metacarpophalangeal joints) (Figs. 6.63A & C).

In forming the extensor expansion, each flexor digitorum tendon divides into a **median band,** which passes to the base of the middle phalanx, and two **lateral bands,** which pass to the base of the distal phalanx (Fig. 6.63D & E). The tendons of the interosseous and lumbrical muscles of the hand join the lateral bands of the extensor expansion (Fig. 6.63).

The **retinacular ligament** is a delicate fibrous band that runs from the proximal phalanx and fibrous digital sheath obliquely across the middle phalanx and two interphalangeal joints (Fig. 6.63C). It joins the extensor expansion to the distal phalanx. During flexion of the distal interphalangeal joint,

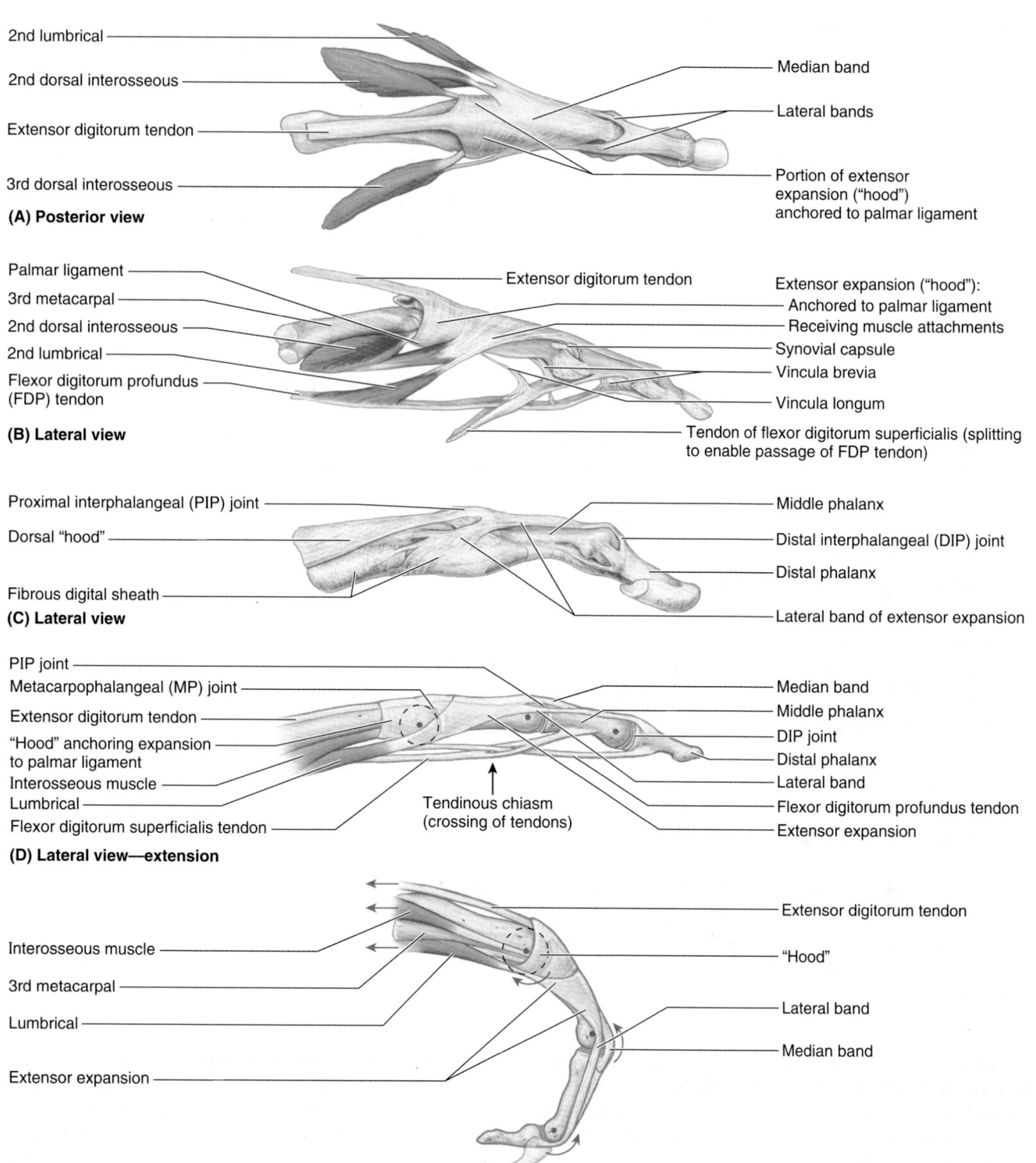

2nd lumbrical
2nd dorsal interosseous
Extensor digitorum tendon
3rd dorsal interosseous
(A) Posterior view

Median band
Lateral bands
Portion of extensor expansion ("hood") anchored to palmar ligament

Palmar ligament
3rd metacarpal
2nd dorsal interosseous
2nd lumbrical
Flexor digitorum profundus (FDP) tendon
(B) Lateral view

Extensor digitorum tendon

Extensor expansion ("hood"):
Anchored to palmar ligament
Receiving muscle attachments
Synovial capsule
Vincula brevia
Vincula longum
Tendon of flexor digitorum superficialis (splitting to enable passage of FDP tendon)

Proximal interphalangeal (PIP) joint
Dorsal "hood"
Fibrous digital sheath
(C) Lateral view

Middle phalanx
Distal interphalangeal (DIP) joint
Distal phalanx
Lateral band of extensor expansion

PIP joint
Metacarpophalangeal (MP) joint
Extensor digitorum tendon
"Hood" anchoring expansion to palmar ligament
Interosseous muscle
Lumbrical
Flexor digitorum superficialis tendon
(D) Lateral view—extension

Tendinous chiasm (crossing of tendons)

Median band
Middle phalanx
DIP joint
Distal phalanx
Lateral band
Flexor digitorum profundus tendon
Extensor expansion

Interosseous muscle
3rd metacarpal
Lumbrical
Extensor expansion
(E) Lateral view—flexion

Extensor digitorum tendon
"Hood"
Lateral band
Median band

FIGURE 6.63. Dorsal digital (extensor) apparatus of 3rd digit. The metacarpal bone and all three phalanges are shown in parts **A, B, D,** and **E;** only the phalanges are shown in part **C. A.** Note the extensor digitorum tendon trifurcating (expanding) into three bands: two lateral bands that unite over the middle phalanx to insert into the base of the distal phalanx, and one median band that inserts into the base of the middle phalanx. **B.** Part of the tendon of the interosseous muscles attaches to the base of the proximal phalanx; the other part contributes to the extensor expansion, attaching primarily to the lateral bands, but also fans out into an aponeurosis. Some of the aponeurotic fibers fuse with the median band, and other fibers arch over it to blend with the aponeurosis arising from the other side. On the radial side of each digit, a lumbrical muscle attaches to the radial lateral band. The dorsal hood consists of a broad band of transversely oriented fibers attached anteriorly to the palmar ligaments of the metacarpophalangeal (MP) joints that encircle the metacarpal head and MP joint, blending with the extensor expansion to keep the apparatus centered over the dorsal aspect of the digit. **C.** Distally, retinacular ligaments extending from the fibrous digital sheath to the lateral bands also help keep the apparatus centered and coordinate movements at the proximal interphalangeal (PIP) and distal interphalangeal (DIP) joints. **D.** Contraction of the extensor digitorum alone results in extension at all joints (including the MP joint in the absence of action by the interossei and lumbricals). **E.** Because of the relationship of the tendons and the lateral bands to the rotational centers of the joints (*red dots* in parts **D** and **E**), simultaneous contraction of the interossei and lumbricals produces flexion at the MP joint but extension at the PIP and DIP joints (the so-called Z-movement).

the retinacular ligament becomes taut and pulls the proximal joint into flexion. Similarly, on extending the proximal joint, the distal joint is pulled by the retinacular ligament into nearly complete extension.

The extensor digitorum acts primarily to extend the proximal phalanges, and through its collateral reinforcements, it secondarily extends the middle and distal phalanges as well. After exerting its traction on the digits, or in the presence of resistance to digital extension, it helps extend the hand at the wrist joint.

To test the extensor digitorum, the forearm is pronated and the fingers are extended. The person attempts to keep the digits extended at the metacarpophalangeal joints as the examiner exerts pressure on the proximal phalanges by attempting to flex them. If acting normally, the extensor digitorum can be palpated in the forearm, and its tendons can be seen and palpated on the dorsum of the hand.

Extensor Digiti Minimi. The **extensor digiti minimi (EDM),** a fusiform slip of muscle, is a partially detached part of the extensor digitorum (Figs. 6.60B, 6.61A & B, and 6.62). The tendon of this extensor of the little finger runs through a separate compartment of the extensor retinaculum, posterior to the distal radio-ulnar joint, within the **tendinous sheath of the extensor digiti minimi.** The tendon then divides into two slips; the lateral one is joined to the tendon of the extensor digitorum, with all three tendons attaching to the dorsal digital expansion of the little finger. After exerting its traction primarily on the 5th digit, it contributes to extension of the hand.

To test the extensor digiti minimi, the little finger is extended against resistance while holding digits 2–4 flexed at the metacarpophalangeal joints.

Extensor Carpi Ulnaris. The **extensor carpi ulnaris (ECU),** a long fusiform muscle located on the medial border of the forearm, has two heads: a humeral head from the common extensor tendon and an ulnar head that arises by a common aponeurosis attached to the posterior border of the ulna and shared by the FCU, FDP, and deep fascia of the forearm. Distally, its tendon runs in a groove between the ulnar head and its styloid process, through a separate compartment of the extensor retinaculum within the **tendinous sheath of the extensor carpi ulnaris.** Acting with the ECRL and ECRB, it extends the hand; acting with the FCU, it adducts the hand. Like the ECRL, it is indispensable when clenching the fist.

To test the extensor carpi ulnaris, the forearm is pronated and the fingers are extended. The extended wrist is then adducted against resistance. If acting normally, the muscle can be seen and palpated in the proximal part of the forearm and its tendon can be felt proximal to the head of the ulna.

Supinator. The **supinator** lies deep in the cubital fossa and, along with the brachialis, forms its floor (Figs. 6.60A & C, 6.61B, and 6.64). Spiraling medially and distally from its continuous, osseofibrous origin, this sheet-like muscle envelops the neck and proximal part of the shaft of the radius. The deep branch of the radial nerve passes between its muscle fibers, separating them into superficial and deep parts, as it passes

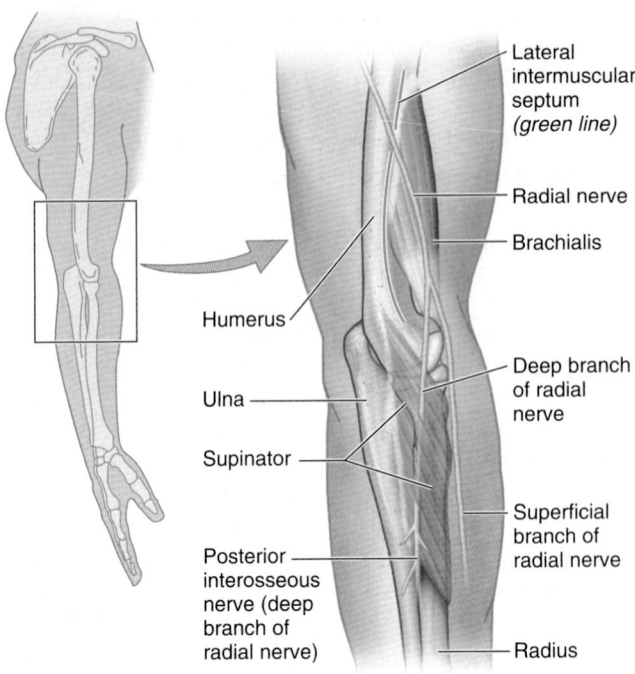

Lateral intermuscular septum *(green line)*

Radial nerve

Brachialis

Humerus

Ulna

Deep branch of radial nerve

Supinator

Superficial branch of radial nerve

Posterior interosseous nerve (deep branch of radial nerve)

Radius

Lateral view

FIGURE 6.64. Relationship of radial nerve to brachialis and supinator muscles. In the cubital fossa, lateral to the brachialis, the radial nerve divides into deep (motor) and superficial (sensory) branches. The deep branch penetrates the supinator muscle and emerges in the posterior compartment of the forearm as the posterior interosseous nerve. It joins the artery of the same name to run in the plane between the superficial and the deep extensors of the forearm.

from the cubital fossa to the posterior part of the arm. As it exits the muscle and joins the posterior interosseous artery, it may be referred to as the posterior interosseous nerve.

The supinator is the prime mover for slow, unopposed supination, especially when the forearm is extended. The biceps brachii also supinates the forearm and is the prime mover during rapid and forceful supination against resistance when the forearm is flexed (e.g., when a right-handed person drives a screw).

The **deep extensors of the forearm** act on the thumb (abductor pollicis longus, extensor pollicis longus, and extensor pollicis brevis) and the index finger (extensor indicis) (Figs. 6.60–6.62; Table 6.11). The three muscles acting on the thumb are deep to the superficial extensors and "crop out" (emerge) from the furrow in the lateral part of the forearm that divides the extensors. Because of this characteristic, they are sometimes referred to as *outcropping muscles of the thumb* (Fig. 6.61A).

Abductor Pollicis Longus. The **abductor pollicis longus (APL)** has a long, fusiform belly that lies just distal to the supinator (Fig. 6.60) and is closely related to the extensor pollicis brevis. Its tendon, and sometimes its belly, is commonly split into two parts, one of which may attach to the trapezium instead of the usual site at the base of the 1st metacarpal. The APL acts with the abductor pollicis brevis during abduction of the thumb and with the extensor pollicis muscles during extension of this digit. Although deeply situated, the APL emerges at the wrist as one of the outcropping muscles. Its tendon passes deep to the extensor retinaculum with the tendon of the extensor pollicis brevis in the common synovial **tendinous sheath of the abductor pollicis longus and extensor pollicis brevis.**

To test the abductor pollicis longus, the thumb is abducted against resistance at the metacarpophalangeal joint. If acting normally, its tendon can be seen and palpated at the lateral side of the *anatomical snuff box* and on the lateral side of the adjacent extensor pollicis brevis tendon.

Extensor Pollicis Brevis. The belly of the **extensor pollicis brevis (EPB),** the fusiform short extensor of the thumb, lies distal to the APL and is partly covered by it. Its tendon lies parallel and immediately medial to that of the APL but extends farther, reaching the base of the proximal phalanx (Fig. 6.62). In continued action after acting to flex the proximal phalanx of the thumb, or acting when that joint is fixed by its antagonists, it helps extend the 1st metacarpal and extend and abduct the hand. When the thumb is fully extended, a hollow called the *anatomical snuff box,* can be seen on the radial aspect of the wrist (Fig. 6.65).

To test the extensor pollicis brevis, the thumb is extended against resistance at the metacarpophalangeal joint. If the EPB is acting normally, the tendon of the muscle can be seen and palpated at the lateral side of the anatomical snuff box and on the medial side of the adjacent APL tendon (Figs. 6.61 and 6.62).

Extensor Pollicis Longus. The **extensor pollicis longus (EPL)** is larger and its tendon is longer than that of the EPB. The tendon passes under the extensor retinaculum in its own tunnel (Fig. 6.60), within the **tendinous sheath of the extensor pollicis longus,** medial to the dorsal tubercle of the radius. It uses the tubercle as a trochlea (pulley) to change its line of pull as it proceeds to the base of the distal phalanx of the thumb. The gap thus created between the long extensor tendons of the thumb is the *anatomical snuff box* (Fig. 6.65). In addition to its main actions (Table 6.11), the EPL also adducts the extended thumb and rotates it laterally.

To test the extensor pollicis longus, the thumb is extended against resistance at the interphalangeal joint. If the EPL is acting normally, the tendon of the muscle can be seen and palpated on the medial side of the anatomical snuff box.

The tendons of the APL and EPB bound the **anatomical snuff box** anteriorly, and the tendon of the EPL bounds it posteriorly (Figs. 6.61, 6.62, and 6.65). The snuff box is visible when the thumb is fully extended; this draws the tendons up and produces a triangular hollow between them. Observe that the:

- *Radial artery* lies in the floor of the snuff box.
- *Radial styloid process* can be palpated proximally and the base of the 1st metacarpal can be palpated distally in the snuff box.
- *Scaphoid* and *trapezium* can be felt in the floor of the snuff box between the radial styloid process and the 1st metacarpal (see the blue box "Fracture of the Scaphoid" on p. 686).

Extensor Indicis. The **extensor indicis** has a narrow, elongated belly that lies medial to and alongside that of the EPL (Figs. 6.61B and 6.62). This muscle confers independence to the index finger in that the extensor indicis may act alone or together with the extensor digitorum to extend the index finger at the proximal interphalangeal joint, as in pointing. It also helps extend the hand.

Arteries of Forearm

The main arteries of the forearm are the *ulnar* and *radial arteries,* which usually arise opposite the neck of the radius in the inferior part of the cubital fossa as terminal branches of the brachial artery (Fig. 6.66). The named arteries of the forearm are illustrated in Figure 6.67 and their origins and courses are described in Table 6.12. The following discussion provides additional details.

ULNAR ARTERY

Pulsations of the **ulnar artery** can be palpated on the lateral side of the FCU tendon, where it lies anterior to the

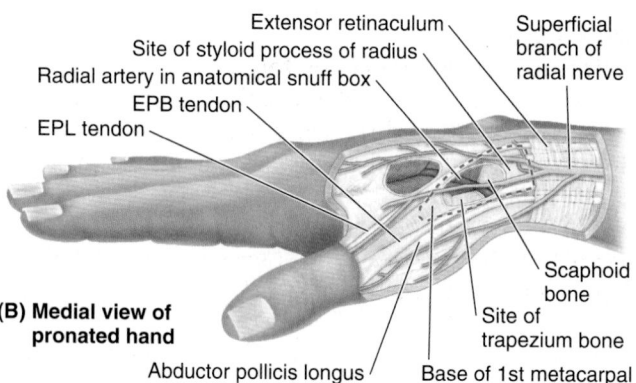

(A) Medial view of pronated hand

Extensor pollicis longus (EPL) tendon

Extensor pollicis brevis (EPB) tendon

Anatomical snuff box

Extensor retinaculum
Site of styloid process of radius
Radial artery in anatomical snuff box
EPB tendon
EPL tendon

Superficial branch of radial nerve

Scaphoid bone

(B) Medial view of pronated hand

Site of trapezium bone

Abductor pollicis longus

Base of 1st metacarpal

FIGURE 6.65. Anatomical snuff box. A. When the thumb is extended, a triangular hollow appears between the tendon of the extensor pollicis longus (EPL) medially and the tendons of the extensor pollicis brevis (EPB) and abductor pollicis longus (APL) laterally. **B.** The floor of the snuff box, formed by the scaphoid and trapezium bones, is crossed by the radial artery as it passes diagonally from the anterior surface of the radius to the dorsal surface of the hand.

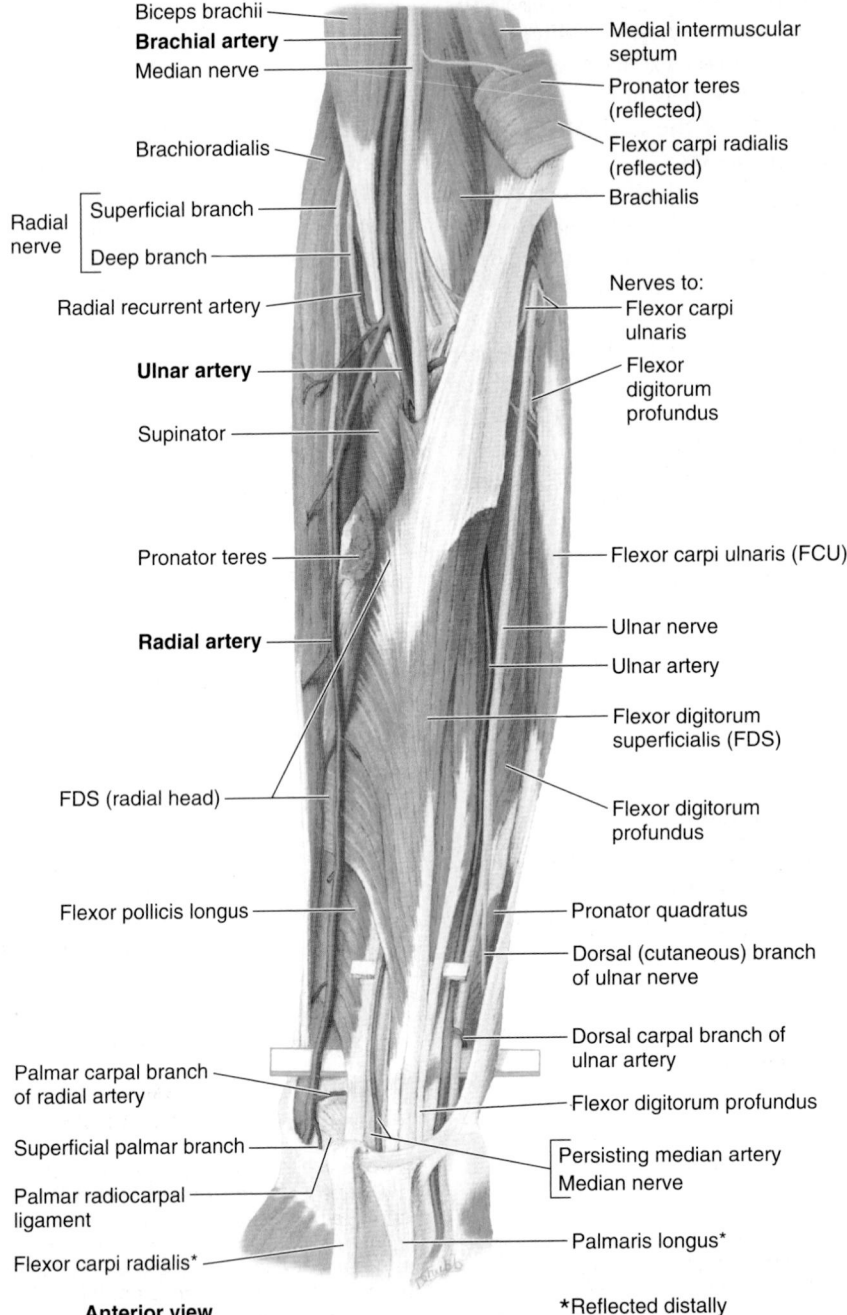

Biceps brachii
Brachial artery
Median nerve
Brachioradialis
Radial nerve — Superficial branch
Radial nerve — Deep branch
Radial recurrent artery
Ulnar artery
Supinator
Pronator teres
Radial artery
FDS (radial head)
Flexor pollicis longus
Palmar carpal branch of radial artery
Superficial palmar branch
Palmar radiocarpal ligament
Flexor carpi radialis*

Medial intermuscular septum
Pronator teres (reflected)
Flexor carpi radialis (reflected)
Brachialis
Nerves to: Flexor carpi ulnaris
Flexor digitorum profundus
Flexor carpi ulnaris (FCU)
Ulnar nerve
Ulnar artery
Flexor digitorum superficialis (FDS)
Flexor digitorum profundus
Pronator quadratus
Dorsal (cutaneous) branch of ulnar nerve
Dorsal carpal branch of ulnar artery
Flexor digitorum profundus
Persisting median artery
Median nerve
Palmaris longus*

Anterior view

*Reflected distally

FIGURE 6.66. Flexor digitorum superficialis and related vasculature. Three muscles of the superficial layer (pronator teres, flexor carpi radialis, and palmaris longus) have been removed, leaving only their attaching ends; the fourth muscle of the layer (the flexor carpi ulnaris) has been retracted medially. The tendinous humeral attachment of the FDS to the medial epicondyle is thick; and the linear attachment to the radius, immediately distal to the radial attachments of the supinator and pronator teres, is thin (Table 6.10). The ulnar artery and median nerve pass between the humeral and the radial heads of the FDS. The artery descends obliquely deep to the FDS to join the ulnar nerve, which descends vertically near the medial border of the FDS (exposed here by splitting a fusion of the FDS and the FCU). A (proximal) probe is elevating the FDS tendons (and median nerve and persisting median artery). A second (distal) probe is elevating all the remaining structures that cross the wrist (radiocarpal) joint anteriorly.

ulnar head. The ulnar nerve is on the medial side of the ulnar artery. Branches of the ulnar artery arising in the forearm participate in the peri-articular anastomoses of the elbow (Fig. 6.67, palmar view) and supply muscles of the medial and central forearm, the common flexor sheath, and the ulnar and median nerves.

• The *anterior* and *posterior ulnar recurrent arteries* anastomose with the inferior and superior ulnar collateral arteries, respectively, thereby participating in the *peri-articular arterial anastomoses of the elbow*. The anterior and posterior arteries may be present as anterior and posterior branches of a (common) ulnar recurrent artery.

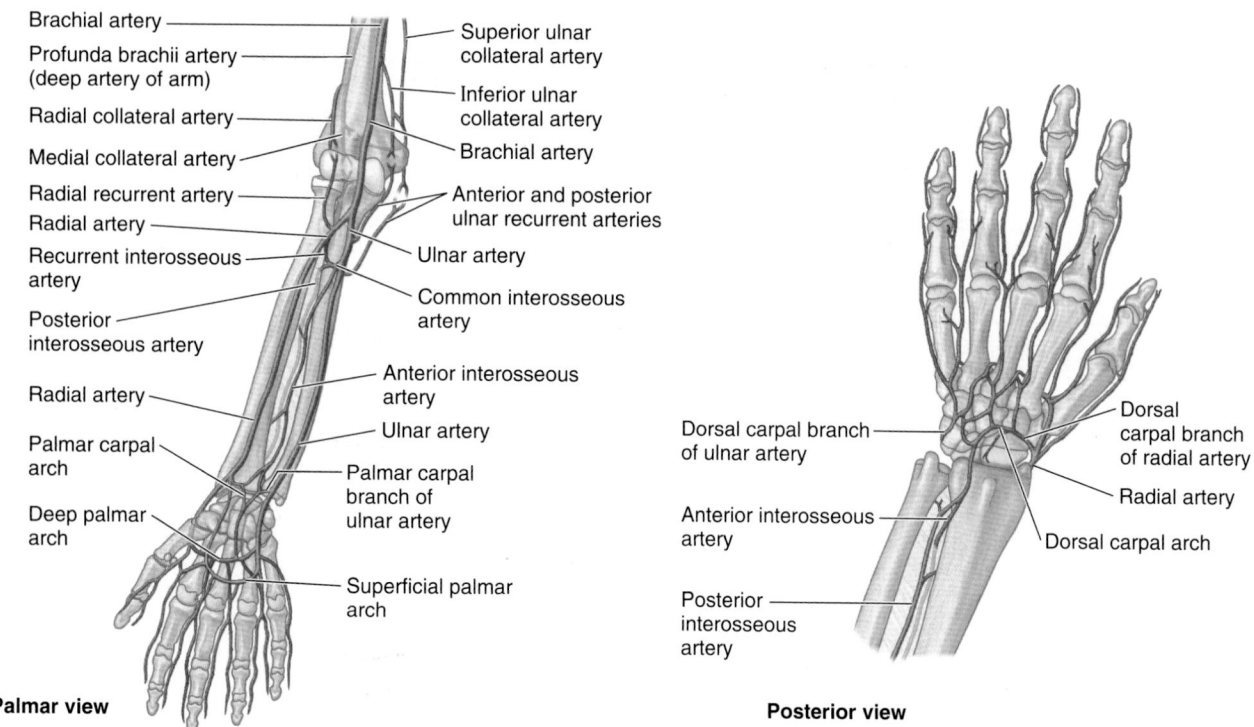

FIGURE 6.67. Arteries of forearm.

TABLE 6.12. ARTERIES OF FOREARM AND WRIST

Artery	Origin	Course in Forearm
Ulnar	As larger terminal branch of brachial artery in cubital fossa	Descends inferomedially and then directly inferiorly, deep to superficial (pronator teres and palmaris longus) and intermediate (flexor digitorum superficialis) layers of flexor muscles to reach medial side of forearm; passes superficial to flexor retinaculum at wrist in ulnar (Guyon) canal to enter hand
Anterior ulnar recurrent artery	Ulnar artery just distal to elbow joint	Passes superiorly between brachialis and pronator teres, supplying both; then anastomoses with inferior ulnar collateral artery anterior to medial epicondyle (Fig. 6.67, palmar view)
Posterior ulnar recurrent artery	Ulnar artery distal to anterior ulnar	Passes superiorly, posterior to medial epicondyle and deep to tendon of flexor carpi ulnaris; then recurrent artery anastomoses with superior ulnar collateral artery
Common interosseous	Ulnar artery in cubital fossa, distal to bifurcation of brachial artery	Passes laterally and deeply, terminating quickly by dividing into anterior and posterior interosseous arteries
Anterior interosseous	As terminal branches of common interosseous artery, between radius and ulna	Passes distally on anterior aspect of interosseous membrane to proximal border of pronator quadratus; pierces membrane and continues distally to join dorsal carpal arch on posterior aspect of interosseous membrane
Posterior interosseous		Passes to posterior aspect of interosseous membrane, giving rise to recurrent interosseous artery; runs distally between superficial and deep extensor muscles, supplying both; replaced distally by anterior interosseous artery
Recurrent interosseous	Posterior interosseous artery, between radius and ulna	Passes superiorly, posterior to proximal radio-ulnar joint and capitulum, to anastomose with middle collateral artery (from deep brachial artery)
Palmar carpal branch	Ulnar artery in distal forearm	Runs across anterior aspect of wrist, deep to tendons of flexor digitorum profundus, to anastomose with the palmar carpal branch of the radial artery, forming palmar carpal arch

TABLE 6.12. ARTERIES OF FOREARM AND WRIST (Continued)

Artery	Origin	Course in Forearm
Dorsal carpal branch	Ulnar artery, proximal to pisiform	Passes across dorsal surface of wrist, deep to extensor tendons, to anastomose with dorsal carpal branch of radial artery, forming dorsal carpal arch
Radial	As smaller terminal branch of brachial artery in cubital fossa	Runs inferolaterally under cover of brachioradialis; lies lateral to flexor carpi radialis tendon in distal forearm; winds around lateral aspect of radius and crosses floor of anatomical snuff box to pierce first dorsal interosseous muscle
Radial recurrent	Lateral side of radial artery, just distal to brachial artery bifurcation	Ascends between brachioradialis and brachialis, supplying both (and elbow joint); then anastomoses with radial collateral artery (from profunda brachii artery)
Palmar carpal branch	Distal radial artery near distal border of pronator quadratus	Runs across anterior wrist deep to flexor tendons to anastomose with the palmar carpal branch of ulnar artery to form palmar carpal arch
Dorsal carpal branch	Distal radial artery in proximal part of snuff box	Runs medially across wrist deep to pollicis and extensor radialis tendons, anastomoses with ulnar dorsal carpal branch forming dorsal carpal arch

- The *common interosseous artery,* a short branch of the ulnar artery, arises in the distal part of the cubital fossa and divides almost immediately into anterior and posterior interosseous arteries.
- The *anterior interosseous artery* passes distally, running directly on the anterior aspect of the interosseous membrane with the anterior interosseous nerve, whereas the *posterior interosseous artery* courses between the superficial and the deep layers of the extensor muscles in the company of the posterior interosseous nerve. The relatively small posterior interosseous artery is the principal artery serving the structures of the middle third of the posterior compartment. Thus it is mostly exhausted in the distal forearm and is replaced by the anterior interosseous artery, which pierces the interosseous membrane near the proximal border of the pronator quadratus.
- Unnamed *muscular branches of the ulnar artery* supply muscles on the medial side of the forearm, mainly those in the flexor–pronator group.

RADIAL ARTERY

The pulsations of the **radial artery** can be felt throughout the forearm, making it useful as an anterolateral demarcation of the flexor and extensor compartments of the forearm. When the brachioradialis is pulled laterally, the entire length of the artery is visible (Figs. 6.66 and 6.67; Table 6.12). The radial artery lies on muscle until it reaches the distal part of the forearm. Here it lies on the anterior surface of the radius and is covered by only skin and fascia, making this an ideal location for checking the radial pulse.

The course of the radial artery in the forearm is represented by a line joining the midpoint of the cubital fossa to a point just medial to the radial styloid process. The radial artery leaves the forearm by winding around the lateral aspect of the wrist and crosses the floor of the anatomical snuff box (Figs. 6.65 and 6.66).

- The *radial recurrent artery* participates in the *periarticular arterial anastomoses around the elbow* by anastomosing with the *radial collateral artery,* a branch of the profunda brachii artery.
- The *palmar* and *dorsal carpal branches of the radial artery* participate in the *peri-articular arterial anastomosis around the wrist* by anastomosing with the corresponding branches of the ulnar artery and terminal branches of the anterior and posterior interosseous arteries, forming the palmar and dorsal carpal arches.
- The unnamed *muscular branches of the radial artery* supply muscles in the adjacent (anterolateral) aspects of both the flexor and the extensor compartments because the radial artery runs along (and demarcates) the anterolateral boundary between the compartments.

Veins of Forearm

In the forearm, as in the arm, there are superficial and deep veins. The superficial veins ascend in the subcutaneous tissue. The deep veins accompany the deep arteries of the forearm.

SUPERFICIAL VEINS

The pattern, common variations, and clinical significance of the superficial veins of the upper limb were discussed earlier in this chapter (p. 689).

DEEP VEINS

Deep veins accompanying arteries are plentiful in the forearm (Fig. 6.68). These accompanying veins (L. *venae comitantes*) arise from the anastomosing **deep venous palmar arch** in the hand. From the lateral side of the

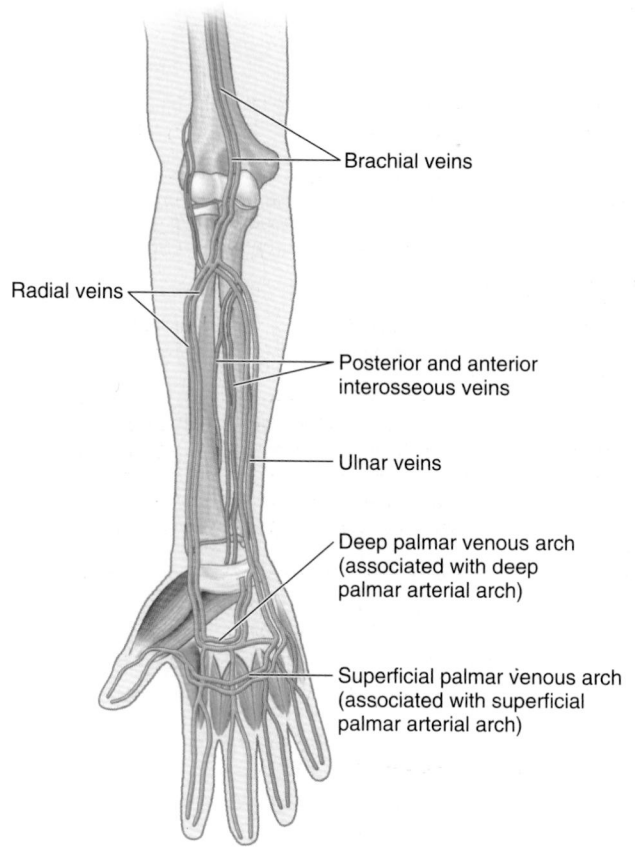

FIGURE 6.68. Deep venous drainage of upper limb.

Brachial veins

Radial veins

Posterior and anterior interosseous veins

Ulnar veins

Deep palmar venous arch (associated with deep palmar arterial arch)

Superficial palmar venous arch (associated with superficial palmar arterial arch)

arch, paired **radial veins** arise and accompany the radial artery; from the medial side, paired **ulnar veins** arise and accompany the ulnar artery. The veins accompanying each artery anastomose freely with each other. The radial and ulnar veins drain the forearm but carry relatively little blood from the hand.

The deep veins ascend in the forearm along the sides of the corresponding arteries, receiving tributaries from veins leaving the muscles with which they are related. Deep veins communicate with the superficial veins. The deep **interosseous veins,** which accompany the interosseous arteries, unite with the accompanying veins of the radial and ulnar arteries. In the cubital fossa the deep veins are connected to the *median cubital vein,* a superficial vein (Fig. 6.55B). These deep cubital veins also unite with the accompanying veins of the brachial artery.

Nerves of Forearm

The nerves of the forearm are the median, ulnar, and radial. The median nerve is the principal nerve of the anterior (flexor–pronator) compartment of the forearm (Figs. 6.57B and 6.69A). Although the radial nerve appears in the cubital region, it soon enters the posterior (extensor–supinator) compartment of the forearm. Besides the cutaneous branches, there are only two nerves of the anterior aspect of the

forearm: the median and ulnar nerves. The named nerves of the forearm are illustrated in Figure 6.69 and their origins and courses are described in Table 6.13. The following sections provide additional details and discuss unnamed branches.

MEDIAN NERVE IN FOREARM

The **median nerve** is the principal nerve of the anterior compartment of the forearm (Figs. 6.69A and 6.70; Table 6.13). It supplies muscular branches directly to the muscles of the superficial and intermediate layers of forearm flexors (except the FCU), and deep muscles (except for the medial [ulnar] half of the FDP) via its branch, the anterior interosseous nerve.

The median nerve has no branches in the arm other than small twigs to the brachial artery. Its major branch in the forearm is the anterior interosseous nerve (Fig. 6.69A, Table 6.13). In addition, the following unnamed branches of the median nerve arise in the forearm:

• *Articular branches.* These branches pass to the elbow joint as the median nerve passes it.
• *Muscular branches.* The *nerve to the pronator teres* usually arises at the elbow and enters the lateral border of the muscle. A broad bundle of nerves pierces the superficial flexor group of muscles and innervates the FCR, the palmaris longus, and the FDS.
• *Anterior interosseous nerve.* This branch runs distally on the interosseous membrane with the anterior interosseous branch of the ulnar artery. After supplying the deep forearm flexors (except the ulnar part of the FDP, which sends tendons to 4th and 5th fingers), it passes deep to and supplies the pronator quadratus, then ends by sending articular branches to the wrist joint.
• *Palmar cutaneous branch of the median nerve.* This branch arises in the forearm, just proximal to the flexor retinaculum, but is distributed to skin of the central part of the palm.

ULNAR NERVE IN FOREARM

Like the median nerve, the *ulnar nerve* does not give rise to branches during its passage through the arm. In the forearm it supplies only one and a half muscles, the FCU (as it enters the forearm by passing between its two heads of proximal attachment) and the ulnar part of the FDP, which sends tendons to the 4th and 5th digits (Fig. 6.69B, Table 6.13). The ulnar nerve and artery emerge from beneath the FCU tendon and become superficial just proximal to the wrist. They pass superficial to the flexor retinaculum and enter the hand by passing through a groove between the pisiform and the hook of the hamate.

A band of fibrous tissue from the flexor retinaculum bridges the groove to form the small **ulnar canal** (Guyon canal) (Fig. 6.70B). The branches of the ulnar nerve arising

(text continues on p. 764)

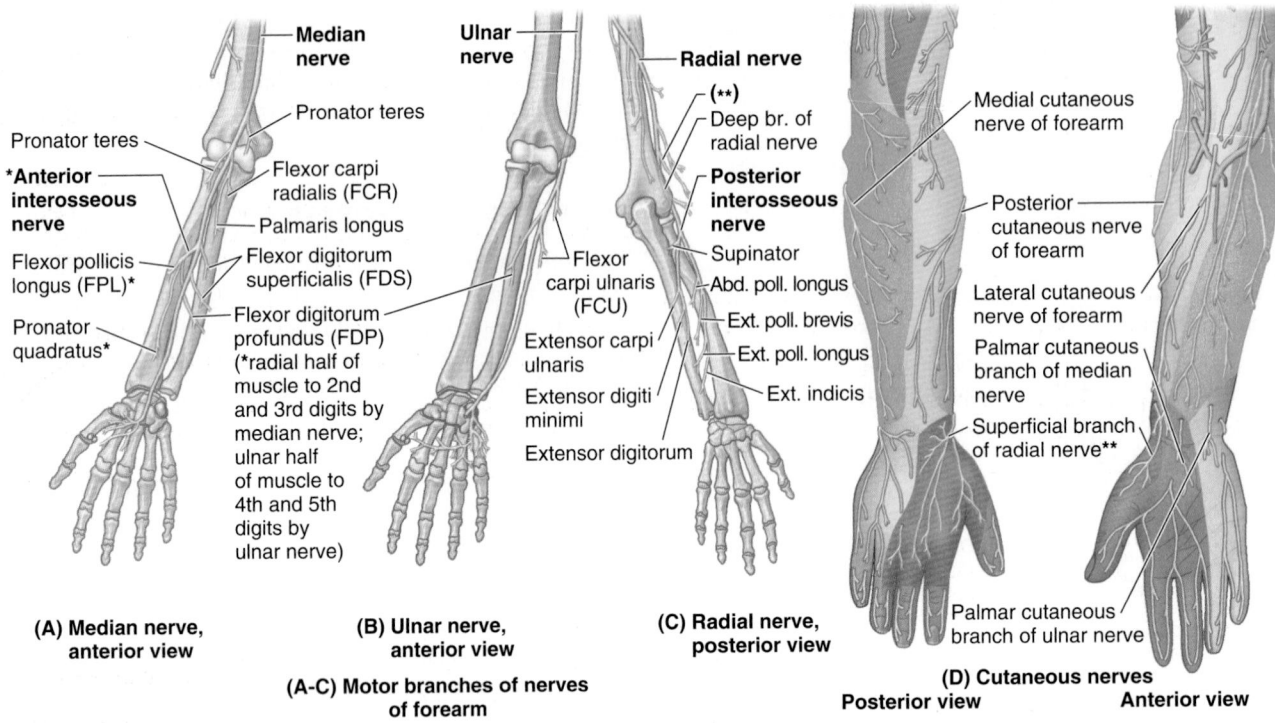

(A) Median nerve, anterior view

(B) Ulnar nerve, anterior view

(A-C) Motor branches of nerves of forearm

(C) Radial nerve, posterior view

(D) Cutaneous nerves
Posterior view   Anterior view

FIGURE 6.69. Nerves of forearm. *Abd. poll. longus,* abductor pollicis longus; *ext. indicis;* extensor indicis; *ext. poll. brevis,* extensor pollicis brevis; *ext. poll. longus;* extensor pollicis longus.

TABLE 6.13. NERVES OF FOREARM

Nerve	Origin	Course in Forearm
Median	By union of lateral root of median nerve (C6 and C7, from lateral cord of brachial plexus) with medial root (C8 and T1) from medial cord)	Enters cubital fossa medial to brachial artery; exits by passing between heads of pronator teres; descends in fascial plane between flexors digitorum superficialis and profundus; runs deep to palmaris longus tendon as it approaches flexor retinaculum to traverse carpal tunnel
Anterior interosseous	Median nerve in distal part of cubital fossa	Descends on anterior aspect of interosseous membrane with artery of same name, between FDP and FPL, to pass deep to pronator quadratus
Palmar cutaneous branch of median nerve	Median nerve in middle to distal forearm, proximal to flexor retinaculum	Passes superficial to flexor reticulum to reach skin of central palm
Ulnar	Larger terminal branch of medial cord of brachial plexus (C8 and T1, often receives fibers from C7)	Enters forearm by passing between heads of flexor carpi ulnaris, after passing posterior to medial epicondyle of humerus; descends forearm between FCU and FDP; becomes superficial in distal forearm
Palmar cutaneous branch of ulnar nerve	Ulnar nerve near middle of forearm	Descends anterior to ulnar artery; perforates deep fascia in distal forearm; runs in subcutaneous tissue to palmar skin medial to axis of 4th digit
Dorsal cutaneous branch of ulnar nerve	Ulnar nerve in distal half of forearm	Passes postero-inferiorly between ulna and flexor carpi ulnaris; enters subcutaneous tissue to supply skin of dorsum medial to axis of 4th digit
Radial	Larger terminal branch of posterior cord of brachial plexus (C5–T1)	Enters cubital fossa between brachioradialis and brachialis; anterior to lateral epicondyle divides into terminal superficial and deep branches
Posterior cutaneous nerve of forearm	Radial nerve, as it traverses radial groove of posterior humerus	Perforates lateral head of triceps; descends along lateral side of arm and posterior aspect of forearm to wrist

(continued)

TABLE 6.13. NERVES OF FOREARM (Continued)

Nerve	Origin	Course in Forearm
Superficial branch of radial nerve	Sensory terminal branch of radial nerve, in cubital fossa	Descends between pronator teres and brachioradialis, emerging from latter to arborize over anatomical snuff box and supply skin of dorsum lateral to axis of 4th digit
Deep branch of radial/posterior interosseous nerve	Motor terminal branch of radial nerve, in cubital fossa	Deep branch exits cubital fossa winding around neck of radius, penetrating and supplying supinator; emerges in posterior compartment of forearm as posterior interosseous; descends on membrane with artery of same name
Lateral cutaneous nerve of forearm	Continuation of musculocutaneous nerve distal to muscular branches	Emerges lateral to biceps brachii on brachialis, running initially with cephalic vein; descends along lateral border of forearm to wrist
Medial cutaneous nerve of forearm	Medial cord of brachial plexus, receiving C8 and T1 fibers	Perforates deep fascia of arm with basilic vein proximal to cubital fossa; descends medial aspect of forearm in subcutaneous tissue to wrist

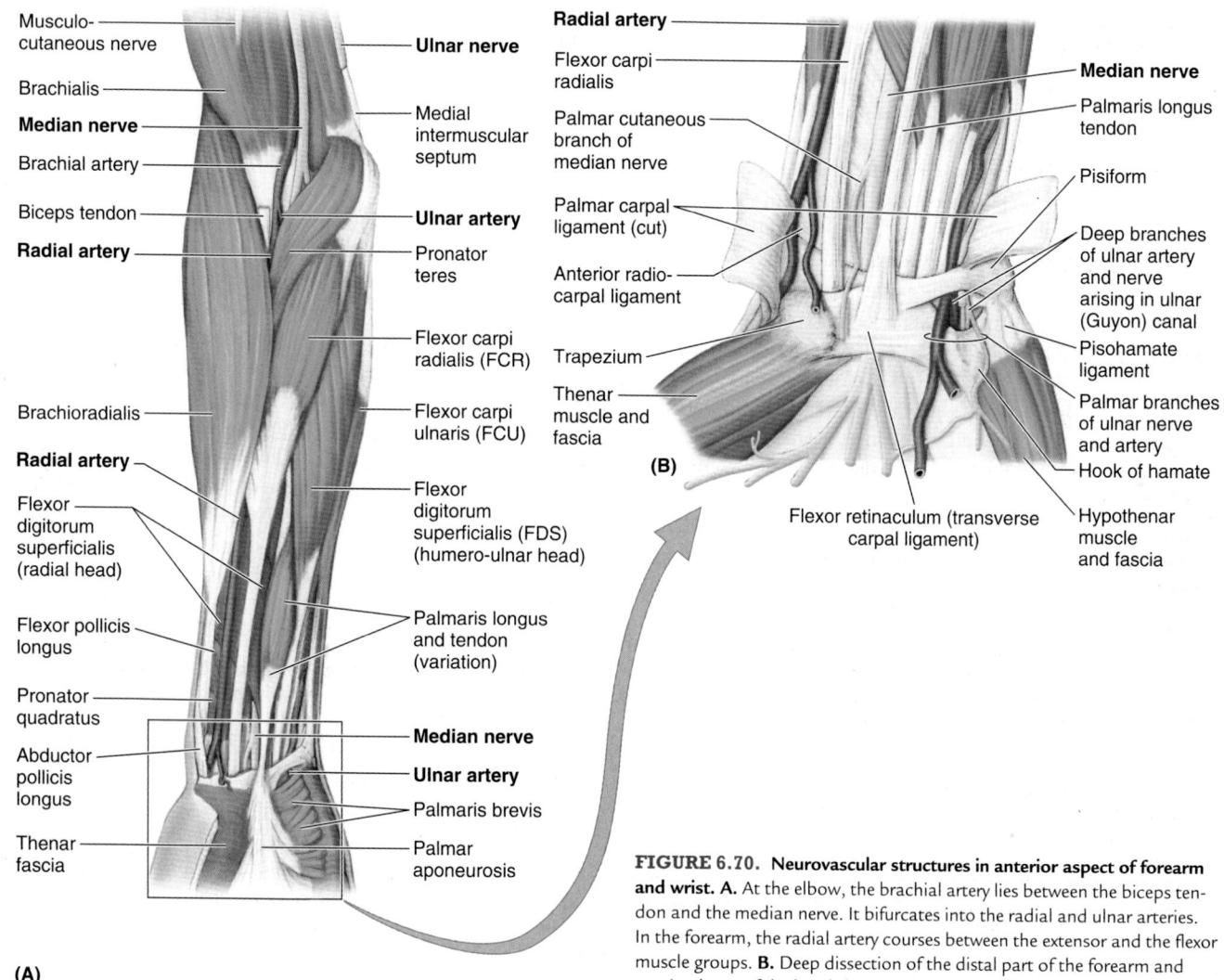

FIGURE 6.70. Neurovascular structures in anterior aspect of forearm and wrist. A. At the elbow, the brachial artery lies between the biceps tendon and the median nerve. It bifurcates into the radial and ulnar arteries. In the forearm, the radial artery courses between the extensor and the flexor muscle groups. **B.** Deep dissection of the distal part of the forearm and proximal part of the hand showing the course of the arteries and nerves.

in the forearm include unnamed muscular and articular branches, and cutaneous branches that pass to the hand:

- *Articular branches* pass to the elbow joint while the nerve is between the olecranon and the medial epicondyle.
- *Muscular branches* supply the FCU and the medial half of the FDP.
- The *palmar* and *dorsal cutaneous branches* arise from the ulnar nerve in the forearm, but their sensory fibers are distributed to the skin of the hand.

RADIAL NERVE IN FOREARM

Unlike the medial and ulnar nerves, the *radial nerve* serves motor and sensory functions in both the arm and the forearm (but only sensory functions in the hand). However, its sensory and motor fibers are distributed in the forearm by two separate branches, the superficial (sensory or cutaneous) and deep radial/posterior interosseous nerve (motor) (Fig. 6.69C & D, Table 6.13). It divides into these terminal branches as it appears in the cubital fossa, anterior to the lateral epicondyle of the humerus, between the brachialis and the brachioradialis (Fig. 6.64). The two branches immediately part company, the deep branch winding laterally around the radius, piercing the supinator en route to the posterior compartment.

The *posterior cutaneous nerve of the forearm* arises from the radial nerve in the posterior compartment of the arm, as it runs along the radial groove of the humerus. Thus it reaches the forearm independent of the radial nerve, descending in the subcutaneous tissue of the posterior aspect of the forearm to the wrist, supplying the skin (Fig. 6.69D).

The *superficial branch of the radial nerve* is also a cutaneous nerve, but it gives rise to articular branches as well. It is distributed to skin on the dorsum of the hand and to a number of joints in the hand, branching soon after it emerges from the overlying brachioradialis and crosses the roof of the anatomical snuff box (Fig. 6.65).

The *deep branch of the radial nerve*, after it pierces the supinator, runs in the fascial plane between superficial and deep extensor muscles in close proximity to the posterior interosseous artery; it is usually referred to as the *posterior interosseous nerve* (Figs. 6.64 and 6.69C). It supplies motor innervation to all the muscles with fleshy bellies located entirely in the posterior compartment of the forearm (distal to the lateral epicondyle of the humerus).

LATERAL AND MEDIAL CUTANEOUS NERVES OF FOREARM

The **lateral cutaneous nerve *of the forearm*** (lateral antebrachial cutaneous nerve) is the continuation of the musculocutaneous nerve after its motor branches have all been given off to the muscles of the anterior compartment of the arm.

The **medial cutaneous nerve of the forearm** (medial antebrachial cutaneous nerve) is an independent branch of the medial cord of the brachial plexus. With the **posterior cutaneous nerve of the forearm** from the radial nerve, each supplying the area of skin indicated by its name, these three nerves provide all the cutaneous innervation of the forearm (Fig. 6.69D). There is no "anterior cutaneous nerve of the forearm." (*Memory device:* This is similar to the brachial plexus, which has lateral, medial, and posterior cords but no anterior cord.)

Although the arteries, veins, and nerves of the forearm have been considered separately, it is important to place them into their anatomical context. Except for the superficial veins, which often course independently in the subcutaneous tissue, these neurovascular structures usually exist as components of neurovascular bundles. These bundles are composed of arteries, veins (in the limbs, usually in the form of accompanying veins), and nerves as well as lymphatic vessels, which are usually surrounded by a neurovascular sheath of varying density.

Surface Anatomy of Forearm

Three bony landmarks are easily palpated at the elbow: the *medial* and *lateral epicondyles* of the humerus and the *olecranon* of the ulna (Fig. 6.71). In the hollow located posterolaterally when the forearm is extended, the *head of the radius* can be palpated distal to the lateral epicondyle. Supinate and pronate your forearm and feel the movement of the radial head. The **posterior border of the ulna** is subcutaneous and can be palpated distally from the olecranon along the entire length of the bone. This landmark demarcates the posteromedial boundary separating the flexor–pronator (anterior) and extensor–supinator (posterior) compartments of the forearm.

The *cubital fossa*, the triangular hollow area on the anterior surface of the elbow, is bounded medially by the prominence formed by the *flexor–pronator group of muscles* that are attached to the medial epicondyle. To estimate the position of these muscles, put your thumb posterior to your medial epicondyle and then place your fingers on your forearm as shown in Figure 6.72A. The *black dot* on the dorsum of the hand indicates the position of the medial epicondyle.

The cubital fossa is bounded laterally by the prominence of the *extensor–supinator group of muscles* attached to the lateral epicondyle (Fig. 6.72B). The *pulsations of the radial artery* can be palpated throughout the forearm as it runs its superficial course from the cubital fossa to the wrist (anterior to the radial styloid process), demarcating the anterolateral boundary separating the flexor–pronator and extensor–supinator compartments of the forearm.

The *head of the ulna* is at its distal end and is easily seen and palpated. It appears as a rounded prominence at the wrist when the hand is pronated. The *ulnar styloid process* can be palpated just distal to the ulnar head. The larger *radial styloid process* can be easily palpated on the lateral side of the wrist when the hand is supinated, particularly when the tendons covering it are relaxed. The radial styloid process is located approximately 1 cm more distal than the ulnar styloid process. This relationship of the styloid processes is important in the diagnosis of certain injuries in the wrist region (e.g., fracture of the distal end of the radius). Proximal to the radial styloid process, the **surfaces of the radius** are palpable for a few centimeters. The lateral surface of the distal half of the radius is easy to palpate.

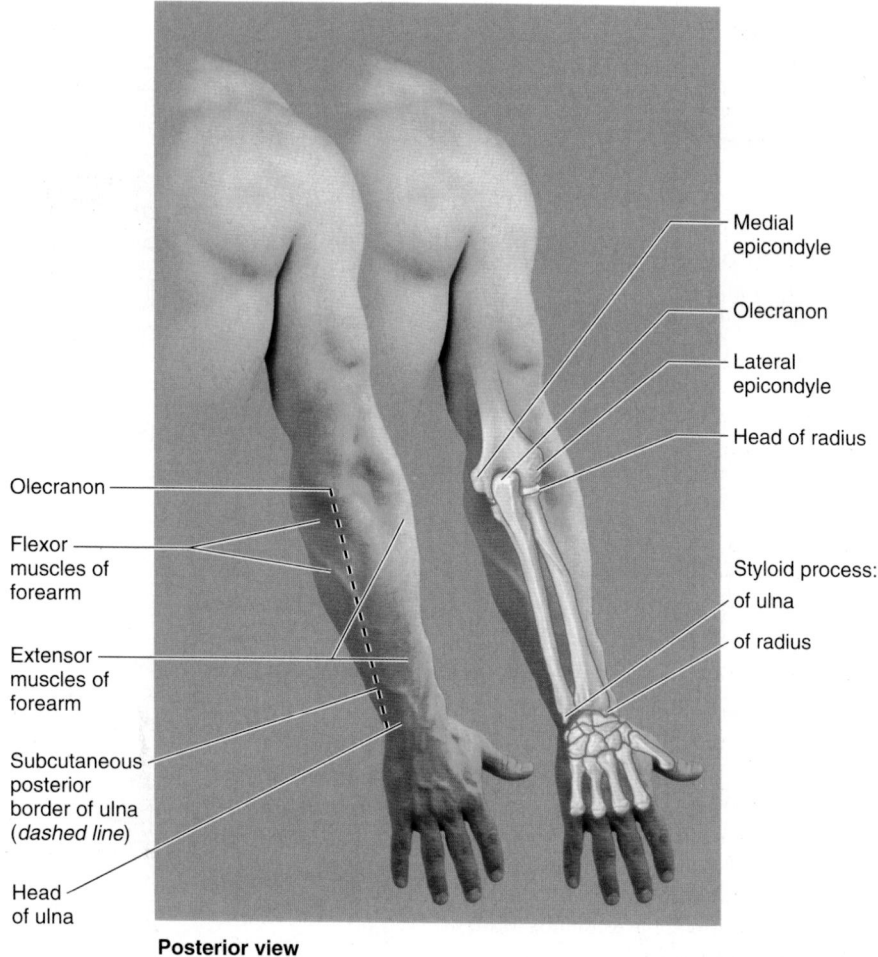

Medial
epicondyle

Olecranon

Lateral
epicondyle

Head of radius

Olecranon

Flexor
muscles of
forearm

Styloid process:
of ulna

of radius

Extensor
muscles of
forearm

Subcutaneous
posterior
border of ulna
(*dashed line*)

Head
of ulna

Posterior view

FIGURE 6.71. Surface anatomy of posterior forearm.

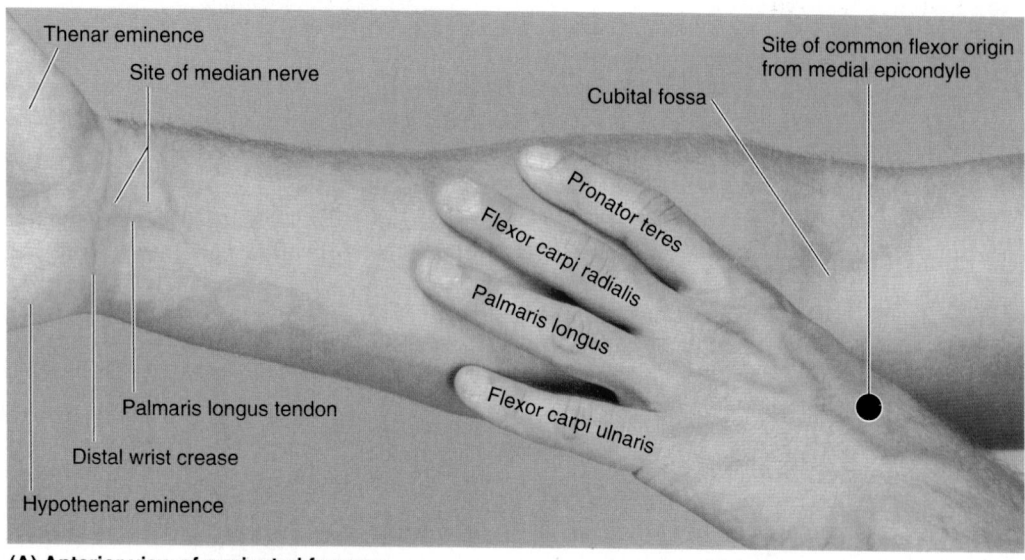

Thenar eminence

Site of median nerve

Site of common flexor origin
from medial epicondyle

Cubital fossa

Pronator teres

Flexor carpi radialis

Palmaris longus

Palmaris longus tendon

Flexor carpi ulnaris

Distal wrist crease

Hypothenar eminence

(A) Anterior view of supinated forearm

FIGURE 6.72. Surface anatomy of anterior forearm.

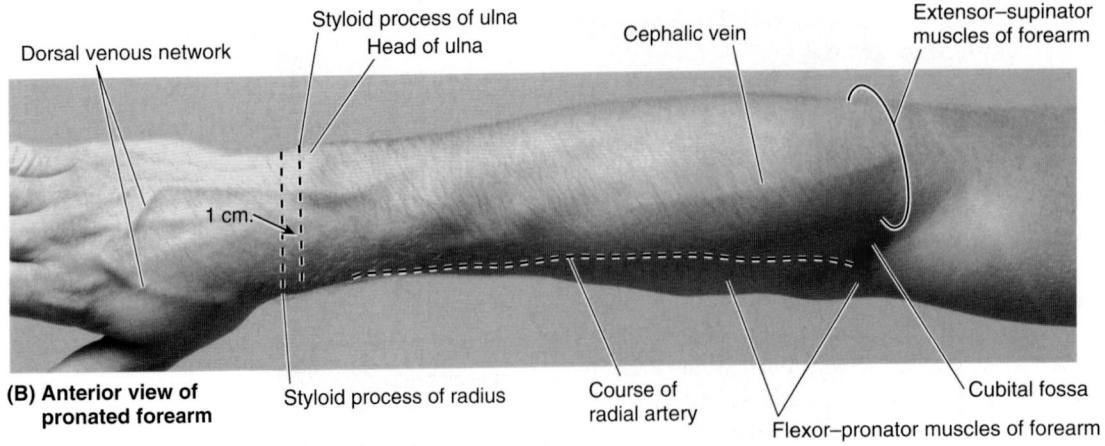

Dorsal venous network
Styloid process of ulna
Head of ulna
Cephalic vein
Extensor–supinator muscles of forearm

1 cm.

(B) Anterior view of pronated forearm
Styloid process of radius
Course of radial artery
Flexor–pronator muscles of forearm
Cubital fossa

FIGURE 6.72. (*Continued*) **Surface anatomy of anterior forearm.**

FOREARM

Elbow Tendinitis or Lateral Epicondylitis

Elbow tendinitis ("tennis elbow") is a painful musculoskeletal condition that may follow repetitive use of the superficial extensor muscles of the forearm. Pain is felt over the lateral epicondyle and radiates down the posterior surface of the forearm. People with elbow tendinitis often feel pain when they open a door or lift a glass. Repeated forceful flexion and extension of the wrist strain the attachment of the common extensor tendon, producing inflammation of the periosteum of the lateral epicondyle (*lateral epicondylitis*).

Mallet or Baseball Finger

Sudden severe tension on a long extensor tendon may avulse part of its attachment to the phalanx. The most common result of the injury is a *mallet* or *baseball finger* (Fig. B6.19A). This deformity results from the distal interphalangeal joint suddenly being forced into extreme flexion (hyperflexion) when, for example, a baseball is miscaught or a finger is jammed into the base pad (Fig. B6.19B). These actions avulse (tear away) the attachment of the tendon to the base of the distal phalanx. As a result, the person cannot extend the distal interphalangeal joint. The resultant deformity bears some resemblance to a mallet.

Fracture of Olecranon

Fracture of the olecranon, called a "fractured elbow" by laypersons, is common because the olecranon is subcutaneous and protrusive. The typical mechanism of injury is a fall on the elbow combined with sudden powerful contraction of the triceps brachii. The fractured olecranon is pulled away by the active and tonic contraction of the triceps

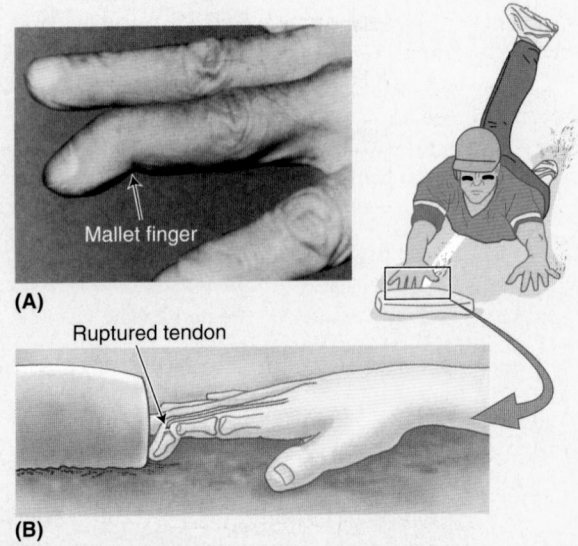

Mallet finger

(A)

Ruptured tendon

(B)

FIGURE B6.19. **Mallet finger. A.** The clinical appearance. **B.** Mechanism of injury.

(Fig. B6.20), and the injury is often considered to be an avulsion fracture (Salter, 1999). Because of the traction produced by the tonus of the triceps on the olecranon fragment, pinning

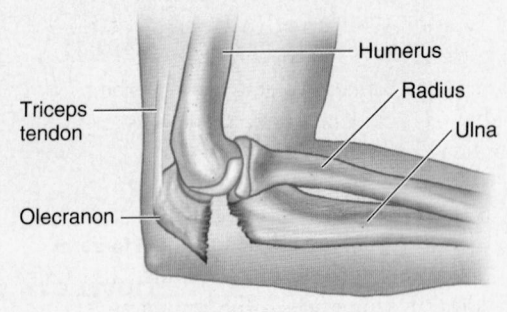

Humerus
Radius
Ulna
Triceps tendon
Olecranon

FIGURE B6.20.

is usually required. Healing occurs slowly, and often a cast must be worn for an extended period of time.

Synovial Cyst of Wrist

Sometimes a non-tender *cystic swelling* appears on the hand, most commonly on the dorsum of the wrist (Fig. B6.21). Usually the cyst is the size of a small grape, but it varies and may be as large as a plum. The thin-walled cyst contains clear mucinous fluid. The cause of the cyst is unknown, but it may result from mucoid degeneration (Salter, 1999). Flexion of the wrist makes the cyst enlarge, and it may be painful. Clinically, this type of swelling is called a "ganglion" (G., swelling or knot). Anatomically, a ganglion refers to a collection of nerve cell bodies (e.g., a spinal ganglion). Synovial cysts are close to and often communicate with the synovial sheaths on the dorsum of the wrist (*purple* in figure). The distal attachment of the ECRB tendon to the base of the 3rd metacarpal is another common site for such a cyst. A cystic swelling of the common flexor synovial sheath on the anterior aspect of the wrist can enlarge enough to produce compression of the median nerve by narrowing the carpal tunnel (*carpal tunnel syndrome*). This syndrome produces pain and paresthesia in the sensory distribution of the median nerve and clumsiness of finger movements (see the blue box "Carpal Tunnel Syndrome" on p. 790).

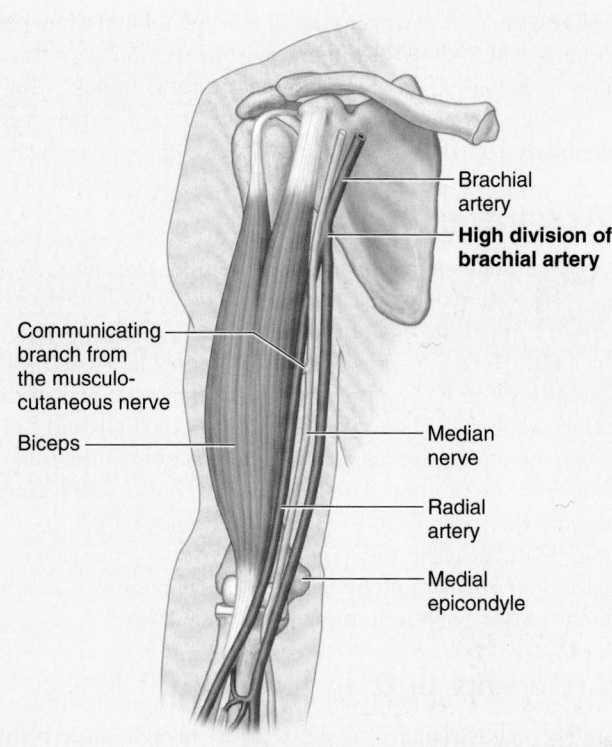

FIGURE B6.22.

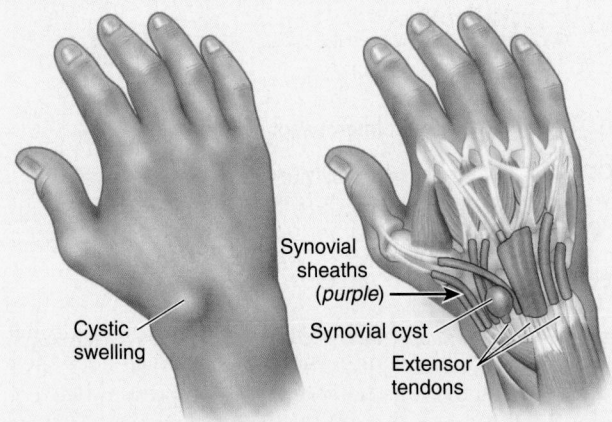
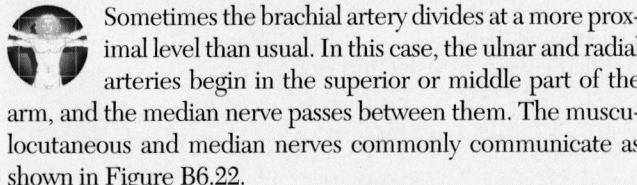

FIGURE B6.21.

High Division of Brachial Artery

Sometimes the brachial artery divides at a more proximal level than usual. In this case, the ulnar and radial arteries begin in the superior or middle part of the arm, and the median nerve passes between them. The musculocutaneous and median nerves commonly communicate as shown in Figure B6.22.

Superficial Ulnar Artery

In approximately 3% of people, the ulnar artery descends superficial to the flexor muscles (Fig. B6.23). Pulsations of a superficial ulnar artery can be felt and

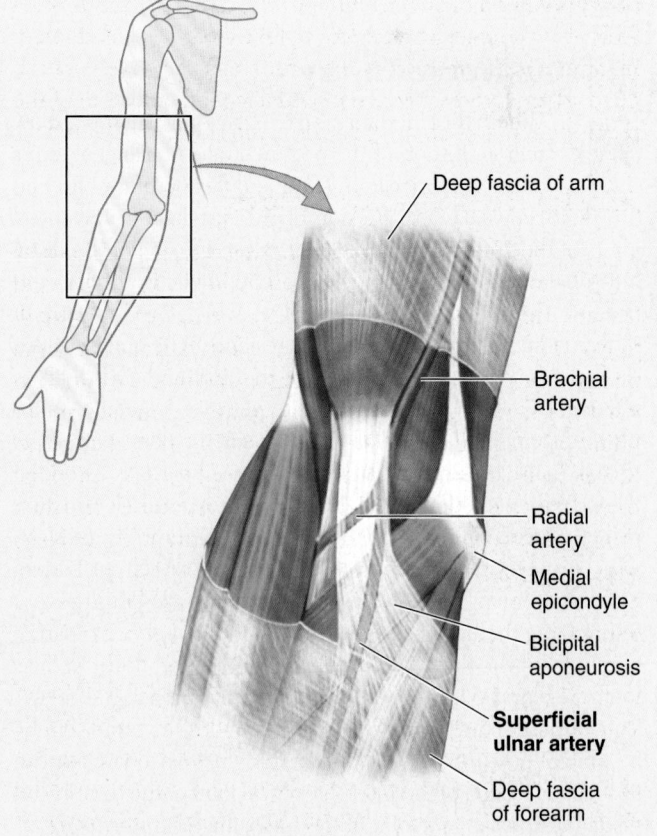

FIGURE B6.23.

may be visible. This variation must be kept in mind when performing venesections for withdrawing blood or making intravenous injections. If an aberrant ulnar artery is mistaken for a vein, it may be damaged and produce bleeding. If certain drugs are injected into the aberrant artery, the result could be fatal.

Measuring Pulse Rate

The common place for measuring the pulse rate is where the radial artery lies on the anterior surface of the distal end of the radius, lateral to the tendon of the FCR. Here the artery is covered by only fascia and skin. The artery can be compressed against the distal end of the radius, where it lies between the tendons of the FCR and APL. When measuring the radial pulse rate, the pulp of the thumb should not be used because it has its own pulse, which could obscure the patient's pulse. If a pulse cannot be felt, try the other wrist because an *aberrant radial artery* on one side may make the pulse difficult to palpate. A radial pulse may also be felt by pressing lightly in the anatomical snuff box.

Variations in Origin of Radial Artery

The origin of the radial artery may be more proximal than usual; it may be a branch of the axillary artery or the brachial artery (Fig. B6.22). Sometimes the radial artery is superficial to the deep fascia instead of deep to it. When a superficial vessel is pulsating near the wrist, it is probably a superficial radial artery. The aberrant vessel is vulnerable to laceration.

Median Nerve Injury

When the median nerve is severed in the elbow region, flexion of the proximal interphalangeal joints of the 1st–3rd digits is lost and flexion of the 4th and 5th digits is weakened. Flexion of the distal interphalangeal joints of the 2nd and 3rd digits is also lost. Flexion of the distal interphalangeal joints of the 4th and 5th digits is not affected because the medial part of the FDP, which produces these movements, is supplied by the ulnar nerve. The ability to flex the metacarpophalangeal joints of the 2nd and 3rd digits is affected because the digital branches of the median nerve supply the 1st and 2nd lumbricals. Thus, when the person attempts to make a fist, the 2nd and 3rd fingers remain partially extended ("hand of benediction") (Fig. B6.24A). Thenar muscle function (function of the muscles at the base of the thumb) is also lost, as in carpal tunnel syndrome (see the blue box "Carpal Tunnel Syndrome" on p. 790). When the anterior interosseous nerve is injured, the thenar muscles are unaffected, but *paresis* (partial paralysis) of the flexor digitorum profundus and flexor pollicis longus occurs. When the person attempts to make the "okay" sign, opposing the tip of the thumb and index finger in a circle, a "pinch" posture of the hand results instead owing to the absence of flexion of the interphalangeal joint of the thumb and distal interphalangeal joint of the index finger (*anterior interosseous syndrome*) (Fig. B6.24B).

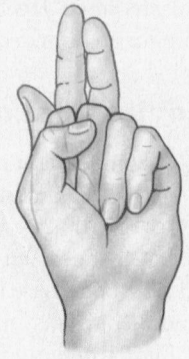

| Inability to flex distal interphalangeal joint of index finger | Inability to flex digits two and three into a compact fist |

(A) Ulnar nerve palsy

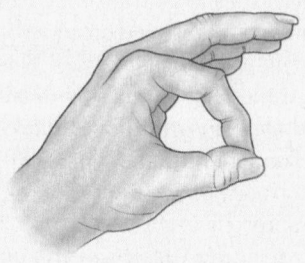
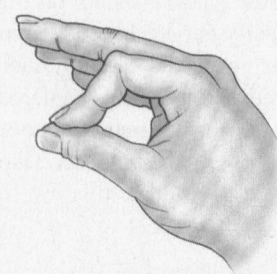

| Normal "ok" sign | Abnormal "pinch" sign |

(B) Anterior interosseous syndrome

FIGURE B6.24. Median nerve injury (palsy). A. Testing for median nerve palsy. **B.** Testing for anterior interosseous syndrome.

Pronator Syndrome

Pronator syndrome, a nerve entrapment syndrome, is caused by compression of the median nerve near the elbow. The nerve may be compressed between the heads of the pronator teres as a result of trauma, muscular hypertrophy, or fibrous bands. Individuals with this syndrome are first seen clinically with pain and tenderness in the proximal aspect of the anterior forearm, and *hypesthesia* (decreased sensation) of palmar aspects of the radial three and half digits and adjacent palm (Fig. B6.25). Symptoms often follow activities that involve repeated pronation.

Communications Between Median and Ulnar Nerves

Occasionally, communications occur between the median and the ulnar nerves in the forearm. These branches are usually represented by slender nerves, but the communications are important clinically because even with a complete lesion of the median nerve, some muscles may

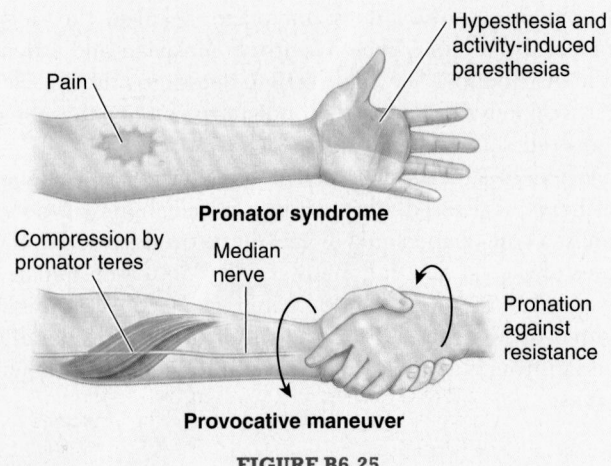

FIGURE B6.25.

not be paralyzed. This may lead to an erroneous conclusion that the median nerve has not been damaged.

Injury of Ulnar Nerve at Elbow and in Forearm

More than 27% of nerve lesions of the upper limb affect the ulnar nerve (Rowland, 2010). Ulnar nerve injuries usually occur in four places: (1) posterior to the medial epicondyle of the humerus, (2) in the cubital tunnel formed by the tendinous arch connecting the humeral and ulnar heads of the FCU, (3) at the wrist, and (4) in the hand.

Ulnar nerve injury occurs most commonly where the nerve passes posterior to the medial epicondyle of the humerus (Fig. B6.26). The injury results when the medial part of the elbow hits a hard surface, fracturing the medial epicondyle ("funny bone"). Any lesion superior to the medial epicondyle will produce paresthesia of the median part of the dorsum of the hand. Compression of the ulnar nerve at the elbow (*cubital tunnel syndrome*) is also common (see the blue box "Cubital Tunnel Syndrome" on p. 770). Ulnar nerve injury usually produces numbness and tingling (*paresthesia*) of the medial part of the palm and the medial one and a half fingers (Fig. B6.27). Pluck your ulnar nerve at the posterior aspect of your elbow with your index finger and you may feel tingling in these fingers. Severe compression may also produce elbow pain that radiates distally. Uncommonly, the ulnar nerve is compressed as it passes through the ulnar canal (see the blue box "Ulnar Canal Syndrome" on p. 792).

Ulnar nerve injury can result in extensive motor and sensory loss to the hand. An injury to the nerve in the distal part of the forearm denervates most intrinsic hand muscles. Power of wrist adduction is impaired, and when an attempt is made to flex the wrist joint, the hand is drawn to the lateral side by the FCR (supplied by the median nerve) in the absence of the "balance" provided by the FCU. After ulnar nerve injury, the person has difficulty making a fist because, in the absence of opposition, the metacarpophalangeal joints become hyperextended, and he or she cannot flex the 4th and 5th digits at

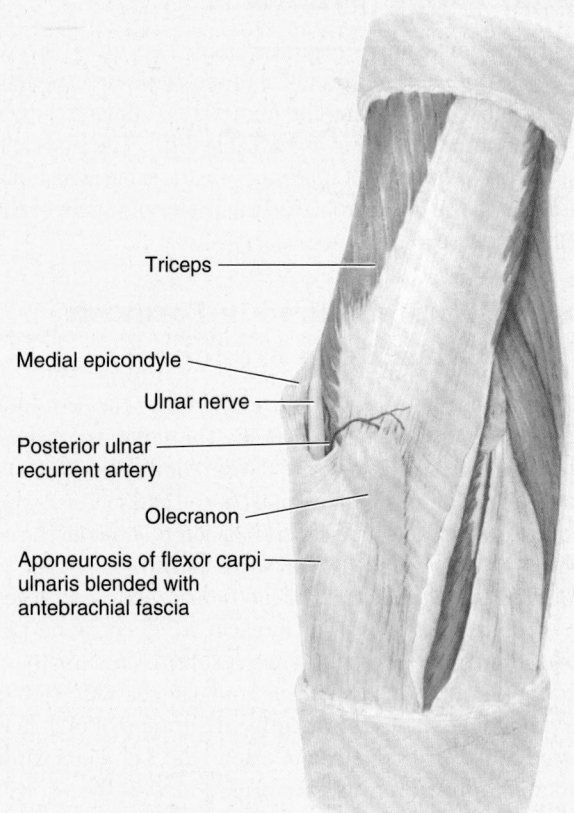

FIGURE B6.26. Vulnerable position of ulnar nerve.

the distal interphalangeal joints when trying to make a fist. Furthermore, the person cannot extend the interphalangeal joints when trying to straighten the fingers. This characteristic appearance of the hand, resulting from a distal lesion of the ulnar nerve, is known as *claw hand (main en griffe)*. The deformity results from atrophy of the interosseous muscles of the hand supplied by the ulnar nerve. The claw is produced by the unopposed action of the extensors and FDP. For a description of ulnar nerve injury at the wrist, see the blue box "Ulnar Canal Syndrome" on p. 792).

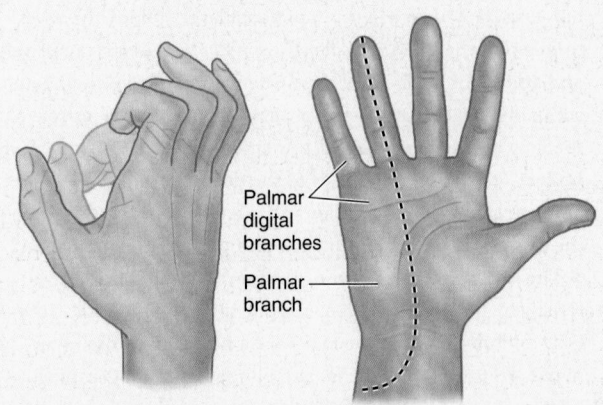

FIGURE B6.27. Claw hand and sensory distribution of ulnar nerve.

Cubital Tunnel Syndrome

The ulnar nerve may be compressed (*ulnar nerve entrapment*) in the cubital tunnel formed by the tendinous arch joining the humeral and ulnar heads of attachment of the FCU (Fig. 6.59; Table 6.10). The signs and symptoms of *cubital tunnel syndrome* are the same as an ulnar nerve lesion in the ulnar groove on the posterior aspect of the medial epicondyle of the humerus.

Injury of Radial Nerve in Forearm (Superficial or Deep Branches)

The radial nerve is usually injured in the arm by a fracture of the humeral shaft. This injury is proximal to the motor branches to the long and short extensors of the wrist from the (common) radial nerve, and so *wrist-drop* is the primary clinical manifestation of an injury at this level (see the blue box "Injury to the Radial Nerve in Arm" on p. 743).

Injury to the deep branch of the radial nerve may occur when wounds of the posterior forearm are deep (penetrating). Severance of the deep branch results in an inability to extend the thumb and the metacarpophalangeal (MP) joints of the other digits. Thus the integrity of the deep branch may be tested by asking the person to extend the MP joints while the examiner provides resistance (Fig. B6.28). If the nerve is intact, the long extensor tendons should appear prominently on the dorsum of the hand, confirming that the extension is occurring at the MP joints rather than at the interphalangeal joints (movements under the control of other nerves).

Loss of sensation does not occur because the deep branch of the radial nerve is entirely muscular and articular in distribution. See Table 6.13 to determine the muscles that are paralyzed (e.g., extensor digitorum) when this nerve is severed.

When the superficial branch of the radial nerve, a cutaneous nerve, is severed, sensory loss is usually minimal. Commonly, a coin-shaped area of anesthesia occurs distal to the bases of the 1st and 2nd metacarpals. The reason the area of sensory loss is less than expected, given the areas highlighted in Figure 6.69D, is the result of the considerable overlap from cutaneous branches of the median and ulnar nerves.

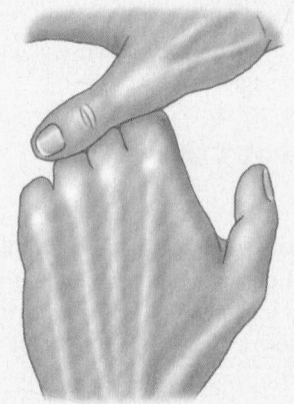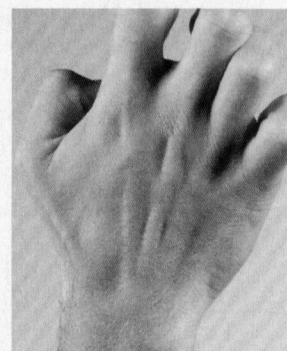

FIGURE B6.28. Testing radial nerve.

The Bottom Line

FOREARM

Muscles of anterior compartment of forearm: The superficial and intermediate muscles of the anterior (flexor–pronator) compartment of the forearm are located anteromedially because they arise mainly from the common flexor attachment (medial epicondyle and supra-epicondylar ridge) of the humerus. ♦ Muscles in the superficial layer "bend" the wrist to position the hand (i.e., flex the wrist when acting exclusively and abduct or adduct the wrist when working with their extensor counterparts) and assist pronation. ♦ The only muscle of the intermediate layer (FDS) primarily flexes the proximal joints of 2nd–5th digits. ♦ Muscles of the deep layer attach to the anterior aspects of the radius and ulna, flex all (but especially the distal) joints of all five digits, and pronate the forearm. ♦ The muscles of the anterior compartment are innervated mostly by the median nerve, but one and a half muscles (the FCU and ulnar half of the FDP) are innervated by the ulnar nerve. ♦ Flexion of the wrist and hand is used for grasping, gripping, and drawing things toward one self. ♦ Pronation is used for positioning the hand to manipulate or pick things up. Both movements are basic protective (defensive) movements.

Muscles of posterior compartment of forearm: The extensor-supinator muscles of the posterior compartment of the forearm are located posterolaterally in the proximal forearm, and are innervated by the radial nerve. ♦ The supinator acts at the radio-ulnar joint, while the remaining muscles extend and abduct the hand at the wrist joint and the thumb. The ECU may also contribute to adduction of the hand. ♦ The extensor muscles become tendinous in the distal forearm, and pass deep to the extensor retinaculum in osseofibrous tunnels. ♦ Tendons passing to the medial 4 digits are involved in complex extensor expansions on the dorsal aspects of the fingers. ♦ Extension ("cocking") of the wrist is important in enabling the flexors of the fingers to grip tightly or make a fist.

Superficial veins and cutaneous nerves of forearm: Well-developed subcutaneous veins course in the subcutaneous tissue of the forearm. These veins are subject to great variation. ◆ Once they have penetrated the deep fascia, cutaneous nerves run independently of the veins in the subcutaneous tissue, where they remain constant in location and size, with lateral, medial, and posterior cutaneous nerves of the forearm supplying the aspects of the forearm described by their names.

Neurovascular bundles of forearm: Three major (radial, median or middle, and ulnar) and two minor (anterior and posterior interosseous) neurovascular bundles occur deep to the antebrachial fascia. ◆ The radial neurovascular bundle—containing the radial artery, accompanying veins, and superficial radial nerve—courses along and defines the border between the anterior and the posterior forearm compartments (the vascular structures serving both) deep to the brachioradialis. ◆ The middle (median nerve and variable median artery and veins) and ulnar (ulnar nerve, artery, and accompanying veins) bundles course in a fascial plane between the intermediate and the deep flexor muscles. The median nerve supplies most muscles in the anterior compartment, many via its anterior interosseous branch, which courses on the interosseous membrane. ◆ The ulnar nerve supplies the one and a half exceptions (FCU and ulnar half of the FDP). ◆ The deep radial nerve penetrates the supinator to join the posterior interosseous artery in the plane between the superficial and the deep extensors. This nerve supplies all the muscles arising in the posterior compartment. ◆ The flexor muscles of the anterior compartment have approximately twice the bulk and strength of the extensor muscles of the posterior compartment. This, and the fact that the flexor aspect of the limb is the more protected aspect, accounts for the major neurovascular structures being located in the anterior compartment, with only the relatively small posterior interosseous vessels and nerve in the posterior compartment.

HAND

The **hand** is the manual part of the upper limb distal to the forearm. The **wrist** is located at the junction of the forearm and hand. Once positioned at the desired height and location relative to the body by movements at the shoulder and elbow, and the direction of action is established by pronation and supination of the forearm, the working position or attitude (tilt) of the hand is adjusted by movement at the wrist joint.

The skeleton of the hand (Fig. 6.9) consists of *carpals* in the wrist, *metacarpals* in the hand proper, and *phalanges* in the digits (fingers). The digits are numbered from one to five, beginning with the thumb: digit 1 is the thumb; digit 2, the index finger; digit 3, the middle finger; digit 4, the ring finger; and digit 5, the little finger. The palmar aspect of the hand features a central concavity that, with the crease proximal to it (over the wrist bones) separates two eminences: a lateral, larger and more prominent **thenar eminence** at the base of the thumb, and a medial, smaller **hypothenar eminence** proximal to the base of the 5th finger (Fig 6.72A).

Because of the importance of manual dexterity in occupational and recreational activities, a good understanding of the structure and function of the hand is essential for all persons involved in maintaining or restoring its activities: free motion, power grasping, precision handling, and pinching.

The **power grip** (palm grasp) refers to forcible motions of the digits acting against the palm; the fingers are wrapped around an object with counterpressure from the thumb—for example, when grasping a cylindrical structure (Fig. 6.73A). The power grip involves the long flexor muscles to the digits (acting at the interphalangeal joints), the intrinsic muscles in the palm (acting at the metacarpophalangeal joints), and the extensors of the wrist (acting at the radiocarpal and midcarpal joints). The "cocking" of the wrist by the extensors increases the distance over which the flexors of the fingers act, producing the same result as a more complete muscular contraction. Conversely, as flexion increases at the wrist, the grip becomes weaker and more insecure.

The **hook grip** is the posture of the hand that is used when carrying a briefcase (Fig. 6.73B). This grip consumes less energy, involving mainly the long flexors of the digits, which are flexed to a varying degree, depending on the size of the object that is grasped.

The **precision handling grip** involves a change in the position of a handled object that requires fine control of the movements of the digits (fingers)—for example, holding a pencil, manipulating a coin, threading a needle, or buttoning a shirt (Fig. 6.73C & D). In a precision grip, the wrist and digits are held firmly by the long flexor and extensor muscles, and the intrinsic hand muscles perform fine movements of the digits.

Pinching refers to compression of something between the thumb and the index finger—for example, handling a teacup or holding a coin on edge (Fig. 6.73E)—or between the thumb and the adjacent two fingers—for example, snapping the fingers.

The **position of rest** is assumed by an inactive hand—for example, when the forearm and hand are laid on a table (Fig. 6.73F). This position is often used when it is necessary to immobilize the wrist and hand in a cast to stabilize a fracture.

Fascia and Compartments of Palm

The *fascia of the palm* is continuous with the antebrachial fascia and the fascia of the dorsum of the hand (Fig. 6.58). The **palmar fascia** is thin over the thenar and hypothenar eminences, as **thenar** and **hypothenar fascia**, respectively (Figs. 6.74A and 6.75A). However, the palmar fascia is thick

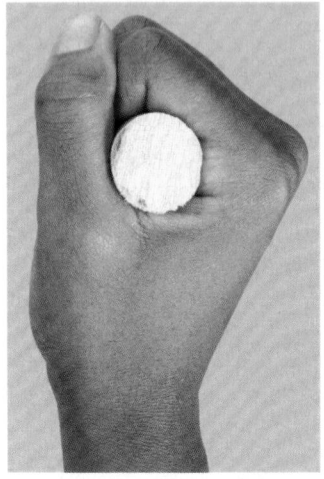

(A) Lateral view; Power grip

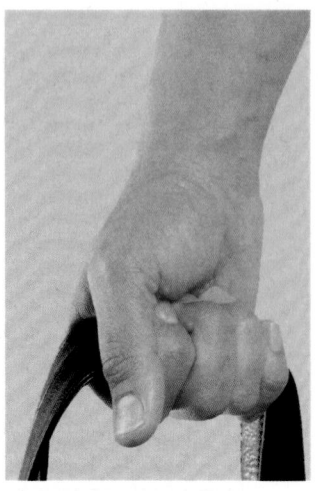

(B) Anteromedial view; Hook grip

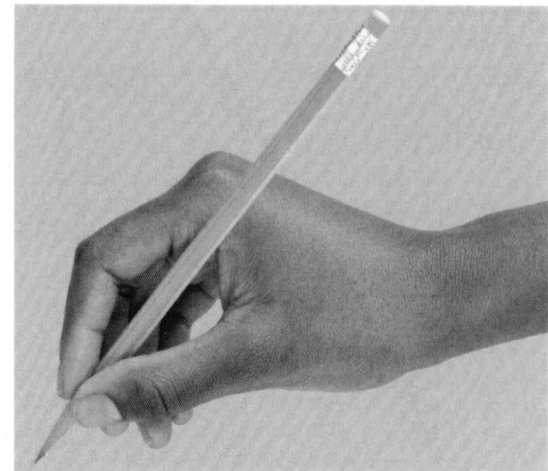

(C) Medial view; Precision handling grip

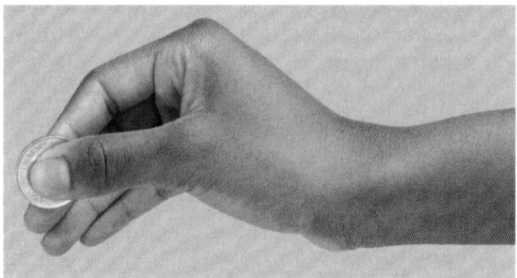

(D) Medial view; Precision grip

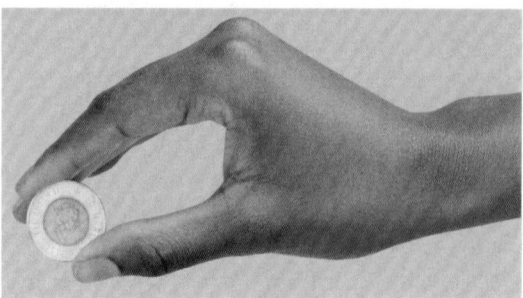

(E) Medial view; Fingertip pinch

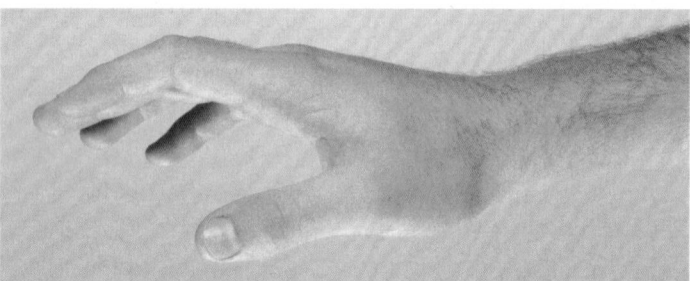

(F) Medial view; Position of rest

(G) Anterior view; Loose grip

(H) Anterior view; Firm grip

FIGURE 6.73. Functional positions of hand. A. In the power grip, when grasping an object, the metacarpophalangeal (MP) and interphalangeal (IP) joints are flexed, but the radiocarpal and midcarpal joints are extended. "Cocking" (extension of) the wrist increases the distance over which the flexor tendons act, increasing tension of the long flexor tendons beyond that produced by maximal contraction of the muscles alone. **B.** The hook grip (flexion of the IP joints of the 2nd–4th digits) resists gravitational (downward) pull with only digital flexion. **C.** The precision grip is used when writing. **D and E.** One uses the precision grip to hold a coin to enable manipulation (**D**) and when pinching an object (**E**). **F.** Casts for fractures are applied most often with the hand and wrist in the position of rest. Note the mild extension of the wrist. **G and H.** When gripping an unattached rod loosely (**G**) or firmly (**H**), the 2nd and 3rd carpometacarpal joints are rigid and stable, but the 4th and 5th are saddle joints permitting flexion and extension. The increased flexion changes the angle of the rod during the firm grip.

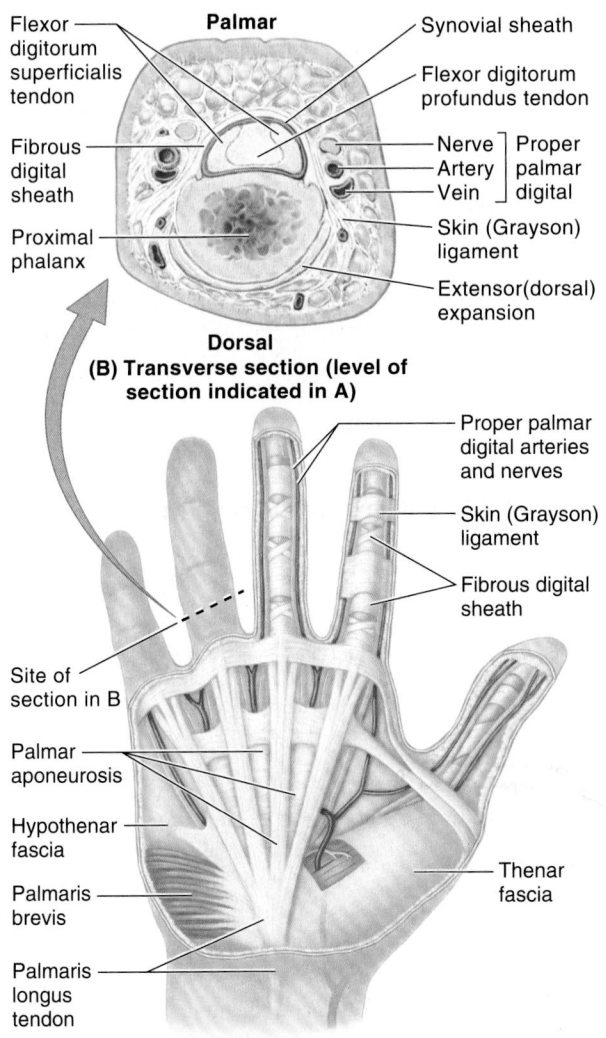

Flexor digitorum superficialis tendon
Palmar
Synovial sheath
Flexor digitorum profundus tendon
Fibrous digital sheath
Nerve ⎤ Proper
Artery ⎥ palmar
Vein ⎦ digital
Proximal phalanx
Skin (Grayson) ligament
Extensor(dorsal) expansion
Dorsal
(B) Transverse section (level of section indicated in A)

Proper palmar digital arteries and nerves
Skin (Grayson) ligament
Fibrous digital sheath
Site of section in B
Palmar aponeurosis
Hypothenar fascia
Palmaris brevis
Palmaris longus tendon
Thenar fascia

(A) Palmar view

FIGURE 6.74. Palmar fascia and fibrous digital sheaths. A. The palmar fascia is continuous with the antebrachial fascia. The thin thenar and hypothenar fascia covers the intrinsic muscles of the thenar and hypothenar eminences, respectively. Between the thenar and the hypothenar muscle masses, the central compartment of the palm is roofed by the thick palmar aponeurosis. **B.** A transverse section of the 4th digit (proximal phalanx level). Within the fibrous digital sheath and proximal to its attachment to the base of the middle phalanx, the FDS tendon has split into two parts to allow continued central passage of the FDP tendon to the distal phalanx.

centrally where it forms the fibrous palmar aponeurosis, and in the fingers where it forms the digital sheaths. The **palmar aponeurosis,** a strong, well-defined part of the deep fascia of the palm, covers the soft tissues and overlies the long flexor tendons. The proximal end or apex of the triangular palmar aponeurosis is continuous with the flexor retinaculum and the palmaris longus tendon.

When the palmaris longus is present, the palmar aponeurosis is the expanded tendon of the palmaris longus. Distal to the apex, the palmar aponeurosis forms four longitudinal digital bands or rays that radiate from the apex and attach distally to the bases of the proximal phalanges and become

continuous with the fibrous digital sheaths (Figs. 6.58 and 6.74). The **fibrous digital sheaths** are ligamentous tubes that enclose the synovial sheaths, the superficial and deep flexor tendons, and the tendon of the FPL in their passage along the palmar aspect of their respective fingers.

A **medial fibrous septum** extends deeply from the medial border of the palmar aponeurosis to the 5th metacarpal (Fig. 6.75A). Medial to this septum is the medial or **hypothenar compartment,** containing the hypothenar muscles and bounded anteriorly by the hypothenar fascia. Similarly, a **lateral fibrous septum** extends deeply from the lateral border of the palmar aponeurosis to the 3rd metacarpal. Lateral to this septum is the lateral or **thenar compartment,** containing the thenar muscles and bounded anteriorly by the thenar fascia.

Between the hypothenar and thenar compartments is the **central compartment,** bounded anteriorly by the palmar aponeurosis and containing the flexor tendons and their sheaths, the lumbricals, the superficial palmar arterial arch, and the digital vessels and nerves.

The deepest muscular plane of the palm is the **adductor compartment** containing the adductor pollicis.

Between the flexor tendons and the fascia covering the deep palmar muscles are two potential spaces, the **thenar space** and the **midpalmar space** (Fig. 6.75). The spaces are bounded by fibrous septa passing from the edges of the palmar aponeurosis to the metacarpals. Between the two spaces is the especially strong *lateral fibrous septum,* which is attached to the 3rd metacarpal. Although most fascial compartments end at the joints, the midpalmar space is continuous with the anterior compartment of the forearm via the carpal tunnel.

Muscles of Hand

The intrinsic muscles of the hand are located in five compartments (Fig. 6.75A):

1. Thenar muscles in the *thenar compartment:* abductor pollicis brevis, flexor pollicis brevis, and opponens pollicis.
2. Adductor pollicis in the *adductor compartment.*
3. Hypothenar muscles in the *hypothenar compartment:* abductor digiti minimi, flexor digiti minimi brevis, and opponens digiti minimi.
4. Short muscles of the hand, the lumbricals, in the *central compartment* with the long flexor tendons.
5. The interossei in separate *interosseous compartments* between the metacarpals.

THENAR MUSCLES

The **thenar muscles** form the *thenar eminence* on the lateral surface of the palm (Fig. 6.72A) and are chiefly responsible for opposition of the thumb. Movement of the thumb is important for the precise activities of the hand. The high degree of freedom of the movements results from the 1st metacarpal being independent, with mobile joints at both

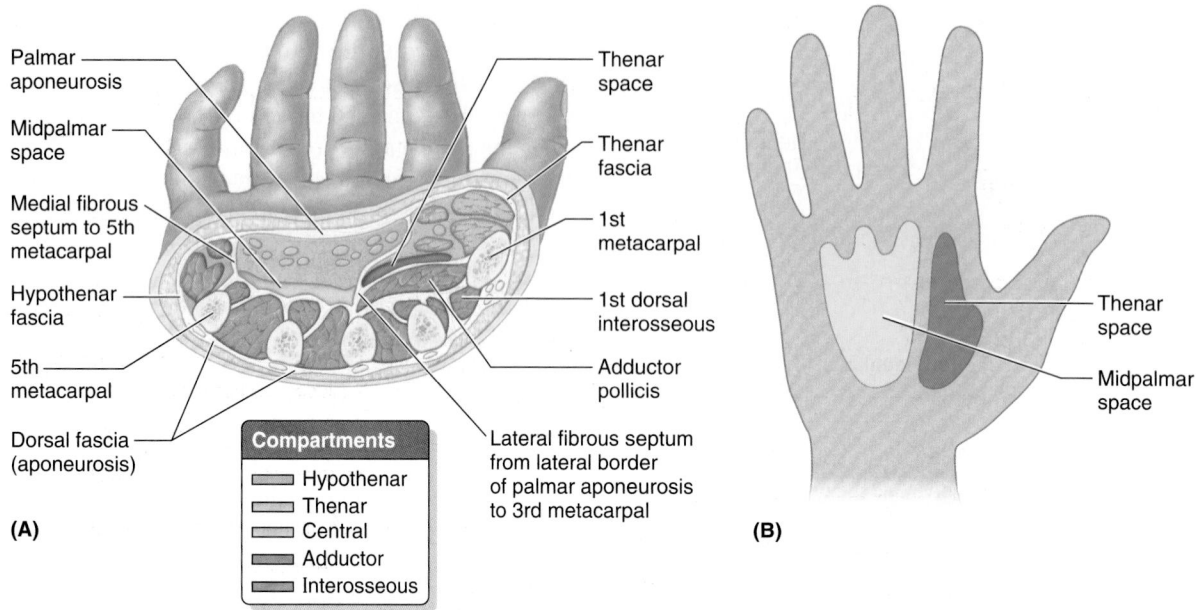

FIGURE 6.75. Compartments, spaces, and fascia of palm. A. Transverse section through the middle of the palm illustrating the fascial compartments of the hand. **B.** Thenar and midpalmar spaces. The midpalmar space underlies the central compartment of the palm and is related distally to the synovial tendon sheaths of the 3rd–5th digits and proximally to the common flexor sheath as it emerges from the carpal tunnel. The thenar space underlies the thenar compartment and is related distally to the synovial tendon sheath of the index finger and proximally to the common flexor sheath distal to the carpal tunnel.

ends. Several muscles are required to control the freedom of thumb movements (Fig. 6.76):

- *Extension:* extensor pollicis longus, extensor pollicis brevis, and abductor pollicis longus.
- *Flexion:* flexor pollicis longus and flexor pollicis brevis.
- *Abduction:* abductor pollicis longus and abductor pollicis brevis.
- *Adduction:* adductor pollicis and 1st dorsal interosseous.
- *Opposition:* opponens pollicis. This movement occurs at the carpometacarpal joint and results in a "cupping" of the palm. Bringing the tip of the thumb into contact with the 5th finger, or any of the other fingers, involves considerably more movement than can be produced by the opponens pollicis alone.

The first four movements of the thumb occur at the carpometacarpal and metacarpophalangeal joints. **Opposition,** a complex movement, begins with the thumb in the extended position and initially involves abduction and medial rotation of the 1st metacarpal (cupping the palm) produced by the action of the opponens pollicis at the carpometacarpal joint and then flexion at the metacarpophalangeal joint (Fig. 6.76). The reinforcing action of the adductor pollicis and FPL increases the pressure that the opposed thumb can exert on the fingertips. In pulp-to-pulp opposition, movements of the finger opposing the thumb are also involved.

The *thenar muscles* are illustrated in Figure 6.77; their attachments are shown in Fig. 6.78A; and their attachments, innervations, and main actions are summarized in Table 6.14.

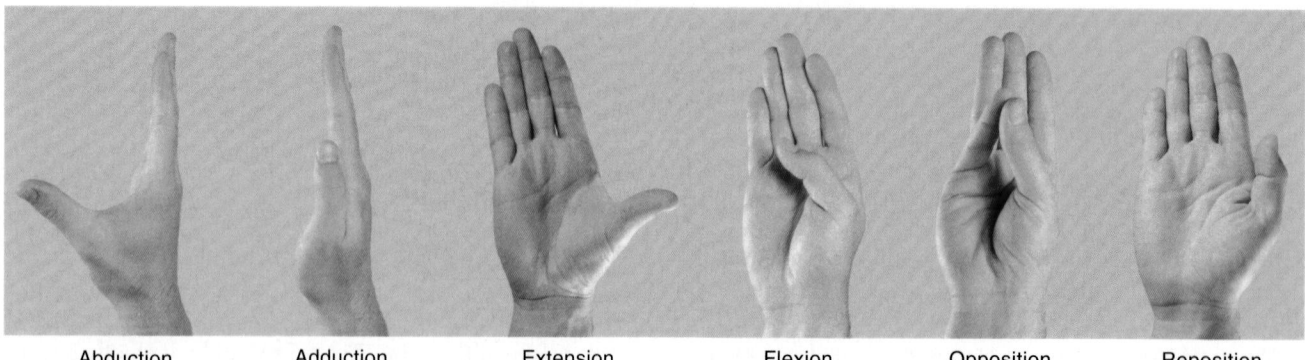

Abduction   Adduction   Extension   Flexion   Opposition   Reposition

FIGURE 6.76. Movements of thumb. The thumb is rotated 90° to the other digits. (This can be confirmed by noting the direction the nail of the thumb faces compared with the nails of the other fingers.) Thus abduction and adduction occur in a sagittal plane and flexion and extension occur in a coronal plane. Opposition, the action bringing the tip of the thumb in contact with the pulps of the other fingers (e.g., with the little finger), is the most complex movement. The components of opposition are abduction and medial rotation at the carpometacarpal joint and flexion of the metacarpophalangeal joint.

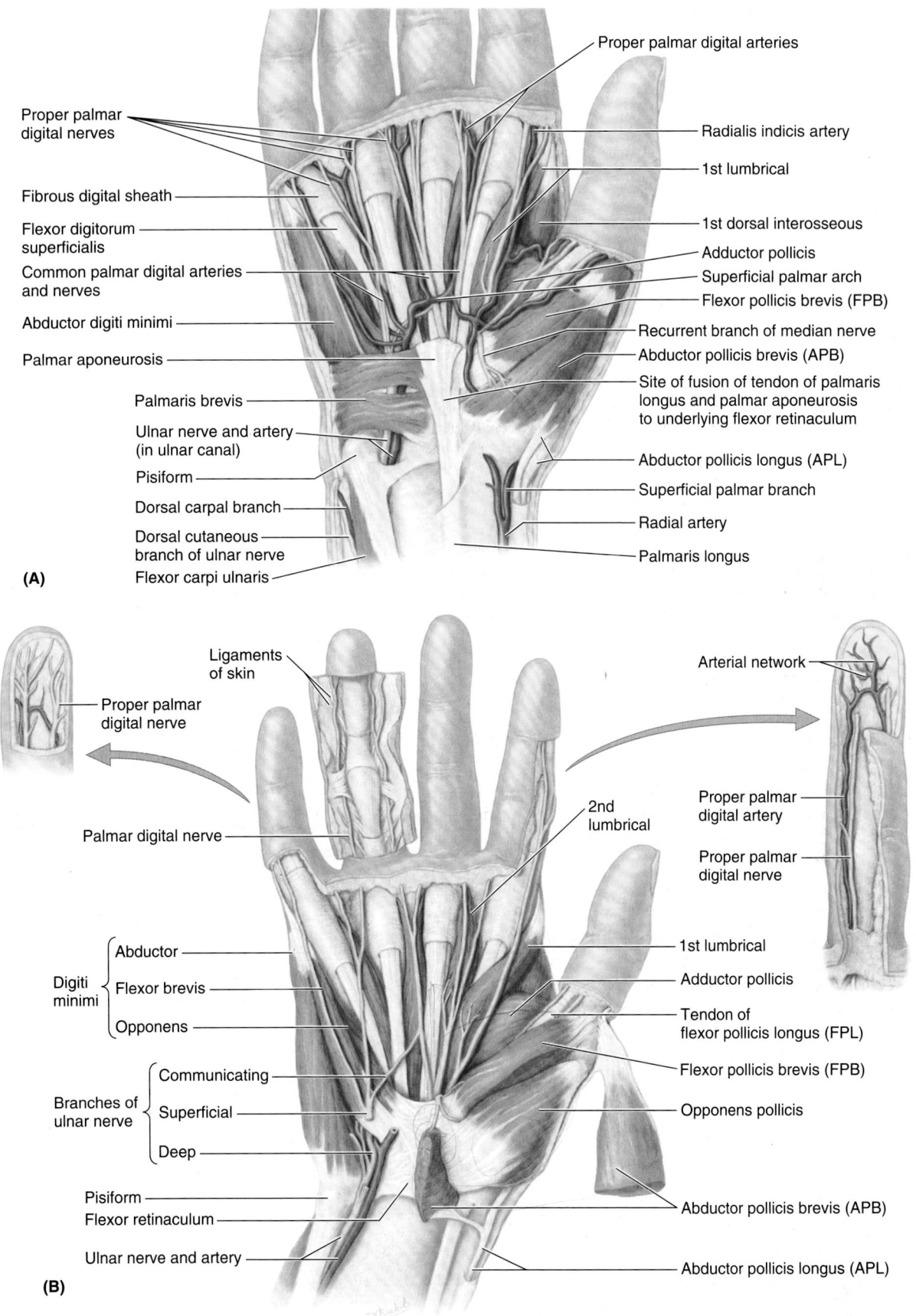

Proper palmar digital arteries

Radialis indicis artery

1st lumbrical

1st dorsal interosseous

Adductor pollicis

Superficial palmar arch

Flexor pollicis brevis (FPB)

Recurrent branch of median nerve

Abductor pollicis brevis (APB)

Site of fusion of tendon of palmaris longus and palmar aponeurosis to underlying flexor retinaculum

Abductor pollicis longus (APL)

Superficial palmar branch

Radial artery

Palmaris longus

Proper palmar digital nerves

Fibrous digital sheath

Flexor digitorum superficialis

Common palmar digital arteries and nerves

Abductor digiti minimi

Palmar aponeurosis

Palmaris brevis

Ulnar nerve and artery (in ulnar canal)

Pisiform

Dorsal carpal branch

Dorsal cutaneous branch of ulnar nerve

Flexor carpi ulnaris

(A)

Ligaments of skin

Proper palmar digital nerve

Palmar digital nerve

Arterial network

2nd lumbrical

Proper palmar digital artery

Proper palmar digital nerve

Digiti minimi
- Abductor
- Flexor brevis
- Opponens

Branches of ulnar nerve
- Communicating
- Superficial
- Deep

1st lumbrical

Adductor pollicis

Tendon of flexor pollicis longus (FPL)

Flexor pollicis brevis (FPB)

Opponens pollicis

Pisiform

Flexor retinaculum

Ulnar nerve and artery

Abductor pollicis brevis (APB)

Abductor pollicis longus (APL)

(B)

FIGURE 6.77. Superficial dissections of right palm. The skin and subcutaneous tissue have been removed, as have most of the palmar aponeurosis and the thenar and hypothenar fasciae. **A.** The superficial palmar arch is located immediately deep to the palmar aponeurosis, superficial to the long flexor tendons. This arterial arch gives rise to the common palmar digital arteries. In the digits, a digital artery (e.g., radialis indicis) and nerve lie on the medial and lateral sides of the fibrous digital sheath. The pisiform bone protects the ulnar nerve and artery as they pass into the palm. **B.** Three thenar and three hypothenar muscles attach to the flexor retinaculum and to the four marginal carpal bones united by the retinaculum.

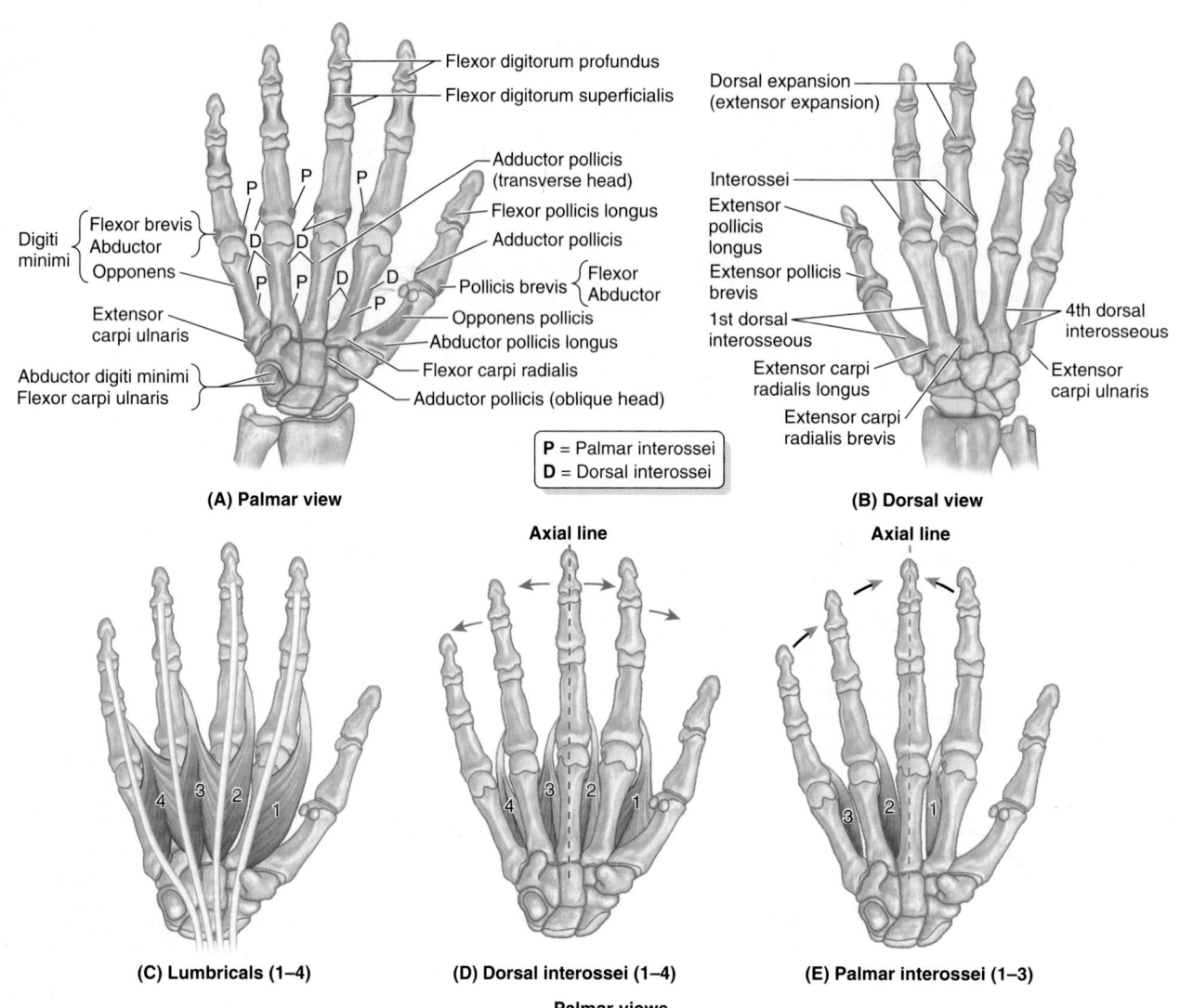

FIGURE 6.78. Attachments of intrinsic muscles of hand and actions of interossei.

TABLE 6.14. INTRINSIC MUSCLES OF HAND

Muscle	Proximal Attachment	Distal Attachment	Innervation[a]	Main Action
Thenar muscles				
Opponens pollicis	Flexor retinaculum and tubercles of scaphoid and trapezium	Lateral side of 1st metacarpal	Recurrent branch of median nerve (**C8**, T1)	To oppose thumb, it draws 1st metacarpal medially to center of palm and rotates it medially
Abductor pollicis brevis		Lateral side of base of proximal phalanx of thumb		Abducts thumb; helps oppose it
Flexor pollicis brevis				Flexes thumb
Superficial head				
Deep head				
Adductor pollicis			Deep branch of ulnar nerve (C8, **T1**)	Adducts thumb toward lateral border of palm
Oblique head	Bases of 2nd and 3rd metacarpals, capitate, and adjacent carpals	Medial side of base of proximal phalanx of thumb		
Transverse head	Anterior surface of shaft of 3rd metacarpal			

(continued)

TABLE 6.14. INTRINSIC MUSCLES OF HAND (Continued)

Muscle	Proximal Attachment	Distal Attachment	Innervation[a]	Main Action
Hypothenar muscles				
Abductor digiti minimi	Pisiform	Medial side of base of proximal phalanx of 5th digit	Deep branch of ulnar nerve (C8, **T1**)	Abducts 5th digit; assists in flexion of its proximal phalanx
Flexor digiti minimi brevis	Hook of hamate and flexor retinaculum			Flexes proximal phalanx of 5th digit
Opponens digiti		Medial border of 5th metacarpal		Draws 5th metacarpal anterior and rotates it, bringing 5th digit into opposition with thumb
Short muscles				
Lumbricals				
1st and 2nd	Lateral two tendons of flexor digitorum profundus (as unipennate muscles)	Lateral sides of extensor expansions of 2nd–5th digits	Median nerve (C8, **T1**)	Flex metacarpophalangeal joints; extend interphalangeal joints of 2nd–5th digits
3rd and 4th	Medial three tendons of flexor digitorum profundus (as bipennate muscles)		Deep branch of ulnar nerve (C8, **T1**)	
Dorsal interossei, 1st–4th	Adjacent sides of two metacarpals (as bipennate muscles)	Bases of proximal phalanges; extensor expansions of 2nd–4th digits		Abduct 2nd–4th digits from axial line; act with lumbricals in flexing metacarpophalangeal joints and extending interphalangeal joints
Palmar interossei, 1st–3rd	Palmar surfaces of 2nd, 4th, and 5th metacarpals (as unipennate muscles)	Bases of proximal phalanges; extensor expansions of 2nd, 4th, and 5th digits		Adduct 2nd, 4th, and 5th digits toward axial line; assist lumbricals in flexing metacarpophalangeal joints and extending interphalangeal joints; extensor expansions of 2nd–4th digits

[a]The spinal cord segmental innervation is indicated (e.g., "**C8**, T1" means that the nerves supplying the opponens pollicis are derived from the eighth cervical segment and first thoracic segment of the spinal cord). Numbers in boldface (**C8**) indicate the main segmental innervation. Damage to one or more of the listed spinal cord segments or to the motor nerve roots arising from them results in paralysis of the muscles concerned.

Abductor Pollicis Brevis. The **abductor pollicis brevis (APB),** the short abductor of the thumb, forms the anterolateral part of the thenar eminence. In addition to abducting the thumb, the APB assists the opponens pollicis during the early stages of opposition by rotating its proximal phalanx slightly medially.

To test the abductor pollicis brevis, abduct the thumb against resistance. If acting normally, the muscle can be seen and palpated.

Flexor Pollicis Brevis. The **flexor pollicis brevis (FPB),** the short flexor of the thumb, is located medial to the APB (Fig. 6.77A). Its two bellies, located on opposite sides of the tendon of the FPL, share (with each other and often with the APB) a common, sesamoid-containing tendon at their distal attachment. The bellies usually differ in their innervation: The larger superficial head of the FPB is innervated by the recurrent branch of the median nerve, whereas the smaller deep head is usually innervated by the deep palmar branch of the ulnar nerve. The FPB flexes the thumb at the carpometacarpal and metacarpophalangeal joints and aids in opposition of the thumb.

To test the flexor pollicis brevis, flex the thumb against resistance. If acting normally, the muscle can be seen and palpated; however, keep in mind that the FPL also flexes the thumb.

Opponens Pollicis. The **opponens pollicis** is a quadrangular muscle that lies deep to the APB and lateral to the FPB (Fig. 6.77B). The opponens pollicis opposes the thumb, the most important thumb movement. It flexes and rotates the 1st metacarpal medially at the carpometacarpal joint during opposition; this movement occurs when picking up an object. During *opposition,* the tip of the thumb is brought into contact with the pulp (pad) of the little finger, as shown in Figure 6.76.

ADDUCTOR POLLICIS

The **adductor pollicis** is located in the adductor compartment of the hand (Fig. 6.75A). The fan-shaped muscle has two heads of origin, which are separated by the radial artery as it enters the palm to form the deep palmar arch (Figs. 6.77A and 6.79). Its tendon usually contains a sesamoid bone. The adductor pollicis adducts the thumb, moving the thumb to

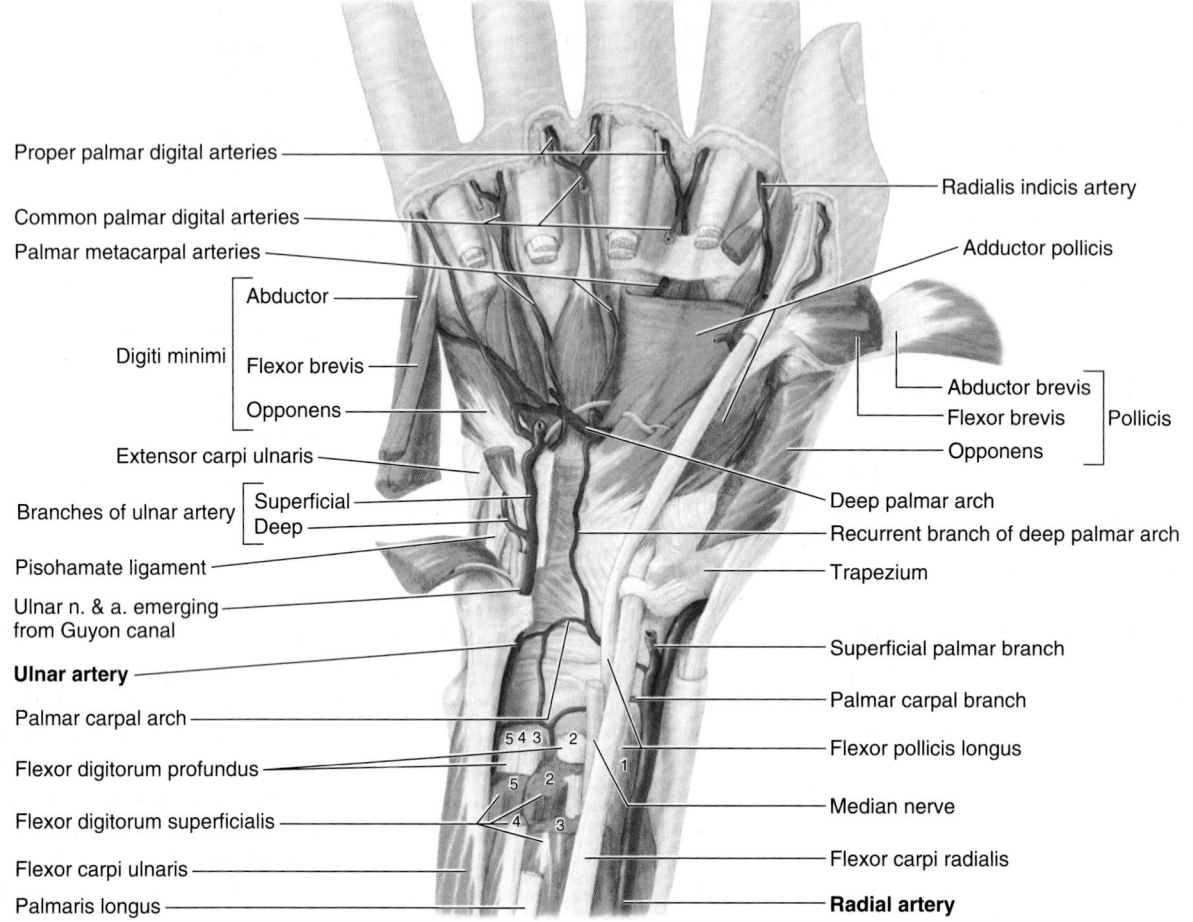

Proper palmar digital arteries

Common palmar digital arteries

Palmar metacarpal arteries

Digiti minimi — Abductor / Flexor brevis / Opponens

Extensor carpi ulnaris

Branches of ulnar artery — Superficial / Deep

Pisohamate ligament

Ulnar n. & a. emerging from Guyon canal

Ulnar artery

Palmar carpal arch

Flexor digitorum profundus

Flexor digitorum superficialis

Flexor carpi ulnaris

Palmaris longus

Radialis indicis artery

Adductor pollicis

Abductor brevis / Flexor brevis — Pollicis / Opponens

Deep palmar arch

Recurrent branch of deep palmar arch

Trapezium

Superficial palmar branch

Palmar carpal branch

Flexor pollicis longus

Median nerve

Flexor carpi radialis

Radial artery

FIGURE 6.79. Muscles and arteries of distal forearm and deep palm. Deep dissection of the palm revealing the anastomosis of the palmar carpal branch of the radial artery with the palmar carpal branch of the ulnar artery to form the palmar carpal arch and the deep palmar arch. The deep palmar arch lies at the level of the bases of the metacarpal bones, 1.5–2 cm proximal to the superficial palmar arch.

the palm of the hand (Fig. 6.76), thereby giving power to the grip (Fig. 6.73G & H).

HYPOTHENAR MUSCLES

The **hypothenar muscles** (abductor digiti minimi, flexor digiti minimi brevis, and opponens digiti minimi) produce the *hypothenar eminence* on the medial side of the palm and move the little finger (Fig. 6.87). These muscles are in the hypothenar compartment with the 5th metacarpal (Figs. 6.75A and 6.77). The attachments are illustrated in Figure 6.78A, and their attachments, innervations, and main actions of the hypothenar muscles are summarized in Table 6.14.

Abductor Digiti Minimi. The **abductor digiti minimi** is the most superficial of the three muscles forming the hypothenar eminence. The abductor digiti minimi abducts the 5th finger and helps flex its proximal phalanx.

Flexor Digiti Minimi Brevis. The **flexor digiti minimi brevis** is variable in size; it lies lateral to the abductor digiti minimi. The flexor digiti minimi brevis flexes the proximal phalanx of the 5th finger at the metacarpophalangeal joint.

Opponens Digiti Minimi. The **opponens digiti minimi** is a quadrangular muscle that lies deep to the abductor and flexor muscles of the 5th finger. The opponens digiti minimi draws the 5th metacarpal anteriorly and rotates it laterally, thereby deepening the hollow of the palm and bringing the 5th finger into opposition with the thumb (Fig. 6.76). Like the opponens pollicis, the opponens digiti minimi acts exclusively at the carpometacarpal joint.

Palmaris Brevis. The **palmaris brevis** is a small, thin muscle in the subcutaneous tissue of the hypothenar eminence (Figs. 6.74A and 6.77A); it is not in the hypothenar compartment. The palmaris brevis wrinkles the skin of the hypothenar eminence and deepens the hollow of the palm, thereby aiding the palmar grip. The palmaris brevis covers and protects the ulnar nerve and artery. It is attached proximally to the medial border of the palmar aponeurosis and to the skin on the medial border of the hand.

SHORT MUSCLES OF HAND

The short muscles of the hand are the lumbricals and interossei (Fig. 6.78C–E; Table 6.14).

Lumbricals. The four slender lumbrical muscles were named because of their worm-like form (L. *lumbricus*, earthworm) (Figs. 6.77B and 6.78C). The lumbricals flex the fingers at the metacarpophalangeal joints and extend the interphalangeal joints.

To test the lumbrical muscles, with the palm facing superiorly the patient is asked to flex the metacarpophalangeal (MP) joints while keeping the interphalangeal joints extended. The examiner uses one finger to apply resistance along the palmar surface of the proximal phalanx of digits 2–5 individually. Resistance may also be applied separately on the dorsal surface of the middle and distal phalanges of digits 2–5 to test extension of the interphalangeal joints, also while flexion of the MP joints is maintained.

Interossei. The four **dorsal interosseous muscles** (dorsal interossei) are located between the metacarpals; the three **palmar interosseous muscles** (palmar interossei) are on the palmar surfaces of the metacarpals in the interosseous compartment of the hand (Fig. 6.75A). The 1st dorsal interosseous muscle is easy to palpate; oppose the thumb firmly against the index finger and it can be easily felt. Some authors describe four palmar interossei; in so doing, they are including the deep head of the FPB because of its similar innervation and placement on the thumb. The four dorsal interossei abduct the fingers, and the three palmar interossei adduct them (Fig. 6.78D & E; Table 6.14).

A mnemonic device is to make acronyms of **d**orsal **ab**duct (DAB) and **p**almar **ad**uct (PAD). Acting together, the dorsal and palmar interossei and the lumbricals produce flexion at the metacarpophalangeal joints and extension of the interphalangeal joints (the so-called Z-movement). This occurs because of their attachment to the lateral bands of the extensor expansions (Fig. 6.63A & B).

Understanding the Z-movement is useful because it is the opposite of claw hand, which occurs in ulnar paralysis when the interossei and the 3rd and 4th lumbricals are incapable of acting together to produce the Z-movement (see the blue box "Injury of Ulnar Nerve" on p. 769).

To test the dorsal interossei, the examiner holds adjacent extended and adducted fingers between thumb and middle finger, providing resistance as the individual attempts to abduct the fingers (the person is asked to "spread the fingers apart") (Fig. 6.80A). *To test the palmar interossei,* a sheet of paper is placed between adjacent fingers. The individual is asked to "keep the fingers together" to prevent the paper from being pulled away by the examiner (Fig. 6.80B).

Long Flexor Tendons and Tendon Sheaths in Hand

The tendons of the FDS and FDP enter the **common flexor sheath** (ulnar bursa) deep to the flexor retinaculum (Fig. 6.81A). The tendons enter the central compartment of the hand and fan out to enter their respective **digital synovial sheaths.** The flexor and digital sheaths enable the tendons to slide freely over

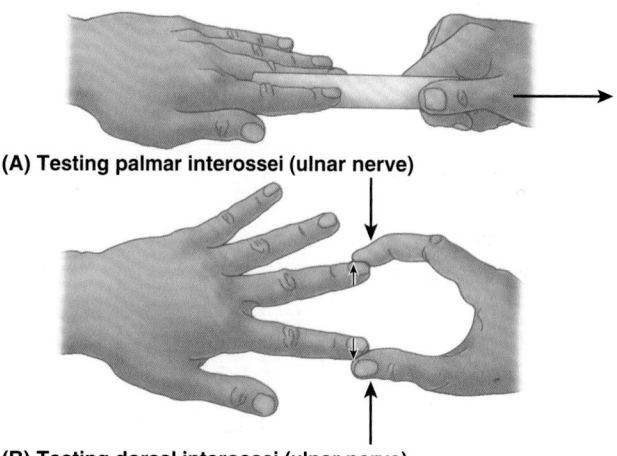

(A) Testing palmar interossei (ulnar nerve)

(B) Testing dorsal interossei (ulnar nerve)

FIGURE 6.80. Testing interossei (ulnar nerve). A. Dorsal interossei. **B.** Palmar interossei.

each other during movements of the fingers. Near the base of the proximal phalanx, the tendon of FDS splits to permit passage of the tendon of FDP; the crossing of the tendons makes up a **tendinous chiasm** (Figs. 6.63D, 6.74B, and 6.81B). The halves of the FDS tendon are attached to the margins of the anterior aspect of the base of the middle phalanx. Distal to the tendinous chiasm, the FDP tendon attaches to the anterior aspect of the base of the distal phalanx (Fig. 6.63D).

The **fibrous digital sheaths** are the strong ligamentous tunnels containing the flexor tendons and their synovial sheaths (Figs. 6.74 and 6.81C & D). The sheaths extend from the heads of the metacarpals to the bases of the distal phalanges. These sheaths prevent the tendons from pulling away from the digits (bowstringing). The fibrous digital sheaths combine with the bones to form **osseofibrous tunnels** through which the tendons pass to reach the digits. The **anular** and **cruciform parts** (often referred to clinically as "pulleys") are thickened reinforcements of the fibrous digital sheaths (Fig. 6.81D).

The long flexor tendons are supplied by small blood vessels that pass within synovial folds (**vincula**) from the periosteum of the phalanges (Fig. 6.63B). The tendon of the FPL passes deep to the flexor retinaculum to the thumb within its own synovial sheath. At the head of the metacarpal, the tendon runs between two *sesamoid bones,* one in the combined tendon of the FPB and APB and the other in the tendon of the adductor pollicis.

Arteries of Hand

Because its function requires it to be placed and held in many different positions, often while grasping or applying pressure, the hand is supplied with an abundance of highly branched and anastomosing arteries so that oxygenated blood is generally available to all parts in all positions. Furthermore, the arteries or their derivatives are relatively superficial, underlying skin that is capable of sweating so that excess heat can be released. To prevent undesirable heat loss in a

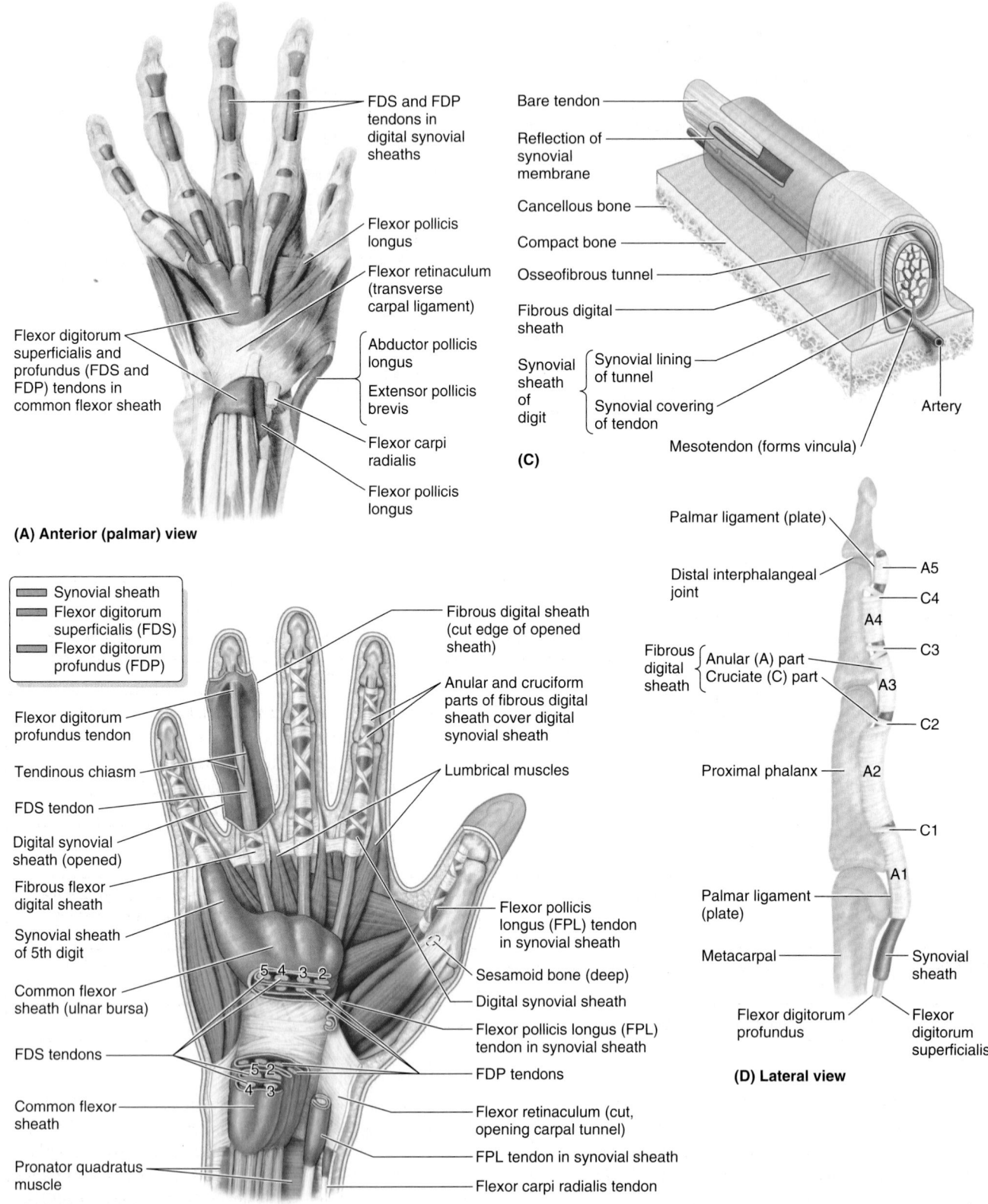

FDS and FDP tendons in digital synovial sheaths

Flexor pollicis longus

Flexor retinaculum (transverse carpal ligament)

Abductor pollicis longus

Extensor pollicis brevis

Flexor carpi radialis

Flexor pollicis longus

Flexor digitorum superficialis and profundus (FDS and FDP) tendons in common flexor sheath

(A) Anterior (palmar) view

Bare tendon

Reflection of synovial membrane

Cancellous bone

Compact bone

Osseofibrous tunnel

Fibrous digital sheath

Synovial sheath of digit { Synovial lining of tunnel / Synovial covering of tendon

Artery

Mesotendon (forms vincula)

(C)

Synovial sheath
Flexor digitorum superficialis (FDS)
Flexor digitorum profundus (FDP)

Flexor digitorum profundus tendon

Tendinous chiasm

FDS tendon

Digital synovial sheath (opened)

Fibrous flexor digital sheath

Synovial sheath of 5th digit

Common flexor sheath (ulnar bursa)

FDS tendons

Common flexor sheath

Pronator quadratus muscle

Fibrous digital sheath (cut edge of opened sheath)

Anular and cruciform parts of fibrous digital sheath cover digital synovial sheath

Lumbrical muscles

Flexor pollicis longus (FPL) tendon in synovial sheath

Sesamoid bone (deep)

Digital synovial sheath

Flexor pollicis longus (FPL) tendon in synovial sheath

FDP tendons

Flexor retinaculum (cut, opening carpal tunnel)

FPL tendon in synovial sheath

Flexor carpi radialis tendon

(B) Anterior (palmar) view

Palmar ligament (plate)

Distal interphalangeal joint

A5

C4

A4

C3

Fibrous digital sheath { Anular (A) part / Cruciate (C) part

A3

C2

Proximal phalanx — A2

C1

A1

Palmar ligament (plate)

Metacarpal

Synovial sheath

Flexor digitorum profundus

Flexor digitorum superficialis

(D) Lateral view

FIGURE 6.81. Flexor tendons, common flexor sheath, fibrous digital sheaths, and synovial sheaths of digits. A. The synovial sheaths of the long flexor tendons to the digits are arranged in two sets: (1) proximal or carpal, posterior to the flexor retinaculum, and (2) distal or digital, within the fibrous sheaths of the digital flexors. **B.** Tendons, tendon bursae, and fibrous digital sheaths. **C.** The structure of an osseofibrous tunnel of a finger, containing a tendon. Within the fibrous sheath, the synovial sheath consists of the (parietal) synovial lining of the tunnel and the (visceral) synovial covering of the tendon. The layers of the synovial sheath are actually separated by only a capillary layer of synovial fluid, which lubricates the synovial surfaces to facilitate gliding of the tendon. **D.** Fibrous digital tendon sheath, demonstrating the anular and cruciate parts ("pulleys").

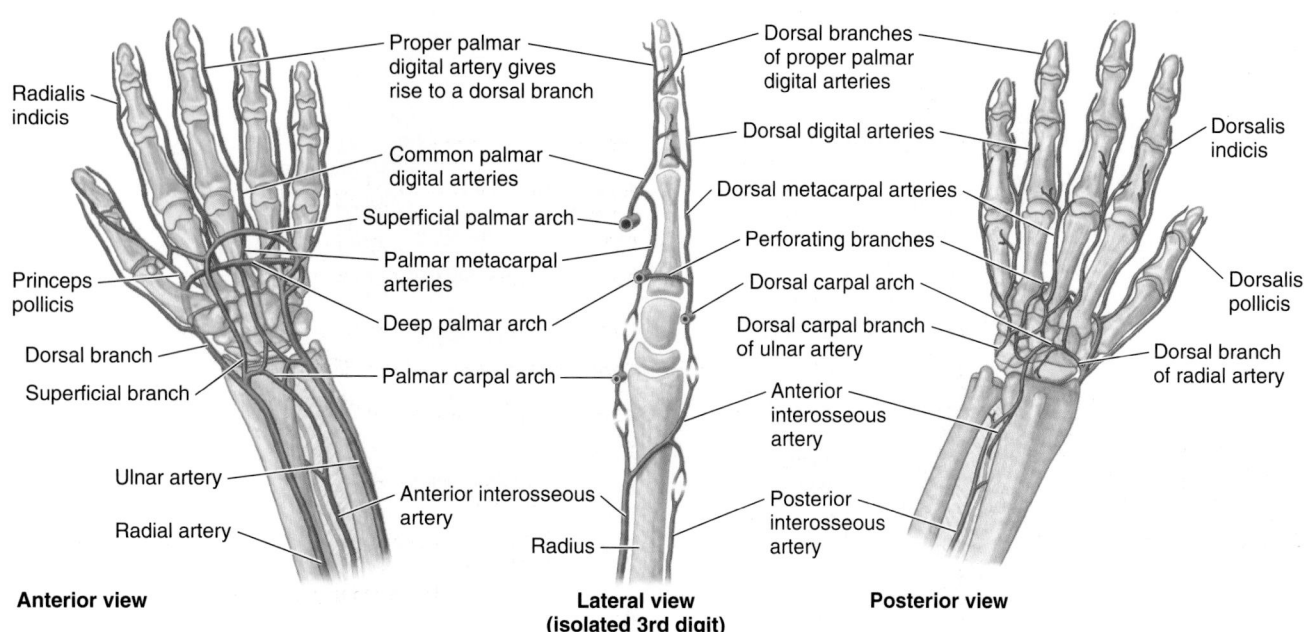

FIGURE 6.82. Arteries of wrist and hand.

TABLE 6.15. ARTERIES OF HAND

Artery	Origin	Course
Superficial palmar arch	Direct continuation of ulnar artery; arch is completed on lateral side by superficial branch of radial artery or another of its branches	Curves laterally deep to palmar aponeurosis and superficial to long flexor tendons; curve of arch lies across palm at level of distal border of extended thumb
Deep palmar arch	Direct continuation of radial artery; arch is completed on medial side by deep branch of ulnar artery	Curves medially, deep to long flexor tendons; is in contact with bases of metacarpals
Common palmar digital	Superficial palmar arch	Pass distally on lumbricals to webbing of digits
Proper palmar digital	Common palmar digital arteries	Run along sides of 2nd–5th digits
Princeps pollicis	Radial artery as it turns into palm	Descends on palmar aspect of 1st metacarpal; divides at base of proximal phalanx into two branches that run along sides of thumb
Radialis indicis	Radial artery but may arise from princeps pollicis artery	Passes along lateral side of index finger to its distal end
Dorsal carpal arch	Radial and ulnar arteries	Arches within fascia on dorsum of hand

cold environment, the arterioles of the hands are capable of reducing blood flow to the surface and to the ends of the fingers. The ulnar and radial arteries and their branches provide all the blood to the hand. The arteries of the hand are illustrated in Figures 6.82 and 6.83, and their origins and courses are described in Table 6.15.

ULNAR ARTERY IN HAND

The **ulnar artery** enters the hand anterior to the flexor retinaculum between the pisiform and the hook of the hamate via the *ulnar canal* (Guyon canal) (Fig. 6.70B). The ulnar artery lies lateral to the ulnar nerve (Fig. 6.77A). The artery divides into two terminal branches, the superficial palmar arch and the deep palmar branch (Figs. 6.82 and 6.83). The **superficial palmar arch,** the main termination of the ulnar artery, gives rise to three **common palmar digital arteries** that anastomose with the **palmar metacarpal arteries** from the deep palmar arch. Each common palmar digital artery divides into a pair of **proper palmar digital arteries,** which run along the adjacent sides of the 2nd–4th digits.

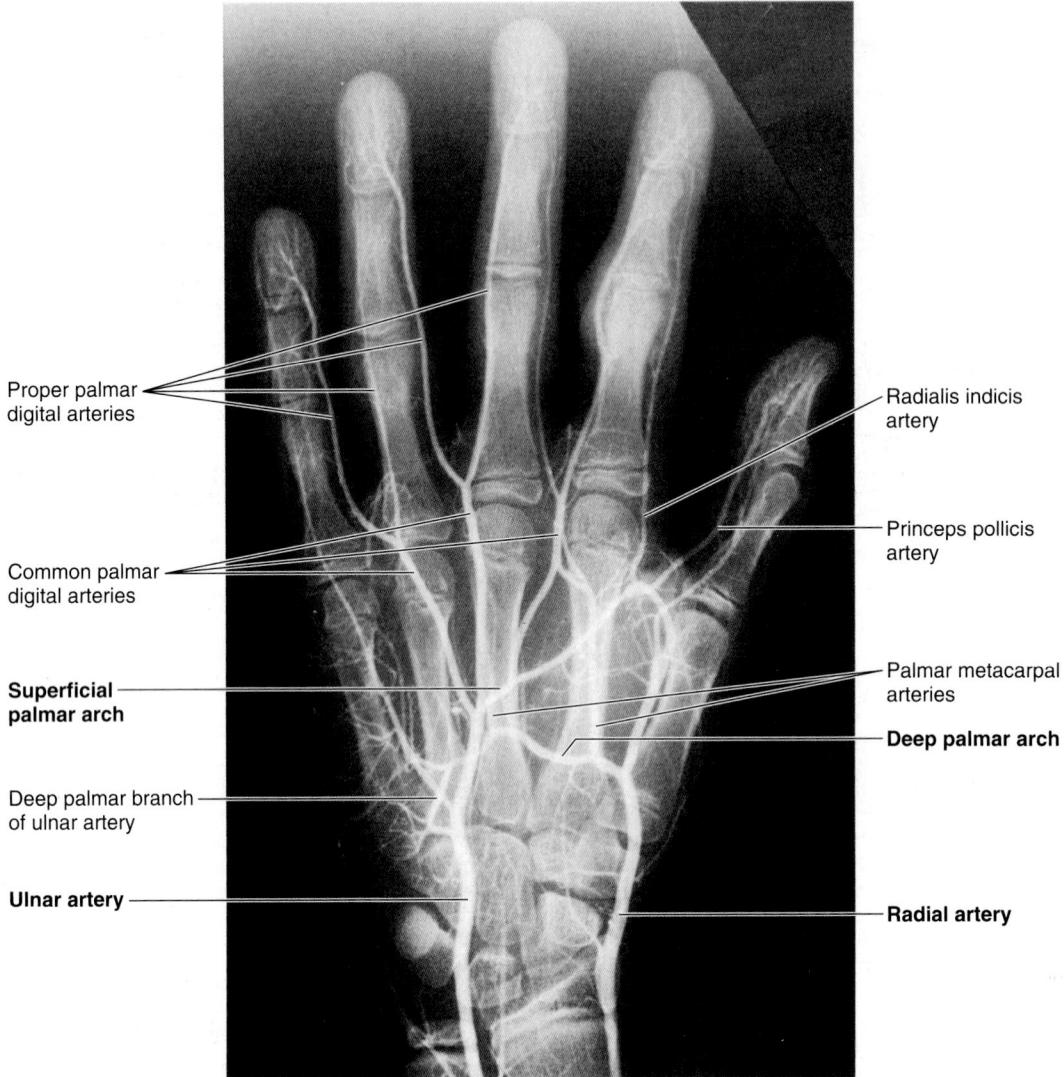

Proper palmar
digital arteries

Common palmar
digital arteries

**Superficial
palmar arch**

Deep palmar branch
of ulnar artery

Ulnar artery

Radialis indicis
artery

Princeps pollicis
artery

Palmar metacarpal
arteries

Deep palmar arch

Radial artery

FIGURE 6.83. Arteriogram of wrist and hand. The carpal bones are fully ossified in this teenage hand, but the epiphysial plates (growth plates) of the long bones remain open. Closure occurs when growth is complete, usually at the end of the teenage years. (Courtesy of Dr. D. Armstrong, University of Toronto, Ontario, Canada.)

RADIAL ARTERY IN HAND

The **radial artery** curves dorsally around the scaphoid and trapezium and crosses the floor of the *anatomical snuff box* (Fig. 6.65). It enters the palm by passing between the heads of the 1st dorsal interosseous muscle and then turns medially, passing between the heads of the adductor pollicis. The radial artery ends by anastomosing with the deep branch of the ulnar artery to form the **deep palmar arch,** which is formed mainly by the radial artery. This arch lies across the metacarpals just distal to their bases (Fig. 6.79). The deep palmar arch gives rise to three *palmar metacarpal arteries* and the *princeps pollicis artery* (Figs. 6.82 and 6.83). The *radialis indicis* artery passes along the lateral side of the index finger. It usually arises from the radial artery, but it may originate from the princeps pollicis.

Veins of Hand

Superficial and *deep venous palmar arches,* associated with the superficial and deep palmar (arterial) arches, drain into the deep veins of the forearm (Fig. 6.68). The dorsal digital veins drain into three dorsal metacarpal veins, which unite to form a *dorsal venous network* (Fig. 6.15A). Superficial to the metacarpus, this network is prolonged proximally on the lateral side as the *cephalic vein.* The *basilic vein* arises from the medial side of the dorsal venous network.

Nerves of Hand

The median, ulnar, and radial nerves supply the hand (Figs. 6.70, 6.77, and 6.84). In addition, branches or communications from the lateral and posterior cutaneous nerves may contribute some fibers that supply the skin of the dorsum of

(*text continues on p. 786*)

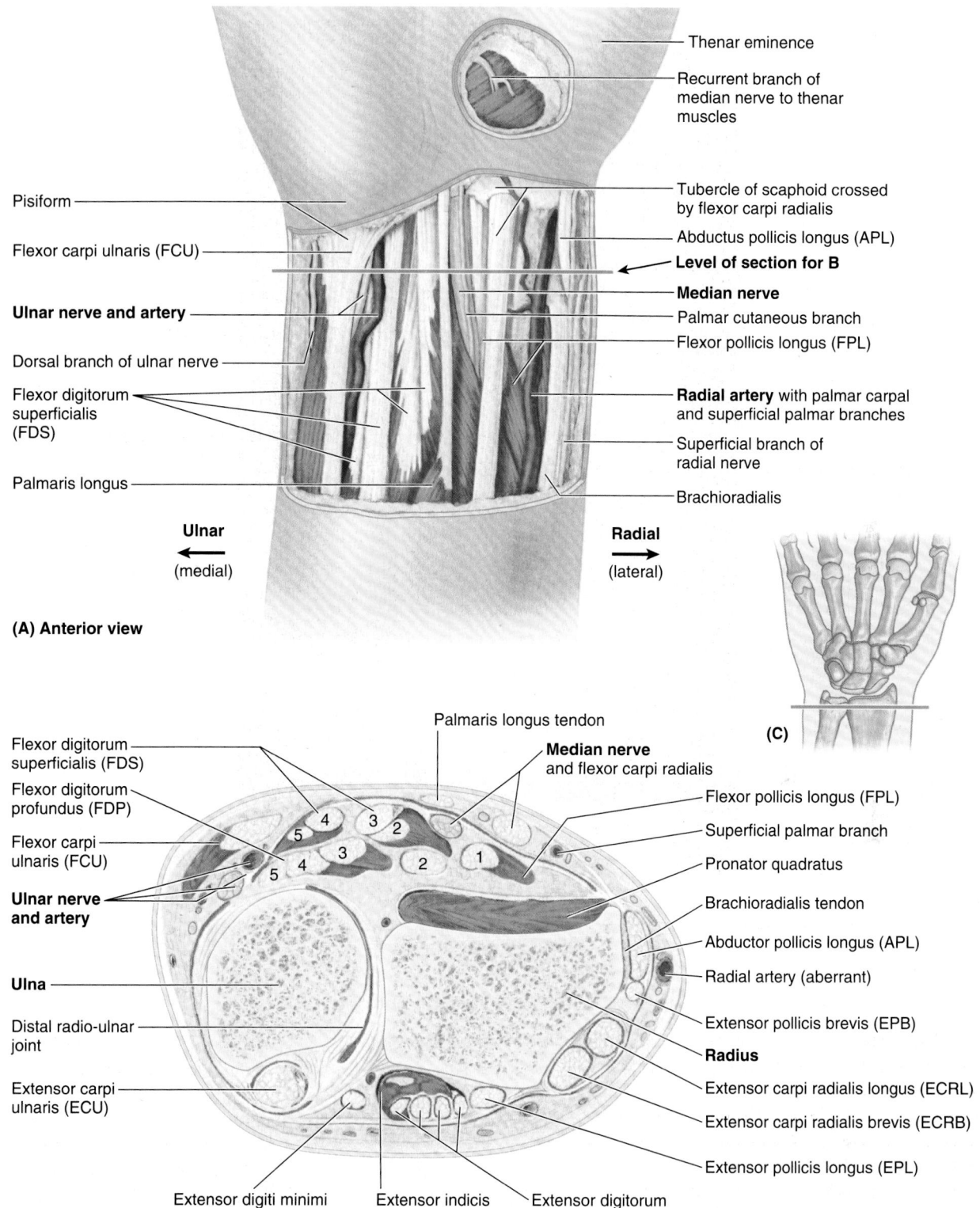

(A) Anterior view

Thenar eminence

Recurrent branch of median nerve to thenar muscles

Tubercle of scaphoid crossed by flexor carpi radialis

Abductus pollicis longus (APL)

Level of section for B

Median nerve

Palmar cutaneous branch

Flexor pollicis longus (FPL)

Radial artery with palmar carpal and superficial palmar branches

Superficial branch of radial nerve

Brachioradialis

Pisiform

Flexor carpi ulnaris (FCU)

Ulnar nerve and artery

Dorsal branch of ulnar nerve

Flexor digitorum superficialis (FDS)

Palmaris longus

Ulnar
(medial)

Radial
(lateral)

(C)

Palmaris longus tendon

Median nerve
and flexor carpi radialis

Flexor pollicis longus (FPL)

Superficial palmar branch

Pronator quadratus

Brachioradialis tendon

Abductor pollicis longus (APL)

Radial artery (aberrant)

Extensor pollicis brevis (EPB)

Radius

Extensor carpi radialis longus (ECRL)

Extensor carpi radialis brevis (ECRB)

Extensor pollicis longus (EPL)

Flexor digitorum superficialis (FDS)

Flexor digitorum profundus (FDP)

Flexor carpi ulnaris (FCU)

Ulnar nerve and artery

Ulna

Distal radio-ulnar joint

Extensor carpi ulnaris (ECU)

Extensor digiti minimi   Extensor indicis   Extensor digitorum

(B) Inferior view of transverse section of distal forearm

FIGURE 6.84. Structures in distal forearm (wrist region). A. A distal skin incision was made along the transverse wrist crease, crossing the pisiform bone. The skin and fasciae are removed proximally, revealing the tendons and neurovascular structures. A circular incision and removal of the skin and thenar fascia reveals the recurrent branch of the median nerve to the thenar muscles, vulnerable to injury when this area is lacerated because of its subcutaneous location. The tendons of the flexor digitorum superficialis and profundus are numbered in **B** according to the digit of insertion. **B.** Transverse section of the distal forearm demonstrating the long flexor and extensor tendons and neurovascular structures en route from forearm to hand. The ulnar nerve and artery are under cover of the flexor carpi ulnaris; therefore, the pulse of the artery cannot be easily detected here. **C.** Orientation drawing indicating the plane of the section shown in part **B.**

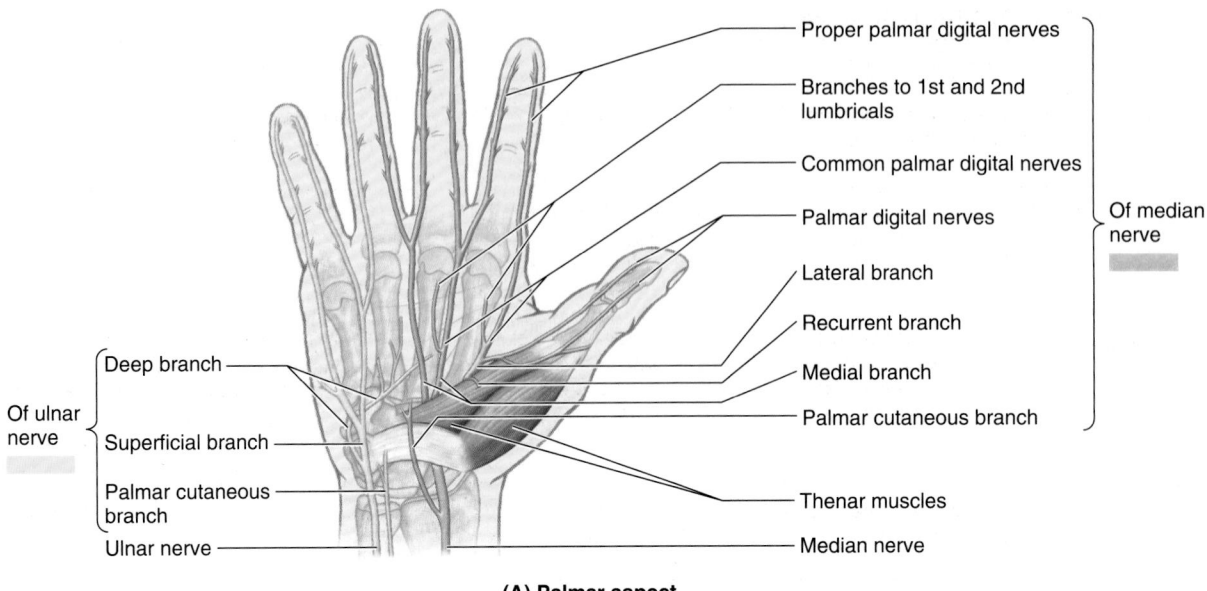

(A) Palmar aspect

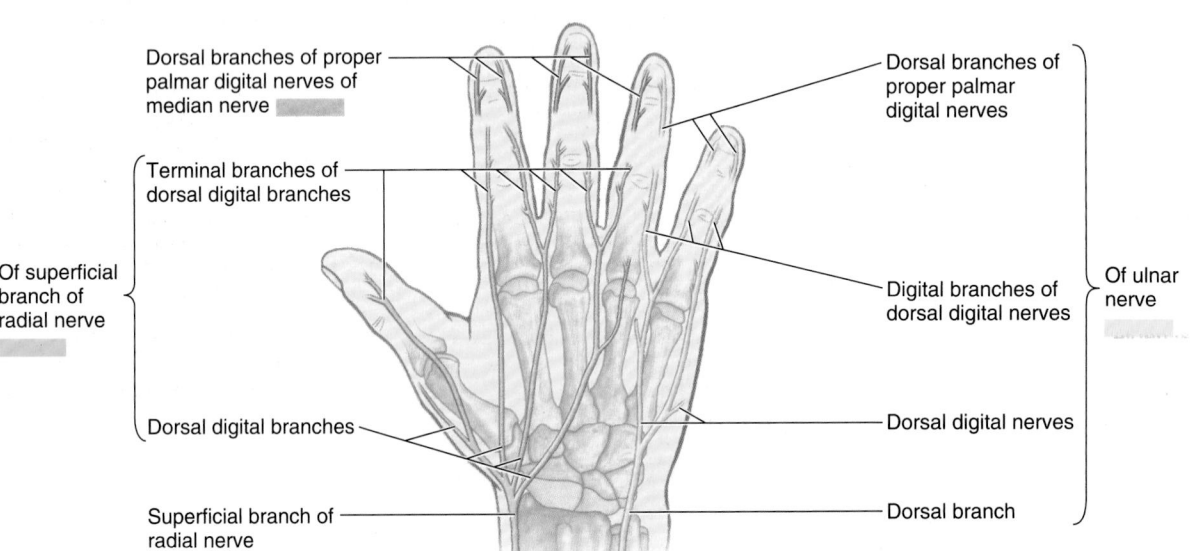

(B) Dorsal aspect

FIGURE 6.85. Branches of nerves to hand.

TABLE 6.16. NERVES OF HAND

Nerve	Origin	Course	Distribution
Median nerve	Arises by two roots, one from lateral cord of brachial plexus (C6, C7 fibers) and one from medial cord (C8, T1 fibers)	Becomes superficial proximal to wrist; passes deep to flexor retinaculum (transverse carpal ligament) as it passes through carpal tunnel to hand	Thenar muscles (except adductor pollicis and deep head of flexor pollicis brevis) and lateral lumbricals (for digits 2 and 3); provides sensation to skin of palmar and distal dorsal aspects of lateral (radial) 3½ digits and adjacent palm
Recurrent (thenar) branch of median nerve	Arises from median nerve as soon as it has passed distal to flexor retinaculum	Loops around distal border of flexor retinaculum; enters thenar muscles	Abductor pollicis brevis; opponens pollicis; superficial head of flexor pollicis brevis
Lateral branch of median nerve	Arises as lateral division of median nerve as it enters palm of hand	Runs laterally to palmar aspect of thumb and radial side of 2nd digit	1st lumbrical; skin of palmar and distal dorsal aspects of thumb and radial half of 2nd digit

(continued)

TABLE 6.16. NERVES OF HAND (Continued)

Nerve	Origin	Course	Distribution
Medial branch of median nerve	Arises as medial division of median nerve as it enters palm of hand	Runs medially to adjacent sides of 2nd–4th digits	2nd lumbrical; skin of palmar and distal dorsal aspects of adjacent sides of 2nd–4th digits
Palmar cutaneous branch of median nerve	Arises from median nerve just proximal to flexor retinaculum	Passes between tendons of palmaris longus and flexor carpi radialis; runs superficial to flexor retinaculum	Skin of central palm
Ulnar nerve	Terminal branch of medial cord of brachial plexus (C8 and T1 fibers; often also receives C7 fibers)	Becomes superficial in distal forearm, passing superficial to flexor retinaculum (transverse carpal ligament) to enter hand	The majority of intrinsic muscles of hand (hypothenar, interosseous, adductor pollicis, and deep head of flexor pollicis brevis, plus the medial lumbricals [for digits 4 and 5]); provides sensation to skin of palmar and distal dorsal aspects of medial (ulnar) 1½ digits and adjacent palm
Palmar cutaneous branch of ulnar nerve	Arises from ulnar nerve near middle of forearm	Descends on ulnar artery and perforates deep fascia in the distal third of forearm	Skin at base of medial palm, overlying the medial carpals
Dorsal branch of ulnar nerve	Arises from ulnar nerve about 5 cm proximal to flexor retinaculum	Passes distally deep to flexor carpi ulnaris, then dorsally to perforate deep fascia and course along medial side of dorsum of hand, dividing into two to three dorsal digital nerves	Skin of medial aspect of dorsum of hand and proximal portions of little and medial half of ring finger (occasionally also adjacent sides of proximal portions of ring and middle fingers)
Superficial branch of ulnar nerve	Arise from ulnar nerve at wrist as they pass between pisiform and hamate bones	Passes palmaris brevis and divides into two common palmar digital nerves	Palmaris brevis and sensation to skin of the palmar and distal dorsal aspects of digit 5 and of the medial (ulnar) side of digit 4 and proximal portion of palm
Deep branch of ulnar nerve		Passes between muscles of hypothenar eminence to pass deeply across palm with deep palmar (arterial) arch	Hypothenar muscles (abductor, flexor, and opponens digiti minimi), lumbricals of digits 4 and 5, all interossei, adductor pollicis, and deep head of flexor pollicis brevis
Radial nerve, superficial branch	Arises from radial nerve in cubital fossa	Courses deep to brachioradialis, emerging from beneath it to pierce the deep fascia lateral to distal radius	Skin of the lateral (radial) half of dorsal aspect of the hand and thumb, the proximal portions of the dorsal aspects of digits 2 and 3, and of the lateral (radial) half of digit 4

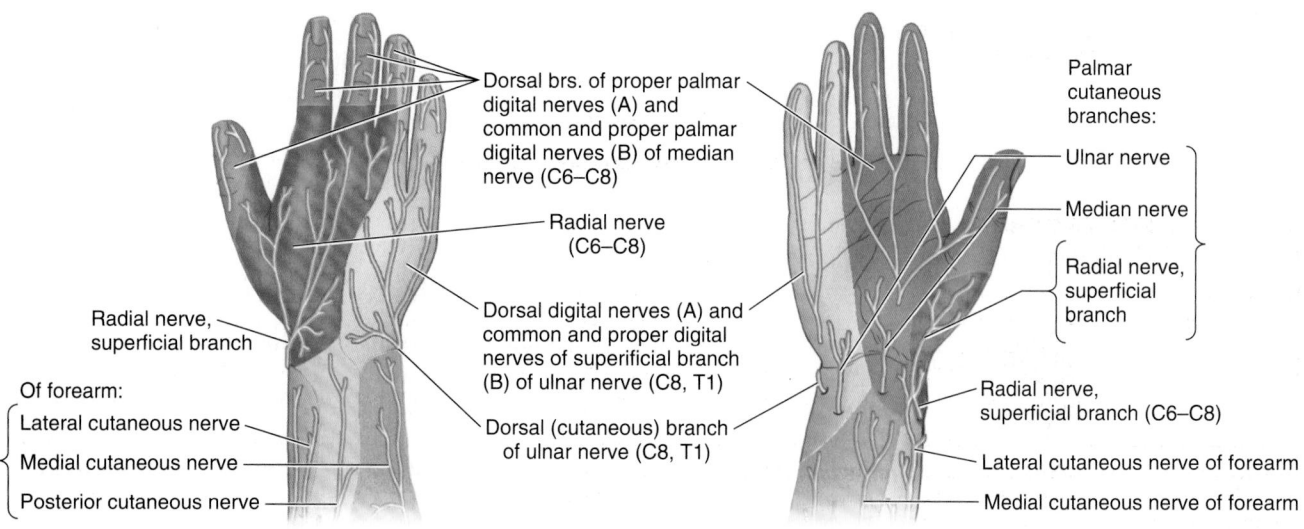

(A) Posterior view   **(B) Anterior view**

FIGURE 6.86. Sensory innervation of wrist and hand. A. Distribution of the peripheral cutaneous nerve to the hand and wrist. **B.** Distribution of the spinal nerve fibers to the hand and wrist (dermatomes).

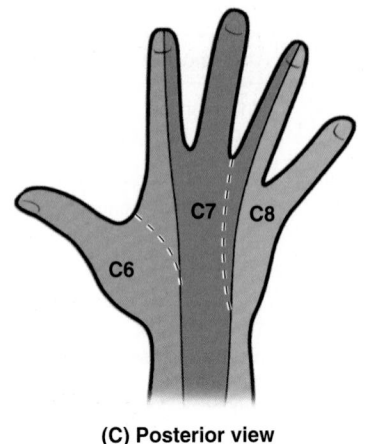

(C) Posterior view

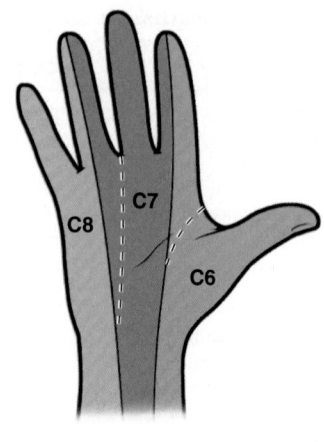

(D) Anterior view

FIGURE 6.86. (*Continued*)

the hand. These nerves and their branches in the hand are illustrated in Figures 6.85 and 6.86A & B, and their origins, courses, and distributions are provided in Table 6.16.

In the hand, these nerves convey sensory fibers from spinal nerves C6–C8 to the skin, so that the C6–C8 dermatomes include the hand (Fig. 6.86C & D). The median and ulnar nerves convey motor fibers from spinal nerve T1 to the hand; the intrinsic muscles of the hand make up myotome T1 (Fig. 6.20F).

MEDIAN NERVE IN HAND

The *median nerve* enters the hand through the carpal tunnel, deep to the flexor retinaculum, along with the nine tendons of the FDS, FDP, and FPL (Fig. 6.84). The **carpal tunnel** is the passageway deep to the flexor retinaculum between the tubercles of the scaphoid and trapezoid bones on the lateral side and the pisiform and hook of the hamate on the medial side (Fig. 6.30A). Distal to the carpal tunnel, the median nerve supplies two and a half thenar muscles and the 1st and 2nd lumbricals (Fig. 6.85A). It also sends sensory fibers to the skin on the entire palmar surface, the sides of the first three digits, the lateral half of the 4th digit, and the dorsum of the distal halves of these digits. Note, however, that the *palmar cutaneous branch of the median nerve,* which supplies the central palm, arises proximal to the flexor retinaculum and passes superficial to it (i.e., it does not pass through the carpal tunnel).

THE ULNAR NERVE IN HAND

The **ulnar nerve** leaves the forearm by emerging from deep to the tendon of the FCU (Figs. 6.77 and 6.84). It continues distally to the wrist via the *ulnar (Guyon) canal* (Fig. 6.70). Here the ulnar nerve is bound by fascia to the anterior surface of the flexor retinaculum as it passes between the pisiform (medially) and the ulnar artery (laterally).

Just proximal to the wrist, the ulnar nerve gives off a **palmar cutaneous branch,** which passes superficial to the flexor retinaculum and palmar aponeurosis and supplies skin on the medial side of the palm (Fig. 6.85A).

The **dorsal cutaneous branch of the ulnar nerve** supplies the medial half of the dorsum of the hand, the 5th finger, and the medial half of the 4th finger (Fig. 6.85B). The ulnar nerve ends at the distal border of the flexor retinaculum by dividing into superficial and deep branches (Fig. 6.77B).

The **superficial branch of the ulnar nerve** supplies cutaneous branches to the anterior surfaces of the medial one and a half digits. The **deep branch of the ulnar nerve** supplies the hypothenar muscles, the medial two lumbricals, the adductor pollicis, the deep head of the FPB, and all the interossei. The deep branch also supplies several joints (wrist, intercarpal, carpometacarpal, and intermetacarpal). The ulnar nerve is often referred to as the *nerve of fine movements* because it innervates most of the intrinsic muscles that are concerned with intricate hand movements (Table 6.16).

RADIAL NERVE IN HAND

The *radial nerve* supplies no hand muscles (Table 6.16). The *superficial branch of the radial nerve* is entirely sensory (Fig. 6.85B). It pierces the deep fascia near the dorsum of the wrist to supply the skin and fascia over the lateral two thirds of the hand, the dorsum of the thumb, and proximal parts of the lateral one and a half digits (Fig. 6.86A).

Surface Anatomy of Hand

The **radial artery pulse,** like other palpable pulses, is a peripheral reflection of cardiac action. The radial pulse rate is measured where the radial artery lies on the anterior surface of the distal end of the radius, lateral to the FCR tendon, which serves as a guide to the artery (Fig. 6.87). Here the artery can be felt pulsating between the tendons of the FCR and the APL and where it can be compressed against the radius.

The *tendons of FCR and palmaris longus* can be palpated anterior to the wrist, a little lateral to its middle, and are usually observed by flexing the closed fist against resistance.

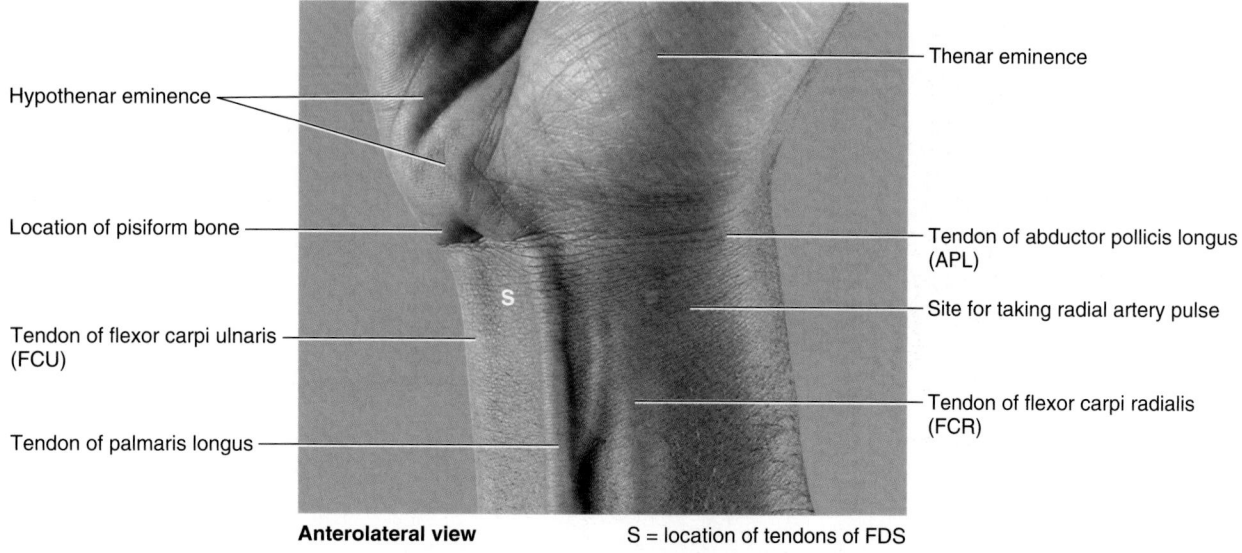

Hypothenar eminence

Location of pisiform bone

Tendon of flexor carpi ulnaris (FCU)

Tendon of palmaris longus

Thenar eminence

Tendon of abductor pollicis longus (APL)

Site for taking radial artery pulse

Tendon of flexor carpi radialis (FCR)

Anterolateral view   S = location of tendons of FDS

FIGURE 6.87. Surface anatomy of anterior wrist region.

The palmaris longus tendon is smaller than the FCR tendon and is not always present. The palmaris longus tendon serves as a guide to the median nerve, which lies deep to it (Fig. 6.84B). The *FCU tendon* can be palpated as it crosses the anterior aspect of the wrist near the medial side and inserts into the pisiform. The FCU tendon serves as a guide to the ulnar nerve and artery.

The *tendons of the FDS* can be palpated as the fingers are alternately flexed and extended. The ulnar pulse is often difficult to palpate. The *tendons of the APL and EPB* indicate the anterior boundary of the *anatomical snuff box* (Fig. 6.88). The *tendon of the EPL* indicates the posterior boundary of the box. The radial artery crosses the floor of the snuff box, where its pulsations may be felt (Fig. 6.65B). The *scaphoid* and, less distinctly, the *trapezium* are palpable in the floor of the snuff box.

The skin covering the dorsum of the hand is thin and loose when the hand is relaxed. Prove this by pinching and pulling folds of your skin here. The looseness of the skin results from the mobility of the subcutaneous tissue and from the relatively few fibrous skin ligaments that are present. Hair is present in this region and on the proximal parts of the digits, especially in men.

If the dorsum of the hand is examined with the wrist extended against resistance and the digits abducted, the *tendons of the extensor digitorum* to the fingers stand out, particularly in thin individuals (Fig. 6.88). These tendons are not visible far beyond the knuckles because they flatten here to form the extensor expansions of the fingers (Fig. 6.63B).

The knuckles that become visible when a fist is made are produced by the heads of the metacarpals. Under the loose subcutaneous tissue and extensor tendons on the dorsum of the hand, the metacarpals can be palpated. A prominent feature of the dorsum of the hand is the *dorsal venous network* (Fig. 6.15A).

The skin on the palm is thick because it must withstand the wear and tear of work and play (Fig. 6.89). It is richly supplied with sweat glands but contains no hair or sebaceous glands.

The *superficial palmar arch* lies across the center of the palm, level with the distal border of the extended thumb. The main part of the arch ends at the *thenar eminence* (Fig. 6.87).

The *deep palmar arch* lies approximately 1 cm proximal to the superficial palmar arch. The palmar skin presents several more or less constant *flexion creases*, where the skin is firmly bound to the deep fascia, that help locate palmar wounds and underlying structures (Fig. 6.89A):

• **Wrist creases—proximal, middle, distal.** The distal wrist crease indicates the proximal border of the flexor retinaculum.
• **Palmar creases—transverse, longitudinal.** The longitudinal creases deepen when the thumb is opposed; the transverse creases deepen when the metacarpophalangeal joints are flexed.
  • **Radial longitudinal crease** (the "life line" of palmistry): partially encircles the *thenar eminence*, formed by the short muscles of the thumb.
  • **Proximal (transverse) palmar crease:** commences on the lateral border of the palm, superficial to the head of the 2nd metacarpal; it extends medially and slightly proximally across the palm, superficial to the bodies of the 3rd–5th metacarpals.
• **Distal (transverse) palmar crease.** The distal palmar crease begins at or near the cleft between the index and the middle fingers; it crosses the palm with a slight convexity, superficial to the head of the 3rd metacarpal and then proximal to the heads of the 4th and 5th metacarpals.

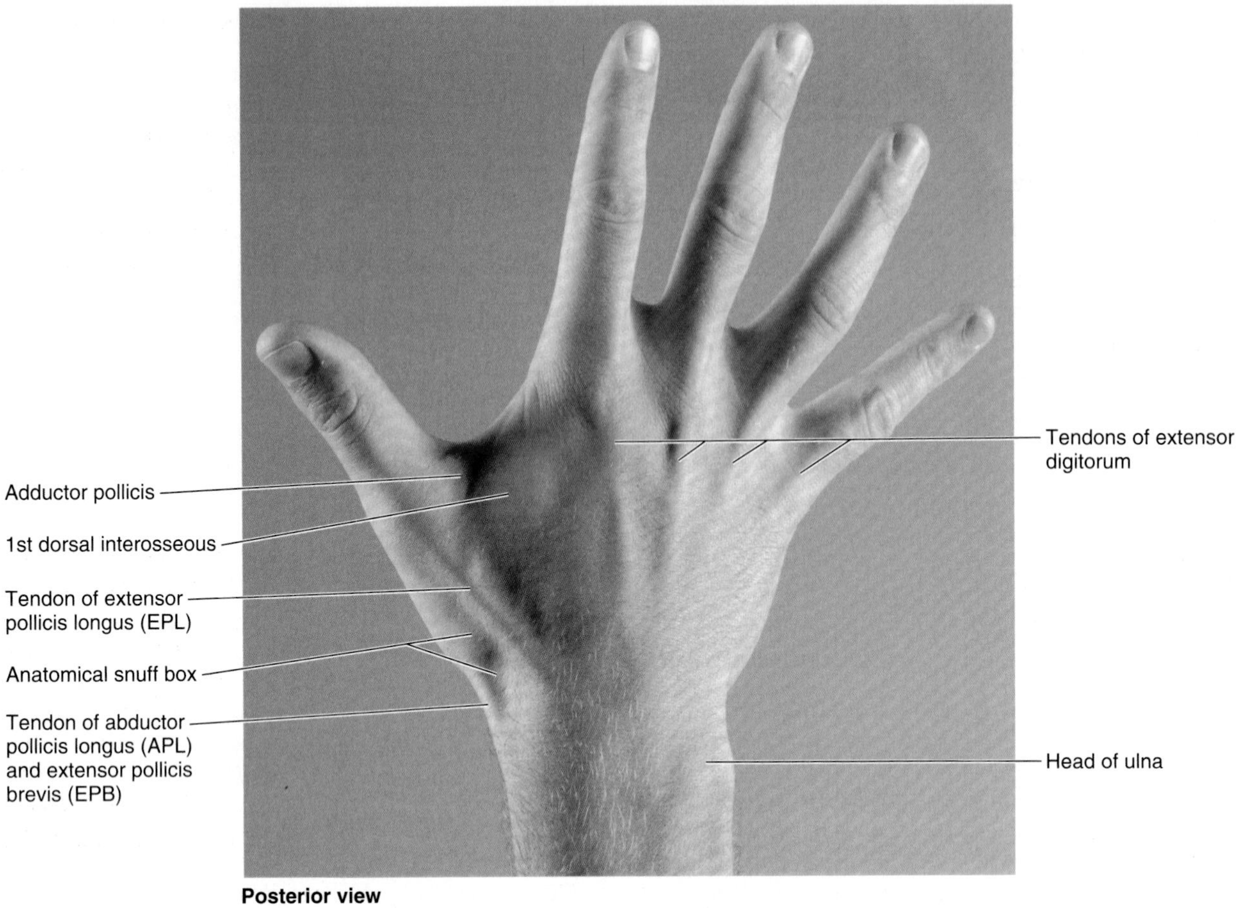

Posterior view

Adductor pollicis

1st dorsal interosseous

Tendon of extensor pollicis longus (EPL)

Anatomical snuff box

Tendon of abductor pollicis longus (APL) and extensor pollicis brevis (EPB)

Tendons of extensor digitorum

Head of ulna

FIGURE 6.88. Surface anatomy of dorsum of hand.

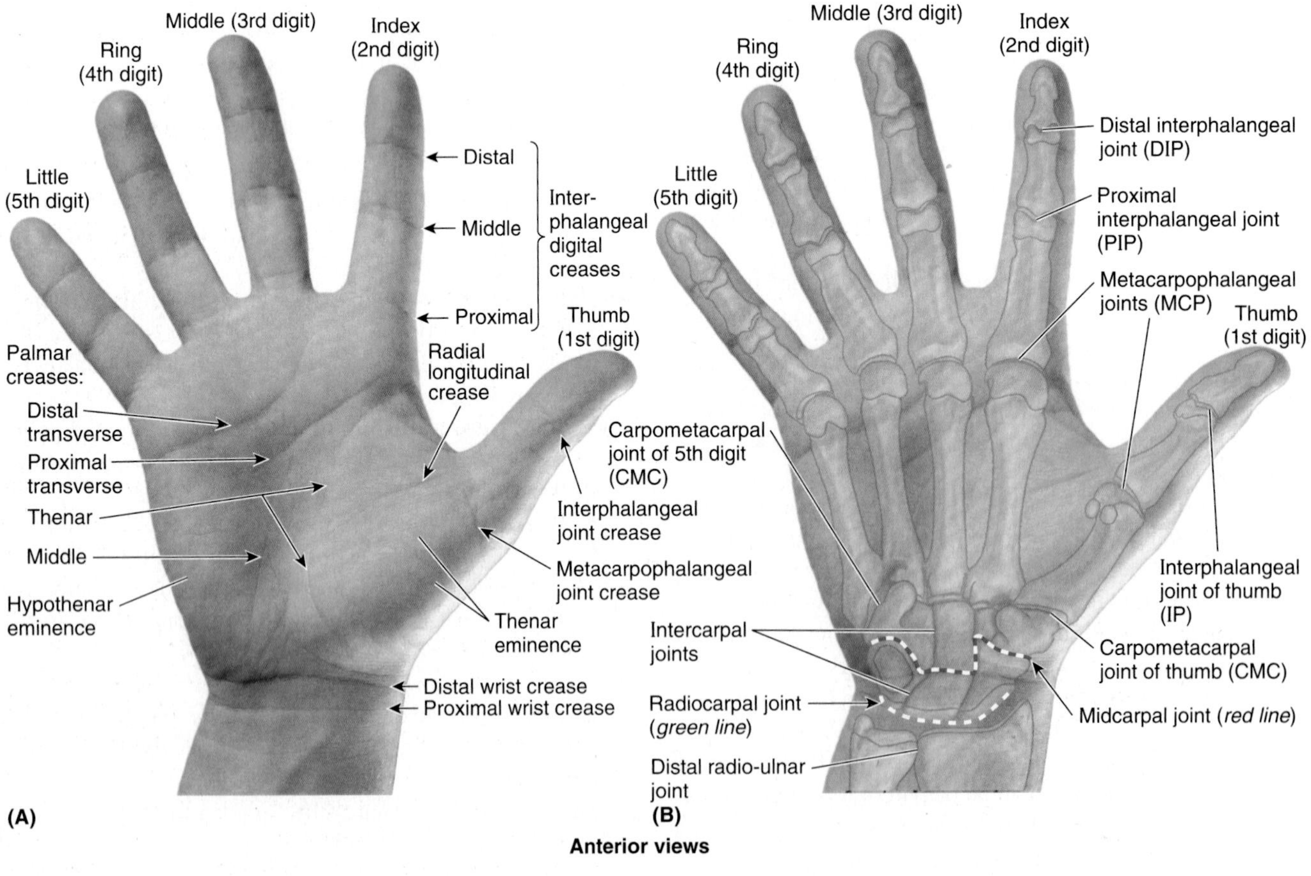

Ring (4th digit)

Middle (3rd digit)

Index (2nd digit)

Little (5th digit)

Distal

Middle

Proximal

Inter-phalangeal digital creases

Thumb (1st digit)

Radial longitudinal crease

Palmar creases:

Distal transverse

Proximal transverse

Thenar

Middle

Hypothenar eminence

Thenar eminence

Interphalangeal joint crease

Metacarpophalangeal joint crease

Distal wrist crease

Proximal wrist crease

(A)

Ring (4th digit)

Middle (3rd digit)

Index (2nd digit)

Little (5th digit)

Distal interphalangeal joint (DIP)

Proximal interphalangeal joint (PIP)

Metacarpophalangeal joints (MCP)

Thumb (1st digit)

Carpometacarpal joint of 5th digit (CMC)

Interphalangeal joint of thumb (IP)

Carpometacarpal joint of thumb (CMC)

Midcarpal joint (*red line*)

Intercarpal joints

Radiocarpal joint (*green line*)

Distal radio-ulnar joint

(B)

Anterior views

FIGURE 6.89. Surface anatomy of palmar aspect of hand.

Each of the medial four fingers usually has three *transverse digital flexion creases:*

- **Proximal digital crease:** located at the root of the finger, approximately 2 cm distal to the metacarpophalangeal joint.
- **Middle digital crease:** lies over the proximal interphalangeal joint.
- **Distal digital crease:** lies over or just proximal to the distal interphalangeal joint.

The thumb, having two phalanges, has only two flexion creases. The proximal digital crease of the thumb crosses obliquely, at or proximal to the 1st metacarpophalangeal joint. The **skin ridges** on the pulp (pads) of the digits, forming the *fingerprints,* are used for identification because of their unique patterns. The physiological function of the skin ridges is to reduce slippage when grasping objects.

HAND

Dupuytren Contracture of Palmar Fascia

 Dupuytren contracture is a disease of the palmar fascia resulting in progressive shortening, thickening, and fibrosis of the palmar fascia and aponeurosis. The fibrous degeneration of the longitudinal bands of the palmar aponeurosis on the medial side of the hand pulls the 4th and 5th fingers into partial flexion at the metacarpophalangeal and proximal interphalangeal joints (Fig. B6.29A).

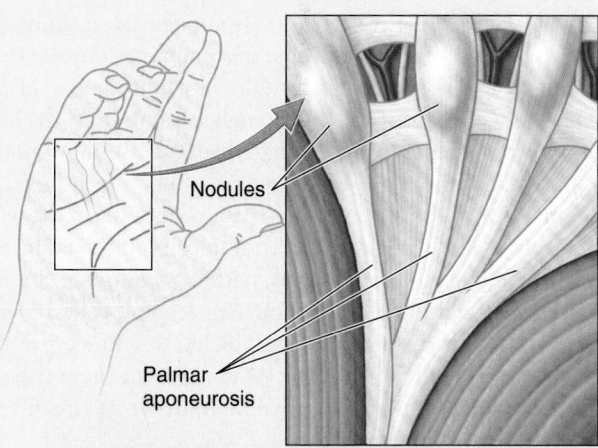

(A) Longitudinal bands of palmar aponeurosis to fibrous digital sheaths of digits four and five nodular and contracted

Nodules

Palmar aponeurosis

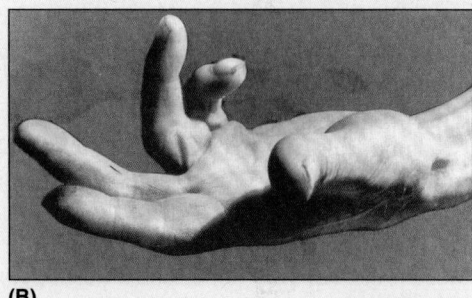

(B)

FIGURE B6.29. Dupuytren contracture.

The contracture is frequently bilateral and is seen in some men > 50 years of age. Its cause is unknown, but evidence points to a hereditary predisposition. The disease first manifests as painless nodular thickenings of the palmar aponeurosis that adhere to the skin. Gradually, progressive contracture of the longitudinal bands produces raised ridges in the palmar skin that extend from the proximal part of the hand to the base of the 4th and 5th fingers (Fig. B6.29B). Treatment of Dupuytren contracture usually involves surgical excision of all fibrotic parts of the palmar fascia to free the fingers (Salter, 1999).

Hand Infections

 Because the palmar fascia is thick and strong, swellings resulting from hand infections usually appear on the dorsum of the hand, where the fascia is thinner. The potential fascial spaces of the palm are important because they may become infected. The fascial spaces determine the extent and direction of the spread of pus formed by these infections.

Depending on the site of infection, pus will accumulate in the thenar, hypothenar, midpalmar, or adductor compartments (Fig. 6.75A). Antibiotic therapy has made infections that spread beyond one of these fascial compartments rare; however, an untreated infection can spread proximally from the midpalmar space through the carpal tunnel into the forearm, anterior to the pronator quadratus and its fascia.

Tenosynovitis

Injuries such as a puncture of a finger by a rusty nail can cause *infection of the digital synovial sheaths* (Fig. 6.81A). When inflammation of the tendon and synovial sheath occurs (*tenosynovitis*), the digit swells and movement becomes painful. Because the tendons of the 2nd, 3rd, and 4th fingers nearly always have separate synovial sheaths, the infection is usually confined to the infected finger. If the infection is untreated, however, the proximal ends of these sheaths may rupture, allowing the infection to spread to the midpalmar space (Fig. 6.75B).

Because the synovial sheath of the little finger is usually continuous with the common flexor sheath (Fig. 6.81B), tenosynovitis in this finger may spread to the common flexor sheath and thus through the palm and carpal tunnel to the anterior forearm, draining into the space between the pronator quadratus and the overlying flexor tendons (*Parona space*). Likewise, tenosynovitis in the thumb may spread via the continuous synovial sheath of the FPL (radial bursa). How far an infection spreads from the fingers depends on variations in their connections with the common flexor sheath.

The tendons of the APL and EPB are in the same tendinous sheath on the dorsum of the wrist. Excessive friction of these tendons on their common sheath results in fibrous thickening of the sheath and stenosis of the osseofibrous tunnel. The excessive friction is caused by repetitive forceful use of the hands during gripping and wringing (e.g., squeezing water out of clothes). This condition, called *Quervain tenovaginitis stenosans*, causes pain in the wrist that radiates proximally to the forearm and distally toward the thumb. Local tenderness is felt over the common flexor sheath on the lateral side of the wrist.

Thickening of a fibrous digital sheath on the palmar aspect of the digit produces stenosis of the osseofibrous tunnel, the result of repetitive forceful use of the fingers. If the tendons of the FDS and FDP enlarge proximal to the tunnel, the person is unable to extend the finger. When the finger is extended passively, a snap is audible. Flexion produces another snap as the thickened tendon moves. This condition is called digital tenovaginitis stenosans (*trigger finger* or snapping finger).

Laceration of Palmar Arches

Bleeding is usually profuse when the palmar (arterial) arches are lacerated. It may not be sufficient to ligate only one forearm artery when the arches are lacerated, because these vessels usually have numerous communications in the forearm and hand and thus bleed from both ends. To obtain a bloodless surgical operating field for treating complicated hand injuries, it may be necessary to compress the brachial artery and its branches proximal to the elbow (e.g., using a pneumatic tourniquet). This procedure prevents blood from reaching the ulnar and radial arteries through the anastomoses around the elbow (Fig. 6.67A).

Ischemia of Digits (Fingers)

Intermittent bilateral attacks of *ischemia of the digits*, marked by cyanosis and often accompanied by paresthesia and pain, is characteristically brought on by cold and emotional stimuli. The condition may result from an anatomical abnormality or an underlying disease. When the cause of the condition is idiopathic (unknown) or primary, it is called *Raynaud syndrome* (disease).

The arteries of the upper limb are innervated by sympathetic nerves. Postsynaptic fibers from the sympathetic ganglia enter nerves that form the brachial plexus and are distributed to the digital arteries through branches arising from the plexus. When treating ischemia resulting from Raynaud syndrome, it may be necessary to perform a cervico-dorsal *presynaptic sympathectomy* (excision of a segment of a sympathetic nerve) to dilate the digital arteries.

Lesions of Median Nerve

Lesions of the median nerve usually occur in two places: the forearm and the wrist. The most common site is where the nerve passes through the carpal tunnel.

CARPAL TUNNEL SYNDROME

Carpal tunnel syndrome results from any lesion that significantly reduces the size of the carpal tunnel (Fig. B6.30A–D) or, more commonly, increases the size of some of the nine structures or their coverings that pass through it (e.g., inflammation of synovial sheaths). Fluid retention, infection, and excessive exercise of the fingers may cause swelling of the tendons or their synovial sheaths. The median nerve is the most sensitive structure in the tunnel. The median nerve has two terminal sensory branches that supply the skin of the hand; hence *paresthesia* (tingling), *hypoesthesia* (diminished sensation), or *anesthesia* (absence of sensation) may occur in the lateral three and a half digits. The palmar cutaneous branch of the median nerve arises proximal to, and does not pass through, the carpal tunnel; thus sensation in the central palm remains unaffected. The nerve also has one terminal motor branch, the recurrent branch, which serves the three thenar muscles (Fig. 6.85A).

Progressive loss of coordination and strength of the thumb (owing to weakness of the APB and opponens pollicis) may occur if the cause of compression is not alleviated. Individuals with carpal tunnel syndrome are unable to oppose their thumbs (Fig. B6.30E) and have difficulty buttoning a shirt or blouse as well as gripping things such as a comb. As the condition progresses, sensory changes radiate into the forearm and axilla. Symptoms of compression can be reproduced by compression of the median nerve with your finger at the wrist for approximately 30 seconds. To relieve both the compression and the resulting symptoms, partial or complete surgical division of the flexor retinaculum, a procedure called *carpal tunnel release*, may be necessary. The incision for carpal tunnel release is made toward the medial side of the wrist and flexor retinaculum to avoid possible injury to the recurrent branch of the median nerve.

TRAUMA TO MEDIAN NERVE

Laceration of the wrist often causes *median nerve injury* because this nerve is relatively close to the surface. In attempted suicides by wrist slashing, the median nerve is commonly injured just proximal to the flexor retinaculum. This results in paralysis of the thenar muscles and the first two lumbricals. Hence opposition of the thumb is not possible and fine control movements of the 2nd and 3rd digits are impaired. Sensation is also lost over the thumb and adjacent two and a half fingers.

Most nerve injuries in the upper limb affect opposition of the thumb (Fig. 6.76). Undoubtedly, injuries to the nerves supplying the intrinsic muscles of the hand, especially the

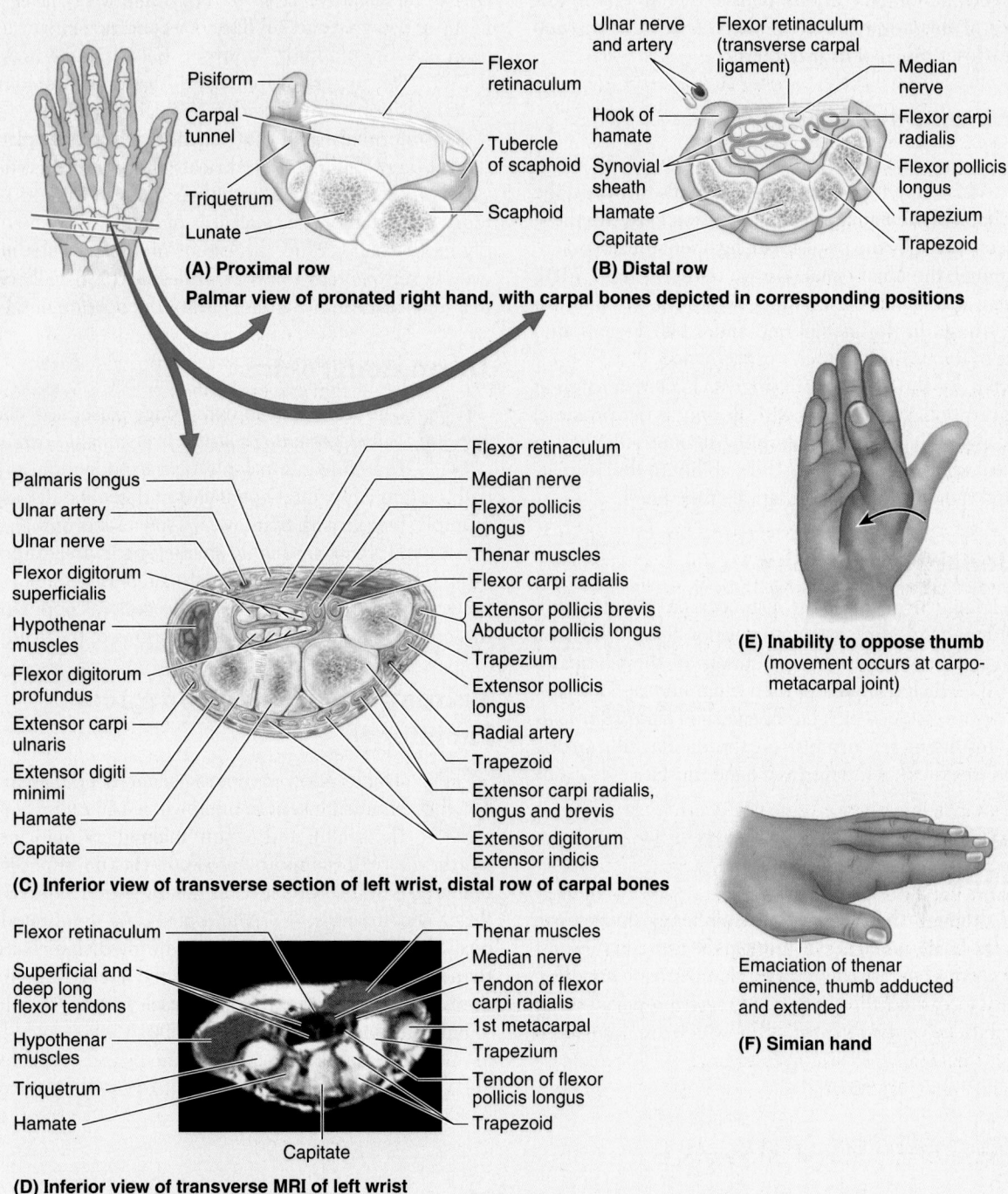

(A) Proximal row

Pisiform
Carpal tunnel
Triquetrum
Lunate
Flexor retinaculum
Tubercle of scaphoid
Scaphoid

(B) Distal row

Ulnar nerve and artery
Flexor retinaculum (transverse carpal ligament)
Median nerve
Hook of hamate
Flexor carpi radialis
Synovial sheath
Flexor pollicis longus
Hamate
Trapezium
Capitate
Trapezoid

Palmar view of pronated right hand, with carpal bones depicted in corresponding positions

Palmaris longus
Ulnar artery
Ulnar nerve
Flexor digitorum superficialis
Hypothenar muscles
Flexor digitorum profundus
Extensor carpi ulnaris
Extensor digiti minimi
Hamate
Capitate

Flexor retinaculum
Median nerve
Flexor pollicis longus
Thenar muscles
Flexor carpi radialis
Extensor pollicis brevis
Abductor pollicis longus
Trapezium
Extensor pollicis longus
Radial artery
Trapezoid
Extensor carpi radialis, longus and brevis
Extensor digitorum
Extensor indicis

(C) Inferior view of transverse section of left wrist, distal row of carpal bones

(E) Inability to oppose thumb
(movement occurs at carpo-metacarpal joint)

Flexor retinaculum
Superficial and deep long flexor tendons
Hypothenar muscles
Triquetrum
Hamate

Thenar muscles
Median nerve
Tendon of flexor carpi radialis
1st metacarpal
Trapezium
Tendon of flexor pollicis longus
Trapezoid
Capitate

(D) Inferior view of transverse MRI of left wrist

Emaciation of thenar eminence, thumb adducted and extended

(F) Simian hand

FIGURE B6.30.

median nerve, have the most severe effects on this complex movement. If the median nerve is severed in the forearm or at the wrist, the thumb cannot be opposed; however, the APL and adductor pollicis (supplied by the posterior interosseous and ulnar nerves, respectively) may imitate opposition, although ineffective.

Median nerve injury resulting from a perforating wound in the elbow region results in loss of flexion of the proximal and distal interphalangeal joints of the 2nd and 3rd digits.

The ability to flex the metacarpophalangeal joints of these fingers is also affected because digital branches of the median nerve supply the 1st and 2nd lumbricals. *Simian hand* (Fig. B6.30F) refers to a deformity in which thumb movements are limited to flexion and extension of the thumb in the plane of the palm. This condition is caused by the inability to oppose and by limited abduction of the thumb. The recurrent branch of the median nerve to the thenar muscles (Fig. 6.84A) lies subcutaneously and may be severed by relatively minor lacerations of the thenar eminence. Severance of this nerve paralyzes the thenar muscles, and the thumb loses much of its usefulness.

Ulnar Canal Syndrome

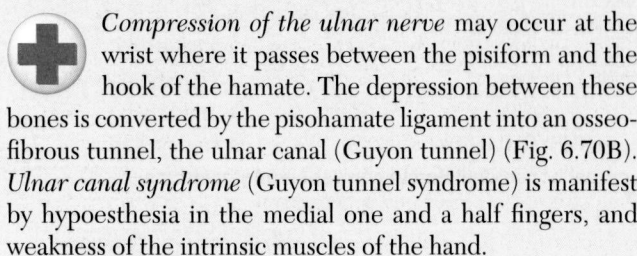

Compression of the ulnar nerve may occur at the wrist where it passes between the pisiform and the hook of the hamate. The depression between these bones is converted by the pisohamate ligament into an osseofibrous tunnel, the ulnar canal (Guyon tunnel) (Fig. 6.70B). *Ulnar canal syndrome* (Guyon tunnel syndrome) is manifest by hypoesthesia in the medial one and a half fingers, and weakness of the intrinsic muscles of the hand.

"Clawing" of the 4th and 5th fingers (hyperextension at the metacarpophalangeal joint with flexion at the proximal interphalangeal joint) may occur, but—in contradistinction to proximal ulnar nerve injury—their ability to flex is unaffected, and there is no radial deviation of the hand.

Handlebar Neuropathy

People who ride long distances on bicycles with their hands in an extended position against the hand grips put pressure on the hooks of their hamates (Fig. 6.70B), which compresses their ulnar nerves. This type of nerve compression, which has been called *handlebar neuropathy,* results in sensory loss on the medial side of the hand, and weakness of the intrinsic hand muscles.

Radial Nerve Injury in Arm and Hand Disability

Although the radial nerve supplies no muscles in the hand, radial nerve injury in the arm can produce serious hand disability. The characteristic handicap is inability to extend the wrist resulting from *paralysis of extensor muscles of the forearm,* all of which are innervated by the radial nerve (Fig. 6.61B; Table 6.11). The hand is flexed at the wrist and lies flaccid, a condition known as *wristdrop* (see the blue box "Injury to Radial Nerve in Arm" on p. 743). The fingers of the relaxed hand also remain in the flexed position at the metacarpophalangeal joints.

The interphalangeal joints can be extended weakly through the action of the intact lumbricals and interossei, which are supplied by the median and ulnar nerves (Table 6.13). The radial nerve has only a small area of exclusive cutaneous supply on the hand. Thus the extent of anesthesia is minimal, even in serious radial nerve injuries, and is usually confined to a small area on the lateral part of the dorsum of the hand.

Dermatoglyphics

The science of studying ridge patterns of the palm, called *dermatoglyphics,* is a valuable extension of the conventional physical examination of people with certain congenital anomalies and genetic diseases. For example, people with trisomy 21 (Down syndrome) have dermatoglyphics that are highly characteristic. In addition, they often have a single transverse palmar crease (Simian crease); however, approximately 1% of the general population has this crease with no other clinical features of the syndrome.

Palmar Wounds and Surgical Incisions

The location of superficial and deep palmar arches should be kept in mind when examining wounds of the palm and when making palmar incisions. Furthermore, it is important to know that the superficial palmar arch is at the same level as the distal end of the common flexor sheath (Figs. 6.77A and 6.81). As mentioned previously, incisions or wounds along the medial surface of the thenar eminence may injure the recurrent branch of the median nerve to the thenar muscles (see the blue box "Trauma to Median Nerve" on p. 790).

The Bottom Line

HAND

Movements: The larger (wider range) and stronger movements of the hand and fingers (grasping, pinching, and pointing) are produced by extrinsic muscles with fleshy bellies located distant from the hand (near the elbow) and long tendons passing into the hand and fingers. ♦ The shorter, more delicate and weaker movements (typing, playing musical instruments, and writing), and positioning of the fingers for the more powerful movements are accomplished largely by the intrinsic muscles.

Organization: The muscles and tendons of the hand are organized into five fascial compartments: two radial compartments (thenar and adductor) that serve the thumb; an ulnar (hypothenar) compartment that serves the little finger; and two more central compartments that serve the medial four digits (a palmar one for the long flexor tendons and lumbricals, and a deep one between the metacarpals for the interossei).

Muscles: The greatest mass of intrinsic muscles is dedicated to the highly mobile thumb. Indeed, when extrinsic muscles are also considered, the thumb has eight muscles producing and controlling the wide array of movements that distinguish the human hand. ♦ The interossei produce multiple movements: the dorsal interossei (and abductors pollicis and digiti minimi) abduct the digits, whereas the palmar interossei (and adductor pollicis) adduct them. Both movements occur at the metacarpophalangeal joints. ♦ Acting together with the lumbricals, the interossei flex the metacarpophalangeal and extend the interphalangeal joints of the medial four digits (the Z-movement)

Vasculature: The vasculature of the hand is characterized by multiple anastomoses between both radial and ulnar vessels and palmar and dorsal vessels. ♦ The arteries of the hand collectively constitute a peri-articular arterial anastomosis around the collective joints of the wrist and hand. Thus blood is generally available to all parts of the hand in all positions as well as while performing functions (gripping or pressing) that might otherwise compromise especially the palmar structures. ♦ The arteries to the digits are also characterized by their ability to

vasoconstrict during exposure to cold to conserve heat and to dilate (while the hand becomes sweaty) to radiate excess heat. ♦ The superficial dorsal venous network is commonly used for administering intravenous fluids.

Innervation: Unlike the dermatomes of the trunk and proximal limbs, the zones of cutaneous innervation and the roles of motor innervation are well defined, as are functional deficits. ♦ In terms of intrinsic structure, the radial nerve is sensory only via its superficial branch to the dorsum of the hand. ♦ The median nerve is most important to the function of the thumb, and sensation from the lateral three and half digits and adjacent palm, whereas the ulnar nerve supplies the remainder. ♦ The intrinsic muscles of the hand constitute the T1 myotome. ♦ The palmar nerves and vessels are dominant, supplying not only the more sensitive and functional palmar aspect, but also the dorsal aspect of the distal part of the digits (nail beds).

JOINTS OF UPPER LIMB

Movement of the pectoral girdle involves the sternoclavicular, acromioclavicular, and glenohumeral joints (Fig. 6.90), usually all moving simultaneously. Functional defects in any of the joints impair movements of the pectoral girdle. Mobility of the scapula is essential for free movement of the upper limb. The clavicle forms a strut (extension) that holds the scapula, and hence the glenohumeral (shoulder) joint, away

from the thorax so it can move freely. The clavicle establishes the radius at which the shoulder (half of the pectoral girdle and glenohumeral joint) rotates at the SC joint. The 15–20° of movement at the AC joint permits positioning of the glenoid cavity that is necessary for arm movements.

When *testing the range of motion of the pectoral girdle,* both scapulothoracic (movement of the scapula on the thoracic wall) and glenohumeral movements must be considered. Although the initial 30° of abduction may occur without

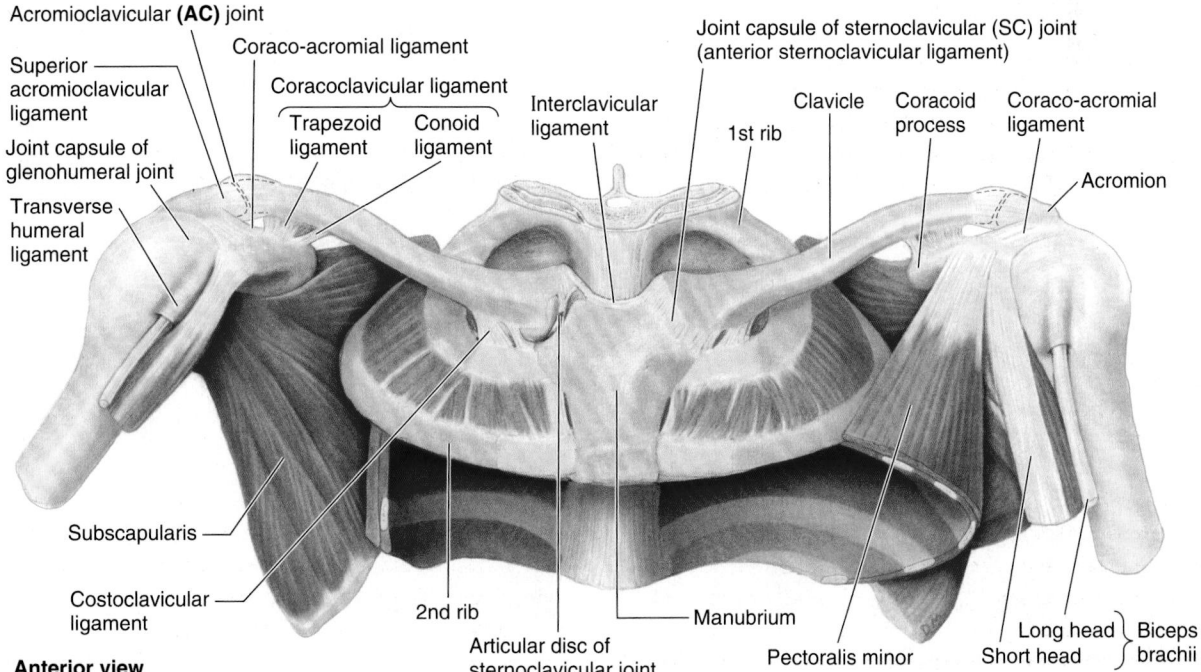

FIGURE 6.90. Pectoral girdle and associated tendons and ligaments. The pectoral girdle is a partial bony ring (incomplete posteriorly) formed by the manubrium of the sternum, the clavicle, and the scapulae. Joints associated with these bones are the sternoclavicular, acromioclavicular, and glenohumeral. The girdle provides for attachment of the superior appendicular skeleton to the axial skeleton and provides the mobile base from which the upper limb operates.

scapular motion, in the overall movement of fully elevating the arm, the movement occurs in a 2:1 ratio: For every 3° of elevation, approximately 2° occurs at the glenohumeral joint and 1° at the physiological scapulothoracic joint.

Hence, when the upper limb has been elevated so that the arm is vertical at the side of the head (180° of arm abduction or flexion), 120° occurred at the glenohumeral joint and 60° occurred at the scapulothoracic joint. This is known as **scapulohumeral rhythm** (Fig.6.92C). The important movements of the pectoral girdle are scapular movements (Table 6.3; p. 698): elevation and depression, protraction (lateral or forward movement of the scapula) and retraction (medial or backward movement of the scapula), and rotation of the scapula.

Sternoclavicular Joint

The **sternoclavicular (SC) joint** is a saddle type of synovial joint but functions as a ball-and-socket joint. The SC joint is divided into two compartments by an *articular disc.* The disc is firmly attached to the *anterior* and *posterior sternoclavicular ligaments,* thickenings of the fibrous layer of the joint capsule, as well as the *interclavicular ligament* (Fig. 6.90).

The great strength of the SC joint is a consequence of these attachments. Thus, although the articular disc serves as a shock absorber of forces transmitted along the clavicle from the upper limb, dislocation of the clavicle is rare, whereas fracture of the clavicle is common.

The SC joint is the only articulation between the upper limb and the axial skeleton, and it can be readily palpated because the sternal end of the clavicle lies superior to the manubrium of the sternum.

ARTICULATION OF STERNOCLAVICULAR JOINT

The sternal end of the clavicle articulates with the manubrium and the 1st costal cartilage. The articular surfaces are covered with fibrocartilage.

JOINT CAPSULE OF STERNOCLAVICULAR JOINT

The *joint capsule* surrounds the SC joint, including the epiphysis at the sternal end of the clavicle. It is attached to the margins of the articular surfaces, including the periphery of the articular disc. A *synovial membrane* lines the internal surface of the *fibrous layer of the joint capsule,* extending to the edges of the articular surfaces.

LIGAMENTS OF STERNOCLAVICULAR JOINT

The strength of the SC joint depends on its ligaments and articular disc. **Anterior** and **posterior sternoclavicular ligaments** reinforce the joint capsule anteriorly and posteriorly. The **interclavicular ligament** strengthens the capsule superiorly. It extends from the sternal end of one clavicle to the sternal end of the other clavicle. In between, it is also attached to the superior border of the manubrium. The **costoclavicular ligament** anchors the inferior surface of the sternal end of the clavicle to the 1st rib and its costal cartilage, limiting elevation of the pectoral girdle.

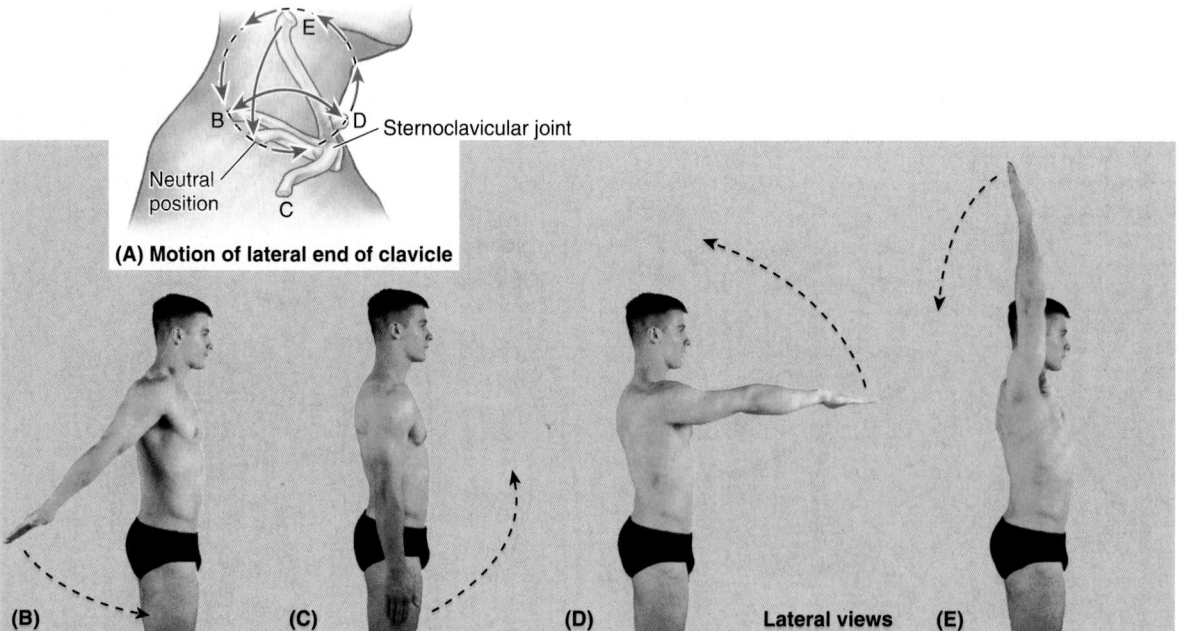

FIGURE 6.91. Movements of upper limb at joints of pectoral girdle. A. Range of motion of lateral end of clavicle permitted by movements at the sternoclavicular joint. *Letters* indicate disposition of the clavicle during the four positions of the limb demonstrated in parts **B–D.** The movements indicated by the double-headed arrows are A↔C, protraction–retraction; A↔D, elevation–depression. **B–E.** Circumduction of upper limb requires coordinated movements of the pectoral girdle and glenohumeral joint. Beginning with extended limb, retracted girdle (**B**); neutral position (**C**); flexed limb, protracted girdle (**D**); and, finally, elevated limb and girdle (**E**).

MOVEMENTS OF STERNOCLAVICULAR JOINT

Although the SC joint is extremely strong, it is significantly mobile to allow movements of the pectoral girdle and upper limb (Figs. 6.91 and 6.92D). During full elevation of the limb, the clavicle is raised to approximately a 60° angle. When ele-

vation is achieved via flexion, it is accompanied by rotation of the clavicle around its longitudinal axis. The SC joint can also be moved anteriorly or posteriorly over a range of up to 25–30°. Although not a typical movement, except perhaps during calisthenics (systematic body exercises), it is capable

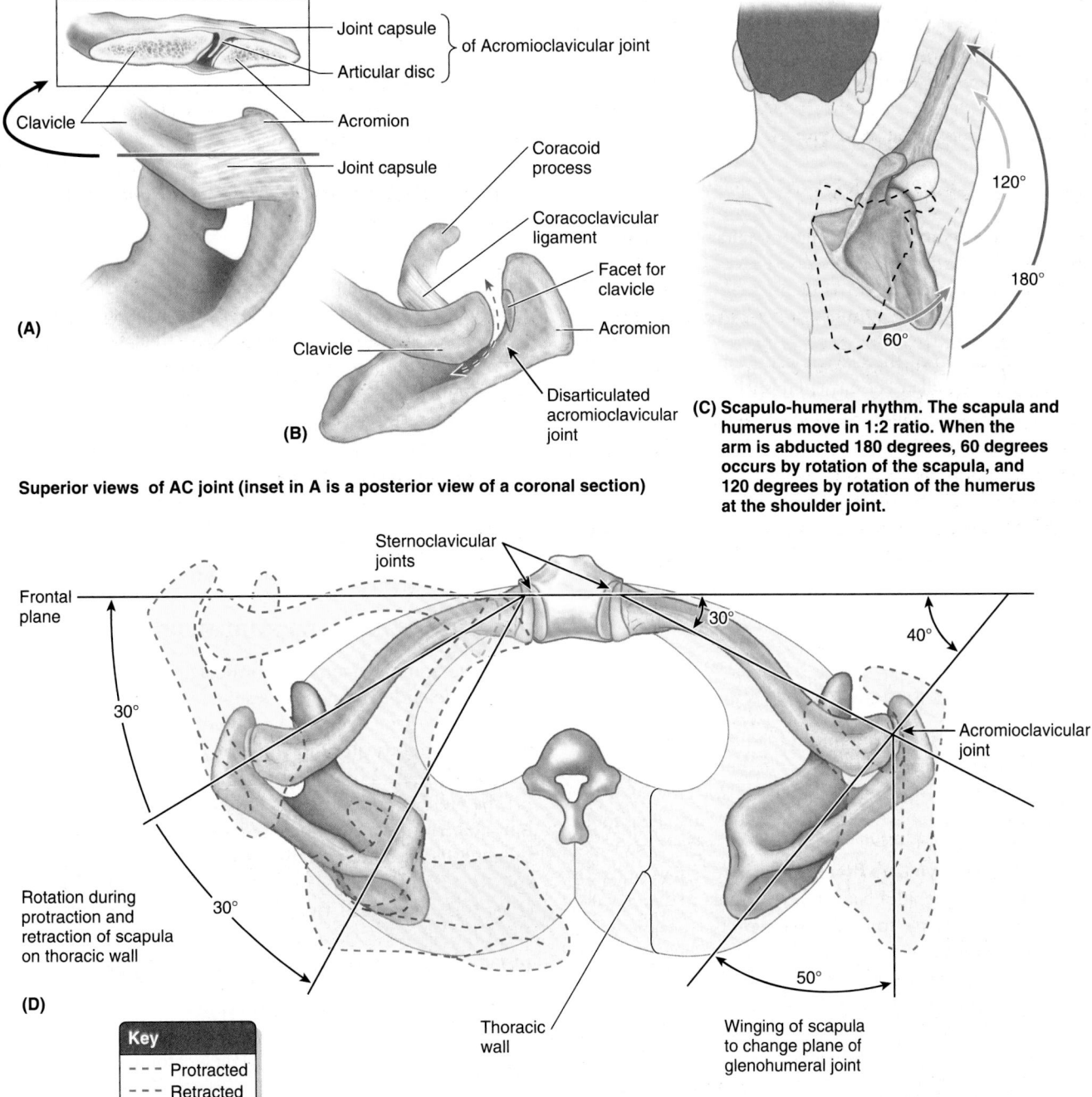

FIGURE 6.92. Acromioclavicular, scapulothoracic, and sternoclavicular joints. A. The right AC joint showing the joint capsule and partial disc (*inset*). **B.** The function of the coracoclavicular ligament. As long as this ligament is intact with the clavicle tethered to the coronoid process, the acromion cannot be driven inferior to the clavicle. The ligament, however, does permit protraction and retraction of the acromion. **C.** Rotation of scapula at "scapulothoracic joint" is an essential component of abduction of the upper limb. **D.** Clavicular movements at the SC and AC joints permit protraction and retraction of the scapula on the thoracic wall (*red and green lines*) and winging of the scapula (*blue line*). Movements of a similar scale occur during elevation, depression, and rotation of the scapula. The latter movements are shown in Table 6.5, which also indicates the muscles specifically responsible for these movements.

of performing these movements sequentially, moving the acromial end along a circular path—a form of *circumduction*.

BLOOD SUPPLY OF STERNOCLAVICULAR JOINT

The SC joint is supplied by the internal thoracic and suprascapular arteries (Fig. 6.39).

NERVE SUPPLY OF STERNOCLAVICULAR JOINT

Branches of the medial supraclavicular nerve and the nerve to the subclavius supply the SC joint (Fig. 6.44 and Table 6.8).

Acromioclavicular Joint

The **acromioclavicular joint** (AC joint) is a plane type of synovial joint, which is located 2–3 cm from the "point" of the shoulder formed by the lateral part of the acromion (Figs. 6.90 and 6.92A).

ARTICULATION OF ACROMIOCLAVICULAR JOINT

The acromial end of the clavicle articulates with the acromion of the scapula. The articular surfaces, covered with fibrocartilage, are separated by an incomplete wedge-shaped *articular disc.*

JOINT CAPSULE OF ACROMIOCLAVICULAR JOINT

The sleeve-like, relatively loose *fibrous layer of the joint capsule* is attached to the margins of the articular surfaces (Fig. 6.92A). A *synovial membrane* lines the fibrous layer. Although relatively weak, the joint capsule is strengthened superiorly by fibers of the trapezius.

LIGAMENTS OF ACROMIOCLAVICULAR JOINT

The **acromioclavicular ligament** is a fibrous band extending from the acromion to the clavicle that strengthens the AC joint superiorly (Figs. 6.90 and 6.93A). However, the integrity of the joint is maintained by extrinsic ligaments, distant from the joint itself.

The **coracoclavicular ligament** is a strong pair of bands that unite the coracoid process of the scapula to the clavicle, anchoring the clavicle to the coracoid process. The coracoclavicular ligament consists of two ligaments, the conoid and trapezoid ligaments, which are often separated by a bursa. The vertical **conoid ligament** is an inverted triangle (cone), which has its apex inferiorly where it is attached to the root of the *coracoid process.* Its wide attachment (base of the triangle) is to the *conoid tubercle* on the inferior surface of the clavicle. The nearly horizontal **trapezoid ligament** is attached to the superior surface of the coracoid process and extends laterally to the trapezoid line on the inferior surface of the clavicle. In addition to augmenting the AC joint, the coracoclavicular ligament provides the means by which the scapula and free limb are (passively) suspended from the clavicular strut.

MOVEMENTS OF ACROMIOCLAVICULAR JOINT

The acromion of the scapula rotates on the acromial end of the clavicle. These movements are associated with motion at the physiological scapulothoracic joint (Figs. 6.25, 6.91, and 6.92; Table 6.5). No muscles connect the articulating bones to move the AC joint; the axio-appendicular muscles that attach to and move the scapula cause the acromion to move on the clavicle.

BLOOD SUPPLY OF ACROMIOCLAVICULAR JOINT

The AC joint is supplied by the suprascapular and thoracoacromial arteries (Fig. 6.39).

NERVE SUPPLY OF ACROMIOCLAVICULAR JOINT

Consistent with Hilton's law (joints are supplied by articular branches of the nerves supplying the muscles that act on the joint), the lateral pectoral and axillary nerves supply the AC joint (Fig. 6.44; Table 6.8). However, consistent with the joint's subcutaneous location, and the fact that no muscles cut across the joint, innervation is also provided to the AC joint by the cutaneous lateral supraclavicular nerve, a manner of innervation more typical of the distal limb.

Glenohumeral Joint

The **glenohumeral (shoulder) joint** is a ball-and-socket type of synovial joint that permits a wide range of movement; however, its mobility makes the joint relatively unstable.

ARTICULATION OF GLENOHUMERAL JOINT

The large, round *humeral head* articulates with the relatively shallow *glenoid cavity* of the scapula (Figs. 6.94 and 6.95), which is deepened slightly but effectively by the ring-like, fibrocartilaginous **glenoid labrum** (L., lip). Both articular surfaces are covered with hyaline cartilage.

The glenoid cavity accepts little more than a third of the humeral head, which is held in the cavity by the tonus of the musculotendinous rotator cuff, or SITS, muscles (supraspinatus, infraspinatus, teres minor, and subscapularis) (Figs. 6.29 and 6.94B; Table 6.6).

JOINT CAPSULE OF GLENOHUMERAL JOINT

The loose *fibrous layer of the joint capsule* surrounds the glenohumeral joint and is attached medially to the margin of the glenoid cavity and laterally to the anatomical neck of the humerus (Fig. 6.95A & B). Superiorly, this part of the capsule encroaches on the root of the coracoid process so that the fibrous layer of the capsule encloses the proximal attachment of the long head of the biceps brachii to the supraglenoid tubercle of scapula within the joint.

The joint capsule has two apertures: (1) an opening between the tubercles of the humerus for passage of the tendon of the long head of the biceps brachii (Fig. 6.93A), and

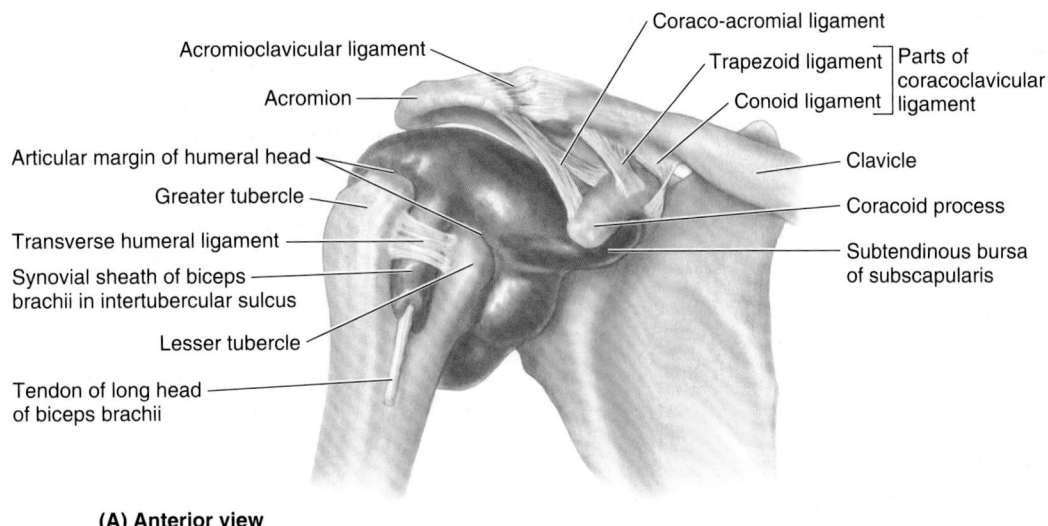

Coraco-acromial ligament

Acromioclavicular ligament

Acromion

Trapezoid ligament ⎤ Parts of
                     ⎦ coracoclavicular
Conoid ligament   ⎦ ligament

Articular margin of humeral head

Greater tubercle

Transverse humeral ligament

Synovial sheath of biceps
brachii in intertubercular sulcus

Lesser tubercle

Tendon of long head
of biceps brachii

Clavicle

Coracoid process

Subtendinous bursa
of subscapularis

(A) Anterior view

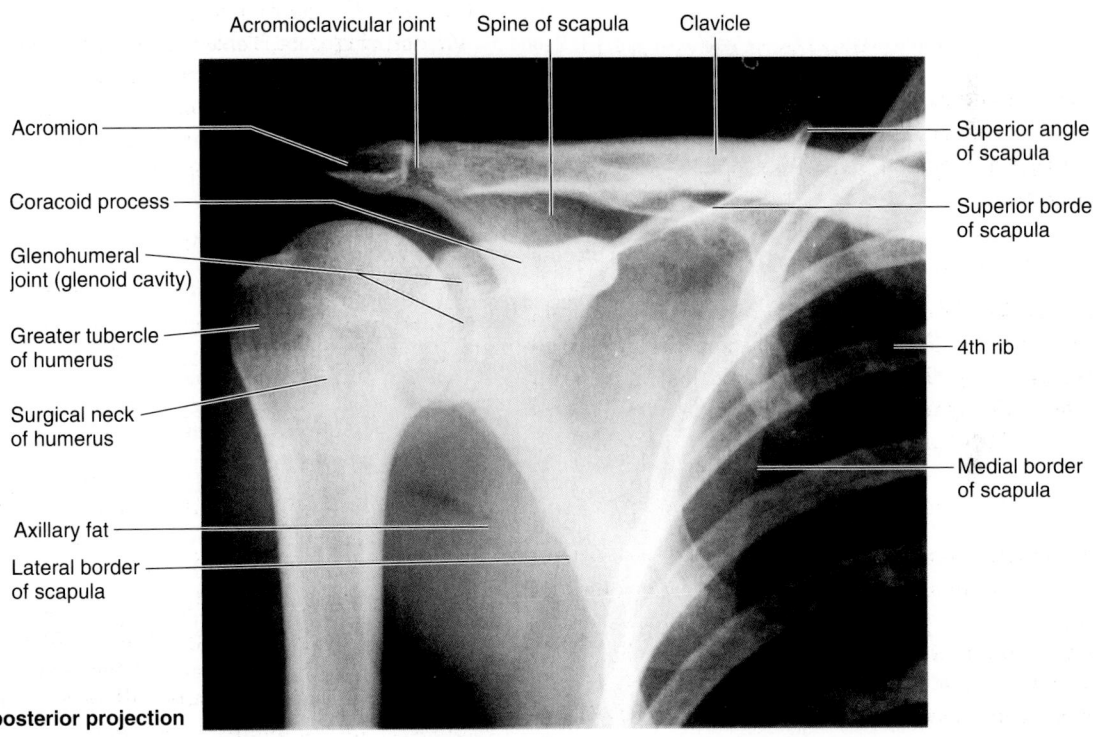

Acromioclavicular joint   Spine of scapula   Clavicle

Acromion

Coracoid process

Glenohumeral
joint (glenoid cavity)

Greater tubercle
of humerus

Surgical neck
of humerus

Axillary fat

Lateral border
of scapula

Superior angle
of scapula

Superior border
of scapula

4th rib

Medial border
of scapula

(B) Anteroposterior projection

FIGURE 6.93. Glenohumeral (shoulder) joint. A. The extent of the synovial membrane of the glenohumeral joint is demonstrated in this specimen in which the articular cavity has been injected with purple latex and the fibrous layer of the joint capsule has been removed. The articular cavity has two extensions: one where it forms a synovial sheath for the tendon of the long head of the biceps brachii in the intertubercular sulcus of the humerus, and the other inferior to the coracoid process where it is continuous with the subscapularis tendon and the margin of the glenoid cavity. The joint capsule and intrinsic ligaments of the AC joint are also seen. **B.** In this radiograph, the head of the humerus and the glenoid cavity overlap, obscuring the joint plane because the scapula does not lie in the coronal plane (therefore the glenoid cavity is oblique, not in a sagittal plane.) (Courtesy of Dr. E. L. Lansdown, Professor of Medical Imaging, University of Toronto, Toronto, Ontario, Canada.)

(2) an opening situated anteriorly, inferior to the coracoid process that allows communication between the *subtendinous bursa of subscapularis* and the synovial cavity of the joint. The inferior part of the joint capsule, the only part not reinforced by the rotator cuff muscles, is its weakest area. Here the capsule is particularly lax and lies in folds when the arm is adducted; however, it becomes taut when the arm is abducted.

The *synovial membrane* lines the internal surface of the fibrous layer of the capsule and reflects from it onto the glenoid labrum and the humerus, as far as the articular margin of the head (Figs. 6.93A, 6.94A, and 6.95A).

The synovial membrane also forms a tubular sheath for the tendon of the long head of the biceps brachii, where it lies in the intertubercular sulcus of the humerus and passes into the joint cavity (Fig. 6.93A).

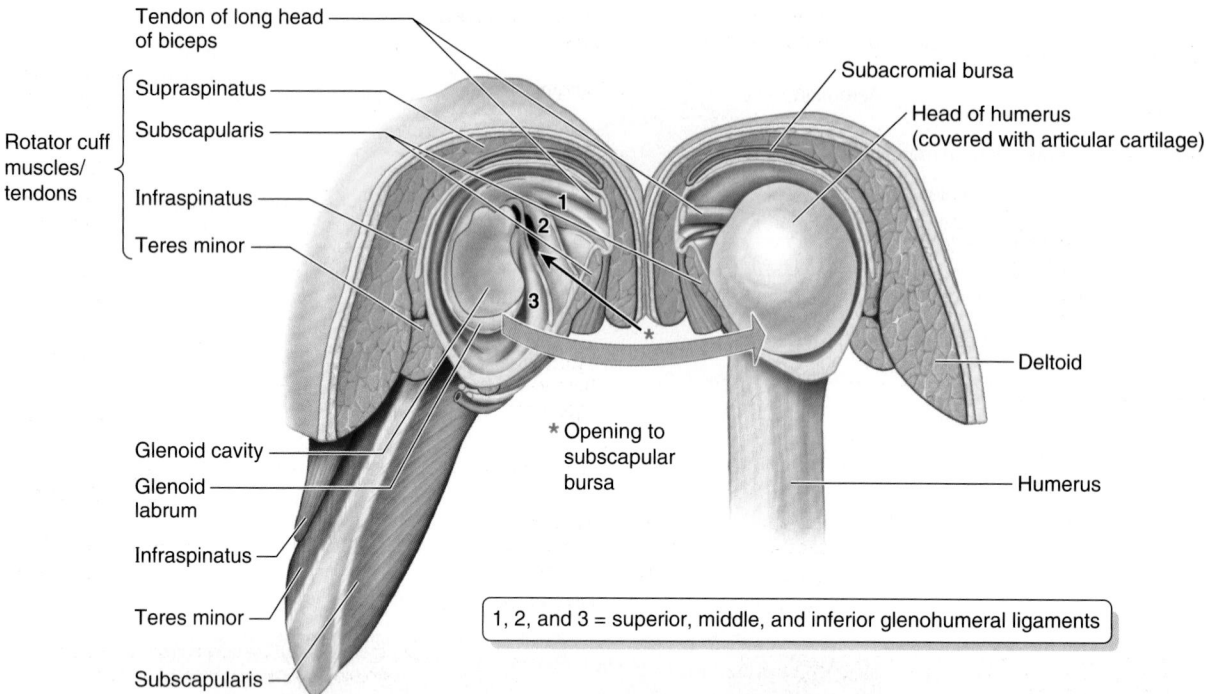

(A) Anterolateral view of glenoid cavity; posteromedial view of humerus

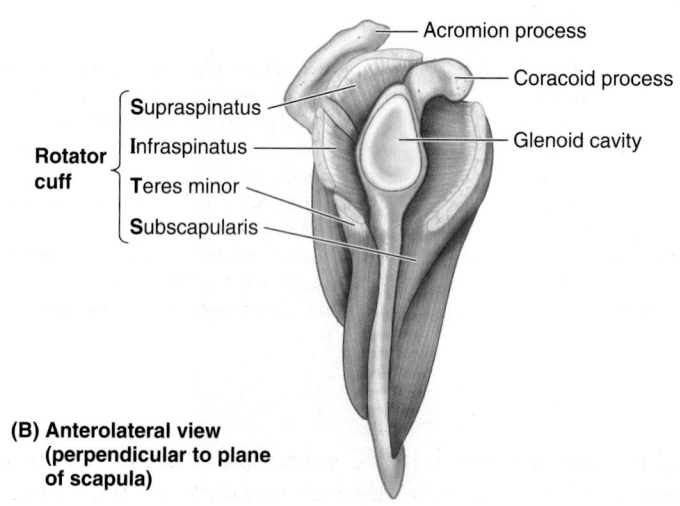

(B) Anterolateral view (perpendicular to plane of scapula)

FIGURE 6.94. Rotator cuff and glenohumeral joint.
A. Dissection of the glenohumeral joint in which the joint capsule was sectioned and the joint opened from its posterior aspect as if it were a book. Four short SITS muscles (supraspinatus, infraspinatus, teres minor, and subscapularis) cross and surround the joint, blending with its capsule. The anterior, internal surface demonstrates the glenohumeral ligaments, which were incised to open the joint. **B.** The SITS muscles of the left rotator cuff are shown as they relate to the scapula and its glenoid cavity. The prime function of these muscles and the musculotendinous rotator cuff is to hold the relatively large head of the humerus in the much smaller and shallow glenoid cavity of the scapula.

LIGAMENTS OF GLENOHUMERAL JOINT

The glenohumeral ligaments, which strengthen the anterior aspect of the joint capsule and the coracohumeral ligament, which strengthens the joint capsule superiorly, are intrinsic ligaments—that is, part of the fibrous layer of the joint capsule (Figs. 6.94A and 6.95B).

The **glenohumeral ligaments** are three fibrous bands, evident only on the internal aspect of the capsule, that reinforce the anterior part of the joint capsule. These ligaments radiate laterally and inferiorly from the glenoid labrum at the supraglenoid tubercle of the scapula and blend distally with the fibrous layer of the capsule as it attaches to the anatomical neck of the humerus.

The **coracohumeral ligament** is a strong broad band that passes from the base of the coracoid process to the anterior aspect of the greater tubercle of the humerus (Fig. 6.95B).

The **transverse humeral ligament** is a broad fibrous band that runs more or less obliquely from the greater to the lesser tubercle of the humerus, bridging over the intertubercular sulcus (Figs. 6.93A and 6.95B). This ligament converts the groove into a canal, which holds the synovial sheath and tendon of the biceps brachii in place during movements of the glenohumeral joint.

The **coraco-acromial arch** is an extrinsic, protective structure formed by the smooth inferior aspect of the

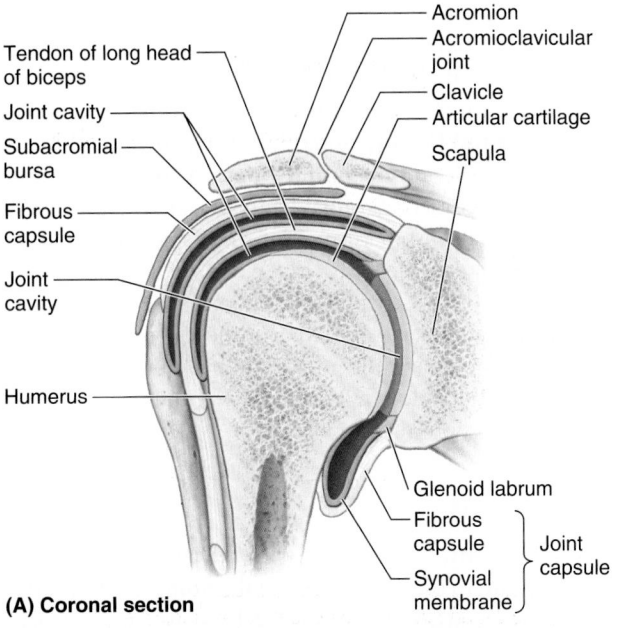

(A) Coronal section

Tendon of long head of biceps
Joint cavity
Subacromial bursa
Fibrous capsule
Joint cavity
Humerus
Acromion
Acromioclavicular joint
Clavicle
Articular cartilage
Scapula
Glenoid labrum
Fibrous capsule
Synovial membrane
} Joint capsule

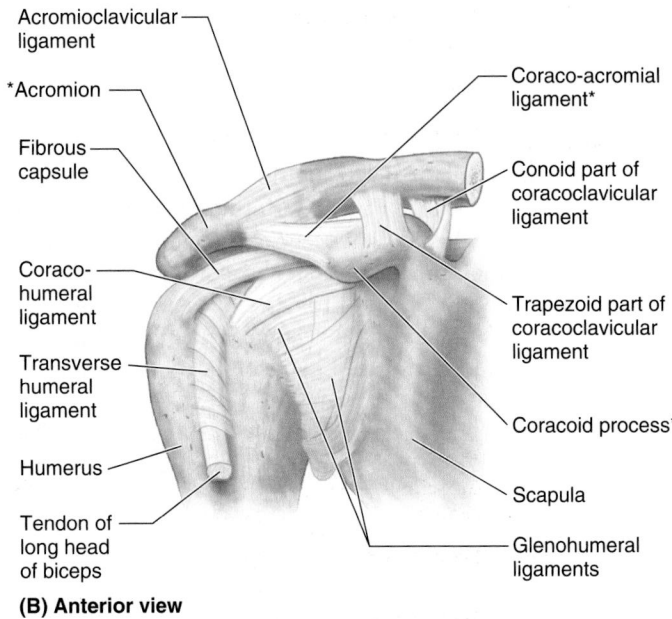

(B) Anterior view

Acromioclavicular ligament
*Acromion
Fibrous capsule
Coraco-humeral ligament
Transverse humeral ligament
Humerus
Tendon of long head of biceps
Coraco-acromial ligament*
Conoid part of coracoclavicular ligament
Trapezoid part of coracoclavicular ligament
Coracoid process*
Scapula
Glenohumeral ligaments

* Coracoid process + coraco-acromial ligament + acromion = **coraco-acromial arch**

FIGURE 6.95. Capsules and ligaments of glenohumeral and acromioclavicular joints. A. The bones, articular surfaces, joint capsule, cavity of the joints, and the subacromial bursa. **B.** The acromioclavicular, coracohumeral, and glenohumeral ligaments. Although shown on the external aspect of the joint capsule, the glenohumeral ligaments are actually a feature observed from the internal aspect of the joint (as in Fig. 6.94A). These ligaments strengthen the anterior aspect of the capsule of the glenohumeral joint, and the coraco-humeral ligament strengthens the capsule superiorly. **C.** Coronal MRI showing the right glenohumeral and AC joints. *A*, acromion; *C*, clavicle; *G*, glenoid cavity; *Gr*, greater tubercle of humerus; *H*, head of humerus; *N*, surgical neck of humerus. (Courtesy of Dr. W. Kucharczyk, Chair of Medical Imaging and Clinical Director of Tri-Hospital Resonance Centre, Toronto, Ontario, Canada.)

acromion and the *coracoid process* of the scapula, with the **coraco-acromial ligament** spanning between them (Fig. 6.95B). This osseoligamentous structure forms a protective arch that overlies the humeral head, preventing its superior displacement from the glenoid cavity. The coraco-acromial arch is so strong that a forceful superior thrust of the humerus will not fracture it; the humeral shaft or clavicle fractures first.

Transmitting force superiorly along the humerus (e.g., when standing at a desk and partly supporting the body with the outstretched limbs), the humeral head presses against the coraco-acromial arch. The supraspinatus muscle passes under this arch and lies deep to the deltoid as its tendon blends with the joint capsule of the glenohumeral joint as part of the rotator cuff (Fig. 6.94).

Movement of the supraspinatus tendon, passing to the greater tubercle of the humerus, is facilitated as it passes

under the arch by the **subacromial bursa** (Fig. 6.95A), which lies between the arch superiorly and the tendon and tubercle inferiorly.

MOVEMENTS OF GLENOHUMERAL JOINT

The glenohumeral joint has more freedom of movement than any other joint in the body. This freedom results from the laxity of its joint capsule and the large size of the humeral head compared with the small size of the glenoid cavity. The glenohumeral joint allows movements around three axes and permits flexion–extension, abduction–adduction, rotation (medial and lateral) of the humerus, and circumduction (Fig. 6.96).

Lateral rotation of the humerus increases the range of abduction. When the arm is abducted without rotation, available articular surface is exhausted and the greater tubercle contacts the *coraco-acromial arch,* preventing further abduction. If the arm is then laterally rotated 180°, the tubercles are rotated posteriorly and more articular surface becomes available to continue elevation.

Circumduction at the glenohumeral joint is an orderly sequence of flexion, abduction, extension, and adduction—or the reverse (Fig. 6.91). Unless performed over a small range, these movements do not occur at the glenohumeral joint in isolation; they are accompanied by movements at the two other joints of the pectoral girdle (SC and AC). Stiffening or fixation of the joints of the pectoral girdle (*ankylosis*) results in a much more restricted range of movement, even if the glenohumeral joint is normal.

MUSCLES MOVING GLENOHUMERAL JOINT

The movements of the glenohumeral joint and the muscles that produce them—the *axio-appendicular muscles,* which may act indirectly on the joint (i.e., act on the pectoral girdle), and the *scapulohumeral muscles,* which act directly on the glenohumeral joint (Tables 6.4 and 6.5)—are illustrated in Figure 6.96 and listed in Table 6.17. Other muscles that serve the glenohumeral joint as shunt muscles, acting to resist dislocation without producing movement at the joint (e.g., when carrying a heavy suitcase), or that maintain the large head of the humerus in the relatively shallow glenoid cavity are also listed in the table.

BLOOD SUPPLY OF GLENOHUMERAL JOINT

The glenohumeral joint is supplied by the *anterior* and *posterior circumflex humeral arteries* and branches of the *suprascapular artery* (Fig. 6.39; Table 6.7).

INNERVATION OF GLENOHUMERAL JOINT

The suprascapular, axillary, and lateral pectoral nerves supply the glenohumeral joint (Table 6.8).

BURSAE AROUND GLENOHUMERAL JOINT

Several **bursae** (sac-like cavities), containing capillary films of *synovial fluid* secreted by the synovial membrane, are situated near the glenohumeral joint. Bursae are located where tendons rub against bone, ligaments, or other tendons, and where skin moves over a bony prominence. The bursae around the glenohumeral joint are of special clinical importance because some of them communicate with the joint cavity (e.g., the subscapular bursa). Consequently, opening a bursa may mean entering the cavity of the glenohumeral joint.

Subtendinous Bursa of Subscapularis. The **subtendinous bursa of subscapularis** is located between the tendon of the subscapularis and the neck of the scapula (Fig. 6.93A). The bursa protects the tendon where it passes inferior to the root of the coracoid process and over the neck of the scapula. It usually communicates with the cavity of the glenohumeral joint through an opening in the fibrous layer of the joint capsule (Fig. 6.94A); thus it is really an extension of the glenohumeral joint cavity.

Subacromial Bursa. Sometimes referred to as the *subdeltoid bursa,* the **subacromial bursa** is located between the acromion, coraco-acromial ligament, and deltoid superiorly and the supraspinatus tendon and joint capsule of the glenohumeral joint inferiorly (Fig. 6.95A). Thus it facilitates movement of the supraspinatus tendon under the coraco-acromial arch and of the deltoid over the joint capsule of the glenohumeral joint and the greater tubercle of the humerus. Its size varies, but it does not normally communicate with the cavity of the glenohumeral joint.

Elbow Joint

The **elbow joint,** a hinge type of synovial joint, is located 2–3 cm inferior to the epicondyles of the humerus (Fig. 6.97).

ARTICULATION OF ELBOW JOINT

The spool-shaped *trochlea* and spheroidal *capitulum* of the humerus articulate with the *trochlear notch* of the ulna and the slightly concave superior aspect of the *head of the radius,* respectively; therefore, there are humero-ulnar and humero-radial articulations. The articular surfaces, covered with hyaline cartilage, are most fully congruent (in contact) when the forearm is in a position midway between pronation and supination and is flexed to a right angle.

JOINT CAPSULE OF ELBOW JOINT

The *fibrous layer of the joint capsule* surrounds the elbow joint (Fig. 6.97A & C). It is attached to the humerus at the margins of the lateral and medial ends of the articular surfaces of the capitulum and trochlea. Anteriorly and posteriorly, it is carried superiorly, proximal to the coronoid and olecranon fossae.

The *synovial membrane* lines the internal surface of the fibrous layer of the capsule and the intracapsular nonarticular parts of the humerus. It is also continuous inferiorly with the synovial membrane of the proximal radio-ulnar joint. The joint capsule is weak anteriorly and posteriorly but is strengthened on each side by collateral ligaments.

LIGAMENTS OF ELBOW JOINT

The collateral ligaments of the elbow joint are strong triangular bands that are medial and lateral thickenings of the fibrous layer of the joint capsule (Figs. 6.97A and 6.98). The lateral, fan-like **radial collateral ligament** extends from the lateral epicondyle of the humerus and blends distally with the **anular ligament of the radius,** which encircles and holds the head of the radius in the radial notch of the ulna, forming the proximal radio-ulnar joint and permitting pronation and supination of the forearm.

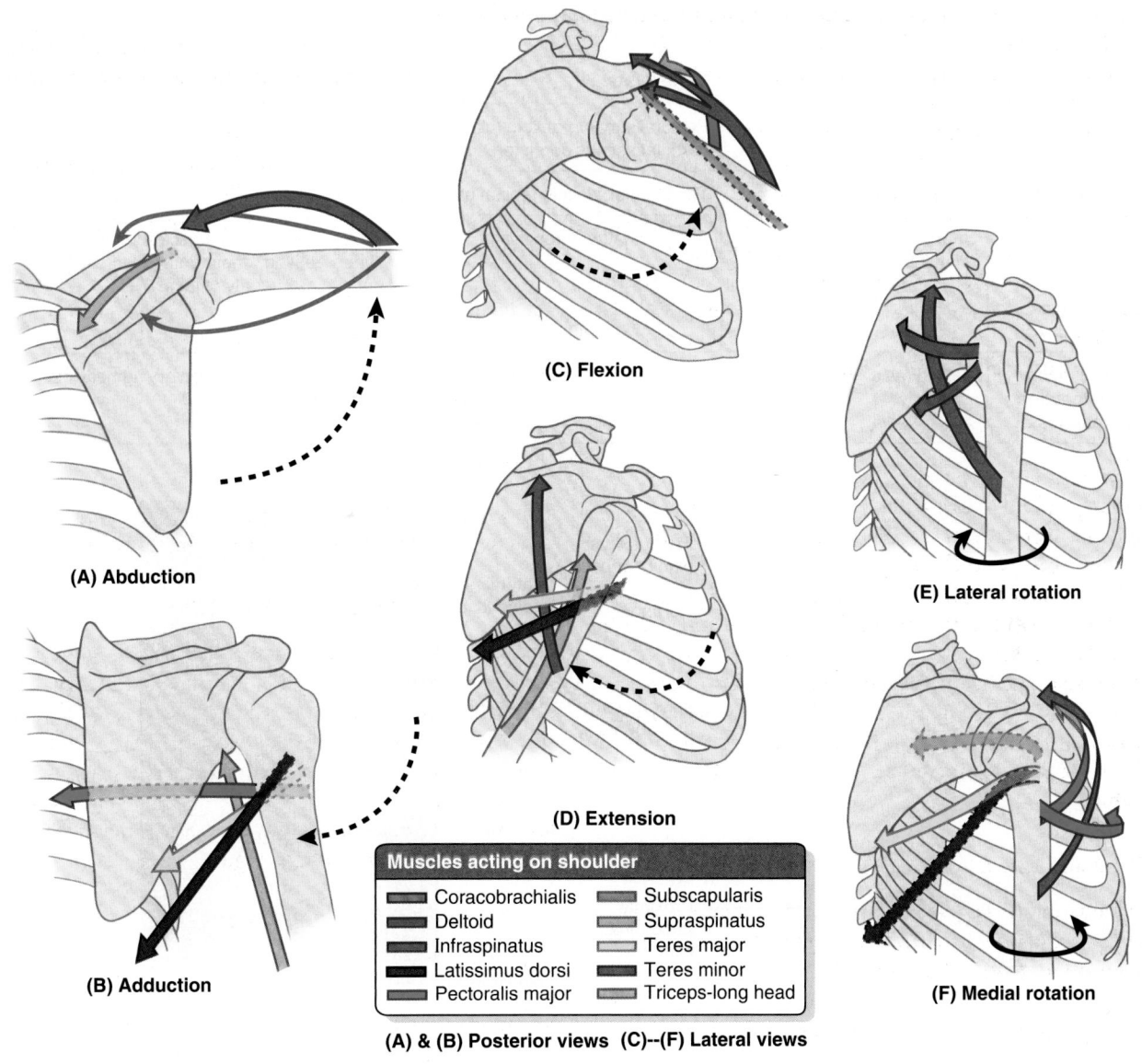

FIGURE 6.96. Movements of glenohumeral joint.

TABLE 6.17. **MOVEMENTS OF GLENOHUMERAL JOINT**

Movement (Function)	Prime Mover(s) (From Pendent Position)	Synergists	Notes
Flexion	Pectoralis major (clavicular head); deltoid (clavicular and anterior acromial parts)	Coracobrachialis (assisted by biceps brachii)	From fully extended position to its own (coronal) plane, sternocostal head of pectoralis major is major force
Extension	Deltoid (spinal part)	Teres major; latissimus dorsi; long head of triceps brachii	Latissimus dorsi, (sternocostal head of pectoralis major, and long head of triceps brachii) act from fully flexed position to their own (coronal) planes
Abduction	Deltoid (as a whole, but especially acromial part)	Supraspinatus	Supraspinatus is particularly important in initiating movement; also, upward rotation of scapula occurs throughout movement, making a significant contribution
Adduction	Pectoralis major; latissimus dorsi	Teres major; long head of triceps brachii	In upright position and in absence of resistance, gravity is prime mover

TABLE 6.17. MOVEMENTS OF GLENOHUMERAL JOINT (Continued)

Movement (Function)	Prime Mover(s) (From Pendent Position)	Synergists	Notes
Medial rotation	Subscapularis	Pectoralis major; deltoid (clavicular part); latissimus dorsi; teres major	With arm elevated, "synergists" become more important than prime movers
Lateral rotation	Infraspinatus	Teres minor; deltoid (spinal part)	
Tensors of articular capsule (to hold head of humerus against the glenoid cavity)	Subscapularis; infraspinatus (simultaneously)	Supraspinatus; teres minor	Rotator cuff (SITS) muscles acting together; when "resting," their tonus adequately maintains integrity of joint
Resisting downward dislocation (shunt muscles)	Deltoid (as a whole)	Long head of triceps brachii; coracobrachialis; short head of biceps brachii	Used especially when carrying heavy objects (suitcases, buckets)

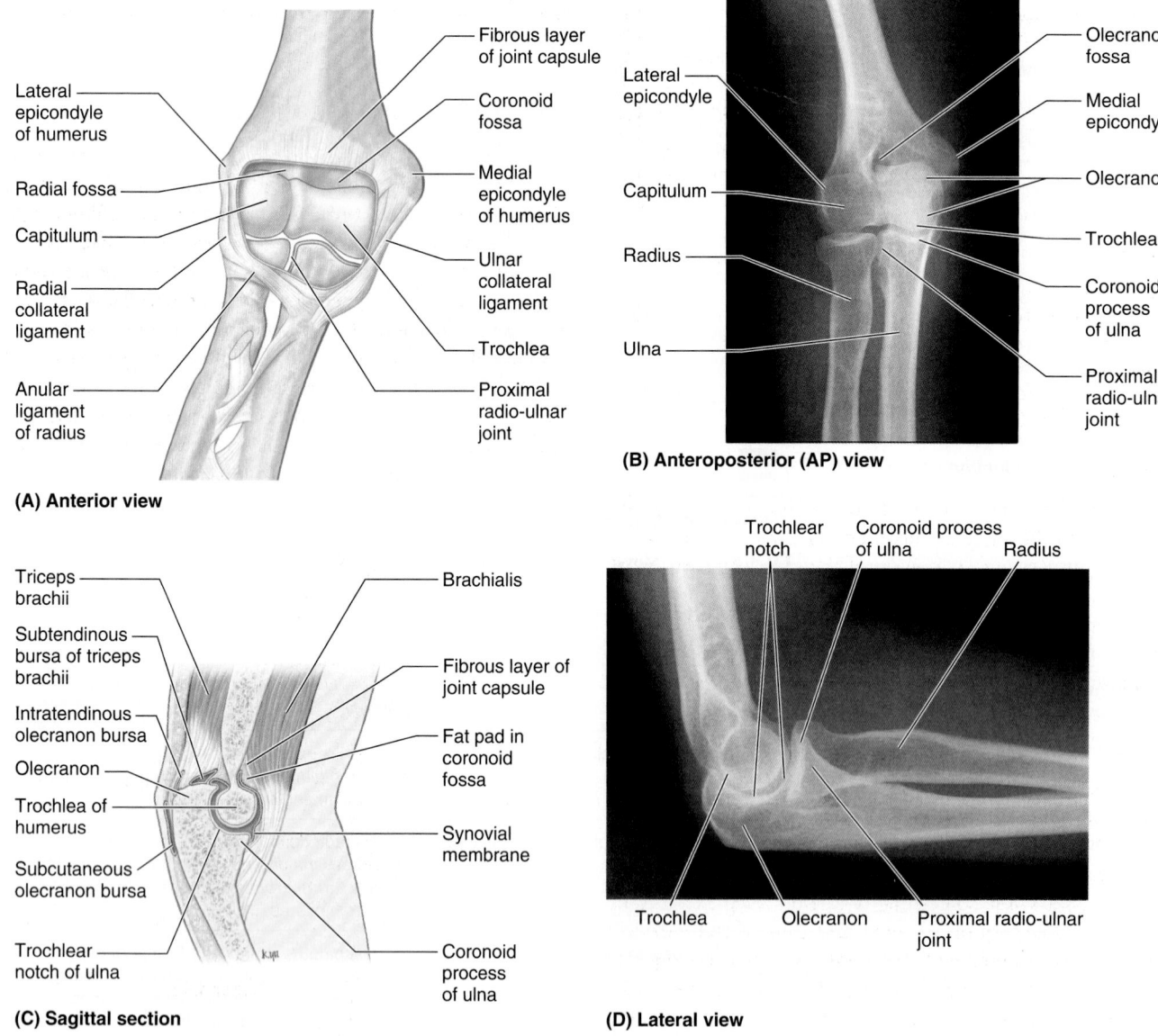

(A) Anterior view

(B) Anteroposterior (AP) view

(C) Sagittal section

(D) Lateral view

FIGURE 6.97. Elbow and proximal radio-ulnar joints. A. The thin anterior aspect of the joint capsule has been removed to reveal the articulating surfaces of the bones inside. The strong collateral ligaments were left intact. **B.** Radiograph of the extended elbow joint. **C.** The fibrous layer and synovial membrane of the joint capsule, the subtendinous and subcutaneous olecranon bursae, and humero-ulnar articulation of the elbow joint. **D.** Radiograph of the flexed elbow joint. (Parts **B** and **D** courtesy of Dr. E. Becker, Associate Professor of Medical Imaging, University of Toronto, Toronto, Ontario, Canada.)

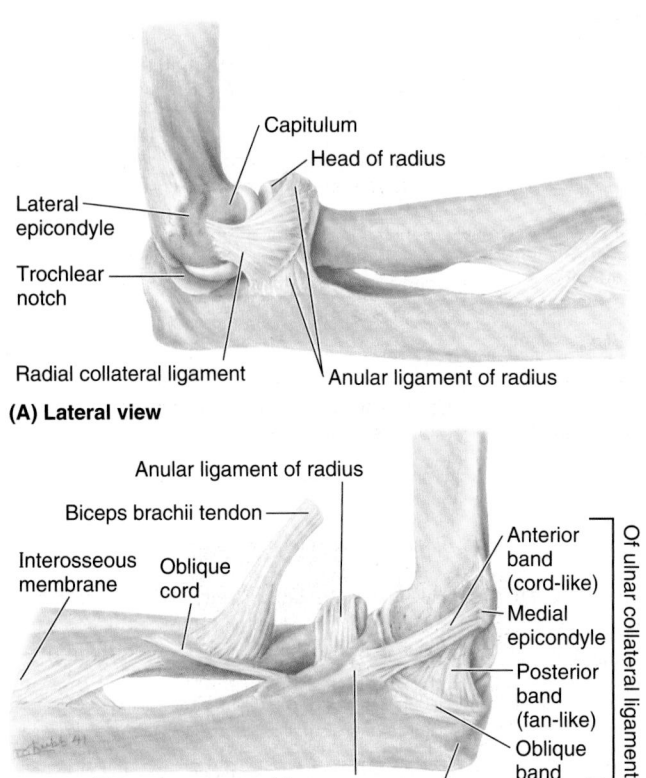

(A) Lateral view

Capitulum
Head of radius
Lateral epicondyle
Trochlear notch
Radial collateral ligament
Anular ligament of radius

(B) Medial view

Anular ligament of radius
Biceps brachii tendon
Interosseous membrane   Oblique cord
Anterior band (cord-like)
Medial epicondyle
Posterior band (fan-like)
Oblique band
Tubercle on coronoid process
Olecranon of ulna
Of ulnar collateral ligament

FIGURE 6.98. Collateral ligaments of elbow joint A. The fan-like radial collateral is attached to the anular ligament of the radius, but its superficial fibers continue on to the ulna. **B.** The ulnar collateral ligament has a strong, round, cord-like anterior band (part), which is taut when the elbow joint is extended, and a weak, fan-like posterior band, which is taut when the joint is flexed. The oblique fibers merely deepen the socket for the trochlea of the humerus.

The medial, triangular **ulnar collateral ligament** extends from the medial epicondyle of the humerus to the coronoid process and olecranon of the ulna and consists of three bands: (1) the *anterior cord-like band* is the strongest, (2) the *posterior fan-like band* is the weakest, and (3) the slender *oblique band* deepens the socket for the trochlea of the humerus.

MOVEMENTS OF ELBOW JOINT

Flexion and extension occur at the elbow joint. The long axis of the fully extended ulna makes an angle of approximately 170° with the long axis of the humerus. This **carrying angle** (Fig. 6.99) is named for the way the forearm angles away from the body when something is carried, such as a pail of water. The obliquity of the ulna and thus of the carrying angle is more pronounced (the angle is approximately 10° more acute) in women than in men. It is teleologically said to enable the swinging limbs to clear the wide female pelvis when walking. In the anatomical position, the elbow is against the waist. The carrying angle disappears when the forearm is pronated.

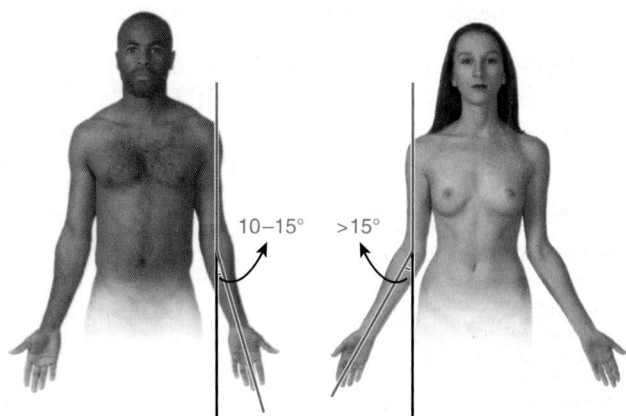

FIGURE 6.99. Carrying angle of elbow joint. This angle is made by the axes of the arm and forearm when the elbow is fully extended. Note that the forearm diverges laterally, forming an angle that is greater in the woman. Teleologically, this is said to allow for clearance of the wider female pelvis as the limbs swing during walking; however, no significant difference exists regarding the function of the elbow.

MUSCLES MOVING ELBOW JOINT

A total of 17 muscles cross the elbow and extend to the forearm and hand, most of which have some potential to affect elbow movement. In turn, their function and efficiency in the other movements they produce are affected by elbow position. *The chief flexors of the elbow joint are the brachialis and biceps brachii* (Fig. 6.100). The brachioradialis can produce rapid flexion in the absence of resistance (even when the chief flexors are paralyzed). Normally, in the presence of resistance, the brachioradialis and pronator teres assist the chief flexors in producing slower flexion. *The chief extensor of the elbow joint is the triceps brachii*, especially the medial head, weakly assisted by the anconeus.

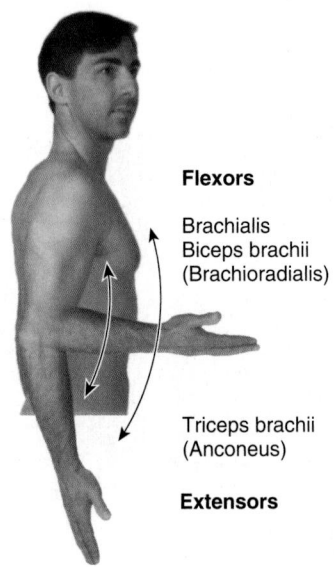

Flexors

Brachialis
Biceps brachii
(Brachioradialis)

Triceps brachii
(Anconeus)

Extensors

FIGURE 6.100. Movements of elbow joint and muscles that produce them.

BLOOD SUPPLY OF ELBOW JOINT

The arteries supplying the elbow joint are derived from the anastomosis around the elbow joint (Fig. 6.51).

NERVE SUPPLY OF ELBOW JOINT

The elbow joint is supplied by the musculocutaneous, radial, and ulnar nerves (Fig. 6.69; Table 6.13).

BURSAE AROUND ELBOW JOINT

Only some of the bursae around the elbow joint are clinically important. The three **olecranon bursae** are (Figs. 6.97C and 6.101) the:

1. **Intratendinous olecranon bursa,** which is sometimes present in the tendon of triceps brachii.
2. **Subtendinous olecranon bursa,** which is located between the olecranon and the triceps tendon, just proximal to its attachment to the olecranon.
3. **Subcutaneous olecranon bursa,** which is located in the subcutaneous connective tissue over the olecranon.

The **bicipitoradial bursa** (biceps bursa) separates the biceps tendon from, and reduces abrasion against, the anterior part of the radial tuberosity.

Proximal Radio-Ulnar Joint

The **proximal (superior) radio-ulnar joint** is a pivot type of synovial joint that allows movement of the head of the radius on the ulna (Figs. 6.97A, B, & D and 6.102).

ARTICULATION OF PROXIMAL RADIO-ULNAR JOINT

The head of the radius articulates with the radial notch of the ulna. The radial head is held in position by the *anular ligament of the radius.*

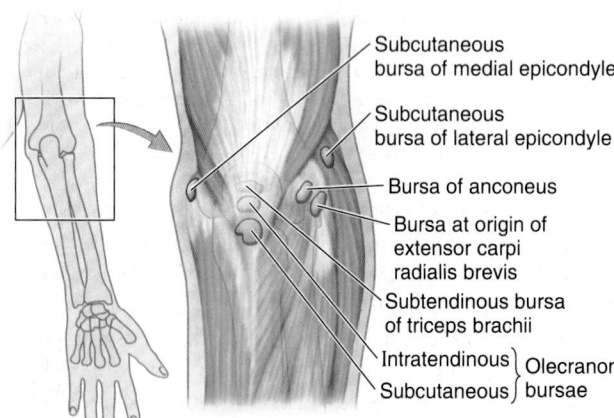

FIGURE 6.101. Bursae around elbow joint. Of the several bursae around the elbow joint, the olecranon bursae are most important clinically. Trauma of these bursae may produce bursitis.

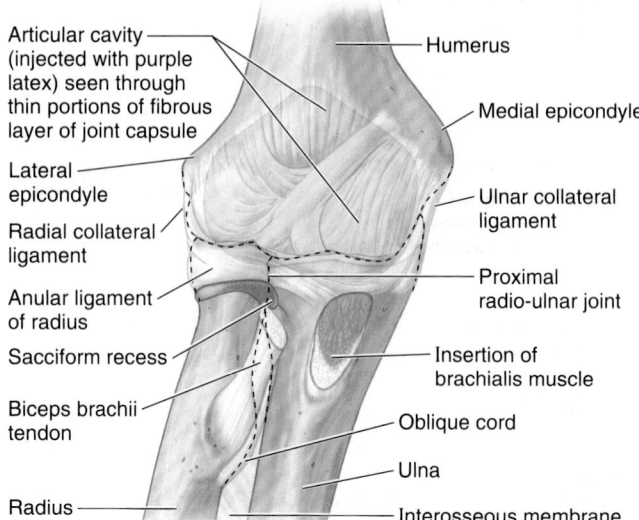

FIGURE 6.102. Proximal radio-ulnar joint. The anular ligament attaches to the radial notch of the ulna, forming a collar around the head of the radius (Fig. 6.103*A*) and creating a pivot type of synovial joint. The articular cavity of the joint is continuous with that of the elbow joint, as demonstrated by the blue latex injected into that space and seen through the thin parts of the fibrous layer of the capsule, including a small area distal to the anular ligament.

JOINT CAPSULE OF PROXIMAL RADIO-ULNAR JOINT

The *fibrous layer of the joint capsule* encloses the joint and is continuous with that of the elbow joint. The *synovial membrane* lines the deep surface of the fibrous layer and non-articulating aspects of the bones. The synovial membrane is an inferior prolongation of the synovial membrane of the elbow joint.

LIGAMENTS OF PROXIMAL RADIO-ULNAR JOINT

The strong *anular ligament*, attached to the ulna anterior and posterior to its radial notch, surrounds the articulating bony surfaces and forms a collar that, with the radial notch, creates a ring that completely encircles the head of the radius (Figs. 6.102, 6.103, and 6.104). The deep surface of the anular ligament is lined with synovial membrane, which continues distally as a **sacciform recess of the proximal radio-ulnar joint** on the neck of the radius. This arrangement allows the radius to rotate within the anular ligament without binding, stretching, or tearing the synovial membrane.

MOVEMENTS OF PROXIMAL RADIO-ULNAR JOINT

During pronation and supination of the forearm, the head of the radius rotates within the collar formed by the anular ligament and the radial notch of the ulna. **Supination** turns the palm anteriorly, or superiorly when the forearm is flexed (Figs. 6.103, 6.105, and 6.106). **Pronation** turns the palm posteriorly, or inferiorly when the forearm is flexed. The axis for these movements passes proximally through the center of the head of the radius, and distally through the site of attachment of the apex of the articular disc to the head (styloid process) of the ulna.

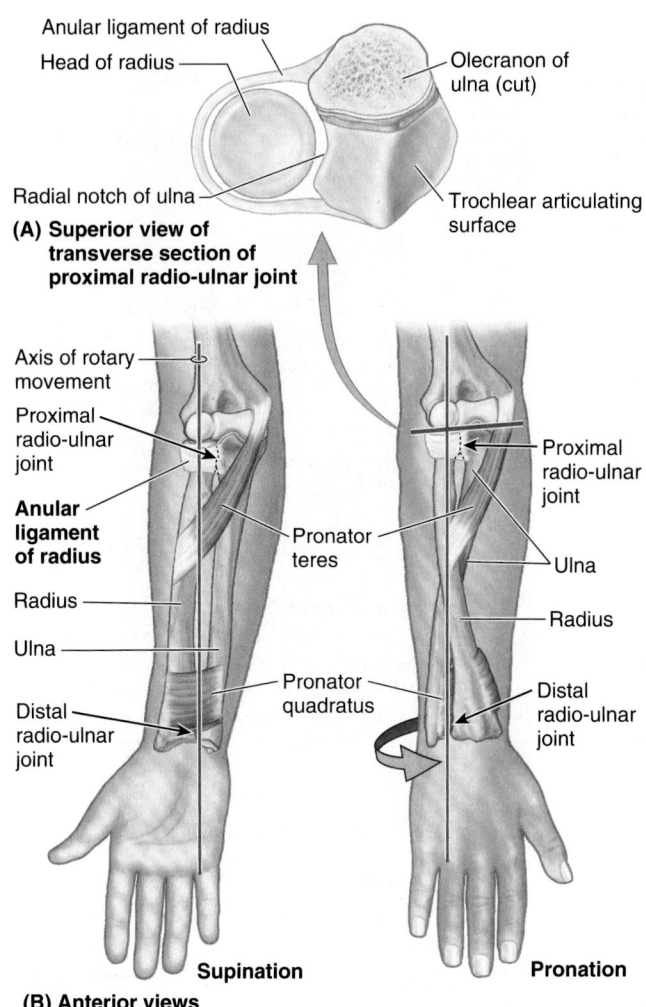

Anular ligament of radius

Head of radius

Olecranon of ulna (cut)

Radial notch of ulna

(A) Superior view of transverse section of proximal radio-ulnar joint

Trochlear articulating surface

Axis of rotary movement

Proximal radio-ulnar joint

Anular ligament of radius

Radius

Ulna

Distal radio-ulnar joint

Pronator teres

Pronator quadratus

Proximal radio-ulnar joint

Ulna

Radius

Distal radio-ulnar joint

Supination   **Pronation**

(B) Anterior views

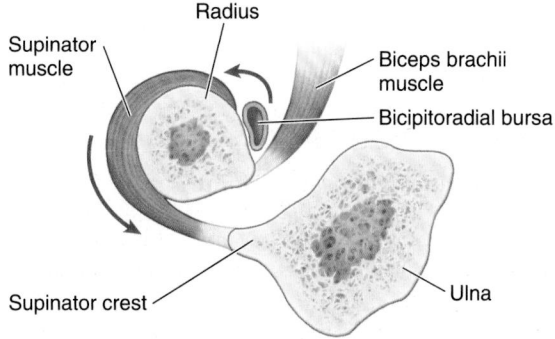

Radius

Supinator muscle

Biceps brachii muscle

Bicipitoradial bursa

Supinator crest

Ulna

(C) Transverse section (inferior view)

FIGURE 6.103. Supination and pronation of forearm. A. The head of the radius rotates in the "socket" formed by the anular ligament and radial notch of the ulna. **B.** Supination is the movement of the forearm that rotates the radius laterally around its longitudinal axis, so that the dorsum of the hand faces posteriorly and the palm faces anteriorly. Pronation is the movement of the forearm, produced by pronators teres and quadratus, that rotates the radius medially around its longitudinal axis, so that the palm of the hand faces posteriorly and its dorsum faces anteriorly (Figs. 6.105 and 6.106). **C.** The actions of the biceps brachii and supinator in producing supination from the pronated position at the radio-ulnar joints.

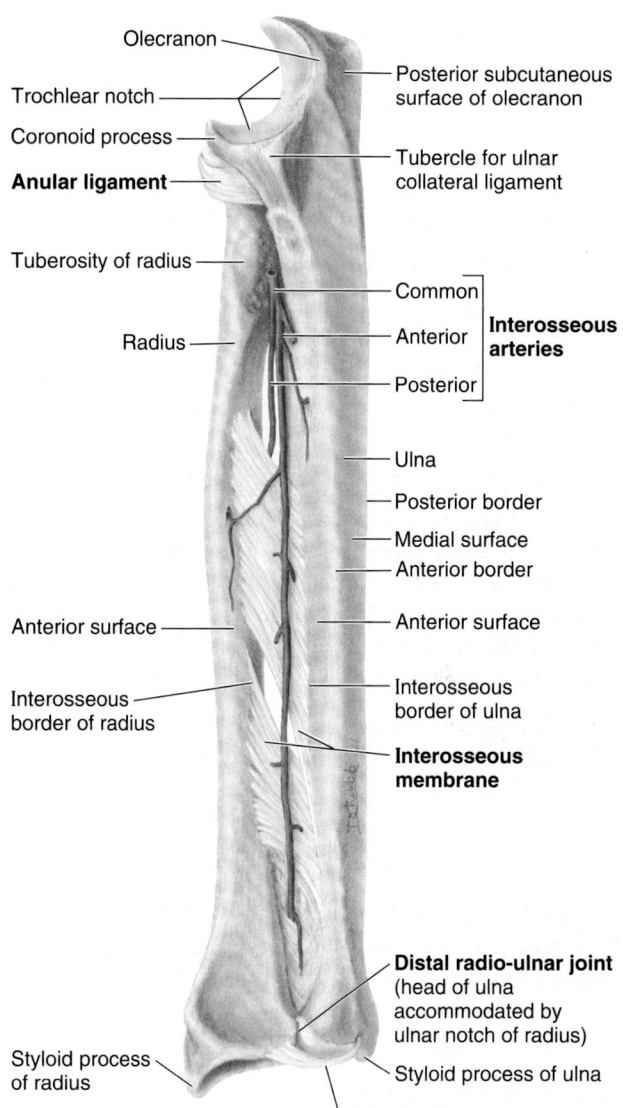

Olecranon

Trochlear notch

Coronoid process

Anular ligament

Tuberosity of radius

Radius

Posterior subcutaneous surface of olecranon

Tubercle for ulnar collateral ligament

Common
Anterior   **Interosseous arteries**
Posterior

Ulna

Posterior border

Medial surface

Anterior border

Anterior surface

Anterior surface

Interosseous border of radius

Interosseous border of ulna

Interosseous membrane

Distal radio-ulnar joint (head of ulna accommodated by ulnar notch of radius)

Styloid process of radius

Styloid process of ulna

Articular disc

Medial view with radius in "resting (midprone) position" (midway between pronation and supination) so that the palm is directed toward the body

FIGURE 6.104. Radio-ulnar ligaments and interosseous arteries. The ligament of the proximal radio-ulnar joint is the anular ligament. The ligament of the distal radio-ulnar joint is the articular disc. The interosseous membrane connects the interosseous margins of the radius and ulna, forming the radio-ulnar syndesmosis. The general direction of the fibers of the interosseous membrane is such that a superior thrust to the hand is received by the radius and is transmitted to the ulna.

During pronation and supination, it is the radius that rotates; its head rotates within the cup-shaped collar formed by the anular ligament and the radial notch on the ulna. Distally, the end of the radius rotates around the head of the ulna. Almost always, supination and pronation are accompanied by synergistic movements of the glenohumeral and elbow joints that produce simultaneous movement of the ulna, except when the elbow is flexed.

MUSCLES MOVING PROXIMAL RADIO-ULNAR JOINT

Supination is produced by the supinator (when resistance is absent), and biceps brachii (when power is required because

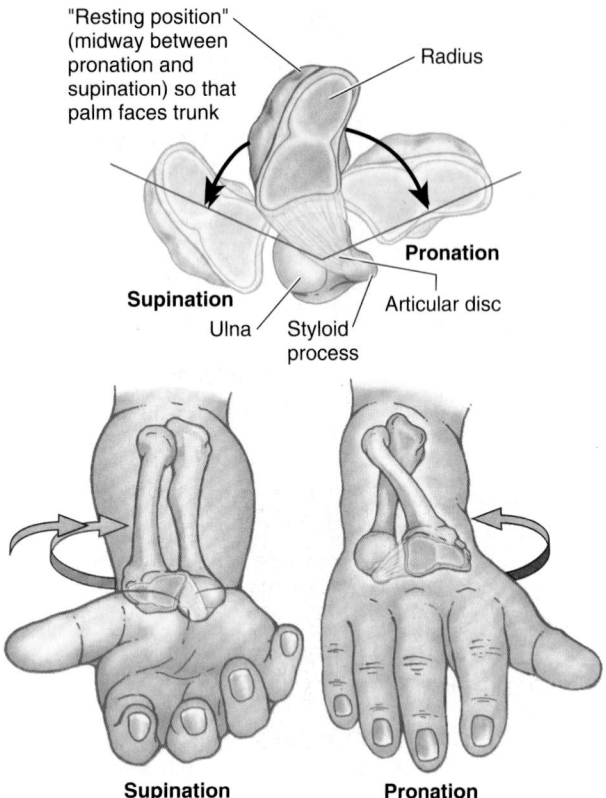

FIGURE 6.105. Movements of distal radio-ulnar joint during supination and pronation of forearm. The distal radio-ulnar joint is the pivot type of synovial joint between the head of the ulna and the ulnar notch of the radius. The inferior end of the radius moves around the relatively fixed end of the ulna during supination and pronation of the hand. The two bones are firmly united distally by the articular disc, referred to clinically as the triangular ligament of the distal radio-ulnar joint. It has a broad attachment to the radius but a narrow attachment to the styloid process of the ulna, which serves as the pivot point for the rotary movement.

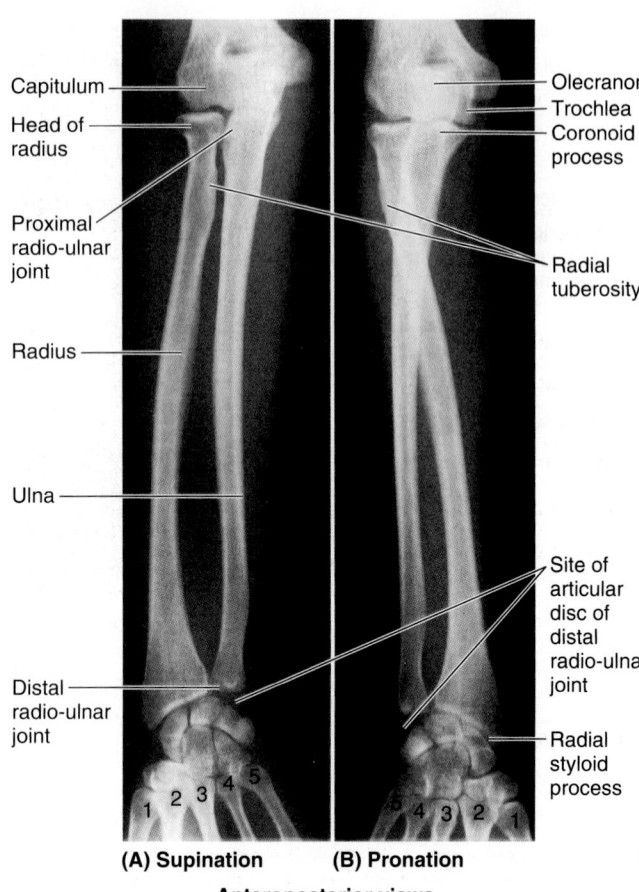

FIGURE 6.106. Radiographs of radio-ulnar joints. A. In the supinated position, the radius and ulna are parallel. **B.** During pronation, the inferior end of the radius moves anteriorly and medially around the inferior end of the ulna, carrying the hand with it; thus in the pronated position, the radius crosses the ulna anteriorly. *1–5*, the metacarpals. (Courtesy of Dr. J. Heslin, Toronto, Ontario, Canada.)

of resistance), with some assistance from the EPL and ECRL (Fig. 6.103C). *Pronation* is produced by the pronator quadratus (primarily), and pronator teres (secondarily) (Fig. 6.103B), with some assistance from the FCR, palmaris longus, and brachioradialis (when the forearm is in the midpronated position).

BLOOD SUPPLY OF PROXIMAL RADIO-ULNAR JOINT

The proximal radio-ulnar joint is supplied by the radial portion of the *peri-articular arterial anastomosis of the elbow joint* (*radial* and *middle collateral arteries* anastomosing with the *radial* and *recurrent interosseous arteries*, respectively) (Fig. 6.67; Table 6.12).

INNERVATION OF PROXIMAL RADIO-ULNAR JOINT

The proximal radio-ulnar joint is supplied mainly by the *musculocutaneous, median, and radial nerves*. Pronation is essentially a function of the median nerve, whereas supination is a function of the musculocutaneous and radial nerves.

Distal Radio-Ulnar Joint

The **distal** (**inferior**) **radio-ulnar joint** is a pivot type of synovial joint (Fig. 6.104). The radius moves around the relatively fixed distal end of the ulna.

ARTICULATION OF DISTAL RADIO-ULNAR JOINT

The rounded head of the ulna articulates with the ulnar notch on the medial side of the distal end of the radius. A fibrocartilaginous, triangular **articular disc of the distal radio-ulnar joint** (sometimes referred to by clinicians as the "triangular ligament") binds the ends of the ulna and radius together, and is the main uniting structure of the joint (Figs. 6.104, 6.105 and 6.107B). The base of the articular disc is attached to the medial edge of the ulnar notch of the radius, and its apex is attached to the lateral side of the base of the styloid process of the ulna. The proximal surface of this disc articulates with the distal aspect of the head of the ulna. Hence, the joint cavity is L-shaped in a coronal section; the vertical bar of the *L* is between the radius and the ulna, and

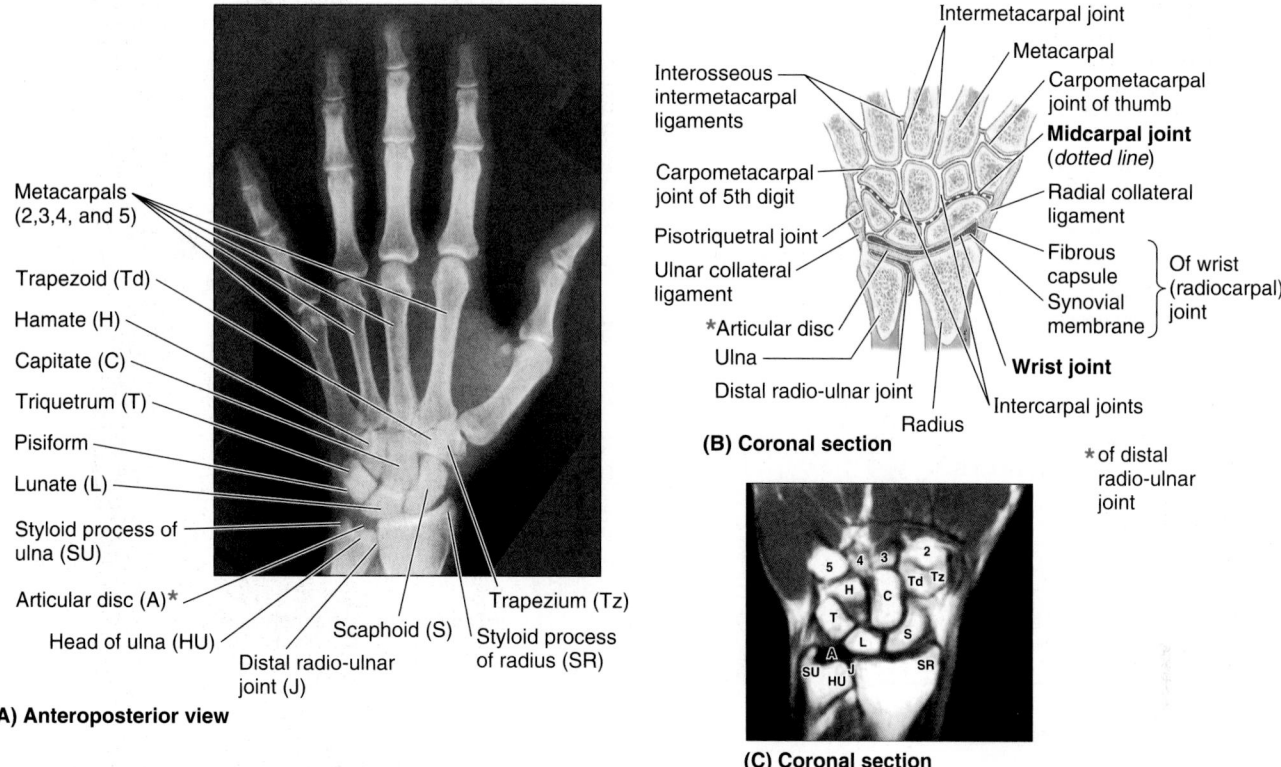

Metacarpals (2,3,4, and 5)
Trapezoid (Td)
Hamate (H)
Capitate (C)
Triquetrum (T)
Pisiform
Lunate (L)
Styloid process of ulna (SU)
Articular disc (A)*
Head of ulna (HU)
Distal radio-ulnar joint (J)
Scaphoid (S)
Trapezium (Tz)
Styloid process of radius (SR)

(A) Anteroposterior view

Intermetacarpal joint
Metacarpal
Carpometacarpal joint of thumb
Midcarpal joint (*dotted line*)
Radial collateral ligament
Fibrous capsule
Synovial membrane } Of wrist (radiocarpal) joint
Wrist joint
Intercarpal joints
Interosseous intermetacarpal ligaments
Carpometacarpal joint of 5th digit
Pisotriquetral joint
Ulnar collateral ligament
*Articular disc
Ulna
Distal radio-ulnar joint
Radius
*of distal radio-ulnar joint

(B) Coronal section

(C) Coronal section

FIGURE 6.107. Bones and joints of wrist and hand. A. In radiographs of the wrist and hand, the "joint space" at the distal end of the ulna appears wide because of the radiolucent articular disc. (Courtesy of Dr. E. L. Lansdown, Professor of Medical Imaging, University of Toronto, Toronto, Ontario, Canada.) **B.** This coronal section of the right hand demonstrates the distal radio-ulnar, wrist, intercarpal, carpometacarpal, and intermetacarpal joints. Although they appear to be continuous when viewed radiographically in parts **A** and **C**, the articular cavities of the distal radio-ulnar and wrist joints are separated by the articular disc of the distal radio-ulnar joint. **C.** Coronal MRI of the wrist. The structures are identified in part **A.** (Courtesy of Dr. W. Kucharczyk, Professor and Neuroradiologist Senior Scientist, Department of Medical Resonance Imaging, University Health Network Toronto, Ontario, Canada.)

the horizontal bar is between the ulna and the articular disc (Figs. 6.107B & C and 6.108A). The articular disc separates the cavity of the distal radio-ulnar joint from the cavity of the wrist joint.

JOINT CAPSULE OF DISTAL RADIO-ULNAR JOINT

The *fibrous layer of the joint capsule* encloses the distal radio-ulnar joint but is deficient superiorly. The *synovial membrane* extends superiorly between the radius and the ulna to form the **sacciform recess of the distal radio-ulnar joint** (Fig. 6.108A). This redundancy of the synovial capsule accommodates the twisting of the capsule that occurs when the distal end of the radius travels around the relatively fixed distal end of the ulna during pronation of the forearm.

LIGAMENTS OF DISTAL RADIO-ULNAR JOINT

Anterior and posterior ligaments strengthen the fibrous layer of the joint capsule of the distal radio-ulnar joint. These relatively weak transverse bands extend from the radius to the ulna across the anterior and posterior surfaces of the joint.

MOVEMENTS OF DISTAL RADIO-ULNAR JOINT

During pronation of the forearm and hand, the distal end of the radius moves (rotates) anteriorly and medially, crossing over the ulna anteriorly (Figs. 6.103, 6.105, and 6.106). During supination, the radius uncrosses from the ulna, its distal end moving (rotating) laterally and posteriorly so the bones become parallel.

MUSCLES MOVING DISTAL RADIO-ULNAR JOINT

The muscles producing movements of the distal radio-ulnar joint are discussed with the proximal radio-ulnar joint (p. 806).

BLOOD SUPPLY OF DISTAL RADIO-ULNAR JOINT

The anterior and posterior *interosseous arteries* supply the distal radio-ulnar joint (Fig. 6.104).

INNERVATION OF DISTAL RADIO-ULNAR JOINT

The anterior and posterior *interosseous nerves* supply the distal radio-ulnar joint.

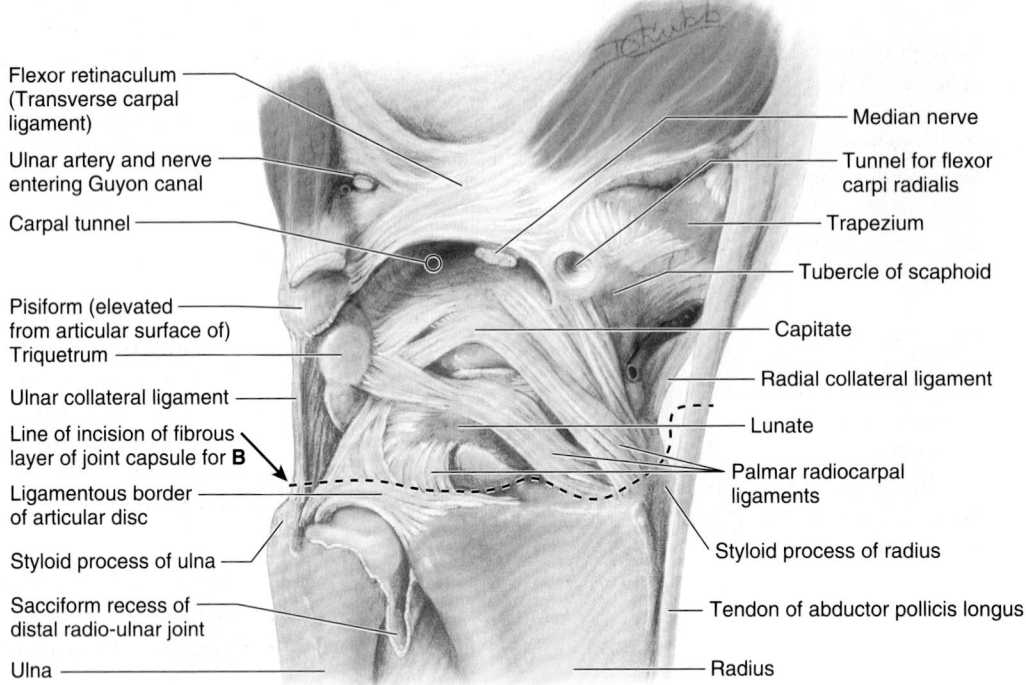

Flexor retinaculum
(Transverse carpal
ligament)

Median nerve

Ulnar artery and nerve
entering Guyon canal

Tunnel for flexor
carpi radialis

Carpal tunnel

Trapezium

Tubercle of scaphoid

Pisiform (elevated
from articular surface of)
Triquetrum

Capitate

Ulnar collateral ligament

Radial collateral ligament

Line of incision of fibrous
layer of joint capsule for **B**

Lunate

Palmar radiocarpal
ligaments

Ligamentous border
of articular disc

Styloid process of ulna

Styloid process of radius

Sacciform recess of
distal radio-ulnar joint

Tendon of abductor pollicis longus

Ulna

Radius

(A) Anterior view

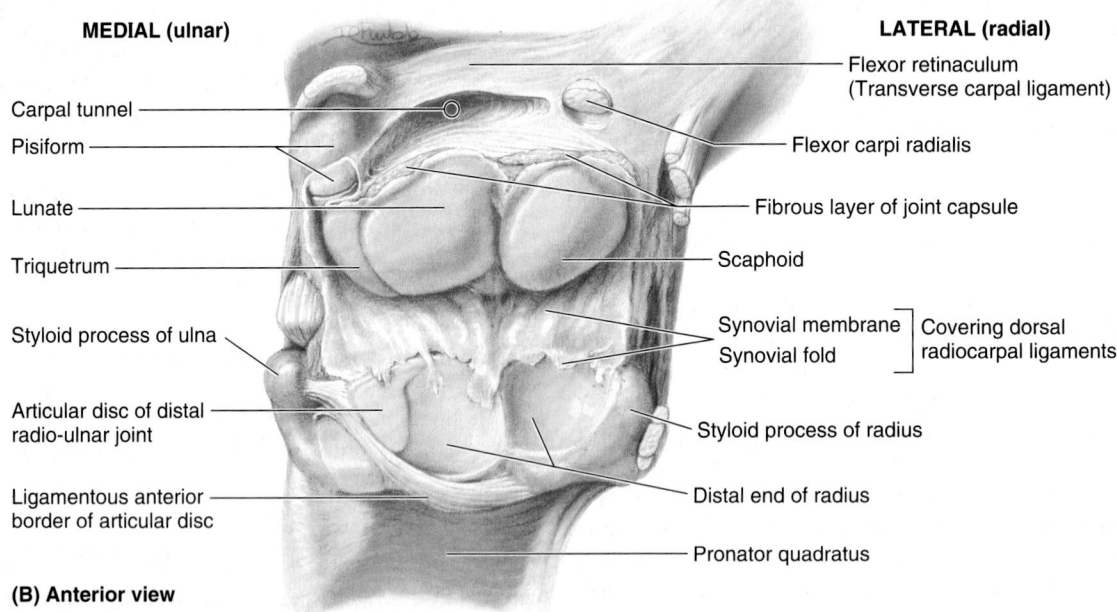

MEDIAL (ulnar)

LATERAL (radial)

Flexor retinaculum
(Transverse carpal ligament)

Carpal tunnel

Flexor carpi radialis

Pisiform

Lunate

Fibrous layer of joint capsule

Triquetrum

Scaphoid

Styloid process of ulna

Synovial membrane   Covering dorsal
Synovial fold   radiocarpal ligaments

Articular disc of distal
radio-ulnar joint

Styloid process of radius

Ligamentous anterior
border of articular disc

Distal end of radius

Pronator quadratus

(B) Anterior view

FIGURE 6.108. Dissection of distal radio-ulnar, radiocarpal, and intercarpal joints. A. The ligaments of these joints are shown. The hand is forcibly extended but the joint is intact. Observe the palmar radiocarpal ligaments, passing from the radius to the two rows of carpal bones. These strong ligaments are directed so that the hand follows the radius during supination. **B.** The joint is opened anteriorly, with the dorsal radiocarpal ligaments serving as a hinge. Observe the nearly equal proximal articular surfaces of the scaphoid and lunate and that the lunate articulates with both the radius and the articular disc. Only during adduction of the wrist does the triquetrum articulate with the articular disc of the distal radio-ulnar joint.

Wrist Joint

The **wrist** (**radiocarpal**) **joint** is a condyloid (ellipsoid) type of synovial joint. The position of the joint is indicated approximately by a line joining the styloid processes of the radius and ulna, or by the proximal wrist crease (Figs. 6.89, 6.106–6.108). The *wrist* (*carpus*), the proximal segment of the hand, is a complex of eight carpal bones, articulating proximally with the forearm via the wrist joint and distally with the five metacarpals.

ARTICULATION OF WRIST JOINT

The ulna does not participate in the wrist joint. The *distal end of the radius* and the *articular disc of the distal radio-ulnar joint* articulate with the *proximal row of carpal bones*, except for the pisiform. The latter bone acts primarily as a sesamoid bone, increasing the leverage of the flexor carpi ulnaris (FCU). The pisiform lies in a plane anterior to the other carpal bones, articulating with only the triquetrum.

JOINT CAPSULE OF WRIST JOINT

The *fibrous layer of the joint capsule* surrounds the wrist joint and is attached to the distal ends of the radius and ulna and the proximal row of carpals (scaphoid, lunate, and triquetrum) (Fig 6.108A & B). The *synovial membrane* lines the internal surface of the fibrous layer of the joint capsule and is attached to the margins of the articular surfaces (Fig. 6.108B). Numerous synovial folds are present.

LIGAMENTS OF WRIST JOINT

The fibrous layer of the joint capsule is strengthened by strong dorsal and palmar radiocarpal ligaments. The **palmar radiocarpal ligaments** pass from the radius to the two rows of carpals (Fig. 6.108A). They are strong and directed so that the hand follows the radius during supination of the forearm. The **dorsal radiocarpal ligaments** take the same direction so that the hand follows the radius during pronation of the forearm.

The joint capsule is also strengthened medially by the **ulnar collateral ligament,** which is attached to the ulnar styloid process and triquetrum (Figs. 6.107B and 6.108A). The joint capsule is also strengthened laterally by the **radial collateral ligament,** which is attached to the radial styloid process and scaphoid.

MOVEMENTS OF WRIST JOINT

The movements at the wrist joint may be augmented by additional smaller movements at the intercarpal and midcarpal joints (Fig. 6.109). The movements are *flexion–extension, abduction–adduction* (radial deviation–ulnar deviation), and *circumduction*. The hand can be flexed on the forearm more than it can be extended; these movements are accompanied (actually, are initiated) by similar movements at the midcarpal joint between the proximal and the distal rows of carpal bones. Adduction of the hand is greater than abduction (Fig. 6.109B). Most adduction occurs at the wrist joint. Abduction from the neutral position occurs at the midcarpal joint. Circumduction of the hand consists of successive flexion, adduction, extension, and abduction.

MUSCLES MOVING WRIST JOINT

Movement at the wrist is produced primarily by the "carpi" muscles of the forearm, the tendons of which extend along the four corners of the wrist (comparing a cross-section of the wrist to a rectangle; Fig. 6.109C) to attach to the bases of the metacarpals. The FCU does so via the *pisohamate ligament* (Fig. 6.110A), a continuation of the FCU tendon if the pisiform is considered a sesamoid bone within the tendon.

- *Flexion of the wrist joint* is produced by the FCR and FCU, with assistance from the flexors of the fingers and thumb, the palmaris longus and the APL (Fig. 6.109C).
- *Extension of the wrist joint* is produced by the ECRL, ECRB, and ECU, with assistance from the extensors of the fingers and thumb.
- *Abduction of the wrist joint* is produced by the APL, FCR, ECRL, and ECRB; it is limited to approximately 15° because of the projecting radial styloid process.
- *Adduction of the wrist joint* is produced by simultaneous contraction of the ECU and FCU.

Most activities require a small amount of wrist flexion; however, tight grip (clenching of the fist) requires extension at the wrist. The mildly extended position is also the most stable and the "resting position."

BLOOD SUPPLY OF WRIST JOINT

The arteries supplying the wrist joint are branches of the *dorsal and palmar carpal arches* (Figs. 6.61A and 6.67).

INNERVATION OF WRIST JOINT

The nerves to the wrist joint are derived from the anterior interosseous branch of the *median nerve,* the posterior interosseous branch of the *radial nerve,* and the dorsal and deep branches of the *ulnar nerve* (Figs. 6.69 and 6.85; Tables 6.13 and 6.16).

Intercarpal Joints

The **intercarpal** (**IC**) **joints,** interconnecting the carpal bones, are plane synovial joints (Fig. 6.107), which may be summarized as follows:

- Joints between the carpal bones of the proximal row.
- Joints between the carpal bones of the distal row.

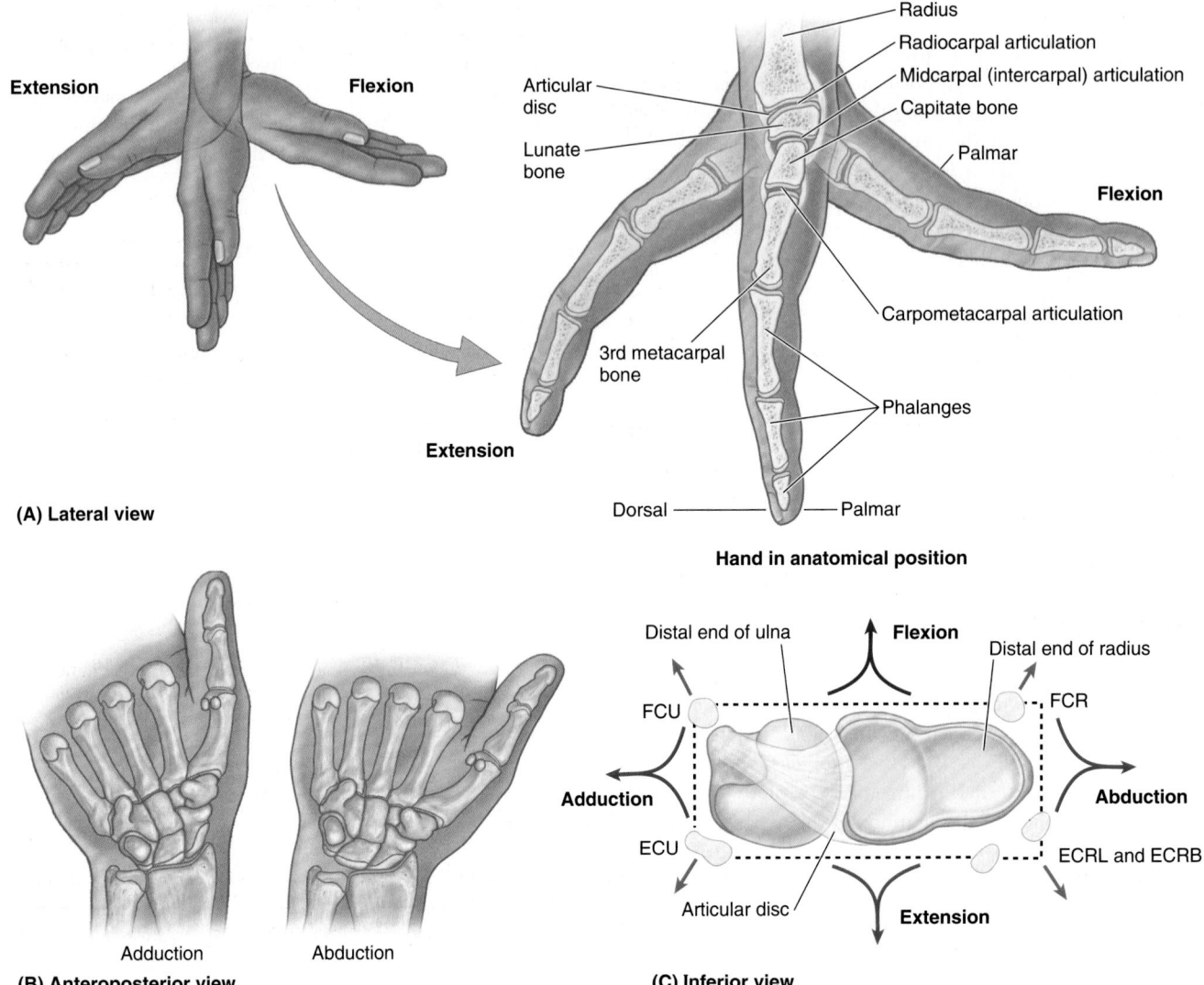

FIGURE 6.109. Movements of wrist. A. In this sagittal section of the wrist and hand during extension and flexion, observe the radiocarpal, midcarpal, and carpometacarpal articulations. Most movement occurs at the radiocarpal joint, with additional movement taking place at the midcarpal joint during full flexion and extension. **B.** Movement at the radiocarpal (RC) and midcarpal (MC) joints during adduction and abduction is demonstrated as seen in posteroanterior radiography. **C.** The *arrows* indicate the direction the hand would move when tendons of the primary ("carpi") muscles acting at the "four corners" of the joint act individually or in unison. *ECRB,* extensor carpi radialis brevis; *ECRL,* extensor carpi radialis longus; *ECU,* extensor carpi ulnaris; *FCR,* flexor carpi radialis; *FCU,* flexor carpi ulnaris.

- The **midcarpal joint,** a complex joint between the proximal and distal rows of carpal bones.
- The **pisotriquetral joint,** formed from the articulation of the pisiform with the palmar surface of the triquetrum.

JOINT CAPSULE OF INTERCARPAL JOINTS

A continuous, common articular cavity is formed by the IC and carpometacarpal joints, with the exception of the **carpometacarpal joint of the thumb,** which is independent. The wrist joint is also independent. The continuity of the articular cavities, or the lack of it, is significant in relation to the spread of infection and to *arthroscopy,* in which a flexible fiberoptic scope is inserted into the articular cavity to view its internal surfaces and features. The *fibrous layer of the joint capsule* surrounds the IC joints, which helps unite the carpals. The *synovial membrane* lines the fibrous layer and is attached to the margins of the articular surfaces of the carpals.

LIGAMENTS OF INTERCARPAL JOINTS

The carpals are united by anterior, posterior, and interosseous ligaments (Figs. 6.108 and 6.110A).

MOVEMENTS OF INTERCARPAL JOINTS

The gliding movements possible between the carpals occur concomitantly with movements at the wrist joint, augment-

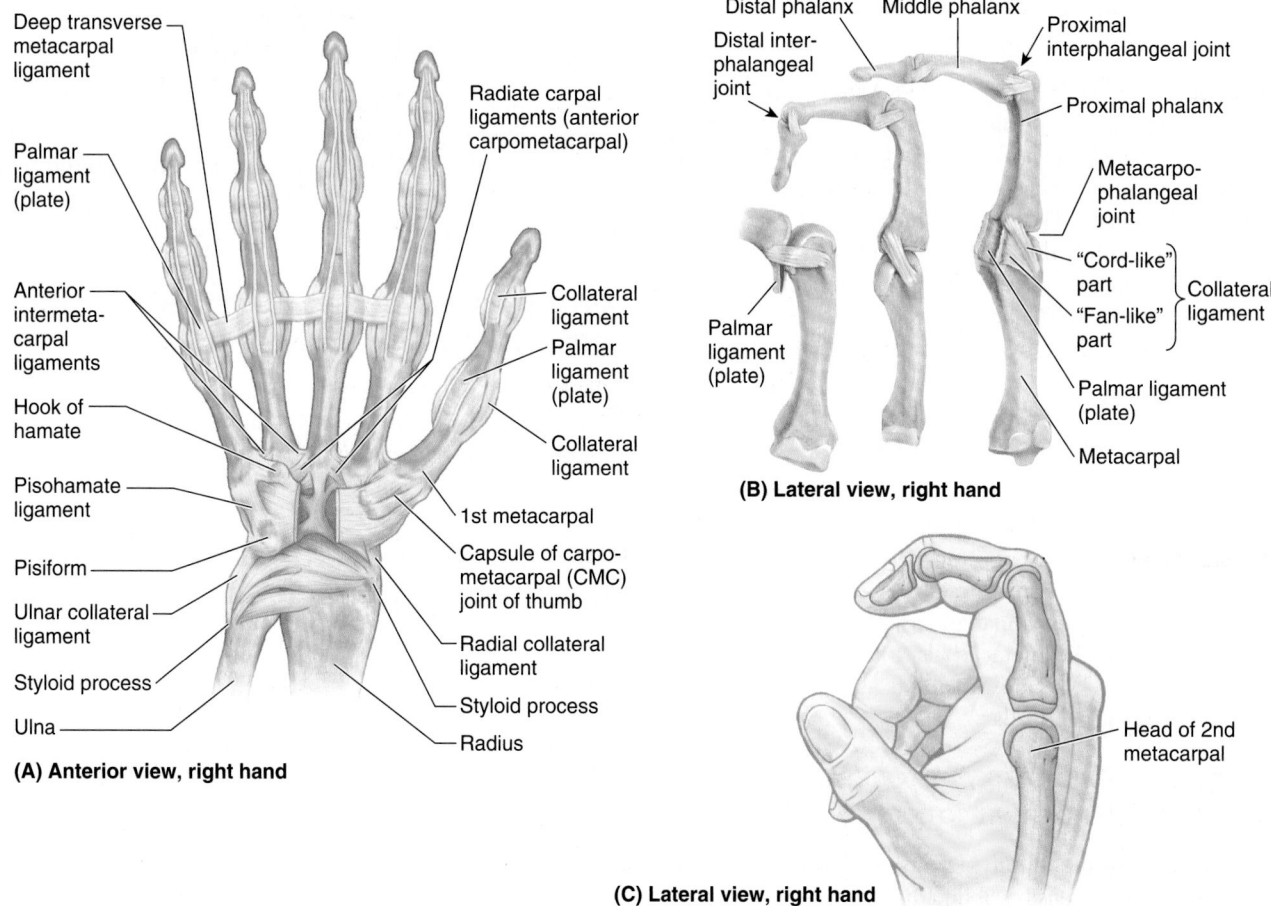

FIGURE 6.110. Joints of hand. A. Palmar ligaments of the radio-ulnar, radiocarpal, intercarpal, carpometacarpal, and interphalangeal joints. **B.** Meta-carpophalangeal and interphalangeal joints. The palmar ligaments (plates) are modifications of the anterior aspect of the MP and IP joint capsules. **C.** The flexed index finger demonstrates its phalanges and the position of the MP and IP joints. The knuckles are formed by the heads of the bones, with the joint plane lying distally.

ing them and increasing the overall range of movement. Flexion and extension of the hand are actually initiated at the *midcarpal joint,* between the proximal and the distal rows of carpals (Figs. 6.107B and 6.109A). Most flexion and adduction occur mainly at the wrist joint, whereas extension and abduction occur primarily at the midcarpal joint. Movements at the other IC joints are small, with the proximal row being more mobile than the distal row.

BLOOD SUPPLY OF INTERCARPAL JOINTS

The arteries supplying the IC joints are derived from the dorsal and palmar carpal arches (Fig. 6.82; Table 6.15).

INNERVATION OF INTERCARPAL JOINTS

The IC joints are supplied by the anterior interosseous branch of the *median nerve* and the dorsal and deep branches of the *ulnar nerve* (Fig. 6.85; Table 6.16).

Carpometacarpal and Intermetacarpal Joints

The **carpometacarpal (CMC)** and **intermetacarpal (IM) joints** are the plane type of synovial joint, except for the **CMC joint of the thumb,** which is a saddle joint (Fig. 6.107).

ARTICULATIONS OF CARPOMETACARPAL AND INTERMETACARPAL JOINTS

The distal surfaces of the carpals of the distal row articulate with the carpal surfaces of the bases of the metacarpals at the CMC joints. The important CMC joint of the thumb is between the trapezium and the base of the 1st metacarpal; it has a separate articular cavity. Like the carpals, adjacent metacarpals articulate with each other; IM joints occur between the radial and ulnar aspects of the bases of the metacarpals.

JOINT CAPSULE OF CARPOMETACARPAL AND INTERMETACARPAL JOINTS

The medial four CMC joints and three IM joints are enclosed by a *common joint capsule* on the palmar and dorsal surfaces. A *common synovial membrane* lines the internal surface of the fibrous layer of the joint capsule, surrounding a common articular cavity. The *fibrous layer of the CMC joint of the thumb* surrounds the joint and is attached to the margins of the articular surfaces. The *synovial membrane* lines the internal surface of the fibrous layer. The looseness of the capsule facilitates free movement of the joint of the thumb.

LIGAMENTS OF CARPOMETACARPAL AND INTERMETACARPAL JOINTS

The bones are united in the region of the joints by **palmar** and **dorsal CMC** and **IM ligaments** (Fig. 6.110A) and by **interosseous IM ligaments** (Fig. 6.107B). In addition, the superficial and deep **transverse metacarpal ligaments** (the former part of the palmar aponeurosis), associated with the distal ends of the metacarpals, play a role in limiting movement at the CMC and IM joints as they limit separation of the metacarpal heads.

MOVEMENTS OF CARPOMETACARPAL AND INTERMETACARPAL JOINTS

The CMC joint of the thumb permits angular movements in any plane (flexion–extension, abduction–adduction, or circumduction) and a restricted amount of axial rotation. Most important, *the movement essential to opposition of the thumb occurs here.* Although the opponens pollicis is the prime mover, all of the hypothenar muscles contribute to opposition.

Almost no movement occurs at the CMC joints of the 2nd and 3rd digits, that of the 4th digit is slightly mobile, and that of the 5th digit is moderately mobile, flexing and rotating slightly during a tight grasp (Fig. 6.73G & H). When the palm of the hand is "cupped" (as during pad-to-pad opposition of thumb and little finger), two thirds of the movement occur at the CMC joint of the thumb, and one third occurs at the CMC and IC joints of the 4th and 5th fingers.

BLOOD SUPPLY OF CARPOMETACARPAL AND INTERMETACARPAL JOINTS

The CMC and IM joints are supplied by *peri-articular arterial anastomoses of the wrist and hand* (dorsal and palmar carpal arches, deep palmar arch, and metacarpal arteries) (Figs. 6.82 and 6.83).

INNERVATION OF CARPOMETACARPAL AND INTERMETACARPAL JOINTS

The CMC and IM joints are supplied by the anterior interosseous branch of the *median nerve,* posterior interosseous branch of the *radial nerve,* and dorsal and deep branches of the *ulnar nerve* (Fig. 6.85).

Metacarpophalangeal and Interphalangeal Joints

The **metacarpophalangeal joints** are the condyloid type of synovial joint that permit movement in two planes: flexion–extension and adduction–abduction. The **interphalangeal joints** are the hinge type of synovial joint that permit flexion–extension only (Fig. 6.110B).

ARTICULATIONS OF METACARPOPHALANGEAL AND INTERPHALANGEAL JOINTS

The heads of the metacarpals articulate with the bases of the proximal phalanges in the MP joints, and the heads of the phalanges articulate with the bases of more distally located phalanges in the IP joints.

JOINT CAPSULES OF METACARPOPHALANGEAL AND INTERPHALANGEAL JOINTS

A *joint capsule* encloses each MC and IP joint with a *synovial membrane* lining a *fibrous layer* that is attached to the margins of each joint.

LIGAMENTS OF METACARPOPHALANGEAL AND INTERPHALANGEAL JOINTS

The fibrous layer of each MC and IP joint capsule is strengthened by two (medial and lateral) **collateral ligaments.** These ligaments have two parts:

- Denser *"cord-like" parts* that pass distally from the heads of the metacarpals and phalanges to the bases of the phalanges (Fig. 6.110A & B).
- Thinner *"fan-like" parts* that pass anteriorly to attach to thick, densely fibrous or fibrocartilaginous plates, the **palmar ligaments** (plates), which form the palmar aspect of the joint capsule.

The fan-like parts of the collateral ligaments cause the palmar ligaments to move like a visor over the underlying metacarpal or phalangeal heads.

The strong cord-like parts of the collateral ligaments of the MP joint, being eccentrically attached to the metacarpal heads, are slack during extension and taut during flexion. As a result, the fingers cannot usually be spread (abducted) when the MP joints are fully flexed. The interphalangeal joints have corresponding ligaments, but the distal ends of the proximal and middle phalanges, being flattened anteroposteriorly and having two small condyles, permit neither adduction or abduction.

The palmar ligaments blend with the fibrous digital sheaths and provide a smooth, longitudinal groove that allows the long flexor ligaments to glide and remain centrally placed as they cross the convexities of the joints. The palmar ligaments of the 2nd–5th MP joints are united by **deep transverse metacarpal ligaments** that hold the heads of the metacarpals together. In addition, the dorsal hood of each extensor apparatus attaches anteriorly to the sides of the palmar plates of the MP joints.

MOVEMENTS OF METACARPOPHALANGEAL AND INTERPHALANGEAL JOINTS

Flexion–extension, abduction–adduction, and circumduction of the 2nd–5th digits occur at the 2nd–5th MP joints. Movement at the MP joint of the thumb is limited to flexion–extension. Only flexion and extension occur at the IP joints.

BLOOD SUPPLY OF METACARPAL AND INTERPHALANGEAL JOINTS

Deep digital arteries that arise from the superficial palmar arches supply the MC and IP joints (Figs. 6.82 and 6.83).

INNERVATION OF METACARPAL AND INTERPHALANGEAL JOINTS

Digital nerves arising from the ulnar and median nerves supply the MC and IP joints (Figs. 6.85A & B).

JOINTS OF UPPER LIMB

Dislocation of Sternoclavicular Joint

 The rarity of *dislocation of the SC joint* attests to its strength, which depends on its ligaments, its disc, and the way forces are generally transmitted along the clavicle. When a blow is received to the acromion of the scapula, or when a force is transmitted to the pectoral girdle during a fall on the outstretched hand, the force of the blow is usually transmitted along the length of the clavicle, i.e., along its long axis. The clavicle may fracture near the junction of its middle and lateral thirds, but it is rare for the SC joint to dislocate. Most dislocations of the SC joint in persons < 25 years of age result from fractures through the epiphysial plate because the epiphysis at the sternal end of the clavicle does not close until 23–25 years of age.

Ankylosis of Sternoclavicular Joint

Movement at the SC joint is critical to movement of the shoulder. When *ankylosis* (stiffening or fixation) of the joint occurs, or is necessary surgically, a section of the center of the clavicle is removed, creating a pseudojoint or "flail" joint to permit scapular movement.

Dislocation of Acromioclavicular Joint

Although its extrinsic coracoclavicular ligament is strong, the AC joint itself is weak and easily injured by a direct blow (Fig. B6.31A-D). In contact sports such as football, soccer, hockey, or the martial arts, it is not uncommon for *dislocation of the AC joint* to result from a hard fall on the shoulder or on the outstretched upper limb. Dislocation of the AC joint can also occur when an ice hockey player is driven into the boards or when a person receives a severe blow to the superolateral part of the back.

An AC joint dislocation, often called a "shoulder separation," is severe when both the AC and coracoclavicular ligaments are torn. When the coracoclavicular ligament tears, the shoulder separates from the clavicle and falls because of the weight of the upper limb. *Rupture of the coracoclavicular ligament* allows the fibrous layer of the joint capsule to be torn so

that the acromion can pass inferior to the acromial end of the clavicle. Dislocation of the AC joint makes the acromion more prominent, and the clavicle may move superior to this process.

Calcific Supraspinatus Tendinitis

 Inflammation and calcification of the subacromial bursa result in pain, tenderness, and limitation of movement of the glenohumeral joint. This condition is also known as *calcific scapulohumeral bursitis*. Deposition of calcium in the supraspinatus tendon is common. This causes increased local pressure that often causes excruciating pain during abduction of the arm; the pain may radiate as far as the hand. The calcium deposit may irritate the overlying subacromial bursa, producing an inflammatory reaction known as *subacromial bursitis* (Fig. B6.32).

As long as the glenohumeral joint is adducted, no pain usually results because in this position the painful lesion is away from the inferior surface of the acromion. In most people, the pain occurs during 50–130° of abduction (*painful arc syndrome*) because during this arc the supraspinatus tendon is in intimate contact with the inferior surface of the acromion. The pain usually develops in males 50 years of age and older after unusual or excessive use of the glenohumeral joint.

Rotator Cuff Injuries

The musculotendinous rotator cuff is commonly injured during repetitive use of the upper limb above the horizontal (e.g., during throwing and racquet sports, swimming, and weightlifting). Recurrent inflammation of the rotator cuff, especially the relatively avascular area of the supraspinatus tendon, is a common cause of shoulder pain and results in tears of the musculotendinous rotator cuff.

Repetitive use of the rotator cuff muscles (e.g., by baseball pitchers) may allow the humeral head and rotator cuff to impinge on the coraco-acromial arch (Fig. 6.95B), producing irritation of the arch and inflammation of the rotator cuff. As a result, *degenerative tendonitis of the rotator cuff* develops. Attrition of the supraspinatus tendon also occurs (Fig. B6.32).

To test for degenerative tendonitis of the rotator cuff, the person is asked to lower the fully abducted limb slowly and

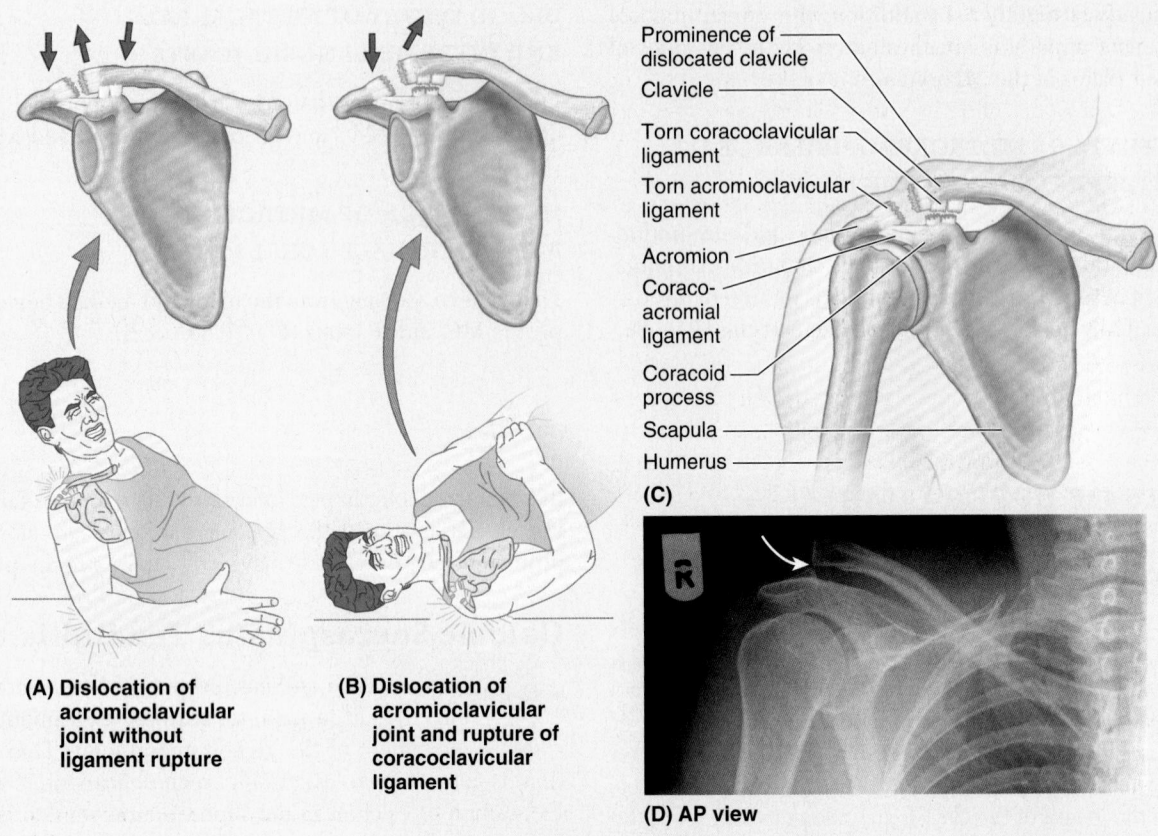

(A) Dislocation of acromioclavicular joint without ligament rupture

(B) Dislocation of acromioclavicular joint and rupture of coracoclavicular ligament

Prominence of dislocated clavicle

Clavicle

Torn coracoclavicular ligament

Torn acromioclavicular ligament

Acromion

Coraco-acromial ligament

Coracoid process

Scapula

Humerus

(C)

(D) AP view

FIGURE B6.31.

smoothly. From approximately 90° abduction, the limb will suddenly drop to the side in an uncontrolled manner if the rotator cuff (especially its supraspinatus part) is diseased and/or torn.

Rotator cuff injuries may also occur during a sudden strain of the muscles, for example, when an older person strains to lift something, such as a window that is stuck. This strain may rupture a previously degenerated musculotendinous rotator cuff. A fall on the shoulder may also tear a previously degenerated rotator cuff. Often the intracapsular part of the tendon of the long head of the biceps brachii becomes frayed (even worn away), leaving it adherent to the intertubercular sulcus. As a result, shoulder stiffness occurs. Because they fuse, the integrity of the fibrous layer of the joint capsule of the glenohumeral joint is usually compromised when the rotator cuff is injured. As a result, the articular cavity communicates with the subacromial bursa. Because the supraspinatus muscle is no longer functional with a complete tear of the rotator cuff, the person cannot initiate abduction of the upper limb. If the arm is passively abducted 15° or more, the person can usually maintain or continue the abduction using the deltoid.

Dislocation of Glenohumeral Joint

Because of its freedom of movement and instability, the glenohumeral joint is commonly dislocated by direct or indirect injury. Because the presence of the coraco-acromial arch and support of the rotator cuff are effective in preventing upward dislocation, most dislocations of the humeral head occur in the downward (inferior)

direction. However, they are described clinically as anterior or (more rarely) posterior dislocations, indicating whether the humeral head has descended anterior or posterior to the infraglenoid tubercle and long head of the triceps. The head ends up lying anterior or posterior to the glenoid cavity.

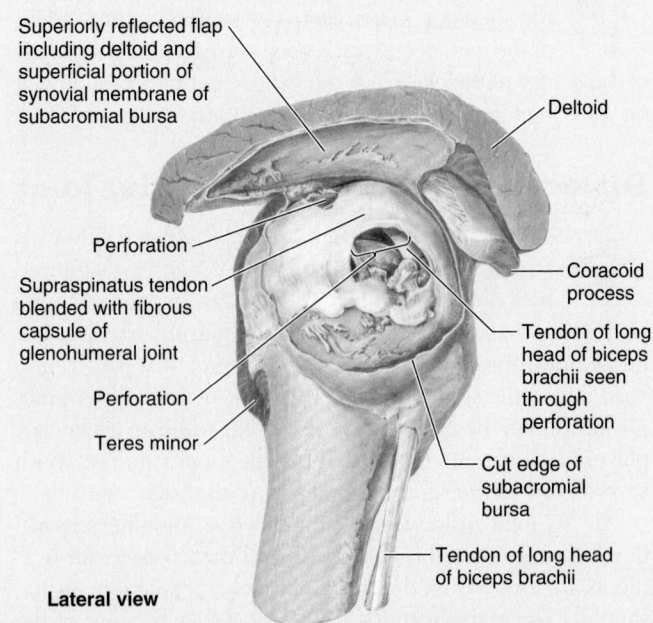

Superiorly reflected flap including deltoid and superficial portion of synovial membrane of subacromial bursa

Deltoid

Perforation

Coracoid process

Supraspinatus tendon blended with fibrous capsule of glenohumeral joint

Tendon of long head of biceps brachii seen through perforation

Perforation

Teres minor

Cut edge of subacromial bursa

Tendon of long head of biceps brachii

Lateral view

FIGURE B6.32. Attrition of the supraspinatus tendon.

Anterior dislocation of the glenohumeral joint occurs most often in young adults, particularly athletes. It is usually caused by excessive extension and lateral rotation of the humerus (Fig. B6.33). The head of the humerus is driven infero-anteriorly, and the fibrous layer of the joint capsule and glenoid labrum may be stripped from the anterior aspect of the glenoid cavity in the process. A hard blow to the humerus when the glenohumeral joint is fully abducted tilts the head of the humerus inferiorly onto the inferior weak part of the joint capsule. This may tear the capsule and dislocate the shoulder so that the humeral head comes to lie inferior to the glenoid cavity and anterior to the infraglenoid tubercle. The strong flexor and adductor muscles of the glenohumeral joint usually subsequently pull the humeral head anterosuperiorly into a subcoracoid position. Unable to use the arm, the person commonly supports it with the other hand.

Inferior dislocation of the glenohumeral joint often occurs after an avulsion fracture of the greater tubercle of the humerus, owing to the absence of the upward and medial pull produced by muscles attaching to the tubercle (see Fig. B6.2A, p. 685).

Axillary Nerve Injury

The axillary nerve may be injured when the gleno-humeral joint dislocates because of its close relation to the inferior part of the joint capsule (Fig. B6.34). The subglenoid displacement of the head of the humerus into the quadrangular space damages the axillary nerve. *Axillary nerve injury* is indicated by paralysis of the deltoid (manifest as an inability to abduct the arm to or above the horizontal level) and loss of sensation in a small area of skin covering the central part of the deltoid (see the blue box "Injury to Axillary Nerve" on p. 710 and Fig. B6.8).

Glenoid Labrum Tears

Tearing of the fibrocartilaginous glenoid labrum commonly occurs in athletes who throw a baseball or football and in those who have shoulder instability

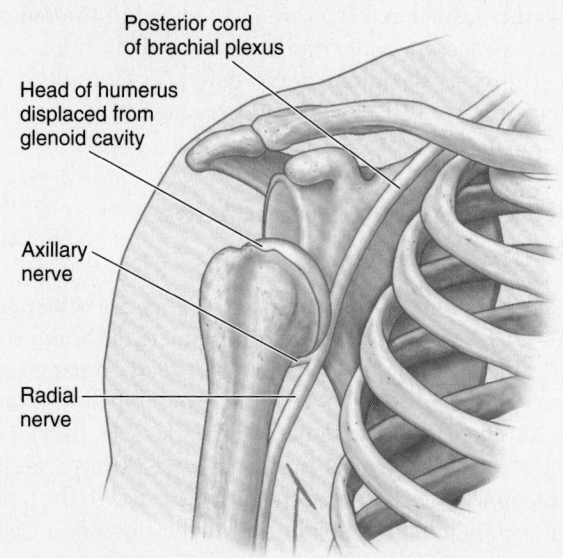

FIGURE B6.33. Dislocation of glenohumeral joint.

Posterior cord of brachial plexus

Head of humerus displaced from glenoid cavity

Axillary nerve

Radial nerve

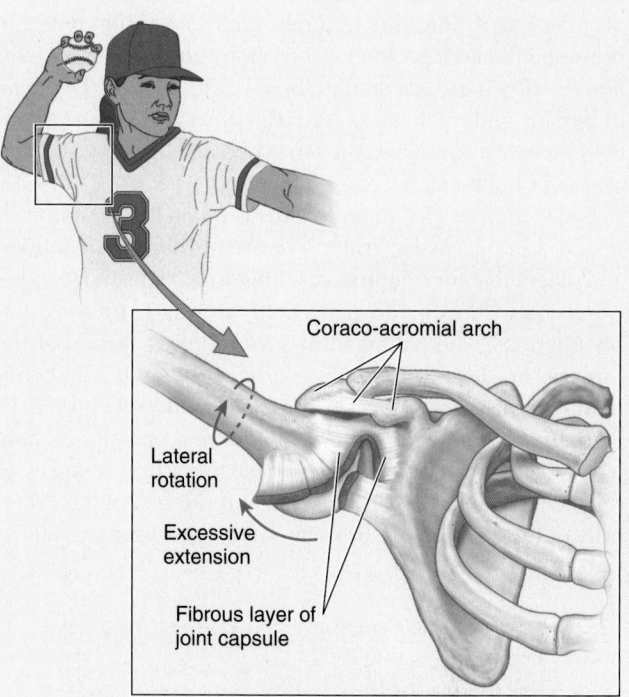

Coraco-acromial arch

Lateral rotation

Excessive extension

Fibrous layer of joint capsule

FIGURE B6.34.

and subluxation (partial dislocation) of the glenohumeral joint. The tear often results from sudden contraction of the biceps or forceful subluxation of the humeral head over the glenoid labrum (Fig. 6.95A). Usually a tear occurs in the anterosuperior part of the labrum. The typical symptom is pain while throwing, especially during the acceleration phase. A sense of popping or snapping may be felt in the glenohumeral joint during abduction and lateral rotation of the arm.

Adhesive Capsulitis of Glenohumeral Joint

Adhesive fibrosis and scarring between the inflamed joint capsule of the glenohumeral joint, rotator cuff, subacromial bursa, and deltoid usually cause *adhesive capsulitis* ("frozen shoulder"), a condition seen in individuals 40–60 years of age. A person with this condition has difficulty abducting the arm and can obtain an apparent abduction of up to 45° by elevating and rotating the scapula. Because of the lack of movement of the glenohumeral joint, strain is placed on the AC joint, which may be painful during other movements (e.g., elevation, or shrugging, of the shoulder). Injuries that may initiate acute capsulitis are glenohumeral dislocations, calcific supraspinatus tendinitis, partial tearing of the rotator cuff, and bicipital tendinitis (Salter, 1999).

Bursitis of Elbow

The subcutaneous olecranon bursa (Figs. 6.97C and 6.101) is exposed to injury during falls on the elbow and infection from abrasions of the skin covering the olecranon. Repeated excessive pressure and friction, as occurs in wrestling, for example, may cause this bursa to become inflamed, producing a friction *subcutaneous olecranon bursitis* (e.g., "student's elbow") (Fig. B6.35). This type of bursitis is also known as "dart thrower's elbow" and "miner's elbow." Occasionally, the bursa becomes infected and the area over the bursa becomes inflamed.

Subtendinous olecranon bursitis is much less common. It results from excessive friction between the triceps tendon and olecranon, for example, resulting from repeated flexion–extension of the forearm, as occurs during certain assembly-line jobs. The pain is most severe during flexion of the forearm because of pressure exerted on the inflamed subtendinous olecranon bursa by the triceps tendon (Fig. 6.101).

Bicipitoradial bursitis (*biceps bursitis*) results in pain when the forearm is pronated because this action compresses the bicipitoradial bursa against the anterior half of the tuberosity of the radius (see "Bursa around elbow joint," p. 804).

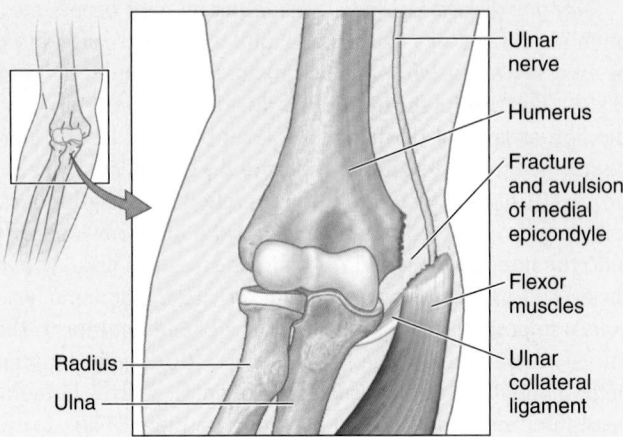

FIGURE B6.36.

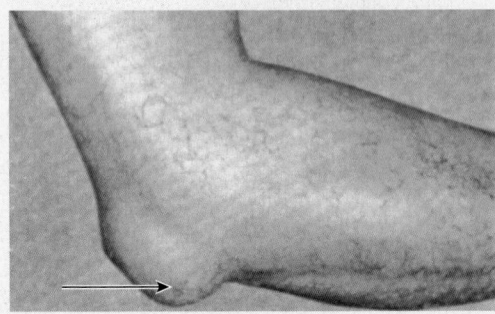

FIGURE B6.35.

Avulsion of Medial Epicondyle

 Avulsion (forced separation) of the medial epicondyle in children can result from a fall that causes severe abduction of the extended elbow, an abnormal movement of this articulation. The resulting traction on the ulnar collateral ligament pulls the medial epicondyle distally (Fig. B6.36). The anatomical basis of the avulsion is that the epiphysis for the medial epicondyle may not fuse with the distal end of the humerus until up to age 20. Usually fusion is complete radiographically at age 14 in females and age 16 in males.

Traction injury of the ulnar nerve is a frequent complication of the abduction type of avulsion of the medial epicondyle of the humerus. The anatomical basis for stretching of the ulnar nerve is that it passes posterior to the medial epicondyle before entering the forearm (see Fig. 6.47A).

Ulnar Collateral Ligament Reconstruction

Rupture, tearing, or stretching of the ulnar collateral ligament (UCL; Fig. 6.107B) are increasingly common injuries related to athletic throwing—primarily baseball pitching (Fig. B6-37A), but also football passing, javelin throwing, and playing water polo. *Reconstruction of the UCL,* known as a "Tommy John procedure" (after the first pitcher to undergo the surgery), involves an autologous transplant of a long tendon from the contralateral forearm or leg (e.g., the palmaris longus or plantaris tendon; Fig. B6-37B). A 10- to 15-cm length of tendon is passed through holes drilled through the medial epicondyle of the humerus and the lateral aspect of the coronoid process of the ulna (Fig. B6-37C–E).

Dislocation of Elbow Joint

Posterior dislocation of the elbow joint may occur when children fall on their hands with their elbows flexed. Dislocations of the elbow may also result from hyperextension or a blow that drives the ulna posterior or posterolateral. The distal end of the humerus is driven through the weak anterior part of the fibrous layer of the joint capsule as the radius and ulna dislocate posteriorly (Fig. B6.38). The ulnar collateral ligament is often torn, and an associated fracture of the head of the radius, coronoid process, or olecranon process of the ulna may occur. Injury to the ulnar nerve may occur, resulting in numbness of the little finger and weakness of flexion and adduction of the wrist.

Subluxation and Dislocation of Radial Head

Preschool children, particularly girls, are vulnerable to transient *subluxation* (incomplete dislocation) of the head of the radius (also called "nursemaid's elbow" and "pulled elbow"). The history of these dislocations is typical. The child is suddenly lifted (jerked) by the upper limb while the forearm is pronated (e.g., lifting a child) (Fig. B6.39A). The child may cry out, refuse to use the limb, and protect their limb by holding it with the elbow flexed and the forearm pronated.

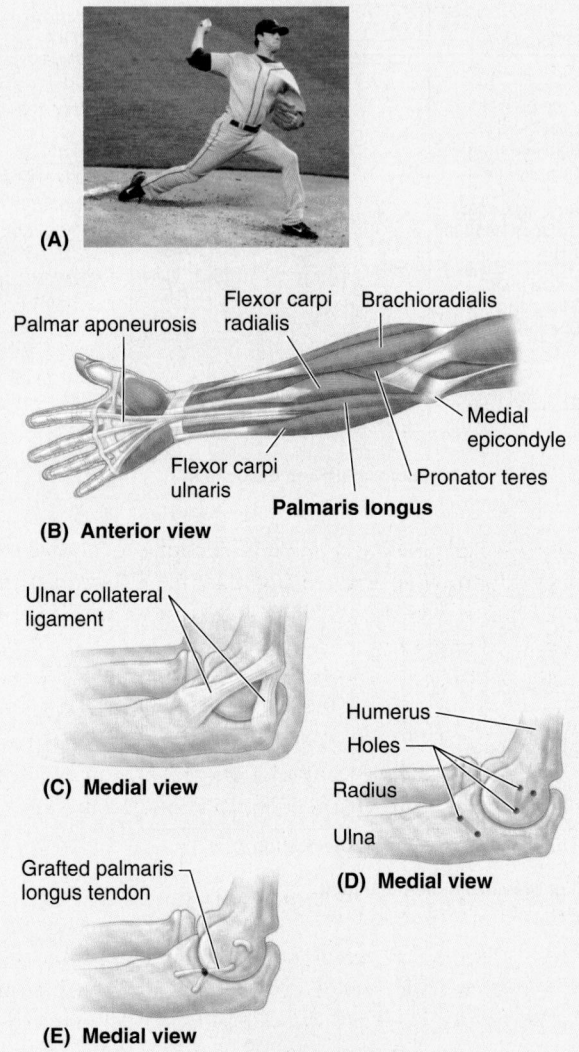

(A)

(B) Anterior view

Palmar aponeurosis — Flexor carpi radialis — Brachioradialis — Medial epicondyle — Pronator teres — **Palmaris longus** — Flexor carpi ulnaris

Ulnar collateral ligament

(C) Medial view

Humerus — Holes — Radius — Ulna

(D) Medial view

Grafted palmaris longus tendon

(E) Medial view

FIGURE B6.37.

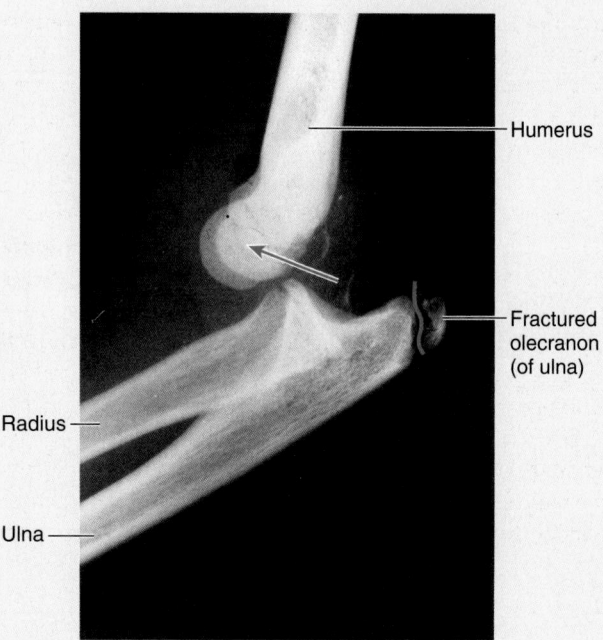

Humerus — Fractured olecranon (of ulna) — Radius — Ulna

FIGURE B6.38. Dislocation of elbow.

Anterior dislocation of the lunate is an uncommon but serious injury that usually results from a fall on the dorsiflexed wrist (Fig. B6.40A). The lunate is pushed out of its place in the floor of the carpal tunnel toward the palmar surface of the wrist. The displaced lunate may compress the median nerve and lead to *carpal tunnel syndrome* (discussed on p. 790). Because of its poor blood supply, *avascular necrosis of the lunate* may occur. In some cases, excision of the lunate may be required. In *degenerative joint disease of the wrist*, surgical fusion of carpals (*arthrodesis*) may be necessary to relieve the severe pain.

Fracture–separation of the distal radial epiphysis is common in children because of frequent falls in which forces are transmitted from the hand to the radius (Fig. B6.40B & C). In a lateral radiograph of a child's wrist, dorsal displacement of the distal radial epiphysis is obvious (Fig. B6.40C). When the epiphysis is placed in its normal position during reduction, the prognosis for normal bone growth is good.

Bull Rider's Thumb

Bull rider's thumb refers to a sprain of the radial collateral ligament, and an avulsion fracture of the lateral part of the proximal phalanx of the thumb. This injury is common in individuals who ride mechanical bulls.

Skier's Thumb

Skier's thumb (historically, game-keeper's thumb) refers to the rupture or chronic laxity of the collateral ligament of the 1st MP joint (Fig. B6.41). The injury results from hyperabduction of the MP joint of the thumb, which occurs when the thumb is held by the ski pole while the rest of the hand hits the ground or enters the snow. In severe injuries, the head of the metacarpal has an avulsion fracture.

The sudden pulling of the upper limb tears the distal attachment of the anular ligament, where it is loosely attached to the neck of the radius. The radial head then moves distally, partially out of the "socket" formed by the anular ligament (Fig. B6.39B). The proximal part of the torn ligament may become trapped between the head of the radius and the capitulum of the humerus.

The source of pain is the pinched anular ligament. Treatment of the subluxation consists of supination of the child's forearm while the elbow is flexed (Salter, 1999). The tear in the anular ligament heals when the limb is placed in a sling for 2 weeks.

Wrist Fractures and Dislocations

Fracture of the distal end of the radius (*Colles fracture*), the most common fracture in people >50 years of age, is discussed in the blue box "Fractures of Radius and Ulna" on p. 685. *Fracture of the scaphoid,* relatively common in young adults, is discussed in the blue box "Fracture of Scaphoid" on p. 686.

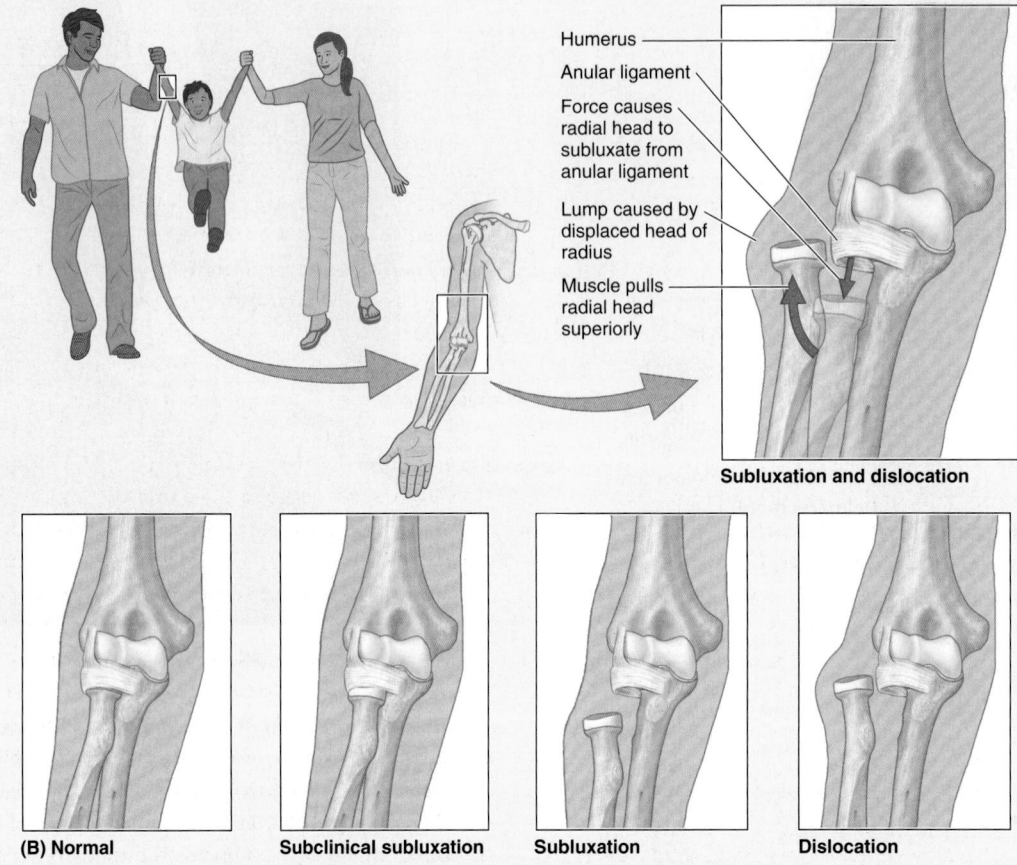

Humerus

Anular ligament

Force causes radial head to subluxate from anular ligament

Lump caused by displaced head of radius

Muscle pulls radial head superiorly

Subluxation and dislocation

(B) Normal   **Subclinical subluxation**   **Subluxation**   **Dislocation**

FIGURE B6.39. Dislocation (subluxation) of proximal radio-ulnar joint.

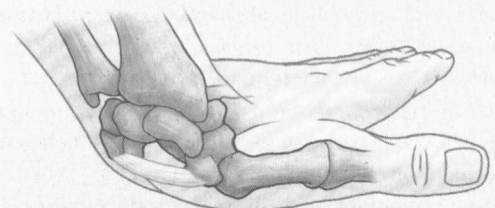

(A) Posterolateral view of pronated limb with wrist extended

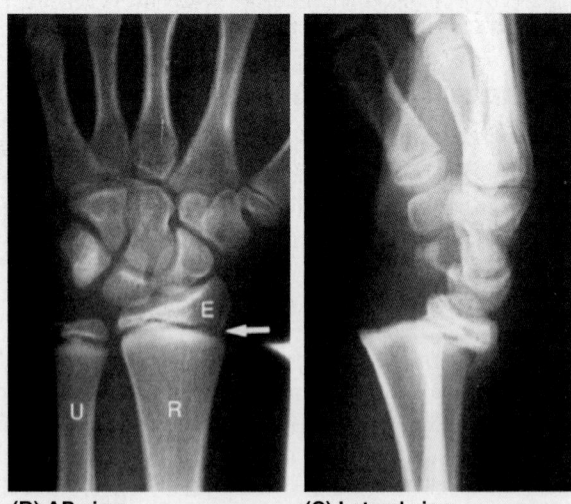

(B) AP view   **(C) Lateral view**

FIGURE B6.40.

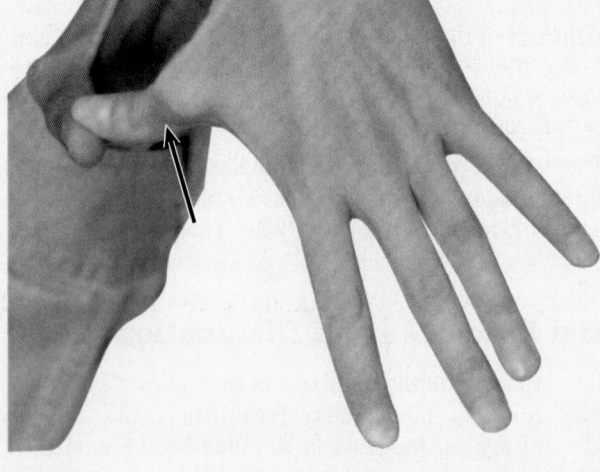

Skier's thumb *(arrow)*

FIGURE B6.41.

The Bottom Line

JOINTS OF UPPER LIMB

Joints of pectoral girdle: The joints of the pectoral girdle are accessory to the glenohumeral joint in positioning the upper limb. ♦ The SC joint links the appendicular skeleton to the axial skeleton. ♦ The SC and AC joints enable the movement at the physiological scapulothoracic joint, where approximately 1° of movement occurs for every 3° of arm movement (scapulohumeral rhythm). In turn, approximately two thirds of the movement at the scapulothoracic joint result from motion at the SC joint, and one third is from motion at the AC joint. ♦ The strength and integrity of the joints of the shoulder complex do not depend on congruity of the articular surfaces. ♦ The integrity of the SC and AC joints results from intrinsic and extrinsic ligaments and the SC articular disc.

Glenohumeral (shoulder) joint: The glenoid cavity of the scapula forms a shallow socket for the relatively large head of the humerus in this ball-and-socket joint; the fossa is deepened only slightly (yet significantly in terms of stability) by the glenoid labrum. ♦ Further, the fibrous capsule is loose to permit the wide range of movement that occurs at this joint. ♦ Integrity of the glenohumeral joint is maintained largely by the tonic and active contraction of the muscles acting across it, particularly the SITS (rotator cuff) muscles. ♦ Degeneration of the rotator cuff is common in advanced age, resulting in pain, limited range and strength of movement, and inflammation of surrounding bursae that develop open communication with the joint cavity.

Elbow joint: Although the elbow joint appears simple because of its primary function as a hinge joint, the fact that it involves the articulation of a single bone proximally with two bones distally, one of which rotates, confers surprising complexity on this compound (three-part) joint. ♦ The hinge movement, the ability to transmit forces, and the high degree of stability of the joint primarily result from the conformation of the articular surfaces of the humero-ulnar joint (i.e., of the trochlear notch of the ulna to the trochlea of the humerus). ♦ The integrity and functions of the humeroradial joint and proximal radio-ulnar joint complex depends primarily on the combined radial collateral and anular ligaments. ♦ The radiohumeral joint is the portion of the elbow joint between the capitulum and the head of the radius.

Radio-ulnar joints: The combined synovial proximal and distal radio-ulnar joints plus the interosseous membrane enable pronation and supination of the forearm. ♦ The anular ligament of the proximal joint, articular disc of the distal joint, and interosseous membrane not only hold the two bones together while permitting the necessary motion between them but (especially the membrane) also transmit forces received from the hand by the radius to the ulna for subsequent transmission to the humerus and pectoral girdle.

Wrist joint: Motion at the wrist moves the entire hand, making a dynamic contribution to a skill or movement, or allowing it to be stabilized in a particular position to maximize the effectiveness of the hand and fingers in manipulating and holding objects. ♦ Complexity as well as flexibility of the wrist results from the number of bones involved. ♦ Extension–flexion, abduction–adduction, and circumduction occur. ♦ Overall, most wrist movement occurs at the wrist (radiocarpal) joint between the radius and articular disc of the distal radio-ulnar joint and the proximal row of carpal bones (primarily the scaphoid and lunate). ♦ However, concomitant movement at the intercarpal (IC) joints (especially the midcarpal IC joint) augments these movements.

Joints of hand: The carpometacarpal (CMC) joints of the four medial fingers, which share a common articular cavity, have limited motion (especially those of the 2nd and 3rd digits), contributing to the stability of the palm as a base from and against which the fingers operate. ♦ Motion occurs at the CMC joints for the 3rd and 4th digits, mostly in association with a tight grip or cupping of the palm, as during opposition. ♦ However, the great mobility of the CMC joint of the thumb, a saddle joint, provides the digit with the major portion of its range of motion and specifically enables opposition. ♦ Therefore, the CMC joint is key to the effectiveness of the human hand. In contrast to the CMC joints, the metacarpophalangeal (MP) joints of the medial four fingers offer considerable freedom of movement (flexion–extension and abduction–adduction), whereas that of the thumb is limited to flexion–extension, as are all interphalangeal joints.

the**Point** **Board-review questions, case studies, and additional resources are available at thePoint.lww.com.**

CHAPTER

7

Head

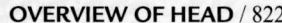

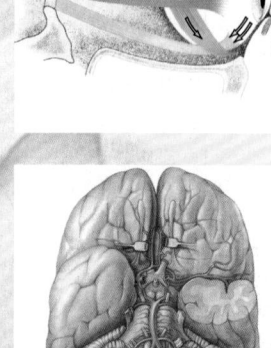

OVERVIEW OF HEAD

The **head** is the superior part of the body that is attached to the trunk by the neck. It is the control and communications center as well as the "loading dock" for the body. It houses the brain; therefore, it is the site of our consciousness: ideas, creativity, imagination, responses, decision making, and memory. It includes special sensory receivers (eyes, ears, mouth, and nose), broadcast devices for voice and expression, and portals for the intake of fuel (food), water, and oxygen and the exhaust of carbon dioxide.

The head consists of the *brain* and its protective coverings (cranial vault and meninges), the *ears,* and the *face.* The face includes openings and passageways, with lubricating glands and valves (seals) to close some of them, the masticatory (chewing) devices, and the orbits that house the visual apparatus. The face also provides our identity as individuals. Disease, malformation, or trauma of structures in the head form the bases of many specialties, including dentistry, maxillofacial surgery, neurology, neuroradiology, neurosurgery, ophthalmology, oral surgery, otology, rhinology, and psychiatry.

CRANIUM

The **cranium** (skull[1]) is the skeleton of the head (Fig. 7.1A). A series of bones form its two parts, the neurocranium and viscerocranium (Fig. 7.1B). The **neurocranium** is the bony case of the brain and its membranous coverings, the cranial meninges. It also contains proximal parts of the cranial nerves and the vasculature of the brain. The neurocranium in adults is formed by a series of eight bones: four singular bones centered on the midline (*frontal, ethmoidal, sphenoidal,* and *occipital*), and two sets of bones occurring as bilateral pairs (*temporal* and *parietal*) (Figs. 7.1A, 7.2A, and 7.3).

The *neurocranium* has a dome-like roof, the **calvaria** (skullcap), and a floor or **cranial base** (basicranium). The bones forming the *calvaria* are primarily flat bones (frontal, parietal, and occipital; see Fig. 7.8A) formed by intramembranous ossification of head mesenchyme from the neural crest. The bones contributing to the *cranial base* are primarily irregular bones with substantial flat portions (sphenoidal and temporal) formed by endochondral ossification of cartilage (*chondrocranium*) or from more than one type of ossification. The *ethmoid bone* is an irregular bone that makes a relatively minor midline contribution

to the neurocranium but is primarily part of the viscerocranium (see Fig. 7.7A). The so-called flat bones and flat portions of the bones forming the neurocranium are actually curved, with convex external and concave internal surfaces.

Most calvarial bones are united by fibrous interlocking **sutures** (Fig. 7.1A & B); however, during childhood, some bones (sphenoid and occipital) are united by hyaline cartilage (**synchondroses**). The spinal cord is continuous with the brain through the *foramen magnum,* a large opening in the cranial base (Fig. 7.1C).

The **viscerocranium** (*facial skeleton*) comprises the facial bones that mainly develop in the mesenchyme of the embryonic pharyngeal arches (Moore et al., 2012). The viscerocranium forms the anterior part of the cranium and consists of the bones surrounding the mouth (upper and lower jaws), nose/nasal cavity, and most of the *orbits* (eye sockets or orbital cavities) (Figs. 7.2 and 7.3).

The *viscerocranium* consists of 15 irregular bones: 3 singular bones centered on or lying in the midline (*mandible, ethmoid,* and *vomer*), and 6 bones occurring as bilateral pairs (*maxillae; inferior nasal conchae;* and *zygomatic, palatine, nasal,* and *lacrimal bones*) (Figs. 7.1A and 7.4A). The maxillae and mandible house the teeth—that is, they provide the sockets and supporting bone for the maxillary and mandibular teeth. The *maxillae* contribute the greatest part of the upper facial skeleton, forming the skeleton of the upper jaw, which is fixed to the cranial base. The *mandible* forms the skeleton of the lower jaw, which is movable because it articulates with the cranial base at the *temporomandibular joints* (Figs. 7.1A and 7.2).

Several bones of the cranium (frontal, temporal, sphenoid, and ethmoid bones) are **pneumatized bones,** which contain **air spaces** (*air cells* or large *sinuses*), presumably to decrease their weight (Fig. 7.5). The total volume of the air spaces in these bones increases with age.

In the *anatomical position,* the cranium is oriented so that the inferior margin of the orbit and the superior margin of the external acoustic opening of the external acoustic meatus of both sides lie in the same horizontal plane (Fig. 7.1A). This standard craniometric reference is the **orbitomeatal plane** (Frankfort horizontal plane).

Facial Aspect of Cranium

Features of the anterior or **facial** (**frontal**) **aspect of the cranium** are the frontal and zygomatic bones, orbits, nasal region, maxillae, and mandible (Figs. 7.2 and 7.3).

The **frontal bone,** specifically its **squamous** (flat) **part,** forms the skeleton of the forehead, articulating inferiorly with the nasal and zygomatic bones. In some adults a *frontal suture* persists; this remnant is called a **metopic suture.** It is in the middle of the **glabella,** the smooth, slightly depressed area between the superciliary arches. The **frontal suture** divides the frontal bones of the fetal cranium (see the blue box "Development of Cranium," p. 839).

---

[1]There is confusion about exactly what the term *skull* means. It may mean the cranium (which includes the mandible), or the part of the cranium excluding the mandible. There has also been confusion because some people have used the term *cranium* for only the neurocranium. The Federative International Committee on Anatomical Terminology (FICAT) has decided to follow the Latin term *cranium* for the skeleton of the head.

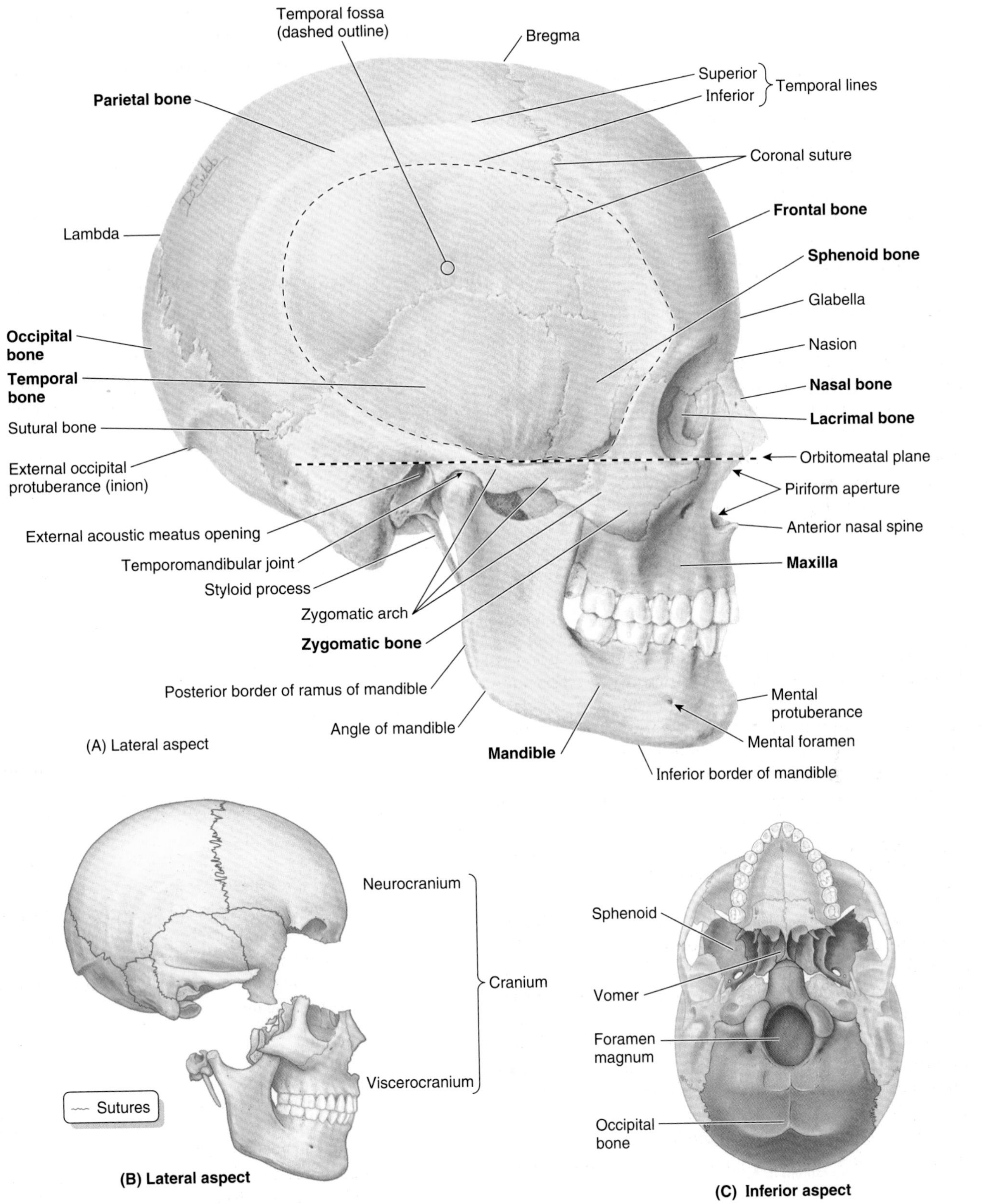

(A) Lateral aspect

(B) Lateral aspect

(C) Inferior aspect

FIGURE 7.1. Adult cranium I. A. In the anatomical position, the inferior margin of the orbit and the superior margin of the external acoustic meatus lie in the same horizontal orbitomeatal (Frankfort horizontal) plane. **B.** The neurocranium and viscerocranium are the two primary functional parts of the cranium. From the lateral aspect, it is apparent that the volume of the neurocranium, housing the brain, is approximately double that of the viscerocranium. **C.** The unpaired sphenoid and occipital bones make substantial contributions to the cranial base. The spinal cord is continuous with the brain through the foramen magnum, the large opening in the basal part of the occipital bone.

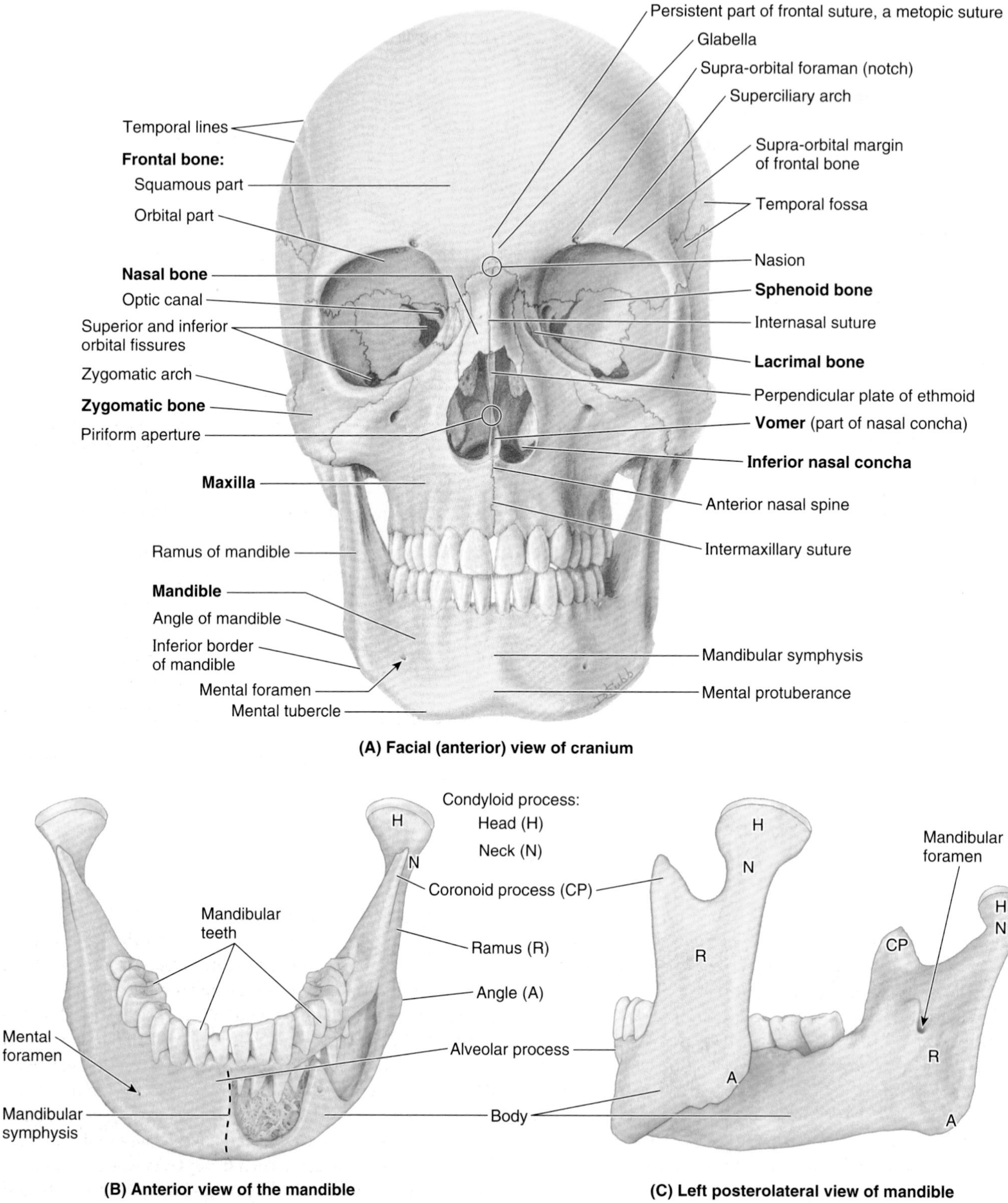

Persistent part of frontal suture, a metopic suture

Glabella

Supra-orbital foraman (notch)

Superciliary arch

Supra-orbital margin
of frontal bone

Temporal fossa

Temporal lines

Frontal bone:

Squamous part

Orbital part

Nasion

Nasal bone

Optic canal

Sphenoid bone

Superior and inferior
orbital fissures

Internasal suture

Lacrimal bone

Zygomatic arch

Zygomatic bone

Perpendicular plate of ethmoid

Piriform aperture

Vomer (part of nasal concha)

Inferior nasal concha

Maxilla

Anterior nasal spine

Ramus of mandible

Intermaxillary suture

Mandible

Angle of mandible

Inferior border
of mandible

Mental foramen

Mental tubercle

Mandibular symphysis

Mental protuberance

(A) Facial (anterior) view of cranium

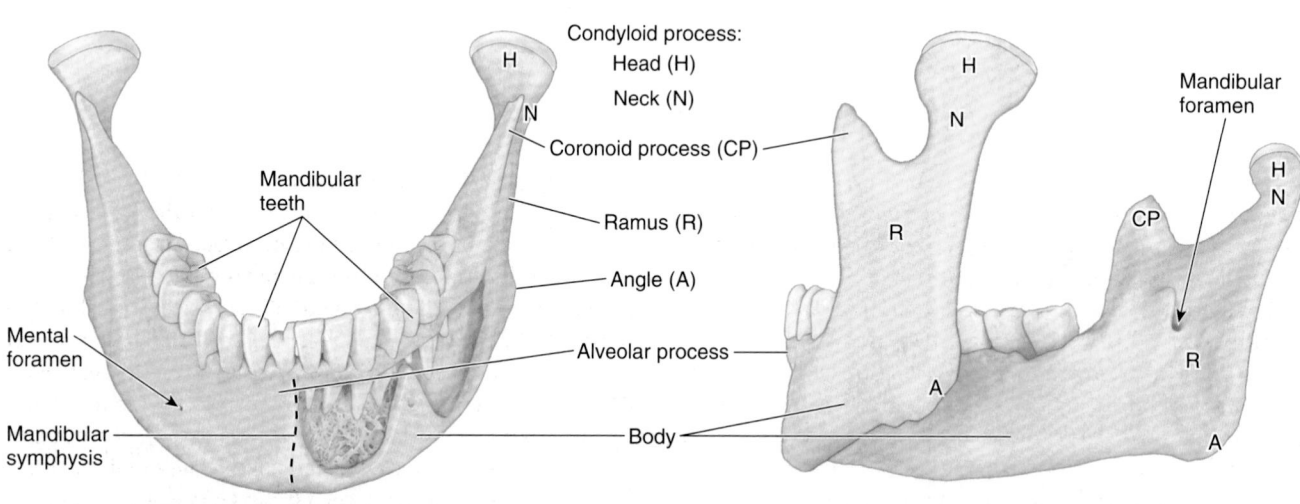

Condyloid process:

Head (H)

Neck (N)

Mandibular
foramen

Mandibular
teeth

Coronoid process (CP)

Ramus (R)

Angle (A)

Mental
foramen

Alveolar process

Mandibular
symphysis

Body

(B) Anterior view of the mandible

(C) Left posterolateral view of mandible

FIGURE 7.2. Adult cranium II. A. The viscerocranium, housing the optical apparatus, nasal cavity, paranasal sinuses, and oral cavity, dominates the facial aspect of the cranium. **B and C.** The mandible is a major component of the viscerocranium, articulating with the remainder of the cranium via the temporomandibular joint. The broad ramus and coronoid process of the mandible provide attachment for powerful muscles capable of generating great force in relationship to biting and chewing (mastication).

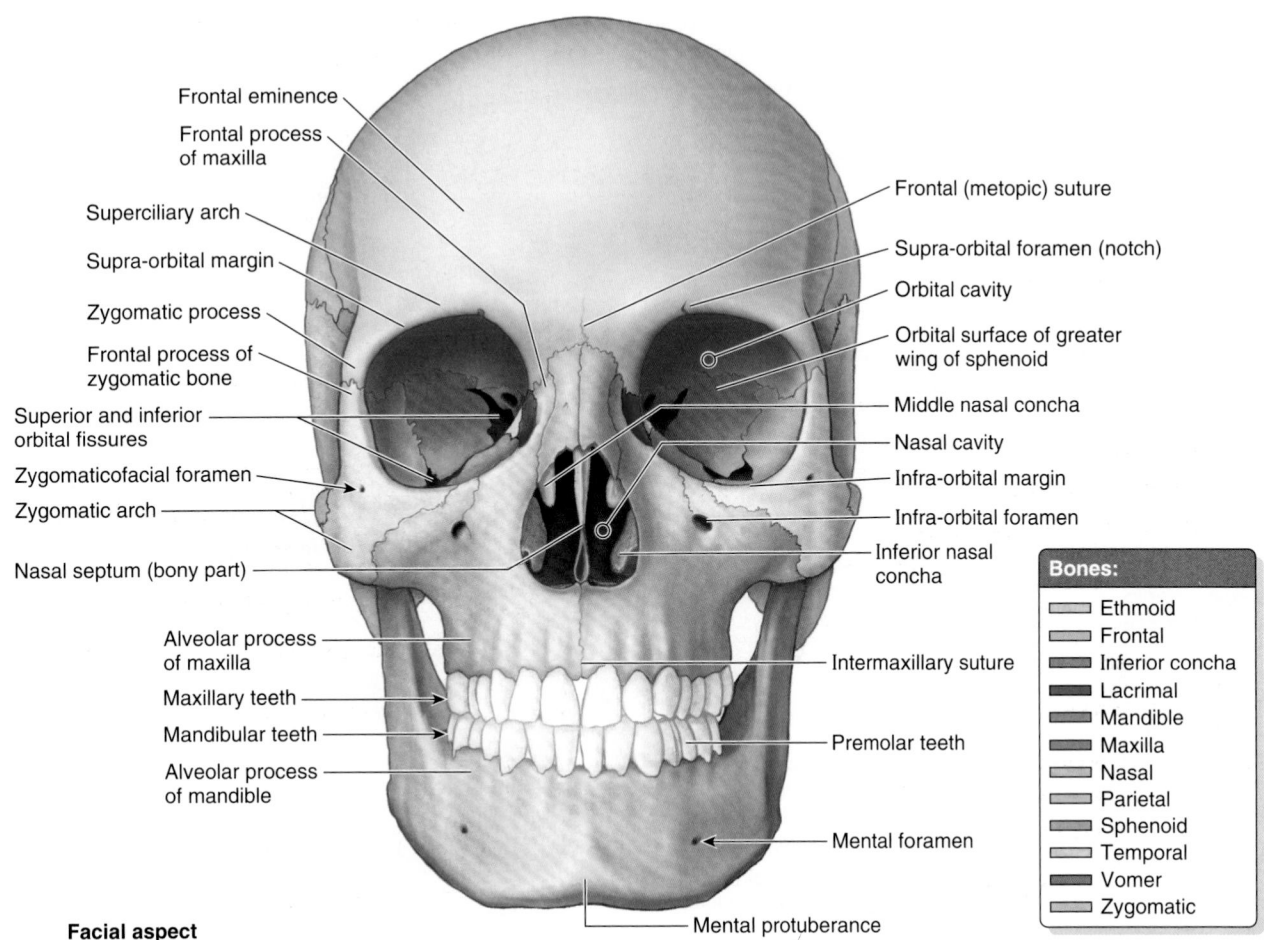

Frontal eminence
Frontal process of maxilla
Superciliary arch
Supra-orbital margin
Zygomatic process
Frontal process of zygomatic bone
Superior and inferior orbital fissures
Zygomaticofacial foramen
Zygomatic arch
Nasal septum (bony part)
Alveolar process of maxilla
Maxillary teeth
Mandibular teeth
Alveolar process of mandible

Frontal (metopic) suture
Supra-orbital foramen (notch)
Orbital cavity
Orbital surface of greater wing of sphenoid
Middle nasal concha
Nasal cavity
Infra-orbital margin
Infra-orbital foramen
Inferior nasal concha
Intermaxillary suture
Premolar teeth
Mental foramen
Mental protuberance

Bones:
- Ethmoid
- Frontal
- Inferior concha
- Lacrimal
- Mandible
- Maxilla
- Nasal
- Parietal
- Sphenoid
- Temporal
- Vomer
- Zygomatic

Facial aspect

FIGURE 7.3. Adult cranium III. The individual bones of the cranium are color coded. The supra-orbital notch, the infra-orbital foramen, and the mental foramen, giving passage to major sensory nerves of the face, are approximately in a vertical line.

The intersection of the frontal and nasal bones is the **nasion** (L. *nasus*, nose), which in most people is related to a distinctly depressed area (bridge of nose) (Figs. 7.1A and 7.2A). The nasion is one of many *craniometric points* that are used radiographically in medicine (or on dry crania in physical anthropology) to make cranial measurements, compare and describe the topography of the cranium, and document abnormal variations (Fig. 7.6; Table 7.1). The frontal bone also articulates with the lacrimal, ethmoid, and sphenoids; a horizontal portion of bone (*orbital part*) forms both the roof of the orbit and part of the floor of the anterior part of the cranial cavity (Fig. 7.3).

The **supra-orbital margin** of the frontal bone, the angular boundary between the squamous and orbital parts, has a **supra-orbital foramen (notch)** in some crania for passage of the supra-orbital nerve and vessels. Just superior to the supra-orbital margin is a ridge, the **superciliary arch,** that extends laterally on each side from the glabella. The prominence of this ridge, deep to the eyebrows, is generally greater in males (Figs. 7.2A and 7.3).

The **zygomatic bones** (cheek bones, malar bones), forming the prominences of the cheeks, lie on the inferolateral sides of the orbits and rest on the maxillae. The anterolateral rims, walls, floor, and much of the infra-orbital margins of the orbits are formed by these quadrilateral bones. A small **zygomaticofacial foramen** pierces the lateral aspect of each bone (Fig. 7.3 and 7.4A). The zygomatic bones articulate with the frontal, sphenoid, and temporal bones and the maxillae.

Inferior to the nasal bones is the pear-shaped **piriform aperture,** the anterior nasal opening in the cranium (Figs. 7.1A and 7.2A). The bony **nasal septum** can be observed through this aperture, dividing the nasal cavity into right and left parts. On the lateral wall of each nasal cavity are curved bony plates, the **nasal conchae** (Figs. 7.2A and 7.3).

The **maxillae** form the upper jaw; their **alveolar processes** include the tooth sockets (alveoli) and constitute the supporting bone for the **maxillary teeth.** The two maxillae are united at the **intermaxillary suture** in the median plane (Fig. 7.2A). The maxillae surround most of the piriform aperture and form the infra-orbital margins medially. They have a

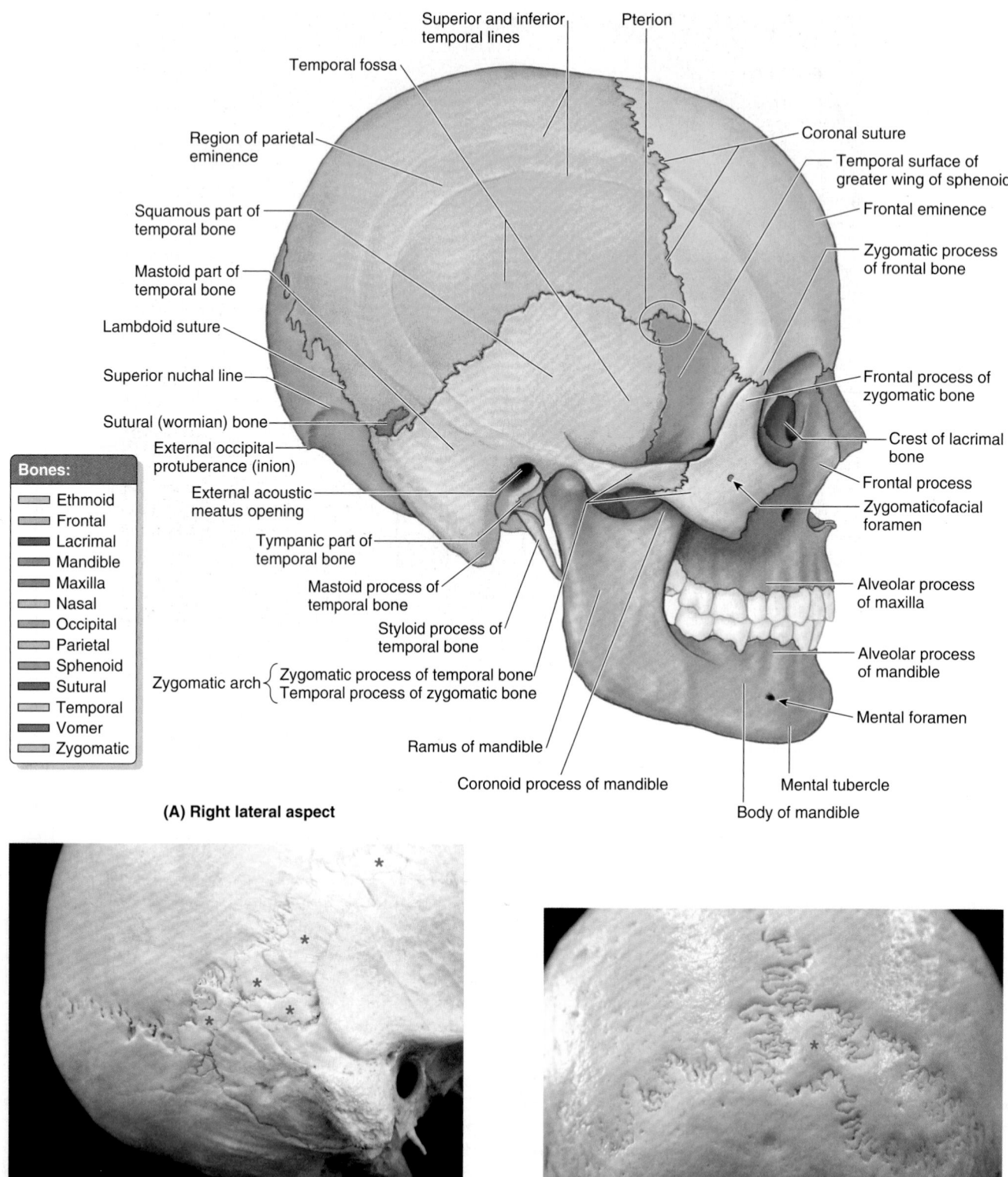

Bones:
- Ethmoid
- Frontal
- Lacrimal
- Mandible
- Maxilla
- Nasal
- Occipital
- Parietal
- Sphenoid
- Sutural
- Temporal
- Vomer
- Zygomatic

Superior and inferior temporal lines

Pterion

Temporal fossa

Region of parietal eminence

Coronal suture

Temporal surface of greater wing of sphenoid

Frontal eminence

Squamous part of temporal bone

Zygomatic process of frontal bone

Mastoid part of temporal bone

Lambdoid suture

Frontal process of zygomatic bone

Superior nuchal line

Crest of lacrimal bone

Sutural (wormian) bone

Frontal process

External occipital protuberance (inion)

Zygomaticofacial foramen

External acoustic meatus opening

Tympanic part of temporal bone

Alveolar process of maxilla

Mastoid process of temporal bone

Styloid process of temporal bone

Alveolar process of mandible

Zygomatic arch { Zygomatic process of temporal bone / Temporal process of zygomatic bone

Mental foramen

Ramus of mandible

Coronoid process of mandible

Mental tubercle

Body of mandible

(A) Right lateral aspect

(B) Right lateral aspect   ⋆ = sutural bones   **(C) Occipital aspect**

FIGURE 7.4. Adult cranium IV. A. The individual bones of the cranium are color coded. Within the temporal fossa, the pterion is a craniometric point at the junction of the greater wing of the sphenoid, the squamous temporal bone, the frontal, and the parietal bones. **B and C.** Sutural bones occurring along the temporoparietal (**B**) and lambdoid (**C**) sutures are shown.

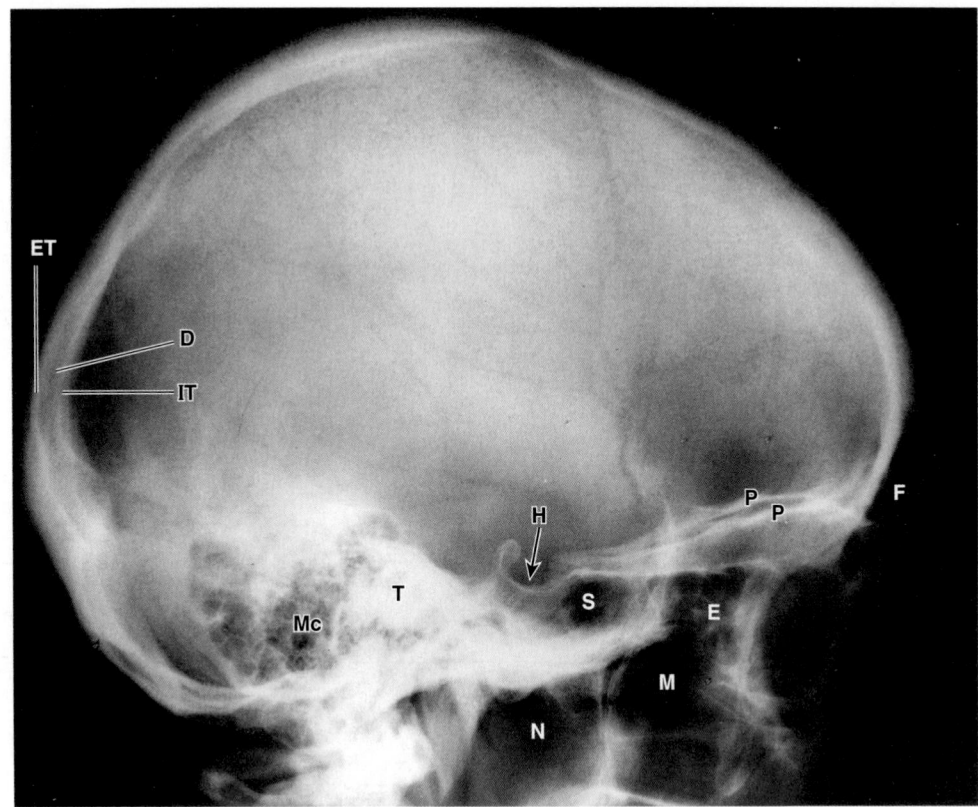

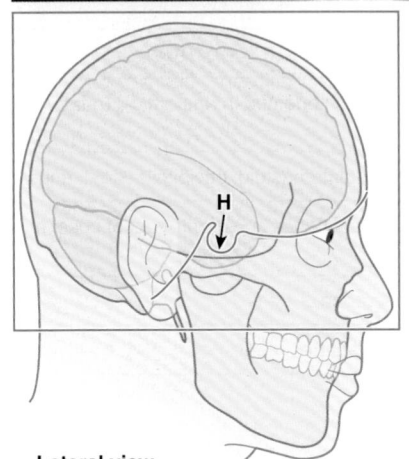

Lateral view

D	Diploë	Mc	Mastoid (air) cells
E	Ethmoid sinus	N	Nasopharynx
ET	External table of bone	P	Orbital part frontal bone
F	Frontal sinus	S	Sphenoidal sinus
H	Hypophysial fossa	T	Petrous part of temporal
IT	Internal table of bone		bone
M	Maxillary sinus		

FIGURE 7.5. Radiograph of cranium. Pneumatized (air-filled) bones contain sinuses or cells that appear as radiolucencies (dark areas) and bear the name of the occupied bone. The right and left orbital parts of the frontal bone are not superimposed; thus the floor of the anterior cranial fossa appears as two lines (*P*). (Courtesy of Dr. E. Becker, Associate Professor of Medical Imaging, University of Toronto, Toronto, Ontario, Canada.)

broad connection with the zygomatic bones laterally and an **infra-orbital foramen** inferior to each orbit for passage of the infra-orbital nerve and vessels (Fig. 7.3).

The **mandible** is a U-shaped bone with an alveolar process that supports the **mandibular teeth.** It consists of a horizontal part, the **body,** and a vertical part, the **ramus** (Fig. 7.2B & C). Inferior to the second premolar teeth are the **mental foramina** for the mental nerves and vessels (Figs. 7.1A, 7.2A & B, and 7.3). The **mental protuberance,** forming the prominence of the chin, is a triangular bony elevation inferior to the **mandibular symphysis** (L. *symphysis menti*), the osseous union where the halves of the infantile mandible fuse (Fig. 7.2A & B).

Lateral Aspect of Cranium

The **lateral aspect of the cranium** is formed by both neurocranium and viscerocranium (Figs. 7.1A & B and 7.4A). The main features of the neurocranial part are the *temporal fossa,* the *external acoustic meatus opening,* and the *mastoid process of the temporal bone.* The main features of the viscerocranial part are the *infratemporal fossa, zygomatic arch,* and lateral aspects of the maxilla and mandible.

The **temporal fossa** is bounded superiorly and posteriorly by the **superior** and **inferior temporal lines,** anteriorly by the frontal and zygomatic bones, and inferiorly by the zygomatic arch (Figs. 7.1A and 7.4A). The superior border

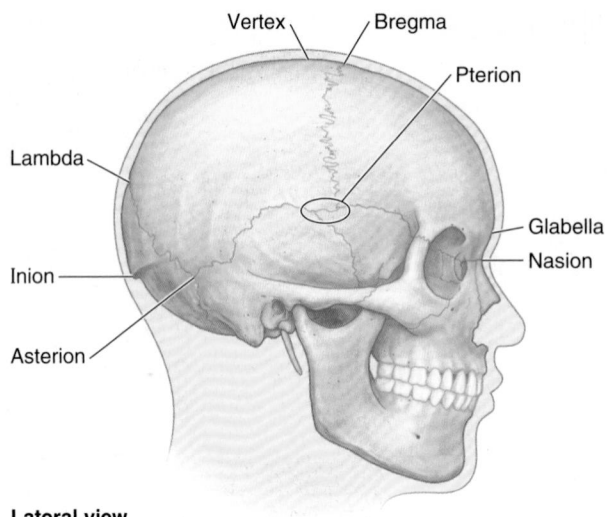

Lateral view

FIGURE 7.6. Craniometric points.

TABLE 7.1. CRANIOMETRIC POINTS OF CRANIUM

Landmark	Shape and Location
Pterion (G. wing)	Junction of greater wing of sphenoid, squamous temporal, frontal, and parietal bones; overlies course of anterior division of middle meningeal artery
Lambda (G. the letter *L*)	Point on calvaria at junction of lambdoid and sagittal sutures
Bregma (G. forepart of head)	Point on calvaria at junction of coronal and sagittal sutures
Vertex (L. whirl, whorl)	Superior point of neurocranium, in middle with cranium oriented in anatomical (orbitomeatal or Frankfort) plane
Asterion (G. *asterios,* starry)	Star shaped; located at junction of three sutures: parietomastoid, occipitomastoid, and lambdoid
Glabella (L. smooth, hairless)	Smooth prominence; most marked in males; on frontal bones superior to root of nose; most anterior projecting part of forehead
Inion (G. back of head)	Most prominent point of external occipital protuberance
Nasion (L. nose)	Point on cranium where frontonasal and internasal sutures meet

of this arch corresponds to the inferior limit of the cerebral hemisphere of the brain. The **zygomatic arch** is formed by the union of the **temporal process of the zygomatic bone** and the **zygomatic process of the temporal bone.**

In the anterior part of the temporal fossa, 3–4 cm superior to the midpoint of the zygomatic arch, is a clinically important area of bone junctions: the **pterion** (G. *pteron,* wing) (Figs. 7.4A and 7.6; Table 7.1). It is usually indicated by an H-shaped formation of sutures that unite the frontal, parietal, sphenoid (greater wing), and temporal bones. Less commonly, the frontal and temporal bones articulate; sometimes all four bones meet at a point.

The **external acoustic meatus opening (pore)** is the entrance to the *external acoustic meatus* (canal), which leads to the tympanic membrane (eardrum) (Fig. 7.4A). The **mastoid process** of the temporal bone is postero-inferior to the external acoustic meatus opening. Anteromedial to the mastoid

process is the *styloid process of the temporal bone,* a slender needle-like, pointed projection. The *infratemporal fossa* is an irregular space inferior and deep to the zygomatic arch, and the mandible and posterior to the maxilla (see Fig. 7.67B).

Occipital Aspect of Cranium

The posterior or **occipital aspect of the cranium** is composed of the **occiput** (L. back of head, the convex posterior protuberance of the **squamous part of the occipital bone**), parts of the parietal bones, and mastoid parts of the temporal bones (Fig. 7.7A).

The **external occipital protuberance,** is usually easily palpable in the median plane; however, occasionally (especially in females) it may be inconspicuous. A craniometric point defined by the tip of the external protuberance is the **inion** (G. nape of neck) (Figs. 7.1A, 7.4A, and 7.6; Table 7.1).

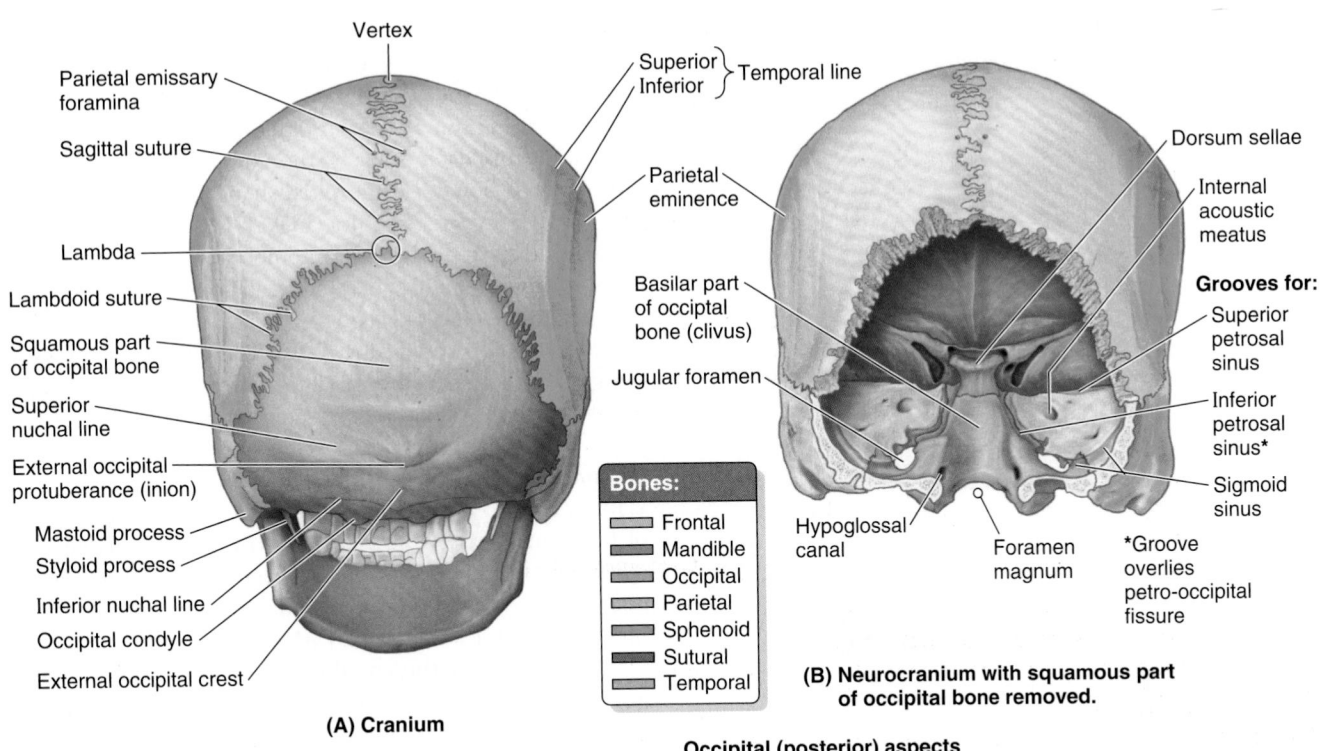

FIGURE 7.7. Adult cranium V: Occipital aspect. A. The posterior aspect of the neurocranium, or occiput, is composed of parts of the parietal bones, the occipital bone, and the mastoid parts of the temporal bones. The sagittal and lambdoid sutures meet at the lambda, which can often be felt as a depression in living persons. **B.** The squamous part of the occipital bone has been removed to expose the anterior part of the posterior cranial fossa.

The **external occipital crest** descends from the protuberance toward the *foramen magnum,* the large opening in the basal part of the occipital bone (Figs. 7.1C, 7.7B, and 7.9).

The **superior nuchal line,** marking the superior limit of the neck, extends laterally from each side of the protuberance; the **inferior nuchal line** is less distinct. In the center of the occiput, **lambda** indicates the junction of the sagittal and the lambdoid sutures (Figs. 7.1A, 7.6, and 7.7A; Table 7.1). Lambda can sometimes be felt as a depression. One or more **sutural bones** (accessory bones) may be located at lambda or near the mastoid process (Fig. 7.4B & C).

Superior Aspect of Cranium

The **superior (vertical) aspect of the cranium,** usually somewhat oval in form, broadens posterolaterally at the **parietal eminences** (Fig. 7.8A). In some people, **frontal eminences** are also visible, giving the calvaria an almost square appearance.

The **coronal suture** separates the frontal and parietal bones (Fig. 7.8A & B), the **sagittal suture** separates the parietal bones, and the **lambdoid suture** separates the parietal and temporal bones from the occipital bone (Fig. 7.8A & C). **Bregma** is the craniometric landmark formed by the intersection of the sagittal and coronal sutures (Figs. 7.6 and 7.8A; Table 7.1). **Vertex,** the most superior point of the calvaria, is near the midpoint of the sagittal suture (Figs. 7.6 and 7.7A).

The **parietal foramen** is a small, inconstant aperture located posteriorly in the parietal bone near the sagittal suture (Fig. 7.8A & C); paired parietal foramina may be present. Most irregular, highly variable foramina that occur in the neurocranium are *emissary foramina* that transmit *emissary veins* connecting scalp veins to the venous sinuses of the dura mater (see "Scalp," p. 843).

External Surface of Cranial Base

The *cranial base* (basicranium) is the inferior portion of the neurocranium (floor of the cranial cavity) and viscerocranium minus the mandible (Fig. 7.9). The **external surface of the cranial base** features the **alveolar arch of the maxillae** (the free border of the alveolar processes surrounding and supporting the maxillary teeth); the palatine processes of the maxillae; and the palatine, sphenoid, vomer, temporal, and occipital bones.

The **hard palate** (bony palate) is formed by the **palatal processes of the maxillae** anteriorly and the **horizontal plates of the palatine bones** posteriorly. The free posterior border of the hard palate projects posteriorly in the median plane as the **posterior nasal spine.** Posterior to the central incisor teeth is the **incisive foramen,** a depression in the midline of the bony palate into which the incisive canals open.

The right and left nasopalatine nerves pass from the nose through a variable number of incisive canals and foramina

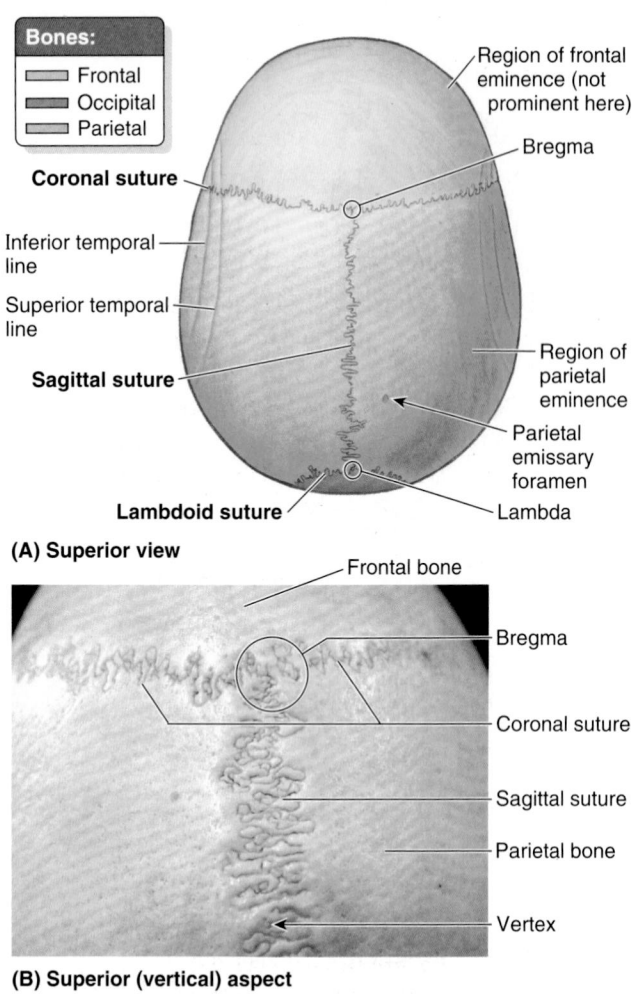

(A) Superior view

Bones:
- Frontal
- Occipital
- Parietal

Region of frontal eminence (not prominent here)

Bregma

Coronal suture

Inferior temporal line

Superior temporal line

Sagittal suture

Region of parietal eminence

Parietal emissary foramen

Lambdoid suture

Lambda

(B) Superior (vertical) aspect

Frontal bone

Bregma

Coronal suture

Sagittal suture

Parietal bone

Vertex

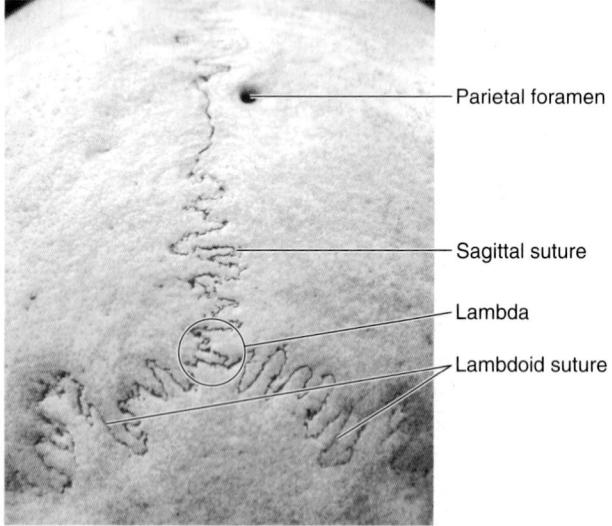

(C) Posterosuperior view

Parietal foramen

Sagittal suture

Lambda

Lambdoid suture

FIGURE 7.8. Adult cranium VI: Calvaria. A. The squamous parts of the frontal and occipital bones, and the paired parietal bones contribute to the calvaria. **B.** The external aspect of the anterior part of the calvaria demonstrates bregma, where the coronal and sagittal sutures meet, and vertex, the superior (topmost) point of the cranium. **C.** This external view demonstrates a prominent, unilateral parietal foramen. Although emissary foramina often occur in this general location, there is much variation.

(they may be bilateral or merged into a single formation). Posterolaterally are the **greater** and **lesser palatine foramina.** Superior to the posterior edge of the palate are two large openings: the **choanae** (posterior nasal apertures), which are separated from each other by the **vomer** (L. plowshare), a flat unpaired bone of trapezoidal shape that forms a major part of the bony nasal septum (Fig. 7.9B).

Wedged between the frontal, temporal, and occipital bones is the **sphenoid,** an irregular unpaired bone that consists of a body and three pairs of processes: greater wings, lesser wings, and pterygoid processes (Fig. 7.10). The **greater** and **lesser wings** of the sphenoid spread laterally from the lateral aspects of the body of the bone. The greater wings have orbital, temporal, and infratemporal surfaces apparent in facial, lateral, and inferior views of the exterior of the cranium (Figs. 7.3, 7.4A, and 7.9A) and cerebral surfaces seen in internal views of the cranial base (Fig. 7.11). The **pterygoid processes,** consisting of **lateral** and **medial pterygoid plates,** extend inferiorly on each side of the sphenoid from the junction of the body and greater wings (Figs. 7.9A and 7.10A & B).

The **groove for the cartilaginous part of the pharyngotympanic (auditory) tube** lies medial to the **spine of the sphenoid,** inferior to the junction of the greater wing of the sphenoid and the **petrous** (L. rock-like) **part of the temporal bone** (Fig. 7.9B). Depressions in the **squamous** (L. flat) **part of the temporal bone,** called the **mandibular fossae,** accommodate the mandibular condyles when the mouth is closed. The cranial base is formed posteriorly by the occipital bone, which articulates with the sphenoid anteriorly.

The four parts of the **occipital bone** are arranged around the *foramen magnum,* the most conspicuous feature of the cranial base. The major structures passing through this large foramen are: the *spinal cord* (where it becomes continuous with the medulla oblongata of the brain); the *meninges* (coverings) of the brain and spinal cord; the *vertebral arteries;* the anterior and posterior *spinal arteries;* and the *spinal accessory nerve* (CN XI). On the lateral parts of the occipital bone are two large protuberances, the **occipital condyles,** by which the cranium articulates with the vertebral column.

The large opening between the occipital bone and the petrous part of the temporal bone is the **jugular foramen,** from which the internal jugular vein (IJV) and several cranial nerves (CN IX–CN XI) emerge from the cranium (Figs. 7.9A and 7.11; Table 7.2). The entrance to the **carotid canal** for the internal carotid artery is just anterior to the jugular foramen (Fig. 7.9B). The *mastoid processes* provide for muscle attachments. The **stylomastoid foramen,** transmitting the facial nerve (CN VII) and stylomastoid artery, lies posterior to the base of the styloid process.

Internal Surface of Cranial Base

The **internal surface of the cranial base** (L. *basis cranii interna*) has three large depressions that lie at different levels: the *anterior, middle,* and *posterior cranial fossae,* which form the bowl-shaped floor of the **cranial cavity** (Fig. 7.12).

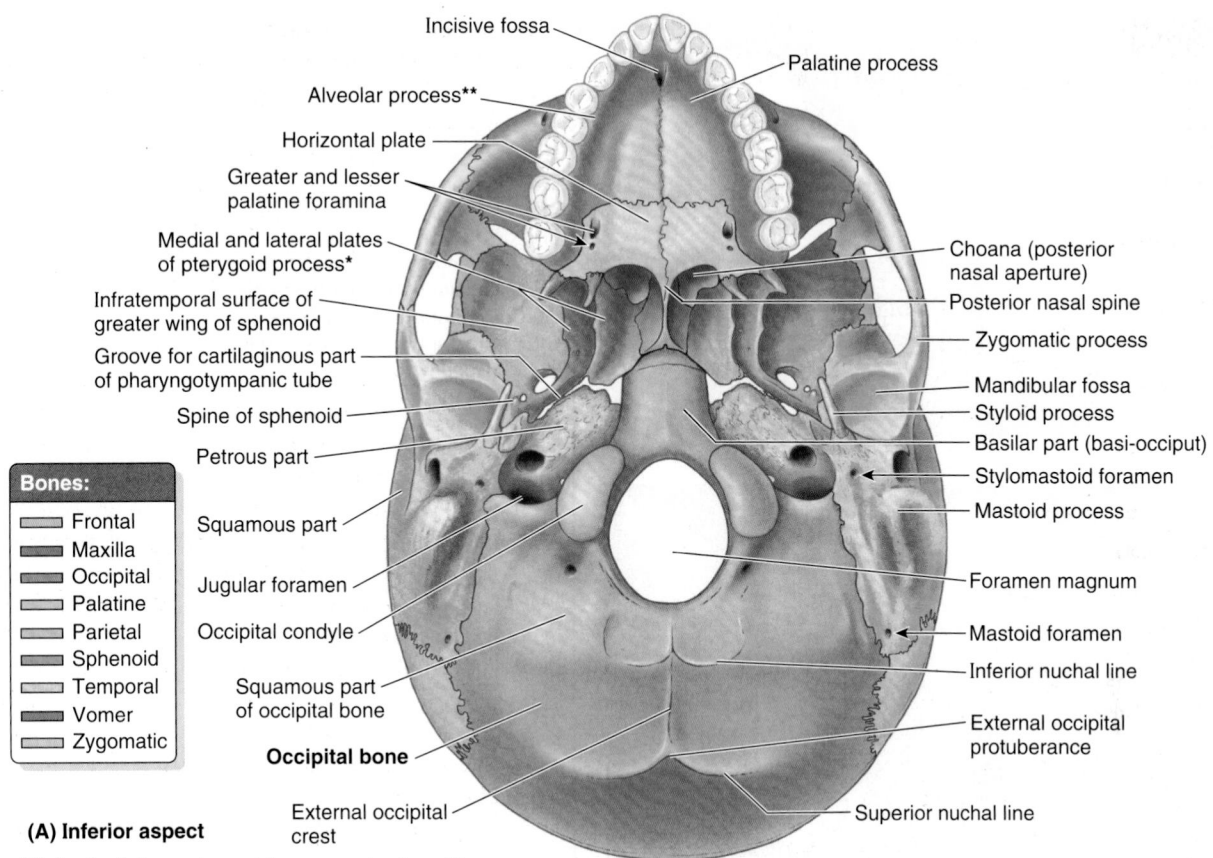

Incisive fossa
Alveolar process**
Horizontal plate
Greater and lesser palatine foramina
Medial and lateral plates of pterygoid process*
Infratemporal surface of greater wing of sphenoid
Groove for cartilaginous part of pharyngotympanic tube
Spine of sphenoid
Petrous part
Squamous part
Jugular foramen
Occipital condyle
Squamous part of occipital bone
Occipital bone

Palatine process
Choana (posterior nasal aperture)
Posterior nasal spine
Zygomatic process
Mandibular fossa
Styloid process
Basilar part (basi-occiput)
Stylomastoid foramen
Mastoid process
Foramen magnum
Mastoid foramen
Inferior nuchal line
External occipital protuberance
Superior nuchal line

Bones:

▭	Frontal
▬	Maxilla
▭	Occipital
▭	Palatine
▭	Parietal
▬	Sphenoid
▭	Temporal
▬	Vomer
▭	Zygomatic

(A) Inferior aspect

External occipital crest

*Collectively form pterygoid process of sphenoid
**The U-shaped (inverted here) ridge formed by the free border of the alveolar processes of the right and left maxillae makes up the alveolar arch

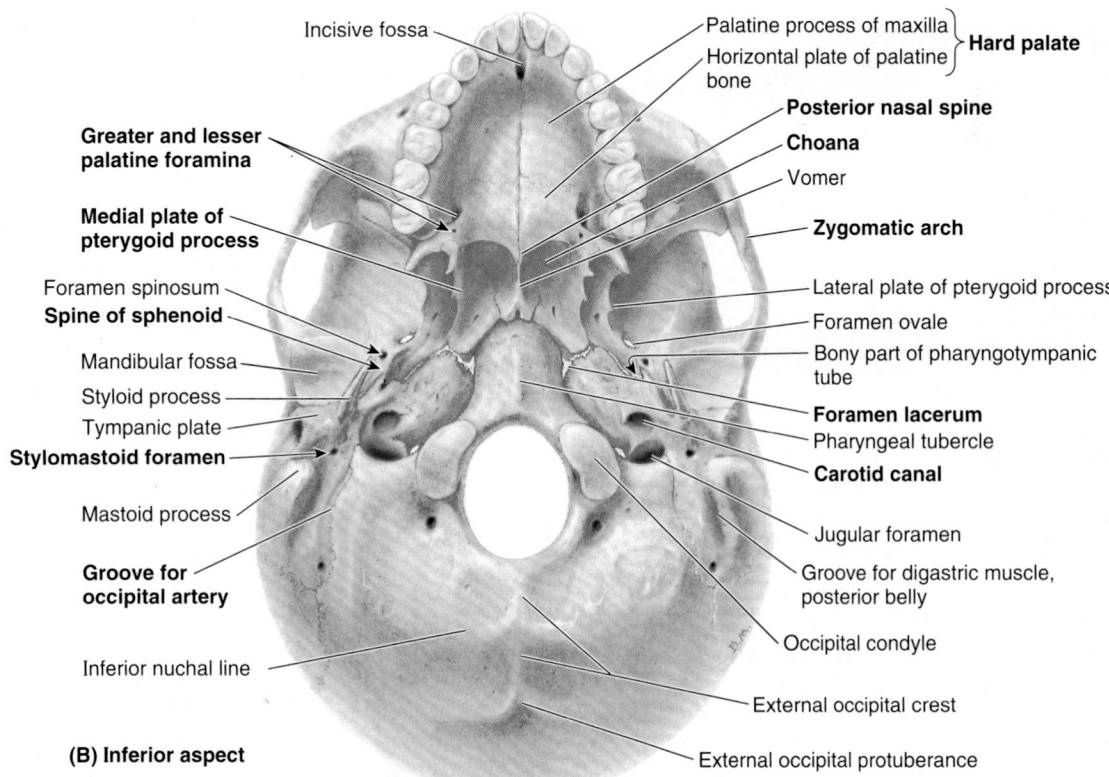

Incisive fossa
Greater and lesser palatine foramina
Medial plate of pterygoid process
Foramen spinosum
Spine of sphenoid
Mandibular fossa
Styloid process
Tympanic plate
Stylomastoid foramen
Mastoid process
Groove for occipital artery
Inferior nuchal line

Palatine process of maxilla ⎫
Horizontal plate of palatine bone ⎬ **Hard palate**
Posterior nasal spine
Choana
Vomer
Zygomatic arch
Lateral plate of pterygoid process
Foramen ovale
Bony part of pharyngotympanic tube
Foramen lacerum
Pharyngeal tubercle
Carotid canal
Jugular foramen
Groove for digastric muscle, posterior belly
Occipital condyle
External occipital crest
External occipital protuberance

(B) Inferior aspect

FIGURE 7.9. Adult cranium VII. External cranial base. A. The contributing bones are color coded. **B.** The foramen magnum is located midway between and on a level with the mastoid processes. The hard palate forms both a part of the roof of the mouth and the floor of the nasal cavity. The large choanae on each side of the vomer make up the posterior entrance to the nasal cavities.

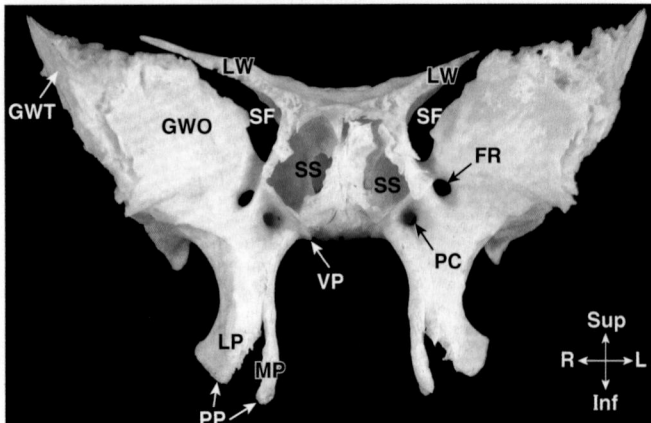

(A) Anterior view

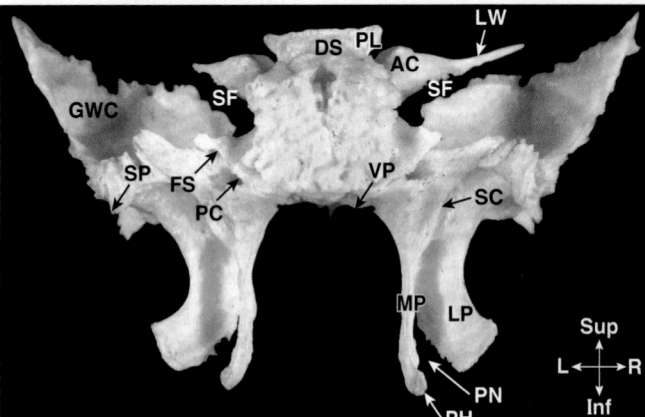

(B) Posterior view

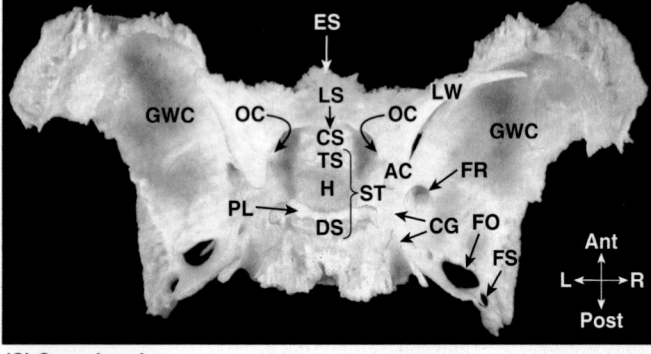

(C) Superior view

Key	
AC	Anterior clinoid process
CG	Carotid sulcus
CS	Prechiasmatic sulcus
DS	Dorsum sellae
ES	Ethmoidal spine
FO	Foramen ovale
FR	Foramen rotundum
FS	Foramen spinosum
GWC	Greater wing (cerebral surface)
GWO	Greater wing (orbital surface)
GWT	Greater wing (temporal surface)
H	Hypophysial fossa
LP	Lateral pterygoid plate
LS	Limbus of sphenoid
LW	Lesser wing
MP	Medial pterygoid plate
OC	Optic canal
PC	Pterygoid canal
PF	Pterygoid fossa
PH	Pterygoid hamulus
PL	Posterior clinoid process
PN	Pterygoid notch
PP	Pterygoid process
SC	Scaphoid fossa
SF	Superior orbital fissure
SP	Spine of sphenoid bone
SS	Sphenoidal sinus (in body of sphenoid)
ST	Sella turcica
TI	Greater wing of sphenoid (Infratemporal surface)
TS	Tuberculum sellae
VP	Vaginal process

FIGURE 7.10. Sphenoid. The unpaired, irregular sphenoid is a pneumatic (air-filled) bone. **A.** Parts of the thin anterior wall of the body of the sphenoid have been chipped off revealing the interior of the sphenoid sinus, which typically is unevenly divided into separate right and left cavities. **B.** The superior orbital fissure is a gap between the lesser and greater wings of the sphenoid. The medial and lateral pterygoid plates are components of the pterygoid processes. **C.** Details of the sella turcica, the midline formation that surrounds the hypophysial fossa, are shown.

The anterior cranial fossa is at the highest level, and the posterior cranial fossa is at the lowest level.

ANTERIOR CRANIAL FOSSA

The inferior and anterior parts of the frontal lobes of the brain occupy the **anterior cranial fossa,** the shallowest of the three cranial fossae (Fig. 7.12B). The fossa is formed by the frontal bone anteriorly, the ethmoid bone in the middle, and the body and lesser wings of the sphenoid posteriorly. The greater part of the fossa is formed by the **orbital parts of the frontal bone,** which support the frontal lobes of the brain and form the roofs of the orbits. This surface shows sinuous impressions (*brain markings*) of the orbital gyri (ridges) of the frontal lobes (Fig. 7.11).

The **frontal crest** is a median bony extension of the frontal bone (Fig. 7.12A). At its base is the **foramen cecum of**

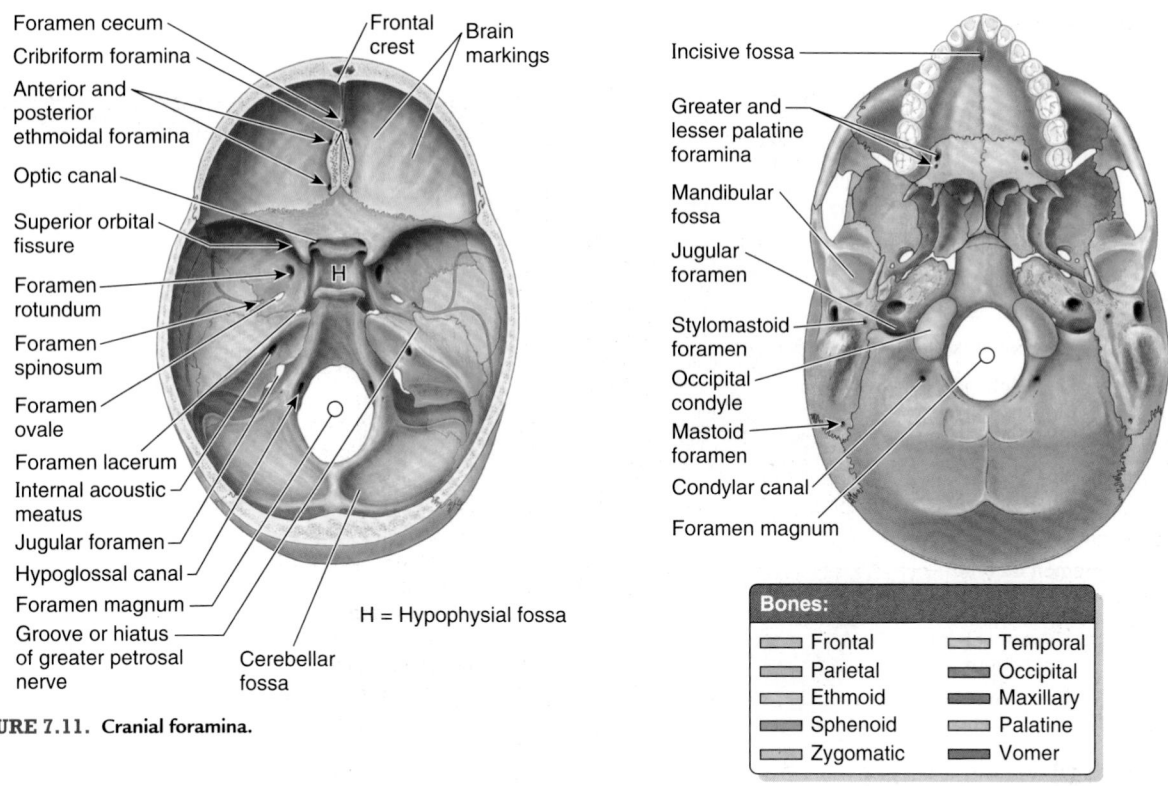

Foramen cecum
Cribriform foramina
Anterior and posterior ethmoidal foramina
Optic canal
Superior orbital fissure
Foramen rotundum
Foramen spinosum
Foramen ovale
Foramen lacerum
Internal acoustic meatus
Jugular foramen
Hypoglossal canal
Foramen magnum
Groove or hiatus of greater petrosal nerve

Frontal crest
Brain markings

H = Hypophysial fossa

Cerebellar fossa

FIGURE 7.11. Cranial foramina.

Incisive fossa
Greater and lesser palatine foramina
Mandibular fossa
Jugular foramen
Stylomastoid foramen
Occipital condyle
Mastoid foramen
Condylar canal
Foramen magnum

Bones:

Frontal	Temporal
Parietal	Occipital
Ethmoid	Maxillary
Sphenoid	Palatine
Zygomatic	Vomer

TABLE 7.2. FORAMINA AND OTHER APERTURES OF CRANIAL FOSSAE AND CONTENTS

Foramina/Apertures	Contents
Anterior cranial fossa	
Foramen cecum	Nasal emissary vein (1% of population)
Cribriform foramina in cribriform plate	Axons of olfactory cells in olfactory epithelium that form olfactory nerves
Anterior and posterior ethmoidal foramina	Vessels and nerves with same names
Middle cranial fossa	
Optic canals	Optic nerves (CN II) and ophthalmic arteries
Superior orbital fissure	Ophthalmic veins; ophthalmic nerve (CN V$_1$); CN III, IV, and VI; and sympathetic fibers
Foramen rotundum	Maxillary nerve (CN V$_2$)
Foramen ovale	Mandibular nerve (CN V$_3$) and accessory meningeal artery
Foramen spinosum	Middle meningeal artery and vein and meningeal branch of CN V$_3$
Foramen lacerum[a]	Deep petrosal nerve and some meningeal arterial branches and small veins
Groove or hiatus of greater petrosal nerve	Greater petrosal nerve and petrosal branch of middle meningeal artery
Posterior cranial fossa	
Foramen magnum	Medulla and meninges, vertebral arteries, CN XI, dural veins, anterior and posterior spinal arteries
Jugular foramen	CN IX, X, and XI; superior bulb of internal jugular vein; inferior petrosal and sigmoid sinuses; and meningeal branches of ascending pharyngeal and occipital arteries
Hypoglossal canal	Hypoglossal nerve (CN XII)
Condylar canal	Emissary vein that passes from sigmoid sinus to vertebral veins in neck
Mastoid foramen	Mastoid emissary vein from sigmoid sinus and meningeal branch of occipital artery

[a] The internal carotid artery and its accompanying sympathetic and venous plexuses actually pass horizontally *across* (rather than vertically through) the area of the foramen lacerum, an artifact of dry crania, which is closed by cartilage in life.

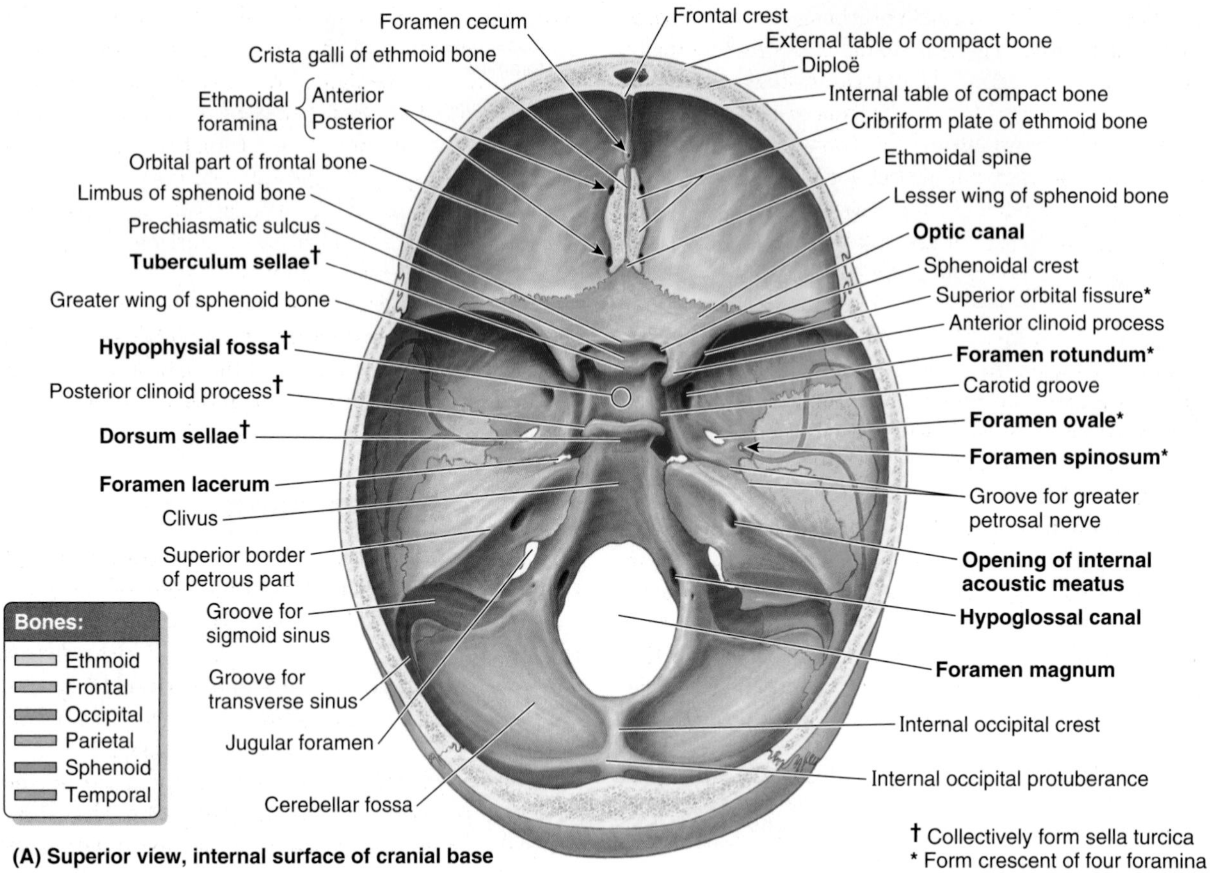

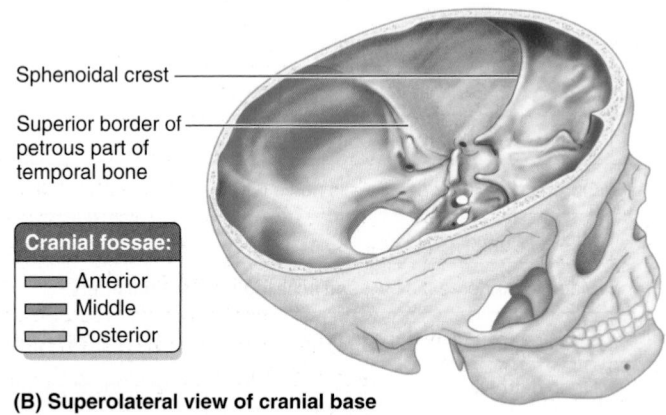

Foramen cecum
Crista galli of ethmoid bone
Ethmoidal {Anterior / Posterior} foramina
Orbital part of frontal bone
Limbus of sphenoid bone
Prechiasmatic sulcus
Tuberculum sellae†
Greater wing of sphenoid bone
Hypophysial fossa†
Posterior clinoid process†
Dorsum sellae†
Foramen lacerum
Clivus
Superior border of petrous part
Groove for sigmoid sinus
Groove for transverse sinus
Jugular foramen
Cerebellar fossa

Frontal crest
External table of compact bone
Diploë
Internal table of compact bone
Cribriform plate of ethmoid bone
Ethmoidal spine
Lesser wing of sphenoid bone
Optic canal
Sphenoidal crest
Superior orbital fissure*
Anterior clinoid process
Foramen rotundum*
Carotid groove
Foramen ovale*
Foramen spinosum*
Groove for greater petrosal nerve
Opening of internal acoustic meatus
Hypoglossal canal
Foramen magnum
Internal occipital crest
Internal occipital protuberance

Bones:
- Ethmoid
- Frontal
- Occipital
- Parietal
- Sphenoid
- Temporal

(A) Superior view, internal surface of cranial base

† Collectively form sella turcica
* Form crescent of four foramina

Sphenoidal crest
Superior border of petrous part of temporal bone

Cranial fossae:
- Anterior
- Middle
- Posterior

(B) Superolateral view of cranial base

FIGURE 7.12. Adult cranium VII. Internal cranial base. A. The internal aspect demonstrates the contributing bones and features. **B.** The floor of the cranial cavity is divisible into three levels (steps): anterior, middle, and posterior cranial fossae.

the frontal bone, which gives passage to vessels during fetal development, but is insignificant postnatally. The **crista galli** (L. cock's comb) is a thick, median ridge of bone posterior to the foramen cecum, which projects superiorly from the ethmoid. On each side of this ridge is the sieve-like **cribriform plate of ethmoid bone.** Its numerous tiny foramina transmit the olfactory nerves (CN I) from the olfactory areas of the nasal cavities to the olfactory bulbs of the brain, which lie on this plate (Fig. 7.12A; Table 7.2).

MIDDLE CRANIAL FOSSA

The butterfly-shaped **middle cranial fossa** has a *central part* composed of the *sella turcica* on the body of the sphenoid and large, depressed *lateral parts* on each side (Fig. 7.12). The middle cranial fossa is postero-inferior to the anterior cranial fossa, separated from it by the sharp *sphenoidal crests* laterally and the *sphenoidal limbus* centrally. The **sphenoidal crests** are formed mostly by the sharp posterior borders of the *lesser* *wings of the sphenoid bones,* which overhang the lateral parts of the fossae anteriorly. The sphenoidal crests end medially in two sharp bony projections, the *anterior clinoid processes.*

A variably prominent ridge, the **limbus of the sphenoid** forms the anterior boundary of the transversely oriented **prechiasmatic sulcus** extending between the right and the left *optic canals.* The bones forming the lateral parts of the fossa are the greater wings of the sphenoid, and squamous parts of the temporal bones laterally, and the petrous parts of the temporal bones posteriorly. The lateral parts of the middle cranial fossa support the temporal lobes of the brain. The boundary between the middle and the posterior cranial fossae is the **superior border of the petrous part of the temporal bone** laterally, and a flat plate of bone, the *dorsum sellae of the sphenoid,* medially.

The **sella turcica** (L. Turkish saddle) is the saddle-like bony formation on the upper surface of the body of the sphenoid, which is surrounded by the **anterior** and **posterior clinoid processes** (Figs. 7.10C and 7.12A). *Clinoid* means "bedpost," and the four processes (two anterior and two posterior) surround the hypophysial fossa, the "bed" of the pituitary gland, like the posts of a four-poster bed. The sella turcica is composed of three parts:

1. The **tuberculum sellae** (horn of saddle): a variable slight to prominent median elevation forming the posterior boundary of the *prechiasmatic sulcus* and the anterior boundary of the hypophysial fossa.
2. The **hypophysial fossa** (pituitary fossa): a median depression (seat of saddle) in the body of the sphenoid that accommodates the *pituitary gland* (L. *hypophysis*).
3. The **dorsum sellae** (back of saddle): a square plate of bone projecting superiorly from the *body of the sphenoid.* It forms the posterior boundary of the sella turcica, and its prominent superolateral angles make up the **posterior clinoid processes.**

On each side of the body of the sphenoid, a *crescent of four foramina* perforate the roots of the cerebral surfaces of the greater wings of the sphenoids (Figs. 7.10C, 7.11, and 7.12A); structures transmitted by the foramina are listed in Table 7.2:

1. **Superior orbital fissure:** Located between the greater and the lesser wings, it opens anteriorly into the orbit (Fig. 7.2A).
2. **Foramen rotundum** (round foramen): Located posterior to the medial end of the superior orbital fissure, it runs a horizontal course to an opening on the anterior aspect of the root of the greater wing of the sphenoid (Figs. 7.10A and 7.11A) into a bony formation between the sphenoid, the maxilla, and the palatine bones, the *pterygopalatine fossa.*
3. **Foramen ovale** (oval foramen): A large foramen posterolateral to the foramen rotundum; it opens inferiorly into the infratemporal fossa (Fig. 7.9B).
4. **Foramen spinosum** (spinous foramen): Located posterolateral to the foramen ovale and opens into the infratemporal fossa in relationship to the *spine of the sphenoid* (Fig. 7.11).

The **foramen lacerum** (lacerated or torn foramen) is not part of the crescent of foramina. This ragged foramen lies posterolateral to the hypophysial fossa, and is an artifact of a dried cranium (Fig. 7.12A). In life, it is closed by a cartilage plate. Only some meningeal arterial branches and small veins are transmitted vertically through the cartilage, completely traversing this foramen. The internal carotid artery and its accompanying sympathetic and venous plexuses pass across the superior aspect of the cartilage (i.e., pass over the foramen), and some nerves traverse it horizontally, passing to a foramen in its anterior boundary.

Extending posteriorly and laterally from the foramen lacerum is a narrow **groove for the greater petrosal nerve** on the anterosuperior surface of the petrous part of the temporal bone. There is also a small groove for the lesser petrosal nerve.

POSTERIOR CRANIAL FOSSA

The **posterior cranial fossa,** the largest and deepest of the three cranial fossae, lodges the cerebellum, pons, and medulla oblongata (Fig. 7.12B). The posterior cranial fossa is formed mostly by the occipital bone, but the *dorsum sellae of the sphenoid* marks its anterior boundary centrally (Fig. 7.12A), and the petrous and mastoid parts of the temporal bones contribute its anterolateral "walls."

From the dorsum sellae there is a marked incline, the **clivus,** in the center of the anterior part of the fossa leading to the *foramen magnum.* Posterior to this large opening, the posterior cranial fossa is partly divided by the **internal occipital crest** into bilateral large concave impressions, the **cerebellar fossae.** The internal occipital crest ends in the **internal occipital protuberance** formed in relationship to the *confluence of the sinuses,* a merging of dural venous sinuses (discussed later on page 867).

Broad grooves show the horizontal course of the *transverse sinus* and the S-shaped *sigmoid sinus.* At the base of the petrous ridge of the temporal bone is the *jugular foramen,* which transmits several cranial nerves in addition to the sigmoid sinus that exits the cranium as the internal jugular vein (IJV) (Fig. 7.11; Table 7.2). Anterosuperior to the jugular foramen is the *internal acoustic meatus* for the facial (CN VII) and vestibulocochlear nerves (CN VIII), and the labyrinthine artery. The **hypoglossal canal** for the hypoglossal nerve (CN XII) is superior to the anterolateral margin of the foramen magnum.

Walls of Cranial Cavity

The **walls of the cranial cavity** vary in thickness in different regions. They are usually thinner in females than in males and are thinner in children and elderly people. The bones tend to be thinnest in areas that are well covered with muscles, such as the squamous part of the temporal bone (Fig. 7.11). Thin areas of bone can be seen radiographically (Fig. 7.5), or by holding a dried cranium up to a bright light.

Most bones of the calvaria consist of **internal** and **external tables of compact bone,** separated by **diploë** (Figs. 7.5 and

7.11). The diploë is cancellous bone containing red bone marrow during life, through which run canals formed by diploic veins. The diploë in a dried calvaria is not red because the protein was removed during preparation of the cranium. The internal table of bone is thinner than the external table, and in some areas there is only a thin plate of compact bone with no diploë.

The bony substance of the cranium is unequally distributed. Relatively thin (but mostly curved) flat bones provide the necessary strength to maintain cavities and protect their contents. However, in addition to housing the brain, the bones of the neurocranium (and processes from them) provide proximal attachment for the strong muscles of mastication that attach distally to the mandible; consequently, high traction forces occur across the nasal cavity and orbits that are sandwiched between. Thus thickened portions of the cranial bones form stronger pillars or *buttresses* that transmit forces, bypassing the orbits and nasal cavity (Fig. 7.13). The main buttresses are the **frontonasal buttress,** extending from the region of the canine teeth between the nasal and the orbital cavities to the central frontal bone, and the **zygomatic arch–lateral orbital margin buttress** from the region of the molars to the lateral frontal and temporal bones. Similarly, **occipital buttresses** transmit forces received lateral to the foramen magnum from the vertebral column. Perhaps to compensate for the denser bone required for these buttresses, some areas of the cranium not as mechanically stressed become pneumatized (air-filled).

Regions of Head

To allow clear communications regarding the location of structures, injuries, or pathologies, the head is divided into regions (Fig. 7.14). The large number of regions into which the relatively small area of the face is divided (eight) is a reflection of both its functional complexity and personal importance, as are annual expenditures for elective aesthetic

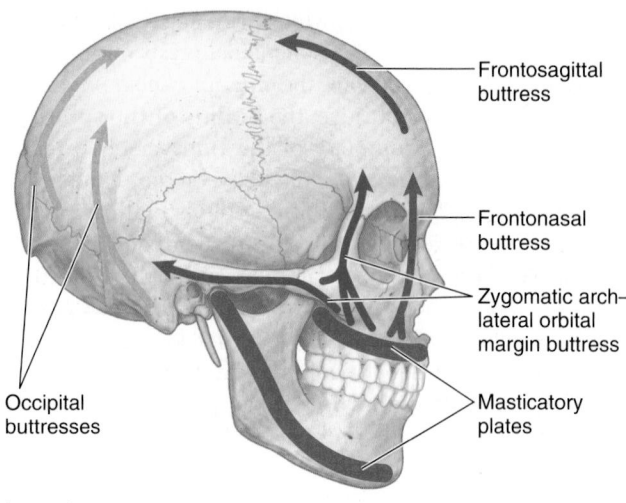

Lateral aspect

FIGURE 7.13. Buttresses of cranium. The buttresses are thicker portions of cranial bone that transmit forces around weaker regions of the cranium.

surgery. With the exception of the **auricular region,** which includes the external ear, the names of the regions of the *neurocranial portion of the head* correspond to the underlying bones or bony features: **frontal, parietal, occipital, temporal,** and **mastoid regions.**

The *viscerocranial portion of the head* includes the **facial region,** which is divided into five bilateral and three median regions related to superficial features (**oral** and **buccal regions**), to deeper soft tissue formations (**parotid region**), and to skeletal features (**orbital, infra-orbital, nasal, zygomatic,** and **mental regions**). The remainder of this chapter discusses several of these regions in detail, as well as some deep regions not represented on the surface (for example, the infratemporal region and pterygopalatine fossa). The surface anatomy of these regions will be discussed with the description of each region.

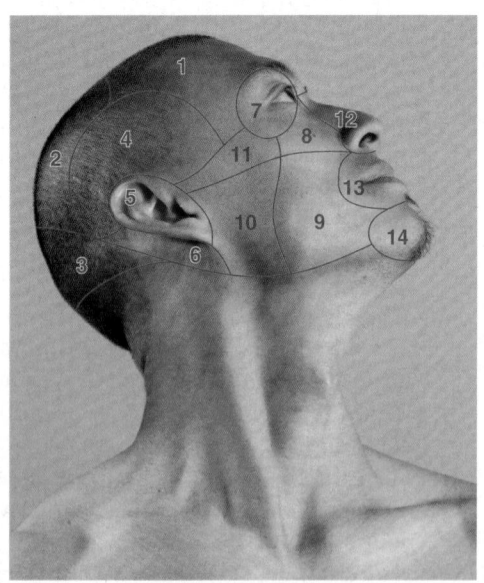

Regions of the head:

1 Frontal region
2 Parietal region
3 Occipital region
4 Temporal region
5 Auricular region
6 Mastoid region

Facial region:
7 Orbital region
8 Infra-orbital region
9 Buccal region
10 Parotid region
11 Zygomatic region
12 Nasal region
13 Oral region
14 Mental region

FIGURE 7.14. Regions of head.

CRANIUM

Head Injuries

Head injuries are a major cause of death and disability. The complications of head injuries include hemorrhage, infection, and injury to the brain (e.g., concussion) and cranial nerves. Disturbance in the level of consciousness is the most common symptom of head injury. Almost 10% of all deaths in the United States are caused by head injuries, and approximately half of traumatic deaths involve the brain (Rowland, 2010). Head injuries occur mostly in young persons between the ages of 15 and 24 years. The major cause of brain injury varies, but motor vehicle and motorcycle accidents are prominent.

Headaches and Facial Pain

Few complaints are more common than *headaches* and *facial pain.* Although usually benign and frequently associated with tension, fatigue, or mild fever, headaches may indicate a serious intracranial problem such as a brain tumor, subarachnoid hemorrhage, or meningitis. *Neuralgias* (G. *algos,* pain) are characterized by severe throbbing or stabbing pain in the course of a nerve caused by a demyelinating lesion. They are a common cause of facial pain. Terms such as *facial neuralgia* describe diffuse painful sensations. Localized aches have specific names, such as *earache* (otalgia) and *toothache* (odontalgia). A sound knowledge of the clinical anatomy of the head helps in understanding the causes of headaches and facial pain.

Injury to Superciliary Arches

The superciliary arches are relatively sharp bony ridges (see Fig. 7.3); consequently, a blow to them (e.g., during boxing) may lacerate the skin and cause bleeding. Bruising of the skin surrounding the orbit causes tissue fluid and blood to accumulate in the surrounding connective tissue, which gravitates into the superior (upper) eyelid and around the eye ("black eye"; see Fig. B7.12).

Malar Flush

The zygomatic bone was once called the malar bone; consequently, you will hear the clinical term *malar flush.* This redness of the skin covering the zygomatic process (malar eminence) is associated with a rise in temperature in various fevers occurring with certain diseases, such as *tuberculosis* and *systemic lupus erythematosus disease.*

Fractures of Maxillae and Associated Bones

Dr. Léon-Clement Le Fort (Paris surgeon and gynecologist, 1829–1893) classified three common variants of fractures of the maxillae (Fig. B7.1):

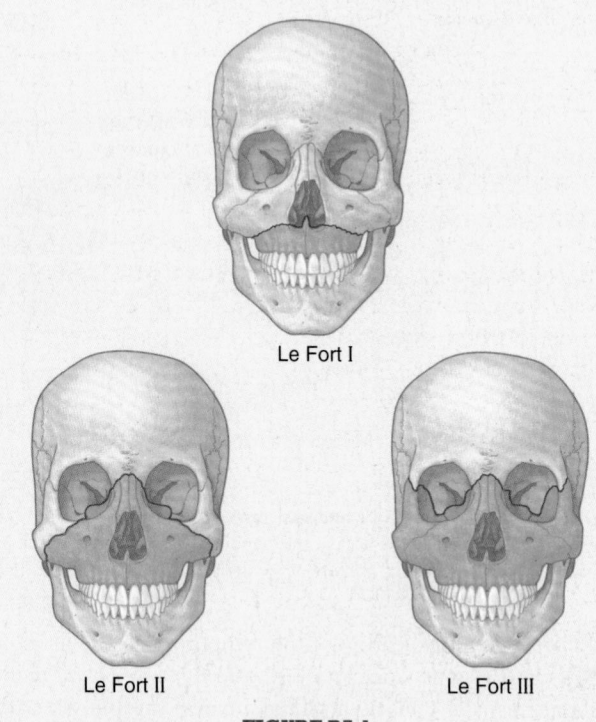

Le Fort I

Le Fort II   Le Fort III

FIGURE B7.1.

- **Le Fort I fracture:** wide variety of horizontal fractures of the maxillae, passing superior to the maxillary alveolar process (i.e., to the roots of the teeth), crossing the bony nasal septum and possibly the pterygoid plates of the sphenoid.
- **Le Fort II fracture:** passes from the posterolateral parts of the maxillary sinuses (cavities in the maxillae) superomedially through the infra-orbital foramina, lacrimals, or ethmoids to the bridge of the nose. As a result, the entire central part of the face, including the hard palate and alveolar processes, is separated from the rest of the cranium.
- **Le Fort III fracture:** horizontal fracture that passes through the superior orbital fissures and the ethmoid and nasal bones and extends laterally through the greater wings of the sphenoid and the frontozygomatic sutures. Concurrent fracturing of the zygomatic arches causes the maxillae and zygomatic bones to separate from the rest of the cranium.

Fractures of Mandible

A broken mandible usually involves two fractures, which frequently occur on opposite sides of the mandible; thus, if one fracture is observed, a search should be made for another. For example, a hard blow to the jaw often fractures the neck and body of the mandible in the region of the opposite canine tooth.

Fractures of the coronoid process of the mandible are uncommon and usually single (Fig. B7.2). *Fractures of the neck of the mandible* are often transverse and may be

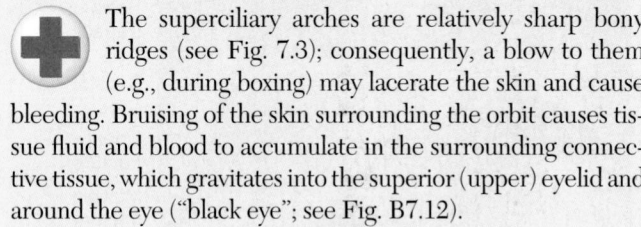

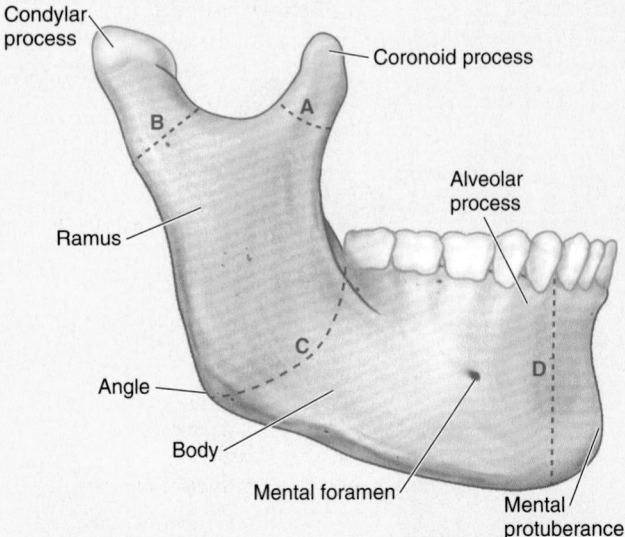

FIGURE B7.2. **Fractures of mandible.** *Line A,* Fracture of the coronoid process; *line B,* fracture of the neck of the mandible; *line C,* fracture of the angle of the mandible; *line D,* fracture of the body of the mandible.

associated with dislocation of the temporomandibular joint (TMJ) on the same side. *Fractures of the angle of the mandible* are usually oblique and may involve the bony socket or alveolus of the 3rd molar tooth (Fig. B7.2, *line C*). *Fractures of the body of the mandible* frequently pass through the socket of a canine tooth (Fig. B7.2, *line D*).

Resorption of Alveolar Bone

Extraction of teeth causes the alveolar bone to resorb in the affected region(s) (Fig. B7.3). Following complete loss or extraction of maxillary teeth, the tooth sockets begin to fill in with bone, and the alveolar process begins to resorb. Similarly, extraction of mandibular teeth causes the bone to resorb. Gradually, the mental foramen lies near the superior border of the body of the mandible (Fig. B7.3A-C). In some cases, the mental foramina disappear, exposing the mental nerves to injury. *Pressure from a dental prosthesis* (e.g., a denture resting on an exposed mental nerve) may produce pain during eating. Loss of all the teeth results in a decrease in the vertical facial dimension and *mandibular prognathism* (overclosure). Deep creases in the facial skin also appear that pass posteriorly from the corners of the mouth.

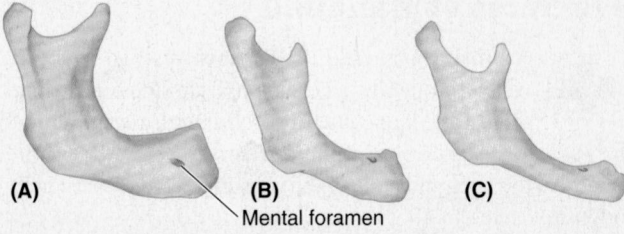

FIGURE B7.3. **Stages of resorption of edentulous (toothless) alveolar bone.**

Fractures of Calvaria

The convexity of the calvaria distributes and thereby usually minimizes the effects of a blow to the head. However, hard blows in thin areas of the calvaria are likely to produce *depressed fractures,* in which a bone fragment is depressed inward, compressing and/or injuring the brain (Fig. B7.4). *Linear calvarial fractures,* the most frequent type, usually occur at the point of impact, but fracture lines often radiate away from it in two or more directions. In *comminuted fractures,* the bone is broken into several pieces. If the area of the calvaria is thick at the site of impact, the bone may bend inward without fracturing; however, a fracture may occur some distance from the site of direct trauma where the calvaria is thinner. In a *contrecoup (counterblow) fracture,* no fracture occurs at the point of impact, but one occurs on the opposite side of the cranium.

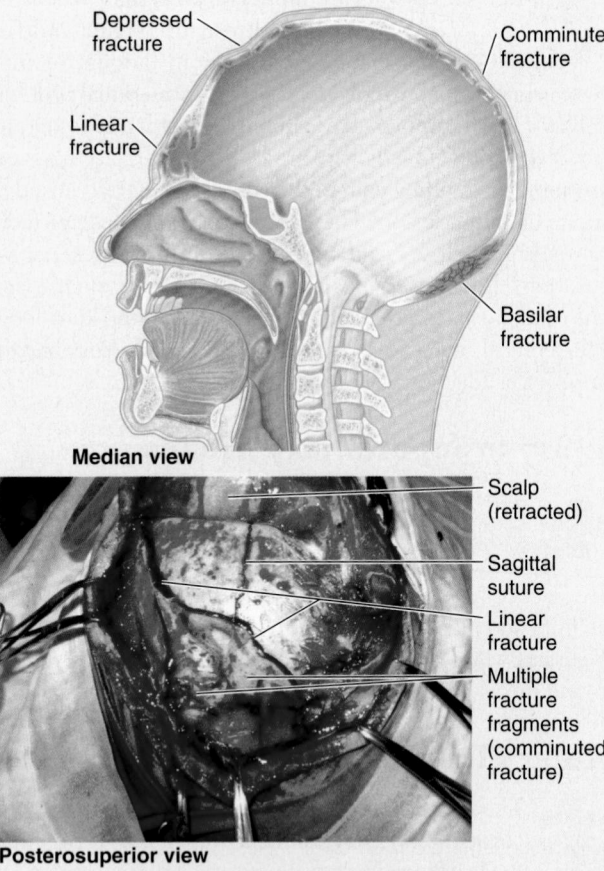

FIGURE B7.4. **Fractures of calvaria.**

Surgical Access to Cranial Cavity: Bone Flaps

Surgeons access the cranial cavity and brain by performing a *craniotomy,* in which a section of the neurocranium, called a bone flap, is elevated or removed (Fig. B7.5). Because the adult pericranium (periosteum of cranium) has poor osteogenic (bone-forming) properties, little regeneration occurs after bone loss (e.g., when pieces of bone are removed during repair of a comminuted cranial fracture). Surgically produced bone flaps are put back into place and wired to other parts of the calvaria, or held in place temporarily with metal plates. Reintegration is most successful when the bone is reflected with its overlying muscle and skin, so that it retains its own blood supply during the procedure and after repositioning. If the bone flap is not replaced (i.e., a permanent plastic or metal plate replaces the flap), the procedure is called a *craniectomy*.

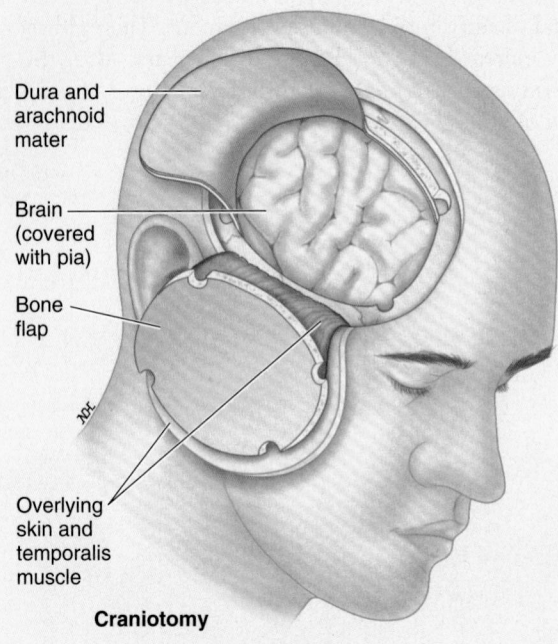

FIGURE B7.5.

Development of Cranium

The bones of the calvaria and some parts of the cranial base develop by *intramembranous ossification*. Most parts of the cranial base develop by *endochondral ossification*. At birth, the bones of the calvaria are smooth and unilaminar; no diploë is present. The frontal and parietal eminences are especially prominent (Fig. B7.6). The cranium of a neonate is disproportionately large compared to other parts of the skeleton; however, the facial aspect is small compared to the *calvaria*, which forms approximately one eighth of the cranium. In the adult, the facial skeleton forms one third of the cranium. The large size of the calvaria in infants results from precocious growth and development of the brain and eyes.

The rudimentary development of the face makes the orbits appear relatively large (Fig. B7.6A). The smallness of the face results from the rudimentary development of the

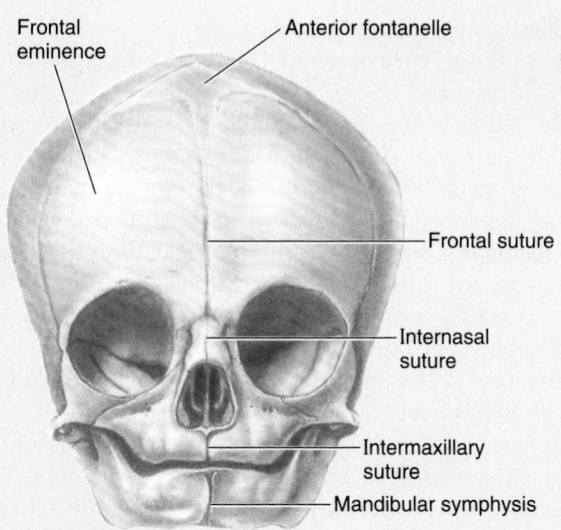

(A) Anterior view

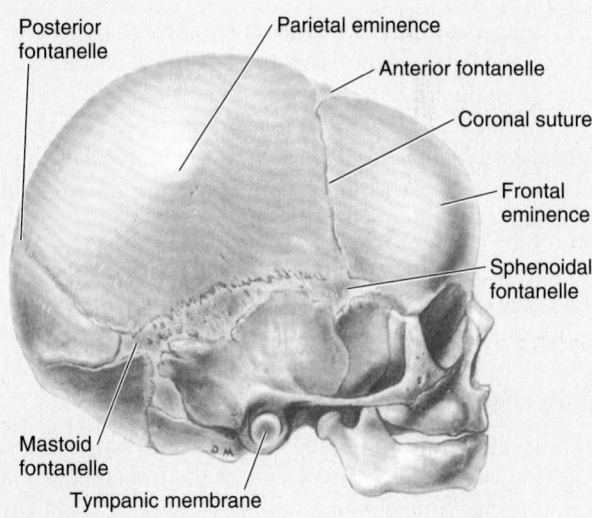

(B) Lateral view

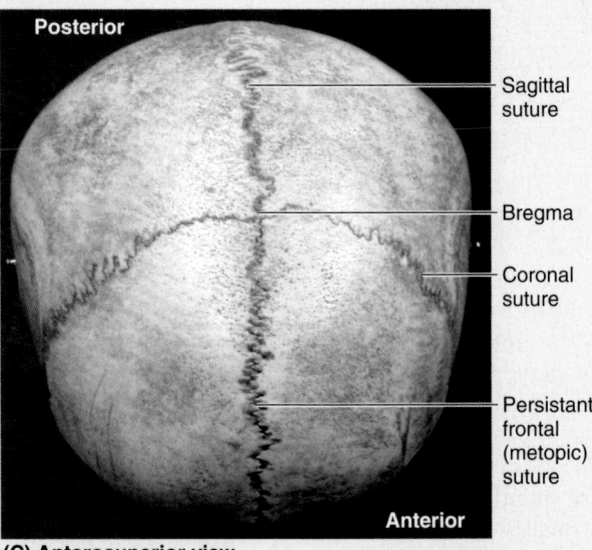

(C) Anterosuperior view

FIGURE B7.6. Cranial development.

maxillae, mandible, and paranasal sinuses (air-filled bone cavities), the absence of erupted teeth, and the small size of the nasal cavities.

The halves of the frontal bone in the neonate are separated by the *frontal suture,* the frontal and parietal bones are separated by the *coronal suture,* and the maxillae and mandibles are separated by the *intermaxillary suture* and *mandibular symphysis* (secondary cartilaginous joint), respectively. There are no mastoid and styloid processes (Figs. B7.6A & B). Because there are no mastoid processes at birth, the facial nerves are close to the surface when they emerge from the stylomastoid foramina. As a result, the facial nerves may be injured by forceps during a difficult delivery or later by an incision posterior to the auricle of the external ear (as for the surgical treatment of mastoiditis or middle ear problems). The mastoid processes form gradually during the 1st year as the sternocleidomastoid muscles complete their development and pull on the petromastoid parts of the temporal bones.

The bones of the calvaria of a neonate are separated by fibrous membranes; the largest occur between the angles (corners) of the flat bones (Fig. B7.6A & B). They include the *anterior* and *posterior fontanelles* and the paired *sphenoidal* and *mastoid fontanelles.* Palpation of the fontanelles during infancy, especially the anterior and posterior ones, enables physicians to determine the:

- Progress of growth of the frontal and parietal bones.
- Degree of hydration of an infant (a depressed fontanelle indicates dehydration).
- Level of intracranial pressure (a bulging fontanelle indicates increased pressure on the brain).

The **anterior fontanelle,** the largest one, is diamond or star shaped; it is bounded by the halves of the frontal bone anteriorly and the parietal bones posteriorly (Fig. B7.6). Thus it is located at the junction of the sagittal, coronal, and frontal sutures, the future site of *bregma* (Fig. 7.6; Table 7.1). By 18 months of age, the surrounding bones have fused, and the anterior fontanelle is no longer clinically palpable.

At birth, the frontal bone consists of two halves. Union of the halves begins in the 2nd year. In most cases, the *frontal suture* is obliterated by the 8th year. However, in approximately 8% of people, a remnant of the frontal suture, the **metopic suture,** persists (Figs. 7.2A and 7.3). Much less frequently, the entire suture remains (Fig. B7.6C). A persistent suture must not be interpreted as a fracture in a radiograph, or other medical image (e.g., a CT scan).

The **posterior fontanelle** is triangular and bounded by the parietal bones anteriorly, and the occipital bone posteriorly. It is located at the junction of the lambdoid and sagittal sutures, the future site of *lambda* (Fig. 7.7A and 7.8C). The posterior fontanelle begins to close during the first few months after birth, and by the end of the 1st year, it is small and no longer clinically palpable. The **sphenoidal** and **mastoid fontanelles,** overlain by the temporalis muscle (Fig. 7.16A), fuse during infancy and are less important clini-

cally than the midline fontanelles. The halves of the mandible fuse early in the 2nd year. The two maxillae and nasal bones usually do not fuse.

The softness of the cranial bones in fetuses and their loose connections at the sutures and fontanelles enable the shape of the cranium to be molded during birth (Fig. B7.7). During passage of the fetus through the birth canal, the halves of the frontal bone become flat, the occipital bone is drawn out, and one parietal bone slightly overrides the other. Within a few days after birth, the shape of the neonatal cranium returns to normal. The resilience of the cranial bones of infants allows them to resist forces that would produce fractures in adults. The fibrous sutures of the calvaria also permit the cranium to enlarge during infancy and childhood. The increase in the size of the calvaria is greatest during the first 2 years, the period of most rapid brain development. The calvaria normally increases in capacity for 15–16 years. After this, the calvaria usually increases slightly in size for 3–4 years as a result of bone thickening.

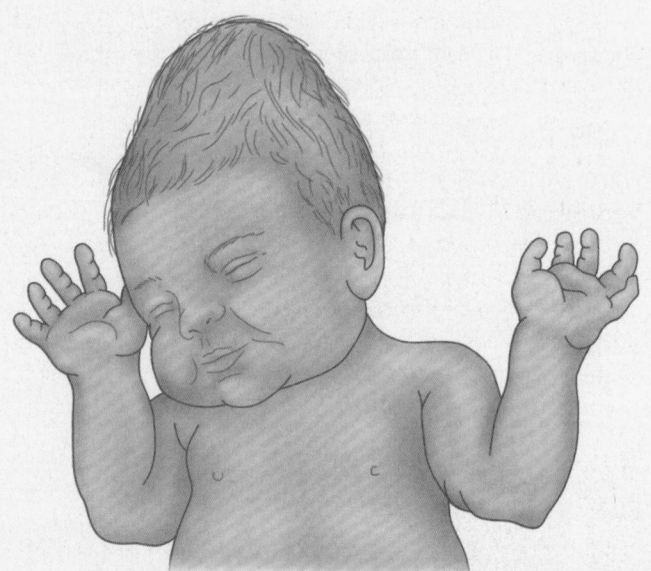

FIGURE B7.7. Molding of calvaria.

Age Changes in Face

The mandible is the most dynamic of our bones; its size and shape and the number of teeth it normally bears, undergo considerable change with age. In the neonate, the mandible consists of two halves united in the median plane by a cartilaginous joint, the *mandibular symphysis.* Union between the halves of the mandible is effected by means of fibrocartilage; this union begins during the 1st year and the halves are fused by the end of the 2nd year. The body of the mandible in neonates is a mere shell lacking alveolar processes; each half enclosing five deciduous teeth. These teeth usually begin to erupt in infants at approximately 6 months of age. The body of the mandible elongates, particularly posterior

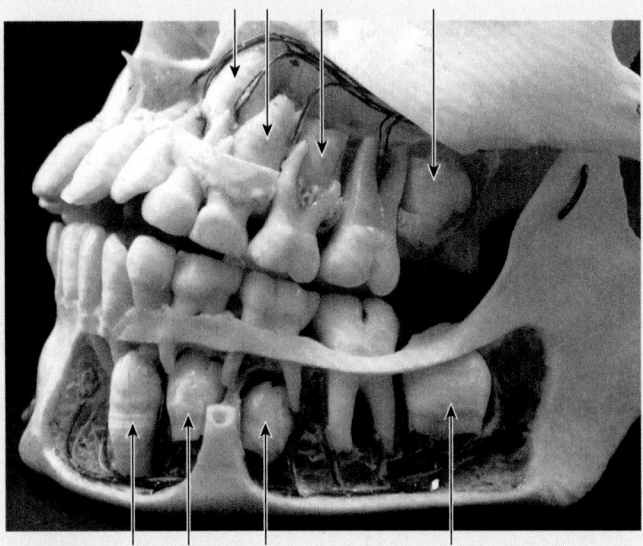

FIGURE B7.8. Left lateral view of dentition. *Arrows,* unerupted permanent teeth.

to the mental foramen (Fig. B7.2), to accommodate this development. Later, eight permanent teeth begin to erupt during the 6th year of life (Fig. B7.8). Eruption of the permanent teeth is not complete until early adulthood.

Rapid growth of the face during infancy and early childhood coincides with the eruption of deciduous teeth. Vertical growth of the upper face results mainly from dentoalveolar development of alveolar bone. These changes are more marked after the permanent teeth erupt. Concurrent enlargement of the frontal and facial regions is associated with the increase in the size of the *paranasal sinuses*, the air-filled extensions of the nasal cavities in certain cranial

bones (Fig. B7.9). Most paranasal sinuses are rudimentary or absent at birth. Growth of the paranasal sinuses is important in altering the shape of the face and in adding resonance to the voice.

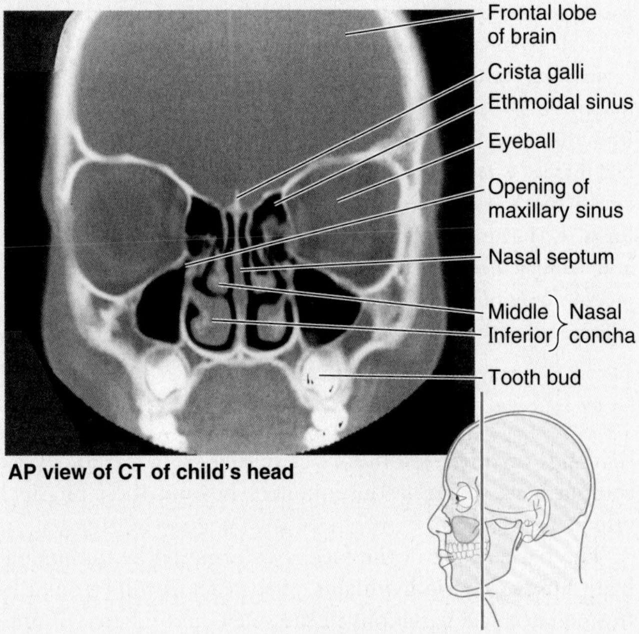

Obliteration of Cranial Sutures

The obliteration of sutures between the bones of the calvaria usually begins between the ages of 30 and 40 years on the internal surface. Approximately 10 years later, the sutures on the external surface obliterate (Fig. B7.10; cf. Fig. 7.8B). Obliteration of sutures usually begins at the bregma and continues sequentially in the sagittal, coronal, and lambdoid sutures. Closure times vary considerably.

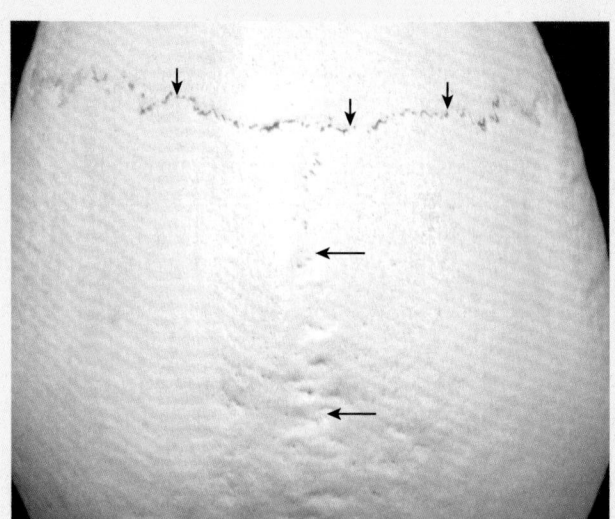

FIGURE B7.10. Obliteration (synostosis) of cranial sutures. *Arrows,* sagittal; *arrowheads,* coronal.

Age Changes in Cranium

As people age, the cranial bones normally become progressively thinner and lighter, and the diploë gradually become filled with a gray gelatinous material. In these individuals, the bone marrow has lost its blood cells and fat, giving it a gelatinous appearance.

Craniosynostosis and Cranial Malformations

Premature closure of the cranial sutures (*primary craniosynostosis*) results in several cranial malformations (Fig. B7.11). The incidence of primary craniosynostosis is approximately 1 per 2000 births (Kliegman et al., 2011). The cause of craniosynostosis is unknown, but genetic factors appear to be important. The prevailing hypothesis is that abnormal development of the cranial base creates exaggerated forces on the *dura mater* (outer covering membrane of the brain) that disrupt normal cranial sutural development. These malformations are more common in males than in

---

- Frontal lobe of brain
- Crista galli
- Ethmoidal sinus
- Eyeball
- Opening of maxillary sinus
- Nasal septum
- Middle ⎱ Nasal
- Inferior ⎰ concha
- Tooth bud

AP view of CT of child's head

FIGURE B7.9.

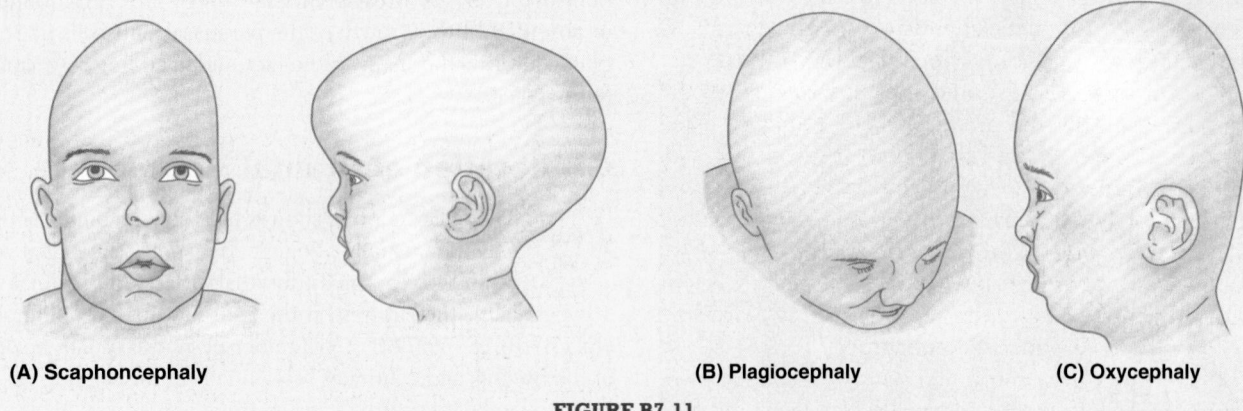

(A) Scaphoncephaly   (B) Plagiocephaly   (C) Oxycephaly

FIGURE B7.11.

females, and are often associated with other skeletal anomalies. The type of malformed cranium that forms depends on which sutures close prematurely.

Premature closure of the sagittal suture, in which the anterior fontanelle is small or absent, results in a long, narrow, wedge-shaped cranium, a condition called *scaphocephaly* (Fig. B7.11A). When *premature closure of the coronal or the lambdoid suture* occurs on one side only, the cranium is twisted and asymmetrical, a condition known as *plagiocephaly* (Fig. B7.11B). *Premature closure of the coronal suture* results in a high, tower-like cranium, called *oxycephaly* or *turricephaly* (Fig. B7.11C). The latter type of cranial malformation is more common in females. Premature closure of sutures usually does not affect brain development.

The Bottom Line

CRANIUM

The cranium is the skeleton of the head, an amalgamation of functional components united to form a single skeletal formation. ♦ The basic functional components include the *neurocranium*, the container of the brain and internal ears, and *viscerocranium*, providing paired orbits, nasal cavities and teeth-bearing plates (alveolar processes) of the oral cavity. ♦ Although some mobility between cranial bones is advantageous during birth, they become fixed together by essentially immovable joints (sutures), allowing independent movement of only the mandible. ♦ Abundant fissures and foramina facilitate communication and passage of neurovascular structures between functional components. ♦ The bony substance of the cranium is unequally distributed. Relatively thin (but mostly curved) flat bones provide the necessary strength to maintain cavities and protect contents. ♦ However, the bones and processes of the neurocranium also provide proximal attachment for the strong muscles of mastication (chewing) that attach distally to the mandible. ♦ The high traction forces generated across the nasal cavity and orbits, sandwiched between the muscle attachments, are resisted by thickened portions of the bones forming stronger pillars or buttresses. ♦ The mostly superficial surface of the cranium provides both visible and palpable landmarks.

Internal features of the cranial base reflect the major formations of the brain that rest on it. ♦ Bony ridges radiating from the centrally located sella turcica divide it into three cranial fossae. ♦ The frontal lobes of the brain lie in the anterior cranial fossa. ♦ The temporal lobes lie in the middle cranial fossa. ♦ The hindbrain, consisting of the pons, cerebellum, and medulla, occupies the posterior cranial fossa, with the medulla continuing through the foramen magnum where it is continuous with the spinal cord.

FACE AND SCALP

Face

The **face** is the anterior aspect of the head from the forehead to the chin and from one ear to the other. The face provides our identity as an individual human. Thus, birth defects scarring, or other alterations resulting from pathology or trauma have marked consequences beyond their physical effects.

The basic shape of the face is determined by the underlying bones. The individuality of the face results primarily from anatomical variations in the shape and relative prominence of the features of the underlying cranium; in the deposition of fatty tissue; in the color and effects of aging on the overlying skin; and in the abundance, nature, and placement of hair on the face and scalp. The relatively large size of the **buccal fat-pads** in infants prevents collapse of the cheeks during sucking and produces their chubby-cheeked appearance. Growth of the facial bones takes longer than those of the calvaria. The ethmoid bone, orbital cavities, and superior parts of the nasal cavities have nearly completed their growth by the 7th year. Expansion of the orbits and growth of the nasal septum carry the maxillae infero-anteriorly. Considerable facial growth occurs during childhood as the paranasal sinuses develop and permanent teeth erupt.

The face plays an important role in communication. Our interactions with others take place largely via the face (including the ears); hence, the term *interface* for a site of interactions. Whereas the shape and features of the face provide our identity, much of our affect on others and their perceptions about us result from the way we use facial muscles to make the slight alterations in the featu... facial expression.

Scalp

The **scalp** consists of skin (normally hair bearing) and subcutaneous tissue that cover the neurocranium from the superior nuchal lines on the occipital bone to the supra-orbital margins of the frontal bone (Figs. 7.3 and 7.4A). Laterally, the scalp extends over the temporal fascia to the zygomatic arches.

The scalp is composed of five layers, the first three of which are connected intimately and move as a unit (e.g., when wrinkling the forehead and moving the scalp). Each letter in the word *scalp* serves as a memory key for one of its five layers (Fig. 7.15A):

1. **S**kin: thin, except in the occipital region, contains many sweat and sebaceous glands and hair follicles. It has an abundant arterial supply and good venous and lymphatic drainage.

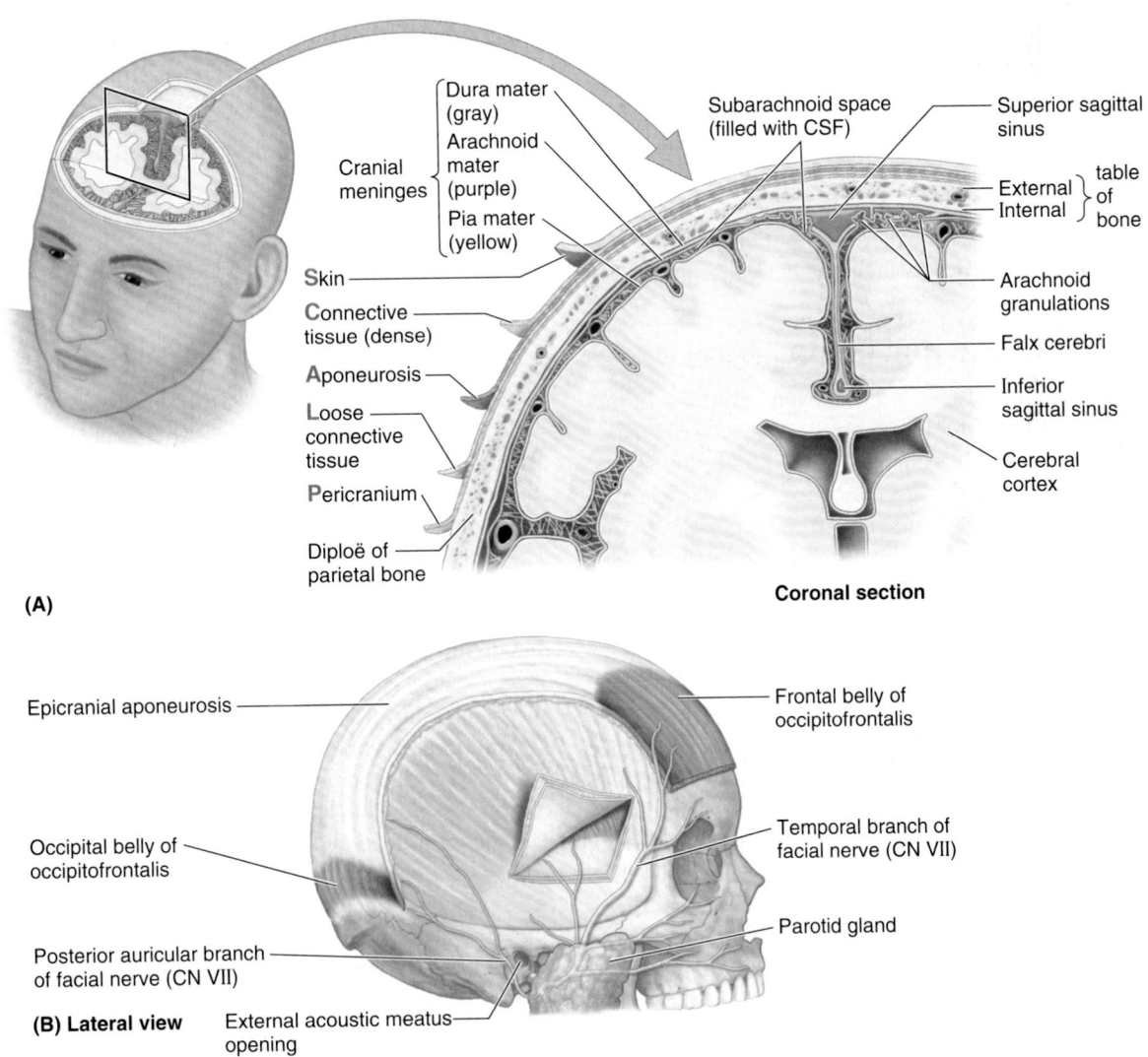

FIGURE 7.15. Layers of scalp, cranium, and meninges. A. The skin is bound tightly to the epicranial aponeurosis, which moves freely over the pericranium and cranium because of the intervening loose connective tissue. *Aponeurosis* refers to the epicranial aponeurosis, the flat intermediate tendon of the occipitofrontalis muscle. The cranial meninges and the subarachnoid (leptomeningeal) space are shown. *CSF,* cerebrospinal fluid. **B.** The occipitofrontalis muscle. Innervation of the two bellies by the posterior auricular and temporal branches of the facial nerve is demonstrated.

2. **C**onnective tissue: forms the thick, dense, richly vascularized subcutaneous layer that is well supplied with cutaneous nerves.

3. **A**poneurosis (**epicranial aponeurosis**): the broad, strong, tendinous sheet that covers the calvaria and serves as the attachment for muscle bellies converging from the forehead and occiput (**occipitofrontalis muscle**) (Fig. 7.15B) and from the temporal bones on each side (**temporoparietalis** and **superior auricular muscles**). Collectively, these structures constitute the musculo-aponeurotic **epicranius.** The *frontal belly of the occipitofrontalis* pulls the scalp anteriorly, wrinkles the forehead, and elevates the eyebrows; the *occipital belly of the occipitofrontalis* pulls the scalp posteriorly, smoothing the skin of the forehead. The superior auricular muscle (actually a specialized posterior part of the temporoparietalis) elevates the auricle of the external ear. All parts of the epicranius (muscle and aponeurosis) are innervated by the facial nerve.

4. **L**oose areolar tissue: a sponge-like layer including potential spaces that may distend with fluid as a result of injury or infection. This layer allows free movement of the **scalp proper** (the first three layers—skin, connective tissue, and epicranial aponeurosis) over the underlying calvaria.

5. **P**ericranium: a dense layer of connective tissue that forms the external periosteum of the neurocranium. It is firmly attached but can be stripped fairly easily from the crania of living persons, except where the pericranium is continuous with the fibrous tissue in the cranial sutures.

Muscles of Face and Scalp

The facial muscles (muscles of facial expression) are in the subcutaneous tissue of the anterior and posterior scalp, face, and neck. They move the skin and change facial expressions to convey mood. Most muscles attach to bone or fascia and produce their effects by pulling the skin. The muscles of the scalp and face are illustrated in Figure 7.16, and their attachments and actions are provided in Table 7.3. Certain muscles and/or muscle groups will be discussed in further detail.

All muscles of facial expression develop from mesoderm in the second pharyngeal arches. A subcutaneous muscular sheet forms during embryonic development that spreads over the neck and face, carrying branches of the nerve of the arch (the facial nerve, CN VII) with it to supply all the muscles formed from the arch (Moore et al., 2012). The muscular sheet differentiates into muscles that surround the facial orifices (mouth, eyes, and nose), serving as sphincter and dilator mechanisms that also produce many facial expressions (Fig. 7.17). Because of their common embryological origin, the platysma and facial muscles are often fused, and their fibers are frequently intermingled.

MUSCLES OF SCALP, FOREHEAD, AND EYEBROWS

The **occipitofrontalis** is a flat digastric muscle, with **occipital** and **frontal bellies** that share a common tendon, the **epicranial aponeurosis** (Figs. 7.15 and 7.16A & B; Table 7.3). Because the aponeurosis is a layer of the scalp, independent contraction of the occipital belly retracts the scalp and contraction of the frontal belly protracts it. Acting simultaneously, the occipital belly, with bony attachments, works as a synergist with the frontal belly, which has no bony attachments, to elevate the eyebrows and produce transverse wrinkles across the forehead. This gives the face a surprised look.

MUSCLES OF MOUTH, LIPS, AND CHEEKS

The lips, shape, and degree of opening of the mouth are important for clear speech. In addition, we add emphasis to our vocal communication with our facial expressions. Several muscles alter the shape of the mouth and lips during speaking as well as during such activities as singing, whistling, and mimicry. The shape of the mouth and lips is controlled by a complex three-dimensional group of muscular slips, which include the following (Fig. 7.16B & C; Table 7.3):

- Elevators, retractors, and evertors of the upper lip.
- Depressors, retractors, and evertors of the lower lip.
- The orbicularis oris, the sphincter around the mouth.
- The buccinator in the cheek.

At rest, the lips are in gentle contact and the teeth are close together.

The **orbicularis oris,** the first of the series of sphincters associated with the alimentary system (digestive tract), encircles the mouth within the lips, controlling entry, and exit through the **oral fissure** (L. *rima oris,* the opening between the lips). The orbicularis oris is important during articulation (speech).

The **buccinator** (L. trumpeter) is a thin, flat, rectangular muscle that attaches laterally to the alveolar processes of the maxillae and mandible, opposite the molar teeth. It also attaches to the **pterygomandibular raphe,** a tendinous thickening of the buccopharyngeal fascia separating and giving origin to the superior pharyngeal constrictor posteriorly. The buccinator occupies a deeper, more medially placed plane than the other facial muscles, passing deep to the mandible so that it is more closely related to the buccal mucosa than to the skin of the face. The buccinator, active in smiling, also keeps the cheek taut, thereby preventing it from folding and being injured during chewing.

Anteriorly, the fibers of the buccinator mingle medially with those of the orbicularis oris, and the tonus of the two muscles compresses the cheeks and lips against the teeth and gums. The tonic contraction of the buccinator, and especially of the orbicularis oris, provides a gentle but continual resistance to the tendency of the teeth to tilt in an outward direction. In the presence of a short upper lip, or retractors that remove this force, crooked or protrusive ("buck") teeth develop.

The orbicularis oris (from the labial aspect) and buccinator (from the buccal aspect) work with the tongue (from the lingual aspect) to keep food between the occlusal surfaces of the teeth during mastication (chewing) and to prevent food from accumulating in the oral vestibule.

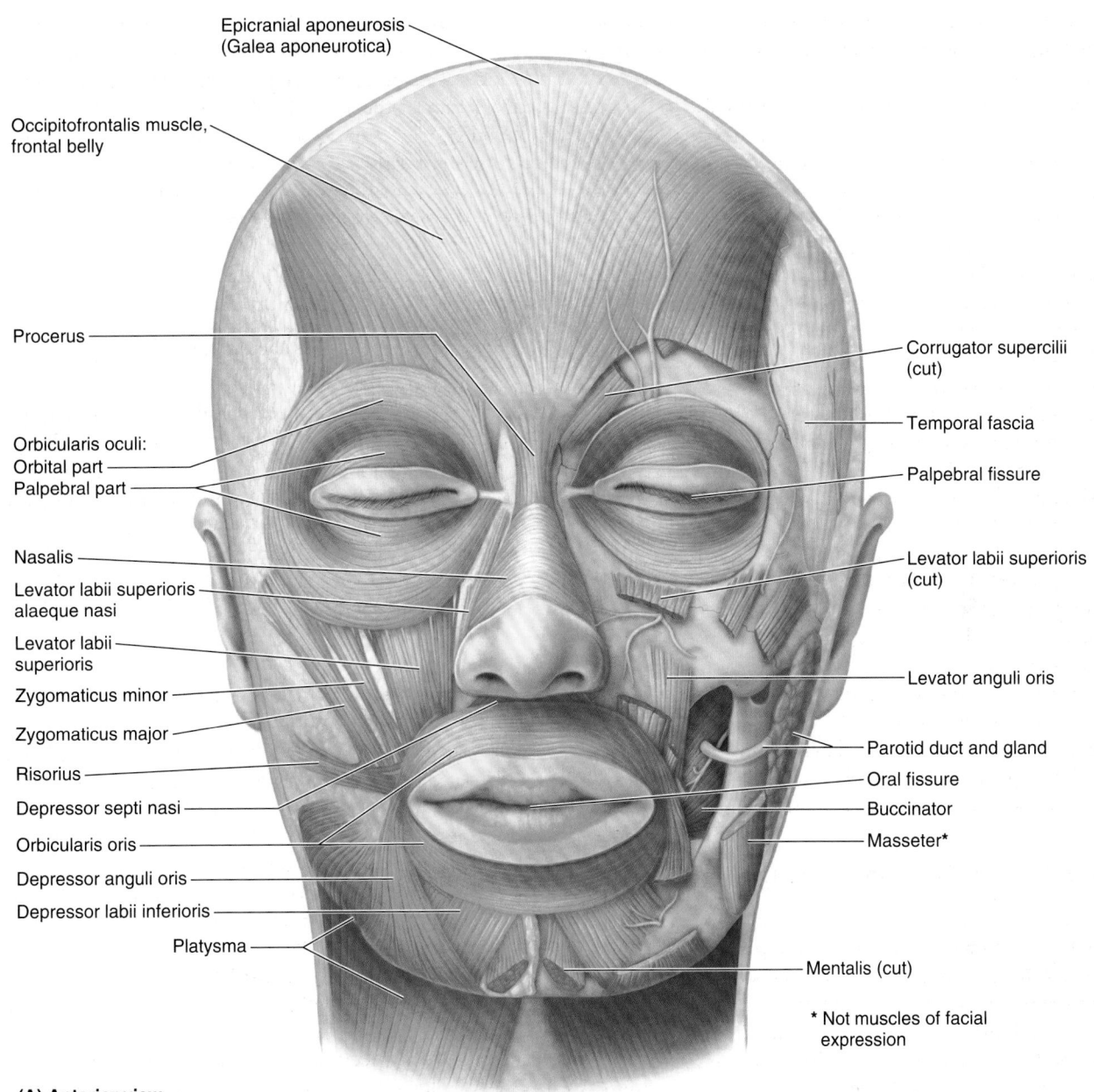

(A) Anterior view

FIGURE 7.16. Muscles of scalp and face.

TABLE 7.3. MUSCLES OF SCALP AND FACE

Muscle[a]	Origin	Insertion	Main Action(s)
Occipitofrontalis			
Front belly[2]	Epicranial aponeurosis	Skin and subcutaneous tissue of eyebrows and forehead	Elevates eyebrows and wrinkles skin of forehead; protracts scalp (indicating surprise or curiosity)
Occipital belly[1]	Lateral two thirds of superior nuchal line	Epicranial aponeurosis	Retracts scalp; increasing effectiveness of frontal belly
Orbicularis oculi (orbital sphincter)[2,3]	Medial orbital margin; medial palpebral ligament; lacrimal bone	Skin around margin of orbit; superior and inferior tarsal plates	Closes eyelids: palpebral part does so gently; orbital part tightly (winking)

[a]All facial muscles are innervated by the facial nerve (CN VII) via its posterior auricular branch (1) or via the temporal (2), zygomatic (3), buccal (4), marginal mandibular (5), or cervical (6) branches of the parotid plexus.

TABLE 7.3. MUSCLES OF SCALP AND FACE (Continued)

Muscle[a]	Origin	Insertion	Main Action(s)
Corrugator supercilii[2]	Medial end of superciliary arch	Skin superior to middle of supra-orbital margin and superciliary arch	Draws eyebrow medially and inferiorly, creating vertical wrinkles above nose (demonstrating concern or worry)
Procerus plus transverse part of nasalis[4]	Fascia aponeurosis covering nasal bone and lateral nasal cartilage	Skin of inferior forehead, between eyebrows	Depresses medial end of eyebrow; wrinkles skin over dorsum of nose (conveying disdain or dislike)
Alar part of nasalis plus levator labii superioris alaeque nasii[4]	Frontal process of maxilla (inferomedial margin of orbit)	Major alar cartilage	Depresses ala laterally, dilating anterior nasal aperture (i.e., "flaring nostrils," as during anger or exertion)
Orbicularis oris (oral sphincter)[4]	Medial maxilla and mandible; deep surface of peri-oral skin; angle of mouth (modiolus)	Mucous membrane of lips	Tonus closes oral fissure; phasic contraction compresses and protrudes lips (kissing) or resists distension (when blowing)
Levator labii superioris[4]	Infra-orbital margin (maxilla)	Skin of upper lip	Part of dilators of mouth; retract (elevate) and/or evert upper lip; deepen nasolabial sulcus (showing sadness)
Zygomaticus minor[4]	Anterior aspect, zygomatic bone		
Buccinator (cheek muscle)[4]	Mandible, alveolar processes of maxilla and mandible, pterygomandibular raphe	Angle of mouth (modiolus); orbicularis oris	Presses cheek against molar teeth; works with tongue to keep food between occlusal surfaces and out of oral vestibule; resists distension (when blowing)
Zygomaticus major[4]	Lateral aspect of zygomatic bone		Part of dilators of mouth; elevate labial commissure—bilaterally to smile (happiness); unilaterally to sneer (disdain)
Levator anguli oris[4]	Infra-orbital maxilla (canine fossa)	Angle of mouth (modiolus)	Part of dilators of mouth; widens oral fissure, as when grinning or grimacing
Risorius[4]	Parotid fascia and buccal skin (highly variable)		Part of dilators of mouth; depresses labial commissure bilaterally to frown (sadness)
Depressor anguli oris[5]	Anterolateral base of mandible		
Depressor labii inferioris[5]	Platysma and anterolateral body of mandible	Skin of lower lip	Part of dilators of mouth; retracts (depresses) and/or everts lower lip (pouting, sadness)
Mentalis[5]	Body of mandible (anterior to roots of inferior incisors)	Skin of chin (mentolabial sulcus)	Elevates and protrudes lower lip; elevates skin of chin (showing doubt)
Platysma[6]	Subcutaneous tissue of infra-clavicular and supraclavicular regions	Base of mandible; skin of cheek and lower lip; angle of mouth (modiolus); orbicularis oris	Depresses mandible (against resistance); tenses skin of inferior face and neck (conveying tension and stress)

[a]All facial muscles are innervated by the facial nerve (CN VII) via its posterior auricular branch (*1*) or via the temporal (*2*), zygomatic (*3*), buccal (*4*), marginal mandibular (*5*), or cervical (*6*) branches of the parotid plexus.

The buccinator also helps the cheeks resist the forces generated by whistling and sucking. The buccinator was given its name because it compresses the cheeks (L. *buccae*) during blowing (e.g., when a musician plays a wind instrument). Some trumpeters (notably the late Dizzy Gillespie) stretch their buccinators and other cheek muscles so much that their cheeks balloon out when they blow forcibly on their instruments.

Several dilator muscles radiate from the lips and angles of the mouth, somewhat like the spokes of a wheel, retracting the various borders of the *oral fissure* collectively, in groups, or individually. Lateral to the angles of the mouth or **commissures of the lips** (the junctions of the upper and lower lips) fibers of as many as nine facial muscles interlace or merge in a highly variable and multiplanar formation called

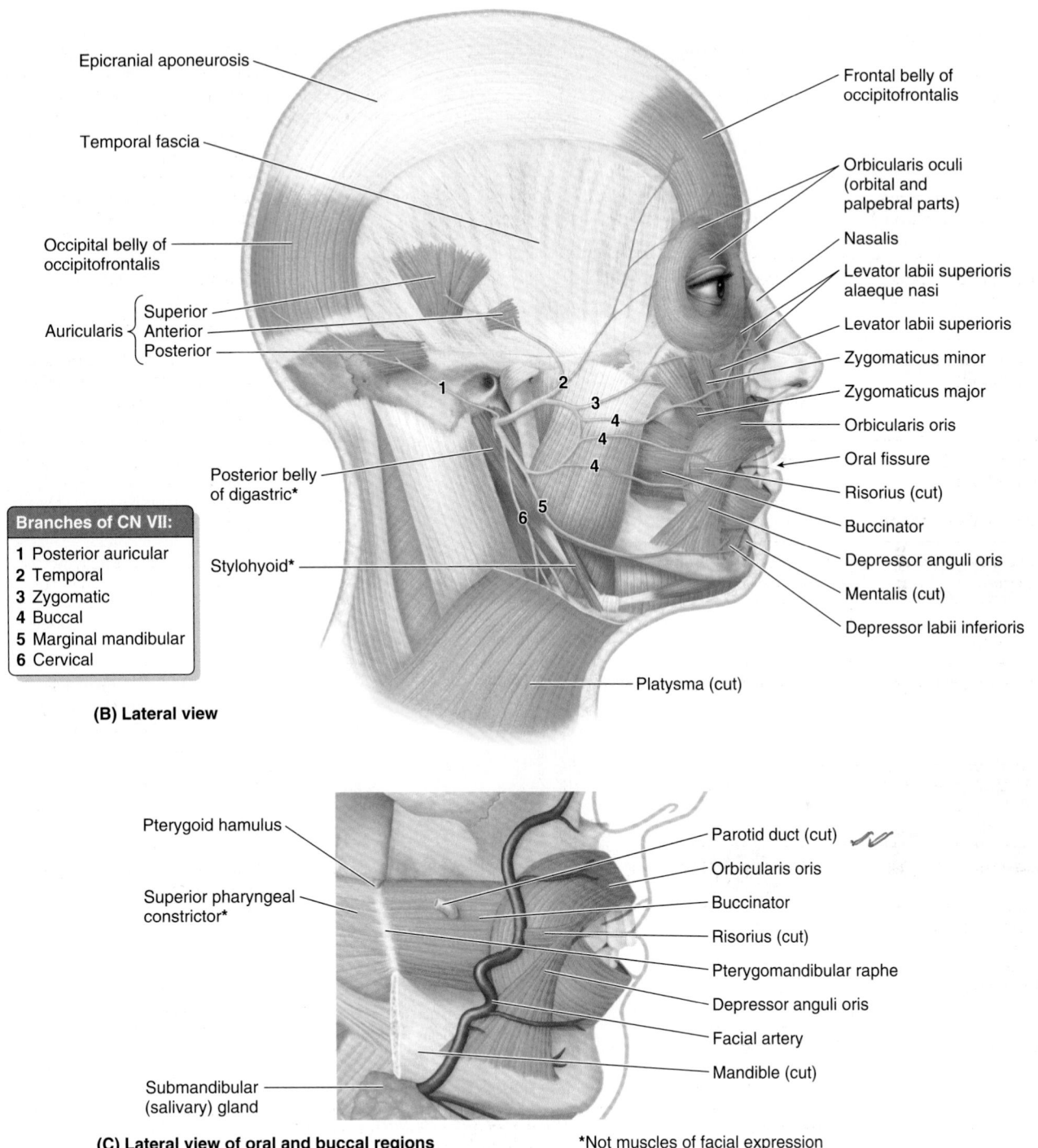

Epicranial aponeurosis

Temporal fascia

Occipital belly of occipitofrontalis

Auricularis { Superior / Anterior / Posterior }

Posterior belly of digastric*

Stylohyoid*

Frontal belly of occipitofrontalis

Orbicularis oculi (orbital and palpebral parts)

Nasalis

Levator labii superioris alaeque nasi

Levator labii superioris

Zygomaticus minor

Zygomaticus major

Orbicularis oris

Oral fissure

Risorius (cut)

Buccinator

Depressor anguli oris

Mentalis (cut)

Depressor labii inferioris

Platysma (cut)

Branches of CN VII:
1 Posterior auricular
2 Temporal
3 Zygomatic
4 Buccal
5 Marginal mandibular
6 Cervical

(B) Lateral view

Pterygoid hamulus

Superior pharyngeal constrictor*

Submandibular (salivary) gland

Parotid duct (cut)

Orbicularis oris

Buccinator

Risorius (cut)

Pterygomandibular raphe

Depressor anguli oris

Facial artery

Mandible (cut)

(C) Lateral view of oral and buccal regions   *Not muscles of facial expression

FIGURE 7.16. *(Continued)*

the **modiolus,** which is largely responsible for the occurrence of dimples in many individuals.

The **platysma** (G. flat plate) is a broad, thin sheet of muscle in the subcutaneous tissue of the neck (Fig. 7.16A & B; Table 7.3). The anterior borders of the two muscles decussate over the chin and blend with the facial muscles. Acting from its superior attachment, the platysma tenses the skin, producing vertical skin ridges, conveying great stress, and releasing pressure on the superficial veins. Acting from

its inferior attachment, the platysma helps depress the mandible and draw the corners of the mouth inferiorly, as in a grimace.

MUSCLES OF ORBITAL OPENING

The function of the eyelids (L. *palpebrae*) is to protect the eyeballs from injury and excessive light. The eyelids also keep the cornea moist by spreading the tears.

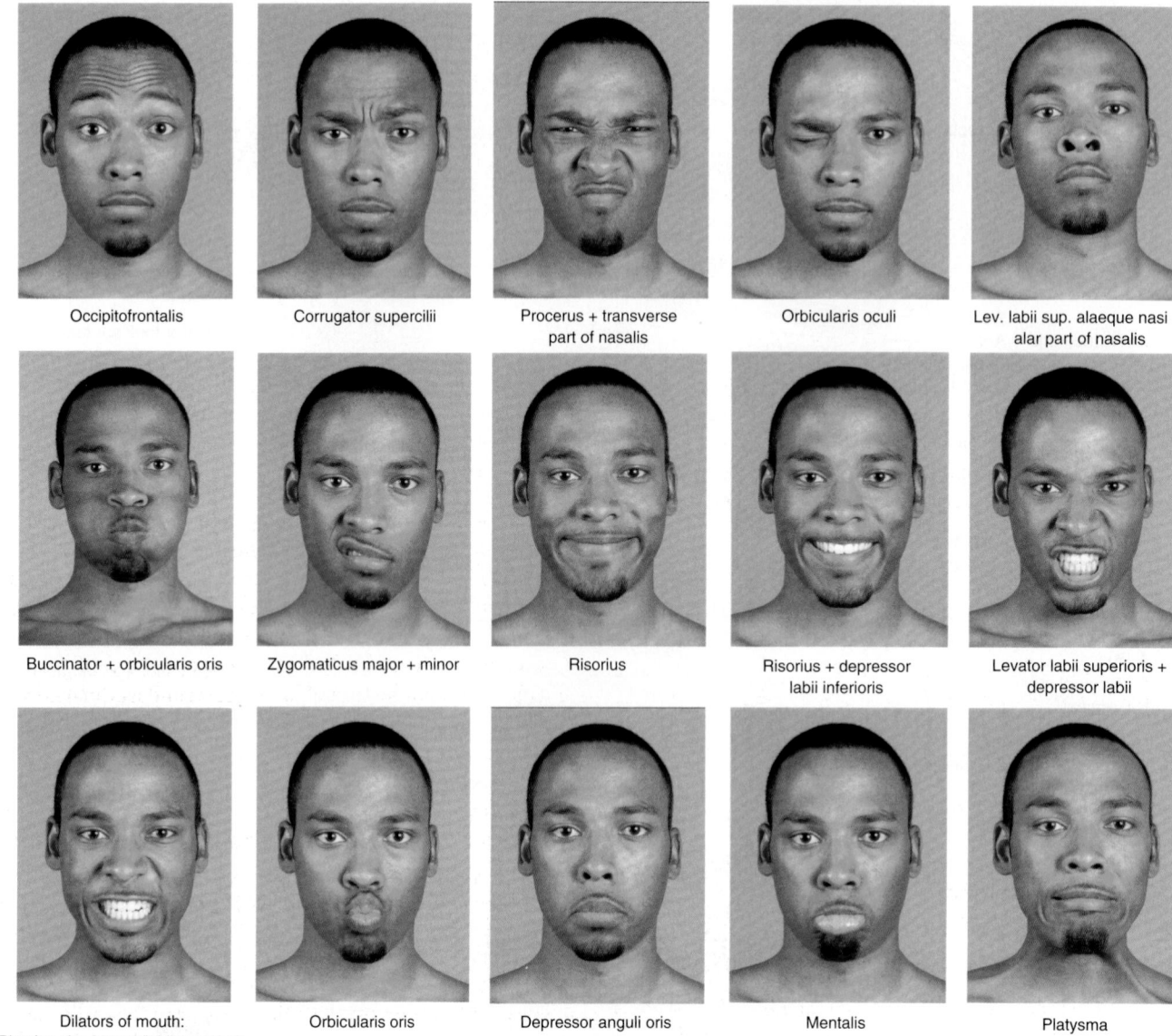

Occipitofrontalis

Corrugator supercilii

Procerus + transverse
part of nasalis

Orbicularis oculi

Lev. labii sup. alaeque nasi +
alar part of nasalis

Buccinator + orbicularis oris

Zygomaticus major + minor

Risorius

Risorius + depressor
labii inferioris

Levator labii superioris +
depressor labii

Dilators of mouth:
Risorius plus levator labii superioris
+ depressor labii inferioris

Orbicularis oris

Depressor anguli oris

Mentalis

Platysma

FIGURE 7.17. Muscles of facial expression in action. These muscles are superficial sphincters and dilators of the orifices of the head. The facial muscles, supplied by the facial nerve (CN VII), are attached to and move the skin of the face, producing many facial expressions.

The **orbicularis oculi** closes the eyelids and wrinkles the forehead vertically (Figs. 7.16A & B and 7.18; Table 7.3). Its fibers sweep in concentric circles around the orbital margin and eyelids. Contraction of these fibers narrows the **palpebral fissure** (aperture between the eyelids) and assists the flow of lacrimal fluid (tears) by bringing the lids together laterally first, closing the palpebral fissure in a lateral to medial direction. The orbicularis oculi muscle consists of three parts:

1. *Palpebral part:* arising from the **medial palpebral ligament** and mostly located within the eyelids, gently closes the eyelids (as in blinking or sleep) to keep the cornea from drying.
2. *Lacrimal part:* passing posterior to the *lacrimal sac,* draws the eyelids medially, aiding drainage of tears.

3. *Orbital part:* overlying the orbital rim and attached to the frontal bone and maxilla medially, tightly closes the eyelids (as in winking or squinting) to protect the eyeballs against glare and dust.

When all three parts of the orbicularis oculi contract, the eyes are firmly closed (Figs. 7.17 and 7.18C).

MUSCLES OF NOSE AND EARS

As demonstrated in the blue box "Flaring of Nostrils" (p. 861), the muscles of the nose may provide evidence of breathing behaviors. Otherwise, although these muscles are functionally important in certain mammals (elephants, tapirs, rabbits,

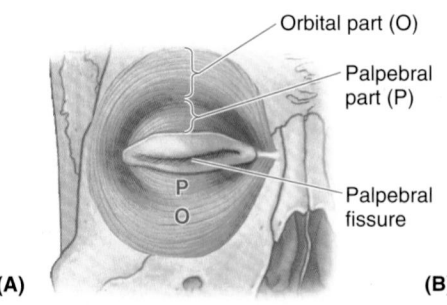
Orbital part (O)
Palpebral part (P)
Palpebral fissure
(A)

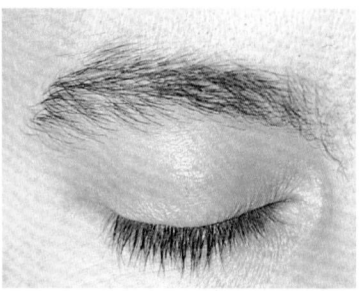
(B)

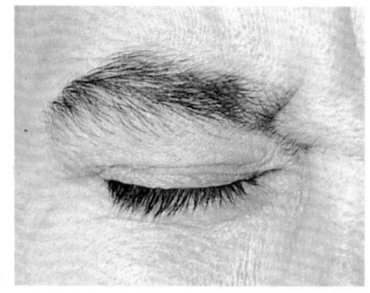
(C)

FIGURE 7.18. Disposition and actions of orbicularis oculi muscle. A. The orbital and palpebral parts of the orbicularis oculi are demonstrated. **B.** The palpebral part gently closes the eyelids. **C.** The orbital part tightly closes the eyelids.

and some diving mammals), they are relatively unimportant in humans, except in terms of facial expression and in the specialized field of aesthetic plastic surgery. The muscles of the ears, important in animals capable of cocking or directing the ears toward the sources of sounds, are even less critical in humans.

Nerves of Face and Scalp

Cutaneous (sensory) innervation of the face and anterosuperior part of the scalp is provided primarily by the *trigeminal nerve* (CN V), whereas motor innervation to the facial muscles is provided by the *facial nerve* (CN VII).

CUTANEOUS NERVES OF FACE AND SCALP

The **trigeminal nerve** (CN V) originates from the lateral surface of the pons of the midbrain by two roots: motor and sensory. These roots are comparable to the motor (anterior) and sensory (posterior) roots of spinal nerves. The sensory root of CN V consists of the central processes of pseudounipolar neurons located in a sensory ganglion (**trigeminal ganglion**) at the distal end of the root, which is bypassed by the multipolar neuronal axons making up the motor root. CN V is the sensory nerve for the face and the motor nerve for the muscles of mastication and several small muscles (Fig. 7.19).

The peripheral processes of the neurons of the trigeminal ganglion constitute three divisions of the nerve: the *ophthalmic nerve* (CN V₁), the *maxillary nerve* (CN V₂), and the sensory component of the *mandibular nerve* (CN V₃). These nerves are named according to their main areas of termination: the eye, maxilla, and mandible, respectively. The first two divisions (ophthalmic and maxillary nerves) are wholly sensory. The mandibular nerve is largely sensory, but it also receives the motor fibers (axons) from the motor root of CN V that mainly supply the muscles of mastication. The cutaneous nerves derived from each division of CN V are illustrated in Figure 7.20, and the origin, course, and distribution of each nerve are listed and described in Table 7.4.

The cutaneous nerves of the neck overlap those of the face. Cutaneous branches of cervical nerves from the *cervical plexus* extend over the posterior aspect of the neck and scalp. The **great auricular nerve** in particular innervates the inferior aspect of the auricle (external ear) and much of

the parotid region of the face (the area overlying the angle of the jaw).

OPHTHALMIC NERVE

The **ophthalmic nerve** (CN V₁), the superior division of the trigeminal nerve, is the smallest of the three divisions of CN V. It arises from the trigeminal ganglion as a wholly sensory nerve and supplies the area of skin derived from the embryonic *frontonasal prominence* (Moore et al., 2012). As CN V₁ enters the orbit through the *superior orbital fissure*, it trifurcates into the frontal, nasociliary, and lacrimal nerves (Fig. 7.19). Except for the external nasal nerve, the cutaneous branches of CN V₁ reach the skin of the face via the orbital opening (Fig. 7.21).

The **frontal nerve,** the largest branch produced by the trifurcation of CN V₁, runs along the roof of the orbit toward the orbital opening, bifurcating approximately midway into the cutaneous **supra-orbital** and **supratrochlear nerves,** distributed to the forehead and scalp (Figs. 7.21 and 7.22).

The **nasociliary nerve,** the intermediate branch of the CN V₁ trifurcation, supplies branches to the eyeball and divides within the orbit into the posterior ethmoidal, anterior ethmoidal, and infratrochlear nerves (Fig. 7.19). The *posterior* and *anterior ethmoidal nerves* leave the orbit, the latter running a circuitous course passing through the cranial and nasal cavities. Its terminal branch, the **external nasal nerve,** is a cutaneous nerve supplying the external nose. The **infratrochlear nerve** is a terminal branch of the nasociliary nerve and its main cutaneous branch.

The **lacrimal nerve,** the smallest branch from the trifurcation of CN V₁, is primarily a cutaneous branch, but it also conveys some secretomotor fibers, sent via a communicating branch, from a ganglion associated with the maxillary nerve for innervation of the lacrimal gland (Figs. 7.20 and 7.21).

MAXILLARY NERVE

The *maxillary nerve* (CN V₂), the intermediate division of the trigeminal nerve, also arises as a wholly sensory nerve (Fig. 7.19A). CN V₂ passes anteriorly from the *trigeminal ganglion* and leaves the cranium through the *foramen rotundum* in the base of the greater wing of the sphenoid.

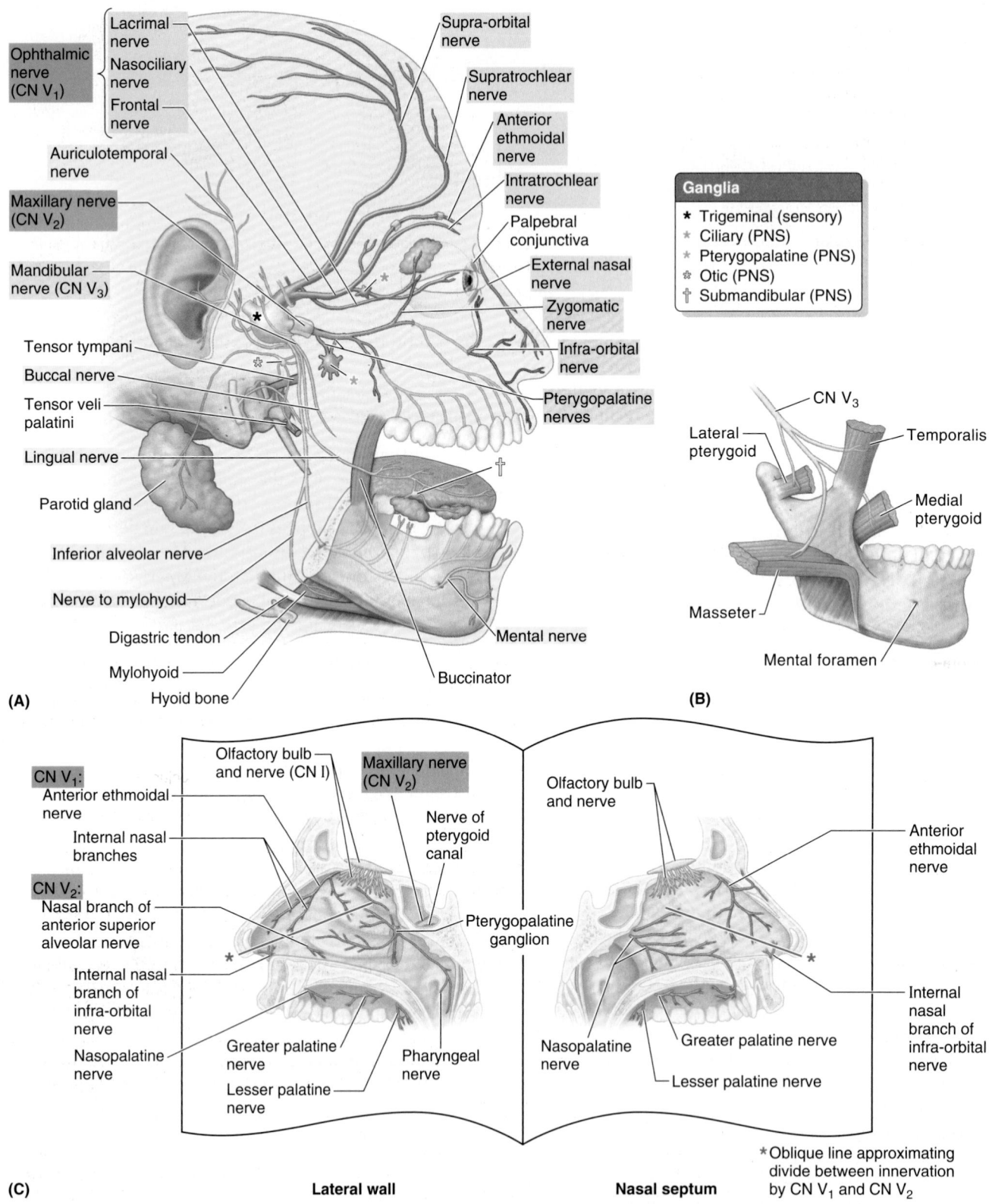

Ophthalmic nerve (CN V₁)
- Lacrimal nerve
- Nasociliary nerve
- Frontal nerve

Auriculotemporal nerve

Maxillary nerve (CN V₂)

Mandibular nerve (CN V₃)

Tensor tympani

Buccal nerve

Tensor veli palatini

Lingual nerve

Parotid gland

Inferior alveolar nerve

Nerve to mylohyoid

Digastric tendon

Mylohyoid

Hyoid bone

Supra-orbital nerve

Supratrochlear nerve

Anterior ethmoidal nerve

Intratrochlear nerve

Palpebral conjunctiva

External nasal nerve

Zygomatic nerve

Infra-orbital nerve

Pterygopalatine nerves

Mental nerve

Buccinator

(A)

Ganglia
- * Trigeminal (sensory)
- * Ciliary (PNS)
- * Pterygopalatine (PNS)
- ✿ Otic (PNS)
- † Submandibular (PNS)

CN V₃

Lateral pterygoid

Temporalis

Medial pterygoid

Masseter

Mental foramen

(B)

CN V₁:
Anterior ethmoidal nerve

Internal nasal branches

CN V₂:
Nasal branch of anterior superior alveolar nerve

Internal nasal branch of infra-orbital nerve

Nasopalatine nerve

Olfactory bulb and nerve (CN I)

Maxillary nerve (CN V₂)

Nerve of pterygoid canal

Pterygopalatine ganglion

Greater palatine nerve

Lesser palatine nerve

Pharyngeal nerve

Olfactory bulb and nerve

Anterior ethmoidal nerve

Internal nasal branch of infra-orbital nerve

Nasopalatine nerve

Greater palatine nerve

Lesser palatine nerve

*Oblique line approximating divide between innervation by CN V₁ and CN V₂

(C)   **Lateral wall**   **Nasal septum**

FIGURE 7.19. Distribution of trigeminal nerve (CN V). A. The three divisions of CN V arise from the trigeminal ganglion. In addition to the trigeminal ganglion, a sensory ganglion (similar to the sensory or dorsal root ganglia of spinal nerves) and four parasympathetic ganglia (three of which are shown here) are associated with the branches of the trigeminal nerve. **B.** Branches of the mandibular nerve (CN V₃) pass to the muscles of mastication. **C.** This "opened book" view of the lateral wall and septum of the right nasal cavity demonstrates superficial and deep distribution of CN V₁, and CN V₂ (and, incidentally, CN I) to the nasal and upper oral cavity, in and near the midline of the head.

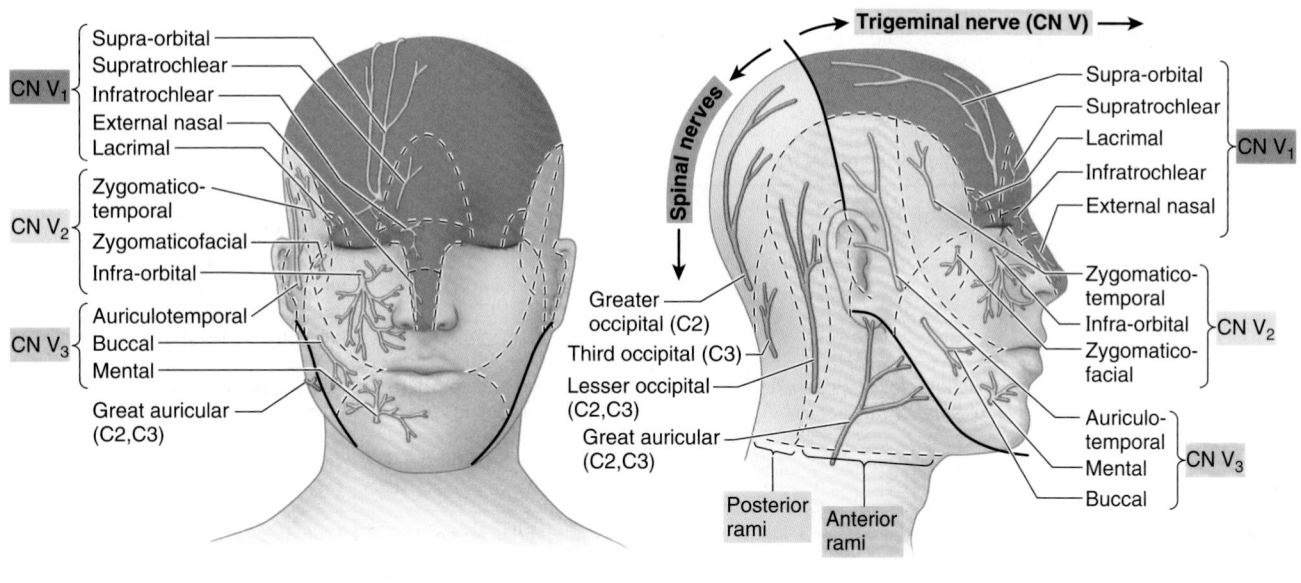

FIGURE 7.20. Cutaneous nerves of face and scalp.

TABLE 7.4. CUTANEOUS NERVES OF FACE AND SCALP

Nerve	Origin	Course	Distribution
Cutaneous nerves derived from ophthalmic nerve (CN V₁)			
Supra-orbital	Largest branch from bifurcation of *frontal nerve,* approximately in middle of orbital roof	Continues anteriorly along roof of orbit, emerging via supra-orbital notch or foramen; ascends forehead, breaking into branches	Mucosa of *frontal sinus;* skin and conjunctiva of middle of *superior eyelid;* skin and pericranium of *anterolateral forehead* and *scalp* to vertex (interauricular line)
Supratrochlear	Smaller branch from bifurcation of *frontal nerve,* approximately in middle of orbital roof	Continues anteromedially along roof of orbit, passing lateral to trochlea and ascending forehead	Skin and conjunctive of medical aspect of *superior eyelid;* skin and pericranium of *anteromedial forehead*
Lacrimal	Smallest branch from trifurcation of *CN V₁* proximal to superior orbital fissure	Runs superolaterally through orbit, receiving secretomotor fibers via a communicating branch from the zygomaticotemporal nerve	*Lacrimal gland* (secretomotor fibers); small area of skin and conjunctive of *lateral part of superior eyelid*
Infratrochlear	Terminal branch (with anterior ethmoidal nerve) of *nasociliary nerve*	Follows medial wall of orbit, passing inferior to trochlea	Skin lateral to *root of nose;* skin and conjunctiva of *eyelids adjacent to medial canthus, lacrimal sac,* and *lacrimal caruncle*
External nasal	Terminal branch of *anterior ethmoidal nerve*	Emerges from nasal cavity by passing between nasal bone and lateral nasal cartilage	Skin of nasal *ala, vestibule,* and *dorsum of nose,* including *apex*
Cutaneous nerves derived from maxillary nerve (CN V₂)			
Infra-orbital	Continuation of *CN V₂* distal to its entrance into the orbit via the inferior orbital fissure	Traverses infra-orbital groove and canal in orbital floor, giving rise to superior alveolar branches; then emerges via infra-orbital foramen, immediately dividing into inferior palpebral, internal and external nasal, and superior labial branches	Mucosa of *maxillary sinus;* premolar, canine, and incisor *maxillary teeth;* skin and conjunctiva of *inferior eyelid;* skin of *cheek, lateral nose,* and antero-inferior *nasal septum;* skin and oral mucosa of *superior lip*
Zygomaticofacial	Smaller terminal branch (with zygomaticotemporal nerve) of *zygomatic nerve*	Traverses zygomaticofacial canal in zygomatic bone at inferolateral angle of orbit	Skin on prominence of *cheek*

TABLE 7.4. CUTANEOUS NERVES OF FACE AND SCALP (Continued)

Nerve	Origin	Course	Distribution
Zygomaticotemporal	Larger terminal branch (with zygomaticofacial nerve) of *zygomatic nerve*	Sends communicating branch to lacrimal nerve in orbit; then passes to temporal fossa via zygomatico-temporal canal in zygomatic bone	Hairless skin *anterior part of temporal fossa*
Cutaneous nerves derived from mandibular nerve (CN V₃)			
Auriculotemporal	In infratemporal fossa via two roots from *posterior trunk of CN V₃ that encircle middle meningeal artery*	Passes posteriorly deep to ramus of mandible and superior deep part of parotid gland, emerging posterior to temporomandibular joint	Skin anterior to auricle and posterior two thirds of *temporal region;* skin of tragus and adjacent helix of *auricle;* skin of roof of *external acoustic meatus;* and skin of superior *tympanic membrane*
Buccal	In infratemporal fossa as sensory branch of *anterior trunk of CN V₃*	Passes between two parts of lateral pterygoid muscle, emerging anteriorly from cover of ramus of mandible and masseter, uniting with buccal branches of facial nerve	Skin and oral mucosa of *cheek* (overlying and deep to anterior part of buccinator); *buccal gingivae* (gums) adjacent to second and third molars
Mental	Terminal branch of *inferior alveolar nerve (CN V₃)*	Emerges from mandibular canal via mental foramen in anterolateral aspect of body of mandible	Skin of *chin* and skin; oral mucosa of *inferior lip*
Cutaneous nerves derived from anterior rami of cervical spinal nerves			
Great auricular	Spinal nerves C2 and C3 via cervical plexus	Ascends vertically across sterno-cleidomastoid, posterior to external jugular vein	Skin overlying angle of mandible and inferior lobe of auricle; parotid sheath
Lesser occipital		Follows posterior border of sterno-cleidomastoid; then ascends posterior to auricle	Scalp posterior to auricle
Cutaneous nerves derived from posterior rami of cervical spinal nerves			
Greater occipital nerve	As medial branch of posterior ramus of spinal nerve C2	Emerges between axis and obliquus capitis inferior; then pierces trapezius	Scalp of occipital region
Third occipital nerve	As lateral branch of posterior ramus of spinal nerve C3	Pierces trapezius	Scalp of lower occipital and suboccipital regions

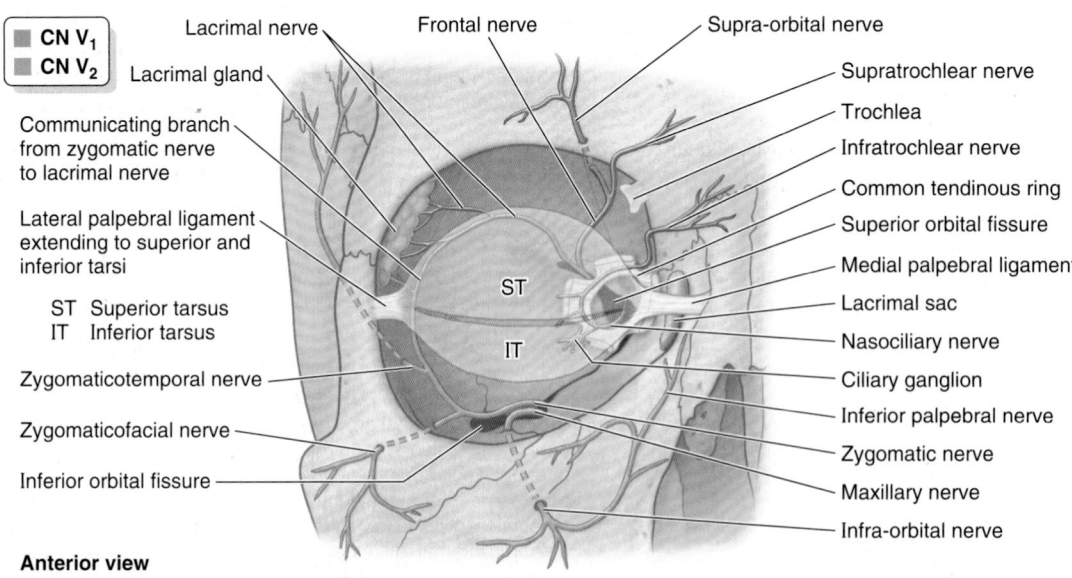

Anterior view

FIGURE 7.21. **Cutaneous nerves of orbital/peri-orbital region.** Cutaneous nerves are shown in relation to the orbital walls and rim and the fibrous skeleton of the eyelids. The skin of the superior eyelid is supplied by branches of the ophthalmic nerve (CN V₁), whereas the inferior eyelid is supplied mainly by branches of the maxillary nerve (CN V₂).

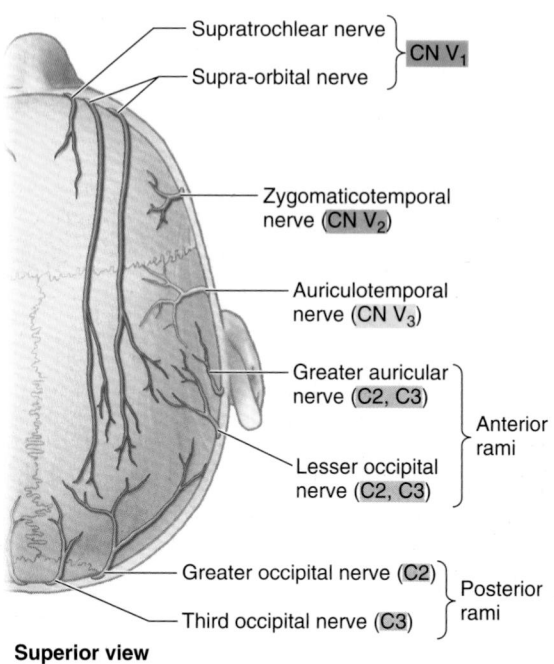

Supratrochlear nerve } CN V₁
Supra-orbital nerve

Zygomaticotemporal nerve (CN V₂)

Auriculotemporal nerve (CN V₃)

Greater auricular nerve (C2, C3) } Anterior rami
Lesser occipital nerve (C2, C3)

Greater occipital nerve (C2) } Posterior rami
Third occipital nerve (C3)

Superior view

FIGURE 7.22. Nerves of scalp. The nerves appear in sequence: CN V₁, CN V₂, CN V₃, anterior rami of C2 and C3, and posterior rami of C2 and C3.

The maxillary nerve enters the *pterygopalatine fossa*, where it gives off branches to the *pterygopalatine ganglion* and continues anteriorly, entering the orbit through the *inferior orbital fissure* (Fig. 7.19). It gives off the zygomatic nerve and passes anteriorly into the *infra-orbital groove and foramen* as the infra-orbital nerve (Fig. 7.21).

The **zygomatic nerve** runs to the lateral wall of the orbit, giving rise to two of the three cutaneous branches of CN V₂, the **zygomaticofacial** and **zygomaticotemporal nerves.** The latter nerve sends a communicating branch conveying secretomotor fibers to the lacrimal nerve. En route to the face, the **infra-orbital nerve** gives off palatine branches, branches to the mucosa of the maxillary sinus, and branches to the posterior teeth. It reaches the skin of the face by traversing the *infra-orbital foramen* on the infra-orbital surface of the maxilla. The three cutaneous branches of the maxillary nerve supply the area of skin derived from the embryonic *maxillary prominences* (Moore et al., 2012).

MANDIBULAR NERVE

The *mandibular nerve* (CN V₃) is the inferior and largest division of the trigeminal nerve (Fig. 7.19A). It is formed by the union of sensory fibers from the sensory ganglion and the motor root of CN V in the *foramen ovale* in the greater wing of the sphenoid, through which CN V₃ emerges from the cranium. CN V₃ has three sensory branches that supply the area of skin derived from the embryonic *mandibular prominence.* It also supplies motor fibers to the muscles of mastication (Fig. 7.19B). CN V₃ is the only division of CN V that carries motor fibers. The major cutaneous branches of CN V₃ are the **auriculotemporal, buccal,** and **mental nerves.** En route to the skin, the auriculotemporal nerve passes deep to the parotid gland, conveying secretomotor fibers to it from a ganglion associated with this division of CN V.

NERVES OF SCALP

Innervation of the scalp anterior to the auricles of the external ears is through branches of all three divisions of CN V, the *trigeminal nerve* (Figs. 7.22, 7.20B, and Table 7.4). Posterior to the auricles, the nerve supply is from spinal cutaneous nerves (C2 and C3).

MOTOR NERVES OF FACE

The motor nerves of the face are the *facial nerve* to the muscles of facial expression and the *motor root of the trigeminal nerve/mandibular nerve* to the muscles of mastication (masseter, temporal, medial, and lateral pterygoids). These nerves also supply some more deeply placed muscles (described later in this chapter in relation to the mouth, middle ear, and neck) (Fig. 7.19A).

FACIAL NERVE

CN VII, the **facial nerve,** has a motor root and a sensory/parasympathetic root (the latter being the *intermediate nerve*). The **motor root of CN VII** supplies the muscles of facial expression, including the superficial muscle of the neck (platysma), auricular muscles, scalp muscles, and certain other muscles derived from mesoderm in the embryonic second pharyngeal arch (Fig. 7.23). Following a circuitous route through the temporal bone, CN VII emerges from the cranium through the *stylomastoid foramen* located between the mastoid and styloid processes (Figs. 7.9B and 7.11). It immediately gives off the **posterior auricular nerve,** which passes posterosuperior to the auricle of the ear to supply the auricularis posterior and occipital belly of the occipitofrontalis muscle (Fig. 7.23A & C).

The main trunk of CN VII runs anteriorly and is engulfed by the parotid gland, in which it forms the **parotid plexus.** This plexus gives rise to the five terminal branches of the facial nerve: *temporal, zygomatic, buccal, marginal mandibular,* and *cervical.* The names of the branches refer to the regions they supply. Specific muscles supplied by each branch are identified in Table 7.4.

The **temporal branch of CN VII** emerges from the superior border of the parotid gland and crosses the zygomatic arch to supply the auricularis superior and auricularis anterior; the frontal belly of the occipitofrontalis; and, most important, the superior part of the orbicularis oculi.

The **zygomatic branch of CN VII** passes via two or three branches superior and mainly inferior to the eye to

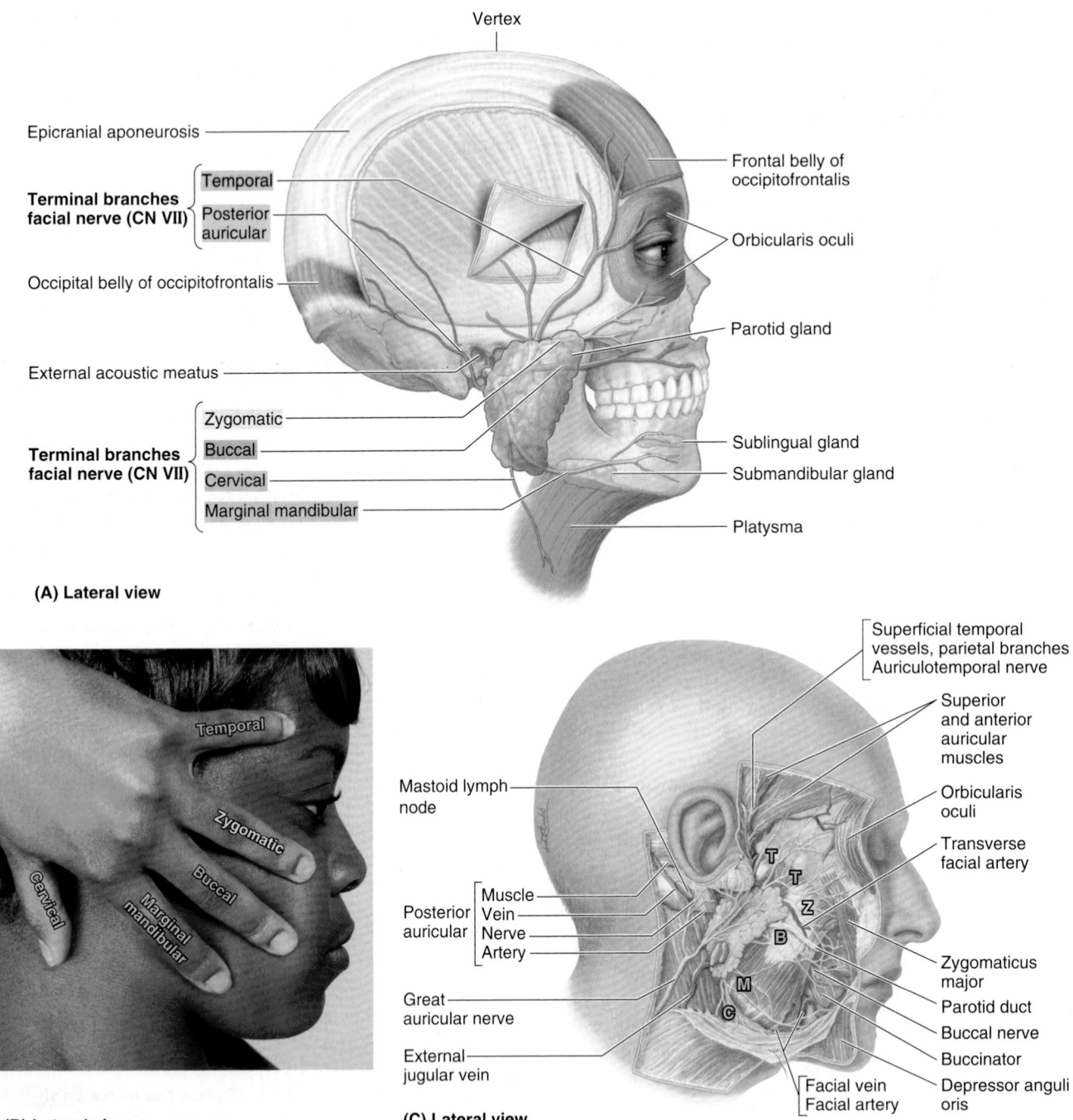

Vertex

Epicranial aponeurosis

Frontal belly of
occipitofrontalis

**Terminal branches
facial nerve (CN VII)**
Temporal
Posterior
auricular

Orbicularis oculi

Occipital belly of occipitofrontalis

Parotid gland

External acoustic meatus

**Terminal branches
facial nerve (CN VII)**
Zygomatic
Buccal
Cervical
Marginal mandibular

Sublingual gland

Submandibular gland

Platysma

(A) Lateral view

Temporal

Zygomatic

Buccal

Cervical

Marginal
mandibular

Superficial temporal
vessels, parietal branches
Auriculotemporal nerve

Superior
and anterior
auricular
muscles

Orbicularis
oculi

Transverse
facial artery

Mastoid lymph
node

Posterior
auricular
Muscle
Vein
Nerve
Artery

Zygomaticus
major

Parotid duct

Buccal nerve

Great
auricular nerve

Buccinator

External
jugular vein

Facial vein
Facial artery

Depressor anguli
oris

(B) Lateral view

(C) Lateral view

FIGURE 7.23. Branches of facial nerve (CN VII). A. The terminal branches of CN VII arise from the parotid plexus within the parotid gland. They emerge from the gland under cover of its lateral surface and radiate in a generally anterior direction across the face. Although intimately related to the parotid gland (and often contacting the submandibular gland via one or more of its lower branches), CN VII does not send nerve fibers to the salivary glands. Two muscles representing the extremes of the distribution of CN VII, the occipitofrontalis and platysma, are also shown. **B.** A simple method for demonstrating and remembering the general course of the five terminal branches of CN VII to the face and neck. **C.** Dissection of the right side of the head showing the great auricular nerve (C2 and C3), which supplies the parotid sheath and skin over the angle of the mandible, and terminal branches of the facial nerve, which supply the muscles of facial expression: *B*, buccal; *C*, cervical; *M*, marginal mandibular; *T*, temporal; *Z*, zygomatic.

supply the inferior part of the orbicularis oculi and other facial muscles inferior to the orbit.

The **buccal branch of CN VII** passes external to the buccinator to supply this muscle and the muscles of the upper lip (upper parts of orbicularis oris and inferior fibers of levator labii superioris).

The **marginal mandibular branch of CN VII** supplies the risorius and muscles of the lower lip and chin. It emerges from the inferior border of the parotid gland and crosses the inferior border of the mandible deep to the platysma to reach the face. In approximately 20% of people, this branch passes inferior to the angle of the mandible.

The **cervical branch of CN VII** passes inferiorly from the inferior border of the parotid gland and runs posterior to the mandible to supply the platysma (Fig. 7.23).

Superficial Vasculature of Face and Scalp

The face is richly supplied by with superficial arteries and external veins, as is evident in blushing and blanching (e.g., becoming pale due to cold). The terminal branches of both arteries and veins anastomose freely, including anastomoses across the midline with contralateral partners.

SUPERFICIAL ARTERIES OF FACE

Most superficial arteries of the face are branches or derivatives of *branches of the external carotid artery,* as illustrated in Figure 7.24. The origin, course, and distribution of these arteries are presented in Table 7.5. The **facial artery** provides the major arterial supply to the face. It arises from the external carotid artery and winds its way to the inferior border of the mandible, just anterior to the masseter (Figs. 7.23C

and 7.24B). The artery lies superficially here, immediately deep to the platysma. The facial artery crosses the mandible, buccinator, and maxilla as it courses over the face to the medial angle (canthus) of the eye, where the superior and inferior eyelids meet (Fig. 7.24B). The facial artery lies deep to the zygomaticus major and levator labii superioris muscles. Near the termination of its sinuous course through the face, the facial artery passes approximately a finger's breadth lateral to the angle of the mouth. The facial artery sends branches to the upper and lower lips (**superior** and **inferior labial arteries**), ascends along the side of the nose, and anastomoses with the dorsal nasal branch of the ophthalmic artery. Distal to the **lateral nasal artery** at the side of the nose, the terminal part of the facial artery is called the **angular artery.**

The **superficial temporal artery** is the smaller terminal branch of the external carotid artery; the other branch is the *maxillary artery.* The superficial temporal artery emerges on the face between the temporomandibular joint (TMJ) and the auricle, enters the temporal fossa, and ends in the scalp by dividing into *frontal* and *parietal branches.* These arterial branches accompany or run in close proximity to the corresponding branches of the auriculotemporal nerve.

The **transverse facial artery** arises from the superficial temporal artery within the parotid gland and crosses the face superficial to the masseter (Figs. 7.23C and 7.24B), approximately a finger's breadth inferior to the zygomatic arch. It divides into numerous branches that supply the parotid gland and duct, the masseter, and the skin of the face. It anastomoses with branches of the facial artery.

In addition to the superficial temporal arteries, several other arteries accompany cutaneous nerves in the face. **Supra-orbital** and **supratrochlear arteries,**

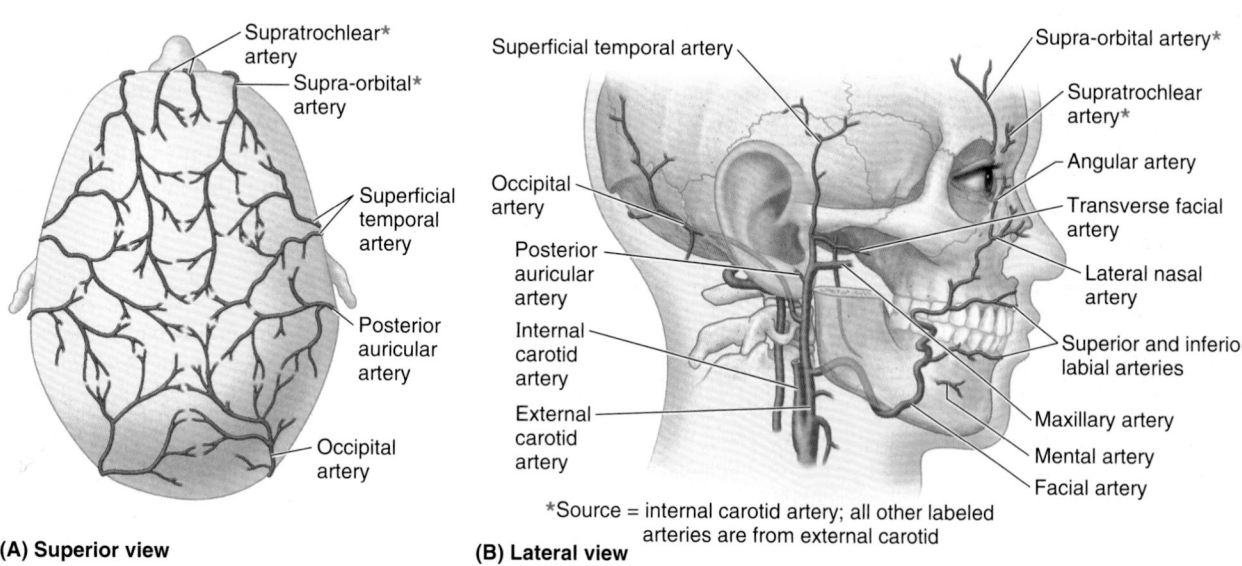

(A) Superior view　　**(B) Lateral view**

*Source = internal carotid artery; all other labeled arteries are from external carotid

FIGURE 7.24. Superficial arteries of face and scalp.

TABLE 7.5. SUPERFICIAL ARTERIES OF FACE AND SCALP

Artery	Origin	Course	Distribution
Facial	External carotid artery	Ascends deep to submandibular gland; winds around inferior border of mandible and enters face	Muscles of facial expression and face
Inferior labial	Facial artery near angle of mouth	Runs medially in lower lip	Lower lip
Superior labial		Runs medially in upper lip	Upper lip and ala (side) and septum of nose
Lateral nasal	Facial artery as it ascends alongside nose	Passes to ala of nose	Skin on ala and dorsum of nose
Angular	Terminal branch of facial artery	Passes to medial angle (canthus) of eye	Superior part of cheek and inferior eyelid
Occipital	External carotid artery	Passes medial to posterior belly of digastric and mastoid process; accompanies occipital nerve in occipital region	Scalp of back of head, as far as vertex
Posterior auricular	External carotid artery	Passes posteriorly, deep to parotid gland, along styloid process between mastoid process and ear	Auricle of ear and scalp posterior to auricle
Superficial temporal	Smaller terminal branch of external carotid artery	Ascends anterior to ear to region and ends in scalp	Facial muscles and skin of temporal frontal and temporal regions
Transverse facial	Superficial temporal artery within parotid gland	Crosses face superficial to masseter and inferior to zygomatic arch	Parotid gland and duct, muscles and skin of face
Mental	Terminal branch of inferior alveolar artery	Emerges from mental foramen and passes to chin	Facial muscles and skin of chin
Supra-orbital[a]	Terminal branch of ophthalmic artery	Passes superiorly from supra-orbital foramen	Muscles and skin of forehead and scalp and superior conjunctiva
Supratrochlear[a]		Passes superiorly from supratrochlear notch	

[a] Source is internal carotid artery.

branches of the ophthalmic artery, accompany nerves of the same name across the eyebrows and forehead (Fig. 7.24; Table 7.5). The supra-orbital artery continues and supplies the anterior scalp to the vertex. The **mental artery,** the only superficial branch derived from the maxillary artery, accompanies the nerve of the same name in the chin.

ARTERIES OF SCALP

The scalp has a rich blood supply (Fig. 7.24A; Table 7.5). The arteries course within layer two of the scalp, the subcutaneous connective tissue layer between the skin and the epicranial aponeurosis. The arteries anastomose freely with adjacent arteries and across the midline with the contralateral artery. The arterial walls are firmly attached to the dense connective tissue in which the arteries are embedded, limiting their ability to constrict when cut. Consequently, bleeding from scalp wounds is profuse.

The arterial supply is from the **external carotid arteries** through the *occipital, posterior auricular,* and *superficial temporal arteries,* and from the *internal carotid arteries* through the *supratrochlear* and *supra-orbital arteries.* The arteries of the scalp supply little blood to the neurocranium, which is supplied primarily by the middle meningeal artery.

EXTERNAL VEINS OF FACE

Most external facial veins are drained by veins that accompany the arteries of the face. As with most superficial veins, they are subject to many variations; a common pattern is shown in Figure 7.25, and Table 7.6 provides details. The venous return from the face is normally superficial, but anastomoses with deep veins, a dural sinus, and venous plexus can provide deep drainage for the valveless veins.

Like veins elsewhere, they have abundant anastomoses that allow drainage to occur by alternate routes during

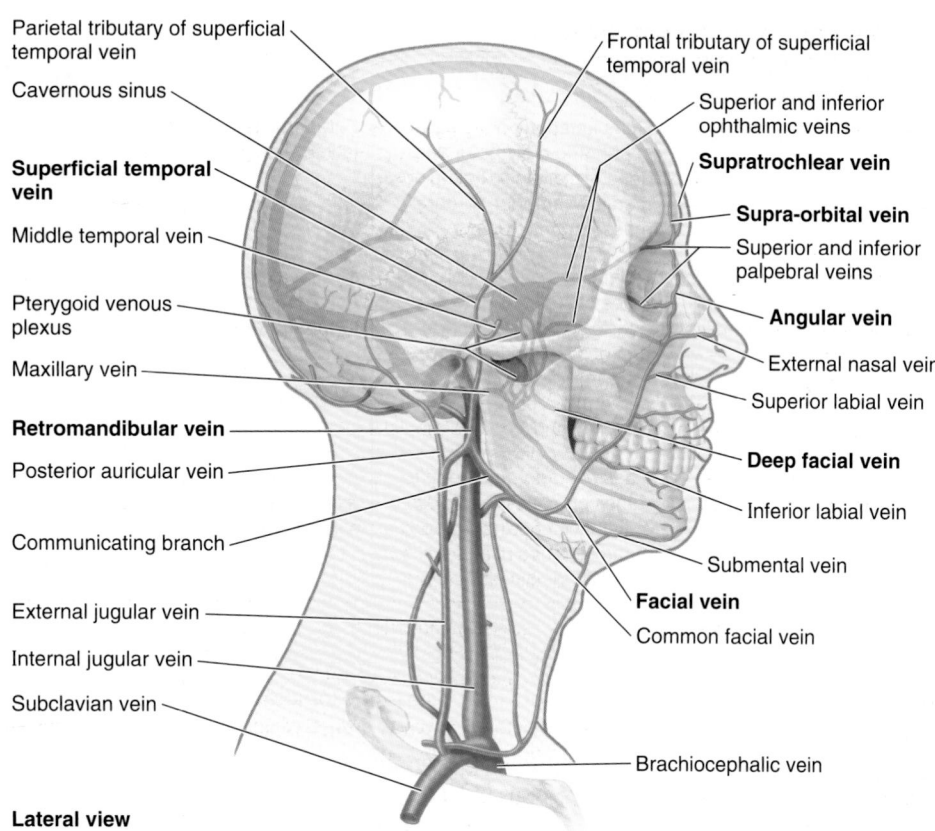

Parietal tributary of superficial temporal vein

Cavernous sinus

Superficial temporal vein

Middle temporal vein

Pterygoid venous plexus

Maxillary vein

Retromandibular vein

Posterior auricular vein

Communicating branch

External jugular vein

Internal jugular vein

Subclavian vein

Frontal tributary of superficial temporal vein

Superior and inferior ophthalmic veins

Supratrochlear vein

Supra-orbital vein

Superior and inferior palpebral veins

Angular vein

External nasal vein

Superior labial vein

Deep facial vein

Inferior labial vein

Submental vein

Facial vein

Common facial vein

Brachiocephalic vein

Lateral view

FIGURE 7.25. Veins of face and scalp.

TABLE 7.6. VEINS OF FACE AND SCALP

Vein	Origin	Course	Termination	Area Drained
Supratrochlear	Begins from venous plexus on forehead and scalp, through which it communicates with frontal branch of superficial temporal vein, its contralateral partner, and supra-orbital vein	Descends near midline of forehead to root of nose, where it joins supra-orbital vein	Angular vein at root of nose	Anterior part of scalp and forehead
Supra-orbital	Begins in forehead by anastomosing with frontal tributary of superficial temporal vein	Passes medially superior to orbit; joins supratrochlear vein; a branch passes through supra-orbital notch and joins with superior ophthalmic vein		
Angular	Begins at root of nose by union of supratrochlear and supra-orbital veins	Descends obliquely along root and side of nose to inferior orbital margin	Becomes facial vein at inferior margin of orbit	Anterior part of scalp and forehead; superior and inferior eyelids and conjunctiva; may receive drainage from cavernous sinus
Facial	Continuation of angular vein past inferior margin of orbit	Descends along lateral border of nose, receiving external nasal and inferior palpebral veins; then passes obliquely across face to cross inferior border of mandible; receives communication from retromandibular vein (after which, it is sometimes called common facial vein)	Internal jugular vein opposite or inferior to level of hyoid bone	Anterior scalp and forehead; eyelids; external nose; anterior cheek; lips; chin; and submandibular gland

TABLE 7.6. VEINS OF FACE AND SCALP (Continued)

Vein	Origin	Course	Termination	Area Drained
Deep facial	Pterygoid venous plexus	Runs anteriorly on maxilla superior to buccinator and deep to masseter, emerging medial to anterior border of masseter onto face	Enters posterior aspect of facial vein	Infratemporal fossa (most areas supplied by maxillary artery)
Superficial temporal	Begins from widespread plexus of veins on side of scalp and along zygomatic arch	Frontal and parietal tributaries unite anterior to the auricle; crosses temporal root of zygomatic arch to pass from temporal region and enter substance of the parotid gland	Joins maxillary vein posterior to neck of mandible to form retromandibular vein	Side of scalp; superficial aspect of temporal muscle; and external ear
Retromandibular	Formed anterior to ear by union of superficial temporal and maxillary veins	Runs posterior and deep to ramus of mandible through substance of parotid gland; communicates at inferior end with facial vein	Unites with posterior auricular vein to form external jugular vein	Parotid gland and masseter muscle

periods of temporary compression. The alternate routes include both superficial pathways (via the facial and retromandibular/external jugular veins) and deep drainage (via the anastomoses with the cavernous sinus, pterygoid venous plexus, and the internal jugular vein).

The **facial veins,** coursing with or parallel to the facial arteries, are *valveless veins* that provide the primary superficial drainage of the face. Tributaries of the facial vein include the **deep facial vein,** which drains the *pterygoid venous plexus* of the infratemporal fossa. Inferior to the margin of the mandible, the facial vein is joined by the anterior (communicating) branch of the retromandibular vein. The facial vein drains directly or indirectly into the internal jugular vein (*IJV*). At the medial angle of the eye, the facial vein communicates with the *superior ophthalmic vein,* which drains into the *cavernous sinus.*

The **retromandibular vein** is a deep vessel of the face formed by the union of the superficial temporal vein and the maxillary vein, the latter draining the pterygoid venous plexus. The retromandibular vein runs posterior to the ramus of the mandible within the substance of the parotid gland, superficial to the external carotid artery and deep to the facial nerve. As it emerges from the inferior pole of the parotid gland, the retromandibular vein divides into an anterior branch that unites with the facial vein and a posterior branch that joins the posterior auricular vein inferior to the parotid gland to form the **external jugular vein.** This vein passes inferiorly and superficially in the neck to empty into the subclavian vein.

VEINS OF SCALP

The venous drainage of the superficial parts of the scalp is through the accompanying veins of the scalp arteries, the **supra-orbital** and **supratrochlear veins.** The **superficial temporal veins** and **posterior auricular veins** drain the scalp anterior and posterior to the auricles, respectively. The posterior auricular vein often receives a *mastoid emissary vein* from the sigmoid sinus, a dural venous sinus (see Fig. 7.33). The **occipital veins** drain the occipital region of the scalp. Venous drainage of deep parts of the scalp in the temporal region is through **deep temporal veins,** which are tributaries of the pterygoid venous plexus.

LYMPHATIC DRAINAGE OF FACE AND SCALP

There are no lymph nodes in the scalp, and except for the parotid/buccal region, there are no lymph nodes in the face. Lymph from the scalp, face, and neck drains into the *superficial ring* (pericervical collar) *of lymph nodes*—submental, submandibular, parotid, mastoid, and occipital—located at the junction of the head and neck (Fig. 7.26A). The lymphatic vessels of the face accompany other facial vessels. Superficial lymphatic vessels accompany veins, and deep lymphatics accompany arteries. All lymphatic vessels from the head and neck drain directly or indirectly into the *deep cervical lymph nodes* (Fig. 7.26B), a chain of nodes mainly located along the IJV in the neck. Lymph from these deep nodes passes to the **jugular lymphatic trunk,** which joins the *thoracic duct* on the left side and the IJV or **brachiocephalic vein** on the right side. A summary of the lymphatic drainage of the face follows:

- Lymph from the lateral part of the face and scalp, including the eyelids, drains to the superficial **parotid lymph nodes.**
- Lymph from the deep parotid nodes drains to the **deep cervical lymph nodes.**
- Lymph from the upper lip and lateral parts of the lower lip drains to the **submandibular lymph nodes.**
- Lymph from the chin and central part of the lower lip drains to the **submental lymph nodes.**

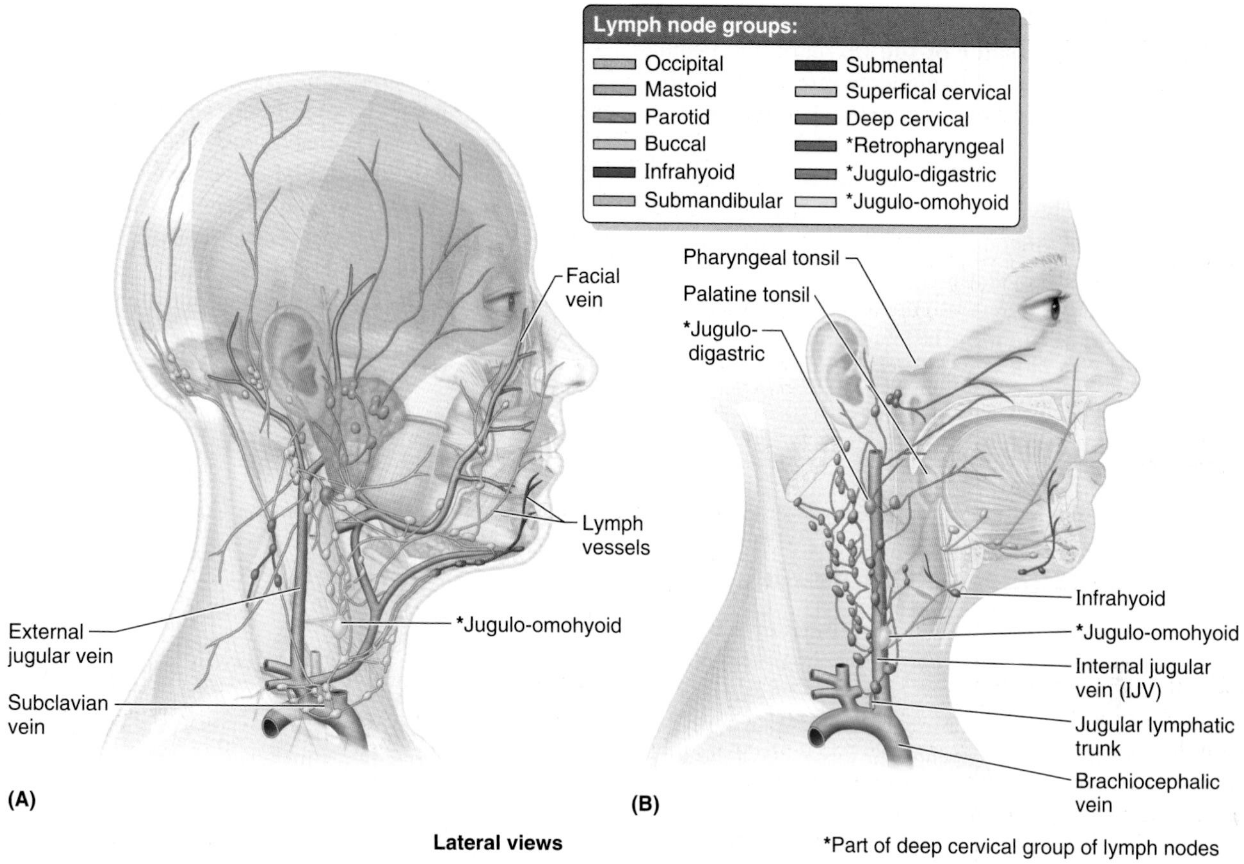

Lymph node groups:

Occipital		Submental	
Mastoid		Superfical cervical	
Parotid		Deep cervical	
Buccal		*Retropharyngeal	
Infrahyoid		*Jugulo-digastric	
Submandibular		*Jugulo-omohyoid	

(A)

Facial vein

Lymph vessels

External jugular vein

Subclavian vein

*Jugulo-omohyoid

Lateral views

(B)

Pharyngeal tonsil

Palatine tonsil

*Jugulo-digastric

Infrahyoid

*Jugulo-omohyoid

Internal jugular vein (IJV)

Jugular lymphatic trunk

Brachiocephalic vein

*Part of deep cervical group of lymph nodes

FIGURE 7.26. Lymphatic drainage of face and scalp. A. Superficial drainage. A pericervical collar of superficial lymph nodes is formed at the junction of the head and neck by the submental, submandibular, parotid, mastoid, and occipital nodes. These nodes initially receive most of the lymph drainage from the face and scalp. **B.** Deep drainage. All lymphatic vessels from the head and neck ultimately drain into the deep cervical lymph nodes, either directly from the tissues or indirectly after passing through an outlying group of nodes.

Surface Anatomy of Face

Despite the apparently infinite variations that enable people to be identified as individuals, the features of the human face are constant (Fig. 7.27). The **eyebrows** (L. *supercilia*) are linear growths of hair overlying the **supra-orbital margin.** The hairless region between the eyebrows overlies the *glabella,* and the prominent ridges that extend laterally on each side above the eyebrows are the *superciliary arches.*

The **eyelids** (L. *palpebrae*) are mobile, musculofibrous folds that overlie the eyeball. They are joined at each end of the **palpebral fissure** between the eyelids at the **medial** and **lateral angles** (canthi) of the eye. The **epicanthal fold (epicanthus)** is a fold of skin that covers the medial angle of the eye in some people, chiefly Asians. The depressions superior and inferior to the eyelids are the **suprapalpebral** and **infrapalpebral sulci.**

The shape of the nose varies remarkably. The external nose presents a prominent *apex* and is continuous with the forehead at the *root of the nose* (bridge). The rounded anterior border between the root and apex is the *dorsum of the nose.* Inferior to the apex, the nasal cavity of each side opens anteriorly by a *naris* (plural = nares), bounded medially by the *nasal septum* and laterally by an *ala* (wing) *of the nose.*

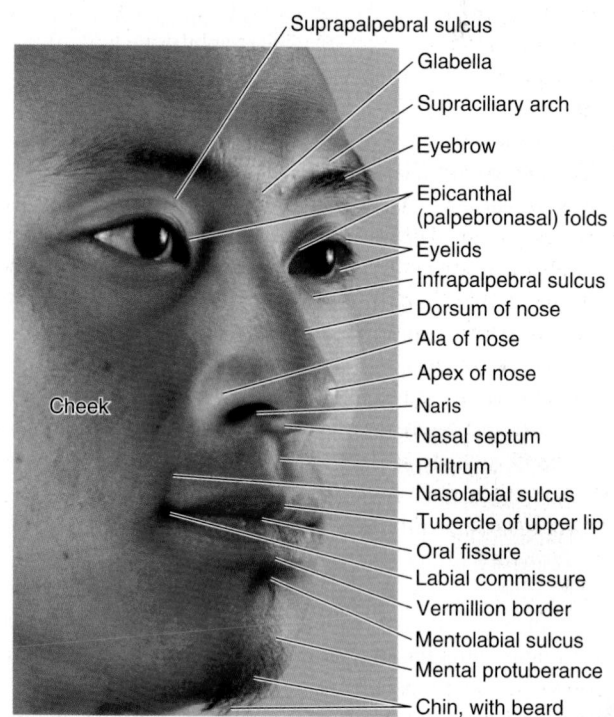

Suprapalpebral sulcus
Glabella
Supraciliary arch
Eyebrow
Epicanthal (palpebronasal) folds
Eyelids
Infrapalpebral sulcus
Dorsum of nose
Ala of nose
Apex of nose
Naris
Nasal septum
Philtrum
Nasolabial sulcus
Tubercle of upper lip
Oral fissure
Labial commissure
Vermillion border
Mentolabial sulcus
Mental protuberance
Chin, with beard

Cheek

FIGURE 7.27. Surface anatomy of face.

The lips surround the opening of the mouth, the *oral fissure.* The **vermillion border of the lip** marks the beginning of the *transitional zone* (commonly referred to as the lip) between the skin and mucous membrane of the lip. The skin of the **transitional zone** is hairless and thin, increasing its sensitivity and causing its color to be different (because of underlying capillary beds) from that of the adjacent skin of the face. The lateral junction of the lips is the **labial commissure;** the angle between the lips, medial to the commissure, that increases as the mouth opens and decreases as it closes is the **angle of the mouth.**

The median part of the upper lip features a **tubercle,** superior to which is a shallow groove, the **philtrum** (G. love charm), extending to the nasal septum. The musculofibrous folds of the lips continue laterally as the **cheek,** which also contains the buccinator muscle and buccal fat-pad. The cheek is separated from the lips by the **nasolabial sulcus,** which runs obliquely between the ala of the nose and the angle of the mouth. These grooves are easiest to observe during smiling. The lower lip is separated from the **mental protuberance** (chin) by the **mentolabial sulcus.** The lips, cheeks, and chin of the mature male grow hair as part of the secondary sex characteristics, the **beard.**

FACE AND SCALP

Facial Lacerations and Incisions

Because the face has no distinct deep fascia and the subcutaneous tissue between the cutaneous attachments of the facial muscles is loose, *facial lacerations* tend to gape (part widely). Consequently, the skin must be carefully sutured to prevent scarring. The looseness of the subcutaneous tissue also enables fluid and blood to accumulate in the loose connective tissue following bruising of the face.

Similarly, facial inflammation causes considerable swelling (e.g., a bee sting on the root of the nose may close both eyes). As a person ages, the skin loses its resiliency (elasticity). As a result, ridges and wrinkles occur in the skin perpendicular to the direction of the facial muscle fibers. Skin incisions along these cleavage or wrinkle lines (Langer lines) heal with minimal scarring (see the blue box "Skin Incisions and Scarring," p. 15).

Scalp Injuries

Because the scalp arteries arising at the sides of the head are well protected by dense connective tissue and anastomose freely, a *partially detached scalp* may be replaced with a reasonable chance of healing as long as one of the vessels supplying the scalp remains intact. During an *attached craniotomy* (surgical removal of a segment of the calvaria with a soft tissue scalp flap to expose the cranial cavity), the incisions are usually made convex and upward, and the superficial temporal artery is included in the tissue flap.

The *scalp proper,* the first three layers of the scalp (see Fig. 7.15A), is often regarded clinically as a single layer because they remain together when a scalp flap is made during a craniotomy and when part of the scalp is torn off (e.g., during an industrial accident). Nerves and vessels of the scalp enter inferiorly and ascend through layer two to the skin. Consequently, surgical pedicle scalp flaps are made so that they remain attached inferiorly to preserve the nerves and vessels, thereby promoting good healing.

The arteries of the scalp supply little blood to the calvaria, which is supplied by the middle meningeal arteries. There-fore, loss of the scalp does not produce necrosis (death) of the calvarial bones.

Scalp Wounds

The epicranial aponeurosis is clinically important. Because of the strength of this aponeurosis, *superficial scalp wounds* do not gape, and the margins of the wound are held together. Furthermore, deep sutures are not necessary when suturing superficial wounds because the epicranial aponeurosis does not allow wide separation of the skin. *Deep scalp wounds* gape widely when the epicranial aponeurosis is lacerated in the coronal plane because of the pull of the frontal and occipital bellies of the occipitofrontalis muscle in opposite directions (anteriorly and posteriorly).

Scalp Infections

The loose connective tissue layer (layer four) of the scalp is the *danger area of the scalp* because pus or blood spreads easily in it. Infection in this layer can also pass into the cranial cavity through small *emissary veins,* which pass through parietal foramina in the calvaria, and reach intracranial structures such as the meninges (Fig. 7.8A & C). An infection cannot pass into the neck because the occipital bellies of the occipitofrontalis muscle attach to the occipital bone and mastoid parts of the temporal bones (Fig. 7.16A). Neither can a scalp infection spread laterally beyond the zygomatic arches because the epicranial aponeurosis is continuous with the temporal fascia that attaches to these arches.

An infection or fluid (e.g., pus or blood) can enter the eyelids and the root of the nose because the occipitofrontalis inserts into the skin and subcutaneous tissue and does not attach to the bone (Fig. 7.16B). The skin of the eyelid is the thinnest of the body and is delicate and sensitive. Because of the loose nature of the subcutaneous tissue within the eyelids, even a relatively slight injury or inflammation may result in an accumulation of fluid, causing the eyelids to swell. Blows to the peri-orbital region usually produce soft tissue damage because the tissues are crushed against the strong and relatively sharp margin. Consequently, "black eyes" (*peri-orbital*

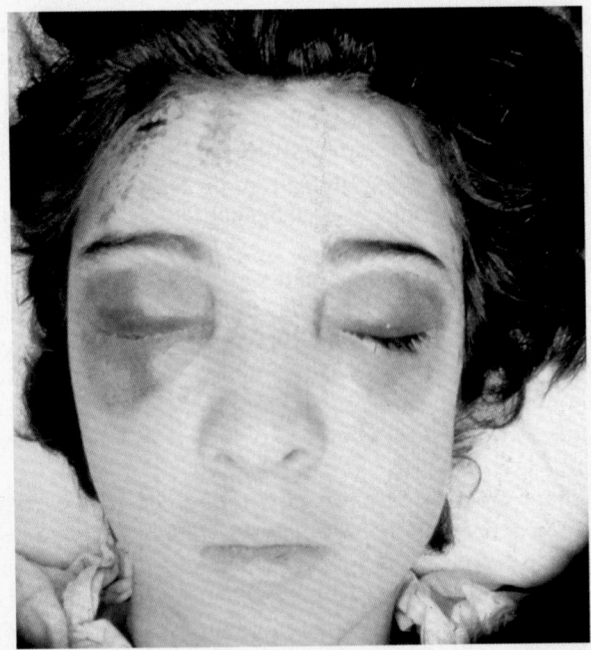

FIGURE B7.12. Ecchymosis (extravasation of blood under the skin).

ecchymosis) can result from an injury to the scalp and/or the forehead (Fig. B7.12). **Ecchymosis** (purple patches), develop as a result of extravasation of blood into the subcutaneous tissue and skin of the eyelids and surrounding regions.

Sebaceous Cysts

 The ducts of sebaceous glands associated with hair follicles in the scalp may become obstructed, resulting in the retention of secretions and the formation of *sebaceous cysts* (pilar cysts). Because they are in the skin, sebaceous cysts move with the scalp.

Cephalhematoma

 Sometimes after a difficult birth, bleeding occurs between the baby's pericranium (layer 5 of scalp; see Fig. 7.15A) and calvaria, usually over one parietal bone. Blood becomes trapped in this area, causing a *cephalhematoma*. This benign condition frequently results from birth trauma that ruptures multiple, minute periosteal arteries that nourish the bones of the calvaria.

Flaring of Nostrils

The actions of the nasalis muscles (Fig. 7.17, center top row) have generally been held as insignificant; however, observant clinicians study their action because of their diagnostic value. For example, true *nasal breathers* can flare their nostrils distinctly. Habitual mouth breathing, caused by chronic nasal obstruction, for example, diminishes and sometimes eliminates the ability

to flare the nostrils. Children who are *chronic mouth breathers* often develop dental malocclusion (improper bite) because the alignment of the teeth is maintained to a large degree by normal periods of occlusion and labial closure. *Anti-snoring devices* have been developed that attach to the nose to flare the nostrils and maintain a more patent air passageway.

Paralysis of Facial Muscles

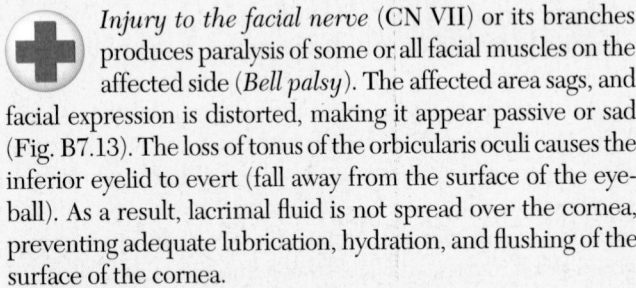 *Injury to the facial nerve* (CN VII) or its branches produces paralysis of some or all facial muscles on the affected side (*Bell palsy*). The affected area sags, and facial expression is distorted, making it appear passive or sad (Fig. B7.13). The loss of tonus of the orbicularis oculi causes the inferior eyelid to evert (fall away from the surface of the eyeball). As a result, lacrimal fluid is not spread over the cornea, preventing adequate lubrication, hydration, and flushing of the surface of the cornea.

This makes the cornea vulnerable to ulceration. A resulting corneal scar can impair vision. If the injury weakens or paralyzes the buccinator and orbicularis oris, food will accumulate in the oral vestibule during chewing, usually requiring continual removal with a finger. When the sphincters or dilators of the mouth are affected, displacement of the mouth (drooping of its corner) is produced by contraction of unopposed contralateral facial muscles and gravity, resulting in food and saliva dribbling out of the side of the mouth. Weakened lip muscles affect speech as a result of an impaired ability to produce labial (*B, M, P,* or *W*) sounds. Affected persons cannot whistle or blow a wind instrument. They frequently dab their eyes and mouth with a handkerchief to wipe the fluid (tears and saliva), which runs from the drooping lid and mouth; the fluid and constant wiping may result in localized skin irritation.

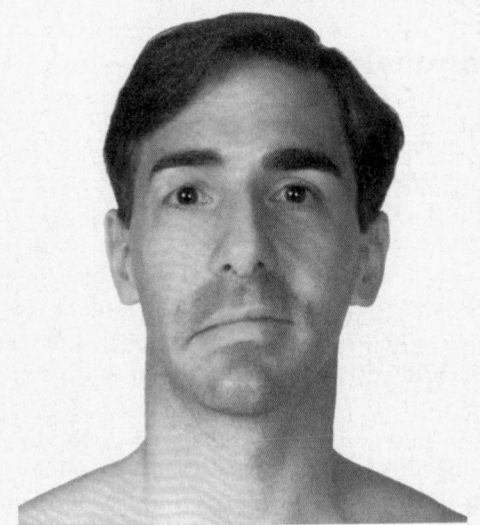

FIGURE B7.13.

Infra-Orbital Nerve Block

 For treating wounds of the upper lip and cheek or, more commonly, for repairing the maxillary incisor teeth, local anesthesia of the inferior part of the face is achieved by infiltration of the infra-orbital nerve with an anesthetic agent. The injection is made in the region of the infra-orbital foramen, by elevating the upper lip and passing the needle through the junction of the oral mucosa and gingiva at the superior aspect of the oral vestibule.

To determine where the infra-orbital nerve emerges, pressure is exerted on the maxilla in the region of the infra-orbital foramen. Too much pressure on the nerve causes considerable pain. Because companion infra-orbital vessels leave the infra-orbital foramen with the nerve, aspiration of the syringe during injection prevents inadvertent injection of anesthetic fluid into a blood vessel. Because the orbit is located just superior to the injection site, a careless injection could result in passage of anesthetic fluid into the orbit, causing temporary paralysis of the extra-ocular muscles.

Mental and Incisive Nerve Blocks

 Occasionally, it is desirable to anesthetize one side of the skin and mucous membrane of the lower lip, and the skin of the chin (e.g., to suture a severe laceration of the lip). Injection of an anesthetic agent into the mental foramen blocks the mental nerve that supplies the skin and mucous membrane of the lower lip from the mental foramen to the midline, including the skin of the chin.

Buccal Nerve Block

 To anesthetize the skin and mucous membrane of the cheek (e.g., to suture a knife wound), an anesthetic injection can be made into the mucosa covering the *retromolar fossa,* a triangular depression posterior to the 3rd mandibular molar tooth between the anterior border of the ramus and the temporal crest.

Trigeminal Neuralgia

 Trigeminal neuralgia or *tic douloureux* is a sensory disorder of the sensory root of CN V that occurs most often in middle-aged and elderly persons. It is characterized by sudden attacks of excruciating, lightening-like jabs of facial pain. A *paroxysm* (sudden sharp pain) can last for 15 minutes or more. The pain may be so intense that the person winces; hence the common term *tic* (twitch). In some cases, the pain may be so severe that psychological changes occur, leading to depression and even suicide attempts.

CN V_2 is most frequently involved, then CN V_3, and least frequently, CN V_1. The paroxysms are often set off by touching the face, brushing the teeth, shaving, drinking, or chewing. The pain is often initiated by touching an especially sensitive *trigger zone*, frequently located around the tip of the nose or the cheek (Haines, 2006). In trigeminal neuralgia, demyelination of axons in the sensory root occurs. In most cases this is caused by pressure of a small aberrant artery (Kiernan, 2008). Often, when the aberrant artery is moved away from the sensory root of CN V, the symptoms disappear. Other scientists believe the condition is caused by a pathological process affecting neurons in the trigeminal ganglion.

Medical or surgical treatment or both are used to alleviate the pain. In cases involving the CN V_2, attempts have been made to block the infra-orbital nerve at the infra-orbital foramen by using alcohol. This treatment usually relieves pain temporarily. The simplest surgical procedure is avulsion or cutting of the branches of the nerve at the infra-orbital foramen.

Other treatments have used *radiofrequency selective ablation of parts of the trigeminal ganglion* by a needle electrode passing through the cheek and foramen ovale. In some cases, it is necessary to section the sensory root for relief of the pain. To prevent regeneration of nerve fibers, the sensory root of the trigeminal nerve may be partially cut between the ganglion and the brainstem (*rhizotomy*). Although the axons may regenerate, they do not do so within the brainstem. Surgeons attempt to differentiate and cut only the sensory fibers to the division of CN V involved.

The same result may be achieved by sectioning the spinal tract of CN V (*tractotomy*). After this operation, the sensation of pain, temperature, and simple (light) touch is lost over the area of skin and mucous membrane supplied by the affected component of the CN V. This loss of sensation may annoy the patient, who may not recognize the presence of food on the lip and cheek or feel it within the mouth on the side of the nerve section, but these disabilities are usually preferable to excruciating pain.

Lesions of Trigeminal Nerve

 Lesions of the entire trigeminal nerve cause widespread anesthesia involving the:

- Corresponding anterior half of the scalp.
- Face, except for an area around the angle of the mandible, the cornea, and the conjunctiva.
- Mucous membranes of the nose, mouth, and anterior part of the tongue.

Paralysis of the muscles of mastication also occurs.

Herpes Zoster Infection of Trigeminal Ganglion

 A *herpes zoster virus infection* may produce a lesion in the cranial ganglia. Involvement of the trigeminal ganglion occurs in approximately 20% of cases (Bernardini, 2010). The infection is characterized by an eruption of groups of vesicles following the course of the affected nerve (e.g., *ophthalmic herpes zoster*). Any division of CN V may be involved, but the ophthalmic division is most commonly affected. Usually, the cornea is involved, often resulting in painful *corneal ulceration* and subsequent *scarring of the cornea.*

Testing Sensory Function of CN V

The sensory function of the trigeminal nerve is tested by asking the person to close his or her eyes and respond when types of touch are felt. For example, a piece of dry gauze is gently stroked across the skin of one side of the face and then to the corresponding position on the other side. The test is then repeated until the skin of the forehead (CN V₁), cheek (CN V₂), and lower jaw (CN V₃) have been tested. The person is asked if one side feels the same as or different from the other side. The testing may then be repeated using warm or cold instruments and the gentle touch of a sharp pin, again alternating sides (Fig. B7.14).

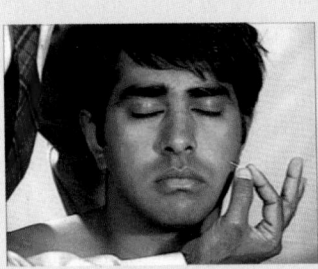 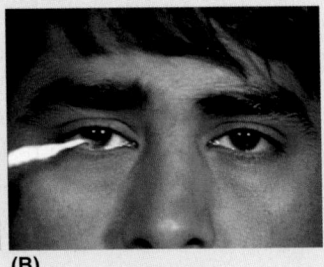

(A) (B)

FIGURE B7.14.

Injuries to Facial Nerve

Injury to branches of the facial nerve causes paralysis of the facial muscles (Bell palsy), with or without loss of taste on the anterior two thirds of the tongue or altered secretion of the lacrimal and salivary glands (see the blue box "Paralysis of Facial Muscles," p. 861). Lesions near the origin of CN VII from the pons of the brain, or proximal to the origin of the greater petrosal nerve (in the region of the geniculate ganglion), result in loss of motor, gustatory (taste), and autonomic functions. Lesions distal to the geniculate ganglion, but proximal to the origin of the chorda tympani nerve, produce the same dysfunction, except that lacrimal secretion is not affected. Lesions near the stylomastoid foramen result in loss of motor function only (i.e., facial paralysis).

Facial nerve palsy has many causes. The most common nontraumatic cause of facial paralysis is *inflammation of the facial nerve* near the stylomastoid foramen (see Fig. 7.9A), often as a result of a viral infection. This produces edema (swelling) and compression of the nerve in the facial canal. Injury of the facial nerve may result from *fracture of the temporal bone;* facial paralysis is evident soon after the injury. If the nerve is completely sectioned, the chances of complete or even partial recovery are remote. Muscular

movement usually improves when the nerve dam[...] ciated with blunt head trauma; however, recover[...] be complete (Rowland, 2010). Facial nerve palsy may be idiopathic (occurring without a known cause), but it often follows exposure to cold, as occurs when riding in a car with a window open.

Facial paralysis may be a complication of surgery; consequently, identification of the facial nerve and its branches is essential during surgery (e.g., *parotidectomy*, removal of a parotid gland). The facial nerve is most distinct as it emerges from the stylomastoid foramen; if necessary, electrical stimulation may be used for confirmation. Facial nerve palsy may also be associated with dental manipulation, vaccination, pregnancy, HIV infection, Lyme disease (inflammatory disorder causing headache and stiff neck), and infections of the middle ear (otitis media). Because the branches of the facial nerve are superficial, they are subject to injury by stab and gunshot wounds, cuts, and injury at birth (Fig. 7.23):

- A *lesion of the zygomatic branch* of CN VII causes paralysis, including loss of tonus of the orbicularis oculi in the inferior eyelid.
- *Paralysis of the buccal branch* of CN VII causes paralysis of the buccinator and superior portion of the orbicularis oris and upper lip muscles.
- *Paralysis of the marginal mandibular branch* of CN VII may occur when an incision is made along the inferior border of the mandible. Injury to this branch (e.g., during a surgical approach to the submandibular gland) causes paralysis of the inferior portion of the orbicularis oris and lower lip muscles.

The consequences of such paralyses are discussed in the blue box "Paralysis of Facial Muscles" on p. 861.

Compression of Facial Artery

The facial artery can be occluded by pressure against the mandible where the vessel crosses it. Because of the numerous anastomoses between the branches of the facial artery and the other arteries of the face, compression of the facial artery on one side does not stop all bleeding from a lacerated facial artery or one of its branches. In lacerations of the lip, pressure must be applied on both sides of the cut to stop the bleeding. In general, facial wounds bleed freely and heal quickly.

Pulses of Arteries of Face and Scalp

The pulses of the superficial temporal and facial arteries may be used for taking the pulse. For example, anesthesiologists at the head of the operating table often take the *temporal pulse* where the superficial temporal artery crosses the zygomatic process just anterior to the auricle. Clench your teeth and palpate the *facial pulse* as the facial artery crosses the inferior border of the mandible immediately anterior to the masseter muscle (Fig. 7.24B).

Stenosis of Internal Carotid Artery

At the medial angle of the eye, an anastomosis occurs between the facial artery, a branch of the external carotid artery, and cutaneous branches of the internal carotid artery. With advancing age, the internal carotid artery may become narrow (stenotic) owing to atherosclerotic thickening of the intima (innermost coat) of the arteries. Because of the arterial anastomosis, intracranial structures such as the brain can receive blood from the connection of the facial artery to the dorsal nasal branch of the ophthalmic artery.

Scalp Lacerations

Scalp lacerations are the most common type of head injury requiring surgical care. These wounds bleed profusely because the arteries entering the periphery of the scalp bleed from both ends owing to abundant anastomoses. The arteries do not retract when lacerated because they are held open by the dense connective tissue in layer two of the scalp. Spasms of the occipitofrontalis muscle can increase gaping of scalp wounds. Bleeding from scalp lacerations can be fatal if not controlled (e.g., by sutures).

Squamous Cell Carcinoma of Lip

Squamous cell carcinoma (cancer) of the lip usually involves the lower lip (Fig. B7.15). Overexposure to sunshine over many years, is a common factor in these cases. Chronic irritation from pipe smoking is also a contributing cause. Cancer cells from the central part of the lower lip, the floor of the mouth, and the apex of the tongue spread to the *submental lymph nodes,* whereas cancer cells from lateral parts of the lower lip drain to the *submandibular lymph nodes.*

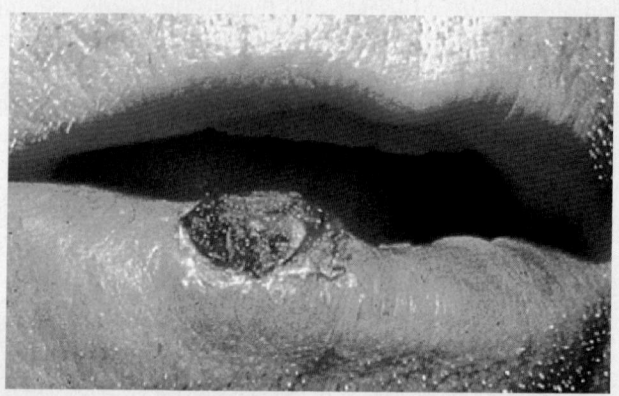

FIGURE B7.15.

The Bottom Line

FACE AND SCALP

The face provides our identity as an individual human. Thus birth or acquired defects have consequences beyond their physical effects. ♦ The individuality of the face results primarily from anatomical variation. ♦ The way in which the facial muscles alter the basic features is critical to communication. ♦ Lips and the shape and degree of opening of the mouth are important components of speech, but emphasis and subtleties of meaning are provided by our facial expressions.

Structure of scalp: The scalp is a somewhat mobile soft tissue mantle covering the calvaria. ♦ The primary subcutaneous component of the scalp is the musculo-aponeurotic epicranius to which the overlying skin is firmly attached, but it is separated from the outer periosteum (pericranium) of the cranium by loose areolar tissue. ♦ The areolar layer enables the mobility of the scalp over the calvaria and permits traumatic separation of the scalp from the cranium. ♦ Attachment of the skin to the epicranial aponeurosis keeps the edges of superficial wounds together, but a wound that also penetrates the epicranial aponeurosis gaps widely. ♦ Blood may collect in the areolar space deep to the aponeurosis after a head injury.

Muscles of face and scalp: The facial muscles play important roles as the dilators and sphincters of the portals of the alimentary (digestive), respiratory, and visual systems (oral and palpebral fissures and nostrils), controlling what enters and some of what exits from our bodies. ♦ Other facial muscles assist the muscles of mastication by keeping food between the teeth during chewing. ♦ Fleshy portions of the face (eyelids and cheeks) form dynamic containing walls for the orbits and oral cavity. ♦ Facial muscles are all derived from the second pharyngeal arch and are therefore supplied by the nerve of this arch, the facial nerve (CN VII). ♦ Facial muscles are subcutaneous, most having a skeletal origin and a cutaneous insertion. ♦ The face lacks the deep fascia present elsewhere in the body.

Innervation of face and scalp: The face is highly sensitive. It receives sensory innervation from the three divisions of the trigeminal nerve (CN V). ♦ The major terminal branches of each division reach the subcutaneous tissue of each side of the face via three foramina that are aligned vertically. ♦ Each division supplies a distinct sensory zone, similar to a dermatome, but without the overlapping of adjacent nerves; therefore, injuries result in distinct and defined areas of paresthesia. ♦ The divisions of CN V supply sensation not only to the superficial skin of the face but also to deep mucosal surfaces of the conjunctival sacs, cornea, nasal cavity, and paranasal sinuses and to the oral cavity and vestibule. ♦ The skin covering the angle of the mandible is innervated by the great auricular nerve, a branch of the cervical plexus. ♦ Eight nerves supply sensation to the scalp via branches arising from all three divisions of CN V, anterior to the auricle of the external ear and branches of cervical spinal nerves posterior to the auricle. ♦ The facial nerve (CN VII) is the motor nerve of the face, supplying all the muscles of facial expression, including the platysma, occipital belly of occipitofrontalis, and auricular muscles that are not part of the face per se. ♦ These muscles receive innervation from CN VII primarily via five branches of the parotid (nerve) plexus.

Vasculature of face and scalp: The face and scalp are highly vascular. The terminal branches of the arteries of the face anastomose freely (including anastomoses across the midline with their contralateral partners). Thus, bleeding from facial lacerations may be diffuse, with the lacerated vessel bleeding from both ends. ♦ Most arteries of the face are branches or derivatives of branches of the external carotid artery; the arteries arising from the internal carotid that supply the forehead are exceptions. ♦ The main artery to the face

is the facial artery. ♦ The arteries of the scalp are firmly embedded in the dense connective tissue overlying the epicranial aponeurosis. Thus, when lacerated, these arteries bleed from both ends, like those of the face, but are less able to constrict or retract than other superficial vessels; therefore, profuse bleeding results.

The veins of the face and scalp generally accompany arteries, providing a primarily superficial venous drainage. ♦ However, they also anastomose with the pterygoid venous plexus and with dural venous sinuses via emissary veins, which provide a potentially dangerous route for the spread of infection. ♦ Most nerves and vessels of the scalp run vertically toward the vertex; thus a horizontal laceration may produce more neurovascular damage than a vertical one.

The lymphatic drainage of most of the face follows the venous drainage to lymph nodes around the base of the anterior part of the head (submandibular, parotid, and superficial cervical nodes). ♦ An exception to this pattern is the lymph drainage of the central part of the lip and chin, which initially drains to the submental lymph nodes. All nodes of the face drain in turn to the deep cervical lymph nodes.

CRANIAL MENINGES

The **cranial meninges** are membranous coverings of the brain that lie immediately internal to the cranium (Figs. 7.15A and 7.28). The cranial meninges:

- Protect the brain.
- Form the supporting framework for arteries, veins, and venous sinuses.
- Enclose a fluid-filled cavity, the subarachnoid space, which is vital to the normal function of the brain.

The meninges are composed of three membranous connective tissue layers (Fig. 7.28A, B, & D):

1. *Dura mater (dura)*: tough, thick external fibrous layer.
2. *Arachnoid mater (arachnoid)*: thin intermediate layer.
3. *Pia mater (pia)*: delicate internal vasculated layer.

The intermediate and internal layers (arachnoid and pia) are continuous membranes that collectively make up the **leptomeninx** (G. slender membrane) (Fig. 7.28B). The arachnoid is separated from the pia by the **subarachnoid (leptomeningeal) space,** which contains **cerebrospinal fluid** (CSF). This fluid-filled space helps maintain the balance of extracellular fluid in the brain. CSF is a clear liquid similar to blood in constitution; it provides nutrients but has less protein and a different ion concentration. CSF is formed by the *choroid plexuses* of the four ventricles of the brain (Fig. 7.28A). This fluid leaves the ventricular system and enters the subarachnoid space between the arachnoid and pia mater, where it cushions and nourishes the brain.

Dura Mater

The **cranial dura mater** (dura), a thick, dense, bilaminar membrane, is also called the *pachymeninx* (G. *pachy,* thick + G. *menix,* membrane) (Fig. 7.28A). It is adherent to the internal table of the calvaria. The two layers of the cranial dura are an *external periosteal layer,* formed by the periosteum covering the internal surface of the calvaria, and an *internal meningeal layer,* a strong fibrous membrane that is continuous at the foramen magnum with the spinal dura covering the spinal cord.

The external **periosteal layer of dura** adheres to the internal surface of the cranium; its attachment is tenacious along the suture lines and in the cranial base (Haines, 2006). The external periosteal layer is continuous at the cranial foramina with the periosteum on the external surface of the calvaria (Fig. 7.28C). This outer layer is not continuous with the dura mater of the spinal cord, which consists of only a meningeal layer.

Except where the dural sinuses and infoldings occur (Fig. 7.28B), the internal meningeal layer is intimately fused with the periosteal layer and cannot be separated from it (Fig. 7.28B & C). The fused external and internal layers of dura over the calvaria can be easily stripped from the cranial bones (e.g., when the calvaria is removed at autopsy). In the cranial base, the two dural layers are firmly attached and difficult to separate from the bones. In life, such separation at the dural–cranial interface occurs only pathologically, creating an actual (blood- or fluid-filled) epidural space.

DURAL INFOLDINGS OR REFLECTIONS

The internal **meningeal layer of dura mater** is a sustentacular (supporting) layer that reflects away from the external

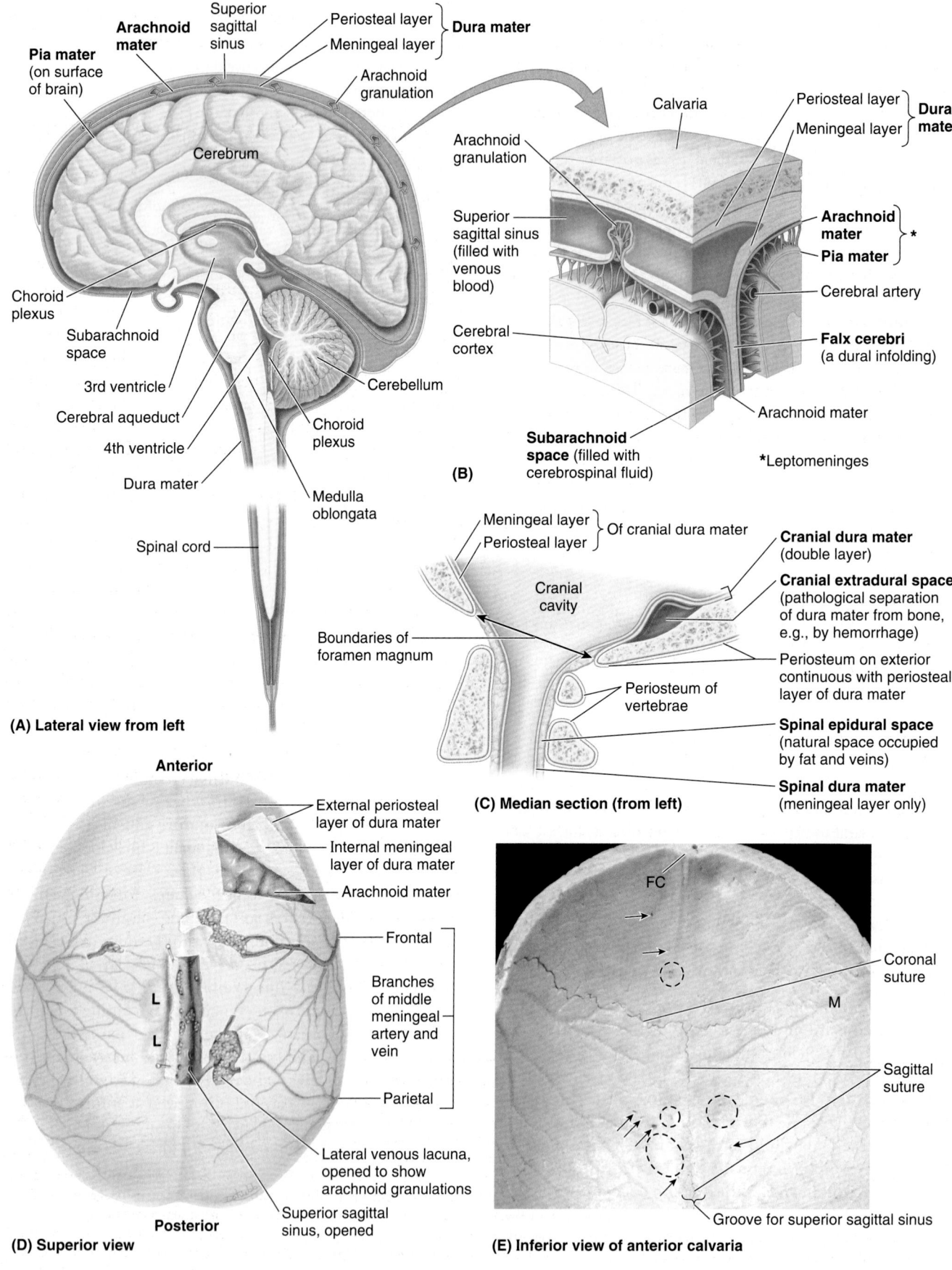

Pia mater (on surface of brain)

Arachnoid mater

Superior sagittal sinus

Periosteal layer

Meningeal layer

Dura mater

Arachnoid granulation

Cerebrum

Choroid plexus

Subarachnoid space

3rd ventricle

Cerebral aqueduct

4th ventricle

Dura mater

Cerebellum

Choroid plexus

Medulla oblongata

Spinal cord

(A) Lateral view from left

Calvaria

Periosteal layer

Meningeal layer

Dura mater

Arachnoid granulation

Superior sagittal sinus (filled with venous blood)

Arachnoid mater

Pia mater

*

Cerebral artery

Cerebral cortex

Falx cerebri (a dural infolding)

Arachnoid mater

Subarachnoid space (filled with cerebrospinal fluid)

*Leptomeninges

(B)

Meningeal layer

Periosteal layer

Of cranial dura mater

Cranial dura mater (double layer)

Cranial cavity

Cranial extradural space (pathological separation of dura mater from bone, e.g., by hemorrhage)

Boundaries of foramen magnum

Periosteum on exterior continuous with periosteal layer of dura mater

Periosteum of vertebrae

Spinal epidural space (natural space occupied by fat and veins)

Spinal dura mater (meningeal layer only)

(C) Median section (from left)

Anterior

External periosteal layer of dura mater

Internal meningeal layer of dura mater

Arachnoid mater

Frontal

Branches of middle meningeal artery and vein

Parietal

L

L

Lateral venous lacuna, opened to show arachnoid granulations

Superior sagittal sinus, opened

Posterior

(D) Superior view

FC

Coronal suture

M

Sagittal suture

Groove for superior sagittal sinus

(E) Inferior view of anterior calvaria

periosteal layer of dura to form dural infoldings (reflections) (Figs. 7.28B and 7.29). The dural infoldings divide the cranial cavity into compartments, forming partial partitions (dural septa) between certain parts of the brain and providing support for other parts. The dural infoldings include the:

- *Falx cerebri* (cerebral falx).
- *Tentorium cerebelli* (cerebellar tentorium).
- *Falx cerebelli* (cerebellar falx).
- *Diaphragma sellae* (sellar diaphragm).

The **falx cerebri** (L. *falx*, sickle-shaped), the largest dural infolding, lies in the **longitudinal cerebral fissure** that separates the right and the left *cerebral hemispheres.* The falx cerebri attaches in the median plane to the internal surface of the calvaria, from the *frontal crest of the frontal bone* and *crista galli of the ethmoid bone* anteriorly to the *internal occipital protuberance* posteriorly (Figs. 7.29A and 7.30). It ends by becoming continuous with the tentorium cerebelli.

The **tentorium cerebelli,** the second largest dural infolding, is a wide crescentic septum that separates the occipital lobes of the cerebral hemispheres from the cerebellum. The tentorium cerebelli attaches rostrally to the clinoid processes of the sphenoid, rostrolaterally to the petrous part of the temporal bone, and posterolaterally to the internal surface of the occipital bone and part of the parietal bone.

The falx cerebri attaches to the tentorium cerebelli and holds it up, giving it a tent-like appearance (L. *tentorium,* tent). The tentorium cerebelli divides the cranial cavity into *supratentorial* and *infratentorial compartments.* The supratentorial compartment is divided into right and left halves by the falx cerebri. The concave anteromedial border of the tentorium cerebelli is free, producing a gap called the **tentorial notch** through which the brainstem (midbrain, pons, and medulla oblongata) extends from the posterior into the middle cranial fossa (Fig. 7.31A & B).

The **falx cerebelli** is a vertical dural infolding that lies inferior to the tentorium cerebelli in the posterior part of the posterior cranial fossa (Figs. 7.29 and 7.30). It is attached to the internal occipital crest and partially separates the cerebellar hemispheres.

The **diaphragma sellae,** the smallest dural infolding, is a circular sheet of dura that is suspended between the clinoid processes forming a partial roof over the hypophysial fossa in the sphenoid (Fig. 7.29A). The diaphragma sellae covers the pituitary gland in this fossa and has an aperture for passage of the infundibulum and hypophysial veins.

DURAL VENOUS SINUSES

The **dural venous sinuses** are endothelium-lined spaces between the periosteal and meningeal layers of the dura. They form where the dural septa attach along the free edge of the falx cerebri and in relation to formations of the cranial floor (Figs. 7.29, 7.31, and 7.32). Large veins from the surface of the brain empty into these sinuses and most of the blood from the brain ultimately drains through them into the IJVs. The **superior sagittal sinus** lies in the convex attached border of the falx cerebri (Fig. 7.29). It begins at the crista galli and ends near the internal occipital protuberance (Fig. 7.30) at the **confluence of sinuses,** a meeting place of the superior sagittal, straight, occipital, and transverse sinuses (Fig. 7.32). The superior sagittal sinus receives the superior cerebral veins and communicates on each side through slit-like openings with the **lateral venous lacunae,** lateral expansions of the superior sagittal sinus (Fig. 7.29D).

Arachnoid granulations (collections of arachnoid villi) are tufted prolongations of the arachnoid that protrude through the meningeal layer of the dura mater into the dural venous sinuses, especially the lateral lacunae and affect transfer of CSF to the venous system (Figs. 7.29B & D and 7.35). Enlarged arachnoid granulations (*pacchionian bodies*) may erode bone, forming pits called **granular foveolae** in the calvaria (Fig. 7.28E). They are usually observed in the vicinity of the superior sagittal, transverse, and some other dural venous sinuses. Arachnoid granulations are structurally adapted for the transport of CSF from the subarachnoid space to the venous system (see "Subarachnoid Cisterns," p. 880).

The **inferior sagittal sinus** is much smaller than the superior sagittal sinus (Fig. 7.29). It runs in the inferior concave free border of the falx cerebri and ends in the straight sinus. The **straight sinus** (L. *sinus rectus*) is formed by the union of the inferior sagittal sinus with the *great cerebral vein.* It runs infero-posteriorly along the line of attachment of the falx cerebri to the tentorium cerebelli, where it joins the confluence of sinuses.

The **transverse sinuses** pass laterally from the confluence of sinuses, forming a groove in the occipital bones and the postero-inferior angles of the parietal bones

---

FIGURE 7.28. Meninges and their relationship to calvaria, brain, and spinal cord. A. The dura mater and subarachnoid space (purple) surround the brain and are continuous with that around the spinal cord. **B.** The two layers of dura separate to form dural venous sinuses, such as the superior sagittal sinus. Arachnoid granulations protrude through the meningeal layer of the dura into the dural venous sinuses and effect transfer of cerebrospinal fluid (CSF) to the venous system. **C.** The normal fat- and vein-filled spinal epidural (extradural) space is not continuous with the potential or pathological cranial epidural space. Cranial dura mater has two layers, whereas spinal dura mater consists of a single layer. **D.** The calvaria has been removed to reveal the external (periosteal layer) of the dura mater. In the median plane, a part of the thick roof of the superior sagittal sinus has been incised and retracted; laterally, parts of the thin roof of two lateral lacunae (*L*) are reflected to demonstrate the abundant arachnoid granulations, which are responsible for absorption of CSF. On the right, an angular flap of dura has been turned anteriorly; the convolutions of the cerebral cortex are visible through the arachnoid mater. **E.** The internal aspect of the calvaria reveals pits (*dotted lines,* granular foveolae) in the frontal and parietal bones, which are produced by enlarged arachnoid granulations or clusters of smaller ones (as in **D**). Multiple small emissary veins pass between the superior sagittal sinus and the veins in the diploë and scalp through small emissary foramina (*arrows*) located on each side of the sagittal suture. The sinuous vascular groove (*M*) on the lateral wall is formed by the frontal branch of the middle meningeal artery. The falx cerebri attaches anteriorly to the frontal crest (*FC*).

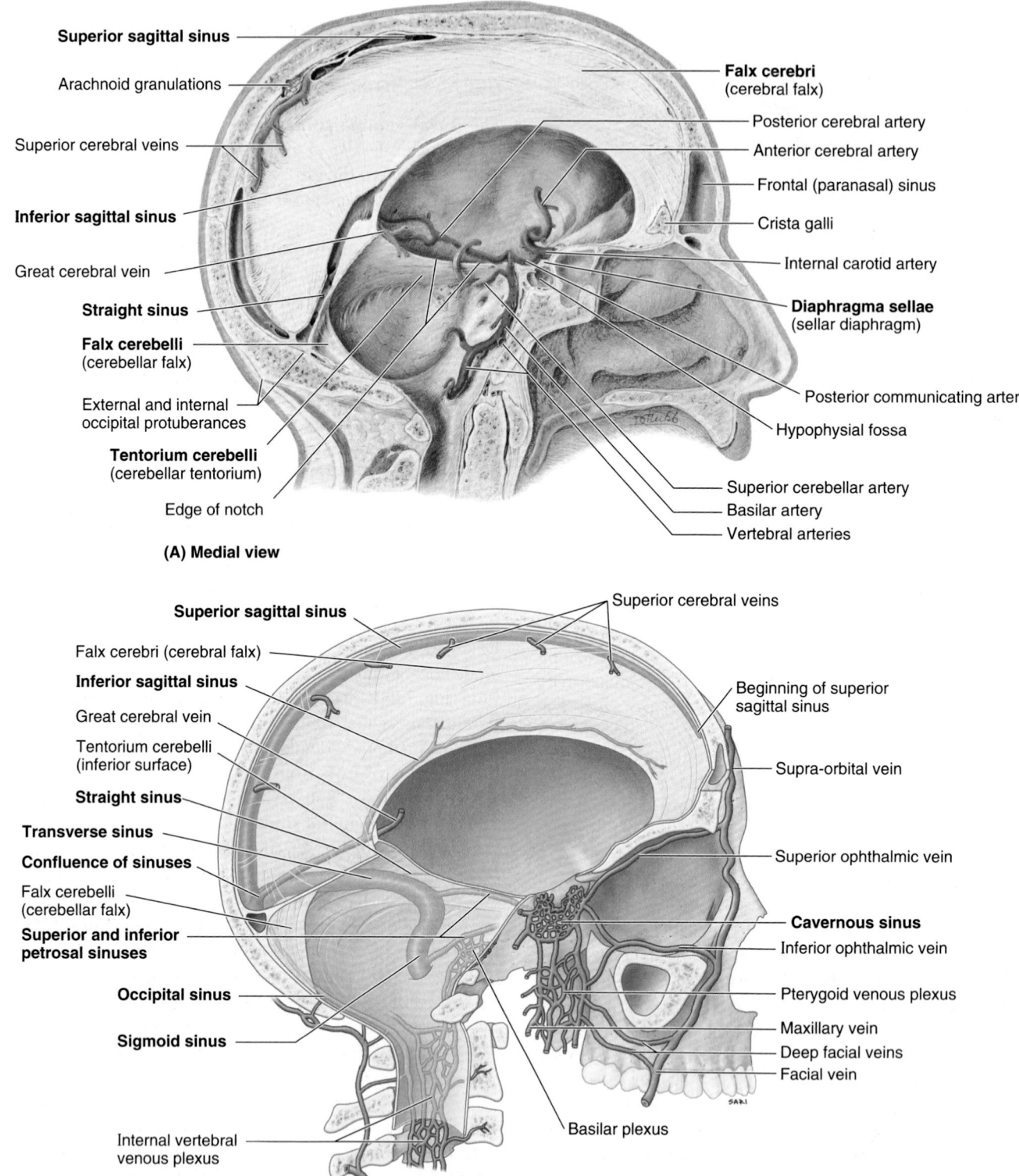

Superior sagittal sinus

Arachnoid granulations

Superior cerebral veins

Inferior sagittal sinus

Great cerebral vein

Straight sinus

Falx cerebelli
(cerebellar falx)

External and internal
occipital protuberances

Tentorium cerebelli
(cerebellar tentorium)

Edge of notch

(A) Medial view

Falx cerebri
(cerebral falx)

Posterior cerebral artery

Anterior cerebral artery

Frontal (paranasal) sinus

Crista galli

Internal carotid artery

Diaphragma sellae
(sellar diaphragm)

Posterior communicating artery

Hypophysial fossa

Superior cerebellar artery

Basilar artery

Vertebral arteries

Superior sagittal sinus

Falx cerebri (cerebral falx)

Inferior sagittal sinus

Great cerebral vein

Tentorium cerebelli
(inferior surface)

Straight sinus

Transverse sinus

Confluence of sinuses

Falx cerebelli
(cerebellar falx)

**Superior and inferior
petrosal sinuses**

Occipital sinus

Sigmoid sinus

Internal vertebral
venous plexus

(B) Medial view

Superior cerebral veins

Beginning of superior
sagittal sinus

Supra-orbital vein

Superior ophthalmic vein

Cavernous sinus

Inferior ophthalmic vein

Pterygoid venous plexus

Maxillary vein

Deep facial veins

Facial vein

Basilar plexus

FIGURE 7.29. Dural infoldings and dural venous sinuses. The left side of the bisected head is shown. **A.** Two sickle-shaped dural folds (septae), the falx cerebri and falx cerebelli, are vertically oriented in the median plane; two roof-like folds, the tentorium cerebelli and the small diaphragma sellae, lie horizontally. **B.** Venous sinuses of the dura mater and their communications are demonstrated.

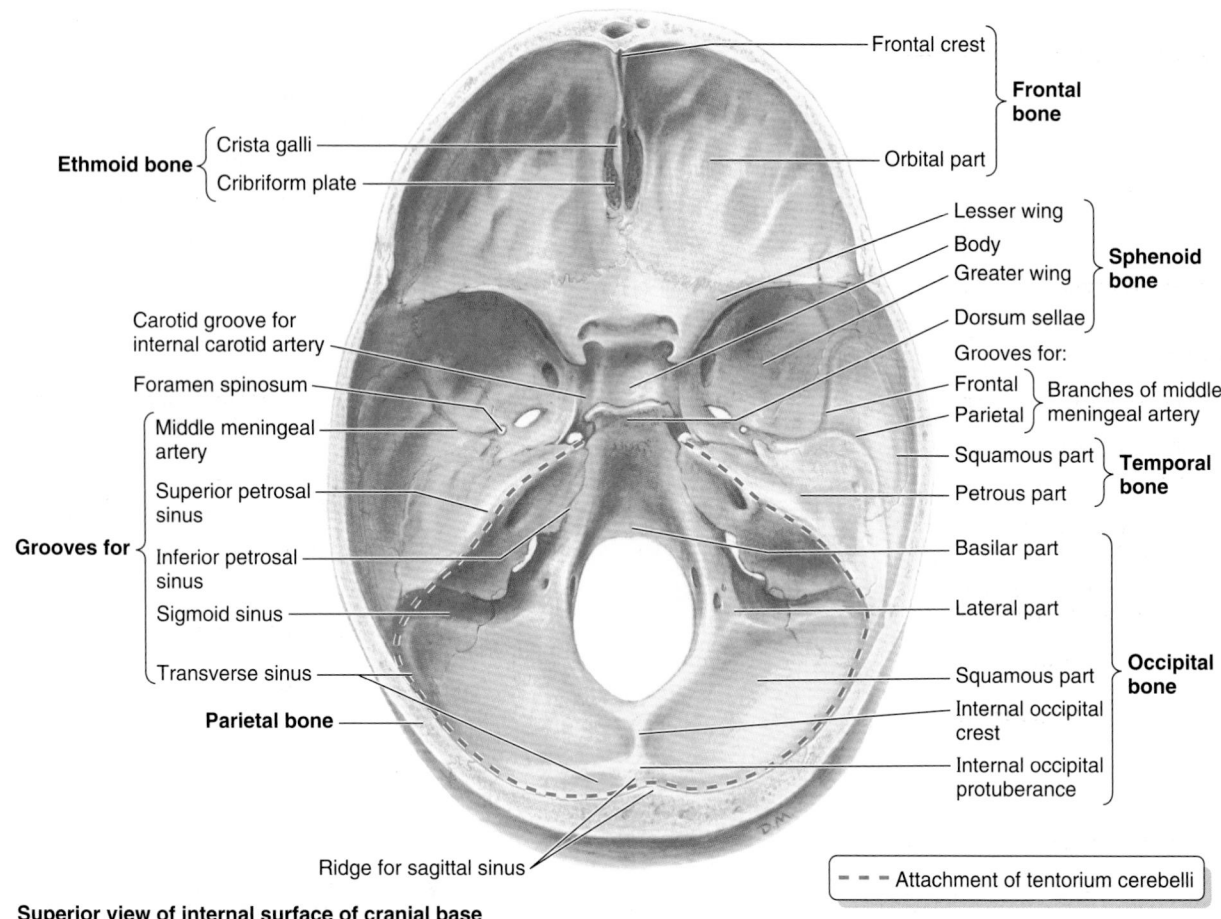

Superior view of internal surface of cranial base

FIGURE 7.30. Interior of base of cranium. The internal occipital protuberance is formed in relationship to the confluence of sinuses (Fig. 7.31A), and grooves are formed in the cranial base by the dural venous sinuses (e.g., the sigmoid sinus). The tentorium cerebelli is attached along the lengths of the transverse and superior petrosal sinuses (*dashed line*).

(Figs. 7.30–7.32). The transverse sinuses course along the posterolateral attached margins of the tentorium cerebelli and then become the sigmoid sinuses as they approach the posterior aspect of the petrous temporal bones. Blood received by the *confluence of sinuses* is drained by the transverse sinuses, but rarely equally. Usually the left sinus is dominant (larger).

The **sigmoid sinuses** follow S-shaped courses in the posterior cranial fossa, forming deep grooves in the temporal and occipital bones. Each sigmoid sinus turns anteriorly and then continues inferiorly as the IJV after traversing the jugular foramen. The **occipital sinus** lies in the attached border of the falx cerebelli and ends superiorly in the confluence of sinuses (Fig. 7.29B). The occipital sinus communicates inferiorly with the *internal vertebral venous plexus* (Figs. 7.29B and 7.33).

The **cavernous sinus,** a large venous plexus, is located on each side of the sella turcica on the upper surface of the body of the sphenoid, which contains the sphenoid (air) sinus (Figs. 7.29B and 7.31). The cavernous sinus consists of a venous plexus of extremely thin-walled veins that extends from the superior orbital fissure anteriorly to the apex of the petrous part of the temporal bone posteriorly. It receives blood from the superior and inferior ophthalmic veins, superficial middle cerebral vein, and sphenoparietal sinus. The venous channels

in these sinuses communicate with each other through venous channels anterior and posterior to the stalk of the pituitary gland—the **intercavernous sinuses** (Fig. 7.31A & B)—and sometimes through veins inferior to the pituitary gland. The cavernous sinuses drain postero-inferiorly through the superior and inferior petrosal sinuses and emissary veins to the *basilar* and *pterygoid plexuses* (Fig. 7.29B).

Inside each cavernous sinus is the **internal carotid artery** with its small branches, surrounded by the carotid plexus of sympathetic nerve(s), and the abducent nerve (CN VI) (Fig. 7.31C). The oculomotor (CN III) and trochlear (CN IV) nerves, plus two of the three divisions of the trigeminal nerve (CN V) are embedded in the lateral wall of the sinus. The artery, carrying warm blood from the body's core, traverses the sinus filled with cooler blood returning from the capillaries of the body's periphery, allowing for heat exchange to conserve energy or cool the arterial blood. This does not appear to be as important in humans as it is in running animals (e.g., horses and cheetahs) in which the carotid artery runs a longer, more tortuous course through the cavernous sinuses, allowing cooling of blood before it enters the brain. Pulsations of the artery within the cavernous sinus are said to promote propulsion of venous blood from the sinus, as does gravity (Standring, 2008).

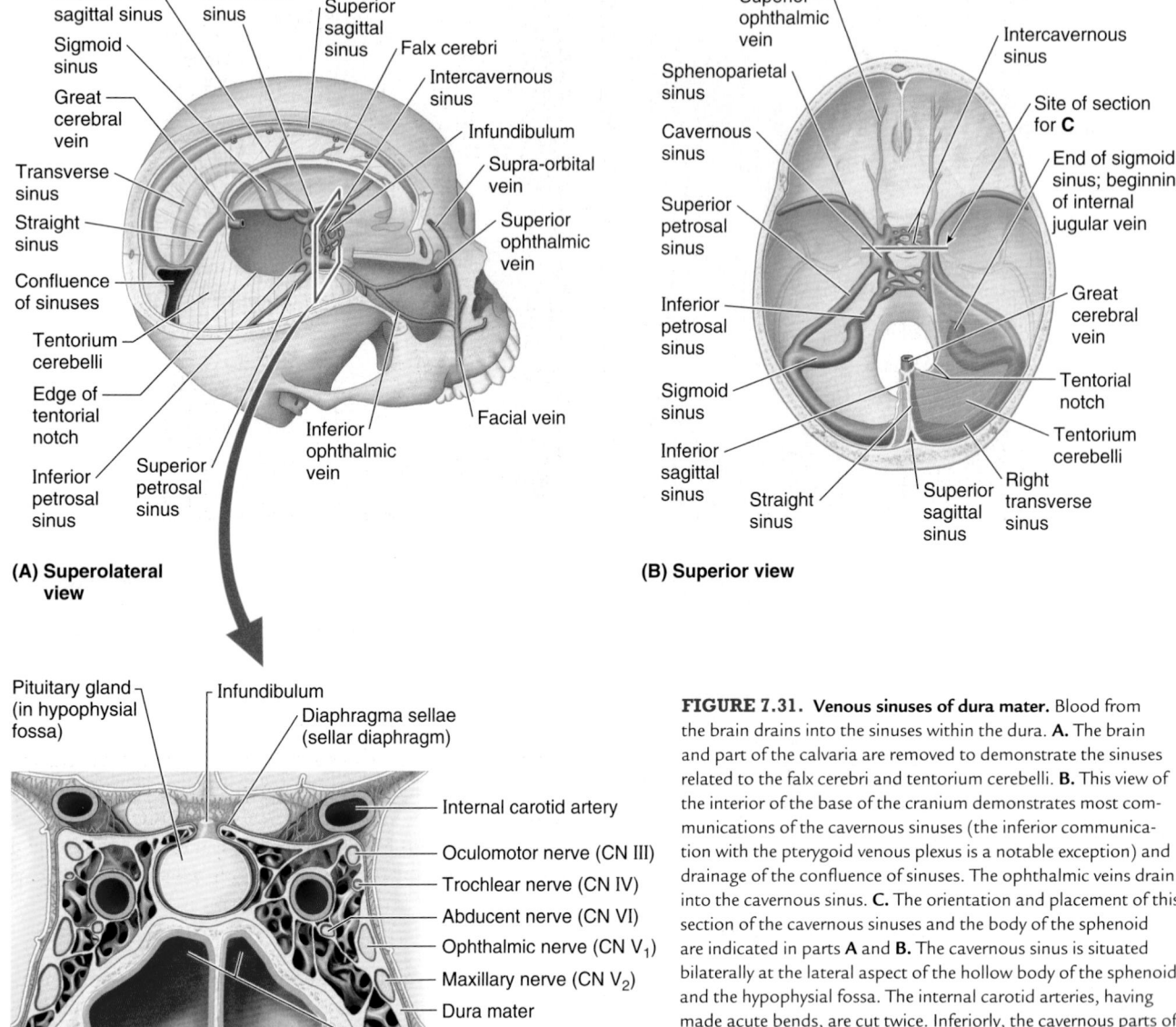

(A) Superolateral view

Inferior sagittal sinus
Cavernous sinus
Superior sagittal sinus
Sigmoid sinus
Falx cerebri
Great cerebral vein
Intercavernous sinus
Transverse sinus
Infundibulum
Straight sinus
Supra-orbital vein
Confluence of sinuses
Superior ophthalmic vein
Tentorium cerebelli
Edge of tentorial notch
Facial vein
Inferior petrosal sinus
Superior petrosal sinus
Inferior ophthalmic vein

(B) Superior view

Superior ophthalmic vein
Intercavernous sinus
Sphenoparietal sinus
Cavernous sinus
Site of section for **C**
Superior petrosal sinus
End of sigmoid sinus; beginning of internal jugular vein
Inferior petrosal sinus
Great cerebral vein
Sigmoid sinus
Tentorial notch
Inferior sagittal sinus
Tentorium cerebelli
Straight sinus
Superior sagittal sinus
Right transverse sinus

Pituitary gland (in hypophysial fossa)
Infundibulum
Diaphragma sellae (sellar diaphragm)
Internal carotid artery
Oculomotor nerve (CN III)
Trochlear nerve (CN IV)
Abducent nerve (CN VI)
Ophthalmic nerve (CN V₁)
Maxillary nerve (CN V₂)
Dura mater
Cavernous sinus
Sphenoidal sinus

(C) Posterior view of coronal section of cavernous sinus

FIGURE 7.31. Venous sinuses of dura mater. Blood from the brain drains into the sinuses within the dura. **A.** The brain and part of the calvaria are removed to demonstrate the sinuses related to the falx cerebri and tentorium cerebelli. **B.** This view of the interior of the base of the cranium demonstrates most communications of the cavernous sinuses (the inferior communication with the pterygoid venous plexus is a notable exception) and drainage of the confluence of sinuses. The ophthalmic veins drain into the cavernous sinus. **C.** The orientation and placement of this section of the cavernous sinuses and the body of the sphenoid are indicated in parts **A** and **B.** The cavernous sinus is situated bilaterally at the lateral aspect of the hollow body of the sphenoid and the hypophysial fossa. The internal carotid arteries, having made acute bends, are cut twice. Inferiorly, the cavernous parts of the arteries are sectioned as they pass anteriorly along the carotid groove toward the acute bend of the artery (some radiologists refer to the bend as the "carotid siphon"); superiorly, the cerebral parts of the arteries are sectioned as they pass posteriorly from the bend to join the cerebral arterial circle.

The **superior petrosal sinuses** run from the posterior ends of the veins making up the cavernous sinus to the transverse sinuses at the site where these sinuses curve inferiorly to form the sigmoid sinuses (Fig. 7.32B). Each superior petrosal sinus lies in the anterolateral attached margin of the tentorium cerebelli, which attaches to the superior border (crest) of the petrous part of the temporal bone (Fig. 7.30).

The **inferior petrosal sinuses** also commence at the posterior end of the cavernous sinus (Fig. 7.31A & B). Each inferior petrosal sinus runs in a groove between the petrous part of the temporal bone and the basilar part of the occipital bone (Fig. 7.30). The inferior petrosal sinuses drain the cavernous sinus directly into the transition of the sigmoid sinus to the IJV at the jugular foramen (Fig. 7.31B). The **basilar plexus** connects the

inferior petrosal sinuses and communicates inferiorly with the *internal vertebral venous plexus* (Figs. 7.29B and 7.33).

Emissary veins connect the dural venous sinuses with veins outside the cranium. Although they are valveless and blood may flow in both directions, flow in the emissary veins is usually away from the brain. The size and number of emissary veins vary; many small ones are unnamed. A **frontal emissary vein** is present in children and some adults. It passes through the foramen cecum of the cranium, connecting the superior sagittal sinus with veins of the frontal sinus and nasal cavities. A **parietal emissary vein,** which may be paired bilaterally, passes through the *parietal foramen* in the calvaria, connecting the superior sagittal sinus with the veins external to it, particularly those in the scalp (see Fig. 7.8A &

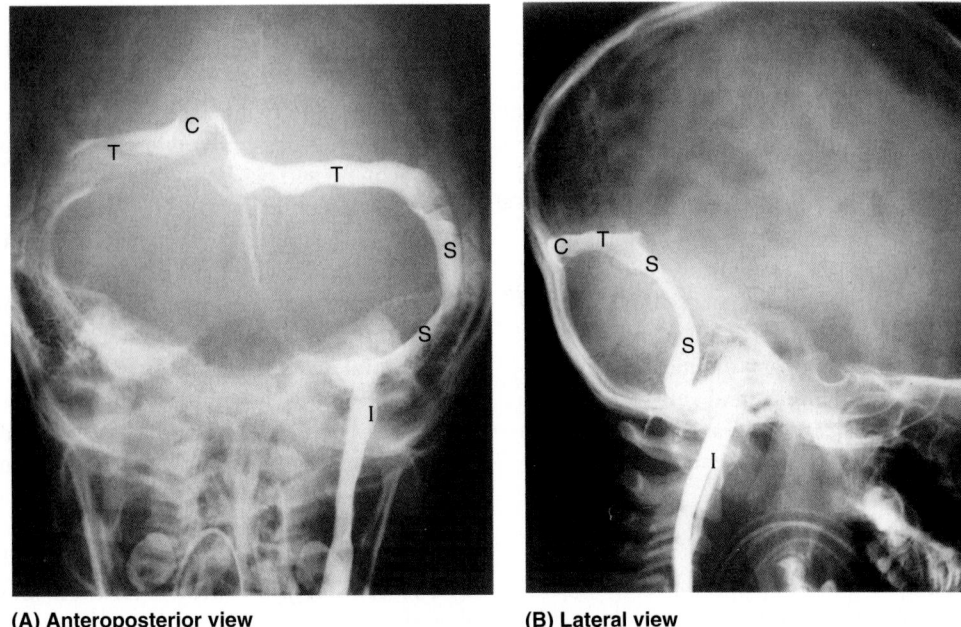

(A) Anteroposterior view   **(B) Lateral view**

FIGURE 7.32. Venograms of dural sinuses. A and B. In these radiographic studies, radiopaque dye injected into the arterial system has circulated through the capillaries of the brain and collected in the dural venous sinuses. *C,* confluence of sinuses; *I,* internal jugular vein; *S,* sigmoid sinus; *T,* transverse sinus. In the AP view (**A**), notice the left-sided dominance in the drainage of the confluence of sinuses. (Courtesy of Dr. D. Armstrong, Associate Professor of Medical Imaging, University of Toronto, Toronto, Ontario, Canada.)

C). A **mastoid emissary vein** passes through the *mastoid foramen* and connects each sigmoid sinus with the *occipital* or *posterior auricular vein* (Fig. 7.33). A **posterior condylar emissary vein** may also be present, passing through the condylar canal, connecting the sigmoid sinus with the *suboccipital venous plexus.*

VASCULATURE OF DURA MATER

The **arteries of the dura** supply more blood to the calvaria than to the dura. The largest of these vessels, the **middle meningeal artery,** is a branch of the maxillary artery (Fig. 7.28D). It enters the floor of the middle cranial fossa through the *foramen spinosum* (Fig. 7.30), runs laterally in the fossa,

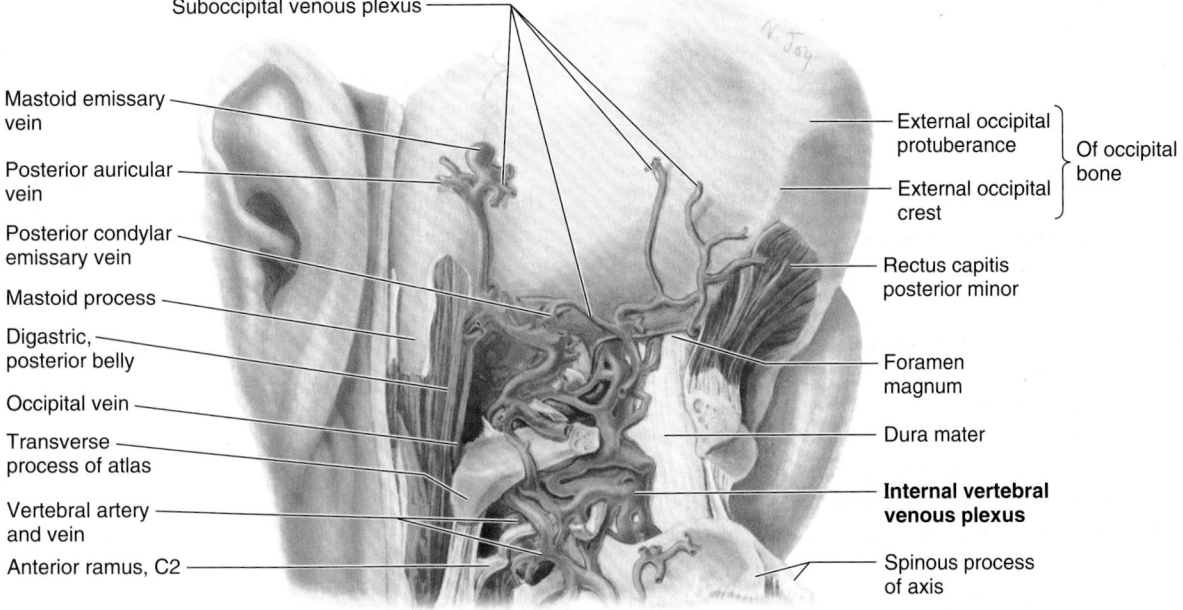

Posterolateral view

FIGURE 7.33. Deep dissection of suboccipital region. The external vertebral venous system has numerous intercommunications and connections, some of which are shown here. Superiorly, the system communicates with the veins of the scalp and the intracranial venous sinuses via the foramen magnum, the mastoid foramina, and the condylar canals. Anteromedially, it passes between the laminae and through the intervertebral foramina to communicate with the internal vertebral venous plexus and veins around the vertebral artery.

and turns supero-anteriorly on the greater wing of the sphenoid, where it divides into anterior and posterior branches (Fig. 7.28D). The **frontal branch of the middle meningeal artery** runs superiorly to the pterion and then curves posteriorly to ascend toward the vertex of the cranium. The **parietal branch of the middle meningeal artery** runs posterosuperiorly and ramifies (breaks up into distributing branches) over the posterior aspect of the cranium. Small areas of dura are supplied by other arteries: meningeal branches of the ophthalmic arteries, branches of the occipital arteries, and small branches of the vertebral arteries.

The **veins of the dura** accompany the meningeal arteries, often in pairs. The **middle meningeal veins** accompany the middle meningeal artery, leave the cranial cavity through the foramen spinosum or foramen ovale, and drain into the *pterygoid venous plexus* (Fig. 7.29B).

NERVE SUPPLY OF DURA MATER

The dura on the floors of the anterior and middle cranial fossa and the roof of the posterior cranial fossa is innervated by meningeal branches arising directly or indirectly from the trigeminal nerve (CN V) (Fig. 7.34). There are three divisions of CN V (CN V_1, CN V_2, and CN V_3), each of which contributes a meningeal branch or branches. The **anterior meningeal branches of the ethmoidal nerves** (CN V_1) and the **meningeal branches of the maxillary** (CN V_2) and **mandibular** (CN V_3) **nerves** supply the dura of the anterior cranial fossa. The latter two nerves also supply the dura of the middle cranial fossa (Fig. 7.34B). The meningeal branches of CN V_2 and CN V_3 are distributed as peri-arterial plexuses, accompanying the branches of the middle meningeal artery (Fig. 7.34A, *inset*).

The dura forming the roof of the posterior cranial fossa (tentorium cerebelli) and posterior part of the falx cerebri is supplied by the **tentorial nerve** (a branch of the ophthalmic nerve), whereas the anterior falx cerebri is innervated by ascending branches of the anterior meningeal branches (Fig. 7.34A). The dura of the floor of the posterior cranial fossa receives sensory fibers from the spinal ganglia of C2 and C3 carried by those spinal nerves or by fibers transferred to and traveling centrally with the vagus (CN X) and hypoglossal (CN XII) nerves. Sensory endings are more numerous in the dura along each side of the superior sagittal sinus and in the tentorium cerebelli than they are in the floor of the cranium.

Pain fibers are most numerous where arteries and veins course in the dura. Pain arising from the dura is generally referred, perceived as a headache arising in cutaneous or mucosal regions supplied by the involved cervical nerve or division of the trigeminal nerve.

Arachnoid Mater and Pia Mater

The *arachnoid mater* and *pia mater* (or simply *arachnoid* and *pia*; together the leptomeninges) develop from a single layer of mesenchyme surrounding the embryonic brain, becoming the parietal part (arachnoid) and visceral part (pia) of the *leptomeninx*. (Fig. 7.35). The derivation of the arachnoid–pia from a single embryonic layer is indicated in the adult by the numerous web-like *arachnoid trabeculae* passing between the arachnoid and pia, which give the arachnoid its name (G. *arachne–*, spider, cobweb + G. *eidos*, resemblance). The trabeculae are composed of flattened, irregularly shaped fibroblasts that bridge the subarachnoid space (Haines, 2006). The arachnoid and pia are in continuity immediately proximal to the exit of each cranial nerve from the dura mater. The **cranial arachnoid mater** contains fibroblasts, collagen fibers, and some elastic fibers. Although thin, the arachnoid is thick enough to be manipulated with forceps. The avascular arachnoid, although closely applied to the meningeal layer of the dura, is not attached to the dura; it is held against the inner surface of the dura by the pressure of the CSF in the subarachnoid space.

The **cranial pia mater** is an even thinner membrane than the arachnoid; it is highly vascularized by a network of fine blood vessels. The pia is difficult to see, but it gives the surface of the brain a shiny appearance. The pia adheres to the surface of the brain and follows all its contours. When the cerebral arteries penetrate the cerebral cortex, the pia follows them for a short distance, forming a **pial coat** and a **peri-arterial space** (Fig. 7.35).

Meningeal Spaces

Of the three meningeal "spaces" commonly mentioned in relation to the cranial meninges, only one exists as a space in the absence of pathology:

- The **dura–cranial interface** (extradural or epidural "space") is not a natural space between the cranium and the external periosteal layer of the dura because the dura is attached to the bones. It becomes an **extradural space** only pathologically—for example, when blood from torn meningeal vessels pushes the periosteum away from the cranium (Fig. 7.28C). The potential or pathological cranial epidural space is not continuous with the **spinal epidural space** (a natural space occupied by epidural fat and a venous plexus) because the former is external to the periosteum lining the cranium, and the latter is internal to the periosteum covering the vertebrae.
- The *dura–arachnoid interface* or junction ("subdural space") is likewise not a natural space between the dura and arachnoid. A space may develop in the dural border cell layer as the result of trauma, such as a hard blow to the head (Haines, 1993, 2006).
- The *subarachnoid space*, between the arachnoid and pia, is a real space that contains CSF, trabecular cells, arteries, and veins.

Although it is commonly stated that the brain "floats" in CSF, the brain is suspended in the CSF-filled subarachnoid space by the arachnoid trabeculae.

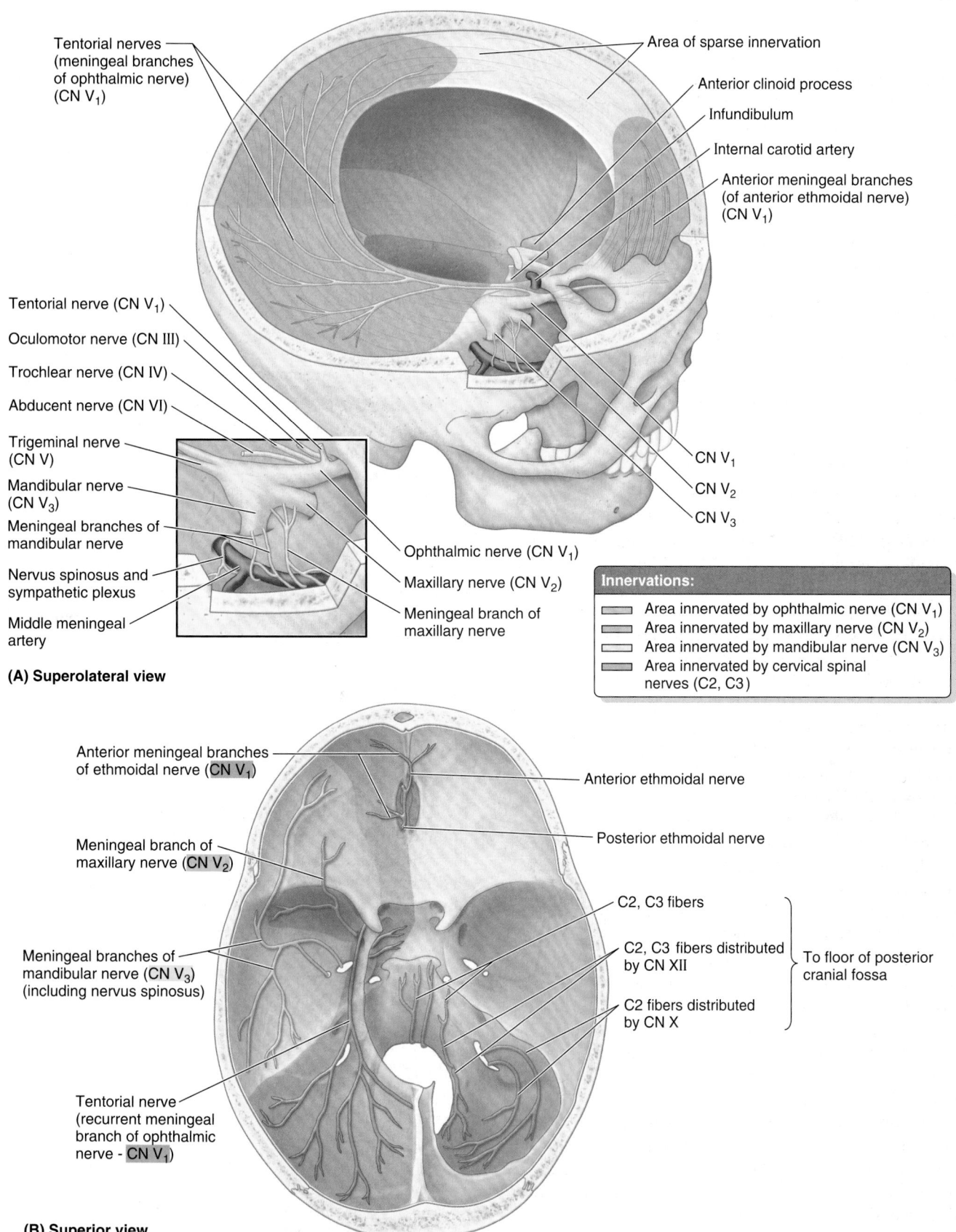

Tentorial nerves
(meningeal branches
of ophthalmic nerve)
(CN V₁)

Area of sparse innervation

Anterior clinoid process

Infundibulum

Internal carotid artery

Anterior meningeal branches
(of anterior ethmoidal nerve)
(CN V₁)

Tentorial nerve (CN V₁)

Oculomotor nerve (CN III)

Trochlear nerve (CN IV)

Abducent nerve (CN VI)

Trigeminal nerve
(CN V)

Mandibular nerve
(CN V₃)

Meningeal branches of
mandibular nerve

Nervus spinosus and
sympathetic plexus

Middle meningeal
artery

Ophthalmic nerve (CN V₁)

Maxillary nerve (CN V₂)

Meningeal branch of
maxillary nerve

CN V₁

CN V₂

CN V₃

(A) Superolateral view

Innervations:

Area innervated by ophthalmic nerve (CN V₁)
Area innervated by maxillary nerve (CN V₂)
Area innervated by mandibular nerve (CN V₃)
Area innervated by cervical spinal
nerves (C2, C3)

Anterior meningeal branches
of ethmoidal nerve (CN V₁)

Anterior ethmoidal nerve

Meningeal branch of
maxillary nerve (CN V₂)

Posterior ethmoidal nerve

C2, C3 fibers

C2, C3 fibers distributed
by CN XII

Meningeal branches of
mandibular nerve (CN V₃)
(including nervus spinosus)

To floor of posterior
cranial fossa

C2 fibers distributed
by CN X

Tentorial nerve
(recurrent meningeal
branch of ophthalmic
nerve - CN V₁)

(B) Superior view

FIGURE 7.34. Innervation of dura mater. A. The right side of the calvaria and brain is removed, and CN V is dissected. The meningeal branches of the maxillary (CN V₂) and mandibular (CN V₃) nerves are distributed to the dura of the lateral part of the anterior and the middle cranial fossae as peri-arterial plexuses that accompany the branches of the middle meningeal artery along with vasomotor sympathetic nerve fibers from the superior cervical ganglion (*inset*). **B.** The internal aspect of the cranial base shows innervation of the dura by branches of the trigeminal and sensory fibers of cervical spinal nerves (C2, C3) passing directly from those nerves or via meningeal branches of the vagus (CN X) and hypoglossal (CN XII) nerves.

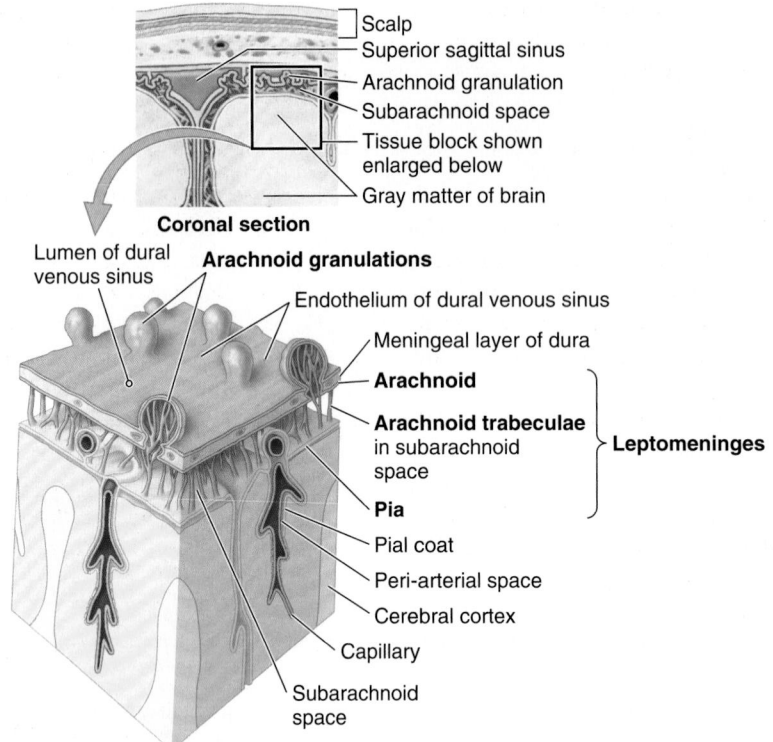

FIGURE 7.35. Leptomeninges. The coronal section (*above*) indicates the site of the tissue block (*below*). The subarachnoid space separates the two layers of the leptomeninges, the arachnoid and pia. CSF pressure keeps the arachnoid apposed to the meningeal layer of dura, and in the region of the superior sagittal sinus and adjacent venous lacunae (Fig. 7.28D), arachnoid granulations project through the dura into the blood-filled dural venous sinus.

CRANIAL CAVITY AND MENINGES

Fracture of Pterion

Fracture of the pterion can be life threatening because it *overlies the frontal branches of the middle meningeal vessels,* which lie in grooves on the internal aspect of the lateral wall of the calvaria (Fig. 7.30). The pterion is two fingers' breadth superior to the zygomatic arch and a thumb's breadth posterior to the frontal process of the zygomatic bone (Fig. B7.16A). A hard blow to the side of the head may fracture the thin bones forming the pterion (Fig. 7.4A), producing a rupture of the frontal branch of the middle meningeal artery or vein crossing the pterion (Fig. B7.16B). The

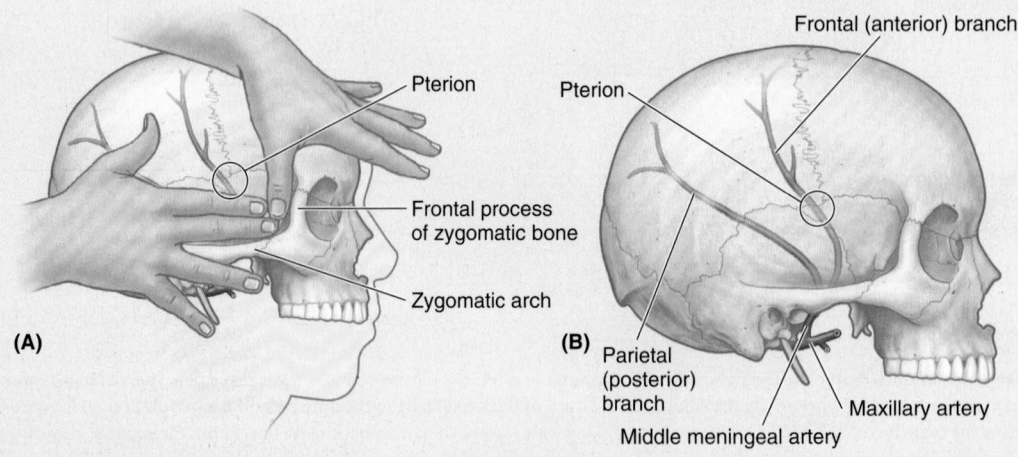

FIGURE B7.16.

resulting *hematoma* exerts pressure on the underlying cerebral cortex (Fig. 7.19A). An untreated *middle meningeal vessel hemorrhage* may cause death in a few hours.

Thrombophlebitis of Facial Vein

The facial vein makes clinically important connections with the cavernous sinus through the superior ophthalmic vein, and the pterygoid venous plexus through the inferior ophthalmic and deep facial veins (Figs. 7.25 and 7.29B; Table 7.6). Because of these connections, an infection of the face may spread to the cavernous sinus and pterygoid venous plexus.

Blood from the medial angle of the eye, nose, and lips usually drains inferiorly through the facial vein, especially when a person is erect. Because the facial vein has no valves, blood may pass through it in the opposite direction. Consequently, venous blood from the face may enter the cavernous sinus. In individuals with *thrombophlebitis of the facial vein*—inflammation of the facial vein with secondary thrombus (clot) formation—pieces of an infected clot may extend into the intracranial venous system and produce *thrombophlebitis of the cavernous sinus.*

Infection of the facial veins spreading to the dural venous sinuses may result from lacerations of the nose or be initiated by squeezing pustules (pimples) on the side of the nose and upper lip. Consequently, the triangular area from the upper lip to the bridge of the nose is considered the *danger triangle of the face* (Fig. B7.17).

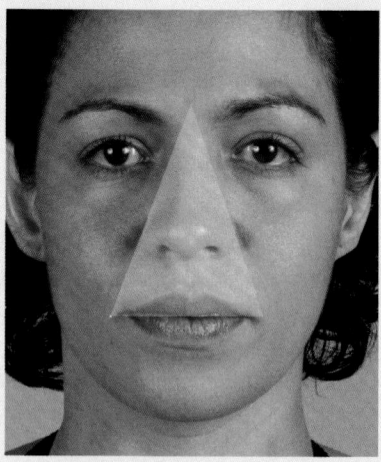

FIGURE B7.17. Danger triangle of face.

Blunt Trauma to Head

A blow to the head can detach the periosteal layer of dura mater from the calvaria without fracturing the cranial bones. In the cranial base, the two dural layers are firmly attached and difficult to separate from the bones. Consequently, a fracture of the cranial base usually tears the dura and results in leakage of CSF. The innermost part of the dura, *the dural border cell layer,* is composed of flattened

fibroblasts that are separated by large extracellular spaces. This layer constitutes a plane of structural weakness at the dura–arachnoid junction (Haines, 2006).

Tentorial Herniation

The tentorial notch is the opening in the tentorium cerebelli for the brainstem, which is slightly larger than is necessary to accommodate the midbrain (Fig. B7.18). Hence, *space-occupying lesions,* such as tumors in the supratentorial compartment, produce increased intracranial pressure, and may cause part of the adjacent temporal lobe of the brain to herniate through the tentorial notch. During *tentorial herniation,* the temporal lobe may be lacerated by the tough tentorium cerebelli, and the oculomotor nerve (CN III) may be stretched, compressed, or both. *Oculomotor lesions* may produce paralysis of the extrinsic eye muscles supplied by CN III.

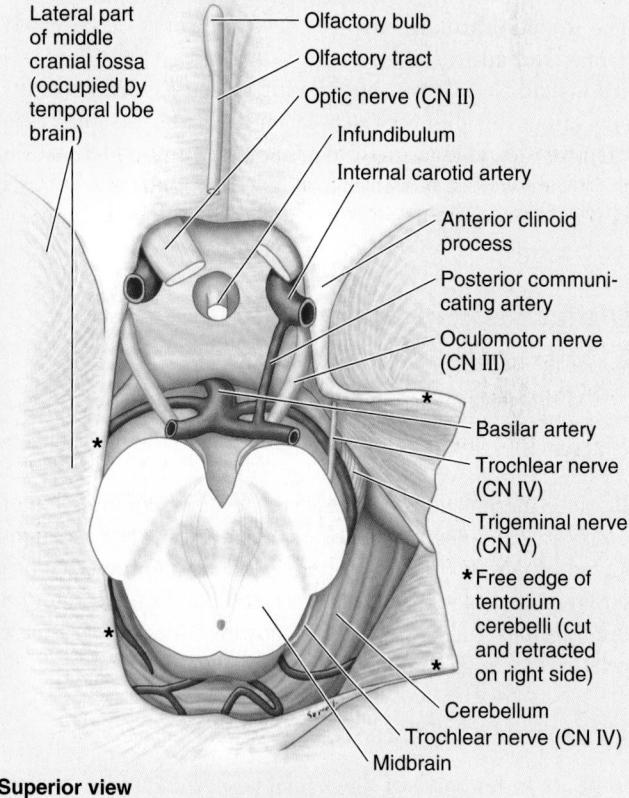

Lateral part of middle cranial fossa (occupied by temporal lobe brain)

Olfactory bulb
Olfactory tract
Optic nerve (CN II)
Infundibulum
Internal carotid artery
Anterior clinoid process
Posterior communicating artery
Oculomotor nerve (CN III)
Basilar artery
Trochlear nerve (CN IV)
Trigeminal nerve (CN V)
*Free edge of tentorium cerebelli (cut and retracted on right side)
Cerebellum
Trochlear nerve (CN IV)
Midbrain

Superior view

FIGURE B7.18.

Bulging of Diaphragma Sellae

Pituitary tumors may extend superiorly through the aperture in the diaphragma sellae, or cause it to bulge. These tumors often expand the diaphragma sellae, producing disturbances in endocrine function early or late (i.e., before or after enlargement of the diaphragma sellae). Superior extension of a tumor may cause visual symptoms owing to pressure on the optic chiasm, the place where the optic nerve fibers cross (Fig. 7.37B, p. 880).

Occlusion of Cerebral Veins and Dural Venous Sinuses

Occlusion of cerebral veins and dural venous sinuses may result from thrombi (clots), thrombophlebitis (venous inflammation), or tumors (e.g., meningiomas). The dural venous sinuses most frequently thrombosed are the transverse, cavernous, and superior sagittal sinuses (Fishman, 2010b).

The facial veins make clinically important connections with the cavernous sinus through the superior ophthalmic veins (Fig. 7.29B). *Cavernous sinus thrombosis* usually results from infections in the orbit, nasal sinuses, and superior part of the face (the danger triangle, Fig. B7.17). In persons with *thrombophlebitis of the facial vein*, pieces of an infected thrombus may extend into the cavernous sinus, producing *thrombophlebitis of the cavernous sinus.* The infection usually involves only one sinus initially, but it may spread to the opposite side through the intercavernous sinuses. Thrombophlebitis of the cavernous sinus may affect the abducent nerve as it traverses the sinus (see Chapter 9), and may also effect the nerves embedded within the lateral wall of the sinus (Fig. 7.31C). Septic thrombosis of the cavernous sinus often results in the development of *acute meningitis.*

Metastasis of Tumor Cells to Dural Venous Sinuses

The basilar and occipital sinuses communicate through the foramen magnum with the internal vertebral venous plexuses (Figs. 7.29B and 7.33). Because these venous channels are valveless, compression of the thorax, abdomen, or pelvis, as occurs during heavy coughing and straining, may force venous blood from these regions into the internal vertebral venous system and from it into the dural venous sinuses. As a result, pus in abscesses and tumor cells in these regions may spread to the vertebrae and brain.

Fractures of Cranial Base

In fractures of the cranial base, the internal carotid artery may be torn, producing an *arteriovenous fistula* within the cavernous sinus. Arterial blood rushes into the cavernous sinus, enlarging it and forcing retrograde blood flow into its venous tributaries, especially the ophthalmic veins. As a result, the eyeball protrudes (*exophthalmos*) and the conjunctiva becomes engorged (*chemosis*). The protruding eyeball pulsates in synchrony with the radial pulse, a phenomenon known as *pulsating exophthalmos.* Because CN III, CN IV, CN V_1, CN V_2, and CN VI lie in or close to the lateral wall of the cavernous sinus, these nerves may also be affected when the sinus is injured (Fig. 7.31C).

Dural Origin of Headaches

The dura is sensitive to pain, especially where it is related to the dural venous sinuses and meningeal arteries (Fig. 7.31A). Consequently, pulling on arteries at the cranial base or veins near the vertex, where they pierce the dura, causes pain. Distension of the scalp or meningeal vessels (or both) is believed to be one cause of headache (Green, 2010).

Many headaches appear to be dural in origin, such as the headache occurring after a lumbar spinal puncture for removal of CSF (see Chapter 4). These headaches are thought to result from stimulation of sensory nerve endings in the dura. When CSF is removed, the brain sags slightly, pulling on the dura; this may also cause a headache. For this reason, patients are asked to keep their heads down after a lumbar puncture to minimize the pull on the dura, reducing the chances of getting a headache.

Leptomeningitis

Leptomeningitis is an inflammation of the leptomeninges (arachnoid and pia) resulting from pathogenic microorganisms. The infection and inflammation are usually confined to the subarachnoid space and the arachnoid–pia (Jubelt, 2010). The bacteria may enter the subarachnoid space through the blood (*septicemia,* or "blood poisoning"), or spread from an infection of the heart, lungs, or other viscera. Microorganisms may also enter the subarachnoid space from a compound cranial fracture or a fracture of the nasal sinuses. *Acute purulent meningitis* can result from infection with almost any pathogenic bacteria (e.g., *meningococcal meningitis*).

Head Injuries and Intracranial Hemorrhage

Extradural (epidural) hemorrhage is arterial in origin. Blood from torn branches of a middle meningeal artery collects between the external periosteal layer of the dura and the calvaria. The extravasated blood strips the dura from the cranium. Usually this follows a hard blow to the head, and forms an *extradural (epidural) hematoma* (Fig. B7.19A & B). Typically, a brief *concussion* (loss of consciousness) occurs, followed by a lucid interval of some hours. Later, drowsiness and coma (profound unconsciousness) occur. Compression of the brain occurs as the blood mass increases, necessitating evacuation of the blood and occlusion of the bleeding vessel(s).

A *dural border hematoma* is commonly called a *subdural hematoma* (Fig. B7.19B); however, this term is a misnomer because there is no naturally occurring space at the dura–arachnoid junction. *Hematomas* at this junction are usually caused by extravasated blood that splits open the dural border cell layer (Haines, 2006). The blood does not collect within a preexisting space, but rather creates a space at the

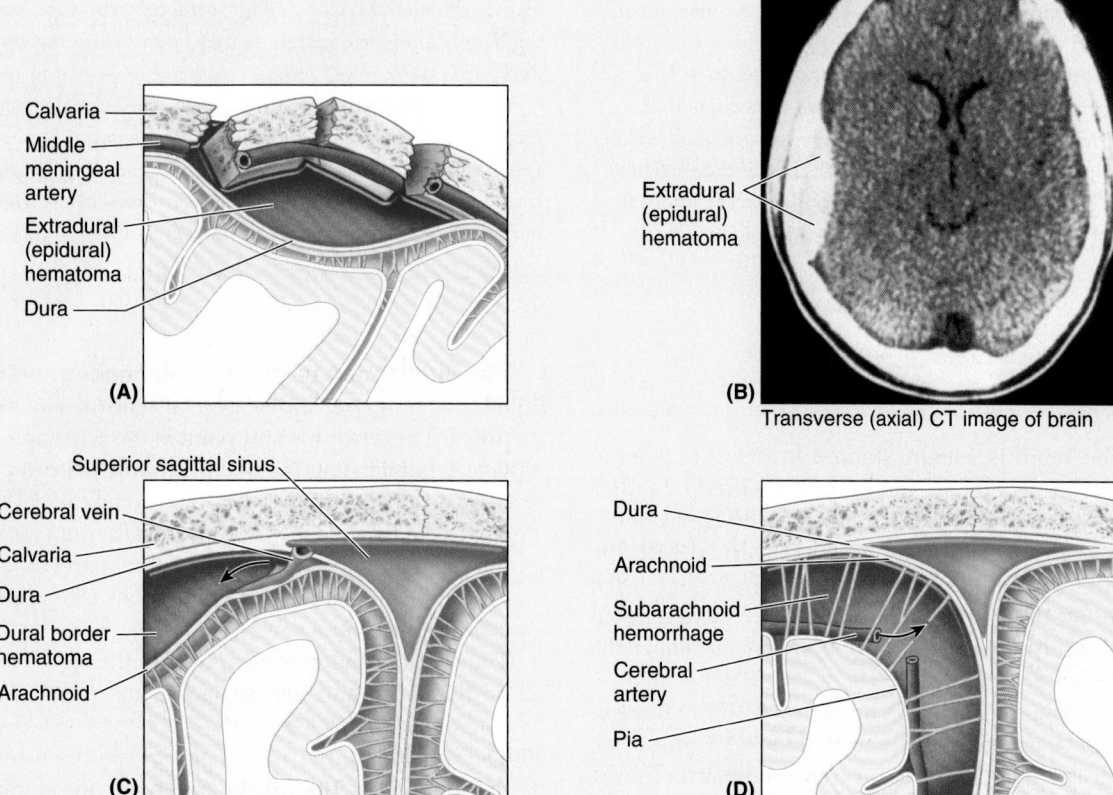

FIGURE B7.19. **Intercranial hemorrhages. A and B.** Extradural (epidural) hemorrhage. **C.** Dural border (subdural) hematoma. **D.** Subarachnoid hemorrhage.

dura–arachnoid junction (Haines, 2006). *Dural border hemorrhage* usually follows a hard blow to the head that jerks the brain inside the cranium and injures it. The precipitating trauma may be trivial or forgotten. Dural border hemorrhage is typically venous in origin and commonly results from tearing a superior cerebral vein as it enters the superior sagittal sinus (Fig. 7.29B) (Haines et al., 1993).

Subarachnoid hemorrhage is an extravasation of blood, usually arterial, into the subarachnoid space (Fig. B7.19C).

Most subarachnoid hemorrhages result from *rupture of a saccular aneurysm* (sac-like dilation on the side of an artery), such as an aneurysm of the internal carotid artery (see the blue box "Strokes," p. 887).

Some subarachnoid hemorrhages are associated with head trauma involving cranial fractures and cerebral lacerations. Bleeding into the subarachnoid space results in meningeal irritation, severe headache, stiff neck, and often loss of consciousness.

The Bottom Line

CRANIAL MENINGES

The cranial meninges consist of three intracranial layers: a substantial, fibrous bilaminar outer layer—the dura—and two continuous, delicate, membranous inner layers—the arachnoid and pia.

Dura mater: The outer (periosteal) lamina of the dura is continuous with the periosteum on the external surface of the cranium, and is intimately applied to the internal surface of the cranial cavity. ♦ The inner (meningeal) lamina is a sustentacular (supporting) layer that more closely reflects

the contours of the brain. ♦ This inner layer separates from the outer layer in certain locations to form dural folds or reflections that penetrate the large fissures between parts of the brain, partially subdividing the cranial cavity into smaller compartments that prevent inertial brain movement. ♦ In separating from the periosteal lamina, intralaminar spaces are created that accommodate the dural venous sinuses, which receive the venous drainage of the brain and mainly drain, in turn, to the internal jugular vein.

BRAIN

Because the brain is usually studied in detail in a separate neuroanatomy course, the brain is covered by only a superficial survey of its gross structure in the typical anatomy course, with attention primarily concerned with the relationship between the brain and its environment—that is, its meningeal coverings, the CSF-filled subarachnoid space, and internal features of its bony encasement (the neurocranium).

Because of their role in the production of CSF (cerebrospinal fluid), the ventricles of the brain and the CSF-producing choroid plexuses found there are also covered. Furthermore, 11 of 12 cranial nerves arise from the brain (see Chapter 9).

Parts of Brain

The **brain** (contained by the neurocranium) is composed of the cerebrum, cerebellum, and brainstem (Fig. 7.36). When the calvaria and dura are removed, **gyri** (folds), **sulci** (grooves), and **fissures** (clefts) of the cerebral cortex are visible through the delicate arachnoid–pia layer. Whereas the gyri and sulci demonstrate much variation, the other features of the brain, including overall brain size, are remarkably consistent from individual to individual.

- The **cerebrum** (L. *brain*) includes the cerebral hemispheres and basal ganglia. The **cerebral hemispheres,** separated by the falx cerebri within the **longitudinal cerebral fissure,** are the dominant features of the brain (Fig. 7.36A-C). Each cerebral hemisphere is divided for descriptive purposes into four lobes, each of which is related to, but the boundaries of which do not correspond to, the overlying bones of the same name. From a superior view, the cerebrum is essentially divided into quarters by the median longitudinal cerebral fissure and the coronal **central sulcus.** The central sulcus separates the **frontal lobes** (anteriorly) from the **parietal lobes** (posteriorly). In a lateral view, these lobes lie superior to the transverse **lateral sulcus** and the temporal lobe inferior to it. The posteriorly placed **occipital lobes** are separated from the parietal and temporal lobes by the plane of the **parieto-occipital sulcus,** visible on the medial surface of the cerebrum in a hemisected brain (Fig. 7.36C).

The anteriormost points of the anteriorly projecting frontal and temporal lobes are the **frontal** and **temporal poles.** The posteriormost point of the posteriorly projecting occipital lobe is the **occipital pole.** The hemispheres occupy the entire supratentorial cranial cavity (Fig. 7.31A & B). The frontal lobes occupy the anterior cranial fossae, the temporal lobes occupy the lateral parts of the middle cranial fossae, and the occipital lobes extend posteriorly over the tentorium cerebelli.

- The **diencephalon** is composed of the epithalamus, dorsal thalamus, and hypothalamus and forms the central core of the brain (Fig. 7.36D).
- The **midbrain,** the rostral part of the brainstem, lies at the junction of the middle and posterior cranial fossae. CN III and IV are associated with the midbrain.
- The **pons** is the part of the brainstem between the midbrain rostrally and the medulla oblongata caudally; it lies in the anterior part of the posterior cranial fossa. CN V is associated with the pons (Fig. 7.36A, C & D).
- The **medulla oblongata** (medulla) is the most caudal subdivision of the brainstem that is continuous with the spinal cord; it lies in the posterior cranial fossa. CN IX, X, and XII are associated with the medulla, whereas CN VI–VIII are associated with the junction of pons and medulla.
- The **cerebellum** is the large brain mass lying posterior to the pons and medulla and inferior to the posterior part of the cerebrum. It lies beneath the tentorium cerebelli in the posterior cranial fossa. It consists of two lateral hemispheres that are united by a narrow middle part, the **vermis.**

Ventricular System of Brain

The ventricular system of the brain consists of two lateral ventricles and the midline 3rd and 4th ventricles connected by the cerebral aqueduct (Figs. 7.37 and 7.38). CSF, largely secreted by the choroid plexuses of the ventricles, fills these brain cavities and the subarachnoid space of the brain and spinal cord.

VENTRICLES OF BRAIN

The **lateral ventricles,** the 1st and 2nd ventricles, are the largest cavities of the ventricular system and occupy large areas of the cerebral hemispheres. Each lateral ventricle

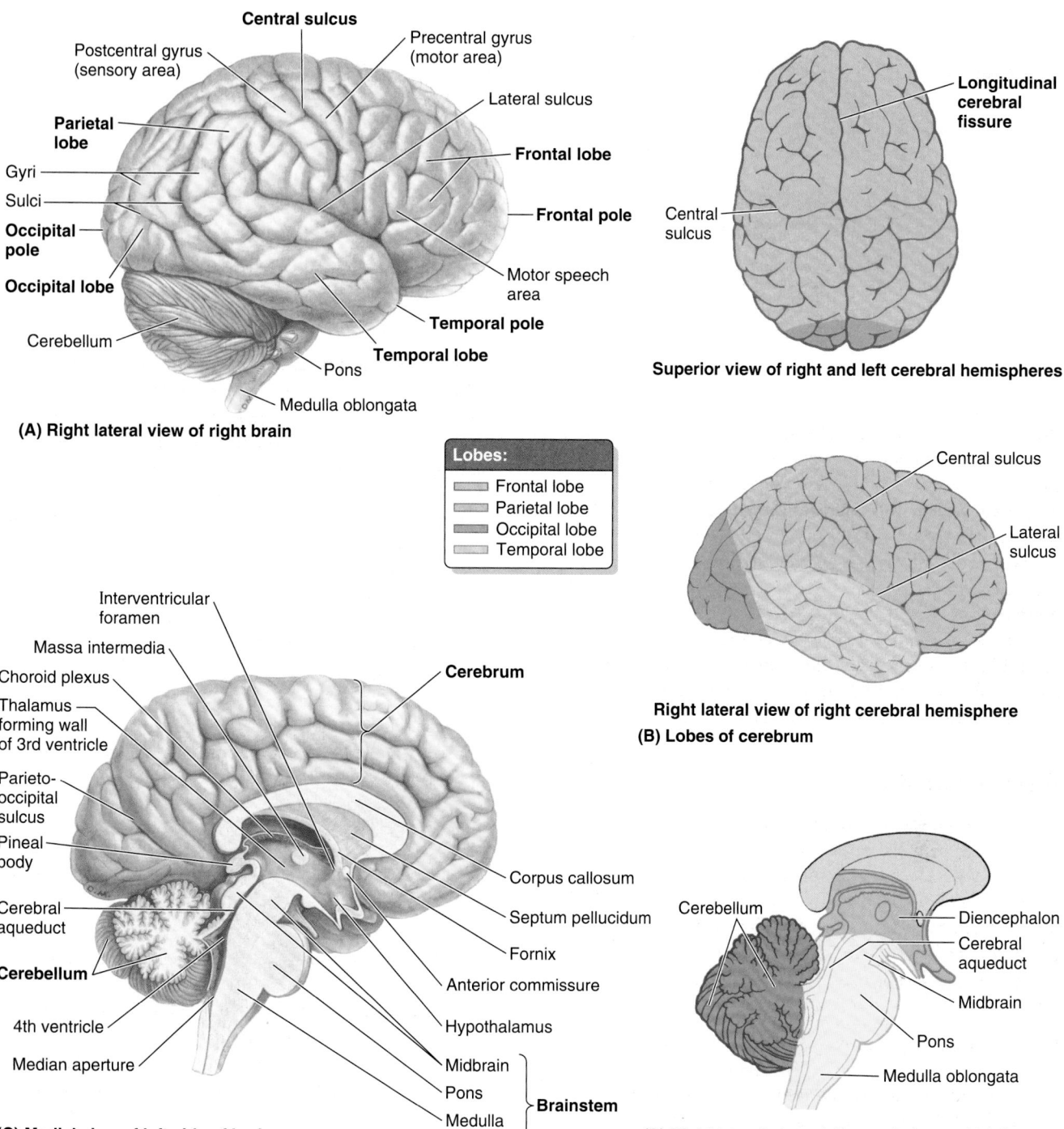

Central sulcus

Postcentral gyrus (sensory area)

Precentral gyrus (motor area)

Lateral sulcus

Parietal lobe

Frontal lobe

Gyri

Sulci

Frontal pole

Occipital pole

Motor speech area

Occipital lobe

Temporal pole

Cerebellum

Temporal lobe

Pons

Medulla oblongata

(A) Right lateral view of right brain

Longitudinal cerebral fissure

Central sulcus

Superior view of right and left cerebral hemispheres

Lobes:
- Frontal lobe
- Parietal lobe
- Occipital lobe
- Temporal lobe

Central sulcus

Lateral sulcus

Right lateral view of right cerebral hemisphere
(B) Lobes of cerebrum

Interventricular foramen

Massa intermedia

Choroid plexus

Cerebrum

Thalamus forming wall of 3rd ventricle

Parieto-occipital sulcus

Pineal body

Cerebral aqueduct

Corpus callosum

Septum pellucidum

Cerebellum

Fornix

Anterior commissure

4th ventricle

Hypothalamus

Median aperture

Midbrain

Pons

Medulla oblongata

Brainstem

(C) Medial view of left side of brain

Cerebellum

Diencephalon

Cerebral aqueduct

Midbrain

Pons

Medulla oblongata

(D) Right lateral view of diencephalon and brainstem

FIGURE 7.36. Structure of brain. A. The cerebral surface features the gyri (folds) and sulci (grooves) of the cerebral cortex. **B.** The lobes of the cerebrum are color coded. Whereas distinct central and lateral sulci demarcate the frontal lobe and anterior boundaries of the parietal and temporal lobes of the cerebrum, the demarcation of the posterior boundaries of the latter and the occipital lobe is less distinct externally. **C.** The medial surface of the cerebrum and deeper parts of the brain (diencephalon and brainstem) are shown after bisection of the brain. The parieto-occipital sulcus demarcating the parietal and occipital lobes is seen on the medial aspect of the cerebrum. **D.** The parts of the brainstem are identified.

opens through an **interventricular foramen** into the 3rd ventricle. The **3rd ventricle,** a slit-like cavity between the right and the left halves of the diencephalon, is continuous postero-inferiorly with the **cerebral aqueduct,** a narrow channel in the midbrain connecting the 3rd and 4th ventricles (Figs. 7.36C and 7.37B).

The pyramid-shaped **4th ventricle** in the posterior part of the pons and medulla extends inferoposteriorly. Inferiorly, it tapers to a narrow channel that continues into the cervical region of the spinal cord as the central canal (Fig. 7.37A). CSF drains into the subarachnoid space from the 4th ventricle through a single **median aperture** and paired **lateral**

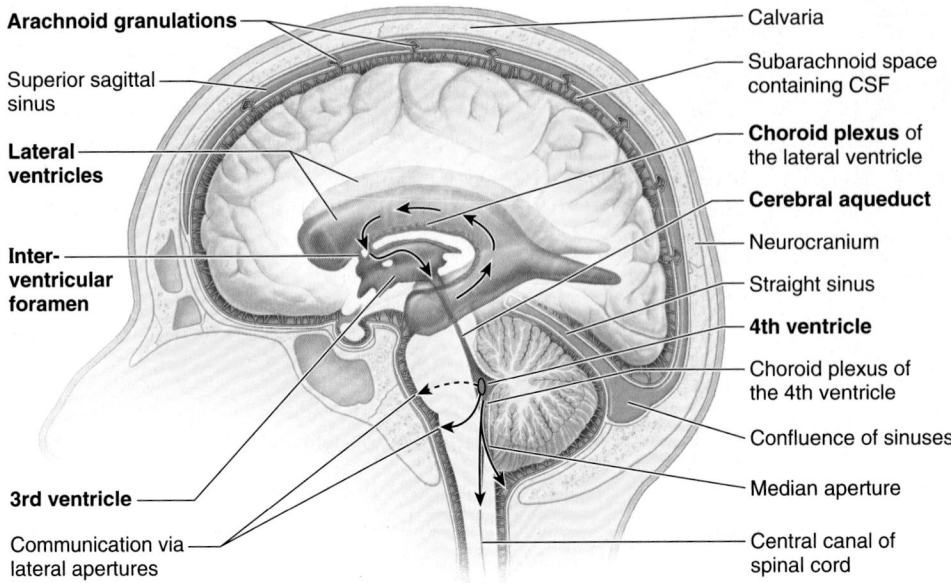

Arachnoid granulations

Superior sagittal sinus

Lateral ventricles

Inter-ventricular foramen

3rd ventricle

Communication via lateral apertures

Calvaria

Subarachnoid space containing CSF

Choroid plexus of the lateral ventricle

Cerebral aqueduct

Neurocranium

Straight sinus

4th ventricle

Choroid plexus of the 4th ventricle

Confluence of sinuses

Median aperture

Central canal of spinal cord

(A) Median section with ventricles viewed from the left

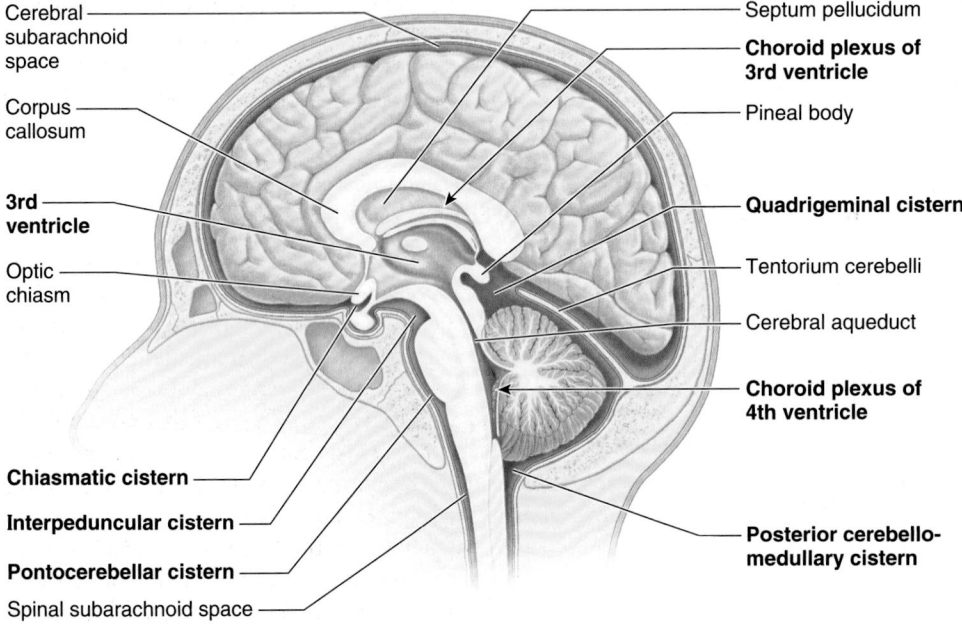

Cerebral subarachnoid space

Corpus callosum

3rd ventricle

Optic chiasm

Chiasmatic cistern

Interpeduncular cistern

Pontocerebellar cistern

Spinal subarachnoid space

Septum pellucidum

Choroid plexus of 3rd ventricle

Pineal body

Quadrigeminal cistern

Tentorium cerebelli

Cerebral aqueduct

Choroid plexus of 4th ventricle

Posterior cerebello-medullary cistern

FIGURE 7.37. Ventricles, subarachnoid spaces, and cisterns. A. The ventricular system and circulation of the CSF are shown. The production of CSF is mainly by the choroid plexuses of the lateral, 3rd, and 4th ventricles. The plexuses of the lateral ventricles are the largest and most important. **B.** Subarachnoid cisterns, expanded regions of the subarachnoid space, contain more substantial amounts of CSF.

(B) Medial view, right half of hemisected head, sectioned to the right of the superior sagittal and straight sinuses

apertures. These apertures are the only means by which CSF enters the subarachnoid space. If they are blocked, CSF accumulates and the ventricles distend, producing compression of the substance of the cerebral hemispheres.

SUBARACHNOID CISTERNS

Although it is not accurate to say that the brain "floats" in the CSF, the brain actually has minimal attachment to the neurocranium. At certain areas on the base of the brain, the arachnoid and pia are widely separated by subarachnoid cisterns (Fig. 7.37B), which contain CSF, and soft tissue structures

that "anchor" the brain, such as the arachnoid trabeculae, vasculature, and in some cases, cranial nerve roots. The cisterns are usually named according to the structures related to them.

Major intracranial subarachnoid cisterns include the:

- **Cerebellomedullary cistern:** the largest of the subarachnoid cisterns, located between the cerebellum and the medulla; receives CSF from the apertures of the 4th ventricle. It is divided into the **posterior cerebellomedullary cistern** (cisterna magna) and the **lateral cerebellomedullary cistern.**

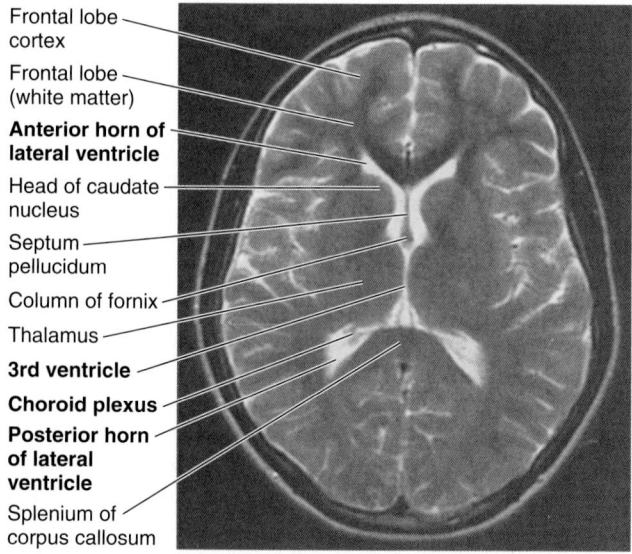

Frontal lobe
cortex

Frontal lobe
(white matter)

**Anterior horn of
lateral ventricle**

Head of caudate
nucleus

Septum
pellucidum

Column of fornix

Thalamus

3rd ventricle

Choroid plexus

**Posterior horn
of lateral
ventricle**

Splenium of
corpus callosum

FIGURE 7.38. Transverse MRI of brain. CSF surrounding the brain, extending into the sulci and fissures, and occupying the ventricles, appears bright white.

- **Pontocerebellar cistern** (pontine cistern): an extensive space ventral to the pons, continuous inferiorly with the spinal subarachnoid space.
- **Interpeduncular cistern** (basal cistern): located in the interpeduncular fossa between the cerebral peduncles of the midbrain.
- **Chiasmatic cistern** (cistern of optic chiasma): inferior and anterior to the *optic chiasm,* the point of crossing or decussation of optic nerve fibers.
- **Quadrigeminal cistern** (cistern of great cerebral vein): located between the posterior part of the corpus callosum and the superior surface of the cerebellum; contains parts of the great cerebral vein.
- **Cisterna ambiens** (ambient cistern): located on the lateral aspect of the midbrain and continuous posteriorly with the quadrigeminal cistern (not illustrated).

SECRETION OF CEREBROSPINAL FLUID

Cerebrospinal fluid (CSF) is secreted (at the rate of 400–500 mL/day) mainly by choroidal epithelial cells (modified ependymal cells) of the **choroid plexuses** in the lateral, 3rd, and 4th ventricles (Figs. 7.36C, 7.37, and 7.38). The choroid plexuses consist of fringes of vascular pia mater (tela choroidea) covered by cuboidal epithelial cells. They are invaginated into the roofs of the 3rd and 4th ventricles and on the floors of the bodies and inferior horns of the lateral ventricles.

CIRCULATION OF CEREBROSPINAL FLUID

CSF leaves the lateral ventricles through the *interventricular foramina* and enters the 3rd ventricle (Fig. 7.37A). From

here, CSF passes through the *cerebral aqueduct* into the 4th ventricle. Some CSF leaves this ventricle through its *median* and *lateral apertures* and enters the *subarachnoid space,* which is continuous around the spinal cord and posterosuperiorly over the cerebellum. However, most CSF flows into the interpeduncular and quadrigeminal cisterns. CSF from the various subarachnoid cisterns flows superiorly through the sulci and fissures on the medial and superolateral surfaces of the cerebral hemispheres. CSF also passes into the extensions of the subarachnoid space around the cranial nerves, the most important of which are the subarachnoid space extensions surrounding the optic nerves (CN II).

ABSORPTION OF CEREBROSPINAL FLUID

The main site of CSF absorption into the venous system is through the **arachnoid granulations** (Figs. 7.35 and 7.37A), especially those that protrude into the superior sagittal sinus and its lateral lacunae (Fig. 7.28D). The subarachnoid space containing CSF extends into the cores of the arachnoid granulations. CSF enters the venous system through two routes: (1) most CSF enters the venous system by transport through the cells of the arachnoid granulations into the dural venous sinuses, and (2) some CSF moves between the cells making up the arachnoid granulations (Corbett et al., 2006).

FUNCTIONS OF CEREBROSPINAL FLUID

Along with the meninges and calvaria, CSF protects the brain by providing a cushion against blows to the head. The CSF in the subarachnoid space provides the buoyancy that prevents the weight of the brain from compressing the cranial nerve roots and blood vessels against the internal surface of the cranium. Because the brain is slightly heavier than the CSF, the gyri on the basal surface of the brain (Fig. 7.42) are in contact with the floor of the cranial cavity when a person is standing erect. In many places at the base of the brain, only the cranial meninges intervene between the brain and cranial bones. In the erect position, the CSF is in the subarachnoid cisterns and sulci on the superior and lateral parts of the brain; therefore, CSF and dura normally separate the superior part of the brain from the calvaria (Fig. 7.37A).

Small, rapidly recurring changes take place in **intracranial pressure** owing to the beating heart; slow recurring changes result from unknown causes. Momentarily large changes in pressure occur during coughing and straining and during changes in position (erect vs. supine). Any change in the volume of the intracranial contents (e.g., a brain tumor, an accumulation of ventricular fluid caused by blockage of the cerebral aqueduct (see Fig. B7.20B), or blood from a ruptured aneurysm) will be reflected by a change in intracranial pressure. This rule is called the **Monro-Kellie doctrine,** which states that the cranial cavity is a closed rigid box and that a change in the quantity of intracranial blood can occur only through the displacement or replacement of CSF.

Arterial Blood Supply to Brain

Although it accounts for only about 2.5% of body weight, the brain receives about one sixth of the cardiac output and one fifth of the oxygen consumed by the body at rest. The *blood supply to the brain* is derived from the internal carotid and vertebral arteries (Fig. 7.39), the terminal branches of which lie in the subarachnoid space. *Venous drainage from the brain* occurs via cerebral and cerebellar veins that drain to the adjacent dural venous sinuses (Fig. 7.29A & B). See also Venous Drainage of Brain on p. 883.

INTERNAL CAROTID ARTERIES

The **internal carotid arteries** arise in the neck from the common carotid arteries (Fig. 7.39). The cervical part of each artery ascends vertically through the neck, without branching, to the cranial base. Each internal carotid artery enters the cranial cavity through the *carotid canal* in the petrous part of the temporal bone. The intracranial course of the internal carotid artery is illustrated and described in Figure 7.40 and demonstrated radiographically in Figure 7.41. In addition to the carotid arteries, the carotid canals contain

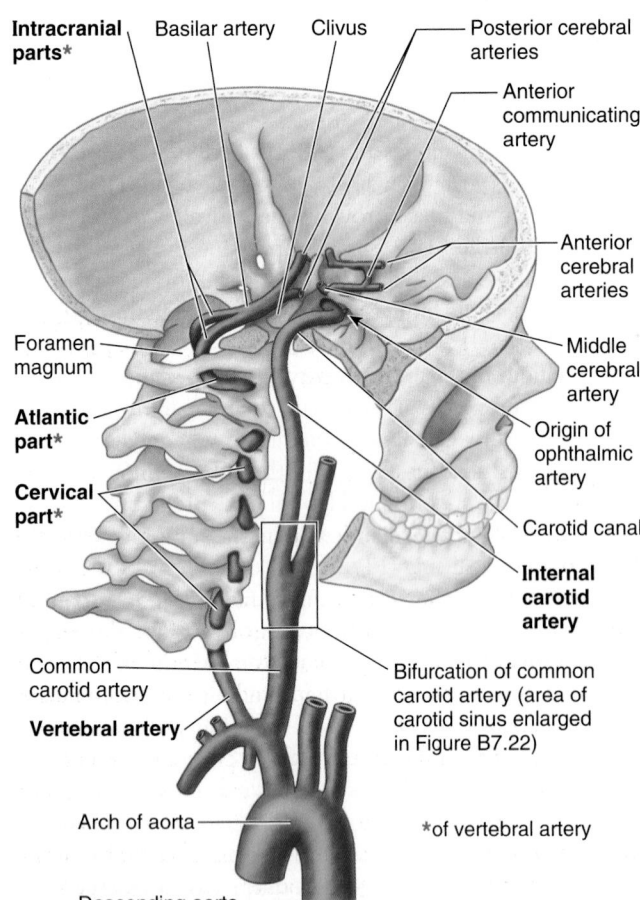

FIGURE 7.39. Arterial supply to brain. The bilaterally paired internal carotid and vertebral arteries deliver an abundant supply of oxygen-rich blood.

venous plexuses and *carotid plexuses of sympathetic nerves* (Fig. 7.40). The internal carotid arteries course anteriorly through the *cavernous sinuses,* with the abducent nerves (CN VI) and in close proximity to the oculomotor (CN III) and trochlear (CN IV) nerves, running in the carotid groove on the side of the body of the sphenoid (Figs. 7.31C and 7.40). The terminal branches of the internal carotid arteries are the **anterior** and **middle cerebral arteries** (Figs. 7.41 and 7.42).

Clinically, the internal carotid arteries and their branches are often referred to as the *anterior circulation of the brain.* The anterior cerebral arteries are connected by the **anterior communicating artery.** Near their termination, the internal carotid arteries are joined to the posterior cerebral arteries by the **posterior communicating arteries,** completing the *cerebral arterial circle* around the *interpeduncular fossa,* the deep depression on the inferior surface of the midbrain between the cerebral peduncles (Figs. 7.42 and 7.43).

VERTEBRAL ARTERIES

The **vertebral arteries** begin in the root of the neck (the prevertebral parts of the vertebral arteries) as the first branches of the first part of the subclavian arteries (Fig. 7.39). The two vertebral arteries are usually unequal in size, the left being larger than the right. The **cervical parts of the vertebral arteries** ascend through the *transverse foramina* of the first six cervical vertebrae. The **atlantic parts of the vertebral arteries** (parts related to the atlas, vertebra C1) perforate the dura and arachnoid and pass through the foramen magnum. The **intracranial parts of the vertebral arteries** unite at the caudal border of the pons to form the *basilar artery* (Figs. 7.29A, 7.39, 7.42 and 7.43C). The vertebrobasilar arterial system and its branches are often referred to clinically as the *posterior circulation of the brain.*

The **basilar artery,** so-named because of its close relationship to the *cranial base,* ascends the *clivus,* the sloping surface from the dorsum sellae to the foramen magnum, through the pontocerebellar cistern to the superior border of the pons. It ends by dividing into the two **posterior cerebral arteries.**

CEREBRAL ARTERIES

In addition to supplying branches to deeper parts of the brain, the cortical branches of each cerebral artery supply a surface and a pole of the cerebrum (Figs. 7.41 and 7.43A & B; Table 7.7). The cortical branches of the:

- **Anterior cerebral artery** supply most of the medial and superior surfaces of the brain and the *frontal pole.*
- **Middle cerebral artery** supply the lateral surface of the brain and the *temporal pole.*
- **Posterior cerebral artery** supply the inferior surface of the brain and the *occipital pole.*

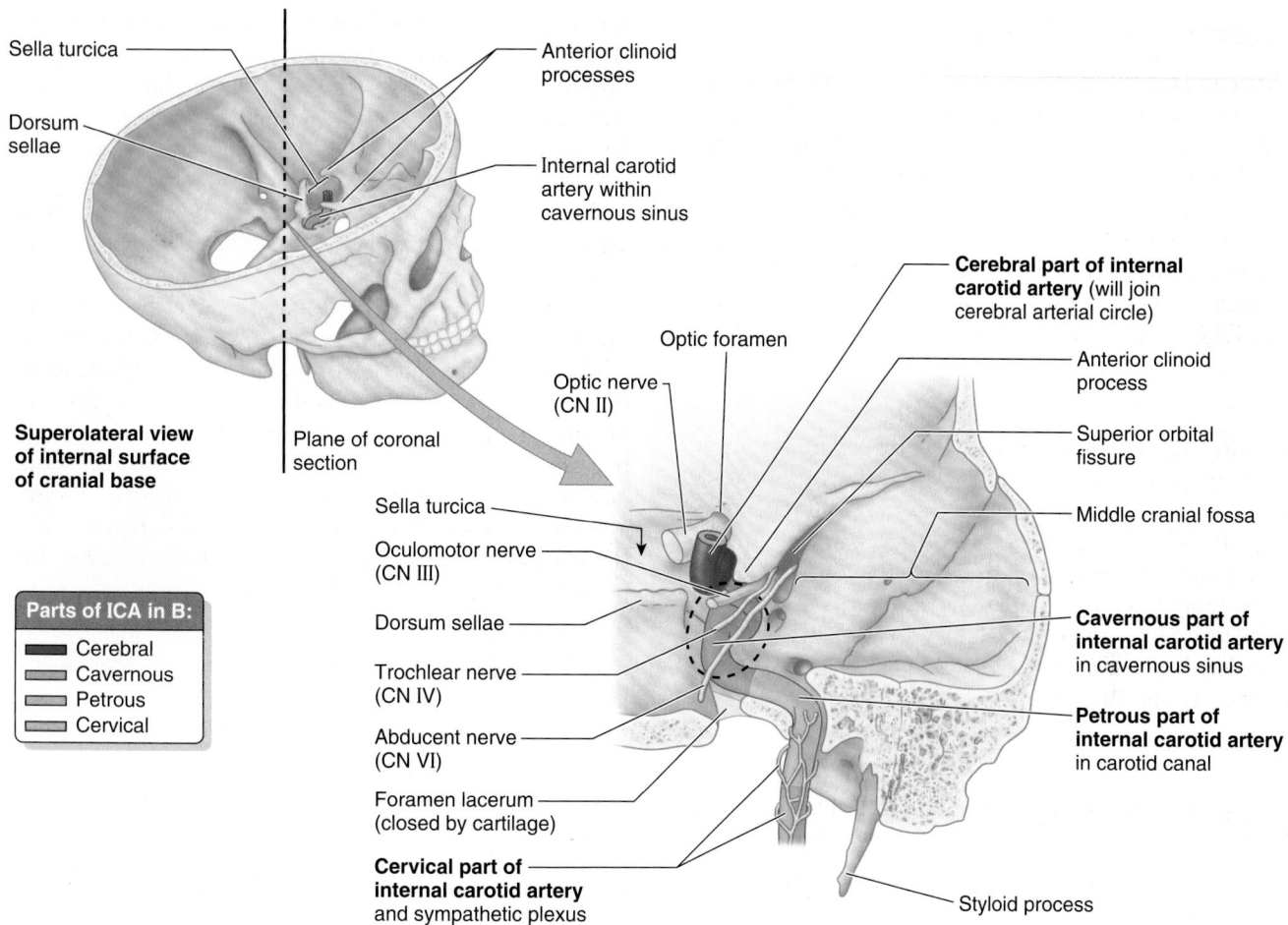

Sella turcica

Dorsum sellae

Anterior clinoid processes

Internal carotid artery within cavernous sinus

Superolateral view of internal surface of cranial base

Plane of coronal section

Optic foramen

Optic nerve (CN II)

Cerebral part of internal carotid artery (will join cerebral arterial circle)

Anterior clinoid process

Superior orbital fissure

Middle cranial fossa

Sella turcica

Oculomotor nerve (CN III)

Dorsum sellae

Trochlear nerve (CN IV)

Abducent nerve (CN VI)

Foramen lacerum (closed by cartilage)

Cervical part of internal carotid artery and sympathetic plexus

Cavernous part of internal carotid artery in cavernous sinus

Petrous part of internal carotid artery in carotid canal

Styloid process

Parts of ICA in B:
Cerebral
Cavernous
Petrous
Cervical

Posterior view (of right side of anterior portion following bisection in coronal plane)

FIGURE 7.40. Course of internal carotid artery. The orientation drawing (*left*) indicates the plane of the coronal section that intersects the carotid canal (*right*). The cervical part of the internal carotid artery ascends vertically in the neck to the entrance of the carotid canal in the petrous temporal bone. The petrous part of the artery turns horizontally and medially in the carotid canal, toward the apex of the petrous temporal bone. It emerges from the canal superior to the foramen lacerum, closed in life by cartilage, and enters the cranial cavity. The artery runs anteriorly across the cartilage; then the cavernous part of the artery runs along the carotid grooves on the lateral side of the body of the sphenoid, traversing the cavernous sinus. Inferior to the anterior clinoid process, the artery makes a 180° turn, its cerebral part heading posteriorly to join the cerebral arterial circle (Figs. 7.42 and 7.43C).

CEREBRAL ARTERIAL CIRCLE

The **cerebral arterial circle** (of Willis) is a roughly pentagon-shaped circle of vessels on the ventral surface of the brain. It is an important anastomosis at the base of the brain between four arteries (two vertebral and two internal carotid arteries) that supply the brain (Figs. 7.42 and 7.43C; Table 7.7). The arterial circle is formed sequentially in an anterior to posterior direction by the:

- Anterior communicating artery.
- Anterior cerebral arteries.
- Internal carotid arteries.
- Posterior communicating arteries.
- Posterior cerebral arteries.

The various components of the cerebral arterial circle give numerous small branches to the brain.

Venous Drainage of Brain

The thin-walled, valveless veins draining the brain pierce the arachnoid and meningeal layers of dura to end in the nearest dural venous sinuses (Figs. 7.28A and 7.29–7.32), which ultimately drain for the most part into the IJVs. The **superior cerebral veins** on the superolateral surface of the brain drain into the superior sagittal sinus; **inferior** and **superficial middle cerebral veins** from the inferior, postero-inferior, and deep aspects of the cerebral hemispheres drain into the straight, transverse, and superior petrosal sinuses. The **great cerebral vein** (of Galen) is a single, midline vein formed inside the brain by the union of two internal cerebral veins; it ends by merging with the inferior sagittal sinus to form the straight sinus (Fig. 7.29). The cerebellum is drained by **superior** and **inferior cerebellar veins,** draining the respective aspect of the cerebellum into the transverse and sigmoid sinuses (Fig. 7.32).

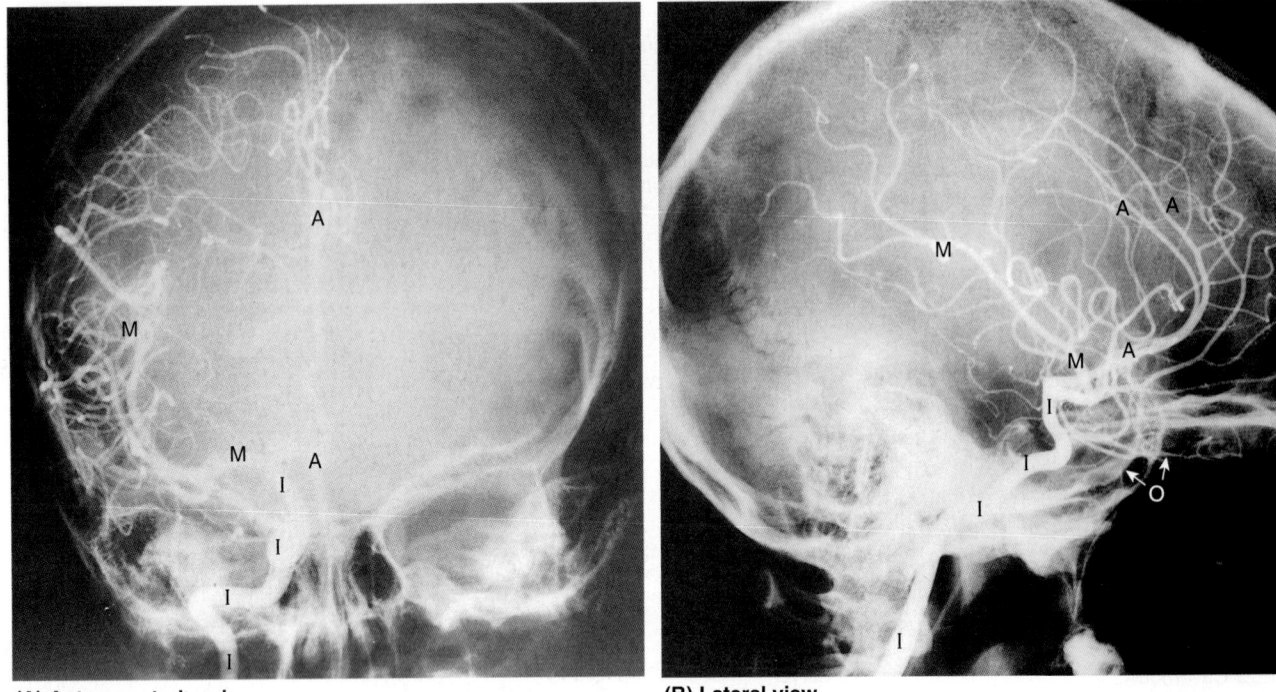

(A) Anteroposterior view   **(B) Lateral view**

FIGURE 7.41. Carotid arteriograms. A and B. Radiopaque dye injected into the carotid arterial system demonstrates unilateral distribution to the brain from the internal carotid artery. *A,* anterior cerebral artery and its branches; *I,* the four parts of the internal carotid artery; *M,* middle cerebral artery and its branches; *O,* ophthalmic artery. (Courtesy of Dr. D. Armstrong, Associate Professor of Medical Imaging, University of Toronto, Ontario, Canada.)

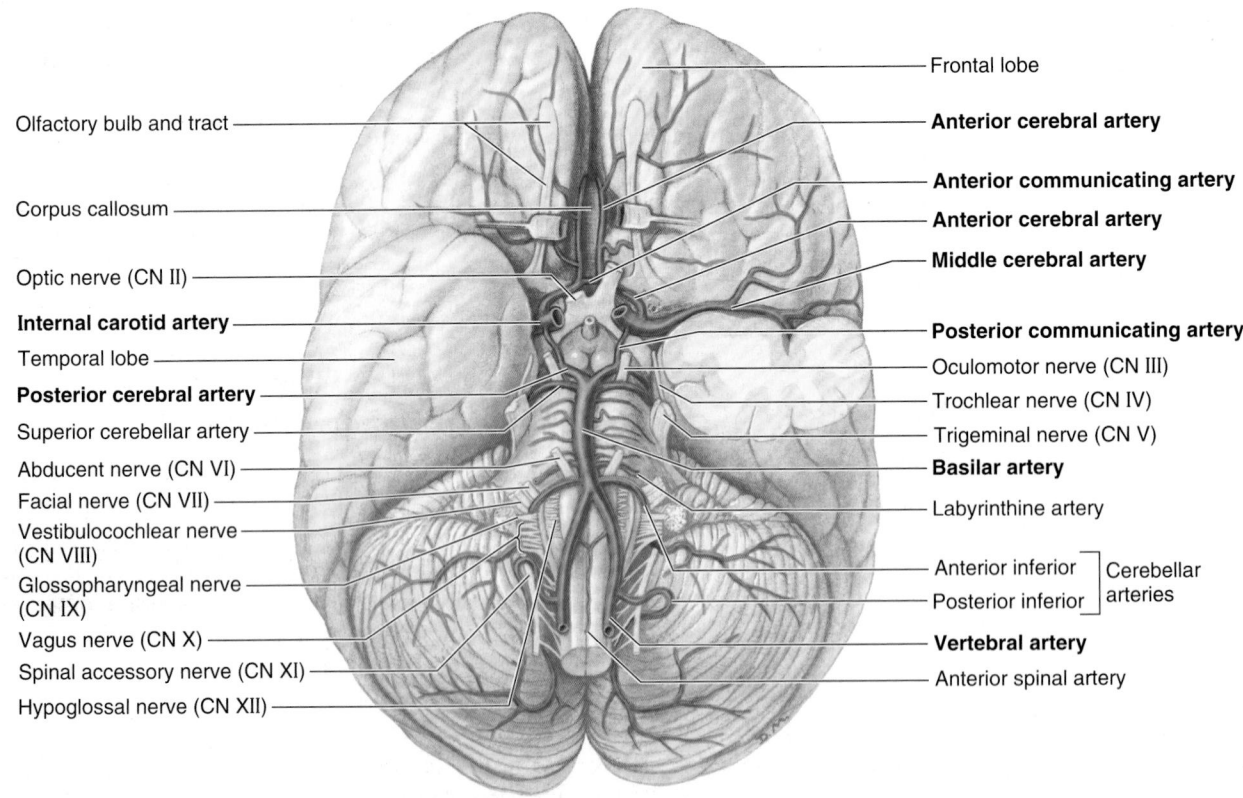

Inferior view

FIGURE 7.42. Base of brain with cerebral arterial circle. The internal carotid and basilar arteries converge, divide, and anastomose to form the cerebral arterial circle (of Willis). The left temporal pole is removed to show the middle cerebral artery in the lateral sulcus of the brain. The frontal lobes are separated to expose the anterior cerebral arteries.

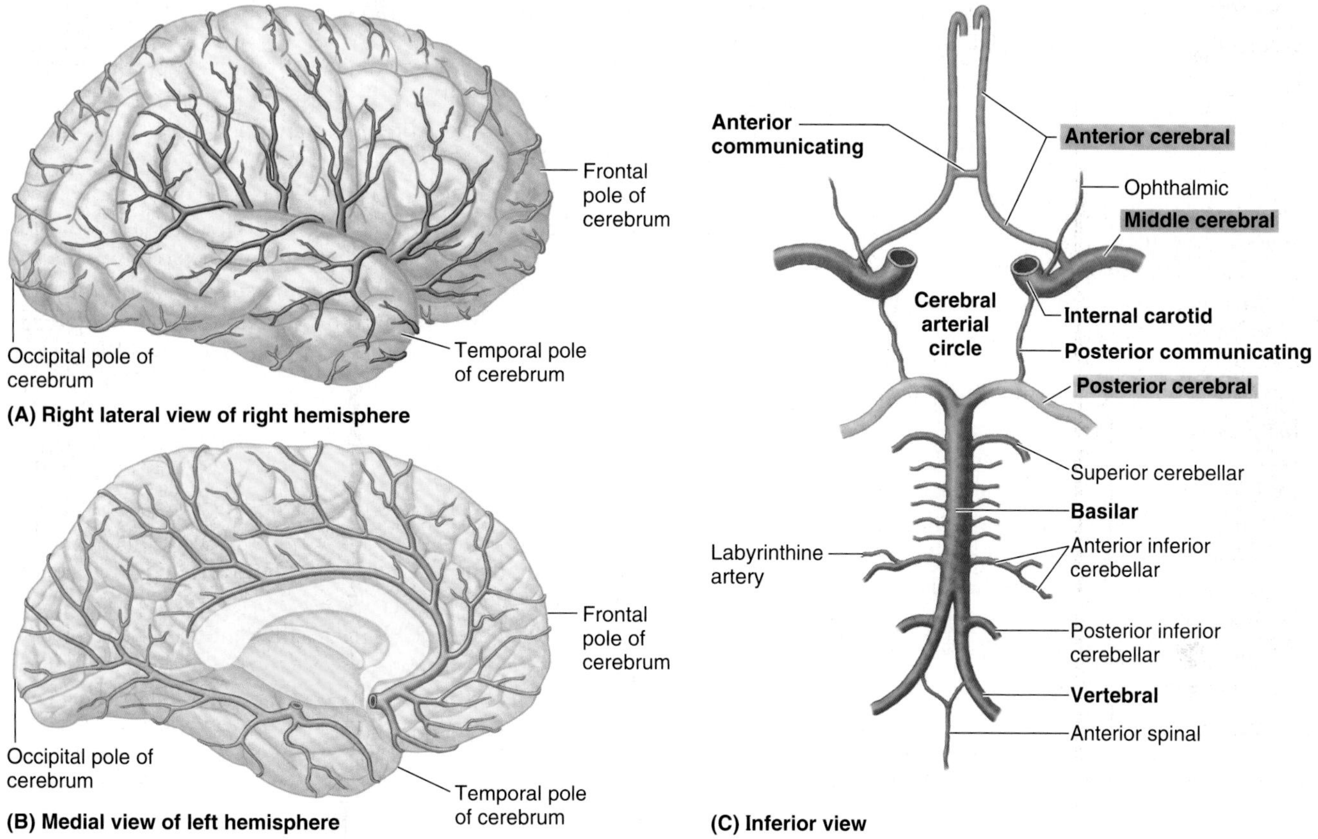

Frontal
pole of
cerebrum

Occipital pole of
cerebrum

Temporal pole
of cerebrum

(A) Right lateral view of right hemisphere

Frontal
pole of
cerebrum

Occipital pole of
cerebrum

Temporal pole
of cerebrum

(B) Medial view of left hemisphere

Anterior
communicating

Anterior cerebral

Ophthalmic

Middle cerebral

Cerebral
arterial
circle

Internal carotid

Posterior communicating

Posterior cerebral

Superior cerebellar

Basilar

Labyrinthine
artery

Anterior inferior
cerebellar

Posterior inferior
cerebellar

Vertebral

Anterior spinal

(C) Inferior view

FIGURE 7.43. Arterial blood supply of cerebrum.

TABLE 7.7. ARTERIAL BLOOD SUPPLY OF CEREBRAL HEMISPHERES

Artery	Origin	Distribution
Internal carotid:	Common carotid artery at superior border of thyroid cartilage	Gives branches to walls of cavernous sinus, pituitary gland, and trigeminal ganglion; provides primary supply to brain
Anterior cerebral	Internal carotid artery	Cerebral hemispheres, except for occipital lobes
Anterior communicating	Anterior cerebral artery	Cerebral arterial circle (of Willis)
Middle cerebral	Continuation of internal carotid artery distal to anterior cerebral artery	Most of lateral surface of cerebral hemispheres
Vertebral:	Subclavian artery	Cranial meninges and cerebellum
Basilar	Formed by union of vertebral arteries	Brainstem, cerebellum, and cerebrum
Posterior cerebral	Terminal branch of basilar artery	Inferior aspect of cerebral hemisphere and occipital lobe
Posterior communicating	Posterior cerebral artery	Optic tract, cerebral peduncle, internal capsule, and thalamus

BRAIN

Cerebral Injuries

Cerebral concussion is an abrupt, brief loss of consciousness immediately after a severe head injury. Consciousness may be lost for only a few seconds, as occurs in most knockdowns during boxing. With a more severe injury, such as that resulting from an automobile accident, consciousness may be lost for hours and even days. If a person recovers consciousness within 6 hr, the long-term outcome is excellent (Rowland, 2010). If the coma lasts longer than 6 hr, brain tissue injury usually occurs.

Professional boxers are especially at risk for *chronic traumatic encephalopathy,* or *"punchdrunk syndrome,"* a brain injury characterized by weakness in the lower limbs, unsteady gait, slowness of muscular movements, tremors of the hands, hesitancy of speech, and slow cerebration (use of one's brain). Brain injuries result from acceleration and deceleration of the head that shears or stretches axons (*diffuse axonal injury*). The sudden stopping of the moving head results in the brain hitting the suddenly stationary cranium. Sometimes concussion occurs without loss of consciousness. The absence of loss of consciousness does not mean that the concussion is any less serious. Over 90% of head injuries are referred to as minor traumatic brain injuries.

Cerebral contusion results from brain trauma in which the pia is stripped from the injured surface of the brain and may be torn, allowing blood to enter the subarachnoid space. The bruising results either from the sudden impact of the still-moving brain against the suddenly stationary cranium, or from the suddenly moving cranium against the still-stationary brain. Cerebral contusion may result in an extended loss of consciousness, but if there is no diffuse axonal injury, brain swelling, or secondary hemorrhage, recovery from a contusion may be excellent (Rowland, 2010).

Cerebral lacerations are often associated with depressed cranial fractures (Fig. B7.4, p. 838) or gunshot wounds. Lacerations result in rupture of blood vessels and bleeding into the brain and subarachnoid space, causing increased intracranial pressure and cerebral compression.

Cerebral compression may be produced by:

- Intracranial collections of blood.
- Obstruction of CSF circulation or absorption.
- Intracranial tumors or abscesses.
- Brain swelling caused by *brain edema,* an increase in brain volume resulting from an increase in water and sodium content (Fishman, 2010a).

Cisternal Puncture

CSF may be obtained from the posterior cerebellomedullary cistern through a *cisternal puncture* for diagnostic or therapeutic purposes. The cerebellomedullary cistern is the site of choice in infants and young children; the lumbar cistern is used most frequently in adults (see Chapter 4, Fig. B4.18). The needle is carefully inserted through the posterior atlanto-occipital membrane into the cistern. The subarachnoid space or the ventricular system may also be entered for measuring or monitoring CSF pressure, injecting antibiotics, or administering contrast media for medical imaging.

Hydrocephalus

Overproduction of CSF, obstruction of CSF flow, or interference with CSF absorption results in excess fluid in the cerebral ventricles and enlargement of the head, a condition called *obstructive hydrocephalus* (Fig. B7.20A). The excess CSF dilates the ventricles, thins the cerebral cortex, and separates the bones of the calvaria in infants. Although an obstruction can occur any place, the blockage usually occurs in the cerebral aqueduct (Fig. B7.20B) or an interventricular foramen. *Aqueductal stenosis* (narrow aqueduct) may be caused by a nearby tumor in the midbrain or by cellular debris following intraventricular hemorrhage or bacterial and fungal infections of the central nervous system (Corbett et al., 2006).

Blockage of CSF circulation results in dilation of the ventricles superior to the point of obstruction and increased pressure on the cerebral hemispheres. This condition squeezes the brain between the ventricular fluid and the calvarial bones. In infants, the internal pressure results in expansion of the brain and calvaria because the sutures and fontanelles are still open. It is possible to produce an artificial

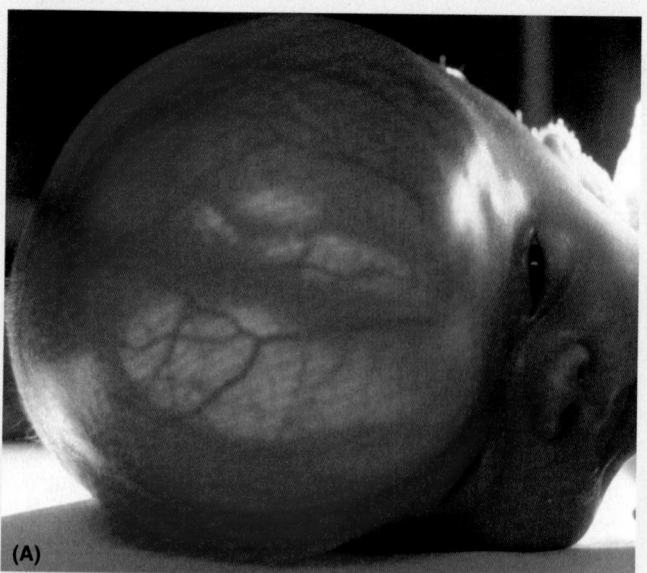

(A)

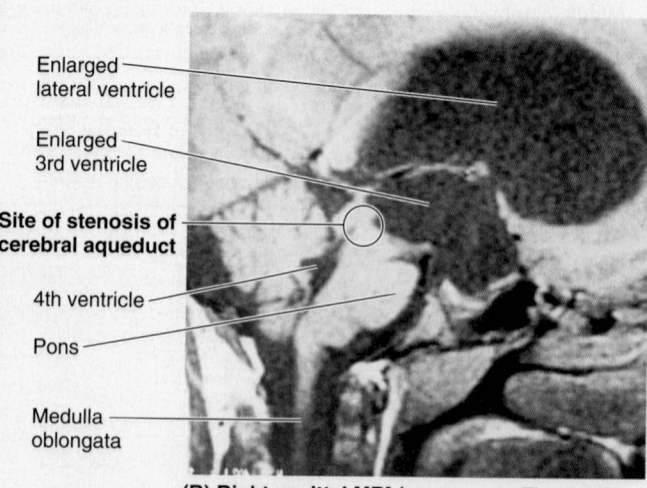

Enlarged lateral ventricle

Enlarged 3rd ventricle

Site of stenosis of cerebral aqueduct

4th ventricle

Pons

Medulla oblongata

(B) Right sagittal MRI (compare to Fig. 7.37)

FIGURE B7.20. Hydrocephalus (A) and aqueductal stenosis (B).

drainage system to bypass the blockage and allow CSF to escape, thereby lessening damage to the brain.

In *communicating hydrocephalus,* the flow of CSF through the ventricles and into the subarachnoid space is not impaired; however, movement of CSF from this space into the venous system is partly or completely blocked. The blockage may be caused by the congenital absence of arachnoid granulations, or the granulations may be blocked by red blood cells as the result of a subarachnoid hemorrhage (Corbett et al., 2006).

Leakage of Cerebrospinal Fluid

 Fractures in the floor of the middle cranial fossa may result in CSF leakage from the external acoustic meatus (*CSF otorrhea*) if the meninges superior to the middle ear are torn and the tympanic membrane is ruptured. Fractures in the floor of the anterior cranial fossa may involve the cribriform plate of the ethmoid (see Fig. 7.12A), resulting in CSF leakage through the nose (*CSF rhinorrhea*). CSF can be distinguished from mucus by testing its glucose level; the glucose level of the CSF reflects that of the blood. CSF otorrhea and rhinorrhea may be the primary indications of a cranial base fracture and increased risk of meningitis because an infection could spread to the meninges from the ear or nose (Rowland, 2010).

Anastomoses of Cerebral Arteries and Cerebral Embolism

Branches of the three cerebral arteries anastomose with each other on the surface of the brain; however, if a cerebral artery is obstructed by a *cerebral embolism* (e.g., a blood clot), these microscopic anastomoses are not capable of providing enough blood for the area of cerebral cortex concerned. Consequently, *cerebral ischemia* and *infarction* occur and an area of necrosis results. Large cerebral emboli occluding major cerebral vessels may cause severe neurologic problems and death.

Variations of Cerebral Arterial Circle

Variations in the size of the vessels forming the cerebral arterial circle are common. The posterior communicating arteries are absent in some individuals; in others there may be two anterior communicating arteries. In approximately 1 in 3 persons, one posterior cerebral artery is a major branch of the internal carotid artery. One of the anterior cerebral arteries is often small in the proximal part of its course; the anterior communicating artery is larger than usual in these individuals. These variations may become clinically significant if emboli or arterial disease occur.

Strokes

An *ischemic stroke* denotes the sudden development of focal neurological deficits that are usually related to *impaired cerebral blood flow.* An ischemic stroke is generally caused by an embolism in a major cerebral artery. Strokes are the most common neurologic disorders affecting adults in the United States (Elkind, 2010); they are more often disabling than fatal. The cardinal feature of a stroke is the sudden onset of neurological symptoms.

The cerebral arterial circle is an important means of collateral circulation in the event of gradual obstruction of one of the major arteries forming the circle. Sudden occlusion, even if only partial, results in neurological deficits. In elderly persons, the anastomoses of the arterial circle are often inadequate when a large artery (e.g., the internal carotid) is occluded, even if the occlusion is gradual (in which case function is impaired at least to some degree). The most common causes of strokes are *spontaneous cerebrovascular accidents,* such as cerebral thrombosis, cerebral hemorrhage, cerebral embolism, and subarachnoid hemorrhage (Rowland, 2010).

Hemorrhagic stroke follows the rupture of an artery or a *saccular aneurysm,* a sac-like dilation on a weak part of the arterial wall (Fig. B7.21A). The most common type of saccular aneurysm is a *berry aneurysm,* occurring in the vessels

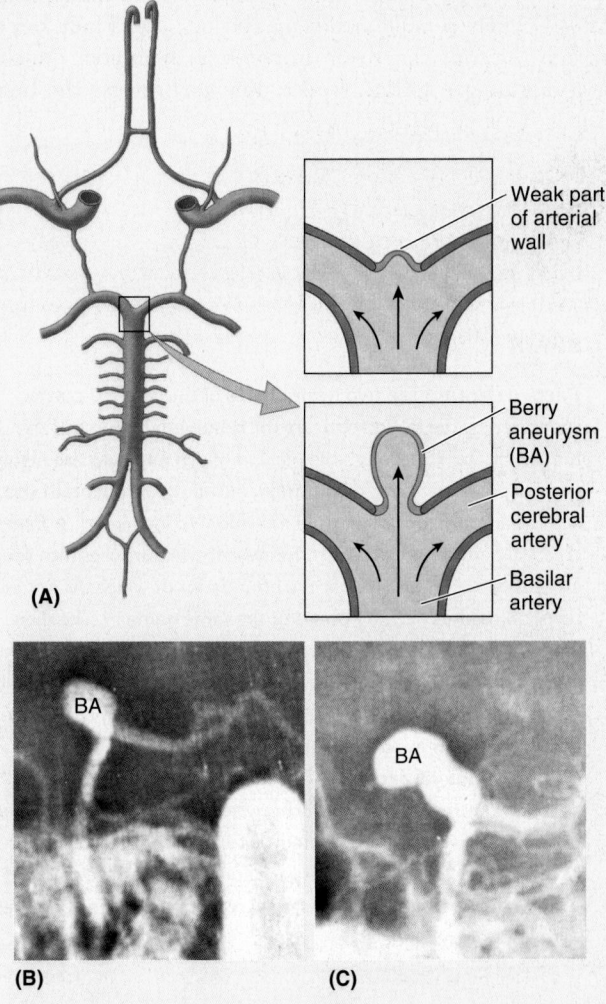

FIGURE B7.21.

of or near the cerebral arterial circle and the medium arteries at the base of the brain (Fig. B7.21B). Aneurysms also occur at the bifurcation of the basilar artery into the posterior cerebral arteries.

In time, especially in individuals with *hypertension* (high blood pressure), the weak part of the wall of the aneurysm expands and may rupture (Fig. B7.21C), allowing blood to enter the subarachnoid space. Sudden *rupture of an aneurysm* usually produces a severe, almost unbearable headache and a stiff neck. These symptoms result from gross bleeding into the subarachnoid space.

Brain Infarction

An *atheromatous plaque* at a bend of an artery (e.g., at the bifurcation of a common carotid artery) results in progressive narrowing (*stenosis*) of the artery, producing increasingly severe neurologic deficits (Fig. B7.22). An *embolus* separates from the plaque and is carried in the blood until it lodges in an artery, usually an intracranial branch that is too small to allow its passage. This event usually results in *acute cortical infarction,* a sudden insufficiency of arterial blood to the brain (e.g., of the left parietal lobes). An interruption of blood supply for 30 sec alters a person's brain metabolism. After 1–2 min, neural function may be lost; after 5 min, lack of oxygen (*anoxia*) can result in cerebral infarction. Quickly restoring oxygen to the blood supply may reverse the brain damage (Elkind, 2010).

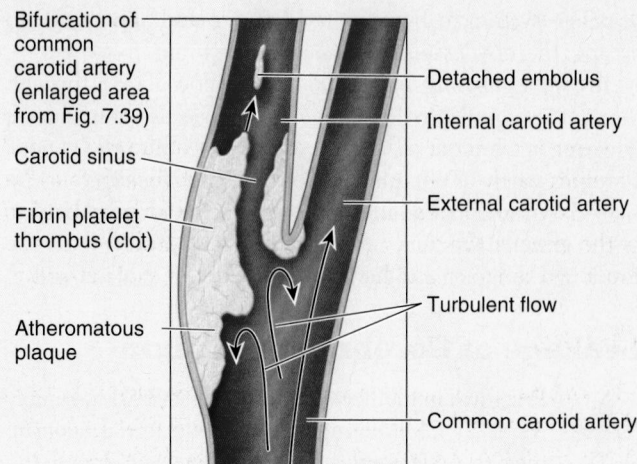

FIGURE B7.22.

Transient Ischemic Attacks

Transient ischemic attacks (TIAs) refer to neurologic symptoms resulting from *ischemia.* Most TIAs last only a few minutes, but some persist for up to an hour. With major *carotid* or *vertebrobasilar stenosis,* the TIA tends to last longer and causes distal closure of intracranial vessels. The symptoms of TIA may be ambiguous: staggering, dizziness, light-headedness, fainting, and *paresthesias.* Persons with TIAs are at increased risk for myocardial infarction and ischemic stroke (Brust, 2010).

The Bottom Line

BRAIN

Parts of brain: The two hemispheres of the cerebral cortex, separated by the falx cerebri, are the dominant features of the human brain. ♦ Although the pattern of gyri and sulci are highly variable, other features of the brain, including overall brain size, are remarkably consistent from individual to individual. ♦ Each cerebral hemisphere is divided for descriptive purposes into four lobes that are related to, but the boundaries of which do not correspond to, the overlying bones of the same name. ♦ The diencephalon forms the central core of the brain, with the midbrain, pons, and medulla oblongata making the brainstem; the medulla is continuous with the spinal cord. ♦ The cerebellum is the subtentorial brain mass occupying the posterior cranial fossa.

Ventricles of brain: Each cerebral hemisphere includes a lateral ventricle in its core; otherwise, the ventricular system of the brain is an unpaired, median formation that communicates with the subarachnoid space surrounding the brain and spinal cord. ♦ Choroid plexuses secrete CSF into the ventricles,

which flows out of them into the subarachnoid space. ♦ CSF is absorbed into the venous system, normally at the same rate at which it is produced, by the arachnoid granulations related to the superior sagittal sinus.

Arterial supply and venous drainage of brain: A continuous supply of oxygen and nutrients is essential for brain function. ♦ The brain receives a dual blood supply from the cerebral branches of the bilaterally paired internal carotid and vertebral arteries. ♦ Anastomoses between these arteries form the cerebral arterial circle. ♦ Anastomoses also occur between the branches of the three cerebral arteries on the surface of the brain. ♦ In adults, if one of the four arteries delivering blood to the brain is blocked, the remaining three are not usually capable of providing adequate collateral circulation; consequently, impaired cerebral blood flow (ischemia) and an ischemic stroke result. ♦ Venous drainage from the brain occurs via the dural venous sinuses and internal jugular veins.

EYE, ORBIT, ORBITAL REGION, AND EYEBALL

The **eye** is the organ of vision and consists of the eyeball and the optic nerve. The *orbit* contains the eyeball and its *accessory visual structures* (L., adnexa oculi). The **orbital region** is the area of the face overlying the orbit and eyeball and includes the upper and lower eyelids and lacrimal apparatus.

Orbits

The **orbits** are bilateral bony cavities in the facial skeleton that resemble hollow, quadrangular pyramids with their bases directed anterolaterally and their apices, posteromedially (Fig. 7.44A). The medial walls of the two orbits, separated by the ethmoidal sinuses and the upper parts of the nasal cavity, are nearly parallel, whereas their lateral walls are approximately at a right (90°) angle.

Consequently, the axes of the orbits (*orbital axes*) diverge at approximately 45°. The *optical axes* (axes of gaze, the direction or line of sight) for the two eyeballs, are parallel, however, and in the anatomical position run directly anteriorly ("looking straight ahead"), the eyeballs being in the *primary position*. The orbits and orbital region anterior to them contain and protect the **eyeballs** and **accessory visual structures** (Fig. 7.45), which include the:

- *Eyelids*, which bound the orbits anteriorly, controlling exposure of the anterior eyeball.

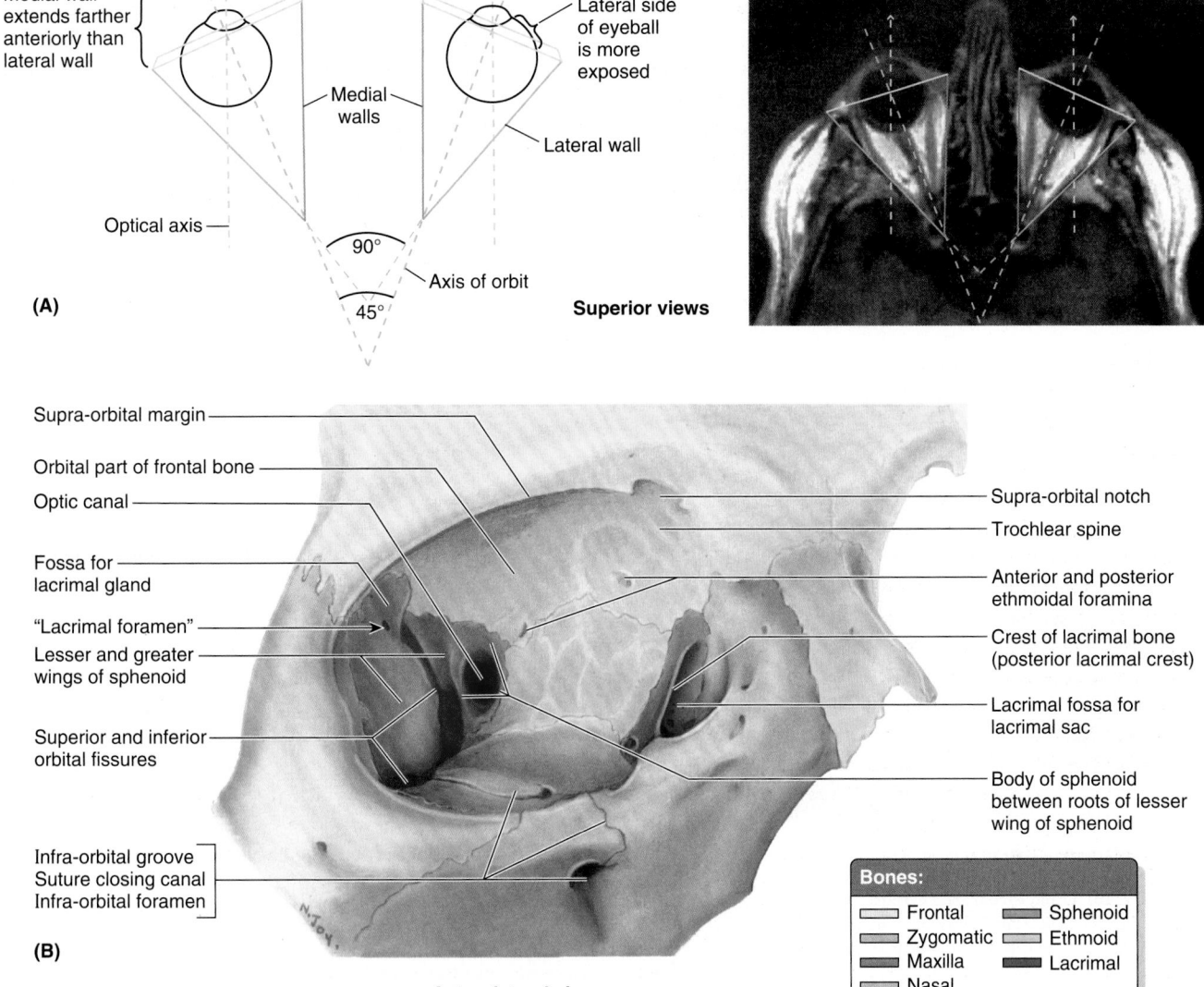

FIGURE 7.44. Orbits and placement of eyeballs within them. A. Note the orbits' disposition relative to each other and to the optical axes (line of gaze). The orbits are separated by ethmoidal cells and the upper nasal cavity and septum. **B.** The bony walls of the orbit are shown. This anterolateral view allows a view of the orbit and apex, which lie in the sagittal plane and are not well seen in an anterior view.

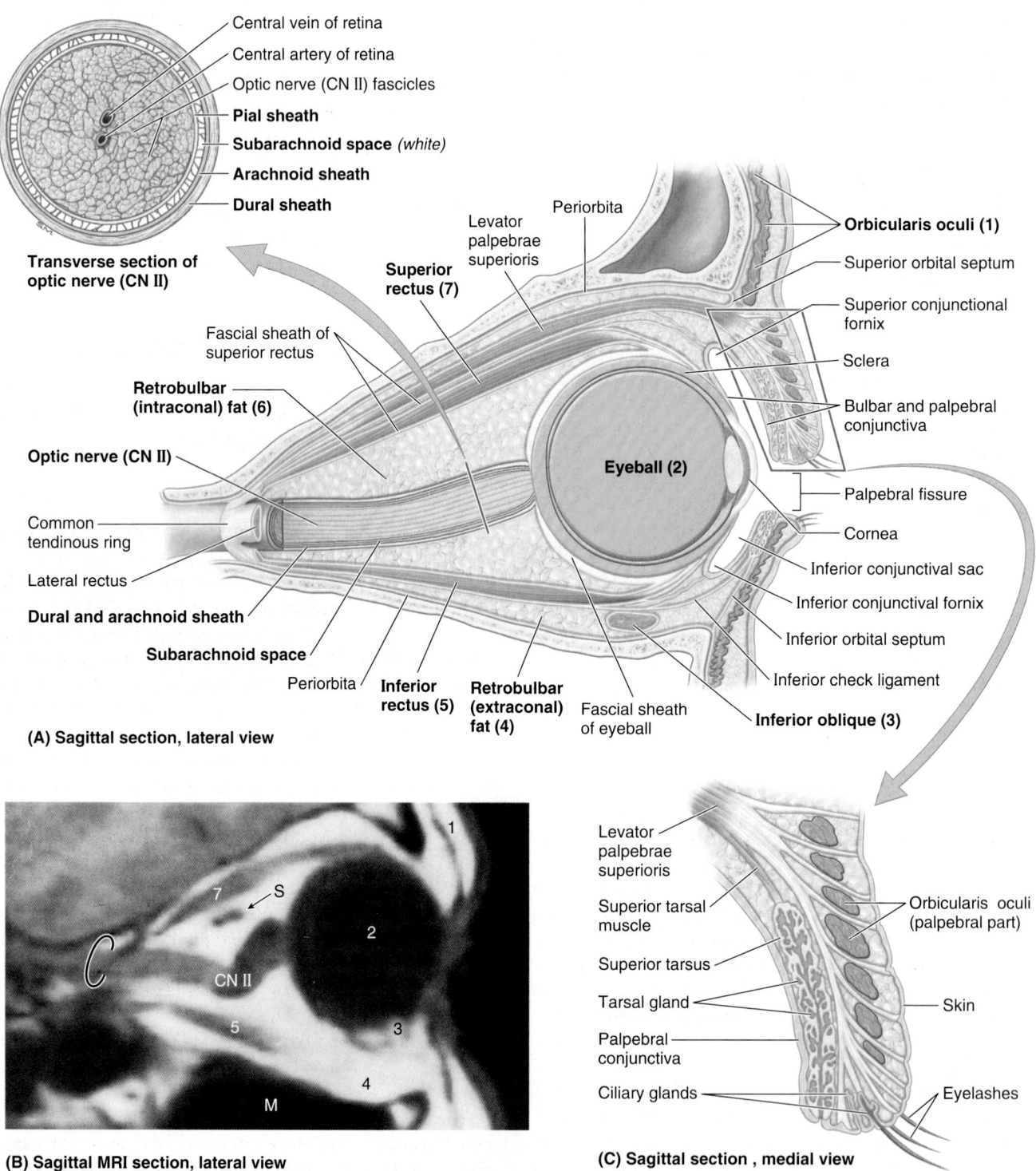

Central vein of retina
Central artery of retina
Optic nerve (CN II) fascicles
Pial sheath
Subarachnoid space (white)
Arachnoid sheath
Dural sheath

**Transverse section of
optic nerve (CN II)**

Periorbita
Levator
palpebrae
superioris
**Superior
rectus (7)**

Orbicularis oculi (1)
Superior orbital septum
Superior conjunctional
fornix
Sclera
Bulbar and palpebral
conjunctiva

Fascial sheath of
superior rectus
**Retrobulbar
(intraconal) fat (6)**

Eyeball (2)

Palpebral fissure

Optic nerve (CN II)

Cornea
Inferior conjunctival sac
Inferior conjunctival fornix
Inferior orbital septum

Common
tendinous ring
Lateral rectus
Dural and arachnoid sheath
Subarachnoid space

Inferior check ligament

Periorbita   **Inferior
rectus (5)**   **Retrobulbar
(extraconal)
fat (4)**   Fascial sheath
of eyeball

Inferior oblique (3)

(A) Sagittal section, lateral view

Levator
palpebrae
superioris

Orbicularis oculi
(palpebral part)

Superior tarsal
muscle
Superior tarsus
Tarsal gland

Skin

Palpebral
conjunctiva
Ciliary glands

Eyelashes

(B) Sagittal MRI section, lateral view

(C) Sagittal section , medial view

FIGURE 7.45. Orbit, eyeball, and eyelids. A. Sagittal section of orbit demonstrating the contents of the orbit. *Inset,* cross-section of optic nerve (CN II).
The subarachnoid space around the optic nerve is continuous with the space between the arachnoid and the pia covering the brain. The numbers refer to
structures labeled in part **B. B.** MRI study providing a sagittal section similar to **A.** *M,* maxillary sinus; *S,* superior ophthalmic vein; *arc,* optic canal. **C.** Detail
of the superior eyelid. The tarsus forms the skeleton of the eyelid and contains tarsal glands. (Part **B** courtesy of Dr. W. Kucharczyk, Professor and Neuroradiologist Senior Scientist, Department of Medical Imaging, University Health Network, Toronto, Ontario, Canada.)

- *Extra-ocular muscles*, which position the eyeballs and raise the superior eyelids.
- *Nerves and vessels* in transit to the eyeballs and muscles.
- *Orbital fascia* surrounding the eyeballs and muscles.
- *Mucous membrane (conjunctiva)* lining the eyelids and anterior aspect of the eyeballs, and most of the *lacrimal apparatus*, which lubricates it.

All space within the orbits not occupied by these structures is filled with **orbital fat;** thus, it forms a matrix in which the structures of the orbit are embedded.

The quadrangular pyramidal **orbit** has a base, four walls, and an apex (Fig. 7.44B):

- The **base of the orbit** is outlined by the **orbital margin,** which surrounds the **orbital opening.** The bone forming the orbital margin is reinforced to afford protection to the orbital contents and provides attachment for the *orbital septum,* a fibrous membrane that extends into the eyelids.
- The **superior wall** (roof) is approximately horizontal and is formed mainly by the *orbital part of the frontal bone,* which separates the orbital cavity from the anterior cranial fossa. Near the apex of the orbit, the superior wall is formed by the *lesser wing of the sphenoid.* Anterolaterally, a shallow depression in the orbital part of the frontal bone, called the **fossa for lacrimal gland** (lacrimal fossa), accommodates the lacrimal gland.
- The **medial walls** of the contralateral orbits are essentially parallel and are formed primarily by the **orbital plate of ethmoid bone,** along with contributions from the *frontal process of the maxilla, lacrimal,* and *sphenoid bones.* Anteriorly, the medial wall is indented by the **lacrimal groove** and **fossa for lacrimal sac;** the *trochlea* (pulley) for the tendon of one of the extra-ocular muscles is located superiorly. Much of the bone forming the medial wall is paper thin; the ethmoid bone is highly pneumatized with ethmoidal cells, often visible through the bone of a dried cranium.
- The **inferior wall** (orbital floor) is formed mainly by the *maxilla* and partly by the *zygomatic* and *palatine bones.* The thin inferior wall is shared by the orbit and maxillary sinus. It slants inferiorly from the apex to the inferior orbital margin. The inferior wall is demarcated from the lateral wall of the orbit by the **inferior orbital fissure,** a gap between the orbital surfaces of the maxilla and the sphenoid.
- The **lateral wall** is formed by the **frontal process of the zygomatic bone** and the *greater wing of the sphenoid.* This is the strongest and thickest wall, which is important because it is most exposed and vulnerable to direct trauma. Its posterior part separates the orbit from the temporal and middle cranial fossae. The lateral walls of the contralateral orbits are nearly perpendicular to each other.
- The **apex of the orbit** is at the **optic canal** in the *lesser wing of the sphenoid* just medial to the *superior orbital fissure.*

The widest part of the orbit corresponds to the equator of the eyeball (Fig. 7.45A), an imaginary line encircling the eyeball equidistant from its anterior and posterior poles. The bones forming the orbit are lined with **periorbita,** the periosteum of the orbit. The periorbita is continuous:

- At the optic canal and superior orbital fissure with the periosteal layer of the dura mater.
- Over the orbital margins and through the inferior orbital fissure with the periosteum covering the external surface of the cranium (pericranium).
- With the orbital septa at the orbital margins.
- With the fascial sheaths of the extra-ocular muscles.
- With the orbital fascia that forms the *fascial sheath of the eyeball.*

Eyelids and Lacrimal Apparatus

The eyelids and lacrimal fluid, secreted by the lacrimal glands, protect the cornea and eyeballs from injury and irritation (e.g., by dust and small particles).

EYELIDS

The **eyelids** are moveable folds that cover the eyeball anteriorly when closed, thereby protecting it from injury and excessive light. They also keep the cornea moist by spreading the lacrimal fluid. The eyelids are covered externally by thin skin and internally by transparent mucous membrane, the **palpebral conjunctiva** (Fig. 7.45A & C). This part of the conjunctiva is reflected onto the eyeball, where it is continuous with the **bulbar conjunctiva.** This part of the conjunctiva is thin and transparent and attaches loosely to the anterior surface of the eyeball. The bulbar conjunctiva, loose and wrinkled over the sclera (where it contains small, visible blood vessels), is adherent to the periphery of the cornea (Fig. 7.46B). The lines of reflection of the palpebral conjunctiva onto the eyeball form deep recesses, the **superior** and **inferior conjunctival fornices** (Figs. 7.45A and 7.46).

The **conjunctival sac** is the space bound by the palpebral and bulbar conjunctivae; it is a closed space when the eyelids are closed, but opens via an anterior aperture, the *palpebral fissure* (L. *rima palpebrae,* the gap between the eyelids), when the eye is open (eyelids are parted) (Fig. 7.45A). The conjunctival sac is a specialized form of mucosal "bursa" that enables the eyelids to move freely over the surface of the eyeball as they open and close.

The superior (upper) and inferior (lower) eyelids are strengthened by dense bands of connective tissue, the **superior** and **inferior tarsi** (singular = **tarsus**), which form the "skeleton" of the eyelids (Figs. 7.45C and 7.47A). Fibers of the palpebral portion of the *orbicularis oculi* (the sphincter of the palpebral fissure) are in the connective tissue superficial to the tarsi and deep to the skin of the eyelids (Fig. 7.45C). Embedded in the tarsi are **tarsal glands** that produce a lipid secretion that lubricates the edges of the eyelids and prevents them from sticking together when they close. The lipid

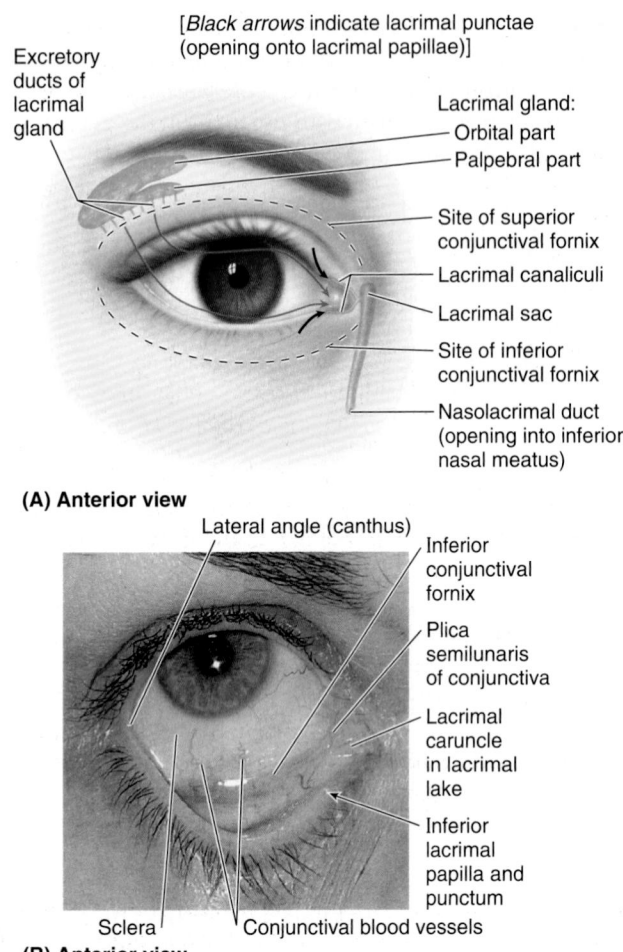

[*Black arrows* indicate lacrimal punctae (opening onto lacrimal papillae)]

Excretory ducts of lacrimal gland

Lacrimal gland:
— Orbital part
— Palpebral part

— Site of superior conjunctival fornix
— Lacrimal canaliculi
— Lacrimal sac
— Site of inferior conjunctival fornix
— Nasolacrimal duct (opening into inferior nasal meatus)

(A) Anterior view

Lateral angle (canthus)

Inferior conjunctival fornix

Plica semilunaris of conjunctiva

Lacrimal caruncle in lacrimal lake

Inferior lacrimal papilla and punctum

Sclera   Conjunctival blood vessels

(B) Anterior view

FIGURE 7.46. Lacrimal apparatus and anterior eyeball. A. The components of the lacrimal apparatus, by which tears flow from the superolateral aspect of the conjunctival sac (*dashed lines*) to the nasal cavity, are demonstrated. **B.** The surface features of the eye are shown. The fibrous outer coat of the eyeball includes the tough white sclera and the central transparent cornea, through which the pigmented iris with its aperture, the pupil, can be seen. The inferior eyelid has been everted to show the reflection of conjunctiva from the anterior surface of the eyeball to the inner surface of the eyelid. The semilunar fold is a vertical fold of conjunctiva near the medial angle, at the lacrimal caruncle.

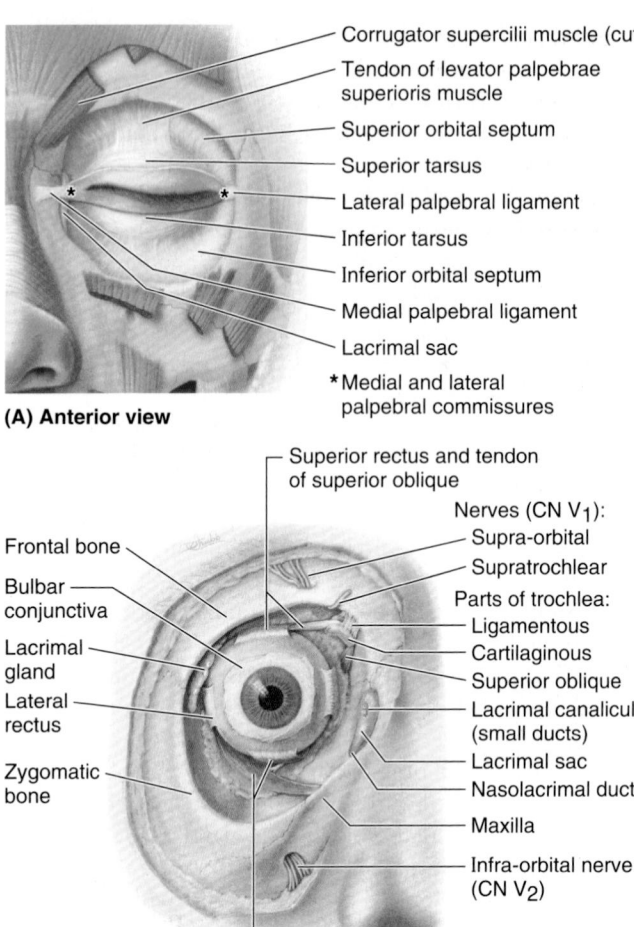

— Corrugator supercilii muscle (cut)
— Tendon of levator palpebrae superioris muscle
— Superior orbital septum
— Superior tarsus
— Lateral palpebral ligament
— Inferior tarsus
— Inferior orbital septum
— Medial palpebral ligament
— Lacrimal sac

(A) Anterior view

* Medial and lateral palpebral commissures

Superior rectus and tendon of superior oblique

Nerves (CN V$_1$):
— Supra-orbital
— Supratrochlear

Frontal bone

Bulbar conjunctiva

Lacrimal gland

Lateral rectus

Zygomatic bone

Parts of trochlea:
— Ligamentous
— Cartilaginous
— Superior oblique
— Lacrimal canaliculi (small ducts)
— Lacrimal sac
— Nasolacrimal duct
— Maxilla
— Infra-orbital nerve (CN V$_2$)

Inferior oblique and inferior rectus

(B) Anterior view

FIGURE 7.47. Skeleton of eyelids, and anterior approach to orbit. A. The superior and inferior tarsi and their attachments are shown. Their ciliary margins are free, but they are attached peripherally to the orbital septum (palpebral fascia in the eyelid). **B.** In this dissection of the orbit, the eyelids, orbital septum, levator palpebrae superioris, and some fat have been removed. Part of the lacrimal gland is seen between the bony orbital wall laterally and the eyeball and lateral rectus muscle medially. Structures receiving lacrimal drainage from the conjunctival sac are seen medially.

secretion also forms a barrier that lacrimal fluid does not cross when produced in normal amounts. When production is excessive, it spills over the barrier onto the cheeks as tears.

The **eyelashes** (L. *cilia*) are in the margins of the eyelids. The large sebaceous glands associated with the eyelashes are **ciliary glands.** The junctions of the superior and inferior eyelids make up the **medial** and **lateral palpebral commissures,** defining the **medial** and **lateral angles of the eye** (G. *kanthos*, corner of eye), or *canthi* (Figs. 7.46B and 7.47A).

Between the nose and the medial angle of the eye is the **medial palpebral ligament,** which connects the tarsi to the medial margin of the orbit (Fig. 7.47A). The orbicularis oculi originates and inserts onto this ligament. A similar **lateral palpebral ligament** attaches the tarsi to the lateral margin of the orbit, but it does not provide for direct muscle attachment.

The **orbital septum** is a fibrous membrane that spans from the tarsi to the margins of the orbit, where it becomes continuous with the periosteum (Figs. 7.45A and 7.47A). It keeps the orbital fat contained and, owing to its continuity with the periorbita, can limit the spread of infection to and from the orbit. The septum constitutes in large part the posterior fascia of the orbicularis oculi muscle.

LACRIMAL APPARATUS

The lacrimal apparatus (Figs. 7.46A and 7.47B) consists of the:

- **Lacrimal gland:** secretes **lacrimal fluid,** a watery physiological saline containing the bacteriocidal enzyme lysozyme. The fluid moistens and lubricates the surfaces of the conjunctiva and cornea and provides some nutrients and dissolved oxygen to the cornea; when produced in excess, the overflowing fluid constitutes tears.

- **Excretory ducts of lacrimal gland:** convey lacrimal fluid from the lacrimal glands to the conjunctival sac (Fig. 7.46A).
- **Lacrimal canaliculi** (L. small canals): commence at a **lacrimal punctum** (opening) on the **lacrimal papilla** near the medial angle of the eye and drain lacrimal fluid from the **lacrimal lake** (L. *lacus lacrimalis;* a triangular space at the medial angle of the eye where the tears collect) to the **lacrimal sac** (dilated superior part of the nasolacrimal duct) (Figs. 7.46A and 7.47B).
- **Nasolacrimal duct:** conveys the lacrimal fluid to the *inferior nasal meatus* (part of the nasal cavity inferior to the inferior nasal concha.

The *lacrimal gland,* almond shaped and approximately 2 cm long, lies in the *fossa for the lacrimal gland* in the superolateral part of each orbit (Figs. 7.44B, 7.46A, and 7.47B). The gland is divided into a superior **orbital** and inferior **palpebral parts** by the lateral expansion of the tendon of the *levator palpebrae superioris* (Fig. 7.46A). **Accessory lacrimal glands** may also be present, sometimes in the middle part of the eyelid, or along the superior or inferior fornices of the conjunctival sac. They are more numerous in the superior eyelid than in the inferior eyelid.

Production of lacrimal fluid is stimulated by parasympathetic impulses from CN VII. It is secreted through 8–12 *excretory ducts,* which open into the lateral part of the *superior conjunctival fornix* of the conjunctival sac. The fluid flows inferiorly within the sac under the influence of gravity. When the cornea becomes dry, the eye blinks. The eyelids come together in a lateral to medial sequence pushing a film of fluid medially over the cornea, somewhat like windshield wipers. In this way, lacrimal fluid, containing foreign material such as dust is pushed toward the medial angle of the eye, accumulating in the lacrimal lake from which it drains by capillary action through the lacrimal puncta and lacrimal canaliculi to the lacrimal sac (Figs. 7.46A & B and 7.47B).

From this sac, the fluid passes to the inferior nasal meatus of the nasal cavity through the nasolacrimal duct. It drains posteriorly across the floor of the nasal cavity to the nasopharynx and is eventually swallowed. In addition to cleansing particles and irritants from the conjunctival sac, lacrimal fluid provides the cornea with nutrients and oxygen.

The **nerve supply of the lacrimal gland** is both sympathetic and parasympathetic (Fig. 7.48). The *presynaptic parasympathetic secretomotor fibers* are conveyed from the facial nerve by the *greater petrosal nerve* and then by the *nerve of the pterygoid canal* to the *pterygopalatine ganglion,* where they synapse with the cell body of the postsynaptic fiber. Vasoconstrictive, postsynaptic sympathetic fibers, brought from the *superior cervical ganglion* by the *internal carotid plexus* and deep petrosal nerve, join the parasympathetic fibers to form the nerve of the pterygoid canal and traverse the pterygopalatine ganglion. The zygomatic nerve (from the maxillary nerve) brings both types of fibers to the lacrimal branch of the ophthalmic nerve, by which they enter the gland (see Chapter 9).

Eyeball

The **eyeball** contains the optical apparatus of the visual system (Fig. 7.45A). It occupies most of the anterior portion of the orbit, suspended by six extrinsic muscles that control its movement, and a fascial *suspensory apparatus.* It measures approximately 25 mm in diameter. All anatomical structures within the eyeball have a circular or spherical arrangement.

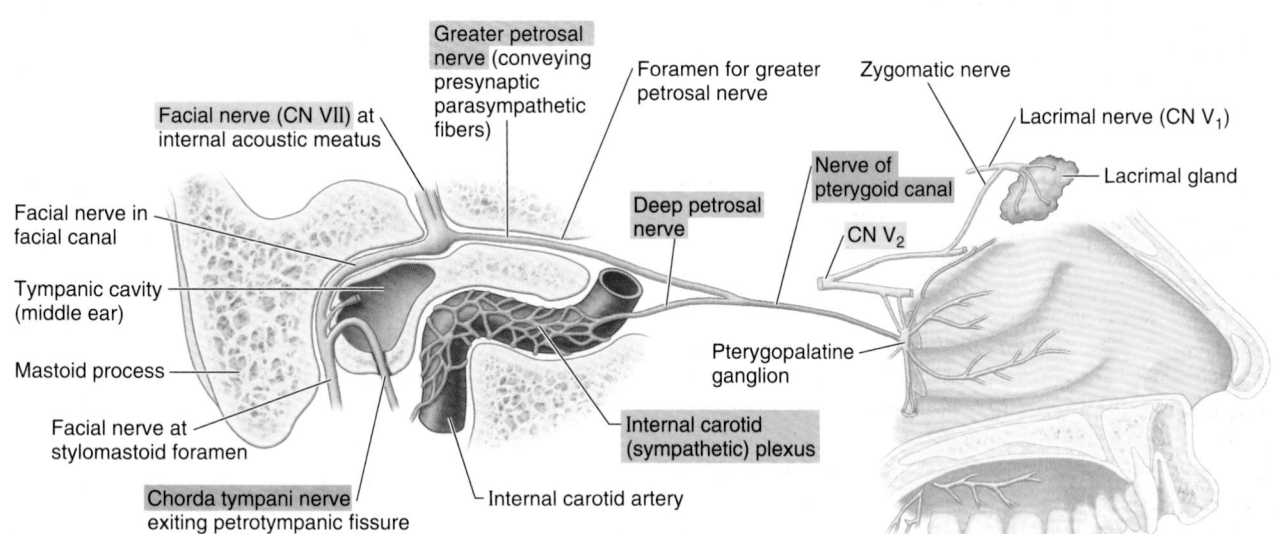

FIGURE 7.48. Innervation of lacrimal gland. The facial nerve (CN VII), greater petrosal nerve, and nerve of pterygoid canal deliver presynaptic parasympathetic fibers to the pterygopalatine ganglion. The synapse between presynaptic and postsynaptic fibers occurs here. The maxillary, infra-orbital, zygomatic, and lacrimal nerves convey the postsynaptic fibers to the gland.

The *eyeball proper* has three layers; however, there is an additional connective tissue layer that surrounds the eyeball, supporting it within the orbit. The connective tissue layer is composed posteriorly of the **fascial sheath of the eyeball** (bulbar fascia or Tenon capsule), which forms the actual socket for the eyeball, and anteriorly of bulbar conjunctiva. The fascial sheath is the most substantial portion of the suspensory apparatus. A very loose connective tissue layer, the **episcleral space** (a potential space) lies between the fascial sheath and the outer layer of the eyeball, facilitating movements of the eyeball within the fascial sheath.

The *three layers of the eyeball* are the (Fig. 7.49):

1. *Fibrous layer* (outer coat), consisting of the *sclera* and *cornea.*
2. *Vascular layer* (middle coat), consisting of the *choroid, ciliary body,* and *iris.*
3. *Inner layer* (inner coat), consisting of the *retina,* which has both *optic* and *non-visual parts.*

FIBROUS LAYER OF EYEBALL

The **fibrous layer of the eyeball** is the external fibrous skeleton of the eyeball, providing shape and resistance. The **sclera** is the tough opaque part of the fibrous layer (coat) of the eyeball, covering the posterior five sixths of the eyeball (Figs. 7.49A and 7.50). It provides attachment for both the extrinsic (extra-ocular) and intrinsic muscles of the eye. The anterior part of the sclera is visible through the transparent bulbar conjunctiva as "the white of the eye" (Fig. 7.46B). The **cornea** is the transparent part of the fibrous layer covering the anterior one sixth of the eyeball (Figs. 7.49A and 7.50). The convexity of the cornea is greater than that of the sclera, and so it appears to protrude from the eyeball when viewed laterally.

The two parts of the fibrous layer differ primarily in terms of the regularity of arrangement of the collagen fibers of which they are composed and the degree of hydration of each. While the sclera is relatively avascular, the cornea is completely avascular, receiving its nourishment from capillary beds around its periphery and fluids on its external and internal surfaces (*lacrimal fluid* and *aqueous humor,* respectively). Lacrimal fluid also provides oxygen absorbed from the air.

The cornea is highly sensitive to touch, its innervation is provided by the ophthalmic nerve (CN V$_1$). Even very small foreign bodies (e.g., dust particles) elicit blinking, flow of tears, and sometimes severe pain. Its nourishment is derived from the capillary beds at its periphery, the *aqueous humor,* and *lacrimal fluid.* The latter also provides oxygen absorbed from air. Drying of the corneal surface may cause ulceration.

The **corneal limbus** is the angle formed by the intersecting curvatures of sclera and cornea at the **corneoscleral junction.** The junction is a 1-mm-wide, gray, translucent circle that includes numerous capillary loops involved in nourishing the avascular cornea.

VASCULAR LAYER OF EYEBALL

The middle **vascular layer of the eyeball** (also called the **uvea** or uveal tract) consists of the choroid, ciliary body, and iris (Fig. 7.49B). The **choroid,** a dark reddish brown layer between the sclera and retina, forms the largest part of the vascular layer of the eyeball and lines most of the sclera (Fig. 7.50A). Within this pigmented and dense vascular bed, larger vessels are located externally (near the sclera). The finest vessels (the **capillary lamina of the choroid,** or *choriocapillaris,* an extensive capillary bed) are innermost, adjacent to the avascular light-sensitive layer of the retina, which it supplies with oxygen and nutrients. Engorged with blood in life (it has the highest perfusion rate per gram of tissue of all vascular beds of the body), this layer is responsible for the "red eye" reflection that occurs in flash photography. The choroid

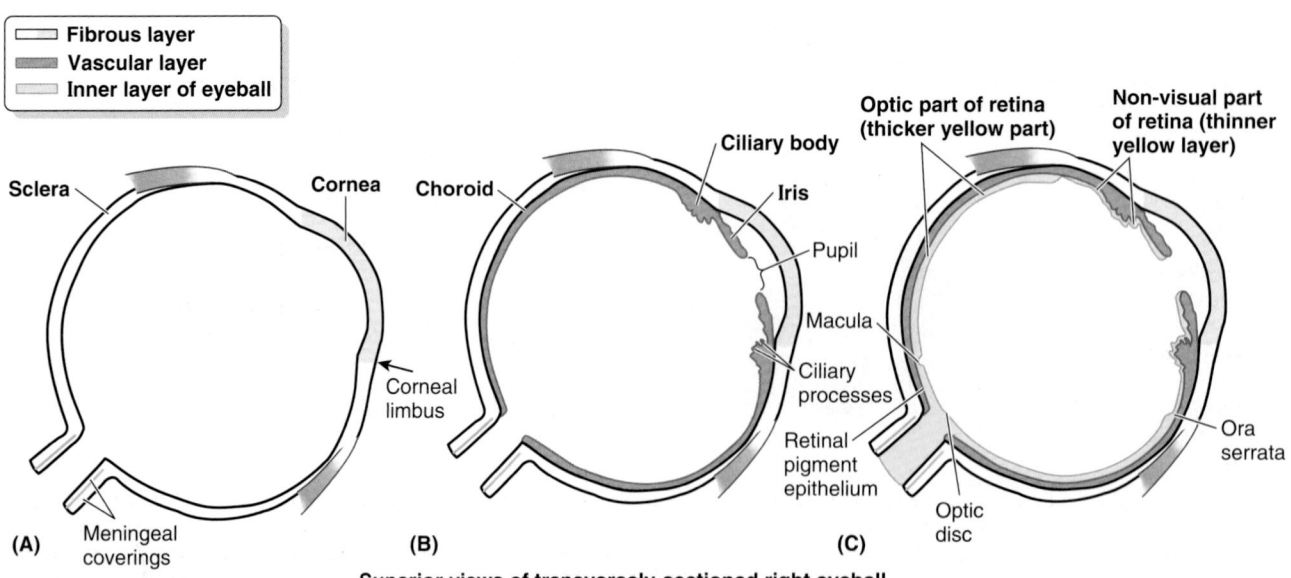

FIGURE 7.49. Layers of eyeball. The three layers are added sequentially. **A.** Outer fibrous layer. **B.** Middle vascular layer. **C.** Inner layer (retina).

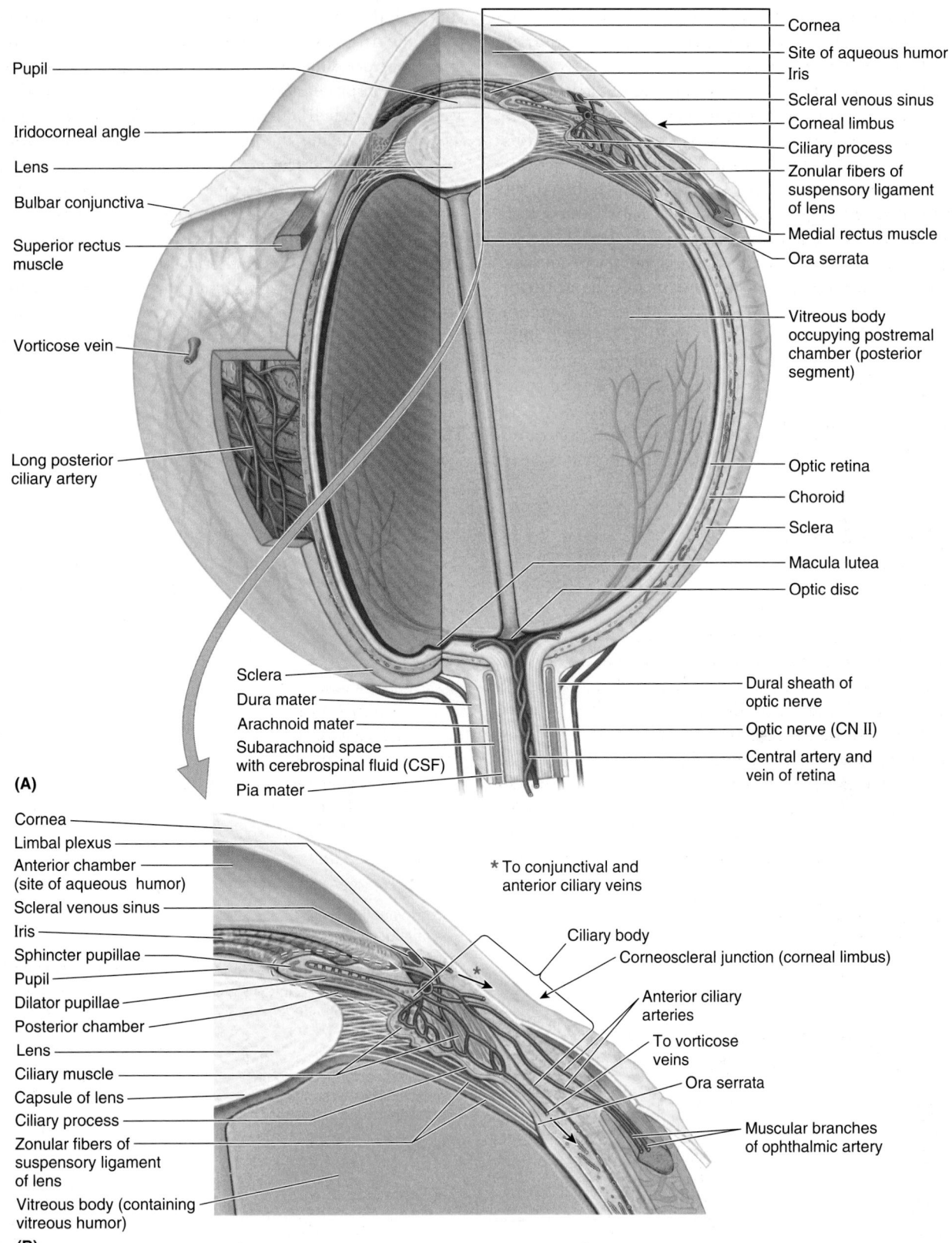

Cornea
Site of aqueous humor
Iris
Scleral venous sinus
Corneal limbus
Ciliary process
Zonular fibers of suspensory ligament of lens
Medial rectus muscle
Ora serrata

Pupil
Iridocorneal angle
Lens
Bulbar conjunctiva
Superior rectus muscle
Vorticose vein
Long posterior ciliary artery

Vitreous body occupying postremal chamber (posterior segment)

Optic retina
Choroid
Sclera
Macula lutea
Optic disc

Sclera
Dura mater
Arachnoid mater
Subarachnoid space with cerebrospinal fluid (CSF)
Pia mater

Dural sheath of optic nerve
Optic nerve (CN II)
Central artery and vein of retina

(A)

Cornea
Limbal plexus
Anterior chamber (site of aqueous humor)
Scleral venous sinus
Iris
Sphincter pupillae
Pupil
Dilator pupillae
Posterior chamber
Lens
Ciliary muscle
Capsule of lens
Ciliary process
Zonular fibers of suspensory ligament of lens
Vitreous body (containing vitreous humor)

* To conjunctival and anterior ciliary veins

Ciliary body
Corneoscleral junction (corneal limbus)
Anterior ciliary arteries
To vorticose veins
Ora serrata
Muscular branches of ophthalmic artery

(B)

FIGURE 7.50. Eyeball with quarter section removed. A. The inner aspect of the optic part of the retina is supplied by the central artery of the retina, whereas the outer, light sensitive aspect is nourished by the capillary lamina of the choroid (Fig. 7.62). The central artery courses through the optic nerve and divides at the optic disc into superior and inferior branches. The branches of the central artery are end arteries that do not anastomose with each other or any other vessel. **B.** The structural details of the ciliary region are shown. The ciliary body is both muscular and vascular, as is the iris; the latter includes two muscles: the sphincter pupillae and dilator pupillae muscles. Venous blood from this region and the aqueous humor in the anterior chamber drain into the scleral venous sinus.

attaches firmly to the pigment layer of the retina, but can easily be stripped from the sclera. The choroid is continuous anteriorly with the ciliary body.

The **ciliary body,** is a ring-like thickening of the layer posterior to the corneoscleral junction, which is muscular as well as vascular (Figs. 7.49B and 7.50). It connects the choroid with the circumference of the iris. The ciliary body provides attachment for the lens. The contraction and relaxation of the circularly arranged smooth muscle of the ciliary body controls the thickness, and therefore the focus, of the lens. Folds on the internal surface of the ciliary body, the **ciliary processes,** secrete *aqueous humor.* Aqueous humor fills the **anterior segment of the eyeball,** the interior of the eyeball anterior to the lens, suspensory ligament, and ciliary body. (Fig. 7.50B).

The **iris,** which literally lies on the anterior surface of the lens, is a thin contractile diaphragm with a central aperture, the **pupil,** for transmitting light (Figs. 7.49B, 7.50, and 7.51A). When a person is awake, the size of the pupil varies continually to regulate the amount of light entering the eye

(Fig. 7.51B). Two involuntary muscles control the size of the pupil: the parasympathetically stimulated, circularly arranged **sphincter pupillae** decreases its diameter (constrict or contracts the pupil, *pupillary miosis*), and the sympathetically stimulated, radially arranged **dilator pupillae** increases its diameter (dilates the pupil). The nature of the pupillary responses is paradoxical: sympathetic responses usually occur immediately, yet it may take up to 20 min for the pupil to dilate in response to low lighting, as in a darkened theater. Parasympathetic responses are typically slower than sympathetic responses, yet parasympathetically stimulated papillary constriction is normally instantaneous. Abnormal sustained pupillary dilation (*mydriasis*) may occur in certain diseases or as a result of trauma or the use of certain drugs.

INNER LAYER OF EYEBALL

The inner layer of the eyeball is the **retina** (Fig. 7.49C and 7.50). It is the sensory neural layer of the eyeball. Grossly,

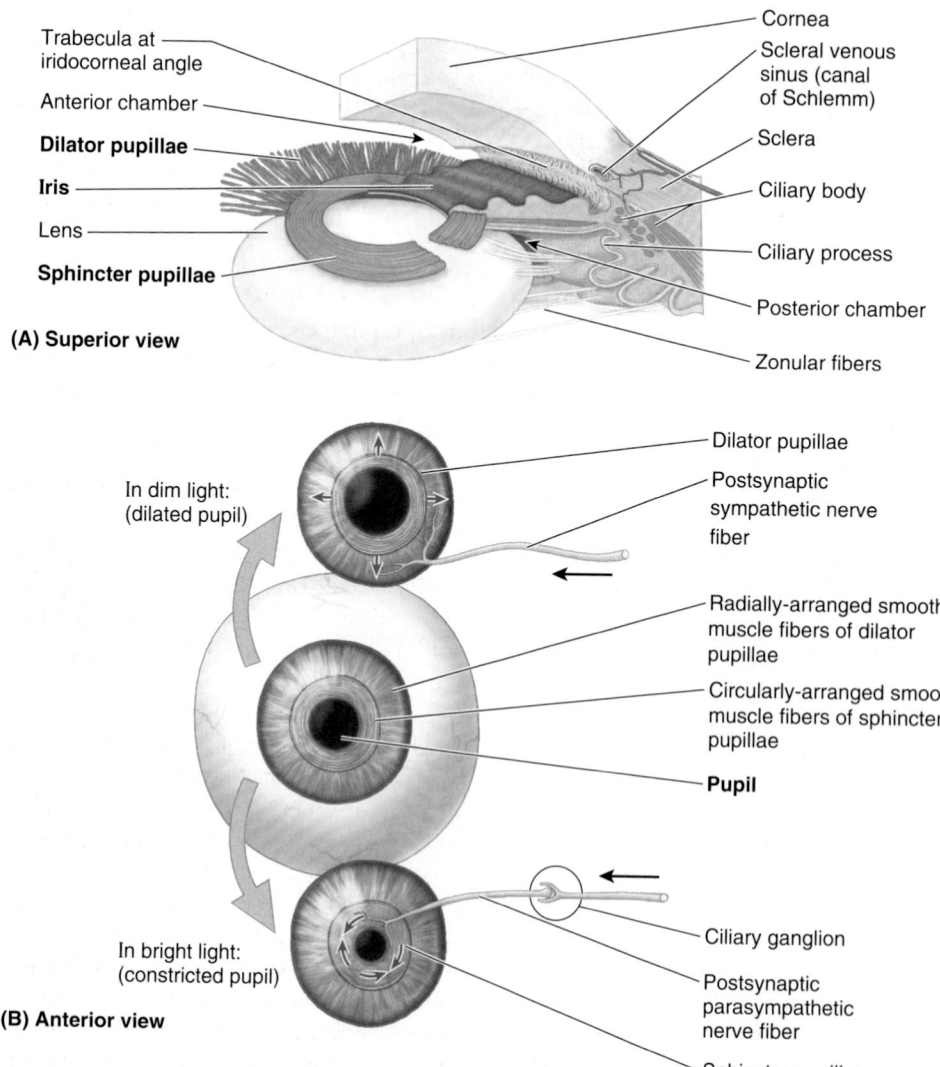

Trabecula at iridocorneal angle
Anterior chamber
Dilator pupillae
Iris
Lens
Sphincter pupillae

(A) Superior view

Cornea
Scleral venous sinus (canal of Schlemm)
Sclera
Ciliary body
Ciliary process
Posterior chamber
Zonular fibers

In dim light: (dilated pupil)

Dilator pupillae
Postsynaptic sympathetic nerve fiber
Radially-arranged smooth muscle fibers of dilator pupillae
Circularly-arranged smooth muscle fibers of sphincter pupillae
Pupil

In bright light: (constricted pupil)

(B) Anterior view

Ciliary ganglion
Postsynaptic parasympathetic nerve fiber
Sphincter pupillae

FIGURE 7.51. Structure and function of the iris. A. Iris dissected *in situ*. The iris separates the anterior and posterior chambers of the anterior segment of the eyeball as it bounds the pupil. B. Dilation and constriction of the pupil. In dim light, sympathetic fibers stimulate dilation of the pupil. In bright light, parasympathetic fibers stimulate constricting the pupil.

the retina consists of two functional parts with distinct locations: an optic part and a non-visual retina. The **optic part of the retina** is sensitive to visual light rays and has two layers: a neural layer and pigmented layer. The **neural layer** is light receptive. The **pigmented layer** consists of a single layer of cells that reinforces the light-absorbing property of the choroid in reducing the scattering of light in the eyeball. The **non-visual retina** is an anterior continuation of the pigmented layer and a layer of supporting cells. The non-visual retina extends over the ciliary body (**ciliary part of retina**) and the posterior surface of the iris (**iridial part of retina**), to the pupillary margin.

Clinically, the internal aspect of the posterior part of the eyeball, where light entering the eyeball is focused, is referred to as the **fundus of the eyeball** (ocular fundus). The retina of the fundus includes a distinctive circular area called the **optic disc** (optic papilla), where the sensory fibers and vessels conveyed by the optic nerve (CN II) enter the eyeball (Figs. 7.49C, 7.50A and 7.52). Because it contains no photoreceptors, the optic disc is insensitive to light. Consequently, this part of the retina is commonly called the *blind spot*.

Just lateral to the optic disc is the **macula of the retina** or macula lutea (L. yellow spot). The yellow color of the macula is apparent only when the retina is examined with red-free light. The macula is a small oval area of the retina with special photoreceptor cones that is specialized for acuity of vision. It is not normally observed with an *ophthalmoscope* (a device for viewing the interior of the eyeball through the pupil). At the center of the macula, a depression, the **fovea centralis** (L., central pit), is the area of most acute vision. The fovea is approximately 1.5 mm in diameter; its center, the **foveola,** does not have the capillary network visible elsewhere deep to the retina.

The optic part of the retina terminates anteriorly along the **ora serrata** (L. serrated edge), the irregular posterior border of the ciliary body (Figs. 7.49C and 7.50A). Except for the cones and rods of the neural layer, the retina is supplied by the **central artery of the retina,** a branch of the ophthalmic artery. The cones and rods of the outer neural layer receive nutrients from the *capillary lamina of the choroid,* or choriocapillaris (discussed in "Vasculature of Orbit" on p. 905). It has the finest vessels of the inner surface of the choroid, against which the retina is pressed. A corresponding system of retinal veins unites to form the **central vein of the retina.**

REFRACTIVE MEDIA AND COMPARTMENTS OF EYEBALL

On their way to the retina, lightwaves pass through the refractive media of the eyeball: cornea, aqueous humor, lens, and vitreous humor (Fig. 7.50A). The *cornea* is the primary refractory medium of the eyeball—that is, it bends light to the greatest degree, focusing an inverted image on the light-sensitive retina of the *fundus of the eyeball.*

The **aqueous humor** (often shortened clinically to "aqueous") occupies the *anterior segment of the eyeball* (Figs. 7.50B and 7.51A). The anterior segment is subdivided by the iris and pupil. The **anterior chamber of the eye** is the space between the cornea anteriorly and the iris/pupil posteriorly. The **posterior chamber of the eye** is between the iris/pupil anteriorly and the lens and ciliary body posteriorly. Aqueous humor is produced in the posterior chamber by the *ciliary processes* of the ciliary body. This clear watery solution provides nutrients for the avascular cornea and lens. After passing through the pupil into the anterior chamber, the aqueous humor drains through a trabecular meshwork at the **iridocorneal angle** into the **scleral venous sinus** (L. *sinus venosus sclerae,* canal of Schlemm) (Fig. 7.51A). The humor is removed by the **limbal plexus,** a network of scleral veins close to the limbus, which drain in turn into both tributaries of the *vorticose* and *anterior ciliary veins* (Fig. 7.50B). Intra-ocular pressure (IOP) is a balance between production and outflow of aqueous humor.

The **lens** is posterior to the iris and anterior to the vitreous humor of the *vitreous body* (Figs. 7.50 and 7.51A). It is a transparent, biconvex structure enclosed in a capsule. The highly elastic **capsule of the lens** is anchored by **zonular fibers** (collectively constituting the **suspensory ligament of the lens**) to the encircling ciliary processes. Although most refraction is produced by the cornea, the convexity of the lens, particularly its anterior surface, constantly varies to fine-tune the focus of near or distant objects on the retina (Fig. 7.53). The isolated unattached lens assumes a nearly spherical shape. In other words, in the absence of external attachment and stretching, it becomes nearly round.

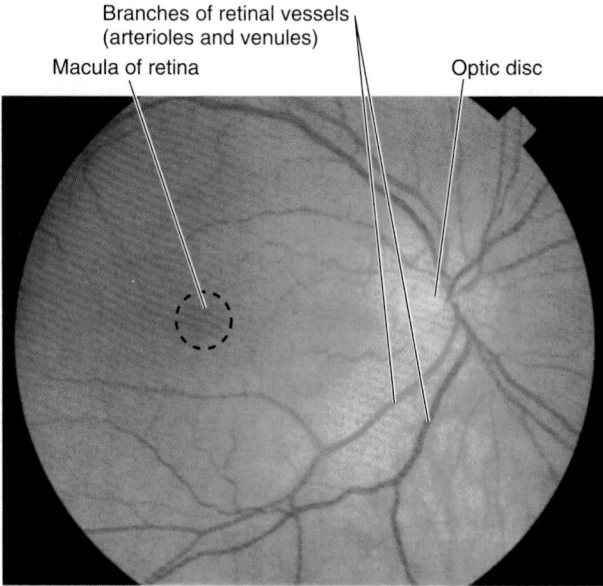

Branches of retinal vessels (arterioles and venules)

Macula of retina

Optic disc

Ophthalmoscopic view

FIGURE 7.52. Fundus of right eyeball. Retinal venules (wider) and retinal arterioles (narrower) radiate from the center of the oval optic disc. The dark area lateral to the disc is the macula. Branches of retinal vessels extend toward this area, but do not reach its center, the fovea centralis—the area of most acute vision.

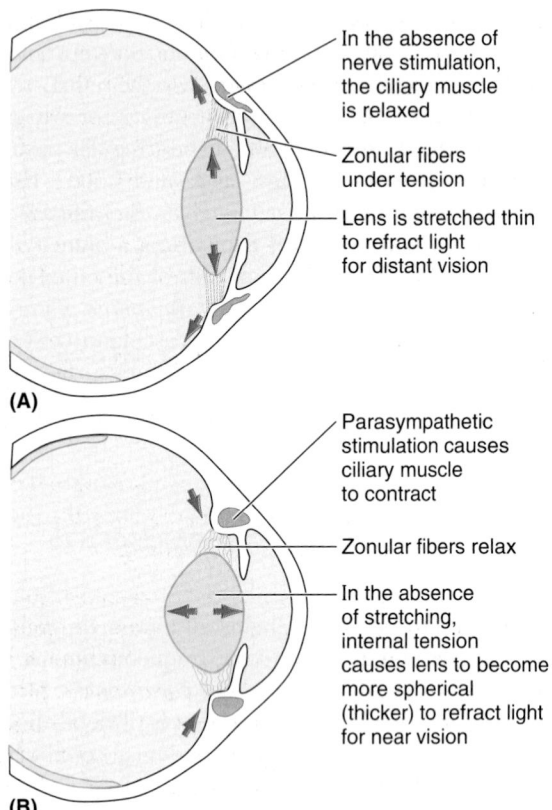

(A)

In the absence of nerve stimulation, the ciliary muscle is relaxed

Zonular fibers under tension

Lens is stretched thin to refract light for distant vision

(B)

Parasympathetic stimulation causes ciliary muscle to contract

Zonular fibers relax

In the absence of stretching, internal tension causes lens to become more spherical (thicker) to refract light for near vision

FIGURE 7.53. Changing lens shape (accommodation). A. Distant vision. **B.** Near vision.

The **ciliary muscle** of the ciliary body changes the shape of the lens. In the absence of nerve stimulation, the diameter of the relaxed muscular ring is larger. The lens suspended within the ring is under tension as its periphery is stretched, causing it to be thinner (less convex). The less convex lens brings more distant objects into focus (far vision). Parasympathetic stimulation via the oculomotor nerve (CN III) causes sphincter-like contraction of the ciliary muscle. The ring becomes smaller, and tension on the lens is reduced. The relaxed lens thickens (becomes more convex), bringing near objects into focus (near vision). The active process of changing the shape of the lens for near vision is called **accommodation.** The thickness of the lens increases with aging so that the ability to accommodate typically becomes restricted after age 40.

The **vitreous humor** is a watery fluid enclosed in the meshes of the **vitreous body,** a transparent jelly-like substance in the posterior four fifths of the eyeball posterior to the lens (**posterior segment of the eyeball,** also called the *postremal* or *vitreous chamber*) (Fig. 7.50A). In addition to transmitting light, the vitreous humor holds the retina in place and supports the lens.

Extra-ocular Muscles of Orbit

The **extra-ocular muscles of the orbit** are the *levator palpebrae superioris,* four *recti* (superior, inferior, medial,

and *lateral*), and two *obliques* (superior and inferior). These seven muscles work together to move the superior eyelids and eyeballs. They are illustrated in Figures 7.54–7.58 and the attachments, nerve supply, and main actions of the orbital muscles, beginning from the primary position, are outlined in Table 7.8. Additional details are provided in the following sections.

LEVATOR PALPEBRAE SUPERIORIS

The **levator palpebrae superioris** broadens into a wide bilaminar aponeurosis as it approaches its distal attachments. The superficial lamina attaches to the skin of the superior eyelid, and the deep lamina to the superior tarsus (Fig. 7.54B). This muscle is opposed most of the time by gravity and is the antagonist of the superior half of the orbicularis oculi, the sphincter of the palpebral fissure. The deep lamina of the distal (palpebral) part of the muscle includes smooth muscle fibers, the **superior tarsal muscle,** that produce additional widening of the palpebral fissure, especially during a sympathetic response (e.g., fright). However, they seem to function continuously (in the absence of a sympathetic response) because an interruption of the sympathetic supply produces a constant *ptosis*—drooping of the upper eyelid.

MOVEMENTS OF EYEBALL

Movements of the eyeball occur as rotations around three *axes—vertical, transverse,* and *anteroposterior* (Fig. 7.54A)—and are described according to the direction of movement of the pupil from the primary position, or of the superior pole of the eyeball from the neutral position. Rotation of the eyeball around the vertical axis moves the pupil medially (toward the midline, **adduction**), or laterally (away from the midline, **abduction**). Rotation around the transverse axis moves the pupil superiorly (**elevation**) or inferiorly (**depression**). Movements around the anteroposterior (AP) axis (corresponding to the axis of gaze in the primary position) move the superior pole of the eyeball medially (**medial rotation,** or intorsion) or laterally (**lateral rotation,** or extorsion). These rotational movements accommodate changes in the tilt of the head. Absence of these movements resulting from nerve lesions contributes to double vision. Movements may occur around the three axes simultaneously, requiring three terms to describe the direction of movement from the primarily position (e.g., the pupil is elevated, adducted, and medially rotated).

RECTI AND OBLIQUE MUSCLES

The four **recti muscles** (L. *rectos,* straight) run anteriorly to the eyeball, arising from a fibrous cuff, the **common tendinous ring,** that surrounds the optic canal and from part of the superior orbital fissure at the apex of the orbit (Figs. 7.54B & C and 7.55A). Structures that enter the orbit through the optic canal and the adjacent part of the fissure lie initially within the cone of recti (Figs. 7.54B & C and 7.55B). The four recti muscles are named for their individual positions relative to the eyeball. Because they mainly run anteriorly to attach to the superior, inferior, medial, and lateral aspects of the eyeball anterior to its equator, the primary actions of the four recti in producing elevation, depression, adduction, and abduction are relatively intuitive.

Several factors make the actions of the obliques and the secondary actions of the superior and inferior recti more challenging to understand. The *apex of the orbit* is medially placed relative to the orbit, so that the *axis of the orbit* does not coincide with the *optical axis* (Figs. 7.44A and 7.54C). Therefore, *when the eye is in the primary position*, the superior rectus (SR) and inferior rectus (IR) muscles also approach the eyeball from its medial side, their line of pull passing medial to the vertical axis. This gives both muscles a secondary action of *adduction*. The SR and IR also extend laterally, passing superior and inferior to the AP axis, respectively, giving the SR a secondary action of *medial rotation*, and the IR a secondary action of *lateral rotation*.

If the gaze is first directed laterally (abducted by the lateral rectus [LR]) so that the line of gaze coincides with the plane of the IR and SR, *the SR produces elevation only* (and is solely responsible for the movement) (Fig. 7.56A), and *the IR produces depression only* (and is likewise solely responsible) (Fig. 7.56B). During a physical examination, the physician directs the patient to follow his or her finger laterally (testing the LR and abducent nerve [CN VI]), then superiorly and inferiorly to isolate and test the function of the SR and IR and the integrity of the oculomotor nerve (CN III), which supplies both (Fig. 7.56E).

The inferior oblique (IO) is the only muscle to originate from the anterior part of the orbit (immediately lateral to the lacrimal fossa) (Fig. 7.47B). The superior oblique (SO) originates from the apex region like the rectus muscles (but super medial to the common tendinous ring) (Fig. 7.55A); however, its tendon traverses the *trochlea* just inside the super medial orbital rim, redirecting its line of pull (Figs. 7.54B & C and 7.55B). Thus, the inserting tendons of the oblique muscles lie in the same oblique vertical plane. When the inserting tendons are viewed anteriorly (Fig. 7.47B) or superiorly (Fig. 7.54C) with the eyeball in the primary position, it can be seen that the tendons of the oblique muscles pass mainly laterally to insert on the lateral half of the eyeball, posterior to its equator. Because they pass inferior and superior to the AP axis as they pass laterally, the IO is the primary lateral rotator, and the SO is the primary medial rotator of the eye. However, in the primary position, the obliques also pass posteriorly across the transverse axis (Fig. 7.54B) and posterior to the vertical axis (Fig. 7.54C), giving the SO a secondary function as a depressor, the IO a secondary function as an elevator, and both muscles a secondary function as abductors (Fig. 7.54B & C).

If the gaze is first directed medially (adducted by the medial rectus [MR]) so that the line of gaze coincides with plane of the inserting tendons of the SO and IO, *the SO*

produces depression only (and is solely responsible for the movement) (Fig. 7.56C) and *the IO produces elevation only* (and is likewise solely responsible) (Fig. 7.56D). During a physical examination, the physician directs the patient to follow his or her finger medially (testing the MR and oculomotor nerve), then inferiorly and superiorly to isolate and test the functions of the SO and IO and the integrity of the trochlear nerve (CN IV), which supplies the SO, and of the inferior division of CN III, which supplies the IO (Fig. 7.56E). In practice:

- The main action of the superior oblique is depression of the pupil in the adducted position (e.g., directing the gaze down the page when the gaze of both eyes is directed medially [*converged*] for reading).
- The main action of the inferior oblique is elevation of the pupil in the adducted position (e.g., directing the gaze up the page during **convergence** for reading).

Although the actions produced by the extra-ocular muscles have been considered individually, all motions require the action of several muscles in the same eye, assisting each other as synergists or opposing each other as antagonists. Muscles that are synergistic for one action may be antagonistic for another. For example, no single muscle can act to elevate the pupil directly from the primary position (Fig. 7.54D). The two elevators (SR and IO) act as synergists to do so. However, these muscles are antagonistic as rotators, and so neutralize each other so that no rotation occurs as they work together to elevate the pupil. Similarly, no single muscle can act to depress the pupil directly from the primary position. The two depressors, the SO and IR, both produce depression when acting alone and also produce opposing actions in terms of adduction–abduction and medial–lateral rotation. However, when the SO and IR act simultaneously, their synergistic actions depress the pupil because their antagonistic actions neutralize each other; therefore, pure depression results.

To direct the gaze, coordination of both eyes must be accomplished by the paired action of contralateral **yoke muscles** (functionally-paired contralateral extra-ocular muscles). For example, in directing the gaze to the right, the right lateral rectus and left medial rectus act as yoke muscles (Fig. 7.59).

SUPPORTING APPARATUS OF EYEBALL

The *fascial sheath of the eyeball* envelops the eyeball, extending posteriorly from the conjunctival fornices to the optic nerve, forming the actual socket for the eyeball (Fig. 7.45A). The cup-like fascial sheath is pierced by the tendons of the extra-ocular muscles and is reflected onto each of them as a tubular *muscle sheath*. The muscle sheaths of the levator and superior rectus muscles are fused; thus, when the gaze is directed superiorly, the superior eyelid is further elevated out of the line of vision. Triangular expansions from the sheaths of the medial and lateral rectus

(continued on p. 903)

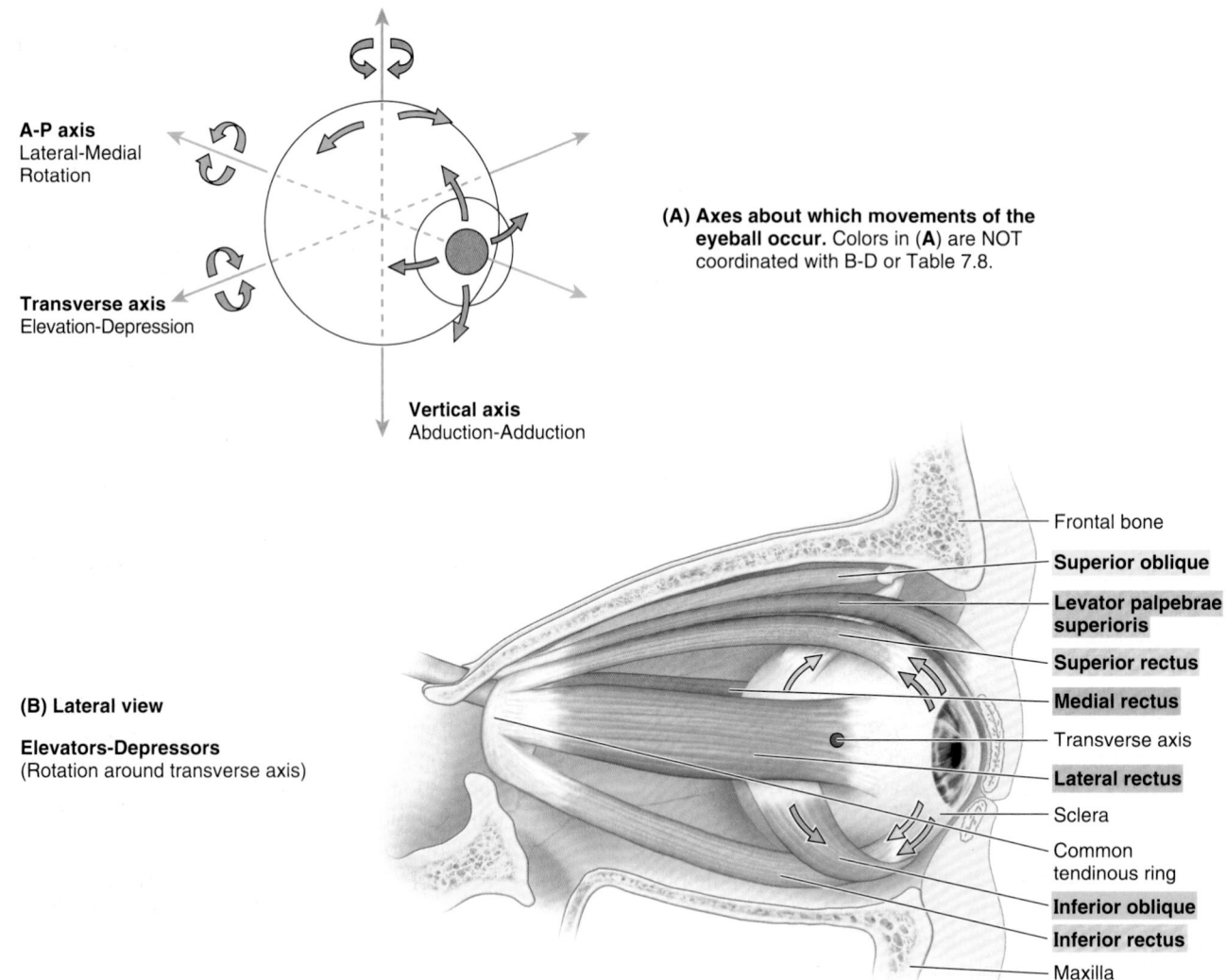

A-P axis
Lateral-Medial Rotation

Transverse axis
Elevation-Depression

Vertical axis
Abduction-Adduction

(A) Axes about which movements of the eyeball occur. Colors in (**A**) are NOT coordinated with B-D or Table 7.8.

(B) Lateral view

Elevators-Depressors
(Rotation around transverse axis)

Frontal bone
Superior oblique
Levator palpebrae superioris
Superior rectus
Medial rectus
Transverse axis
Lateral rectus
Sclera
Common tendinous ring
Inferior oblique
Inferior rectus
Maxilla

FIGURE 7.54. Extra-ocular muscles and their movements. A. Axes around which movements of the eyeball occur. **B.** Position of muscles in right orbit. *Arrows,* movements of the eyeball around the transverse axis.

TABLE 7.8. EXTRA-OCULAR MUSCLES OF ORBIT

Muscle	Origin	Insertion	Innervation	Main Action[a]
Levator palpebrae superioris	Lesser wing of sphenoid bone, superior and anterior to optic canal	Superior tarsus and skin of superior eyelid	Oculomotor nerve (CN III); deep layer (superior tarsal muscle) is supplied by sympathetic fibers	Elevates superior eyelid
Superior oblique (SO)	Body of sphenoid bone	Its tendon passes through a fibrous ring or trochlea, changes its direction, and inserts into sclera deep to superior rectus muscle	Trochlear nerve (CN IV)	Abducts, depresses, and medially rotates eyeball
Inferior oblique (IO)	Anterior part of floor of orbit	Sclera deep to lateral rectus muscle	Oculomotor nerve (CN III)	Abducts, elevates, and laterally rotates eyeball
Superior rectus (SR)	Common tendinous ring	Sclera just posterior to corneo-scleral junction		Elevates, adducts, and rotates eyeball medially
Inferior rectus (IR)				Depresses, adducts, and rotates eyeball laterally
Medial rectus (MR)				Adducts eyeball
Lateral rectus (LR)			Abducent nerve (CN VI)	Abducts eyeball

[a]The actions described are for muscles acting alone, starting from the primary position (gaze directed anteriorly). In fact, muscles rarely act independently and almost always work together in synergistic and antagonistic groups. Clinical testing requires maneuvers to isolate muscle actions. Only the actions of the medial and lateral rectus are tested, starting from the primary position (Fig. 7.56E).

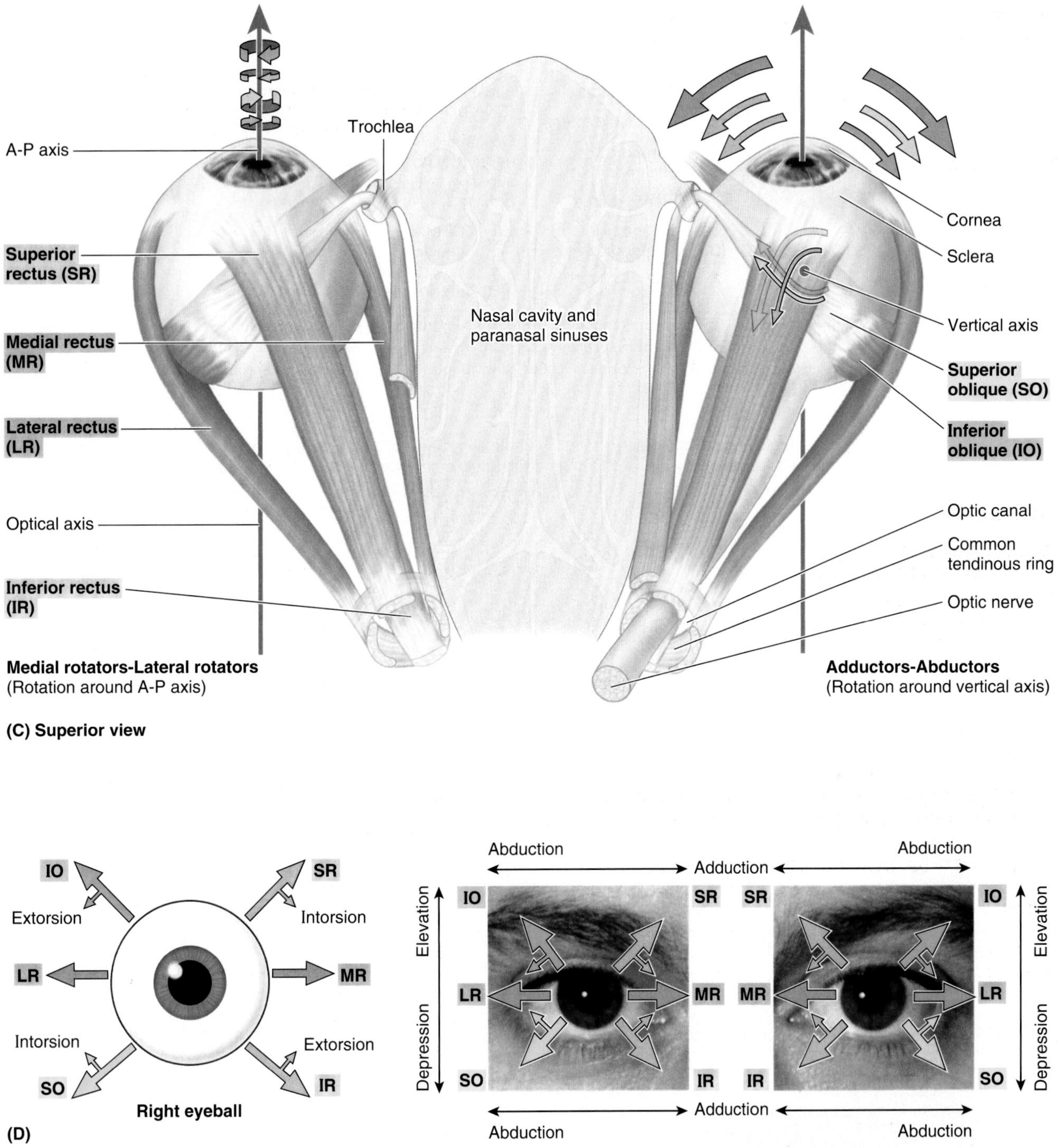

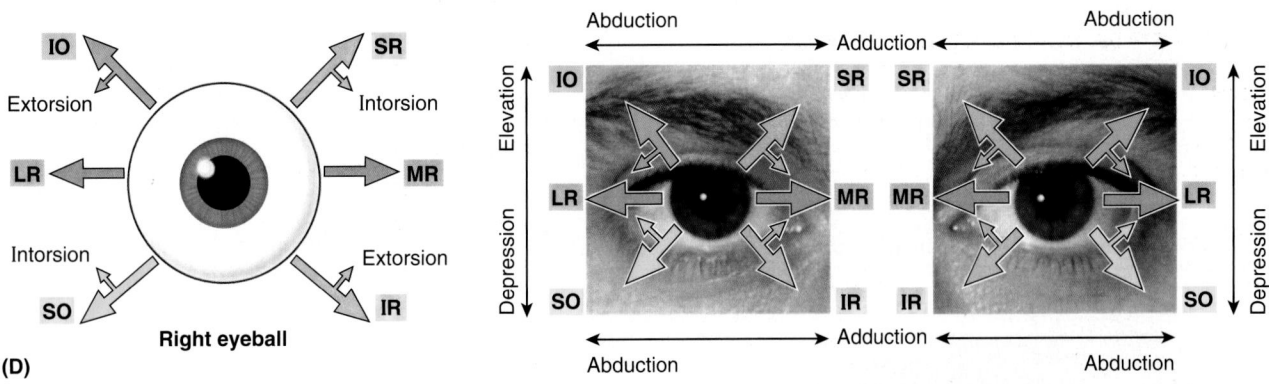

FIGURE 7.54. *(Continued)* **C.** Position of muscles in right and left orbits. *Arrows at left,* movements of the eyeball around the AP axis; *arrows at right,* movements of the eyeball around the vertical axis. To understand the actions produced by muscles starting from the primary position, it is necessary to observe the placement and line of pull of the muscle relative to the axes about which the movements occur. **D.** Unilateral and bilateral demonstration of extra-ocular muscle actions, starting from the primary position. For movements in any of the six cardinal directions (*large arrows*) the indicated muscle is the prime mover. Movements in directions between large arrows requires synergistic actions by the adjacent muscles. For example, direct elevation requires the synergistic actions of IO and SR; direct depression requires synergistic action of SO and IR. *Small arrows,* muscles producing rotational movements around the AP axis. Coordinated action of the contralateral yoke muscles is required to direct the gaze. For example, in directing the gaze to the right, the right LR and left MR are yoke muscles.

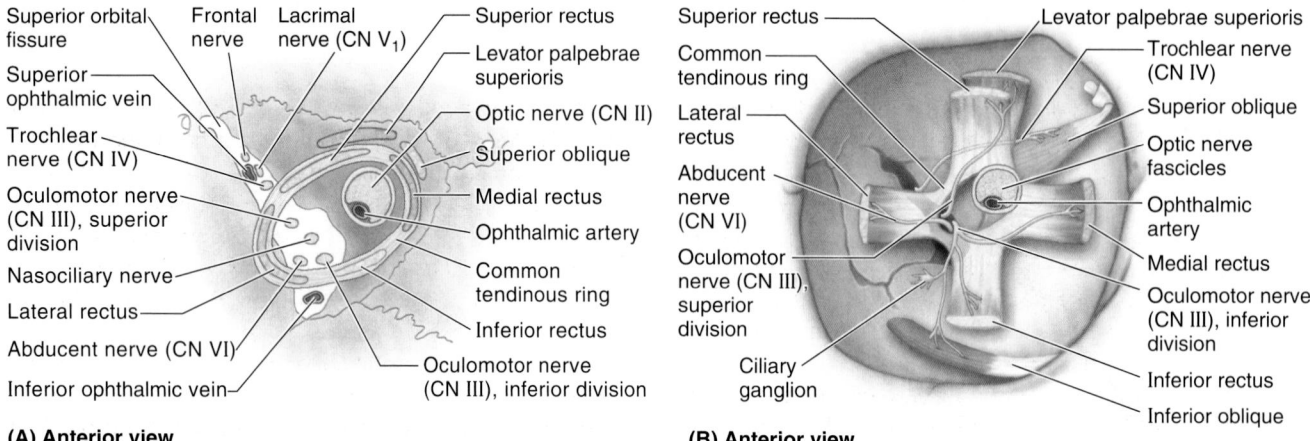

(A) Anterior view

Superior orbital fissure
Frontal nerve
Lacrimal nerve (CN V₁)
Superior ophthalmic vein
Trochlear nerve (CN IV)
Oculomotor nerve (CN III), superior division
Nasociliary nerve
Lateral rectus
Abducent nerve (CN VI)
Inferior ophthalmic vein

Superior rectus
Levator palpebrae superioris
Optic nerve (CN II)
Superior oblique
Medial rectus
Ophthalmic artery
Common tendinous ring
Inferior rectus
Oculomotor nerve (CN III), inferior division

(B) Anterior view

Superior rectus
Common tendinous ring
Lateral rectus
Abducent nerve (CN VI)
Oculomotor nerve (CN III), superior division
Ciliary ganglion

Levator palpebrae superioris
Trochlear nerve (CN IV)
Superior oblique
Optic nerve fascicles
Ophthalmic artery
Medial rectus
Oculomotor nerve (CN III), inferior division
Inferior rectus
Inferior oblique

FIGURE 7.55. Relationship at apex of orbit. A. The common tendinous ring is formed by the origin of the four recti muscles and encircles the optic sheath of CN II, the superior and inferior divisions of CN III, the nasociliary nerve (CN V₁), and CN VI. The nerves supplying the extra-ocular muscles enter the orbit through the superior orbital fissure: oculomotor (CN III), trochlear (CN IV), and abducent (CN VI). **B.** Structures (minus membranous fascia and fat) after enucleation (excision) of the eyeball.

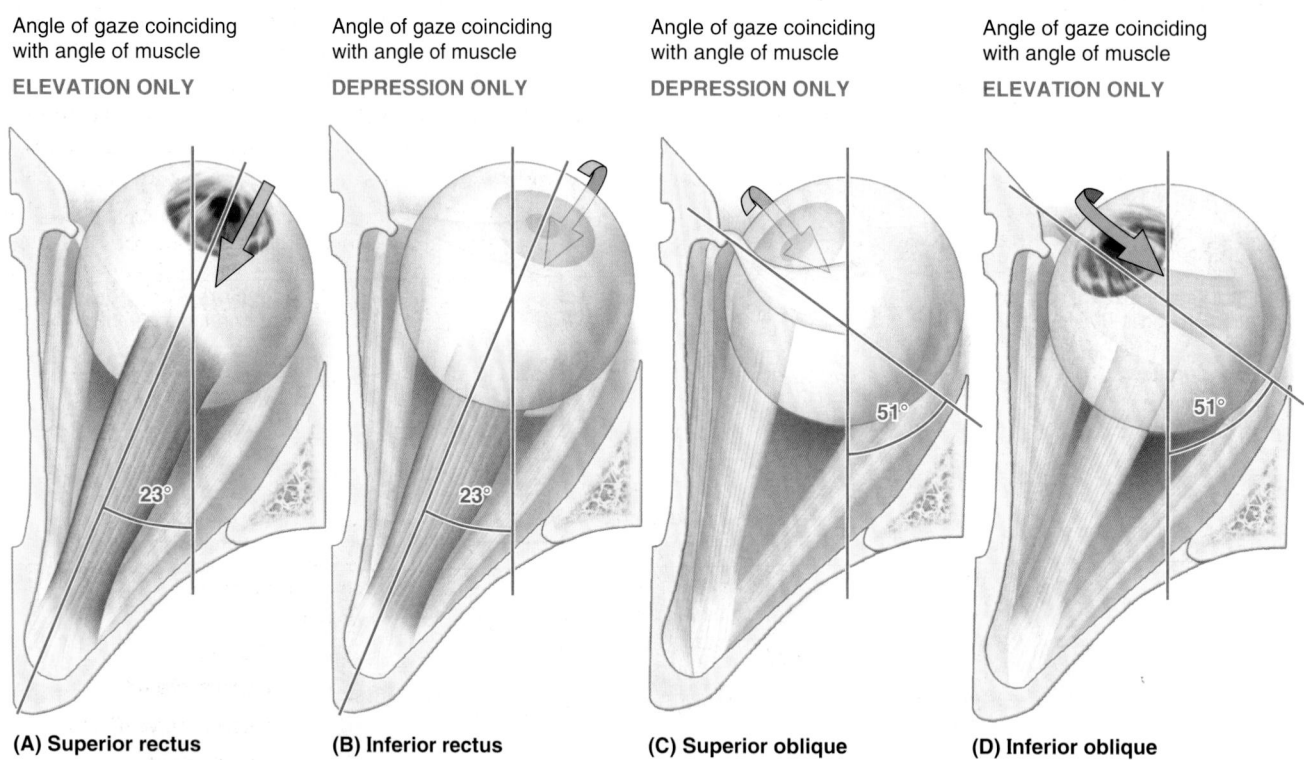

Angle of gaze coinciding with angle of muscle
ELEVATION ONLY

Angle of gaze coinciding with angle of muscle
DEPRESSION ONLY

Angle of gaze coinciding with angle of muscle
DEPRESSION ONLY

Angle of gaze coinciding with angle of muscle
ELEVATION ONLY

23°

23°

51°

51°

(A) Superior rectus

(B) Inferior rectus

(C) Superior oblique

(D) Inferior oblique

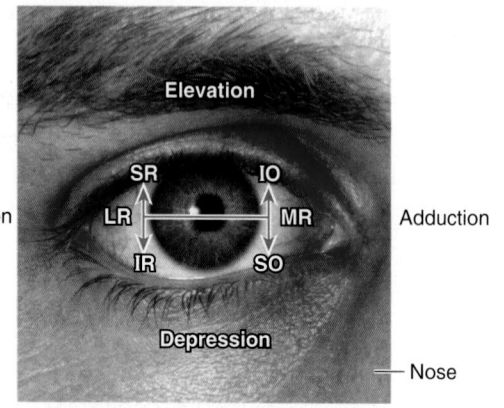

Elevation

Abduction

SR   IO
LR   MR
IR   SO

Adduction

Depression

Nose

(E) Actions of muscles of orbit as tested clinically (Right Eye)

FIGURE 7.56. Clinical testing of extra-ocular muscles. A and B. When the eye is initially abducted by LR, only the rectus muscles can produce elevation and depression. **C and D.** When the eye is initially abducted by MR, only the oblique muscles can produce elevation and depression. **E.** Following movements of the examiner's finger, the pupil is moved in an extended H-pattern to isolate and test individual extra-ocular muscles and the integrity of their nerves.

muscles, called the **medial** and **lateral check ligaments,** are attached to the lacrimal and zygomatic bones, respectively. These ligaments limit abduction and adduction. A blending of the check ligaments with the fascia of the inferior rectus and inferior oblique muscles forms a hammock-like sling, the **suspensory ligament of the eyeball.** A similar **inferior check ligament** from the fascial sheath of the inferior rectus retracts the inferior eyelid when the gaze is directed downward (Fig. 7.45A). Collectively, the check ligaments act with the oblique muscles and the **retrobulbar fat** to resist the posterior pull on the eyeball produced by the rectus muscles. In diseases or starvation that reduce the retrobulbar fat, the eyeball is retracted into the orbit (*inophthalmos*).

Nerves of Orbit

The large **optic nerves** convey purely sensory nerves that transmit impulses generated by optical stimuli (Figs. 7.45A and 7.50A). They are cranial nerves (CN II) by convention, but develop as paired anterior extensions of the forebrain and are actually central nervous system (CNS) fiber tracts formed of second-order neurons. The optic nerves begin at the **lamina cribrosa of the sclera,** where the unmyelinated nerve fibers pierce the sclera and become myelinated, posterior to the *optic disc.* They exit the orbits via the optic canals. Throughout their course in the orbit, the optic

nerves are surrounded by extensions of the *cranial meninges* and *subarachnoid space,* the latter occupied by a thin layer of *CSF* (Fig. 7.45A, *inset*). The intra-orbital extensions of the cranial dura and arachnoid constitute the **optic nerve sheath,** which becomes continuous anteriorly with the fascial sheath of the eyeball and the sclera. A layer of pia mater covers the surface of the optic nerve within the sheath.

In addition to the optic nerve (CN II), the nerves of the orbit include those that enter through the *superior orbital fissure* and supply the ocular muscles: **oculomotor** (CN III), **trochlear** (CN IV), and **abducent** (CN VI) nerves (Figs. 7.55 and 7.57). A memory device for the innervation of the extra-ocular muscles moving the eyeball is similar to a chemical formula: $LR_6SO_4AO_3$ (**l**ateral **r**ectus, CN **VI**; **s**uperior **o**blique, CN **IV**; **a**ll **o**thers, CN **III**). The trochlear and abducent nerves pass directly to the single muscle supplied by each nerve. The oculomotor nerve divides into a superior and an inferior division. The superior division supplies the superior rectus and levator palpebrae superioris. The inferior division supplies the medial and inferior rectus and inferior oblique and carries presynaptic parasympathetic fibers to the ciliary ganglion (Fig. 7.58). The movements are stimulated by the oculomotor, trochlear, and abducent nerves, starting from the primary position in the right and left orbits, and produce binocular vision, demonstrated in Fig. 7.59.

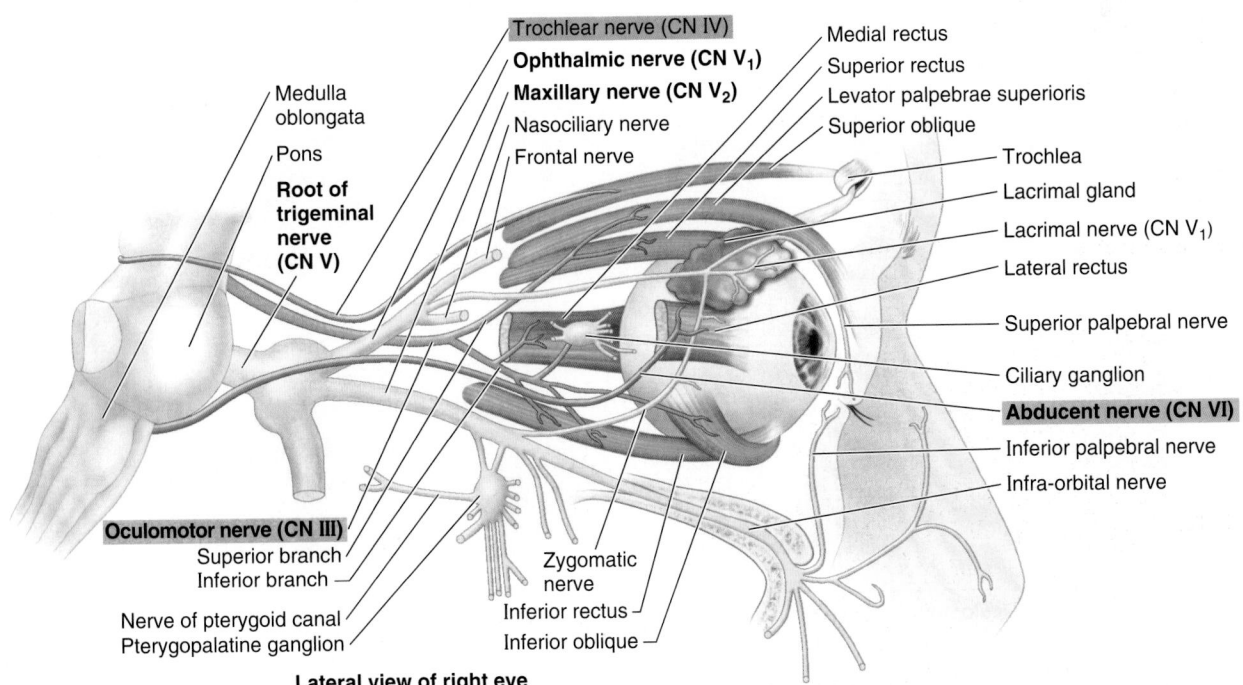

Lateral view of right eye

FIGURE 7.57. Nerves of orbit. Three cranial nerves (CN III, IV, and VI) supply the seven voluntary extra-ocular muscles. CN IV supplies the superior oblique, CN VI supplies the lateral rectus, and CN III supplies the remaining five muscles. The CN III also brings presynaptic parasympathetic fibers to the ciliary ganglion. The trigeminal nerve (CN V) supplies sensory fibers to the orbit, orbital region, and eyeball.

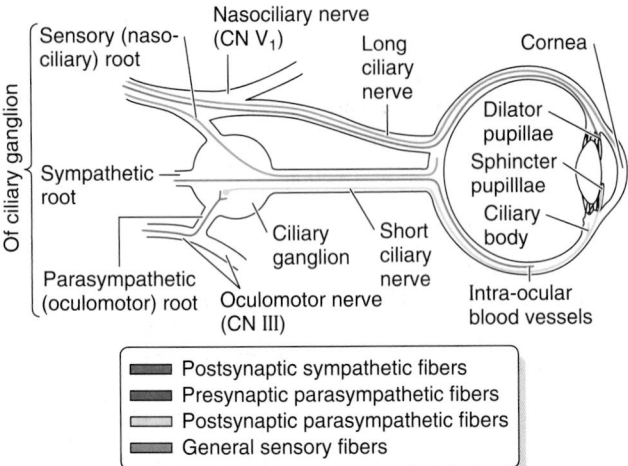

FIGURE 7.58. Distribution of nerve fibers to ciliary ganglion and eyeball. The ciliary ganglion receives three types of nerve fibers from three separate sources. All parasympathetic innervation but only some of the sensory and sympathetic innervation to the eyeball traverses the ganglion. Sympathetic and sensory fibers in the long ciliary nerve bypass the ganglion.

The three terminal branches of the *ophthalmic nerve*, CN V₁ (the frontal, nasociliary, and lacrimal nerves), pass through the superior orbital fissure and supply structures related to the anterior orbit (e.g., lacrimal gland and eyelids), face, and scalp (Fig. 7.60). The cutaneous branches of CN V₁ (lacrimal, frontal, and infratrochlear nerves) are described in "Cutaneous Nerves of Face and Scalp" (p. 849) and in Table 7.5.

The **ciliary ganglion** is a small group of postsynaptic parasympathetic nerve cell bodies associated with CN V₁. It is located between the optic nerve and the lateral rectus toward the posterior limit of the orbit. The ganglion receives nerve fibers from three sources (Fig. 7.58):

- Sensory fibers from CN V₁ via the **sensory** or **nasociliary root of the ciliary ganglion**.
- Presynaptic parasympathetic fibers from CN III via the **parasympathetic** or **oculomotor root of the ciliary ganglion**.
- Postsynaptic sympathetic fibers from the *internal carotid plexus* via the **sympathetic root of the ciliary ganglion**.

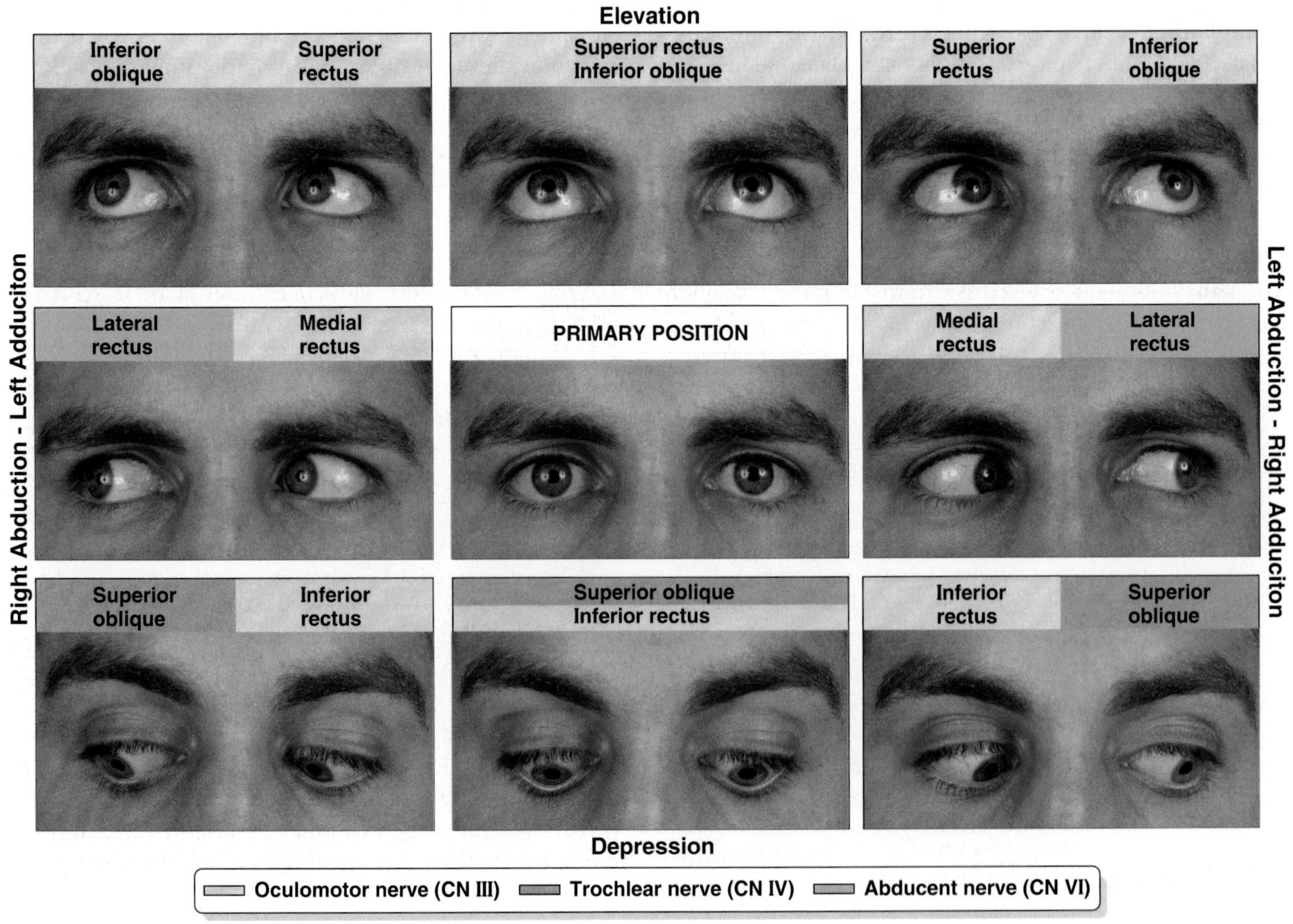

FIGURE 7.59. Binocular movements and muscles producing them. All movements start from the primary position.

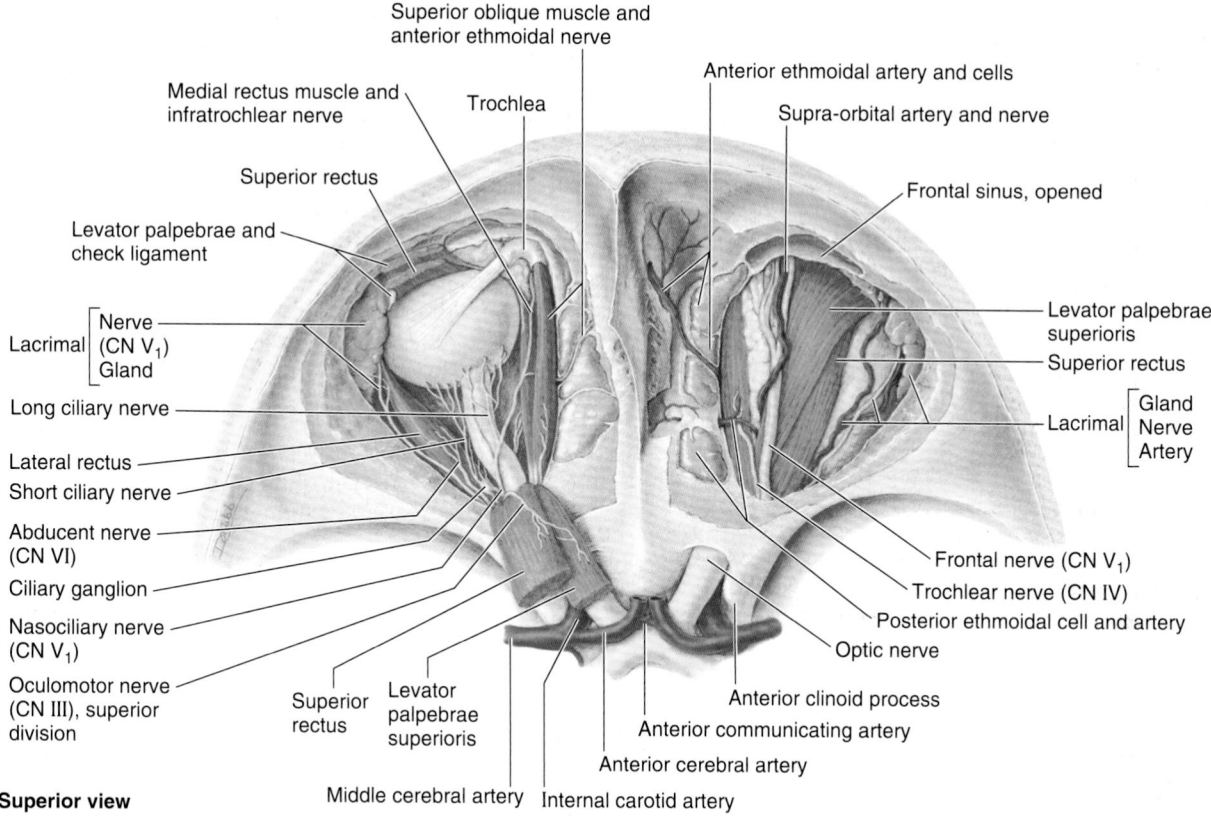

Superior oblique muscle and
anterior ethmoidal nerve

Medial rectus muscle and
infratrochlear nerve

Trochlea

Anterior ethmoidal artery and cells

Supra-orbital artery and nerve

Superior rectus

Frontal sinus, opened

Levator palpebrae and
check ligament

Levator palpebrae
superioris

Superior rectus

Lacrimal [Nerve (CN V₁) Gland]

Lacrimal [Gland Nerve Artery]

Long ciliary nerve

Lateral rectus

Short ciliary nerve

Abducent nerve
(CN VI)

Ciliary ganglion

Frontal nerve (CN V₁)

Trochlear nerve (CN IV)

Posterior ethmoidal cell and artery

Nasociliary nerve
(CN V₁)

Optic nerve

Oculomotor nerve
(CN III), superior
division

Superior
rectus

Levator
palpebrae
superioris

Anterior clinoid process

Anterior communicating artery

Anterior cerebral artery

Superior view

Middle cerebral artery   Internal carotid artery

FIGURE 7.60. Dissection of orbit. In this superior approach, the orbital part of the frontal bone has been removed. On the right side, three nerves applied to the roof of the orbit (trochlear, frontal, and lacrimal) are evident. On the left side, the levator palpebrae superioris and superior rectus have been cut and reflected and the orbital fat removed to demonstrate the nerves that traverse the intraconal fat.

The **short ciliary nerves** arise from the ciliary ganglion and are considered to be branches of CN V₁ (Figs. 7.58 and 7.60). They carry parasympathetic and sympathetic fibers to the ciliary body and iris. The short ciliary nerves consist of postsynaptic parasympathetic fibers originating in the ciliary ganglion, afferent fibers from the nasociliary nerve that pass through the ganglion, and postsynaptic sympathetic fibers that also pass through it. **Long ciliary nerves,** branches of the nasociliary nerve (CN V₁) that pass to the eyeball, bypassing the ciliary ganglion, convey postsynaptic sympathetic fibers to the dilator pupillae and afferent fibers from the iris and cornea.

The *posterior* and *anterior ethmoidal nerves,* branches of the nasociliary nerve arising in the orbit, exit via openings in the medial wall of the orbit to supply the mucous membrane of the sphenoidal and ethmoidal sinuses and the nasal cavities, as well as the dura of the anterior cranial fossa.

Vasculature of Orbit

ARTERIES OF ORBIT

The blood supply of the orbit is mainly from the **ophthalmic artery,** a branch of the internal carotid artery (Fig. 7.61; Table 7.9); the **infra-orbital artery,** from the

external carotid artery, also contributes blood to structures related to the orbital floor. The **central artery of the retina,** a branch of the ophthalmic artery arising inferior to the optic nerve, pierces the sheath of the optic nerve and runs within the nerve to the eyeball, emerging at the optic disc (Fig. 7.45A, *inset*). Its branches spread over the internal surface of the retina (Figs. 7.52 and 7.62). The terminal branches are *end arteries (arterioles)*, which provide the only blood supply to the internal aspect of the retina.

The external aspect of the retina is also supplied by the *capillary lamina of the choroid (choriocapillaris)*. Of the eight or so *posterior ciliary arteries* (also branches of the ophthalmic artery), six **short posterior ciliary arteries** directly supply the choroid, which nourishes the outer non-vascular layer of the retina. Two **long posterior ciliary arteries,** one on each side of the eyeball, pass between the sclera and the choroid to anastomose with the **anterior ciliary arteries** (continuations of the **muscular branches of the ophthalmic artery** to the rectus muscles) to supply the ciliary plexus.

VEINS OF ORBIT

Venous drainage of the orbit is through the **superior** and **inferior ophthalmic veins,** which pass through the superior orbital fissure and enter the cavernous sinus (Fig. 7.63).

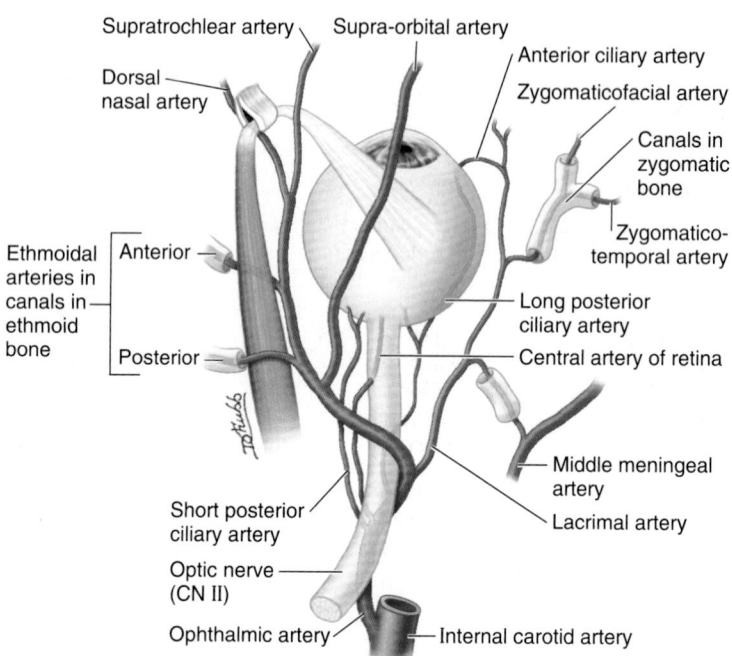

FIGURE 7.61. Arteries of orbit.

TABLE 7.9. ARTERIES OF ORBIT

Artery	Origin	Course and Distribution
Ophthalmic	Internal carotid artery	Traverses optic foramen to reach orbital cavity
Central artery of retina	Ophthalmic artery	Pierces dural sheath of optic nerve and runs to eyeball; branches from center of optic disc; supplies optic retina (except cones and rods)
Supra-orbital		Passes superiorly and posteriorly from supra-orbital foramen to supply forehead and scalp
Supratrochlear		Passes from supra-orbital margin to forehead and scalp
Lacrimal		Passes along superior border of lateral rectus muscle to supply lacrimal gland, conjunctiva, and eyelids
Dorsal nasal		Courses along dorsal aspect of nose and supplies its surface
Short posterior ciliaries		Pierce sclera at periphery of optic nerve to supply choroid, which in turn supplies cones and rods of optic retina
Long posterior ciliaries		Pierce sclera to supply ciliary body and iris
Posterior ethmoidal		Passes through posterior ethmoidal foramen to posterior ethmoidal cells
Anterior ethmoidal		Passes through anterior ethmoidal foramen to anterior cranial fossa; supplies anterior and middle ethmoidal cells, frontal sinus, nasal cavity, and skin on dorsum of nose
Anterior ciliary	Muscular (rectus) branches of ophthalmic artery	Pierces sclera at attachments of rectus muscles and forms network in iris and ciliary body
Infra-orbital	Third part of maxillary artery	Passes along infra-orbital groove and foramen to face

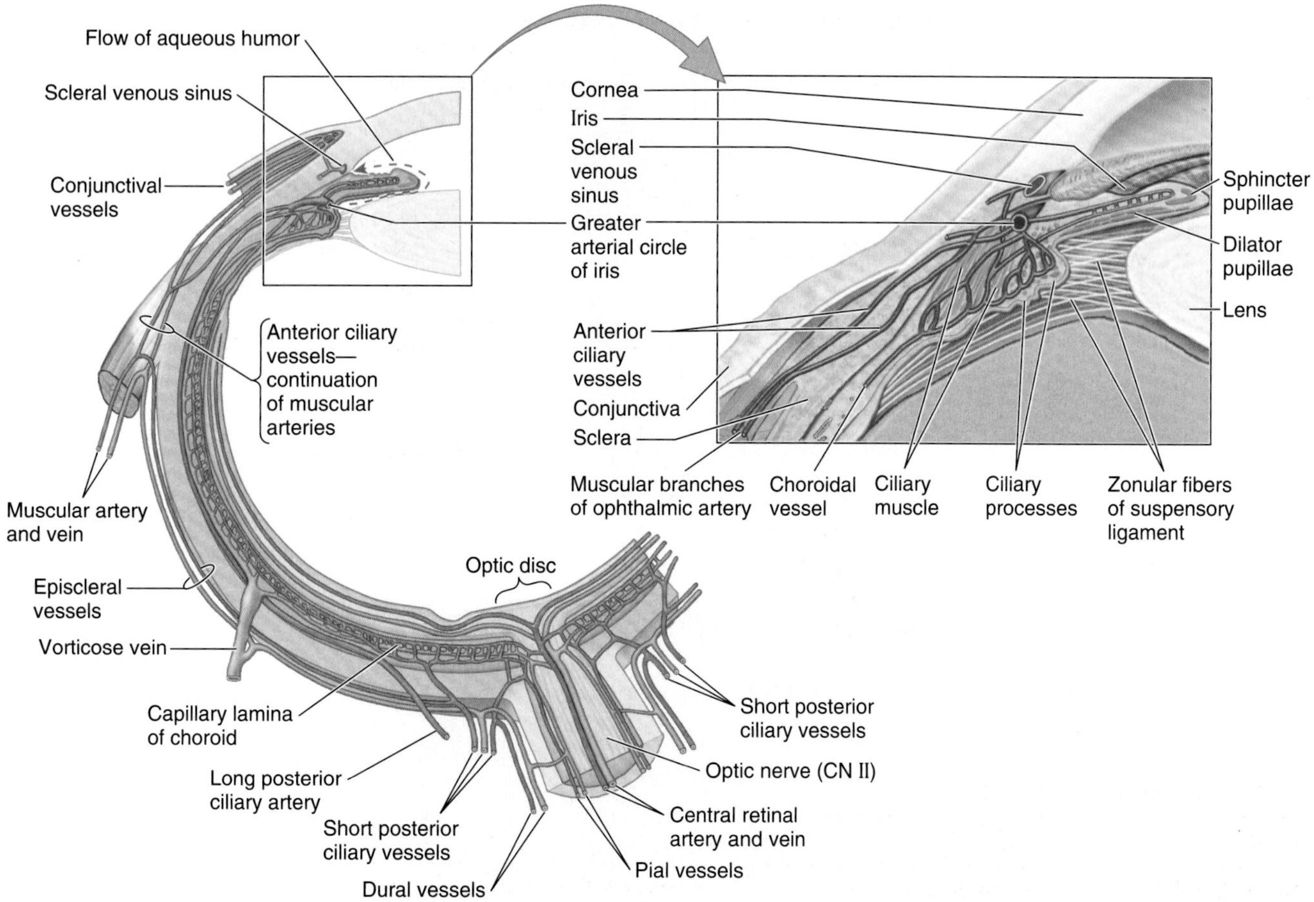

Flow of aqueous humor

Scleral venous sinus

Conjunctival vessels

Cornea

Iris

Scleral venous sinus

Greater arterial circle of iris

Sphincter pupillae

Dilator pupillae

Lens

Anterior ciliary vessels— continuation of muscular arteries

Anterior ciliary vessels

Conjunctiva

Sclera

Muscular branches of ophthalmic artery

Choroidal vessel

Ciliary muscle

Ciliary processes

Zonular fibers of suspensory ligament

Muscular artery and vein

Episcleral vessels

Vorticose vein

Optic disc

Capillary lamina of choroid

Long posterior ciliary artery

Short posterior ciliary vessels

Dural vessels

Short posterior ciliary vessels

Optic nerve (CN II)

Central retinal artery and vein

Pial vessels

FIGURE 7.62. Partial horizontal section of right eyeball. The artery supplying the inner part of the retina (central retinal artery) and the choroid, which in turn nourishes the outer non-vascular layer of the retina, are shown. The choroid is arranged so that the supplying vessels and larger choroidal vessels are externally placed, and the smallest vessels (the capillary lamina) are most internal, adjacent to the non-vascular layer of the retina. The vorticose vein (one of four to five) drains venous blood from the choroid into the posterior ciliary and ophthalmic veins. The scleral venous sinus returns the aqueous humor, secreted into the anterior chamber by the ciliary processes, to the venous circulation.

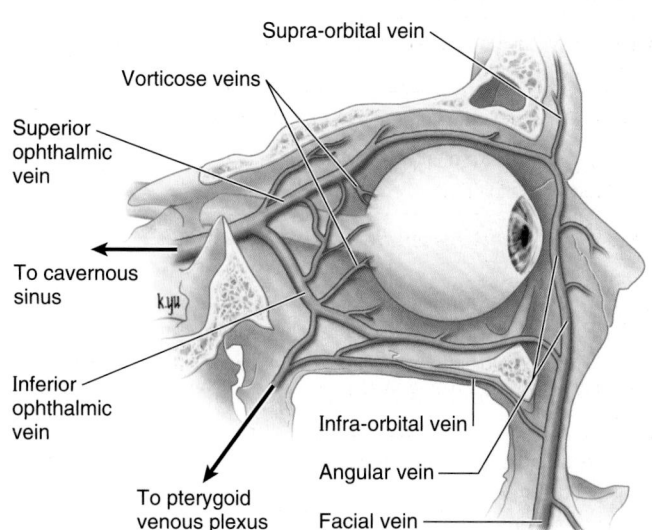

Supra-orbital vein

Vorticose veins

Superior ophthalmic vein

To cavernous sinus

Inferior ophthalmic vein

Infra-orbital vein

Angular vein

To pterygoid venous plexus

Facial vein

FIGURE 7.63. Ophthalmic veins. The superior ophthalmic vein empties into the cavernous sinus, and the inferior ophthalmic vein empties into the pterygoid venous plexus. They communicate with the facial and supra-orbital veins anteriorly and each other posteriorly. The superior ophthalmic vein accompanies the ophthalmic artery and its branches.

The **central vein of the retina** (Fig. 7.62) usually enters the cavernous sinus directly, but it may join one of the ophthalmic veins. The vortex, or **vorticose veins,** from the vascular layer of the eyeball drain into the inferior ophthalmic vein. The **scleral venous sinus** is a vascular structure encircling the anterior chamber of the eyeball through which the aqueous humor is returned to the blood circulation.

Surface Anatomy of Eye and Lacrimal Apparatus

For a description of the surface anatomy of the eyelids, see "Surface Anatomy of Face" (p. 859). The anterior part of the *sclera* (the "white" of the eye) is covered by the transparent *bulbar conjunctiva,* which contains minute but apparent conjunctival blood vessels (Fig. 7.64B). When irritated, the vessels may enlarge noticeably, and the bulbar conjunctiva may take on a distinctly pink appearance when inflamed ("red" eyes). The normal tough, opaque sclera often appears slightly blue in infants and children and commonly has a yellow hue in many older people.

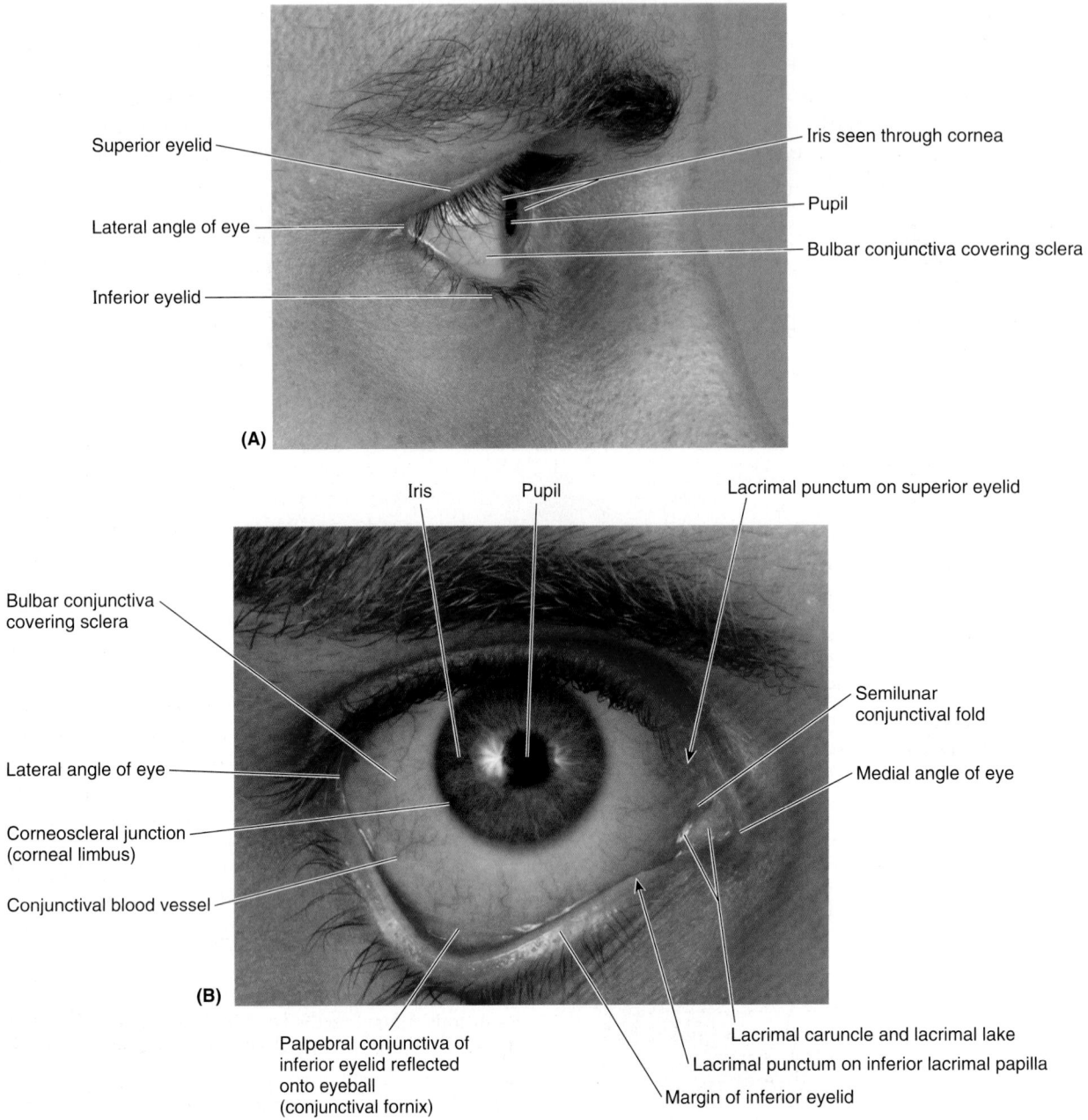

FIGURE 7.64. Surface anatomy of the eye (A) and lacrimal apparatus (B).

The anterior transparent part of the eye is the *cornea*, which is continuous with the sclera at its margins. In a lateral view (Fig. 7.64A), most of the visible part of the eyeball protrudes slightly through the *palpebral fissure*. It is apparent that the cornea has a greater curvature (convexity) than that of the rest of the eyeball (the part covered by sclera); thus a shallow angle occurs at the *corneoscleral junction*, the *corneal limbus* (Fig. 7.64B). The prominence of the cornea also makes movements of the eyeball apparent when the eyelids are closed.

The dark circular opening through which light enters the eyeball, the *pupil*, is surrounded by the *iris* (plural = irides), a circular pigmented diaphragm. The relative size of the pupil and iris varies with the brightness of the entering light; however, the size of the contralateral pupils and irides should be uniform.

Normally, when the eyes are open and the gaze is directed anteriorly, the superior part of the cornea and iris are covered by the edge of the *superior eyelid,* and the inferior part of the cornea and iris are fully exposed above the *inferior eyelid,* usually exposing a narrow rim of sclera. Even slight variations in the position of the eyeballs are noticeable, causing a change in facial expression to a surprised look when the superior eyelid is elevated (as occurs in *exophthalmos,* or protrusion of the eyeballs, caused by hyperthyroidism), or a sleepy appearance (as occurs when the superior eyelid droops, *ptosis,* owing to an absence of sympathetic innervation in Horner syndrome).

The *bulbar conjunctiva* is reflected from the sclera onto the deep surface of the eyelid. The *palpebral conjunctiva* is normally red and vascular and, with experience, can provide some assessment of hemoglobin levels. It is commonly examined in cases of suspected *anemia*, a blood condition commonly manifested by pallor (paleness) of the mucous membranes. When the superior eyelid is everted, the size and extent of the enclosed *superior tarsus* can be appreciated, and commonly the *tarsal glands* can be distinguished through the palpebral conjunctiva as slightly yellow vertical stripes. Under close examination, the openings of these glands (approximately 20 per eyelid) can be seen on the margins of the eyelids, posterior to the two to three rows of emerging cilia or *eyelashes*. As the bulbar conjunctiva is continuous with the anterior epithelium of the cornea and the palpebral conjunctiva, it forms the *conjunctival sac*. The palpebral fissure is the "mouth," or anterior aperture, of the conjunctival sac.

In the *medial angle of the eye*, a reddish shallow reservoir of tears, the *lacrimal lake*, can be observed. Within the lake is the *lacrimal caruncle*, a small mound of moist modified skin. Lateral to the caruncle is a *semilunar conjunctival fold*, which slightly overlaps the eyeball. The semilunar fold is a rudiment of the nictitating membrane of birds and reptiles. When the edges of the eyelids are everted, a small pit, the *lacrimal punctum*, is visible at its medial end on the summit of a small elevation, the *lacrimal papilla*.

ORBITAL REGION, ORBIT, AND EYEBALL

Fractures of Orbit

The orbital margin is strong to protect the orbital content. However, when the blows are powerful enough and the impact is directly on the bony rim, the resulting fractures usually occur at the three sutures between the bones forming the orbital margin. Because of the thinness of the medial and inferior walls of the orbit, a blow to the eye may fracture the orbital walls while the margin remains intact (Fig. B7.23). Indirect traumatic injury that displaces the orbital walls is called a "blowout" fracture. Fractures of the medial wall may involve the ethmoidal and sphenoidal sinuses, whereas fractures of the inferior wall (orbital floor) may involve the maxillary sinus.

Although the superior wall is stronger than the medial and inferior walls, it is thin enough to be translucent and may be readily penetrated. Thus a sharp object may pass through it and enter the frontal lobe of the brain.

Orbital fractures often result in intra-orbital bleeding, which exerts pressure on the eyeball, causing *exophthalmos* (protrusion of the eyeball). Any trauma to the eye may affect adjacent structures—for example, bleeding into the maxillary sinus, displacement of maxillary teeth, and fracture of nasal bones resulting in hemorrhage, airway obstruction, and infection that could spread to the cavernous sinus through the ophthalmic vein.

Orbital Tumors

Because of the closeness of the optic nerve to the sphenoidal and posterior ethmoidal sinuses, a malignant tumor in these sinuses may erode the thin bony walls of the orbit and compress the optic nerve and orbital contents. Tumors in the orbit produce *exophthalmos*. The easiest entrance to the orbital cavity for a tumor in the middle cranial fossa is through the superior orbital fissure; tumors in the temporal or infratemporal fossa gain access to this cavity through the inferior orbital fissure. Although the lateral wall of the orbit is nearly as long as the medial wall because it extends laterally and anteriorly, it does not reach as far anteriorly as the medial wall does, which occupies essentially a sagittal plane (Fig. 7.44A). Nearly 2.5 cm of the eyeball is exposed when the pupil is turned medially as far as possible. This is why the lateral side affords a good approach for operations on the eyeball.

Injury to Nerves Supplying Eyelids

Because it supplies the levator palpebrae superioris, a lesion of the oculomotor nerve causes paralysis of the muscle, and the superior eyelid droops (*ptosis*). Damage to the facial nerve involves paralysis of the orbicularis oculi, preventing the eyelids from closing fully. Normal rapid protective blinking of the eye is also lost.

The loss of tonus of the muscle in the inferior eyelid causes the lid to fall away (evert) from the surface of the eyeball, leading to drying of the cornea. This leaves the eyeball unprotected from dust and small particles. Thus, irritation of the unprotected eyeball results in excessive but inefficient lacrimation (tear formation). *Excessive lacrimal fluid* also forms when the lacrimal drainage apparatus is obstructed, thereby preventing the fluid from reaching the inferior part of the eyeball. People often dab their eyes constantly to wipe the tears, resulting in further irritation.

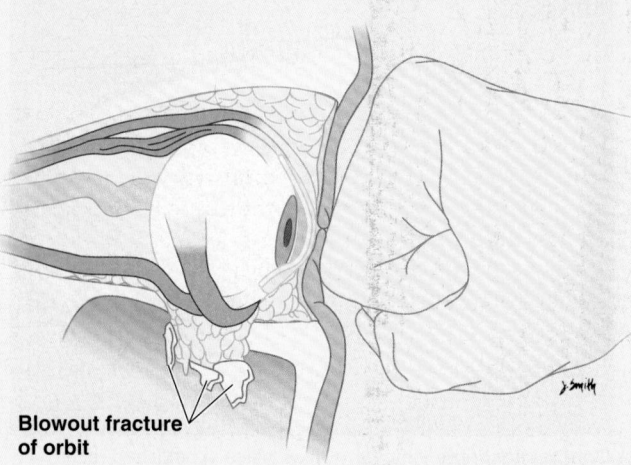

Blowout fracture of orbit

FIGURE B7.23.

Inflammation of Palpebral Glands

Any of the glands in the eyelid may become inflamed and swollen from infection or obstruction of their ducts. If the ducts of the ciliary glands are obstructed, a painful red suppurative (pus-producing) swelling, a *sty* (*hordeolum*), develops on the eyelid. *Cysts of the sebaceous glands* of the eyelid, called *chalazia,* may also form. Obstruction of a tarsal gland produces inflammation, a *tarsal chalazion,* that protrudes toward the eyeball and rubs against it as the eyelids blink.

Hyperemia of Conjunctiva

The conjunctiva is colorless, except when its vessels are dilated and congested ("bloodshot eyes"). Hyperemia of the conjunctiva is caused by local irritation (e.g., from dust, chlorine, or smoke). An inflamed conjunctiva, *conjunctivitis* ("pinkeye"), is a common contagious infection of the eye.

Subconjunctival Hemorrhages

Subconjunctival hemorrhages are common and are manifested by bright or dark red patches deep to and within the bulbar conjunctiva. The hemorrhages may result from injury or inflammation. A blow to the eye, excessively hard blowing of the nose, and paroxysms of coughing or violent sneezing can cause hemorrhages resulting from rupture of small subconjunctival capillaries.

Development of Retina

The retina and optic nerve develop from the **optic cup,** an outgrowth of the embryonic forebrain, the **optic vesicle** (Fig. B7.24A). As it evaginates from the forebrain (Fig. B7.24B), the optic vesicle carries the developing meninges with it. Hence the optic nerve is invested with cranial meninges and an extension of the subarachnoid space (Fig. B7.24C). The central artery and vein of the retina cross the subarachnoid space and run within the distal part of the optic nerve. The pigment cell layer of the retina develops from the outer layer of the optic cup, and the neural layer develops from the inner layer of the cup.

Retinal Detachment

The layers of the developing retina are separated in the embryo by an *intraretinal space* (Fig. B7.24B). During the early fetal period, the layers fuse, obliterating this space. Although the pigment cell layer becomes firmly fixed to the choroid, its attachment to the neural layer is not firm. Consequently, detachment of the retina may follow a blow to the eye (Fig. B7.25). A detached retina usually results from seepage of fluid between the neural and pigment cell layers of the retina, perhaps days or even weeks after trauma to the eye. Persons with a retinal detachment may complain of flashes of light or specks floating in front of the eye.

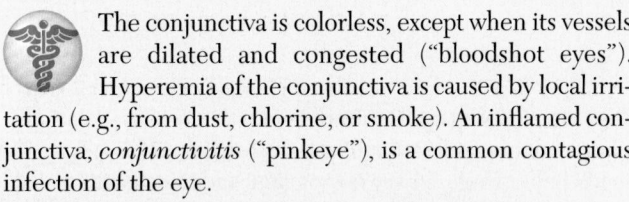
(A)

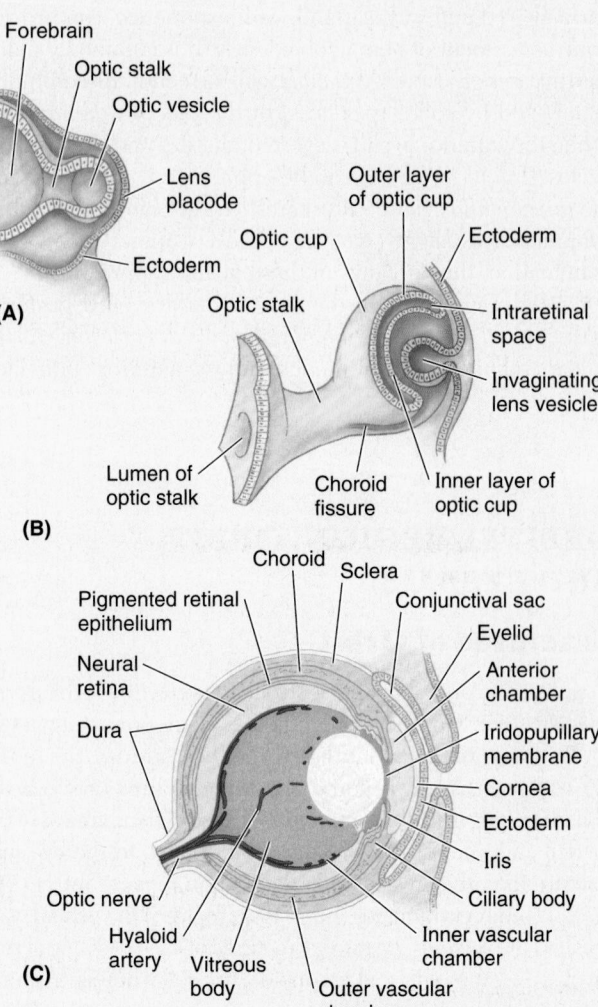
(B)
(C)

FIGURE B7.24.

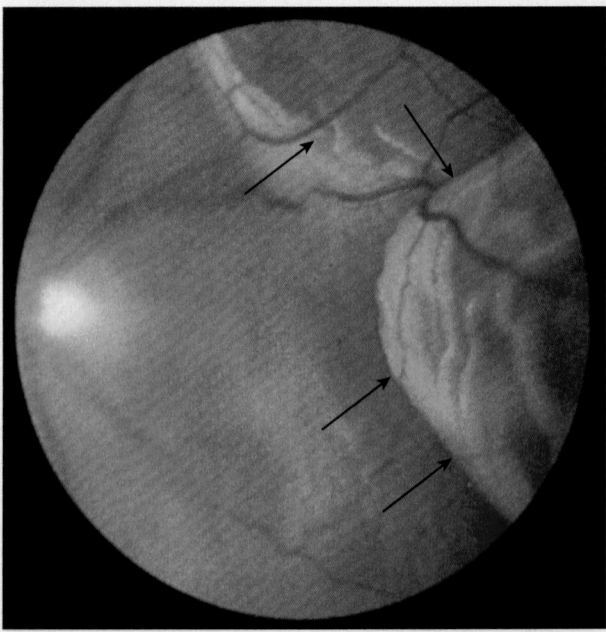

Ophthalmoscopic view (*arrows,* wrinkles in detached retina)

FIGURE B7.25.

Pupillary Light Reflex

The *pupillary light reflex* is tested using a penlight during a neurological examination. This reflex, involving CN II (afferent limb) and CN III (efferent limb), is the rapid constriction of the pupil in response to light. When light enters one eye, both pupils constrict because each retina sends fibers into the optic tracts of both sides. The sphincter pupillae muscle is innervated by parasympathetic fibers; consequently, interruption of these fibers causes dilation of the pupil because of the unopposed action of the sympathetically innervated dilator pupillae muscle. The first sign of *compression of the oculomotor nerve* is ipsilateral slowness of the pupillary response to light.

Uveitis

Uveitis, inflammation of the vascular layer of the eyeball (uvea), may progress to severe visual impairment and blindness if the inflammation is not treated by a specialist in ophthalmology.

Ophthalmoscopy

Physicians use an *ophthalmoscope* (funduscope) to view the fundus of the eyeball (Fig. 7.52). The retinal arteries and veins radiate over the fundus from the optic disc. The pale, oval disc appears on the medial side with the retinal vessels radiating from its center. Pulsation of the retinal arteries is usually visible. Centrally, at the posterior pole of the eyeball, the macula appears darker than the reddish hue of surrounding areas of the retina because the black melanin pigment in the choroid and pigment cell layer is not screened by capillary blood.

Papilledema

An increase in CSF pressure slows venous return from the retina, causing *edema of the retina* (fluid accumulation). The edema is viewed during ophthalmoscopy as swelling of the optic disc, a condition called *papilledema*. Normally, the disc is flat and does not form a papilla. Papilledema results from increased intracranial pressure and increased CSF pressure in the extension of the subarachnoid space around the optic nerve (Fig. 7.50A).

Presbyopia and Cataracts

As people age, their lenses become harder and more flattened. These changes gradually reduce the focusing power of the lenses, a condition known as *presbyopia* (G. *presbyos*, old). Some people also experience a loss of transparency (cloudiness) of the lens from areas of opaqueness (*cataracts*). *Cataract extraction* combined with an *intra-ocular lens implant* has become a common operation. An extracapsular cataract extraction involves removing the lens but leaving the capsule of the lens intact to receive a synthetic intra-ocular

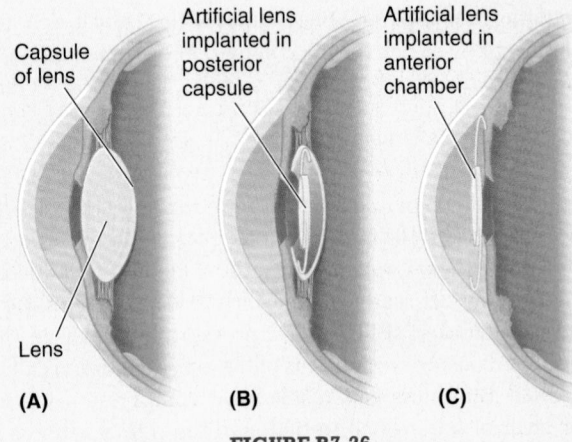

FIGURE B7.26.

lens (Fig. B7.26A & B). Intracapsular lens extraction involves removing the lens and lens capsule, and implanting a synthetic intra-ocular lens in the anterior chamber (Fig. B7.26C).

Coloboma of Iris

The absence of a section of iris (Fig. B7.27) may result from a birth defect, in which the choroid (retinal) fissure fails to close properly (Fig. B7.24B), from

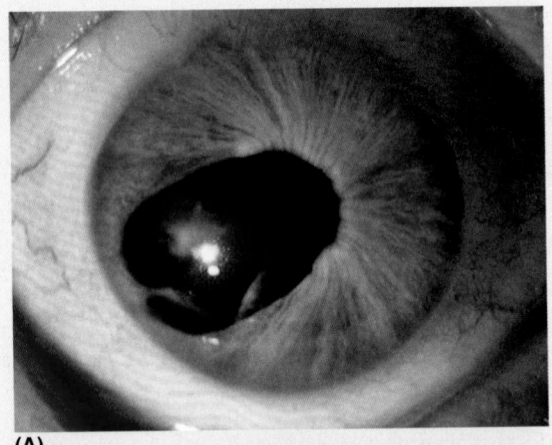

(A)

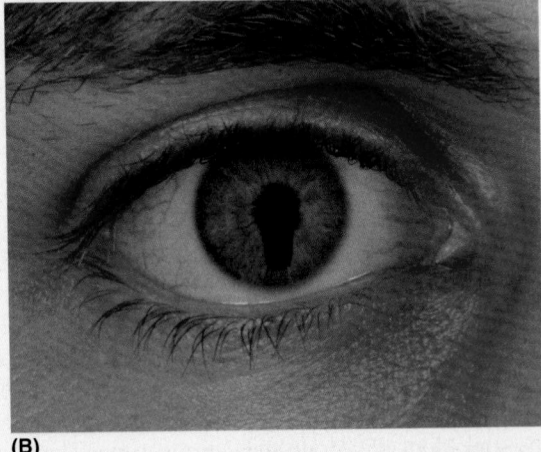

(B)

FIGURE B7.27.

penetrating or non-penetrating injuries to the eyeball, or a surgical iridectomy. When the iris is injured in such a manner, the iridial fissure does not heal.

Glaucoma

Outflow of aqueous humor through the scleral venous sinus into the blood circulation must occur at the same rate at which the aqueous is produced. If the outflow decreases significantly because the outflow pathway is blocked (Fig. B7.28), pressure builds up in the anterior and posterior chambers of the eye, a condition called *glaucoma*. Blindness can result from compression of the inner layer of the eyeball (retina) and the retinal arteries if aqueous humor production is not reduced to maintain normal intra-ocular pressure.

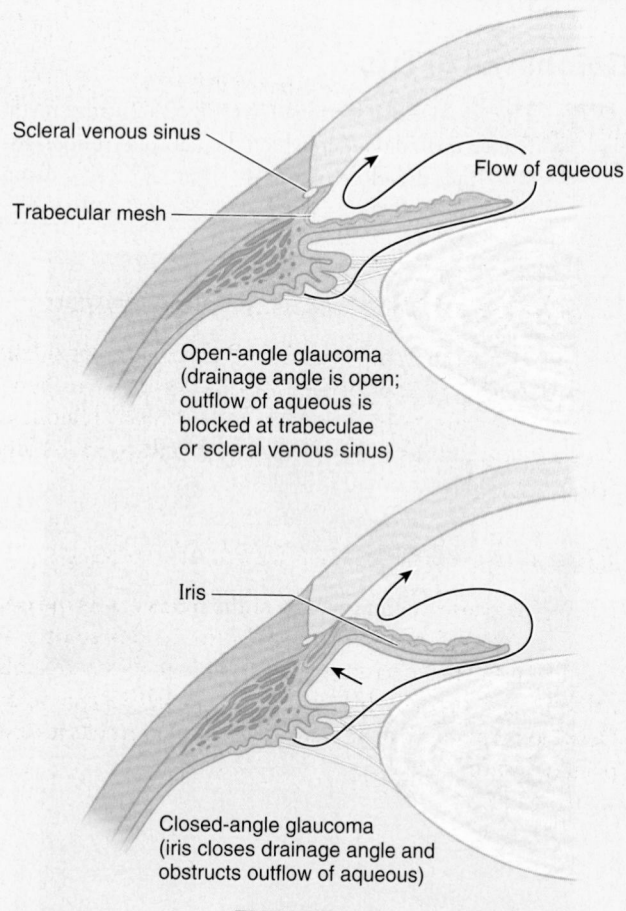

Scleral venous sinus

Trabecular mesh

Flow of aqueous

Open-angle glaucoma (drainage angle is open; outflow of aqueous is blocked at trabeculae or scleral venous sinus)

Iris

Closed-angle glaucoma (iris closes drainage angle and obstructs outflow of aqueous)

FIGURE B7.28.

Hemorrhage into Anterior Chamber

Hemorrhage within the anterior chamber of the eyeball (*hyphema*) usually results from blunt trauma to the eyeball, such as from a squash or racquet ball or a hockey stick (Fig. B7.29). Initially, the anterior chamber is tinged red, but blood soon accumulates in this chamber. The

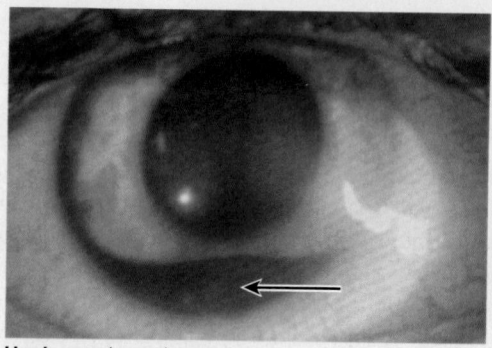

Hyphema *(arrow)*

FIGURE B7.29.

initial hemorrhage usually stops in a few days and recovery is usually good.

Artificial Eye

The fascial sheath of the eyeball forms a socket for an artificial eye when the eyeball is removed (*enucleated*). After this operation, the eye muscles cannot retract too far because their fascial sheaths remain attached to the fascial sheath of the eyeball. Thus, some coordinated movement of a properly fitted artificial eyeball is possible. Because the suspensory ligament supports the eyeball (Fig. 7.62), it is preserved when surgical removal of the bony floor of the orbit is performed (e.g., during the removal of a tumor).

Corneal Reflex

During a neurological examination, the examiner touches the cornea with a wisp of cotton (Fig. B7.14, p. 863). A normal (positive) response is a blink. Absence of a blink suggests a lesion of CN V_1; a lesion of CN VII (the motor nerve to the orbicularis oculi) may also impair this reflex. The examiner must be certain to touch the cornea (not just the sclera) to evoke the reflex. The presence of a contact lens may hamper or abolish the ability to evoke this reflex.

Corneal Abrasions and Lacerations

Foreign objects such as sand or metal filings (particles) produce *corneal abrasions* that cause sudden, stabbing pain in the eyeball and tears. Opening and closing the eyelids is also painful. *Corneal lacerations* are caused by sharp objects such as a tree branch, fingernails, or the corner of a page of a book.

Corneal Ulcers and Transplants

Damage to the sensory innervation of the cornea from CN V_1 leaves the cornea vulnerable to injury by foreign particles. People with corneal lesions (scarred or opaque corneas) may receive *corneal transplants* from donors or implants of non-reactive plastic material.

Horner Syndrome

Horner syndrome results from interruption of a cervical sympathetic trunk and is manifest by the absence of sympathetically stimulated functions on the ipsilateral side of the head. The syndrome includes the following signs: constriction of the pupil (miosis), drooping of the superior eyelid (ptosis), redness and increased temperature of the skin (vasodilation), and absence of sweating (anhidrosis). Constriction of the pupil occurs because the parasympathetically stimulated sphincter of the pupil is unopposed. The ptosis is a consequence of paralysis of the smooth muscle fibers interdigitated with the aponeurosis of the levator palpebrae superioris that collectively constitute the superior tarsal muscle, supplied by sympathetic fibers.

Paralysis of Extra-ocular Muscles/ Palsies of Orbital Nerves

One or more extra-ocular muscles may be paralyzed by disease in the brainstem or by a head injury, resulting in *diplopia* (double vision). Paralysis of a muscle is apparent by the limitation of movement of the eyeball in the field of action of the muscle and by the production of two images when one attempts to use the muscle.

OCULOMOTOR NERVE PALSY

Complete *oculomotor nerve palsy* affects most of the ocular muscles, the levator palpebrae superioris, and the sphincter pupillae. The superior eyelid droops and cannot be raised voluntarily because of the unopposed activity of the orbicularis oculi (supplied by the facial nerve) (Fig. B7.30A). The pupil is also fully dilated and non-reactive because of the unopposed dilator pupillae. The pupil is fully abducted and depressed ("down and out") because of the unopposed activity of the lateral rectus and superior oblique, respectively.

ABDUCENT NERVE PALSY

When the abducent nerve (CN VI) supplying only the lateral rectus is paralyzed, the individual cannot abduct the pupil on the affected side (abducent nerve palsy or paralysis). The pupil is fully adducted by the unopposed pull of the medial rectus (Fig. B7.30B).

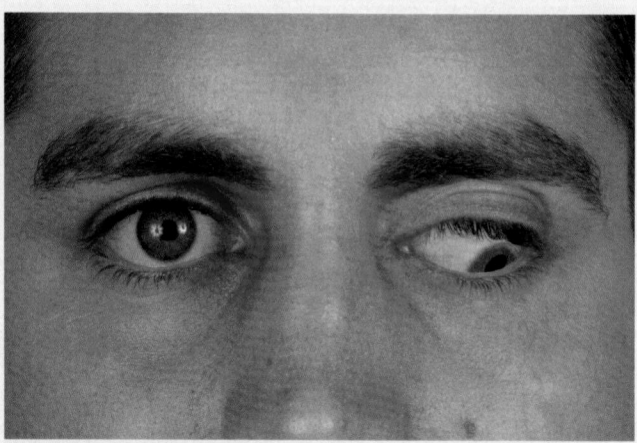

(A) Oculomotor paralysis

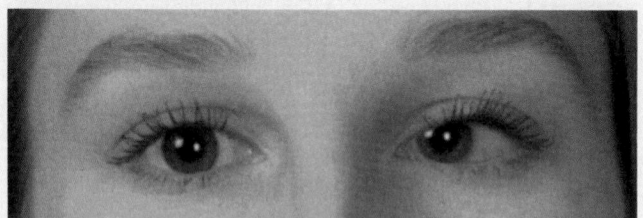

(B) Abducent paralysis

FIGURE B7.30.

Blockage of Central Artery of Retina

Because terminal branches of the central artery of the retina are end arteries, obstruction of them by an embolus results in instant and total blindness. Blockage of the artery is usually unilateral and occurs in older people.

Blockage of Central Vein of Retina

Because the central vein of the retina enters the cavernous sinus, *thrombophlebitis* of this sinus may result in the passage of a thrombus to the central retinal vein and produce blockage of the small retinal veins. Occlusion of a branch of the central vein usually results in slow, painless loss of vision.

The Bottom Line

ORBITAL REGION, ORBIT, AND EYEBALL

Orbits: The orbits are pyramidal cavities, with bases directed anteriorly and apices posteriorly, that house the eyeballs and accessory visual structures. ♦ The medial walls of the contralateral orbits are parallel, and the lateral walls are perpendicular to each other. ♦ The margins and lateral walls of the orbits, being most vulnerable to direct trauma, are strong.

♦ The superior wall (roof) and inferior wall (floor) are shared with the anterior cranial fossa and the maxillary sinus, respectively, and much of the paper-thin medial wall is common to the ethmoidal cells. ♦ The medial wall and floor are thus vulnerable to the spread of disease processes from the paranasal sinuses and to blowout fractures when blunt force is applied to the orbital contents, suddenly increasing intraorbital pressure. ♦ The optic canal and superior orbital fissure at the apex of the orbit are the primary paths by which structures enter and exit the orbits.

Accessory visual structures: The eyelids and lacrimal apparatus are protective devices for the eyeball. ♦ The conjunctival sac is a special form of mucosal bursa, which enables the eyelids to move over the surface of the eyeball as they open and close, spreading the moistening and lubricating film of lacrimal fluid within the sac. ♦ The fluid is secreted into the lateral superior fornix of the sac and is spread by gravity and blinking across the anterior eyeball, cleansing and providing the cornea with nutrients and oxygen as it is pushed toward the medial angle of the eye. ♦ The fluid and contained irritants accumulate in the lacrimal lake. ♦ They are drained from here by capillary action through superior and inferior lacrimal puncta into lacrimal canaliculi that pass to the lacrimal sac. ♦ The sac drains via the nasolacrimal duct into the nasal cavity, where the fluid flows posteriorly and is eventually swallowed. ♦ Although the conjunctival sac opens anteriorly via the palpebral fissure, the watery lacrimal fluid will not cross the lipid barrier secreted by the tarsal glands onto the margins of the fissure, unless it is produced in excess, as when crying.

Eyeball: The eyeball contains the optical apparatus of the visual system. ♦ It has a trilaminar construction, with (1) a supporting outer fibrous layer, consisting of the opaque sclera and transparent anterior cornea; (2) a middle vascular layer, consisting of the choroid (largely concerned with providing nourishment to the cones and rods of the retina), the ciliary body (producer of the aqueous humor and adjuster of the lens), and the iris (protector of the retina); and (3) an inner layer, consisting of optic and non-visual parts of the retina. ♦ The cornea is the major refractive component of the eyeball, with focusing adjustments made by the lens. ♦ Parasympathetic stimulation of the ciliary body reduces tension on the lens, allowing it to thicken for near vision. ♦ Relaxation of the ciliary body in the absence of stimulation stretches the lens, making it thinner for far vision. ♦ Parasympathetic stimulation also constricts the sphincter of the iris, which closes the pupil in response to bright light. ♦ Sympathetic stimulation of the dilator of the iris opens the pupil to admit more light. ♦ The anterior segment of the eyeball is filled with aqueous humor, produced by the ciliary processes in the posterior chamber. ♦ The aqueous humor passes through the pupil into the anterior chamber and is absorbed into the venous circulation at the scleral venous sinus. ♦ The posterior segment or vitreous chamber is filled with vitreous humor, which maintains the shape of the eye, transmits light, and holds the retina in place against the choroid.

Extra-ocular muscles: There are seven extra-ocular muscles: four recti, two obliques, and a levator of the superior eyelid. ♦ Six muscles originate from the apex of the orbit, and the four rectus muscles arise from a common tendinous ring. ♦ Only the inferior oblique arises anteriorly in the orbit. The levator palpebrae superioris elevates the superior eyelid. ♦ Associated smooth muscle (superior tarsal muscle) widens the palpebral fissure even more during sympathetic responses; ptosis results from the absence of sympathetic innervation to the head (Horner syndrome). ♦ When the eyes are adducted (converged) as for close reading, the superior and inferior obliques produce depression and elevation, respectively, directing the gaze down or up the page. ♦ Coordination of the contralateral extra-ocular muscles as yoke muscles is necessary to direct the gaze in a particular direction.

Nerves of orbit: All muscles of the orbit are supplied by CN III, except for the superior oblique and lateral rectus, which are supplied by CN IV and VI, respectively. ♦ Memory device: $LR_6SO_4AO_3$.

Vasculature of orbit: Extra-ocular circulation is provided mainly by the ophthalmic (internal carotid) and infra-orbital (external carotid) arteries, the latter supplying structures near the orbital floor. ♦ Superior and inferior ophthalmic veins drain anteriorly to the facial vein, posteriorly to the cavernous sinus, and inferiorly to the pterygoid venous plexus. ♦ Intraocular circulation is exclusively from the ophthalmic artery, with the central retinal artery supplying all of the retina except the layer of cones and rods, which is nourished by the capillary lamina of the choroid. ♦ The ciliary-iridial structures receive blood from anterior ciliary arteries (from the rectus muscle branches of the ophthalmic artery) and two long posterior ciliary arteries. ♦ Multiple short posterior ciliary arteries supply the choroid. ♦ Superior and inferior vorticose veins drain the eyeballs to the respective ophthalmic veins.

PAROTID AND TEMPORAL REGIONS, INFRATEMPORAL FOSSA, AND TEMPOROMANDIBULAR JOINT

Parotid Region

The **parotid region** is the posterolateral part of the facial region (Fig. 7.23A), bounded by the:

- Zygomatic arch superiorly.
- External ear and anterior border of the sternocleidomastoid posteriorly.
- Ramus of the mandible medially.
- Anterior border of the masseter muscle anteriorly.
- Angle and inferior border of the mandible inferiorly.

The parotid region includes the parotid gland and duct, the parotid plexus of the facial nerve (CN VII), the retromandibular vein, the external carotid artery, and the masseter muscle.

PAROTID GLAND

The **parotid gland** is the largest of three paired salivary glands. From a functional viewpoint, it would seem logical to discuss all three glands simultaneously in association with the anatomy of the mouth. However, from an anatomical viewpoint, particularly in dissection courses, the parotid gland is usually examined with or immediately subsequent to the dissection of the face to expose the facial nerve. Although the parotid plexus of the facial nerve (CN VII) is embedded within the parotid gland, the branches extending from the gland to innervate the muscles of facial expression are encountered during the dissection of the face, and were discussed and illustrated on p. 853. Dissection of the parotid region must be completed before dissection of the infratemporal region and muscles of mastication or the carotid triangle of the neck. The *submandibular gland* is encountered primarily during dissection of the submandibular triangle of the neck, and the *sublingual glands* when dissecting the floor of the mouth.

The parotid gland is enclosed within a tough, unyielding, fascial capsule, the **parotid sheath** (capsule), derived from the *investing layer of deep cervical fascia* (Figs. 7.65, 8.4 and 8.16). The parotid gland has an irregular shape because the area occupied by the gland, the **parotid bed,** is antero-inferior to the external acoustic meatus, where it is wedged between the ramus of the mandible and the mastoid process (Figs. 7.23A & C and 7.65). Fatty tissue between the lobes

of the gland confers the flexibility the gland must have to accommodate the motion of the mandible. The apex of the parotid gland is posterior to the angle of the mandible, and its base is related to the zygomatic arch. The subcutaneous lateral surface of the parotid gland is almost flat.

The **parotid duct** passes horizontally from the anterior edge of the gland (Fig. 7.65). At the anterior border of the masseter, the duct turns medially, pierces the buccinator, and enters the oral cavity through a small orifice opposite the 2nd maxillary molar tooth. Embedded within the substance of the parotid gland, from superficial to deep, are the *parotid plexus of the facial nerve* (CN VII) and its branches (Figs. 7.23A & C and 7.65), the *retromandibular vein,* and the *external carotid artery.* On the parotid sheath and within the gland are *parotid lymph nodes.*

INNERVATION OF PAROTID GLAND AND RELATED STRUCTURES

Although the parotid plexus of CN VII is embedded within it, CN VII does not provide innervation to the gland. The *auriculotemporal nerve,* a branch of CN V_3, is closely related to the parotid gland and passes superior to it with the

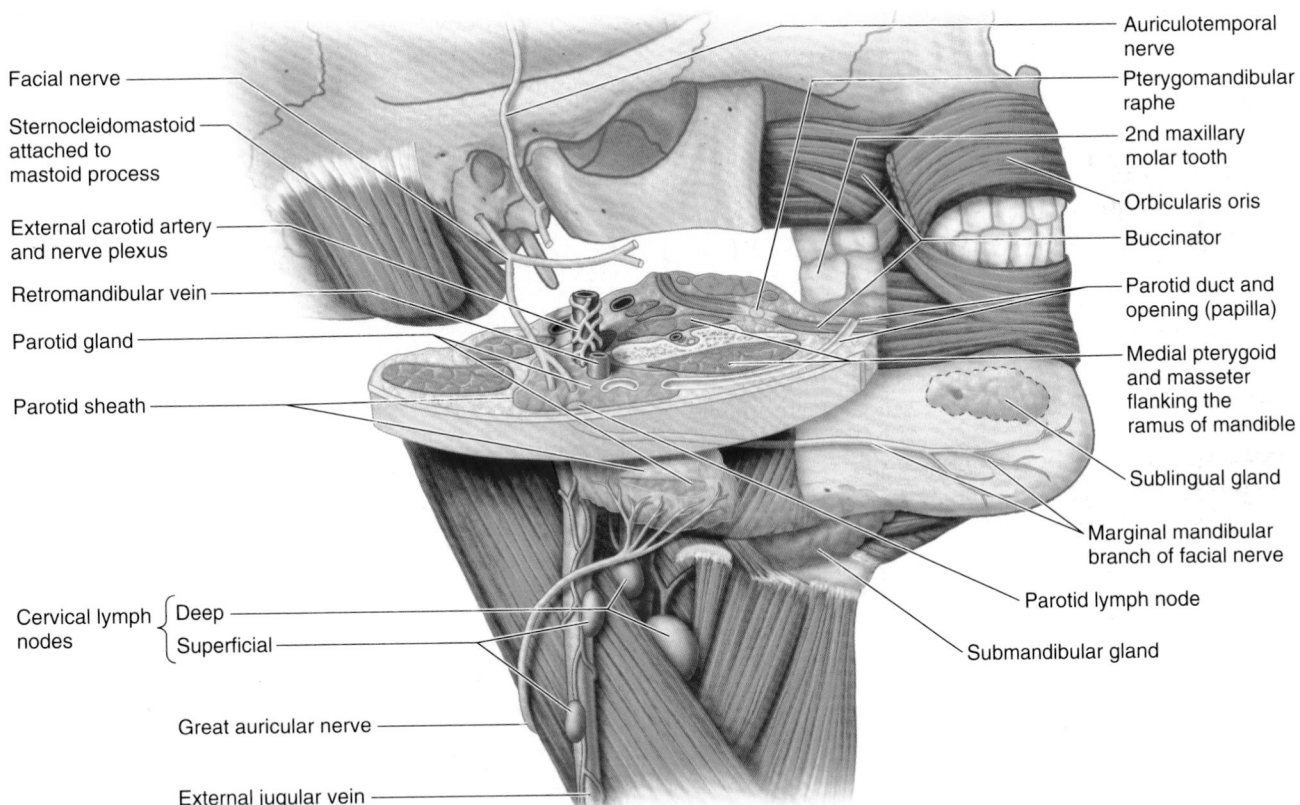

FIGURE 7.65. Relationships of parotid gland. A transverse slice through the bed of the parotid gland demonstrates the relationship of the gland to the surrounding structures. The gland passes deeply between the ramus of the mandible, flanked by the muscles of mastication anteriorly and the mastoid process and sternocleidomastoid muscle posteriorly. The dimensions of the parotid bed change with movements of the mandible. The external carotid artery and peri-arterial plexus, retromandibular vein, and parotid plexus of the facial nerve (CN VII) are embedded within the gland itself. The parotid duct turns medially at the anterior border of the masseter muscle and pierces the buccinator muscle.

superficial temporal vessels. The auriculotemporal nerve and the **great auricular nerve,** a branch of the cervical plexus composed of fibers from C2 and C3 spinal nerves, innervates the parotid sheath (Fig. 7.65) as well as the overlying skin.

The parasympathetic component of the **glossopharyngeal nerve** (CN IX) supplies presynpatic secretory fibers to the **otic ganglion** (Fig. 7.66). The postsynaptic parasympathetic fibers are conveyed from the ganglion to the gland by the auriculotemporal nerve. Stimulation of the parasympathetic fibers produces a thin, watery saliva. Sympathetic fibers are derived from the cervical ganglia through the **external carotid nerve plexus** on the external carotid artery (Fig. 7.65). The vasomotor activity of these fibers may reduce secretion from the gland. Sensory nerve fibers pass to the gland through the great auricular and auriculotemporal nerves.

Temporal Region

The **temporal region** of the head includes the lateral area of the scalp and the deeper soft tissues overlying the temporal fossa of the cranium, superior to the zygomatic arch (Figs. 7.14 and 7.67A, *inset*). The **temporal fossa,** occupied primarily by the upper portion of the *temporalis muscle,* is bounded (Figs. 7.1A and 7.67A):

- Posteriorly and superiorly by the temporal lines.
- Anteriorly by the frontal and zygomatic bones.
- Laterally by the zygomatic arch.
- Inferiorly by the infratemporal crest (Fig. 7.67B).

The *floor of the temporal fossa* is formed by parts of the four bones that form the *pterion:* frontal, parietal, temporal, and greater wing of the sphenoid. The fan-shaped *temporalis muscle* arises from the bony floor and overlying **temporal fascia** (Fig. 7.68), which forms the *roof of the temporal fossa.* This tough fascia covers the temporalis, attaching superiorly to the *superior temporal line.* Inferiorly, the fascia splits into two layers, which attach to the lateral and medial surfaces of the zygomatic arch. The temporal fascia also tethers the zygomatic arch superiorly. When the powerful masseter muscle,

which is attached to the inferior border of the arch, contracts and exerts a strong downward pull on the zygomatic arch, the temporal fascia provides resistance.

Infratemporal Fossa

The **infratemporal fossa** is an irregularly shaped space deep and inferior to the zygomatic arch, deep to the ramus of the mandible, and posterior to the maxilla (Fig. 7.67B). It communicates with the temporal fossa through the interval between (deep to) the zygomatic arch and (superficial to) the cranial bones.

The *boundaries of the infratemporal fossa* are as follows (Fig. 7.67):

- Laterally: the ramus of the mandible.
- Medially: the lateral pterygoid plate.
- Anteriorly: the posterior aspect of the maxilla.
- Posteriorly: the tympanic plate and the mastoid and styloid processes of the temporal bone.
- Superiorly: the inferior (infratemporal) surface of the greater wing of the sphenoid.
- Inferiorly: where the medial pterygoid muscle attaches to the mandible near its angle (see Fig. 7.72D).

The *infratemporal fossa contains* the (Figs. 7.68-7.70):

- Inferior part of the temporalis muscle.
- Lateral and medial pterygoid muscles.
- Maxillary artery.
- Pterygoid venous plexus.
- Mandibular, inferior alveolar, lingual, buccal, and chorda tympani nerves
- Otic ganglion (see Fig. 7.75).

The parotid and temporal regions and the infratemporal fossa collectively include the *temporomandibular joint* and the *muscles of mastication* that produce its movements.

TEMPOROMANDIBULAR JOINT

The **temporomandibular joint** (**TMJ**) is a modified hinge type of synovial joint, permitting gliding (translation) and a small degree of rotation (pivoting) in addition to flexion (elevation) and extension (depression) movements typical for hinge joints. The bony articular surfaces involved are the *mandibular fossa* and *articular tubercle of the temporal bone* superiorly, and the *head of the mandible* inferiorly (Figs. 7.9B and 7.69A–D). The loose *fibrous layer of the joint capsule* attaches to the margins of the articular cartilage on the temporal bone and around the neck of the mandible (Figs. 7.69E and 7.70A & C). The two bony articular surfaces are completely separated by intervening fibrocartilage, the **articular disc of the TMJ,** attached at its periphery to the internal aspect of the fibrous capsule. This creates separate **superior** and **inferior articular cavities,** or compartments, lined by separate **superior** and **inferior synovial membranes** (Figs. 7.69A & B and 7.70B & C).

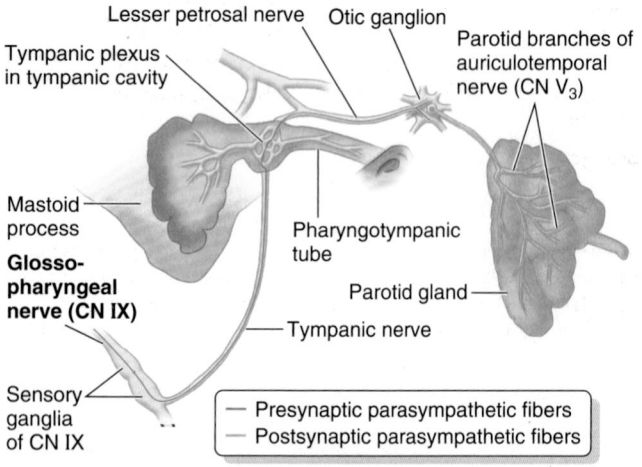

FIGURE 7.66. Innervation of parotid gland.

Lesser petrosal nerve   Otic ganglion

Tympanic plexus in tympanic cavity

Parotid branches of auriculotemporal nerve (CN V₃)

Mastoid process

Glossopharyngeal nerve (CN IX)

Pharyngotympanic tube

Parotid gland

Tympanic nerve

Sensory ganglia of CN IX

— Presynaptic parasympathetic fibers
— Postsynaptic parasympathetic fibers

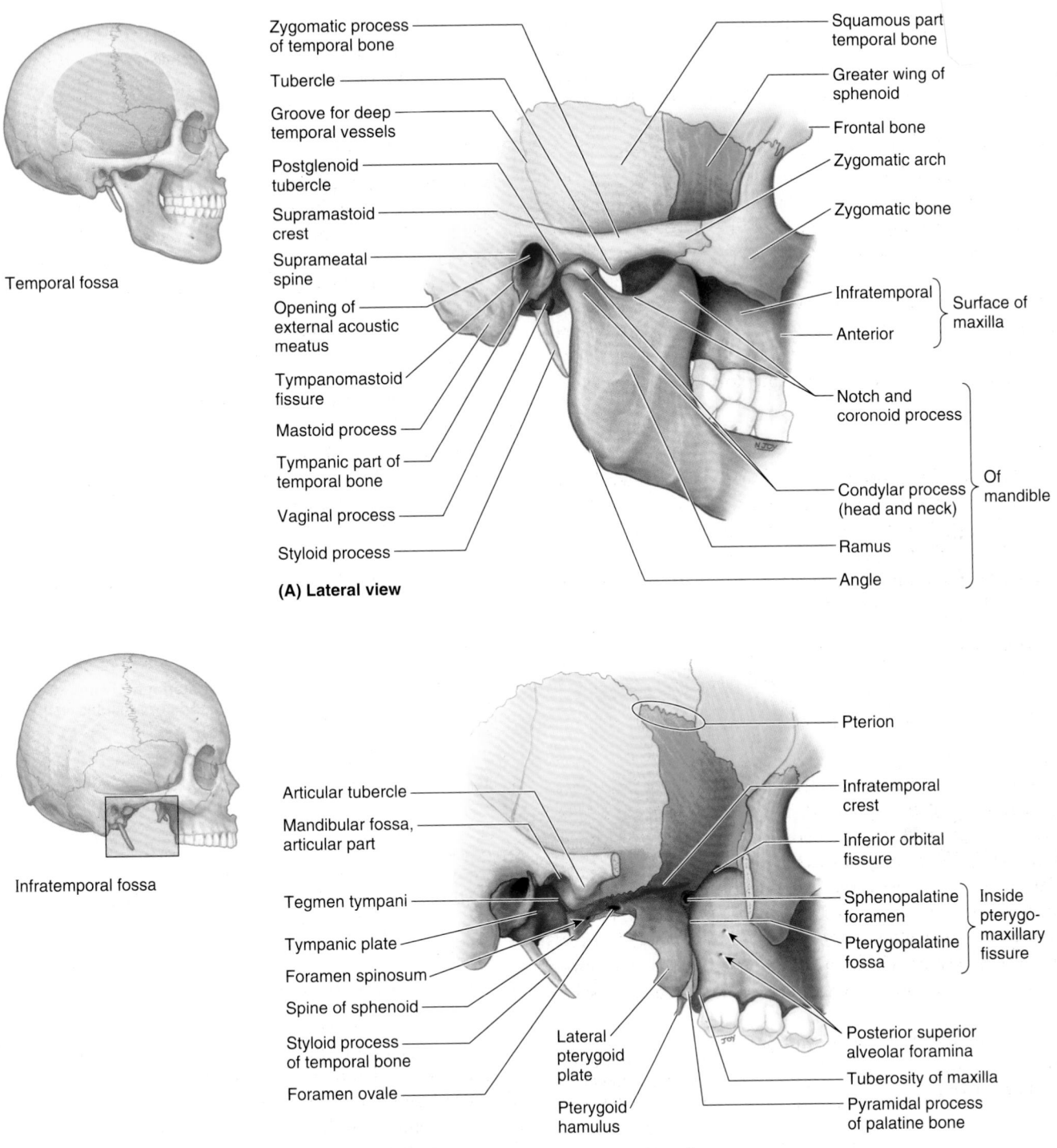

Temporal fossa

Zygomatic process
of temporal bone

Tubercle

Groove for deep
temporal vessels

Postglenoid
tubercle

Supramastoid
crest

Suprameatal
spine

Opening of
external acoustic
meatus

Tympanomastoid
fissure

Mastoid process

Tympanic part of
temporal bone

Vaginal process

Styloid process

Squamous part
temporal bone

Greater wing of
sphenoid

Frontal bone

Zygomatic arch

Zygomatic bone

Infratemporal ⎱ Surface of
                   ⎰ maxilla
Anterior

Notch and
coronoid process

Condylar process
(head and neck) ⎰ Of
                     ⎱ mandible
Ramus

Angle

(A) Lateral view

Infratemporal fossa

Articular tubercle

Mandibular fossa,
articular part

Tegmen tympani

Tympanic plate

Foramen spinosum

Spine of sphenoid

Styloid process
of temporal bone

Foramen ovale

Lateral
pterygoid
plate

Pterygoid
hamulus

Pterion

Infratemporal
crest

Inferior orbital
fissure

Sphenopalatine ⎱ Inside
foramen   ⎰ pterygo-
                     maxillary
Pterygopalatine   fissure
fossa

Posterior superior
alveolar foramina

Tuberosity of maxilla

Pyramidal process
of palatine bone

**(B) Lateral view following removal of zygomatic arch
and ramus of mandible**

FIGURE 7.67. Bony boundaries of temporal and infratemporal fossae. A. The lateral wall of the infratemporal fossa is formed by the ramus of the mandible. The space is deep to the zygomatic arch and is traversed by the temporal muscle and the deep temporal nerves and vessels. Through this interval, the temporal fossa communicates inferiorly with the infratemporal fossa. **B.** The roof and three walls of the infratemporal fossa are shown. The fossa is an irregularly shaped space posterior to the maxilla (anterior wall). The roof of the fossa is formed by the infratemporal surface of the greater wing of the sphenoid. The medial wall is formed by the lateral pterygoid plate; and the posterior wall is formed by the tympanic plate, styloid process, and mastoid process of the temporal bone. The infratemporal fossa communicates with the pterygopalatine fossa through the pterygomaxillary fissure.

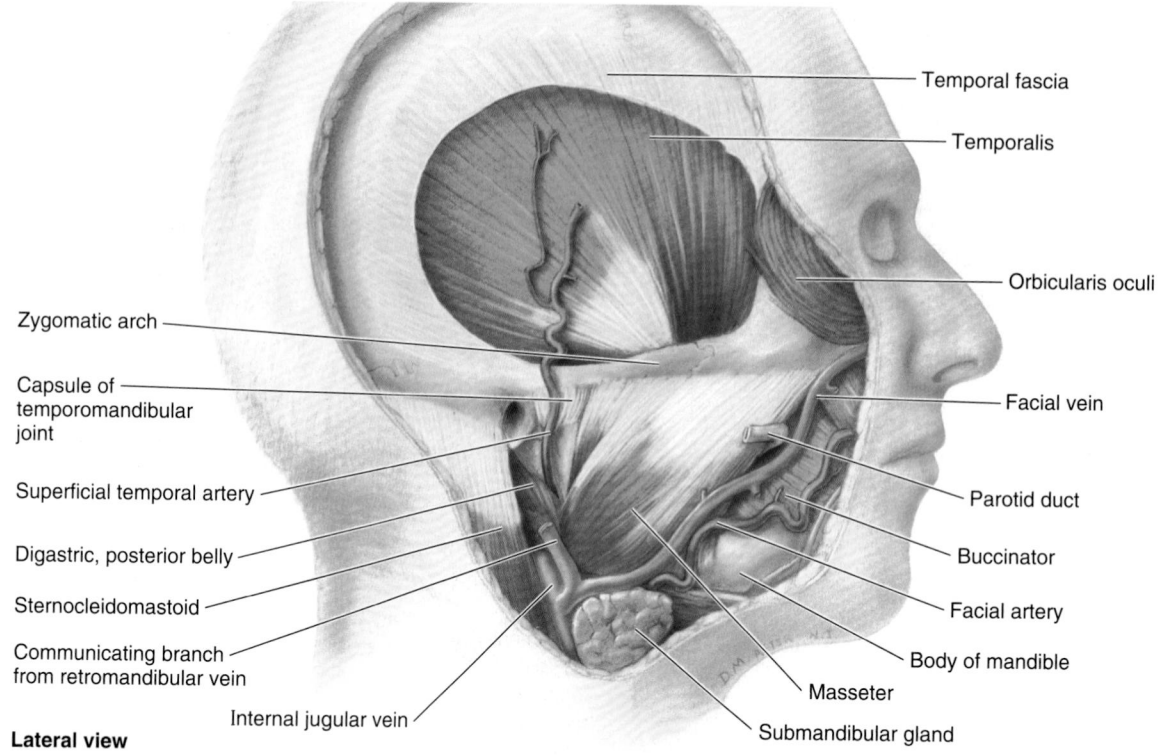

Temporal fascia

Temporalis

Orbicularis oculi

Zygomatic arch

Capsule of temporomandibular joint

Superficial temporal artery

Digastric, posterior belly

Sternocleidomastoid

Communicating branch from retromandibular vein

Internal jugular vein

Lateral view

Facial vein

Parotid duct

Buccinator

Facial artery

Body of mandible

Masseter

Submandibular gland

FIGURE 7.68. Dissections of temporal and infratemporal regions. In this superficial dissection of the great muscles on the side of the cranium, the parotid gland and most of the temporal fascia have been removed. The temporal and masseter muscles are both supplied by the trigeminal nerve (CN V) and both close the jaw. The facial artery passes deep to the submandibular gland, whereas the facial vein passes superficial to it.

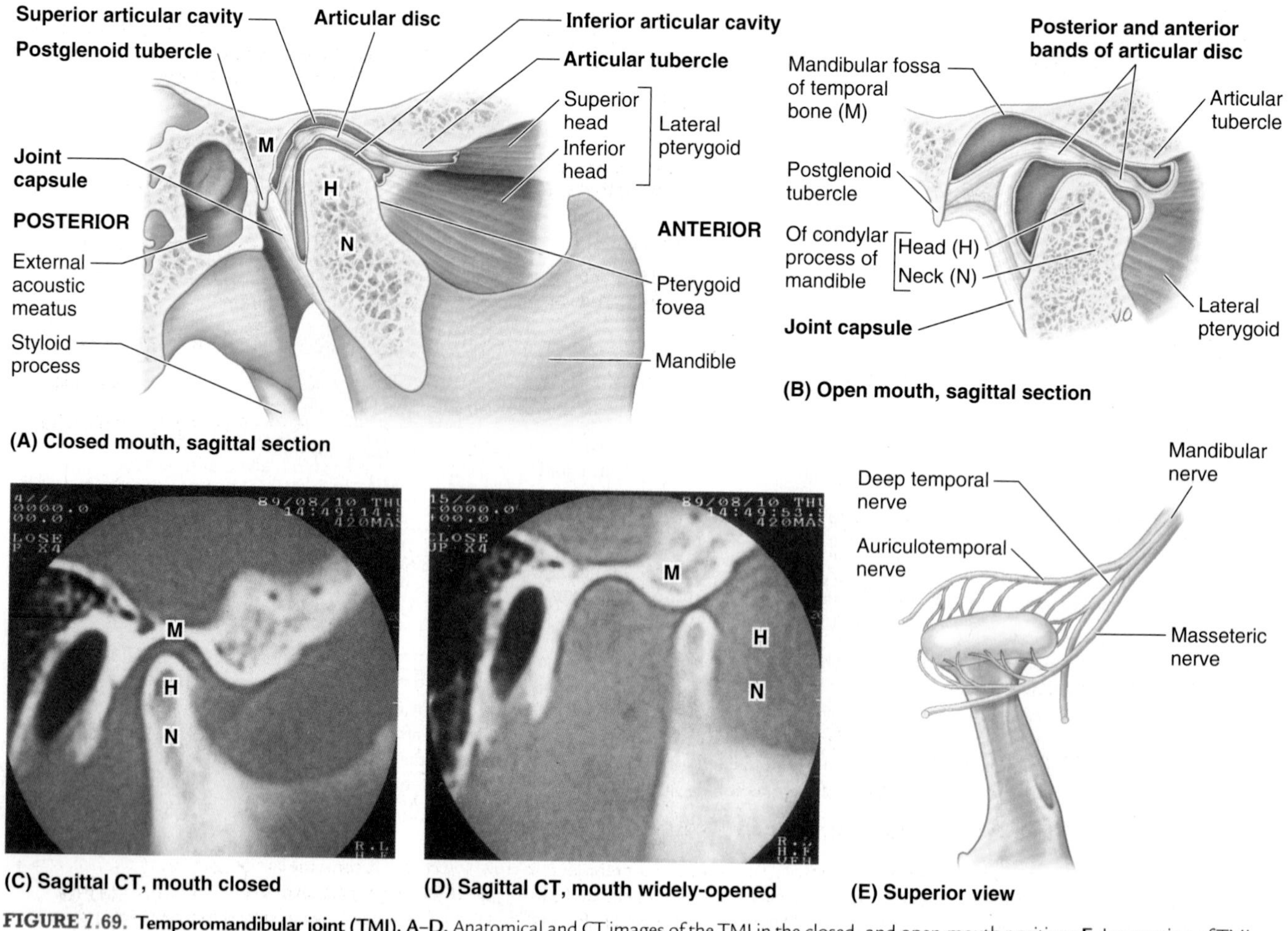

Superior articular cavity

Postglenoid tubercle

Articular disc

Inferior articular cavity

Articular tubercle

Superior head
Inferior head
} Lateral pterygoid

Joint capsule

POSTERIOR

External acoustic meatus

Styloid process

ANTERIOR

Pterygoid fovea

Mandible

(A) Closed mouth, sagittal section

Posterior and anterior bands of articular disc

Mandibular fossa of temporal bone (M)

Articular tubercle

Postglenoid tubercle

Of condylar process of mandible
{ Head (H)
Neck (N) }

Joint capsule

Lateral pterygoid

(B) Open mouth, sagittal section

(C) Sagittal CT, mouth closed

(D) Sagittal CT, mouth widely-opened

Deep temporal nerve

Auriculotemporal nerve

Mandibular nerve

Masseteric nerve

(E) Superior view

FIGURE 7.69. Temporomandibular joint (TMJ). A–D. Anatomical and CT images of the TMJ in the closed- and open-mouth positions **E.** Innervation of TMJ.

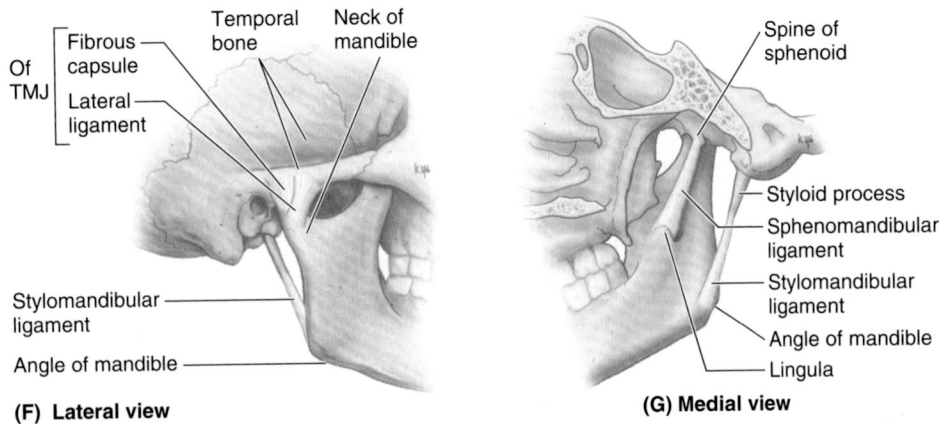

(F) Lateral view

(G) Medial view

In (F) Lateral view:
- Of TMJ: Fibrous capsule, Lateral ligament
- Temporal bone
- Neck of mandible
- Stylomandibular ligament
- Angle of mandible

In (G) Medial view:
- Spine of sphenoid
- Styloid process
- Sphenomandibular ligament
- Stylomandibular ligament
- Angle of mandible
- Lingula

FIGURE 7.69. *(Continued)* **F and G.** The TMJ and the extrinsic stylomandibular and sphenomandibular ligaments are shown. The sphenomandibular ligament passively bears the weight of the lower jaw and is the "swinging hinge" of the mandible, permitting protrusion and retrusion as well as elevation and depression.

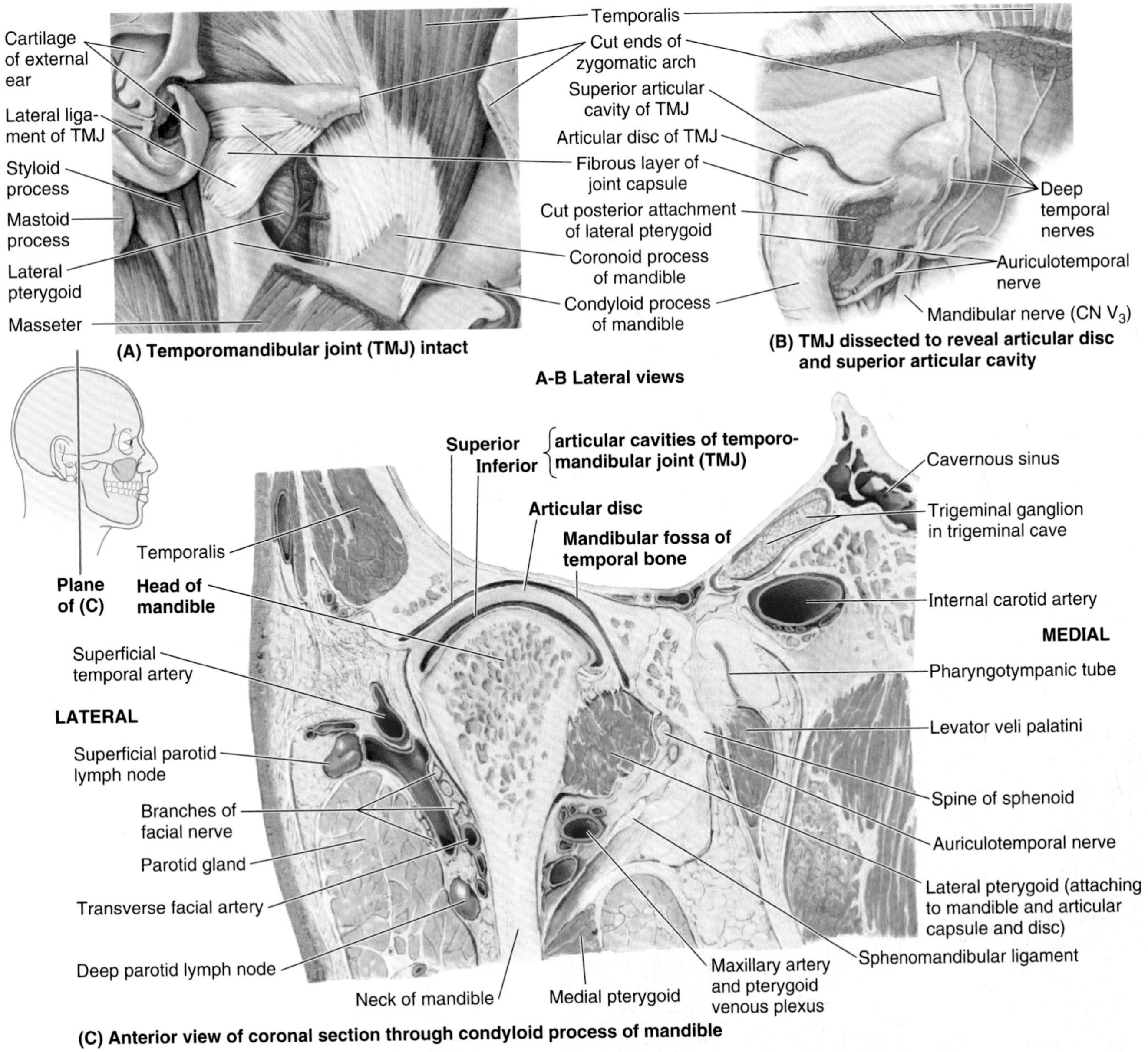

(A) Temporomandibular joint (TMJ) intact

Labels: Cartilage of external ear, Lateral ligament of TMJ, Styloid process, Mastoid process, Lateral pterygoid, Masseter, Temporalis, Coronoid process of mandible, Condyloid process of mandible

(B) TMJ dissected to reveal articular disc and superior articular cavity

Labels: Temporalis, Cut ends of zygomatic arch, Superior articular cavity of TMJ, Articular disc of TMJ, Fibrous layer of joint capsule, Cut posterior attachment of lateral pterygoid, Deep temporal nerves, Auriculotemporal nerve, Mandibular nerve (CN V₃)

A-B Lateral views

(C) Anterior view of coronal section through condyloid process of mandible

Labels: Temporalis, Plane of (C), Head of mandible, Superficial temporal artery, LATERAL, Superficial parotid lymph node, Branches of facial nerve, Parotid gland, Transverse facial artery, Deep parotid lymph node, Neck of mandible, Medial pterygoid, Superior / Inferior articular cavities of temporomandibular joint (TMJ), Articular disc, Mandibular fossa of temporal bone, Cavernous sinus, Trigeminal ganglion in trigeminal cave, Internal carotid artery, MEDIAL, Pharyngotympanic tube, Levator veli palatini, Spine of sphenoid, Auriculotemporal nerve, Lateral pterygoid (attaching to mandible and articular capsule and disc), Sphenomandibular ligament, Maxillary artery and pterygoid venous plexus

FIGURE 7.70. Dissections and coronal sections of TMJ. A. The fibrous layer of the joint capsule is thickened to form the lateral ligament of the TMJ, which with the postglenoid tubercle, prevents excessive posterior displacement of the head of the mandible. **B.** The upper portion of the fibrous capsule has been removed, demonstrating the superior compartment of the TMJ between the mandibular fossa and the articular disc. The auriculotemporal nerve provides articular branches to the joint. **C.** Coronal section of the right TMJ demonstrating the articular disc dividing the joint cavity into superior and inferior compartments.

The gliding movements of protrusion and retrusion (translation) occur between the temporal bone and the articular disc (superior cavity) (Fig. 7.71); the hinge movements of depression and elevation and the rotational or pivoting movements occur in the inferior compartment. A thickened part of the joint capsule forms the intrinsic **lateral ligament of the TMJ** (Figs. 7.69E and 7.70A), which strengthens the joint laterally and, with the **postglenoid tubercle** (Fig. 7.69A), acts to prevent posterior dislocation of the joint.

Two extrinsic ligaments and the lateral ligament connect the mandible to the cranium. The **stylomandibular ligament,** which is actually a thickening of the fibrous capsule of the parotid gland, runs from the styloid process to the angle of the mandible (Fig. 7.69E & F). It does not contribute significantly to the strength of the joint. The **sphenomandibular ligament** runs from the spine of the sphenoid to the lingula of the mandible (Figs. 7.69F and 7.70C). It is the primary passive support of the mandible, although the tonus of the muscles of mastication usually bears the mandible's weight. However, the sphenomandibular ligaments serve as a "swinging hinge" for the mandible, serving both as a fulcrum and as a check ligament for the movements of the mandible at the TMJs.

The movements of the mandible at the TMJs are shown in Figure 7.71, and the muscles (or forces) producing the movements are summarized in Table 7.10. When the mouth is closed and at rest, the heads of the mandible are held in the retracted position in the mandibular fossae, and the chin is elevated by the tonus of the retractors and elevators of the mandible (Figs. 7.69A & C, 7.70B & C, and 7.71A). When sleeping in the supine or sitting position (head upright), as one enters a state of deep sleep, the tonic contraction relaxes and gravity causes depression of the mandible (the mouth falls open).

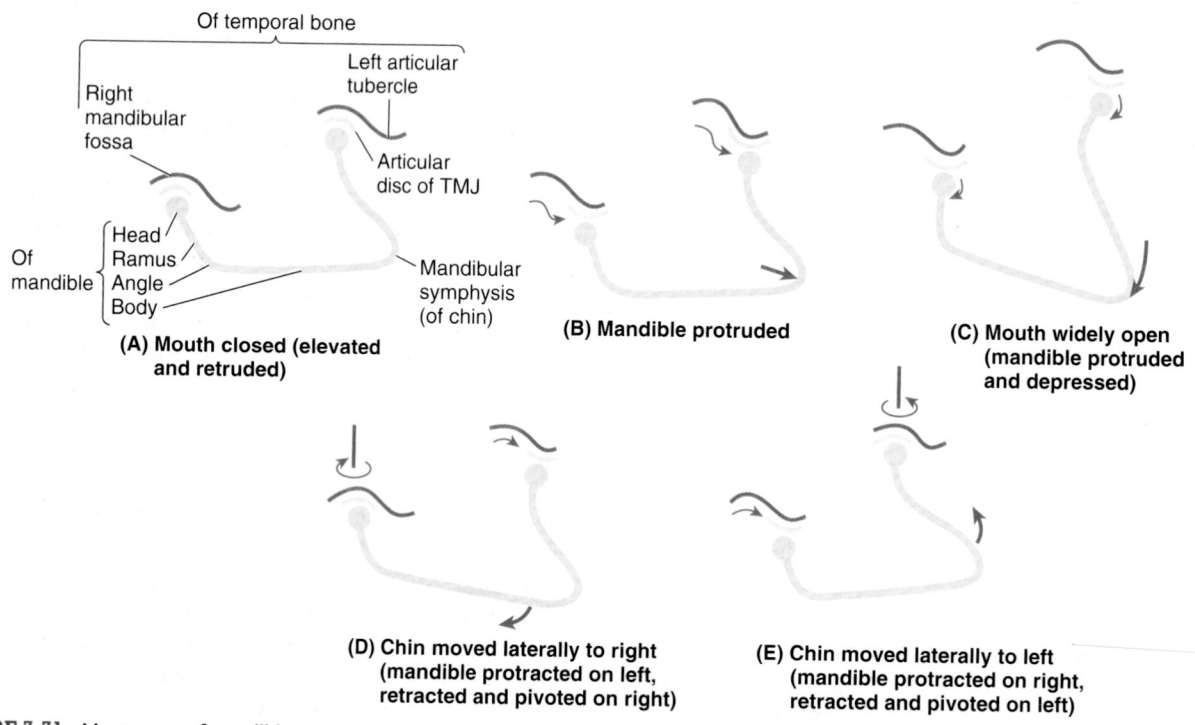

FIGURE 7.71. Movements of mandible consequent to movement at TMJs.

TABLE 7.10. MOVEMENTS OF TEMPOROMANDIBULAR JOINT

Movements of Mandible	Muscle(s)
Elevation (close mouth)	Temporalis, masseter, and medial pterygoid
Depression (open mouth)	Lateral pterygoid, suprahyoid, and infrahyoid muscles[a]
Protrusion (protrude chin)	Lateral pterygoid, masseter, and medial pterygoid[b]
Retrusion (retrude chin)	Temporalis (posterior oblique and near horizontal fibers) and masseter
Lateral movements (grinding and chewing)	Temporalis of same side, pterygoids of opposite side, and masseter

[a] The prime mover is normally gravity; these muscles are mainly active against resistance.

[b] The lateral pterygoid is the prime mover here, with minor secondary roles played by the masseter and medial pterygoid.

To enable more than a small amount of depression of the mandible—that is, to open the mouth wider than just to separate the upper and lower teeth—the head of the mandible and articular disc must move anteriorly on the articular surface until the head lies inferior to the articular tubercle (a movement referred to as "translation" by dentists) (Fig. 7.71B). If this occurs without depression, the chin protrudes. Most often, the mandible is depressed (the mouth is opened) as the head of the mandible and articular disc glide toward the articular tubercle, with full depression possible only when the heads and discs are fully protracted (Figs. 7.69B & D and 7.71C). If protraction of head and disc occurs unilaterally, the contralateral head rotates (pivots) on the inferior surface of the articular disc in the retracted position, permitting simple side-to-side chewing or grinding movements over a small range (Fig. 7.71D & E). During protrusion and retrusion of the mandible, the head and articular disc slide anteriorly and posteriorly on the articular surface of the temporal bone, with both sides moving together (Fig. 7.71A & B).

MUSCLES OF MASTICATION

TMJ movements are produced chiefly by the **muscles of mastication.** These four muscles (**temporal, masseter,** and **medial** and **lateral pterygoid muscles**) develop from the mesoderm of the embryonic first pharyngeal arch; consequently, they are all innervated by the nerve of that arch, the (*motor root of the*) *mandibular nerve* (CN V₃). The muscles of mastication are shown in isolation in Figure 7.72 and in situ in Figures 7.68 and 7.74; their attachments, details concerning their innervation, and their main actions are described in Table 7.11. In addition to the movements listed, studies indicate that the superior head of the lateral pterygoid muscle is active during the retraction movement produced by the posterior fibers of the temporalis. Traction is applied to the articular disc so that it is not pushed posteriorly ahead of the retracting mandible.

Generally, depression of the mandible is produced by gravity. The *suprahyoid* and *infrahyoid muscles* are strap-like muscles on each side of the neck (Fig. 7.72E; Table 7.11). They are primarily used to raise and depress the hyoid bone and larynx, respectively—for example, during swallowing (see Chapter 8). Indirectly they can also help depress the mandible, especially when opening the mouth suddenly, against resistance, or when inverted (e.g., standing on one's head). The platysma can be similarly used.

NEUROVASCULATURE OF INFRATEMPORAL FOSSA

The **maxillary artery** is the larger of the two terminal branches of the external carotid artery. It arises posterior to the neck of the mandible and is divided into three parts based on its relation to the lateral pterygoid muscle. The three parts of the maxillary artery and their branches are illustrated in isolation in Figure 7.73, and their courses and distributions are listed in Table 7.12. Relationships of the maxillary artery and many of its branches are shown in Figure 7.74.

The **pterygoid venous plexus** is located partly between the temporalis and pterygoid muscles (Fig. 7.25). It is the venous equivalent of most of the maxillary artery—that is, most of the veins that accompany the branches of the maxillary artery drain into this plexus. The plexus anastomoses anteriorly with the facial vein via the deep facial vein and superiorly with the cavernous sinus via emissary veins. The extensive nature and volume of the pterygoid venous plexus is difficult to appreciate in the cadaver, in which it is usually drained of blood.

The **mandibular nerve** arises from the trigeminal ganglion in the middle cranial fossa. It immediately receives the motor root of the trigeminal nerve and descends through the foramen ovale into the infratemporal fossa (Fig. 7.75). The branches of CN V₃ are the auriculotemporal, inferior alveolar, lingual, and buccal nerves. Branches of CN V₃ also supply the four muscles of mastication but not the buccinator, which is supplied by the facial nerve.

The **auriculotemporal nerve** encircles the middle meningeal artery and divides into numerous branches, the largest of which passes posteriorly, medial to the neck of the mandible, and supplies sensory fibers to the auricle and temporal region. The auriculotemporal nerve also sends articular (sensory) fibers to the TMJ (Fig. 7.69E). It conveys postsynaptic parasympathetic secretomotor fibers from the *otic ganglion* to the parotid gland.

The **inferior alveolar nerve** enters the mandibular foramen and passes through the mandibular canal, forming the *inferior dental plexus,* which sends branches to all mandibular teeth on its side. Another branch of the plexus, the *mental nerve,* passes through the mental foramen and supplies the skin and mucous membrane of the lower lip, the skin of the chin, and the vestibular gingiva of the mandibular incisor teeth.

The **lingual nerve** lies anterior to the inferior alveolar nerve (Fig. 7.74). It is sensory to the anterior two thirds of the tongue, the floor of the mouth, and the lingual gingivae. It enters the mouth between the medial pterygoid muscle and the ramus of the mandible and passes anteriorly under cover of the oral mucosa, medial and inferior to the 3rd molar tooth. The **chorda tympani nerve,** a branch of CN VII carrying taste fibers from the anterior two thirds of the tongue, joins the lingual nerve in the infratemporal fossa (Fig. 7.74B). The chorda tympani also carries secretomotor fibers for the submandibular and sublingual salivary glands.

The **otic ganglion** (parasympathetic) is located in the infratemporal fossa, just inferior to the foramen ovale, medial to CN V₃ and posterior to the medial pterygoid muscle (Fig. 7.75). Presynaptic parasympathetic fibers, derived mainly from the glossopharyngeal nerve, synapse in the otic ganglion (Fig. 7.66). Postsynaptic parasympathetic fibers, which are secretory to the parotid gland, pass from the otic ganglion to this gland through the auriculotemporal nerve.

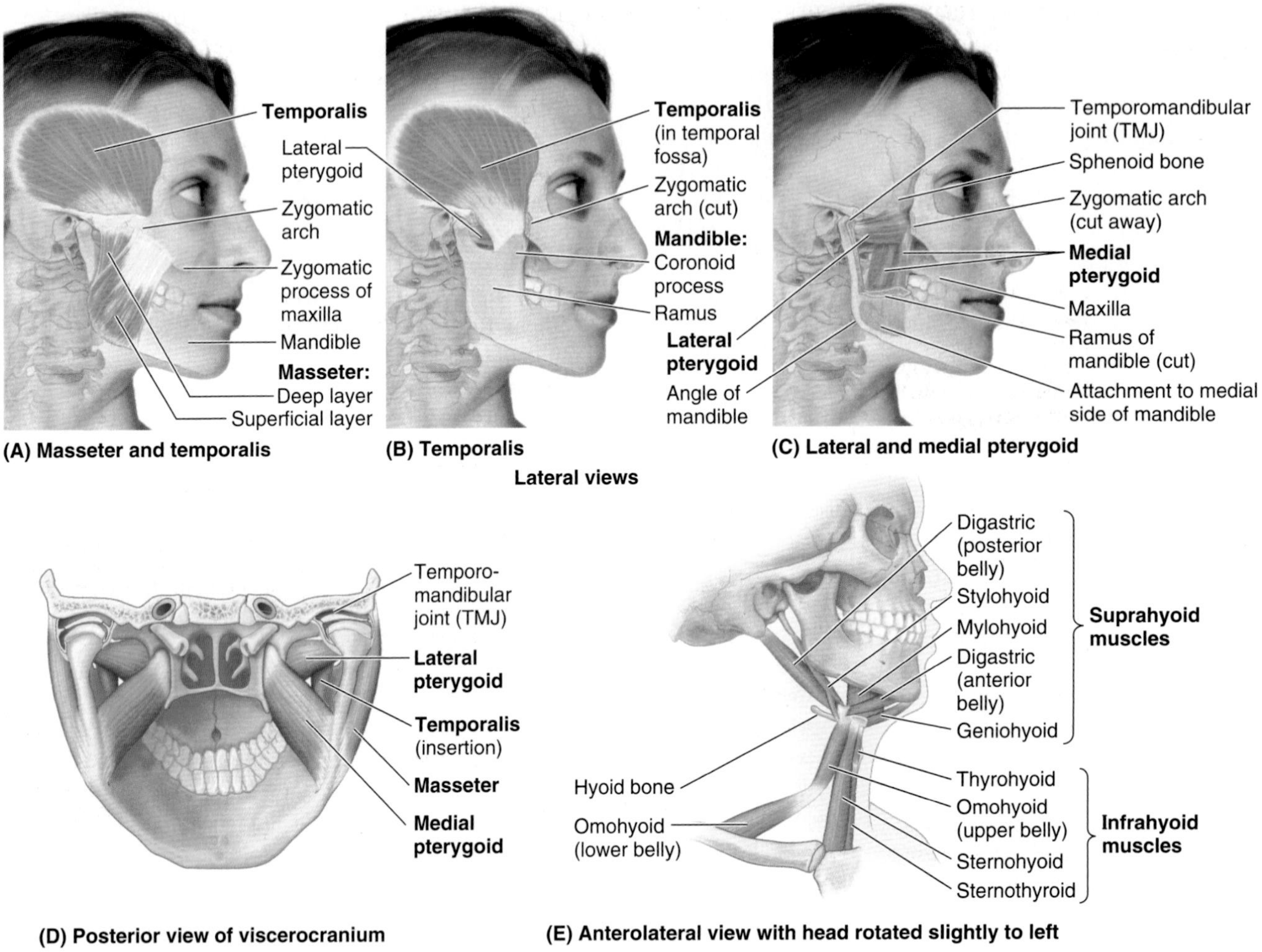

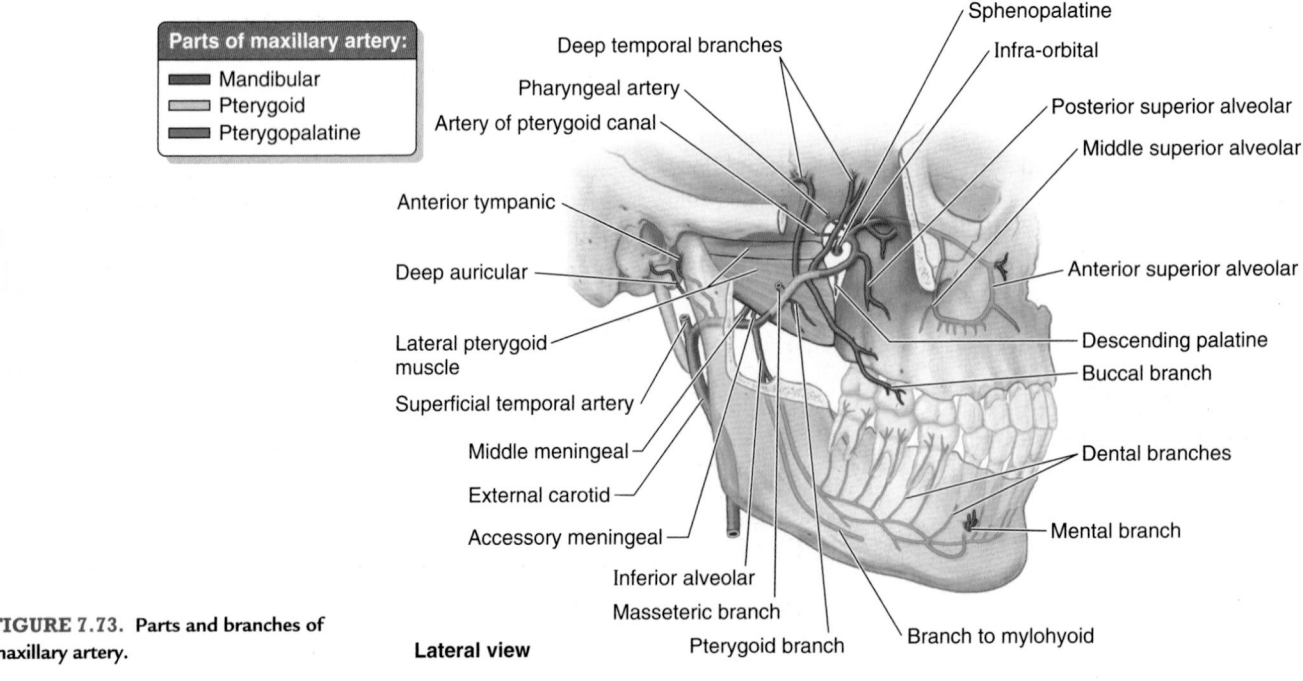

(A) Masseter and temporalis

Temporalis
Lateral pterygoid
Zygomatic arch
Zygomatic process of maxilla
Mandible
Masseter:
Deep layer
Superficial layer

(B) Temporalis

Temporalis (in temporal fossa)
Zygomatic arch (cut)
Mandible:
Coronoid process
Ramus
Lateral pterygoid
Angle of mandible

(C) Lateral and medial pterygoid

Temporomandibular joint (TMJ)
Sphenoid bone
Zygomatic arch (cut away)
Medial pterygoid
Maxilla
Ramus of mandible (cut)
Attachment to medial side of mandible

Lateral views

(D) Posterior view of viscerocranium

Temporo-mandibular joint (TMJ)
Lateral pterygoid
Temporalis (insertion)
Masseter
Medial pterygoid

(E) Anterolateral view with head rotated slightly to left

Digastric (posterior belly)
Stylohyoid
Mylohyoid
Digastric (anterior belly)
Geniohyoid
Suprahyoid muscles
Hyoid bone
Omohyoid (lower belly)
Thyrohyoid
Omohyoid (upper belly)
Sternohyoid
Sternothyroid
Infrahyoid muscles

FIGURE 7.72. Muscles acting on mandible/TMJ.

Parts of maxillary artery:
Mandibular
Pterygoid
Pterygopalatine

Deep temporal branches
Pharyngeal artery
Artery of pterygoid canal
Anterior tympanic
Deep auricular
Lateral pterygoid muscle
Superficial temporal artery
Middle meningeal
External carotid
Accessory meningeal
Inferior alveolar
Masseteric branch
Pterygoid branch

Sphenopalatine
Infra-orbital
Posterior superior alveolar
Middle superior alveolar
Anterior superior alveolar
Descending palatine
Buccal branch
Dental branches
Mental branch
Branch to mylohyoid

FIGURE 7.73. Parts and branches of maxillary artery.

Lateral view

TABLE 7.11. MUSCLES ACTING ON MANDIBLE/TEMPOROMANDIBULAR JOINT

Muscle(s)	Proximal Attachment	Distal Attachment	Innervation		Action on Mandible
Muscles of mastication:					
Temporalis	Triangular muscle with broad attachment to floor of temporal fossa and deep surface of temporal fascia	Narrow attachment to tip and medial surface of coronoid process and anterior border of ramus of mandible	Anterior trunk of mandibular nerve (CN V₃)	Via deep temporal branches	Elevates mandible, closing jaws; posterior, more horizontal fibers are 1° retractors of mandible
Masseter	Quadrate muscle attaching to inferior border and medial surface of maxillary process of zygomatic bone and the zygomatic arch	Angle and lateral surface of ramus of mandible		Via masseteric nerve	Elevates mandible, closing jaws; superficial fibers make limited contribution to protrusion of mandible
Lateral pterygoid	Triangular two-headed muscle from (1) infratemporal surface and crest of greater wing of sphenoid and (2) lateral surface of lateral pterygoid plate	Superior head attaches primarily to joint capsule and articular disc of TMJ; inferior head attaches primarily to pterygoid fovea on anteromedial aspect of neck of condyloid process of mandible		Via lateral pterygoid nerve	Acting bilaterally, protracts mandible and depresses chin; acting unilaterally, swings jaw toward contralateral side; alternate unilateral contraction produces larger lateral chewing movements
Medial pterygoid	Quadrangular two-headed muscle from (1) medial surface of lateral pterygoid plate and pyramidal process of palatine bone and (2) tuberosity of maxilla	Medial surface of ramus of mandible, inferior to mandibular foramen; in essence, a "mirror image" of ipsilateral masseter, two muscles flanking ramus	Anterior trunk of mandibular nerve (CN V₃)	Via medial pterygoid nerve	Acts synergistically with masseter to elevate mandible; contributes to protrusion; alternate unilateral activity produces smaller grinding movements
Suprahyoid muscles:					
Digastric	Base of cranium	Hyoid bone	Facial and mandibular nerves		Depresses mandible against resistance when infrahyoid muscles fix or depress hyoid bone
Stylohyoid	Styloid process		Facial nerve		
Mylohyoid	Medial body of mandible		Mandibular nerve		
Geniohyoid	Anterior body of mandible		Nerve to geniohyoid (C1–C2)		
Infrahyoid muscles:					
Omohyoid	Scapula	Hyoid bone	Ansa cervicalis from cervical plexus (C1–C3)		Fixes or depresses hyoid bone
Sternohyoid	Manubrium of sternum				
Sternothyroid		Thyroid cartilage			
Thyrohyoid	Thyroid cartilage	Hyoid bone	C1 (via hypoglossal n.–CN XII)		
Muscle of facial expression:					
Platysma	Subcutaneous tissue of infraclavicular and supraclavicular regions	Base of mandible, skin of cheek and lower lip, angle of mouth (modiolus), and orbicularis oris	Cervical branch of facial nerve (CN VII)		Depresses mandible against resistance

TABLE 7.12. PARTS AND BRANCHES OF MAXILLARY ARTERY

Part	Course	Branches	Distribution
First (mandibular)	Proximal (posterior) to lateral pterygoid muscle; runs horizontally, deep (medial) to neck of condylar process of mandible and lateral to stylomandibular ligament	Deep auricular artery	Supplies external acoustic meatus, external tympanic membrane, and temporomandibular joint
		Anterior tympanic artery	Supplies internal aspect of tympanic membrane
		Middle meningeal artery	Enters cranial cavity via foramen spinosum to supply periosteum, bone, red bone marrow, dura mater of lateral wall and calvaria of neurocranium, trigeminal ganglion, facial nerve and geniculate ganglion, tympanic cavity, and tensor tympani muscle
		Accessory meningeal artery	Enters cranial cavity via foramen ovale; its distribution is mainly extracranial to muscles of infra-temporal fossa, sphenoid bone, mandibular nerve, and otic ganglion
		Inferior alveolar artery	Descends to enter mandibular canal of mandible via mandibular foramen; supplies mandible, mandibular teeth, chin, mylohyoid muscle
Second (pterygoid)	Adjacent (superficial or deep) to lateral pterygoid muscle; ascends obliquely anterosuperiorly, medial to temporalis muscle	Masseteric artery	Traverses mandibular notch, supplying temporomandibular joint and masseter muscle
		Deep temporal arteries	Anterior and posterior arteries ascend between temporalis muscle and bone of temporal fossa, supplying mainly muscle
		Pterygoid branches	Irregular in number and origin; supply pterygoid muscle
		Buccal artery	Runs antero-inferiorly with buccal nerve to supply buccal fat-pad, buccinator, and buccal oral mucosa
Third (pterygoid-palatine)	Distal (anteromedial) to lateral pterygoid muscle; passes between heads of lateral pterygoid and through pterygomaxillary fissure into pterygopalatine fossa	Posterior superior alveolar artery	Descends on maxilla's infratemporal surface with branches traversing alveolar canals to supply maxillary molar and premolar teeth, adjacent gingiva, and mucous membrane of maxillary sinus
		Infra-orbital artery	Traverses inferior orbital fissure, infra-orbital groove, canal, and foramen; supplies inferior oblique and rectus muscles, lacrimal sac, maxillary canines and incisors teeth, mucous membrane of maxillary sinus, and skin of infra-orbital region of face
		Artery of pterygoid canal	Passes posteriorly through pterygoid canal; supplies mucosa of upper pharynx, pharyngotympanic tube, and tympanic cavity
		Pharyngeal branch	Passes through palatovaginal canal to supply mucosa of nasal roof, nasopharynx, sphenoidal air sinus, and pharyngotympanic tube
		Descending palatine artery	Descends through palatine canal, dividing into greater and lesser palatine arteries to mucosa and glands of hard and soft palate
		Sphenopalatine artery	Terminal branch of maxillary artery, traverses sphenopalatine foramen to supply walls and septum of nasal cavity; frontal, ethmoidal, sphenoid, and maxillary sinuses; and anteriormost palate

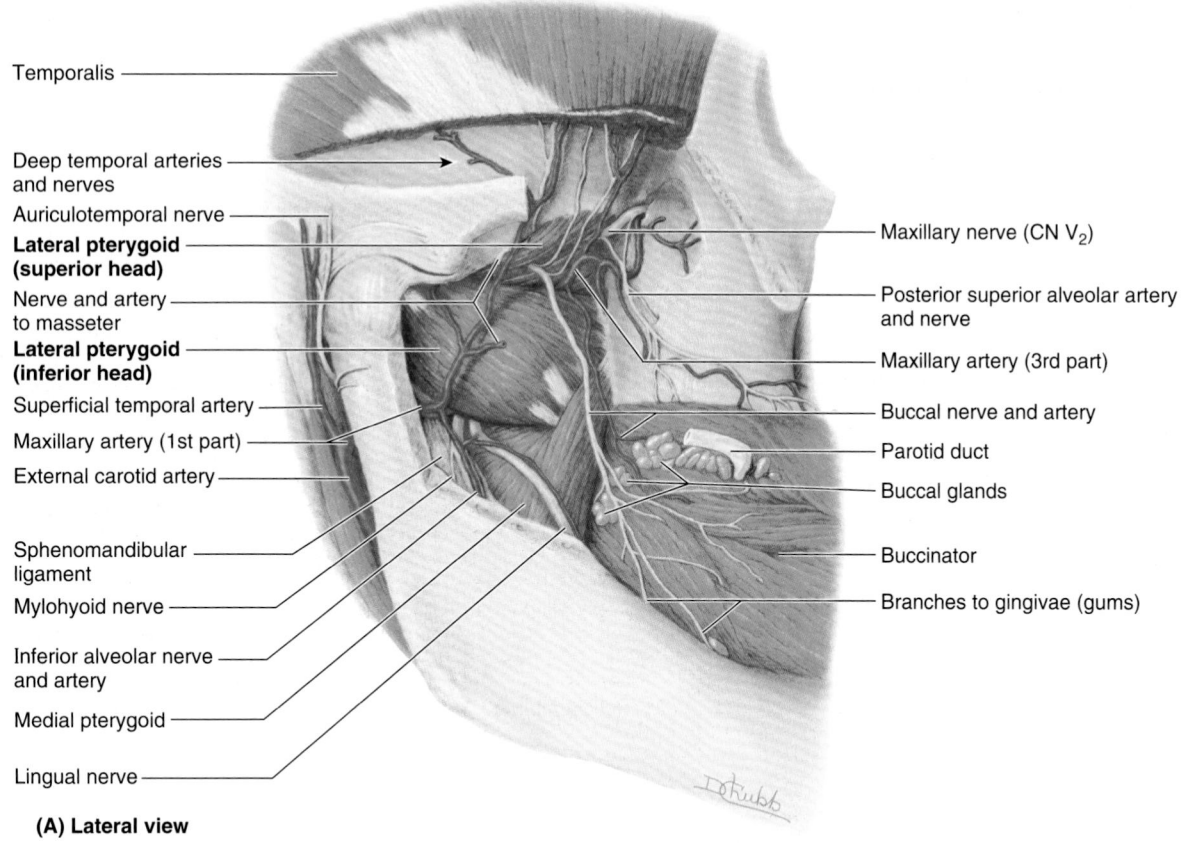

Temporalis

Deep temporal arteries
and nerves

Auriculotemporal nerve

**Lateral pterygoid
(superior head)**

Nerve and artery
to masseter

**Lateral pterygoid
(inferior head)**

Superficial temporal artery

Maxillary artery (1st part)

External carotid artery

Sphenomandibular
ligament

Mylohyoid nerve

Inferior alveolar nerve
and artery

Medial pterygoid

Lingual nerve

Maxillary nerve (CN V₂)

Posterior superior alveolar artery
and nerve

Maxillary artery (3rd part)

Buccal nerve and artery

Parotid duct

Buccal glands

Buccinator

Branches to gingivae (gums)

(A) Lateral view

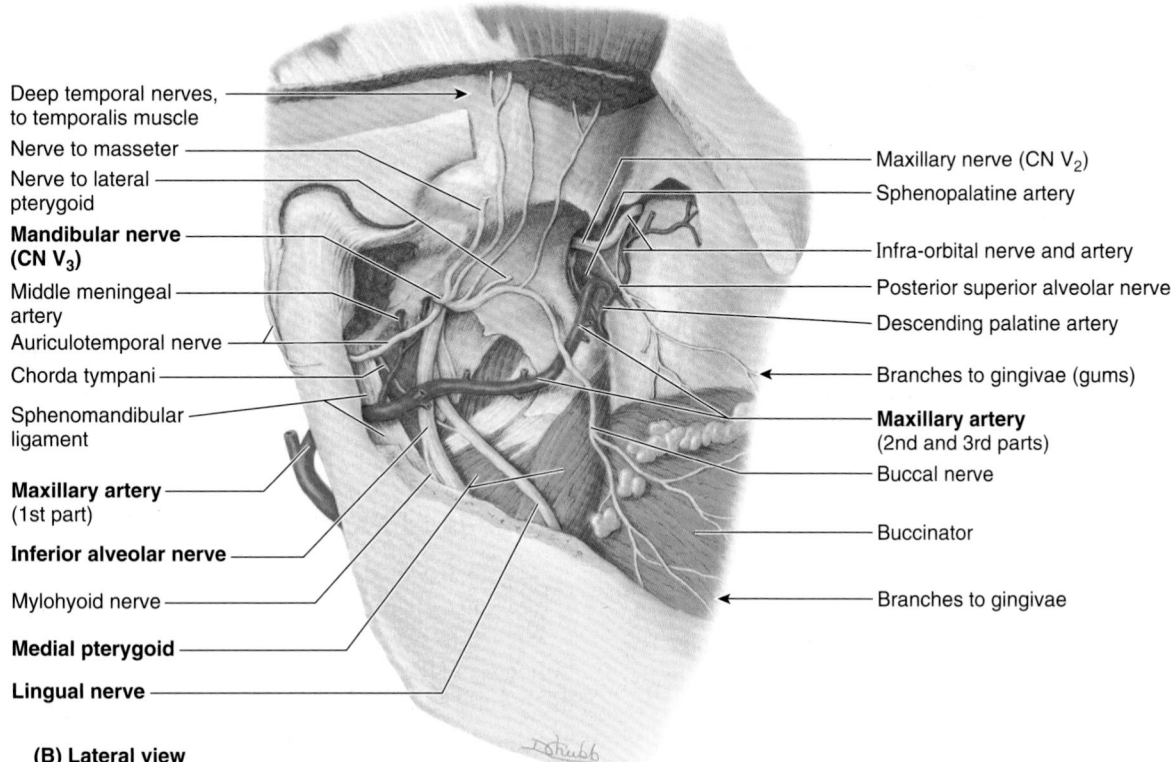

Deep temporal nerves,
to temporalis muscle

Nerve to masseter

Nerve to lateral
pterygoid

**Mandibular nerve
(CN V₃)**

Middle meningeal
artery

Auriculotemporal nerve

Chorda tympani

Sphenomandibular
ligament

**Maxillary artery
(1st part)**

Inferior alveolar nerve

Mylohyoid nerve

Medial pterygoid

Lingual nerve

Maxillary nerve (CN V₂)

Sphenopalatine artery

Infra-orbital nerve and artery

Posterior superior alveolar nerve

Descending palatine artery

Branches to gingivae (gums)

Maxillary artery
(2nd and 3rd parts)

Buccal nerve

Buccinator

Branches to gingivae

(B) Lateral view

FIGURE 7.74. Dissections of infratemporal region. A. In this superficial dissection, most of the zygomatic arch and attached masseter, the coronoid process and adjacent parts of the ramus of the mandible, and the inferior half of the temporal muscle have been removed. The first part of the maxillary artery, the larger of the two end branches of the external carotid, run anteriorly, deep to the neck of the mandible and then pass deeply between the lateral and the medial pterygoid muscles. **B.** In this deep dissection, more of the ramus of the mandible, the lateral pterygoid muscle, and most branches of the maxillary artery have been removed. Branches of the mandibular nerve (CN V₃), including the auriculotemporal nerve, and the second part of the maxillary artery pass between the sphenomandibular ligament and the neck of the mandible.

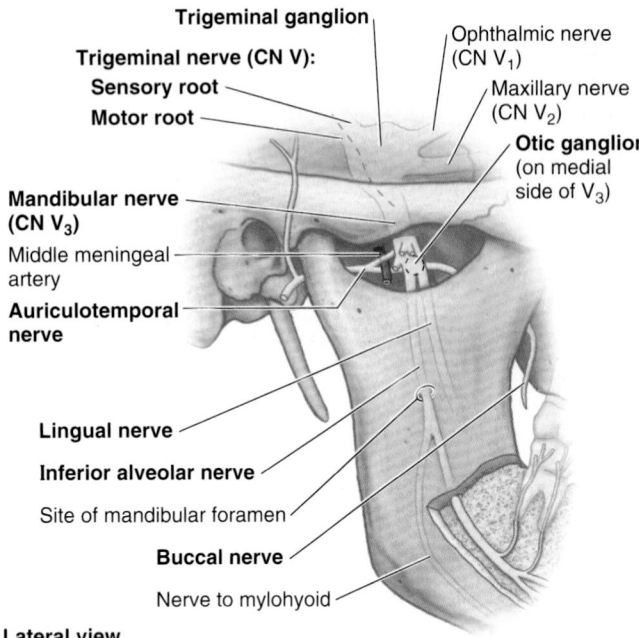

FIGURE 7.75. Nerves of infratemporal fossa.

PAROTID AND TEMPORAL REGIONS, INFRATEMPORAL FOSSA, AND TEMPOROMANDIBULAR JOINT

Parotidectomy

 About 80% of salivary gland tumors occur in the parotid glands. Most tumors of the parotid glands are benign, but most salivary gland cancers begin in the parotid glands. Surgical excision of the parotid gland (*parotidectomy*) is often performed as part of the treatment. Because the parotid plexus of CN VII is embedded in the parotid gland, the plexus and its branches are in jeopardy during surgery (see Fig. 7.23A & C). An important step in parotidectomy is the identification, dissection, isolation, and preservation of the facial nerve. A superficial portion of the gland (often erroneously referred to as a "lobe") is removed, after which the parotid plexus, which occupies a distinct plane within the gland, can be retracted to enable dissection of the deep portion of the gland. The parotid gland makes a substantial contribution to the posterolateral contour of the face, the extent of its contribution being especially evident after it has been surgically removed. See the blue box "Paralysis of Facial Muscles" (p. 861) for a discussion of the functional consequences of injury to the facial nerve.

Infection of Parotid Gland

 The parotid gland may become infected by infectious agents that pass through the bloodstream, as occurs in *mumps,* an acute communicable viral disease. Infection of the gland causes inflammation (*parotiditis*) and swelling of the gland. Severe pain occurs because the parotid sheath limits

swelling. Often the pain is worse during chewing because the enlarged gland is wrapped around the posterior border of the ramus of the mandible and is compressed against the mastoid process of the temporal bone when the mouth is opened. The mumps virus may also cause *inflammation of the parotid duct,* producing *redness of the parotid papilla,* the small projection at the opening of the duct into the superior oral vestibule (Fig. 7.65, p. 915). Because the pain produced by mumps may be confused with a toothache, redness of the papilla is often an early sign that the disease involves the parotid gland and not a tooth.

Parotid gland disease often causes pain in the auricle and external acoustic meatus of the external ear, the temporal region, and TMJ because the auriculotemporal and great auricular nerves, from which the parotid gland and sheath receives sensory fibers, also supplies sensory fibers to the skin over the temporal fossa and auricle.

Abscess in Parotid Gland

A bacterial infection localized in the parotid gland usually produces an abscess (pus formation). The infection could result from extremely poor dental hygiene, and the infection could spread to the gland through the parotid ducts. Physicians and dentists must determine whether a swelling of the cheek results from infection of the parotid gland or from an abscess of dental origin.

Sialography of Parotid Duct

A radiopaque fluid can be injected into the duct system of the parotid gland through a cannula inserted through the orifice of the parotid duct in the mucous membrane of the cheek. This technique (*sialography*) is followed by radiography of the gland. *Parotid sialograms* (G. *sialon*, saliva + G. *grapho*, to write) demonstrate parts of the parotid duct system that may be displaced or dilated by disease.

Blockage of Parotid Duct

The parotid duct may be blocked by a calcified deposit, called a *sialolith* or *calculus* (L. pebble). The resulting pain in the parotid gland is made worse by eating. Sucking a lemon slice is painful because of the buildup of saliva in the proximal part of the blocked parotid duct.

Accessory Parotid Gland

Sometimes an accessory parotid gland lies on the masseter muscle between the parotid duct and the zygomatic arch. Several ducts open from this accessory gland into the parotid duct.

Mandibular Nerve Block

To produce a mandibular nerve block, an anesthetic agent is injected near the mandibular nerve where it enters the infratemporal fossa (Fig. 7.67B). In the extra-oral approach, the needle passes through the mandibular notch of the ramus of the mandible into the infratemporal fossa. The injection usually anesthetizes the auriculotemporal, inferior alveolar, lingual, and buccal branches of CN V₃.

Inferior Alveolar Nerve Block

An inferior alveolar nerve block anesthetizes the inferior alveolar nerve, a branch of CN V₃. The site of the anesthetic injection is around the *mandibular foramen*, the opening into the mandibular canal on the medial aspect of the ramus of the mandible (Fig. 7.75). This canal gives passage to the inferior alveolar nerve, artery, and vein. When this nerve block is successful, all mandibular teeth are anesthetized to the median plane. The skin and mucous membrane of the lower lip, the labial alveolar mucosa and gingivae, and the skin of the chin are also anesthetized because they are supplied by the mental nerve, a branch of the inferior alveolar nerve (Fig. 7.79A). There are possible problems associated with an inferior alveolar nerve block, such as injection of the anesthetic into the parotid gland or the medial pterygoid muscle. This would affect ability to open the mouth (*pterygoid trismus*)

Dislocation of TMJ

Sometimes during yawning or taking a large bite, excessive contraction of the lateral pterygoids may cause the heads of the mandible to dislocate anteriorly (pass anterior to the articular tubercles) (Fig. B7.31). In this position, the mandible remains depressed and the person

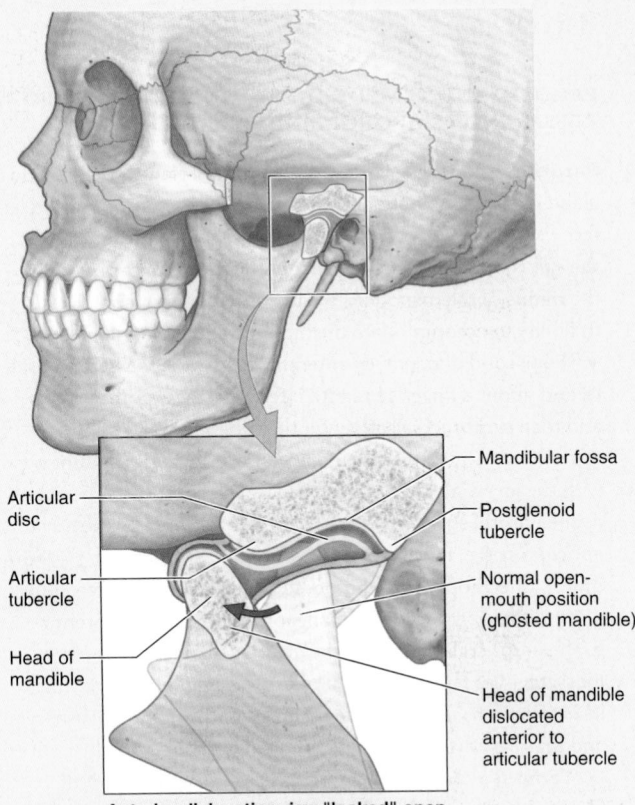

Anterior dislocation: jaw "locked" open

FIGURE B7.31. Dislocation of TMJ.

is unable to close his or her mouth. Most common, a sideways blow to the chin by a clenched hand (fist) when the mouth is open dislocates the TMJ on the side that received the blow. Dislocation of the TMJ may also accompany fractures of the mandible. Posterior dislocation is uncommon, being resisted by the presence of the postglenoid tubercle and the strong intrinsic lateral ligament. Usually in falls on or direct blows to the chin, the neck of the mandible fractures before dislocation occurs. Because of the close relationship of the facial and auriculotemporal nerves to the TMJ, care must be taken during surgical procedures to preserve both the branches of the facial nerve overlying it and the articular branches of the auriculotemporal nerve that enter the posterior part of the joint. Injury to articular branches of the auriculotemporal nerve supplying the TMJ, associated with traumatic dislocation and rupture of the articular capsule and lateral ligament, leads to laxity and instability of the TMJ.

Arthritis of TMJ

The TMJ may become inflamed from degenerative arthritis, for example. Abnormal function of the TMJ may result in structural problems such as dental occlusion and joint clicking (*crepitus*). The clicking is thought to result from delayed anterior disc movements during mandibular depression and elevation.

The Bottom Line

PAROTID AND TEMPORAL REGIONS, INTRATEMPORAL FOSSA, AND TEMPOROMANDIBULAR JOINT

Parotid region: The largest of the salivary glands, the parotid gland makes a substantial contribution to the contour of the face. ♦ Occupying a complex space anterior to the auricle of the ear, the gland straddles most of the posterior aspect of the ramus of the mandible. ♦ Fatty tissue in the gland gives it flexibility to accommodate the motions of the mandible.
♦ The parotid duct passes anteriorly across the masseter, parallel and about a finger's breadth inferior to the zygomatic arch, and then turns medially to enter the superior oral vestibule opposite the 2nd maxillary molar. ♦ Parotid fascia, continuous with the investing layer of deep cervical fascia, invests the gland as the parotid sheath. ♦ The sheath is innervated by the great auricular nerve, but the gland receives parasympathetic secretomotor innervation from the glossopharyngeal nerve via a complex route involving the otic ganglion. ♦ Medial and anterior to the parotid gland, one of the muscles of mastication—the masseter—lies lateral to the ramus of the mandible, receiving its innervation via masseteric branches of the mandibular nerve and maxillary artery that traverse the mandibular notch.

Temporal and infratemporal fossa: The temporal fossa and its inferior continuation deep to the zygomatic arch and ramus of the mandible, the infratemporal fossa, are largely occupied by derivatives of the embryonic first pharyngeal arch: three of the four muscles of mastication (temporalis muscle and two pterygoid muscles) and the nerve that conveys motor fibers to them, the mandibular nerve (CN V$_3$).

TMJ and muscles of mastication: The TMJ is a hinge joint, modified by the presence of an articular disc that intervenes between the mandibular head and the articular surfaces of the temporal bone. ♦ Gliding movements between the mandibular fossa and articular eminence occur in the upper compartment and are produced by the lateral pterygoid (protraction) and posterior fibers of the temporalis (retraction). ♦ The mandible must be protracted for full opening of the mouth. ♦ Hinge and pivoting movements occur in the lower compartment and are produced by gravity (depression) and three of the four muscles of mastication (elevation): masseter, medial pterygoid, and anterior portion of the temporalis.

Neurovasculature of infratemporal fossa: Also contained in the infratemporal fossa is the second part of the maxillary artery and its venous equivalent, the pterygoid venous plexus. ♦ Adjacent cranial compartments communicate with the fossae, and neurovascular structures pass to and from the fossae via bony passages, including the (1) foramen ovale, through which the mandibular nerve enters from the middle cranial fossa; (2) foramen spinosum, through which the middle meningeal artery enters and the meningeal branch of CN V$_3$ returns to the middle cranial fossa; (3) pterygomaxillary fissure, through which the maxillary artery passes into the pterygopalatine fossa for further distribution; (4) inferior orbital fissure, through which the inferior ophthalmic veins drain to the pterygoid venous plexus; and (5) mandibular foramen, through which the inferior alveolar nerve passes to mandibular canal for distribution to the mandible and teeth.

ORAL REGION

The **oral region** includes the oral cavity, teeth, gingivae, tongue, palate, and the region of the palatine tonsils. The oral cavity is where food is ingested and prepared for digestion in the stomach and small intestine. Food is chewed by the teeth, and saliva from the salivary glands facilitates the formation of a manageable *food bolus* (L. lump). *Deglutition* (swallowing) is voluntarily initiated in the oral cavity. The voluntary phase of the process pushes the bolus from the oral cavity into the pharynx, the expanded part of the alimentary (digestive) system, where the involuntary (automatic) phase of swallowing occurs.

Oral Cavity

The **oral cavity** (**mouth**) consists of two parts: the *oral vestibule* and the *oral cavity proper* (Fig. 7.76). It is in the oral cavity that food and drinks are tasted and where mastication (chewing) and lingual manipulation of food occur. The **oral vestibule** is the slit-like space between the teeth and gingivae (gums) and the lips and cheeks. The vestibule communicates with the exterior through the **oral fissure** (opening). The size of the fissure is controlled by the peri-oral muscles, such as the orbicularis oris (the sphincter of the oral fissure), the buccinator, risorius, and depressors and elevators of the lips (dilators of the fissure).

The **oral cavity proper** is the space between the upper and the lower **dental arches** or arcades (maxillary and mandibular alveolar arches and the teeth they bear). The oral cavity is limited laterally and anteriorly by the dental arches. The *roof of the oral cavity* is formed by the palate. Posteriorly, the oral cavity communicates with the oropharynx (oral part of the pharynx). When the mouth is closed and at rest, the oral cavity is fully occupied by the tongue.

Lips, Cheeks, and Gingivae

LIPS AND CHEEKS

The **lips** are mobile, musculofibrous folds surrounding the mouth, extending from the *nasolabial sulci* and *nares* laterally, and superiorly to the *mentolabial sulcus* inferiorly

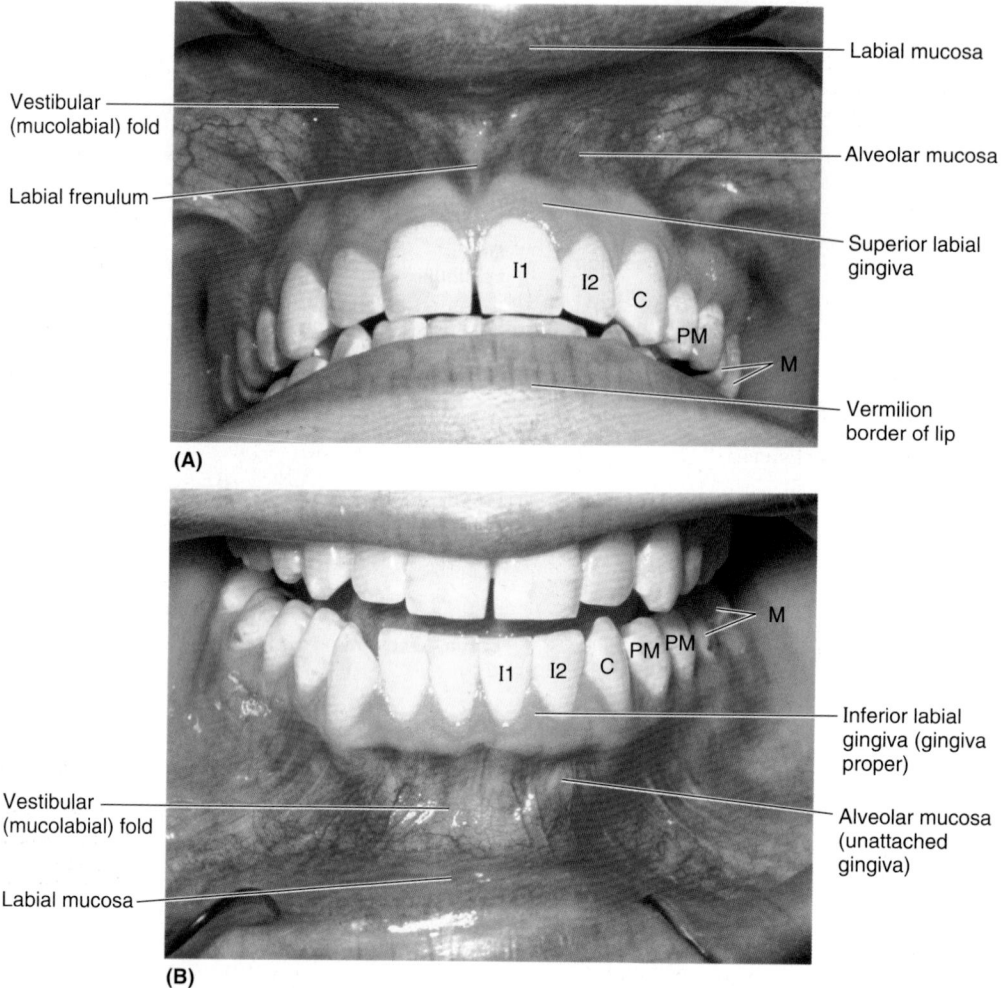

Vestibular (mucolabial) fold

Labial frenulum

Labial mucosa

Alveolar mucosa

Superior labial gingiva

Vermilion border of lip

I1   I2   C   PM   M

(A)

M   PM PM   C   I2   I1

Inferior labial gingiva (gingiva proper)

Alveolar mucosa (unattached gingiva)

Vestibular (mucolabial) fold

Labial mucosa

(B)

FIGURE 7.78. Oral vestibule and gingivae. A. The vestibule and gingivae of the maxilla are shown. **B.** The vestibule and gingivae of the mandible are shown. As the alveolar mucosa approaches the necks of the teeth, it changes in texture and color to become the gingiva proper. (Courtesy of Dr. B. Liebgott, Professor, Division of Anatomy, Department of Surgery, University of Toronto, Toronto, Ontario, Canada.)

The principal muscles of the cheeks are the buccinators (Figs. 7.76). Numerous small **buccal glands** lie between the mucous membrane and the buccinators (Fig. 7.74A). Superficial to the buccinators are encapsulated collections of fat; these *buccal fat-pads* are proportionately much larger in infants, presumably to reinforce the cheeks and keep them from collapsing during sucking. The cheeks are supplied by buccal branches of the maxillary artery and innervated by buccal branches of the mandibular nerve.

GINGIVAE

The **gingivae** (gums) are composed of fibrous tissue covered with mucous membrane. The **gingiva proper** (attached gingiva) is firmly attached to the alveolar processes of the mandible and maxilla and the necks of the teeth (Figs. 7.76 and 7.78). The gingiva proper adjacent to the tongue is the superior and inferior lingual gingivae, and that adjacent to the lips and cheeks is the **maxillary** and **mandibular labial** or **buccal gingiva,** respectively. The gingiva proper is normally

pink, stippled, and keratinizing. The **alveolar mucosa** (unattached gingiva) is normally shiny red and non-keratinizing. The nerves and vessels supplying the gingiva, underlying alveolar bone, and *periodontium* (which surrounds the root[s] of a tooth, anchoring it to the tooth socket), are presented in Fig. 7.79A & C.

Teeth

The *chief functions of teeth* are to:

- Incise (cut), reduce, and mix food material with saliva during mastication (chewing).
- Help sustain themselves in the tooth sockets by assisting the development and protection of the tissues that support them.
- Participate in articulation (distinct connected speech).

The teeth are set in the *tooth sockets* and are used in mastication and in assisting in articulation. A tooth is identified and described on the basis of whether it is **deciduous** (primary)

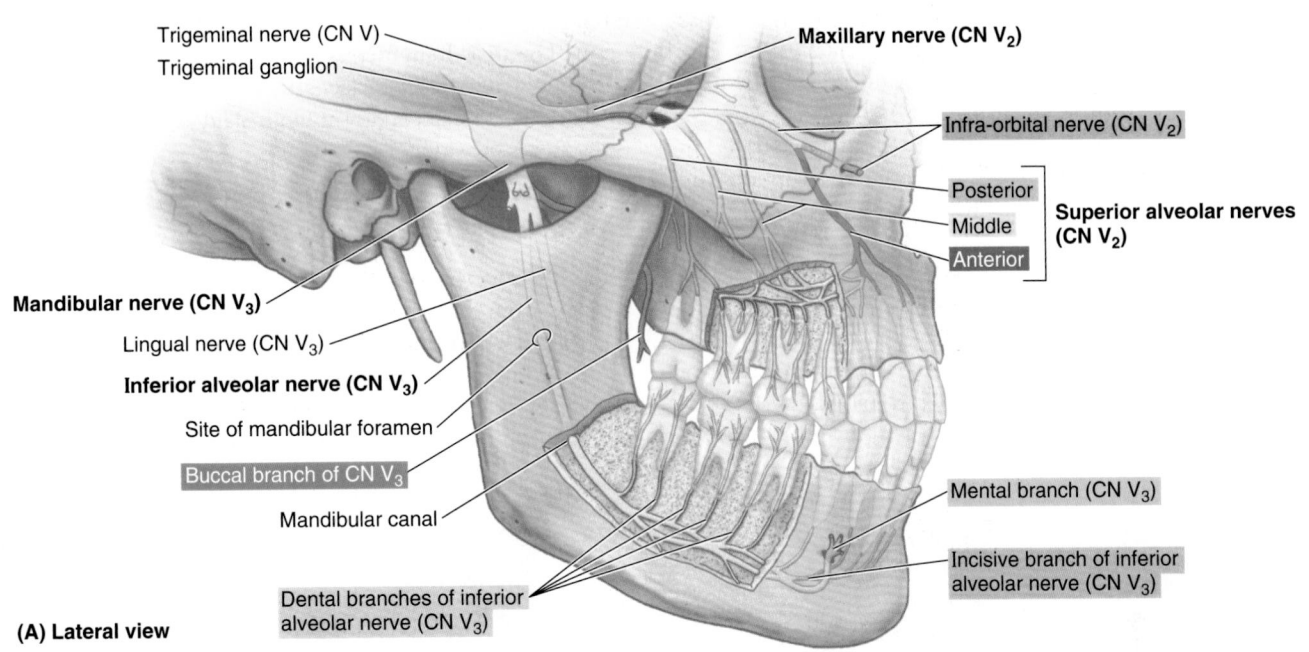

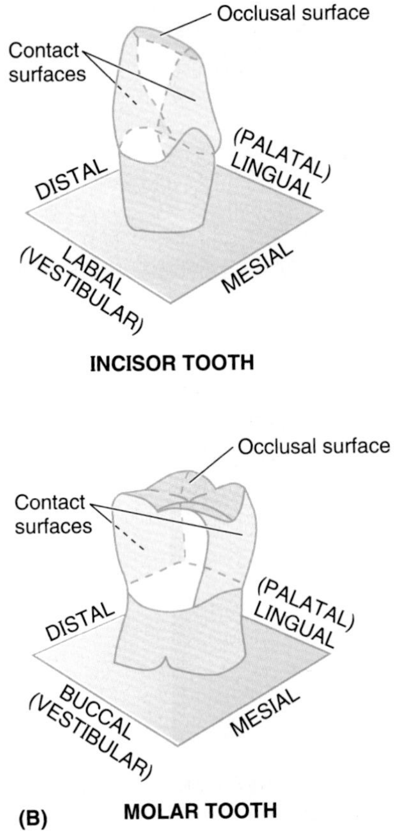

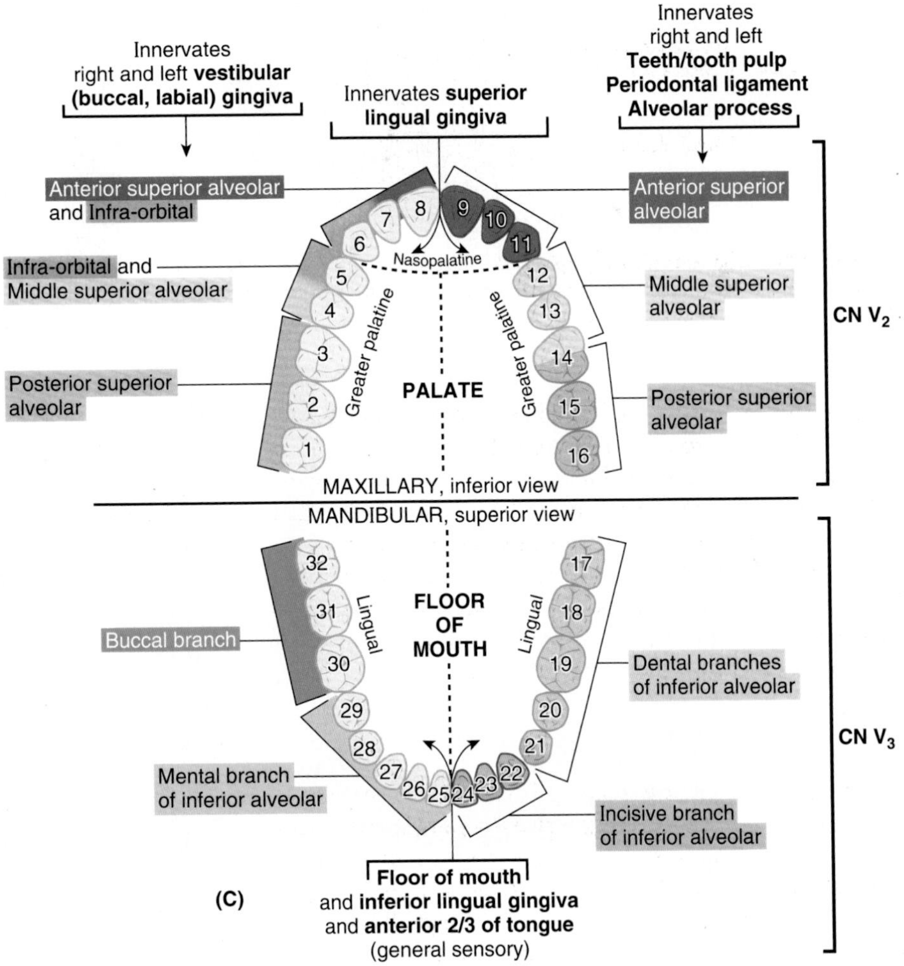

(A) Lateral view

- Trigeminal nerve (CN V)
- Trigeminal ganglion
- **Maxillary nerve (CN V₂)**
- Infra-orbital nerve (CN V₂)
- Posterior
- Middle
- Anterior
- **Superior alveolar nerves (CN V₂)**
- **Mandibular nerve (CN V₃)**
- Lingual nerve (CN V₃)
- **Inferior alveolar nerve (CN V₃)**
- Site of mandibular foramen
- Buccal branch of CN V₃
- Mandibular canal
- Dental branches of inferior alveolar nerve (CN V₃)
- Mental branch (CN V₃)
- Incisive branch of inferior alveolar nerve (CN V₃)

(B)

INCISOR TOOTH
- Occlusal surface
- Contact surfaces
- DISTAL
- (PALATAL) LINGUAL
- LABIAL (VESTIBULAR)
- MESIAL

MOLAR TOOTH
- Occlusal surface
- Contact surfaces
- DISTAL
- (PALATAL) LINGUAL
- BUCCAL (VESTIBULAR)
- MESIAL

(C)

Innervates right and left **vestibular (buccal, labial) gingiva**

Innervates **superior lingual gingiva**

Innervates right and left **Teeth/tooth pulp Periodontal ligament Alveolar process**

- Anterior superior alveolar and Infra-orbital
- Infra-orbital and Middle superior alveolar
- Posterior superior alveolar
- Nasopalatine
- Greater palatine
- **PALATE**
- Greater palatine
- Anterior superior alveolar
- Middle superior alveolar
- Posterior superior alveolar

CN V₂

MAXILLARY, inferior view

MANDIBULAR, superior view

- Buccal branch
- Lingual
- **FLOOR OF MOUTH**
- Lingual
- Mental branch of inferior alveolar
- Dental branches of inferior alveolar
- Incisive branch of inferior alveolar

CN V₃

Floor of mouth and **inferior lingual gingiva** and **anterior 2/3 of tongue** (general sensory)

Teeth numbers: 1–16 (maxillary), 17–32 (mandibular)

FIGURE 7.79. Innervation of teeth and gingiva. A. Superior and inferior alveolar nerves. **B.** Surfaces of an incisor and molar tooth. **C.** Innervation of the mouth and teeth.

or **permanent** (secondary), the type of tooth, and its proximity to the midline or front of the mouth (e.g., medial and lateral incisors; the 1st molar is anterior to the 2nd).

Children have 20 deciduous teeth; adults normally have 32 permanent teeth (Fig. 7.80A & C). The usual ages of the eruption ("cutting") of these teeth are demonstrated in Figure 7.81 and listed in Table 7.13. Before eruption, the developing teeth reside in the alveolar arches as **tooth buds** (Fig. 7.80B).

The types of teeth are identified by their characteristics: **incisors,** thin cutting edges; **canines,** single prominent cones; **premolars** (bicuspids), two cusps; and **molars,** three or more cusps (Fig. 7.80A & C). The **vestibular surface** (labial or buccal) of each tooth is directed outwardly, and the **lingual surface** is directed inwardly (Fig. 7.79B). As used

in clinical (dental) practice, the **mesial surface** of a tooth is directed toward the median plane of the facial part of the cranium. The **distal surface** is directed away from this plane; both mesial and distal surfaces are *contact surfaces*—that is, surfaces that contact adjacent teeth. The masticatory surface is the **occlusal surface.**

PARTS AND STRUCTURE OF TEETH

A tooth has a crown, neck, and root (Fig. 7.82). The **crown** projects from the gingiva. The **neck** is between the crown and root. The **root** is fixed in the tooth socket by the *periodontium* (connective tissue surrounding roots); the number of roots varies. Most of the tooth is composed of **dentine** (L. *dentinium*), which is covered by **enamel** over the crown

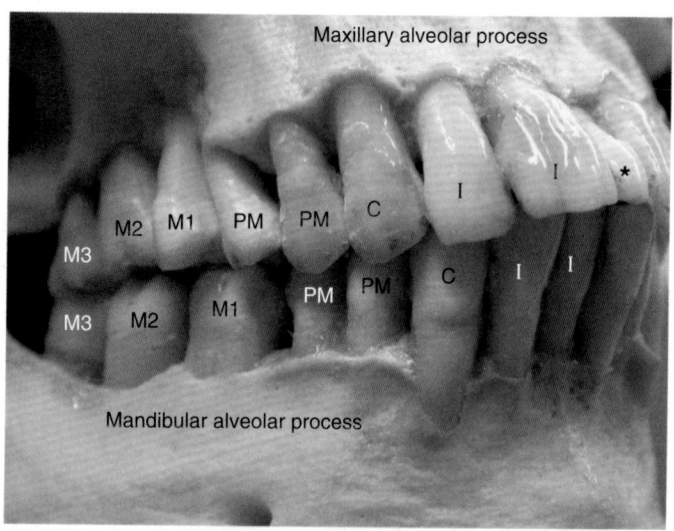

(A) Right anterolateral view

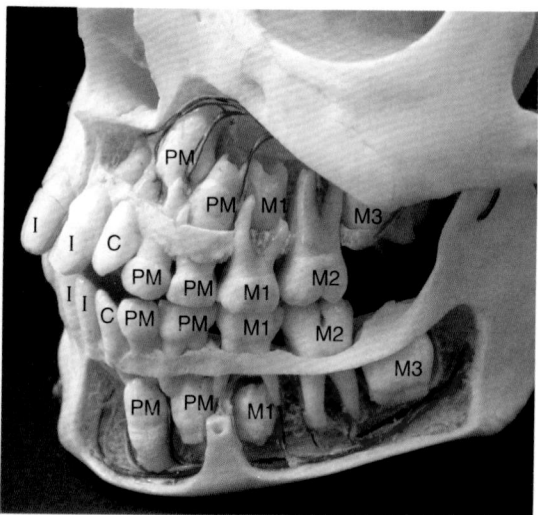

(B) Left anterolateral view

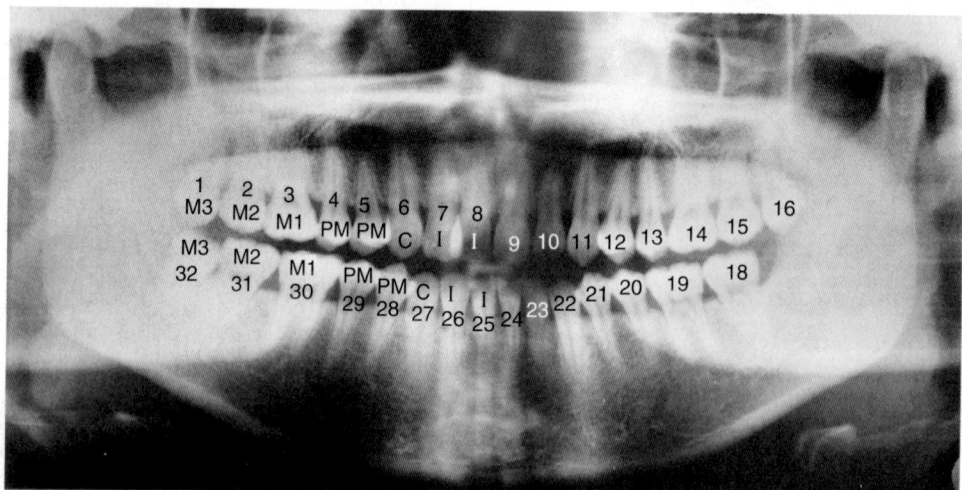

(C) Panoramic view

FIGURE 7.80. Secondary dentition. A. The teeth are shown in occlusion. There is a supernumerary midline tooth (mesiodens) in this specimen (*). **B.** Maxillary and mandibular jaws of a child acquiring secondary dentition are shown. The alveolar processes are carved to reveal the roots of the teeth and tooth buds. **C.** A pantomographic radiograph of an adult mandible and maxilla is shown. The left lower 3rd molar is not present. *I,* incisor; *C,* canine; *PM,* premolar; *M1, M2,* and *M3,* 1st, 2nd, and 3rd molars. (Part **C** courtesy of M. J. Pharoah, Associate Professor of Dental Radiology, Faculty of Dentistry, University of Toronto, Toronto, Ontario, Canada.)

TABLE 7.13A. DECIDUOUS TEETH

Deciduous Teeth	Central Incisor	Lateral Incisor	Canine	1st Molar	2nd Molar
Eruption (months)[a]	6–8	8–10	16–20	12–16	20–24
Shedding (years)	6–7	7–8	10–12	9–11	10–12

[a] In some normal infants, the first teeth (medial incisors) may not erupt until 12–13 months of age.

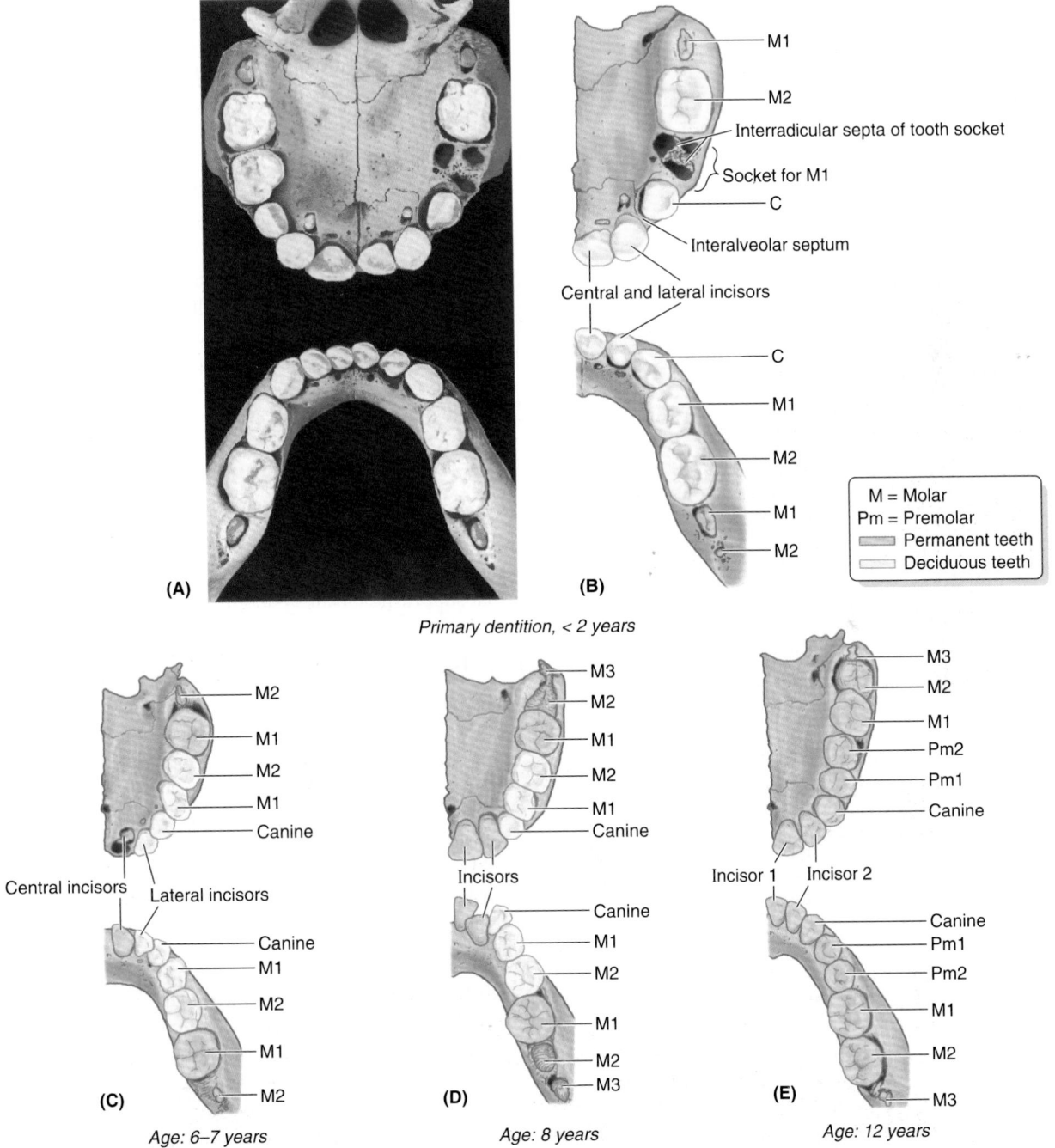

(A)

M1
M2
Interradicular septa of tooth socket
Socket for M1
C
Interalveolar septum
Central and lateral incisors
C
M1
M2
M1
M2

M = Molar
Pm = Premolar
Permanent teeth
Deciduous teeth

(B)

Primary dentition, < 2 years

M2
M1
M2
M1
Canine
Central incisors
Lateral incisors
Canine
M1
M2
M1
M2

(C)

Age: 6–7 years

M3
M2
M1
M2
M1
Canine
Incisors
Canine
M1
M2
M1
M2
M3

(D)

Age: 8 years

M3
M2
M1
Pm2
Pm1
Canine
Incisor 1   Incisor 2
Canine
Pm1
Pm2
M1
M2
M3

(E)

Age: 12 years

FIGURE 7.81. Primary dentition (deciduous teeth) and eruption of permanent teeth.

TABLE 7.13B. PERMANENT TEETH

Permanent Teeth	Central Incisor	Lateral Incisor	Canine	1st Premolar	2nd Premolar	1st Molar	2nd Molar	3rd Molar
Eruption (years)	7–8	8–9	10–12	10–11	11–12	6–7	12	13–25

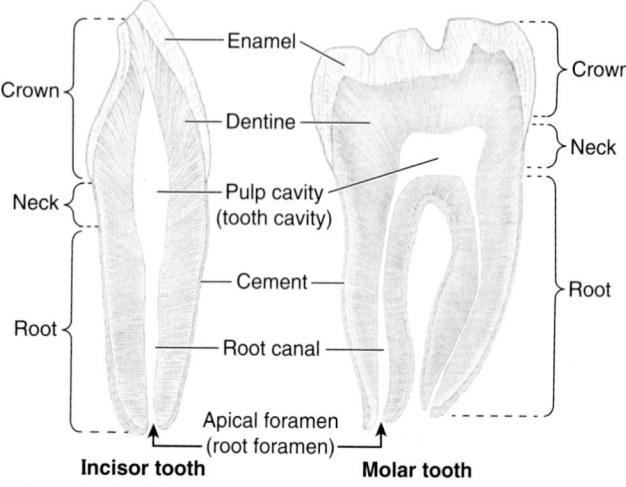

(A) Longitudinal section

Crown
Neck
Root

Enamel
Dentine
Pulp cavity (tooth cavity)
Cement
Root canal
Apical foramen (root foramen)

Incisor tooth

Crown
Neck
Root

Molar tooth

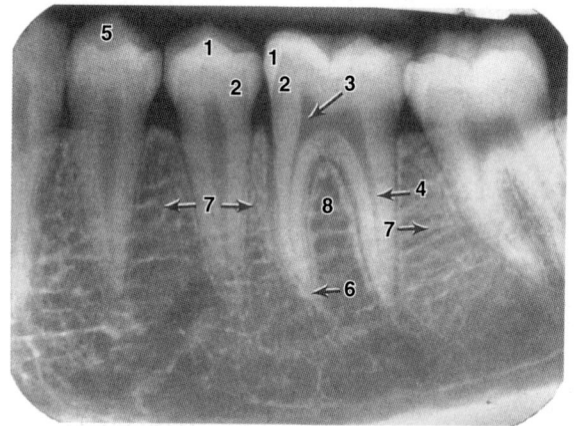

(B) Lateral radiograph

1 Enamel	**2** Dentine	**3** Pulp cavity
4 Root canal	**5** Buccal cusp	**6** Root apex
7 Interalveolar septa (alveolar bone)		
8 Interradicular septum (alveolar bone)		

FIGURE 7.82. Sections of teeth. A. An incisor and a molar are shown. In living people, the pulp cavity is a hollow space within the crown and neck of the tooth containing connective tissue, blood vessels, and nerves. The cavity narrows down to the root canal in a single-rooted tooth or to one canal per root of a multirooted tooth. The vessels and nerves enter or leave through the apical foramen. **B.** Bite-wing radiograph of maxillary premolar and molar teeth demonstrating features shown and described in part **A.**

and **cement** (L. *cementum*) over the root. The **pulp cavity** contains connective tissue, blood vessels, and nerves. The **root canal** (pulp canal) transmits the nerves and vessels to and from the pulp cavity through the **apical foramen.**

The **tooth sockets** are in the *alveolar processes* of the maxillae and mandible (Fig. 7.80A); they are the skeletal features that display the greatest change during a lifetime (Fig. 7.81B). Adjacent sockets are separated by **interalveolar septa;** within the socket, the roots of teeth with more than one root are separated by **interradicular septa** (Figs. 7.81B and 7.82B). The bone of the socket has a thin cortex separated from the adjacent labial and lingual cortices by a

variable amount of trabeculated bone. The labial wall of the socket is particularly thin over the incisor teeth; the reverse is true for the molars, where the lingual wall is thinner. Thus the labial surface commonly is broken to extract incisors and the lingual surface is broken to extract molars.

The roots of the teeth are connected to the bone of the alveolus by a springy suspension forming a special type of fibrous joint called a **dento-alveolar syndesmosis** or **gomphosis.** The **periodontium** (periodontal membrane) is composed of collagenous fibers that extend between the cement of the root and the periosteum of the alveolus. It is abundantly supplied with tactile, pressoreceptive nerve endings, lymph capillaries, and glomerular blood vessels that act as hydraulic cushioning to curb axial masticatory pressure. *Pressoreceptive nerve endings* are capable of receiving changes in pressure as stimuli.

VASCULATURE OF TEETH

The **superior** and **inferior alveolar arteries,** branches of the maxillary artery, supply the maxillary and mandibular teeth, respectively (Figs. 7.73 and 7.74A; Table 7.12). The **alveolar veins** have the same names and distribution accompany the arteries. **Lymphatic vessels** from the teeth and gingivae pass mainly to the *submandibular lymph nodes* (Fig. 7.77).

INNERVATION OF TEETH

The *nerves supplying the teeth* are illustrated in Figure 7.79A. The named branches of the *superior* (CN V_2) and *inferior* (CN V_3) *alveolar nerves* give rise to **dental plexuses** that supply the maxillary and mandibular teeth.

Palate

The **palate** forms the arched roof of the mouth and the floor of the nasal cavities (Fig. 7.83). It separates the oral cavity from the nasal cavities and the nasopharynx, the part of the pharynx superior to the soft palate. The superior (nasal) surface of the palate is covered with respiratory mucosa, and the inferior (oral) surface is covered with oral mucosa, densely packed with glands. The palate consists of two regions: the hard palate anteriorly and the soft palate posteriorly.

HARD PALATE

The **hard palate** is vaulted (concave); this space is mostly filled by the tongue when it is at rest. The anterior two thirds of the palate has a bony skeleton formed by the palatine processes of the maxillae and the horizontal plates of the palatine bones (Fig. 7.84A). The **incisive fossa** is a depression in the midline of the bony palate posterior to the central incisor teeth into which the incisive canals open. The nasopalatine nerves pass from the nose through a variable number of incisive canals and foramina that open into the incisive fossa (Fig. 7.87B).

Medial to the 3rd molar tooth, the *greater palatine foramen* pierces the lateral border of the bony palate (Fig. 7.84A). The *greater palatine vessels and nerve* emerge from this

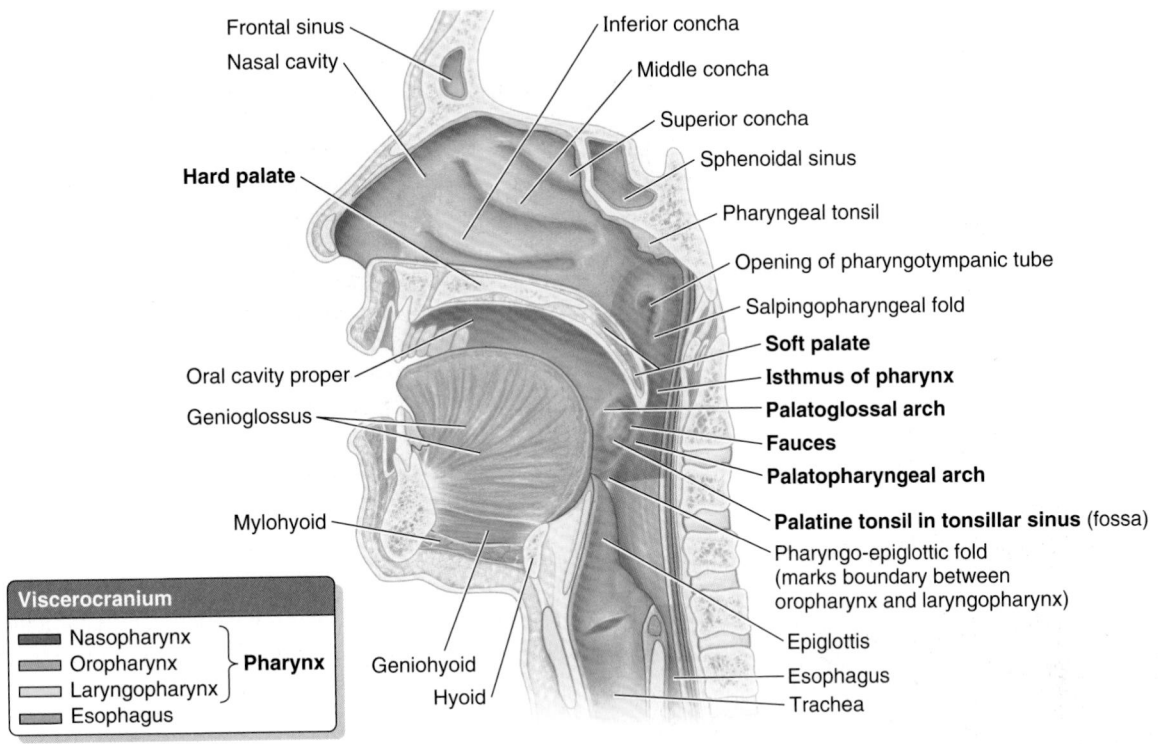

Medial view of right half of viscerocranium

FIGURE 7.83. Median section of head and neck. The airway and food passageways cross in the pharynx. The soft palate acts as a valve, elevating to seal the pharyngeal isthmus connecting the nasal cavity and nasopharynx with the oral cavity and oropharynx.

foramen and run anteriorly on the palate. The *lesser palatine foramina* posterior to the greater palatine foramen pierce the pyramidal process of the palatine bone. These foramina transmit the *lesser palatine nerves and vessels* to the soft palate and adjacent structures (Fig. 7.87).

SOFT PALATE

The **soft palate** is the movable posterior third of the palate and is suspended from the posterior border of the hard palate (Figs. 7.83 and 7.84B). The soft palate has no bony skeleton; however, its anterior *aponeurotic part* is strengthened by the **palatine aponeurosis,** which attaches to the posterior edge of the hard palate. The aponeurosis is thick anteriorly and thin posteriorly, where it blends with a posterior *muscular part*. Postero-inferiorly, the soft palate has a curved free margin from which hangs a conical process, the **uvula.**

When a person swallows, the soft palate initially is tensed to allow the tongue to press against it, squeezing the bolus of food to the back of the mouth. The soft palate is then elevated posteriorly and superiorly against the wall of the pharynx, thereby preventing passage of food into the nasal cavity.

Laterally, the soft palate is continuous with the wall of the pharynx and is joined to the tongue and pharynx by the **palatoglossal** and **palatopharyngeal arches,** respectively (Fig. 7.83). A few taste buds are located in the epithelium covering the oral surface of the soft palate, the posterior wall of the oropharynx, and the epiglottis.

The **fauces** (L. throat) is the space between the oral cavity and the pharynx. The fauces is bounded superiorly by the soft palate, inferiorly by the root of the tongue, and laterally by the **pillars of the fauces,** the *palatoglossal* and *palatopharyngeal arches*. The **isthmus of the fauces** is the short constricted space that establishes the connection between the oral cavity proper and oropharynx. The isthmus is bounded anteriorly by the palatoglossal folds and posteriorly by the palatopharyngeal folds. The **palatine tonsils,** often referred to as "the tonsils," are masses of lymphoid tissue, one on each side of the oropharynx. Each tonsil is in a **tonsillar sinus (fossa),** bounded by the palatoglossal and palatopharyngeal arches and the tongue.

SUPERFICIAL FEATURES OF PALATE

The mucosa of the hard palate is tightly bound to the underlying bone (Fig. 7.85A); consequently, submucous injections here are extremely painful. The *superior lingual gingiva,* the part of the gingiva covering the lingual surface of the teeth and the alveolar process, is continuous with the mucosa of the palate; therefore, injection of an anesthetic agent into the gingiva of a tooth anesthetizes the adjacent palatal mucosa.

Deep to the mucosa are mucus-secreting **palatine glands** (Fig. 7.85B). The openings of the ducts of these glands give the palatine mucosa a pitted (orange-peel) appearance. In the midline, posterior to the maxillary incisor

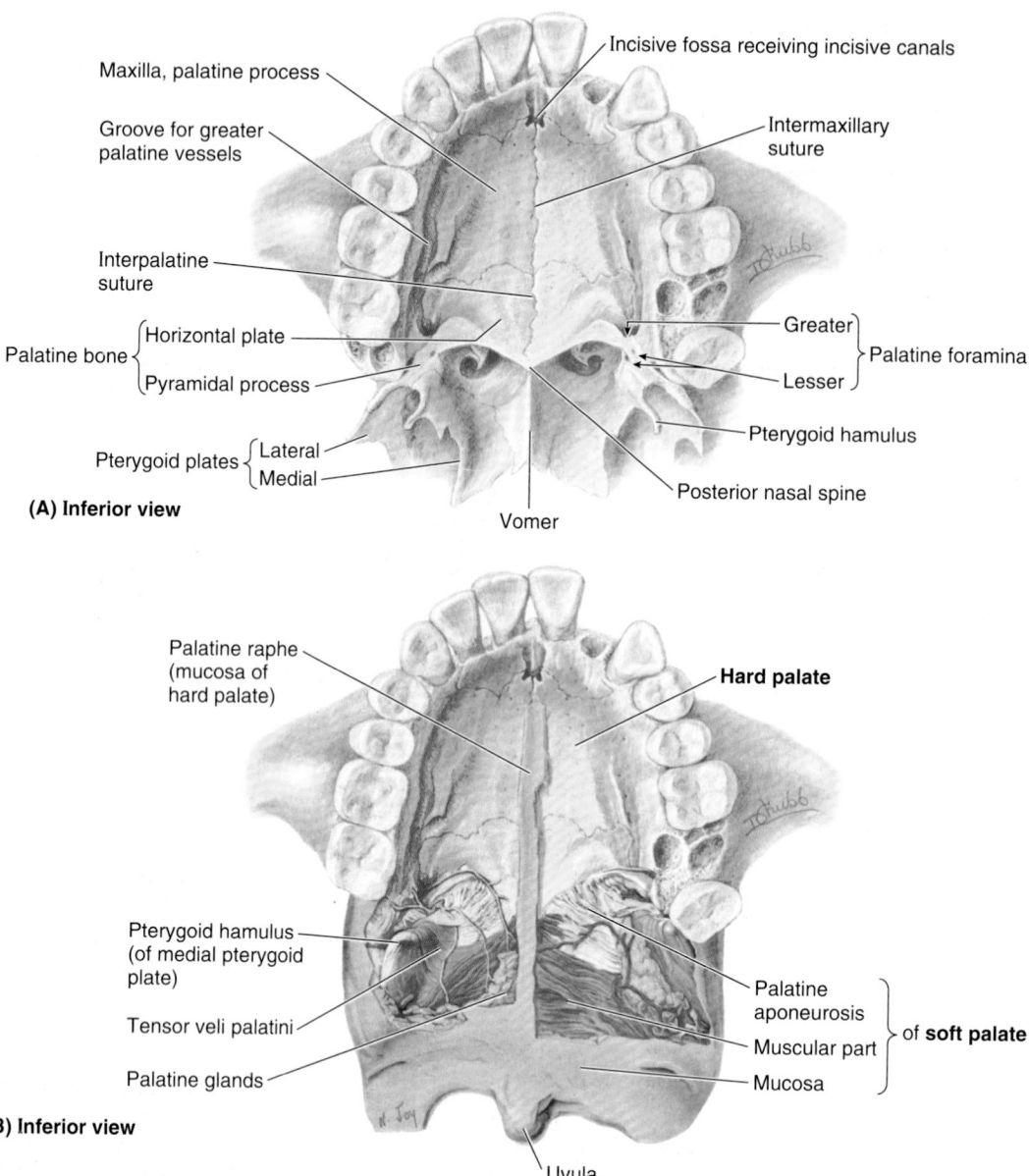

FIGURE 7.84. Palate. The bones and structures of the hard palate and soft palate are shown. The palate has bony (**A**), aponeurotic, and muscular parts (**B**). The mucosa has been removed on each side of the palatine raphe in part **B**. The palatine aponeurosis is formed by the merging of the flattened tendons of the right and left tensor veli palatini muscles. Before they become flattened, each tendon uses the pterygoid hamulus as a trochlea or pulley, redirecting its line of pull approximately 90°.

teeth, is the **incisive papilla.** This elevation of the mucosa lies directly anterior to the underlying *incisive fossa.*

Radiating laterally from the incisive papilla are several parallel **transverse palatine folds** or rugae (Fig. 7.85). These folds assist with manipulation of food during mastication. Passing posteriorly in the midline of the palate from the incisive papilla is a narrow whitish streak, the **palatine raphe.** It may present as a ridge anteriorly and a groove posteriorly. The palatine raphe marks the site of fusion of the embryonic palatal processes (palatal shelves) (Moore et al. 2012). You can feel the transverse palatine folds and the palatine raphe with your tongue.

MUSCLES OF SOFT PALATE

The soft palate may be elevated so that it is in contact with the posterior wall of the pharynx. This closes the **isthmus of the pharynx,** requiring that one breathes through the mouth. The soft palate may also be drawn inferiorly so that it is in contact with the posterior part of the tongue. This closes the *isthmus of the fauces,* so that expired air passes through the nose (even when the mouth is open) and prevents substances in the oral cavity from passing to the pharynx. Tensing the soft palate pulls it tight at an intermediate level so that the tongue may push against it, compressing masticated food and propelling it into the pharynx for swallowing.

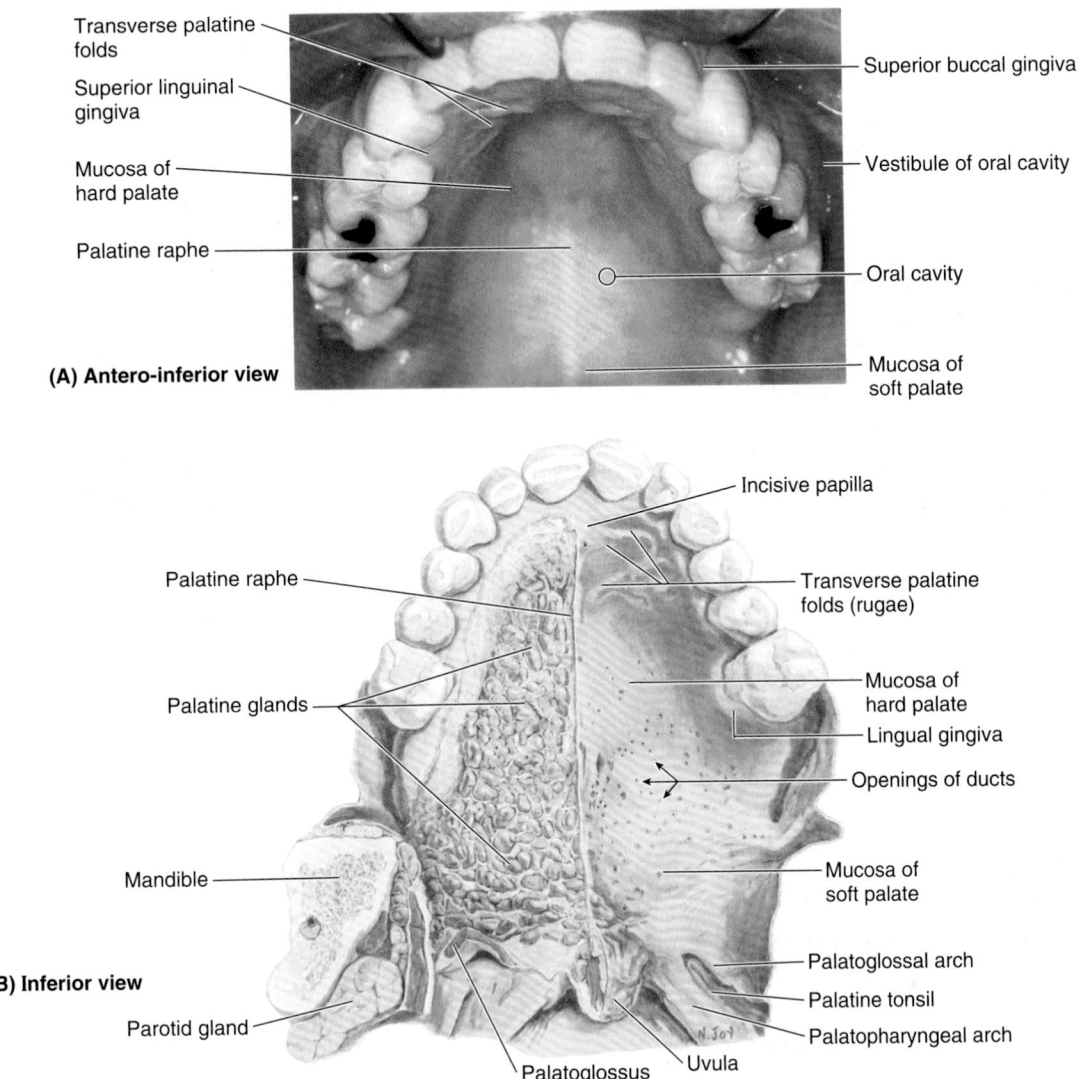

Transverse palatine folds

Superior linguinal gingiva

Mucosa of hard palate

Palatine raphe

Superior buccal gingiva

Vestibule of oral cavity

Oral cavity

Mucosa of soft palate

(A) Antero-inferior view

Incisive papilla

Palatine raphe

Transverse palatine folds (rugae)

Mucosa of hard palate

Lingual gingiva

Palatine glands

Openings of ducts

Mucosa of soft palate

Mandible

Palatoglossal arch

Palatine tonsil

Palatopharyngeal arch

(B) Inferior view

Parotid gland

Palatoglossus

Uvula

FIGURE 7.85. Maxillary teeth and palate. A. The maxillary teeth and the mucosa covering the hard palate in a living person are shown. **B.** The mucous membrane and glands of the palate are demonstrated. The orifices of the ducts of the palatine glands give the mucous membrane an orange-skin appearance. The palatine glands form a thick layer in the soft palate and a thin one in the hard palate; they and are absent in the region of the incisive fossa and the anterior part of the palatine raphe. (Part **A** Courtesy of Dr. B. Liebgott, Professor, Division of Anatomy, Department of Surgery, University of Toronto, Toronto, Ontario, Canada.)

The five muscles of the soft palate arise from the base of the cranium and descend to the palate. The muscles of the soft palate are illustrated in Figure 7.86 and their attachments, nerve supply, and actions are described in Table 7.14. Note that the direction of pull of the belly of the *tensor veli palatini* is redirected approximately 90° because its tendon uses the pterygoid hamulus as a pulley or trochlea, allowing it to pull horizontally on the aponeurosis (Figs. 7.84B and 7.86).

VASCULATURE AND INNERVATION OF PALATE

The palate has a rich blood supply, chiefly from the **greater palatine artery** on each side, a branch of the descending palatine artery (Fig. 7.87). The greater palatine artery passes through the greater palatine foramen and runs anteromedially. The **lesser palatine artery,** a smaller branch of the

descending palatine artery, enters the palate through the lesser palatine foramen and anastomoses with the **ascending palatine artery,** a branch of the facial artery (Fig. 7.87B). The **veins of the palate** are tributaries of the *pterygoid venous plexus.*

The *sensory nerves of the palate* are branches of the maxillary nerve (CNV₂), which branch from the *pterygopalatine ganglion* (Fig. 7.87A). The **greater palatine nerve** supplies the gingivae, mucous membrane, and glands of most of the hard palate. The **nasopalatine nerve** supplies the mucous membrane of the anterior part of the hard palate (Fig. 7.87B). The **lesser palatine nerves** supply the soft palate. The palatine nerves accompany the arteries through the greater and lesser palatine foramina, respectively. Except for the tensor veli palatini supplied by CN V₃, all muscles of the soft palate are supplied through the *pharyngeal plexus of nerves* (see Chapter 8).

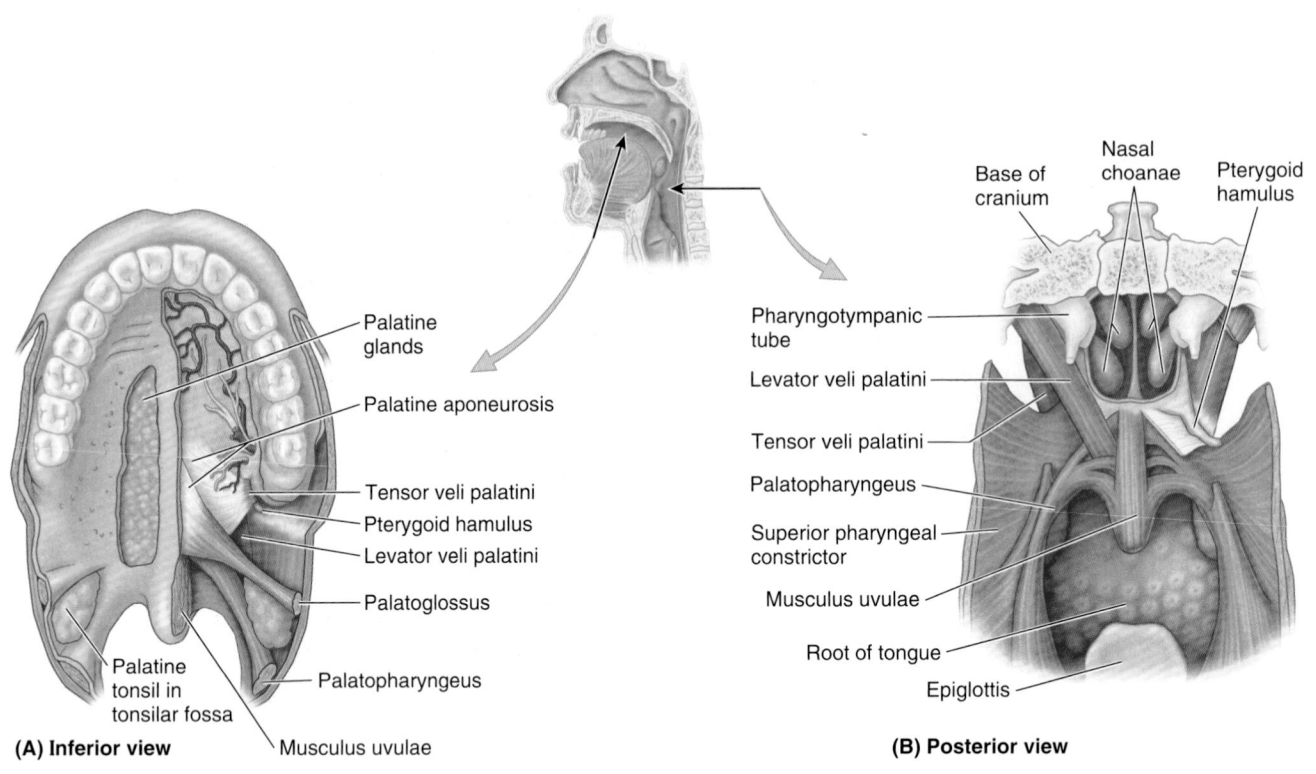

(A) Inferior view

(B) Posterior view

FIGURE 7.86. Muscles of soft palate.

TABLE 7.14. MUSCLES OF SOFT PALATE

Muscle	Superior Attachment	Inferior Attachment	Innervation	Main Action
Tensor veli palatini	Scaphoid fossa of medial pterygoid plate, spine of sphenoid bone, and cartilage of pharyngotympanic tube	Palatine aponeurosis	Medial pterygoid nerve (a branch of mandibular nerve, CN V₃) via otic ganglion	Tenses soft palate and opens mouth of pharyngotympanic tube during swallowing and yawning
Levator veli palatini	Cartilage of pharyngotym-panic tube and petrous part of temporal bone		Pharyngeal branch of vagus nerve (CN X) via pharyngeal plexus	Elevates soft palate during swallowing and yawning
Palatoglossus	Palatine aponeurosis	Side of tongue		Elevates posterior part of tongue and draws soft palate onto tongue
Palatopharyngeus	Hard palate and palatine aponeurosis	Lateral wall of pharynx		Tenses soft palate and pulls walls of pharynx superiorly, anteriorly, and medially during swallowing
Musculus uvulae	Posterior nasal spine and palatine aponeurosis	Mucosa of uvula		Shortens uvula and pulls it superiorly

Tongue

The **tongue** (L. *lingua;* G. *glossa*) is a mobile muscular organ covered with mucous membrane. It can assume a variety of shapes and positions. It is partly in the oral cavity and partly in the oropharynx. The tongue's main functions are articulation (forming words during speaking) and squeezing food into the oropharynx as part of deglutition (swallowing). The tongue is also involved with mastication, taste, and oral cleansing.

PARTS AND SURFACES OF TONGUE

The tongue has a root, body, and apex (Fig. 7.88A). The **root of the tongue** is the attached posterior portion, extending

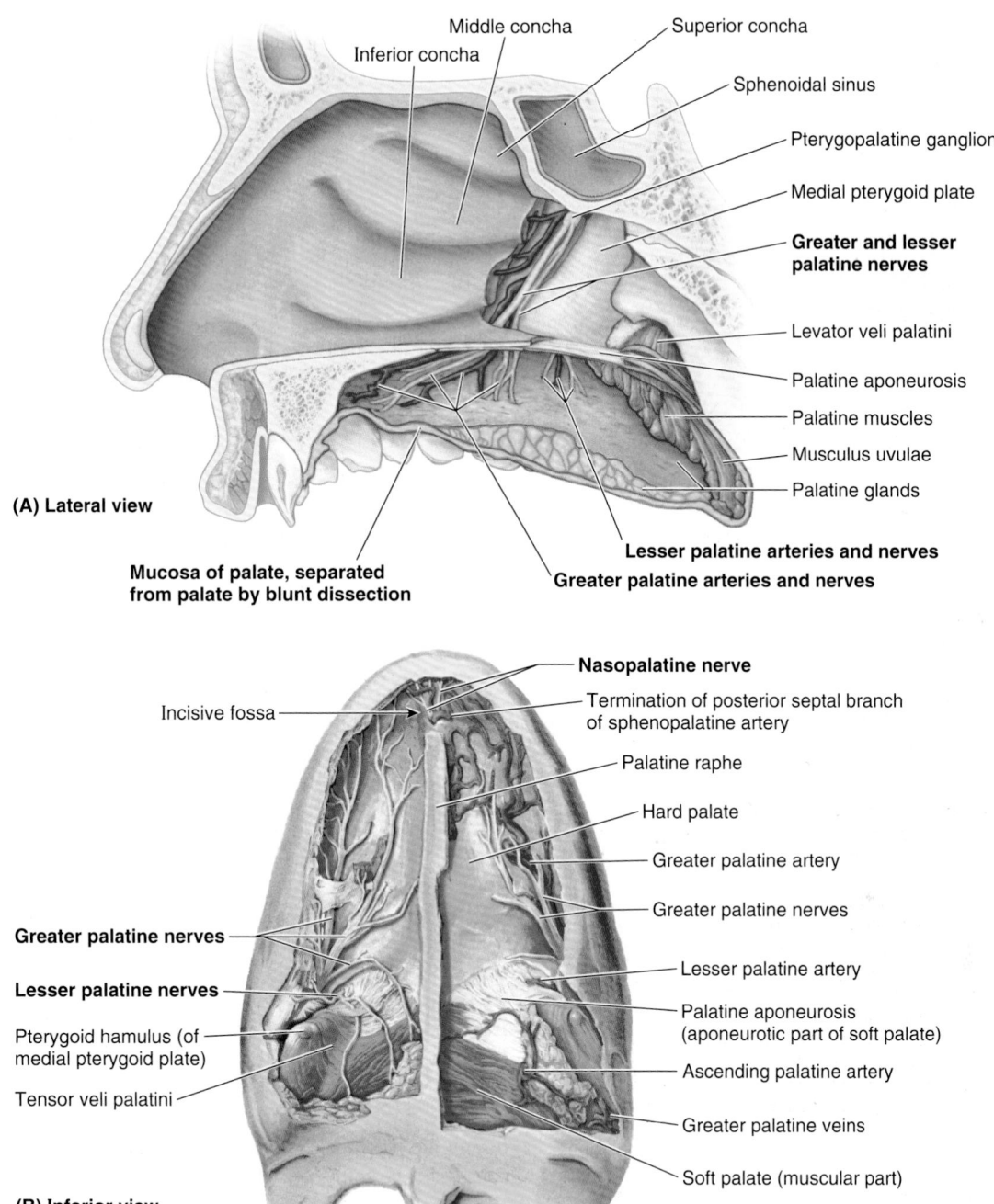

(A) Lateral view

Middle concha

Inferior concha

Superior concha

Sphenoidal sinus

Pterygopalatine ganglion

Medial pterygoid plate

Greater and lesser palatine nerves

Levator veli palatini

Palatine aponeurosis

Palatine muscles

Musculus uvulae

Palatine glands

Mucosa of palate, separated from palate by blunt dissection

Lesser palatine arteries and nerves

Greater palatine arteries and nerves

Incisive fossa

Nasopalatine nerve

Termination of posterior septal branch of sphenopalatine artery

Palatine raphe

Hard palate

Greater palatine artery

Greater palatine nerves

Lesser palatine artery

Palatine aponeurosis (aponeurotic part of soft palate)

Ascending palatine artery

Greater palatine veins

Soft palate (muscular part)

Uvula

Greater palatine nerves

Lesser palatine nerves

Pterygoid hamulus (of medial pterygoid plate)

Tensor veli palatini

(B) Inferior view

FIGURE 7.87. Nerves and vessels of palate. A. In this dissection of the posterior part of the lateral wall of the nasal cavity and the palate, the mucous membrane of the palate, containing a layer of mucous glands, has been separated from the hard and soft regions of the palate by blunt dissection. The posterior ends of the middle and inferior nasal conchae are cut through; these and the mucoperiosteum are pulled off the side wall of the nose as far as the posterior border of the medial pterygoid plate. The perpendicular plate of the palatine bone is broken through to expose the palatine nerves and arteries descending from the pterygopalatine fossa in the palatine canal. **B.** The nerves and vessels of an edentulous palate are shown. The mucosa has been removed on each side of the palatine raphe, demonstrating a branch of the greater palatine nerve on each side and the artery on the lateral side. There are four palatine arteries, two on the hard palate (greater palatine and the terminal branch of posterior nasal septal/sphenopalatine artery) and two on the soft palate (lesser palatine and ascending palatine).

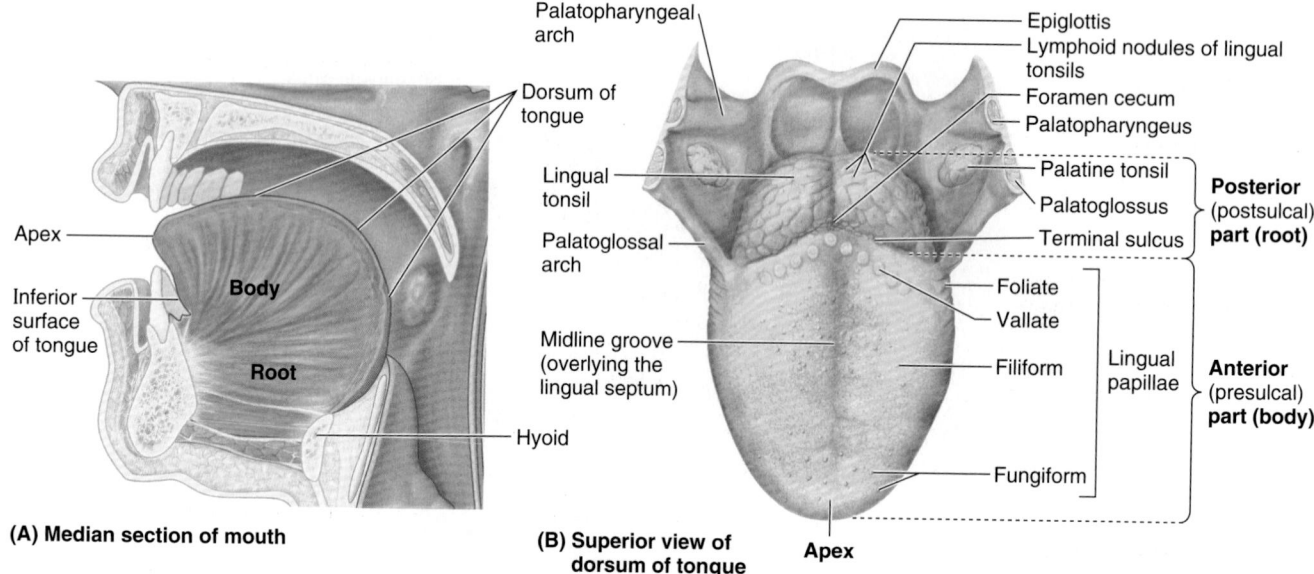

FIGURE 7.88. **Parts and features of tongue.** The anterior free part constituting the majority of the mass of the tongue is the body. The posterior attached portion is the root. The anterior (two thirds) and posterior (third) parts of the dorsum of the tongue are separated by the terminal sulcus (groove) and foramen cecum. *Brackets,* indicate parts of the dorsum of the tongue and do not embrace specific labels.

between the mandible, hyoid, and the nearly vertical posterior surface of the tongue. The **body of the tongue** is the anterior, approximately two thirds of the tongue between root and apex. The **apex** (tip) of the tongue is the anterior end of the body, which rests against the incisor teeth. The body and apex of the tongue are extremely mobile.

The tongue features two surfaces. The more extensive, superior and posterior surface is the *dorsum of the tongue* (commonly referred to as the "top" of the tongue). The *inferior surface of the tongue* (commonly referred to as its "underside") usually rests against the floor of the mouth. The margin of the tongue separating the two surfaces is related on each side to the lingual gingivae and lateral teeth. The **dorsum of the tongue** is characterized by a V-shaped groove, the **terminal sulcus of the tongue,** the angle of which points posteriorly to the *foramen cecum* (Fig. 7.88B). This small pit, frequently absent, is the non-functional remnant of the proximal part of the embryonic thyroglossal duct from which the thyroid gland developed. The terminal sulcus divides the dorsum of the tongue transversely into a presulcal **anterior part** in the oral cavity proper and a postsulcal **posterior part** in the oropharynx.

A **midline groove** divides the anterior part of the tongue into right and left parts. The *mucosa of the anterior part* of the tongue is relatively thin and closely attached to the underlying muscle. It has a rough texture because of numerous small **lingual papillae:**

• **Vallate papillae:** large and flat topped, lie directly anterior to the terminal sulcus and are arranged in a V-shaped row. They are surrounded by deep circular trenches, the walls of which are studded with *taste buds.* The ducts of the serous glands of the tongue open into the trenches.
• **Foliate papillae:** small lateral folds of the lingual mucosa. They are poorly developed in humans.

• **Filiform papillae:** long and numerous, contain afferent nerve endings that are sensitive to touch. These scaly, conical projections are pinkish gray and are arranged in V-shaped rows that are parallel to the terminal sulcus, except at the apex, where they tend to be arranged transversely.
• **Fungiform papillae:** mushroom shaped pink or red spots scattered among the filiform papillae but most numerous at the apex and margins of the tongue.

The vallate, foliate, and most of the fungiform papillae contain taste receptors in the taste buds.

The *mucosa of the posterior part* of the tongue is thick and freely movable. It has no lingual papillae, but the underlying **lymphoid nodules** give this part of the tongue an irregular, cobblestone appearance. The lymphoid nodules are known collectively as the **lingual tonsil.** The pharyngeal part of the tongue constitutes the anterior wall of the oropharynx and can be inspected only with a mirror or downward pressure on the tongue with a tongue depressor.

The **inferior surface of the tongue** is covered with a thin, transparent mucous membrane (Fig. 7.89). This surface is connected to the floor of the mouth by a midline fold called the **frenulum of the tongue.** The frenulum allows the anterior part of the tongue to move freely. On each side of the frenulum, a deep lingual vein is visible through the thin mucous membrane. A **sublingual caruncle** (papilla) is present on each side of the base of the lingual frenulum that includes the *opening of the submandibular duct* from the submandibular salivary gland.

MUSCLES OF TONGUE

The tongue is essentially a mass of muscles that is mostly covered by mucosa (mucous membrane—Fig. 7.90; Table 7.15).

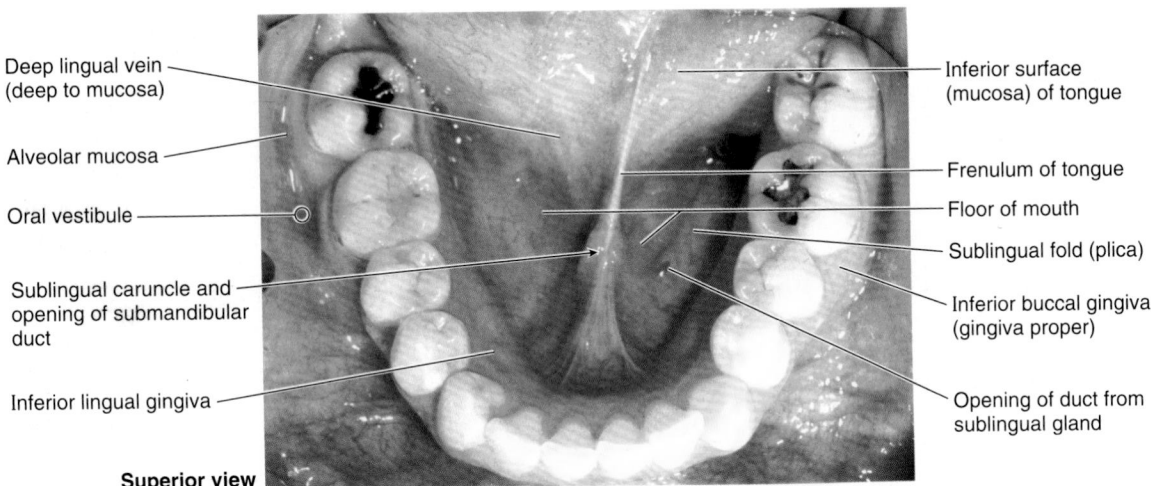

Deep lingual vein
(deep to mucosa)

Alveolar mucosa

Oral vestibule

Sublingual caruncle and
opening of submandibular
duct

Inferior lingual gingiva

Inferior surface
(mucosa) of tongue

Frenulum of tongue

Floor of mouth

Sublingual fold (plica)

Inferior buccal gingiva
(gingiva proper)

Opening of duct from
sublingual gland

Superior view

FIGURE 7.89. Floor of mouth and oral vestibule. The tongue is elevated and retracted superiorly. (Courtesy of Dr. B. Liebgott, Professor, Division of Anatomy, Department of Surgery, University of Toronto, Toronto, Ontario, Canada.)

As in the orbit, it is traditional to provide descriptions of the actions of tongue muscles ascribing (1) a single action to a specific muscle, or (2) implying that a particular movement is the consequence of a single muscle acting. This approach facilitates learning, but greatly oversimplifies the actions of the tongue. The muscles of the tongue do not act in isolation, and some muscles perform multiple actions. Parts of a single muscle are capable of acting independently, producing different, even antagonistic actions. In general, *extrinsic muscles* **alter the position of the tongue**, and *intrinsic muscles* **alter its shape**. The four intrinsic and four extrinsic muscles in each half of the tongue are separated by a median fibrous **lingual septum,** which merges posteriorly with the lingual aponeurosis.

Extrinsic Muscles of Tongue. The **extrinsic muscles of the tongue** (genioglossus, hyoglossus, styloglossus, and palatoglossus) originate outside the tongue and attach to it. They mainly move the tongue but they can alter its shape as well. They are illustrated in Figure 7.90 and their shape, position, attachments, and main actions are described in Table 7.15.

Intrinsic Muscles of Tongue. The superior and inferior longitudinal, transverse, and vertical muscles are confined to the tongue. They have their attachments entirely within the tongue and are not attached to bone. They are illustrated in Figure 7.90, and their shape, position, attachments, and main actions are described in Table 7.15. The **superior** and **inferior longitudinal muscles** act together to make the tongue short and thick and to retract the protruded tongue. The **transverse** and **vertical muscles** act simultaneously to make the tongue long and narrow, which may push the tongue against the incisor teeth or protrude the tongue from the open mouth (especially when acting with the posterior inferior part of the genioglossus).

INNERVATION OF TONGUE

All muscles of the tongue, except the *palatoglossus*, receive motor innervation from CN XII, the **hypoglossal nerve**

(Fig. 7.91). Palatoglossus is a palatine muscle supplied by the *pharyngeal plexus* (see Fig. 8.46A, p. 1035). For general sensation (touch and temperature), the mucosa of the anterior two thirds of the tongue is supplied by the *lingual nerve*, a branch of CN V_3 (Fig. 7.91, 7.95, and 7.96). For special sensation (taste), this part of the tongue, except for the vallate papillae, is supplied the *chorda tympani nerve*, a branch of CN VII. The chorda tympani joins the lingual nerve in the infratemporal fossa and runs anteriorly in its sheath. The mucosa of the posterior third of the tongue and the vallate papillae are supplied by the lingual branch of the *glossopharyngeal nerve* (CN IX) for both general and special sensation (Fig. 7.91). Twigs of the **internal laryngeal nerve,** a branch of the vagus nerve (CN X), supply mostly general but some special sensation to a small area of the tongue just anterior to the epiglottis. These mostly sensory nerves also carry **parasympathetic secretomotor fibers** to serous glands in the tongue.

There are *four basic taste sensations: sweet, salty, sour,* and *bitter*. Sweetness is detected at the apex, saltiness at the lateral margins, and sourness and bitterness at the posterior part of the tongue. All other "tastes" expressed by gourmets are olfactory (smell and aroma).

VASCULATURE OF TONGUE

The *arteries of the tongue* are derived from the **lingual artery,** which arises from the *external carotid artery* (Fig. 7.92). On entering the tongue, the lingual artery passes deep to the hyoglossus muscle. The **dorsal lingual arteries** supply the root of the tongue; the **deep lingual arteries** supply the lingual body. The deep lingual arteries communicate with each other near the apex of the tongue. The dorsal lingual arteries are prevented from communicating by the *lingual septum* (Fig. 7.90C).

The *veins of the tongue* are the **dorsal lingual veins,** which accompany the lingual artery; the **deep lingual veins,** which begin at the apex of the tongue, run posteriorly beside the lingual frenulum to join the **sublingual vein** (Fig. 7.93).

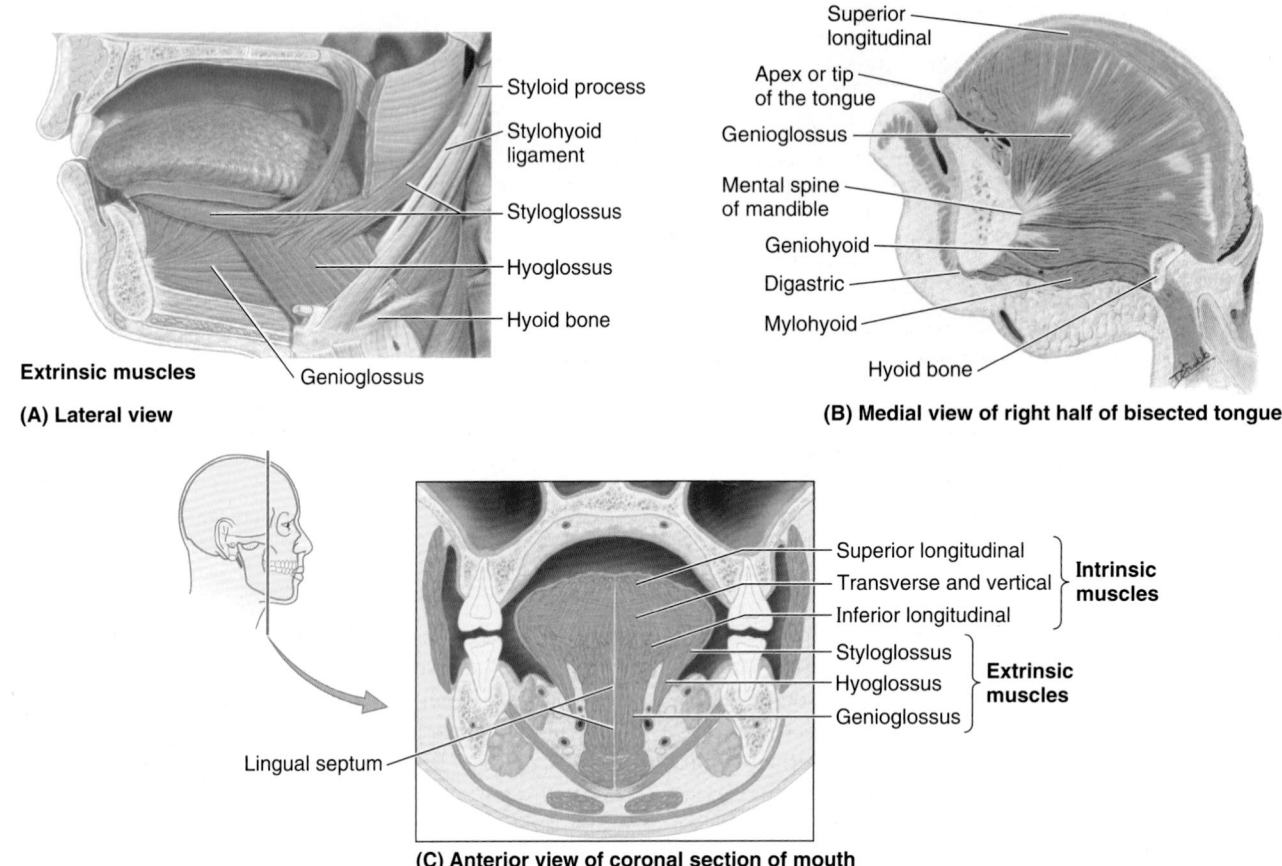

Extrinsic muscles

(A) Lateral view

Styloid process
Stylohyoid ligament
Styloglossus
Hyoglossus
Hyoid bone
Genioglossus

(B) Medial view of right half of bisected tongue

Superior longitudinal
Apex or tip of the tongue
Genioglossus
Mental spine of mandible
Geniohyoid
Digastric
Mylohyoid
Hyoid bone

Superior longitudinal
Transverse and vertical   } **Intrinsic muscles**
Inferior longitudinal
Styloglossus
Hyoglossus   } **Extrinsic muscles**
Genioglossus

Lingual septum

(C) Anterior view of coronal section of mouth

FIGURE 7.90. Muscles of tongue.

TABLE 7.15. MUSCLES OF TONGUE

Muscle	Shape and Position	Proximal Attachment	Distal Attachment	Main Action(s)
Extrinsic muscles of tongue[a]				
Genioglossus	Fan-shaped muscle; constitutes bulk of tongue	Via a short tendon from superior part of mental spine of mandible	Entire dorsum of tongue; inferiormost and posteriormost fibers attach to body of hyoid bone	Bilateral activity depresses tongue, especially central part, creating a longitudinal furrow; posterior part pulls tongue anteriorly for protrusion;[a] most anterior part retracts apex of protruded tongue; unilateral contraction deviates ("wags") tongue to contralateral side
Hyoglossus	Thin, quadrilateral muscle	Body and greater horn of hyoid bone	Inferior aspects of lateral part of tongue	Depresses tongue, especially pulling its sides inferiorly; helps shorten (retrude) tongue
Styloglossus	Small, short triangular muscle	Anterior border of distal styloid process; stylohyoid ligament	Sides of tongue posteriorly, interdigitating with hyoglossus	Retrudes tongue and curls (elevates) its sides, working with genioglossus to form a central trough during swallowing
Palatoglossus[b]	Narrow crescent-shaped palatine muscle; forms posterior column of isthmus of fauces	Palatine aponeurosis of soft palate	Enters posterolateral tongue transversely, blending with intrinsic transverse muscles	Capable of elevating posterior tongue or depressing soft palate; most commonly acts to constrict isthmus of fauces

[a] Except for palatoglossus, the muscles of the tongue are innervated by the hypoglossal nerve (CN XII).
[b] Actually a palatine muscle, the palatoglossus is innervated by the vagus nerve (CN X).

TABLE 7.15. MUSCLES OF TONGUE (Continued)

Muscle	Shape and Position	Proximal Attachment	Distal Attachment	Main Action(s)
Intrinsic muscles of tongue[a]				
Superior longitudinal	Thin layer deep to mucous membrane of dorsum	Submucosal fibrous layer and median fibrous septum	Margins of tongue and mucous membrane	Curls tongue longitudinally upward, elevating apex and sides of tongue; shortens (retrudes) tongue
Inferior longitudinal	Narrow band close to inferior surface	Root of tongue and body of hyoid bone	Apex of tongue	Curls tongue longitudinally downward, depressing apex; shortens (retrudes) tongue
Transverse	Deep to superior longitudinal muscle	Median fibrous septum	Fibrous tissue at lateral lingual margins	Narrows and elongates (protrudes) tongue[c]
Vertical	Fibers intersect transverse muscle	Submucosal fibrous layer of dorsum of tongue	Inferior surface of borders of tongue	Flattens and broadens tongue[c]

[c] Act simultaneously to protrude tongue.

The sublingual veins in elderly people are often varicose (enlarged and tortuous). Some or all of them may drain into the IJV, or they may do so indirectly, joining first to form a **lingual vein** that accompanies the initial part of the lingual artery.

The **lymphatic drainage of the tongue** is exceptional. Most of the lymphatic drainage converges toward and follows the venous drainage; however, lymph from the tip of the tongue, frenulum, and central lower lip runs an independent course (Fig. 7.94). *Lymph from the tongue takes four routes:*

1. Lymph from the *root* drains bilaterally into the **superior deep cervical lymph nodes.**
2. Lymph from the *medial part of the body* drains bilaterally and directly to the **inferior deep cervical lymph nodes.**
3. Lymph from the right and left *lateral parts of body* drains to the **submandibular lymph nodes** on the ipsilateral side.
4. The *apex and frenulum* drain to the **submental lymph nodes,** the medial portion draining bilaterally.

All lymph from the tongue ultimately drains to the deep cervical nodes, and passes via the jugular venous trunks into the venous system at the right and left venous angles.

Salivary Glands

The **salivary glands** are the parotid, submandibular, and sublingual glands (Fig. 7.95). The clear, tasteless, odorless

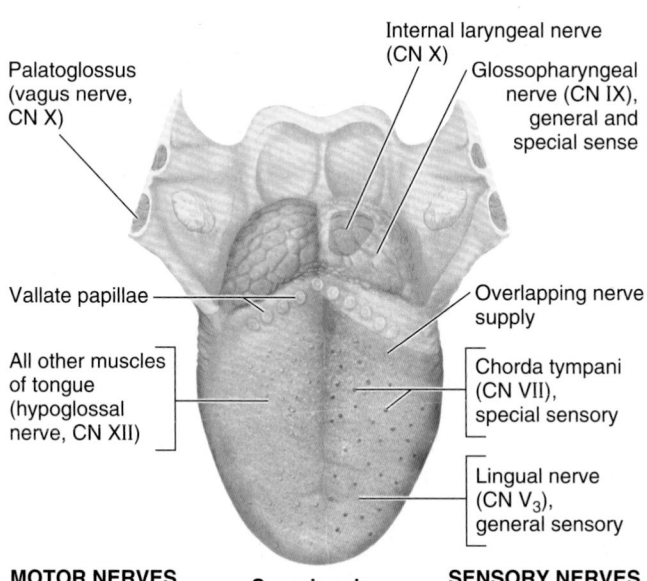

FIGURE 7.91. Nerve supply to parts of tongue.

MOTOR NERVES — Palatoglossus (vagus nerve, CN X); Vallate papillae; All other muscles of tongue (hypoglossal nerve, CN XII); Superior view

SENSORY NERVES — Internal laryngeal nerve (CN X); Glossopharyngeal nerve (CN IX), general and special sense; Overlapping nerve supply; Chorda tympani (CN VII), special sensory; Lingual nerve (CN V₃), general sensory

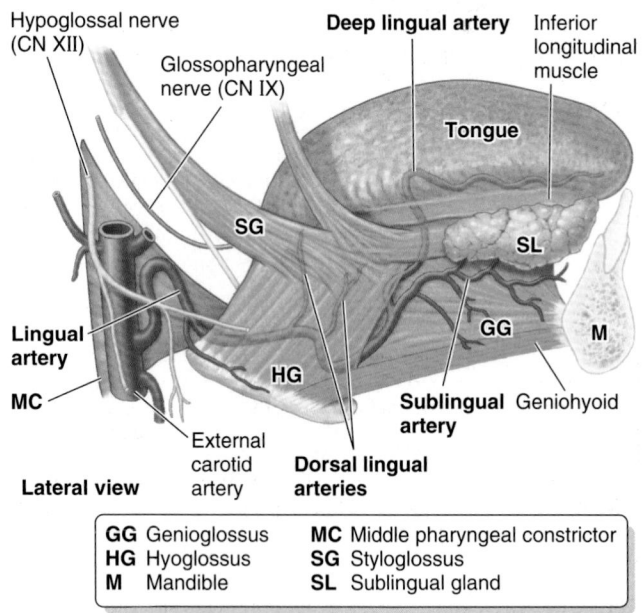

FIGURE 7.92. Blood supply of tongue. The main artery to the tongue is the lingual, a branch of the external carotid artery. The dorsal lingual arteries provide the blood supply to the root of the tongue and a branch to the palatine tonsil. The deep lingual arteries supply the body of the tongue. The sublingual arteries provide the blood supply to the floor of the mouth, including the sublingual glands.

Labels in figure: Hypoglossal nerve (CN XII); Glossopharyngeal nerve (CN IX); Deep lingual artery; Inferior longitudinal muscle; Tongue; SG; SL; Lingual artery; MC; GG; M; HG; External carotid artery; Dorsal lingual arteries; Sublingual artery; Geniohyoid; Lateral view

GG Genioglossus	MC Middle pharyngeal constrictor
HG Hyoglossus	SG Styloglossus
M Mandible	SL Sublingual gland

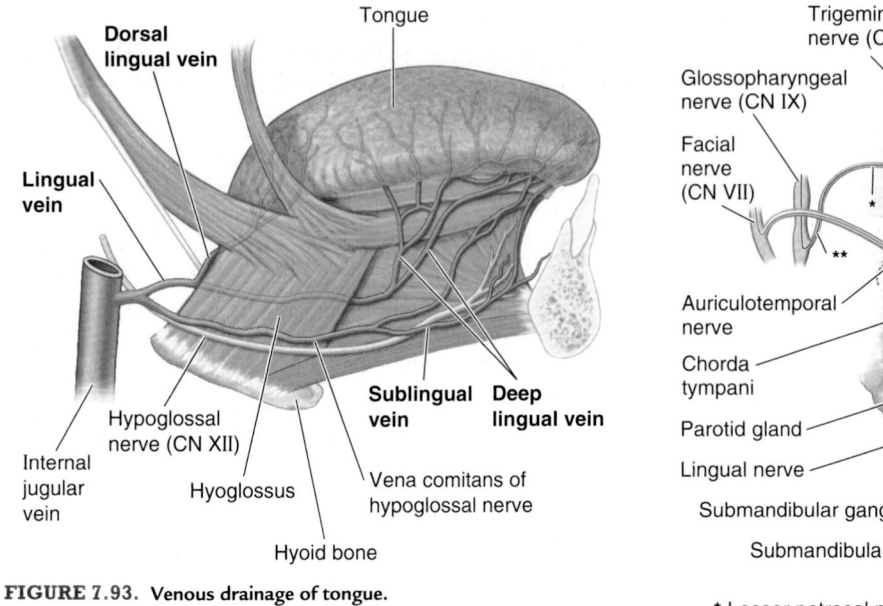

FIGURE 7.93. Venous drainage of tongue.

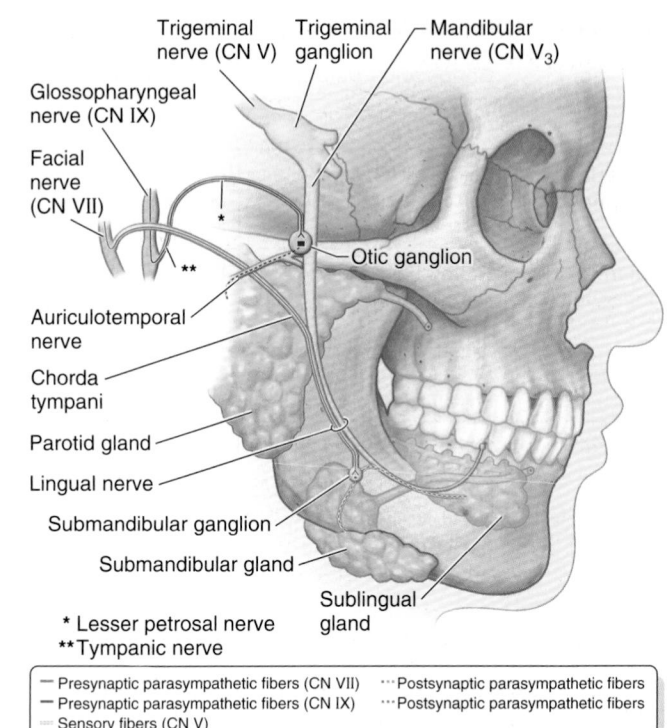

* Lesser petrosal nerve
**Tympanic nerve

— Presynaptic parasympathetic fibers (CN VII)	···Postsynaptic parasympathetic fibers
— Presynaptic parasympathetic fibers (CN IX)	···Postsynaptic parasympathetic fibers
Sensory fibers (CN V)	

Postsynaptic sympathetic fibers from superior cervical ganglion travel with arteries to glands in peri-arterial plexuses

FIGURE 7.95. Innervation of salivary glands.

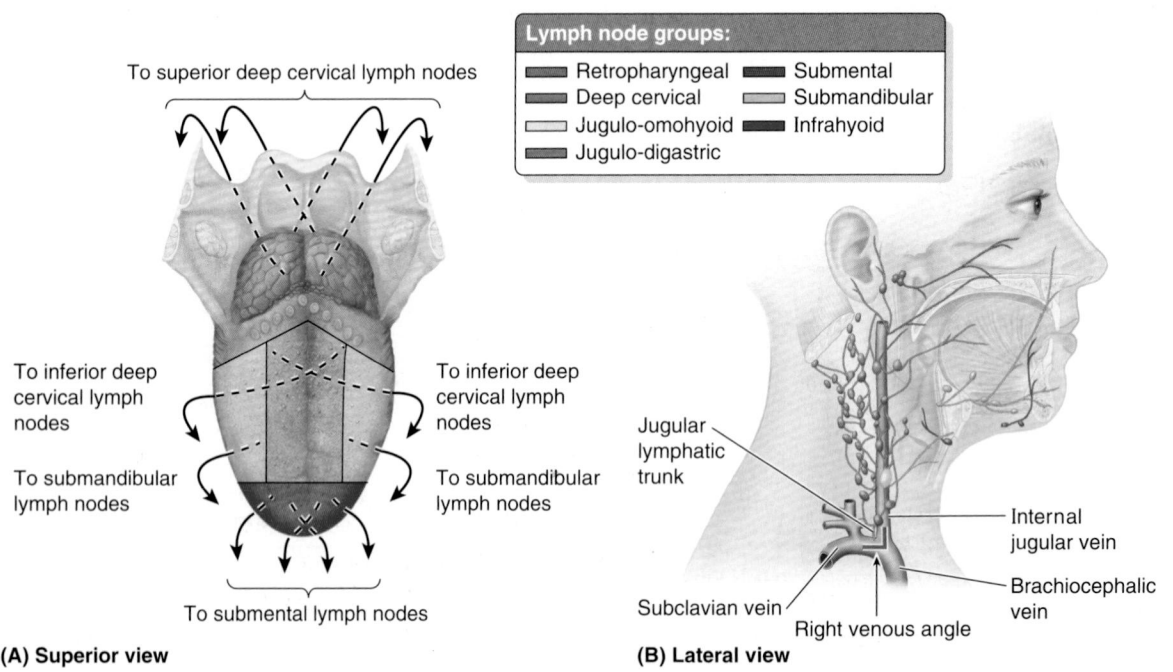

Lymph node groups:

- Retropharyngeal
- Deep cervical
- Jugulo-omohyoid
- Jugulo-digastric
- Submental
- Submandibular
- Infrahyoid

(A) Superior view

(B) Lateral view

FIGURE 7.94. Lymphatic drainage of tongue. **A.** The dorsum of the tongue is shown. **B.** Lymph drains to the submental, submandibular, and superior and inferior deep cervical lymph nodes, including the jugulodigastric and jugulo-omohyoid nodes. Extensive communications occur across the midline of the tongue.

viscid fluid, **saliva,** secreted by these glands and the mucous glands of the oral cavity:

- Keeps the mucous membrane of the mouth moist.
- Lubricates the food during mastication.
- Begins the digestion of starches.
- Serves as an intrinsic "mouthwash."
- Plays significant roles in the prevention of tooth decay and in the ability to taste.

In addition to the main salivary glands, small **accessory salivary glands** are scattered over the palate, lips, cheeks, tonsils, and tongue. The *parotid glands*, the largest of the three paired salivary glands, were discussed earlier in this chapter (p. 914). The parotid glands are located lateral and posterior to the rami of the mandible and masseter muscles, within unyielding fibrous sheaths. The parotid glands drain anteriorly via single ducts that enter the oral vestibule opposite the second maxillary molar teeth (see Fig. 7.65).

SUBMANDIBULAR GLANDS

The **submandibular glands** lie along the body of the mandible, partly superior and partly inferior to the posterior half of the mandible, and partly superficial and partly deep to the mylohyoid muscle (Fig. 7.96). The **submandibular duct,** approximately 5 cm long, arises from the portion of the gland that lies between the mylohyoid and hyoglossus muscles. Passing from lateral to medial, the *lingual nerve* loops under the duct that runs anteriorly, opening by one to three orifices on a small sublingual papilla beside the base of the lingual frenulum (Fig. 7.96B). The orifices of the submandibular

ducts are visible, and saliva can often be seen trickling from them (or spraying from them during yawning). The **arterial supply of the submandibular glands** is from the **submental arteries** (Fig. 7.92). The **veins** accompany the arteries. The **lymphatic vessels** of the glands end in the *deep cervical lymph nodes*, particularly the *jugulo-omohyoid node* (Fig. 7.94B).

The submandibular glands are supplied by presynaptic parasympathetic secretomotor fibers conveyed from the facial nerve to the lingual nerve by the chorda tympani nerve, which synapse with postsynaptic neurons in the submandibular ganglion (Fig. 7.95). The latter fibers accompany arteries to reach the gland, along with vasoconstrictive postsynaptic sympathetic fibers from the superior cervical ganglion.

SUBLINGUAL GLANDS

The **sublingual glands** are the smallest and most deeply situated of the salivary glands (Fig. 7.96). Each almond-shaped gland lies in the floor of the mouth between the mandible and the genioglossus muscle. The glands from each side unite to form a horseshoe-shaped mass around the connective tissue core of the lingual frenulum. Numerous small **sublingual ducts** open into the floor of the mouth along the sublingual folds. The **arterial supply of the sublingual glands** is from the *sublingual* and *submental arteries*, branches of the lingual and facial arteries, respectively (Fig. 7.92). The **nerves of the glands** accompany those of the submandibular gland. Presynaptic parasympathetic secretomotor fibers are conveyed by the facial, chorda tympani, and lingual nerves to synapse in the submandibular ganglion (Fig. 7.95).

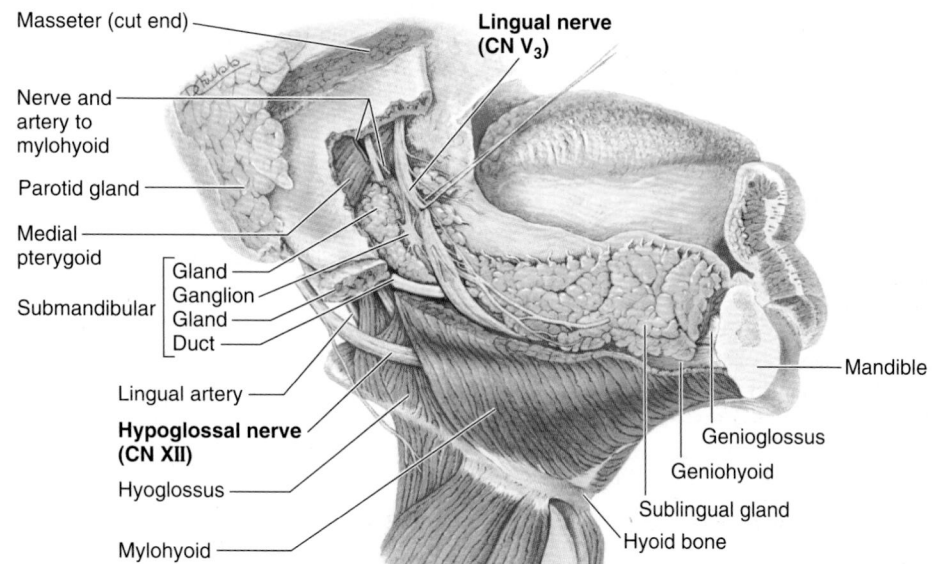

(A) Right lateral view

FIGURE 7.96. **Parotid, submandibular, and sublingual salivary glands. A.** The body and parts of the ramus of the mandible have been removed. The parotid gland contacts the deep part of the submandibular gland posteriorly. Fine ducts passing from the superior border of the sublingual gland open on the sublingual fold.

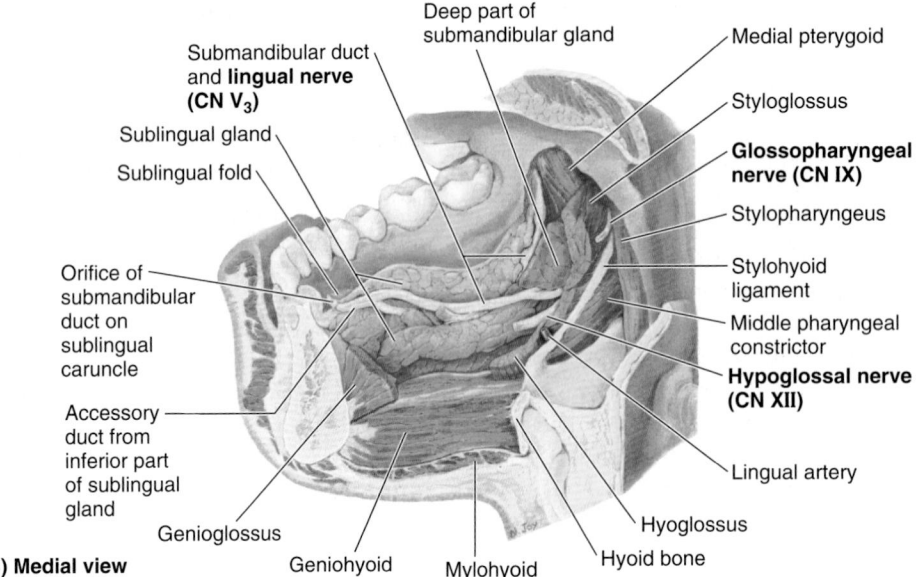

Deep part of submandibular gland

Submandibular duct and **lingual nerve (CN V₃)**

Sublingual gland

Sublingual fold

Orifice of submandibular duct on sublingual caruncle

Accessory duct from inferior part of sublingual gland

Medial pterygoid

Styloglossus

Glossopharyngeal nerve (CN IX)

Stylopharyngeus

Stylohyoid ligament

Middle pharyngeal constrictor

Hypoglossal nerve (CN XII)

Lingual artery

Hyoglossus

Hyoid bone

Genioglossus   Geniohyoid   Mylohyoid

(B) Medial view

FIGURE 7.96. *(Continued)* **B.** The right sublingual and submandibular glands and the floor of the mouth are shown; the tongue has been excised. The orifice of the duct of the submandibular gland is visible at the anterior end of the sublingual fold. The submandibular duct adheres to the medial side of the sublingual gland; here it is receiving, as it sometimes does, a large accessory duct from the inferior part of the sublingual gland. The sublingual carunculae (*) are bilateral papillae flanking the frenulum of the tongue, each bearing the opening of the ipsilateral submandibular duct.

ORAL REGION

Cleft Lip

Cleft lip (*harelip*, a misnomer) is a birth defect (usually of the upper lip) that occurs in 1 of 1000 births; 60–80% of affected infants are males. The clefts vary from a small notch in the transitional zone of the lip and vermilion border to a notch that extends through the lip into the nose (Fig. B7.32). In severe cases, the cleft extends deeper and is continuous with a cleft in the palate. Cleft lip may be unilateral or bilateral (Moore et al., 2012).

Cyanosis of Lips

The lips, like fingers, have an abundant, relatively superficial arterial blood flow. Because of this, they can lose a disproportionate amount of body heat when exposed to a cold environment. Both lips are provided with sympathetically innervated arteriovenous anastomoses, capable of redirecting a considerable portion of the blood back to the body core, reducing heat loss while producing cyanosis of the lips and fingers. *Cyanosis*, a dark bluish or purplish coloration of the lips and mucous membranes, results from deficient oxygenation of capillary blood and is a sign of many pathologic conditions. The common blue discoloration of the lips owing to cold exposure does not indicate pathology; instead, it results from the decreased blood flow in the capillary beds supplied by the superior and inferior labial arteries and the increased extraction of oxygen. Simple warming restores the normal coloring of the lips.

Large Labial Frenulum

An excessively large superior labial frenulum in children may cause a space between the central incisor teeth. Resection of the frenulum and the underlying connective tissue (*frenulectomy*) between the incisors allows approximation of the teeth, which may require an orthodontic appliance ("brace"). A large lower labial frenulum in adults may

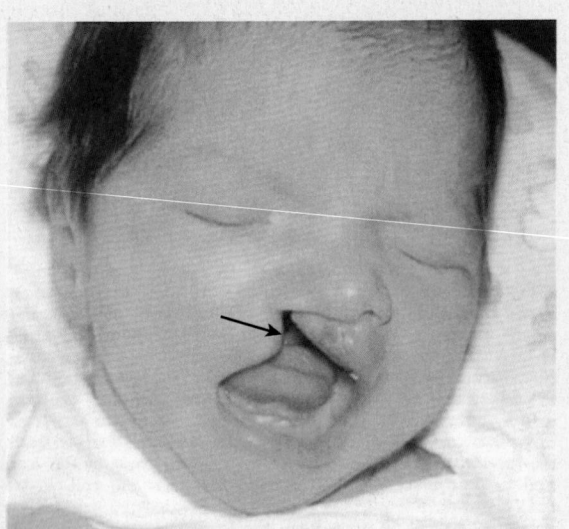

FIGURE B7.32. Unilateral cleft lip (*arrow*).

pull on the labial gingiva and contribute to *gingival recession,* which results in an abnormal exposure of the roots of the teeth.

Gingivitis

Improper oral hygiene results in food and bacterial deposits in tooth and gingival crevices that may cause inflammation of the gingivae (*gingivitis*). The gingivae swell and redden as a result. If untreated, the disease spreads to other supporting structures, including alveolar bone, producing *periodontitis* (inflammation and destruction of bone and periodontium). *Dento-alveolar abscesses* (collections of pus resulting from death of inflamed tissues) may drain to the oral cavity and lips.

Dental Caries, Pulpitis, and Tooth Abscesses

Acid, enzymes, or both produced by oral bacteria may break down (decay) the hard tissues of a tooth. This results in the formation of *dental caries* (*cavities*) (Fig. B7.33A). Neglected dental caries eventually invade and inflame tissues in the pulp cavity (Fig. B7.33B). Invasion of the pulp by a deep carious lesion results in infection and irritation of the tissues (*pulpitis*). Because the pulp cavity is a rigid space, the swollen tissues cause considerable pain (*toothache*). If untreated, the small vessels in the root canal may die from the pressure of the swollen tissue, and the infected material may pass through the apical canal and foramen into the periodontal tissues (Fig. B7.33C). An infective process develops and spreads through the root canal to the alveolar bone, producing an *abscess* (*peri-apical disease*). If untreated, loss of the tooth may occur with an abscess remaining (Fig. B7.33D). Treatment involves removal of the decayed tissue and restoration of the anatomy of the tooth with prosthetic dental material (commonly referred to as a "filling") (Fig. B7.33E).

Pus from an abscess of a maxillary molar tooth may extend into the nasal cavity or the maxillary sinus. The roots of the maxillary molar teeth are closely related to the floor of this sinus. As a consequence, infection of the pulp cavity may also cause sinusitis, or sinusitis may stimulate nerves entering the teeth and simulate a toothache. The roots of mandibular teeth are closely related to the mandibular canal (Fig. B7.33E), and abscess formation may compress the nerve causing pain that may be referred to (perceived as coming from) more anterior teeth.

Supernumerary Teeth (Hyperdontia)

Supernumerary teeth are teeth present in addition to the normal complement (number) of teeth. They may be single, multiple, unilateral or bilateral, erupted or unerupted, and in one or both maxillary and mandibular alveolar arches (Fig. B7.34). They may occur in both deciduous and permanent dentitions, but more commonly occur in the latter. The presence of a single supernumerary (accessory) tooth is

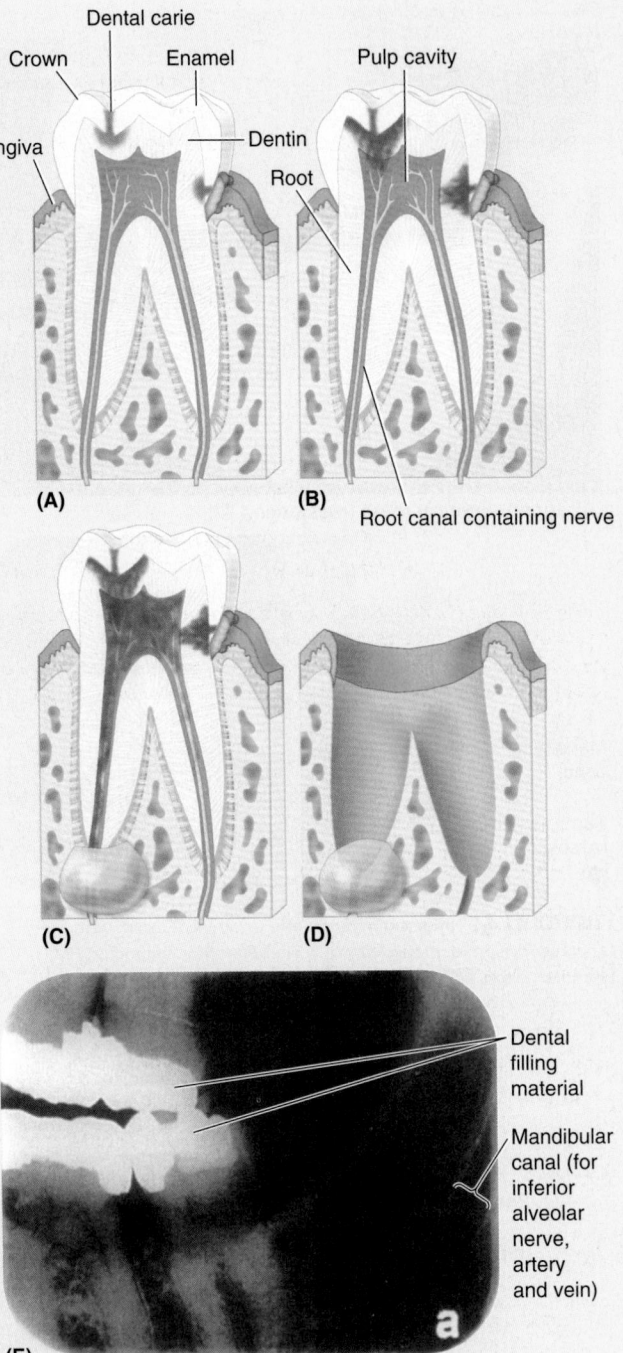

FIGURE B7.33. Dental caries and peri-apical disease.

usually seen in the anterior maxilla. The most common supernumerary tooth is a *mesiodens,* which is a malformed, peg-like tooth that occurs between the maxillary central incisor teeth (Fig. B7.34A). A *supernumerary tooth* occurs in addition to the normal number but resembles the size, shape, or placement of normal teeth. An *accessory tooth* does not resemble the form or disposition of a normal tooth (Fig. B7.34B).

Multiple supernumerary teeth are rare in individuals with no other associated diseases or syndromes, such as cleft lip or

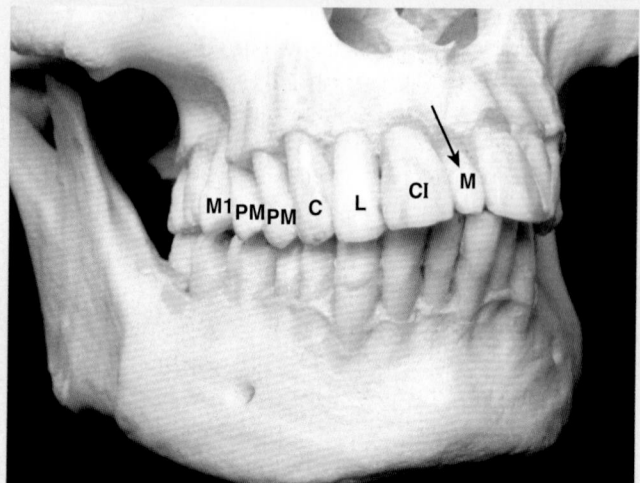

(A) Supernumerary tooth (mesiodens–M)

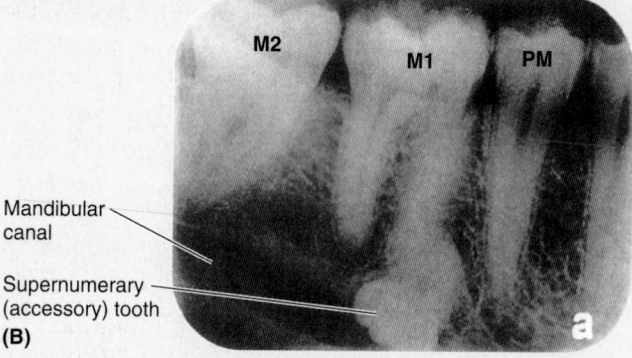

Mandibular canal

Supernumerary (accessory) tooth

(B)

FIGURE B7.34. Supernumerary teeth. *C,* canine; *CI,* central incisor; *LI,* lateral incisor; *M,* mesiodens; *M1,* 1st molar; *M2,* 2nd molar; *PM,* premolar; *arrow,* supernumerary tooth.

palate, or cranial dysplasia (malformation). The supernumerary teeth can cause problems for the eruption and alignment of normal dentition, and are usually surgically extracted.

Extraction of Teeth

Sometimes it is not practical to restore a tooth because of extreme tooth destruction. The only alternative is tooth extraction. A tooth may lose its blood supply as a result of trauma. The blow to the tooth disrupts the blood vessels entering and leaving the apical foramen. It is not always possible to save the tooth. Supernumerary teeth are also extracted.

The *lingual nerve* is closely related to the medial aspect of the 3rd molar teeth; therefore, caution is taken to avoid injuring this nerve during their extraction. Damage to this nerve results in altered sensation to the ipsilateral side of the tongue.

Unerupted 3rd molars are common dental problems; these teeth are the last to erupt, usually when people are in their late teens or early 20s. Often there is not enough room for these molars to erupt, and they become lodged (impacted) under or against the 2nd molar teeth (Fig. B7.35, insets). If impacted 3rd molars become painful, they are usually removed. When doing so, the surgeon takes care not to injure the alveolar nerves (Figs. 7.79A and B7.33E).

Dental Implants

Following extraction of a tooth, or fracture of a tooth at the neck, a prosthetic crown may be placed on an abutment (metal peg) inserted into a metal socket

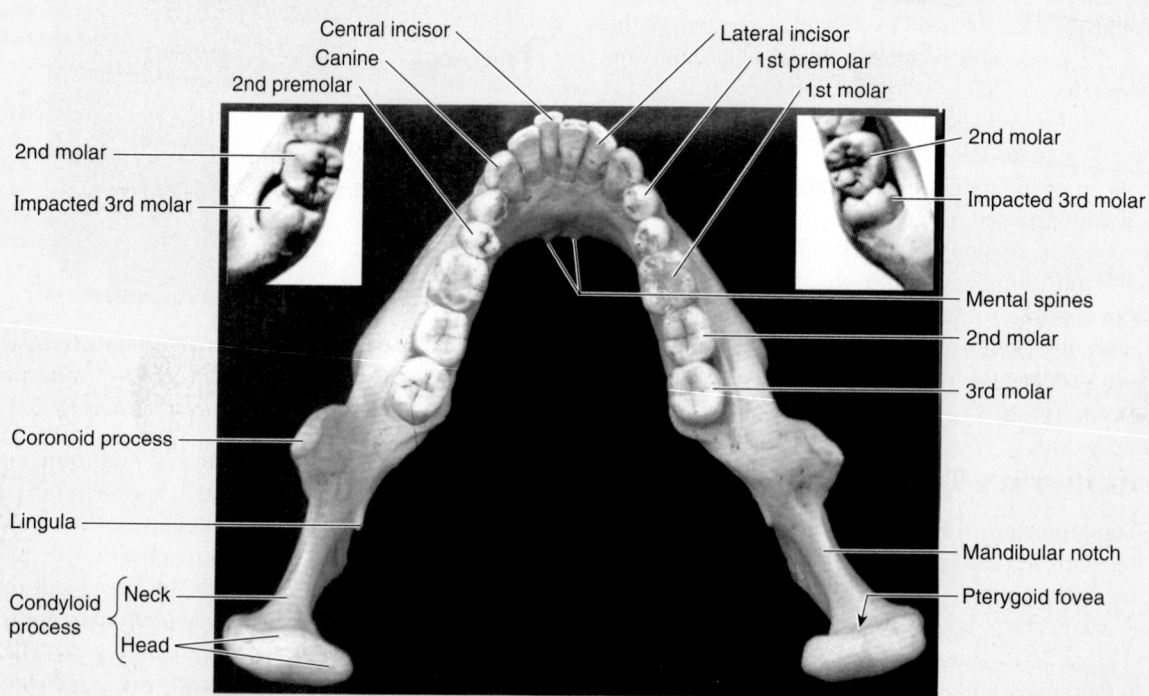

FIGURE B7.35. Normal adult mandible with full dentition. *Insets,* impacted 3rd molars.

surgically implanted into the alveolar bone (Fig. B7.36). A procedure to augment the alveolar bone with calf or cadaveric bone may be required before the socket can be implanted. A waiting period of several months may be necessary to allow bone growth around the implanted socket before the abutment and prosthetic crown are mounted.

molar teeth. This nerve block anesthetizes all the palatal mucosa and lingual gingivae posterior to the maxillary canine teeth and the underlying bone of the palate. Branches of the greater palatine arteries should be avoided. The anesthetic should be injected slowly to prevent stripping of the mucosa from the hard palate.

Cleft Palate

Cleft palate, with or without cleft lip, occurs in approximately 1 of 2500 births and is more common in females than in males (Moore et al., 2012). The cleft may involve only the uvula, giving it a fishtail appearance, or it may extend through the soft and hard regions of the palate (Fig. B7.37). In severe cases associated with cleft lip, the cleft palate extends through the alveolar processes of the maxillae and the lips on both sides. The embryological basis of cleft palate is failure of mesenchymal masses in the lateral palatine processes to meet and fuse with each other, with the nasal septum, and/or with the posterior margin of the median palatine process.

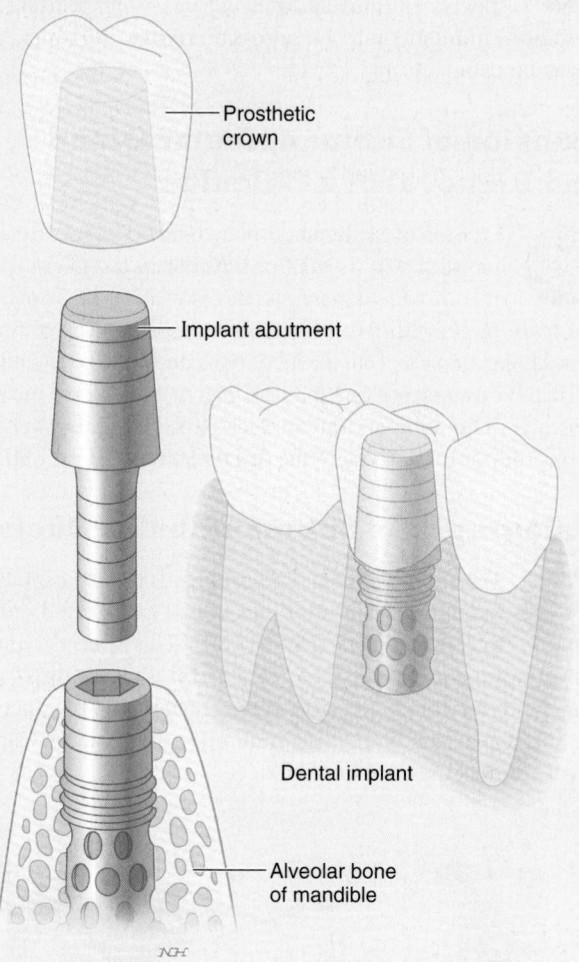

FIGURE B7.36. Dental implants.

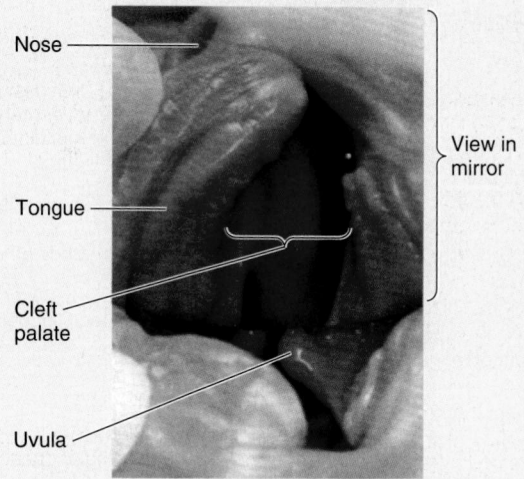

FIGURE B7.37. Bilateral cleft palate.

Nasopalatine Block

The nasopalatine nerves can be anesthetized by injecting anesthetic into the incisive fossa in the hard palate. The needle is inserted immediately posterior to the incisive papilla. Both nerves are anesthetized by the same injection where they emerge through the incisive fossa (Fig. 7.87B). The affected tissues are the palatal mucosa, the lingual gingivae and alveolar bone of the six anterior maxillary teeth, and the hard palate.

Greater Palatine Block

The greater palatine nerve can be anesthetized by injecting anesthetic into the greater palatine foramen. The nerve emerges between the 2nd and the 3rd

Gag Reflex

It is possible to touch the anterior part of the tongue without feeling discomfort; however, when the posterior part is touched, the individual gags. CN IX and CN X are responsible for the muscular contraction of each side of the pharynx. Glossopharyngeal branches provide the afferent limb of the gag reflex.

Paralysis of Genioglossus

When the genioglossus muscle is paralyzed, the tongue has a tendency to fall posteriorly, obstructing the airway and presenting the risk of suffocation. Total relaxation of the genioglossus muscles occurs during general anesthesia; therefore, an airway is inserted in an anesthetized person to prevent the tongue from relapsing.

Injury to Hypoglossal Nerve

 Trauma, such as a fractured mandible, may injure the hypoglossal nerve (CN XII), resulting in paralysis and eventual atrophy of one side of the tongue. The tongue deviates to the paralyzed side during protrusion because of the action of the unaffected genioglossus muscle on the other side.

Sublingual Absorption of Drugs

 For quick absorption of a drug, for instance, when nitroglycerin is used as a vasodilator in persons with *angina pectoris;* the pill or spray is put under the tongue where it dissolves and enters the deep lingual veins in <1 min (Figs. 7.89 and B7.38).

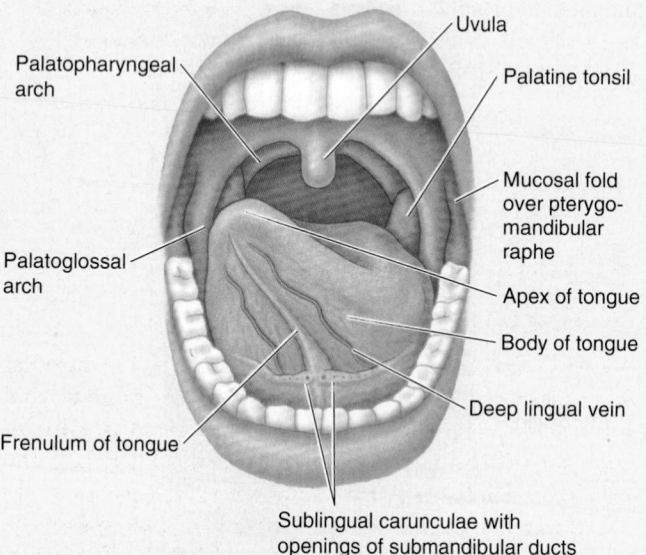

FIGURE B7.38.

Lingual Carcinoma

 A *lingual carcinoma* in the posterior part of the tongue metastasizes to the superior deep cervical lymph nodes on both sides, whereas a tumor in the anterior part usually does not metastasize to the inferior deep cervical lymph nodes until late in the disease. Because the nodes are closely related to the IJV, metastases from the tongue

may be distributed through the submental and submandibular regions and along the IJVs in the neck (Fig. 7.94).

Frenectomy

An overly large frenulum of the tongue (tongue-tie) interferes with tongue movements and may affect speech. In unusual cases, a *frenectomy* (cutting the frenulum) in infants may be necessary to free the tongue for normal movements and speech.

Excision of Submandibular Gland and Removal of a Calculus

Excision of a submandibular gland because of a calculus (stone) in its duct or a tumor in the gland is not uncommon. Skin incision is made at least 2.5 cm inferior to the angle of the mandible to avoid injury to the marginal mandibular branch of the facial nerve (see Fig. 7.65). Caution must also be taken not to injure the lingual nerve when incising the duct. The submandibular duct passes directly over the nerve inferior to the neck of the 3rd molar tooth (Fig. 7.96).

Sialography of Submandibular Ducts

The submandibular salivary glands may be examined radiographically after injection of a contrast medium into their ducts (Fig. B7.39). This special type of radiograph (*sialogram*) demonstrates the salivary ducts and some secretory units. Because of the small size of the ducts of the sublingual glands and their multiplicity, one cannot usually inject contrast medium into the ducts.

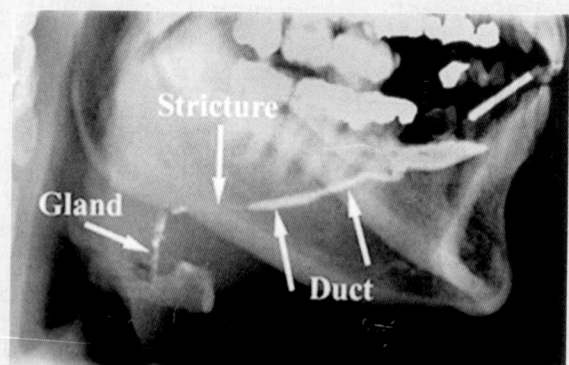

Sialogram of submandibular duct and gland. Lateral view.

FIGURE B7.39. Sialogram of submandibular duct and gland.

The Bottom Line

ORAL REGION

Oral cavity: The oral cavity (mouth) is the primary portal of the alimentary system, and a secondary portal for the respiratory system, especially important for speech in the latter case.

♦ The oral cavity extends from the oral fissure to the oropharyngeal isthmus. ♦ The oral cavity is divided by the upper and lower jaws and their dental arches into a superficial oral vestibule (between the lips and cheeks and the gingival and teeth) and a deeper oral cavity proper (internal to the jaws and dental arcades). ♦ The oral cavity (and specifically the oral vestibule) is bounded by the lips and cheeks, which are flexible dynamic musculofibrous folds containing muscles, neurovasculature, and mucosal glands, covered superficially with skin and deeply with oral mucosa. ♦ The cheeks also include buccal fat-pads.

Teeth: The strong alveolar parts of the maxilla and mandible contain, in sequence, two sets of teeth (20 deciduous and 32 permanent teeth). ♦ The crowns of the teeth project from the gingiva, and the roots are anchored in tooth sockets by periodontium. ♦ The maxilla, its teeth, gingivae, and adjacent vestibule are supplied by branches of the maxillary nerve (CN V$_2$), the alveolar arteries, and accompanying veins. ♦ The same features of the mandible are supplied by the mandibular nerve (CN V$_3$) and inferior alveolar vessels.

Palate: The roof of the oral cavity proper is formed by the hard (anterior two thirds) and soft (posterior one third) palates, the latter being a controlled flap that allows or limits communication with the nasal cavity. ♦ The mucosa of the hard palate includes abundant palatine glands. ♦ Branches of the maxillary (greater and lesser palatine arteries) and facial (ascending palatine artery) arteries supply the palate; its venous blood drains to the pterygoid plexus. The palate receives sensory innervation from the maxillary nerve (CN V$_2$); the muscles of the soft palate receive motor innervation from the pharyngeal plexus (CN X) plus a branch from the mandibular nerve (CN V$_3$) for the tensor veli palatini.

Tongue: The tongue is a mass of striated muscle, innervated by CN XII, and covered with a specialized mucosa textured with lingual papillae. ♦ It occupies most of the oral cavity when the mouth is closed. ♦ Its extrinsic muscles primarily control its placement, whereas its intrinsic muscles primarily control its shape, for manipulation of food during chewing, swallowing, and speech. ♦ It is highly sensitive, with four cranial nerves contributing sensory fibers to it. ♦ The terminal sulcus divides it into an anterior two thirds, receiving general sensation from the lingual nerve (CN V$_3$) and taste fibers from CN VII, and a posterior third receiving all sensory innervation from CN IX. ♦ Adjacent to the epiglottis, CN X provides general and special sensory innervation.

Salivary glands: Salivary glands secrete saliva to initiate digestion by facilitating chewing and swallowing. ♦ The parotid gland, the largest, receives parasympathetic innervation from CN IX via the otic ganglion. ♦ The submandibular and sublingual glands receive parasympathetic innervation from CN VII by way of the chorda tympani nerve, lingual nerve, and submandibular ganglion. Their ducts open into the oral cavity under the tongue.

PTERYGOPALATINE FOSSA

The **pterygopalatine fossa** is a small pyramidal space inferior to the apex of the orbit and medial to the infratemporal fossa (Fig. 7.97). It lies between the *pterygoid process of the sphenoid* posteriorly and the rounded posterior aspect of the maxilla anteriorly. The fragile *perpendicular plate of the palatine bone* forms its medial wall. The incomplete roof of the pterygopalatine fossa is formed by the a medial continuation of the *infratemporal surface of the greater wing of the sphenoid.* The floor of the pterygopalatine fossa is formed by the *pyramidal process of the palatine bone.* Its superior larger end opens anterosuperiorly into the *inferior orbital fissure;* its inferior end narrows, continuing as the *greater and lesser palatine canals.* The pterygopalatine fossa communicates through many passageways, distributing and receiving nerves and vessels to and from most of the major compartments of the viscerocranium (Fig. 7.98A).

The **contents of the pterygopalatine fossa** (Fig. 7.98B & C) are the:

- Terminal (pterygopalatine or third) part of the maxillary artery, and the initial parts of its branches, and accompanying veins (tributaries of the pterygoid venous plexus).
- Maxillary nerve (CN V$_2$), with which the pterygopalatine ganglion is associated. Branches arising from the ganglion within the fossa are considered to be branches of the maxillary nerve.
- Neurovascular sheaths of the vessels and nerves and a fatty matrix occupy all remaining space.

Pterygopalatine Part of Maxillary Artery

The *maxillary artery,* a terminal branch of the external carotid artery, passes anteriorly through the infratemporal fossa, as described previously on p. 921. The **pterygopalatine part of the maxillary artery,** its third part (i.e., the part located anterior to the lateral pterygoid), passes medially through the *pterygomaxillary fissure* and enters the pterygopalatine fossa (Figs. 7.98B and 7.99A). The artery lies anterior to the pterygopalatine ganglion and gives rise to branches that accompany all nerves entering and exiting the fossa, sharing the same names with many (Table 7.12).

Maxillary Nerve

The **maxillary nerve** runs anteriorly through the *foramen rotundum,* which enters posterior wall of the fossa (Figs. 7.98C, 7.99B and 7.100C). Within the pterygopalatine fossa, the maxillary nerve gives off the *zygomatic nerve,* which divides into zygomaticofacial and zygomaticotemporal nerves (Figs. 7.99B and 7.100A). These nerves emerge from the zygomatic bone through cranial foramina of the same name and supply general sensation to the lateral region of the cheek and temple. The **zygomaticotemporal nerve** also gives rise to a communicating branch, which conveys postsynaptic parasympathetic secretomotor fibers to the lacrimal gland by way of the heretofore purely sensory lacrimal nerve from CN V$_1$ (Fig. 7.100A & B).

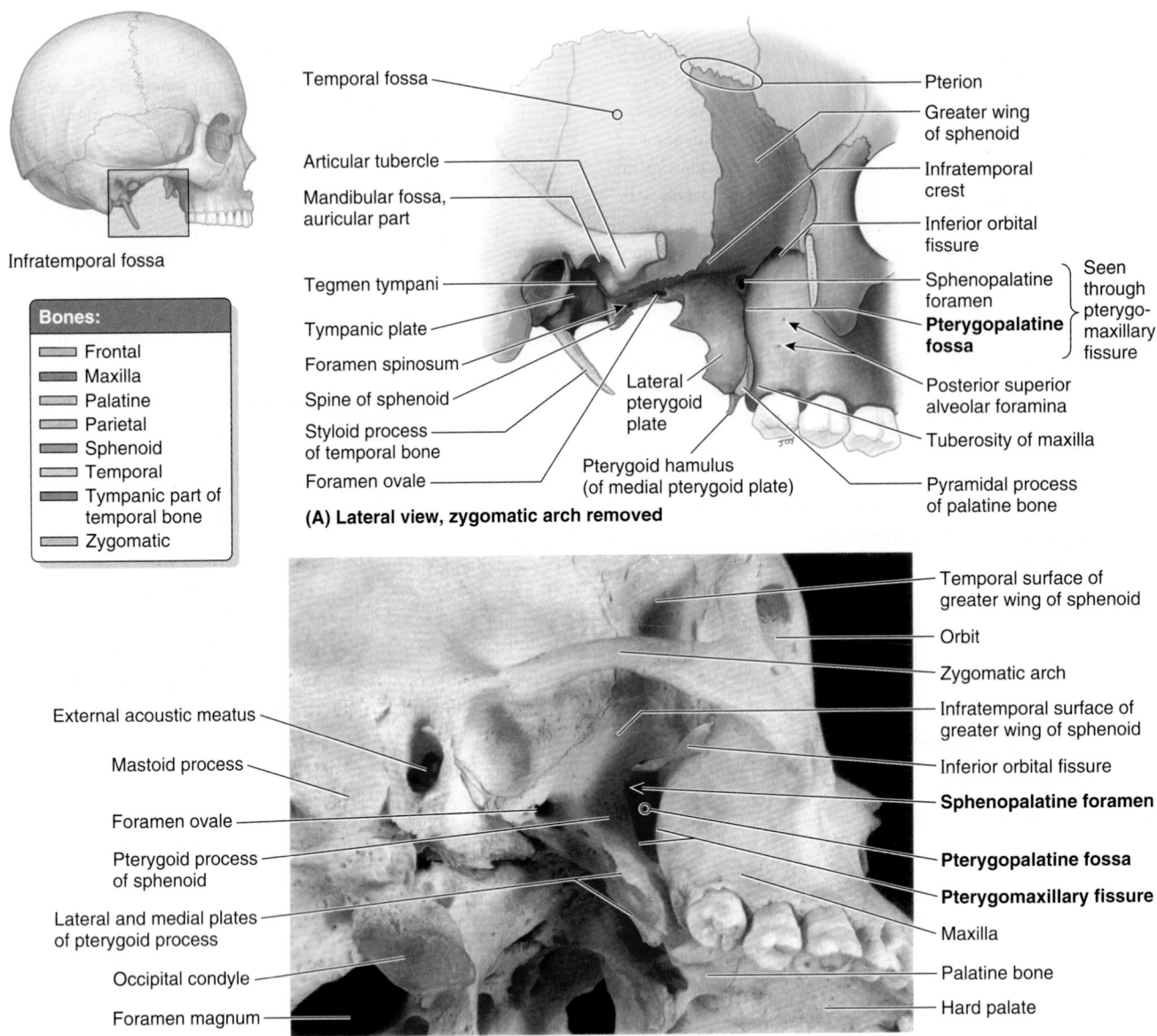

Infratemporal fossa

Bones:
- Frontal
- Maxilla
- Palatine
- Parietal
- Sphenoid
- Temporal
- Tympanic part of temporal bone
- Zygomatic

(A) Lateral view, zygomatic arch removed

(B) Inferolateral and slightly posterior view, looking into infratemporal and pterygopalatine fossae

FIGURE 7.97. Temporal, infratemporal, and pterygopalatine fossae. The pterygopalatine fossa is seen medial to the infratemporal fossa through the pterygomaxillary fissure, between the pterygoid process and the maxilla. The sphenopalatine foramen is an opening into the nasal cavity at the top of the palatine bone.

While in the pterygopalatine fossa, the maxillary nerve also gives off the two *ganglionic branches to the pterygopalatine ganglion* (sensory roots of the pterygopalatine ganglion) that suspend the parasympathetic **pterygopalatine ganglion** in the superior part of the pterygopalatine fossa (Figs. 7.98C and 7.100A). The pterygopalatine nerves convey general sensory fibers of the maxillary nerve, which pass through the pterygopalatine ganglion without synapsing to supply the nose, palate, and pharynx (Fig. 7.100C). The maxillary nerve leaves the pterygopalatine fossa through the inferior orbital fissure, after which it is known as the *infra-orbital nerve* (Figs. 7.98C and 7.99B).

The **parasympathetic fibers to the pterygopalatine ganglion** come from the facial nerve by way of its first branch, the *greater petrosal nerve* (Figs. 7.98C and 7.100A & B). This

nerve joins the *deep petrosal nerve* as it passes through the foramen lacerum to form the **nerve of the pterygoid canal,** which passes anteriorly through this canal to the pterygopalatine fossa. The parasympathetic fibers of the greater petrosal nerve synapse in the pterygopalatine ganglion.

The **deep petrosal nerve** is a sympathetic nerve arising from the *internal carotid peri-arterial plexus* as the artery exits the carotid canal (Figs. 7.98C and 7.100A & C). It conveys postsynaptic fibers from nerve cell bodies in the *superior cervical sympathetic ganglion* to the pterygopalatine ganglion by joining the *nerve of the pterygoid canal.* The fibers do not synapse in the ganglion but pass directly through it into the branches (of CN V₂) arising from it (Fig. 7.100C). The postsynaptic sympathetic fibers pass to the palatine glands, and the mucosal glands of the nasal cavity and superior pharynx.

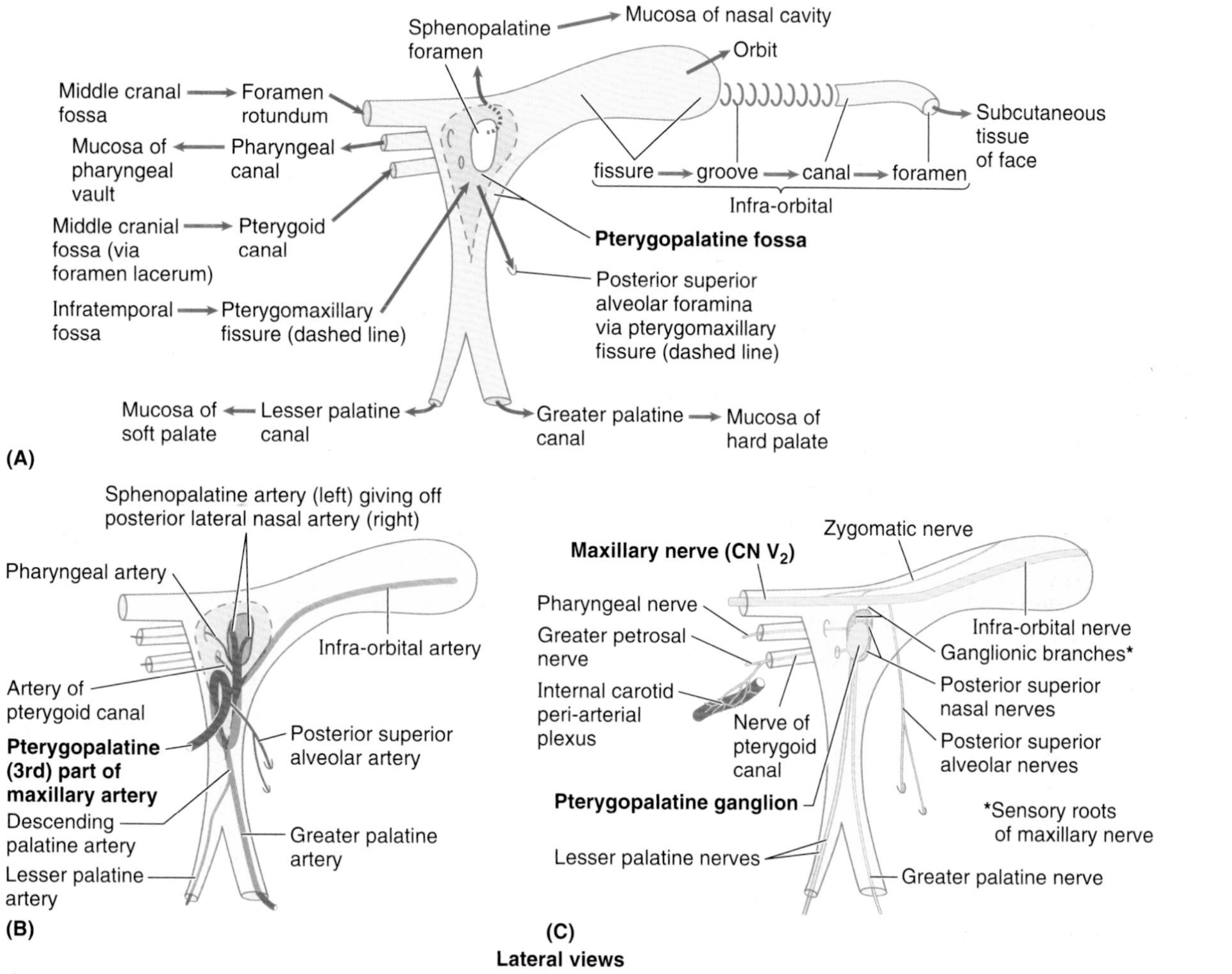

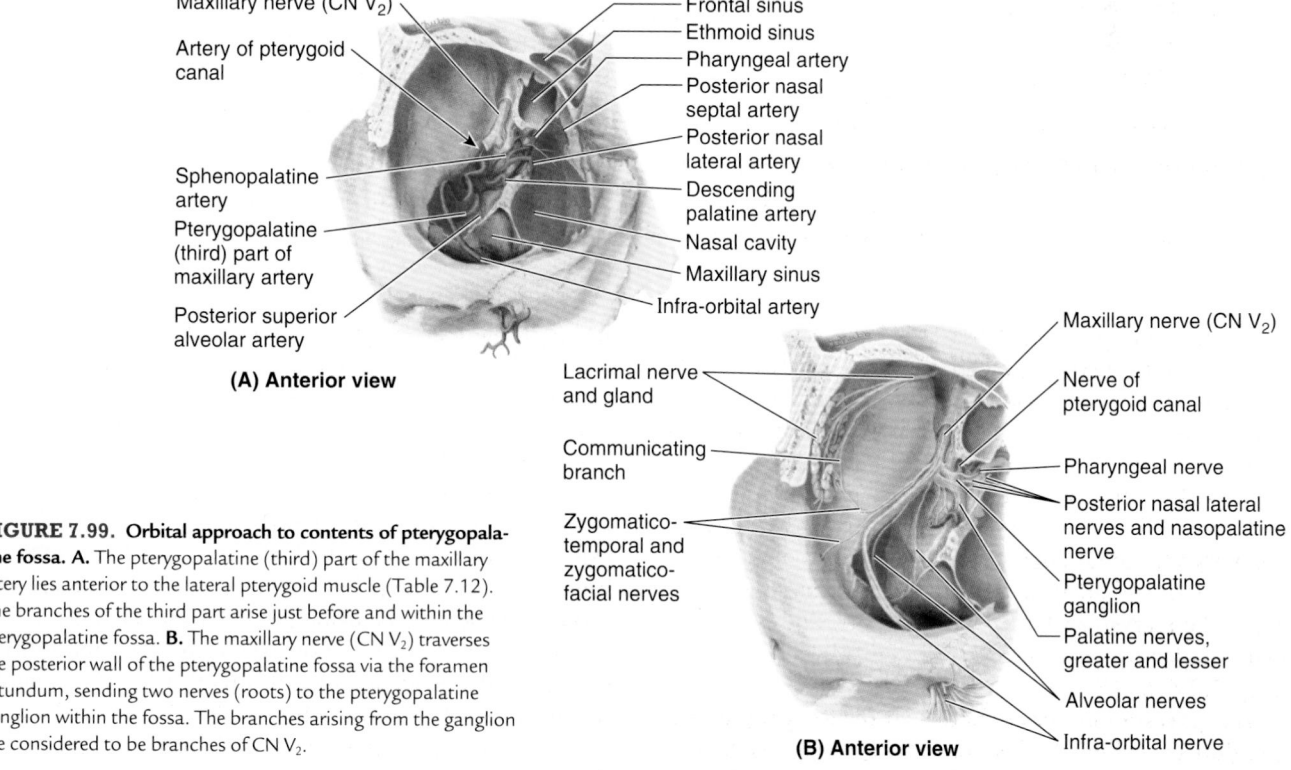

FIGURE 7.98. Pterygopalatine fossa—communications and contents. A. Communications of the pterygopalatine fossa and the passageways by which structures enter and exit fossae are shown. **B.** The distribution of branches of the pterygopalatine part of the maxillary artery is demonstrated. **C.** Branches of the maxillary nerve and pterygopalatine ganglion enter and exit the fossa.

(A)

Sphenopalatine foramen → Mucosa of nasal cavity

Middle cranal fossa → Foramen rotundum → Orbit

Mucosa of pharyngeal vault ← Pharyngeal canal ← Pharyngeal canal

Middle cranial fossa (via foramen lacerum) → Pterygoid canal

Infratemporal fossa → Pterygomaxillary fissure (dashed line)

fissure → groove → canal → foramen → Subcutaneous tissue of face

Infra-orbital

Pterygopalatine fossa

Posterior superior alveolar foramina via pterygomaxillary fissure (dashed line)

Mucosa of soft palate ← Lesser palatine canal

Greater palatine canal → Mucosa of hard palate

(B)

Sphenopalatine artery (left) giving off posterior lateral nasal artery (right)

Pharyngeal artery

Infra-orbital artery

Artery of pterygoid canal

Pterygopalatine (3rd) part of maxillary artery

Posterior superior alveolar artery

Descending palatine artery

Greater palatine artery

Lesser palatine artery

(C)

Maxillary nerve (CN V₂) / **Maxillary nerve (CN V$_2$)**

Zygomatic nerve

Pharyngeal nerve

Greater petrosal nerve

Internal carotid peri-arterial plexus

Infra-orbital nerve

Ganglionic branches*

Posterior superior nasal nerves

Nerve of pterygoid canal

Posterior superior alveolar nerves

Pterygopalatine ganglion

Lesser palatine nerves

*Sensory roots of maxillary nerve

Greater palatine nerve

Lateral views

FIGURE 7.99. Orbital approach to contents of pterygopalatine fossa. A. The pterygopalatine (third) part of the maxillary artery lies anterior to the lateral pterygoid muscle (Table 7.12). The branches of the third part arise just before and within the pterygopalatine fossa. **B.** The maxillary nerve (CN V$_2$) traverses the posterior wall of the pterygopalatine fossa via the foramen rotundum, sending two nerves (roots) to the pterygopalatine ganglion within the fossa. The branches arising from the ganglion are considered to be branches of CN V$_2$.

Maxillary nerve (CN V$_2$)

Artery of pterygoid canal

Frontal sinus

Ethmoid sinus

Pharyngeal artery

Posterior nasal septal artery

Posterior nasal lateral artery

Descending palatine artery

Nasal cavity

Maxillary sinus

Infra-orbital artery

Sphenopalatine artery

Pterygopalatine (third) part of maxillary artery

Posterior superior alveolar artery

(A) Anterior view

Lacrimal nerve and gland

Communicating branch

Zygomatico-temporal and zygomatico-facial nerves

Maxillary nerve (CN V$_2$)

Nerve of pterygoid canal

Pharyngeal nerve

Posterior nasal lateral nerves and nasopalatine nerve

Pterygopalatine ganglion

Palatine nerves, greater and lesser

Alveolar nerves

Infra-orbital nerve

(B) Anterior view

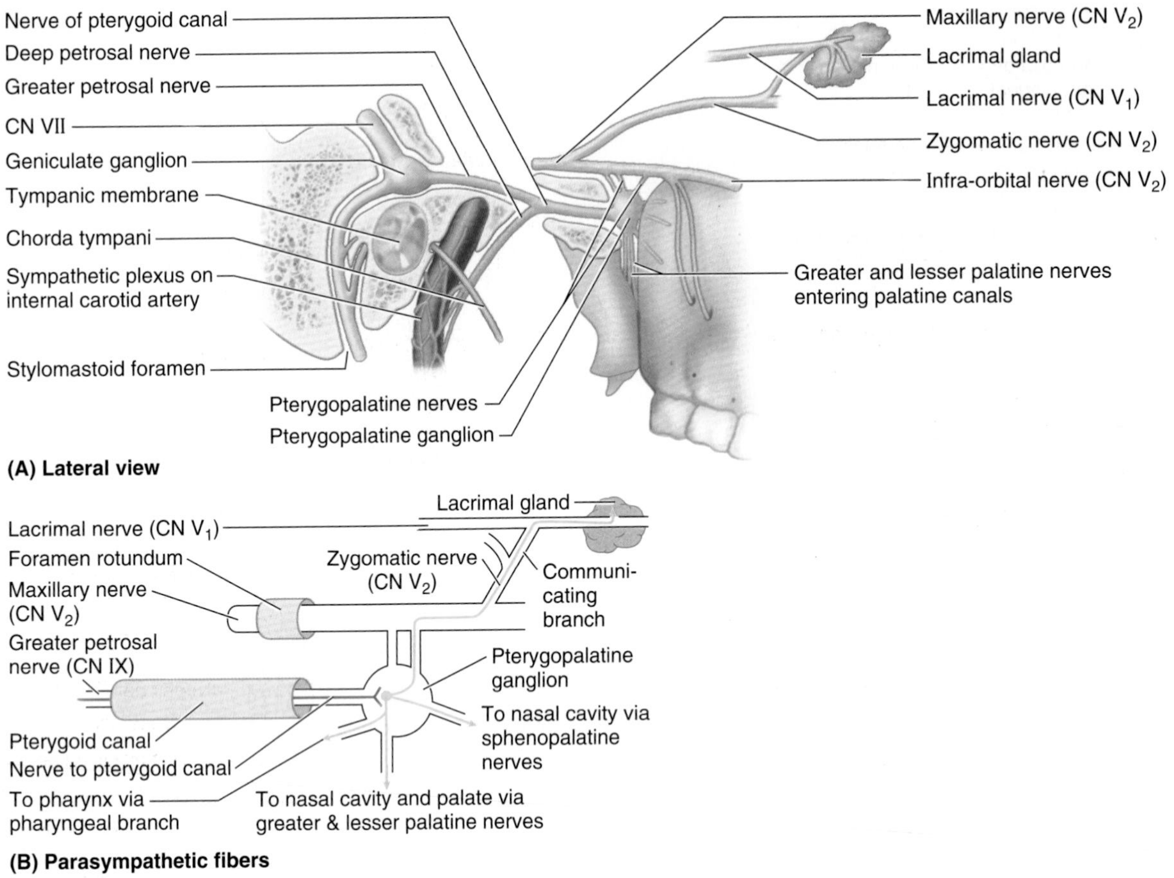

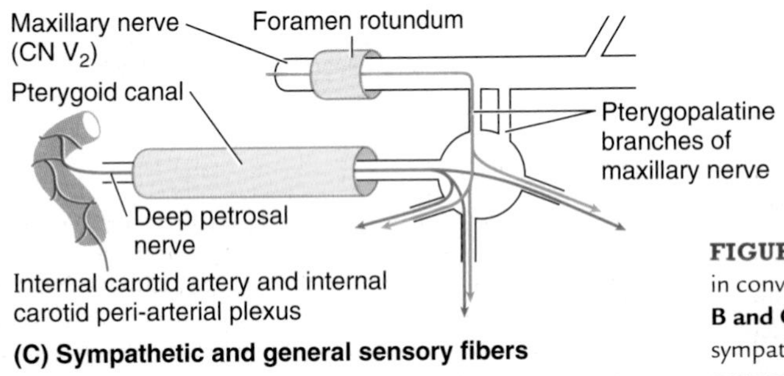

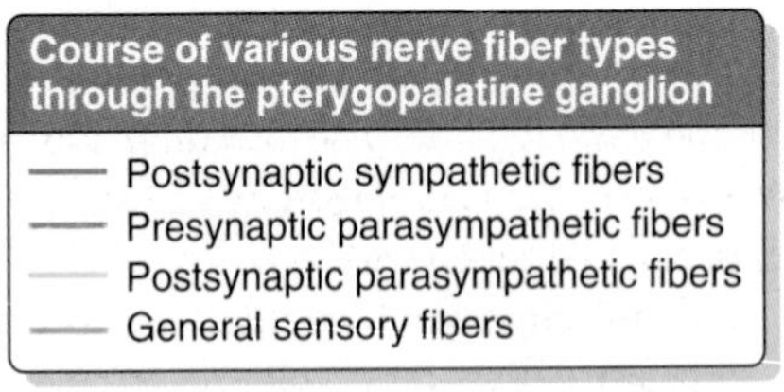

Nerve of pterygoid canal
Deep petrosal nerve
Greater petrosal nerve
CN VII
Geniculate ganglion
Tympanic membrane
Chorda tympani
Sympathetic plexus on internal carotid artery

Stylomastoid foramen

Pterygopalatine nerves
Pterygopalatine ganglion

Maxillary nerve (CN V₂)
Lacrimal gland
Lacrimal nerve (CN V₁)
Zygomatic nerve (CN V₂)
Infra-orbital nerve (CN V₂)

Greater and lesser palatine nerves entering palatine canals

(A) Lateral view

Lacrimal nerve (CN V₁)
Foramen rotundum
Maxillary nerve (CN V₂)
Greater petrosal nerve (CN IX)

Pterygoid canal
Nerve to pterygoid canal
To pharynx via pharyngeal branch

Lacrimal gland
Zygomatic nerve (CN V₂)
Communi-cating branch
Pterygopalatine ganglion
To nasal cavity via sphenopalatine nerves
To nasal cavity and palate via greater & lesser palatine nerves

(B) Parasympathetic fibers

Maxillary nerve (CN V₂)
Foramen rotundum
Pterygoid canal
Pterygopalatine branches of maxillary nerve

Deep petrosal nerve
Internal carotid artery and internal carotid peri-arterial plexus

(C) Sympathetic and general sensory fibers

Course of various nerve fiber types through the pterygopalatine ganglion

— Postsynaptic sympathetic fibers
— Presynaptic parasympathetic fibers
— Postsynaptic parasympathetic fibers
— General sensory fibers

FIGURE 7.100. Pterygopalatine ganglion. A. Nerves involved in conveying nerve fibers to and from the ganglion are shown. **B and C.** The nerve of the pterygoid canal conveys presynaptic parasympathetic fibers from the facial nerve (via its branch, the greater petrosal nerve) to the ganglion, where they will synapse with postsynaptic fibers. The nerve of the pterygoid canal also brings postsynaptic sympathetic fibers to the ganglion from the internal carotid plexus (via the deep petrosal nerve). Sensory fibers reach the ganglion via pterygopalatine branches of the maxillary nerve (CN V₂). Secretomotor postsynaptic parasympathetic and vasoconstrictive postsynaptic sympathetic fibers are distributed to the lacrimal, nasal, palatine, and pharyngeal glands. Similarly, sensory fibers are distributed to the mucosa of the nasal cavity, palate, and uppermost pharynx.

PTERYGOPALATINE FOSSA

Transantral Approach to Pterygopalatine Fossa

 Surgical access to the deeply placed pterygopalatine fossa is gained through the maxillary sinus. After elevating the upper lip, the maxillary gingiva and anterior wall of the sinus are transversed to enter the sinus. The posterior wall is then chipped away as needed to open the anterior wall of the pterygopalatine fossa. In the case of chronic *epistaxis* (nosebleed), the third part of the maxillary artery may be ligated in the fossa to control the bleeding.

PTERYGOPALATINE FOSSA

The pterygopalatine fossa is a major distributing center for branches of the maxillary nerve and the pterygopalatine (third) part of the maxillary artery. ♦ It is located between, and has communications with, the infratemporal fossa, nasal cavity, orbit, middle cranial fossa, pharyngeal vault, maxillary sinus, and oral cavity (palate). ♦ The contents of the pterygopalatine fossa are the maxillary nerve (CN V$_2$), the parasympathetic pterygopalatine ganglion, the third part of the maxillary artery and accompanying veins, and a surrounding fatty matrix.

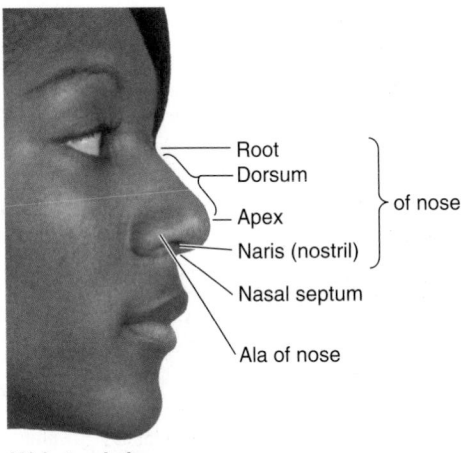

(A) Lateral view

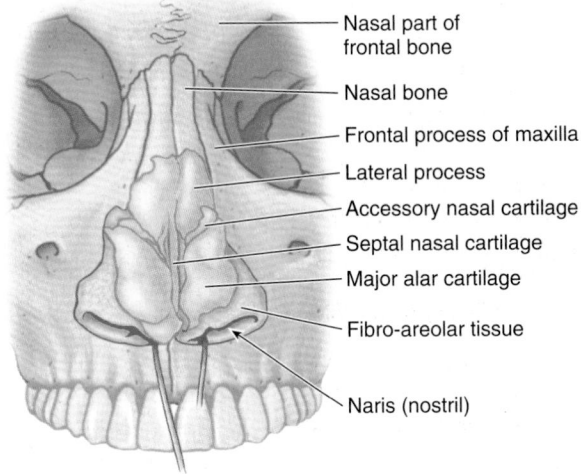

(B) Anterior view

FIGURE 7.101. External nose. A. The surface anatomy of the nose is shown. The nose is attached to the forehead by its root. The rounded border between the apex and the root is the dorsum. **B.** The cartilages of the nose are retracted inferiorly to expose the sesamoid cartilages. The lateral nasal cartilages are fixed by sutures to the nasal bones and are continuous with the septal cartilage.

NOSE

The **nose** is the part of the respiratory tract superior to the hard palate and contains the peripheral organ of smell. It includes the external nose and nasal cavity, which is divided into right and left cavities by the *nasal septum* (Fig. 7.101A). The functions of the nose include olfaction (smelling), respiration (breathing), filtration of dust, humidification of inspired air, and reception and elimination of secretions from the paranasal sinuses and nasolacrimal ducts.

External Nose

The **external nose** is the visible portion that projects from the face; its skeleton is mainly cartilaginous (Fig. 7.101B). Noses vary considerably in size and shape, mainly because of differences in these cartilages. The **dorsum of the nose** extends from the **root of the nose** to the **apex** (tip) **of the nose.** The inferior surface of the nose is pierced by two piriform (L. pear-shaped) openings, the **nares** (nostrils, anterior nasal apertures), which are bound laterally by the **alae** (wings) **of the nose.** The superior bony part of the nose, including its root, is covered by thin skin.

The skin over the cartilages of the nose is covered with thicker skin, which contains many sebaceous glands. The skin extends into the **vestibule of the nose** (Fig. 7.103A), where it has a variable number of stiff hairs (*vibrissae*). Because they are usually moist, these hairs filter dust particles from air entering the nasal cavity. The junction of the skin and mucous membrane is beyond the hair-bearing area.

SKELETON OF EXTERNAL NOSE

The supporting skeleton of the nose is composed of bone and hyaline cartilage. The **bony part of the nose** (Figs. 7.101B and 7.102) consists of the *nasal bones, frontal processes of the maxillae, the nasal part of the frontal bone* and its *nasal spine,* and the bony parts of the nasal septum. The **cartilaginous part of the nose** consists of five main cartilages: two lateral cartilages, two alar cartilages, and one septal cartilage. The U-shaped **alar cartilages** are free and movable; they dilate or constrict the nares when the muscles acting on the nose contract.

NASAL SEPTUM

The *nasal septum* divides the chamber of the nose into two *nasal cavities.* The septum has a bony part and a soft mobile cartilaginous part. The main components of the nasal septum are the perpendicular plate of the ethmoid, the vomer, and the septal cartilage. The thin **perpendicular plate of the ethmoid bone,** forming the superior part of the nasal septum, descends from the *cribriform plate* and is continued superior to this plate as the *crista galli.* The **vomer,** a thin

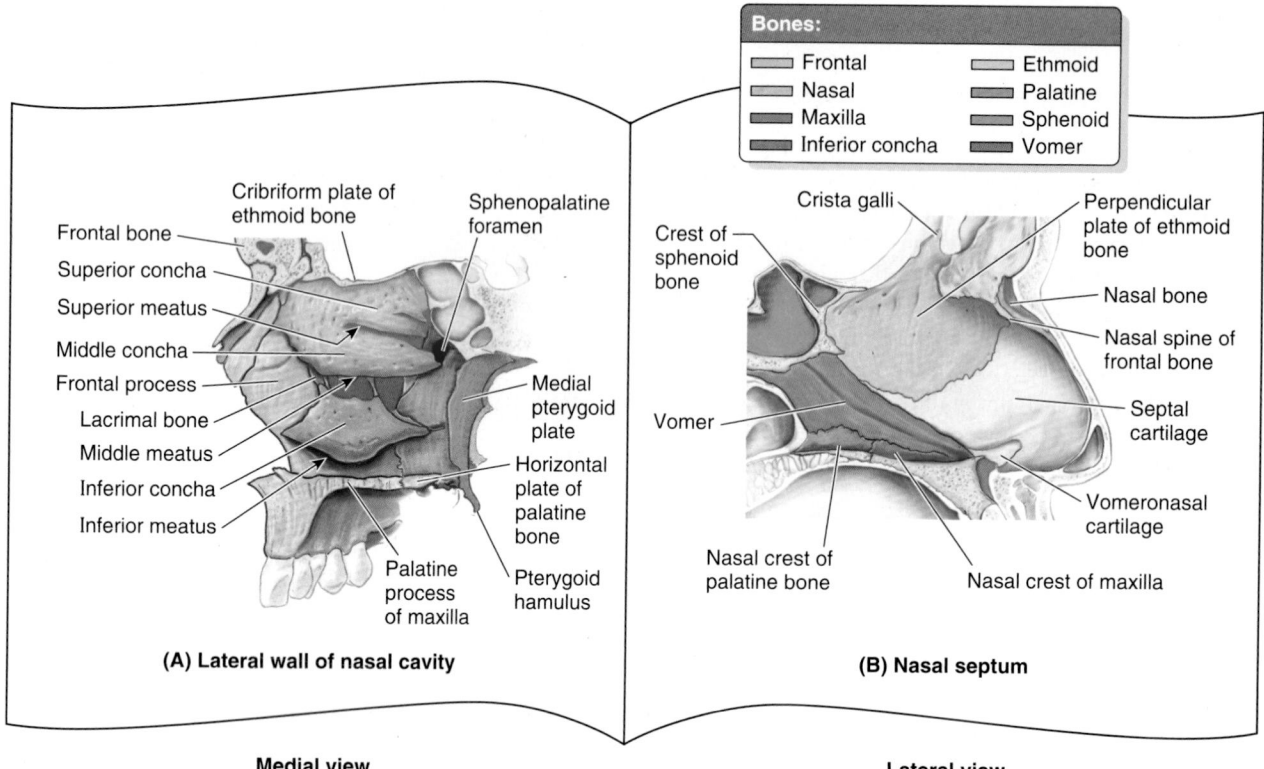

Bones:
- Frontal
- Nasal
- Maxilla
- Inferior concha
- Ethmoid
- Palatine
- Sphenoid
- Vomer

(A) Lateral wall of nasal cavity

Cribriform plate of ethmoid bone
Sphenopalatine foramen
Frontal bone
Superior concha
Superior meatus
Middle concha
Frontal process
Lacrimal bone
Middle meatus
Inferior concha
Inferior meatus
Medial pterygoid plate
Horizontal plate of palatine bone
Palatine process of maxilla
Pterygoid hamulus

Medial view

(B) Nasal septum

Crista galli
Crest of sphenoid bone
Perpendicular plate of ethmoid bone
Nasal bone
Nasal spine of frontal bone
Vomer
Septal cartilage
Vomeronasal cartilage
Nasal crest of palatine bone
Nasal crest of maxilla

Lateral view

FIGURE 7.102. Lateral and medial (septal) walls of right side of nasal cavity. The walls are separated and shown as adjacent pages of a book. The medial view shows the right lateral wall of the nasal cavity, and the lateral view shows the nasal septum. The nasal septum has a hard (bony) part located deeply (posteriorly) where it is protected and a soft or mobile part located superficially (anteriorly) mostly in the more vulnerable external nose.

flat bone, forms the postero-inferior part of the nasal septum, with some contribution from the nasal crests of the maxillary and palatine bones. The **septal cartilage** has a tongue-and-groove articulation with the edges of the bony septum.

Nasal Cavities

The term *nasal cavity* refers to either the entire cavity or to the right or left half, depending on the context. The nasal cavity is entered anteriorly through the *nares* (nostrils). It opens posteriorly into the *nasopharynx* through the *choanae* (Fig. 7.9). Mucosa lines the nasal cavity, except for the *nasal vestibule,* which is lined with skin (Fig. 7.103A).

The **nasal mucosa** is firmly bound to the periosteum and perichondrium of the supporting bones and cartilages of the nose. The mucosa is continuous with the lining of all the chambers with which the nasal cavities communicate: the nasopharynx posteriorly, the paranasal sinuses superiorly and laterally, and the lacrimal sac and conjunctiva superiorly. The inferior two thirds of the nasal mucosa is the respiratory area, and the superior one third is the olfactory area (Fig. 7.106B). Air passing over the **respiratory area** is warmed and moistened before it passes through the rest of the upper respiratory tract to the lungs. The **olfactory area** contains the peripheral organ of smell; sniffing draws air to the area.

BOUNDARIES OF NASAL CAVITIES

The nasal cavities have a roof, floor, and medial and lateral walls.

- The *roof of the nasal cavities* is curved and narrow, except at its posterior end, where the hollow *body of the sphenoid* forms the roof. It is divided into three parts (frontonasal, ethmoidal, and sphenoidal) named from the bones forming each part (Fig. 7.102).
- The *floor of the nasal cavities* is wider than the roof and is formed by the *palatine processes of the maxilla* and the *horizontal plates of the palatine bone.*
- The *medial wall of the nasal cavities* is formed by the nasal septum.
- The *lateral walls of the nasal cavities* are irregular owing to three bony plates, the *nasal conchae,* which project inferiorly, somewhat like louvers (Figs. 7.102A, 7.103, and 7.108).

FEATURES OF NASAL CAVITIES

The **nasal conchae** (superior, middle, and inferior) curve inferomedially, hanging like louvers or short curtains from the lateral wall. The conchae (L. shells) or turbinates of many mammals (especially running mammals and those existing in extreme environments) are highly convoluted, scroll-like structures that offer a vast surface area for heat exchange. In both

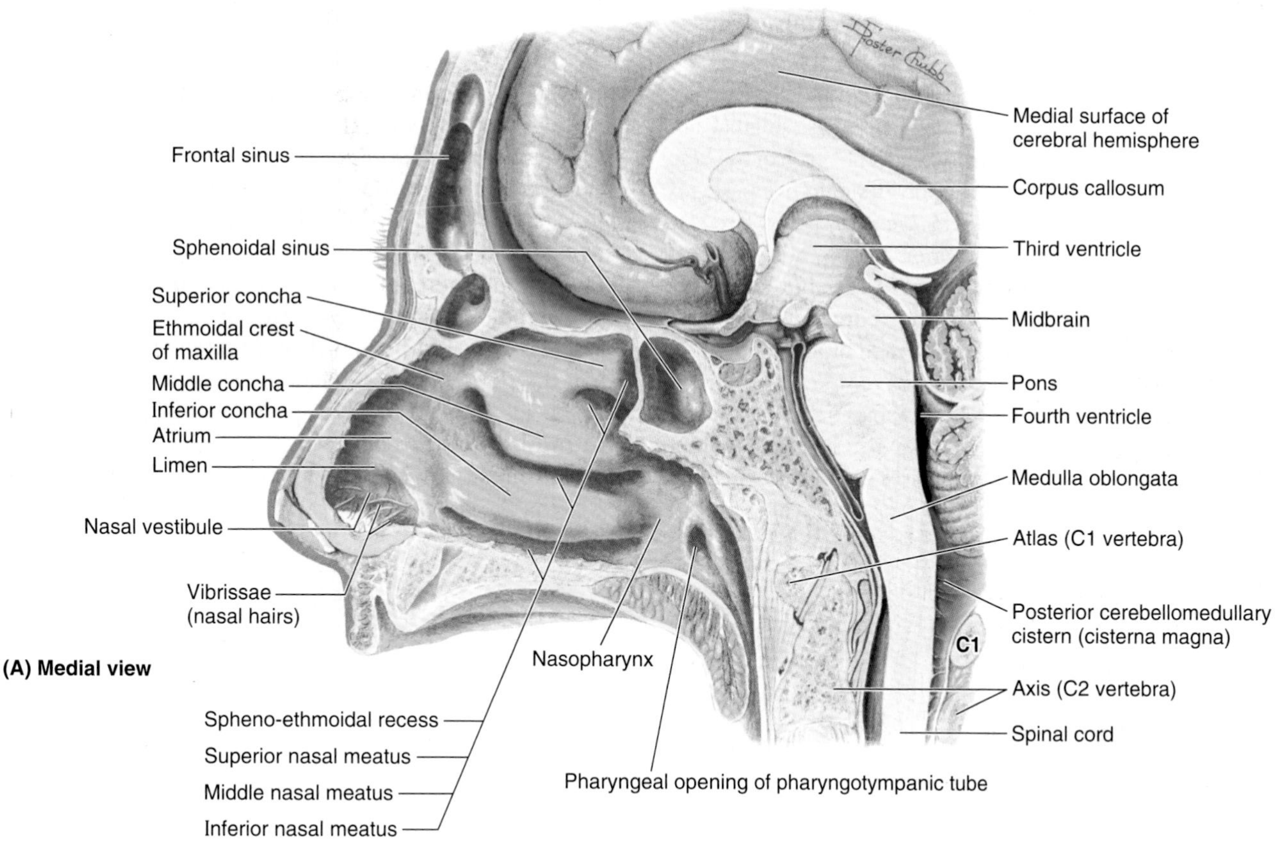

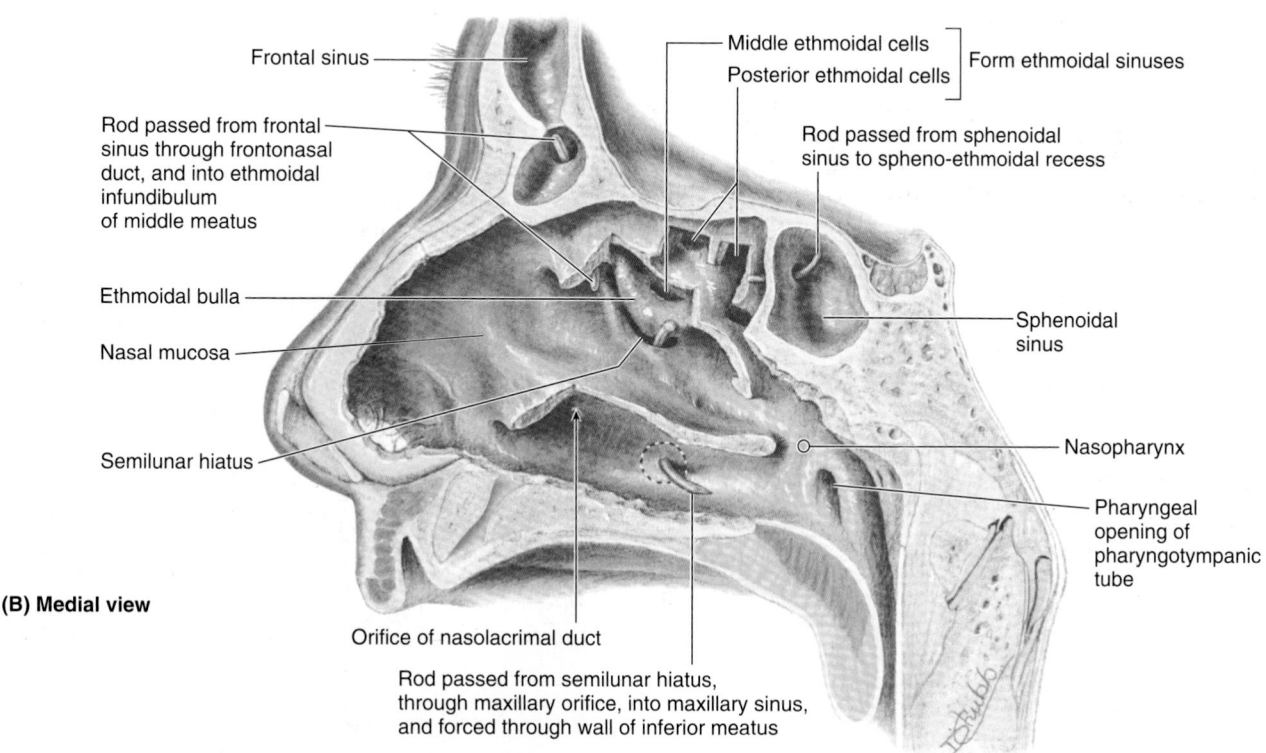

(A) Medial view

Frontal sinus

Sphenoidal sinus

Superior concha

Ethmoidal crest
of maxilla

Middle concha

Inferior concha

Atrium

Limen

Nasal vestibule

Vibrissae
(nasal hairs)

Spheno-ethmoidal recess

Superior nasal meatus

Middle nasal meatus

Inferior nasal meatus

Nasopharynx

Pharyngeal opening of pharyngotympanic tube

Medial surface of
cerebral hemisphere

Corpus callosum

Third ventricle

Midbrain

Pons

Fourth ventricle

Medulla oblongata

Atlas (C1 vertebra)

Posterior cerebellomedullary
cistern (cisterna magna)

Axis (C2 vertebra)

Spinal cord

C1

(B) Medial view

Frontal sinus

Rod passed from frontal
sinus through frontonasal
duct, and into ethmoidal
infundibulum
of middle meatus

Ethmoidal bulla

Nasal mucosa

Semilunar hiatus

Orifice of nasolacrimal duct

Rod passed from semilunar hiatus,
through maxillary orifice, into maxillary sinus,
and forced through wall of inferior meatus

Middle ethmoidal cells

Posterior ethmoidal cells

Form ethmoidal sinuses

Rod passed from sphenoidal
sinus to spheno-ethmoidal recess

Sphenoidal
sinus

Nasopharynx

Pharyngeal
opening of
pharyngotympanic
tube

FIGURE 7.103. Lateral wall of nasal cavity of right half of head. A. The inferior and middle conchae, curving medially and inferiorly from the lateral wall, divide the wall into three nearly equal parts and cover the inferior and middle meatus, respectively. The superior concha is small and anterior to the sphenoidal sinus and the middle concha has an angled inferior border and ends inferior to the sphenoidal sinus. The inferior concha has a slightly curved inferior border and ends inferior to the middle concha approximately 1 cm anterior to the orifice of the pharyngotympanic tube (approximately the width of the medial pterygoid plate). **B.** This dissection of the lateral wall of the nasal cavity shows the communications through the lateral wall of the nasal cavity. Parts of the superior, middle, and inferior conchae are cut away. The sphenoidal sinus occupies the body of the sphenoid bone; its orifice, superior to the middle of its anterior wall, opens into the spheno-ethmoidal recess. The orifices of posterior, middle, and anterior ethmoidal cells open into the superior meatus, middle meatus, and semilunar hiatus, respectively.

humans with simple plate-like nasal conchae and animals with complex turbinates, a recess or **nasal meatus** (singular and plural; passage(s) in the nasal cavity) underlies each of the bony formations. The nasal cavity is thus divided into five passages: a posterosuperiorly placed *spheno-ethmoidal recess*, three laterally located *nasal meatus* (superior, middle, and inferior), and a medially placed *common nasal meatus* into which the four lateral passages open. The **inferior concha** is the longest and broadest of the conchae and is formed by an independent bone (of the same name, inferior concha) covered by a mucous membrane that contains large vascular spaces that can enlarge to control the caliber of the nasal cavity. The **middle** and **superior conchae** are medial processes of the ethmoid bone.

When infected or irritated, the mucosa covering the conchae may swell rapidly, blocking the nasal passage(s) on that side.

The **spheno-ethmoidal recess,** lying superoposterior to the superior concha, receives the opening of the *sphenoidal sinus,* an air-filled cavity in the body of the sphenoid. The **superior nasal meatus** is a narrow passage between the superior and the middle nasal conchae into which the posterior ethmoidal sinuses open by one or more orifices (Fig. 7.103A). The **middle nasal meatus** is longer and deeper than the superior one. The anterosuperior part of this passage leads into a funnel-shaped opening, the **ethmoidal infundibulum,** through which it communicates with the frontal sinus (Fig. 7.104). The passage that leads inferiorly

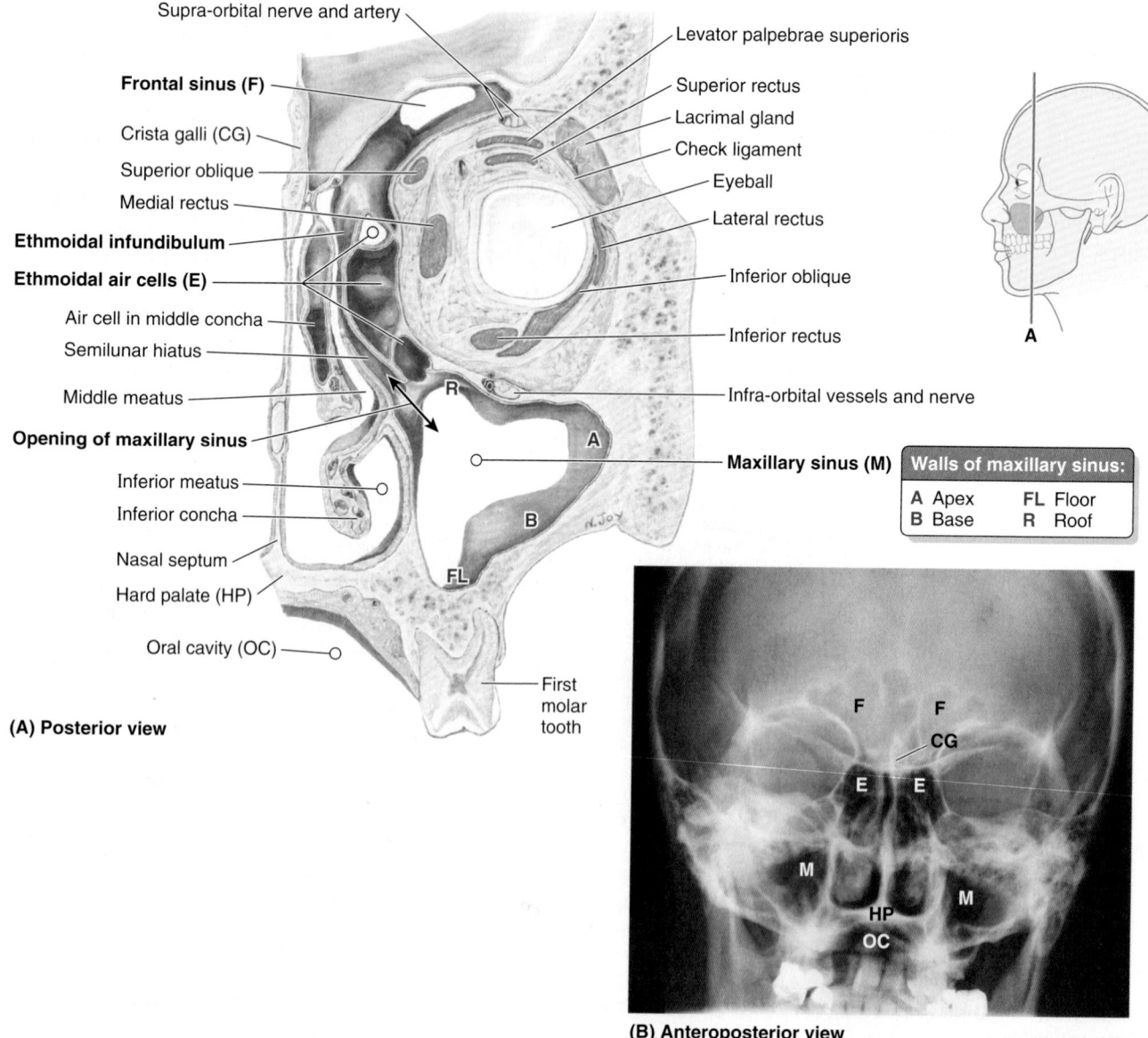

Walls of maxillary sinus:

A	Apex	FL	Floor
B	Base	R	Roof

(A) Posterior view

(B) Anteroposterior view

FIGURE 7.104. Coronal section of right half of head. A. The orientation drawing illustrates the plane of the section. Observe the relationship of the orbit, nasal cavity, and paranasal sinuses. The orbital contents, including the four recti and the fascia uniting them, form a circle (a cone when viewed in three dimensions) around the posterior part (fundus) of the eyeball. **B.** Radiograph of cranium demonstrating nasal cavity and paranasal sinuses. Letters refer to structures labeled in part **A.**

from each frontal sinus to the infundibulum is the *fronto-nasal duct* (Fig. 7.103B). The **semilunar hiatus** (L. *hiatus semilunaris*) is a semicircular groove into which the frontal sinus opens. The **ethmoidal bulla** (L. bubble), a rounded elevation located superior to the semilunar hiatus, is visible when the middle concha is removed. The bulla is formed by middle ethmoidal cells that form the *ethmoidal sinuses*.

The **inferior nasal meatus** is a horizontal passage infero-lateral to the inferior nasal concha. The *nasolacrimal duct*, which drains tears from the lacrimal sac, opens into the anterior part of this meatus (see Fig. 7.46A, p. 892). The **common nasal meatus** is the medial part of the nasal cavity between the conchae and the nasal septum, into which the lateral recesses and meatus open.

Vasculature and Innervation of Nose

The *arterial supply of the medial and lateral walls of the nasal cavity* (Fig. 7.105) is from five sources:

1. *Anterior ethmoidal artery* (from the ophthalmic artery).
2. *Posterior ethmoidal artery* (from the ophthalmic artery).
3. *Sphenopalatine artery* (from the maxillary artery).
4. *Greater palatine artery* (from the maxillary artery).
5. *Septal branch of the superior labial artery* (from the facial artery).

The first three arteries divide into lateral and medial (septal) branches. The greater palatine artery reaches the septum via the incisive canal through the anterior hard palate. The

anterior part of the nasal septum is the site of an anastomotic arterial plexus involving all five arteries supplying the septum (*Kiesselbach area*). The external nose also receives blood from first and fifth arteries listed, plus nasal branches of the infra-orbital artery and the lateral nasal branches of the facial artery.

A rich **submucosal venous plexus,** deep to the nasal mucosa, provides *venous drainage of the nose via* the spheno-palatine, facial, and ophthalmic veins. The plexus is an important part of the body's thermoregulatory system, exchanging heat and warming air before it enters the lungs. Venous blood from the external nose drains mostly into the facial vein via the angular and lateral nasal veins (see Fig. 7.25). However, recall that it lies within the "danger area" of the face because of communications with the *cavernous (dural venous) sinus* (see the blue box "Thrombophlebitis of Facial Vein," p. 875).

Regarding its *nerve supply of the nose,* the nasal mucosa can be divided into postero-inferior and anterosuperior portions by an oblique line passing approximately through the anterior nasal spine and the spheno-ethmoidal recess (Fig. 7.106). The nerve supply of the postero-inferior portion of the nasal mucosa is chiefly from the maxillary nerve, by way of the *nasopalatine nerve* to the nasal septum, and posterior superior lateral nasal and inferior lateral nasal branches of the *greater palatine nerve* to the lateral wall. The nerve supply of the anterosuperior portion is from the ophthalmic nerve (CN V$_1$) by way of the **anterior** and **posterior ethmoidal nerves,** branches of the nasociliary nerve. Most of the external nose (dorsum and apex) is also supplied by CN V$_1$ (via the infratrochlear nerve and the external nasal branch of the

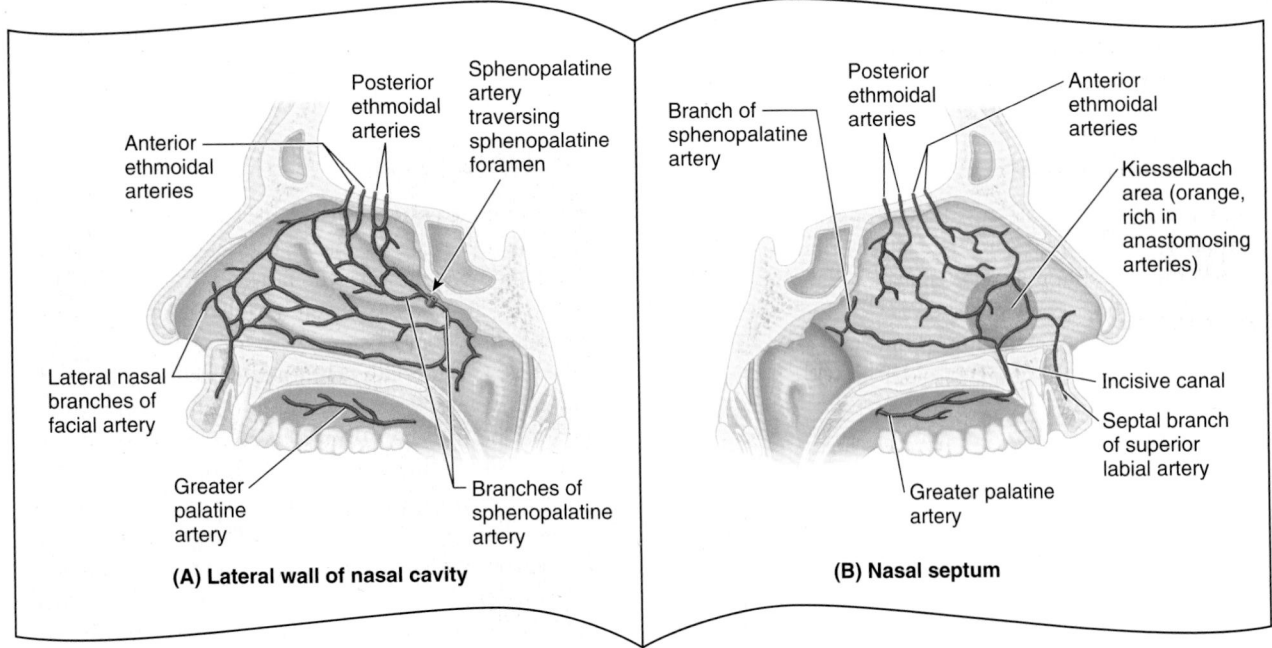

Open-book view

FIGURE 7.105. Arterial supply of nasal cavity. An open-book view of the lateral and medial walls of the right side of the nasal cavity is shown. The left "page" shows the lateral wall of the nasal cavity. The sphenopalatine artery (a branch of the maxillary) and the anterior ethmoidal artery (a branch of the oph-thalmic) are the most important arteries to the nasal cavity. The right "page" shows the nasal septum. An anastomosis of four to five named arteries supplying the septum occurs in the antero-inferior portion of the nasal septum (Kiesselbach area, *orange*), an area commonly involved in chronic epistaxis (nosebleeds).

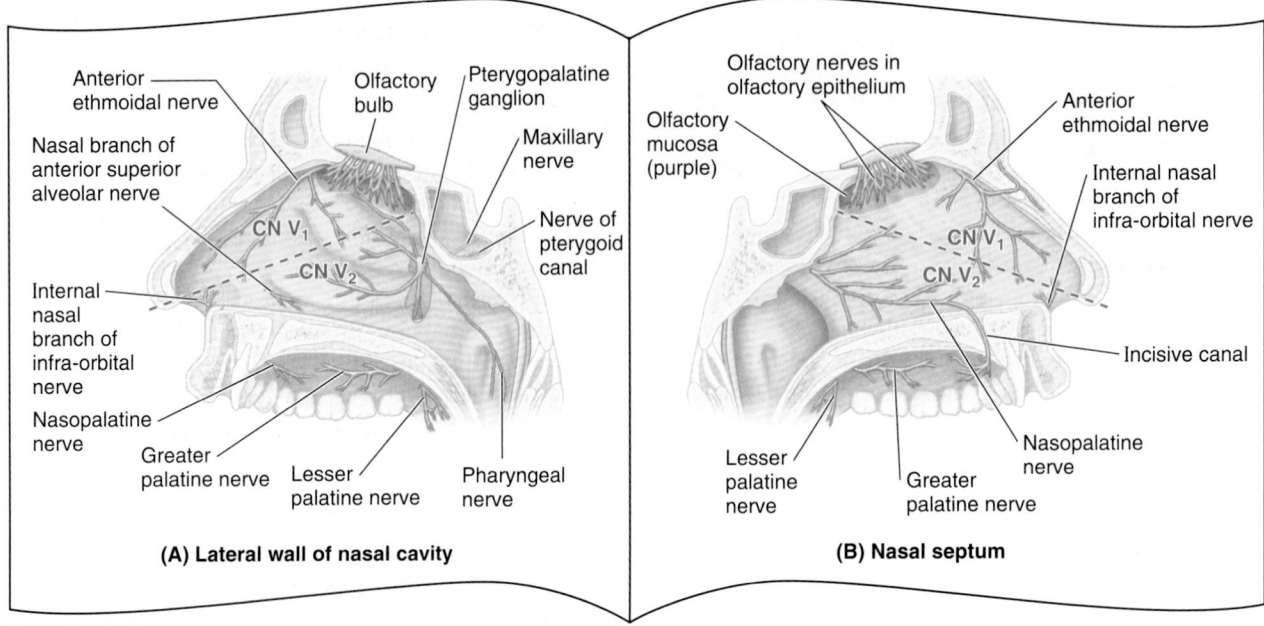

Open-book view

FIGURE 7.106. Innervation of nasal cavity. An open-book view of the lateral and medial (septal) walls of the right side of the nasal cavity is shown. A *dashed line* extrapolated approximately from the spheno-ethmoidal recess to the apex of the nose demarcates the territories of the ophthalmic (CN V₁) and maxillary (CN V₂) nerves for supplying general sensation to both the lateral wall and the nasal septum. The olfactory nerve (CN I) is distributed to the olfactory mucosa superior to the level of the superior concha on both the lateral wall and the nasal septum.

anterior ethmoidal nerve), but the alae of the nose are supplied by the nasal branches of the infra-orbital nerve (CN V₂). The **olfactory nerves,** concerned with smell, arise from cells in the **olfactory epithelium** in the superior part of the lateral and septal walls of the nasal cavity. The central processes of these cells (forming the olfactory nerve) pass through the *cribriform plate* and end in the **olfactory bulb,** the rostral expansion of the **olfactory tract** (Fig. 7.102A).

Paranasal Sinuses

The **paranasal sinuses** are air-filled extensions of the respiratory part of the nasal cavity into the following cranial bones: frontal, ethmoid, sphenoid, and maxilla. They are named according to the bones in which they are located. The sinuses continue to invade the surrounding bone, and marked extensions are common in the crania of older individuals.

FRONTAL SINUSES

The **right** and **left frontal sinuses** are between the outer and inner tables of the frontal bone, posterior to the superciliary arches and the root of the nose (Figs. 7.103, 7.104, and 7.107). Frontal sinuses are usually detectable in children by 7 years of age. The right and left sinuses each drain through a **frontonasal duct** into the *ethmoidal infundibulum,* which opens into the *semilunar hiatus* of the middle nasal meatus. The frontal sinuses are innervated by branches of the *supra-orbital nerves* (CN V₁).

The right and left frontal sinuses are rarely of equal size, and the septum between them is not usually situated entirely

in the median plane. The frontal sinuses vary in size from approximately 5 mm to large spaces extending laterally into the greater wings of the sphenoid. Often a frontal sinus has two parts: a vertical part in the squamous part of the frontal bone, and a horizontal part in the orbital part of the frontal bone. One or both parts may be large or small. When the supra-orbital part is large, its roof forms the floor of the anterior cranial fossa and its floor forms the roof of the orbit.

ETHMOIDAL CELLS

The **ethmoidal cells** (**sinuses**) are small invaginations of the mucous membrane of the middle and superior nasal meatus into the ethmoid bone between the nasal cavity and the orbit (Figs. 7.104, 7.107, and 7.108). The ethmoidal cells usually are not visible in plain radiographs before 2 years of age but are recognizable in CT scans. The **anterior ethmoidal cells** drain directly or indirectly into the middle nasal meatus through the ethmoidal infundibulum. The **middle ethmoidal cells** open directly into the middle meatus and are sometimes called "bullar cells" because they form the *ethmoidal bulla,* a swelling on the superior border of the semilunar hiatus (Fig. 7.103B). The **posterior ethmoidal cells** open directly into the superior meatus. The ethmoidal cells are supplied by the anterior and posterior ethmoidal branches of the *nasociliary nerves* (CN V₁) (Figs. 7.19 and 7.106).

SPHENOIDAL SINUSES

The **sphenoidal sinuses** are located in the body of the sphenoid, but they may extend into the wings of this bone

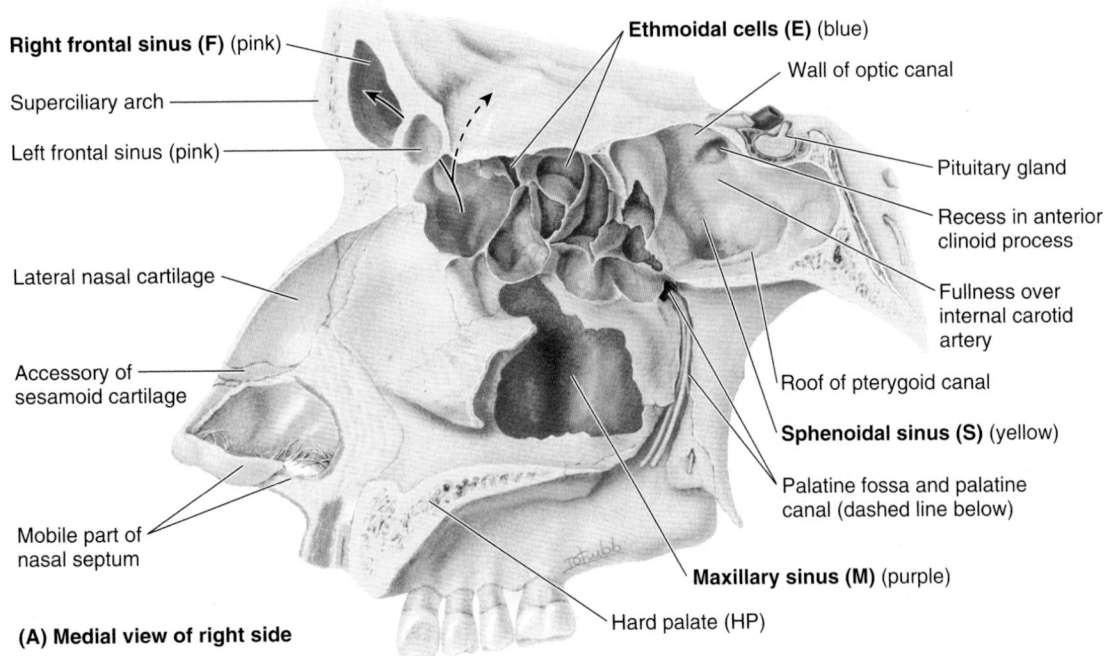

Right frontal sinus (F) (pink)

Superciliary arch

Left frontal sinus (pink)

Lateral nasal cartilage

Accessory of sesamoid cartilage

Mobile part of nasal septum

(A) Medial view of right side

Ethmoidal cells (E) (blue)

Wall of optic canal

Pituitary gland

Recess in anterior clinoid process

Fullness over internal carotid artery

Roof of pterygoid canal

Sphenoidal sinus (S) (yellow)

Palatine fossa and palatine canal (dashed line below)

Maxillary sinus (M) (purple)

Hard palate (HP)

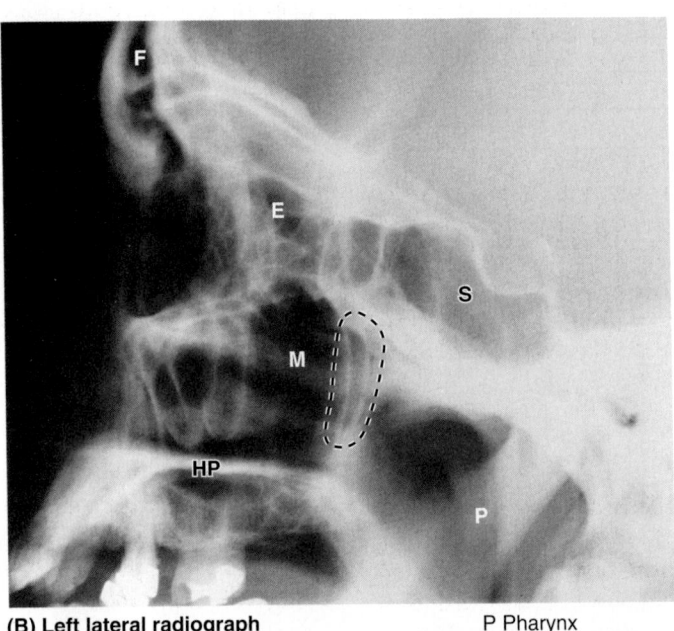

(B) Left lateral radiograph   P Pharynx

FIGURE 7.107. Paranasal sinuses I. A. The paranasal sinuses of the right side have been opened from a nasal approach and color coded. An anterior ethmoidal cell (*pink*) is invading the diploë of the frontal bone to become a frontal sinus. An offshoot (*broken arrow*) invades the orbital plate of the frontal bone. The sphenoidal sinus in this specimen is extensive, extending (1) posteriorly, inferior to the pituitary gland, to the clivus; (2) laterally, inferior to the optic nerve (CN II), into the anterior clinoid process; and (3) inferior to the pterygoid process but leaving the pterygoid canal and rising as a ridge on the floor of the sinus. The maxillary sinus is pyramidal. **B.** Radiograph of cranium demonstrating air densities (dark areas) associated with paranasal sinuses, nasal cavity, oral cavity, and pharynx. The letters are defined in part **A.**

(Figs. 7.103 and 7.107). They are unevenly divided and separated by a bony septum. Because of this extensive pneumatization (formation of air cells), the body of the sphenoid is fragile. Only thin plates of bone separate the sinuses from several important structures: the optic nerves and optic chiasm, the pituitary gland, the internal carotid arteries, and the cavernous sinuses. The sphenoidal sinuses are derived from a posterior ethmoidal cell that begins to invade the sphenoid at approximately 2 years of age. In some people, several posterior ethmoidal cells invade the sphenoid, giving rise to multiple sphenoidal sinuses that open separately into the *spheno-ethmoidal recess* (Fig. 7.103A). The posterior ethmoidal arteries and the posterior ethmoidal nerves that accompany the arteries supply the sphenoidal sinuses (Fig. 7.105).

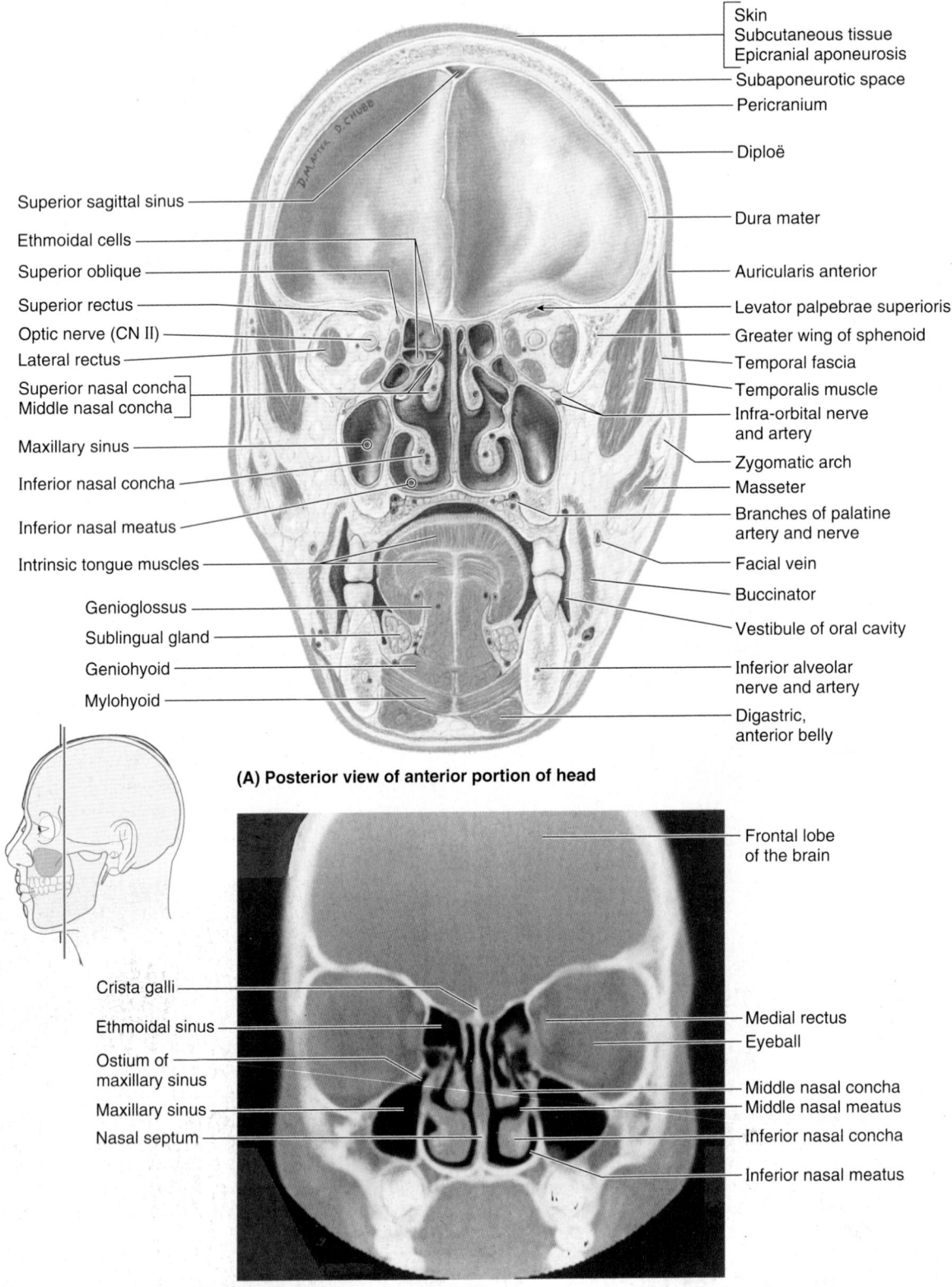

(A) Posterior view of anterior portion of head

Skin
Subcutaneous tissue
Epicranial aponeurosis
Subaponeurotic space
Pericranium

Diploë

Dura mater

Auricularis anterior

Levator palpebrae superioris
Greater wing of sphenoid
Temporal fascia
Temporalis muscle
Infra-orbital nerve
and artery
Zygomatic arch
Masseter
Branches of palatine
artery and nerve
Facial vein
Buccinator
Vestibule of oral cavity
Inferior alveolar
nerve and artery
Digastric,
anterior belly

Superior sagittal sinus
Ethmoidal cells
Superior oblique
Superior rectus
Optic nerve (CN II)
Lateral rectus
Superior nasal concha
Middle nasal concha
Maxillary sinus
Inferior nasal concha
Inferior nasal meatus
Intrinsic tongue muscles
Genioglossus
Sublingual gland
Geniohyoid
Mylohyoid

(B) Anterior view of coronal CT scan

Frontal lobe
of the brain

Crista galli
Ethmoidal sinus
Ostium of
maxillary sinus
Maxillary sinus
Nasal septum

Medial rectus
Eyeball

Middle nasal concha
Middle nasal meatus
Inferior nasal concha
Inferior nasal meatus

FIGURE 7.108. Paranasal sinuses II. The orientation drawing shows the plane of the section shown in both parts. **A.** The ethmoid bone occupies a central position, with its horizontal component forming the central part of the anterior cranial fossa superiorly and the roof of the nasal cavity inferiorly. The ethmoidal cells give attachment to the superior and middle concha and form part of the medial wall of the orbit; the perpendicular plate of the ethmoid forms part of the nasal septum. The maxillary sinus forms the inferior part of the lateral wall of the nose and shares a common wall with the orbit. The middle concha shelters the semilunar hiatus into which the maxillary ostium opens (*arrow*). **B.** The CT scan demonstrates air-filled cavities shown in anatomical section in part **A.** (Courtesy of Dr. D. Armstrong, Associate Professor of Medical Imaging, University of Toronto, Toronto, Ontario, Canada.)

MAXILLARY SINUSES

The **maxillary sinuses** are the largest of the paranasal sinuses. They occupy the bodies of the maxillae and communicate with the middle nasal meatus (Figs. 7.104, 7.107, and 7.108).

- The **apex** of the maxillary sinus extends toward and often into the zygomatic bone.
- The **base** of the maxillary sinus forms the inferior part of the lateral wall of the nasal cavity.
- The **roof** of the maxillary sinus is formed by the floor of the orbit.
- The **floor** of the maxillary sinus is formed by the alveolar part of the maxilla. The roots of the maxillary teeth,

particularly the first two molars, often pr█████ elevations in the floor of the sinus.

Each maxillary sinus drains by one or more openings, the **maxillary ostium (ostia)**, into the middle nasal meatus of the nasal cavity by way of the semilunar hiatus.

The **arterial supply of the maxillary sinus** is mainly from superior alveolar branches of the **maxillary artery** (Fig. 7.73; Table 7.12); however, branches of the *descending* and *greater palatine arteries* supply the floor of the sinus (Figs. 7.98B). **Innervation of the maxillary sinus** is from the anterior, middle, and posterior **superior alveolar nerves,** which are branches of the maxillary nerve (Fig. 7.79A).

THE NOSE

Nasal Fractures

Because of the prominence of the nose, fractures of the nasal bones are common in automobile accidents and contact sports (unless face guards are worn). Fractures usually result in deformation of the nose, particularly when a lateral force is applied by someone's elbow, for example;

epistaxis (bleeding from the nose) usually occurs. In severe fractures, disruption of the bones and cartilages results in displacement of the nose. When the injury results from a direct blow, the cribriform plate of the ethmoid bone may also fracture.

Deviation of Nasal Septum

The nasal septum is usually deviated to one side or the other (Fig. B7.40). This could be the result of a birth injury, but more often the deviation occurs during adolescence and adulthood from trauma (e.g., during a fist fight). Sometimes the deviation is so severe that the nasal septum is in contact with the lateral wall of the nasal cavity and often obstructs breathing or exacerbates snoring. The deviation can be corrected surgically.

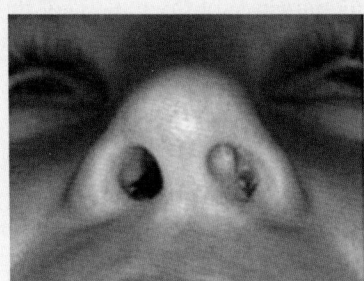

Nasal septum deviated to left side

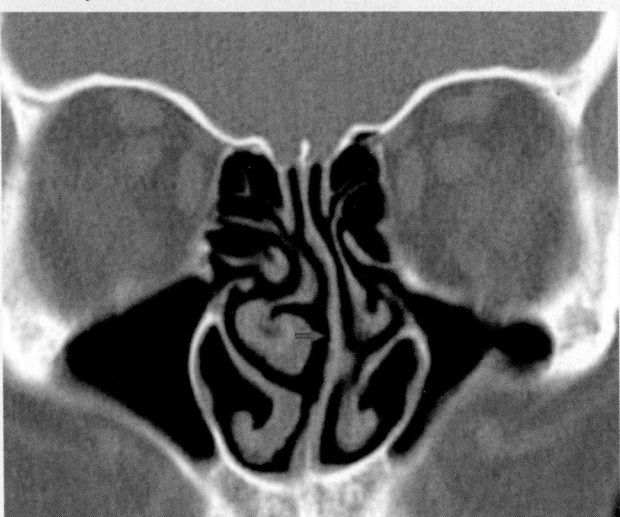

Anterior view, CT scan

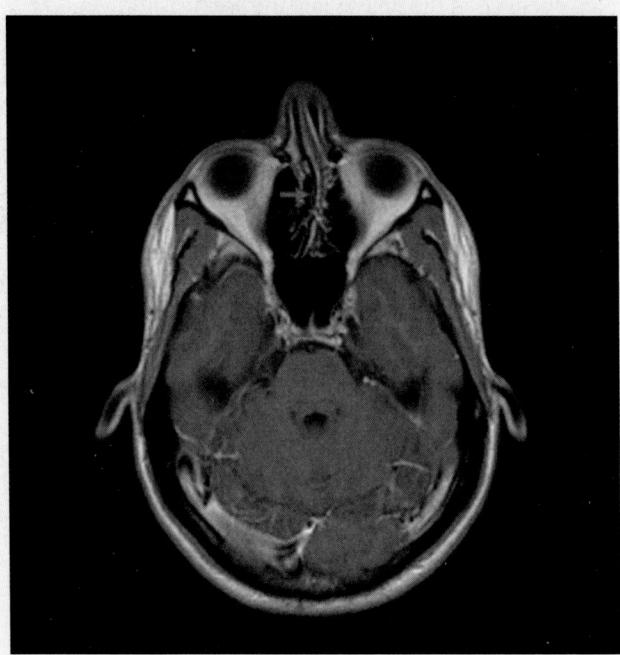

Inferior view, MRI

FIGURE B7.40. Deviated nasal septum.

Rhinitis

The nasal mucosa becomes swollen and inflamed (*rhinitis*) during severe upper respiratory infections and allergic reactions (e.g., hayfever). Swelling of the mucosa occurs readily because of its vascularity. Infections of the nasal cavities may spread to the:

- Anterior cranial fossa through the cribriform plate.
- Nasopharynx and retropharyngeal soft tissues.
- Middle ear through the *pharyngotympanic tube* (auditory tube), which connects the tympanic cavity and nasopharynx.
- Paranasal sinuses.
- Lacrimal apparatus and conjunctiva.

Epistaxis

Epistaxis (nosebleed) is relatively common because of the rich blood supply to the nasal mucosa. In most cases, the cause is trauma and the bleeding is from an area in the anterior third of the nose (Kiesselbach area—Fig. 7.105B). Epistaxis is also associated with infections and hypertension. Spurting of blood from the nose results from rupture of arteries. Mild epistaxis may also result from nose picking, which tears veins in the vestibule of the nose.

Sinusitis

Because the paranasal sinuses are continuous with the nasal cavities through apertures that open into them, infection may spread from the nasal cavities, producing inflammation and swelling of the mucosa of the sinuses (*sinusitis*) and local pain. Sometimes several sinuses are inflamed (*pansinusitis*), and the swelling of the mucosa may block one or more openings of the sinuses into the nasal cavities.

Infection of Ethmoidal Cells

If nasal drainage is blocked, infections of the ethmoidal cells may break through the fragile medial wall of the orbit. Severe infections from this source may cause blindness because some posterior ethmoidal cells lie close to the optic canal, which gives passage to the optic nerve and ophthalmic artery. Spread of infection from these cells could also affect the dural sheath of the optic nerve, causing *optic neuritis*.

Infection of Maxillary Sinuses

The maxillary sinuses are the most commonly infected, probably because their ostia (openings) are commonly small and are located high on their superomedial walls (Fig. 7.108). When the mucous membrane of the sinus is congested, the maxillary ostia are often obstructed. Because of the high location of the ostia, when the head is erect it is impossible for the sinuses to drain until they are full. Because the ostia of the right and left sinuses lie on the medial sides (i.e., are directed toward each other), when lying on one's side only the upper sinus (e.g., the right sinus if lying on the left side) drains. A cold or allergy involving both sinuses can result in nights of rolling from side to side in an attempt to keep the sinuses drained. A maxillary sinus can be cannulated and drained by passing a cannula from the naris through the maxillary ostium into the sinus.

Relationship of Teeth to Maxillary Sinus

The close proximity of the three maxillary molar teeth to the floor of the maxillary sinus poses potentially serious problems. During removal of a maxillary molar tooth, a fracture of a root of the tooth may occur. If proper retrieval methods are not used, a piece of the root may be driven superiorly into the maxillary sinus. A communication may be created between the oral cavity and the maxillary sinus as a result, and an infection may occur. Because the superior alveolar nerves (branches of the maxillary nerve) supply both the maxillary teeth and the mucous membrane of the maxillary sinuses, inflammation of the mucosa of the sinus is frequently accompanied by a sensation of toothache in the molar teeth.

Transillumination of Sinuses

Transillumination of the maxillary sinuses is performed in a darkened room. A bright light is placed in the patient's mouth on one side of the hard palate or firmly against the cheek (an in Fig. B7.41A). The light passes through the maxillary sinus and appears as a crescent-shaped, dull glow inferior to the orbit. If a sinus contains excess fluid, a mass, or a thickened mucosa, the glow is decreased. The frontal sinuses can also be transilluminated by directing the light superiorly under the medial aspect of the eyebrow, normally producing a glow superior to the orbit (Fig. B7.41B). Because of the great variation in the development of the sinuses, the pattern and extent of sinus illumination differs from person to person (Swartz, 2009).

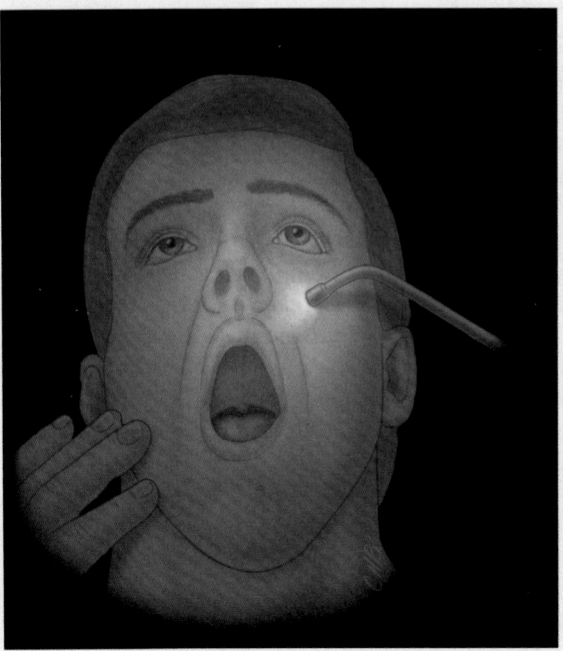

(A) Transillumination of maxillary sinus

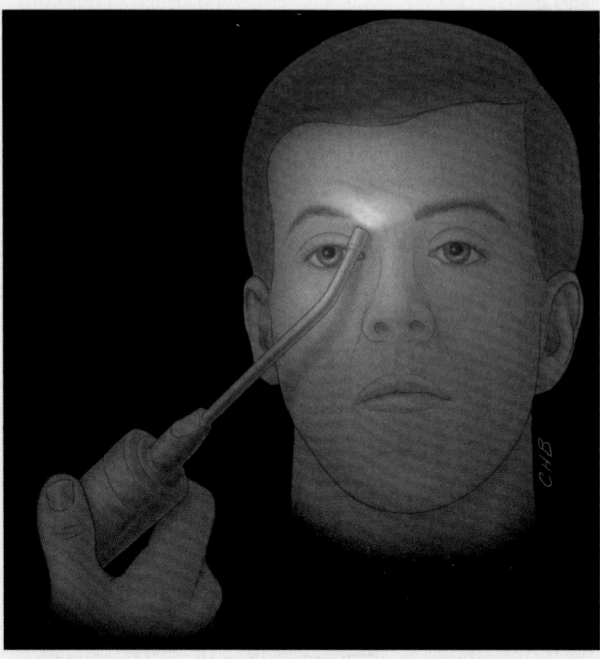

(B) Transillumination of frontal sinus

FIGURE B7.41.

The Bottom Line

THE NOSE

The nose is the ventilation system that traverses the head, enabling the flow of air between the external environment and the lower respiratory system (lungs). ♦ As air is drawn in through the nose, its chemistry is sampled (smell and taste augmentation), and it is warmed, humidified, and filtered for the lungs. As it exits, heat and moisture are released with it. ♦ The nose also provides a drainage route for mucus and lacrimal fluid.

Skeleton of Nose: Opening anteriorly via the nares, the nasal cavity is subdivided by a median nasal septum. ♦ The protruding external nose and anterior septum benefit from the flexibility provided by a cartilaginous skeleton, reducing the potential for nasal fractures. ♦ Except for the septum and floor, the walls of the nasal cavity are highly pneumatized by the paranasal sinuses, and its lateral walls bear conchae.

Nasal cavities: Both the sinuses and conchae increase the secretory surface area for exchange of moisture and heat. ♦ Essentially, all surfaces are covered with thick, vascular, secretory mucosa, the anterosuperior portion of which

(including that of most of the paranasal sinuses) is supplied by the ophthalmic artery and nerve (CN V₁), and the postero-inferior portion (including that of the maxillary sinus) by the maxillary artery and nerve (CN V₂). ♦ The mucosa of the roof and adjacent areas of walls and septum also receive special sensory innervation from the olfactory nerve (CN I). ♦ Posteriorly, the nasal cavity is continuous with the nasopharynx via the choanae; the soft palate serves as a valve or gate controlling access to and from the nasal passageway. ♦ The bone and mucosa of the lateral walls of this passageway are perforated by openings of the nasolacrimal ducts, the paranasal sinuses and the pharyngotympanic tube. ♦ Only the bone is perforated by the pterygopalatine foramen, providing passage of neurovascular structures into the nasal mucosa.

Paranasal sinuses: The paranasal sinuses are named for the bones they occupy. ♦ The maxillary sinus is the largest. ♦ Most sinuses open into the middle nasal meatus, but the sphenoidal sinuses enter the spheno-ethmoidal recess.

EAR

The **ear**—the organ of hearing and equilibrium (balance)—is divided into the external, middle, and internal ear (Fig. 7.109). The external ear and middle ear are mainly concerned with the transfer of sound to the internal ear, which contains the organ for equilibrium as well as for hearing. The *tympanic membrane* separates the external ear from the middle ear. The *pharyngotympanic tube* joins the middle ear to the nasopharynx.

External Ear

The **external ear** is composed of the shell-like *auricle (pinna)*, which collects sound, and the *external acoustic meatus* (ear canal), which conducts sound to the tympanic membrane.

AURICLE

The **auricle** (L. *auris*, ear) is composed of an irregularly shaped plate of elastic cartilage that is covered by thin skin (Fig. 7.110). The auricle has several depressions and elevations. The **concha of the auricle** is the deepest depression.

The elevated margin of the auricle is the **helix.** The other depressions and elevations are identified in Figure 7.110. The non-cartilaginous **lobule** (lobe) consists of fibrous tissue, fat, and blood vessels. It is easily pierced for taking small blood samples and inserting earrings. The **tragus** (G. *tragos,* goat; alluding to the hairs that tend to grow from this formation, like a goat's beard) is a tongue-like projection overlapping the opening of the external acoustic meatus. The **arterial supply** to the auricle is derived mainly from the *posterior auricular* and *superficial temporal arteries* (Fig. 7.111A).

The main **nerves to the skin of the auricle** are the great auricular and auriculotemporal nerves. The **great auricular nerve** supplies the cranial (medial) surface (commonly called the "back of the ear") and the posterior part (helix, antihelix, and lobule) of the lateral surface ("front"). The *auriculotemporal nerve,* a branch of CN V$_3$, supplies the skin of the auricle anterior to the external acoustic meatus (Figs. 7.109 and 7.111A). Minor contributions of embryological significance are made to the skin of the concha and its eminence by the vagus and facial nerves.

The **lymphatic drainage** of the auricle is as follows: the lateral surface of the superior half of the auricle drains to the *superficial parotid lymph nodes* (Fig. 7.111B); the cranial surface of the superior half of the auricle drains to the

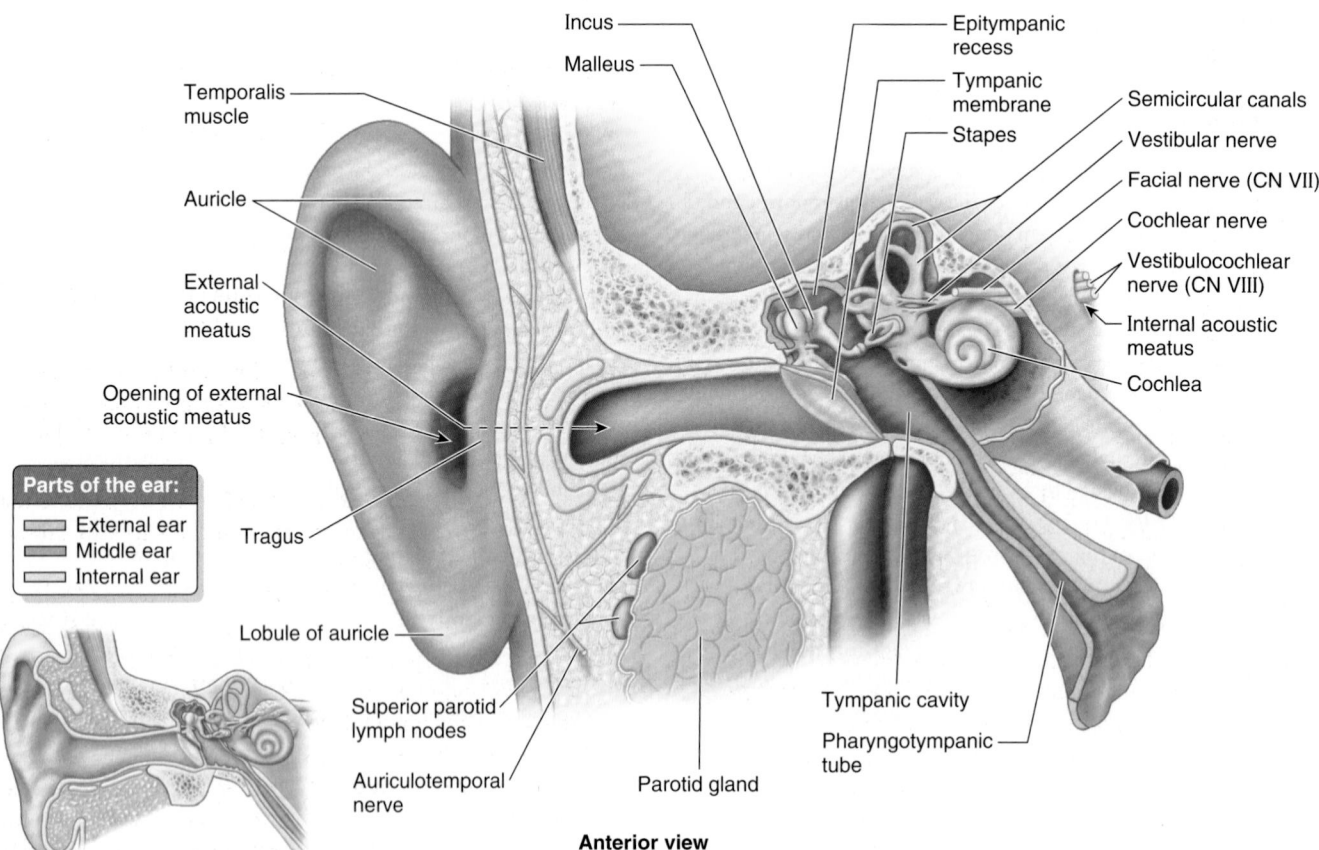

Parts of the ear:
- External ear
- Middle ear
- Internal ear

Labels: Temporalis muscle; Auricle; External acoustic meatus; Opening of external acoustic meatus; Tragus; Lobule of auricle; Superior parotid lymph nodes; Auriculotemporal nerve; Parotid gland; Incus; Malleus; Epitympanic recess; Tympanic membrane; Stapes; Semicircular canals; Vestibular nerve; Facial nerve (CN VII); Cochlear nerve; Vestibulocochlear nerve (CN VIII); Internal acoustic meatus; Cochlea; Tympanic cavity; Pharyngotympanic tube

Anterior view

FIGURE 7.109. Parts of ear. A coronal section of the ear, with accompanying orientation figure, demonstrates that the ear has three parts: external, middle, and internal. The external ear consists of the auricle and external acoustic meatus. The middle ear is an air space in which the auditory ossicles are located. The internal ear contains the membranous labyrinth; its chief divisions are the cochlear labyrinth and the vestibular labyrinth.

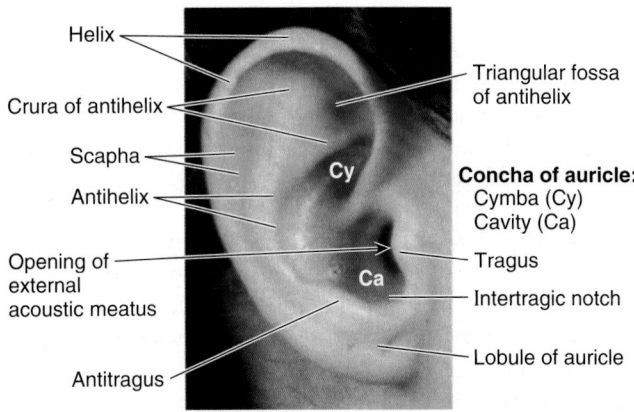

Helix
Crura of antihelix
Scapha
Antihelix
Opening of external acoustic meatus
Antitragus
Triangular fossa of antihelix
Concha of auricle: Cymba (Cy) Cavity (Ca)
Tragus
Intertragic notch
Lobule of auricle
Cy
Ca

FIGURE 7.110. External ear. The parts of the auricle commonly used in clinical descriptions are labeled. The external ear includes the auricle and external acoustic meatus.

mastoid lymph nodes and *deep cervical lymph nodes;* and the remainder of the auricle, including the lobule, drains into the **superficial cervical lymph nodes.**

EXTERNAL ACOUSTIC MEATUS

The **external acoustic meatus** is an ear canal that leads inward through the tympanic part of the temporal bone from the auricle to the tympanic membrane, a distance of 2–3 cm in adults (Fig. 7.109). The lateral third of this slightly S-shaped canal is cartilaginous and is lined with skin that is continuous with the auricular skin. The medial two thirds of the meatus is bony and lined with thin skin that is continuous with the external layer of the tympanic membrane. The ceruminous and sebaceous glands in the subcutaneous tissue of the cartilaginous part of the meatus produce *cerumen* (earwax).

The **tympanic membrane,** approximately 1 cm in diameter, is a thin, oval semitransparent membrane at the medial end of the external acoustic meatus (Figs. 7.109 and 7.112). This membrane forms a partition between the external acoustic meatus and the tympanic cavity of the middle ear. The tympanic membrane is covered with thin skin externally and mucous membrane of the middle ear internally. Viewed through an *otoscope,* the tympanic membrane has a concavity toward the external acoustic meatus with a shallow, cone-like central depression, the peak of which is the **umbo** (Fig. 7.112A) (see the blue box "Otoscopic Examination," p. 977). The central axis of the tympanic membrane passes perpendicularly through the umbo like the handle of an umbrella, running anteriorly and inferiorly as it runs laterally. Thus, the tympanic membrane is oriented like a mini radar or satellite dish positioned to receive signals coming from the ground in front and to the side of the head.

Superior to the lateral process of the *malleus* (one of the small ear bones, or *auditory ossicles,* of the middle ear), the membrane is thin and is called the **pars flaccida** (flaccid part). It lacks the radial and circular fibers present in the remainder of the membrane, called the **pars tensa** (tense part). The flaccid part forms the lateral wall of the superior recess of the tympanic cavity.

The tympanic membrane moves in response to air vibrations that pass to it through the external acoustic meatus. Movements of the membrane are transmitted by the auditory ossicles through the middle ear to the internal ear (Fig. 7.109). The external surface of the tympanic membrane is supplied mainly by the *auriculotemporal nerve* (Fig. 7.111A), a branch of CN V$_3$. Some innervation is supplied by a small *auricular branch of the vagus* (CN X). The internal surface of the tympanic membrane is supplied by the glossopharyngeal nerve (CN IX).

Middle Ear

The **tympanic cavity** or **cavity of the middle ear** is the narrow air-filled chamber in the petrous part of the temporal bone (Fig. 7.109 and 7.113). The cavity has two parts: the **tympanic cavity proper,** the space directly internal to the tympanic membrane, and the **epitympanic recess,** the space superior to the membrane. The tympanic cavity is connected anteromedially with the nasopharynx by the *pharyngotympanic tube* and posterosuperiorly with the mastoid cells through the *mastoid antrum* (Figs. 7.113A and 7.114). The tympanic cavity is lined with mucous membrane that is continuous with the lining of the pharyngotympanic tube, mastoid cells, and mastoid antrum.

The contents of the middle ear are the:

- Auditory ossicles (malleus, incus, and stapes).
- Stapedius and tensor tympani muscles.
- Chorda tympani nerve, a branch of CN VII (Fig. 7.114).
- Tympanic plexus of nerves.

WALLS OF TYMPANIC CAVITY

The middle ear is shaped like a lozenge or narrow box with concave sides. It has six walls (Fig. 7.114B).

1. The **tegmental wall** (**roof**) is formed by a thin plate of bone, the *tegmen tympani,* which separates the tympanic cavity from the dura mater on the floor of the middle cranial fossa.
2. The **jugular wall** (**floor**) is formed by a layer of bone that separates the tympanic cavity from the superior bulb of the internal jugular vein.
3. The **membranous (lateral) wall** is formed almost entirely by the peaked convexity of the *tympanic membrane;* superiorly it is formed by the lateral bony wall of the *epitympanic recess.* The handle of the malleus is attached to the tympanic membrane, and its head extends into the epitympanic recess.
4. The **labyrinthine (medial) wall** (**medial wall**) separates the tympanic cavity from the internal ear. It also features the *promontory of the labyrinthine wall,* formed by the initial part (basal turn) of the cochlea, and the *oval* and *round windows,* which, in a dry cranium, communicate with the internal ear.

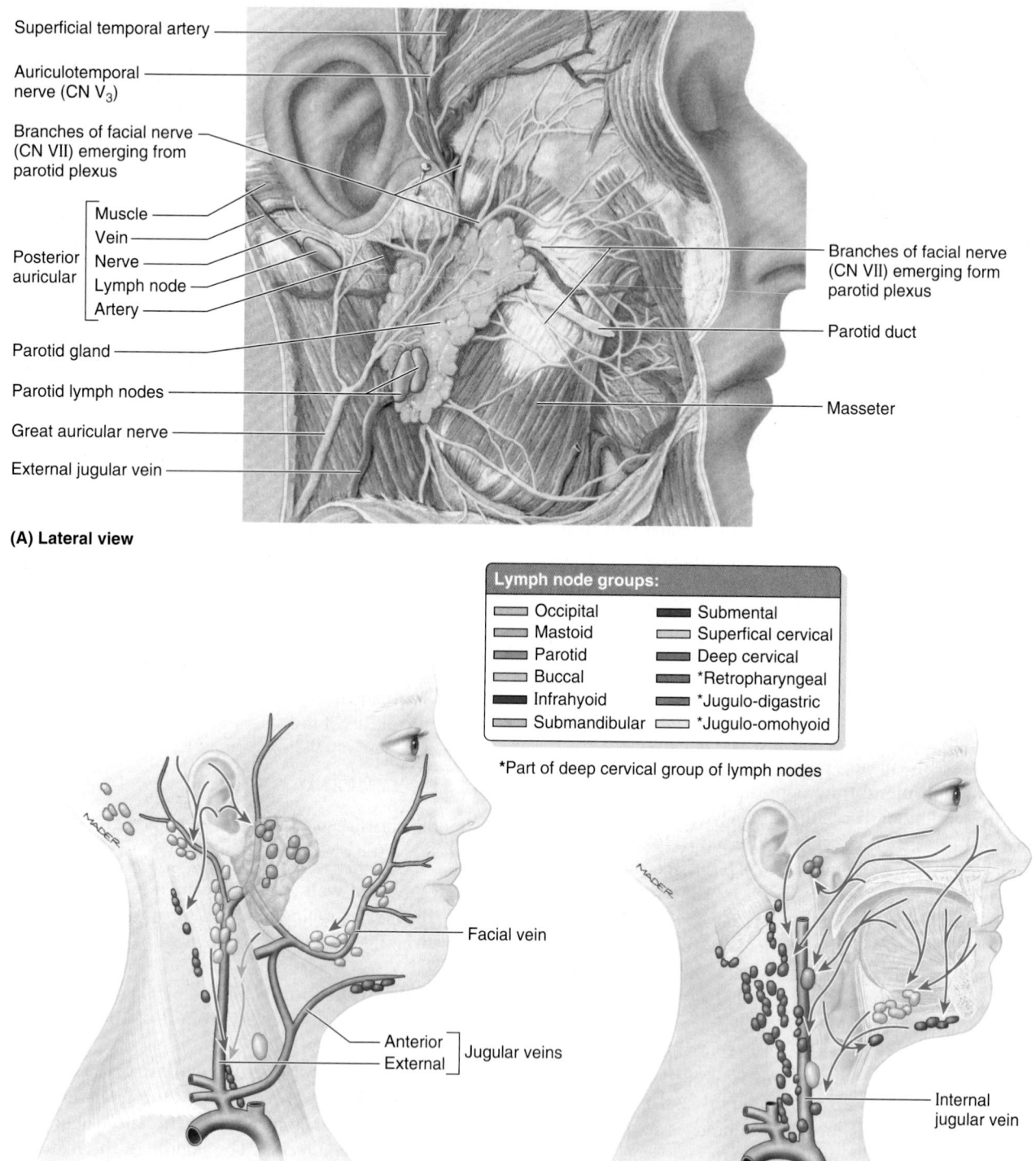

Superficial temporal artery

Auriculotemporal nerve (CN V₃)

Branches of facial nerve (CN VII) emerging from parotid plexus

Posterior auricular
- Muscle
- Vein
- Nerve
- Lymph node
- Artery

Parotid gland

Parotid lymph nodes

Great auricular nerve

External jugular vein

Branches of facial nerve (CN VII) emerging form parotid plexus

Parotid duct

Masseter

(A) Lateral view

Lymph node groups:

Occipital	Submental
Mastoid	Superfical cervical
Parotid	Deep cervical
Buccal	*Retropharyngeal
Infrahyoid	*Jugulo-digastric
Submandibular	*Jugulo-omohyoid

*Part of deep cervical group of lymph nodes

Facial vein

Anterior | Jugular veins
External

Internal jugular vein

(B)

FIGURE 7.111. Dissection of face and lymphatic drainage of head. A. The posterior auricular and superficial temporal arteries and veins and the great auricular and auriculotemporal nerves provide the circulation and innervation of the external ear. **B.** Lymphatic drainage is to the parotid lymph nodes and the mastoid and superficial cervical lymph nodes, all which drain to the deep cervical nodes.

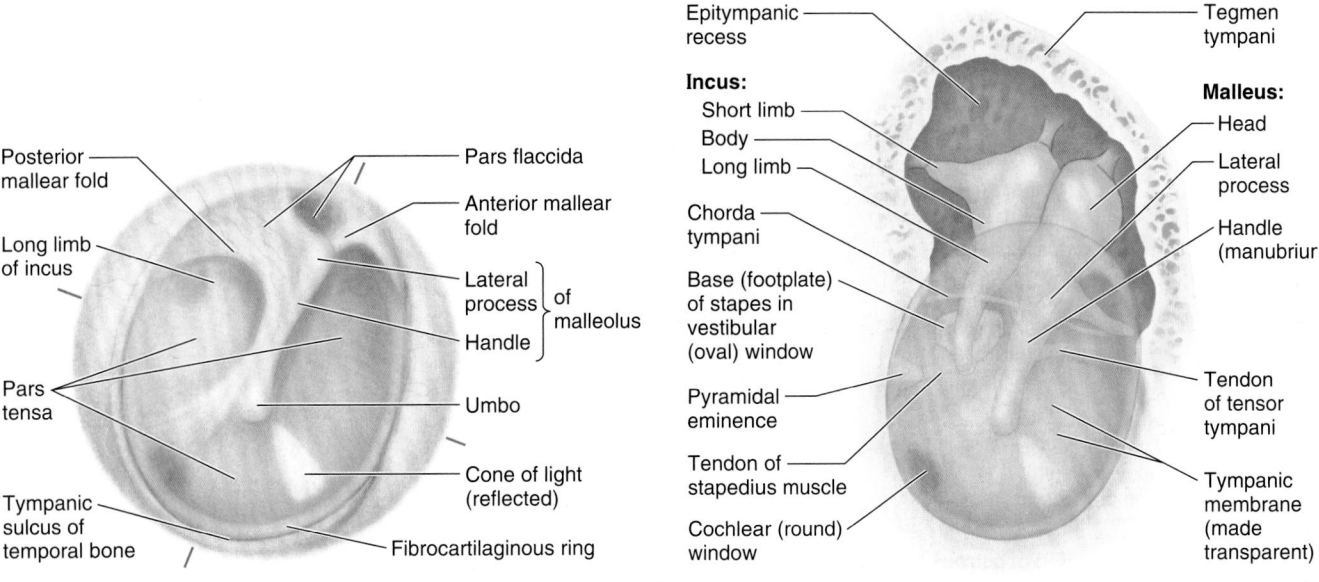

Epitympanic recess

Tegmen tympani

Incus:
- Short limb
- Body
- Long limb

Chorda tympani

Base (footplate) of stapes in vestibular (oval) window

Pyramidal eminence

Tendon of stapedius muscle

Cochlear (round) window

Malleus:
- Head
- Lateral process
- Handle (manubriur

Tendon of tensor tympani

Tympanic membrane (made transparent)

Posterior mallear fold

Pars flaccida

Anterior mallear fold

Long limb of incus

Lateral process | of malleolus
Handle |

Pars tensa

Umbo

Cone of light (reflected)

Tympanic sulcus of temporal bone

Fibrocartilaginous ring

(A) Otoscopic view of right tympanic membrane

(B) Ossicles of ear seen through tympanic membrane

FIGURE 7.112. Tympanic membrane and lateral approach to tympanic cavity. A. An otoscopic view of the right tympanic membrane is demonstrated. The *cone of light* is a reflection of the light of the otoscope. **B.** The tympanic membrane has been rendered semitransparent and the lateral wall of the epitympanic recess has been removed to demonstrate the ossicles of the middle ear in situ.

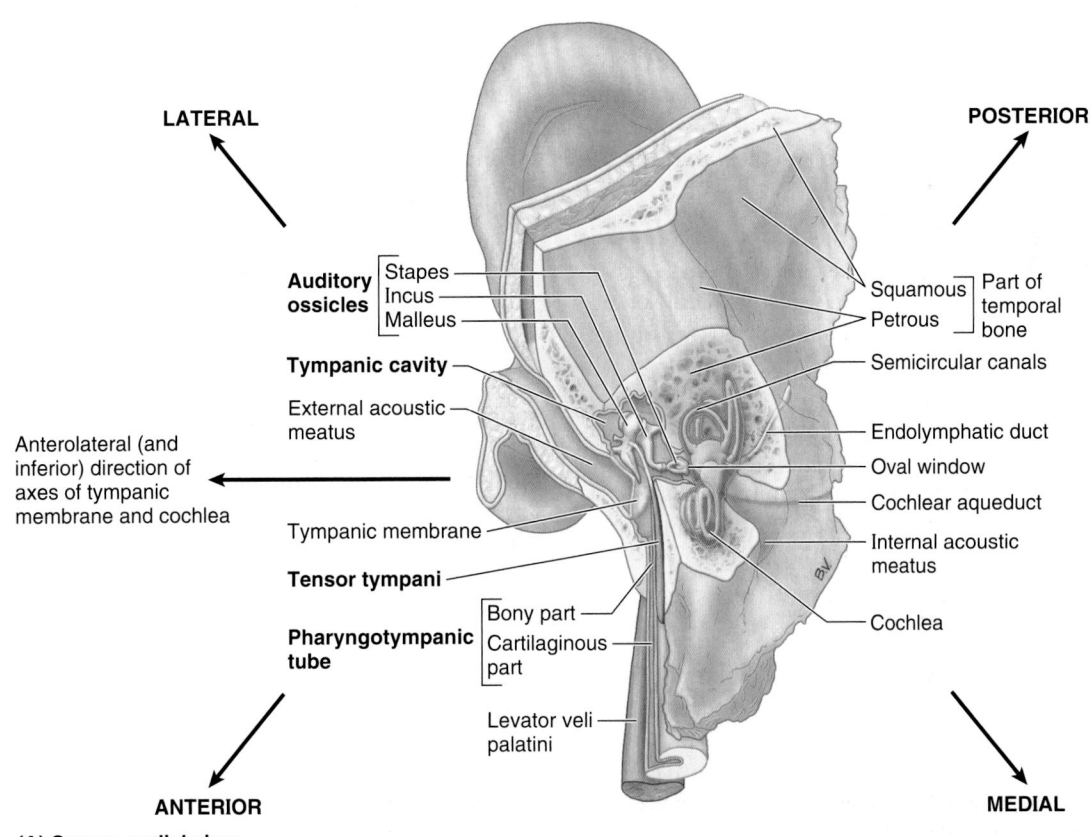

LATERAL

POSTERIOR

Auditory ossicles
- Stapes
- Incus
- Malleus

Tympanic cavity

External acoustic meatus

Anterolateral (and inferior) direction of axes of tympanic membrane and cochlea

Tympanic membrane

Tensor tympani

Pharyngotympanic tube
- Bony part
- Cartilaginous part

Levator veli palatini

Squamous | Part of temporal bone
Petrous |

Semicircular canals

Endolymphatic duct

Oval window

Cochlear aqueduct

Internal acoustic meatus

Cochlea

ANTERIOR

MEDIAL

(A) Superomedial view

FIGURE 7.113. General scheme and orientation of components of ear. A. The ear is shown in situ. The external acoustic meatus runs lateral to medial; the axis of the tympanic membrane and the axis about which the cochlea winds runs inferiorly and anteriorly as it proceeds laterally. The long axes of the bony and membranous labyrinths and of the pharyngotympanic tube and parallel tensor tympani and levator palatini muscles lie perpendicular to those of the tympanic membrane and cochlea (i.e., they run inferiorly and anteriorly as they proceed medially).

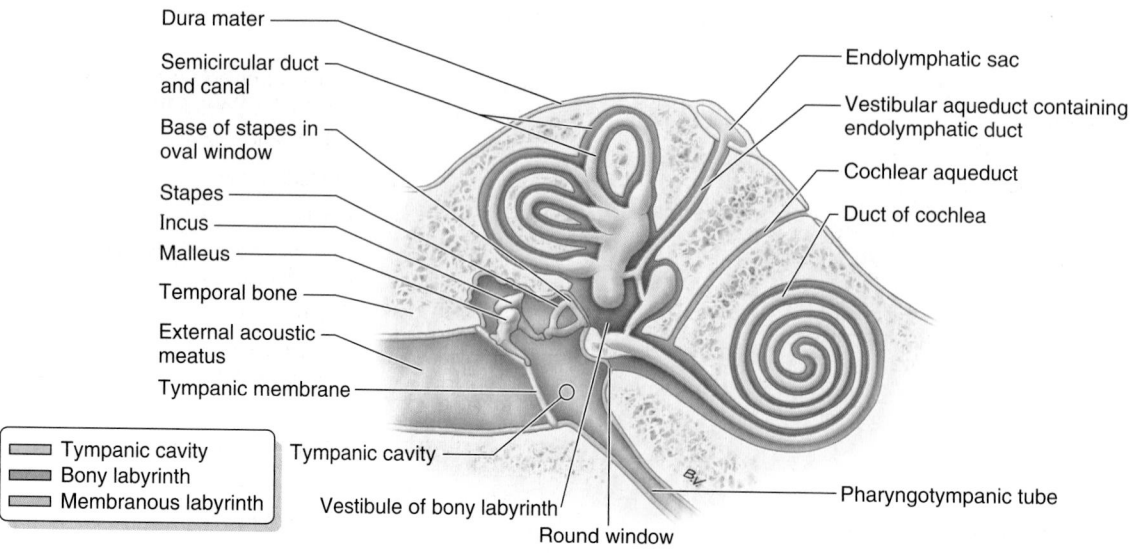

Dura mater
Semicircular duct and canal
Base of stapes in oval window
Stapes
Incus
Malleus
Temporal bone
External acoustic meatus
Tympanic membrane

Endolymphatic sac
Vestibular aqueduct containing endolymphatic duct
Cochlear aqueduct
Duct of cochlea

Tympanic cavity
Bony labyrinth
Membranous labyrinth

Tympanic cavity
Vestibule of bony labyrinth
Round window
Pharyngotympanic tube

(B) Oblique section of petrous temporal bone

FIGURE 7.113. *(Continued)* **B.** The middle and internal parts of the ear are shown. The middle ear lies between the tympanic membrane and the internal ear. Three auditory ossicles stretch from the lateral to the medial wall of the tympanic cavity. The pharyngotympanic tube is a communication between the anterior wall of the tympanic cavity and the lateral wall of the nasopharynx. The internal ear is composed of a closed system of membranous tubes and bulbs, the membranous labyrinth, which is filled with a fluid called endolymph *(orange)* and bathed in surrounding fluid called perilymph *(purple)*.

5. The **mastoid wall (posterior wall)** features an opening in its superior part, the **aditus** (L. access) **to the mastoid antrum,** connecting the tympanic cavity to the mastoid cells; the canal for the facial nerve descends between the posterior wall and the antrum, medial to the aditus.
6. The anterior **carotid wall** separates the tympanic cavity from the carotid canal; superiorly, it has the **opening of the pharyngotympanic tube** and the **canal for the tensor tympani.**

The **mastoid antrum** is a cavity in the mastoid process of the temporal bone (Fig. 7.114A). The antrum (L. from G., cave), like the tympanic cavity, is separated from the middle cranial fossa by a thin plate of the temporal bone, called the **tegmen tympani.** This structure forms the *tegmental wall* (roof) for the ear cavities and is also part of the floor of the lateral part of the middle cranial fossa. The mastoid antrum is the common cavity into which the mastoid cells open. The antrum and mastoid cells are lined by mucous membrane that is continuous with the lining of the middle ear. Antero-inferiorly, the antrum is related to the canal for the facial nerve.

PHARYNGOTYMPANIC TUBE

The **pharyngotympanic tube** (auditory tube) connects the tympanic cavity to the nasopharynx, where it opens posterior to the inferior nasal meatus (Fig. 7.113). The posterolateral third of the tube is bony, and the remainder is cartilaginous. The pharyngotympanic tube is lined by mucous membrane that is continuous posteriorly with that of the tympanic cavity and anteriorly with that of the nasopharynx.

The *function of the pharyngotympanic tube* is to equalize pressure in the middle ear with the atmospheric pressure, thereby allowing free movement of the tympanic membrane.

By allowing air to enter and leave the tympanic cavity, this tube balances the pressure on both sides of the membrane. Because the walls of the cartilaginous part of the tube are normally in apposition, the tube must be actively opened. The tube is opened by the expanding girth of the belly of the *levator veli palatini* as it contracts longitudinally, pushing against one wall while the *tensor veli palatini* pulls on the other. Because these are muscles of the soft palate, equalizing pressure ("popping the eardrums") is commonly associated with activities such as yawning and swallowing.

The *arteries of the pharyngotympanic tube* are derived from the *ascending pharyngeal artery*, a branch of the external carotid artery, and the *middle meningeal artery* and *artery of the pterygoid canal*, branches of the maxillary artery (Fig. 7.115; Table 7.12).

The *veins of the pharyngotympanic tube* drain into the pterygoid venous plexus. **Lymphatic drainage** of the tube is to the *deep cervical lymph nodes* (Fig. 7.111B).

The **nerves of the pharyngotympanic tube** arise from the *tympanic plexus* (Fig. 7.114B), which is formed by fibers of the glossopharyngeal nerve (CN IX). Anteriorly, the tube also receives fibers from the *pterygopalatine ganglion* (Fig. 7.106A).

AUDITORY OSSICLES

The **auditory ossicles** form a mobile *chain of small bones* across the tympanic cavity from the tympanic membrane to the **oval window** (L. *fenestra vestibuli*), an oval opening on the labyrinthine wall of the tympanic cavity leading to the *vestibule of the bony labyrinth* (Figs. 7.113B and 7.116A). These ossicles are the first bones to be fully ossified during development and are essentially mature at birth. The bone from which they are formed is exceptionally dense (hard).

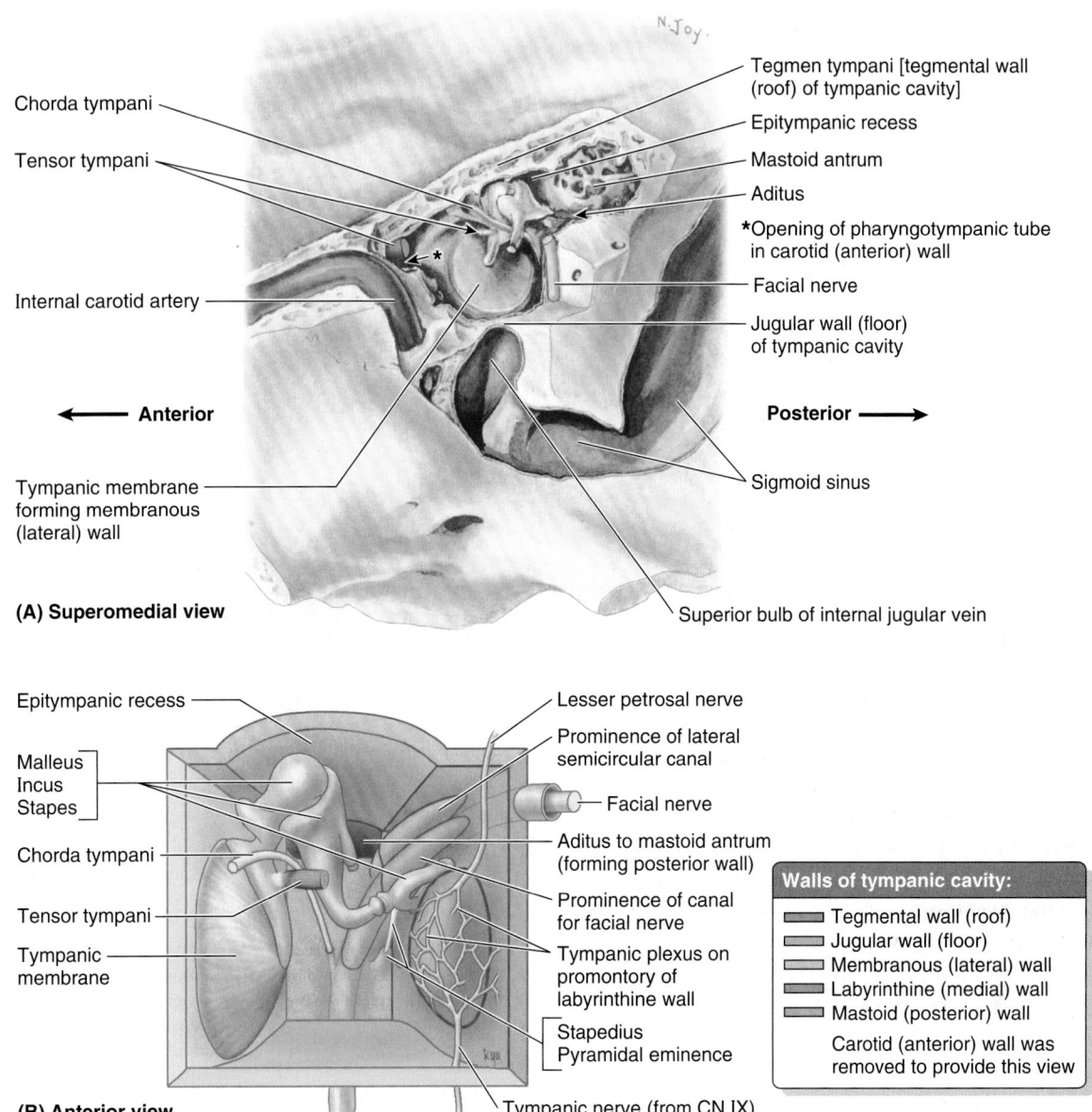

Chorda tympani

Tensor tympani

Internal carotid artery

← **Anterior**

Tympanic membrane
forming membranous
(lateral) wall

(A) Superomedial view

Tegmen tympani [tegmental wall
(roof) of tympanic cavity]

Epitympanic recess

Mastoid antrum

Aditus

*Opening of pharyngotympanic tube
in carotid (anterior) wall

Facial nerve

Jugular wall (floor)
of tympanic cavity

Posterior →

Sigmoid sinus

Superior bulb of internal jugular vein

Epitympanic recess

Malleus
Incus
Stapes

Chorda tympani

Tensor tympani

Tympanic
membrane

(B) Anterior view

Lesser petrosal nerve

Prominence of lateral
semicircular canal

Facial nerve

Aditus to mastoid antrum
(forming posterior wall)

Prominence of canal
for facial nerve

Tympanic plexus on
promontory of
labyrinthine wall

Stapedius
Pyramidal eminence

Tympanic nerve (from CN IX)

Walls of tympanic cavity:

- Tegmental wall (roof)
- Jugular wall (floor)
- Membranous (lateral) wall
- Labyrinthine (medial) wall
- Mastoid (posterior) wall
- Carotid (anterior) wall was removed to provide this view

FIGURE 7.114. Walls of tympanic cavity. A. This specimen was dissected with a drill from the medial aspect. The tegmen tympani, forming the roof of the tympanic cavity and the mastoid antrum, is fairly thick in this specimen; usually it is extremely thin. The internal carotid artery is the main relation of the anterior wall, the internal jugular vein is the main relation of the floor, and the facial nerve (CN VII) is a main feature of the posterior wall. The chorda tympani passes between the malleus and the incus. **B.** In this view of the middle ear, the carotid (anterior) wall of the tympanic cavity has been removed. The tympanic membrane forms most of the membranous (lateral) wall; superior to it is the epitympanic recess, in which are housed the larger parts of the malleus and incus. Branches of the tympanic plexus provide innervation to the mucosa of the middle ear and adjacent pharyngotympanic tube; but one branch, the lesser petrosal nerve, is conveying presynaptic parasympathetic fibers to the otic ganglion for secretomotor innervation of the parotid gland.

The ossicles are covered with the mucous membrane lining the tympanic cavity; but unlike other bones, they lack a surrounding layer of osteogenic periosteum.

Malleus. The **malleus** (L. a hammer) attaches to the tympanic membrane. The rounded superior **head of the malleus** lies in the epitympanic recess (Fig. 7.116B). The **neck of the malleus** lies against the flaccid part of the tympanic membrane, and the **handle of the malleus** is embedded in the tympanic membrane, with its tip at the umbo; thus the

malleus moves with the membrane. The head of the malleus articulates with the incus; the tendon of the tensor tympani inserts into its handle near the neck. The *chorda tympani* crosses the medial surface of the neck of the malleus. The malleus functions as a lever, with the longer of its two processes and its handle attached to the tympanic membrane.

Incus. The **incus** (L. an anvil) is located between the malleus and the stapes and articulates with them. It has a body and two limbs. Its large **body** lies in the epitympanic recess (Fig. 7.116A),

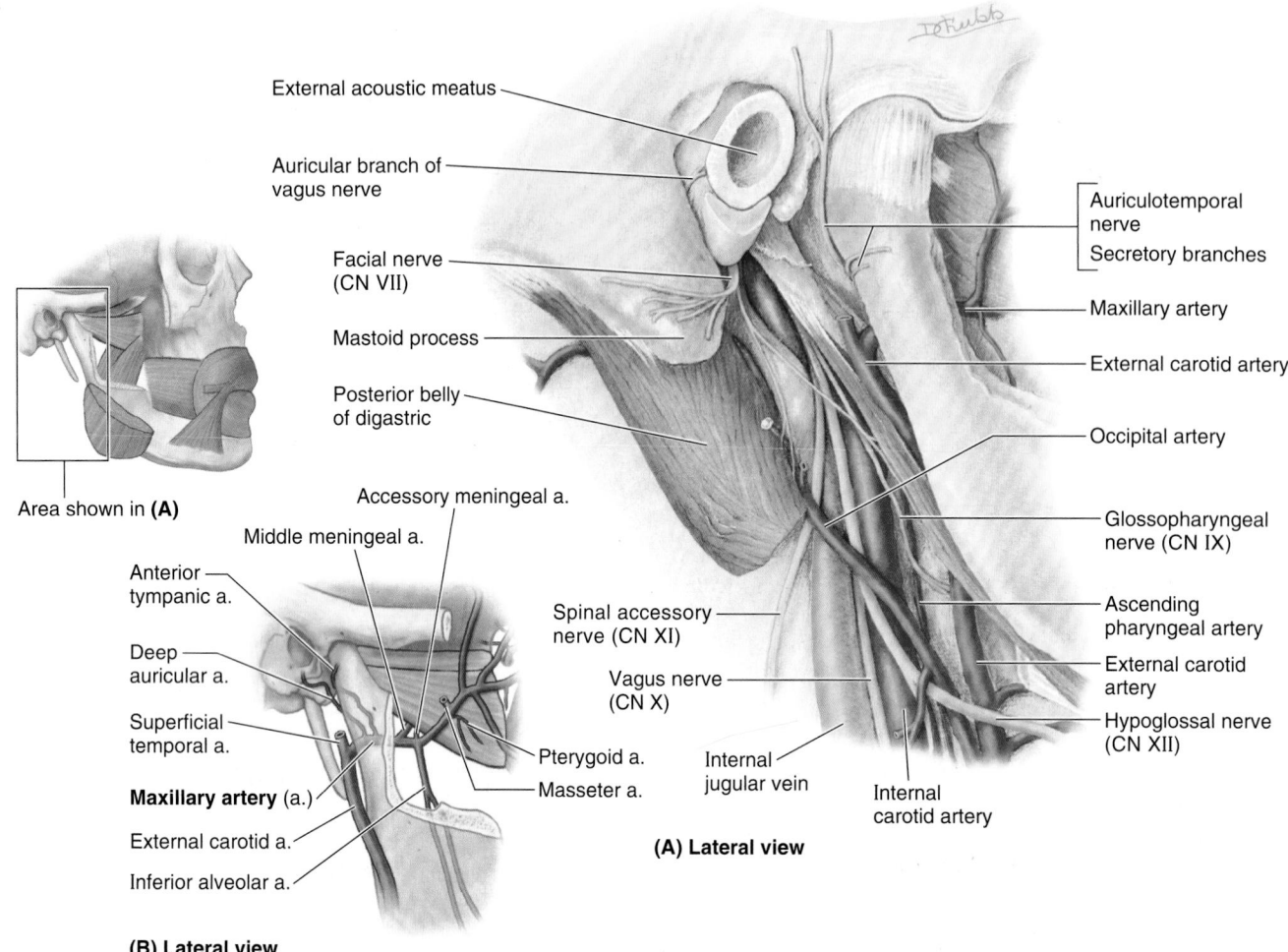

FIGURE 7.115. Neurovascular structures in vicinity of external and middle ear. A. In this dissection of structures deep to parotid bed, the facial nerve (CN VII), the posterior belly of the digastric muscle, and the nerve to it are retracted. The deeply placed ascending pharyngeal artery is the only medial branch of the external carotid artery. It supplies the pharynx, palatine tonsil, pharyngotympanic tube, and the medial wall of the tympanic cavity before it terminates by sending meningeal branches to the cranial cavity. **B.** Maxillary artery and its branches. The branches of the first (mandibular) part supply the external acoustic meatus and tympanic membrane. The middle meningeal artery sends branches to the pharyngotympanic tube before entering the cranium through the foramen spinosum.

where it articulates with the head of the malleus (Fig. 7.116C). The **long limb** lies parallel to the handle of the malleus, and its interior end articulates with the stapes by way of the **lenticular process**, a medially directed projection. The **short limb** is connected by a ligament to the posterior wall of the tympanic cavity.

Stapes. The **stapes** (L. stirrup) is the smallest ossicle. It has a head, two limbs, and a base (Fig. 7.116D). Its **head**, directed laterally, articulates with the incus (Fig. 7.116A). The **base** (footplate) of the stapes fits into the *oval window* on the medial wall of the tympanic cavity. The oval base is attached to the margins of the oval window. The base of the stapes is considerably smaller than the tympanic membrane; as a result, the vibratory force of the stapes is increased approximately 10 times over that of the tympanic membrane. Consequently, the auditory ossicles increase the force but decrease the amplitude of the vibrations transmitted from the tympanic membrane through the ossicles to the internal ear (see Fig. 7.120).

Muscles Associated with Auditory Ossicles. Two muscles dampen or resist movements of the auditory ossicles; one also dampens movements (vibration) of the tympanic membrane. The **tensor tympani** is a short muscle that arises from the superior surface of the cartilaginous part of the pharyngotympanic tube, the greater wing of the sphenoid, and the petrous part of the temporal bone (Figs. 7.113A and 7.114). The muscle inserts into the handle of the malleus. The tensor tympani pulls the handle medially, tensing the tympanic membrane and reducing the amplitude of its oscillations. This action tends to prevent damage to the internal ear when one is exposed to loud sounds. The tensor tympani is supplied by the mandibular nerve (CN V_3).

The **stapedius** is a tiny muscle inside the **pyramidal eminence** (pyramid), a hollow, cone-shaped prominence on the posterior wall of the tympanic cavity (Figs. 7.112B and 7.114B). Its tendon enters the tympanic cavity by emerging

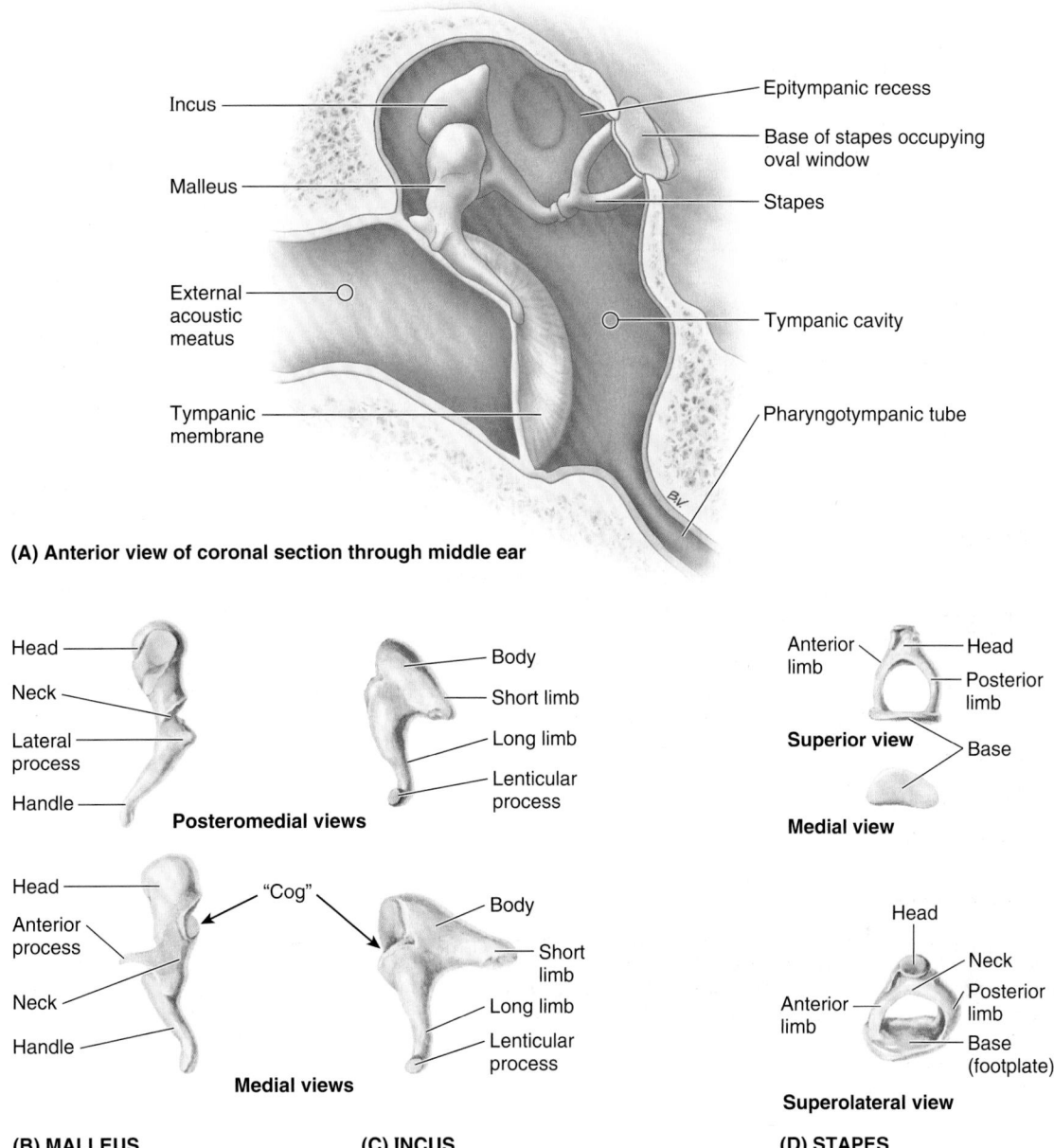

(A) Anterior view of coronal section through middle ear

Posteromedial views

Medial views

Superior view

Medial view

Superolateral view

(B) MALLEUS   **(C) INCUS**   **(D) STAPES**

FIGURE 7.116. Auditory ossicles. A. The ossicles in situ are shown. **B–E.** The isolated ossicles are shown.

from a pinpoint foramen in the apex of the eminence and inserts on the neck of the stapes. The stapedius pulls the stapes posteriorly and tilts its base in the *oval window*, thereby tightening the anular ligament and reducing the oscillatory range. It also prevents excessive movement of the stapes. The nerve to the stapedius arises from the facial nerve (CN VII).

Internal Ear

The **internal ear** contains the **vestibulocochlear organ** concerned with the reception of sound and the maintenance of balance. Buried in the petrous part of the temporal bone (Figs. 7.113 and 7.117A), the internal ear consists of the sacs and ducts of the membranous labyrinth. The *membranous*

labyrinth, containing *endolymph*, is suspended within the perilymph-filled *bony labyrinth*, either by delicate filaments similar to the filaments of arachnoid mater that traverse the subarachnoid space or by the substantial spiral ligament. It does not float. These fluids are involved in stimulating the end organs for balance and hearing, respectively.

BONY LABYRINTH

The **bony labyrinth** is a series of cavities (cochlea, vestibule, and semicircular canals) contained within the otic capsule of the petrous part of the temporal bone (Figs. 7.113A and 7.117B). The **otic capsule** is made of bone that is denser than the remainder of the petrous temporal bone and can

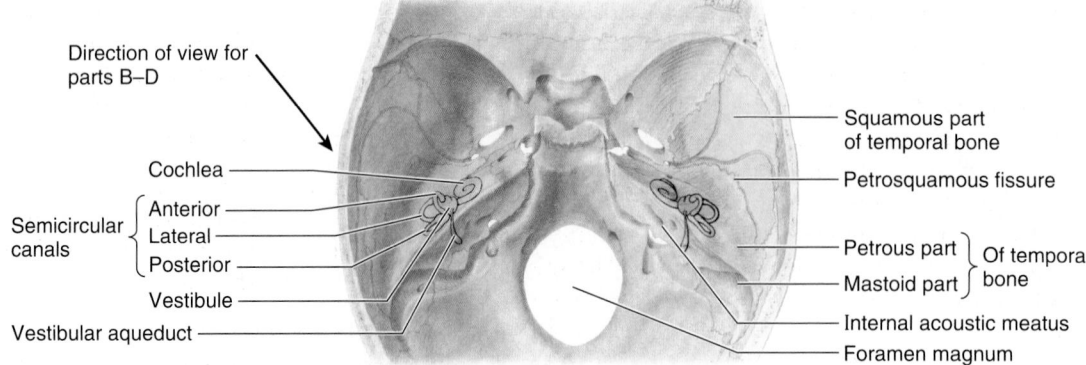

(A) Superior view of internal surface of cranial base

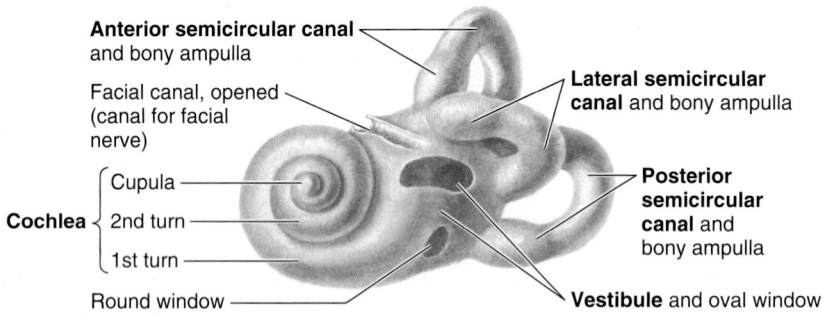

(B) Anterolateral view of left otic capsule

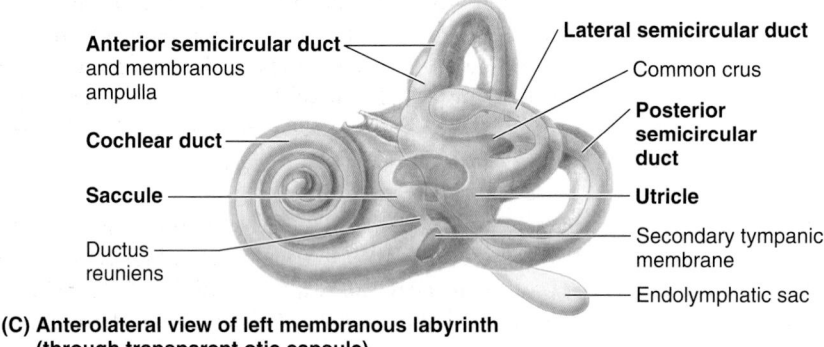

(C) Anterolateral view of left membranous labyrinth (through transparent otic capsule)

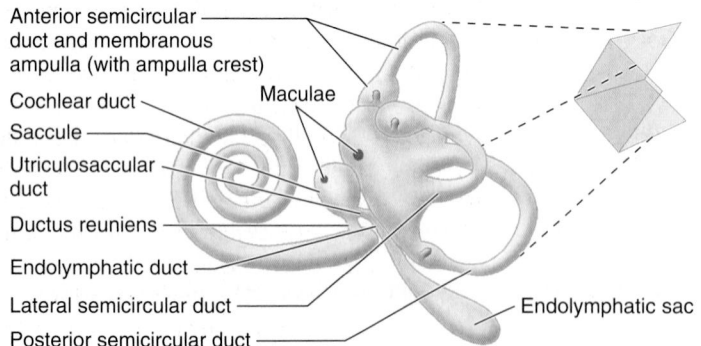

(D) Anterolateral view of left membranous labyrinth

FIGURE 7.117. Bony and membranous labyrinths of internal ear. A. This view of the interior of the base of the cranium shows the temporal bone and the location of the bony labyrinth. **B.** The walls of the bony labyrinth have been carved out of the petrous temporal bone. **C.** A similar view of the bony labyrinth occupied by perilymph and the membranous labyrinth is shown. **D.** The membranous labyrinth, shown after removal from the bony labyrinth, is a closed system of ducts and chambers filled with endolymph and bathed by perilymph. It has three parts: the cochlear duct, which occupies the cochlea; the saccule and utricle, which occupy the vestibule; and the three semicircular ducts, which occupy the semicircular canals. The utricle communicates with the saccule through the utriculosaccular duct. The lateral semicircular duct lies in the horizontal plane and is more horizontal than it appears in this drawing.

be isolated (carved) from it using a dental drill. The otic capsule is often erroneously illustrated and identified as being the bony labyrinth. However, *the bony labyrinth is the fluid-filled space*, which is surrounded by the otic capsule, and is most accurately represented by a cast of the otic capsule after removal of the surrounding bone.

Cochlea. The **cochlea** is the shell-shaped part of the bony labyrinth that contains the **cochlear duct** (Fig. 7.117C), the part of the internal ear concerned with hearing. The **spiral canal of the cochlea** begins at the vestibule and makes 2.5 turns around a bony core, the **modiolus** (Fig. 7.118), the cone-shaped core of spongy bone about which the spiral canal of the cochlea turns. The modiolus contains canals for blood vessels and for distribution of the branches of the cochlear nerve. The apex of the cone-shaped modiolus, like the axis of the tympanic membrane, is directed laterally, anteriorly, and inferiorly. The large basal turn of the cochlea produces the *promontory of the labyrinthine wall* of the tympanic cavity (Fig. 7.114B). At the basal turn, the bony labyrinth communicates with the subarachnoid space superior to the jugular foramen through the **cochlear aqueduct** (Fig. 7.113B). It also features the **round window** (L. *fenestra cochleae*), closed by the **secondary tympanic membrane** (Fig. 7.117B & C).

Vestibule of Bony Labyrinth. The **vestibule of the bony labyrinth** is a small oval chamber (approximately 5 mm long) that contains the **utricle** and **saccule** (Fig. 7.117C) and parts of the balancing apparatus (vestibular labyrinth). The vestibule features the *oval window* on its lateral wall, occupied by the base of the stapes. The vestibule is continuous with the bony cochlea anteriorly, the semicircular canals posteriorly, and the posterior cranial fossa by the **vestibular aqueduct** (Fig. 7.113B). The aqueduct extends to the posterior surface of the petrous part of the temporal bone, where it opens posterolateral to the *internal acoustic meatus* (Fig. 7.117A). The vestibular aqueduct transmits the **endolymphatic duct** (Figs. 7.113B and 7.117D) and two small blood vessels.

Semicircular Canals. The **semicircular canals (anterior, posterior**, and **lateral)** communicate with the vestibule of the bony labyrinth (Fig. 7.117B). The canals lie posterosuperior to the vestibule into which they open; they are set at right angles to each other. The canals occupy three planes in space. Each semicircular canal forms approximately two thirds of a circle, and is approximately 1.5 mm in diameter, except at one end where there is a swelling, the **bony ampulla.** The canals have only five openings into the vestibule because the anterior and posterior canals have one limb common to both. Lodged within the canals are the *semicircular ducts* (Fig. 7.117C & D).

MEMBRANOUS LABYRINTH

The **membranous labyrinth** consists of a series of communicating sacs and ducts that are suspended in the bony labyrinth (Fig. 7.117C). The labyrinth contains **endolymph,** a watery fluid similar in composition to intracellular fluid, thus differing in composition from the surrounding **perilymph** (which is like extracellular fluid) that fills the remainder of the bony labyrinth. The membranous labyrinth—composed of two divisions, the *vestibular labyrinth* and the *cochlear labyrinth*—consists of more parts than does the bony labyrinth:

- **Vestibular labyrinth:** utricle and saccule, two small communicating sacs in the vestibule of the bony labyrinth.
- Three semicircular ducts in the semicircular canals.
- **Cochlear labyrinth:** cochlear duct in the cochlea.

The **spiral ligament,** a spiral thickening of the periosteal lining of the cochlear canal, secures the *cochlear duct* to the spiral canal of the cochlea (Fig. 7.118). The remainder of the membranous labyrinth is suspended by delicate filaments that traverse the perilymph.

The **semicircular ducts** open into the **utricle** through five openings, reflective of the way the surrounding semicircular canals open into the vestibule. The utricle communicates with the saccule through the **utriculosaccular duct,** from which the *endolymphatic duct* arises (Fig. 7.117D). The **saccule** is continuous with the cochlear duct through the **ductus reuniens,** a uniting duct. The utricle and saccule have specialized areas of sensory epithelium called **maculae.** The **macula of the utricle** (L. *macula utriculi*) is in the floor of the utricle, parallel with the base of the cranium, whereas the **macula of the saccule** (L. *macula sacculi*) is vertically placed on the medial wall of the saccule. The **hair cells in the maculae** are innervated by fibers of the vestibular division of the **vestibulocochlear nerve.** The primary sensory neurons are in the **vestibular ganglia** (Fig. 7.119), which are in the internal acoustic meatus.

The *endolymphatic duct* traverses the *vestibular aqueduct* (Fig. 7.113B) and emerges through the bone of the posterior cranial fossa, where it expands into a blind pouch called the **endolymphatic sac** (Figs. 7.113B, 7.117C and 7.119). The endolymphatic sac is located under the dura mater on the posterior surface of the petrous part of the temporal bone. The sac is a storage reservoir for excess endolymph, formed by the blood capillaries in the membranous labyrinth.

Semicircular Ducts. Each semicircular duct has an **ampulla** at one end containing a sensory area, the **ampullary crest** (L. *crista ampullari*) (Fig. 7.119). The crests are sensors for recording movements of the endolymph in the ampulla resulting from rotation of the head in the plane of the duct. The **hair cells of the crests**, like those of the maculae, stimulate primary sensory neurons, whose cell bodies are in the vestibular ganglia.

Cochlear Duct. The **cochlear duct** is a spiral tube, closed at one end and triangular in cross section. The duct is firmly suspended across the cochlear canal between the *spiral ligament* on the external wall of the cochlear canal (Fig. 7.118) and the **osseous spiral lamina** of the modiolus. Spanning the spiral canal in this manner, the endolymph-filled cochlear duct divides the perilymph-filled spiral canal into two channels that are continuous at the apex of the cochlea at the **helicotrema,** a semilunar communication at the apex of the cochlea.

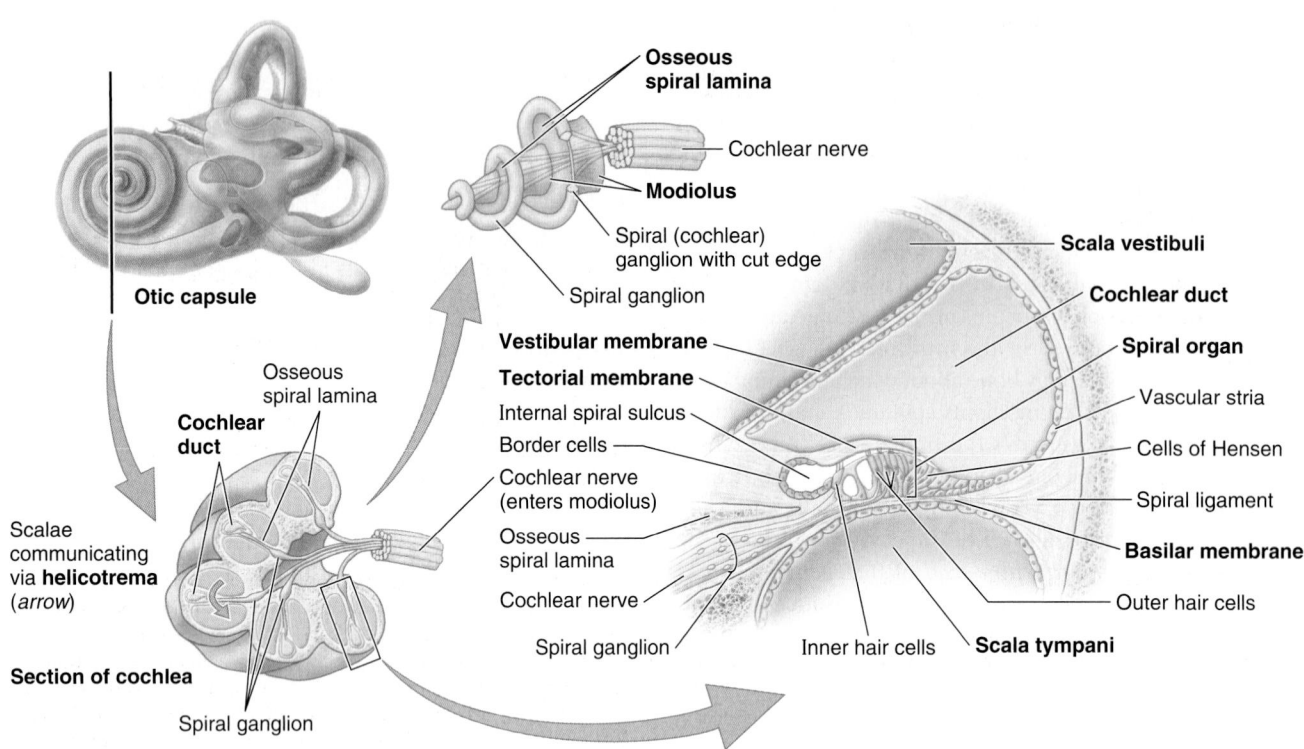

FIGURE 7.118. Structure of cochlea. The cochlea has been sectioned along the axis about which the cochlea winds (see *orientation figure*). An isolated, cone-like, bony core of the cochlea, the modiolus, is shown after the turns of the cochlea are removed, leaving only the spiral lamina winding around it like the thread of a screw. Details of the area enclosed in the rectangle are also shown.

Waves of hydraulic pressure created in the perilymph of the vestibule by the vibrations of the base of the stapes ascend to the apex of the cochlea by one channel, the **scala vestibuli** (Fig. 7.120). The pressure waves then pass through the helicotrema and descend back to the basal turn of the cochlea by the other channel, the **scala tympani.** Here, the pressure waves again become vibrations, this time of the *secondary tympanic membrane* in the round window, and the energy initially received by the (primary) tympanic membrane is finally dissipated into the air of the tympanic cavity.

The roof of the cochlear duct is formed by the **vestibular membrane.** The floor of the duct is also formed by part of the duct, the **basilar membrane,** plus the outer edge of the osseous spiral lamina. The receptor of auditory stimuli is the **spiral organ** (of Corti), situated on the basilar membrane (Fig. 7.118). It is overlaid by the gelatinous **tectorial membrane.**

The spiral organ contains hair cells, the tips of which are embedded in the tectorial membrane. The organ is stimulated to respond by deformation of the cochlear duct induced by the hydraulic pressure waves in the perilymph,

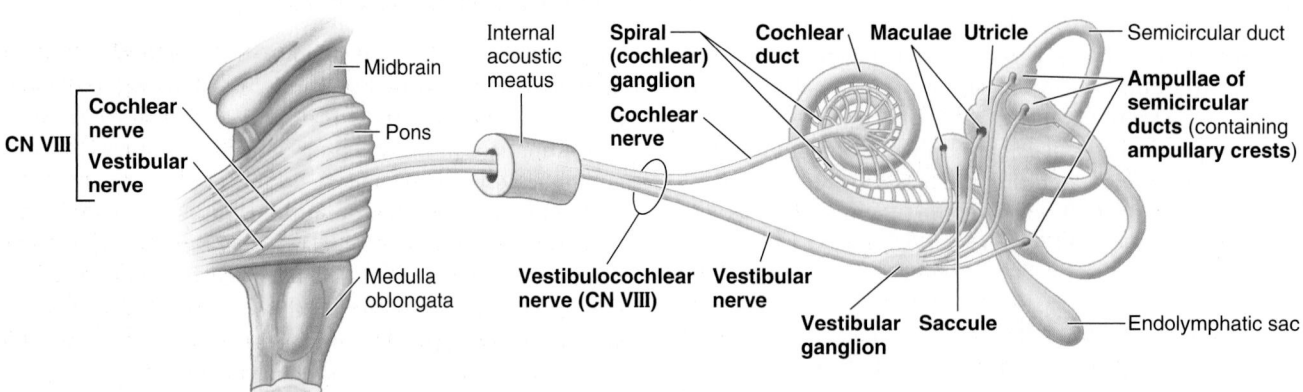

FIGURE 7.119. Vestibulocochlear nerve. CN VIII has two parts: the cochlear nerve (the nerve of hearing) and the vestibular nerve (the nerve of balance). The cell bodies of the scensory fibers that make up the two parts of this nerve constitute the spiral and vestibular ganglia.

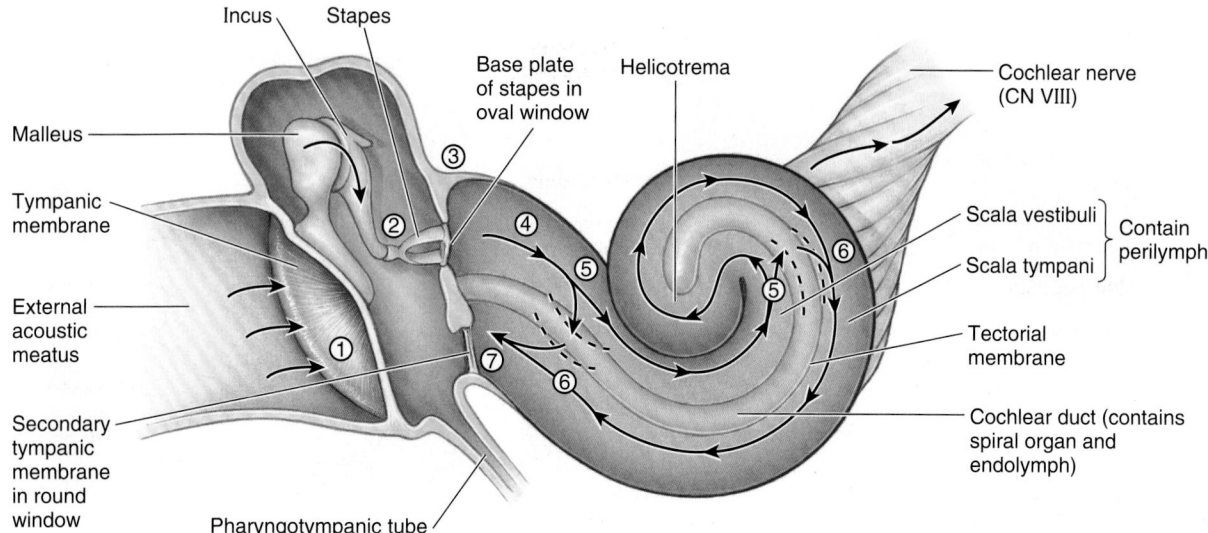

FIGURE 7.120. Sound transmission through the ear. The cochlea is depicted schematically as if consisting of a single coil to demonstrate the transmission of sound stimuli through the ear. *1,* Sound waves entering the external ear strike the tympanic membrane, causing it to vibrate. *2,* Vibrations initiated at the tympanic membrane are transmitted through the ossicles of the middle ear and their articulations. *3,* The base of the stapes vibrates with increased strength and decreased amplitude in the oval window. *4,* Vibrations of the base of the stapes create pressure waves in the perilymph of the scala vestibuli. *5,* Pressure waves in the scala vestibuli cause displacement of the basilar membrane of the cochlear duct. Short waves (high pitch) cause displacement near the oval window; longer waves (low pitch) cause more distant displacement, nearer to the helicotrema at the apex of the cochlea. Movement of the basilar membrane bends the hair cells of the spiral organ. Neurotransmitter is released, stimulating action potentials conveyed by the cochlear nerve to the brain. *6,* Vibrations are transferred across the cochlear duct to the perilymph of the scala tympani. *7,* Pressure waves in the perilymph are dissipated (dampened) by the secondary tympanic membrane at the round window into the air of the tympanic cavity.

which ascend and descend in the surrounding scalae vestibuli and tympani.

INTERNAL ACOUSTIC MEATUS

The **internal acoustic meatus** is a narrow canal that runs laterally for approximately 1 cm within the petrous part of the temporal bone (Fig. 7.117A). The **internal acoustic meatus** **opening** is in the posteromedial part of this bone, in line with the external acoustic meatus. The internal acoustic meatus is closed laterally by a thin, perforated plate of bone that separates it from the internal ear. Through this plate pass the facial nerve (CN VII), the vestibulocochlear nerve (CN VIII) and its divisions, and blood vessels. The vestibulocochlear nerve divides near the lateral end of the internal acoustic meatus into two parts: a cochlear nerve and a vestibular nerve (Fig. 7.119).

EAR

External Ear Injury

Bleeding within the auricle resulting from trauma may produce an *auricular hematoma.* A localized collection of blood forms between the perichondrium and auricular cartilage, causing distortion of the contours of the auricle. As the hematoma enlarges, it compromises the blood supply to the cartilage. If untreated (e.g., by aspiration of blood), fibrosis (formation of fibrous tissue) develops in the overlying skin, forming a deformed auricle (e.g., the cauliflower or boxer's ear of some professional boxers and wrestlers).

Otoscopic Examination

Examination of the external acoustic meatus and tympanic membrane begins by straightening the meatus. In adults, the helix is grasped and pulled posterosuperiorly (up, out, and back). These movements reduce the curvature of the external acoustic meatus, facilitating insertion of the *otoscope* (Fig. B7.42A). The meatus is relatively short in infants; therefore, extra care must be exercised to prevent injury to the tympanic membrane. The meatus is straightened in infants by pulling the auricle inferoposteriorly (down and back). The examination also provides a clue to tenderness, which can indicate inflammation of the auricle and/or the meatus.

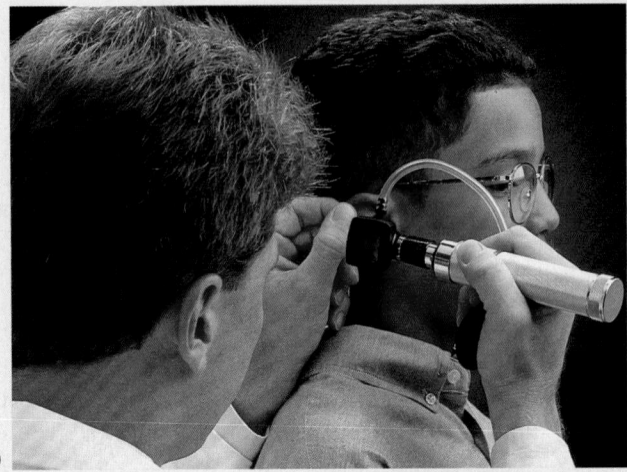
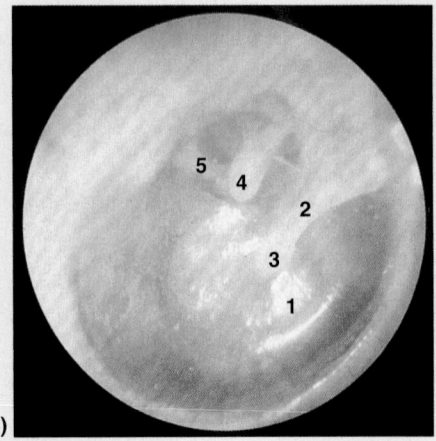

(A)   (B)

FIGURE B7.42. **A.** Otoscopic examination. **B.** Normal tympanic membrane. *1,* cone of light; *2,* handle of malleus; *3,* umbo; *4,* long limb of incus; *5,* posterior limb of stapes.

The tympanic membrane is normally translucent and pearly gray (Fig. B7.42B). The handle of the malleus is usually visible near the center of the membrane (the umbo). From the inferior end of the handle, a bright *cone of light* is reflected from the otoscope's illuminator. This *light reflex* is visible radiating antero-inferiorly in the healthy ear.

Acute Otitis Externa

 Otitis externa is an inflammation of the external acoustic meatus. The infection often develops in swimmers who do not dry their meatus (ear canals) after swimming and/or use ear drops. The inflammation may also be the result of a bacterial infection of the skin lining the meatus. The affected individual complains of itching and pain in the external ear. Pulling the auricle or applying pressure on the tragus increases the pain.

Otitis Media

An earache and a bulging red tympanic membrane may indicate pus or fluid in the middle ear, a sign of *otitis media* (Fig. B7.43A). Infection of the middle ear is often secondary to upper respiratory infections. Inflammation and swelling of the mucous membrane lining the tympanic cavity may cause partial or complete blockage of the pharyngotympanic tube (Fig. 7.109). The tympanic membrane becomes red and bulges, and the person may complain of "ear popping." An amber-colored bloody fluid may be observed through the tympanic membrane. If untreated, otitis media may produce impaired hearing owing to scarring of the auditory ossicles, limiting their ability to move in response to sound.

Perforation of Tympanic Membrane

Perforation of the tympanic membrane ("ruptured eardrum") may result from otitis media and is one of several causes of middle ear deafness. Perforation may also result from foreign bodies in the external acoustic meatus, trauma, or excessive pressure (e.g., during scuba diving). Minor ruptures of the tympanic membrane often heal spontaneously. Large ruptures usually require surgical repair. Because the superior half of the tympanic membrane is much more vascular than the inferior half, incisions to release pus from a middle ear abscess (*myringotomy*), for example, are made postero-inferiorly through the membrane

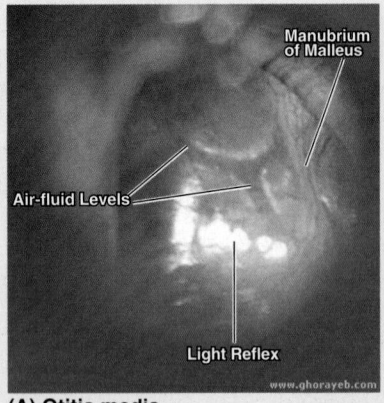

(A) Otitis media

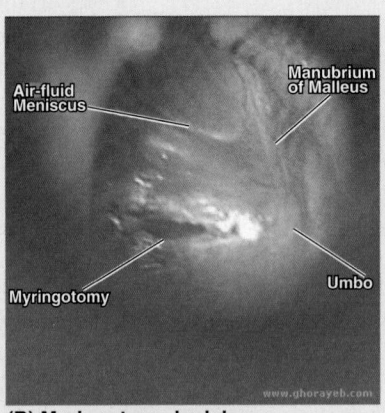

(B) Myringotomy incision

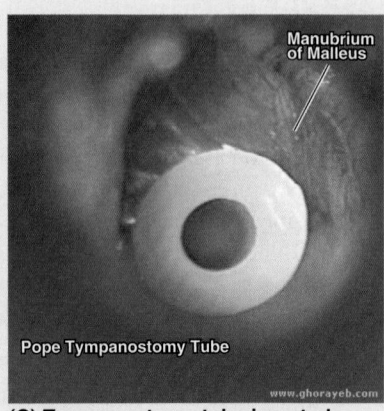

(C) Tympanostomy tube inserted

FIGURE B7.43.

(Fig. B7.43B). This incision also avoids injury to the chorda tympani nerve and auditory ossicles. In persons with chronic middle ear infections, myringotomy may be followed by insertion of *tympanostomy or pressure-equalization* (PE) *tubes* in the incision to enable drainage of effusion and ventilation of pressure (Fig. B7.43C).

Mastoiditis

Infections of the mastoid antrum and mastoid cells (*mastoiditis*) result from a middle ear infection that causes inflammation of the mastoid process (Fig. B7.44). Infections may spread superiorly into the middle cranial fossa through the petrosquamous fissure in children and cause *osteomyelitis* (bone infection) of the tegmen tympani. Since the advent of antibiotics, mastoiditis is uncommon. During operations for mastoiditis, surgeons are conscious of the course of the facial nerve to avoid injuring it. One point of access to the tympanic cavity is through the mastoid antrum. In children, only a thin plate of bone must be removed from the lateral wall of the antrum to expose the tympanic cavity. In adults, bone must be penetrated for 15 mm or more. At present, most mastoidectomies are endaural (i.e., performed through the posterior wall of the external acoustic meatus).

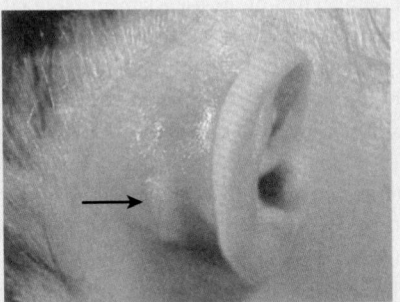

FIGURE B7.44. Mastoiditis (ruptured retro-auricular abscess).

Blockage of Pharyngotympanic Tube

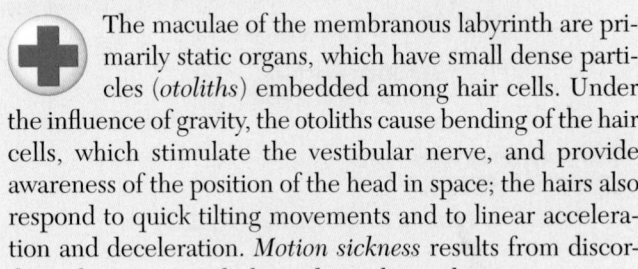
The pharyngotympanic tube forms a route for an infection to pass from the nasopharynx to the tympanic cavity. This tube is easily blocked by swelling of its mucous membrane, even as a result of mild infections (e.g., a cold), because the walls of its cartilaginous part are normally already in apposition. When the pharyngotympanic tube is occluded, residual air in the tympanic cavity is usually absorbed into the mucosal blood vessels, resulting in lower pressure in the tympanic cavity, retraction of the tympanic membrane, and interference with its free movement. Finally, hearing is affected.

Paralysis of Stapedius

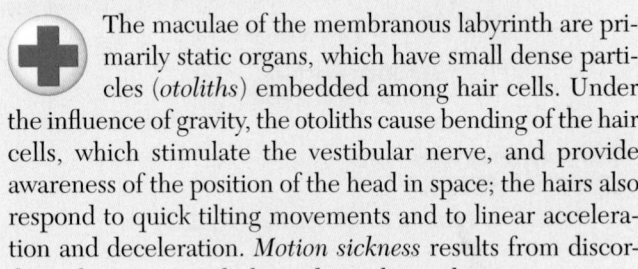
The tympanic muscles have a protective action in that they dampen large vibrations of the tympanic membrane resulting from loud noises. Paralysis of the stapedius (e.g., resulting from a lesion of the facial nerve) is associated with excessive acuteness of hearing, called *hyperacusis* or *hyperacusia*. This condition results from uninhibited movements of the stapes.

Motion Sickness

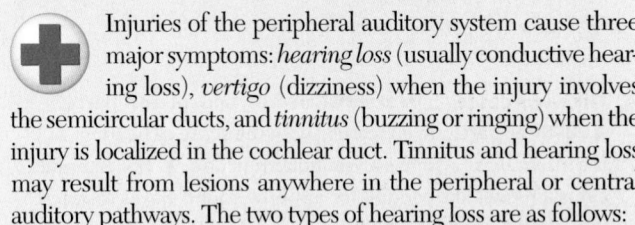
The maculae of the membranous labyrinth are primarily static organs, which have small dense particles (*otoliths*) embedded among hair cells. Under the influence of gravity, the otoliths cause bending of the hair cells, which stimulate the vestibular nerve, and provide awareness of the position of the head in space; the hairs also respond to quick tilting movements and to linear acceleration and deceleration. *Motion sickness* results from discordance between vestibular and visual stimulation.

Dizziness and Hearing Loss

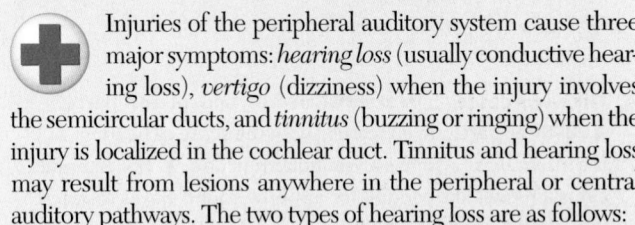
Injuries of the peripheral auditory system cause three major symptoms: *hearing loss* (usually conductive hearing loss), *vertigo* (dizziness) when the injury involves the semicircular ducts, and *tinnitus* (buzzing or ringing) when the injury is localized in the cochlear duct. Tinnitus and hearing loss may result from lesions anywhere in the peripheral or central auditory pathways. The two types of hearing loss are as follows:

- *Conductive hearing loss:* resulting from anything in the external or middle ear that interferes with conduction of sound or movement of the oval or round windows. People with this type of hearing loss often speak with a soft voice because, to them, their own voices sound louder than background sounds. This type of hearing loss may be improved surgically or by use of a hearing aid device.
- *Sensorineural hearing loss:* resulting from defects in the pathway from cochlea to brain: defects of the cochlea, cochlear nerve, brainstem, or cortical connections. *Cochlear implants* are one approach employed to restore sound perception when the hair cells of the spiral organ have been damaged (Fig. B7.45). Sound received by a small external microphone are transmitted to an implanted receiver that sends electrical impulses to the cochlea, stimulating the cochlear nerve. Hearing remains relatively crude but enables perception of rhythm and intensity of sounds.

Ménière Syndrome

Ménière syndrome is related to excess endolymph production or *blockage of the endolymphatic duct* (Fig. 7.113B) and is characterized by recurrent attacks of tinnitus, hearing loss, and vertigo. These symptoms are accompanied by a sense of pressure in the ear, distortion of

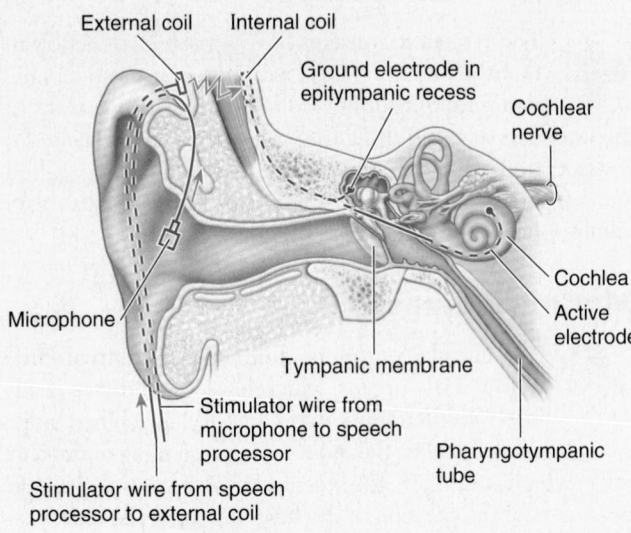

FIGURE B7.45. Cochlear implant.

sounds, and sensitivity to noises (Storper, 2010). A characteristic sign is ballooning of the cochlear duct, utricle, and saccule caused by an increase in endolymphatic volume.

High Tone Deafness

 Persistent exposure to excessively loud sounds causes degenerative changes in the spiral organ, resulting in high tone deafness. This type of hearing loss commonly occurs in workers who are exposed to loud noises and do not wear protective earmuffs (e.g., individuals working for long periods around jet engines).

Otic Barotrauma

 Injury caused to the ear by an imbalance in pressure between ambient (surrounding) air and the air in the middle ear is called *otic barotrauma*. This type of injury usually occurs in fliers and divers.

The Bottom Line

The ear is divided into external, middle, and internal parts.
♦ All three parts are concerned with the sense of hearing, but the internal ear also has a vestibular function. ♦ The external ear is a funnel-like conduit for airborne sound waves to reach the middle ear. ♦ The protruding auricle and lateral part of the external acoustic meatus have an elastic cartilage skeleton that allows flexibility. ♦ The primary sensory innervation of the external ear is provided by CN V and CN X. ♦ The tympanic membrane responds to airborne sound waves, converting them to vibrations transmitted by the solid medium of the ossicles of the middle ear. ♦ Because its entire lateral wall is formed by a thin membrane, the middle ear (tympanic cavity) is a pressure-sensitive space, ventilated by means of the pharyngotympanic tube. ♦ The mucosa lining both the cavity and tube is innervated by CN IX. ♦ At the oval window, between the middle and internal ears, the solid medium vibrations of the ossicles are converted to fluid-borne sonar waves. ♦ The internal ear consists of a delicate and complex membranous labyrinth filled with fluid that resembles intracellular fluid (endolymph), suspended within a bony cave otherwise occupied by extracellular fluid (perilymph). ♦ Although much larger and slightly less complex, the architecture of the bony labyrinth is a reflection of that of the membranous labyrinth. ♦ The posterior portion of the bony labyrinth takes the form of three semicircular canals and ducts; the ampulla of each of the ducts contains an ampullary crest that is sensitive to motion of the head. ♦ The central bony vestibule contains a membranous utricle and saccule, each provided with a macula to monitor the position of the head relative to the line of gravitational pull. ♦ The neuroepithelial crests and maculae are innervated by the vestibular portion of CN VIII. ♦ The anterior portion of the internal ear contains a membranous cochlear duct, suspended between two limbs of continuous pathway for the sonar waves that are conducted by the perilymph; the duct and perilymphatic channels spiral through the 2.5 turns of the bony cochlea. ♦ Deformation of the spiral organ within the cochlear duct by the sonar waves stimulates impulses conducted by the cochlear part of CN VIII for the sense of hearing.

Neck

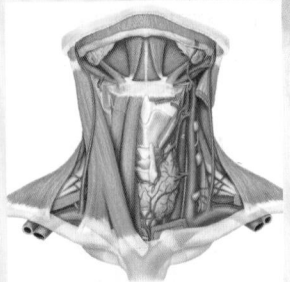

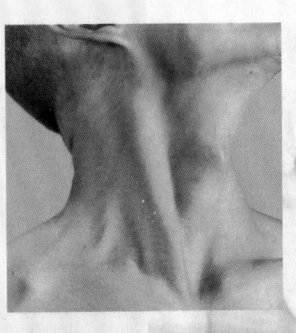

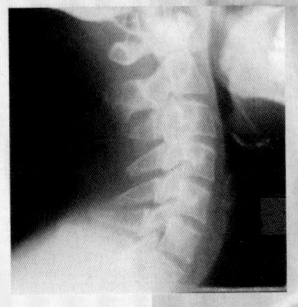

OVERVIEW

The **neck** is the transitional area between the base of the cranium superiorly and the clavicles inferiorly. The neck joins the head to the trunk and limbs, serving as a major conduit for structures passing between them. In addition, several important organs with unique functions are located here: the larynx and the thyroid and parathyroid glands, for example.

The neck is relatively slender to allow the flexibility necessary to position the head to maximize the efficiency of its sensory organs (mainly the eyeballs but also the ears, mouth, and nose). Thus, many important structures are crowded together in the neck, such as muscles, glands, arteries, veins, nerves, lymphatics, trachea, esophagus, and vertebrae. Consequently, the neck is a well-known region of vulnerability. Further, several vital structures, including the trachea, esophagus, and thyroid gland, lack the bony protection afforded other parts of the systems to which these structures belong.

The main arterial blood flow to the head and neck (the *carotid arteries*) and the principal venous drainage (the *jugular veins*) lie anterolaterally in the neck (Fig. 8.1). Carotid/jugular blood vessels are the major structures commonly injured in penetrating wounds of the neck. The *brachial plexuses of nerves* originate in the neck and pass inferolaterally to enter the axillae and continue into and supply the upper limbs.

In the middle of the anterior aspect of the neck is the *thyroid cartilage,* the largest of the cartilages of the larynx, and the trachea. Lymph from structures in the head and neck drains into cervical lymph nodes.

BONES OF NECK

The skeleton of the neck is formed by the cervical vertebrae, hyoid bone, manubrium of the sternum, and clavicles (Figs. 8.2 and 8.3). These bones are parts of the axial skeleton except the clavicles, which are part of the appendicular skeleton.

Cervical Vertebrae

Seven *cervical vertebrae* form the cervical region of the vertebral column, which encloses the spinal cord and meninges. The stacked, centrally placed vertebral bodies support the head, and the intervertebral (IV) articulations—especially the craniovertebral joints at its superior end—provide the flexibility necessary to allow positioning of the head.

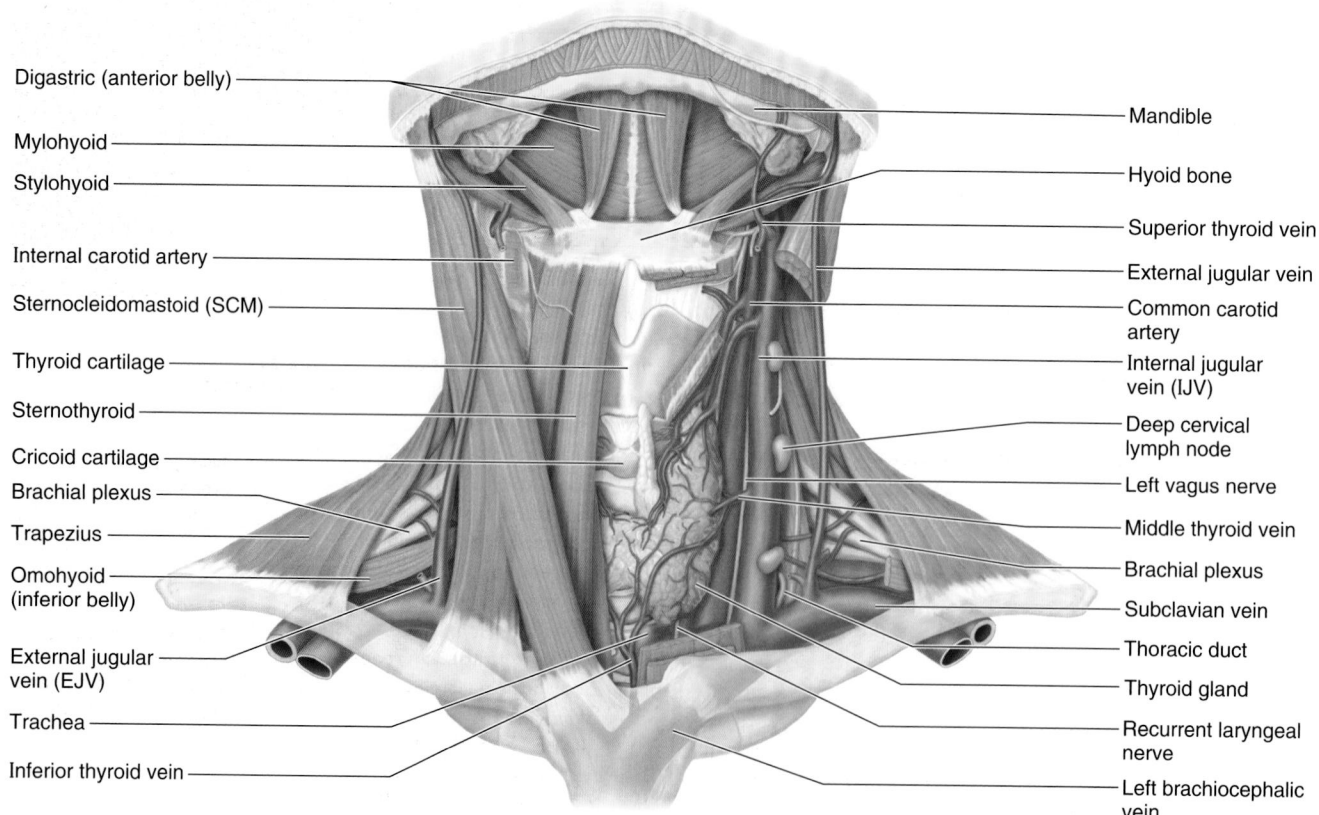

Digastric (anterior belly)
Mylohyoid
Stylohyoid
Internal carotid artery
Sternocleidomastoid (SCM)
Thyroid cartilage
Sternothyroid
Cricoid cartilage
Brachial plexus
Trapezius
Omohyoid (inferior belly)
External jugular vein (EJV)
Trachea
Inferior thyroid vein

Mandible
Hyoid bone
Superior thyroid vein
External jugular vein
Common carotid artery
Internal jugular vein (IJV)
Deep cervical lymph node
Left vagus nerve
Middle thyroid vein
Brachial plexus
Subclavian vein
Thoracic duct
Thyroid gland
Recurrent laryngeal nerve
Left brachiocephalic vein

Anterior view

FIGURE 8.1. Dissection of anterior neck. The fascia has been removed and the muscles on the left side have been reflected to show the hyoid bone, thyroid gland, and structures related to the carotid sheath: carotid artery, internal jugular vein (IJV), vagus nerve (CN X), and deep cervical lymph nodes.

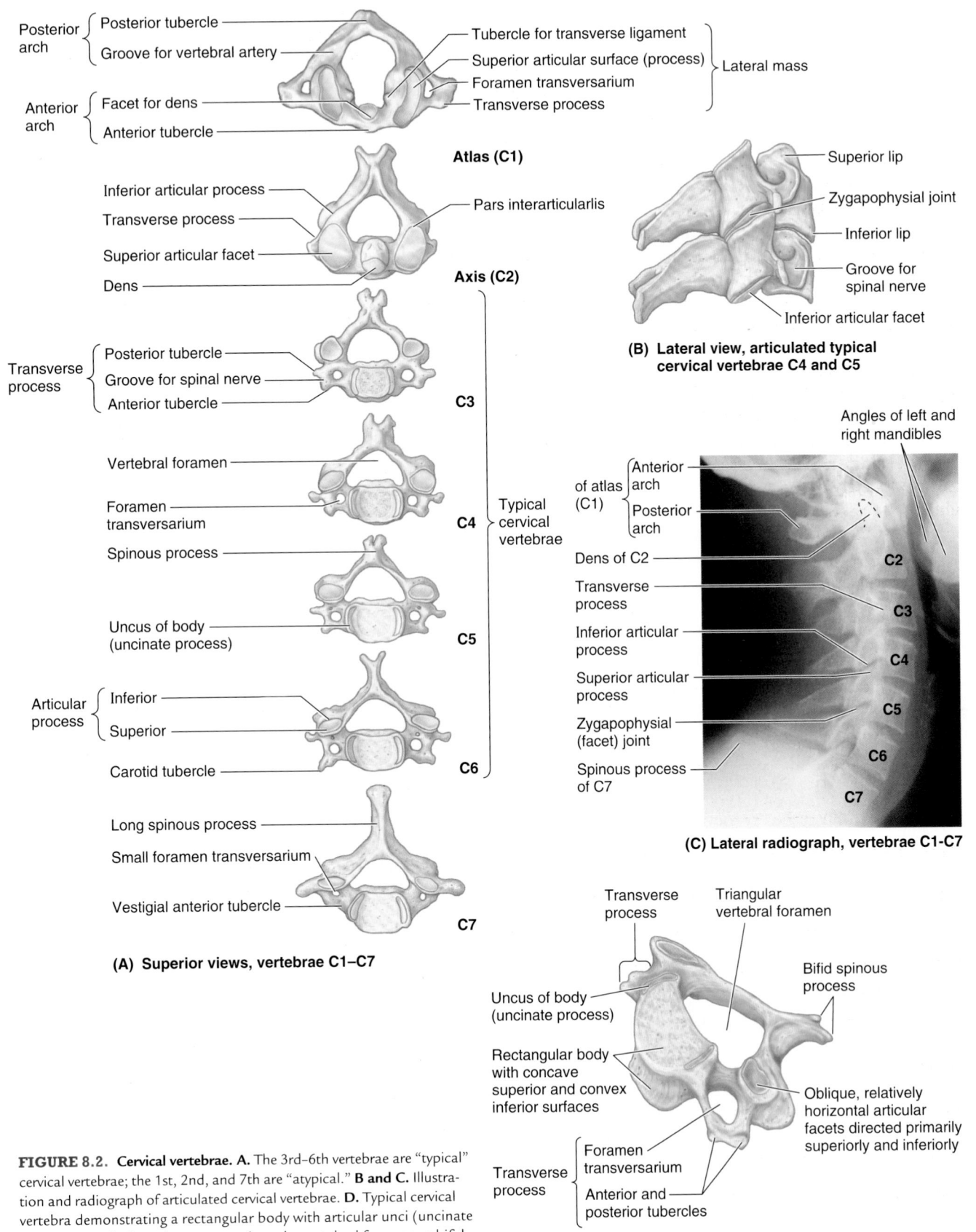

Posterior arch
{ Posterior tubercle
  Groove for vertebral artery }

Tubercle for transverse ligament
Superior articular surface (process)
Foramen transversarium
Transverse process } Lateral mass

Anterior arch
{ Facet for dens
  Anterior tubercle }

Atlas (C1)

Inferior articular process
Transverse process
Superior articular facet
Dens

Pars interarticularis

Axis (C2)

Transverse process
{ Posterior tubercle
  Groove for spinal nerve
  Anterior tubercle }

C3

Vertebral foramen
Foramen transversarium

C4

Spinous process

Uncus of body (uncinate process)

C5

Articular process
{ Inferior
  Superior }

Carotid tubercle

C6

} Typical cervical vertebrae

Long spinous process
Small foramen transversarium
Vestigial anterior tubercle

C7

(A) Superior views, vertebrae C1–C7

Superior lip
Zygapophysial joint
Inferior lip
Groove for spinal nerve
Inferior articular facet

(B) Lateral view, articulated typical cervical vertebrae C4 and C5

Angles of left and right mandibles

of atlas (C1)
{ Anterior arch
  Posterior arch }

Dens of C2
Transverse process
Inferior articular process
Superior articular process
Zygapophysial (facet) joint
Spinous process of C7

C2
C3
C4
C5
C6
C7

(C) Lateral radiograph, vertebrae C1-C7

Transverse process
Triangular vertebral foramen
Bifid spinous process
Uncus of body (uncinate process)
Rectangular body with concave superior and convex inferior surfaces
Oblique, relatively horizontal articular facets directed primarily superiorly and inferiorly
Foramen transversarium
Transverse process
Anterior and posterior tubercles

(D) Superior oblique view of typical cervical vertebra

FIGURE 8.2. Cervical vertebrae. A. The 3rd–6th vertebrae are "typical" cervical vertebrae; the 1st, 2nd, and 7th are "atypical." **B and C.** Illustration and radiograph of articulated cervical vertebrae. **D.** Typical cervical vertebra demonstrating a rectangular body with articular unci (uncinate processes) on its lateral aspects, a triangular vertebral foramen, a bifid spinous process, and foramina transversaria.

The cervical vertebrae, cervical IV joints, and movement of the cervical region of the vertebral column are described with the back (Chapter 4); therefore, only a brief review follows.

The four *typical cervical vertebrae (3rd–6th)* have the following characteristics (Fig. 8.2A & D):

- The vertebral body is small and longer from side to side than anteroposteriorly; the superior surface is concave, and the inferior surface is convex.
- The vertebral foramen is large and triangular.
- The transverse processes of all cervical vertebrae (typical or atypical) include **foramina transversaria** for the vertebral vessels (the vertebral veins and, except for vertebra C7, the vertebral arteries).
- The superior facets of the articular processes are directed superoposteriorly, and the inferior facets are directed inferoposteriorly.
- Their spinous processes are short and, in individuals of European heritage, bifid.

There are three *atypical cervical vertebrae (C1, C2, and C7)* (Fig. 8.2A):

- The *C1 vertebra or atlas:* a ring-like, kidney-shaped bone lacking a spinous process or body and consisting of two lateral masses connected by anterior and posterior arches. Its concave superior articular facets receive the occipital condyles.
- The *C2 vertebra or axis:* a peg-like *dens* (odontoid process) projects superiorly from its body.
- The *vertebra prominens (C7):* so-named because of its long spinous process, which is not bifid. Its transverse processes are large, but its foramina transversaria are small.

Hyoid Bone

The mobile **hyoid bone** (or simply, the **hyoid**), lies in the anterior part of the neck at the level of the C3 vertebra in the angle between the mandible and the thyroid cartilage (Fig. 8.3). The hyoid is suspended by muscles that connect it to the mandible, styloid processes, thyroid cartilage, manubrium of the sternum, and scapulae.

The hyoid is unique among bones for its isolation from the remainder of the skeleton. The U-shaped hyoid derives its name from the Greek word *hyoeidçs,* meaning "shaped like the letter upsilon," the 20th letter in the Greek alphabet. The hyoid does not articulate with any other bone. It is suspended from the styloid processes of the temporal bones by the *stylohyoid ligaments* (Fig. 8.3A) and is firmly bound to the thyroid cartilage. The hyoid consists of a body and greater and lesser horns (L. *cornua*). Functionally, the hyoid serves as an attachment for anterior neck muscles and a prop to keep the airway open.

The **body of the hyoid,** its middle part, faces anteriorly and is approximately 2.5 cm wide and 1 cm thick (Fig. 8.3B & C). Its anterior convex surface projects anterosuperiorly; its posterior concave surface projects postero-inferiorly. Each end of its body is united to a **greater horn** that projects posterosuperiorly and laterally from the body. In young people, the greater horns are united to the body by fibrocartilage. In older people, the horns are usually united by bone. Each **lesser horn** is a small bony projection from the superior part of the body of the hyoid near its union with the greater horn. It is connected to the body of the hyoid by fibrous tissue and sometimes to the greater horn by a synovial joint. The lesser horn projects superoposteriorly toward the styloid process; it may be partly or completely cartilaginous in some adults.

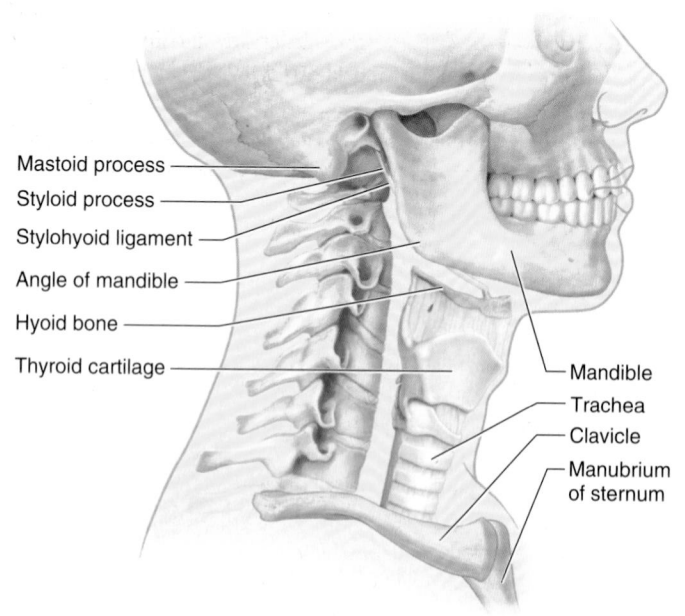

(A) Lateral view

Mastoid process
Styloid process
Stylohyoid ligament
Angle of mandible
Hyoid bone
Thyroid cartilage
Mandible
Trachea
Clavicle
Manubrium of sternum

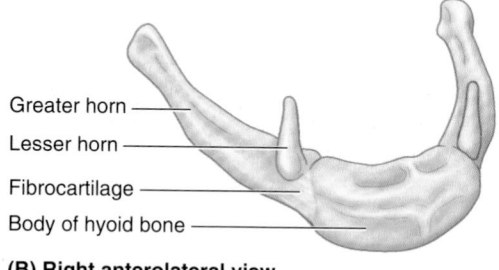

Greater horn
Lesser horn
Fibrocartilage
Body of hyoid bone

(B) Right anterolateral view

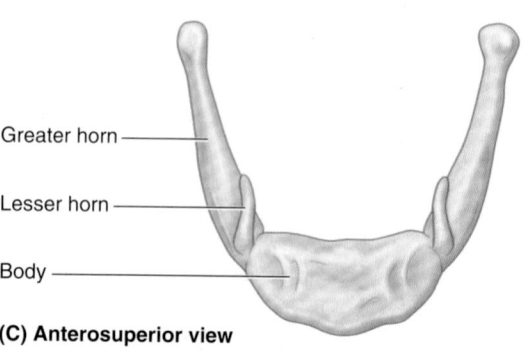

Greater horn
Lesser horn
Body

(C) Anterosuperior view

FIGURE 8.3. Bones and cartilages of neck. A. The bony and cartilaginous landmarks of the neck are the vertebrae, mastoid and styloid processes, angles of the mandible, hyoid bone, thyroid cartilage, clavicle, and manubrium of the sternum. **B and C.** The hyoid bone and its features are demonstrated.

BONES OF NECK

Cervical Pain

✚ *Cervical pain* (neck pain) has several causes, including inflamed lymph nodes, muscle strain, and protruding intervertebral (IV) discs. Enlarged cervical lymph nodes may indicate a malignant tumor in the head; however, the primary cancer may be in the thorax or abdomen because the neck connects the head to the trunk (e.g., lung cancer may metastasize through the neck to the cranium). Most chronic cervical pain is caused by bony abnormalities (e.g., *cervical osteoarthritis*) or by trauma. Cervical pain is usually affected by movement of the head and neck, and it may be exaggerated during coughing or sneezing, for example.

Injuries of Cervical Vertebral Column

✚ *Fractures and dislocations of the cervical vertebra* may injure the spinal cord and/or the vertebral arteries and sympathetic plexuses passing through the foramina transversaria. See the blue boxes "Dislocation of Cervical Vertebrae" (p. 457), "Fracture and Dislocation of Atlas" (p. 458), and "Fracture and Dislocation of Axis" (p. 459).

Fracture of Hyoid Bone

✚ *Fracture of the hyoid* (or of the styloid processes of the temporal bone; see Chapter 7), occurs in people who are manually strangled by compression of the throat. This results in depression of the body of the hyoid onto the thyroid cartilage. Inability to elevate the hyoid and move it anteriorly beneath the tongue makes swallowing and maintenance of the separation of the alimentary and respiratory tracts difficult and may result in *aspiration pneumonia*.

FASCIA OF NECK

Structures in the neck are surrounded by a layer of subcutaneous tissue (superficial fascia) and are compartmentalized by layers of deep cervical fascia. The fascial planes determine the direction an infection may spread in the neck.

Cervical Subcutaneous Tissue and Platysma

The **cervical subcutaneous tissue** (superficial cervical fascia) is a layer of fatty connective tissue that lies between the dermis of the skin and the investing layer of deep cervical fascia (Fig. 8.4A). The cervical subcutaneous tissue is usually thinner than in other regions, especially anteriorly. It contains cutaneous nerves, blood and lymphatic vessels, superficial lymph nodes, and variable amounts of fat. Anterolaterally, it contains the platysma (Fig. 8.4B).

PLATYSMA

The **platysma** (G. flat plate) is a broad, thin sheet of muscle in the subcutaneous tissue of the neck (Figs. 8.4B, 8.5, and Fig. 8.7A). Like other facial and scalp muscles, the platysma develops from a continuous sheet of musculature derived from mesenchyme in the 2nd pharyngeal arch of the embryo and is supplied by branches of the facial nerve, CN VII. The external jugular vein (EJV), descending from the angle of the mandible to the middle of the clavicle (Fig. 8.1), and the main cutaneous nerves of the neck are deep to the platysma.

The platysma covers the anterolateral aspect of the neck. Its fibers arise in the deep fascia covering the superior parts of the deltoid and pectoralis major muscles and sweep superomedially over the clavicle to the inferior border of the mandible. The anterior borders of the two muscles decussate over the chin and blend with the facial muscles. Inferiorly, the fibers diverge, leaving a gap anterior to the larynx and trachea (Fig. 8.5). Much variation exists in terms of the continuity (completeness) of this muscular sheet, which often occurs as isolated slips. The platysma is supplied by the cervical branch of CN VII.

The Bottom Line

BONES OF NECK

Cervical vertebrae: The neck is a movable "connecting stalk" with a segmented axial skeleton. ◆ The stacked, centrally placed vertebral bodies support the head. ◆ The IV articulations—especially the craniovertebral joints at its superior end—provide the flexibility necessary to allow positioning of the head to maximize the use of its sensory organs. ◆ Multiple processes of the vertebrae provide both the attachments and the leverage necessary to move the head into and maintain those positions. ◆ The foramina of the cervical vertebrae provide protective passage for the spinal cord and the vertebral arteries that nourish the bones and are a major component of the brain's blood supply. ◆ The vertebrae provide little protection for other structures of the neck.

Hyoid bone: Unique in terms of its isolation from the rest of the skeleton, the U-shaped hyoid is suspended between the body of the mandible superiorly and the manubrium of the sternum inferiorly. ◆ The hyoid provides a movable base for the tongue and attachment for the middle part of the pharynx. ◆ The hyoid also maintains the patency of the pharynx, required for swallowing and respiration.

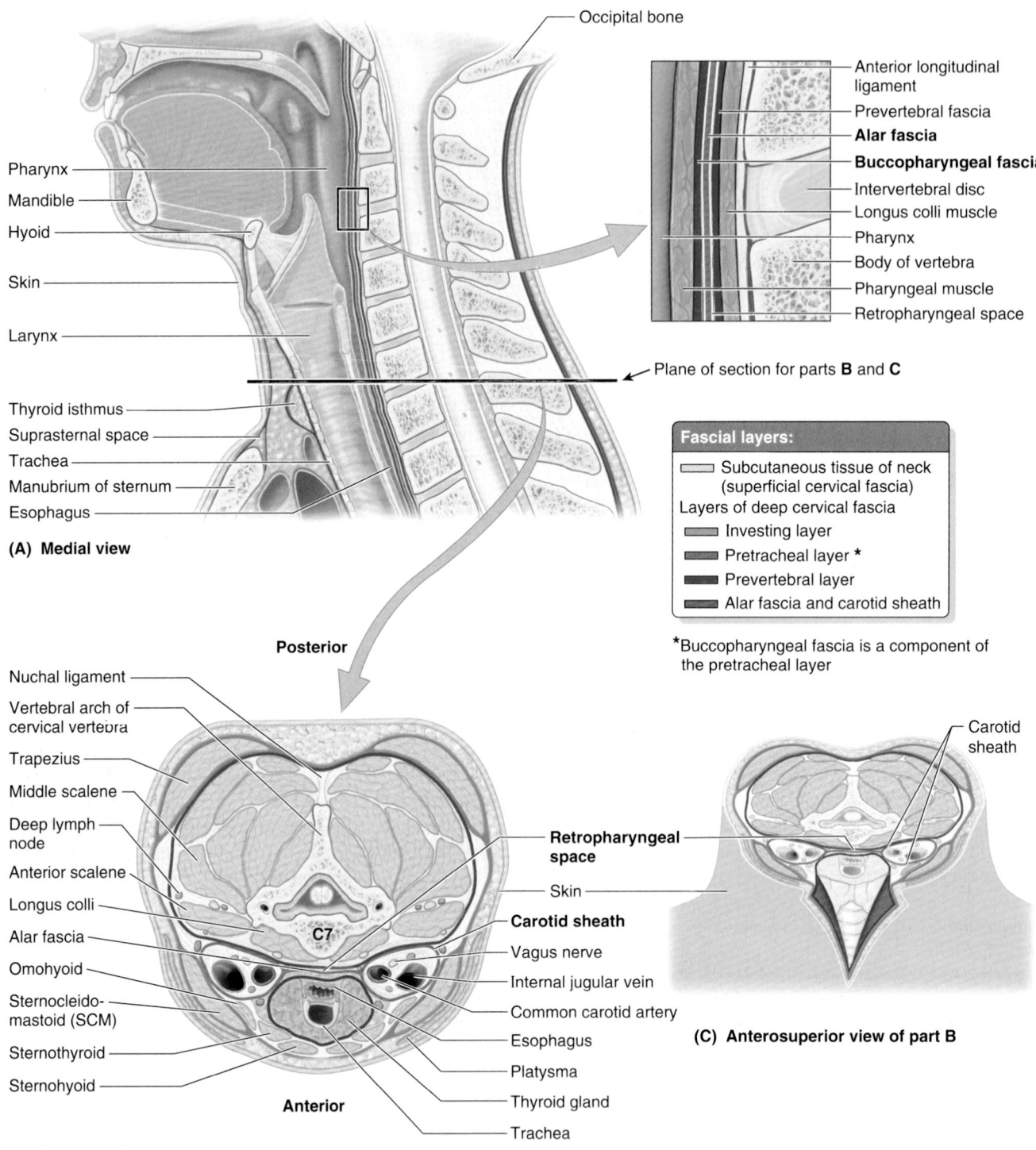

(A) Medial view

Occipital bone

Pharynx

Mandible

Hyoid

Skin

Larynx

Thyroid isthmus

Suprasternal space

Trachea

Manubrium of sternum

Esophagus

Anterior longitudinal ligament

Prevertebral fascia

Alar fascia

Buccopharyngeal fascia*

Intervertebral disc

Longus colli muscle

Pharynx

Body of vertebra

Pharyngeal muscle

Retropharyngeal space

Plane of section for parts **B** and **C**

Fascial layers:

Subcutaneous tissue of neck (superficial cervical fascia)

Layers of deep cervical fascia

Investing layer

Pretracheal layer *

Prevertebral layer

Alar fascia and carotid sheath

*Buccopharyngeal fascia is a component of the pretracheal layer

Posterior

Nuchal ligament

Vertebral arch of cervical vertebra

Trapezius

Middle scalene

Deep lymph node

Anterior scalene

Longus colli

Alar fascia

Omohyoid

Sternocleido-mastoid (SCM)

Sternothyroid

Sternohyoid

C7

Anterior

Retropharyngeal space

Skin

Carotid sheath

Vagus nerve

Internal jugular vein

Common carotid artery

Esophagus

Platysma

Thyroid gland

Trachea

Carotid sheath

(C) Anterosuperior view of part B

(B) Superior view of transverse section (at level C7 vertebra)

FIGURE 8.4. Sections of head and neck demonstrating cervical fascia. A. The right half of the head and neck have been sectioned in the median plane. The *detail* illustrates the fascia in the retropharyngeal region. **B.** This transverse section of the neck passes through the isthmus of the thyroid gland at the C7 vertebral level, as indicated in part **A.** The outermost layer of deep cervical fascia, the investing layer, splits to enclose the trapezius and sternocleidomastoid (SCM) at the four corners of the neck. The investing layer and its embedded muscles surround two main fascial columns. The pretracheal (visceral) layer encloses muscles and viscera in the anterior neck; the prevertebral (musculoskeletal) layer encircles the vertebral column and associated muscles. The carotid sheaths are neurovascular conduits related to both fascial columns. **C.** The fascial compartments of the neck are shown to demonstrate an anterior midline approach to the thyroid gland. Although the larynx, trachea, and thyroid gland are nearly subcutaneous in the midline, two layers of deep cervical fascia (the investing and pretracheal layers) must be incised to reach them.

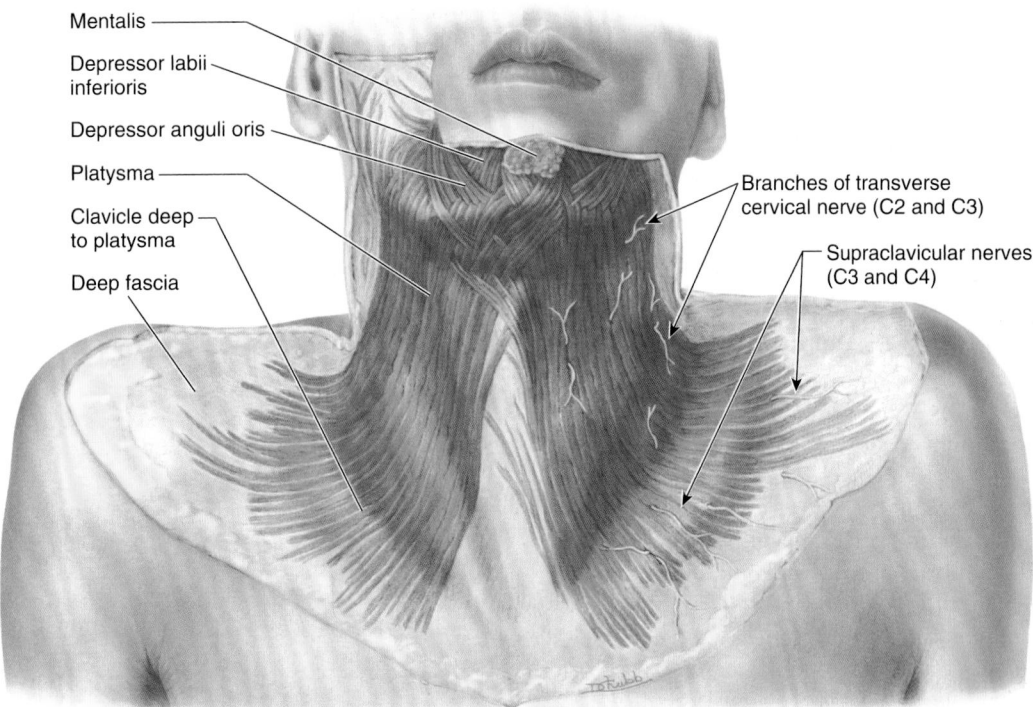

Mentalis
Depressor labii inferioris
Depressor anguli oris
Platysma
Clavicle deep to platysma
Deep fascia

Branches of transverse cervical nerve (C2 and C3)
Supraclavicular nerves (C3 and C4)

FIGURE 8.5. Platysma. The thin platysma muscle spreads subcutaneously like a sheet, passes over the clavicles, and is pierced by cutaneous nerves. Much variation is present in the continuity of this muscular sheet.

Acting from its superior attachment to the mandible, *the platysma tenses the skin*, producing vertical skin ridges and releasing pressure on the superficial veins (Table 8.2). Men commonly use actions of the platysma when shaving their necks and when easing tight collars. Acting from its inferior attachment, the platysma helps depress the mandible and draw the corners of the mouth inferiorly, as in a grimace. As a muscle of facial expression, the platysma serves to convey tension or stress.

Deep Cervical Fascia

The **deep cervical fascia** consists of three fascial layers (sheaths): *investing, pretracheal,* and *prevertebral* (Fig. 8.4A & B). These layers support the cervical viscera (e.g., thyroid gland), muscles, vessels, and deep lymph nodes. The deep cervical fascia also condenses around the common carotid arteries, internal jugular veins (IJVs), and vagus nerves to form the *carotid sheath* (Fig. 8.4B & C).

These three fascial layers form natural cleavage planes through which tissues may be separated during surgery, and they limit the spread of abscesses (collections of pus) resulting from infections. The deep cervical fascial layers also afford the slipperiness that allows structures in the neck to move and pass over one another without difficulty, such as when swallowing and turning the head and neck.

INVESTING LAYER OF DEEP CERVICAL FASCIA

The **investing layer of deep cervical fascia,** the most superficial deep fascial layer, surrounds the entire neck deep to the skin and subcutaneous tissue. At the "four corners" of the neck, it splits into superficial and deep layers to enclose (invest) the *trapezius* and *sternocleidomastoid (SCM) muscles* (Fig. 8.4B & C). These muscles are derived from the same embryonic sheet of muscle and are innervated by the same nerve (CN XI). They have essentially continuous attachments to the cranial base superiorly and to the scapular spine, acromion, and clavicle inferiorly.

Superiorly, the investing layer of deep cervical fascia attaches to the:

- Superior nuchal lines of the occipital bone.
- Mastoid processes of the temporal bones.
- Zygomatic arches.
- Inferior border of the mandible.
- Hyoid bone.
- Spinous processes of the cervical vertebrae.

Just inferior to its attachment to the mandible, the investing layer of deep fascia splits to enclose the submandibular gland; posterior to the mandible, it splits to form the fibrous capsule of the parotid gland. The *stylomandibular ligament* is a thickened modification of this fascial layer (see Fig. 7.69).

Inferiorly, the investing layer of deep cervical fascia attaches to the manubrium of the sternum, clavicles, and acromions and spines of the scapulae. The investing layer of deep cervical fascia is continuous posteriorly with the periosteum covering the C7 spinous process, and with the *nuchal ligament* (L. *ligamentum nuchae*), a triangular membrane that forms a median fibrous septum between the muscles of the two sides of the neck (Fig. 8.4B).

Inferiorly between the sternal heads of the SCMs and just superior to the manubrium, the investing layer of deep cervical fascia remains divided into two layers to enclose the SCM; one layer attaches to the anterior and the other to the posterior surface of the manubrium. A **suprasternal space** lies between these layers (Fig. 8.4A). It encloses the inferior ends of the anterior jugular veins, the jugular venous arch, fat, and a few deep lymph nodes.

PRETRACHEAL LAYER OF DEEP CERVICAL FASCIA

The thin **pretracheal layer of deep cervical fascia** is limited to the anterior part of the neck (Fig. 8.4). It extends inferiorly from the hyoid into the thorax, where it blends with the fibrous pericardium covering the heart. The pretracheal layer of fascia includes a thin *muscular part*, which encloses the infrahyoid muscles, and a *visceral part*, which encloses the thyroid gland, trachea, and esophagus, and is continuous posteriorly and superiorly with the **buccopharyngeal fascia** of the pharynx. The pretracheal layer of deep fascia blends laterally with the carotid sheaths. Superior to the hyoid, a thickening of the pretracheal fascia forms a pulley or trochlea through which the *intermediate tendon of the digastric muscle* passes, suspending the hyoid. By wrapping around the lateral border of the intermediate tendon of the omohyoid, the pretracheal layer also tethers the two-bellied *omohyoid muscle,* redirecting the course of the muscle between the bellies.

PREVERTEBRAL LAYER OF DEEP CERVICAL FASCIA

The **prevertebral layer of deep cervical fascia** forms a tubular sheath for the vertebral column and the muscles associated with it, such as the *longus colli* and *longus capitis* anteriorly, the *scalenes* laterally, and the *deep cervical muscles* posteriorly (Fig. 8.4A & B).

The prevertebral layer of deep fascia is fixed to the cranial base superiorly. Inferiorly, it blends with the *endothoracic fascia* peripherally and fuses with the *anterior longitudinal ligament* centrally at approximately the T3 vertebra (see Chapter 4) (Fig. 8.4A). The prevertebral fascia extends laterally as the *axillary sheath* (Chapter 6), which surrounds the axillary vessels and brachial plexus. The cervical parts of the sympathetic trunks are embedded in the prevertebral layer of deep cervical fascia.

Carotid Sheath. The **carotid sheath** is a tubular fascial investment that extends from the cranial base to the root of the neck. This sheath blends anteriorly with the investing and pretracheal layers of fascia and posteriorly with the prevertebral layer of fascia (Fig. 8.4B & C). The carotid sheath contains the:

- Common and internal carotid arteries.
- Internal jugular vein.
- Vagus nerve (CN X).
- Some deep cervical lymph nodes.
- Carotid sinus nerve.
- Sympathetic nerve fibers (carotid peri-arterial plexuses).

The carotid sheath and pretracheal fascia communicate freely with the mediastinum of the thorax inferiorly and the cranial cavity superiorly. These communications represent potential pathways for the spread of infection and extravasated blood.

Retropharyngeal Space. The **retropharyngeal space** is the largest and most important interfascial space in the neck (Fig. 8.4A-C). It is a potential space that consists of loose connective tissue between the visceral part of the prevertebral layer of deep cervical fascia and the buccopharyngeal fascia surrounding the pharynx superficially. Inferiorly, the buccopharyngeal fascia is continuous with the pretracheal layer of deep cervical fascia.

The **alar fascia** forms a further subdivision of the retropharyngeal space. This thin layer is attached along the midline of the buccopharyngeal fascia from the cranium to the level of the C7 vertebra. From this attachment, it extends laterally and terminates in the carotid sheath. The retropharyngeal space permits movement of the pharynx, esophagus, larynx, and trachea relative to the vertebral column during swallowing. This space is closed superiorly by the cranial base and on each side by the carotid sheath. It opens inferiorly into the superior mediastinum (see Chapter 1).

CERVICAL FASCIA

Paralysis of Platysma

Paralysis of the platysma, resulting from injury to the cervical branch of the facial nerve (Fig. 8.8), causes the skin to fall away from the neck in slack folds. Consequently, during surgical dissections of the neck, extra care is necessary to preserve the cervical branch of the facial nerve. When suturing wounds of the neck, surgeons carefully suture the skin and edges of the platysma. If this is not done, the skin wound will be distracted (pulled in different directions) by the contracting platysma muscle fibers, and an ugly scar may develop.

Spread of Infections in Neck

The investing layer of deep cervical fascia helps prevent the *spread of abscesses (purulent infections)* caused by tissue destruction. If an infection occurs between the investing layer of deep cervical fascia and the muscular part of the pretracheal fascia surrounding the infrahyoid muscles, the infection will usually not spread beyond the superior edge of the manubrium of the sternum. If, however, the infection occurs between the investing fascia and the visceral part of pretracheal fascia, it can spread into the thoracic cavity anterior to the pericardium.

Pus from an abscess posterior to the prevertebral layer of deep cervical fascia may extend laterally in the neck and form a swelling posterior to the SCM. The pus may perforate the prevertebral layer of deep cervical fascia and enter the retropharyngeal space, producing a bulge in the pharynx (*retropharyngeal abscess*). This abscess may cause difficulty in swallowing (*dysphagia*) and speaking (*dysarthria*).

Infections in the head may also spread inferiorly posterior to the esophagus and enter the posterior mediastinum, or it may spread anterior to the trachea and enter the anterior mediastinum. Infections in the retropharyngeal space may also extend inferiorly into the superior mediastinum. Similarly, air from a ruptured trachea, bronchus, or esophagus (*pneumomediastinum*) can pass superiorly in the neck.

The Bottom Line

CERVICAL FASCIA

Cervical subcutaneous tissue and platysma: The subcutaneous tissue (superficial cervical fascia) is usually thinner in the neck than in other regions, especially anteriorly. ♦ It contains the platysma, a muscle of facial expression.

Deep cervical fascia: Like deep fascia elsewhere, the function of the deep cervical fascia is (1) to provide containment of muscles and viscera in compartments with varying degrees of rigidity, (2) to provide the slipperiness that allows structures to slide over each other, and (3) to serve as a conduit for the passage of neurovascular structures. ♦ Two major fascial compartments of the neck are separated by the retropharyngeal space. ♦ Anteriorly, the pretracheal fascia surrounds the cervical viscera and extrinsic musculature associated with it (suprahyoid and infrahyoid muscles). ♦ Posteriorly, the prevertebral fascia surrounds the musculoskeletal elements of the neck associated with and including the cervical vertebrae. ♦ These two fascial compartments are contained within the third and most superficial layer of deep cervical fascia, the investing layer, which includes the superficial muscles (trapezius and SCM). ♦ The investing fascia attaches to the cranium superiorly and the pectoral girdle inferiorly. ♦ Lying anterolateral at the common junctions of these three layers are the major neurovascular conduits, the carotid sheaths.
♦ The superior and inferior boundaries and continuities of these fascial layers, compartments, and interfascial spaces establish pathways for the spread of infection, fluid, gas, or tumors.

SUPERFICIAL STRUCTURES OF NECK: CERVICAL REGIONS

To allow clear communication regarding the location of structures, injuries, or pathologies, the neck is divided into regions (Fig. 8.6; Table 8.1). Between the cranium (mandible anteriorly and occipital bone posteriorly) and clavicles, the neck is divided into four major regions based on the usually visible and/or palpable borders of the large and relatively superficial SCM and trapezius muscles, which are contained within the outermost (investing) layer of deep cervical fascia.

Sternocleidomastoid Region

The **sternocleidomastoid (SCM) muscle** is a key muscular landmark in the neck, forming the **sternocleidomastoid region.** The SCM visibly divides each side of the neck into the *anterior* and *lateral cervical regions* (anterior and posterior triangles). The SCM is a broad, strap-like muscle that has two heads: the rounded tendon of the **sternal head** attaches to the manubrium, and the thick, fleshy **clavicular head** attaches to the superior surface of the medial third of the clavicle (Figs. 8.6 and 8.7; Table 8.2).

The two heads of the SCM are separated inferiorly by a space, visible superficially as a small triangular depression, the **lesser supraclavicular fossa** (Fig. 8.6B). The heads join superiorly as they pass obliquely upward toward the cranium. The superior attachment of the SCM is the mastoid process of the temporal bone and the superior nuchal line of the occipital bone. The investing layer of deep cervical fascia splits to form a sheath for the SCM (Fig. 8.4B).

The SCMs produce movement at the craniovertebral joints, the cervical intervertebral joints, or at both (Fig. 8.7; Table 8.2). The cranial attachments of the SCMs lie posterior to the axis of the atlanto-occipital (AO) joints. Starting from the anatomical position, with tonic contraction maintaining the position of the cervical vertebral column, bilateral contraction of the SCMs (especially their more posterior fibers) will cause extension of the head at the AO joints, elevating the chin (Fig. 8.7D).

Acting bilaterally, the SCMs can also flex the neck. They can do this in two different ways:

1. If initially the head is flexed anteriorly at the AO joints by the prevertebral muscles (and/or the suprahyoid and infrahyoid muscles) against resistance, the SCMs (especially the anterior fibers) flex the entire cervical vertebral

(text continues on p. 992)

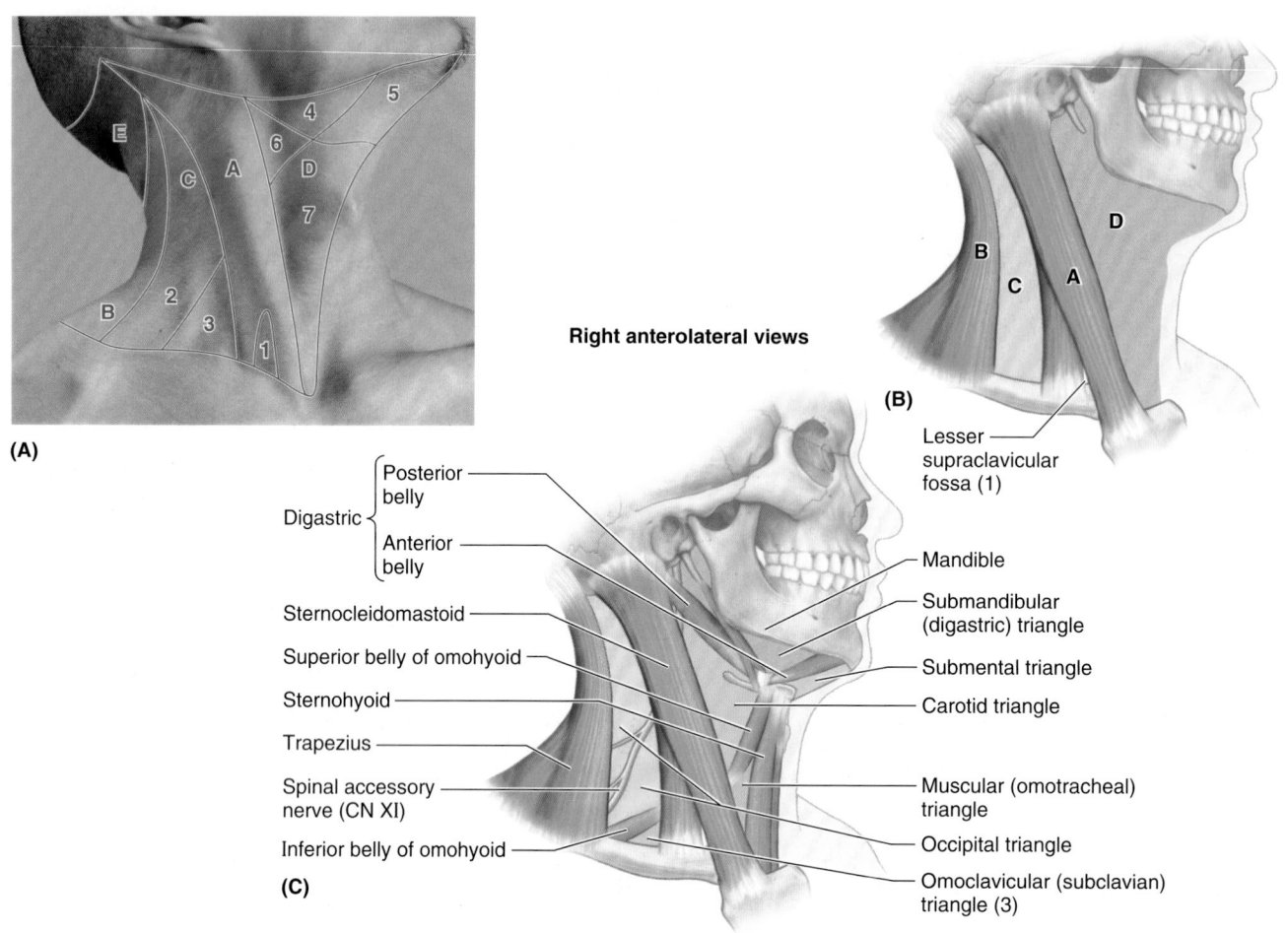

Right anterolateral views

FIGURE 8.6. Cervical regions and triangles.

TABLE 8.1. CERVICAL REGIONS/TRIANGLES AND CONTENTS

Region[a]	Main Contents and Underlying Structures
Sternocleidomastoid region (A)	Sternocleidomastoid muscle; superior part of external jugular vein; greater auricular nerve; transverse cervical nerve
Lesser supraclavicular fossa (1)	Inferior part of internal jugular vein
Posterior cervical region (B)	Trapezius muscle; cutaneous branches of posterior rami of cervical spinal nerves; suboccipital region or triangle (E) lies deep to superior part of this region
Lateral cervical region (posterior triangle) (C)	
Occipital triangle (2)	Part of external jugular vein; posterior branches of cervical plexus of nerves; spinal accessory nerve (CN XI);[b] trunks of brachial plexus; cervicodorsal trunk; cervical lymph node
Omoclavicular (subclavian) triangle (3)	Subclavian artery (third part); part of subclavian vein (sometimes); suprascapular artery; supraclavicular lymph nodes
Anterior cervical region (anterior triangle) (D)	
Submandibular (digastric) triangle (4)	Submandibular gland almost fills triangle; submandibular lymph nodes; hypoglossal nerve (CN XII); mylohyoid nerve; parts of facial artery and vein
Submental triangle (5)	Submental lymph nodes and small veins that unite to form anterior jugular vein
Carotid triangle (6)	Carotid sheath containing common carotid artery and its branches; internal jugular vein and its tributaries; vagus nerve; external carotid artery and some of its branches; hypoglossal nerve (CN XII) and superior root of ansa cervicalis; spinal accessory nerve (CN XI);[b] thyroid gland, larynx, and pharynx; deep cervical lymph nodes; branches of cervical plexus
Muscular (omotracheal) triangle (7)	Sternothyroid and sternohyoid muscles; thyroid and parathyroid glands

[a]Letters and numbers in parentheses refer to Figure 8.6A & B.
[b]The spinal accessory nerve (CN XI) refers to the traditional "spinal root of CN XI." The traditional "cranial root" is now considered part of the vagus nerve (CN X) (Lachman et al., 2002).

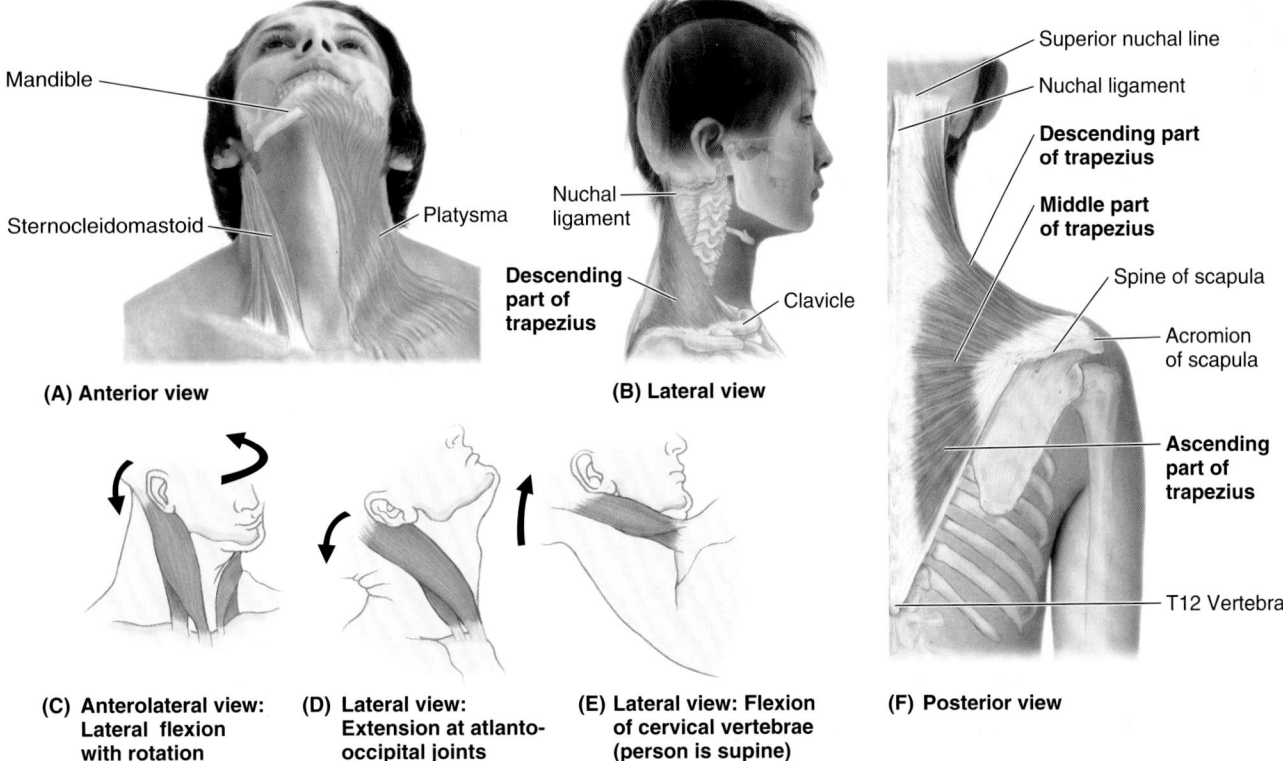

FIGURE 8.7. Muscles of neck.

(A) Anterior view

Mandible

Sternocleidomastoid

Platysma

(B) Lateral view

Nuchal ligament

Descending part of trapezius

Clavicle

(C) Anterolateral view: Lateral flexion with rotation

(D) Lateral view: Extension at atlanto-occipital joints

(E) Lateral view: Flexion of cervical vertebrae (person is supine)

(F) Posterior view

Superior nuchal line

Nuchal ligament

Descending part of trapezius

Middle part of trapezius

Spine of scapula

Acromion of scapula

Ascending part of trapezius

T12 Vertebra

TABLE 8.2. CUTANEOUS AND SUPERFICIAL MUSCLES OF NECK

Muscle	Superior Attachment	Inferior Attachment	Innervation	Main Action(s)
Platysma	Inferior border of mandible, skin, and subcutaneous tissues of lower face	Fascia covering superior parts of pectoralis major and deltoid muscles	Cervical branch of facial nerve (CN VII)	Draws corners of mouth inferiorly and widens it as in expressions of sadness and fright; draws skin of neck superiorly when teeth are clenched
Sternocleido-mastoid (SCM)	Lateral surface of mastoid process of temporal bone and lateral half of superior nuchal line	*Sternal head:* anterior surface of manubrium of sternum *Clavicular head:* superior surface of medial third of clavicle	Spinal accessory nerve (CN XI, motor); C2 and C3 nerves (pain and proprioception)	*Unilateral contraction:* tilts head to same side (i.e., laterally flexes neck) and rotates it so face is turned superiorly toward opposite side (Fig. 8.6C) *Bilateral contraction:* (1) extends neck at atlanto-occipital joints (Fig. 8.6D), (2) flexes cervical vertebrae so that chin approaches manubrium (Fig. 8.6E), or (3) extends superior cervical vertebrae while flexing inferior vertebrae so chin is thrust forward with head kept level With cervical vertebrae fixed, may elevate manubrium and medial ends of clavicles, assisting pump-handle action of deep respiration (see Chapter 1)
Trapezius	Medial third of superior nuchal line, external occipital protuberance, nuchal ligament, spinous processes of C7–T12 vertebrae, and lumbar and sacral spinous processes	Lateral third of clavicle, acromion, and spine of scapula	Spinal accessory nerve (CN XI; motor); C2 and C3 nerves (pain and proprioception)	Elevates, retracts, and rotates scapulae superiorly *Descending (superior) fibers:* elevate scapulae/shoulders, maintain level of shoulders against gravity or resistance *Transverse (middle) fibers:* retract scapulae *Ascending (inferior) fibers:* depress scapulae/shoulders *Descending and ascending fibers together:* rotate spinous process of scapulae superiorly With shoulders fixed, bilateral contraction extends neck; unilateral contraction produces lateral flexion to same side

hat the chin approaches the manubrium Iowever, gravity is usually the prime mover ... ... movement when standing erect.

2. Acting antagonistically with the extensors of the neck (i.e., the deep cervical muscles), bilateral contraction of the SCMs can flex the lower neck while producing limited extension at the AO joint and upper neck, protruding the chin while keeping the head level. Such flexion movements also occur when lifting the head off the ground while lying supine (with gravity providing the resistance in place of the deep cervical muscles).

It is probable that most of the time smaller synergistic muscles and/or eccentric contraction (controlled relaxation of the muscle, gradually yielding to gravity; see the Introduction) are involved in initiating flexion or extension, with the SCMs providing the power and range of movement once initiated.

Acting unilaterally, the SCM laterally flexes the neck (bends the neck sideways) and rotates the head so the ear approaches the shoulder of the ipsilateral (same) side while elevating and rotating the chin toward the contralateral (opposite) side. If the head and neck are fixed, *bilateral contraction* of the SCMs elevates the clavicles and manubrium and thus the anterior ribs. In this way, the SCMs act as accessory muscles of respiration in assisting production of the pump-handle movement of the thoracic wall (see p. 83).

To test the SCM, the head is turned to the opposite side against resistance (hand against chin). If it is acting normally, the SCM can be seen and palpated.

Posterior Cervical Region

The region posterior to the anterior borders of (i.e., corresponding to the area of) the trapezius is the **posterior cervical region** (Fig. 8.6; Table 8.1). The *suboccipital region* is deep to the superior part of this region (see Fig. 8.9, and p. 492). The **trapezius** is a large, flat triangular muscle that covers the posterolateral aspect of the neck and thorax (Fig. 8.7F). The trapezius is a:

- Superficial muscle of the back (see Chapter 4).
- Posterior axio-appendicular muscle, that acts on the pectoral girdle (see Chapter 6).
- Cervical muscle, that can produce movement of the cranium.

The trapezius attaches the pectoral girdle to the cranium and the vertebral column and assists in suspending it. Its attachments, nerve supply, and main actions are described in Table 8.2. The skin of the posterior cervical region is innervated in a segmental pattern by the posterior rami of cervical spinal nerves that pierce, but do not innervate, the trapezius (see Fig. 4.28, and p. 483).

To test the trapezius, the shoulder is shrugged against resistance. If the muscle is acting normally, its superior border can be seen and palpated. If the trapezius is paralyzed, the shoulder droops; however, the combined actions of the levator scapulae and superior fibers of the serratus anterior help support the shoulder and may compensate for the paralysis to some degree (see Chapter 6).

Lateral Cervical Region

The **lateral cervical region** (posterior triangle) is bounded (Figs. 8.6 and 8.8):

- Anteriorly by the posterior border of the SCM.
- Posteriorly by the anterior border of the trapezius.
- Inferiorly by the middle third of the clavicle between the trapezius and the SCM.
- By an *apex,* where the SCM and trapezius meet on the superior nuchal line of the occipital bone.
- By a *roof,* formed by the investing layer of deep cervical fascia.
- By a *floor,* formed by muscles covered by the prevertebral layer of deep cervical fascia.

The lateral cervical region wraps around the lateral surface of the neck like a spiral. The region is covered by skin and subcutaneous tissue containing the platysma.

MUSCLES IN LATERAL CERVICAL REGION

The floor of the lateral cervical region is usually formed by the prevertebral fascia overlying four muscles (Fig. 8.9): splenius capitis, levator scapulae, middle scalene (L. *scalenus medius*), and posterior scalene (L. *scalenus posterior*). Sometimes the inferior part of the anterior scalene (L. *scalenus anterior*) appears in the inferomedial angle of the lateral cervical region, where it is usually hidden by the SCM. An occasional offshoot of the anterior scalene, the *smallest scalene* (L. *scalenus minimus*), passes posterior to the subclavian artery to attach to the 1st rib (Agur and Dalley, 2013).

For a more precise localization of structures, the lateral cervical region is divided into a large occipital triangle superiorly and a small omoclavicular triangle inferiorly by the *inferior belly of the omohyoid* (Table 8.1).

- The **occipital triangle** is so called because the *occipital artery* appears in its apex (Figs. 8.8 and 8.10). The most important nerve crossing the occipital triangle is the *spinal accessory nerve* (CN XI).
- The **omoclavicular** (subclavian) **triangle** is indicated on the surface of the neck by the *supraclavicular fossa.* The inferior part of the EJV crosses this triangle superficially; the *subclavian artery* lies deep in it (Figs. 8.8 and 8.10). These vessels are separated by the investing layer of deep cervical fascia. Because the third part of the subclavian artery is located in this region, the omoclavicular triangle is often called the subclavian triangle (Fig. 8.6).

ARTERIES IN LATERAL CERVICAL REGION

The arteries in the lateral cervical region include the lateral branches of the thyrocervical trunk, the third part of

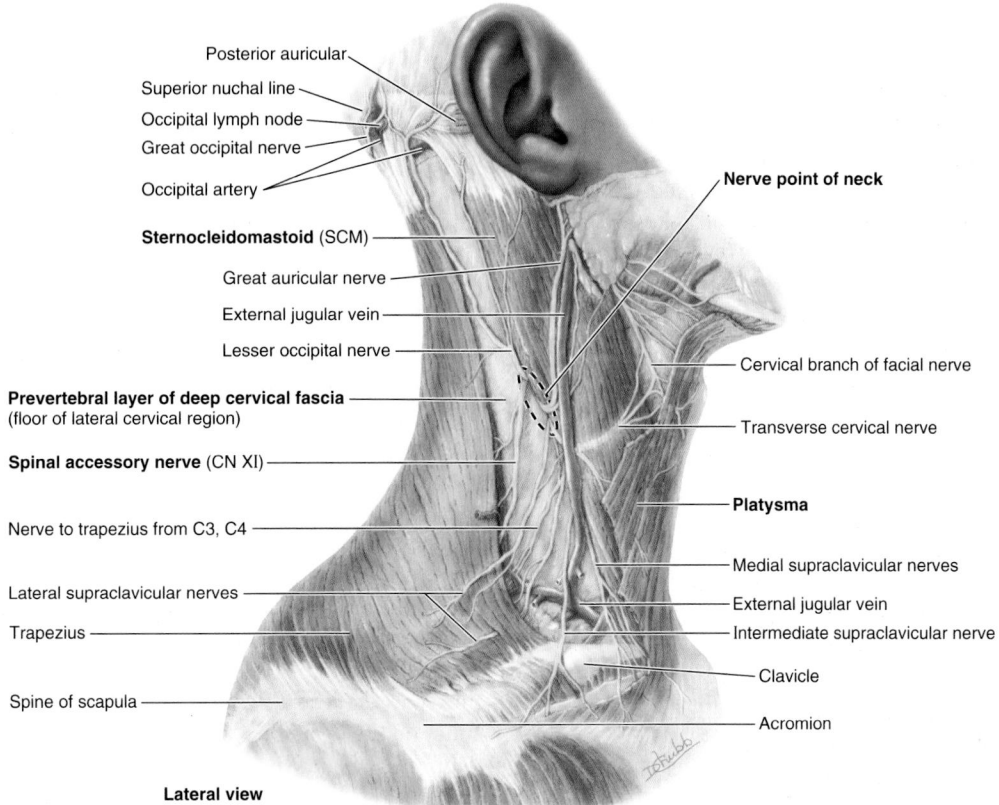

Posterior auricular
Superior nuchal line
Occipital lymph node
Great occipital nerve
Occipital artery

Sternocleidomastoid (SCM)

Great auricular nerve
External jugular vein
Lesser occipital nerve

Prevertebral layer of deep cervical fascia
(floor of lateral cervical region)

Spinal accessory nerve (CN XI)

Nerve to trapezius from C3, C4

Lateral supraclavicular nerves
Trapezius

Spine of scapula

Nerve point of neck

Cervical branch of facial nerve

Transverse cervical nerve

Platysma

Medial supraclavicular nerves
External jugular vein
Intermediate supraclavicular nerve
Clavicle
Acromion

Lateral view

FIGURE 8.8. Superficial dissection of lateral cervical region. The subcutaneous tissue and the investing layer of deep fascia have been removed, sparing most of the platysma and the cutaneous nerves. Between the trapezius (in the posterior cervical region) and the SCM, the prevertebral layer of deep cervical fascia forms the floor of the lateral cervical region. The spinal accessory nerve (CN XI) is the only motor nerve superficial to this fascia.

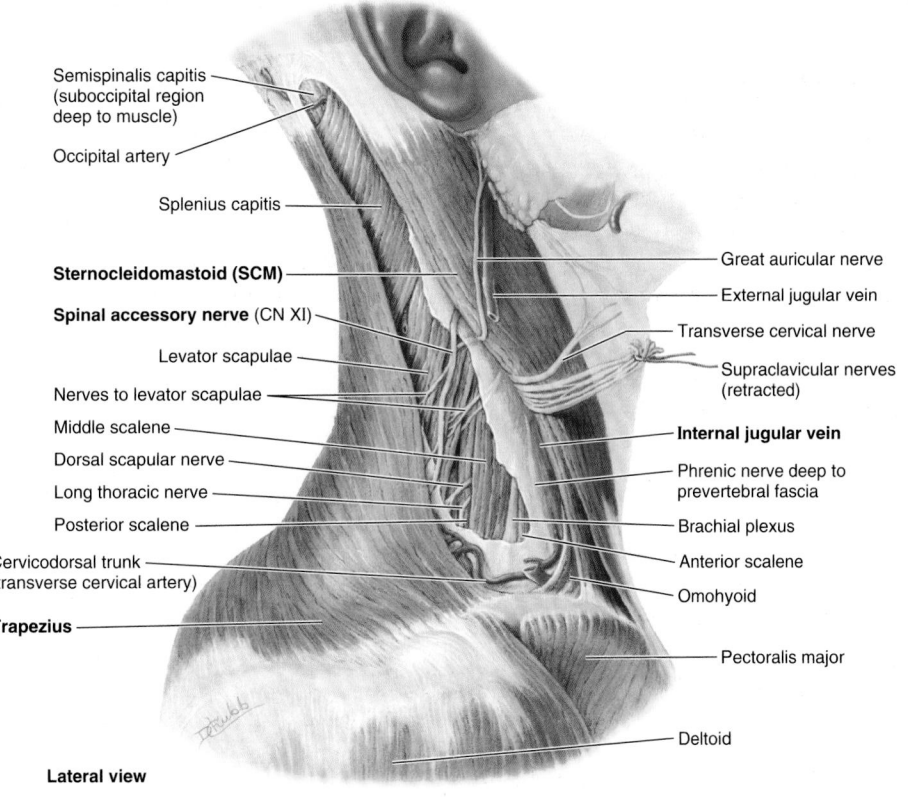

Semispinalis capitis
(suboccipital region
deep to muscle)

Occipital artery

Splenius capitis

Sternocleidomastoid (SCM)

Spinal accessory nerve (CN XI)

Levator scapulae

Nerves to levator scapulae
Middle scalene
Dorsal scapular nerve
Long thoracic nerve
Posterior scalene
Cervicodorsal trunk
(transverse cervical artery)

Trapezius

Great auricular nerve
External jugular vein
Transverse cervical nerve
Supraclavicular nerves
(retracted)

Internal jugular vein

Phrenic nerve deep to
prevertebral fascia
Brachial plexus
Anterior scalene
Omohyoid

Pectoralis major

Deltoid

Lateral view

FIGURE 8.9. Deep dissection of lateral cervical region. The investing layer of the deep cervical fascia has been removed. Although the spinal accessory nerve (CN XI) is superficial to it, the brachial plexus and motor nerves of the cervical plexus run deep to the prevertebral layer of deep cervical fascia that covers the floor of the triangle.

the subclavian artery, and part of the occipital artery. The *thyrocervical trunk,* a branch of the subclavian artery (Figs. 8.9–8.11) most commonly gives rise to a suprascapular artery and a cervicodorsal trunk from its lateral aspect; its terminal branches are the ascending cervical and inferior thyroid artery (discussed on p. 995).

The **suprascapular artery** passes inferolaterally across the anterior scalene muscle and phrenic nerve (Fig. 8.10). It then crosses the third part of the subclavian artery and the cords of the brachial plexus. It then passes posterior to the clavicle to supply muscles on the posterior aspect of the scapula. Alternately, the suprascapular artery may arise directly from the third part of the subclavian artery.

Also arising laterally, the **cervicodorsal trunk** (Weiglein et al., 2005), sometimes known as the transverse cervical artery, further bifurcates into the *superficial cervical artery* (superficial branch of transverse cervical artery) and the *dorsal scapular artery* (deep branch of transverse cervical artery). These branches run superficially and laterally across the phrenic nerve and anterior scalene muscle, 2–3 cm superior to the clavicle. They then cross or pass through the *trunks of the brachial plexus,* supplying branches to

their *vasa nervorum* (blood vessels of nerves). The **superficial cervical artery** passes deep (anterior) to the trapezius accompanying the spinal accessory nerve (CN XI). The **dorsal scapular artery** may arise independently, directly from the third (or, less often, the second) part of the subclavian artery. When it is a branch of the subclavian, the dorsal scapular artery passes laterally through the trunks of the brachial plexus, anterior to the middle scalene. Regardless of its origin, its distal portion runs deep to the levator scapulae and rhomboid muscles, supplying both and participating in the arterial anastomoses around the scapula (Chapter 6). The *occipital artery,* a branch of the external carotid artery, enters the lateral cervical region at its apex and ascends over the head to supply the posterior half of the scalp (Fig. 8.9).

The **subclavian artery** supplies blood to the upper limb. The third part begins approximately a finger's breadth superior to the clavicle, opposite the lateral border of the anterior scalene muscle. It is hidden in the inferior part of the lateral cervical region, posterosuperior to the subclavian vein. The third part of the artery is the longest and most superficial part. It lies on the 1st rib, and its pulsations can be felt by applying deep pressure in the omoclavicular triangle. The artery is in

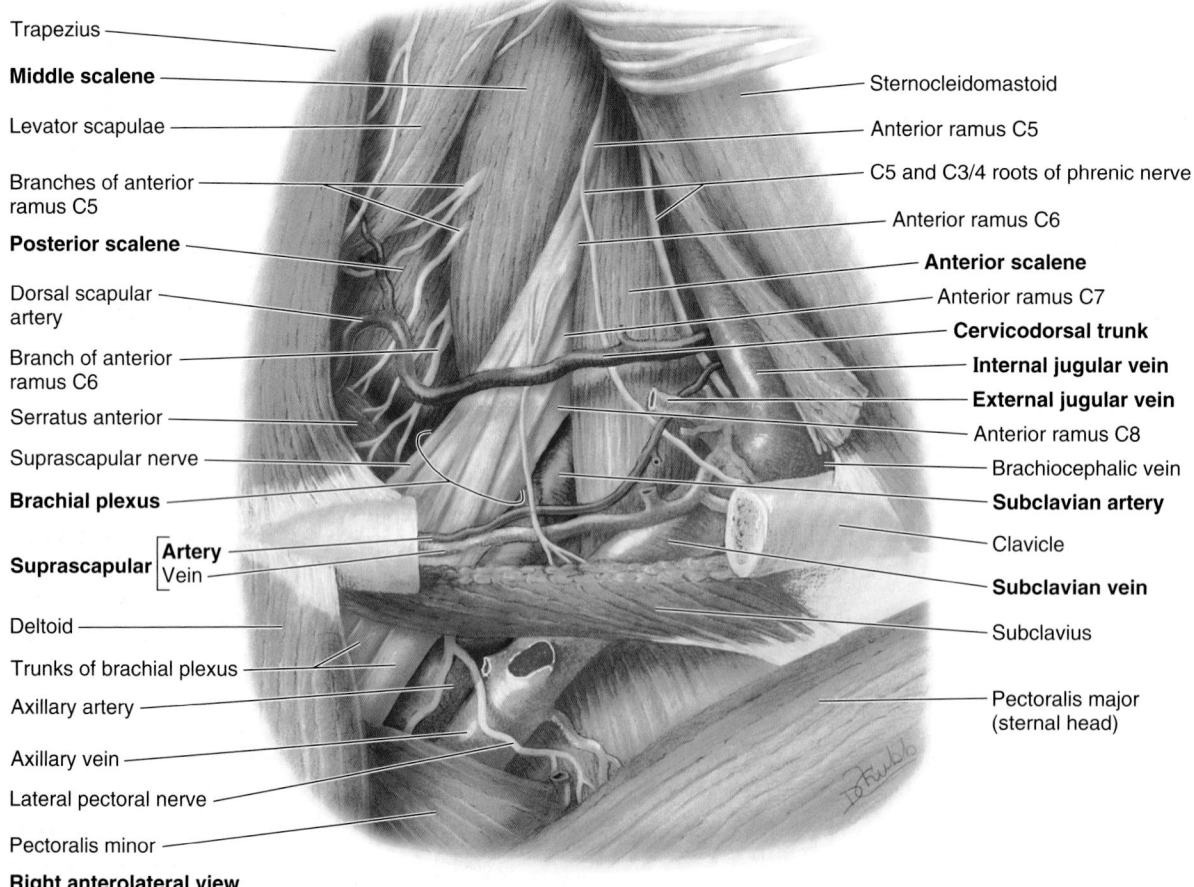

Trapezius
Middle scalene
Levator scapulae
Branches of anterior ramus C5
Posterior scalene
Dorsal scapular artery
Branch of anterior ramus C6
Serratus anterior
Suprascapular nerve
Brachial plexus
Suprascapular [Artery / Vein]
Deltoid
Trunks of brachial plexus
Axillary artery
Axillary vein
Lateral pectoral nerve
Pectoralis minor
Right anterolateral view

Sternocleidomastoid
Anterior ramus C5
C5 and C3/4 roots of phrenic nerve
Anterior ramus C6
Anterior scalene
Anterior ramus C7
Cervicodorsal trunk
Internal jugular vein
External jugular vein
Anterior ramus C8
Brachiocephalic vein
Subclavian artery
Clavicle
Subclavian vein
Subclavius
Pectoralis major (sternal head)

FIGURE 8.10. Deeper dissection of inferior part of lateral cervical region. All fascia, the omohyoid muscle, and the clavicular head of the pectoralis major have been removed to reveal the subclavian vein and third part of the subclavian artery. The internal jugular vein, deep to the SCM, is not in the lateral cervical region but is close to it. The brachial plexus of nerves and subclavian vessels pass to the upper limb, the name of the vessels changing to *axillary* inferior to the clavicle at the lateral border of the 1st rib.

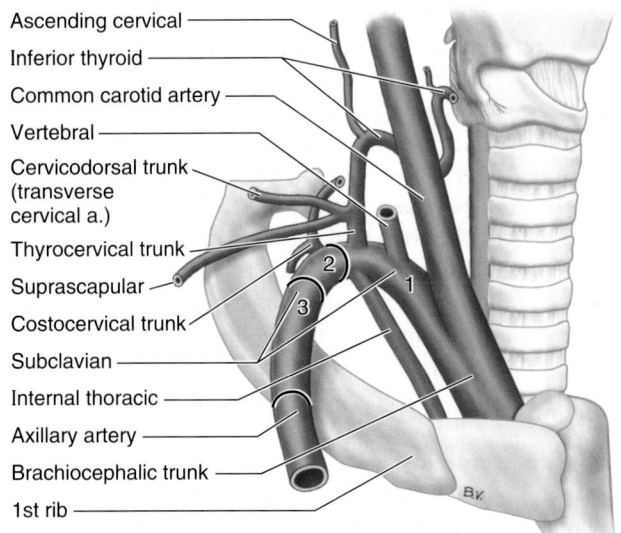

Ascending cervical
Inferior thyroid
Common carotid artery
Vertebral
Cervicodorsal trunk
(transverse
cervical a.)
Thyrocervical trunk
Suprascapular
Costocervical trunk
Subclavian
Internal thoracic
Axillary artery
Brachiocephalic trunk
1st rib

Lateral view of right side

FIGURE 8.11. Subclavian artery: its parts and branches. There are three parts of the subclavian artery: medial (*1*), posterior (*2*), and lateral (*3*) to the anterior scalene muscle. The cervicodorsal trunk (transverse cervical artery) and suprascapular artery occasionally arise directly (or via a common trunk) from the second or third parts of the subclavian artery instead of directly from the thyrocervical trunk via a common trunk, as shown here, or independently.

contact with the 1st rib as it passes posterior to the anterior scalene muscle; consequently, compression of the subclavian artery against this rib can control bleeding in the upper limb. The inferior trunk of the brachial plexus lies directly posterior to the third part of the artery. The branches that occasionally arise from the third part (suprascapular artery, dorsal scapular artery) are aberrant forms of more typical patterns

in which they arise elsewhere (from the thyrocervical trunk via a cervicodorsal trunk in particular).

VEINS IN LATERAL CERVICAL REGION

The **external jugular vein** (**EJV**) begins near the angle of the mandible (just inferior to the auricle) by the union of the posterior division of the *retromandibular vein* with the *posterior auricular vein* (Fig. 8.12). The EJV crosses the SCM obliquely, deep to the platysma, and enters the antero-inferior part of the lateral cervical region (Fig. 8.8). It then pierces the investing layer of deep cervical fascia, which forms the roof of this region, at the posterior border of the SCM. The EJV descends to the inferior part of the lateral cervical region and terminates in the subclavian vein (Figs. 8.10 and 8.12). It drains most of the scalp and side of the face.

The **subclavian vein,** the major venous channel draining the upper limb, curves through the inferior part of the lateral cervical region. It passes anterior to the anterior scalene muscle and phrenic nerve and unites at the medial border of the muscle with the IJV to form the **brachiocephalic vein,** posterior to the medial end of the clavicle. Just superior to the clavicle, the EJV receives the *cervicodorsal, suprascapular*, and *anterior jugular veins*.

NERVES OF LATERAL CERVICAL REGION

The **spinal accessory nerve** (**CN XI**) passes deep to the SCM, supplying it before entering the lateral cervical region at or inferior to the junction of the superior and middle thirds of the posterior border of the SCM (Figs. 8.8 and 8.9). The nerve passes postero-inferiorly, within or deep to the investing layer of deep cervical fascia, running on the levator scapulae from which it is separated by the prevertebral layer of fascia. CN XI then disappears deep to the anterior border of

FIGURE 8.12. Superficial veins of neck. The superficial temporal and maxillary veins merge, forming the retromandibular vein, the posterior division of which unites with the posterior auricular vein to form the EJV. The facial vein receives the anterior division of the retromandibular vein before emptying into the internal jugular vein, deep to the SCM. The anterior jugular veins may lie superficial or deep to the investing layer of the deep cervical fascia.

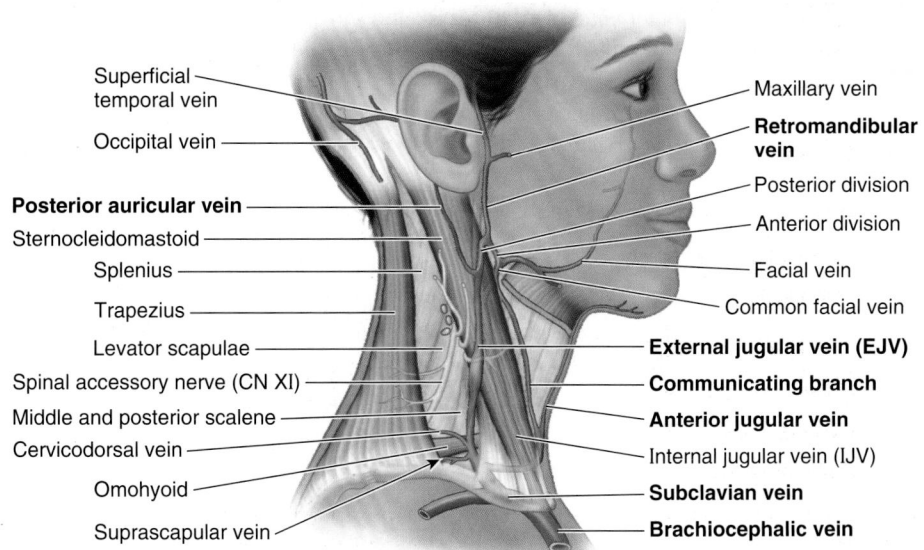

Superficial
temporal vein
Occipital vein
Posterior auricular vein
Sternocleidomastoid
Splenius
Trapezius
Levator scapulae
Spinal accessory nerve (CN XI)
Middle and posterior scalene
Cervicodorsal vein
Omohyoid
Suprascapular vein

Maxillary vein
Retromandibular vein
Posterior division
Anterior division
Facial vein
Common facial vein
External jugular vein (EJV)
Communicating branch
Anterior jugular vein
Internal jugular vein (IJV)
Subclavian vein
Brachiocephalic vein

Lateral view

he junction of its superior two thirds with its

The **roots of the brachial plexus** (anterior rami of C5–C8 and T1) appear between the anterior and the middle scalene muscles (Fig. 8.10). The five rami unite to form the *three trunks of the brachial plexus,* which descend inferolaterally through the lateral cervical region. The plexus then passes between the 1st rib, clavicle, and superior border of the scapula (the *cervico-axillary canal*) to enter the axilla, providing innervation for most of the upper limb (see Chapter 6).

The **suprascapular nerve,** which arises from the superior trunk of the *brachial plexus* (not cervical plexus), runs laterally across the lateral cervical region to supply the supraspinatus and infraspinatus muscles on the posterior aspect of the scapula. It also sends articular branches to the glenohumeral joint.

The *anterior rami of C1–C4* make up the **roots of the cervical plexus** (Fig. 8.13). The **cervical plexus** consists of an irregular series of (primary) nerve loops and the branches that arise from the loops. Each participating ramus, except the first, divides into ascending and descending branches that unite with the branches of the adjacent spinal nerve to form the loops. The cervical plexus lies anteromedial to the levator scapulae and middle scalene muscles and deep to the SCM. The superficial branches of the plexus that initially pass posteriorly are cutaneous (sensory) branches (Fig. 8.13A, C, & D). The deep branches passing anteromedially are motor branches, including the *roots of the phrenic nerve* (to the diaphragm) and the *ansa cervicalis* (Fig. 8.13A & B).

The **superior root of the ansa cervicalis,** conveying fibers from spinal nerves C1 and C2, briefly joins and then descends from the hypoglossal nerve (CN XII) as it traverses the lateral cervical region (Fig. 8.13). The **inferior root of the ansa cervicalis** arises from a loop between spinal nerves C2 and C3. The superior and inferior roots of the ansa cervicalis unite, forming a secondary loop, the **ansa cervicalis,** consisting of fibers from the C1–C3 spinal nerves, which branch from the secondary loop to supply the infrahyoid muscles, including the *omohyoid, sternothyroid,* and *sternohyoid* (Figs. 8.13, 8.14A, and 8.15). The fourth infrahyoid muscle, the *thyrohyoid,* receives C1 fibers, which descend independently from the hypoglossal nerve, distal to the superior root of the ansa cervicalis (**nerve to thyrohyoid**) (Figs. 8.13A & B and 8.14B).

Cutaneous branches of the cervical plexus emerge around the middle of the posterior border of the SCM, often called the **nerve point of the neck** (Fig. 8.8), and supply the skin of the neck, superolateral thoracic wall, and scalp between the auricle and the external occipital protuberance (Fig. 8.13 A, C, & D). Close to their origin, the roots of the cervical plexus receive gray rami communicantes, most of which descend from the large *superior cervical ganglion* in the superior part of the neck.

Branches of cervical plexus arising from the nerve loop between the anterior rami of C2 and C3 are the:

- **Lesser occipital nerve** (C2): supplies the skin of the neck and scalp posterosuperior to the auricle.

- **Great auricular nerve** (C2 and C3): ascends vertically across the oblique SCM to the inferior pole of the parotid gland, where it divides to supply the skin over—and the sheath surrounding—the gland, the mastoid process, and both surfaces of the auricle and an area of skin extending from the angle of the mandible to the mastoid process.

- **Transverse cervical nerve** (C2 and C3): supplies the skin covering the anterior cervical region. It curves around the middle of the posterior border of the SCM inferior to the great auricular nerve and passes anteriorly and horizontally across it deep to the EJV and platysma, dividing into superior and inferior branches.

The branches of the cervical plexus arising from the nerve loop formed between the anterior rami of C3–C4 are the:

- **Supraclavicular nerves** (C3 and C4): emerge as a common trunk under cover of the SCM, sending small branches to the skin of the neck that cross the clavicle and supply the skin over the shoulder.

In addition to the ansa cervicalis and phrenic nerves arising from the loops of the plexus, **deep motor branches of the cervical plexus** include branches arising from the roots that supply the rhomboids (dorsal scapular nerve; C4 and C5), serratus anterior (long thoracic nerve; C5–C7), and nearby prevertebral muscles.

The **phrenic nerves** originate chiefly from the C4 nerve but receive contributions from the C3 and C5 nerves (Figs. 8.10 and 8.13A). The phrenic nerves contain motor, sensory, and sympathetic nerve fibers. These nerves provide the sole motor supply to the diaphragm as well as sensation to its central part. In the thorax, each phrenic nerve supplies the mediastinal pleura and pericardium (see Chapter 1). Receiving variable communicating fibers in the neck from the cervical sympathetic ganglia or their branches, each phrenic nerve forms at the superior part of the lateral border of the anterior scalene muscle at the level of the superior border of the thyroid cartilage. The phrenic nerve descends obliquely with the IJV across the anterior scalene, deep to the prevertebral layer of deep cervical fascia and the transverse cervical and suprascapular arteries.

On the left, the phrenic nerve crosses anterior to the first part of the subclavian artery; *on the right,* it lies on the anterior scalene muscle and crosses anterior to the second part of the subclavian artery. On both sides, the phrenic nerve runs posterior to the subclavian vein and anterior to the internal thoracic artery as it enters the thorax.

The contribution of the C5 nerve to the phrenic nerve may be derived from an **accessory phrenic nerve** (Fig. 8.10). Frequently, it is a branch of the nerve to the subclavius. If present, the accessory phrenic nerve lies lateral to the main nerve and descends posterior and sometimes anterior to the subclavian vein. The accessory phrenic nerve joins the phrenic nerve either in the root of the neck or in the thorax.

LYMPH NODES IN LATERAL CERVICAL REGION

Lymph from superficial tissues in the lateral cervical region enters the **superficial cervical lymph nodes** that lie along

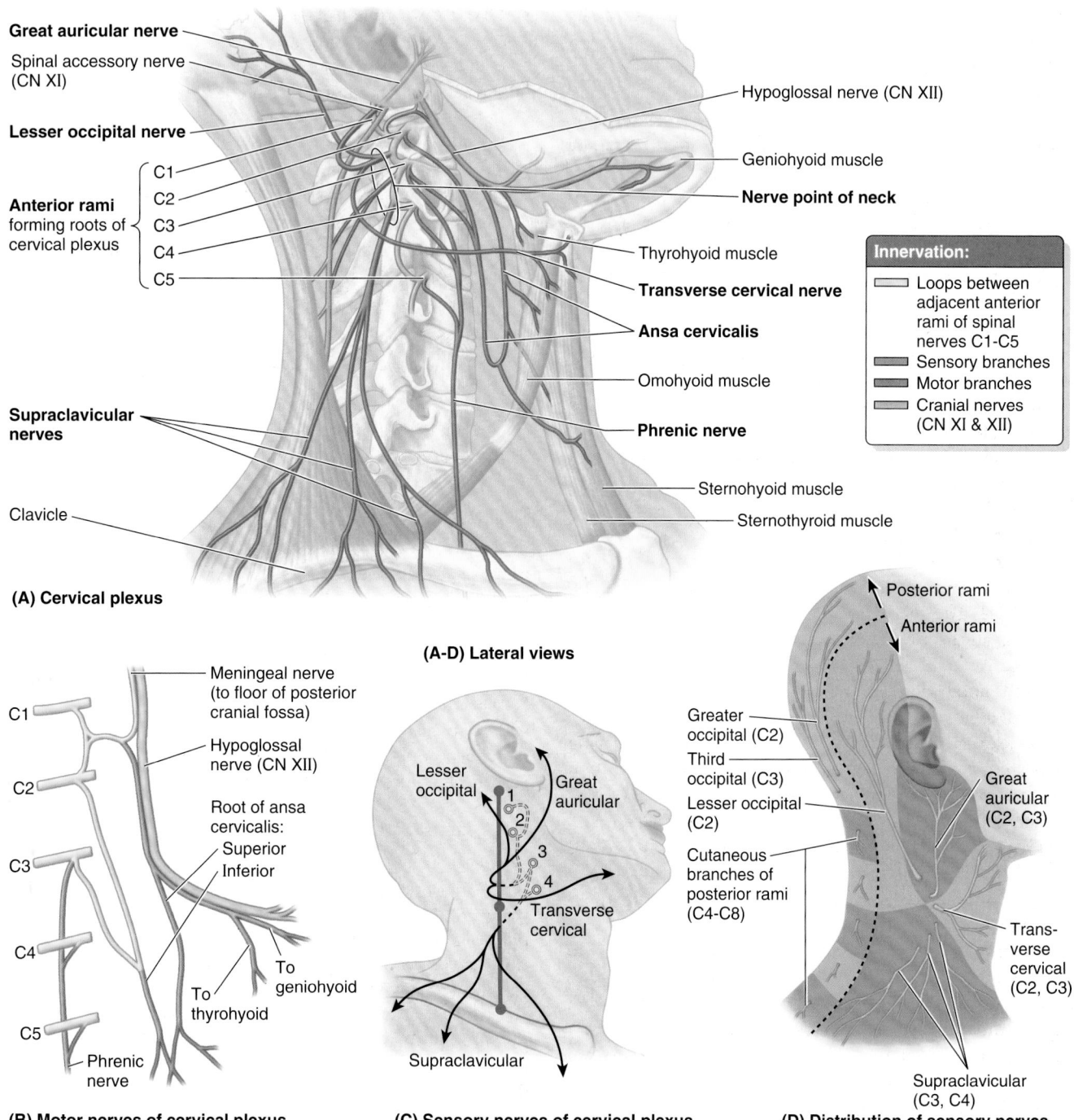

(A) Cervical plexus

(A–D) Lateral views

(B) Motor nerves of cervical plexus **(C) Sensory nerves of cervical plexus** **(D) Distribution of sensory nerves**

FIGURE 8.13. Cervical plexus of nerves. A–C. The plexus consists of nerve loops formed between the adjacent anterior rami of the first four cervical nerves and the receiving gray rami communicantes from the superior cervical sympathetic ganglion (not shown here) (Fig. 8.25A). Motor (**B**) and sensory nerves (**C**) arise from the loops of the plexus. The ansa cervicalis (**A, B**) is a second-level loop, the superior limb of which arises from the loop between the C1 and the C2 vertebrae but travels initially with the hypoglossal nerve (CN XII), which is not part of the cervical plexus. **D.** The areas of skin innervated by the sensory (cutaneous) nerves of the cervical plexus (derived from anterior rami) and by the posterior rami of cervical spinal nerves are shown.

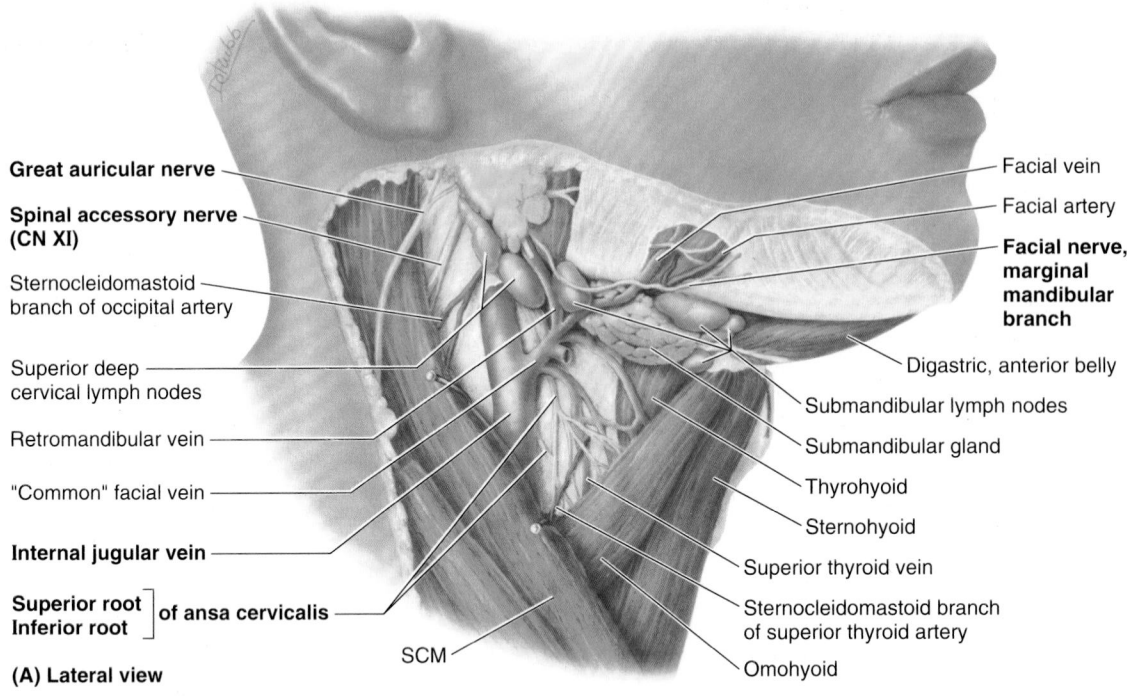

Great auricular nerve

Spinal accessory nerve
(CN XI)

Sternocleidomastoid
branch of occipital artery

Superior deep
cervical lymph nodes

Retromandibular vein

"Common" facial vein

Internal jugular vein

Superior root
Inferior root ⎤ of ansa cervicalis

(A) Lateral view

Facial vein

Facial artery

Facial nerve,
marginal
mandibular
branch

Digastric, anterior belly

Submandibular lymph nodes

Submandibular gland

Thyrohyoid

Sternohyoid

Superior thyroid vein

Sternocleidomastoid branch
of superior thyroid artery

Omohyoid

SCM

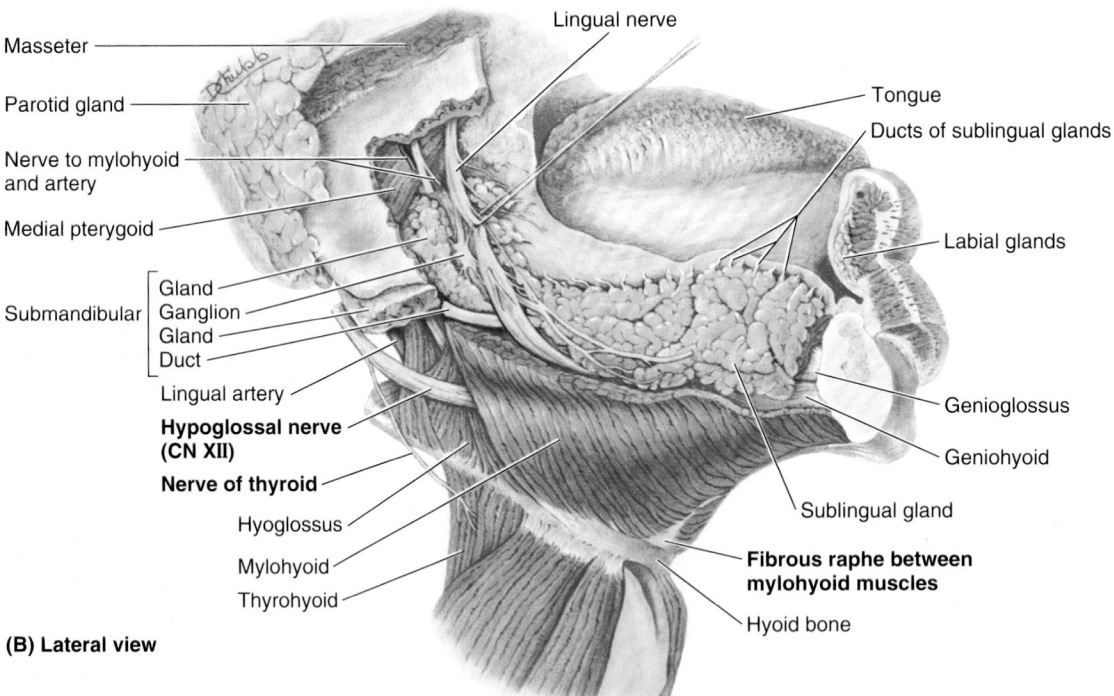

Masseter

Parotid gland

Nerve to mylohyoid
and artery

Medial pterygoid

Submandibular ⎡ Gland
⎢ Ganglion
⎢ Gland
⎣ Duct

Lingual artery

Hypoglossal nerve
(CN XII)

Nerve of thyroid

Hyoglossus

Mylohyoid

Thyrohyoid

(B) Lateral view

Lingual nerve

Tongue

Ducts of sublingual glands

Labial glands

Genioglossus

Geniohyoid

Sublingual gland

Fibrous raphe between
mylohyoid muscles

Hyoid bone

FIGURE 8.14. Dissections of anterior cervical and suprahyoid regions. A. This superficial dissection of the neck displays the submandibular gland and lymph nodes. **B.** In this dissection of the suprahyoid region, the right half of the mandible and the superior part of the mylohyoid muscle have been removed. The cut surface of the mylohyoid becomes progressively thinner as it is traced anteriorly.

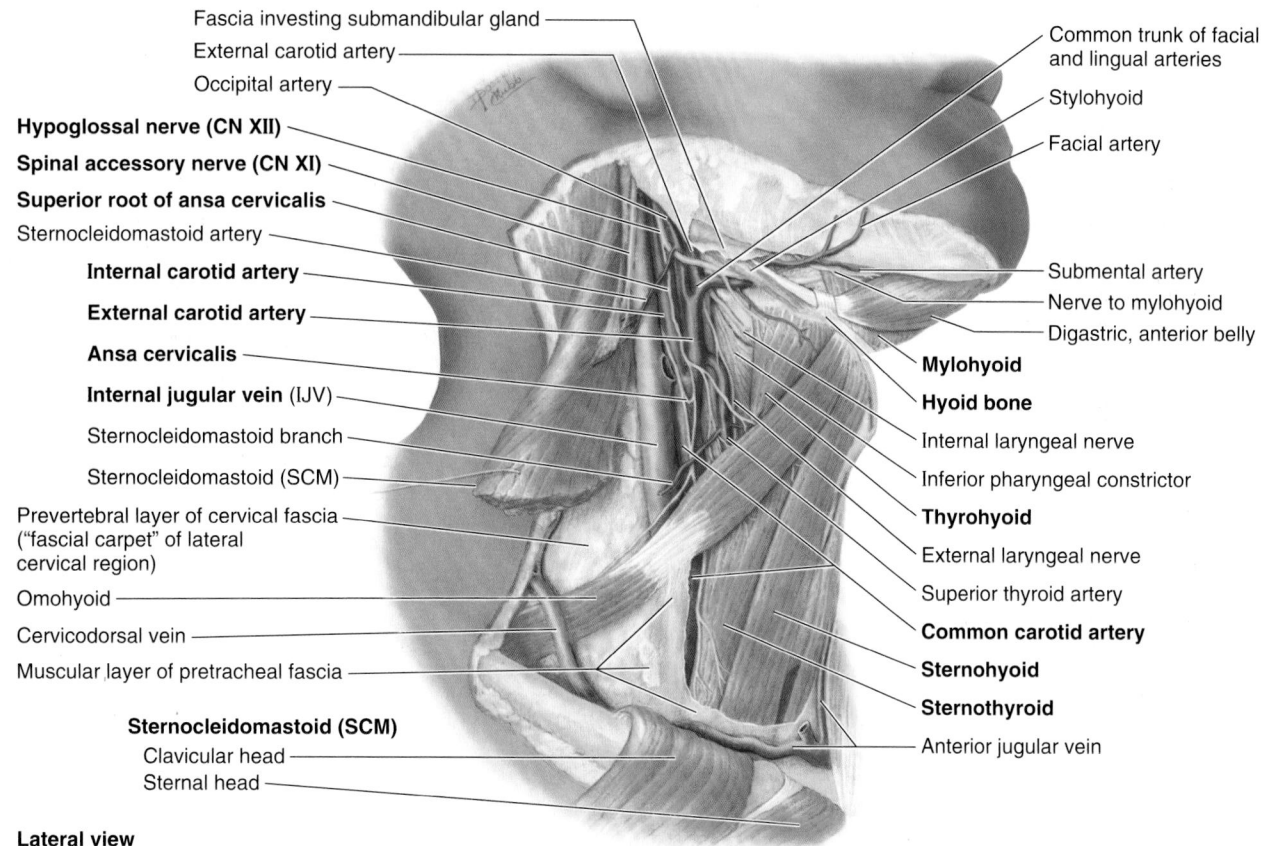

Fascia investing submandibular gland
External carotid artery
Occipital artery
Hypoglossal nerve (CN XII)
Spinal accessory nerve (CN XI)
Superior root of ansa cervicalis
Sternocleidomastoid artery
Internal carotid artery
External carotid artery
Ansa cervicalis
Internal jugular vein (IJV)
Sternocleidomastoid branch
Sternocleidomastoid (SCM)
Prevertebral layer of cervical fascia ("fascial carpet" of lateral cervical region)
Omohyoid
Cervicodorsal vein
Muscular layer of pretracheal fascia

Sternocleidomastoid (SCM)
Clavicular head
Sternal head

Common trunk of facial and lingual arteries
Stylohyoid
Facial artery
Submental artery
Nerve to mylohyoid
Digastric, anterior belly
Mylohyoid
Hyoid bone
Internal laryngeal nerve
Inferior pharyngeal constrictor
Thyrohyoid
External laryngeal nerve
Superior thyroid artery
Common carotid artery
Sternohyoid
Sternothyroid
Anterior jugular vein

Lateral view

FIGURE 8.15. Deep dissection of anterior cervical region. The common facial vein and its tributaries have been removed, revealing arteries and nerves, including the ansa cervicalis and its branches to the infrahyoid muscles. The facial and lingual arteries in this person arise by a common trunk that passes deep to the stylohyoid and digastric muscles to enter the submandibular triangle.

the EJV superficial to the SCM. Efferent vessels from these nodes drain into the **deep cervical lymph nodes,** which form a chain along the course of the IJV embedded in the fascia of the carotid sheath (Figs. 8.4B and 8.14A).

Anterior Cervical Region

The **anterior cervical region** (anterior triangle) (Table 8.1) has the following:

- An *anterior boundary* formed by the median line of the neck.
- A *posterior boundary* formed by the anterior border of the SCM.
- A *superior boundary* formed by the inferior border of the mandible.
- An *apex* located at the jugular notch in the manubrium.
- A *roof* formed by subcutaneous tissue containing the platysma.
- A *floor* formed by the pharynx, larynx, and thyroid gland.

For more precise localization of structures, the anterior cervical region is subdivided into four smaller triangles by the digastric and omohyoid muscles: the unpaired submental triangle and three small paired triangles—submandibular, carotid, and muscular.

The **submental triangle,** inferior to the chin, is a suprahyoid area bounded inferiorly by the body of the hyoid and laterally by the right and left anterior bellies of the digastric muscles. The floor of the submental triangle is formed by the two mylohyoid muscles, which meet in a median **fibrous raphe** (Fig. 8.14B). The apex of the submental triangle is at the *mandibular symphysis,* the site of union of the halves of the mandible during infancy. The base of the submental triangle is formed by the hyoid (Fig. 8.16). This triangle contains several small **submental lymph nodes** and small veins that unite to form the *anterior jugular vein* (Fig. 8.15).

The **submandibular triangle** is a glandular area between the inferior border of the mandible and the anterior and posterior bellies of the digastric muscle (Fig. 8.14A). The floor of the submandibular triangle is formed by the mylohyoid and hyoglossus muscles, and the middle pharyngeal constrictor. The **submandibular gland** nearly fills this triangle (Fig. 8.12B). (Because of its functional association with the mouth as well as its anatomical association with the floor of the mouth, the gland is discussed in Chapter 7, p. 945.)

Submandibular lymph nodes lie on each side of the submandibular gland and along the inferior border of the mandible (Fig. 8.14A). The *hypoglossal nerve* (CN XII) provides motor innervation to the intrinsic and extrinsic muscles

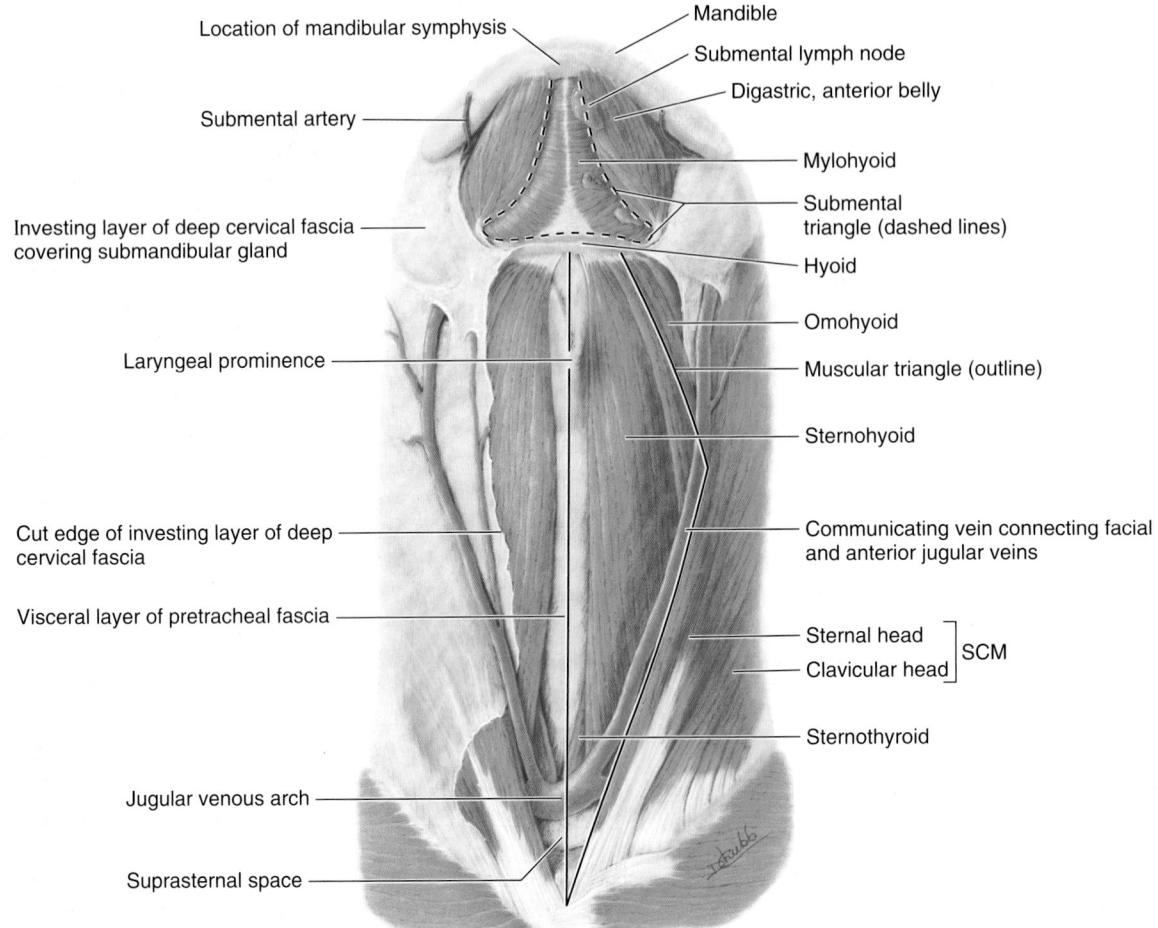

FIGURE 8.16. Superficial dissection of anterior cervical region. The submental triangle is bounded inferiorly by the body of the hyoid and laterally by the right and left anterior bellies of the digastric muscles. The floor of this triangle is formed by the two mylohyoid muscles and the raphe between them (not distinct here; see Fig. 8.14B). The muscular triangle is bounded by the superior belly of the omohyoid, the anterior border of the SCM and the midline.

of the tongue. It passes into the submandibular triangle, as does the *nerve to the mylohyoid* (a branch of CN V_3, which also supplies the anterior belly of the digastric), parts of the *facial artery* and vein, and the *submental artery* (a branch of the facial artery) (Figs. 8.14 and 8.15).

The **carotid triangle** is a vascular area bounded by the superior belly of the omohyoid, the posterior belly of the digastric, and the anterior border of the SCM (Figs. 8.6, 8.14A and 8.15). This triangle is important because the *common carotid artery* ascends into it. Its pulse can be auscultated or palpated by compressing it lightly against the transverse processes of the cervical vertebrae. At the level of the superior border of the thyroid cartilage, the common carotid artery divides into the *internal* and *external carotid arteries* (Figs. 8.15, 8.17, and Fig. 8.19). Located within the carotid triangle are the following:

- **Carotid sinus:** a dilation of the proximal part of the internal carotid artery (Fig. 8.17), which may involve the common carotid artery. Innervated principally by the glossopharyngeal nerve (CN IX) through the **carotid sinus nerve** as well as by the vagus nerve (CN X), it is a *baroreceptor* (pressoreceptor) that reacts to changes in arterial blood pressure.

- **Carotid body:** a small, reddish brown ovoid mass of tissue in life that lies in a septum on the medial (deep) side of the bifurcation of the common carotid artery in close relation to the carotid sinus. Supplied mainly by the carotid sinus nerve (CN IX) and by CN X, it is a *chemoreceptor* that monitors the level of oxygen in the blood. It is stimulated by low levels of oxygen and initiates a reflex that increases the rate and depth of respiration, cardiac rate, and blood pressure.

The neurovascular structures in the carotid triangle are surrounded by the *carotid sheath:* the carotid arteries medially, the *IJV* laterally, and the *vagus nerve* posteriorly (Fig 8.4B & C). Superiorly, the common carotid is replaced by the internal carotid artery. The *ansa cervicalis* usually lies on (or is embedded in) the anterolateral aspect of the sheath (Fig. 8.15). Many *deep cervical lymph nodes* lie along the carotid sheath and the IJV.

The **muscular triangle** is bounded by the superior belly of the omohyoid muscle, the anterior border of the SCM, and the median plane of the neck (Figs. 8.6 and 8.16). This triangle contains the *infrahyoid muscles* and viscera (e.g., the thyroid and parathyroid glands).

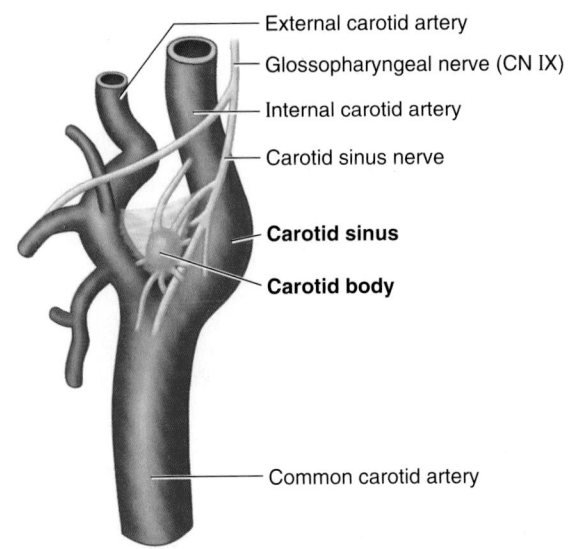

External carotid artery

Glossopharyngeal nerve (CN IX)

Internal carotid artery

Carotid sinus nerve

Carotid sinus

Carotid body

Common carotid artery

Medial view of right carotid artery

FIGURE 8.17. Carotid body and carotid sinus. This small epithelioid body lies within the bifurcation of the common carotid artery. The carotid sinus and the associated network of sensory fibers of the glossopharyngeal nerve (CN IX) are also shown.

MUSCLES IN ANTERIOR CERVICAL REGION

In the anterolateral part of the neck, the *hyoid* provides attachments for the suprahyoid muscles superior to it and the infrahyoid muscles inferior to it. These **hyoid muscles** steady or move the hyoid and larynx (Figs. 8.15, 8.16, and 8.18). For descriptive purposes, they are divided into suprahyoid and infrahyoid muscles, the attachments, innervation, and main actions of which are presented in Table 8.3.

The **suprahyoid muscles** are superior to the hyoid and connect it to the cranium (Figs. 8.14–8.16 and 8.18; Table 8.3). The suprahyoid group of muscles includes the *mylohyoid, geniohyoid, stylohyoid,* and *digastric muscles.* As a group, these muscles constitute the substance of the floor of the mouth, supporting the hyoid in providing a base from which the tongue functions and elevating the hyoid and larynx in relation to swallowing and tone production. Each **digastric muscle** has two bellies, joined by an **intermediate tendon** that descends toward the hyoid. A **fibrous sling** derived from the pretracheal layer of deep cervical fascia allows the tendon to slide anteriorly and posteriorly as it connects this tendon to the body and greater horn of the hyoid.

The difference in nerve supply between the anterior and the posterior bellies of the digastric muscles results from their different embryological origin from the 1st and 2nd pharyngeal arches, respectively. CN V supplies derivatives of the 1st arch, and CN VII supplies those of the 2nd arch.

The **infrahyoid muscles,** often called *strap muscles* because of their ribbon-like appearance, are inferior to the hyoid (Figs. 8.14 and 8.18; Table 8.3). These four muscles anchor the hyoid, sternum, clavicle, and scapula and depress

the hyoid and larynx during swallowing and speaking. They also work with the suprahyoid muscles to steady the hyoid, providing a firm base for the tongue. The infrahyoid group of muscles are arranged in two planes: a *superficial plane,* made up of the sternohyoid and omohyoid, and a *deep plane,* composed of the sternothyroid and thyrohyoid.

Like the digastric, the omohyoid has two bellies (superior and inferior) united by an *intermediate tendon.* The fascial sling for the intermediate tendon connects to the clavicle.

The **sternothyroid** is wider than the **sternohyoid,** under which it lies. The sternothyroid covers the lateral lobe of the thyroid gland. Its attachment to the *oblique line* of the lamina of the thyroid cartilage immediately superior to the gland limits upward extension of an enlarged thyroid (see the blue box "Enlargement of Thyroid Gland" on p. 1042). The **thyrohyoid** appears to be the continuation of the sternothyroid muscle, running superiorly from the oblique line of the thyroid cartilage to the hyoid.

ARTERIES IN ANTERIOR CERVICAL REGION

The anterior cervical region contains the **carotid system of arteries,** consisting of the common carotid artery and its terminal branches, the internal and external carotid arteries. It also contains the IJV, its tributaries, and the anterior jugular veins (Figs. 8.19 and Fig. 8.20). The common carotid artery and one of its terminal branches, the *external carotid artery,* are the main arterial vessels in the carotid triangle. Branches of the external carotid (e.g., the superior thyroid artery) also originate in the carotid triangle. Each *common carotid artery* ascends within the *carotid sheath* with the IJV and vagus nerve to the level of the superior border of the thyroid cartilage. Here, each common carotid artery terminates by dividing into the internal and external carotid arteries. The *internal carotid artery* has no branches in the neck; the external carotid has several.

The **right common carotid artery** begins at the bifurcation of the *brachiocephalic trunk.* The right subclavian artery is the other branch of this trunk. From the arch of the aorta, the **left common carotid artery** ascends into the neck. Consequently, the left common carotid has a course of approximately 2 cm in the superior mediastinum before entering the neck. ⟶ *left common carotid artery & left subclavian do not come from the same trunk → they branch out from the aortic arch independently*

The **internal carotid arteries** are direct continuations of the common carotids superior to the origin of the external carotid artery, at the level of the superior border of the thyroid cartilage. The proximal part of each internal carotid artery is the site of the *carotid sinus,* discussed earlier on p. 1000 (Figs. 8.17 and 8.19). The *carotid body* is located in the cleft between the internal and the external carotid arteries. The internal carotid arteries enter the cranium through the *carotid canals* in the petrous parts of the temporal bones and become the main arteries of the brain and structures in the orbits (see Chapter 7). No named branches arise from the internal carotid arteries in the neck.

The **external carotid arteries** supply most structures external to the cranium; the orbit and the part of the forehead

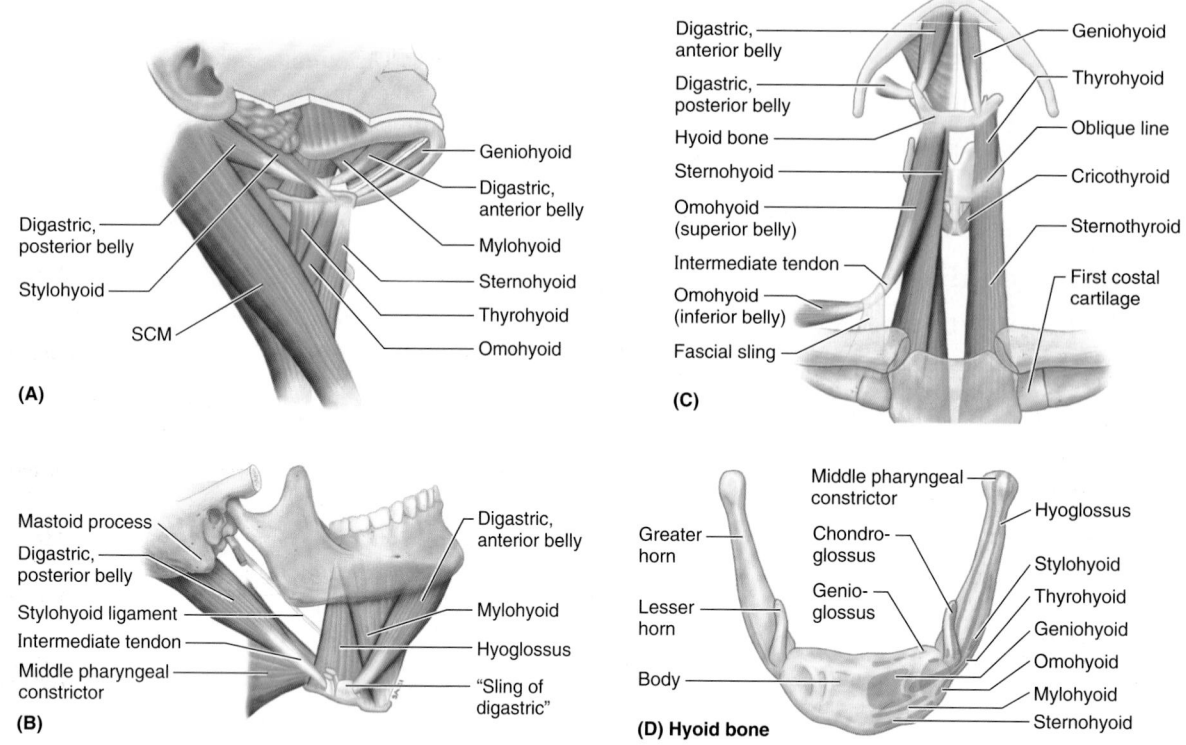

FIGURE 8.18. Muscles of anterior cervical region.

TABLE 8.3. MUSCLES OF ANTERIOR CERVICAL REGION (EXTRINSIC MUSCLES OF LARYNX)

Muscle	Origin	Insertion	Innervation	Main Action(s)
Suprahyoid muscles				
Mylohyoid	Mylohyoid line of mandible	Mylohyoid raphe and body of hyoid	Nerve to mylohyoid, a branch of inferior alveolar nerve (from mandibular nerve, CN V₃)	Elevates hyoid, floor of mouth, and tongue during swallowing and speaking
Geniohyoid	Inferior mental spine of mandible	Body of hyoid	C1 via hypoglossal nerve (CN XII)	Pulls hyoid anterosuperiorly; shortens floor of mouth; widens pharynx
Stylohyoid	Styloid process of temporal bone		Stylohyoid (preparotid) branch of facial nerve (CN VII)	Elevates and retracts hyoid, thus elongating floor of mouth
Digastric	*Anterior belly:* digastric fossa of mandible	Intermediate tendon to body and greater horn of hyoid	Nerve to mylohyoid, a branch of inferior alveolar nerve	Working with infrahyoid muscles, depresses mandible against resistance; elevates and steadies hyoid during swallowing and speaking
	Posterior belly: mastoid notch of temporal bone		Digastric (preparotid) branch of facial nerve (CN VII)	
Infrahyoid muscles				
Sternohyoid	Manubrium of sternum and medial end of clavicle	Body of hyoid	C1–C3 by a branch of ansa cervicalis	Depresses hyoid after elevation during swallowing
Omohyoid	Superior border of scapula near suprascapular notch	Inferior border of hyoid		Depresses, retracts, and steadies hyoid
Sternothyroid	Posterior surface of manubrium of sternum	Oblique line of thyroid cartilage	C2 and C3 by a branch of ansa cervicalis	Depresses hyoid and larynx
Thyrohyoid	Oblique line of thyroid cartilage	Inferior border of body and greater horn of hyoid	C1 via hypoglossal nerve (CN XII)	Depresses hyoid and elevates larynx

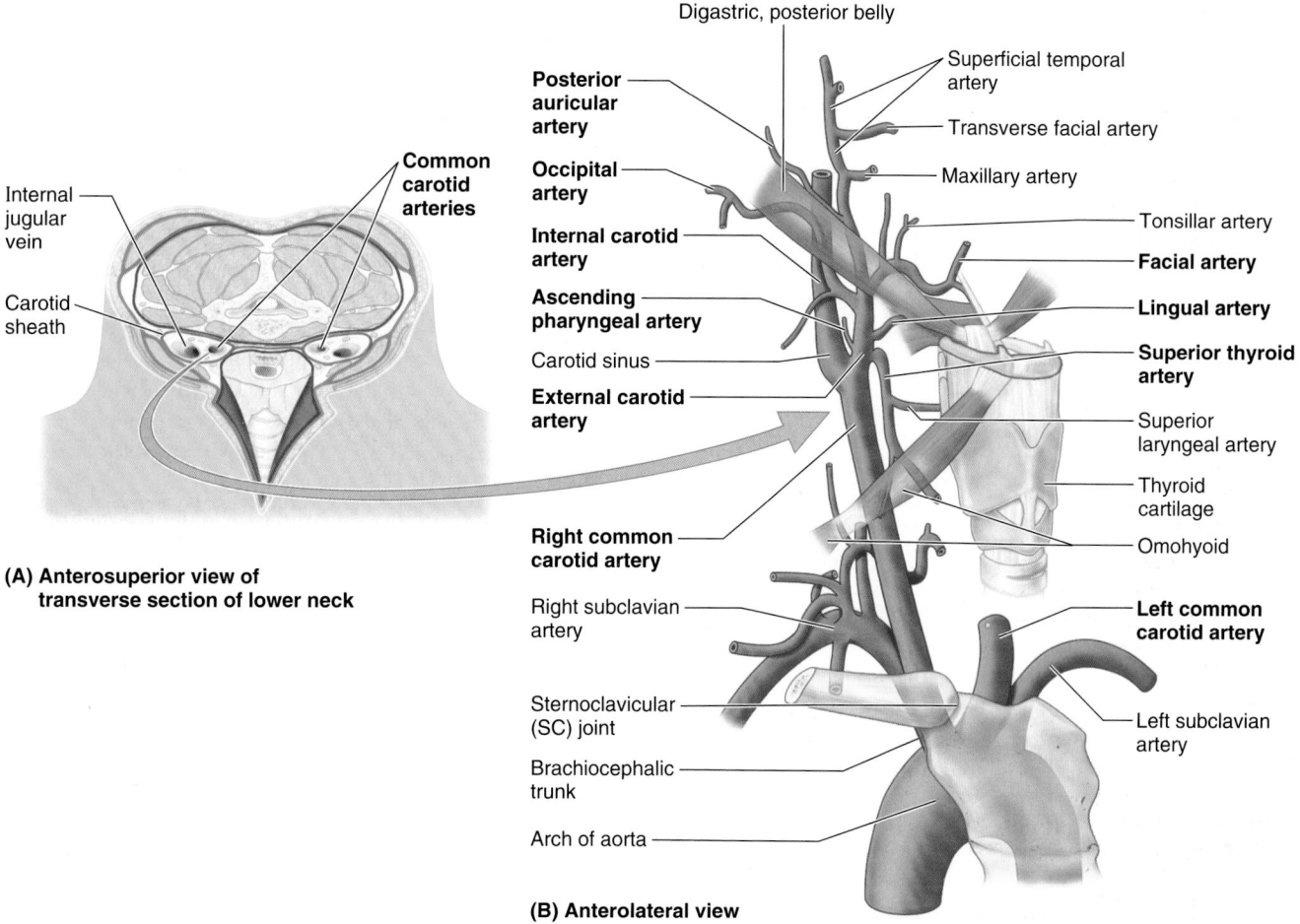

FIGURE 8.19. Subclavian and carotid arteries and their branches. A. The positions of the carotid sheaths are demonstrated. **B.** The muscles (posterior belly of the digastric and omohyoid muscles) indicate the superior and inferior boundaries of the carotid triangle.

and scalp supplied by the supra-orbital artery are the major exceptions. There is some deep distribution as well (e.g., via the middle meningeal artery). Each external carotid artery runs posterosuperiorly to the region between the neck of the mandible and the lobule of the auricle, where it is embedded in the parotid gland, and terminates by dividing into two branches, the *maxillary artery* and the *superficial temporal artery* (Fig. 8.19). Before these terminal branches, six arteries arise from the external carotid artery:

1. **Ascending pharyngeal artery:** arises as the first or second branch of the external carotid artery and is its only medial branch. It ascends on the pharynx deep (medial) to the internal carotid artery and sends branches to the pharynx, prevertebral muscles, middle ear, and cranial meninges.

2. **Occipital artery:** arises from the posterior aspect of the external carotid artery, superior to the origin of the facial artery. It passes posteriorly, immediately medial and parallel to the attachment of the posterior belly of the digastric muscle in the **occipital groove** in the temporal bone, and ends by dividing into numerous branches in the posterior part of the scalp. During its course, it passes superficial to the internal carotid artery and CN IX–CN XI.

3. **Posterior auricular artery:** a small posterior branch of the external carotid artery, which is usually the last preterminal branch. It ascends posteriorly between the external acoustic meatus and mastoid process to supply the adjacent muscles; parotid gland; facial nerve; and structures in the temporal bone, auricle, and scalp.

4. **Superior thyroid artery:** the most inferior of the three anterior branches of the external carotid artery, runs antero-inferiorly deep to the infrahyoid muscles to reach the thyroid gland. In addition to supplying this gland, it gives off branches to the infrahyoid muscles and SCM and gives rise to the *superior laryngeal artery,* supplying the larynx.

5. **Lingual artery:** arises from the anterior aspect of the external carotid artery, where it lies on the middle pharyngeal constrictor. It arches supero-anteriorly and passes deep to the hypoglossal nerve (CN XII), the stylohyoid muscle, and the posterior belly of the digastric muscle. It disappears deep to the hyoglossus muscle, giving branches to the posterior tongue. It then turns superiorly at the anterior border of this muscle, bifurcating into the deep lingual and sublingual arteries.

6. **Facial artery:** arises anteriorly from the external carotid artery, either in common with the lingual artery or immediately superior to it (Figs. 8.15 and 8.19). After giving rise to the *ascending palatine artery* and a *tonsillar artery*, the facial artery passes superiorly under cover of the digastric and stylohyoid muscles and the angle of the mandible. It loops anteriorly and enters a deep groove in and supplies the submandibular gland. It then gives rise to the *submental artery* to the floor of the mouth and hooks around the middle of the inferior border of the mandible to enter the face.

Memory device for the six branches of the carotid artery: 1-2-3—one branch arises medially (ascending pharyngeal), two branches arise posteriorly (occipital and posterior auricular), and three branches arise anteriorly (superior thyroid, lingual, and facial).

VEINS IN ANTERIOR CERVICAL REGION

Most veins in the anterior cervical region are tributaries of the IJV, typically the largest vein in the neck (Figs. 8.15 and 8.20). The IJV drains blood from the brain, anterior face, cervical viscera, and deep muscles of the neck. It commences at the *jugular foramen* in the posterior cranial fossa as the direct continuation of the sigmoid sinus (see Chapter 7).

From a dilation at its origin, the **superior bulb of the IJV,** the vein descends in the *carotid sheath* (Fig. 8.19A), accompanying the internal carotid artery superior to the carotid bifurcation and the common carotid artery and vagus nerve inferiorly (Fig. 8.20). The vein lies laterally within the carotid sheath, with the nerve located posteriorly.

The *cervical sympathetic trunk* lies posterior to the carotid sheath. Although closely related, the trunk is not within the sheath; instead, it is embedded in the prevertebral layer of deep cervical fascia. The IJV leaves the anterior cervical region by passing deep to the SCM. The inferior end of the vein passes deep to the gap between the sternal and clavicular heads of this muscle. Posterior to the sternal end of the clavicle, the IJV merges with the subclavian vein to form the *brachiocephalic vein.* The inferior end of the IJV dilates to form the **inferior bulb of the IJV.** This bulb has a bicuspid valve that permits blood to flow toward the heart while preventing backflow into the vein, as might occur if inverted (e.g., standing on one's head or when intrathoracic pressure is increased).

The tributaries of the IJV are the inferior petrosal sinus and the facial and lingual (often by a common trunk), pharyngeal, and superior and middle thyroid veins. The **occipital vein** usually drains into the *suboccipital venous plexus,* drained by the deep cervical vein and the vertebral vein, but it may drain into the IJV.

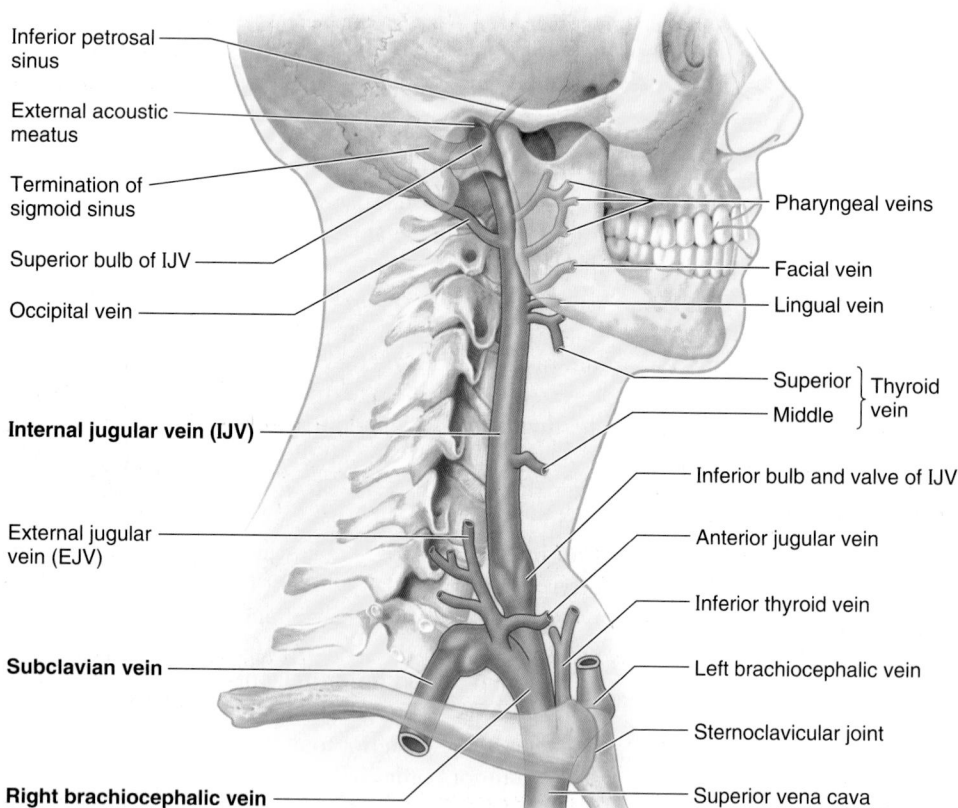

Inferior petrosal sinus
External acoustic meatus
Termination of sigmoid sinus
Superior bulb of IJV
Occipital vein
Internal jugular vein (IJV)
External jugular vein (EJV)
Subclavian vein
Right brachiocephalic vein

Pharyngeal veins
Facial vein
Lingual vein
Superior ⎱ Thyroid
Middle ⎰ vein
Inferior bulb and valve of IJV
Anterior jugular vein
Inferior thyroid vein
Left brachiocephalic vein
Sternoclavicular joint
Superior vena cava

FIGURE 8.20. Internal jugular vein. The IJV is the main venous structure in the neck. It originates as a continuation of the S-shaped sigmoid (dural venous) sinus. As it descends in the neck, it is contained in the carotid sheath. It terminates at the T1 vertebral level, superior to the sternoclavicular joint, by uniting with the subclavian vein to form the brachiocephalic vein. A large valve near its termination prevents reflux of blood into the vein.

The **inferior petrosal sinus** leaves the cranium through the jugular foramen and enters the superior bulb of the IJV (Fig. 8.20). The *facial vein* empties into the IJV opposite or just inferior to the level of the hyoid. The facial vein may receive the superior thyroid, lingual, or sublingual veins. The *lingual veins* form a single vein from the tongue, which empties into the IJV at the level of origin of the lingual artery. The *pharyngeal veins* arise from the venous plexus on the pharyngeal wall and empty into the IJV approximately at the level of the angle of the mandible. The superior and middle thyroid veins leave the thyroid gland and drain into the IJV.

NERVES IN ANTERIOR CERVICAL REGION

Several nerves, including branches of cranial nerves, are located in the anterior cervical region.

- **Transverse cervical nerve** (C2 and C3): supplies the skin covering the anterior cervical region. This nerve was discussed with the cervical plexus earlier in this chapter (Figs. 8.8 and 8.13).
- **Hypoglossal nerve** (CN XII): the motor nerve of the tongue, enters the submandibular triangle deep to the posterior belly of the digastric muscle to supply the intrinsic and four of the five extrinsic muscles of the tongue (Figs. 8.13A, 8.15, and 8.21). The nerve passes between the external carotid and jugular vessels and gives off the superior root of the ansa cervicalis and then a branch to the geniohyoid muscle (Fig. 8.13). In both cases, the branch conveys only fibers from the C1 spinal nerve, which joined its proximal part; no hypoglossal fibers are conveyed in these branches (see Chapter 9 for details).
- Branches of the **glossopharyngeal** (CN IX) and **vagus** (CN X) nerves: in the submandibular and carotid triangles

(Figs. 8.15 and 8.21). CN IX is primarily related to the tongue and pharynx. In the neck, CN X gives rise to pharyngeal, laryngeal, and cardiac branches.

Surface Anatomy of Cervical Regions and Triangles of Neck

The skin of the neck is thin and pliable. The subcutaneous tissue contains the *platysma,* a thin sheet of striated muscle that ascends to the face (Figs. 8.5 and 8.22A). Its fibers can be observed, especially in thin people, by asking them to contract the platysma muscles (e.g., by pretending to ease a tight collar).

The *SCM* is the key muscular landmark of the neck. It defines the *sternocleidomastoid region* and divides the neck into anterior and lateral cervical regions (Fig. 8.22B). This broad bulging muscle is easy to observe and palpate throughout its length as it passes superolaterally from the *sternum* and *clavicle.* Its superior attachment to the *mastoid process* is palpable posterior to the lobule of the auricle. The SCM can be made to stand out by asking the person to rotate the face toward the contralateral side and elevate the chin. In this contracted state, the anterior and posterior borders of the muscle are clearly defined.

The *jugular notch* of the manubrium forms the inferior boundary of the fossa between the sternal heads of the SCMs (Fig. 8.22C & D). The *suprasternal space* and *jugular venous arch* are located superior to this notch (Fig. 8.16). The *lesser supraclavicular fossa,* between the sternal and the clavicular heads of the SCM, overlies the inferior end of the IJV (Fig. 8.22B & D). It can be entered here by a needle or catheter (see "Internal Jugular Vein Puncture," p. 1011).

The *EJV* runs vertically across the SCM toward the *angle of the mandible* (Fig. 8.22D). It may be prominent, especially

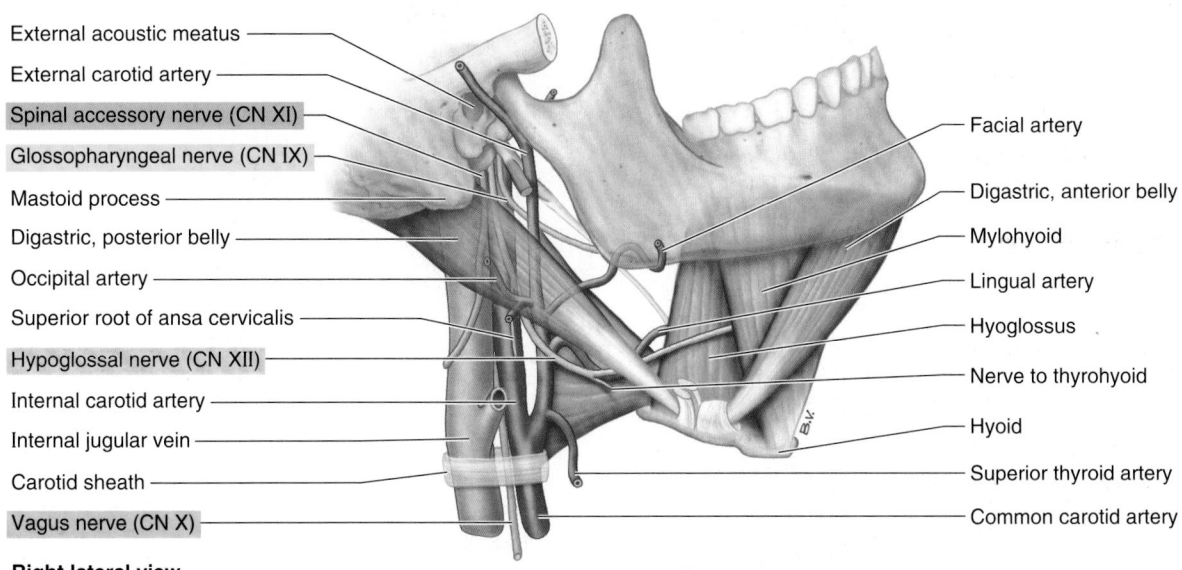

External acoustic meatus

External carotid artery

Spinal accessory nerve (CN XI)

Glossopharyngeal nerve (CN IX)

Mastoid process

Digastric, posterior belly

Occipital artery

Superior root of ansa cervicalis

Hypoglossal nerve (CN XII)

Internal carotid artery

Internal jugular vein

Carotid sheath

Vagus nerve (CN X)

Facial artery

Digastric, anterior belly

Mylohyoid

Lingual artery

Hyoglossus

Nerve to thyrohyoid

Hyoid

Superior thyroid artery

Common carotid artery

Right lateral view

FIGURE 8.21. **Relationships of nerves and vessels to suprahyoid muscles of anterior cervical region.** The posterior belly of the digastric muscle, running from the mastoid process to the hyoid, holds a superficial and key position in the neck.

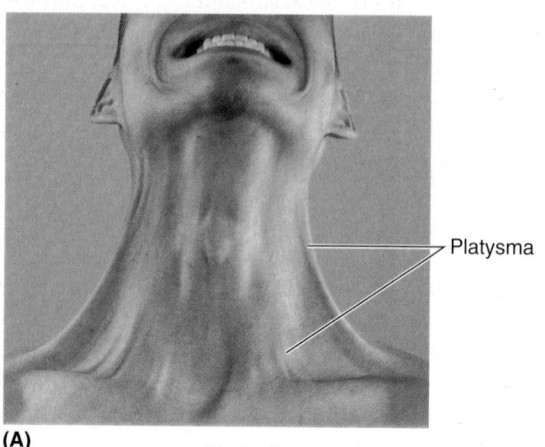

Platysma

(A)

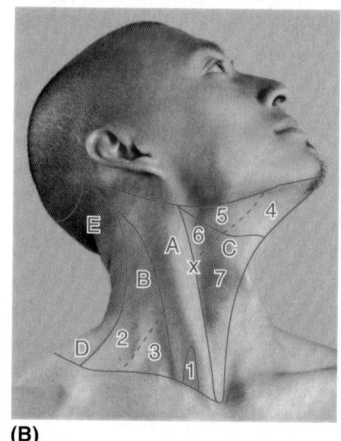

Key for (B):

A Sternocleidomastoid region
B Lateral cervical region
C Anterior cervical region
D Posterior cervical region
E Suboccipital region
1 Lesser supraclavicular fossa
2 Occipital triangle
3 Omoclavicular triangle
4 Submental triangle
5 Submandibular triangle
6 Carotid triangle
7 Muscular triangle
X Carotid bifurcation
   (take pulse inferior to
   this point)

(B)

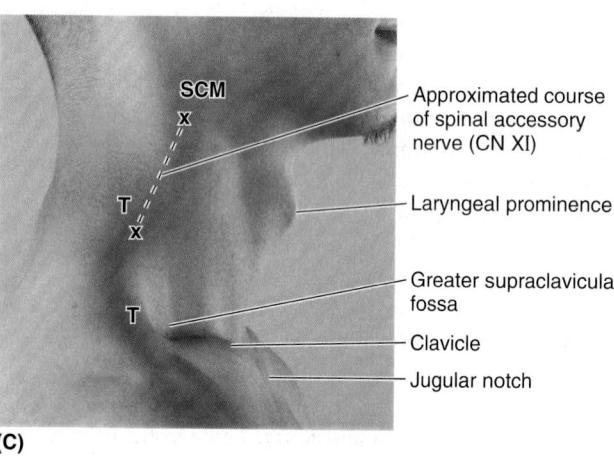

SCM
x

T
x

T

Approximated course
of spinal accessory
nerve (CN XI)

Laryngeal prominence

Greater supraclavicular
fossa

Clavicle

Jugular notch

(C)

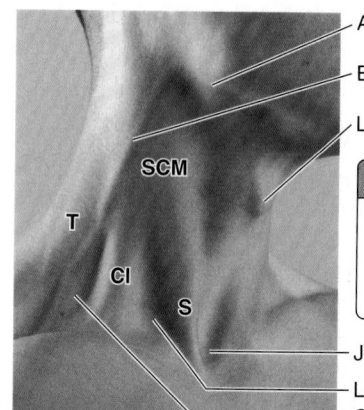

Angle of mandible

External jugular vein

Laryngeal prominence

Key for (C) & (D):

T   Trapezius (ant. border)
SCM Sternocleidomastoid
Cl   Clavicular head
S   Sternal head

SCM

T

Cl

S

Jugular notch

Lesser ⎫
Greater ⎭ Supraclavicular fossae

(D)

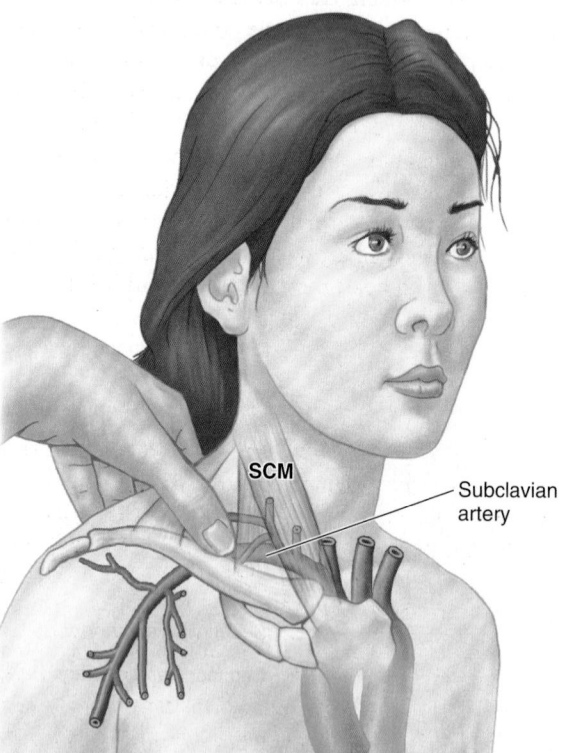

SCM

Subclavian
artery

(E)

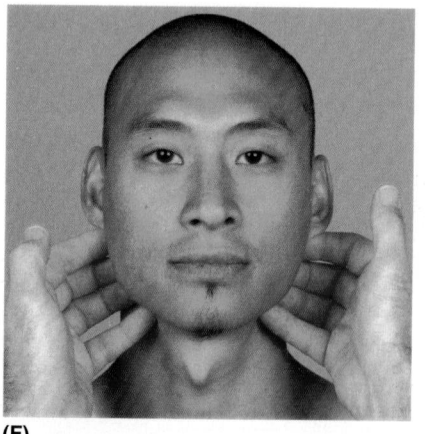

(F)

FIGURE 8.22. Surface anatomy of neck. A. Contraction of the platysma.
B. The regions (A–E) and triangles (2–7) of the neck. **C.** The course of the
spinal accessory nerve (CN XI). **D.** The landmarks of the anterolateral neck.
E. Checking the subclavian artery pulse. **F.** Palpation of the submandibular
lymph.

if distended by asking the person to take a deep breath and hold it, expiring against resistance (*Valsalva maneuver*) or by using gentle pressure on the inferior part of the vein. These actions impede venous return to the right side of the heart. The EJV is less obvious in children and middle-aged women because their subcutaneous tissues tend to be thicker than those in men.

The great auricular nerve parallels the EJV, approximately a finger's breadth posterior to the vein. Deep to the superior half of the SCM is the cervical plexus, and deep to the inferior half of the SCM are the IJV, common carotid artery, and vagus nerve in the carotid sheath (Fig. 8.21).

The *trapezius,* which defines the *posterior cervical region,* can be observed and palpated by asking the person to shrug the shoulders against resistance (Fig. 8.22B–D). Superiorly, inferior to its attachment to the *external occipital protuberance,* the muscle overlies the *suboccipital region* (see Fig. 4.37).

The inferior belly of the *omohyoid muscle* can just barely be seen and palpated as it passes superomedially across the inferior part of the *lateral cervical region.* Easiest to observe in thin people, the omohyoid muscle can often be seen contracting when they are speaking.

Just inferior to the inferior belly of the omohyoid is the *greater supraclavicular fossa,* the depression overlying the *omoclavicular triangle* (Fig. 8.22C & D). The third part of the subclavian artery passes through this triangle before coursing posterior to the clavicle and across the 1st rib. The greater supraclavicular fossa is clinically important because **subclavian arterial pulsations** can be palpated here in most people. The course of the *subclavian artery* in the neck is represented by a curved line from the *sternoclavicular (SC) joint* to the midpoint of the clavicle. To feel subclavian pulsations, press inferoposteriorly (down and back) immediately posterior to the junction of the medial and middle thirds of the clavicle (Fig. 8.22E). This is the pressure point for the subclavian artery; firmer pressure, compressing the artery against the 1st rib, can occlude the artery when hemorrhage is occurring distally in the upper limb.

The chief contents of the larger *occipital triangle,* superior to the omohyoid muscle, are the spinal accessory nerve (CN XI); cutaneous branches of cervical nerves C2, C3, and C4; and cervical lymph nodes. Because of the vulnerability and frequency of iatrogenic injury of the spinal accessory nerve, it is important to be able to estimate the location of CN XI in the lateral cervical region. Its course can be approximated by a line that intersects the junction of the superior and middle thirds of the posterior border of the SCM and the junction of the middle and lower thirds of the anterior border of the trapezius (Fig. 8.22C).

The cervical viscera and carotid arteries and their branches are approached surgically through the *anterior cervical region,* between the anterior border of the SCM and the midline (Fig. 8.22B). Of the four smaller triangles into which this region is subdivided, the submandibular and carotid triangles are especially important clinically.

The *submandibular gland* nearly fills the *submandibular triangle.* It is palpable as a soft mass inferior to the body of the mandible, especially when the apex of the tongue is forced against the maxillary incisor teeth. The *submandibular lymph nodes* lie superficial to the gland (Fig. 8.14A). These nodes receive lymph from the face inferior to the eye and from the mouth. If enlarged, these nodes can be palpated by moving the fingertips from the angle of the mandible along its inferior border (Figs. 8.22D and F). If continued until the fingers meet under the chin, enlarged *submental lymph nodes* can be palpated in the *submental triangle* (Fig. 8.22B).

The carotid arterial system is located in the *carotid triangle.* This area is important for surgical approaches to the carotid sheath containing the common carotid artery, IJV, and vagus nerve (Figs. 8.15 and 8.21). The carotid triangle also contains the hypoglossal nerve (CN XII) and cervical sympathetic trunk. The *carotid sheath* can be marked out by a line joining the SC joint to a point midway between the mastoid process and the angle of the mandible. The **carotid pulse** can be palpated by placing the index and 3rd fingers on the thyroid cartilage and pointing them posterolaterally between the trachea and SCM. The pulse is palpable just medial to the SCM. The palpation is performed low in the neck to avoid pressure on the *carotid sinus,* which could cause a reflex drop in blood pressure and heart rate (Figs. 8.17 and 8.22B).

SUPERFICIAL STRUCTURES OF NECK: CERVICAL REGIONS

Congenital Torticollis

Torticollis (L. *tortus,* twisted + L. *collum,* neck) is a contraction or shortening of the cervical muscles that produces twisting of the neck and slanting of the head. The most common type of torticollis (wry neck) results from a *fibrous tissue tumor* (L. *fibromatosis colli*) that develops in the SCM before or shortly after birth. The lesion, like a normal unilateral SCM contraction, causes the head to tilt toward, and the face to turn away from, the affected side (Fig. B8.1). When torticollis occurs prenatally, the abnormal position of the infant's head usually necessitates a breech delivery.

Occasionally, the SCM is injured when an infant's head is pulled too much during a difficult birth, tearing its fibers (*muscular torticollis*) (Kliegman et al., 2011). A *hematoma* (localized mass of extravasated blood) occurs that may develop into a fibrotic mass that entraps a branch of the spinal accessory nerve (CN XI) and thus denervates part of the SCM. The stiffness and twisting of the neck results from fibrosis and shortening of the SCM. Surgical release of the SCM from its inferior attachments to the manubrium and

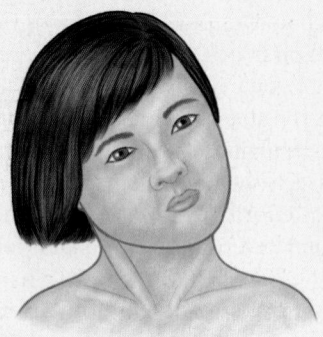

FIGURE B8.1.

clavicle inferior to the level of CN XI may be necessary to enable the person to hold and rotate the head normally.

Spasmodic Torticollis

Cervical dystonia (abnormal tonicity of the cervical muscles), commonly known as *spasmodic torticollis*, usually begins in adulthood. It may involve any bilateral combination of lateral neck muscles, especially the SCM and trapezius. Characteristics of this disorder are sustained turning, tilting, flexing, or extending of the neck. Shifting the head laterally or anteriorly can occur involuntarily (Fahn et al., 2010). The shoulder is usually elevated and displaced anteriorly on the side to which the chin turns.

Subclavian Vein Puncture

The right or left subclavian vein is often the point of entry to the venous system for *central line placement*, such as a Swan-Ganz catheter. Central lines are inserted to administer parenteral (venous nutritional) fluids

and medications and to measure central venous pressure. In an infraclavicular subclavian vein approach, the administrator places the thumb of one hand on the middle part of the clavicle and the index finger on the jugular notch in the manubrium (Fig. B8.2). The needle punctures the skin inferior to the thumb (middle of the clavicle) and is advanced medially toward the tip of the index finger (jugular notch) until the tip enters the right venous angle, posterior to the sternoclavicular joint. Here the internal jugular and subclavian veins merge to form the brachiocephalic vein. If the needle is not inserted carefully, it may puncture the pleura and lung, resulting in *pneumothorax*. Furthermore, if the needle is inserted too far posteriorly, it may enter the subclavian artery. When the needle has been inserted correctly, a soft, flexible catheter is inserted into the subclavian vein, using the needle as a guide.

Right Cardiac Catheterization

For *right cardiac catheterization* (to take measurements of pressures in the right chambers of the heart), puncture of the IJV can be used to introduce a catheter through the right brachiocephalic vein into the superior vena cava (SVC) and the right side of the heart. Although the preferred route is through the IJV or subclavian vein, it may be necessary in some patients to use the EJV. This vein is not ideal for catheterization because its angle of junction with the subclavian vein makes passage of the catheter difficult.

Prominence of External Jugular Vein

The EJV may serve as an "internal barometer." When venous pressure is in the normal range, the EJV is usually visible above the clavicle for only a

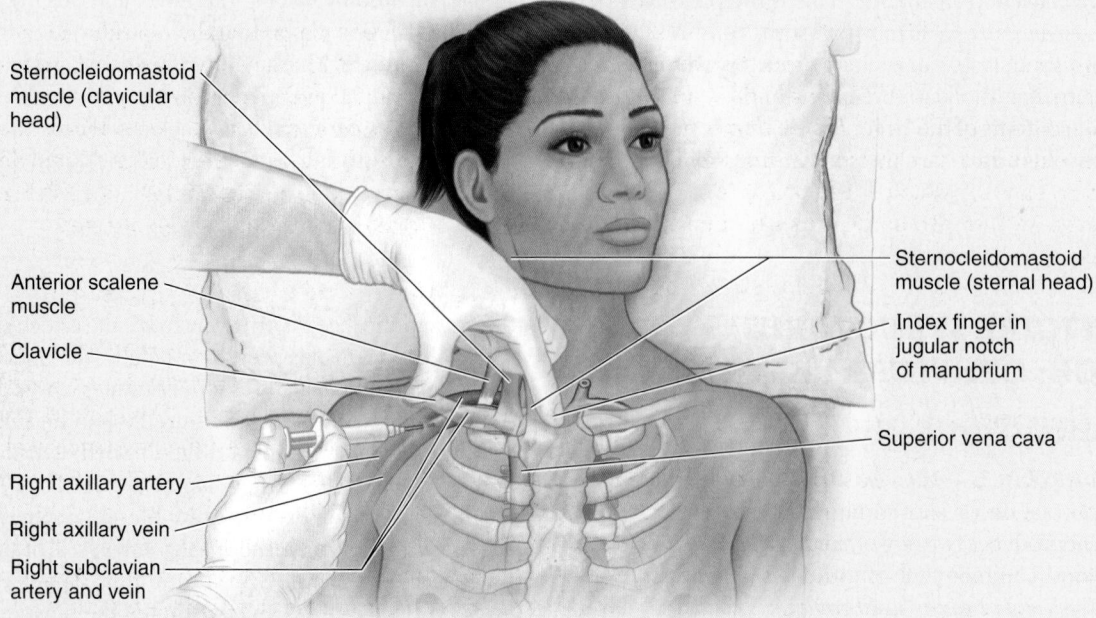

Sternocleidomastoid muscle (clavicular head)

Anterior scalene muscle

Clavicle

Right axillary artery

Right axillary vein

Right subclavian artery and vein

Sternocleidomastoid muscle (sternal head)

Index finger in jugular notch of manubrium

Superior vena cava

FIGURE B8.2. Subclavian vein puncture.

short distance. However, when venous pressure rises (e.g., as in heart failure), the vein is prominent throughout its course along the side of the neck. Consequently, routine observation of the EJVs during physical examinations may give diagnostic signs of heart failure, SVC obstruction, enlarged supraclavicular lymph nodes, or increased intrathoracic pressure.

Severance of External Jugular Vein

 If the *EJV is severed* along the posterior border of the SCM, where it pierces the roof of the lateral cervical region (e.g., by a knife slash), its lumen is held open by the tough investing layer of deep cervical fascia, and the negative intrathoracic pressure air will suck air into the vein. This action produces a churning noise in the thorax and cyanosis (a bluish discoloration of the skin and mucous membranes resulting from an excessive concentration of reduced hemoglobin in the blood). A *venous air embolism* produced in this way will fill the right side of the heart with froth, which nearly stops blood flow through it, resulting in *dyspnea* (shortness of breath). The application of firm pressure to the severed jugular vein until it can be sutured will stop the bleeding and entry of air into the blood.

Lesions of Spinal Accessory Nerve (CN XI)

Lesions of the spinal accessory nerve (CN XI) are uncommon. This nerve may be damaged by:

- Penetrating trauma, such as a stab or bullet wound.
- Surgical procedures in the lateral cervical region.
- Tumors at the cranial base or cancerous cervical lymph nodes.
- Fractures of the jugular foramen, where CN XI leaves the cranium.

Although contraction of one SCM turns the head to one side, a unilateral lesion of CN XI usually does not produce an abnormal position of the head. However, people with CN XI damage usually have weakness in turning the head to the opposite side against resistance. Lesions of the CN XI produce weakness and atrophy of the trapezius, impairing neck movements.

Unilateral paralysis of the trapezius is evident by the patient's inability to elevate and retract the shoulder and by difficulty in elevating the upper limb superior to the horizontal level. The normal prominence in the neck produced by the trapezius is also reduced. *Drooping of the shoulder* is an obvious sign of CN XI injury. During extensive surgical dissections in the lateral cervical region—for example, during removal of cancerous lymph nodes—the surgeon isolates CN XI to preserve it, if possible. An awareness of the superficial location of this nerve during superficial procedures in the lateral cervical region is important because CN XI is the most commonly iatrogenic nerve injury (G. *iatros,* physician or surgeon).

Severance of Phrenic Nerve, Phrenic Nerve Block, and Phrenic Nerve Crush

 Severance of a phrenic nerve results in paralysis of the corresponding half of the diaphragm (see the blue box "Paralysis of the Diaphragm" on p. 85). A *phrenic nerve block* produces a short period of paralysis of the diaphragm on one side (e.g., for a lung operation). The anesthetic is injected around the nerve where it lies on the anterior surface of the middle third of the anterior scalene muscle. A surgical *phrenic nerve crush* (e.g., compressing the nerve injuriously with forceps) produces a longer period of paralysis (sometimes for weeks after surgical repair of a diaphragmatic hernia). If an accessory phrenic nerve is present, it must also be crushed to produce complete paralysis of the hemidiaphragm.

Nerve Blocks in Lateral Cervical Region

 For regional anesthesia before neck surgery, a *cervical plexus block* inhibits nerve impulse conduction. The anesthetic agent is injected at several points along the posterior border of the SCM, mainly at the junction of its superior and middle thirds, the *nerve point of the neck* (Figs. 8.8 and 8.13A). Because the phrenic nerve supplying half the diaphragm is usually paralyzed by a cervical nerve block, this procedure is not performed on persons with pulmonary or cardiac disease. For anesthesia of the upper limb, the anesthetic agent in a *supraclavicular brachial plexus block* is injected around the supraclavicular part of the brachial plexus. The main injection site is superior to the midpoint of the clavicle.

Injury to Suprascapular Nerve

 The suprascapular nerve is vulnerable to injury in fractures of the middle third of the clavicle. *Injury to the suprascapular nerve* results in loss of lateral rotation of the humerus at the glenohumeral joint. Consequently the relaxed limb rotates medially into the *waiter's tip position* (see Fig. B6.12B, p. 729). The ability to initiate abduction of the limb is also affected.

Ligation of External Carotid Artery

 Ligation of an external carotid artery is sometimes necessary to control bleeding from one of its relatively inaccessible branches. This procedure decreases blood flow through the artery and its branches but does not eliminate it. Blood flows in a retrograde (backward) direction into the artery from the external carotid artery on the other side through communications between its branches (e.g., those in the face and scalp) and across the midline. When the external carotid or subclavian arteries are ligated, the descending branch of the occipital artery provides the main collateral circulation, anastomosing with the vertebral and deep cervical arteries.

Surgical Dissection of Carotid Triangle

The carotid triangle provides an important surgical approach to the carotid system of arteries. It also provides access to the IJV, the vagus and hypoglossal nerves, and the cervical sympathetic trunk. Damage or compression of the vagus and/or recurrent laryngeal nerves during *surgical dissection of the carotid triangle* may produce an alteration in the voice because these nerves supply laryngeal muscles.

Carotid Occlusion and Endarterectomy

Atherosclerotic thickening of the intima of the internal carotid artery may obstruct blood flow. Symptoms resulting from this obstruction depend on the degree of obstruction and the amount of collateral blood flow to the brain and structures in the orbit from other arteries. A partial occlusion of the internal carotid may cause a *transient ischemic attack* (TIA), a sudden focal loss of neurological function (e.g., dizziness and disorientation) that disappears within 24 hr. Arterial occlusion may also cause a *minor stroke*, a loss of neurological function such as weakness or sensory loss on one side of the body that exceeds 24 hr but disappears within 3 weeks.

Obstruction of blood flow can be observed in a *Doppler color study* (Fig. B8.3A). A Doppler is a diagnostic instrument that emits an ultrasonic beam and detects its reflection from moving fluid (blood) in a manner that distinguishes the fluid from the static surrounding tissue, providing information about its pressure, velocity, and turbulence. *Carotid occlusion*, causing stenosis (narrowing) in otherwise healthy persons (Fig. B8.3B), can be relieved by opening the artery at its origin and stripping off the atherosclerotic plaque with the intima. This procedure is called *carotid endarterectomy.* After the operation, drugs that inhibit clot formation are administered until the endothelium has regrown. Because of the relations of the internal carotid artery, there is risk of cranial nerve injury during the procedure involving one or more of the following nerves: CN IX, CN X (or its branch, the superior laryngeal nerve), CN XI, or CN XII (Fig. 8.21).

Carotid Pulse

The *carotid pulse* ("neck pulse") is easily felt by palpating the common carotid artery in the side of the neck, where it lies in a groove between the trachea and the infrahyoid muscles (Fig. 8.15). It is usually easily palpated just deep to the anterior border of the SCM at the level of the superior border of the thyroid cartilage. It is routinely checked during *cardiopulmonary resuscitation* (CPR). Absence of a carotid pulse indicates cardiac arrest.

Carotid Sinus Hypersensitivity

In people with *carotid sinus hypersensitivity* (exceptional responsiveness of the carotid sinuses in various types of vascular disease), external pressure on the carotid artery may cause slowing of the heart rate, a fall in

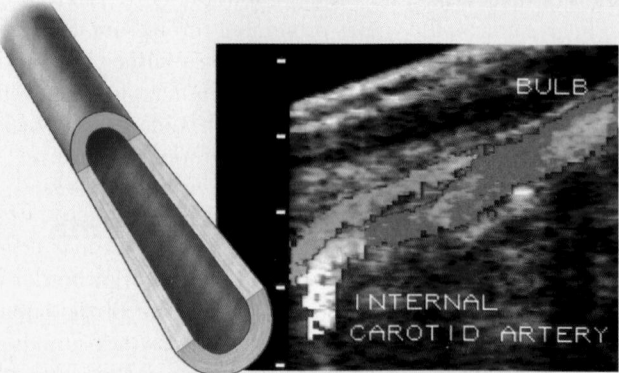

(A) Doppler color flow study of normal internal carotid artery

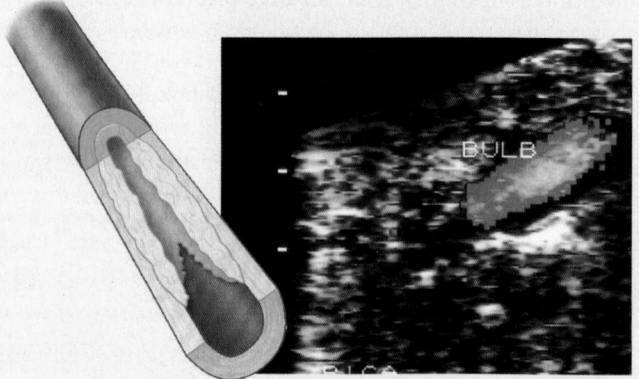

(B) Doppler color flow study of occluded carotid artery

FIGURE B8.3.

blood pressure, and cardiac ischemia resulting in fainting (*syncope*). In all forms of syncope, symptoms result from a sudden and critical decrease in cerebral perfusion (Hirsch et al, 2010). Consequently, this method of checking the pulse is not recommended for people with cardiac or vascular disease. Alternate sites, such as the radial artery at the wrist, should be used to check pulse rate in people with carotid sinus hypersensitivity.

Role of Carotid Bodies

The *carotid bodies* are in an ideal position to monitor the oxygen content of blood before it reaches the brain (Fig. 8.17). A decrease in Po_2 (partial pressure of oxygen), as occurs at high altitudes or in pulmonary disease, activates the aortic and carotid chemoreceptors, increasing alveolar ventilation. The carotid bodies also respond to increased carbon dioxide (CO_2) tension or free hydrogen ions in the blood. The glossopharyngeal nerve (CN IX, perhaps with involvement of the vagus nerve) conducts the information centrally, resulting in reflexive stimulation of the respiratory centers of the brain that increase the depth and rate of breathing. The pulse rate and blood pressure also increase. With increased ventilation and circulation, more oxygen is taken in and the concentration of CO_2 is reduced accordingly.

Internal Jugular Pulse

Although pulsations are most commonly associated with arteries, *pulsations of the (IJV)* can provide information about heart activity corresponding to electrocardiogram (ECG) recordings and right atrial pressure. The IJV pulse is not palpable in the same manner as arterial pulses; however, the vein's pulsations are transmitted through the surrounding tissue and may be observed beneath the SCM superior to the medial end of the clavicle.

Because there are no valves in the brachiocephalic vein or the superior vena cava, a wave of contraction passes up these vessels to the inferior bulb of the IJV. The pulsations are especially visible when the person's head is inferior to the lower limbs (*Trendelenburg position*). The internal jugular pulse increases considerably in conditions such as mitral valve disease (see Chapter 1), which increases pressure in the pulmonary circulation and right side of the heart. The right IJV runs a straighter, more direct course to the right atrium than does the left; thus it is the one that is examined (Swartz, 2009).

Internal Jugular Vein Puncture

A needle and catheter may be inserted into the IJV for diagnostic or therapeutic purposes. The right IJV is preferable because it is usually larger and straighter. During this procedure, the clinician palpates the common carotid artery and inserts the needle into the IJV just lateral to it at a 30° angle, aiming at the apex of the triangle between the sternal and clavicular heads of the SCM, the lesser supraclavicular fossa (Fig. B8.4). The needle is then directed inferolaterally toward the ipsilateral nipple.

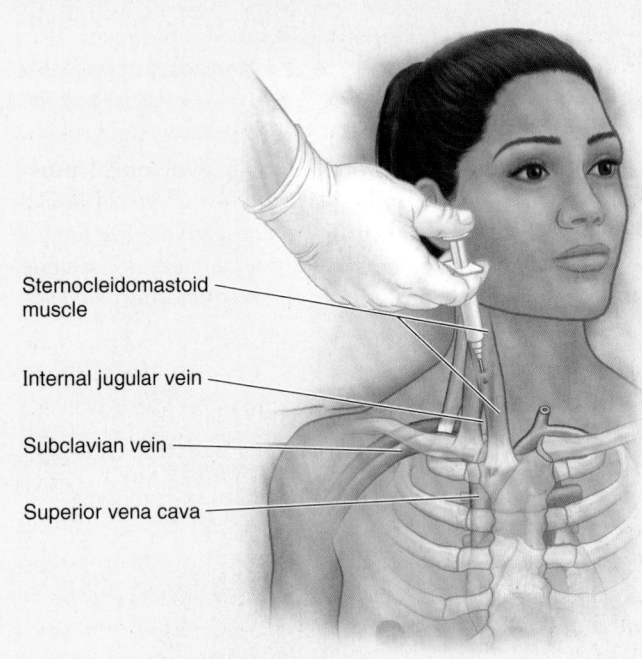

Sternocleidomastoid muscle

Internal jugular vein

Subclavian vein

Superior vena cava

FIGURE B8.4. **Internal jugular vein puncture.**

The Bottom Line

SUPERFICIAL STRUCTURES OF NECK: CERVICAL REGIONS

Sternocleidomastoid (SCM) and trapezius: The SCM and trapezius muscles share their origins from a common embryologic source, innervation by the spinal accessory nerve (CN XI), enclosure by the investing layer of deep cervical fascia, a linear superior attachment to the cranial base, and an inferior attachment to the pectoral girdle. ◆ Their superficial masses and palpable borders provide the basis for describing the regions of the neck. ◆ The SCM produces multiple movements of the head and neck. ◆ The trapezius causes multiple movements of the scapula, depending on whether the muscles act unilaterally or bilaterally, and independently or in conjunction with concentric or eccentric contraction of other muscles.

Lateral cervical region: The lateral cervical region is bounded by the SCM, trapezius, and middle third of the clavicle, with a muscular floor formed by the lateral deep cervical muscles. ◆ It is subdivided by the diagonally placed inferior belly of the omohyoid. ◆ Most apparent within the superior occipital triangle is the lower half of the external jugular vein. ◆ Most important clinically is the superficially located spinal accessory nerve (CN XI). ◆ In the inferior and much smaller omoclavicular triangle, the brachial plexus emerges between the middle and anterior scalene muscles, the latter of which is crossed anteriorly by the phrenic nerve. ◆ Superior to the brachial plexus, and in the same plane, is the cervical plexus. ◆ The cutaneous branches of this plexus emerge from the midpoint of the posterior border of the SCM and radiate toward the scalp, auricle, anterior neck, and shoulder.

Anterior cervical region: The anterior cervical region is inferior to the body of the mandible, extending anteriorly from the SCM to the midline. ◆ The bellies of the digastric, the anterior belly of the omohyoid, and the hyoid subdivide the region into smaller triangles. ◆ The submental triangle is superficial to the floor of the mouth. ◆ The submandibular triangle, superior to the digastric bellies, is occupied by the submandibular salivary gland and submandibular lymph nodes. ◆ The facial artery, coursing within this triangle, is palpable as it emerges from it and crosses the body of the mandible. ◆ The carotid triangle, between the posterior belly of the digastric, superior belly of the omohyoid, and SCM, includes much of the carotid sheath and related structures, including the bifurcation of the common carotid, the carotid sinus and body, and the initial branches of the external carotid artery. ◆ The muscular triangle is formed and occupied by the infrahyoid muscles.

DEEP STRUCTURES OF NECK

The **deep structures of the neck** are the prevertebral muscles, located posterior to the cervical viscera and anterolateral to the cervical vertebral column and the viscera extending through the superior thoracic aperture, at the inferiormost part or the *root of the neck*.

Prevertebral Muscles

The anterior and lateral vertebral or **prevertebral muscles** are deep to prevertebral layer of deep cervical fascia. The **anterior vertebral muscles,** consisting of the longus colli and capitis, rectus capitis anterior, and anterior scalene muscles, lie directly posterior to the *retropharyngeal space*

(Fig. 8.4A & B) and medial to the neurovascular plane of the cervical and brachial plexuses and subclavian artery. The **lateral vertebral muscles,** consisting of the rectus capitis lateralis, splenius capitis, levator scapulae, and middle and posterior scalene muscles, lie posterior to this neurovascular plane and (except for the highly placed rectus capitis lateralis) form the floor of the lateral cervical region. These muscles are illustrated in Figure 8.23; their attachments, innervation, and main actions are given in Table 8.4.

Root of Neck

The **root of the neck** is the junctional area between the thorax and neck (Fig. 8.24A). It is located on the cervical side of the *superior thoracic aperture*, through which pass all

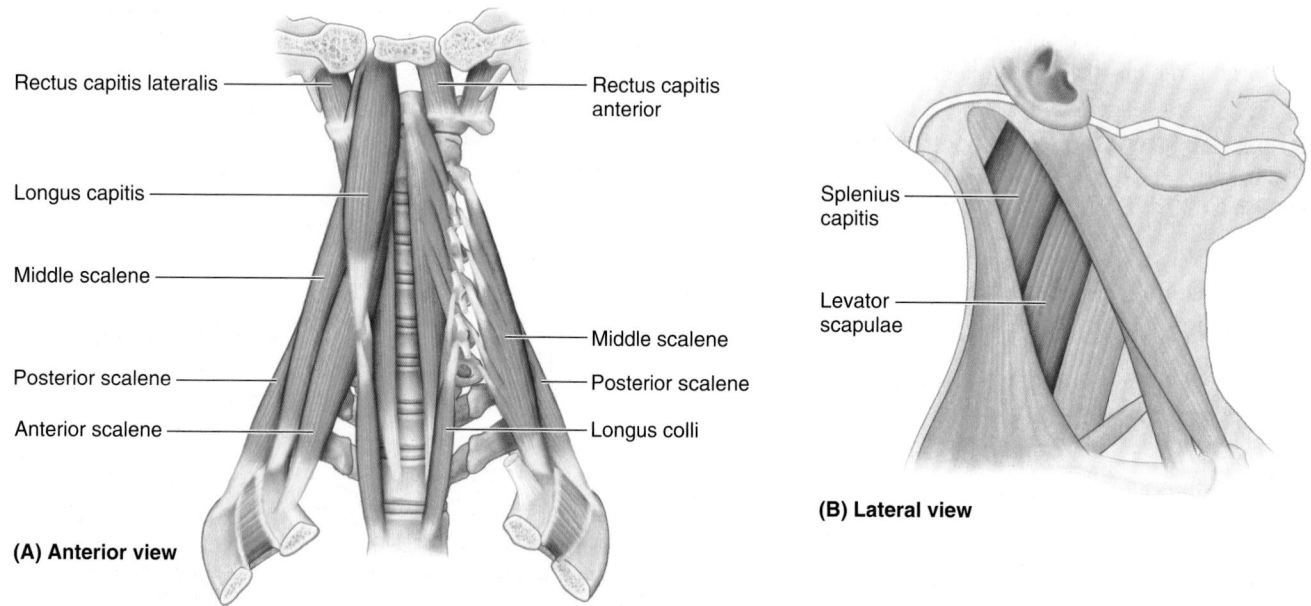

Rectus capitis lateralis — Longus capitis — Middle scalene — Posterior scalene — Anterior scalene — Rectus capitis anterior — Middle scalene — Posterior scalene — Longus colli

(A) Anterior view

Splenius capitis — Levator scapulae

(B) Lateral view

FIGURE 8.23. Prevertebral muscles.

TABLE 8.4. **PREVERTEBRAL MUSCLES**

Muscle	Superior Attachment	Inferior Attachment	Innervation	Main Action(s)
Anterior vertebral muscles				
Longus colli	Anterior tubercle of C1 vertebra (atlas); bodies of C1–C3 and transverse processes of C3–C6 vertebrae	Bodies of C5–T3 vertebrae; transverse processes of C3–C5 vertebrae	Anterior rami of C2–C6 spinal nerves	Flexes neck with rotation (torsion) to opposite side if acting unilaterally[a]
Longus capitis	Basilar part of occipital bone	Anterior tubercles of C3–C6 transverse processes	Anterior rami of C1–C3 spinal nerves	Flex head[b]
Rectus capitis anterior	Base of cranium, just anterior to occipital condyle	Anterior surface of lateral mass of atlas (C1 vertebra)	Branches from loop between C1 and C2 spinal nerves	
Anterior scalene	Transverse processes of C3–C6 vertebrae	1st rib	Cervical spinal nerves C4–C6	

TABLE 8.4. PREVERTEBRAL MUSCLES (Continued)

Muscle	Superior Attachment	Inferior Attachment	Innervation	Main Action(s)
Lateral vertebral muscles				
Rectus capitis lateralis	Jugular process of occipital bone	Transverse process of atlas (C1 vertebra)	Branches from loop between C1 and C2 spinal nerves	Flexes head and helps stabilize it[b]
Splenius capitis	Inferior half of nuchal ligament and spinous processes of superior six thoracic vertebrae	Lateral aspect of mastoid process and lateral third of superior nuchal line	Posterior rami of middle cervical spinal nerves	Laterally flexes and rotates head and neck to same side; acting bilaterally, extends head and neck[c]
Levator scapulae	Posterior tubercles of transverse processes C2–C6 vertebrae	Superior part of medial border of scapula	Dorsal scapular nerve C5 and cervical spinal nerves C3 and C4	Downward rotation of scapula and tilts its glenoid cavity inferiorly by rotating scapula
Middle scalene	Posterior tubercles of transverse processes of C5–C7 vertebrae	Superior surface of 1st rib; posterior to groove for subclavian artery	Anterior rami of cervical spinal nerves	Flexes neck laterally; elevates 1st rib during forced inspiration[a]
Posterior scalene		External border of 2nd rib	Anterior rami of cervical spinal nerves C7 and C8	Flexes neck laterally; elevates 2nd rib during forced inspiration[a]

[a]Flexion of neck = anterior (or lateral) bending of cervical vertebrae C2–C7.

[b]Flexion of head = anterior (or lateral) bending of the head relative to the vertebral column at the atlanto-occipital joints.

[c]Rotation of the head occurs at the atlanto-axial joints.

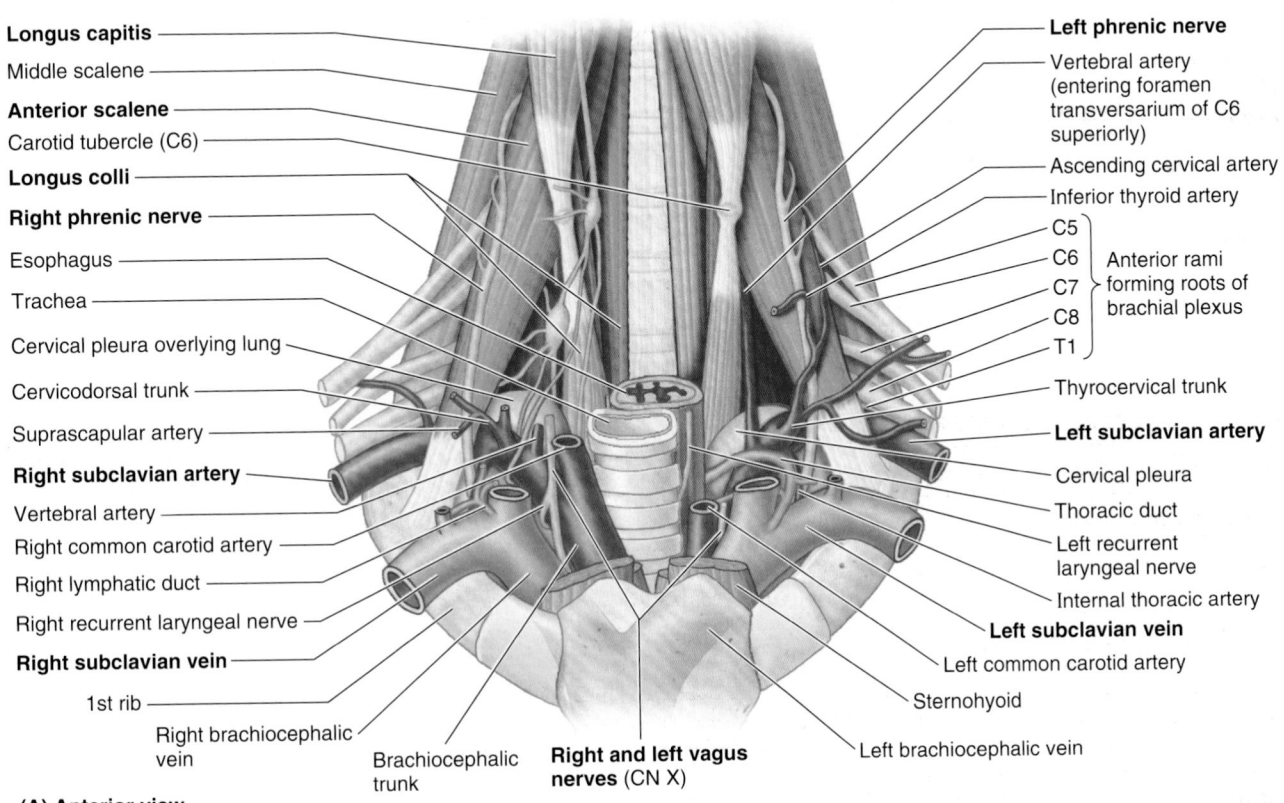

(A) Anterior view

FIGURE 8.24. Root of neck and prevertebral region. A. A dissection of the root of the neck is shown. The brachial plexus and the third part of the subclavian artery emerge between the anterior and the middle scalene muscles. The brachiocephalic veins, the first parts of the subclavian arteries, and the internal thoracic arteries arising from the subclavian arteries are closely related to the cervical pleura (cupula). The thoracic duct terminates in the root of the neck as it enters the left venous angle.

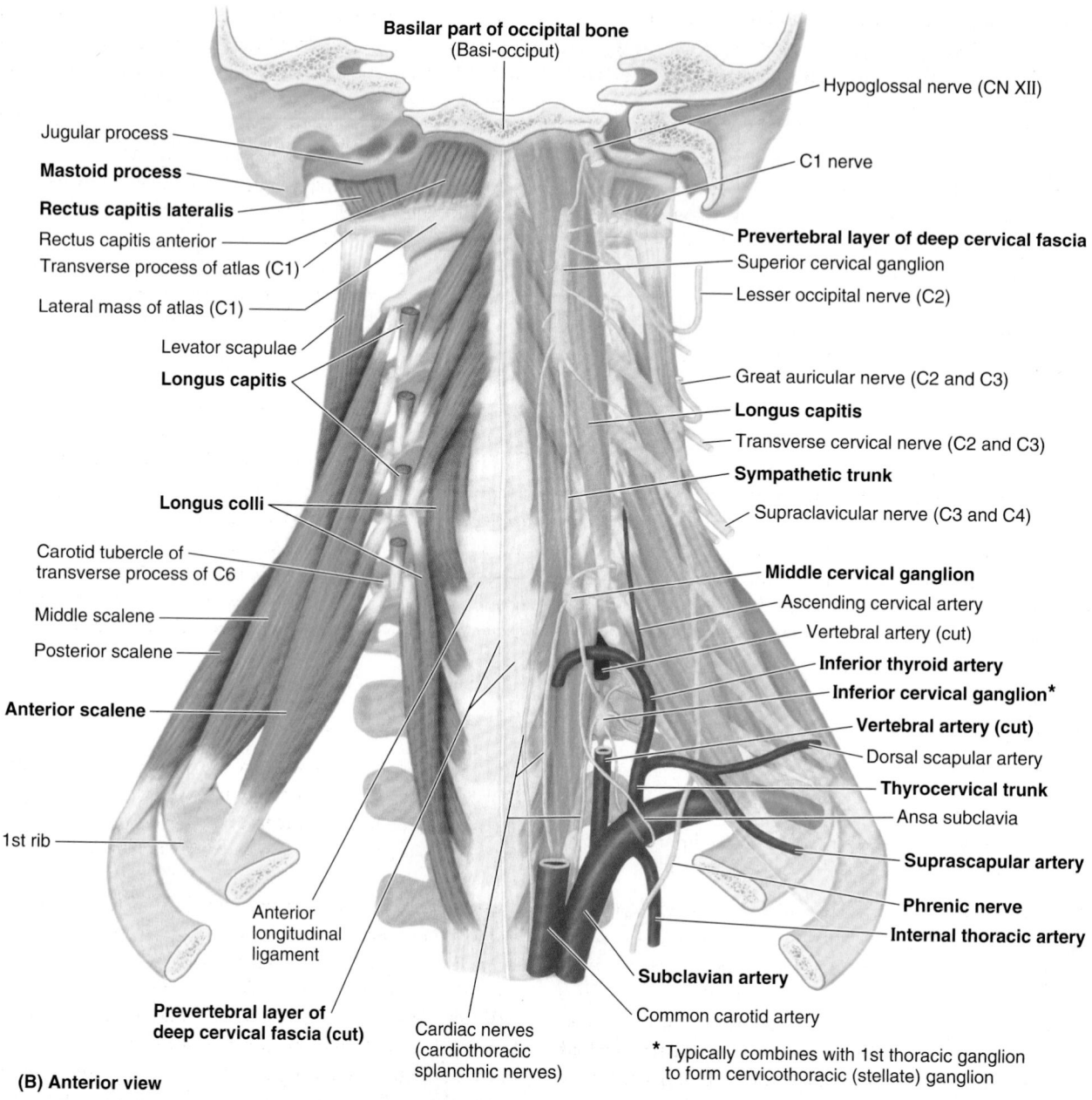

Basilar part of occipital bone
(Basi-occiput)

Hypoglossal nerve (CN XII)

Jugular process

Mastoid process

Rectus capitis lateralis

Rectus capitis anterior

Transverse process of atlas (C1)

Lateral mass of atlas (C1)

Levator scapulae

Longus capitis

Longus colli

Carotid tubercle of
transverse process of C6

Middle scalene

Posterior scalene

Anterior scalene

1st rib

**Prevertebral layer of
deep cervical fascia (cut)**

Anterior
longitudinal
ligament

Cardiac nerves
(cardiothoracic
splanchnic nerves)

C1 nerve

Prevertebral layer of deep cervical fascia

Superior cervical ganglion

Lesser occipital nerve (C2)

Great auricular nerve (C2 and C3)

Longus capitis

Transverse cervical nerve (C2 and C3)

Sympathetic trunk

Supraclavicular nerve (C3 and C4)

Middle cervical ganglion

Ascending cervical artery

Vertebral artery (cut)

Inferior thyroid artery

Inferior cervical ganglion*

Vertebral artery (cut)

Dorsal scapular artery

Thyrocervical trunk

Ansa subclavia

Suprascapular artery

Phrenic nerve

Internal thoracic artery

Subclavian artery

Common carotid artery

* Typically combines with 1st thoracic ganglion
to form cervicothoracic (stellate) ganglion

(B) Anterior view

FIGURE 8.24. (*Continued*) **Root of neck and prevertebral region. B.** In this dissection of the prevertebral region and root of the neck, the prevertebral layer of the deep cervical fascia and the arteries and nerves have been removed from the right side; the longus capitis muscle has been excised on the right side. The cervical plexus of nerves, arising from the anterior rami of C1–C4; the brachial plexus of nerves, arising from the anterior rami of C5–C8 and T1; and branches of the subclavian artery are visible on the left side.

structures going from the thorax to the head or upper limb and vice versa (see Fig. 1.7, p. 79). The inferior boundary of the root of the neck is the *superior thoracic aperture,* formed laterally by the 1st pair of ribs and their costal cartilages, anteriorly by the manubrium of the sternum, and posteriorly by the body of T1 vertebra. The visceral structures in the root of the neck are described in "Viscera of Neck" (p. 1018). Only the neurovascular elements of the root of the neck are described here.

ARTERIES IN ROOT OF NECK

The *brachiocephalic trunk* is covered anteriorly by the right sternohyoid and sternothyroid muscles; it is the largest branch of the arch of the aorta (Fig. 8.24). It arises in the midline from the beginning of the arch of the aorta, posterior to the manubrium. It passes superolaterally to the right where it divides into the right common carotid and right subclavian arteries posterior to the sternoclavicular (SC) joint. The brachiocephalic trunk usually has no preterminal branches.

The **subclavian arteries** supply the upper limbs; they also send branches to the neck and brain (Figs. 8.19 and 8.24). The **right subclavian artery** arises from the brachiocephalic trunk. The **left subclavian artery** arises from the arch of the aorta, about 1 cm distal to the left common carotid artery. The **left vagus nerve** runs parallel to the first part of the artery (Fig. 8.24A). Although the subclavian arteries of the two sides have different origins, their courses in the neck begin posterior to the respective SC joints as they ascend through the superior thoracic aperture and enter the root of the neck.

The subclavian arteries arch superolaterally, reaching an apex as they pass posterior to the *anterior scalene muscles.* As they begin to descend, they disappear posterior to the middle of the clavicles. As the subclavian arteries cross the outer margin of the first ribs, their name changes; they become the axillary arteries. Three parts of each subclavian artery are described relative to the anterior scalene: the first part is medial to the muscle, the second part is posterior to it, and the third part is lateral to it (Figs. 8.11 and 8.24B). The cervical pleurae, apices of the lung, and sympathetic trunks lie posterior to the first part of the arteries. The third part of the subclavian artery was discussed previously in this chapter.

The branches of the subclavian arteries are the:

- *Vertebral artery, internal thoracic artery,* and *thyrocervical trunk* from the first part of the subclavian artery.
- *Costocervical trunk* from the second part of the subclavian artery.
- *Dorsal scapular artery,* often arising from the third part of the subclavian artery.

The **cervical part of the vertebral artery** arises from the first part of the subclavian artery and ascends in the pyramidal space formed between the scalene and longus muscles (colli and capitis) (Fig. 8.24). At the apex of this space, the artery passes deeply to course through the foramina transversaria of vertebrae C1–C6. This is the **vertebral part of the vertebral artery.** Occasionally, the vertebral artery may enter a foramen more superior than vertebra C6. In approximately 5% of people, the left vertebral artery arises from the arch of the aorta.

The **suboccipital part of the vertebral artery** courses in a groove on the posterior arch of the atlas before it enters the cranial cavity through the foramen magnum. The **cranial part of the vertebral artery** supplies branches to the medulla and spinal cord, parts of the cerebellum, and the dura of the posterior cranial fossa. At the inferior border of the pons of the brainstem, the vertebral arteries join to form the *basilar artery,* which participates in the formation of the cerebral arterial circle (see Chapter 7, p. 883).

The **internal thoracic artery** arises from the anteroinferior aspect of the subclavian artery and passes inferomedially into the thorax. The cervical part of the internal thoracic artery has no branches; its thoracic distribution is described in Chapter 1 (Figs. 1.14 and 1.15A, pp. 89–90).

The **thyrocervical trunk** arises from the anterosuperior aspect of the first part of the subclavian artery, near the medial border of the anterior scalene muscle. It has four branches, the largest and most important of which is the **inferior thyroid artery,** the primary visceral artery of the neck, supplying the larynx, trachea, esophagus, and thyroid and parathyroid glands, as well as adjacent muscles. The other branches of the thyrocervical trunk are the ascending cervical and suprascapular arteries, and the cervicodorsal trunk (transverse cervical artery). The branches of the cervicodorsal artery were discussed previously, with the lateral cervical region (p. 994). The terminal branches of the thyrocervical trunk are the inferior thyroid and ascending cervical arteries. The latter is a small artery that sends muscular branches to the lateral muscles of the upper neck and spinal branches into the intervertebral foramina.

The **costocervical trunk** arises from the posterior aspect of the second part of the subclavian artery (posterior to the anterior scalene on the right side [Fig. 8.11] and usually just medial to this muscle on the left side). The trunk passes posterosuperiorly and divides into the superior intercostal and deep cervical arteries, which supply the first two intercostal spaces and the posterior deep cervical muscles, respectively.

VEINS IN ROOT OF NECK

Two large veins terminating in the root of the neck are the *EJV,* draining blood received mostly from the scalp and face, and the variable **anterior jugular vein (AJV),** usually the smallest of the jugular veins (Figs. 8.15 and 8.20). The AJV typically arises near the hyoid from the confluence of superficial submandibular veins. The AJV descends either in the subcutaneous tissue or deep to the investing layer of deep cervical fascia between the anterior median line and the anterior border of the SCM. At the root of the neck, the AJV turns laterally, posterior to the SCM, and opens into the termination of the EJV or into the subclavian vein. Superior to the manubrium, the right and left AJVs commonly unite across the midline to form the **jugular venous arch** in the suprasternal space (Fig. 8.16).

The **subclavian vein,** the continuation of the axillary vein, begins at the lateral border of the 1st rib and ends when it unites with the IJV (Fig. 8.24A). The subclavian vein passes over the 1st rib anterior to the scalene tubercle parallel to the subclavian artery, but it is separated from it by the anterior scalene muscle. It usually has only one named tributary, the *EJV* (Fig. 8.20).

The IJV ends posterior to the medial end of the clavicle by uniting with the subclavian vein to form the brachiocephalic vein. This union is commonly referred to as the **venous angle** and is the site where the *thoracic duct* (left side) and the *right lymphatic trunk* (right side) drain lymph collected throughout the body into the venous circulation (see Fig. 8.48). Throughout its course, the IJV is enclosed by the *carotid sheath* (Fig. 8.21).

NERVES IN ROOT OF NECK

There are three pairs of major nerves in the root of the neck: (1) vagus nerves, (2) phrenic nerves, and (3) sympathetic trunks.

Vagus Nerves (CN X). After its exit from the jugular foramen, each vagus nerve passes inferiorly in the neck within the posterior part of the carotid sheath in the angle between the IJV and common carotid artery (Figs. 8.21 and 8.25). The **right vagus nerve** passes anterior to the first part of the subclavian artery and posterior to the brachiocephalic vein and SC joint to enter the thorax. The **left vagus nerve**

descends between the left common carotid and left subclavian arteries and posterior to the SC joint to enter the thorax.

The **recurrent laryngeal nerves** arise from the vagus nerves in the inferior part of the neck (Fig. 8.25). The nerves of the two sides have essentially the same distribution; however, they loop around different structures and at different levels on the two sides. The **right recurrent laryngeal nerve** loops inferior to the right subclavian artery at approximately the T1–T2 vertebral level. The **left recurrent laryngeal nerve** loops inferior to the arch of the aorta at approximately the T4–T5 vertebral level. After looping, the

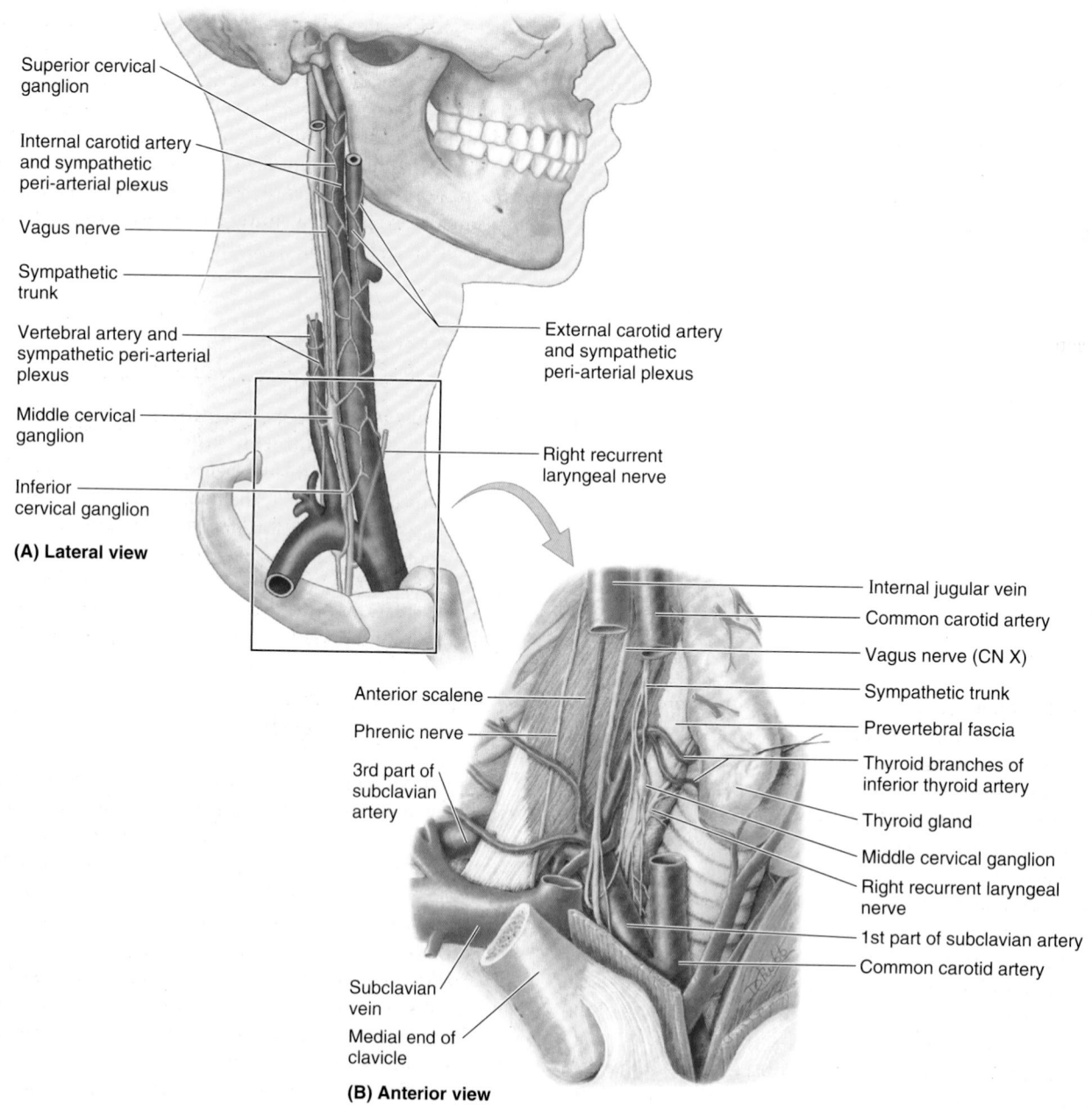

(A) Lateral view

(B) Anterior view

FIGURE 8.25. **Nerves in neck. A.** The cervical sympathetic trunk and ganglia, the carotid arteries, and the sympathetic peri-arterial plexuses surrounding them are shown. **B.** In this view of the root of the neck (right side), the clavicle is removed and sections are taken from the common carotid artery and IJV. The right lobe of the thyroid gland is retracted to reveal the right recurrent laryngeal nerve and middle cervical (sympathetic) ganglion.

recurrent laryngeal nerves ascend superiorly to the postero-medial aspect of the thyroid gland (Figs. 8.24, 8.26B, and 8.27), where they ascend in the *tracheo-esophageal groove*, supplying both trachea and esophagus and all the intrinsic muscles of the larynx except the cricothyroid.

The **cardiac branches of CN X** originate in the neck (Fig. 8.24B) as well as in the thorax and convey presynaptic parasympathetic and visceral afferent fibers to the cardiac plexus of nerves (see Chapter 1, p.150 and Fig. 1.68C).

Phrenic Nerves. The phrenic nerves are formed at the lateral borders of the anterior scalene muscles (Figs. 8.24A and 8.25B), mainly from the C4 nerve with contributions from C3 and C5. The phrenic nerves descend anterior to the anterior scalene muscles under cover of the IJVs and the SCMs. They pass under the prevertebral layer of deep cervical fascia, between the subclavian arteries and veins, and proceed to the thorax to supply the diaphragm. The phrenic nerves are important because, in addition to their sensory distribution, they provide the sole motor supply to their own half of the diaphragm (see Chapter 2, p. 307 for details).

Sympathetic Trunks. The **cervical portion of the sympathetic trunks** lie anterolateral to the vertebral column, extending superiorly to the level of the C1 vertebra or cranial base (Figs. 8.24B and 8.25). The sympathetic trunks receive no white rami communicantes in the neck (recall that white rami associated with cervical spinal nerves). The cervical portion of the trunks includes three **cervical sympathetic ganglia:** superior, middle, and inferior. These ganglia receive presynaptic fibers conveyed to the trunk by the superior thoracic spinal nerves and their associated white rami communicantes, which then ascend through the sympathetic trunk to the ganglia. After synapsing with the postsynaptic neuron in the cervical sympathetic ganglia, postsynaptic neurons send fibers to the:

1. Cervical spinal nerves via *gray rami communicantes.*
2. Thoracic viscera via *cardiopulmonary splanchnic nerves.*
3. Head and viscera of the neck via *cephalic arterial branches* (rami).

The latter fibers accompany arteries as *sympathetic peri-arterial nerve plexuses*, especially the vertebral and internal and external carotid arteries (Fig. 8.25).

In approximately 80% of people, the **inferior cervical ganglion** fuses with the first thoracic ganglion to form the large **cervicothoracic ganglion (stellate ganglion).** This star-shaped (L. *stella*, a star) ganglion lies anterior to the transverse process of the C7 vertebra, just superior to the neck of the 1st rib on each side and posterior to the origin of the vertebral artery (Fig. 8.24B). Some postsynaptic fibers from the ganglion pass via gray rami communicantes to the anterior rami of the C7 and C8 spinal nerves (roots of the brachial plexus), and others pass to the heart via the *inferior cervical cardiac nerve* (a cardiopulmonary splanchnic nerve), which passes along the trachea to the deep *cardiac plexus.* Other fibers pass via arterial branches to contribute to the sympathetic peri-arterial nerve plexus around the vertebral artery running into the cranial cavity (Fig. 8.25A).

The **middle cervical ganglion,** the smallest of the three ganglia, is occasionally absent. When present, it lies on the anterior aspect of the inferior thyroid artery at the level of the cricoid cartilage and the transverse process of C6 vertebra, just anterior to the vertebral artery (Figs. 8.25 and 8.27). Postsynaptic fibers pass from the ganglion via gray rami communicantes to the anterior rami of the C5 and C6 spinal nerves, via a middle cervical cardiac (cardiopulmonary splanchnic) nerve to the heart and via arterial branches to form the peri-arterial plexuses to the thyroid gland.

The **superior cervical ganglion** is at the level of the C1 and C2 vertebrae (Figs. 8.24B and 8.25A). Because of its large size, it forms a good landmark for locating the sympathetic trunk, but it may need to be distinguished from a large sensory (nodose) ganglion of the vagus (CN X) when present. Postsynaptic fibers pass from it by means of cephalic arterial branches to form the internal carotid sympathetic plexus and then enter the cranial cavity (Fig. 8.25). This ganglion also sends arterial branches to the external carotid artery and gray rami to the anterior rami of the superior four cervical spinal nerves. Other postsynaptic fibers pass from it to the cardiac plexus of nerves via a **superior cervical cardiac** (cardiopulmonary splanchnic) **nerve** (see Chapter 1, p. 150).

DEEP STRUCTURES OF NECK

Cervicothoracic Ganglion Block

Anesthetic injected around the large cervico-thoracic ganglion blocks transmission of stimuli through the cervical and superior thoracic ganglia. This *ganglion block* may relieve vascular spasms involving the brain and upper limb. It is also useful when deciding if a surgical resection of the ganglion would be beneficial to a person with excess vasoconstriction in the ipsilateral limb.

Lesion of Cervical Sympathetic Trunk

A *lesion of a cervical sympathetic trunk* in the neck results in a sympathetic disturbance called *Horner syndrome,* which is characterized by:

- Contraction of the pupil (*miosis*), resulting from paralysis of the dilator pupillae muscle (see Chapter 7).
- Drooping of the superior eyelid (*ptosis*), resulting from paralysis of the smooth (tarsal) muscle intermingled with the striated muscle of the levator palpebrae superioris.
- Sinking in of the eye (*enophthalmos*), possibly caused by paralysis of the rudimentary smooth (orbital) muscle in the floor of the orbit.
- Vasodilation and absence of sweating on the face and neck (*anhydrosis*), caused by lack of a sympathetic (vasoconstrictive) nerve supply to the blood vessels and sweat glands.

om Line

RES OF NECK

Prevertebral muscles: The prevertebral muscles, deep to the prevertebral layer of deep cervical fascia, are divided into anterior and lateral vertebral muscles by the neurovascular plane of the cervical and brachial plexuses and subclavian artery. ♦ The anterior vertebral muscles flex the head and neck; however, this movement is normally produced by gravity in conjunction with eccentric contraction of the extensors of the neck. ♦ Thus, the anterior vertebral muscles are called into action mainly when this movement is performed against resistance, probably initiating the movement while the strength of the movement is produced by the SCM. ♦ The lateral vertebral muscles laterally flex the neck, participate in rotation of the neck, and fix or elevate the superiormost ribs during forced inspiration.

Root of the neck: The branches of the arch of the aorta bifurcate and/or traverse the root of the neck, with the branches of the subclavian artery arising here also. ♦ The internal jugular and subclavian veins converge at the root of the neck to form the brachiocephalic veins. ♦ The major lymphatic trunks (right lymphatic duct and thoracic duct) enter the venous angles formed by the convergence of these veins. ♦ The phrenic and vagus nerves enter the thorax by passing anterior to the subclavian arteries and posterior to the brachiocephalic veins. ♦ The sympathetic trunks and recurrent laryngeal nerves traverse the root of the neck posterior to the arteries, as do the visceral structures (trachea and esophagus). ♦ The cervical portion of the sympathetic trunks include three cervical sympathetic ganglia (inferior, middle, and superior), in which presynaptic fibers from the superior thoracic spinal cord synapse with postsynaptic neurons. ♦ These neurons send fibers to the cervical spinal nerves, via gray rami communicantes; to the head and viscera of the neck, via cephalic arterial branches and peri-arterial plexuses; and to the thoracic viscera, via cardiac (cardiopulmonary splanchnic) nerves.

VISCERA OF NECK

The cervical viscera are disposed in three layers, named for their primary function (Fig. 8.26). Superficial to deep, they are the:

1. *Endocrine layer:* the thyroid and parathyroid glands.
2. *Respiratory layer:* the larynx and trachea.
3. *Alimentary layer:* the pharynx and esophagus.

Endocrine Layer of Cervical Viscera

The viscera of the **endocrine layer** are part of the body's endocrine system of ductless, hormone-secreting glands. The *thyroid gland* is the body's largest endocrine gland. It produces *thyroid hormone,* which controls the rate of metabolism, and *calcitonin,* a hormone controlling calcium metabolism. The thyroid gland affects all areas of the body except itself and the spleen, testes, and uterus. The hormone produced by the *parathyroid glands, parathormone* (PTH), controls the metabolism of phosphorus and calcium in the blood. The parathyroid glands target the skeleton, kidneys, and intestine.

THYROID GLAND

The **thyroid gland** lies deep to the sternothyroid and sternohyoid muscles, located anteriorly in the neck at the level of the C5–T1 vertebrae (Fig. 8.26). It consists primarily of right and left **lobes,** anterolateral to the larynx and trachea. A relatively thin **isthmus** unites the lobes over the trachea, usually anterior to the second and third tracheal rings. The thyroid gland is surrounded by a thin **fibrous capsule,** which sends septa deeply into the gland. Dense connective tissue attaches the capsule to the cricoid cartilage and superior tracheal rings. External to the capsule is a loose sheath formed by the visceral portion of the pretracheal layer of deep cervical fascia.

Arteries of Thyroid Gland. The highly vascular thyroid gland is supplied by the superior and inferior *thyroid arteries* (Figs. 8.26B and 8.27). These vessels lie between the fibrous capsule and the loose fascial sheath. Usually the first branches of the external carotid arteries, the **superior thyroid arteries,** descend to the superior poles of the gland, pierce the pretracheal layer of deep cervical fascia, and divide into anterior and posterior branches supplying mainly the anterosuperior aspect of the gland.

The **inferior thyroid arteries,** the largest branches of the thyrocervical trunks arising from the subclavian arteries, run superomedially posterior to the carotid sheaths to reach the posterior aspect of the *thyroid gland.* They divide into several branches that pierce the pretracheal layer of the deep cervical fascia and supply the posteroinferior aspect, including the **inferior poles of the gland.** The right and left superior and inferior thyroid arteries anastomose extensively within the gland, ensuring its supply while providing potential collateral circulation between the subclavian and external carotid arteries.

In approximately 10% of people, a small, unpaired **thyroid ima artery** (L. *arteria thyroidea ima*) arises from the brachiocephalic trunk (see the blue box "Thyroid Ima Artery" on p. 1040); however, it may arise from the arch of the aorta or from the right common carotid, subclavian, or internal thoracic arteries. When present, this small artery ascends on the anterior surface of the trachea, supplying small branches to it. The artery then continues to the isthmus of the thyroid gland, where it divides and supplies it.

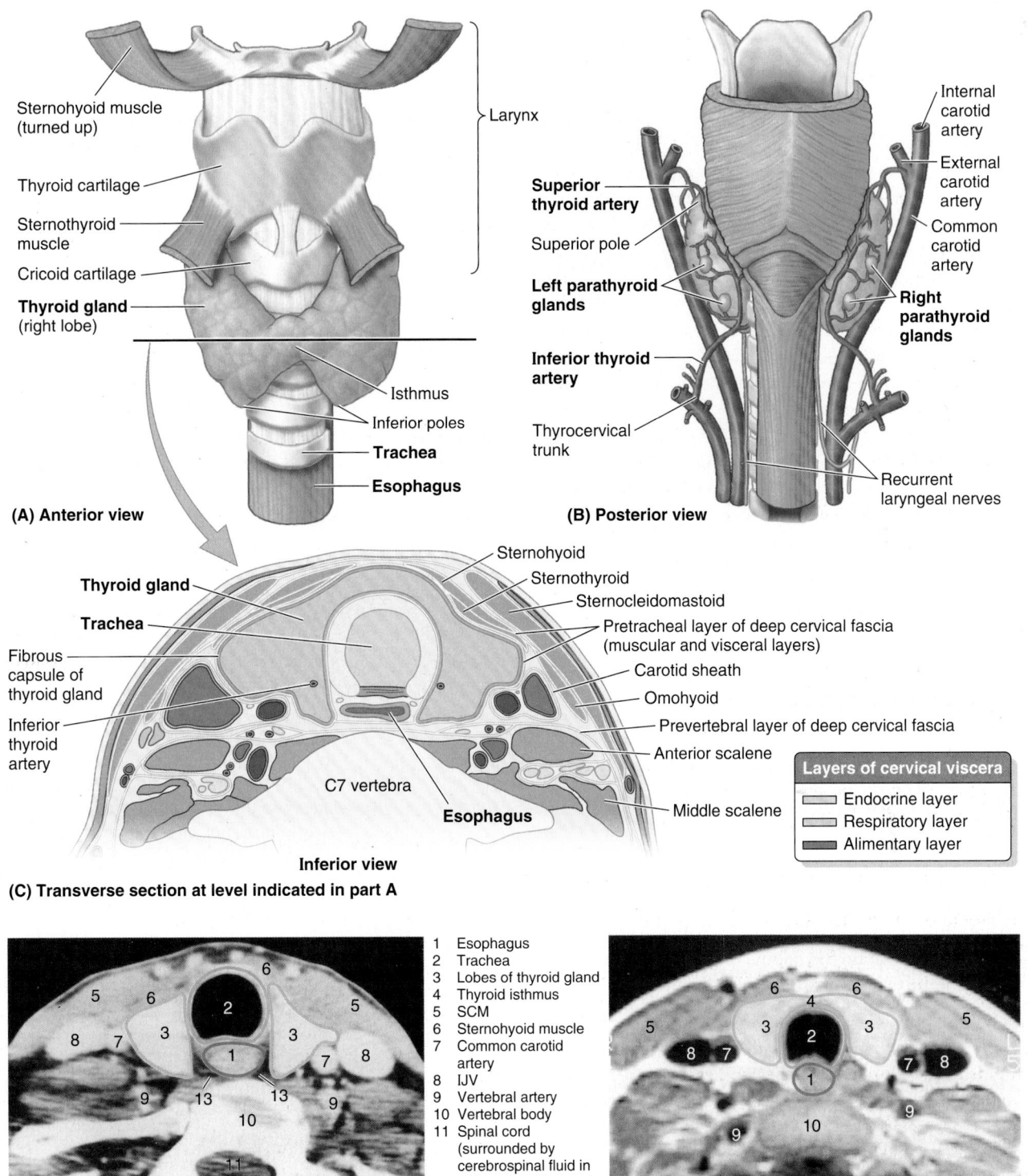

FIGURE 8.26. Relationships of thyroid gland. A. The sternothyroid muscles have been cut to expose the lobes of the normal thyroid gland. The isthmus lies anterior to the second and third tracheal rings. **B.** The parathyroid glands are usually embedded in the fibrous capsule on the posterior surface of the thyroid gland. **C–E.** The functional layers of the cervical viscera are indicated. The levels of the imaging studies are close to that shown in part **C.** (Part **D** courtesy of Dr. M. Keller, Medical Imaging, University of Toronto, Toronto, Ontario, Canada. Part **E** Courtesy of Dr. W. Kucharczyk, Professor and Neuroradiologist Senior Scientist, Department of Medical Resonance Imaging, University Health Network, Toronto, Ontario, Canada.)

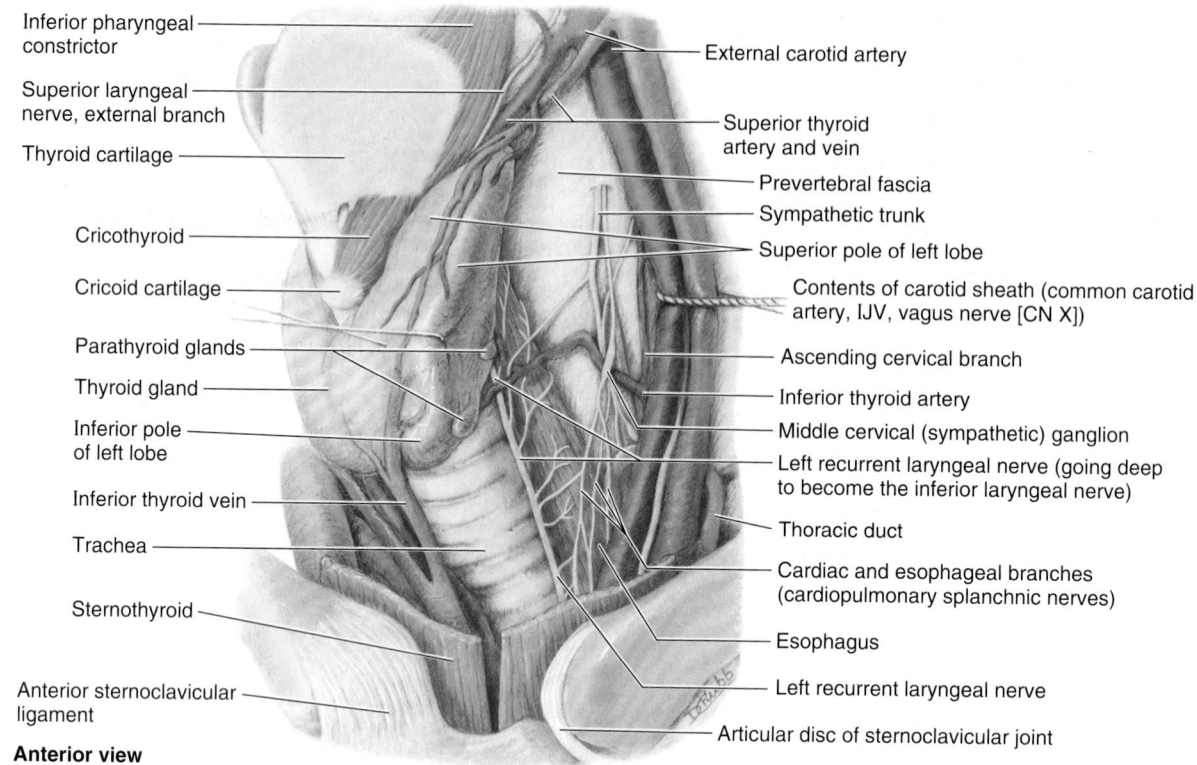

Inferior pharyngeal constrictor

Superior laryngeal nerve, external branch

Thyroid cartilage

Cricothyroid

Cricoid cartilage

Parathyroid glands

Thyroid gland

Inferior pole of left lobe

Inferior thyroid vein

Trachea

Sternothyroid

Anterior sternoclavicular ligament

Anterior view

External carotid artery

Superior thyroid artery and vein

Prevertebral fascia

Sympathetic trunk

Superior pole of left lobe

Contents of carotid sheath (common carotid artery, IJV, vagus nerve [CN X])

Ascending cervical branch

Inferior thyroid artery

Middle cervical (sympathetic) ganglion

Left recurrent laryngeal nerve (going deep to become the inferior laryngeal nerve)

Thoracic duct

Cardiac and esophageal branches (cardiopulmonary splanchnic nerves)

Esophagus

Left recurrent laryngeal nerve

Articular disc of sternoclavicular joint

FIGURE 8.27. Dissection of left side of root of neck. The viscera (thyroid gland, trachea, and esophagus) are retracted to the right, and the contents of the left carotid sheath are retracted to the left. The middle thyroid vein, severed to allow such retraction, is not apparent. The left parathyroid glands on the posterior aspect of the left lobe of the thyroid gland are exposed. The recurrent laryngeal nerve ascends beside the trachea, in the angle between the trachea and the esophagus. The thoracic duct passes laterally, posterior to the contents of the carotid sheath as the thyrocervical trunk passes medially.

Veins of Thyroid Gland. Three pairs of thyroid veins usually form a **thyroid plexus of veins** on the anterior surface of the thyroid gland and anterior to the trachea (Figs. 8.27 and 8.28). The **superior thyroid veins** accompany the superior thyroid arteries; they drain the **superior poles** of the thyroid gland; the **middle thyroid veins** do not accompany but run essentially parallel courses with the inferior thyroid arteries; they drain the middle of the lobes. The usually independent **inferior thyroid veins** drain the inferior poles. The superior and middle thyroid veins drain into the IJVs; the inferior thyroid veins drain into the brachiocephalic veins posterior to the manubrium.

Lymphatic Drainage of Thyroid Gland. The lymphatic vessels of this gland run in the interlobular connective tissue, usually near the arteries; they communicate with a capsular network of lymphatic vessels. From here, the vessels pass initially to **prelaryngeal, pretracheal,** and **paratracheal lymph nodes.** The prelaryngeal nodes drain in turn to the superior deep cervical lymph nodes, and the pretracheal and paratracheal lymph nodes drain to the inferior deep cervical nodes (Fig. 8.29). Laterally, lymphatic vessels located along the superior thyroid veins pass directly to the inferior deep cervical lymph nodes. Some lymphatic vessels may drain into the *brachiocephalic lymph nodes* or the *thoracic duct* (Fig. 8.27).

Nerves of Thyroid Gland. The nerves of the thyroid gland are derived from the *superior, middle, and inferior cervical* (sympathetic) *ganglia* (Figs. 8.25 and 8.27). They

reach the gland through the *cardiac* and *superior* and *inferior thyroid peri-arterial plexuses* that accompany the thyroid arteries. These fibers are vasomotor, not secretomotor. They cause constriction of blood vessels. Endocrine secretion from the thyroid gland is hormonally regulated by the pituitary gland.

PARATHYROID GLANDS

The small flattened, oval **parathyroid glands** usually lie external to the thyroid capsule on the medial half of the posterior surface of each lobe of the thyroid gland, inside its sheath (Figs. 8.26B, 8.27 and 8.30A). The **superior parathyroid glands** usually lie slightly more than 1 cm superior to the point of entry of the inferior thyroid arteries into the thyroid gland. The **inferior parathyroid glands** usually lie slightly more than 1 cm inferior to the arterial entry point (Skandalakis et al., 1995). Most people have four parathyroid glands. Approximately 5% of people have more; some have only two glands. The superior parathyroid glands, more constant in position than the inferior ones, are usually at the level of the inferior border of the cricoid cartilage. The inferior parathyroid glands are usually near the inferior poles of the thyroid gland, but they may lie in various positions (Fig. 8.30B). In 1–5% of people, an inferior parathyroid gland is deep in the superior mediastinum (Norton and Wells, 1994).

Vessels of Parathyroid Glands. Because the *inferior thyroid arteries* provide the primary blood supply to the posterior

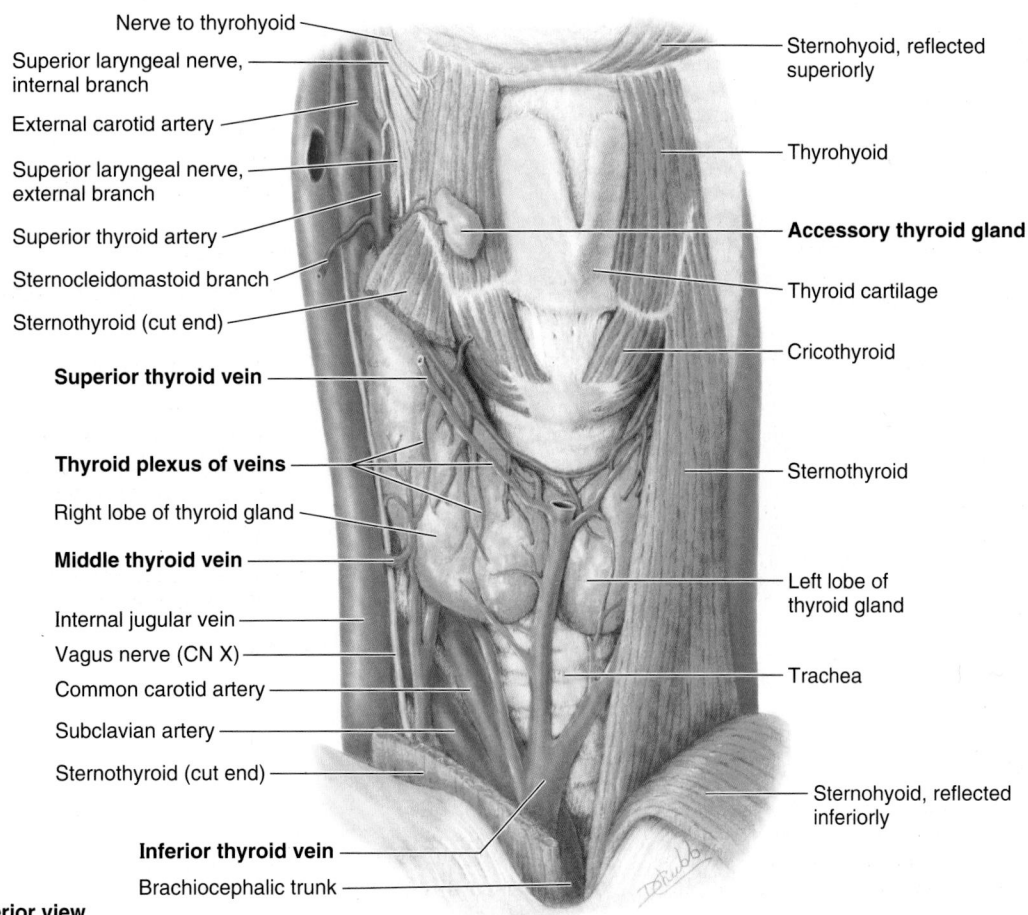

Nerve to thyrohyoid

Superior laryngeal nerve, internal branch

External carotid artery

Superior laryngeal nerve, external branch

Superior thyroid artery

Sternocleidomastoid branch

Sternothyroid (cut end)

Superior thyroid vein

Thyroid plexus of veins

Right lobe of thyroid gland

Middle thyroid vein

Internal jugular vein

Vagus nerve (CN X)

Common carotid artery

Subclavian artery

Sternothyroid (cut end)

Inferior thyroid vein

Brachiocephalic trunk

Sternohyoid, reflected superiorly

Thyrohyoid

Accessory thyroid gland

Thyroid cartilage

Cricothyroid

Sternothyroid

Left lobe of thyroid gland

Trachea

Sternohyoid, reflected inferiorly

Anterior view

FIGURE 8.28. Thyroid gland. A dissection of the anterior aspect of the neck is shown. In this specimen, there is a small accessory thyroid gland on the right, lying on the thyrohyoid muscle, lateral to the thyroid cartilage. The superior thyroid artery is distributed primarily to the anterosuperior portion of the gland.

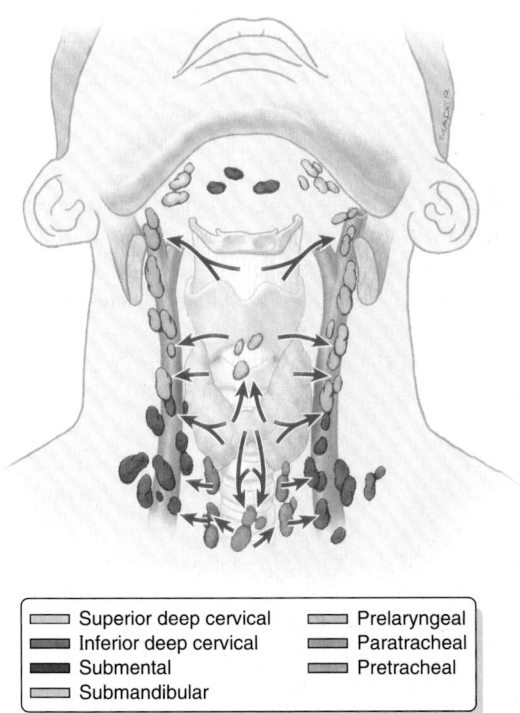

Superior deep cervical
Inferior deep cervical
Submental
Submandibular
Prelaryngeal
Paratracheal
Pretracheal

FIGURE 8.29. Lymphatic drainage of thyroid gland, larynx, and trachea. The *arrows* indicate the direction of lymph flow.

aspect of the thyroid gland where the parathyroid glands are located, branches of these arteries usually supply these glands (Figs. 8.26B and 8.30A). However, they may also be supplied by branches from the superior thyroid arteries; thyroid ima artery; or laryngeal, tracheal, and esophageal arteries. **Parathyroid veins** drain into the *thyroid plexus of veins* of the thyroid gland and trachea (Fig. 8.28). *Lymphatic vessels* from the parathyroid glands drain with those from the thyroid gland into deep cervical lymph nodes and paratracheal lymph nodes (Fig. 8.29).

Nerves of Parathyroid Glands. The nerve supply of the parathyroid glands is abundant; it is derived from thyroid branches of the cervical (sympathetic) ganglia (Fig. 8.25). Like the nerves to the thyroid, they are vasomotor rather than secretomotor because these glands are hormonally regulated.

Respiratory Layer of Cervical Viscera

The viscera of the **respiratory layer,** the *larynx* and *trachea,* contribute to the respiratory functions of the body. The main functions of the cervical respiratory viscera are as follows:

- Routing air and food into the respiratory tract and esophagus, respectively.
- Providing a patent airway and a means of sealing it off temporarily (a "valve").
- Producing voice.

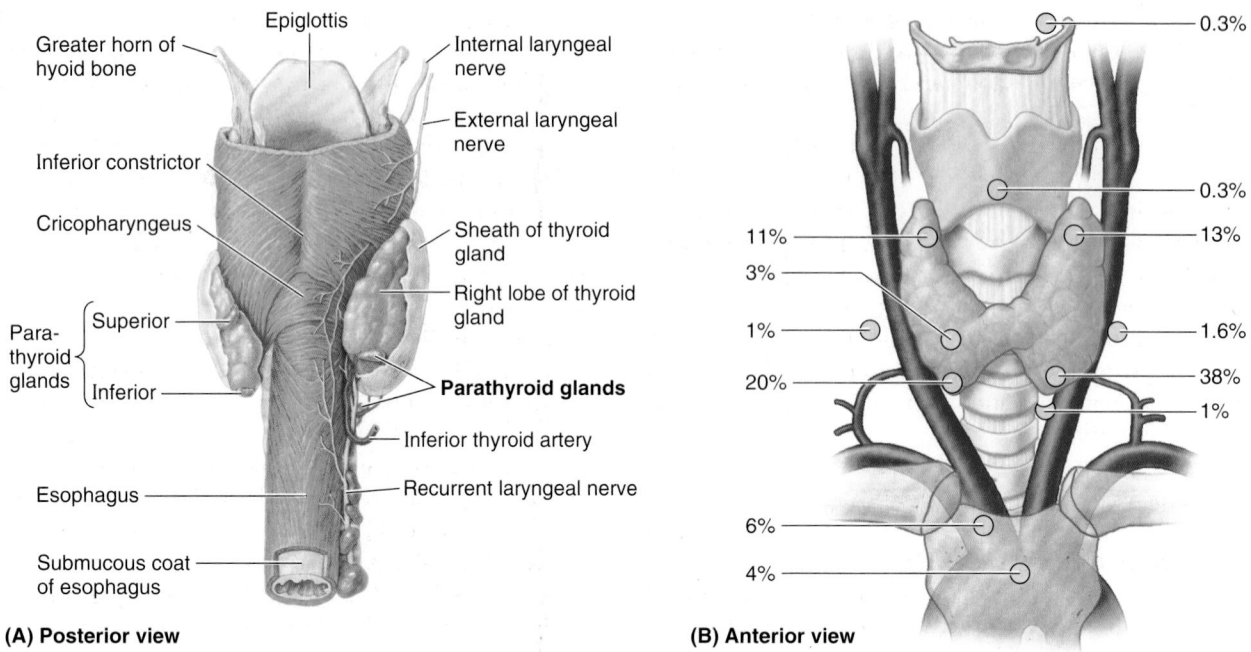

FIGURE 8.30. Thyroid and parathyroid glands. A. The thyroid sheath has been dissected from the posterior surface of the thyroid gland to reveal the three embedded parathyroid glands. Both parathyroid glands on the right side are rather low, and the inferior gland is inferior to the thyroid gland. **B.** Sites and frequencies of aberrant parathyroid glandular tissue are shown.

LARYNX

The **larynx** is the complex organ of voice production (the "voice box") composed of nine cartilages connected by membranes and ligaments and containing the *vocal folds* ("cords"). The larynx is located in the anterior neck at the level of the bodies of C3–C6 vertebrae (Fig. 8.31). It connects the inferior part of the pharynx (oropharynx) with the trachea. Although most commonly known for its role as the phonating mechanism for voice production, its most vital function is to guard the air passages, especially during swallowing when it serves as the "sphincter" or "valve" of the lower respiratory tract, thus maintaining a patent airway.

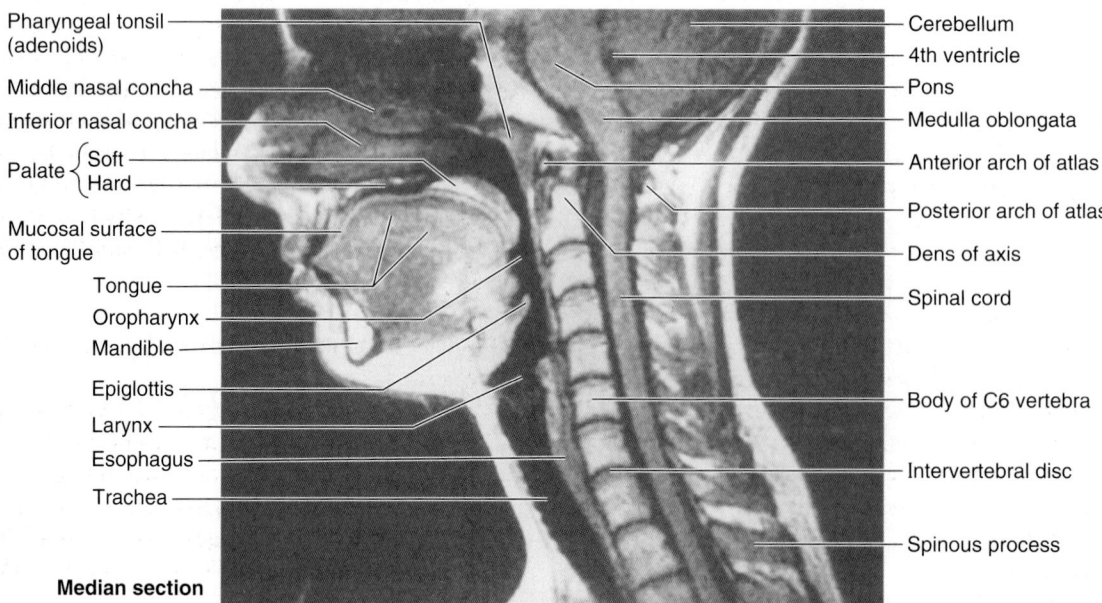

FIGURE 8.31. Median MRI of head and neck. Because the air and food passages share the oropharynx, separation of food and air must occur to continue into the trachea (anterior) and esophagus (posterior). (Courtesy of Dr. W. Kucharczyk, University Health Network, Toronto, Ontario, Canada.)

Laryngeal Skeleton. The laryngeal skeleton *consists of nine cartilages:* three are single (thyroid, cricoid, and epiglottic), and three are paired (arytenoid, corniculate, and cuneiform) (Fig. 8.32A & B).

The **thyroid cartilage** is the largest of the cartilages; its superior border lies opposite the C4 vertebra. The inferior two thirds of its two plate-like **laminae** fuse anteriorly in the median plane to form the **laryngeal prominence** (Fig. 8.32A & D). This projection ("Adam's apple") is well marked in men but seldom visible in women. Superior to this prominence, the laminae diverge to form a V-shaped **superior thyroid notch.** The less distinct **inferior thyroid notch** is a shallow indentation in the middle of the inferior border of the cartilage.

The posterior border of each lamina projects superiorly as the **superior horn** and inferiorly as the **inferior horn.** The superior border and superior horns attach to the hyoid by the **thyrohyoid membrane** (Fig. 8.32A & B). The thick median part of this membrane is the **median thyrohyoid ligament;** its lateral parts are the **lateral thyrohyoid ligaments.**

The inferior horns articulate with the lateral surfaces of the cricoid cartilage at the **cricothyroid joints** (Fig. 8.32B). The main movements at these joints are rotation and gliding of the thyroid cartilage, which result in changes in the length of the vocal folds. The **cricoid cartilage** is shaped like a signet ring with its band facing anteriorly. This ring-like opening of the cartilage fits an average finger. The posterior (signet) part of the cricoid is the **lamina,** and the anterior (band) part is the **arch** (Fig. 8.32A). Although much smaller than the thyroid cartilage, the cricoid cartilage is thicker and stronger and is the only complete ring of cartilage to encircle any part of the airway. It attaches to the inferior margin of the thyroid cartilage by the **median cricothyroid ligament** and to the first tracheal ring by the **cricotracheal ligament.** Where the larynx is closest to the skin and most accessible, the median cricothyroid ligament may be felt as a soft spot during palpation inferior to the thyroid cartilage.

The **arytenoid cartilages** are paired, three-sided pyramidal cartilages that articulate with the lateral parts of the superior border of the cricoid cartilage lamina (Fig. 8.32B). Each cartilage has an apex superiorly, a vocal process anteriorly, and a large muscular process that projects laterally from its base. The **apex** bears the corniculate cartilage and attaches to the ary-epiglottic fold. The **vocal process** provides the posterior attachment for the vocal ligament, and the muscular process serves as a lever to which the posterior and lateral crico-arytenoid muscles are attached. The **crico-arytenoid joints,** located between the bases of the arytenoid cartilages and the superolateral surfaces of the lamina of the cricoid cartilage (Fig. 8.32B), permit the arytenoid cartilages to slide toward or away from one to another, to tilt anteriorly and posteriorly, and rotate. These movements are important in approximating, tensing, and relaxing the vocal folds.

The elastic **vocal ligaments** extend from the junction of the laminae of the thyroid cartilage anteriorly to the vocal process of the arytenoid cartilage posteriorly (Fig. 8.32E).

The vocal ligaments make up the submucosal skeleton of the vocal folds. These ligaments are the thickened, free superior border of the **conus elasticus** or **cricovocal membrane.** The parts of the membrane extending laterally between the vocal folds and the superior border of the cricoid are the **lateral cricothyroid ligaments.** The fibro-elastic conus elasticus blends anteriorly with the *median cricothyroid ligament.* The conus elasticus and overlying mucosa close the tracheal inlet except for the central **rima glottidis** (opening between the vocal folds).

The **epiglottic cartilage,** consisting of elastic cartilage, gives flexibility to the **epiglottis,** a heart-shaped cartilage covered with mucous membrane (Fig. 8.32B). Situated posterior to the root of the tongue and the hyoid and anterior to the **laryngeal inlet,** the epiglottic cartilage forms the superior part of the anterior wall and the superior margin of the inlet. Its broad superior end is free. Its tapered inferior end, the **stalk of the epiglottis,** is attached to the angle formed by the thyroid laminae by the **thyro-epiglottic ligament** (Fig. 8.32E). The **hyo-epiglottic ligament** attaches the anterior surface of the epiglottic cartilage to the hyoid (Fig. 8.33). The **quadrangular membrane** (Figs. 8.32B and 8.34) is a thin, submucosal sheet of connective tissue that extends between the lateral aspects of the arytenoid and epiglottic cartilages. Its free inferior margin constitutes the **vestibular ligament,** which is covered loosely by mucosa to form the **vestibular fold** (Fig. 8.34). This fold lies superior to the vocal fold and extends from the thyroid cartilage to the arytenoid cartilage. The free superior margin of the quadrangular membrane forms the **ary-epiglottic ligament,** which is covered with mucosa to form the **ary-epiglottic fold.** The **corniculate** and **cuneiform cartilages** appear as small nodules in the posterior part of the ary-epiglottic folds. The corniculate cartilages attach to the apices of the arytenoid cartilages; the cuneiform cartilages do not directly attach to other cartilages. The quadrangular membrane and conus elasticus are the superior and inferior parts of the submucosal **fibro-elastic membrane of the larynx.**

Interior of Larynx. The **laryngeal cavity** extends from the *laryngeal inlet,* through which it communicates with the *laryngopharynx,* to the level of the inferior border of the cricoid cartilage. Here the laryngeal cavity is continuous with the cavity of the trachea (Figs. 8.34 and 8.35A & B). The laryngeal cavity includes the:

- **Laryngeal vestibule:** between the laryngeal inlet and the vestibular folds.
- **Middle part of the laryngeal cavity:** the central cavity (airway) between the vestibular and vocal folds.
- **Laryngeal ventricle:** recesses extending laterally from the middle part of the laryngeal cavity between vestibular and vocal folds. The **laryngeal saccule** is a blind pocket opening into each ventricle that is lined with mucosal glands.
- **Infraglottic cavity:** the inferior cavity of the larynx between the vocal folds and the inferior border of the

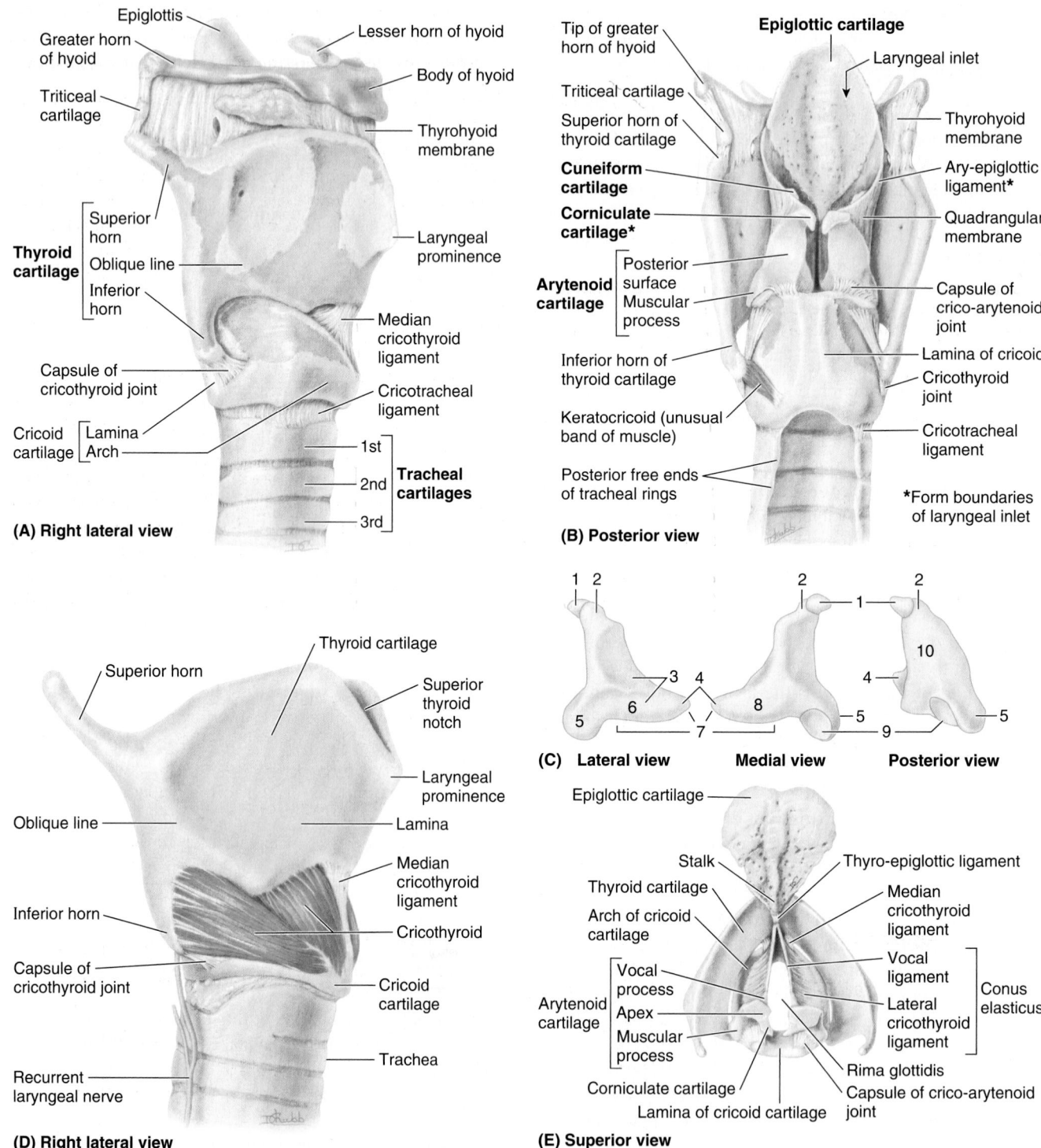

(A) Right lateral view

Epiglottis
Greater horn of hyoid
Triticeal cartilage
Thyroid cartilage
Superior horn
Oblique line
Inferior horn
Capsule of cricothyroid joint
Cricoid cartilage
Lamina
Arch

Lesser horn of hyoid
Body of hyoid
Thyrohyoid membrane
Laryngeal prominence
Median cricothyroid ligament
Cricotracheal ligament
1st
2nd
3rd
Tracheal cartilages

(B) Posterior view

Tip of greater horn of hyoid
Triticeal cartilage
Superior horn of thyroid cartilage
Cuneiform cartilage
Corniculate cartilage*
Arytenoid cartilage
Posterior surface
Muscular process
Inferior horn of thyroid cartilage
Keratocricoid (unusual band of muscle)
Posterior free ends of tracheal rings

Epiglottic cartilage
Laryngeal inlet
Thyrohyoid membrane
Ary-epiglottic ligament*
Quadrangular membrane
Capsule of crico-arytenoid joint
Lamina of cricoid
Cricothyroid joint
Cricotracheal ligament
*Form boundaries of laryngeal inlet

(C)   Lateral view   Medial view   Posterior view

1 2   2   2
3 4   1   10
6   8   4
5   5   5
7   9

(D) Right lateral view

Superior horn
Oblique line
Inferior horn
Capsule of cricothyroid joint
Recurrent laryngeal nerve

Thyroid cartilage
Superior thyroid notch
Laryngeal prominence
Lamina
Median cricothyroid ligament
Cricothyroid
Cricoid cartilage
Trachea

(E) Superior view

Epiglottic cartilage
Stalk
Thyroid cartilage
Arch of cricoid cartilage
Arytenoid cartilage
Vocal process
Apex
Muscular process
Corniculate cartilage
Lamina of cricoid cartilage

Thyro-epiglottic ligament
Median cricothyroid ligament
Vocal ligament
Lateral cricothyroid ligament
Conus elasticus
Rima glottidis
Capsule of crico-arytenoid joint

FIGURE 8.32. Skeleton of larynx. A. Although firmly connected to it, the hyoid is not part of the larynx. The larynx extends vertically from the tip of the heart-shaped epiglottis to the inferior border of the cricoid cartilage. **B.** The thyroid cartilage shields the smaller cartilages of the larynx, and the hyoid shields the superior part of the epiglottic cartilage. **C.** Three views of an isolated arytenoid cartilage are shown. *1,* Corniculate cartilage; *2,* apex of arytenoid cartilage; *3,* anterolateral surface; *4,* vocal process (projects anteriorly, provides attachment for vocal ligament); *5,* muscular process (projects laterally, for attachment of posterior and lateral crico-arytenoid muscles); *6,* oblong fovea (for attachment of thyro-arytenoid muscle); *7,* base; *8,* medial surface; *9,* articular surface; *10,* posterior surface. **D.** The thyroid cartilage and cricothyroid muscle are shown. This muscle produces movement at the cricothyroid joint. **E.** The epiglottic cartilage is pitted for mucous glands, and its stalk is attached by the thyro-epiglottic ligament to the angle of the thyroid cartilage superior to the vocal ligaments. The vocal ligament, which forms the skeleton of the vocal fold, extends from the vocal process of the arytenoid cartilage to the "angle" of the thyroid cartilage, and there joins its fellow inferior to the thyro-epiglottic ligament.

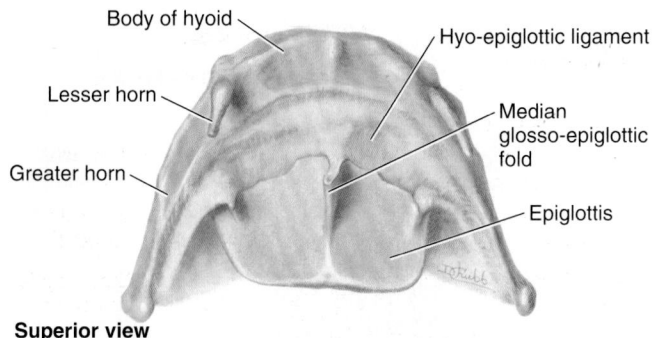

Superior view

FIGURE 8.33. Epiglottis and hyo-epiglottic ligament. The epiglottis is a leaf-shaped plate of elastic fibrocartilage, which is covered with mucous membrane (*pink*) and is attached anteriorly to the hyoid by the hyo-epiglottic ligament (*blue*). The epiglottis serves as a diverter valve over the superior aperture of the larynx during swallowing.

cricoid cartilage, where it is continuous with the lumen of the trachea.

The **vocal folds** control sound production (Figs. 8.35 and 8.36). The apex of each wedge-shaped fold projects medially into the laryngeal cavity. Each vocal fold contains a:

- *Vocal ligament*, consisting of thickened elastic tissue that is the medial free edge of the conus elasticus (Figs. 8.32E and 8.34).

- *Vocalis muscle*, composed of exceptionally fine muscle fibers immediately lateral to and terminating at intervals relative to the length of the vocal ligaments (Fig. 8.35A).

The vocal folds are the sharp-edged folds of mucous membrane overlying and incorporating the vocal ligaments and the thyro-arytenoid muscles. They are the source of the sounds (tone) that come from the larynx. These folds produce audible vibrations when their free margins are closely (but not tightly) apposed during phonation, and air is forcibly expired intermittently (Fig. 8.36C). The vocal folds also serve as the main inspiratory sphincter of the larynx when they are tightly closed. Complete adduction of the folds forms an effective sphincter that prevents entry of air.

The **glottis** (the vocal apparatus of the larynx) makes up the vocal folds and processes, together with the **rima glottidis,** the aperture between the vocal folds (Fig. 8.35C). The shape of the rima (L. slit) varies according to the position of the vocal folds (Fig. 8.36). During ordinary breathing, the rima is narrow and wedge shaped; during forced respiration, it is wide and trapezoidal in shape. The rima glottidis is slit-like when the vocal folds are closely approximated during phonation. Variation in the tension and length of the vocal folds, in the width of the rima glottidis, and in the intensity of the expiratory effort produces changes in the pitch of the

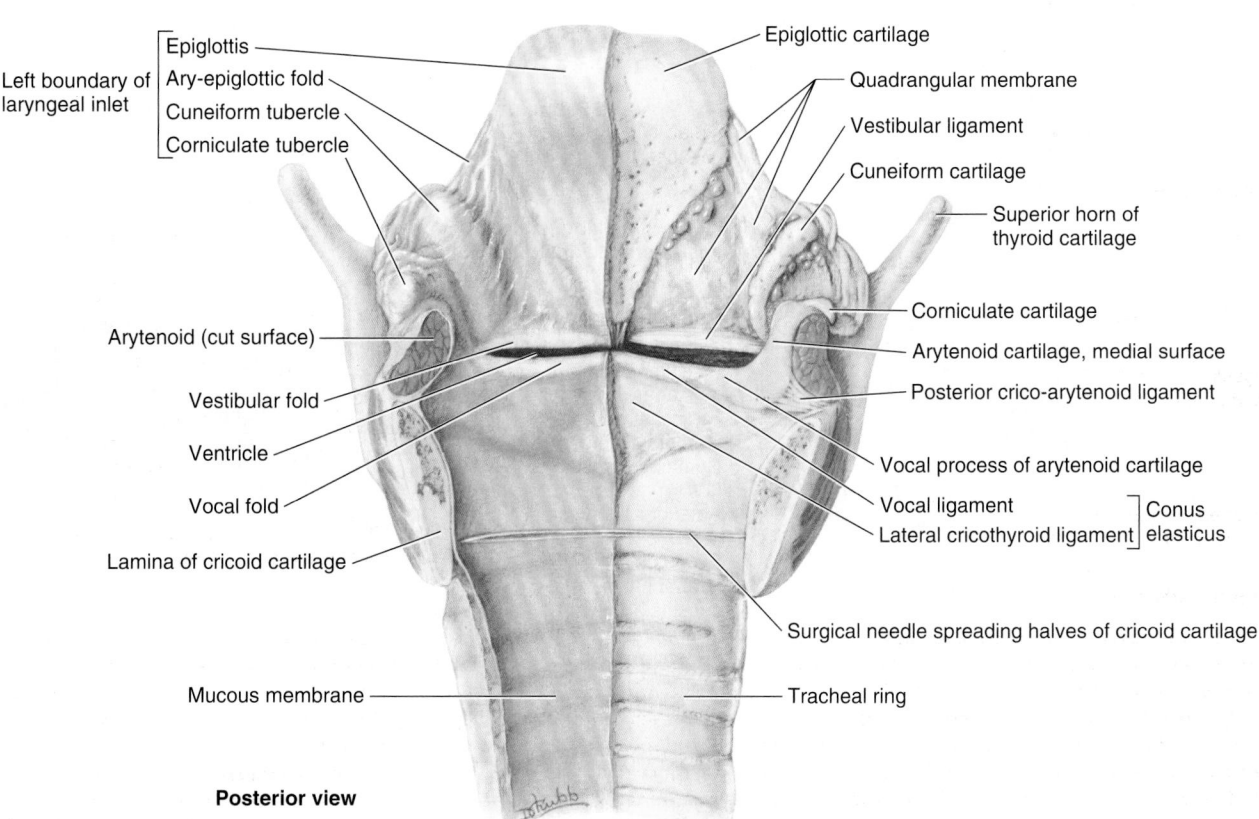

Posterior view

FIGURE 8.34. Interior of larynx. The posterior wall of the larynx is split in the median plane, and the two sides are spread apart and held in place by a surgical needle. On the left side, the mucous membrane is intact. On the right side, the mucous and submucous coats are peeled off, and the skeletal coat—consisting of cartilages, ligaments, and the fibro-elastic membrane—is uncovered.

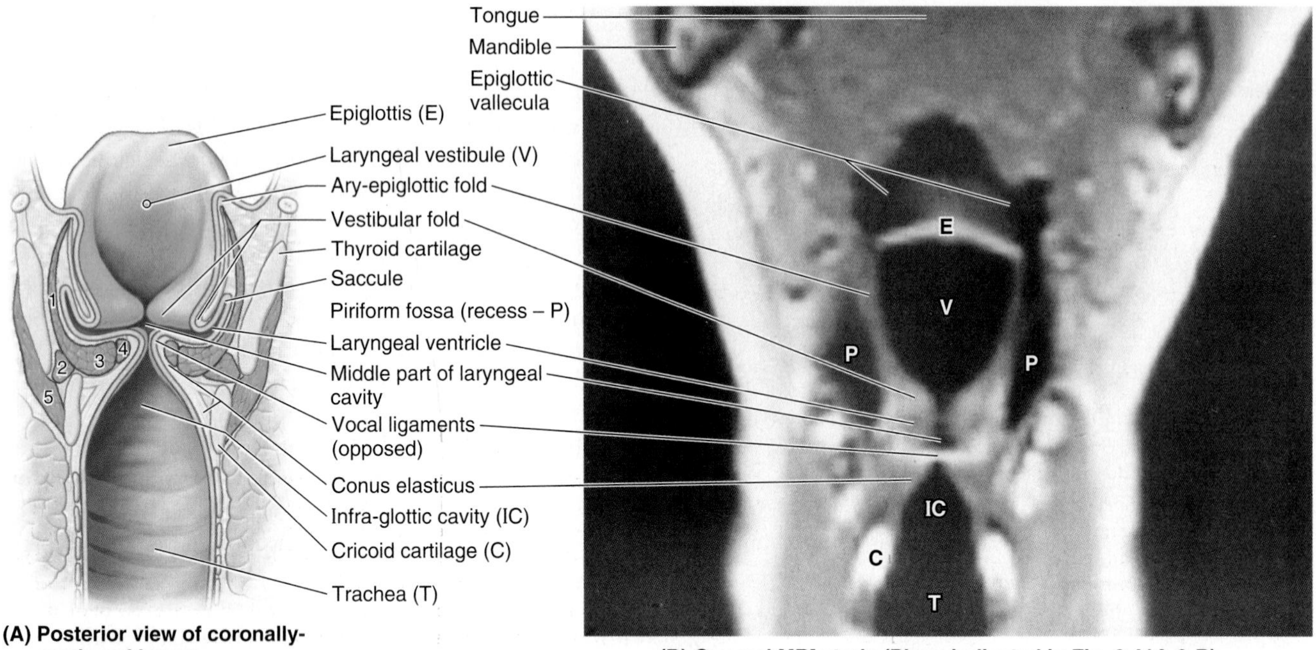

Tongue
Mandible
Epiglottic vallecula

Epiglottis (E)
Laryngeal vestibule (V)
Ary-epiglottic fold
Vestibular fold
Thyroid cartilage
Saccule
Piriform fossa (recess – P)
Laryngeal ventricle
Middle part of laryngeal cavity
Vocal ligaments (opposed)
Conus elasticus
Infra-glottic cavity (IC)
Cricoid cartilage (C)
Trachea (T)

(A) Posterior view of coronally-sectioned larynx

(B) Coronal MRI study (Plane indicated in Fig. 8.41A & B)

Key for (A):
1 Ary-epiglottic muscle
2 Lateral crico-arytenoid
3 Thyro-arytenoid
4 Vocalis
5 Cricothyroid

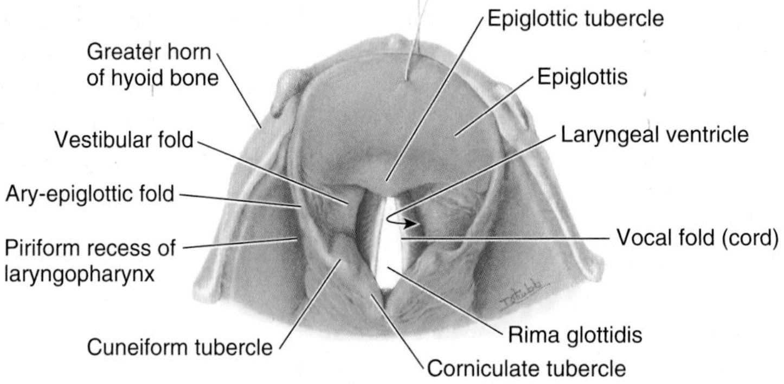

Greater horn of hyoid bone
Vestibular fold
Ary-epiglottic fold
Piriform recess of laryngopharynx
Cuneiform tubercle

Epiglottic tubercle
Epiglottis
Laryngeal ventricle
Vocal fold (cord)
Rima glottidis
Corniculate tubercle

(C) Posterosuperior view, looking antero-inferiorly through laryngeal vestibule and rima glottidis

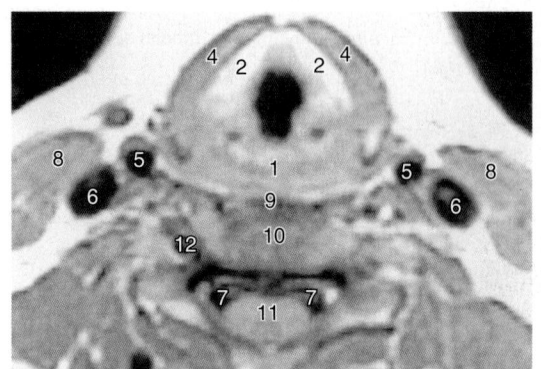

1 Esophagus
2 Thyroid cartilage
3 Lamina of cricoid cartilage
4 Sternothyroid and sternohyoid muscles
5 Common carotid artery
6 IJV
7 Anterior root of spinal nerve
8 SCM
9 Inferior pharyngeal constrictor
10 Vertebral body
11 Spinal cord
12 Vertebral artery

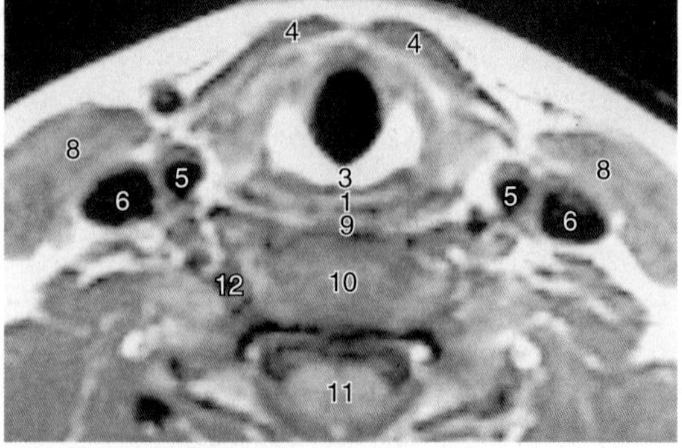

(D) Transverse MRI study of layrnx transecting thyroid cartilage

(E) Transverse MRI study of larynx transecting cricoid cartilage

FIGURE 8.35. Folds and compartments of larynx. A. This coronal section shows the compartments of the larynx: the vestibule, middle compartment with left and right ventricles, and the infraglottic cavity. **B.** This MRI study shows the epiglottic valleculae of the oropharynx, piriform fossae of the laryngopharynx, and vestibular and vocal folds of the larynx. **C.** The rima glottidis (the space between the vocal folds) is visible through the laryngeal inlet and vestibule. The laryngeal inlet is bounded (1) anteriorly by the free curved edge of the epiglottis; (2) posteriorly by the arytenoid cartilages, the corniculate cartilages that cap them, and the interarytenoid fold that unites them; and (3) on each side by the ary-epiglottic fold that contains the superior end of the cuneiform cartilage. **D and E.** The planes of these transverse studies, oriented in the same direction as part **C,** pass superior (**D**) and inferior (**E**) to the rima glottidis. (MRI studies courtesy of Dr. W. Kucharczyk, University Health Network, Toronto, Ontario, Canada.)

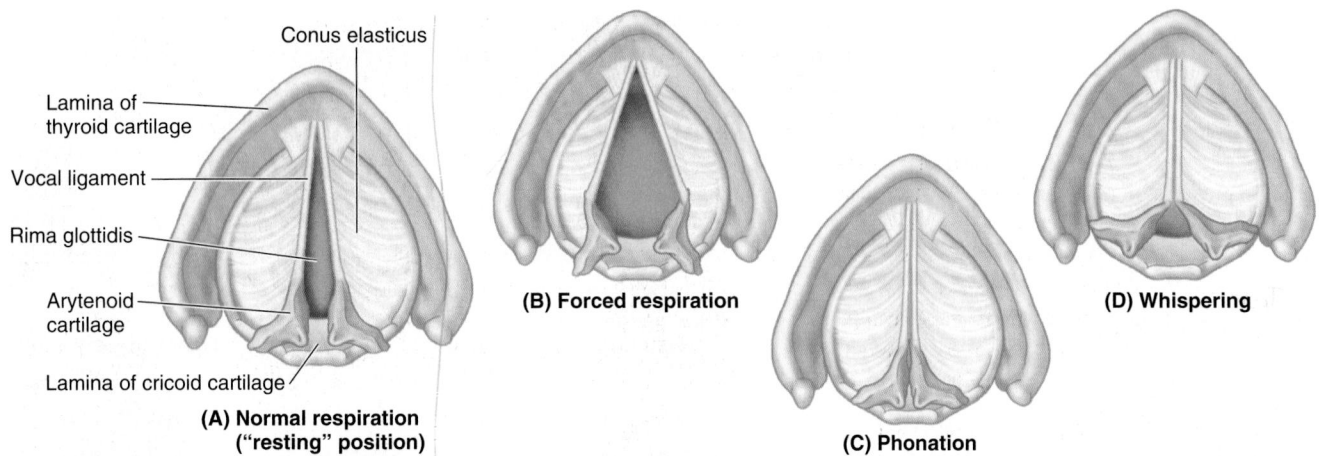

Conus elasticus

Lamina of thyroid cartilage

Vocal ligament

Rima glottidis

Arytenoid cartilage

Lamina of cricoid cartilage

(A) Normal respiration ("resting" position)

(B) Forced respiration

(C) Phonation

(D) Whispering

FIGURE 8.36. **Variations in shape of rima glottidis. A.** The shape of the rima glottidis, the aperture between the vocal folds, varies according to the position of the vocal folds. During normal respiration, the laryngeal muscles are relaxed and the rima glottidis assumes a narrow, slit-like position. **B.** During a deep inhalation, the vocal ligaments are abducted by contraction of the posterior crico-arytenoid muscles, opening the rima glottidis widely into an inverted kite shape. **C.** During phonation, the arytenoid muscles adduct the arytenoid cartilages at the same time that the lateral crico-arytenoid muscles moderately adduct. Air forced between the adducted vocal ligaments produces tone. Stronger contraction of the same muscles seals the rima glottidis (Valsalva maneuver). **D.** During whispering, the vocal ligaments are strongly adducted by the lateral crico-arytenoid muscles, but the relaxed arytenoid muscles allow air to pass between the arytenoid cartilages (intercartilaginous part of rima glottidis), which is modified into toneless speech. No tone is produced.

voice. The lower range of pitch of the voice of postpubertal males results from the greater length of the vocal folds.

The *vestibular folds*, extending between the thyroid and the arytenoid cartilages (Figs. 8.34 and 8.35), play little or no part in voice production; they are protective in function. They consist of two thick folds of mucous membrane enclosing the *vestibular ligaments*. The space between these ligaments is the **rima vestibuli.** The lateral recesses between the vocal and the vestibular folds are the *laryngeal ventricles*.

Laryngeal Muscles. The laryngeal muscles are divided into extrinsic and intrinsic groups.

- **Extrinsic laryngeal muscles** move *the larynx as a whole* (Fig. 8.18; Table 8.3). The *infrahyoid muscles* are depressors of the hyoid and larynx, whereas the *suprahyoid muscles* (and the *stylopharyngeus*, a pharyngeal muscle discussed later in this chapter) are elevators of the hyoid and larynx.
- **Intrinsic laryngeal muscles** move *the laryngeal components*, altering the length and tension of the vocal folds and the size and shape of the rima glottidis (Fig. 8.36). All but one of the intrinsic muscles of the larynx are supplied by the *recurrent laryngeal nerve* (Figs. 8.37, 8.39, and 8.40), a branch of CN X. The cricothyroid is supplied by the external laryngeal nerve, one of the two terminal branches of the *superior laryngeal nerve*.

The actions of the intrinsic laryngeal muscles are easiest to understand when they are considered as functional groups: adductors and abductors, sphincters, and tensors and relaxers. The intrinsic muscles are illustrated in situ in Figures 8.35D & E and 8.38; their attachments, innervation, and main actions are summarized in Table 8.5.

- *Adductors and abductors:* These muscles move the vocal folds to open and close the rima glottidis. The principal adductors are the **lateral crico-arytenoid muscles,**

which pull the muscular processes anteriorly, rotating the arytenoid cartilages so that their vocal processes swing medially. When this action is combined with that of the **transverse** and **oblique arytenoid muscles,** which pull the arytenoid cartilages together, air pushed through the rima glottidis causes vibrations of the vocal ligaments

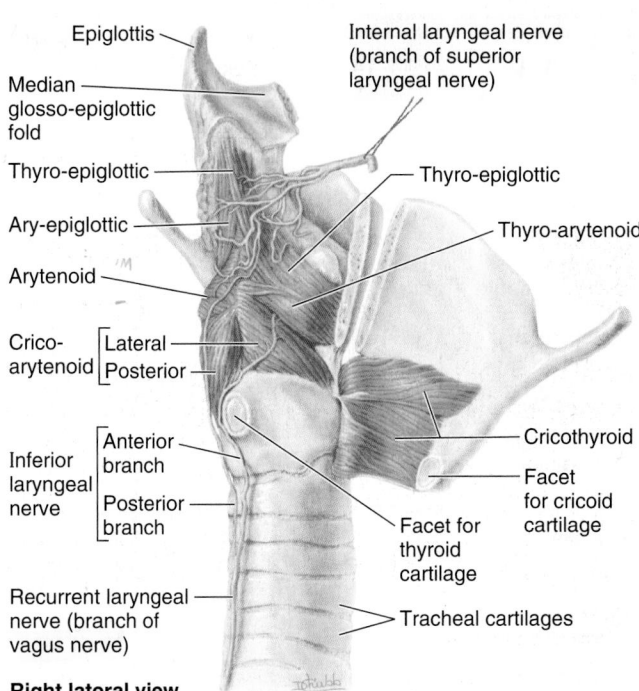

Epiglottis

Median glosso-epiglottic fold

Thyro-epiglottic

Ary-epiglottic

Arytenoid

Crico-arytenoid [Lateral / Posterior]

Inferior laryngeal nerve [Anterior branch / Posterior branch]

Recurrent laryngeal nerve (branch of vagus nerve)

Internal laryngeal nerve (branch of superior laryngeal nerve)

Thyro-epiglottic

Thyro-arytenoid

Cricothyroid

Facet for cricoid cartilage

Facet for thyroid cartilage

Tracheal cartilages

Right lateral view

FIGURE 8.37. Muscles and nerves of larynx and cricothyroid joint. The thyroid cartilage is sawn through to the right of the median plane. The cricothyroid joint is disarticulated, and the right lamina of the thyroid cartilage is turned anteriorly (like opening a book), stripping the cricothyroid muscles off the arch of the cricoid cartilage.

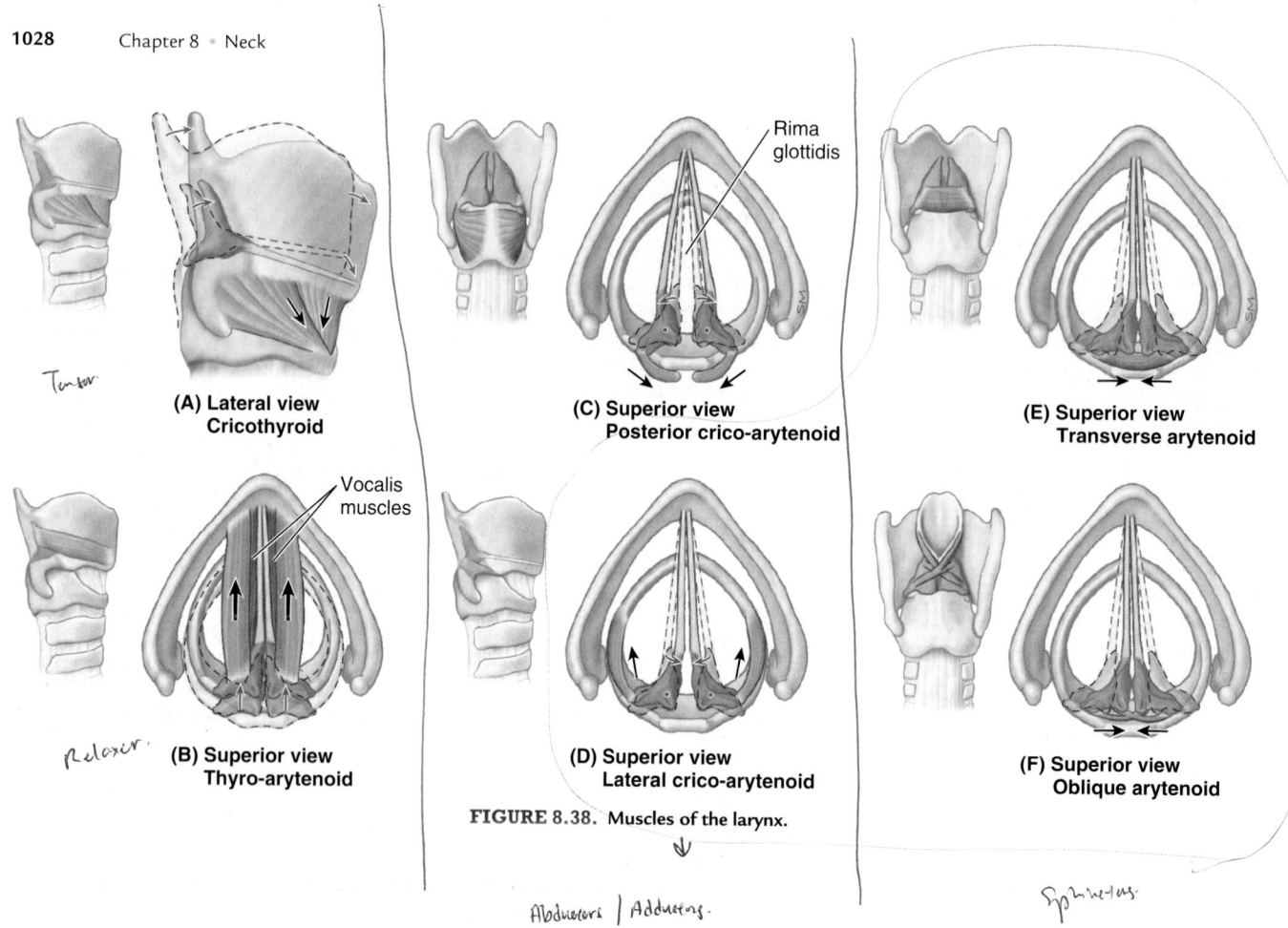

(A) Lateral view
Cricothyroid

Tensor.

Relaxer.

(B) Superior view
Thyro-arytenoid

Vocalis
muscles

Rima
glottidis

(C) Superior view
Posterior crico-arytenoid

(D) Superior view
Lateral crico-arytenoid

(E) Superior view
Transverse arytenoid

(F) Superior view
Oblique arytenoid

FIGURE 8.38. Muscles of the larynx.

Abductors / Adductors.

Sphincters.

TABLE 8.5. MUSCLES OF LARYNX

Muscle	Origin	Insertion	Innervation	Main Action(s)
Cricothyroid	Anterolateral part of cricoid cartilage	Inferior margin and inferior horn of thyroid cartilage	External laryngeal nerve (from CN X)	Stretches and tenses vocal ligament
Thyro-arytenoid[a]	Lower half of posterior aspect of angle of thyroid laminae and cricothyroid ligament	Anterolateral arytenoid surface	Inferior laryngeal nerve (terminal part of recurrent laryngeal nerve, from CN X – see Fig. 8.37)	Relaxes vocal ligament
Posterior crico-arytenoid	Posterior surface of lamina of cricoid cartilage	Muscular ~~Vocal~~ process of arytenoid cartilage		Abducts vocal folds
Lateral crico-arytenoid	Arch of cricoid cartilage			Adducts vocal folds (inter-ligamentous portion)
Transverse and oblique arytenoids[b]	One arytenoid cartilage	Contralateral arytenoid cartilage		Adduct arytenoid cartilages (adducting intercartilaginous portion of vocal folds, closing posterior rima glottidis)
Vocalis[c]	Lateral surface of vocal process of arytenoid cartilage	Ipsilateral vocal ligament		Relaxes posterior vocal ligament while maintaining (or increasing) tension of anterior part

[a]Superior fibers of the thyro-arytenoid muscles pass into the ary-epiglottic fold, and some of them reach the epiglottic cartilage. These fibers constitute the thyro-epiglottic muscle, which widens the laryngeal inlet.

[b]Some fibers of the oblique arytenoid muscles continue as ary-epiglottic muscles (Fig. 8.39).

[c]This slender muscle slip lies medial to and is composed of fibers finer than those of the thyro-arytenoid muscle.

(phonation). When the vocal ligaments are adducted, but the transverse arytenoid muscles do not act, the arytenoid cartilages remain apart and air may bypass the ligaments. This is the position of whispering when the breath is modified into voice in the absence of tone. The sole abductors are the **posterior crico-arytenoid muscles,** which pull the muscular processes posteriorly, rotating the vocal processes laterally and thus widening the rima glottidis.

- *Sphincters:* The combined actions of most of the muscles of the laryngeal inlet result in a sphincteric action that closes the laryngeal inlet as a protective mechanism during swallowing. Contraction of the *lateral crico-arytenoids, transverse* and *oblique arytenoids,* and *ary-epiglottic muscles* brings the ary-epiglottic folds together and pulls the arytenoid cartilages toward the epiglottis. This action occurs reflexively in response to the presence of liquid or particles approaching or within the laryngeal vestibule. It is perhaps our strongest reflex, diminishing only after loss of consciousness, as in drowning.
- *Tensors:* The principal tensors are the **cricothyroid muscles,** which tilt or pull the prominence or angle of the

thyroid cartilage anteriorly and inferiorly toward the arch of the cricoid cartilage. This increases the distance between the thyroid prominence and the arytenoid cartilages. Because the anterior ends of the vocal ligaments attach to the posterior aspect of the prominence, the vocal ligaments elongate and tighten, raising the pitch of the voice.

- *Relaxers:* The principal muscles in this group are the **thyro-arytenoid muscles,** which pull the arytenoid cartilages anteriorly, toward the thyroid angle (prominence), thereby relaxing the vocal ligaments to lower the pitch of the voice.

The **vocalis muscles** lie medial to the thyro-arytenoid muscles and lateral to the vocal ligaments within the vocal folds. The vocalis muscles produce minute adjustments of the vocal ligaments, selectively tensing and relaxing the anterior and posterior parts, respectively, of the vocal folds during animated speech and singing.

Arteries of Larynx. The laryngeal arteries, branches of the superior and inferior thyroid arteries, supply the larynx (Fig. 8.39). The **superior laryngeal artery** accompanies the internal branch of the superior laryngeal nerve through

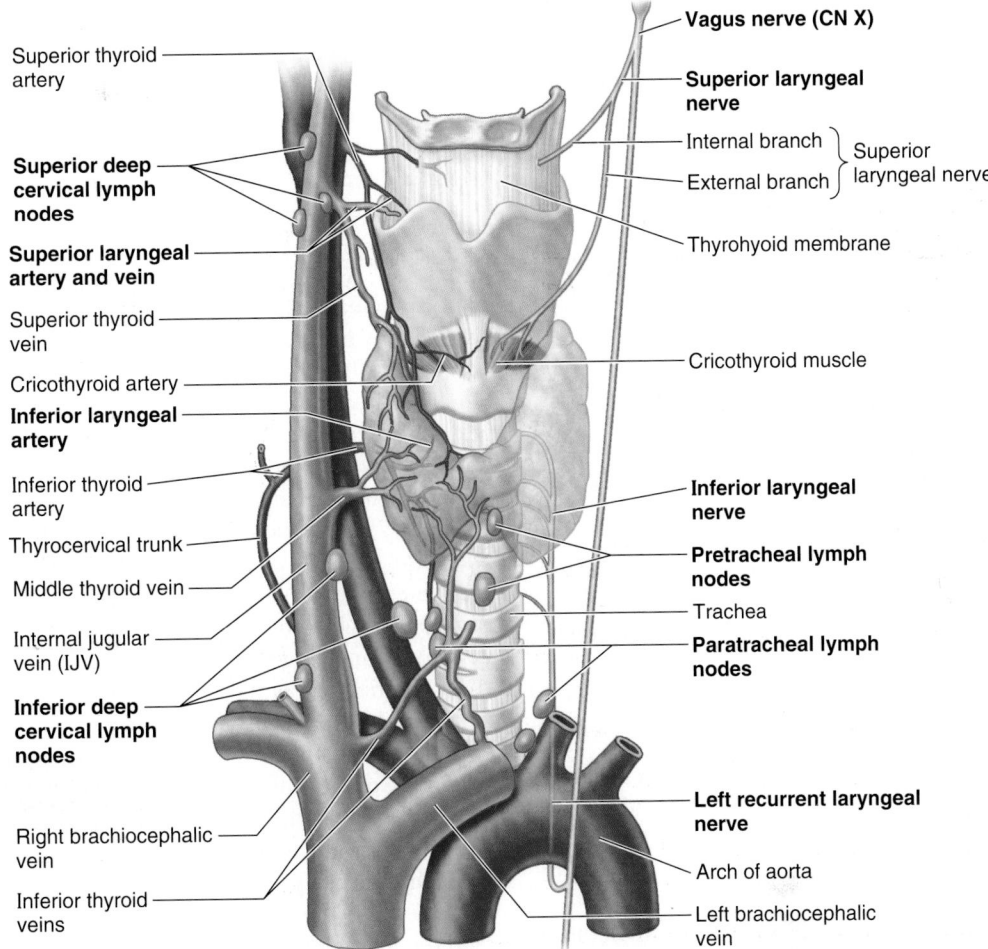

FIGURE 8.39. Vessels, nerves, and lymph nodes of larynx. The superior and inferior thyroid arteries give rise to the superior and inferior laryngeal arteries, respectively; they anastomose with each other. The laryngeal nerves are derived from the vagus (CN X) through the internal and external branches of the superior laryngeal nerve and the inferior laryngeal nerve from the recurrent laryngeal nerve. The left recurrent laryngeal nerve passes inferior to the arch of the aorta.

the thyrohyoid membrane and branches to supply the internal surface of the larynx. The **cricothyroid artery,** a small branch of the superior thyroid artery, supplies the cricothyroid muscle. The **inferior laryngeal artery,** a branch of the inferior thyroid artery, accompanies the *inferior laryngeal nerve* (terminal part of the recurrent laryngeal nerve) and supplies the mucous membrane and muscles in the inferior part of the larynx.

Veins of Larynx. The laryngeal veins accompany the laryngeal arteries. The **superior laryngeal vein** usually joins the superior thyroid vein and through it drains into the IJV (Fig. 8.39). The **inferior laryngeal vein** joins the inferior thyroid vein or the venous plexus of veins on the anterior aspect of the trachea, which empties into the left brachiocephalic vein.

Lymphatics of Larynx. The laryngeal lymphatic vessels superior to the vocal folds accompany the superior laryngeal artery through the thyrohyoid membrane and drain into the **superior deep cervical lymph nodes.** The lymphatic vessels inferior to the vocal folds drain into the *pretracheal* or *paratracheal lymph nodes,* which drain into the **inferior deep cervical lymph nodes** (Fig. 8.39).

Nerves of Larynx. The nerves of the larynx are the superior and inferior laryngeal branches of the vagus nerves (CN X). The **superior laryngeal nerve** arises from the **inferior vagal ganglion** at the superior end of the carotid triangle (Fig. 8.40). The nerve divides into two terminal branches within the carotid sheath: the internal laryngeal nerve (sensory and autonomic) and the external laryngeal nerve (motor).

The **internal laryngeal nerve,** the larger of the terminal branches of the superior laryngeal nerve, pierces the thyrohyoid membrane with the superior laryngeal artery, supplying sensory fibers to the laryngeal mucous membrane of the laryngeal vestibule and middle laryngeal cavity, including the superior surface of the vocal folds. The **external laryngeal nerve,** the smaller terminal branch of the superior laryngeal nerve, descends posterior to the sternothyroid muscle in company with the superior thyroid artery. At first, the external laryngeal nerve lies on the inferior pharyngeal constrictor; it then pierces the muscle, contributing to its innervation (with the pharyngeal plexus), and continues to supply the cricothyroid muscle.

The **inferior laryngeal nerve,** the continuation of the recurrent laryngeal nerve (a branch of the vagus nerve), enters the larynx by passing deep to the inferior border of the inferior pharyngeal constrictor and medial to the lamina of the thyroid cartilage (Figs. 8.37, 8.39, and 8.40). It divides into anterior and posterior branches, which accompany the inferior laryngeal artery into the larynx. The anterior branch supplies the lateral crico-arytenoid, thyro-arytenoid, vocalis, ary-epiglottic, and thyro-epiglottic muscles. The posterior branch supplies the posterior crico-arytenoid and transverse and oblique arytenoid muscles. Because it supplies all the intrinsic muscles except the cricothyroid, the inferior laryngeal nerve is the primary motor nerve of the larynx. However, it also provides sensory fibers to the mucosa of the infraglottic cavity.

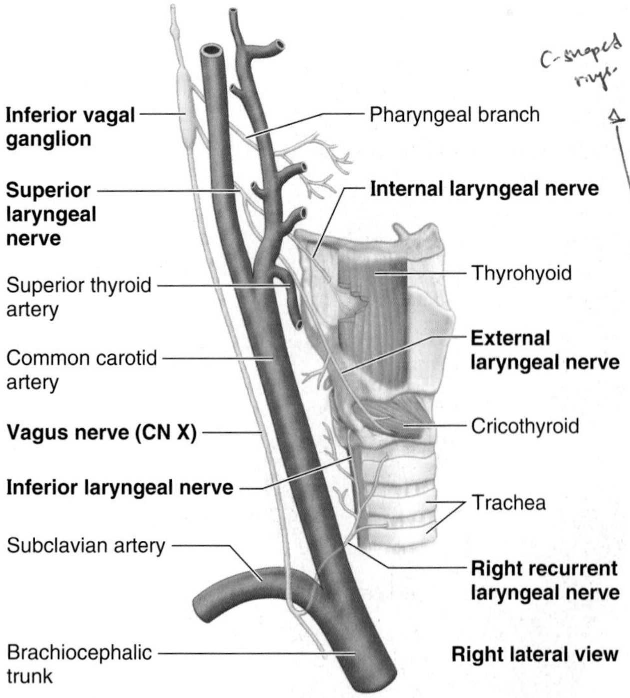

TRACHEA

The **trachea,** extending from the larynx into the thorax, terminates inferiorly as it divides into right and left main bronchi. It transports air to and from the lungs, and its epithelium propels debris-laden mucus toward the pharynx for expulsion from the mouth. The trachea is a fibrocartilaginous tube, supported by incomplete cartilaginous **tracheal cartilages** (rings), that occupies a median position in the neck (Fig. 8.37). The tracheal cartilages keep the trachea patent; they are deficient posteriorly where the trachea is adjacent to the esophagus. The posterior gaps in the tracheal rings are spanned by the involuntary **trachealis muscle,** smooth muscle connecting the ends of the rings (Fig. 8.41). Hence the posterior wall of the trachea is flat.

In adults, the trachea is approximately 2.5 cm in diameter, whereas in infants it has the diameter of a pencil. The trachea extends from the inferior end of the larynx at the level of the C6 vertebra. It ends at the level of the sternal angle or the T4–T5 IV disc, where it divides into the right and left main bronchi (see Chapter 1, p. 114).

Lateral to the trachea are the common carotid arteries and the lobes of the thyroid gland (Fig. 8.39). Inferior to the isthmus of the thyroid gland are the jugular venous arch and the inferior thyroid veins (Fig. 8.16). The brachiocephalic trunk is related to the right side of the trachea in the root of

FIGURE 8.40. Laryngeal branches of right vagus nerve (CN X). The nerves of the larynx are the internal and external branches of the superior laryngeal nerve and the inferior laryngeal nerve from the recurrent laryngeal nerve. The right recurrent laryngeal nerve passes inferior to the right subclavian artery.

Labels: Inferior vagal ganglion; Superior laryngeal nerve; Superior thyroid artery; Common carotid artery; Vagus nerve (CN X); Inferior laryngeal nerve; Subclavian artery; Brachiocephalic trunk; Pharyngeal branch; Internal laryngeal nerve; Thyrohyoid; External laryngeal nerve; Cricothyroid; Trachea; Right recurrent laryngeal nerve; Right lateral view; C-shaped rings

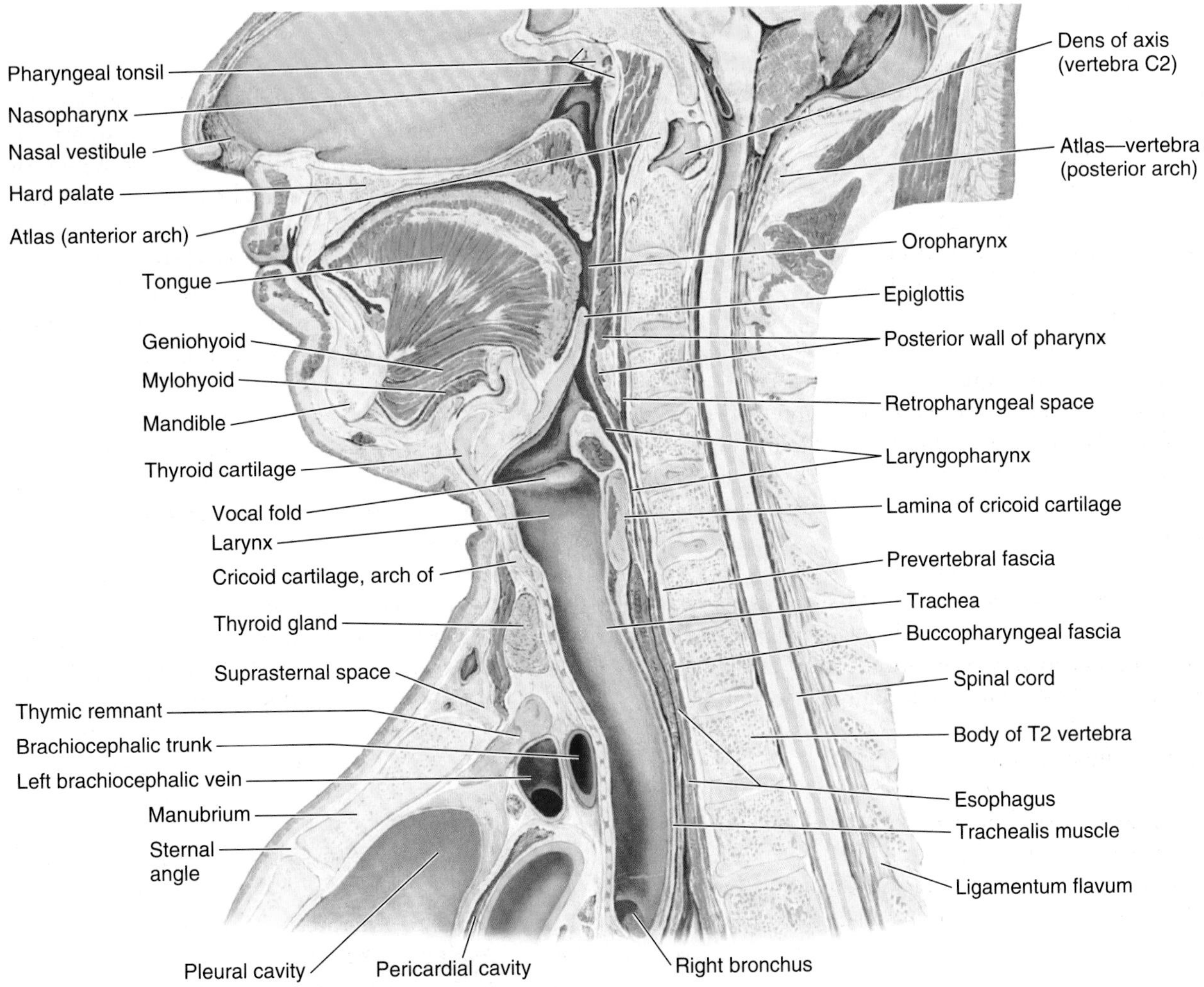

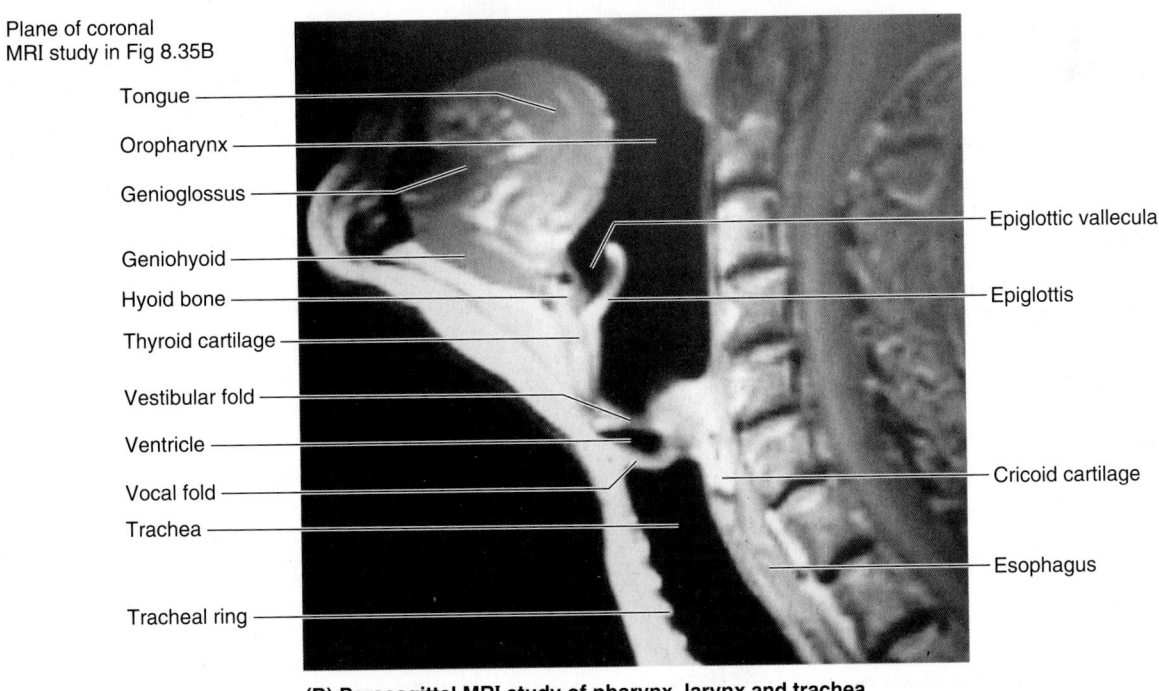

Pharyngeal tonsil

Nasopharynx

Nasal vestibule

Hard palate

Atlas (anterior arch)

Tongue

Geniohyoid

Mylohyoid

Mandible

Thyroid cartilage

Vocal fold

Larynx

Cricoid cartilage, arch of

Thyroid gland

Suprasternal space

Thymic remnant

Brachiocephalic trunk

Left brachiocephalic vein

Manubrium

Sternal angle

Pleural cavity

Pericardial cavity

Right bronchus

Dens of axis (vertebra C2)

Atlas—vertebra C1 (posterior arch)

Oropharynx

Epiglottis

Posterior wall of pharynx

Retropharyngeal space

Laryngopharynx

Lamina of cricoid cartilage

Prevertebral fascia

Trachea

Buccopharyngeal fascia

Spinal cord

Body of T2 vertebra

Esophagus

Trachealis muscle

Ligamentum flavum

(A) Medial view of right half of bisected head and neck (median section)

Plane of coronal MRI study in Fig 8.35B

Tongue

Oropharynx

Genioglossus

Geniohyoid

Hyoid bone

Thyroid cartilage

Vestibular fold

Ventricle

Vocal fold

Trachea

Tracheal ring

Epiglottic vallecula

Epiglottis

Cricoid cartilage

Esophagus

(B) Parasagittal MRI study of pharynx, larynx and trachea

FIGURE 8.41. Median sections of head and neck. A. The pharynx extends from the cranial base to the level of the cricoid cartilage (body of C6 vertebra or the C6–C7 IV disc, as shown here), where it is continuous with the esophagus. **B.** This sagittal section does not demonstrate the continuities of the upper respiratory tract because the soft palate is elevated, closing off the nasopharynx, and the plane of section passes through the vestibular and vocal folds to the side of the rima glottidis. (Part **B** Courtesy of Dr. W. Kucharczyk, University Health Network, Toronto, Ontario, Canada.)

a of the trachea from the midline, appar-
r radiographically, often signals the pres-
ical process. Tracheal trauma often affects
ent esophagus.

Alimentary Layer of Cervical Viscera

In the **alimentary layer,** cervical viscera take part in the digestive functions of the body. Although the pharynx conducts air to the larynx, trachea, and lungs, the pharyngeal constrictors direct (and the *epiglottis* deflects) food to the esophagus. The esophagus, also involved in food propulsion, is the beginning of the *alimentary canal* (digestive tract).

PHARYNX

The **pharynx** is the superior expanded part of the alimentary system posterior to the nasal and oral cavities, extending inferiorly past the larynx (Figs. 8.41 and 8.42). The pharynx extends from the *cranial base* to the inferior border of the cricoid cartilage anteriorly and the inferior border of the C6 vertebra posteriorly. The pharynx is widest (approximately 5 cm) opposite the hyoid and narrowest (approximately 1.5 cm) at its inferior end, where it is continuous with the esophagus. The flat posterior wall of the pharynx lies against the prevertebral layer of deep cervical fascia.

Interior of Pharynx. The pharynx is divided into three parts:

- *Nasopharynx:* posterior to the nose and superior to the soft palate.
- *Oropharynx:* posterior to the mouth.
- *Laryngopharynx:* posterior to the larynx.

The **nasopharynx** has a respiratory function; it is the posterior extension of the nasal cavities (Figs. 8.41–8.43). The nose opens into the nasopharynx through two **choanae** (paired openings between the nasal cavity and the nasopharynx). The roof and posterior wall of the nasopharynx form a continuous surface that lies inferior to the body of the sphenoid bone and the basilar part of the occipital bone (Fig. 8.42).

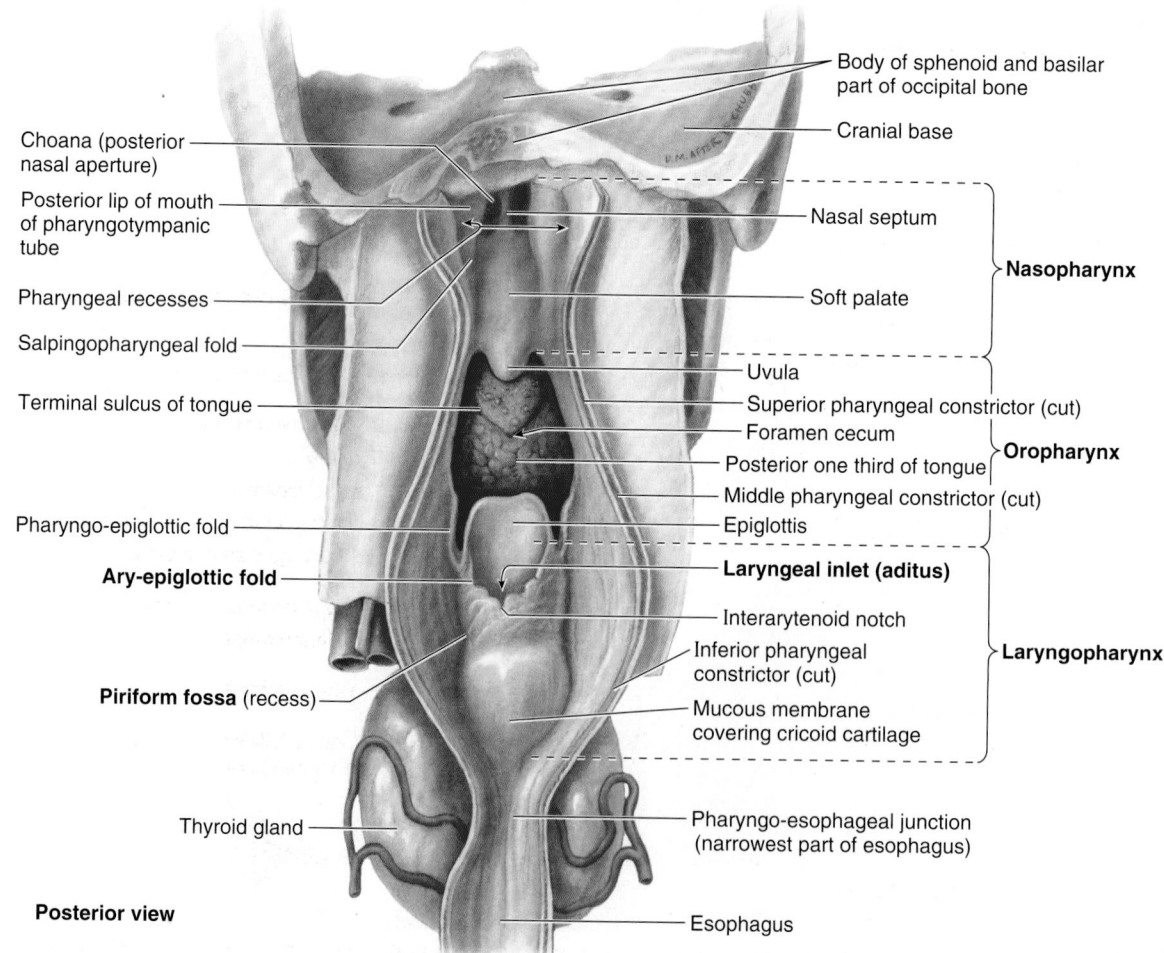

FIGURE 8.42. Anterior wall of pharynx. In this dissection, the posterior wall has been incised along the midline and spread apart. Openings in the anterior wall communicate with the nasal, oral, and laryngeal cavities. On each side of the laryngeal inlet, separated from it by the ary-epiglottic fold, a piriform fossa (recess) is formed by the invagination of the larynx into the anterior wall of the laryngopharynx.

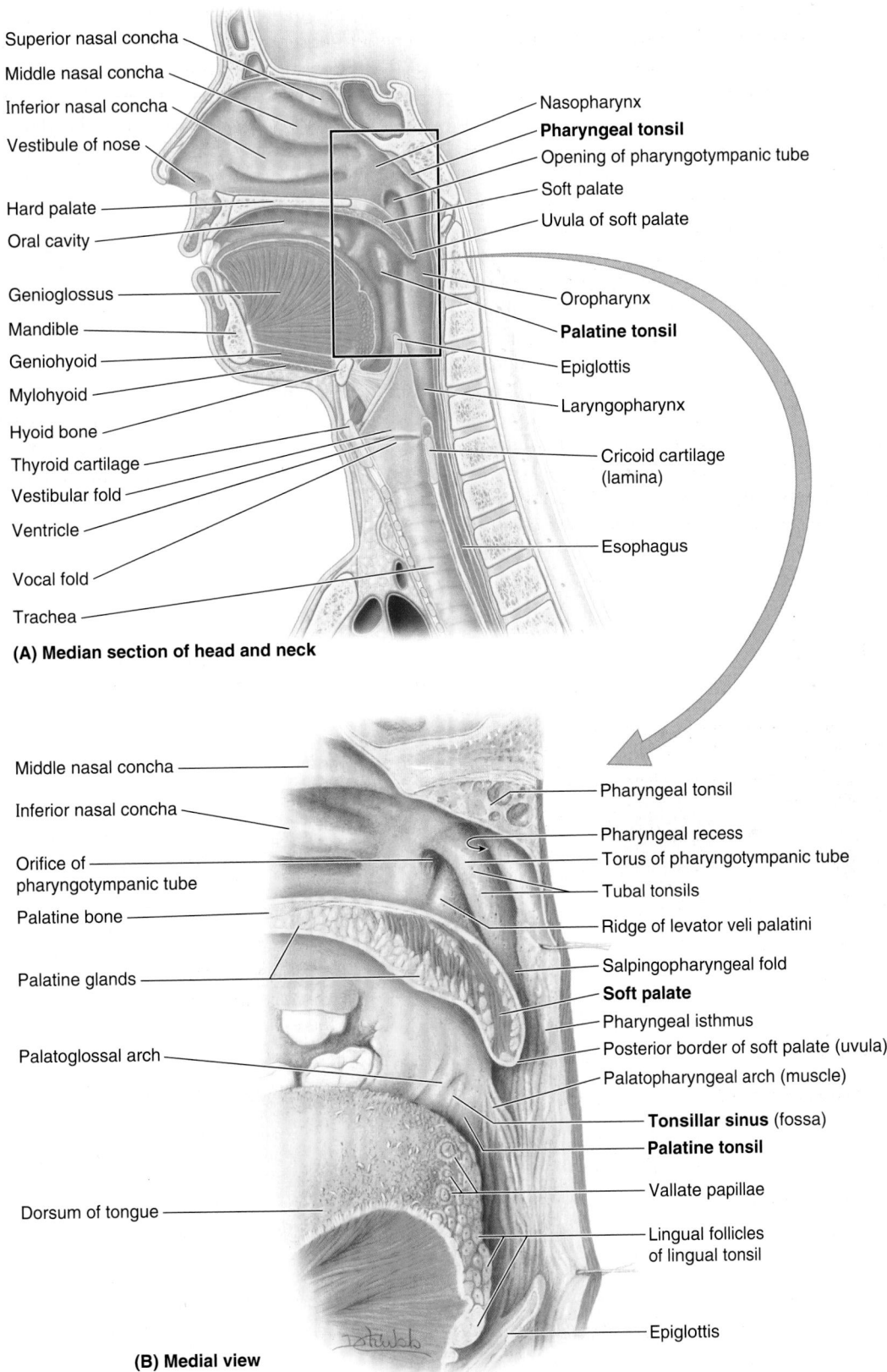

Superior nasal concha
Middle nasal concha
Inferior nasal concha
Vestibule of nose
Hard palate
Oral cavity
Genioglossus
Mandible
Geniohyoid
Mylohyoid
Hyoid bone
Thyroid cartilage
Vestibular fold
Ventricle
Vocal fold
Trachea

Nasopharynx
Pharyngeal tonsil
Opening of pharyngotympanic tube
Soft palate
Uvula of soft palate
Oropharynx
Palatine tonsil
Epiglottis
Laryngopharynx
Cricoid cartilage (lamina)
Esophagus

(A) Median section of head and neck

Middle nasal concha
Inferior nasal concha
Orifice of pharyngotympanic tube
Palatine bone
Palatine glands
Palatoglossal arch
Dorsum of tongue

Pharyngeal tonsil
Pharyngeal recess
Torus of pharyngotympanic tube
Tubal tonsils
Ridge of levator veli palatini
Salpingopharyngeal fold
Soft palate
Pharyngeal isthmus
Posterior border of soft palate (uvula)
Palatopharyngeal arch (muscle)
Tonsillar sinus (fossa)
Palatine tonsil
Vallate papillae
Lingual follicles of lingual tonsil
Epiglottis

(B) Medial view

FIGURE 8.43. Internal aspect of lateral wall of pharynx. A. The upper respiratory passages and alimentary canal in the right half of a bisected head and neck are shown. The *rectangle* indicates the location of the section shown in part **B. B.** A closer view of the nasopharynx and oropharynx, which are separated anteriorly by the soft palate, is provided. The posterior border of the soft palate forms the anterior margin of the pharyngeal isthmus through which the two spaces communicate posteriorly.

Neck

1033

...mphoid tissue in the pharynx forms an *ring* around the superior part of the phar- in this chapter, p. 1038). The lymphoid in certain regions to form masses called *tonsils*. The **pharyngeal tonsil** (commonly called the adenoid when enlarged) is in the mucous membrane of the roof and posterior wall of the nasopharynx (Figs. 8.41A and 8.43). Extending inferiorly from the medial end of the pharyngotympanic tube is a vertical fold of mucous membrane, the **salpingopharyngeal fold** (Figs. 8.42 and 8.43B). It covers the salpingopharyngeus muscle, which opens the pharyngeal orifice of the pharyngotympanic tube during swallowing. The collection of lymphoid tissue in the submucosa of the pharynx near the nasopharyngeal opening, or orifice of the pharyngotympanic tube, is the **tubal tonsils** (Fig. 8.43B). Posterior to the **torus of the pharyngotympanic tube** and the salpingopharyngeal fold is a slit-like lateral projection of the pharynx, the **pharyngeal recess,** which extends laterally and posteriorly.

The **oropharynx** has a digestive function. It is bounded by the soft palate superiorly, the base of the tongue inferiorly, and the palatoglossal and palatopharyngeal arches laterally (Figs. 8.43 and 8.44A). It extends from the soft palate to the superior border of the epiglottis.

Deglutition (swallowing) is the complex process that transfers a food bolus from the mouth through the pharynx and esophagus into the stomach. Solid food is masticated (chewed) and mixed with saliva to form a soft bolus (mass) that is easier to swallow. Deglutition occurs in three stages:

- **Stage 1:** voluntary; the bolus is compressed against the palate and pushed from the mouth into the oropharynx, mainly by movements of the muscles of the tongue and soft palate (Fig. 8.45A & B).
- **Stage 2:** involuntary and rapid; the soft palate is elevated, sealing off the nasopharynx from the oropharynx and laryngopharynx (Fig. 8.45C). The pharynx widens and shortens to receive the bolus of food as the suprahyoid muscles and longitudinal pharyngeal muscles contract, elevating the larynx.
- **Stage 3:** involuntary; sequential contraction of all three pharyngeal constrictor muscles creates a peristaltic ridge that forces the food bolus inferiorly into the esophagus (Fig. 8.45B-D).

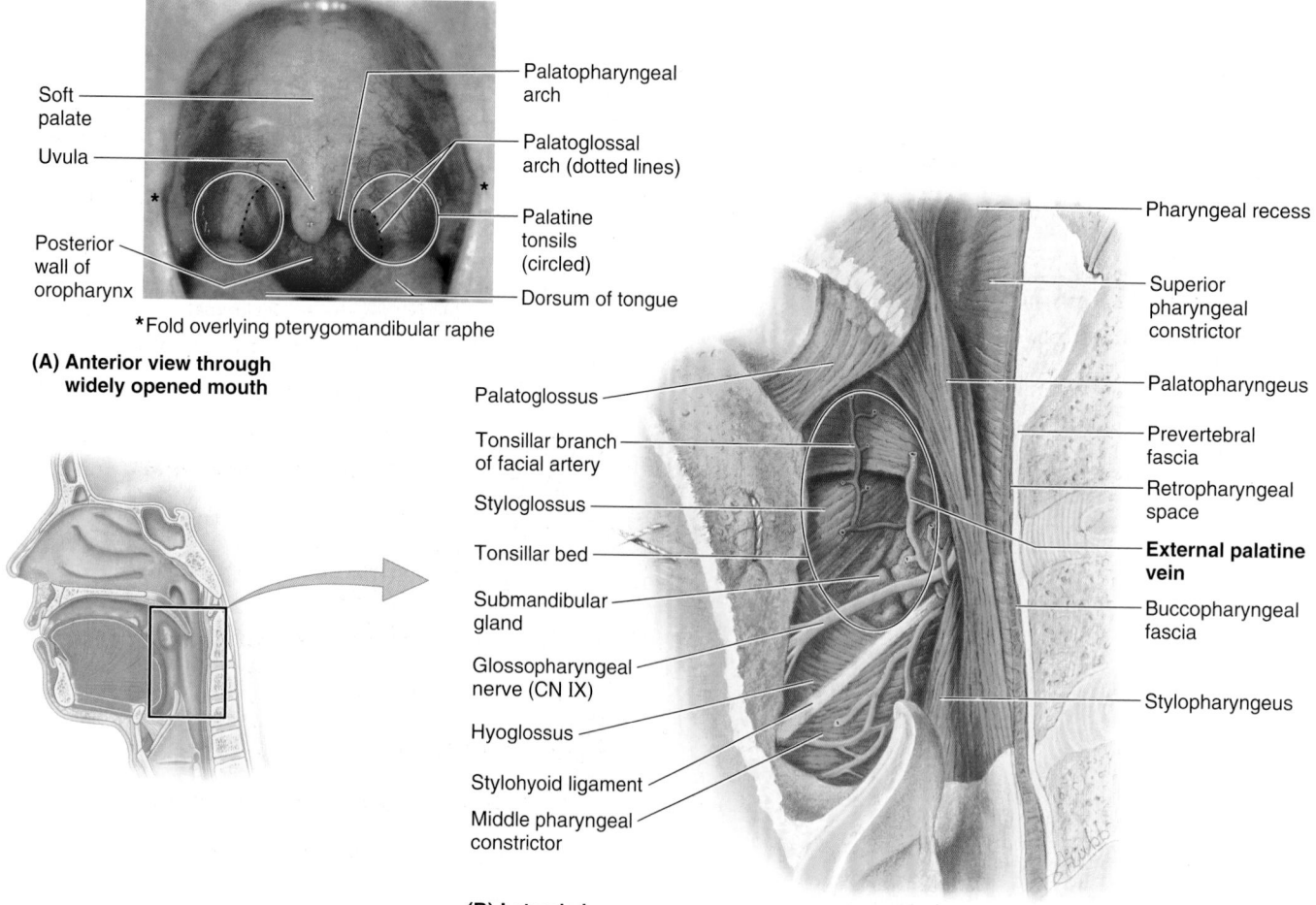

FIGURE 8.44. Oral cavity and tonsillar bed. A. The oral cavity and palatine tonsils in a young child, with the mouth wide open and the tongue protruding as far as possible. The uvula is a muscular projection from the posterior edge of the soft palate. **B.** In this deep dissection of the tonsillar bed, the palatine tonsil has been removed. The tongue is pulled anteriorly, and the inferior (lingual) attachment of the superior pharyngeal constrictor muscle is cut away.

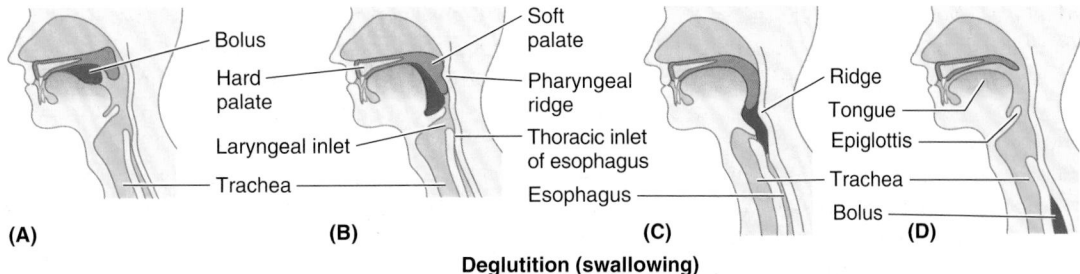

Deglutition (swallowing)

FIGURE 8.45. Deglutition. A. The bolus of food is squeezed to the back of the mouth by pushing the tongue against the palate. **B.** The nasopharynx is sealed off and the larynx is elevated, enlarging the pharynx to receive food. **C.** The pharyngeal sphincters contract sequentially, creating a "peristaltic ridge," squeezing food into the esophagus. The epiglottis deflects the bolus from but does not close the inlet to the larynx and trachea. **D.** The bolus of food moves down the esophagus by peristaltic contractions.

The **palatine tonsils** are collections of lymphoid tissue on each side of the oropharynx in the interval between the palatine arches (Figs. 8.43 and 8.44A). The tonsil does not fill the **tonsillar sinus** (fossa) between the palatoglossal and palatopharyngeal arches in adults. The submucosal **tonsillar bed,** in which the palatine tonsil lies, is between these arches (Fig. 8.44B). The tonsillar bed is formed by the superior pharyngeal constrictor and the thin, fibrous sheet of **pharyngobasilar fascia** (Fig. 8.46A & B). This fascia blends with the periosteum of the cranial base and defines the limits of the pharyngeal wall in its superior part.

The **laryngopharynx** lies posterior to the larynx (Figs. 8.41A and 8.43), extending from the superior border of the epiglottis and the pharyngo-epiglottic folds to the inferior

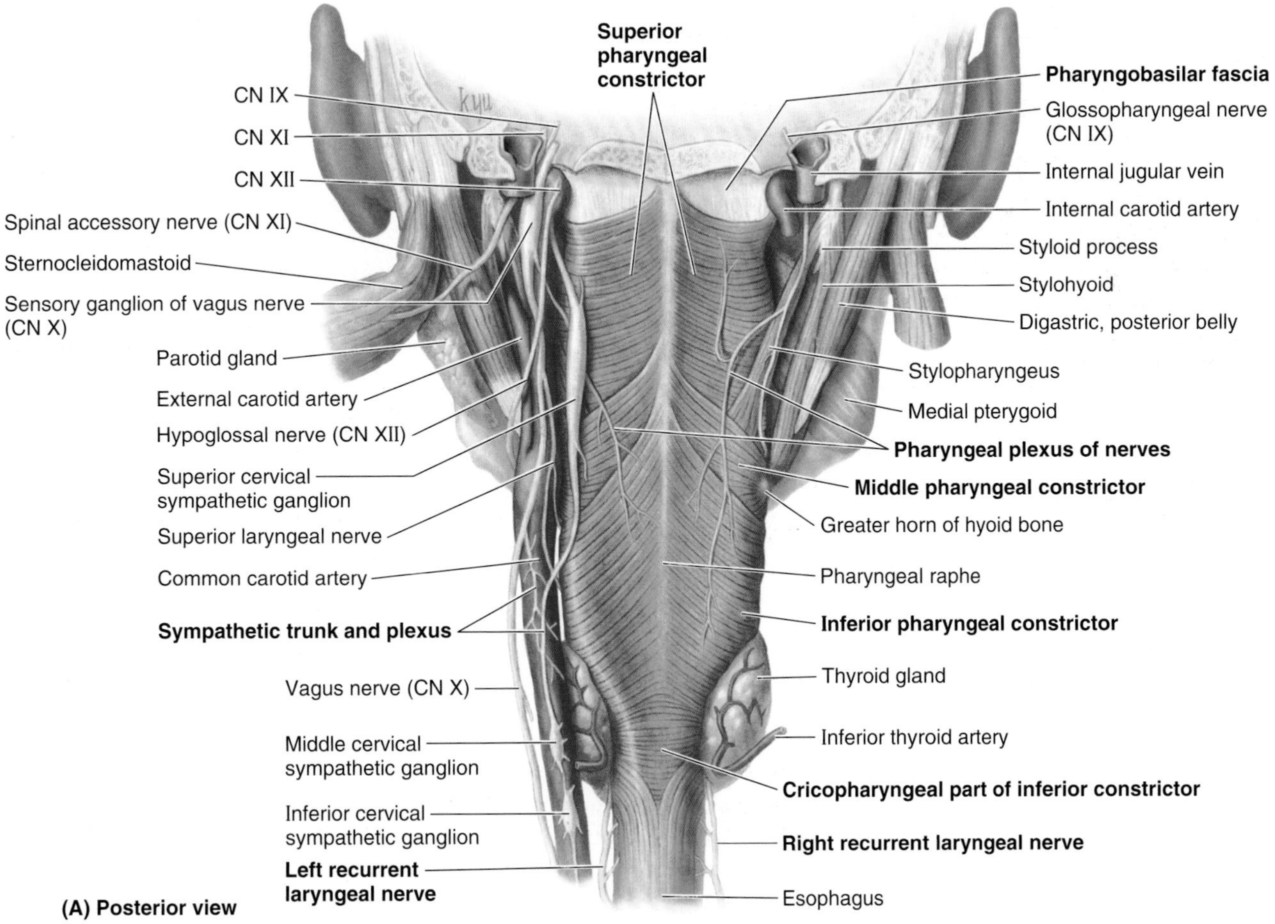

FIGURE 8.46. Pharynx and cranial nerves. A. This dissection shows the posterior aspect of the pharynx and associated structures. The buccopharyngeal fascia has been removed. Of the three pharyngeal constrictor muscles, the inferior muscle overlaps the middle one and the middle one overlaps the superior one. All three muscles form a common median pharyngeal raphe posteriorly.

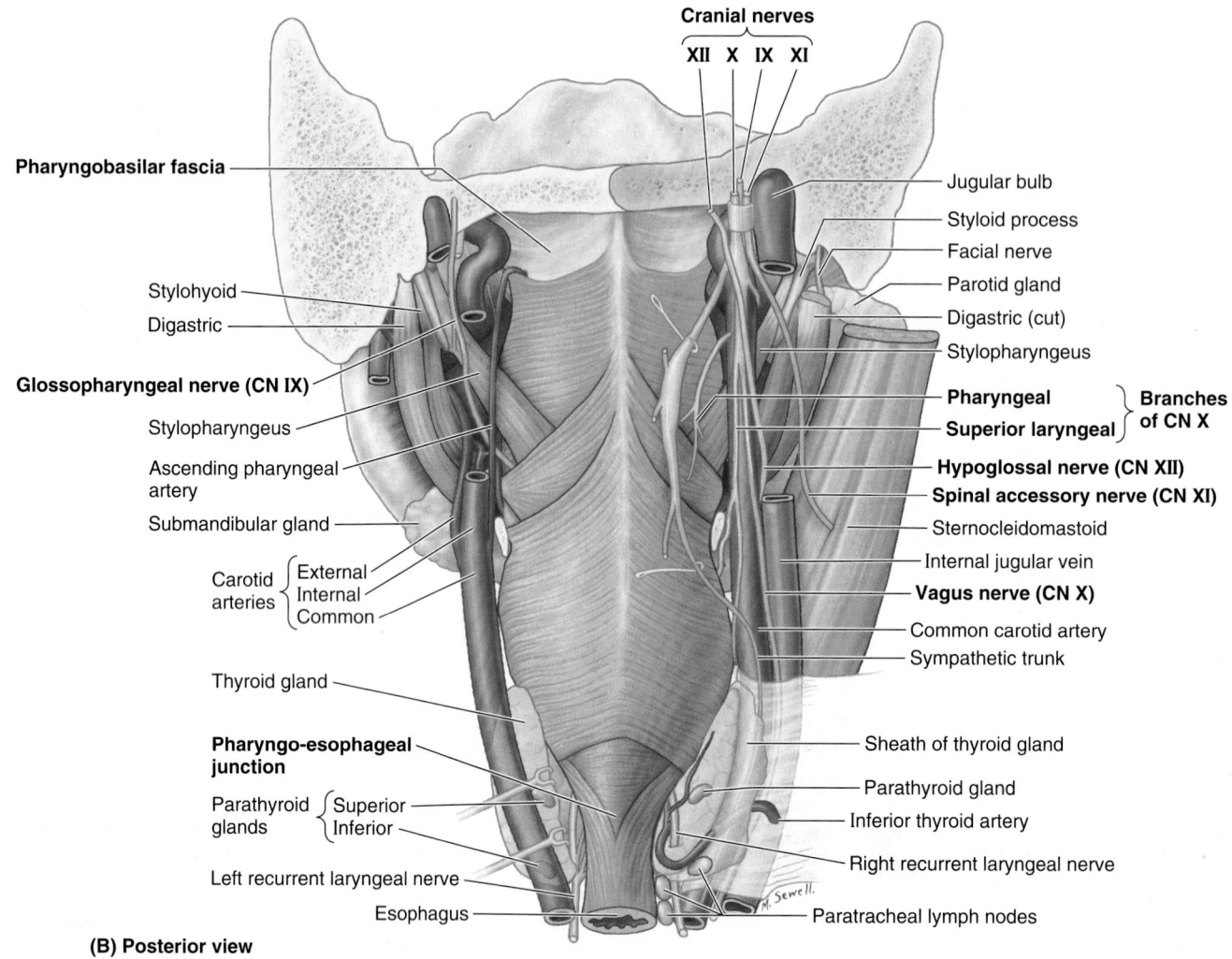

Cranial nerves

XII X IX XI

Pharyngobasilar fascia

Stylohyoid

Digastric

Glossopharyngeal nerve (CN IX)

Stylopharyngeus

Ascending pharyngeal artery

Submandibular gland

Carotid arteries { External / Internal / Common }

Thyroid gland

Pharyngo-esophageal junction

Parathyroid glands { Superior / Inferior }

Left recurrent laryngeal nerve

Esophagus

Jugular bulb

Styloid process

Facial nerve

Parotid gland

Digastric (cut)

Stylopharyngeus

Pharyngeal } Branches

Superior laryngeal } of CN X

Hypoglossal nerve (CN XII)

Spinal accessory nerve (CN XI)

Sternocleidomastoid

Internal jugular vein

Vagus nerve (CN X)

Common carotid artery

Sympathetic trunk

Sheath of thyroid gland

Parathyroid gland

Inferior thyroid artery

Right recurrent laryngeal nerve

Paratracheal lymph nodes

(B) Posterior view

FIGURE 8.46. (*Continued*) **Pharynx and cranial nerves. B.** The narrowest and least distensible part of the alimentary tract is the pharyngo-esophageal junction, where the laryngopharynx becomes the esophagus.

border of the cricoid cartilage, where it narrows and becomes continuous with the esophagus. Posteriorly, the laryngopharynx is related to the bodies of the C4–C6 vertebrae. Its posterior and lateral walls are formed by the *middle* and *inferior pharyngeal constrictor muscles* (Fig. 8.46A). Internally, the wall is formed by the *palatopharyngeus* and *stylopharyngeus muscles*. The laryngopharynx communicates with the larynx through the **laryngeal inlet** on its anterior wall (Fig. 8.42).

The **piriform fossa** (recess) is a small depression of the laryngopharyngeal cavity on either side of the laryngeal inlet. This mucosa-lined fossa is separated from the laryngeal inlet by the **ary-epiglottic fold.** Laterally, the piriform fossa is bounded by the medial surfaces of the thyroid cartilage and the *thyrohyoid membrane* (Fig. 8.39). Branches of the internal laryngeal and recurrent laryngeal nerves lie deep to the mucous membrane of the piriform fossa and are vulnerable to injury when a foreign body lodges in the fossa.

Pharyngeal Muscles. The wall of the pharynx is exceptional for the alimentary tract, having a muscular layer composed entirely of *voluntary muscle*, arranged with longitudinal muscles

internal to a circular layer of muscles. Most of the alimentary tract is composed of smooth muscle, with a layer of longitudinal muscle external to a circular layer. The external circular layer of pharyngeal muscles consists of three **pharyngeal constrictors: superior, middle,** and **inferior** (Figs. 8.44 and 8.46A & B). The internal longitudinal muscles consists of the **palatopharyngeus, stylopharyngeus,** and **salpingopharyngeus.** These muscles elevate the larynx and shorten the pharynx during swallowing and speaking. The pharyngeal muscles are illustrated in Figure 8.47, and their attachments, nerve supply, and actions of the pharyngeal muscles are described in Table 8.6.

The pharyngeal constrictors have a strong internal fascial lining, the *pharyngobasilar fascia* (Fig. 8.46B), and a thin external fascial lining, the *buccopharyngeal fascia* (Fig. 8.41A). Inferiorly, the buccopharyngeal fascia blends with the pretracheal layer of the *deep cervical fascia*. The pharyngeal constrictors contract involuntarily so that contraction takes place sequentially from the superior to the inferior end of the pharynx, propelling food into the esophagus. All three pharyngeal constrictors are supplied by the *pharyngeal*

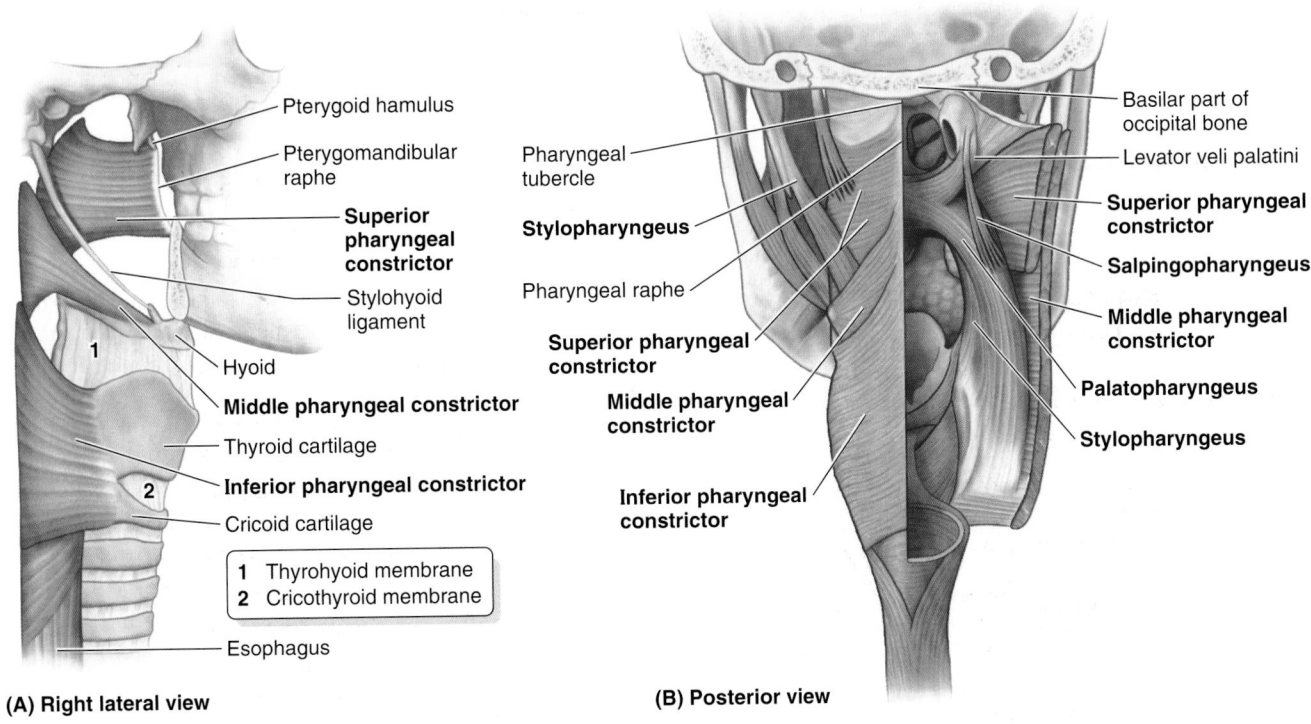

(A) Right lateral view   (B) Posterior view

FIGURE 8.47. Muscles of pharynx.

TABLE 8.6. MUSCLES OF PHARYNX

Muscle	Origin	Insertion	Innervation	Main Action(s)
External layer				
Superior pharyngeal constrictor	Pterygoid hamulus, pterygomandibular raphe, posterior end of mylohyoid line of mandible, and side of tongue	Pharyngeal tubercle on basilar part of occipital bone	Pharyngeal branch of vagus (CN X) and pharyngeal plexus	Constrict walls of pharynx during swallowing
Middle pharyngeal constrictor	Stylohyoid ligament and greater and lesser horns of hyoid	Pharyngeal raphe	Pharyngeal branch of vagus (CN X) and pharyngeal plexus, plus branches of external and recurrent laryngeal nerves of vagus	
Inferior pharyngeal constrictor	Oblique line of thyroid cartilage and side of cricoid cartilage	Cricopharyngeal part encircles pharyngo-esophageal junction without forming a raphe		
Internal layer				
Palatopharyngeus	Hard palate and palatine aponeurosis	Posterior border of lamina of thyroid cartilage and side of pharynx and esophagus	Pharyngeal branch of vagus (CN X) and pharyngeal plexus	Elevate (shorten and widen) pharynx and larynx during swallowing and speaking
Salpingo-pharyngeus	Cartilaginous part of pharyngotympanic tube	Blends with palatopharyngeus		
Stylopharyngeus	Styloid process of temporal bone	Posterior and superior borders of thyroid cartilage with palatopharyngeus	Glossopharyngeal nerve (CN IX)	

s that is formed by pharyngeal branches of glossopharyngeal nerves and by sympathetic ...n the superior cervical ganglion (Fig. 8.46A; Table 8.6). The pharyngeal plexus lies on the lateral wall of the pharynx, mainly on the middle pharyngeal constrictor.

The overlapping of the pharyngeal constrictor muscles leaves four gaps in the musculature for structures to enter or leave the pharynx (Fig. 8.47):

1. Superior to the superior pharyngeal constrictor, the levator veli palatini, pharyngotympanic tube, and ascending palatine artery pass through a *gap between the superior pharyngeal constrictor and the cranium.* It is here that the pharyngobasilar fascia blends with the buccopharyngeal fascia to form, with the mucous membrane, the thin wall of the pharyngeal recess (Fig. 8.42).
2. A *gap between the superior and middle pharyngeal constrictors* forms a passageway that allows the stylopharyngeus, glossopharyngeal nerve, and stylohyoid ligament to pass to the internal aspect of the pharyngeal wall (Fig 8.47).
3. A *gap between the middle and inferior pharyngeal constrictors* allows the internal laryngeal nerve and superior laryngeal artery and vein to pass to the larynx.
4. A *gap inferior to the inferior pharyngeal constrictor* allows the recurrent laryngeal nerve and inferior laryngeal artery to pass superiorly into the larynx.

Vessels of Pharynx. A branch of the facial artery, the **tonsillar artery** (Fig. 8.44B) passes through the superior pharyngeal constrictor muscle and enters the inferior pole of the palatine tonsil. The tonsil also receives arterial twigs from the ascending palatine, lingual, descending palatine, and ascending pharyngeal arteries. The large **external palatine vein** (paratonsillar vein) descends from the soft palate and passes close to the lateral surface of the tonsil before it enters the pharyngeal venous plexus.

The **tonsillar lymphatic vessels** pass laterally and inferiorly to the lymph nodes near the angle of the mandible and the **jugulodigastric node,** referred to as the *tonsillar node* because of its frequent enlargement when the tonsil is inflamed (*tonsillitis*) (Fig. 8.48). The palatine, lingual, and pharyngeal tonsils form the **pharyngeal lymphatic** (tonsillar) **ring,** an incomplete circular band of lymphoid tissue around the superior part of the pharynx (Fig. 8.49). The antero-inferior part of the ring is formed by the **lingual tonsil** in the posterior part of the tongue. Lateral parts of the ring are formed by the palatine and tubal tonsils, and posterior and superior parts are formed by the pharyngeal tonsil.

Pharyngeal Nerves. The nerve supply to the pharynx (motor and most of sensory) derives from the **pharyngeal plexus of nerves** (Fig. 8.46A). Motor fibers in the plexus are derived from the vagus nerve (CN X) via its pharyngeal branch or branches. They supply all muscles of the pharynx and soft palate, except the stylopharyngeus (supplied by CN IX) and the tensor veli palatini (supplied by CN V₃). The inferior pharyngeal constrictor also receives some motor fibers from the external and recurrent laryngeal branches of the vagus. Sensory fibers in

the plexus are derived from the glossopharyngeal nerve. They are distributed to all three parts of the pharynx. In addition, the mucous membrane of the anterior and superior nasopharynx receives innervation from the maxillary nerve (CN V₂). The **tonsillar nerves** are derived from the *tonsillar plexus of nerves* formed by branches of the glossopharyngeal and vagus nerves.

ESOPHAGUS

The *esophagus* is a muscular tube that connects the pharynx to the stomach. It begins in the neck where it is continuous with the laryngopharynx at the **pharyngo-esophageal junction** (Figs. 8.42 and 8.46B). The esophagus consists of striated (voluntary) muscle in its upper third, smooth (involuntary) muscle in its lower third, and a mixture of striated and smooth muscle in between.

Its first part, the **cervical esophagus,** is part of the voluntary upper third. It begins immediately posterior to, and at the level of, the inferior border of the cricoid cartilage in the median plane. This is the level of the C6 vertebra.

Externally, the pharyngo-esophageal junction appears as a constriction produced by the **cricopharyngeal part of the inferior pharyngeal constrictor muscle** (the superior esophageal sphincter) and is the narrowest part of the esophagus. The cervical esophagus inclines slightly to the left as it descends and enters the superior mediastinum via the superior thoracic aperture, where it becomes the thoracic esophagus.

When the esophagus is empty, it is a slit-like lumen. When a food bolus descends in it, the lumen expands, eliciting reflex peristalsis in the inferior two thirds of the esophagus. The cervical esophagus lies between the trachea and the cervical vertebral column (Figs. 8.41 and 8.43A). It is attached to the trachea by loose connective tissue. The *recurrent laryngeal nerves* lie in or near the **tracheo-esophageal grooves** between the trachea and esophagus (Fig. 8.46). On the right of the esophagus is the right lobe of the *thyroid gland* and the right *carotid sheath* and its contents.

The esophagus is in contact with the cervical pleura at the root of the neck. On the left is the left lobe of the thyroid gland and the left carotid sheath. The thoracic duct adheres to the left side of the esophagus and lies between the pleura and the esophagus. For details concerning the thoracic and abdominal regions of the esophagus, see Chapters 1 and 2.

Vessels of Cervical Esophagus. The arteries to the cervical esophagus are branches of the *inferior thyroid arteries.* Each artery gives off ascending and descending branches that anastomose with each other and across the midline. Veins from the cervical esophagus are tributaries of the *inferior thyroid veins.* Lymphatic vessels of the cervical part of the esophagus drain into the *paratracheal lymph nodes* and *inferior deep cervical lymph nodes* (Fig. 8.48).

Nerves of Cervical Esophagus. The nerve supply to the esophagus is somatic motor and sensory to the upper half and parasympathetic (vagal), sympathetic, and visceral sensory to the lower half. The cervical esophagus receives somatic fibers via branches from the *recurrent laryngeal nerves* and vasomotor fibers from the *cervical sympathetic trunks* through the plexus around the inferior thyroid artery (Fig. 8.46).

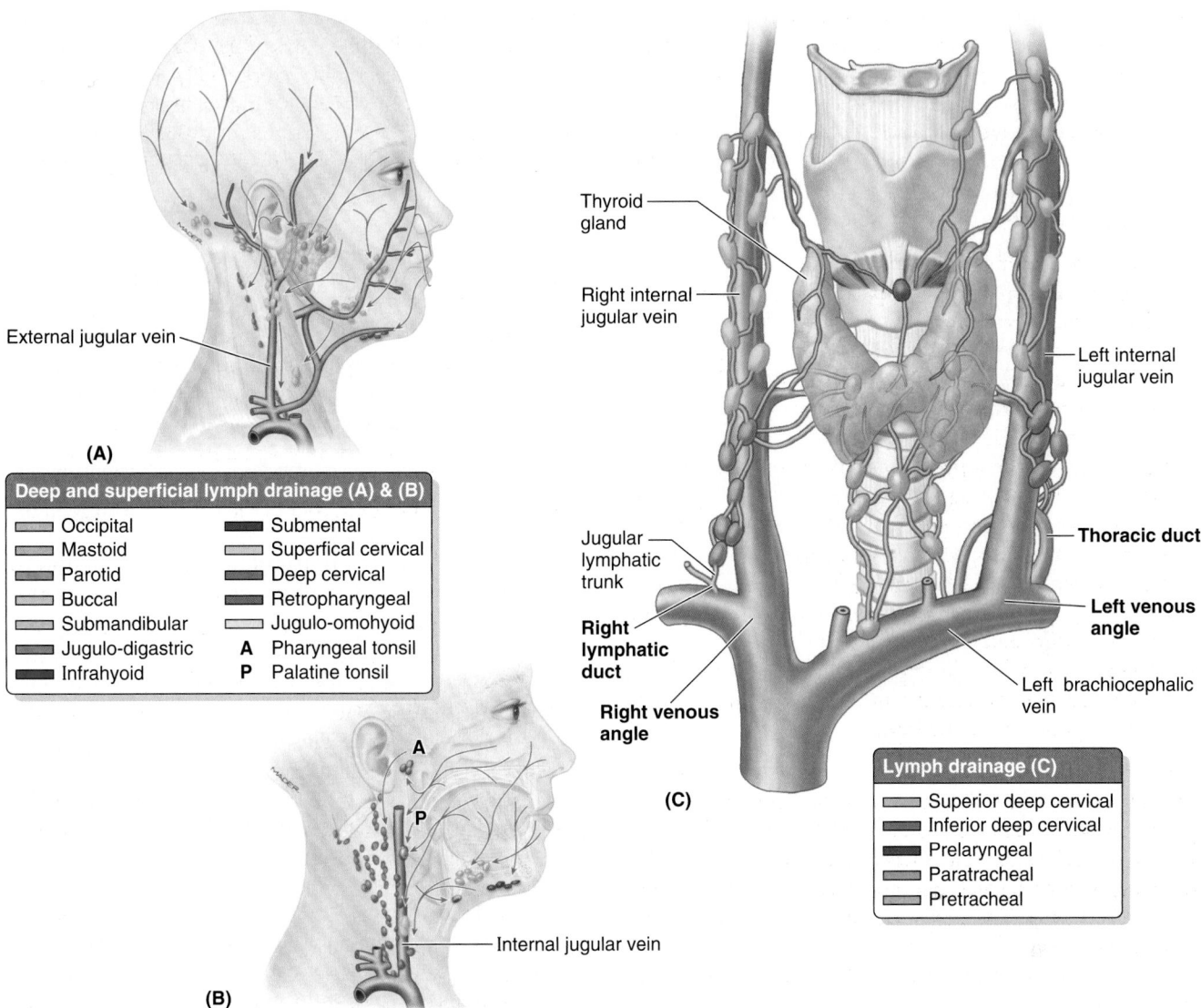

FIGURE 8.48. Lymphatic drainage of head and neck. A and B. The pathways of the superficial and deep lymphatic drainages are shown, respectively. **C.** The lymph nodes, lymphatic trunks, and thoracic duct are shown.

Surface Anatomy of Endocrine and Respiratory Layers of Cervical Viscera

The neck of an infant is short; therefore, the cervical viscera are located more superiorly in infants than in adults. The cervical viscera do not reach their final levels until after the 7th year. The elongation of the neck is accompanied by growth changes in the skin. Consequently, a midline incision in the inferior neck of an infant results in a scar that will lie over the superior part of the sternum as a child.

The U-shaped *hyoid bone* lies in the anterior part of the neck in the deep angle between the mandible and the thyroid cartilage at the level of the C3 vertebra (Fig. 8.50). Swallow, and the hyoid will move under your fingers when they are placed at the angle between the chin and anterior neck. The greater horn of one side of the hyoid is palpable only when the greater horn on the opposite side is steadied.

The *laryngeal prominence* is produced by the meeting of the laminae of the thyroid cartilage at an acute angle in the anterior midline. This *thyroid angle*, most acute in postpubertal males, forms the laryngeal prominence ("Adam's apple"), which is palpable and frequently visible. During palpation of the prominence, it can be felt to recede on swallowing. The vocal folds are at the level of the middle of the laryngeal prominence.

The *cricoid cartilage* can be felt inferior to the laryngeal prominence at the level of the C6 vertebra. Extend your neck as far as possible and run your finger over the laryngeal prominence. As your finger passes inferiorly from the prominence, feel the *cricothyroid ligament,* the site for a *needle cricothyrotomy* or *coniotomy* (see the blue box "Aspiration of Foreign Bodies and Heimlich Maneuver" on p. 1044). After your finger passes over the arch of the cricoid cartilage, note that your fingertip sinks in because the arch of the cartilage projects farther anteriorly than the rings of the

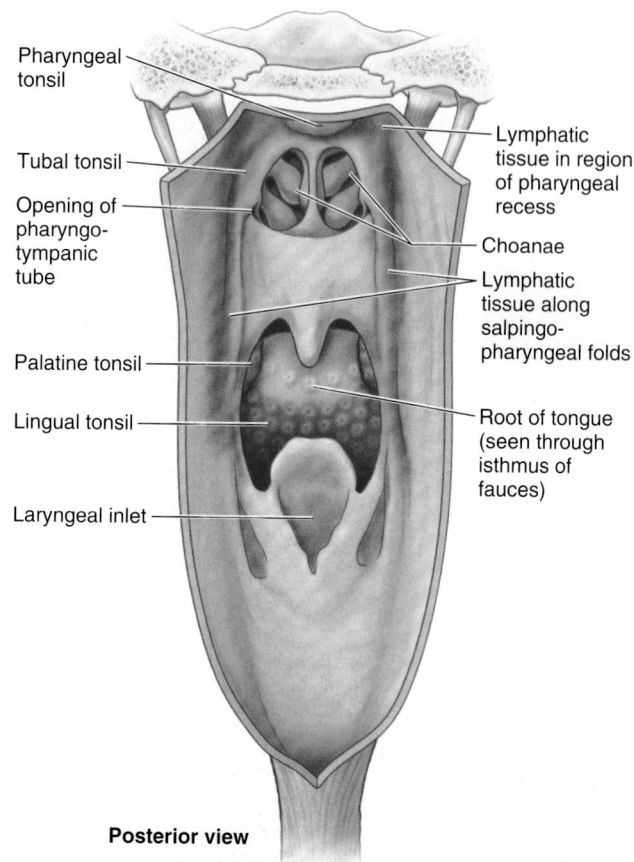

Pharyngeal tonsil

Tubal tonsil

Opening of pharyngo-tympanic tube

Palatine tonsil

Lingual tonsil

Laryngeal inlet

Lymphatic tissue in region of pharyngeal recess

Choanae

Lymphatic tissue along salpingo-pharyngeal folds

Root of tongue (seen through isthmus of fauces)

Posterior view

FIGURE 8.49. Lymphoid tissue in tongue and pharynx. The pharyngeal lymphatic (tonsillar) ring (*pink*) around the superior pharynx is formed of the pharyngeal, tubal, palatine, and lingual tonsils.

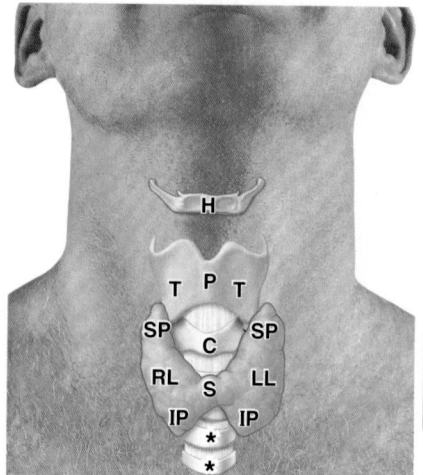

C	Cricoid cartilage
H	Hyoid bone
IP	Inferior pole of thyroid gland
LL	Left lobe of thyroid gland
P	Laryngeal prominence
RL	Right lobe of thyroid gland
S	Isthmus
SP	Superior pole of thyroid gland
T	Thyroid cartilage
*	Tracheal rings

FIGURE 8.50. Surface anatomy of endocrine and respiratory layers of neck.

trachea. The cricoid cartilage, a key landmark in the neck, indicates the:

- Level of the C6 vertebra.
- Site where the carotid artery can be compressed against the transverse process of the C6 vertebra.
- Junction of the larynx and trachea.
- Joining of the pharynx and esophagus.
- Point where the recurrent laryngeal nerve enters the larynx.
- Site that is approximately 3 cm superior to the isthmus of the thyroid gland.

The **first tracheal cartilage** is broader than the others and is palpable (Fig. 8.32A). The second through fourth cartilages cannot be felt because the thyroid isthmus connecting the right and left lobes of the thyroid covers them.

The *thyroid gland* may be palpated by anterior or posterior approaches (i.e., standing in front of or behind the person). Place your fingertips anterior (for the isthmus) or immediately lateral (for the lobes) to the trachea and then direct the person to swallow (see Bickley, 2009, for details). Although both approaches to examining the thyroid are performed, the posterior approach usually allows better palpation, but the anterior approach allows observation. A perfectly normal thyroid gland may not be visible or distinctly palpable in some females, except during menstruation or pregnancy. The normal gland has the consistency of muscle tissue.

The *isthmus of the thyroid gland* lies immediately inferior to the cricoid cartilage; it extends approximately 1.25 cm on either side of the midline. It can usually be felt by placing the fingertips of one hand on the midline below the cricoid arch and then asking the person to swallow. The isthmus will be felt moving up and then down. The apex of each *lobe of the thyroid gland* extends superiorly to the middle of the lamina of the thyroid cartilage (Fig. 8.50).

The surface anatomy of the posterior aspect of the neck is described in Chapter 4 (p. 492). Key points are the following:

- The *spinous processes of the C6 and C7 vertebrae* are palpable and visible, especially when the neck is flexed.
- The *transverse processes of the C1, C6, and C7 vertebrae* are palpable.
- The *tubercles of the C1 vertebra* can be palpated by deep pressure posteroinferior to the tips of the mastoid processes.

VISCERA OF NECK

Thyroid Ima Artery

In approximately 10% of people, a small, unpaired *thyroid ima artery* (L. *arteria thyroidea ima*) arises from the brachiocephalic trunk (Fig. B8.5); however, it may arise from the arch of the aorta or from the right common carotid, subclavian, or internal thoracic arteries. This small ima artery ascends on the anterior surface of the trachea to the isthmus of the thyroid gland, supplying branches to both structures. The possible presence of this artery must be considered when performing procedures in the midline of the

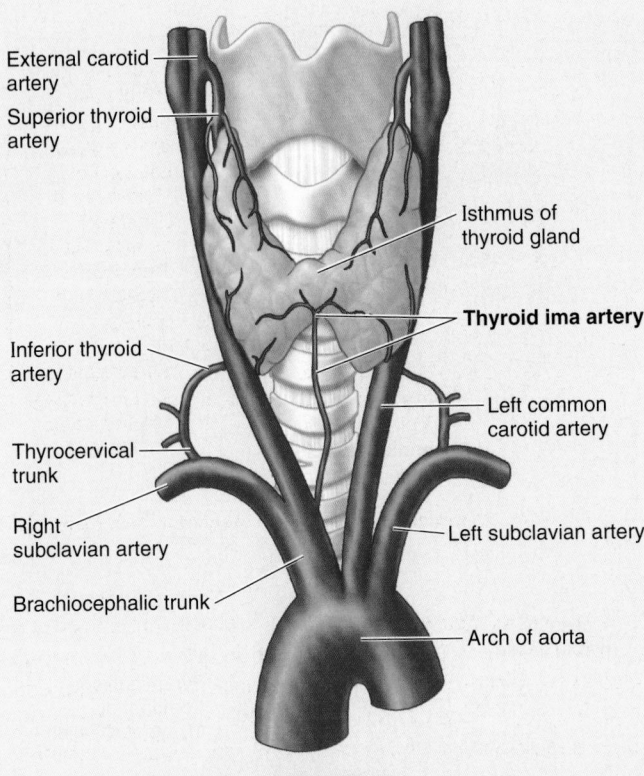

FIGURE B8.5.

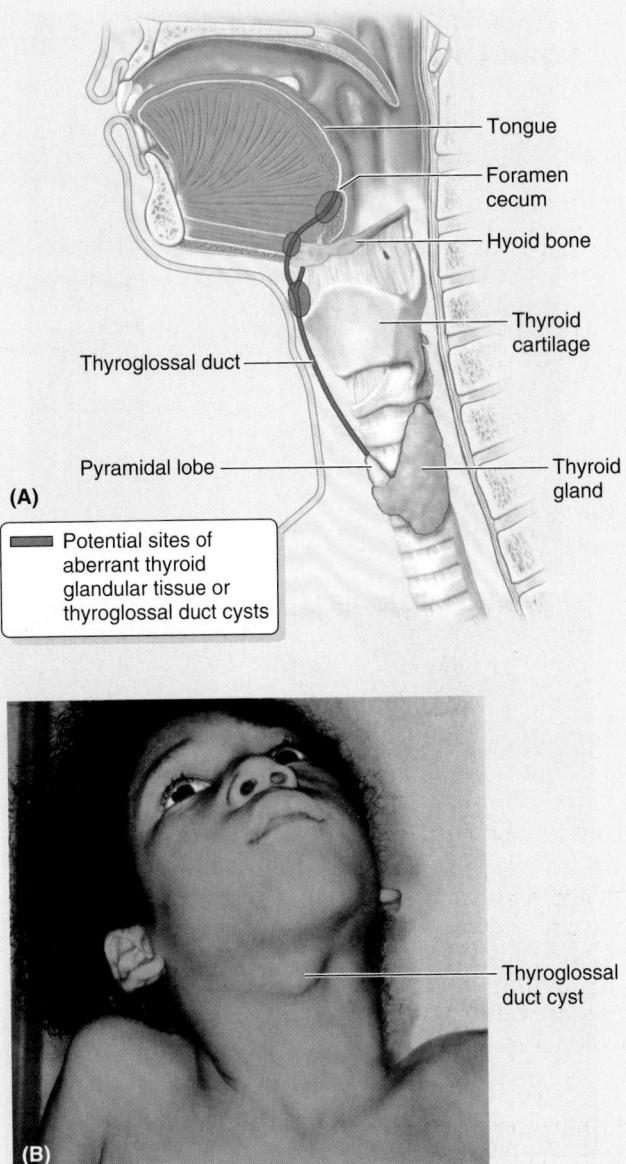

(A)

Potential sites of aberrant thyroid glandular tissue or thyroglossal duct cysts

FIGURE B8.6. **A.** Thyroglossal duct vestiges. **B.** Child with thyroglossal duct cyst.

neck inferior to the isthmus, because it is a potential source of bleeding (see the blue box "Tracheostomy," p. 1045).

Thyroglossal Duct Cysts

Development of the thyroid gland begins in the floor of the embryonic pharynx at the site indicated by a small pit, the *foramen cecum,* in the dorsum of the postnatal tongue (Chapter 7, p. 940). Subsequently, the developing gland relocates from the tongue into the neck, passing anterior to the hyoid and thyroid cartilages to reach its final position anterolateral to the superior part of the trachea (Moore et al., 2012). During this relocation, the thyroid gland is attached to the foramen cecum by the **thyroglossal duct.** This duct normally disappears but remnants of epithelium may remain and form a *thyroglossal duct cyst* at any point along the path of its descent (Fig. B8.6A). The cyst is usually in the neck, close or just inferior to the hyoid, and forms a swelling in the anterior part of the neck. Surgical excision of the cyst may be necessary. Most thyroglossal duct cysts are in the neck, close or just inferior to the body of the hyoid (Fig. B8.6B).

Aberrant Thyroid Gland

Aberrant thyroid glandular tissue may be found anywhere along the path of the embryonic thyroglossal duct. Although uncommon, the thyroglossal duct carrying thyroid-forming tissue at its distal end may fail to relocate to its definitive position in the neck. Aberrant thyroid tissue may be in the root of the tongue, just posterior to the foramen cecum, resulting in a **lingual thyroid gland,** or in the neck, at

or just inferior to the hyoid (Fig. B8.7A). Cystic remnants of the thyroglossal duct may be differentiated from an undescended thyroid by radioisotope scanning (Fig. B8.7B). As a rule, an ectopic thyroid gland in the median plane of the neck is the only thyroid tissue present. Occasionally, thyroid glandular tissue is associated with a thyroglossal duct cyst. Therefore, it is important to differentiate between an ectopic thyroid gland and a thyroglossal duct cyst when excising a cyst. Failure to do so may result in a *total thyroidectomy,* leaving the person permanently dependent on thyroid medication (Leung et al., 1995).

Accessory Thyroid Glandular Tissue

Portions of the thyroglossal duct may persist to form thyroid tissue. *Accessory thyroid glandular tissue* may appear anywhere along the embryonic course

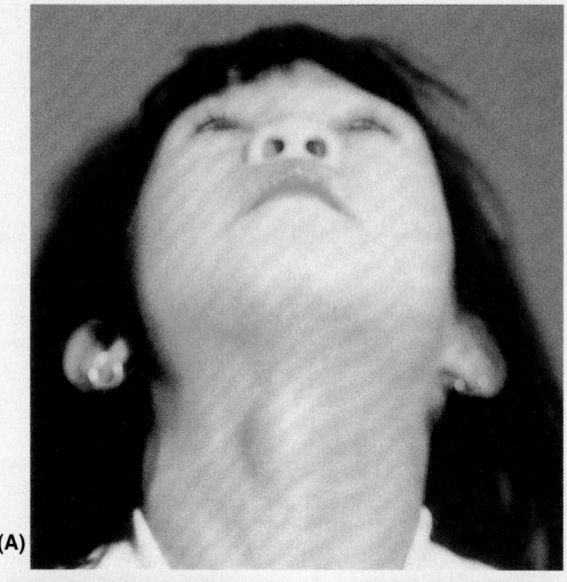

(A)

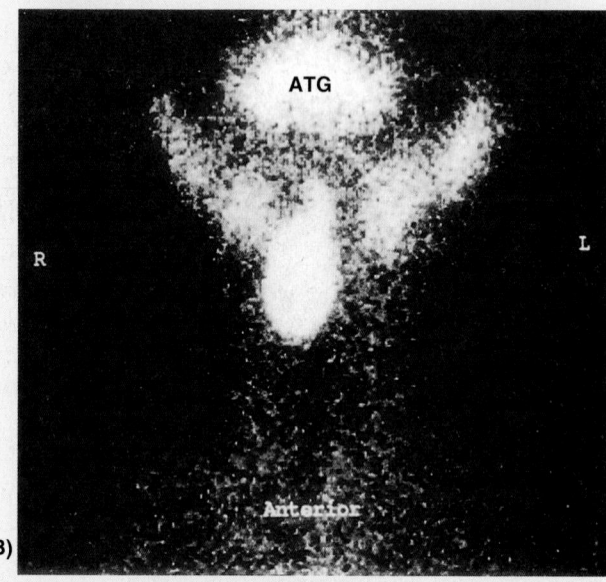

ATG

R   L

Anterior

(B)

FIGURE B8.7. Aberrant thyroid glandular tissue. A. Aberrant tissue inferior to hyoid bone. **B.** Radioisotope scan demonstrating presence of aberrant thyroid glandular tissue (*ATG*). Glandular tissue in the typical position is present in irregularly shaped masses making up small tapering lobes and a large isthmus.

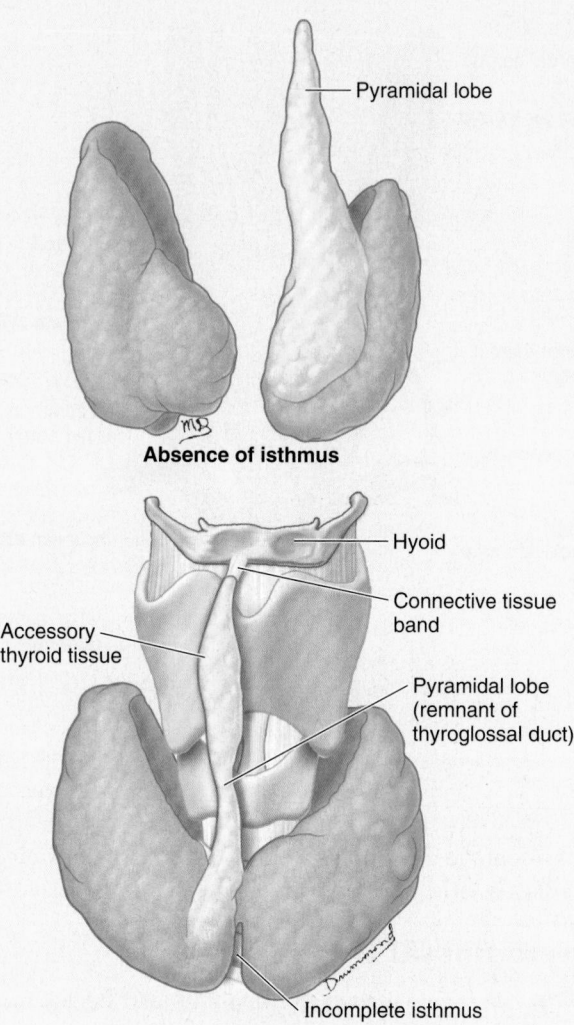

Pyramidal lobe

Absence of isthmus

Hyoid

Connective tissue band

Accessory thyroid tissue

Pyramidal lobe (remnant of thyroglossal duct)

Incomplete isthmus

FIGURE B8.8.

of the thyroglossal duct (e.g., in the thymus inferior to the thyroid gland or in the thorax). An *accessory thyroid gland* may develop in the neck lateral to the thyroid cartilage; it usually lies on the thyrohyoid muscle (Fig. 8.28). Although the accessory gland may be functional, it is often of insufficient size to maintain normal function if the thyroid gland is removed.

Pyramidal Lobe of Thyroid Gland

 Approximately 50% of thyroid glands have a *pyramidal lobe*. This lobe, which varies in size, extends superiorly from the isthmus of the thyroid gland,

usually to the left of the median plane; the isthmus may be incomplete or absent (Fig. B8.8). A band of connective tissue, often containing accessory thyroid tissue, may continue from the apex of the pyramidal lobe to the hyoid. This narrow lobe and connective tissue band develop from remnants of the epithelium and connective tissue of the thyroglossal duct.

Enlargement of Thyroid Gland

A non-neoplastic, noninflammatory *enlargement of the thyroid gland*, other than the variable enlargement that may occur during menstruation and pregnancy, is called a *goiter*, which results from a lack of iodine. It is common in parts of the world where the soil and water are deficient in iodine. The enlarged gland causes a swelling in the neck that may compress the trachea, esophagus, and recurrent laryngeal nerves (Fig. B8.9). When the gland enlarges, it may do so anteriorly, posteriorly, inferiorly, or laterally. It cannot move superiorly because of the superior attachments of the overlying sternothyroid and sternohyoid muscles (Table 8.3). Substernal extension of a goiter is also common.

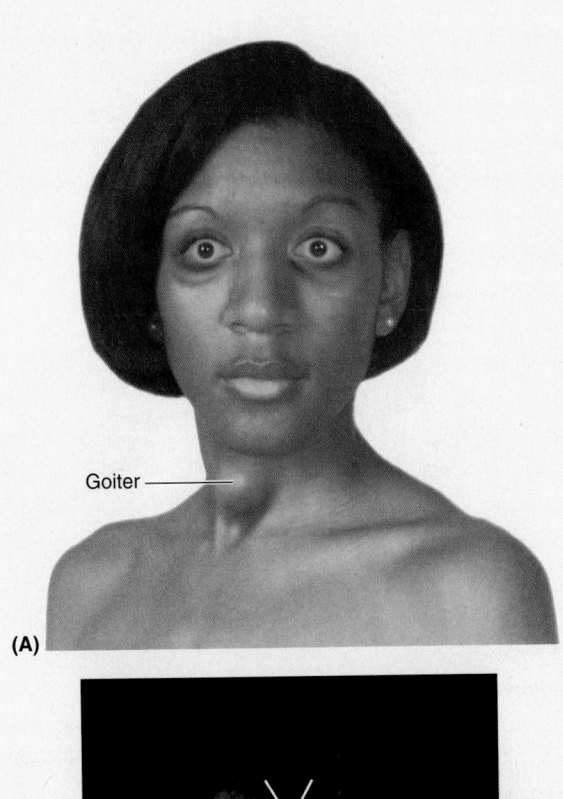

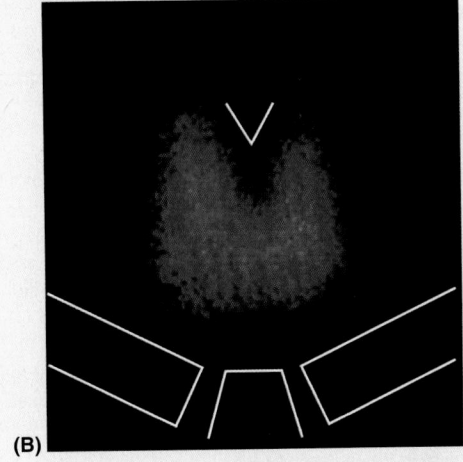

FIGURE B8.9. Enlarged thyroid. A. Individual with a goiter. **B.** Scintigram showing a diffuse, enlarged thyroid gland.

Thyroidectomy

 Excision of a malignant tumor of the thyroid gland, or other surgical procedure, sometimes necessitates removal of part or all of the gland (*hemithyroidectomy* or *thyroidectomy*). In the surgical treatment of hyperthyroidism, the posterior part of each lobe of the enlarged thyroid is usually preserved, a procedure called *near-total thyroidectomy,* to protect the recurrent and superior laryngeal nerves and to spare the parathyroid glands. Postoperative hemorrhage after thyroid gland surgery may compress the trachea, making breathing difficult. The blood collects within the fibrous capsule of the gland.

Injury to Recurrent Laryngeal Nerves

The risk of *injury to the recurrent laryngeal nerves* is ever present during neck surgery. Near the inferior pole of the thyroid gland, the right recurrent

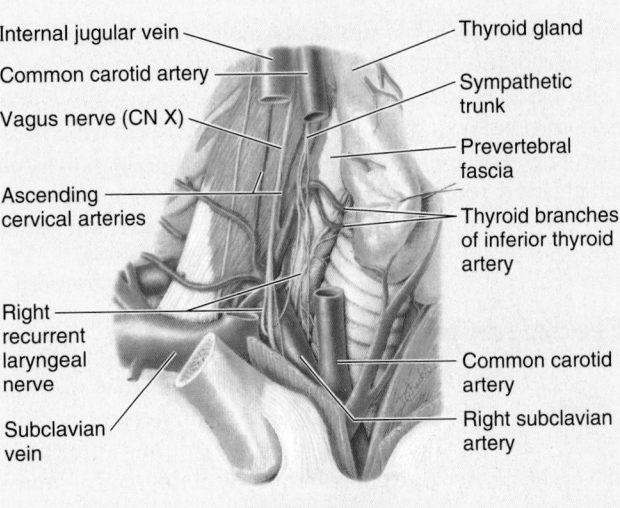

FIGURE B8.10.

laryngeal nerve is intimately related to the inferior thyroid artery and its branches (Fig. B8.10). This nerve may cross anterior or posterior to branches of the artery, or it may pass between them. Because of this close relationship, the inferior thyroid artery is ligated some distance lateral to the thyroid gland, where it is not close to the nerve. Although the danger of injuring the left recurrent laryngeal nerve during surgery is not as great, owing to its more vertical ascent from the superior mediastinum, the artery and nerve are also closely associated near the inferior pole of the thyroid gland (Fig. 8.27). Hoarseness is the usual sign of unilateral recurrent nerve injury; however, temporary *aphonia* or disturbance of phonation (voice production) and laryngeal spasm may occur. These signs usually result from bruising the recurrent laryngeal nerves during surgery or from the pressure of accumulated blood and serous exudate after the operation.

Inadvertent Removal of Parathyroid Glands

The variable position of the parathyroid glands, especially the inferior ones, puts them in danger of being damaged or removed during surgical procedures in the neck. The superior parathyroid glands may be as far superior as the thyroid cartilage, and the inferior glands may be as far inferior as the superior mediastinum (Fig. 8.30B). The aberrant sites of these glands are of concern when searching for abnormal parathyroid glands, as may be necessary in treating *parathyroid adenoma,* an ordinarily benign tumor of epithelial tissue associated with *hyperparathyroidism.*

Atrophy or inadvertent surgical removal of all the parathyroid glands results in *tetany,* a severe neurologic syndrome characterized by muscle twitches and cramps. The generalized spasms are caused by decreased serum calcium levels. Because laryngeal and respiratory muscles are involved, failure to respond immediately with appropriate therapy can result in death. To safeguard these glands during thyroidectomy, surgeons usually preserve the posterior part of the lobes of the thyroid gland.

In cases in which it is necessary to remove the whole thyroid gland (e.g., because of malignant disease), the parathyroid glands are carefully isolated with their blood vessels intact before removal of the thyroid gland. Parathyroid tissue may also be transplanted, usually to the arm, so it will not be damaged by subsequent surgery or radiation therapy.

Fractures of Laryngeal Skeleton

Laryngeal fractures may result from blows received in sports, such as kick boxing and hockey, or from compression by a shoulder strap during an automobile accident. Because of the frequency of this type of injury, most goalies in ice hockey and catchers in baseball have protective guards hanging from their masks that cover their larynges. Laryngeal fractures produce submucous hemorrhage and edema, respiratory obstruction, hoarseness, and sometimes a temporary inability to speak.

Laryngoscopy

Laryngoscopy is the procedure used to examine the interior of the larynx. The larynx may be examined visually by *indirect laryngoscopy* using a laryngeal mirror (Fig. B8.11A). The anterior part of the tongue is gently pulled from the oral cavity to minimize the extent to which the posterior part of the tongue covers the epiglottis and laryngeal inlet. Because the rima vestibuli is larger than the rima glottidis during normal respiration, the vestibular folds and vocal folds are visible during a laryngoscopic examination (Fig. B8.11B). The larynx can also be viewed by *direct laryngoscopy,* using a tubular endoscopic instrument, a laryngoscope. A *laryngoscope* is a tube or flexible fiber optic endoscope equipped with electrical lighting for examining or operating on the interior of the larynx through the mouth. The vestibular folds normally appear pink, whereas the vocal folds are usually pearly white.

Valsalva Maneuver

The sphincteric actions of the vestibular and vocal folds are important during the *Valsalva maneuver,* any forced expiratory effort against a closed airway, such as a cough, sneeze, or strain during a bowel movement or weight lifting. The vestibular and vocal folds abduct widely as the lungs inflate during deep inspiration. In the Valsalva maneuver, both the vestibular and vocal folds are tightly adducted at the end of deep inspiration. The anterolateral abdominal muscles then contract strongly to increase the intrathoracic and intra-abdominal pressures. The relaxed diaphragm passively transmits the increased abdominopelvic pressure to the thoracic cavity. Because high intrathoracic pressure impedes venous return to the right atrium, the Valsalva maneuver is used to study cardiovascular effects of

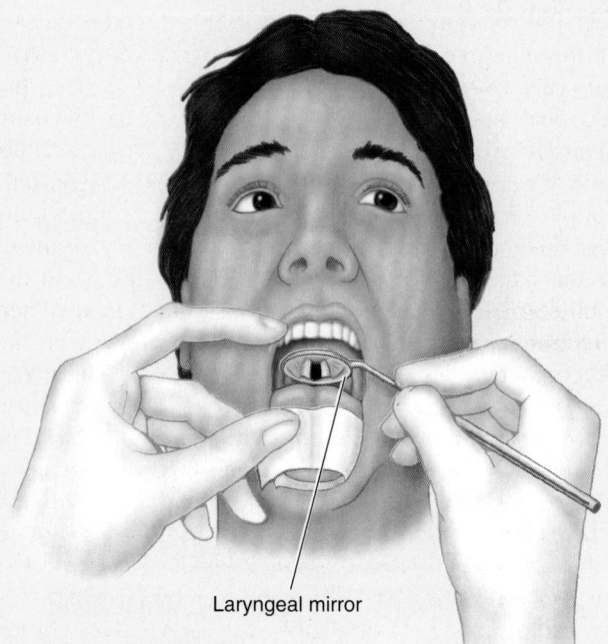

Laryngeal mirror

(A) Indirect laryngoscopy

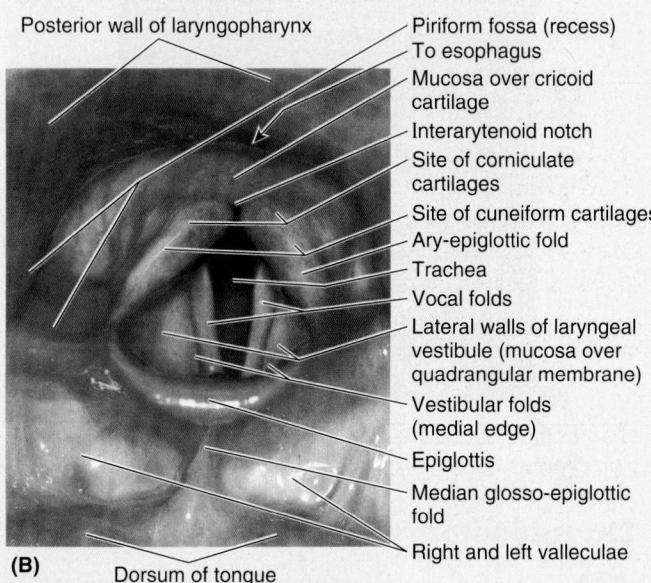

Posterior wall of laryngopharynx — Piriform fossa (recess) — To esophagus — Mucosa over cricoid cartilage — Interarytenoid notch — Site of corniculate cartilages — Site of cuneiform cartilages — Ary-epiglottic fold — Trachea — Vocal folds — Lateral walls of laryngeal vestibule (mucosa over quadrangular membrane) — Vestibular folds (medial edge) — Epiglottis — Median glosso-epiglottic fold — Right and left valleculae

(B) Dorsum of tongue

FIGURE B8.11.

raised peripheral venous pressure and decreased cardiac filling and cardiac output.

Aspiration of Foreign Bodies and Heimlich Maneuver

A foreign object, such as a piece of steak, may accidentally *aspirate* (be inhaled into the airways) through the laryngeal inlet into the vestibule of the larynx, where it becomes trapped superior to the vestibular folds. When a foreign object enters the vestibule of the larynx, the laryngeal muscles go into spasm, tensing the vocal folds. The rima glottidis closes and no air enters the trachea. The resulting blockage may completely seal off the larynx (*laryngeal obstruction*) and choke the person, leaving the individual speechless because the larynx is blocked. *Asphyxiation* occurs, and the person will die in approximately 5 min from lack of oxygen if the obstruction is not removed.

A person who is choking will cough in an attempt to dislodge the object. The vestibular folds are part of the protective mechanism that closes the larynx. The mucosa of the vestibule is sensitive to foreign objects such as food. When an object passes through the laryngeal inlet and contacts the vestibular epithelium, violent coughing occurs. Emergency therapy must be given to open the airway. The procedure used depends on the condition of the person, the facilities available, and the experience of the person giving first aid.

Because the lungs still contain air, sudden compression of the abdomen (*Heimlich maneuver*) causes the diaphragm to elevate and compress the lungs, expelling air from the trachea into the larynx. This maneuver usually dislodges the food or other material from the larynx. To perform the Heimlich maneuver, the person giving first aid uses subdiaphragmatic abdominal thrusts to expel the foreign object from the larynx. First, the closed fist, with the base of the palm facing inward, is placed on the victim's abdomen between the umbilicus and the xiphoid process of the sternum (Fig. B8.12). The fist is grasped by the other hand and forcefully thrust inward and superiorly, forcing the diaphragm superiorly. This action forces air from the lungs and creates an artificial cough that usually expels the foreign object. Several abdominal thrusts may be necessary to remove the obstruction in the larynx.

In extreme cases, experienced persons (e.g., physicians) insert a large-bore needle through the cricothyroid ligament (*needle cricothyrotomy*, or "coniotomy") to permit fast entry of air. Later, a *surgical cricothyrotomy* may be performed, which involves an incision through the skin and cricothyroid ligament and insertion of a small *tracheostomy tube* into the trachea (Fig. B8.13).

Tracheostomy

 A transverse incision through the skin of the neck and anterior wall of the trachea, *tracheostomy* establishes an airway in patients with upper airway obstruction or respiratory failure (Fig. B8.13). The infrahyoid muscles are retracted laterally, and the isthmus of the thyroid gland is either divided or retracted superiorly. An opening is made in the trachea between the first and second tracheal rings or through the second through fourth rings. A *tracheostomy tube* is then inserted into the trachea and secured. To avoid complications during a tracheostomy, the following anatomical relationships are important:

- The *inferior thyroid veins* arise from a venous plexus on the thyroid gland and descend anterior to the trachea.
- A small *thyroid ima artery* is present in approximately 10% of people; it ascends from the brachiocephalic trunk or the arch of the aorta to the isthmus of the thyroid gland.
- The *left brachiocephalic vein,* jugular venous arch, and pleurae may be encountered, particularly in infants and children.
- The *thymus* covers the inferior part of the trachea in infants and children.
- The trachea is small, mobile, and soft in infants, making it easy to cut through its posterior wall and damage the esophagus.

Injury to Laryngeal Nerves

Because the inferior laryngeal nerve, the continuation of the recurrent laryngeal nerve, innervates the muscles moving the vocal fold, *paralysis of the vocal fold* results when *injury to laryngeal nerves* occurs. The voice is poor initially because the paralyzed vocal fold cannot adduct to meet the normal vocal fold. Within weeks, the contralateral fold crosses the midline when its muscles act to compensate. When bilateral paralysis of the vocal folds occurs, the voice is almost absent because the vocal folds are motionless in a position that is slightly narrower than the usually neutral respiratory position. They cannot be adducted for phonation, nor can they be abducted for increased respiration, resulting in *stridor* (high pitched, noisy respiration) often accompanied by anxiety similar to that accompanying an asthmatic episode.

In progressive lesions of the recurrent laryngeal nerve, abduction of the vocal ligaments is lost before adduction; conversely, during recovery, adduction returns before abduction. Hoarseness is the common symptom of serious disorders of the larynx, such as carcinoma of the vocal folds.

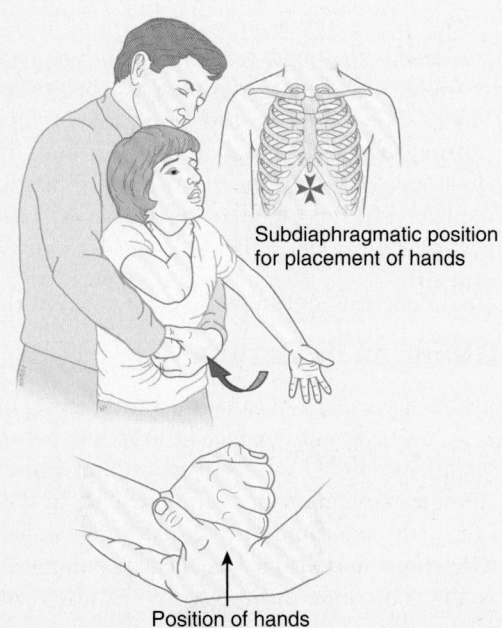

Subdiaphragmatic position for placement of hands

Position of hands

FIGURE B8.12. **Heimlich maneuver.**

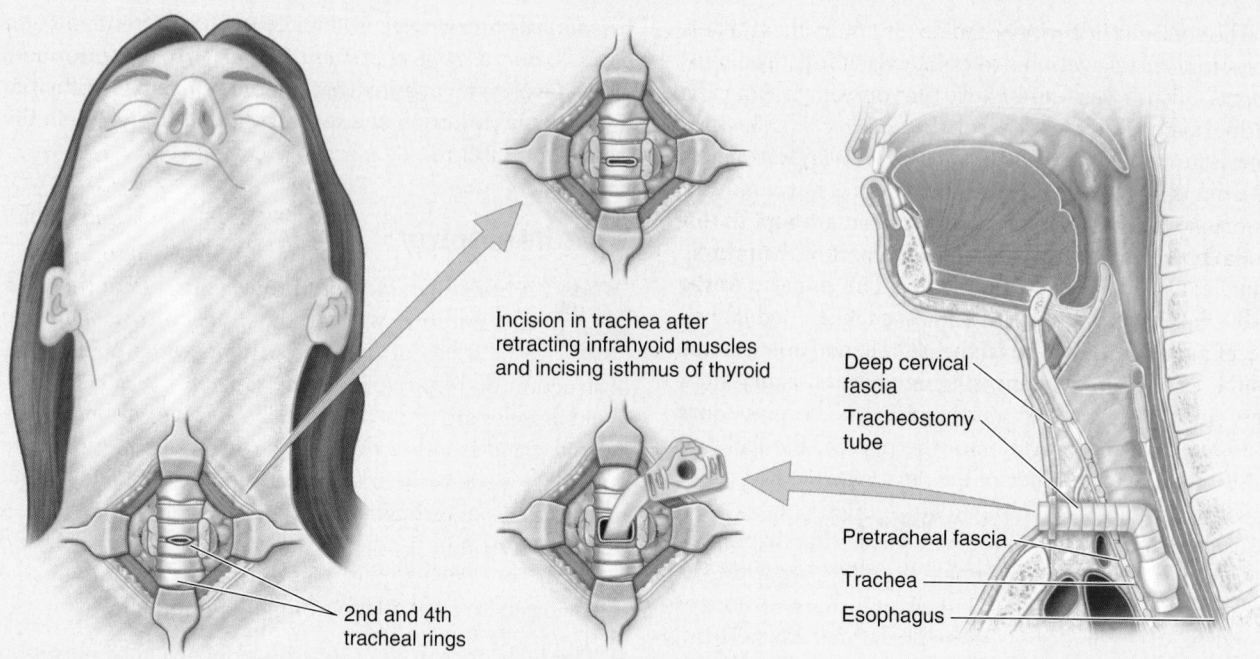

Incision in trachea after
retracting infrahyoid muscles
and incising isthmus of thyroid

Deep cervical
fascia

Tracheostomy
tube

Pretracheal fascia

Trachea

Esophagus

2nd and 4th
tracheal rings

FIGURE B8.13. Tracheostomy.

Paralysis of the superior laryngeal nerve causes anesthesia of the superior laryngeal mucosa. As a result, the protective mechanism designed to keep foreign bodies out of the larynx is inactive, and foreign bodies can easily enter the larynx. Injury to the external branch of the superior laryngeal nerve results in a voice that is monotonous in character because the paralyzed cricothyroid muscle supplied by it is unable to vary the length and tension of the vocal fold (Table 8.5, p. 1028). Such an injury may be unnoticed in individuals who do not usually employ a wide range of tone in their speech, but it may be critical to singers or public speakers.

To avoid injury to the external branch of the superior laryngeal nerve (e.g., during thyroidectomy), the superior thyroid artery is ligated and sectioned more superior to the gland, where it is not as closely related to the nerve. Because an enlarged thyroid gland (goiter) may itself cause impaired innervation of the larynx by compressing the laryngeal nerves, the vocal folds are examined by laryngoscopy before an operation in this area. In this way, damage to the larynx or its nerves resulting from a surgical mishap may be distinguished from a pre-existing injury resulting from nerve compression.

Superior Laryngeal Nerve Block

A *superior laryngeal nerve block* is often administered with endotracheal intubation in the conscious patient. This technique is used for peroral endoscopy, transesophageal echocardiography, and laryngeal and esophageal instrumentation. The needle is inserted midway between the thyroid cartilage and the hyoid, 1–5 cm anterior to the greater horn of the hyoid. The needle passes through the thyrohyoid membrane and the anesthetic agent bathes the internal laryngeal nerve, the larger terminal branch of the superior laryngeal nerve. Anesthesia of the laryngeal mucosa occurs superior to the vocal folds and includes the superior surface of these folds.

Cancer of Larynx

The incidence of *cancer of the larynx* is high in individuals who smoke cigarettes or chew tobacco. Most persons present with persistent hoarseness, often associated with *otalgia* (earache) and *dysphagia* (difficulty in swallowing). Enlarged pretracheal or paratracheal lymph nodes may indicate the presence of laryngeal cancer. *Laryngectomy* (removal of the larynx) may be performed in severe cases of cancer. Vocal rehabilitation can be accomplished by the use of an electrolarynx, a tracheoesophageal prosthesis, or esophageal speech (regurgitation of ingested air).

Age Changes in Larynx

The larynx grows steadily until approximately 3 years of age, after which little growth occurs until approximately 12 years of age. Before puberty, no major laryngeal sex differences exist. Owing to the presence of testosterone at puberty in males, the walls of the larynx strengthen, and the laryngeal cavity enlarges. There is only a slight increase in the size of the larynx of most girls. In boys, all of the laryngeal cartilages enlarge and the laryngeal prominence becomes conspicuous in most males. The anteroposterior diameter of the rima glottidis almost doubles its prepubescent measurement in males, the vocal folds lengthening and thickening proportionately and abruptly. This growth accounts for the voice changes that occur in males: The pitch typically becomes an octave lower.

The pitch of the voice of *eunuchs*, males whose testes have not developed (*agonadal males*) or have been surgically removed (e.g., cancerous testes), does not become lower without administration of male hormones. The thyroid, cricoid, and most of the arytenoid cartilages often ossify as age advances, commencing at approximately 25 years of age in the thyroid cartilage. By 65 years of age, the cartilages are frequently visible in radiographs.

Foreign Bodies in Laryngopharynx

When food passes through the laryngopharynx during swallowing, some of it enters the piriform fossae. *Foreign bodies* (e.g., a chicken bone or fishbone) *entering the pharynx* may lodge in this recess. If the object is sharp, it may pierce the mucous membrane and injure the internal laryngeal nerve.

The superior laryngeal nerve and its internal laryngeal branch are also vulnerable to injury during removal of the object if the instrument used to remove the foreign body accidentally pierces the mucous membrane. Injury to these nerves may result in anesthesia of the laryngeal mucous membrane as far inferiorly as the vocal folds. Young children swallow a variety of objects, most of which reach the stomach and pass through the alimentary tract without difficulty. In some cases, the foreign body stops at the inferior end of the laryngopharynx, its narrowest part. A medical image such as a radiograph or a CT scan will reveal the presence of a radiopaque foreign body. Foreign bodies in the pharynx are often removed under direct vision through a pharyngoscope.

Sinus Tract from Piriform Fossa

Although uncommon, a *sinus tract* may pass from the piriform fossa to the thyroid gland, becoming a potential site for recurring thyroiditis (inflammation of the thyroid gland). This sinus tract apparently develops from a remnant of the thyroglossal duct that adheres to the developing laryngopharynx. Removal of this sinus tract essentially involves a partial thyroidectomy because the piriform fossa lies deep to the superior pole of the gland (Scher and Richtsmeier, 1994).

Tonsillectomy

Tonsillectomy (removal of the tonsils) is performed by dissecting the palatine tonsil from the tonsillar bed or by a guillotine or snare operation. Each procedure involves removal of the tonsil and

surrounding connective tissue (Fig. B8.14). Because of the rich blood supply of the tonsil, bleeding commonly arises from the large *external palatine vein* (Fig. 8.44B) or, less commonly, from the tonsillar artery or other arterial twigs. The glossopharyngeal nerve (CN IX) accompanies the tonsillar artery on the lateral wall of the pharynx. Because this wall is thin, the nerve is vulnerable to injury. The internal carotid artery is especially vulnerable when it is tortuous and lies directly lateral to the tonsil (Fig. 8.46B).

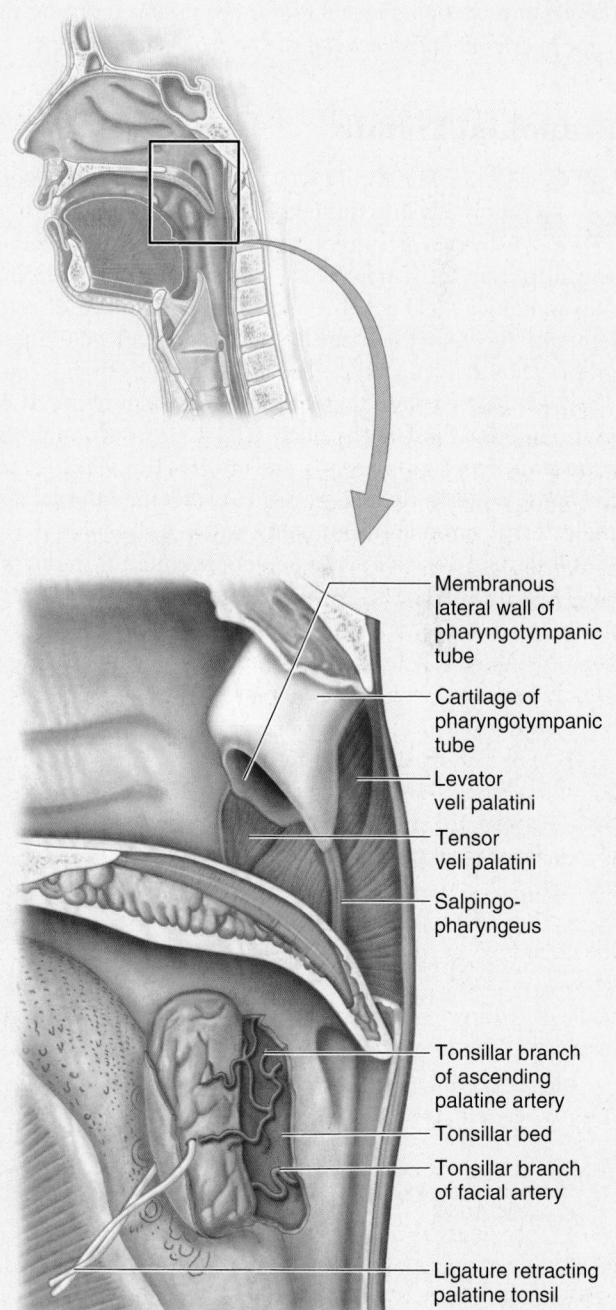

- Membranous lateral wall of pharyngotympanic tube
- Cartilage of pharyngotympanic tube
- Levator veli palatini
- Tensor veli palatini
- Salpingo-pharyngeus
- Tonsillar branch of ascending palatine artery
- Tonsillar bed
- Tonsillar branch of facial artery
- Ligature retracting palatine tonsil

FIGURE B8.14. Tonsillectomy.

Adenoiditis

Inflammation of the pharyngeal tonsils (adenoids—Fig. 8.43) is called *adenoiditis*, which can obstruct the passage of air from the nasal cavities through the choanae into the nasopharynx, making mouth breathing necessary. Infection from the enlarged pharyngeal tonsils may spread to the tubal tonsils, causing swelling and closure of the pharyngotympanic tubes. Impairment of hearing may result from nasal obstruction and blockage of the pharyngotympanic tubes. Infection spreading from the nasopharynx to the middle ear causes *otitis media* (middle ear infection), which may produce temporary or permanent hearing loss. Sometimes the palatine and pharyngeal tonsils are removed during the same operation (*tonsillectomy and adenoidectomy*; T&A).

Branchial Fistula

A *branchial fistula* is an abnormal canal that opens internally into the tonsillar sinus (fossa) and externally on the side of the neck (Fig. B8.15A). Saliva may drip from the fistula, which may become infected. This uncommon cervical canal results from persistence of remnants of the 2nd pharyngeal pouch and 2nd pharyngeal groove (Moore et al., 2012). The fistula ascends from its cervical opening, usually along the anterior border of the SCM in the inferior third of the neck. It first passes through the subcutaneous tissue, platysma, and fascia of the neck to enter the carotid sheath. It then passes between the internal and the external carotid arteries on its way to its opening in the tonsillar sinus. Its course can be demonstrated by radiography (Fig. B8.15B).

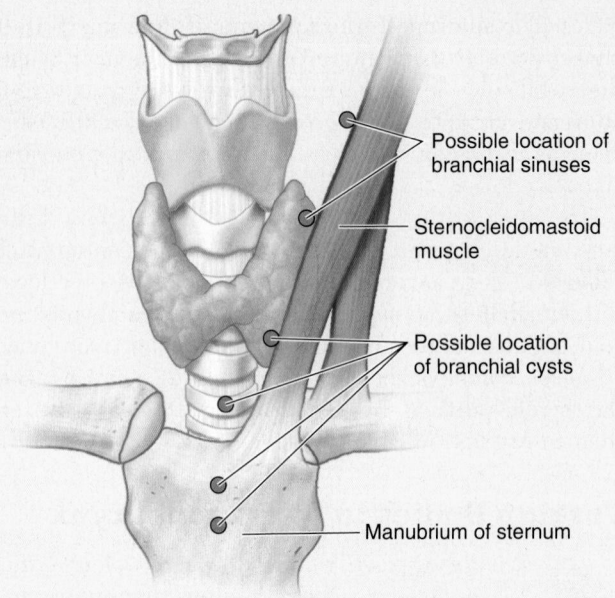

FIGURE B8.16. Branchial sinuses.

Branchial Sinuses and Cysts

When the embryonic cervical sinus fails to disappear, it may retain its connection with the lateral surface of the neck by a *branchial sinus,* a narrow canal. The opening of the sinus may be anywhere along the anterior border of the SCM (Fig. B8.16). If a remnant of the cervical sinus is not connected with the surface, it may form a *branchial cyst* (lateral cervical cyst), usually located just inferior to the angle of the mandible. Although branchial cysts may be present in infants and children,

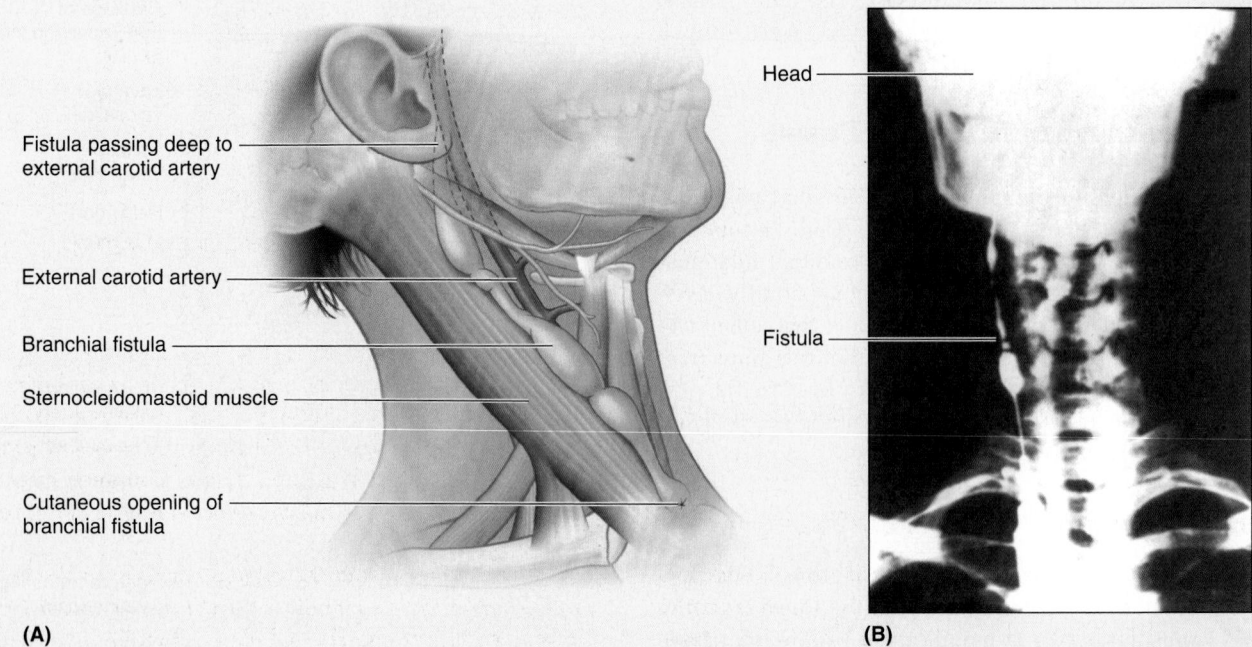

FIGURE B8.15. Branchial fistula.

they may not enlarge and become visible until early adulthood. The sinus and cyst are usually excised. The cyst passes close to the hypoglossal, glossopharyngeal, and spinal accessory nerves (Fig. 8.46A). Therefore, care must be taken to avoid damage to these nerves during removal of the cyst.

Esophageal Injuries

Esophageal injuries are the rarest kinds of penetrating neck trauma; however, they cause most complications after a surgical procedure or other treatment. Most esophageal injuries occur in conjunction with airway injuries because the airway lies anterior to the esophagus and provides some protection to it. Esophageal injuries are often occult (hidden), which makes the injury difficult to detect, especially if it is isolated. Unrecognized esophageal perforation is fatal in nearly all non-operative cases and in approximately 50% of surgical occurances (Sinkinson, 1991).

Tracheo-Esophageal Fistula

The most common birth defect of the esophagus is *tracheo-esophageal fistula (TEF)*. Usually, it is combined with some form of esophageal atresia. In the most common type of TEF (approximately 90% of cases), the superior part of the esophagus ends in a blind pouch and the inferior part communicates with the trachea (Fig. B8.17A). In these cases, the pouch fills with mucus, which the infant aspirates. In some cases, the superior esophagus communicates with the trachea and the inferior esophagus joins the stomach (Fig. B8.17C), but sometimes it does not, producing TEF with esophageal atresia (Fig. B8.17B). TEFs result from failures in partitioning of the esophagus and trachea (Moore et al., 2012).

Esophageal Cancer

The most common presenting complaint of *esophageal cancer* is *dysphagia* (difficulty in swallowing), which is not usually recognized until the lumen is reduced by 30–50%. *Esophagoscopy* is a common diagnostic tool for observing these cancers. Painful swallowing in some patients suggests extension of the tumor to peri-esophageal tissues. Enlargement of the inferior deep cervical lymph nodes also suggests *esophageal cancer*. Compression of the recurrent laryngeal nerves by an esophageal tumor produces hoarseness.

Zones of Penetrating Neck Trauma

Three zones are common clinical guides to the seriousness of neck trauma (Fig. B8.18). The zones give physicians an understanding of the structures that are at risk with penetrating neck injuries.

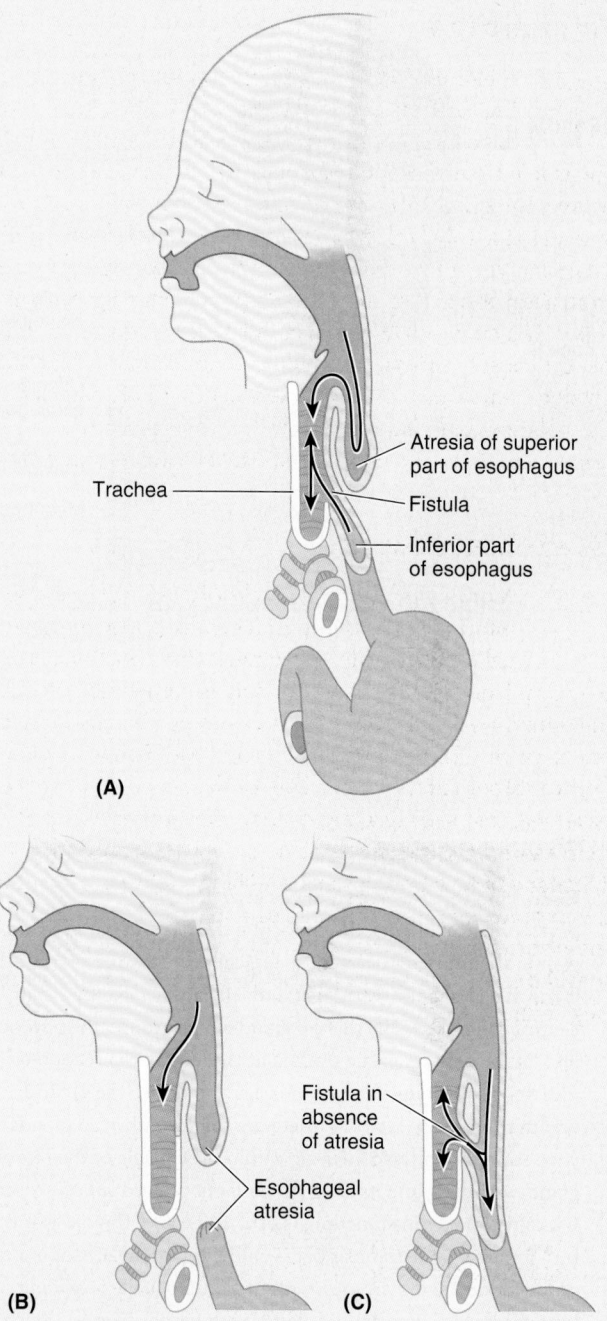

FIGURE B8.17. Tracheo-esophageal fistulae (TEF).

- **Zone I:** includes the root of the neck and extends from the clavicles and the manubrium to the level of the inferior border of the cricoid cartilage. Structures at risk are the cervical pleurae, apices of lungs, thyroid and parathyroid glands, trachea, esophagus, common carotid arteries, jugular veins, and the cervical region of the vertebral column.
- **Zone II:** extends from the cricoid cartilage to the level of the angles of the mandible. Structures at risk are the superior poles of the thyroid gland, thyroid and cricoid cartilages, larynx, laryngopharynx, carotid arteries, jugular

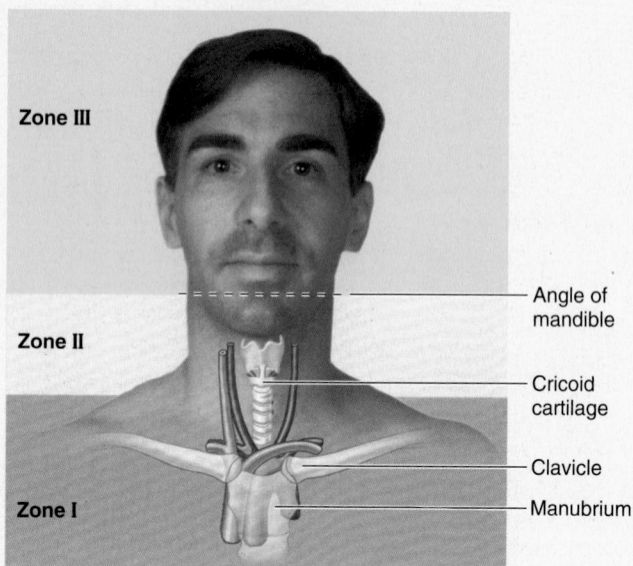

Angle of mandible

Cricoid cartilage

Clavicle

Manubrium

FIGURE B8.18. Zones of penetrating neck trauma.

veins, esophagus, and cervical region of the vertebral column.

• **Zone III:** occurs at the angles of the mandibles superiorly. Structures at risk are the salivary glands, oral and nasal cavities, oropharynx, and nasopharynx.

Injuries in zones I and III obstruct the airway and have the greatest risk for **morbidity** (complications following surgical procedures and other treatments) and **mortality** (a fatal outcome) because the injured structures are difficult to visualize and repair and the vascular damage is difficult to control. Injuries in zone II are most common; however, morbidity and mortality are lower because physicians can control vascular damage by direct pressure and surgeons can visualize and treat injured structures more easily than they can in the other zones.

The Bottom Line

VISCERA OF NECK

Endocrine layer of cervical viscera: Despite different developmental origins, the endocrine thyroid and parathyroid glands are intimately related. ♦ Typically, the thyroid gland is roughly H-shaped, with right and left lobes linked by a thin central isthmus. ♦ The thyroid gland wraps around the anterior and lateral aspects of the trachea at the level of the second to fourth tracheal rings; the isthmus lies anterior to the second and third rings. ♦ Typically, there are four parathyroid glands (two superior and two inferior) within the capsule of the thyroid gland or in the gland itself. ♦ An abundant blood supply, essential to the endocrine function, is provided to the thyroid gland by a four-way anastomosis between the right and left superior and inferior thyroid arteries, with the latter usually providing branches to the parathyroid glands. ♦ Superior thyroid veins accompany the arteries of the same name, draining the area they supply. ♦ Unaccompanied middle and inferior thyroid veins drain the inferior part of the thyroid gland: The superior and middle thyroid veins drain to the IJV, whereas the usually singular inferior thyroid vein enters the left brachiocephalic vein. ♦ Vasomotor nerves course along the arteries, but the glands are regulated hormonally rather than by secretomotor nerve fibers. ♦ Lymphatic vessels pass directly to the deep cervical lymph nodes or via nodes associated with the larynx and trachea.

  Respiratory layer of cervical viscera: The larynx is the superior end of the lower respiratory tract, modified to regulate entry into or close off the lower respiratory tract. ♦ The larynx also modifies the exit of air from the tract to produce

tone for vocalization. ♦ With the diaphragm, it regulates intra-abdominal pressure through air retention and the force and suddenness by which air exits the tract (e.g., exhaling verses coughing or sneezing). ♦ The larynx consists of a cartilaginous articulating skeleton joined by ligaments, membranes, and muscles, lined with mucous membrane. ♦ All the laryngeal muscles except one (posterior crico-arytenoid) participate in closure of the rima glottidis. ♦ Active opening of the rima is required only during deep inspiration. ♦ Otherwise, opening occurs passively by the tidal flow of air, with the other muscles controlling the amount and nature of resistance provided at the rima glottidis to produce tone and control its pitch. ♦ In addition to intrinsically produced movements between its components, extrinsic musculature (hyoid muscles) can move the entire larynx for swallowing and to modify pitch further. ♦ The internal laryngeal nerve, a branch of the superior laryngeal nerve, is the sensory nerve of the larynx. ♦ The recurrent laryngeal nerve (via its terminal branch, the inferior laryngeal nerve) is the motor nerve, which supplies all muscles of the larynx, with one exception. ♦ The external laryngeal nerve, a smaller branch of the superior laryngeal nerve, supplies the cricothyroid muscle. ♦ The trachea is the median fibrocartilaginous tube extending between the cricoid cartilage at the C6 vertebral level and its bifurcation into main bronchi at the T4–T5 IV disc level (level of sternal angle).

  Alimentary layer of cervical viscera: Although generally considered part of the alimentary tract, the pharynx is shared with the respiratory system. ♦ The superior, non-collapsible nasopharynx is exclusively respiratory, and the air and food pathways cross within the oropharynx and laryngopharynx. ♦ The contractile pharynx is unique within the alimentary tract in being constructed of voluntary muscle with the circular layer (pharyngeal constrictors) external to longitudinal muscle, the stylopharyngeus, palatopharyngeus, and salpingopharyngeus. ♦ The flat posterior wall of the pharynx, abutting the musculoskeletal neck at the retropharyngeal space, is without openings; however, its anterior wall includes openings to the nose, mouth, and larynx. These openings determine the three segments of the pharynx. ♦ The soft palate serves as a flap valve regulating access to or from the nasopharynx and oropharynx, whereas the larynx is the "valve" ultimately separating food and air before they enter the esophagus and trachea, respectively. ♦ The superior two openings of the pharynx, which connect

to the external environment, are encircled by a ring of lymphoid (tonsillar) tissue. ♦ Gaps in the submucosal lateral wall, between attachments of the pharyngeal constrictor muscles, permit the passage of slip-like longitudinal muscles and neurovascular elements. ♦ Innervation of the pharynx is from the pharyngeal nerve plexus, with the vagus providing the motor fibers and the glossopharyngeal providing sensory fibers. ♦ At the level of the cricoid cartilage (C6 vertebral level), a relatively abrupt change is made to the muscular pattern more typical of the alimentary tract. ♦ The cricopharyngeal part of the inferior pharyngeal constrictor, the most inferior part of the external circular layer, forms the superior esophageal sphincter. ♦ Immediately inferior, as the outer muscular layer becomes longitudinal, the esophagus begins. ♦ Also at approximately this point, sensory and motor innervation is transferred to the recurrent laryngeal nerves. ♦ The cervical esophagus is composed of voluntary muscle.

LYMPHATICS OF NECK

Most superficial tissues in the neck are drained by lymphatic vessels that enter the **superficial cervical lymph nodes,** which are located along the course of the EJV. Lymph from these nodes, like lymph from all of the head and neck, drains

into **inferior deep cervical lymph nodes** (Figs. 8.48 and 8.51). The specific group of inferior deep cervical nodes involved here descends across the lateral cervical region with the spinal accessory nerve (CN XI).

Most lymph from the six to eight lymph nodes then drains into the *supraclavicular group of nodes,* which accompany

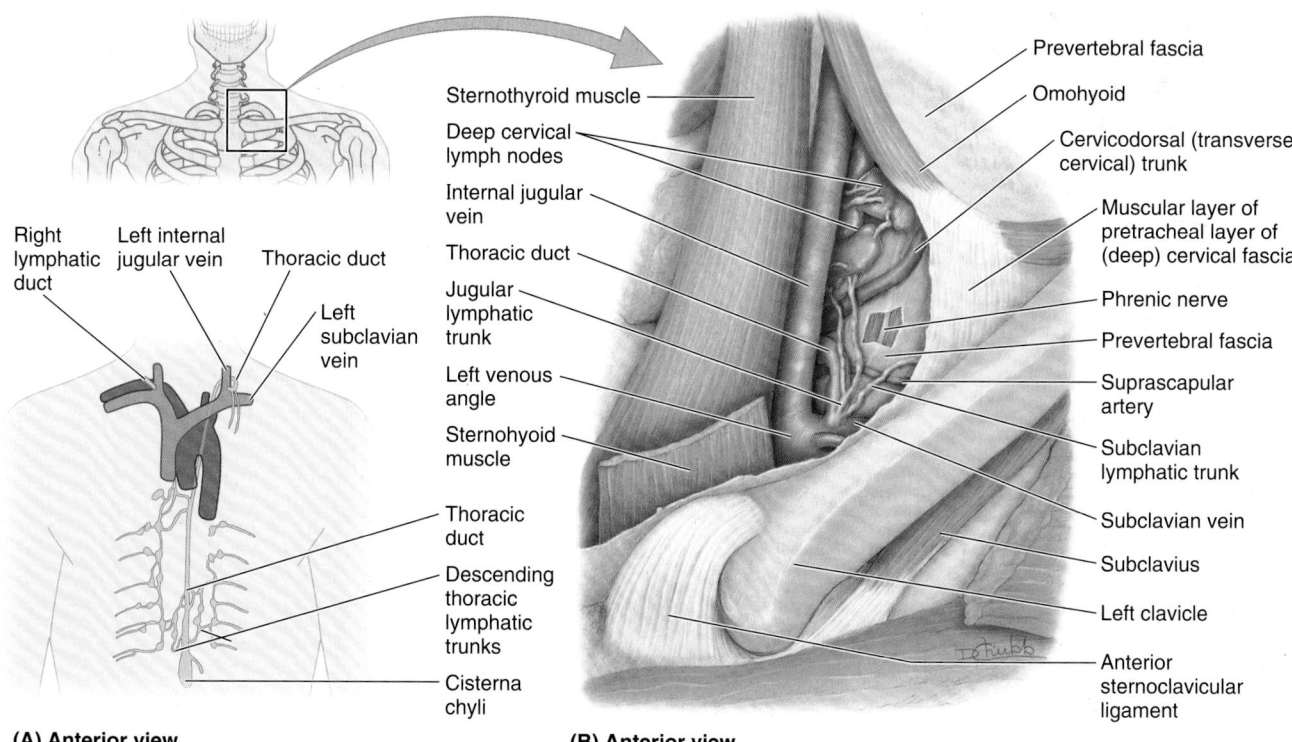

(A) Anterior view

Right lymphatic duct
Left internal jugular vein
Thoracic duct
Left subclavian vein
Thoracic duct
Descending thoracic lymphatic trunks
Cisterna chyli

(B) Anterior view

Sternothyroid muscle
Deep cervical lymph nodes
Internal jugular vein
Thoracic duct
Jugular lymphatic trunk
Left venous angle
Sternohyoid muscle

Prevertebral fascia
Omohyoid
Cervicodorsal (transverse cervical) trunk
Muscular layer of pretracheal layer of (deep) cervical fascia
Phrenic nerve
Prevertebral fascia
Suprascapular artery
Subclavian lymphatic trunk
Subclavian vein
Subclavius
Left clavicle
Anterior sternoclavicular ligament

FIGURE 8.51. Lymphatic vessels in root of neck. A. This overview demonstrates the course of the thoracic duct and site of the termination of the thoracic and right lymphatic ducts. **B.** This dissection of the left side shows the deep cervical lymph nodes and the termination of the thoracic duct at the junction of the subclavian and internal jugular veins (left venous angle). The cervicodorsal trunk is often called the transverse cervical artery.

the cervicodorsal trunk. The main group of deep cervical lymph nodes forms a chain along the IJV, mostly under cover of the SCM. Other deep cervical nodes include the prelaryngeal, pretracheal, paratracheal, and retropharyngeal nodes. Efferent lymphatic vessels from the deep cervical nodes join to form the **jugular lymphatic trunks,** which usually join the thoracic duct on the left side and enter the junction of the internal jugular and subclavian veins (right venous angle) directly or via a short right lymphatic duct on the right.

The **thoracic duct** passes superiorly through the superior thoracic aperture along the left border of the esophagus. It arches laterally in the root of the neck, posterior to the carotid sheath and anterior to the sympathetic trunk and vertebral and subclavian arteries (Fig. 8.51B). The thoracic duct enters the left brachiocephalic vein at the junction of the subclavian and IJVs (*left venous angle*). When the right jugular, subclavian, and bronchomediastinal lymphatic trunks unite to form a *right lymphatic duct*, it enters the right venous angle as the thoracic duct does on the left (Fig. 8.51A). Often, however, these lymphatic trunks enter the venous system independently in the region of the right venous angle.

LYMPHATICS IN NECK
Radical Neck Dissections

Radical neck dissections are performed when cancer invades the cervical lymphatics. During the procedure, the deep cervical lymph nodes and tissues around them are removed as completely as possible. The major arteries, brachial plexus, CN X, and phrenic nerve are preserved; however, most cutaneous branches of the cervical plexus are removed. The aim of the dissection is to remove all tissue that bears lymph nodes in one piece. The deep cervical lymph nodes, particularly those located along the cervicodorsal trunk, may be involved in the spread of cancer from the thorax and abdomen. Because their enlargement may give the first clue to cancer in these regions, they are often referred to as the *cervical sentinel lymph nodes.*

thePoint: **Board-review questions, case studies, and additional resources are available at thePoint.lww.com.**

Summary of Cranial Nerves

9 CHAPTER

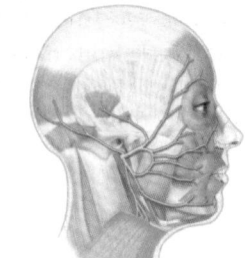

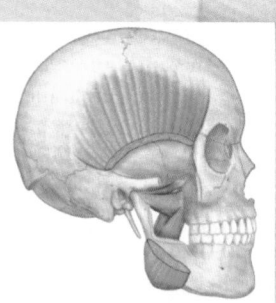

The regional aspects of the cranial nerves are described in the preceding chapters, especially those for the head and neck. This chapter summarizes all of the cranial nerves, largely in figures and tables. Figures 9.1–9.3 and Tables 9.1 and 9.2 summarize specific cranial nerves. Figure 9.4 and Table 9.3 summarize the cranial parasympathetic ganglia, their location, sympathetic and parasympathetic roots, and main distribution.

OVERVIEW

Like spinal nerves, **cranial nerves** are bundles of sensory or motor fibers that innervate muscles or glands, carry impulses from sensory receptors, or have a combination of motor and sensory fibers. They are called cranial nerves because they emerge through foramina or fissures in the cranium and are covered by tubular sheaths derived from the cranial meninges. There are 12 pairs of cranial nerves, which are numbered I–XII, from rostral to caudal (Figs. 9.1–9.3). Their names reflect their general distribution or function.

Cranial nerves carry one or more of the following five main functional components (Fig. 9.3, Table 9.1).

- Motor (efferent) fibers.
  1. *Motor fibers to voluntary* (striated) *muscle.* These include the somatic motor (general somatic efferent) axons. On the basis of the embryologic/phylogenetic derivation of certain muscles of the head and neck,[1] some motor fibers conveyed by cranial nerves to striated muscle have traditionally been classified as "special visceral." When appropriate, these fibers are designated somatic (branchial) motor, referring to the muscle tissue derived from the pharyngeal arches in the embryo (e.g., muscles of mastication).
  2. *Motor fibers involved in innervating involuntary* (smooth) *muscles or glands.* These include visceral motor (general visceral efferent) axons that constitute the cranial outflow of the parasympathetic division of the autonomic nervous system (ANS). The presynaptic (preganglionic) fibers emerge from the brain and synapse outside the central nervous system (CNS) in a parasympathetic ganglion. The postsynaptic (postganglionic) fibers continue to innervate smooth muscles and glands (e.g., the sphincter pupillae and lacrimal gland).
- Sensory (afferent) fibers.
  3. *Fibers transmitting general sensation* (e.g., touch, pressure, heat, cold, etc.) *from the skin and mucous membranes.* These include somatic sensory (general somatic afferent) fibers, mainly carried by CN V, but also by CN VII, CN IX, and CN X.
  4. *Fibers conveying sensation from the viscera.* These include visceral sensory (general visceral afferent) fibers conveying information from the carotid body and sinus (see Fig. 8.17), pharynx, larynx, trachea, bronchi, lungs, heart, and gastrointestinal tract.
  5. *Fibers transmitting unique sensations.* These include special sensory fibers conveying taste and smell (special visceral afferent fibers) and those serving the special senses of vision, hearing, and balance (special somatic afferent fibers).

Some cranial nerves are purely sensory, others are considered purely motor, and several are mixed. CN III, CN IV, CN VI, CN XI, CN XII, and the motor root of CN V are considered to be pure motor nerves that appear to have evolved from primordial anterior roots. However, a small number of sensory fibers for proprioception (nonvisual perception of movement and position) are also present in these nerves, the cell bodies of which are probably located in the mesencephalic nucleus of CN V. The sensory root of CN V is purely a somatic (general) sensory nerve. Four cranial nerves (CN III, CN VII, CN IX, and CN X) contain presynaptic parasympathetic (visceral motor) axons as they emerge from the brainstem. CN V, CN VII, CN IX, and CN X are mixed nerves with both somatic (branchial) motor and somatic (general) sensory components, and each nerve supplies derivatives of a different pharyngeal arch.

The fibers of cranial nerves connect centrally to **cranial nerve nuclei**—groups of neurons in which sensory or afferent fibers terminate and from which motor or efferent fibers originate (Fig. 9.5, see p. 1061). Except for CN I and CN II, which involve extensions of the forebrain, the nuclei of the cranial nerves are located in the brainstem. Nuclei of similar functional components (e.g., somatic or visceral motor, or somatic or visceral sensory) are generally aligned into functional columns in the brainstem.

OLFACTORY NERVE (CN I)

Function: Special sensory (special visceral afferent)—that is, the special sense of smell. "*Olfaction* is the sensation of odors that results from the detection of odorous substances aerosolized in the environment" (Simpson, 2006).

The cell bodies of olfactory receptor neurons are located in the **olfactory organ** (the olfactory part of the nasal mucosa or olfactory area), which is located in the roof of the nasal cavity, and along the nasal septum and medial wall of the superior nasal concha (Fig. 9.6, see p. 1062). **Olfactory receptor neurons** are both receptors and conductors. The apical surfaces of the neurons possess fine **olfactory cilia**, bathed by a film of watery mucus secreted by the **olfactory glands** of the epithelium. The olfactory cilia are stimulated by molecules of an odiferous gas dissolved in the fluid.

The basal surfaces of the bipolar olfactory receptor neurons of the nasal cavity of one side give rise to central processes that are collected into approximately 20 **olfactory nerves** (L. *fila olfactoria*), constituting the right or left **olfactory nerve (CN I)**.

---

[1] Historically, the sternocleidomastoid and trapezius have been classified as branchiomeric muscles; students may see them classified as such in other references.

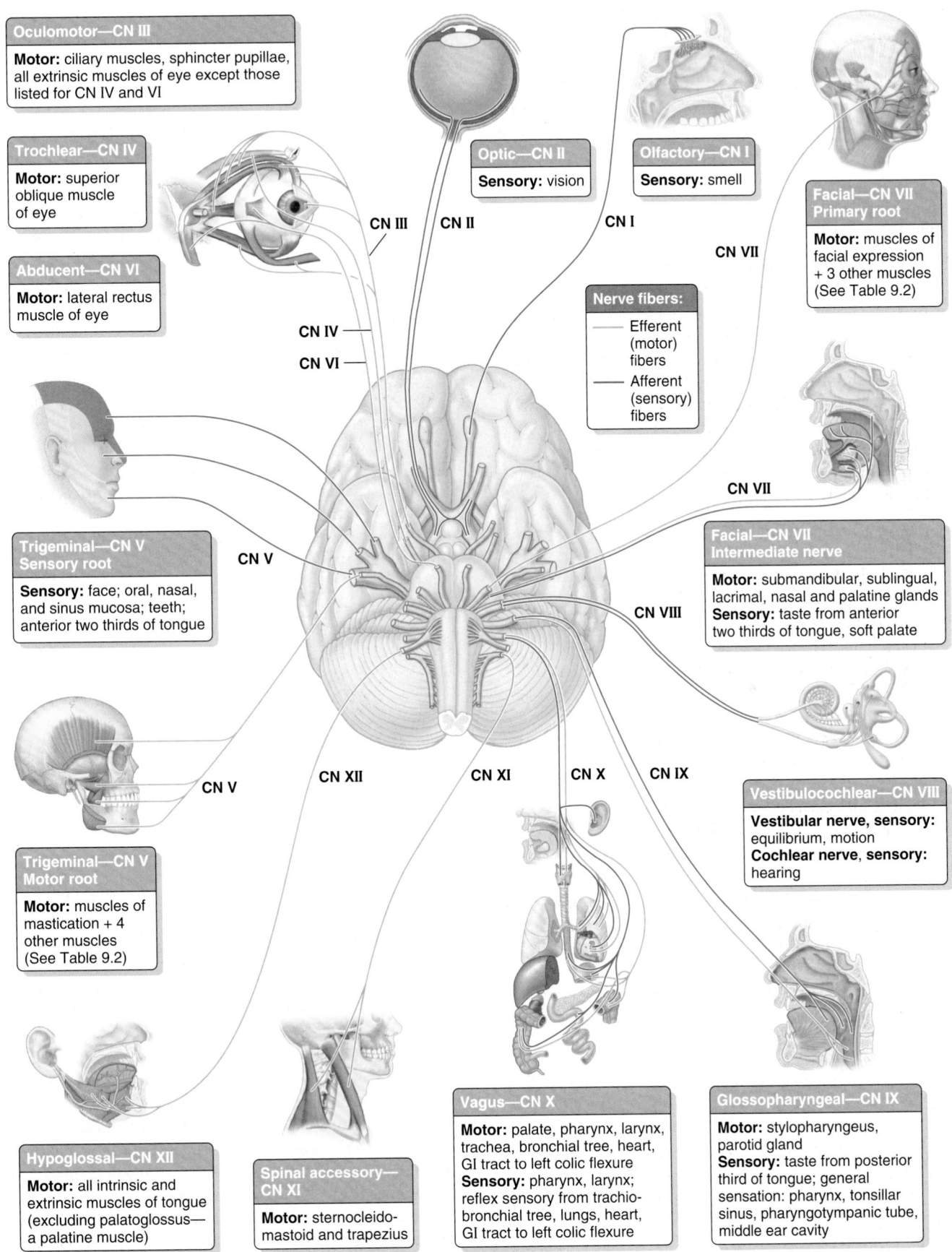

Oculomotor—CN III

Motor: ciliary muscles, sphincter pupillae, all extrinsic muscles of eye except those listed for CN IV and VI

Trochlear—CN IV

Motor: superior oblique muscle of eye

Abducent—CN VI

Motor: lateral rectus muscle of eye

Optic—CN II

Sensory: vision

Olfactory—CN I

Sensory: smell

Facial—CN VII Primary root

Motor: muscles of facial expression + 3 other muscles (See Table 9.2)

Nerve fibers:
— Efferent (motor) fibers
— Afferent (sensory) fibers

Trigeminal—CN V Sensory root

Sensory: face; oral, nasal, and sinus mucosa; teeth; anterior two thirds of tongue

Facial—CN VII Intermediate nerve

Motor: submandibular, sublingual, lacrimal, nasal and palatine glands
Sensory: taste from anterior two thirds of tongue, soft palate

Trigeminal—CN V Motor root

Motor: muscles of mastication + 4 other muscles (See Table 9.2)

Vestibulocochlear—CN VIII

Vestibular nerve, sensory: equilibrium, motion
Cochlear nerve, sensory: hearing

Hypoglossal—CN XII

Motor: all intrinsic and extrinsic muscles of tongue (excluding palatoglossus—a palatine muscle)

Spinal accessory—CN XI

Motor: sternocleidomastoid and trapezius

Vagus—CN X

Motor: palate, pharynx, larynx, trachea, bronchial tree, heart, GI tract to left colic flexure
Sensory: pharynx, larynx; reflex sensory from trachiobronchial tree, lungs, heart, GI tract to left colic flexure

Glossopharyngeal—CN IX

Motor: stylopharyngeus, parotid gland
Sensory: taste from posterior third of tongue; general sensation: pharynx, tonsillar sinus, pharyngotympanic tube, middle ear cavity

FIGURE 9.1. Summary of cranial nerves.

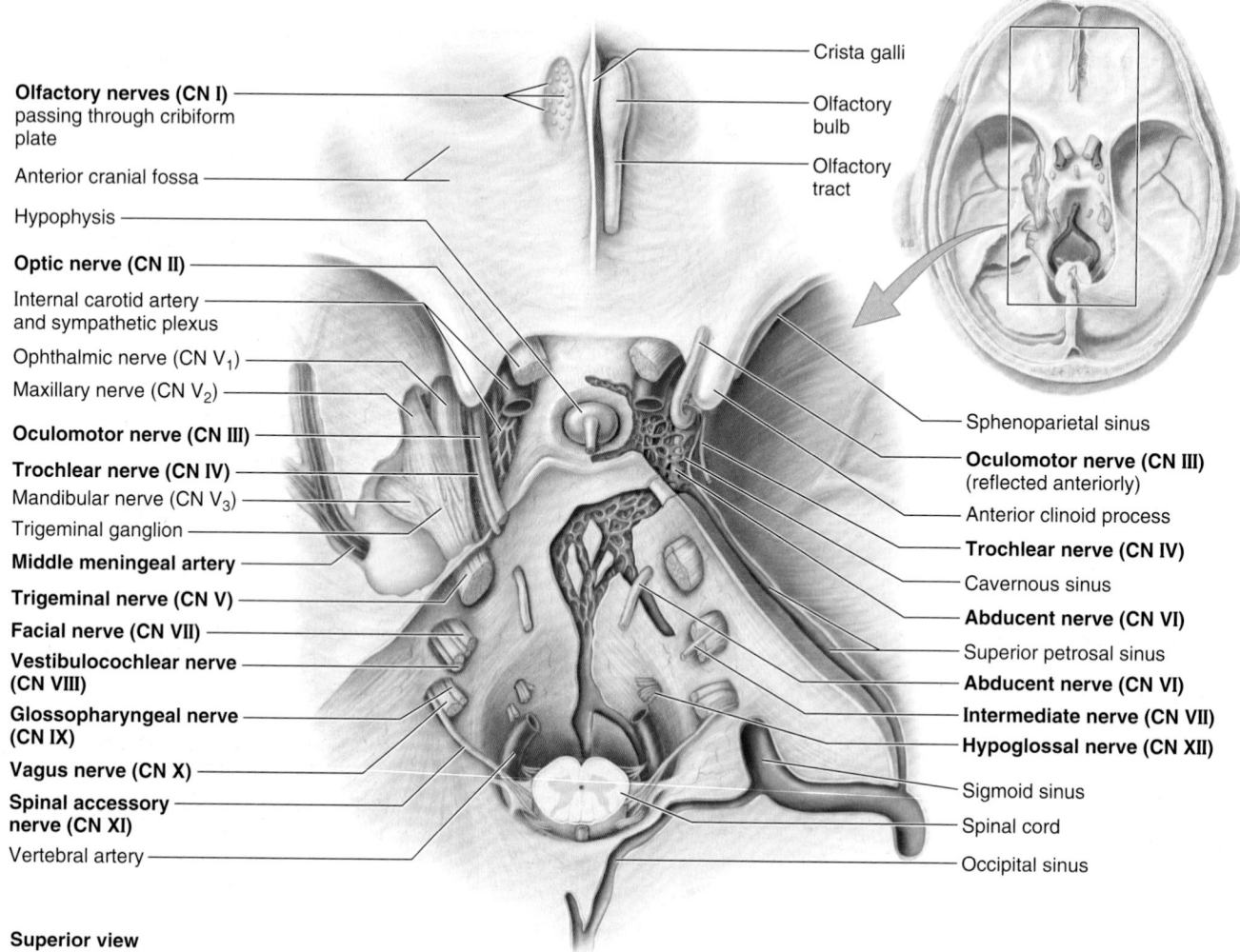

- Crista galli
- Olfactory bulb
- Olfactory tract

Olfactory nerves (CN I)
passing through cribiform plate

Anterior cranial fossa

Hypophysis

Optic nerve (CN II)

Internal carotid artery and sympathetic plexus

Ophthalmic nerve (CN V$_1$)

Maxillary nerve (CN V$_2$)

Oculomotor nerve (CN III)

Trochlear nerve (CN IV)

Mandibular nerve (CN V$_3$)

Trigeminal ganglion

Middle meningeal artery

Trigeminal nerve (CN V)

Facial nerve (CN VII)

Vestibulocochlear nerve (CN VIII)

Glossopharyngeal nerve (CN IX)

Vagus nerve (CN X)

Spinal accessory nerve (CN XI)

Vertebral artery

- Sphenoparietal sinus
- **Oculomotor nerve (CN III)** (reflected anteriorly)
- Anterior clinoid process
- **Trochlear nerve (CN IV)**
- Cavernous sinus
- **Abducent nerve (CN VI)**
- Superior petrosal sinus
- **Abducent nerve (CN VI)**
- **Intermediate nerve (CN VII)**
- **Hypoglossal nerve (CN XII)**
- Sigmoid sinus
- Spinal cord
- Occipital sinus

Superior view

FIGURE 9.2. Cranial nerves in relation to internal aspect of cranial base. The tentorium cerebelli has been removed and the venous sinuses have been opened on the right side. The dural roof of the trigeminal cave has been removed on the left side and CN V$_1$, CN III, and CN IV have been dissected from the lateral wall of the cavernous sinus.

They pass through tiny foramina in the *cribriform plate of the ethmoid bone*, surrounded by sleeves of dura mater and arachnoid mater, and enter the olfactory bulb in the anterior cranial fossa (Figs. 9.2 and 9.3). The **olfactory bulb** lies in contact with the inferior or orbital surface of the frontal lobe of the cerebral hemisphere. The olfactory nerve fibers synapse with **mitral cells** in the olfactory bulb. The axons of these secondary neurons form the **olfactory tract.** The olfactory bulbs and tracts are anterior extensions of the forebrain.

Each olfactory tract divides into lateral and medial **olfactory striae** (distinct fiber bands). The lateral olfactory stria terminates in the piriform cortex of the anterior part of the temporal lobe, and the medial olfactory stria projects through the anterior commissure to contralateral olfactory structures. The olfactory nerves are the only cranial nerves to enter the cerebrum directly.

The Bottom Line

OLFACTORY NERVE

♦ The olfactory nerves (CN I) have sensory fibers concerned with the special sense of smell. ♦ The olfactory receptor neurons are in the olfactory epithelium (olfactory mucosa) in the roof of the nasal cavity. ♦ The central processes of the olfactory receptor neurons ascend through foramina in the cribriform plate of the ethmoid bone to reach the olfactory bulbs in the anterior cranial fossa. These nerves synapse on neurons in the bulbs, and the processes of these neurons follow the olfactory tracts to the primary and associated areas of the cerebral cortex.

(text continued on p. 1061)

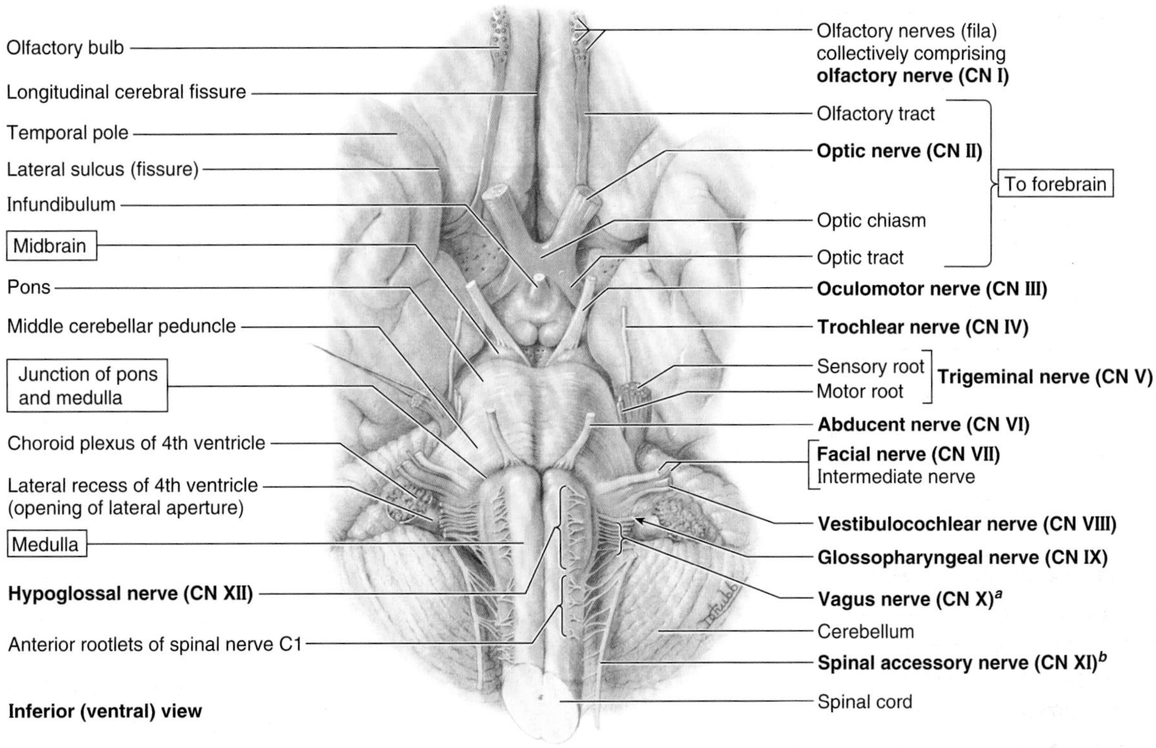

Olfactory bulb
Longitudinal cerebral fissure
Temporal pole
Lateral sulcus (fissure)
Infundibulum
Midbrain
Pons
Middle cerebellar peduncle
Junction of pons and medulla
Choroid plexus of 4th ventricle
Lateral recess of 4th ventricle (opening of lateral aperture)
Medulla
Hypoglossal nerve (CN XII)
Anterior rootlets of spinal nerve C1
Inferior (ventral) view
Olfactory nerves (fila) collectively comprising olfactory nerve (CN I)
Olfactory tract
Optic nerve (CN II)
To forebrain
Optic chiasm
Optic tract
Oculomotor nerve (CN III)
Trochlear nerve (CN IV)
Sensory root
Motor root
Trigeminal nerve (CN V)
Abducent nerve (CN VI)
Facial nerve (CN VII)
Intermediate nerve
Vestibulocochlear nerve (CN VIII)
Glossopharyngeal nerve (CN IX)
Vagus nerve (CN X)a
Cerebellum
Spinal accessory nerve (CN XI)b
Spinal cord

FIGURE 9.3. Superficial origins of cranial nerves from brain and spinal cord (except for CN IV, which arises from the posterior aspect of the midbrain). [a]The traditional "cranial root of the accessory nerve" is considered here as part of the vagus nerve. [b]The spinal accessory nerve as listed here refers to only the traditional "spinal root of the accessory nerve."

TABLE 9.1. CRANIAL NERVES: ATTACHMENT TO CENTRAL NERVOUS SYSTEM, GENERAL FUNCTIONS, AND DISTRIBUTION

Cranial Nerve Number	Cranial Nerve Name	Part of Central Nervous System from Which Nerve(s) Enter(s) or Emerge(s)		General Functional Types of Fibers[1]		General Distribution
I	**Olfactory**	Forebrain (prosencephalon)	Cerebral hemispheres (telencephalon)	Special sensory		Olfactory mucosa of nose
II	**Optic**		Diencephalon			Retina of eye
III	**Oculomotor**	Midbrain (mesencephalon)		Motor[2]		Intra-ocular and four extra-ocular muscles
IV	**Trochlear**	Midbrain				One extra-ocular muscle (superior oblique)
V	**Trigeminal**	Brainstem	Pons (metencephalon)	Mixed	Motor root	Derivatives of frontonasal process and 1st pharyngeal arch
					Sensory root	
VI	**Abducent (abducens)**		Junction between pons and medulla	Motor[2]		One extra-ocular muscle
VII	**Facial**			Mixed	Primary root	Derivatives of 2nd pharyngeal arch
					Intermediate nerve	
VIII	**Vestibulocochlear**			Special sensory[5]		Internal ear
IX	**Glossopharyngeal**		Medulla (myelencephalon)	Mixed		Derivatives of 3rd pharyngeal arch
X	**Vagus**					Derivatives of 4th pharyngeal arch
XI	**Spinal accessory**	Superior spinal cord		Motor[3]		Superficial layer of neck
XII	**Hypoglossal**	Brainstem	Medulla (myelencephalon)	Motor[4]		Muscles of tongue

[1]Note that the colors in this column match those of the nerves in Figure 9.3.
[2]The presence and function of proprioceptive afferent fibers to the extra-ocular muscles is controversial.
[3]Cranial nerve XI is purely motor as it leaves the CNS, but gains pain and proprioceptive fibers from the cervical plexus in the lateral cervical region (posterior triangle) of the neck.
[4]Cranial nerve XII is purely motor as it leaves the CNS; pathways for proprioception associated with the tongue are unknown and may involve the lingual and glossopharyngeal nerves, and cervical spinal nerves that communicate with CN XII.
[5]The cochlear part of CN VIII, traditionally considered "purely sensory," actually conveys some efferent fibers that appear to modulate sensory sensitivity.

TABLE 9.2. SUMMARY OF CRANIAL NERVES

Nerve	Components	Location of Nerve Cell Bodies	Cranial Exit	Main Action(s)
Olfactory (CN I)	Special sensory	Olfactory epithelium (olfactory cells)	Foramina in cribriform plate of ethmoid bone	Smell from nasal mucosa of roof of each nasal cavity and superior sides of nasal septum and superior concha
Optic (CN II)	Special sensory	Retina (ganglion cells)	Optic canal	Vision from retina
Oculomotor (CN III)	Somatic motor	Midbrain	Superior orbital fissure	Motor to superior, inferior, and medial recti, inferior oblique, and levator palpebrae superioris muscles that raise superior eyelids and direct gaze superiorly, inferiorly, and medially
	Visceral motor	*Presynaptic:* midbrain *Postsynaptic:* ciliary ganglion		Parasympathetic innervation to sphincter pupillae and ciliary muscles that constrict pupil and accommodate-lens of eye
Trochlear (CN IV)	Somatic motor	Midbrain		Motor to superior oblique that assists in directing gaze inferolaterally (or inferiorly from the adducted position)
Trigeminal (CN V) **Ophthalmic (CN V₁)**	Somatic (general) sensory	Trigeminal ganglion	Superior orbital fissure	Sensation from cornea, skin of forehead, scalp, eyelids, nose, and mucosa of nasal cavity and paranasal sinuses
Maxillary (CN V₂)			Foramen rotundum	Sensation from skin of face over maxilla, including upper lip, maxillary teeth, mucosa of nose, maxillary sinuses, and palate
Mandibular (CN V₃)			Foramen ovale	Sensation from skin of face overlying mandible including lower lip, mandibular teeth, temporomandibular joint, mucosa of mouth and anterior two thirds of tongue
	Somatic (branchial) motor	Pons		Motor to muscles of mastication, mylohyoid, anterior belly of digastric, tensor veli palatini, and tensor tympani
Abducent (CN VI) (L. *abducens*)	Somatic motor	Pons	Superior orbital fissure	Motor to lateral rectus that directs gaze laterally
Facial (CN VII)	Somatic (branchial) motor	Pons	Internal acoustic meatus; facial canal; stylomastoid foramen	Motor to muscles of facial expression and scalp; also supplies stapedius of middle ear, stylohyoid, and posterior belly of digastric
	Special sensory	Geniculate ganglion		Taste from anterior two thirds of tongue and palate
	Visceral motor	*Presynaptic:* pons *Postsynaptic:* pterygopalatine ganglion; submandibular ganglion		Parasympathetic innervation of submandibular and sublingual salivary glands, lacrimal gland, and glands of nose and palate

TABLE 9.2. SUMMARY OF CRANIAL NERVES (Continued)

Nerve	Components	Location of Nerve Cell Bodies	Cranial Exit	Main Action(s)
Vestibulocochlear (CN VIII)	Special sensory	Vestibular ganglion	Internal acoustic meatus	Vestibular sensation from semicircular ducts, utricle, and saccule related to position and movements of head
Vestibular				
Cochlear		Spiral ganglion		Hearing from spiral organ
Glossopharyngeal (CN IX)	Somatic (branchial) motor	Medulla	Jugular foramen	Motor to stylopharyngeus that assists with swallowing
	Visceral motor	*Presynaptic:* medulla *Postsynaptic:* otic ganglion		Parasympathetic innervation to parotid gland
	Visceral sensory	Superior ganglion		Visceral sensation from parotid gland, carotid body and sinus, pharynx, and middle ear
	Special sensory	Inferior ganglion		Taste from posterior third of tongue
	Somatic (general) sensory			Cutaneous sensation from external ear
Vagus (CN X)	Somatic (branchial) motor	Medulla		Motor to muscles of pharynx (except stylopharyngeus), intrinsic muscles of larynx, muscles of palate (except tensor veli palatini), and striated muscle in superior two thirds of esophagus
	Visceral motor	*Presynaptic:* medulla *Postsynaptic:* neurons in, on, or near viscera		Parasympathetic innervation to smooth muscle and glands of trachea, bronchi, digestive tract, coronary arteries, and nodes of conduction system of heart
	Visceral sensory	Inferior ganglion		Visceral sensation from base of tongue, pharynx, larynx, trachea, bronchi, heart, esophagus, stomach, and intestine to left colic flexure
	Special sensory	Inferior ganglion		Taste from epiglottis and palate
	Somatic (general) sensory	Superior ganglion		Sensation from auricle, external acoustic meatus, and dura mater of posterior cranial fossa
Spinal accessory (CN XI)	Somatic motor	Spinal cord		Motor to sternocleidomastoid and trapezius
Hypoglossal (CN XII)	Somatic motor	Medulla	Hypoglossal canal	Motor to intrinsic and extrinsic muscles of tongue (except palatoglossus)

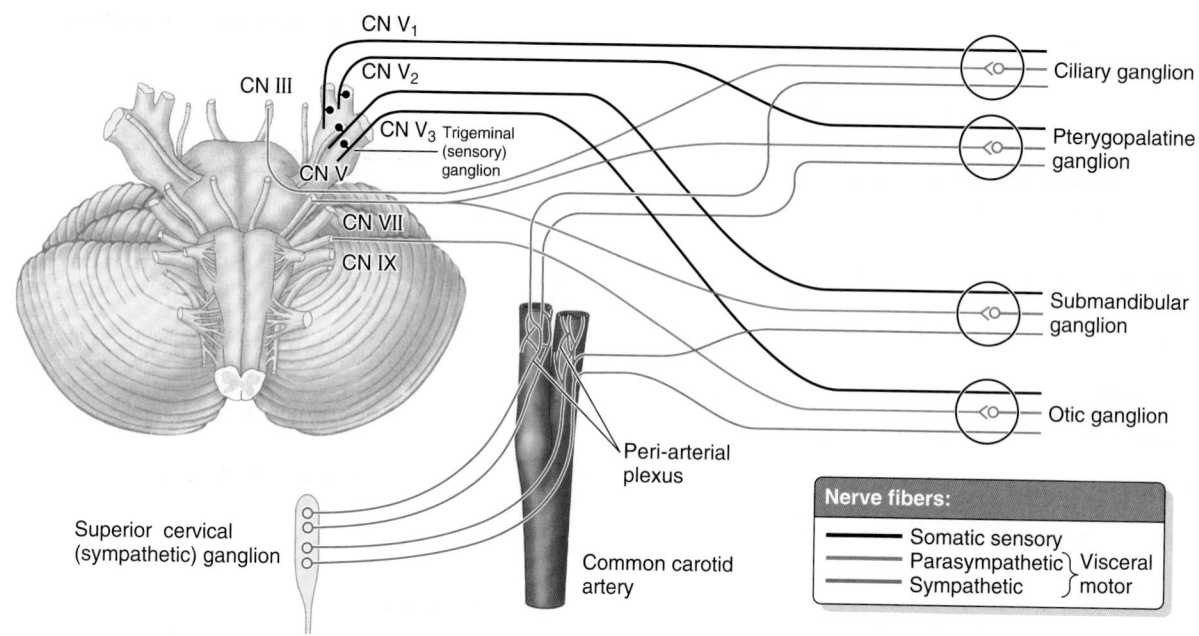

FIGURE 9.4. Summary of cranial parasympathetic ganglia.

TABLE 9.3. CRANIAL PARASYMPATHETIC GANGLIA: LOCATION, PARASYMPATHETIC AND SYMPATHETIC ROOTS, AND MAIN DISTRIBUTION

Ganglion	Location	Parasympathetic Root	Sympathetic Root	Main Distribution
Ciliary	Between optic nerve and lateral rectus, close to apex of orbit	Inferior branch of oculomotor nerve (CN III)	Branches from internal carotid plexus in cavernous sinus	Parasympathetic postsynaptic fibers from ciliary ganglion pass to ciliary muscle and sphincter pupillae of iris; sympathetic postsynaptic fibers from superior cervical ganglion pass to dilator pupillae and blood vessels of eye
Pterygopalatine	In pterygopalatine fossa, where it is suspended by ganglionic branches of maxillary nerve (sensory roots of pterygopalatine ganglion); just anterior to opening of pterygoid canal and inferior to CN V₂	Greater petrosal nerve from facial nerve (CN VII) via nerve of pterygoid canal	Deep petrosal nerve, a branch of internal carotid plexus that is a continuation of postsynaptic fibers of cervical sympathetic trunk; fibers from superior cervical ganglion pass through pterygopalatine ganglion and enter branches of CN V₂	Parasympathetic postganglionic (secretomotor) fibers from pterygopalatine ganglion innervate lacrimal gland via zygomatic branch of CN V₂; sympathetic postsynaptic fibers from superior cervical ganglion accompany branches of pterygopalatine nerve that are distributed to blood vessels of nasal cavity, palate, and superior parts of pharynx
Otic	Between tensor veli palatini and mandibular nerve (CN V₃); lies inferior to foramen ovale of sphenoid bone	Tympanic nerve from glossopharyngeal nerve (CN IX); continues from tympanic plexus as lesser petrosal nerve	Fibers from superior cervical ganglion come from plexus on middle meningeal artery	Parasympathetic postsynaptic fibers from otic ganglion are distributed to parotid gland via auriculotemporal nerve (branch of CN V₃); sympathetic postsynaptic fibers from superior cervical ganglion pass to parotid gland and supply its blood vessels
Submandibular	Suspended from lingual nerve by two ganglionic branches (sensory roots); lies on surface of hyoglossus muscle inferior to submandibular duct	Parasympathetic fibers join facial nerve (CN VII) and leave it in its chorda tympani branch, which unites with lingual nerve	Sympathetic fibers from superior cervical ganglion via plexus on facial artery chorda tympani	Parasympathetic postsynaptic (secretomotor) fibers from submandibular ganglion are distributed to sublingual and submandibular glands; sympathetic fibers from superior cervical ganglion supply sublingual and submandibular glands

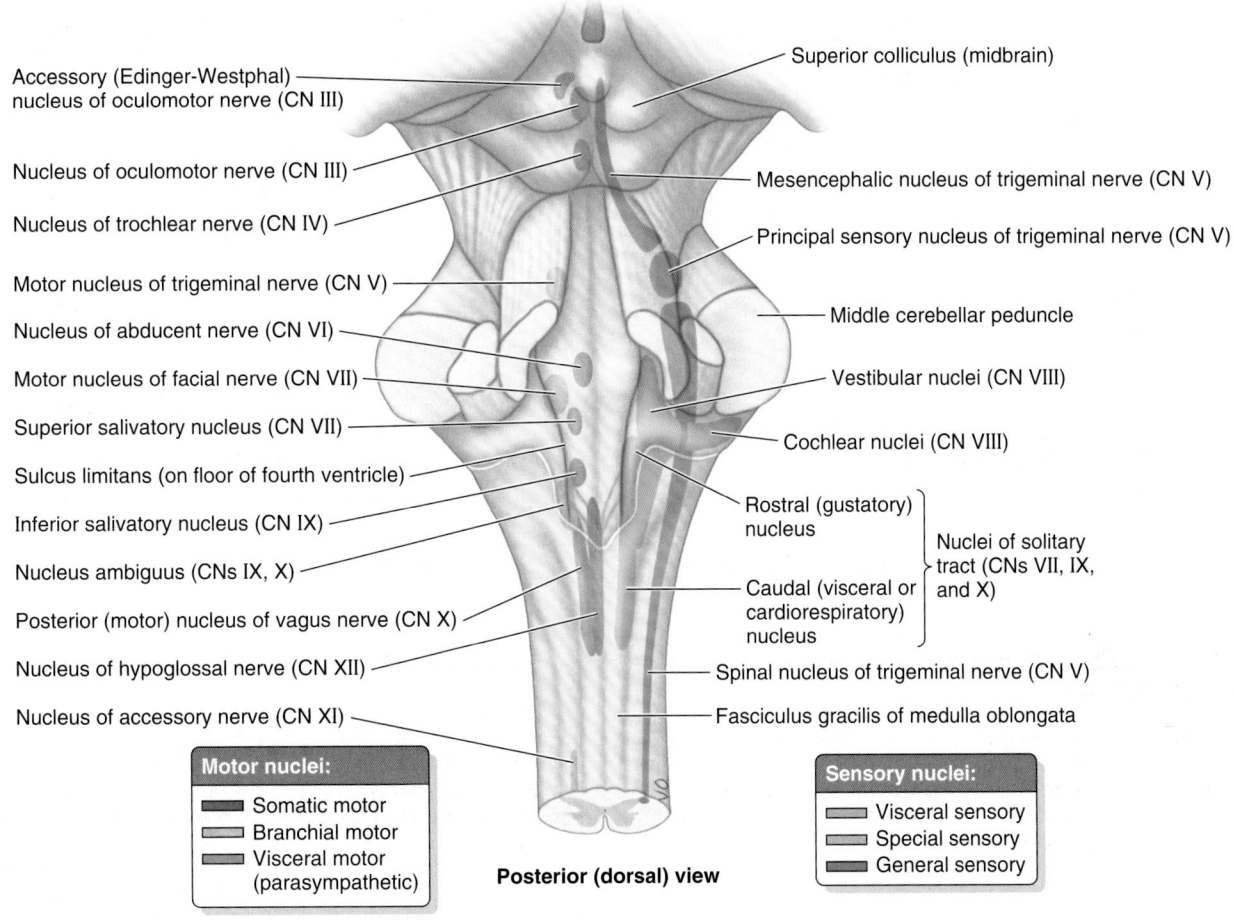

Accessory (Edinger-Westphal) nucleus of oculomotor nerve (CN III)

Nucleus of oculomotor nerve (CN III)

Nucleus of trochlear nerve (CN IV)

Motor nucleus of trigeminal nerve (CN V)

Nucleus of abducent nerve (CN VI)

Motor nucleus of facial nerve (CN VII)

Superior salivatory nucleus (CN VII)

Sulcus limitans (on floor of fourth ventricle)

Inferior salivatory nucleus (CN IX)

Nucleus ambiguus (CNs IX, X)

Posterior (motor) nucleus of vagus nerve (CN X)

Nucleus of hypoglossal nerve (CN XII)

Nucleus of accessory nerve (CN XI)

Superior colliculus (midbrain)

Mesencephalic nucleus of trigeminal nerve (CN V)

Principal sensory nucleus of trigeminal nerve (CN V)

Middle cerebellar peduncle

Vestibular nuclei (CN VIII)

Cochlear nuclei (CN VIII)

Rostral (gustatory) nucleus

Caudal (visceral or cardiorespiratory) nucleus

Nuclei of solitary tract (CNs VII, IX, and X)

Spinal nucleus of trigeminal nerve (CN V)

Fasciculus gracilis of medulla oblongata

Motor nuclei:
- Somatic motor
- Branchial motor
- Visceral motor (parasympathetic)

Sensory nuclei:
- Visceral sensory
- Special sensory
- General sensory

Posterior (dorsal) view

FIGURE 9.5. Cranial nerve nuclei.

OPTIC NERVE (CN II)

Function: Special sensory (special somatic afferent)—that is, the special sense of vision.

Although officially nerves by convention, the **optic nerves** (CN II) develop in a completely different manner from the other cranial nerves. The structures involved in receiving and transmitting optical stimuli (the optical fibers and neural retina, together with the pigmented epithelium of the eyeball) develop as evaginations of the diencephalon. The optic nerves are paired, anterior extensions of the forebrain (diencephalon) and are therefore actually CNS fiber tracts formed by axons of **retinal ganglion cells** (Moore et al., 2012). In other words, they are third-order neurons, with their cell bodies located in the retina (Fig. 9.7B).

The optic nerves are surrounded by extensions of the cranial meninges and subarachnoid space, which is filled with cerebrospinal fluid (CSF). The meninges extend all the way to the eyeball. The central artery and vein of the retina traverse the meningeal layers and course in the anterior part of the optic nerve. Cranial nerve II begins where the unmyelinated axons of retinal ganglion cells pierce the sclera (the opaque part of the external fibrous coat of the eyeball) and become myelinated, deep to the *optic disc.*

The nerve passes posteromedially in the orbit, exiting through the *optic canal* to enter the middle cranial fossa, where it forms the **optic chiasm** (Fig. 9.7A). Here, fibers from the nasal (medial) half of each retina decussate in the chiasm and join uncrossed fibers from the temporal (lateral) half of the retina to form the **optic tract.**

The partial crossing of optic nerve fibers in the chiasm is a requirement for binocular vision, allowing depth-of-field perception (three-dimensional vision). Thus fibers from the right halves of both retinas form the right optic tract. The decussation of nerve fibers in the chiasm results in the right optic tract conveying impulses from the left visual field and vice versa. The **visual field** is what is seen by a person who has both eyes wide open and who is looking straight ahead. Most fibers in the optic tracts terminate in the **lateral geniculate bodies** of the thalamus. From these nuclei, axons are relayed to the visual cortices of the occipital lobes of the brain.

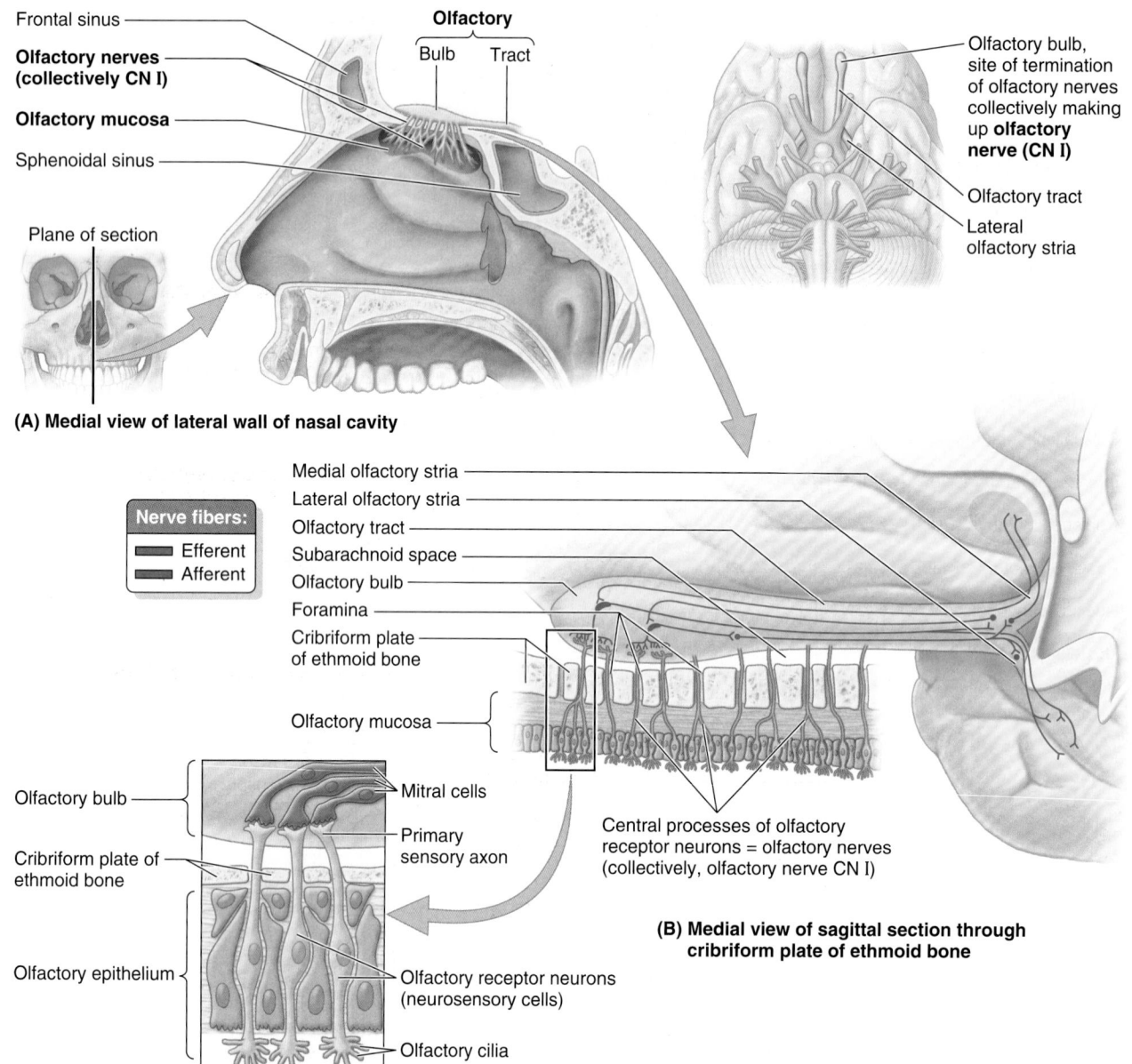

(A) Medial view of lateral wall of nasal cavity

(B) Medial view of sagittal section through cribriform plate of ethmoid bone

FIGURE 9.6. Olfactory system. A. This sagittal section through the nasal cavity shows the relationship of the olfactory mucosa to the olfactory bulb. **B.** The bodies of the olfactory receptor neurons are in the olfactory epithelium. These bundles of axons are collectively called the olfactory nerve (CN I).

The Bottom Line

OPTIC NERVE

♦ The optic nerves (CN II) have sensory fibers concerned with the special sense of vision. ♦ The optic nerve fibers arise from ganglion cells in the retina. ♦ The nerve fibers exit the orbit via the optic canals; fibers from the nasal half of the retina cross to the contralateral side at the optic chiasm. ♦ The nerve fibers then pass via the optic tracts to the geniculate bodies of the thalamus, where they synapse on neurons whose processes form the optic radiations to the primary visual cortex of the occipital lobe.

OCULOMOTOR NERVE (CN III)

Functions: Somatic motor (general somatic efferent) and visceral motor (general visceral efferent–parasympathetic).

Nuclei: There are two oculomotor nuclei, each serving one of the functional components of the nerve. The somatic **motor nucleus of the oculomotor nerve** is in the midbrain (Fig. 9.5). The visceral motor (parasympathetic) **accessory (Edinger-Westphal) nucleus of the oculomotor nerve** lies dorsal to the rostral two thirds of the somatic motor nucleus (Haines, 2006).

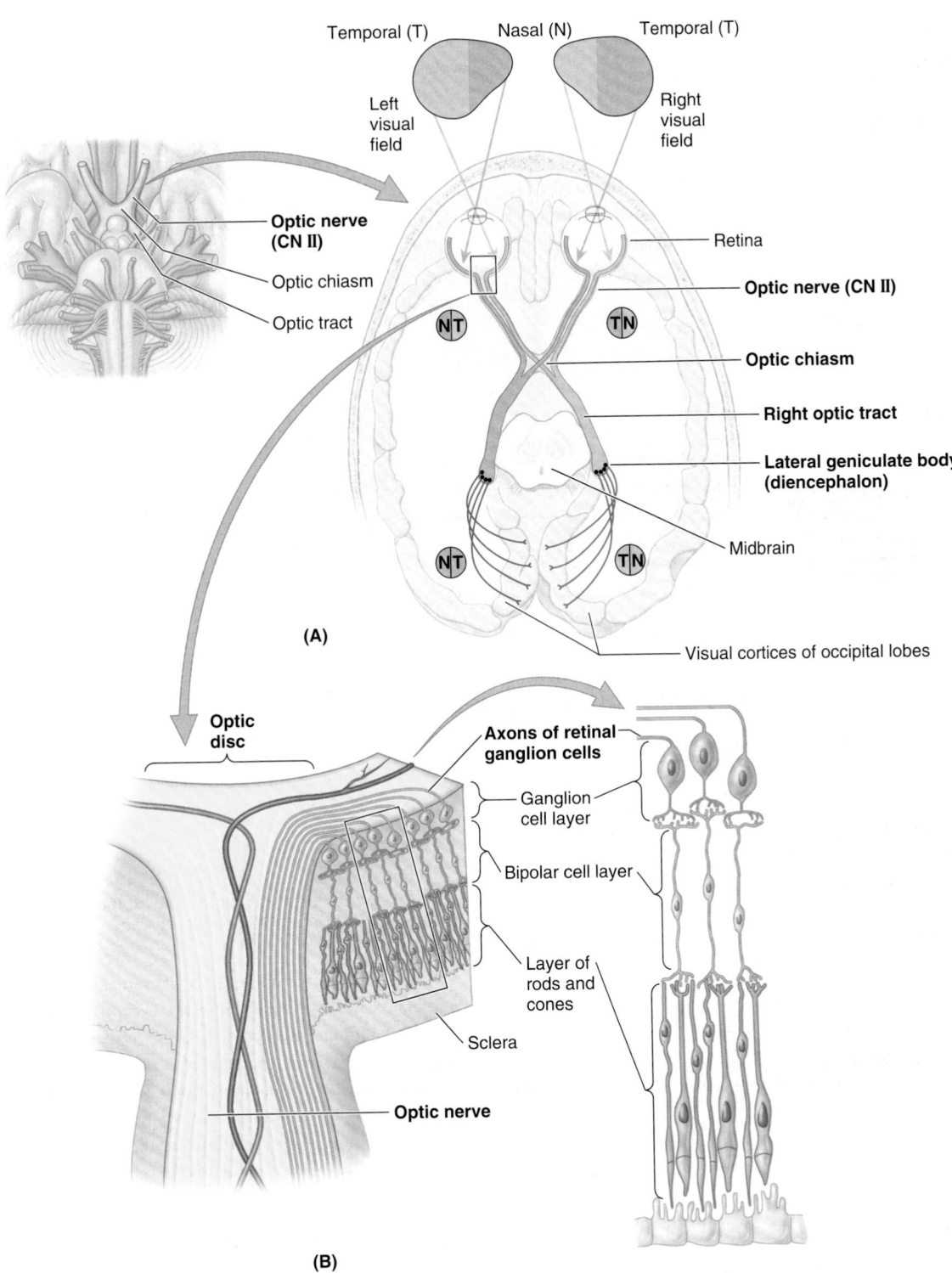

FIGURE 9.7. Visual system. A. The origin, course, and distribution of the visual pathway are shown. The axons of retinal ganglionic neurons convey visual information to the lateral geniculate body of the diencephalon (thalamus) through the optic nerve (CN II) and optic tract. Fibers from the lateral geniculate body project to the visual cortices of the occipital lobes. The axons of the ganglion cells of the nasal halves of the retinas cross in the optic chiasm; those from the temporal halves do not cross. **B.** The visual pathway begins with photoreceptor cells (rods and cones) in the retina. The responses of the photoreceptors are transmitted by bipolar cells (neurons with two processes) to ganglion cells in the ganglion cell layer of the retina. The central processes of this third-order neuron are the fibers conducted by the optic nerves.

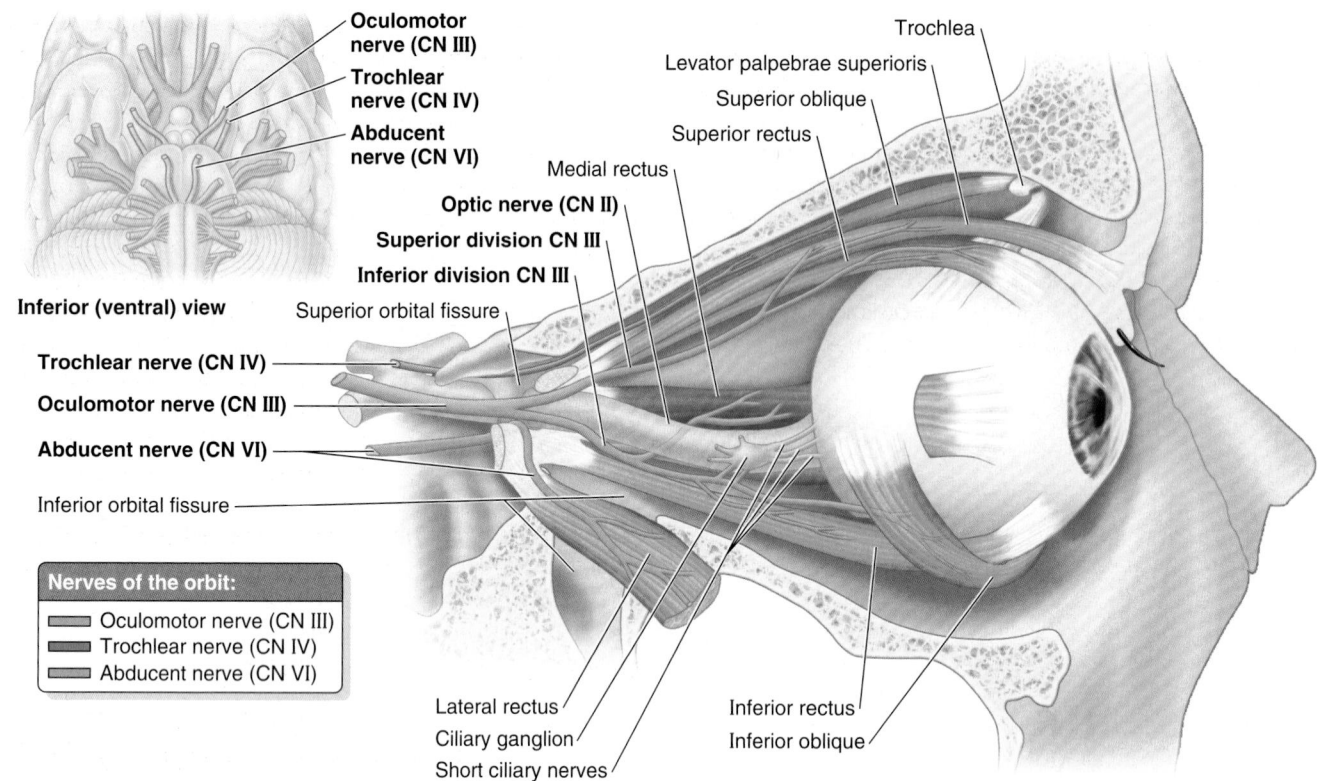

FIGURE 9.8. Distribution of oculomotor (CN III), trochlear (CN IV), and abducent (CN VI) nerves. CN IV supplies the superior oblique, CN VI supplies the lateral rectus, and CN III supplies five striated extra-ocular muscles (levator palpebrae superioris, superior rectus, medial rectus, inferior rectus, and inferior oblique) and two intra-ocular muscles (ciliary muscle and sphincter pupillae muscle—not shown; see Chapter 7, p. 896).

The oculomotor nerve (CN III) provides the following (Fig. 9.8):

- Motor to the striated muscle of four of the six extra-ocular muscles (*superior, medial,* and *inferior recti* and *inferior oblique*) and superior eyelid (L. *levator palpebrae superioris*); hence the nerve's name.
- Parasympathetic through the ciliary ganglion to the smooth muscle of the sphincter pupillae, which causes constriction of the pupil and ciliary muscle, which produces accommodation (allowing the lens to become more rounded) for near vision.

CN III is the chief motor nerve to the ocular and extra-ocular muscles. It emerges from the midbrain, pierces the dura mater lateral to the sellar diaphragm roofing over the hypophysis, and then runs through the roof and lateral wall of the *cavernous sinus.* CN III leaves the cranial cavity and enters the orbit through the *superior orbital fissure.* Within this fissure, CN III divides into a **superior division** (which supplies the superior rectus and levator palpebrae superioris) and an **inferior division** (which supplies the inferior and medial rectus and inferior oblique). The inferior division also carries presynaptic parasympathetic (visceral efferent) fibers to the *ciliary ganglion,* where they synapse (Fig. 9.4; Table 9.3). Postsynaptic fibers from this ganglion pass to the eyeball in the *short ciliary nerves* to innervate the ciliary body and sphincter pupillae (see Chapter 7, p. 896).

The Bottom Line

OCULOMOTOR NERVE

♦ The oculomotor nerves (CN III) send somatic motor fibers to all extra-ocular muscles, except the superior oblique and lateral rectus. ♦ These nerves also send presynaptic parasympathetic fibers to the ciliary ganglion for innervation of the ciliary body and sphincter pupillae. ♦ These nerves originate from the brainstem, emerging medial to the cerebral peduncles, and run in the lateral wall of the cavernous sinus. ♦ These nerves enter the orbit through the superior orbital fissures and divide into superior and inferior branches.

TROCHLEAR NERVE (CN IV)

Functions: Somatic motor (general somatic efferent) to one extra-ocular muscle (superior oblique).

Nucleus: The **nucleus of the trochlear nerve** is located in the midbrain, immediately caudal to the oculomotor nucleus (Fig. 9.5).

The **trochlear nerve** (CN IV) is the smallest cranial nerve. It emerges from the posterior (dorsal) surface of the midbrain (the only cranial nerve to do so), passing anteriorly around the brainstem. It has the longest *intracranial (subarachnoid) course* of the cranial nerves. The trochlear nerve pierces the dura mater at the margin of the tentorium cerebelli, and passes anteriorly in the lateral wall of the cavernous sinus (Fig. 9.8). CN IV then continues through the superior orbital fissure into the orbit, where it supplies the superior oblique—the only extra-ocular muscle that uses a pulley, or trochlea, to redirect its line of action (hence the nerve's name).

The Bottom Line

TROCHLEAR NERVE

♦ The trochlear nerves (CN IV) supply somatic motor fibers to the superior oblique muscles, which abduct, depress, and medially rotate the pupil. ♦ The trochlear nerves emerge from the posterior aspect of the brainstem. ♦ The nerves run a long intracranial course, passing around the brainstem to enter the dura mater in the free edge of the tentorium cerebelli close to the posterior clinoid process. ♦ The nerves then run in the lateral wall of the cavernous sinus, entering the orbit via the superior orbital fissures.

TRIGEMINAL NERVE (CN V)

Functions: Somatic (general) sensory and somatic (branchial) motor to derivatives of the 1st pharyngeal arch.

Nuclei: There are four trigeminal nuclei (Fig. 9.5)—one motor (**motor nucleus of trigeminal nerve**), and three sensory (**mesencephalic, principal sensory and spinal nuclei of trigeminal nerve**).

The **trigeminal nerve** (CN V) is the largest cranial nerve (if the atypical optic nerve is excluded). It emerges from the lateral aspect of the pons by a large sensory root and a small motor root (Fig. 9.3). The roots of CN V are comparable to the posterior and anterior roots of spinal nerves. CN V is the principal somatic (general) sensory nerve for the head (face, teeth, mouth, nasal cavity, and dura mater of the cranial cavity). The large **sensory root of CN V** is composed mainly of the central processes of the pseudounipolar neurons that make up the sensory **trigeminal ganglion** (Fig. 9.9). The ganglion is flattened and crescent shaped (hence its unofficial name, semilunar ganglion), and is housed within a dural recess (**trigeminal cave**) lateral to the cavernous sinus.

The peripheral processes of the ganglionic neurons form three nerves or divisions: *ophthalmic nerve (CN V₁), maxillary nerve (CN V₂),* and sensory component of the *mandibular nerve (CN V₃).* Maps of the zones of cutaneous innervation by the three divisions resemble the dermatome maps for cutaneous innervation by spinal nerves (Fig. 9.9A). Unlike spinal nerve dermatomes, however, there is little overlap in innervation by the divisions; lesions of a single nerve result in clearly demarcated areas of numbness.

The fibers of the **motor root of CN V** pass inferior to the trigeminal ganglion along the floor of the trigeminal cave, bypassing the ganglion (just as the anterior roots of spinal nerves bypass the spinal sensory ganglia). They are distributed exclusively via the mandibular nerve (CN V₃), blending with the sensory fibers as the nerve traverses the foramen ovale in the cranium. Branches pass to the muscles of mastication, mylohyoid, anterior belly of the digastric, tensor veli palatini, and tensor tympani, which are derived from the 1st pharyngeal arch.

Although CN V conveys no presynaptic parasympathetic fibers from the CNS, all four parasympathetic ganglia are associated with the divisions of CN V. Postsynaptic parasympathetic fibers from the ganglia join branches of CN V and are carried to their destinations along with the CN V sensory and motor fibers (Fig. 9.9; Table 9.3).

Ophthalmic Nerve (CN V₁)

In contrast to the other two CN V divisions, CN V₁ is not a branchial nerve (i.e., it does not supply pharyngeal arch derivatives). It serves structures derived from the paraxial mesoderm of the embryonic frontonasal process. The ophthalmic nerve's association with the other CN V divisions is a secondary occurrence. The somatic (general) sensory fibers of CN V₁ are distributed to skin and mucous membranes and conjunctiva of the front of the head and nose (Fig. 9.9).

Testing CN V₁: The integrity of this division is tested by checking the corneal reflex—touching the cornea, which is also supplied by CN V₁, with a wisp of cotton will evoke a reflexive blink if the nerve is functional (Table 9.4).

Maxillary Nerve (CN V₂)

CN V₂ innervates derivatives of the maxillary prominence of the 1st pharyngeal arch. Exiting the cranial cavity via the foramen rotundum, its somatic (general) sensory fibers are generally distributed to skin and mucous membranes associated with the upper jaw. The pterygopalatine (parasympathetic) ganglion is associated with this division of CN V, involved in innervating the lacrimal gland and glands of the nose and palate.

Mandibular Nerve (CN V₃)

CN V₃ innervates derivatives of the mandibular prominence of the 1st pharyngeal arch. CN V₃ is the only division of CN V to convey somatic (branchial) motor fibers, distributed to the striated muscle derived from mandibular prominence mesoderm, primarily the muscles of mastication. Two parasympathetic ganglia, the otic and submandibular, are associated with this division of CN V; both are concerned with the innervation of salivary glands.

Tables 9.1 and 9.2 provide a general summary of CN V. Table 9.4 summarizes the branches of the three divisions.

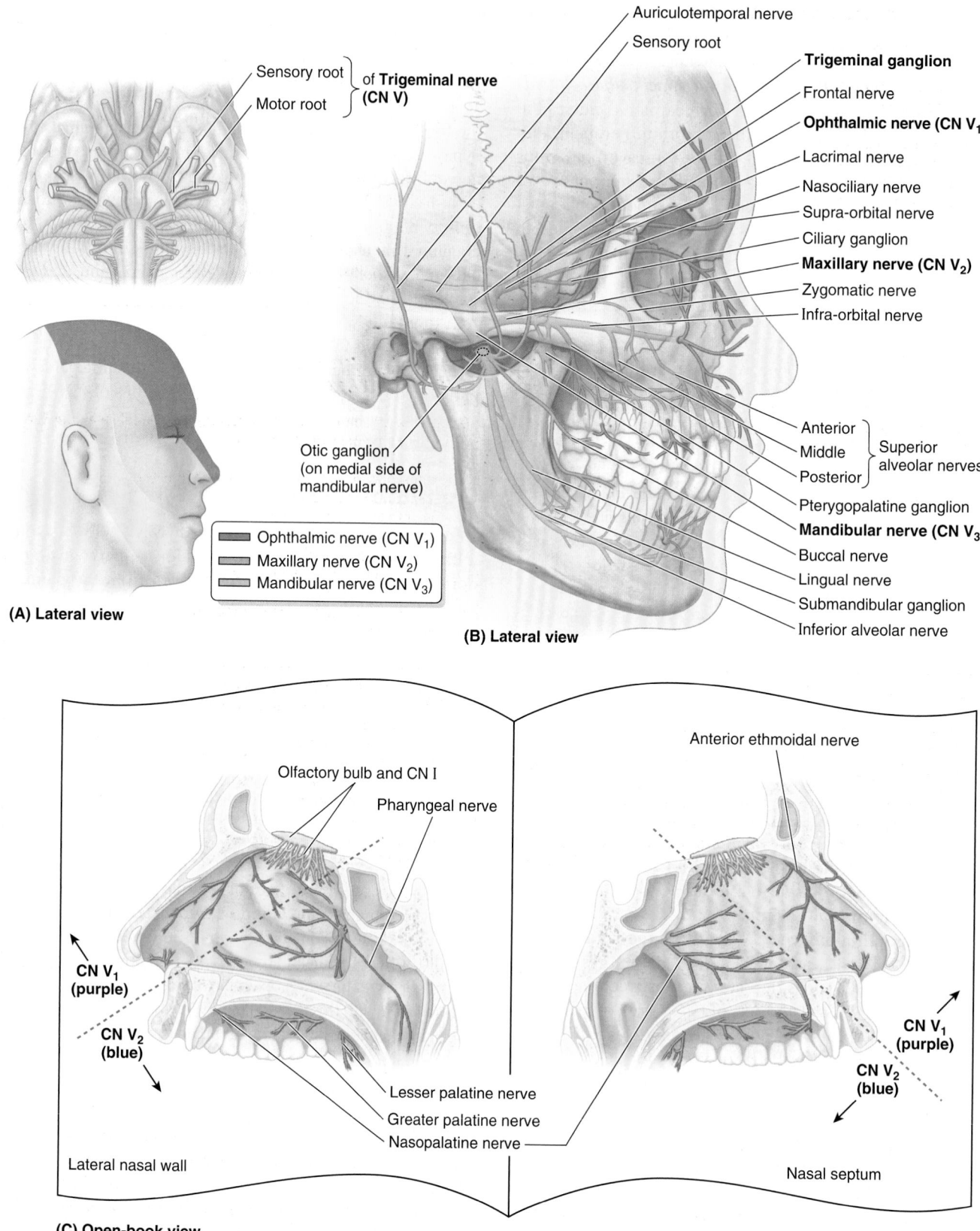

(A) Lateral view

Sensory root }
Motor root } of **Trigeminal nerve (CN V)**

Ophthalmic nerve (CN V₁)
Maxillary nerve (CN V₂)
Mandibular nerve (CN V₃)

Otic ganglion (on medial side of mandibular nerve)

Auriculotemporal nerve
Sensory root
Trigeminal ganglion
Frontal nerve
Ophthalmic nerve (CN V₁)
Lacrimal nerve
Nasociliary nerve
Supra-orbital nerve
Ciliary ganglion
Maxillary nerve (CN V₂)
Zygomatic nerve
Infra-orbital nerve
Anterior } Superior
Middle } alveolar nerves
Posterior }
Pterygopalatine ganglion
Mandibular nerve (CN V₃)
Buccal nerve
Lingual nerve
Submandibular ganglion
Inferior alveolar nerve

(B) Lateral view

Olfactory bulb and CN I
Pharyngeal nerve
Anterior ethmoidal nerve

CN V₁ (purple)
CN V₂ (blue)

CN V₁ (purple)
CN V₂ (blue)

Lesser palatine nerve
Greater palatine nerve
Nasopalatine nerve

Lateral nasal wall
Nasal septum

(C) Open-book view

FIGURE 9.9. Distribution of trigeminal nerve (CN V). A. Cutaneous (sensory) zones are innervated by the three divisions of the trigeminal nerve. **B.** Each cranial nerve division supplies skin and mucous membranes and sends a branch to the dura of the anterior and middle cranial fossae. Each division is associated with one or two parasympathetic ganglia and delivers the postsynaptic parasympathetic fibers from that ganglion: CN V₁ for the ciliary ganglion, CN V₂ for the pterygopalatine ganglion, and CN V₃ for the submandibular and otic ganglia. **C.** This open-book view shows the innervation of the lateral wall and septum of the nasal cavity and palate. CN V₁ supplies the anterosuperior portions of the cavity, and CN V₂, the postero-inferior portions and the palate.

TABLE 9.4. SUMMARY OF DIVISIONS OF TRIGEMINAL NERVE (CN V)

Divisions/Distributions	Branches

Ophthalmic nerve (CN V₁)
Sensory only
Passes through superior orbital fissure into orbit
Supplies cornea; superior conjunctiva; mucosa of anterosuperior nasal cavity; frontal, ethmoidal, and sphenoidal sinuses; anterior and supratentorial dura mater; skin of dorsum of external nose; superior eyelid; forehead; and anterior scalp
	Tentorial nerve (a meningeal branch)
	Lacrimal nerve
	  Communicating branch from zygomatic nerve
	Frontal nerve
	  Supra-orbital nerve
	  Supratrochlear nerve
	Nasociliary nerve
	  Sensory root of ciliary ganglion
	  Short ciliary nerves
	  Long ciliary nerves
	  Anterior and posterior ethmoidal nerves
	  Infratrochlear nerves

Maxillary nerve (CN V₂)
Sensory only
Passes through foramen rotundum to enter pterygopalatine fossa
Supplies dura mater of anterior part of middle cranial fossa; conjunctiva of inferior eyelid; mucosa of postero-inferior nasal cavity, maxillary sinus, palate and anterior part of superior oral vestibule; maxillary teeth; and skin of lateral external nose, inferior eyelid, anterior cheek, and upper lip
	Meningeal branch
	Zygomatic nerve
	  Zygomaticofacial branch
	  Zygomaticotemporal branch
	  Communicating branch to lacrimal nerve
	Ganglionic branches to (sensory root of) pterygopalatine ganglion
	Posterior superior alveolar branches
	Infra-orbital nerve
	  Anterior and middle superior alveolar branches
	  Superior labial branches
	  Inferior palpebral branches
	  External nasal branches
	Greater palatine nerves
	  Posterior inferior lateral nasal nerves
	Lesser palatine nerves
	Posterior superior lateral nasal branches
	Nasopalatine nerve
	Pharyngeal nerve

Mandibular nerve (CN V₃)
Sensory and motor
Passes through foramen ovale into infratemporal fossa
Supplies sensory innervation to mucosa of anterior two thirds of tongue, floor of mouth, and posterior and anterior inferior oral vestibule; mandibular teeth; and skin of lower lip, buccal, parotid, and temporal regions of face; and external ear (auricle, upper external acoustic meatus, and tympanic membrane)
Supplies motor innervation to 4 muscles of mastication, mylohyoid, anterior belly of digastric, tensor veli palatini, and tensor tympani
	Somatic (general) sensory branches
	  Meningeal branch (nervus spinosum)
	  Buccal nerve
	  Auriculotemporal nerve
	  Lingual nerve
	  Inferior alveolar nerve
	    Inferior dental plexus
	    Mental nerve
	Somatic (branchial) motor branches
	  Masseteric nerve
	  Deep temporal nerves
	  Nerves to medial and lateral pterygoid
	  Nerve to mylohyoid (and anterior belly of digastric)
	  Nerve to tensor veli palatini
	  Nerve to tensor tympani

The Bottom Line

TRIGEMINAL NERVE

♦ The trigeminal nerve (CN V) supplies somatic motor fibers to the muscles of mastication, mylohyoid, anterior belly of the digastric, tensor tympani, and tensor veli palatini muscles. ♦ It also distributes postsynaptic parasympathetic fibers of the head to their destinations. ♦ CN V is sensory to the dura of the anterior and middle cranial fossae, skin of the face, teeth, gingiva, mucous membrane of the nasal cavity, paranasal sinuses, and mouth. ♦ CN V originates from the lateral surface of the pons by two roots: motor and sensory. ♦ These roots cross the medial part of the crest of the petrous part of the temporal bone and enter the trigeminal cave of the dura mater lateral to the body of the sphenoid and cavernous sinus. ♦ The sensory root leads to the trigeminal ganglion; the motor root runs parallel to the sensory root, then bypasses the ganglion and becomes part of the mandibular nerve (CN V₃).

ABDUCENT NERVE (CN VI)

Functions: Somatic motor (general somatic efferent) to one extra-ocular muscle, the *lateral rectus.*

Nucleus: The **abducent nucleus** is in the pons near the median plane (Fig. 9.5).

The **abducent nerve** (CN VI) emerges from the brainstem between the pons and the medulla and traverses the pontine cistern of the subarachnoid space, the right and left nerves straddling the basilar artery (Figs. 9.3 and 9.8). Each abducent nerve then pierces the dura to run the *longest intradural course* within the cranial cavity of all the cranial nerves—that is, its point of entry into the dura mater covering the clivus is the most distant from its exit from the cranium via the superior orbital fissure. During its intradural course, it bends sharply over the crest of the petrous part of the temporal bone and then courses through the cavernous sinus, surrounded by the venous blood in the same manner as the internal carotid artery, which it parallels in the sinus. CN VI traverses the common tendinous ring (*L. anulus tendineus communis*) as it enters the orbit (see Chapter 7), running on and penetrating the medial surface of the lateral rectus, which abducts the pupil.

The Bottom Line

ABDUCENT NERVE

♦ The abducent nerves (CN VI) supply somatic motor fibers to the lateral rectus muscles of the eyeballs. ♦ The nerves originate from the pons, pierce the dura mater on the clivus, traverse the cavernous sinuses and superior orbital fissures, and enter the orbits.

FACIAL NERVE (CN VII)

Functions: *Sensory*—special sensory (taste) and somatic (general) sensory. *Motor*—somatic (branchial) motor and visceral (parasympathetic) motor. It also carries proprioceptive fibers from the muscles it innervates.

Nuclei: The motor nucleus of the facial nerve is a branchiomotor nucleus in the ventrolateral part of the pons (Fig. 9.5). The cell bodies of the primary sensory neurons are in the geniculate ganglion (Fig. 9.10B). The central processes of those concerned with taste end in the nuclei of the solitary tract in the medulla. The processes of those concerned with general sensations (pain, touch, and thermal) from around the external ear end in the *spinal nucleus of the trigeminal nerve* (Fig. 9.5).

The **facial nerve** (CN VII) emerges from the junction of the pons and medulla as two divisions: the primary root and the intermediate nerve. The larger **primary root** (facial nerve proper) innervates the muscles of facial expression, and the smaller **intermediate nerve** (L. *nervus intermedius*) carries taste, parasympathetic, and somatic sensory fibers. During its course, CN VII traverses the posterior cranial fossa, internal acoustic meatus, facial canal, stylomastoid foramen of the temporal bone, and parotid gland. After traversing the internal acoustic meatus, the nerve proceeds a short distance anteriorly within the temporal bone and then turns abruptly posteriorly to course along the medial wall of the tympanic cavity. The sharp bend, the **geniculum of the facial nerve** (L. *genu,* knee) is the site of the **geniculate ganglion,** the sensory ganglion of CN VII (Fig. 9.10). While traversing the temporal bone within the facial canal, CN VII gives rise to the:

- Greater petrosal nerve.
- Nerve to the stapedius.
- Chorda tympani nerve.

Then, after running the *longest intra-osseous course* of any cranial nerve, CN VII emerges from the cranium via the *stylomastoid foramen;* gives off the posterior auricular branch; enters the parotid gland; and forms the *parotid plexus,* which gives rise to the following five terminal motor branches: temporal, zygomatic, buccal, marginal mandibular, and cervical.

Somatic (Branchial) Motor

As the nerve of the 2nd pharyngeal arch, the facial nerve supplies striated muscles derived from its mesoderm, mainly the muscles of facial expression and auricular muscles. It also supplies the posterior bellies of the digastric, stylohyoid, and stapedius muscles.

Visceral (Parasympathetic) Motor

The visceral (parasympathetic) motor distribution of the facial nerve is presented in Figure 9.11. CN VII provides presynaptic parasympathetic fibers to the *pterygopalatine ganglion* for innervation of the lacrimal glands and to the *submandibular ganglion* for innervation of the sublingual and submandibular salivary glands. The pterygopalatine ganglion is associated with the maxillary nerve (CN V$_2$), which distributes its postsynaptic fibers, whereas the submandibular ganglion is associated with the mandibular nerve (CN V$_3$). The main features of the parasympathetic ganglia supplied by the facial nerve and other cranial nerves are summarized in Figure 9.4 and Table 9.3. Parasympathetic fibers synapse in these ganglia, whereas sympathetic and other fibers pass through them.

Somatic (General) Sensory

Some fibers from the geniculate ganglion supply a small area of the skin of the concha of the auricle, close to external acoustic meatus.

Special Sensory (Taste)

Fibers carried by the chorda tympani join the *lingual nerve* of CN V$_3$ to convey taste sensation from the anterior two thirds of the tongue and soft palate (Fig. 9.10).

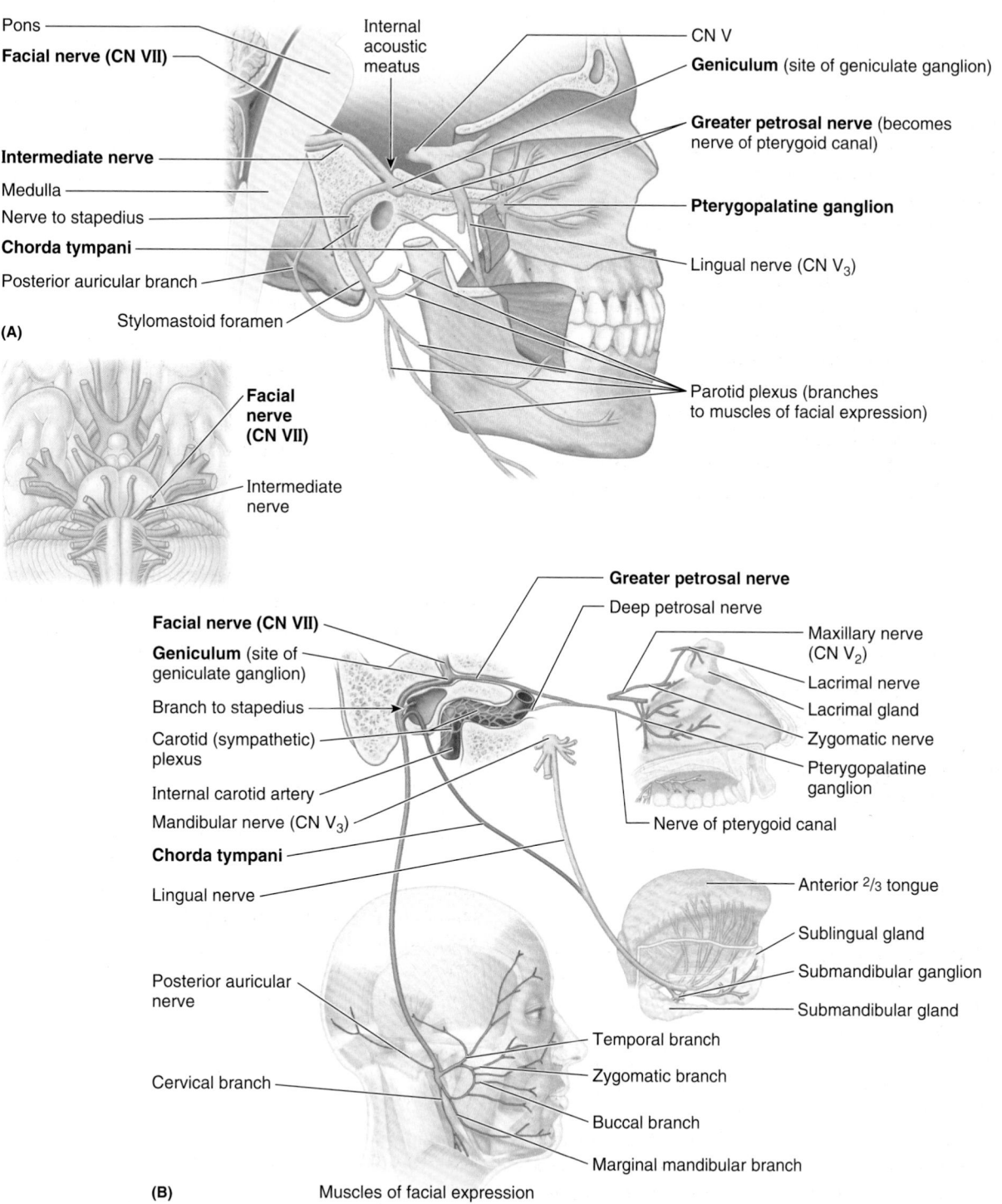

FIGURE 9.10. Distribution of facial nerve (CN VII). A. The facial nerve in situ demonstrates its intraosseous course and branches. **B.** The distribution of the facial nerve fibers is shown. Observe that CN VII supplies (1) somatic (branchial) motor innervation (*blue*) to derivatives of the 2nd pharyngeal arch (muscles of facial expression, including the auricular and occipitofrontalis muscles plus the stapedius and posterior bellies of the digastric and stylohyoid); (2) special sensory (taste) and presynaptic parasympathetic (secretomotor) fibers (*green*) to the anterior tongue and submandibular ganglion via the chorda tympani; and (3) presynaptic parasympathetic (secretomotor) fibers (*purple*) to the pterygopalatine ganglion via the greater petrosal nerve.

(A) Parasympathetic (visceral motor) to lacrimal gland

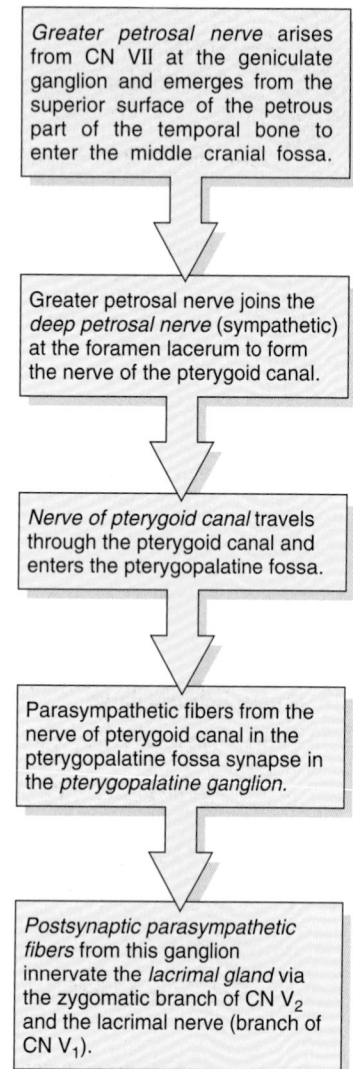

Greater petrosal nerve arises from CN VII at the geniculate ganglion and emerges from the superior surface of the petrous part of the temporal bone to enter the middle cranial fossa.

Greater petrosal nerve joins the *deep petrosal nerve* (sympathetic) at the foramen lacerum to form the nerve of the pterygoid canal.

Nerve of pterygoid canal travels through the pterygoid canal and enters the pterygopalatine fossa.

Parasympathetic fibers from the nerve of pterygoid canal in the pterygopalatine fossa synapse in the *pterygopalatine ganglion.*

Postsynaptic parasympathetic fibers from this ganglion innervate the *lacrimal gland* via the zygomatic branch of CN V_2 and the lacrimal nerve (branch of CN V_1).

(B) Parasympathetic (visceral motor) to submandibular and sublingual glands

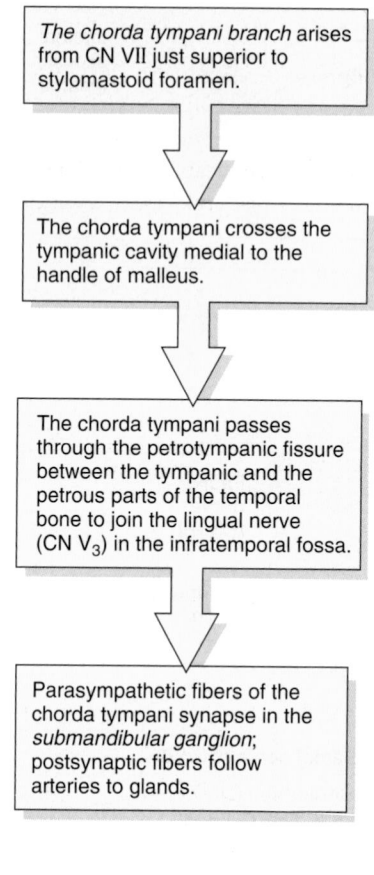

The chorda tympani branch arises from CN VII just superior to stylomastoid foramen.

The chorda tympani crosses the tympanic cavity medial to the handle of malleus.

The chorda tympani passes through the petrotympanic fissure between the tympanic and the petrous parts of the temporal bone to join the lingual nerve (CN V_3) in the infratemporal fossa.

Parasympathetic fibers of the chorda tympani synapse in the *submandibular ganglion*; postsynaptic fibers follow arteries to glands.

FIGURE 9.11. Parasympathetic innervation involving CN VII. A. Innervation of the lacrimal gland. **B.** Innervation of the submandibular and sublingual glands.

The Bottom Line

FACIAL NERVE

♦ The facial nerves (CN VII) supply motor fibers to the sta-pedius, posterior belly of the digastric, stylohyoid, facial, and scalp muscles. ♦ They also supply presynaptic parasympa-thetic fibers via the intermediate nerve (smaller root of CN VII) destined for the pterygopalatine and submandibular ganglia via the greater petrosal nerve and chorda tympani, respectively. ♦ CN VII is sensory to some skin of the external acoustic meatus and, via the intermediate nerve, is sensory to taste from the anterior two thirds of the tongue and the soft palate. ♦ CN VII originates from the posterior border of the pons and runs through the internal acoustic meatus and facial canal in the petrous part of the temporal bone. ♦ CN VII exits via the stylomastoid foramen; its main trunk forms the intraparotid nerve plexus.

VESTIBULOCOCHLEAR NERVE (CN VIII)

Functions: Special sensory (special somatic afferent)—that is, special sensations of hearing, equilibrium, and motion (acceleration/deceleration).

Nuclei: Vestibular nuclei are located at the junction of the pons and medulla in the lateral part of the floor of the 4th ventricle; the **cochlear nuclei**, anterior and posterior, are in the medulla (Fig. 9.5).

The **vestibulocochlear nerve** (CN VIII) emerges from the junction of the pons and medulla and enters the *internal acoustic meatus* (Figs. 9.2 and 9.3). Here it separates into the vestibular and cochlear nerves (Fig. 9.12).

- The **vestibular nerve** is composed of the central processes of bipolar neurons in the **vestibular ganglion.**

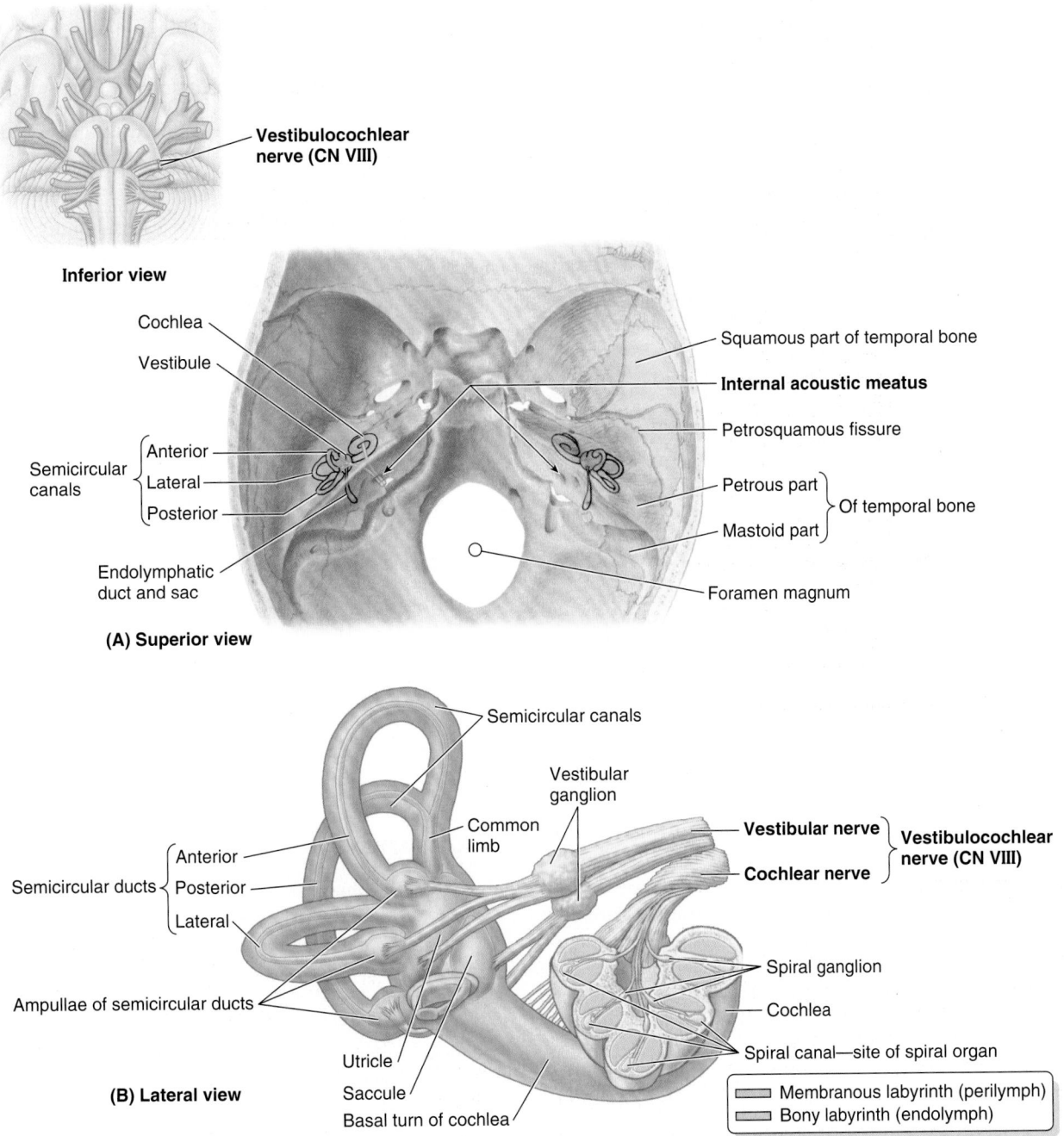

FIGURE 9.12. Vestibulocochlear nerve (CN VIII). A. The internal surface of the cranial base shows the location of the bony labyrinth of the internal ear within the temporal bone. **B.** This view of the bony and membranous labyrinths shows (1) innervation of the cochlea by the cochlear nerve of CN VIII for the sense of hearing and (2) innervation of the vestibular apparatus by the vestibular nerve of CN VIII for equilibrium and motion.

The peripheral processes of the neurons extend to the *maculae of the utricle and saccule* (sensitive to linear acceleration and the pull of gravity relative to the position of the head) and to the cristae of the ampullae of the semicircular ducts (sensitive to rotational acceleration).

- The **cochlear nerve** is composed of the central processes of bipolar neurons in the **spiral ganglion;** the peripheral processes of the neurons extend to the spiral organ for the sense of hearing.

Within the internal acoustic meatus, the two divisions of CN VIII are accompanied by the primary root and intermediate nerve of CN VII and the labyrinthine artery (see Chapter 7).

The Bottom Line

VESTIBULOCOCHLEAR NERVE

♦ The vestibulocochlear nerves (CN VIII) carry fibers concerned with the special senses of hearing, equilibrium, and motion. ♦ The nerves originate from the grooves between the pons and medulla. ♦ They run through the internal acoustic meatus and divide into the cochlear and vestibular nerves. ♦ The cochlear nerve is sensory to the spiral organ (for the sense of hearing). ♦ The vestibular nerve is sensory to the cristae of the ampullae of the semicircular ducts and the maculae of the saccule and utricle (for the sense of equilibration and motion).

GLOSSOPHARYNGEAL NERVE (CN IX)

Functions: *Sensory*—somatic (general) sensory, special sensory (taste), and visceral sensory. *Motor*—somatic (branchial) motor, and visceral (parasympathetic) motor for derivatives of the 3rd pharyngeal arch.

Nuclei: Four nuclei in the medulla send or receive fibers via CN IX: two motor (*nucleus ambiguus* and **inferior salivary nucleus**) and two sensory (*sensory nuclei of the trigeminal nerve* [CN V] and *nuclei of the solitary tract*). Three of these nuclei (in italics) are shared with CN X (Fig. 9.5).

The **glossopharyngeal nerve** (CN IX) emerges from the lateral aspect of the medulla and passes anterolaterally to leave the cranium through the anterior aspect of the *jugular foramen* (Figs. 9.13 and 9.14). At this foramen are the sensory **superior** and **inferior ganglia of CN IX,** which contain the pseudounipolar cell bodies for the afferent components of the nerve. CN IX follows the *stylopharyngeus,* the only muscle the nerve supplies, and passes between the superior and middle pharyngeal constrictor muscles to reach the oropharynx and tongue. It contributes sensory fibers to the *pharyngeal plexus of nerves.* CN IX is afferent from the tongue and pharynx (hence its name) and efferent to the stylopharyngeus and parotid gland.

Somatic (Branchial) Motor

Motor fibers pass to one muscle, the stylopharyngeus, derived from the 3rd pharyngeal arch.

Visceral (Parasympathetic) Motor

Following a circuitous route initially involving the tympanic nerve, presynaptic parasympathetic fibers are provided to the *otic ganglion* for innervation of the parotid gland (Fig. 9.15). The otic ganglion is associated with the mandibular nerve (CN V_3), branches of which convey the postsynaptic parasympathetic fibers to the parotid gland.

Somatic (General) Sensory

The general sensory branches of CN IX are as follows (Fig. 9.13):

- The *tympanic nerve.*
- The *carotid sinus nerve* to the carotid sinus, a baro- (presso-) receptor sensitive to changes in blood pressure, and the carotid body, a chemoreceptor sensitive to blood gas (oxygen and carbon dioxide levels).
- The *pharyngeal, tonsillar,* and *lingual nerves* to the mucosa of the oropharynx and isthmus of the fauces (L., throat), including palatine tonsil, soft palate, and posterior third of the tongue. In addition to general sensation (touch, pain, temperature), tactile (actual or threatened) stimuli determined to be unusual or unpleasant here may evoke the gag reflex or even vomiting.

Special Sensory (Taste)

Taste fibers are conveyed from the posterior third of the tongue to the sensory inferior ganglia of CN IX (Fig. 9.14). Details of the distribution of CN IX are outlined in Figure 9.13.

The Bottom Line

GLOSSOPHARYNGEAL NERVE

♦ The glossopharyngeal nerves (CN IX) send somatic motor fibers to the stylopharyngeus and visceral motor (presynaptic parasympathetic) fibers to the otic ganglion for innervation of the parotid gland. ♦ They also send sensory fibers to the posterior third of the tongue (including taste), pharynx, tympanic cavity, pharyngotympanic cavity, carotid body, and carotid sinus. ♦ The nerves originate from the rostral end of the medulla and exit from the cranium via the jugular foramina. ♦ They pass between the superior and the middle pharyngeal constrictors to the tonsillar sinus and enter the posterior third of the tongue.

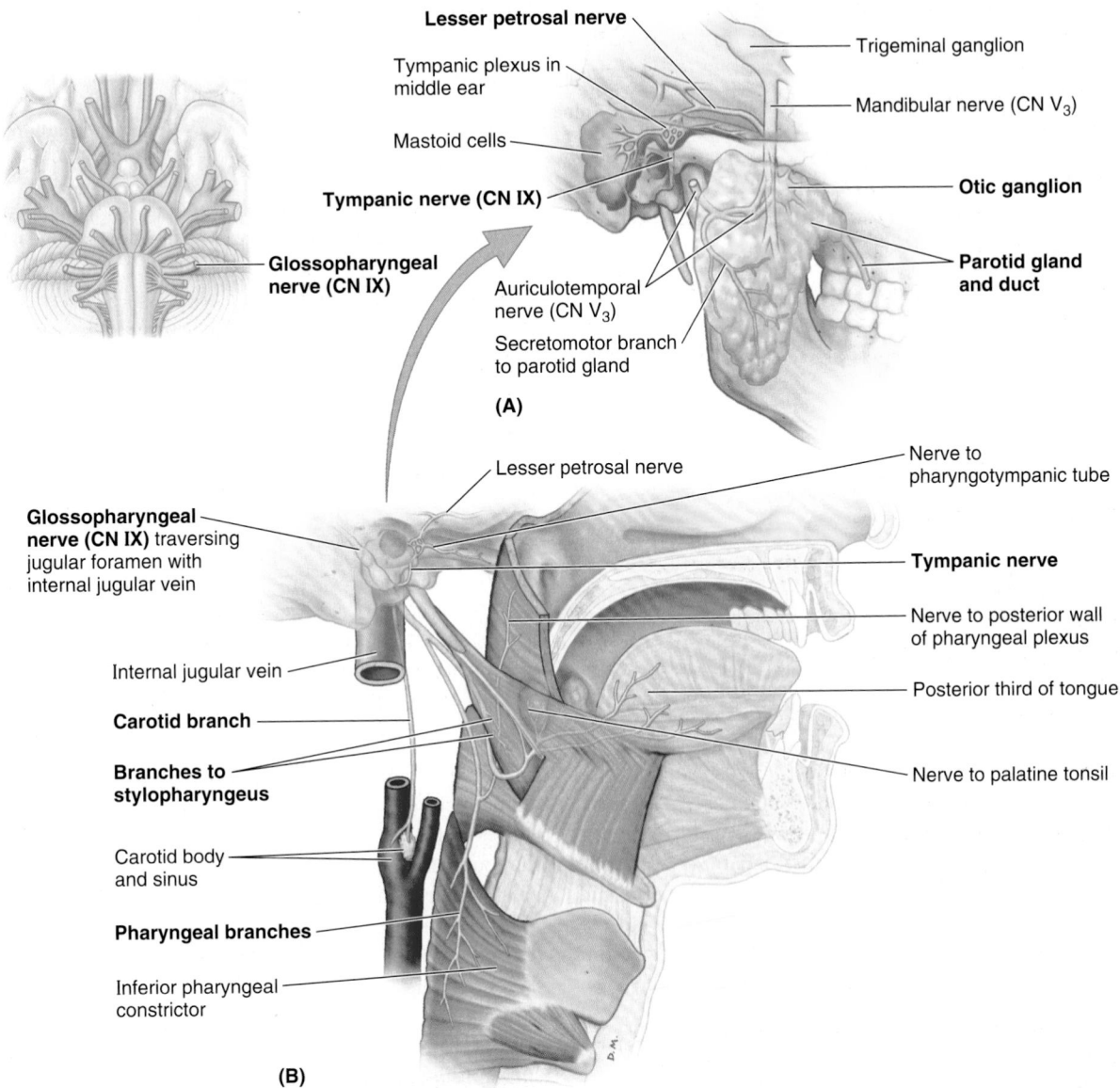

FIGURE 9.13. Distribution of glossopharyngeal nerve (CN IX). A. CN IX is motor to one striated pharyngeal muscle, the stylopharyngeus. It also carries sensory fibers from the carotid body and carotid sinus, conveying information about blood pressure and gas levels as well as somatic (general) sensation from the internal ear, pharynx, and fauces and taste from the posterior tongue. **B.** The parasympathetic component of CN IX supplies presynaptic secretory fibers to the otic ganglion; postsynaptic fibers pass to the parotid gland via the auriculotemporal nerve (CN V₃).

VAGUS NERVE (CN X)

Functions: *Sensory*—somatic (general) sensory, special sensory (taste), visceral sensory. *Motor*—somatic (branchial) motor and visceral (parasympathetic) motor.

- Somatic (general) sensory from the inferior pharynx, and larynx.
- Visceral sensory from the thoracic and abdominal organs.
- Taste and somatic (general) sensation from the root of the tongue and taste buds on the epiglottis. Branches of the internal laryngeal nerve (a branch of CN X) supply a small area, mostly somatic (general) sensory, but also some special sensation (taste).

- Somatic (branchial) motor to the soft palate; pharynx; intrinsic laryngeal muscles (phonation); and a nominal extrinsic tongue muscle, the palatoglossus, which is actually a palatine muscle based on its derivation and innervation.
- Proprioceptive to the muscles listed above.
- Visceral (parasympathetic) motor to thoracic and abdominal viscera.

Nuclei: *Sensory*—sensory nucleus of the trigeminal nerve (somatic sensory) and nuclei of the solitary tract (taste and visceral sensory). *Motor*—nucleus ambiguus (somatic [branchial] motor) and **dorsal vagal nucleus** (visceral [parasympathetic] motor) (Fig. 9.5).

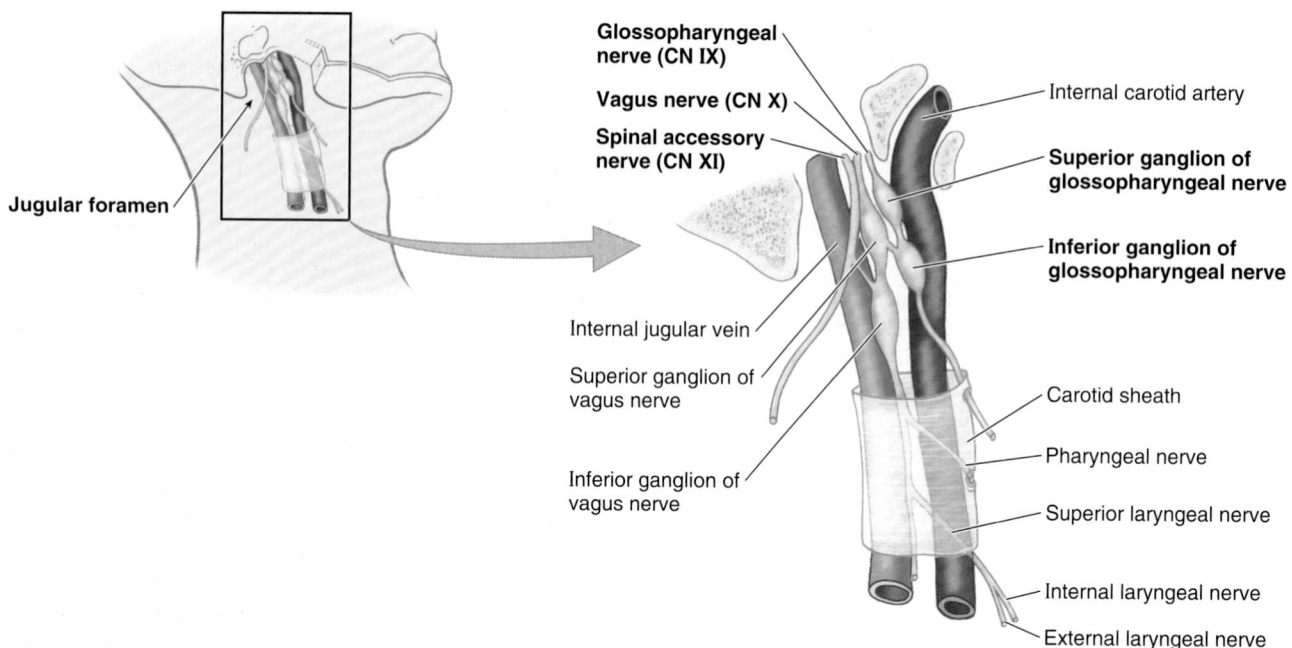

FIGURE 9.14. **Relationship of structures traversing jugular foramen.** CN IX, CN X, and CN XI are, in numerical order, anterior to the internal jugular vein as they traverse the foramen. They are immediately posterior to the internal carotid artery as they emerge from it. The superior and inferior sensory ganglia of CN IX and CN X are seen as thickenings of those nerves immediately inferior to their exit from the cranium.

The **vagus nerve** (CN X) has the longest course and most extensive distribution of all the cranial nerves, most of which is outside of (inferior to) the head. The term *vagus* is derived from the Latin word *vagary*, meaning "wandering." CN X was so called because of its extensive distribution (Table 9.5). It arises by a series of rootlets from the lateral aspect of the medulla that merge and leave the cranium through the *jugular foramen* positioned between CN IX and CN XI (Figs. 9.14 and 9.16).

What was formerly called the "cranial root of the accessory nerve" is actually a part of CN X (Fig. 9.17). CN X has a *superior ganglion* in the jugular foramen that is mainly concerned with the general sensory component of the nerve. Inferior to the foramen is an *inferior ganglion* (nodose ganglion) concerned with the visceral and special sensory components of the nerve (Fig. 9.14). In the region of the superior ganglion are connections to CN IX and the superior cervical (sympathetic) ganglion. CN X continues inferiorly in the *carotid sheath* to the root of the neck (see Chapter 8), supplying branches to the palate, pharynx, and larynx (Fig. 9.16; Table 9.5).

The courses of the vagi are asymmetrical in the thorax, a consequence of rotation of the midgut during development (see Chapters 1 and 2). CN X supplies branches to the heart, bronchi, and lungs. The vagi form **anterior** and **posterior vagal trunks** that are continuations of the *esophageal plexus* surrounding the esophagus, which is also joined by branches of the sympathetic trunks. The trunks pass with the esophagus through the diaphragm into the abdomen, where the vagal trunks break up into branches that innervate the stomach and intestinal tract as far as the left colic flexure.

The Bottom Line

VAGUS NERVE

♦ The vagus nerves (CN X) supply motor fibers to the voluntary muscles of the larynx and superior esophagus.
♦ They also send visceral motor (presynaptic parasympathetic) fibers to the involuntary muscles and glands of: (1) the tracheobronchial tree and esophagus via the pulmonary and esophageal plexuses, (2) to the heart via the cardiac plexus, and (3) to the alimentary tract as far as the left colic flexure via the vagal trunks. ♦ The vagus nerves also send sensory fibers to the pharynx, larynx, and reflex afferents from these same areas (1-3 above). ♦ They originate via 8–10 rootlets from the lateral sides of the medulla of the brainstem. They enter the superior mediastinum posterior to the sternoclavicular joints and brachiocephalic veins.
♦ The nerves give rise to right and left recurrent nerves and then, from the esophageal plexus, reform as anterior and posterior vagal trunks, which continue into the abdomen.

Parasympathetic (visceral motor)

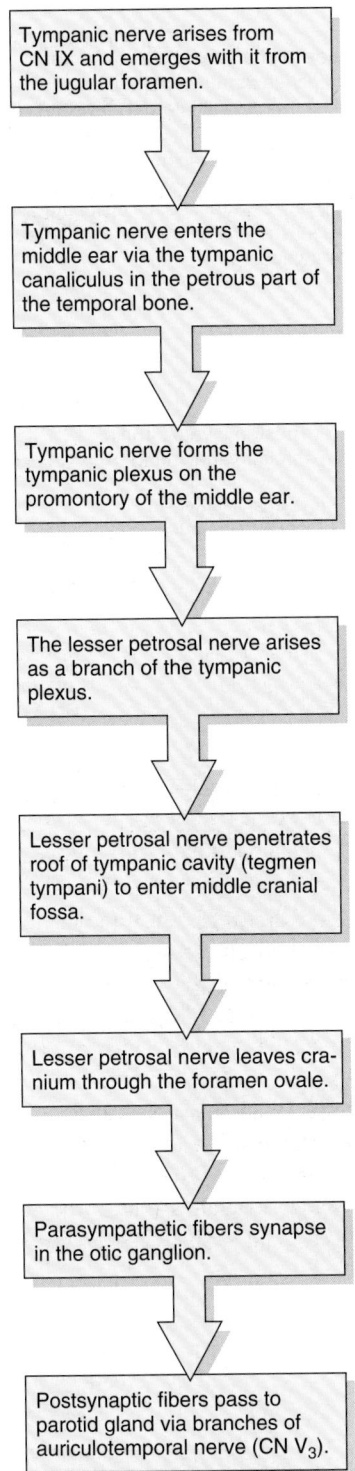

Tympanic nerve arises from CN IX and emerges with it from the jugular foramen.

Tympanic nerve enters the middle ear via the tympanic canaliculus in the petrous part of the temporal bone.

Tympanic nerve forms the tympanic plexus on the promontory of the middle ear.

The lesser petrosal nerve arises as a branch of the tympanic plexus.

Lesser petrosal nerve penetrates roof of tympanic cavity (tegmen tympani) to enter middle cranial fossa.

Lesser petrosal nerve leaves cranium through the foramen ovale.

Parasympathetic fibers synapse in the otic ganglion.

Postsynaptic fibers pass to parotid gland via branches of auriculotemporal nerve (CN V₃).

FIGURE 9.15. Parasympathetic innervation involving the glossopharyngeal nerve (CN IX). CN IX sends presynaptic parasympathetic (secretomotor) fibers to the otic ganglion via a convoluted route; postsynaptic fibers pass from the ganglion to the parotid gland via the auriculotemporal nerve (Fig. 9.13B).

SPINAL ACCESSORY NERVE (CN XI)

Functions: Somatic motor to the striated sternocleidomastoid and trapezius muscles.

Nuclei: The spinal accessory nerve arises from the **nucleus of the spinal accessory nerve**, a column of anterior horn motor neurons in the superior five or six cervical segments of the spinal cord (Fig. 9.5).

The traditional "cranial root" of CN XI is actually a part of CN X (Lachman et al., 2002). It may be united for a short distance with the **spinal accessory nerve** (CN XI) (Fig. 9.17). CN XI emerges as a series of rootlets from the first five or six cervical segments of the spinal cord. It joins the CN X temporarily as they pass through the *jugular foramen*, separating again as they exit (Fig. 9.14). CN XI descends along the internal carotid artery, penetrates and innervates the sternocleidomastoid, and emerges from the muscle near the middle of its posterior border. Next, CN XI crosses the posterior cervical region and passes deep to the superior border of the trapezius to descend on its deep surface, providing multiple branches to the muscle. Branches of the cervical plexus conveying sensory fibers from spinal nerves C2–C4 join the spinal accessory nerve in the posterior cervical region, providing these muscles with pain and proprioceptive fibers.

The Bottom Line

SPINAL ACCESSORY NERVE

♦ The spinal accessory nerves (CN XI) supply somatic motor fibers to the sternocleidomastoid and trapezius muscles. ♦ The nerves arise as rootlets from the sides of the spinal cord in the superior five or six cervical segments. ♦ They ascend into the cranial cavity via the foramen magnum and exit through the jugular foramina, crossing the lateral cervical region, where pain and proprioceptive fibers from the cervical plexus join the nerves.

HYPOGLOSSAL NERVE (CN XII)

Functions: Somatic motor to the intrinsic and extrinsic muscles of the tongue (G. *glossa*)—styloglossus, hyoglossus, and genioglossus.

The **hypoglossal nerve** (CN XII) arises as a purely motor nerve by several rootlets from the medulla and leaves the cranium through the *hypoglossal canal* (Figs. 9.2 and 9.3). After exiting the cranial cavity, CN XII is joined by a branch or branches of the cervical plexus conveying general somatic motor fibers from C1 and C2 spinal nerves and somatic (general) sensory fibers from the spinal ganglion of C2 (Fig. 9.18). These spinal nerve fibers "hitch a ride" with CN XII to reach

(text continues on p. 1078)

TABLE 9.5. SUMMARY OF VAGUS NERVE (CN X)

Divisions (Parts)	Branches
Cranial Vagi arise by a series of rootlets from medulla (includes traditional cranial root of CN XI)	Meningeal branch to dura mater (sensory; actually fibers of C2 spinal ganglion neurons that hitch a ride with vagus nerve) Auricular branch
Cervical Exit cranium/enter neck through jugular foramen; right and left vagus nerves enter carotid sheaths and continue to root of neck	Pharyngeal branches to pharyngeal plexus (motor) Cervical cardiac branches (parasympathetic, visceral afferent) Superior laryngeal nerve (mixed) internal (sensory) and external (motor) branches Right recurrent laryngeal nerve (mixed)
Thoracic Vagi enter thorax through superior thoracic aperture; left vagus contributes to anterior esophageal plexus; right vagus to posterior plexus; form anterior and posterior vagal trunks	Left recurrent laryngeal nerve (mixed; all distal branches convey parasympathetic and visceral afferent fibers for reflex stimuli) Thoracic cardiac branches Pulmonary branches Esophageal plexus
Abdominal Anterior and posterior vagal trunks enter abdomen through esophageal hiatus in diaphragm; distribute asymmetrically	Esophageal branches Gastric branches Hepatic branches Celiac branches (from posterior vagal trunk) Pyloric branch (from anterior vagal trunk) Renal branches Intestinal branches (to left colic flexure)

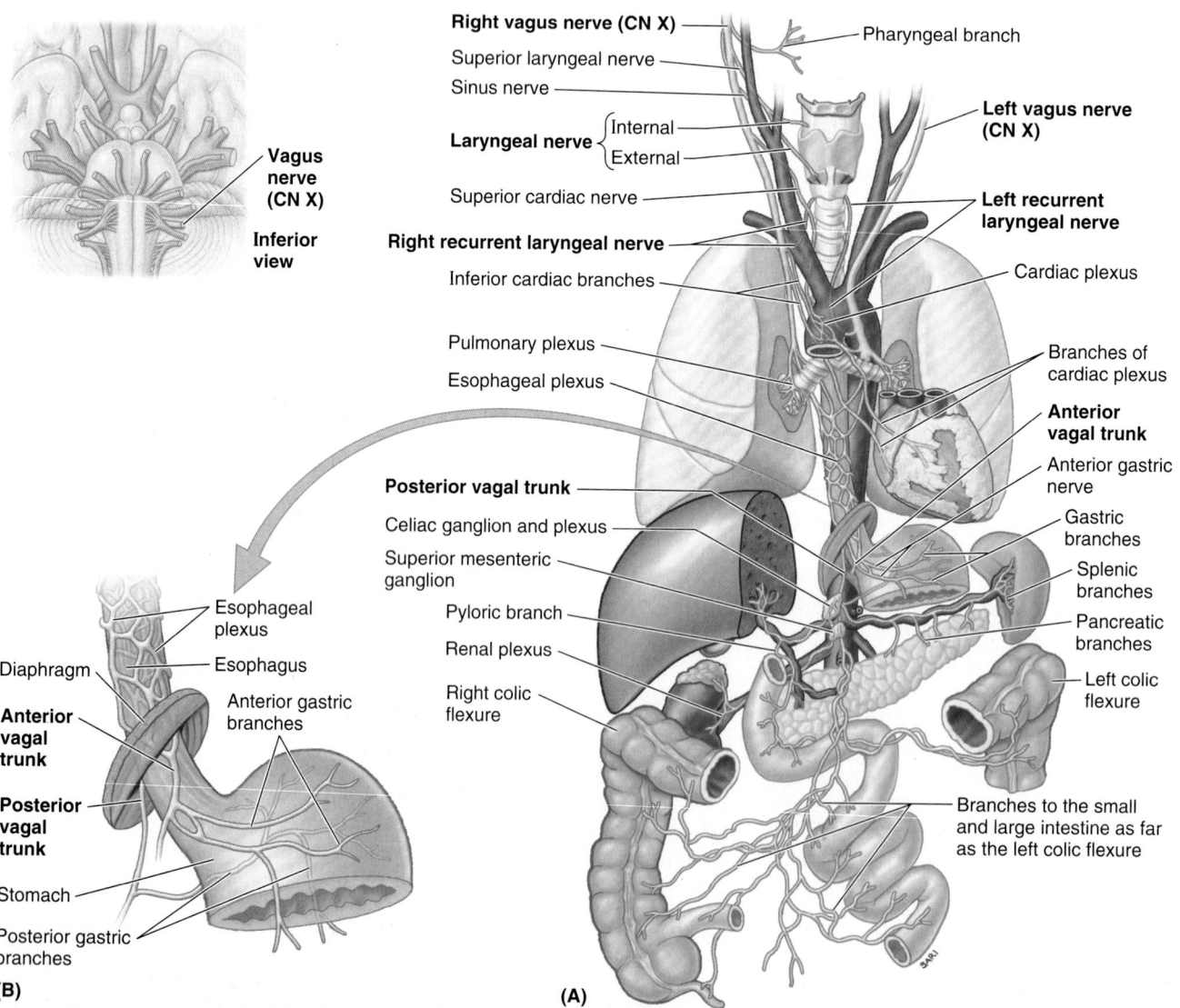

FIGURE 9.16. Distribution of vagus nerve (CN X). After supplying the palatine, pharyngeal, and laryngeal branches, CN X descends into the thorax. The recurrent laryngeal nerves ascend to the larynx, the left from a more inferior (thoracic) level. In the abdomen, the anterior and posterior vagal trunks demonstrate further asymmetry as they supply the terminal esophagus, stomach, and intestinal tract as far distally as the left colic flexure.

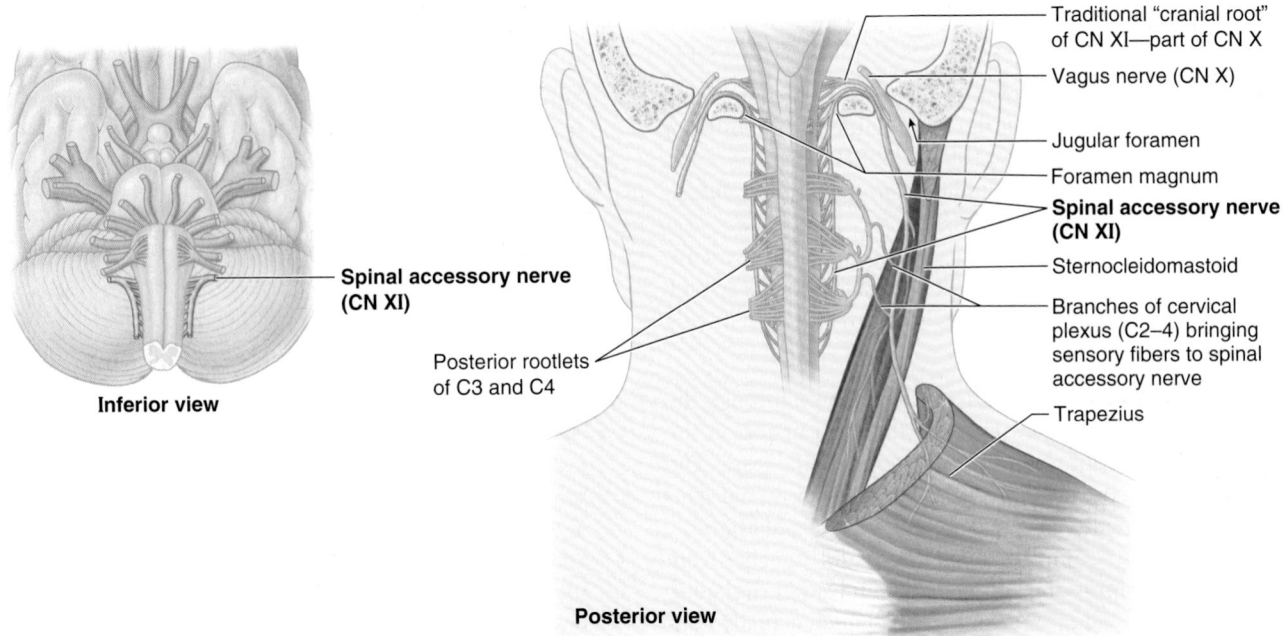

Inferior view

Traditional "cranial root" of CN XI—part of CN X

Vagus nerve (CN X)

Jugular foramen

Foramen magnum

Spinal accessory nerve (CN XI)

Sternocleidomastoid

Branches of cervical plexus (C2–4) bringing sensory fibers to spinal accessory nerve

Trapezius

Spinal accessory nerve (CN XI)

Posterior rootlets of C3 and C4

Posterior view

FIGURE 9.17. Distribution of spinal accessory nerve (CN XI).

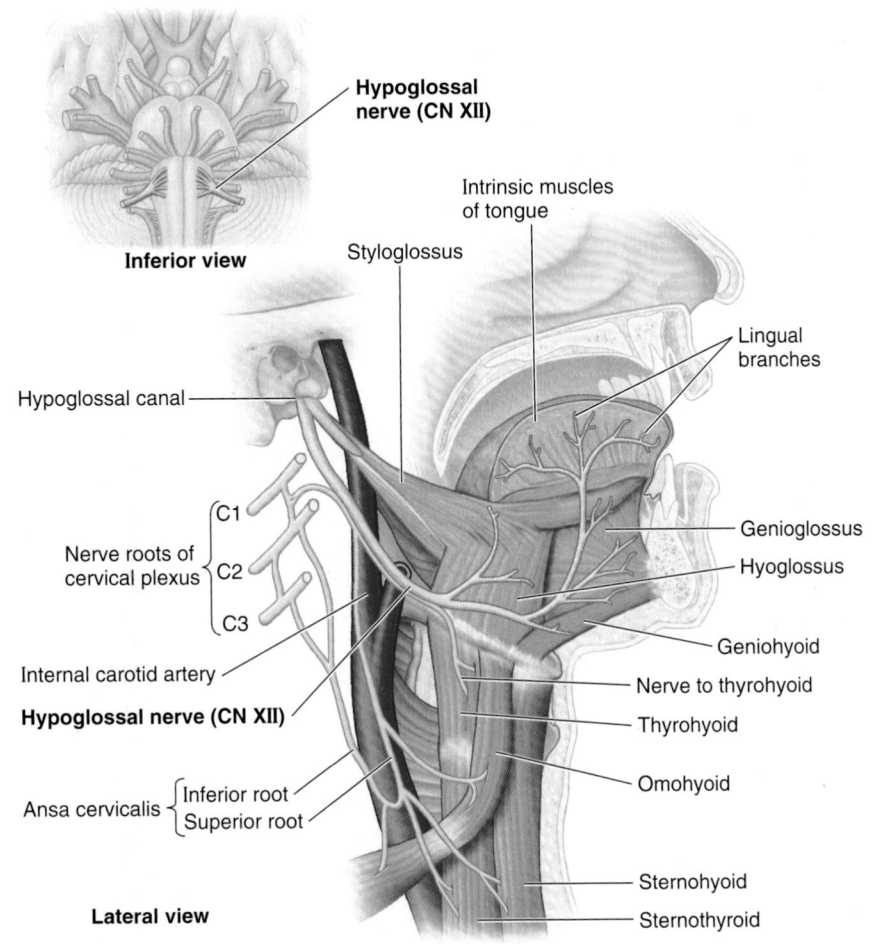

Hypoglossal nerve (CN XII)

Inferior view

Intrinsic muscles of tongue

Styloglossus

Lingual branches

Hypoglossal canal

Nerve roots of cervical plexus { C1, C2, C3 }

Internal carotid artery

Hypoglossal nerve (CN XII)

Ansa cervicalis { Inferior root, Superior root }

Genioglossus

Hyoglossus

Geniohyoid

Nerve to thyrohyoid

Thyrohyoid

Omohyoid

Sternohyoid

Sternothyroid

Lateral view

FIGURE 9.18. Distribution of hypoglossal nerve (CN XII). CN XII leaves the cranium through the hypoglossal canal and passes deep to the mandible to enter the tongue, where it supplies all intrinsic and extrinsic lingual muscles, except the palatoglossus. CN XII is joined immediately distal to the hypoglossal canal by a branch conveying fibers from the C1 and C2 loop of the cervical plexus. These fibers hitch a ride with CN XII, leaving it as the superior root of the ansa cervicalis and the nerve to the thyrohyoid muscle. Cervical spinal nerves, not CN XII, supply the infrahyoid muscles.

1078   Chapter 9 • Summary of Cranial Nerves

the hyoid muscles, with some of the sensory fibers passing retrograde along it to reach the dura mater of the posterior cranial fossa (see Fig. 8.13B, p. 997). CN XII passes inferiorly medial to the angle of the mandible and then curves anteriorly to enter the tongue (Fig. 9.18).

CN XII ends in many branches that supply all the extrinsic muscles of the tongue, except the palatoglossus (which is actually a palatine muscle). CN XII has the following branches:

- A **meningeal branch** returns to the cranium through the hypoglossal canal and innervates the dura mater on the floor and posterior wall of the posterior cranial fossa. The nerve fibers conveyed are from the sensory spinal ganglion of spinal nerve C2 and are not hypoglossal fibers.
- The **superior root of the ansa cervicalis** branches from CN XII to supply the infrahyoid muscles (sternohyoid, sternothyroid, and omohyoid). This branch actually conveys only fibers from the cervical plexus (the loop between the anterior rami of C1 and C2) that joined the nerve outside the cranial cavity, not hypoglossal fibers (Fig. 9.18).

Some fibers continue past the origin of the superior root to reach the thyrohyoid muscle.
- Terminal **lingual branches** supply the styloglossus, hyoglossus, genioglossus, and intrinsic muscles of the tongue.

The Bottom Line

HYPOGLOSSAL NERVE

♦ The hypoglossal nerves (CN XII) supply somatic motor fibers to the intrinsic and extrinsic muscles of the tongue, except the palatoglossus (a palatine muscle). ♦ They arise by several rootlets between the pyramids and the olives of the medulla. ♦ They pass through the hypoglossal canals and run inferiorly and anteriorly, passing medial to the angles of the mandible and between the mylohyoid and the hyoglossus to reach the muscles of the tongue.

CRANIAL NERVES

Cranial Nerve Injuries

 Table 9.6 summarizes some common cranial nerve injuries, indicating the type or site of lesions and abnormal findings that result. *Injury to the cranial nerves* is a frequent complication of a fracture in the base of the cranium. Furthermore, excessive movement of the brain within the cranium may tear or bruise cranial nerve fibers, especially those of CN I. Paralysis of cranial nerves as a result of trauma can usually be detected as soon as the patient's state of consciousness permits (Brannagan et al., 2010); however, in some people, the paralysis may not be evident for several days.

Because of their location within the confined cranial cavity, relatively fixed positions, and sometimes close relationships to bony or vascular formations, the intracranial portions of certain cranial nerves are also susceptible to compression owing to a tumor or aneurysm. In such cases, the onset of symptoms usually occurs gradually, and the effects depend on the extent of the pressure exerted. Because of their close relationship to the cavernous sinus, CN III, CN IV, CN V$_1$ and especially CN VI are susceptible to compression or injury related to pathologies (infections, thrombophlebitis) affecting the sinus.

OLFACTORY NERVE

Anosmia—Loss of Smell

The *loss of smell (anosmia)* is frequently associated with upper respiratory infections, sinus disease, and head trauma. Loss of olfactory fibers

occurs with aging. Consequently, elderly people often have reduced acuity of the sensation of smell, resulting from progressive reduction in the number of olfactory receptor neurons in the olfactory epithelium. The chief complaint of most people with anosmia is the loss or alteration of taste; however, clinical studies reveal that in all but a few people, the dysfunction is in the olfactory system (Simpson, 2006). The reason is that most people confuse taste with flavor. Transitory olfactory impairment occurs as a result of viral or *allergic rhinitis*—inflammation of the nasal mucous membrane.

To test the sense of smell, the person is blindfolded and asked to identify common odors, such as freshly ground coffee placed near the external nares (nostrils). One naris is occluded and the eyes are closed. Because anosmia is usually unilateral, each naris is tested separately. If the loss of smell is unilateral, the person may not be aware of it without clinical testing.

Injury to the nasal mucosa, olfactory nerve fibers, olfactory bulbs, or olfactory tracts may also impair smell. In severe head injuries, the olfactory bulbs may be torn away from the olfactory nerves, or some olfactory nerve fibers may be torn as they pass through a *fractured cribriform plate.* If all the nerve bundles on one side are torn, a complete loss of smell will occur on that side; consequently, anosmia may be a clue to a fracture of the cranial base and cerebrospinal fluid *rhinorrhea* (leakage of the fluid through the nose).

A tumor and/or abscess in the frontal lobe of the brain or a tumor of the meninges (*meningioma*) in the anterior cranial fossa may also cause anosmia by compressing the olfactory bulb and/or tract (Bruce et al., 2010).

TABLE 9.6. SUMMARY OF CRANIAL NERVE LESIONS

Nerve	Types(s) and/or Site(s) of Lesion	Abnormal Finding(s)
CN I	Fracture of cribriform plate	Anosmia (loss of smell); cerebrospinal fluid rhinorrhea
CN II	Direct trauma to orbit or eyeball; fracture involving optic canal	Loss of pupillary constriction
	Pressure on optic pathway; laceration or intracerebral clot in temporal, parietal, or occipital lobes of brain	Visual field defects
CN III	Pressure from herniating uncus on nerve; fracture involving cavernous sinus; aneurysms	Dilated pupil; ptosis; eye turns down and out; pupillary reflex on side of lesion will be lost
CN IV	Stretching of nerve during its course around brainstem; fracture of orbit	Inability to look down when eye is adducted
CN V	Injury to terminal branches (particularly CN V_2) in roof of maxillary sinus; pathological processes affecting trigeminal ganglion	Loss of pain and touch sensations; paraesthesia; masseter and temporalis muscles do not contract; deviation of mandible to side of lesion when mouth is opened
CN VI	Base of brain or fracture involving cavernous sinus or orbit	Eye fails to move laterally; diplopia on lateral gaze
CN VII	Laceration or contusion in parotid region	Paralysis of facial muscles; eye remains open; angle of mouth droops; forehead does not wrinkle
	Fracture of temporal bone	As above, plus associated involvement of cochlear nerve and chorda tympani; dry cornea; loss of taste on anterior two thirds of tongue
	Intracranial hematoma ("stroke")	Forehead wrinkles because of bilateral innervation of frontalis muscle; otherwise paralysis of contralateral facial muscles
CN VIII	Tumor of nerve (acoustic neuroma)	Progressive unilateral hearing loss; tinnitus (noises in ear)
CN IX	Brainstem lesion or deep laceration of neck	Loss of taste on posterior third of tongue; loss of sensation on affected side of soft palate
CN X	Brainstem lesion or deep laceration of neck	Sagging of soft palate; deviation of uvula to normal side; hoarseness owing to paralysis of vocal fold
CN XI	Laceration of neck	Paralysis of sternocleidomastoid and descending fibers of trapezius; drooping of shoulder
CN XII	Neck laceration; fractures of cranial base	Protruded tongue deviates toward affected side; moderate dysarthria (disturbance of articulation)

Olfactory Hallucinations

Occasionally *olfactory hallucinations* (false perceptions of smell) may accompany lesions in the temporal lobe of the cerebral hemisphere. A lesion that irritates the lateral olfactory area (deep to the uncus) may cause *temporal lobe epilepsy* or "uncinate fits," which are characterized by imaginary disagreeable odors and involuntary movements of the lips and tongue.

OPTIC NERVE

Demyelinating Diseases and Optic Nerves

Because the optic nerves are actually CNS tracts, the myelin sheath that surrounds the sensory fibers from the point at which the fibers penetrate the sclera is formed by oligodendrocytes (glial cells) rather than by neurolemma (Schwann) cells, as in other cranial or spinal nerves of the peripheral nervous system (PNS). Consequently, the optic nerves are susceptible to the effects of *demyelinating diseases* of the CNS, such as *multiple sclerosis* (MS), which usually do not affect other nerves of the PNS.

Optic Neuritis

Optic neuritis refers to lesions of the optic nerve that cause diminution of visual acuity, with or without changes in peripheral fields of vision (Brannagan et al., 2010). Optic neuritis may be caused by inflammatory, degenerative, demyelinating, or toxic disorders. The optic disc appears pale and smaller than usual on ophthalmoscopic examination. Many toxic substances (e.g., methyl and ethyl alcohol, tobacco, lead, and mercury) may also injure the optic nerve.

Right monocular blindness

Visual fields

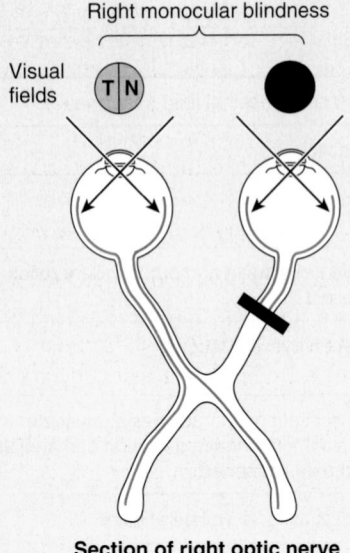

Section of right optic nerve

Bitemporal hemianopsia

Visual fields

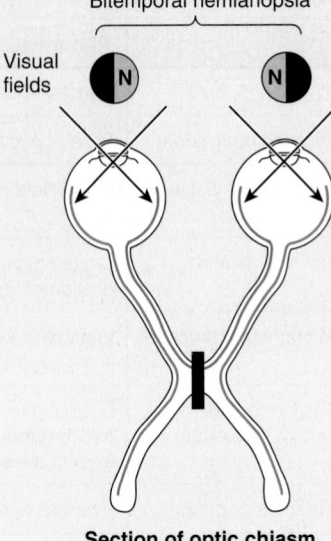

Section of optic chiasm

Left homonymous hemianopsia

Visual fields

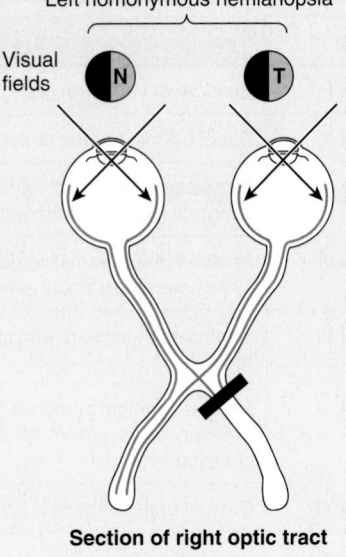

Section of right optic tract

FIGURE B9.1

Visual Field Defects

Visual field defects result from lesions that affect different parts of the visual pathway. The type of defect depends on where the pathway is interrupted (Fig. B9.1):

- Complete section of an optic nerve results in blindness in the temporal (*T*) and nasal (*N*) visual fields of the ipsilateral eye (depicted in *black*).
- Complete section of the optic chiasm reduces peripheral vision and results in *bitemporal hemianopsia*, the loss of vision of one half of the visual field of both eyes.
- Complete section of the right optic tract at the midline eliminates vision from the left temporal and right nasal visual fields. A lesion of the right or left optic tract causes a contralateral homonymous hemianopsia, indicating that visual loss is in similar fields. This defect is the most common form of visual field loss and is often observed in patients with strokes (Swartz, 2009).

Defects of vision caused by compression of the optic pathway, as may result from tumors of the pituitary gland or berry aneurysms of the internal carotid arteries (see Chapter 7), may produce only part of the visual losses described here. Patients may not be aware of changes in their visual fields until late in the course of disease, because lesions affecting the visual pathway often develop insidiously.

OCULOMOTOR NERVE

Injury to Oculomotor Nerve

A lesion of CN III results in *ipsilateral oculomotor palsy*, summarized in Table 9.6 and discussed in detail in Chapter 7, p. 913.

Compression of Oculomotor Nerve

Rapidly increasing intracranial pressure (e.g., resulting from an extradural hematoma) often compresses CN III against the crest of the petrous part of the temporal bone. Because autonomic fibers in CN III are superficial, they are affected first. As a result, the pupil dilates progressively on the injured side. Consequently, the first sign of *CN III compression* is ipsilateral slowness of the pupillary response to light.

Aneurysm of Posterior Cerebral or Superior Cerebellar Artery

An *aneurysm of a posterior cerebral or superior cerebellar artery* may also exert pressure on CN III as it passes between these vessels. The effects of this pressure depend on its severity. Because CN III lies in the lateral wall of the cavernous sinus, injuries or infections of the sinus may also affect this nerve.

TROCHLEAR NERVE

CN IV is rarely paralyzed alone. *Lesions of the trochlear nerve* or its nucleus cause paralysis of the superior oblique and impair the ability to turn the affected eyeball inferomedially. CN IV may be torn when there are severe head injuries because of its long intracranial course. The characteristic sign of trochlear nerve injury is *diplopia* (double vision) when looking down. Diplopia occurs because the superior oblique normally assists the inferior rectus in depressing the pupil (directing the gaze downward) and is the only muscle to do so when the pupil is adducted. In addition, because the superior oblique is the primary muscle producing intorsion of the eyeball, the primary muscle producing extorsion (the inferior oblique) is unopposed when the superior oblique is paralyzed. Thus the direction of gaze and rotation of the eyeball about its anteroposterior axis is different for the two eyes when an attempt is made to look downward, and especially when looking downward and medially. The person can compensate for the diplopia by inclining the head anteriorly and laterally toward the side of the normal eye.

TRIGEMINAL NERVE

Injury to Trigeminal Nerve

 CN V may be injured by trauma, tumors, aneurysms, or meningeal infections (Brannagan et al., 2010). It may be involved occasionally in poliomyelitis and generalized polyneuropathy, a disease process involving several nerves. The sensory and motor nuclei in the pons and medulla may be destroyed by intramedullary tumors or vascular lesions. An isolated lesion of the spinal trigeminal tract also may occur with multiple sclerosis (MS). *Injury to CN V* causes the following:

- Paralysis of the muscles of mastication with deviation of the mandible toward the side of the lesion (Table 9.6).
- Loss of the ability to appreciate soft tactile, thermal, or painful sensations in the face.
- Loss of corneal reflex (blinking in response to the cornea being touched) and the sneezing reflex (stimulated by irritants to clear the respiratory tract).

Common causes of facial numbness are dental trauma, *herpes zoster ophthalmicus* (infection caused by a herpesvirus), cranial trauma, head and neck tumors, intracranial tumors, and idiopathic trigeminal neuropathy (a nerve disease of unknown cause). *Trigeminal neuralgia* (tic douloureux), the principal disease affecting the sensory root of CN V, produces excruciating, episodic pain that is usually restricted to the areas supplied by the maxillary and/or mandibular divisions of this nerve. (See detailed discussion in Chapter 7, p. 862.)

Dental Anesthesia

Anesthetic agents are commonly administered by injection to block pain during dental procedures. CN V is of great importance in the practice of dentistry because it is the sensory nerve of the head, serving the teeth and mucosa of the oral cavity. Because the superior alveolar nerves (branches of CN V_2) are not accessible, the maxillary teeth are locally anesthetized by injecting the agent into the tissues surrounding the roots of the teeth and allowing the solution to infiltrate the tissue to reach the terminal (dental) nerve branches that enter the roots. By contrast, the inferior alveolar nerve (CN V_3) is readily accessible and is probably anesthetized more frequently than any other nerve. This procedure is discussed in the blue box "Inferior Alveolar Nerve Block" in Chapter 7, p. 927.

ABDUCENT NERVE

Because CN VI has a long intradural course, it is often stretched when intracranial pressure rises, partly because of the sharp bend it makes over the crest of the petrous part of the temporal bone after entering

the dura. A space-occupying lesion, such as a brain tumor, may compress CN VI, causing paralysis of the lateral rectus. Complete *paralysis of CN VI* causes medial deviation of the affected eye—that is, it is fully adducted owing to the unopposed action of the medial rectus, leaving the person unable to abduct the eye. *Diplopia* is present in all ranges of movement of the eyeball, except on gazing to the side opposite the lesion. Paralysis of CN VI may also result from:

- An aneurysm of the cerebral arterial circle (at the base of the brain) (see Chapter 7, p. 887).
- Pressure from an atherosclerotic internal carotid artery in the cavernous sinus, where CN VI is closely related to this artery.
- Septic thrombosis of the sinus subsequent to infection in the nasal cavities and/or paranasal sinuses.

FACIAL NERVE

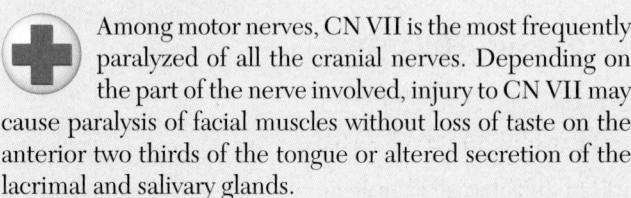 Among motor nerves, CN VII is the most frequently paralyzed of all the cranial nerves. Depending on the part of the nerve involved, injury to CN VII may cause paralysis of facial muscles without loss of taste on the anterior two thirds of the tongue or altered secretion of the lacrimal and salivary glands.

A *lesion of CN VII near its origin* or near the geniculate ganglion is accompanied by loss of motor, gustatory (taste), and autonomic functions. The motor paralysis of facial muscles involves superior and inferior parts of the face on the ipsilateral side.

A *central lesion of CN VII* (lesion of the CNS) results in paralysis of muscles in the inferior face on the contralateral side; consequently, forehead wrinkling is not visibly impaired because it is innervated bilaterally. Lesions between the geniculate ganglion and the origin of the chorda tympani produce the same effects as that resulting from injury near the ganglion, except that lacrimal secretion is not affected. Because it passes through the facial canal in the temporal bone, CN VII is vulnerable to compression when a viral infection produces inflammation (viral neuritis) and swelling of the nerve just before it emerges from the stylomastoid foramen.

Because the branches of CN VII are superficial, they are subject to injury from knife and gunshot wounds, cuts, and birth injury. Damage to CN VII is common with fracture of the temporal bone and is usually detectable immediately after the injury. CN VII may also be affected by tumors of the brain and cranium, aneurysms, meningeal infections, and herpes viruses. Although injuries to CN VII cause paralysis of facial muscles, sensory loss in the small area of skin on the posteromedial surface of the auricle and around the opening of the external acoustic meatus is rare. Similarly, hearing is not usually impaired, but the ear may become more sensitive to low tones when the stapedius (supplied by CN VII) is paralyzed; this muscle dampens vibration of the stapes (see Chapter 7).

Bell palsy is a unilateral facial paralysis of sudden onset resulting from a lesion of CN VII. This syndrome is illustrated and discussed in detail in Chapter 7, p. 861.

VESTIBULOCOCHLEAR NERVE

Injuries to Vestibulocochlear Nerve

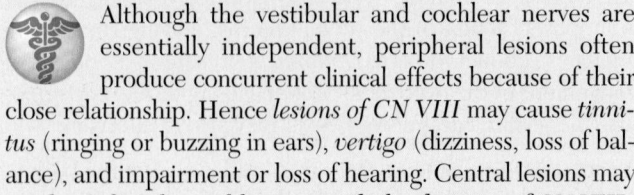 Although the vestibular and cochlear nerves are essentially independent, peripheral lesions often produce concurrent clinical effects because of their close relationship. Hence *lesions of CN VIII* may cause *tinnitus* (ringing or buzzing in ears), *vertigo* (dizziness, loss of balance), and impairment or loss of hearing. Central lesions may involve either the cochlear or vestibular divisions of CN VIII.

Deafness

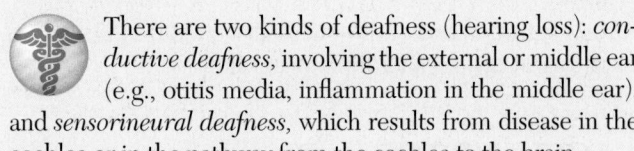 There are two kinds of deafness (hearing loss): *conductive deafness*, involving the external or middle ear (e.g., otitis media, inflammation in the middle ear), and *sensorineural deafness*, which results from disease in the cochlea or in the pathway from the cochlea to the brain.

Acoustic Neuroma

An *acoustic neuroma* (neurofibroma) is a slow-growing benign tumor of the neurolemma (Schwann) cells. The tumor begins in the vestibular nerve while it is in the internal acoustic meatus. The early symptom of an acoustic neuroma is usually loss of hearing. Dysequilibrium (derangement of the sense of equilibrium) and tinnitus occur in approximately 70% of patients (Bruce et al., 2010).

Trauma and Vertigo

People with head trauma often experience headache, dizziness, vertigo, and other features of post-traumatic injury. *Vertigo* is a hallucination of movement involving the person or the environment (Wazen, 2010). It often involves a spinning sensation but may be felt as a swaying back and forth or falling. These symptoms, often accompanied by nausea and vomiting, are usually related to a peripheral vestibular nerve lesion.

GLOSSOPHARYNGEAL NERVE

Lesions of Glossopharyngeal Nerve

Isolated *lesions of CN IX* or its nuclei are uncommon and are not associated with perceptible disability (Brannagan et al., 2010). Taste is absent on the posterior third of the tongue, and the gag reflex is absent on the side of the lesion. Ipsilateral weakness may produce a noticeable change in swallowing.

Injuries of CN IX resulting from infection or tumors are usually accompanied by signs of involvement of adjacent nerves. Because CN IX, CN X, and CN XI pass through the jugular foramen, tumors in this region produce multiple cranial nerve palsies, called the *jugular foramen syndrome*. Pain in the distribution of CN IX may be associated with involvement of the nerve in a tumor in the neck.

Glossopharyngeal Neuralgia

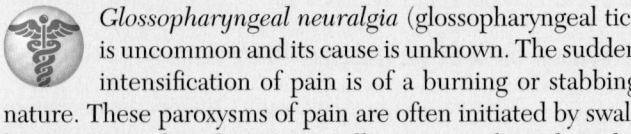 *Glossopharyngeal neuralgia* (glossopharyngeal tic) is uncommon and its cause is unknown. The sudden intensification of pain is of a burning or stabbing nature. These paroxysms of pain are often initiated by swallowing, protruding the tongue, talking, or touching the palatine tonsil (Brannagan et al., 2010). Pain paroxysms occur during eating when trigger areas are stimulated.

VAGUS NERVE

Isolated *lesions of CN X* are uncommon. Injury to pharyngeal branches of CN X results in *dysphagia* (difficulty in swallowing). Lesions of the superior laryngeal nerve produce anesthesia of the superior part of the larynx and paralysis of the cricothyroid muscle (see Chapter 8). The voice is weak and tires easily. Injury of a recurrent laryngeal nerve may be caused by aneurysms of the arch of the aorta and may occur during neck operations. Injury of the recurrent laryngeal nerve causes hoarseness and *dysphonia* (difficulty in speaking) because of paralysis of the vocal folds (cords). Paralysis of both recurrent laryngeal nerves causes *aphonia* (loss of voice) and *inspiratory stridor* (a harsh, high pitched respiratory sound). Paralysis of recurrent laryngeal nerves usually results from cancer of the larynx and thyroid gland and/or from injury during surgery on the thyroid gland, neck, esophagus, heart, and lungs. Because of its longer course, lesions of the left recurrent laryngeal nerve are more common than those of the right. Proximal lesions of CN X also affect the pharyngeal and superior laryngeal nerves, causing difficulty in swallowing and speaking.

SPINAL ACCESSORY NERVE

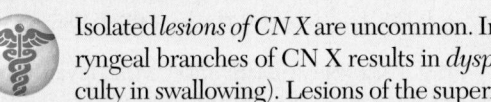 Because of its nearly subcutaneous passage through the posterior cervical region, *iatrogenic* (physician-caused) *injury of CN XI* may occur during surgical procedures such as lymph node biopsy, cannulation of the internal jugular vein, and carotid endarterectomy (see Chapter 8).

HYPOGLOSSAL NERVE

Injury to CN XII paralyzes the ipsilateral half of the tongue. After some time, the tongue atrophies, making it appear shrunken and wrinkled. When the tongue is protruded, its apex deviates toward the paralyzed side because of the unopposed action of the genioglossus muscle on the normal side of the tongue (see Chapter 7).

A References and Suggested Reading

INTRODUCTION

Bergman RA, Thompson SA, Afifi AK, Saadeh FA: *Compendium of Human Anatomic Variation: Text, Atlas, and World Literature.* Baltimore, Urban & Schwarzenberg, 1988. This useful source has been updated and is available from the Virtual Hospital's Web site *Illustrated Encyclopedia of Human Anatomic Variation* at www.vh.org/Providers/Textbooks/AnatomicVariants/AnatomyHP.html, 2008.

Federative Committee on Anatomical Terminology: *Terminologia Anatomica: International Anatomical Nomenclature.* Stuttgart, Thieme, 1998.

Haines DE (ed): *Fundamental Neuroscience for Basic and Clinical Applications,* 3rd ed. New York, Churchill Livingstone (Elsevier), 2006.

Hutchins JB, Naftel JP, Ard MD: The cell biology of neurons and glia. *In* Haines DE (ed): *Fundamental Neuroscience,* 4th ed. Saunders/Elsevier, 2012.

Keegan JJ, Garrett FD: The segmental distribution of the cutaneous nerves in the limbs of man. *Anat Rec* 102:409, 1948.

Kliegman RM, Stanton BMD, St. Geme J, Schor NF and Behrman RE (eds): *Nelson Textbook of Pediatrics,* 19th ed. Philadelphia, Saunders/Elsevier, 2011.

Kumar V, Abbas AK, Fausto N, Aster J: *Robbins and Cotran Pathologic Basis of Disease,* 8th ed. Philadelphia, Saunders/Elsevier, 2009.

Maklad A, Quinn T, Fritsch B: Intracranial distribution of the sympathetic system in mice: DiI tracing and immunocytochemical labeling. *Anat Rec* 263:99, 2001.

Marieb E and Hoehn K: *Human Anatomy and Physiology,* 9th ed. Menlo Park, CA, Benjamin/Cummings, 2012.

Moore KL, Persaud TVN and Torchia MG: *The Developing Human: Clinically Oriented Embryology,* 9th ed. Philadelphia, Saunders/Elsevier, 2012.

O'Rahilly R: Making planes plain. *Clin Anat* 10:129, 1997.

Ross MH, Pawlina W: *Histology. A Text and Atlas,* 6th ed. Baltimore, Lippincott Williams & Wilkins, 2011.

Salter RB: *Textbook of Disorders and Injuries of the Musculoskeletal System,* 3rd ed. Baltimore, Lippincott Williams & Wilkins, 1998.

Stedman's Medical Dictionary, 28th ed. Baltimore, Lippincott Williams & Wilkins, 2006.

Swartz MH: *Textbook of Physical Diagnosis, History and Examination,* 6th ed. Philadelphia, Saunders/Elsevier, 2009.

Willis MC: *Medical Terminology: The Language of Health Care,* 2nd ed. Baltimore, Lippincott Williams & Wilkins, 2005.

Wilson-Pauwels L, Stewart PA, Akesson E: *Autonomic Nerves: Basic Science, Clinical Aspects, Case Studies.* Hamilton, ON, Decker, 1997.

CHAPTER 1

Bergman RA, Thompson SA, Afifi AK, Saadeh FA: *Compendium of Human Anatomic Variation: Text, Atlas, and World Literature.* Baltimore, Urban & Schwarzenberg, 1988. This useful source has been updated and is available from the Virtual Hospital's Web site Illustrated Encyclopedia of Human Anatomic Variation at www.vh.org/Providers/Textbooks/AnatomicVariants/AnatomyHP.html (accessed May 2004).

Goroll AH and Mulley AG: *Primary Care Medicine: Office Evaluation and Management of the Adult Patient,* 6th ed. Philadelphia: Lippincott Williams & Wilkins, 2009.

Greene MF, Creasy RK, Resnik R, Iams JD, Lockwood CJ and Moore T: *Creasy and Resnik's Maternal-Fetal Medicine,* 6th ed. Philadelphia, Saunders/Elsevier, 2008.

Hardy SGP, Naftel JP: Viscerosensory pathways. *In* Haines DE (ed): *Fundamental Neuroscience for Basic and Clinical Applications,* 3rd ed. New York, Churchill Livingstone (Elsevier), 2006.

Kliegman RM, Stanton BMD, St. Geme J, Schor NF and Behrman RE (eds): *Nelson Textbook of Pediatrics,* 19th ed. Philadelphia, Saunders/Elsevier, 2011.

Kumar V, Abbas AK, Aster JC, Fausto N: *Robbins & Cotran Pathological Basis of Disease,* 8th ed. Philadelphia, Saunders/Elsevier, 2009.

Marx J, Hockberger R, and Walls R: *Rosen's Emergency Medicine: Concepts and Clinical Practice,* 7th ed. St. Louis, Mosby (Elsevier), 2009.

Moore KL, Persaud TVN and Torchia MG: *The Developing Human. Clinically Oriented Embryology,* 9th ed. Philadelphia, Saunders/Elsevier, 2012.

Rowland LP, Pedley TA (ed): *Merritt's Neurology,* 12th ed. Baltimore, Lippincott Williams & Wilkins, 2009.

Shields TW, LoCicero J, Reed CE, and Feins RH: General Thoracic Surgery, Vol. 1. Baltimore, Lippincott Williams & Wilkins, 2009.

Slaby FJ, McCune SK, Summers RW: *Gross Anatomy in the Practice of Medicine.* Philadelphia: Lea & Febiger, 1994.

Standring S (Editor-in-Chief): *Gray's Anatomy: The Anatomical Basis of Medicine and Surgery,* 40th British ed. London, Churchill Livingstone/Elsevier, 2008.

Swartz MH: *Textbook of Physical Diagnosis, History and Examination,* 6th ed. Philadelphia, Saunders/Elsevier, 2009.

Torrent-Guasp F, Buckberg GD, Clemente C, Cox JL, Coghlan HC, Gharib M: The structure and function of the helical heart and its buttress wrapping. I. The normal macroscopic structure of the heart. *Semin Thoracic Cardiovasc Surg* 13:301–319, 2001.

Vilensky JA, Baltes M, Weikel L, Fortin JD, Fourie LJ: Serratus posterior muscles: Anatomy, clinical relevance, and function. *Clin Anat* 14:237–241, 2001.

Wilson-Pauwels L, Stewart PA, Akesson EJ: *Autonomic Nerves—Basic Science, Clinical Aspects, Case Studies.* Hamilton, ON, Canada, Decker, 1997.

CHAPTER 2

Agur AMR, Dalley AF: *Grant's Atlas of Anatomy,* 13th ed. Baltimore, Lippincott Williams & Wilkins, 2013.

Bickley LS, Szilagyi PG: *Bates' Guide to Physical Examination and History Taking,* 10th ed. Baltimore, Lippincott Williams & Wilkins, 2009.

Cantlie J: On a new arrangement of the right and left lobes of the liver. *J Anat Physiol (Lond)* 32:iv, 1898.

Cheng YF, Huang TL, Chen CL, et al.: Variations of the middle and inferior right hepatic vein: Application in hepatectomy. *J Clin Ultrasound* 25:175, 1997.

Ellis H and Mahadevan: *Clinical Anatomy,* 12th ed. Blackwell Scientific, London 2010.

Fruchaud H: Anatomie chirurgicales des hernies de l'aine. Paris, Doin, 1956. [Cited in Skandalakis LJ, Gadacz TR, Mansberger AR Jr, Mitchell WE Jr, Colborn GL, Skandalakis JE: *Modern Hernia Repair: The Embryological and Anatomical Basis of Surgery.* New York, Parthenon, 1996.]

Kliegman RM, Stanton BMD, St. Geme J, Schor NF and Behrman RE (eds): *Nelson Textbook of Pediatrics,* 19th ed. Philadelphia, Saunders, 2011.

Kumar V, Abbas AK, Fausto N: *Robbins & Cotran Pathological Basis of Disease,* 8th ed. Philadelphia, Saunders/Elsevier, 2009.

Magee DF, Dalley AF: *Digestion and the Structure and Function of the Gut* [Karger Continuing Education Series, Vol. 8]. Basel, Karger, 1986.

Moore KL, Persaud TVN and Torchia MG: *The Developing Human. Clinically Oriented Embryology,* 9th ed. Philadelphia, Saunders/ Elsevier, 2012.

Moore KL, Persaud TVN, Shiota K: *Color Atlas of Clinical Embryology,* 2nd ed. Philadelphia, Saunders, 2000.

Rosse C, Gaddum-Rosse P: *Hollinshead's Textbook of Anatomy,* 5th ed. Philadelphia, Lippincott-Raven, 1997.

Sabiston DC Jr, Lyerly H (eds): *Sabiston Essentials of Surgery,* 2nd ed. Philadelphia, Saunders, 1994.

Skandalakis JE, Skandalakis PN, Skandalakis LJ: *Surgical Anatomy and Technique. A Pocket Manual.* 3rd ed. New York, Springer-Verlag, 2009.

Skandalakis LJ, Gadacz TR, Mansberger AR Jr, Mitchell WE Jr, Colborn GL, Skandalakis JE: *Modern Hernia Repair. The Embryology of Anatomical Basis of Surgery.* New York, Parthenon, 1996.

Standring S (ed.): *Gray's Anatomy: The Anatomical Basis of Clinical Practice,* 40th British ed. New York, Churchill Livingstone, 2008.

Swartz MH: *Textbook of Physical Diagnosis, History and Examination,* 6th ed. Philadelphia, Saunders, 2009.

Townsend CM, Beauchamp RD, Evers BM, Mattox KL: *Sabiston Textbook of Surgery,* 19th ed. Philadelphia, Saunders/Elsevier, 2012.

CHAPTER 3

Ashton-Miller JA, DeLancey JOL: Functional Anatomy of Female Pelvic Floor. *Annals of NY Academy of Science,* 11001:266–296 (2007).

Copeland LJ (ed): *Textbook of Gynecology,* 2nd ed. Philadelphia, Saunders, 2000.

DeLancey JOL: Anatomic aspects of vaginal eversion after hysterectomy. *Am J Obstet Gynecol* 166:1717–1728 (1992).

Federative International Committee on Anatomical Terminology [FICAT]: *Terminologia Anatomica: International Anatomical Nomenclature.* Stuttgart, Thieme, 1998.

Gabbe SG, Niebyl JR, Galan HL, Jauniaux ERM, Landon MB, Simpson JL, Driscoll DA: *Obstetrics—Normal and Problem Pregnancies,* 6th ed. Saunders/Elsevier, 2012.

Krebs H-B: Premalignant lesions of the cervix. *In* Copeland LJ (ed): *Textbook of Gynecology,* 2nd ed. Philadelphia, Saunders, 2000.

Moore KL, Persaud TVN and Torchia MG: *The Developing Human. Clinically Oriented Embryology,* 9th ed. Philadelphia, Saunders/Elsevier, 2012.

Morris M, Burke TW: Cervical cancer. *In* Copeland LJ (ed): *Textbook of Gynecology,* 2nd ed. Philadelphia, Saunders, 2000.

Myers RP, Cahill DR, Devine RM, King BF: Anatomy of radical prostatectomy as defined by magnetic resonance imaging. *J Urol* 159:2148, 1998a.

Myers RP, King BF, Cahill DR: Deep perineal "space" as defined by magnetic resonance imaging. *Clin Anat* 11:132, 1998b.

Oelrich TM: The urethral sphincter muscle in the male. *Am J Anat* 158:229, 1980.

Oelrich TM: The striated urogenital sphincter muscle in the female. *Anat Rec* 205:223, 1983.

Wendell-Smith CP: Muscles and fasciae of the pelvis. *In* Williams PL, Bannister LH, Berry MM, Collins P, Dussek JE, Fergusson MWJ (eds): *Gray's Anatomy, The Anatomical Basis of Medicine and Surgery,* 38th ed. Edinburgh, Churchill-Livingstone, 1995.

CHAPTER 4

Bergman RA, Thompson SA, Afifi AK, Saadeh FA: *Compendium of Human Anatomic Variation. Text, Atlas, and World Literature.* Baltimore, Urban & Schwarzenberg, 1988. This useful source has been updated and is available from the Virtual Hospital's Web site *Illustrated Encyclopedia of Human Anatomic Variation* at www.vh.org/Providers/ Textbooks/AnatomicVariants/AnatomyHP.html, 2008.

Bogduk N: *Clinical and Radiological Anatomy of the Lumbar Spine and Sacrum,* 4th ed. London: Churchill Livingstone (Elsevier), 2005.

Bogduk N, Macintosh JE: Applied anatomy of the thoracolumbar fascia. *Spine* 9:164, 1984.

Buxton DF, Peck D: Neuromuscular spindles relative to joint movement complexities. *Clin Anat* 2:211, 1989.

Crockard HA, Heilman AE, Stevens JM: Progressive myelopathy secondary to odontoid fractures: Clinical, radiological, and surgical features. *J Neurosurg* 78:579, 1993.

Duray SM, Morter HB, Smith FJ: Morphological variation in cervical spinous processes: potential applications in the forensic identification of race from the skeleton. *J Forensic Sci* 44(5):937–944, 1999.

Dvorak J, Schneider E, Saldinger P, Rahn B: Biomechanics of the craniovertebral region: The alar and transverse ligaments. *J Orthop Res* 6:452, 1988.

Greer M: Cerebral and spinal malformations. *In* Rowland LP, Tedley TA (ed): *Merritt's Textbook of Neurology,* 12th ed. Baltimore, Lippincott Williams & Wilkins, 2009.

Haines DE (ed): *Fundamental Neuroscience for Basic and Clinical Applications,* 3rd ed. New York, Churchill Livingstone (Elsevier), 2006.

Mercer S, Bogduk N: The ligaments and anulus fibrosus of human adult cervical intervertebral discs. *Spine* 24:619–628, 1999.

Moore KL, Persaud TVN and Torchia MG: *The Developing Human: Clinically Oriented Embryology,* 9th ed. Philadelphia, Saunders, 2012.

Rickenbacher J, Landolt AM, Theiler K: *Applied Anatomy of the Back.* New York: Springer Verlag, 1985.

Standring S (ed.): *Gray's Anatomy,* 40th British ed. New York, Churchill Livingstone, 2008.

Swartz MH: *Textbook of Physical Diagnosis. History and Examination,* 6th ed. Philadelphia, Saunders/Elsevier, 2009.

Vilensky JA, Baltes M, Weikel L, Fortin JD, Fourie LJ: Serratus posterior muscles: Anatomy, clinical relevance, and function. *Clin Anat* 14:237, 2001.

Yochum TR, Rowe LJ: *Essentials of Skeletal Radiology,* 3rd ed. Baltimore, Lippincott Williams & Wilkins, 2004.

CHAPTER 5

Anderson MK, Hall SJ, Martin M: *Sports Injury Management,* 2nd ed. Baltimore, Lippincott Williams & Wilkins, 2000.

Clay JH, Pounds DM: *Basic Clinical Message Therapy: Integrating Anatomy and Treatment,* 2nd ed. Baltimore, Lippincott Williams & Wilkins, 2008.

Foerster O: The dermatomes in man. *Brain* 56:1, 1933.

Ger R, Sedlin E: The accessory soleus muscle. *Clin Orthop* 116: 200, 1976.

Hamill J, Knutzen KM: *Biomechanical Basis of Human Movement,* 3rd ed. Baltimore, Lippincott Williams & Wilkins, 2008.

Jenkins DB: *Hollinshead's Functional Anatomy of the Limbs and Back,* 9th ed. Philadelphia, Saunders/Elsevier, 2008.

Kapandji IA: *The Physiology of the Joints, Vol. 2. Lower Limb,* 5th ed. Edinburgh, Churchill Livingstone, 1987.

Keegan JJ, Garrett FD: The segmental distribution of the cutaneous nerves in the limbs of man. *Anat Rec* 102:409, 1948.

Kendall FP, McCreary EK, Provance PG, Rodgers M, Romani W: *Muscles: Testing and Function with Posture and Gait,* 5th ed. Baltimore, Lippincott Williams & Wilkins, 2005.

Markhede G, Stener G: Function after removal of various hip and thigh muscles for extirpation of tumors. *Acta Orthop Scand* 52:373, 1981.

Moore KL, Persaud TVN and Torchia MG: *The Developing Human. Clinically Oriented Embryology,* 9th ed. Philadelphia, Saunders, 2012.

Palastanga N, Soames RW: *Anatomy and Human Movement,* 6th ed. Edinburgh, Churchill Livingstone/Elsevier, 2011.

Rancho Los Amigos National Rehabilitation Center Pathokinesiology Service and Physical Therapy Department: *Observational Gait Analysis.* Downey, CA, Los Amigos Research and Education Institute, Inc., 2001.

Rose J, Gamble JG: *Human Walking,* 3rd ed. Baltimore, Lippincott Williams & Wilkins, 2005.

Salter RB: *Textbook of Disorders and Injuries of the Musculoskeletal System,* 3rd ed. Baltimore, Lippincott Williams & Wilkins, 1999.

Soderberg GL: *Kinesiology: Application to Pathological Motion.* Baltimore: Williams & Wilkins, 1986.

Standring S (ed.): *Gray's Anatomy: The Anatomical Basis of Clinical Practice*, 40th British ed. New York, Churchill Livingstone, 2008.

Swartz MH: *Textbook of Physical Diagnosis*, 6th ed. Philadelphia, Saunders, 2009.

CHAPTER 6

Anderson MK, Hall SJ, Martin, M: *Sports Injury Management*, 2nd ed. Baltimore, Lippincott Williams & Wilkins, 2000.

Bergman RA, Thompson SA, Afifi AK, Saadeh FA: *Compendium of Human Anatomic Variation: Text, Atlas, and World Literature.* Baltimore, Urban & Schwarzenberg, 1988. This useful source has been updated and is available from the Virtual Hospital's website *Illustrated Encyclopedia of Human Anatomic Variation* at www.vh.org/Providers/Textbooks/AnatomicVariants/AnatomyHP.html.

Foerster O: The dermatomes in man. *Brain* 56:1, 1933.

Ger R, Abrahams P, Olson T: *Essentials of Clinical Anatomy*, 2nd ed. New York, Parthenon, 1996.

Halpern BC: Shoulder injuries. *In* Birrer RB, O'Connor FG (eds): *Sports Medicine for the Primary Care Physician*, 3rd ed. Boca Raton, FL, CRC Press, 2004.

Hamill J, Knutzen KM: *Biomechanical Basis of Human Movement*, 3rd ed. Baltimore, Lippincott Williams & Wilkins, 2008.

Keegan JJ, Garrett FD: The segmental distribution of the cutaneous nerves in the limbs of man. *Anat Rec* 102:409, 1948.

Leonard LJ (Chair), Educational Affairs Committee, American Association of Clinical Anatomists: The clinical anatomy of several invasive procedures. *Clin Anat* 12:43, 1999.

Moore KL, Persaud TVN and Torchia MG: *The Developing Human: Clinically Oriented Embryology*, 9th ed. Philadelphia, Saunders, 2012.

Rowland LP, Tedley TA (ed): *Merritt's Neurology*, 12th ed. Baltimore: Lippincott Williams & Wilkins, 2010.

Salter RB: *Textbook of Disorders and Injuries of the Musculoskeletal System*, 3rd ed. Baltimore, Lippincott Williams & Wilkins, 1999.

CHAPTER 7

Bernardini GL: Focal infections. *In* Rowland LP, Tedley TA (ed): *Merritt's Neurology*, 12th ed. Baltimore, Lippincott Williams & Wilkins, 2009.

Brust JCM: Coma. *In* Rowland LP, Tedley TA (ed): *Merritt's Neurology*, 12th ed. Baltimore, Lippincott Williams & Wilkins, 2010.

Corbett JJ, Haines DE, Ard MD, Lancon JA: The ventricles, choroid plexus, and cerebrospinal fluid. *In* Haines DE (ed): *Fundamental Neuroscience for Basic and Clinical Applications*, 3rd ed. New York, Churchill Livingstone (Elsevier), 2006.

Elkind MSV, Sacco RL: Pathogenesis, classification, and epidemiology of cerebrovascular disease. *In* Rowland LP, Pedley TA (ed): *Merritt's Neurology*, 12th ed. Baltimore, Lippincott Williams & Wilkins, 2010.

Fishman RA: Brain edema and disorders of intracranial pressure. *In* Rowland LP (ed): *Merritt's Neurology*, 11th ed. Baltimore, Lippincott Williams & Wilkins, 2005a.

Fishman RA: Cerebral veins and sinuses. *In* Rowland LP, Tedley TA (ed): *Merritt's Neurology*, 11th ed. Baltimore, Lippincott Williams & Wilkins, 2005b.

Green MW: Headache. *In* Rowland LP, Pedley TA (ed): *Merritt's Neurology*, 12th ed. Baltimore, Lippincott Williams & Wilkins, 2010.

Haines DE (ed): *Fundamental Neuroscience for Basic and Clinical Applications*, 3rd ed. New York, Churchill Livingstone (Elsevier), 2006.

Haines DE: *Neuroanatomy: An Atlas of Structures, Sections, and Systems*, 8th ed. Baltimore, Lippincott Williams & Wilkins, 2011.

Haines DE, Harkey HL, Al-Mefty O: The "subdural" space: A new look at an outdated concept. *Neurosurgery* 32:111, 1993.

Jubelt B: Bacterial infections. *In* Rowland LP (ed): *Merritt's Neurology*, 12th ed. Baltimore, Lippincott Williams & Wilkins, 2010.

Kiernan JA: *Barr's The Human Nervous System: An Anatomical Viewpoint*, 9th ed. Baltimore, Lippincott Williams & Wilkins, 2008.

Kliegman RM, Stanton BF, St. Geme JW, Schor NF, Behrman RE (eds): *Nelson Textbook of Pediatrics*, 19th ed. Philadelphia, Saunders (Elsevier), 2011.

Moore KL, Persaud TVN and Torchia MG: *The Developing Human: Clinically Oriented Embryology*, 9th ed. Philadelphia, Saunders (Elsevier), 2012.

Olson TR, Abrahams PR, Ger R: *Ger's Essentials of Clinical Anatomy*, 3rd ed. New York, Cambridge University Press, 2009.

Rowland LP, Pedley TA (ed): *Merritt's Neurology*, 12th ed. Baltimore, Lippincott Williams & Wilkins, 2010.

Standring S (ed.): *Gray's Anatomy: The Anatomical Basis of Clinical Practice*, 39th British ed. Edinburgh, UK, Churchill Livingstone, 2004.

Storper IS: Ménière disease. *In* Rowland LP, Pedley TA (ed): *Merritt's Neurology*, 12th ed. Baltimore, Lippincott Williams & Wilkins, 2010.

Swartz MH: *Textbook of Physical Diagnosis. History and Examination*, 6th ed. Philadelphia, Saunders (Elsevier), 2009.

CHAPTER 8

Agur AMR, Dalley AF: *Grant's Atlas of Anatomy*, 13th ed. Baltimore, Lippincott Williams & Wilkins, 2013.

Bickley LS: *Bates' Guide to Physical Examination and History Taking*, 10th ed. Baltimore, Lippincott Williams & Wilkins, 2009.

Elkind MSV, Sacco RL: Pathogenesis, classification, and epidemiology of cerebrovascular disease. *In* Rowland LP, Pedley TA (ed): *Merritt's Textbook of Neurology*, 12th ed. Baltimore, Lippincott Williams & Wilkins, 2010.

Fahn S, Bressman SB: Dystonia. *In* Rowland LP, Pedley TA (ed): *Merritt's Textbook of Neurology*, 12th ed. Baltimore, Lippincott Williams & Wilkins, 2010.

Hirsch LJ, Pedley TA: Syncope, seizures and their mimics. *In* Rowland LP (ed): *Merritt's Textbook of Neurology*, 12th ed. Baltimore, Lippincott Williams & Wilkins, 2010.

Kliegman RM, Stanton BF, St. Geme JW, Schor NF, Behrman RE (eds): *Nelson Textbook of Pediatrics*, 19th ed. Philadelphia, Saunders/Elsevier, 2011.

Lachman N, Acland RD, Rosse C: Anatomical evidence for the absence of a morphologically distinct cranial root of the accessory nerve in man. *Clin Anat* 15:4–10, 2002.

Leung AKC, Wong AL, Robson WLLM: Ectopic thyroid gland simulating a thyroglossal duct cyst: A case report. *Can J Surg* 38:87, 1995.

Moore KL, Persaud TVN, Torchia MG: *The Developing Human: Clinically Oriented Embryology*, 9th ed. Philadelphia, Saunders, 2012.

Norton JA, Wells SA Jr: The parathyroid glands. *In* Sabiston DC Jr, Lyerly HK (eds): *Sabiston Essentials of Surgery*, 2nd ed. Philadelphia, Saunders, 1994.

Olson TR, Abrahams P, Ger R: *Essentials of Clinical Anatomy*, 3rd ed. New York, Cambridge University Press, 2009.

Rowland LP, Pedley TA (ed): *Merritt's Textbook of Neurology*, 12th ed. Baltimore: Lippincott Williams & Wilkins, 2010A.

Rowland LP, Pedley TA: Diagnosis of pain and paresthesias. *In* Rowland LP (ed): *Merritt's Textbook of Neurology*, 12th ed. Baltimore, Lippincott Williams & Wilkins, 2010B.

Sabiston DC Jr, Lyerly H: *Sabiston Essentials of Surgery*, 2nd ed. Philadelphia, Saunders, 1994.

Scher RL, Richtsmeier WJ: Otolaryngology: Head and neck surgery. *In* Sabiston DC Jr, Lyerly HK (eds): *Sabiston Essentials of Surgery*, 2nd ed. Philadelphia, Saunders, 1994.

Sinkinson CA: The continuing saga of penetrating neck injuries. *Emerg Med* 12:135, 1991.

Skandalakis JE, Skandalakis PN, Skandalakis LJ: *Surgical Anatomy and Technique. A Pocket Manual*. 3rd ed., New York, Springer-Verlag, 2009.

Standring S (ed.): *Gray's Anatomy: The Anatomical Basis of Clinical Practice*, 40th British ed. New York, Churchill Livingstone/Elsevier, 2008.

Swartz MH: *Textbook of Physical Diagnosis: History and Examination*, 6th ed. Philadelphia, Saunders, 2009.

Weiglein AH, Morrigl B, Schalk C, et al.: Arteries in the posterior cervical triangle in man. *Clin Anat* 18:533–557, 2005.